MATHEMATICAL PHYSICS

ABOUT THE AUTHORS

H K Dass is recipient of British Scholarship in 1971-72 for Specialist Study in **ADVANCED MATHEMATICS** at University of Hull, England. He is M.Sc. (Mathematics) First Division, college topper, and has been teaching Mathematics to Engineering students for more than 50 years and for all these year has remained on of the most popular faculty among students.
He is an established and prolific author of more than 30 books for engineering and 15 for school and enjoys immense popularity not only in India but also in our neighbouring countries and many East Asian and African countries.

Dr. Rama Verma is Gold Medalist M.Sc. (Mathematics) and Ph.D. At present she is Associate Professor, Mata Sundri College, Delhi University, Delhi.

MATHEMATICAL PHYSICS

H K DASS
M.Sc. (Mathematics)
Diploma in Specialist Studies
University of Hull
England, UK

Dr. RAMA VERMA
M.Sc. (Gold Medalist), Ph.D.
Associate Professor
Mata Sundri College
University of Delhi

Dr. RAJNISH VERMA
Fellow IETE, MBA
B.E. Electronics Engg. DCE / DTU
Advisory Board – Global Research Conferences
Consultant (Retd.) - TCS Ltd.
Ex. DGM - CMC Ltd.

S Chand And Company Limited
(ISO 9001 Certified Company)

S Chand And Company Limited
(ISO 9001 Certified Company)
Head Office: D-92, Sector–2, Noida – 201301, U.P. (India), Ph. 91-120-4682700
Registered Office: A-27, 2nd Floor, Mohan Co-operative Industrial Estate, New Delhi – 110 044, Phone: 011-49731800
www.**schandpublishing.com;** e-mail: **info@schandpublishing.com**

Marketing Offices:

Chennai	:	Ph: 23632120; chennai@schandpublishing.com
Guwahati	:	Ph: 2738811, 2735640; guwahati@schandpublishing.com
Hyderabad	:	Ph: 40186018; hyderabad@schandpublishing.com
Jalandhar	:	Ph: 4645630; jalandhar@schandpublishing.com
Kolkata	:	Ph: 23357458, 23353914; kolkata@schandpublishing.com
Lucknow	:	Ph: 4003633; lucknow@schandpublishing.com
Mumbai	:	Ph: 25000297; mumbai@schandpublishing.com
Patna	:	Ph: 2260011; patna@schandpublishing.com

First Edition 1997
Subsequent Editions and Reprints 2003, 2004, 2005, 2007, 2008 (Twice), 2009 (Twice), 2010, 2011 (Twice) 2013, 2014, Seventh Revised Edition 2014; Reprints 2015, 2016
Eighth Revised Edition 2018

Low Priced Students' Paperback Edition (LPSPE) 2019, Reprints 2020, 2021 (Four times), 2022 (Twice), 2022 (Twice)

Reprint 2025

ISBN: 978-93-528-3722-9 **Product Code:** H6MAP61PHYS10ENAH19L

PRINTED IN INDIA

By Vikas Publishing House Private Limited, Plot 20/4, Site-IV, Industrial Area Sahibabad, Ghaziabad – 201 010 and Published by S Chand And Company Limited, A-27, 2nd Floor, Mohan Co-operative Industrial Estate, New Delhi – 110 044.

PREFACE TO EIGHTH REVISED EDITION

The interest shown in the textbook by the professors and students has encouraged me to revise this edition.

A hallmark of this text has been our effort to eliminate errors of any kind. Add to that a robust Index (book-end) and the availability of useful formulae and solved question papers on the S Chand Publishing website **(https://www.schandpublishing.com/books/higher-education/physics/mathematical-physics/9789352533572/)**, the students have all the resources needed to find success in Mathematical Physics.

Keeping no stone unturned, the textbook contains 49 chapters on all major topics in an exhaustive endeavour to cover syllabuses of all major universities.

The author thanks everyone at S Chand And Company Limited in their efforts to publish this new and revised edition.

H K DASS

D-1/87 Janakpuri
New-Delhi–110058
Tele: 011-28525078
Mob: +919350055078
Email: **hk_dass@yahoo.com**

PREFACE

This is my first effort to write a book on Mathematical Physics.

The chief aim of this book is to meet the requirements of students of B.Sc. Honours (Physics) and M.Sc. of various Indian Universities.

The subject-matter is presented in a very systematic and logical manner. Every endeavour has been made to make the content simple and lucid as far as possible.

While every effort has been made to present the material correctly, no attempt has been made to be absolutely rigorous. The subject matter has been so arranged that even an average student can understand how to apply the mathematical operations to the problems of Physics.

All valuable suggestions for the improvement of the book will be highly appreciated and gratefully acknowledged.

I am thankful to Shri Rajendra Kumar Gupta, Managing Director, Shri Ravindra Kumar Gupta, Director and other members of the staff of the Publishers, M/s S. Chand & Co. Ltd., New Delhi without whose co-operation, it would not have been possible to put this book in such a fine format and that too in record time.

D-1/87, Janakpuri **H K DASS**
New Delhi-110 058

CONTENTS

UNIT – I

UNIT – II

14. Cauchy–Euler Equations, Method of Variation of Parameters 371–395

15. Differential Equation of other Types 396–420

UNIT - III

UNIT - IV

32. Abstract Vector Spaces 873–887

33. Vectors in R^n 888–922

47. Inverse Laplace Transforms (Solution of differential equations) 1334–1378

48. Dirac-Delta Function 1379–1381

49. Tensor Analysis 1382–1410

For Useful Formulae and Solved Question Papers please go to :
https://www.schandpublishing.com/books/higher-education/physics/mathematical-physics/9789352533572/

Or

Scan  for immediate access

UNIT – I

Review of Vector Algebra

1.1 VECTORS

A vector is a quantity having both magnitude and direction such as force, velocity acceleration, displacement etc.

Definition of Polar and Axial vectors. *(Delhi University, April 2010)*

Polar vectors:

The vectors associated with a linear directional effect are called polar vectors.

The examples of polar vectors are force, acceleration, linear velocity, linear momentum etc.

Axial vectors

The vectors associated with rotation about an axis are called axial vectors.

The examples of axial vectors are: torque, angular velocity, angular momentum etc.

1.2 ADDITION OF VECTORS

Let $\vec{a}$ and $\vec{b}$ be two given vectors

$\vec{OA} = \vec{a}$ and $\vec{AB} = \vec{b}$ then vector $\vec{OB}$ is called the sum of $\vec{a}$ and $\vec{b}$.

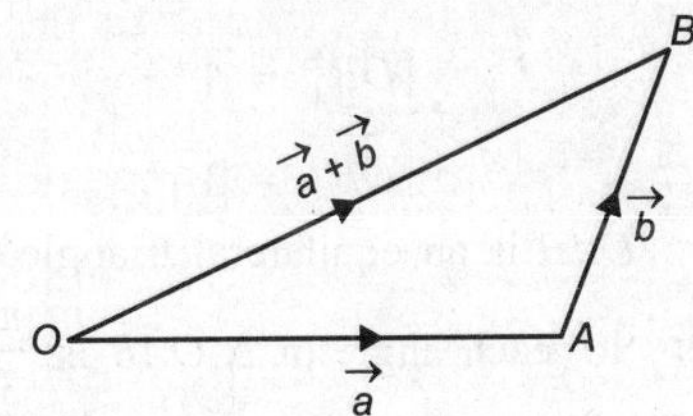

Symbolically

$$\vec{OA} + \vec{AB} = \vec{OB}$$

$$\vec{a} + \vec{b} = \vec{OB}$$

1.3 RECTANGULAR RESOLUTION OF A VECTOR

Let OX, OY, OZ be the three rectangular axes. Let $\hat{i}, \hat{j}, \hat{k}$ be three unit vectors and parallel to three axes.

If $\vec{OP} = \hat{r}$ and the co-ordinates of P be (x, y, z)

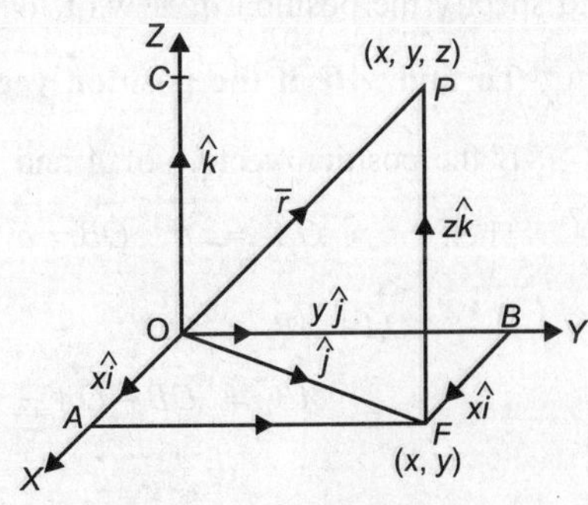

$$\vec{OA} = x\hat{i}, \quad \vec{OB} = y\hat{j} \quad \text{and} \quad \vec{OC} = z\hat{k}$$

$$\vec{OP} = \vec{OF} + \vec{FP}$$

$$\Rightarrow \quad \vec{OP} = (\vec{OA} + \vec{AF}) + \vec{FP}$$

$$\Rightarrow \quad \vec{OP} = \vec{OA} + \vec{OB} + \vec{OC}$$

$$\vec{r} = x\hat{i} + y\hat{j} + z\hat{k}$$

$$\Rightarrow \quad OP^2 = OF^2 + FP^2$$
$$= (OA^2 + AF^2) + FP^2$$
$$= OA^2 + OB^2 + OC^2$$
$$= x^2 + y^2 + z^2$$
$$OP = \sqrt{x^2 + y^2 + z^2}$$
$$|\vec{r}| = \sqrt{x^2 + y^2 + z^2}$$

1.4 UNIT VECTOR

Let a vector be $x\hat{i} + y\hat{j} + z\hat{k}$.

$$\text{Unit vector} = \frac{x\hat{i} + y\hat{j} + z\hat{k}}{\sqrt{x^2 + y^2 + z^2}}$$

Example 1. *If $\vec{a}$ and $\vec{b}$ be two unit vectors and α be the angle between them, then find the value of α such that $\vec{a} + \vec{b}$ is a unit vector.*

Solution. Let $\vec{OA} = \vec{a}$ be a unit vector and $\vec{AB} = \vec{b}$ is another unit vector and α be the angle between $\vec{a}$ and $\vec{b}$.

If $\vec{OB} = \vec{c} = \vec{a} + \vec{b}$ is also a unit vector then, we have

$$|\vec{OA}| = 1$$
$$|\vec{OB}| = 1$$
$$|\vec{OB}| = 1$$

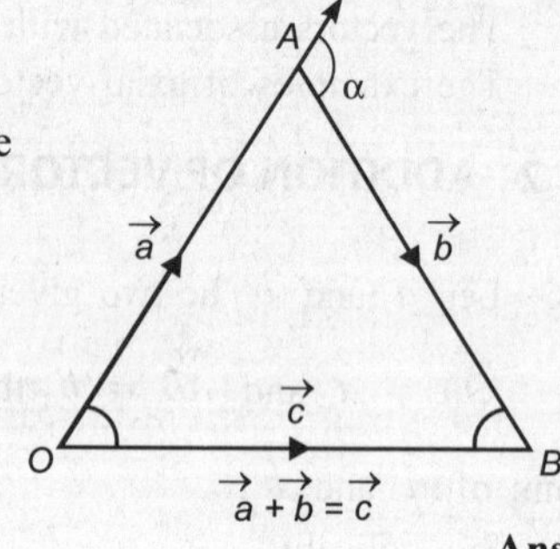

OAB is an equilateral triangle.

So, each angle of $\Delta\ OAB$ is $\frac{\pi}{3}$ **Ans.**

Hence, $\alpha = \pi - \frac{\pi}{3} - \frac{\pi}{3} = \frac{\pi}{3}$

1.5 POSITION VECTOR OF A POINT

The position vector of a point A with respect to origin O is the vector $\vec{OA}$ which is used to specify the position of A w.r.t. O.

To find $\vec{AB}$ if the position vectors of the point A and point B are given.

If the position vectors of A and B are $\vec{a}$ and $\vec{b}$. Let the origin be O.

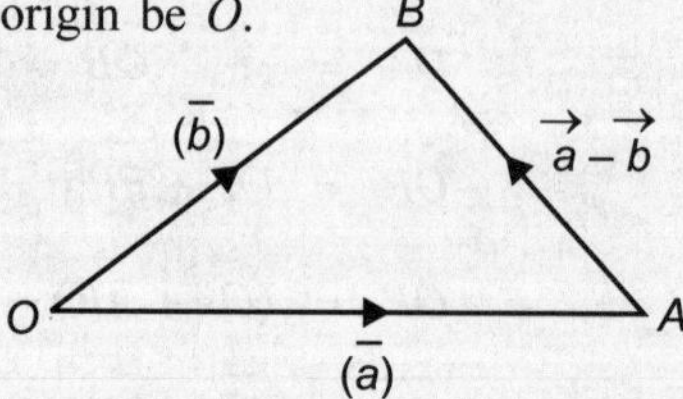

Then $\quad \vec{OA} = \vec{a}, \ \vec{OB} = \vec{b}$

$$\vec{OA} + \vec{AB} = \vec{OB}$$
$$\vec{AB} = \vec{OB} - \vec{OA}$$

$$\Rightarrow \quad \vec{AB} = \vec{b} - \vec{a}$$

$$\vec{AB} = \textbf{Position vector of } \boldsymbol{B} - \textbf{Position vector of } \boldsymbol{A}$$

Example 2. *If A and* B *are (3, 4, 5) and (6, 8, 9), find* $\vec{AB}$.

Solution.
$$\vec{AB} = \text{Position vector of } B - \text{Position vector of } A$$
$$= (6\hat{i} + 8\hat{j} + 9\hat{k}) - (3\hat{i} + 4\hat{j} + 5\hat{k})$$
$$= 3\hat{i} + 4\hat{j} + 4\hat{k}$$
Ans.

1.6 RATIO FORMULA

To find the position vector of the point which divides the line joining two given points. Let A and B be two points and a point C divides AB in the ratio of $m : n$.

Let O be the origin, then

$$\vec{OA} = \vec{a}, \quad \text{and} \quad \vec{OB} = \bar{b}, \quad \vec{OC} = ?$$

$$\vec{OC} = \vec{OA} + \vec{AC}$$

$$= \vec{OA} + \frac{m}{m+n}\vec{AB} \quad \left(\therefore AC = \frac{m}{m+n} AB\right)$$

$$= \vec{a} + \frac{m}{m+n}.(\vec{b} - \vec{a}) \quad (\therefore \vec{AB} = \vec{b} - \vec{a})$$

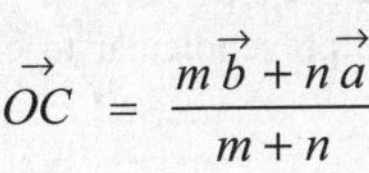

$$\vec{OC} = \frac{m\vec{b} + n\vec{a}}{m+n}$$

Cor. If $m = n = 1$, then C will be the mid-point, and

$$\vec{OC} = \frac{\vec{a} + \vec{b}}{2}$$

1.7 PRODUCT OF TWO VECTORS

The product of two vectors results in two different ways, the one is a number and the other is vector. So, there are two types of product of two vectors, namely scalar product and vector product. They are written as $\vec{a} \,.\, \vec{b}$ and $\vec{a} \times \vec{b}$.

1.8 SCALAR, OR DOT PRODUCT

The scalar, or dot product of two vectors $\vec{a}$ and $\vec{b}$ is defined to be $|\vec{a}||\vec{b}| \cos\theta$ *i.e.*, scalar where θ is the angle between $\vec{a}$ and $\vec{b}$.

Symbolically, $\vec{\boldsymbol{a}} \,.\, \vec{\boldsymbol{b}} = |\vec{\boldsymbol{a}}||\vec{\boldsymbol{b}}| \cos\theta$

Due to a dot between $\vec{a}$ and $\vec{b}$ this product is also called dot product.

The scalar product is commutative

To Prove. $\vec{a}\cdot\vec{b} = \vec{b}\cdot\vec{a}$

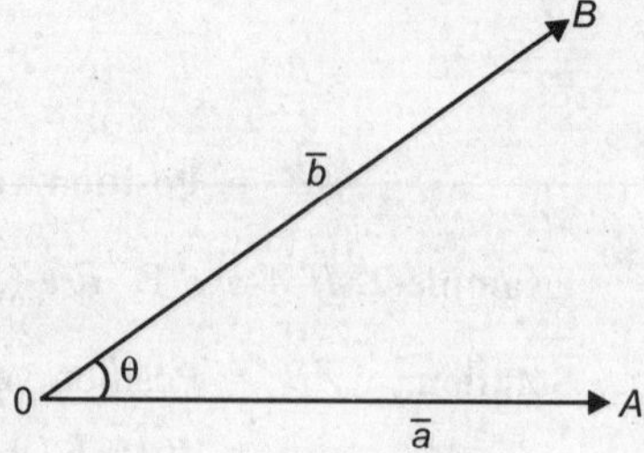

Proof.
$$\vec{b}\cdot\vec{a} = \left|\vec{b}\right|\left|\vec{a}\right|\cos(-\theta)$$
$$= \left|\vec{a}\right|\left|\vec{b}\right|\cos\theta$$
$$= \vec{a}\cdot\vec{b} \qquad \textbf{Proved.}$$

Geometrical interpretation. The scalar product of two vectors is the product of one vector and the length of the projection of the other in the direction of the first.

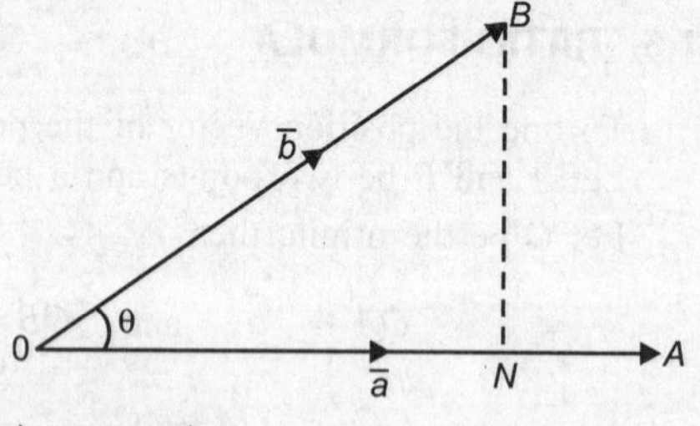

Let $\vec{OA} = \vec{a}$ and $\vec{OB} = \vec{b}$

then
$$\vec{a}\cdot\vec{b} = (OA)\,.\,(OB)\cos\theta$$
$$= OA\,.\,OB\,.\,\frac{ON}{OB}$$
$$= OA\,.\,ON$$
$$= (\text{Length of } \vec{a})\ (\text{projection of } \vec{b} \text{ along } \vec{a})$$

1.9 USEFUL RESULTS

$$\hat{i}\cdot\hat{i} = (1)(1)\cos 0° = 1 \qquad \text{Similarly, } \hat{j}\cdot\hat{j} = 1, \quad \hat{k}\cdot\hat{k} = 1$$
$$\hat{i}\cdot\hat{j} = (1)(1)\cos 90° = 0 \qquad \text{Similarly, } \hat{j}\cdot\hat{k} = 0, \quad \hat{k}\cdot\hat{i} = 0$$

Note. If the dot product of two vectors is zero then vectors are prependicular to each other.

1.10 WORK DONE AS A SCALAR PRODUCT

If a constant force F acting on a particle displaces it from A to B then,

Work done = (component of F along AB). Displacement
$$= F\cos\theta.\ AB$$
$$= \vec{F}\cdot\vec{AB}$$

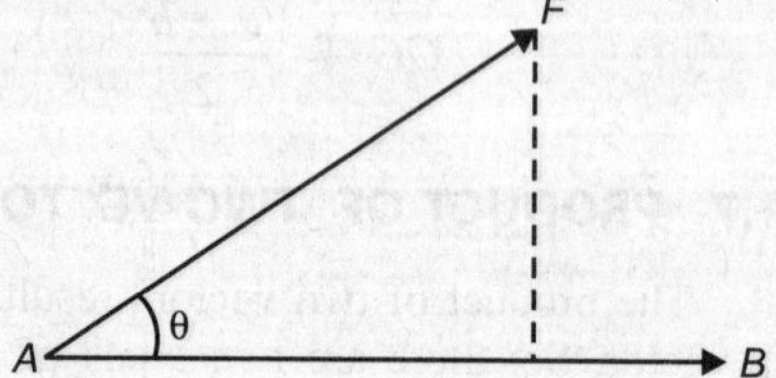

Work done = Force . Displacement

1.11 VECTOR PRODUCT OR CROSS PRODUCT

1. **The vector, or cross product of two vectors $\vec{a}$ and $\vec{a}$ is defined to be a vector such that**

(*i*) Its magnitude is $\left|\vec{a}\right|\left|\vec{b}\right|\sin\theta$, where θ is the angle between $\vec{a}$ and $\vec{b}$.

(*ii*) Its direction is perpendicular to both vectors $\vec{a}$ and $\vec{b}$.

(*iii*) It forms with a right handed system.

Let $\hat{\eta}$ be a unit vector perpendicular to both the vectors $\vec{a}$ and $\vec{b}$.

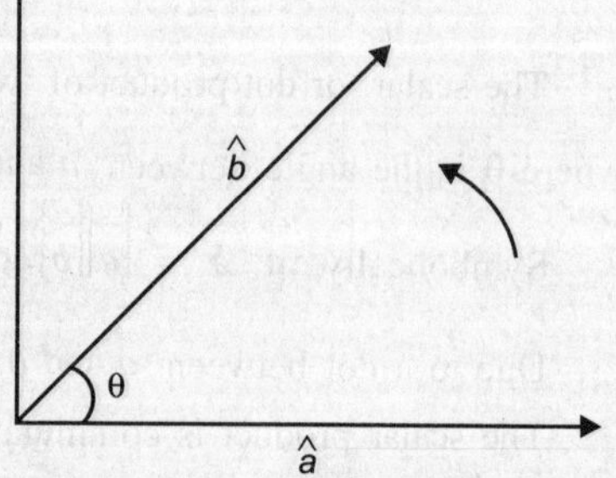

$$\vec{a}\times\vec{b} = |\vec{a}||\vec{b}|\sin\theta.\hat{\eta}$$

2. Useful results

Since $\hat{i}, \hat{j}, \hat{k}$ are three mutually perpendicular unit vectors, then

$$\hat{i}\times\hat{i} = \hat{j}\times\hat{j} = \hat{k}\times\hat{k} = 0$$

$$\hat{i}\times\hat{j} = -\hat{j}\times\hat{i} = \hat{k} \qquad \hat{j}\times\hat{i} = -\hat{i}\times\hat{j}$$

$$\hat{j}\times\hat{k} = -\hat{k}\times\hat{j} = \hat{i} \qquad \text{and} \qquad \hat{k}\times\hat{j} = -\hat{j}\times\hat{k}$$

$$\hat{k}\times\hat{i} = -\hat{i}\times\hat{k} = \hat{j} \qquad \hat{i}\times\hat{k} = -\hat{k}\times\hat{i}$$

1.12 VECTOR PRODUCT EXPRESSED AS A DETERMINANT

If $\vec{a} = a_1\hat{i} + a_2\hat{j} + a_3\hat{k}$

$\vec{b} = b_1\hat{i} + b_2\hat{j} + b_3\hat{k}$

$$\vec{a}\times\vec{b} = (a_1\hat{i} + a_2\hat{j} + a_3\hat{k})\times(b_1\hat{i} + b_2\hat{j} + b_3\hat{k})$$

$$= a_1b_1\,(\hat{i}\times\hat{i}) + a_1b_2\,(\hat{i}\times\hat{j}) + a_1b_3\,(\hat{i}\times\hat{k}) + a_2b_1\,(\hat{j}\times\hat{i}) + a_2b_2\,(\hat{j}\times\hat{j})$$
$$+ a_2b_3(\hat{j}\times\hat{k}) + a_3b_1(\hat{k}\times\hat{i}) + a_3b_2(\hat{k}\times\hat{j}) + a_3b_3(\hat{k}\times\hat{k})$$

$$= a_1b_2\hat{k} - a_1b_3\hat{j} - a_2b_1\hat{k} + a_2b_3\hat{i} + a_3b_1\hat{j} - a_3b_2\hat{i}$$

$$= (a_2b_3 - a_3b_2)\hat{i} - (a_1b_3 - a_3b_1)\hat{j} + (a_1b_2 - a_2b_1)\hat{k}$$

$$\vec{a}\times\vec{b} = \begin{vmatrix} \hat{i} & \hat{j} & \hat{k} \\ a_1 & a_2 & a_3 \\ b_1 & b_2 & b_3 \end{vmatrix}$$

1.13 AREA OF PARALLELOGRAM

Example 3. *Find the area of a parallelogram whose adjacent sides are* $\hat{i} - 2\hat{j} + 3\hat{k}$ and $2\hat{i} + \hat{j} - 4\hat{k}$

Solution. Vector area of || gm $= \begin{vmatrix} \hat{i} & \hat{j} & \hat{k} \\ 1 & -2 & 3 \\ 2 & 1 & -4 \end{vmatrix}$

$$= (8-3)\hat{i} - (-4-6)\hat{j} + (1+4)\hat{k} = 5\hat{i} + 10\hat{j} + 5\hat{k}$$

Area of parallelogram $= \sqrt{(5)^2 + (10)^2 + (5)^2} = 5\sqrt{6}$ **Ans.**

1.14 MOMENT OF A FORCE

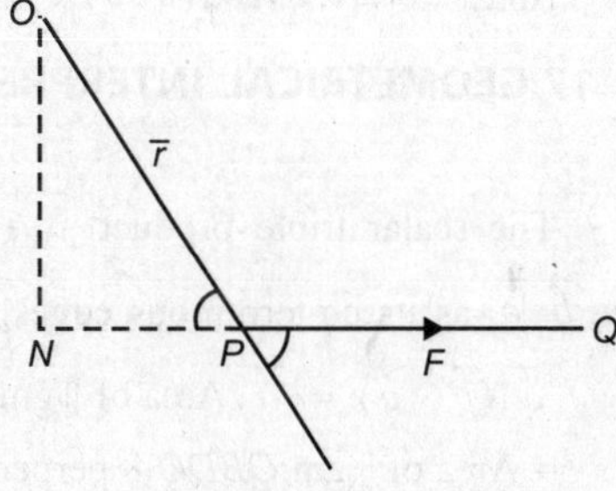

Let a force F $(\vec{PQ})$ act at a point P.

Moment of $\vec{F}$ about O

= Product of force F and perpendicular distance (ON. $\hat{\eta}$)

= (PQ) $(ON)(\hat{\eta})$ = (PQ) (OP) $\sin\theta$ $(\hat{\eta})$

= $\vec{OP}\times\vec{PQ}$

$\Rightarrow$ $\vec{M} = \vec{r}\times\vec{F}$

1.15 ANGULAR VELOCITY

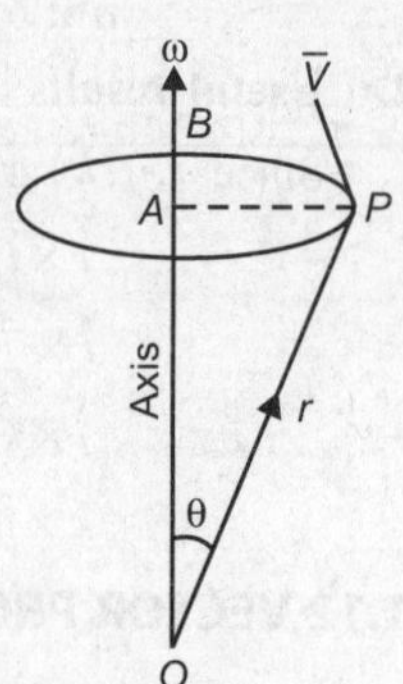

Let a rigid body be rotating about the axis OA with the angular velocity ω which is a vector and its magnitude is ω radians per second and its direction is parallel to the axis of rotation OA.

Let P be any point on the body such that $\overline{OP} = \vec{r}$ and $\angle AOP = \theta$ and $AP \perp OA$. Let the velocity of P be V.

Let $\hat{\eta}$ be a unit vector perpendicular to $\vec{\omega}$ and $\vec{r}$.

$$\vec{\omega} \times \vec{r} = (\omega\, r \sin\theta)\ \hat{\eta} = (\omega\, AP)\ \hat{\eta} = (\text{Speed of } P)\ \hat{\eta}$$

$$= \text{Velocity of } P \perp \text{ to } \vec{\omega} \text{ and } r$$

Hence $\boxed{\vec{V} = \vec{\omega} \times \vec{r}}$

1.16 SCALAR TRIPLE PRODUCT

Let $\vec{a}, \vec{b}, \vec{c}$ be three vectors then their dot product is written as $\vec{a}\,.(\vec{b} \times \vec{c})$ or $[\vec{a}\ \vec{b}\ \vec{c}]$.

If $\vec{a} = a_1\hat{i} + a_2\hat{j} + a_3\hat{k},\ \vec{b} = b_1\hat{i} + b_2\hat{j} + b_3\hat{k}$, and $\vec{c} = c_1\hat{i} + c_2\hat{j} + c_3\hat{k}$

$$\vec{a}.(\vec{b} \times \vec{c}) = (a_1\hat{i} + a_2\hat{j} + a_3\hat{k}).[(b_1\hat{i} + b_2\hat{j} + b_3\hat{k}) \times (c_1\hat{i} + c_2\hat{j} + c_3\hat{k})]$$

$$= (a_1\hat{i} + a_2\hat{j} + a_3\hat{k}).[(b_2c_3 - b_3c_2)\hat{i} + (b_3c_1 - b_1c_3)\hat{j} + (b_1c_2 - b_2c_1)\hat{k}]$$

$$= a_1\,(b_2c_3 - b_3c_2) + a_2\,(b_3c_1 - b_1c_3) + a_3\,(b_1c_2 - b_2c_1)$$

$$= \begin{vmatrix} a_1 & a_2 & a_3 \\ b_1 & b_2 & b_3 \\ c_1 & c_2 & c_3 \end{vmatrix}$$

Similarly, $\vec{b}\,.(\vec{c} \times \vec{a})$ and $\vec{c}\,.(\vec{a} \times \vec{b})$ have the same value.

$$\therefore\quad \vec{a}\,.(\vec{b} \times \vec{c}) = \vec{b}\,.(\vec{c} \times \vec{a}) = \vec{c}\,.(\vec{a} \times \vec{b})$$

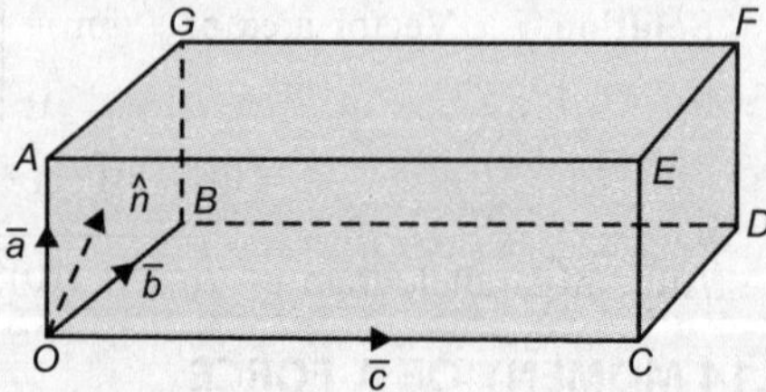

The value of the product depends upon the cyclic order of the vector, but is independent of the position of the dot and cross. These may be interchanged.

The value of the product changes if the order is non-cyclic.

Note. $\vec{a} \times (\vec{b}\,.\,\vec{c})$ and $(\vec{a}\,.\,\vec{b}) \times \vec{c}$ are meaningless.

1.17 GEOMETRICAL INTERPRETATION

The scalar triple product $\vec{a}\,.(\vec{b} \times \vec{c})$ represents the volume of the parallelopiped having $\vec{a}, \vec{b}, \vec{c}$ as its co-terminous edges.

$$\vec{a}\,.(\vec{b} \times \vec{c}) = \vec{a}\,.\,\text{Area of} \parallel \text{gm } OBDC\ \hat{\eta}$$

= Area of || gm $OBDC$ × perpendicular distance between the parallel faces $OBDC$ and $AEFG$.
= Volume of the parallelopiped

Note. (1) If $\vec{a}.(\vec{b}\times\vec{c}) = 0$, then $\vec{a}, \vec{b}, \vec{c}$ are coplanar.

(2) Volume of tetrahedron $\frac{1}{6}(\vec{a}\ \vec{b}\ \vec{c})$.

Example 4. *Find the volume of parallelopiped if*

$$\vec{a} = -3\hat{i} + 7\hat{j} + 5\hat{k},\quad \vec{b} = -3\hat{i} + 7\hat{j} - 3\hat{k}, \text{ and } \vec{c} = 7\hat{i} - 5\hat{j} - 3\hat{k}$$

are the three co-terminous edges of the parallelopiped.

Solution.

$$\text{Volume} = \vec{a}.(\vec{b}\times\vec{c})$$

$$= \begin{vmatrix} -3 & 7 & 5 \\ -3 & 7 & -3 \\ 7 & -5 & -3 \end{vmatrix} = -3(-21-15) - 7(9+21) + 5(15-49)$$

$$= 108 - 210 - 170 = -272$$

Volume = 272 cube units. **Ans.**

Example 5. *Show that the volume of the tetrahedron having* $\vec{A}+\vec{B}, \vec{B}+\vec{C}, \vec{C}+\vec{A}$ *as concurrent edges is twice the volume of the tetrahendron having* $\vec{A}, \vec{B}, \vec{C}$ *as concurrent edges.*

Solution. Volume of tetrahendron $= \frac{1}{6}(\vec{A}+\vec{B}).[(\vec{B}+\vec{C})\times(\vec{C}+\vec{A})]$

$$= \frac{1}{6}(\vec{A}+\vec{B}).[\vec{B}\times\vec{C} + \vec{B}\times\vec{A} + \vec{C}\times\vec{C} + \vec{C}\times\vec{A}] \qquad [\vec{C}\times\vec{C} = 0]$$

$$= \frac{1}{6}(\vec{A}+\vec{B}).(\vec{B}\times\vec{C} + \vec{B}\times\vec{A} + \vec{C}\times\vec{A})$$

$$= \frac{1}{6}[\vec{A}.(\vec{B}\times\vec{C}) + \vec{A}.(\vec{B}\times\vec{A}) + \vec{A}.(\vec{C}\times\vec{A}) + \vec{B}.(\vec{B}\times\vec{C}) + \vec{B}.(\vec{B}\times\vec{A}) + \vec{B}.(\vec{C}\times\vec{A})]$$

$$= \frac{1}{6}[\vec{A}.(\vec{B}\times\vec{C}) + \vec{B}.(\vec{C}\times\vec{A})] = \frac{1}{3}\vec{A}.(\vec{B}\times\vec{C})$$

$$= 2\times\frac{1}{6}[\vec{A}\ \vec{B}\ \vec{C}]$$

= 2 Volume of tetrahedron having $\vec{A}, \vec{B}, \vec{C}$, as concurrent edges. **Proved.**

Example 6. *Using vectors, establish Cramer's rule for solving these following linear equations*

$$a_1x + b_1y + c_1z = d_1,\ a_2x + b_2y + c_2z = d_2,\ a_3x + b_3y + c_3z = d_3$$

Solution. Here, we have,

$$a_1x + b_1y + c_1z = d_1 \qquad ...(1)$$

$$a_2x + b_2y + c_2z = d_2 \qquad ...(2)$$

$$a_3x + b_3y + c_3z = d_3 \qquad ...(3)$$

Multiplying equations (1), (2) and (3) by $\hat{i}, \hat{j}$ and $\hat{k}$ respectively, we get

$$(a_1x + b_1y + c_1z)\,\hat{i} = d_1\hat{i} \qquad ...(4)$$

$$(a_2x + b_2y + c_2z)\,\hat{j} = d_2\hat{j} \qquad ...(5)$$

$$(a_3x + b_3y + c_3z)\,\hat{k} = d_3\hat{k} \qquad ...(6)$$

Adding (4), (5) and (6), we have

$$(a_1\hat{i}+a_2\hat{j}+a_3\hat{k})\,x+(b_1\hat{i}+b_2\hat{j}+b_3\hat{k})\,y+(c_1\hat{i}+c_2\hat{j}+c_3\hat{k})\,z=d_1\hat{i}+d_2\hat{j}+d_3\hat{k}$$

$$(\vec{a}\,x+\vec{b}\,y+\vec{c}\,z) = \overline{d} \qquad ...(7)$$

$$[\vec{a}=a_1\hat{i}+a_2\hat{j}+a_3\hat{k},\ \vec{b} = b_1\hat{i}+b_2\hat{j}+b_3\hat{k},\ \text{and}\ \vec{c}=c_1\hat{i}+c_2\hat{j}+c_3\hat{k}\,]$$

Take scalar product of (7) with $(\vec{b}\times\vec{c})$

$$\vec{a}\,.(\vec{b}\times\vec{c})\,x+\vec{b}\,.(\vec{b}\times\vec{c})\,y+\vec{c}\,.(\vec{b}\times\vec{c})\,z=\overline{d}\,.(\vec{b}\times\vec{c})$$

$$\Rightarrow \qquad (\vec{a}\,\vec{b}\,\vec{c})\,x = (\vec{d}\,\vec{b}\,\vec{c}) \quad \Rightarrow x = \frac{(\vec{d}\,\vec{b}\,\vec{c})}{(\vec{a}\,\vec{b}\,\vec{c})}$$

Similarly, $\qquad (\vec{b}\,\vec{c}\,\vec{a})\,y = (\vec{d}\,\vec{c}\,\vec{a}) \quad \Rightarrow y = \dfrac{(\vec{d}\,\vec{c}\,\vec{a})}{(\vec{b}\,\vec{c}\,\vec{a})}$

$$(\vec{c}\,\vec{a}\,\vec{b})\,z = (\vec{d}\,\vec{a}\,\vec{b}) \quad \Rightarrow z = \frac{(\vec{d}\,\vec{a}\,\vec{b})}{(\vec{c}\,\vec{a}\,\vec{b})}$$

Thus $\qquad x = \dfrac{(\vec{d}\,\vec{b}\,\vec{c})}{(\vec{a}\,\vec{b}\,\vec{c})}, \qquad y=\dfrac{(\vec{d}\,\vec{c}\,\vec{a})}{(\vec{b}\,\vec{c}\,\vec{a})}, \qquad z=\dfrac{(\vec{d}\,\vec{a}\,\vec{b})}{(\vec{c}\,\vec{a}\,\vec{b})}$ **Proved.**

EXERCISE 1.1

1. Find the volume of the parallelopiped with adjacent sides.

$$\overline{OA} = 3\hat{i}-\hat{j},\quad \overline{OB}=\hat{j}+2\hat{k},\ \text{and}\ \overline{OC}=\hat{i}+5\hat{j}+4\hat{k}$$

extending from the origin of co-ordinates O. **Ans.** 20

2. Find the volume of the tetrahedron whose vertices are the points A (2, –1, –3), B (4, 1, 3) C (3, 2, –1) and D (1, 4, 2). **Ans.** $7\frac{1}{3}$

3. Choose y in order that the vectors $\vec{a}=7\hat{i}+y\hat{j}+\hat{k},\ \vec{b}=3\hat{i}+2\hat{j}+\hat{k}$,

$\vec{c}=5\hat{i}+3\hat{j}+\hat{k}$ are linearly dependent. **Ans.** $y = 4$

4. Prove that

$$[\vec{a}+\vec{b},\ \vec{b}+\vec{c},\ \vec{c}+\vec{a}]=2[\vec{a}\ \vec{b}\ \vec{c}]$$

1.18 COPLANARITY QUESTIONS

Example 7. *Find the volume of tetrahedron having vertices*

$$(-\hat{j}-\hat{k}),\ (4\hat{i}+5\hat{j}+q\hat{k}),\ (3\hat{i}+9\hat{j}+4\hat{k})\ \text{and}\ 4(-\hat{i}+\hat{j}+\hat{k}).$$

Also find the value of q for which these four points are coplanar.

Solution. Let $\vec{A} = -\hat{j}-\hat{k},\ \overline{B}=4\hat{i}+5\hat{j}+q\hat{k},\ \overline{C}=3\hat{i}+9\hat{j}+4\hat{k},\ \overline{D}=4(-\hat{i}+\hat{j}+\hat{k})$

$$\overline{AB} = \vec{B}-\vec{A}=4\hat{i}+5\hat{j}+q\hat{k}-(-\hat{j}-\hat{k})=4\hat{i}+6\hat{j}+(q+1)\hat{k}$$

$$\overline{AC} = \vec{C}-\vec{A}=(3\hat{i}+9\hat{j}+4\hat{k})-(-\hat{j}-\hat{k})=3\hat{i}+10\hat{j}+5\hat{k}$$

$$\overline{AD} = \vec{D}-\vec{A}=4(-\hat{i}+\hat{j}+\hat{k})-(-\hat{j}-\hat{k})=-4\hat{i}+5\hat{j}+5\hat{k}$$

$$\text{Volume of the tetrahedron} = \frac{1}{6}[\overline{AB}\ \overline{AC}\ \overline{AD}]$$

$$= \frac{1}{6}\begin{vmatrix} 4 & 6 & q+1 \\ 3 & 10 & 5 \\ -4 & 5 & 5 \end{vmatrix} = \frac{1}{6}\{4(50-25)-6(15+20)+(q+1)(15+40)\}$$

$$= \frac{1}{6}\{100-210+55(q+1)\} = \frac{1}{6}(-110+55+55q)$$

$$= \frac{1}{6}(-55+55q) = \frac{55}{6}(q-1)$$

If four points A, B, C and D are coplanar, then $(\overline{AB}\ \overline{AC}\ \overline{AD}) = 0$

i.e., Volume of the tetrahedron = 0

$$\Rightarrow \quad \frac{55}{6}(q-1) = 0 \quad \Rightarrow \quad q = 1$$ **Ans.**

Example 8. *Find m so that the vectors*

$2\hat{i}-4\hat{j}+5\hat{k};\ \hat{i}-m\hat{j}+\hat{k}$, *and* $3\hat{i}+2\hat{j}-5\hat{k}$ *are coplanar.*

Solution. Let
$$\vec{a} = 2\hat{i}-4\hat{j}+5\hat{k}$$
$$\vec{b} = \hat{i}-m\hat{j}+\hat{k}$$
$$\vec{c} = 3\hat{i}+2\hat{j}-5\hat{k}$$

$\vec{a}, \vec{b}$ and $\vec{c}$ are coplanar if $\vec{a}.(\vec{b}\times\vec{c})=0$

$$\vec{a}.(\vec{b}\times\vec{c}) = \begin{vmatrix} 2 & -4 & 5 \\ 1 & -m & 1 \\ 3 & 2 & -5 \end{vmatrix} = 0$$

$$0 = 2(5m-2) + 4(-5-3) + 5(2+3m)$$
$$0 = 10m - 4 - 32 + 10 + 15m$$
$$0 = 25m - 26$$
$$25m = 26$$
$$\therefore \quad m = \frac{26}{25}.$$ **Ans.**

Example 9. *Show that the vectors*

$$(5\vec{a}+6\vec{b}+7\vec{c}),(7\vec{a}-8\vec{b}+9\vec{c}) \text{ and } (3\vec{a}+20\vec{b}+5\vec{c})$$

are coplanar, $\vec{a},\vec{b},\vec{c}$ *being three non-collinear vectors.*

Solution. Let $\vec{\alpha} = 5\bar{a}+6\bar{b}+7\bar{c}$

$$\vec{\beta} = 7\bar{a} - 8\bar{b} + 9\bar{c}$$

$$\vec{\gamma} = 3\bar{a} + 20\bar{b} + 5\bar{c}$$

$$\vec{\alpha}.(\vec{\beta} \times \vec{\gamma}) = \begin{vmatrix} 5 & 6 & 7 \\ 7 & -8 & 9 \\ 3 & 20 & 5 \end{vmatrix}$$

$$= 5\,(-40 - 180) - 6\,(35 - 27) + 7\,(140 + 24)$$

$$= -5 \times 220 - 6 \times 8 + 7 \times 164$$

$$= -1100 - 48 + 1148 = 0$$

Hence $\vec{\alpha}, \vec{\beta}$ and $\vec{\gamma}$ are coplanar. **Proved.**

Example 10. *If four points whose position vectors are $\bar{a}, \bar{b}, \bar{c}, \bar{d}$ are coplanar, show that*

$$[\vec{a}\;\vec{b}\;\vec{c}] = [\vec{a}\;\vec{d}\;\vec{b}] + [\vec{a}\;\vec{d}\;\vec{c}] + [\vec{d}\;\vec{b}\;\vec{c}]$$ (*Nagpur University, Summer 2005*)

Solution. Let A, B, C, D be four points whose position vectors are $\vec{a}, \vec{b}, \vec{c}, \vec{d}$.

$$\vec{AD} = \vec{d} - \vec{a}, \quad \vec{BD} = \vec{d} - \vec{b} \quad \text{and} \quad \vec{CD} = \vec{d} - \vec{c}$$

If $\vec{AD}, \vec{BD}, \vec{CD}$ are coplanar, then

$$\vec{AD}.(\vec{BD} \times \vec{CD}) = 0$$

$$\Rightarrow \quad (\vec{d} - \vec{a}).[(\vec{d} - \vec{b}) \times (\vec{d} - \vec{c})] = 0$$

$$\Rightarrow \quad (\vec{d} - \vec{a}).[\vec{d} \times \vec{d} - \vec{d} \times \vec{c} - \vec{b} \times \vec{d} + \vec{b} \times \vec{c}] = 0$$

$$\Rightarrow \quad (\vec{d} - \vec{a}).[-\vec{d} \times \vec{c} - \vec{b} \times \vec{d} + \vec{b} \times \vec{c}] = 0$$

$$\Rightarrow -\vec{d}.(\vec{d} \times \vec{c}) - \vec{d}.(\vec{b} \times \vec{d}) + \vec{d}.(\vec{b} \times \vec{c}) + \vec{a}.(\vec{d} \times \vec{c}) + \vec{a}.(\vec{b} \times \vec{d}) - \vec{a}.(\vec{b} \times \vec{c}) = 0$$

$$\Rightarrow -0 + 0 + [\vec{d}\;\vec{b}\;\vec{c}] + [\vec{a}\;\vec{d}\;\vec{c}] + [\vec{a}\;\vec{b}\;\vec{d}] - [\vec{a}\;\vec{b}\;\vec{c}] = 0$$

$$\Rightarrow \quad [\vec{a}\;\vec{b}\;\vec{c}] = [\vec{a}\;\vec{b}\;\vec{d}] + [\vec{a}\;\vec{d}\;\vec{c}] + [\vec{d}\;\vec{b}\;\vec{c}]$$ **Proved.**

Example 11. *Assuming that $\vec{a}.(\vec{b} \times \vec{c}) \neq 0$ and $\vec{d} = x\vec{a} + y\vec{b} + z\vec{c}$.*

Find the values of x, y and z.

Solution. We have, $\vec{d} = x\vec{a} + y\vec{b} + z\vec{c}$ ...(1)

Taking dot product of (1) with $\vec{b} \times \vec{c}$, we get

$$[\vec{d}\;\vec{b}\;\vec{c}] = \vec{d}.(\vec{b} \times \vec{c}) = (x\vec{a} + y\vec{b} + z\vec{c}).(\vec{b} \times \vec{c})$$

$$= x\,\{\vec{a}.(\vec{b} \times \vec{c})\} + y\,\{\vec{b}.(\vec{b} \times \vec{c})\} + z\,\{\vec{c}.(\vec{b} \times \vec{c})\} = x[\vec{a}\;\vec{b}\;\vec{c}] + 0 + 0$$

$$\therefore \qquad x = \frac{[\vec{d}\,\vec{b}\,\vec{c}]}{[\vec{a}\,\vec{b}\,\vec{c}]}$$

Similarly taking dot product of (1) with $(\vec{c} \times \vec{a})$ and $(\vec{a} \times \vec{b})$

Separately, we get

$$y = \frac{[\vec{d}\,\vec{c}\,\vec{a}]}{[\vec{a}\,\vec{b}\,\vec{c}]} \text{ and } z = \frac{[\vec{d}\,\vec{a}\,\vec{b}]}{[\vec{a}\,\vec{b}\,\vec{c}]}$$ **Ans.**

EXERCISE 1.2

1. Determine λ such that

   $\bar{a} = \hat{i} + \hat{j} + \hat{k}, \bar{b} = 2\hat{i} - 4\hat{k}$, and $\bar{c} = \hat{i} + \lambda\hat{j} + 3\hat{k}$ are coplanar. **Ans.** $\lambda = 5/3$

2. Show that the four points

   $-6\hat{i} + 3\hat{j} + 2\hat{k}, 3\hat{i} - 2\hat{j} + 4\hat{k}, 5\hat{i} + 7\hat{j} + 3\hat{k}$ and $-13\hat{i} + 17\hat{j} - \hat{k}$ are coplanar.

3. Find the constant a such that the vectors

   $2\hat{i} - \hat{j} + \hat{k}, \hat{i} + 2\hat{j} - 3\hat{k}$, and $3\hat{i} + a\hat{j} + 5\hat{k}$ are coplanar. **Ans.** -4

4. Prove that four points

   $4\hat{i} + 5\hat{j} + \hat{k}, -(\hat{j} + \hat{k}), 3\hat{i} + 9\hat{j} + 4\hat{k}, 4(-\hat{i} + \hat{j} + \hat{k})$ are coplanar.

5. If the vectors $\vec{a}, \vec{b}$ and $\vec{c}$ are coplanar, show that

$$\begin{vmatrix} \vec{a} & \vec{b} & \vec{c} \\ \vec{a}.\vec{a} & \vec{a}.\vec{b} & \vec{a}.\vec{c} \\ \vec{b}.\vec{a} & \vec{b}.\vec{b} & \vec{b}.\vec{c} \end{vmatrix} = 0$$

1.19 VECTOR PRODUCT OF THREE VECTORS

Let $\vec{a}, \vec{b}$ and $\vec{c}$ be three vectors then their vector product is written as $\vec{a} \times (\vec{b} \times \vec{c})$.

Let $\vec{a} = a_1\hat{i} + a_2\hat{j} + a_3\hat{k},$ *(DU, IIrem 2014)*

$$\vec{b} = b_1\hat{i} + b_2\hat{j} + b_3\hat{k},$$

$$\vec{c} = c_1\hat{i} + c_2\hat{j} + c_3\hat{k}$$

$$\vec{a} \times (\vec{b} \times \vec{c}) = (a_1\hat{i} + a_2\hat{j} + a_3\hat{k}) \times (b_1\hat{i} + b_2\hat{j} + b_3\hat{k}) \times (c_1\hat{i} + c_2\hat{j} + c_3\hat{k})$$

$$= (a_1\hat{i} + a_2\hat{j} + a_3\hat{k}) \times [(b_2c_3 - b_3c_2)\hat{i} + (b_3c_1 - b_1c_3)\hat{j} + (b_1c_2 - b_2c_1)\hat{k}]$$

$$= [a_2(b_1c_2 - b_2c_1) - a_3(b_3c_1 - b_1c_3)]\hat{i} + [a_3(b_2c_3 - b_3c_2) - a_1(b_1c_2 - b_2c_1)]\hat{j}$$

$$+ [a_1(b_3c_1 - b_1c_3) - a_2(b_2c_3 - b_3c_2)]\hat{k}$$

$$= (a_1c_1 + a_2c_2 + a_3c_3)(b_1\hat{i} + b_2\hat{j} + b_3\hat{k}) - (a_1b_1 + a_2b_2 + a_3b_3)(c_1\hat{i} + c_2\hat{j} + c_3\hat{k})$$

$$= (\vec{a}\cdot\vec{c})\,\vec{b} - (\vec{a}\cdot\vec{b})\,\vec{c}.$$ **Ans.**

Example 12. *Let* $\vec{a} = \hat{i} + \hat{j} - \hat{k},\quad \vec{b} = \hat{i} - \hat{j} + \hat{k},\quad \vec{c} = \hat{i} - \hat{j} - \hat{k}.$

Find the vector $\vec{a} \times (\vec{b} \times \vec{c})$.

Solution. $\vec{a} \times (\vec{b} \times \vec{c}) = (\vec{a}\cdot\vec{c})\,\vec{b} - (\vec{a}\cdot\vec{b})\,\vec{c}$

$$= [(\hat{i} + \hat{j} - \hat{k})\cdot(\hat{i} - \hat{j} - \hat{k})](\hat{i} - \hat{j} + \hat{k}) - [(\hat{i} + \hat{j} - \hat{k})\cdot(\hat{i} - \hat{j} + \hat{k})](\hat{i} - \hat{j} - \hat{k})$$

$$= (1 - 1 + 1)(\hat{i} - \hat{j} + \hat{k}) - [(1 - 1 - 1)(\hat{i} - \hat{j} - \hat{k})]$$

$$= (\hat{i} - \hat{j} + \hat{k}) + (\hat{i} - \hat{j} - \hat{k}) = 2\hat{i} - 2\hat{j}$$ **Ans.**

Example 13. *Prove that :*

$$\vec{a} \times (\vec{b} \times \vec{c}) + \vec{b} \times (\vec{c} \times \vec{a}) + \vec{c} \times (\vec{a} \times \vec{b}) = 0$$ (*Nagpur University, Winter 2008*)

Solution. Here, we have

$$\vec{a} \times (\vec{b} \times \vec{c}) + \vec{b} \times (\vec{c} \times \vec{a}) + \vec{c} \times (\vec{a} \times \vec{b})$$

$$= [(\vec{a}\cdot\vec{c})\,\vec{b} - (\vec{a}\cdot\vec{b})\,\vec{c}] + [(\vec{b}\cdot\vec{a})\,\vec{c} - (\vec{b}\cdot\vec{c})\,\vec{a}] + [(\vec{c}\cdot\vec{b})\,\vec{a} - (\vec{c}\cdot\vec{a})\,\vec{b}]$$

$$= [(\vec{b}\cdot\vec{a})\,\vec{c} - (\vec{a}\cdot\vec{b})\,\vec{c}] + [(\vec{c}\cdot\vec{b})\,\vec{a} - (\vec{b}\cdot\vec{c})\,\vec{a}] + [(\vec{a}\cdot\vec{c})\,\vec{b} - (\vec{c}\cdot\vec{a})\,\vec{b}]$$

$$= [(\vec{a}\cdot\vec{b})\,\vec{c} - (\vec{a}\cdot\vec{b})\,\vec{c}] + [(\vec{b}\cdot\vec{c})\,\vec{a} - (\vec{b}\cdot\vec{c})\,\vec{a}] + [(\vec{c}\cdot\vec{a})\,\vec{b} - (\vec{c}\cdot\vec{a})\,\vec{b}]$$

$$= 0 + 0 + 0 = 0$$ **Proved.**

Example 14. *Prove that :*

$$\hat{i} \times (\vec{a} \times \hat{i}) + \hat{j} \times (\vec{a} \times \hat{j}) + \hat{k} \times (\vec{a} \times \hat{k}) = 2\,\vec{a}$$

Solution. Let $\vec{a} = a_1\hat{i} + a_2\hat{j} + a_3\hat{k}$

Now, L.H.S. $= \hat{i} \times (\vec{a} \times \hat{i}) + \hat{j} \times (\vec{a} \times \hat{j}) + \hat{k} \times (\vec{a} \times \hat{k})$

$$= \hat{i} \times \left[(a_1\hat{i} + a_2\hat{j} + a_3\hat{k}) \times \hat{i}\right] + \hat{j} \times \left[(a_1\hat{i} + a_2\hat{j} + a_3\hat{k}) \times \hat{j}\right] + \hat{k} \times \left[(a_1\hat{i} + a_2\hat{j} + a_3\hat{k}) \times \hat{k}\right]$$

$$= \hat{i} \times \left[a_1(\hat{i} \times \hat{i}) + a_2(\hat{j} \times \hat{i}) + a_3(\hat{k} \times \hat{i})\right] + \hat{j} \times \left[a_1(\hat{i} \times \hat{j}) + a_2(\hat{j} \times \hat{j}) + a_3(\hat{k} \times \hat{j})\right] + \hat{k} \times \left[a_1(\hat{i} \times \hat{k}) + a_2(\hat{j} \times \hat{k}) + a_3(\hat{k} \times \hat{k})\right]$$

$$= \hat{i} \times \left[0 - a_2\hat{k} + a_3\hat{j}\right] + \hat{j} \times \left[a_1\hat{k} + 0 - a_3\hat{i}\right] + \hat{k} \times \left[-a_1\hat{j} + a_2\hat{i} + 0\right]$$

$$= -a_2(\hat{i} \times \hat{k}) + a_3(\hat{i} \times \hat{j}) + a_1(\hat{j} \times \hat{k}) - a_3(\hat{j} \times \hat{i}) - a_1(\hat{k} \times \hat{j}) + a_2(\hat{k} \times \hat{i})$$

$$= a_2 \hat{j} + a_3 \hat{k} + a_1 \hat{i} + a_3 \hat{k} + a_1 \hat{i} + a_2 \hat{j} = 2a_1 \hat{i} + 2a_2 \hat{j} + 2a_3 \hat{k}$$

$$= 2(a_1 \hat{i} + a_2 \hat{j} + a_3 \hat{k}) = 2\vec{a}$$

Proved.

Example 15. *Show that for any scalar λ, the vectors $\vec{x}, \vec{y}$ given by*

$$\vec{x} = \lambda \vec{a} + \frac{q(\vec{a} \times \vec{b})}{a^2},\ \vec{y} = \frac{(1 - p\lambda)}{q}\vec{a} - \frac{p(\vec{a} \times \vec{b})}{a^2}$$ *satisfy the equations*

$$p\vec{x} + q\vec{y} = \vec{a} \text{ and } \vec{x} \times \vec{y} = \vec{b}.$$

Solution. The given equations are

$$p\vec{x} + q\vec{y} = \vec{a} \qquad ...(1)$$

$$\vec{x} \times \vec{y} = \vec{b} \qquad ...(2)$$

Multiplying equation (1) vectorially by $\vec{x}$, we get

$$\vec{x} \times (p\vec{x} + q\vec{y}) = \vec{x} \times \vec{a}$$

$$p(\vec{x} \times \vec{x}) + q(\vec{x} \times \vec{y}) = \vec{x} \times \vec{a}$$

$$q(\vec{x} \times \vec{y}) = \vec{x} \times \vec{a}, \qquad \text{as } \vec{x} \times \vec{x} = 0$$

$$\vec{x} \times \vec{a} = q\vec{b}, \qquad [\text{From (2) } \vec{x} \times \vec{y} = \vec{b}] \quad ...(3)$$

Multiplying (3) vectorially by $\vec{a}$, we have

$$\vec{a} \times (\vec{x} \times \vec{a}) = \vec{a} \times q\vec{b}$$

$$(\vec{a} \cdot \vec{a})\vec{x} - (\vec{a} \cdot \vec{x})\vec{a} = q(\vec{a} \times \vec{b})$$

$$a^2\vec{x} - (\vec{a} \cdot \vec{x})\vec{a} = q(\vec{a} \times \vec{b}) \quad \Rightarrow \quad a^2\vec{x} = (\vec{a} \cdot \vec{x})\vec{a} + q(\vec{a} \times \vec{b})$$

$$\vec{x} = \frac{(\vec{a} \cdot \vec{x})\vec{a}}{a^2} + \frac{q(\vec{a} \times \vec{b})}{a^2}$$

$$\vec{x} = \lambda \vec{a} + \frac{q(\vec{a} \times \vec{b})}{a^2} \qquad \text{where } \lambda = \frac{\vec{a} \cdot \vec{x}}{a^2}$$

Substituting the value of $\vec{x}$ in (1), we get $p\left\{\lambda \vec{a} + \frac{q(\vec{a} \times \vec{b})}{a^2}\right\} + q\vec{y} = \vec{a}$

$$q\vec{y} = \vec{a} - p\left\{\lambda \vec{a} + \frac{q(\vec{a} \times \vec{b})}{a^2}\right\}$$

Proved.

$$\vec{y} = \frac{(1 - p\lambda)\vec{a}}{q} - \frac{p(\vec{a} \times \vec{b})}{a^2}$$

EXERCISE 1.3

1. Show that $\vec{a} \times (\vec{b} \times \vec{a}) = (\vec{a} \times \vec{b}) \times \vec{a}$

2. Write the correct answer

(*a*) $(\vec{A} \times \vec{B}) \times \vec{C}$ lies in the plane of

(*i*) $\vec{A}$ and B (*ii*) $\vec{B}$ and $\vec{C}$ (*iii*) $\vec{C}$ and $\vec{A}$ **Ans.** (*ii*)

(*b*) The value of $\vec{a} \cdot \left((\vec{b} + \vec{c}) \times (\vec{a} + \vec{b} + \vec{c}) \right)$ is

(*i*) Zero (*ii*) $[\vec{a}, \vec{b}, \vec{c}] + [\vec{b}, \vec{c}, \vec{a}]$

(*iii*) $[\vec{a}, \vec{b}, \vec{c}]$ (*iv*) None of these **Ans.** (*i*)

1.20 SCALAR PRODUCT OF FOUR VECTORS

Prove the identity

$$(\vec{a} \times \vec{b}) \cdot (\vec{c} \times \vec{d}) = (\vec{a} \cdot \vec{c})(\vec{b} \cdot \vec{d}) - (\vec{a} \cdot \vec{d})(\vec{b} \cdot \vec{c})$$

Proof. $(\vec{a} \times \vec{b}) \cdot (\vec{c} \times \vec{d}) = (\vec{a} \times \vec{b}) \cdot \vec{r}$

$= \vec{a} \cdot (\vec{b} \times \vec{r})$ dot and cross can be interchanged. Put $\vec{c} \times \vec{d} = \vec{r}$

$$= \vec{a} \cdot [\vec{b} \times (\vec{c} \times \vec{d})] = \vec{a} \cdot [(\vec{b} \cdot \vec{d})\, \vec{c} - (\vec{b} \cdot \vec{c})\, \vec{d}]$$

$$= (\vec{a} \cdot \vec{c})(\vec{b} \cdot \vec{d}) - (\vec{a} \cdot \vec{d})(\vec{b} \cdot \vec{c})$$

$$= \begin{vmatrix} \vec{a} \cdot \vec{c} & \vec{a} \cdot \vec{d} \\ \vec{b} \cdot \vec{c} & \vec{b} \cdot \vec{d} \end{vmatrix}$$

Proved.

EXERCISE 1.4

1. If $\vec{a} = 2i + 3j - k$, $\vec{b} = -i + 2j - 4k$, $\vec{c} = i + j + k$, find $(\vec{a} \times \vec{b}) \cdot (\vec{a} \times \vec{c})$. **Ans.** –74

2. Prove that $(\vec{a} \times \vec{b}) \cdot (\vec{a} \times \vec{c}) = a^2 (\vec{b} \cdot \vec{c}) - (\vec{a} \cdot \bar{b})(\vec{a} \cdot \vec{c})$.

1.21 VECTOR PRODUCT OF FOUR VECTORS

Let $\vec{a}, \vec{b}, \vec{c}$ and $\vec{d}$ be four vectors then their vector product is written as

$$(\vec{a} \times \vec{b}) \times (\vec{c} \times \vec{d})$$

Now, $(\vec{a} \times \vec{b}) \times (\vec{c} \times \vec{d}) = \vec{r} \times (\vec{c} \times \vec{d})$ [Put $\vec{a} \times \vec{b} = \vec{r}$]

$$= (\vec{r} \cdot \vec{d})\, \vec{c} - (\vec{r} \cdot \vec{c})\, \vec{d}$$

$$= [(\vec{a} \times \vec{b}) \cdot \vec{d}]\,\vec{c} - [(\vec{a} \times \vec{b}) \cdot \vec{c}]\,\vec{d}$$

$$= [\vec{a}\;\vec{b}\;\vec{d}]\,\vec{c} - [\vec{a}\;\vec{b}\;\vec{c}]\,\vec{d}$$

$\therefore\ (\vec{a} \times \vec{b}) \times (\vec{c} \times \vec{d})$ lies in the plane of $\vec{c}$ and $\vec{d}$. ...(1)

Again, $(\vec{a} \times \vec{b}) \times (\vec{c} \times \vec{d}) = (\vec{a} \times \vec{b}) \times \vec{s}$ [Put = $\vec{c} \times \vec{d} = \vec{s}$]

$$= -\,\vec{s} \times (\vec{a} \times \vec{b}) = -(\vec{s} \cdot \vec{b})\,\vec{a} + (\vec{s} \cdot \vec{a})\,\vec{b}$$

$$= -[(\vec{c} \times \vec{d}) \cdot \vec{b}]\,\vec{a} + [(\vec{c} \times \vec{d}) \cdot \vec{a}]\,\vec{b} = -[(\vec{b}\;\vec{c}\;\vec{d}]\,\vec{a} + [\vec{a}\;\vec{c}\;\vec{d}]\,\vec{b}$$

$\therefore\ (\vec{a} \times \vec{b}) \times (\vec{c} \times \vec{d})$ lies in the plane of $\vec{a}$ and $\vec{b}$. ...(2)

Geometrical interpretation : From (1) and (2) we conclude that $(\vec{a} \times \vec{b}) \times (\vec{c} \times \vec{d})$ is a vector parallel to the line of intersection of the plane containing $\vec{a}, \vec{b}$ and plane containing $\vec{c}, \vec{d}$.

Example 16. *Show that*

$$(\vec{B} \times \vec{C}) \times (\vec{A} \times \vec{D}) + (\vec{C} \times \vec{A}) \times (\vec{B} \times \vec{D}) + (\vec{A} \times \vec{B}) \times (\vec{C} \times \vec{D}) = -2\,(\vec{A}\;\vec{B}\;\vec{C})\,\vec{D}$$

Solution. L.H.S. = $(\vec{B} \times \vec{C}) \times (\vec{A} \times \vec{D}) + (\vec{C} \times \vec{A}) \times (\vec{B} \times \vec{D}) + (\vec{A} \times \vec{B}) \times (\vec{C} \times \vec{D})$

$$= [(\vec{B}\;\vec{C}\;\vec{D})\,\vec{A} - (\vec{B}\;\vec{C}\;\vec{A})\,\vec{D}] + [(\vec{C}\;\vec{A}\;\vec{D})\,\vec{B} - (\vec{C}\;\vec{A}\;\vec{B})\,\vec{D}] + [(-\vec{B}\;\vec{C}\;\vec{D})\,\vec{A} + (\vec{A}\;\vec{C}\;\vec{D})\,\vec{B}]$$

$$= (\vec{B}\;\vec{C}\;\vec{D})\,\vec{A} - (\vec{B}\;\vec{C}\;\vec{D})\,\vec{A} + (\vec{C}\;\vec{A}\;\vec{D})\,\vec{B} + (\vec{A}\;\vec{C}\;\vec{D})\,\vec{B} - (\vec{B}\;\vec{C}\;\vec{A})\,\vec{D} - (\vec{C}\;\vec{A}\;\vec{B})\,\vec{D}$$

$$= -(\vec{A}\;\vec{C}\;\vec{D})\,\vec{B} + (\vec{A}\;\vec{C}\;\vec{D})\,\vec{B} - (\vec{A}\,\vec{B}\;\vec{C})\,\vec{D} - (\vec{A}\;\vec{B}\;\vec{C})\,\vec{D}$$

$= -2\,(\vec{A}\,\vec{B}\;\vec{C})\,\vec{D}$ = R.H.S. **Proved.**

Example 17. *Prove that*

$$(\vec{a} \times \vec{b}) \cdot \left\{(\vec{b} \times \vec{c}) \times (\vec{c} \times \vec{a})\right\} = [\vec{a} \cdot (\vec{b} \times \vec{c})]^2$$

Solution. L.H.S. = $(\vec{a} \times \vec{b}) \cdot \left\{(\vec{b} \times \vec{c}) \times (\vec{c} \times \vec{a})\right\}$ ...(1)

By applying the formula of vector triple product on $(\vec{b} \times \vec{c}) \times (\vec{c} \times \vec{a})$ in (1), we get

$$\text{L.H.S.} = (\vec{a} \times \vec{b}) \cdot \left[\left\{(\vec{b}\;\vec{c}\;\vec{a})\,\vec{c} - (\vec{b}\;\vec{c}\;\vec{c})\,\vec{a}\right\}\right]$$

$$= (\vec{a} \times \vec{b}) \cdot [(\vec{a}\;\vec{b}\;\vec{c})\,\vec{c} - 0\,(\vec{a})] = (\vec{a}\;\vec{b}\;\vec{c})[(\vec{a} \times \vec{b}) \cdot \vec{c}]$$

$= [\vec{a}\,(\vec{b} \times \vec{c})][\vec{a} \cdot (\vec{b} \times \vec{c})] = [\vec{a} \cdot (\vec{b} \times \vec{c})]^2$ = R.H.S. **Proved.**

Example 18. *Prove that:*

$$\left[\left(\vec{A}\times\vec{B}\right)\times\vec{C}\right]\times\vec{D}+\left[\left(\vec{B}\times\vec{A}\right)\times\vec{D}\right]\times\vec{C}+\left[\left(\vec{C}\times\vec{D}\right)\times\vec{A}\right]\times\vec{B}+\left[\left(\vec{D}\times\vec{C}\right)\times\vec{B}\right]\times\vec{A}=0$$

(Delhi University April, 2010)

Solution.

$$\text{L.H.S.}=\left[\left(\vec{A}\times\vec{B}\right)\times\vec{C}\right]\times\vec{D}+\left[\left(\vec{B}\times\vec{A}\right)\times\vec{D}\right]\times\vec{C}+\left[\left(\vec{C}\times\vec{D}\right)\times\vec{A}\right]\times\vec{B}+\left[\left(\vec{D}\times\vec{C}\right)\times\vec{B}\right]\times\vec{A}$$

$$=\left[\left(\vec{A}.\vec{C}\right).\vec{B}-\left(\vec{B}.\vec{C}\right)\vec{A}\right]\times\vec{D}+\left[\left(\vec{B}.\vec{D}\right).\vec{A}-\left(\vec{A}.\vec{D}\right)\vec{B}\right]\times\vec{C}$$

$$+\left[\left(\vec{C}.\vec{A}\right).\vec{D}-\left(\vec{D}.\vec{A}\right)\vec{C}\right]\times\vec{B}+\left[\left(\vec{D}.\vec{B}\right).\vec{C}-\left(\vec{C}.\vec{B}\right)\vec{D}\right]\times\vec{A}$$

$$\left[\left(\vec{a}\times\vec{b}\right)\times\vec{c}=\left(\vec{a}.\vec{c}\right)\vec{b}-\left(\vec{b}.\vec{c}\right)\vec{a}\right]$$

$$=\left(\vec{A}.\vec{C}\right)\left(\vec{B}\times\vec{D}\right)-\left(\vec{B}.\vec{C}\right)\left(\vec{A}\times\vec{D}\right)+\left(\vec{B}.\vec{D}\right)\left(\vec{A}\times\vec{C}\right)-\left(\vec{A}.\vec{D}\right)\left(\vec{B}\times\vec{C}\right)$$

$$\left(\vec{C}.\vec{A}\right)\left(\vec{D}\times\vec{B}\right)-\left(\vec{D}.\vec{A}\right)\left(\vec{C}\times\vec{B}\right)+\left(\vec{D}.\vec{B}\right)\left(\vec{C}\times\vec{A}\right)-\left(\vec{C}.\vec{B}\right)\left(\vec{D}\times\vec{A}\right)$$

$$=\left(\vec{A}.\vec{C}\right)\left[\left(\vec{B}\times\vec{D}\right)+\left(\vec{D}\times\vec{B}\right)\right]-\left(\vec{B}.\vec{C}\right)\left[\left(\vec{A}\times\vec{D}\right)+\left(\vec{D}\times\vec{A}\right)\right]$$

$$+\left(\vec{B}.\vec{D}\right)\left[\left(\vec{A}\times\vec{C}\right)+\left(\vec{C}\times\vec{A}\right)\right]-\left(\vec{A}.\vec{D}\right)\left[\left(\vec{B}\times\vec{C}\right)+\left(\vec{C}\times\vec{B}\right)\right]\quad\left[\because\vec{a}\times\vec{b}+\vec{b}\times\vec{a}=0\right]$$

$$=0$$

Proved.

EXERCISE 1.5

Show that:

1. $(\vec{b}\times\vec{c})\times(\vec{c}\times\vec{a})=\vec{c}\,(\vec{a}\;\vec{b}\;\vec{c})$ when $(\vec{a}\;\vec{b}\;\vec{c})$ stands for scalar triple product.
2. $[\vec{b}\times\vec{c},\quad\vec{c}\times\vec{a},\quad\vec{a}\times\vec{b})]=[\vec{a}\;\vec{b}\;\vec{c}]^2$
3. $\vec{d}\,[\vec{a}\times\{\vec{b}\times(\vec{c}\times\vec{d})\}]=[(\vec{b}\,.\,\vec{d})[\vec{a}\,.(\vec{c}\times\vec{d})]$
4. $\vec{a}\,[\vec{a}\times[\vec{a}\times(\vec{a}\times\vec{b})]=a^2\,(\vec{b}\times\vec{a})$
5. $[(\vec{a}\times\vec{b})\times(\vec{a}\times\vec{c})].\,\vec{d}=(\vec{a}\,.\,\vec{d})[\vec{a}\;\vec{b}\;\vec{c}]$
6. $2a^2=\left|\vec{a}\times\hat{i}\right|^2+\left|\vec{a}\times\hat{j}\right|^2+\left|\vec{a}\times\hat{k}\right|^2$
7. $\vec{a}\times\vec{b}=[(\hat{i}\times\vec{a}).\,\vec{b}]\,\hat{i}+[(\hat{j}\times\vec{a}).\,\vec{b}]\,\hat{j}+[(\hat{k}\times\vec{a}).\,\vec{b}]\,\hat{k}$
8. $\vec{p}\times[(\vec{a}\times\vec{q})\times(\vec{b}\times\vec{r})]+\vec{q}\times[(\vec{a}\times\vec{r})\times(\vec{b}\times\vec{p})]+\vec{r}\times[(\vec{a}\times\vec{p})\times(\vec{b}\times\vec{q})]=0$

CHAPTER

Differentiation of Vectors

(Point function, gradient, Divergence and Curl of a Vector and their Physical Interpretations)

2.1 VECTOR FUNCTION

If vector r is a function of a scalar variable t, then we write

$$\vec{r} = \vec{r}(t)$$

If a particle is moving along a curved path then the position vector $\vec{r}$ of the particle is a function of t. If the component of $f(t)$ along x-axis, y-axis, z-axis are $f_1(t), f_2(t), f_3(t)$ respectively. Then,

$$\overline{f(t)} = f_1(t)\hat{i} + f_2(t)\hat{j} + f_3(t)\hat{k}$$

2.2 DIFFERENTIATION OF VECTORS

Let O be the origin and P be the position of a moving particle at time t.

Let $\qquad \overrightarrow{OP} = \vec{r}$

Let Q be the position of the particle at the time $t + \delta t$ and the position vector of Q is $\overrightarrow{OQ} = \vec{r} + \delta\vec{r}$

$$\overrightarrow{PQ} = \overrightarrow{OQ} - \overrightarrow{OP}$$

$$= (\vec{r} + \delta\vec{r}) - \vec{r} = \delta\vec{r}$$

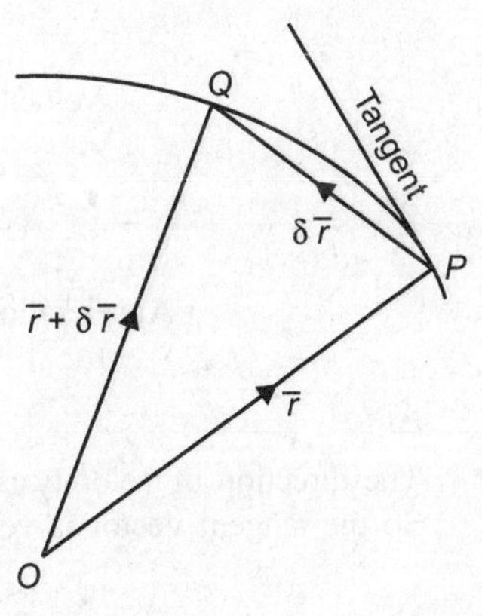

$\dfrac{\delta\vec{r}}{\delta t}$ is a vector. As $\delta t \to 0$, Q tends to P and the chord becomes the tangent at P.

We define $\qquad \dfrac{d\vec{r}}{dt} = \lim\limits_{\delta t \to 0} \dfrac{\delta\vec{r}}{\delta t}$, then

$\dfrac{d\vec{r}}{dt}$ is a vector in the direction of the *tangent* at P.

$\dfrac{d\vec{r}}{dt}$ is also called the differential coefficient of $\vec{r}$ with respect to 't'.

Similarly, $\dfrac{d^2\vec{r}}{dt^2}$ is the second order derivative of $\vec{r}$.

$\frac{d\vec{r}}{dt}$ gives the velocity of the particle at P, which is along the tangent to its path. Also

$\frac{d^2\vec{r}}{dt^2}$ gives the *acceleration* of the particle at P.

2.3 FORMULAE OF DIFFERENTIATION

(*i*) $\frac{d}{dt}(\vec{F}+\vec{G})=\frac{d\vec{F}}{dt}+\frac{d\vec{G}}{dt}$ (*ii*) $\frac{d}{dt}(\vec{F}\phi)=\frac{d\vec{F}}{dt}\phi+\vec{F}\frac{d\phi}{dt}$ (*U.P. I semester, Dec. 2005*)

(*iii*) $\frac{d}{dt}(\vec{F}.\vec{G})=\vec{F}.\frac{d\vec{G}}{dt}+\frac{d\vec{F}}{dt}.\overline{G}$ (*iv*) $\frac{d}{dt}(\vec{F}\times\vec{G})=\vec{F}\times\frac{d\vec{G}}{dt}+\frac{d\vec{F}}{dt}\times\vec{G}$

(*v*) $\frac{d}{dt}[\vec{a}\,\vec{b}\,\vec{c}]=\left[\frac{d\vec{a}}{dt}\vec{b}\,\vec{c}\right]+\left[\vec{a}\frac{d\vec{b}}{dt}\vec{c}\right]+\left[\vec{a}\,\vec{b}\frac{d\vec{c}}{dt}\right]$

(*vi*) $\frac{d}{dt}[\vec{a}\times(\vec{b}\times\vec{c})]=\frac{d\vec{a}}{dt}\times(\vec{b}\times\vec{c})+\vec{a}\times\left(\frac{d\vec{b}}{dt}\times\vec{c}\right)+\vec{a}\times\left(\vec{b}\times\frac{d\vec{c}}{dt}\right)$

The order of the functions $\vec{F},\vec{G}$ is not to be changed.

Example 1. *A particle moves along the curve* $\overline{r}=(t^3-4t)\hat{i}+(t^2+4t)\hat{j}+(8t^2-3t^3)\hat{k}$, *where t is the time. Find the magnitude of the tangential components of its acceleration at t = 2.* (*Nagpur University, Summer 2005*)

Solution. We have, $\vec{r} = (t^3-4t)\hat{i}+(t^2+4t)\hat{j}+(8t^2-3t^3)\hat{k}$

$$\text{Velocity} = \frac{d\vec{r}}{dt}=(3t^2-4)\hat{i}+(2t+4)\hat{j}+(16t-9t^2)\hat{k}$$

At $t = 2$, Velocity $= 8\hat{i}+8\hat{j}-4\hat{k}$

$$\text{Acceleration} = \vec{a} = \frac{d^2\vec{r}}{dt^2}=6t\,\hat{i}+2\hat{j}+(16-18t)\hat{k}$$

At $t = 2$ $\vec{a}=12\hat{i}+2\hat{j}-20\hat{k}$

The direction of velocity is along tangent.
So the tangent vector is velocity.

Unit tangent vector, $\hat{T} = \frac{\vec{v}}{|v|}=\frac{8\hat{i}+8\hat{j}-4\hat{k}}{\sqrt{64+64+16}}=\frac{8\hat{i}+8\hat{j}-4\hat{k}}{12}=\frac{2\hat{i}+2\hat{j}-\hat{k}}{3}$

Tangential component of acceleration, $a_t = \overline{a}.\hat{T}$

$$= (12\hat{i}+2\hat{j}-20\hat{k}).\frac{2\hat{i}+2\hat{j}-\hat{k}}{3}=\frac{24+4+20}{3}=\frac{48}{3}=16$$ **Ans.**

Example 2. *At any point of the curve x = 3 cos t, y = 3 sin t, z = 4 t, find*

(*i*) *Tangent vector* (*ii*) *Unit tangent vector* (*R.G.P.V., Bhopal, II Semester June 2007*)

Solution. We have, $x = 3\cos t$, $y = 3\sin t$, $z = 4t$

$$\vec{r} = x\hat{i} + y\hat{j} + z\hat{k} \quad \Rightarrow \quad \vec{r} = (3\cos t)\hat{i} + (3\sin t)\hat{j} + (4t)\hat{k}$$

(*i*) $$\frac{d\vec{r}}{dt} = (-3\sin t)\hat{i} + (3\cos t)\hat{j} + 4\hat{k}$$

which is the required tangent vector.

Magnitude of tangent vector $= \sqrt{(-3\sin t)^2 + (3\cos t)^2 + (4)^2} = 5.$

(*ii*) Unit tangent vector $= \frac{1}{5}(-3\sin t\,\hat{i} + 3\cos t\,\hat{j} + 4\hat{k})$ **Ans.**

Example 3. *Show that* $\frac{d}{dt}\left[\vec{u}\,\frac{d\vec{u}}{dt}\,\frac{d^2\vec{u}}{dt^2}\right] = \left[\vec{u}\,\frac{d\vec{u}}{dt}\,\frac{d^3\vec{u}}{dt^3}\right]$

Solution. We know that $\frac{d}{dt}(x\,y\,z) = \frac{dx}{dt}yz + x\frac{dy}{dt}z + xy\frac{dz}{dt}$

$$\frac{d}{dt}\left[\vec{u}\,\frac{d\vec{u}}{dt}\,\frac{d^2\vec{u}}{dt^2}\right] = \left[\frac{d\vec{u}}{dt}\,\frac{d\vec{u}}{dt}\,\frac{d^2\vec{u}}{dt}\right] + \left[\vec{u}\,\frac{d^2\vec{u}}{dt^2}\,\frac{d^2\vec{u}}{dt^2}\right] + \left[\vec{u}\,\frac{d\vec{u}}{dt}\,\frac{d^3\vec{u}}{dt^3}\right]$$

$$= 0 + 0 + \left[\vec{u}\,\frac{d\vec{u}}{dt}\,\frac{d^3\vec{u}}{dt^3}\right] = \left[\vec{u}\,\frac{d\vec{u}}{dt}\,\frac{d^3\vec{u}}{dt^3}\right]$$ **Proved.**

Example 4. *If* $\frac{d\vec{a}}{dt} = \vec{u} \times \vec{a}$ *and* $\frac{d\vec{b}}{dt} = \vec{u} \times \vec{b}$ *then prove that* $\frac{d}{dt}[\vec{a} \times \vec{b}] = \vec{u} \times (\vec{a} \times \vec{b})$

(*M.U. 2009*)

Solution. We have,

$$\frac{d}{dt}[\vec{a} \times \vec{b}] = \vec{a} \times \frac{d\vec{b}}{dt} + \frac{d\vec{a}}{dt} \times \vec{b} = \vec{a} \times (\vec{u} \times \vec{b}) + (\vec{u} \times \vec{a}) \times \vec{b}$$

$$= \vec{a} \times (\vec{u} \times \vec{b}) - \vec{b} \times (\vec{u} \times \vec{a})$$

$$= (\vec{a}.\vec{b})\,\vec{u} - (\vec{a}.\vec{u})\,\vec{b} - [(\vec{b}.\vec{a})\,\vec{u} - (\vec{b}.\vec{u})\,\vec{a}]$$

(Vector triple product)

$$= (\vec{a}.\vec{b})\,\vec{u} - (\vec{u}.\vec{a})\,\vec{b} - (\vec{a}.\vec{b})\,\vec{u} + (\vec{u}.\vec{b})\,\vec{a}$$

$$= (\vec{u}.\vec{b})\,\vec{a} - (\vec{u}.\vec{a})\,\vec{b}$$

$$= \vec{u} \times (\vec{a} \times \vec{b})$$ **Proved.**

Example 5. *Find the angle between the surface* $x^2 + y^2 + z^2 = 9$ *and* $z = x^2 + y^2 - 3$ *at* $(2, -1, 2)$. (*M.D.U. Dec. 2009*)

Solution. Here, we have

$$x^2 + y^2 + z^2 = 9 \quad \text{...(1)}$$

$$z = x^2 + y^2 - 3 \quad \text{...(2)}$$

Normal to (1) $\eta_1 = \Delta(x^2 + y^2 + z^2 - 9)$

$$= \left(\hat{i}\frac{\partial}{\partial x} + \hat{j}\frac{\partial}{\partial y} + \hat{k}\frac{\partial}{\partial z}\right)(x^2 + y^2 + z^2 - 9) = 2x\hat{i} + 2y\hat{j} + 2z\hat{k}$$

Normal to (1) at (2, –1, 2), $\eta_1 = 4\hat{i} - 2\hat{j} + 4\hat{k}$ ...(3)

Normal to (2), $\eta_2 = \Delta(z - x^2 - y^2 + 3)$

$$= \left(\hat{i}\frac{\partial}{\partial x} + \hat{j}\frac{\partial}{\partial y} + \hat{k}\frac{\partial}{\partial z}\right)(z - x^2 - y^2 + 3) = -2x\hat{i} - 2y\hat{j} + \hat{k}$$

Normal to (2) at (2, – 1, 2), $\eta_2 = -4\hat{i} + 2\hat{j} + \hat{k}$ ...(4)

$$\eta_1 . \eta_2 = |\eta_1||\eta_2| \cos\theta$$

$$\cos\theta = \frac{\eta_1.\eta_2}{|\eta_1||\eta_2|} = \frac{(4\hat{i} - 2\hat{j} + 4\hat{k}).(-4\hat{i} + 2\hat{j} + \hat{k})}{|4\hat{i} - 2\hat{j} + 4\hat{k}||-4\hat{i} + 2\hat{j} + \hat{k}|} = \frac{-16 - 4 + 4}{\sqrt{16+4+16}\sqrt{16+4+1}}$$

$$= \frac{-16}{6\sqrt{21}} = \frac{-8}{3\sqrt{21}}$$

$$\theta = \cos^{-1}\left(\frac{-8}{3\sqrt{21}}\right)$$

Hence the angle between (1) and (2) $\cos^{-1}\left(\frac{-8}{3\sqrt{21}}\right)$ **Ans.**

EXERCISE 2.1

1. The coordinates of a moving particle are given by $x = 4t - \frac{t^2}{2}$ and $y = 3 + 6t - \frac{t^3}{6}$. Find the velocity and acceleration of the particle when $t = 2$ secs. **Ans.** 4.47, 2.24
2. A particle moves along the curve
$$x = 2t^2, \quad y = t^2 - 4t \quad \text{and} \quad z = 3t - 5$$
where t is the time. Find the components of its velocity and acceleration at time $t = 1$, in the direction $\hat{i} - 3\hat{j} + 2\hat{k}$. **Ans.** $\frac{8\sqrt{14}}{7}, -\frac{\sqrt{14}}{7}$
3. Find the unit tangent and unit normal vector at $t = 2$ on the curve $x = t^2 - 1$, $y = 4t - 3$, $z = 2t^2 - 6t$ where t is any variable. **Ans.** $\frac{1}{3}(2\hat{i} + 2\hat{j} + \hat{k})$, $\frac{1}{3\sqrt{5}}(2\hat{i} + 2\hat{k})$
4. Prove that $\frac{d}{dt}(\vec{F} \times \vec{G}) = \vec{F} \times \frac{d\vec{G}}{dt} + \frac{d\vec{F}}{dt} \times \vec{G}$
5. Find the angle between the tangents to the curve $\bar{r} = t^2\hat{i} + 2t\hat{j} - t^3\hat{k}$, at the points $t = \pm 1$.
Ans. $\cos^{-1}\left(\frac{9}{17}\right)$
6. If the surface $5x^2 - 2byz = 9x$ be orthogonal to the surface $4x^2y + z^3 = 4$ at the point (1, –1, 2) then b is equal to
(*a*) 0 (*b*) 1 (*c*) 2 (*d*) 3 (*AMIETE, Dec. 2009*) **Ans.** (*b*)

2.4 SCALAR AND VECTOR POINT FUNCTIONS

Point function. A variable quantity whose value at any point in a region of space depends upon the position of the point, is called a *point function*. There are two types of point functions.

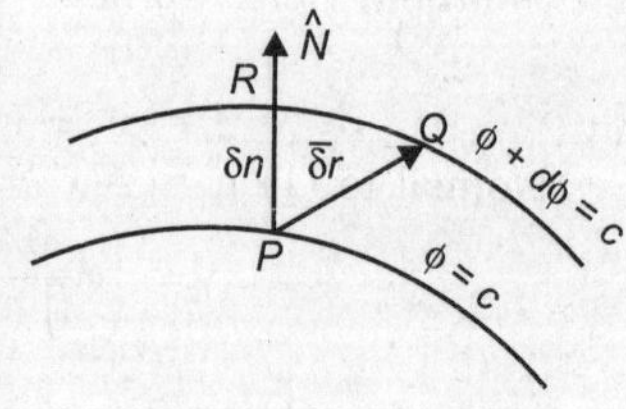

(*i*) **Scalar point function.** If to each point $P(x, y, z)$ of a region R in space there corresponds a unique scalar $f(P)$, then f is called a scalar point function. *For example*, the temperature distribution in a heated body, density of a body and potential due to gravity are the examples of a scalar point function.

(*ii*) **Vector point function.** If to each point $P(x, y, z)$ of a region R in space there corresponds a unique vector $f(P)$, then f is called a *vector point function.* The velocity of a moving fluid, gravitational force are the examples of vector point function.

Vector Differential Operator Del *i.e.* ∇

The vector differential operator Del is denoted by ∇. It is defined as

$$\nabla = \hat{i}\frac{\partial}{\partial x} + \hat{j}\frac{\partial}{\partial y} + \hat{k}\frac{\partial}{\partial z}$$

2.5 GRADIENT OF A SCALAR FUNCTION

If $\phi(x, y, z)$ be a scalar function then $\hat{i}\frac{\partial \phi}{\partial x} + \hat{j}\frac{\partial \phi}{\partial y} + \hat{k}\frac{\partial \phi}{\partial z}$ is called the gradient of the scalar function ϕ.

And is denoted by grad ϕ.

Thus,
$$\text{grad } \phi = \hat{i}\frac{\partial \phi}{\partial x} + \hat{j}\frac{\partial \phi}{\partial y} + \hat{k}\frac{\partial \phi}{\partial z}$$

$$\text{gard } \phi = \left(\hat{i}\frac{\partial}{\partial x} + \hat{j}\frac{\partial}{\partial y} + \hat{k}\frac{\partial}{\partial z}\right)\phi(x, y, z)$$

$$\text{gard } \phi = \nabla \phi \qquad (\nabla \text{ is read del or nebla})$$

2.6 GEOMETRICAL MEANING OF GRADIENT, NORMAL

(*U.P. Ist Semester, Dec 2006*)

If a surface $\phi(x, y, z) = c$ passes through a point P. The value of the function at each point on the surface is the same as at P. Then such a surface is called a *level surface* through P. *For example*, If $\phi(x, y, z)$ represents potential at the point P, then *equipotential surface* $\phi(x, y, z) = c$ is a *level surface.*

Two level surfaces can not intersect.

Let the level surface pass through the point P at which the value of the function is ϕ. Consider another level surface passing through Q, where the value of the function is $\phi + d\phi$.

Let $\bar{r}$ and $\bar{r} + \delta\bar{r}$ be the position vector of P and Q then $\overrightarrow{PQ} = \delta\bar{r}$

$$\nabla\phi . d\bar{r} = \left(\hat{i}\frac{\partial \phi}{\partial x} + \hat{j}\frac{\partial \phi}{\partial y} + \hat{k}\frac{\partial \phi}{\partial z}\right).(\hat{i}\, dx + \hat{j}\, dy + \hat{k}\, dz)$$

$$= \frac{\partial \phi}{\partial x}dx + \frac{\partial \phi}{\partial y}dy + \frac{\partial \phi}{\partial z}dz = d\phi \qquad \text{...(1)}$$

If Q lies on the level surface of P, then $d\phi = 0$

Equation (1) becomes $\nabla\phi . dr = 0$. Then $\overline{\nabla}\phi$ is ⊥ to $d\bar{r}$ (tangent).

Hence, $\nabla\phi$ is **normal** to the surface $\phi(x, y, z) = c$

Let $\nabla\phi = |\nabla\phi|\,\widehat{N}$, where $\widehat{N}$ is a unit normal vector. Let δn be the perpendicular distance between two level surfaces through P and R. Then the rate of change of ϕ in the direction of the normal to the surface through P is $\frac{\partial \phi}{\partial n}$.

$$\frac{d\phi}{dn} = \lim_{\delta n \to 0}\frac{\delta\phi}{\delta n} = \lim_{\delta n \to 0}\frac{\nabla\phi . d\vec{r}}{\delta n}$$

$$= \lim_{\delta n \to 0} \frac{|\nabla\phi|\,\hat{N}.d\vec{r}}{\delta n} \qquad \left\{\begin{array}{l}\hat{N}.\vec{\delta r} = |\vec{N}||\vec{\delta r}|\cos\theta \\ = |\vec{\delta r}|\cos\theta = \delta n\end{array}\right\}$$

$$= \lim_{\delta n \to 0} \frac{|\nabla\phi|\,\delta n}{\delta n} = |\nabla\phi|$$

$$\therefore \qquad |\nabla\phi| = \frac{\partial\phi}{\partial n}$$

Hence, gradient ϕ is a vector normal to the surface $\phi = c$ and has a magnitude equal to the rate of change of ϕ along this normal.

2.7 NORMAL AND DIRECTIONAL DERIVATIVE

(*i*) **Normal.** If $\phi(x, y, z) = c$ represents a family of surfaces for different values of the constant c. On differentiating ϕ, we get $d\phi = 0$

But $\qquad d\phi = \nabla\phi.\,d\vec{r} \quad \text{so} \quad \nabla\phi.dr = 0$

The scalar product of two vectors $\nabla\phi$ and $d\vec{r}$ being zero, $\nabla\phi$ and $d\vec{r}$ are perpendicular to each other. $d\vec{r}$ is in the direction of tangent to the given surface.

Thus $\nabla\phi$ is a vector *normal* to the surface $\phi(x, y, z) = c$.

(*ii*) **Directional derivative.** The component of $\nabla\phi$ in the direction of a vector $\vec{d}$ is equal to $\nabla\phi.\hat{d}$ and is called the directional derivative of ϕ in the direction of $\vec{d}$.

$$\frac{\partial\phi}{\partial r} = \lim_{\delta r \to 0} \frac{\delta\phi}{\delta r} \qquad \text{where, } \delta r = PQ$$

$\frac{\partial\phi}{\partial r}$ is called the *directional derivative* of ϕ at P in the direction of PQ.

Let a unit vector along PQ be $\hat{N}'$.

$$\frac{\delta n}{\delta r} = \cos\theta \Rightarrow \delta r = \frac{\delta n}{\cos\theta} = \frac{\delta n}{\hat{N}\cdot\hat{N}'} \qquad ...(1)$$

Now $\qquad \frac{\partial\phi}{\partial r} = \lim_{\delta r \to 0}\left[\frac{\delta\phi}{\dfrac{\delta n}{\hat{N}.\hat{N}'}}\right] = \hat{N}.\hat{N}'\frac{\partial\phi}{\partial n} \qquad \left[\text{From (1), } \delta r = \frac{\delta n}{\hat{N}.\hat{N}'}\right]$

$$= \hat{N}'.\hat{N}|\nabla\phi| = \hat{N}'.\nabla\phi \qquad (\because \hat{N}|\nabla\phi| = \nabla\phi)$$

Hence, $\frac{\partial\phi}{\partial r}$, directional derivative is the component of $\nabla\phi$ in the direction $\hat{N}'$.

$$\frac{\partial\phi}{\partial r} = \hat{N}'.\nabla\phi = |\nabla\phi|\cos\theta \le |\nabla\phi|$$

Hence, $\nabla\phi$ is the maximum rate of change of ϕ.

Example 6. *For the vector field (i)* $\vec{A} = m\hat{i}$ *and (ii)* $\vec{A} = m\vec{r}$. *Find* $\nabla.\vec{A}$ *and* $\nabla \times \vec{A}$. *Draw the sketch in each case.* (*Gujarat, I Semester, Jan. 2009*)

Solution. (*i*) Vector $\vec{A} = m\hat{i}$ is represented in the figure. (*i*).

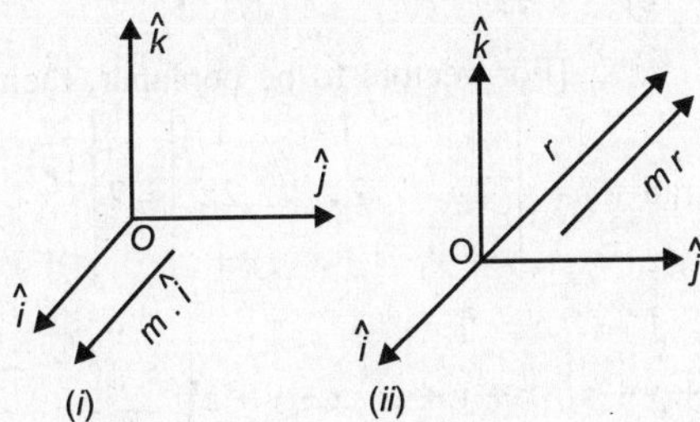

(*ii*) $\vec{A} = m\vec{r}$ is represented in the figure. (*ii*).

(*iii*) $\nabla.\vec{A} = \left(\hat{i}\frac{\partial}{\partial x} + \hat{j}\frac{\partial}{\partial y} + \hat{k}\frac{\partial}{\partial z}\right)m.(x\hat{i} + y\hat{j} + z\hat{k}) = m + m + m = 3m$ 0 1 2 3 Number line

$\nabla.\vec{A} = 3\,m$ is represented on the number line at 3.

(*iv*) $\nabla \times \vec{A} = \left(\hat{i}\frac{\partial}{\partial x} + \hat{j}\frac{\partial}{\partial y} + \hat{k}\frac{\partial}{\partial z}\right) \times m\,(x\hat{i} + y\hat{j} + z\hat{k})$

$$= m\begin{vmatrix} \hat{i} & \hat{j} & \hat{k} \\ \frac{\partial}{\partial x} & \frac{\partial}{\partial y} & \frac{\partial}{\partial z} \\ x & y & z \end{vmatrix} = 0$$

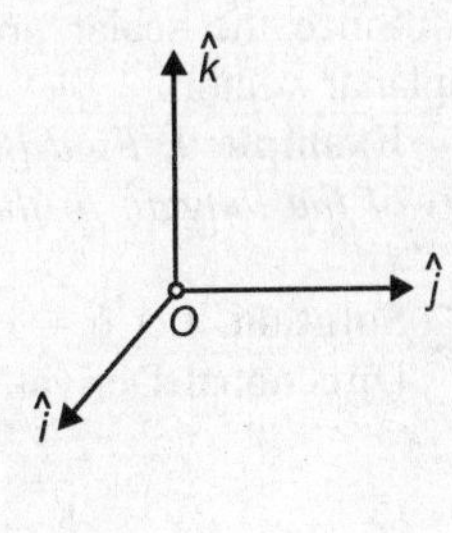

are represented in the adjoining figure.

Example 7. *If* $\phi = 3x^2y - y^3z^2$; *find grad* ϕ *at the point* (1, –2, –1).

(*AMIETE, June 2009, U.P., I Semester, Dec. 2006*)

Solution. grad $\phi = \nabla\phi$

$$= \left(\hat{i}\frac{\partial}{\partial x} + \hat{j}\frac{\partial}{\partial y} + \hat{k}\frac{\partial}{\partial z}\right)(3x^2y - y^3z^2)$$

$$= \hat{i}\frac{\partial}{\partial x}(3x^2y - y^3z^2) + \hat{j}\frac{\partial}{\partial y}(3x^2y - y^3z^2) + \hat{k}\frac{\partial}{\partial z}(3x^2y - y^3z^2)$$

$$= \hat{i}\,(6xy) + \hat{j}\,(3x^2 - 3y^2z^2) + \hat{k}\,(-2y^3z)$$

$$\text{grad } \phi \text{ at } (1, -2, -1) = \hat{i}\,(6)(1)(-2) + \hat{j}\,[(3)(1) - 3(4)(1)] + \hat{k}\,(-2)(-8)(-1)$$

$$= -12\hat{i} - 9\hat{j} - 16\hat{k}$$ **Ans.**

Example 8. *If* $u = x + y + z$, $v = x^2 + y^2 + z^2$, $w = yz + zx + xy$ *prove that grad u, grad v and grad w are coplanar vectors.*

Solution. We have,

$$\text{grad } u = \left(\hat{i}\frac{\partial}{\partial x} + \hat{j}\frac{\partial}{\partial y} + \hat{k}\frac{\partial}{\partial z}\right)(x + y + z) = \hat{i} + \hat{j} + \hat{k}$$

$$\text{grad } v = \left(\hat{i}\frac{\partial}{\partial x} + \hat{j}\frac{\partial}{\partial y} + \hat{k}\frac{\partial}{\partial z}\right)(x^2 + y^2 + z^2) = 2x\hat{i} + 2y\hat{j} + 2z\hat{k}$$

$$\text{grad } w = \left(\hat{i}\frac{\partial}{\partial x} + \hat{j}\frac{\partial}{\partial y} + \hat{k}\frac{\partial}{\partial z}\right)(yz + zx + xy) = \hat{i}(z+y) + \hat{j}(z+x) + \hat{k}(y+x)$$

[For vectors to be coplanar, their scalar triple product is 0]

$$\text{Now, grad } u.(\text{grad } v \times \text{grad } w) = \begin{vmatrix} 1 & 1 & 1 \\ 2x & 2y & 2z \\ z+y & z+x & y+x \end{vmatrix} = 2\begin{vmatrix} 1 & 1 & 1 \\ x & y & z \\ z+y & z+x & y+x \end{vmatrix}$$

$$= 2\begin{vmatrix} 1 & 1 & 1 \\ x+y+z & x+y+z & x+y+z \\ z+y & z+x & y+x \end{vmatrix} \qquad [\text{Applying } R_2 \to R_2 + R_3]$$

$$= 2(x+y+z)\begin{vmatrix} 1 & 1 & 1 \\ 1 & 1 & 1 \\ y+z & z+x & x+y \end{vmatrix} = 0 \qquad [\text{Since } R_1 = R_2]$$

Since the scalar product of grad u, grad v and grad w are zero, hence these vectors are coplanar vectors.

Proved.

Example 9. *Find the directional derivative of $x^2y^2z^2$ at the point (1, 1, –1) in the direction of the tangent to the curve $x = e^t$, $y = \sin 2t + 1$, $z = 1 - \cos t$ at $t = 0$.*

(Nagpur University, Summer 2005)

Solution. Let $\phi = x^2\, y^2\, z^2$

Directional Derivative of ϕ

$$= \nabla\phi = \left(\hat{i}\frac{\partial}{\partial x} + \hat{j}\frac{\partial}{\partial y} + \hat{k}\frac{\partial}{\partial z}\right)(x^2y^2z^2)$$

$$\nabla\phi = 2xy^2z^2\,\hat{i} + 2yx^2z^2\,\hat{j} + 2zx^2y^2\,\hat{k}$$

Directional Derivative of ϕ at (1, 1, –1)

$$= 2(1)(1)^2(-1)^2\,\hat{i} + 2(1)(1)^2(-1)^2\,\hat{j} + 2(-1)(1)^2(1)^2\,\hat{k}$$

$$= 2\,\hat{i} + 2\,\hat{j} - 2\,\hat{k} \qquad \text{...(1)}$$

$$\vec{r} = x\,\hat{i} + y\,\hat{j} + z\,\hat{k} = e^t\,\hat{i} + (\sin 2t + 1)\,\hat{j} + (1 - \cos t)\,\hat{k}$$

$$\text{Tangent vector, } \quad \vec{T} = \frac{d\vec{r}}{dt} = e^t\,\hat{i} + 2\cos 2t\,\hat{j} + \sin t\,\hat{k}$$

$$\text{Tangent (at } t = 0) = e^0\,\hat{i} + 2(\cos 0)\,\hat{j} + (\sin 0)\,\hat{k} = \hat{i} + 2\,\hat{j} \qquad \text{...(2)}$$

$$\text{Required directional derivative along tangent} = (2\hat{i} + 2\hat{j} - 2\hat{k})\frac{(\hat{i} + 2\hat{j})}{\sqrt{1+4}} \qquad [\text{From (1), (2)}]$$

$$= \frac{2+4+0}{\sqrt{5}} = \frac{6}{\sqrt{5}} \qquad \textbf{Ans.}$$

Example 10. *Find the unit normal to the surface $xy^3z^2 = 4$ at (–1, –1, 2).* *(M.U. 2008)*

Solution. Let $\phi(x, y, z) = xy^3z^2 - 4$

We know that $\nabla\phi$ is the vector normal to the surface $\phi(x, y, z) = x\,y^3z^2 - 4$.

$$\text{Normal vector} = \nabla\phi = \hat{i}\frac{\partial\phi}{\partial x} + \hat{j}\frac{\partial\phi}{\partial y} + \hat{k}\frac{\partial\phi}{\partial z}$$

Now $= \hat{i}\frac{\partial}{\partial x}(xy^3z^2) + \hat{j}\frac{\partial}{\partial y}(xy^3z^2) + \hat{k}\frac{\partial}{\partial z}(xy^3z^2)$

$\Rightarrow$ Normal vector $= y^3z^2\hat{i} + 3xy^2z^2\hat{j} + 2xy^3z\hat{k}$

Normal vector at (–1, –1, 2) $= -4\hat{i} - 12\hat{j} + 4\hat{k}$

Unit vector normal to the surface at (–1, –1, 2).

$$= \frac{\nabla\phi}{|\nabla\phi|} = \frac{-4\hat{i} - 12\hat{j} + 4\hat{k}}{\sqrt{16+144+16}} = -\frac{1}{\sqrt{11}}(\hat{i} + 3\hat{j} - \hat{k}) \quad \textbf{Ans.}$$

Example 11. *Find the unit normal to the surface:*

$$x^2 + y^2 = z$$ *(Delhi University, 2010)*

at a point (1, 2, 5)

Solution. Let $\phi = x^2 + y^2 - z$

$$\text{Gradient } \phi = \nabla\phi = \left(\hat{i}\frac{\partial}{\partial x} + \hat{j}\frac{\partial}{\partial y} + \hat{k}\frac{\partial}{\partial z}\right)(x^2 + y^2 - z) = 2x\hat{i} + 2y\hat{j} - \hat{k}$$

$$(\text{Gradient } \phi)_{1,\,2,\,5} = 2\hat{i} + 4\hat{j} - \hat{k}$$

$$\text{Unit normal vector} = \frac{\Delta\phi}{|\Delta\phi|} = \frac{2\hat{i} + 4\hat{j} - \hat{k}}{\sqrt{4+16+1}} = \frac{2}{\sqrt{21}}\hat{i} + \frac{4}{\sqrt{21}}\hat{j} - \frac{\hat{k}}{\sqrt{21}} \quad \textbf{Ans.}$$

Example 12. *Find the rate of change of $\phi = xyz$ in the direction normal to the surface $x^2y + y^2x + yz^2 = 3$ at the point (1, 1, 1).*

Solution. Rate of change of $\phi = \nabla\phi$

$$= \left(\hat{i}\frac{\partial}{\partial x} + \hat{j}\frac{\partial}{\partial y} + \hat{k}\frac{\partial}{\partial z}\right)(x\,yz) = \hat{i}\,yz + \hat{j}\,xz + \hat{k}\,xy$$

Rate of change of ϕ at (1, 1, 1) $= (\hat{i} + \hat{j} + \hat{k})$

Normal to the surface $\Psi = x^2y + y^2x + yz^2 - 3$ is given as

$$\nabla\Psi = \left(\hat{i}\frac{\partial}{\partial x} + \hat{j}\frac{\partial}{\partial y} + \hat{k}\frac{\partial}{\partial z}\right)(x^2y + y^2x + yz^2 - 3)$$

$$= \hat{i}(2xy + y^2) + \hat{j}(x^2 + 2xy + z^2) + \hat{k}\,2yz$$

$$(\nabla\psi)_{(1,\,1,\,1)} = 3\hat{i} + 4\hat{j} + 2\hat{k}$$

$$\text{Unit normal} = \frac{3\hat{i} + 4\hat{j} + 2\hat{k}}{\sqrt{9+16+4}}$$

Required rate of change of $\phi = (\hat{i} + \hat{j} + \hat{k}).\dfrac{(3\hat{i} + 4\hat{j} + 2\hat{k})}{\sqrt{9+16+4}} = \dfrac{3+4+2}{\sqrt{29}} = \dfrac{9}{\sqrt{29}}$ **Ans.**

Example 13. *Find the constants m and n such that the surface $m\,x^2 - 2nyz = (m + 4)x$ will be orthogonal to the surface $4x^2y + z^3 = 4$ at the point (1, –1, 2).*

(M.D.U. Dec. 2009)

Solution. The point P (1, –1, 2) lies on both surfaces. As this point lies in

$$mx^2 - 2nyz = (m+4)x, \text{ so we have}$$

$$m - 2n(-2) = (m+4)$$

$$\Rightarrow \quad m + 4n = m + 4 \quad \Rightarrow \quad n = 1$$

$$\therefore \quad \text{Let } \phi_1 = mx^2 - 2yz - (m+4)x \text{ and } \phi_2 = 4x^2y + z^3 - 4$$

$$\text{Normal to } \phi_1 = \nabla\phi_1$$

$$= \left(\hat{i}\frac{\partial}{\partial x} + \hat{j}\frac{\partial}{\partial y} + \hat{k}\frac{\partial}{\partial z}\right)[mx^2 - 2yz - (m+4)x]$$

$$= \hat{i}(2mx - m - 4) - 2z\,\hat{j} - 2y\,\hat{k}$$

$$\text{Normal to } \phi_1 \text{ at } (1, -1, 2) = \hat{i}(2m - m - 4) - 4\hat{j} + 2\hat{k} = (m-4)\hat{i} - 4\hat{j} + 2\hat{k}$$

$$\text{Normal to } \phi_2 = \nabla\phi_2$$

$$= \left(\hat{i}\frac{\partial}{\partial x} + \hat{j}\frac{\partial}{\partial y} + \hat{k}\frac{\partial}{\partial z}\right)(4x^2y + z^3 - 4) = \hat{i}\,8xy + 4x^2\,\hat{j} + 3z^2\,\hat{k}$$

$$\text{Normal to } \phi_2 \text{ at } (1, -1, 2) = -8\hat{i} + 4\hat{j} + 12\hat{k}$$

Sinec ϕ_1 and ϕ_2 are orthogonal, then normals are perpendicular to each other.

$$\nabla\phi_1 . \nabla\phi_2 = 0$$

$$\Rightarrow \quad [(m-4)\hat{i} - 4\hat{j} + 2\hat{k}].[-8\hat{i} + 4\hat{j} + 12\hat{k}] = 0$$

$$\Rightarrow \quad -8(m-4) - 16 + 24 = 0$$

$$\Rightarrow \quad m - 4 = -2 + 3 \quad \Rightarrow \quad m = 5$$

Hence $m = 5$, $n = 1$ **Ans.**

Example 14. *Find the values of constants* λ *and* μ *so that the surfaces* $\lambda x^2 - \mu yz = (\lambda + 2)x$, $4x^2y + z^3 = 4$ *intersect orthogonally at the point (1, – 1, 2).*

(*AMIETE, II Sem., Dec. 2010, June 2009*)

Solution. Here, we have

$$\lambda x^2 - \mu yz = (\lambda + 2)\,x \qquad ...(1)$$

$$4x^2\,y + z^3 = 4 \qquad ...(2)$$

$$\text{Normal to the surface (1)}, = \nabla[\lambda x^2 - \mu yz - (\lambda+2)x]$$

$$= \left[\hat{i}\frac{\partial}{\partial x} + \hat{j}\frac{\partial}{\partial y} + \hat{k}\frac{\partial}{\partial z}\right][\lambda x^2 - \mu yz - (\lambda+2)x]$$

$$= \hat{i}\,(2\lambda x - \lambda - 2) + \hat{j}\,(-\mu z) + \hat{k}\,(-\mu y)$$

$$\text{Normal at } (1, -1, 2) = \hat{i}\;(2\lambda - \lambda - 2) - \hat{j}\;(2\mu) + \hat{k}\,\mu \qquad ...(3)$$

$$= \hat{i}\;(\lambda - 2) - \hat{j}\,(2\mu) + \hat{k}\,\mu$$

Normal at the surface (2)

$$= \left(\hat{i}\frac{\partial}{\partial x} + \hat{j}\frac{\partial}{\partial y} + \hat{k}\frac{\partial}{\partial z}\right)(4x^2y + z^3 - 4)$$

$$= \hat{i}\;(8xy) + \hat{j}\;(4x^2) + \hat{k}\;(3z^2)$$

$$\text{Normal at the point } (1, -1, 2) = -8\hat{i} + 4\hat{j} + 12\hat{k} \qquad ...(4)$$

Since (3) and (4) are orthogonal so

$$[\hat{i}(\lambda-2)-\hat{j}(2\mu)+\hat{k}\mu].[-8\hat{i}+4\hat{j}+12\hat{k}]=0$$

$$-8(\lambda-2)-4(2\mu)+12\mu=0 \quad \Rightarrow \quad -8\lambda+16-8\mu+12\mu=0$$

$$-8\lambda+4\mu+16=0 \quad \Rightarrow \quad 4(-2\lambda+\mu+4)=0$$

$$\Rightarrow \quad 2\lambda-\mu=4 \qquad ...(5)$$

Point (1, – 1, 2) will satisfy (1)

$$\therefore \quad \lambda(1)^2-\mu(-1)(2)=(\lambda+2)(1) \Rightarrow \lambda+2\mu=\lambda+2 \Rightarrow \mu=1$$

Putting $\mu = 1$ in (5), we get

$$2\lambda-1=4 \quad \Rightarrow \quad \lambda=\frac{5}{2}$$

Hence $\lambda=\frac{5}{2}$ and $\mu=1$ **Ans.**

Example 15. *Find the directional derivative of $\frac{1}{r}$ in the direction $\bar{r}$ where $\bar{r}=x\hat{i}+y\hat{j}+z\hat{k}$.*

(Nagpur University, Summer 2004, U.P., I Semester, Winter 2005)

Solution. Here, $\phi(x, y, z) = \frac{1}{r}=\frac{1}{\sqrt{x^2+y^2+z^2}}=(x^2+y^2+z^2)^{-\frac{1}{2}}$

$$\text{Now} \quad \nabla\left(\frac{1}{r}\right) = \left(\hat{i}\frac{\partial}{\partial x}+\hat{j}\frac{\partial}{\partial y}+\hat{k}\frac{\partial}{\partial z}\right)(x^2+y^2+z^2)^{-\frac{1}{2}}$$

$$= \frac{\partial}{\partial x}(x^2+y^2+z^2)^{-\frac{1}{2}}\hat{i}+\frac{\partial}{\partial y}(x^2+y^2+z^2)^{-\frac{1}{2}}\hat{j}+\frac{\partial}{\partial z}(x^2+y^2+z^2)^{-\frac{1}{2}}\hat{k}$$

$$= \left\{-\frac{1}{2}(x^2+y^2+z^2)^{-\frac{3}{2}}2x\right\}\hat{i}+\left(-\frac{1}{2}(x^2+y^2+z^2)^{-\frac{3}{2}}2y\right)\hat{j}$$

$$+\left\{-\frac{1}{2}(x^2+y^2+z^2)^{-\frac{3}{2}}2z\right\}\hat{k}$$

$$= \frac{-(x\hat{i}+y\hat{j}+z\hat{k})}{(x^2+y^2+z^2)^{3/2}} \qquad ...(1)$$

and $\hat{r}$ = unit vector in the direction of $x\hat{i}+y\hat{j}+z\hat{k}$

$$= \frac{x\hat{i}+y\hat{j}+z\hat{k}}{\sqrt{x^2+y^2+z^2}} \qquad ...(2)$$

So, the required directional derivative

$$= \nabla\phi.\hat{r} = -\frac{x\hat{i}+y\hat{j}+z\hat{k}}{(x^2+y^2+z^2)^{3/2}}\cdot\frac{x\hat{i}+y\hat{j}+z\hat{k}}{(x^2+y^2+z^2)^{1/2}}=\frac{x^2+y^2+z^2}{(x^2+y^2+z^2)^2} \quad \text{[From (1), (2)]}$$

$$= \frac{1}{x^2+y^2+z^2}=\frac{1}{r^2}$$ **Ans.**

Example 16. *Find the direction in which the directional derivative of $\phi(x, y) = \frac{x^2+y^2}{xy}$ at (1, 1) is zero and hence find out component of velocity of the vector $\bar{r}=(t^3+1)\hat{i}+t^2\hat{j}$ in* *the same direction at t = 1.*

Solution. Directional derivative $= \nabla\phi = \left(\hat{i}\frac{\partial}{\partial x}+\hat{j}\frac{\partial}{\partial y}+\hat{k}\frac{\partial}{\partial z}\right)\left(\frac{x^2+y^2}{xy}\right)$

$$= \hat{i}\left[\frac{xy.2x-(x^2+y^2)y}{x^2y^2}\right]+\hat{j}\left[\frac{xy.2y-x(y^2+x^2)}{x^2y^2}\right]$$

$$= \hat{i}\left[\frac{x^2y-y^3}{x^2y^2}\right]+\hat{j}\left[\frac{xy^2-x^3}{x^2y^2}\right]$$

Directional Derivative at (1, 1) $= \hat{i}\,0+\hat{j}\,0=0$

Since $(\nabla\phi)_{(1,\,1)} = 0$, the directional derivative of ϕ at (1, 1) is zero in any direction.

Again $\quad \bar{r} = (t^3+1)\hat{i}+t^2\hat{j}$

Velocity, $\quad \bar{v} = \frac{d\bar{r}}{dt}=3t^2\hat{i}+2t\hat{j}$

Velocity (at $t = 1$) $= 3\hat{i}+2\hat{j}$

The component of velocity in the same direction of velocity

$$= (3\hat{i}+2\hat{j}).\left(\frac{3\hat{i}+2\hat{j}}{\sqrt{9+4}}\right)=\frac{9+4}{\sqrt{13}}=\sqrt{13}$$ **Ans.**

Example 17. *Find the directional derivative of* $\phi\,(x, y, z) = x^2yz + 4xz^2$ *at (1, –2, 1) in the direction of* $2\hat{i}-\hat{j}-2\hat{k}$. *Find the greatest rate of increase of* ϕ.

(Uttarakhand, I Semester, Dec. 2006)

Solution. Here, $\quad \phi\,(x, y, z) = x^2yz + 4xz^2$

Now, $\quad \nabla\phi = \left(\hat{i}\frac{\partial}{\partial x}+\hat{j}\frac{\partial}{\partial y}+\hat{k}\frac{\partial}{\partial z}\right)(x^2yz+4xz^2)$

$$= (2xyz+4z^2)\hat{i}+(x^2z)\hat{j}+(x^2y+8xz)\hat{k}$$

$\nabla\phi$ at (1, – 2, 1) $= \{2(1)(-2)(1)+4(1)^2\}\hat{i}+(1\times 1)\hat{j}+\{1(-2)+8(1)(1)\}\hat{k}$

$$= (-4+4)\hat{i}+\hat{j}+(-2+8)\hat{k}=\hat{j}+6\hat{k}$$

Let $\quad \hat{a}$ = unit vector $= \frac{2\hat{i}-\hat{j}-2\hat{k}}{\sqrt{4+1+4}}=\frac{1}{3}(2\hat{i}-\hat{j}-2\hat{k})$

So, the required directional derivative at (1, –2, 1)

$$= \nabla\phi.\hat{a}=(\hat{j}+6\hat{k}).\frac{1}{3}(2\hat{i}-\hat{j}-2\hat{k}) = \frac{1}{3}(-1-12)=\frac{-13}{3}$$

Greatest rate of increase of $\phi = \left|\hat{j}+6\hat{k}\right|=\sqrt{1+36}=\sqrt{37}$ **Ans.**

Example 18. *Find the directional derivative of the function* $\phi = x^2 - y^2 + 2z^2$ *at the point P (1, 2, 3) in the direction of the line PQ where Q is the point (5, 0, 4).*

(AMIETE, Dec. 2010, Nagpur University, Summer 2008)

Solution. Directional derivative $= \nabla\phi$

$$= \left(\hat{i}\frac{\partial}{\partial x}+\hat{j}\frac{\partial}{\partial y}+\hat{k}\frac{\partial}{\partial z}\right)(x^2-y^2+2z^2)=2x\hat{i}-2y\hat{j}+4z\hat{k}$$

Directional Derivative at the point P (1, 2, 3) $= 2\hat{i} - 4\hat{j} + 12\hat{k}$ ...(1)

$$\overline{PQ} = \overline{Q} - \overline{P} = (5\hat{i} + 4\hat{k}) - (\hat{i} + 2\hat{j} + 3\hat{k}) = 4\hat{i} - 2\hat{j} + \hat{k} \quad \text{...(2)}$$

Directional Derivative along $PQ = (2\hat{i} - 4\hat{j} + 12\hat{k}).\dfrac{(4\hat{i} - 2\hat{j} + \hat{k})}{\sqrt{16+4+1}}$ [From (1) and (2)]

$$= \frac{8+8+12}{\sqrt{21}} = \frac{28}{\sqrt{21}} \qquad \textbf{Ans.}$$

Example 19. *Find the directional derivative of* $\phi = 4\, e^{2x-y+z}$ *at the point (1, 1, –1) in the directional towards the point (–3, 5, 6).*

Solution. Directional derivative $= \nabla\phi$

$$= \left(\hat{i}\frac{\partial}{\partial x} + \hat{j}\frac{\partial}{\partial y} + \hat{k}\frac{\partial}{\partial z}\right)(4e^{2x-y+z})$$

$$= 4[\hat{i}\, 2e^{2x-y+z} - \hat{j}\, e^{2x-y+z} + \hat{k}\, e^{2x-y+z}] = 4[2\hat{i} - \hat{j} + \hat{k}]\, e^{2x-y+z}$$

Directional Derivative at (1, 1, –1)

$$= 4[2\hat{i} - \hat{j} + \hat{k}]e^{2-1-1} = 4[2\hat{i} - \hat{j} + \hat{k}] \quad \text{...(1)}$$

Direction of Directional Derivative

$$= (-3\hat{i} + 5\hat{j} + 6\hat{k}) - (\hat{i} + \hat{j} - \hat{k}) = -4\hat{i} + 4\hat{j} + 7\hat{k} \quad \text{...(2)}$$

Directional Derivative in the direction of $(-4\hat{i} + 4\hat{j} + 7\hat{k})$

$$= (8\hat{i} - 4\hat{j} + 4\hat{k}).\frac{(-4\hat{i} + 4\hat{j} + 7\hat{k})}{\sqrt{16+16+49}} \qquad \text{[From (1) and (2)]}$$

$$= \frac{1}{9}[-32 - 16 + 28] = -\frac{20}{9} \qquad \textbf{Ans.}$$

Example 20. *For the function* $\phi(x, y) = \dfrac{x}{x^2 + y^2}$, *find the magnitude of the directional derivative along a line making an angle 30° with the positive x-axis at (0, 2).*

Solution. Directional derivative $= \vec{\nabla}\phi$

$$= \left(\hat{i}\frac{\partial}{\partial x} + \hat{j}\frac{\partial}{\partial y} + \hat{k}\frac{\partial}{\partial z}\right)\frac{x}{x^2+y^2} = \hat{i}\left(\frac{1}{x^2+y^2} - \frac{x(2x)}{(x^2+y^2)^2}\right) - \hat{j}\frac{x(2y)}{(x^2+y^2)^2}$$

$$= \hat{i}\frac{y^2 - x^2}{(x^2+y^2)^2} - \hat{j}\frac{2xy}{(x^2+y^2)^2}$$

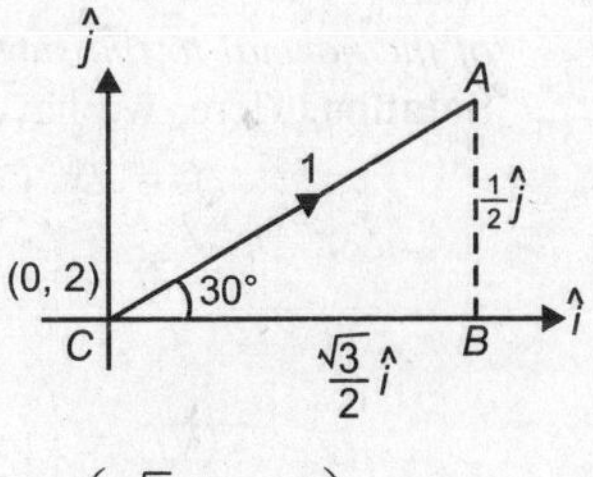

Directional derivative at the point (0, 2)

$$= \hat{i}\frac{4-0}{(0+4)^2} - \hat{j}\frac{2(0)(2)}{(0+4)^2} = \frac{\hat{i}}{4}$$

$\hat{j}$, A, 1, $\frac{1}{2}\hat{j}$, (0, 2), 30°, C, B, $\hat{i}$, $\frac{\sqrt{3}}{2}\hat{i}$

Directional derivative at the point (0, 2) in the direction $\overrightarrow{CA}$ *i.e.* $\left(\dfrac{\sqrt{3}}{2}\hat{i} + \dfrac{1}{2}\hat{j}\right)$

$$= \frac{\hat{i}}{4}\cdot\left(\frac{\sqrt{3}}{2}\hat{i}+\frac{1}{2}\hat{j}\right) \qquad \left\{\begin{aligned} \overrightarrow{CA} = \overrightarrow{OB} + \overrightarrow{BA} &= \hat{i}\cos 30° + \hat{j}\sin 30° \\ &= \left(\frac{\sqrt{3}}{2}\hat{i}+\frac{1}{2}\hat{j}\right)\end{aligned}\right\}$$

$$= \frac{\sqrt{3}}{8}$$ **Ans.**

Example 21. *Find the directional derivative of $\vec{V}^2$, where $\vec{V} = xy^2\hat{i} + zy^2\hat{j} + xz^2\hat{k}$, at the point (2, 0, 3) in the direction of the outward normal to the sphere $x^2 + y^2 + z^2 = 14$ at the point (3, 2, 1).* (*A.M.I.E.T.E., Dec. 2007*)

Solution. $V^2 = \vec{V}.\vec{V}$

$$= (xy^2\hat{i} + zy^2\hat{j} + xz^2\hat{k}).(xy^2\hat{i} + zy^2\hat{j} + xz^2\hat{k}) = x^2y^4 + z^2y^4 + x^2z^4$$

Directional derivative $= \bar{\nabla} V^2$

$$= \left(\hat{i}\frac{\partial}{\partial x} + \hat{j}\frac{\partial}{\partial y} + \hat{k}\frac{\partial}{\partial z}\right)(x^2y^4 + z^2y^4 + x^2z^4)$$

$$= (2xy^4 + 2xz^4)\hat{i} + (4x^2y^3 + 4y^3z^2)\hat{j} + (2y^4z + 4x^2z^3)\hat{k}$$

Directional derivative at (2, 0, 3) $= (0 + 2\times2\times81)\hat{i} + (0+0)\hat{j} + (0 + 4\times4\times27)\hat{k}$

$$= 324\hat{i} + 432\hat{k} = 108(3\hat{i} + 4\hat{k}) \qquad \text{...(1)}$$

Normal to $x^2 + y^2 + z^2 - 14 = \nabla\phi$

$$= \left(\hat{i}\frac{\partial}{\partial x} + \hat{j}\frac{\partial}{\partial y} + \hat{k}\frac{\partial}{\partial z}\right)(x^2 + y^2 + z^2 - 14)$$

$$= (2x\hat{i} + 2y\hat{j} + 2z\hat{k})$$

Normal vector at (3, 2, 1) $= 6\hat{i} + 4\hat{j} + 2\hat{k}$ ...(2)

Unit normal vector $= \dfrac{6\hat{i} + 4\hat{j} + 2\hat{k}}{\sqrt{36+16+4}} = \dfrac{2(3\hat{i} + 2\hat{j} + \hat{k})}{2\sqrt{14}} = \dfrac{3\hat{i} + 2\hat{j} + \hat{k}}{\sqrt{14}}$ [From (1), (2)]

Directional derivative along the normal $= 108(3\hat{i} + 4\hat{k}).\dfrac{3\hat{i} + 2\hat{j} + \hat{k}}{\sqrt{14}}$.

$$= \frac{108\times(9+4)}{\sqrt{14}} = \frac{1404}{\sqrt{14}}$$ **Ans.**

Example 22. *Find the directional derivative of ∇ (∇f) at the point (1, – 2, 1) in the direction of the normal to the surface $xy^2z = 3x + z^2$, where $f = 2x^3y^2z^4$.* (*U.P., I Semester, Dec 2008*)

Solution. Here, we have

$$f = 2x^3y^2z^4$$

$$\nabla f = \left(\hat{i}\frac{\partial}{\partial x} + \hat{j}\frac{\partial}{\partial y} + \hat{k}\frac{\partial}{\partial z}\right)(2x^3y^2z^4) = 6x^2y^2z^4\hat{i} + 4x^3yz^4\hat{j} + 8x^3y^2z^3\hat{k}$$

$$\nabla(\nabla f) = \left(\hat{i}\frac{\partial}{\partial x} + \hat{j}\frac{\partial}{\partial y} + \hat{k}\frac{\partial}{\partial z}\right)(6x^2y^2z^4\hat{i} + 4x^3yz^4\hat{j} + 8x^3y^2z^3\hat{k})$$

$$= 12xy^2z^4 + 4x^3z^4 + 24x^3y^2z^2$$

Directional derivative of $\nabla(\nabla f)$

$$= \left(\hat{i}\frac{\partial}{\partial x} + \hat{j}\frac{\partial}{\partial y} + \hat{k}\frac{\partial}{\partial z}\right)(12xy^2z^4 + 4x^3z^4 + 24x^3y^2z^2)$$

$$= (12y^2z^4 + 12x^2z^4 + 72x^2y^2z^2)\hat{i} + (24xyz^4 + 48x^3yz^2)\hat{j}$$

$$+ (48xy^2z^3 + 16x^3z^3 + 48x^3y^2z)\hat{k}$$

Directional derivative at (1, – 2, 1) = (48 + 12 + 288) $\hat{i}$ + (– 48 – 96) $\hat{j}$ + (192 + 16 + 192) $\hat{k}$

$$= 348\hat{i} - 144\hat{j} + 400\hat{k}$$

Normal to $(xy^2z - 3x - z^2) = \nabla(xy^2z - 3x - z^2)$

$$= \left(\hat{i}\frac{\partial}{\partial x} + \hat{j}\frac{\partial}{\partial y} + \hat{k}\frac{\partial}{\partial z}\right)(xy^2z - 3x - z^2)$$

$$= (y^2z - 3)\hat{i} + (2xyz)\hat{j} + (xy^2 - 2z)\hat{k}$$

Normal at (1, – 2, 1) = $\hat{i} - 4\hat{j} + 2\hat{k}$

Unit Normal Vector = $\dfrac{\hat{i} - 4\hat{j} + 2\hat{k}}{\sqrt{1+16+4}} = \dfrac{1}{\sqrt{21}}(\hat{i} - 4\hat{j} + 2\hat{k})$

Directional derivative in the direction of normal

$$= (348\hat{i} - 144\hat{j} + 400\hat{k})\cdot\frac{1}{\sqrt{21}}(\hat{i} - 4\hat{j} + 2\hat{k})$$

$$= \frac{1}{\sqrt{21}}(348 + 576 + 800) = \frac{1724}{\sqrt{21}}$$ **Ans.**

Example 23. *If the directional derivative of* $\phi = a\,x^2\,y + b\,y^2\,z + c\,z^2\,x$ *at the point (1, 1, 1) has maximum magnitude 15 in the direction parallel to the line* $\dfrac{x-1}{2} = \dfrac{y-3}{-2} = \dfrac{z}{1}$, *find the values of a, b and c.*

Solution. Given $\phi = a\,x^2\,y + b\,y^2\,z + c\,z^2\,x$

$$\therefore \quad \overline{\nabla}\phi = \left(\hat{i}\frac{\partial}{\partial x} + \hat{j}\frac{\partial}{\partial y} + \hat{k}\frac{\partial}{\partial z}\right)(a\,x^2\,y + b\,y^2\,z + c\,z^2\,x)$$

$$= \hat{i}(2axy + cz^2) + \hat{j}(ax^2 + 2byz) + \hat{k}(by^2 + 2czx)$$

$\overline{\nabla}\phi$ at the point (1, 1, 1) = $\hat{i}(2a + c) + \hat{j}(a + 2b) + \hat{k}(b + 2c)$ ...(1)

We know that the maximum value of the directional derivative is in the direction of $\overline{\nabla}\phi$.

i.e. $|\nabla\phi| = 15 \Rightarrow (2a + c)^2 + (2b + a)^2 + (2c + b)^2 = (15)^2$

But, the directional derivative is given to be maximum parallel to the line

$\dfrac{x-1}{2} = \dfrac{y-3}{-2} = \dfrac{z}{1}$ *i.e.*, parallel to the vector $2\hat{i} - 2\hat{j} + \hat{k}$. ...(2)

$$\hat{i}(2a+c) + \hat{j}(a+2b) + \hat{k}(b+2c) = 15\frac{(2\hat{i} - 2\hat{j} + \hat{k})}{\sqrt{4+4+1}}$$

$$\hat{i}(2a+c) + \hat{j}(a+2b) + \hat{k}(b+2c) = 10\hat{i} - 10\hat{j} + 5\hat{k}$$

Equating the coefficients of corresponding of corresponding terms, we

$2a + c = 10$ ...(1)

$a + 2b = -10$ ...(2)

$b + 2c = 5$ ...(3)

On solving (1), (2) and (3), we get

$$a=\frac{20}{9};\ jb=-\frac{55}{9};\ jc=\frac{50}{9}$$ **Ans.**

Example 24. *If* $\bar{r}=x\hat{i}+y\hat{j}+z\hat{k}$, *show that:*

(i) grad $r=\frac{\vec{r}}{r}$ (ii) grad $\left(\frac{1}{r}\right)=-\frac{\vec{r}}{r^3}$.

Solution. (i) $\bar{r}=x\hat{i}+y\hat{j}+z\hat{k} \Rightarrow r=\sqrt{x^2+y^2+z^2} \Rightarrow r^2=x^2+y^2+z^2$

$$\therefore \quad 2r\frac{\partial r}{\partial x}=2x \quad\Rightarrow\quad \frac{\partial r}{\partial x}=\frac{x}{r}$$

Similarly, $\frac{\partial r}{\partial y}=\frac{y}{r}$ and $\frac{\partial r}{\partial z}=\frac{z}{r}$

$$\text{grad } r=\nabla r=\left(\hat{i}\frac{\partial}{\partial x}+\hat{j}\frac{\partial}{\partial y}+\hat{k}\frac{\partial}{\partial z}\right)r=\hat{i}\frac{\partial r}{\partial x}+\hat{j}\frac{\partial r}{\partial y}+\hat{k}\frac{\partial r}{\partial z}$$

$$=\hat{i}\frac{x}{r}+\hat{j}\frac{y}{r}+\hat{k}\frac{z}{r}=\frac{x\hat{i}+y\hat{j}+z\hat{k}}{r}=\frac{\bar{r}}{r}$$ **Proved.**

(ii) $$\text{grad}\left(\frac{1}{r}\right)=\nabla\left(\frac{1}{r}\right)=\left(\hat{i}\frac{\partial}{\partial x}+\hat{j}\frac{\partial}{\partial y}+\hat{k}\frac{\partial}{\partial z}\right)\left(\frac{1}{r}\right)=\hat{i}\frac{\partial}{\partial x}\left(\frac{1}{r}\right)+\hat{j}\frac{\partial}{\partial y}\left(\frac{1}{r}\right)+\hat{k}\frac{\partial}{\partial z}\left(\frac{1}{r}\right)$$

$$=\hat{i}\left(-\frac{1}{r^2}\frac{\partial r}{\partial x}\right)+\hat{j}\left(\frac{-1}{r^2}\frac{\partial r}{\partial y}\right)+\hat{k}\left(-\frac{1}{r^2}\frac{\partial r}{\partial z}\right)$$

$$=\hat{i}\left(-\frac{1}{r^2}\frac{x}{r}\right)+\hat{j}\left(-\frac{1}{r^2}\frac{y}{r}\right)+\hat{k}\left(-\frac{1}{r^2}\frac{z}{r}\right)=-\frac{x\hat{i}+y\hat{j}+z\hat{k}}{r^3}=-\frac{\bar{r}}{r^3}$$ **Proved.**

Example 25. *Prove that* $\nabla^2 f(r)=f''(r)+\frac{2}{r}f'(r)$. *(K. University, Dec. 2008)*

Solution. $$\nabla f(r)=\left(i\frac{\partial}{\partial x}+j\frac{\partial}{\partial y}+k\frac{\partial}{\partial z}\right)f(r)$$

$$\left[r^2=x^2+y^2+z^2 \Rightarrow 2r\frac{\partial r}{\partial x}=2x \Rightarrow \frac{\partial r}{\partial x}=\frac{x}{r},\quad \frac{\partial r}{\partial y}=\frac{y}{r} \text{ and } \frac{\partial r}{\partial z}=\frac{z}{r}\right]$$

$$=\hat{i}f'(r)\frac{\partial r}{\partial x}+\hat{j}f'(r)\frac{\partial r}{\partial y}+\hat{k}f'(r)\frac{\partial r}{\partial z}=f'(r)\left[\hat{i}\frac{x}{r}+\hat{j}\frac{y}{r}+\hat{k}\frac{z}{r}\right]$$

$$=f'(r)\frac{x\hat{i}+y\hat{j}+z\hat{k}}{r}$$

$$\nabla^2 f(r)=\nabla[\nabla f(r)]=\left(\hat{i}\frac{\partial}{\partial x}+\hat{j}\frac{\partial}{\partial y}+\hat{k}\frac{\partial}{\partial z}\right)\left[f'(r)\frac{x\hat{i}+y\hat{j}+z\hat{k}}{r}\right]$$

$$= \frac{\partial}{\partial x}\left[f'(r)\frac{x}{r}\right]+\frac{\partial}{\partial y}\left[f'(r)\frac{y}{r}\right]+\frac{\partial}{\partial z}\left[f'(r)\frac{z}{r}\right]$$

$$= \left(f''(r)\frac{\partial r}{\partial x}\right)\left(\frac{x}{r}\right)+f'(r)\frac{r.1-x\dfrac{\partial r}{\partial x}}{r^2}+\left(f''(r)\frac{\partial r}{\partial y}\right)\left(\frac{y}{r}\right)+$$

$$f'(r)\frac{r.1-y\dfrac{\partial r}{\partial y}}{r^2}+\left(f''(r)\frac{\partial r}{\partial z}\right)\left(\frac{z}{r}\right)+f'(r)\frac{r.1-z\dfrac{\partial r}{\partial z}}{r^2}$$

$$= \left(f''(r)\frac{x}{r}\right)\left(\frac{x}{r}\right)+f'(r)\frac{r-\dfrac{x^2}{r}}{r^2}+\left(f''(r)\frac{y}{r}\right)\left(\frac{y}{r}\right)+f'(r)\frac{r-\dfrac{y^2}{r}}{r^2}$$

$$+\left(f''(r)\frac{z}{r}\right)\left(\frac{z}{r}\right)+f'(r)\frac{r-\dfrac{z^2}{r}}{r^2}$$

$$= \left(f''(r)\frac{x}{r}\right)\left(\frac{x}{r}\right)+f'(r)\frac{r^2-x^2}{r^3}+\left(f''(r)\frac{y}{r}\right)+f'(r)\frac{r^2-y^2}{r^3}$$

$$+\left(f''(r)\frac{z}{r}\right)\left(\frac{z}{r}\right)+f'(r)\frac{r^2-z^2}{r^3}$$

$$= \left(f''(r)\frac{x^2}{r^2}\right)\left(\frac{x}{r}\right)+f'(r)\frac{r^2-x^2}{r^3}+\left(f''(r)\frac{y}{r}\right)+f'(r)\frac{r^2-y^2}{r^3}$$

$$= f''(r)\left[\frac{x^2}{r^2}+\frac{y^2}{r^2}+\frac{z^2}{r^2}\right]+f'(r)\left[\frac{y^2+z^2}{r^3}+\frac{z^2+x^2}{r^3}+\frac{x^2+y^2}{r^3}\right]$$

$$= f''(r)\frac{x^2+y^2+x^2}{r^2}+f'(r)\frac{2(x^2+y^2+z^2)}{r^3} = f''(r)\frac{r^2}{r^2}+f'(r)\frac{2r^2}{r^3}$$

$$= f''(r)+f'(r)\frac{2}{r}$$ **Ans.**

EXERCISE 2.2

1. Evaluate grad ϕ if $\phi = \log(x^2+y^2+z^2)$ **Ans.** $\dfrac{2(x\hat{i}+y\hat{j}+z\hat{k})}{x^2+y^2+z^2}$

2. Find a unit normal vector to the surface $x^2+y^2+z^2 = 5$ at the point (0, 1, 2). **Ans.** $\dfrac{1}{\sqrt{5}}(\hat{j}+2\hat{k})$

   (AMIETE, June 2010)

3. Calculate the directional derivative of the function $\phi(x, y, z) = xy^2 + yz^3$ at the point (1, –1, 1) in the direction of (3, 1, –1) *(A.M.I.E.T.E. Winter 2009)* **Ans.** $\dfrac{5}{\sqrt{11}}$

4. Find the direction in which the directional derivative of $f(x, y) = (x^2 - y^2)/xy$ at (1, 1) is zero.

   Ans. $\dfrac{\hat{i}+\hat{j}}{\sqrt{2}}$

5. Find the directional derivative of the scalar function of $(x, y, z) = xyz$ in the direction of the outer normal to the surface $z = xy$ at the point (3, 1, 3). **Ans.** $\frac{27}{\sqrt{11}}$

6. The temperature of the points in space is given by $T(x, y, z) = x^2 + y^2 - z$. A mosquito located at (1, 1, 2) desires to fly in such a direction that it will get warm as soon as possible. In what direction should it move? **Ans.** $\frac{1}{3}(2\hat{i} + 2\hat{j} - \hat{k})$

7. If $\phi(x, y, z) = 3xz^2y - y^3z^2$, find grad ϕ at the point (1, –2, –1) **Ans.** $-(16\hat{i} + 9\hat{j} + 4\hat{k})$

8. Find a unit vector normal to the surface $x^2y + 2xz = 4$ at the point (2, –2, 3).

Ans. $\frac{1}{3}(-\hat{i} + 2\hat{j} + 2\hat{k})$

9. What is the greatest rate of increase of the function $u = xyz^2$ at the point (1, 0, 3)? **Ans.** 9

10. If θ is the acute angle between the surfaces $xyz^2 = 3x + z^2$ and $3x^2 - y^2 + 2z = 1$ at the point (1, –2, 1) show that $\cos\theta = 3/7\sqrt{6}$.

11. Find the values of constants a, b, c so that the maximum value of the directional derivative of $\phi = axy^2 + byz + cz^2x^3$ at (1, 2, –1) has a maximum magnitude 64 in the direction parallel to the axis of z. **Ans.** $a = 6, b = 24, c = -8$

12. The position vector of a particle at time t is $R = \cos(t-1)\, i + \sinh(t-1)\, j + at^2 k$. If at $t = 1$, the acceleration of the particle be perpendicular to its position vector, then a is equal to

(*a*) 0 (*b*) 1 (*c*) $\frac{1}{2}$ (*d*) $\frac{1}{\sqrt{2}}$ (*AMIETE, Dec. 2009*) **Ans.** (*d*)

2.8 DIVERGENCE OF A VECTOR FUNCTION

The divergence of a vector point function $\vec{F}$ is denoted by *div F* and is defined as below.

Let $\vec{F} = F_1\hat{i} + F_2\hat{j} + F_3\hat{k}$

$$\operatorname{div}\vec{F} = \vec{\nabla}.\vec{F} = \left(\hat{i}\frac{\partial}{\partial x} + \hat{j}\frac{\partial}{\partial y} + \hat{k}\frac{\partial}{\partial z}\right)(\hat{i}F_1 + \hat{j}F_2 + \hat{k}F_3) = \frac{\partial F_1}{\partial x} + \frac{\partial F_2}{\partial y} + \frac{\partial F_3}{\partial z}$$

It is evident that div F is scalar function.

2.9 PHYSICAL INTERPRETATION OF DIVERGENCE

Let us consider the case of a fluid flow. Consider a small rectangular parallelopiped of dimensions dx, dy, dz parallel to x,y and z axes respectively.

Let $\vec{V} = V_x\hat{i} + V_y\hat{j} + V_z\hat{k}$ be the velocity of the fluid at $P(x, y, z)$.

∴ Mass of fluid flowing in through the face *ABCD* in unit time

$$= \text{Velocity} \times \text{Area of the face} = V_x\,(dy\,dz)$$

Mass of fluid flowing out across the face *PQRS* per unit time

$$= V_{x+dx}\,(dy\,dz)$$

$$= \left(V_x + \frac{\partial V_x}{\partial x}dx\right)(dy\,dz)$$

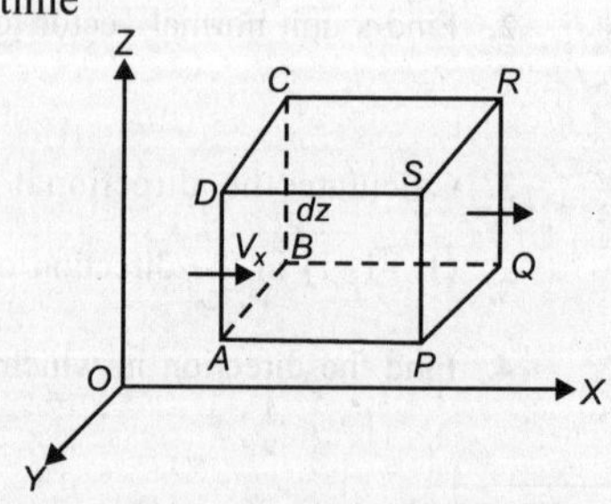

Net decrease in mass of fluid in the parallelopiped corresponding to the flow along x-axis per unit time

$$= V_x\,dy\,dz - \left(V_x + \frac{\partial V_x}{\partial x}dx\right)dy\,dz$$

$$= -\frac{\partial V_x}{\partial x} dx\, dy\, dz \qquad \text{(Minus sign shows decrease)}$$

Similarly, the decrease in mass of fluid to the flow along y-axis $= \frac{\partial V_y}{\partial y} dx\, dy\, dz$

and the decrease in mass of fluid to the flow along z-axis $= \frac{\partial V_z}{\partial z} dx\, dy\, dz$

Total decrease of the amount of fluid per unit time $= \left(\frac{\partial V_x}{\partial x} + \frac{\partial V_y}{\partial y} + \frac{\partial V_z}{\partial z}\right) dx\, dy\, dz$

Thus the rate of loss of fluid per unit volume $= \frac{\partial V_x}{\partial x} + \frac{\partial V_y}{\partial y} + \frac{\partial V_z}{\partial z}$

$$= \left(\hat{i}\frac{\partial}{\partial x} + \hat{j}\frac{\partial}{\partial y} + \hat{k}\frac{\partial}{\partial z}\right).(\hat{i}V_x + \hat{j}V_y + \hat{k}V_z) = \overline{\nabla}.\overline{V} = \text{div}\,\overline{V}$$

If the fluid is compressible, there can be no gain or loss in the volume element. Hence

$$\text{div}\,\vec{V} = 0 \qquad ...(1)$$

and V is called a *Solenoidal* vector function.
Equation (1) is also called the *equation of continuity or conservation of mass.*

Example 26. *If* $\vec{v} = \frac{x\hat{i} + y\hat{j} + z\hat{k}}{\sqrt{x^2 + y^2 + z^2}}$, *find the value of div* $\overline{v}$.

Solution. We have, $\vec{v} = \frac{x\hat{i} + y\hat{j} + z\hat{k}}{\sqrt{x^2 + y^2 + z^2}}$

$$\text{div}\,\vec{v} = \vec{\nabla}.\vec{v} = \left(\hat{i}\frac{\partial}{\partial x} + \hat{j}\frac{\partial}{\partial y} + \hat{k}\frac{\partial}{\partial z}\right).\left(\frac{x\hat{i} + y\hat{j} + z\hat{k}}{(x^2 + y^2 + z^2)^{1/2}}\right)$$

$$= \frac{\partial}{\partial x}\frac{x}{(x^2 + y^2 + z^2)^{1/2}} + \frac{\partial}{\partial y}\frac{y}{(x^2 + y^2 + z^2)^{1/2}} + \frac{\partial}{\partial z}\frac{z}{(x^2 + y^2 + z^2)^{1/2}}$$

$$= \frac{\left[(x^2 + y^2 + z^2)^{1/2} - x.\frac{1}{2}(x^2 + y^2 + z^2)^{-\frac{1}{2}}.2x\right]}{(x^2 + y^2 + z^2)} + \frac{\left[(x^2 + y^2 + z^2)^{\frac{1}{2}} - y.\frac{1}{2}(x^2 + y^2 + z^2)^{-\frac{1}{2}} \times 2y\right]}{(x^2 + y^2 + z^2)}$$

$$+ \frac{\left[(x^2 + y^2 + z^2)^{1/2} - z.\frac{1}{2}(x^2 + y^2 + z^2)^{-1/2}.2z\right]}{(x^2 + y^2 + z^2)}$$

$$= \frac{(x^2 + y^2 + z^2) - x^2}{(x^2 + y^2 + z^2)^{3/2}} + \frac{(x^2 + y^2 + z^2) - y^2}{(x^2 + y^2 + z^2)^{3/2}} + \frac{(x^2 + y^2 + z^2) - z^2}{(x^2 + y^2 + z^2)^{3/2}}$$

$$= \frac{y^2 + z^2 + x^2 + z^2 + x^2 + y^2}{(x^2 + y^2 + z^2)^{3/2}} = \frac{2(x^2 + y^2 + z^2)}{(x^2 + y^2 + z^2)^{3/2}} = \frac{2}{\sqrt{(x^2 + y^2 + z^2)}} \qquad \textbf{Ans.}$$

Example 27. *If $u = x^2 + y^2 + z^2$, and $\overline{r} = x\hat{i} + y\hat{j} + z\hat{k}$, then find div $(u\overline{r})$ in terms of u.*

Solution. $\quad \text{div}\ (u\vec{r}) = \left(\hat{i}\frac{\partial}{\partial x} + \hat{j}\frac{\partial}{\partial y} + \hat{k}\frac{\partial}{\partial z}\right).[(x^2+y^2+z^2)(x\hat{i}+y\hat{j}+z\hat{k})]$

$$= \left(\hat{i}\frac{\partial}{\partial x} + \hat{j}\frac{\partial}{\partial y} + \hat{k}\frac{\partial}{\partial z}\right).[(x^2+y^2+z^2)x\hat{i}+(x^2+y^2+z^2)y\hat{j}+(x^2+y^2+z^2)z\hat{k}]$$

$$= \frac{\partial}{\partial x}(x^3+xy^2+xz^2) + \frac{\partial}{\partial y}(x^2y+y^3+yz^2) + \frac{\partial}{\partial z}(x^2z+y^2z+z^3)$$

$$= (3x^2+y^2+z^2) + (x^2+3y^2+z^2) + (x^2+y^2+3z^2) = 5(x^2+y^2+z^2) = 5u \qquad \textbf{Ans.}$$

Example 28. *Find the value of n for which the vector $r^n\vec{r}$ is solenoidal, where $\overline{r} = x\hat{i} + y\hat{j} + z\hat{k}$.*

Solution. Divergence $(\vec{F}) = \vec{\nabla}.\vec{F} = \vec{\nabla}.r^n\vec{r} = \nabla.(x^2+y^2+z^2)^{n/2}(x\hat{i}+y\hat{j}+z\hat{k})$

$$= \left[\hat{i}\frac{\partial}{\partial x} + \hat{j}\frac{\partial}{\partial y} + \hat{k}\frac{\partial}{\partial z}\right].[(x^2+y^2+z^2)^{n/2}x\hat{i}+(x^2+y^2+z^2)^{n/2}y\hat{j}+(x^2+y^2+z^2)^{n/2}z\hat{k}]$$

$$= \frac{n}{2}(x^2+y^2+z^2)^{n/2-1}(2x^2) + (x^2+y^2+z^2)^{n/2} + \frac{n}{2}(x^2+y^2+z^2)^{n/2-1}(2y^2)$$

$$+ (x^2+y^2+z^2)^{n/2} + \frac{n}{2}(x^2+y^2+z^2)^{n/2-1}(2z^2) + (x^2+y^2+z^2)^{n/2}$$

$$= n(x^2+y^2+z^2)^{n/2-1}(x^2+y^2+z^2) + 3(x^2+y^2+z^2)^{n/2}$$

$$= n(x^2+y^2+z^2)^{n/2} + 3(x^2+y^2+z^2)^{n/2} = (n+3)(x^2+y^2+z^2)^{n/2}$$

If $r^n\vec{r}$ is solenoidal, then $(n+3)(x^2+y^2+z^2)^{n/2} = 0$ or $n+3=0$ or $n=-3$. **Ans.**

Example 29. *Show that* $\nabla\left[\frac{(\vec{a}.\vec{r})}{r^n}\right] = \frac{\vec{a}}{r^n} - \frac{n(\vec{a}.\vec{r})\vec{r}}{r^{n+2}}$. *(M.U. 2005)*

Solution. We have, $\quad \frac{\vec{a}.\vec{r}}{r^n} = \frac{(a_1\hat{i}+a_2\hat{j}+a_3\hat{k}).(x\hat{i}+y\hat{j}+z\hat{k})}{r^n} = \frac{a_1x+a_2y+a_3z}{r^n}$

Let $\quad \phi = \frac{\vec{a}.\vec{r}}{r^n} = \frac{a_1x+a_2y+a_3z}{r^n}$

$$\therefore \quad \frac{\partial\phi}{\partial x} = \frac{r^n.a_1 - (a_1x+a_2y+a_3z)\,nr^{n-1}(\partial r/\partial x)}{r^{2n}}$$

But $r^2 = x^2+y^2+z^2 \quad \Rightarrow \quad 2r\frac{\partial r}{\partial x} = 2x \quad \Rightarrow \quad \frac{\partial r}{\partial x} = \frac{x}{r}$

$$\therefore \quad \frac{\partial\phi}{\partial x} = \frac{a_1r^n - (a_1x+a_2y+a_3z).n\,r^{n-2}.x}{r^{2n}} = \frac{a_1}{r^n} - \frac{n(a_1x+a_2y+a_3z)x}{r^{n+2}}$$

$$\therefore \quad \nabla\phi = \frac{\partial\phi}{\partial x}\hat{i} + \frac{\partial\phi}{\partial y}\hat{j} + \frac{\partial\phi}{\partial z}\hat{k}$$

$$= \frac{1}{r^n}(a_1\hat{i}+a_2\hat{j}+a_3\hat{k}) - \frac{n}{r^{n+2}}[(a_1x+a_2y+a_3z)(x\hat{i}+y\hat{j}+z\hat{k})] = \frac{\overline{a}}{r^n} - \frac{n}{r^{n+2}}(\overline{a}.\overline{r})\overline{r}$$

Example 30. *Determine* $\vec{\nabla}\cdot\left(\dfrac{\vec{r}}{r^n}\right)$, $n > 0$ *and* $\vec{r} = x\hat{i} + y\hat{j} + z\hat{k}$ (*DU, 2012*)

Solution. $$\vec{\nabla}\cdot\left(\frac{\vec{r}}{r^n}\right) = \left(\hat{i}\frac{\partial}{\partial x}+\hat{j}\frac{\partial}{\partial y}+\hat{k}\frac{\partial}{\partial z}\right)\left[\frac{x\hat{i}+y\hat{j}+z\hat{k}}{(x^2+y^2+z^2)^{\frac{n}{2}}}\right]$$

$$= \hat{i}\cdot\frac{\partial}{\partial x}\frac{x\hat{i}}{(x^2+y^2+z^2)^{\frac{n}{2}}}+\hat{j}\cdot\frac{\partial}{\partial y}\frac{y\hat{j}}{(x^2+y^2+z^2)^{\frac{n}{2}}}+\hat{k}\cdot\frac{\partial}{\partial z}\frac{z\hat{k}}{(x^2+y^2+z^2)^{\frac{n}{2}}}$$

$$= \frac{\partial}{\partial x}\frac{x}{(x^2+y^2+z^2)^{\frac{n}{2}}}+\frac{\partial}{\partial y}\frac{y}{(x^2+y^2+z^2)^{\frac{n}{2}}}+\frac{\partial}{\partial z}\frac{z}{(x^2+y^2+z^2)^{\frac{n}{2}}}$$

$$= \frac{(x^2+y^2+z^2)^{\frac{n}{2}}\cdot 1 - x\frac{n}{2}(x^2+y^2+z^2)^{\frac{n}{2}-1}(2x)}{(x^2+y^2+z^2)^n}+\frac{(x^2+y^2+z^2)^{\frac{n}{2}}\cdot 1 - y\cdot\frac{n}{2}(x^2+y^2+z^2)^{\frac{n}{2}-1}(2y)}{(x^2+y^2+z^2)^n}$$

$$+\frac{(x^2+y^2+z^2)^{\frac{n}{2}}\cdot 1 - z\cdot\frac{n}{2}(x^2+y^2+z^2)^{\frac{n}{2}-1}(2z)}{(x^2+y^2+z^2)^n}$$

$$= \frac{(x^2+y^2+z^2)-nx^2}{(x^2+y^2+z^2)^{\frac{n}{2}+1}}+\frac{(x^2+y^2+z^2)-ny^2}{(x^2+y^2+z^2)^{\frac{n}{2}+1}}+\frac{(x^2+y^2+z^2)-nz^2}{(x^2+y^2+z^2)^{\frac{n}{2}+1}}$$

$$= \frac{x^2-nx^2+y^2+z^2}{(x^2+y^2+z^2)^{\frac{n}{2}+1}}+\frac{x^2+y^2-ny^2+z^2}{(x^2+y^2+z^2)^{\frac{n}{2}+1}}+\frac{x^2+y^2+z^2-nz^2}{(x^2+y^2+z^2)^{\frac{n}{2}+1}}$$

$$= \frac{(1-n)x^2+y^2+z^2}{(x^2+y^2+z^2)^{\frac{n}{2}+1}}+\frac{x^2+(1-n)y^2+z^2}{(x^2+y^2+z^2)^{\frac{n}{2}+1}}+\frac{x^2+y^2+(1-n)z^2}{(x^2+y^2+z^2)^{\frac{n}{2}+1}}$$

$$= \frac{(3-n)(x^2+y^2+z^2)}{(x^2+y^2+z^2)^{\frac{n}{2}+1}} = \frac{3-n}{(x^2+y^2+z^2)^{\frac{n}{2}}} = \frac{3-n}{r^n}$$ **Ans.**

Example 31. *Find the directional derivative of div* $(\vec{u})$ *at the point (1, 2, 2) in the direction of the outer normal of the sphere* $x^2 + y^2 + z^2 = 9$ *for* $\vec{u} = x^4\hat{i} + y^4\hat{j} + z^4\hat{k}$.

Solution. $\text{div}(\vec{u}) = \nabla.\vec{u}$

$$= \left(\hat{i}\frac{\partial}{\partial x}+\hat{j}\frac{\partial}{\partial y}+\hat{k}\frac{\partial}{\partial z}\right).(x^4\hat{i}+y^4\hat{j}+z^4\hat{k}) = 4x^3+4y^3+4z^3$$

Outer normal of the sphere $= \nabla(x^2 + y^2 + z^2 - 9)$

$$= \left(\hat{i}\frac{\partial}{\partial x}+\hat{j}\frac{\partial}{\partial y}+\hat{k}\frac{\partial}{\partial z}\right)(x^2+y^2+z^2-9) = 2x\hat{i}+2y\hat{j}+2z\hat{k}$$

Outer normal of the sphere at (1, 2, 2) $= 2\hat{i}+4\hat{j}+4\hat{k}$ ...(1)

Directional derivative $= \vec{\nabla}(4x^3 + 4y^3 + 4z^3)$

$$= \left(\hat{i}\frac{\partial}{\partial x} + \hat{j}\frac{\partial}{\partial y} + \hat{k}\frac{\partial}{\partial z}\right)(4x^3 + 4y^3 + 4z^3) = 12x^2\,\hat{i} + 12y^2\,\hat{j} + 12z^2\,\hat{k}$$

Directional derivative at (1, 2, 2) $= 12\hat{i} + 48\hat{j} + 48\hat{k}$ ...(2)

Directional derivative along the outer normal $= (12\hat{i} + 48\hat{j} + 48\hat{k}).\dfrac{2\hat{i} + 4\hat{j} + 4\hat{k}}{\sqrt{4+16+16}}$

[From (1), (2)]

$$= \frac{24 + 192 + 192}{6} = 68$$ **Ans.**

Example 32. *Show that div (grad r^n) = $n(n+1)r^{n-2}$, where*

$$r = \sqrt{x^2 + y^2 + z^2}$$

Hence, show that $\nabla^2\left(\dfrac{1}{r}\right) = 0.$ *(AMIETE, Dec. 2010)*

Solution. grad $(r^n) = \hat{i}\dfrac{\partial}{\partial x}r^n + \hat{j}\dfrac{\partial}{\partial y}r^n + \hat{k}\dfrac{\partial}{\partial z}r^n$ by definition

$$= \hat{i}\,n\,r^{n-1}\frac{\partial r}{\partial x} + \hat{j}\,n\,r^{n-1}\frac{\partial r}{\partial y} + \hat{k}.n\,r^{n-1}\frac{\partial r}{\partial z} = n\,r^{n-1}\left[\hat{i}\frac{\partial r}{\partial x} + \hat{j}\frac{\partial r}{\partial y} + \hat{k}\frac{\partial r}{\partial z}\right]$$

$$= n\,r^{n-1}\left[\hat{i}\left(\frac{x}{r}\right) + \hat{j}\left(\frac{y}{r}\right) + \hat{k}\left(\frac{z}{r}\right)\right] = nr^{n-2}(x\hat{i} + y\hat{j} + z\hat{k}) = nr^{n-2}\overline{r}.$$

$$\left[\because r^2 = x^2 + y^2 + z^2 \Rightarrow 2r\frac{\partial r}{\partial x} = 2x \Rightarrow \frac{\partial r}{\partial x} = \frac{x}{r} \text{ etc.}\right]$$

Thus, grad $(r^n) = n\,r^{n-2}\,x\hat{i} + n\,r^{n-2}\,y\hat{j} + n\,r^{n-2}\,z\hat{k}$ ...(1)

$\therefore$ div grad r^n = div $[n\,r^{n-2}\,x\hat{i} + n\,r^{n-2}\,y\hat{j} + n\,r^{n-2}\,z\hat{k}]$

$$= \left(\hat{i}\frac{\partial}{\partial x} + \hat{j}\frac{\partial}{\partial y} + \hat{k}\frac{\partial}{\partial z}\right).(nr^{n-2}\,x\hat{i} + nr^{n-2}\,y\hat{j} + nr^{n-2}\,z\hat{k}) \quad \text{[From (1)]}$$

$$= \frac{\partial}{\partial x}(n\,r^{n-2}x) + \frac{\partial}{\partial y}(n\,r^{n-2}y) + \frac{\partial}{\partial z}(n\,r^{n-2}z) \quad \text{(By definition)}$$

$$= \left(n\,r^{n-2} + nx\,(n-2)\,r^{n-3}\frac{\partial r}{\partial x}\right) + \left(n\,r^{n-2} + ny(n-2)\,r^{n-3}\frac{\partial r}{\partial y}\right) + \left(n\,r^{n-2} + nz\,(n-2)\,r^{n-3}\frac{\partial r}{\partial z}\right)$$

$$= 3n\,r^{n-2} + n(n-2)r^{n-3}\left[x\frac{\partial r}{\partial x} + y\frac{\partial r}{\partial y} + z\frac{\partial r}{\partial z}\right]$$

$$= 3n\,r^{n-2} + n(n-2)r^{n-3}\left[x\left(\frac{x}{r}\right) + y\left(\frac{y}{r}\right) + z\left(\frac{z}{r}\right)\right]$$

$$\left[\because r^2 = x^2 + y^2 + z^2 \Rightarrow 2r\frac{\partial r}{\partial x} = 2x \Rightarrow \frac{\partial r}{\partial x} = \frac{x}{r} \text{ etc.}\right]$$

$$= 3nr^{n-2} + n(n-2)r^{n-4}[x^2 + y^2 + z^2]$$

$$= 3nr^{n-2} + n(n-2)r^{n-4}.r^2 \qquad (\because r^2 = x^2 + y^2 + z^2)$$

$$= r^{n-2}[3n + n^2 - 2n] = r^{n-2}(n^2 + n) = n(n+1)r^{n-2}$$

If we put $n = -1$

$$\text{div grad } (r^{-1}) = -1(-1+1)r^{-1-2}$$

$$\Rightarrow \quad \nabla^2\left(\frac{1}{r}\right) = 0$$

EXERCISE 2.3

1. If $r = x\hat{i} + y\hat{j} + z\hat{k}$ and $r = |\vec{r}|$, show that (*i*) div $\left(\frac{\vec{r}}{|\vec{r}|^3}\right) = 0$,
   (*ii*) div $(r\,\phi) = 3\phi + r$ grad ϕ.
2. Show that the vector $V = (x+3y)\hat{i} + (y-3z)\hat{j} + (x-2z)\hat{k}$ is solenoidal.
3. Show that $\nabla.(\phi A) = \nabla\phi.A + \phi(\nabla.A)$
4. If ρ, ϕ, z are cylindrical coordinates, show that grad $(\log \rho)$ and grad ϕ are solenoidal vectors.
5. Obtain the expression for $\nabla^2 f$ in spherical coordinates from their corresponding expression in orthogonal curvilinear coordinates.
6. If $\vec{r} = x\hat{i} + y\hat{j} + z\hat{k}$, and $r = |r|$ find div $\left(\frac{r}{r^2}\right)$. (*U.P. I Sem., Dec. 2006*) **Ans.** $\frac{1}{r^2}$

Prove the following:

7. $\nabla.(\nabla\phi) = \nabla^2\phi$
8. $\vec{\nabla} \times \frac{(\vec{A} \times \vec{R})}{r^n} = \frac{(2-n)\vec{A}}{r^n} + \frac{n(\vec{A}.\vec{R})\vec{R}}{r^{n+2}}, r = |\vec{R}|$
9. div $(f\nabla g)$ – div $(g\nabla f) = f\nabla^2 g - g\nabla^2 f$

2.10 CURL

(*U.P., I semester, Dec. 2006*)

The curl of a vector point function F is defined as below

$$\text{curl } \vec{F} = \vec{\nabla} \times \vec{F} \qquad (\vec{F} = F_1\hat{i} + F_2\hat{j} + F_3\hat{k})$$

$$= \left(\hat{i}\frac{\partial}{\partial x} + \hat{j}\frac{\partial}{\partial y} + \hat{k}\frac{\partial}{\partial z}\right) \times (F_1\hat{i} + F_2\hat{j} + F_3\hat{k})$$

$$= \begin{vmatrix} \hat{i} & \hat{j} & \hat{k} \\ \frac{\partial}{\partial x} & \frac{\partial}{\partial y} & \frac{\partial}{\partial z} \\ F_1 & F_2 & F_3 \end{vmatrix} = \hat{i}\left(\frac{\partial F_3}{\partial y} - \frac{\partial F_2}{\partial z}\right) - \hat{j}\left(\frac{\partial F_3}{\partial x} - \frac{\partial F_1}{\partial z}\right) + \hat{k}\left(\frac{\partial F_2}{\partial x} - \frac{\partial F_1}{\partial y}\right)$$

Curl $\vec{F}$ is a vector quantity.

2.11 PHYSICAL MEANING OF CURL

(*Delhi University April, 2010, M.D.U., Dec. 2009, U.P. I Semester, Winter 2009*)

We know that $\vec{V} = \vec{\omega} \times \vec{r}$, where ω is the angular velocity, $\vec{V}$ *is the linear velocity and* $\vec{r}$ *is the position vector of a point on the rotating body.*

$$\text{Curl } \vec{V} = \vec{\nabla} \times \vec{V} \qquad \left[\begin{array}{l}\vec{\omega} = \omega_1 \hat{i} + \omega_2 \hat{j} + \omega_3 \hat{k} \\ \vec{r} = x\hat{i} + y\hat{j} + z\hat{k}\end{array}\right]$$

$$= \vec{\nabla} \times (\vec{\omega} \times \vec{r}) = \vec{\nabla} \times [(\omega_1 \hat{i} + \omega_2 \hat{j} + \omega_3 \hat{k}) \times (x\hat{i} + y\hat{j} + z\hat{k})]$$

$$= \vec{\nabla} \times \begin{vmatrix} \hat{i} & \hat{j} & \hat{k} \\ \omega_1 & \omega_2 & \omega_3 \\ x & y & z \end{vmatrix} = \vec{\nabla} \times [(\omega_2 z - \omega_3 y)\hat{i} - (\omega_1 z - \omega_3 x)\hat{j} + (\omega_1 y - \omega_2 x)\hat{k}]$$

$$= \left(\hat{i}\frac{\partial}{\partial x} + \hat{j}\frac{\partial}{\partial y} + \hat{k}\frac{\partial}{\partial z}\right) \times [(\omega_2 z - \omega_3 y)\hat{i} - (\omega_1 z - \omega_3 x)\hat{j} + (\omega_1 y - \omega_2 x)\hat{k}]$$

$$= \begin{vmatrix} \hat{i} & \hat{j} & \hat{k} \\ \dfrac{\partial}{\partial x} & \dfrac{\partial}{\partial y} & \dfrac{\partial}{\partial z} \\ \omega_2 z - \omega_3 y & \omega_3 x - \omega_1 z & \omega_1 y - \omega_2 x \end{vmatrix}$$

$$= (\omega_1 + \omega_1)\hat{i} - (-\omega_2 - \omega_2)\hat{j} + (\omega_3 + \omega_3)\hat{k} = 2(\omega_1 \hat{i} + \omega_2 \hat{j} + \omega_3 \hat{k}) = 2\omega$$

Curl $\vec{V} = 2\omega$ which shows that curl of a vector field is connected with rotational properties of the vector field and justifies the name *rotation* used for curl.

If Curl $\overline{F} = 0$, the field F is termed as *irrotational.*

Example 33. *Find the divergence and curl of* $\vec{v} = (xyz)\hat{i} + (3x^2 y)\hat{j} + (xz^2 - y^2 z)\hat{k}$ *at (2, –1, 1)*

Solution. Here, we have

$$\vec{v} = (xyz)\hat{i} + (3x^2 y)\hat{j} + (xz^2 - y^2 z)\hat{k}$$

$$\text{Div. } \vec{v} = \nabla . \vec{v}$$

$$\text{Div } \vec{v} = \frac{\partial}{\partial x}(xyz) + \frac{\partial}{\partial y}(3x^2 y) + \frac{\partial}{\partial z}(xz^2 - y^2 z)$$

$$= yz + 3x^2 + 2xz - y^2 = -1 + 12 + 4 - 1 = 14 \text{ at } (2, -1, 1)$$

$$\text{Curl } \overline{v} = \begin{vmatrix} \hat{i} & \hat{j} & \hat{k} \\ \dfrac{\partial}{\partial x} & \dfrac{\partial}{\partial y} & \dfrac{\partial}{\partial z} \\ xyz & 3x^2 y & xz^2 - y^2 z \end{vmatrix} = -2yz\,\hat{i} - (z^2 - xy)\hat{j} + (6xy - xz)\hat{k}$$

$$= -2yz\,\hat{i} + (xy - z^2)\hat{j} + (6xy - xz)\hat{k}$$

Curl at (2, –1, 1)

$$= -2(-1)(1)\hat{i} + \{(2)(-1) - 1\}\hat{j} + \{6(2)(-1) - 2(1)\}\hat{k}$$

$$= 2\hat{i} - 3\hat{j} - 14\hat{k}$$

Ans.

Example 34. *If* $\vec{V} = \dfrac{x\hat{i} + y\hat{j} + z\hat{k}}{\sqrt{x^2 + y^2 + z^2}}$, *find the value of curl* $\vec{V}$.

Solution.

$$\text{Curl } \vec{V} = \vec{\nabla} \times \vec{V}$$

$$= \left(\hat{i}\frac{\partial}{\partial x} + \hat{j}\frac{\partial}{\partial y} + \hat{k}\frac{\partial}{\partial z}\right) \times \left(\frac{x\hat{i} + y\hat{j} + z\hat{k}}{(x^2 + y^2 + z^2)^{1/2}}\right)$$

$$= \begin{vmatrix} \hat{i} & \hat{j} & \hat{k} \\ \dfrac{\partial}{\partial x} & \dfrac{\partial}{\partial y} & \dfrac{\partial}{\partial z} \\ \dfrac{x}{(x^2 + y^2 + z^2)^{1/2}} & \dfrac{y}{(x^2 + y^2 + z^2)^{1/2}} & \dfrac{z}{(x^2 + y^2 + z^2)^{1/2}} \end{vmatrix}$$

$$= \hat{i}\left[\frac{\partial}{\partial y}\left(\frac{z}{(x^2 + y^2 + z^2)^{1/2}}\right) - \frac{\partial}{\partial z}\left(\frac{y}{(x^2 + y^2 + z^2)^{1/2}}\right)\right] - \hat{j}\left[\frac{\partial}{\partial x}\left(\frac{z}{(x^2 + y^2 + z^2)^{1/2}}\right)\right.$$

$$\left. - \frac{\partial}{\partial z}\left(\frac{x}{(x^2 + y^2 + z^2)^{1/2}}\right)\right] + \hat{k}\left[\frac{\partial}{\partial x}\left(\frac{y}{(x^2 + y^2 + z^2)^{1/2}}\right) - \frac{\partial}{\partial y}\left(\frac{x}{(x^2 + y^2 + z^2)^{1/2}}\right)\right]$$

$$= \hat{i}\left[\frac{-yz}{(x^2+y^2+z^2)^{3/2}} + \frac{y.z}{(x^2+y^2+z^2)^{3/2}}\right] - \hat{j}\left[\frac{-zx}{(x^2+y^2+z^2)^{3/2}} + \frac{zx}{(x^2+y^2+z^2)^{3/2}}\right]$$

$$+ \hat{k}\left[\frac{-xy}{(x^2+y^2+z^2)^{3/2}} + \frac{xy}{(x^2+y^2+z^2)^{3/2}}\right] = 0 \quad \textbf{Ans.}$$

Example 35. *Prove that* $(y^2 - z^2 + 3yz - 2x)\hat{i} + (3xz + 2xy)\hat{j} + (3xy - 2xz + 2z)\hat{k}$ *is both solenoidal and irrotational.* *(U.P., I Sem, Dec. 2008)*

Solution. Let $\vec{F} = (y^2 - z^2 + 3yz - 2x)\hat{i} + (3xz + 2xy)\hat{j} + (3xy - 2xz + 2z)\hat{k}$

For solenoidal, we have to prove $\vec{\nabla}.\vec{F} = 0$.

Now, $$\vec{\nabla}.\vec{F} = \left[\hat{i}\frac{\partial}{\partial x} + \hat{j}\frac{\partial}{\partial y} + \hat{k}\frac{\partial}{\partial z}\right]\cdot\left[(y^2 - z^2 + 3yz - 2x)\hat{i} + (3xz + 2xy)\hat{j} + (3xy - 2xz + 2z)\hat{k}\right]$$

$$= -2 + 2x - 2x + 2 = 0$$

Thus, $\vec{F}$ is solenoidal. For irrotational, we have to prove Curl $\overline{F} = 0$.

Now, $$\text{Curl } \vec{F} = \begin{vmatrix} \hat{i} & \hat{j} & \hat{k} \\ \dfrac{\partial}{\partial x} & \dfrac{\partial}{\partial y} & \dfrac{\partial}{\partial z} \\ y^2 - z^2 + 3yz - 2x & 3xz + 2xy & 3xy - 2xz + 2z \end{vmatrix}$$

$$= (3x - 3x)\hat{i} - (-2z + 3y - 3y + 2z)\hat{j} + (3z + 2y - 2y - 3z)k$$

$$= 0\hat{i} + 0\hat{j} + 0\hat{k} = 0$$

Thus, $\vec{F}$ is irrotational.

Hence, $\vec{F}$ is both solenoidal and irrotational. **Proved.**

Example 36. *Determine the constants a and b such that the curl of vector*

$$\overline{A} = (2xy - 3yz)\hat{i} + (x^2 + axz - 4z^2)\hat{j} - (3xy + byz)\hat{k} \text{ is zero.}$$

(*U.P. I Semester, Dec 2008*)

Solution. Curl $A = \left(\hat{i}\frac{\partial}{\partial x} + \hat{j}\frac{\partial}{\partial y} + \hat{k}\frac{\partial}{\partial z}\right) \times [(2xy - 3yz)\hat{i} + (x^2 + axz - 4z^2)\hat{j} - (3xy + byz)\hat{k}]$

$$= \begin{vmatrix} \hat{i} & \hat{j} & \hat{k} \\ \frac{\partial}{\partial x} & \frac{\partial}{\partial y} & \frac{\partial}{\partial z} \\ 2xy - 3yz & x^2 + axz - 4z^2 & -3xy - byz \end{vmatrix}$$

$$= [-3x - bz - ax + 8z]\hat{i} - [-3y + 3y]\hat{j} + [2x + az - 2x - 3z]\hat{k}$$

$$= [-x(3 + a) + z(8 - b)]\hat{i} + 6y\hat{j} + z(3 + a)\hat{k}] = 0 \quad \text{(given)}$$

i.e., $3 + a = 0$ and $8 - b = 0$,

$a = -3, \quad b = 8$ **Ans.**

Example 37. *If a vector field is given by*

$\vec{F} = (x^2 - y^2 + x)\hat{i} - (2xy + y)\hat{j}$. *Is this field irrotational ? If so, find its scalar potential.*

(*U.P. I Semester, Dec 2009*)

Solution. Here, we have

$$\vec{F} = (x^2 - y^2 + x)\hat{i} - (2xy + y)\hat{j}$$

$$\text{Curl } F = \nabla \times \vec{F}$$

$$= \left(\hat{i}\frac{\partial}{\partial x} + \hat{j}\frac{\partial}{\partial y} + \hat{k}\frac{\partial}{\partial z}\right) \times (x^2 - y^2 + x)\hat{i} - (2xy + y)\hat{j}$$

$$= \begin{vmatrix} \hat{i} & \hat{j} & \hat{k} \\ \frac{\partial}{\partial x} & \frac{\partial}{\partial y} & \frac{\partial}{\partial z} \\ x^2 - y^2 + x & -2xy - y & 0 \end{vmatrix} = \hat{i}(0 - 0) - \hat{j}(0 - 0) + \hat{k}(-2y + 2y) = 0$$

Hence, vector field $\vec{F}$ is irrotational.

To find the scalar potential function ϕ

$$\vec{F} = \nabla \phi$$

$$d\phi = \frac{\partial \phi}{\partial x}dx + \frac{\partial \phi}{\partial y}dy + \frac{\partial \phi}{\partial z}dz = \left|\hat{i}\frac{\partial \phi}{\partial x} + \hat{j}\frac{\partial \phi}{\partial y} + \hat{k}\frac{\partial \phi}{\partial z}\right| \cdot (\hat{i}\, dx + \hat{j}\, dy + \hat{k}\, dz)$$

$$= \left(\hat{i}\frac{\partial}{\partial x} + \hat{j}\frac{\partial}{\partial y} + \hat{k}\frac{\partial}{\partial z}\right)\phi \cdot (d\vec{r}) = \nabla\phi \cdot d\vec{r} = \vec{F} \cdot d\vec{r}$$

$$= [(x^2 - y^2 + x)\hat{i} - (2xy + y)\hat{j}] \cdot (\hat{i}\, dx + \hat{j}\, dy + \hat{k}\, dz)$$

$$= (x^2 - y^2 + x)dx - (2xy + y)dy.$$

$$\phi = \int [(x^2 - y^2 + x)dx - (2xy + y)dy] + c$$

$$= \int \left[x^2\, dx + x\, dx - y\, dy - y^2\, dx - 2\, xy\, dy\right] + c = \frac{x^3}{3} + \frac{x^2}{2} - \frac{y^2}{2} - xy^2 + c$$

Hence, the scalar potential is $\frac{x^3}{3} + \frac{x^2}{2} - \frac{y^2}{2} - xy^2 + c$ **Ans.**

Example 38. *Find the scalar potential function f for* $\vec{A} = y^2\hat{i} + 2xy\hat{j} - z^2\hat{k}$.

(*Gujarat, I Semester, Jan. 2009*)

Solution. We have, $\vec{A} = y^2\hat{i} + 2xy\hat{j} - z^2\hat{k}$

$$\text{Curl } \vec{A} = \nabla \times \vec{A} = \left(\hat{i}\frac{\partial}{\partial x} + \hat{j}\frac{\partial}{\partial y} + \hat{k}\frac{\partial}{\partial z}\right) \times (y^2\hat{i} + 2xy\hat{j} - z^2\hat{k})$$

$$= \begin{vmatrix} \hat{i} & \hat{j} & \hat{k} \\ \frac{\partial}{\partial x} & \frac{\partial}{\partial y} & \frac{\partial}{\partial z} \\ y^2 & 2xy & -z^2 \end{vmatrix} = \hat{i}(0) - \hat{j}(0) + \hat{k}(2y - 2y) = 0$$

Hence, $\vec{A}$ is irrotational. To find the scalar potential function *f*.

$$\vec{A} = \nabla f$$

$$df = \frac{\partial f}{\partial x}dx + \frac{\partial f}{\partial y}dy + \frac{\partial f}{\partial z}dz = \left(\hat{i}\frac{\partial f}{\partial x} + \hat{j}\frac{\partial f}{\partial y} + \hat{k}\frac{\partial f}{\partial z}\right).(\hat{i}\, dx + \hat{j}\, dy + \hat{k}\, dz)$$

$$= \left(\hat{i}\frac{\partial}{\partial x} + \hat{j}\frac{\partial}{\partial y} + \hat{k}\frac{\partial}{\partial z}\right) f.dr = \nabla f.d\vec{r} = \vec{A}.dr \qquad (A = \nabla f)$$

$$= (y^2\hat{i} + 2xy\hat{j} - z^2\hat{k}).(\hat{i}\, dx + \hat{j}\, dy + \hat{k}\, dz)$$

$$= y^2\, dx + 2xy\, dy - z^2\, dz = d\,(xy^2) - z^2\, dz$$

$$f = \int d\,(xy^2) - \int z^2\, dz = xy^2 - \frac{z^3}{3} + C$$ **Ans.**

Example 39. *A vector field is given by* $\vec{A} = (x^2 + xy^2)\hat{i} + (y^2 + x^2y)\hat{j}$. *Show that the field is irrotational and find the scalar potential.*

Solution. $\vec{A}$ is irrotational if curl $\vec{A} = 0$

$$\text{Curl } \vec{A} = \nabla \times \vec{A} = \begin{vmatrix} \hat{i} & \hat{j} & \hat{k} \\ \frac{\partial}{\partial x} & \frac{\partial}{\partial y} & \frac{\partial}{\partial z} \\ x^2 + xy^2 & y^2 + x^2 y & 0 \end{vmatrix} = \hat{i}(0-0) - \hat{j}(0-0) + \hat{k}(2xy - 2xy) = 0$$

Hence, $\vec{A}$ is irrotational. If ϕ is the scalar potential, then

$$\vec{A} = \text{grad } \phi$$

$$d\phi = \frac{\partial \phi}{\partial x} dx + \frac{\partial \phi}{\partial y} dy + \frac{\partial \phi}{\partial z} dz \qquad \text{[Total differential coefficient]}$$

$$= \left(\hat{i} \frac{\partial \phi}{\partial x} + \hat{j} \frac{\partial \phi}{\partial y} + \hat{k} \frac{\partial \phi}{\partial z} \right) . (\hat{i}\, dx + \hat{j}\, dy + \hat{k}\, dz) = \text{grad } \phi . dr$$

$$= \vec{A}.dr = [(x^2 + xy^2)\hat{i} + (y^2 + x^2 y)\hat{j}].(\hat{i}\, dx + \hat{j}\, dy + \hat{k}\, dz)$$

$$= (x^2 + xy^2)\, dx + (y^2 + x^2 y)\, dy = x^2\, dx + y^2\, dy + (x\, dx) y^2 + (x^2)\,(y\, dy)$$

$$\phi = \int x^2 dx + \int y^2 dy + \int [(x\, dx)\, y^2 + (x^2)\,(y\, dy)] = \frac{x^3}{3} + \frac{y^3}{3} + \frac{x^2 y^2}{2} + c \qquad \textbf{Ans.}$$

Example 40. *Show that* $\vec{V}(x, y, z) = 2x\,y\,z\,\hat{i} + (x^2 z + 2y)\,\hat{j} + x^2 y\,\hat{k}$ *is irrotational and find a scalar function u(x, y, z) such that* $\vec{V}$ = *grad (u).*

Solution. $\vec{V}\,(x, y, z) = 2x\,y\,z\,\hat{i} + (x^2 z + 2y)\,\hat{j} + x^2 y\,\hat{k}$

$$\text{Curl } \vec{V} = \left(\hat{i} \frac{\partial}{\partial x} + \hat{j} \frac{\partial}{\partial y} + \hat{k} \frac{\partial}{\partial z} \right) \times [2x\,y\,z\,\hat{i} + (x^2 z + 2\,y)\,\hat{j} + x^2 y\,\hat{k}]$$

$$= \begin{vmatrix} \hat{i} & \hat{j} & \hat{k} \\ \frac{\partial}{\partial x} & \frac{\partial}{\partial y} & \frac{\partial}{\partial z} \\ 2x\,y\,z & x^2 z + 2\,y & x^2 y \end{vmatrix}$$

$$= (x^2 - x^2)\,\hat{i} - (2xy - 2xy)\,\hat{j} + (2xz - 2xz)\,\hat{k} = 0$$

Hence, $\vec{V}$ (x, y, z) is irrotational.

To find corresponding scalar function u, consider the following relations given

$$\vec{V} = \text{grad } (u)$$

or

$$\vec{V} = \vec{\nabla}(u) \qquad ...(1)$$

$$du = \frac{\partial u}{\partial x} dx + \frac{\partial u}{\partial y} dy + \frac{\partial u}{\partial z} dz \qquad \text{(Total differential coefficient)}$$

$$= \left(\hat{i} \frac{\partial u}{\partial x} + \hat{j} \frac{\partial u}{\partial y} + \hat{k} \frac{\partial u}{\partial z} \right) . (\hat{i}\, dx + \hat{j}\, dy + \hat{k}\, dz)$$

$$= \vec{\nabla} u . d\vec{r} = \vec{V} . d\vec{r} \qquad \text{[From (1)]}$$

$$= [2xyz\hat{i} + (x^2 z + 2y)\hat{j} + x^2 y\hat{k}].(\hat{i}\,dx + \hat{j}\,dy + \hat{k}\,dz)$$

$$= 2\,x\,y\,z\,dx + (x^2 z + 2y)\,dy + x^2 y\,dz$$

$$= y(2x\,z\,dx + x^2\,dz) + (x^2 z)\,dy + 2y\,dy$$

$$= [yd\,(x^2 z) + (x^2 z)\,dy] + 2y\,dy = d(x^2 yz) + 2y\,dy$$

Integrating, we get $u = x^2 yz + y^2$ **Ans.**

Example 41. *A fluid motion is given by* $\vec{v} = (y+z)\hat{i} + (z+x)\hat{j} + (x+y)\hat{k}$. *Show that the motion is irrotational and hence find the velocity potential.*

(*A.M.I.E.T.E., Dec.2007, Uttarakhand, I Semester 2006; U.P.,*)

Solution. Curl $\vec{v} = \nabla \times \vec{v}$

$$= \left(\hat{i}\frac{\partial}{\partial x} + \hat{j}\frac{\partial}{\partial y} + \hat{k}\frac{\partial}{\partial z}\right) \times [(y+z)\hat{i} + (z+x)\hat{j} + (x+y)\hat{k}]$$

$$= \begin{vmatrix} \hat{i} & \hat{j} & \hat{k} \\ \dfrac{\partial}{\partial x} & \dfrac{\partial}{\partial y} & \dfrac{\partial}{\partial z} \\ y+z & z+x & x+y \end{vmatrix} = (1-1)\hat{i} - (1-1)\hat{j} + (1-1)\hat{k} = 0$$

Hence, $\vec{v}$ is irrotational.

To find the corresponding velocity potential ϕ, consider the following relation.

$$\bar{v} = \nabla\phi$$

$$d\phi = \frac{\partial \phi}{\partial x}dx + \frac{\partial \phi}{\partial y}dy + \frac{\partial \phi}{\partial z}dz \qquad \text{[Total Differential coefficient]}$$

$$= \left(\hat{i}\frac{\partial \phi}{\partial x} + \hat{j}\frac{\partial \phi}{\partial y} + \hat{k}\frac{\partial \phi}{\partial z}\right).(\hat{i}\,dx + \hat{j}\,dy + \hat{k}\,dz)$$

$$= \left(\hat{i}\frac{\partial}{\partial x} + \hat{j}\frac{\partial}{\partial y} + \hat{k}\frac{\partial}{\partial z}\right)\phi . d\vec{r} = \nabla\phi . d\vec{r} = \vec{v} . d\vec{r}$$

$$= [(y+z)\hat{i} + (z+x)\hat{j} + (x+y)\hat{k}].(\hat{i}\,dx + \hat{j}\,dy + \hat{k}\,dz)$$

$$= (y+z)\,dx + (z+x)\,dy + (x+y)\,dz$$

$$= y\,dx + z\,dx + z\,dy + x\,dy + x\,dz + y\,dz$$

$$\phi = \int (y\,dx + x\,dy) + \int (z\,dy + y\,dz) + \int (z\,dx + x\,dz)$$

$$\phi = xy + yz + zx + c$$

Velocity potential $= xy + yz + zx + c$ **Ans.**

Example 42. *A fluid motion is given by*

$\vec{v} = (y \sin z - \sin x)\hat{i} + (x \sin z + 2yz)\hat{j} + (xy \cos z + y^2)\hat{k}$ *is the motion irrotational? If so, find the velocity potential.*

Solution. Curl $\vec{v} = \vec{\nabla} \times \vec{v}$

$$= \left(\hat{i}\frac{\partial}{\partial x} + \hat{j}\frac{\partial}{\partial y} + \hat{k}\frac{\partial}{\partial z}\right) \times (y \sin z - \sin x)\hat{i} + (x \sin z + 2yz)\hat{j} + (xy \cos z + y^2)\hat{k}$$

$$= \begin{vmatrix} \hat{i} & \hat{j} & \hat{k} \\ \frac{\partial}{\partial x} & \frac{\partial}{\partial y} & \frac{\partial}{\partial z} \\ y\sin z - \sin x & x\sin z + 2yz & xy\cos z + y^2 \end{vmatrix}$$

$$= (x\cos z + 2y - x\cos z - 2y)\hat{i} - [y\cos z - y\cos z]\hat{j} + (\sin z - \sin z)\hat{k} = 0$$

Hence, the motion is irrotational.

So, $\bar{v} = \bar{\nabla}\phi$ where ϕ is called velocity potential.

$$d\phi = \frac{\partial \phi}{\partial x}dx + \frac{\partial \phi}{\partial y}dy + \frac{\partial \phi}{\partial z}dz \qquad \text{[Total differential coefficient]}$$

$$= \left(\hat{i}\frac{\partial \phi}{\partial x} + \hat{j}\frac{\partial \phi}{\partial y} + \hat{k}\frac{\partial \phi}{\partial z}\right).(\hat{i}\,dx + \hat{j}\,dy + \hat{k}\,dz) = \vec{\nabla}\phi.d\vec{r} = \vec{v}.d\vec{r}$$

$$= [(y\sin z - \sin x)\hat{i} + (x\sin z + 2yz)\hat{j} + (xy\cos z + y^2)\hat{k}].[\hat{i}\,dx + \hat{j}\,dy + \hat{k}\,dz]$$

$$= (y\sin z - \sin x)\,dx + (x\sin z + 2\,y\,z)\,dy + (x\,y\cos z + y^2)\,dz$$

$$= (y\sin z\,dx + x\,dy\sin z + x\,y\cos z\,dz) - \sin x\,dx + (2\,y\,z\,dy + y^2\,dz)$$

$$= d\,(x\,y\sin z) + d\,(\cos x) + d\,(y^2\,z)$$

$$\phi = \int d\,(xy\sin z) + \int d\,(\cos x) + \int d(y^2 z)$$

$$\phi = xy\sin z + \cos x + y^2 z + c$$

Hence, Velocity potential $= xy\sin z + \cos x + y^2 z + c$. **Ans.**

Example 43. *Prove that* $\vec{F} = r^2\vec{r}$ *is conservative and find the scalar potential* ϕ *such that*

$$\vec{F} = \nabla\phi.$$

Solution. Given $\vec{F} = r^2\vec{r} = r^2(x\hat{i} + y\hat{j} + z\hat{k}) = r^2\,x\hat{i} + r^2\,y\hat{j} + r^2\,z\hat{k}$

Consider $\vec{\nabla}\times\vec{F} = \begin{vmatrix} \hat{i} & \hat{j} & \hat{k} \\ \frac{\partial}{\partial x} & \frac{\partial}{\partial y} & \frac{\partial}{\partial z} \\ r^2 x & r^2 y & r^2 z \end{vmatrix}$

$$= \hat{i}\left[\frac{\partial}{\partial y}r^2 z - \frac{\partial}{\partial z}r^2 y\right] - \hat{j}\left[\frac{\partial}{\partial x}r^2 z - \frac{\partial}{\partial z}r^2 x\right] + \hat{k}\left[\frac{\partial}{\partial x}r^2 y - \frac{\partial}{\partial y}r^2 x\right]$$

$$= \hat{i}\left[2rz\frac{\partial r}{\partial y} - 2ry\frac{\partial r}{\partial z}\right] - \hat{j}\left[2rz\frac{\partial r}{\partial x} - 2rx\frac{\partial r}{\partial z}\right] + \hat{k}\left[2ry\frac{\partial r}{\partial x} - 2rx\frac{\partial r}{\partial y}\right]$$

$$\left[\text{But } r^2 = x^2 + y^2 + z^2, \frac{\partial r}{\partial x} = \frac{x}{r}, \frac{\partial r}{\partial y} = \frac{y}{r}, \frac{\partial r}{\partial z} = \frac{z}{r}\right]$$

$$= \hat{i}\left[2rz\frac{y}{r} - 2ry\frac{z}{r}\right] - \hat{j}\left[2rz\frac{x}{r} - 2rx\frac{z}{r}\right] + \hat{k}\left[2ry\frac{x}{r} - 2rx\frac{y}{r}\right]$$

$$= \hat{i}(2yz - 2yz) - \hat{j}(2zx - 2zx) + \hat{k}(2xy - 2xy) = 0\hat{i} - 0\hat{j} + 0\hat{k} = 0$$

$$\therefore \quad \nabla\times\vec{F} = 0$$

$\therefore \quad \vec{F}$ is irrotational $\quad \therefore \quad F$ is conservative.

Consider scalar potential ϕ such that $\vec{F} = \nabla\phi$.

$$d\phi = \frac{\partial\phi}{\partial x}dx + \frac{\partial\phi}{\partial y}dy + \frac{\partial\phi}{\partial z}dz \qquad \text{[Total differential coefficient]}$$

$$= \left(\hat{i}\frac{\partial\phi}{\partial x} + \hat{j}\frac{\partial\phi}{\partial y} + \hat{k}\frac{\partial\phi}{\partial z}\right).(\hat{i}\,dx + \hat{j}\,dy + \hat{k}\,dz)$$

$$= \left(\hat{i}\frac{\partial}{\partial x} + \hat{j}\frac{\partial}{\partial y} + \hat{k}\frac{\partial}{\partial z}\right)\phi.(\hat{i}\,dx + \hat{j}\,dy + \hat{k}\,dz) = \nabla\phi.(\hat{i}\,dx + \hat{j}\,dy + \hat{k}\,dz)$$

$$= \vec{F}.(\hat{i}\,dx + \hat{j}\,dy + \hat{k}\,dz) = r^2\,\vec{r}.(\hat{i}\,dx + \hat{j}\,dy + \hat{k}\,dz) \qquad (\nabla\phi = \vec{F})$$

$$= (x^2 + y^2 + z^2)(\hat{i}\,x + \hat{j}\,y + \hat{k}\,z).(\hat{i}\,dx + \hat{j}\,dy + \hat{k}\,dz)$$

$$= (x^2 + y^2 + z^2)\,(x\,dx + y\,dy + z\,dz)$$

$$= x^3\,dx + y^3\,dy + z^3\,dz + (x\,dx)\,y^2 + (x^2)\,(y\,dy) + (x\,dx)z^2 + z^2\,(y\,dy) + x^2\,(z\,dz) + y^2\,(z\,dz)$$

$$\phi = \int x^3\,dx + \int y^3\,dy + \int z^3\,dz + \int[(x\,dx)y^2 + (y\,dy)x^2] + \int[(x\,dx)z^2 + (z\,dz)x^2] + \int[(y\,dy)z^2 + (z\,dz)y^2]$$

$$= \frac{x^4}{4} + \frac{y^4}{4} + \frac{z^4}{4} + \frac{1}{2}x^2y^2 + \frac{1}{2}x^2z^2 + \frac{1}{2}y^2z^2 + c$$

$$= \frac{1}{4}\,(x^4 + y^4 + z^4 + 2x^2y^2 + 2x^2z^2 + 2y^2\,z^2) + c$$ **Ans.**

Example 44. *Show that the vector field* $\vec{F} = \dfrac{\vec{r}}{|\vec{r}|^3}$ *is irrotational as well as solenoidal. Find the scalar potential.* (*Nagpur University, Summer 2008*)

Solution. $F = \dfrac{\vec{r}}{|\vec{r}|^3} = \dfrac{x\,\hat{i} + y\,\hat{j} + z\,\hat{k}}{(x^2 + y^2 + z^2)^{3/2}}$

$$\text{Curl } \vec{F} = \vec{\nabla} \times \vec{F} = \left(\hat{i}\frac{\partial}{\partial x} + \hat{j}\frac{\partial}{\partial y} + \hat{k}\frac{\partial}{\partial z}\right) \times \left(\frac{x\,\hat{i} + y\,\hat{j} + z\,\hat{k}}{(x^2 + y^2 + z^2)^{3/2}}\right)$$

$$= \begin{vmatrix} \hat{i} & \hat{j} & \hat{k} \\ \dfrac{\partial}{\partial x} & \dfrac{\partial}{\partial y} & \dfrac{\partial}{\partial z} \\ \dfrac{x}{(x^2 + y^2 + z^2)^{3/2}} & \dfrac{y}{(x^2 + y^2 + z^2)^{3/2}} & \dfrac{z}{(x^2 + y^2 + z^2)^{3/2}} \end{vmatrix}$$

$$= \hat{i}\left[\frac{-3}{2}\,\frac{2yz}{(x^2 + y^2 + z^2)^{5/2}} + \frac{3}{2}\,\frac{2yz}{(x^2 + y^2 + z^2)^{5/2}}\right] - \hat{j}\left[\frac{-3}{2}\,\frac{2xz}{(x^2 + y^2 + z^2)^{5/2}} - \left(-\frac{3}{2}\right)\frac{2xz}{(x^2 + y^2 + z^2)^{5/2}}\right]$$

$$+\hat{k}\left[-\frac{3}{2}\frac{2xy}{(x^2+y^2+z^2)^{5/2}}-\left(-\frac{3}{2}\right)\frac{2xy}{(x^2+y^2+z^2)^{5/2}}\right]=0$$

Hence, $\vec{F}$ is irrotational.

$$\Rightarrow \qquad \vec{F} = \vec{\nabla}\phi, \text{ where } \phi \text{ is called scalar potential}$$

$$d\phi = \frac{\partial\phi}{\partial x}dx+\frac{\partial\phi}{\partial y}dy+\frac{\partial\phi}{\partial z}dz \qquad \text{[Total differential coefficient]}$$

$$= \left(\hat{i}\frac{\partial\phi}{\partial x}+\hat{j}\frac{\partial\phi}{\partial y}+\hat{k}\frac{\partial\phi}{\partial z}\right).(\hat{i}\,dx+\hat{j}\,dy+\hat{k}\,dz) = \vec{\nabla}\phi.d\vec{r} = \vec{F}.d\vec{r}$$

$$= \frac{x\hat{i}+y\hat{j}+z\hat{k}}{(x^2+y^2+z^2)^{3/2}}.(\hat{i}\,dx+\hat{j}\,dy+\hat{k}\,dz) = \frac{x\,dx+y\,dy+z\,dz}{(x^2+y^2+z^2)^{3/2}}$$

$$\phi = \frac{1}{2}\int\frac{2x\,dx+2y\,dy+2z\,dz}{(x^2+y^2+z^2)^{3/2}}$$

$$= \frac{1}{2}\left(-\frac{2}{1}\right)(x^2+y^2+z^2)^{-\frac{1}{2}} = -\frac{1}{(x^2+y^2+z^2)^{\frac{1}{2}}} = -\frac{1}{|\vec{r}|} \qquad \textbf{Ans.}$$

Now, $\quad \text{Div}\,\vec{F} = \vec{\nabla}.\vec{F} = \left(\hat{i}\frac{\partial}{\partial x}+\hat{j}\frac{\partial}{\partial y}+\hat{k}\frac{\partial}{\partial z}\right).\frac{x\hat{i}+y\hat{j}+z\hat{k}}{(x^2+y^2+z^2)^{3/2}}$

$$= \frac{\partial}{\partial x}\frac{x}{(x^2+y^2+z^2)^{3/2}}+\frac{\partial}{\partial y}\frac{y}{(x^2+y^2+z^2)^{3/2}}+\frac{\partial}{\partial z}\frac{z}{(x^2+y^2+z^2)^{3/2}}$$

$$= \frac{(x^2+y^2+z^2)^{3/2}\,(1)-x\left(\frac{3}{2}\right)(x^2+y^2+z^2)^{1/2}\,(2x)}{(x^2+y^2+z^2)^3}$$

$$+\frac{(x^2+y^2+z^2)^{3/2}\,(1)-y\left(\frac{3}{2}\right)(x^2+y^2+z^2)^{1/2}\,(2y)}{(x^2+y^2+z^2)^3}$$

$$+\frac{(x^2+y^2+z^2)^{3/2}\,(1)-z\left(\frac{3}{2}\right)(x^2+y^2+z^2)^{1/2}\,(2z)}{(x^2+y^2+z^2)^3}$$

$$= \frac{(x^2+y^2+z^2)^{1/2}}{(x^2+y^2+z^2)^3}\,[x^2+y^2+z^2-3x^2+x^2+y^2+z^2-3y^2+x^2+y^2+z^2-3z^2] = 0$$

Hence, $\vec{F}$ is solenoidal. **Proved.**

Example 45. *Show that* $\vec{A}=(6xy+z^3)\hat{i}+(3x^2-z)\hat{j}+(3xz^2-y)\hat{k}$ *is irrotational.*

Find ϕ *such that* $\vec{A}=\vec{\nabla}\phi$. *(DU, 2012)*

Solution. We have

$$\vec{A} = (6xy+z^3)\hat{i}+(3x^2-z)\hat{j}+(3xz^2-y)\hat{k}\,;$$

$$\vec{\nabla}\times\vec{A} = \nabla\times[(6xy+z^3)\hat{i}+(3x^2-z)\hat{j}+(3xz^2-y)\hat{k}]$$

$$= \begin{vmatrix} \hat{i} & \hat{j} & \hat{k} \\ \dfrac{\partial}{\partial x} & \dfrac{\partial}{\partial y} & \dfrac{\partial}{\partial z} \\ 6xy+z^3 & 3x^2-z & 3xz^2-y \end{vmatrix} = \hat{i}(-1+1)-\hat{j}(3z^2-3z^2)+\hat{k}(6x-6x) = \hat{i}(0)-\hat{j}(0)+\hat{k}(0)=0$$

Hence, $\bar{A}$ is irrotational. $\Rightarrow \bar{A}=\nabla\phi$, where ϕ is called scalar potential.

$$d\phi = \frac{\partial\phi}{\partial x}dx+\frac{\partial\phi}{\partial y}dy+\frac{\partial\phi}{\partial z}dz = \left(\hat{i}\frac{\partial\phi}{\partial x}+\hat{j}\frac{\partial\phi}{\partial y}+\hat{k}\frac{\partial\phi}{\partial z}\right)\cdot(\hat{i}\,dx+\hat{j}\,dy+\hat{k}dz) = \nabla\phi\cdot dr = A\,.\,dr$$

$$= [(6xy+z^3)\hat{i}+(3x^2-z)\hat{j}+(3xz^2-y)\hat{k}]\cdot[\hat{i}dx+\hat{j}dy+\hat{k}dz]$$

$$= (6xy+z^3)dx+(3x^2-z)\,dy+(3xz^2-y)\,dz$$

$$= (6xy\,dx+3x^2\,dy)-(ydz+zdy)+(z^3dx+3xz^2dz)$$

$$\phi = \int(6xy\,dx+3x^2dy)-\int(y\,dz+zdy)+\int(z^3\,dx+3xz^2dz) = 3x^2y-yz+xz^3+C$$ **Ans.**

Example 46. *Find the constants a, b, c, so that*

$$\vec{F} = (x+2y+az)\hat{i}+(bx-3y-z)\hat{j}+(4x+cy+2z)\hat{k} \quad ...(1)$$

is irrotational and hence find function ϕ such that $\vec{F} = \nabla\phi$.

(*Nagpur University, Summer 2005, R.G.P.V., Bhopal 2009*)

Solution. We have,

$$\therefore \quad \nabla\times\vec{F} = \begin{vmatrix} \hat{i} & \hat{j} & \hat{k} \\ \dfrac{\partial}{\partial x} & \dfrac{\partial}{\partial y} & \dfrac{\partial}{\partial z} \\ (x+2y+az) & (bx-3y-z) & (4x+cy+2z) \end{vmatrix}$$

$$= (c+1)\hat{i}-(4-a)\hat{j}+(b-2)\hat{k}$$

As $\vec{F}$ is irrotational, $\nabla\times\vec{F}=\vec{0}$

i.e., $(c+1)\hat{i}-(4-a)\hat{j}+(b-2)\hat{k}=0\hat{i}+0\hat{j}+0\hat{k}$

$\therefore \quad c+1=0, \quad 4-a=0 \quad$ and $\quad b-2=0$

i.e., $\quad a=4, \quad b=2, \quad c=-1$

Putting the values of a, b, c in (1), we get

$$\vec{F} = (x+2y+4z)\hat{i}+(2x-3y-z)\hat{j}+(4x-y+2z)\hat{k}$$

Now we have to find ϕ such that $\vec{F} = \nabla\phi$

We know that

$$d\phi = \frac{\partial\phi}{\partial x}dx+\frac{\partial\phi}{\partial y}dy+\frac{\partial\phi}{\partial z}dz \quad \text{[Total differential coefficient]}$$

$$= \left(\hat{i}\frac{\partial\phi}{\partial x}+\hat{j}\frac{\partial\phi}{\partial y}+\hat{k}\frac{\partial\phi}{\partial z}\right).(\hat{i}\,dx+\hat{j}\,dy+\hat{k}\,dz)$$

$$= \left(\hat{i}\frac{\partial}{\partial x}+\hat{j}\frac{\partial}{\partial y}+\hat{k}\frac{\partial}{\partial z}\right)\phi.(\hat{i}\,dx+\hat{j}\,dy+\hat{k}\,dz) = \nabla\phi.(\hat{i}\,dx+\hat{j}\,dy+\hat{k}\,dz)$$

$$= \bar{F}.(\hat{i}\,dx + \hat{j}\,dy + \hat{k}\,dz)$$

$$= [(x+2y+4z)\hat{i} + (2x-3y-z)\hat{j} + (4x-y+2z)\hat{k})].(\hat{i}\,dx + \hat{j}\,dy + \hat{k}\,dz)$$

$$= (x + 2y + 4z)\,dx + (2x - 3y - z)\,dy + (4x - y + 2z)\,dz$$

$$= x\,dx - 3y\,dy + 2z\,dz + (2y\,dx + 2x\,dy) + (4z\,dx + 4x\,dz) + (-z\,dy - y\,dz)$$

$$\phi = \int x\,dx - 3\int y\,dy + 2\int z\,dz + \int (2y\,dx + 2x\,dy) + \int (4z\,dx + 4x\,dz) - \int (z\,dy + y\,dz)$$

$$= \frac{x^2}{2} - \frac{3y^2}{2} + z^2 + 2xy + 4zx - yz + c$$ **Ans.**

Example 47. *Let $\vec{V}$ (x, y, z) be a differentiable vector function and $\phi(x, y, z)$ be a scalar function. Derive an expression for div $(\phi\vec{V})$ in terms of ϕ . $\vec{V}$, div $\vec{V}$ and $\nabla\phi$.*

Solution. Let $\vec{V} = V_1\hat{i} + V_2\hat{j} + V_3\hat{k}$

$$\text{div }(\phi\vec{V}) = \vec{\nabla}.(\phi\vec{F})$$

$$= \left(\hat{i}\frac{\partial}{\partial x} + \hat{j}\frac{\partial}{\partial y} + \hat{k}\frac{\partial}{\partial z}\right).[\phi V_1\hat{i} + \phi V_2\hat{j} + \phi V_3\hat{k}] = \frac{\partial}{\partial x}(\phi V_1) + \frac{\partial}{\partial y}(\phi V_2) + \frac{\partial}{\partial z}(\phi V_3)$$

$$= \left(\phi\frac{\partial V_1}{\partial x} + \frac{\partial\phi}{\partial x}V_1\right) + \left(\phi\frac{\partial V_2}{\partial y} + \frac{\partial\phi}{\partial y}V_2\right) + \left(\phi\frac{\partial V_3}{\partial z} + \frac{\partial\phi}{\partial z}V_3\right)$$

$$= \phi\left(\frac{\partial V_1}{\partial x} + \frac{\partial V_2}{\partial y} + \frac{\partial V_3}{\partial z}\right) + \left(\frac{\partial\phi}{\partial x}V_1 + \frac{\partial\phi}{\partial y}V_2 + \frac{\partial\phi}{\partial z}V_3\right)$$

$$= \phi\left(\hat{i}\frac{\partial}{\partial x} + \hat{j}\frac{\partial}{\partial y} + \hat{k}\frac{\partial}{\partial z}\right).(V_1\hat{i} + V_2\hat{j} + V_3\hat{k})$$

$$+ \left(\hat{i}\frac{\partial\phi}{\partial x} + \hat{j}\frac{\partial\phi}{\partial y} + \hat{k}\frac{\partial\phi}{\partial z}\right).(V_1\hat{i} + V_2\hat{j} + V_3\hat{k})$$

$$= \phi(\nabla.\vec{V}) + (\vec{\nabla}\phi).\vec{V} = \phi\,(\text{div }\vec{V}) + (\text{grad }\phi).\,\vec{V}$$ **Ans.**

Example 48. *If $\vec{A}$ is a constant vector and $\vec{R} = x\hat{i} + y\hat{j} + z\hat{k}$, then prove that*

$$Curl\left[\left(\vec{A}.\vec{R}\right)\vec{R}\right] = \vec{A} \times \vec{R}$$ *(K. University, Dec. 2009)*

Solution. Let $\vec{A} = A_1\hat{i} + A_2\hat{j} + A_3\hat{k}$, $\vec{R} = x\hat{i} + y\hat{j} + z\hat{k}$

$$\vec{A}.\vec{R} = (A_1\hat{i} + A_2\hat{j} + A_3\hat{k}).(x\hat{i} + y\hat{j} + z\hat{k}) = A_1x + A_2y + A_3z$$

$$[\vec{A}.\vec{R}]\vec{R} = (A_1x + A_2y + A_3z)\;(x\hat{i} + y\hat{j} + z\hat{k})$$

$$= (A_1x^2 + A_2xy + A_3zx)\;\hat{i} + (A_1xy + A_2y^2 + A_3yz)\;\hat{j} + (A_1xz + A_2yz + A_3z^2)\;\hat{k}$$

$$\text{Curl}\left[(\vec{A}.\vec{R})\vec{R}\right] = \begin{vmatrix} \hat{i} & \hat{j} & \hat{k} \\ \frac{\partial}{\partial x} & \frac{\partial}{\partial y} & \frac{\partial}{\partial z} \\ A_1x^2 + A_2xy + A_3zx & A_1xy + A_2y^2 + A_3yz & A_1xz + A_2yz + A_3z^2 \end{vmatrix}$$

$$= (A_2 z - A_3 y)\ \hat{i}\ - (A_1 z - A_3 x)\ \hat{j}\ + [A_1 y - A_2 x]\ \hat{k} \qquad \text{... (1)}$$

$$\text{R.H.S.} = \vec{A} \times \vec{R}$$

$$= (A_1 \hat{i} + A_2 \hat{j} + A_3 k) \times (x\hat{i} + y\hat{j} + z\hat{k})$$

$$= \begin{vmatrix} \hat{i} & \hat{j} & \hat{k} \\ A_1 & A_2 & A_3 \\ x & y & z \end{vmatrix}$$

$$= (A_2 z - A_3 y)\ \hat{i}\ - (A_1 z - A_3 x)\ \hat{j}\ + (A_1 y - A_2 x)\ \hat{k}$$

$$= \text{L.H.S.}$$

[From (1)] **Proved.**

Example 49. *Suppose that* $\vec{U}, \vec{V}$ *and f are continuously differentiable fields then Prove that, div* $(\vec{U} \times \vec{V}) = \vec{V}.curl\,\vec{U} - \vec{U}.curl\,\vec{V}$. *(GBTU, Dec. 2012, M.U., 2005)*

Solution. Let $\vec{U} = u_1 \hat{i} + u_2 \hat{j} + u_3 \hat{k},\ \vec{V} = v_1 \hat{i} + v_2 \hat{j} + v_3 \hat{k}$

$$\vec{U} \times \vec{V} = \begin{vmatrix} \hat{i} & \hat{j} & \hat{k} \\ u_1 & u_2 & u_3 \\ v_1 & v_2 & v_3 \end{vmatrix}$$

$$= (u_2 v_3 - u_3 v_2)\hat{i} - (u_1 v_3 - u_3 v_1)\hat{j} + (u_1 v_2 - u_2 v_1)\hat{k}$$

$$\text{div}\ (\vec{U} \times \vec{V}) = \left(\hat{i}\frac{\partial}{\partial x} + \hat{j}\frac{\partial}{\partial y} + \hat{k}\frac{\partial}{\partial z}\right).[(u_2 v_3 - u_3 v_2)\hat{i} - (u_1 v_3 - u_3 v_1)\hat{j} + (u_1 v_2 - u_2 v_1)\hat{k}]$$

$$= \frac{\partial}{\partial x}(u_2 v_3 - u_3 v_2) + \frac{\partial}{\partial y}(-u_1 v_3 + u_3 v_1) + \frac{\partial}{\partial z}(u_1 v_2 - u_2 v_1)$$

$$= \left[u_2 \frac{\partial v_3}{\partial x} + v_3 \frac{\partial u_2}{\partial x} - u_3 \frac{\partial v_2}{\partial x} - v_2 \frac{\partial u_3}{\partial x}\right] + \left[-u_1 \frac{\partial v_3}{\partial y} - v_3 \frac{\partial u_1}{\partial y} + u_3 \frac{\partial v_1}{\partial y} + v_1 \frac{\partial u_3}{\partial y}\right]$$
$$+ \left[u_1 \frac{\partial v_2}{\partial z} + v_2 \frac{\partial u_1}{\partial z} - u_2 \frac{\partial v_1}{\partial z} - v_1 \frac{\partial u_2}{\partial z}\right]$$

$$= v_1\left(\frac{\partial u_3}{\partial y} - \frac{\partial u_2}{\partial z}\right) + v_2\left(-\frac{\partial u_3}{\partial x} + \frac{\partial u_1}{\partial z}\right) + v_3\left(\frac{\partial u_2}{\partial x} - \frac{\partial u_1}{\partial y}\right)$$
$$+ u_1\left(-\frac{\partial v_3}{\partial y} + \frac{\partial v_2}{\partial z}\right) + u_2\left(\frac{\partial v_3}{\partial x} - \frac{\partial v_1}{\partial z}\right) + u_3\left(\frac{\partial v_1}{\partial y} - \frac{\partial v_2}{\partial x}\right)$$

$$= (v_1 \hat{i} + v_2 \hat{j} + v_3 \hat{k}).\left[\hat{i}\left(\frac{\partial u_3}{\partial y} - \frac{\partial u_2}{\partial z}\right) + \hat{j}\left(\frac{\partial u_1}{\partial z} - \frac{\partial u_3}{\partial x}\right) + \hat{k}\left(\frac{\partial u_2}{\partial x} - \frac{\partial u_1}{\partial y}\right)\right]$$
$$-(u_1 \hat{i} + u_2 \hat{j} + u_3 \hat{k}).\left[\hat{i}\left(-\frac{\partial v_2}{\partial z} + \frac{\partial v_3}{\partial y}\right) + \hat{j}\left(-\frac{\partial v_3}{\partial x} + \frac{\partial v_1}{\partial z}\right) + \hat{k}\left(-\frac{\partial v_1}{\partial y} + \frac{\partial v_2}{\partial x}\right)\right]$$

$$= \text{div.}(\vec{U} \times \vec{V}) = \vec{V}.\text{curl}\,\vec{U} - \vec{U}.\text{curl}\,\vec{V}$$

Proved

Example 50. *Prove that*

$$\vec{\nabla} \times (\vec{F} \times \vec{G}) = \vec{F}(\vec{\nabla}.\vec{G}) - \vec{G}(\vec{\nabla}.\vec{F}) + (\vec{G}.\vec{\nabla})\vec{F} - (\vec{F}.\vec{\nabla})\vec{G}$$

(M.U. 2004, 2005)

Solution. $\vec{\nabla}\times(\vec{F}\times\vec{G}) = \Sigma\,\hat{i}\times\dfrac{\partial}{\partial x}(\vec{F}\times\vec{G})$

$$= \Sigma\,\hat{i}\times\left(\frac{\partial F}{\partial x}\times\vec{G}+\vec{F}\times\frac{\partial G}{\partial x}\right)=\Sigma\,\hat{i}\times\left(\frac{\partial F}{\partial x}\times\vec{G}\right)+\Sigma\,\hat{i}\times\left(\vec{F}\times\frac{\partial G}{\partial x}\right)$$

$$= \Sigma\left[(\hat{i}.\vec{G})\frac{\partial F}{\partial x}-\left(\hat{i}\frac{\partial F}{\partial x}\right)\vec{G}\right]+\Sigma\left[\left(\hat{i}\frac{\partial G}{\partial x}\right)\vec{F}-(\hat{i}.\vec{F})\frac{\partial G}{\partial x}\right]$$

$$= \Sigma\,(\vec{G}.\hat{i})\frac{\partial F}{\partial x}-\vec{G}\Sigma\left(\hat{i}.\frac{\partial F}{\partial x}\right)+\vec{F}\Sigma\left(\hat{i}.\frac{\partial G}{\partial x}\right)-\Sigma(\vec{F}.\hat{i})\frac{\partial G}{\partial x}$$

$$= \vec{F}\left(\Sigma\,\hat{i}\,\frac{\partial G}{\partial x}\right)-\vec{G}\,\Sigma\left(\hat{i}.\frac{\partial F}{\partial x}\right)+\Sigma(\vec{G}.\hat{i})\frac{\partial F}{\partial x}-\Sigma(\vec{F}.\hat{i})\frac{\partial G}{\partial x}$$

$$= \vec{F}\,(\vec{\nabla}\vec{G})-\vec{G}\,(\vec{\nabla}.\vec{F})+(\vec{G}.\vec{\nabla})\vec{F}-(\vec{F}.\vec{\nabla})\vec{G}$$ **Proved**

Example 51. *Prove that, for every field $\vec{V}$; div curl $\vec{V} = 0$.*

(*AMIETE, Sem II, June 2010*)

Solution. Let $V = V_1\,\hat{i}+V_2\,\hat{j}+V_3\,\hat{k}$

$$\text{div }\overline{(\text{curl }\vec{V})} = \vec{\nabla}.(\vec{\nabla}\times\vec{V})$$

$$= \vec{\nabla}.\begin{vmatrix}\hat{i} & \hat{j} & \hat{k}\\ \dfrac{\partial}{\partial x} & \dfrac{\partial}{\partial y} & \dfrac{\partial}{\partial z}\\ V_1 & V_2 & V_3\end{vmatrix}$$

$$= \left(\hat{i}\frac{\partial}{\partial x}+\hat{j}\frac{\partial}{\partial y}+\hat{k}\frac{\partial}{\partial z}\right).\left[\hat{i}\left(\frac{\partial V_3}{\partial y}-\frac{\partial V_2}{\partial z}\right)-\hat{j}\left(\frac{\partial V_3}{\partial x}-\frac{\partial V_1}{\partial z}\right)+\hat{k}\left(\frac{\partial V_2}{\partial x}-\frac{\partial V_1}{\partial y}\right)\right]$$

$$= \frac{\partial}{\partial x}\left(\frac{\partial V_3}{\partial y}-\frac{\partial V_2}{\partial z}\right)-\frac{\partial}{\partial y}\left(\frac{\partial V_3}{\partial x}-\frac{\partial V_1}{\partial z}\right)+\frac{\partial}{\partial z}\left(\frac{\partial V_2}{\partial x}-\frac{\partial V_1}{\partial y}\right)$$

$$= \frac{\partial^2 V_3}{\partial x\,\partial y}-\frac{\partial^2 V_2}{\partial x\,\partial z}-\frac{\partial^2 V_3}{\partial y\,\partial x}+\frac{\partial^2 V_1}{\partial y\,\partial z}+\frac{\partial^2 V_2}{\partial z\,\partial x}-\frac{\partial^2 V_1}{\partial z\,\partial y}$$

$$= \left(\frac{\partial^2 V_1}{\partial y\,\partial z}-\frac{\partial^2 V_1}{\partial z\,\partial y}\right)+\left(\frac{\partial^2 V_2}{\partial z\,\partial x}-\frac{\partial^2 V_2}{\partial x\,\partial z}\right)+\left(\frac{\partial^2 V_3}{\partial x\,\partial y}-\frac{\partial^2 V_3}{\partial y\,\partial x}\right)=0$$ **Ans.**

Example 52. *If $\vec{a}$ is a constant vector, show that*

$$\vec{a}\times(\vec{\nabla}\times\vec{r}) = \vec{\nabla}(\vec{a}.\vec{r})-(\vec{a}.\vec{\nabla})\vec{r}.$$ (*U.P., Ist Semester, Dec. 2007*)

Solution. $\vec{a} = a_1\,\hat{i}+a_2\,\hat{j}+a_3\,\hat{k},\quad \vec{r}=r_1\,\hat{i}+r_2\,\hat{j}+r_3\,\hat{k}$

$$\vec{\nabla}\times\vec{r} = \begin{vmatrix}\hat{i} & \hat{j} & \hat{k}\\ \dfrac{\partial}{\partial x} & \dfrac{\partial}{\partial y} & \dfrac{\partial}{\partial z}\\ r_1 & r_2 & r_3\end{vmatrix}=\left(\frac{\partial r_3}{\partial y}-\frac{\partial r_2}{\partial z}\right)\hat{i}-\left(\frac{\partial r_3}{\partial x}-\frac{\partial r_1}{\partial z}\right)\hat{j}+\left(\frac{\partial r_2}{\partial x}-\frac{\partial r_1}{\partial y}\right)\hat{k}$$

$$\vec{a} \times (\vec{\nabla} \times \vec{r}) = \begin{vmatrix} \hat{i} & \hat{j} & \hat{k} \\ a_1 & a_2 & a_3 \\ \frac{\partial r_3}{\partial y} - \frac{\partial r_2}{\partial z} & -\frac{\partial r_3}{\partial x} + \frac{\partial r_1}{\partial z} & \frac{\partial r_2}{\partial x} - \frac{\partial r_1}{\partial y} \end{vmatrix}$$

$$= \left[\left(a_2 \frac{\partial r_2}{\partial x} - a_2 \frac{\partial r_1}{\partial y}\right) - \left(-a_3 \frac{\partial r_3}{\partial x} + a_3 \frac{\partial r_1}{\partial z}\right)\right]\hat{i} - \left[a_1 \frac{\partial r_2}{\partial x} - a_1 \frac{\partial r_1}{\partial y} - a_3 \frac{\partial r_3}{\partial y} + a_3 \frac{\partial r_2}{\partial z}\right]\hat{j}$$

$$+ \left[-a_1 \frac{\partial r_3}{\partial x} + a_1 \frac{\partial r_1}{\partial z} - a_2 \frac{\partial r_3}{\partial y} + a_2 \frac{\partial r_2}{\partial z}\right]\hat{k}$$

$$= \left[\left(a_1 \hat{i} \frac{\partial r_1}{\partial x} + a_2 \hat{i} \frac{\partial r_2}{\partial x} + a_3 \hat{i} \frac{\partial r_3}{\partial x}\right) + \left(a_1 \hat{j} \frac{\partial r_1}{\partial y} + a_2 \hat{j} \frac{\partial r_2}{\partial y} + a_3 \hat{j} \frac{\partial r_3}{\partial y}\right)\right.$$

$$\left. + \left(a_1 \hat{k} \frac{\partial r_1}{\partial z} + a_2 \hat{k} \frac{\partial r_2}{\partial z} + a_3 \hat{k} \frac{\partial r_3}{\partial z}\right)\right] - \left[\left(a_1 \hat{i} \frac{\partial r_1}{\partial x} + a_1 \hat{j} \frac{\partial r_2}{\partial x} + a_1 \hat{k} \frac{\partial r_3}{\partial x}\right)\right.$$

$$\left. + \left(a_2 \hat{i} \frac{\partial r_1}{\partial y} + a_2 \hat{j} \frac{\partial r_2}{\partial y} + a_2 \hat{k} \frac{\partial r_3}{\partial y}\right) + \left(a_3 \hat{i} \frac{\partial r_1}{\partial z} + a_3 \hat{j} \frac{\partial r_2}{\partial z} + a_3 \hat{k} \frac{\partial r_3}{\partial z}\right)\right]$$

$$= \left(\hat{i} \frac{\partial}{\partial x} + \hat{j} \frac{\partial}{\partial y} + \hat{k} \frac{\partial}{\partial z}\right)(a_1 r_1 + a_2 r_2 + a_3 r_3) - \left[a_1 \frac{\partial}{\partial x} + a_2 \frac{\partial}{\partial y} + a_3 \frac{\partial}{\partial z}\right](r_1 \hat{i} + r_2 \hat{j} + r_3 \hat{k})$$

$$= \nabla(\vec{a} \cdot \vec{r}) - (\vec{a} \cdot \vec{\nabla})\vec{r}$$ **Proved.**

Example 53. *If r is the distance of a point (x, y, z) from the origin, prove that* $Curl\left(\hat{k} \times grad \frac{1}{r}\right) + grad\left(\hat{k}\ grad \frac{1}{r}\right) = 0$*, where k is the unit vector in the direction OZ.*

Solution. $r^2 = (x-0)^2 + (y-0)^2 + (z-0)^2 = x^2 + y^2 + z^2$

$$\Rightarrow \quad \frac{1}{r} = (x^2 + y^2 + z^2)^{-1/2}$$

$$\text{grad } \frac{1}{r} = \vec{\nabla} \frac{1}{r} = \left(\hat{i} \frac{\partial}{\partial x} + \hat{j} \frac{\partial}{\partial y} + \hat{k} \frac{\partial}{\partial z}\right)(x^2 + y^2 + z^2)^{-1/2}$$

$$= -\frac{1}{2}(x^2 + y^2 + z^2)^{-3/2}(2x\hat{i} + 2y\hat{j} + 2z\hat{k})$$

$$= -(x^2 + y^2 + z^2)^{-3/2}(x\hat{i} + y\hat{j} + z\hat{k})$$

$$\hat{k} \times \text{grad } \frac{1}{r} = \hat{k} \times [-(x^2 + y^2 + z^2)^{-3/2}(x\hat{i} + y\hat{j} + z\hat{k})]$$

$$= -(x^2 + y^2 + z^2)^{-3/2}(x\hat{j} - y\hat{i})$$

$$\text{curl}\left(\hat{k} \times \text{grad } \frac{1}{r}\right) = \vec{\nabla} \times \left(\hat{k} \times \text{grad } \frac{1}{r}\right)$$

$$= \left(\hat{i} \frac{\partial}{\partial x} + \hat{j} \frac{\partial}{\partial y} + \hat{k} \frac{\partial}{\partial z}\right) \times [-(x^2 + y^2 + z^2)^{-3/2}(x\hat{j} - y\hat{i})]$$

$$= \begin{vmatrix} \hat{i} & \hat{j} & \hat{k} \\ \dfrac{\partial}{\partial x} & \dfrac{\partial}{\partial y} & \dfrac{\partial}{\partial z} \\ \dfrac{y}{(x^2+y^2+z^2)^{3/2}} & \dfrac{-x}{(x^2+y^2+z^2)^{3/2}} & 0 \end{vmatrix}$$

$$= -\left(-\frac{3}{2}\right)\frac{(-x)\,(2z)}{(x^2+y^2+z^2)^{5/2}}\hat{i} + -\frac{3}{2}\frac{y\,(2z)}{(x^2+y^2+z^2)^{5/2}}\hat{j} + \left[-\frac{3}{2}\frac{(-x)(2x)}{(x^2+y^2+z^2)^{5/2}}\right.$$

$$\left. - \frac{1}{(x^2+y^2+z^2)^{3/2}} - \frac{(-3/2)\,(y)\,(2y)}{(x^2+y^2+z^2)^{5/2}} - \frac{1}{(x^2+y^2+z^2)^{3/2}}\right]\hat{k}$$

$$= \frac{-3xz}{(x^2+y^2+z^2)^{5/2}}\hat{i} - \frac{3yz}{(x^2+y^2+z^2)^{5/2}}\hat{j} + \frac{(3x^2-x^2-y^2-z^2+3y^2-x^2-y^2-z^2)}{(x^2+y^2+z^2)^{5/2}}\hat{k}$$

$$= \frac{-3xz\,\hat{i} - 3yz\,\hat{j} + (x^2+y^2-2z^2)\hat{k}}{(x^2+y^2+z^2)^{5/2}} \qquad \text{...(1)}$$

$$\hat{k}\ \text{grad}\ \frac{1}{r} = \hat{k}\,[-(x^2+y^2+z^2)^{-3/2}\ (x\,\hat{i} + y\,\hat{j} + z\,\hat{k})] = \frac{-z}{(x^2+y^2+z^2)^{3/2}}$$

$$\text{grad}\left(\hat{k}\ \text{grad}\,\frac{1}{r}\right) = \left(\hat{i}\frac{\partial}{\partial x} + \hat{j}\frac{\partial}{\partial y} + \hat{k}\frac{\partial}{\partial z}\right)\frac{-z}{(x^2+y^2+z^2)^{3/2}}$$

$$= -\frac{3}{2}\frac{\hat{i}(-z)(2x)}{(x^2+y^2+z^2)^{5/2}} + -\frac{3}{2}\frac{\hat{j}(-z)(2y)}{(x^2+y^2+z^2)^{5/2}}$$

$$+ \left[-\frac{3}{2}\frac{(-z)(2z)}{(x^2+y^2+z^2)^{5/2}} - \frac{1}{(x^2+y^2+z^2)^{3/2}}\right]\hat{k}$$

$$= \frac{3xz\,\hat{i} + 3yz\,\hat{j} + (3z^2-x^2-y^2-z^2)\hat{k}}{(x^2+y^2+z^2)^{5/2}} = \frac{3xz\,\hat{i} + 3yz\,\hat{j} - (x^2+y^2-2z^2)\hat{k}}{(x^2+y^2+z^2)^{5/2}} \qquad \text{...(2)}$$

Adding (1) and (2), we get

$$\text{Curl}\left(\hat{k} \times \text{grad}\,\frac{1}{r}\right) + \text{grad}\left(\hat{k}\ \text{grad}\,\frac{1}{r}\right) = 0$$ **Proved.**

Example 54. *Prove that* $\nabla \times \left(\dfrac{\vec{a} \times \vec{r}}{r^n}\right) = \dfrac{(2-n)\,\vec{a}}{r^n} + \dfrac{n(\vec{a}\,.\,\vec{r})\,\vec{r}}{r^{n+2}}$.

(*M.U. 2009, 2005; AMIETE, II Sem. June 2010*)

Solution. We have,

$$\frac{\vec{a} \times \vec{r}}{r^n} = \frac{1}{r^n}\begin{vmatrix} \hat{i} & \hat{j} & \hat{k} \\ a_1 & a_2 & a_3 \\ x & y & z \end{vmatrix}$$

$$= \frac{1}{r^n}(a_2z - a_3y)\hat{i} + \frac{1}{r^n}(a_3x - a_1z)\hat{j} + \frac{1}{r^n}(a_1y - a_2x)\hat{k}$$

$$\nabla \times \frac{(\vec{a} \times \vec{r})}{r^n} = \begin{vmatrix} \hat{i} & \hat{j} & \hat{k} \\ \frac{\partial}{\partial x} & \frac{\partial}{\partial y} & \frac{\partial}{\partial z} \\ \frac{a_2z - a_3y}{r^n} & \frac{a_3x - a_1z}{r^n} & \frac{a_1y - a_2x}{r^n} \end{vmatrix}$$

$$= \hat{i}\left[\frac{\partial}{\partial y}\left(\frac{a_1y - a_2x}{r^n}\right) - \frac{\partial}{\partial z}\left(\frac{a_3x - a_1z}{r^n}\right)\right] - \hat{j}\left[\frac{\partial}{\partial x}\left(\frac{a_1y - a_2x}{r^n}\right) - \frac{\partial}{\partial z}\left(\frac{a_2z - a_3y}{r^n}\right)\right]$$
$$+ \hat{k}\left[\frac{\partial}{\partial x}\left(\frac{a_3x - a_1z}{r^n}\right) - \frac{\partial}{\partial y}\left(\frac{a_2z - a_3y}{r^n}\right)\right]$$

Now, $\quad r^2 = x^2 + y^2 + z^2 \Rightarrow 2r\frac{\partial r}{\partial x} = 2x \quad \Rightarrow \quad \frac{\partial r}{\partial x} = \frac{x}{r}$

Similarly, $\quad \frac{\partial r}{\partial y} = \frac{y}{r}, \quad \frac{\partial r}{\partial z} = \frac{z}{r}$

$$\therefore \nabla \times \left(\frac{\vec{a} \times \vec{r}}{r^n}\right) = \hat{i}\left[\left\{-nr^{-n-1}\cdot\left(\frac{y}{r}\right)(a_1y - a_2x) + \frac{1}{r^n}a_1\right\}\right.$$
$$\left. - \left\{-nr^{-n-1}\left(\frac{z}{r}\right)(a_3x - a_1z) + \frac{1}{r^n}(-a_1)\right\}\right] + \text{two similar terms}$$

$$= \hat{i}\left[-\frac{n}{r^{n+2}}(a_1y^2 - a_2xy) + \frac{a_1}{r^n} + \frac{n}{r^{n+2}}(a_3xz - a_1z^2) + \frac{a_1}{r^n}\right] + \text{two similar terms}$$

$$= \hat{i}\left[\frac{2a_1}{r^n} - \frac{n}{r^{n+2}}a_1(y^2 + z^2) + \frac{n}{r^{n+2}}(a_2xy + a_3xz)\right] + \text{two similar terms}$$

Adding and subtracting $\frac{n}{r^{n+2}}a_1x^2$ to third and from second term, we get

$$\vec{\nabla} \times \left(\frac{\vec{a} \times \vec{r}}{r^n}\right) = \hat{i}\left[\frac{2a_1}{r^n} - \frac{na_1}{r^{n+2}}(x^2 + y^2 + z^2) + \frac{n}{r^{n+2}}(a_1x^2 + a_2xy + a_3xz)\right]$$
$$+ \text{two similar terms}$$

$$= \hat{i}\left[\frac{2a_1}{r^n} - \frac{na_1}{r^{n+2}}r^2 + \frac{n}{r^{n+2}}x(a_1x + a_2y + a_3z)\right] + \text{two similar terms}$$

$$= \hat{i}\left[\frac{2a_1}{r^n} - \frac{na_1}{r^n} + \frac{n}{r^{n+2}}x(a_1x + a_2y + a_3z)\right] + \hat{j}\left[\frac{2a_2}{r^n} - \frac{na_2}{r^n} + \frac{n}{r^{n+2}}y(a_2y + a_3z + a_1x)\right]$$
$$+ \hat{k}\left[\frac{2a_3}{r^n} - \frac{na_3}{r^n} + \frac{n}{r^{n+2}}z(a_3z + a_1x + a_2y)\right]$$

$$= \frac{2}{r^n}(a_1\hat{i} + a_2\hat{j} + a_3\hat{k}) - \frac{n}{r^n}(a_1\hat{i} + a_2\hat{j} + a_3\hat{k}) + \frac{n}{r^{n+2}}(a_1x + a_2y + a_3z)(x\hat{i} + y\hat{j} + z\hat{k})$$

$$= \frac{2-n}{r^n}(a_1\hat{i}+a_2\hat{j}+a_3\hat{k})+\frac{n}{r^{n+2}}(a_1x+a_2y+a_3z)(x\hat{i}+y\hat{j}+z\hat{k})$$

$$= \frac{2-n}{r^n}\vec{a}+\frac{n}{r^{n+2}}(\vec{a}.\vec{r})\vec{r}$$ **Proved.**

Example 55. *If f and g are two scalar point functions, prove that*

$$div\ (f\,\nabla g) = f\,\nabla^2 g + \nabla f\,\nabla g.$$

Solution. We have, $\nabla g = \frac{\partial g}{\partial x}\hat{i}+\frac{\partial g}{\partial y}\hat{j}+\frac{\partial g}{\partial z}\hat{k}$

$$\Rightarrow \quad f\,\nabla g = f\frac{\partial g}{\partial x}\hat{i}+f\frac{\partial g}{\partial y}\hat{j}+f\frac{\partial g}{\partial z}\hat{k}$$

$$\Rightarrow \quad \text{div}\ (f\,\nabla g) = \frac{\partial}{\partial x}\left(f\frac{\partial g}{\partial x}\right)+\frac{\partial}{\partial y}\left(f\frac{\partial g}{\partial y}\right)+\frac{\partial}{\partial z}\left(f\frac{\partial g}{\partial z}\right)$$

$$= f\left(\frac{\partial^2 g}{\partial x^2}+\frac{\partial^2 g}{\partial y^2}+\frac{\partial^2 g}{\partial z^2}\right)+\left(\frac{\partial f}{\partial x}\frac{\partial g}{\partial x}+\frac{\partial f}{\partial y}\frac{\partial g}{\partial y}+\frac{\partial f}{\partial z}\frac{\partial g}{\partial z}\right)$$

$$= f\left(\frac{\partial^2}{\partial x^2}+\frac{\partial^2}{\partial y^2}+\frac{\partial^2}{\partial z^2}\right)g+\left(\frac{\partial f}{\partial x}\hat{i}+\frac{\partial f}{\partial y}\hat{j}+\frac{\partial f}{\partial z}\hat{k}\right).\left(\frac{\partial g}{\partial x}\hat{i}+\frac{\partial g}{\partial y}\hat{j}+\frac{\partial g}{\partial z}\hat{k}\right)$$

$$= f\,\nabla^2 g+\nabla f.\nabla g$$ **Proved.**

Example 56. *Establish the reation curl curl* $\vec{f}=\vec{\nabla}\ div\ \vec{f}-\nabla^2\vec{f}$.

Solution. Let $\vec{f} = f_1\hat{i}+f_2\hat{j}+f_3\hat{k}$, then by definition,

$$\text{Curl } \vec{f} = \begin{vmatrix} \hat{i} & \hat{j} & \hat{k} \\ \frac{\partial}{\partial x} & \frac{\partial}{\partial y} & \frac{\partial}{\partial z} \\ f_1 & f_2 & f_3 \end{vmatrix} = \left(\frac{\partial f_3}{\partial y}-\frac{\partial f_2}{\partial z}\right)\hat{i}+\left(\frac{\partial f_1}{\partial z}-\frac{\partial f_3}{\partial x}\right)\hat{j}+\left(\frac{\partial f_2}{\partial x}-\frac{\partial f_1}{\partial y}\right)\hat{k}$$

$$\therefore \text{ Curl curl } \vec{f} = \begin{vmatrix} \hat{i} & \hat{j} & \hat{k} \\ \frac{\partial}{\partial x} & \frac{\partial}{\partial y} & \frac{\partial}{\partial z} \\ \frac{\partial f_3}{\partial y}-\frac{\partial f_2}{\partial z} & \frac{\partial f_1}{\partial z}-\frac{\partial f_3}{\partial x} & \frac{\partial f_2}{\partial x}-\frac{\partial f_1}{\partial y} \end{vmatrix} = \left[\frac{\partial}{\partial y}\left(\frac{\partial f_2}{\partial x}-\frac{\partial f_1}{\partial y}\right)-\frac{\partial}{\partial z}\left(\frac{\partial f_1}{\partial z}-\frac{\partial f_3}{\partial x}\right)\right]\hat{i}$$

$$-\left[\frac{\partial}{\partial x}\left(\frac{\partial f_2}{\partial x}-\frac{\partial f_1}{\partial y}\right)-\frac{\partial}{\partial z}\left(\frac{\partial f_3}{\partial y}-\frac{\partial f_2}{\partial z}\right)\right]\hat{j}+\left[\frac{\partial}{\partial x}\left(\frac{\partial f_1}{\partial z}-\frac{\partial f_3}{\partial x}\right)-\frac{\partial}{\partial y}\left(\frac{\partial f_3}{\partial y}-\frac{\partial f_2}{\partial z}\right)\right]\hat{k}$$

$$= \left[\frac{\partial^2 f_2}{\partial y\partial x}-\frac{\partial^2 f_1}{\partial y^2}-\frac{\partial^2 f_1}{\partial z^2}+\frac{\partial^2 f_3}{\partial z\partial x}\right]\hat{i}$$

$$-\left[\frac{\partial^2 f_2}{\partial x^2}-\frac{\partial^2 f_1}{\partial x\partial y}-\frac{\partial^2 f_3}{\partial z\partial y}+\frac{\partial^2 f_2}{\partial z^2}\right]\hat{j}+\left[\frac{\partial^2 f_1}{\partial x\partial z}-\frac{\partial^2 f_3}{\partial x^2}-\frac{\partial^2 f_3}{\partial y^2}+\frac{\partial^2 f_2}{\partial y\partial z}\right]\hat{k}$$

$$= \left[\frac{\partial^2 f_1}{\partial x^2} + \frac{\partial^2 f_2}{\partial y\,\partial x} + \frac{\partial^2 f_3}{\partial z\,\partial x} - \left(\frac{\partial^2 f_1}{\partial x^2} + \frac{\partial^2 f_1}{\partial y^2} + \frac{\partial^2 f_1}{\partial z^2}\right)\right]\hat{i}$$

$$- \left[\frac{\partial^2 f_1}{\partial x\,\partial z} + \frac{\partial^2 f_2}{\partial y^2} + \frac{\partial^2 f_3}{\partial x\,\partial y} - \left(\frac{\partial^2 f_2}{\partial x^2} + \frac{\partial^2 f_2}{\partial y^2} + \frac{\partial^2 f_2}{\partial z^2}\right)\right]\hat{j}$$

$$+ \left[\frac{\partial^2 f_1}{\partial x\,\partial z} - \frac{\partial^2 f_1}{\partial y^2} + \frac{\partial^2 f_2}{\partial y\,\partial z} - \left(\frac{\partial^2 f_3}{\partial x^2} + \frac{\partial^2 f_3}{\partial y^2} + \frac{\partial^2 f_3}{\partial z^2}\right)\right]\hat{k}$$

$$= \left[\frac{\partial}{\partial x}\left(\frac{\partial f_1}{\partial x} + \frac{\partial f_2}{\partial y} + \frac{\partial f_3}{\partial z}\right) - \left(\frac{\partial^2}{\partial x^2} + \frac{\partial^2}{\partial y^2} + \frac{\partial^2}{\partial z^2}\right) f_1\right]\hat{i}$$

$$+ \left[\frac{\partial}{\partial y}\left(\frac{\partial f_2}{\partial x} + \frac{\partial f_2}{\partial y} + \frac{\partial f_2}{\partial z}\right) - \left(\frac{\partial^2}{\partial x^2} + \frac{\partial^2}{\partial y^2} + \frac{\partial^2}{\partial z^2}\right) f_2\right]\hat{j}$$

$$+ \left[\frac{\partial}{\partial z}\left(\frac{\partial f_3}{\partial x} + \frac{\partial f_3}{\partial y} + \frac{\partial f_3}{\partial z}\right) - \left(\frac{\partial^2}{\partial x^2} + \frac{\partial^2}{\partial y^2} + \frac{\partial^2}{\partial z^2}\right) f_3\right]\hat{k}$$

$$= \left(\hat{i}\frac{\partial}{\partial x} + \hat{j}\frac{\partial}{\partial y} + \hat{k}\frac{\partial}{\partial z}\right)(\nabla f_1 + \nabla f_2 + \nabla f_3)$$

$$= \frac{\partial}{\partial x}(\text{div}\,\vec{f} - \nabla^2 f_1)\,\hat{i} + \frac{\partial}{\partial y}(\text{div}\,\vec{f} - \nabla^2 f_2)\,\hat{j} + \frac{\partial}{\partial z}(\text{div}\,\vec{f} - \nabla^2 f_3)\,\hat{k}$$

$$= \left(\hat{i}\frac{\partial}{\partial x} + \hat{j}\frac{\partial}{\partial y} + \hat{k}\frac{\partial}{\partial z}\right)\text{div}\,\vec{f} - \nabla^2[f_1\,\hat{i} + f_2\,\hat{j} + f_3\,\hat{k}] = \nabla(\nabla.\vec{f}) - \nabla^2\,\vec{f}$$

$$= \text{grad div}\;\vec{f} - \nabla^2\,\vec{f}$$

Proved.

Example 57. *For a solenoidal vector $\vec{F}$, show that curl curl curl curl $\vec{F} = \nabla^4\,\vec{F}$.*

(M.D.U., Dec. 2009)

Solution. Since vector $\vec{F}$ is solenoidal, so div $\vec{F} = 0$ ... (1)

We know that curl curl $\vec{F}$ = grad div $\vec{F} - \nabla^2\,\vec{F}$ ... (2)

Using (1) in (2), grad div $\vec{F}$ = grad (0) = 0 ... (3)

On putting the value of grad div F in (2), we get

curl curl $\vec{F} = -\nabla^2\,\vec{F}$ ... (4)

Now, curl curl curl curl $\vec{F}$ = curl curl $(-\nabla^2\,\vec{F})$ [Using (4)]

$= -$ curl curl $(-\nabla^2\,\vec{F}) = -$ [grad div $(\nabla^2\,\vec{F}) - \nabla^2\,(\nabla^2\,\vec{F})$] [Using (2)]

$= -$ grad $(\nabla.\nabla^2\,\vec{F}) + \nabla^2\,(\nabla^2\,\vec{F}) = -$ grad $(\nabla^2\,\nabla.\vec{F}) + \nabla^4\,\vec{F}$ $[\nabla.\vec{F} = 0]$

$= 0 + \nabla^4\,\vec{F} = \nabla^4\,\vec{F}$ [Using (1)] **Proved.**

EXERCISE 2.4

1. Find the divergence and curl of the vector field $V = (x^2 - y^2)\hat{i} + 2xy\,\hat{j} + (y^2 - xy)\,\hat{k}$.

Ans. Divergence = $4x$, Curl = $(2y - x)\hat{i} + y\hat{j} + 4y\hat{k}$

2. If a is constant vector and r is the radius vector, prove that

(i) $\nabla(\vec{a}.\vec{r}) = \vec{a}$ (ii) $\text{div}\,(\vec{r} \times \vec{a}) = 0$ (iii) $\text{curl}\,(\vec{r} \times \vec{a}) = -2\vec{a}$

where $\vec{r} = x\hat{i} + y\hat{j} + z\hat{k}$ and $\vec{a} = a_1\hat{i} + a_2\hat{j} + a_3\hat{k}$.

3. Prove that:

$\nabla(A.B) = (A.\nabla)B + (B.\nabla)A + A \times (\nabla \times B) + B \times (\nabla \times A)$ *(R.G.P.V. Bhopal, June 2004)*

4. If $F = (x + y + 1)\hat{i} + \hat{j} - (x + y)\hat{k}$. show that $F.\text{curl}\,F = 0$.

(R.G.P.V. Bhopal, Feb. 2006, June 2004)

Prove that

5. $\vec{\nabla} \times (\phi\vec{F}) = (\vec{\nabla}\phi) \times \vec{F} + \phi(\vec{\nabla} \times \vec{F})$

6. $\nabla.(\vec{F} \times \vec{G}) = \vec{G}.(\nabla \times \vec{F}) - \vec{F}.(\nabla \times \vec{G})$ *(Delhi University, April 2010)*

7. Prove that curl $(\vec{a} \times \vec{r}) = 2\vec{a}$

8. Prove that Div. (curl $\vec{v}$) = $\nabla.(\nabla \times \vec{v}) = 0$

9. If $V = e^{xyz}\,(\hat{i} + \hat{j} + \hat{k})$, find curl V **Ans.** $e^{xyz}\,[x\,(z - y)\hat{i} + y\,(x - z)\hat{j} + z\,(y - x)\hat{k}]$

10. If $\vec{u} = \dfrac{\vec{r}}{r^2}$, then evaluate curl $\vec{u}$ **Ans.** 0

11. Evaluate curlgrad r^m where $\vec{r} = |\vec{r}| = |x\hat{i} + y\hat{j} + z\hat{k}|$ **Ans.** 0

12. Find div $\vec{F}$ and curl F where F = grad $(x^3 + y^3 + z^3 - 3xyz)$.

Ans. div $\vec{F} = 6(x + y + z)$, curl $\vec{F} = 0$

13. Find out values of a, b, c for which $\vec{v} = (x + y + az)\hat{i} + (bx + 3y - z)\hat{j} + (3x + cy + z)\hat{k}$ is irrotational. **Ans.** $a = 3, b = 1, c = -1$

Choose the correct alternative:

14. The magnitude of the vector drawn in a direction perpendicular to the surface $x^2 + 2y^2 + z^2 = 7$ at the point $(1, -1, 2)$ is

(i) $\dfrac{2}{3}$ (ii) $\dfrac{3}{2}$ (iii) 3 (iv) 6 **Ans.** (iv)

15. A unit normal to $x^2 + y^2 + z^2 = 5$ at $(0, 1, 2)$ is equal to

(i) $\dfrac{1}{\sqrt{5}}(\hat{i} + \hat{j} + \hat{k})$ (ii) $\dfrac{1}{\sqrt{5}}(\hat{i} + \hat{j} - \hat{k})$ (iii) $\dfrac{1}{\sqrt{5}}(\hat{j} + 2\hat{k})$ (iv) $\dfrac{1}{\sqrt{5}}(\hat{i} - \hat{j} + \hat{k})$

(A.M.I.E.T.E., Dec. 2008) **Ans.** (iii)

16. The directional derivative of $\phi = x\,y\,z$ at the point $(1, 1, 1)$ in the direction $\hat{i}$ is:

(i) -1 (ii) $-\dfrac{1}{3}$ (iii) 1 (iv) $\dfrac{1}{3}$ **Ans.** (iii)

(R.G.P.V. Bhopal, II Sem., June 2007)

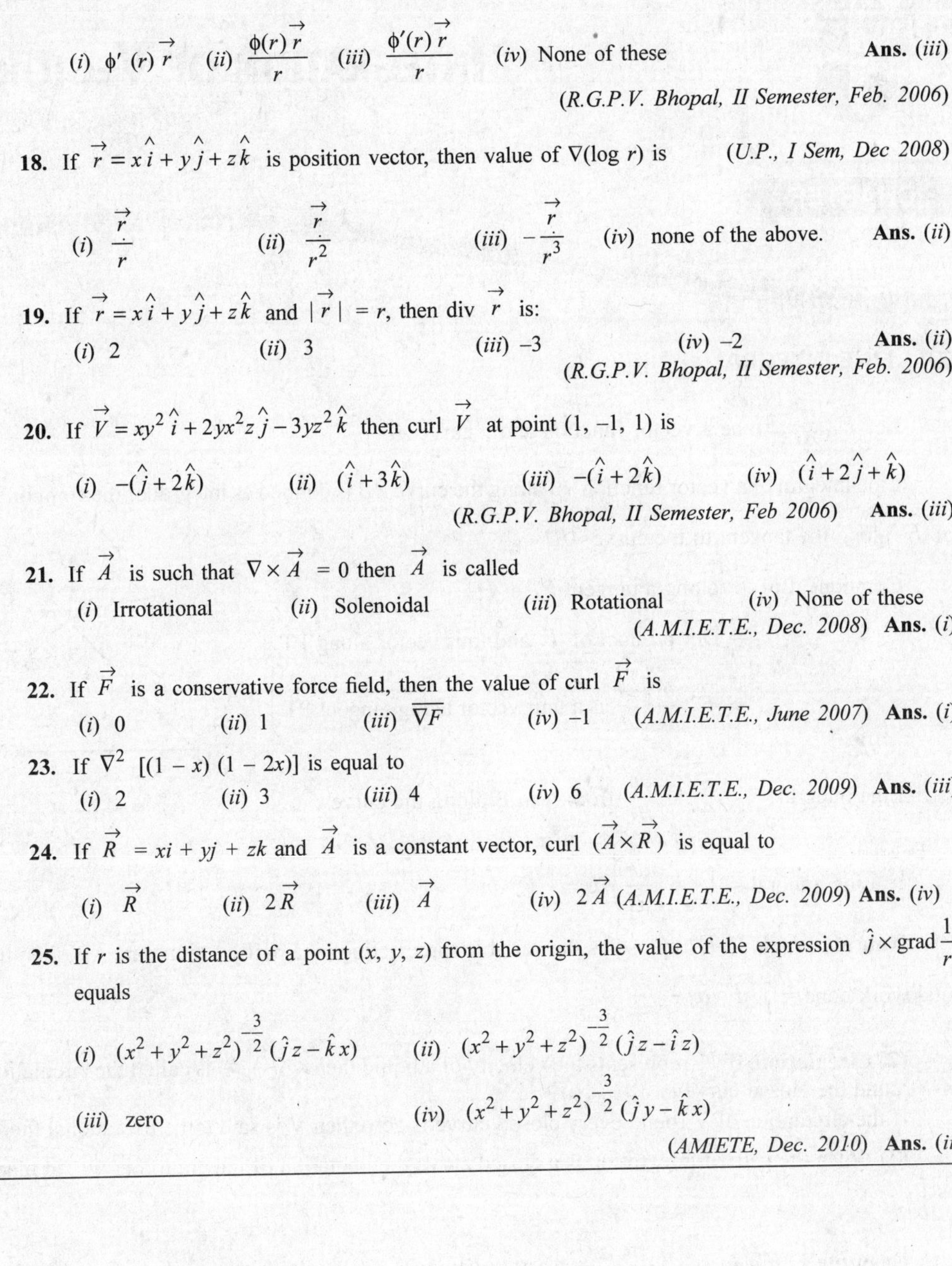

17. If $\vec{r} = x\hat{i} + y\hat{j} + z\hat{k}$ and $r = |\vec{r}|$ then $\nabla\phi(r)$ is:

(*i*) $\phi'(r)\vec{r}$ (*ii*) $\frac{\phi(r)\vec{r}}{r}$ (*iii*) $\frac{\phi'(r)\vec{r}}{r}$ (*iv*) None of these **Ans.** (*iii*)

(*R.G.P.V. Bhopal, II Semester, Feb. 2006*)

18. If $\vec{r} = x\hat{i} + y\hat{j} + z\hat{k}$ is position vector, then value of $\nabla(\log r)$ is (*U.P., I Sem, Dec 2008*)

(*i*) $\frac{\vec{r}}{r}$ (*ii*) $\frac{\vec{r}}{r^2}$ (*iii*) $-\frac{\vec{r}}{r^3}$ (*iv*) none of the above. **Ans.** (*ii*)

19. If $\vec{r} = x\hat{i} + y\hat{j} + z\hat{k}$ and $|\vec{r}| = r$, then div $\vec{r}$ is:

(*i*) 2 (*ii*) 3 (*iii*) –3 (*iv*) –2 **Ans.** (*ii*)

(*R.G.P.V. Bhopal, II Semester, Feb. 2006*)

20. If $\vec{V} = xy^2\hat{i} + 2yx^2z\hat{j} - 3yz^2\hat{k}$ then curl $\vec{V}$ at point (1, –1, 1) is

(*i*) $-(\hat{j} + 2\hat{k})$ (*ii*) $(\hat{i} + 3\hat{k})$ (*iii*) $-(\hat{i} + 2\hat{k})$ (*iv*) $(\hat{i} + 2\hat{j} + \hat{k})$

(*R.G.P.V. Bhopal, II Semester, Feb 2006*) **Ans.** (*iii*)

21. If $\vec{A}$ is such that $\nabla \times \vec{A} = 0$ then $\vec{A}$ is called

(*i*) Irrotational (*ii*) Solenoidal (*iii*) Rotational (*iv*) None of these

(*A.M.I.E.T.E., Dec. 2008*) **Ans.** (*i*)

22. If $\vec{F}$ is a conservative force field, then the value of curl $\vec{F}$ is

(*i*) 0 (*ii*) 1 (*iii*) $\overline{\nabla F}$ (*iv*) –1 (*A.M.I.E.T.E., June 2007*) **Ans.** (*i*)

23. If ∇^2 [(1 – *x*) (1 – 2*x*)] is equal to

(*i*) 2 (*ii*) 3 (*iii*) 4 (*iv*) 6 (*A.M.I.E.T.E., Dec. 2009*) **Ans.** (*iii*)

24. If $\vec{R}$ = *xi* + *yj* + *zk* and $\vec{A}$ is a constant vector, curl $(\vec{A} \times \vec{R})$ is equal to

(*i*) $\vec{R}$ (*ii*) $2\vec{R}$ (*iii*) $\vec{A}$ (*iv*) $2\vec{A}$ (*A.M.I.E.T.E., Dec. 2009*) **Ans.** (*iv*)

25. If *r* is the distance of a point (*x*, *y*, *z*) from the origin, the value of the expression $\hat{j} \times \text{grad}\,\frac{1}{r}$ equals

(*i*) $(x^2 + y^2 + z^2)^{-\frac{3}{2}}(\hat{j}z - \hat{k}x)$ (*ii*) $(x^2 + y^2 + z^2)^{-\frac{3}{2}}(\hat{j}z - \hat{i}z)$

(*iii*) zero (*iv*) $(x^2 + y^2 + z^2)^{-\frac{3}{2}}(\hat{j}y - \hat{k}x)$

(*AMIETE, Dec. 2010*) **Ans.** (*ii*)

CHAPTER

INTEGRATION OF VECTORS

3.1 LINE INTEGRAL

Let $\vec{F}(x, y, z)$ be a vector function and a curve AB.

Line integral of a vector function $\vec{F}$ along the curve AB is defined as integral of the component of $\vec{F}$ along the tangent to the curve AB.

Component of $\vec{F}$ along a tangent PT at P

$$= \text{Dot product of } \vec{F} \text{ and unit vector along PT}$$

$$= \vec{F} \cdot \frac{\vec{dr}}{ds} \left(\frac{\vec{dr}}{ds} \text{ is a unit vector along tangent PT} \right)$$

$$\text{Line integral} = \sum \vec{F} \cdot \frac{\vec{dr}}{ds} \text{ from A to B along the curve}$$

$$\therefore \text{Line integral} = \int_c \left(\vec{F} \cdot \frac{\vec{dr}}{ds} \right) ds = \int_c \vec{F} \cdot \vec{dr}$$

Note (1) Work. If $\overline{F}$ represents the variable force acting on a particle along arc AB, then the total work done $= \int_A^B \vec{F} \cdot \vec{dr}$

(2) Circulation. If $\vec{V}$ represents the velocity of a liquid then $\oint_c \vec{V} \cdot \vec{dr}$ is called the circulation of V round the closed curve c.

If the circulation of V round every closed curve is zero then V is said to be irrotational there.

(3) When the path of integration is a closed curve then notation of integration is $\oint$ in place of $\int$.

Example 1. *If a force $\vec{F} = 2x^2 y\hat{i} + 3\,xy\hat{j}$ displaces a particle in the xy-plane from (0, 0) to (1, 4) along a curve $y = 4\,x^2$. Find the work done.*

Solution. Work done $= \int_c \vec{F} . \vec{dr}$

$$\left[\begin{aligned} \vec{r} &= x\hat{i} + y\hat{j} \\ \vec{dr} &= dx\,\hat{i} + dy\,\hat{j} \end{aligned} \right]$$

$$= \int_c (2x^2 y\hat{i} + 3xy\,\hat{j}).(dx\hat{i} + dy\,\hat{j})$$

$$= \int_c (2x^2 y\,dx + 3xy\,dy)$$

Putting the values of y and dy, we get $\begin{pmatrix} y = 4x^2 \\ dy = 8x\,dx \end{pmatrix}$

$$= \int_0^1 \cdot [2x^2 (4x^2)\,dx + 3x(4x^2)\,8x\,dx]$$

$$= 104\int_0^1 x^4\,dx = 104\left(\frac{x^5}{5}\right)_0^1 = \frac{104}{5}$$ **Ans.**

Example 2. *Evaluate* $\int_C \vec{F}.\vec{dr}$ *where* $\vec{F} = x^2\hat{i} + xy\hat{j}$ *and C is the boundary of the square in the plane z = 0 and bounded by the lines x = 0, y = 0, x = a and y = a.*

Solution. $\int_C \vec{F}.\vec{d}\,\vec{r} = \int_{OA} \vec{F}.\vec{dr} + \int_{AB} \vec{F}.\vec{dr} + \int_{BC} \vec{F}.\vec{dr} + \int_{CO} \vec{F}.\vec{dr}$

Here $\vec{r} = x\hat{i} + y\hat{j},\ \vec{d}\,r = dx\hat{i} + dy\hat{j}, \vec{F} = x^2\hat{i} + xy\,\hat{j}$

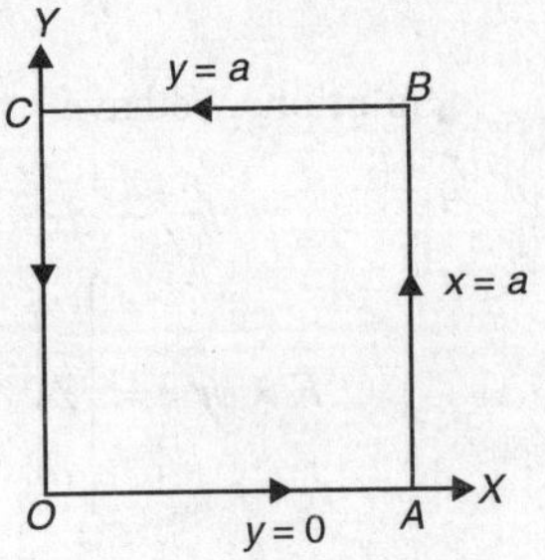

$$\vec{F}.\vec{dr} = x^2dx + xydy \quad ...(1)$$

On OA, $y = 0$ $\therefore\ \vec{F}.\vec{dr} = x^2dx$

$$\int_{OA} \vec{F}.\vec{dr} = \int_0^a x^2dx = \left[\frac{x^3}{3}\right]_0^a = \frac{a^3}{3} \quad ...(2)$$

On AB, $x = a$ $\therefore$ $dx = 0$

(1) becomes

$$\therefore\ \vec{F}.\vec{dr} = aydy$$

$$\int_{AB} \vec{F}.\vec{dr} = \int_0^a aydy = a\left[\frac{y^2}{2}\right]_0^a = \frac{a^3}{2} \quad ...(3)$$

On BC, $y = a$ $\therefore dy = 0$

$\Rightarrow$ (1) becomes $\vec{F}.\vec{dr} = x^2dx$

$$\int_{BC} \vec{F}.\vec{dr} = \int_a^0 x^2dx = \left[\frac{x^3}{3}\right]_a^0 = \frac{-a^3}{3} \quad ...(4)$$

On CO, $x = 0$, $\therefore\ \vec{F}.\vec{dr} = 0$

(1) becomes

$$\int_{CO} \vec{F}.\vec{dr} = 0 \quad ...(5)$$

On adding (2), (3), (4) and (5), we get $\int_C \vec{F}.\vec{dr} = \frac{a^3}{3} + \frac{a^3}{2} - \frac{a^3}{3} + 0 = \frac{a^3}{2}$ **Ans.**

Example 3. *A vector field is given by*

$\vec{F} = (2y + 3)\,\hat{i} + xz\hat{j} + (yz - x)\,\hat{k}$. *Evaluate* $\int_C \vec{F}.\vec{dr}$ *along the path c is* $x = 2t$,

$y = t,\ z = t^3$ *from* $t = 0$ *to* $t = 1$.

Solution. $\int_C \vec{F}.\vec{dr} = \int_C (2y+3)\,dx + (xz)\,dy + (yz-x)\,dz$

$$\left[\begin{array}{lll} \text{Since } x = 2t & y = t & z = t^3 \\ \therefore \dfrac{dx}{dt} = 2 & \dfrac{dy}{dt} = 1 & \dfrac{dz}{dt} = 3t^2 \end{array}\right]$$

$$= \int_0^1 (2t+3)(2\,dt) + (2t)(t^3)\,dt + (t^4 - 2t)(3t^2\,dt) = \int_0^1 (4t + 6 + 2t^4 + 3t^6 - 6t^3)\,dt$$

$$= \left[4\frac{t^2}{2} + 6t + \frac{2}{5}t^5 + \frac{3}{7}t^7 - \frac{6}{4}t^4\right]_0^1 = \left[2t^2 + 6t + \frac{2}{5}t^5 + \frac{3}{7}t^7 - \frac{3}{2}t^4\right]_0^1$$

$$= 2 + 6 + \frac{2}{5} + \frac{3}{7} - \frac{3}{2} = 7.32857.$$

Ans.

Example 4. *If* $\overline{F} = 2y\hat{i} - z\hat{j} + x\hat{k}$, *evaluate* $\int_C \vec{F} \times \vec{dr}$ *along the curve*

$x = \cos t,\ y = \sin t,\ z = 2\cos t$ *from* $t = 0$ *to* $t = \dfrac{\pi}{2}$.

Solution. We have, $\vec{r} = x\hat{i} + y\hat{j} + z\hat{k}$

$$\vec{dr} = dx\,\hat{i} + dy\,\hat{j} + dz\,\hat{k}$$

$$\vec{F} \times \vec{dr} = \begin{vmatrix} \hat{i} & \hat{j} & \hat{k} \\ 2y & -z & x \\ dx & dy & dz \end{vmatrix}$$

$$= (-zdz - xdy)\,\hat{i} - (2y\,dz - xdx)\,\hat{j} + (2ydy + zdx)\,\hat{k}$$

$$= [-2\cos t\,(-2\sin t)\,dt - \cos t\,(\cos t)\,dt]\,\hat{i}$$

$$-[\,2\sin t\,(-2\sin t)\,dt - \cos t\,(-\sin t)\,dt\,]\,\hat{j}$$

$$+\ [\,2\sin t\,(\cos t)\,dt + 2\cos t\,(-\sin t)\,dt\,]\,\hat{k}$$

$$= [\,(4\cos t\sin t - \cos^2 t)\,\hat{i} + (4\sin^2 t - \cos t\sin t)\,\hat{j}\,]\,dt$$

$$\therefore \int_C \vec{F} \times \vec{dr} = \int_0^{\frac{\pi}{2}} [\,(4\cos t\sin t - \cos^2 t)\,\hat{i} + (4\sin^2 t - \cos t\sin t)\,\hat{j}\,]\,dt$$

$$= \int_0^{\frac{\pi}{2}} \left[\left\{2\sin 2t - \frac{\cos 2t + 1}{2}\right\}\hat{i}\right] dt + \int_0^{\frac{\pi}{2}} \left[\left\{2(1 - \cos 2t) - \frac{1}{2}\sin 2t\right\}\hat{j}\right] dt$$

$$= \left[-\cos 2t - \frac{1}{4}\sin 2t - \frac{1}{2}t\right]_0^{\frac{\pi}{2}} \hat{i} + \left[2t - \sin 2t + \frac{1}{4}\cos 2t\right]_0^{\frac{\pi}{2}} \hat{j}$$

$$= \left[-\cos\pi - \frac{1}{4}\sin\pi - \frac{1}{2}\left(\frac{\pi}{2}\right) + \cos 0 + \frac{1}{4}\sin 0 + \frac{1}{2}(0)\right]\hat{i} +$$

$$\left[\pi - \sin\pi + \frac{1}{4}\cos\pi - 0 + \sin 0 - \frac{1}{4}\cos 0\right]\hat{j}$$

$$= \left[1-0-\frac{\pi}{4}+1+0\right]\hat{i}+\left[\pi-0-\frac{1}{4}+0-\frac{1}{4}\right]\hat{j} = \left(2-\frac{\pi}{4}\right)\hat{i}+\left(\pi-\frac{1}{2}\right)\hat{j}$$ **Ans.**

Example 5. *The acceleration of a particle at time t is given by*

$$\vec{a} = 18\cos 3t\,\hat{i} - 8\sin 2t\,\hat{j} + 6t\,\hat{k}.$$

If the velocity $\vec{v}$ *and displacement* $\vec{r}$ *be zero at* $t = 0$, *find* $\vec{v}$ *and* $\vec{r}$ *at any point t.*

Solution. Here, $\vec{a} = \dfrac{d^2\vec{r}}{dt^2} = 18\cos 3t\,\hat{i} - 8\sin 2t\,\hat{j} + 6t\,\hat{k}.$

On integrating, we have

$$\vec{v} = \frac{d\vec{r}}{dt} = \hat{i}\int 18\cos 3t\,dt + \hat{j}\int -8\sin 2t\,dt + \hat{k}\int 6t\,dt$$

$$\Rightarrow \quad \vec{v} = 6\sin 3t\,\hat{i} + 4\cos 2t\,\hat{j} + 3t^2\,\hat{k} + \vec{c} \qquad ...(1)$$

At $\quad t = 0, \vec{v} = \vec{0}$

Putting $t = 0$ and $\vec{v} = 0$ in (1), we get

$$\vec{0} = 4\hat{j} + \vec{c} \Rightarrow \vec{c} = -4\hat{j}$$

$$\therefore \quad \vec{v} = \frac{d\vec{r}}{dt} = 6\sin 3t\,\hat{i} + 4(\cos 2t - 1)\,\hat{j} + 3t^2\hat{k}$$

Again integrating, we have

$$\vec{r} = \hat{i}\int 6\sin 3t\,dt + \hat{j}\int 4(\cos 2t - 1)\,dt + \hat{k}\int 3t^2\,dt$$

$$\Rightarrow \quad \vec{r} = -2\cos 3t\,\hat{i} + (2\sin 2t - 4t)\,\hat{j} + t^3\,\hat{k} + \vec{c_1} \qquad ...(2)$$

At, $\quad t = 0, \vec{r} = 0$

Putting $t = 0$ and $\vec{r} = 0$ in (2), we get

$$\therefore \quad \vec{0} = -2\hat{i} + \vec{C_1} \Rightarrow \vec{C_1} = 2\hat{i}$$

Hence, $\quad \vec{r} = 2(1-\cos 3t)\,\hat{i} + 2(\sin 2t - 2t)\,\hat{j} + t^3\hat{k}$ **Ans.**

Example 6. *If* $\vec{A} = (3x^2 + 6y)\,\hat{i} - 14yz\hat{j} + 20xz^2\,\hat{k}$, *evaluate the line integral* $\oint \vec{A}.\vec{dr}$ *from* (0, 0, 0) to (1, 1, 1) *along the curve C.*

$x = t, y = t^2, z = t^3.$ *(Uttarakhand, I Semester, Dec. 2006)*

Solution. We have,

$$\int_C \vec{A}.\vec{dr} = \int_C [(3x^2+6y)\,\hat{i} - 14yz\hat{j} + 20xz^2\hat{k}].[\hat{i}\,dx + \hat{j}\,dy + \hat{k}\,dz]$$

$$= \int_C [(3x^2+6y)\,dx - 14yzdy + 20xz^2dz]$$

If $x = t, y = t^2, z = t^3$, then points (0, 0, 0) and (1, 1, 1) correspond to $t = 0$ and $t = 1$ respectively.

Now, $\int_C \vec{A}.\vec{dr} = \int_{t=0}^{t=1} [(3t^2 + 6t^2)\,d(t) - 14t^2\,t^3\,d(t^2) + 20t\,(t^3)^2\,d(t^3)]$

$$= \int_{t=0}^{t=1} [9t^2\,dt - 14t^5 \,.\, 2t\,dt + 20\,t^7 \,.\, 3t^2\,dt] = \int_0^1 (9t^2 - 28t^6 + 60t^9)\,dt$$

$$= \left[9\left(\frac{t^3}{3}\right) - 28\left(\frac{t^7}{7}\right) + 60\left(\frac{t^{10}}{10}\right)\right]_0^1 = 3 - 4 + 6 = 5 \qquad \textbf{Ans.}$$

Example 7. *Compute* $\int_c \vec{F} \,.\, \vec{dr}$, *where* $\vec{F} = \dfrac{\hat{i}y - \hat{j}x}{x^2 + y^2}$ *and c is the circle* $x^2 + y^2 = 1$ *traversed counter clockwise.*

Solution. $\vec{r} = \hat{i}\,x + \hat{j}\,y + \hat{k}\,z,\ d\overline{r} = \hat{i}\,dx + \hat{j}\,dy + \hat{k}\,dz$

$$\int_c \vec{F} \,.\, d\vec{r} = \int_c \frac{\hat{i}y - \hat{j}x}{x^2 + y^2} \cdot (\hat{i}dx + \hat{j}dy + \hat{k}dz)$$

$$= \int_c \frac{ydx - xdy}{x^2 + y^2} = \int_c (ydx - xdy) \qquad ...(1)\ [\because\ x^2 + y^2 = 1]$$

Parametric equation of the circle are $x = \cos\theta$, $y = \sin\theta$.

Putting $x = \cos\theta$, $y = \sin\theta$, $dx = -\sin\theta\, d\theta$, $dy = \cos\theta\, d\theta$ in (1), we get

$$\int_C \vec{F} d\vec{r} = \int_0^{2\pi} \sin\theta\,(-\sin\theta\, d\theta) - \cos\theta\,(\cos\theta\, d\theta)$$

$$= -\int_0^{2\pi} (\sin^2\theta + \cos^2\theta)\, d\theta = -\int_0^{2\pi} d\theta = -(\theta)_0^{2\pi} = -2\pi \qquad \textbf{Ans.}$$

Example 8. *Show that the vector field* $\vec{F} = 2x(y^2 + z^3)\hat{i} + 2x^2y\hat{j} + 3x^2z^2\hat{k}$ *is conservative. Find its scalar potential and the work done in moving a particle from* (–1, 2, 1) to (2, 3, 4).

(*A.M.I.E.T.E. June 2010, 2009*)

Solution. Here, we have

$$\vec{F} = 2x(y^2 + z^3)\,\hat{i} + 2x^2y\,\hat{j} + 3x^2z^2\hat{k}$$

$$\text{Curl } \vec{F} = \nabla \times \vec{F}$$

$$= \begin{vmatrix} \hat{i} & \hat{j} & \hat{k} \\ \dfrac{\partial}{\partial x} & \dfrac{\partial}{\partial y} & \dfrac{\partial}{\partial z} \\ 2x(y^2 + z^3) & 2x^2y & 3x^2z^2 \end{vmatrix} = (0-0)i - (6xz^2 - 6xz^2)\hat{j} + (4xy - 4xy)\hat{k} = 0$$

Hence, vector field $\vec{F}$ is irrotational.

To find the scalar potential function ϕ

$$\vec{F} = \vec{\nabla}\phi$$

$$d\phi = \frac{\partial\phi}{\partial x}dx + \frac{\partial\phi}{\partial y}dy + \frac{\partial\phi}{\partial z}dz = \left(\hat{i}\frac{\partial\phi}{\partial x} + \hat{j}\frac{\partial\phi}{\partial y} + \hat{k}\frac{\partial\phi}{\partial z}\right) . \left(\hat{i}dx + \hat{j}dy + \hat{k}dz\right)$$

$$= \left(\hat{i}\frac{\partial}{\partial x} + \hat{j}\frac{\partial}{\partial y} + \hat{k}\frac{\partial}{\partial z}\right)\phi . \left(d\vec{r}\right) = \nabla\phi . d\vec{r} = \vec{F} \,.\, d\vec{r}$$

$$= [2x(y^2 + z^3)\hat{i} + 2x^2 y\hat{j} + 3x^2 z^2 \hat{k}] (\hat{i}dx + \hat{j}dy + \hat{k}dz)$$

$$= 2x(y^2 + z^3)dx + 2x^2 y\, dy + 3x^2 z^2 dz$$

$$\phi = \int \left[2x(y^2 + z^3)dx + 2x^2 y dy + 3x^2 z^2 dz\right] + C$$

$$\int (2xy^2 dx + 2x^2 y dy) + (2xz^3 dx + 3x^2 z^2 dz) + C = x^2y^2 + x^2z^3 + C$$

Hence, the scalar potential is $x^2y^2 + x^2z^3 + C$

Now, for conservative field

$$\text{Work done} = \int_{(-1,2,1)}^{(2,3,4)} \vec{F}.d\vec{r} = \int_{(-1,2,1)}^{(2,3,4)} d\phi = [\phi]_{(-1,2,1)}^{(2,3,4)} = \left[x^2y^2 + x^2z^3\right]_{(-1,2,1)}^{(2,3,4)}$$

$$= (36 + 256) - (4 - 1) = 287$$ **Ans.**

Example 9. *A vector field is given by* $\vec{F} = (\sin y)\,\hat{i} + x\,(1 + \cos y)\,\hat{j}$*. Evaluate the line integral over a circular path* $x^2 + y^2 = a^2$, $z = 0$.

Solution. We have,

$$\text{Work done} = \int_C \vec{F}.\vec{dr}$$

$$= \int_C [\,(\sin y)\,\hat{i} + x\,(1 + \cos y)\,\hat{j}].[dx\hat{i} + dy\hat{j}] \quad (\because z = 0 \text{ hence } dz = 0)$$

$$\Rightarrow \quad \int_C \vec{F}.\vec{dr} = \int_C \sin y\, dx + x\,(1 + \cos y)\, dy = \int_C (\sin y\, dx + x\cos y\, dy + x\,dy)$$

$$= \int_C d\,(x \sin y) + \int_C x\, dy$$

(where d is differential operator).

The parametric equations of given path

$x^2 + y^2 = a^2$ are $x = a\cos\theta$, $y = a\sin\theta$,

Where θ varies from 0 to 2π

$$\therefore \quad \int_C \vec{F}.\vec{dr} = \int_0^{2\pi} d\,[\,a\cos\theta \sin(a\sin\theta)\,] + \int_0^{2\pi} a\cos\theta\,.\,a\cos\theta\, d\,\theta$$

$$= \int_0^{2\pi} d\,[a\cos\theta\sin(a\sin\theta)\,] + \int_0^{2\pi} a^2\cos^2\theta\,.\,d\,\theta$$

$$= [a\cos\theta \sin(a\sin\theta)\,]_0^{2\pi} + \int_0^{2\pi} a^2\cos^2\theta\, d\,\theta$$

$$= 0 + a^2 \int_0^{2\pi} \left(\frac{1 + \cos 2\theta}{2}\right) d\theta = \frac{a^2}{2}\left[\theta + \frac{\sin 2\theta}{2}\right]_0^{2\pi}$$

$$= \frac{a^2}{2}.\,2\pi = \pi\, a^2$$ **Ans.**

Example 10. *Determine whether the line integral*

$\int (2xyz^2)\, dx + (x^2z^2 + z\cos yz)\, dy + (2x^2yz + y\cos yz)\, dz$ *is independent of the path of integration ? If so, then evaluate it from (1, 0, 1) to* $\left(0, \frac{\pi}{2}, 1\right)$.

Solution. $\int_c (2xyz^2)\,dx + (x^2z^2 + z\cos yz)\,dy + (2x^2yz + y\cos yz)\,dz$

$$= \int_c [(2xyz^2\hat{i}) + (x^2z^2 + z\cos yz)\,\hat{j} + (2x^2yz + y\cos yz)\,\hat{k}].(\hat{i}dx + \hat{j}dy + \hat{k}dz)$$

$$= \int_c \vec{F}\cdot\vec{dr}$$

This integral is independent of path of integration if

$$\vec{F} = \nabla\phi \Rightarrow \nabla\times\vec{F} = 0$$

$$\nabla\times\overline{F} = \begin{vmatrix} \hat{i} & \hat{j} & \hat{k} \\ \dfrac{\partial}{\partial x} & \dfrac{\partial}{\partial y} & \dfrac{\partial}{\partial z} \\ 2xyz^2 & x^2z^2 + z\cos yz & 2x^2yz + y\cos yz \end{vmatrix}$$

$$= (2x^2z + \cos yz - yz\sin yz - 2x^2z - \cos yz + yz\sin yz)\,\hat{i} - (4xyz - 4xyz)\,\hat{j} + (2xz^2 - 2xz^2)\,\hat{k} = 0$$

Hence, the line integral is independent of path.

$$d\phi = \frac{\partial\phi}{\partial x}dx + \frac{\partial\phi}{\partial y}dy + \frac{\partial\phi}{\partial z}dz \qquad \text{(Total differentiation)}$$

$$= \left(\hat{i}\frac{\partial\phi}{\partial x} + \hat{j}\frac{\partial\phi}{\partial y} + \hat{k}\frac{\partial\phi}{\partial z}\right)\cdot(\hat{i}dx + \hat{j}dy + \hat{k}dz) = \nabla\phi\cdot dr = \vec{F}\cdot\vec{d}r$$

$$= [(2xyz^2)\,\hat{i} + (x^2z^2 + z\cos yz)\,\hat{j} + (2x^2yz + y\cos yz)\hat{k}].\ (\hat{i}dx + \hat{j}dy + \hat{k}dz)$$

$$= 2xyz^2\,dx + (x^2z^2 + z\cos yz)\,dy + (2x^2yz + y\cos yz)\,dz$$

$$= [(2x\,dx)\,yz^2 + x^2\,(dy)\,z^2 + x^2y\,(2z\,dz)] + [(\cos yz\,dy)\,z + (\cos yz\,dz)\,y]$$

$$= d\,(x^2yz^2) + d\,(\sin yz)$$

$$\phi = \int d\,(x^2yz^2) + \int d\,(\sin yz) = x^2yz^2 + \sin yz$$

$$[\phi]_A^B = \phi\,(B) - \phi\,(A)$$

$$= [x^2yz^2 + \sin yz]_{(0,\frac{\pi}{2},1)} - [x^2yz^2 + \sin yz]_{(1,0,1)} = \left[0 + \sin\left(\frac{\pi}{2}\times 1\right)\right] - [0+0]$$

$$= 1$$

Ans.

EXERCISE 3.1

1. Find the work done by a force $y\hat{i} + x\hat{j}$ which displaces a particle from origin to a point $(\hat{i} + \hat{j})$. **Ans.** 1

2. Find the work done when a force $\overline{F} = (x^2 - y^2 + x)\,\hat{i} - (2xy + y)\,\hat{j}$ moves a particle from origin to (1, 1) along a parabola $y^2 = x$. **Ans.** $\dfrac{2}{3}$

3. Show that $\vec{V} = (2xy + z^3)\,\hat{i} + x^2\,\hat{j} + 3xz^2\,\hat{k}$ is a conservative field. Find its scalar potential ϕ such that $\vec{V}$ = grad ϕ. Find the work done by the force $\vec{V}$ in moving a particle from (1, – 2, 1) to (3, 1, 4). **Ans.** $x^2y + xz^3$, 202

4. Show that the line integral $\int_c (2xy + 3)\,dx + (x^2 - 4z)\,dy - 4y\,dz$

   where c is any path joining (0, 0, 0) to (1, – 1, 3) does not depend on the path c and evaluate the line integral. **Ans.** 14

5. Find the work done in moving a particle once round the ellipse $\frac{x^2}{25}+\frac{y^2}{16}=1$, $z = 0$, under the field of force given by $F = (2x - y + z)\ \hat{i} + (x + y - z^2)\ \hat{j} + (3x - 2y + 4z)\ \hat{k}$. Is the field of force conservative?

Ans. 40 π, No

6. If $\vec{\nabla}\phi = (y^2 - 2xyz^3)\ \hat{i} + (3 + 2xy - x^2z^3)\ \hat{j} + (z^3 - 3x^2yz^2)\ \hat{k}$, find φ. **Ans.** $3y+\frac{z^4}{4}+xy^2-x^2yz^3$

7. $\int_C \vec{F}.d\vec{R}$ is independent of the path joining any two point if it is. *(A.M.I.E.T.E., June 2010)*

(*i*) irrotational field (*ii*) solenoidal field (*iii*) rotational field (*iv*) vector field. **Ans.** (*i*)

3.2 SURFACE INTEGRAL

A surface $r = f(u, v)$ is called smooth if $f(u, v)$ posses continous first order partial derivative.

Let $\vec{F}$ be a vector function and S be the given surface.

Surface integral of a vector function $\vec{F}$ over the surface S is defined as the integral of the components of $\vec{F}$ along the normal to the surface.

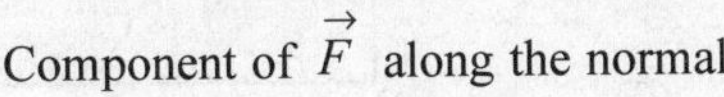

Component of $\vec{F}$ along the normal

$= \vec{F}.\hat{n}$, where n is the unit normal vector to an element ds and

$$\hat{n} = \frac{\text{grad } f}{|\text{grad } f|} \qquad ds = \frac{dx\,dy}{(\hat{n}\cdot\hat{k})}$$

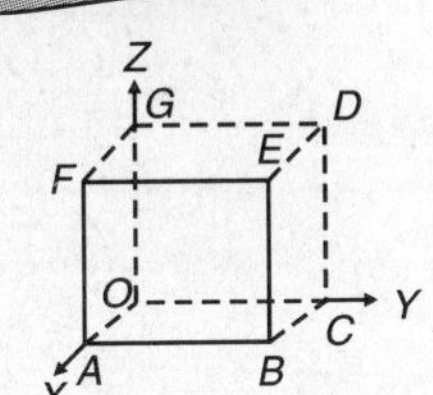

Surface integral of F over S

$$= \Sigma\, \vec{F}\cdot\hat{n} \qquad = \iint_S (\vec{F}\cdot\hat{n})\, ds$$

Note. (1) Flux $= \iint_S (\vec{F}\cdot\hat{n})\, ds$ where, $\overline{F}$ represents the velocity of a liquid.

If $\iint_S (\vec{F}\cdot\hat{n})ds = 0$, then $\vec{F}$ is said to be a *solenoidal* vector point function.

Example 11. *Evaluate* $\iint_S \vec{A}.\hat{n}\, ds$ *where* $\vec{A} = (x + y^2)\,\hat{i} - 2x\hat{j} + 2yz\hat{k}$ *and S is the surface of the plane* $2x + y + 2z = 6$ *in the first octant.*

Solution. A vector normal to the surface "S" is given by

$$\nabla\,(2x + y + 2z) = \left(\hat{i}\frac{\partial}{\partial x}+\hat{j}\frac{\partial}{\partial y}+\hat{k}\frac{\partial}{\partial z}\right)(2x + y + 2z) = 2\hat{i} + \hat{j} + 2\hat{k}$$

And $\hat{n}$ = a unit vector normal to surface S

$$= \frac{2\hat{i} + \hat{j} + 2\hat{k}}{\sqrt{4+1+4}} = \frac{2}{3}\hat{i}+\frac{1}{3}\hat{j}+\frac{2}{3}\hat{k}$$

$$\hat{k}.\hat{n} = \hat{k}.\left(\frac{2}{3}\hat{i}+\frac{1}{3}\hat{j}+\frac{2}{3}\hat{k}\right) = \frac{2}{3}$$

$$\therefore \quad \iint_S \overline{A}.\hat{n}ds = \iint_R \overline{A}.\hat{n}\frac{dx\,dy}{\hat{k}.\hat{n}}$$

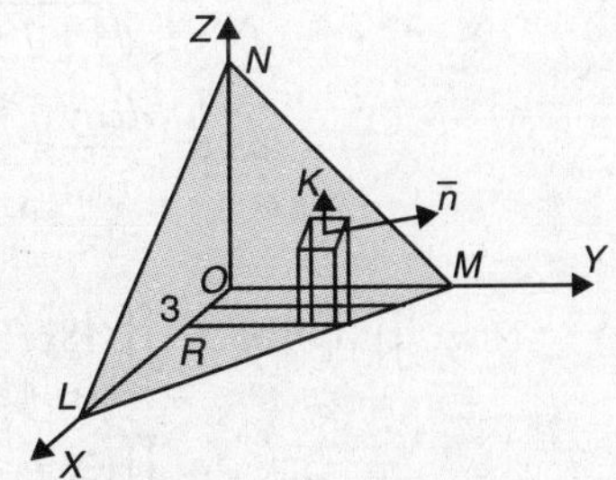

Where R is the projection of S.

Now, $\vec{A}.\hat{n} = [(x+y^2)\hat{i} - 2x\hat{j} + 2yz\hat{k}].\left(\frac{2}{3}\hat{i} + \frac{1}{3}\hat{j} + \frac{2}{3}\hat{k}\right)$

$$= \frac{2}{3}(x+y^2) - \frac{2}{3}x + \frac{4}{3}yz = \frac{2}{3}y^2 + \frac{4}{3}yz \quad ...(1)$$

Putting the value of z in (1), we get

$$\vec{A}.\hat{n} = \frac{2}{3}y^2 + \frac{4}{3}y\left(\frac{6-2x-y}{2}\right) \qquad \left(\because \text{ on the plane } 2x+y+2z=6, \; z = \frac{(6-2x-y)}{2}\right)$$

$$\vec{A}.\hat{n} = \frac{2}{3}y = \frac{4}{3}y(3-x) \quad ...(2)$$

Hence, $\iint_S \vec{A}.\hat{n}\,ds = \iint_R \overline{A}.\overline{n}\frac{dx\,dy}{|\hat{k}.\overline{n}|}$ ...(3)

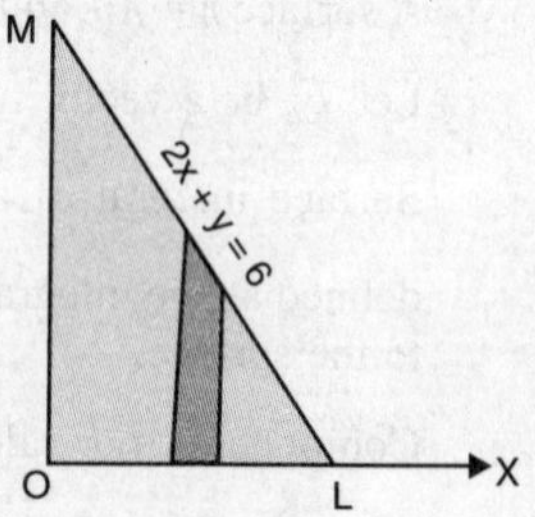

Putting the value of $\vec{A}.\hat{n}$ from (2) in (3), we get

$$\iint_S \vec{A}.\hat{n}\,ds = \iint_R \frac{4}{3}y(3-x).\frac{3}{2}\,dx\,dy = \int_0^3 \int_0^{6-2x} 2y(3-x)\,dy\,dx$$

$$= \int_0^3 2(3-x)\left[\frac{y^2}{2}\right]_0^{6-2x} dx$$

$$= \int_0^3 (3-x)(6-2x)^2\,dx = 4\int_0^3 (3-x)^3\,dx$$

$$= 4.\left[\frac{(3-x)^4}{4(-1)}\right]_0^3 = -(0-81) = 81$$ **Ans.**

Example 12. *Evaluate* $\iint_S \vec{A}.\hat{n}\,dS$, *where* $\vec{A} = 18z\hat{i} - 12\hat{j} + 3y\hat{k}$ *and S is the part of the plane* $2x + 3y + 6z = 12$ *included in the first octant.* (*Uttarakhand, I semester, Dec. 2006*)

Solution. Here, $\vec{A} = 18z\hat{i} - 12\hat{j} + 3y\hat{k}$

Given surface $f(x, y, z) = 2x + 3y + 6z - 12$

Normal vector $= \nabla f = \left(\hat{i}\frac{\partial}{\partial x} + \hat{j}\frac{\partial}{\partial y} + \hat{k}\frac{\partial}{\partial z}\right)(2x+3y+6z-12) = 2\hat{i} + 3\hat{j} + 6\hat{k}$

$\hat{n}$ = unit normal vector at any point (x, y, z) of $2x + 3y + 6z = 12$

$$= \frac{2\hat{i}+3\hat{j}+6\hat{k}}{\sqrt{4+9+36}} = \frac{1}{7}(2\hat{i}+3\hat{j}+6\hat{k})$$

$$dS = \frac{dx\,dy}{\hat{n}.\hat{k}} = \frac{dx\,dy}{\frac{1}{7}(2\hat{i}+3\hat{j}+6\hat{k}).\hat{k}} = \frac{dx\,dy}{\frac{6}{7}} = \frac{7}{6}\,dx\,dy$$

Now, $\iint \vec{A}.\hat{n}\,dS = \iint (18z\hat{i} - 12\hat{j} + 3y\hat{k}).\frac{1}{7}(2\hat{i}+3\hat{j}+6\hat{k})\frac{7}{6}\,dx\,dy$

$$= \iint (36z - 36 + 18y)\frac{dx\,dy}{6} = \iint (6z - 6 + 3y)\,dx\,dy$$

Putting the value of $6z = 12 - 2x - 3y$, we get

$$= \int_0^6 \int_0^{\frac{1}{3}(12-2x)} (12 - 2x - 3y - 6 + 3y)\, dx\, dy$$

$$= \int_0^6 \int_0^{\frac{1}{3}(12-2x)} (6 - 2x)\, dx\, dy$$

$$= \int_0^6 (6 - 2x)\, dx \int_0^{\frac{1}{3}(12-2x)} dy$$

$$= \int_0^6 (6 - 2x)\, dx\, (y)_0^{\frac{1}{3}(12-2x)}$$

$$= \int_0^6 (6 - 2x) \frac{1}{3}(12 - 2x)\, dx = \frac{1}{3}\int_0^6 (4x^2 - 36x + 72)\, dx$$

$$= \frac{1}{3}\left[\frac{4x^3}{3} - 18x^2 + 72x\right]_0^6 = \frac{1}{3}[4 \times 36 \times 2 - 18 \times 36 + 72 \times 6] = \frac{72}{3}[4 - 9 + 6] = 24 \text{ **Ans.**}$$

Example 13. *Evaluate* $\iint_S (yz\hat{i} + zx\hat{j} + xy\hat{k})\cdot \vec{ds}$ *where S is the surface of the sphere* $x^2 + y^2 + z^2 = a^2$ *in the first octant.* *(U.P., I Semester, Dec. 2004)*

Solution. Here, $\phi = x^2 + y^2 + z^2 - a^2$

Vector normal to the surface $= \nabla \phi = \hat{i}\frac{\partial \phi}{\partial x} + \hat{j}\frac{\partial \phi}{\partial y} + \hat{k}\frac{\partial \phi}{\partial z}$

$$= \left(\hat{i}\frac{\partial}{\partial x} + \hat{j}\frac{\partial}{\partial y} + \hat{k}\frac{\partial}{\partial z}\right)(x^2 + y^2 + z^2 - a^2) = 2x\hat{i} + 2y\hat{j} + 2z\hat{k}$$

$$\hat{n} = \frac{\nabla\phi}{|\nabla\phi|} = \frac{2x\hat{i} + 2y\hat{j} + 2z\hat{k}}{\sqrt{4x^2 + 4y^2 + 4z^2}} = \frac{x\hat{i} + y\hat{j} + z\hat{k}}{\sqrt{x^2 + y^2 + z^2}}$$

$$= \frac{x\hat{i} + y\hat{j} + z\hat{k}}{a} \qquad [\because x^2 + y^2 + z^2 = a^2]$$

Here, $\vec{F} = yz\hat{i} + zx\hat{j} + xy\hat{k}$

$$\vec{F}\cdot\hat{n} = (yz\hat{i} + zx\hat{j} + xy\hat{k})\cdot\left(\frac{x\hat{i} + y\hat{j} + z\hat{k}}{a}\right) = \frac{3xyz}{a} \qquad \left[ds = \frac{dx\,dy}{\hat{n}\,\hat{k}} = \frac{dx\,dy}{\frac{z}{a}}\right]$$

Now, $\iint_S F\cdot\hat{n}\, ds = \iint_S (\vec{F}\cdot\hat{n})\frac{dx\,dy}{|\hat{k}\,.\,\hat{n}|} = \int_0^a \int_0^{\sqrt{a^2 - x^2}} \frac{3xyz\, dx\, dy}{a\left(\frac{z}{a}\right)}$

$$= 3\int_0^a \int_0^{\sqrt{a^2 - x^2}} xy\, dy\, dx = 3\int_0^a x\left(\frac{y^2}{2}\right)_0^{\sqrt{a^2 - x^2}} dx$$

$$= \frac{3}{2}\int_0^a x(a^2 - x^2)\, dx = \frac{3}{2}\left(\frac{a^2x^2}{2} - \frac{x^4}{4}\right)_0^a = \frac{3}{2}\left(\frac{a^4}{2} - \frac{a^4}{4}\right) = \frac{3a^4}{8}. \quad \textbf{Ans.}$$

Example 14. *Show that* $\iint_S \vec{F}\cdot\hat{n}\, ds = \frac{3}{2}$, *where* $\vec{F} = 4xz\hat{i} - y^2\hat{j} + yz\hat{k}$ *and S is the surface of the cube bounded by the planes,*

$x = 0, x = 1, y = 0, y = 1, z = 0, z = 1.$

S.No.	Surface	Outward normal	ds	
1	OABC	$-k$	$dx\,dy$	$z = 0$
2	DEFG	k	$dx\,dy$	$z = 1$
3	OAGF	$-j$	$dx\,dz$	$y = 0$
4	BCED	j	$dx\,dz$	$y = 1$
5	ABDG	i	$dy\,dz$	$x = 1$
6	OCEF	$-i$	$dy\,dz$	$x = 0$

Solution. $\iint_S \vec{F}\cdot\hat{n}\,ds = \iint_{OABC} \vec{F}\cdot\hat{n}\,ds$

$+\iint_{DEFG} \vec{F}\cdot\hat{n}\,ds + \iint_{OAGF} \vec{F}\cdot\hat{n}\,ds$

$+\iint_{BCED} \vec{F}\cdot\hat{n}\,ds + \iint_{ABDG} \vec{F}\cdot\hat{n}\,ds$

$+\iint_{OCEF} \vec{F}\cdot\hat{n}\,ds$ ...(1)

Now, $\iint_{OABC} \vec{F}\cdot n\,ds = \iint_{OABC} (4\,xz\hat{i} - y^2\,\hat{j} + yz\,\hat{k})\,(-k)\,dx\,dy = \int_0^1\int_0^1 -yz\,dx\,dy = 0$ (as $z = 0$)

$\iint_{DEFG} (4\,xz\hat{i} - y^2\,\hat{j} + yz\,\hat{k})\cdot\hat{k}\,dx\,dy$

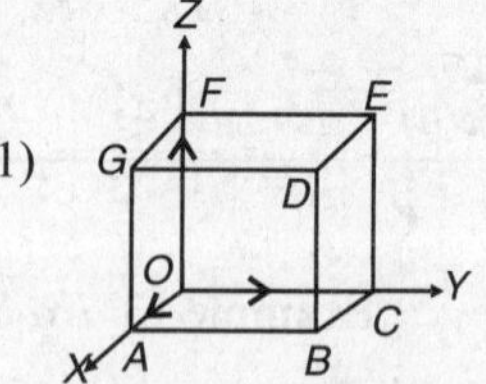

$$= \iint_{DEFG} yz\,dx\,dy = \int_0^1\int_0^1 y\,(1)\,dx\,dy \qquad (\text{as } z = 1)$$

$$= \int_0^1 dx\left[\frac{y^2}{2}\right]_0^1 = [x]_0^1\,\frac{1}{2} = \frac{1}{2}$$

$\iint_{OAGF} (4xz\,\hat{i} - y^2\hat{j} + yz\,\hat{k})\cdot(-j)\,dx\,dz = \iint_{OAGF} y^2\,dx\,dz = 0$ (as $y = 0$)

$\iint_{BCED} (4xz\,\hat{i} - y^2\hat{j} + yz\,\hat{k})\cdot\hat{j}\,dx\,dz = \iint_{BCED} (-y^2)\,dx\,dz$

$$= -\int_0^1 dx\int_0^1 dz = -(x)_0^1\,(z)_0^1 = -1 \qquad (\text{as } y = 1)$$

$\iint_{ABDG} (4xz\hat{i} - y^2\hat{j} + yz\hat{k})\cdot\hat{i}\,dy\,dz = \iint 4xz\,dy\,dz = \int_0^1\int_0^1 4\,(1)\,z\,dy\,dz$ (as $x = 1$)

$$= 4\,(y)_0^1\left(\frac{z^2}{2}\right)_0^1 = 4\,(1)\left(\frac{1}{2}\right) = 2$$

$\iint_{OCEF} (4xz\,\hat{i} - y^2\,\hat{j} + yz\,\hat{k})\,(-\hat{i})\,dy\,dz = \int_0^1\int_0^1 -4xz\,dy\,dz = 0$ (as $x = 0$)

On putting these values in (1), we get

$\iint_S \overline{F}\cdot\hat{n}\,ds = 0 + \frac{1}{2} + 0 - 1 + 2 + 0 = \frac{3}{2}$ **Proved**

EXERCISE 3.2

1. Evaluate $\iint_S \vec{A}\,.\,\hat{n}\,ds$, where $\vec{A} = z\hat{i} + x\hat{j} - 3y^2z\hat{k}$ and S is the surface of the cylinder $x^2 + y^2 = 16$ included in the first octant between $z = 0$ and $z = 5$. **Ans.** 90

2. If $\vec{r} = t\hat{i} - t^2\hat{j} + (t-1)\,\hat{k}$ and $\vec{S} = 2t^2\hat{i} + 6t\hat{k}$, evaluate $\int_0^2 \vec{r}\cdot\vec{S}\,dt$. **Ans.** 12

3. Evaluate $\iint_S \vec{F}\cdot\hat{n}\,ds$, where, $F = 2yx\hat{i} - yz\hat{j} + x^2\hat{k}$ over the surface S of the cube bounded by the coordinate planes and planes $x = a, y = a$ and $z = a$. **Ans.** $\frac{1}{2}a^4$

4. If $\vec{F} = 2y\hat{i} - 3\hat{j} + x^2\hat{k}$ and S is the surface of the parabolic cylinder $y^2 = 8x$ in the first octant bounded by the planes $y = 4$, and $z = 6$, then evaluate $\iint_S \vec{F}\cdot\hat{n}\,dS$. **Ans.** 132

3.3 VOLUME INTEGRAL

Let $\vec{F}$ be a vector point function and volume V enclosed by a closed surface.

The volume integral $= \iiint_V \vec{F}\, dv$

Example 15. *If* $\vec{F} = 2z\hat{i} - x\hat{j} + y\hat{k}$, *evaluate* $\iiint_V \vec{F}\, dv$ *where, v is the region bounded by the surfaces*

$$x = 0,\ y = 0,\ x = 2,\ y = 4,\ z = x^2,\ z = 2.$$

Solution. $\iiint_V \vec{F}\, dv = \iiint (2z\hat{i} - x\hat{j} + y\hat{k})\, dx\, dy\, dz$

$$= \int_0^2 dx \int_0^4 dy \int_{x^2}^2 (2z\hat{i} - x\hat{j} + y\hat{k})\, dz \quad = \int_0^2 dx \int_0^4 dy\, [z^2\hat{i} - xz\hat{j} + yz\hat{k}]_{x^2}^2$$

$$= \int_0^2 dx \int_0^4 dy\, [4\hat{i} - 2x\hat{j} + 2y\hat{k} - x^4\hat{i} + x^3\hat{j} - x^2 y\hat{k}]$$

$$= \int_0^2 dx \left[4y\hat{i} - 2xy\hat{j} + y^2\hat{k} - x^4 y\hat{i} + x^3 y\hat{j} - \frac{x^2y^2}{2}\hat{k}\right]_0^4$$

$$= \int_0^2 (16\hat{i} - 8x\hat{j} + 16\hat{k} - 4x^4\hat{i} + 4x^3\hat{j} - 8x^2\hat{k})\, dx$$

$$= \left[16x\hat{i} - 4x^2\hat{j} + 16x\hat{k} - \frac{4x^5}{5}\hat{i} + x^4\hat{j} - \frac{8x^3}{3}\hat{k}\right]_0^2$$

$$= 32\hat{i} - 16\hat{j} + 32\hat{k} - \frac{128}{5}\hat{i} + 16\hat{j} - \frac{64}{3}\hat{k} = \frac{32\hat{i}}{5} + \frac{32\hat{k}}{3} = \frac{32}{15}(3\hat{i} + 5\hat{k})$$ **Ans.**

EXERCISE 3.3

1. If $\vec{F} = (2x^2 - 3z)\hat{i} - 2xy\hat{j} - 4x\hat{k}$, then evaluate $\iiint_V \nabla \vec{F}\, dV$, where V is bounded by the plane $x = 0, y = 0, z = 0$ and $2x + 2y + z = 4$. **Ans.** $\frac{8}{3}$

2. Evaluate $\iiint_V \phi\, dV$, where $\phi = 45\, x^2 y$ and V is the closed region bounded by the planes $4x + 2y + z = 8,\ x = 0, y = 0, z = 0$ **Ans.** 128

3. If $\vec{F} = (2x^2 - 3z)\, \hat{i} - 2xy\, \hat{j} - 4x\hat{k}$, then evaluate $\iiint_V \nabla \times \vec{F}\, dV$, where V is the closed region bounded by the planes $x = 0, y = 0, z = 0$ and $2x + 2y + z = 4$. **Ans.** $\frac{8}{3}(\hat{j} - \hat{k})$

4. Evaluate $\iiint_V (2x + y)\, dV$, where V is closed region bounded by the cylinder $z = 4 - x^2$ and the planes $x = 0, y = 0, y = 2$ and $z = 0$. **Ans.** $\frac{80}{3}$

5. If $\vec{F} = 2xz\hat{i} - x\hat{j} + y^2\hat{k}$, evaluate $\iiint \vec{F}\, dV$ over the region bounded by the surfaces $x = 0$, $y = 0$, $y = 6$ and $z = x^2$, $z = 4$. **Ans.** $(16\hat{i} - 3\hat{j} + 48\hat{k})$ *(Delhi University, 2010)*

3.4 GREEN'S THEOREM (FOR A PLANE)

Statement. *If* $\phi(x, y), \psi(x, y), \frac{\partial\phi}{\partial y}$ *and* $\frac{\partial\psi}{\partial x}$ *be continuous functions over a region R bounded by simple closed curve C in x – y plane, then* (*MTU Dec. 2012*)

$$\oint_C (\phi\, dx + \psi\, dy) = \iint_R \left(\frac{\partial\psi}{\partial x} - \frac{\partial\phi}{\partial y}\right) dx\, dy$$ (*AMIETE, June 2010, U.P., I Semester, Dec. 2007*)

Proof. Let the curve C be divided into two curves C_1 (*ABC*) and C_2 (*CDA*).

Let the equation of the curve C_1 (*ABC*) be $y = y_1(x)$ and equation of the curve C_2 (*CDA*) be $y = y_2(x)$.

Let us see the value of

$$\iint_R \frac{\partial\phi}{\partial y} dx\, dy = \int_{x=a}^{x=c}\left[\int_{y=y_1(x)}^{y=y_2(x)} \frac{\partial\phi}{\partial y} dy\right] dx = \int_a^c [\phi(x, y)]_{y=y_1(x)}^{y=y_2(x)} dx$$

$$= \int_a^c [\phi(x, y_2) - \phi(x, y_1)]\, dx = -\int_c^a \phi(x, y_2)\, dx - \int_a^c \phi(x, y_1)\, dx$$

$$= -\left[\int_c^a \phi(x, y_2)\, dx + \int_a^c \phi(x, y_1)\, dx\right]$$

$$= -\left[\int_{c_2} \phi(x, y)\, dx + \int_{c_1} \phi(x, y)\, dx\right] = -\oint_c \phi(x, y)\, dx$$

Thus, $$\oint_c \phi\, dx = -\iint_R \frac{\partial\phi}{\partial y} dx\, dy \qquad ...(1)$$

Similarly, it can be shown that

$$\oint_c \psi\, dy = \iint \frac{\partial\psi}{\partial x} dx\, dy \qquad ...(2)$$

On adding (1) and (2), we get

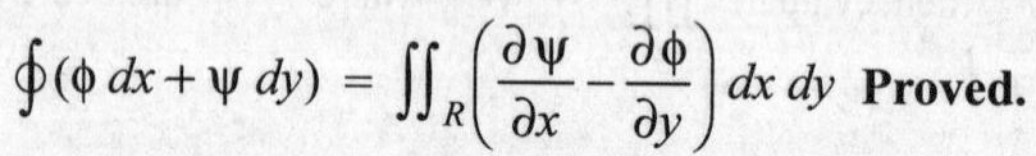

$$\oint(\phi\, dx + \psi\, dy) = \iint_R \left(\frac{\partial\psi}{\partial x} - \frac{\partial\phi}{\partial y}\right) dx\, dy$$ **Proved.**

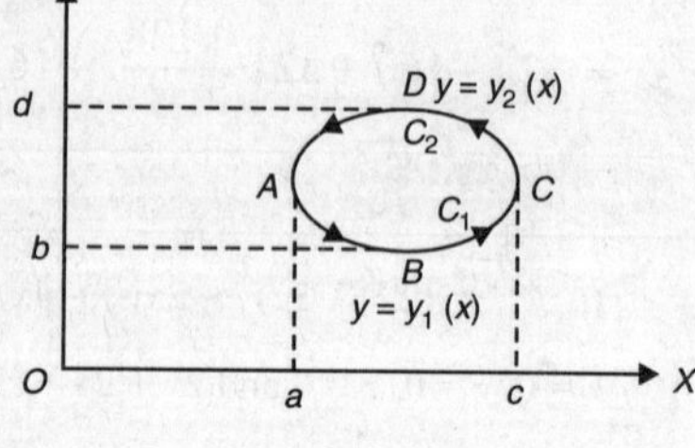

Note. Green's Theorem in vector form

$$\int_c \vec{F} \cdot d\vec{r} = \iint_R (\nabla \times \vec{F}) \cdot \hat{k}\, dR$$

where, $\vec{F} = \phi\,\hat{i} + \psi\,\hat{j}, \bar{r} = x\hat{i} + y\hat{j}, \hat{k}$ is a unit vector along z-axis and $dR = dx\, dy$.

Example 16. *A vector field* $\vec{F}$ *is given by* $\vec{F} = \sin y\hat{i} + x(1 + \cos y)\,\hat{j}$.

Evaluate the line integral $\int_C \vec{F} \cdot d\vec{r}$ *where C is the circular path given by* $x^2 + y^2 = a^2$.

Solution. $\vec{F} = \sin y\hat{i} + x(1 + \cos y)\,\hat{j}$

$$\int_C \vec{F} \cdot d\vec{r} = \int_C [\sin y\hat{i} + x(1 + \cos y)\,\hat{j}] \cdot (\hat{i}dx + \hat{j}dy) = \int_C \sin y\, dx + x(1 + \cos y)\, dy$$

On applying Green's Theorem, we have

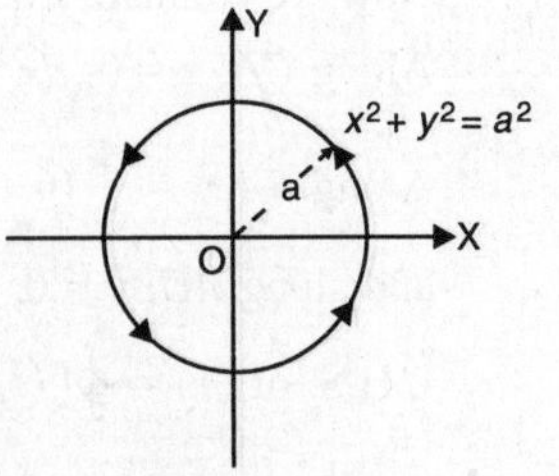

$$\oint_c (\phi\, dx + \psi\, dy) = \iint_s \left(\frac{\partial \psi}{\partial x} - \frac{\partial \phi}{\partial y}\right) dx\, dy$$

$$= \iint_s [(1 + \cos y) - \cos y]\, dx\, dy$$

where s is the circular plane surface of radius a.

$$= \iint_s dx\, dy = \text{Area of circle} = \pi a^2. \quad \textbf{Ans.}$$

Example 17. *Using Green's Theorem, evaluate* $\int_c (x^2\, y dx + x^2\, dy)$, *where c is the boundary described counter clockwise of the triangle with vertices (0, 0), (1, 0), (1, 1).*

Solution. By Green's Theorem, we have

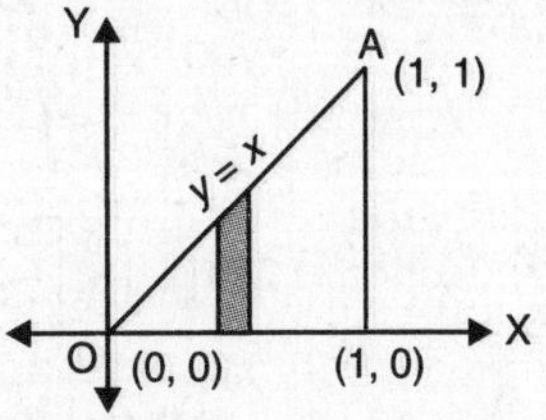

$$\int_c (\phi\, dx + \psi\, dy) = \iint_R \left(\frac{\partial \psi}{\partial x} - \frac{\partial \phi}{\partial y}\right) dx\, dy$$

$$\int_c (x^2 y\, dx + x^2\, dy) = \iint_R (2x - x^2)\, dx\, dy$$

$$= \int_0^1 (2x - x^2)\, dx \int_0^x dy = \int_0^1 (2x - x^2)\, dx\, [y]_0^x$$

$$= \int_0^1 (2x - x^2)(x)\, dx = \int_0^1 (2x^2 - x^3)\, dx = \left(\frac{2x^3}{3} - \frac{x^4}{4}\right)_0^1$$

$$= \left(\frac{2}{3} - \frac{1}{4}\right) = \frac{5}{12}$$ **Ans.**

Example 18. *State and verify Green's Theorem in the plane for* $\oint (3x^2 - 8y^2)\, dx + (4y - 6xy)\, dy$ *where C is the boundary of the region bounded by* $x \geq 0$, $y \leq 0$ *and* $2x - 3y = 6$.

(*Uttarakhand, I Semester, Dec. 2006*)

Solution. Statement: See Article 3.4 on page 72.

Here the closed curve *C* consists of straight lines *OB*, *BA* and *AO*, where coordinates of *A* and *B* are (3, 0) and (0, – 2) respectively. Let *R* be the region bounded by *C*.

Then by Green's Theorem in plane, we have

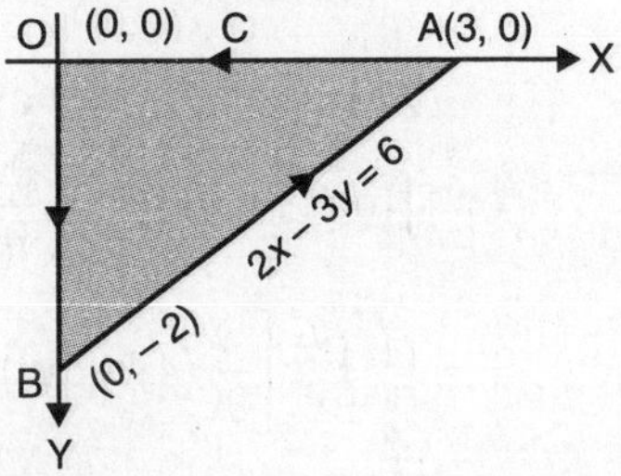

$$\oint [(3x^2 - 8y^2)\, dx + (4y - 6xy)\, dy]$$

$$= \iint_R \left[\frac{\partial}{\partial x}(4y - 6xy) - \frac{\partial}{\partial y}(3x^2 - 8y^2)\right] dx\, dy \quad ..(1)$$

$$= \iint_R (-6y + 16y)\, dx\, dy = \iint_R 10y\, dx\, dy$$

$$= 10 \int_0^3 dx \int_{\frac{1}{3}(2x-6)}^0 y\, dy = 10 \int_0^3 dx \left[\frac{y^2}{2}\right]_{\frac{1}{3}(2x-6)}^0 = -\frac{5}{9} \int_0^3 dx\, (2x - 6)^2$$

$$= -\frac{5}{9}\left[\frac{(2x-6)^3}{3 \times 2}\right]_0^3 = -\frac{5}{54}(0 + 6)^3 = -\frac{5}{54}(216) = -20 \quad ...(2)$$

Now we evaluate L.H.S. of (1) along OB, BA and AO.

Along OB, $x = 0$, $dx = 0$ and y varies form 0 to –2.

Along BA, $x = \frac{1}{2}(6+3y)$, $dx = \frac{3}{2}\, dy$ and y varies from –2 to 0.

and along AO, $y = 0$, $dy = 0$ and x varies from 3 to 0.

L.H.S. of (1) $= \oint[(3x^2 - 8y^2)\,dx + (4y - 6xy)\,dy]$

$$= \int_{OB}[(3x^2 - 8y^2)\,dx + (4y - 6xy)\,dy] + \int_{BA}[(3x^2 - 8y^2)\,dx + (4x - 6xy)\,dy]$$

$$+ \int_{AO}[(3x^2 - 8y^2)\,dx + (4y - 6xy)\,dy]$$

$$= \int_0^{-2} 4y\,dy + \int_{-2}^{0}\left[\frac{3}{4}(6+3y)^2 - 8y^2\right]\left(\frac{3}{2}\,dy\right) + [4y - 3(6+3y)\,y]\,dy + \int_3^0 3x^2\,dx$$

$$= [2y^2]_0^{-2} + \int_{-2}^{0}\left[\frac{9}{8}(6+3y)^2 - 12y^2 + 4y - 18y - 9y^2\right]dy + (x^3)_3^0$$

$$= 2[4] + \int_{-2}^{0}\left[\frac{9}{8}(6+3y)^2 - 21y^2 - 14y\right]dy + (0 - 27)$$

$$= 8 + \left[\frac{9}{8}\frac{(6+3y)^3}{3\times 3} - 7y^3 - 7y^2\right]_{-2}^{0} - 27 = -19 + \left[\frac{216}{8} + 7(-2)^3 + 7(-2)^2\right]$$

$$= -19 + 27 - 56 + 28 = -20 \qquad \text{...(3)}$$

With the help of (2) and (3), we find that (1) is true and so Green's Theorem is verified.

Example 19. *Verify Green's Theorem in the plane for*

$$\oint_C (3x^2 - 8y^2)\,dx + (4y - 6xy)\,dy$$

Where C is the boundary of the region defined by

$y = \sqrt{x}$, *and* $y = x^2$ *(K.University, Dec. 2008)*

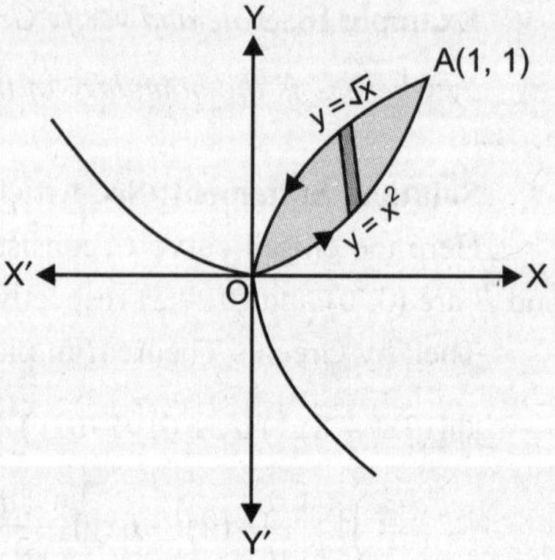

Solution. Here we have,

$$\oint_C \left(3x^2 - 8y^2\right)dx + (4y - 6xy)dy$$

By Green's Theorem, we have

$$\int_C (\phi\,dx + \psi\,dy) = \iint_S \left(\frac{\partial \psi}{\partial x} - \frac{\partial \phi}{\partial y}\right) dx\,dy$$

$$= \int_0^1 \int_{x^2}^{\sqrt{x}} \left[\frac{\partial}{\partial x}(4y - 6xy) - \frac{\partial}{\partial y}\left(3x^2 - 8y^2\right)\right] dx\,dy$$

$$= \int_0^1 \int_{x^2}^{\sqrt{x}} (-6y + 16y)\,dx\,dy \;\; 10\int_0^1 \int_{x^2}^{\sqrt{x}} y\,dx\,dy = 10\int_0^1 dx\left(\frac{y^2}{2}\right)_{x^2}^{\sqrt{x}} = \frac{10}{2}\int_0^1 dx\,(x - x^4)$$

$$= 5\left(\frac{x^2}{2} - \frac{x^5}{5}\right)_0^1 = 5\left(\frac{1}{2} - \frac{1}{5}\right) = 5\left(\frac{3}{10}\right) = \frac{3}{2}$$

Ans.

Verify By line integral

$$\int_c \phi\,dx + \psi\,dy = \int_c (3x^2 - 8y^2)\,dx\,(4y - 6xy)\,dy$$

$$= \int_0^1 (3x^2 - 8x^4)\,dx + (4x^2 - 6x^3)\,2x\,dx + \int_1^0 (3y^4 - 8y^2)\,2y dy + (4y - 6y^3)\,dy$$

$$= \int_0^1 (3x^2 - 8x^4 + 8x^3 - 12x^4)\,dx + \int_0^1 6y^5 - 16y^3 + 4y - 6y^3)\,dy$$

$$= \left[x^3 - \frac{8}{5}x^5 + 2x^4 - \frac{12}{5}x^5\right]_0^1 + \left[y^6 - 4y^4 + 2y^2 - \frac{3}{2}y^4\right]_1^0$$

$$= \left[1 - \frac{8}{5} + 2\frac{-12}{5}\right] + \left[0 - \left(1 - 4 + 2\frac{-3}{2}\right)\right]$$

$$= \left[\frac{5 - 8 + 10 - 12}{5}\right] + \left[-\left(\frac{2 - 8 + 4 - 3}{2}\right)\right]$$

$= -1 + 5/2 = 3/2 =$ L.H.S. = R.H.S.

Hence $\int_c \phi\,dx + \psi\,dy = \iint_s \left(\frac{\partial\psi}{\partial x} - \frac{\partial\phi}{\partial y}\right) dx dy$ proved.

Example 20. *Apply Green's Theorem to evaluate* $\int_C [(2x^2 - y^2)\,dx + (x^2 + y^2)\,dy]$, *where C is the boundary of the area enclosed by the x-axis and the upper half of circle* $x^2 + y^2 = a^2$.

(M.D.U. Dec. 2009, U.P., I Sem., Dec. 2004)

Solution. $\int_C [(2x^2 - y^2)\,dx + (x^2 + y^2)\,dy]$

By Green's Theorem, we've $\int_C (\phi\,dx + \psi\,dy) = \iint_S \left(\frac{\partial\psi}{\partial x} - \frac{\partial\phi}{\partial y}\right) dx\,dy$

$$= \int_{-a}^{a}\int_0^{\sqrt{a^2-x^2}} \left[\frac{\partial}{\partial x}(x^2 + y^2) - \frac{\partial}{\partial y}(2x^2 - y^2)\right] dx\,dy$$

$$= \int_{-a}^{a}\int_0^{\sqrt{a^2-x^2}} (2x + 2y)\,dx\,dy = 2\int_{-a}^{a} dx \int_0^{\sqrt{a^2-x^2}} (x + y)\,dy$$

$$= 2\int_{-a}^{a} dx \left(xy + \frac{y^2}{2}\right)_0^{\sqrt{a^2-x^2}} = 2\int_{-a}^{a}\left(x\sqrt{a^2 - x^2} + \frac{a^2 - x^2}{2}\right) dx$$

$$= 2\int_{-a}^{a} x\sqrt{a^2 - x^2}\,dx + \int_{-a}^{a}(a^2 - x^2)\,dx \qquad \left[\begin{array}{ll}\int_{-a}^{a} f(x)\,dx = 2\int_0^a f(x)\,dx, & f \text{ is even} \\ \qquad = 0, & f \text{ is odd}\end{array}\right]$$

$$= 0 + 2\int_0^a (a^2 - x^2)\,dx = 2\left(a^2x - \frac{x^3}{3}\right)_0^a = 2\left(a^3 - \frac{a^3}{3}\right) = \frac{4a^3}{3}$$ **Ans.**

Example 21. *Evaluate* $\oint_C -\frac{y}{x^2 + y^2}\,dx + \frac{x}{x^2 + y^2}\,dy$, *where* $C = C_1 \cup C_2$ *with* $C_1 : x^2 + y^2 = 1$ *and* $C_2 : x = \pm 2, y = \pm 2$.

(Gujarat, I Semester, Jan 2009)

Solution. $\oint_C -\frac{y}{x^2+y^2}\,dx + \frac{x}{x^2+y^2}\,dy$

$$= \iint\left(\frac{\partial}{\partial x}\frac{x}{x^2+y^2} + \frac{\partial}{\partial y}\frac{y}{x^2+y^2}\right)dx\,dy$$

$$= \iint\left[\frac{(x^2+y^2)1-2x(x)}{(x^2+y^2)^2} + \frac{(x^2+y^2)1-2y(y)}{(x^2+y^2)^2}\right]dx\,dy$$

$$= \iint\left[\frac{x^2+y^2-2x^2}{(x^2+y^2)^2} + \frac{x^2+y^2-2y^2}{(x^2+y^2)^2}\right]dx\,dy$$

$$= \iint\left[\frac{y^2-x^2}{(x^2+y^2)^2} + \frac{x^2-y^2}{(x^2+y^2)^2}\right]dx\,dy = \iint\frac{0}{(x^2+y^2)^2}\,dx\,dy = 0$$ **Ans.**

3.5 AREA OF THE PLANE REGION BY GREEN'S THEOREM

Proof. We know that

$$\int_C M dx + N dy = \iint_A\left(\frac{\partial N}{\partial x} - \frac{\partial M}{\partial y}\right)dx\,dy \qquad ...(1)$$

On putting $N = x \; x\left(\frac{\partial N}{\partial x} = 1\right)$ and $M = -y\left(\frac{\partial M}{\partial y} = -1\right)$ in (1), we get

$$\int_C -y\,dx + x\,dy = \iint_A [1-(-1)]\,dx\,dy = 2\iint dx\,dy = 2A$$

$$\text{Area} = \frac{1}{2}\int_C (x\,dy - y\,dx)$$

Example 22. *Using Green's theorem, find the area of the region in the first quadrant bounded by the curves*

$$y = x,\ y = \frac{1}{x},\ y = \frac{x}{4}$$ *(U.P. I, Semester, Dec. 2008)*

Solution. By Green's Theorem Area A of the region bounded by a closed curve C is given by

$$A = \frac{1}{2}\oint_C (xdy - ydx)$$

Here, C consists of the curves $C_1 : y = \frac{x}{4}$, $C_2 : y = \frac{1}{x}$

and $C_3 : y = x$ So

$$\left[A = \frac{1}{2}\oint_C = \frac{1}{2}\left[\int_{C_1} + \int_{C_2} + \int_{C_3}\right] = \frac{1}{2}(I_1 + I_2 + I_3)\right]$$

Along $C_1 : y = \frac{x}{4},\ dy = \frac{1}{4}dx,\ x : 0 \text{ to } 2$

$$I_1 = \int_{C_1}(xdy - ydx) = \int_{C_1}\left(x\frac{1}{4}dx - \frac{x}{4}dx\right) = 0$$

Along $C_2 : y = \frac{1}{x},\ dy = -\frac{1}{x^2}dx,\ x : 2 \text{ to } 1$

$$I_2 = \int_{C_2} (xdy - ydx) = \int_2^1 \left[x\left(-\frac{1}{x^2}\right) dx - \frac{1}{2} dx \right] = [-2 \log x]_2^1 = 1 \log 2$$

Along $C_3 : y = x,\ dy = dx\ ;\ x : 1 \text{ to } 0\ ;$

$$I_3 = \int_{C_3} (xdy - ydx) = \int (xdx - xdx) = 0$$

$$A = \frac{1}{2}(I_1 + I_2 + I_3) = \frac{1}{2}(0 + 2\log 2 + 0) = \log 2$$ **Ans.**

EXERCISE 3.4

1. Evaluate $\int_c [(3x^2 - 6yz)\,dx + (2y + 3xz)\,dy + (1 - 4xyz^2)\,dz]$ from (0, 0, 0) to (1, 1, 1) along the path c given by the straight line from (0, 0, 0) to (0, 0, 1) then to (0, 1, 1) and then to (1, 1, 1). **Ans.** -3

2. Verify Green's Theorem in plane for $\int_C (x^2 + 2xy)\,dx + (y^2 + x^3 y)\,dy$, where c is a square with the vertices P (0, 0), Q (1, 0), R (1, 1) and S (0, 1). **Ans.** $-\frac{1}{2}$

3. Verify Green's Theorem for $\int_c (x^2 - 2xy)\,dx + (x^2 y + 3)\,dy$ around the boundary c of the region $y^2 = 8x$ and $x = 2$.

4. Use Green's Theorem in a plane to evaluate the integral $\int_c [(2x^2 - y^2)\,dx + (x^2 + y^2)\,dy]$, where c is the boundary in the xy-plane of the area enclosed by the x-axis and the semi-circle $x^2 + y^2 = 1$ in the upper half xy-plane. **Ans.** $\frac{4}{3}$

5. Apply Green's Theoem to evaluate $\int_c [(y - \sin x)\,dy + \cos x\,dy]$, where c is the plane triangle enclosed by the lines $y = 0$, $x = \frac{\pi}{2}$ and $y = \frac{2x}{\pi}$. **Ans.** $-\frac{\pi^2 + 8}{4\pi}$

6. Either directly or by Green's Therorem, evaluate the line integral $\int_c e^{-x} (\cos y\,dx - \sin y\,dy)$, where c is the rectangle with vertices (0, 0), (π, 0,), $\left(\pi, \frac{\pi}{2}\right)$ and $\left(0, \frac{\pi}{2}\right)$. **Ans.** $2(1 - e^{-\pi})$

   (AMIETE II Sem June 2010)

7. Verify the Green's Theorem to evaluate the line integral $\int_c (2y^2\,dx + 3x\,dy)$, where c is the boundary of the closed region bounded by $y = x$ and $y = x^2$.

   (U.P., I Semester, Dec. 2005, AMIETE Summer 2004) **Ans.** $-\frac{7}{30}$

8. Evaluate $\iint_s \overline{F} . \hat{n}.ds$, where $\vec{F} = xy\,\hat{i} - x^2\,\hat{j} + (x + z)\,\hat{k}$ and s is the region of the plane $2x + 2y + z = 6$ in the first octant. **Ans.** $\frac{27}{4}$

9. Verify Green's Theorem for $\int_C [(xy + y^2)\,dx + x^2 dy]$ where C is the boundary by $y = x$ and $y = x^2$.

   (AMIETE, June 2010) **Ans.** $-\frac{1}{20}$

3.6 STOKE'S THEOREM (Relation between Line Integral and Surface Integral)

(*Uttarakhand, I Sem. 2008, U.P., Ist Semester, Dec. 2006*)

Statement. Surface integral of the component of curl $\overline{F}$ along the normal to the surface S, taken over the surface S bounded by curve C is equal to the line integral of the vector point function $\vec{F}$ taken along the closed curve C.

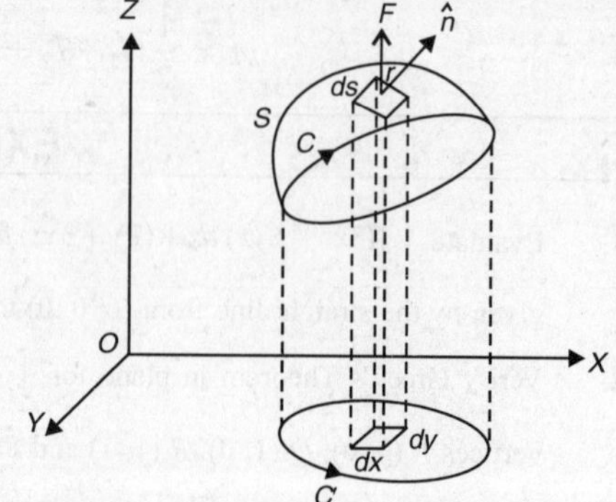

Mathematically

$$\oint \vec{F}.d\,\overline{r} = \iint_S \text{curl } \vec{F}\cdot\hat{n}\, ds$$

where $\hat{n} = \cos\alpha\,\hat{i} + \cos\beta\,\hat{j} + \cos\gamma\,\hat{k}$ is a unit external normal to any surface ds,

Proof. Let
$$\vec{r} = x\hat{i} + y\hat{j} + z\hat{k}$$
$$d\vec{r} = \hat{i}\,dx + \hat{j}\,dy + \hat{k}\,dz$$
$$F = F_1\,\hat{i} + F_2\,\hat{j} + F_3\,\hat{k}$$

On putting the values of $\vec{F}\cdot d\,\vec{r}$ in the statement of the theorem

$$\oint_c (F_1\,\hat{i} + F_2\,\hat{j} + F_3\,\hat{k})\cdot(\hat{i}\,dx + \hat{j}\,dy + \hat{k}\,dz)$$

$$= \iint_S \left(i\frac{\partial}{\partial x} + j\frac{\partial}{\partial y} + k\frac{\partial}{\partial z}\right) \times (F_1\,\hat{i} + F_2\,\hat{j} + F3\,\hat{k}).\,(\cos\alpha\,\hat{i} + \cos\beta\,\hat{j} + \cos\gamma\,\hat{k})\,ds$$

$$\oint (F_1\,dx + F_2\,dy + F_3\,dz) = \iint_S \left[\left(\frac{\partial F_3}{\partial y} - \frac{\partial F_2}{\partial z}\right)\hat{i} + \left(\frac{\partial F_1}{\partial z} - \frac{\partial F_3}{\partial x}\right)\hat{j} + \left(\frac{\partial F_2}{\partial x} - \frac{\partial F_1}{\partial y}\right)\hat{k}\right].$$
$$(\hat{i}\cos\alpha + \hat{j}\cos\beta + \hat{k}\cos\gamma)\,ds$$

$$= \iint_S \left[\left(\frac{\partial F_3}{\partial y} - \frac{\partial F_2}{\partial z}\right)\cos\alpha + \left(\frac{\partial F_1}{\partial z} - \frac{\partial F_3}{\partial x}\right)\cos\beta + \left(\frac{\partial F_2}{\partial x} - \frac{\partial F_1}{\partial y}\right)\cos\gamma\right] ds \qquad ...(1)$$

Let us first prove

$$\oint_c F_1\,dx = \iint_S \left[\left(\frac{\partial F_1}{\partial z}\cos\beta - \frac{\partial F_1}{\partial y}\cos\gamma\right)\right] ds \qquad ...(2)$$

Let the equation of the surface S be $z = g(x, y)$. The projection of the surface on $x - y$ plane is region R.

$$\oint_c F_1(x, y, z)\,dx = \oint_c F_1[x, y, g(x, y)]\,dx$$

$$= -\iint_R \frac{\partial}{\partial y} F_1(x, y, g)\,dx\,dy \qquad \text{[By Green's Theorem]}$$

$$= -\iint_R \left(\frac{\partial F_1}{\partial y} + \frac{\partial F_1}{\partial z}\frac{\partial g}{\partial y}\right) dx\,dy \qquad ...(3)$$

The direction consines of the normal to the surface $z = g(x, y)$ are given by

$$\frac{\cos\alpha}{\dfrac{-\partial g}{\partial x}} = \frac{\cos\beta}{\dfrac{-\partial g}{\partial y}} = \frac{\cos\gamma}{1}$$

And $dx\,dy$ = projection of ds on the xy-plane = $ds\cos\gamma$

Putting the values of ds in R.H.S. of (2)

$$\iint_S \left(\frac{\partial F_1}{\partial z}\cos\beta - \frac{\partial F_1}{\partial y}\cos\gamma\right) ds = \iint_R \left(\frac{\partial F_1}{\partial z}\cos\beta - \frac{\partial F_1}{\partial y}\cos\gamma\right)\frac{dx\,dy}{\cos\gamma}$$

$$= \iint_R \left(\frac{\partial F_1}{\partial z}\frac{\cos\beta}{\cos\gamma} - \frac{\partial F_1}{\partial y}\right) dx\,dy = \iint_R \left(\frac{\partial F_1}{\partial z}\left(-\frac{\partial g}{\partial y}\right) - \frac{\partial F_1}{\partial y}\right) dx\,dy$$

$$= -\iint_R \left(\frac{\partial F_1}{\partial y} + \frac{\partial F_1}{\partial z}\frac{\partial g}{\partial y}\right) dx\,dy \qquad ...(4)$$

From (3) and (4), we get

$$\oint_c F_1\,dx = \iint_S \left(\frac{\partial F_1}{\partial z}\cos\beta - \frac{\partial F_1}{\partial y}\cos\gamma\right) ds \qquad ...(5)$$

Similarly, $$\oint_c F_2\,dy = \iint_S \left(\frac{\partial F_2}{\partial x}\cos\gamma - \frac{\partial F_2}{\partial z}\cos\alpha\right) ds \qquad ...(6)$$

and $$\oint_c F_3\,dz = \iint_S \left(\frac{\partial F_3}{\partial y}\cos\alpha - \frac{\partial F_3}{\partial x}\cos\beta\right) ds \qquad ...(7)$$

On adding (5), (6) and (7), we get

$$\oint_c (F_1\,dx + F_2\,dy + F_3\,dz) = \iint_S \left(\frac{\partial F_1}{\partial z}\cos\beta - \frac{\partial F_1}{\partial y}\cos\gamma + \frac{\partial F_2}{\partial x}\cos\gamma - \frac{\partial F_2}{\partial z}\cos\alpha + \frac{\partial F_3}{\partial y}\cos\alpha - \frac{\partial F_3}{\partial x}\cos\beta\right) ds$$ **Proved.**

3.7 ANOTHER METHOD OF PROVING STOKE'S THEOREM

The circulation of vector F around a closed curve C is equal to the flux of the curve of the vector through the surface S bounded by the curve C.

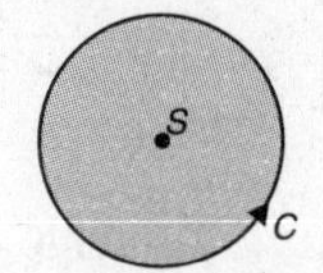

$$\oint_c \overline{F}\cdot d\overline{r} = \iint_S curl\,\vec{F}\cdot\hat{n}\,d\vec{s} = \iint_S curl\,\vec{F}\cdot d\vec{S}$$

Proof : The projection of any curved surface over xy-plane can be treated as kernal of the surface integral over actual surface

Now, $$\iint_S (\nabla\times\vec{F})\cdot\hat{k}\,d\vec{S} = \iint_S (\nabla\times\vec{F})\cdot(\vec{i}\times\vec{j})\,dx\,dy \qquad [\hat{k} = \hat{i}\times\hat{j}]$$

$$= \iint_S [(\nabla\cdot\hat{i})(\vec{F}\cdot\hat{j}) - (\nabla\cdot\hat{j})(\vec{F}\cdot\hat{i})]\,dx\,dy = \iint_S \left[\frac{\partial}{\partial x}(F_y) - \frac{\partial}{\partial y}(F_x)\right] dx\,dy$$

$$= \iint_S [F_x\,dx + F_y\,dy] \quad \text{[By Green's theorem]}$$

$$= \iint_S [\hat{i}\,F_x + \hat{j}\,F_y]\cdot(\hat{i}\,dx + \hat{j}\,dy) = \oint_c \vec{F}\,.\,\vec{dr}$$

$$\iint_S curl\,\vec{F}\cdot\hat{n}\,dS = \oint_c \vec{F}\,.\,\vec{d}\,r.$$

where, $\vec{F} = F_x\,\hat{i} + F_y\,\hat{j} + F_z\,\hat{k}$ and $\vec{dr} = dx\,\hat{i} + dy\,\hat{j} + dz\,\hat{k}$

Example 23. *Evaluate by Stokes theorem* $\oint_C (yz\,dx + zx\,dy + xy\,dz)$ *where C is the curve* $x^2 + y^2 = 1,\ z = y^2$. *(M.D.U., Dec 2009)*

Solution. Here we have $\oint yz\,dx + zx\,dy + xy\,dz$

$$= \int (yz\hat{i} + zx\hat{j} + xy\hat{k}).(\hat{i}dx + \hat{j}dy + kdz) = \int (yz\,dx + zx\,dy + xy\,dz)$$

$$\text{Curl } \overline{F} = \begin{vmatrix} \hat{i} & \hat{j} & \hat{k} \\ \dfrac{\partial}{\partial x} & \dfrac{\partial}{\partial y} & \dfrac{\partial}{\partial z} \\ yz & zx & xy \end{vmatrix} = (x - x)\ \hat{i} + (y - y)\ \hat{j} + (z - z)\ \hat{k} = 0$$

$$= \oint \vec{F}.d\vec{r} = \iint curl\ F.\hat{\eta} ds = 0$$ **Ans.**

Example 24. *Using Stoke's theorem or otherwise, evaluate*

$$\int_c [(2x - y)\,dx - yz^2 dy - y^2 z\,dz]$$

where c is the circle $x^2 + y^2 = 1$, corresponding to the sursface of sphere of unit radius.

Solution. $\int_c [(2x - y)\,dx - yz^2 dy - y^2 z\,dz]$

$$= \int_c [(2x - y)\,\hat{i} - yz^2\,\hat{j} - y^2 z\,\hat{k}] \cdot (\hat{i}\,dx + \hat{j}\,dy + \hat{k}\,dz)$$

By Stoke's theorem $\oint \vec{F} \cdot d\vec{r} = \iint_S \text{Curl } \vec{F} \cdot \hat{n}\,ds$ ...(1)

$$\text{Curl } \vec{F} = \nabla \times \vec{F} = \begin{vmatrix} \hat{i} & \hat{j} & \hat{k} \\ \dfrac{\partial}{\partial x} & \dfrac{\partial}{\partial y} & \dfrac{\partial}{\partial z} \\ 2x - y & -yz^2 & -y^2 z \end{vmatrix}$$

x² + y² = 1
O 1

$$= (-2yz + 2yz)\hat{i} - (0 - 0)\,\hat{j} + (0 + 1)\hat{k} = \hat{k}$$

Putting the value of curl $\vec{F}$ in (1), we get

$$\oint \vec{f}.d\vec{r} = \iint \hat{k} \cdot \hat{n}\,ds = \iint \hat{k} \cdot \hat{n} \frac{dx\,dy}{\hat{n} \cdot \hat{k}} = \iint dx\,dy = \text{Area of the circle} = \pi \left[\because ds = \frac{dx\,dy}{(\hat{n} \cdot \hat{k})} \right]$$

Example 25. *Evaluate* $\int_C \vec{F}.d\vec{r}$, *where* $F(x, y, z) = -y^2\hat{i} + x\hat{j} + z^2\hat{k}$ *and C is the curve of intersection of the plane y + z = 2 and the cylinder $x^2 + y^2 = 1$. (Gujarat, I sem. Jan. 2009)*

Solution. $\oint_C \vec{F}.\vec{dr} = \iint_S \text{curl } \vec{F}.\hat{n}\,ds = \iint_S \text{curl}\,(-y^2\,\hat{i} + x\,\hat{j} + z^2\,\hat{k})\,\hat{n}\,ds$ ...(1)

$$F(x, y, z) = -y^2\,\hat{i} + x\,\hat{j} + z^2\,\hat{k}$$ *(By Stoke's Theorem)*

$$\text{Curl } \vec{F} = \begin{vmatrix} \hat{i} & \hat{j} & \hat{k} \\ \dfrac{\partial}{\partial x} & \dfrac{\partial}{\partial y} & \dfrac{\partial}{\partial z} \\ -y^2 & x & z^2 \end{vmatrix}$$

$$= \hat{i}\,(0 - 0) - \hat{j}\,(0 - 0) + \hat{k}\,(1 + 2y) = (1 + 2y)\,\hat{k}$$

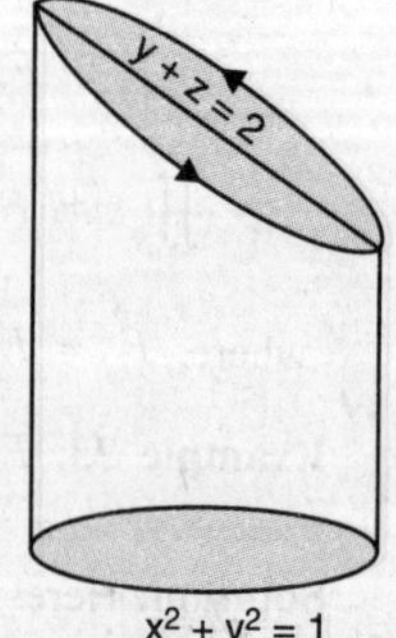

Normal vector

$$= \nabla\vec{F} = \left(\hat{i}\frac{\partial}{\partial x} + \hat{j}\frac{\partial}{\partial y} + \hat{k}\frac{\partial}{\partial z}\right)(y + z - 2) = \hat{j} + \hat{k}$$

Unit normal vector $\hat{n} = \dfrac{\hat{j} + \hat{k}}{\sqrt{2}}$

$$ds = \frac{dx\,dy}{\hat{\eta}.\hat{k}}$$

On putting the values of curl $\vec{F}, \hat{n}$ and ds in (1), we get

$$\int_C \vec{F}.\vec{dr} = \iint_S (1+2y)\,\hat{k}.\frac{\hat{j}+\hat{k}}{\sqrt{2}}\frac{dx\,dy}{\left(\frac{\hat{j}+\hat{k}}{\sqrt{2}}\right).\hat{k}}$$

$$= \iint \frac{1+2y}{\sqrt{2}}\frac{dx\,dy}{\frac{1}{\sqrt{2}}} = \iint (1+2y)\,dx\,dy = \int_0^{2\pi}\int_0^1 (1+2r\sin\theta)\,r\,d\theta\,dr$$

$$= \int_0^{2\pi}\int_0^1 (r+2r^2\sin\theta)\,d\theta\,dr$$

$$= \int_0^{2\pi} d\theta\left[\frac{r^2}{2}+\frac{2r^3}{3}\sin\theta\right]_0^1 = \int_0^{2\pi}\left[\frac{1}{2}+\frac{2}{3}\sin\theta\right]d\theta$$

$$= \left[\frac{\theta}{2}-\frac{2}{3}\cos\theta\right]_0^{2\pi} = \left(\pi-\frac{2}{3}-0+\frac{2}{3}\right) = \pi \qquad \textbf{Ans.}$$

Example 26. *Apply Stoke's Theorem to find the value of*

$$\int_c (y\,dx + z\,dy + x\,dz)$$

where c is the curve of intersection of $x^2 + y^2 + z^2 = a^2$ *and* $x + z = a$.

Solution. $\int_c (y\,dx + z\,dy + x\,dz)$

$$= \int_c (y\hat{i}+z\hat{j}+x\hat{k})\cdot(\hat{i}\,dx+\hat{j}\,dy+\hat{k}\,dz) = \int_C (y\hat{i}+z\hat{j}+x\hat{k})\cdot\vec{dr}$$

$$= \iint_S \text{curl}\,(y\hat{i}+z\hat{j}+x\hat{k})\cdot\hat{n}\,ds \qquad \text{(By Stoke's Theorem)}$$

$$= \iint_S \left(\hat{i}\frac{\partial}{\partial x}+\hat{j}\frac{\partial}{\partial y}+\hat{k}\frac{\partial}{\partial z}\right)\times(y\hat{i}+z\hat{j}+x\hat{k})\cdot\hat{n}\,ds = \iint_S -(\hat{i}+\hat{j}+\hat{k})\cdot\hat{n}\,ds \quad ...(1)$$

where S is the circle formed by the intersection of $x^2 + y^2 + z^2 = a^2$ and $x + z = a$.

$$\hat{n} = \frac{\nabla\phi}{|\nabla\phi|} = \frac{\left(\hat{i}\frac{\partial}{\partial x}+\hat{j}\frac{\partial}{\partial y}+\hat{k}\frac{\partial}{\partial z}\right)(x+z-a)}{|\nabla\phi|} = \frac{\hat{i}+\hat{k}}{\sqrt{1+1}}$$

$$\therefore \qquad \hat{n} = \frac{\hat{i}}{\sqrt{2}}+\frac{\hat{k}}{\sqrt{2}}$$

Putting the value of $\hat{n}$ in (1), we have

$$= \iint_S -(\hat{i}+\hat{j}+\hat{k})\cdot\left(\frac{\hat{i}}{\sqrt{2}}+\frac{\hat{k}}{\sqrt{2}}\right)ds$$

$$= \iint_S -\left(\frac{1}{\sqrt{2}}+\frac{1}{\sqrt{2}}\right)ds \qquad \left[\text{Use } r^2 = a^2 - p^2 = a^2 - \frac{a^2}{2} = \frac{a^2}{2}\right]$$

$$= \frac{-2}{\sqrt{2}}\iint_S ds = \frac{-2}{\sqrt{2}}\pi\left(\frac{a}{\sqrt{2}}\right)^2 = -\frac{\pi a^2}{\sqrt{2}} \qquad \textbf{Ans.}$$

Example 27. *Use Stoke's Theorem to evaluate* $\int_c \vec{v}\cdot\vec{dr}$, *where* $\vec{v} = y^2\hat{i} + xy\hat{j} + xz\hat{k}$, *and c is the bounding curve of the hemisphere* $x^2 + y^2 + z^2 = 9$, $z > 0$, *oriented in the positive direction.*

Solution. By Stoke's theorem

$$\int_c \vec{v}\cdot\vec{dr} = \iint_S (\text{curl } \vec{v})\cdot\hat{n}\,ds = \iint_S (\nabla\times\vec{v})\cdot\hat{n}\,ds$$

$$\nabla\times\vec{v} = \begin{vmatrix} \hat{i} & \hat{j} & \hat{k} \\ \frac{\partial}{\partial x} & \frac{\partial}{\partial y} & \frac{\partial}{\partial z} \\ y^2 & xy & xz \end{vmatrix} = (0-0)\,\hat{i} - (z-0)\,\hat{j} + (y-2y)\,\hat{k} = -z\hat{j} - y\hat{k}$$

$$\hat{n} = \frac{\nabla\phi}{|\nabla\phi|} = \frac{\left(i\frac{\partial}{\partial x} + j\frac{\partial}{\partial y} + k\frac{\partial}{\partial z}\right)(x^2+y^2+z^2-9)}{|\nabla\phi|}$$

$$= \frac{2x\hat{i} + 2y\hat{j} + 2z\hat{k}}{\sqrt{4x^2+4y^2+4z^2}} = \frac{x\hat{i} + y\hat{j} + z\hat{k}}{\sqrt{x^2+y^2+z^2}} = \frac{x\hat{i} + y\hat{j} + z\hat{k}}{3}$$

$$(\nabla\times\vec{v})\cdot\hat{n} = (-z\hat{j} - y\hat{k})\cdot\frac{x\hat{i} + y\hat{j} + z\hat{k}}{3} = \frac{-yz-yz}{3} = \frac{-2yz}{3}$$

$$\hat{n}\cdot\hat{k}\,ds = dx\,dy \Rightarrow \frac{x\hat{i} + y\hat{j} + z\hat{k}}{3}.\hat{k}\;ds = dx\,dy \Rightarrow \frac{z}{3}ds = dx\,dy$$

$$\therefore \quad ds = \frac{3}{z}dx\,dy$$

$$\iint_S (\nabla\times\vec{v})\cdot\hat{n}\,ds = \iint\left(\frac{-2yz}{3}\right)\left(\frac{3}{z}dx\,dy\right) = -\iint 2y\,dx\,dy$$

$$= -\iint 2r\sin\theta\, r\,d\theta\,dr = -2\int_0^{2\pi}\sin\theta\,d\theta\int_0^3 r^2\,dr$$

$$= -2(-\cos\theta)_0^{2\pi}\cdot\left[\frac{r^3}{3}\right]_0^3 = -2(-1+1)\,9 = 0 \textbf{ Ans.}$$

Y, r = 3, O, X

Example 28. *Evaluate the surface integral* $\iint_S \text{curl}\,\vec{F}.\hat{n}\,dS$ *by transforming it into a line integral, S being that part of the surface of the paraboloid* $z = 1 - x^2 - y^2$ *for which* $z \geq 0$ *and* $\vec{F} = y\,\hat{i} + z\,\hat{j} + x\,\hat{k}$. *(K. University, Dec. 2008)*

Solution.
$$\vec{\nabla}\times\vec{F} = \begin{vmatrix} \hat{i} & \hat{j} & \hat{k} \\ \frac{\partial}{\partial x} & \frac{\partial}{\partial y} & \frac{\partial}{\partial z} \\ y & z & x \end{vmatrix} = -\hat{i} - \hat{j} - \hat{k}$$

Obviously $\hat{n} = \hat{k}$.

Therefore $(\nabla\times\vec{F})\cdot\hat{n} = (-\hat{i} - \hat{j} - \hat{k}).\,\hat{k} = -1$

Hence $$\iint_S (\nabla\times\vec{F})\cdot\hat{n}\,ds = \iint_S (-1)\,dx\,dy = -\iint_S dx\,dy$$

$$= -\pi\,(1)^2 = -\pi.$$ (Area of circle $= \pi r^2$) **Ans.**

Example 29. *Evaluate* $\oint_C \vec{F}\cdot \vec{dr}$ *by Stoke's Theorem, where* $\vec{F} = y^2\,\hat{i} + x^2\,\hat{j} - (x+z)\,\hat{k}$ *and C is the boundary of triangle with vertices at (0, 0, 0), (1, 0, 0) and (1, 1, 0).*

Solution. We have, curl $\vec{F} = \nabla \times \vec{F}$

$$= \begin{vmatrix} \hat{i} & \hat{j} & \hat{k} \\ \dfrac{\partial}{\partial x} & \dfrac{\partial}{\partial y} & \dfrac{\partial}{\partial z} \\ y^2 & x^2 & -(x+z) \end{vmatrix} = 0.\hat{i} + \hat{j} + 2\,(x-y)\,\hat{k}.$$

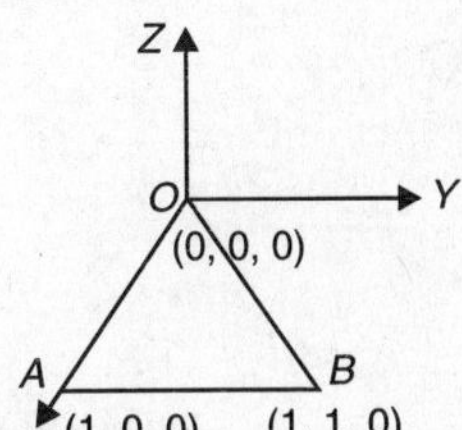

We observe that z co-ordinate of each vertex of the triangle is zero. Therefore, the triangle lies in the xy-plane.

$\therefore \quad \hat{n} = \hat{k}$

$\therefore \quad$ curl $\vec{F}\cdot\hat{n} = [\hat{j} + 2\,(x-y)\hat{k}].\hat{k} = 2\,(x-y).$

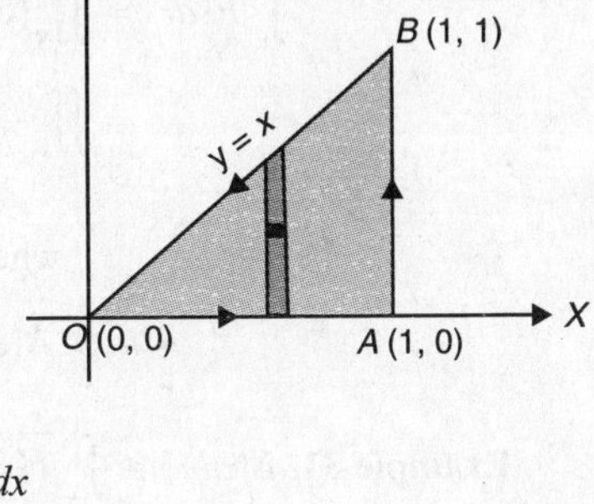

In the figure, only xy-plane is considered.

The equation of the line OB is $y = x$

By Stoke's theorem, we have

$$\oint_C \vec{F}\cdot\vec{dr} = \iint_S (\text{curl } \vec{F}\cdot\hat{n})\, ds$$

$$= \int_{x=0}^{1}\int_{y=0}^{x} 2(x-y)\,dx\,dy = 2\int_0^1 \left[xy - \frac{y^2}{2}\right]_0^x dx$$

$$= 2\int_0^1 \left[x^2 - \frac{x^2}{2}\right] dx = 2\int_0^1 \frac{x^2}{2}\,dx = \int_0^1 x^2\,dx = \left[\frac{x^3}{3}\right]_0^1 = \frac{1}{3}. \qquad \textbf{Ans.}$$

Example 30. *Use the Stoke's Theorem to evaluate*

$$\int_C [(x+2y)\,dx + (x-z)\,dy + (y-z)\,dz]$$

where c is the boundary of the triangle with vertices (2, 0, 0), (0, 3, 0) and (0, 0, 6) oriented in the anti-clockwise direction.

Solution.

$$\int_C \vec{F}\cdot\vec{dr} = \int_C [(x+2y)\,dx + (x-z)\,dy + (y-z)\,dz]$$

$$= \int_C [(x+2y)\,\hat{i} + (x-z)\,\hat{j} + (y-z)\,\hat{k}].[\hat{i}\,dx + \hat{j}dy + \hat{k}dz]$$

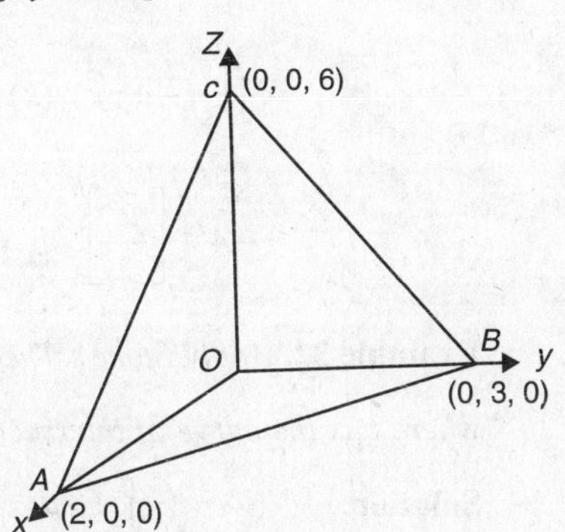

$$\therefore \quad \vec{F} = (x+2y)\,\hat{i} + (x-z)\,\hat{j} + (y-z)\,\hat{k}$$

$$\nabla\times\vec{F} = \begin{vmatrix} \hat{i} & \hat{j} & \hat{k} \\ \dfrac{\partial}{\partial x} & \dfrac{\partial}{\partial y} & \dfrac{\partial}{\partial z} \\ x+2y & x-z & y-z \end{vmatrix}$$

$$= (1+1)\,\hat{i} - (0-0)\,\hat{j} + (1-2)\,\hat{k} = 2\hat{i} - \hat{k}$$

S is the surface of the plane $\dfrac{x}{2} + \dfrac{y}{3} + \dfrac{z}{6} = 1,$

$\hat{n}$ is the normal to the plane *ABC*.

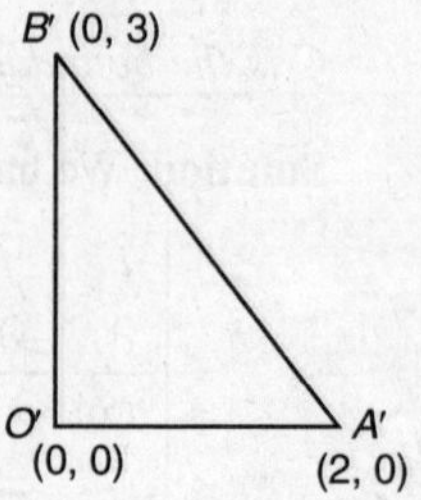

$$\text{Normal Vector} = \nabla\phi = \left[\hat{i}\frac{\partial}{\partial x} + \hat{j}\frac{\partial}{\partial y} + \hat{k}\frac{\partial}{\partial z}\right]\left[\frac{x}{2} + \frac{y}{3} + \frac{z}{6} - 1\right]$$

$$= \frac{\hat{i}}{2} + \frac{\hat{j}}{3} + \frac{\hat{k}}{6} = \frac{1}{6}(3\hat{i} + 2\hat{j} + \hat{k})$$

$$\hat{n} = \frac{\frac{1}{6}(3\hat{i} + 2\hat{j} + \hat{k})}{\frac{1}{6}\sqrt{9+4+1}} = \frac{1}{\sqrt{14}}(3\hat{i} + 2\hat{j} + \hat{k})$$

B′ (0, 3)

O′ (0, 0)   A′ (2, 0)

$$(\nabla \times \vec{F})\cdot\hat{n} = (2\hat{i} - \hat{k}).\frac{1}{\sqrt{14}}(3\hat{i} + 2\hat{j} + \hat{k}) = \frac{1}{\sqrt{14}}(6-1) = \frac{5}{\sqrt{14}}$$

$$\int_C \vec{F}\cdot\vec{dr} = \iint_S (\text{curl } \vec{F})\cdot\hat{n}\, ds$$

$$= \iint_s \frac{5}{\sqrt{14}}\, ds = \frac{5}{\sqrt{14}}\iint_R \frac{dx\, dy}{\hat{k}.\frac{1}{\sqrt{14}}(3\hat{i} + 2\hat{j} + \hat{k})} = 5\iint_R dx\, dy$$

where R is the projection of S on the $x\,y$-plane *i.e.* triangle $O'A'B'$.

$$= 5.\ \text{Area of triangle } O'A'B' = \frac{5}{2}(2\times 3) = 15$$ **Ans.**

Example 31. *Evaluate* $\oint_C \vec{F}\cdot\vec{dr}$ *by Stoke's Theorem, where* $\vec{F} = (x^2 + y^2)\,\hat{i} - 2\,xy\,\hat{j}$ *and C is the boundary of the rectangle* $x = \pm\, a$, $y = 0$ *and* $y = b$.

Solution. Since the z co-ordinate of each vertex of the given rectangle is zero, hence the given rectangle must lie in the xy-plane.

Here, the co-ordinates of *A, B, C* and *D* are $(a, 0)$, (a, b), $(-a, b)$ and $(-a, 0)$ respectively.

$$\therefore \quad \text{Curl } \overline{F} = \begin{vmatrix} \hat{i} & \hat{j} & \hat{k} \\ \frac{\partial}{\partial x} & \frac{\partial}{\partial y} & \frac{\partial}{\partial z} \\ x^2 + y^2 & -2xy & 0 \end{vmatrix} = -4\,y\,k$$

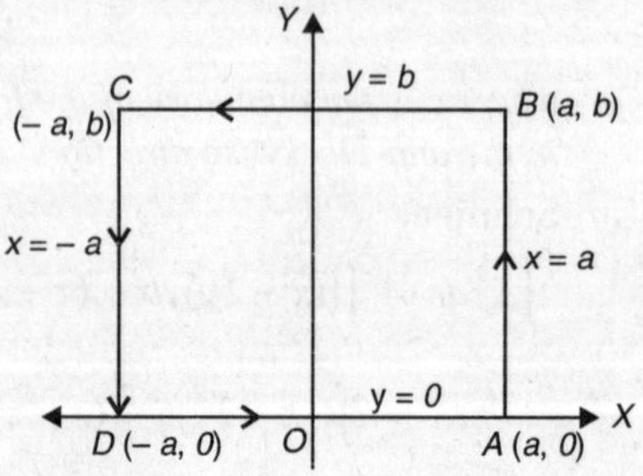

Y, C (− a, b), y = b, B (a, b), x = − a, x = a, y = 0, X, D (− a, 0), O, A (a, 0)

Here, $\hat{n} = \hat{k}$, so by Stoke's theorem, we've

$$\oint_C \vec{F}\cdot\vec{dr} = \iint_S \text{curl } \overline{F}\cdot\hat{n}\, d\,s$$

$$= \iint_S (-4\,y\hat{k})\cdot(\hat{k})\,d\,x\,d\,y = -4\int_{x=-a}^{a}\int_{y=0}^{b} y\, dx\, dy$$

$$= -4\int_{-a}^{a}\left[\frac{y^2}{2}\right]_0^b dx = -2\,b^2\int_{-a}^{a} dx = -4\,a\,b^2$$ **Ans.**

Example 32. *Apply Stoke's Theorem to calculate* $\int_c 4\,y\,dx + 2\,z\,dy + 6\,y\,dz$ *where c is the curve of intersection of* $x^2 + y^2 + z^2 = 6\,z$ *and* $z = x + 3$.

Solution. $$\int_c \vec{F}\cdot\vec{dr} = \int_c 4\,y\,dx + 2\,z\,dy + 6\,y\,dz$$

$$= \int_c (4\,y\hat{i} + 2\,z\hat{j} + 6\,y\hat{k})\cdot(\hat{i}dx + \hat{j}dy + \hat{k}dz)$$

$$\vec{F} = 4\,y\hat{i} + 2\,z\hat{j} + 6\,y\hat{k}$$

$$\nabla\times\vec{F} = \begin{vmatrix} \hat{i} & \hat{j} & \hat{k} \\ \frac{\partial}{\partial x} & \frac{\partial}{\partial y} & \frac{\partial}{\partial z} \\ 4y & 2z & 6y \end{vmatrix} \begin{matrix} = (6-2)\,\hat{i} - (0-0)\,\hat{j} + (0-4)\,\hat{k} \\ = 4\hat{i} - 4\,\hat{k} \end{matrix}$$

S is the surface of the circle $x^2 + y^2 + z^2 = 6z$, $z = x + 3$, $\hat{n}$ is normal to the plane $x - z + 3 = 0$

$$\hat{n} = \frac{\nabla\phi}{|\nabla\phi|} = \frac{\left(\hat{i}\frac{\partial}{\partial x} + \hat{j}\frac{\partial}{\partial y} + \hat{k}\frac{\partial}{\partial z}\right)(x - z + 3)}{|\nabla\phi|} = \frac{\hat{i} - \hat{k}}{\sqrt{1+1}} = \frac{\hat{i} - \hat{k}}{\sqrt{2}}$$

$$(\nabla\times F)\cdot\hat{n} = (4\,\hat{i} - 4\,\hat{k})\cdot\frac{\hat{i} - \hat{k}}{\sqrt{2}} = \frac{4+4}{\sqrt{2}} = 4\sqrt{2}$$

$$\int_c \vec{F}\cdot\vec{dr} = \iint_S (\text{curl } F)\cdot\hat{n}\,ds = \iint_S 4\sqrt{2}\,(dx\,dz) = 4\sqrt{2}\ \text{(area of circle)}$$

Centre of the sphere $x^2 + y^2 + (z-3)^2 = 9$, $(0, 0, 3)$ lies on the plane $z = x + 3$. It means that the given circle is a great circle of sphere, where radius of the circle is equal to the radius of the sphere.

Radius of circle = 3, Area = $\pi\,(3)^2 = 9\,\pi$

$$\iint_S (\nabla\times F)\cdot\hat{n}\,ds = 4\sqrt{2}(9\pi) = 36\sqrt{2}\,\pi$$ **Ans.**

Example 33. *Verify Stoke's Theorem for the function* $\overline{F} = z\hat{i} + x\hat{j} + y\hat{k}$, *where C is the unit circle in xy-plane bounding the hemisphere* $z = \sqrt{(1-x^2-y^2)}$.

Solution. Here $\overline{F} = z\hat{i} + x\hat{j} + y\hat{k}$. ...(1)

Also, $\overline{r} = x\hat{i} + y\hat{j} + z\hat{k} \Rightarrow \overline{dr} = d\,x\hat{i} + dy\,\hat{j} + dz\,\hat{k}$.

$\therefore$ $\overline{F}\cdot\overline{dr} = z\,dx + x\,dy + y\,dz$.

$\therefore$ $\oint_C \overline{F}\cdot\overline{dr} = \oint_C (z\,dx + x\,dy + y\,dz)$. ...(2)

On the circle C, $x^2 + y^2 = 1$, $z = 0$ on the xy-plane. Hence on C, we have $z = 0$ so that $dz = 0$. Hence (2) reduces to

$$\oint_C \overline{F}\cdot\overline{dr} = \oint_C x\,dy. \qquad ...(3)$$

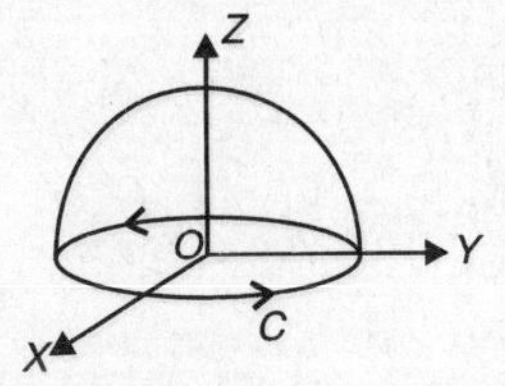

Now the parametric equations of C, *i.e.*, $x^2 + y^2 = 1$ are

$$x = \cos\phi,\ y = \sin\phi. \qquad ...(4)$$

$$dy = \cos\phi,\ d\phi.$$

Using (4), (3) reduces to $\oint_C \vec{F}\cdot\vec{dr} = \int_{\phi=0}^{2\pi} \cos\phi\cos\phi\,d\phi = \int_0^{2\pi} \frac{1+\cos 2\phi}{2}\,d\phi$

$$= \frac{1}{2}\left[\phi + \frac{\sin 2\phi}{2}\right]_0^{2\pi} = \pi \qquad ...(5)$$

Let $P(x, y, z)$ be any point on the surface of the hemisphere $x^2 + y^2 + z^2 = 1$, O origin is the centre of the sphere.

Radius $= OP = x\hat{i} + y\hat{j} + z\hat{k}$

Normal $= x\hat{i} + y\hat{j} + z\hat{k}$

$$\hat{n} = \frac{x\hat{i} + y\hat{j} + z\hat{k}}{\sqrt{x^2 + y^2 + z^2}} = x\hat{i} + y\hat{j} + z\hat{k}$$

(Radius is $\perp$ to tangent *i.e.* Radius is normal)

$x = \sin\theta\cos\phi,\ y = \sin\theta\sin\phi,\ z = \cos\theta$ ...(6)

$\hat{n} = \sin\theta\cos\phi\ \hat{i} + \sin\theta\sin\phi\ \hat{j} + \cos\theta\ \hat{k}$

Also, $$\text{Curl } \vec{F} = \begin{vmatrix} \hat{i} & \hat{j} & \hat{k} \\ \partial/\partial x & \partial/\partial y & \partial/\partial z \\ z & x & y \end{vmatrix} = \hat{i} + \hat{j} + \hat{k}$$ ...(7)

$$\text{Curl } \vec{F}\cdot\hat{n} = (\hat{i} + \hat{j} + \hat{k}).(\sin\theta\cos\phi\,\hat{i} + \sin\theta\sin\phi\,\hat{j} + \sin\theta\,\hat{k})$$

$$= \sin\theta\cos\phi + \sin\theta\sin\phi + \cos\theta$$

$$\therefore \quad \iint_s \text{Curl } \vec{F}\cdot\hat{n}\, dS = \int_{\theta=0}^{\pi/2}\int_{\phi=0}^{2\pi} (\hat{i} + \hat{j} + \hat{k})$$

$$(\sin\theta\cos\phi\ \hat{i} + \sin\theta\sin\phi\ \hat{j} + \cos\theta\ \hat{k})\sin\theta\, d\theta\, d\phi$$

$$= \int_{\theta=0}^{\pi/2} \sin\theta\, d\theta \int_{\phi=0}^{2\pi} (\sin\theta\cos\phi + \sin\theta\sin\phi + \cos\theta)\, d\phi$$

[$\because$ dS = Elementary area on hemisphere = $\sin\theta\, d\theta\, d\phi$]

$$= \int_0^{\pi/2} \sin\theta\, d\theta\,[\sin\theta\sin\phi + \sin\theta(-\cos\phi) + \phi\cos\theta]_0^{2\pi} = \int_0^{\pi/2} \sin\theta\, d\theta$$

$$= \int_0^{\pi/2} (0 + 0 + 2\pi\sin\theta\cos\theta)\, d\theta = \pi\int_0^{\pi/2} \sin 2\theta\, d\theta \quad = \pi\left[-\frac{\cos 2\theta}{2}\right]_0^{\pi/2}$$

$= -(\pi/2)\,[-1-1] = \pi.$ ...(8)

From (5) and (8), $\oint_C \vec{F}\cdot\vec{dr} = \iint_S \text{curl } \overline{F}\cdot\hat{n}\, dS,$ which verifies Stokes's theorem.

Example 34. *Verify Stoke's theorem for the vector field* $\overline{F} = (2x - y)\hat{i} - yz^2\,\hat{j} - y^2 z\,\hat{k}$ *over the upper half of the surface* $x^2 + y^2 + z^2 = 1$ *bounded by its projection on xy- plane.*

Solution. Let S be the upper half surface of the sphere $x^2 + y^2 + z^2 = 1$. The boundary C or S is a circle in the xy plane of radius unity and centre O. The equation of C are $x^2 + y^2 = 1$, $z = 0$ whose parametric form is

$x = \cos t,\ y = \sin t,\ z = 0,\ 0 < t < 2\pi$

$$\int_C \vec{F}.\vec{dr} = \int_C [(2x - y)\,\hat{i} - yz^2\,\hat{j} - y^2 z\hat{k}].[\hat{i}\,dx + \hat{j}\,dy + \hat{k}\,dz]$$

$$= \int_C [(2x - y)\,dx - yz^2\,dy - y^2 z\,dz]$$

$$= \int_C (2x - y)\,dx, \text{ since on } C,\ z = 0 \text{ and } 2z = 0$$

$$= \int_0^{2\pi} (2\cos t - \sin t)\frac{dx}{dt}\,dt = \int_0^{2\pi} (2\cos t - \sin t)(-\sin t)\,dt$$

$$= \int_0^{2\pi} (-\sin 2t + \sin^2 t)\,dt = \int_0^{2\pi}\left(-\sin 2t + \frac{1 - \cos 2t}{2}\right)dt$$

$$= \left[\frac{\cos 2t}{2} + \frac{t}{2} - \frac{\sin 2t}{4}\right]_0^{2\pi} = \frac{1}{2} + \pi - \frac{1}{2} = \pi$$ ...(1)

$$\text{Curl } \vec{F} = \begin{vmatrix} \hat{i} & \hat{j} & \hat{k} \\ \dfrac{\partial}{\partial x} & \dfrac{\partial}{\partial y} & \dfrac{\partial}{\partial z} \\ 2x-y & -yz^2 & -y^2z \end{vmatrix} = (-2yz+2yz)\,\hat{i} + (0-0)\,\hat{j} + (0+1)\,\hat{k} = \hat{k}$$

$$\text{Curl } \vec{F}.\hat{n} = \hat{k}.\hat{n} = \hat{n}.\hat{k}$$

$$\iint_S Curl\ \vec{F}.\hat{n}\,ds = \iint_S \hat{n}.\hat{k}\,ds = \iint_R \hat{n}.\hat{k}.\frac{dx}{\hat{n}}.\frac{dy}{\hat{k}}$$

Where R is the projection of S on xy-plane.

$$= \int_{-1}^{1}\int_{-\sqrt{1-x^2}}^{\sqrt{1-x^2}} dx\,dy$$

$$= \int_{-1}^{1} 2\sqrt{1-x^2}\,dx = 4\int_0^1 \sqrt{1-x^2}\,dx$$

$$= 4\left[\frac{x}{2}\sqrt{1-x^2} + \frac{1}{2}\sin^{-1}x\right]_0^1 = 4\left[\frac{1}{2}.\frac{\pi}{2}\right] = \pi \qquad ...(2)$$

From (1) and (2), we have

$$\therefore \quad \int_C \vec{F}.\vec{dr} = \iint \text{Curl } \overline{F}.\hat{n}\,ds \text{ which is the Stoke's theorem.}$$ **Ans.**

Example 35. *Verify Stoke's Theorem for*

$$\vec{F} = (x^2+y-4)\,\hat{i} + 3\,xy\hat{j} + (2\,xz+z^2)\,\hat{k}$$

over the surface of hemisphere $x^2 + y^2 + z^2 = 16$ *above the xy-plane.*

Solution. $\int_c \vec{F}\cdot\vec{dr}$, where c is the boundary of the circle $x^2 + y^2 + z^2 = 16$ (bounding the hemispherical surface)

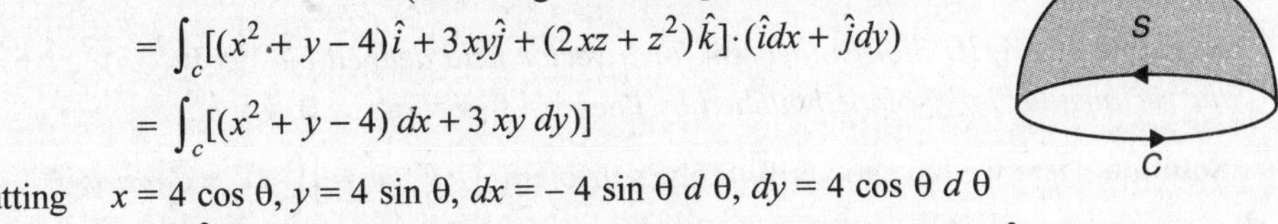

$$= \int_c [(x^2+y-4)\hat{i} + 3xy\hat{j} + (2xz+z^2)\hat{k}]\cdot(\hat{i}dx + \hat{j}dy)$$

$$= \int_c [(x^2+y-4)\,dx + 3\,xy\,dy)]$$

Putting $x = 4\cos\theta$, $y = 4\sin\theta$, $dx = -4\sin\theta\,d\theta$, $dy = 4\cos\theta\,d\theta$

$$= \int_0^{2\pi} [(16\cos^2\theta + 4\sin\theta - 4)\,(-4\sin\theta\,d\theta) + (192\sin\theta\cos^2\theta\,d\theta)]$$

$$= 16\int_0^{2\pi} [-4\cos^2\theta\sin\theta - \sin^2\theta + \sin\theta + 12\sin\theta\cos^2\theta]\,d\theta$$

$$= 16\int_0^{2\pi} (8\sin\theta\cos^2\theta - \sin^2\theta + \sin\theta)\,d\theta$$

$$= -16\int_0^{2\pi} \sin^2\theta\,d\theta \qquad \left\{\begin{aligned} &\int_0^{2\pi} \sin^n\theta\cos\theta\,d\theta = 0 \\ &\int_0^{2\pi} \cos^n\theta\sin\theta\,d\theta = 0 \end{aligned}\right\}$$

$$= -16\times 4\int_0^{\frac{\pi}{2}} \sin^2\theta\,d\theta = -64\left(\frac{1}{2}\frac{\pi}{2}\right) = -16\pi.$$

To evaluate surface integral

$$\nabla\times\vec{F} = \begin{vmatrix} \hat{i} & \hat{j} & \hat{k} \\ \dfrac{\partial}{\partial x} & \dfrac{\partial}{\partial y} & \dfrac{\partial}{\partial z} \\ x^2+y-4 & 3xy & 2xz+z^2 \end{vmatrix}$$

$$= (0-0)\,\hat{i} - (2\,z-0)\,\hat{j} + (3\,y-1)\,\hat{k} = -2\,z\hat{j} + (3\,y-1)\,\hat{k}$$

$$\hat{n} = \frac{\nabla\phi}{|\nabla\phi|} = \frac{\left(\hat{i}\frac{\partial}{\partial x} + \hat{j}\frac{\partial}{\partial y} + \hat{k}\frac{\partial}{\partial z}\right)(x^2+y^2+z^2-16)}{|\nabla\phi|}$$

$$= \frac{2x\hat{i}+2y\hat{j}+2z\hat{k}}{\sqrt{4x^2+4y^2+4z^2}} = \frac{x\hat{i}+y\hat{j}+z\hat{k}}{\sqrt{x^2+y^2+z^2}} = \frac{x\hat{i}+y\hat{j}+z\hat{k}}{4}$$

$$(\nabla\times\vec{F})\cdot\hat{n} = [-2z\hat{j}+(3y-1)\hat{k}]\cdot\frac{x\hat{i}+y\hat{j}+z\hat{k}}{4} = \frac{-2yz+(3y-1)z}{4}$$

$$\hat{k}\cdot\hat{n}\cdot ds = dx\,dy \Rightarrow \frac{x\hat{i}+y\hat{j}+z\hat{k}}{4}\,.\,k\,ds = dx\,dy \Rightarrow \frac{z}{4}ds = dx\,dy$$

$$\therefore \quad ds = \frac{4}{z}\,dx\,dy$$

$$\iint(\nabla\times F)\cdot\hat{n}\,ds = \iint\frac{-2yz+(3y-1)z}{4}\left(\frac{4}{z}\,dx\,dy\right)$$

$$= \iint[-2y+(3y-1)]\,dx\,dy = \iint(y-1)\,dx\,dy$$

On putting $x = r\cos\theta$, $y = r\sin\theta$, $dx\,dy = r\,d\theta\,dr$, we get

$$= \iint(r\sin\theta-1)\,r\,d\theta\,dr = \int_0^{2\pi}d\theta\int_0^4(r^2\sin\theta - r)\,dr$$

$$= \int_0^{2\pi}d\theta\left(\frac{r^3}{3}\sin\theta - \frac{r^2}{2}\right)_0^4 = \int_0^{2\pi}d\theta\left(\frac{64}{3}\sin\theta - 8\right)$$

$$= \left(-\frac{64}{3}\cos\theta - 8\theta\right)_0^{2\pi} = \frac{-64}{3} - 16\pi + \frac{64}{3} = -16\pi$$

The line integral is equal to the surface integral, hence Stoke's Theorem is verified. **Proved.**

Example 36. *Verify Stoke's theorem for a vector field defined by* $\vec{F} = (x^2 - y^2)\hat{i} + 2xy\hat{j}$ *in the rectangular in xy-plane bounded by lines* $x = 0, x = a, y = 0, y = b$.

Solution. Here we have to verify Stoke's theorem $\int_C \vec{F}\,.\,\vec{dr} = \iint_S (\nabla\times\bar{F})\,.\,\hat{n}\,ds$

Where 'C' be the boundary of rectangle ($ABCO$) and S be the surface enclosed by curve C.

$$\vec{F} = (x^2-y^2)\hat{i} + (2xy)\hat{j}$$

$$\vec{F}\,.\,\vec{dr} = [(x^2-y^2)\hat{i} + 2xy\hat{j}]\,.\,[\hat{i}\,dx + \hat{j}\,dy]$$

$$\Rightarrow \quad \vec{F}\,.\,\vec{dr} = (x^2+y^2)\,dx + 2xy\,dy \qquad ...(1)$$

$$\text{Now,} \quad \int_C \vec{F}\,.\,\vec{dr} = \int_{OA} F\,.\,\vec{dr} + \int_{AB}\vec{F}\,.\,\vec{dr} + \int_{BC}\vec{F}\,.\,\vec{dr} + \int_{CO}\vec{F}\,.\,\vec{dr} \qquad ...(2)$$

Along OA, put $y = 0$ so that $k\,dy = 0$ in (1) and $\vec{F}\,.\,\vec{dr} = x^2\,dx$,
Where x is from 0 to a.

$$\therefore \quad \int_{OA}\vec{F}\,.\,\vec{dr} = \int_0^a x^2dx = \left[\frac{x^3}{3}\right]_0^a = \frac{a^3}{3} \qquad ...(3)$$

Along AB, put $x = a$ so that $dx = 0$ in (1), we get $\vec{F}\,.\,\vec{dr} = 2ay\,dy$
Where y is from 0 to b.

$$\therefore \quad \int_{AB}\vec{F}\,.\,\vec{dr} = \int_0^b 2ay\,dy = [ay^2]_0^b = ab^2 \qquad ...(4)$$

Along BC, put $y = b$ and $dy = 0$ in (1) we get $\overline{F}\,.\,\overline{dr} = (x^2 - b^2)\,dx$, where x is from a to 0.

$$\therefore \quad \int_{BC} \vec{F}\,.\,\vec{dr} = \int_a^0 (x^2 - b^2)\,dx = \left[\frac{x^3}{3} - b^2 x\right]_a^0 = \frac{-a^3}{3} + b^2 a \quad ...(5)$$

Along CO, put $x = 0$ and $dx = 0$ in (1), we get $\vec{F}\,.\,\vec{dr} = 0$

$$\therefore \quad \int_{CO} \vec{F}\,.\,\vec{dr} = 0 \quad ...(6)$$

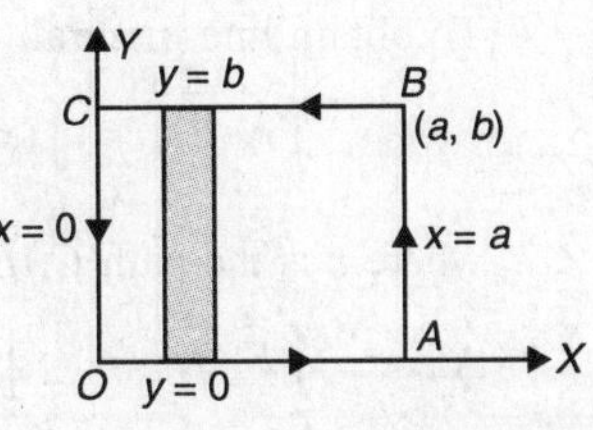

Putting the values of integrals (3), (4), (5) and (6) in (2), we get

$$\int_C \vec{F}\,.\,\vec{dr} = \frac{a^3}{3} + ab^2 - \frac{a^3}{3} + ab^2 + 0 = 2ab^2 \quad ...(7)$$

Now we have to evaluate R.H.S. of Stoke's Theorem *i.e.* $\iint_S (\nabla \times \vec{F})\,.\,\hat{n}\,ds$

We have,

$$\nabla \times \vec{F} = \begin{vmatrix} \hat{i} & \hat{j} & \hat{k} \\ \dfrac{\partial}{\partial x} & \dfrac{\partial}{\partial y} & \dfrac{\partial}{\partial z} \\ x^2 - y^2 & 2xy & 0 \end{vmatrix} = (2y + 2y)\,\hat{k} = 4y\,\hat{k}$$

Also the unit vector normal to the surface S in outward direction is $\hat{n} = k$

($\because$ z-axis is normal to surface S)

Also in xy-plane $ds = dx\,dy$

$$\therefore \quad \iint_S (\nabla \times \vec{F})\,.\,\hat{n}\,.\,ds = \iint_R 4y\,\hat{k}\,.\,\hat{k}\,dx\,dy = \iint_R 4y\,dx\,dy.$$

Where R be the region of the surface S.

Consider a strip parallel to y-axis. This strip starts on line y = 0 (*i.e.* x-axis) and end on the line $y = b$, We move this strip from $x = 0$ (y-axis) to $x = a$ to cover complete region R.

$$\therefore \quad \iint_S (\nabla \times \vec{F})\,.\,\hat{n}\,.\,ds = \int_0^a \left[\int_0^b 4y\,dy\right] dx = \int_0^a [2y^2]_0^b\,dx$$

$$= \int_0^a 2b^2\,dx = 2b^2\,[x]_0^a = 2ab^2 \quad ...(8)$$

$\therefore$ From (7) and (8), we get

$\int_C \vec{F}\,.\,\vec{dr} = \iint_S (\nabla \times \overline{F})\,.\,\hat{n}\,ds$ and hence the Stoke's theorem is verified.

Example 37. *Verify Stoke's Theorem for the function*

$$\vec{F} = x^2\hat{i} - xy\hat{j}$$

integrated round the square in the plane $z = 0$ and bounded by the lines

$$x = 0,\ y = 0,\ x = a,\ y = a.$$

Solution. We have, $\vec{F} = x^2\hat{i} - xy\hat{j}$

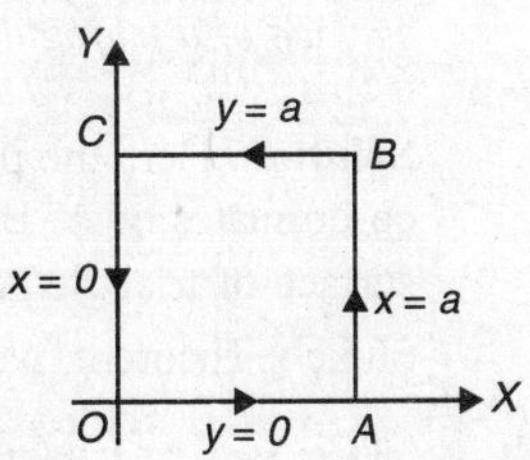

$$\nabla \times \vec{F} = \begin{vmatrix} \hat{i} & \hat{j} & \hat{k} \\ \dfrac{\partial}{\partial x} & \dfrac{\partial}{\partial y} & \dfrac{\partial}{\partial z} \\ x^2 & -xy & 0 \end{vmatrix}$$

$$= (0 - 0)\,\hat{i} - (0 - 0)\,\hat{j} + (-y - 0)\,\hat{k} = -y\hat{k}$$

$$(\hat{n} \perp \text{ to } xy \text{ plane } i.e.\ \hat{k})$$

$$\iint_S (\nabla \times \vec{F}) \cdot \hat{n}\, ds = \iint_S (-yk) \cdot k\, dx\, dy$$

$$= \int_0^a dx \int_0^a -y dy = \int_0^a dx \left[-\frac{y^2}{2} \right]_0^a = -\frac{a^2}{2} (x)_0^a = -\frac{a^3}{2} \qquad ...(1)$$

To obtain line integral

$$\int_C \vec{F} \cdot \vec{dr} = \int_C (x^2 \hat{i} - xy\hat{j}) \cdot (\hat{i}\, dx + \hat{j}\, dy) = \int_C (x^2\, dx - xy\, dy)$$

where c is the path $OABCO$ as shown in the figure.

Also, $$\int_C \vec{F} \cdot \vec{dr} = \int_{OABCO} \vec{F} \cdot dr = \int_{OA} \vec{F} \cdot dr + \int_{AB} \vec{F} \cdot dr + \int_{BC} \vec{F} \cdot dr + \int_{CO} \vec{F} \cdot dr \qquad ...(2)$$

line	Eq. of line		Lower limit	Upper limit
OA	$y = 0$	$dy = 0$	$x = 0$	$x = a$
AB	$x = a$	$dx = 0$	$y = 0$	$y = a$
BC	$y = a$	$dy = 0$	$x = 0$	$x = a$
CO	$x = 0$	$dx = 0$	$y = 0$	$y = a$

Along OA, $y = 0$, $dy = 0$

$$\int_{OA} \vec{F} \cdot \vec{dr} = \int_{OA} (x^2 dx - xydy)$$

$$= \int_0^a x^2 dx = \left[\frac{x^3}{3} \right]_0^a = \frac{a^3}{3}$$

Along AB, $x = a$, $dx = 0$

$$\int_{AB} \vec{F} \cdot \vec{dr} = \int_{AB} (x^2\, dx - x\, y\, d\, y)$$

$$= \int_0^a -a\, y\, d\, y = -a \left[\frac{y^2}{2} \right]_0^a = -\frac{a^3}{2}$$

Along BC, $y = a$, $dy = 0$

$$\int_{BC} \vec{F} \cdot \vec{dr} = \int_{BC} (x^2 dx - xy\, dy) = \int_0^a x^2 dx = \left[\frac{x^3}{3} \right]_0^a = -\frac{a^3}{3}$$

Along CO, $x = 0$, $dx = 0$

$$\int_{CO} \vec{F} \cdot \vec{dr} = \int_{CO} (x^2 dx - xy\, dy) = 0$$

Putting the values of these integrals in (2), we have

$$\int_C \vec{F} \cdot \vec{dr} = \frac{a^3}{3} - \frac{a^3}{2} - \frac{a^3}{3} + 0 = -\frac{a^3}{2} \qquad ...(3)$$

From (1) and (3), $\iint_S (\nabla \times \vec{F}) \cdot \hat{n}\, ds = \int_C \vec{F} \cdot \vec{dr}$

Hence, Stoke's Theorem is verified. **Ans.**

Example 38. *Verify Stoke's Theorem for* $\vec{F} = (x + y)\,\hat{i} + (2x - z)\,\hat{j} + (y + z)\,\hat{k}$ *for the surface of a triangular lamina with vertices (2, 0, 0), (0, 3, 0) and (0, 0, 6).*

(Nagpur University 2004, K. U. Dec. 2009, 2008, A.M.I.E.T.E.)

Solution. Here the path of integration c consists of the straight lines AB, BC, CA where the co-ordinates of A, B, C and (2, 0, 0), (0, 3, 0) and (0, 0, 6) respectively. Let S be the plane surface of triangle ABC bounded by C. Let $\hat{n}$ be unit normal vector to surface S. Then by Stoke's Theorem, we must have

$$\oint_c \vec{F} \cdot \vec{dr} = \iint_s \text{curl } \vec{F} \cdot \hat{n}\, ds \qquad ...(1)$$

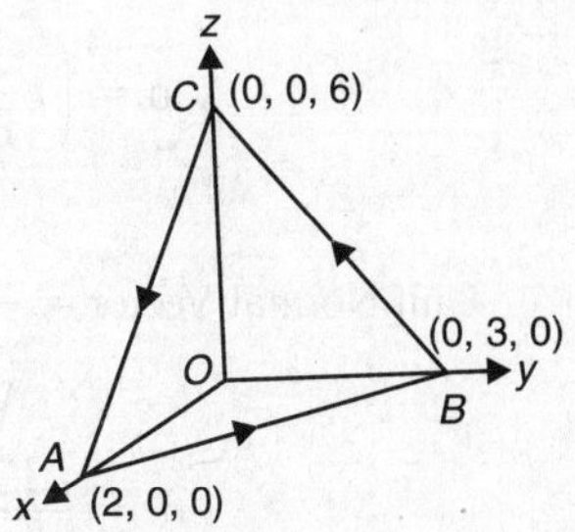

L.H.S. of (1) $= \int_{ABC} \vec{F}\cdot\vec{dr} = \int_{AB} \vec{F}\cdot\vec{dr} + \int_{BC} \vec{F}\cdot\vec{dr} + \int_{CA} \vec{F}\cdot\vec{dr}$

z
C (0, 0, 6)
(0, 3, 0)
O
B
y
A
x (2, 0, 0)

Along line AB, $z = 0$, equation of AB is $\frac{x}{2}+\frac{y}{3}=1$

$$\Rightarrow \qquad y = \frac{3}{2}(2-x),\ dy = -\frac{3}{2}dx$$

At A, $x = 2$, At B, $x = 0$, $\bar{r} = x\hat{i} + y\hat{j}$

$$\int_{AB} \vec{F}\cdot\vec{dr} = \int_{AB}[(x+y)\hat{i} + 2x\hat{j} + y\hat{k}]\cdot(\hat{i}dx + \hat{j}dy)$$

$$= \int_{AB}(x+y)\,dx + 2x\,dy$$

line	Eq. of line		Lower limit	Upper limit
AB	$\frac{x}{2}+\frac{y}{3}=1$, $z=0$	$dy=-\frac{3}{2}dx$	At A $x=2$	At B $x=0$
BC	$\frac{y}{3}+\frac{z}{6}=1$, $z=0$	$dz=-2dy$	At B $y=3$	At C $y=0$
CA	$\frac{x}{2}+\frac{z}{6}=1$, $y=0$	$dz=-3dx$	At C $x=0$	At A $X=2$

$$= \int_{AB}\left(x+3-\frac{3x}{2}\right)dx + 2x\left(-\frac{3}{2}dx\right)$$

$$= \int_2^0\left(-\frac{7x}{2}+3\right)dx = \left(-\frac{7x^2}{4}+3x\right)_2^0$$

$= (7-6) = +1$

Along line BC, $x = 0$, Equation of BC is $\frac{y}{3}+\frac{z}{6} = 1$ or $z = 6-2y$, $dz = -2dy$

At B, $y = 3$, At C, $y = 0$, $\vec{r} = y\hat{j} + z\hat{k}$

$$\int_{BC} \vec{F}\cdot\vec{dr} = \int_{BC}[y\hat{i} + z\hat{j} + (y+z)\hat{k}]\cdot(\hat{j}dy + \hat{k}dz) = \int_{BC} -z\,dy + (y+z)\,dz$$

$$= \int_3^0(-6+2y)\,dy + (y+6-2y)(-2dy)$$

$$= \int_3^0(4y-18)\,dy = (2y^2-18y)_3^0 = 36$$

Along line CA, $y = 0$, Eq. of CA, $\frac{x}{2}+\frac{z}{6} = 1$ or $z = 6-3x$, $dz = -3dx$

At C, $x = 0$, at A, $x = 2$, $\vec{r} = x\hat{i} + z\hat{k}$

$$\int_{CA} \vec{F}\cdot\vec{dr} = \int_{CA}[x\hat{i} + (2x-z)\,\hat{j} + z\hat{k}]\cdot[dx\hat{i} + dz\hat{k}] = \int_{CA}(x\,dx + z\,dz)$$

$$= \int_0^2 x\,dx + (6-3x)(-3dx) = \int_0^2(10x-18)\,dx = [5x^2-18x]_0^2 = -16$$

$$\text{L.H.S. of (1)} = \int_{ABC} \vec{F}\cdot\vec{dr} = \int_{AB} \vec{F}\cdot\vec{dr} + \int_{BC} \vec{F}\cdot\vec{dr} + \int_{CA} \vec{F}\cdot\vec{dr} = 1+36-16 = 21 \qquad ...(2)$$

$$\text{Curl } \bar{F} = \nabla\times\bar{F} = \left(\hat{i}\frac{\partial}{\partial x} + \hat{j}\frac{\partial}{\partial y} + \hat{k}\frac{\partial}{\partial z}\right)\times[(x+y)\,\hat{i} + (2x-z)\,\hat{j} + (y+z)\,k]$$

$$= \begin{vmatrix} \hat{i} & \hat{j} & \hat{k} \\ \frac{\partial}{\partial x} & \frac{\partial}{\partial y} & \frac{\partial}{\partial z} \\ x+y & 2x-z & y+z \end{vmatrix} = (1+1)\,\hat{i} - (0-0)\,\hat{j} + (2-1)\,\hat{k} = 2\hat{i} + \hat{k}$$

Equation of the plane of ABC is $\frac{x}{2}+\frac{y}{3}+\frac{z}{6}=1$

Normal to the plane ABC is

$$\nabla\phi = \left(\hat{i}\frac{\partial}{\partial x}+\hat{j}\frac{\partial}{\partial y}+\hat{k}\frac{\partial}{\partial z}\right)\left(\frac{x}{2}+\frac{y}{3}+\frac{z}{6}-1\right)=\frac{\hat{i}}{2}+\frac{\hat{j}}{3}+\frac{\hat{k}}{6}$$

$$\text{Unit Normal Vector} = \frac{\frac{\hat{i}}{2}+\frac{\hat{j}}{3}+\frac{\hat{k}}{6}}{\sqrt{\frac{1}{4}+\frac{1}{9}+\frac{1}{36}}}$$

$$\hat{n} = \frac{1}{\sqrt{14}}(3\hat{i}+2\hat{j}+\hat{k})$$

$$\text{R.H.S. of (1)} = \iint_s \text{curl}\,\vec{F}\cdot\hat{n}\,ds = \iint_s (2\hat{i}+\hat{k})\cdot\frac{1}{\sqrt{14}}(3\hat{i}+2\hat{j}+\hat{k})\frac{dx\,dy}{\frac{1}{\sqrt{14}}(3\hat{i}+2\hat{j}+\hat{k}).\hat{k}}$$

$$= \iint_s \frac{(6+1)}{\sqrt{14}}\frac{dx\,dy}{\frac{1}{\sqrt{14}}} = 7\iint dx\,dy = 7 \text{ Area of } \Delta \text{ OAB}$$

$$= 7\left(\frac{1}{2}\times 2\times 3\right) = 21 \qquad ...(3)$$

with the help of (2) and (3) we find (1) is true and so Stoke's *Theorem* is verified.

Example 39. *Verify Stoke's Theorem for*

$$\vec{F} = (y-z+2)\,\hat{i} + (yz+4)\,\hat{j} - (xz)\,\hat{k}$$

over the surface of a cube x = 0, y = 0, z = 0, x = 2, y = 2, z = 2 above the XOY plane (open the bottom). *(DU, I Sem. 2012)*

Solution. Consider the surface of the cube as shown in the figure. Bounding path is *OABCO* shown by arrows.

$$\int_c \vec{F}\cdot\vec{d}r = \int[(y-z+2)\,\hat{i}+(yz+4)\,\hat{j}-(xz)\,\hat{k}]\cdot(\hat{i}dx+\hat{j}dy+\hat{k}dz)$$

$$= \int_c (y-z+2)\,dx+(yz+4)\,dy - xzdz$$

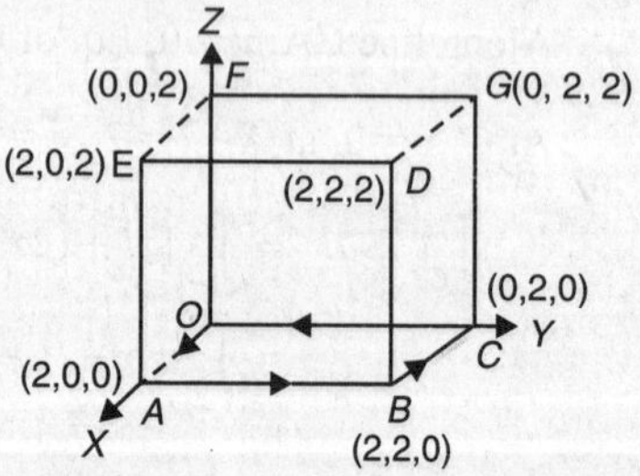

$$\int_c \vec{F}\cdot\vec{dr} = \int_{OA}\vec{F}\cdot\vec{dr}+\int_{AB}\vec{F}\cdot\vec{dr}+\int_{BC}\vec{F}\cdot\vec{dr}+\int_{CO}\vec{F}\cdot\vec{dr} \quad ...(1)$$

Z, (0,0,2) F, G(0, 2, 2), (2,0,2) E, (2,2,2) D, O, (0,2,0), Y, C, (2,0,0), A, B, X, (2,2,0)

(1) Along *OA*, $y = 0$, $dy = 0$, $z = 0$, $dz = 0$

$$\int_{OA}\vec{F}\cdot\vec{dr} = \int_0^2 2\,dx = [2x]_0^2 = 4$$

(2) Along *AB*, $x = 2$, $dx = 0$, $z = 0$, $dz = 0$

$$\int_{AB}\vec{F}\cdot\vec{dr} = \int_0^2 4\,dy = 4\,(y)_0^2 = 8$$

(3) Along *BC*, $y = 2$, $dy = 0$, $z = 0$, $dz = 0$

$$\int_{BC}\vec{F}\cdot\vec{dr} = \int_2^0 (2-0+2)\,dx = (4x)_2^0 = -8$$

(4) Along *CO*, $x = 0$, $dx = 0$, $z = 0$, $dz = 0$

$$\int_{CO}\vec{F}\cdot\vec{dr} = \int_2^0 (y-0+2)\times 0+(0+4)dy - 0$$

	line	Equ. of line		Lower limit	Upper limit	$\bar{F}.\bar{d}r$
1	*OA*	$y=0$ $z=0$	$dy=0$ $dz=0$	$x=0$	$x=2$	$2\,dx$
2	*AB*	$x=2$ $z=0$	$dx=0$ $dz=0$	$y=0$	$y=2$	$4\,dy$
3	*BC*	$y=2$ $z=0$	$dy=0$ $dz=0$	$x=2$	$x=0$	$4\,dx$
4	*CO*	$x=2$ $z=0$	$dx=0$ $dz=0$	$y=2$	$y=0$	$4\,dy$

$$= 4\int dy = 4\,(y)_2^0 = -8$$

On putting the values of these integrals in (1), we get

$$\int_c \vec{F}\cdot\vec{dr} = 4 + 8 - 8 - 8 = -4$$

To obtain surface integral

$$\nabla\times\vec{F} = \begin{vmatrix} \hat{i} & \hat{j} & \hat{k} \\ \dfrac{\partial}{\partial x} & \dfrac{\partial}{\partial y} & \dfrac{\partial}{\partial z} \\ y - z + 2 & yz + 4 & -xz \end{vmatrix}$$

$$= (0 - y)\,\hat{i} - (-z + 1)\,\hat{j} + (0 - 1)\,\hat{k} = -y\hat{i} + (z - 1)\,\hat{j} - \hat{k}$$

Here we have to integrate over the five surfaces, *ABDE, OCGF, BCGD, OAEF, DEFG.*

	Surface	Outward normal	ds	
1	*ABDE*	$\hat{i}$	$dy\,dz$	$x = 2$
2	*OCGF*	$-\hat{i}$	$dy\,dz$	$x = 0$
3	*BCGD*	$\hat{j}$	$dx\,dz$	$y = 2$
4	*OAEF*	$-\hat{j}$	$dx\,dz$	$y = 0$
5	*DEFG*	$\hat{k}$	$dx\,dy$	$z = 2$

Over the surface *ABDE* ($x = 2$), $\hat{n} = i,\ ds = dy\,dz$

$$\iint(\nabla\times\vec{F})\cdot\hat{n}\,ds = \iint[-yi + (z-1)\,j - k]\cdot i\,dx\,dz = \iint -y\,dy\,dz$$

$$= \iint_R [F_3\,(x, y, z)]_{z=f_1(x,y)}^{z=f_2(x,y)}\,dx\,dy$$

$$= -\int_0^2 y\,dy\int_0^2 dz = -\left[\frac{y^2}{2}\right]_0^2 [z]_0^2 = -4$$

Over the surface OCGF ($x = 0$), $\hat{n} = -i,\ ds = dy\,dz$

$$\iint(\nabla\times\vec{F})\cdot\hat{n}\,ds = \iint[-y\hat{i} + (z-1)\,\hat{j} - \hat{k}]\cdot(-\hat{i})\,dy\,dz$$

$$= \iint y\,dy\,dz = \int_0^2 y\,dy\int_0^2 dz = 2\left[\frac{y^2}{2}\right]_0^2 = 4$$

(3) Over the surface *BCGD*, ($y = 2$), $\hat{n} = j,\ ds = dx\,dz$

$$\iint(\nabla\times\vec{F})\cdot\hat{n}\,ds = \iint[-y\hat{i} + (z-1)\,\hat{j} - \hat{k}]\cdot\hat{j}\,dx\,dz$$

$$= -\iint(z-1)\,dx\,dz = \int_0^2 dx\int_0^2(z-1)\,dz = -(x)_0^2\left(\frac{z^2}{2} - z\right)_0^2 = 0$$

(4) Over the surface *OAEF*, ($y = 0$), $\hat{n} = -\hat{j}$, $ds = dx\,dz$

$$\iint(\nabla\times\vec{F})\cdot\hat{n}\,ds = \iint[-y\hat{i} + (z-1)\hat{j} - \hat{k}]\cdot(-\hat{j})\,dx\,dz$$

$$= -\iint(z-1)\,dx\,dz = -\int_0^2 dx\int_0^2(z-1)\,dz = -(x)_0^2\left(\frac{z^2}{2} - z\right)_0^2 = 0$$

(5) Over the surface *DEFG,* ($z = 2$), $\hat{n} = k,\ ds = dx\,dy$

$$\iint(\nabla\times\vec{F})\cdot\hat{n}\,ds = \iint[-y\hat{i} + (z-1)\hat{j} - \hat{k}]\cdot\hat{k}\,dx\,dy = -\iint dx\,dy$$

$$= -\int_0^2 dx\int_0^2 dy = -[x]_0^2\,[y]_0^2 = -4$$

Total surface integral = – 4 + 4 + 0 + 0 – 4 = – 4

Thus $\iint_S \text{curl}\,\vec{F}\cdot\hat{n}\,ds = \int_c \vec{F}\cdot\vec{dr} = -4$

which verifies Stoke's Theorem. **Ans.**

EXERCISE 3.5

1. Use the Stoke's Theorem to evaluate $\int_C y^2 dx + xy\, dy + xz\, dz,$
where C is the bounding curve of the hemisphere $x^2 + y^2 + z^2 = 1$, $z \geq 0$, oriented in the positive direction. **Ans.** 0

2. Evaluate the integral for $\int_C y^2 dx + z^2 dy + x^2 dz,$ where C is the triangular closed path joining the points (0, 0, 0), (0, a, 0) and (0, 0, a) by transforming the integral to surface integral using Stoke's Theorem. **Ans.** $\frac{a^3}{3}$.

3. Verify Stoke's Theorem for $\vec{A} = 3y\hat{i} - xz\hat{j} + yz^2\hat{k},$ where S is the surface of the paraboloid $2z = x^2 + y^2$ bounded by $z = 2$ and c is its boundary traversed in the clockwise direction. **Ans.** -20π

4. Evaluate $\int_C \vec{F} \cdot d\vec{r}$ where $\vec{F} = y\hat{i} + xz^3\hat{j} - zy^3\hat{k},$ C is the circl $x^2 + y^2 = 4$, $z = 1.5$ **Ans.** $\frac{19}{2}\pi$

5. Verify Stoke's Theorem for the vector field
$\vec{F} = (2y + z)\,\hat{i} + (x - z)\,\hat{j} + (y - x)\,\hat{k}$
over the portion of the plane $x + y + z = 1$ cut off by the co-ordinate planes.

6. Evaluate $\int_c \vec{F} \cdot dr$ by Stoke's Theorem for $\vec{F} = yz\,\hat{i} + zx\,\hat{j} + xy\,k$ and C is the curve of intersection of $x^2 + y^2 = 1$ and $y = z^2$. **Ans.** 0

7. If $\vec{F} = (x - z)\,\hat{i} + (x^3 + yz)\,\hat{j} + 3xy^2\,\hat{k}$ and S is the surface of the cone $z = a - \sqrt{(x^2 + y^2)}$ above the xy-plane, show that $\iint_s \text{curl}\,\vec{F} \cdot dS = 3\pi a^4 / 4$.

8. If $\vec{F} = 3y\hat{i} - xy\hat{j} + yz2\hat{k}$ and S is the surface of the paraboloid $2z = x^2 + y^2$ bounded by $z = 2$, show by using Stoke's Theorem that $\iint_s (\nabla \times \vec{F}) \cdot dS = 20\pi$.

9. If $\vec{F} = (y^2 + z^2 - x^2)\,\hat{i} + (z^2 + x^2 - y^2)\,\hat{j} + (x^2 + y^2 - z^2)\,\hat{k},$ evaluate $\int \text{curl}\,\vec{F} \cdot \hat{n}\, ds$ integrated over the portion of the surface $x^2 + y^2 - 2ax + az = 0$ above the plane $z = 0$ and verify Stoke's Theorem; where $\hat{n}$ is unit vector normal to the surface. **Ans.** $2\pi a^3$

10. Evaluate by using Stoke's Theorem $\int_C [\sin z\, dx - \cos x\, dy + \sin y\, dz]$ where C is the boundary of rectangle $0 \leq x \leq \pi$, $0 \leq 1$, $y \leq 1$, $z = 3$. *(DU, I Sem. 2012, AMIETE, June 2010)* **Ans.** -2

3.8 GAUSS'S THEOREM OF DIVERGENCE

(Relation between surface integral and volume integral)

(GBTU, 2012 U.P., Ist Semester, Jan., 2011, Dec, 2006)

Statement. The surface integral of the normal component of a vector function F taken around a closed surface S is equal to the integral of the divergence of F taken over the volume V enclosed by the surface S.

Mathematically

$$\iint_S \vec{F} . \hat{n}\, ds = \iiint_V div\, \vec{F} dv$$

Proof. Let $\vec{F} = F_1\hat{i} + F_2\hat{j} + F_3\hat{k}$.

Putting the values of $\vec{F}, \hat{n}$ in the statement of the divergence theorem, we have

$$\iint_S F_1\,\hat{i} + F_2\,\hat{j} + F_3\,\hat{k}\cdot\hat{n}\,ds = \iiint_V \left(\hat{i}\frac{\partial}{\partial x} + \hat{j}\frac{\partial}{\partial y} + \hat{k}\frac{\partial}{\partial z}\right)(F_1\,\hat{i} + F_2\,\hat{j} + F_3\,\hat{k})\,dx\,dy\,dz.$$

$$= \iiint_V \left(\frac{\partial F_1}{\partial x} + \frac{\partial F_2}{\partial y} + \frac{\partial F_3}{\partial z}\right) dx\,dy\,dz \qquad ...(1)$$

We require to prove (1).

Let us first evaluate $\iiint_V \frac{\partial F_3}{\partial z}\,dx\,dy\,dz$.

$$\iiint_V \frac{\partial F_3}{\partial z}\,dx\,dy\,dz = \iint_R \left[\int_{z=f_1(x,y)}^{z=f_2(x,y)} \frac{\partial F_3}{\partial z}\,dz\right] dx\,dy$$

$$= \iint_R [F_3\,(x, y, f_2) - F_3\,(x, y, f_1)]\,dx\,dy \qquad ...(2)$$

For the upper part of the surface *i.e.* S_2, we have

$$dx\,dy = ds_2\cos r_2 = \hat{n}_2.\hat{k}\,ds_2$$

Again for the lower part of the surface *i.e.* S_1, we have,

$$dx\,dy = -\cos r_1\,ds_1 = \hat{n}_1.\hat{k}\,ds_1$$

$$\iint_R F_3\,(x, y, f_2)\,dx\,dy = \iint_{S_2} F_3\,\hat{n}_2\cdot\hat{k}\,ds_2$$

and $$\iint_R F_3\,(x, y, f_1)\,dx\,dy = -\iint_{S_1} F_3\,\hat{n}_1\cdot\hat{k}\,ds_1$$

Putting these values in (2), we have

$$\iiint_V \frac{\partial F_3}{\partial z}\,dv = \iint_{S_2} F_3\,\hat{n}_2\cdot\hat{k}\,ds_2 + \iint_{S_1} F_3\,\hat{n}_1\cdot\hat{k}\,ds_1 = \iint_S F_3\,\hat{n}\cdot\hat{k}\,ds \qquad ...(3)$$

Similarly, it can be shown that

$$\iiint_V \frac{\partial F_2}{\partial y}\,dv = \iint_S F_2\,\hat{n}\cdot\hat{j}\,ds \qquad ...(4)$$

$$\iiint_V \frac{\partial F_1}{\partial x}\,dv = \iint_S F_1\,\hat{n}\cdot\hat{i}\,ds \qquad ...(5)$$

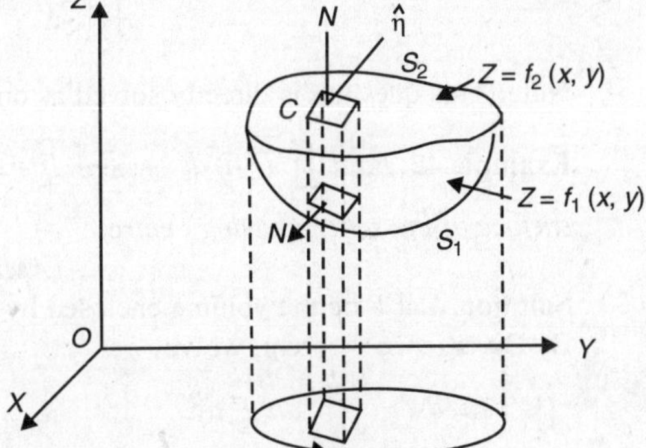

Adding (3), (4) & (5), we have

$$\iiint_V \left(\frac{\partial F_1}{\partial x} + \frac{\partial F_2}{\partial y} + \frac{\partial F_3}{\partial z}\right) dv$$

$$= \iint_S (F_1\,\hat{i} + F_2\,\hat{j} + F_3\,\hat{k})\cdot\hat{n}\cdot ds$$

$$\Rightarrow \iiint_V (\nabla\cdot\overline{F})\,dv = \iint_S \overline{F}\cdot\hat{n}\cdot ds$$ **Proved.**

Example 40. *State Gauss's Divergence theorem* $\iint_S \vec{F}.\hat{n}\,ds = \iiint Div\,\vec{F}\,dv$ *where S is the surface of the sphere* $x^2 + y^2 + z^2 = 16$ *and* $\vec{F} = 3x\,\hat{i} + 4y\,\hat{j} + 5z\,\hat{k}$.

Solution. Statement of Gauss's Divergence theorem is given in Art 3.8 on page 94.

Thus by Gauss's divergence theorem,

$$\iint_S \vec{F}.\hat{n}\,ds = \iiint_v \nabla.\vec{F}\,dv \qquad \text{Here } \vec{F} = 3x\,\hat{i} + 4y\,\hat{j} + 5z\,\hat{k}$$

$$\nabla.\vec{F} = \left(\hat{i}\frac{\partial}{\partial x} + \hat{j}\frac{\partial}{\partial y} + \hat{k}\frac{\partial}{\partial z}\right).(3x\hat{i} + 4y\hat{j} + 5z\hat{k})$$

$$\nabla . \vec{F} = 3 + 4 + 5 = 14$$

Putting the value of $\nabla . F$, we get

$$\iint_S \vec{F} . \hat{n}\, ds = \iiint_v 12 . dv = 12\, v \qquad \text{where } v \text{ is volume of a sphere}$$

$$= 12 \frac{4}{3} \pi (4)^3 = 1024\pi \qquad \textbf{Ans.}$$

Example 41. *Evaluate* $\iint_S \vec{F} . \hat{n}\, ds$ *where* $\vec{F} = 4xz\,\hat{i} - y^2\,\hat{j} + yz\,\hat{k}$ *and S is the surface of the cube bounded by x = 0, x = 1, y = 0, y = 1, z = 0, z = 1.* *(U.P., Ist Semester, 2009)*

Solution. By Divergence theorem,

$$\iint_S \vec{F} . \hat{n}\, ds = \iiint_v (\nabla . \vec{F})\, dv$$

$$= \iiint_v \left(\hat{i} \frac{\partial}{\partial x} + \hat{j} \frac{\partial}{\partial y} + \hat{k} \frac{\partial}{\partial z} \right) . (4xz\,\hat{i} - y^2\,\hat{j} + yz\,\hat{k})\, dv$$

$$= \iiint_v \left[\frac{\partial}{\partial x}(4xz) + \frac{\partial}{\partial y}(-y^2) + \frac{\partial}{\partial z}(yz) \right] dx\, dy\, dz$$

$$= \iiint_v (4z - 2y + y)\, dx\, dy\, dz$$

$$= \iiint_v (4z - y)\, dx\, dy\, dz = \int_0^1 \int_0^1 \left(\frac{4z^2}{2} - yz \right)_0^1 dx\, dy$$

$$= \int_0^1 \int_0^1 (2z^2 - yz)_0^1\, dx\, dy = \int_0^1 \int_0^1 (2 - y)\, dx\, dy$$

$$= \int_0^1 \left(2y - \frac{y^2}{2} \right)_0^1 dx = \frac{3}{2} \int_0^1 dx = \frac{3}{2} [x]_0^1 = \frac{3}{2}(1) = \frac{3}{2} \qquad \textbf{Ans.}$$

Note: This question is directly solved as on example 14 on Page 69.

Example 42. *Find* $\iint \vec{F} \cdot \hat{n} \cdot ds$, *where* $\vec{F} = (2x + 3z)\,\hat{i} - (xz + y)\,\hat{j} + (y^2 + 2z)\,\hat{k}$ *and S is the surface of the sphere having centre (3, – 1, 2) and radius 3.*

(AMIETE, Dec. 2010, U.P., I Semester, Winter 2005)

Solution. Let V be the volume enclosed by the surface S.
By Divergence theorem, we've

$$\iint_S \vec{F} \cdot \hat{n} \cdot ds = \iiint_V div\, \vec{F}\, dv.$$

Now, $$div\, \vec{F} = \frac{\partial}{\partial x}(2x + 3z) + \frac{\partial}{\partial y}[-(xz + y)] + \frac{\partial}{\partial z}(y^2 + 2z) = 2 - 1 + 2 = 3$$

$$\therefore \quad \iint_S \vec{F} \cdot \hat{n} \cdot ds = \iiint_V 3\, dv = 3 \iiint_V dv = 3V.$$

Again V is the volume of a sphere of radius 3. Therefore

$$V = \frac{4}{3} \pi r^3 = \frac{4}{3} \pi (3)^3 = 36\, \pi.$$

$$\therefore \quad \iint_S \vec{F} \cdot \hat{n} \cdot ds = 3V = 3 \times 36\, \pi = 108\, \pi \qquad \textbf{Ans.}$$

Example 43. *Use Divergence Theorem to evaluate* $\iint_S \vec{A}\cdot \vec{ds}$,

where $\vec{A} = x^3\hat{i} + y^3\hat{j} + z^3\hat{k}$ *and S is the surface of the sphere* $x^2 + y^2 + z^2 = a^2$.

(*AMIETE, Dec. 2009*)

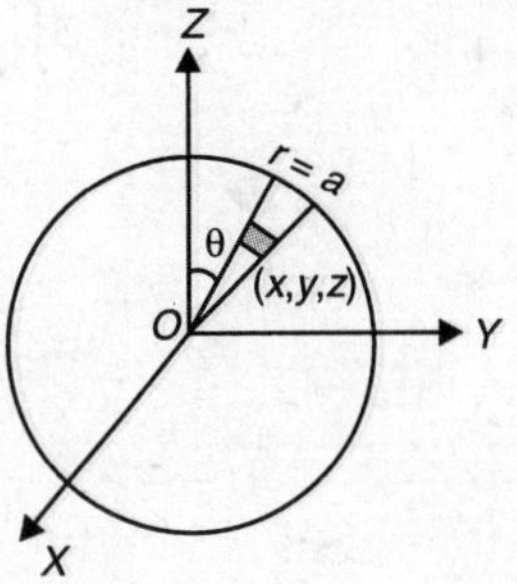

Solution. $\iint_S \vec{A}\cdot\vec{ds} = \iiint_v \text{div}\,\vec{A}\,dV$

$$= \iiint_v \left(\hat{i}\frac{\partial}{\partial x} + \hat{j}\frac{\partial}{\partial y} + \hat{k}\frac{\partial}{\partial z}\right)\cdot(x^3\hat{i} + y^3\hat{j} + z^3\hat{k})\,dV$$

$$= \iiint_v (3x^2 + 3y^2 + 3z^2)\,dV = 3\iiint_v (x^2 + y^2 + z^2)\,dV$$

On putting $x = r\sin\theta\cos\phi,\ y = r\sin\theta\sin\phi,\ z = r\cos\theta$, we get

$$= 3\iiint_V r^2\,(r^2\sin\theta\,dr\,d\theta\,d\phi) = 3\times 8\int_0^{\frac{\pi}{2}} d\phi\int_0^{\frac{\pi}{2}}\sin\theta\,d\theta\int_0^a r^4\,dr$$

$$= 24\,(\phi)_0^{\frac{\pi}{2}}\,(-\cos\theta)_0^{\frac{\pi}{2}}\left(\frac{r^5}{5}\right)_0^a = 24\left(\frac{\pi}{2}\right)(-0+1)\left(\frac{a^5}{5}\right) = \frac{12\pi a^5}{5}$$ **Ans.**

Example 44. *Use divergence Theorem to show that*

$$\iint_S \nabla(x^2 + y^2 + z^2)\,d\vec{s} = 6\,V$$

where S is any closed surface enclosing volume V.

Solution. Here $\nabla(x^2 + y^2 + z^2) = \left(\hat{i}\frac{\partial}{\partial x} + \hat{j}\frac{\partial}{\partial y} + \hat{k}\frac{\partial}{\partial z}\right)\cdot(x^2 + y^2 + z^2)$

$$= 2x\,\hat{i} + 2y\,\hat{j} + 2z\,\hat{k} = 2\,(x\,\hat{i} + y\,\hat{j} + z\,\hat{k})$$

$$\therefore \quad \iint_S \nabla(x^2 + y^2 + z^2)\cdot ds = \iint_S \nabla(x^2 + y^2 + z^2)\cdot\hat{n}\,ds$$

$\hat{n}$ being outward drawn unit normal vector to S

$$= \iint_S 2\,(x\hat{i} + y\,\hat{j} + z\hat{k})\cdot\hat{n}\,ds$$

$$= 2\iiint_V div\,(x\hat{i} + y\,\hat{j} + z\hat{k})\,dv \qquad ...(1)$$

(By Divergence Theorem)
(V being volume enclosed by S)

Now, div. $(x\,\hat{i} + y\,\hat{j} + z\hat{k}) = \left(\hat{i}\frac{\partial}{\partial x} + \hat{j}\frac{\partial}{\partial y} + \hat{k}\frac{\partial}{\partial z}\right)\cdot(x\hat{i} + y\,\hat{j} + z\,\hat{k})$

$$= \frac{\partial x}{\partial x} + \frac{\partial y}{\partial y} + \frac{\partial z}{\partial z} = 3 \qquad ...(2)$$

From (1) & (2), we have

$$\iint \nabla(x^2 + y^2 + z^2)\cdot dS = 2\iiint_V 3\,dv = 6\iiint_V dv = 6\,V$$ **Proved.**

Example 45. *Evaluate* $\iint_S (y^2z^2\hat{i} + z^2x^2\hat{j} + z^2y^2\hat{k})\,\hat{n}\,dS$, *where S is the part of the sphere* $x^2 + y^2 + z^2 = 1$ *above the xy-plane and bounded by this plane.*

Solution. Let V be the volume enclosed by the surface S. Then by divergence Theorem, we have

$$\iint_S (y^2z^2\hat{i} + z^2x^2\hat{j} + z^2y^2\hat{k})\,\hat{n}\,dS = \iiint_V div\,(y^2z^2\hat{i} + z^2x^2\hat{j} + z^2y^2\hat{k})\,dV$$

$$= \iiint_V \left[\frac{\partial}{\partial x}(y^2z^2) + \frac{\partial}{\partial y}(z^2x^2) + \frac{\partial}{\partial z}(z^2y^2)\right]dV = = \iint_V 2z\,y^2\,dV = 2\iint_V zy^2\,dV$$

Changing to spherical polar coordinates by putting

$$x = r\sin\theta\cos\phi, \quad y = r\sin\theta\sin\phi, \quad z = r\cos\theta,\ dV = r^2\sin\theta\, dr\, d\theta\, d\phi$$

To cover V, the limits of r will be 0 to 1, those of θ will be 0 to $\frac{\pi}{2}$ and those of ϕ will be 0 to 2π.

$$\therefore \quad 2\iiint_V zy^2\, dV = 2\int_0^{2\pi}\int_0^{\pi/2}\int_0^1 (r\cos\theta)(r^2\sin^2\theta\sin^2\phi)\, r^2\sin\theta\cdot dr\, d\theta\, d\phi$$

$$= 2\int_0^{2\pi}\int_0^{\pi/2}\int_0^1 r^5\sin^3\theta\cos\theta\sin^2\phi\, dr\, d\theta\, d\phi$$

$$= 2\int_0^{2\pi}\int_0^{\frac{\pi}{2}} \sin^3\theta\cos\theta\sin^2\phi\left[\frac{r^6}{6}\right]_0^1 d\theta\, d\phi$$

$$= \frac{2}{6}\int_0^{2\pi}\sin^2\phi\cdot\frac{2}{4.2}\, d\phi = \frac{1}{12}\int_0^{2\pi}\sin^2\phi\, d\phi = \frac{\pi}{12} \qquad \textbf{Ans.}$$

Example 46. *Use Divergence Theorem to evaluate* $\iint_S \vec{F}\cdot d\vec{S}$ *where* $\vec{F} = 4x\hat{i} - 2y^2\hat{j} + z^2\hat{k}$ *and S is the surface bounding the region* $x^2 + y^2 = 4$, $z = 0$ *and* $z = 3$.

Solution. By Divergence Theorem,

$$\iint_S \vec{F}\cdot dS = \iiint_V div\,\vec{F}\, dV$$

$$= \iiint_V \left(\hat{i}\frac{\partial}{\partial x} + \hat{j}\frac{\partial}{\partial y} + \hat{k}\frac{\partial}{\partial z}\right)\cdot(4x\hat{i} - 2y^2\hat{j} + z^2\hat{k})\, dV$$

$$= \iiint_V (4 - 4y + 2z)\, dx\, dy\, dz$$

$$= \iint dx\, dy\int_0^3 (4 - 4y + 2z)dz = \iint dx\, dy\,[4z - 4yz + z^2]_0^3$$

$$= \iint (12 - 12y + 9)\, dx\, dy = \iint (21 - 12y)\, dx\, dy$$

Let us put $x = r\cos\theta$, $y = r\sin\theta$

$$= \iint (21 - 12r\sin\theta)\, r\, d\theta\, dr = \int_0^{2\pi} d\theta\int_0^2 (21r - 12r^2\sin\theta)\, dr$$

$$= \int_0^{2\pi} d\theta\left[\frac{21r^2}{2} - 4r^3\sin\theta\right]_0^2 = \int_0^{2\pi} d\theta\,(42 - 32\sin\theta) = (42\theta + 32\cos\theta)_0^{2\pi}$$

$$= 84\pi + 32 - 32 = 84\pi \qquad \textbf{Ans.}$$

Example 47. *Apply the Divergence Theorem to compute* $\iint_s \vec{u}\cdot\hat{n}\, ds$, *where s is the surface of the cylinder* $x^2 + y^2 = a^2$ *bounded by the planes* $z = 0$, $z = b$ *and where* $u = \hat{i}x - \hat{j}y + \hat{k}z$.

Solution. By Gauss's Divergence Theorem

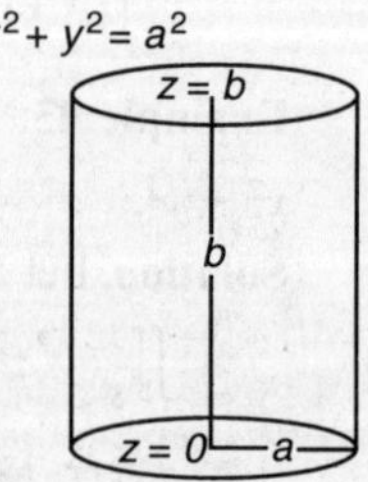

$$\iint \vec{u}\cdot\hat{n}ds = \iiint_V (\nabla\cdot\vec{u})\, dv$$

$$= \iiint_V \left(\hat{i}\frac{\partial}{\partial x} + \hat{j}\frac{\partial}{\partial y} + \hat{k}\frac{\partial}{\partial z}\right)\cdot(\hat{i}x - \hat{j}y + \hat{k}z)dv$$

$$= \iiint_V \left(\frac{\partial x}{\partial x} - \frac{\partial y}{\partial y} + \frac{\partial z}{\partial z}\right) dv = \iiint_V (1 - 1 + 1)\, dv$$

$$= \iiint_V dv = \iiint_V dx\, dy\, dz = \text{Volume of the cylinder} = \pi a^2 b \quad \textbf{Ans.}$$

Example 48. *Apply Divergence Theorem to evaluate* $\iint_S \vec{F}\,.\,\hat{n}\,ds$, *where* $\vec{F} = 4x^3\hat{\imath} - x^2 y\,\hat{\jmath} + x^2 z\hat{k}$ *and S is the surface of the cylinder* $x^2 + y^2 = a^2$ *bounded by the planes* $z = 0$ *and* $z = b$.

(U.P. Ist Semester, Dec. 2006)

Solution. We have,

$$\vec{F} = 4x^3\hat{\imath} - x^2 y\,\hat{\jmath} + x^2 z\hat{k}$$

$$\therefore \quad div\,\overline{F} = \left(\hat{\imath}\frac{\partial}{\partial x} + \hat{\jmath}\frac{\partial}{\partial y} + \hat{k}\frac{\partial}{\partial z}\right)\cdot(4x^3\hat{\imath} - x^2 y\hat{\jmath} + x^2 z\hat{k})$$

$$= \frac{\partial}{\partial x}(4x^3) + \frac{\partial}{\partial y}(-x^2 y) + \frac{\partial}{\partial z}(x^2 z) = 12x^2 - x^2 + x^2 = 12\,x^2$$

Now, $$\iiint_V div\,\vec{F}\,dV = 12\int_{x=-a}^{a}\int_{y=-\sqrt{a^2-x^2}}^{\sqrt{a^2-x^2}}\int_{z=0}^{b} x^2\,dz\,dy\,dx$$

$$= 12\int_{x=-a}^{a}\int_{y=-\sqrt{a^2-x^2}}^{\sqrt{a^2-x^2}} x^2\,(z)_0^b\,dy\,dx = 12\,b\int_{-a}^{a} x^2 (y)_{-\sqrt{a^2-x^2}}^{\sqrt{a^2-x^2}}\,dx$$

$$= 12\,b\int_{-a}^{a} x^2 .2\sqrt{a^2 - x^2}\,dx \qquad = 24\,b\int_{-a}^{a} x^2\sqrt{a^2 - x^2}\,dx$$

$$= 48\,b\int_0^a x^2\sqrt{a^2 - x^2}\,dx \qquad [\text{Put } x = a\sin\theta,\ dx = a\cos\theta\,d\theta]$$

$$= 48\,b\int_0^{\pi/2} a^2\sin^2\theta\,a\cos\theta\,a\cos\theta\,d\theta$$

$$= 48\,ba^4\int_0^{\pi/2}\sin^2\theta\cdot\cos^2\theta\,d\theta = 48\,ba^4\,\frac{\overline{\left|\frac{3}{2}\right.}\,\overline{\left|\frac{3}{2}\right.}}{2\overline{\left|3\right.}}$$

$$= 48\,ba^4\,\frac{\frac{1}{2}\sqrt{\pi}\cdot\frac{1}{2}\sqrt{\pi}}{2\cdot 2} = 3\,b\,a^4\pi$$ **Ans.**

Example 49. *Evaluate surface integral* $\iint_S \vec{F}\cdot\hat{n}\,ds$, *where* $\vec{F} = (x^2 + y^2 + z^2)\,(\hat{\imath} + \hat{\jmath} + \hat{k})$, *S is the surface of the tetrahedron* $x = 0, y = 0, z = 0, x + y + z = 2$ *and n is the unit normal in the outward direction to the closed surface S.*

Solution. By Divergence theorem

$$\iint_S \vec{F}\cdot\hat{n}\,ds = \iiint_V \text{div}\,\vec{F}\cdot dv$$

where S is the surface of tetrahedron $x = 0, y = 0, z = 0, x + y + z = 2$

$$= \iiint_V\left(\hat{\imath}\frac{\partial}{\partial x} + \hat{\jmath}\frac{\partial}{\partial y} + \hat{k}\frac{\partial}{\partial z}\right)\cdot(x^2 + y^2 + z^2)(\hat{\imath} + \hat{\jmath} + \hat{k})\,dv$$

$$= \iiint_V (2x + 2y + 2z)\,dv$$

$$= 2\iiint_V (x + y + z)\,dx\,dy\,dz$$

$$= 2\int_0^2 dx\int_0^{2-x} dy\int_0^{2-x-y}(x + y + z)\,dz$$

$$= 2\int_0^2 dx\int_0^{2-x} dy\left(xz + yz + \frac{z^2}{2}\right)_0^{2-x-y}$$

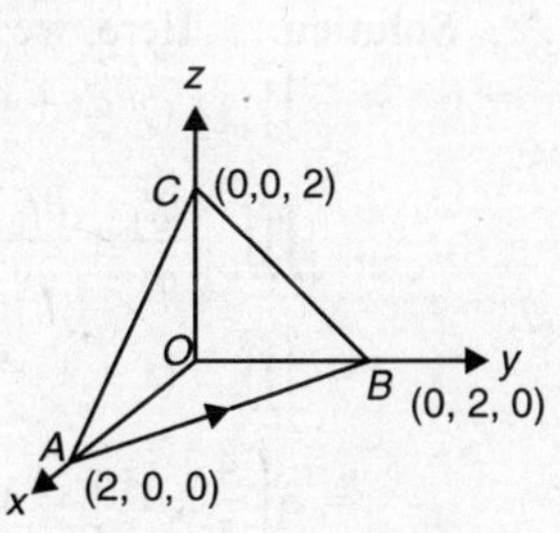

$$= 2\int_0^2 dx \int_0^{2-x} dy\left(2x - x^2 - xy + 2y - xy - y^2 + \frac{(2-x-y)^2}{2}\right]$$

$$= 2\int_0^2 dx\left[2xy - x^2y - xy^2 + y^2 - \frac{y^3}{3} - \frac{(2-x-y)^3}{6}\right]_0^{2-x}$$

$$= 2\int_0^2 dx\left[2x(2-x) - x^2(2-x) - x(2-x)^2 + (2-x)^2 - \frac{(2-x)^3}{3} + \frac{(2-x)^3}{6}\right]$$

$$= 2\int_0^2\left(4x - 2x^2 - 2x^2 + x^3 - 4x + 4x^2 - x^3 + (2-x)^2 - \frac{(2-x)^3}{3} + \frac{(2-x)^3}{6}\right] dx$$

$$= 2\left[2x^2 - \frac{4x^3}{3} + \frac{x^4}{4} - 2x^2 + \frac{4x^3}{3} - \frac{x^4}{4} - \frac{(2-x)^3}{3} + \frac{(2-x)^4}{12} - \frac{(2-x)^4}{24}\right]_0^2$$

$$= 2\left[-\frac{(2-x)^3}{3} + \frac{(2-x)^4}{12} - \frac{(2-x)^4}{24}\right]_0^2 = 2\left[\frac{8}{3} - \frac{16}{12} + \frac{16}{24}\right] = 4$$ **Ans.**

Example 50. *Use the Divergence Theorem to evaluate*

$$\iint_S (x\,dy\,dz + y\,dz\,dx + z\,dx\,dy)$$

where S is the portion of the plane x + 2 y + 3 z = 6 which lies in the first Octant.

Solution. $\iint_S (f_1\,dy\,dz + f_2\,dx\,dz + f_3\,dx\,dy)$

$$= \iiint_V\left(\frac{\partial f_1}{\partial x} + \frac{\partial f_2}{\partial y} + \frac{\partial f_3}{\partial z}\right) dx\,dy\,dz$$

where S is a closed surface bounding a volume V.

$\therefore$ $\iint_S (x\,dy\,dz + y\,dz\,dx + z\,dx\,dy)$

$$= \iiint_V\left[\frac{\partial x}{\partial x} + \frac{\partial y}{\partial y} + \frac{\partial z}{\partial z}\right] dx\,dy\,dz$$

$$= \iiint_V (1+1+1)\,dx\,dy\,dz = 3\iiint_V dx\,dy\,dz$$

= 3 (Volume of tetrahedron *OABC*)

$$= 3[(\tfrac{1}{3}\text{ Area of the base } \Delta\, OAB) \times \text{height } OC]$$

$$= 3\left[\frac{1}{3}\left(\frac{1}{2}\times 6\times 3\right)\times 2\right] = 18$$ **Ans.**

Example 51. *Use Divergence Theorem to evaluate :* $\iint (x\,dy\,dz + y\,dz\,dx + z\,dx\,dy)$ *over the surface of a sphere radius a.* (*K. University, Dec. 2009*)

Solution. Here, we have

$$\iint_S [x\,dy\,dz + y\,dx\,dz + z\,dx\,dy]$$

$$= \iiint_V\left(\frac{\partial f_1}{\partial x} + \frac{\partial f_2}{\partial y} + \frac{\partial f_3}{\partial z}\right) dx\,dy\,dz = \iiint_V\left(\frac{\partial x}{\partial x} + \frac{\partial y}{\partial y} + \frac{\partial z}{\partial z}\right) dx\,dy\,dz$$

$$= \iiint_V (1+1+1)\,dx\,dy\,dz = 3\text{ (volume of the sphere)}$$

$$= 3\left(\frac{4}{3}\pi a^3\right) = 4\pi a^3$$ **Ans.**

Example 52. *Using the divergence theorem, evaluate the surface integral* $\iint_S (yz\,dy\,dz + zx\,dz\,dx + xy\,dy\,dx)$ *where* $S : x^2 + y^2 + z^2 = 4.$

(AMIETE, Dec. 2010, UP, I Sem., Dec 2008)

Solution. $\iint_S (f_1\,dy\,dz + f_2\,dx\,dz + f_3\,dx\,dy)$

$$= \iiint_v \left(\frac{\partial f_1}{\partial x} + \frac{\partial f_2}{\partial y} + \frac{\partial f_3}{\partial z}\right) dx\,dy\,dz$$

where S is closed surface bounding a volume V.

$$\therefore \quad \iint_S (yz\,dy\,dz + zx\,dx\,dz + xy\,dx\,dy)$$

$$= \iiint_v \left(\frac{\partial (yz)}{\partial x} + \frac{\partial (zx)}{\partial y} + \frac{\partial (xy)}{\partial z}\right) dx\,dy\,dz = \iiint_v (0+0+0)\,dx\,dy\,dz = 0 \quad \textbf{Ans.}$$

Example 53. *Evaluate* $\iint_S xz^2\,dy\,dz + (x^2 y - z^3)\,dz\,dx + (2xy + y^2 z)\,dx\,dy$ *where S is the surface of hemispherical region bounded by*

$$z = \sqrt{a^2 - x^2 - y^2} \quad \textit{and } z = 0.$$

Solution. $\iint_S (f_1\,dy\,dz + f_2\,dz\,dx + f_3\,dx\,dy) = \iiint_V \left(\frac{\partial f_1}{\partial x} + \frac{\partial f_2}{\partial y} + \frac{\partial f_3}{\partial z}\right) dx\,dy\,dz$

where S is a closed surface bounding a volume V.

$$\therefore \quad \iint_S xz^2\,dy\,dz + (x^2 y - z^3)\,dz\,dx + (2xy + y^2 z)\,dx\,dy$$

$$= \iiint_V \left[\frac{\partial}{\partial x}(xz^2) + \frac{\partial}{\partial y}(x^2 y - z^3) + \frac{\partial}{\partial z}(2xy + y^2 z)\right] dx\,dy\,dz$$

(Here V is the volume of hemisphere)

$$= \iiint_V (z^2 + x^2 + y^2)\,dx\,dy\,dz$$

Let $x = r\sin\theta\cos\phi,\ y = r\sin\theta\sin\phi,\ z = r\cos\theta$

$$= \iiint r^2 (r^2 \sin\theta\,dr\,d\theta\,d\phi) = \int_0^{2\pi} d\phi \int_0^{\frac{\pi}{2}} \sin\theta\,d\theta \int_0^a r^4\,dr$$

$$= (\phi)_0^{2\pi} (-\cos\theta)_0^{\pi/2} \left(\frac{r^5}{5}\right)_0^a = 2\pi(-0+1)\frac{a^5}{5} = \frac{2\pi a^5}{5} \quad \textbf{Ans.}$$

Example 54. *Evaluate* $\iint_S \vec{F}\cdot\hat{n}\,ds$ *over the entire surface of the region above the xy-plane bounded by the cone* $z^2 = x^2 + y^2$ *and the plane* $z = 4$, *if* $F = 4xz\,\hat{i} + xyz^2\,\hat{j} + 3z\,\hat{k}$.

Solution. If V is the volume enclosed by S, then V is bounded by the surfaces $z = 0$, $z = 4$, $z^2 = x^2 + y^2$.

By divergence theorem, we have

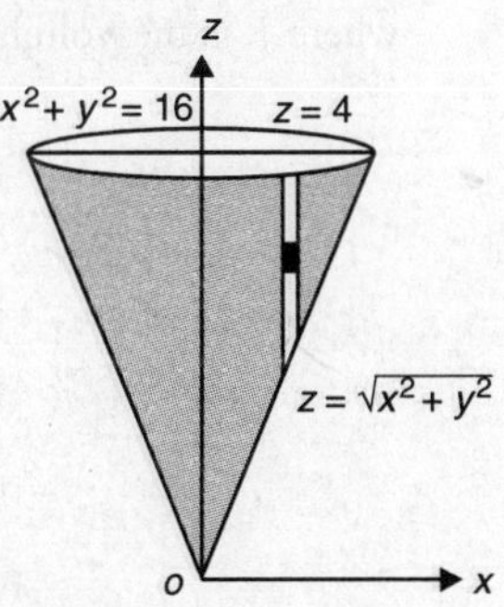

$$\iint_S \overline{F}\cdot\hat{n}\,ds = \iiint_V \operatorname{div}\vec{F}\,dx\,dy\,dz$$

$$= \iiint_V \left[\frac{\partial}{\partial x}(4xz) + \frac{\partial}{\partial y}(xyz^2) + \frac{\partial}{\partial z}(3z)\right] dx\,dy\,dz$$

$$= \iiint_V (4z + xz^2 + 3)\,dx\,dy\,dz$$

Limits of z are $\sqrt{x^2 + y^2}$ and 4.

$$\iiint_{\sqrt{x^2+y^2}}^{4} (4z + xz^2 + 3)\, dz\, dy\, dx = \iint \left[2z^2 + \frac{xz^3}{3} + 3z \right]_{\sqrt{x^2+y^2}}^{4} dy\, dx$$

$$= \iint \left[\left(32 + \frac{64x}{3} + 12 \right) - \{ 2(x^2+y^2) + \frac{x(x^2+y^2)^{3/2}}{3} + 3\sqrt{x^2+y^2} \} \right] dy\, dx$$

$$= \iint \left(44 + \frac{64x}{3} - 2(x^2+y^2) - \frac{x(x^2+y^2)^{3/2}}{3} - 3\sqrt{x^2+y^2} \right) dy\, dx$$

Putting $x = r\cos\theta$ and $y = r\sin\theta$, we have

$$= \iint \left(44 + \frac{64 r\cos\theta}{3} - 2r^2 - \frac{r\cos\theta\, r^3}{3} - 3r \right) r\, d\theta\, dr$$

Limits of r are 0 to 4.

and limits of θ are 0 to 2π.

$$= \int_0^{2\pi} \int_0^4 \left(44r + \frac{64 r^2 \cos\theta}{3} - 2r^3 - \frac{r^5 \cos\theta}{3} - 3r^2 \right) d\theta\, dr$$

$$= \int_0^{2\pi} \left[22r^2 + \frac{64 \times r^3 \cos\theta}{9} - \frac{r^4}{2} - \frac{r^6}{18}\cos\theta - r^3 \right]_0^4 d\theta$$

$$= \int_0^{2\pi} \left[22(4)^2 + \frac{64 \times (4)^3 \cos\theta}{9} - \frac{(4)^4}{2} - \frac{(4)^6}{18}\cos\theta - (4)^3 \right] d\theta$$

$$= \int_0^{2\pi} \left[352 + \frac{64 \times 64}{9}\cos\theta - 128 - \frac{2048}{9}\cos\theta - 64 \right] d\theta$$

$$= \int_0^{2\pi} \left[160 + \left(\frac{64 \times 64}{9} - \frac{2048}{9} \right) \cos\theta \right] d\theta$$

$$= \left[160\,\theta + \left(\frac{64 \times 64}{9} - \frac{2048}{9} \right) \sin\theta \right]_0^{2\pi} = 160\,(2\pi) + \left(\frac{64 \times 64}{9} - \frac{2048}{9} \right) \sin 2\pi$$

$$= 320\,\pi$$ **Ans.**

Example 55. *The vector field* $\vec{F} = x^2\hat{i} + z\hat{j} + yz\hat{k}$ *is defined over the volume of the cuboid given by* $0 \le x \le a$, $0 \le y \le b$, $0 \le z \le c$, *enclosing the surface S. Evaluate the surface integral*

$$\iint_S \vec{F} \cdot \vec{ds}$$

Solution. By Divergence Theorem, we have

$$\iint_S (x^2\hat{i} + z\hat{j} + yz\hat{k}) \cdot ds = \iiint_v div\,(x^2\hat{i} + z\hat{j} + yz\hat{k})\, dv,$$

where V is the volume of the cuboid enclosing the surface S.

$$= \iiint_v \left(\hat{i}\frac{\partial}{\partial x} + \hat{j}\frac{\partial}{\partial y} + \hat{k}\frac{\partial}{\partial z} \right) \cdot (x^2\hat{i} + z\hat{j} + yz\hat{k})\, dv$$

$$= \iiint_v \left\{ \frac{\partial}{\partial x}(x^2) + \frac{\partial}{\partial y}(z) + \frac{\partial}{\partial z}(yz) \right\} dx\, dy\, dz$$

$$= \int_{x=0}^{a} \int_{y=0}^{b} \int_{z=0}^{c} (2x + y)\, dx\, dy\, dz = \int_0^a dx \int_0^b dy \int_0^c (2x + y)\, dz$$

$$= \int_0^a dx \int_0^b [2xz + yz]_0^c\, dy = \int_0^a dx \int_0^b (2xc + yc)\, dy$$

$$= c\int_0^a dx \int_0^b (2x+y)\,dy = c\int_0^a \left[2xy+\frac{y^2}{2}\right]_0^b dx = c\int_0^a \left(2bx+\frac{b^2}{2}\right) dx$$

$$= c\left[\frac{2bx^2}{2}+\frac{b^2x}{2}\right]_0^a = c\left[a^2b+\frac{ab^2}{2}\right] = abc\left(a+\frac{b}{2}\right) \qquad \textbf{Ans.}$$

Example 56. *Verify the divergence Theorem for the function* $\overline{F} = 2x^2yi - y^2j + 4xz^2k$ *taken over the region in the first octant bounded by* $y^2+z^2=9$ *and* $x=2$.

Solution. $\iiint_V \nabla\cdot\vec{F}\,dV = \iiint\left(\hat{i}\frac{\partial}{\partial x}+\hat{j}\frac{\partial}{\partial y}+\hat{k}\frac{\partial}{\partial z}\right)\cdot(2x^2y\hat{i}-y^2\hat{j}+4xz^2\hat{k})\,dV$

$$= \iiint(4xy-2y+8xz)\,dx\,dy\,dz = \int_0^2 dx\int_0^3 dy\int_0^{\sqrt{9-y^2}}(4xy-2y+8xz)\,dz$$

$$= \int_0^2 dx\int_0^3 dy\,(4xyz-2yz+4xz^2)_0^{\sqrt{9-y^2}}$$

$$= \int_0^2 dx\int_0^3 [4xy\sqrt{9-y^2}-2y\sqrt{9-y^2}+4x(9-y^2)]\,dy$$

$$= \int_0^2 dx\left[-\frac{4x}{2}\frac{2}{3}(9-y^2)^{3/2}+\frac{2}{3}(9-y^2)^{3/2}+36xy-\frac{4xy^3}{3}\right]_0^3$$

$$= \int_0^2(0+0+108x-36x+36x-18)\,dx = \int_0^2(108x-18)\,dx = \left[108\frac{x^2}{2}-18x\right]_0^2$$

$$= 216-36 = 180 \qquad ...(1)$$

Here $\iint_S \vec{F}\cdot\hat{n}\,ds = \iint_{OABC}\vec{F}\cdot\hat{n}\,ds+\iint_{OCE}\vec{F}\cdot\hat{n}\,ds+\iint_{OADE}\vec{F}\cdot\hat{n}\,ds+\iint_{ABD}\vec{F}\cdot\hat{n}\,ds+\iint_{BDEC}\vec{F}\cdot\hat{n}\,ds$

$$\iint_{BDEC}\overline{F}\cdot\hat{n}\,ds = \iint_{BDEC}(2x^2y\hat{i}-y^2\hat{j}+4xz^2\hat{k})\,.\,\hat{n}\,ds$$

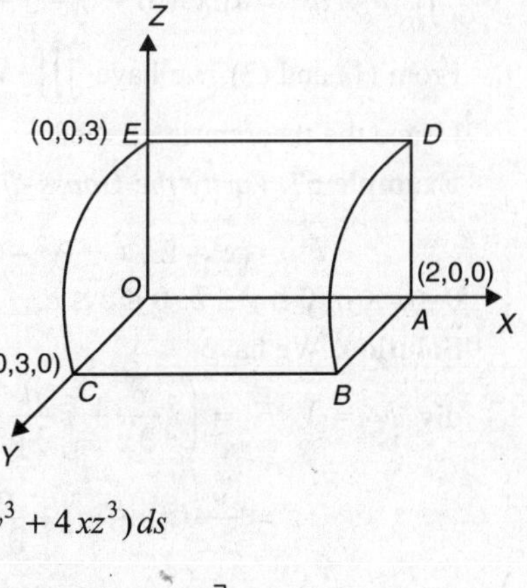

Z

(0,0,3) E D

O (2,0,0)

A X

(0,3,0) C B

Y

Normal vector

$$= \nabla\phi = \left(\hat{i}\frac{\partial}{\partial x}+\hat{j}\frac{\partial}{\partial y}+\hat{k}\frac{\partial}{\partial z}\right)(y^2+z^2-9)$$

$$= 2y\hat{j}+2z\hat{k}$$

Unit normal vector $= \hat{n} = \dfrac{2y\hat{j}+2z\hat{k}}{\sqrt{4y^2+4z^2}} = \dfrac{y\hat{j}+z\hat{k}}{\sqrt{y^2+z^2}}$

$$= \frac{y\hat{j}+z\hat{k}}{\sqrt{9}} = \frac{y\hat{j}+z\hat{k}}{3}$$

$$\iint_{BDEC}(2x^2y\hat{i}-y^2\hat{j}+4xz^2k)\cdot\frac{y\hat{j}+zk}{3}\,ds = \frac{1}{3}\iint_{BDEC}(-y^3+4xz^3)\,ds$$

$$\left[dx\,dy = ds\,(\hat{n}\cdot k) = ds\left(\frac{y\hat{j}+z\hat{k}}{3}\cdot\hat{k}\right) = ds\,\frac{z}{3}\text{ or } ds = \frac{dx\,dy}{\frac{z}{3}}\right]$$

$$= \frac{1}{3}\iint_{BDEC}(-y^3+4xz^3)\frac{dxdy}{\frac{z}{3}} = \int_0^2 dx\int_0^3\left(-\frac{y^3}{z}+4xz^2\right)dy \qquad \begin{pmatrix} y=3\sin\theta,\\ z=3\cos\theta \end{pmatrix}$$

$$= \int_0^2 dx\int_0^{\frac{\pi}{2}}\left[\frac{-27\sin^3\theta}{3\cos\theta}+4x(9\cos^2\theta)\right]$$

$$= \int_0^2 dx\left(-27\times\frac{2}{3}+108\,x\times\frac{2}{3}\right) = \int_0^2(-18+72\,x)\,dx$$

$$= \left[-18x+36x^2\right]_0^2 = 108 \qquad ...(2)$$

$$\iint_{OABC} \vec{F}\cdot\hat{n}\,ds = \iint_{OABC}(2x^2y\hat{i}-y^2\hat{j}+4xz^2\hat{k})\cdot(-\hat{k})\,ds$$

$$= \iint_{OABC} 4xz^2\,ds = 0 \qquad ...(3) \text{ because in } OABC\ xy\text{-plane, } z=0$$

$$\iint_{OADE} \vec{F}\cdot\hat{n}\,ds = \iint_{OADE}(2x^2y\hat{i}-y^2\hat{j}+4xz^2\hat{k})\cdot(-\hat{j})\,ds = \iint_{OADE} y^2\,ds = 0 \qquad ...(4)$$

because in *OADE* xz-plane, $y = 0$

$$\iint_{OCE} \vec{F}\cdot\hat{n}\,ds = \iint_{OCE}(2x^2y\hat{i}-y^2\hat{j}+4xz^2\hat{k})\cdot(-\hat{i})\,ds = \iint_{OCE} -2x^2y\,ds = 0 \qquad ...(5)$$

because in *OCE* yz-plane, $x = 0$

$$\iint_{ABD} \vec{F}\cdot\hat{n}\,ds = \iint_{ABD}(2x^2y\hat{i}-y^2\hat{j}+4xz^2\hat{k})\cdot(\hat{i})\,ds = \iint_{ABD} 2x^2y\,ds$$

$$= \iint 2x^2y\,dy\,dz = \int_0^3 dz\int_0^{\sqrt{9-z^2}} 2\,(2)^2\,y\,dy \qquad \text{because in } ABD \text{ plane, } x=2$$

$$= 8\int_0^3 dz\left[\frac{y^2}{2}\right]_0^{\sqrt{9-z^2}} = 4\int_0^3 dz\,(9-z^2) = 4\left[9z-\frac{z^3}{3}\right]_0^3 = 4\,[27-9] = 72 \qquad ...(6)$$

On adding (2), (3), (4), (5) and (6), we get

$$\iint_S \vec{F}\cdot\hat{n}\,ds = 108+0+0+0+72 = 180 \qquad ...(7)$$

From (1) and (7), we have $\iiint_V \nabla\cdot\vec{F}\,dV = \iint_S \vec{F}\cdot\hat{n}\,ds$

Hence the theorem is verified.

Example 57. *Verify the Gauss divergence Theorem for*

$\vec{F} = (x^2-yz)\,\hat{i} + (y^2-zx)\,\hat{j} + (z^2-xy)\,\hat{k}$ *taken over the rectangular parallelopiped* $0 \le x \le a,\ 0 \le y \le b,\ 0 \le z \le c$.

Solution. We have

$$\text{div } \vec{F} = \nabla\cdot\vec{F} = \left(\hat{i}\frac{\partial}{\partial x}+\hat{j}\frac{\partial}{\partial y}+\hat{k}\frac{\partial}{\partial z}\right)\cdot[(x^2-yz)\,\hat{i}+(y^2-zx)\,\hat{j}+(z^2-xy)\,\hat{k}]$$

$$= \frac{\partial}{\partial x}(x^2-yz)+\frac{\partial}{\partial y}(y^2-zx)+\frac{\partial}{\partial z}(z^2-xy) = 2x+2y+2z$$

$$\therefore \text{Volume integral} = \iiint_V \nabla\cdot\vec{F}\,dV = \iiint_V 2\,(x+y+z)\,dV$$

$$= 2\int_{x=0}^{a}\int_{y=0}^{b}\int_{z=0}^{c}(x+y+z)\,dx\,dy\,dz = 2\int_0^a dx\int_0^b dy\int_0^c (x+y+z)\,dz$$

$$= 2\int_0^a dx\int_0^b dy\left(xz+yz+\frac{z^2}{2}\right)_0^c = 2\int_0^a dx\int_0^b dy\left(cx+cy+\frac{c^2}{2}\right)$$

$$= 2\int_0^a dx\left(c\,xy+c\,\frac{y^2}{2}+\frac{c^2y}{2}\right)_0^b = 2\int_0^a dx\left(bcx+\frac{b^2c}{2}+\frac{bc^2}{2}\right)$$

$$= 2\left[\frac{bc\,x^2}{2} + \frac{b^2\,cx}{2} + \frac{bc^2 x}{2}\right]_0^a = [a^2bc + ab^2c + abc^2]$$

$$= abc\,(a + b + c) \qquad \text{...(A)}$$

To evaluate $\iint_S \vec{F}\cdot\hat{n}\,ds,$ where S consists of six plane surfaces.

$$\iint_S \vec{F}\cdot\hat{n}\,ds = \iint_{OABC} \vec{F}\cdot\hat{n}\,ds + \iint_{DEFG} \vec{F}\cdot\hat{n}\,ds + \iint_{OAFG} \vec{F}\cdot\hat{n}\,ds$$

$$+ \iint_{BCDE} \vec{F}\cdot\hat{n}\,ds + \iint_{ABEF} \vec{F}\cdot\hat{n}\,ds + \iint_{OCDG} \vec{F}\cdot\hat{n}\,ds$$

$$\iint_{OABC} \vec{F}\cdot\hat{n}\,ds = \iint_{OABC} \{(x^2 - yz)\hat{i} + (y^2 - xz)\,\hat{j} + (z^2 - xy)\hat{k}\}\,(-\hat{k})\,dx\,dy$$

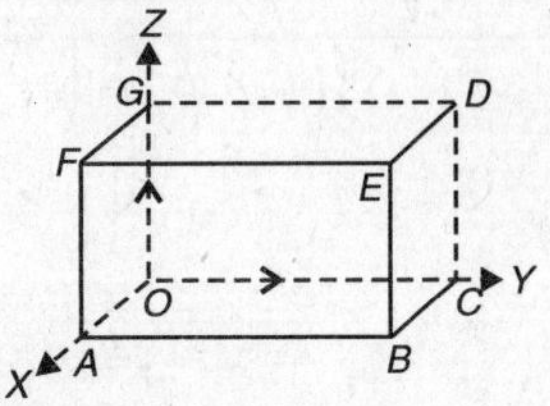

$$= -\int_0^a\int_0^b (z^2 - xy)\,dx\,dy$$

$$= -\int_0^a\int_0^b (0 - xy)\,dx\,dy = \frac{a^2\,b^2}{4} \qquad \text{...(1)}$$

$$\iint_{DEFG} \vec{F}\cdot\hat{n}\,ds = \iint_{DEFG} \{(x^2 - yz)\hat{i} + (y^2 - xz)\,\hat{j} + (z^2 - xy)\hat{k}\}\,(\hat{k})\,dx\,dy$$

$$= \int_0^a\int_0^b (z^2 - xy)\,dx\,dy = \int_0^a\int_0^b (c^2 - xy)\,dx\,dy$$

$$= \int_0^a \left[c^2 y - \frac{xy^2}{2}\right]_0^b dx = \int_0^a \left(c^2 b - \frac{x\,b^2}{2}\right) dx$$

$$= \left[c^2 bx - \frac{x^2\,b^2}{4}\right]_0^a = abc^2 - \frac{a^2\,b^2}{4} \qquad \text{...(2)}$$

S. No.	Surface	Outward Normal	ds	
1	OABC	$-k$	$dx\,dy$	$z = 0$
2	DEFG	k	$dx\,dy$	$z = c$
3	OAFG	$-j$	$dx\,dz$	$y = 0$
4	BCDE	j	$dx\,dz$	$y = b$
5	ABEF	i	$dy\,dz$	$x = a$
6	OCDG	$-i$	$dy\,dz$	$x = 0$

$$\iint_{OAFG} \vec{F}\cdot\hat{n}\,ds = \iint_{OAFG} \{(x^2 - yz)\hat{i} + (y^2 - zx)\,\hat{j} + (z^2 - xy)\,\hat{k}\}\,(-\hat{j})\,dx\,dz$$

$$= -\iint_{OAFG} (y^2 - zx)\,dx\,dz$$

$$= -\int_0^a dx \int_0^c (0 - zx)\,dz = \int_0^a dx\left(\frac{x\,z^2}{2}\right)_0^c = \int_0^a \frac{x\,c^2}{2}\,dx = \left[\frac{x^2\,c^2}{4}\right]_0^a = \frac{a^2\,c^2}{4} \qquad \text{...(3)}$$

$$\iint_{BCDE} \vec{F}\cdot\hat{n}\,ds = \iint \{(x^2 - y\,z)\hat{i} + (y^2 - zx)\,\hat{j} + (z^2 - xy)\hat{k}\}\cdot\hat{j}\,dx\,dz = \iint_{BCDE} (y^2 - xz)\,dx\,dz$$

$$= -\int_0^a dx \int_0^c (b^2 - x\,z)\,dz = \int_0^a \left(b^2\,z - \frac{x\,z^2}{2}\right)_0^c dx = \int_0^a \left(b^2\,c - \frac{x\,c^2}{2}\right) dx$$

$$= \left[b^2\,c\,x - \frac{x^2\,c^2}{4}\right]_0^a = a\,b^2\,c - \frac{a^2 c^2}{4} \qquad \text{...(4)}$$

$$\iint_{ABEF} \vec{F}\cdot\hat{n}\,ds = \iint_{ABEF} \{(x^2 - y\,z)\hat{i} + (y^2 - xz)\,\hat{j} + (z^2 - x\,y)\hat{k}\}\cdot\hat{i}\,dy\,dz$$

$$= \iint_{ABEF} (x^2 - y\,z)\,dy\,dz = \int_0^b dy \int_0^c (a^2 - y\,z)\,d\,z = \int_0^b dy\left(a^2 z - \frac{y\,z^2}{2}\right)_0^c$$

$$= \int_0^b \left(a^2 c - \frac{y c^2}{2} \right) dy = \left[a^2 c y - \frac{y^2 c^2}{4} \right]_0^b = a^2 b c - \frac{b^2 c^2}{4} \qquad ...(5)$$

$$\iint_{OCDG} \vec{F} \cdot \hat{n}\, ds = \iint_{OCDG} \{(x^2 - yz)\hat{i} + (y^2 - zx)\hat{j} + (z^2 - xy)\hat{k}\} \cdot (-\hat{i})\, dy\, dz$$

$$= \int_0^b \int_0^c (x^2 - yz)\, dy\, dz = -\int_0^b dy \int_0^c (-yz)\, dz = -\int_0^b dy \left[\frac{-y z^2}{2} \right]_0^c$$

$$= \int_0^b \frac{y c^2}{2} dy = \left[\frac{y^2 c^2}{4} \right]_0^b = \frac{b^2 c^2}{4} \qquad ...(6)$$

Adding (1), (2), (3), (4), (5) and (6), we get

$$\iint \vec{F} \cdot \hat{n}\, ds = \left(\frac{a^2 b^2}{4} \right) + \left(abc^2 - \frac{a^2 b^2}{4} \right) + \left(\frac{a^2 c^2}{4} \right) + \left(a b^2 c - \frac{a^2 c^2}{4} \right) + \left(\frac{b^2 c^2}{4} \right) + \left(a^2 b c - \frac{b^2 c^2}{4} \right)$$

$$= abc^2 + ab^2c + a^2bc$$

$$= abc\,(a + b + c) \qquad ...(B)$$

From (A) and (B), Gauss divergence Theorem is verified. **Verified.**

Example 58. *Verify Divergence Theorem, given that* $\vec{F} = 4xz\hat{i} - y^2\hat{j} + yz\,\hat{k}$ *and S is the surface of the cube bounded by the planes* $x = 0, x = 1, y = 0, y = 1, z = 0, z = 1$.

Solution. $\nabla \cdot \vec{F} = \left(\hat{i} \frac{\partial}{\partial x} + \hat{j} \frac{\partial}{\partial y} + \hat{k} \frac{\partial}{\partial z} \right) \cdot (4\, zx\hat{i} - y^2\hat{j} + yz\hat{k})$

$$= 4z - 2y + y$$

$$= 4z - y$$

Volume Integral $= \iiint \nabla \cdot \vec{F}\, dv$

$$= \iiint (4z - y)\, dx\, dy\, dz$$

$$= \int_0^1 dx \int_0^1 dy \int_0^1 (4z - y)\, dz$$

$$= \int_0^1 dx \int_0^1 dy\, (2z^2 - yz)_0^1 = \int_0^1 dx \int_0^1 dy\, (2 - y)$$

$$= \int_0^1 dx \left(2y - \frac{y^2}{2} \right)_0^1 = \int_0^1 dx \left(2 - \frac{1}{2} \right) = \frac{3}{2} \int_0^1 dx = \frac{3}{2}(x)_0^1 = \frac{3}{2} \qquad ...(1)$$

To evaluate $\iint_S \vec{F} \cdot \hat{n}\, ds$, where S consists of six plane surfaces.

(*i*) Over the face *OABC*, $z = 0$, $dz = 0$, $\hat{n} = -\hat{k}$, $ds = dx\, dy$

$$\iint \vec{F} \cdot \hat{n}\, ds = \int_0^1 \int_0^1 (-y^2\hat{j}) \cdot (-\hat{k})\, dx\, dy = 0$$

(*ii*) Over the face *BCDE*, $y = 1$, $dy = 0$

$$\hat{n} = -\hat{j},\ ds = dx\, dz$$

$$\iint \vec{F} \cdot \hat{n}\, ds = \int_0^1 \int_0^1 (4\, xz\hat{i} - \hat{j} + z\hat{k}) \cdot (\hat{j})\, dx\, dz$$

$$= \int_0^1 \int_0^1 -\, dx dz$$

S. No.				$\hat{n}$	ds
1	OABC	$z = 0$	$dz = 0$	$-\hat{k}$	$dx\, dy$
2	BCDE	$y = 1$	$dy = 0$	$\hat{j}$	$dx\, dz$
3	DEFG	$z = 1$	$dz = 0$	$\hat{k}$	dx dy
4	OCDG	$x = 0$	$dx = 0$	$-\hat{i}$	$dy\, dz$
5	AOGF	$y = 0$	$dy = 0$	$-\hat{j}$	$dx\, dz$
6	ABEF	$x = 1$	$dx = 0$	$\hat{i}$	$dy\, dz$

$$= -\int_0^1 dx \int_0^1 dz = -(x)_0^1 (z)_0^1 = -(1)(1) = -1$$

(*iii*) Over the face *DEFG*, $z = 1$, $dz = 0$, $\hat{n} = \hat{k}$, $ds = dx\, dy$

$$\iint \vec{F}\cdot\hat{n}\, ds = \int_0^1\int_0^1 [4x(1) - y^2\hat{j} + y(1)\hat{k}]\cdot(\hat{k})\, dx\, dy$$

$$= \int_0^1\int_0^1 y\, dx\, dy = \int_0^1 dx \int_0^1 y\, dy = (x)_0^1 \left(\frac{y^2}{2}\right)_0^1 = \frac{1}{2}$$

(*iv*) Over the face *OCDG*, $x = 0$, $dx = 0$, $\hat{n} = -\hat{i}$, $ds = dy\, dz$

$$\iint \vec{F}\cdot\hat{n}\, ds = \int_0^1\int_0^1 (0\hat{i} - y^2\hat{j} + yz\hat{k})\cdot(-\hat{i})\, dy\, dz = 0$$

(*v*) Over the face *AOGF*, $y = 0$, $dy = 0$, $\hat{n} = -\hat{j}$, $ds = dx\, dz$

$$\iint \vec{F}\cdot\hat{n}\, ds = \int_0^1\int_0^1 (4xz\hat{i})\cdot(-\hat{j})\, dx\, dz = 0$$

(*vi*) Over the face *ABEF*, $x = 1$, $dx = 0$, $\hat{n} = \hat{i}$, $ds = dy\, dz$

$$\iint \vec{F}\cdot\hat{n}\, ds = \int_0^1\int_0^1 [(4z\hat{i} - y^2\hat{j} + yz\hat{k})\cdot(\hat{i})]\, dy\, dz = \int_0^1\int_0^1 4z\, dy\, dz$$

$$= \int_0^1 dy \int_0^1 4z\, dz = \int_0^1 dy\, (2z^2)_0^1 = 2\int_0^1 dy = 2(y)_0^1 = 2$$

On adding we see that over the whole surface

$$\iint \vec{F}\cdot\hat{n}\, ds = \left(0 - 1 + \frac{1}{2} + 0 + 0 + 2\right) = \frac{3}{2} \quad ...(2)$$

From (1) and (2), we have $\iiint_V \nabla\cdot\vec{F}\, dv = \iint_S \vec{F}\cdot\hat{n}\, ds$ **Verified.**

Example 60. *Evaluate integral*:

$$I = \oiint_S (x^3 dy\, dz + x^2 y\, dz\, dx + x^2 z\, dx\, dy)$$ (*Delhi University, April 2010*)

Where s is the surface bounded by $z = 0$, $z = 4$ *and* $x^2 + y^2 = 9$.

Solution. We have,

$$I = \oiint_S (x^3 dy\, dz + x^2 y\, dz\, dx + x^2 z\, dx\, dy)$$

Where S is the surface bounded by $z = 0$, $z = 4$ and $x^2 + y^2 = 9$

$$I = \oiint_S (x^3\hat{i} + x^2 y\hat{j} + x^2 z\hat{k}).\, d\vec{s}$$

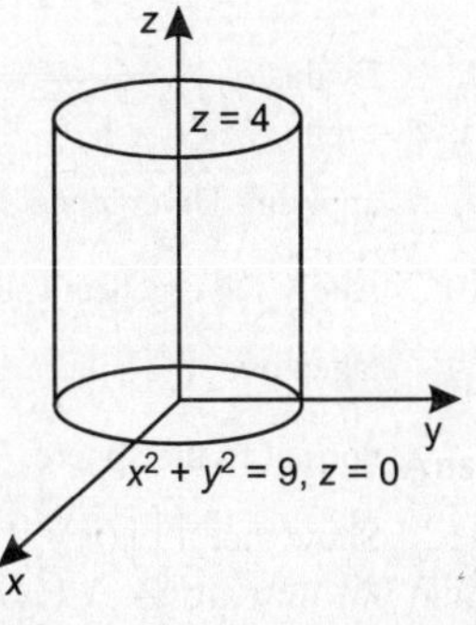

Apply Gauss's divergence theorem

$$\oiint_S \vec{A}.\, d\vec{s} = \oiiint_V (\nabla.\vec{A})\, dv$$

$$\therefore \quad \vec{A} = x^3\hat{i} + x^2 y\hat{j} + x^2 z\hat{k}$$

$$\nabla.\vec{A} = 3x^2 + x^2 + x^2 = 5x^2$$

$$\therefore \quad \oiint_S \vec{A}.\, d\vec{S} = \oiiint_V (\nabla.\vec{A})\, dV = \oiiint_V 5x^2\, dV$$

Convert into cylinder co-ordinate system.

$$x = r\cos\theta,\; dV = r\, dr d\theta dz$$

$$\iiint_V 5x^2\, dV = 5\int_0^3\int_0^{\pi}\int_0^4 r^2\cos^2\theta\, r dr\, d\theta\, dz$$

$$= 5\int_0^3 r^3 dr \int_0^{\pi}\cos^2\theta\, d\theta \int_0^4 dz = 5\left[\frac{r^4}{4}\right]_0^3\left(\frac{\pi}{2}\right)(4) = \frac{405\pi}{2}$$ **Ans.**

EXERCISE 3.6

1. Use Divergence Theorem to evaluate $\iint_s (y^2z^2\hat{i} + z^2x^2\hat{j} + x^2y^2\hat{k}).\overline{ds}$,

   where S is the upper part of the sphere $x^2 + y^2 + z^2 = 9$ above xy- plane. **Ans.** $\frac{243\pi}{8}$

2. Evaluate $\iint_S (\nabla \times \vec{F}) . ds$, where S is the surface of the paraboloid $x^2 + y^2 + z = 4$ above the xy-plane and

   $\vec{F} = (x^2 + y - 4)\,\hat{i} + 3\,xy\hat{j} + (2\,xz + z^2)\,\hat{k}$. **Ans.** -4π

3. Evaluate $\iint_s [xz^2\, dy\, dz + (x^2y - z^3)\, dz\, dx + (2xy + y^2z)\, dx\, dy]$, where S is the surface enclosing a region

   bounded by hemisphere $x^2 + y^2 + z^2 = 4$ above XY-plane. **Ans.** $\frac{64\pi}{5}$

4. Verify Divergence Theorem for $\vec{F} = x^2\hat{i} + z\hat{j} + yz\hat{k}$, taken over the cube bounded by **Ans.** $\frac{3}{2}$

   $x = 0, x = 1, y = 0, y = 1, z = 0, z = 1$.

5. Evaluate $\iint_S (2xy\,\hat{i} + yz^2\,\hat{j} + xz\,\hat{k})\cdot\vec{ds}$ over the surface of the region bounded by

   $x = 0, y = 0, y = 3, z = 0$ and $x + 2\,z = 6$ **Ans.** $\frac{351}{2}$

6. Verify Divergence Theorem for $\vec{F} = (x + y^2)\,\hat{i} - 2\,x\hat{j} + 2\,yz\hat{k}$ and the volume of a tetrahedron bounded by co-ordinate planes and the plane $2\,x + y + 2\,z = 6$.

7. Verify Divergence Theorem for the function $\vec{F} = y\hat{i} + x\hat{j} + z^2\,\hat{k}$ over the region bounded by

   $x^2 + y^2 = 9, \quad z = 0$ and $z = 2$.

8. Evaluate the integral $\iint_s (z^2 - x)\, dy\, dz - xy\, dx\, dz + 3z\, dx\, dy$, where S is the surface of closed region bounded by $z = 4 - y^2$ and planes $x = 0$, $x = 3$, $z = 0$ by transforming it with the help of Divergence Theorem to a triple integral. **Ans.** 16

9. Evaluate $\iint_s \frac{ds}{\sqrt{a^2x^2 + b^2y^2 + c^2z^2}}$ over the closed surface of the ellipsoid $ax^2 + by^2 + cz^2 = 1$ by

   applying Divergence Theorem. **Ans.** $\frac{4\pi}{\sqrt{(a\,b\,c)}}$

10. Apply Divergence Theorem to evaluate $\iint (l\,x^2 + m\,y^2 + n\,z^2)\, ds$

    taken over the sphere $(x - a)^2 + (y - b)^2 + (z - c)^2 = r^2$, l, m, n being the direction cosines of the external normal to the sphere. *(AMIETE June 2010, 2009)* **Ans.** $\frac{8\pi}{3}(a + b + c)\, r^3$

11. Show that $\iiint_V (u\,\nabla\cdot\vec{V} + \nabla u\cdot\vec{V})\, dv = \iint_s u\vec{V}\cdot ds$.

12. If $E = \text{grad}\,\phi$ and $D^2\phi = 4\,\pi\,\rho$, prove that $\iint_S \vec{E}\cdot\vec{n}\, ds = -4\,\pi\iiint_V \rho\, dv$

    where $\vec{n}$ is the outward unit normal vector, while dS and dV are respectively surface and volume elements.

Pick up the correct option from the following:

13. If $\vec{F}$ is the velocity of a fluid particle then $\int_C \vec{F}.\vec{dr}$ represents.

(*a*) Work done (*b*) Circulation (*c*) Flux (*d*) Conservative field.

(*U.P. Ist Semester, Dec 2009*) **Ans.** (*c*)

14. If $\vec{f} = ax\,\vec{i} + by\,\vec{j} + cz\,\vec{k}$, *a, b, c,* constants, then $\iint \vec{f}.\vec{dS}$ where *S* is the surface of a unit sphere is

(*a*) $\frac{\pi}{3}(a+b+c)$ (*b*) $\frac{4}{3}\pi(a+b+c)$ (*c*) $2\pi\,(a+b+c)$ (*d*) $\pi\,(a+b+c)$

(*U.P., Ist Semester, 2009*) **Ans.** (*b*)

15. A force field $\vec{F}$ is said to be conservative if

(*a*) Curl $\vec{F} = 0$ (*b*) grad $\overline{F} = 0$ (*c*) Div $\overline{F} = 0$ (*d*) Curl (grad $\overline{F}$) = 0

(*AMIETE, Dec. 2006*) **Ans.** (*a*)

16. The line integral $\int_c x^2\,dx + y^2\,dy$, where *C* is the boundary of the region $x^2 + y^2 < a^2$ equals

(*a*) 0, (*b*) *a* (*c*) $\pi\,a^2$ (*d*) $\frac{1}{2}\pi\,a^2$

(*AMIETE, Dec. 2006*) **Ans.** (*b*)

3.9 DEDUCTIONS FROM GAUSS DIVERGENCE THEOREM

Gauss's Law. Gauss's law of electrostatics states that the total electric flux through a closed surface is equal to $\frac{1}{\epsilon_0}$ times the total charge enclosed by the surface.

In mathematical form it may be expressed as

$$\iiint_V E.\,dS = \begin{cases} \frac{q}{\epsilon_0}, & \text{if charge } q \text{ lies inside the closed surface} \\ 0, & \text{if charge } q \text{ lies outside the closed surface} \end{cases}$$

where ϵ_0 is the absolute permitivity of free space.

Proof. Let a point charge *q* be placed at the origin. The electric field strength due to point charge at any point having position vector $\vec{r}$ is given by

$$E = \frac{1}{4\pi\epsilon_0}\,\frac{q\,\vec{r}}{r^3} \qquad \ldots (1)$$

Case (*a*): When origin *O* lies outside the closed surface *S*:

We know Gauss divergence theorem,

$$\iint_S E\,.\,ds = \iiint_V \text{div } E\,dV = \frac{q}{4\pi\epsilon_0} = \iiint_V \text{div}\left(\frac{\vec{r}}{r^3}\right) dV \qquad \text{From (1)}$$

$$= \frac{q}{4\pi\,\epsilon_0} = \iiint_V \nabla.\left(\frac{\vec{r}}{r^3}\right) dV = \frac{q}{4\pi\,\epsilon_0}\iiint_V \left[\left(\nabla.\frac{1}{r^3}\right)\vec{r} + \frac{1}{r^3}\nabla.\vec{r}\right] dV$$

$$= \frac{q}{4\pi\,\epsilon_0}\iiint_V \left[-\frac{3}{r^5}\,\vec{r}.\vec{r} + \frac{3}{r^3}\right] dV = 0$$

$$(\because \text{div } r^m = mr^{m-2}\,\vec{r},\ \text{div } \vec{r} = 3,\ \vec{r}\,.\,\vec{r} = r^2)$$

Hence, the total electric flux $\left(\iint_S E\,.\,dS\right)$ through closed surface is zero when the surface does not include the origin i.e., charge where the integrands are not defined. This is the second part of Gauss's law.

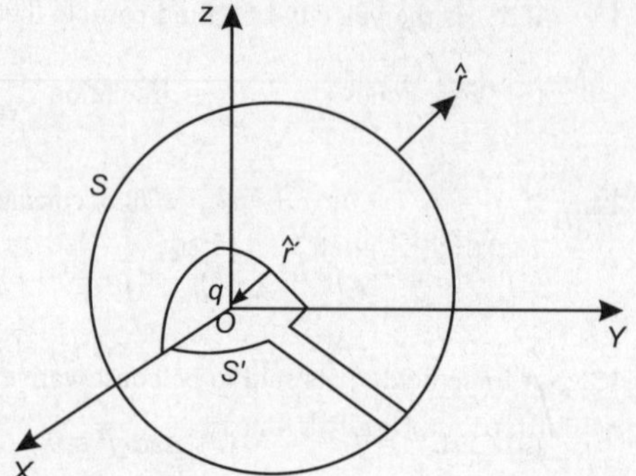

Case (*b*): When origin O lies inside the closed surface S.

We surround the origin with a small sphere S' of radius a. Imagine the volume outside the outer surface S and inside the surface S' connected by a small hole. This joins surface S and S', combining them into one single simply connected closed surface. As the radius of the imaginary hole may be made vanishingly small, there is no additional contributes to the surface integral. Gauss divergence theorem is applicable between S and S'.

$$\iint_{S+S'} E \cdot dS = \iint_{S'} E \cdot dS + \iint_{S'} E \cdot dS' = \iiint_V \text{div } E\, dV$$

$$\text{i.e.} \quad \iint_S E.dS + \iint_{S'} E \cdot dS'$$

On putting the value of E from (1), we get

$$\frac{q}{4\pi\epsilon_0}\iiint_V \nabla \cdot \left(\frac{\vec{r}}{r^3}\right) dV = 0 \text{ since } r \neq 0 \text{ between } S \text{ and } S'.$$

$$\text{Thus,} \quad \iint_S E \cdot dS = - - \iint_{S'} E \cdot dS'$$

On putting the value of E from (1), we get

$$= -\frac{q}{4\pi\epsilon_0}\iint_{S'}\frac{\vec{r} \cdot dS'}{r^3} \qquad \text{... (3)}$$

The integral on R.H.S. is evaluated as follows:

On surface S', $|r| = a$, $dS' = \hat{r}'\ dS'$ and $\dfrac{dS'}{a^2} = d\Omega$ is an element of solid angle subtended by the surface dS' at O and $\hat{r}'$ is the unit vector along the outward drawn normal on surface dS'. Obviously, the outward drawn normal $\hat{r}'$ is in the negative radial direction, $\hat{r} = -\hat{r}$ i.e. $\hat{r} \cdot \hat{r}' = -1$.

$$\Rightarrow \quad \iint_S \frac{\vec{r} \cdot dS'}{r} = \iint_{S'} \frac{a\hat{r} \cdot (\hat{r}')\, dS'}{a^3} = \iint_{S'} \frac{a\hat{r} \cdot r'\,(a^2\, d\Omega)}{a^3}. = \iint_{S'} d\Omega = 4\pi$$

Putting the value of this integral in (3), we get $\iiint_S E \cdot dS = -\dfrac{q}{4\pi\epsilon_0}(-4\pi) = \dfrac{q}{\epsilon_0}$ ... (4)

So the first part is proved.

Hence the Gauss's law is proved.

2. Gauss's law in differential form. Let volume V enclosed by surface S in electric field of varying charge density ρ, then by Gauss's law of electrostatics

$$\iint_S E.dS = \frac{1}{\epsilon_0} \times \text{charge enclosed within surface } S = \frac{1}{\epsilon_0}\iiint_V \rho dV \qquad \text{... (5)}$$

On applying Gauss divergence theorem to electric field E, we get

$$\iint_S E \cdot dS = \iiint_V \text{div } E\, dv \qquad \text{... (6)}$$

Comparing equations (5) and (6), we get

$$\iiint_V \text{div } E\, dV = \frac{1}{\epsilon_0} \iiint_V \rho dV$$

or $$\iiint_V \left(\text{div } E - \frac{\rho}{\epsilon_0} \right) dV = 0$$

As the volume, over which integration takes place, is arbitrary, therefore, we must have

$$\text{div } E - \frac{\rho}{\epsilon_0} = 0$$

i.e. $$\text{div } E = \frac{\rho}{\epsilon_0} \qquad \text{... (7)}$$

This is the different form of Gauss's law.

3. Poisson's equation and Laplace's equation:

If E and ϕ are the electric field strength and electric potential respectively, then

$$E = -\text{ grad } \phi$$

Replacing E by – grad ϕ in equation (7), we get

$$\text{div grad } \phi = -\frac{\rho}{\epsilon_0}$$

$$\Rightarrow \qquad \nabla . \nabla \phi = -\frac{\rho}{\epsilon_0} \qquad \text{... (8)}$$

The operator div grad $= \nabla . \nabla = \nabla^2 \left(\frac{\partial^2}{\partial x^2} + \frac{\partial^2}{\partial y^2} + \frac{\partial^2}{\partial z^2} \right)$ is called the Laplacian. (8) is written as

$$\nabla^2 \phi = -\frac{\rho}{\epsilon_0} \qquad (9)$$

This is Poisson's equation.

In the charge free region, volume charge density $\rho = 0$; so that equation (9) becomes

$$\nabla^2 \phi = 0 \qquad \text{... (10)}$$

This is called the Laplace's equation.

3.10 HELMHOLTZ THEOREM

Statement. Any general vector V (for which div $V \neq 0$ and curl $V \neq 0$) may be expressed as the sum of two parts, one of which is irrotational and the other is solenoidal.

Proof. As div $V \neq 0$ and curl $V \neq 0$, let

$$\text{div } V = \nabla . V = s \text{ and curl } V = \nabla \times V = c \qquad \text{... (1)}$$

where s and c represent charge and current densities respectively, both vanishing at infinity.

To show that the given vector V is written as

$$V = -\text{ grad } \phi + \text{curl } A = -\nabla\phi + \nabla \times A \qquad \text{... (2)}$$

$\nabla\phi$ being *irrotational* and $\nabla \times A$ being **solenoidal.**

If r_1 i.e., (x_1, y_1, z_1) and r_2 i.e., (x_2, y_2, z_2), indicate the source and field points respectively, then

$$r_{12} = | r_2 - r_1 | = [(x_2 - x_1)^2 + (y_2 - y_1)^2 + (z_2 - z_1)^2]^{1/2} \qquad \text{... (3)}$$

In particular ϕ and A corresponding to *scalar* and *vector potentials* respectively are given by

$$\phi(\vec{r_2}) = \frac{1}{4\pi}\iiint \frac{s(\vec{r_1})}{r_{12}}\, dv_1 \qquad \text{... (4)}$$

$$A(\vec{r_2}) = \frac{1}{4\pi}\iiint \frac{c(\vec{r_1})}{r_{12}}\, dv_1 \qquad \text{... (5)}$$

where dv_1 is an element of volume around source point $\vec{r_1}$. Of course s (charge density) and c (current density) must vanish sufficiently rapidly at large distances.

Equation (2) represents V in the desired form, resolved into irrorational and solenoidal parts. With the scalar and vector potentials ϕ and V given by equations (4) and (5) if we show that V satisfies the conditions given by equation (1), then the validity of equation (2) will be proved thus verifying Helmholtz theorem.

By the divergence of V

$$\text{div} = \text{div}(-\text{grad}\,\phi + \text{curl}\,A) = -\,\text{div grad}\,\phi + \text{div curl}\,A$$

$$= -\nabla.\nabla\phi \qquad [\because \text{div curl A} = 0]$$

Thus $\text{div } V = -\nabla^2\phi = -\nabla^2 \frac{1}{4\pi}\iiint \frac{s(r_1)}{r_{12}}\, dv_1$ ... (6) [From (4)]

The Laplacian operator ∇^2 operates on the field coordinates (x_2, y_2, z_2) and so commutes with the integration with respect to (x_1, y_1, z_1); so that we have

$$\text{div } V = \frac{1}{-4\pi}\iiint s(r_1)\,\nabla_2{}^2\left(\frac{1}{r_{12}}\right) dv_1 \qquad \text{... (7)}$$

In Art. 3.9, we have shown in the development of Gauss's law that

$$\iiint \nabla.\left(\frac{\vec{r}}{r^3}\right) dv = \iiint \nabla^2\left(\frac{1}{r}\right) dv = \begin{cases} 0 \\ 4\pi \end{cases} \qquad \text{... (8)}$$

0 if the volume did not include the origin and 4π if the volume included the origin ($r = 0$). In terms of Dirac delta function this result is expressed as

$$\nabla^2\left(\frac{1}{r}\right) = -\,4\pi\,\delta(r) \qquad \text{... (9a)}$$

Here the source has shifted from origin to the position $\vec{r} = \vec{r_1}$ and $r_{12} = |\vec{r_2} - \vec{r_1}|$, we have

$$\nabla^2\left(\frac{1}{r_{12}}\right) = 4\pi\,\delta(\vec{r_2} - \vec{r_1}) \qquad \text{... (9b)}$$

The Dirac delta function has its assigned properties.

$$\left.\begin{aligned} &\delta(\vec{r_2} - \vec{r_1}) = 0,\ \vec{r_1} \neq \vec{r_2} \\ &\iiint f(\vec{r_1})\,\delta(\vec{r_1} - \vec{r_2})\, d\,v_1 = f(\vec{r_2}) \end{aligned}\right\} \qquad \begin{aligned} &\text{...(10a)} \\ &\text{...(10b)} \end{aligned}$$

Since the differentiation of $r_{12}{}^{-1}$ twice with respect to (x_1, y_1, z_1) is the same as the differentiation twice with respect to (x_2, y_2, z_2); we have

$$\nabla_1{}^2\left(\frac{1}{r_{12}}\right) = \nabla_2{}^2\left(\frac{1}{r_{12}}\right) = -\,4\pi\,\delta(\vec{r_1} - \vec{r_2}) = -\,4\pi\,\delta(\vec{r_2} - \vec{r_1}) \qquad \text{... (11)}$$

Using equation (11), equation (7) may be rewritten as

$$\text{div } V = -\frac{1}{4\pi}.\iiint s(\vec{r_1})\,\nabla_1^{\,2}\left(\frac{1}{r_{12}}\right)d\,v_1$$

$$= \frac{1}{4\pi}.\iiint s(\vec{r_1})(-4\pi)\,\delta(\vec{r_1}-\vec{r_2})\,dv_1 = s(\vec{r_2}) \qquad \text{... (12)} \quad \text{[Using (11)]}$$

This equation shows that the assumed form of V and that of scalar potential f are it complete agreement with the given divergence [eqn. (1)]; thus first condition is proved.

To prove the second condition [*i.e.*, curl $V = c\,(\vec{r_2})$, let us take the curl of equation (2), i.e.

$$\text{Curl V} = -\text{ curl grad }\phi + \text{curl curl A}$$

$$= \text{curl curl } A \qquad \text{(since curl grad } \phi = 0)$$

$$= \text{grad div } A - \nabla^2 A \qquad \text{... (13)}$$

From (5) the first term grad div A leads to

$$\text{grad div A} = \nabla\nabla . A = \iiint c\,(\vec{r_1}).\nabla_2\,\nabla_2\left(\frac{1}{r_{12}}\right)dv_1 \qquad \text{... (14)}$$

Again replacing the second derivatives with respect to (x_2, y_2, z_2) by second derivatives of (x_1, y_1, z_1), we integrate each component of (14) by parts.

$$\nabla\nabla . A\big|_x = \iiint c\,(r_1).\nabla_1\frac{\partial}{\partial x_1}\left(\frac{1}{r_{12}}\right)dv_1$$

$$= \iiint \nabla_1 . c\,(\vec{r_1})\frac{\partial}{\partial x_1}\left(\frac{1}{r_{12}}\right)dv_1 - \iiint\left[\nabla_1 . c\,(\vec{r_1})\right]\frac{\partial}{\partial x_1}\left(\frac{1}{r_{12}}\right)dv_1 \qquad \text{... (15)}$$

The second integral vanishes because the circulation density c is solenoidal. The first integral may be transformed to a surface integral by Gauss divergence theorem. If c is bounded in space or vanishes faster than r^{-1} for large r, so that the integral in (15) exists, then be choosing a sufficiently large surface, the first integral on the R.H.S. of equation (15) also vansihes.

$$\text{div } A = \nabla . A = 0, \text{ equation (13) now reduce to}$$

$$\text{curl } V = \nabla^2 A$$

$$= \iiint c\,(\vec{r_1})\,\nabla_2^{\,2}\left(\frac{1}{r_{12}}\right)dv_1 \qquad \text{... (16)}$$

This is exactly like equation (7) except that the scalar s $(\vec{r_1})$ is replaced by the vector circulation density $c\,(\vec{r_1})$. Introducing the Dirac delta function, as before, to carry out the integration it is seen that

$$\text{curl } V = c\,(\vec{r_2}).$$

This is same as second condition of equation (1).

Thus our assumed from of V and that of vector potential A given by equation (5) are agreement with equation (1) specifying the curl of V.

This completes the proof of **Helmholtz' theorem.**

ORTHOGONAL CURVILINEAR COORDINATES

4.1 CURVILINEAR COORDINATES

Let the rectangular cartesian coordinates of a point P in space be (x, y, z) . Now we introduce one more system of coordinates.

Let $x = X(u_1, u_2, u_3), y = Y(u_1, u_2, u_3), z = Z(u_1, u_2, u_3),$ ... (1)

On solving (1), we get, the values of u, v, w in terms of x, y, z.

$u = U(x, y, z), v = V(x, y, z), w = W(x, y, z)$ ... (2)

Then (u, v, w) are called curvilinear coordinates of a point (x, y, z).

The surfaces $u = c_1, v = c_2, w = c_3$ are called **co-ordinate surfaces**. When it is taken in pairs these co-ordinate surfaces intersect each other in curves called co-ordinate curves.

(*i*) u-curve is given by $v = c_2$, $w = c_3$

(*ii*) v-curve is given by $w = c_3, u = c_1$

(*iii*) w-curve is given by $u = c_1, v = c_2$

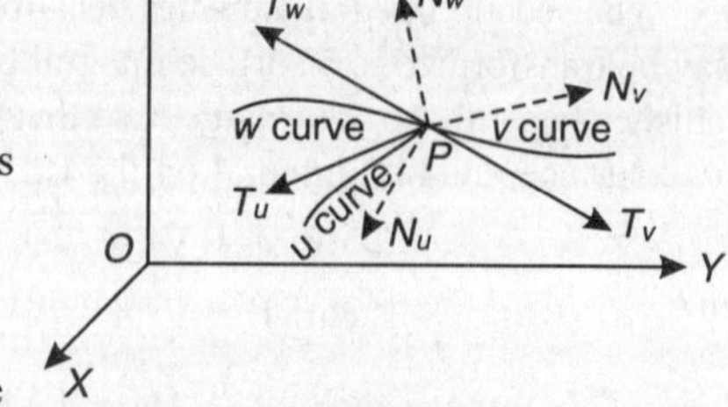

The coordinate axes are determined by the tangents PT_1, PT_2, PT_3 to the co-ordinate curves at the point P. The directions of these co-ordinates axes depend on the chosen point P of space.

If at every point P (x, y, z) , the coordinate axes are mutually perpendicular, then u, v, w are called orthogonal curvilinear coordinates of P.

Let *Tu, Tv, Tw* form a right-handed system of *unit vectors* tangents to the coordinate curves at P. Then, we have

$$\left.\begin{aligned} &\overline{T}_u.\overline{T}_u = \overline{T}_v.\overline{T}_v = \overline{T}_w.\overline{T}_w = 1 \\ &\overline{T}_u.\overline{T}_v = \overline{T}_v.\overline{T}_w = \overline{T}_w.\overline{T}_u = 0 \\ &\overline{T}_u \times \overline{T}_u = \overline{T}_v \times \overline{T}_v = \overline{T}_w \times \overline{T}_w = 0 \\ &\overline{T}_u \times \overline{T}_v = \overline{T}_w, \overline{T}_v \times \overline{T}_w = \overline{T}_u, \ \overline{T}_w \times \overline{T}_u = \overline{T}_v \end{aligned}\right\} \quad ...(3)$$

We now define three numbers h_1, h_2, h_3 known as scalar factors or material coefficients as follows.

$$h_1 = \left|\frac{\partial \overline{r}}{\partial u}\right|, \ h_2 = \left|\frac{\partial \overline{r}}{\partial v}\right|, \ h_3 = \left|\frac{\partial \overline{r}}{\partial w}\right| \quad ...(4)$$

The position vector $\overline{r}$ is given by

$$\bar{r} = x\hat{i} + y\hat{j} + z\hat{k}$$
$$= X(u, v, w)\ \hat{i}\ + Y(u, v, w)\ \hat{j}\ + Z(u, v, w)\ \hat{k}$$
$$= \bar{r}\ (u, v, w)$$

The tangent vector to u-curve at P is $\frac{\partial \bar{r}}{\partial u}$. Since, T_u is the unit vector in this direction,we have

$$\frac{\partial \bar{r}}{\partial u} = \left|\frac{\partial \bar{r}}{\partial u}\right| \bar{T}_u = h_1 \bar{T}_u \qquad \text{[Using (1)]}$$

$$\frac{d\bar{r}}{du} = h_1 \bar{T}_u,\ \frac{d\bar{r}}{dv} = h_2 \bar{T}_v,\ \frac{d\bar{r}}{dw} = h_3 \bar{T}_w \qquad \text{... (5)}$$

4.2 DIFFERENTIAL OF AN ARC LENGTH

$$(ds)^2 = h_1\ (du)^2 + h_2\ (dv)^2 + h_3\ (dw)^2$$

Proof. $$d\bar{r} = \frac{\partial \bar{r}}{\partial u} du + \frac{\partial \bar{r}}{\partial v} dv + \frac{\partial \bar{r}}{\partial w} dw$$

From (5), we have $$d\bar{r} = h_1 du \bar{T}_u + h_2 dv \bar{T}_v + h_3 dw \bar{T}_w \qquad \text{... (1)}$$

Then the differential of an arc length, ds, is given by

$$(ds)^2 = d\bar{r}.d\bar{r}$$
$$= (h_1 du \bar{T}_u + h_2 dv \bar{T}_v + h_3 dw \bar{T}_w).(h_1 du \bar{T}_u + h_2 dv \bar{T}_v + h_3 dw \bar{T}_w)$$
$$= h_1^2 (du)^2 + h_2^2 (dv)^2 + h_3^2 (dw)^2 \qquad \text{... (2)}$$

which is known as quadratic differential form.

4.3 GEOMETRICAL SIGNIFICANCE OF h_1, h_2, h_3

Let the element of arc ds is directed along u–curve so that $dv = dw = 0$. Then differential length of arc ds along u–curve is given by

$$ds_1 = h_1\ du$$

Similarly, $$ds_2 = h_2\ dv$$
$$ds_3 = h_3\ dw$$

Consider an infinitesimal parallelopiped with vertex at P.

Lengths of edges of parallelopiped are $h_1\ du,\ h_2\ dv,\ h_3\ dw$

Area of the faces of the parallelopiped are $h_1\ h_2\ du\ dv,\ h_2\ h_3\ dv\ dw,\ h_3\ h_1\ dw\ du$.

Volume of the parallelopiped is $h_1\ h_2\ h_3\ du\ dv\ dw$.

4.4 DIFFERENTIAL OPERATOR

Let $\phi(x, y, z)$ be continuously differentiable scalar point function of u, v, w

Gradient ϕ *(DU, I Sem. 2012)*

$$\textbf{grad}\ \phi = \nabla\phi = \frac{T_u}{h_1}\frac{\partial f}{\partial u} + \frac{T_v}{h_2}\frac{\partial f}{\partial v} + \frac{T_u}{h_3}\frac{\partial f}{\partial w}$$

Proof. grad $\phi = \nabla\phi = k_1 \bar{T}_u + k_2 \bar{T}_v + k_3 \bar{T}_w$ ... (1)

where $\bar{T}_u, \bar{T}_v, \bar{T}_w$ are unit vectors along curvilinear co-ordinates axes, k_1, k_2, k_3 are scalars.

We know that

$$d\overline{r} = h_1 du\, \overline{T}_u + h_2\, dv\, \overline{T}_v + h_3\, dw\, \overline{T}_w \quad \text{.... (2)}$$

$$d\phi = \nabla\phi . d\overline{r}$$

$$= k_1 h_1 du + k_2 h_2 dv + k_3 h_3 dw \quad \text{... (3)}$$

But $$d\phi = \frac{\partial\phi}{\partial u} du + \frac{\partial\phi}{\partial v} dv + \frac{\partial\phi}{\partial w} dw \quad \text{.... (4)}$$

Comparing (3) and (4), we have

$$k_1 h_1 = \frac{\partial\phi}{\partial u}, \quad k_2 h_2 = \frac{\partial\phi}{\partial v}, \quad k_3 h_3 = \frac{\partial\phi}{\partial w}$$

$$k_1 = \frac{1}{h_1}\frac{\partial\phi}{\partial u}, \quad k_2 = \frac{1}{h_2}\frac{\partial\phi}{\partial v}, \quad k_3 = \frac{1}{h_3}\frac{\partial\phi}{\partial w}$$

From (1), grad $\phi = \nabla\phi$

$$= \frac{1}{h_1}\frac{\partial\phi}{\partial u}\overline{T}_u + \frac{1}{h_2}\frac{\partial\phi}{\partial v}\overline{T}_v + \frac{1}{h_3}\frac{\partial\phi}{\partial w}\overline{T}_w \quad \text{.... (5)}$$

Again $$\text{grad}\,\phi = \left(\frac{\overline{T}_u}{h_1}\frac{\partial}{\partial u} + \frac{\overline{T}_v}{h_2}\frac{\partial}{\partial v} + \frac{\overline{T}_w}{h_3}\frac{\partial}{\partial w}\right)\phi$$

$\therefore$ $$\nabla = \frac{\overline{T}_u}{h_1}\frac{\partial}{\partial u} + \frac{\overline{T}_v}{h_2}\frac{\partial}{\partial v} + \frac{\overline{T}_w}{h_3}\frac{\partial}{\partial w} \quad \text{... (6)}$$

Note: From (6)

$$\nabla u = \frac{\overline{T}_u}{h_1}\frac{\partial u}{\partial u} + \frac{\overline{T}_v}{h_2}\frac{\partial u}{\partial v} + \frac{\overline{T}_w}{h_3}\frac{\partial u}{\partial w}$$

or $$\nabla u = \frac{\overline{T}_u}{h_1}, \quad \nabla v = \frac{\overline{T}_v}{h_2}, \quad \nabla w = \frac{\overline{T}_w}{h_3} \quad \text{... (7)} \quad \text{(others are equal to zero)}$$

where $\nabla u, \nabla v, \nabla w$ are normal to the surfaces $u = c_1$, $v = c_2$ and $w = c_3$ respectively.

From (5) we know that

$$\nabla\overline{\phi} = \frac{\partial\phi}{\partial u}\frac{\overline{T}_u}{h_1} + \frac{\partial\phi}{\partial v}\frac{\overline{T}_v}{h_2} + \frac{\partial\phi}{\partial w}\frac{\overline{T}_w}{h_3}$$

$$\nabla\overline{\phi} = \frac{\partial\phi}{\partial u}\nabla u + \frac{\partial\phi}{\partial v}\nabla v + \frac{\partial\phi}{\partial w}\nabla w \quad \text{...(8)}$$

4.5 DIVERGENCE

(Delhi University, April 2010)

Let $\phi(u, v, w)$ be a vector point function of orthogonal curvilinear co-ordinates u, v, w.

Divergence $$\overline{f} = \nabla.\overline{f} = \frac{1}{h_1 h_2 h_3}\left[\frac{\partial}{\partial u}(h_2 h_3 f_1) + \frac{\partial}{\partial v}(h_3 h_1 f_2) + \frac{\partial}{\partial w}(h_1 h_2 f_3)\right]$$

Proof. Let $$\overline{f} = f_1\overline{T}_u + f_2\overline{T}_v + f_3\overline{T}_w$$

$$= f_1\overline{T}_v \times \overline{T}_w + f_2\overline{T}_w \times \overline{T}_u + f_3\overline{T}_u \times \overline{T}_v \quad \text{[From (3) of Art. 4.1]}$$

$$= f_1 h_2 h_3 \nabla v \times \nabla w + f_2 h_3 h_1 \nabla w \times \nabla u + f_3 h_1 h_2 \nabla u \times \nabla v$$

Divergence $$f = \nabla.\overline{f} = \nabla[f_1 h_2 h_3 \nabla v \times \nabla w + f_2 h_3 h_1 \nabla w \times \nabla u + f_3 h_1 h_2 \nabla u \times \nabla v]$$

$$= \nabla.(f_1 h_2 h_3 \nabla v \times \nabla w) + \nabla.[f_2 h_3 h_1 \nabla w \times \nabla u] + \nabla.[f_3 h_1 h_2 \nabla u \times \nabla v]$$

$$\nabla.(f_1 h_2 h_3 \nabla v \times \nabla w) = f_1 h_2 h_3 \nabla.(\nabla v \times \nabla w) + (\nabla v \times \nabla w).\nabla(f_1 h_2 h_3)$$

$$= f_1h_2h_3[\text{curl } \nabla v\cdot\nabla w - \text{curl}\,\nabla w\cdot\nabla v] + [\nabla v\times\nabla w]\cdot\nabla(f_1h_2h_3)$$

$$= f_1h_2h_3[\text{curl grad } v\cdot\nabla w - \textit{curl grad } w\cdot\nabla v] + [\nabla v\times\nabla w]\cdot\nabla(f_1h_2h_3)$$

(curl grad $\phi = 0$)

$$= (\nabla v\times\nabla w)\cdot\nabla(f_1\,h_2\,h_3)$$

$$= (\nabla v\times\nabla w)\cdot\left[\frac{\partial}{\partial u}(f_1h_2h_3)\nabla u_1 + \frac{\partial}{\partial v}(f_1h_2h_3)\nabla v + \frac{\partial}{\partial w}(f_1h_2h_3)\nabla u_3\right]$$

$$= (\nabla v\times\nabla w)\cdot\nabla u\frac{\partial}{\partial u}(f_1h_2h_3), \qquad \begin{bmatrix}\nabla v\times\nabla w\cdot\nabla v = 0\\ \nabla v\times\nabla w\cdot\nabla w = 0\end{bmatrix}$$

$$= \frac{1}{h_1h_2h_3}(T_v\times T_w)\cdot T_u\frac{\partial}{\partial u}(f_1\,h_2\,h_3)$$

$$= \frac{1}{h_1h_2h_3}\frac{\partial}{\partial u}(f_1\,h_2\,h_3) \qquad \{(T_v\times T_w)\cdot T_u = T_u\cdot T_u = 1\}$$

By symmetry

$$\nabla\cdot(f_2\,h_3\,h_1\,\nabla w\times\nabla u) = \frac{1}{h_1\,h_2\,h_3}\frac{\partial}{\partial v}(f_2\,h_3\,h_1)$$

and $$\nabla\cdot(f_3\,h_1\,h_2\,\nabla u\times\nabla v) = \frac{1}{h_1\,h_2\,h_3}\frac{\partial}{\partial w}(f_3\,h_1\,h_2)$$

On substituting the values

$$\text{Div}.\overline{f} = \nabla\cdot\overline{f} = \frac{1}{h_1\,h_2\,h_3}\left[\frac{\partial}{\partial u}(f_1\,h_2\,h_3) + \frac{\partial}{\partial v}(f_2\,h_3\,h_1) + \frac{\partial}{\partial w}(f_3\,h_1\,h_2)\right]$$

4.6 CURL

$$\text{curl } \overline{f} = \nabla\times\overline{f} = \begin{vmatrix}\dfrac{\overline{T}_u}{h_2h_3} & \dfrac{\overline{T}_v}{h_1h_3} & \dfrac{\overline{T}_w}{h_1h_2}\\ \dfrac{\partial}{\partial u} & \dfrac{\partial}{\partial v} & \dfrac{\partial}{\partial w}\\ h_1f_1 & h_2f_2 & h_3f_3\end{vmatrix}, \text{ where } \overline{f} = f_1\,T_u + f_2\,T_v + f_3\,T_w$$

Proof. $\overline{f} = f_1\,\overline{T}_u + f_2\,\overline{T}_v + f_3\,\overline{T}_w$ (*DU, I sem 2012*)

$$\overline{f} = h_1f_1\nabla u + h_2f_2\nabla v + h_3f_3\,\nabla w$$

$$\text{curl } \overline{f} = \nabla\times\overline{f}$$

$$= \nabla\times(h_1\,f_1\nabla u) + \nabla v\times(h_2\,f_2\nabla v) + \nabla\times(h_3\,f_3\nabla w)$$

$$\nabla\times(h_1\,f_1\nabla u) = \nabla(h_1\,f_1)\times\nabla u + h_1\,f_1 \text{ curl grad} u$$

$$= \nabla(h_1\,f_1)\times\nabla u$$

$$= \left[\frac{\partial}{\partial u}(h_1f_1)\nabla u + \frac{\partial}{\partial v}(h_1f_1)\nabla v + \frac{\partial}{\partial w}(h_1f_1)\nabla w\right]\times\nabla u$$

$$= \frac{\partial}{\partial v}(h_1 f_1)\nabla v \times \nabla u + \frac{\partial}{\partial w}(h_1 f_1)\nabla w \times \nabla u \qquad [\nabla u \times \nabla v = 0]$$

$$= \frac{1}{h_1 h_2}\frac{\partial(h_1 f_1)}{\partial v}\overline{T}_v \times \overline{T}_u + \frac{1}{h_3 h_1}\frac{\partial(h_1 f_1)}{\partial w}\overline{T}_w \times \overline{T}_u$$

$$= -\frac{1}{h_1 h_2}\frac{\partial(h_1 f_1)}{\partial v}\overline{T}_w + \frac{1}{h_3 h_1}\frac{\partial(h_1 f_1)}{\partial w}\overline{T}_v$$

$$\nabla \times (h_1 f_1 \nabla u) = \frac{\overline{T}_v}{h_3 h_1}\frac{\partial(h_1 f_1)}{\partial w} - \frac{\overline{T}_w}{h_1 h_2}\frac{\partial(h_1 f_1)}{\partial v} \in$$

Similarly, $\nabla \times (h_2 f_2 \nabla v) = \frac{\overline{T}_w}{h_1 h_2}\frac{\partial(h_2 f_2)}{\partial u} - \frac{\overline{T}_u}{h_2 h_3}\frac{\partial(h_2 f_2)}{\partial w}$

and $\nabla \times (h_3 f_3 \nabla w) = \frac{\overline{T}_u}{h_2 h_3}\frac{\partial(h_3 f_3)}{\partial v} - \frac{\overline{T}_v}{h_3 h_1}\frac{\partial(h_3 f_3)}{\partial u}$

Substituting these values, we get

$$\text{curl}\,\overline{f} = \nabla \times \overline{f} = \frac{\overline{T}_v}{h_3 h_1}\frac{\partial}{\partial w}(h_1 f_1) - \frac{\overline{T}_w}{h_1 h_2}\frac{\partial}{\partial v}(h_1 f_1) + \frac{\overline{T}_w}{h_1 h_2}\frac{\partial}{\partial u}(h_2 f_2)$$

$$- \frac{\overline{T}_u}{h_2 h_3}\frac{\partial(h_2 f_2)}{\partial w} + \frac{\overline{T}_u}{h_2 h_3}\frac{\partial(h_3 f_3)}{\partial v} - \frac{\overline{T}_v}{h_3 h_1}\frac{\partial(h_3 f_3)}{\partial u}$$

$$= \frac{\overline{T}_u}{h_2 h_3}\left[\frac{\partial}{\partial v}(h_3 f_3) - \frac{\partial}{\partial w}(h_2 f_2)\right] + \frac{\overline{T}_v}{h_3 h_1}\left[\frac{\partial}{\partial w}(h_1 f_1) - \frac{\partial}{\partial u}(h_3 f_3)\right] + \frac{\overline{T}_w}{h_1 h_2}\left[\frac{\partial(h_2 f_2)}{\partial u} - \frac{\partial(h_1 f_1)}{\partial v}\right]$$

$$= \begin{vmatrix} \frac{\overline{T}_u}{h_2 h_3} & \frac{\overline{T}_v}{h_3 h_1} & \frac{\overline{T}_w}{h_1 h_2} \\ \frac{\partial}{\partial u} & \frac{\partial}{\partial v} & \frac{\partial}{\partial w} \\ h_1 f_1 & h_2 f_2 & h_3 f_3 \end{vmatrix}$$

4.7 LAPLACIAN OPERATOR ∇^2

$$\nabla^2 \phi = \nabla \,.\, (\nabla \phi)$$

$$= \nabla \cdot \left[\frac{1}{h_1}\frac{\partial \phi}{\partial u}\overline{T}_u + \frac{1}{h_2}\frac{\partial \phi}{\partial v}\overline{T}_v + \frac{1}{h_3}\frac{\partial \phi}{\partial w}\overline{T}_w\right]$$

$$= \frac{1}{h_1 h_2 h_3}\nabla \cdot \left[h_2 h_3 \frac{\partial \phi}{\partial u}\overline{T}_u + h_3 h_1 \frac{\partial \phi}{\partial v}\overline{T}_v + h_1 h_2 \frac{\partial \phi}{\partial w}\overline{T}_w\right]$$

$$= \frac{1}{h_1 h_2 h_3}\left[\frac{\partial}{\partial u}\left(\frac{h_2 h_3}{h_1}\frac{\partial \phi}{\partial u}\right) + \frac{\partial}{\partial v}\left(\frac{h_3 h_1}{h_2}\frac{\partial \phi}{\partial v}\right) + \frac{\partial}{\partial w}\left(\frac{h_1 h_2}{h_3}\frac{\partial \phi}{\partial w}\right)\right]$$

4.8 CYLINDRICAL (POLAR) CO-ORDINATES

Let P be a point (x, y, z) in the space. Let ρ, ϕ, z denote the projection OQ of OP on x-y plane, the angle which OQ makes with x-axis and perpendicular PQ on x-y plane. Then cylindrical coordinates of P are (ρ, ϕ, z) and so here, we get

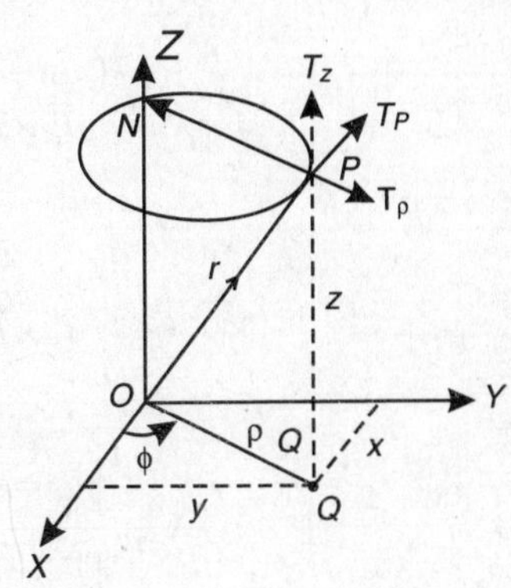

$$\mu_1 = \rho,\ u_2 = \phi,\ u_3 = z$$

$$x = \rho\cos\phi,\ y = \rho\sin\phi,\ z = z$$

On squaring and adding, we get

$$\rho = (x^2 + y^2 + z^2)^{1/2}$$

$$\phi = \tan^{-1}\left(\frac{y}{x}\right)$$

$$z = z$$

Then the co-ordinate surfaces are given by

$$\rho = c_1 \Rightarrow x^2 + y^2 = c_1^2$$

Cylinders co-axial with z-axis

(*i*) $\phi = c_2$, *i.e.*, $y = (\tan c_2)\,x$, *i.e.*, planes through the z-axis.

(*ii*) $z = c_3$, *i.e.*, planes perpendicular to the z-axis.

The point P is the point of intersection of these surfaces.

Let the unit vectors T_u, T_v, T_w be denoted by T_p, T_ϕ, T_z respectively in cylindrical co-ordinates.

Let $\bar{r}$ be the position vector of P. Then we have

$$\bar{r} = x\hat{i} + y\hat{j} + z\hat{k}$$

$$\bar{r} = (\rho\cos\phi)\hat{i} + (\rho\sin\phi)\hat{j} + z\hat{k}$$

$$\frac{\partial\bar{r}}{\partial\rho} = \cos\phi\hat{i} + \sin\phi\,\hat{j}$$

$$\frac{\partial\bar{r}}{\partial\phi} = -\rho\sin\phi\hat{i} + \rho\cos\phi\, j$$

$$\frac{\partial\bar{r}}{\partial z} = \hat{k}$$

$$h_1T_u = \frac{\partial\bar{r}}{\partial u},\ h_2T_v = \frac{\partial\bar{r}}{\partial v},\quad h_3T_w = \frac{\partial\bar{r}}{\partial w}$$

But $$h_1T_u = \frac{\partial\bar{r}}{\partial u},\ h_2T_v = \frac{\partial\bar{r}}{\partial v},\quad h_3T_w = \frac{\partial\bar{r}}{\partial w}$$

Here $$h_1T_\rho = \frac{\partial\bar{r}}{\partial\rho},\ h_2T_\phi = \frac{\partial\bar{r}}{\partial\phi},\quad h_3T_z = \frac{\partial\bar{r}}{\partial z}$$

$$\bar{T}_\rho = \cos\phi\hat{i} + \sin\phi\,\hat{j},\ \bar{T}_\phi = -\sin\phi\hat{i} + \cos\phi\,\hat{j},\ \bar{T}_z = \hat{k}$$

$$\bar{T}_\rho.\bar{T}_\phi = \bar{T}_\phi.\bar{T}_z = \bar{T}_z.\bar{T}_\rho = 0$$

It shows that $\bar{T}_\rho, \bar{T}_\phi, \bar{T}_z$ are mutually perpendicular and hence cylindrical co-ordinates are orthogonal curvilinear co-ordinates.

$$\bar{T}_\rho \times \bar{T}_\phi = \begin{vmatrix} \hat{i} & \hat{j} & \hat{k} \\ \cos\phi & \sin\phi & 0 \\ -\sin\phi & \cos\phi & 0 \end{vmatrix} = \hat{k} = \bar{T}_z$$

$$\overline{T}_\phi \times \overline{T}_z = \begin{vmatrix} \hat{i} & \hat{j} & \hat{k} \\ -\sin\phi & \cos\phi & 0 \\ 0 & 0 & 1 \end{vmatrix} = \cos\phi\hat{i} + \sin\phi\hat{j} = \overline{T}_\rho$$

$$\overline{T}_z \times \overline{T}_\rho = \begin{vmatrix} \hat{i} & \hat{j} & \hat{k} \\ 0 & 0 & 1 \\ \cos\phi & \sin\phi & 0 \end{vmatrix} = -\sin\phi\hat{i} + \cos\phi\,\hat{j} = \overline{T}_\phi$$

$$\left|\overline{T}_\rho\right| = \sqrt{(\cos^2\phi + \sin^2\phi)} = 1$$

$$\left|\overline{T}_\phi\right| = \sqrt{(\sin^2\phi + \cos^2\phi)} = 1$$

$$\left|\overline{T}_z\right| = 1$$

Hence, it follows, that $\overline{T}_\rho$, $\overline{T}_\phi$, $\overline{T}_z$ forms an orthogonal right handed basis.

Now, $$\text{grad } f = \nabla f = \frac{\partial f}{\partial \rho}\overline{T}_\rho + \frac{1}{\rho}\frac{\partial f}{\partial \phi}\overline{T}_\phi + \frac{\partial f}{\partial z}\overline{T}_z$$

$$\nabla^2 f = \frac{1}{\rho}\frac{\partial}{\partial \rho}\left(\partial\frac{\partial f}{\partial \rho}\right) + \frac{1}{\rho^2}\frac{\partial^2 f}{\partial \phi^2} + \frac{\partial^2 f}{\partial z^2}$$

$$div.\overline{F} = \nabla.\overline{F} = \frac{1}{\rho}\partial\frac{(\partial F_\rho)}{\partial \rho} + \frac{1}{\rho}\frac{\partial F_\phi}{\partial \phi} + \frac{\partial F_z}{\partial z}$$

$$\text{Curl}\,\overline{F} = \nabla \times \overline{F} = \frac{1}{\rho}\begin{vmatrix} \overline{T}_\rho & \rho\overline{T}_\phi & \overline{T}_z \\ \frac{\partial}{\partial \rho} & \frac{\partial}{\partial \phi} & \frac{\partial}{\partial z} \\ F_\rho & \rho F_\phi & F_z \end{vmatrix}$$

where $$\overline{F} = F_\rho\overline{T}_\rho + F_\phi\overline{T}_\phi + F_z\overline{T}_z$$

In curvilinear co-ordinates,

$$(ds)^2 = h_1^2(du_1)^2 + h_2^2(du_2)^2 + h_3^2(du_3)^2$$

In cylindrical co-ordinates,

$$(ds)^2 = (d\rho)^2 + (\rho d\phi)^2 + (dz)^2$$

4.9 SPHERICAL POLAR CO-ORDINATES *(DU, I Sem. 2012)*

Let $P(x, y, z)$ be any point in space. Let r, θ, ϕ respectively denote the distance OP of P from the origin, the angle which OP makes with z-axis and the angle between the projection OM of OP on xy-plane and the x-axis.

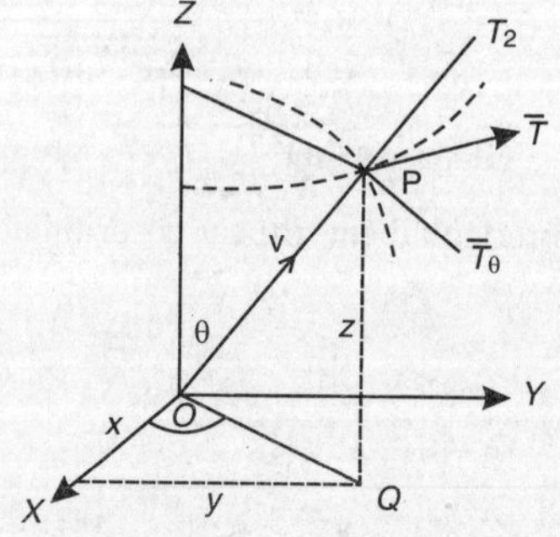

So, we have

$$u_1 = r,\ u_2 = \theta,\ u_3 = \phi$$

Again, $x = r\sin\theta\cos\phi,\ y = r\sin\theta\sin\phi,\ z = r\cos\theta$

where $r \geq 0,\ \theta \geq \pi,\ \phi \geq 2\pi$

The co-ordinates surfaces are given by

(*i*) $r = c_1,\ x^2 + y^2 + z^2 = c_1^2$

(*ii*) $\theta = c_2,\ x^2 + y^2 = (\tan^2 c_2) z^2$

(*iii*) $\phi = c_3,\ y = (\tan c_3)\, x$

The point P is the point of intersection of these surfaces. Let the unit vectors be denoted by T_r, T_θ, T_ϕ in spherical co-ordinates.

$$\bar{r} = x\hat{i} + y\hat{j} + z\hat{k}$$

$$\bar{r} = r\sin\theta(\cos\phi\hat{i} + \sin\phi\hat{j}) + r\cos\theta\hat{k}$$

$$\frac{\partial \bar{r}}{\partial r} = \sin\theta(\cos\phi\hat{i} + \sin\phi\hat{j}) + \cos\theta\hat{k}$$

$$\frac{\partial \bar{r}}{\partial \theta} = r\cos\theta(\cos\phi\hat{i} + \sin\phi\hat{j}) - r\sin\theta\hat{k}$$

$$\frac{\partial \bar{r}}{\partial \phi} = r\sin\theta(-\sin\phi\hat{i} + \cos\phi\hat{j})$$

$$h_1 = \left|\frac{\partial \bar{r}}{\partial r}\right| = 1,\ h_2 = \left|\frac{\partial \bar{r}}{\partial \theta}\right| = r,\ h_3 = \left|\frac{\partial \bar{r}}{\partial \phi}\right| = r\sin\theta$$

But $$h_1\bar{T}_r = \frac{\partial \bar{r}}{\partial u},\ h_2\bar{T}_\theta = \frac{\partial \bar{r}}{\partial v},\ h_3\bar{T}_\phi = \frac{\partial \bar{r}}{\partial w}$$

$$\bar{T}_r = \sin\theta(\cos\phi\hat{i} + \sin\phi\hat{j}) + \cos\theta\hat{k}$$

$$\bar{T}_\theta = \cos\theta(\cos\phi\hat{i} + \sin\phi\hat{j}) - \sin\theta\hat{k}$$

$$\bar{T}_\phi = -\sin\phi\hat{i} + \cos\phi\,\hat{j}$$

$$\bar{T}_r \cdot \bar{T}_\theta = \bar{T}_\theta \cdot \bar{T}_\phi = \bar{T}_\phi \cdot \bar{T}_r = 0$$

It shows that $\bar{T}_r, \bar{T}_\theta, \bar{T}_\phi$ are mutually perpendicular.

$$T_r \times T_\theta = \begin{vmatrix} \hat{i} & \hat{j} & \hat{k} \\ \sin\theta\cos\phi & \sin\theta\sin\phi & \cos\theta \\ \cos\theta\cos\phi & \cos\theta\sin\phi & -\sin\theta \end{vmatrix}$$

$$= -\sin\phi\hat{i} + \cos\phi\hat{j} = \bar{T}_\phi$$

$$\bar{T}_\theta \times \bar{T}_\phi = \begin{vmatrix} \hat{i} & \hat{j} & \hat{k} \\ \cos\theta\cos\phi & \cos\theta\sin\phi & -\sin\theta \\ -\sin\phi & \cos\phi & 0 \end{vmatrix}$$

$$= \sin\theta(\cos\phi\hat{i} + \sin\phi\hat{j}) + \cos\theta\hat{k} = \bar{T}_r$$

$$\bar{T}_\phi \times \bar{T}_r = \begin{vmatrix} \hat{i} & \hat{j} & \hat{k} \\ -\sin\theta & \cos\phi & 0 \\ \sin\theta\cos\phi & \sin\theta\sin\phi & \cos\theta \end{vmatrix}$$

$$= \cos\theta(\cos\phi\hat{i} + \sin\phi\hat{j}) - \sin\theta\hat{k} = \bar{T}_\phi$$

$$|\bar{T}_r| = \sqrt{\sin^2\theta\cos^2\phi + \sin^2\theta\sin^2\phi + \cos^2\theta} = 1$$

$$|\bar{T}_\theta| = \sqrt{\cos^2\theta\cos^2\phi + \cos^2\theta\sin^2\phi + \sin^2\theta} = 1$$

$$|\bar{T}_\phi| = \sqrt{\sin^2\phi + \cos^2\phi} = 1$$

$$\text{grad } f = \frac{\partial f}{\partial r}\bar{T}_r + \frac{1}{r}\frac{\partial f}{d\theta}\bar{T}_\theta + \frac{1}{r\sin\theta}\frac{\partial f}{\partial\phi}\bar{T}_\phi$$

$$\nabla^2 f = \frac{1}{r^2}\frac{\partial}{\partial r}\left(r^2\frac{\partial f}{\partial r}\right) + \frac{1}{r^2\sin\theta}\frac{\partial}{\partial\theta}\left(\sin\theta\frac{\partial f}{\partial\theta}\right) + \frac{1}{r^2\sin^2\theta}\frac{\partial^2 f}{\partial\phi^2}$$

$$\text{div.}\bar{F} = \nabla\cdot\bar{F} = \frac{1}{r^2}\frac{\partial}{\partial r}(r^2 F_r) + \frac{1}{r\sin\theta}\frac{\partial}{\partial\theta}(\sin\theta F\theta) + \frac{1}{r\sin\theta}\frac{\partial F_\phi}{\partial\phi}$$

$$\text{Curl } \bar{F} = \nabla\times\bar{F} = \frac{1}{r^2\sin\theta}\begin{vmatrix} \bar{T}_r & r\bar{T}_\theta & r\sin\theta T_\phi \\ \frac{\partial}{\partial r} & \frac{\partial}{\partial\theta} & \frac{\partial}{\partial\phi} \\ F_r & rF_\theta & r\sin\theta F_\phi \end{vmatrix}$$

$$|ds|^2 = (dr)^2 + (rd\theta)^2 + (r\sin\theta\, d\phi)^2$$

Example 1. *If u, v, w are orthogonal curvilinear co-ordinates, show that* $\frac{\partial\bar{r}}{\partial u}, \frac{\partial\bar{r}}{\partial v}, \frac{\partial\bar{r}}{\partial w}$ *and* $\nabla u, \nabla v, \nabla w$ *are reciprocal system of vectors.*

Solution. We know that

$$\nabla u = \frac{\bar{T}_u}{h_1},\ \nabla v = \frac{\bar{T}_v}{h_2},\ \nabla w = \frac{\bar{T}_w}{h_3}$$

$$(\nabla u, \nabla v, \nabla w) = (\nabla u\times\nabla v)\cdot\nabla w$$

$$= \left(\frac{\bar{T}_u}{h_1}\times\frac{\bar{T}_v}{h_2}\right)\cdot\frac{\bar{T}_w}{h_3}$$

$$= \frac{1}{h_1 h_2 h_3}\left[\bar{T}_w\cdot\bar{T}_w\right] = \frac{1}{h_1 h_2 h_3}$$

$$\frac{\partial\bar{r}}{\partial u} = h_1 T_u,\ \frac{\partial\bar{r}}{\partial v} = h_2 T_v,\ \frac{\partial\bar{r}}{\partial w} = h_3 T_w$$

Now,
$$\frac{\nabla u\times\nabla w}{[\nabla u\nabla v\nabla w]} = \frac{\left(\frac{1}{h_2}\right)\bar{T}_v\times\left(\frac{1}{h_3}\right)\bar{T}_w}{\frac{1}{h_1 h_2 h_3}} = h_1\bar{T}_u \qquad [\bar{T}_v\times\bar{T}_w = \bar{T}_u]$$

$$= \frac{\partial\bar{r}}{\partial u}$$

Similarly,
$$\frac{\partial\bar{r}}{\partial v} = \frac{\nabla T_w\times\nabla T_u}{[\nabla u, \nabla v, \nabla w]},\ \frac{\partial\bar{r}}{\partial w} = \frac{\nabla u_1\times\nabla u_2}{[\nabla u, \nabla v, \nabla w]}$$

This shows that $\frac{\partial\bar{r}}{\partial u}, \frac{\partial\bar{r}}{\partial v}, \frac{\partial\bar{r}}{\partial w}$ and $\nabla u, \nabla v, \nabla w$ form reciprocal system of vectors. **Proved.**

Example 2. *Transform the wave equation*

$$\frac{\partial^2 u}{\partial r^2} = c^2\nabla^2 u \text{ in spherical co-ordinates if } u \text{ is independent of } \phi.$$

Solution. Since u is independent of ϕ, so $\dfrac{\partial^2 u}{\partial \phi^2} = 0$

$$\frac{\partial^2 u}{\partial t^2} = c^2\left[\frac{1}{r^2}\frac{\partial}{\partial r}\left(r^2\frac{\partial u}{\partial r}\right) + \frac{1}{r^2 \sin\theta}\frac{\partial}{\partial \theta}\left(\sin\theta\frac{\partial u}{\partial \theta}\right)\right] \quad \textbf{Ans.}$$

Example 3. *Show that the spherical co-ordinate system $((\overline{T}_r, \overline{T}_\theta, \overline{T}_\phi))$ is self-reciprocal.*

Solution. We know that

$$\overline{T}_r = \sin\theta(\cos\phi\hat{i} + \sin\phi\hat{j}) + \cos\theta\,\hat{k}$$

$$\overline{T}_\theta = \cos\theta(\cos\phi\hat{i} + \sin\phi\hat{j}) + \sin\theta\,\hat{k}$$

$$\overline{T}_\phi = -\sin\phi\,\hat{i} + \cos\phi\,\hat{j}$$

Let $\quad \overline{T}_r = \dfrac{\overline{T}_\theta' \times \overline{T}_\phi}{[T_r T_\theta T_\phi]}, \overline{T}_\theta' = \dfrac{\overline{T}_\phi \times \overline{T}_r}{[T_r T_\theta T_\phi]}, \overline{T}_\phi' = \dfrac{\overline{T}_r \times \overline{T}_\theta}{[T_r T_\theta T_\phi]}$

be the reciprocal vectors of $\overline{T}_r,\ \overline{T}_\theta,\ \overline{T}_\phi$.

$$[\overline{T}_r \overline{T}_\theta \overline{T}_\phi] = [\overline{T}_r \times \overline{T}_\theta]\cdot\overline{T}_\phi = \overline{T}_\phi \cdot \overline{T}_\phi = 1$$

$$\overline{T}_r' = \frac{T_r}{1} = \overline{T}_r$$

similarly $\quad \overline{T}_\theta' = \overline{T}_\theta,\ \overline{T}_\phi' = \overline{T}_\phi$

Hence, $(\overline{T}_r, \overline{T}_\theta, \overline{T}_\phi)$ is a self-reciprocal base. **Proved.**

4.10 TRANSFORMATION OF CYLINDRICAL POLAR CO-ORDINATES INTO $\hat{i}, \hat{j}, \hat{k}$

Solution. We know that

$$T_\rho = \cos\phi\hat{i} + \sin\phi\hat{j} + 0\,\hat{k}$$

$$T_\phi = -\sin\phi\hat{i} + \cos\phi\hat{j} + 0\,\hat{k}$$

$$T_z = 0\hat{i} + 0\,\hat{j} + \hat{k}$$

The above equations are rewritten in matrix form.

$$\begin{bmatrix}\overline{T}_\rho \\ \overline{T}_\phi \\ \overline{T}_z\end{bmatrix} = \begin{bmatrix}\cos\phi & \sin\phi & 0 \\ -\sin\phi & \cos\phi & 0 \\ 0 & 0 & 1\end{bmatrix}\begin{bmatrix}\hat{i} \\ \hat{j} \\ \hat{k}\end{bmatrix}$$

On inverting, we get

$$\begin{bmatrix}\hat{i} \\ \hat{j} \\ \hat{k}\end{bmatrix} = \begin{bmatrix}\cos\phi & -\sin\phi & 0 \\ \sin\phi & \cos\phi & 0 \\ 0 & 0 & 1\end{bmatrix}\begin{bmatrix}\overline{T}_\rho \\ \overline{T}_\phi \\ \overline{T}_z\end{bmatrix}$$

4.11 CONVERSION OF SPHERICAL POLAR CO-ORDINATES (r, θ, Φ) INTO $\hat{i}, \hat{j}, \hat{k}$

Solution. We know that

$$\overline{T}_r = \sin\theta\cos\phi\hat{i} + \sin\theta\sin\phi\hat{j} + \cos\phi\,\hat{k}$$

$$\overline{T}_\theta = \cos\theta\cos\phi\hat{i} + \cos\theta\sin\phi\hat{j} - \sin\theta\,\hat{k}$$

$$\overline{T}_\phi = -\sin\phi\hat{i} + \cos\phi\hat{j} + 0\hat{k}$$

The equations are written in matrix form

$$\begin{bmatrix} \overline{T}_r \\ \overline{T}_\theta \\ \overline{T}_\phi \end{bmatrix} = \begin{vmatrix} \sin\theta\cos\phi & \sin\theta\sin\phi & \cos\phi \\ \cos\theta\cos\phi & \cos\theta\sin\phi & -\sin\theta \\ -\sin\phi & \cos\phi & 0 \end{vmatrix} \begin{bmatrix} \hat{i} \\ \hat{j} \\ \hat{k} \end{bmatrix}$$

On inverting, we get

$$\begin{bmatrix} \hat{i} \\ \hat{j} \\ \hat{k} \end{bmatrix} = \begin{vmatrix} \sin\theta\cos\phi & \cos\theta\cos\phi & -\sin\phi \\ \sin\theta\sin\phi & \cos\theta\sin\phi & \cos\phi \\ \cos\phi & -\sin\theta & 0 \end{vmatrix} \begin{bmatrix} \overline{T}_r \\ \overline{T}_\theta \\ \overline{T}_\phi \end{bmatrix}$$

4.12 RELATION BETWEEN CYLINDRICAL AND SPHERICAL CO-ORDINATES

Solution. We know that

$$\begin{bmatrix} \hat{i} \\ \hat{j} \\ \hat{k} \end{bmatrix} = \begin{bmatrix} \cos\phi & -\sin\phi & 0 \\ \sin\phi & \cos\phi & 0 \\ 0 & 0 & 1 \end{bmatrix} \begin{bmatrix} T_\rho \\ T_\phi \\ T_z \end{bmatrix} \qquad \text{(cylindrical co-oridinates)}$$

$$\begin{bmatrix} \hat{i} \\ \hat{j} \\ \hat{k} \end{bmatrix} = \begin{bmatrix} \sin\theta\cos\phi & \cos\theta\cos\phi & -\sin\phi \\ \sin\theta\sin\phi & \cos\theta\sin\phi & \cos\phi \\ \cos\theta & -\sin\phi & 0 \end{bmatrix} \begin{bmatrix} \overline{T}_r \\ \overline{T}_\theta \\ \overline{T}_\phi \end{bmatrix} \qquad \text{(spherical co-ordinates)}$$

Equating, we get

$$\begin{bmatrix} \cos\phi & -\sin\phi & 0 \\ \sin\phi & \cos\phi & 0 \\ 0 & 0 & 1 \end{bmatrix} \begin{bmatrix} T_\rho \\ T_\phi \\ T_z \end{bmatrix} = \begin{bmatrix} \sin\theta\cos\phi & \cos\theta\cos\phi & -\sin\phi \\ \sin\theta\sin\phi & \cos\theta\sin\phi & \cos\phi \\ \cos\theta & -\sin\phi & 0 \end{bmatrix} \begin{bmatrix} \overline{T}_r \\ \overline{T}_\theta \\ \overline{T}_\phi \end{bmatrix}$$

$$\begin{bmatrix} \overline{T}_\rho \\ \overline{T}_\phi \\ \overline{T}_z \end{bmatrix} = \begin{bmatrix} \cos\phi & -\sin\phi & 0 \\ \sin\phi & \cos\phi & 0 \\ 0 & 0 & 1 \end{bmatrix}^{-1} \begin{bmatrix} \sin\theta\cos\phi & \cos\theta\cos\phi & -\sin\phi \\ \sin\theta\sin\phi & \cos\theta\sin\phi & \cos\phi \\ \cos\theta & -\sin\phi & 0 \end{bmatrix} \begin{bmatrix} \overline{T}_r \\ \overline{T}_\theta \\ \overline{T}_\phi \end{bmatrix}$$

$$= \begin{bmatrix} \sin\theta & 0 & \cos\theta \\ \cos\theta & 0 & -\sin\theta \\ 0 & 1 & 0 \end{bmatrix} \begin{bmatrix} \overline{T}_r \\ \overline{T}_\theta \\ \overline{T}_\phi \end{bmatrix} \qquad \textbf{Ans.}$$

Example 4. *Express* $z\hat{i} - 2\hat{j} + y\hat{k}$ *in cylindrical co-ordinates.*

Solution. $x = r\cos\phi$, $y = r\sin\phi$, $z = z$

$$\overline{R} = x\hat{i} + y\hat{j} + z\hat{k}$$

$$\overline{R} = \rho\cos\phi\,\hat{i} + \rho\sin\phi\,\hat{j} + z\hat{k}$$

If T_ρ, T_ϕ, T_z be the unit vectors at P in the directions of the tangents to the ρ, ϕ, z curves respectively, then

$$\overline{T}_\rho = \frac{\dfrac{\partial \overline{R}}{\partial \rho}}{\left|\dfrac{\partial \overline{R}}{\partial \rho}\right|} = \frac{\cos\phi\hat{i} + \sin\phi\hat{j}}{\sqrt{\cos^2\phi + \sin^2\phi}} = \cos\phi\,\hat{i} + \sin\phi\,\hat{j}$$

$$\overline{T}_\phi = \frac{\dfrac{\partial \overline{R}}{\partial \phi}}{\left|\dfrac{\partial \overline{R}}{\partial \phi}\right|} = \frac{-\rho\sin\phi\hat{i} + \rho\cos\phi\,\hat{j}}{\sqrt{(-\rho\sin\phi)^2 + (\rho\cos\phi)^2}} = -\sin\phi\,\hat{i} + \sin\phi\,\hat{j}$$

$$\overline{T}_z = \frac{\dfrac{\partial \overline{R}}{\partial z}}{\left|\dfrac{\partial \overline{R}}{\partial z}\right|} = \hat{k}$$

$$\overline{f} = z\hat{i} - 2x\hat{j} + y\hat{k}$$

$$\overline{f} = z\hat{i} - 2\rho\cos\phi\,\hat{j} + \rho\sin\phi\,\hat{k}$$

$$f_1 = \overline{f}\cdot T_\rho = (z\hat{i} - 2\rho\cos\phi\,\hat{j} + \rho\sin\phi\,\hat{k})\cdot(\cos\phi\hat{i} + \sin\phi\,\hat{j})$$

$$= z\cos\phi - 2\rho\sin\phi\cos\phi$$

$$f_2 = \overline{f}\cdot\overline{T}_\phi = (z\hat{i} - 2\rho\cos\phi\hat{j} + \rho\sin\phi\hat{k})\cdot(-\sin\phi\hat{i} + \cos\phi\hat{j})$$

$$= -z\sin\phi - 2\rho\cos^2\phi$$

$$f_3 = \overline{f}\cdot\overline{T}_z = (z\hat{i} - 2\rho\cos\phi\hat{j} + \rho\sin\phi\,\hat{k})\cdot\hat{k}$$

$$= \rho\sin\phi$$

$$\overline{f} = f_1\overline{T}_\rho + f_2\overline{T}_\phi + f_3\overline{T}_z, \text{ where } f_1 = z\cos\phi - 2\rho\sin\phi\cos\phi,$$

$$f_2 = -z\sin\phi - 2\rho\cos^2\phi,\ f_3 = \rho\sin\phi$$ **Ans.**

Example 5. *Express* $x\hat{i} + 2y\hat{j} + yz\hat{k}$ *in spherical polar co-ordinates.*

Solution. $x = r\sin\theta\cos\phi,\ y = r\sin\theta\sin\phi,\ z = r\cos\theta$

$$\overline{R} = x\hat{i} + y\hat{j} + z\hat{k}$$

$$\overline{R} = r\sin\theta\cos\phi\hat{i} + r\sin\theta\sin\phi\hat{j} + r\cos\theta\,\hat{k}$$

In spherical co-ordinates, T_r, T_θ, T_ϕ be the unit vectors along the tangents to r, θ, ϕ curves respectively, then

$$\overline{T}_r = \frac{\partial\overline{R}/\partial r}{\left|\dfrac{\partial\overline{R}}{\partial r}\right|} = \frac{\sin\theta\cos\phi\hat{i} + \sin\theta\sin\phi\hat{j} + \cos\theta\,\hat{k}}{\sqrt{(\sin\theta\cos\phi)^2 + (\sin\theta\sin\phi)^2 + \cos^2\theta}}$$

$$= \sin\theta\cos\phi\hat{i} + \sin\theta\sin\phi\hat{j} + \cos\theta\,\hat{k}$$

$$\overline{T}_\theta = \frac{\dfrac{\partial\overline{R}}{\partial\theta}}{\left|\dfrac{\partial\overline{R}}{\partial\theta}\right|} = \frac{r\cos\theta\cos\phi\hat{i} + r\cos\theta\sin\phi\hat{j} - r\sin\theta\,\hat{k}}{\sqrt{(r\cos\theta\cos\phi)^2 + (r\cos\theta\sin\phi)^2 + (-r\sin\theta)^2}}$$

$$= \cos\theta\cos\phi\hat{i} + \cos\theta\sin\phi\hat{j} - \sin\theta\hat{k}$$

$$\overline{T}_\phi = \frac{\dfrac{\partial \overline{R}}{\partial \phi}}{\left|\dfrac{\partial \overline{R}}{\partial \phi}\right|} = \frac{-r\sin\theta\sin\phi\hat{i} + r\sin\theta\cos\phi\,\hat{j}}{\sqrt{(-r\sin\theta\sin\phi)^2 + (r\sin\theta\cos\phi)^2}}$$

$$= -\sin\phi\;\hat{i} + \cos\phi\;\hat{j}$$

$$\overline{f} = x\hat{i} + 2y\hat{j} + yz\hat{k}$$

$$= r\sin\theta\cos\phi\hat{i} + 2r\sin\theta\sin\phi\hat{j} + r^2\sin\theta\sin\phi\cos\theta\hat{k}$$

$$f_1 = \overline{f}\cdot\overline{T}_r = [r\sin\theta\cos\phi\hat{i} + 2r\sin\theta\sin\phi\hat{j} + r^2\sin\theta\sin\phi\cos\theta\hat{k}]$$

$$[\sin\theta\cos\phi\hat{i} + \sin\theta\sin\phi\,\hat{j} - \cos\theta\hat{k}]$$

$$= r\sin^2\theta\cos^2\phi + 2r\sin^2\theta\sin^2\phi + r^2\sin\theta\sin\phi\cos^2\theta$$

$$= r\sin^2\theta(\cos^2\phi + \sin^2\phi) - r\sin^2\theta\sin^2\phi + r^2\sin\theta\sin\phi\cos^2\theta$$

$$= r\sin^2\theta + r\sin^2\theta\sin^2\phi + r^2\sin\theta\sin\phi\cos^2\theta$$

$$= r\sin^2\theta(1 + \sin^2\phi) + r^2\sin\theta\cos^2\theta\sin\phi$$

$$f_2 = \overline{f}\cdot\overline{T}_\theta = [r\sin\theta\cos\phi\hat{i} + 2r\sin\theta\sin\phi\,\hat{j} + r^2\sin\theta\sin\phi\cos\theta\hat{k}]$$

$$[\cos\theta\cos\phi\hat{i} + \cos\theta\sin\phi\hat{j} - \sin\theta\hat{k}]$$

$$= r\sin\theta\cos\theta\cos^2\phi + 2r\sin\theta\cos\theta\sin^2\phi - r^2\sin^2\theta\sin\phi\cos\theta$$

$$= r\sin\theta\cos\theta(1 + \sin^2\phi) - r^2\sin^2\theta\cos\theta\sin\phi$$

$$f_3 = \overline{f}\cdot T_\phi = [r\sin\theta\cos\phi\hat{i} + 2r\sin\theta\sin\phi\hat{j} + r^2\sin\theta\sin\phi\cos\theta\hat{k}]\;[-\sin\phi\hat{i} + \cos\phi\hat{j}]$$

$$= -r\sin\theta\sin\phi\cos\phi + 2r\sin\theta\sin\phi\cos\phi$$

$$= r\sin\theta\sin\phi\cos\phi$$

$$\overline{f} = f_1\overline{T}_r + f_2\overline{T}_\theta + f_3\overline{T}_\phi$$

where
$$f_1 = r\sin^2\theta(1 + \sin^2\phi) + r^2\sin\theta\cos^2\theta\sin\phi$$

$$f_2 = r\sin\theta\cos\theta(1 + \sin^2\phi) - r^2\sin^2\theta\cos\theta\sin\phi$$

$$f_3 = r\sin\theta\sin\phi\cos\phi$$

Ans.

EXERCISE 4.1

1. Express $2y\hat{i} - z\hat{j} + 3x\hat{k}$ in spherical co-ordinates. **Ans.** $\overline{f} = f_1T_r + f_2T_\theta + f_3T_\phi$, where

$$f_1 = 2r^2\sin^2\theta\sin\phi\cos\phi - r\sin\theta\cos\theta\sin\phi + 3r\sin\theta\cos\theta\cos\phi$$

$$f_2 = 2r\sin\theta\cos\theta\sin\phi\cos\phi - r\cos^2\theta\sin\phi - 3r\sin^2\theta\cos\phi$$

$$f_3 = -2r\sin\theta\sin^2\phi - r\cos\theta\cos\phi$$

2. Express $f = 3y\hat{i} + x^2\hat{j} - x^2\hat{k}$ in terms of cylindrical polar co-ordinates.

Ans. $f = \rho\sin\phi\cos\phi(3 + R\cos\phi)\,T_\rho + R(R\cos^3\phi - 3\sin^2\phi)\,T_\phi - z^2k$

3. Transform the function $f = \rho\,T_p + \rho\,T_\phi$ from cylindrical to cartesian co-ordinates.

Ans. $(x - y)\;\hat{i} + (x + y)\;\hat{j}$

4. Transform $f = \frac{1}{2} T_r$ from spherical to cartesian system. **Ans.** $\frac{(x\hat{i} + y\hat{j} + z\hat{k})}{x^2 + y^2 + z^2}$

5. If ρ, ϕ, z are cylindrical co-ordinates, show that $\nabla(\log \rho)$ and $\nabla\phi$ are solenoidal vectors.

6. If (r, θ, ϕ) are spherical polar co-ordinates, show that grad $(\cos\theta) \times$ grad ϕ = grad $\left(\frac{1}{r}\right)$.

7. Prove that the spherical polar co-ordinates system is orthogonal.

8. The parametric representation of the surface is given by $\bar{r} = (w\cos\theta)\hat{i} + (w\sin\theta)\hat{j} + w^2\hat{k}$. Find the equation of the surface in the cartesian form. **Ans.** $x^2 + y^2 = z$

9. Starting from the principles, derive an expression for divergence of a vector in orthogonal curvilinear coordinates. *(Delhi University, April 2010)*

DOUBLE INTEGRALS

5.1 DOUBLE INTEGRATION

We know that

$$\int_a^b f(x)\,dx = \lim_{\substack{n\to\infty \\ \delta x\to 0}} [f(x_1)\,\delta x_1 + f(x_2)\,\delta x_2 + f(x_3)\,\delta x_3 + ... + f(x_n)\,\delta\, x_n]$$

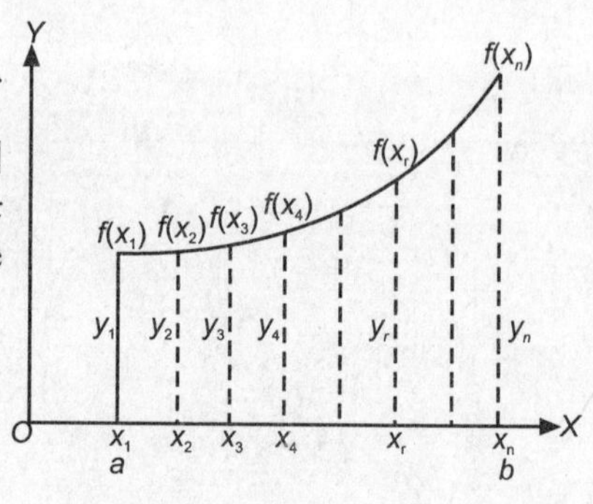

Let us consider a function $f(x, y)$ of two variable x and y defined in the finite region A of xy-plane. Divide the region A into elementary areas.

$$\delta A_1,\ \delta A_2,\ \delta A_3,\ ......\ \delta A_n$$

Then $$\iint_A f(x,y)\,dA$$

$$= \lim_{\substack{n\to\infty \\ \delta A\to 0}} \left[f(x_1,y_1)\,\delta A_1 + f(x_2,y_2)\,\delta A_2 + ... + f(x_n,y_n)\,\delta A_n\right]$$

5.2 EVALUATION OF DOUBLE INTEGRAL

Double integral over region A may be evaluated by two successive integrations.

If A is described as $f_1(x) \le y \le f_2(x)$, $[y_1 \le y \le y_2]$

and $$a \le x \le b,$$

Then $$\iint_A f(x,y)\,dA = \int_a^b \int_{y_1}^{y_2} f(x,y)\,dx\,dy$$

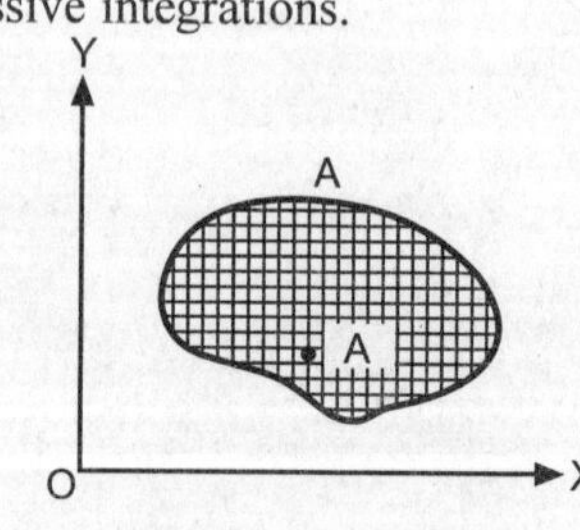

(1) First Method

$$\iint_A f(x,y)\,dA = \int_a^b \boxed{\int_{y_1}^{y_2} f(x,y)\,dy}\,dx$$

$f(x, y)$ is first integratred with respect to y treating x as constant between the limits a and b.

In the region we take an elementary area $\delta x \delta y$. Then integration w.r.t y (x keeping constant). converts small rectangle $\delta x \delta y$ into a strip PQ ($y\ \delta x$). While the integration of the result w.r.t. x corresponding to the sliding to the strip PQ, from AD to BC covering the while region $ABCD$.

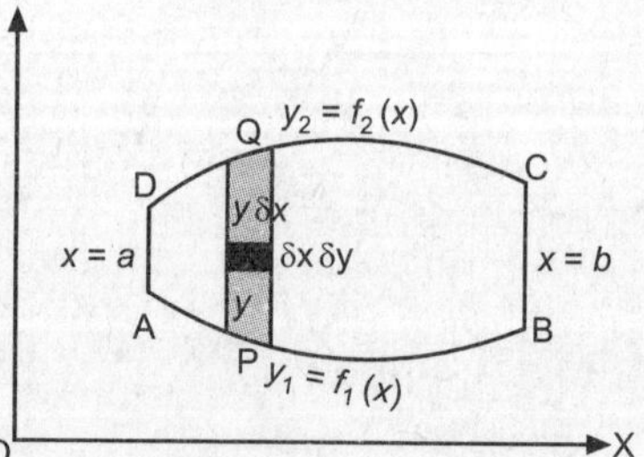

Second method

$$\iint_A f(x,y)\,dxdy = \int_c^d \boxed{\int_{x_1}^{x_2} f(x,y)\,dx}\,dy$$

Here $f(x,y)$ is first integrated w.r.t x keeping y constant between the limits x_1 and x_2 and then the resulting expression is integrated with respect to y between the limits c and d.

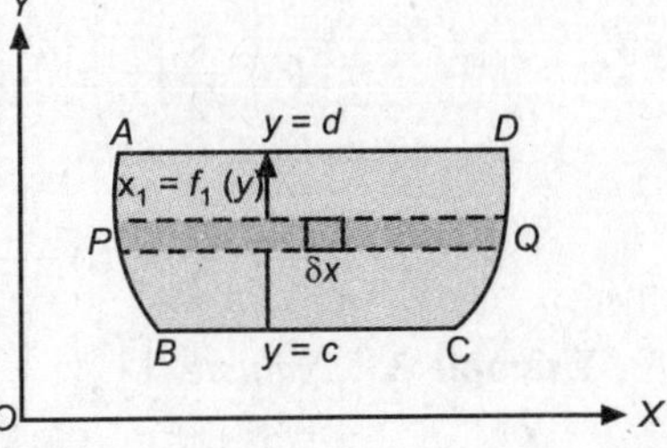

Take a small area $\delta x \delta y$. The integration w.r.t. x between the limits x_1, x_2 keeping y fixed indicates that integration is done, along PQ. Then the integration of result w.r.t y corresponds to sliding the strips PQ from BC to AD covering the whole region $ABCD$.

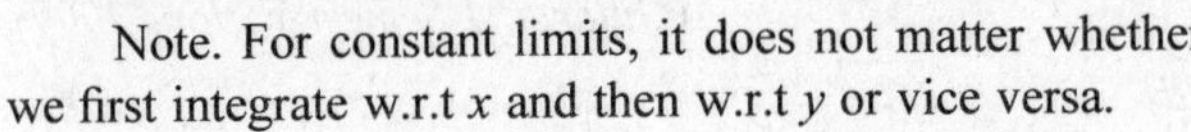
Note. For constant limits, it does not matter whether we first integrate w.r.t x and then w.r.t y or vice versa.

Example 1. *Evaluate* $\int_0^1 \int_0^x (x^2 + y^2)\, dA$, *where dA indicates small area in xy-plane.*

(Gujarat, I Semester, Jan. 2009)

Solution. Let $\quad I = \int_0^1 \int_0^x (x^2 + y^2)\, dy\, dx = \int_0^1 \left[x^2 y + \frac{y^3}{3} \right]_0^x dx$

$$= \int_0^1 \left[x^2 (x - 0) + \frac{1}{3} (x^3 - 0) \right] dx = \int_0^1 \left[x^3 + \frac{x^3}{3} \right] dx$$

$$= \int_0^1 \frac{4}{3} x^3\, dx = \frac{4}{3} \left[\frac{x^4}{4} \right]_0^1 = \frac{1}{3}[1 - 0] = \frac{1}{3} \text{ sq. units.}$$ **Ans.**

Example 2. *Evaluate* $\int_{-1}^{1} \int_0^{1-x} x^{1/3}\, y^{-1/2}\, (1 - x - y)^{1/2}\, dy\, dx$.

Solution. Here, we have

$$I = \int_{-1}^{1} \int_0^{1-x} x^{1/3}\, y^{-1/2}\, (1 - x - y)^{1/2}\, dy\, dx \qquad ...(1)$$

Putting $(1 - x) = c$ in (1), we get

$$I = \int_{-1}^{1} x^{1/3}\, dx \int_0^c y^{-1/2}\, (c - y)^{1/2}\, dy \qquad ...(2)$$

Again putting $y = ct \Rightarrow dy = c\, dt$ in (2), we get

$$I = \int_{-1}^{1} x^{\frac{1}{3}}\, dx \int_0^1 c^{-\frac{1}{2}}\, t^{-\frac{1}{2}}\, (c - ct)^{\frac{1}{2}}\, c\, dt$$

$$= \int_{-1}^{1} x^{1/3}\, dx \int_0^1 c^{-1/2}\, t^{-1/2}\, c^{1/2}\, (1 - t)^{1/2}\, c\, dt$$

$$= \int_{-1}^{1} c\, x^{1/3}\, dx \int_0^1 t^{-1/2} (1 - t)^{1/2}\, dt = \int_{-1}^{1} c\, x^{1/3}\, dx \int_0^1 t^{1/2 - 1}\, (1 - t)^{3/2 - 1}\, dt$$

$$= \int_{-1}^{1} c\, x^{\frac{1}{3}}\, dx\, \beta\left(\frac{1}{2}, \frac{3}{2} \right) \qquad \left[\int_0^1 x^{l-1} (1 - x)^{m-1}\, dx = \beta(l, m) \right]$$

$$= \int_{-1}^{1} c\, x^{1/3}\, dx \frac{\left|\frac{1}{2}\right. \left|\frac{3}{2}\right.}{\left|\frac{1}{2} + \frac{3}{2}\right.} = \int_{-1}^{1} c\, x^{1/3}\, dx \frac{\left|\frac{1}{2}\right. . \frac{1}{2} \left|\frac{1}{2}\right.}{\left|2\right.} = \int_{-1}^{1} c\, x^{1/3}\, dx \frac{\sqrt{\pi}\, \frac{1}{2} \sqrt{\pi}}{1}$$

$$= \int_{-1}^{1} c\, x^{1/3}\, \frac{\pi}{2}\, dx = \frac{\pi}{2} \int_{-1}^{1} x^{1/3} . c\, dx$$

Putting the value of c, we get

$$I = \frac{\pi}{2}\int_{-1}^{1} x^{1/3}(1-x)\,dx = \frac{\pi}{2}\int_{-1}^{1}(x^{1/3}-x^{4/3})\,dx = \frac{\pi}{2}\left[\frac{x^{4/3}}{\frac{4}{3}} - \frac{x^{7/3}}{\frac{7}{3}}\right]_{-1}^{1}$$

$$= \frac{\pi}{2}\left[\frac{3}{4}(1) - \frac{3}{7}(1) - \frac{3}{4}(-1) + \frac{3}{7}(-1)\right] = \frac{\pi}{2}\left[\frac{9}{14}\right] = \frac{9\pi}{28} \qquad \textbf{Ans.}$$

Example 3. *Evaluate* $\iint_R (x+y)\,dy\,dx$, *R is the region bounded by* $x = 0$, $x = 2$, $y = x$, $y = x + 2$. (*Gujarat, I Semester, Jan. 2009*)

Solution. Let $I = \iint_R (x+y)\,dy\,dx$

The limits are $x = 0$, $x = 2$, $y = x$ and $y = x + 2$

$$I = \int_0^2 dx \int_x^{x+2}(x+y)\,dy = \int_0^2 \left[xy + \frac{y^2}{2}\right]_x^{x+2} dx$$

$$= \int_0^2 \left[x(x+2) + \frac{1}{2}(x+2)^2 - x^2 - \frac{x^2}{2}\right] dx$$

$$= \int_0^2 \left[x^2 + 2x + \frac{1}{2}(x^2+4x+4) - x^2 - \frac{x^2}{2}\right] dx$$

$$= \int_0^2 [2x + 2x + 2]\,dx$$

$$= 2\int_0^2 (2x+1)\,dx = 2[x^2+x]_0^2 = 2\,[4+2] = 12 \quad \textbf{Ans.}$$

Example 4. *Evaluate* $\iint_R xy\,dx\,dy$

where R is the quadrant of the circle $x^2 + y^2 = a^2$ *where* $x \geq 0$ *and* $y \geq 0$.

Solution. Let the region of integration be the first quadrant of the circle *OAB*.

$$\iint_R xy\,dx\,dy \;(x^2+y^2=a^2,\, y=\sqrt{a^2-x^2})$$

First we integrate w.r.t. y and then w.r.t. x.

The limits for y are 0 and $\sqrt{a^2-x^2}$ and for x, 0 to a.

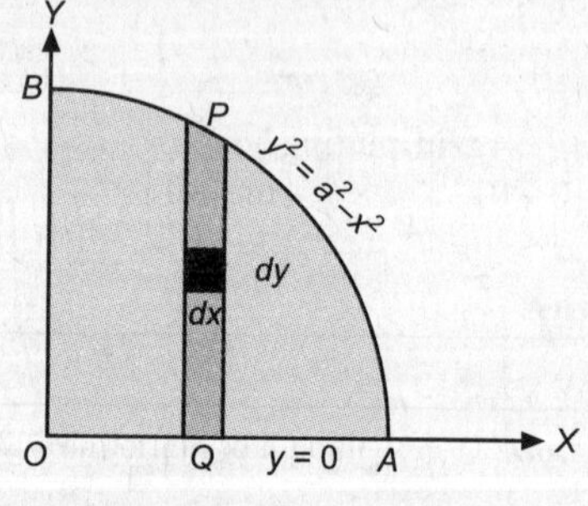

$$= \int_0^a x\,dx \int_0^{\sqrt{a^2-x^2}} y\,dy = \int_0^a x\,dx \left[\frac{y^2}{2}\right]_0^{\sqrt{a^2-x^2}}$$

$$= \frac{1}{2}\int_0^a x(a^2-x^2)\,dx = \frac{1}{2}\left[\frac{a^2x^2}{2} - \frac{x^4}{4}\right]_0^a = \frac{a^4}{8} \quad \textbf{Ans.}$$

Example 5. *Evaluate* $\iint_s \sqrt{xy-y^2}\,dy\,dx$,

where S is a triangle with vertices (0, 0), (10, 1) and (1, 1).

Solution. Let the vertices of a triangle *OBA* be (0, 0) (10, 1) and (1, 1).

Equation of *OA* is $x = y$.

Equation of *OB* is $x = 10\,y$.

The region of ΔOBA, given by the limits

$y \leq x \leq 10\,y$ and $0 \leq y \leq 1$.

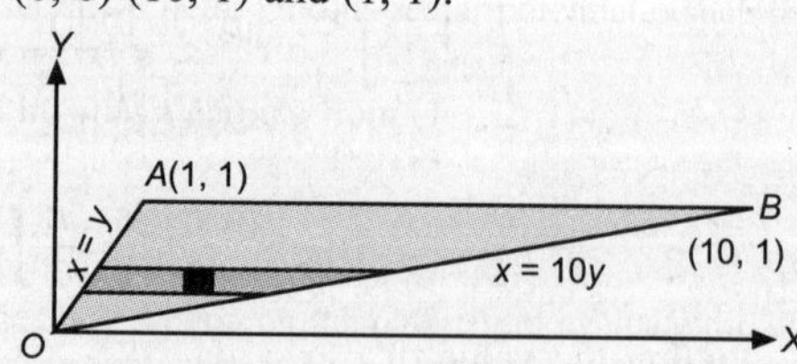

$$\iint_s \sqrt{xy-y^2}\,dy\,dx = \int_0^1 dy \int_y^{10y} (xy-y^2)^{1/2}\,dx$$

$$= \int_0^1 dy \left[\frac{2}{3}\frac{1}{y}(xy - y^2)^{3/2}\right]_y^{10y} = \int_0^1 \frac{2}{3}\frac{1}{y}(9y^2)^{3/2}\, dy = 18\int_0^1 y^2\, dy$$

$$= 18\left[\frac{y^3}{3}\right]_0^1 = \frac{18}{3} = 6$$ **Ans.**

Example 6. *Evaluate* $\iint_A x^2\, dx\, dy$, *where A is the region in the first quadrant bounded by the hyperbola xy = 16 and the lines y = x, y = 0 and x = 8.*

Solution. The line OP, $y = x$ and the curve PS, $xy = 16$ intersect at (4, 4).
The line SN, $x = 8$ intersects the hyperbola at S (8, 2). $y = 0$ is x-axis.
The area A is shown shaded.

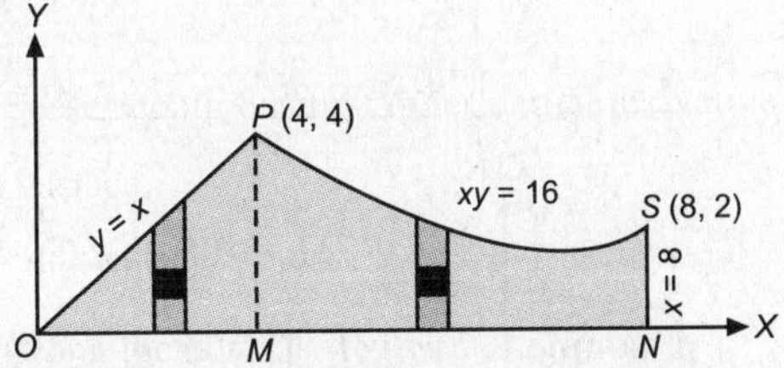

Divide the area in to two part by PM perpendicular to OX.
For the area OMP, y varies from 0 to x, and then x varies from 0 to 4.
For the area PMNS, y -series from 0 to $16/x$ and then x varies from 4 to 8.

$$\therefore \quad \iint_A x^2\, dx\, dy = \int_0^4\int_0^x x^2\, dx\, dy + \int_4^8\int_0^{16/x} x^2\, dx\, dy$$

$$= \int_0^4 x^2\, dx \int_0^x dy + \int_4^8 x^2\, dx \int_0^{16/x} dy = \int_0^4 x^2\,[y]_0^x\, dx + \int_4^8 x^2\,[y]_0^{16/x}\, dx$$

$$= \int_0^4 x^3\, dx + \int_4^8 16\, x\, dx = \left[\frac{x^4}{4}\right]_0^4 + 16\left[\frac{x^2}{2}\right]_4^8 = 64 + 8\,(8^2 - 4^2) = 64 + 384 = 448.$$ **Ans.**

Example 7. *Evaluate* $\iint (x+y)^2\, dx\, dy$ *over the area bounded by the ellipse* $\frac{x^2}{a^2} + \frac{y^2}{b^2} = 1$

Solution. For the ellipse $\frac{x^2}{a^2} + \frac{y^2}{b^2} = 1$

$$\Rightarrow \quad \frac{y}{b} = \pm\sqrt{1 - \frac{x^2}{a^2}} \quad \Rightarrow \quad y = \pm\frac{b}{a}\sqrt{a^2 - x^2}$$

$\therefore$ The region of integration can be expressed as

$-a \le x \le a$ and $-\frac{b}{a}\sqrt{a^2 - x^2} \le y \le \frac{b}{a}\sqrt{a^2 - x^2}$

$$\therefore \quad \iint (x+y)^2\, dx\, dy = \iint (x^2 + y^2 + 2xy)\, dx\, dy$$

$$= \int_{-a}^{a}\int_{(-b/a)\sqrt{a^2-x^2}}^{b/a\sqrt{a^2-x^2}} (x^2 + y^2 + 2xy)\, dx\, dy$$

$$= \int_{-a}^{a}\int_{(-b/a)\sqrt{a^2-x^2}}^{b/a\sqrt{a^2-x^2}} (x^2 + y^2)\, dx\, dy + \int_{-a}^{a}\int_{(-b/a)\sqrt{a^2-x^2}}^{b/a\sqrt{a^2-x^2}} 2xy\, dy\, dx$$

$$= \int_{-a}^{a}\int_0^{b/a\sqrt{a^2-x^2}} 2\,(x^2 + y^2)\, dy\, dx + 0$$

[Since $(x^2 + y^2)$ is an even function of y and $2xy$ is an odd function of y]

$$= \int_{-a}^{a}\left[2\left(x^2 y + \frac{y^3}{3}\right)\right]_0^{\left(\frac{b}{a}\right)\sqrt{a^2-x^2}} dx$$

$$= 2\int_{-a}^{a}\left[x^2 \times \frac{b}{a}\sqrt{a^2 - x^2} + \frac{1}{3}\frac{b^3}{a^3}(a^2 - x^2)^{3/2}\right] dx$$

$$= 4\int_{0}^{a}\left[\frac{b}{a}x^2\sqrt{a^2 - x^2} + \frac{b^3}{3a^3}(a^2 - x^2)^{3/2}\right] dx$$

[On putting $x = a \sin \theta$ and $dx = a \cos \theta \, d\theta$]

$$= 4\int_{0}^{\frac{\pi}{2}}\left(\frac{b}{a}.a^2 \sin^2\theta .a\cos\theta + \frac{b^3}{3a^3}a^3\cos^3\theta\right)\times a\cos\theta\, d\theta$$

$$= 4\int_{0}^{\frac{\pi}{2}}\left(a^3 b\sin^2\theta\cos^2\theta + \frac{ab^3}{3}\cos^4\theta\right) d\theta$$

$$= 4\left[a^3 b.\frac{1}{4}.\frac{1}{2}.\frac{\pi}{2} + \frac{ab^3}{3}.\frac{3}{4}.\frac{1}{2}.\frac{\pi}{2}\right]$$

$$= \frac{\pi}{4}(a^3 b + ab^3) = \frac{\pi}{4}ab\,(a^2 + b^2)$$ **Ans.**

Example 8. *Evaluate* $\iint (x^2 + y^2)\, dx\, dy$ *throughout the area enclosed by the curves* $y = 4x$, $x + y = 3$, $y = 0$ *and* $y = 2$.

Solution. Let OC represent $y = 4x$; BD, $x + y = 3$; OB, $y = 0$, and CD, $y = 2$. The given integral is to be evaluated over the area A of the trapezium $OCDB$.

Area $OCDB$ consists of area OCE, area $ECDF$ and area FDB.

The co-ordinates of C, D and B are $\left(\frac{1}{2}, 2\right)$ (1, 2) and (3, 0) respectively.

$$\therefore \quad \iint_A (x^2 + y^2)\, dy\, dx$$

$$= \iint_{OCE}(x^2 + y^2)\, dy\, dx + \iint_{ECDE}(x^2 + y^2)\, dy\, dx + \iint_{FDB}(x^2 + y^2)\, dy\, dx$$

$$= \underbrace{\int_0^{1/2} dx\int_0^{4x}(x^2 + y^2)\, dy}_{I_1} + \underbrace{\int_{1/2}^{1} dx\int_0^{2}(x^2 + y^2)\, dy}_{I_2} + \underbrace{\int_1^{3} dx\int_0^{3-x}(x^2 + y^2)\, dy}_{I_3}$$

Now, $$I_1 = \int_0^{1/2} dx\int_0^{4x}(x^2 + y^2)\, dy = \int_0^{1/2}\left[x^2 y + \frac{y^3}{3}\right]_0^{4x} dx = \int_0^{1/2}\frac{76}{3}x^3\, dx$$

$$= \frac{76}{3}\int_0^{1/2} x^3\, dx = \frac{76}{3}\left[\frac{x^4}{4}\right]_0^{1/2} = \frac{76}{3}\left[\frac{1}{4}.\frac{1}{16}\right] = \frac{19}{48}$$

$$I_2 = \int_{1/2}^{1} dx\int_0^{2}(x^2 + y^2)\, dy = \int_{1/2}^{1}\left[x^2 y + \frac{y^3}{3}\right]_0^{2} dx = \int_{1/2}^{1}\left(2x^2 + \frac{8}{3}\right) dx$$

$$= \left[\frac{2x^3}{3} + \frac{8}{3}x\right]_{1/2}^{1} = \left[\left(\frac{2}{3} + \frac{8}{3}\right) - \left(\frac{2}{3}.\frac{1}{8} + \frac{8}{3}.\frac{1}{2}\right)\right] = \frac{23}{12}$$

$$I_3 = \int_1^{3} dx\int_0^{3-x}(x^2 + y^2)\, dy = \int_1^{3}\left[x^2 y + \frac{y^3}{3}\right]_0^{3-x} dx = \int_1^{3}\left[x^2(3 - x) + \frac{(3 - x)^3}{3}\right] dx$$

$$= \int_1^3 \left[3x^2 - x^3 + \frac{(3-x)^3}{3}\right] dx = \left[x^3 - \frac{x^4}{4} - \frac{(3-x)^4}{3\times 4}\right]_1^3$$

$$= \left[27 - \frac{81}{4} - 0 - 1 + \frac{1}{4} + \frac{16}{12}\right] = \frac{22}{3}$$

$$\therefore \quad \int_A\int (x^2 + y^2)\, dy\, dx = I_1 + I_2 + I_3 = \frac{19}{48} + \frac{23}{12} + \frac{22}{3} = \frac{463}{48} = 9\frac{31}{48}. \qquad \textbf{Ans.}$$

EXERCISE 5.1

Evaluate

1. $\int_0^2\int_0^{x^2} e^{\frac{y}{x}}\, dy\, dx$ **Ans.** $e^2 - 1$

2. $\int_0^a\int_0^{\sqrt{ay}} xy\, dx\, dy$ **Ans.** $\frac{a^4}{6}$

3. $\int_0^a\int_0^{\sqrt{a^2-y^2}} dx\, dy$ **Ans.** $\frac{\pi a^2}{4}$

4. $\int_0^1\int_{y^2}^{y} (1 + xy^2)\, dx\, dy$ **Ans.** $\frac{41}{210}$

5. $\int_0^{2a}\int_0^{\sqrt{2ax-x}} xy\, dy\, dx$ **Ans.** $\frac{2a^4}{3}$

6. $\int_0^{2a}\int_0^{\sqrt{2ax-x^2}} x^2\, dy\, dx$ **Ans.** $\frac{5\pi a^4}{8}$

7. $\int_0^{a}\int_0^{\sqrt{a^2-x^2}} \sqrt{a^2 - x^2 - y^2}\, dy\, dx$ **Ans.** $\frac{\pi a^3}{4}$

8. $\int_0^{1}\int_0^{\sqrt{\frac{1}{2}(1-y^2)}} \frac{dx\, dy}{\sqrt{1 - x^2 - y^2}}$ **Ans.** $\frac{\pi}{4}$

9. $\int_0^{a}\int_0^{\sqrt{a^2-x^2}} \frac{dx\, dy}{(1+e^y)\sqrt{a^2 - x^2 - y^2}}$ **Ans.** $\frac{\pi}{2}\log\frac{2e^a}{1+e^a}$

10. $\int_0^{a}\int_y^{a} \frac{x\, dx\, dy}{\sqrt{x^2 + y^2}}$ **Ans.** $\frac{a^2}{2}\log(\sqrt{2} + 1)$

11 $\int_{x=0}^{1}\int_{y=0}^{2} (x^2 + 3xy^2)\, dx\, dy$ *(A.M.I.E.T.E., June 2009)* **Ans.** $\frac{14}{3}$

12. $\iint_A (5 - 2x - y)\, dx\, dy$, where A is given by $y = 0$, $x + 2y = 3$, $x = y^2$. **Ans.** $\frac{217}{60}$

13. $\iint_A xy\, dx\, dy$, where A is given by $x^2 + y^2 - 2x = 0$, $y^2 = 2x$, $y = x$. **Ans.** $\frac{7}{12}$

14. $\iint_A \sqrt{4x^2 - y^2}\, dx\, dy$, where A is the triangle given by $y = 0$, $y = x$ and $x = 1$. **Ans.** $\frac{1}{3}\left(\frac{\pi}{3} + \frac{\sqrt{3}}{2}\right)$

15. $\iint_R x^2\, dx\, dy$, where R is the two-dimensional region bounded by the curves $y = x$ and $y = x^2$. **Ans.** $\frac{1}{20}$

16. $\iint_A \sqrt{xy(1 + x - y)}\, dx\, dy$ where A is the area bounded by $x = 0$, $y = 0$ and $x + y = 1$. **Ans.** $\frac{2\pi}{105}$

5.3 EVALUATION OF DOUBLE INTEGRALS IN POLAR CO-ORDINATES

We have to evalaute $\int_{\theta_1}^{\theta_2}\int_{r_1(\theta)}^{r_2(\theta)} f(r, \theta)\, dr\, d\theta$ over the region bounded by the straight lines $\theta = \theta_1$ and $\theta = \theta_2$ and the curves $r = r_1(\theta)$ and $r = r_2(\theta)$. We first integrate with respect to r between the limits $r = r_1(\theta)$ and $r = r_2(\theta)$ and taking θ as constant. Then the resulting expression is integrated with respect to θ between the limits $\theta = \theta_1$ and $\theta = \theta_2$.

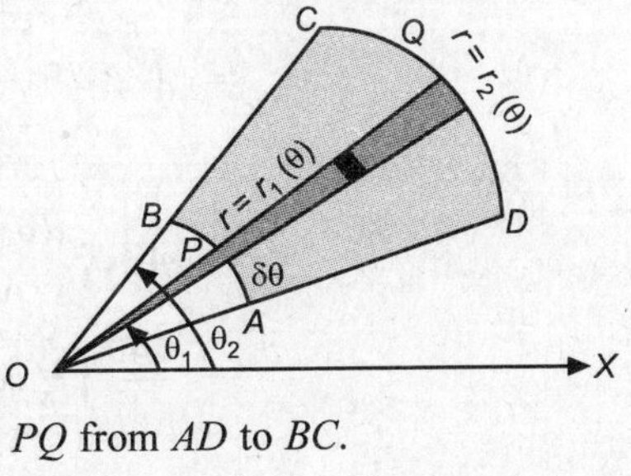

The area of integration is $ABCD$. On integrating first with respect to r, the strip extends from P to Q and the integration with respect to θ means the rotation ot this strip PQ from AD to BC.

Example 9. *Evaluate* $\int_0^{\pi/2}\left[\int_0^{a\cos\theta} r\sqrt{a^2-r^2}\,dr\right]d\theta.$

Solution. $I = \int_0^{\pi/2}\left[\int_0^{a\cos\theta} r\sqrt{a^2-r^2}\,dr\right]d\theta. = \int_0^{\frac{\pi}{2}}\left[\int_0^{a\cos\theta} -\frac{1}{2}(a^2-r^2)^{\frac{1}{2}}(-2r)\,dr\right]d\theta.$

$$= \int_0^{\pi/2}\left[-\frac{1}{2}\cdot\frac{(a^2-r^2)^{3/2}}{3/2}\right]_0^{a\cos\theta} d\theta. = -\frac{1}{3}\int_0^{\frac{\pi}{2}}\left[(a^2-a^2\cos^2\theta)^{\frac{3}{2}}-(a^2)^{\frac{3}{2}}\right]d\theta.$$

$$= -\frac{1}{3}\int_0^{\frac{\pi}{2}}\left[a^2(1-\cos^2\theta)^{\frac{3}{2}}-a^3\right]d\theta.$$

$$= -\frac{1}{3}\int_0^{\pi/2}(a^3\sin^3\theta - a^3)\,d\theta = -\frac{a^3}{3}\left[\frac{2}{3}-\frac{\pi}{2}\right] = \frac{a^3}{18}(3\pi-4).$$ **Ans.**

Example 10. *Evaluate* $\int_0^{\frac{\pi}{2}}\int_{a(1-\cos\theta)}^{a} r^2\,dr\,d\theta$

Solution. $\int_0^{\frac{\pi}{2}} d\theta \int_{a(1-\cos\theta)}^{a} r^2\,dr$

$$= \int_0^{\frac{\pi}{2}} d\theta\left[\frac{r^3}{3}\right]_{a(1-\cos\theta)}^{a} = \int_0^{\frac{\pi}{2}} d\theta\left(\frac{a^3}{3}-\frac{a^3(1-\cos\theta)^3}{3}\right)$$

$$= \frac{a^3}{3}\int_0^{\frac{\pi}{2}}[1-(1-\cos\theta)^3]\,d\theta = \frac{a^3}{3}\int_0^{\frac{\pi}{2}}[1-(1-3\cos\theta+3\cos^2\theta-\cos^3\theta]\,d\theta$$

$$= \frac{a^3}{3}\int_0^{\frac{\pi}{2}}(3\cos\theta-3\cos^2\theta+\cos^3\theta)\,d\theta$$

$$= \frac{a^3}{3}\left[[3\sin\theta]_0^{\frac{\pi}{2}} - 3\frac{1}{2}\frac{\pi}{2}+\frac{2}{3.1}\right] = \frac{a^3}{3}\left[3-\frac{3\pi}{4}+\frac{2}{3}\right] = \frac{a^3}{36}[44-9\pi]$$ **Ans.**

Example 11. *Evaluate* $\iint r^3\,dr\,d\theta$*, over the area bounded between the circles* $r = 2\cos\theta$ *and* $r = 4\cos\theta$. *(K. University, Dec. 2008)*

Solution. Here, we have

$r = 2\cos\theta$ (circle) ... (1)

$r = 4\cos\theta$ (circle) ... (2)

Let $I = \iint r^3\,dr\,d\theta$ ... (3)

$$= \int_{-\frac{\pi}{2}}^{\frac{\pi}{2}} d\theta \int_{2\cos\theta}^{4\cos\theta} r^3\,dr$$

$$= \int_{-\frac{\pi}{2}}^{\frac{\pi}{2}} d\theta\left[\frac{r^4}{4}\right]_{2\cos\theta}^{4\cos\theta}$$

$$= \frac{1}{4}\int_{-\frac{\pi}{2}}^{\frac{\pi}{2}} d\theta\,(256\cos^4\theta - 16\cos^4\theta)$$

$$= \frac{240}{4}\int_{-\frac{\pi}{2}}^{\frac{\pi}{2}}\cos^4\theta\,d\theta = 120\int_0^{\frac{\pi}{2}}\cos^4\theta\,d\theta$$ [$\cos^4\theta$ is an even function]

$$= 120\frac{\boxed{\frac{4+1}{2}}\boxed{\frac{0+1}{2}}}{2\boxed{\frac{4+1+0+1}{2}}} = 60\frac{\boxed{\frac{5}{2}}\boxed{\frac{1}{2}}}{\boxed{3}}$$

$$= 60\left[\frac{3}{2}\left(\frac{1}{2}\right)\boxed{\frac{1}{2}}\right]\boxed{\frac{1}{2}} = \frac{45}{2}(\sqrt{\pi})(\sqrt{\pi}) = \frac{45\pi}{2}$$ **Ans.**

Example 12. *Transform the integral to cartesian form and hence evaluate* $\int_0^{\pi}\int_0^{a} r^3 \sin\theta\cos\theta \, dr\, d\theta$.

Solution. Here, we have

$$\int_0^{\pi}\int_0^{a} r^3 \sin\theta\cos\theta \, dr\, d\theta \qquad ...(1)$$

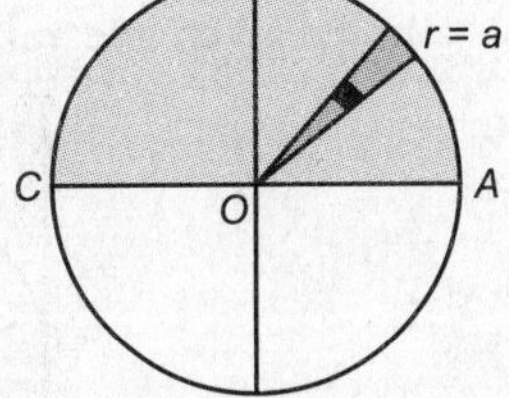

Here the region *i.e.*, semicircle ABC of integration is bounded by $r = 0$, *i.e.*, x-axis.

$r = a$ *i.e.*, circle, $\theta = 0$ and $\theta = \pi$ *i.e.*, x-axis in the second quadrant.

$$\int\int (r\sin\theta)(r\cos\theta)(r\, d\theta\, dr)$$

Putting $x = r\cos\theta$, $y = r\sin\theta$, $dx\, dy = r\, d\theta\, dr$ in (1), we get

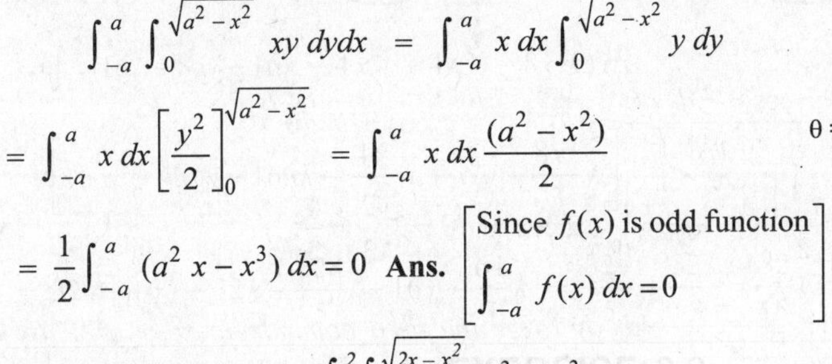

$$\int_{-a}^{a}\int_0^{\sqrt{a^2-x^2}} xy\, dydx = \int_{-a}^{a} x\, dx\int_0^{\sqrt{a^2-x^2}} y\, dy$$

$$= \int_{-a}^{a} x\, dx\left[\frac{y^2}{2}\right]_0^{\sqrt{a^2-x^2}} = \int_{-a}^{a} x\, dx\,\frac{(a^2-x^2)}{2}$$

$$= \frac{1}{2}\int_{-a}^{a}(a^2x - x^3)\, dx = 0 \quad \textbf{Ans.} \quad \left[\begin{array}{l}\text{Since } f(x) \text{ is odd function}\\ \int_{-a}^{a} f(x)\, dx = 0\end{array}\right]$$

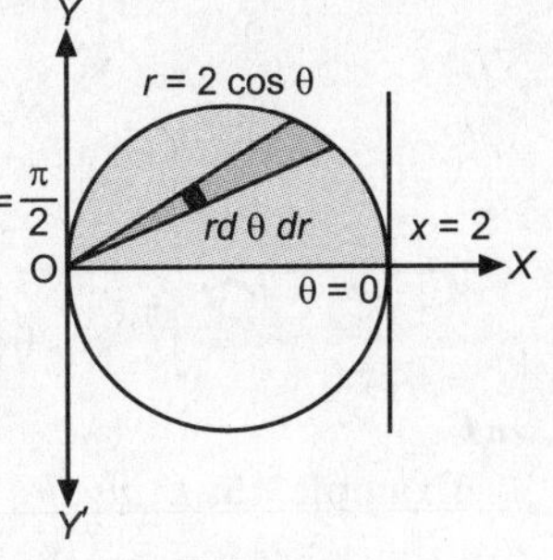

Example 13. *Evaluate* $\int_0^2\int_0^{\sqrt{2x-x^2}}(x^2+y^2)\, dydx$

Solution. $\int_0^2\int_0^{\sqrt{2x-x^2}}(x^2+y^2)\, dy\, dx$

Limits of $\quad y = \sqrt{2x-x^2} \Rightarrow y^2 = 2x - x^2 \Rightarrow x^2 + y^2 - 2x = 0 \qquad ...(1)$

(1) represents a circle whose centre is (1, 0) and radius = 1.

Lower limit of y is 0 *i.e.*, x-axis.

Region of integration is upper half circle.

Let us convert (1) into polar co-ordinate by putting

$$x = r\cos\theta,\ y = r\sin\theta$$

$$r^2 - 2r\cos\theta = 0 \Rightarrow r = 2\cos\theta$$

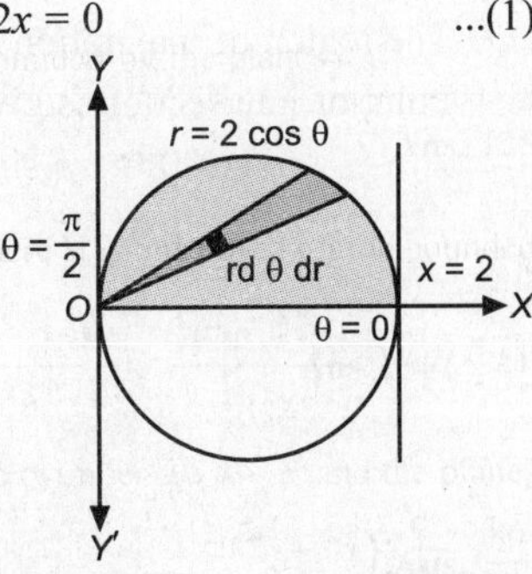

Limits of r are 0 to $2\cos\theta$

Limits of θ are 0 to $\frac{\pi}{2}$

$$\int_0^2\int_0^{\sqrt{2x-x^2}}(x^2+y^2)\, dy\, dx = \int_0^{\frac{\pi}{2}}\int_0^{2\cos\theta} r^2(r\, d\theta\, dr) = \int_0^{\frac{\pi}{2}} d\theta\int_0^{2\cos\theta} r^3\, dr = \int_0^{\frac{\pi}{2}} d\theta\left[\frac{r^4}{4}\right]_0^{2\cos\theta}$$

$$= 4\int_0^{\frac{\pi}{2}}\cos^4\theta\, d\theta = 4\times\frac{3\times1\times\pi}{4\times2\times2} = \frac{3\pi}{4}$$ **Ans.**

Example 14. *Evaluate* $\int_0^2 \int_0^{\sqrt{2x-x^2}} \dfrac{x\,dy\,dx}{\sqrt{x^2+y^2}}$ *by changing to polar coordinates.*

Solution. In the given integral, y varies from 0 to $\sqrt{2x-x^2}$ and x varies from 0 to 2.

$$y = \sqrt{2x - x^2}$$

$$\Rightarrow \quad y^2 = 2x - x^2$$

$$\Rightarrow \quad x^2 + y^2 = 2x$$

In polar co-ordinates, we have $r^2 = 2r\cos\theta \;\Rightarrow\; r = 2\cos\theta$.

$\therefore$ For the region of integarion, r varies from 0 to $2\cos\theta$ and θ varies from 0 to $\dfrac{\pi}{2}$.

In the given integral, replacing x by $r\cos\theta$, y by $r\sin\theta$, $dy\,dx$ by $r\,dr\,d\theta$, we have

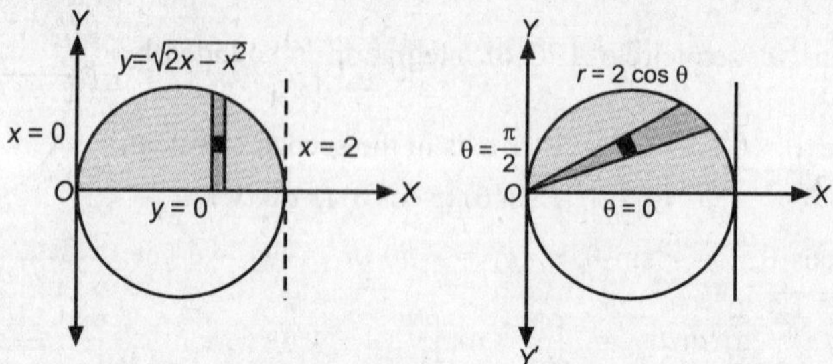

$$I = \int_0^{\pi/2}\int_0^{2\cos\theta} \frac{r\cos\theta . r\,dr\,d\theta}{r} = \int_0^{\pi/2}\int_0^{2\cos\theta} r\cos\theta\,dr\,d\theta$$

$$= \int_0^{\pi/2} \cos\theta \left[\frac{r^2}{2}\right]_0^{2\cos\theta} d\theta = \int_0^{\pi/2} 2\cos^3\theta\,d\theta = 2.\frac{2}{3} = \frac{4}{3}. \qquad \textbf{Ans.}$$

Example 15. *Evaluate* $\iint \dfrac{(x^2+y^2)^2}{x^2y^2}\,dx\,dy$ *over the area common to* $x^2 + y^2 = ax$ *and* $x^2 + y^2 = by$, $a, b > 0$. *(M.U. II Semester 2008)*

Solution. The boundary of area of integration are

$$x^2 + y^2 = ax \;\Rightarrow\; r^2 = ar\cos\theta \;\Rightarrow\; r = a\cos\theta$$

and
$$x^2 + y^2 = by \;\Rightarrow\; r^2 = br\sin\theta \;\Rightarrow\; r = b\sin\theta$$

The region of integration is bounded by $r = a\cos\theta$ and $r = b\sin\theta$.

Point of intersection is given by

$$ra\cos\theta = rb\sin\theta$$

$$\tan\theta = \frac{a}{b} \;\Rightarrow\; \theta = \tan^{-1}\frac{a}{b}$$

And
$$\frac{(x^2+y^2)^2}{x^2y^2} = \frac{r^4}{r^4\sin^2\theta\cos^2\theta} = \frac{1}{\sin^2\theta\cos^2\theta}$$

$$dx\,dy = r\,dr\,d\theta$$

$$\iint \frac{(x^2+y^2)^2}{x^2y^2}\,dx\,dy = \int_0^{\tan^{-1}\frac{a}{b}}\int_0^{b\sin\theta} \frac{1}{\sin^2\theta\cos^2\theta}\,r\,dr\,d\theta$$

$$+ \int_{\tan^{-1}\frac{a}{b}}^{\frac{\pi}{2}}\int_0^{a\cos\theta} \frac{1}{\sin^2\theta\cos^2\theta}\,r\,dr\,d\theta$$

$$= I_1 + I_2 \text{ (say)}$$

$$I_1 = \int_0^{\tan^{-1}\frac{a}{b}} \int_0^{b\sin\theta} \frac{1}{\sin^2\theta\cos^2\theta} r\,dr\,d\theta = \int_0^{\tan^{-1}\frac{a}{b}} \frac{1}{\sin^2\theta\cos^2\theta} d\theta \left[\frac{r^2}{2}\right]_0^{b\sin\theta}$$

$$= \int_0^{\tan^{-1}\frac{a}{b}} \frac{1}{\sin^2\theta\cos^2\theta}\left(\frac{b^2\sin^2\theta}{2}\right) d\theta = \frac{1}{2}b^2 \int_0^{\tan^{-1}\frac{a}{b}} \sec^2\theta\,d\theta = \frac{b^2}{2}[\tan\theta]_0^{\tan^{-1}\frac{a}{b}}$$

$$= \frac{1}{2}b^2 \tan\left(\tan^{-1}\frac{a}{b}\right) = \frac{1}{2}b^2\frac{a}{b} = \frac{ab}{2}.$$

$$I_2 = \int_{\tan^{-1}\frac{a}{b}}^{\frac{\pi}{2}} \int_0^{r = a\cos\theta} \frac{d\theta}{\sin^2\theta\cos^2\theta} r\,dr$$

$$= \int_{\tan^{-1}\frac{a}{b}}^{\frac{\pi}{2}} \frac{d\theta}{\sin^2\theta\cos^2\theta}\left[\frac{r^2}{2}\right]_0^{a\cos\theta} = \int_{\tan^{-1}\frac{a}{b}}^{\frac{\pi}{2}} \frac{d\theta}{\sin^2\theta\cos^2\theta}\left(\frac{a^2\cos^2\theta}{2}\right)$$

$$= \frac{a^2}{2}\int_{\tan^{-1}\frac{a}{b}}^{\frac{\pi}{2}} \frac{d\theta}{\sin^2\theta} = \frac{a^2}{2}\int_{\tan^{-1}\frac{a}{b}}^{\frac{\pi}{2}} \operatorname{cosec}^2\theta\,d\theta = \frac{a^2}{2}[-\cot\theta]_{\tan^{-1}\frac{a}{b}}^{\frac{\pi}{2}}$$

$$= -\frac{a^2}{2}\left[\cot\frac{\pi}{2} - \cot\left(\tan^{-1}\frac{a}{b}\right)\right] = -\frac{a^2}{2}\left[0 - \frac{b}{a}\right] = \frac{ab}{2}$$

$$\iint \frac{(x^2+y^2)^2}{x^2y^2}\,dx\,dy = I_1 + I_2 = \frac{ab}{2} + \frac{ab}{2} = ab \qquad \textbf{Ans.}$$

Example 16. *Evaluate :* $\int_0^\infty\int_0^\infty e^{-(x^2+y^2)}\,dx\,dy$ *by changing to polar co-ordinates.*

Hence, show that $\int_0^\infty e^{-x^2}\,dx = \frac{\sqrt{\pi}}{2}$ *(AMIETE, June 2010)*

Solution. From the limit of integration, we find that the first integration is along a vertical strip extending from $y = 0$ to $y = \infty$. The strip slides from $x = 0$ and goes to $x = \infty$. Thus the region of integration is the whole of first quadrant.

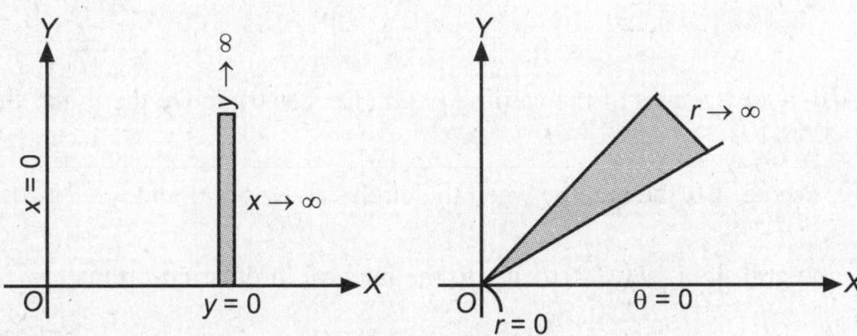

This region can be covered by radial strips extending from $r = 0$ to $r = \infty$. The strip starts from $\theta = 0$ and goes upto $\theta = \pi/2$.

Hence, $\int_0^\infty\int_0^\infty e^{-(x^2+y^2)}\,dx\,dy = \int_0^{\pi/2}\int_0^\infty e^{-r^2}\,r\,dr\,d\theta$

$$= -\frac{1}{2}\int_0^{\pi/2}\int_0^\infty(-2r)\,e^{-r^2}\,dr\,d\theta = -\frac{1}{2}\int_0^{\pi/2}[e^{-r^2}]_0^\infty\,d\theta$$

$$= -\frac{1}{2}\int_0^{\pi/2}(0-1)\,d\theta = \frac{1}{2}\int_0^{\pi/2}1\,d\theta = \frac{\pi}{4} \qquad \textbf{Ans.}$$

Let $$I = \int_0^\infty e^{-x^2}\,dx \qquad ...(1)$$

Also, $$I^1 = \int_0^\infty e^{-y^2}\, dy \qquad \text{...(2) [Property of definite integrals]}$$

Multiplying (1) and (2), we get

$$I^2 = \int_0^\infty \int_0^\infty e^{-(x^2+y^2)}\, dx\, dy = \frac{\pi}{4} \qquad \text{[As obtained above]}$$

$$\Rightarrow \qquad I = \sqrt{\frac{\pi}{4}} = \frac{\sqrt{\pi}}{2}$$ **Proved.**

EXERCISE 5.2

Evaluate the following:

1. $\int_0^\pi \int_0^{a(1-\cos\theta)} 2\pi r^2 \sin\theta\, d\theta\, dr$ **Ans.** $\frac{8}{3}\pi a^3$

2. $\int_0^\pi \int_0^{a(1+\cos\theta)} r^2 \cos\theta\, dr\, d\theta$ **Ans.** $\frac{5}{8}\pi a^3$

3. $\iint_A \frac{r\, dr\, d\theta}{\sqrt{r^2+a^2}}$ where A is a loop of $r^2 = a^2 \cos 2\theta$ **Ans.** $2a - \frac{\pi a}{2}$

4. $\iint_A r^2 \sin\theta\, d\theta\, dr$ where A is $r = 2a\cos\theta$ above initial line.

5. Calculate the integral $\iint \frac{(x-y)^2}{x^2+y^2}\, dx\, dy$ over the circle $x^2 + y^2 \le 1$. **Ans.** $\pi - 2$

6. $\int (x^2+y^2)\, x\, dx\, dy$ over the possitive quadrant of the circle $x^2 + y^2 = a^2$ by changing to polar coordinates.

Ans. $\frac{a^2}{5}$

7. $\iint_R \sqrt{x^2+y^2}\, dx\, dy$ by changing to polar coordinates, R is the region in the xy-plane bounded by the circles $x^2 + y^2 = 4$ *(AMIETE, Dec. 2009)* **Ans.** $\frac{38\pi}{3}$

8. Convert into polar coordinates

$\int_0^{2a} \int_0^{2ax-x^2} dx\, dy$ **Ans.** $\int_0^{\pi/2} \int_0^{2a\cos\theta} r\, d\theta\, dr$

9. $\iint r \sin\theta\, dr\, d\theta$ over the area of the cardiod $r = a(1 + \cos\theta)$ above the initial line. **Ans.** $\frac{5}{8}\pi a^3$

10. $\int\int_A x^2\, dr\, d\theta$, where A is the area between the circles $r = a\cos\theta$ and $r = 2a\cos\theta$. **Ans.** $\frac{28a^3}{9}$

11. Transform the integral $\int_0^1 \int_0^x f(x, y)\, dy\, dx$ to the integral in polar co-ordinates.

Ans. $\int_0^{\pi/4} \int_0^{\sec\theta} f(r, \theta)\, r\, d\theta\, dr$

5.4 CHANGE OF ORDER OF INTEGRATION

On changing the order of integration, the limits of integration change. To find the new limits, we draw the rough sketch of the region of integration.

Some of the problems connected with double integrals, which seem to be complicated, can be made easy to handle by a change in the order of integration.

Example 17. *Evaluate* $\int_0^a \int_y^a \frac{x}{x^2+y^2}\, dx\, dy$ *by changing the order of integration.*

(AMIETE, June 2010, Nagpur University, Summer 2008)

Solution. Here we have

$$I = \int_0^a \int_y^a \frac{x}{x^2 + y^2}\, dxdy$$

Here $\quad x = a, x = y, y = 0$ and $y = a$

The area of integration is OAB.

On changing the order of integration Lower limit of $y = 0$ and upper limit is $y = x$.

Lower limit of $x = 0$ and upper limit is $x = a$.

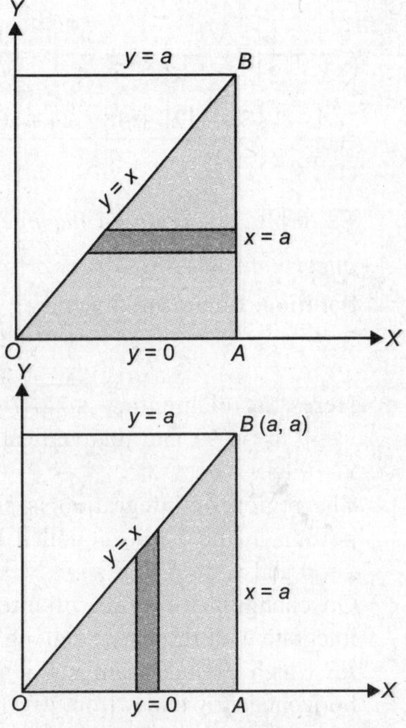

$$I = \int_0^a xdx = \int_0^{y=x} \frac{1}{x^2 + y^2} dy$$

$$= \int_0^a xdx \left[\frac{1}{x}\tan^{-1}\frac{y}{x}\right]_0^{y=x}$$

$$= \int_0^a \frac{x}{x} dx \left(\tan^{-1}\frac{x}{x} - \tan^{-1} 0\right)$$

$$= \int_0^a dx \left(\frac{\pi}{4}\right) \quad = \frac{\pi}{4}[x]_0^a = \frac{a\pi}{4} \quad \textbf{Ans.}$$

Example 18. *Change the order of integration in* $I = \int_0^1 \int_{x^2}^{2-x} xy\, dx\, dy$ *and hence evaluate the same.*

(*A.M.I.E.T.E., June 2010, 2009*)

Solution. $I = \int_0^1 \int_{x^2}^{2-x} xy\, dxdy$

The region of integration is shown by shaded portion in the figure bounded by parabola $y = x^2$ and the line $y = 2 - x$.

The point of intersection of the parabola $y = x^2$ and the line $y = 2 - x$ is B (1, 1).

In the figure below (left) we have taken a strip parallel to y-axis and the order of integration is

$$\int_0^1 x\, dx \int_{x^2}^{2-x} y\, dy$$

In the second figure above we have taken a strip parallel to x-axis in the area OBC and second strip in the area ABC. The limits of x in the area OBC are 0 and $\sqrt{y}$ and the limits of x in the area ABC are 0 and $2 - y$.

$$= \int_0^1 y\, dy \int_0^{\sqrt{y}} x\, dx + \int_1^2 y dy \int_0^{2-y} x\, dx = \int_0^1 y\, dy \left[\frac{x^2}{2}\right]_1^{\sqrt{y}} + \int_1^{\sqrt{z}} y\, dy \left[\frac{x^2}{2}\right]_0^{2-y}$$

$$= \frac{1}{2}\int_0^1 y^2\, dy + \frac{1}{2}\int_1^2 y\,(2 - y)^2\, dy = \frac{1}{2}\left[\frac{y^3}{3}\right]_0^1 + \frac{1}{2}\int_1^2 (4y - 4y^2 + y^3)\, dy$$

$$= \frac{1}{6}+\frac{1}{2}\left[2y^2-\frac{4}{3}y^3+\frac{y^4}{4}\right]_1^2 = \frac{1}{6}+\frac{1}{2}\left[8-\frac{32}{3}+4-2+\frac{4}{3}-\frac{1}{4}\right]$$

$$= \frac{1}{6}+\frac{1}{2}\left[\frac{96-128+48-24+16-3}{12}\right] = \frac{1}{6}+\frac{5}{24} = \frac{9}{24} = \frac{3}{8}$$ **Ans.**

Example 19. *Evaluate the integral* $\int_0^\infty \int_0^x x \exp\left(-\frac{x^2}{y}\right) dx\, dy$ *by changing the order of integration* (*U.P. I Semester Dec., 2005*)

Solution. Limits are given

$$y = 0 \text{ and } y = x$$
$$x = 0 \text{ and } x = \infty$$

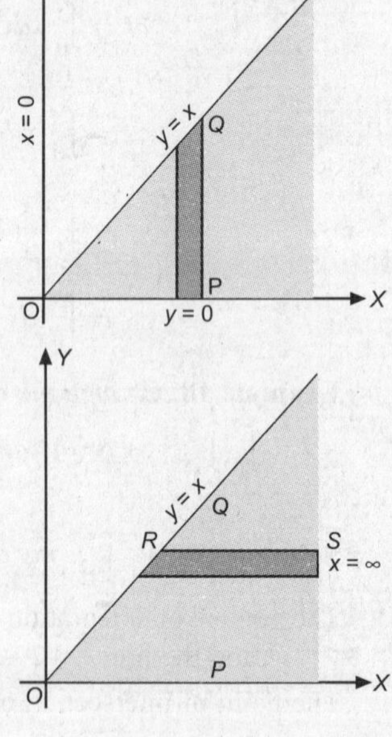

Here, the elementary strip PQ extends from $y = 0$ to $y = x$ and this vertical strip slides from $x = 0$ to $x = \infty$.

The region of integration is shown by shaded portion in the figure bounded by $y = 0$, $y = x$, $x = 0$ and $x = \infty$.

On changing the order of integration, we first integrate with respect to x along a horizontal strip RS which extends from $x = y$ to $x = \infty$ and this horizontal strip slides from $y = 0$ to $y = \infty$ to cover the given region of integration.

New limits :

$$x = y \quad \text{and} \quad x = \infty$$
$$y = 0 \quad \text{and} \quad y = \infty$$

We first integrate with respect to x.

Thus,

$$\int_0^\infty dy \int_y^\infty x e^{-\frac{x^2}{y}} dx = \int_0^\infty dy \int_y^\infty -\frac{y}{2}\left(-\frac{2x}{y} e^{-\frac{x^2}{y}}\right) dx$$

$$= \int_0^\infty dy \left[-\frac{y}{2} e^{-\frac{x^2}{y}}\right]_y^\infty = \int_0^\infty dy \left[0+\frac{y}{2} e^{-\frac{y^2}{y}}\right] = \int_0^\infty \frac{y}{2} e^{-y}\, dy$$

$$= \left[\frac{y}{2}(-e^{-y}) - \left(\frac{1}{2}\right)(e^{-y})\right]_0^\infty$$ (Integrating by parts)

$$= \left[(0-0) - \left(0-\frac{1}{2}\right)\right] = \frac{1}{2}$$ **Ans.**

Example 20. *Evaluate the double integral.*

$\int_0^1 \int_x^1 \sin(y^2)\, dy\, dx$ (*B.P.U.T.; I Semester 2008*)

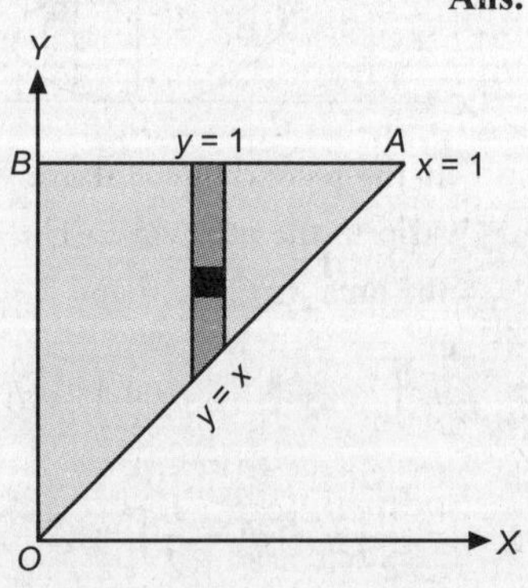

Solution. Here, we have $\int_0^1 \int_x^1 \sin(y^2)\, dy\, dx$

The region OAB integration is bounded by the straight lines $y = x$, $x = 0$ and $y = 1$. A strip is drawn parallel to y axis. y varies from x to 1 and x varies from 0 to 1.

In the given problem the integration is first w.r.t. y and then w.r.t. x. But by this way the evaluation of integral is difficult.

So we change the order of integration. Now we will integrate first w.r.t. x and then w.r.t. y. Here we draw a strip parallel to x axis. On this strip x varies from 0 to y and y varies from 0 to 1.

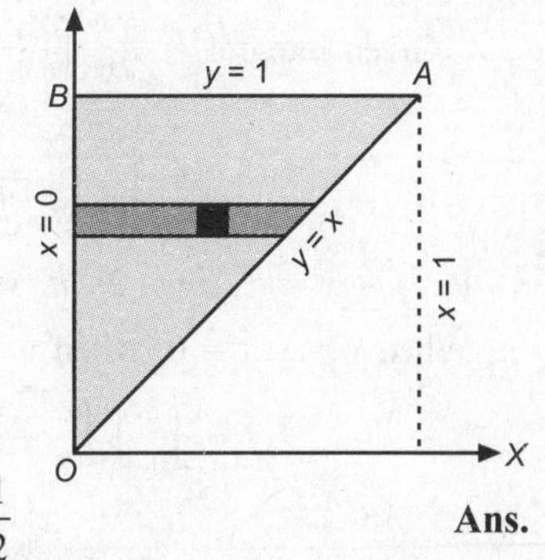

Hence, $\int_0^1 dx \int_x^1 \sin(y^2)\, dy$

$$= \int_0^1 \sin(y^2)\, dy \int_0^y dx = \int_0^1 \sin(y^2)\, dy\, [x]_0^y$$

$$= \int_0^1 \sin y^2 . (y)\, dy = \left[\frac{-\cos y^2}{2}\right]_0^1 - \left(\frac{\cos 1}{2} - \frac{1}{2}\right) = \frac{1}{2} - \cos\frac{1}{2} \qquad \textbf{Ans.}$$

Example 21. *Change the order of the integration*

$$\int_0^\infty \int_0^x e^{-xy}\, y\, dy\, dx$$

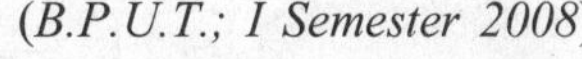
(*B.P.U.T.; I Semester 2008*)

Solution. Here, we have

$$\int_0^\infty \int_0^x e^{-xy}\, y\, dy\, dx$$

Here the region OAB of integration is bounded by $y = 0$ (x-axis), $y = x$ (a straight line), $x = 0$, *i.e.*, y axis. A strip is drawn parallel to y-axis, y varies 0 to x and x varies 0 to ∞.

On changing the order of integration, first we integrate w.r.t. x and then w.r.t. y.

A strip is drawn parallel to x-axis. On this strip x varies from y to ∞ and y varies from 0 to ∞.

Hence $\int_0^\infty \int_0^x e^{-xy}\, y\, dy\, dx = \int_0^\infty y\, dy \int_y^\infty e^{-xy} dx$

$$= \int_0^\infty y\, dy \left(\frac{e^{-xy}}{-y}\right)_y^\infty$$

$$= \int_0^\infty \frac{y\, dy}{-y}[0 - e^{-y^2}]$$

$$= \int_0^\infty e^{-y^2}\, dy = \frac{1}{2}\sqrt{\pi} \quad \textbf{Ans.}$$

Example 22. *Change the order of integration and evaluate:*

$$\int_0^a \int_0^y \frac{dx\, dy}{\sqrt{(a^2 + x^2)(a - y)(y - x)}}$$

Solution. Here we have

$$\int_0^a \int_0^y \frac{dx\, dy}{\sqrt{(a^2 + x^2)(a - y)(y - x)}}$$

The region of integration is bounded by $x = 0$ (y axis) $x = y$ (straight line), $y = 0$ (x-axis), $y = a$ (straight line).

Here we integrate first w.r.t. x and then y.

On changing the order of integration we have to integrate first w.r.t. y and then x.

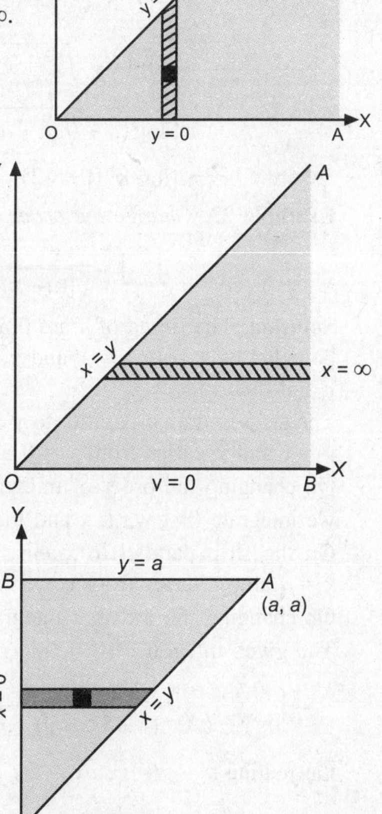

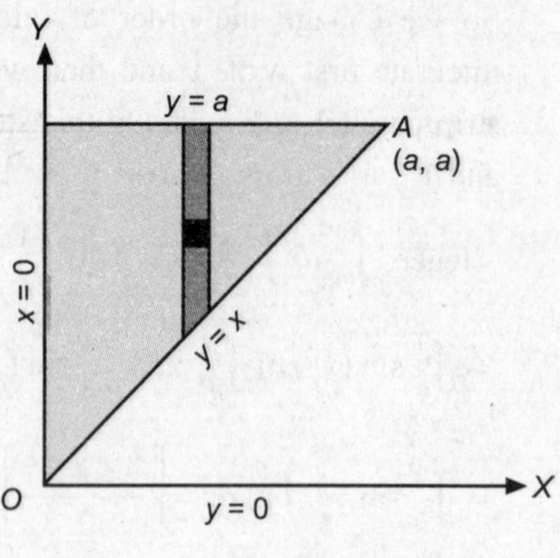

Given integral $= \int_0^a \int_0^y \frac{dx\,dy}{\sqrt{(a^2+x^2)(a-y)(y-x)}}$

$$= \int_0^a \int_x^a \frac{dy\,dx}{\sqrt{(a^2+x^2)(a-y)(y-x)}}$$

Put $y - x = t^2$, $dy = 2t\,dt$

When $y = x$, $t = 0$, when $y = a$, $t = \sqrt{a-x}$

$$\therefore \quad I = \int_0^a \int_0^{\sqrt{a-x}} \frac{dx}{\sqrt{a^2+x^2}} \cdot \frac{2t\,dt}{\sqrt{(a-x)-t^2}}$$

$$= 2\int_0^a \frac{dx}{\sqrt{a^2+x^2}} \left[\sin^{-1}\frac{t}{\sqrt{a-x}}\right]_0^{\sqrt{a-x}}$$

$$= 2\int_0^a \frac{1}{\sqrt{a^2+x^2}}[\sin^{-1}1 - \sin^{-1}0]\,dx$$

$$= 2\int_0^a \frac{1}{\sqrt{a^2+x^2}} \cdot \frac{\pi}{2}\,dx$$

$$= 2\left(\frac{\pi}{2}\right)\int_0^a \frac{1}{\sqrt{a^2+x^2}}\,dx$$

$$= \pi\left[\log\left(x+\sqrt{a^2+x^2}\right)\right]_0^a$$

$$= \pi\left[\log\left(a+\sqrt{a^2+a^2}\right) - \log\left(0+\sqrt{a^2+0}\right)\right] = \pi\,[\log\,(a+\sqrt{2}a) - \log a]$$

$$= \pi\,[\log a\,(1+\sqrt{2}) - \log a] = \pi\,\log\,(1+\sqrt{2})$$ **Ans.**

Example 23. *Change the order of integration and evaluate*

$$\int_0^a \int_0^x \frac{\sin y\,dy\,dx}{\sqrt{[(a-x)(x-y)](4-5\cos y)^2}}$$

Solution. The limits of y are 0 and x, that of x are 0 and a. The area OAB of integration is bounded by $y = 0$, $y = x$ and $x = 0$ and $x = a$. The given function is integrated first w.r.t. y and then x.

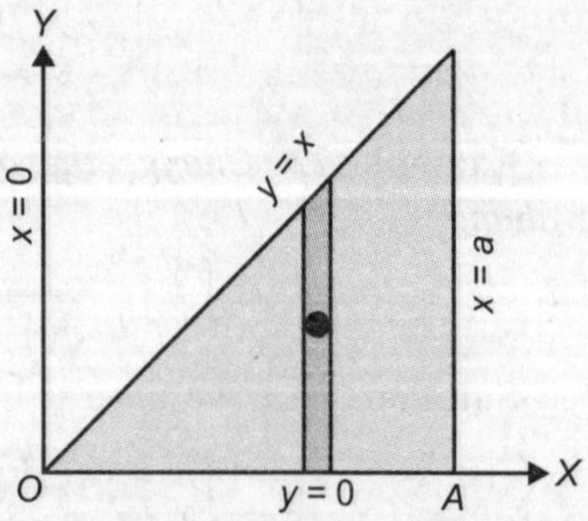

The strip is drawn parallel to y-axis and varies $y = 0$ and $y = x$ and x varies from $x = 0$ and $x = a$.

On changing the order of integration we integrate first w.r.t. x and then y.

On the strip parallel to x-axis, x varies from $x = y$ to $x = a$ and y varies from $y = 0$ to $y = a$.

On changing the order of integration.

The given integral

$$= \int_{y=0}^{y=a} \int_{x=y}^{x=a} \frac{\sin y}{(4-5\cos y)} \cdot \frac{dx}{\sqrt{[(a-x)(x-y)]}}\,dy \quad ...(1)$$

On putting $x - y = t^2$, $dx = 2\,t\,dt$ in (1), we get

$$= \int_0^a \int_0^{\sqrt{a-y}} \frac{\sin y}{(4-5\cos y)} \cdot \frac{2t\,dt}{\sqrt{[(a-y-t^2)]\,t}}\,dy$$

$$= \int_0^a \frac{\sin y\, dy}{(4-5\cos y)} \int_0^{\sqrt{a-y}} \frac{2t\, dt}{\sqrt{[(a-y)-t^2]\,t}}$$

$$\left[\begin{array}{l}\text{Put } x-y=t^2 \;\therefore\; dx = 2t\,dt \\ \text{When } x = y, \text{ then } t=0; \\ \text{when } x=a, \text{ then } t=\sqrt{a-y}\end{array}\right]$$

$$= \int_0^a \frac{\sin y\, dy}{(4-5\cos y)} \int_0^{\sqrt{a-y}} \frac{2dt}{\sqrt{(a-y)-t^2}}$$

$$= 2\int_0^a \frac{\sin y\, dy}{4-5\cos y} \cdot \left[\sin^{-1}\left(\frac{t}{\sqrt{a-y}}\right)\right]_0^{\sqrt{a-y}} dy$$

$$= 2\int_0^a \frac{\sin y}{4-5\cos y}\, dy \left(\sin^{-1}\frac{\sqrt{a-y}}{\sqrt{a-y}} - \sin^{-1} 0\right)$$

$$= 2\int_0^a \frac{\sin y}{4-5\cos y}\, dy\, [\sin^{-1}(1)] = 2 \cdot \frac{\pi}{2}\int_0^a \frac{\sin y\, dy}{4-5\cos y}$$

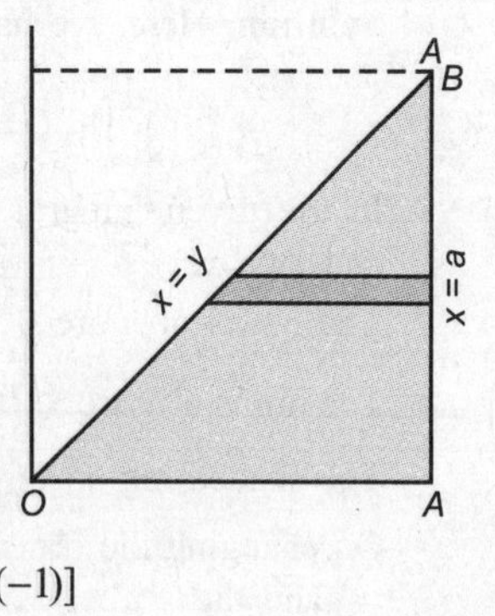

$$= \pi\left[\frac{1}{5}\log(4-5\cos y)\right]_0^a = \frac{\pi}{5}[\log(4-5\cos a) - \log(-1)]$$

$$= \frac{\pi}{5}\log\frac{(4-5\cos a)}{-1} = \frac{\pi}{5}\log(5\cos a - 4)$$ **Ans.**

Example 24. *Change the order of integration and evaluate* $\int_0^{a/\sqrt{2}} \int_x^{\sqrt{a^2-x^2}} y^2\, dA$

(Gujarat, I Semester, Jan,. 2009)

Solution. We have,

$$\int_0^{a/\sqrt{2}} \int_x^{\sqrt{a^2-x^2}} y^2\, dA$$

Here the limits are

$$x = 0$$
$$x = \frac{a}{\sqrt{2}}$$
$$y = x$$
$$y = \sqrt{a^2-x^2} \Rightarrow x^2+y^2=a^2$$

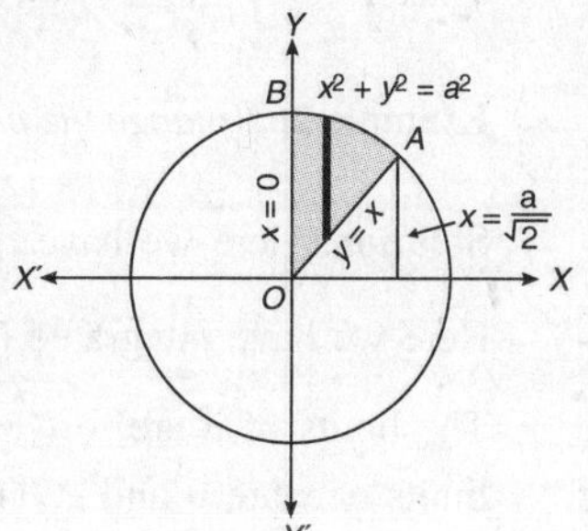

Area of integration is shaded area OAB in the figure.

On changing the order of integration we integrate first w.r.t. 'x' and then w.r.t. 'y'

In this way we have to divide the area of integration in two parts OAC and ABC.

$$\int_0^{\frac{a}{\sqrt{2}}} dx \int_x^{\sqrt{a^2-x^2}} y^2 dy$$

$$= \int_0^{\frac{a}{\sqrt{2}}} y^2 dy \int_0^y dx + \int_{\frac{a}{\sqrt{2}}}^a y^2 dy \int_0^{\sqrt{a^2-y^2}} dx$$

$$= \int_0^{\frac{a}{\sqrt{2}}} y^2 dy (x)_0^y + \int_{\frac{a}{\sqrt{2}}}^a y^2 dy (x)_0^{\sqrt{a^2-y^2}}.$$

$$= \int_0^{\frac{a}{2}} y^2\, dy (y) + \int_{\frac{a}{\sqrt{2}}}^a y^2\, dy \sqrt{a^2-y^2} \quad (\text{Put } y = a\sin\theta)$$

$$= \int_0^{\frac{a}{\sqrt{2}}} y^3\, dy + \int_{\frac{\pi}{4}}^{\frac{\pi}{2}} a^2 \sin^2\theta (a\cos\theta) a\cos\theta\, d\theta = \left[\frac{y^4}{4}\right]_0^{\frac{a}{\sqrt{2}}} + a^4 \int_{\frac{\pi}{4}}^{\frac{\pi}{2}} \sin^2\theta\cos^2\theta\, d\theta$$

$$= \frac{\frac{a^4}{4}}{4} + a^4 \int_{\frac{\pi}{4}}^{\frac{\pi}{2}} \sin^2\theta(1-\sin^2\theta)\, d\theta = \frac{a^4}{16} + a^4\left(\frac{\pi}{32}\right) == a^4\left(\frac{\pi+2}{32}\right)$$ **Ans.**

Example 25. *Change the order of integration*

$$\int_0^1 \int_{\sqrt{2x-x^2}}^{1+\sqrt{1-x^2}} f(x,y)\,dy\,dx \quad (M.U.\ II\ Semester\ 2009)$$

Solution. Here, we have

$$\int_0^1 \int_{\sqrt{2x-x^2}}^{1+\sqrt{1-x^2}} f(x,y)\,dy\,dx$$

In the given integral problem it is integrated first w.r.t. y and then x.

The limits of x are 0 and 1.

The limits of y are $\sqrt{2x-x^2}$ (circle) and $1+\sqrt{1-x^2}$ (circle).

The shaded region of integration is bounded by two circles.

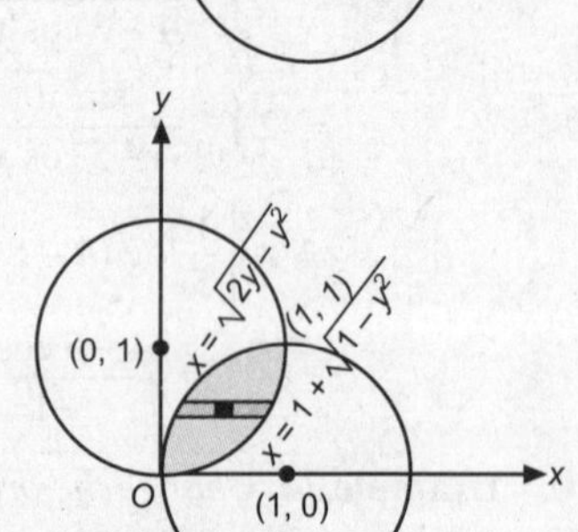

On changing the order of integration we integrate first w.r.t. 'x' and then 'y'.

The limits of y are 0 and 1.

The limits of x are $1+\sqrt{1-y^2}$ and $\sqrt{2y-y^2}$.

Thus $\displaystyle\int_0^1 dx \int_{\sqrt{2x-x^2}}^{1+\sqrt{1-x^2}} f(x,y)\,dy = \int_0^1 dy \int_{\sqrt{2y-y^2}}^{1+\sqrt{1-y^2}} f(x,y)\,dx$

Example 26. *Change the order of integration* $\displaystyle\int_0^a \int_{\sqrt{a^2-x^2}}^{x+3a} f(x,y)\,dx\,dy$. *(M.U., II Sem. 2008)*

Solution. Here we have $\displaystyle\int_0^a \int_{\sqrt{a^2-x^2}}^{x+3a} f(x,y)\,dx dy \quad ...(1)$

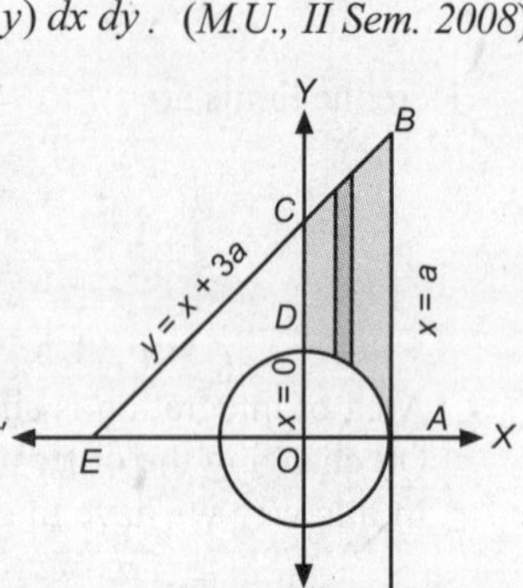

Here we have integrated (1) first w.r.t. 'y' and then x.

The limits of y are $\sqrt{a^2-x^2}$ (circle) and $x + 3a$ and the limits of x are 0 and a. The shaded portion $ABCDA$ of the region of the integration is bounded by $y = \sqrt{a^2-x^2}$ (circle), $y = x + 3a$ (straight line) $x = 0$ (y-axis) and $x = a$ (a straight line).

On changing the order of integration we have to integrate (1) w.r.t. to x first and then y.

For this way we have to divide the region $ABCDA$ of integration into three parts AFD, $DFGC$ and BCG.

One part is $AFDA$ in which the limits of x are $\sqrt{a^2-y^2}$ and $x = a$ and the limits of y are 0 and a.

Second part is $FGCD$ in which the limits of $x = 0$ and $x = a$ and the limits of y are a and $3a$.

In third part $BCGB$ the limits of x are $x = y - 3a$ and $x = a$ and the limits of y are $3a$ and $4a$.

Hence, $\displaystyle\int_0^a \int_{\sqrt{a^2-x^2}}^{x+3a} f(x,y)\,dy\,dx$

$$= \int_0^a \int_{\sqrt{a^2-y^2}}^{a} f(x,y)\,dx\,dy + \int_a^{3a} \int_0^a f(x,y)\,dx\,dy + \int_{3a}^{4a} \int_{y-3a}^{a} f(x,y)\,dx\,dy \qquad \textbf{Ans.}$$

Example 27. *Change the order of integration in the double integral*

Solution. Limits are given as $\int_0^{2a}\int_{\sqrt{2ax-x^2}}^{\sqrt{2ax}} V\,dx\,dy$

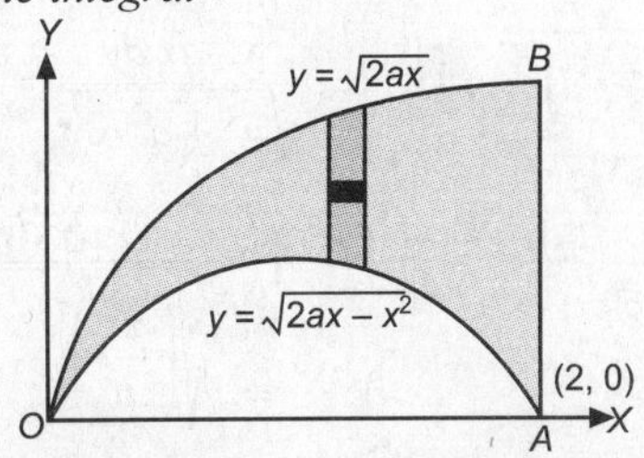

$$x = 0,\ x = 2a$$

$$y = \sqrt{2ax}$$

and $y = \sqrt{2ax - x^2} \quad \Rightarrow \quad y^2 = 2ax - x^2$

and $\qquad (x-a)^2 + y^2 = a^2$

The area of integration is the shaded portion *OAB*. On changing the order of integration first we have to integrate w.r.t. *x*, The area of integration has three portions *BCE*, *ODE* and *ACD*.

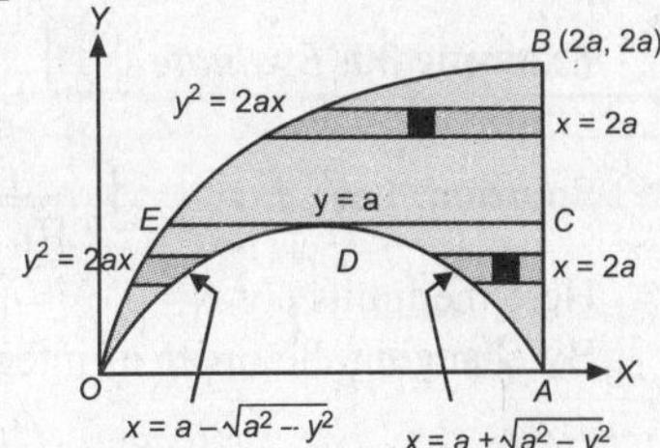

$$\int_0^{2a} dx \int_{\sqrt{2ax-x^2}}^{\sqrt{2ax}} V\,dy$$

$$= \int_a^{2a} dy \int_{y^2/2a}^{2a} V\,dx + \int_0^a dy \int_{y^2/2a}^{a+\sqrt{a^2-y^2}} V\,dx$$

$$+ \int_0^a dy \int_{a+\sqrt{a^2-y^2}}^{2a} V\,dx \qquad \textbf{Ans.}$$

Example 28. *Changing the order of integration of* $\int_0^\infty\int_0^\infty e^{-xy}\sin nx\,dx\,dy$ *show that*

$$\int_0^\infty \frac{\sin nx}{x}\,dx = \frac{\pi}{2}.$$

(*AMIETE, Dec. 2010*)

Solution. The region of integration is bounded by $x = 0, x = \infty, y = 0, y = \infty$, *i.e.*, first quadrant.

$$\int_0^\infty\int_0^\infty e^{-xy}\sin nx\,dx\,dy = \int_0^\infty dy \int_0^\infty e^{-xy}\sin nx\,dx$$

$$= \int_0^\infty dy\left[\frac{e^{-xy}}{n^2+y^2}\{-y\sin x - n\cos nx\}\right]_0^\infty$$

$$= \int_0^\infty dy\left[0 + \frac{n}{n^2+y^2}\right] = \int_0^\infty \frac{n}{n^2+y^2}\,dy = \left[\tan^{-1}\left(\frac{y}{n}\right)\right]_0^\infty = \frac{\pi}{2} \quad \text{...(1)}$$

On changing the order of integration

$$\int_0^\infty\int_0^\infty e^{-xy}\sin nx\,dx\,dy = \int_0^\infty \sin nx\,dx \int_0^\infty e^{-xy}\,dy$$

$$= \int_0^\infty \sin nx\,dx\left[\frac{e^{-xy}}{-x}\right]_0^\infty = \int_0^\infty \frac{\sin nx\,dx}{x}\left[-\frac{1}{e^{xy}}\right]_0^\infty$$

$$= \int_0^\infty \frac{\sin nx}{x}\,dx\,[-0+1] = \int_0^\infty \frac{\sin nx}{x}\,dx \quad \text{...(2)}$$

From (1) and (2), $\int_0^\infty \frac{\sin nx}{x}\,dx = \frac{\pi}{2}$ **Proved.**

Example 29. *Change order of integration and hence evaluate:*

$$\int_0^a\int_{\sqrt{ax}}^a \frac{y^2\,dx\,dy}{\sqrt{y^4 - a^2x^2}}$$

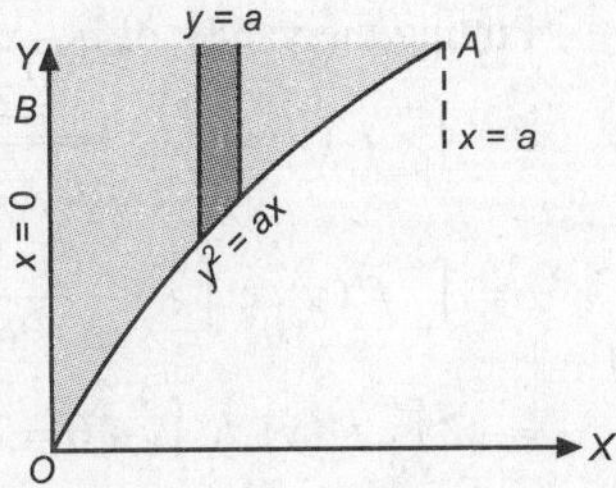

Solution. The given limits show that the area of integration lies between $y^2 = ax$, $y = a$, $x = 0$ and $x = a$. We can consider it as lying between $y = 0$, $y = a$, $x = 0$ and $x = y^2/a$ by changing the order of integration. Hence, the given integral.

$$\int_{x=0}^{a}\int_{y=\sqrt{ax}}^{a}\frac{y^2\,dx\,dy}{\sqrt{y^4-a^2x^2}}=\int_{y=0}^{a}\int_{x=0}^{\frac{y^2}{a}}\frac{y^2\,dy\,dx}{\sqrt{y^4-a^2x^2}}$$

$$=\frac{1}{a}\int_0^a\int_0^{\frac{y^2}{a}}\frac{y^2\,dy\,dx}{\sqrt{\left(\frac{y^2}{a}\right)^2-x^2}}\qquad=\frac{1}{a}\int_a^a y^2\left[\sin^{-1}\left(\frac{ax}{y^2}\right)\right]_0^{y^2/a}dy$$

$$=\frac{1}{a}\int_0^a y^2\,[\sin^{-1}(1)-\sin^{-1}(0)]\,dy=\frac{\pi}{2a}\int_0^a y^2\,dy=\frac{\pi}{2a}\left(\frac{y^3}{3}\right)_0^a=\frac{\pi}{6a}(a^3)=\frac{\pi a^2}{6}.\ \textbf{Ans.}$$

Example 30. *Evaluate* $\int_0^a\int_0^x\frac{f'(y)\,dy\,dx}{[(a-x)(x-y)]^{1/2}}$

Solution. Let $I=\int_0^a\int_0^x\frac{f'(y)\,dy\,dx}{[(a-x)(x-y)]^{1/2}}$

Here the limits are $x = 0$, $x = a$ and $y = 0$, $y = x$. Evidently the region of integration is $OABO$. By changing the order of integration, we have

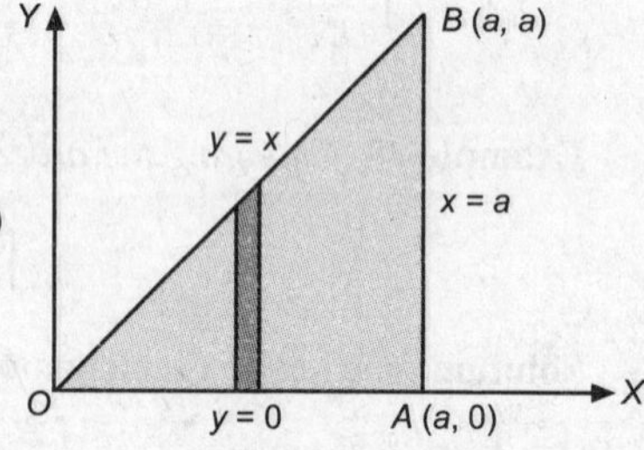

$$I=\int_0^a\int_y^a\frac{f'(y)\,dy\,dx}{[(a-x)(x-y)]^{1/2}}$$

$$=\int_0^a f'(y)\,dy\int_y^a\frac{dx}{\sqrt{(a-x)(x-y)}}\quad ...(1)$$

Let us find the values of $(a - x)$ and $(x - y)$ for (1)

Putting $x = a\cos^2\theta + y\sin^2\theta$

We have $\quad a - x = a - a\cos^2\theta - y\sin^2\theta$

$= a(1-\cos^2\theta) - y\sin^2\theta$

$a - x = a\sin^2\theta - y\sin^2\theta = (a-y)\sin^2\theta \quad ...(2)$

$\Rightarrow\quad -dx = 2(a-y)\sin\theta\cos\theta\,d\theta$, keeping y constant.

Also, $\quad x - y = a\cos^2\theta + y\sin^2\theta - y$

$= a\cos^2\theta - y(1-\sin^2\theta)$

$= a\cos^2\theta - y\cos^2\theta$

$= (a-y)\cos^2\theta \quad ...(3)$

$dx = -2(a-y)\sin\theta\cos\theta\,d\theta \quad ...(4)$

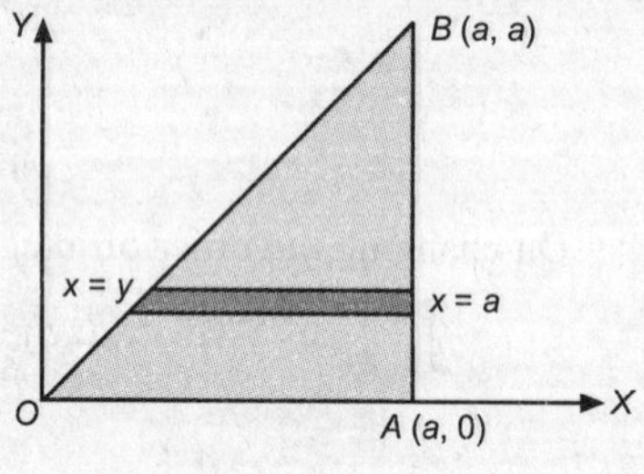

when $\quad x = y$ then $x - y = 0$

Upper limt $\quad x = a$

$x - y = (a-y)\cos^2\theta$

$a - y = (a-y)\cos^2\theta$

$\Rightarrow\quad \cos^2\theta = 1 \Rightarrow \theta = 0$

lower limit $\quad x = y$

$x - y = (a-y)\cos^2\theta$

$x - x = (a-y)\cos^2\theta \;\Rightarrow\; 0 = \cos^2\theta \;\Rightarrow\; \theta=\frac{\pi}{2}$

Putting the values of $a - x$, $x - y$ and dx from (2), (3) and (4) respectively in (1), we get

$$I=\int_0^a f'(y)\,dy\int_{\frac{\pi}{2}}^0\frac{-2(a-y)\sin\theta\cos\theta}{\sqrt{(a-y)\sin^2\theta.(a-y)\cos^2\theta}}\,d\theta$$

$$=\int_0^a f'(y)\,dy\int_{\frac{\pi}{2}}^0\frac{-2(a-y)\sin\theta\cos\theta}{(a-y)\sin\theta\cos\theta}\,d\theta$$

$$=-2\int_0^a f'(y)\,dy\int_{\frac{\pi}{2}}^0 d\theta=2\int_0^a f'(y)\,[\theta]_0^{\pi/2}=2\int_0^a f'(y)\,dy.\frac{\pi}{2}=2[f(y)]_0^a.\frac{\pi}{2}=[f(a)-f(0)]\,\pi$$

EXERCISE 5.3

Change the order of integration and hence evaluate the following:

1. $\int_0^a \int_0^x \frac{\cos y\, dy}{\sqrt{(a-x)(a-y)}}\, dx$ **Ans.** (a) $\int_0^a dy \int_y^a \frac{\cos y\, dx}{\sqrt{(a-x)(a-y)}}$ (b) $2 \sin a$.

2. $\int_0^{2a} \int_{\frac{x^2}{4a}}^{3a-x} (x^2+y^2)\, dy\, dx$ **Ans.** (a) $\int_0^a dy \int_0^{2\sqrt{ay}} (x^2+y^2)\, dx + \int_a^{3a} dy \int_0^{3a-y} (x^2+y^2)\, dx$ (b) $\frac{314\, a^4}{35}$.

3. $\int_0^1 \int_{x^2}^x (x^2+y^2)^{-1/2}\, dy\, dx$ **Ans.** $\int_0^1 dy \int_y^{\sqrt{y}} (x^2+y^2)^{-1/2}\, dx$.

4. $\int_0^a \int_{\sqrt{a^2-y^2}}^{y+a} f(x, y)\, dx\, dy$ **Ans.** $\int_0^a dx \int_{\sqrt{a^2-x^2}}^a f(x, y)\, dy + \int_a^{2a} dx \int_{x-a}^a f(x, y)\, dy$

5. $\int_{-a}^a \int_0^{\sqrt{a^2-y^2}} f(x, y)\, dx\, dy$ **Ans.** $\int_0^a dx \int_{-\sqrt{a^2-x^2}}^{\sqrt{a^2-x^2}} f(x, y)\, dy$

6. $\int_0^1 \int_x^{2-x} \frac{x}{y}\, dy\, dx$ **Ans.** $\int_0^1 \frac{dy}{y} \int_0^y x dx + \int_1^2 \frac{dy}{y} \int_0^{2-y} x dx$; $\log \frac{4}{e}$

7. $\int_0^b \int_y^a \frac{x\, dy\, dx}{x^2+y^2}$; $a > b$ **Ans.** $\int_0^b x dx \int_0^x \frac{dy}{x^2+y^2} + \int_0^a x dx \int_0^b \frac{dy}{x^2+y^2}$2

8. $\int_0^a \int_0^{bx/a} x\, dy\, dx$ **Ans.** (a) $\int_0^b dy \int_{ay/b}^a x\, dx$ (b) $\frac{1}{3} a^2 b$

9. $\int_0^5 \int_{2-x}^{2+x} f(x, y)\, dx\, dy$ **Ans.** $\int_0^2 dy \int_{2-y}^5 f(x, y)\, dx + \int_2^7 dy \int_{y-2}^5 f(x, y)\, dx$

10. $\int_0^\infty \int_{-y}^y (y^2-x^2)\, e^{-y}\, dx\, dy$ **Ans.** $\int_{-\infty}^\infty dx \int_{-x}^x (y^2-x^2)\, e^{-y}\, dy$,8

11. $\int_{y=0}^1 \int_{x=\sqrt{y}}^{2-y} \times yx\, dy$ *(A.M.I.E.T.E., June 2009)*

12. $\int_0^a \int_{\frac{x^2}{a}}^{2a-x} xy\, dx\, dy$ *(U.P. I Semester, Dec., 2007)* **Ans.** $\int_0^a \int_0^{\sqrt{ay}} xy\, dx\, dy + \int_0^{2a-y} xy\, dx\, dy$, $\frac{3a^2}{8}$

13. $\int_0^a \int_{a-\sqrt{a^2-y^2}}^{a+\sqrt{a^2-y^2}} xy\, dx\, dy$ **Ans.** $\int_0^{2a} x\, dx \int_0^{\sqrt{a^2-(x-a)^2}} y\, dy$, $\frac{2}{3} a^4$

[**Hint:** Put $x = a \sin^2 \theta \Rightarrow dx = 2\, a \sin \theta \cos \theta\, d\theta$]

14. $\int_0^1 \int_{-1}^{1-y} x^{1/3}\, y^{-1/2}\, (1-x-y)^{1/2}\, dx\, dy$ **Ans.** $\int_{-1}^1 x^{\frac{1}{3}}\, dx \int_0^{1-x} y^{-\frac{1}{2}} (1-x-y)^{\frac{1}{2}}\, dy$, $-\frac{3\pi}{7}$

15. $\int_0^{2a} dx \int_0^{\frac{x^2}{4a}} (x+y)^3\, dy$ **Ans.** $\int_0^a dy \int_{\sqrt{4ay}}^{2a} (x+y)^3\, dx$

16. $\int_0^1 \int_0^y (x^2+y^2)\, dx\, dy + \int_1^2 \int_0^{2-y} (x^2+y^2)\, dx\, dy$ **Ans.** $\int_0^1 dx \int_x^{2-x} (x^2+y^2)\, dy$, $\frac{5}{3}$

17. $\int_0^a \int_0^{\sqrt{a^2-x^2}} y^2 \sqrt{x^2+y^2}\, dx\, dy$ by changing into polar coordinates. **Ans.** $\frac{\pi\, a^5}{20}$

(U.P., I Semester, Dec. 2007)

18. $\int_0^1 \int_1^2 \frac{1}{x^2+y^2}\, dx\, dy + \int_0^2 \int_y^2 \frac{1}{x^2+y^2}\, dx\, dy = \int_R \int \frac{1}{x^2+y^2}\, dy\, dx$

Recognise the region R of integration on the R.H.S. and then evaluate the integral on the right in the order indicated. **Ans.** Region R is $x = 0$, $x = y$, $y = 1$ and $y = 2$, $\frac{\pi}{4} \log 2$.

19. Express as single integral and evaluate :

$$\int_0^{\frac{a}{\sqrt{2}}}\int_0^x x\,dx\,dy+\int_{\frac{a}{\sqrt{2}}}^{a}\int_0^{\sqrt{a^2-x^2}} x\,dx\,dy \qquad \textbf{Ans. } \int_0^{\frac{a}{\sqrt{2}}} dy\int_y^{\sqrt{a^2-y^2}} x\,dx,\ \frac{5a^3}{6\sqrt{2}}$$

20. Express as single integral and evaluate :

$$\int_0^1\int_0^y (x^2+y^2)\,dx\,dy+\int_1^2\int_0^{2-y}(x^2+y^2)\,dx\,dy \qquad \textbf{Ans. } \int_0^1 dx\int_x^{2-x}(x^2+y^2)\,dy,\ \frac{5}{3}$$

21. If $\iint_R f(x, y)\,dx\,dy$, where R is the circle $x^2 + y^2 = a^2$, is R equivalent to the Polar integral.

$$\left[\textbf{Ans. } \int_0^{2\pi}\int_0^{a}(r,\theta)\,r\,dr\,d\theta.\right]$$

5.5 CHANGE OF VARIABLES

Sometimes the problems of double integration can be solved easily by change of independent variables. Let the double integral as be $\iint_R f(x, y)\,dx\,dy$. It is to be changed by the new variables u, v.

The relation of x, y with u, v are given as $x = f(u, v)$, $y = \Psi(u, v)$. Then the double integration is converted into.

$$\iint_{R'} f\,\{\phi(u, v), \Psi(u, v)\}\,|J|\,du\,dv, \text{ where}$$

$$dx\,dy = |J|\,du\,dv = \frac{\partial(x, y)}{\partial(u, v)}\,du\,dv = \begin{vmatrix} \dfrac{\partial x}{\partial u} & \dfrac{\partial x}{\partial v} \\ \dfrac{\partial y}{\partial u} & \dfrac{\partial y}{\partial v} \end{vmatrix}\,du\,dv$$

Example 31. *Evaluate* $\iint_R (x+y)^2\,dx\,dy$, *where R is the parallelogram in the xy-plane with vertices (1, 0), (3, 1), (2, 2), (0, 1), using the transformation* $u = x + y$ *and* $v = x - 2y$.

Solution. The region of integration is a parallelogram $ABCD$, where A (1, 0), B (3, 1), C (2, 2) and D (0, 1) in xy-plane.

The new region of integration is a rectangle $A'B'C'D'$ in uv-plane

xy-plane	$A \equiv (x, y)$ $A \equiv (1, 0)$	$B \equiv (x, y)$ $B \equiv (3, 1)$	$C \equiv (x, y)$ $C \equiv (2, 2)$	$D \equiv (x, y)$ $D \equiv (0, 1)$
uv-plane	$A' \equiv (u, v)$ $A' \equiv (x + y, x - 2y)$ $A' \equiv (1 + 0, 1 - 2 \times 0)$ $A' \equiv (1, 1)$	$B' \equiv (u, v)$ $B' \equiv (x + y, x - 2y)$ $B' \equiv (3 + 1, 3 - 2 \times 1)$ $B' \equiv (4, 1)$	$C' \equiv (u, v)$ $C' \equiv (x + y, x - 2y)$ $C' \equiv (2 + 2, 2 - 2 \times 2)$ $C' \equiv (4, -2)$	$D' \equiv (u, v)$ $D' \equiv (x + y, x - 2y)$ $D' \equiv (0 + 1, 0 - 2 \times 1)$ $D' \equiv (1, -2)$

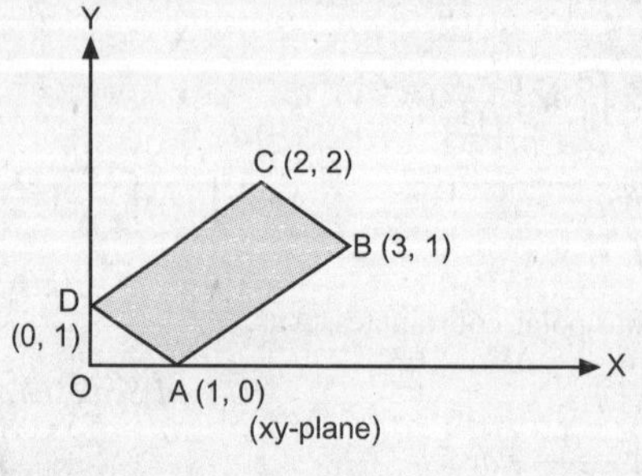

(xy-plane)

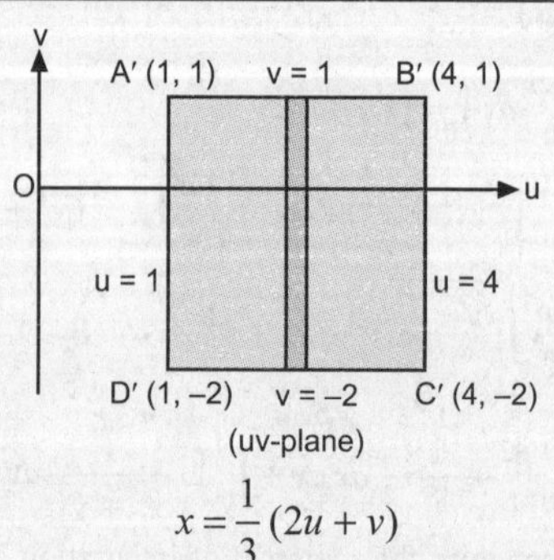

(uv-plane)

and $\begin{Bmatrix} u = x + y \\ v = x - 2y \end{Bmatrix} \Rightarrow \quad x = \dfrac{1}{3}(2u + v)$

and $\quad y = \dfrac{1}{3}(u - v)$

$$J = \frac{\partial(x, y)}{\partial(u, v)} = \begin{vmatrix} \frac{\partial x}{\partial u} & \frac{\partial x}{\partial v} \\ \frac{\partial y}{\partial u} & \frac{\partial y}{\partial v} \end{vmatrix} = \begin{vmatrix} \frac{2}{3} & \frac{1}{3} \\ \frac{1}{3} & -\frac{1}{3} \end{vmatrix} = -\frac{1}{3}$$

$$dx\, dy = |J|\, du\, dv = \frac{1}{3}\, du\, dv$$

$$\iint_R (x+y)^2\, dx\, dy = \int_{-2}^{1}\int_{1}^{4} u^2 . \frac{1}{3}\, du\, dv = \int_{-2}^{1} \frac{1}{3}\left[\frac{u^3}{3}\right]_1^4 dv = \int_{-2}^{1} 7\, dv = 7\,[v]_{-2}^{1} = 7 \times 3 = 21 \textbf{ Ans.}$$

Example 32. *Transform* $\int_0^{\pi/2}\int_0^{\pi/2} \sqrt{\frac{\sin\phi}{\sin\theta}}\, d\phi\, d\theta$ *by the substitution* $x = \sin\phi\cos\theta$, $y = \sin\phi\sin\theta$ *and show that its value is* π.

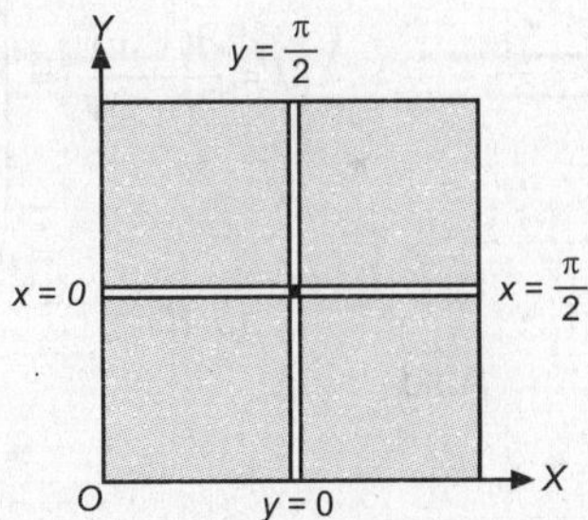

Solution. $x = \sin\phi\cos\theta$, $y = \sin\phi\sin\theta$.

$\Rightarrow \quad x^2 + y^2 = \sin^2\phi$ and $\frac{y}{x} = \tan\theta$

limit of θ are 0 and $\frac{\pi}{2}$.

Also, the limits of ϕ are 0 and $\frac{\pi}{2}$.

$$x^2 + y^2 = \sin^2\phi = \sin^2\frac{\pi}{2} = 1$$

limits of x are 0 to $\sqrt{1-y^2}$

limits of y are 0 to 1.

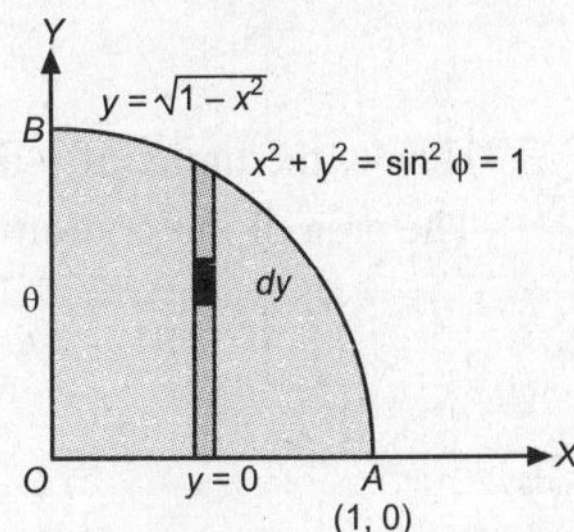

Now, $\quad d\phi\, d\theta = \frac{\partial(\phi,\theta)}{\partial(x, y)}\, dx\, dy = \begin{vmatrix} \frac{\partial\phi}{\partial x} & \frac{\partial\theta}{\partial x} \\ \frac{\partial\phi}{\partial y} & \frac{\partial\theta}{\partial y} \end{vmatrix} dx\, dy$

$$= \begin{vmatrix} \frac{x}{\sin\phi\cos\phi} & \frac{-y\cos^2\theta}{x^2} \\ \frac{y}{\sin\phi\cos\phi} & \frac{\cos^2\theta}{x} \end{vmatrix} dx\, dy = \left[\frac{\cos^2\theta}{\sin\phi\cos\phi} + \frac{y^2\cos^2\theta}{x^2\sin\phi\cos\phi}\right] dx\, dy$$

$$= \frac{(x^2+y^2)\cos^2\theta}{x^2\sin\phi\cos\phi}\, dx\, dy = \frac{\sin^2\phi\cos^2\theta}{(\sin^2\phi\cos^2\theta)\sin\phi\cos\phi}\, dx\, dy = \frac{1}{\sin\phi\cos\phi}\, dx\, dy$$

Again, $\quad \iint \sqrt{\left(\frac{\sin\phi}{\sin\theta}\right)}\, d\phi\, d\theta = \iint \sqrt{\left\{\left(\frac{\sin\phi}{\sin\theta}\right)\right\}}\, \frac{dx\, dy}{\sin\phi\cos\phi}$

$$= \iint \frac{dx\, dy}{\cos\phi\sqrt{\sin\phi\sin\theta}} = \iint \frac{dx\, dy}{\sqrt{1-\sin^2\phi}\sqrt{\sin\phi\sin\theta}}$$

$$= \int_0^1\int_0^{\sqrt{1-y^2}} \frac{1}{\sqrt{1-(x^2+y^2)}\sqrt{y}}\, dx\, dy \qquad \left[\begin{matrix} x^2+y^2 = \sin^2\phi \\ y = \sin\phi\sin\theta \end{matrix}\right]$$

$$I = \int_0^1\int_0^{\sqrt{1-y^2}} \frac{dx\, dy}{\sqrt{(y)}\sqrt{(1-y^2)-x^2}} = \int_0^1 \frac{dy}{\sqrt{y}}\int_0^{\sqrt{1-y^2}} \frac{dx}{\sqrt{\{(1-y^2)-x^2\}}}$$

$$= \int_0^1 \frac{1}{\sqrt{(y)}} \left\{ \sin^{-1} \frac{x}{\sqrt{(1-y^2)}} \right\}_0^{\sqrt{(1-y^2)}} dy = \int_0^1 \frac{1}{\sqrt{(y)}} \{\sin^{-1} 1 - \sin^{-1} 0\}\, dy$$

$$= \frac{\pi}{2} \{2\sqrt{(y)}\}_0^1 = \pi$$

Proved.

Example 33. *Using the transformation $x + y = u$, $y = uv$ show that*

$$\int_0^1 \int_0^{1-x} e^{y/(x+y)}\, dy\, dx = \frac{1}{2}(e-1)$$

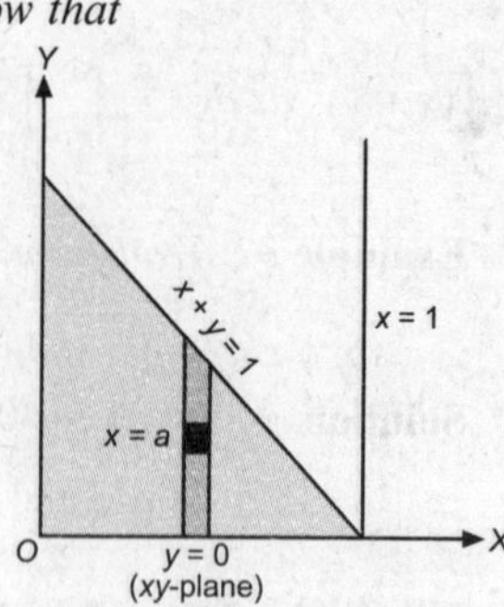

Solution. Since, $x = u(1 - v)$, $y = uv$

$$\therefore \quad J = \frac{\partial(x,y)}{\partial(u,v)} = \begin{vmatrix} \frac{\partial x}{\partial u} & \frac{\partial x}{\partial v} \\ \frac{\partial y}{\partial u} & \frac{\partial y}{\partial v} \end{vmatrix} = \begin{vmatrix} 1-v & -u \\ v & u \end{vmatrix}$$

$$= u - uv + uv = u$$

$$\therefore \quad dx\, dy = |J|\, du\, dv = u\, du\, dv$$

Also

$$x = 0 \Rightarrow u(1 - v) = 0$$
$$\Rightarrow u = 0, v = 1$$
$$y = 0 \Rightarrow uv = 0$$
$$\Rightarrow u = 0, v = 0$$
$$x + y = \Rightarrow u = 1$$

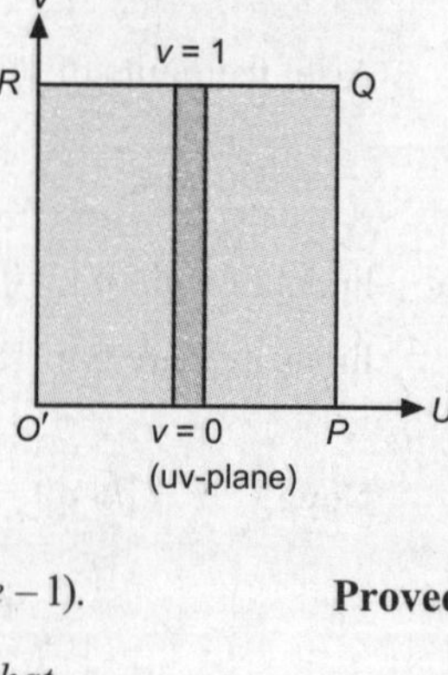

Hence, the limits of u are 0 to 1 and the limits of v are 0 to 1.

The area of integration is $O'PQR$ in uv-plance.

$$\therefore \quad \int_0^1 \int_0^{1-x} e^{\frac{y}{x+y}}\, dydx = \int_0^1 \int_0^1 e^{uv/u}\, |J|\, du\, dv$$

$$= \int_0^1 \int_0^1 ue^v\, du\, dv = \left(\frac{u^2}{2}\right)_0^1 (e^v)_0^1 = \frac{1}{2}(e-1).$$

Proved.

Example 34. *Using the transformation $x + y = u$, $y = uv$, show that*

$$\iint [xy(1-x-y)]^{1/2}\, dx\, dy = \frac{2\pi}{105},$$ *integration being taken over the area of the tringle bounded by the lines $x = 0$, $y = 0$, $x + y = 1$.*

Solution. $\iint [xy(1-x-y)]^{1/2}\, dxdy$ ...(1)

$$x + y = u \text{ or } x = u - y = u - uv,$$

$$dxdy = \frac{\partial(x,y)}{\partial(u,v)}\, du\, dv = \begin{vmatrix} \frac{\partial x}{\partial u} & \frac{\partial x}{\partial v} \\ \frac{\partial y}{\partial u} & \frac{\partial y}{\partial v} \end{vmatrix} dudv$$

$$dxdy = \begin{vmatrix} 1-v & -u \\ v & u \end{vmatrix} du\, dv = u\, du\, dv.$$

$$x = 0 \Rightarrow u(1-v) = 0$$
$$\Rightarrow u = 0, v = 1$$
$$y = 0 \Rightarrow uv = 0$$
$$\Rightarrow u = 0, v = 0$$

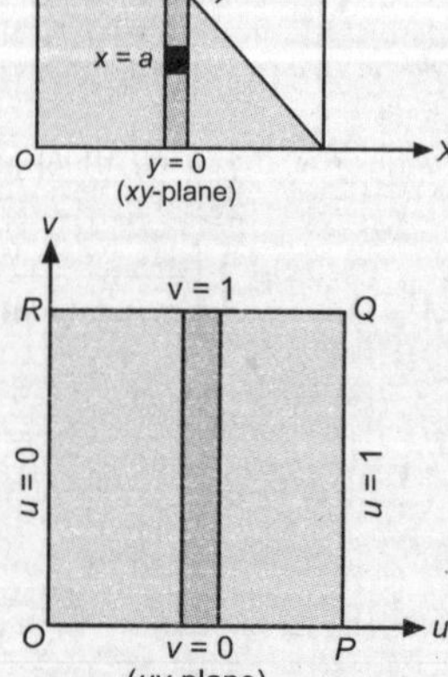

$x + y = 1 \quad \Rightarrow \quad u = 1$

Hence, the limits of u are from 0 to 1 and the limits of v are from 0 to 1.

The area of integration is a square $OPQR$ in uv-plane.

On putting $x = u - uv$, $y = uv$, $dx\,dy = u\,du\,dv$ in (1), we get

$$\iint (u - uv)^{1/2} (uv)^{1/2} (1 - v)^{1/2}\, u\, du\, dv$$

$$= \int_0^1 u^2 (1-u)^{1/2}\, du \int_0^1 v^{1/2} (1-u)^{1/2}\, dv = \left(\frac{\left|\overline{3}\right. \left|\overline{\frac{3}{2}}\right.}{\left|\overline{\frac{9}{2}}\right.} \times \frac{\left|\overline{\frac{3}{2}}\right. \left|\overline{\frac{3}{2}}\right.}{\left|\overline{3}\right.} \right)$$

$$= \frac{2.\left|\overline{\frac{3}{2}}\right.}{\frac{7}{2}.\frac{5}{2}.\frac{3}{2}\left|\overline{\frac{3}{2}}\right.} \times \frac{\frac{1}{2}\left|\overline{\frac{1}{2}}\right..\frac{1}{2}\left|\overline{\frac{1}{2}}\right.}{2} = \frac{1}{\frac{7}{2}.\frac{5}{2}.\frac{3}{2}} \times \frac{\frac{1}{2}\sqrt{\pi}.\frac{1}{2}\sqrt{\pi}}{1} = \frac{2\pi}{105}$$ **Ans.**

Example 35. *Using the transformation* $x + y = u$, $y = v$, *evaluate* $\int_0^{\pi}\int_0^{\pi} |\cos(x + y)|\, dx\, dy$.

Solution. Let $\quad I = \int_0^{\pi}\int_0^{\pi} |\cos(x + y)|\, dx\, dy$

$$\left\{\begin{matrix} x + y = u \\ y = v \end{matrix}\right\} \Rightarrow \begin{matrix} x = u - v \\ y = v \end{matrix}$$

$$J = \frac{\partial(x, y)}{\partial(u, v)} = \begin{vmatrix} \frac{\partial x}{\partial u} & \frac{\partial x}{\partial v} \\ \frac{\partial y}{\partial u} & \frac{\partial y}{\partial v} \end{vmatrix} = \begin{vmatrix} 1 & -1 \\ 0 & 1 \end{vmatrix} = 1$$

$$dx\, dy = |J|\, du\, dv = du\, dv$$

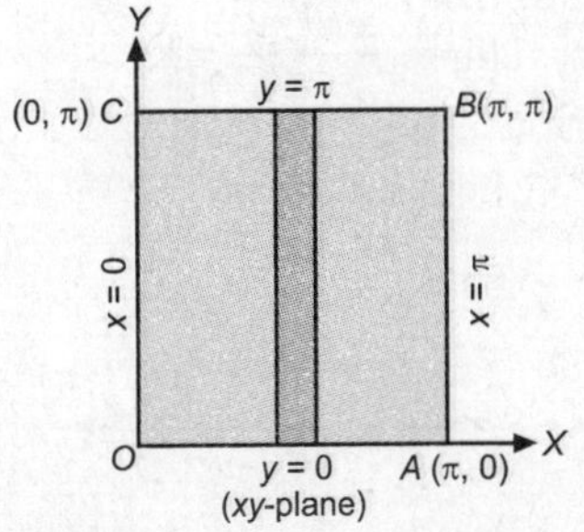

Y
$y = \pi$
(0, π) C
B(π, π)
$x = 0$
$x = \pi$
O
$y = 0$
A (π, 0)
X
(xy-plane)

The area of integration is a square OABC in xy-plane, where O is (0, 0), $A \equiv (\pi, 0)$, $B \equiv (\pi, \pi)$, $C \equiv (0, \pi)$)

xy-plane	$O \equiv (x, y)$	$A \equiv (x, y)$	$B \equiv (x, y)$	$C \equiv (x, y)$
	$O \equiv (1, 0)$	$A \equiv (\pi, 0)$	$B \equiv (2, 2)$	$C \equiv (0, 1)$
	$O' \equiv (u, v)$	$A' \equiv (u, v)$	$B' \equiv (u, v)$	$C' \equiv (u, v)$
uv-plane	$O' \equiv (x + y, y)$	$A' \equiv (x + y, y)$	$B' \equiv (x + y, y)$	$C' \equiv (x + y, y)$
	$O' \equiv (0 + 0, 0)$	$A' \equiv (\pi + 0, 0)$	$B' \equiv (\pi + \pi, \pi)$	$C' \equiv (0 + \pi, \pi)$
	$O' \equiv (0, 0)$	$A' \equiv (\pi, 0)$	$B' \equiv (2\pi, \pi)$	$C' \equiv (\pi, \pi)$

The new area of integration is $O'A'B'C'$ in uv-plane where $O'(0, 0)$, $A'(\pi, 0)$, $B'(2\pi, \pi)$, $C'(\pi, \pi)$.

Equation of $O'C'$ is $v = u$

Equation of $A'B'$ is $v - 0 = \dfrac{0 - \pi}{\pi - 2\pi}(u - \pi)$

$\Rightarrow v = u - \pi \Rightarrow u = v + \pi$

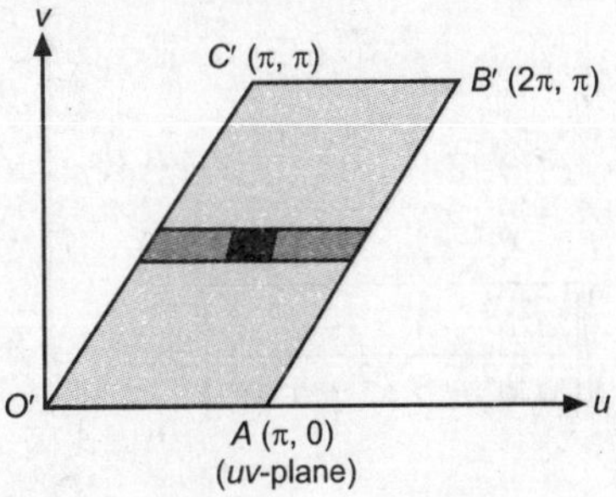

v
C' (π, π)
B' (2π, π)
O'
u
A (π, 0)
(uv-plane)

Now, $I = \int_0^{\pi}\int_0^{\pi} |\cos(x + y)|\, dx\, dy$

$$= \int_0^{\pi}\int_v^{\pi + v} |\cos u|\, du\, dv$$

$$= \int_0^{\pi} dv \int_v^{\pi + v} |\cos u|\, du$$

$$= \int_0^{\pi} dv \left[\int_v^{\pi/2} |\cos u|\, du + \int_{\pi/2}^{\pi} |\cos u|\, du + \int_{\pi}^{\pi + v} |\cos u|\, du\right]$$

$$= \int_0^{\pi} dv \left[\int_v^{\pi/2} \cos u \, du + \int_{\pi/2}^{\pi} - \cos u \, du + \int_{\pi}^{\pi+v} - \cos u \, du \right]$$

$$= \int_0^{\pi} dv \left[(\sin u)_v^{\pi/2} - (\sin u)_{\pi/2}^{\pi} - (\sin u)_{\pi}^{\pi+v} \right]$$

$$= \int_0^{\pi} dv \, [1 - \sin v) - (0 - 1) - |\sin(\pi + v) - 0|]$$

$$= \int_0^{\pi} dv \, [(1 - \sin v) + 1 - \{- \sin v - 0|]$$

$$= \int_0^{\pi} dv \, (2) = 2\int_0^{\pi} dv = 2(v)_0^{\pi} = 2(\pi - 0) = 2\pi$$

Interval	cos u	$\lvert \cos u \rvert$
$\left(v, \frac{\pi}{2}\right)$	+ ve	$\cos u$
$\left(\frac{\pi}{2}, \pi\right)$	– ve	$-\cos u$
$(\pi, \pi + v)$	– ve	$-\cos u$

Example 36. *Evaluate* $\iint \sqrt{\dfrac{a^2b^2 - b^2x^2 - a^2y^2}{a^2b^2 + b^2x^2 + a^2y^2}} \, dx \, dy$ *where R is the region bounded by the ellipse* $\dfrac{x^2}{a^2} + \dfrac{y^2}{b^2} = 1$*, using polar coordinates.* *(M.U. II Semester, 2009)*

Solution. The region of integration is bounded by the ellipse $\dfrac{x^2}{a^2} + \dfrac{y^2}{b^2} = 1$.

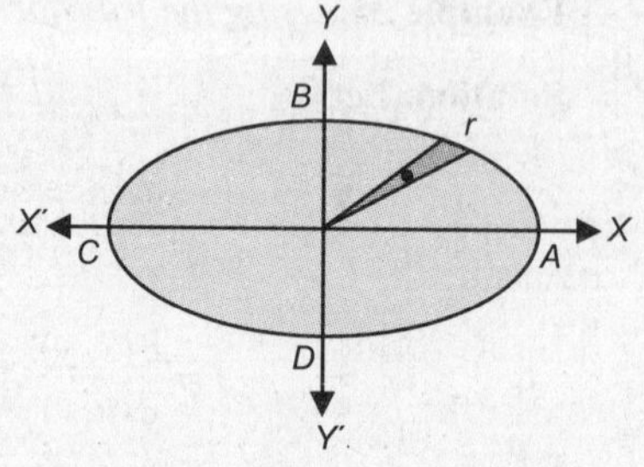

Let us substitute $\dfrac{x}{a} = r\cos\theta$

$\Rightarrow \quad x = a\, r\cos\theta$

$\dfrac{y}{b} = r\sin\theta$

$\Rightarrow \quad y = b\, r\sin\theta$

$$\left[\begin{aligned} \frac{x^2}{a^2} + \frac{y^2}{b^2} &= 1 \\ r^2\cos^2\theta + r^2\sin^2\theta &= 1 \\ r^2(\cos^2\theta + \sin^2\theta) &= 1 \\ r^2 &= 1 \\ r &= 1 \end{aligned}\right]$$

$$J = \frac{\partial(x, y)}{\partial(r, \theta)} = \begin{vmatrix} \dfrac{\partial x}{\partial r} & \dfrac{\partial x}{\partial \theta} \\ \dfrac{\partial y}{\partial r} & \dfrac{\partial y}{\partial \theta} \end{vmatrix}$$

$$= \begin{vmatrix} a\cos\theta & -ar\sin\theta \\ b\sin\theta & br\cos\theta \end{vmatrix}$$

$$= abr\cos^2\theta + abr\sin^2\theta$$

$$= abr(\cos^2\theta + \sin^2\theta)$$

$$= abr$$

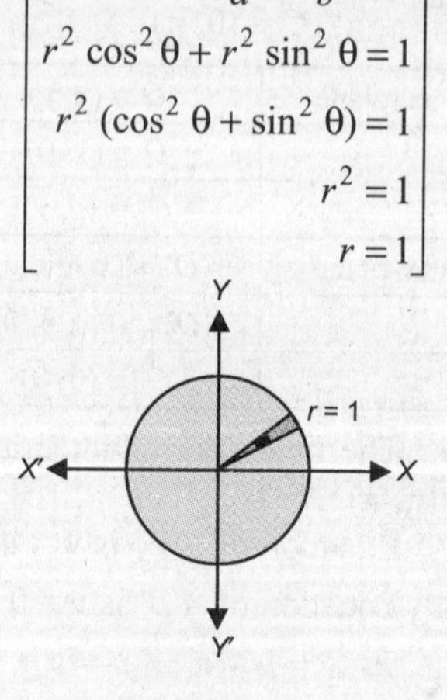

$$dx\,dy = J\,dr\,d\theta$$

$$= abr\,dr\,d\theta$$

$$\iint \sqrt{\left[\frac{a^2b^2 - b^2x^2 - a^2y^2}{a^2b^2 + b^2x^2 + a^2y^2}\right]} dx\,dy = 4\int_0^{\frac{\pi}{2}}\int_0^1 \sqrt{\left[\frac{a^2b^2 - b^2a^2r^2\cos^2\theta - a^2b^2r^2\sin^2\theta}{a^2b^2 + b^2a^2r^2\cos^2\theta + a^2b^2r^2\sin^2\theta}\right]}\, abr\,dr\,d\theta$$

$$= 4\int_0^{\frac{\pi}{2}}\int_0^1 \sqrt{\frac{a^2b^2 - a^2b^2r^2(\cos^2\theta + \sin^2\theta)}{a^2b^2 + a^2b^2r^2(\cos^2\theta + \sin^2\theta)}}\, abr\,dr\,d\theta$$

$$= 4\int_0^{\frac{\pi}{2}} \int_0^1 \sqrt{\frac{a^2b^2 - a^2b^2r^2}{a^2b^2 + a^2b^2r^2}}\, abr\, dr\, d\theta$$

$$= 4\int_0^{\frac{\pi}{2}} \int_0^1 \sqrt{\frac{a^2b^2(1-r^2)}{a^2b^2(1+r^2)}}\, abr\, dr\, d\theta = 4\int_0^{\frac{\pi}{2}} \int_0^1 \sqrt{\frac{1-r^2}{1+r^2}}\, abr\, dr\, d\theta$$

$$= 4\int_0^{\frac{\pi}{2}} \int_0^1 \sqrt{\frac{(1-r^2)(1-r^2)}{(1+r^2)(1-r^2)}}\, abr\, dr\, d\theta = 4ab\int_0^{\frac{\pi}{2}} \int_0^1 \frac{1-r^2}{\sqrt{1-r^4}}\, r\, dr\, d\theta \quad [\text{Put } r^2 = \sin t]$$

$$= 4ab\int_0^{\frac{\pi}{2}} \int_0^{\frac{\pi}{2}} \frac{1-\sin t}{\cos t}\cdot\frac{1}{2}\cos t\, dt\, d\theta = 2ab\int_0^{\frac{\pi}{2}} \int_0^{\frac{\pi}{2}} (1-\sin t)\, dt\, d\theta$$

$$= 2ab\int_0^{\frac{\pi}{2}} [t+\cos t]_0^{\pi/2}\, d\theta = 2ab\int_0^{\frac{\pi}{2}} \left[\frac{\pi}{2}-1\right] d\theta = 2ab\left(\frac{\pi}{2}-1\right)\int_0^{\frac{\pi}{2}} d\theta$$

$$= 2ab\left(\frac{\pi}{2}-1\right)[\theta]_0^{\pi/2} = \pi ab\left(\frac{\pi}{2}-1\right).$$ **Ans.**

Example 37. *Using the transformation* $u = x + y$ and $v = 2x - y$,

evaluate the integral $\iint_R (x+y)\, dx\, dy$,

Where R is the surface of the parallelogram $x + y = 1$ *and* $x + y = 3$, $2x - y = 0$ *and* $2x - y = 4$.

(Delhi University, April 2010)

Solution. We have, $\iint_R (x+y)\, dx\, dy$,

$u = x + y$ and $v = 2x - y$

$$\Rightarrow \quad x = \frac{u+v}{3} \text{ and } y = \frac{2u-v}{3}$$

$$dx\, dy = \frac{\partial(x,y)}{\partial(u,v)}\, du\, dv = \begin{vmatrix} \frac{\partial x}{\partial u} & \frac{\partial x}{\partial v} \\ \frac{\partial y}{\partial u} & \frac{\partial y}{\partial v} \end{vmatrix} du\, dv$$

$$= \begin{vmatrix} \frac{1}{3} & \frac{1}{3} \\ \frac{2}{3} & -\frac{1}{3} \end{vmatrix} du\, dv = \left(-\frac{1}{9}-\frac{2}{9}\right) du\, dv = -\frac{1}{3}\, du\, dv$$

Lower Limit of $u = 1$ and upper limit of $u = 3$

Lower Limit of $v = 0$ and upper limit of $v = 4$

On putting $x + y = 4$, we get

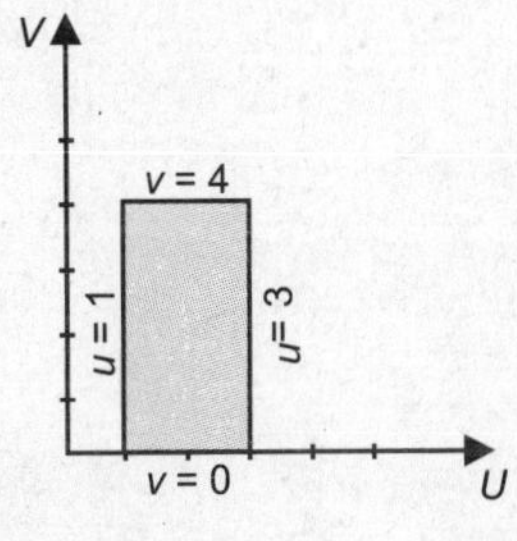

$$\iint u\left(-\frac{1}{3}\right) du\, dv = -\frac{1}{3}\int_1^3 u\, dv \int_0^4 dv$$

$$= -\frac{1}{3}\left[\frac{u^2}{2}\right]_1^3 [v]_0^4 = -\frac{1}{3}\left(\frac{9}{2}-\frac{1}{2}\right)(4-0)$$

$$= -\frac{1}{3}\times 4\times 4 = -\frac{16}{3}$$ **Ans.**

EXERCISE 5.4

1. Evaluate $\int_0^\infty \int_0^\infty e^{-(x+y)} \sin\left(\frac{\pi y}{x+y}\right) dx\, dy$ by means of the transformation $u = x + y$, $v = y$ from (x, y) to (u, v) **Ans.** $\frac{1}{\pi}$

2. Using the transformation $x + y = u$, $y = uv$, show that $\int_0^1 \int_0^{1-x} \frac{y}{e^{x+y}} dy\, dx = \frac{1}{2}(e-1)$

3. Using the transformation $u = x - y$, $v = x + y$, prove that $\iint_R \cos\frac{x-y}{x+y} dx\, dy = \frac{1}{2}\sin 1$ where R is bounded by $x = 0$, $y = 0$, $x + y = 1$

$$\left[\textbf{Hint}: x = \frac{1}{2}(u+v),\ y = \frac{1}{2}(v-u) \text{ so that } |J| = \frac{1}{2}\right]$$

APPLICATION OF THE DOUBLE INTEGRALS

(AREA, CENTRE OF GRAVITY, MASS, VOLUME)

6.1 INTRODUCTION

In this chapter, we will study how to find out area, centre of gravity, mass of lamina and the volume by revolving the area.

6.2 AREA IN CARTESIAN CO-ORDINATES

Let the curves *AB* and *CD* be $y_1 = f_1(x)$ and $y_2 = f_2(x)$.

Let the ordinates *AD* and *BC* be $x = a$ and $x = b$.

So the area enclosed by the two curves $y_1 = f_1(x)$ and $y_2 = f_2(x)$ and $x = a$ and $x = b$ is *ABCD*.

Let $P(x, y)$ and $Q(x + \delta x, y + \delta y)$ be two neighbouring points, then the area of the small rectangle $PQ = \delta x.\ \delta y$.

Area of the vertical strip $= \lim\limits_{\partial y \to 0} \sum\limits_{y_1}^{y_2} \delta x\, \delta y = \delta x \int_{y_1}^{y_2} dy\, x$ the width of the strip is constant throughout.

If we add all the strips from $x = a$ to $x = b$, we get

$$\text{The area } ABCD = \lim_{\partial x \to 0} \sum_{a}^{b} \delta x \int_{y_1}^{y_2} dy = \int_a^b dx \int_{y_1}^{y_2} dy$$

$$\boxed{\textbf{Area} = \int_a^b \int_{y_1}^{y_2} dx\, dy}$$

Example 1. *Find the area bounded by the parabola $y^2 = 4ax$ and its latus rectum.*

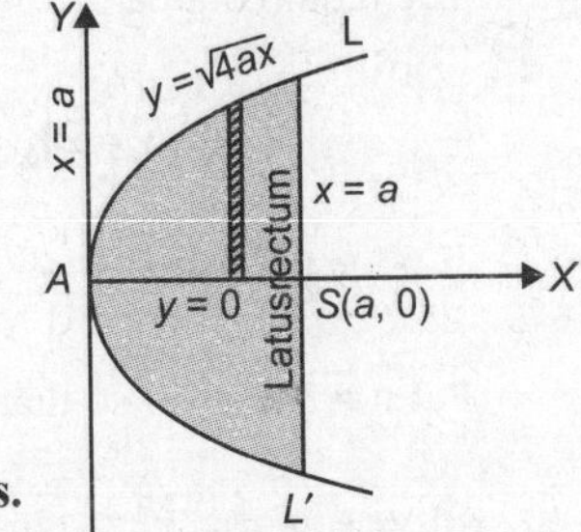

Solution. Required area $= 2$ [(area (ASL)]

$$= 2\int_0^a \int_0^{2\sqrt{ax}} dy\, dx$$

$$= 2\int_0^a 2\sqrt{ax}\, dx$$

$$= 4\sqrt{a}\left(\frac{x^{3/2}}{3/2}\right)_0^a = \frac{8a^2}{3} \qquad \textbf{Ans.}$$

Example 2. *Find the area between the parabolas $y^2 = 4\,ax$ and $x^2 = 4\,ay$.*

Solution.

$$y^2 = 4ax \qquad \text{...(1)}$$

$$x^2 = 4ay \qquad \text{...(2)}$$

On solving the equations (1) and (2) we get the point of intersection $(4a, 4a)$

Divide the area into horizontal strips of width δy, x varies from $P, \dfrac{y^2}{4a}$ to $Q, \sqrt{4\,ay}$ and then y varies from $O(y = 0)$ to $A\,(y = 4a)$.

$\therefore$ The required area $= \int_0^{4a} dy \int_{y^2/4a}^{\sqrt{4ay}} dx$

$$= \int_0^{4a} dy\,[x]_{y^2/4a}^{\sqrt{4ay}} = \int_0^{4a} dy\left[\sqrt{4ay} - \frac{y^2}{4a}\right] = \left[\sqrt{4a}\,\frac{y^{3/2}}{\frac{3}{2}} - \frac{y^3}{12\,a}\right]_0^{4a}$$

$$= \left[\frac{4\sqrt{a}}{3}(4a)^{3/2} - \frac{(4a)^3}{12a}\right] = \left[\frac{32}{3}a^2 - \frac{16}{3}a^2\right] = \frac{16}{3}a^2 \qquad \textbf{Ans.}$$

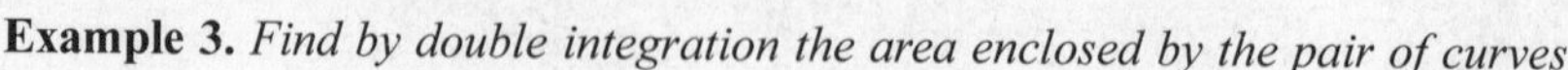

Example 3. *Find by double integration the area enclosed by the pair of curves*

$$y = 2 - x \text{ and } y^2 = 2(2 - x)$$

Solution. $y = 2 - x$ ...(1)

$y^2 = 2\,(2 - x)$ ...(2)

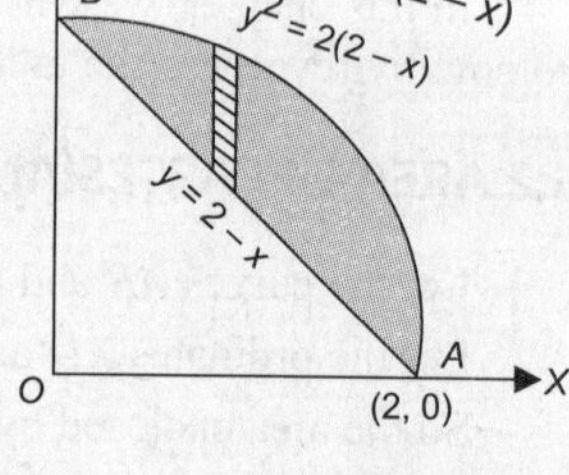

On solving the equations (1) and (2), we get the points of intersection (2, 0) and (0, 2).

$$A = \int\int dx\,dy$$

The required area $= \int_0^2 dx \int_{2-x}^{\sqrt{2(2-x)}} dy$

$$= \int_0^2 dx\,[y]_{2-x}^{\sqrt{2(2-x)}} = \int_0^2 dx\,[\sqrt{4 - 2x} - 2 + x] = \left[\frac{2}{3 \times -2}(4 - 2x)^{3/2} - 2x + \frac{x^2}{2}\right]_0^2$$

$$= \left[-\frac{1}{3}(4 - 2x)^{3/2} - 2x + \frac{x^2}{2}\right]_0^2 = \left(-4 + \frac{4}{2}\right) + \frac{8}{3} = \frac{2}{3} \qquad \textbf{Ans.}$$

Example 4. *By double integration, find the whole area of the curve* $a^2x^2 = y^3\,(2a - y)$.

Solution. $a^2\,x^2 = y^3\,(2a - y)$ ...(1)

The area enclosed by the curve (1) is the shaded portion of the figure

Take a horizontal strip extending from $x = 0$ to $x = \dfrac{y^{3/2}\sqrt{(2a - y)}}{a}$.

The required area

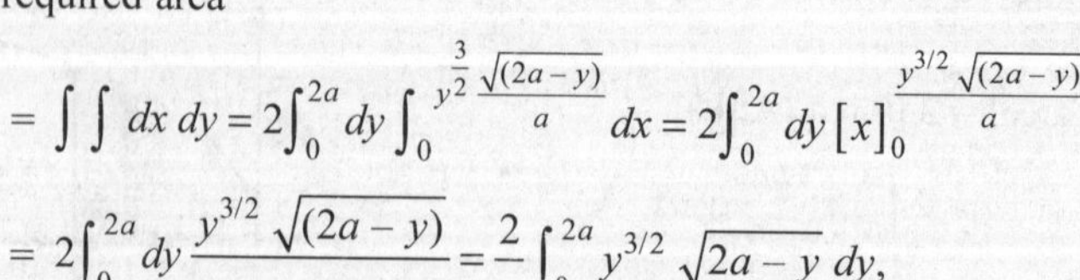

$$= \int\int dx\,dy = 2\int_0^{2a} dy \int_0^{\frac{y^{\frac{3}{2}}\sqrt{(2a-y)}}{a}} dx = 2\int_0^{2a} dy\,[x]_0^{\frac{y^{3/2}\sqrt{(2a-y)}}{a}}$$

$$= 2\int_0^{2a} dy\,\frac{y^{3/2}\sqrt{(2a - y)}}{a} = \frac{2}{a}\int_0^{2a} y^{3/2}\,\sqrt{2a - y}\,dy,$$

Put $y = 2\,a\,\sin^2\theta$ so that $dy = 4\,a\,\sin\theta\,\cos\theta\,d\,\theta$

$$\text{A} = \frac{2}{a}\int_0^{2a} y^{3/2}\,\sqrt{2a - y}\,dy = \frac{2}{a}\int_0^{\frac{\pi}{2}} (2a\sin^2\theta)^{\frac{3}{2}}\,\sqrt{2a - 2a\sin^2\theta}\,[4a\sin\theta\cos\theta\,d\theta]$$

$$= \frac{2}{a}(2a)^{3/2}\,\sqrt{2a}.\,(4a)\int_0^{\frac{\pi}{2}} \sin^3\theta\cos\theta.\,\sin\theta\cos\theta\,d\theta$$

$$= 32a^2 \int_0^{\frac{\pi}{2}} \sin^4\theta \cos^2\theta \, d\theta = 32a^2 \frac{\left|\frac{5}{2}\right|\frac{3}{2}}{2\left|4\right.}$$

$$= 32a^2 \frac{\frac{3}{2}\times\frac{1}{2}\times\left|\frac{1}{2}\times\frac{1}{2}\right|\frac{1}{2}}{2\times3\times2\times1} = 32a^2 \frac{\frac{3}{2}\times\frac{1}{2}\sqrt{\pi}\times\frac{1}{2}\sqrt{\pi}}{12} = \pi a^2 \text{ Ans.}$$

Example 5. *Find the mass of the area bounded by the curves $y^2 = x$ and $y = x^3$, if $\rho = \mu\,(x^2 + y^2)$.*

(Nagpur University, Summer 2008)

Solution. Here we have

$y^2 = x$ and $y = x^3$

Area bounded by the curves $= \iint dx\, dy$

Mass of the area

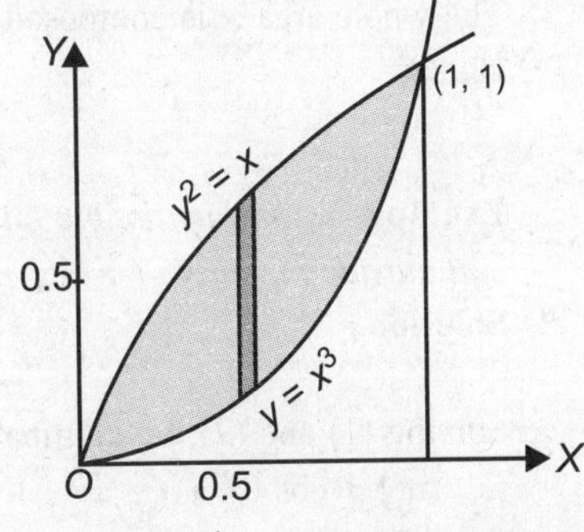

$$= \iint \mu(x^2 + y^2)\, dx\, dy = \mu \int_0^1 \int_{x^3}^{\sqrt{x}} (x^2 + y^2)\, dx\, dy$$

$$= \mu \int_0^1 dx \int_{x^3}^{\sqrt{x}} (x^2 + y^2)\, dy = \mu \int_0^1 dx \left[x^2 y + \frac{y^3}{3}\right]_{x^3}^{\sqrt{x}}$$

$$= \mu \int_0^1 \left[\left(x^{\frac{5}{2}} + \frac{x^{\frac{3}{2}}}{3}\right) - \left(x^5 + \frac{x^9}{3}\right)\right] dx = \mu\left[\frac{2}{7}x^{\frac{7}{2}} + \frac{2}{15}x^{\frac{5}{2}} - \frac{x^6}{6} - \frac{x^{10}}{30}\right]_0^1 = \mu\left[\frac{2}{7} + \frac{2}{15} - \frac{1}{6} - \frac{1}{30}\right]$$

$$= \frac{\mu}{210}[60 + 28 - 35 - 7] = \frac{46}{210}\mu = \frac{23}{105}\mu \qquad \text{Ans}$$

EXERCISE 6.1

Use double integration in the following questions:

1. Find the area bounded by $y = x - 2$ and $y^2 = 2x + 4$. **Ans.** 18.
2. Find the area between the circle $x^2 + y^2 = a^2$ and the line $x + y = a$ in the first quadrant. **Ans.** $(\pi - 2)a^2/4$
3. Find the area of a plate in the form of quadrant of the ellipse $\frac{x^2}{a^2} + \frac{y^2}{b^2} = 1$. **Ans.** $\frac{\pi ab}{4}$
4. Find the area included between the curves $y^2 = 4a(x + a)$ and $y^2 = 4b(b - x)$. **Ans.** $\frac{8\sqrt{ab}}{3}$
5. Find the area bounded by (*a*) $y^2 = 4 - x$ and $y^2 = x$. **Ans.** $\frac{16\sqrt{2}}{3}$

   (*b*) $x - 2y + 4 = 0,\ x + y - 5 = 0,\ y = 0$ **Ans.** $\frac{27}{2}$
6. Find the area enclosed by the leminscate $r^2 = a^2 \cos 2\theta$. **Ans.** a^2
7. Find the area common to the circles $x^2 + y^2 = a^2$ and $x^2 + y^2 = 2ax$. **Ans.** $\left[\frac{\pi}{3} - \frac{\sqrt{3}}{4}\right]a^2$
8. Find the area included between the curves $y = x^2 - 6x + 3$ and $y = 2x + 9$. **Ans.** $\frac{88\sqrt{22}}{3}$
9. Determine the area of region bounded by the curves $xy = 2,\ 4y = x^2,\ y = 4$. **Ans.** $\frac{28}{3} - 4\log 2$

6.3 AREA IN POLAR CO-ORDINATES

$$\text{Area} = \int\int r\, d\theta\, dr$$

Let us consider the area enclosed by the curve $r = f(\theta)$.

Let $P(r, \theta)$, $Q(r + \delta r, \theta + \delta\theta)$ be two neighbouring points.

Draw ares PL and QM, radii r and $r + \delta r$.

$$PL = r\delta\theta,\ PM = \delta r$$

Area of rectangle $\quad PLQM = PL \times PM$

$$= (r\delta\theta)\,\delta r = r\,\delta\theta\,\delta r.$$

The whole area A is composed of such small rectangles.

Hence,

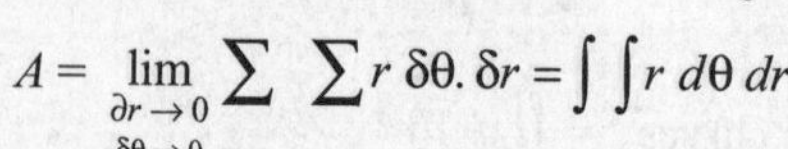

$$A = \lim_{\substack{\partial r \to 0 \\ \delta\theta \to 0}} \sum \sum r\,\delta\theta.\,\delta r = \int\int r\, d\theta\, dr$$

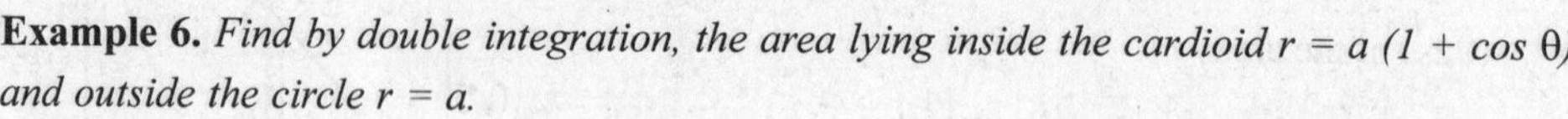

Example 6. *Find by double integration, the area lying inside the cardioid $r = a(1 + \cos\theta)$ and outside the circle $r = a$.*

Solution.
$$r = a(1 + \cos\theta) \qquad \text{...(1)}$$
$$r = a \qquad \text{...(2)}$$

Solving (1) and (2), by eliminating r, we get

$$a(1 + \cos\theta) = a \quad \Rightarrow \quad 1 + \cos\theta = 1$$

$$\cos\theta = 0 \quad \Rightarrow \quad \theta = -\frac{\pi}{2} \text{ or } \frac{\pi}{2}$$

limits of r are a and $a(1 + \cos\theta)$

limits of θ are $-\dfrac{\pi}{2}$ to $\dfrac{\pi}{2}$

Required area = Area $ABCDA$

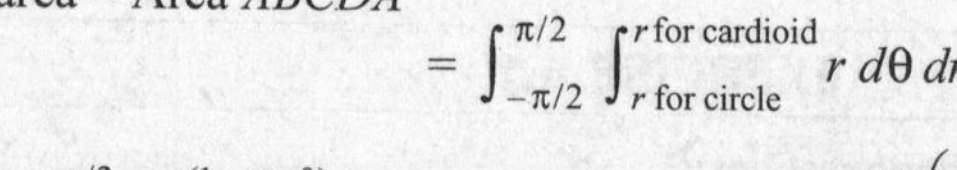

$$= \int_{-\pi/2}^{\pi/2} \int_{r \text{ for circle}}^{r \text{ for cardioid}} r\, d\theta\, dr$$

$$= \int_{-\pi/2}^{\pi/2} \int_{a}^{a(1+\cos\theta)} r\, d\theta\, dr \qquad = \int_{-\pi/2}^{\pi/2} \left(\frac{r^2}{2}\right)_a^{a(1+\cos\theta)} d\theta$$

$$= \frac{a^2}{2} \int_{-\pi/2}^{\pi/2} [(1 + \cos\theta)^2 - 1]\, d\theta \qquad = \frac{a^2}{2} \int_{-\pi/2}^{\pi/2} (\cos^2\theta + 2\cos\theta)\, d\theta$$

$$= a^2 \int_0^{\pi/2} (\cos^2\theta + 2\cos\theta)\, d\theta \qquad = a^2 \left[\int_0^{\pi/2} \cos^2\theta\, d\theta + 2\int_0^{\pi/2} \cos\theta\, d\theta\right]$$

$$= a^2 \left[\frac{\pi}{4} + 2(\sin\theta)_0^{\pi/2}\right] = a^2 \left[\frac{\pi}{4} + 2\right] \quad = \frac{a^2}{4}(\pi + 8) \qquad \textbf{Ans.}$$

Example 7. *Find by double integration, the area lying inside the circle $r = a\sin\theta$ and outside the cardioid $r = a(1 - \cos\theta)$.*

Solution. We have,

$$r = a\sin\theta \qquad \text{...(1)}$$
$$r = a(1 - \cos\theta) \qquad \text{...(2)}$$

Solving (1) and (2) by eliminating r, we have

$$\sin\theta = 1 - \cos\theta \Rightarrow \sin\theta + \cos\theta = 1$$

Squaring above, we get

$$\sin^2\theta + \cos^2\theta + 2\sin\theta\cos\theta = 1$$

$$\Rightarrow \quad 1 + \sin 2\theta = 1 \Rightarrow \sin 2\theta = 0 \Rightarrow 2\theta = 0 \text{ or } \pi \Rightarrow \theta = 0 \text{ or } \frac{\pi}{2}$$

The required area is shaded portion in the fig.

Limits of r are $a\,(1-\cos\theta)$ and $a\sin\theta$, limits of θ are 0 and $\frac{\pi}{2}$.

$$\text{Required area} = \int_0^{\frac{\pi}{2}}\int_{a(1-\cos\theta)}^{a\sin\theta} r\,dr\,d\theta$$

$$= \int_0^{\frac{\pi}{2}}\left[\frac{r^2}{2}\right]_{a(1-\cos\theta)}^{a\sin\theta} d\theta = \frac{1}{2}\int_0^{\pi/2} a^2\,[\sin^2\theta - (1-\cos\theta)^2]\,d\theta$$

$$= \frac{a^2}{2}\int_0^{\pi/2}(\sin^2\theta - 1 - \cos^2\theta + 2\cos\theta]\,d\theta$$

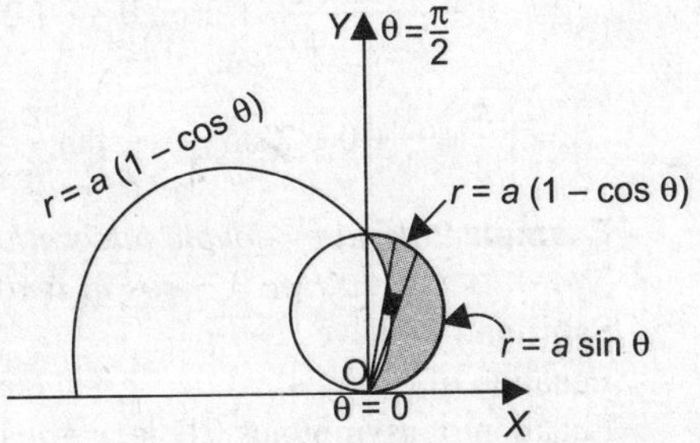

$$= \frac{a^2}{2}\int_0^{\pi/2}(-2\cos^2\theta + 2\cos\theta)\,d\theta$$

$$= \frac{a^2}{2}\left[\int_0^{\pi/2} -2\cos^2\theta\,d\theta + \int_0^{\pi/2} 2\cos\theta\,d\theta\right]$$

$$= \frac{a^2}{2}\left[\left(-2.\frac{\pi}{4}\right) + 2\,(\sin\theta)_0^{\pi/2}\right]$$

$$= \frac{a^2}{2}\left[-\frac{\pi}{2} + 2\left(\sin\frac{\pi}{2} - \sin 0\right)\right] = \frac{a^2}{2}\left[-\frac{\pi}{2}+2\right] = a^2\left(1-\frac{\pi}{4}\right) \qquad \textbf{Ans.}$$

Example 8. *Find by double integration, the area lying inside a cardioid $r = 1 + \cos\theta$ and outside the parabola $r\,(1 + \cos\theta) = 1$.*

Solution . We have,

$$r = 1 + \cos\theta \qquad ...(1)$$

$$r\,(1 + \cos\theta) = 1 \qquad ...(2)$$

Solving (1) and (2), we get

$$(1 + \cos\theta)\,(1 + \cos\theta) = 1$$

$$(1 + \cos\theta)^2 = 1$$

$$1 + \cos\theta = 1$$

$$\cos\theta = 0 \Rightarrow \theta = \pm\frac{\pi}{2}$$

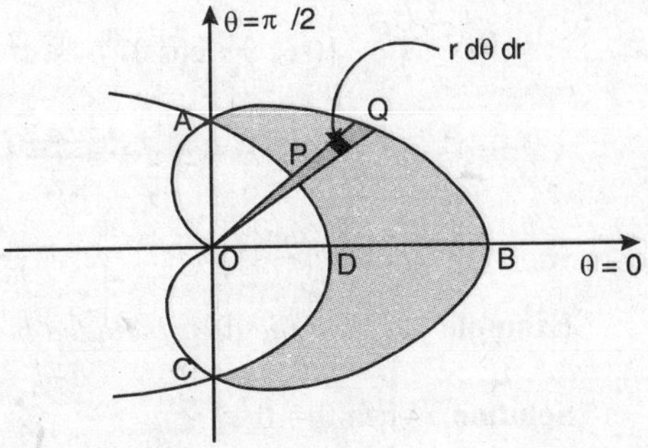

limits of r are $1 + \cos\theta$ and $\frac{1}{1+\cos\theta}$

limits of θ are $-\frac{\pi}{2}$ to $\frac{\pi}{2}$.

Required area = Area *ADCBA* (Shaded portion)

$$= \int_{-\pi/2}^{\pi/2}\int_{\frac{1}{1+\cos\theta}}^{1+\cos\theta} r\,dr\,d\theta = \int_{\frac{-\pi}{2}}^{\frac{\pi}{2}}\left(\frac{r^2}{2}\right)_{\frac{1}{1+\cos\theta}}^{1+\cos\theta} d\theta = \frac{1}{2}\int_{-\pi/2}^{\pi/2}\left[(1+\cos\theta)^2 - \frac{1}{(1+\cos\theta)^2}\right]d\theta$$

$$= \frac{1}{2}\int_{-\pi/2}^{\pi/2}\left[(1+\cos^2\theta + 2\cos\theta) - \frac{1}{\left(2\cos^2\frac{\theta}{2}\right)^2}\right]d\theta$$

$$= 2\times\frac{1}{2}\int_0^{\pi/2}\left[(1+\cos^2\theta + 2\cos\theta) - \frac{1}{4}\sec^4\frac{\theta}{2}\right]d\theta$$

$$= \int_0^{\pi/2}\left[(1+\cos^2\theta+2\cos\theta)-\frac{1}{4}\left(1+\tan^2\frac{\theta}{2}\right)\sec^2\frac{\theta}{2}\right]d\theta$$

$$= \int_0^{\pi/2}\left[\left(1+\frac{1+\cos 2\theta}{2}+2\cos\theta\right)-\frac{1}{4}\left(1+\tan^2\frac{\theta}{2}\right)\sec^2\frac{\theta}{2}\right]d\theta$$

$$= \int_0^{\pi/2}\left[1+\frac{1}{2}+\frac{\cos 2\theta}{2}+2\cos\theta-\frac{1}{4}\left(\sec^2\frac{\theta}{2}+\tan^2\frac{\theta}{2}\times\sec^2\frac{\theta}{2}\right)\right]d\theta$$

$$= \left[\theta+\frac{\theta}{2}+\frac{\sin 2\theta}{4}+2\sin\theta-\frac{1}{4}\left(2\tan\frac{\theta}{2}+\frac{2}{3}\tan^3\frac{\theta}{2}\right)\right]_0^{\frac{\pi}{2}}$$

$$= \left[\frac{\pi}{2}+\frac{\pi}{4}+0+2\sin\frac{\pi}{2}-\frac{1}{2}\tan\frac{\pi}{4}-\frac{1}{6}\tan^3\frac{\pi}{4}\right]=\left[\frac{3\pi}{4}+2-\frac{1}{2}-\frac{1}{6}\right]=\left[\frac{3\pi}{4}+\frac{4}{3}\right]$$ **Ans.**

Example 9. *Find by double integration the area included between the curves $r = a(\sec\theta + \cos\theta)$ and its asymptotes.*

Solution. We have,

Equation of curve is, $r = a(\sec\theta + \cos\theta)$ ...(1)

Equation of asymptotes CD is, $r = a\sec\theta$ ...(2)

limits of r are $a(\sec\theta + \cos\theta)$ and $a\sec\theta$

limits of θ are $-\frac{\pi}{2}$ and $\frac{\pi}{2}$

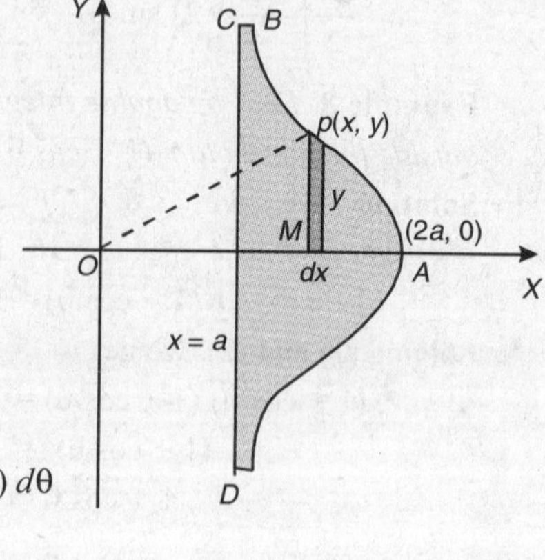

Required area = Shaded portion of the figure

$$= \int_{-\pi/2}^{\pi/2}\int_{a\sec\theta}^{a(\sec\theta+\cos\theta)} r\,dr\,d\theta$$

$$= \int_{-\pi/2}^{\pi/2}\left(\frac{r^2}{2}\right)_{a\sec\theta}^{a(\sec\theta+\cos\theta)} d\theta$$

$$= \frac{a^2}{2}\int_{-\pi/2}^{\pi/2}[(\sec\theta+\cos\theta)^2-\sec^2\theta]\,d\theta$$

$$= \frac{a^2}{2}\int_{-\pi/2}^{\pi/2}(\cos^2\theta+2)\,d\theta = \frac{2a^2}{2}\int_0^{\pi/2}(\cos^2\theta+2)\,d\theta$$

$$= a^2\left(\frac{1}{2}.\frac{\pi}{2}+2.\frac{\pi}{2}\right)=a^2\left(\frac{\pi}{4}+\pi\right)=\frac{5\pi}{4}a^2.$$ **Ans.**

Example 10. *Find the area included between the curve $x = a(\theta - \sin\theta)$, $y = a(1-\cos\theta)$ and its base.*

Solution. When $\theta = 0$

$x = a(0-\sin 0) = a\times 0 = 0$

$y = a(1-\cos 0) = a(1-1) = a\times 0 = 0$

When $\theta = 2\pi$

$x = a(2\pi - \sin 2\pi) = a(2\pi - 0) = 2a\pi$

$y = a(1-\cos 2\pi) = a(1-1) = a\times 0 = 0$

Therefore, the limits for θ are from 0 to 2π.

Required area = area $OBAO$

$$= \int_0^{2\pi} y\frac{dx}{d\theta}\,d\theta = \int_0^{2\pi} a(1-\cos\theta)\,a(1-\cos\theta)\,d\theta$$

$$= a^2\int_0^{2\pi}(1-\cos\theta)^2\,d\theta = 4a^2\int_0^{2\pi}\sin^4\frac{\theta}{2}\,d\theta$$

$$= 8a^2 \int_0^{\pi} \sin^4 \phi \, d\phi \qquad \left[\text{Put } \frac{\theta}{2} = \phi \Rightarrow d\theta = 2d\phi\right]$$

$$= 8a^2 . 2\int_0^{\pi/2} \sin^4 \phi \, d\phi = 3\pi a^2$$ **Ans.**

EXERCISE 6.2

1. Find the area of cardioid $r = a(1 + \cos \theta)$. **Ans.** $\frac{3\pi a^2}{2}$
2. Find the area of the curve $r^2 = a^2 \cos 2\theta$. **Ans.** a^2
3. Find the area enclosed by the curve $r = 2\, a \cos \theta$ **Ans.** πa^2
4. Find the area enclosed by the curve $r = 3 + 2 \cos \theta$. **Ans.** $11\,\pi$
5. Find the area enclosed by the curve
   $r^3 = a^2 \cos^2\theta + b^2 \sin^2\theta$. **Ans.** $\frac{\pi}{2}(a^2 + b^2)$
6. Show that the area of the region included between the cardioides $r = a(1 + \cos \theta)$ and $r = a\,(1 - \cos \theta)$ is $\frac{a^2}{2}(3\pi - 8)$.
7. Find the area outside the circle $r = 2$ and inside the cardioid $r = 2(1 + \cos \theta)$. **Ans.** $(\pi + 8)$
8. Find the area inside the circle $r = 2a \cos \theta$ and outside the circle $r = a$. **Ans.** $2a^2\left(\frac{\pi}{3} + \frac{\sqrt{3}}{4}\right)$
9. Find the area inside the circle $r = 4 \sin \theta$ and outside the lemniscate $r^2 = 8 \cos 2\,\theta$.
   Ans. $\left(\frac{8}{3}\pi + 4\sqrt{3} - 4\right)$

6.4 VOLUME OF SOLID BY ROTATION OF AN AREA (DOUBLE INTEGRAL)

When the area enclosed by a curve $y = f(x)$ is revolved about an axis, a solid is generated, we have to find out the volume of solid generated.

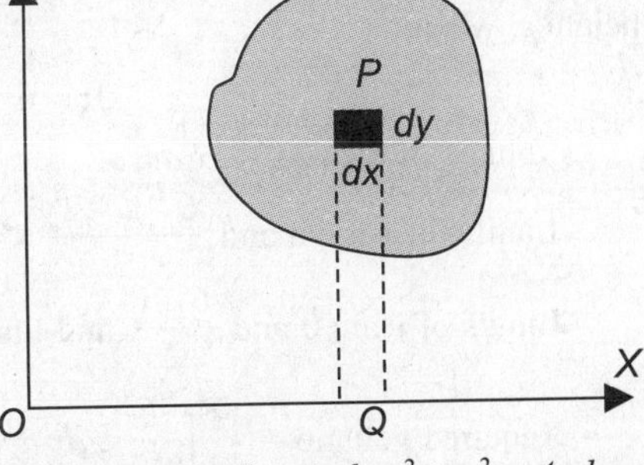

Volume of the solid generated about x-axis

$$= \int_a^b \int_{y_1(x)}^{y_2(x)} 2\pi \, PQ \, dx \, dy$$

Example 11. *Find the volume of the torus generated by revolving the circle $x^2 + y^2 = 4$ about the line $x = 3$.*

Solution. $x^2 + y^2 = 4$

$$V = \int \int (2\pi \, PQ) \, dx \, dy = 2\pi \int \int (3 - x) \, dx \, dy$$

$$= 2\pi \int_{-2}^{+2} dx \int_{-\sqrt{4-x^2}}^{+\sqrt{4-x^2}} (3 - x) \, dy$$

$$= 2\pi \int_{-2}^{+2} dx \, (3y - x\,y)_{-\sqrt{4-x^2}}^{+\sqrt{4-x^2}}$$

$$= 2\pi \int_{-2}^{+2} dx \, [3\sqrt{4-x^2} - x\sqrt{4-x^2} + 3\sqrt{4-x^2} - x\sqrt{4-x^2}]$$

$$= 4\pi \, [3\int_{-2}^{2} \sqrt{4-x^2} - x\sqrt{4-x^2}] \, dx = 4\pi \left[3\frac{x}{2}\sqrt{4-x^2} + 3 \times \frac{4}{2} \sin^{-1}\frac{x}{2} + \frac{1}{3}(4-x^2)^{3/2}\right]_{-2}^{2}$$

$$= 4\pi\left[6\times\frac{\pi}{2}+6\times\frac{\pi}{2}\right]=24\,\pi^2$$ **Ans.**

Example 12. *Calculate by double integration the volume generated by the revolution of the cardioid $r = a\,(1 - \cos\theta)$ about its axis.* *(AMIETE, June 2010)*

Solution. $r = a\,(1 - \cos\theta)$

$$V = 2\pi\int\int y\,dx\,dy \Rightarrow V = 2\pi\int\int (r\,d\theta\,dr)\,y$$

$$= 2\pi\int d\theta\int r\,dr\,(r\sin\theta)$$

$$= 2\pi\int_0^{\pi}\sin\theta\,d\theta\int_0^{a(1-\cos\theta)} r^2\,dr$$

$$= 2\pi\int_0^{\pi}\sin\theta\,d\theta\left[\frac{r^3}{3}\right]_0^{a\,(1-\cos\theta)} = \frac{2\pi}{3}\int_0^{\pi} a^3\,(1-\cos\theta)^3\sin\theta\,d\theta$$

$$= \frac{2\pi\,a^3}{3}\left[\frac{(1-\cos\theta)^4}{4}\right]_0^{\pi} = \frac{2\pi\,a^3}{12}[16] = \frac{8}{3}\pi\,a^3$$ **Ans.**

Example 13. *A pyramid is bounded by the three co-ordinate planes and the plane $x + 2y + 3z = 6$. Compute this volume by double integration.*

Solution. $x + 2y + 3z = 6$ ...(1)

$x = 0,\ y = 0,\ z = 0$ are co-ordinate planes.

The line of intersection of plane (1) and xy plane $(z = 0)$ is

$x + 2y = 6$ ...(2)

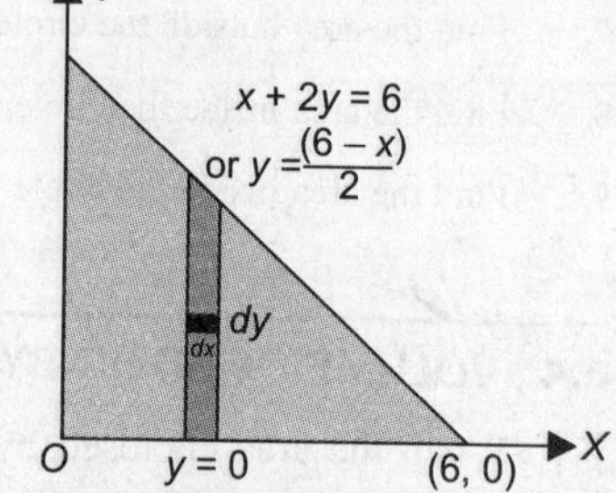

The base of the pyramid may be taken to be the triangle bounded by x-axis, y-axis and the line (2).

An elementary area on the base is $dx\,dy$.

Consider the elementary rod standing on this area and having height z, where

$$3z = 6 - x - 2y \text{ or } z = \frac{6-x-2y}{3}$$

Volume of the rod = $dx\,dy$,

Limits for z are 0 and $\dfrac{6-x-2y}{3}$.

Limits of y are 0 and $\dfrac{6-x}{2}$ and limits of x are 0 and 6.

$$\text{Required volume} = \int_0^6\int_0^{\frac{6-x}{2}} z\,dx\,dy = \int_0^6 dx\int_0^{\frac{6-x}{2}}\frac{6-x-2y}{3}\,dy$$

$$= \frac{1}{3}\int_0^6 dx\left(6y - xy - y^2\right)_0^{\frac{6-x}{2}} = \frac{1}{3}\int_0^6\left(\frac{6(6-x)}{2} - \frac{x(6-x)}{2} - \left(\frac{6-x}{2}\right)^2\right)dx$$

$$= \frac{1}{3}\int_0^6\left(\frac{36-6x}{2} - \frac{6x-x^2}{2} - \frac{36+x^2-12x}{4}\right)dx$$

$$= \frac{1}{12}\int_0^6 (72 - 12x - 12x + 2x^2 - 36 - x^2 + 12x)\,dx$$

$$= \frac{1}{12}\int_0^6 (x^2 - 12x + 36)\,dx = \frac{1}{12}\left[\frac{x^3}{3} - \frac{12x^2}{2} + 36x\right]_0^6$$

$$= \frac{1}{12}[72 - 216 + 216] = 6$$ **Ans.**

EXERCISE 6.3

1. Find the volume of the sphere $x^2 + y^2 + z^2 = a^2$ by revolving area of the circle $x^2 + y^2 = a^2$. **Ans.** $\frac{4}{3}\pi a^3$

6.5 CENTRE OF GRAVITY

$$\bar{x} = \frac{\iint \rho\, x\, dx\, dy}{\iint \rho\, dx\, dy},\ \bar{y} = \frac{\iint \rho\, y\, dx\, dy}{\iint \rho\, dx\, dy}$$

Example 14. *Find the position of the C.G. of a semi-circular lamina of radius a if its density varies as the square of the distance from the diameter.* (*AMIETE, Dec. 2010*)

Solution. Let the bounding diameter be as the x-axis and a line perpendicular to the diameter and passing through the centre is y-axis. Equation of the circle is $x^2 + y^2 = a^2$. By symmetry $\bar{x} = 0$.

$$\bar{y} = \frac{\iint y\, \rho\, dx\, dy}{\iint \rho\, dx\, dy} = \frac{\iint (\lambda\, y^2)\, y\, dx\, dy}{\iint (\lambda\, y^2)\, dx\, dy} = \frac{\int_{-a}^{a} dx \int_0^{\sqrt{a^2 - x^2}} y^3\, dy}{\int_{-a}^{a} dx \int_0^{\sqrt{a^2 - x^2}} y^2\, dy}$$

$$= \frac{\int_{-a}^{a} dx \left[\frac{y^4}{4}\right]_0^{\sqrt{a^2 - x^2}}}{\int_{-a}^{a} dx \left(\frac{y^3}{3}\right)_0^{\sqrt{a^2 - x^2}}} = \frac{3\int_{-a}^{a} (a^2 - x^2)^2\, dx}{4\int_{-a}^{a} (a^2 - x^2)^{3/2}\, dx} \quad \text{[Put } x = a \sin\theta\ \ dx = a\cos\theta\, d\theta$$

$$= \frac{3\int_{-\frac{\pi}{2}}^{\frac{\pi}{2}} (a^2 - a^2 \sin^2\theta)^2\, a\cos\theta\, d\theta}{4\int_{-\frac{\pi}{2}}^{\frac{\pi}{2}} (a^2 - a^2 \sin^2\theta)^{3/2}\, a\cos\theta\, d\theta} = \frac{3\int_{-\frac{\pi}{2}}^{\frac{\pi}{2}} a^5 \cos^5\theta\, d\theta}{4\int_{-\frac{\pi}{2}}^{\frac{\pi}{2}} a^4 \cos^4\theta\, d\theta}$$

$$= \frac{3a}{4}\,\frac{\frac{4\times 2}{5\times 3}}{\frac{3\times 1}{4\times 2}\,\frac{\pi}{2}} = \left(\frac{3a}{4}\right)\left(\frac{8}{15}\right)\left(\frac{16}{3\pi}\right) = \frac{32a}{15\pi}$$

Hence C.G. is $\left(0, \frac{32a}{15\pi}\right)$ **Ans.**

Example 15. *Find C.G. of the area in the positive quadrant of the curve*

$$x^{2/3} + y^{2/3} = a^{2/3}.$$

Solution. For C.G. of area; $\bar{x} = \frac{\iint x\, dx\, dy}{\iint dx\, dy},\ \bar{y} = \frac{\iint y\, dx\, dy}{\iint dx\, dy}$

$$\bar{x} = \frac{\int_0^a x\, dx \int_0^{(a^{2/3} - x^{2/3})^{3/2}} dy}{\int_0^a dx \int_0^{(a^{2/3} - x^{2/3})^{3/2}} dy} = \frac{\int_0^a x\, dx\, [y]_0^{(a^{2/3} - x^{2/3})^{3/2}}}{\int_0^a dx\, [y]_0^{(a^{2/3} - x^{2/3})^{3/2}}} \quad \text{[Put } x = a\cos^3\theta]$$

$$= \frac{\int_0^a x\, dx\, (a^{2/3} - x^{2/3})^{3/2}}{\int_0^a dx\, (a^{2/3} - x^{2/3})^{3/2}} = \frac{\int_{\frac{\pi}{2}}^{0} a\cos^3\theta\, (a^{2/3} - a^{2/3}\cos^2\theta)^{3/2}\, (-3a\cos^2\theta\sin\theta\, d\theta)}{\int_{\frac{\pi}{2}}^{0} (a^{2/3} - a^{2/3}\cos^2\theta)^{3/2}\, (-3a\cos^2\theta\sin\theta\, d\theta)}$$

$$= \frac{\int_0^{\frac{\pi}{2}} 3a^3 \cos^3\theta \sin^3\theta \cos^2\theta \sin\theta \, d\theta}{\int_0^{\frac{\pi}{2}} 3a^2 \sin^3\theta \cos^2\theta \sin\theta \, d\theta} = \frac{a\int_0^{\frac{\pi}{2}} \sin^4\theta \cos^5\theta \, d\theta}{\int_0^{\frac{\pi}{2}} \sin^4\theta \cos^2\theta \, d\theta} = \frac{\dfrac{\Gamma\frac{5}{2}\,\Gamma\frac{6}{2}}{2\,\Gamma\frac{11}{2}}\,a}{\dfrac{\Gamma\frac{5}{2}\,\Gamma\frac{3}{2}}{2\,\Gamma 4}}$$

$$= \frac{\Gamma 3\,\Gamma 4\, a}{\Gamma\frac{3}{2}\,\Gamma\frac{11}{2}} = \frac{(2)\,(6)\,a}{\frac{1}{2}\cdot\frac{9}{2}\cdot\frac{7}{2}\cdot\frac{5}{2}\cdot\frac{3}{2}\cdot\frac{1}{2}\,\pi} = \frac{256\,a}{315\,\pi}, \quad \text{Similarly,} \quad \overline{y} = \frac{256\,a}{315\,\pi}$$

Hence, C.G. of the area is $\left(\dfrac{256\,a}{315\,\pi}, \dfrac{256\,a}{315\,\pi}\right)$. **Ans.**

Example 16. *Find by double integration, the centre of gravity of the area of the cardioid $r = a\,(1 + \cos\theta)$.*

Solution. Let $(\overline{x}, \overline{y})$ be the C.G. the cardioid

By Symmetry, $\overline{y} = 0$.

$$\overline{x} = \frac{\iint_A x\,dx\,dy}{\iint_A dx\,dy} = \frac{\iint_A x\,dx\,dy}{\iint_A dx\,dy}$$

$$= \frac{\int_{-\pi}^{\pi}\int_0^{a(1+\cos\theta)} (r\cos\theta)\,(r\,d\theta\,dr)}{\int_{-\pi}^{\pi}\int_0^{a(1+\cos\theta)} r\,d\theta\,dr} = \frac{\int_{-\pi}^{\pi}\cos\theta\,d\theta\int_0^{a\,(1+\cos\theta)} r^2\,dr}{\int_{-\pi}^{\pi} d\theta\int_0^{a(1+\cos\theta)} r\,dr}$$

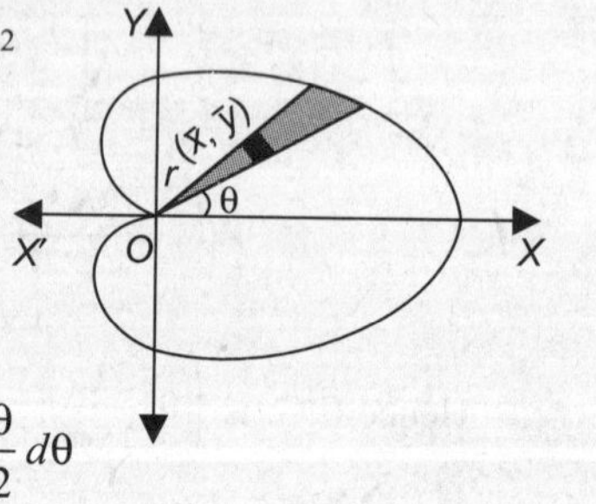

$$= \frac{\int_{-\pi}^{\pi}\cos\theta\,d\theta\left[\frac{r^3}{3}\right]_0^{a\,(1+\cos\theta)}}{\int_{-\pi}^{\pi} d\theta\left(\frac{r^2}{2}\right)_0^{a\,(1+\cos\theta)}} = \frac{\int_{-\pi}^{\pi}\cos\theta\,d\theta.\,\frac{a^3}{3}\,(1+\cos\theta)^3}{\int_{-\pi}^{\pi} d\theta\,\frac{a^2}{2}\,(1+\cos\theta)^2}$$

$$= \frac{\frac{a^3}{3}\int_{-\pi}^{\pi}\left(2\cos^2\frac{\theta}{2}-1\right)\left(1+2\cos^2\frac{\theta}{2}-1\right)^3 d\theta}{\frac{a^2}{2}\int_{-\pi}^{\pi}\left(1+2\cos^2\frac{\theta}{2}-1\right)^2 d\theta}$$

$$= \frac{a^3}{3}\int_{-\pi}^{\pi}\left(2\cos^2\frac{\theta}{2}-1\right)\left(8\cos^6\frac{\theta}{2}\right)d\theta \div \frac{a^2}{2}\int_{-\pi}^{\pi} 4\cos^4\frac{\theta}{2}\,d\theta$$

$$= \frac{8a^3}{3}\int_{-\pi}^{\pi}\left(2\cos^8\frac{\theta}{2}-\cos^6\frac{\theta}{2}\right)d\theta \div 2a^2\int_{-\pi}^{\pi}\cos^4\frac{\theta}{2}\,d\theta$$

$$= \frac{2\times 8a^3}{3}\int_0^{\pi}\left(2\cos^8\frac{\theta}{2}-\cos^6\frac{\theta}{2}\right)d\theta \div 4a^2\int_0^{\pi}\cos^4\frac{\theta}{2}\,d\theta$$

$$= \frac{16a^3}{3}\int_0^{\pi/2}(2\cos^8 t-\cos^6 t)\,(2\,dt) \div 4a^2\int_0^{\pi/2}\cos^4 t\,(2\,dt) \qquad \left[\text{Put } \frac{\theta}{2} = t \;\; d\theta = 2\,dt\right]$$

$$= \frac{32\,a^3}{3}\left[\frac{2\times 7\times 5\times 3\times 1}{8\times 6\times 4\times 2}\,\frac{\pi}{2} - \frac{5\times 3\times 1}{6\times 4\times 2}\,\frac{\pi}{2}\right] \div 8a^2\left(\frac{3\times 1}{4\times 2}\,\frac{\pi}{2}\right)$$

$$= \frac{32\,a^3}{3}\left(\frac{35\pi}{128}-\frac{5\pi}{32}\right) \div 8a^2\left(\frac{3\pi}{16}\right) = \frac{32a^3}{3}\times\frac{15\pi}{128}\times\frac{16}{8a^2\times 3\pi} = \frac{5}{6}a$$

Ans.

Example 17. *OA is the diameter of semicircular disc, the density at any point varies its distance from O. Find the position of entire of gravity given that OA = a.*

Solution. $\rho \propto r \Rightarrow \rho = kr$, where k is constant.

r varies from $r = 0$ to $r = a\cos\theta$ and θ varies from $\theta = 0$ to $\theta = \frac{\pi}{2}$.

If $(\bar{x}, \bar{y})$ are the coordinates of centre of gravity, then

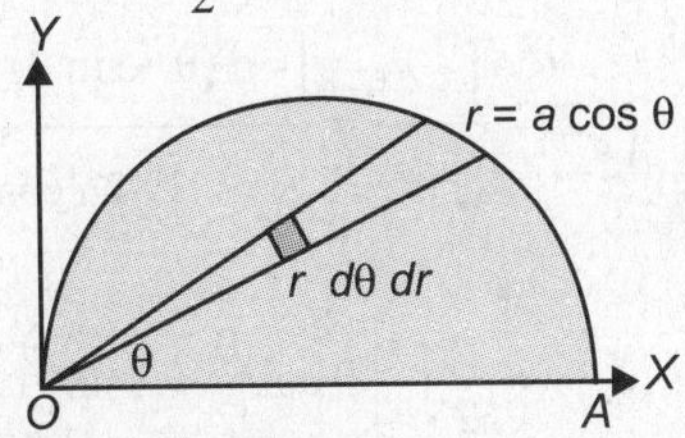

Y, r = a cos θ, r dθ dr, θ, O, A, X

$$\bar{x} = \frac{\iint x\rho\, dA}{\iint \rho\, dA} = \frac{\int_0^{\frac{\pi}{2}}\int_0^{a\cos\theta}(r\cos\theta)(kr)(r\,d\theta dr)}{\int_0^{\frac{\pi}{2}}\int_0^{a\cos\theta}(kr)(r\,d\theta dr)}$$

$$= \frac{\int_0^{\frac{\pi}{2}}\cos\theta\, d\theta \int_0^{a\cos\theta} r^3\, dr}{\int_0^{\pi/2}\int_0^{a\cos\theta} r^2\, dr\, d\theta} = \frac{\int_0^{\frac{\pi}{2}}\cos\theta\, d\theta\left[\frac{r^4}{4}\right]_0^{a\cos\theta}}{\int_0^{\frac{\pi}{2}}\left[\frac{r^3}{3}\right]_0^{a\cos\theta} d\theta} = \frac{\frac{1}{4}\int_0^{\frac{\pi}{2}}\cos\theta\, a(a^4\cos^4\theta)}{\frac{1}{3}\int_0^{\frac{\pi}{2}} d\theta(a^3\cos^3\theta)}$$

$$= 3\frac{a^4}{4a^3}\frac{\int_0^{\frac{\pi}{2}}\cos^5\theta\, d\theta}{\int_0^{\frac{\pi}{2}}\cos^3\theta\, d\theta} = \frac{3a}{4}\left[\frac{\frac{4\times 2}{5\times 3}}{\frac{2}{3}}\right] = \frac{3a}{4}\left(\frac{4}{5}\right) = \frac{3a}{5}$$

$$\bar{y} = \frac{\iint y\rho\, dA}{\iint \rho\, dA} = \frac{\int_0^{\frac{\pi}{2}}\int_0^{a\cos\theta}(r\sin\theta)kr(r\,d\theta\,dr)}{\int_0^{\frac{\pi}{2}}\int_0^{a\cos\theta}(kr)\,r\,d\theta\,dr},\quad \rho = kr$$

$$= = \frac{k\int_0^{\frac{\pi}{2}}\sin\theta\, d\theta\int_0^{a\cos\theta} r^3\, dr}{k\int_0^{\frac{\pi}{2}} d\theta\int_0^{a\cos\theta} r^2\, dr} = \frac{\int_0^{\frac{\pi}{2}}\sin\theta\, d\theta\left[\frac{r^4}{4}\right]_0^{a\cos\theta}}{\int_0^{\frac{\pi}{2}} d\theta\left[\frac{r^3}{3}\right]_0^{a\cos\theta}}$$

$$= \frac{\frac{1}{4}\int_0^{\frac{\pi}{2}}\sin\theta\, d\theta(a^4\cos^4\theta)}{\frac{1}{3}\int_0^{\frac{\pi}{2}} d\theta(a^3\cos^3\theta)} = \frac{3a^4\int_0^{\frac{\pi}{2}}\cos^4\theta(\sin\theta\, d\theta)}{4a^3\int_0^{\frac{\pi}{2}}\cos^3\theta\, d\theta}$$

$$= \frac{3a}{4}\frac{\left[-\frac{\cos^5\theta}{5}\right]_0^{\frac{\pi}{2}}}{\frac{2}{3}} = \frac{3}{4}\times\frac{3}{2}a\left[-0+\frac{1}{5}\right] = \frac{9a}{40}$$ Hence C.G is at $\left(\frac{3a}{5}, \frac{9a}{40}\right)$. **Ans.**

6.6 CENTRE OF GRAVITY OF AN ARC

Example 18. *Find the C.G. of the arc of the curve*

$x = a(\theta + \sin\theta)$, $y = a(1 - \cos\theta)$ in the positive quadrant.

Solution. We know that, $\bar{x} = \frac{\int x\,ds}{\int ds}, \bar{y} = \frac{\int y\,ds}{\int ds}$

Now, $ds = \sqrt{\left(\frac{dx}{d\theta}\right)^2 + \left(\frac{dy}{d\theta}\right)^2}\, d\theta$

$$= \sqrt{\{a^2 (1+\cos\theta)^2 + a^2 \sin^2\theta\}}\, d\theta = a\sqrt{1+2\cos\theta+\cos^2\theta+\sin^2\theta}\, d\theta$$

$$= a\sqrt{1+2\cos\theta+1}\, d\theta = a\sqrt{2(1+\cos\theta)}\, d\theta = a\sqrt{4\cos^2\frac{\theta}{2}}\, d\theta = 2a\cos\frac{\theta}{2}\, d\theta$$

$$\overline{x} = \frac{\int x\, dx}{\int ds} = \frac{\int_0^\pi a(\theta+\sin\theta)\, 2a\cos\frac{\theta}{2}\, d\theta}{\int_0^\pi 2a\cos\frac{\theta}{2}\, d\theta} = \frac{a\int_0^\pi\left(\theta+2\sin\frac{\theta}{2}\cos\frac{\theta}{2}\, d\theta\right)}{\left[2\sin\frac{\pi}{2}\right]_0^\pi} \quad \left[\text{Put } \frac{\theta}{2} = t\ d\theta = 2\, dt\right]$$

$$= \frac{a}{2}\int_0^\pi\left[\theta\cos\frac{\theta}{2} + 2\sin\frac{\theta}{2}\cos^2\frac{\theta}{2}\right] d\theta = \frac{a}{2}\int_0^{\frac{\pi}{2}} (2t\cos t + 2\sin t\cos^2 t)\, 2\, dt$$

$$= 2a\left[t\sin t + \cos t - \frac{\cos^3 t}{3}\right]_0^{\frac{\pi}{2}} = 2a\left[\frac{\pi}{2} - 1 + \frac{1}{3}\right] = a\left[\pi - \frac{4}{3}\right]$$

$$\overline{y} = \frac{\int y\, ds}{\int ds} = \frac{\int_0^\pi a(1-\cos\theta)\, 2a\cos\frac{\theta}{2}\, d\theta}{\int_0^\pi 2a\cos\frac{\theta}{2}\, d\theta} = \frac{a\int_0^\pi 2\sin^2\frac{\theta}{2}\cos\frac{\theta}{2}\, d\theta}{\int_0^\pi \cos\frac{\theta}{2}\, d\theta}$$

$$= \frac{a\, r\left[\sin^3\frac{\theta}{2}\right]_0^\pi}{3\left[2\sin\frac{\theta}{2}\right]_0^\pi} = \frac{4a}{3\times 2} = \frac{2a}{3} \quad \text{Hence, C.G. of the arc is } \left[a\left(\pi - \frac{4}{3}\right), \frac{2a}{3}\right] \qquad \textbf{Ans.}$$

EXERCISE 6.4

1. Find the centre of gravity of the area bounded by the parabola $y^2 = x$ and the line $x + y = 2$. **Ans.** $\left(\frac{8}{5}, -\frac{1}{2}\right)$

2. Find the centroid of the tetrahedron bounded by the coordinate planes and the plane $x + y + z = 1$, the density at any point varying as its distance from the plane $z = 0$. **Ans.** $\left(\frac{1}{5}, \frac{1}{5}, \frac{2}{5}\right)$

3. Find the centroid of the area enclosed by the parabola $y^2 = 4\, ax$, the axis of x and latus rectum. **Ans.** $\left(\frac{3a}{20}, \frac{3a}{16}\right)$

4. Find the centroid of the loop of curve $r^2 = a^2 \cos 2\theta$. **Ans.** $\left(\frac{\pi a\sqrt{2}}{8}, 0\right)$

5. Find the centroid of solid formed by revolving about the x-axis that part of the area of the ellipse $\frac{x^2}{a^2} + \frac{y^2}{b^2} = 1$ which lies in the first quadrant. **Ans.** $\left(\frac{3a}{8}, 0\right)$

6. Find the average density of the sphere of radius a whose density at a distance r from the centre of the sphere is $\rho = \rho_0\left[1 + k\frac{r^3}{a^3}\right]$. $\rho_0\left(1 + \frac{k}{2}\right)$

7. The density at a point on a circular lamina varies as the distance from a point O on the circumference. Show that the C.G. divides the diameter through O in the ratio 3 : 2.

TRIPLE INTEGRATION

7.1 INTRODUCTION

In this chapter we will learn triple integration. Before that we will discuss the following coordinate systems.

1. Cartesian coordinates

2. Spherical coordinates

3. Cylindrical coordinates

1. Cartesian Coordinates

Take a point O in the space draw three mutually perpendicular lines through O. O is known as origin and these three lines are known as x-axis, y-axis, z-axis. There are three coordinate planes.

1. ***xy-plane*** **:** The plane passing through x-axis and y-axis is known as xy-plane.

2. ***yz-plane*** **:** The plane passing through y-axis and z-axis is known as yz-plane.

3. ***zx-plane*** **:** The plane passing through z-axis and x-axis is known as zx-plane.

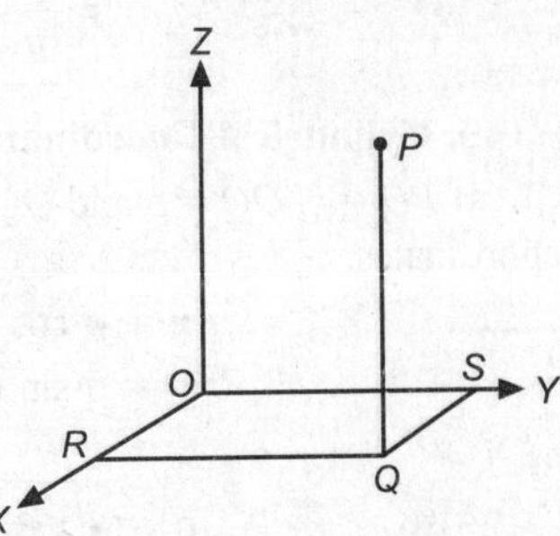

Consider a point P in the space draw perpendicular PQ to xy-plane. PQ is known as z-coordinate; from Q draw perpendicular lines QR and QS to x-axis and y-axis respectively. QS is x-coordinate, QR is y-coordinate and small element of the solid in cartesian coordinates is $dxdydz$.

2. Spherical Coordinates

In the adjoining figure join OP, OP is denoted by r, θ is called the angle between OP *i.e.* r and z-axis.

ϕ is the angle between x-axis and OQ.

Then the coordinates (r, θ, ϕ) of a point P are known as spherical coordinates.

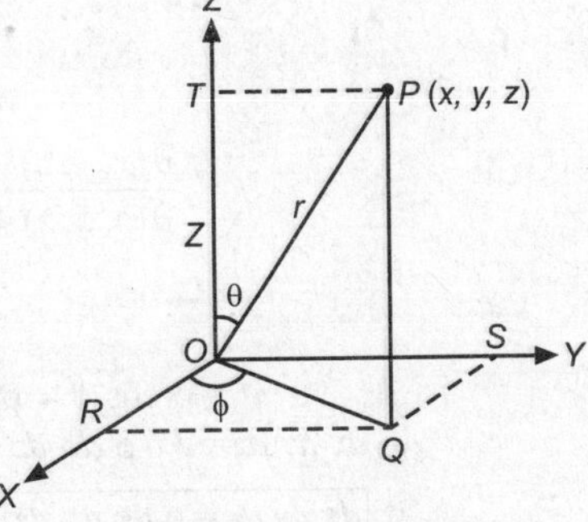

$$Z = PQ = OT = OP \cos\theta$$

$$= r\cos\theta$$

$$PT = OP \sin\theta$$

$$= r\sin\theta$$

$$\Rightarrow \quad OQ = r\sin\theta$$

In right angled triangle QRO, $\angle R$ is right angle.

$$x = OR = OQ \cos\phi = r\sin\theta\cos\phi$$

Again in right angled triangle QSO

$$y = OQ\sin\phi = r\sin\theta\sin\phi$$

$$\boxed{\begin{aligned} x &= r\sin\theta\cos\Phi \\ y &= r\sin\theta\sin\Phi \\ z &= r\cos\theta \end{aligned}}$$

Also,

$$r^2 = x^2 + y^2 + z^2$$

$$\Phi = \tan^{-1}\frac{y}{x}$$

$$\theta = \tan^{-1}\left(\frac{\sqrt{x^2+y^2}}{z}\right)$$

If

$$J = \frac{\partial(x, y, z)}{\partial(r, \theta, \phi)} \qquad \begin{bmatrix} x = r\sin\theta\cos\phi \\ y = r\sin\theta\sin\phi \\ z = r\cos\theta \end{bmatrix}$$

$$= \begin{vmatrix} \frac{\partial x}{\partial r} & \frac{\partial x}{\partial \theta} & \frac{\partial x}{\partial \phi} \\ \frac{\partial y}{\partial r} & \frac{\partial y}{\partial \theta} & \frac{\partial y}{\partial \phi} \\ \frac{\partial z}{\partial r} & \frac{\partial z}{\partial \theta} & \frac{\partial z}{\partial \phi} \end{vmatrix} = r^2\sin\theta$$

$$dx\,dy\,dz = J\,dr\,d\theta\,d\phi$$

$$\boxed{dx\,dy\,dz = r^2\sin\theta\,dr\,d\theta\,d\phi}$$

3. Cylindrical Coordinate System

If $PQ = z$, $OQ = \rho$ and OQ makes angle Φ with the x-axis then (z, ρ, Φ) are called cylindrical coordinates.

$$x = \rho\cos\Phi$$

$$y = \rho\sin\Phi$$

$$z = z$$

Also

$$\rho = \sqrt{x^2+y^2}$$

$$\Phi = \tan^{-1}\left(\frac{y}{x}\right)$$

$$z = z$$

$$J = \frac{\partial(x, y, z)}{\partial(\rho, \phi, z)} = \begin{vmatrix} \frac{\partial x}{\partial \rho} & \frac{\partial x}{\partial \phi} & \frac{\partial x}{\partial z} \\ \frac{\partial y}{\partial \rho} & \frac{\partial y}{\partial \phi} & \frac{\partial y}{\partial z} \\ \frac{\partial z}{\partial \rho} & \frac{\partial z}{\partial \phi} & \frac{\partial z}{\partial z} \end{vmatrix} = \begin{vmatrix} \cos\phi & -\rho\sin\phi & 0 \\ \sin\phi & \rho\cos\phi & 0 \\ 0 & 0 & 1 \end{vmatrix}$$

$$= \rho\cos^2\phi + \rho\sin^2\phi = \rho(\cos^2\phi + \sin^2\phi) = \rho$$

$$dx\,dy\,dz = J\,d\rho\,d\phi\,dz$$

$$\boxed{dx\,dy\,dz = \rho\,d\rho\,d\phi\,dz}$$

7.2 TRIPLE INTEGRATION

Let a function $f(x, y, z)$ be a continuous at every point of a finite region S of three dimensional space. Consider n sub-spaces $\delta S_1, \delta S_2, \delta S_3, \ldots \delta S_n$ of the space S.

If (x_r, y_r, z_r) be a point in the rth subspace.

The limit of the sum $\sum_{r=1}^{n} f(x_r, y_r, z_r)\, \delta S_r$, as $n \to \infty$, $\delta S_r \to 0$ is known as the triple integral of $f(x, y, z)$ over the space S.

Symbolically, it is denoted by

$$\iiint_S f(x, y, z)\, dS$$

It can be calculated as $\int_{x_1}^{x_2}\int_{y_1}^{y_2}\int_{z_1}^{z_2} f(x, y, z)\ dx\ dy\ dz$. First we integrate with respect to z treating x, y as constant between the limits z_1 and z_2. The resulting expression (function of x, y) is integrated with respect to y keeping x as constant between the limits y_1 and y_2. At the end we integrate the resulting expression (function of x only) within the limits x_1 and x_2.

$$\boxed{\int_{x_1=a}^{x_2=b} \Psi(x)\, dx \boxed{\int_{y_1=\phi_1(x)}^{y_2=\phi_2(x)} \phi(x, y)\, dy \boxed{\int_{z_1=f_1(x,y)}^{z_2=f_2(x,y)} f(x, y, z)\, dz}}}$$

First we integrate from inner most integral w.r.t. z, then we integrate with respect to y and finally the outer most with respect to x.

But the above order of integration is immaterial provided the limits change accordingly.

Example 1. *Evaluate* $\iiint_R (x + y + z)\ dx\ dy\ dz$, *where* $R : 0 \le x \le 1,\ 1 \le y \le 2,\ 2 \le z \le 3$.

Solution. $\int_0^1 dx\int_1^2 dy\int_2^3 (x+y+z)\, dz \ = \ \int_0^1 dx\int_1^2 dy\left[\frac{(x+y+z)^2}{2}\right]_2^3$

$$= \frac{1}{2}\int_0^1 dx\int_1^2 dy\,[(x+y+3)^2 - (x+y+2)^2] = \frac{1}{2}\int_0^1 dx\int_1^2 (2x+2y+5)\,.1.dy$$

$$= \frac{1}{2}\int_0^1 dx\left[\frac{(2x+2y+5)^2}{4}\right]_1^2 = \frac{1}{8}\int_0^1 dx\,[(2x+4+5)^2 - (2x+2+5)^2]$$

$$= \frac{1}{8}\int_0^1 (4x+16)\,.\,2\, dx = \int_0^1 (x+4)\, dx = \left[\frac{x^2}{2} + 4x\right]_0^1 = \frac{1}{2} + 4 = \frac{9}{2} \qquad \textbf{Ans.}$$

Example 2. *Evaluate the integral* : $\int_0^{\log 2}\int_0^x\int_0^{x+\log y} e^{x+y+z}\, dz\, dy\, dx.$

Solution. $\int_0^{\log 2}\int_0^x\int_0^{x+\log y} e^{x+y+z}\, dz\, dy\, dx.$

$$= \int_0^{\log 2} e^x dx\int_0^x e^y dy\int_0^{x+\log y} e^z dz = \int_0^{\log 2} e^x\, dx\int_0^x e^y dy (e^z)_0^{x+\log y}$$

$$= \int_0^{\log 2} e^x dx\int_0^x e^y dy\,(e^{x+\log y} - 1) = \int_0^{\log 2} e^x\, dx\int_0^x e^y dy\,(e^{\log y}\,.\,e^x - 1)$$

$$= \int_0^{\log 2} e^x dx\int_0^x e^y\,(ye^x - 1)\, dy \ = \ \int_0^{\log 2} e^x dx\left[(ye^x - 1)\, e^y - \int e^x\,.\,e^y\, dy\right]_0^x$$

$$= \int_0^{\log 2} e^x dx\left[(ye^x - 1)\, e^y - e^{x+y}\right]_0^x = \int_0^{\log 2} e^x dx\,[(xe^x - 1)\, e^x - e^{2x} + 1 + e^x]$$

$$= \int_0^{\log 2} e^x dx\,[xe^{2x} - e^x - e^{2x} + 1 + e^x] = \int_0^{\log 2} (xe^{3x} - e^{3x} + e^x)\, dx$$

$$= \left[x\frac{e^{3x}}{3} - \int 1.\frac{e^{3x}}{3}\,dx - \frac{e^{3x}}{3} + e^x\right]_0^{\log 2} = \left[\frac{x}{3}e^{3x} - \frac{e^{3x}}{9} - \frac{e^{3x}}{3} + e^x\right]_0^{\log 2}$$

$$= \frac{\log 2}{3}e^{3\log 2} - \frac{e^{3\log 2}}{9} - \frac{e^{3\log 2}}{3} + e^{\log 2} + \frac{1}{9} + \frac{1}{3} - 1$$

$$= \frac{\log 2}{3}e^{\log 2^3} - \frac{e^{\log 2^3}}{9} - \frac{e^{\log 2^3}}{3} + e^{\log 2} + \frac{1}{9} + \frac{1}{3} - 1$$

$$= \frac{8}{3}\log 2 - \frac{8}{9} - \frac{8}{3} + 2 + \frac{1}{9} + \frac{1}{3} - 1 = \frac{8}{3}\log 2 - \frac{19}{9}$$ **Ans.**

Example 3. *Evaluate* $\int_0^{\log 2}\int_0^x\int_0^{x+y} e^{x+y+z}\,dx\,dy\,dz.$ *(M.U. II Semester, 2005)*

Solution. $I = \int_0^{\log 2}\int_0^x e^{x+y}\left[e^z\right]_0^{x+y} dx\,dy$

$$= \int_0^{\log 2}\int_0^x e^{x+y}(e^{x+y} - 1)\,dx\,dy = \int_0^{\log 2}\int_0^x \left[e^{2(x+y)} - e^{(x+y)}\right] dx\,dy$$

$$= \int_0^{\log 2}\left[e^{2x}.\frac{e^{2y}}{2} - e^x.e^y\right]_0^x dx = \int_0^{\log 2}\left(\frac{e^{4x}}{2} - e^{2x} - \frac{e^{2x}}{2} + e^x\right) dx$$

$$= \left[\frac{e^{4x}}{8} - \frac{e^{2x}}{2} - \frac{e^{2x}}{4} + e^x\right]_0^{\log 2} = \left[\frac{e^{4\log 2}}{8} - \frac{e^{2\log 2}}{2} - \frac{e^{2\log 2}}{4} + e^{\log 2}\right] - \left(\frac{1}{8} - \frac{1}{2} - \frac{1}{4} + 1\right)$$

$$= \left(\frac{e^{\log 16}}{8} - \frac{e^{\log 4}}{2} - \frac{e^{\log 4}}{4} + e^{\log 2}\right) - \left(\frac{1}{8} - \frac{1}{2} - \frac{1}{4} + 1\right)$$

$$= \left(\frac{16}{8} - \frac{4}{2} - \frac{4}{4} + 2\right) - \left(\frac{1}{8} - \frac{1}{2} - \frac{1}{4} + 1\right) = \frac{5}{8}$$ **Ans.**

Example 4. *Evaluate* $\iiint_R (x^2 + y^2 + z^2)\,dz\,dy\,dx$

where R denotes the region bounded by $x = 0,\ y = 0,\ z = 0$ *and* $x + y + z = a,\ (a > 0)$

Solution. $\iiint_R (x^2 + y^2 + z^2)\,dx\,dy\,dz$

$x + y + z = a$ or $z = a - x - y$

Upper limit of $z = a - x - y$

On x-y plane, $x + y + z = a$ becomes $x + y = a$ as shown in the figure.

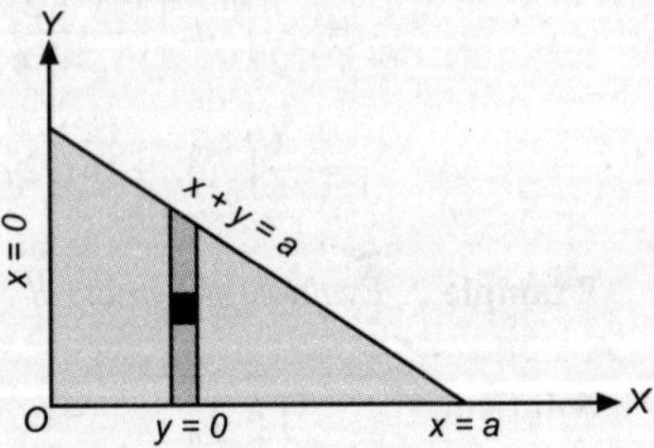

Upper limit of $y = a - x$

Upper limit of $x = a$

$$= \int_{x=0}^{a} dx \int_{y=0}^{a-x} dy \int_{z=0}^{a-x-y} (x^2 + y^2 + z^2)\,dz = \int_0^a dx \int_0^{a-x} dy\left(x^2 z + y^2 z + \frac{z^3}{3}\right)_0^{a-x-y}$$

$$= \int_0^a dx \int_0^{a-x} dy\left[x^2(a - x - y) + y^2(a - x - y) + \frac{(a - x - y)^3}{3}\right]$$

$$= \int_0^a dx \int_0^{a-x}\left[x^2(a - x) - x^2 y + (a - x)\,y^2 - y^3 + \frac{(a - x - y)^3}{3}\right] dy$$

$$= \int_0^a dx\left[x^2(a - x)\,y - \frac{x^2 y^2}{2} + (a - x)\frac{y^3}{3} - \frac{y^4}{4} - \frac{(a - x - y)^4}{12}\right]_0^{a-x}$$

$$= \int_0^a dx\left[x^2(a-x)^2 - \frac{x^2}{2}(a-x)^2 + (a-x)\frac{(a-x)^3}{3} - \frac{(a-x)^4}{4} + \frac{(a-x)^4}{12}\right]$$

$$= \int_0^a \left[\frac{x^2}{2}(a-x)^2 + \frac{(a-x)^4}{6}\right]dx = \int_0^a \left[\frac{1}{2}(a^2x^2 - 2ax^3 + x^4) + \frac{(a-x)^4}{6}\right]dx$$

$$= \left[\frac{1}{2}a^2\frac{x^3}{3} - \frac{ax^4}{4} + \frac{x^5}{10} - \frac{(a-x)^5}{30}\right]_0^a = \frac{a^5}{6} - \frac{a^5}{4} + \frac{a^5}{10} + \frac{a^5}{30} = \frac{a^5}{20}$$ **Ans.**

Example 5. *Compute* $\iiint \frac{dx\,dy\,dz}{(x+y+z+1)^3}$ *if the region of integration is bounded by the coordinate planes and the plane $x + y + z = 1$.* (*M.U., II Semester 2007, 2006*)

Solution. Let the given region be R, then R is expressed as

$$0 \le z \le 1 - x - y, \quad 0 \le y \le 1 - x, \quad 0 \le x \le 1.$$

$$\iiint_R \frac{dz\,dy\,dx}{(x+y+z+1)^3} = \int_0^1 dx \int_0^{1-x} dy \int_0^{1-x-y} \frac{dz}{(x+y+z+1)^3}$$

$$= \int_0^1 dx \int_0^{1-x} dy \left[\frac{1}{-2(x+y+z+1)^2}\right]_0^{1-x-y}$$

$$= -\frac{1}{2}\int_0^1 dx \int_0^{1-x} dy \left[\frac{1}{(x+y+1-x-y+1)^2} - \frac{1}{(x+y+1)^2}\right]$$

$$= -\frac{1}{2}\int_0^1 dx \int_0^{1-x} \left[\frac{1}{4} - \frac{1}{(x+y+1)^2}\right]dy = -\frac{1}{2}\int_0^1 dx\left[\frac{y}{4} + \frac{1}{x+y+1}\right]_0^{1-x}$$

$$= -\frac{1}{2}\int_0^1 dx\left[\frac{1-x}{4} + \frac{1}{x+1+1-x} - \frac{1}{x+1}\right] = -\frac{1}{2}\int_0^1 \left[\frac{1-x}{4} + \frac{1}{2} - \frac{1}{x+1}\right]dx$$

$$= -\frac{1}{2}\left[-\frac{(1-x)^2}{8} + \frac{x}{2} - \log(x+1)\right]_0^1 = -\frac{1}{2}\left[\frac{1}{2} - \log 2 + \frac{1}{8}\right] = -\frac{1}{2}\left[\frac{5}{8} - \log 2\right]$$

$$= \frac{1}{2}\log 2 - \frac{5}{16}$$ **Ans.**

Example 6. *Compute* $\iiint_V x^2\,dx\,dy\,dz$ *over volume of tetrahedron bounded by*

$x = 0,\ y = 0,\ z = 0$ *and* $\frac{x}{a} + \frac{y}{b} + \frac{z}{c} = 1.$ (*M.U. II Semester, 2008*)

Solution. Here, we have

$$I = \iiint_V x^2\,dx\,dy\,dz \quad \text{...(1)}$$

Where V is bounded by $x = 0$, $y = 0$, $z = 0$ and $\frac{x}{a} + \frac{y}{b} + \frac{z}{c} = 1$

Putting $\frac{x}{a} = u$, $\frac{y}{b} = v$, $\frac{z}{c} = w$ so that $dx = a\,du$, $dy = b\,dv$, $dz = c\,dw$

in (1), we get

$$I = \int_0^1 \int_0^{1-u} \int_0^{1-u-v} a^2u^2\ (adu)\,(bdv)\,(cdw)$$

$$= a^3bc\int_0^1 u^2du \int_0^{1-u} dv \int_0^{1-u-v} dw = a^3bc\int_0^1 u^2\,du \int_0^{1-u} dv\,[w]_0^{1-u-v}$$

$$= a^3bc\int_0^1 u^2du \int_0^{1-u} (1-u-v)\,dv = a^3bc\int_0^1 u^2du\left[v - uv - \frac{v^2}{2}\right]_0^{1-u}$$

$$= a^3bc\int_0^1 u^2\left[1 - u - u(1-u) - \frac{(1-u)^2}{2}\right]du$$

$$= a^3bc\int_0^1 u^2\left[1-u-u+u^2-\frac{1}{2}-\frac{u^2}{2}+u\right]du$$

$$= a^3bc\int_0^1 u^2\left(\frac{1}{2}-u+\frac{1}{2}u^2\right)du = a^3bc\int_0^1\left(\frac{u^2}{2}-u^3+\frac{u^4}{2}\right)du$$

$$= a^3bc\left[\frac{1}{2}.\frac{u^3}{3}-\frac{u^4}{4}+\frac{1}{2}.\frac{u^5}{5}\right]_0^1 = a^3bc\left[\frac{1}{6}-\frac{1}{4}+\frac{1}{10}\right]=a^3bc\left(\frac{1}{60}\right)=\frac{a^3bc}{60}$$ **Ans.**

Example 7. *Evaluate* $\iiint x^2yz\ dx\ dy\ dz$ *throughout the volume bounded by the planes* $x = 0, y = 0, z = 0, \frac{x}{a}+\frac{y}{b}+\frac{z}{c}=1.$

Solution. Here, we have

$$I = \iiint x^2yz\ dx\ dy\ dz \quad ...(1)$$

Putting $x = au, \quad y = bv, \quad z = cw$

$dx = a\ du, \quad dy = b\ dv, \quad dz = c\ dw$ in (1), we get

$$u + v + w = 1 \quad I = \iiint a^2bc\,u^2\,v\,w\,a\,bc\ du\ dv\ dw$$

Limits are for $u = 0, 1$ for $v = 0, 1 - u$ and for $w = 0, 1 - u - v$

$$I = \int_{u=o}^{1}\int_{v=0}^{1-u}\int_{w=0}^{1-u-v} a^3b^2c^2\,u^2vw\ du\ dv\ dw = \int_0^1\int_0^{1-u} a^3b^2c^2\,u^2v\left[\frac{w^2}{2}\right]_0^{1-u-v} du\ dv$$

$$= \frac{a^3b^2c^2}{2}\int_0^1\int_0^{1-u} u^2v(1-u-v)^2\ du\ dv$$

$$= \frac{a^3b^2c^2}{2}\int_0^1\int_0^{1-u} u^2v\left[(1-u)^2-2(1-u)v+v^2\right]du\ dv$$

$$= \frac{a^3b^2c^2}{2}\int_0^1\int_0^{1-u} u^2[(1-u)^2v-2(1-u)v^2+v^3]\,du\ dv$$

$$= \frac{a^3b^2c^2}{2}\int_0^1 u^2\left[(1-u)^2\frac{v^2}{2}-2(1-u)\frac{v^3}{3}+\frac{v^4}{4}\right]_0^{1-u} du$$

$$= \frac{a^3b^2c^2}{2}\int_0^1 u^2\left[\frac{(1-u)^4}{2}-\frac{2(1-u)^4}{3}+\frac{(1-u)^4}{4}\right]du$$

$$= \frac{a^3b^2c^2}{2}\int_0^1 \frac{u^2(1-u)^4}{12}\,du = \frac{a^3b^2c^2}{24}\int_0^1 u^{3-1}(1-u)^{5-1}\,du$$

$$= \frac{a^3b^2c^2}{24}\beta(3,5) = \frac{a^3b^2c^2}{24}.\frac{\overline{|3}\,\overline{|5}}{\overline{|8}} = \frac{a^3b^2c^2}{24}.\left(\frac{2!\,4!}{7!}\right)=\frac{a^3\,b^2\,c^2}{2520}.$$ **Ans.**

7.3 INTEGRATION BY CHANGE OF CARTESIAN COORDINATES INTO SPHERICAL COORDINATES

Sometime it becomes easy to integrate by changing the cartesian coordinates into spherical coordinates.

The relations between the cartesian and spherical polar co-ordinates of a point are given by the relations

$$x = r\sin\theta\cos\phi$$
$$y = r\sin\theta\sin\phi$$

$$z = r\cos\theta$$
$$dx\,dy\,dz = |J|\,dr\,d\theta\,d\phi$$
$$= r^2\sin\theta\,dr\,d\theta\,d\phi$$

Note. 1. Spherical coordinates are very useful if the expression $x^2 + y^2 + z^2$ is involved in the problem.

2. In a sphere $x^2 + y^2 + z^2 = a^2$ the limits of r are 0 and a and limits of θ are 0, π and that of ϕ are 0 and 2π.

Example 8. *Evaluate the integral* $\iiint (x^2 + y^2 + z^2)\; dx\,dy\,dz$ *taken over the volume enclosed by the sphere* $x^2 + y^2 + z^2 = 1$.

Solution. Let us convert the given integral into spherical polar co-ordinates. By putting

$x = r\sin\theta\cos\phi$; $y = r\sin\theta\sin\phi$; $z = r\cos\theta$

$$\iiint (x^2+y^2+z^2)\; dx\,dy\,dz = \int_0^{2\pi}\int_0^{\pi}\int_0^1 r^2(r^2\sin\theta\,dr\,d\theta\,d\phi)$$

$$= \int_0^{2\pi} d\phi\int_0^{\pi}\sin\theta\,d\theta\int_0^1 r^4 dr = \int_0^{2\pi} d\phi\int_0^{\pi}\sin\theta\,d\theta\left(\frac{r^5}{5}\right)_0^1 = \frac{1}{5}\int_0^{2\pi} d\phi\,[-\cos\theta]_0^{\pi} = \frac{2}{5}\int_0^{2\pi} d\phi$$

$$= \frac{2}{5}(\phi)_0^{2\pi} = \frac{4\pi}{5}$$ **Ans.**

Example 9. *Evaluate* $\iiint (x^2 + y^2 + z^2)\; dx\,dy\,dz$ *over the first octant of the sphere* $x^2 + y^2 + z^2 = a^2$. *(M.U. II Semester 2007)*

Solution. Here, we have

$$I = \iiint (x^2 + y^2 + z^2)\,dx\,dy\,dz \qquad ...(1)$$

Putting $x = r\sin\theta\cos\phi$, $y = r\sin\theta\sin\phi$, $z = r\cos\theta$ and $dx\,dy\,dz = r^2\sin\theta\,dr\,d\theta\,d\phi$ in (1), we get

Limits of r are 0, a for θ are 0, $\frac{\pi}{2}$ for ϕ are 0, $\frac{\pi}{2}$.

$$I = \int_0^{\frac{\pi}{2}}\int_0^{\frac{\pi}{2}}\int_0^a r^2.\,r^2\sin\theta\,dr\,d\theta\,d\phi = \int_0^{\frac{\pi}{2}} d\phi\int_0^{\frac{\pi}{2}}\sin\theta\,d\theta\int_0^a r^4\,dr$$

$$\left(\begin{aligned} x^2+y^2+z^2 &= r^2\sin^2\theta\cos^2\phi + r^2\sin^2\theta\sin^2\phi + r^2\cos^2\theta \\ &= r^2\sin^2\theta + r^2\cos^2\theta = r^2 \end{aligned}\right)$$

$$= [\phi]_0^{\pi/2}\,[-\cos\theta]_0^{\pi/2}\left[\frac{r^5}{5}\right]_0^a = \frac{\pi}{2}.(1).\frac{a^5}{5} = \pi.\frac{a^5}{10}.$$ **Ans.**

Example 10. *Evaluate* $\iiint \frac{dx\,dy\,dz}{x^2+y^2+z^2}$ *throughout the volume of the sphere* $x^2 + y^2 + z^2 = a^2$.

Solution. Here, we have

$$I = \iiint \frac{dx\,dy\,dz}{x^2+y^2+z^2} \qquad ...(1)$$

Putting $x = r\sin\theta\cos\phi$, $y = r\sin\theta\sin\phi$, $z = r\cos\theta$ and $dx\,dy\,dz = r^2\sin\theta\,dr\,d\theta\,d\phi$ in (1), we get

The limits of r are 0 and a, for θ are 0 and $\frac{\pi}{2}$ for ϕ are 0 and $\frac{\pi}{2}$ in first octant.

$$I = 8\int_0^{\frac{\pi}{2}}\int_0^{\frac{\pi}{2}}\int_0^a \frac{r^2\sin\theta\,dr\,d\theta\,d\phi}{r^2} \qquad [\text{Sphere } x^2+y^2+z^2 \text{ lies in 8 octants}]$$

$$I = 8\int_0^{\frac{\pi}{2}} d\phi \int_0^{\frac{\pi}{2}} \sin\theta\, d\theta \int_0^a dr = 8[\phi]_0^{\pi/2}[-\cos\theta]_0^{\pi/2}[r]_0^a = 8\left(\frac{\pi}{2} - 0\right)(0+1)(a+0)$$

$$= 8\frac{\pi}{2}.1.a = 4\pi a$$ **Ans.**

Example 11. *Evaluate* $\iiint \frac{z^2 dx\, dy\, dz}{x^2+y^2+z^2}$ *over the volume of the sphere* $x^2+y^2+z^2 = 2$.

(M.U. II Semester 2005)

Solution. Here, we have

$$I = \iiint \frac{z^2 dx\, dy\, dz}{x^2+y^2+z^2} \qquad ...(1)$$

Putting $x = r\sin\theta\cos\phi$, $y = r\sin\theta\sin\phi$, $z = r\cos\theta$, $dx\, dy\, dz = r^2 \sin\theta\, dr\, d\theta\, d\phi$ in (1), we get

[The limits r, θ and ϕ over the first octant of $x^2+y^2+z^2 = r^2$ are 0, $\sqrt{2}$; 0, $\frac{\pi}{2}$ and 0, $\frac{\pi}{2}$].

$$I = 8\int_0^{\frac{\pi}{2}}\int_0^{\frac{\pi}{2}}\int_0^{\sqrt{2}} \frac{r^4\cos^2\theta\sin\theta}{r^2}\, dr\, d\theta\, d\phi = 8\int_0^{\frac{\pi}{2}} d\phi \int_0^{\frac{\pi}{2}} \cos^2\theta\sin\theta\, d\theta . \int_0^{\sqrt{2}} r^2 dr$$

$$= 8[\phi]_0^{\pi/2}\left[-\frac{\cos^3\theta}{3}\right]_0^{\frac{\pi}{2}}\left[\frac{r^3}{3}\right]_0^{\sqrt{2}} = 8\frac{\pi}{2}.\frac{1}{3}.\frac{2\sqrt{2}}{3} = \frac{8\pi\sqrt{2}}{9}.$$ **Ans.**

Example 12. *Evaluate* $\iiint xyz\ dx\, dy\, dz$ *over the positive octant of the sphere* $x^2+y^2+z^2 = a^2$.

Solution. Here, we have

$$I = \iiint xyz\, dx\, dy\, dz \text{ over the first quadrant} \qquad ...(1)$$

Putting $x = r\sin\theta\cos\phi$, $y = r\sin\theta\sin\phi$, $z = r\cos\theta$, $dx\, dy\, dz = r^2\sin\theta\, dr\, d\theta\, d\phi$ in (1), we get

$$I = \int_0^{\frac{\pi}{2}}\int_0^{\frac{\pi}{2}}\int_0^a (r\sin\theta\cos\phi)(r\sin\theta\sin\phi)(r\cos\theta)(r^2\sin\theta\, d\theta\, dr\, d\phi)$$

$$I = \int_0^{\frac{\pi}{2}}\int_0^{\frac{\pi}{2}}\int_0^a r^5\sin^3\theta\cos\theta\sin\phi\cos\phi\, dr\, d\theta\, d\phi$$

$$= \int_0^{\frac{\pi}{2}} \sin\phi\cos\phi\, d\phi \int_0^{\frac{\pi}{2}} \sin^3\theta\cos\theta\, d\theta \int_0^a r^5 dr$$

$$= \left[\frac{\sin^2\phi}{2}\right]_0^{\frac{\pi}{2}}\left[\frac{\sin^4\theta}{4}\right]_0^{\frac{\pi}{2}}\left[\frac{r^6}{6}\right]_0^a = \frac{1}{2}.\frac{1}{4}.\frac{a^6}{6} = \frac{a^6}{48}.$$ **Ans.**

[In first octant of $x^2+y^2+z^2 = a^2$
Limits of $r = 0$ and a
for $\theta = 0$ and $\frac{\pi}{2}$
for $\phi = 0$ and $\frac{\pi}{2}$.]

Example 13. *Evaluate* $\iiint \sqrt{x^2+y^2}\ dx\, dy\, dz$ *over the volume bounded by the right circular cone* $x^2+y^2 = z^2$, $z > 0$ *and the planes* $z = 0$ *and* $z = 1$.

Solution. Here, we have

$$I = \iiint \sqrt{x^2+y^2}\ dx\, dy\, dz \qquad ...(1)$$

Putting $x = r\cos\theta$, $y = r\sin\theta$, $z = z$ (cylindrical coordinates),
$dx\, dy\, dz = r.dr\, d\theta\, dz$ in (1), we get

[Now limits for r are 0 to 1 for θ are 0 to 2π for z are r to 1.]

$$I = \int_0^1 r^2 dr \int_0^{2\pi} d\theta \int_r^1 dz$$

$$= \int_0^1 r^2 dr \, [\theta]_0^{2\pi} \, [z]_r^1$$

$$= \int_0^1 r^2 \,.\, 2\pi \,.\, (1-r)\, dr$$

$$= 2\pi \left[\frac{r^3}{3} - \frac{r^4}{4}\right]_0^1 = \frac{2\pi}{12} = \frac{\pi}{6} \qquad \textbf{Ans.}$$

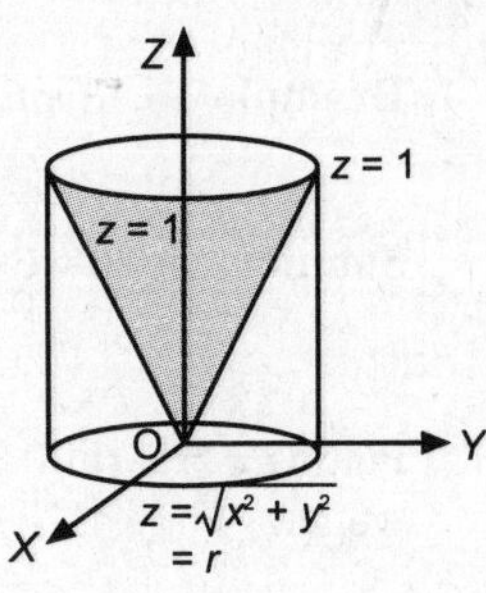

Example 14. *Using cylindrical co-ordinates evaluate:*

$$\iiint_V \sqrt{x^2 + y^2}\; dx\, dy\, dz$$

where V is the region bounded by $z = x^2 + y^2$ *and* $z = 8 - (x^2 + y^2)$ *(Delhi University, April 2010)*

Solution. We have,

$$\iiint_V \sqrt{x^2 + y^2}\; dx\, dy\, dz$$

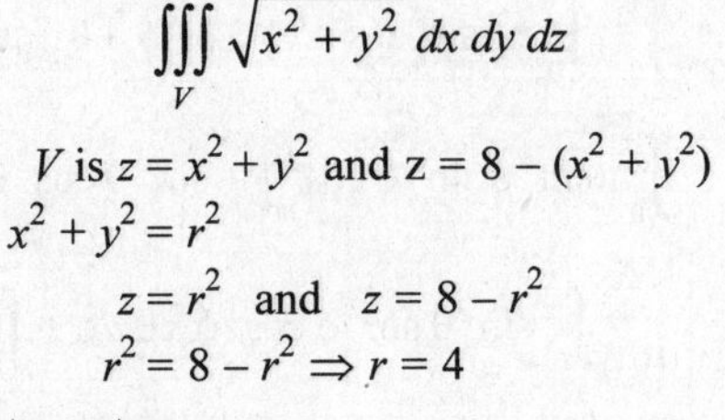

V is $z = x^2 + y^2$ and $z = 8 - (x^2 + y^2)$

Put $\quad x^2 + y^2 = r^2$

$\therefore \quad z = r^2$ and $z = 8 - r^2$

$\Rightarrow \quad r^2 = 8 - r^2 \Rightarrow r = 4$

In cylindrical co-ordinate system

$$\iiint_V \sqrt{r^2}\; dv = \iiint_V r.\, r dr d\theta dz$$

$$= \int_0^4 r^2 dr \int_0^{\pi} d\theta \int_0^8 dz = \left[\frac{r^3}{3}\right]_0^4 (\pi)\,(8) = \frac{512\,\pi}{3} \qquad \textbf{Ans.}$$

Example 15. *Evaluate* $\iiint xyz\,(x^2 + y^2 + z^2)\; dx\; dy\; dz$ *over the first octant of the sphere* $x^2 + y^2 + z^2 = a^2$. *(M.U. II Semester 2009, 2005)*

Solution. Here, we have

$$I = \iiint xyz\,(x^2 + y^2 + z^2)\;\; dx\; dy\; dz \qquad ...(1)$$

Putting $x = r\sin\theta\cos\phi$, $y = r\sin\theta\sin\phi$, $z = r\cos\theta$, $dx\;dy\;dz = r^2\sin\theta\; dr\; d\theta\; d\phi$ in (1), we get

$$\left(\begin{aligned} x^2 + y^2 + z^2 &= r^2\sin^2\theta\cos^2\phi + r^2\sin^2\theta\sin^2\phi + r^2\cos^2\theta \\ &= r^2\sin^2\theta + r^2\cos^2\theta = r^2 \end{aligned}\right)$$

$$\left[\text{Limits of } r,\ \theta \text{ and } \phi \text{ in the first octant are } 0, a\ ;\ 0, \frac{\pi}{2} \text{ and } 0, \frac{\pi}{2}\right]$$

$$I = \int_0^{\frac{\pi}{2}} \int_0^{\frac{\pi}{2}} \int_0^a r^3 \sin^2\theta \cos\theta \sin\phi \cos\phi\, r^2 \,.\, r^2 \sin\theta\; dr\; d\theta\; d\phi$$

$$= \int_0^{\frac{\pi}{2}} \sin\phi\cos\phi\; d\phi \int_0^{\frac{\pi}{2}} \sin^3\theta\cos\theta\; d\theta \int_0^a r^7 dr$$

$$= \left[\frac{\sin^2\phi}{2}\right]_0^{\frac{\pi}{2}} \left[\frac{\sin^4\theta}{4}\right]_0^{\frac{\pi}{2}} \left[\frac{r^8}{8}\right]_0^a = \frac{1}{2}\cdot\frac{1}{4}\cdot\frac{a^8}{8} = \frac{a^8}{64} \qquad \textbf{Ans.}$$

Example 16. *Evaluate* $\iiint (x^2y^2 + y^2z^2 + z^2x^2)\ dx\ dy\ dz$ *over the volume of the sphere* $x^2 + y^2 + z^2 = a^2$.

Solution. Here, we have

$$I = \iiint (x^2y^2 + y^2z^2 + z^2x^2)\ dx\ dy\ dz \qquad ...(1)$$

Putting $x = r \sin\theta \cos\phi$, $y = r \sin\theta \sin\phi$, $z = r\cos\theta$, $dx\ dy\ dz = r^2 \sin\theta\ dr\ d\theta\ d\phi$, in (1), we get

$$\left[\text{In first octant the limits of } r \text{ are } 0, a \text{ for } \theta \text{ are } 0, \frac{\pi}{2} \text{ for } \phi \text{ are } 0 \text{ and } \frac{\pi}{2}\right]$$

$$I = 8\int_0^{\frac{\pi}{2}}\int_0^{\frac{\pi}{2}}\int_0^{a} (r^4 \sin^4\theta \sin^2\phi\ \cos^2\phi + r^4 \sin^2\theta \cos^2\theta \sin^2\phi + r^4 \sin^2\theta \cos^2\theta \cos^2\phi)\ .\ r^2 \sin\theta\ dr\ d\theta\ d\phi$$

$$= 8\int_0^{\frac{\pi}{2}}\int_0^{\frac{\pi}{2}}\int_0^{a} r^6(\sin^4\theta \sin^2\phi\ \cos^2\phi + \sin^2\theta \cos^2\theta) \sin\theta\ dr\ d\theta\ d\phi$$

$$= 8\int_0^{\frac{\pi}{2}}\int_0^{\frac{\pi}{2}} (\sin^4\theta \sin^2\phi \cos^2\phi + \sin^2\theta \cos^2\theta)\ .\ \sin\theta\ d\theta\ d\phi \int_0^a r^6 dr$$

$$= \frac{8a^7}{7}\left[\int_0^{\frac{\pi}{2}}\int_0^{\frac{\pi}{2}} \sin^5\theta \sin^2\phi \cos^2\phi\ d\theta\ d\phi + \int_0^{\frac{\pi}{2}}\int_0^{\frac{\pi}{2}} \sin^3\theta \cos^2\theta\ d\theta\ d\phi\right]$$

$$= \frac{8a^7}{7}\left[\int_0^{\frac{\pi}{2}} \sin^5\theta\ d\theta \int_0^{\frac{\pi}{2}} \sin^2\phi \cos^2\phi\ d\phi + \int_0^{\frac{\pi}{2}} \sin^3\theta \cos^2\theta\ d\theta \int_0^{\frac{\pi}{2}} d\phi\right]$$

$$= \frac{8a^7}{7}\left[\frac{1}{2}\cdot\frac{2!\left\lceil\frac{1}{2}\right.}{\frac{5}{2}\cdot\frac{3}{2}\cdot\frac{1}{2}\left\lceil\frac{1}{2}\right.}\cdot\frac{1}{2}\cdot\frac{\frac{1}{2}\cdot\left\lceil\frac{1}{2}\right.\cdot\frac{1}{2}\cdot\left\lceil\frac{1}{2}\right.}{2!} + \frac{1}{2}\cdot\frac{1!\cdot\frac{1}{2}\left\lceil\frac{1}{2}\right.}{\frac{5}{2}\cdot\frac{3}{2}\cdot\frac{1}{2}\left\lceil\frac{1}{2}\right.}\cdot\frac{\pi}{2}\right]$$

$$= \frac{8a^7}{7}\left[\frac{1}{2}\cdot\frac{2!}{\frac{5}{2}\cdot\frac{3}{2}\cdot\frac{1}{2}}\cdot\frac{1}{2}\cdot\frac{\frac{1}{2}\cdot\sqrt{\pi}\cdot\frac{1}{2}\cdot\sqrt{\pi}}{2!} + \frac{1}{2}\cdot\frac{1!\cdot\frac{1}{2}}{\frac{5}{2}\cdot\frac{3}{2}\cdot\frac{1}{2}}\cdot\frac{\pi}{2}\right]$$

$$= \frac{8a^7}{7}\left[\frac{8}{15}\cdot\frac{1}{16}\cdot\pi + \frac{1}{15}\cdot\pi\right] = \frac{8a^7}{7}\left[\frac{1}{30} + \frac{1}{15}\right]\pi = \frac{8a^7}{7}\cdot\frac{1}{10}\pi = \frac{4a^7\pi}{35} \quad \textbf{Ans.}$$

EXERCISE 7.1

Evaluate the following :

1. $\int_{-1}^{1}\int_{-2}^{2}\int_{-3}^{3} dx\ dy\ dz$ **Ans.** 48

2. $\int_0^4\int_0^x\int_0^{x+y} z\ dz\ dy\ dx$ **Ans.** 70

3. $\int_1^2\int_0^1\int_{-1}^{1} (x^2 + y^2 + z^2)\ dx\ dy\ dz$ **Ans.** 6

4. $\int_0^1\int_0^1\int_0^1 (x^2 + y^2 + z^2)\ dz\ dy\ dx$ *(AMIETE, June 2006)* **Ans.** 1

5. $\int_{-1}^{1}\int_0^z\int_{x-z}^{x+z} (x - y + z)\ dx\ dy\ dz$ **Ans.** 0

6. $\iiint_R (x - y - z)\ dx\ dy\ dz$, where $R : 1 \le x \le 2;\ 2 \le y \le 3;\ 1 \le z \le 3$ **Ans.** 2

7. $\int_0^2 \int_1^3 \int_1^2 xy^2 z\ dx\ dy\ dz$ (*AMIETE, Dec. 2007*) **Ans.** 26

8. $\int_0^1 dx \int_0^2 dy \int_1^2 x^2\ yz\ dz$ **Ans.** 1

9. $\iiint x^2 yz\ dx\ dy\ dz$ throughout the volume bounded by $x = 0,\ y = 0,\ z = 0,\ x + y + z = 1.$

Ans. $\frac{1}{2520}$

10. $\int_0^1 \int_0^{1-x} \int_0^{1-x^2-y^2} dz\ dy\ dx$ **Ans.** $\frac{1}{3}$

11. $\int_1^e \int_1^{\log y} \int_1^{e^x} \log z\ dz\ dx\ dy$ **Ans.** $\frac{1}{2}(e^2 - 8e + 13)$

12. $\iiint_T y\ dx\ dy\ dz$, where T is the region bounded by the surfaces $x = y^2,\ x = y + 2,\ 4z = x^2 + y^2$ and $z = y + 3.$ (*AMIETE Dec. 2008*) **Ans.** $\frac{92}{15}$

13. $\int_0^2 \int_0^x \int_0^{2x+2y} e^{x+y+z}\ dz\ dy\ dx$ **Ans.** $\frac{1}{3}\left[\frac{e^{12}}{6} - \frac{e^6}{3} - \frac{1}{6} + \frac{1}{3}\right] - \frac{1}{2}[e^4 - 1] + [e^2 - 1]$

14. $\iiint_V (2x + y)\ dV$

where V is the closed region bounded by the cylinder $z = 4 - x^2$ and the planes $x = 0,\ y = 0,\ y = 2$ and $z = 0.$ (*Delhi University, April 2010*) **Ans.** $\frac{3^2}{3}$

15. $\iiint (x + y + z)\ dx\ dy\ dz$ over the tetrahedron bounded by the planes $x = 0,\ y = 0,\ z = 0$ and $x + y + z = 1.$ **Ans.** $\frac{1}{8}$

16. $\int_0^a \int_0^{a-x} \int_0^{a-x-y} x^2\ dx\ dy\ dz$ **Ans.** $\frac{a^5}{60}$

17. $\int_{-2}^2 \int_{-\sqrt{(4-x^2)/2}}^{\sqrt{(4-x^2)/2}} \int_{x^2+3y^2}^{8-x^2-y^2} dz\ dy\ dx$ **Ans.** $8\sqrt{2}\pi$

18. $\int_{-1}^1 \int_0^z \int_{x-z}^{x+z} (x + y + z)\ dz\ dx\ dy$ **Ans.** 0

19. $\int_0^2 \int_0^y \int_{x-y}^{x+y} (x + y + z)\ dx\ dy\ dz$ **Ans.** 16

20. $\iiint \sqrt{1 - \frac{x^2}{a^2} - \frac{y^2}{b^2} - \frac{z^2}{c^2}}\ dx\ dy\ dz$ throughout the volume of the ellipsoid $\frac{x^2}{a^2} + \frac{y^2}{b^2} + \frac{z^2}{c^2} = 1.$

Ans. $\frac{\pi^2}{4} abc$

21. $\iiint \sqrt{\frac{x^2}{a^2} + \frac{y^2}{b^2} + \frac{z^2}{c^2}}\ dx\ dy\ dz$ over the volume of the ellipsoid $\frac{x^2}{a^2} + \frac{y^2}{b^2} + \frac{z^2}{c^2} = 1.$ **Ans.** $\frac{4\pi}{3} abc$

22. $\iiint x^{l-1} y^{m-1} z^{n-1}\, dx\, dy\, dz$

throughout the volume of the tetrahedron $x \geq 0, y \geq 0, z \geq 0, x + y + z \leq 1$.

Ans. $\dfrac{1}{(l+m+n)} \cdot \dfrac{\lceil l \lceil m \lceil n}{\lceil l+m+n}$

23. $\iiint \dfrac{dx\, dy\, dz}{\sqrt{1 - x^2 - y^2 - z^2}}$ taken throughout the volume of the sphere $x^2 + y^2 + z^2 = 1$, lying in the first octant.

Ans. $\dfrac{\pi^2}{8}$

24. $\displaystyle\int_0^{\pi} 2d\theta \int_0^{a(1+\cos\theta)} r\, dr \int_0^h \left[1 - \frac{r}{a(1+\cos\theta)}\right] dz$

Ans. $\dfrac{\pi a^2}{2} h$

25. $\displaystyle\int_0^{\pi/2} \int_0^{a\sin\theta} \int_0^{(a^2 - r^2)/a} r\, d\theta\, dr\, dz$

Ans. $\dfrac{5a^3}{64}$

26. $\iiint z^2\, dx\, dy\, dz$ over the volume common to the sphere $x^2 + y^2 + z^2 = a^2$ and the cylinder $x^2 + y^2 + z^2 = ax$.

Ans. $\dfrac{2a^5\pi}{15}$

27. $\displaystyle\iiint_V \frac{dx\, dy\, dz}{(1 + x^2 + y^2 + z^2)^2}$ where V is the volume in the first octant.

Ans. $\dfrac{\pi^2}{8}$

28. $\iiint \dfrac{dx\, dy\, dz}{(x^2 + y^2 + z^2)^{3/2}}$ over the volume bounded by the spheres $x^2 + y^2 + z^2 = 16$ and $x^2 + y^2 + z^2 = 25$.

Ans. $4\pi \log (5/4)$

29. $\displaystyle\iiint_T z^2\, dx\, dy\, dz$ over the volume bounded by the cylinder $x^2 + y^2 = a^2$ and the paraboloid $x^2 + y^2 = z$ and the plane $z = 0$.

Ans. $\dfrac{\pi a^8}{12}$

30. $\displaystyle\iiint_T z\, dx\, dy\, dz$, where T is region bounded by the cone $x^2 \tan^2 \alpha + y^2 \tan^2 \beta = z^2$ and the planes $z = 0$ to $z = h$ in the first octant.

(AMIETE, Dec. 2009)

31. Find the components of a vector $\vec{A} = x^2\hat{i} + xyz\hat{j} + y^2\hat{k}$ in cylindrical coordinates.

(Delhi University, April 2010)

APPLICATION OF TRIPLE INTEGRATION

8.1 INTRODUCTION

In this chapter we will discuss how to find out volume, surface area, mass, C.G., moment of Inertia of solids and centre of pressure of fluids.

8.2 VOLUME = $\iiint dx\, dy\, dz$.

The elementary volume δv is $\delta x \,.\, \delta y \,.\, \delta z$ and therefore the volume of the whole solid is obtained by evaluating the triple integral.

$$\delta V = \delta x\ \delta y\ \delta z$$

$$\boxed{V = \iiint dx\, dy\, dz.}$$

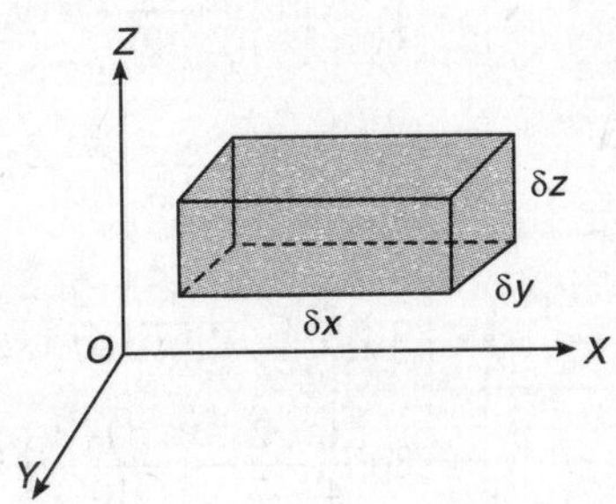

Note : (*i*) Mass = volume × density

$$= \iiint \rho\, dx\, dy\, dz \text{ if } \rho \text{ is the density.}$$

(*ii*) In cylindrical co-ordinaties, we have

$$\boxed{V = \iiint_V r\, dr\, d\theta\, dz}$$

(*iii*) In spherical polar co-ordinates, we have

$$\boxed{V = \iiint_V r^2 \sin\theta dr\, d\theta\, d\varphi}$$

Example 1. *Find the volume of the tetrahedron bounded by the planes $x = 0$, $y = 0$, $z = 0$ and $x + y + z = a$.* (*M.U. II Semester, 2005*)

Solution. Here, we have a solid which is bounded by $x = 0$, $y = 0$, $z = 0$ and $x + y + z = a$ planes.

The limits of z are 0 and $a - x - y$, the limits of y are 0 and $a - x$,

the limits of x are 0 and a.

$$V = \int_{x=0}^{a} \int_{y=0}^{a-x} \int_{z=0}^{a-x-y} dx\, dy\, dz = \int_{x=0}^{a} \int_{y=0}^{a-x} [z]_0^{a-x-y}\, dx\, dy$$

$$= \int_{x=0}^{a} \int_{y=0}^{a-x} (a - x - y)\, dx\, dy$$

$$= \int_{x=0}^{a} \left[ay - xy - \frac{y^2}{2}\right]_0^{a-x} dx$$

$$= \int_0^a \left[a(a-x) - x(a-x) - \frac{(a-x)^2}{2}\right] dx$$

$$= \int_0^a \left[a^2 - ax - ax + x^2 - \frac{a^2}{2} + ax - \frac{x^2}{2}\right] dx$$

$$= \int_0^a \left(\frac{a^2}{2} - ax + \frac{x^2}{2}\right) dx$$

$$= \left[\frac{a^2}{2}.x - \frac{ax^2}{2} + \frac{x^3}{6}\right]_0^a = a^3\left(\frac{1}{2} - \frac{1}{2} + \frac{1}{6}\right) = \frac{a^3}{6}. \quad \textbf{Ans.}$$

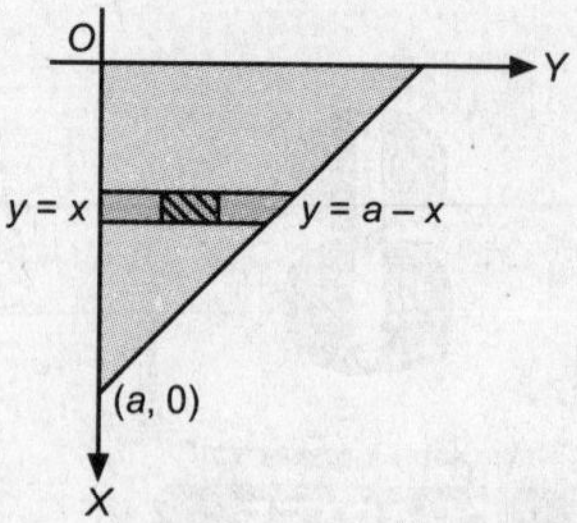

O, Y, $y = x$, $y = a - x$, (a, 0), X

Example 2. *Find the volume of the cylindrical column standing on the area common to the parabolas $y^2 = x$, $x^2 = y$ and cut off by the surface $z = 12 + y - x^2$.*

Solution. We have,

$$y^2 = x$$
$$x^2 = y$$
$$z = 12 + y - x^2$$

$$V = \int_0^1 dx \int_{x^2}^{\sqrt{x}} dy \int_0^{12+y-x^2} dz = \int_0^1 dx \int_{x^2}^{\sqrt{x}} (12 + y - x^2)\, dy$$

$$= \int_0^1 dx \left(12y + \frac{y^2}{2} - x^2 y\right)_{x^2}^{\sqrt{x}}$$

$$= \int_0^1 \left(12\sqrt{x} + \frac{x}{2} - x^{5/2} - 12x^2 - \frac{x^4}{2} + x^4\right) dx$$

$$= \left[\frac{2}{3} \times 12x^{3/2} + \frac{x^2}{4} - \frac{2}{7}x^{7/2} - 4x^3 - \frac{x^5}{10} + \frac{x^5}{5}\right]_0^1$$

$$= 8 + \frac{1}{4} - \frac{2}{7} - 4 - \frac{1}{10} + \frac{1}{5} = 4 + \frac{1}{4} - \frac{2}{7} - \frac{1}{10} + \frac{1}{5} = \frac{560 + 35 - 40 - 14 + 28}{140} = \frac{569}{140} \quad \textbf{Ans.}$$

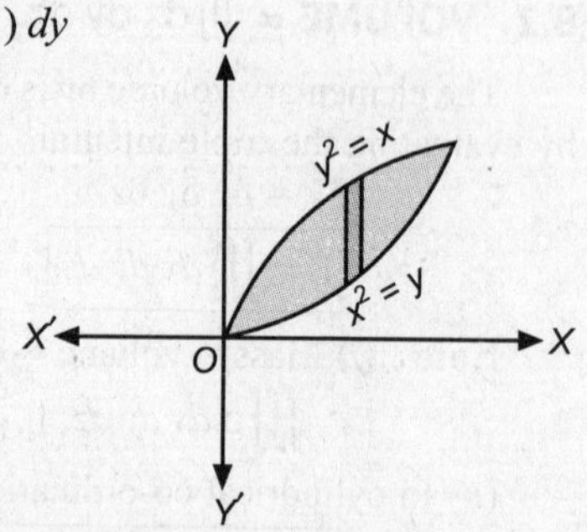

Y, $y^2 = x$, $x^2 = y$, X', O, X, Y'

Example 3. *A triangular prism is formed by planes whose equations are $ay = bx$, $y = 0$ and $x = a$. Find the volume of the prism between the planes $z = 0$ and surface $z = c + xy$.*

(*U.P., Ist Semester, 2009*)

Solution. Required volume $= \displaystyle\int_0^a \int_0^{\frac{bx}{a}} \int_0^{c+xy} dz\, dy\, dx$

$$= \int_0^a \int_0^{\frac{bx}{a}} (c + xy)\, dy\, dx$$

$$= \int_0^a \left(cy + \frac{xy^2}{2}\right)_0^{\frac{bx}{a}} dx$$

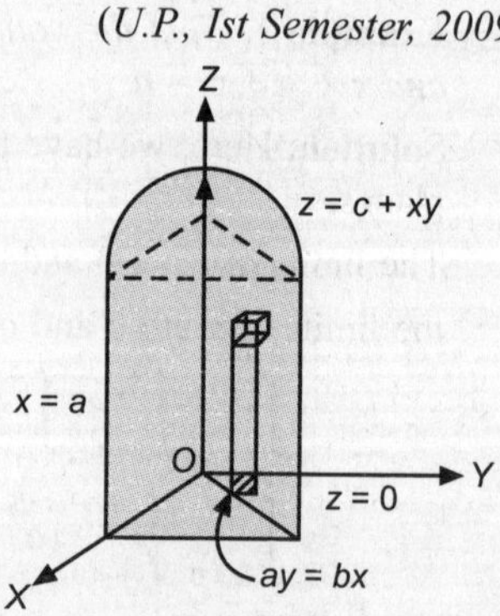

Z, $z = c + xy$, $x = a$, O, Y, $z = 0$, X, $ay = bx$

$$= \int_0^a \left(\frac{cbx}{a} + \frac{b^2}{2a^2}x^3\right) dx = \frac{bc}{a}\left(\frac{x^2}{2}\right)_0^a + \frac{b^2}{2a^2}\left(\frac{x^4}{4}\right)_0^a$$

$$= \frac{abc}{2} + \frac{b^2a^2}{8} = \frac{ab}{8}(4c + ab) \quad \textbf{Ans.}$$

8.3 VOLUME OF SOLID BOUNDED BY SPHERE OR BY CYLINDER

We use spherical coordinates (r, θ, ϕ) and the cylindrical coordinates are (ρ, ϕ, z) and the relations are $x = \rho \cos \phi$, $y = \rho \sin \phi$.

Example 4. *Find the volume of a solid bounded by the spherical surface* $x^2 + y^2 + z^2 = 4a^2$ *and the cylinder* $x^2 + y^2 - 2ay = 0$.

Solution.
$$x^2 + y^2 + z^2 = 4a^2 \qquad ...(1)$$
$$x^2 + y^2 - 2ay = 0 \qquad ...(2)$$

Considering the section in the positive quadrant of the xy-plane and taking z to be positive (that is volume above the xy-plane) and changing to polar co-ordinates, (1) becomes

Y, (0, a), X′, O, X, Y′

$$r^2 + z^2 = 4a^2 \quad \Rightarrow \quad z^2 = 4a^2 - r^2$$
$$\therefore \quad z = \sqrt{4a^2 - r^2}$$

(2) becomes $r^2 - 2ar \sin \theta = 0 \Rightarrow r = 2a \sin \theta$

$$\text{Volume} = \iiint dx\, dy\, dz$$
$$= 4 \int_0^{\pi/2} d\theta \int_0^{2a \sin \theta} r\, dr \int_0^{\sqrt{4a^2 - r^2}} dz \qquad \text{(Cylindrical coordinates)}$$
$$= 4 \int_0^{\pi/2} d\theta \int_0^{2a \sin \theta} r\, dr [z]_0^{\sqrt{4a^2 - r^2}} = 4 \int_0^{\pi/2} d\theta \int_0^{2a \sin \theta} r\, dr \,.\, \sqrt{4a^2 - r^2}$$
$$= 4 \int_0^{\pi/2} d\theta \left[-\frac{1}{3}(4a^2 - r^2)^{3/2} \right]_0^{2a \sin \theta} = \frac{4}{3} \int_0^{\pi/2} \left[-(4a^2 - 4a^2 \sin^2 \theta)^{3/2} + 8a^3 \right] d\theta$$
$$= \frac{4}{3} \int_0^{\pi/2} (-8a^3 \cos^3 \theta + 8a^3)\, d\theta = \frac{8 \times 4a^3}{3} \int_0^{\pi/2} (1 - \cos^3 \theta)\, d\theta$$
$$= \frac{32a^3}{3} \int_0^{\pi/2} \left(1 - \frac{1}{4} \cos 3\theta - \frac{3}{4} \cos \theta \right) d\theta$$
$$= \frac{32a^3}{3} \left[\theta - \frac{1}{12} \sin 3\theta - \frac{3}{4} \sin \theta \right]_0^{\pi/2} = \frac{32a^3}{3} \left(\frac{\pi}{2} + \frac{1}{12} - \frac{3}{4} \right) = \frac{32a^3}{3} \left[\frac{\pi}{2} - \frac{2}{3} \right] \quad \textbf{Ans.}$$

Example 5. *Find the volume enclosed by the solid*
$$\left(\frac{x}{a}\right)^{2/3} + \left(\frac{y}{b}\right)^{2/3} + \left(\frac{z}{c}\right)^{2/3} = 1$$

Solution. The equation of the solid is
$$\left(\frac{x}{a}\right)^{2/3} + \left(\frac{y}{b}\right)^{2/3} + \left(\frac{z}{c}\right)^{2/3} = 1$$

Putting
$$\left(\frac{x}{a}\right)^{1/3} = u \quad \Rightarrow \quad x = a\, u^3 \quad \Rightarrow \quad dx = 3\, au^2\, du$$
$$\left(\frac{y}{b}\right)^{1/3} = v \quad \Rightarrow \quad y = b\, v^3 \quad \Rightarrow \quad dy = 3\, bv^2\, dv$$
$$\left(\frac{z}{c}\right)^{1/3} = w \quad \Rightarrow \quad z = c\, w^3 \quad \Rightarrow \quad dz = 3\, c\, w^2\, dw$$

The equation of the solid becomes

$$u^2 + v^2 + w^2 = 1 \qquad \text{...(1)}$$

$$V = \iiint dx\, dy\, dz \qquad \text{...(2)}$$

On putting the values of dx, dy and dz in (2), we get

$$V = \iiint 27abc\, u^2 v^2 w^2 du\, dv\, dw \qquad \text{...(3)}$$

(1) represents a sphere.

Let us use spherical coordinates.

$$u = r \sin\theta \cos\phi, \qquad v = r \sin\theta \sin\phi,$$

$$w = r \cos\theta, \qquad du\, dv\, dw = r^2 \sin\theta\, dr\, d\theta\, d\phi$$

On substituting spherical coordinates in (3), we have

$$V = 27abc\,.\,8 \int_{r=0}^{1} \int_{\phi=0}^{\pi/2} \int_{\theta=0}^{\pi/2} r^2 \sin^2\theta \cos^2\phi \,.\, r^2 \sin^2\theta \sin^2\phi \,.\, r^2 \cos^2\theta \,.\, r^2 \sin\theta\, dr\, d\theta\, d\phi$$

$$= 216\, abc \int_{r=0}^{1} r^8\, dr \int_{\phi=0}^{\pi/2} \sin^2\phi \cos^2\phi\, d\phi \int_{\theta=0}^{\pi/2} \sin^5\theta \cos^2\theta\, d\theta$$

$$= 216 abc \left[\frac{r^9}{9}\right]_0^1 . \left(\frac{\Gamma\left(\frac{3}{2}\right)\Gamma\left(\frac{3}{2}\right)}{2\,\Gamma(3)}\right)\left(\frac{\Gamma(3)\,\Gamma\left(\frac{3}{2}\right)}{2\,\Gamma\left(\frac{9}{2}\right)}\right) = 24\, abc\,.\frac{1}{2}.\frac{\Gamma\left(\frac{3}{2}\right)\Gamma\left(\frac{3}{2}\right)}{\Gamma(3)}.\frac{1}{2}.\frac{\Gamma(3)\,\Gamma\left(\frac{3}{2}\right)}{\Gamma\left(\frac{9}{2}\right)}$$

$$= 6abc\,.\frac{\left[\left(\frac{1}{2}\right)\Gamma\left(\frac{1}{2}\right)\right]^2}{2!}.\frac{2!\,\Gamma\left(\frac{3}{2}\right)}{\left(\frac{7}{2}\right)\left(\frac{5}{2}\right)\frac{3}{2}\,\Gamma\left(\frac{3}{2}\right)} = 6abc\,.\frac{1}{4}.\pi\frac{1}{\left(\frac{7}{2}\right)\left(\frac{5}{2}\right)\left(\frac{3}{2}\right)} = \frac{4}{35}\, abc\, \pi \qquad \textbf{Ans.}$$

Example 6. *Find the volume bounded above by the sphere $x^2 + y^2 + z^2 = a^2$ and below by the cone $x^2 + y^2 = z^2$.*

Solution. The equation of the sphere is $x^2 + y^2 + z^2 = a^2$ ...(1)

and that of the cone is $x^2 + y^2 = z^2$ ...(2)

In polar coordinates $x = r \sin\theta \cos\phi$, $y = r \sin\theta \sin\phi$, $z = r \cos\theta$

The equation (1) in polar co-ordinates is

$$(r \sin\theta \cos\phi)^2 + (r \sin\theta \sin\phi)^2 + (r \cos\theta)^2 = a^2$$

$$\Rightarrow \quad r^2 \sin^2\theta \cos^2\phi + r^2 \sin^2\theta \sin^2\phi + r^2 \cos^2\theta = a^2$$

$$\Rightarrow \quad r^2 \sin^2\theta (\cos^2\phi + \sin^2\phi) + r^2 \cos^2\theta = a^2$$

$$\Rightarrow \quad r^2 \sin^2\theta + r^2 \cos^2\theta = a^2$$

$$\Rightarrow \quad r^2 (\sin^2\theta + \cos^2\theta) = a^2$$

$$\Rightarrow \quad r^2 = a^2 \quad \Rightarrow \quad r = a$$

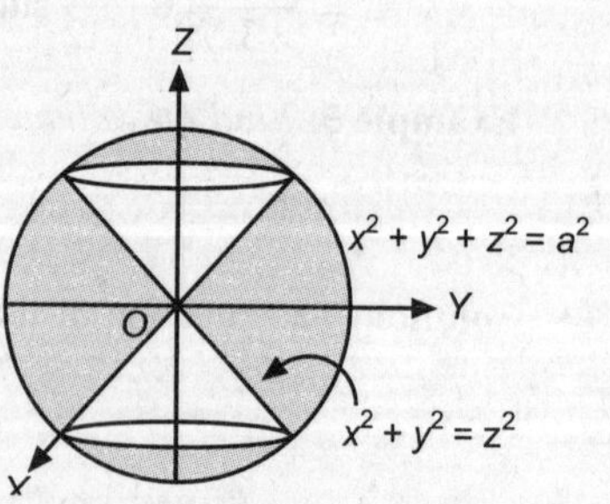

The equation (2) in polar co-ordinates is

$$(r \sin\theta \cos\phi)^2 + (r \sin\theta \sin\phi)^2 = (r \cos\theta)^2$$

$$\Rightarrow \quad r^2 \sin^2\theta (\cos^2\phi + \sin^2\phi) = r^2 \cos^2\theta \Rightarrow r^2 \sin^2\theta = r^2 \cos^2\theta$$

$$\Rightarrow \quad \tan^2\theta = 1 \quad \Rightarrow \tan\theta = 1 \Rightarrow \quad \theta = \pm\frac{\pi}{4}$$

Thus equations (1) and (2) in polar coordinates are respectively,

$$r = a \quad \text{and} \quad \theta = \pm\frac{\pi}{4}$$

The volume in the first octant is one fourth only.

Limits in the first octant : r varies 0 to a, θ from 0 to $\frac{\pi}{4}$ and ϕ from 0 to $\frac{\pi}{2}$.

The required volume lies between $x^2 + y^2 + z^2 = a^2$ and $x^2 + y^2 = z^2$.

$$V = 4\int_0^{\frac{\pi}{2}}\int_0^{\frac{\pi}{4}}\int_0^a r^2 \sin\theta \, dr\, d\theta\, d\phi = 4\int_0^{\frac{\pi}{2}} d\phi \int_0^{\frac{\pi}{4}} \sin\theta\, d\theta \left[\frac{r^3}{3}\right]_0^a$$

$$= 4\int_0^{\frac{\pi}{2}} d\phi \int_0^{\frac{\pi}{4}} \sin\theta\, d\theta \,.\, \frac{a^3}{3} = \frac{4a^3}{3}\int_0^{\frac{\pi}{2}} d\phi \left[-\cos\theta\right]_0^{\frac{\pi}{4}} = \frac{4a^3}{3}(\phi)_0^{\frac{\pi}{2}}\left[-\frac{1}{\sqrt{2}}+1\right]$$

$$= \frac{2}{3}\pi a^3\left(1-\frac{1}{\sqrt{2}}\right)$$ **Ans.**

8.4 VOLUME OF SOLID BOUNDED BY CYLINDER OR CONE

We use cylindrical coordinates (r, θ, z).

Example 7. *Find the volume of the solid bounded by the parabolic $y^2 + z^2 = 4x$ and the plane $x = 5$.*

Solution. $y^2 + z^2 = 4x$, $x = 5$

$$V = \int_0^5 dx \int_{-2\sqrt{x}}^{2\sqrt{x}} dy \int_{-\sqrt{4x-y^2}}^{\sqrt{4x-y^2}} dz = 4\int_0^5 dx \int_0^{2\sqrt{x}} dy \int_0^{\sqrt{4x-y^2}} dz$$

$$= 4\int_0^5 dx \int_0^{2\sqrt{x}} dy\, [z]_0^{\sqrt{4x-y^2}} = 4\int_0^5 dx \int_0^{2\sqrt{x}} dy\, \sqrt{4x-y^2}$$

$$= 4\int_0^5 dx \left[\frac{y}{2}\sqrt{4x-y^2} + \frac{4x}{2}\sin^{-1}\frac{y}{2\sqrt{x}}\right]_0^{2\sqrt{x}} = 4\int_0^5 \left[0 + 2x\left(\frac{\pi}{2}\right)\right] dx = 4\pi\int_0^5 x\, dx$$

$$= 4\pi\left[\frac{x^2}{2}\right]_0^5 = 50\pi$$ **Ans.**

Example 8. *Calculate the volume of the solid bounded by the following surfaces :*

$z = 0$, $x^2 + y^2 = 1$, $x + y + z = 3$

Solution. $x^2 + y^2 = 1$ ...(1)

$x + y + z = 3$ ...(2)

$z = 0$ ...(3)

$$\text{Required Volume} = \iiint dx\, dy\, dz = \iint dx\, dy\, [z]_0^{3-x-y} = \iint (3-x-y)\, dx\, dy$$

On putting $x = r\cos\theta$, $y = r\sin\theta$, $dx\, dy = r\, d\theta\, dr$, we get

$$= \iint (3 - r\cos\theta - r\sin\theta)\, r\, d\theta\, dr = \int_0^{2\pi} d\theta \int_0^1 (3r - r^2\cos\theta - r^2\sin\theta)\, dr$$

$$= \int_0^{2\pi} d\theta \left(\frac{3r^2}{2} - \frac{r^3}{3}\cos\theta - \frac{r^3}{3}\sin\theta\right)_0^1 = \int_0^{2\pi}\left(\frac{3}{2} - \frac{1}{3}\cos\theta - \frac{1}{3}\sin\theta\right) d\theta$$

$$= \left[\frac{3}{2}\theta - \frac{1}{3}\sin\theta + \frac{1}{3}\cos\theta\right]_0^{2\pi} = 3\pi - \frac{1}{3}\sin 2\pi + \frac{1}{3}\cos 2\pi - \frac{1}{3} = 3\pi$$ **Ans.**

Example 9. *Find the volume bounded by the cylinder $x^2 + y^2 = 4$ and the planes $y + z = 4$ and $z = 0$.*

Solution. $x^2 + y^2 = 4 \Rightarrow y = \pm\sqrt{4-x^2}$

$$y + z = 4 \quad \Rightarrow \quad z = 4 - y \text{ and } z = 0$$

x varies from -2 to $+2$.

$$V = \iiint dx\, dy\, dz$$

$$= \int_{-2}^{2} dx \int_{-\sqrt{4-x^2}}^{\sqrt{4-x^2}} dy \int_{0}^{4-y} dz$$

$$= \int_{-2}^{2} dx \int_{-\sqrt{4-x^2}}^{\sqrt{4-x^2}} dy\, [z]_0^{4-y}$$

$$= \int_{-2}^{2} dx \int_{-\sqrt{4-x^2}}^{\sqrt{4-x^2}} dy\, (4-y) = \int_{-2}^{2} dx \left[4y - \frac{y^2}{2}\right]_{-\sqrt{4-x^2}}^{\sqrt{4-x^2}}$$

$$= \int_{-2}^{2} dx \left[4\sqrt{4-x^2} - \frac{1}{2}(4-x^2) + 4\sqrt{4-x^2} + \frac{1}{2}(4-x^2)\right]$$

$$= 8\int_{-2}^{2} \sqrt{4-x^2}\, dx = 8\left[\frac{x}{2}\sqrt{4-x^2} + \frac{4}{2}\sin^{-1}\frac{x}{2}\right]_{-2}^{2} = 16\pi \qquad \textbf{Ans.}$$

Example 10. *Find the volume in the first octant bounded by the cylinder $x^2 + y^2 = 2$ and the planes $z = x + y$, $y = x$, $z = 0$ and $x = 0$.* (*M.U. II Semester 2005*)

Solution. Here, we have the solid bounded by

$$x^2 + y^2 = 2 \text{ (cylinder)}$$

$$(\text{or } \quad r^2 = 2)$$

$$z = x + y \quad \Rightarrow \quad z = r(\cos\theta + \sin\theta) \qquad \text{(plane)}$$

$$y = x \quad \Rightarrow \quad r\sin\theta = r\cos\theta \qquad \text{(plane)}$$

$$\Rightarrow \quad \tan\theta = 1 \quad \Rightarrow \quad \theta = \frac{\pi}{4}$$

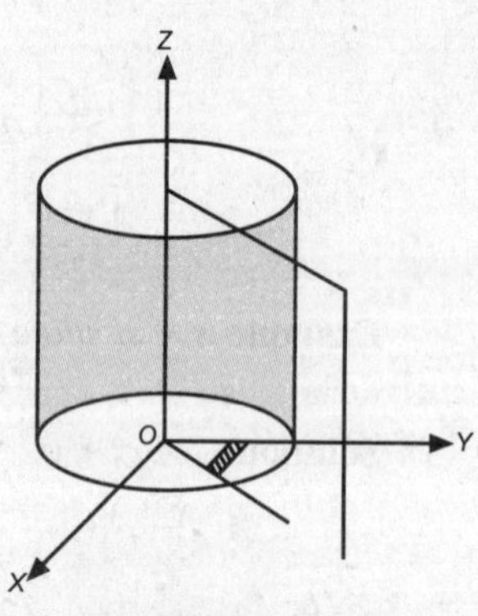

$$x = 0 \Rightarrow r\cos\theta = 0 \Rightarrow \cos\theta = 0 \Rightarrow \theta = \frac{\pi}{2}$$

z varies from 0 to $r(\cos\theta + \sin\theta)$

r varies from 0 to $\sqrt{2}$

θ varies from $\frac{\pi}{4}$ to $\frac{\pi}{2}$

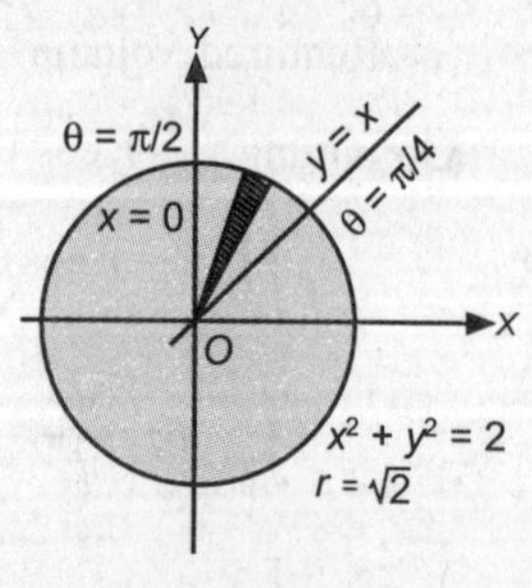

$$\therefore \quad V = \int_{\theta=\pi/4}^{\pi/2} \int_{r=0}^{\sqrt{2}} \int_{z=0}^{r(\cos\theta+\sin\theta)} r\, dr\, d\theta\, dz$$

$$= \int_{\theta=\pi/4}^{\pi/2} \int_{r=0}^{\sqrt{2}} r[z]_0^{r(\cos\theta+\sin\theta)}\, dr\, d\theta$$

$$= \int_{\theta=\pi/4}^{\pi/2} \int_{r=0}^{\sqrt{2}} r^2(\cos\theta + \sin\theta)\, dr\, d\theta$$

$$= \int_{\theta=\pi/4}^{\pi/2} (\cos\theta + \sin\theta)\left[\frac{r^3}{3}\right]_0^{\sqrt{2}} d\theta = \frac{2\sqrt{2}}{3}\int_{\theta=\pi/4}^{\pi/2} (\cos\theta + \sin\theta)\, d\theta$$

$$= \frac{2\sqrt{2}}{3}[\sin\theta - \cos\theta]_{\pi/4}^{\pi/2} = \frac{2\sqrt{2}}{3}\left[(1-0) - \left(\frac{1}{\sqrt{2}} - \frac{1}{\sqrt{2}}\right)\right] = \frac{2\sqrt{2}}{3} \qquad \textbf{Ans.}$$

Example 11. *Show that the volume of the wedge intercepted between the cylinder $x^2 + y^2 = 2ax$ and planes $z = mx$, $z = nx$ is $\pi (m - n) a^3$.*

Solution. The equation of the cylinder is $x^2 + y^2 = 2\,a\,x$

we convert the cartesian coordinates into cylindrical coordinates.

$$x = r\cos\theta$$

$$y = r\sin\theta$$

$$x^2 + y^2 = 2ax \Rightarrow r^2 = 2ar\cos\theta$$

$$\Rightarrow \quad r = 2a\cos\theta$$

r varies from 0 to $2a\cos\theta$

θ varies from $-\dfrac{\pi}{2}$ to $-\dfrac{\pi}{2}$

and $\quad z$ varies from $z = nx$ $(z = nr\cos\theta)$ to $z = m\,x$ $(z = m\,r\cos\theta)$

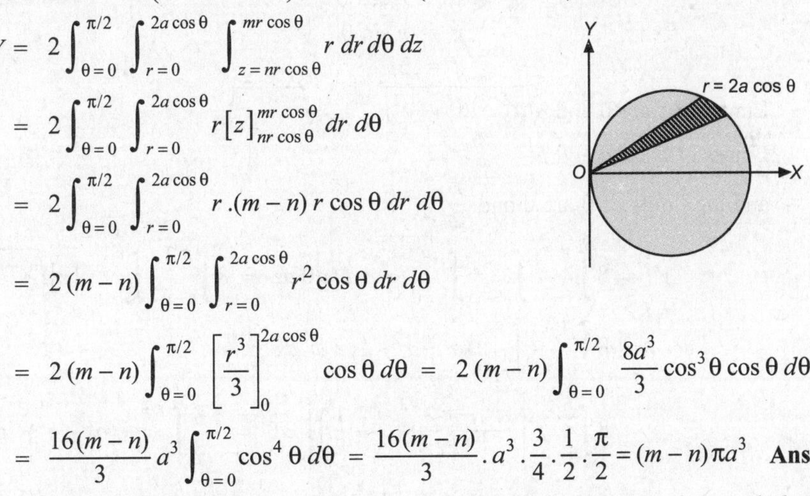

$$V = 2\int_{\theta=0}^{\pi/2}\int_{r=0}^{2a\cos\theta}\int_{z=nr\cos\theta}^{mr\cos\theta} r\,dr\,d\theta\,dz$$

$$= 2\int_{\theta=0}^{\pi/2}\int_{r=0}^{2a\cos\theta} r\left[z\right]_{nr\cos\theta}^{mr\cos\theta} dr\,d\theta$$

$$= 2\int_{\theta=0}^{\pi/2}\int_{r=0}^{2a\cos\theta} r\,.(m-n)\,r\cos\theta\,dr\,d\theta$$

$$= 2(m-n)\int_{\theta=0}^{\pi/2}\int_{r=0}^{2a\cos\theta} r^2\cos\theta\,dr\,d\theta$$

$$= 2(m-n)\int_{\theta=0}^{\pi/2}\left[\frac{r^3}{3}\right]_0^{2a\cos\theta}\cos\theta\,d\theta = 2(m-n)\int_{\theta=0}^{\pi/2}\frac{8a^3}{3}\cos^3\theta\cos\theta\,d\theta$$

$$= \frac{16(m-n)}{3}a^3\int_{\theta=0}^{\pi/2}\cos^4\theta\,d\theta = \frac{16(m-n)}{3}.a^3.\frac{3}{4}.\frac{1}{2}.\frac{\pi}{2} = (m-n)\pi a^3 \quad \textbf{Ans.}$$

Example 12. *Find by triple integration, the volume of the region bounded by the paraboloid $a\,z = x^2 + y^2$ and the cylinder $x^2 + y^2 = R^2$.* (*U.P. I Semester Dec. 2008*)

Solution. The volume bounded by the paraboloid and the cylinder is shown as shaded portion of the figure.

Transforming the given equations to the polar form, by substituting $x = r\cos\theta$, $y = r\sin\theta$

We get the equation of the cylinder

$$x^2 + y^2 = R^2 \text{ as } r = R$$

and that of the paraboloid as

$$a\,z = r^2 \Rightarrow z = \frac{r^2}{a}$$

In the figure, only one fourth of the common volume is shown. Thus in the common region, z varies from 0 to $\dfrac{r^2}{a}$ and r and ϕ vary on the circle $r = R$ (In the xy or r, θ plane).

The variation of r is from 0 to R and that of ϕ is 0 to $\dfrac{\pi}{2}$.

Required volume $V = 4\displaystyle\int_0^{\pi/2}\int_0^R\int_0^{\frac{r^2}{a}} r\,dz\,dr\,d\theta$ (cylindrical coordinates)

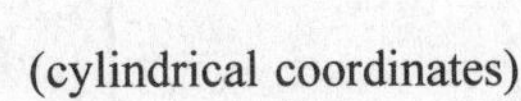

$$= 4\int_0^{\pi/2}\int_0^R r[z]_0^{r^2/a}\,dr\,d\theta = 4\int_0^{\pi/2}\int_0^R r\frac{r^2}{a}\,dr\,d\theta = 4\int_0^{\pi/2}\left[\frac{r^4}{4a}\right]_0^R d\theta$$

$$= \frac{1}{a}\int_0^{\pi/2} R^4\,d\theta = \frac{R^4}{a}.\frac{\pi}{2} = \frac{\pi R^4}{2a}$$ **Ans.**

Example 13. *A cylindrical hole of radius b is bored through a sphere of radius a. Find the volume of the remaining solid.*

Solution. Let the equation of the sphere be

$$x^2 + y^2 + z^2 = a^2$$

Now, we will solve this problem using cylindrical coordinates

$$x = r\cos\theta$$
$$y = r\sin\theta$$
$$z = z$$

Limits of z are 0 and $\sqrt{a^2-(x^2+y^2)}$ *i.e.,* $\sqrt{a^2-r^2}$

Limits of r are a and b.

and the limits of θ are 0 and $\frac{\pi}{2}$

$$V = 8\int_{\theta=0}^{\pi/2}\int_{r=b}^{a}\int_{z=0}^{\sqrt{a^2-r^2}} r\,dr\,d\theta\,dz = 8\int_{\theta=0}^{\pi/2}\int_{r=b}^{a}[z]_0^{\sqrt{a^2-r^2}}\,r\,dr\,d\theta$$

$$= 8\int_{\theta=0}^{\pi/2}\int_{r=b}^{a}(a^2-r^2)^{1/2}.r\,dr\,d\theta$$

$$= 8\int_{\theta=0}^{\pi/2}\left[\frac{(a^2-r^2)^{3/2}}{3/2}.\left(-\frac{1}{2}\right)\right]_b^a d\theta = -\frac{8}{3}\int_0^{\pi/2} -(a^2-b^2)^{\frac{3}{2}}\,d\theta$$

$$= \frac{8}{3}(a^2-b^2)^{\frac{3}{2}}[\theta]_0^{\pi/2} = \frac{4\pi}{3}(a^2-b^2)^{\frac{3}{2}}$$ **Ans.**

Example 14. *Find the volume cut off from the sphere* $x^2 + y^2 + z^2 = a^2$ *by the cylinder* $x^2 + y^2 = ax$.

Solution. Cylindrical co-ordinates

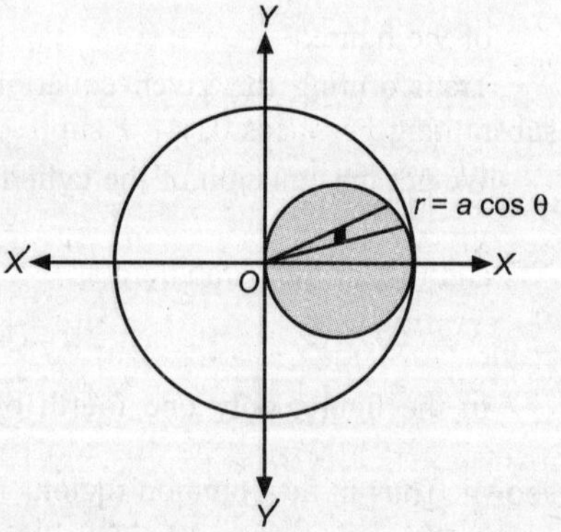

$$x = r\cos\theta\ ;\ y = r\sin\theta$$

$$x^2+y^2+z^2 = a^2 \Rightarrow r^2+z^2 = a^2$$

$$x^2+y^2 = ax \Rightarrow r^2 = ar\cos\theta \Rightarrow r = a\cos\theta$$

$$\text{Volume} = \iiint dx\,dy\,dz = \iiint (r\,d\theta\,dr)\,dz$$

$$= \int_{-\pi/2}^{\pi/2} d\theta\int_0^{a\cos\theta} r\,dr\int_{-\sqrt{a^2-r^2}}^{\sqrt{a^2-r^2}} dz$$

$$= 4\int_0^{\pi/2} d\theta\int_0^{a\cos\theta} r\,dr\int_0^{\sqrt{a^2-r^2}} dz$$

$$= 4\int_0^{\pi/2} d\theta\int_0^{a\cos\theta} r\,dr[z]_0^{\sqrt{a^2-r^2}} = 4\int_0^{\pi/2} d\theta\int_0^{a\cos\theta} r\,dr\sqrt{a^2-r^2}$$

$$= 4\int_0^{\pi/2} d\theta\left[\left(-\frac{1}{2}\right)\frac{2}{3}(a^2-r^2)^{\frac{3}{2}}\right]_0^{a\cos\theta}$$

$$= -\frac{4}{3}\int_0^{\frac{\pi}{2}} d\theta\left[\left(a^2 - a^2\cos^2\theta\right)^{\frac{3}{2}} - a^3\right] = -\frac{4}{3}\int_0^{\pi/2} d\theta\left[a^3\left(1-\cos^2\theta\right)^{3/2} - a^3\right]$$

$$= \frac{4}{3}\int_0^{\pi/2}(a^3 - a^3\sin^3\theta)\,d\theta = \frac{4a^3}{3}\int_0^{\pi/2}(1-\sin^3\theta)\,d\theta = \frac{4a^3}{3}\left[\frac{\pi}{2}-\frac{2}{3}\right]$$ **Ans.**

Example 15. *Find the volume common to the cylinders*

$x^2 + y^2 = a^2$ *and* $x^2 + z^2 = a^2$

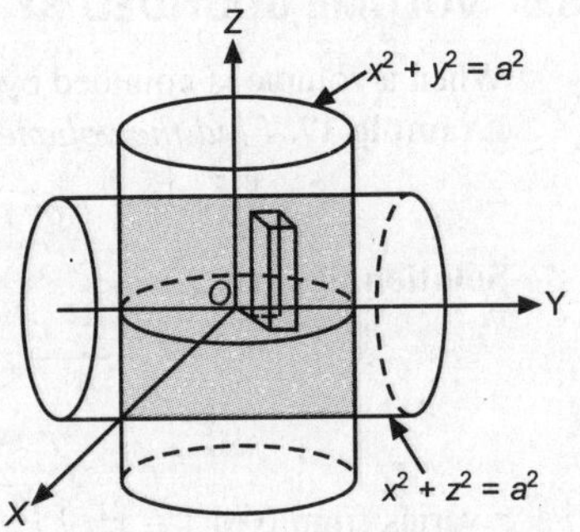

Solution. $x^2 + y^2 = a^2$; $x^2 + z^2 = a^2$

(*i*) z varies from $-\sqrt{a^2 - x^2}$ to $\sqrt{a^2 - x^2}$

(*ii*) y varies from $-\sqrt{a^2 - x^2}$ to $\sqrt{a^2 - x^2}$

(*ii*) x varies from $-a$ to a.

Required volume $= \int_{-a}^{+a} dx \int_{-\sqrt{a^2-x^2}}^{\sqrt{a^2-x^2}} dy \int_{-\sqrt{a^2-x^2}}^{\sqrt{a^2-x^2}} dz$

$$= \int_{-a}^{+a} dx \int_{-\sqrt{a^2-x^2}}^{\sqrt{a^2-x^2}} dy\,[z]_{-\sqrt{a^2-x^2}}^{\sqrt{a^2-x^2}}$$

$$= \int_{-a}^{+a} dx \int_{-\sqrt{a^2-x^2}}^{\sqrt{a^2-x^2}} 2\sqrt{a^2-x^2}\,dy = 2\int_{-a}^{+a} dx\,\sqrt{a^2-x^2}\,[y]_{-\sqrt{a^2-x^2}}^{\sqrt{a^2-x^2}}$$

$$= 2\int_{-a}^{+a} dx\,\sqrt{a^2-x^2}\left[\sqrt{a^2-x^2}+\sqrt{a^2-x^2}\right]$$

$$= 4\int_{-a}^{+a}(a^2-x^2)\,dx = 4\left[a^2x - \frac{x^3}{3}\right]_{-a}^{+a} = 4\left[a^3 - \frac{a^3}{3} + a^3 - \frac{a^3}{3}\right] = \frac{16a^3}{3}$$

Ans.

Example 16. *Find by triple integration the volume of a solid bounded by the sphere* $x^2 + y^2 + z^2 = 4$ *and the paraboloid* $x^2 + y^2 = 3z$.

Solution. $x^2 + y^2 + z^2 = 4 \Rightarrow z = \sqrt{4 - x^2 - y^2}$ ...(1)

and $x^2 + y^2 = 3z \Rightarrow z = \dfrac{x^2+y^2}{3}$ ...(2)

Limits of z are $\sqrt{4-x^2-y^2}$ and $\dfrac{x^2+y^2}{3}$

Area of cross-section at the intersection of (1) and (2) is given by

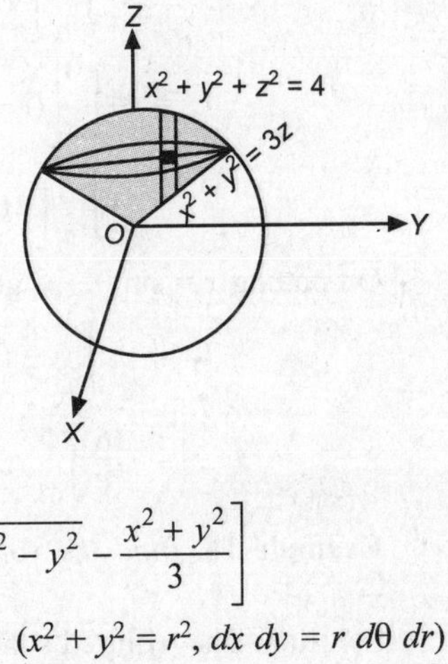

$$4 - x^2 - y^2 = \left(\frac{x^2+y^2}{3}\right)^2$$

$\Rightarrow \quad (x^2+y^2)^2 + 9(x^2+y^2) - 36 = 0$

$\Rightarrow \quad (x^2+y^2-3)(x^2+y^2+12) = 0$

$\Rightarrow \quad x^2 + y^2 = 3$, a circle in the area of cross-section

$$V = \iiint dx\,dy\,dz = \iint dx\,dy\int_{\frac{x^2+y^2}{3}}^{\sqrt{4-x^2-y^2}} dz$$

$$= \iint dx\,dy\,[z]_{\frac{x^2+y^2}{3}}^{\sqrt{4-x^2-y^2}} = \iint dx\,dy\left[\sqrt{4-x^2-y^2} - \frac{x^2+y^2}{3}\right]$$

$(x^2 + y^2 = r^2,\ dx\,dy = r\,d\theta\,dr)$

Limits of r are 0 to $\sqrt{3}$ and that of θ are from 0 to 2π.

$$V = \int_0^{2\pi}\int_0^{\sqrt{3}}\left(\sqrt{4-r^2}-\frac{r^2}{3}\right) r\,d\theta\,dr = \int_0^{2\pi} d\theta \int_0^{\sqrt{3}}\left[r\left(4-r^2\right)^{\frac{1}{2}}-\frac{r^3}{3}\right] dr.$$

$$= (\theta)_0^{2\pi}\left[-\frac{1}{2}\times\frac{2}{3}\left(4-r^2\right)^{\frac{3}{2}}-\frac{r^4}{12}\right]_0^{\sqrt{3}} = 2\pi\left[-\frac{1}{3}-\frac{3}{4}+\frac{8}{3}\right]=\frac{19\pi}{6}$$ **Ans.**

8.5 VOLUME BOUNDED BY A PARABOLOID

When a volume is bounded by a paraboloid it is convenient to use cartesian coordinates again.

Example 17. *Find the volume cut off from the paraboloid*

$x^2+\frac{y^2}{4}+z = 1$ by the plane $z = 0$. (*M.U. II Semester 2005*)

Solution. We have

$$x^2+\frac{y^2}{4}+z = 1 \quad \text{(Paraboloid)} \quad ...(1)$$

$$z = 0 \quad (x\text{-}y \text{ plane}) \quad ...(2)$$

z varies from 0 to $1-x^2-\frac{y^2}{4}$

y varies from $-2\sqrt{1-x^2}$ to $2\sqrt{1-x^2}$

x varies from -1 to 1.

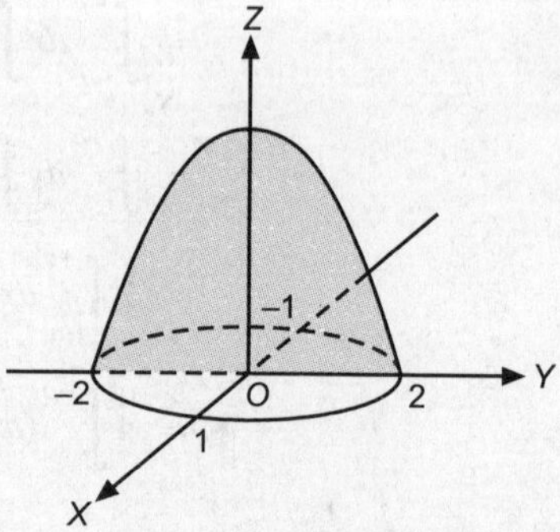

$$V = \iiint dx\,dy\,dz$$

$$= \int_{-1}^{1} dx \int_{-2\sqrt{1-x^2}}^{2\sqrt{1-x^2}} dy \int_0^{\left(1-x^2-\frac{y^2}{4}\right)} dz$$

$$= \int_{-1}^{1}\int_{-2\sqrt{1-x^2}}^{2\sqrt{1-x^2}}\left(1-x^2-\frac{y^2}{4}\right) dx\,dy$$

$$= 4\int_0^1\int_0^{2\sqrt{1-x^2}}\left(1-x^2-\frac{y^2}{4}\right) dx\,dy$$

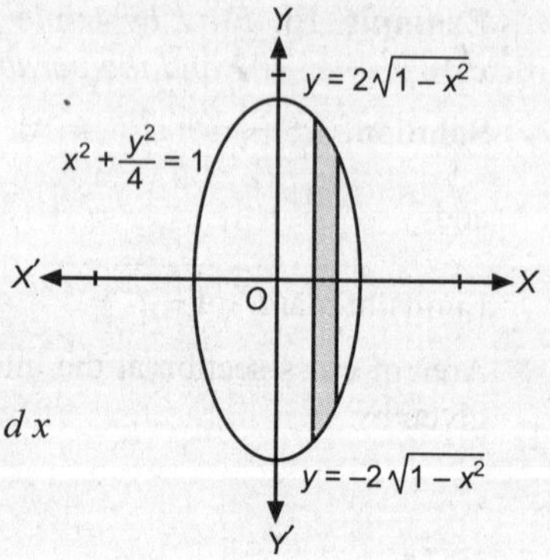

$$= 4\int_0^1\left[(1-x^2)\,y-\frac{y^3}{12}\right]_0^{2\sqrt{1-x^2}} .\,dx$$

$$= 4\int_0^1\left[(1-x^2).\,2\sqrt{1-x^2}-\frac{8}{12}(1-x^2)^{3/2}\right] dx$$

$$= 4\int_0^1\left[2\,(1-x^2)^{3/2}-\frac{2}{3}(1-x^2)^{3/2}\right] dx$$

On putting $x = \sin\theta$, we get

$$V = 4\int_0^1 \frac{4}{3}(1-x^2)^{3/2}\,dx = \frac{16}{3}\int_0^{\pi/2}(1-\sin^2\theta)^{3/2}\cos\theta\,d\theta$$

$$= \frac{16}{3}\int_0^{\pi/2}\cos^4\theta\,d\theta = \frac{16}{3}.\frac{3}{4}.\frac{1}{2}.\frac{\pi}{2} = \pi$$ **Ans.**

Z
-1
-2 O 2 Y
1
X
Y
$y=2\sqrt{1-x^2}$
$x^2+\frac{y^2}{4}=1$
X' O X
$y=-2\sqrt{1-x^2}$
Y'

Example 18. *Find the volume bounded by the paraboloid $x^2 + y^2 = az$ and the cylinder $x^2 + y^2 = a^2$.* (*M.U. II Semester 2007*)

Solution. The required solid is bounded by a cylinder

$$x^2 + y^2 = a^2 \quad \Rightarrow \quad r^2 = a^2$$

and the paraboloid $x^2 + y^2 = a\,z \Rightarrow r^2 = az \Rightarrow z = \dfrac{r^2}{a}$

z varies from 0 to $\dfrac{r^2}{a}$

r varies from 0 to a

and θ varies from 0 to $\dfrac{\pi}{2}$

and the solid lies in four octants.

$$V = 4\int_{\theta=0}^{\pi/2}\int_{r=0}^{a}\int_{z=0}^{r^2/a} r\,dr\,d\theta\,dz$$

$$= 4\int_{\theta=0}^{\pi/2}\int_{r=0}^{a} r\left[z\right]_{z=0}^{r^2/a} dr\,d\theta = 4\int_{\theta=0}^{\pi/2}\int_{r=0}^{a}\frac{r^3}{a}\,dr\,d\theta = \frac{4}{a}\int_{\theta=0}^{\pi/2}\left[\frac{r^4}{4}\right]_0^a d\theta$$

$$= \frac{4}{a}\int_{\theta=0}^{\pi/2}\left[\frac{a^4}{4}\right] d\theta = a^3\int_0^{\pi/2} d\theta = a^3\left[\theta\right]_0^{\pi/2} = \frac{\pi}{2}a^3 \qquad \textbf{Ans.}$$

Example 19. *Find the volume bounded by the surfaces*

$$z = 4 - x^2 - \frac{1}{4}y^2 \quad \text{and} \quad z = 3x^2 + \frac{y^2}{4}$$

Solution. Here, we have

$$z + x^2 + \frac{y^2}{4} = 4 \qquad \text{(Paraboloid)} \qquad ...(1)$$

$$z = 3x^2 + \frac{y^2}{4} \qquad \text{(Paraboloid)} \qquad ...(2)$$

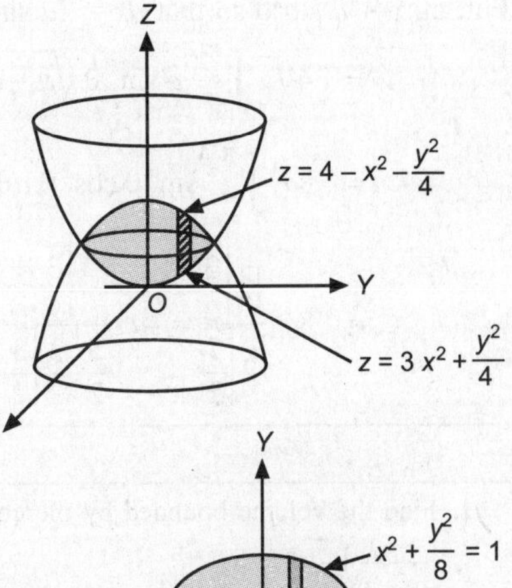

Let us find out the equation of the section intersected by (1) and (2).
Solving them, we get

$$4 - x^2 - \frac{1}{4}y^2 = 3x^2 + \frac{y^2}{4}$$

$$\Rightarrow \qquad 4x^2 + \frac{y^2}{2} = 4 \quad i.e. \quad x^2 + \frac{y^2}{8} = 1$$

The cross-section is an ellipse in four octants

$$V = 4\int_{x=0}^{1}\int_{y=0}^{\sqrt{8(1-x^2)}}\int_{3x^2+(y^2/4)}^{4-x^2-(y^2/4)} dz\,dy\,dx$$

$$= 4\int_{x=0}^{1}\int_{y=0}^{\sqrt{8(1-x^2)}}\left(4 - x^2 - \frac{y^2}{4} - 3x^2 - \frac{y^2}{4}\right) dy\,dx$$

$$= 4\int_{x=0}^{1}\int_{y=0}^{\sqrt{8(1-x^2)}}\left(4 - 4x^2 - \frac{y^2}{2}\right) dy\,dx$$

$$= 4\int_{x=0}^{1}\left[4y - 4x^2y - \frac{y^3}{6}\right]_0^{\sqrt{8(1-x^2)}} dx = 4\int_0^1\left[4(1-x^2)\,y - \frac{y^3}{6}\right]_0^{\sqrt{8(1-x^2)}} dx$$

$$= 4\int_0^1 \left[4(1-x^2).\sqrt{8}.\sqrt{1-x^2} - \frac{1}{6}8\sqrt{8}\,(1-x^2)^{3/2}\right] dx = 4\int_0^1 \left(4\sqrt{8} - \frac{4\sqrt{8}}{3}\right)(1-x^2)^{3/2} dx$$

$$= \frac{64\sqrt{2}}{3}\int_0^1 (1-x^2)^{3/2}\,dx \qquad [\text{Put } x = \sin\theta \;\Rightarrow\; dx = \cos\theta\, d\theta]$$

$$= \frac{64\sqrt{2}}{3}\int_0^{\pi/2} \cos^4\theta\, d\theta = \frac{64\sqrt{2}}{3}.\frac{3}{4}.\frac{1}{2}.\frac{\pi}{2} = 4\sqrt{2}\,.\pi$$ **Ans.**

Example 20. *Find the volume enclosed between the cylinders $x^2 + y^2 = a\,x$, and $z^2 = a\,x$.*

Solution. Here, we have $x^2 + y^2 = ax$ ...(1)

$z^2 = ax$ ...(2)

$$V = \iiint dx\,dy\,dz$$

$$= \int_0^a dx \int_{-\sqrt{ax-x^2}}^{\sqrt{ax-x^2}} dy \int_{-\sqrt{ax}}^{\sqrt{ax}} dz = 2\int_0^a dx \int_{-\sqrt{ax-x^2}}^{\sqrt{ax-x^2}} dy \int_0^{\sqrt{ax}} dz$$

$$= 2\int_0^a dx \int_{-\sqrt{ax-x^2}}^{\sqrt{ax-x^2}} dy\,(z)_0^{\sqrt{ax}} = 2\int_0^a dx \int_{-\sqrt{ax-x^2}}^{\sqrt{ax-x^2}} dy\,\sqrt{ax} = 2\int_0^a \sqrt{ax}\,dx\,[y]_{-\sqrt{ax-x^2}}^{\sqrt{ax-x^2}}$$

$$= 2\int_0^a \sqrt{ax}\,dx\left(2\sqrt{ax-x^2}\right) = 4\sqrt{a}\int_0^a x\sqrt{a-x}\,dx$$

Putting $x = a\sin^2\theta$ so that $dx = 2a\sin\theta\cos\theta\, d\theta$, we get

$$V = 4\sqrt{a}\int_0^{\pi/2} a\sin^2\theta\sqrt{a - a\sin^2\theta}\,.\,2a\sin\theta\cos\theta\, d\theta$$

$$= 8a^3\int_0^{\pi/2} \sin^3\theta\cos^2\theta\, d\theta$$

$$= 8a^3\,\frac{\left|\!\overline{2}\right.\left|\!\overline{\tfrac{3}{2}}\right.}{2\left|\!\overline{\tfrac{7}{2}}\right.} = 4a^3\,\frac{\left|\!\overline{\tfrac{3}{2}}\right.}{\frac{5}{2}.\frac{3}{2}\left|\!\overline{\tfrac{3}{2}}\right.} = \frac{16a^3}{15}$$ **Ans.**

EXERCISE 8.1

1. Find the volume bounded by the coordinate planes and the plane.
   $\frac{x}{a}+\frac{y}{b}+\frac{z}{c}=1$ **Ans.** $\frac{abc}{6}$
2. Find the volume bounded by the cylinders $y^2 = x$ and $x^2 = y$ between the planes $z = 0$ and $x + y + z = 2$. **Ans.** $\frac{11}{30}$
3. Find the volume bounded by the co-ordinate planes and the plane.
   $l\,x + m\,y + n\,z = 1$ **Ans.** $\frac{1}{6\,l\,m\,n}$
4. Find the volume of the sphere $x^2 + y^2 + z^2 = a^2$ by triple integration. (*AMIETE, June 2009*) **Ans.** $\frac{4}{3}\pi a^3$
5. Find the volume of the ellipsoid $\frac{x^2}{a^2}+\frac{y^2}{b^2}+\frac{z^2}{c^2}=1$ **Ans.** $\frac{4\pi\,a\,b\,c}{3}$
6. Find the volume bounded by the cylinder $x^2 + y^2 = a^2$ and the planes $y + z = 2a$ and $z = 0$.
   (*M.U. II Semester 06*) **Ans.** $2\pi a^3$

7. Find the volume bounded by the cylinder $x^2 + y^2 = a^2$ and the planes $z = 0$ and $y + z = b$.
Ans. $\pi a^2 b$

8. Find the volume of the region bounded by $z = x^2 + y^2$, $z = 0$, $x = -a$, $x = a$ and $y = -a$, $y = a$.
Ans. $\frac{8}{3}a^4$

9. Find the volume enclosed by the cylinder $x^2 + y^2 = 9$ and the planes $x + z = 5$ and $z = 0$.
Ans. $45\pi - 36$

10. Compute the volume of the solid bounded by $x^2 + y^2 = z$, $z = 2x$. **Ans.** 2π

11. Find the volume cut from the paraboloid $4z = x^2 + y^2$ by plane $z = 4$.
(U.P. I Semester, Dec. 2005) **Ans.** 32π

12. By using triple integration find the volume cut off from the sphere $x^2 + y^2 + z^2 = 16$ by the plane $z = 0$ and the cylinder $x^2 + y^2 = 4x$. **Ans.** $\frac{64}{9}(3\pi - 4)$

13. The sphere $x^2 + y^2 + z^2 = a^2$ is pierced by the cylinder $x^2 + y^2 = a^2(x^2 - y^2)$.

Prove that the volume of the sphere that lies inside the cylinder is $\frac{8}{3}\left[\frac{\pi}{4} + \frac{5}{3} - \frac{4\sqrt{2}}{3}\right]a^3$.

14. Find the volume of the solid bounded by the surfaces $z = 0$, $3z = x^2 + y^2$ and $x^2 + y^2 = 9$.
(A.M.I.E.T.E., Summer 2005) **Ans.** $\frac{27\pi}{2}$

15. Obtain the volume bounded by the surface $z = c\left(1 - \frac{x}{a}\right)\left(1 - \frac{y}{b}\right)$ and a quadrant of the elliptic cylinder $\frac{x^2}{a^2} + \frac{y^2}{b^2} = 1$, $z > 0$ and where $a, b > 0$. *(A.M.I.E.T.E., Dec. 2005)* **Ans.** $\pi\, a\, b\, c$

16. Find the volume of the paraboloid $x^2 + y^2 = 4z$ cut off by the plane $z = 4$. **Ans.** 32π

17. Find the volume bounded by the cone $z^2 = x^2 + y^2$ and the paraboloid $z = x^2 + y^2$. **Ans.** $\frac{\pi}{6}$

18. Find the volume enclosed by the cylinders $x^2 + y^2 = 2ax$ and $z^2 = 2ax$. **Ans.** $\frac{128a^3}{15}$

19. Find the volume of the solid bounded by the plane $z = 0$, the paraboloid $z = x^2 + y^2 + 2$ and the cylinder $x^2 + y^2 = 4$. **Ans.** 16π

8.6 SURFACE AREA

Let $z = f(x,y)$ be the surface S. Let its projection on the x-y plane be the region A. Consider an element $\delta x . \delta y$ in the region A. Erect a cylinder on the element $\delta x . \delta y$ having its generator parallel to OZ and meeting the surface S in an element of area δs.

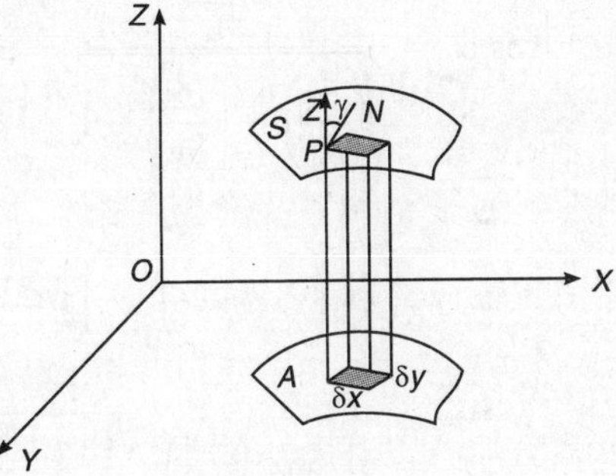

$$\therefore \quad \delta x\, \delta y = \delta s \cos \gamma,$$

Where γ is the angle between the xy-plane and the tangent plane to S at P, *i.e.*, it is the angle between the Z-axis and the normal to S at P.

The direction cosines of the normal to the surface $F(x, y, z) = 0$ are proportional to

$$\frac{\partial F}{\partial x}, \frac{\partial F}{\partial y}, \frac{\partial F}{\partial z}$$

$\therefore$ The direction of the normal to S $[F = f(x, y) - z]$ are proportional to $-\frac{\partial z}{\partial x}, -\frac{\partial z}{\partial y}, 1$ and those of the Z-axis are 0, 0 , 1.

$$\text{Direction cosines} = \frac{-\frac{\partial z}{\partial x}}{\sqrt{\left(\frac{\partial z}{\partial x}\right)^2 + \left(\frac{\partial z}{\partial y}\right)^2 + 1}}, \frac{-\frac{\partial z}{\partial y}}{\sqrt{\left(\frac{\partial z}{\partial x}\right)^2 + \left(\frac{\partial z}{\partial y}\right)^2 + 1}}, \frac{1}{\sqrt{\left(\frac{\partial z}{\partial x}\right)^2 + \left(\frac{\partial z}{\partial y}\right)^2 + 1}},$$

Hence
$$\cos\gamma = \frac{1}{\sqrt{\left[\left(\frac{\partial z}{\partial x}\right)^2 + \left(\frac{\partial z}{\partial y}\right)^2 + 1\right]}} \quad (\cos\theta = l_1 l_2 + m_1 m_2 + n_1 n_2)$$

$$\delta S = \frac{\delta x\, \delta y}{\cos\gamma} = \sqrt{\left[\left(\frac{\partial z}{\partial x}\right)^2 + \left(\frac{\partial z}{\partial y}\right)^2 + 1\right]}\, \delta x\, \delta y\,; \quad S = \iint_A \sqrt{\left[\left(\frac{\partial z}{\partial x}\right)^2 + \left(\frac{\partial z}{\partial y}\right)^2 + 1\right]}\, dx\, dy$$

Example 21. $\iint_s 6xy\, ds$, *where S is the portion of the plane x + y + z = 1 that lies infront of the yz-plane.* (*Gujarat, I Semester, Jan. 2009*)

Solution. Here, we have

$$I = \iint_s 6xy\, ds$$

$$x + y + z = 1, z = 1 - x - y \qquad \text{... (1)}$$

$$\Rightarrow \quad \frac{\partial z}{\partial x} = -1 \quad \Rightarrow \quad \frac{\partial z}{\partial y} = -1$$

The direction ratio of the normal to the plane (1) are $-\frac{\partial z}{\partial x}, = -1$

Direction cosines of the normal to the surface

$$\frac{-\frac{\partial z}{\partial x}}{\sqrt{\left(\frac{\partial z}{\partial x}\right)^2 + \left(\frac{\partial z}{\partial y}\right)^2 + 1}}, \frac{-\frac{\partial z}{\partial y}}{\sqrt{\left(\frac{\partial z}{\partial x}\right)^2 + \left(\frac{\partial z}{\partial y}\right)^2 + 1}}, \frac{1}{\sqrt{\left(\frac{\partial z}{\partial x}\right)^2 + \left(\frac{\partial z}{\partial y}\right)^2 + 1}}$$

Direction cosines of the normal (x-axis) to the plane 1, 0, 0

$$\cos\alpha = \frac{-\frac{\partial z}{\partial x}}{\sqrt{\left(\frac{\partial z}{\partial x}\right)^2 + \left(\frac{\partial z}{\partial y}\right)^2 + 1}} \quad \Rightarrow \delta s = \frac{\delta x\, \delta y}{\cos\alpha} = \frac{\sqrt{\left(\frac{\partial z}{\partial x}\right)^2 + \left(\frac{\partial z}{\partial y}\right)^2 + 1}}{-\frac{\partial z}{\partial x}} \delta x\, \delta y$$

$$I = \iint 6xy\, ds = 6\int x dx \int y \frac{\sqrt{\left(\frac{\partial z}{\partial x}\right)^2 + \left(\frac{\partial z}{\partial s}\right)^2 + 1}}{-\frac{\partial z}{\partial x}} dy$$

$$= 6\int_0^1 x dx \int_0^{1-x} y \frac{\sqrt{1+1+1}}{-1} dy = 6\int_0^1 x\, dx \left[\frac{y^2}{2}\right]_0^{1-x} (-\sqrt{3})$$

$$= -3\sqrt{3}\int_0^{1-x} x\, dx\, (1-x)^2 = -3\sqrt{3}\int_0^1 (x^3 - 2x^2 + x)\, dx$$

$$= -3\sqrt{3}\left(\frac{x^4}{4} - \frac{2x^3}{3} + \frac{x^2}{2}\right)_0^1 = -3\sqrt{3}\left[\frac{1}{4} - \frac{2}{3} + \frac{1}{2}\right] = -\frac{3\sqrt{3}}{12} = -\frac{\sqrt{3}}{4}$$

Surface Area $= \dfrac{\sqrt{3}}{4}$ **Ans.**

Example 22. *Find the surface area of the cylinder* $x^2 + z^2 = 4$ *inside the cylinder* $x^2 + y^2 = 4$.

Solution. $x^2 + y^2 = 4$ $x^2 + x^2 = 4$

$$2x + 2z\frac{\partial z}{\partial x} = 0 \text{ or } \frac{\partial z}{\partial x} = -\frac{x}{z},\ \frac{\partial z}{\partial y} = 0$$

$$\left(\frac{\partial z}{\partial x}\right)^2 + \left(\frac{\partial z}{\partial y}\right)^2 + 1 = \frac{x^2}{z^2} + 1 = \frac{x^2 + z^2}{z^2} = \frac{4}{4 - x^2}$$

Hence, the required surface area

$$= 8\int_0^2\int_0^{\sqrt{4-x^2}} \sqrt{\left[\left(\frac{\partial z}{\partial x}\right)^2 + \left(\frac{\partial z}{\partial y}\right)^2 + 1\right]}\, dx\, dy$$

$$= 8\int_0^2\int_0^{\sqrt{4-x^2}} \frac{2}{\sqrt{4-x^2}}\, dx\, dy = 16\int_0^2 \frac{1}{\sqrt{4-x^2}}[y]_0^{\sqrt{4-x^2}}\, dx = 16\int_0^2 \frac{1}{\sqrt{4-x^2}}[\sqrt{4-x^2}]\, dx$$

$$= 16\int_0^2 dx = 16(x)_0^2 = 32$$ **Ans.**

Example 23. *Find the surface area of the sphere* $x^2 + y^2 + z^2 = 9$ *lying inside the cylinder* $x^2 + y^2 = 3y$.

Solution.
$$x^2 + y^2 + z^2 = 9$$
$$2x + 2z\frac{\partial z}{\partial x} = 0,\quad \frac{\partial z}{\partial x} = -\frac{x}{z}$$
$$2y + 2z\frac{\partial z}{\partial y} = 0,\quad \frac{\partial z}{\partial y} = -\frac{y}{z}$$

$$\left[\left(\frac{\partial z}{\partial x}\right)^2 + \left(\frac{\partial z}{\partial y}\right)^2 + 1\right] = \frac{x^2}{z^2} + \frac{y^2}{z^2} + 1 = \frac{x^2 + y^2 + z^2}{z^2} = \frac{9}{9 - x^2 - y^2} = \frac{9}{9 - r^2} \quad \begin{bmatrix} x = r\cos\theta \\ y = r\sin\theta \end{bmatrix}$$

$$x^2 + y^2 = 3y \text{ or } r^2 = 3r\sin\theta \text{ or } r = 3\sin\theta.$$

Hence, the required surface area

$$= \iint \sqrt{\left[\left(\frac{\partial z}{\partial x}\right)^2 + \left(\frac{\partial z}{\partial y}\right)^2 + 1\right]}\, dx\, dy = 4\int_0^{\pi/2}\int_0^{3\sin\theta} \frac{3}{\sqrt{9 - r^2}}\, r\, d\theta\, dr = 12\int_0^{\pi/2} d\theta \int_0^{3\sin\theta} \frac{r\, dr}{\sqrt{9 - r^2}}$$

$$= 12\int_0^{\pi/2} d\theta\,[-\sqrt{9 - r^2}]_0^{3\sin\theta} = 12\int_0^{\pi/2} [-\sqrt{9 - 9\sin^2\theta} + 3]\, d\theta$$

$$= 36\int_0^{\pi/2} (-\cos\theta + 1)\, d\theta = 36(-\sin\theta + \theta)_0^{\pi/2} = 36\left(-1 + \frac{\pi}{2}\right) = 18(\pi - 2)$$ **Ans.**

Example 24. *Find the surface area of the section of the cylinder* $x^2 + y^2 = a^2$ *made by the plane* $x + y + z = a$.

Solution.
$$x^2 + y^2 = a^2 \quad \text{... (1)}$$
$$x + y + z = a \quad \text{... (2)}$$

The projection of the surface area on *xy*-plane is a circle

$x^2 + y^2 = a^2$

$$1 + \frac{\partial z}{\partial x} = 0 \quad \text{or} \quad \frac{\partial z}{\partial x} = -1$$
$$1 + \frac{\partial z}{\partial y} = 0 \quad \text{or} \quad \frac{\partial z}{\partial y} = -1$$

$$\sqrt{\left(\frac{\partial z}{\partial x}\right)^2+\left(\frac{\partial z}{\partial y}\right)^2+1} = \sqrt{(-1)^2+(-1)^2+1}=\sqrt{3}$$

Hence the required surface area

$$= 4\int_0^a\int_0^{\sqrt{a^2-x^2}}\sqrt{\left(\frac{\partial z}{\partial x}\right)^2+\left(\frac{\partial z}{\partial y}\right)^2+1}\,dx\,dy = 4\int_0^a\int_0^{\sqrt{a^2-x^2}}\sqrt{3}\,dx\cdot dy$$

$$= 4\sqrt{3}\int_0^a [y]_0^{\sqrt{a^2-x^2}}\,dx = 4\sqrt{3}\int_0^a\sqrt{a^2-x^2}\,dx$$

$$= 4\sqrt{3}\left[\frac{x}{2}\sqrt{a^2-x^2}+\frac{a^2}{2}\sin^{-1}\frac{x}{a}\right]_0^a = 4\sqrt{3}\left[0+\frac{a^2}{2}\frac{\pi}{2}\right] = 4\sqrt{3}\left(\frac{a^2\pi}{4}\right)=\sqrt{3}\,\pi\,a^2 \quad \textbf{Ans.}$$

Example 25. *Find the area of that part of the surface of the paraboloid $y^2 + z^2 = 2\,ax$, which lies between the cylinder, $y^2 = ax$ and the plane $x = a$.*

Solution.

$$y^2+z^2=2\,ax \quad \text{... (1)}$$

$$y^2=ax \quad \text{... (2)}$$

$$x=a \quad \text{... (3)}$$

Differentiating (1), we get

$$2\,z\frac{\partial z}{\partial x} = 2\,a,\quad \frac{\partial z}{\partial x}=\frac{a}{z}$$

$$2y+2\,z\frac{\partial z}{\partial y} = 0,\quad \frac{\partial z}{\partial y}=-\frac{y}{z}$$

$$\left(\frac{\partial z}{\partial x}\right)^2+\left(\frac{\partial z}{\partial y}\right)^2+1 = \frac{a^2}{z^2}+\frac{y^2}{z^2}+1=\frac{a^2+y^2}{z^2}+1 \qquad \left[\begin{array}{l}y^2+z^2=2\,ax\\ z^2=2\,ax-y^2\end{array}\right]$$

$$= \frac{a^2+y^2}{2\,a\,x-y^2}+1=\frac{a^2+y^2+2\,a\,x-y^2}{2\,a\,x-y^2}=\frac{a^2+2\,a\,x}{2\,a\,x-y^2}$$

$$S = \int_0^a\int_{-\sqrt{ax}}^{\sqrt{ax}}\sqrt{\left(\frac{\partial z}{\partial x}\right)^2+\left(\frac{\partial z}{\partial y}\right)^2+1}\,dx\,dy = \int_0^a\int_{-\sqrt{ax}}^{\sqrt{ax}}\sqrt{\frac{a^2+2\,ax}{2\,ax-y^2}}\,dx\,dy \qquad \left[\begin{array}{c}y^2=ax\\ y=\pm\sqrt{ax}\end{array}\right]$$

$$= \sqrt{a}\int_0^a\int_{-\sqrt{ax}}^{\sqrt{ax}}\sqrt{\frac{a+2\,x}{2\,ax-y^2}}\,dx\,dy = \sqrt{a}\int_0^a\sqrt{a+2\,x}\,dx\int_{-\sqrt{ax}}^{\sqrt{ax}}\frac{1}{\sqrt{2\,ax-y^2}}\,dy$$

$$= \sqrt{a}\int_0^a\sqrt{a+2\,x}\,dx\left[\sin^{-1}\frac{y}{\sqrt{2ax}}\right]_{-\sqrt{ax}}^{\sqrt{ax}}$$

$$= \sqrt{a}\int_0^a\sqrt{a+2\,x}\,dx\left[\sin^{-1}\frac{1}{\sqrt{2}}-\sin^{-1}\left(-\frac{1}{\sqrt{2}}\right)\right] = \sqrt{a}\int_0^a\sqrt{a+2\,x}\,dx\left[\frac{\pi}{4}+\left(\frac{\pi}{4}\right)\right]$$

$$= \sqrt{a}\,\frac{\pi}{2}\int_0^a\sqrt{a+2\,x}\,dx=\frac{\pi}{2}\cdot\frac{\sqrt{a}}{2}\cdot\frac{2}{3}[(a+2x)^{3/2}]_0^a$$

$$= \frac{\pi\sqrt{a}}{6}[(3a)^{3/2}-a^{3/2}]=\frac{\pi\,a^2}{6}[3\sqrt{3}-1] \qquad \textbf{Ans.}$$

EXERCISE 8.2

1. Find the surface area of sphere $x^2 + y^2 + z^2 = 16$. **Ans.** 64π
2. Find the surface area of the portion of the cylinder $x^2 + y^2 = 4y$ lying inside the sphere $x^2 + y^2 + z^2 = 16$. **Ans.** 64.
3. Show that the area of surfaces $cz = xy$ intercepted by the cylinder $x^2 + y^2 = b^2$ is $\iint_A \frac{\sqrt{c^2 + x^2 + y^2}}{c}\, dx\, dy$, where A is the area of the circle $x^2 + y^2 = b^2, z = 0$ **Ans.** $\frac{2\pi}{3c}\left[(c^2 + b^2)^{\frac{1}{2}} - c^2\right]$
4. Find the area of the portion of the sphere $x^2 + y^2 + z^2 = a^2$ lying inside the cylinder $x^2 + y^2 = ax$. **Ans.** $2(\pi - 2)a^2$
5. Find the area of the surface of the cone $z^2 = 3(x^2 + y^2)$ cut out by the paraboloid $z = x^2 + y^2$ using surface integral. **Ans.** 6π

8.7 CALCULATION OF MASS

We have,

$$\text{Volume} = \iiint_V dx\, dy\, dz \qquad \text{Density} = \rho = f(x, y, z)$$

$$[\text{Density} = \text{Mass per unit volume}] \qquad \text{Mass} = \text{Volume} \times \text{Density}$$

$$\text{Mass} = \iiint_V dx\, dy\, dz \qquad \boxed{Mass = \iiint_V f(x, y, z)\, dx\, dy\, dz}$$

Example 26. *Find the mass of a plate which is formed by the co-ordinate planes and the plane $\frac{x}{a} + \frac{y}{b} + \frac{z}{c} = 1$, the density is given by $\rho = k\,x\,y\,z$.* (U.P., I Semester, Dec., 2003)

Solution. The plate is bounded by the planes $x = 0, y = 0, z = 0$ and $\frac{x}{a} + \frac{y}{b} + \frac{z}{c} = 1$.

$$\text{Mass} = \iiint dx\, dy\, dz\, \rho = \int_0^c \int_0^{b\left(1-\frac{z}{c}\right)} \int_0^{a\left(1-\frac{y}{b}-\frac{z}{c}\right)} dx\, dy\, dz\, (k\,xyz)$$

$$= k\int_0^c z\, dz \int_0^{b\left(1-\frac{z}{c}\right)} y\, dy \int_0^{a\left(1-\frac{y}{b}-\frac{z}{c}\right)} x\, dx = k\int_0^c z\, dz \int_0^{b\left(1-\frac{z}{c}\right)} y\, dy \left(\frac{x^2}{2}\right)_0^{a\left(1-\frac{y}{b}-\frac{z}{c}\right)}$$

$$= k\int_0^c z\, dz \int_0^{b\left(1-\frac{z}{c}\right)} y\, dy \frac{a^2}{2}\left(1-\frac{y}{b}-\frac{z}{c}\right)^2 = \frac{k\,a^2}{2}\int_0^c z\, dz \int_0^{b\left(1-\frac{z}{c}\right)} y\left[\left(1-\frac{z}{c}\right)-\frac{y}{b}\right]^2 dy$$

$$= \frac{k\,a^2}{2}\int_0^c z\, dz \int_0^{b\left(1-\frac{z}{c}\right)} \left[y\left(1-\frac{z}{c}\right)^2 + \frac{y^3}{b^2} - \frac{2y^2}{b}\left(1-\frac{z}{c}\right)\right] dy$$

$$= \frac{k\,a^2}{2}\int_0^c z\, dz \left[\frac{y^2}{2}\left(1-\frac{z}{c}\right)^2 + \frac{y^4}{4b^2} - \frac{2y^3}{3b}\left(1-\frac{z}{c}\right)\right]_0^{b\left(1-\frac{z}{c}\right)}$$

$$= \frac{k\,a^2}{2}\int_0^c z\, dz \left[\frac{b^2}{2}\left(1-\frac{z}{c}\right)^4 + \frac{b^4}{4b^2}\left(1-\frac{z}{c}\right)^4 - \frac{2}{3}\cdot\frac{b^3}{b}\left(1-\frac{z}{c}\right)^4\right]$$

$$= \frac{k\,a^2}{2}\int_0^c z\left[\frac{b^2}{2} + \frac{b^2}{4} - \frac{2b^2}{3}\right]\left(1-\frac{z}{c}\right)^4 dz = \frac{k\,a^2}{2}\frac{b^2}{12}\int_0^z \left(1-\frac{z}{c}\right)^4 dz \qquad [\text{Put } z = c\sin^2\theta]$$

$$= \frac{k\,a^2\,b^2}{24}\int_0^{\frac{\pi}{2}} c\sin^2\theta\,(1-\sin^2\theta)^4\,(2c\sin\theta\cos\theta\, d\theta)$$

$$= \frac{k^2\, a^2\, b^2\, c^2}{12} \int_0^{\pi/2} \sin^2\theta\,(\cos^8\theta) \sin\theta \cos\theta\, d\theta \;=\; \frac{k^2\, a^2\, b^2\, c^2}{12} \int_0^{\pi/2} \sin^3\theta \cos^9\theta\, d\theta$$

$$= \frac{k^2\, a^2\, b^2\, c^2}{12} \frac{\Gamma\!\left(\frac{3+1}{2}\right)\Gamma\!\left(\frac{9+1}{2}\right)}{2\,\Gamma\!\left(\frac{3+9+2}{2}\right)} = \frac{k\, a^2\, b^2\, c^2}{12} \cdot \frac{\Gamma 2\, \Gamma 5}{2\,\Gamma 7} = \frac{k\, a^2\, b^2\, c^2}{12} \frac{(1)\,(\Gamma 5)}{2 \times 6 \times 5\,\Gamma 5} = \frac{k\, a^2\, b^2\, c^2}{720} \quad \textbf{Ans.}$$

Example 27. *Find the mass of a plate in the shape of the curve* $\left(\frac{x}{a}\right)^{2/3} + \left(\frac{y}{b}\right)^{2/3} = 1$, *the density being given by* $\rho = \mu\, xy$.

Solution. Let the required mass be M which is four times the mass in the first quadrant.

From the equation of the curve $\left(\frac{y}{b}\right)^{2/3} = 1 - \left(\frac{x}{a}\right)^{2/3} \quad \Rightarrow y = b\left[1 - \left(\frac{x}{a}\right)^{2/3}\right]^{3/2} = y_1$

For the region *OAB*, x varies from 0 to a and y varies from 0 to y_1.

$$\therefore\; M = 4\int_0^a \int_0^{y_1} \rho\, dy\, dx = 4\int_0^a \int_0^{y_1} \mu\, xy\, dy\, dx$$

$$= 4\int_0^a \mu x \cdot \left[\frac{y^2}{2}\right]_0^{y_1} dx = 2\mu \int_0^a x y_1^2\, dx$$

$$= 2\mu \int_0^a x b^2 \left[1 - \left(\frac{x}{a}\right)^{2/3}\right]^3 dx$$

Put $x = a \sin^3\theta$, then $dx = 3a \sin^2\theta \cos\theta\, d\theta$

When $x = 0,\ \theta = 0$; when $x = a,\ \theta = \frac{\pi}{2}$

$$\therefore\; M = 2\mu b^2 \int_0^{\pi/2} a \sin^3\theta\,(1 - \sin^2\theta)^3 \,.\, 3a \sin^2\theta \cos\theta\, d\theta$$

$$= 6\mu a^2 b^2 \int_0^{\pi/2} \sin^5\theta \cos^7\theta\, d\theta \;=\; 6\mu a^2 b^2 \cdot \frac{4 \cdot 2 \cdot 6 \cdot 4 \cdot 2}{12 \cdot 10 \cdot 8 \cdot 6 \cdot 4 \cdot 2} = \frac{\mu a^2 b^2}{20} \quad \textbf{Ans.}$$

Example 28. *Find the mass of a lamina in the form of the cardioid* $r = a\,(1 + \cos\theta)$ *whose density at any point varies as the square of its distance from the initial line.*

(Nagpur University, Winter 2005)

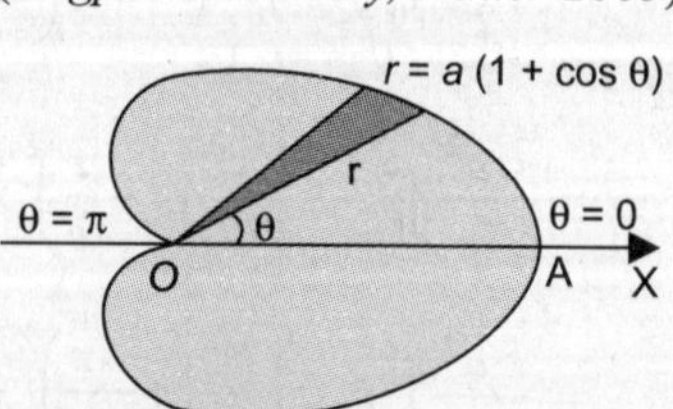

Solution. Let the required mass be M which is twice the mass above the initial line.

Since the distance of any point (r, θ) from the initial line is $r \sin\theta$, the density at (r, θ) is given by $r = m\,(r \sin\theta)^2 = \mu\, r^2 \sin^2\theta$.

For the region above the initial line, θ varies from 0 to π and r varies from 0 to $a(1 + \cos\theta)$.

$$\therefore \quad M = 2\int_0^{\pi} \int_0^{a(1+\cos\theta)} \rho\, r\, dr\, d\theta$$

$$= 2\int_0^{\pi} \int_0^{a(1+\cos\theta)} \mu\, r^3 \sin^2\theta\, dr\, d\theta \;=\; 2\int_0^{\pi} \mu \sin^2\theta \,.\left[\frac{r^4}{4}\right]_0^{a(1+\cos\theta)} d\theta$$

$$= \frac{\mu a^4}{2}\int_0^{\pi}\left(2\sin\frac{\theta}{2}\cos\frac{\theta}{2}\right)^2\cdot\left(2\cos^2\frac{\theta}{2}\right)^4 d\theta = 32\,\mu a^4\int_0^{\pi}\sin^2\frac{\theta}{2}\cos^{10}\frac{\theta}{2}\,d\theta$$

Put $\frac{\theta}{2} = t$, then $d\theta = 2dt$; when $\theta = 0$, $t = 0$; when $\theta = \pi$, $t = \frac{\pi}{2}$

$$\therefore \quad M = 32\,\mu a^4\int_0^{\pi} 2\sin^2 t\cos^{10} t\,dt = 64\,\mu\,a^4\cdot\frac{1.9.7.5\,3.1}{12.10.8.6.4.2}\cdot\frac{\pi}{2} = \frac{21}{32}\mu\pi a^4$$ **Ans.**

8.8 CENTRE OF GRAVITY

$$\bar{x} = \frac{\iiint x\rho\,dx\,dy\,dz}{\iiint \rho\,dx\,dy\,dz},\quad \bar{y} = \frac{\iiint y\rho\,dx\,dy\,dz}{\iiint \rho\,dx\,dy\,dz},\quad \bar{z} = \frac{\iiint z\rho\,dx\,dy\,dz}{\iiint \rho\,dx\,dy\,dz}$$

Example 29. *Find the co-ordinates of the centre of gravity of the positive octant of the sphere $x^2 + y^2 + z^2 = a^2$, density being given = k xyz.*

Solution. $$\bar{x} = \frac{\iiint_V x\rho\,dx\,dy\,dz}{\iiint_V \rho\,dx\,dy\,dz} = \frac{\iiint z\rho\,dx\,dy\,dz}{\iiint \rho\,dx\,dy\,dz} = \frac{\iiint_V x^2yz\,dx\,dy\,dz}{\iiint_V xyz\,dx\,dy\,dz}$$

Converting into polar co-ordinates, $x = r\sin\theta\cos\phi$, $y = r\sin\theta\sin\phi$, $z = r\cos\theta$,

$dx\,dy\,dz = r^2\sin\theta\,dr\,d\theta\,d\phi$

$$\bar{x} = \frac{\int_0^{\pi/2}\int_0^{\pi/2}\int_0^{a}(r\sin\theta\cos\phi)^2\,(r\sin\theta\sin\phi)\,(r\cos\theta)\,(r^2\sin\theta\,dr\,d\theta\,d\phi)}{\int_0^{\pi/2}\int_0^{\pi/2}\int_0^{a}(r\sin\theta\cos\phi)\,(r\sin\theta\sin\phi)\,(r\cos\theta)\,(r^2\sin\theta\,dr\,d\theta\,d\phi)}$$

$$= \frac{\int_0^{\pi/2}\int_0^{\pi/2}\int_0^{a} r^6\sin^4\theta\cos\theta\sin\phi\cos^2\phi\,dr\,d\theta\,d\phi}{\int_0^{\pi/2}\int_0^{\pi/2}\int_0^{a} r^5\sin^3\theta\cos\theta\sin\phi\cos\phi\,dr\,d\theta\,d\phi}$$

$$= \frac{\int_0^{\pi/2}\sin\phi\cos^2\phi\,d\phi\int_0^{\pi/2}\sin^4\theta\cos\theta\,d\theta\int_0^{a} r^6\,dr}{\int_0^{\pi/2}\sin\phi\cos\phi\,d\phi\int_0^{\pi/2}\sin^3\theta\cos\theta\,d\theta\int_0^{a} r^5\,dr}$$

$$= \frac{\left[-\frac{\cos^3\phi}{3}\right]_0^{\pi/2}\left[\frac{\sin^5\theta}{5}\right]_0^{\pi/2}\left[\frac{r^7}{7}\right]_0^{a}}{-\left[\frac{\cos^2\phi}{2}\right]_0^{\pi/2}\left[\frac{\sin^4\theta}{4}\right]_0^{\pi/2}\left[\frac{r^6}{6}\right]_0^{a}} = \frac{\left(\frac{1}{3}\right)\left(\frac{1}{5}\right)\left(\frac{a^7}{7}\right)}{\left(\frac{1}{2}\right)\left(\frac{1}{4}\right)\left(\frac{a^6}{6}\right)} = \frac{16\,a}{35}$$

Similarly, $\bar{y} = \bar{x} = \frac{16\,a}{35}$

Hence, C.G. is $\left(\frac{16\,a}{35}, \frac{16\,a}{35}, \frac{16\,a}{35}\right)$ **Ans.**

8.9 MOMENT OF INERTIA OF A SOLID

Let the mass of an element of a solid of volume V be $\rho\delta x\delta y\delta z$.

Perpendicular distance of this element from the x-axis $= \sqrt{y^2 + z^2}$

M.I. of this element about the x-axis $= \rho\,\delta x\,\delta y\,\delta z\,\sqrt{y^2 + z^2}$

M.I. of the solid about x-axis $= \iiint_V \rho\,(y^2 + z^2)\,dx\,dy\,dz$

M.I. of the solid about y-axis $= \iiint_V \rho\,(x^2 + z^2)\,dx\,dy\,dz$

M.I. of the solid about z-axis $= \iiint_V \rho\,(x^2 + y^2)\,dx\,dy\,dz$

The Perpendicular Axes Theorem

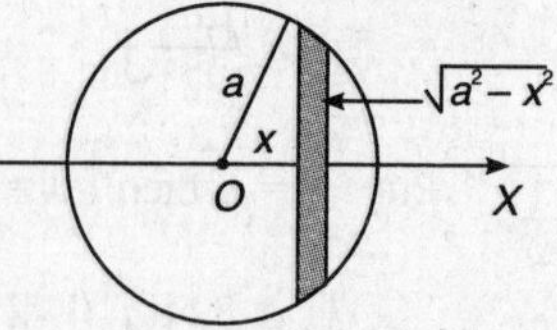

If I_{ox} and I_{oy} be the moments of inertia of a lamina about x-axis and y-axis respectively and I_{oz} be the moment of inertia of the lamina about an axis perpendicular to the lamina and passing through the point of intersection of the axes OX and OY.

$$I_{OZ} = I_{OX} + I_{OY}$$

The Parallel Axes Theorem

M.I. of a lamina about an axis in the plane of the lamina equals the sum of the moment of inertia about a parallel centroidal axis in the plane of lamina together with the product of the mass of the lamina and square of the distance between the two axes.

$$I_{AB} = I_{XX} + M\bar{y}^2$$

Example 30. *Find M.I. of a sphere about diameter.*

Solution. Let a circular disc of δx thickness be perpendicular to the given diameter XX' at a distance x from it.

The radius of the disc= $\sqrt{a^2 - x^2}$

Mass of the disc= $\rho\, \pi\, (a^2 - x^2)$

Moment of inertia of the disc about a diameter perpendicular on it

$$= \frac{1}{2} MR^2 = \frac{1}{2}[\rho\, \pi\, (a^2 - x^2)]\,(a^2 - x^2) = \frac{1}{2}\rho\, \pi\, (a^2 - x^2)^2$$

$$\text{M.I. of the sphere} = \int_{-a}^{a} \frac{1}{2}\rho\, \pi\, (a^2 - x^2)^2\, dx = 2\left(\frac{1}{2}\rho\, \pi\right)\int_0^a [a^4 - 2\, a^2\, x^2 + x^4]\, dx$$

$$= \rho\, \pi \left[a^4 x - \frac{2\, a^2\, x^3}{3} + \frac{x^5}{5}\right]_0^a = \rho\, \pi\left[a^5 - \frac{2\, a^5}{3} + \frac{a^5}{5}\right]$$

$$= \frac{8}{15}\pi\, \rho\, a^5 = \frac{2}{5}\left(\frac{4\,\pi}{3}a^3\, \rho\right)a^2 = \frac{2}{5}Ma^2$$ **Ans.**

Example 31. *The mass of a solid right circular cylinder of radius a and height h* is *M. Find the moment of inertia of the cylinder about (i) its axis (ii) a line through its centre of gravity perpendicular to its axis (iii) any diameter through its base.*

Solution. *To find M.I. about OX. Consider a disc at a distance x from O at the base.*

M.I. *of the about OX,*

$$= \frac{(\pi\, a^2\, \rho\, dx)\, a^2}{2} = \frac{\pi\, \rho\, a^4\, dx}{2}$$

(*i*) M.I. of the cylinder about OX

$$\int_0^h \frac{\pi\, \rho\, a^4 dx}{2} = \frac{\pi\, \rho\, a^4}{2}(x)_0^h = \frac{\pi\, \rho\, a^4\, h}{2} = (\pi\, a^2\, h)\,\rho \cdot \frac{a^2}{2} = \frac{M\, a^2}{2}$$

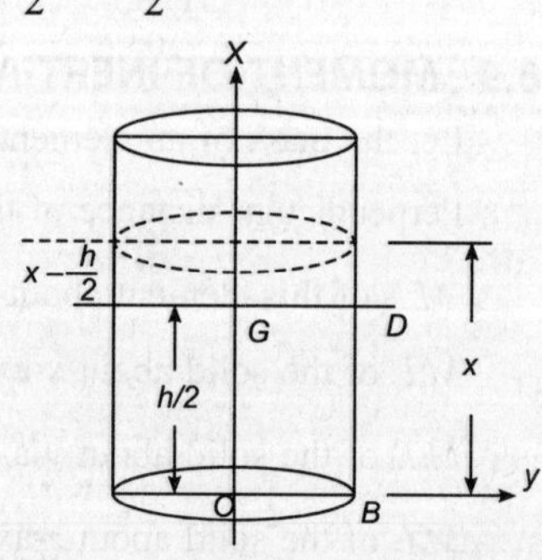

(*ii*) *M.I.* of the disc about a line through *C.G.* and perpendicular to *OX*.

$$I_{OX} + I_{OY} = I_{OZ}$$

$$I_{OX} + I_{OX} = I_{OZ}$$

$$I_{OX} = \frac{1}{2} I_{OZ}$$

M.I. of the disc about a line through

$$C.G. = \frac{1}{2}\left(\frac{M\, a^2}{2}\right) = \frac{M\, a^2}{4}$$

M.I. of the disc about the diameter $= \left(\dfrac{\pi\, a^2\, \rho\, dx}{4}\right) a^2$

M.I. of the disc about line $GD = \dfrac{\pi\, a^2\, \rho\, dx}{4} + (\pi\, a^2\, \rho\, dx)\left(x - \dfrac{h}{2}\right)^2$

Hence, *M.I.* of cylinder about $GD = \displaystyle\int_0^h \frac{\pi\, a^2\, \rho}{4}\, dx + \int_0^h (\pi\, a^2\, \rho\, dx)\left(x - \frac{h}{2}\right)^2$

$$= \frac{\pi\, a^2\, \rho}{4}(x)_0^h + \left[\frac{\pi\, a^2\, \rho}{4}\left(x - \frac{h}{2}\right)^3\right]_0^h = \frac{\pi\, a^2\, \rho\, h}{4} + \left[\frac{\pi\, a^2\, \rho}{3}\left(\frac{h}{2}\right)^3 + \frac{\pi\, a^2\, \rho}{3}\left(\frac{h}{2}\right)^3\right]$$

$$= \frac{\pi\, a^2\, \rho\, h}{4} + \frac{\pi\, a^2\, \rho\, h^3}{12} = \frac{M\, a^2}{4} + \frac{M\, h^2}{12}$$

(*iii*) *M.I.* of cylinder about line OB (through) base

$$I_{OB} \quad = I_G + M\left(\frac{h}{2}\right)^2 = \frac{M\, a^2}{4} + \frac{M\, h^2}{12} + \frac{M\, h^2}{4} = \frac{M\, a^2}{4} + \frac{M\, h^2}{3}$$ **Ans.**

Example 32. *Find the moment of inertia and radius of gyration about z-axis of the region in the first octant bounded by* $\dfrac{x}{a} + \dfrac{y}{b} + \dfrac{z}{c} = 1$.

Solution. Let r be the density.

M.I. of tetrahedron about z-axis

$$= \iiint (\rho\, dx\, dy\, dz)\,(x^2 + y^2)$$

$$= \rho \int_0^a dx \int_0^{b\left(1-\frac{x}{a}\right)} (x^2 + y^2)\, dy \int_0^{c\left(1-\frac{x}{a}-\frac{y}{b}\right)} dz = \rho \int_0^a dx \int_0^{b\left(1-\frac{x}{a}\right)} (x^2 + y^2)\, dy\, (z)_0^{c\left(1-\frac{x}{a}-\frac{y}{b}\right)}$$

$$= \rho \int_0^a dx \int_0^{b\left(1-\frac{x}{a}\right)} (x^2 + y^2)\, dy\; c\left(1 - \frac{x}{a} - \frac{y}{b}\right)$$

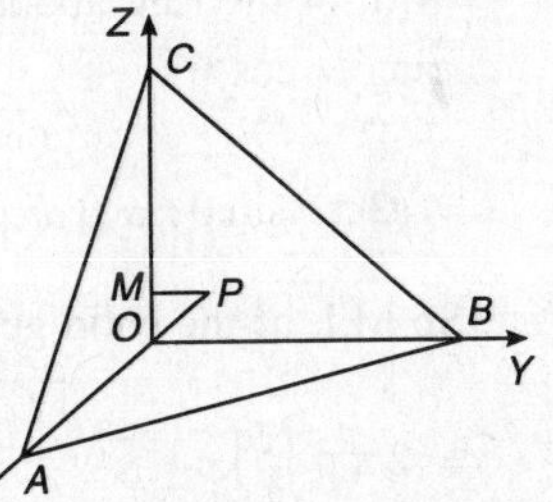

$$= \rho c \int_0^a dx \int_0^{b\left(1-\frac{x}{a}\right)} \left[x^2\left(1 - \frac{x}{a}\right) - \frac{x^2 y}{b} + y^2\left(1 - \frac{x}{a}\right) - \frac{y^3}{b}\right] dy$$

$$= \rho c \int_0^a dx \left[x^2\left(1 - \frac{x}{a}\right) y - \frac{x^2 y^2}{2\, b} + \frac{y^3}{3}\left(1 - \frac{x}{a}\right) - \frac{y^4}{4\, b}\right]_0^{b\left(1-\frac{x}{a}\right)}$$

$$= \rho c \int_0^a dx \left[x^2\left(1 - \frac{x}{a}\right) b\left(1 - \frac{x}{a}\right) - \frac{x^2}{2\, b} b^2\left(1 - \frac{x}{a}\right)^2 + \frac{b^3}{3}\left(1 - \frac{x}{a}\right)^3\left(1 - \frac{x}{a}\right) - \frac{b^4}{4\, b}\left(1 - \frac{x}{a}\right)^4\right]$$

$$= b\rho c \int_0^a \left[x^2\left(1 - \frac{x}{a}\right)^2 - \frac{x^2}{2}\left(1 - \frac{x}{a}\right)^2 - \frac{b^2}{3}\left(1 - \frac{x}{a}\right)^4 - \frac{b^2}{4}\left(1 - \frac{x}{a}\right)^4\right] dx$$

$$= \rho bc \int_0^a \left[\frac{x^2}{2}\left(1 - \frac{x}{a}\right)^2 + \frac{b^2}{12}\left(1 - \frac{x}{a}\right)^4\right] dx$$

$$= \rho bc \int_0^a \left[\frac{1}{2}\left(x^2 - \frac{2\, x^3}{a} + \frac{x^4}{a^2}\right) + \frac{b^2}{12}\left(1 - \frac{4\, x}{a} + \frac{6\, x^2}{a^2} - \frac{4\, x^3}{a^3} + \frac{x^4}{a^4}\right)\right] dx$$

$$= \rho bc \left[\frac{1}{2}\left(\frac{x^3}{3} - \frac{x^4}{2\, a} + \frac{x^5}{5\, a^2}\right) + \frac{b^2}{12}\left(x - \frac{2\, x^2}{a} + \frac{6\, x^3}{3a^2} - \frac{x^4}{a^3} + \frac{x^5}{5a^4}\right)\right]_0^a$$

$$= \rho bc \int_0^a \left[\frac{1}{2}\left(\frac{a^3}{3} - \frac{a^3}{2} + \frac{a^3}{5}\right) + \frac{b^2}{12}\left(a - 2a + 2a - a + \frac{a}{5}\right)\right]$$

$$= \rho bc \left[\frac{a^3}{60} + \frac{ab^2}{60}\right] = \rho \frac{abc}{60}(a^2 + b^2)$$

Radius of gyration $= \sqrt{\dfrac{M.I.}{Mass}} = \dfrac{\sqrt{\dfrac{\rho abc}{60}(a^2+b^2)}}{\dfrac{\rho abc}{6}} = \sqrt{\dfrac{1}{10}(a^2+b^2)}$ **Ans.**

Example 33. *A solid body of density ρ is in the shape of the solid formed by revolution of the cardioid r = a (1 + cos θ) about the initial line. Show that its moment of inertia about a straight line through the pole perpendicular to the initial line is* $\left(\frac{352}{105}\right)\pi \rho a^5$.

Solution. $r = a(1 + \cos\theta)$

Consider an elementary area $r\,d\theta\,dr$. This area when revolved about OX generates a circular ring of radius $r \sin\theta$.

Its mass $= (2\pi r \sin\theta)(r\,d\theta\,dr\,\rho)$.

($M.I.$ of a ring about a diameter $= \dfrac{Ma^2}{2}$)

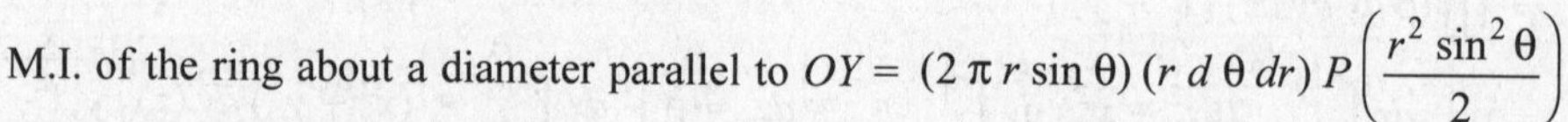

M.I. of the ring about a diameter parallel to $OY = (2\pi r \sin\theta)(r\,d\theta\,dr)\,P\left(\dfrac{r^2\sin^2\theta}{2}\right)$

M.I. of the ring about OY = M.I. of the ring about a diameter parallel to OY + mass of the ring $(r\cos\theta)^2$

$$= (2\pi r \sin\theta\, r\, d\theta\, dr\, \rho)\left(\frac{r^2\sin^2\theta}{2} + r^2\cos^2\theta\right)$$

So M.I. of the solid generated by revolution about OY

$$= 2\pi\rho \int_0^\pi \int_a^{a(1+\cos\theta)} r^4 \sin\theta \left(\frac{\sin^2\theta}{2} + \cos^2\theta\right) d\theta\, dr$$

$$= \pi\rho \int_0^\pi \sin\theta\,(1+\cos^2\theta)\, d\theta \int_a^{a(1+\cos\theta)} r^4\, dr = \pi\rho \int_0^\pi \sin\theta\,(1+\cos^2\theta)\, d\theta \left(\frac{r^5}{5}\right)_0^{a(1+\cos\theta)}$$

$$= \pi\rho \int_0^\pi \sin\theta\,(1+\cos^2\theta)\, d\theta\, \frac{a^5(1+\cos\theta)^5}{5} = \pi\rho\frac{a^5}{5}\int_0^\pi (1+\cos^2\theta)(1+\cos\theta)^5 \sin\theta\, d\theta$$

$$= \frac{\pi\rho a^5}{5}\int_0^\pi \left[1 + \left(2\cos^2\frac{\theta}{2} - 1\right)^2\right]\left(2\cos^2\frac{\theta}{5}\right)^5 2\sin\frac{\theta}{2}\cos\frac{\theta}{2}\, d\theta$$

Putting $\left(\frac{\theta}{2} = t,\ d\theta = 2\,dt\right) = \dfrac{265\pi\rho a^5}{5}\displaystyle\int_0^{\pi/2}[\cos^{11} t \sin t + 2\cos^{15} t \sin t - 2\cos^{13} t \sin t]\, dt$

$$= \frac{265\pi\rho a^5}{5}\left[-\frac{\cos^{12} t}{12} - 2\frac{\cos^{16} t}{16} + 2\frac{\cos^{14} t}{14}\right]_0^{\pi/2} = \frac{352\pi\rho a^5}{105}$$ **Proved.**

8.10 CENTRE OF PRESSURE

The centre of pressure of a plane area immersed in a fluid is the point at which the resultant force acts on the area.

Consider a plane area A immersed vertically in a homogeneous liquid. Let x-axis be the line of intersection of the plane with the free surface. Any line in this plane and perpendicular to x-axis is the *y-axis*.

Let P be the pressure at the point (x, y). Then the pressure on elementary area $\delta x\ \delta y$ is $P\ \delta x\ \delta y$. Let $(\overline{x}, \overline{y})$ be the centre of pressure. Taking moment about y-axis.

$$\overline{x} \cdot \iint_A P\,dx\,dy = \iint_A Px\,dx\,dy$$

$$\overline{x} = \frac{\iint_A Px\,dx\,dy}{\iint_A P\,dx\,dy}$$

Similarly, $$\overline{y} = \frac{\iint_A Py\,dx\,dy}{\iint_A P\,dx\,dy}$$

Example 34. *A uniform semi-circular lamina is immersed in a fluid with its plane vertical and its bounding diameter on the free surface. If the density at any point of the fluid varies as the depth of the point below the free surface, find the position of the centre of pressure of the lamina.*

Solution. Let the semi-circular lamina be

$$x^2 + y^2 = a^2$$

By symmetry its centre of pressure lies on OY.

Let ky be the density of the fluid.

$$\overline{y} = \frac{\iint_A Py\,dx\,dy}{\iint_A P\,dx\,dy} = \frac{\iint_A (\rho y)\,y\,dx\,dy}{\iint_A (\rho y)\,dx\,dy} \qquad (\because \rho = ky)$$

$$= \frac{\iint_A (ky \,.\, y)\,y\,dx\,dy}{\iint_A (ky \,.\, y)\,dx\,dy} = \frac{\iint_A y^3\,dx\,dy}{\iint_A y^2\,dx\,dy} = \frac{\int_{-a}^{a} dx \int_0^{\sqrt{a^2-x^2}} y^3\,dy}{\int_{-a}^{a} dx \int_0^{\sqrt{a^2-x^2}} y^2\,dy}$$

$$= \frac{\int_{-a}^{a} dx \left[\frac{y^4}{4}\right]_0^{\sqrt{a^2-x^2}}}{\int_{-a}^{a} dx \left[\frac{y^3}{3}\right]_0^{\sqrt{a^2-x^2}}} = \frac{3}{4}\,\frac{\int_{-a}^{a} dx\,(a^2-x^2)^2}{\int_{-a}^{a} dx\,(a^2-x^2)^{3/2}}$$

$$= \frac{3}{4}\,\frac{\int_{-\pi/2}^{\pi/2} (a\cos\theta\,d\theta)\,(a^2 - a^2\sin^2\theta)^2}{\int_{-\pi/2}^{\pi/2} (a\cos\theta\,d\theta)\,(a^2 - a^2\sin^2\theta)^{3/2}} \qquad (\text{Put } x = a\sin\theta)$$

$$= \frac{3\,a}{4}\,\frac{\int_{-\pi/2}^{\pi/2} \cos^5\theta\,d\theta}{\int_{-\pi/2}^{\pi/2} \cos^4\theta\,d\theta}$$

$$= \frac{3\,a}{4}\,\frac{2\int_0^{\pi/2} \cos^5\theta\,d\theta}{2\int_0^{\pi/2} \cos^4\theta\,d\theta} = \frac{3\,a}{4}\,\frac{\frac{4\times 2}{5\times 3}}{\frac{3\times 1}{4\times 2}\,\frac{\pi}{2}} = \frac{32\,a}{15\,\pi} \qquad \textbf{Ans.}$$

EXERCISE 8.3

1. Find the mass of the solid bounded by the ellipsoid $\frac{x^2}{a^2}+\frac{y^2}{b^2}+\frac{z^2}{c^2}=1$ and the co-ordinate planes, where the density at any point $P(x, y, z)$ is $k\,xyz$. **Ans.** P
2. If the density at a point varies as the square of the distance of the point from XOY plane, find the mass of the volume common to the sphere $x^2 + y^2 + z^2 = a^2$ and cylinder $x^2 + y^2 = ax$. **Ans.** $\frac{4k}{15}a^5\left(\frac{\pi}{2}-\frac{8}{15}\right)$
3. Find the mass of the plate in the form of one loop of leminscate $r^2 = a^2 \sin 2\theta$, where $\rho = k\,r^2$. **Ans.** $\frac{k\pi a^4}{16}$
4. Find the mass of the plate which is inside the circle $r = 2a\cos\theta$ and outside the circle $r = a$, if the density varies as the distance from the pole.
5. Find the mass of a lamina in the form of the cardioid $r = a(1 + \cos\theta)$ whose density at any point varies as the square of its distance from the initial line. **Ans.** $\frac{21\pi k a^4}{32}$
6. Find the centroid of the region in the first octant bounded by $\frac{x}{a}+\frac{y}{b}+\frac{z}{c}=1$. **Ans.** $\left(\frac{a}{4},\frac{b}{4},\frac{c}{4}\right)$
7. Find the centroid of the region bounded by $z = 4 - x^2 - y^2$ and xy-plane. **Ans.** $\left(0,0,\frac{4}{3}\right)$
8. Find the position of $C.G.$. of the volume intercepted between the parallelepiped $x^2+y^2 = a(a - z)$ and the plane $z = 0$. **Ans.** $\left(0,0,\frac{a}{3}\right)$
9. A solid is cut off the cylinder $x^2 + y^2 = a^2$ by the plane $z = 0$ and that part of the olane $z = mx$ for which z is positive. The density of the solid cut off at any point varies as the height of the point above plane $z = 0$. Find $C.G.$ of the solid. **Ans.** $\bar{z}=\frac{64\,ma}{45\pi}$
10. If an area is bounded by two concentric semi-circles with their common bounding diameter in a free surface, prove that the depth of the centre of pressure is $\frac{3\pi}{16}\frac{(a+b)(a^2+b^2)}{a^2+ab+b^2}$
11. An ellipse $\frac{x^2}{a^2}+\frac{y^2}{b^2}=1$ is immersed vertically in a fluid with its major axis horizontal. If its centre be at depth h, find the depth of its centre of pressure. **Ans.** $h+\frac{b^2}{4h}$
12. A horizontal boiler has a flat bottom and its ends are plane and semi-circular. If it is just full of water, show that the depth of centre of pressure of either end is 0.7 × total depth approximately.
13. A quadrant of a circle of radius a is just immersed vertically in a homogeneous liquid with one edge in the surface. Determine the co-ordinates of the centre of pressure. **Ans.** $\left(\frac{3a}{8},\frac{3\pi a}{16}\right)$
14. Find the product of inertia of an equilateral triangle about two perpendicular axes in its plane at a vertex, one of the axes being along a side.
15. Find the $M.I.$ of a right circular cylinder of radius a and height h about axis if density varies as distance from the axis. **Ans.** $\frac{2}{5}k\,\delta\,a^5 h$

16. Compute the moment of inertia of a right circular cone whose altitude is h and base radius r, about (*i*) the axis of symmetry (*ii*) the diameter of the base. **Ans.** (*i*) $\frac{\pi h r^4}{10}$ (*ii*) $\frac{\pi h r^2}{60}(2h^2+3r^2)$

17. Find the moment of inertia for the area of the cardioid $r = a(1 - \cos\theta)$ relative to the pole.

Ans. $\frac{35\pi a^4}{16}$

18. Find the M.I. about the line $\theta = \frac{\pi}{2}$ of the area enclosed by $r = a(1 + \cos\theta)$.

19. Find the moment of inertia of the uniform solid in the form of octant of the ellipsoid $\frac{x^2}{a^2}+\frac{y^2}{b^2}+\frac{z^2}{c^2}=1$ about OX **Ans.** $\frac{M}{5}(b^2+c^2)$

20. Prove that the moment of inertia of the area included between the curves $y^2 = 4ax$ and $x^2 = 4ay$ about the x-axis is $\frac{144}{35} M a^2$, where M is the mass of area included between the curves.

21. A solid body of density p is the shape of solid formed by revolution of the cardioid $r = a(1 + \cos\theta)$ about the initial line. Show that its moment of inertia about a straight line through the pole perpendicular to the initial line is $\left(\frac{352}{105}\right)\pi l a^5$.

22. Find the product of inertia of a disc in the form of a quadrant of a circle of radius 'a' about bounding radii. **Ans.** $\rho\frac{a^4}{4}$

23. Show that the principal axes at the origin of the triangle enclosed by $x = 0$, $y = 0$, $\frac{x}{a}+\frac{y}{b}=1$ are inclined at angles α and $\alpha+\frac{\pi}{2}$ to the x-axis, where $a = \frac{1}{2}\tan^{-1}\left(\frac{ab}{a^2-b^2}\right)$

Choose the correct answer:

24. The triple integral $\iiint_T dx\,dy\,dz$ gives

(*i*) Volume of region T (*ii*) Surface area of region T

(*ii*) Area of region T (*iv*) Density of region T.

Ans. (*i*)

25. The volume of the solid under the surface $az = x^2 + y^2$ and whose base R is the circle $x^2 + y^2 = a^2$ is given as

(*i*) $\frac{\pi}{2a}$ (*ii*) $\frac{\pi a^3}{2}$ **Ans.** (*ii*)

(*iii*) $\frac{4}{3}\pi a^3$ (*iv*) None of the above. [*U.P., I. Sem. Dec. 2008*]

GAMMA, BETA FUNCTION

9.1 GAMMA FUNCTION

(*U.P. I Semester Dec. 2007*)

$$\int_0^\infty e^{-x} x^{n-1} dx \qquad (n > 0)$$

is called gamma function of *n*. It is also written as $\overline{|n} = \int_0^\infty e^{-x} x^{n-1} dx$

$$\boxed{\int_0^\infty e^{-x} x^{n-1} dx = \overline{|n}}$$

Example 1. *Prove that* $\overline{|1} = 1$

Solution. We know that, $\overline{|n} = \int_0^\infty e^{-x} x^{n-1} dx$

Put $n = 1$, $\overline{|1} = \int_0^\infty e^{-x} x^{1-1} dx = \int_0^\infty e^{-x} dx = \left[\frac{e^{-x}}{-1}\right]_0^\infty = 1$ **Proved.**

Example 2. *Prove that*

(*i*) $\overline{|n+1} = n\overline{|n}$ (*ii*) $\overline{|n+1} = n!$ ***(Reduction formula)***

Solution.

(*i*) We know that, $\overline{|n} = \int_0^\infty x^{n-1} e^{-x} dx$ ...(1)

Integrating by parts, we have

$$\overline{|n} = \left[x^{n-1}\frac{e^{-x}}{-1}\right]_0^\infty - (n-1)\int_0^\infty x^{n-2}\frac{e^{-x}}{-1} dx$$

$$= \lim_{x\to 0}\left\{\left(1 - \frac{x}{1!} + \frac{x^2}{2!} - \ldots + (-1)^n\frac{x^n}{n!} + \ldots + \infty\right)x^{n-1}\right\} + (n-1)\int_0^\infty x^{n-2} e^{-x} dx$$

$$= 0 + (n-1)\int_0^\infty x^{n-2} e^{-x} dx$$

$\therefore$ $\overline{|n} = (n-1)\overline{|n-1}$ ...(2)

$\boxed{\overline{|n+1} = n\overline{|n}}$ Replacing *n* by $(n + 1)$ **Proved.**

(*ii*) Replacing *n* by $n - 1$ in (2), we get

$$\overline{|n-1} = (n-2)\overline{|n-2}$$

Putting the value $\Gamma(n-1)$ in (2), we get

$$\Gamma(n)=(n-1)(n-2)\Gamma(n-2)$$

Similarly, $$\Gamma(n)=(n-1)(n-2)\ldots\ldots 3.2.1\,\Gamma(1) \qquad \ldots(3)$$

Putting the value of $\Gamma(1)$ in (3), we have

$$\Gamma(n)=(n-1)(n-2)\ldots\ldots 3.2.1.1$$

$$\Gamma(n)=(n-1)\,!$$

Replacing n by $n+1$, we have $\Gamma(n+1)=n\,!$ **Proved.**

Example 3. *Evaluate* $\Gamma\left(\frac{-5}{2}\right)$ (*MTU 2014*)

Solution. $\Gamma\left(-\frac{1}{2}+1\right)=-\frac{1}{2}\Gamma\left(-\frac{1}{2}\right)$ $\qquad \Gamma(n+1)=n\Gamma(n)$

$$\Gamma\left(\frac{1}{2}\right)=-\frac{1}{2}\Gamma\left(-\frac{3}{2}+1\right) \quad\Rightarrow\quad \sqrt{\pi}=\left(-\frac{1}{2}\right)\left(-\frac{3}{2}\right)\Gamma\left(-\frac{3}{2}\right)$$

$$\Rightarrow \qquad \sqrt{\pi}=\left(-\frac{1}{2}\right)\left(-\frac{3}{2}\right)\Gamma\left(-\frac{5}{2}+1\right) \quad\Rightarrow\quad \sqrt{\pi}=\left(-\frac{1}{2}\right)\left(-\frac{3}{2}\right)\left(-\frac{5}{2}\right)\Gamma\left(-\frac{5}{2}\right)$$

$$\sqrt{\pi}=-\frac{15}{8}\Gamma\left(-\frac{5}{2}\right)$$

$$\Gamma\left(-\frac{5}{2}\right)=-\frac{8}{15}\sqrt{\pi} \qquad \textbf{Ans.}$$

Example 4. *Evaluate* $\int_0^\infty \sqrt[4]{x}\, e^{-\sqrt{x}}\,dx$

Solution. Let $\quad I=\int_0^\infty x^{1/4}\, e^{-\sqrt{x}}\,dx \qquad \ldots(1)$

Putting $\sqrt{x}=t \;\Rightarrow\; x=t^2$ so that $dx=2t\,dt$ in (1), we get

$$I=\int_0^\infty t^{1/2}\,e^{-t}\,2t\,dt=2\int_0^\infty t^{3/2}e^{-t}dt=2\int_0^\infty t^{\frac{5}{2}-1}\,e^{-t}\,dt$$

$$=2\Gamma\left(\frac{5}{2}\right) \qquad \text{[By definition]}$$

$$=2\cdot\frac{3}{2}\Gamma\left(\frac{3}{2}\right)=2\cdot\frac{3}{2}\cdot\frac{1}{2}\Gamma\left(\frac{1}{2}\right)=\frac{3}{2}\sqrt{\pi} \qquad \textbf{Ans.}$$

Example 5. *Evaluate* $\int_0^\infty \sqrt{x}\, e^{-\sqrt[3]{x}}\,dx.$

Solution. Let $I=\int_0^\infty \sqrt{x}\, e^{-\sqrt[3]{x}}\,dx \qquad \ldots(1)$

Putting $\sqrt[3]{x}=t \;\Rightarrow\; x=t^3$ so that $dx=3\,t^2\,dt$ in (1), we get

$$I=\int_0^\infty t^{3/2}\,e^{-t}\,3t^2\,dt=3\int_0^\infty t^{7/2}e^{-t}dt=3\int_0^\infty t^{\frac{9}{2}-1}\,e^{-t}\,dt=3\Gamma\left(\frac{9}{2}\right)=3\cdot\frac{7}{2}\cdot\frac{5}{2}\cdot\frac{3}{2}\cdot\frac{1}{2}\Gamma\left(\frac{1}{2}\right)=\frac{315}{16}\sqrt{\pi}$$

Ans.

Example 6. *Evaluate* $\int_0^\infty x^{n-1}\, e^{-h^2x^2}\,dx.$

Solution. Let $\qquad I=\int_0^\infty x^{n-1}\, e^{-h^2x^2}\,dx \qquad \ldots(1)$

Putting $t = h^2 x^2 \quad \Rightarrow \quad x = \frac{\sqrt{t}}{h}$ so that $dx = \frac{dt}{2h\sqrt{t}}$, we get

$$I = \int_0^\infty \left(\frac{\sqrt{t}}{h}\right)^{n-1} e^{-t} \frac{dt}{2h\sqrt{t}}$$

$$= \frac{1}{2h^n}\int_0^\infty t^{\frac{n-1}{2}} e^{-t} \frac{dt}{\sqrt{t}} = \frac{1}{2h^n}\int_0^\infty t^{\frac{n-2}{2}} e^{-t} dt = \frac{1}{2h^n}\left\lceil\frac{n}{2}\right. \qquad \textbf{Ans.}$$

Example 7. *Evaluate* $\int_0^\infty \frac{x^a}{a^x} dx$, *hence show that* $\int_0^\infty \frac{x^7}{7^x} dx = \frac{7!}{(\log 7)^8}$ $\quad (a > 1)$

Solution. Here, we have $\int_0^\infty \frac{x^a}{a^x} dx$ ...(1)

Putting $\quad a^x = e^t \Rightarrow x \log a = t \Rightarrow x = \frac{t}{\log a}, \Rightarrow \quad dx = \frac{dt}{\log a}$ in (1), we have

$$\int_0^\infty \frac{x^a}{a^x} dx = \int_0^\infty \left(\frac{t}{\log a}\right)^a e^{-t} \frac{dt}{\log a} = \frac{1}{(\log a)^{a+1}}\int_0^\infty e^{-t} t^a dt = \frac{1}{(\log a)^{a+1}}\int_0^\infty t^{(a+1)-1} e^{-t} dt$$

$$= \frac{1}{(\log a)^{a+1}}\left\lceil a+1\right.$$

On putting $a = 7$, we get $\quad \int_0^\infty \frac{x^7}{7^x} dx = \frac{7!}{(\log 7)^8}$ **Ans.**

9.2 PROVE THAT

$$\int_0^1 x^m (\log x)^n dx = \frac{(-1)^n}{(m+1)^{n+1}}\left\lceil n+1\right.$$

Proof : Put $\log x = -t$ so that $x = e^{-t} \Rightarrow dx = -e^{-t} dt$

$$\therefore \qquad x^m = e^{-mt}$$

$$(\log x)^n = (-t)^n$$

Now, $\quad \int_0^1 x^m (\log x)^n dx = \int_\infty^0 e^{-mt} (-t)^n (-e^{-t})\, dt = \int_0^\infty (-1)^n e^{-mt-t} t^n dt$

Putting $(m + 1)\, t = u$ so that $(m + 1)\, dt = du$, we get

$$\therefore \quad I = \int_0^\infty (-1)^n e^{-u} . \frac{u^n}{(m+1)^n} \frac{du}{(m+1)}$$

$$= \frac{(-1)^n}{(m+1)^{n+1}}\int_0^\infty e^{-u} u^n du$$

$$= \frac{(-1)^n}{(m+1)^{n+1}}\int_0^\infty e^{-u}\, u^{(n+1)-1}\, du = \frac{(-1)^n}{(m+1)^{n+1}}\left\lceil n+1\right. \qquad \textbf{Proved.}$$

Example 8. *Prove that* $\int_0^1 (x \log x)^4 dx = \frac{4!}{5^5}$ *(M.U. II Semester, 2009)*

Solution. We know that

$$\int_0^1 x^m (\log x)^n dx = \frac{(-1)^n}{(m+1)^{n+1}} \cdot \left\lceil n+1\right. \qquad \text{...(1) [From Art 9.2]}$$

Putting $\quad m = n = 4$ in (1), we get

$$\int_0^1 x^4(\log x)^4 dx = \frac{(-1)^4}{(4+1)^{4+1}}\Gamma(4+1) = \frac{\Gamma 5}{5^5} = \frac{4!}{5^5}$$ **Proved.**

Example 9. *Evaluate* $\int_0^1 \frac{dx}{\sqrt{-\log x}}$

Solution. Let $-\log x = y \Rightarrow \log x = -y \Rightarrow e^{-y} = x$ so that $dx = -e^{-y}\,dy$

$$\int_0^1 \frac{dx}{\sqrt{-\log x}} = \int_0^1 \frac{-e^{-y}dy}{\sqrt{y}} = \int_0^\infty y^{-\frac{1}{2}} e^{-y}\,dy = \Gamma\frac{1}{2} = \sqrt{\pi}$$ **Ans.**

Example 10. *Evaluate* $\int_0^1 x^{n-1}\left[\log_e\left(\frac{1}{x}\right)\right]^{m-1} \cdot dx$

Solution: Put $\log_e \frac{1}{x} = t$ or $x = e^{-t}$ $\therefore dx = -e^{-t}\,dt$

$$\int_0^1 x^{n-1}\left[\log_e\left(\frac{1}{x}\right)\right]^{m-1} dx = \int_\infty^0 \left(e^{-t}\right)^{n-1}[t]^{m-1}\left(-e^{-t}dt\right) = \int_0^\infty e^{-nt}t^{m-1}\,dt$$

Putting $nt = u \Rightarrow t = \frac{u}{n}$ so that $dt = \frac{du}{n}$

$$= \int_0^\infty e^{-u}\left(\frac{u}{n}\right)^{m-1}\frac{du}{n} = \frac{1}{n^m}\int_0^\infty e^{-u}u^{m-1}du = \frac{1}{n^m}\Gamma m$$ **Ans.**

9.3 TRANSFORMATION OF GAMMA FUNCTION

Prove that (*i*) $\int_0^\infty e^{-ky}y^{n-1}dy = \frac{\Gamma n}{k^n}$ (*AMIETE, Dec. 2010*) (*ii*) $\Gamma\frac{1}{2} = \sqrt{\pi}$

(*iii*) $\int_0^1 \left(\log\frac{1}{y}\right)^{n-1} dy = \Gamma n$ (*iv*) $\Gamma n = \frac{1}{n}\int_0^\infty e^{-x^{\frac{1}{n}}}dx$

Solution: We know that $\Gamma n = \int_0^\infty x^{n-1}e^{-x}dx$ ...(1)

(*i*) Replace x by $k\,y$, so that $dx = k\,dy$; then (1) becomes

$$\Gamma n = \int_0^\infty (k\,y)^{n-1}e^{-ky}\,k\,dy, \quad \Gamma n = k^n\int_0^\infty e^{-ky}y^{n-1}\,dy$$

$\therefore$ $$\boxed{\int_0^\infty e^{-ky}y^{n-1}\,dy = \frac{\Gamma n}{k^n}}$$ ...(2) **Proved.**

(*ii*) Replace x^n by y, so that $n\,x^{n-1}\,dx = dy$ in (1), then

$$\Gamma n = \int_0^\infty y^{\frac{n-1}{n}}e^{-y^{1/n}}\frac{dy}{nx^{n-1}} = \int_0^\infty y^{\frac{n-1}{n}}e^{-y^{1/n}}\frac{dy}{ny^{\frac{n-1}{n}}} = \frac{1}{n}\int_0^\infty e^{-y^{1/n}}dy$$

When $n = \frac{1}{2}$, $$\Gamma\frac{1}{2} = \frac{1}{\frac{1}{2}}\int_0^\infty e^{-y^2}dy = 2\left[\frac{1}{2}\sqrt{\pi}\right]$$

$$\boxed{\Gamma\frac{1}{2} = \sqrt{\pi}}$$ **Proved.**

(*iii*) Putting $e^{-x} = y$, so that $-e^{-x}\,dx = dy$ and $-x = \log y$, $x = \log\frac{1}{y}$, (1) becomes

$$\Gamma n = -\int_1^0 \left(\log\frac{1}{y}\right)^{n-1} y \cdot \frac{dy}{e^{-x}} = \int_0^1 \left(\log\frac{1}{y}\right)^{n-1} y \cdot \frac{dy}{y} = \int_0^1 \left(\log\frac{1}{y}\right)^{n-1} dy$$ **Proved.**

(*iv*) We know that, $\Gamma(n) = \int_0^\infty e^{-x} . x^{n-1}\, dx$ ...(1)

Putting $x^n = y \Rightarrow x = y^{1/n}$ so that $dx = \frac{1}{n} y^{\frac{1}{n}-1}\, dy$ in (1), we get

$$\Gamma n = \int_0^\infty e^{-y^{1/n}}\, y^{\frac{n-1}{n}} . \frac{1}{n} y^{\frac{1}{n}-1}\, dy = \frac{1}{n}\int_0^\infty e^{-y^{\frac{1}{n}}}\, dy$$

$$\Gamma n = \frac{1}{n}\int_0^\infty e^{-x^{\frac{1}{n}}}\, dx.$$

Proved.

EXERCISE 9.1

Evaluate:

1. (*i*) $\left|\overline{-\frac{3}{2}}\right.$ (*ii*) $\left|\overline{-\frac{15}{2}}\right.$ (*iii*) $\left|\overline{\frac{7}{2}}\right.$ (*iv*) $\left|\overline{0}\right.$

Ans. (*i*) $\frac{4}{3}\sqrt{\pi}$ (*ii*) $\frac{2^8\sqrt{\pi}}{15\times 13\times 11\times 9\times 7\times 5\times 3}$ (*iii*) $\frac{15\sqrt{\pi}}{8}$ (*iv*) ∞

2. $\int_0^\infty \sqrt{x}\, e^{-x} dx$ **Ans.** $\left|\overline{\frac{3}{2}}\right.$ **3.** $\int_0^\infty x^4\, e^{-x^2} dx$ **Ans.** $\frac{3\sqrt{\pi}}{8}$

4. $\int_0^\infty e^{-h^2x^2} dx$ **Ans.** $\frac{\sqrt{\pi}}{2h}$

5. $\int_0^\infty\int_0^\infty e^{-\left(ax^2+by^2\right)} x^{2m-1} y^{2n-1} dx\, dy,\quad a,b,m,n > 0$ **Ans.** $\frac{\left|\overline{m}\right.\left|\overline{n}\right.}{4\, a^m b^n}$

6. $\int_0^1 (x\log x)^3\, dx$ **Ans.** $-\frac{3}{128}$ **7.** $\int_0^1 \frac{dx}{\sqrt{x\log\frac{1}{x}}}$ **Ans.** $\sqrt{2\pi}$

8. Prove that 1.3.5.... (2 n – 1) = $\frac{2^n\left|\overline{n+\frac{1}{2}}\right.}{\sqrt{\pi}}$ **9.** $\int_0^\infty e^{-y^{1/m}}\, dy = m\left|\overline{m}\right.$

10. $\int_0^\infty\int_0^\infty e^{-(ax^2+by^2)}\, x^{2m-1}\, y^{2n-1}\, dx\, dy = \frac{\Gamma(m)\,\Gamma(n)}{4a^m b^n}$, where a, b, m, n are positive.

11. $\int_0^{\pi/2} \frac{d\theta}{(a\cos^4\theta + b\sin^4\theta)} = \frac{(\Gamma 1/4)^2}{4(ab)^{1/4}\sqrt{\pi}}$ **[Hint.** Put $\tan\theta = t$ then $bt^4 = az$**]**

9.4 BETA FUNCTION

(*DU, I Sem. 2012, U.P. I Semester Dec. 2007*)

$$\int_0^1 x^{l-1}(1-x)^{m-1} dx \qquad (l > 0,\ m > 0)$$

is called the Beta function of l, m. It is also written as $\beta\,(l, m) = \int_0^1 x^{l-1}(1-x)^{m-1} dx$

9.5 EVALUATION OF BETA FUNCTION

$$\beta\,(l, m) = \frac{\left|\overline{l}\right.\left|\overline{m}\right.}{\left|\overline{l+m}\right.}$$

Solution. We have, $\beta\,(l, m) = \int_0^1 x^{l-1}(1-x)^{m-1} dx = \int_0^1 (1-x)^{m-1} x^{l-1} dx$

Integrating by parts, we have

$$= \left[(1-x)^{m-1}\frac{x^l}{l}\right]_0^1 + (m-1)\int_0^1 (1-x)^{m-2}\left(\frac{x^l}{l}\right) dx = \frac{(m-1)}{l}\int_0^1 (1-x)^{m-2}\, x^l dx$$

Again integrating by parts, we get

$$= \frac{(m-1)(m-2)}{l(l+1)}\int_0^1 (1-x)^{m-3}\, x^{l+1} dx = \frac{(m-1)(m-2)...2.1}{l\,(l+1).....(l+m-2)}\int_0^1 x^{l+m-2} dx$$

$$= \frac{(m-1)(m-2)...2.1}{l\,(l+1).....(l+m-2)}\left[\frac{x^{l+m-1}}{l+m-1}\right]_0^1 = \frac{(m-1)(m-2)...2.1}{l(l+1).....(l+m-2)(l+m-1)}$$

$$= \frac{(m-1)!}{l(l+1).....(l+m-2)(l+m-1)} \times \frac{(l-1)\,(l-2).....1}{(l-1)\,(l-2)\,.....1}$$

$$= \frac{(m-1)!\ (l-1)!}{1.2....(l-2)(l-1).l(l+1)...(l+m-2)(l+m-1)} = \frac{(l-1)!\,(m-1)!}{(l+m-1)!} = \frac{\boxed{l}\,\boxed{m}}{\boxed{l+m}}$$

And if only l is positive integer and not m then

$$\boxed{\beta(l, m) = \frac{\boxed{l}\,\boxed{m}}{\boxed{l+m}}}$$ **Ans.**

9.6 A PROPERTY OF BETA FUNCTION

$$\beta\ (l,\ m) = \beta(m,\ l\ )$$

Solution. We have

$$\beta\ (l,\ m) = \int_0^1 x^{l-1}(1-x)^{m-1} dx \qquad \left[\int_0^a f(x)dx = \int_0^a f(a-x)dx\right]$$

$$= \int_0^1 (1-x)^{l-1}\left[1-(1-x)\right]^{m-1} dx = \int_0^1 (1-x)^{l-1} x^{m-1} dx$$

$$= \int_0^1 x^{m-1}(1-x)^{l-1} dx$$

l and m are interchanged.

$$\boxed{\beta(l, m) = \beta(m, l)}$$ **Proved.**

Example 11. *Evaluate* $\int_0^1 x^4\left(1-\sqrt{x}\right)^5 dx$

Solution. Let $\sqrt{x} = t \Rightarrow x = t^2$ so that $dx = 2\ t\ dt$

$$\int_0^1 x^4\left(1-\sqrt{x}\right)^5 dx = \int_0^1 (t^2)^4\,(1-t)^5(2\,t\,dt)$$

$$= 2\int_0^1 t^9\,(1-t)^5\,dt = 2\,\beta\,(10,6) = 2\frac{\boxed{10}\,\boxed{6}}{\boxed{16}} = 2\,.\,\frac{9!\,5!}{(15)!}$$

$$= 2.\frac{5!}{10\times11\times12\times13\times14\times15} = \frac{2\times1\times2\times3\times4\times5}{10\times11\times12\times13\times14\times15} = \frac{1}{11\times13\times7\times15} = \frac{1}{15015}$$ **Ans.**

Example 12. *Evaluate* $\int_0^1 (1-x^3)^{-\frac{1}{2}} dx$

Solution. Let $x^3 = y \Rightarrow x = y^{1/3}$ so that $dx = \frac{1}{3}y^{-\frac{2}{3}}\,dy$

$$\int_0^1 (1-x^3)^{-\frac{1}{2}} dx = \int_0^1 (1-y)^{-\frac{1}{2}}\left(\frac{1}{3}y^{-\frac{2}{3}}\,dy\right)$$

$$= \frac{1}{3}\int_0^1 y^{-\frac{2}{3}}(1-y)^{-\frac{1}{2}} dy = \frac{1}{3}\beta\left(\frac{1}{3},\frac{1}{2}\right) = \frac{1}{3}\frac{\boxed{\frac{1}{3}}\,\boxed{\frac{1}{2}}}{\boxed{\frac{5}{6}}}$$ **Ans.**

Example 13. *Determine the value of* $\beta\left(\frac{1}{2},\frac{1}{2}\right)$. *(Delhi University, April 2010)*

Solution. We know that $\beta(m,n)=\dfrac{\overline{|m}\,\overline{|n}}{\overline{|m+n}}$ ... (1)

Putting $m=n=\frac{1}{2}$ in (1), we get

$$\beta\left(\frac{1}{2},\frac{1}{2}\right)=\frac{\overline{\left|\frac{1}{2}\right.}\,\overline{\left|\frac{1}{2}\right.}}{\overline{\left|\frac{1}{2}+\frac{1}{2}\right.}}=\pi \qquad \left[\overline{\left|\frac{1}{2}\right.}=\sqrt{\pi},\ \overline{|1}=1\right]$$ **Ans.**

9.7 TRANSFORMATION OF BETA FUNCTION

We know that

$$\beta(l,\,m)\;=\;\int_0^1 x^{l-1}(1-x)^{m-1}\,dx\,,$$

Putting $x=\dfrac{1}{1+y}$ so that $dx=-\dfrac{1}{(1+y)^2}dy$ and $1-x=\dfrac{y}{1+y}$ in (1), we get

$$\beta(l,\,m)=\int_{\infty}^{0}\left(\frac{1}{1+y}\right)^{l-1}\left(\frac{y}{1+y}\right)^{m-1}\left[-\frac{1}{(1+y)^2}dy\right]=\int_0^{\infty}\frac{y^{m-1}}{(1+y)^{l+m}}dy$$

Since l, m can be interchanged in $\beta(l,\,m)$,

$$\beta(l,\,m)\;=\;\int_0^{\infty}\frac{y^{l-1}}{(1+y)^{m+l}}dy\quad\Rightarrow\quad\beta(l,\,m)=\int_0^{\infty}\frac{x^{l-1}}{(1+x)^{m+l}}dx \qquad ...(1)$$

Example 14. *Evaluate* $\displaystyle\int_0^1\frac{x^{m-1}+x^{n-1}}{(1+x)^{m+n}}dx$

Solution. We know that

$$\beta(m,\,n)=\int_0^{\infty}\frac{x^{m-1}}{(1+x)^{m+n}}dx\quad\Rightarrow\quad\int_0^{\infty}\frac{x^{m-1}}{(1+x)^{m+n}}dx=\beta(m,n)$$

$$\Rightarrow\quad\int_0^1\frac{x^{m-1}}{(1+x)^{m+n}}dx+\int_1^{\infty}\frac{x^{m-1}}{(1+x)^{m+n}}dx\;=\;\beta(m,\,n) \qquad ...(1)$$

Consider $$\int_1^{\infty}\frac{x^{m-1}}{(1+x)^{m+n}}dx=\int_1^{0}\frac{\left(\frac{1}{t}\right)^{m-1}}{\left(1+\frac{1}{t}\right)^{m+n}}\left(-\frac{1}{t^2}dt\right)=\int_0^1\frac{\left(\frac{1}{t}\right)^{m-1}\frac{1}{t^2}}{\left(\frac{1}{t}\right)^{m+n}(t+1)^{m+n}}dt \qquad \left(\text{Put } x=\frac{1}{t}\right)$$

$$=\int_0^1\frac{t^{n-1}}{(1+t)^{m+n}}dt=\int_0^1\frac{x^{n-1}}{(1+x)^{m+n}}dx$$

Putting the value of $\displaystyle\int_1^{\infty}\frac{x^{m-1}}{(1+x)^{m+n}}dx$ in (1), we get

$$\int_0^1\frac{x^{m-1}}{(1+x)^{m+n}}dx+\int_0^1\frac{x^{n-1}}{(1+x)^{m+n}}dx\;=\;\beta(m,\,n)$$

$$\Rightarrow\qquad\int_0^1\frac{x^{m-1}+x^{n-1}}{(1+x)^{m+n}}dx\;=\;\beta(m,\,n)$$ **Ans.**

9.8 RELATION BETWEEN BETA AND GAMMA FUNCTIONS (*DU, I Sem. 2012*)

We know that, $\Gamma l = \int_0^\infty e^{-x}\, x^{l-1} dx, \quad \Rightarrow \quad \Gamma l = \int_0^\infty e^{-y} y^{l-1} dy$ [Put $zx = y$]

$$\frac{\Gamma l}{z^l} = \int_0^\infty e^{-zx}\, x^{l-1} dx$$

$$\Gamma l = \int_0^\infty z^l\, e^{-zx} x^{l-1} dx$$

Multiplying both sides by $e^{-z}\, z^{m-1}$, we have

$$\Gamma l . e^{-z} . z^{m-1} = \int_0^\infty e^{-z} . z^{m-1} . z^l . e^{-zx} . x^{l-1} dx$$

$$\Gamma l . e^{-z} . z^{m-1} = \int_0^\infty e^{-(1+x)z} . z^{l+m-1} . x^{l-1} dx$$

Integrating both sides w.r.t. 'z', we get

$$\int_0^\infty \Gamma l\, e^{-z} z^{m-1}\, dz = \int_0^\infty \int_0^\infty e^{-(1+x)z} z^{l+m-1} x^{l-1}\, dx\, dz$$

$$\Gamma l\, \Gamma m = \int_0^\infty x^{l-1}\, dx \int_0^\infty e^{-(1+x)z} z^{l+m-1} dz$$

$$= \int_0^\infty x^{l-1} dx \cdot \frac{\Gamma (l+m)}{(1+x)^{l+m}} \quad \text{[From (1), Art 9.7]}$$

$$\Gamma l\, \Gamma m = \Gamma (l+m) \int_0^\infty \frac{x^{l-1}}{(1+x)^{l+m}} dx = \Gamma (l+m) . \beta\,(l, m)$$

$$\therefore \quad \boxed{\beta(l, m) = \frac{\Gamma l\, \Gamma m}{\Gamma (l+m)}}$$

This is the required relation.

9.9. SHOW THAT

$$\boxed{\int_0^{\frac{\pi}{2}} sin^p\, \theta\, cos^q\, \theta d\theta = \frac{\Gamma\left(\frac{p+1}{2}\right) \Gamma\left(\frac{q+1}{2}\right)}{2\,\Gamma\left(\frac{p+q+2}{2}\right)}}$$

Solution. We know that

$$\beta\ (m,\ n) = \int_0^1 x^{m-1} (1-x)^{n-1}\ dx \quad ...(1)$$

Putting $x = \sin^2\, \theta,\ dx = 2\, \sin\, \theta\, \cos\, \theta\ d\theta$

and $1 - x = 1 - \sin^2\, \theta = \cos^2\, \theta$

Then (1) becomes

$$\beta\ (m,n) = \int_0^{\frac{\pi}{2}} \sin^{2m-2} \theta \cos^{2n-2} \theta\ \ 2 \sin\theta \cos\theta\, d\theta$$

or

$$\frac{\Gamma m\, \Gamma n}{\Gamma (m+n)} = 2\int_0^{\frac{\pi}{2}} \sin^{2m-1} \theta \cos^{2n-1} \theta\, d\,\theta$$

Putting $2\, m - 1 = p, \quad i.e., \quad m = \dfrac{p+1}{2}$

and $\qquad 2n-1 = q, \quad i.e., \quad n = \dfrac{q+1}{2}$

$$\frac{\Gamma\left(\frac{p+1}{2}\right)\Gamma\left(\frac{q+1}{2}\right)}{\Gamma\left(\frac{p+q+2}{2}\right)} = 2\int_0^{\frac{\pi}{2}} \sin^p\theta\cdot\cos^q\theta\,d\theta$$

$$\boxed{\int_0^{\frac{\pi}{2}} \sin^p\theta\cos^q\theta\,d\theta = \frac{\Gamma\left(\frac{p+1}{2}\right)\Gamma\left(\frac{q+1}{2}\right)}{2\,\Gamma\left(\frac{p+q+2}{2}\right)}}$$

Proved.

Example 15. *Prove that* $\displaystyle\int_0^{\frac{\pi}{2}} \sqrt{\sin\theta}\,d\theta \times \int_0^{\frac{\pi}{2}} \frac{1}{\sqrt{\sin\theta}}\,d\theta = \pi$ *(AMIETE, June 2009)*

Solution. L.H.S. $\displaystyle = \int_0^{\frac{\pi}{2}} \sqrt{\sin\theta}\,d\theta \times \int_0^{\frac{\pi}{2}} \frac{1}{\sqrt{\sin\theta}}\,d\theta$

$$= \int_0^{\frac{\pi}{2}} \sin^{\frac{1}{2}}\theta\cos^0\theta\,d\theta \times \int_0^{\frac{\pi}{2}} \sin^{-\frac{1}{2}}\theta\cos^0\theta\,d\theta$$

$$= \frac{\Gamma\left(\frac{\frac{1}{2}+1}{2}\right)\Gamma\left(\frac{0+1}{2}\right)}{2\,\Gamma\left(\frac{3}{4}+\frac{1}{2}\right)} \times \frac{\Gamma\left(\frac{-\frac{1}{2}+1}{2}\right)\Gamma\left(\frac{0+1}{2}\right)}{2\,\Gamma\left(\frac{1}{4}+\frac{1}{2}\right)} = \frac{\Gamma\left(\frac{3}{4}\right)\Gamma\left(\frac{1}{2}\right)}{2\,\Gamma\left(\frac{5}{4}\right)} \times \frac{\Gamma\left(\frac{1}{4}\right)\Gamma\left(\frac{1}{2}\right)}{2\,\Gamma\left(\frac{3}{4}\right)}$$

$$= \frac{\Gamma\left(\frac{1}{2}\right)\Gamma\left(\frac{1}{2}\right)\Gamma\left(\frac{1}{4}\right)}{4\,\Gamma\left(\frac{5}{4}\right)} = \frac{(\sqrt{\pi})(\sqrt{\pi})\,\Gamma\left(\frac{1}{4}\right)}{4\cdot\frac{1}{4}\Gamma\left(\frac{1}{4}\right)} = \pi = \text{R.H.S.}$$

Proved.

Example 16. *Prove that* $\displaystyle\int_0^1 \frac{x^2}{\sqrt{(1-x^4)}}\,dx \times \int_0^1 \frac{dx}{\sqrt{(1+x^4)}} = \frac{\pi}{4\sqrt{2}}$ *(AMIETE, Dec. 2009)*

Solution. Here, we have $\displaystyle\int_0^1 \frac{x^2}{\sqrt{1-x^4}}\,dx \times \int_0^1 \frac{dx}{\sqrt{1+x^4}}$ ... (1)

Let $\displaystyle I_1 = \int_0^1 \frac{x^2}{\sqrt{1-x^4}}\,dx$

$$\left[\begin{array}{l}\text{Put } x^2 = \sin\theta \Rightarrow 2x\,dx = \cos\theta\,d\theta \\ x = \sqrt{\sin\theta} \Rightarrow dx = \frac{1}{2}(\sin\theta)^{-\frac{1}{2}}\cos\theta\,d\theta\end{array}\right]$$

$$= \int_0^{\frac{\pi}{2}} \frac{\sin\theta}{\sqrt{1-\sin^2\theta}}\,\frac{1}{2}(\sin\theta)^{-\frac{1}{2}}\cos\theta\,d\theta.$$

$$= \frac{1}{2}\int_0^{\frac{\pi}{2}} \frac{(\sin\theta)^{1-\frac{1}{2}}}{\cos\theta}.\cos\theta\,d\theta = \frac{1}{2}\int_0^{\frac{\pi}{2}} (\sin\theta)^{\frac{1}{2}}(\cos\theta)^0\,d\theta$$

$$= \frac{1}{2}\left(\frac{\Gamma\left(\frac{\frac{1}{2}+1}{2}\right)\Gamma\left(\frac{0+1}{2}\right)}{2\,\Gamma\left(\frac{3}{4}+\frac{1}{2}\right)}\right) = \frac{1}{4}\,\frac{\Gamma\left(\frac{3}{4}\right)\Gamma\left(\frac{1}{2}\right)}{\Gamma\left(\frac{5}{4}\right)} \qquad \text{... (2)}$$

Let $\quad I_2 = \int_0^1 \frac{dx}{\sqrt{1+x^4}}$ $\qquad \left[\begin{array}{l} \text{Put } x^2 = \tan\theta \Rightarrow \quad x = \sqrt{\tan\theta} \\ \Rightarrow dx = \frac{1}{2}(\tan\theta)^{-\frac{1}{2}} \sec^2\theta\, d\theta \end{array}\right]$

$$= \frac{1}{2}\int_0^{\frac{\pi}{4}} \frac{(\tan\theta)^{-\frac{1}{2}} \sec^2\theta}{\sqrt{1+\tan^2\theta}}\, d\theta = \frac{1}{2}\int_0^{\frac{\pi}{4}} (\tan\theta)^{-\frac{1}{2}} \sec\theta\, d\theta$$

$$= \frac{1}{2}\int_0^{\frac{\pi}{4}} (\sin\theta)^{-\frac{1}{2}} . (\cos\theta)^{-\frac{1}{2}}\, d\theta = \frac{1}{2}\int_0^{\frac{\pi}{4}} \sqrt{\frac{2}{2\sin\theta\cos\theta}}\, d\theta$$

$$= \frac{1}{\sqrt{2}}\int_0^{\frac{\pi}{2}} \frac{d\theta}{\sqrt{\sin 2\theta}} \qquad \text{Put } 2\theta = t \quad \Rightarrow \quad d\theta = \frac{dt}{2}$$

$$= \frac{1}{2\sqrt{2}}\int_0^{\frac{\pi}{2}} \frac{dt}{\sqrt{\sin t}} = \frac{1}{2\sqrt{2}}\int_0^{\frac{\pi}{2}} (\sin t)^{-\frac{1}{2}} (\cos\theta)^{\circ}\, dt$$

$$= \frac{1}{2\sqrt{2}} \left(\frac{\Gamma\left(\frac{1-\frac{1}{2}}{2}\right)\Gamma\left(\frac{0+1}{2}\right)}{2\,\Gamma\left(\frac{1}{4}+\frac{1}{2}\right)} \right) = \frac{1}{4\sqrt{2}} \frac{\Gamma\left(\frac{1}{4}\right)\Gamma\left(\frac{1}{2}\right)}{\Gamma\left(\frac{3}{4}\right)} \qquad \text{... (3)}$$

Putting the value in (1) from equation (1) and (2), we get

$$\int_0^1 \frac{x^2 dx}{\sqrt{1-x^4}} \times \int_0^1 \frac{dx}{\sqrt{1-x^4}} = \frac{1}{4}\frac{\Gamma\left(\frac{3}{4}\right)\Gamma\left(\frac{1}{2}\right)}{\Gamma\left(\frac{5}{4}\right)} \times \frac{1}{4\sqrt{2}}\frac{\Gamma\left(\frac{1}{4}\right)\Gamma\left(\frac{1}{2}\right)}{\Gamma\left(\frac{3}{4}\right)} = \frac{1}{16\sqrt{2}}\frac{\Gamma\left(\frac{1}{2}\right)\Gamma\left(\frac{1}{2}\right)\Gamma\left(\frac{1}{4}\right)}{\Gamma\left(\frac{5}{4}\right)} = \frac{1}{16\sqrt{2}}\frac{\pi\,\Gamma\left(\frac{1}{4}\right)}{\frac{1}{4}\Gamma\left(\frac{1}{4}\right)} = \frac{\pi}{4\sqrt{2}}$$

Proved.

Example 17. *Prove that* $\int_0^1 \frac{dx}{\sqrt{1-x^4}} = \frac{\sqrt{\pi}}{4}\frac{\Gamma\left(\frac{1}{4}\right)}{\Gamma\left(\frac{3}{4}\right)}$ *(AMIETE, June 2010)*

Solution. Here, we have $\int_0^1 \frac{dx}{\sqrt{1-x^4}}$, $\qquad$ Put $x^2 = \sin\theta \quad \Rightarrow \quad x = \sqrt{\sin\theta}$

$$\Rightarrow \quad dx = \frac{1}{2}(\sin\theta)^{-\frac{1}{2}} \cos\theta . d\theta$$

$$= \frac{1}{2}\int_0^{\frac{\pi}{2}} \frac{(\sin\theta)^{-\frac{1}{2}}\cos\theta}{\sqrt{1-\sin^2\theta}} = \frac{1}{2}\int_0^{\frac{\pi}{2}} \frac{(\sin\theta)^{-\frac{1}{2}}\cos\theta}{\cos\theta}\, d\theta$$

$$= \frac{1}{2}\int_0^{\frac{\pi}{2}} (\sin\theta)^{-\frac{1}{2}} (\cos\theta)^0\, d\theta = \frac{1}{2}\left(\frac{\Gamma\left(\frac{-\frac{1}{2}+1}{2}\right)\Gamma\left(\frac{0+1}{2}\right)}{2\,\Gamma\left(\frac{1}{4}+\frac{1}{2}\right)} \right) = \frac{1}{4}\frac{\Gamma\left(\frac{1}{4}\right)\Gamma\left(\frac{1}{2}\right)}{\Gamma\left(\frac{3}{4}\right)} = \frac{\sqrt{\pi}}{4}\frac{\Gamma\left(\frac{1}{4}\right)}{\Gamma\left(\frac{3}{4}\right)} \qquad \textbf{Proved.}$$

Example 18. *Find the value of* $\Gamma\left(\frac{1}{2}\right)$.

Solution. We have already solved this problem in Art. 9.3 (*ii*) Transformation of the Gamma Function.

Now, by **Second method:** We know that,

$$\int_0^{\frac{\pi}{2}} \sin^p\theta \cdot \cos^q\theta\, d\theta = \frac{\left\lceil\frac{p+1}{2}\right.\left\lceil\frac{q+1}{2}\right.}{2\left\lceil\frac{p+q+2}{2}\right.}$$

Putting $p = q = 0$,we get $\int_0^{\frac{\pi}{2}} d\theta = \frac{\left\lceil\frac{1}{2}\right.\left\lceil\frac{1}{2}\right.}{2\lceil 1}$ $\Rightarrow$ $[\theta]_0^{\pi/2} = \frac{1}{2}\left(\left\lceil\frac{1}{2}\right.\right)^2$ $\Rightarrow$ $\frac{\pi}{2} = \frac{1}{2}\left(\left\lceil\frac{1}{2}\right.\right)^2$

$\Rightarrow$ $\left(\left\lceil\frac{1}{2}\right.\right)^2 = \pi$ $\Rightarrow$ $\left\lceil\frac{1}{2}\right. = \sqrt{\pi}$ **Ans.**

Example 19. *Show that*

$$\int_0^{\frac{\pi}{2}} \sqrt{\cot\theta}\, d\theta = \frac{1}{2}\left\lceil\frac{1}{4}\right.\left\lceil\frac{3}{4}\right.$$

Solution. We know that

$$\int_0^{\frac{\pi}{2}} \sin^p\theta \cdot \cos^q\theta\, d\theta = \frac{\left\lceil\frac{p+1}{2}\right.\left\lceil\frac{q+1}{2}\right.}{2\left\lceil\frac{p+q+2}{2}\right.} \quad ...(1)$$

$$\int_0^{\frac{\pi}{2}} \sqrt{\cot\theta}\, d\theta = \int_0^{\frac{\pi}{2}} \frac{\cos^{1/2}\theta}{\sin^{1/2}\theta}\, d\theta = \int_0^{\frac{\pi}{2}} \sin^{-1/2}\theta \cos^{1/2}\theta\, d\theta$$

On applying formula (1), we have

$$\int_0^{\frac{\pi}{2}} \sqrt{\cot\theta}\, d\theta = \frac{\left\lceil\frac{-\frac{1}{2}+1}{2}\right.\left\lceil\frac{\frac{1}{2}+1}{2}\right.}{2\left\lceil\frac{-\frac{1}{2}+\frac{1}{2}+2}{2}\right.} = \frac{\left\lceil\frac{1}{4}\right.\left\lceil\frac{3}{4}\right.}{2\lceil 1} = \frac{1}{2}\left\lceil\frac{1}{4}\right.\left\lceil\frac{3}{4}\right.$$ **Proved.**

Example 20. *Using Beta and Gamma functions, evaluate*

$$\int_0^1 \left(\frac{x^3}{1-x^3}\right)^{\frac{1}{2}} dx$$

Solution.
$$\int_0^1 \left(\frac{x^3}{1-x^3}\right)^{\frac{1}{2}} dx \quad ...(1)$$

Putting $x^3 = \sin^2\theta$, so that $x = \sin^{\frac{2}{3}}\theta$, $dx = \frac{2}{3}\sin^{-\frac{1}{3}}\theta\cos\theta\, d\theta$ in (1), we get

$$\int_0^1 \left(\frac{x^3}{1-x^3}\right)^{\frac{1}{2}} dx = \int_0^{\pi/2} \left(\frac{\sin^2\theta}{1-\sin^2\theta}\right)^{\frac{1}{2}} \frac{2}{3}\sin^{-\frac{1}{3}}\theta\cos\theta\, d\theta$$

$$= \frac{2}{3}\int_0^{\pi/2} \left(\frac{\sin\theta}{\cos\theta}\right) \sin^{-\frac{1}{3}}\theta\cos\theta\, d\theta = \frac{2}{3}\int_0^{\pi/2} \sin^{\frac{2}{3}}\theta\, d\theta$$

$$= \frac{2}{3} \frac{\left|\frac{\frac{2}{3}+1}{2}\right|\frac{0+1}{2}}{\left|\frac{\frac{2}{3}+1+1}{2}\right.} = \frac{2}{3}\frac{\left|\frac{5}{6}\right|\frac{1}{2}}{\left|\frac{4}{3}\right.} = \frac{2}{3}\frac{\sqrt{\pi}\left|\frac{5}{6}\right.}{\left|\frac{4}{3}\right.}$$ **Ans.**

Example 21. *Evaluate* $\int_{-1}^{+1} (1+x)^{p-1}(1-x)^{q-1}\, dx$.

Solution. Put $x = \cos 2\theta$, then $dx = -2 \sin 2\theta \, d\theta$

$$\int_{-1}^{+1} (1+x)^{p-1}(1-x)^{q-1} dx = \int_{\frac{\pi}{2}}^{0} (1+\cos 2\theta)^{p-1}(1-\cos 2\theta)^{q-1}(-2\sin 2\theta\, d\theta)$$

$$= \int_{\frac{\pi}{2}}^{0} \left(1+2\cos^2\theta - 1\right)^{p-1}\left(1-1+2\sin^2\theta\right)^{q-1}(-4\sin\theta\cos\theta\, d\theta)$$

$$= 4\int_0^{\frac{\pi}{2}} 2^{p-1}\cos^{2p-2}\theta \,.\, 2^{q-1}\sin^{2q-2}\theta \,.\, \sin\theta\cos\theta\, d\theta = 2^{p+q}\int_0^{\infty} \sin^{2q-1}\theta\cos^{2p-1}\theta\, d\theta$$

$$= 2^{p+q}\frac{\left|\frac{2q}{2}\right|\frac{2p}{2}}{2\left|\frac{2p+2q}{2}\right.} = 2^{p+q-1}\frac{\left|p\right|q}{\left|p+q\right.}$$ **Ans.**

Example 22. *Show that* $\left|n\right|1-n = \frac{\pi}{\sin n\pi}$ *(0 < n < 1) (Delhi University, 2010)*

Solution. We know that

$$\beta(m, n) = \int_0^{\infty} \frac{x^{n-1}}{(1+x)^{m+n}} dx$$ [From (1), Art 9.7]

$$\frac{\left|m\right|n}{\left|m+n\right.} = \int_0^{\infty} \frac{x^{n-1}}{(1+x)^{m+n}} dx$$

Putting $m + n = 1$ or $m = 1 - n$, we get

$$\frac{\left|1-n\right|n}{\left|1\right.} = \int_0^{\infty} \frac{x^{n-1}}{(1+x)^1} dx$$

$$\left|1-n\right|n = \int_0^{\infty} \frac{x^{n-1}}{1+x} dx \qquad \left[\int_0^{\infty} \frac{x^{n-1}}{1+x} dx = \frac{\pi}{\sin n\pi}\right]$$

$$\Rightarrow \quad \left|n\right|1-n = \frac{\pi}{\sin n\pi}$$ **Proved.**

Example 23. *Assuming* $\left|n\right|1-n = \pi \operatorname{cosec} n\pi$. $0 < n < 1$,

show that $\int_0^{\infty} \frac{x^{p-1}}{1+x} dx = \left(\frac{\pi}{\sin p\pi}\right)$. $0 < p < 1$ *(U.P., I Semester, Dec 2009)*

Solution: Here, we have

$$\Rightarrow \quad \frac{\pi}{\sin n\pi} = \left|n\right|1-n \qquad ...(1)$$

We know that $\beta(m, n) = \int_0^{\infty} \frac{x^{n-1}}{(1+x)^{m+n}} dx$

$$\frac{\overline{|m}\,\overline{|n}}{\overline{|m+n}} = \int_0^\infty \frac{x^{n-1}}{(1+x)^{m+n}}\,dx$$

Putting $m + n = 1$, $m = 1 - n$

$$\frac{\overline{|1-n}\,\overline{|n}}{\overline{|1}} = \int_0^\infty \frac{x^{n-1}}{(1+x)^1}\,dx$$

$$\overline{|1-n}\,\overline{|n} = \int_0^\infty \frac{x^{n-1}}{1+x}\,dx$$

$$\pi \operatorname{cosec} n\pi = \int_0^\infty \frac{x^{n-1}}{1+x}\,dx \qquad [p = n]$$

$$= \int_0^\infty \frac{x^{p-1}}{1+x}\,dx = \frac{\pi}{\sin p\pi}$$

Proved

Example 24. *Prove that* $\overline{\left|\left(\frac{1}{4}\right)\right.}\,\overline{\left|\left(\frac{3}{4}\right)\right.} = \pi\sqrt{2}$

Solution. Putting $n = \frac{1}{4}$ in result of example 22, we obtain

$$\overline{\left|\left(\frac{1}{4}\right)\right.}\,\overline{\left|\left(1-\frac{1}{4}\right)\right.} = \frac{\pi}{\sin\frac{\pi}{4}}$$

$$\Rightarrow \quad \overline{\left|\left(\frac{1}{4}\right)\right.}\,\overline{\left|\left(\frac{3}{4}\right)\right.} = \frac{\pi}{\left(\frac{1}{\sqrt{2}}\right)} \quad \Rightarrow \quad \overline{\left|\left(\frac{1}{4}\right)\right.}\,\overline{\left|\left(\frac{3}{4}\right)\right.} = \pi\sqrt{2}$$

Proved.

Example 25. *Evaluate* $\int_0^1 \frac{dx}{(1-x^n)^{\frac{1}{n}}}$

Solution. Let $x^n = \sin^2\theta$ or $x = \sin^{2/n}\theta$

So that
$$dx = \frac{2}{n}\sin^{2/n-1}\theta\cos\theta\,d\theta$$

$$\int_0^1 \frac{dx}{(1-x^n)^{\frac{1}{n}}} = \int_0^{\frac{\pi}{2}} \frac{\frac{2}{n}\sin^{2/n-1}\theta\cos\theta}{(1-\sin^2\theta)^{1/n}}\,d\theta = \frac{2}{n}\int_0^{\frac{\pi}{2}} \frac{\sin^{2/n-1}\theta\cos\theta}{(\cos^2\theta)^{1/n}}\,d\theta$$

$$= \frac{2}{n}\int_0^{\frac{\pi}{2}} \sin^{2/n-1}\theta\cos^{1-2/n}\theta\,d\theta$$

$$= \frac{2}{n}\,\frac{\overline{\left|\frac{\frac{2}{n}-1+1}{2}\right.}\,\overline{\left|\frac{1-\frac{2}{n}+1}{2}\right.}}{2\,\overline{\left|\frac{\frac{2}{n}-1+1+2-\frac{2}{n}}{2}\right.}} = \frac{1}{n}\,\frac{\overline{\left|\frac{1}{n}\right.}\,\overline{\left|\frac{n-1}{n}\right.}}{\overline{|1}} \qquad \left(\because \overline{\left|\frac{1}{n}\right.}\,\overline{\left|1-\frac{1}{n}\right.} = \frac{\pi}{\sin\frac{\pi}{n}}\right)$$

$$= \frac{\pi}{n\sin\frac{\pi}{n}}$$

Ans.

Example 26. *Show that* $\int_0^a \frac{dx}{\sqrt[n]{(a^n - x^n)}} = \frac{\pi}{n}\,\text{cosec}\left(\frac{\pi}{n}\right)$, *where* $n > 1$.*(M.U. II Semester 2009)*

Solution. Let $x^n = a^n \sin^2\theta \qquad \Rightarrow \qquad x = a\sin^{\frac{2}{n}}\theta$

So that $\qquad dx = \frac{2a}{n}\sin^{\frac{2}{n}-1}\theta\cos\theta\, d\theta$

$$\therefore \quad \int_0^a \frac{dx}{\sqrt[n]{a^n - x^n}} = \int_0^{\frac{\pi}{2}} \frac{a \times \frac{2}{n}\sin^{\frac{2}{n}-1}\theta\cos\theta}{(a^n - a^n\sin^2\theta)^{\frac{1}{n}}}\, d\theta = \frac{2}{n}\int_0^{\frac{\pi}{2}} \frac{\sin^{\frac{2}{n}-1}\theta\cos\theta}{\cos^{\frac{2}{n}}\theta}\, d\theta$$

$$= \frac{2}{n}\int_0^{\frac{\pi}{2}} \sin^{\frac{2}{n}-1}\theta\cos^{1-\frac{2}{n}}\theta\, d\theta = \frac{2}{n}\,\frac{\Gamma\left(\frac{\frac{2}{n}-1+1}{2}\right)\Gamma\left(\frac{1-\frac{2}{n}+1}{2}\right)}{2\,\Gamma\left(\frac{\frac{2}{n}-1+1+2-\frac{2}{n}}{2}\right)}$$

$$= \frac{1}{n}\,\frac{\Gamma\left(\frac{1}{n}\right)\Gamma\left(\frac{n-1}{n}\right)}{\Gamma(1)} \qquad \left[\Gamma\left(\frac{1}{n}\right)\Gamma\left(1-\frac{1}{n}\right) = \frac{\pi}{\sin\frac{\pi}{n}}\right]$$

$$= \frac{\pi}{n\sin\frac{\pi}{n}} = \frac{\pi}{n}\,\text{cosec}\left(\frac{\pi}{n}\right)$$ **Proved.**

Example 27. *Show that* $\int_0^{\frac{\pi}{2}} \tan^P\theta\, d\theta = \frac{\pi}{2}\sec\frac{P\pi}{2}$ *and indicate the restriction on the values of P.*

Solution. $\int_0^{\frac{\pi}{2}} \tan^P\theta\, d\theta = \int_0^{\frac{\pi}{2}} \sin^P\theta\cos^{-P}\theta\, d\theta$

$$= \frac{\Gamma\left(\frac{P+1}{2}\right)\Gamma\left(\frac{-P+1}{2}\right)}{2\,\Gamma\left(\frac{P+1-P+1}{2}\right)} \qquad \begin{bmatrix} 1-P>0 \\ 1>P \end{bmatrix}$$

$$= \frac{\Gamma\left(\frac{P+1}{2}\right)\Gamma\left(\frac{-P+1}{2}\right)}{2\,\Gamma(1)} \qquad \begin{bmatrix} 1+P>0 \\ P>-1 \end{bmatrix}$$

$$= \frac{1}{2}\Gamma\left(\frac{P+1}{2}\right)\Gamma\left(\frac{-P+1}{2}\right)$$

$$= \frac{1}{2}\,\frac{\pi}{\sin\frac{P+1}{2}\pi} = \frac{1}{2}\,\frac{\pi}{\cos\frac{P\pi}{2}} = \frac{\pi}{2}\sec\frac{P\pi}{2} \qquad (\because 1 > P > -1)$$ **Proved.**

9.10 DUPLICATION FORMULA

$$\sqrt{m}\,\sqrt{m+\frac{1}{2}} = \frac{\sqrt{\pi}}{2^{2m-1}}\sqrt{2m}, \text{ where } m \text{ is positive.}$$

Hence show that $\beta(m, m) = 2^{1-2m}\beta\left(m, \frac{1}{2}\right)$ (*DU, I Sem. 2012, AMIETE, Dec. 2010*)

Proof. We know that $\dfrac{\sqrt{\frac{p+1}{2}}\sqrt{\frac{q+1}{2}}}{2\sqrt{\frac{p+q+2}{2}}} = \int_0^{\frac{\pi}{2}} \sin^p\theta \cos^q\theta\, d\theta$

Putting $q = p$, we get $\dfrac{\sqrt{\frac{p+1}{2}}\sqrt{\frac{p+1}{2}}}{2\sqrt{p+1}}$

$$= \int_0^{\frac{\pi}{2}} \sin^p\theta \cos^p\theta\, d\theta = \int_0^{\frac{\pi}{2}} (\sin\theta \cos\theta)^p\, d\theta$$

$$= \int_0^{\frac{\pi}{2}} \frac{1}{2^p}(2\sin\theta\cos\theta)^p\, d\theta = \frac{1}{2^p}\int_0^{\frac{\pi}{2}} (\sin 2\theta)^p\, d\theta \qquad \left[\text{Putting } 2\theta = t \Rightarrow d\theta = \frac{dt}{2}\right]$$

$$= \frac{1}{2^p}\int_0^{\pi} \sin^p t\,\frac{dt}{2} = \frac{1}{2^p}\cdot\frac{1}{2}\cdot 2\int_0^{\frac{\pi}{2}} \sin^p t\, dt = \frac{1}{2^p}\int_0^{\frac{\pi}{2}} \sin^p t \cos^0 t\, dt$$

$$= \frac{1}{2^p}\frac{\sqrt{\frac{p+1}{2}}\sqrt{\frac{0+1}{2}}}{2\sqrt{\frac{p+2}{2}}}$$

$$= \frac{1}{2^p}\frac{\sqrt{\frac{p+1}{2}}\sqrt{\frac{1}{2}}}{2\sqrt{\frac{p+2}{2}}}$$

$$\Rightarrow \frac{\sqrt{\frac{p+1}{2}}}{\sqrt{p+1}} = \frac{1}{2^p}\frac{\sqrt{\frac{1}{2}}}{\sqrt{\frac{p+2}{2}}} \qquad \Rightarrow \qquad \frac{\sqrt{\frac{p+1}{2}}}{\sqrt{p+1}} = \frac{1}{2^p}\frac{\sqrt{\pi}}{\sqrt{\frac{p+2}{2}}}$$

Take $\dfrac{p+1}{2} = m \quad \Rightarrow \quad p = 2m - 1$

$$\Rightarrow \qquad \frac{\sqrt{m}}{\sqrt{2m}} = \frac{1}{2^{2m-1}}\frac{\sqrt{\pi}}{\sqrt{\frac{2m+1}{2}}} \qquad ...(1)$$

$$\sqrt{m}\,\sqrt{m+\frac{1}{2}} = \frac{\sqrt{\pi}}{2^{2m-1}}\sqrt{2m}$$ **Proved.**

Multiplying both sides of (1) by $\sqrt{m}$, we have

$$\frac{\Gamma m\,\Gamma m}{\Gamma 2m} = 2^{1-2m}\,\frac{\Gamma\frac{1}{2}\,\Gamma m}{\Gamma\left(m+\frac{1}{2}\right)}$$

$$\boxed{\beta(m, m) = 2^{1-2m}\,\beta\left(m, \frac{1}{2}\right)}$$ **Proved.**

Example 28. *For α β function, show that*

$\beta(p, q) = \beta(p + 1, q) + \beta(p, q + 1)$ *(U.P., Ist Semester, Dec 2008)*

Solution. $\beta(p + 1, q) + \beta(p, q + 1) = \int_0^1 x^p (1-x)^{q-1}\,dx + \int_0^1 x^{p-1}(1-x)^q\,dx$

$= \int_0^1 x^{p-1}(1-x)^{q-1}\,[x + 1 - x]\,dx$ (Taking common)

$= \int_0^1 x^{p-1}(1-x)^{q-1}\,dx = \beta(p, q)$ **Proved.**

Example 29. *Prove that* $\beta(m, m) \times \beta\left(m+\frac{1}{2}, m+\frac{1}{2}\right) = \frac{\pi}{m}2^{1-4m}$ *(M.U. II, Semester, 2008)*

Solution. L.H.S. $= \beta(m, m) \times \beta\left(m+\frac{1}{2}, m+\frac{1}{2}\right) = \frac{\Gamma m\,\Gamma m}{\Gamma 2m} \times \frac{\Gamma\left(m+\frac{1}{2}\right)\Gamma\left(m+\frac{1}{2}\right)}{\Gamma(2m+1)}$

$= \frac{(\Gamma m)^2}{\Gamma 2m} \times \frac{\left(\Gamma\left(m+\frac{1}{2}\right)\right)^2}{2m\,\Gamma 2m}$ $[\because \Gamma(2m+1) = 2m\,\Gamma 2m]$

$= \left(\frac{\Gamma m\,\Gamma\left(m+\frac{1}{2}\right)}{\Gamma 2m}\right)^2 \cdot \frac{1}{2m} = \left(\frac{\sqrt{\pi}}{2^{2m-1}}\right)^2 \frac{1}{2m}$ $\left[\text{By Duplication formula} \quad 2^{2m-1}\cdot\Gamma m\cdot\Gamma\left(m+\frac{1}{2}\right) = \sqrt{\pi}\cdot\sqrt{2m}\right]$

$= \frac{\pi}{2^{4m-2}}\cdot\frac{1}{2m}\;\frac{\pi}{(2^{4m-1})m}$ $\qquad \frac{\Gamma m\,\Gamma\left(m+\frac{1}{2}\right)}{\Gamma 2m} = \frac{\sqrt{\pi}}{2^{2m-1}}$

$= \frac{\pi}{m}\cdot 2^{1-4m}$

= R.H.S. **Proved.**

9.11 TO SHOW THAT

$$\Gamma\left(\frac{1}{n}\right)\Gamma\left(\frac{2}{n}\right)\Gamma\left(\frac{3}{n}\right)\cdots\cdots\Gamma\left(\frac{n-1}{n}\right) = \frac{(2\pi)^{\left(\frac{n-1}{2}\right)}}{n^{1/2}}$$

where n is a positive integer than one.

Proof. Let $P = \Gamma\left(\frac{1}{n}\right)\Gamma\left(\frac{2}{n}\right)\Gamma\left(\frac{3}{n}\right)\cdots\cdots\Gamma\left(\frac{n-2}{n}\right)\Gamma\left(\frac{n-1}{n}\right)$

$= \Gamma\left(\frac{1}{n}\right)\Gamma\left(\frac{2}{n}\right)\Gamma\left(\frac{3}{n}\right)\cdots\cdots\Gamma\left(1-\frac{2}{n}\right)\Gamma\left(1-\frac{1}{n}\right)$ ...(1)

Writing the value of P in the reverse order, we have

$$P = \left|\overline{\left(1-\frac{1}{n}\right)}\right. \left|\overline{\left(1-\frac{2}{n}\right)}\right. \ldots\ldots \left|\overline{\left(\frac{3}{n}\right)}\right. \left|\overline{\left(\frac{2}{n}\right)}\right. \left|\overline{\left(\frac{1}{n}\right)}\right. \quad \ldots(2)$$

Multiplying (1) and (2), we get

$$P^2 = \left(\left|\overline{\left(\frac{1}{n}\right)}\right. \left|\overline{\left(1-\frac{1}{n}\right)}\right.\right) \left(\left|\overline{\left(\frac{2}{n}\right)}\right. \left|\overline{\left(1-\frac{2}{n}\right)}\right.\right) \ldots\ldots \left(\left|\overline{\left(1-\frac{2}{n}\right)}\right. \left|\overline{\left(\frac{2}{n}\right)}\right.\right) \left(\left|\overline{\left(1-\frac{1}{n}\right)}\right. \left|\overline{\left(\frac{1}{n}\right)}\right.\right)$$

$$P^2 = \frac{\pi}{\sin\left(\frac{\pi}{n}\right)} \cdot \frac{\pi}{\sin\left(\frac{2\pi}{n}\right)} \cdot \frac{\pi}{\sin\left(\frac{3\pi}{n}\right)} \ldots\ldots \frac{\pi}{\sin\frac{(n-1)\pi}{n}} \qquad \left[\because \Gamma(n)\,\Gamma(1-n) = \frac{\pi}{\sin n\pi}\right]$$

$$\Rightarrow \quad P^2 = \frac{\pi^{n-1}}{\sin\left(\frac{\pi}{n}\right) \sin\left(\frac{2\pi}{n}\right) \sin\left(\frac{3\pi}{n}\right) \ldots\ldots \sin\left\{\frac{(n-1)\pi}{n}\right\}} \quad \ldots(3)$$

But from Trigonometry, we know that

$$\frac{\sin n\,\theta}{\sin\theta} = 2^{n-1} \sin\left(\theta + \frac{\pi}{n}\right) \sin\left(\theta + \frac{2\pi}{n}\right) \ldots\ldots\ldots \sin\left\{\theta + \frac{(n-1)\pi}{n}\right\} \quad \ldots(4)$$

Take Limit as $\theta \to 0$,

$$\underset{\theta\to 0}{\text{Lt}} \frac{\sin n\theta}{\sin\theta} = n = \underset{\theta\to 0}{\text{Lt}} \left(n \cdot \frac{\sin n\theta}{n\,\theta} \cdot \frac{\theta}{\sin\theta}\right)$$

On putting this limit in (4), we get

$$n = 2^{n-1} \sin\frac{\pi}{n} \sin\frac{2\pi}{n} \sin\frac{3\pi}{n} \ldots \sin\frac{(n-1)\pi}{n} \qquad \text{[From (4)]}$$

$$\Rightarrow \qquad \sin\frac{\pi}{n} \cdot \sin\frac{2\pi}{n} \cdot \sin\frac{3\pi}{n} \ldots \sin\frac{(n-1)\pi}{n} = \frac{n}{2^{n-1}}$$

Substituting this in equation (3), we obtain

$$P^2 = \frac{\pi^{n-1}}{\left(\frac{n}{2^{n-1}}\right)} = \frac{(2\pi)^{n-1}}{n} \qquad \therefore \quad P = \frac{(2\pi)^{(n-1)/2}}{n^{1/2}}$$

$$\Rightarrow \qquad \boxed{\left|\overline{\left(\frac{1}{n}\right)}\right. \left|\overline{\left(\frac{2}{n}\right)}\right. \left|\overline{\left(\frac{3}{n}\right)}\right. \ldots\ldots \left|\overline{\left(\frac{n-1}{n}\right)}\right. = \frac{(2\pi)^{(n-1)/2}}{n^{1/2}}}$$

9.12 TO SHOW THAT

(*i*) $\int_0^\infty e^{-ax}\, x^{n-1} \cos bx\, dx = \dfrac{\Gamma(n) \cos n\,\theta}{(a^2+b^2)^{n/2}}$

(*ii*) $\int_0^\infty e^{-ax}\, x^{n-1} \sin bx\, dx = \dfrac{\Gamma(n) \sin n\,\theta}{(a^2+b^2)^{n/2}}$ where $\theta = \tan^{-1}\left(\dfrac{b}{a}\right)$

Proof. We know that $\int_0^\infty e^{-ax} . x^{n-1}\, dx = \dfrac{\Gamma(n)}{a^n}$, where a, n are positive.

Put $ax = z$ so that $\qquad dx = \dfrac{dz}{a}$

$$\therefore \qquad \int_0^\infty e^{-ax}\, x^{n-1}\, dx = \int_0^\infty e^{-z} \left(\frac{z}{a}\right)^{n-1} . \frac{dz}{a} = \int_0^\infty e^{-z} \left(\frac{z}{a}\right)^{n-1} . \frac{dz}{a}$$

Replacing a by $(a + ib)$, we have

$$\int_0^\infty e^{-(a+ib)x}\, x^{n-1}\, dx = \frac{\Gamma n}{(a+ib)^n} \quad \ldots(1)$$

Now $$e^{-(a+ib)x} = e^{-ax} \cdot e^{-ibx} = e^{-ax}(\cos bx - i \sin bx)$$

Putting the value of $e^{-(a+ib)x}$ in (1), we get

$$\int_0^\infty e^{-ax}(\cos bx - i\sin bx)\, x^{n-1}\, dx = \frac{\Gamma n}{(a+ib)^n} \quad ...(2)$$

Putting $a = r\cos\theta$ and $b = r\sin\theta$ so that $r^2 = a^2 + b^2$ and $\theta = \tan^{-1}\frac{b}{a}$

$$(a+ib)^n = (r\cos\theta + ir\sin\theta)^n = r^n(\cos\theta + i\sin\theta)^n$$

$$(a+ib)^n = r^n(\cos n\theta + i\sin n\theta) \qquad \text{[De Moivre's Theorm]}$$

Putting the value of $(a+ib)^n$ in (2), we have

$$\begin{aligned}\int_0^\infty e^{-ax}(\cos bx - i\sin bx)\, x^{n-1}\, dx &= \frac{\Gamma n}{r^n(\cos n\theta + i\sin n\theta)}\\ &= \frac{\Gamma n}{r^n}(\cos n\theta + i\sin n\theta)^{-1}\\ &= \frac{\Gamma n}{r^n}(\cos n\theta - i\sin n\theta)\end{aligned}$$

Now, equating real and imaginary parts on the two sides, we get

(*i*) $$\int_0^\infty e^{-ax}\, x^{n-1}\cos bx\, dx = \frac{\Gamma n}{r^n}\cos n\theta$$

(*ii*) and $$\int_0^\infty e^{-ax}\, x^{n-1}\sin bx\, dx = \frac{\Gamma n}{r^n}\sin n\theta$$

where $r^2 = a^2 + b^2$ and $\theta = \tan^{-1}\frac{b}{a}$. **Proved.**

Example 30. *Evaluate:*

(*i*) $\int_0^\infty \cos x^2\, dx$ (*ii*) $\int_{-\infty}^\infty \cos\frac{\pi x^2}{2}\, dx.$

Solution. (*i*) We know that

$$\int_0^\infty e^{-ax}.x^{n-1}\cos bx\, dx = \frac{\Gamma n \cos n\theta}{(a^2+b^2)^{n/2}}, \text{ where } \theta = \tan^{-1}\left(\frac{b}{a}\right)$$

Put $a = 0$, $$\int_0^\infty x^{n-1}\cos bx\, dx = \frac{\Gamma n}{b^n}\cos\frac{n\pi}{2} \qquad \left[\theta = \tan^{-1}\left(\frac{b}{0}\right) = \tan^{-1}(\infty) = \frac{\pi}{2}\right]$$

Put $x^n = z$ so that $x^{n-1}\, dx = \frac{dz}{n}$ and $x = z^{1/n}$

then, $$\int_0^\infty \cos bz^{1/n}\, dz = \frac{n\,\Gamma n}{b^n}\cos\frac{n\pi}{2}$$

$\Rightarrow$ $$\int_0^\infty \cos(bx^{1/n})\, dx = \frac{\Gamma(n+1)}{b^n}\cos\frac{n\pi}{2} \quad ...(1)$$

Here $b = 1$, $n = \frac{1}{2}$

$\therefore$ $$\int_0^\infty \cos x^2\, dx = \Gamma\left(\frac{3}{2}\right)\cos\frac{\pi}{4} = \frac{\sqrt{\pi}}{2}\cdot\frac{1}{\sqrt{2}} = \frac{\sqrt{\pi}}{2\sqrt{2}} \qquad \textbf{Ans.}$$

(*ii*) $$I = \int_{-\infty}^\infty \cos\frac{\pi x^2}{2}\, dx = 2\int_0^\infty \cos\frac{\pi x^2}{2}\, dx \qquad [f(-x) = f(x)] \quad ...(2)$$

Putting $b = \frac{\pi}{2}$ and $n = \frac{1}{2}$ in equation (1), we get

$$\int_0^\infty \cos\left(\frac{\pi}{2}x^2\right) dx = \frac{\Gamma\left(\frac{3}{2}\right)}{\left(\frac{\pi}{2}\right)^{1/2}} \cos\frac{\pi}{4}$$

$\therefore$ From (2), $$I = 2\frac{\Gamma\left(\frac{3}{2}\right)}{\left(\frac{\pi}{2}\right)^{1/2}} \cos\frac{\pi}{4} = 2.\frac{1}{2}\sqrt{\pi}.\frac{\sqrt{2}}{\sqrt{\pi}}.\frac{1}{\sqrt{2}} = 1.$$ **Ans.**

Example 31. *Evaluate:* $\int_0^1 \log\Gamma(x)\,dx$

Solution. Let $$I = \int_0^1 \log\Gamma(x)\,dx \quad ...(1)$$

$$= \int_0^1 \log\Gamma(1-x)\,dx \quad ....(2) \quad \left[\int_0^a f(x)\,dx = \int_0^a f(a-x)\,dx\right]$$

Adding (1) and (2), we get

$$2I = \int_0^1 (\log\Gamma(x) + \log\Gamma(1-x))\,dx$$

$$= \int_0^1 \log(\Gamma(x)\,\Gamma(1-x))\,dx = \int_0^1 \log\left(\frac{\pi}{\sin\pi x}\right) dx \quad \left[\overline{|x}\,\overline{|1-x} = \frac{\pi}{\sin\pi x}\right]$$

$$= \int_0^1 (\log\pi - \log\sin\pi x)\,dx = \int_0^1 \log\pi\,dx - \int_0^1 \log\sin\pi x\,dx$$

$$= I_1 - I_2 \quad ...(3)$$

where $$I_1 = \int_0^1 \log\pi\,dx = \log\pi$$

$$I_2 = \int_0^1 \log\sin\pi x\,dx \quad \left[\text{Put } \pi x = t \Rightarrow dx = \frac{1}{\pi}dt\right]$$

$$= \int_0^\pi \log\sin t\left(\frac{dt}{\pi}\right) = \frac{1}{\pi}\cdot 2\int_0^{\pi/2} \log\sin t\,dt = \frac{2}{\pi}\left(-\frac{\pi}{2}\log 2\right) = -\log 2$$

From (3), $$2I = \log\pi + \log 2 = \log 2\pi$$

$$I = \frac{1}{2}\log 2\pi.$$ **Ans.**

Example 32. *Prove that* $\int_0^\infty x^2 e^{-x^4}\,dx \times \int_0^\infty \frac{e^{-x^2}}{\sqrt{x}}\,dx = \frac{\pi}{4\sqrt{2}}$ *(M.U., II Semester, 2008)*

Solution. Here, we have

$$\int_0^\infty x^2 e^{-x^4}\,dx \quad ...(1)$$

Putting $x^4 = t \Rightarrow x = t^{\frac{1}{4}}$, $dx = \frac{1}{4}t^{-\frac{3}{4}}\,dt$ in (1), we get

$$\int_0^\infty x^2 e^{-x4} dx = \int_0^\infty t^{\frac{1}{2}} e^{-t}\left(\frac{1}{4}t^{\frac{-3}{4}}\,dt\right) = \frac{1}{4}\int_0^\infty e^{-t} t^{-\frac{1}{4}}\,dt = \frac{1}{4}\int_0^\infty e^{-t}\,t^{\frac{3}{4}-1}\,dt = \frac{1}{4}\overline{\left|\frac{3}{4}\right.} \quad ...(2)$$

Now, we calculate the value of $\int_0^\infty \frac{e^{-x^2}}{\sqrt{x}}\,dx$ ...(3)

Putting $x^2 = t$, $x = t^{\frac{1}{2}} \Rightarrow dx = \frac{1}{2}t^{-\frac{1}{2}}\,dt$ in (2), we get

$$\int_0^\infty \frac{e^{-x2}}{\sqrt{x}}\,dx = \int_0^\infty e^{-t} t^{-\frac{1}{4}}\left(\frac{1}{2}t^{\frac{-1}{2}}\,dt\right) = \frac{1}{2}\int_0^\infty e^{-t} t^{-\frac{3}{4}}\,dt = \frac{1}{2}\int_0^\infty e^{-t}\,t^{\frac{1}{4}-1}\,dt = \frac{1}{2}\left|\frac{1}{4}\right. \qquad \text{...(4)}$$

On multiplying (2) and (4), we get

$$\int_0^\infty x^2 e^{-x^4}dx.\int_0^\infty \frac{e^{-x^2}}{\sqrt{x}}\,dx = \left(\frac{1}{4}\left|\frac{3}{4}\right.\right)\left(\frac{1}{2}\left|\frac{1}{4}\right.\right) = \frac{1}{8}\left|\frac{3}{4}\right.\left|\frac{1}{4}\right.$$

$$= \frac{1}{8}(\pi\sqrt{2}) = \frac{\pi}{4\sqrt{2}} \qquad \textbf{Proved.}$$

Diplication formula

$$2^{2m-1}\left|m\right.\left|m+\frac{1}{2}\right. = \sqrt{2}\left|2m\right.$$

Putting $m = \frac{1}{4}$, we get

$$2^{-\frac{1}{2}}\left|\frac{1}{4}\right.\left|\frac{3}{4}\right. = \sqrt{\pi}\left|\frac{1}{2}\right.$$

$$\left|\frac{1}{4}\right.\left|\frac{3}{4}\right. = \sqrt{\pi}\left|\pi\right.(2)^{\frac{1}{2}}$$

$$= \pi\sqrt{2}$$

Example 33. *Show that*

$$\int_0^1 \frac{x^{m-1}(1-x)^{n-1}}{(a+x)^{m+n}}\,dx = \frac{\beta(m,n)}{a^n(1+a)^m}$$

Solution. Put $\frac{x}{a+x} = \frac{t}{a+1}$

$$(a+1)\,x = t\,(a+x) \Rightarrow x = \frac{a\,t}{a+1-t}$$

$$dx = \frac{(a+1-t)\,a\,dt - a\,t\,(-dt)}{(a+1-t)^2} = \frac{(a^2+a-at+at)}{(a+1-t)^2}dt = \frac{a(a+1)}{(a+1-t)^2}dt$$

$$z\int_0^1 \frac{x^{m-1}(1-x)^{n-1}}{(a+x)^{m+n}}\,dx = \int_0^1 \frac{\left(\frac{at}{a+1-t}\right)^{m-1}\cdot\left(1-\frac{at}{a+1-t}\right)^{n-1}}{\left(a+\frac{at}{a+1-t}\right)^{m+n}}\cdot\frac{a(a+1)}{(a+1-t)^2}dt$$

$$= \int_0^1 \frac{(at)^{m-1}(a+1-t-at)^{n-1}}{(a^2+a-at+at)^{m+n}}a\,(a+1)\,dt$$

$$= \int_0^1 \frac{a^{m-1}t^{m-1}(a+1)^{n-1}(1-t)^{n-1}}{a^{m+n}(a+1)^{m+n}}a\,(a+1)\,dt$$

$$= \frac{1}{a^n(a+1)^m}\int_0^1 t^{m-1}(1-t)^{n-1}\,dt = \frac{1}{a^n(a+1)^m}\beta\,(m,n) \qquad \textbf{Proved.}$$

Example 34. *Prove that* $\int_0^\infty \frac{x^4(1+x^5)}{(1+x)^{15}}\,dx = \frac{1}{5005}$. *(M.U. II Semester, 2008)*

Solution. Let $I = \int_0^\infty \frac{x^4(1+x^5)}{(1+x)^{15}}dx$

$$\Rightarrow \qquad I = \int_0^\infty \frac{x^4}{(1+x)^{15}}\,dx + \int_0^\infty \frac{x^9}{(1+x)^{15}}\,dx = I_1 + I_2 \qquad \text{...(1)}$$

Now, put $\quad x = \dfrac{t}{1-t}$, when $\quad x = 0, \quad t = 0$; when $\quad x = \infty, \quad t = 1$

$$1 + x = 1 + \frac{t}{1-t} = \frac{1}{1-t} \quad \Rightarrow \quad dx = \frac{dt}{(1-t)^2}$$

$$\therefore \qquad I_1 = \int_0^1 \left(\frac{t}{1-t}\right)^4 .(1-t)^{15} . \frac{1}{(1-t)^2} dt = \int_0^1 t^4 (1+t)^9 \, dt = \beta\,(5, 10) \qquad ...(2)$$

and
$$I_2 = \int_0^1 \left(\frac{t}{1-t}\right)^9 .(1-t)^{15} . \frac{dt}{(1-t)^2} = \int_0^1 t^9 (1-t)^4 \, dt = \beta\,(10, 5) \qquad ...(3)$$

$$\therefore \qquad I = I_1 + I_2$$

$$= \beta\,(5, 10) + \beta\,(10, 5) \qquad \text{[Using (2) and (3)]}$$

$$= \beta\,(5, 10) + \beta\,(5, 10) \qquad [\because \beta(m, n) = \beta(n, m)]$$

$$= 2\,\beta\,(5, 10) = \frac{2\,\Gamma(5)\,\Gamma(10)}{\Gamma(15)} = \frac{2.4!.9!}{14!}$$

$$= \frac{2 \times 4 \times 3 \times 2 \times 1 \times 9!}{14 \times 13 \times 12 \times 11 \times 10 \times 9!} = \frac{1}{7 \times 13 \times 11 \times 5} = \frac{1}{5005} \qquad \textbf{Proved.}$$

Example 35. *Prove that*

$$\int_0^1 \frac{x^3 - 2x^4 + x^5}{(1+x)^7} dx = \frac{7}{60}. \qquad (\textit{M.U., II Semester, 2008})$$

Solution. Let
$$I = \int_0^1 \frac{x^3 - 2x^4 + x^5}{(1+x)^7} dx$$

$$\Rightarrow \qquad I = \int_0^1 \frac{x^3}{(1+x)^7} dx - 2\int_0^1 \frac{x^4}{(1+x)^7} dx + \int_0^1 \frac{x^5}{(1+x)^7} dx$$

$$\Rightarrow \qquad I = I_1 - 2\,I_2 + I_3 \qquad ...(1)$$

Now put $x = \dfrac{t}{1-t}$ so that $1 + x = 1 + \dfrac{t}{1-t} \Rightarrow 1 + x = \dfrac{1}{1-t} \Rightarrow dx = \dfrac{d\,t}{(1-t)^2}$

$$\therefore \qquad I_1 = \int_0^1 \left(\frac{t}{1-t}\right)^3 .(1-t)^7 . \frac{1}{(1-t)^2} d\,t$$

$$= \int_0^1 t^3 (1-t)^2 \, d\,t = \int_0^1 t^{4-1} (1-t)^{3-1} \, d\,t = \beta\,(4, 3) \qquad ...(2)$$

and
$$I_2 = \int_0^1 \left(\frac{t}{1-t}\right)^4 (1-t)^7 . \frac{1}{(1-t)^2} d\,t$$

$$= \int_0^1 t^4 (1-t) \, d\,t = \int_0^1 t^{5-1} (1-t)^{2-1} \, d\,t = \beta\,(5, 2) \qquad ...(3)$$

Also,
$$I_3 = \int_0^1 \left(\frac{t}{1-t}\right)^5 (1-t)^7 \frac{d\,t}{(1-t)^2} = \int_0^1 t^5 (1-t)^0 \, d\,t = \int_0^1 t^{6-1} (1-t)^{1-1} \, d\,t = \beta\,(6, 1) \;...(4)$$

Putting the values of I_1, I_2 and I_3 in (1), we get

$$I = \beta\,(4, 3) - 2\,\beta\,(5, 2) + \beta\,(6, 1) = \frac{\Gamma(4)\,\Gamma(3)}{\Gamma(4+3)} - 2\frac{\Gamma(5)\,\Gamma(2)}{\Gamma(5+2)} + \frac{\Gamma(6)\,\Gamma(1)}{\Gamma(6+1)}$$

$$= \frac{\overline{|4}\,\overline{|3}}{\overline{|7}} - 2\frac{\overline{|5}\,\overline{|2}}{\overline{|7}} + \frac{\overline{|6}\,\overline{|1}}{\overline{|7}} = \frac{\overline{|3}}{4\times5\times6} - 2\frac{1}{5\times6} + \frac{1}{6}$$

$$= \frac{1}{60} - \frac{1}{15} + \frac{1}{6} = \frac{1-4+10}{60} = \frac{7}{60}$$ **Ans.**

EXERCISE 9.2

Prove that

1. $\int_0^{\frac{\pi}{2}} \sin^2\theta \cos^4\theta \, d\theta = \frac{\pi}{32}$ **2.** $\int_0^{\frac{\pi}{2}} \sin^6\theta \, d\theta = \frac{5\pi}{32}$ *(Delhi, University, April 2010)*

3. $\int_0^{\frac{\pi}{2}} \sin^{m-1}(2\theta)\, d\theta = \frac{2^{m-1}}{m!}\left(\overline{\left|\frac{m}{2}\right.}\right)^2$

4. $\int_0^{\pi} \sin^5 x\,(1-\cos x)^3\, dx = \frac{32}{21}$

5. $\int_0^{2} x\,(8-x^3)^{\frac{1}{3}}\, dx = \frac{8}{9}\,\overline{\left|\frac{1}{3}\right.}\,\overline{\left|\frac{2}{3}\right.}$

6. $\int_0^{2\pi} x\sqrt{2ax-x^2}\, dx = \frac{\pi a^3}{2}$

7. $\beta(m+1, n) = \frac{m}{m+n}\beta(m, n)$ **8.** $\beta(m, n+1) = \frac{n}{m+n}\beta(m, n)$

9. $\int_0^1 \sqrt{x}\,\sqrt[3]{1-x^2}\, dx = \frac{\overline{\left|\frac{3}{4}\right.}\,\overline{\left|\frac{4}{3}\right.}}{2\,\overline{\left|\frac{5}{12}\right.}}$ **10.** $\int_0^1 (1-x^n)^{-\frac{1}{2}}\, dx = \frac{\overline{\left|\frac{1}{n}\right.}\,\overline{\left|\frac{1}{2}\right.}}{n\,\overline{\left|\frac{n+2}{2n}\right.}}$

11. $\int_0^1 (1-x^{1/n})^m\, dx = \frac{\overline{|m}\,\overline{|n}}{\overline{|m+n}}$ **12.** $\int_1^{\infty} \frac{dx}{x^{P+1}(x-1)^q} = \beta(P+q, 1-q)$ if $-P < q < 1$

13. $\int_0^1 x^m(1-x^n)^P\, dx = \frac{1}{n}\frac{\overline{\left|\frac{m+1}{2}\right.}\,\overline{|P+1}}{\overline{\left|\frac{m+1}{n}+P+1\right.}}$ **14.** $\int_a^b (x-a)^m(b-x)^n\, dx = (b-a)^{m+n+1}.\beta(m+1, n+1)$

15. $\int_a^b (x-a)^{m-1}(b-x)^{n-1}\, dx = (b-a)^{m+n-1}.\beta(m, n)$

16. $\int_3^7 \sqrt[4]{(x-3)(7-x)}\, dx = \frac{2\left(\overline{\left|\frac{1}{4}\right.}\right)^2}{3\sqrt{\pi}}$ **[Hint.** Put $x = 4t + 3$] **17.** $\int_0^{\frac{\pi}{2}} \frac{d\theta}{\sqrt{1-\frac{1}{2}\sin^2\theta}} = \frac{\left(\overline{\left|\frac{1}{4}\right.}\right)^2}{4\sqrt{\pi}}$

18. If $\int_0^{\infty} e^{-x}x^{n-1}\, dx = I_n$ for $n > 0$ find $\frac{I_{n+1}}{I_n}$ **Ans.** n

Show that :

19. $\int_0^{\infty} \sqrt{x}e^{-x^2}\, dx \times \int_0^{\infty} \frac{e^{-x^2}}{\sqrt{x}}\, dx = \frac{\pi}{2\sqrt{2}}$ **20.** $\int_0^1 \frac{x^2 dx}{\sqrt{1-x^4}} \int_0^1 \frac{dx}{\sqrt{1+x^4}} = \frac{\pi}{4\sqrt{2}}$

21. $\int_0^{\infty} x^2 e^{-x^4}\, dx \times \int_0^{\infty} e^{-x^4}\, dx = \frac{\pi}{8\sqrt{2}}$ **22.** $\int_0^{\pi/2} \sin^3 x \cos^{5/2} x\, dx = \frac{8}{77}$

23. $\frac{\beta(p, q+1)}{q} = \frac{\beta(p+1, q)}{p} = \frac{\beta(p, q)}{p+q}$

9.13 DOUBLE INTEGRATION

Example 36. *Evaluate* $\iint_A \frac{dx\,dy}{\sqrt{xy}}$ *using the substitutions*

$$x = \frac{u}{1+v^2}, \quad y = \frac{uv}{1+v^2}$$

where A is bounded by $x^2 + y^2 - x = 0,\ y = 0,\ y > 0.$

Solution. Here $\sqrt{xy} = \sqrt{\left(\frac{u}{1+v^2}\right)\left(\frac{uv}{1+v^2}\right)} = \frac{u\sqrt{v}}{1+v^2}$

$$dx\ dy = \left|\frac{\partial(x,y)}{\partial(u,v)}\right| du\,dv$$

$$= \begin{vmatrix} \frac{\partial x}{\partial u} & \frac{\partial x}{\partial v} \\ \frac{\partial y}{\partial u} & \frac{\partial y}{\partial v} \end{vmatrix} du\,dv = \begin{vmatrix} \frac{1}{1+v^2} & -\frac{2uv}{(1+v^2)^2} \\ \frac{v}{1+v^2} & \frac{u(1-v^2)}{(1+v^2)^2} \end{vmatrix} du\,dv$$

$$= \left[\frac{u(1-v^2)}{(1+v^2)^3} + \frac{2uv^2}{(1+v^2)^3}\right] du\,dv = \left[\frac{u - uv^2 + 2uv^2}{(1+v^2)^3}\right] du\,dv$$

$$= \frac{u(1+v^2)}{(1+v^2)^3} du\,dv = \frac{u}{(1+v^2)^2} du\,dv$$

Also the circle $x^2 + y^2 - x = 0$ is transformed into

$$\frac{u^2}{(1+v^2)^2} + \frac{u^2v^2}{(1+v^2)^2} - \frac{u}{1+v^2} = 0 \Rightarrow \frac{u^2(1+v^2)}{(1+v^2)^2} - \frac{u}{1+v^2} = 0$$

$$\Rightarrow \quad \frac{u^2}{1+v^2} - \frac{u}{1+v^2} = 0 \Rightarrow u^2 - u = 0 \Rightarrow u(u-1) = 0 \Rightarrow u = 0,\ u = 1$$

Further $\quad y = 0 \Rightarrow \frac{uv}{1+v^2} = 0 \Rightarrow u = 0, v = 0$

and $y > 0 \Rightarrow uv > 0$ either both u and v are positive or both negative.

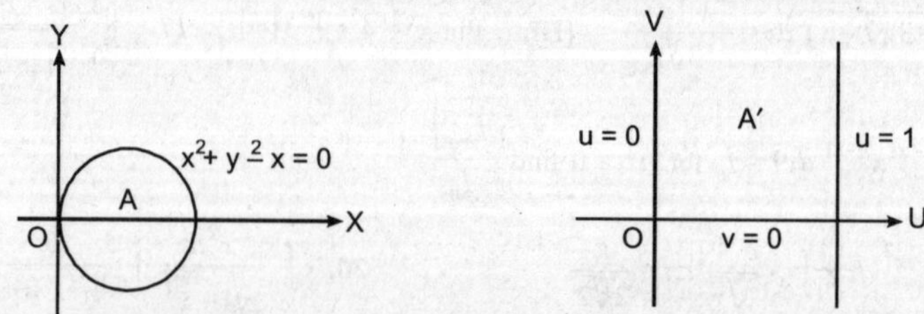

The area A, *i.e.*, $x^2 + y^2 - x = 0$ is transformed into A' bounded by $u = 0$, $v = 0$ and $u = 1$ and $v = \infty$.

$$\iint \frac{dx\,dy}{\sqrt{xy}} = \int_0^1\int_0^\infty \frac{\frac{u}{(1+v^2)^2}}{\frac{u\sqrt{v}}{1+v^2}} du\,dv = \int_0^1\int_0^\infty \frac{1}{\sqrt{v}(1+v^2)} dv\,du$$

On putting $v = \tan\theta$, $dv = \sec^2\theta\, d\theta$

$$= \int_0^1 \int_0^{\frac{\pi}{2}} \frac{\sec^2\theta}{\sqrt{\tan\theta}\,(1+\tan^2\theta)}\, d\theta\, du = \int_0^1 du \int_0^{\frac{\pi}{2}} \sqrt{\frac{\cos\theta}{\sin\theta}}\, d\theta = \int_0^1 du \int_0^{\frac{\pi}{2}} \sin^{-\frac{1}{2}}\theta \cos^{\frac{1}{2}}\theta\, d\theta$$

Duplication formula $\Gamma(m)\,\Gamma\!\left(m+\frac{1}{2}\right) = \frac{\sqrt{\pi}}{2^{2m-1}}\,\Gamma(2m)$

$$= \int_0^1 du \frac{\Gamma\!\left(\frac{-\frac{1}{2}+1}{2}\right)}{\Gamma(2)} \frac{\Gamma\!\left(\frac{\frac{1}{2}+1}{2}\right)}{\Gamma(2)} = \frac{1}{2}\int_0^1 du\, \Gamma\!\left(\frac{1}{4}\right)\Gamma\!\left(\frac{3}{4}\right) = \frac{1}{2}\int_0^1 du \left[\frac{\sqrt{\pi}}{2^{\frac{-1}{2}}}\,\Gamma\!\left(\frac{1}{2}\right)\right]$$

$$= \frac{1}{2}\int_0^1 du\, \sqrt{2}\,\sqrt{\pi}\,\square\sqrt{\pi} = \frac{\pi}{\sqrt{2}}[u]_0^1 = \frac{\pi}{\sqrt{2}}$$ **Ans.**

Example 37. *Prove that*

$$\iint_D x^{l-1}y^{m-1}\,dx\,dy = \frac{\Gamma(l)\,\Gamma(m)}{\Gamma(l+m+1)}\,h^{l+m}$$

where D is the domain $x \geq 0$, $y \geq 0$ and $x + y \leq h$.

Solution. Putting $x = X h$ and $y = Yh$, $dx\, dy = h^2\, dX\, dY$

$$\iint_D x^{l-1}y^{m-1}\,dx\,dy = \iint_{D'} (X h)^{l-1}(Y h)^{m-1} h^2 dX\, dY$$

where D' is the domain

$$X \geq 0, \qquad Y \geq 0, \qquad X + Y \leq 1$$

$$= h^{l+m}\int_0^1\int_0^{1-X} X^{l-1}Y^{m-1}dX\,dY = h^{l+m}\int_0^1 X^{l-1}dX\int_0^{1-X} Y^{m-1}dY$$

$$= h^{l+m}\int_0^1 X^{l-1}dX\left[\frac{Y^m}{m}\right]_0^{1-X} = \frac{h^{l+m}}{m}\int_0^1 X^{l-1}(1-X)^m\, dX$$

$$= \frac{h^{l+m}}{m}\beta(l, m+1) = \frac{h^{l+m}}{m}\frac{\Gamma(l)\,\Gamma(m+1)}{\Gamma(l+m+1)} = \frac{h^{l+m}}{m}\frac{m\,\Gamma(l)\,\Gamma(m)}{\Gamma(l+m+1)} = h^{l+m}\frac{\Gamma(l)\,\Gamma(m)}{\Gamma(l+m+1)}$$ **Proved.**

9.14 DIRICHLET'S INTEGRAL (Triple Integration)

If l, m, n are all positive, then the triple integral

$$\iiint_V x^{l-1}y^{m-1}z^{n-1}\,dx\,dy\,dz = \frac{\Gamma(l)\,\Gamma(m)\,\Gamma(n)}{\Gamma(l+m+n+1)}$$

where V is the region $x \geq 0$, $y \geq 0$, $z \geq 0$ and $0 \leq x + y + z \leq 1$.

Proof. Putting $y + z \leq 1 - x = h$. Then $z \leq h - y$

$$\iiint_V x^{l-1}y^{m-1}z^{n-1}\,dx\,dy\,dz = \int_0^1 x^{l-1}dx\int_0^{1-x} y^{m-1}dy\int_0^{1-x-y} z^{n-1}\,dz$$

$$= \int_0^1 x^{l-1}dx\left[\int_0^h\int_0^{h-y} y^{m-1}z^{n-1}\,dy\,dz\right] \qquad [\text{Put } x = h]$$

$$= \int_0^1 x^{l-1}\,dx\left[\frac{\Gamma(m)\,\Gamma(n)}{\Gamma(m+n+1)}h^{m+n}\right] = \int_0^1 x^{l-1}dx\left[\frac{\Gamma(m)\,\Gamma(n)}{\Gamma(m+n+1)}h^{m+n}\right]$$

$$= \frac{\Gamma(m)\,\Gamma(n)}{\Gamma(m+n+1)}\int_0^1 x^{l-1}(1-x)^{m+n}dx = \frac{\Gamma(m)\,\Gamma(n)}{\Gamma(m+n+1)}\beta(l, m+n+1)$$

$$= \frac{\Gamma(m)\,\Gamma(n)}{\Gamma(m+n+1)}\frac{\Gamma(l)\,\Gamma(m+n+1)}{\Gamma(l+m+n+1)}$$

$$\iiint_V x^{l-1} y^{m-1} z^{n-1} dx\, dy\, dz = \frac{\overline{|l}\ \overline{|m}\ \overline{|n}}{\overline{|l+m+n+1}}$$

Note. $$\iiint_V x^{l-1} y^{m-1} z^{n-1} dx\, dy\, dz = \frac{\overline{|l}\ \overline{|m}\ \overline{|n}}{\overline{|l+m+n+1}} h^{l+m+n}$$

where V is the domain, $x \geq 0$, $y \geq 0$, $z \geq 0$ and $x + y + z \leq h$.

Corollary: Dirichlet's theorem for *n* variables, the theorem states that

$$\iiint \ldots \int x_1^{l_1 - 1}\, x_2^{l_2 - 1} \ldots x_n^{l_n - 1}\, dx_1\, dx_2\, dx_3 \ldots dx_n = \frac{\overline{|l_1}\ \overline{|l_2}\ \overline{|l_3} \ldots \overline{|l_n}}{\overline{|1 + l_1 + l_2 + \ldots + l_n}} h^{l_1 + l_2 + \ldots + l_n}$$

9.15 LIOUVILLE'S EXTENSION OF DIRICHLET THEOREM

If the variables *x, y, z* are all positive such that $h_1 < x + y + z < h_2$, then

$$\iiint f(x+y+z)\, x^{l-1}\, y^{m-1} z^{n-1}\, dx\, dy\, dz = \frac{\overline{|l}\ \overline{|m}\ \overline{|n}}{\overline{|l+m+n}} \int_{h_1}^{h_2} f(u)\, u^{l+m+n-1}\, du$$

Proof. By Dirichlet Theorem, we have

$$I = \iiint X^{l-1}\, Y^{m-1}\, Z^{n-1}\, dX\, dY\, dZ = \frac{\overline{|l}\ \overline{|m}\ \overline{|n}}{\overline{|1+l+m+n}} \qquad \ldots(1)$$

Under the condition $x + y + z \leq u \Rightarrow \frac{x}{u} + \frac{y}{u} + \frac{z}{u} \leq 1$

Putting $X = \frac{x}{u}$, $Y = \frac{y}{u}$ and $Z = \frac{z}{u}$

so that $dX = \frac{dx}{u}$, $dY = \frac{dy}{u}$, $dZ = \frac{dz}{u}$ in (1), we get

$$\iiint \left(\frac{x}{u}\right)^{l-1} \left(\frac{y}{u}\right)^{m-1} \left(\frac{z}{u}\right)^{n-1} \frac{dx}{u} \frac{dy}{u} \frac{dz}{u} = \frac{\overline{|l}\ \overline{|m}\ \overline{|n}}{\overline{|1+l+m+n}}$$

$$\iiint x^{l-1}\, y^{m-1}\, z^{n-1}\, dx\, dy\, dz = u^{l+m+n} \frac{\overline{|l}\ \overline{|m}\ \overline{|n}}{\overline{|1+l+m+n}}$$

Similarly, if $x + y + z \leq u + \delta u$ then

$$\iiint x^{l-1}\, y^{m-1}\, z^{n-1}\, dx\, dy\, dz = (u + \delta u)^{l+m+n} \frac{\overline{|l}\ \overline{|m}\ \overline{|n}}{\overline{|1+l+m+n}}$$

Hence, value of the integral I extended to all such values of the variables as make the sum of the variables lie between u and $u + \delta u$ is given by

$$I = (u + \delta u)^{l+m+n} \frac{\overline{|l}\ \overline{|m}\ \overline{|n}}{\overline{|1+l+m+n}} - u^{l+m+n} \frac{\overline{|l}\ \overline{|m}\ \overline{|n}}{\overline{|1+l+m+n}}$$

$$\iiint x^{l-1}\, y^{m-1}\, z^{n-1}\, dx\, dy\, dz = \frac{\overline{|l}\ \overline{|m}\ \overline{|n}}{\overline{|1+l+m+n}} [(u + \delta u)^{l+m+n} - u^{l+m+n}]$$

$$= \frac{\overline{|l}\ \overline{|m}\ \overline{|n}}{\overline{|1+l+m+n}} u^{l+m+n} \left[\left(1 + \frac{\delta u}{u}\right)^{l+m+n} - 1\right]$$

$$= \frac{\overline{|l}\,\overline{|m}\,\overline{|n}}{\overline{|1+l+m+n}} u^{l+m+n}\left[1+(l+m+n)\frac{\delta u}{u}+\ldots\ldots-1\right]$$

$$= \frac{\overline{|l}\,\overline{|m}\,\overline{|n}}{\overline{|1+l+m+n}} u^{l+m+n}(l+m+n)\frac{\delta u}{u} = \frac{\overline{|l}\,\overline{|m}\,\overline{|n}}{\overline{|l+m+n}} u^{l+m+n-1}\,\delta u$$

Let us consider $\iiint f(x+y+z)\,x^{l-1}\,y^{m-1}\,z^{n-1}\,dx\,dy\,dz$

Under the condition $h_1 \le x+y+z \le h_2$

When $x + y + z$ lies between u and $u + \delta u$, the value of $f(x + y + z)$ can only differ from $f(u)$ by a small quantity of the same order as δu. Hence, neglecting square of δu, the part of the integral

$$\iiint f(x+y+z)\,x^{l-1}\,y^{m-1}\,z^{n-1}\,dx\,dy\,dz = \frac{\overline{|l}\,\overline{|m}\,\overline{|n}}{\overline{|l+m+n}} f(u)u^{l+m+n-1}\,\delta u$$

(supposing the sum of variables to be between u and $u + \delta u$)

So $$\boxed{\iiint f(x+y+z)\,x^{l-1}\,y^{m-1}\,z^{n-1}\,dx\,dy\,dz = \frac{\overline{|l}\,\overline{|m}\,\overline{|n}}{\overline{|(l+m+n)}}\int_{h_1}^{h_2} f(u)\,u^{l+m+n-1}\,du}$$

Example 38. *Show that* $\iiint \frac{dx\,dy\,dz}{(x+y+z+1)^3} = \frac{1}{2}\log 2 - \frac{5}{16}$, *the integral being taken throughout the volume bounded by planes* $x = 0,\ y = 0,\ z = 0,\ x + y + z = 1.$

Solution. By Liouville's theorem when $0 < x + y + z < 1$

$$\iiint \frac{dx\,dy\,dz}{(x+y+z+1)^3} = \iiint \frac{x^{1-1}\,y^{1-1}\,z^{1-1}\,dx\,dy\,dz}{(x+y+z+1)^3} \qquad (0 \le x+y+z \le 1)$$

$$= \frac{\overline{|1}\,\overline{|1}\,\overline{|1}}{\overline{|1+1+1}}\int_0^1 \frac{1}{(u+1)^3} u^{3-1}\,du = \frac{1}{2}\int_0^1 \frac{u^2}{(u+1)^3}\,du$$

$$= \frac{1}{2}\int_0^1 \left[\frac{1}{u+1} - \frac{2}{(u+1)^2} + \frac{1}{(u+1)^3}\right] du \qquad \text{(Partial fractions)}$$

$$= \frac{1}{2}\left[\log(u+1) + \frac{2}{u+1} - \frac{1}{2(u+1)^2}\right]_0^1$$

$$= \frac{1}{2}\left[\log 2 + 2\left(\frac{1}{2}-1\right) - \left(\frac{1}{8}-\frac{1}{2}\right)\right] = \frac{1}{2}\log 2 - \frac{5}{16} \qquad \textbf{Proved.}$$

Example 39. *Find the value of* $\iiint \log(x+y+z)\,dx\,dy\,dz$ *the integral extending over all positive and zero values of x, y, z subject to the condition* $o < x + y + z < 1.$

Solution. By Liouville's theorem when $0 < x + y + z < 1$

$\iiint \log(x+y+z)\,dx\,dy\,dz$

$$= \iiint \log(x+y+z)\,x^{1-1}\,y^{1-1}\,z^{1-1}\,dx\,dy\,dz = \frac{\overline{|1}\,\overline{|1}\,\overline{|1}}{\overline{|1+1+1}}\int_0^1 (\log u)u^{1+1+1-1}\,du$$

$$= \frac{1}{\overline{|3}}\int_0^1 u^2 \log u\,du = \frac{1}{2}\left[\log u\left(\frac{u^3}{3}\right) - \frac{1}{3}\frac{u^3}{3}\right]_0^1 = \frac{1}{2}\left(-\frac{1}{9}\right) = -\frac{1}{18} \qquad \textbf{Ans.}$$

Example 40. *Evaluate* $\iiint \dfrac{dx_1\, dx_2 \ldots dx_n}{\sqrt{1 - x_1^2 - x_2^2 \ldots - x_n^2}}$, *integral being extended to all positive values of the variables for which the expression is real.*

Solution. $\sqrt{1 - x_1^2 - x_2^2 \ldots x_n^2}$ is real only when $0 < x_1^2 + x_2^2 \ldots x_n^2 < 1$

Hence, the given integral is extended for all positive values of the variables $x_1, x_2 \ldots$ and x_n such that $0 < x_1^2 + x_2^2 + \ldots + x_n^2 < 1$

Let us now put $x_1^2 = u_1$ *i.e.* $x_1 = u_1^{\frac{1}{2}}$ so that $dx_1 = \frac{1}{2} u_1^{-\frac{1}{2}}\, du_1$

$$x_2^2 = u_2 \text{ i.e. } x_2 = u_2^{\frac{1}{2}} \text{ so that } dx_2 = \frac{1}{2}\, u_2^{-\frac{1}{2}}\, du_2$$

$$x_n^2 = u_n \text{ i.e. } x_n = u_n^{\frac{1}{2}} \text{ so that } dx_n = \frac{1}{2} u_n^{\frac{1}{2}}\, du_n$$

Making these substitutions, the given condition becomes $0 < u_1 + u_2 + \ldots + u_n < 1$.

Hence, the required integral becomes

$$= \frac{1}{2^n} \iiint \frac{u_1^{-\frac{1}{2}} . u_2^{-\frac{1}{2}} \ldots u_n^{-\frac{1}{2}} . du_1 . du_2 \ldots du_n}{\sqrt{1 - u_1 - u_2 \ldots - u_n}}$$

$$= \frac{1}{2^n} \iiint \frac{u_1^{\frac{1}{2}-1} . u_2^{\frac{1}{2}-1} \ldots u_n^{\frac{1}{2}-1} . du_1 . du_2 \ldots du_n}{\sqrt{1 - u_1 - u_2 \ldots - u_n}}$$

$$= \frac{1}{2^n} \frac{\Gamma\left(\frac{1}{2}\right) . \Gamma\left(\frac{1}{2}\right) \ldots \Gamma\left(\frac{1}{2}\right)}{\Gamma\left(\frac{1}{2} + \frac{1}{2} + \ldots + \frac{1}{2}\right)} \int_0^1 \frac{1}{\sqrt{1-u}} . u^{\left(\frac{1}{2}+\frac{1}{2}+\ldots+\frac{1}{2}\right)-1}\, du$$

By Liouville's Extension of Dirichlet's Therorem

$$= \frac{1}{2^n} \frac{\left(\Gamma\left(\frac{1}{2}\right)\right)^n}{\Gamma\left(\frac{n}{2}\right)} \int_0^1 \frac{1}{\sqrt{1-u}} . u^{\frac{n}{2}-1}\, du$$

$$= \frac{1}{2^n} \frac{(\sqrt{\pi})^n}{\Gamma\left(\frac{n}{2}\right)} \int_0^{\frac{\pi}{2}} \frac{1}{\sqrt{1 - \sin^2\theta}} . (\sin^2\theta)^{\frac{n}{2}-1}\; 2 \sin\theta \cos\theta\, d\theta \qquad (\text{Put } u = \sin^2\theta)$$

$$= \frac{1}{2^{n-1}} \frac{(\sqrt{\pi})^n}{\Gamma\left(\frac{n}{2}\right)} \int_0^{\frac{\pi}{2}} \sin^{n-1}\theta\, (\cos\theta)^\circ\, d\theta = \frac{1}{2^{n-1}} \frac{\pi^{\frac{n}{2}}}{\Gamma\left(\frac{n}{2}\right)} \frac{\Gamma\left(\frac{n}{2}\right)\Gamma\left(\frac{1}{2}\right)}{2\,\Gamma\left(\frac{n+1}{2}\right)} = \frac{1}{2^n} \frac{\pi^{\frac{n+1}{2}}}{\Gamma\left(\frac{n+1}{2}\right)} \qquad \textbf{Ans.}$$

Example 41. *Evaluate* $\iiint \frac{\sqrt{1-x^2-y^2-z^2}}{1+x^2+y^2+z^2}\,dx\,dy\,dz,$ *integral being taken over all positive values of x, y, z such that* $0 \le x^2+y^2+z^2 \le 1$.

Solution. Putting $x^2 = u, y^2 = v, z^2 = w$ so that $u+v+w \le 1$

Also,
$$x = \sqrt{u} \Rightarrow dx = \frac{1}{2\sqrt{u}}\,du$$
$$y = \sqrt{v} \Rightarrow dy = \frac{1}{2\sqrt{v}}\,dv$$
$$z = \sqrt{w} \Rightarrow dz = \frac{1}{2\sqrt{w}}\,dw$$

$\therefore$ The given integral

$$= \iiint \sqrt{\frac{1-(u+v+w)}{1+(u+v+w)}}\,\frac{du\,dv\,dw}{8\sqrt{uv\,w}}$$

$$= \frac{1}{8}\iiint u^{1/2-1}\,v^{1/2-1}\,w^{1/2-1}\sqrt{\frac{1-(u+v+w)}{1+(u+v+w)}}\,du\,dv\,dw$$

$$= \frac{1}{8}\frac{\left|\overline{\left(\frac{1}{2}\right)}\right.\left|\overline{\left(\frac{1}{2}\right)}\right.\left|\overline{\left(\frac{1}{2}\right)}\right.}{\left|\overline{\left(\frac{1}{2}+\frac{1}{2}+\frac{1}{2}\right)}\right.}\int_0^1 \sqrt{\frac{1-u}{1+u}}\,u^{\frac{1}{2}+\frac{1}{2}+\frac{1}{2}-1}\,du \quad \text{[Using Liouville's extension]}$$

$$= \frac{1}{8}\frac{\left[\left|\overline{\left(\frac{1}{2}\right)}\right.\right]^3}{\frac{1}{2}\left|\overline{\left(\frac{1}{2}\right)}\right.}\int_0^1 \frac{(1-u)}{\sqrt{1-u^2}}\,u^{1/2}\,du = \frac{\pi}{4}\int_0^1 \frac{(1-\sqrt{t})}{\sqrt{1-t}}\,t^{1/4}\,\frac{dt}{2\sqrt{t}} \qquad \text{where } u^2 = t$$

$$= \frac{\pi}{8}\int_0^1 \frac{(1-\sqrt{t})\,t^{-1/4}}{\sqrt{1-t}}\,dt = \frac{\pi}{8}\left[\int_0^1 t^{\frac{3}{4}-1}(1-t)^{1/2-1}\,dt - \int_0^1 t^{5/4-1}(1-t)^{1/2-1}\,dt\right]$$

$$= \frac{\pi}{8}\left[\beta\left(\frac{3}{4},\frac{1}{2}\right) - \beta\left(\frac{5}{4},\frac{1}{2}\right)\right]$$ **Ans.**

Example 42. *Find the area and the mass contained m the first quadrant enclosed by the curve* $\left(\frac{x}{a}\right)^\alpha + \left(\frac{y}{b}\right)^\beta = 1$ *where* $\alpha > 0$, $\beta > 0$ *given that density at any point p (xy) is k* $\sqrt{xy}$.

(U.P. I Semester 2008)

Solution. Here, we have $\left(\frac{x}{a}\right)^\alpha + \left(\frac{y}{b}\right)^\beta = 1$

Put $\left(\frac{x}{a}\right)^\alpha = \cos^2 t$ and $\left(\frac{y}{b}\right)^\beta = \sin^2 t$

$\Rightarrow \quad x = a\cos^{\frac{2}{\alpha}} t \qquad \Rightarrow y = b\sin^{\frac{2}{\beta}} t$

$$dx = \frac{2}{\alpha}a\cos^{\frac{2}{\alpha}-1} t\,(-\sin t)dt$$

Y
X′
X
O
Y′

$$\text{Area} = \int y\,dx = \int_0^{\frac{\pi}{2}} (b \sin^{\frac{2}{\beta}} t) \left(-\frac{2a}{\alpha} \cos^{\frac{2}{\alpha}-1} t \sin t \right) dt$$

$$= \frac{-2ab}{\alpha} \int_0^{\frac{\pi}{2}} \sin^{\frac{2}{\beta}+1} t \cos^{\frac{2}{\alpha}-1} t\, dt = \frac{-2ab}{\alpha} \frac{\Gamma\left(\frac{\frac{2}{\beta}+1+1}{2}\right) \Gamma\left(\frac{\frac{2}{\alpha}-1+1}{2}\right)}{2\,\Gamma\left(\frac{\beta+1}{\beta}+\frac{1}{\alpha}\right)}$$

$$= \frac{-ab}{\alpha} \frac{\Gamma\left(1+\frac{1}{\beta}\right) \Gamma\left(\frac{1}{\alpha}\right)}{\Gamma\left(\frac{\alpha+\beta+\alpha\beta}{\alpha\beta}\right)} = -ab \frac{\frac{1}{\alpha}\Gamma\left(\frac{1}{\alpha}\right) \Gamma\left(1+\frac{1}{\beta}\right)}{\Gamma\left(\frac{\alpha+\beta+\alpha\beta}{\alpha\beta}\right)} = -ab \frac{\Gamma\left(\frac{1}{\alpha}+1\right) \Gamma\left(1+\frac{1}{\beta}\right)}{\Gamma\left(\frac{\alpha+\beta+\alpha\beta}{\alpha\beta}\right)}$$

$$\text{Required area} = \frac{\Gamma\left(\frac{\alpha+1}{\alpha}\right) \Gamma\left(\frac{\beta+1}{\beta}\right)}{\Gamma\left(\frac{\alpha+\beta+\alpha\beta}{\alpha\beta}\right)} \qquad \textbf{Ans.}$$

$$\text{Density} = k\sqrt{xy}$$

$$\text{Mass} = \text{Area} \times \text{Density} = \int \left[y\,dx\, k\sqrt{xy} \right] = 4k \int_0^a x^{\frac{1}{2}} y^{\frac{3}{2}}\, dx$$

$$= 4k \int_0^{\frac{\pi}{2}} (a \cos^{\frac{2}{\alpha}} t)^{\frac{1}{2}} (b \sin^{\frac{2}{\beta}} t)^{\frac{3}{2}} \frac{2}{\alpha} a \cos^{\frac{2}{\alpha}-1} t\, (-\sin t)\, dt$$

$$= 4k \int_0^{\frac{\pi}{2}} a^{\frac{1}{2}} \cos^{\frac{1}{\alpha}} t\, b^{\frac{3}{2}} \sin^{\frac{3}{\beta}} t \frac{2}{\alpha} a \cos^{\frac{2}{\alpha}-1} t\, (-\sin t\, dt)$$

$$= 4k\, a^{\frac{1}{2}+1} b^{\frac{3}{2}} \cdot \frac{2}{\alpha} \int_0^{\frac{\pi}{2}} \sin^{\frac{3}{\beta}+1} t \cos^{\frac{1}{\alpha}+\frac{2}{\alpha}-1} t\, dt$$

$$= \frac{8k}{\alpha} a^{\frac{3}{2}} b^{\frac{3}{2}} \int_0^{\frac{\pi}{2}} \sin^{\frac{3}{\beta}+1} t \cos^{\frac{3}{\alpha}-1} t\, dt$$

$$= \frac{8k}{\alpha} a^{\frac{3}{2}} b^{\frac{3}{2}} \frac{\Gamma\left(\frac{\frac{3}{\beta}+1+1}{2}\right) \Gamma\left(\frac{\frac{3}{\alpha}-1+1}{2}\right)}{2\,\Gamma\left(\frac{\frac{3}{\beta}+1+1+\frac{3}{\alpha}-1+1}{2}\right)} = \frac{8k}{\alpha} a^{\frac{3}{2}} b^{\frac{3}{2}} \frac{\Gamma\left(\frac{3+2\beta}{2\beta}\right) \Gamma\left(\frac{3}{2\alpha}\right)}{2\,\Gamma\left(\frac{3}{2\alpha}+\frac{3}{2\beta}+1\right)} \qquad \textbf{Ans.}$$

Example 43. *Find the mass of an octant of the ellipsoid* $\frac{x^2}{a^2}+\frac{y^2}{b^2}+\frac{z^2}{c^2}=1$, *the density at any point being* $\rho = k\,x\,y\,z$. (*U.P. , I Semester, 2009*)

Solution. $\quad \text{Mass} = \iiint \rho\, dv = \iiint (k\,x\,y\,z)\, dx\, dy\, dz$

$$= k \iiint (x\,dx)(y\,dy)(z\,dz) \qquad \text{...(1)}$$

Putting $\quad \frac{x^2}{a^2} = u,\ \frac{y^2}{b^2} = v,\ \frac{z^2}{c^2} = w$ and $u + v + w = 1$

so that $\dfrac{2x\,dx}{a^2} = du, \dfrac{2y\,dy}{b^2} = dv, \dfrac{2z\,dz}{c^2} = dw$

$$\text{Mass} = k\iiint\left(\frac{a^2du}{2}\right)\left(\frac{b^2dv}{2}\right)\left(\frac{c^2dw}{2}\right)$$

$$= \frac{ka^2b^2c^2}{8}\iiint du\,dv\,dw, \qquad \text{where } u+v+w \le 1.$$

$$= \frac{ka^2b^2c^2}{8}\iiint u^{1-1}\,v^{1-1}\,w^{1-1}\,du\,dv\,dw$$

$$= \frac{ka^2b^2c^2}{8}\frac{\overline{|1}\,\overline{|1}\,\overline{|1}}{\overline{|3+1}} = \frac{ka^2b^2c^2}{8\times 6} = \frac{ka^2b^2c^2}{48}$$ **Ans.**

Example 44. *Find the mass of a solid* $\left(\dfrac{x}{a}\right)^p + \left(\dfrac{y}{b}\right)^q + \left(\dfrac{z}{c}\right)^r = 1$, *the density at any point being* $\rho = kx^{l-1}y^{m-1}z^{n-1}$, *where x, y, z are all positive.*

Solution. Here, we have

$$\left(\frac{x}{a}\right)^p + \left(\frac{y}{b}\right)^q + \left(\frac{z}{c}\right)^r = 1$$

Put $\dfrac{x}{a} = u \Rightarrow x = au \Rightarrow dx = a\,du$

$\dfrac{y}{b} = v \Rightarrow y = bv \Rightarrow dy = b\,dv$

$\dfrac{z}{c} = w \Rightarrow z = cw \Rightarrow dz = c\,dw$

$$\text{Density} = k\,x^{l-1}\,y^{m-1}\,z^{n-1}$$

$$\text{Mass} = \text{Volume} \times \text{Density}$$

$$= \iiint dx\,dy\,dz.\,kx^{l-1}y^{m-1}z^{n-1}$$

$$= k\iiint (au)^{l-1}\,(bv)^{m-1}\,(cw)^{n-1}\;adu.bdv.cdw$$

$$= k\,a^l\,b^m\,c^n\iiint u^{l-1}\,v^{m-1}\,w^{n-1}\,du\,dv\,dw$$

$$= k.a^l.\,b^m\,c^n\,\frac{\overline{|l}\,\overline{|m}\,\overline{|n}}{\overline{|l+m+n+1}}$$ **Ans.**

Example 45. *Evaluate* $I = \iiint_V x^{\alpha-1}\,y^{\beta-1}\,z^{\gamma-1}\,dx\,dy\,dz$, *where V is the region in the first cotenant bonded by sphere* $x^2+y^2+z^2 = 1$ *and the coordinate planes.*

Solution. Let $x^2 = u \Rightarrow x = \sqrt{u} \quad \therefore \quad dx = \dfrac{1}{2\sqrt{u}}du$

$y^2 = v \Rightarrow y = \sqrt{v} \quad \therefore \quad dy = \dfrac{1}{2\sqrt{v}}dv$

$z^2 = w \Rightarrow z = \sqrt{w} \quad \therefore \quad dz = \dfrac{1}{2\sqrt{w}}dw$

Then, $u+v+w = 1$. Also, $u \ge 0, v \ge 0, w \ge 0$.

$$I = \iiint_V (\sqrt{u})^{\alpha-1}\,(\sqrt{v})^{\beta-1}\,(\sqrt{w})^{\gamma-1}\,\frac{du}{2\sqrt{u}}.\frac{dv}{2\sqrt{v}}.\frac{dw}{2\sqrt{w}}$$

$$= \frac{1}{8}\iiint u^{(\alpha/2)-1}\,v^{(\beta/2)-1}\,w^{(\gamma/2)-1}\,du\,dv\,dw = \frac{1}{8}\,\frac{\Gamma(\alpha/2)\,\Gamma(\beta/2)\,\Gamma(\gamma/2)}{\overline{|(\alpha/2)+(\beta/2)+(\gamma/2)+1}}$$ **Ans.**

EXERCISE 9.3

Evaluate:

1. $\iiint e^{x+y+z}\, dx\, dy\, dz$ taken over the positive octant such that $x + y + z \le 1$. **Ans.** $\dfrac{e-2}{2}$

2. $\iiint \dfrac{dx\,dy\,dz}{(a^2-x^2-y^2-z^2)}$ for all positive values of the variables for which the expression is real.

[**Hint.** $a^2 - x^2 - y^2 - z^2 > 0 \quad \Rightarrow \quad 0 < x^2 + y^2 + z^2 < a^2$] **Ans.** $\dfrac{\pi^2 a^2}{8}$

3. $\iiint_R (x+y+z+1)^2\, dx\, dy\, dz$ where R is defined by $x \ge 0,\ y \ge 0,\ z \ge 0,\ x + y + z \le 1$. **Ans.** $\dfrac{31}{60}$

4. $\iiint x^{-\frac{1}{2}} y^{-\frac{1}{2}} z^{-\frac{1}{2}} (1-x-y-z)^{\frac{1}{2}}\, dx\, dy\, dz,\ x + y + z \le 1,\ x > 0,\ y > 0,\ z > 0$ **Ans.** $\dfrac{\pi^2}{4}$

5. Show that $\iiint \dfrac{dx\, dy\, dz}{(x+y+z+1)^2} = \dfrac{3}{4} - \log 2$, the integral being taken throughout the volume bounded by the planes $x = 0,\ y = 0,\ z = 0$ and $x + y + z = 1$.

Choose the correct alternative:

6. $\left|\!\overline{\dfrac{1}{2}}\right.$ is equal to

(*i*) π (*ii*) $\dfrac{1}{2!}$ (*iii*) $\sqrt{\pi}$ (*iv*) $\dfrac{\pi}{2}$ **Ans.** (*iii*)

(*R.G.P.V. Bhopal, I Semester, June 2006*)

7. $\beta\,(m, n) =$

(*i*) $\int_0^{\pi} \sin^{2m-1}\theta \cos^{2n-1}\theta\, d\theta$ (*ii*) $\int_0^{\frac{\pi}{2}} \sin^{2m-1}\theta \cos^{2n-1}\theta\, d\theta$

(*iii*) $2\int_0^{\frac{\pi}{2}} \sin^{2m-1}\theta \cos^{2n-1}\theta\, d\theta$ (*iv*) $\int_0^{\frac{\pi}{2}} \sin^{m}\theta \cos^{n}\theta\, d\theta$ **Ans.** (*iii*)

(*R.G.P.V. Bhopal, I Semester Dec. 2006*)

8. What is the relation between Beta and Gamma functions?

(*i*) $\beta\,(m, l) = \dfrac{\overline{|m} + \overline{|l}}{\overline{|l+m}}$ (*ii*) $\beta\,(l, m) = \dfrac{\overline{|l}\ \overline{|m}}{\overline{|l+m}}$

(*iii*) $\beta\,(l, m) = \dfrac{\overline{|l} + \overline{|m}}{\overline{|m+l}}$ (*iv*) $\beta\,(l, m) = \dfrac{\overline{|l}\ \overline{|m}}{\overline{|m}\,\overline{|l}}$ **Ans.** (*ii*)

(*R.G.P.V. Bhopal, I Semester Dec. 2006*)

9. The value of $\overline{|3.5}$

(*i*) $\dfrac{15}{4}\sqrt{\pi}$ (*ii*) $\dfrac{15}{8}\sqrt{\pi}$ (*iii*) $\dfrac{15}{2}\sqrt{\pi}$ (*iv*) $\dfrac{15}{16}\sqrt{\pi}$ (*AMIETE, June 2010*) **Ans.** (*ii*)

10. The value of $\int_{-\infty}^{\infty} e^{-x^2}\, dx$ is

(*i*) $\dfrac{\sqrt{\pi}}{2}$ (*ii*) $\left|\!\overline{\dfrac{\pi}{2}}\right.$ (*iii*) $\sqrt{\pi}$ (*iv*) $\dfrac{\pi}{\sqrt{2}}$

(*AMIETE, June 2010*) **Ans.** (*iii*)

11. In terms of Beta function $\int_0^{\frac{\pi}{2}} \sin^7\theta \sqrt{\cos\theta}\, d\theta$ is

(*i*) $\dfrac{1}{2}\beta\left(4, \dfrac{3}{4}\right)$ (*ii*) $\dfrac{1}{2}\beta\left(2, \dfrac{1}{4}\right)$ (*iii*) $\dfrac{1}{2}\beta\left(3, \dfrac{3}{4}\right)$ (*iv*) None of these

(*AMIETE, June 2010*) **Ans.** (*i*)

12. If l, m, n are all positive, then the triple integral

$$\iiint_V x^{l-1} y^{m-1} z^{n-1}\, dx\, dy\, dz = \frac{\overline{|l}\,\overline{|m}\,\overline{|n}}{\overline{|l+m+n}}$$

where V is the region $x \geq 0$, $y \geq 0$, $z \geq 0$ and $x + y + z \leq 1$. (*U.P., Ist Semester, 2005*) **Ans.** True

9.16 ELLIPTIC INTEGRALS

Draw a circle with AA' (diameter) the major axis of the ellipse $\frac{x^2}{a^2}+\frac{y^2}{b^2}=1$. This circle is called the auxiliary circle $x^2+y^2=a^2$. The co-ordinates of a point P on the ellipse are (a sin ϕ, b cos ϕ)

$x = a \sin\phi$, $y = b\cos\phi$ is the parametric equation of the ellipse.

Now the length of the arc BP of the ellipse

B (a cos φ l₁ sin φ)

A′ O A

$$= \int_0^{\phi} \sqrt{\left\{\left(\frac{dx}{d\phi}\right)^2 + \left(\frac{dy}{d\phi}\right)^2\right\}}\, d\phi \qquad \{b^2 = a^2(1-e^2)\}$$

$$= \int_0^{\phi} \sqrt{(a^2\cos^2\phi + b^2\sin^2\phi)}\; d\phi$$

$$= \int_0^{\phi} \sqrt{\{a^2\cos^2\phi + (a^2 - a^2e^2)\sin^2\phi\}}\, d\phi = \boldsymbol{a\int_0^{\phi}\sqrt{(1-e^2\sin^2 j)}\, d\phi}$$

where e is the eccentricity of the ellipse.

This integral cannot be evaluated in the form of the elementary function. It defines a new function, called *elliptic function.* This integral is called the elliptic integrals as it is derived from the determination of the Perimeter of the ellipse. This integral cannot be evaluated by standard methods of integration. First the integrand $\sqrt{1-e^2\sin^2\phi}$ is expanded as power series and then is integrated term by term.

9.17 DEFINITION AND PROPERTY

Elliptic integral of first kind $= F(k,\phi) = \int_0^{\phi} \frac{1}{\sqrt{(1-k^2\sin^2\phi)}}\, d\phi \qquad k^2 < 1$

Elliptic integral of second kind $= E(k,\phi) = \int_0^{\phi} \sqrt{(1-k^2\sin^2\phi)}\, d\phi \qquad k^2 < 1$

Here k is known as modulus and ϕ amplitude.

The following results are easy to prove

$$F(0,\phi) = E(0,\phi) = \phi$$

$$F(1,\phi) = \log(\tan\phi + \sec\phi)$$

$$E(1,\phi) = \sin\phi$$

If $\phi = \frac{\pi}{2}$ is the upper limit of the integral then the integral is called *complete elliptic integral* as under:

$$F(k) = \int_0^{\frac{\pi}{2}} \frac{d\phi}{\sqrt{1-k^2 \sin^2 \phi}} \qquad \text{... (1)}$$

and
$$E(k) = \int_0^{\frac{\pi}{2}} \sqrt{1-k^2 \sin^2 \phi}\, d\phi \qquad \text{... (2)}$$

These integrals can be evaluated by expanding the integrand in binomial series and integrating term by term.

$$(1-k^2 \sin^2 \phi)^{-1/2} = \frac{k^2}{2}\sin^2\phi + \frac{1.3}{2.4}k^4 \sin^4\phi + ...$$

$$F(k, \phi) \int_0^{\phi} \frac{d\phi}{\sqrt{(1-k^2\sin^2\phi)}} = \phi + \frac{k^2}{2}\int_0^{\phi} \sin^2\phi\, d\phi + \frac{1.3}{2.4}k^4 \int_0^{\phi} \sin^4\phi\, d\phi + ... \qquad \text{...(3)}$$

which can be evaluated by the *Reduction Formula*

$$\int_0^{\phi} \sin^n \phi\, d\phi = -\frac{\sin^{n-1}\phi\cos\phi}{n} + \frac{n-1}{n}\int_0^{\phi} \sin^{n-2}\phi\, d\phi$$

From (3), we get

$$K(k) = \int_0^{\frac{\pi}{2}} \frac{d\phi}{\sqrt{(1-k^2\sin^2\phi)}} = \frac{\pi}{2} + \frac{k^2}{2}\left(\frac{1}{2}\frac{\pi}{2}\right) + \frac{1.3}{2.4}k^4\left(\frac{3.1}{4.2}\frac{\pi}{2}\right) + ...$$

or
$$K(k) = \frac{\pi}{2}\left[1 + \frac{k^2}{4} + \frac{9k^4}{64} + ...\right]$$

If $k = \sin 10'$

$$K = \frac{\pi}{2}[1 + 0.00754 + 0.00012 + ...] = 1.5828$$

The elliptic integrals are periodic functios with a period π.

$$F(k, \phi + P\pi) = PF(k, \pi) + F(k, \phi), P = 0, 1, 2......$$
$$E(k, \phi + P\pi) = PE(k, \pi) + E(k, \phi), P = 0, 1, 2......$$
$$F(k, \phi + P\pi) = 2PF(k) + F(k, \phi)$$
$$E(k, \phi + P\pi) = 2PE(k) + E(k, \phi)$$

If we substitute $\sin\phi = x,\ d\phi = \dfrac{dx}{\sqrt{1-x^2}}$ in (1) and (2), we have

$$F_1(k, x) = \int_0^x \frac{dx}{\sqrt{\{(1-x^2)(1-k^2x^2)\}}}$$

$$E_1(k, x) = \int_0^x \sqrt{\left(\frac{1-k^2x^2}{1-x^2}\right)}dx$$

These are known as Jacobi's form of elliptic integrals.

Example 46. *Express* $\int_0^{\frac{\pi}{2}} \sqrt{\cos x}\, dx$ *in terms of elliptic integrals.*

Solution. Substitute $\cos x = \cos^2\phi$

so that
$$x = \cos^{-1}\cos^2\phi,\ dx = \frac{2\cos\phi\sin\phi\, d\phi}{\sqrt{1-\cos^4\phi}} = \frac{2\cos\phi\, d\phi}{\sqrt{1+\cos^2\phi}}$$

$$\int_0^{\frac{\pi}{2}} \sqrt{\cos x}\, dx = \int_0^{\frac{\pi}{2}} \frac{2\cos^2 \phi d\phi}{\sqrt{1+\cos^2 \phi}} = 2\int_0^{\frac{\pi}{2}} \frac{(1+\cos^2 \phi)-1}{\sqrt{1+\cos^2 \phi}} d\phi$$

$$= 2\left\{\int_0^{\frac{\pi}{2}} \sqrt{1+\cos^2 \phi}\, d\phi - \int_0^{\pi/2} \frac{1}{\sqrt{1+\cos^2 \phi}} d\phi\right\}$$

$$= 2\left\{\int_0^{\frac{\pi}{2}} \sqrt{(2-\sin^2 \phi}\, d\phi - \int_0^{\frac{\pi}{2}} \frac{1}{\sqrt{(2-\sin^2 \phi)}} d\phi\right\}$$

$$= 2\sqrt{2}\int_0^{\frac{\pi}{2}} \sqrt{\left(1-\frac{1}{2}\sin^2 \phi\right)}\, d\phi - \frac{2}{\sqrt{2}}\int_0^{\frac{\pi}{2}} \frac{1}{\sqrt{\left(1-\frac{1}{2}\sin^2 \phi\right)}} d\phi$$

$$= 2\sqrt{2}\, E\left(\frac{1}{\sqrt{2}}\right) - \sqrt{2} K\left(\frac{1}{\sqrt{2}}\right) \qquad \textbf{Ans.}$$

Example 47. *Express* $\int_0^{\frac{\pi}{2}} \frac{dx}{\sqrt{(2-\cos x)}}$ *in terms of elliptic integrals.*

Solution. $\int_0^{\frac{\pi}{2}} \frac{dx}{\sqrt{(2-\cos x)}} = \int_0^{\frac{\pi}{2}} \frac{dx}{\sqrt{\left\{2-\left(2\cos^2 \frac{x}{2}-1\right)\right\}}}$

$$= \int_0^{\frac{\pi}{2}} \frac{dx}{\sqrt{\left(3-2\cos^2 \frac{x}{2}\right)}} = \frac{1}{\sqrt{3}}\int_0^{\frac{\pi}{2}} \frac{dx}{\sqrt{\left(1-\frac{2}{3}\cos^2 \frac{x}{2}\right)}}$$

On Putting $x = \pi - 2\phi$, so that $dx = -2\, d\phi$

$$= \frac{1}{\sqrt{3}}\int_{\frac{\pi}{2}}^{\frac{\pi}{4}} \frac{-2d\phi}{\sqrt{1-\frac{2}{3}\cos^2\left(\frac{\pi}{2}-\phi\right)}} = \frac{-2}{\sqrt{3}}\int_{\frac{\pi}{2}}^{\frac{\pi}{4}} \frac{d\phi}{\sqrt{\left(1-\frac{2}{3}\sin^2 \phi\right)}}$$

$$= \frac{2}{\sqrt{3}}\left[\int_0^{\frac{\pi}{2}} \frac{d\phi}{\sqrt{\left(1-\frac{2}{3}\sin^2 \phi\right)}} - \int_0^{\frac{\pi}{4}} \frac{d\phi}{\sqrt{\left(1-\frac{2}{3}\sin^2 \phi\right)}}\right] = \frac{2}{\sqrt{3}}\left[K\left(\frac{\sqrt{2}}{3}\right) - F\left(\frac{\sqrt{2}}{3}, \frac{\pi}{4}\right)\right] \quad \textbf{Ans.}$$

Example 48. *Show that* $\int_0^{\frac{a}{2}} \frac{dx}{\sqrt{(2ax-x^2)(a^2-x^2)}} = \frac{2}{3a} K\left(\frac{1}{3}\right)$

Solution. On substituting $x = \frac{a}{2}(1-\sin\theta)$ so that $dx = -\frac{a}{2}\, 2\cos\theta\, d\theta$

Upper limit, $x = \frac{a}{2}, \frac{a}{2} = \frac{a}{2}(1-\sin\theta) \quad \Rightarrow \quad \theta = 0$

Lower limit, $x = 0,\ 0 = \frac{a}{2}(1-\sin\theta) \quad \Rightarrow \quad \theta = \frac{\pi}{2}$

$$2ax - x^2 = (2a)\frac{a}{2}\ (1 - \sin\theta) - \frac{a^2}{4}\ (1 - \sin\theta)^2 = \frac{a^2}{4}\ [4 - 4\sin\theta - 1 + 2\sin\theta - \sin^2\theta]$$

$$= \frac{a^2}{4}(3 - 2\sin\theta - \sin^2\theta) = \frac{a^2}{4}\ (1 - \sin\theta)\ (3 + \sin\theta)$$

$$a^2 - x^2 = a^2 - \frac{a^2}{4}\ (1 - \sin\theta)^2 = \frac{a^2}{4}\ [4 - 1 - \sin^2\theta + 2\sin\theta] = \frac{a^2}{4}\ [3 + 2\sin\theta - \sin^2\theta]$$

$$= \frac{a^2}{4}(1 + \sin\theta)(3 - \sin\theta)$$

$$\int_0^{\frac{a}{2}} \frac{dx}{\sqrt{(2ax - x^2)(a^2 - x^2)}} = \int_{\frac{\pi}{2}}^{0} \frac{-\frac{a}{2}\cos\theta\, d\theta}{\sqrt{\frac{a^2}{4}(1 - \sin\theta)(3 + \sin\theta)\frac{a^2}{4}(1 + \sin\theta)(3 - \sin\theta)}}$$

$$= \int_0^{\frac{\pi}{2}} \frac{\cos\theta\, d\theta}{\frac{a}{2}\sqrt{(1 - \sin^2\theta)(9 - \sin^2\theta)}} = \frac{2}{a}\int_0^{\frac{\pi}{2}} \frac{d\theta}{\sqrt{9 - \sin^2\theta}}$$

$$= \frac{2}{3a}\int_0^{\frac{\pi}{2}} \frac{d\theta}{\sqrt{\left(1 - \left(\frac{1}{3}\right)^2 \sin^2\theta\right)}} = \frac{2}{3a}K\left(\frac{1}{3}\right)$$ **Proved.**

EXERCISE 9.4

Show that

1. $\displaystyle\int_0^{\pi} \frac{dx}{\sqrt{(1 - k^2\sin^2\phi)}} = \frac{1}{k}F\left(\frac{1}{k}, x\right)$

2. $\displaystyle\int_0^{\frac{\pi}{2}} \frac{dx}{\sqrt{(1 + 3\sin^2 x)}} = \frac{1}{2}K\left(\frac{\sqrt{3}}{2}\right)$

3. $\displaystyle\int_0^{\frac{\pi}{6}} \frac{dx}{\sqrt{\sin x}} = \sqrt{2}\left[K\left(\frac{1}{\sqrt{2}}\right) - F\left(\frac{1}{\sqrt{2}}, \frac{\pi}{4}\right)\right]$

4. $\displaystyle\int_0^{\phi} \frac{\sin\phi}{\sqrt{(1 - k^2\sin^2\phi)}} d\phi = \frac{1}{k^2}[F(k, \phi) - E(k, \phi)]$

5. $\displaystyle\int_0^{1} \frac{dx}{\sqrt{(1 - x^4)}} = \frac{1}{\sqrt{2}}K\left(\frac{1}{\sqrt{2}}\right)$

6. $\displaystyle\int_0^{\phi} \sqrt{(1 - k^2\sin^2\phi)}\, d\phi = \left(\frac{1}{k} - k\right)F\left(\frac{1}{k}, x\right) + kE\left(\frac{1}{k}, x\right)$ and $k\sin\phi < 1$ **[Hint.** Put $\sin x = k\sin\phi$]

9.18 ERROR FUNCTION

1. $\displaystyle\frac{2}{\sqrt{\pi}}\int_0^{x} e^{-t^2} dt$ is called error function of x and is also written as $erf(x)$.

2. $\displaystyle\frac{2}{\sqrt{\pi}}\int_x^{\infty} e^{-t^2} dt$ is called complementary error function of x and is also written as $erf_c(x)$.

3. Important formula.

$$\int_0^{\infty} e^{-t^2} dt = \frac{\sqrt{\pi}}{2}$$

Example 49. *Prove that* $erf(0) = 0$

Solution. $$erf(x) = \frac{2}{\sqrt{\pi}}\int_0^x e^{-t^2}\,dt$$

$$erf(0) = \frac{2}{\sqrt{\pi}}\int_0^0 e^{-t^2}\,dt = 0$$ **Proved.**

Example 50. *Prove that* $erf(\infty) = 1$

Solution. $$erf(x) = \frac{2}{\sqrt{\pi}}\int_0^x e^{-t^2}\,dt$$

$$erf(\infty) = \frac{2}{\sqrt{\pi}}\int_0^\infty e^{-t^2}\,dt = \frac{2}{\sqrt{\pi}}\frac{\sqrt{\pi}}{2} = 1$$ **Proved.**

Example 51. *Prove that* $erf(x) + erf_c(x) = 1$

Solution. $$erf(x) + erf_c(x) = \frac{2}{\sqrt{\pi}}\int_0^x e^{-t^2}\,dt + \frac{2}{\sqrt{\pi}}\int_x^\infty e^{-t^2}\,dt = \frac{2}{\sqrt{\pi}}\left[\int_0^x e^{-t^2}\,dt + \int_x^\infty e^{-t^2}\,dt\right]$$

$$= \frac{2}{\sqrt{\pi}}\int_0^\infty e^{-t^2}\,dt = \frac{2}{\sqrt{\pi}}\frac{\sqrt{\pi}}{2} = 1$$ **Proved.**

Example 52. *Prove that* $erf(-x) = -erf(x)$

Solution. $$erf(x) = \frac{2}{\sqrt{\pi}}\int_0^x e^{-t^2}\,dt$$

$$erf(-x) = \frac{2}{\sqrt{\pi}}\int_0^{-x} e^{-t^2}\,dt$$ [Put $t = -u$]

$$= \frac{2}{\sqrt{\pi}}\int_0^x e^{-u^2}(-du) = -\frac{2}{\sqrt{\pi}}\int_0^x e^{-u^2}(du) = -erf(x)$$ **Proved.**

Example 53. *Show that* $\int_a^b e^{-x^2}\,dx = \frac{\sqrt{\pi}}{2}[erf(b) - erf(a)]$

Solution. $$\frac{\sqrt{\pi}}{2}[erf(b) - erf(a)]$$

$$= \frac{\sqrt{\pi}}{2}\left[\frac{2}{\sqrt{\pi}}\int_0^b e^{-t^2}\,dt - \frac{2}{\sqrt{\pi}}\int_0^a e^{-t^2}\,dt\right] = \int_0^b e^{-t^2}\,dt - \int_0^a e^{-t^2}\,dt$$

$$= \int_0^b e^{-t^2}\,dt + \int_a^0 e^{-t^2}\,dt = \int_a^b e^{-t^2}\,dt = \int_a^b e^{-x^2}\,dx$$ **Proved.**

Example 54. *Show that*

$$\int_0^\infty e^{-x^2-2bx}\,dx = \frac{\sqrt{\pi}}{2}e^{b^2}\,[1 - erf(b)]$$

Solution. $$\int_0^\infty e^{-x^2-2bx}\,dx = \int_0^\infty e^{-x^2-2bx-b^2+b^2}\,dx = \int_0^\infty e^{-(x+b)^2}\cdot e^{b^2}\,dx$$

$$= e^{b^2}\int_b^\infty e^{-t^2}\,dt$$ [Put $x + b = t$]

$$= e^{b^2}\left[\int_0^\infty e^{-t^2}\,dt - \int_0^b e^{-t^2}\,dt\right]$$

$$= e^{b^2}\left[\frac{\sqrt{\pi}}{2}-\frac{\sqrt{\pi}}{2}erf(b)\right] = e^{b^2}\frac{\sqrt{\pi}}{2}[1-erf(b)]$$ **Proved.**

Example 55. *Prove that*

$$\frac{d}{dx}[erf_c(\alpha\, x)] = \frac{-2a}{\sqrt{\pi}}e^{-\alpha^2 x^2}$$

Solution. $$\frac{d}{dx}[erf_c(\alpha\, x)] = \frac{d}{dx}\left[\frac{2}{\sqrt{\pi}}\int_{\alpha\, x}^{\infty} e^{-t^2}\,dt\right]$$

On applying the rule of differentiation under integral sign, we get

$$= \frac{2}{\sqrt{\pi}}\left[\int_{\alpha\, x}^{\infty}\left(\frac{\partial}{\partial x}e^{-t^2}\right)dt + \frac{d}{dx}(\infty)e^{-\infty} - \frac{d}{dx}(\alpha\, x)e^{-\alpha^2 x^2}\right]$$

$$= \frac{2}{\sqrt{\pi}}\left[0+0-\alpha . e^{-\alpha^2 x^2}\right] = -\frac{2\alpha}{\sqrt{\pi}}e^{-\alpha^2 x^2}$$ **Proved.**

EXERCISE 9.5

Prove that

1. $erf_c(x) + erf_c(-x) = 2$
2. $erf_c(-x) = 1 + erf_c(x)$
3. $erf(x)\frac{2}{\sqrt{\pi}}\left[x - \frac{x^3}{3} + \frac{1}{2!}\frac{x^5}{5} - \frac{1}{3!}\frac{x^7}{7} + \ldots\ldots\right]$
4. $\int_0^{\infty} e^{-(x+a)^2}\,dx = \frac{\sqrt{\pi}}{2}[1-erf(a)]$
5. $\int_0^t erf_c(a\,x)\,dx = t\,erf_c(at) - \frac{e^{-a^2t^2}}{a\sqrt{\pi}} + \frac{1}{a\sqrt{\pi}}$
6. $\frac{d}{dx}[erf(a\,x)] = \frac{2a}{\sqrt{\pi}}e^{-a^2x^2}$
7. $\frac{d}{dx}[erf(\sqrt{x})] = \frac{e^{-x}}{\sqrt{\pi x}}$
8. $\frac{d}{dx}[erf(\sqrt{x})] = \frac{2}{\sqrt{\pi}}e^{-x^2}$

9.19 DIFFERENTIATION UNDER THE INTEGRAL SIGN

The value of a definite integral $\int_a^b f(x,\alpha)dx$ is a function of α (parameter), $F(\alpha)$ say. To find $F'(\alpha)$, first we have to evaluate the integral $\int_a^b f(x,\alpha)\,dx$ and then differentiate $F(\alpha)$ w.r.t. α. However, it is not always possible to evaluate the integral and then to find its derivative. Such problems are solved by reversing the order of the integration and differentiation *i.e.*, first differentiate $f(x,\alpha)$ partially w.r.t. "α" and then integrate it.

9.20 LEIBNITZ'S RULE

If $f(x,\alpha)$ and $\frac{\partial f(x,\alpha)}{\partial\alpha}$ be continuous functions of x and α, then

$$\frac{d}{d\alpha}\left[\int_a^b f(x,\alpha)\,dx\right] = \int_a^b \frac{\partial f(x,\alpha)}{\partial\alpha}dx\,.$$

Proof. Let $$\int_a^b f(x,\alpha)\,dx = F(\alpha)$$

then $$F(\alpha+\delta\alpha)=\int_a^b f(x,\alpha+\delta\alpha)\,dx$$

Hence $$F(\alpha+\delta)-F(\alpha)=\int_a^b f(x,\alpha+\delta\alpha)\,dx-\int_a^b f(x,\alpha)dx$$

$$=\int_a^b [f(x,\alpha+\delta\alpha)-f(x,\alpha)]\,dx$$

$$\frac{F(\alpha+\delta\alpha)-F(\alpha)}{\delta\alpha}=\int_a^b \frac{f(x,\alpha+\delta\alpha)-f(x,\alpha)}{\delta\alpha}dx$$

Taking limits of both sides as $\delta\alpha\to 0$, we have

$$\frac{\partial F}{d\alpha}=\int_a^b \frac{\partial f(x,\alpha)}{\partial\alpha}dx$$

The above formula is useful for evaluating definite integrals which are otherwise impossible to evaluate.

Example 56. *Evaluate* $\displaystyle\int_0^\infty \frac{\tan^{-1}(ax)}{x(1+x^2)}dx$

Solution. Let $$I=\int_0^\infty \frac{\tan^{-1}(ax)}{x(1+x^2)}dx$$

$\therefore$ $$\frac{dI}{da}=\int_0^\infty \frac{\partial}{\partial a}\frac{\tan^{-1}(ax)}{x(1+x^2)}dx \qquad ...(1)$$

$$=\int_0^\infty \frac{1}{x(1+x^2)}\cdot\frac{x}{1+a^2x^2}dx=\int_0^\infty \frac{1}{(1+x^2)(1+a^2x^2)}dx$$

Breaking the integrand into partial fractions,

$$=\int_0^\infty \frac{1}{1-a^2}\left[\frac{1}{1+x^2}-\frac{a^2}{1+a^2x^2}\right]dx=\frac{1}{1-a^2}\left[\tan^{-1}x-a\tan^{-1}ax\right]_0^\infty$$

$$=\frac{1}{1-a^2}\left[\frac{\pi}{2}-a\frac{\pi}{2}\right]=\frac{\pi}{2}\frac{1-a}{1-a^2}=\frac{\pi}{2}\cdot\frac{1}{1+a}$$

Now, integrating with respect to "a", we get $I=\dfrac{\pi}{2}\log(1+a)+c$ ...(2)

From (1), when $a=0$, then $I=0$

Putting $a=0$ and $I=0$ in (2), we get $c=0$

Hence (2) gives $\quad I=\dfrac{\pi}{2}\log(1+a)$ **Ans.**

Example 57. *Evaluate* $\displaystyle\int_0^\infty \frac{e^{-x}}{x}\left(a-\frac{1}{x}+\frac{1}{x}e^{-ax}\right)dx$

using the rule of differentiation under the sign of integration.

Solution. Let $$I=\int_0^\infty \frac{e^{-x}}{x}\left(a-\frac{1}{x}+\frac{1}{x}e^{-ax}\right)dx \qquad ...(1)$$

$$\frac{dI}{da}=\int_0^\infty \frac{\partial}{\partial a}\left[\frac{e^{-x}}{x}\left(a-\frac{1}{x}+\frac{1}{x}e^{-ax}\right)\right]dx=\int_0^\infty \frac{e^{-x}}{x}\left(1-0-\frac{x}{x}e^{-ax}\right)dx$$

$$=\int_0^\infty \frac{e^{-x}}{x}(1-e^{-ax})dx \qquad ...(2)$$

$$\frac{d^2I}{da^2}=\int_0^\infty \frac{\partial}{\partial a}\left[\frac{e^{-x}}{x}(1-e^{-ax})\right]dx=\int_0^\infty \frac{e^{-x}}{x}\left(xe^{-ax}\right)dx=\int_0^\infty e^{-(a+1)x}dx$$

$$=\left[\frac{e^{-(a+1)x}}{-(a+1)}\right]_0^\infty=\left[0+\frac{1}{a+1}\right]=\frac{1}{a+1} \qquad ...(3)$$

Integrating w.r.t.a, we have $\dfrac{dI}{da}=\log(a+1)+c_1$ ...(4)

Putting $a=0$ in (2), we get $\dfrac{dI}{da}=0$

Putting $a=0$ and $\dfrac{dI}{da}=0$ in (4), we get $c_1=0$

From (4), $\dfrac{dI}{da}=\log(a+1)$

$$I=\int\log(a+1)\,da=\log(a+1)\cdot a-\int\frac{a}{a+1}da=a\log(a+1)-\int\left(1-\frac{1}{a+1}\right)da$$

$$I=a\log(a+1)-a+\log(a+1)+c_2$$

$$I=(a+1)\log(a+1)-a+c_2 \qquad ...(5)$$

Putting $a=0$ in (1), we get $I=0$

Putting $a=0, I=0$ in (5), we get

$$0=c_2$$

Hence (5) gives

$$I=(a+1)\log(a+1)-a$$ **Ans.**

Example 58. *Evaluate the integral*

$$\int_0^\infty \frac{e^{-x}\sin bx}{x}dx$$

Solution. Let $I=\displaystyle\int_0^\infty \frac{e^{-x}\sin bx}{x}dx$ ... (1)

$$\frac{dI}{db}=\int_0^\infty \frac{\partial}{\partial b}\left(\frac{e^{-x}\sin bx}{x}\right)dx=\int_0^\infty \frac{e^{-x}\cdot x\cos bx}{x}dx=\int_0^\infty e^{-x}\cos bx\,dx$$

[We know that $\int e^{ax}\cos bx\,dx=\dfrac{e^{ax}}{a^2+b^2}(a\cos bx+b\sin bx)$]

$$=\left[\frac{e^{-x}}{1+b^2}(-\cos bx+b\sin bx)\right]_0^\infty$$

$$\frac{dI}{db}=\frac{1}{1+b^2} \qquad ....(2)$$

Integrating both sides of (2) w.r.t. 'b', we have $I = \tan^{-1} b + c$ ...(3)

On putting $b = 0$ in (1), we have $I = 0$

On putting $b = 0, I = 0$ in (3), we get $c = 0$

Hence (3) gives

$$I = \tan^{-1} b$$

or $$\int_0^\infty \frac{e^{-x} \sin bx}{x} dx = \tan^{-1} b$$ **Ans.**

Example 59. *Find the value of* $\int_0^\pi \frac{dx}{a + b\cos x}$ *(when* $a > 0, \ |b| < a$*)*

and deduce that $\int_0^\pi \frac{dx}{(a + b\cos x)^2} = \frac{\pi a}{(a^2 - b^2)^{3/2}}$

Solution. Let $I = \int_0^\pi \frac{dx}{a + b\cos x} = \int_0^\pi \frac{dx}{a\left(\cos^2 \frac{x}{2} + \sin^2 \frac{\pi}{2}\right) + b\left(\cos^2 \frac{x}{2} - \sin^2 \frac{x}{2}\right)}$

$$= \int_0^\pi \frac{dx}{(a+b)\cos^2 \frac{x}{2} + (a-b)\sin^2 \frac{x}{2}} = \frac{1}{a-b}\int_0^\infty \frac{\sec^2 \frac{x}{2} dx}{\frac{a+b}{a-b} + \tan^2 \frac{x}{2}}$$

$$= \frac{2}{a-b}\sqrt{\frac{a-b}{a+b}}\left[\tan^{-1}\left\{\tan\frac{x}{2}.\sqrt{\frac{a-b}{a+b}}\right\}\right]_0^\pi = \frac{2}{a-b}\sqrt{\frac{a-b}{a+b}}\left[\tan^{-1}\infty - \tan^{-1} 0\right]$$

Now differentiating both sides w.r.t. 'a' we get

$$\frac{dI}{da} = -\frac{1}{2}\frac{2\pi a}{(a^2 - b^2)^{3/2}} \quad \text{or} \quad \int_0^\pi \frac{\partial}{\partial a}\left(\frac{1}{a + b\cos . x}\right) dx = -\frac{1}{2}\frac{2\pi a}{(a^2 - b^2)^{3/2}}$$

$$\Rightarrow \quad \int_0^\pi \frac{-1}{(a + b\cos x)^2} dx = \frac{-\pi a}{(a^2 - b^2)^{3/2}}$$

$$\Rightarrow \quad \int_0^\pi \frac{1}{(a + b\cos x)^2} dx = \frac{\pi a}{(a^2 - b^2)^{3/2}}$$ **Ans.**

EXERCISE 9.6

Prove that:

1. $\int_0^\infty \frac{1 - e^{-ax}}{x} e^{-x} dx = \log(1 + a)$

2. $\int_0^\infty \frac{e^{-ax} - e^{-bx}}{x} dx = \log\frac{b}{a}$

3. $\int_0^\infty \frac{e^{-ax}\text{sinx}}{x} dx = \cot^{-1} \alpha$ and hence deduce that $\int_0^\infty \frac{\sin x}{x} dx = \frac{\pi}{2}$

4. $\int_0^1 \frac{x^a - x^b}{\log x} dx = \log\frac{a+1}{b+1}$

5. $\int_0^\infty \frac{\cos \lambda x}{x}(e^{-ax} - e^{-bx}) dx = \frac{1}{2}\log\frac{b^2 + \lambda^2}{a^2 + \lambda^2}$, $(a > 0, \ b > 0)$

6. $\int_0^\infty e^{-bx^2} \cos 2ax\, dx = \frac{1}{2}\frac{\sqrt{\pi}}{b} e^{-a^2/b}$ $(b > 0)$ Assume $\int_0^\infty e^{-x^2} dx = \frac{\sqrt{\pi}}{2}$

Evaluate the following.

7. $\int_0^{\frac{\pi}{2}} \log(\alpha^2 \cos^2\theta + \beta^2 \sin^2\theta)\, d\theta$ $(\alpha > 0, \beta > 0)$ **Ans.** $\pi \log \frac{\alpha+\beta}{2}$

8. $\int_0^\infty \frac{\log(1+a^2x^2)}{1+b^2x^2} dx$ **Ans.** $\frac{\pi}{l} \log \frac{a+b}{b}$

9. $\int_0^{\frac{\pi}{2}} \log\left(\frac{a+b\sin\theta}{a-b\sin\theta}\right) \cdot \frac{d\theta}{\sin\theta}$ **Ans.** $\pi \sin^{-1}\frac{b}{a}$

10. $\int_0^{\frac{\pi}{2}} \frac{\log(1+\cos\alpha \cdot \cos x)}{\cos x} dx$ **Ans.** $\frac{1}{2}\left(\frac{\pi^2}{4} - \alpha^2\right)$

Prove that

11. $\int_0^{\frac{\pi}{2}} \frac{\log(1+y\sin^2 x)}{\sin^2 x} dx = \pi[\sqrt{(1+y)} - 1]$ When $y > 1$.

12. $\int_0^{\frac{\pi}{2}} \frac{dx}{(a^2 + \sin^2 x + b^2 \cos^2 x)^2} = \frac{\pi(a^2+b^2)}{4a^3b^3}$

9.21 RULE OF DIFFERENTIATION UNDER THE INTEGRAL SIGN WHEN THE LIMITS OF INTEGRATION ARE FUNCTIONS OF THE PARAMETER

If $f(x, \alpha)$, $\frac{\partial f(x,\alpha)}{\partial \alpha}$ be continuous functions of x and α, then

$$\frac{d}{d\alpha}\left\{\int_{\phi(\alpha)}^{\psi(\alpha)} f(x,\alpha)dx\right\} = \int_{\phi(\alpha)}^{\psi(\alpha)} \frac{\partial f(x,\alpha)}{\partial \alpha} dx - \frac{d\phi}{d\alpha} f[\phi(\alpha),\alpha] + \frac{d\psi}{d\alpha} f[\psi(\alpha),\alpha]$$

Example 60. *Verify the rule of differentiation under the sign of integration for* $\int_0^{a^2} \tan^{-1}\frac{x}{a} dx$

Solution. Let $I = \int_0^{a^2} \tan^{-1}\frac{x}{a} dx$

$$\frac{d}{d\alpha}\left\{\int_{\phi(\alpha)}^{\psi(\alpha)} f(x,\alpha)dx\right\} = \int_{\phi(\alpha)}^{\psi(\alpha)} \frac{\partial f(x,\alpha)}{\partial \alpha} dx - \frac{d\phi}{d\alpha} f[\phi(\alpha),\alpha] + \frac{d\psi}{d\alpha} f[\psi(\alpha),\alpha]$$

$$\frac{dI}{da} = \int_0^{a^2} \left[\frac{\partial}{\partial a}\left(\tan^{-1}\frac{x}{a}\right)\right] dx - 0 + 2a\left[\tan^{-1}\frac{a^2}{a}\right]$$

$$= \int_0^{a^2} \frac{1}{1+\frac{x^2}{a^2}}\left(-\frac{x}{a^2}\right) dx + 2a\tan^{-1} a = \int_0^{a^2} -\frac{x}{a^2+x^2} dx + 2a\tan^{-1} a$$

$$= \left[-\frac{1}{2}\log(a^2+x^2)\right]_0^{a^2} + 2a\tan^{-1} a = -\frac{1}{2}\log(a^2+a^4) + \frac{1}{2}\log a^2 + 2a\tan^{-1} a$$

$$= -\frac{1}{2}\log\frac{a^2+a^4}{a^2} + 2a\tan^{-1}a = -\frac{1}{2}\log(a^2+1) + 2a\tan^{-1}a \qquad ...(1)$$

Now integration by parts

$$I = \int_0^{a^2} \tan^{-1}\frac{x}{a}\cdot 1\,dx = \left[\left(\tan^{-1}\frac{x}{a}\right)x\right]_0^{a^2} - \int_0^{a^2} \frac{1}{1+\frac{x^2}{a^2}}\frac{1}{a}\cdot x\,dx$$

$$= \left[a^2\tan^{-1}\frac{a^2}{a}\right] - \int_0^{a^2}\frac{ax}{a^2+x^2}dx = a^2\tan^{-1}a - \frac{a}{2}[\log(x^2+a^2)]_0^{a^2}$$

$$= a^2\tan^{-1}a - \frac{a}{2}[\log(a^4+a^2) - \log a^2] = a^2\tan^{-1}a - \frac{a}{2}\log\frac{a^4+a^2}{a^2}$$

$$= a^2\tan^{-1}a - \frac{a}{2}\log(a^2+1)$$

$$\frac{dI}{da} = \left[a^2\frac{1}{1+a^2} + 2a\tan^{-1}a\right] - \left[\frac{a}{2}\frac{2a}{a^2+1} + \frac{1}{2}\log(a^2+1)\right]$$

$$= \frac{a^2}{1+a^2} + 2a\tan^{-1}a - \frac{a^2}{a^2+1} - \frac{1}{2}\log(a^2+1) = 2a\tan^{-1}a - \frac{1}{2}\log(a^2+1) \qquad ...(2)$$

From (1) and (2), the rule is verified.

Example 61. *Evaluate* $\int_0^{\alpha}\frac{\log(1+\alpha x)}{1+x^2}dx$ *and hence show that*

$$\int_0^1 \frac{\log(1+x)}{1+x^2}dx = \frac{\pi}{8}\log_e 2$$

Solution. Let $\quad I = \int_0^{\alpha}\frac{\log(1+\alpha x)}{1+x^2}dx \qquad ...(1)$

$$\left[\frac{d}{d\alpha}\left\{\int_{\phi(\alpha)}^{\psi(\alpha)} f(x,\alpha)dx\right\} = \int_{\phi(\alpha)}^{\psi(\alpha)}\frac{\partial f(x,\alpha)}{\partial\alpha}dx - \frac{d\phi}{d\alpha}f[\phi(\alpha),\alpha] + \frac{d\psi}{d\alpha}f[\psi(\alpha),\alpha]\right]$$

$$\frac{dI}{d\alpha} = \int_0^{\alpha}\frac{\partial}{\partial\alpha}\left\{\frac{\log(1+\alpha x)}{1+x^2}\right\}dx + \frac{d\alpha}{d\alpha}f(\alpha,\alpha) = \int_0^{\alpha}\frac{x}{(1+x^2)(1+\alpha x)}dx + \frac{\log(1+\alpha^2)}{1+\alpha^2}$$

Converting into partial fractions,

$$= -\frac{\alpha}{1+\alpha^2}\int_0^{\alpha}\frac{dx}{1+\alpha x} + \frac{1}{2(1+\alpha^2)}\int_0^{\alpha}\frac{2x}{1+x^2}dx + \frac{\alpha}{1+\alpha^2}\int_0^{\alpha}\frac{dx}{1+x^2} + \frac{\log(1+\alpha^2)}{1+\alpha^2}$$

$$= -\frac{\alpha}{1+\alpha^2}\left[\frac{1}{\alpha}\log(1+\alpha x)\right]_0^{\alpha} + \frac{1}{2(1+\alpha^2)}[\log(1+x^2)]_0^{\alpha} + \frac{\alpha}{1+\alpha^2}[\tan^{-1}x]_0^{\alpha} + \frac{\log(1+\alpha^2)}{1+\alpha^2}$$

$$= -\frac{1}{1+\alpha^2}\log(1+\alpha^2) + \frac{\log(1+\alpha^2)}{2(1+\alpha^2)} + \frac{\alpha}{1+\alpha^2}\tan^{-1}\alpha + \frac{\log(1+\alpha^2)}{1+\alpha^2}$$

$$= \frac{\log(1+\alpha^2)}{2(1+\alpha^2)} + \frac{\alpha}{1+\alpha^2}\tan^{-1}\alpha$$

On integrating both sides w.r.t. α, we have

$$I=\frac{1}{2}\int \log(1+\alpha^2)\cdot\frac{1}{1+\alpha^2}d\alpha+\int\frac{\alpha\tan^{-1}\alpha}{1+\alpha^2}d\alpha+c$$

$$I=\frac{1}{2}\log(1+\alpha^2)\cdot\tan^{-1}\alpha-\frac{1}{2}\int\frac{2\alpha}{1+\alpha^2}\cdot\tan^{-1}\alpha\,d\alpha+\int\frac{\alpha\tan^{-1}\alpha}{1+\alpha^2}d\alpha$$

$$I=\frac{1}{2}\log(1+\alpha^2)\cdot\tan^{-1}\alpha+c \qquad \text{... (2)}$$

From (1), when $\alpha = 0$, then $I = 0$. From (2), when $\alpha = 0$, $I = 0$, then $c = 0$

Hence (2) gives

$$I=\frac{1}{2}\log(1+\alpha^2)\cdot\tan^{-1}\alpha \qquad \text{...(3)}$$

On putting $\alpha = 1$ in (3), we have

$$\int_0^1\frac{\log(1+x)}{1+x^2}dx=\frac{1}{2}\log(1+1)\cdot\tan^{-1}(1)=\frac{1}{2}(\log_e 2)\frac{\pi}{4}=\frac{\pi}{8}\log_e 2$$ **Ans.**

Example 62. *Evaluate* $\displaystyle\int_{\frac{\pi}{6a}}^{\frac{\pi}{2a}}\frac{\sin ax}{x}dx$

Solution. Let $\displaystyle I=\int_{\frac{\pi}{6a}}^{\frac{\pi}{2a}}\frac{\sin ax}{x}dx \qquad \text{...(1)}$

$$\left[\frac{d}{d\alpha}\left\{\int_{\phi(\alpha)}^{\psi(\alpha)}f(x,\alpha)dx\right\}=\int_{\phi(\alpha)}^{\psi(\alpha)}\frac{\partial f(x,\alpha)}{\partial\alpha}dx-\frac{d\phi}{d\alpha}f[\phi(\alpha),\alpha]+\frac{d\psi}{d\alpha}f[\psi(\alpha),\alpha]\right]$$

$$\frac{dI}{da}=\int_{\frac{\pi}{6a}}^{\frac{\pi}{2a}}\frac{\partial}{\partial a}\left(\frac{\sin ax}{x}\right)dx-\frac{d}{da}\left(\frac{\pi}{6a}\right)\cdot\frac{\left[\sin a\cdot\frac{\pi}{6a}\right]}{\frac{\pi}{6a}}+\frac{d}{da}\left(\frac{\pi}{2a}\right)\cdot\frac{\sin a\cdot\frac{\pi}{2a}}{\frac{\pi}{2a}}$$

$$=\int_{\frac{\pi}{6a}}^{\frac{\pi}{2a}}\cos ax\,dx+\frac{1}{2a}-\frac{1}{a}=\left[\frac{\sin ax}{a}\right]_{\frac{\pi}{6a}}^{\frac{\pi}{2a}}-\frac{1}{2a}$$

$$=\frac{1}{a}\left[\sin a\cdot\frac{\pi}{2a}-\sin a\cdot\frac{\pi}{6a}\right]-\frac{1}{2a}$$

$$=\frac{1}{a}\left[\sin\frac{\pi}{2}-\sin\frac{\pi}{6}\right]-\frac{1}{2a}=\frac{1}{a}\left[1-\frac{1}{2}\right]-\frac{1}{2a}=\frac{1}{2a}-\frac{1}{2a}=0$$

Integrating we have $\quad I = \text{Constant}$ **Ans.**

Example 63. *If* $y=\int_0^x f(t)\sin[k(x-t)]\,dt$, *prove that y satisfies the differential equation*

$$\frac{d^2y}{dx^2}+k^2y=kf(x)$$

Solution. $$y = \int_0^x f(t) \sin [k (x - t)]\, dt$$

$$\frac{dy}{dx} = \int_0^x \frac{\partial}{\partial x} [f(t) \sin\{k (x - t\}]dt - 0\ \frac{d}{dx}(x) \cdot f(x) \sin\{k (x - x)\}$$

$$= \int_0^x f(t)\, k \cos\{k (x - t)\} \cdot dt = k \int_0^x f(t) \cos\{k (x - t)\}\, dt$$

Again applying the same rule

$$\frac{d^2y}{dx^2} = k\left[\int_0^x \frac{\partial}{\partial x}\{f(t) \cos k(x - t)\} dt - 0 + \frac{d}{dx}(x) f(x) \cos k(x - x)\right]$$

$$= -k^2 \int_0^x f(t) \sin [k (x - t)]\, dt + k f(x) = -k^2 y + k f(x)$$

$$\Rightarrow \qquad \frac{d^2y}{dx^2} + k^2 y = k f(x)$$ **Proved.**

Example 64. *Using differentition under integral sign, evaluate* $\int_0^1 \frac{x^\alpha - 1}{\log x} dx, \alpha \geq 0$

(AMIETE, June 2010)

Solution. Here, we have

$$I = \int_0^1 \frac{x^\alpha - 1}{\log x} dx \qquad \text{... (1)}$$

$$\left[\frac{d}{d\alpha}\left\{\int_{\phi(\alpha)}^{\psi(\alpha)} f(x, \alpha)\, dx\right\} = \int_{\phi(\alpha)}^{\psi(\alpha)} \frac{\partial f(x,\alpha)}{\partial \alpha} dx - \frac{d\phi}{d\alpha} f[\phi(\alpha), \alpha] + \frac{d\psi}{d\alpha} f[\psi(\alpha), \alpha]\right]$$

$$\frac{dI}{d\alpha} = \int_0^1 \frac{\partial}{\partial \alpha}\left(\frac{x^\alpha - 1}{\log x}\right) dx - \frac{d(0)}{d(\alpha)} f(\theta, \alpha) + \frac{d(1)}{d\alpha} f(1, \alpha)$$

$$= \int_0^1 \frac{x^\alpha \log x}{\log x} dx - 0 + 0 = \int_0^1 x^\alpha\, dx$$

$$= \left[\frac{x^{\alpha+1}}{\alpha + 1}\right]_0^1 = \frac{1}{\alpha + 1}$$

$$\frac{dI}{d\alpha} = \frac{1}{\alpha + 1}$$

Now, integrating both sides *w.r.t.* α, we get

$$I = \int \frac{d\alpha}{\alpha + 1} = \log(\alpha + 1) + C \qquad \text{... (2)}$$

From (1), when $\alpha = 0$, then $I = 0$

Putting $\alpha = 0$ and $I = 0$ in (2), we get $C = 0$

Hence (2) becomes $I = \log(\alpha + 1)$ **Ans.**

CHAPTER 10

THEORY OF ERRORS

10.1 NUMBERS

There are two types of numbers

(*i*) Exact

(*ii*) Approximate

For example; Exact numbers are 1, 3, 5, 7, 10, $\frac{5}{2}$, 6.23.

Approximate numbers are $\frac{4}{3} = 1.3333..........$

$\sqrt{2} = 1.414213..........$

$\pi = 3.141592..........$

The value of the left hand side can not be expressed by a finite number of digits.

Approximate value of $\frac{4}{3} = 1.3333$

App. value of $\sqrt{2} = 1.4142$

and the app. value of $\pi = 3.1416$

10.2 SIGNIFICANT FIGURES

The digits used to express a number are called significant digits (figures).

8123, 3.187, 0.8725, contains 4 significant figures. While the numbers 0.0163, 0.00127, 0.000365 and 0.0000345 contain only three significant figures (digits).

Since zeroes before decimal and after decimal only helps to fix the position of decimal point. Similarly, the numbers 52000 and 8700.00 have two significant figures only.

10.3 ROUNDING OFF

These are number with larger number of digits.

For example; $\frac{22}{7} = 3.14285143$

In practice it is convenient to limit such number as 3.14 or 3.143.

The dropping of the digits is called rounding off.

Rule: (1) To round of a number to *n* significant numbers ignore all the digits to the right of *n*th digit if there is some digit ignore it.

(2) Less than half a unit leave this unit.

(3) Greater than half unit is taken as full unit.

(4) Exactly half unit is taken as one unit in the case of odd numbers *i.e.*, increased the odd number by one. If nth number is even, then nth number should not be changed.

5.783 to 5.78	7.767 to 7.77
15.976 to 15.9	95767 to 95800
8.4365 to 8.44	87.656 to 87.6

Also the numbers 7.284359, 15.864651, 9.464762 rounded off to four places of decimals at 7.2844, 15.8646, 9.4648 respectively.

10.4 TYPES OF ERRORS

(1) Absolute Errors

The error is defined as a quantity which is added to true value in order to obtain the measured value.

True value + Error = Measured value/observed value.

Correction. The error with sign changed is called correction.

Measured value + Correction = True value.

If x is the true value and X' is approximate value then $|X - X'|$ is called the absolute error.

(2) **Relative Error**

$$\text{Relative error} = \left|\frac{X - X'}{X}\right|$$

(3) **Percentage Error**

$$\text{Percentage error} = \frac{100\,|X - X'|}{X}$$

(4) **Inherent Error**

Errors which are already in data for calculation of a problem before its solution are called inherent error. Such error arrise due to limitation of mathematical tables or the digital computer.

(5) **Rounding off the errors**

Such error arrise by the process of rounding off the numbers. Such errors are un avoidable most of the calculation.

(6) **Truncation Error**

Truncation error are caused by using approximate result on replacing an infinite series.

For example ; if $e^x = 1 + x + \frac{x^2}{2!} + \frac{x^3}{3!} + \frac{x^4}{4!} + ... \infty = X$ (say)

is replaced by $1 + x + \frac{x^2}{2!} + \frac{x^3}{3!} + \frac{x^4}{4!} = X'$ (say)

then the truncation error $= X - X'$

Note : (1) If a number is correct to n decimal places then the error is $= \frac{1}{2} 10^{-n}$.

For example **:** If the number is $\sqrt{2} = 1.1414$ correct to four decimal places, then the error $= \frac{1}{2} \times 10^{-4}$

(2) If the first significant figure of a number is λ and the number is correct to n significant figures, then the relative error is less than $\dfrac{1}{\lambda \times 10^{n-1}}$.

Verification. 974.16 is correct to five significant figure.

Here, $\lambda = 9, n = 5$

$$\text{Absolute error} = \frac{0.01}{2} = 0.005$$

$$\text{Relative error} < \frac{0.005}{974.16} = \frac{5}{974160} = \frac{1}{2 \times 97416}$$

$$< \frac{1}{2 \times 90000} = \frac{1}{2 \times 9 \times 10^4}$$

$$< \frac{1}{9 \times 10^4}$$

$$\textit{i.e.,} \quad \frac{1}{\lambda \times 10^{n-1}}.$$

Example 1. *Round off the numbers 754126 and 16.73117 to four significant figures. Compute absolute error relative error and percentile error.*

Solution. Number rounded off to 4 significant figure equal to 754100

$$\text{Absolute error} = |X - X'| = |754126 - 754100| = |26| = 26$$

$$\text{Relative error} = \left|\frac{X - X'}{X}\right| = \frac{26}{754126} = 3.45 \times 10^{-5}$$

$$\text{Percentile error} = \frac{|X - X'|}{X} \times 100$$

$$= 3.45 \times 10^{-5} \times 100 = 3.45 \times 10^{-3}$$

(*ii*) Number rounded off to four significant figure is 16.73

$$\text{Absolute error} = |X - X'| = |16.73117 - 16.73| = 0.00117$$

$$\text{Relative error} = \left|\frac{X - X'}{X}\right| = \frac{0.00117}{16.73117} = 6.99 \times 10^{-5}$$

$$\text{Percentile error} = \left|\frac{X - X'}{X}\right| \times 100 = 6.99 \times 10^{-5} \times 100 = 6.99 \times 10^{-3}$$ **Ans.**

EXERCISE 10.1

Round off the following numbers correct to three significant figures :

1. 0.0031614 **Ans.** 0.00316

2. 16.132102 **Ans.** 16.1

3. 0.30617 **Ans.** 0.306

4. 2945567 **Ans.** 2940000

5. 45.56735 **Ans.** 45.6

6. 5.26521 **Ans.** 5.26

7. Find the relative error if $\frac{1}{3}$ is approximated to 0.334. **Ans.** 0.002

8. Find the percentage error if 625.483 is approximated to three significant figures. **Ans.** 0.077

9. $\sqrt{29} = 5.385$ and $\pi = 3.317$ correct to four significant figures. Find the relative errors in their sum and difference. **Ans.** 1.149×10^{-4}, 4.836×10^{-4}

10.5 ERROR DUE TO APPROXIMATION OF THE FUNCTION

Let $z = f(x, y)$ be a function of two variables x and y.

If δx, δy be the errors in x and y, then the error in z is given by $z + \delta z = f(x + \delta x, y + \delta y)$.

Expanding $f(x, y)$ by Taylor's series, we get

$$z + \delta z = f(x, y) + \left(\frac{\partial f}{\partial x}\delta x + \frac{\partial f}{\partial y}\delta y\right) + \text{terms involving higher powers of } \delta x \text{ and } \delta y. \quad ...(1)$$

If δx and δy be so small that their squares and higher powers can be neglected, then (1) can be written as

$$dz = \frac{\partial z}{\partial x}\delta x + \frac{\partial z}{\partial y}\delta y \quad \text{(app.)}$$

In general, if $z = f(x_1, x_2, ... x_n)$ and there are errors in $x_1, x_2 ..... x_n$, then

$$\delta z = \frac{\partial z}{\partial x_1}dx_1 + \frac{\partial z}{\partial x_2}dx_2 + \frac{\partial z}{\partial x_3}dx_3 + ... + \frac{\partial z}{\partial x_n}dx_n.$$

Example 2. *If* $u = \dfrac{5x^3y^4}{z^5}$ *and errors in x, y, z be 0.001, and compute the relative maximum error when x =1, y = 1, z = 1.*

Solution.

$$u = \frac{5x^3y^4}{z^5} \quad ...(1)$$

$$\delta x = \delta y = \delta z = 0.001$$

and $$x = y = z = 1$$

Differentiating (1) partially, with respect to 'x', we get

$$\frac{\delta u}{\delta x} = \frac{15x^2y^4}{z^5}, \quad \frac{\delta u}{\delta y} = \frac{20x^3y^3}{z^5}, \quad \frac{\delta u}{\delta z} = -\frac{25x^3y^4}{z^6}$$

Now, we know that

$$\delta u = \frac{\partial u}{\partial x}\delta x + \frac{\partial u}{\partial y}\delta y + \frac{\partial u}{\partial z}\delta z$$

$$= \frac{15x^2y^4}{z^5}\delta x + \frac{20x^3y^3}{z^5}\delta y - \frac{25x^3y^4}{z^6}\delta z$$

The error being maximum

$$(\delta u)_{max} = \left|\frac{15x^2y^4}{z^5}\delta x\right| + \left|\frac{20x^3y^3}{z^5}\delta y\right| + \left|\frac{25x^3y^4}{z^6}\delta z\right|$$

$$= \left|\frac{15(1)(1)}{(1)}(0.001)\right| + \left|\frac{20(1)(1)}{(1)}(0.001)\right| + \left|\frac{25(1)(1)}{(1)}(0.001)\right|$$

$$= 0.015 + 0.020 + 0.025 = 0.06$$

$$\text{Relative error} = \frac{(\delta u)_{max}}{u} = \frac{0.06}{5} = 0.012 \quad \textbf{Ans.}$$

Example 3. *Find the maximum error in magnitude in the approximation* $f(x, y) = x^2 - xy + \frac{1}{2}y^2 + 3$ *over the rectangle* $R : |x - 3| < 0.01$ *and* $|y - 2| < 0.01$.

Solution. Here, we have

$$f(x, y) = x^2 - xy + \frac{1}{2}y^2 + 3$$

$$\frac{\partial f}{\partial x} = 2x - y, \quad \frac{\partial f}{\partial y} = -x + y.$$

We know that

$$\text{Maximum } \delta f = \left|\frac{\partial f}{\partial x}\delta x\right| + \left|\frac{\partial f}{\partial y}\delta y\right|$$

$$= |(2x - y)\delta x| + |(-x + y)\delta y|$$

$$= |(2 \times 3 - 2)(0.01)| + |(-3 + 2)\, 0.01|$$

$$= 4(0.01) + |-0.01| = 0.05$$ **Ans.**

10.6 ERROR IN A SERIES APPROXIMATION

By Taylor series of one variable

$$f(x) = f(a + \overline{x - a}) = f(a) + (x - a)f'(a) + \frac{(x-a)^2}{2!}f''(a) + \ldots + \frac{(x-a)^{n-1}}{(n-1)!}f^{n-1}(a) + R_n(x)$$

Here $R_n(x) = \frac{(x-a)^n}{n!}f^n(\theta),\ a < \theta < x$

For a convergent series $R_n(x) \to 0$ as $n \to \infty$.

Approximate value of series = First n terms of the series.

We can find the number of terms for a particular desired accuracy.

Example 4. *Correct to five places of decimal at* $x = 1$ *find the number of terms of the approximate series of* e^x.

Solution. We know that

$$e^x = 1 + x + \frac{x^2}{2!} + \frac{x^3}{3!} + \ldots + \frac{x^{n-1}}{(n-1)!} + R_n(x)$$

Here $$R_n(x) = \frac{x^n}{n!}e^\theta, 0 < \theta < x$$

Maximum absolute error at $(\theta = x) = \frac{x^n}{n!}f^n(x) = \frac{x^n}{n!}e^x$

Maximum error at $(x = 1) = \frac{1}{n!}$ $\qquad (x = 1)$

Maximum error correct to five decimal places $\frac{1}{n!} < \frac{1}{2}10^{-5}$.

$$\Rightarrow \quad n! > 2 \times 10^5 \qquad (8! = 40320)$$

$$8! > 2 \times 10^5 \Rightarrow n = 8$$

Hence there are 8 terms in order that the sum is correct to five places of decimal. **Ans.**

EXERCISE 10.2

1. Find the number of term of the approximated series of e^x correct to six decimal places.

   Ans. $n = 10$

2. Find the number of terms in the approximated series of log $(1 + x)$ at $x = 1$, (log 2) to six decimal places.

   Ans. $n = 0$

3. The fractional error in the measurement of x is 0.001. What is the corresponding error in expansion of e^x.

4. If $R = \frac{4xy^2}{z^3}$ and errors in x, y, z be 0.001, show that the maximum relative error at $x = y = z = 1$ is 0.006.

10.7 ORDER OF APPROXIMATION

$$\text{Function} = f(h)$$

$$\text{Approximate value of function} = \phi(x)$$

$$\text{Error} = E(h^n)$$

$$|f(h) - \phi(h)| \le E\,|h^n|$$

$$\text{Order of error} = O(h^n)$$

$$f(h) = \phi(h) + O(h^n)$$

Example 5. *Write down with fifth order of approximation of* $\frac{1}{1-h}$.

Solution. We know that

$$\frac{1}{1-h} = (1-h)^{-1} = 1 + h + h^2 + h^3 + h^4 + h^5 + h^6 + h^7 + ...$$

$$= 1 + h + h^2 + h^3 + h^4 + 0(h^5)$$

Example 6. *Write down the seventh order of approximation of* $\sin|h|$.

Solution. We know that

$$\sin|h| = h - \frac{h^3}{3!} + \frac{h^5}{5!} - \frac{h^7}{7!} + \frac{h^9}{9!} + ...$$

$\sin|h|$ with seventh order of approximation

$$\sin(h) = h - \frac{h^3}{3!} + \frac{h^5}{5!} + O(h^7)$$ **Ans.**

10.8 MOST PROBABLE VALUE AND RESIDUAL

Let true value of a quantity be X.

Their approximate values are $X_1, X_2, X_3, ..., X_n$.

and the corresponding probable errors are $x_1, x_2, x_3, ... x_n$.

$$x_1 = X_1 - X, x_2 = X_2 - X, x_3 = X_3 - X, ..., x_n = X_n - X$$

In fact we cannot get true value of a quantity due to random errors. For practical purposes we take a probable value $\bar{X}$ of a quantity in place of true value. The probable value $\bar{X}$ is the average of $X_1, X_2, X_3, ..., X_n$.
$$\left[\bar{X} = \frac{X_1 + X_2 + X_3 + ... + X_n}{n}\right]$$

We define the residual by.

$$d_1 = X_1 - \bar{X}, d_2 = X_2 - \bar{X}, d_3 = X_3 - \bar{X} \,..... \, d_n = X_n - \bar{X}.$$

$d_1, d_2, d_3 \,.....\, d_n$ are the residual error and $x_1, x_2, x_3, \,.....\, x_n$ are the probable error.

10.9 GAUSSIAN ERROR

Errors and residuals are neither systematic nor constants but equally likely to be positive or negative.

Small errors are more frequent than large ones.

Very large errors don't occur at all.

Under these conditions the errors follow the law of probability given by normal distribution.

$$y = \frac{1}{\sigma\sqrt{2\pi}} e^{-\frac{(x-\mu)^2}{2\sigma^2}}$$

$$y = \frac{1}{\sigma\sqrt{2\pi}} e^{-\frac{x^2}{2\sigma^2}} \qquad (\mu = 0)$$

On putting $y = N$, $h^2 = \frac{1}{2\sigma^2}$

$$N = \frac{h}{\sqrt{\pi}} e^{-h^2x^2}$$

This given the relative number of measurements of N having error x and h is called precision index.

On plotting a graph between N and x, we get the Gaussian error curve.

10.10 THEORETICAL DISTRIBUTIONS

(1) **Binomial Distribution** $(q + p)^n$

$$P(r) = {}^nC_r p^r q^{n-r}$$

$$\text{Mean} = np$$

$$\text{S.D.} = \sqrt{npq}$$

$$\text{Variance} = npq$$

Mode = Most probable of success = $(n + 1)p$

Recurrence relation, $P(r+1) = \frac{n-r}{r+1}\left(\frac{p}{q}\right).P(r).$

(2) **Poisson's distribution**

$$P(r) = \frac{e^{-m}\, m^r}{r!}$$

$$\text{Mean} = m$$

$$\text{S.D.} = \sqrt{m}$$

$$\text{Variance} = m$$

Mode = $[m]$ = Integral part of m $\qquad m - 1 \le r \le m$

Recurrence relation $P(r+1) = \frac{m}{r+1} P(r).$

(3) **Normal distribution**

$$f(x) = \frac{1}{\sigma\sqrt{2\pi}} e^{-\frac{(x-\mu)^2}{2\sigma^2}}$$

$$\text{Mean} = \mu$$

$$\text{Standard deviation} = \sigma$$

$$\text{Medium} = 0$$

$$\text{Modal ordinate} = \frac{1}{\sigma\sqrt{2\pi}}.$$

EXERCISE 10.3

1. Explain the meaning of the terms mean and standard deviation of a term.
2. Calculate the mean deviation and standard deviation of the series
   $a, a + d, a + 2d, \ldots\ldots, a + 2nd$
3. Explain what do you mean by binomial distribution. Find its mean and standard deviation.
4. Define Poisson's distribution. Discuss its importance in physics.
5. Calculate mean and standard deviation of Poisson's distribution.
6. Define probability density function for the normal distribution.
7. Define binomial and normal probability distribution and compare them.
8. Assuming that N is large, show that the error in writing $\sigma_{\bar{x}} = \frac{\sigma}{\sqrt{n}}$ is approximately $\frac{50(n-1)}{N}$ percent of the value of $\sigma_{\bar{x}}$.
9. State and prove the normal law of errors and find an expression of the measure of precision and the probable error of the arithmetic mean *(D.U. May 2010)*.
10. Derive the normal law of errors and calculate the probable error of an observation.

Chapter 11: Fourier Series

11.1 PERIODIC FUNCTIONS

If the value of each ordinate $f(t)$ repeats itself at equal intervals in the abscissa, then $f(t)$ is said to be a periodic function.

If $f(t) = f(t+T) = f(t+2T) = \ldots$ then T is called the period of the function $f(t)$.

For example:

The period of $\sin x$, $\cos x$, $\sec x$, and $\operatorname{cosec} x$ is 2π.

The period of $\tan x$ and $\cot x$ is π.

$\sin x = \sin(x + 2\pi) = \sin(x + 4\pi) = \ldots$ so $\sin x$ is a periodic function with the period 2π.

$$\sin 5x = \sin(5x + 2\pi) = \sin 5\left(x + \frac{2\pi}{5}\right), \qquad \text{Period} = \frac{2\pi}{5}$$

$$\cos 3x = \cos(3x + 2\pi) = \cos 3\left(x + \frac{2\pi}{3}\right), \qquad \text{Period} = \frac{2\pi}{3}$$

$$\cos\frac{2n\pi x}{k} = \cos\left(\frac{2n\pi x}{k} + 2\pi\right) = \cos\frac{2n\pi}{k}\left(x + \frac{2\pi k}{2n\pi}\right)$$

$$= \cos\frac{2n\pi}{k}\left(x + \frac{k}{n}\right), \qquad \text{Period} = \frac{k}{n}$$

$$\tan 2x = \tan(2x + \pi) = \tan 2\left(x + \frac{\pi}{2}\right), \qquad \text{Period} = \frac{\pi}{2}$$

This is also called sinusoidal periodic function.

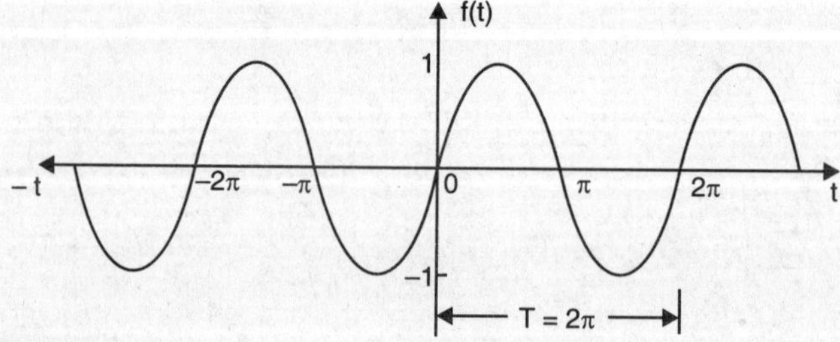

11.2 FOURIER SERIES

Here we will express a non-sinusoidal periodic function into a fundamental and its harmonics. A series of sines and cosines of an angle and its multiples of the form

$$\frac{a_0}{2} + a_1 \cos x + a_2 \cos 2x + a_3 \cos 3x + ... + a_n \cos nx + ...$$
$$+ b_1 \sin x + b_2 \sin 2x + b_3 \sin 3x + ... + b_n \sin nx + ...$$
$$= \frac{a_0}{2} + \sum_{n=1}^{\infty} a_n \cos nx + \sum_{n=1}^{\infty} b_n \sin nx.$$

is called the Fourier series, where $a_0, a_1, a_2, ... a_n, ... b_1, b_2, b_3 ... b_n ...$ are constants.

A periodic function $f(x)$ can be expanded in a Fourier Series. The series consists of the following:

(*i*) A constant term a_0 (called d.c. component in electrical work).

(*ii*) A component at the fundamental frequency determined by the values of a_1, b_1.

(*iii*) Components of the harmonics (multiples of the fundamental frequency) determined by a_2, $,a_3 ... b_2, b_3 ....$ And $a_0, a_1, a_2 ... ,b_1, b_2 ...$ are known as *Fourier coefficients* or Fourier constants.

Note. (1) When the function and its derivatives are continuous then the function can be expanded in powers of x by Maclaurin's theorem.

(2) But by Fourier series we can expand continuous and discontinuous both types of functions under certain conditions.

11.3 DIRICHLET'S CONDITIONS FOR A FOURIER SERIES (*D.U. April, 2010*)

If the function $f(x)$ for the interval $(-\pi, \pi)$

(1) is single-valued **(2)** is bounded

(3) has at most a finite number of maxima and minima.

(4) has only a finite number of discontinuous

(5) is $f(x + 2\pi) = f(x)$ for values of x outside $[-\pi, \pi]$, then

$$S_P(x) = \frac{a_0}{2} + \sum_{n=1}^{P} a_n \cos nx + \sum_{n=1}^{P} b_n \sin nx$$

converges to $f(x)$ as $P \to \infty$ at values of x for which $f(x)$ is continuous and the sum of the series is equal to $\frac{1}{2}[f(x+0) + f(x-0)]$ at points of discontinuity.

11.4 ADVANTAGES OF FOURIER SERIES

1. Discontinuous function can be represented by Fourier series. Although derivatives of the discontinuous functions do not exist. (This is not true for Taylor's series).

2. The Fourier series is useful in expanding the periodic functions since outside the closed interval, there exists a periodic extension of the function.

3. Expansion of an oscillating function by Fourier series gives all modes of oscillation (fundamental and all overtones) which is extremely useful in physics.

4. Fourier series of a discontinuous function is not uniformly convergent at all points.

5. Term by term integration of a convergent Fourier series is always valid, and it may be valid if the series is not convergent. However, term by term, differentiation of a Fourier series is not valid in most cases.

11.5 USEFUL INTEGRALS

The following integrals are useful in Fourier Series.

(*i*) $\int_0^{2\pi} \sin nx \, dx = 0$ (*ii*) $\int_0^{2\pi} \cos nx \, dx = 0$

(*iii*) $\int_0^{2\pi} \sin^2 nx \, dx = \pi$ (*iv*) $\int_0^{2\pi} \cos^2 nx \, dx = \pi$

(v) $\int_0^{2\pi} \sin nx \cdot \sin mx \, dx = 0$ (vi) $\int_0^{2\pi} \cos nx \cos mx \, dx = 0$

(vii) $\int_0^{2\pi} \sin nx \cdot \cos mx \, dx = 0$ (viii) $\int_0^{2\pi} \sin nx \cdot \cos nx \, dx = 0$

(ix) $[uv]_1 = uv_1 - u'v_2 + u''v_3 - u'''v_4 + \ldots$

where $v_1 = \int v \, dx, v_2 = \int v_1 \, dx$ and so on $u' = \frac{du}{dx}$, $u'' = \frac{d^2u}{dx^2}$ and so on

(x) $\sin n\pi = 0$, $\cos n\pi = (-1)^n$ where $n \in I$

11.6 DETERMINATION OF FOURIER COEFFICIENTS (EULER'S FORMULAE)

$$f(x) = \frac{a_0}{2} + a_1 \cos x + a_2 \cos 2x + \ldots + a_n \cos nx + \ldots$$
$$+ b_1 \sin x + b_2 \sin 2x + \ldots + b_n \sin nx + \ldots \quad \ldots(1)$$

(i) **To find a_0 :** Integrate both sides of (1) from $x = 0$ to $x = 2\pi$.

$$\int_0^{2\pi} f(x)\,dx = \frac{a_0}{2}\int_0^{2\pi} dx + a_1 \int_0^{2\pi} \cos x\,dx + a_2 \int_0^{2\pi} \cos 2x\,dx + \ldots + a_n \int_0^{2\pi} \cos nx\,dx + \ldots$$
$$+ b_1 \int_0^{2\pi} \sin x\,dx + b_2 \int_0^{2\pi} \sin 2x\,dx + \ldots + b_n \int_0^{2\pi} \sin nx\,dx + \ldots$$

$$= \frac{a_0}{2}\int_0^{2\pi} dx$$ (other integrals = 0 by formulae (i) and (ii) of Art 11.5)

$$\int_0^{2\pi} f(x)\,dx = \frac{a_0}{2} 2\pi \quad \Rightarrow \quad \boxed{a_0 = \frac{1}{\pi}\int_0^{2\pi} f(x)\,dx} \quad \ldots(2)$$

(ii) **To find a_n :** Multiply each side of (1) by $\cos nx$ and integrate from $x = 0$ to $x = 2\pi$.

$$\int_0^{2\pi} f(x) \cos nx\,dx = \frac{a_0}{2}\int_0^{2\pi} \cos nx\,dx + a_1 \int_0^{2\pi} \cos x \cos nx\,dx + \ldots + a_n \int_0^{2\pi} \cos^2 nx\,dx \ldots$$
$$+ b_1 \int_0^{2\pi} \sin x \cos nx\,dx + b_2 \int_0^{2\pi} \sin 2x \cos nx\,dx + \ldots$$

$$= a_n \int_0^{2\pi} \cos^2 nx\,dx = a_n \pi,$$ (Other integrals = 0, by formulae Art. 11.5)

$$\therefore \quad \boxed{a_n = \frac{1}{\pi}\int_0^{2\pi} f(x) \cos nx\,dx} \quad \ldots(3)$$

By taking $n = 1, 2 \ldots$ we can find the values of $a_1, a_2 \ldots$.

(iii) **To find b_n :** Multiply each side of (1) by $\sin nx$ and integrate from $x = 0$ to $x = 2\pi$.

$$\int_0^{2\pi} f(x) \sin nx\,dx = \frac{a_0}{2}\int_0^{2\pi} \sin nx\,dx + a_1 \int_0^{2\pi} \cos x \sin nx\,dx + \ldots + a_n \int_0^{2\pi} \cos nx \sin nx\,dx + \ldots$$
$$\ldots + b_1 \int_0^{2\pi} \sin x \sin nx\,dx + \ldots + b_n \int_0^{2\pi} \sin^2 nx\,dx + \ldots$$

$$= b_n \int_0^{2\pi} \sin^2 nx\,dx$$

$$= b_n \pi$$ (All other integrals = 0, Article No. 11.5)

$$\therefore \quad \boxed{b_n = \frac{1}{\pi}\int_0^{2\pi} f(x) \sin nx\,dx} \quad \ldots(4)$$

Note : To get similar formula of a_0, $\frac{1}{2}$ has been written with a_0 in Fourier series.

Example 1. *Find the Fourier series representing*

$$f(x) = x,\ 0 < x < 2\pi$$

and sketch its graph from $x = -4\pi$ *to* $x = 4\pi$.

Solution. Let $f(x) = \frac{a_0}{2} + a_1 \cos x + a_2 \cos 2x + ... + b_1 \sin x + b_2 \sin 2x + ...$ ...(1)

Hence $a_0 = \frac{1}{\pi}\int_0^{2\pi} f(x)\,dx = \frac{1}{\pi}\int_0^{2\pi} x\,dx = \frac{1}{\pi}\left[\frac{x^2}{2}\right]_0^{2\pi} = 2\pi$

$$a_n = \frac{1}{\pi}\int_0^{2\pi} f(x)\cos nx\,dx = \frac{1}{\pi}\int_0^{2\pi} x\cos nx\,dx$$

$$= \frac{1}{\pi}\left[x\frac{\sin nx}{n} - 1.\left(-\frac{\cos nx}{n^2}\right)\right]_0^{2\pi} = \frac{1}{\pi}\left[\frac{\cos 2n\pi}{n^2} - \frac{1}{n^2}\right] = \frac{1}{n^2\pi}(1-1) = 0$$

$$b_n = \frac{1}{\pi}\int_0^{2\pi} f(x)\sin nx\,dx = \frac{1}{\pi}\int_0^{2\pi} x\sin nx\,dx$$

$$= \frac{1}{\pi}\left[x\left(-\frac{\cos nx}{n}\right) - 1.\left(\frac{-\sin nx}{n^2}\right)\right]_0^{2\pi} = \frac{1}{\pi}\left[\frac{-2\pi\cos 2n\pi}{n}\right] = -\frac{2}{n}$$

Substituting the values of $a_0, a_1, a_2 ..., b_1, b_2 ...$ in (1), we get

$$x = \pi - 2\left[\sin x + \frac{1}{2}\sin 2x + \frac{1}{3}\sin 3x + ...\right]$$ **Ans.**

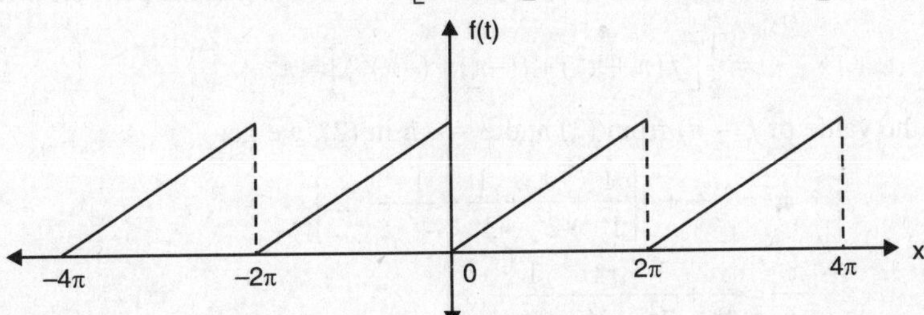

Example 2. *Given that* $f(x) = x + x^2$ *for* $-\pi < x < \pi$, *find the Fourier expression of* $f(x)$.

Deduce that $\frac{\pi^2}{6} = 1 + \frac{1}{2^2} + \frac{1}{3^2} + \frac{1}{4^2} + ...$

(DU, I Sem. 2012, U.P., II Semester, Uttarakhand, June 2009)

Solution. Let $x + x^2 = \frac{a_0}{2} + a_1\cos x + a_2\cos 2x + ... + b_1\sin x + b_2\sin 2x + ...$ ...(1)

$$a_0 = \frac{1}{\pi}\int_{-\pi}^{\pi} f(x)\,dx = \frac{1}{\pi}\int_{-\pi}^{\pi}(x + x^2)\,dx$$

$$= \frac{1}{\pi}\left[\frac{x^2}{2} + \frac{x^3}{3}\right]_{-\pi}^{\pi} = \frac{1}{\pi}\left[\frac{\pi^2}{2} + \frac{\pi^3}{3} - \frac{\pi^2}{2} + \frac{\pi^3}{3}\right] = \frac{2\pi^2}{3}$$

$$a_n = \frac{1}{\pi}\int_{-\pi}^{\pi} f(x)\cos nx\,dx = \int_{-\pi}^{\pi}(x + x^2)\cos nx\,dx$$

$$= \frac{1}{\pi}\left[(x+x^2)\frac{\sin nx}{n} - (2x+1)\frac{(-\cos nx)}{n^2} + (2)\left(-\frac{\sin nx}{n^3}\right)\right]_{-\pi}^{\pi}$$

$$= \frac{1}{\pi}\left[(2\pi+1)\frac{\cos n\pi}{n^2} - (-2\pi+1)\frac{\cos n\pi}{n^2}\right] = \frac{1}{\pi}\left[4\pi\frac{\cos n\pi}{n^2}\right] = \frac{4(-1)^n}{n^2}$$

$$b_n = \frac{1}{\pi}\int_{-\pi}^{\pi} f(x)\sin nx\,dx = \frac{1}{\pi}\int_{-\pi}^{\pi}(x+x^2)\sin nx\,dx$$

$$= \frac{1}{\pi}\left[(x+x^2)\left(-\frac{\cos nx}{n}\right) - (2x+1)\left(\frac{-\sin nx}{n^2}\right) + 2\frac{\cos nx}{n^3}\right]_{-\pi}^{\pi}$$

$$= \frac{1}{\pi}\left[-(\pi+\pi^2)\frac{\cos n\pi}{n} + 2\frac{\cos n\pi}{n^3} + (-\pi+\pi^2)\frac{\cos n\pi}{n} - 2\frac{\cos n\pi}{n^3}\right]$$

$$= \frac{1}{\pi}\left[-\frac{2\pi}{n}\cos n\pi\right] = -\frac{2}{n}(-1)^n$$

Substituting the values of $a_0, a_1, a_2, a_3, b_1, b_2, b_3$, in (1), we get

$$x + x^2 = \frac{\pi^2}{3} + 4\left[-\cos x + \frac{1}{2^2}\cos 2x - \frac{1}{3^2}\cos 3x + \ldots\right]$$

$$-2\left[-\sin x + \frac{1}{2}\sin 2x - \frac{1}{3}\sin 3x + \ldots\right] \quad \ldots(2)$$

Here $f(x) = x + x^2$ is valid for all values of x between $-\pi$ and π but not at the end points $-\pi$ and π due to open interval.

$$f(-x) = \frac{1}{2}[f(-\pi-0) + f(-\pi+0)]$$

$$= \frac{1}{2}[f(\pi-0) + f(-\pi+0)] \qquad [f(x) \text{ is periodic with period } 2\pi]$$

$$= \frac{1}{2}\left[f(\pi+\pi^2) + \{(-\pi)+(-\pi)^2\}\right] = \pi^2 \quad \ldots(3)$$

Putting the value of $f(-\pi)$ from (3) and $x = -\pi$ in (2), we get

$$\pi^2 = \frac{\pi^2}{3} + 4\left[\frac{1}{1^2} + \frac{1}{2^2} + \frac{1}{3^2} + \frac{1}{4^2} + \ldots\right]$$

$$\Rightarrow \quad \frac{\pi^2}{6} = \frac{1}{1^2} + \frac{1}{2^2} + \frac{1}{3^2} + \frac{1}{4^2} + \ldots\ldots$$

Proved

Example 3. *Find the Fourier series expansion for* $f(x) = x + \frac{x^2}{4}, -\pi \le x \le \pi$

(U.P. II Semester, 2009)

Solution. Let $x + \frac{x^2}{4} = \frac{a_0}{2} + a_1\cos x + a_2\cos 2x + \ldots\ldots + b_1\sin x + b_2\sin 2x + \ldots\ldots$

where
$$a_0 = \frac{1}{\pi}\int_{-\pi}^{\pi} f(x)\,dx = \frac{1}{\pi}\int_{-\pi}^{\pi}\left(x + \frac{x^2}{4}\right)dx$$

$$= \frac{1}{\pi}\left[\frac{x^2}{2} + \frac{x^3}{12}\right]_{-\pi}^{\pi} = \frac{1}{\pi}\left[\frac{\pi^2}{2} + \frac{\pi^3}{12} - \frac{\pi^2}{2} + \frac{\pi^3}{12}\right] = \frac{1}{\pi}\left[\frac{2\pi^3}{12}\right] = \frac{\pi^2}{6}$$

$$a_n = \frac{1}{\pi}\int_{-\pi}^{\pi} f(x)\cos nx\,dx = \frac{1}{\pi}\int_{-\pi}^{\pi}\left(x + \frac{x^2}{4}\right)\cos nx\,dx$$

$$= \frac{1}{\pi}\left[\left(x + \frac{x^2}{4}\right)\left(\frac{\sin nx}{n}\right) - \left(1 + \frac{2x}{4}\right)\left(\frac{-\cos nx}{n^2}\right) + \frac{1}{2}\left(-\frac{\sin nx}{n^3}\right)\right]_{-\pi}^{\pi}$$

$$= \frac{1}{\pi}\left[\left(\pi+\frac{\pi^2}{4}\right)\left(\frac{\sin n\pi}{n}\right)+\left(1+\frac{2\pi}{4}\right)\left(\frac{\cos n\pi}{n^2}\right)-\frac{1}{2}\left(\frac{\sin n\pi}{n^3}\right)-\left(-\pi+\frac{\pi^2}{4}\right)\left(\frac{\sin(-n\pi)}{n}\right)\right.$$

$$\left.-\left(1-\frac{2\pi}{4}\right)\frac{\cos(-n\pi)}{n^2}+\frac{1}{2}\frac{\sin(-n\pi)}{n^3}\right]$$

$$= \frac{1}{\pi}\left[\left(1+\frac{\pi}{2}\right)\frac{(-1)^n}{n^2}-\left(1-\frac{\pi}{2}\right)\frac{(-1)^n}{n^2}\right] = \frac{(-1)^n}{n^2\pi}\left[1+\frac{\pi}{2}-1+\frac{\pi}{2}\right] = \frac{(-1)^n}{n^2\pi}(\pi) = \frac{(-1)^n}{n^2}$$

$$a_1 = -1, \quad a_2 = \frac{1}{4}, \quad a_3 = -\frac{1}{9}, \quad a_4 = \frac{1}{16} \quad \ldots\ldots\ldots\ldots$$

$$b_n = \frac{1}{\pi}\int_{-\pi}^{\pi} f(x)\sin nx\,dx = \frac{1}{\pi}\int_{-\pi}^{\pi}\left(x+\frac{x^2}{4}\right)\sin nx\,dx$$

$$= \frac{1}{\pi}\int_{-\pi}^{\pi} x\sin nx\,dx + \frac{1}{\pi}\int_{-\pi}^{\pi}\frac{x^2}{4}\sin nx\,dx$$

Even function   Odd function

$$= \frac{2}{\pi}\int_0^{\pi} x\sin nx\,dx + 0$$

$$= \frac{2}{\pi}\left[x\left(\frac{\cos nx}{n}\right)-(1)\left(\frac{-\sin nx}{n^2}\right)\right]_0^{\pi}$$

$$= \frac{2}{\pi}\left[-\pi\frac{\cos n\pi}{n}+\frac{\sin n\pi}{n^2}\right]$$

$$= \frac{2}{\pi}\left[-\pi\frac{(-1)^n}{n}\right] = -\frac{2(-1)^n}{n} = \frac{2}{n}(-1)^{n+1}$$

$$b_1 = \frac{2}{1}, \quad b_2 = -1, \quad b_3 = \frac{2}{3}, \quad b_4 = -\frac{1}{2} \ldots\ldots$$

Hence, Fourier series of the given function is.

$$f(x) = \frac{a_0}{2}+\sum_{n=1}^{\infty} a_n\cos nx+\sum_{n=1}^{\infty} b_n\sin nx = \frac{\pi^2}{12}+\sum\frac{(-1)^n}{n^2}\cos nx-\sum_{n=1}^{\infty}\frac{2}{n}(-1)^{n+1}\sin nx$$

$$f(x) = \frac{\pi^2}{12}-\cos x+\frac{1}{4}\cos 2x-\frac{1}{9}\cos 3x+\frac{1}{16}\cos 4x+\ldots$$

$$+2\sin x-\sin 2x+\frac{2}{3}\sin 3x-\frac{1}{2}\sin 4x+\ldots$$ **Ans.**

EXERCISE 11.1

1. Find a Fourier series to represent, $f(x) = \pi - x$ for $0 < x < 2\pi$.

Ans. $2\left[\sin x+\frac{1}{2}\sin 2x+\frac{1}{3}\sin 3x+\ldots+\frac{1}{n}\sin nx+\ldots\right]$

2. Find a Fourier series to represent $x - x^2$ from $x = -\pi$ to π and show that

$$\frac{\pi^2}{12} = \frac{1}{1^2}-\frac{1}{2^2}+\frac{1}{3^2}-\frac{1}{4^2}+\ldots$$

Ans. $-\frac{\pi^2}{3}+4\left[\frac{\cos x}{1^2}-\frac{\cos 2x}{2^2}+\frac{\cos 3x}{3^2}-\frac{\cos 4x}{4^2}+\ldots\right]+2\left[\frac{\sin x}{1}-\frac{\sin 2x}{2}+\frac{\sin 3x}{3}-\frac{\sin 4x}{4}+\ldots\right]$

3. Find a Fourier series to represent the function $f(x) = e^x$, for $-\pi < x < \pi$ and hence derive a series for $\frac{\pi}{\sinh \pi}$.

Ans. $\frac{2\sinh\pi}{\pi}\left[\frac{1}{2}-\frac{1}{1^2+1}\cos x+\frac{1}{2^2+1}\cos 2x-\frac{1}{3^2+1}\cos 3x+\ldots\right.$

$\left.+\frac{1}{1^2+1}\sin x-\frac{2}{2^2+1}\sin 2x-\frac{3}{3^2+1}\sin 3x \ldots,\right]$, $\frac{\pi}{\sinh\pi}=1+2\left[-\frac{1}{2}+\frac{1}{5}-\frac{1}{10}+\ldots\right]$

4. Obtain the Fourier series for $f(x) = e^{-x}$ in the interval $0 \le x < 2\pi$.

Ans. $\frac{1-e^{-2\pi}}{\pi}\left[\frac{1}{2}+\frac{1}{2}\cos x+\frac{1}{5}\cos 2x+\frac{1}{10}\cos 3x+\frac{1}{2}\sin x+\frac{2}{5}\sin 2x+\frac{3}{10}\sin 3x+\ldots\right]$

5. If $f(x) = \left(\frac{\pi - x}{2}\right)^2, 0 < x < 2\pi$, show that $f(x) = \frac{\pi^2}{12}+\sum_{n=1}^{\infty}\frac{\cos nx}{n^2}$

6. Prove that $x^2 = \frac{\pi^2}{3}+4\sum_{n=1}^{\infty}(-1)^n\frac{\cos nx}{n^2}, -\pi < x < \pi$. Hence show that

(*i*) $\Sigma\frac{1}{(2n-1)^2}=\frac{\pi^2}{8}$ (*ii*) $\Sigma\frac{1}{n^4}=\frac{\pi^4}{90}$

7. If $f(x)$ is a periodic function defined over a period $(0, 2\pi)$ by $f(x) = \frac{(3x^2-6x\pi+2\pi^2)}{12}$.

Prove that $f(x) = \sum_{n=1}^{\infty}\frac{\cos nx}{n^2}$ and hence show that $\frac{\pi^2}{6}=1+\frac{1}{2^2}+\frac{1}{3^2}+\ldots$

11.7 FOURIER SERIES FOR DISCONTINUOUS FUNCTIONS

Let the function $f(x)$ be defined by

$$f(x) = \begin{cases} f_1(x), & c < x < x_0 \\ f_2(x), & x_0 < x < c + 2\pi, \end{cases}$$

where x_0 is the point of discontinuity in the interval $(c, c + 2\pi)$.

In such cases also, we obtain the Fourier series for $f(x)$ in the usual way. The values of a_0, a_n, b_n are evaluated by

$$a_0 = \frac{1}{\pi}\left[\int_c^{x_0} f_1(x)\,dx + \int_{x_0}^{c+2\pi} f_2(x)\,dx\right];$$

$$a_n = \frac{1}{\pi}\left[\int_c^{x_0} f_1(x)\cos nx dx + \int_{x_0}^{c+2\pi} f_2(x)\cos nx dx\right]$$

$$b_n = \frac{1}{\pi}\left[\int_c^{x_0} f_1(x)\sin nx dx + \int_{x_0}^{c+2\pi} f_2(x)\cos nx dx\right]$$

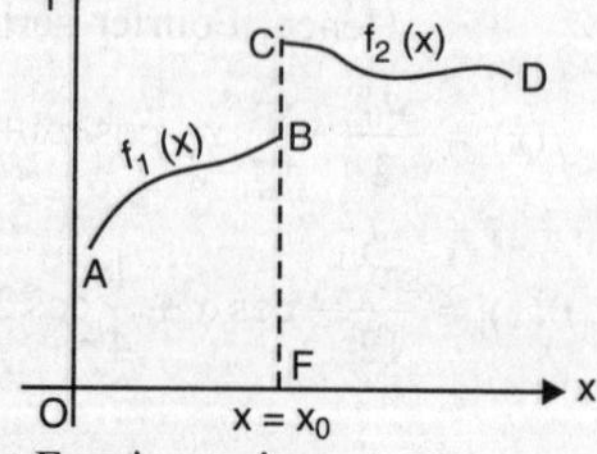

If $x = x_0$ is the point of finite discontinuity, then the sum of the Fourier series

$$= \frac{1}{2}\left[\lim_{h\to 0} f(x_0 - h) + \lim_{h\to 0} f(x_0 + h)\right]$$

$$= \frac{1}{2}\left[f(x_0 - 0) + f(x_0 + 0)\right] = \frac{1}{2}(FB + FC)$$

Remarks.

1. It may be seen from the graph, that at a point of finite discontinuity $x = x_0$, there is a finite jump equal to BC in the value of the function $f(x)$ at $x = x_0$.

2. A given function $f(x)$ may be defined by different formulae in different regions. Such types of functions are quite common in Fourier Series.

3. At a point of discontinuity the sum of the series is equal to the mean of the limits on the right and left.

11.8 FUNCTION DEFINED IN TWO OR MORE SUB-RANGES

Example 4. *Find the fourier series to represent the function f (x) given by :*

$$f(x) = \begin{cases} -k \text{ for } -\pi < x < 0 \\ k \text{ for } 0 < x < \pi \end{cases}$$ *Hence show that :*

$$1 - \frac{1}{3} + \frac{1}{5} - \frac{1}{7} + ... = \frac{\pi}{4}.$$ *(U.P. II Semester 2010)*

Solution. $f(x) = \begin{cases} -k, & -\pi < x < 0 \\ k, & 0 < x < \pi \end{cases}$

$$a_0 = \frac{1}{\pi}\int_{-\pi}^{\pi} f(x)\, dx = \frac{1}{\pi}\left[\int_{-\pi}^{0} -k\, dx + \int_0^{\pi} k\, dx\right] = \frac{1}{\pi}\left[[-kx]_{-\pi}^{0} + [kx]_0^{\pi}\right]$$

$$= \frac{1}{\pi} k\, [0 - \pi + \pi - 0] = 0$$

$$a_n = \frac{1}{\pi}\int_{-\pi}^{\pi} f(x) \cos nx\, dx = \frac{1}{\pi}\left[\int_{-\pi}^{0} -k \cos nx\, dx + \int_0^{\pi} k \cos nx\, dx\right]$$

$$= \frac{1}{\pi} k \left[-\left\{\frac{\sin nx}{n}\right\}_{-\pi}^{0} + \left\{\frac{\sin nx}{n}\right\}_0^{\pi}\right] = \frac{1}{\pi} k\, [-0 + 0] = 0$$

$$b_n = \frac{1}{\pi}\int_{-\pi}^{\pi} f(x) \sin nx\, dx = \frac{1}{\pi}\left[\int_{-\pi}^{0} -k \sin nx\, dx + \int_0^{\pi} k \sin nx\, dx\right]$$

$$= \frac{1}{\pi} k \left[\left\{\frac{\cos nx}{n}\right\}_{-\pi}^{0} - \left\{\frac{\cos nx}{n}\right\}_0^{\pi}\right]$$

$$= \frac{1}{\pi} k \left[\frac{1}{n} - \frac{(-1)^n}{n} - \frac{(-1)^n}{n} + \frac{1}{n}\right] = \frac{1}{\pi} k \left[\frac{2}{n} - \frac{2(-1)^n}{n}\right]$$

If n is even $b_n = 0$

If n is odd $b_n = \dfrac{4k}{n\pi}$

$$f(x) = \frac{a_0}{2} + \sum_{n=1}^{\infty} a_n \cos nx + \sum_{n=1}^{\infty} b_n \sin nx$$

$$f(x) = \sum_{n=1}^{\infty} b_n \sin nx = b_1 \sin x + b_2 \sin 2x + b_3 \sin 3x + b_4 \sin 4x + ...$$

Thus required Fourier sine series is

$$f(x) = \frac{4k}{\pi} \sin x + \frac{4k}{3\pi} \sin 3x + \frac{4k}{5\pi} \sin 5x + ...$$

$$\Rightarrow \quad f(x) = \frac{4k}{\pi}\left[\sin x + \frac{1}{3} \sin 3x + \frac{1}{5} \sin 5x + ...\right] \qquad ...(1)$$

Putting $x = \dfrac{\pi}{2}$ in (1), we get

$$k = \frac{4k}{\pi}\left[\sin \frac{\pi}{2} + \frac{1}{3} \sin \frac{3\pi}{2} + \frac{1}{5} \sin \frac{5\pi}{2} + ...\right]$$

$$\Rightarrow \quad 1 = \frac{4}{\pi}\left[1 + \frac{1}{3}(-1) + \frac{1}{5}(1) + \frac{1}{7}(-1) + ...\right]$$

$$= \frac{4}{\pi}\left[1 - \frac{1}{3} + \frac{1}{5} - \frac{1}{7} + ...\right] \Rightarrow \frac{\pi}{4} = 1 - \frac{1}{3} + \frac{1}{5} - \frac{1}{7} + ...$$ **Proved**

Example 5. *Find the Fourier series of the function*

$$f(x) = \begin{cases} -1 & \text{for} \quad -\pi < x < -\dfrac{\pi}{2} \\ 0 & \text{for} \quad -\dfrac{\pi}{2} < x < \dfrac{\pi}{2} \\ +1 & \text{for} \quad \dfrac{\pi}{2} < x < \pi. \end{cases}$$

Solution. Let $f(x) = \frac{a_0}{2} + a_1 \cos x + a_2 \cos 2x + ... + b_1 \sin x + b_2 \sin 2x + ...$ ...(1)

$$a_0 = \frac{1}{\pi}\int_{-\pi}^{\pi} f(x)\,dx = \frac{1}{\pi}\int_{-\pi}^{-\pi/2}(-1)\,dx + \frac{1}{\pi}\int_{-\pi/2}^{\pi/2} 0\,dx + \frac{1}{\pi}\int_{\pi/2}^{\pi} 1\,dx$$

$$= \frac{1}{\pi}[-x]_{-\pi}^{-\pi/2} + \frac{1}{\pi}[x]_{\pi/2}^{\pi} = \frac{1}{\pi}\left[\frac{\pi}{2} - \pi + \pi - \frac{\pi}{2}\right] = 0$$

$$a_n = \frac{1}{\pi}\int_{-\pi}^{\pi} f(x)\cos nx\,dx$$

$$= \frac{1}{\pi}\int_{-\pi}^{-\pi/2}(-1)\cos nx\,dx + \frac{1}{\pi}\int_{-\pi/2}^{\pi/2}(0)\cos nx\,dx + \frac{1}{\pi}\int_{\pi/2}^{\pi}(1)\cos nx\,dx$$

$$= -\frac{1}{\pi}\left[\frac{\sin nx}{n}\right]_{-\pi}^{-\pi/2} + \frac{1}{\pi}\left[\frac{\sin nx}{n}\right]_{\pi/2}^{\pi}$$

$$= -\frac{1}{\pi}\left[-\frac{\sin\frac{n\pi}{2}}{n}\right] + \frac{1}{\pi}\left[-\frac{\sin\frac{n\pi}{2}}{n}\right] = 0$$

$$b_n = \frac{1}{\pi}\int_{-\pi}^{\pi} f(x)\sin nx\,dx$$

$$= \frac{1}{\pi}\int_{-\pi}^{-\pi/2}(-1)\sin nx\,dx + \frac{1}{\pi}\int_{-\pi/2}^{\pi/2}(0)\sin nx\,dx$$

$$+ \frac{1}{\pi}\int_{\pi/2}^{\pi}(1)\sin nx\,dx = \frac{1}{\pi}\left[\frac{\cos nx}{n}\right]_{-\pi}^{-\pi/2} - \frac{1}{\pi}\left[\frac{\cos nx}{n}\right]_{\pi/2}^{\pi}$$

$$= \frac{1}{n\pi}\left[\cos\frac{n\pi}{2} - \cos n\pi\right] - \frac{1}{n\pi}\left(\cos n\pi - \cos\frac{n\pi}{2}\right) = \frac{2}{n\pi}\left[\cos\frac{n\pi}{2} - \cos n\pi\right]$$

$$b_1 = \frac{2}{\pi}, \qquad b_2 = -\frac{2}{\pi}, \qquad b_3 = \frac{2}{3\pi}$$

Putting the values of a_0, a_n, b_n in (1), we get

$$f(x) = \frac{1}{\pi}\left[2\sin x - 2\sin 2x + \frac{2}{3}\sin 3x + ...\right]$$ **Ans.**

11.9 DISCONTINUOUS FUNCTIONS

At a point of discontinuity, Fourier series gives the value of $f(x)$ as the arithmetic mean of left and right limits.

At the point of discontinuity, $x = c$

$$\text{At } x = c,\ f(x) = \frac{1}{2}[f(c-0) + f(c+0)]$$

Example 6. *Find the Fourier series for $f(x)$, if* $f(x) = \begin{cases} -\pi, & -\pi < x < 0 \\ x, & 0 < x < \pi \end{cases}$

Deduce that $\dfrac{1}{1^2} + \dfrac{1}{3^2} + \dfrac{1}{5^2} + \ldots = \dfrac{\pi^2}{8}$

Solution. Let $f(x) = \dfrac{a_0}{2} + a_1 \cos x + a_2 \cos 2x + \ldots + a_n \cos nx + \ldots$

$$+ b_1 \sin x + b_2 \sin 2x + \ldots + b_n \sin nx + \ldots \qquad \ldots(1)$$

$$a_0 = \frac{1}{\pi}\int_{-\pi}^{\pi} f(x)\,dx$$

Then $a_0 = \dfrac{1}{\pi}\left[\int_{-\pi}^{0}(-\pi)\,dx + \int_0^{\pi} x\,dx\right] = \dfrac{1}{\pi}\left[-\pi\,(x)_{-\pi}^{0} + (x^2/2)_0^{\pi}\right] = \dfrac{1}{\pi}(-\pi^2 + \pi^2/2) = -\dfrac{\pi}{2};$

$$a_n = \frac{1}{\pi}\int_{-\pi}^{\pi} f(x)\cos nx\,dx$$

$$a_n = \frac{1}{\pi}\left[\int_{-\pi}^{0}(-\pi)\cos nx\,dx + \int_0^{\pi} x\cos nx\,dx\right] = \frac{1}{\pi}\left[-\pi\left(\frac{\sin nx}{n}\right)_{-\pi}^{0} + \left(\frac{x\sin nx}{n} + \frac{\cos nx}{n^2}\right)_0^{\pi}\right]$$

$$= \frac{1}{\pi}\left[0 + \frac{1}{n^2}\cos n\pi - \frac{1}{n^2}\right] = \frac{1}{\pi n^2}(\cos n\pi - 1)$$

$$b_n = \frac{1}{\pi}\int_{-\pi}^{\pi} f(x)\sin nx\,dx$$

$$= \frac{1}{\pi}\left[\int_{-\pi}^{0}(-\pi)\sin nx\,dx + \int_0^{\pi} x\sin nx\,dx\right]$$

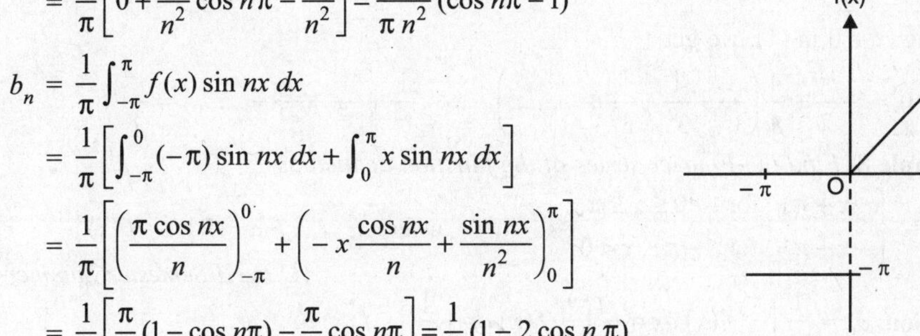

$$= \frac{1}{\pi}\left[\left(\frac{\pi\cos nx}{n}\right)_{-\pi}^{0} + \left(-x\frac{\cos nx}{n} + \frac{\sin nx}{n^2}\right)_0^{\pi}\right]$$

$$= \frac{1}{\pi}\left[\frac{\pi}{n}(1 - \cos n\pi) - \frac{\pi}{n}\cos n\pi\right] = \frac{1}{n}(1 - 2\cos n\pi)$$

$$f(x) = -\frac{\pi}{4} - \frac{2}{\pi}\left(\cos x + \frac{\cos 3x}{3^2} + \frac{\cos 5x}{5^2} + \ldots\right) + 3\sin x - \frac{\sin 2x}{2} + \frac{3\sin 3x}{3} - \frac{\sin 4x}{4} \qquad \ldots(2)$$

Putting $x = 0$ in (2), we get $f(0) = -\dfrac{\pi}{4} - \dfrac{2}{\pi}\left(1 + \dfrac{1}{3^2} + \dfrac{1}{5^2} + \ldots\infty\right)$ ...(3)

Now, $f(x)$ is discontinuous at $x = 0$.

But $f(0 - 0) = -\pi$ and $f(0 + 0) = 0$

$$\therefore\quad f(0) = \frac{1}{2}\,[f(0-0) + f(0+0)] = -\pi/2 \qquad \ldots(4)$$

From (3) and (4), we get

$$-\frac{\pi}{2} = -\frac{\pi}{4} - \frac{2}{\pi}\left[\frac{1}{1^2} + \frac{1}{3^2} + \frac{1}{5^2} + \ldots\right] \Rightarrow \frac{\pi^2}{8} = \frac{1}{1^2} + \frac{1}{3^2} + \frac{1}{5^2} + \ldots$$

Proved.

Example 7. *Obtain Fourier Series of the function*

$$f(x) = \begin{cases} x, & -\pi < x < 0 \\ -x, & 0 < x < \pi \end{cases}$$

and hence show that

$$\frac{1}{1^2} + \frac{1}{3^2} + \frac{1}{5^2} + \ldots\ldots\ldots\ldots = \frac{\pi^2}{8}$$

(U.P., II Semester, June 2008)

Solution. We have, $f(x) = \begin{cases} x, & -\pi < x < 0 \\ -x, & 0 < x < \pi \end{cases}$

Here $f(x)$ is an even function so $b_n = 0$

$$a_0 = \frac{2}{\pi}\int_0^{\pi} f(x)\,dx = \frac{2}{\pi}\int_0^{\pi} -x\,dx = -\frac{2}{\pi}\left[\frac{x^2}{2}\right]_0^{\pi} = -\frac{2}{\pi}\left[\frac{\pi^2}{2}\right] = -\pi$$

$$a_n = \frac{2}{\pi}\int_0^{\pi} f(x)\cos nx\,dx = \frac{2}{\pi}\int_0^{\pi} -x\cos nx\,dx = -\frac{2}{\pi}\left[x\frac{\sin nx}{n} + \frac{\cos nx}{n^2}\right]_0^{\pi}$$

$$= -\frac{2}{\pi}\left[\frac{(-1)^n}{n^2} - \frac{1}{n^2}\right] = \frac{2}{\pi n^2}[1-(-1)^n] = \begin{cases} 0, & n \text{ is even} \\ \dfrac{4}{\pi n^2}, & n \text{ is odd} \end{cases}$$

Fourier series

$$f(x) = -\frac{\pi}{2} + \frac{4}{\pi}\left[\frac{\cos x}{1^2} + \frac{\cos 3x}{3^2} + \frac{\cos 5x}{5^2} + \ldots\right] \qquad \ldots(1) \qquad \textbf{Ans.}$$

Now $f(x)$ is discontinuous at $x = 0$.

At $x = 0$, the point of discontinuity

$$f(0-0) = 0 \text{ and } f(0+0) = 0$$

$$f(0) = \frac{1}{2}[f(0-0) + f(0+0)] = \frac{1}{2}(0+0) = 0$$

Putting $x = 0$ in (1), we get

$$0 = -\frac{\pi}{2} + \frac{4}{\pi}\left(\frac{1}{1^2} + \frac{1}{3^2} + \frac{1}{5^2} + \ldots\ldots\ldots\right) \Rightarrow \frac{1}{1^2} + \frac{1}{3^2} + \frac{1}{5^2} + \ldots\ldots\ldots = \frac{\pi^2}{8} \qquad \textbf{Ans.}$$

Example 8. *Find the Fourier series of the function defined as*

$$f(x) = \begin{cases} x+\pi, & \text{for } \quad 0 \le x \le \pi, \\ -x-\pi, & \text{for } \quad -\pi \le x < 0 \end{cases} \qquad \textit{where} \qquad f(x+2\pi) = f(x).$$

(*U.P., II Semester Summer 2006*)

Solution. $a_0 = \frac{1}{\pi}\int_{-\pi}^{\pi} f(x)\,dx = \frac{1}{\pi}\int_{-\pi}^{0} f(x)\,dx + \frac{1}{\pi}\int_0^{\pi} f(x)\,dx$

$$= \frac{1}{\pi}\int_{-\pi}^{0}(-x-\pi)\,dx + \frac{1}{\pi}\int_0^{\pi}(x+\pi)\,dx = \frac{1}{\pi}\left(-\frac{x^2}{2} - \pi x\right)_{-\pi}^{0} + \frac{1}{\pi}\left(\frac{x^2}{2} + \pi x\right)_0^{\pi}$$

$$= \frac{1}{\pi}\left(\frac{\pi^2}{2} - \pi^2\right) + \frac{1}{\pi}\left(\frac{\pi^2}{2} + \pi^2\right) = \pi\left(\frac{1}{2} - 1\right) + \pi\left(\frac{1}{2} + 1\right) = \pi$$

$$a_n = \frac{1}{\pi}\int_{-\pi}^{0} f(x)\cos nx\,dx = \frac{1}{\pi}\int_{-\pi}^{0} f(x)\cos nx\,dx + \frac{1}{\pi}\int_0^{\pi} f(x)\cos nx\,dx$$

$$= \frac{1}{\pi}\int_{-\pi}^{0}(-x-\pi)\cos nx\,dx + \frac{1}{\pi}\int_0^{\pi}(x+\pi)\cos nx\,dx$$

$$= \frac{1}{\pi}\left[(-x-\pi)\frac{\sin nx}{n} - (-1)\left\{-\frac{\cos nx}{n^2}\right\}\right]_{-\pi}^{0} + \frac{1}{\pi}\left[(x+\pi)\frac{\sin nx}{n} - (-1)\left\{-\frac{\cos nx}{n^2}\right\}\right]_0^{\pi}$$

$$= \frac{1}{\pi}\left[-\frac{1}{n^2} + \frac{(-1)^n}{n^2}\right] + \frac{1}{\pi}\left[\frac{(-1)^n}{n^2} - \frac{1}{n^2}\right] = \frac{2}{n^2\pi}[(-1)^n - 1] = \frac{-4}{n^2\pi}, \text{ if } n \text{ is odd.}$$

$= 0$, if n is even.

$$b_n = \frac{1}{\pi}\int_{-\pi}^{\pi} f(x)\sin nx\,dx = \frac{1}{\pi}\int_{-\pi}^{0} f(x)\sin nx\,dx + \frac{1}{\pi}\int_0^{\pi} f(x)\sin nx\,dx$$

$$= \frac{1}{\pi}\int_{-\pi}^{0}(-x-\pi)\sin nx\,dx + \frac{1}{\pi}\int_0^{\pi}(x+\pi)\sin nx\,dx$$

$$= \frac{1}{\pi}\left[(-x-\pi)\left(-\frac{\cos nx}{n}\right)-(-1)\left(-\frac{\sin nx}{n^2}\right)\right]_{-\pi}^{0}$$

$$+\frac{1}{\pi}\left[(x+\pi)\left(-\frac{\cos nx}{n}\right)-(-1)\left(-\frac{\sin nx}{n^2}\right)\right]_{0}^{\pi}$$

$$= \frac{1}{\pi}\left[\frac{\pi}{n}\right]+\frac{1}{\pi}\left[-\frac{2\pi}{n}(-1)^n+\frac{\pi}{n}\right]=\frac{1}{n}[(1)-2(-1)^n+(1)]=\frac{2}{n}[1-(-1)^n]$$

$$= \frac{4}{n}, \quad \text{if } n \text{ is odd.}$$

$$= 0, \quad \text{if } n \text{ is even.}$$

Fourier series is

$$f(x) = \frac{a_0}{2}+a_1\cos x+a_2\cos 2x+\ldots+b_1\sin x+b_2\sin 2x+\ldots$$

$$f(x) = \frac{\pi}{2}-\frac{4}{\pi}\left(\frac{\cos x}{1^2}+\frac{\cos 3x}{3^2}+\ldots\right)+4\left(\frac{\sin x}{1}+\frac{\sin 3x}{3}+\ldots\right)$$ **Ans.**

EXERCISE 11.2

1. Find the Fourier series of the function

$$f(x)=\begin{cases}-1 & \text{for } -\pi<x<0\\ 1 & \text{for } 0<x<\pi\end{cases}$$ (*U.P. II Semester 2005*)

where $f(x+2\pi)=f(x)$. **Ans.** $\frac{4}{\pi}\left[\frac{1}{1}\sin x+\frac{1}{3}\sin 3x+\frac{1}{5}\sin 5x+\frac{1}{7}\sin 7x+\ldots\right]$

2. Find the Fourier series of the function

$$f(x)=\begin{cases}0 & \text{for } -\pi\le x\le 0\\ 1 & \text{for } 0<x<\frac{\pi}{2}\\ 0 & \text{for } \frac{\pi}{2}\le x\le\pi\end{cases}$$

Ans. $\frac{1}{4}+\frac{1}{\pi}\left[\frac{\cos x}{1}-\frac{\cos 3x}{3}+\frac{\cos 5x}{5}+\ldots+\frac{\sin x}{1}+\frac{\sin 2x}{2}+\frac{\sin 3x}{3}+\ldots\right]$

3. Obtain a Fourier series to represent the following periodic function

$$f(x)=\begin{cases}0, & \text{when } 0<x<\pi\\ 1, & \text{when } \pi<x<2\pi\end{cases}$$ **Ans.** $\frac{1}{2}-\frac{2}{\pi}\left(\sin x+\frac{1}{3}\sin 3x+\frac{1}{5}\sin 5x+\ldots\right)$

4. Find the Fourier expansion of the function defined in a single period by the relations.

$$f(x)=\begin{cases}1 & \text{for } 0<x<\pi\\ 2 & \text{for } \pi<x<2\pi\end{cases}$$

and from it deduce that $\frac{\pi}{4}=1-\frac{1}{3}+\frac{1}{5}-\frac{1}{7}+\ldots$ **Ans.** $\frac{3}{2}-\frac{2}{\pi}\left(\sin x+\frac{1}{3}\sin 3x+\frac{1}{5}\sin 5x+\ldots\right)$

5. Find a Fourier series to represent the function

$$f(x)=\begin{cases}0 & \text{for } -\pi<x\le 0\\ \frac{1}{4}\pi x & \text{for } 0<x<\pi\end{cases}$$ and hence deduce that $\frac{\pi^2}{8}=\frac{1}{1^2}+\frac{1}{3^2}+\frac{1}{5^2}+\ldots$

Ans. $\frac{\pi^2}{16}+\sum_{n=1}^{\infty}\left(\frac{[(-1)^n-1]}{4n^2}\cos nx-\frac{(-1)^n\pi}{4n}\sin nx+\ldots\right)$

6. Find the Fourier series for $f(x)$, if

$$f(x) = \begin{cases} -\pi & \text{for } -\pi < x \le 0 \\ x & \text{for } 0 < x < 0 \\ \dfrac{-\pi}{2} & \text{for } x = 0 \end{cases}$$

Deduce that $\dfrac{1}{1^2}+\dfrac{1}{3^2}+\dfrac{1}{5^2}+\ldots=\dfrac{\pi^2}{8}$

Ans. $-\dfrac{\pi}{4}-\dfrac{2}{\pi}\left(\cos x+\dfrac{\cos 3x}{3^2}+\dfrac{\cos 5x}{5^2}+\ldots\right)+3\sin x-\dfrac{1}{2}\sin 2x+\dfrac{3}{3}\sin 3x-\dfrac{1}{4}\sin 4x+\ldots$

7. Obtain a Fourier series to represent the function

$f(x) = |x| \quad$ for $\quad -\pi < x < \pi$

and hence deduce $\dfrac{\pi^2}{8}=\dfrac{1}{1^2}+\dfrac{1}{3^2}+\dfrac{1}{5^2}+\ldots$ **Ans.** $\dfrac{\pi}{2}-\dfrac{4}{\pi}\left[\cos x+\dfrac{1}{3^2}\cos 3x+\dfrac{1}{5^2}\cos 5x+\ldots\right]$

8. Expand as a Fourier series, the function $f(x)$ defined as

$$f(x) = \begin{cases} \pi + x, & \text{for } -\pi < x < -\dfrac{\pi}{2} \\ \dfrac{\pi}{2}, & \text{for } -\dfrac{\pi}{2} < x < \dfrac{\pi}{2} \\ \pi - x, & \text{for } \dfrac{\pi}{2} < x < \pi \end{cases}$$

Ans. $\dfrac{3\pi}{8}+\dfrac{2}{\pi}\left[\dfrac{1}{1^2}\cos x-\dfrac{2}{2^2}\cos 2x+\dfrac{1}{3^2}\cos 3x+\ldots\right]$

9. Obtain a Fourier series to represent the function

$f(x) = |\sin x| \;$ for $\; -\pi < x < \pi$ **Hint** $f(x)=\begin{Bmatrix} -\sin x & \text{for} & -\pi < x < 0 \\ \sin x & \text{for} & 0 < x < \pi \end{Bmatrix}$

Ans. $\dfrac{2}{\pi}-\dfrac{4}{\pi}\left[\dfrac{1}{3}\cos 2x+\dfrac{1}{15}\cos 4x+\dfrac{1}{35}\cos 6x+\ldots\right]$

10. An alternating current after passing through a rectifier has the form

$$i = \begin{cases} I\sin\theta & \text{for } 0 < \theta < \pi \\ 0 & \text{for } 0 < \theta < \pi \end{cases}$$

Find the Fourier series of the function.

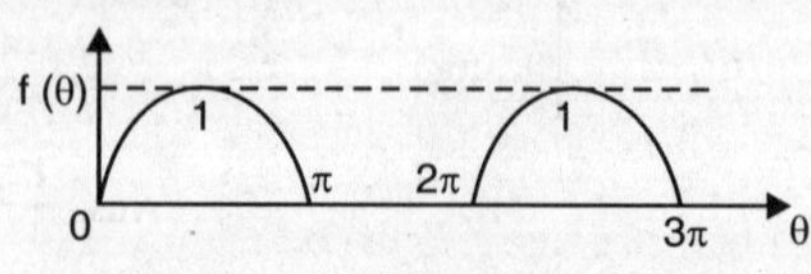

Ans. $\dfrac{I}{\pi}-\dfrac{2I}{\pi}\left(\dfrac{\cos 2\theta}{3}+\dfrac{\cos 4\theta}{15}+\ldots\right)+\dfrac{I}{2}\sin\theta$

11. If $f(x) = \begin{cases} 0 & \text{for } -\pi < x < 0 \\ \sin x & \text{for } 0 < x < \pi \end{cases}$

Prove that $f(x) = \dfrac{1}{\pi}+\dfrac{\sin x}{2}-\dfrac{2}{\pi}\sum_{m=1}^{\infty}\dfrac{\cos 2mx}{4m^2-1}$. Hence show that $\dfrac{1}{1.3}-\dfrac{1}{3.5}+\dfrac{1}{5.7}\ldots\infty=\dfrac{1}{4}(\pi-2)$

11.10 EVEN FUNCTION AND ODD FUNCTION

(*a*) Even Function

A function $f(x)$ is said to be even (or symmetric) function if, $f(-x) = f(x)$

The graph of such a function is symmetrical with respect to y-axis [$f(x)$ axis]. Here y-axis is a mirror for the reflection of the curve.

The area under such a curve from $-\pi$ to π is double the area from 0 to π.

$$\therefore \qquad \int_{-\pi}^{\pi} f(x)\,dx = 2\int_0^{\pi} f(x)\,dx$$

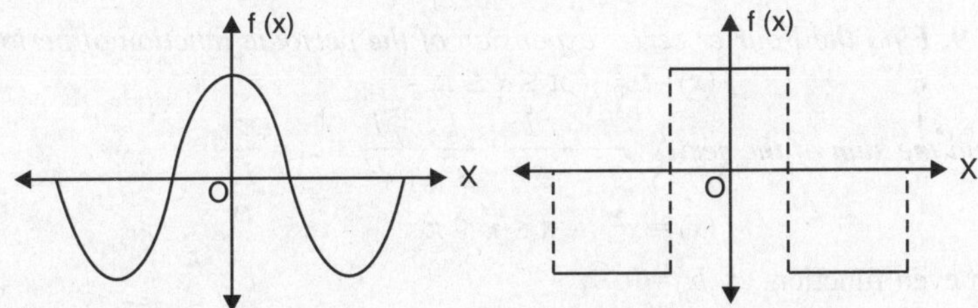

(*b*) **Odd Function**

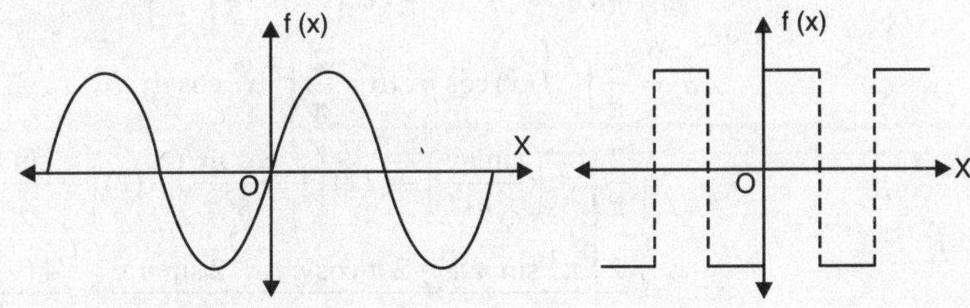

A function $f(x)$ is called odd (or skew symmetric) function if

$$f(-x) = -f(x)$$

Here the area under the curve from $-\pi$ to π is zero.

$$\int_{-\pi}^{\pi} f(x)\,dx = 0$$

Expansion of an even function:

$$a_0 = \frac{1}{\pi}\int_{-\pi}^{\pi} f(x)\,dx = \frac{2}{\pi}\int_0^{\pi} f(x)\,dx$$

$$a_n = \frac{1}{\pi}\int_{-\pi}^{\pi} f(x)\cos nx\,dx = \frac{2}{\pi}\int_0^{\pi} f(x)\cos nx\,dx$$

As $f(x)$ and $\cos nx$ are both even functions, therefore, the product of $f(x).\cos nx$ is also an even function.

$$b_n = \frac{1}{\pi}\int_{-\pi}^{\pi} f(x)\sin nx\,dx = 0$$

As $\sin nx$ is an odd function so $f(x).\sin nx$ is also an odd function. We need not to calculate b_n. It saves our labour a lot.

The series of the even function will contain only cosine terms. (*U.P. II Semester 2010*)

Expansion of an odd function :

$$a_0 = \frac{1}{\pi}\int_{-\pi}^{\pi} f(x)\,dx = 0$$

$$a_n = \frac{1}{\pi}\int_{-\pi}^{\pi} f(x)\cos nx\,dx = 0 \qquad [f(x).\cos nx \text{ is odd function}]$$

$$b_n = \frac{1}{\pi}\int_{-\pi}^{\pi} f(x)\sin nx\,dx = \frac{2}{\pi}\int_0^{\pi} f(x)\sin nx\,dx$$

$$[f(x).\sin nx \text{ is even function}]$$

The series of the odd function will contain only sine terms.

The function shown below is neither odd nor even so it contains both sine and cosine terms

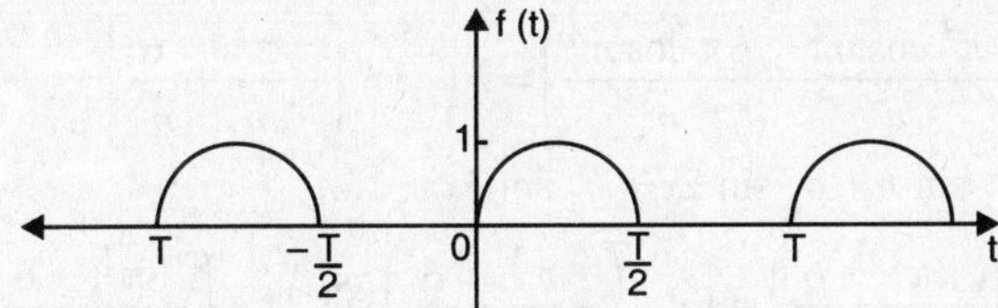

Example 9. *Find the Fourier series expansion of the periodic function of period 2π*

$$f(x) = x^2, \; -\pi \le x \le \pi.$$

Hence, find the sum of the series $\frac{1}{1^2} - \frac{1}{2^2} + \frac{1}{3^2} - \frac{1}{4^2} + ... = \frac{\pi^2}{12}$

Solution. $f(x) = x^2, \; -\pi \le x \le \pi$

This is an even function. $\therefore b_n = 0$ $[f(-x) = f(x)]$

$$a_0 = \frac{2}{\pi}\int_0^{\pi} f(x)\,dx = \frac{2}{\pi}\int_0^{\pi} x^2\,dx = \frac{2}{\pi}\left[\frac{x^3}{3}\right]_0^{\pi} = \frac{2\pi^2}{3}$$

$$a_n = \frac{2}{\pi}\int_0^{\pi} f(x)\cos nx\,dx = \frac{2}{\pi}\int_0^{\pi} x^2 \cos nx\,dx$$

$$= \frac{2}{\pi}\left[x^2\left(\frac{\sin nx}{n}\right) - (2x)\left(-\frac{\cos nx}{n^2}\right) + (2)\left(-\frac{\sin nx}{n^3}\right)\right]_0^{\pi}$$

$$= \frac{2}{\pi}\left[\frac{\pi^2 \sin n\pi}{n} + \frac{2\pi\cos n\pi}{n^2} - \frac{2\sin n\pi}{n^3}\right] = \frac{4(-1)^n}{n^2}$$

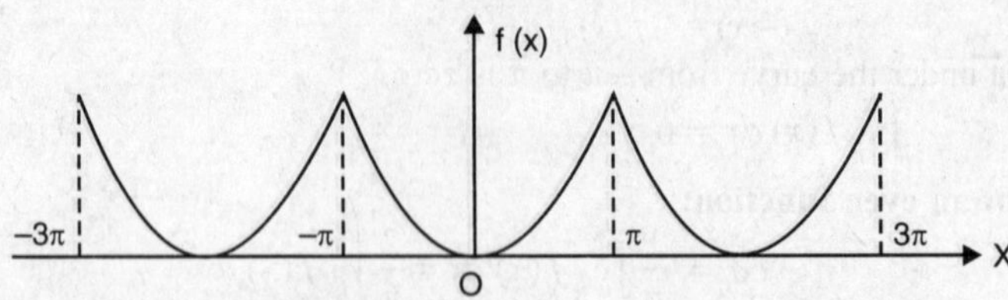

Fourier series is $f(x) = \frac{a_0}{2} + a_1\cos x + a_2\cos 2x + a_3\cos 3x + ... + a_n\cos nx + ...$

$$x^2 = \frac{\pi^2}{3} - 4\left[\frac{\cos x}{1^2} - \frac{\cos 2x}{2^2} + \frac{\cos 3x}{3^2} - \frac{\cos 4x}{4^2} + ...\right]$$

On putting $x = 0$, we have

$$0 = \frac{\pi^2}{3} - 4\left[\frac{1}{1^2} - \frac{1}{2^2} + \frac{1}{3^2} - \frac{1}{4^2} ...\right]$$

$$\Rightarrow \quad \frac{1}{1^2} - \frac{1}{2^2} + \frac{1}{3^2} - \frac{1}{4^2} ... = \frac{\pi^2}{12}$$ **Ans.**

Example 10. *Obtain a Fourier expression for*

$$f(x) = x^3 \quad for \quad -\pi < x < \pi.$$

Solution. $f(x) = x^3$ is an odd function. $[f(-x) = -f(x)]$

$\therefore$ $a_0 = 0$ and $a_n = 0$

$$b_n = \frac{2}{\pi}\int_0^{\pi} f(x)\sin nx\,dx = \frac{2}{\pi}\int_0^{\pi} x^3 \sin nx\,dx, \quad \left[\int uv = uv_1 - u'v_2 + u''v_3 - u'''v_4 + ...\right]$$

$$= \frac{2}{\pi}\left[x^3\left(-\frac{\cos nx}{n}\right) - 3x^2\left(-\frac{\sin nx}{n^2}\right) + 6x\left(\frac{\cos nx}{n^3}\right) - 6\left(\frac{\sin nx}{n^4}\right)\right]_0^{\pi}$$

$$= \frac{2}{\pi}\left[-\frac{\pi^3\cos n\pi}{n} + \frac{6\pi\cos n\pi}{n^3}\right] = 2.(-1)^n\left[-\frac{\pi^2}{n} + \frac{6}{n^3}\right]$$

$$f(x) = b_1\sin x + b_2\sin 2x + b_3\sin 3x + ...$$

$$\therefore \quad x^3 = 2\left[-\left(-\frac{\pi^2}{1} + \frac{6}{1^3}\right)\sin x + \left(-\frac{\pi^2}{2} + \frac{6}{2^3}\right)\sin 2x - \left(-\frac{\pi^2}{3} + \frac{6}{3^3}\right)\sin 3x + ...\right]$$ **Ans.**

Example 11. *Expand the function $f(x) = x \sin x$, as a Fourier series in the interval $-\pi \le x \le \pi$.*

Hence deduce that $\frac{1}{1.3} - \frac{1}{3.5} + \frac{1}{5.7} - \frac{1}{7.9} + \ldots = \frac{\pi - 2}{4}$

(*MDU, Dec. 2010, U.P., II Sem., Summer 2008, Uttarakhand, II Sem., June 2007*)

Solution. $f(x) = x \sin x$

$$a_0 = \frac{2}{\pi}\int_0^{\pi} f(x)\,dx = \frac{2}{\pi}\int_0^{\pi} x \sin x\,dx \qquad \text{(Here } x \sin x \text{ is an even function)}$$

$$= \frac{2}{\pi}[x(-\cos x) - (1)(-\sin x)]_0^{\pi} = \frac{2}{\pi}(\pi) = 2$$

$$a_n = \frac{2}{\pi}\int_0^{\pi} f(x)\cos nx\,dx = \frac{2}{\pi}\int_0^{\pi} x \sin x \cos nx\,dx$$

$$= \frac{1}{\pi}\int_0^{\pi} x\,\{\sin(n+1)x - \sin(n-1)x\}\,dx$$

$$= \frac{1}{\pi}\int_0^{\pi} x \sin(n+1)x\,dx - \frac{1}{\pi}\int_0^{\pi} x \sin(n-1)x\,dx$$

$$= \frac{1}{\pi}\left[x\left(-\frac{\cos(n+1)x}{n+1}\right)_0^{\pi} - (1)\left\{-\frac{\sin(n+1)x}{(n+1)^2}\right\}\right]_0^{\pi}$$

$$-\frac{1}{\pi}\left[x\left(-\frac{\cos(n-1)x}{(n-1)}\right) - (1)\left\{-\frac{\sin(n-1)x}{(n-1)^2}\right\}\right]_0^{\pi}$$

$$= \frac{1}{\pi}\left[-\pi\frac{(-1)^{(n+1)}}{n+1} + 0\right] - \frac{1}{\pi}\left[-\pi\frac{(-1)^{(n-1)}}{n-1} - 0\right]$$

$$= -\frac{(-1)^{n+1}}{n+1} + \frac{(-1)^{n-1}}{n-1} = (-1)^{n+1}\left[-\frac{1}{n+1} + \frac{1}{n-1}\right] = \frac{2(-1)^{n+1}}{n^2 - 1}$$

$$a_1 = \frac{2}{\pi}\int_0^{\pi} x \sin x \cos x\,dx = \frac{1}{\pi}\int_0^{\pi} x \sin 2x\,dx$$

$$= \frac{1}{\pi}\left[x\left(-\frac{\cos 2x}{2}\right) - (1)\left(\frac{-\sin 2x}{4}\right)\right]_0^{\pi} = \frac{1}{\pi}\left[-\frac{\pi}{2}\right] = -\frac{1}{2}$$

$$b_n = 0 \qquad \text{[As } x \sin x \sin nx \text{ is an odd function]}$$

Hence $f(x) = 1 - \frac{1}{2}\cos x + 2\sum_{n=2}^{\infty}\frac{(-1)^{n+1}}{n^2 - 1}\cos nx = 1 - \frac{1}{2}\cos x + 2\sum_{n=2}^{\infty}\frac{(-1)^{n+1}}{(n-1)(n+1)}\cos nx$

$$x \sin x = 1 + 2\left[-\frac{1}{4}\cos x - \frac{1}{1.3}\cos 2x + \frac{1}{2.4}\cos 3x - \frac{1}{3.5}\cos 4x + \ldots\right] \qquad \ldots(1)$$

Putting $x = \frac{\pi}{2}$ in (1) we get $\qquad \frac{\pi}{2} = 1 + 2\left\{\frac{1}{1.3} - \frac{1}{3.5} + \frac{1}{5.7} - \ldots\right\}$

or $\quad \frac{\pi}{4} = \frac{1}{2} + \frac{1}{1.3} - \frac{1}{3.5} + \frac{1}{5.7} - \ldots \quad \Rightarrow \quad \frac{\pi}{4} - \frac{1}{2} = \frac{1}{1.3} - \frac{1}{3.5} + \frac{1}{5.7} - \ldots$

$\Rightarrow \quad \frac{\pi - 2}{4} = \frac{1}{1.3} - \frac{1}{3.5} + \frac{1}{5.7} - \ldots$ **Proved.**

Example 12. *Find the Fourier Series expansion for the function*

$f(x) = x \cos x, -\pi < x < \pi.$

Solution. Since $x \cos x$ is an odd function therefore, $a_0 = a_n = 0$.

Let $x \cos x = \Sigma\, b_n \sin bx$, where

$$b_n = \frac{2}{\pi}\int_0^{\pi} x \cos x . \sin nx\, dx, \quad = \frac{1}{\pi}\int_0^{\pi} x\,\{\sin (n+1)\, x + \sin (n-1)\, x\}\, dx$$

$$= \frac{1}{\pi}\int_0^{\pi} x \sin (n+1)\, x\, dx + \frac{1}{\pi}\int_0^{\pi} x \sin (n-1)\, x\, dx$$

$$= \frac{1}{\pi}\left[x\left(\frac{-\cos (n+1)\, x}{n+1}\right) + \frac{\sin (n+1)\, x}{(n+1)^2}\right]_0^{\pi} + \frac{1}{\pi}\left[-x\frac{\cos (n-1)\, x}{n-1} + \frac{\sin (n-1)\, x}{(n-1)^2}\right]_0^{\pi}$$

$$= \frac{1}{\pi}\left[x\cdot\left\{-\frac{\cos (n+1)\, x}{(n+1)} - \frac{\cos (n-1)\, x}{(n-1)}\right\} + 1\cdot\left\{\frac{\sin (n+1)\, x}{(n+1)^2} + \frac{\sin (n-1)\, x}{(n-1)^2}\right\}\right]_0^{\pi}$$

$$= \frac{1}{\pi}\left[\pi .\left\{-\frac{\cos (n+1)\,\pi}{(n+1)} - \frac{\cos (n-1)\,\pi}{(n-1)}\right\}\right]$$

$$\Rightarrow \quad b_n = \left\{-\frac{(-1)^{n+1}}{n+1} - \frac{(-1)^{n-1}}{n-1}\right\}, \; n \neq 1$$

$$b_n = -(-1)^{n+1}\left[\frac{1}{n+1} + \frac{1}{n-1}\right]$$

$$= -\left\{\frac{1}{(n+1)} + \frac{1}{(n-1)}\right\} = \frac{-2n}{n^2-1}, \qquad \text{If } n \text{ is odd; } n \neq 1.$$

$$\text{But } b_n = \left\{\frac{1}{n+1} + \frac{1}{n-1}\right\} = \frac{2n}{n^2-1}, \qquad \text{If } n \text{ is even; } n \neq 1$$

If $n = 1$, then $b_1 = \frac{2}{\pi}\int_0^{\pi} x \cos x . \sin x\, dx = \frac{1}{\pi}\int_0^{\pi} x . \sin 2x\, dx$

$$= \frac{1}{\pi}\left[x .\left(-\frac{\cos 2x}{2}\right) - 1\left(-\frac{\sin 2x}{4}\right)\right]_0^{\pi} = \frac{1}{\pi}\left[\pi\left(-\frac{1}{2}\right)\right] = -\frac{1}{2}$$

$\therefore\; x \cos x = b_1 \sin x + b_2 \sin 2x + b_3 \sin 3x + \ldots$

$$= -\frac{1}{2}\sin x + \frac{4 \sin 2x}{2^2-1} - \frac{6 \sin 3x}{3^2-1} + \ldots$$ **Ans.**

11.11 HALF-RANGE SERIES, PERIOD 0 TO π

The given function is defined in the interval $(0, \pi)$ and it is immaterial whatever the function may be outside the interval $(0, \pi)$. To get the series of cosines only we assume that $f(x)$ is an even function in the interval $(-\pi, \pi)$.

$$\boldsymbol{a_n = \frac{2}{\pi}\int_0^{\pi} f(x) \cos nx\, dx \;\text{ and }\; b_n = 0}$$

To expand $f(x)$ as a sine series we extend the function in the interval $(-\pi, \pi)$ as an odd function.

$$\boldsymbol{b_n = \frac{2}{\pi}\int_0^{\pi} f(x) \sin nx\, dx \;\text{ and }\; a_n = 0}$$

Example 13. *Represent the following function by a Fourier sine series:*

$$f(t)=\begin{cases} t, & 0<t\le\dfrac{\pi}{2} \\ \dfrac{\pi}{2}, & \dfrac{\pi}{2}<t\le\pi \end{cases}$$

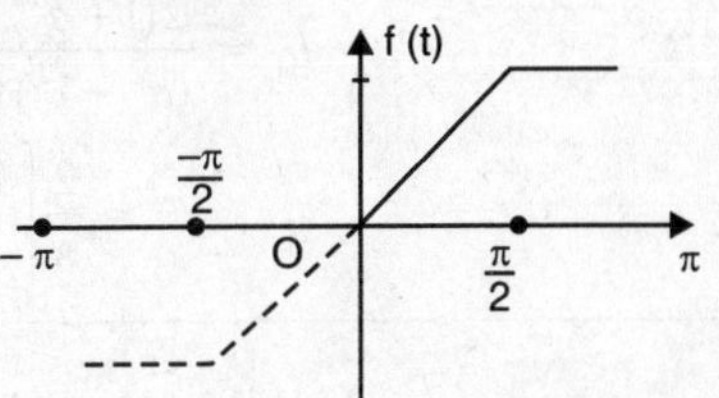

Solution. $b_n=\dfrac{2}{\pi}\displaystyle\int_0^{\pi} f(t)\sin nt\,dt$

$$=\frac{2}{\pi}\int_0^{\pi/2} t\sin nt\,dt+\frac{2}{\pi}\int_{\pi/2}^{\pi}\frac{\pi}{2}\sin nt\,dt$$

$$=\frac{2}{\pi}\left[t\left(-\frac{\cos nt}{n}\right)-(1)\left(-\frac{\sin nt}{n^2}\right)\right]_0^{\pi/2}+\frac{2}{\pi}\frac{\pi}{2}\left[-\frac{\cos nt}{n}\right]_{\pi/2}^{\pi}$$

$$=\frac{2}{\pi}\left[-\frac{\pi}{2}\frac{\cos\frac{n\pi}{2}}{n}+\frac{\sin\frac{n\pi}{2}}{n^2}\right]+\left[-\frac{\cos n\pi}{n}+\frac{\cos\frac{n\pi}{2}}{n}\right]$$

$$b_1=\frac{2}{\pi}\left[-\frac{\pi}{2}\cos\frac{\pi}{2}+\sin\frac{\pi}{2}\right]+\left[-\cos\pi+\cos\frac{\pi}{2}\right]=\frac{2}{\pi}[0+1]+[1]=\frac{2}{\pi}+1$$

$$b_2=\frac{2}{\pi}\left[-\frac{\pi}{2}\frac{\cos\pi}{2}+\frac{\sin\pi}{2^2}\right]+\left[-\frac{\cos 2\pi}{2}+\frac{\cos\pi}{2}\right]=\frac{2}{\pi}\left[-\frac{\pi}{2}\frac{(-1)}{2}+0\right]+\left[-\frac{1}{2}-\frac{1}{2}\right]$$

$$=\frac{2}{\pi}\left[\frac{\pi}{4}\right]-1=\frac{1}{2}-1=-\frac{1}{2}$$

$$b_3=\frac{2}{\pi}\left[-\frac{\pi}{2}\frac{\cos\frac{3\pi}{2}}{3}+\frac{\sin\frac{3\pi}{2}}{3^2}\right]+\left[-\frac{\cos 3\pi}{3}+\frac{\cos\frac{3\pi}{2}}{3}\right]$$

$$=\frac{2}{\pi}\left[-\frac{\pi}{2}(0)-\frac{1}{9}\right]+\left[\frac{1}{3}+0\right]=-\frac{2}{9\pi}+\frac{1}{3}$$

$$f(t)=\left(\frac{2}{\pi}+1\right)\sin t-\frac{1}{2}\sin 2t+\left(-\frac{2}{9\pi}+\frac{1}{3}\right)\sin 3t+\ldots$$ **Ans.**

Example 14. *Find the Fourier sine series for the function*

$$f(x)=e^{ax}\ \text{for}\ 0<x<\pi$$

where a is constant.

Solution. $b_n=\dfrac{2}{\pi}\displaystyle\int_0^{\pi} e^{ax}\sin nx\,dx$ $\qquad\left[\displaystyle\int e^{ax}\sin bx\,dx=\frac{e^{ax}}{a^2+b^2}[a\sin bx-b\cos bx]\right]$

$$=\frac{2}{\pi}\left[\frac{e^{ax}}{a^2+n^2}(a\sin nx-n\cos nx)\right]_0^{\pi}$$

$$=\frac{2}{\pi}\left[\frac{e^{a\pi}}{a^2+n^2}(a\sin n\pi-n\cos n\pi)+\frac{n}{a^2+n^2}\right]$$

$$=\frac{2}{\pi}\frac{n}{a^2+n^2}\left[-(-1)^n e^{a\pi}+1\right]=\frac{2n}{(a^2+n^2)\pi}[1-(-1)^n e^{a\pi}]$$

$$b_1 = \frac{2(1+e^{a\pi})}{(a^2+1^2)\pi}, \qquad b_2 = \frac{2.2.(1-e^{a\pi})}{.(a^2+2^2)\pi}$$

$$e^{ax} = \frac{2}{\pi}\left[\frac{1+e^{a\pi}}{a^2+1^2}\sin x + \frac{2(1-e^{a\pi})}{a^2+2^2}\sin 2x + \ldots\right]$$ **Ans.**

EXERCISE 11.3

1. Find the Fourier cosine series for the function

$$f(x) = \begin{cases} 1 & \text{for} \quad 0 < x < \frac{\pi}{2} \\ 0 & \text{for} \quad \frac{\pi}{2} < x < \pi. \end{cases}$$

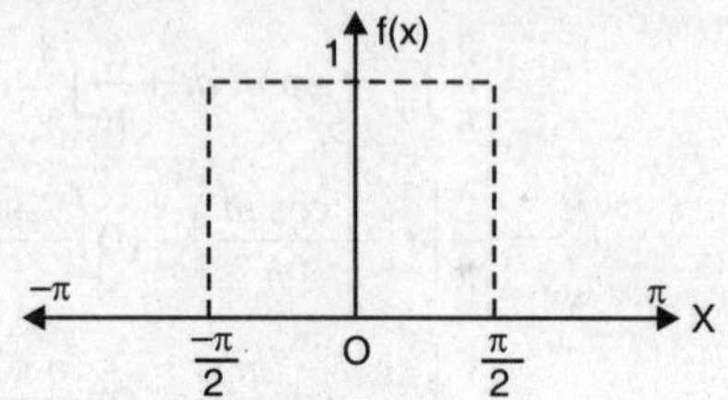

Ans. $\frac{1}{2} + \frac{2}{\pi}\left[\cos x - \frac{1}{3}\cos 3x + \frac{1}{5}\cos 5x - \ldots\right]$

2. Find a series of cosine of multiples of x which will represent $f(x)$ in $(0, \pi)$ where

$$f(x) = \begin{cases} 0, & 0 < x < \frac{\pi}{2} \\ \frac{\pi}{2}, & \frac{\pi}{2} < x < \pi \end{cases}$$

Deduce that $1 - \frac{1}{3} + \frac{1}{5} - \frac{1}{7} + \ldots \infty = \frac{\pi}{4}$ **Ans.** $\frac{\pi}{4} - \cos x + \frac{1}{3}\cos 3x - \frac{1}{5}\cos 5x + \ldots$

3. Express $f(x) = x$ as a sine series in $0 < x < \pi$. **Ans.** $2\left[\sin x - \frac{1}{2}\sin 2x + \frac{1}{3}\sin 3x - \ldots\right]$

4. Find the cosine series for $f(x) = \pi - x$ in the interval $0 < x < \pi$.

(DU, I Sem. 2012) **Ans.** $\frac{\pi}{2} + \frac{4}{\pi}\left(\cos x + \frac{\cos 3x}{3^2} + \frac{\cos 5x}{5^2} + \ldots\right)$

5. If $f(x) = \begin{cases} 0 & \text{for} \quad 0 < x < \frac{\pi}{2} \\ \pi - x, & \text{for} \quad \frac{\pi}{2} < x < \pi \end{cases}$

Show that: (*i*) $f(x) = \frac{4}{\pi}\left(\sin x - \frac{1}{3^2}\sin 3x + \frac{1}{5^2}\sin 5x - \ldots\right)$

(*ii*) $f(x) = \frac{\pi}{4} - \frac{2}{\pi}\left(\frac{1}{1^2}\cos 2x + \frac{1}{3^2}\cos 6x + \frac{1}{5^2}\cos 10x + \ldots\right)$

6. Obtain the half-range cosine series for $f(x) = x^2$ in $0 < x < \pi$.

Ans. $\frac{\pi^2}{3} - \frac{4}{\pi}\left(\cos x - \frac{1}{2^2}\cos 2x + \frac{1}{3^2}\cos 3x - \ldots\right)$

7. Find (*i*) sine series and (*ii*) cosine series for the function

$f(x) = e^x$ for $0 < x < \pi$.

Ans. (*i*) $\frac{2}{\pi}\sum_1^{\infty} n\left[\frac{1-(-1)^n e^{\pi}}{n^2+1}\right]\sin nx$ (*ii*) $\frac{e^{\pi}-1}{\pi} - \frac{2}{\pi}\sum_1^{\infty}\frac{1-(-1)^n e^{\pi}}{n^2+1}\cos nx$

8. If $f(x) = x + 1$, for $0 < x < \pi$, find its Fourier (*i*) sine series (*ii*) cosine series. Hence deduce that

(*i*) $1 - \frac{1}{3} + \frac{1}{5} - \frac{1}{7} + \ldots = \frac{\pi}{4}$ (*ii*) $1 + \frac{1}{3^2} + \frac{1}{5^2} + \frac{1}{7^2} + \ldots = \frac{\pi^2}{8}$

Ans. (*i*) $\frac{2}{\pi}\left[(\pi+2)\sin x - \frac{\pi}{2}\sin 2x + \frac{1}{3}(\pi+2)\sin 3x - \frac{\pi}{4}\sin 4x + \ldots\right]$

(*ii*) $\frac{\pi}{2} + 1 - \frac{4}{\pi}\left(\cos x + \frac{\cos 3x}{3^2} + \frac{\cos 5x}{5^2} + \ldots\right)$

9. Find the Fourier series expansion of the function
$f(x) = \cos(sx), -\pi \le x \le \pi$
where s is a fraction. Hence, show that $\cot\theta = \frac{1}{\theta} + \frac{2\theta}{\theta^2 - \pi^2} + \frac{2\theta}{\theta^2 - 4\pi^2} + \ldots$

Ans. $\frac{\sin \pi x}{\pi s} + \frac{1}{\pi} \Sigma \left(\frac{\sin(s\pi + n\pi)}{s+n} + \frac{\sin(s\pi - n\pi)}{s-n} \right) \cos nx$

11.12 CHANGE OF INTERVAL AND FUNCTIONS HAVING ARBITRARY PERIOD

In electrical engineering problems, the period of the function is not always 2π but T or $2c$. This period must be converted to the length 2π. The independent variable x is also to be changed proportionally.

Let the function $f(x)$ be defined in the interval $(-c, c)$. Now we want to change the function to the period of 2π so that we can use the formulae of a_n, b_n as discussed in Article 11.6.

$\because$ $2c$ is the interval for the variable x.

$\therefore$ 1 is the interval for the variable $= \frac{x}{2c}$

$\therefore$ 2π is the interval for the variable $= \frac{x\,2\pi}{2c} = \frac{\pi x}{c}$

so put $$z = \frac{\pi x}{c} \text{ or } x = \frac{zc}{\pi}$$

Thus the function $f(x)$ of period $2c$ is transformed to the function

$f\left(\frac{cz}{\pi}\right)$ or $F(z)$ of period 2π.

$F(z)$ can be expanded in the Fourier series :

$$F(z) = f\left(\frac{cz}{\pi}\right) = \frac{a_0}{2} + a_1 \cos z + a_2 \cos 2z + \ldots + b_1 \sin z + b_2 \sin 2z + \ldots$$

where $$a_0 = \frac{1}{\pi}\int_0^{2\pi} F(z)\,dz = \frac{1}{\pi}\int_0^{2\pi} f\left(\frac{cz}{\pi}\right) dz$$

$$= \frac{1}{\pi}\int_0^{2c} f(x)\, d\left(\frac{\pi x}{c}\right) = \frac{1}{c}\int_0^{2c} f(x)\,dx \qquad \left(\text{put } z = \frac{\pi x}{c}\right)$$

$$\boxed{a_0 = \frac{1}{c}\int_0^{2c} f(x)\,dx}$$

$$a_n = \frac{1}{\pi}\int_0^{2\pi} F(z)\cos nz\,dz = \frac{1}{\pi}\int_0^{2\pi} f\left(\frac{cz}{\pi}\right)\cos nz\,dz$$

$$= \frac{1}{\pi}\int_0^{2c} f(x)\cos\frac{n\pi x}{c}\, d\left(\frac{\pi x}{c}\right) = \frac{1}{c}\int_0^{2c} f(x)\cos\frac{n\pi x}{c}\,dx. \qquad \left[\text{Put } z = \frac{\pi x}{c}\right]$$

$$\boxed{a_n = \frac{1}{c}\int_0^{2c} f(x)\cos\frac{n\pi x}{c}\,dx}$$

Similarly, $$\boxed{b_n = \frac{1}{c}\int_0^{2c} f(x)\sin\frac{n\pi x}{c}\,dx}$$

Example 15. *Find the Fourier series corresponding to the function $f(x)$ defined in $(-2, 2)$ as follows*

$$f(x) = \begin{cases} 2 & in \quad -2 \le x \le 0 \\ x & in \quad 0 < x < 2 \end{cases}$$

Solution. Here the interval is $(-2, 2)$ and $c = 2$

$$a_0 = \frac{1}{c}\int_{-c}^{c} f(x)\,dx = \frac{1}{2}\left[\int_{-2}^{0} 2.dx + \int_0^2 x.dx\right]$$

$$= \frac{1}{2}\left[[2x]_{-2}^{0} + \left(\frac{x^2}{2}\right)_0^2\right] = \frac{1}{2}[4+2] = 3$$

$$a_n = \frac{1}{c}\int_{-c}^{c} f(x)\cos\left(\frac{n\pi x}{c}\right) dx = \frac{1}{2}\left[\int_{-2}^{0} 2\cos\frac{n\pi x}{2}\,dx + \int_0^2 x\cos\frac{n\pi x}{2}\,dx\right]$$

$$= \frac{1}{2}\left[\frac{4}{n\pi}\left(\sin\frac{n\pi x}{2}\right)_{-2}^{0} + \left(x\frac{2}{n\pi}\sin\frac{n\pi x}{2} + \frac{4}{n^2\pi^2}\cos\frac{n\pi x}{2}\right)_0^2\right]$$

$$= \frac{1}{2}\left[\frac{4}{n^2\pi^2}\cos n\pi - \frac{4}{n^2\pi^2}\right] = \frac{2}{n^2\pi^2}[(-1)^n - 1]$$

$$= -\frac{4}{n^2\pi^2}, \qquad \text{when } n \text{ is odd}$$

$$= 0, \qquad \text{when } n \text{ is even.}$$

$$b_n = \frac{1}{c}\int_{-c}^{c} f(x)\sin\frac{n\pi x}{c}\,dx = \frac{1}{2}\int_{-2}^{0} 2\sin\frac{n\pi x}{2}\,dx + \frac{1}{2}\int_0^2 x\sin\frac{n\pi x}{2}\,dx$$

$$= \frac{1}{2}\left[2\left(-\frac{2}{n\pi}\cos\frac{n\pi x}{2}\right)\right]_{-2}^{0} + \frac{1}{2}\left[x\left(-\frac{2}{n\pi}\cos\frac{n\pi x}{2}\right) + (1)\frac{4}{n^2\pi^2}\sin\frac{n\pi x}{2}\right]_0^2$$

$$= \frac{1}{2}\left[\frac{4}{n\pi} + \frac{4}{n\pi}\cos n\pi\right] + \frac{1}{2}\left[-\frac{4}{n\pi}\cos n\pi + \frac{4}{n^2\pi^2}\sin n\pi\right] = \frac{1}{2}\left[-\frac{4}{n\pi}\right] = -\frac{2}{n\pi}$$

$$f(x) = \frac{a_0}{2} + a_1\cos\frac{\pi x}{c} + a_2\cos\frac{2\pi x}{c} + a_3\cos\frac{3\pi x}{c} + \ldots$$

$$+ b_1\sin\frac{\pi x}{c} + b_2\sin\frac{2\pi x}{c} + b_3\sin\frac{3\pi x}{c} + \ldots$$

$$= \frac{3}{2} - \frac{4}{\pi^2}\left\{\frac{1}{1^2}\cos\frac{\pi x}{2} + \frac{1}{3^2}\cos\frac{3\pi x}{2} + \ldots\right\}$$

$$- \frac{2}{\pi}\left\{\frac{1}{1}\sin\frac{\pi x}{2} + \frac{1}{2}\sin\frac{2\pi x}{2} + \frac{1}{3}\sin\frac{3\pi x}{2} + \ldots\right\}$$ **Ans.**

Example 16. *A periodic function of period 4 is defined as*

$$f(x) = |x|, \; -2 < x < 2.$$

Find its Fourier series expansion.

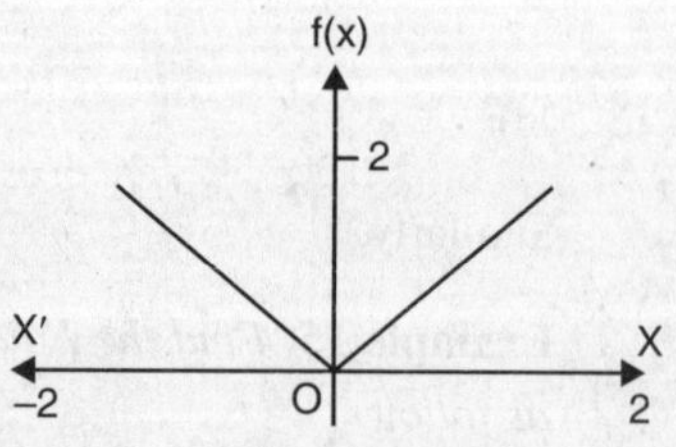

Solution.

$$f(x) = |x| \; -2 < x < 2$$

$$f(x) = \begin{cases} x, & 0 < x < 2 \\ -x, & -2 < x < 0 \end{cases}$$

$$a_0 = \frac{1}{c}\int_{-c}^{c} f(x)\,dx = \frac{1}{2}\int_0^2 x\,dx + \frac{1}{2}\int_{-2}^{0}(-x)\,dx$$

$$= \frac{1}{2}\left[\frac{x^2}{2}\right]_0^2 + \frac{1}{2}\left[\frac{-x^2}{2}\right]_{-2}^0 = \frac{1}{4}(4-0) + \frac{1}{4}(0+4) = 2$$

$$a_n = \frac{1}{c}\int_{-c}^{c} f(x)\cos\frac{n\pi x}{c}\,dx = \frac{1}{2}\int_0^2 x\cos\frac{n\pi x}{2}\,dx + \frac{1}{2}\int_{-2}^0 (-x)\cos\frac{n\pi x}{2}\,dx$$

$$= \frac{1}{2}\left[x\left(\frac{2}{n\pi}\sin\frac{n\pi x}{2}\right) - (1)\left(-\frac{4}{n^2\pi^2}\cos\frac{n\pi x}{2}\right)\right]_0^2$$

$$+ \frac{1}{2}\left[(-x)\left(\frac{2}{n\pi}\sin\frac{n\pi x}{2}\right) - (-1)\left(-\frac{4}{n^2\pi^2}\right)\cos\frac{n\pi x}{2}\right]_{-2}^0$$

$$= \frac{1}{2}\left[0 + \frac{4}{n^2\pi^2}(-1)^n - \frac{4}{n^2\pi^2}\right] + \frac{1}{2}\left[0 - \frac{4}{n^2\pi^2} + \frac{4}{n^2\pi^2}(-1)^n\right]$$

$$= \frac{1}{2}\frac{4}{n^2\pi^2}[(-1)^n - 1 - 1 + (-1)^n] = \frac{4}{n^2\pi^2}[(-1)^n - 1]$$

$$= -\frac{8}{n^2\pi^2} \qquad \text{(If } n \text{ is odd.)}$$

$$= 0 \qquad \text{(If } n \text{ is even)}$$

$b_n = 0$ as $f(x)$ is even function.

Fourier series is

$$f(x) = \frac{a_0}{2} + a_1\cos\frac{\pi x}{c} + a_2\cos\frac{2\pi x}{c} + \ldots + b_1\sin\frac{\pi x}{c} + b_2\sin\frac{2\pi x}{c} + \ldots$$

$$f(x) = 1 - \frac{8}{\pi^2}\left[\frac{\cos\frac{\pi x}{2}}{1^2} + \frac{\cos\frac{3\pi x}{2}}{3^2} + \frac{\cos\frac{5\pi x}{2}}{5^2} + \ldots\right]$$ **Ans.**

Example 17. *Prove that*

$$\frac{l}{2} - x = \frac{l}{\pi}\sum_{1}^{\infty}\frac{1}{n}\sin\frac{2n\pi x}{l}, \quad 0 < x < l.$$

Solution. $f(x) = \frac{1}{2} - x$

$$a_0 = \frac{1}{l/2}\int_0^l f(x)\,dx = \frac{2}{l}\int_0^l\left(\frac{1}{2} - x\right)dx = \frac{2}{l}\left[\frac{lx}{2} - \frac{x^2}{2}\right]_0^l = 0$$

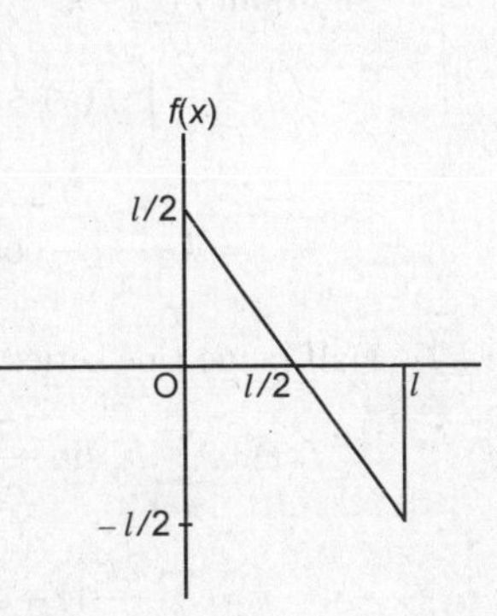

$$a_n = \frac{1}{l/2}\int_0^l f(x)\cos\frac{n\pi x}{l/2}\,dx = \frac{2}{l}\int_0^l\left(\frac{1}{2} - x\right)\cos\frac{2n\pi x}{l}\,dx$$

$$= \frac{2}{l}\left[\left(\frac{l}{2} - x\right)\frac{l}{2n\pi}\sin\frac{2n\pi x}{l} + (-1)\frac{l^2}{4n^2\pi^2}\cos\frac{2n\pi x}{l}\right]_0^l$$

$$= \frac{2}{l}\left[0 - \frac{l^2}{4n^2\pi^2}\cos 2n\pi + \frac{l^2}{4n^2\pi^2}\right]$$

$$= \frac{2}{l}\frac{l^2}{4n^2\pi^2}(-\cos 2n\pi + 1) = \frac{l}{2n^2\pi^2}(-1+1) = 0$$

$$b_n = \frac{1}{l/2}\int_0^l f(x)\sin\frac{n\pi x}{l/2}\,dx = \frac{2}{l}\int_0^l\left(\frac{1}{2}-x\right)\sin\frac{2n\pi x}{l}\,dx$$

$$= \frac{2}{l}\left[\left(\frac{1}{2}-x\right)\left(-\frac{1}{2n\pi}\cos\frac{2n\pi x}{l}\right) - (-1)\left(-\frac{l^2}{4n^2\pi^2}\sin\frac{2n\pi x}{l}\right)\right]_0^l$$

$$= \frac{2}{l}\left[\frac{l}{2}\frac{l}{2n\pi}\cos 2n\pi - 0 + \frac{l}{2}\cdot\frac{l}{2n\pi}(1)\right] = \frac{2}{l}\left[\frac{l^2}{2n\pi}\right] = \frac{l}{n\pi}$$

Fourier series is

$$f(x) = \frac{a_0}{2} + a_1\cos\frac{\pi x}{l/2} + a_2\cos\frac{2\pi x}{l/2} + a_3\cos\frac{3\pi x}{l/2} + \ldots$$

$$+ b_1\sin\frac{\pi x}{l/2} + b_2\sin\frac{2\pi x}{l/2} + b_3\sin\frac{3\pi x}{l/2} + \ldots$$

$$\frac{l}{2} - x = \frac{l}{\pi}\sin\frac{2\pi x}{l} + \frac{l}{2\pi}\sin\frac{4\pi x}{l} + \frac{l}{3\pi}\sin\frac{6\pi x}{l} + \ldots$$

$$= \frac{l}{\pi}\sum_1^\infty\frac{1}{n}\sin\frac{2n\pi x}{l}$$

Proved.

11.13 HALF PERIOD SERIES

Cosine series: $f(x) = \frac{a_0}{2} + a_1\cos\frac{\pi x}{c} + a_2\cos\frac{2\pi x}{c} + \ldots + a_n\cos\frac{n\pi x}{c} + \ldots$

where $a_0 = \frac{2}{c}\int_0^c f(x)\,dx,\ a_n = \frac{2}{c}\int_0^c f(x)\cos\frac{n\pi x}{c}\,dx$

Sine series: $f(x) = b_1\sin\frac{\pi x}{c} + b_2\sin\frac{2\pi x}{c} + \ldots + b_n\sin\frac{n\pi x}{c} + \ldots$

where $b_n = \frac{2}{c}\int_0^c f(x)\sin\frac{n\pi x}{c}\,dx.$

Example 18. *Expand for f (x) = k for 0 < x < 2 in a half range sine series.*

(U.P., II Semester, June 2007)

Solution. $f(x) = k$

$$b_n = \frac{2}{c}\int_0^c f(x)\cdot\sin\frac{n\pi x}{c}\,dx \text{. In half range (0, 2)} = \frac{2}{2}\int_0^2 k\sin\frac{n\pi x}{2}\,dx$$

$$= k\frac{2}{n\pi}\left(-\cos\frac{n\pi x}{2}\right)_0^2 = \frac{2k}{n\pi}[-\cos n\pi + 1]$$

Half range sine series is

$$f(x) = \sum_{n=1}^\infty b_n\sin\frac{n\pi x}{2}$$

$$k = \sum\frac{2k}{n\pi}[1-\cos n\pi]\sin\frac{n\pi x}{2}$$

Ans.

Example 19. *Obtain the half-range sine series for the function $f(x) = x^2$ in the interval $0 < x < 3$.*

Solution. We know that half range sine series is given by $f(x) = \Sigma b_n \sin nx$

Where $$b_n = \frac{2}{c}\int_0^c f(x)\sin\frac{n\pi x}{c}\,dx \text{ in the half-range } (0, c).$$

Here, we have half range $0 < x < 3$ and $f(x) = x^2$

$$b_n = \frac{2}{3}\int_0^3 x^2 \sin\frac{n\pi x}{3}\,dx$$

$$= \frac{2}{3}\left[x^2\left(\frac{3}{n\pi}\right)\left(-\cos\frac{n\pi x}{3}\right) + 2x\left(\frac{3}{n\pi}\right)\left(\frac{3}{n\pi}\right)\sin\frac{n\pi x}{3} - 2\left(\frac{3}{n\pi}\right)\left(\frac{3}{n\pi}\right)\left(\frac{3}{n\pi}\right)\left(-\cos\frac{n\pi x}{3}\right)\right]_0^3$$

$$= \frac{2}{3}\left[\left\{-\frac{27}{n\pi}(-1)^n - \frac{54}{n^3\pi^3}(-1)^n\right\} + \frac{54}{n^3\pi^3}\right]$$

$$\Rightarrow \quad b_n = \frac{2}{3}\left[\frac{54}{n^3\pi^3}\{1-(-1)^n\} - \frac{27}{n\pi}(-1)^n\right] \Rightarrow b_n = \frac{2}{3}\left[\frac{108}{n^3\pi^3} + \frac{27}{n\pi}\right] \text{ when } n \text{ is odd}$$

And $b_n = -\dfrac{18}{n\pi}$ when n is even

$\therefore$ Half range sine series

$$f(x) = b_1 \sin x + b_2 \sin 2x + b_3 \sin 3x + \ldots$$

$$= \frac{2}{3}\left[\frac{108}{\pi^3}\left(\frac{\sin x}{1^3} + \frac{\sin 3x}{3^3} + \frac{\sin 5x}{5^3} + \ldots\right) + \frac{27}{\pi}\left(\frac{\sin x}{1} + \frac{\sin 3x}{3} + \frac{\sin 5x}{5} + \ldots\right)\right.$$
$$\left. - \frac{18}{\pi}\left(\frac{\sin 2x}{2} + \frac{\sin 4x}{4} + \ldots\right)\right] \quad \textbf{Ans.}$$

Example 20. *Expand $f(x) = e^x$ in a cosine series over (0, 1).*

Solution. Here, we have $f(x) = e^x$ and $c = 1$

$$a_0 = \frac{2}{c}\int_0^c f(x)\,dx = \frac{2}{1}\int_0^1 e^x\,dx = 2(e-1)$$

$$a_n = \frac{2}{c}\int_0^c f(x)\cos\frac{n\pi x}{c}\,dx = \frac{2}{1}\int_0^1 e^x \cos\frac{n\pi x}{1}\,dx$$

$$= 2\left[\frac{e^x}{n^2\pi^2+1}(n\pi \sin n\pi x + \cos n\pi x)\right]_0^1$$

$$= 2\left[\frac{e^1}{n^2\pi^2+1}(n\pi \sin n\pi + \cos n\pi) - \frac{1}{n^2\pi^2+1}\right]$$

$$= \frac{2}{n^2\pi^2+1}[(-1)^n e - 1]$$

$$f(x) = \frac{a_0}{2} + a_1 \cos\pi x + a_2 \cos 2\pi x + a_3 \cos 3\pi x + \ldots$$

$$e^x = e - 1 + 2\left[\frac{-e-1}{\pi^2+1}\cos\pi x + \frac{e-1}{4\pi^2+1}\cos 2\pi x + \frac{-e-1}{9\pi^2+1}\cos 3\pi x + \ldots\right] \quad \textbf{Ans.}$$

Example 21. *Find the Fourier half-range cosine series of the function*

$$f(t) = \begin{cases} 2t, & 0 < t < 1 \\ 2(2-t), & 1 < t < 2 \end{cases}$$

(U.P., II Semester, Summer 2007, 2006)

Solution. $f(t) = \begin{cases} 2t, & 0 < t < 1 \\ 2(2-t), & 1 < t < 2 \end{cases}$

Let $$f(t) = \frac{a_0}{2} + a_1 \cos \frac{\pi t}{c} + a_2 \cos \frac{2\pi t}{c} + a_3 \cos \frac{3\pi t}{c} + \ldots$$

$$+ b_1 \sin \frac{\pi t}{c} + b_2 \sin \frac{2\pi t}{c} + b_3 \sin \frac{3\pi t}{c} + \ldots \quad \ldots(1)$$

Here, $c = 2$, because it is half range series.

Hence, $a_0 = \frac{2}{c}\int_0^c f(t)\, dt = \frac{2}{2}\int_0^1 2t\, dt + \frac{2}{2}\int_1^2 2(2-t)\, dt$

$$= [t^2]_0^1 + \left[2\left(2t - \frac{t^2}{2}\right)\right]_1^2 = 1 + [4t - t^2]_1^2 = 1 + (8 - 4 - 4 + 1) = 2$$

$$a_n = \frac{2}{c}\int_0^c f(t) \cos \frac{n\pi t}{c}\, dt = \frac{2}{2}\int_0^1 2t \cos \frac{n\pi t}{2}\, dt + \frac{2}{2}\int_1^2 2(2-t) \cos \frac{n\pi t}{2}\, dt$$

$$= \left[2t\left(\frac{2}{n\pi} \sin \frac{n\pi t}{2}\right) - (2)\left(-\frac{4}{n^2\pi^2} \cos \frac{n\pi t}{2}\right)\right]_0^1$$

$$+ \left[(4-2t)\left(\frac{2}{n\pi} \sin \frac{n\pi t}{2}\right) - (-2)\left(-\frac{4}{n^2\pi^2} \cos \frac{n\pi t}{2}\right)\right]_1^2$$

$$= \left[\frac{4}{n\pi} \sin \frac{n\pi}{2} + \frac{8}{n^2\pi^2} \cos \frac{n\pi}{2} - \frac{8}{n^2\pi^2}\right] + \left[0 - \frac{8}{n^2\pi^2} \cos n\pi - \frac{4}{n\pi} \sin \frac{n\pi}{2} + \frac{8}{n^2\pi^2} \cos \frac{n\pi}{2}\right]$$

$$= \frac{16}{n^2\pi^2} \cos \frac{n\pi}{2} - \frac{8}{n^2\pi^2} - \frac{8}{n^2\pi^2} \cos n\pi = \frac{8}{n^2\pi^2}\left[2 \cos \frac{n\pi}{2} - 1 - \cos n\pi\right]$$

$$f(t) = 1 + \frac{8}{\pi^2} \sum_{n=1}^{\infty} \frac{1}{n^2}\left[2 \cos \frac{n\pi}{2} - 1 \cos n\pi\right] \cos \frac{n\pi t}{2}$$ **Ans.**

Example 22. *Obtain the Fourier cosine series expansion of the periodic function defined by* $f(t) = \sin\left(\frac{\pi t}{l}\right)$, $0 < t < l$

Solution. We have, $f(t) = \sin\left(\frac{\pi t}{l}\right)$, $0 < t < l$

$$a_0 = \frac{2}{l}\int_0^l \sin\left(\frac{\pi t}{l}\right) dt = \frac{2}{l}\left(-\frac{l}{\pi} \cos \frac{\pi t}{l}\right)_0^l = -\frac{2}{\pi}(\cos \pi - \cos 0) = -\frac{2}{\pi}(-1-1) = \frac{4}{\pi}$$

$$a_n = \frac{2}{l}\int_0^l \sin\left(\frac{\pi t}{l}\right) \cos \frac{n\pi t}{l}\, dt = \frac{1}{l}\int_0^l \left[\sin\left(\frac{\pi t}{l} + \frac{n\pi t}{l}\right) - \sin\left(\frac{n\pi t}{l} - \frac{\pi t}{l}\right)\right] dt$$

$$= \frac{1}{l}\int_0^l \sin (n+1)\frac{\pi t}{l}\, dt - \frac{1}{l}\int_0^l \sin (n-1)\frac{\pi t}{l}\, dt$$

$$= \frac{1}{l}\left[-\frac{l}{(n+1)\pi}\cos\frac{(n+1)\pi t}{l}\right]_0^l - \frac{1}{l}\left[\frac{l}{(n-1)\pi}\cos\frac{(n-1)\pi t}{l}\right]_0^l$$

$$= \frac{-1}{(n+1)\pi}[\cos(n+1)\pi - \cos 0] + \frac{1}{(n-1)\pi}[\cos(n-1)\pi - \cos 0]$$

$$= \frac{1}{(n+1)\pi}[(-1)^{n+1} - 1] + \frac{1}{(n-1)\pi}[(-1)^{n-1} - 1]$$

$$= (-1)^{n+1}\left[\frac{1}{(n+1)\pi} + \frac{1}{(n-1)\pi}\right] + \frac{1}{(n+1)\pi} - \frac{1}{(n-1)\pi}$$

$$= (-1)^{n+1}\frac{2}{(n^2-1)\pi} - \frac{2}{(n^2-1)\pi} = \frac{2}{(n^2-1)\pi}[(-1)^{n+1} - 1]$$

$$= \frac{-4}{(n^2-1)\pi}, \quad \text{when } n \text{ is even}$$

$$= 0, \quad \text{when } n \text{ is odd.}$$

The above formula for finding the value of a_1 is not applicable.

$$a_1 = \frac{2}{l}\int_0^l \sin\frac{\pi t}{l}\cos\frac{\pi t}{l}\,dt = \frac{1}{l}\int_0^l \sin\frac{2\pi t}{l}\,dt$$

$$= \frac{1}{l}\left(-\frac{l}{2\pi}\cos\frac{2\pi t}{l}\right)_0^l = -\frac{1}{2\pi l}(\cos 2\pi - \cos 0) = -\frac{1}{2\pi l}(1-1) = 0$$

$$f(t) = \frac{a_0}{2} + a_1\cos\frac{\pi t}{l} + a_2\cos\frac{2\pi t}{l} + a_3\cos\frac{3\pi t}{l} + a_4\cos\frac{4\pi t}{l} + \ldots$$

$$= \frac{2}{\pi} - \frac{4}{\pi}\left[\frac{1}{3}\cos\frac{2\pi t}{l} + \frac{1}{15}\cos\frac{4\pi t}{l} + \frac{1}{35}\cos\frac{6\pi t}{l} + \ldots\right]$$ **Ans.**

Example 23. *Find the Fourier series expansion of the output of Half Wave Rectifier*

(*Delhi University, April 2010*)

$$f(t) = \begin{cases} 0, & -\pi < wt < 0 \\ E_0 \sin wt, & 0 < wt < \pi \end{cases}$$

Solution. We have, $f(t) = \begin{cases} 0, & -\pi < wt < 0 \\ E_0 \sin wt, & 0 < wt < \pi \end{cases}$

$$a_0 = \frac{1}{2\pi}\int_{-\pi}^{\pi} f(t)\,d(wt) = \frac{1}{2\pi}\left[\int_{-\pi}^{0} 0\,d(wt) + \int_0^{\pi} E_0\sin(wt)\,d(wt)\right]$$

$$= \frac{1}{2\pi}\int_0^{\pi} E_0\sin(wt)\,d(wt) = \frac{E_0}{\pi} \quad \ldots(1) \qquad \text{and } a_n = \frac{1}{\pi}\left[\int_{-\pi}^{\pi} f(t)\cos nwt\,d(wt)\right]$$

$$= \frac{1}{\pi}\left[\int_{-\pi}^{0} 0\,d(wT) + \int_0^{\pi} E_0\sin wt\cos nwt\,d(wt)\right] = \frac{E_0}{\pi}\int_0^{\pi}\sin wt\cos n\,wt\,d(wt)$$

$$= \frac{1}{2}\frac{E_0}{\pi}\int_0^{\pi}[(\sin(n+1)wt - \sin(n-1)wt]\,d(wt) = \frac{E_0}{2\pi}\left[-\frac{\cos(n+1)wt}{n+1} + \frac{\cos(n-1)wt}{n-1}\right]_0^{\pi}$$

$$= \frac{E_0}{2\pi}\left[\left\{-\frac{(-1)^{n+1}}{n+1}+\frac{(-1)^{n-1}}{n-1}\right\}-\left\{-\frac{1}{n+1}+\frac{1}{n-1}\right\}\right] = \begin{cases} \dfrac{2E_0}{\pi(n^2-1)} & \text{for } n \text{ is even} \\ 0, & \text{for } n \text{ is odd} \quad (n \neq 1) \end{cases}$$

$$b_n = \frac{E_0}{\pi}\int_0^{\pi} \sin wt \sin nwt \, d(wt) = \frac{E_0}{2\pi}\int_0^{\pi}[\cos(n-1)wt - \cos(n+1)wt]\, d(wt)$$

$$= \frac{E_0}{2\pi}\left[\frac{\sin(n-1)wt}{n-1}-\frac{\sin(n+1)wt}{n+1}\right]_0^{\pi} = \frac{E_0}{2\pi}\left[\frac{\sin(n-1)\pi}{n-1}-\frac{\sin(n+1)\pi}{n+1}\right]$$ **Ans.**

$$= \frac{E_0 \sin(n-1)\pi}{2\pi}\left[\frac{1}{n-1}-\frac{1}{n+1}\right] = \frac{E_0 \sin(n-1)\pi}{(n^2-1)\pi} = 0 \qquad (n \neq 1)$$

$$b_1 = \frac{E_0}{\pi}\int_0^{\pi} \sin wt \sin wt \, d(wt) = \frac{E_0}{\pi}\int_0^{\pi} \sin^2 wt \, d(wt) = \frac{E_0}{2\pi}\int_0^{\pi}(1-\cos 2wt)\, d(wt)$$

$$= \frac{E_0}{\pi}\int_0^{\pi}\left[wt - \frac{\sin 2wt}{2}\right]_0^{\pi} = \frac{E_0}{2\pi}[\pi] = \frac{E_0}{2}$$

Hence $f(t) = a_0 + b_1 \sin wt + \sum_{n=2,3....}^{n} a_n \cos nwt$

$$\Rightarrow f(t) = \frac{E_0}{\pi} + \frac{E_0}{2}\sin wt - \frac{2E_0}{\pi}\left[\frac{\cos 2wt}{2^2-1}+\frac{\cos 4wt}{4^2-1}+\ldots\right]$$ **Ans.**

Example 24. *Obtain a cosine series expansion of the function $f(x) = 1 + x$ valid in the interval $0 \le x \le 2$ and hence deduce that*:

$$\frac{1}{1^2}+\frac{1}{3^2}+\frac{1}{5^2}+\ldots\ldots = \frac{\pi^2}{8}$$ (*Delhi University, April 2010*)

Solution. Let $f(x) = \frac{a_0}{2} + \sum_n a_n \cos\frac{nx\pi}{2}$ as $f(x)$ is a cosine series

$$\therefore \quad a_0 = \frac{2}{2}\int_0^2 f(x)\, dx = 1\int_0^2 (1+x)dx = \left[x + \frac{x^2}{2}\right]_0^2 = 4$$

and $$a_n = \frac{2}{2}\int_0^2 f(x)\cos\frac{nx\pi}{2}\, dx = \int_0^2 (1+x)\cos\frac{nx\pi}{2}\, dx$$

$$= \left[\frac{2(1+x)\sin\frac{nx\pi}{2}}{\pi n}\right]_0^2 - 2\int_0^2 \frac{\sin\frac{nx\pi}{2}}{\pi n}\, dx = 0 + 4\left[\frac{\cos\frac{nx\pi}{2}}{n^2\pi^2}\right]_0^2 = \frac{4}{\pi^2 n^2}[\cos n\pi - 1]$$

$$a_n = \begin{cases} -\dfrac{8}{n^2\pi^2}, & \text{for } n \text{ odd} \\ 0, & \text{for } n \text{ even} \end{cases} \qquad f(x) = 2 - \frac{8}{\pi^2}\left[\frac{\cos\frac{x\pi}{2}}{1^2}+\frac{\cos\frac{3x\pi}{2}}{3^2}+\ldots\ldots\right]$$

$$1 + x = 2 - \frac{8}{\pi^2}\left[\frac{\cos\frac{x\pi}{2}}{1^2}+\frac{\cos\frac{3x\pi}{2}}{3^2}+\ldots\ldots\right]$$

Put $x = 2$

$$3 = 2 + \frac{8}{\pi^2}\left[\frac{1}{1^2} + \frac{1}{3^2} + \frac{1}{5^2} + \cdots\cdots\right] \qquad \left[\begin{array}{l}\cos m\pi = -1 \\ \text{if } m \text{ is odd}\end{array}\right]$$

$$\Rightarrow \qquad \frac{\pi^2}{8} = \frac{1}{1^2} + \frac{1}{3^2} + \frac{1}{5^2} + \cdots\cdots \qquad \textbf{Proved.}$$

Example 25. *Find the Fourier cosine series expansion of the periodic function of period 1*

$$f(x) = \begin{cases} \dfrac{1}{2} + x, & -\dfrac{1}{2} < x \le 0 \\ \dfrac{1}{2} - x, & 0 < x < \dfrac{1}{2} \end{cases}$$

Solution. Let $f(x) = \dfrac{a_0}{2} + a_1 \cos \dfrac{\pi x}{c} + a_2 \cos \dfrac{2\pi x}{c} + \ldots$ ...(1)

as $f(x)$ is a cosine series.

Here $2c = 1 \Rightarrow c = \dfrac{1}{2}$

$$a_0 = \frac{1}{c}\int_{-c}^{c} f(x)\,dx = \frac{1}{1/2}\int_{-1/2}^{0}\left(\frac{1}{2} + x\right)dx + \frac{1}{1/2}\int_{0}^{1/2}\left(\frac{1}{2} - x\right)dx$$

$$= 2\left[\frac{x}{2} + \frac{x^2}{2}\right]_{-1/2}^{0} + 2\left[\frac{x}{2} - \frac{x^2}{2}\right]_{0}^{1/2} = 2\left[\frac{1}{4} - \frac{1}{8}\right] + 2\left[\frac{1}{4} - \frac{1}{8}\right] = \frac{1}{2}$$

$$a_n = \frac{1}{c}\int_{-c}^{c} f(x)\cos\frac{n\pi x}{c}\,dx$$

$$= \frac{1}{1/2}\int_{-1/2}^{0}\left(\frac{1}{2} + x\right)\cos\frac{n\pi x}{1/2}\,dx + \frac{1}{1/2}\int_{0}^{1/2}\left(\frac{1}{2} - x\right)\cos\frac{n\pi x}{1/2}\,dx$$

$$= 2\int_{-1/2}^{0}\left(\frac{1}{2} + x\right)\cos 2n\pi x\,dx + 2\int_{0}^{1/2}\left(\frac{1}{2} - x\right)\cos 2n\pi x\,dx$$

$$= 2\left[\left(\frac{1}{2} + x\right)\frac{\sin 2n\pi x}{2n\pi} - (1)\left(-\frac{\cos 2n\pi x}{4n^2\pi^2}\right)\right]_{-1/2}^{0}$$

$$+ 2\left[\left(\frac{1}{2} - x\right)\frac{\sin 2n\pi x}{2n\pi} - (-1)\left(\frac{-\cos 2n\pi x}{4n^2\pi^2}\right)\right]_{0}^{1/2}$$

$$= 2\left[0 + \frac{1}{4n^2\pi^2} - \frac{(-1)^n}{4n^2\pi^2}\right] + 2\left[0 - \frac{(-1)^n}{4n^2\pi^2} + \frac{1}{4n^2\pi^2}\right] = \frac{1}{\pi^2}\left[\frac{1}{n^2} - \frac{(-1)^n}{n^2}\right]$$

$$= \frac{2}{n^2\pi^2} \qquad \text{(if } n \text{ is odd)}$$

$$= 0 \qquad \text{(if } n \text{ is even)}$$

Substituting the values of $a_0, a_1, a_2, a_3, \ldots$ in (1), we have

$$f(x) = \frac{1}{4} + \frac{2}{\pi^2}\left[\frac{\cos 2\pi x}{1^2} + \frac{\cos 6\pi x}{3^2} + \frac{\cos 10\pi x}{5^2} + \ldots\right] \qquad \textbf{Ans.}$$

Example 26. *Let* $f(x) = \begin{cases} wx, & \text{where } 0 \le x \le \dfrac{l}{2} \\ w(l-x), & \text{where } \dfrac{l}{2} \le x \le l \end{cases}$

Show that $$f(x) = \frac{4wl}{\pi^2}\sum_{n=0}^{\infty}\frac{(-1)^n}{(2n+1)^2}\sin\frac{(2n+1)\pi x}{l}$$

Hence, obtain the sum of the series

$$1+\frac{1}{3^2}+\frac{1}{5^2}+\ldots$$

Solution. Half range sine series is

$$f(x) = \sum_{n=1}^{\infty} b_n \sin\frac{n\pi x}{l} \qquad \ldots(1)$$

where, $b_n = \dfrac{2}{l}\displaystyle\int_0^l f(x)\sin\frac{n\pi x}{l}\,dx$

$$b_n = \frac{2}{l}\left[\int_0^{\frac{l}{2}} f(x)\sin\frac{n\pi x}{l}\,dx + \int_{\frac{l}{2}}^{l} f(x)\sin\frac{n\pi x}{l}\,dx\right]$$

$$= \frac{2}{l}\left[\int_0^{\frac{l}{2}} wx\sin\frac{n\pi x}{l}\,dx + \int_{\frac{l}{2}}^{l} w(l-x)\sin\frac{n\pi x}{l}\,dx\right]$$

$$= \frac{2}{l}\left[\left\{wx\frac{\left(-\cos\frac{n\pi x}{l}\right)}{\left(\frac{n\pi}{l}\right)} - w\frac{\left(-\sin\frac{n\pi x}{l}\right)}{\left(\frac{n\pi}{l}\right)^2}\right\}_0^{\frac{l}{2}} + \left\{w(l-x)\frac{\left(-\cos\frac{n\pi x}{l}\right)}{\left(\frac{n\pi}{l}\right)} - w(-1)\frac{\left(-\sin\frac{n\pi x}{l}\right)}{\left(\frac{n\pi}{l}\right)^2}\right\}_{\frac{l}{2}}^{l}\right]$$

$$= \frac{2}{l}\left[\left\{\frac{wl}{2}\frac{\left(-\cos\frac{n\pi}{2}\right)}{\left(\frac{n\pi}{l}\right)} + \frac{w\sin\frac{n\pi}{2}}{\left(\frac{n\pi}{l}\right)^2} - 0 - 0\right\} - \left\{\frac{wl}{2}\frac{\left(-\cos\frac{n\pi}{2}\right)}{\left(\frac{n\pi}{l}\right)} - \frac{w\sin\frac{n\pi}{2}}{\left(\frac{n\pi}{l}\right)^2}\right\}\right]$$

$$= \frac{2w}{l}\left[-\frac{l^2}{2n\pi}\cos\frac{n\pi}{2} + \frac{l^2}{n^2\pi^2}\sin\frac{n\pi}{2} + \frac{l^2}{2n\pi}\cos\frac{n\pi}{2} + \frac{l^2}{n^2\pi^2}\sin\frac{n\pi}{2}\right]$$

$$= \frac{2w}{l}\left[\frac{l^2}{n^2\pi^2}\sin\frac{n\pi}{2} + \frac{l^2}{n^2\pi^2}\sin\frac{n\pi}{2}\right] = \frac{4wl^2}{ln^2\pi^2}\sin\frac{n\pi}{2}$$

$$b_n = \frac{4wl}{n^2\pi^2}\sin\frac{n\pi}{2}, \text{ when } n \text{ is odd.}$$

$$b_n = 0, \quad \text{when } n \text{ is even.}$$

$$\left[\begin{aligned} &\sin(2n+1)\frac{\pi}{2} \\ &= \sin\left(n\pi+\frac{\pi}{2}\right) = \cos n\pi \\ &= (-1)^n \end{aligned}\right]$$

Now, putting the value of b_n in (1), we get

$$f(x) = \sum_{n=1}^{\infty}\frac{4wl}{n^2\pi^2}\sin\frac{n\pi}{2}\sin\frac{n\pi x}{l}, \text{ when } n \text{ is odd}$$

$$f(x) = \frac{4wl}{\pi^2}\sum_{n=0}^{\infty}\frac{(-1)^n}{(2n+1)^2}\sin\frac{(2n+1)\pi x}{l} \qquad (n \to 2n+1)$$

$$f(x) = \frac{4wl}{\pi^2}\left[\frac{1}{1^2}\sin\frac{\pi x}{l} - \frac{1}{3^2}\sin\frac{3\pi x}{l} + \frac{1}{5^2}\sin\frac{5\pi x}{l} + \ldots\right] \qquad \ldots(2)$$

Putting $x = \frac{l}{2}$, $f(x) = wx$ and $f\left(\frac{l}{2}\right) = \frac{wl}{2}$ in (2), we get

$$\frac{wl}{2} = \frac{4wl}{\pi^2}\left[\frac{1}{1^2}\sin\frac{\pi}{2} - \frac{1}{3^2}\sin\frac{3\pi}{2} + \frac{1}{5^2}\sin\frac{5\pi}{2} + \ldots\right]$$

$$\Rightarrow \frac{\pi^2}{8} = \frac{1}{1^2} + \frac{1}{3^2} + \frac{1}{5^2} + \ldots$$ **Ans.**

Example 27. *Find the half period sine series for f (x) given in the range (0, l) by the graph OPQ as shown in figure.* *(U.P. II semester, 2009)*

Solution. The equation of line *OP* is $y = \frac{kx}{\frac{l}{2}} \Rightarrow y = \frac{2kx}{l}$

and the equation of the line *PQ* is $y - 0 = \frac{0-k}{l - \frac{l}{2}}(x - l)$

$$\Rightarrow \quad y = \frac{-2kx}{l} + 2k$$

$f(x)$ is the half period

$$f(x) = \begin{cases} \frac{2kx}{l}, & 0 < x < \frac{l}{2} \\ -\frac{2kx}{l} + 2k, & \frac{l}{2} < x < l \end{cases}$$

$f(x)$ is half period series. It is to be expanded as sine series.

Here, $a_0 = 0$ and $a_n = 0$

$$b_n = \frac{2}{l}\int_0^{\frac{l}{2}} f(x)\sin\frac{n\pi x}{l}\,dx + \frac{2}{l}\int_{\frac{l}{2}}^{l} f(x)\sin\frac{n\pi x}{l}\,dx$$

$$= \frac{2}{l}\int_0^{\frac{l}{2}} \frac{2kx}{l}\sin\frac{n\pi x}{l}\,dx + \frac{2}{l}\int_{\frac{l}{2}}^{l}\left(\frac{-2kx}{l} + 2k\right)\sin\frac{n\pi x}{l}\,dx$$

$$= \frac{4k}{l^2}\int_0^{\frac{l}{2}} x\sin\frac{n\pi x}{l}\,dx + \frac{4k}{l^2}\int_{\frac{l}{2}}^{l}(-x + 1)\sin\frac{n\pi x}{l}\,dx$$

$$= \frac{4k}{l^2}\left[x\left(\frac{-l}{n\pi}\cos\frac{n\pi x}{l}\right) - \left(\frac{-l^2}{n^2\pi^2}\sin\frac{n\pi x}{l}\right)\right]_0^{\frac{l}{2}} + \frac{4k}{l^2}\left[(-x + l)\left(\frac{-l}{n\pi}\cos\frac{n\pi x}{l}\right) - (-1)\left(\frac{-l^2}{n^2\pi^2}\sin\frac{n\pi x}{l}\right)\right]_{\frac{l}{2}}^{l}$$

$$= \frac{4k}{l^2}\left[-\frac{l}{2}\left(\frac{l}{n\pi}\right)\cos\frac{n\pi\frac{l}{2}}{l} + \frac{l^2}{n^2\pi^2}\sin\frac{n\pi\frac{l}{2}}{l}\right] + \frac{4k}{l^2}\left[(-l+l)\left(-\frac{l}{n\pi}\cos n\pi\right) - \frac{l^2}{n^2\pi^2}\sin n\pi\right.$$
$$\left. - \left(-\frac{l}{2} + l\right)\left(-\frac{1}{n\pi}\cos\frac{n\pi\frac{1}{2}}{l}\right) + \frac{l^2}{n^2\pi^2}\sin\frac{n\pi\frac{1}{2}}{l}\right]$$

$$= \frac{4k}{l^2}\left[-\frac{l^2}{2n\pi}\cos\frac{n\pi}{2} + \frac{l^2}{n^2\pi^2}\sin\frac{n\pi}{2}\right] + \frac{4k}{l^2}\left[0 - 0 + \frac{l^2}{2n\pi}\cos\frac{n\pi}{2} + \frac{l^2}{n^2\pi^2}\sin\frac{n\pi}{2}\right]$$

$$= \frac{4k}{l^2}\left(\frac{l^2}{2n\pi}\right)\left[-\cos\frac{n\pi}{2}+\frac{2}{n\pi}\sin\frac{n\pi}{2}\right]+\frac{4k}{l^2}\left[\frac{l^2}{2n\pi}\cos\frac{n\pi}{2}+\frac{l^2}{n^2\pi^2}\sin\frac{n\pi}{2}\right]$$

$$= \frac{4k}{l^2}\left(\frac{l^2}{2n\pi}\right)\left[-\cos\frac{n\pi}{2}+\frac{2}{n\pi}\sin\frac{n\pi}{2}+\cos\frac{n\pi}{2}+\frac{2}{n\pi}\sin\frac{n\pi}{2}\right]$$

$$= \frac{2k}{n\pi}\left[\frac{4}{n\pi}\sin\frac{n\pi}{2}\right]=\frac{8k}{n^2\pi^2}\sin\frac{n\pi}{2}$$

Hence, Fourier series of $f(x)$ is

$$f(x) = \frac{8k}{\pi^2}\sum_{n=1}^{\infty}\frac{1}{n^2}\sin\frac{n\pi}{2}\sin\frac{n\pi x}{l}$$ **Ans.**

EXERCISE 11.4

1. Find the Fourier series to represent $f(x)$, where

$$f(x) = \begin{cases} -a, & -c<x<0 \\ a, & 0<x<c \end{cases}$$ **Ans.** $\frac{4a}{\pi}\left[\sin\frac{\pi x}{c}+\frac{1}{3}\sin\frac{3\pi x}{c}+\frac{1}{5}\sin\frac{5\pi x}{c}+...\right]$

2. Find the half-range sine series for the function

$f(x) = 2x - 1 \quad 0<x<1$ **Ans.** $-\frac{2}{\pi}\left[\sin 2\pi x+\frac{1}{2}\sin 4\pi x+\frac{1}{3}\sin 6\pi x+...\right]$

3. Express $f(x) = x$ as a cosine, half range series in $0 < x < 2$.

Ans. $1-\frac{8}{\pi^2}\left[\frac{1}{1^2}\cos\frac{\pi x}{2}+\frac{1}{3^2}\cos\frac{3\pi x}{2}+\frac{1}{5^2}\cos\frac{5\pi x}{2}+...\right]$

4. Find the Fourier series of the function

$$f(x) = \begin{cases} -2 & \text{for } -4<x<-2 \\ x & \text{for } -2<x<2 \\ 2 & \text{for } 2<x<4 \end{cases}$$

Ans. $\frac{4}{\pi}+\frac{8}{\pi^2}\sin\frac{\pi x}{4}-\frac{2}{\pi}\sin\frac{2\pi x}{4}+\left(\frac{4}{3\pi}-\frac{8}{3^2\pi}\right)\sin\frac{3\pi x}{4}-\frac{2}{2\pi}\sin\frac{4\pi x}{4}+...$

5. Find the Fourier series to represent
$f(x) = x^2 - 2$ from $-2 < x < 2$.

Ans. $-\frac{2}{3}-\frac{16}{\pi^2}\left[\cos\frac{\pi x}{2^2}-\frac{1}{4}\cos\pi x+\frac{1}{9}\cos\frac{3\pi x}{2}+...\right]$

6. If $f(x) = e^{-x}$, $-c < x < c$, show that

$$f(x) = (e^c - e^{-c})\left\{\frac{1}{2c}-c\left(\frac{1}{c^2+\pi^2}\cos\frac{\pi x}{c}-\frac{1}{c^2+4\pi^2}\cos\frac{2\pi x}{c}+...\right)\right.$$
$$\left.-\pi\left(\frac{1}{c^2+\pi^2}\sin\frac{\pi x}{c}-\frac{2}{c^2+4\pi^2}\sin\frac{2\pi x}{c}...\right)\right\}$$ *(MDU, Dec. 2010)*

7. A periodic square wave has a period 4. The function generating the square is

$$f(t) = \begin{cases} 0 & \text{for } -2<t<-1 \\ k & \text{for } -1<t<1 \\ 0 & \text{for } 1<t<2 \end{cases}$$

Find the Fourier series of the function. **Ans.** $f(t)=\frac{k}{2}+\frac{2k}{\pi}\left[\cos\frac{\pi t}{2}-\frac{1}{3}\cos\frac{3\pi t}{2}+...\right]$

8. Find a Fourier series to represent x^2 in the interval $(-l, l)$.

Ans. $\frac{l^2}{3}-\frac{4l^2}{\pi^2}\left[\cos\pi x-\frac{\cos\pi x}{2^2}+\frac{\cos 3\pi x}{3^2}+....\right]$

11.14 PARSEVAL'S FORMULA

$$\int_{-c}^{c}[f(x)]^2\,dx = c\left\{\frac{1}{2}a_0^2+\sum_{n=1}^{\infty}(a_n^2+b_n^2)\right\}$$

We know that $f(x) = \dfrac{a_0}{2}+\displaystyle\sum_{n=1}^{\infty}\left(a_n\cos\frac{n\pi x}{c}+b_n\sin\frac{n\pi x}{c}\right)$ ...(1)

Multiplying (1) by $f(x)$, we get

$$[f(x)]^2 = \frac{a_0}{2}f(x)+\sum_{n=1}^{\infty}a_n f(x)\cos\frac{n\pi x}{c}+\sum_{n=1}^{\infty}b_n f(x)\sin\frac{n\pi x}{c} \quad ..(2)$$

Intergrating term by term from $-c$ to c, we have

$$\int_{-c}^{c}[f(x)]^2\,dx = \frac{a_0}{2}\int_{-c}^{c}f(x)\,dx+\sum_{n=1}^{\infty}a_n\int_{-c}^{c}f(x)\cos\frac{n\pi x}{c}\,dx$$

$$+\sum_{n=1}^{\infty}b_n\int_{-c}^{c}f(x)\sin\frac{n\pi x}{c}\,dx \quad ...(3)$$

In article 11.12, we have the following results

$$\int_{-c}^{c}f(x)\,dx = c\,a_0$$

$$\int_{-c}^{c}f(x)\cos\frac{n\pi x}{c}\,dx = c\,a_n$$

$$\int_{-c}^{c}f(x)\sin\frac{n\pi x}{c}\,dx = c\,b_n$$

On putting these integrals in (3), we get

$$\int_{-c}^{c}[f(x)]^2\,dx = c\frac{a_0^2}{2}+\sum_{n=1}^{\infty}c\,a_n^2+\sum_{n=1}^{\infty}c\,b_n^2 = c\left[\frac{a_0^2}{2}+\sum_{n=1}^{\infty}(a_n^2+b_n^2)\right]$$

This is the Parseval's formula.

Note. 1. If $0 < x < 2c$, then $\displaystyle\int_0^{2c}[f(x)]^2\,dx = c\left[\frac{a_0^2}{2}+\sum_{n=1}^{\infty}(a_n^2+b_n^2)\right]$

2. If $0 < x < c$ (Half range consine series), $\displaystyle\int_0^{c}[f(x)]^2\,dx = \frac{c}{2}\left[\frac{a_0^2}{2}+\sum_{n=1}^{\infty}a_n^2\right]$

3. If $0 < x < c$ (Half range sine series), $\displaystyle\int_0^{c}[f(x)]^2\,dx = \frac{c}{2}\left[\sum_{n=1}^{\infty}b_n^2\right]$

4. R.M.S. $= \left\{\dfrac{\int_a^b[f(x)]^2\,dx}{b-a}\right\}^{\frac{1}{2}}$

Example 28. *By using the series for $f(x) = 1$ in $0 < x < \pi$, show that*

$$\frac{\pi^2}{8} = 1+\frac{1}{3^2}+\frac{1}{5^2}+\frac{1}{7^2}+\ldots$$

Solution. Sine series is $f(x) = \Sigma\, b_n \sin nx$

$$b_n = \frac{2}{\pi}\int_0^{\pi}f(x)\sin nx\,dx$$

$$= \frac{2}{\pi}\int_0^{\pi}(1)\sin nx\,dx = \frac{2}{\pi}\left(\frac{-\cos nx}{n}\right)_0^{\pi} = \frac{-2}{n\pi}[\cos n\pi-1] = \frac{-2}{n\pi}[(-1)^n-1]$$

$$= \frac{4}{n\pi} \text{ if } n \text{ is odd } = 0$$

if n is even

Then the sine series is

$$1 = \frac{4}{\pi}\sin x + \frac{4}{3\pi}\sin 3x + \frac{4}{5\pi}\sin 5x + \frac{4}{7\pi}\sin 7x + ...$$

$$\int_0^c [f(x)]^2\, dx = \frac{c}{2}\left[b_1^2 + b_2^2 + b_3^2 + b_4^2 + b_5^2 + ...\right]$$

$$\int_0^\pi (1)^2\, dx = \frac{\pi}{2}\left[\left(\frac{4}{\pi}\right)^2 + \left(\frac{4}{3\pi}\right)^2 + \left(\frac{4}{5\pi}\right)^2 + \left(\frac{4}{7\pi}\right)^2 + ...\right]$$

$$[x]_0^\pi = \left(\frac{\pi}{2}\right)\left(\frac{16}{\pi^2}\right)\left[1 + \frac{1}{3^2} + \frac{1}{5^2} + \frac{1}{7^2} + ...\right]$$

$$\pi = \frac{\pi}{2}\left(\frac{16}{\pi^2}\right)\left[1 + \frac{1}{3^2} + \frac{1}{5^2} + \frac{1}{7^2} + ...\right]$$

$$\frac{\pi^2}{8} = 1 + \frac{1}{3^2} + \frac{1}{5^2} + \frac{1}{7^2} + ...$$ **Proved.**

Example 29. *If* $f(x) = \begin{cases} \pi x, & 0 < x < 1 \\ \pi(2-x), & 1 < x < 2 \end{cases}$

using half range cosine series, show that

$$\frac{1}{1^4} + \frac{1}{3^4} + \frac{1}{5^4} + ... = \frac{\pi^4}{96}$$

Solution. Half range consine series is

$$f(x) = \frac{a_0}{2} + \Sigma\, a_n \cos\frac{n\pi x}{c}$$

where $a_0 = \frac{2}{c}\int_0^c f(x)\, dx = \frac{2}{2}\left[\int_0^1 \pi x\, dx + \int_1^2 \pi(2-x)\, dx\right]$

$$= \pi\left(\frac{x^2}{2}\right)_0^1 + \pi\left(2x - \frac{x^2}{2}\right)_1^2 = \frac{\pi}{2} + \pi\left[(4-2) - \left(2 - \frac{1}{2}\right)\right] = \pi$$

$$a_n = \frac{2}{c}\int_0^c f(x)\cos\frac{n\pi x}{c}\, dx$$

$$= \frac{2}{2}\left[\int_0^1 \pi x\cos\frac{n\pi x}{2}\, dx + \int_1^2 \pi(2-x)\cos\frac{n\pi x}{2}\, dx\right]$$

$$= \pi\left[\frac{x\,\sin\frac{n\pi x}{2}}{\frac{n\pi}{2}} - \left(\frac{-\cos\frac{n\pi x}{2}}{\frac{n^2\pi^2}{4}}\right)\right]_0^1 + \pi\left[(2-x)\frac{\sin\frac{n\pi x}{2}}{\frac{n\pi}{2}} - (-1)\left(\frac{-\cos\frac{n\pi x}{2}}{\frac{n^2\pi^2}{4}}\right)\right]_1^2$$

$$= \pi\left[\frac{2}{n\pi}\sin\frac{n\pi}{2} + \frac{4}{n^2\pi^2}\cos\frac{n\pi}{2} - \frac{4}{n^2\pi^2}\right] + \pi\left[0 - \frac{4}{n^2\pi^2}\cos n\pi - \frac{2}{n\pi}\sin\frac{n\pi}{2} + \frac{4}{n^2\pi^2}\cos\frac{n\pi}{2}\right]$$

$$= \pi\left[\frac{8}{n^2\pi^2}\cos\frac{n\pi}{2} - \frac{4}{n^2\pi^2} - \frac{4}{n^2\pi^2}\cos n\pi\right] = \frac{4}{n^2\pi}\left[2\cos\frac{n\pi}{2} - 1 - \cos n\pi\right]$$

$$a_1 = 0,\ a_2 = \frac{-4}{\pi},\ a_3 = 0,\ a_4 = 0,\ a_5 = 0,\ a_6 = \frac{-4}{9\pi}\ \ldots$$

$$\int_0^c [f(x)]^2\,dx = \frac{c}{2}\left[\frac{a_0^2}{2} + a_1^2 + a_2^2 + a_3^2 + \ldots\right]$$

$$\int_0^1 (\pi x)^2\,dx + \int_1^2 \pi^2 (2-x)^2\,dx = \frac{2}{2}\left[\frac{\pi^2}{2} + \frac{16}{\pi^2} + \frac{16}{81\pi^2} + \ldots\right]$$

$$\pi^2\left[\frac{x^3}{3}\right]_0^1 - \pi^2\left[\frac{(2-x)^3}{3}\right]_1^2 = \frac{\pi^2}{2} + \frac{16}{\pi^2} + \frac{16}{81\pi^2} + \ldots$$

$$\frac{\pi^2}{3} - \pi^2\left(0 - \frac{1}{3}\right) = \frac{\pi^2}{2} + \frac{16}{\pi^2}\left[1 + \frac{1}{81} + \ldots\right]$$

$$\frac{2\pi^2}{3} - \frac{\pi^2}{2} = \frac{16}{\pi^2}\left[1 + \frac{1}{3^4} + \frac{1}{5^4} + \ldots\right]$$

$$\frac{\pi^2}{6} = \frac{16}{\pi^2}\left[1 + \frac{1}{3^4} + \frac{1}{5^4} + \ldots\right]$$

$$\frac{\pi^4}{96} = 1 + \frac{1}{3^4} + \frac{1}{5^4} + \ldots$$ **Ans.**

Example 30. *Prove that for $0 < x < \pi$*

(a) $x(\pi - x) = \dfrac{\pi^2}{6} - 4\left[\dfrac{\cos 2x}{2^2} + \dfrac{\cos 4x}{4^2} + \dfrac{\cos 6x}{3^2} + \ldots\right]$

(b) $x(\pi - x) = \dfrac{8}{\pi}\left[\dfrac{\sin x}{1^2} + \dfrac{\sin 3x}{3^2} + \dfrac{\sin 5x}{5^2} + \ldots\right]$

Deduce from (a) and (b) respectively that

(c) $\displaystyle\sum_{n=1}^{\infty} \frac{1}{n^4} = \frac{\pi^4}{90}$

(d) $\displaystyle\sum_{n=1}^{\infty} \frac{1}{n^6} = \frac{\pi^6}{945}$

Solution. (a) Half range cosine series

$$a_0 = \frac{2}{\pi}\int_0^\pi x(\pi - x) = \frac{2}{\pi}\left[\frac{\pi x^2}{2} - \frac{x^3}{3}\right]_0^\pi = \frac{2}{\pi}\left[\frac{\pi^3}{2} - \frac{\pi^3}{3}\right] = \frac{\pi^2}{3}$$

$$a_n = \frac{2}{\pi}\int_0^\pi x(\pi - x)\cos nx\,dx$$

$$= \frac{2}{\pi}\left[(\pi x - x^2)\frac{\sin nx}{n} - (\pi - 2x)\left(\frac{-\cos nx}{n^2}\right) + (-2)\left(\frac{-\sin nx}{n^3}\right)\right]_0^\pi$$

$$= \frac{2}{\pi}\left[0 - \frac{\pi(-1)^n}{n^2} + 0 - \frac{\pi}{n^2}\right] = \frac{2}{\pi}\left(\frac{\pi}{n^2}\right)[-(-1)^n - 1]$$

$$= -\frac{4}{n^2}$$ (when n is even)

$$= 0$$ (when n is odd)

Hence, $x(\pi - x) = \dfrac{\pi^2}{6} - 4\left[\dfrac{\cos 2x}{2^2} + \dfrac{\cos 4x}{4^2} + \ldots\right]$

By Parseval's formula $\frac{2}{\pi}\int_0^{\pi} x^2\,(\pi - x)^2\,dx = \frac{a_0^2}{2} + \Sigma a_n^2$

$$\frac{2}{\pi}\int_0^{\pi} (\pi^2 x^2 - 2\pi x^3 + x^4)\,dx = \frac{1}{2}\left(\frac{\pi^4}{9}\right) + 16\left[\frac{1}{2^4} + \frac{1}{4^4} + \frac{1}{6^4} + \ldots\right]$$

$$\frac{2}{\pi}\left[\frac{\pi^2 x^3}{3} - \frac{2\pi x^4}{4} + \frac{x^5}{5}\right]_0^{\pi} = \frac{\pi^4}{18} + \left[\frac{1}{1^4} + \frac{1}{2^4} + \frac{1}{3^4} + \ldots\right]$$

$$-\frac{2}{\pi}\left[\frac{\pi^5}{3} - \frac{2\pi^5}{4} + \frac{\pi^5}{5}\right] = \frac{\pi^4}{18} + \left[\frac{1}{1^4} + \frac{1}{2^4} + \frac{1}{3^4} + \ldots\right]$$

$$\frac{\pi^4}{15} = \frac{\pi^4}{18} + \sum_{n=1}^{\infty}\frac{1}{n^4} \quad \text{or} \quad \sum_{n=1}^{\infty}\frac{1}{n^4} = \frac{\pi^4}{90}$$

(*b*) Half range sine series

$$b_n = \frac{2}{\pi}\int_0^{\pi} x\,(\pi - x)\sin nx\,dx$$

$$= \frac{2}{\pi}\left[(\pi x - x^2)\left(\frac{-\cos nx}{n}\right) - (\pi - 2x)\left(\frac{-\sin nx}{n^2}\right) + (-2)\frac{\cos nx}{n^3}\right]_0^{\pi}$$

$$= \frac{2}{\pi}\left[-2\frac{(-1)^n}{n^3} + \frac{2}{n^3}\right] = \frac{4}{\pi n^3}[-(-1)^n + 1]$$

$$= \begin{cases} \frac{8}{n^3\pi}, & \text{(when } n \text{ is odd)} \\ 0, & \text{(when } n \text{ is odd)} \end{cases}$$

$$\therefore \quad x\,(\pi - x) = \frac{8}{\pi}\left[\frac{\sin x}{1^3} + \frac{\sin 3x}{3^3} + \frac{\sin 5x}{5^3} + \ldots\right]$$

By Parseval's formula

$$\frac{2}{\pi}\int_0^{\pi} x^2\,(\pi - x)^2\,dx = \Sigma\, b_n^2$$

$$\frac{\pi^4}{15} = \frac{64}{\pi^2}\left[\frac{1}{1^6} + \frac{1}{3^6} + \frac{1}{5^6} + \ldots\right]$$

$$\frac{\pi^6}{960} = \frac{1}{1^6} + \frac{1}{3^6} + \frac{1}{5^6} + \ldots$$

$$\text{Let } S = \frac{1}{1^6} + \frac{1}{2^6} + \frac{1}{3^6} + \frac{1}{4^6} + \ldots = \left(\frac{1}{1^6} + \frac{1}{3^6} + \frac{1}{5^6}\ldots\right) + \left(\frac{1}{2^6} + \frac{1}{4^6} + \frac{1}{6^6} + \ldots\right)$$

$$S = \frac{\pi^6}{960} + \left(\frac{1}{2^6} + \frac{1}{4^6} + \frac{1}{6^6} + \ldots\right) = \frac{\pi^6}{960} + \frac{1}{2^6}\left[\frac{1}{1^6} + \frac{1}{2^6} + \frac{1}{3^6} + \ldots\right]$$

$$S = \frac{\pi^6}{960} + \frac{S}{64}$$

$$S - \frac{S}{64} = \frac{\pi^6}{960} \quad \Rightarrow \quad \frac{63}{64}S = \frac{\pi^6}{960}$$

$$S = \frac{\pi^6}{960} \times \frac{64}{63} = \frac{\pi^6}{945}$$

$$\sum_{n=1}^{\infty}\frac{1}{n^6} = \frac{\pi^6}{945}$$

Proved.

EXERCISE 11.5

1. Prove that in $0 < x < c$,

$$x = \frac{c}{2} - \frac{4c}{\pi^2}\left(\cos\frac{\pi x}{c} + \frac{1}{3^2}\cos\frac{3\pi x}{c} + \frac{1}{5^2}\cos\frac{5\pi x}{c} + \ldots\right) \text{ and deduce that}$$

(i) $\frac{1}{1^4} + \frac{1}{3^4} + \frac{1}{5^4} + \ldots = \frac{\pi^4}{96}$ (ii) $\frac{1}{1^4} + \frac{1}{2^4} + \frac{1}{3^4} + \frac{1}{4^4} + \ldots = \frac{\pi^4}{90}$

11.15 FOURIER SERIES IN COMPLEX FORM

Fourier series of a function $f(x)$ of period $2l$ is

$$f(x) = \frac{a_0}{2} + a_1\cos\frac{\pi x}{l} + a_2\cos\frac{2\pi x}{l} + \ldots + a_n\cos\frac{n\pi x}{l} + \ldots$$
$$+ b_1\sin\frac{\pi x}{l} + b_2\sin\frac{2\pi x}{l} + \ldots + b_n\sin\frac{n\pi x}{l} + \ldots \quad \ldots(1)$$

We know that $\cos x = \dfrac{e^{ix} + e^{-ix}}{2}$ and $\sin x = \dfrac{e^{ix} - e^{-ix}}{2i}$

On putting the values of cos x and sin x in (1), we get

$$f(x) = \frac{a_0}{2} + a_1\frac{e^{\frac{i\pi x}{l}} + e^{\frac{-i\pi x}{l}}}{2} + a_2\frac{e^{\frac{2i\pi x}{l}} + e^{\frac{-2i\pi x}{l}}}{2} + \ldots + b_1\frac{e^{\frac{i\pi x}{l}} - e^{\frac{-i\pi x}{l}}}{2i} + b_2\frac{e^{\frac{2i\pi x}{l}} - e^{\frac{-2i\pi x}{l}}}{2i} + \ldots$$

$$= \frac{a_0}{2} + (a_1 - ib_1)e^{\frac{i\pi x}{l}} + (a_2 - ib_2)e^{\frac{2i\pi x}{l}} + \ldots + (a_1 + ib_1)e^{\frac{-i\pi x}{l}} + (a_2 + ib_2)e^{\frac{-2i\pi x}{l}} + \ldots$$

$$= c_0 + c_1 e^{\frac{i\pi x}{l}} + c_2 e^{\frac{2i\pi x}{l}} + \ldots + c_{-1}e^{\frac{-i\pi x}{l}} + c_{-2}e^{-\frac{2i\pi x}{l}} + \ldots$$

$$= c_0 + \sum_{n=1}^{\infty} c_n e^{\frac{in\pi x}{l}} + \sum_{n=1}^{\infty} c_{-n} e^{\frac{-in\pi x}{l}}$$

$$c_n = \frac{1}{2}(a_n - ib_n),\ c_{-n} = \frac{1}{2}(a_n + ib_n)$$

where $c_0 = \dfrac{a_0}{2} = \dfrac{1}{2}\dfrac{1}{l}\displaystyle\int_0^{2l} f(x)\,dx$

$$c_n = \frac{1}{2}\left[\frac{1}{l}\int_0^{2l} f(x)\cos\frac{n\pi x}{l}\,dx - \frac{i}{l}\int_0^{2l} f(x)\sin\frac{n\pi x}{l}\,dx\right]$$

$$\Rightarrow \quad c_n = \frac{1}{2}\frac{1}{l}\int_0^{2l} f(x)\left\{\cos\frac{n\pi x}{l} - i\sin\frac{n\pi x}{l}\right\}dx$$

$$c_n = \frac{1}{2l}\int_0^{2l} f(x)\,e^{\frac{-in\pi x}{l}}\,dx,$$

$$c_{-n} = \frac{1}{2l}\int_0^{2l} f(x)\,e^{\frac{in\pi x}{l}}\,dx$$

Example 31. *Obtain the complex form of the Fourier series of the function*

$$f(x) = \begin{cases} 0, & -\pi \le x \le 0 \\ 1, & 0 \le x \le \pi \end{cases}$$

Solution. $c_0 = \dfrac{1}{2\pi}\displaystyle\int_0^{\pi} dx = \dfrac{1}{2}$

$$c_n = \frac{1}{2\pi}\int_{-\pi}^{\pi} f(x)\, e^{-inx}\, dx$$

$$= \frac{1}{2\pi}\left[\int_{-\pi}^{0} 0\cdot e^{-inx}\, dx + \int_{0}^{\pi} 1\cdot e^{-inx}\, dx\right] = \frac{1}{2\pi}\int_{0}^{\pi} e^{-inx}\, dx = \frac{1}{2\pi}\left[\frac{e^{-inx}}{-in}\right]_0^{\pi}$$

$$= -\frac{1}{2n\pi i}[e^{-inx} - 1] = -\frac{1}{2n\pi i}[\cos n\pi - i\sin n\pi - 1] = -\frac{1}{2n\pi i}[(-1)^n - 0 - 1]$$

$$= \begin{cases} \dfrac{1}{in\pi}, & n \text{ is odd} \\ 0 \quad , & n \text{ is even} \end{cases}$$

$$f(x) = \frac{1}{2} + \frac{1}{i\pi}\left[\frac{e^{ix}}{1} + \frac{e^{3ix}}{3} + \frac{e^{5ix}}{5} + \ldots\right] + \frac{1}{i\pi}\left[\frac{e^{-ix}}{-1} + \frac{e^{-3ix}}{-3} + \frac{e^{-5ix}}{-5} + \ldots\right]$$

$$= \frac{1}{2} + \frac{1}{i\pi}\left[\left(e^{ix} - e^{-ix}\right) + \frac{1}{3}\left(e^{3ix} - e^{-3ix}\right) + \frac{1}{5}\left(e^{5ix} - e^{-5ix}\right) + \ldots\right]$$

$$= \frac{1}{2} + \frac{2}{\pi}\left[\sin x + \frac{1}{3}\sin 3x + \frac{1}{5}\sin 5x + \ldots\ldots\right]$$ **Ans.**

EXERCISE 11.6

Find the complex form of the Fourier series of

1. $f(x) = e^{-x},\ -1 \le x \le 1.$ **Ans.** $\displaystyle\sum_{n=-\infty}^{\infty} \frac{(-1)^n (1 - in\pi)}{1 + n^2\pi^2} \sinh 1 . e^{in\pi x}$

2. $f(x) = e^{ax},\ -l < x < l$ **Ans.** $\displaystyle\frac{2}{\pi} - \frac{2}{\pi}\left[\frac{e^{2it} + e^{-2it}}{1\cdot 3} + \frac{e^{4it} + e^{-4it}}{3\cdot 5} + \frac{e^{6it} + e^{-6it}}{5-7} + \ldots\right]$

3. $f(x) = \cos ax,\ -\pi < x < \pi$ **Ans.** $\displaystyle\frac{a}{\pi}\sin a\pi \sum_{-\infty}^{\infty} \frac{(-1)^n e^{inx}}{a^2 - n^2}$

11.16 PRACTICAL HARMONIC ANALYSIS

Some times the function is not given by a formula, but by a graph or by a table of corresponding values. The process of finding the Fourier series for a function given by such values of the function and independent variable is known as Harmonic Analysis. The Fourier constants are evaluated by the following formulae :

(1) $$a_0 = \frac{1}{\pi}\int_0^{2\pi} f(x)\, dx$$

$$= 2\frac{1}{2\pi - 0}\int_0^{2\pi} f(x)\, dx \qquad \left[\text{Mean} = \frac{1}{b-a}\int_a^b f(x)\, dx\right]$$

$\Rightarrow$ **a_0 = 2 [mean value of $f(x)$ in $(0, 2\pi)$]**

(2) $$a_n = \frac{1}{\pi}\int_0^{2\pi} f(x)\cos nx\, dx = 2\frac{1}{2\pi - 0}\int_0^{2\pi} f(x)\cos nx\, dx$$

= 2 [mean value of $f(x)\cos nx$ in $(0, 2\pi)$]

(3) $$b_n = \frac{1}{\pi}\int_0^{2\pi} f(x)\sin nx\, dx = 2\frac{1}{2\pi - 0}\int_0^{2\pi} f(x)\sin nx\, dx$$

= 2 [mean value of $f(x)\sin nx$ in $(0, 2\pi)$]

Fundamental or first harmonic. The term $(a_1 \cos x + b_1 \sin x)$ in Fourier series is called the fundamental or first harmonic.

Second harmonic. The term $(a_2 \cos 2x + b_2 \sin 2x)$ in Fourier series is called the second harmonic and so on.

Example 32. *Find the Fourier series as far as the second harmonic to represent the function given by table below :*

x	0°	30°	60°	90°	120°	150°	180°	210°	240°	270°	300°	330°
$f(x)$	2.34	3.01	3.69	4.15	3.69	2.20	0.83	0.51	0.88	1.09	1.19	1.64

Solution.

$x°$	$\sin x$	$\sin 2x$	$\cos x$	$\cos 2x$	$f(x)$	$f(x)$ $\sin x$	$f(x)$ $\sin 2x$	$f(x)$ $\cos x$	$f(x)$ $\cos 2x$
0°	0	0	1	1	2.34	0	0	2.340	2.340
30°	0.50	0.87	0.87	0.50	3.01	1.505	2.619	2.619	1.505
60°	0.87	0.87	0.50	– 0.50	3.69	3.210	3.210	1.845	– 1.845
90°	1.00	0	0	– 1.00	4.15	4.150	0	0	– 4.150
120°	0.87	– 0.87	– 0.50	– 0.50	3.69	3.210	– 3.210	– 1.845	– 1.845
150°	0.50	– 0.87	– 0.87	0.50	2.20	1.100	– 1.914	– 1.914	1.100
180°	0	0	– 1	1.00	0.83	0	0	– 0.830	0.830
210°	– 0.50	0.87	– 0.87	0.50	0.51	– 0.255	0.444	– 0.444	0.255
240°	– 0.87	0.87	– 0.50	– 0.50	0.88	– 0.766	0.766	–0.440	– 0.440
270°	– 1.00	0	0	– 1.00	1.09	– 1.090	0	0	– 1.090
300°	– 0.87	– 0.87	0.50	– 0.50	1.19	– 1.035	– 1.035	0.595	– 0.595
330°	– 0.50	– 0.87	0.87	0.50	1.64	– 0.820	– 1.427	1.427	0.820
					25.22	9.209	– 0.547	3.353	– 3.115

$$a_0 = 2 \times \text{Mean of } f(x) = 2 \times \frac{25.22}{12} = 4.203$$

$$a_1 = 2 \times \text{Mean of } f(x) \cos x = 2 \times \frac{3.353}{12} = 0.559$$

$$a_2 = 2 \times \text{Mean of } f(x) \cos 2x = 2 \times \frac{-3.115}{12} = -0.519$$

$$b_1 = 2 \times \text{Mean of } f(x) \sin x = 2 \times \frac{9.209}{12} = 1.535$$

$$b_2 = 2 \times \text{Mean of } f(x) \sin 2x = 2 \times \frac{-0.547}{12} = -0.091$$

Fourier series is

$$f(x) = \frac{a_0}{2} + a_1 \cos x + a_2 \cos 2x + \ldots + b_1 \sin x + b_2 \sin 2x + \ldots$$

$$= 2.1015 + 0.559 \cos x - 0.519 \cos 2x + \ldots + 1.535 \sin x - 0.091 \sin 2x + \ldots$$ **Ans.**

Example 33. *A machine completes its cycle of operations every time as certain pulley completes a revolution. The displacement $f(x)$ of a point on a certain portion of the machine is given in the table given below for twelve positions of the pulley, x being the angle in degree turned through by the pulley. Find a Fourier series to represent $f(x)$ for all values of x.*

x	30°	60°	90°	120°	150°	180°	210°	240°	270°	300°	330°	360°
$f(x)$	7.976	8.026	7.204	5.676	3.674	1.764	0.552	0.262	0.904	2.492	4.736	6.824

Solution.

x	$\sin x$	$\sin 2x$	$\sin 3x$	$\cos x$	$\cos 2x$	$\cos 3x$	$f(x)$	$f(x) \times \sin x$	$f(x) \times \sin 2x$	$f(x) \times \sin 3x$	$f(x) \times \cos x$	$f(x) \times \cos 2x$	$f(x) \times \cos 3x$
30°	0.50	0.87	1	0.87	0.50	0	7.976	3.988	6.939	7.976	6.939	3.988	0
60°	0.87	0.87	0	0.50	– 0.50	– 1	8.026	6.983	6.983	0	4.013	4.013	– 8.026
90°	1.00	0	– 1	0	– 1	0	7.204	7.204	0	– 7.204	0	– 7.204	0
120°	0.87	– 0.87	0	– 0.50	– 0.50	1	5.676	4.938	– 4.939	0	– 2.838	– 2.838	5.676
150°	0.50	– 0.87	1	– 0.87	0.50	0	3.674	1.837	– 3.196	– 3.196	– 3.196	1.837	0
180°	0	0	0	– 1	1	– 1	1.764	0	0	– 1.764	– 1.764	1.764	– 1.764
210°	– 0.50	0.87	– 1	– 0.87	0.50	0	0.552	– 0.276	0.480	0.480	–0.480	0.276	0
240°	– 0.87	0.87	0	– 0.50	– 0.50	1	0.262	– 0.228	0.228	– 0.131	– 0.131	0.131	0.262
270°	– 1.00	0	1	0	– 1.00	0	0.904	– 0.904	0	0	0	– 0.904	0
300°	– 0.87	– 0.87	0	0.50	– 0.50	– 1	2.492	– 2.168	– 2.168	1.246	1.246	–1.296	– 2.492
330°	– 0.50	– 0.87	– 1	0.87	0.50	0	4.736	–2.368	– 4.120	4.120	4.120	2.368	0
360°	0	0	0	1	1	1	6.824	0	0	0	6.824	6.824	6.824
						Σ	50.09	19.206	0.207	0.062	14.733	0.721	0.460

$$a_0 = 2 \times \text{Mean value of } f(x) \qquad = 2 \times \frac{50.09}{12} = 8.34$$

$$a_1 = 2 \times \text{Mean value of } f(x) \cos x = 2 \times \frac{14.733}{12} = 2.45$$

$$a_2 = 2 \times \text{Mean value of } f(x) \cos 2x = 2 \times \frac{0.721}{12} = 0.12$$

$$a_3 = 2 \times \text{Mean value of } f(x) \cos 3x = 2 \times \frac{0.460}{12} = 0.08$$

$$b_1 = 2 \times \text{Mean value of } f(x) \sin x = 2 \times \frac{19.206}{12} = 3.16$$

$$b_2 = 2 \times \text{Mean value of } f(x) \sin 2x = 2 \times \frac{0.207}{12} = 0.03$$

$$b_3 = 2 \times \text{Mean value of } f(x) \sin 3x = 2 \times \frac{0.062}{12} = 0.01$$

Fourier series is

$$f(x) = \frac{a_0}{2} + a_1 \cos x + a_2 \cos 2x + a_3 \cos 3x + \ldots + b_1 \sin x + b_2 \sin 2x + b_3 \sin 3x + \ldots$$

$$= 4.17 + 2.45 \cos x + 0.12 \cos 2x + 0.08 \cos 3x + \ldots$$

$$+ 3.16 \sin x + 0.03 \sin 2x + 0.01 \sin 3x + \ldots$$ **Ans.**

Example 34. *Obtain the constant term and the coefficient of the first sine and cosine terms in the Fourier series of $f(x)$ as given in the following table.*

x	0	1	2	3	4	5
$f(x)$	9	18	24	28	26	20

Solution.

x	$\frac{x\pi}{3}$	$\sin\frac{\pi x}{3}$	$\cos\frac{\pi x}{3}$	$f(x)$	$f(x)\sin\frac{\pi x}{3}$	$f(x)\cos\frac{\pi x}{3}$
0	0	0	1.0	9	0	9
1	$\frac{\pi}{3}$	0.87	0.5	18	15.66	9
2	$\frac{2\pi}{3}$	0.87	– 0.5	24	20.88	– 12
3	$\frac{3\pi}{3}$	0	– 1.0	28	0	– 28
4	$\frac{4\pi}{3}$	– 0.87	– 0.5	26	– 22.62	– 13
5	$\frac{5\pi}{3}$	– 0.87	0.5	20	– 17.4	10
				$\Sigma = 125$	$\Sigma = -3.468$	$\Sigma = 25$

$$a_0 = 2 \text{ Mean value of } f(x) = 2 \times \frac{125}{6} = 41.67$$

$$a_1 = 2 \text{ Mean value of } f(x) \cos\frac{\pi x}{3} = 2 \times \frac{-25}{6} = -8.33$$

$$b_1 = 2 \text{ Mean value of } f(x) \sin\frac{\pi x}{3} = 2 \times \frac{-3.48}{6} = -1.16$$

Fourier series is $f(x) = \frac{a_0}{2} + a_1 \cos\frac{\pi x}{3} + \ldots + b_1 \sin\frac{\pi x}{3} + \ldots$

$$= 20.84 - 8.33 \cos\frac{\pi x}{3} + \ldots - 1.16 \sin\frac{\pi x}{3} + \ldots$$ **Ans.**

EXERCISE 11.7

1. In a machine the displacement $f(x)$ of a given point is given for a certain angle $x°$ as follows:

$x°$	0°	30°	60°	90°	120°	150°	180°	210°	240°	270°	300°	330°
$f(x)$	7.9	8.0	7.2	5.6	3.6	1.7	0.5	0.2	0.9	2.5	4.7	6.8

Find the coefficient of sin 2 x in the Fourier series representing the above variations. **Ans.** – 0.072

2. The displacement $f(x)$ of a part of a machine is tabulated with corresponding angular moment 'x' of the crank. Express $f(x)$ as a Fourier series upto third harmonic.

$x°$	0°	30°	60°	90°	120°	150°	180°	210°	240°	270°	300°	330°
$f(x)$	1.80	1.10	0.30	0.16	0.50	1.30	2.16	1.25	1.30	1.52	1.76	2.00

Ans. $f(x) = 1.26 + 0.04 \cos x + 0.53 \cos 2x - 0.1 \cos 3x + \ldots$
$- 0.63 \sin x - 0.23 \sin 2x + 0.085 \sin 3x + \ldots$

3. Fourier coefficient 'a_0' in Fourier series expansion of a function represents the:
(*i*) maximum value of the function (*ii*) 2 mean value of the function
(*iii*) minimum value of the function (*iv*) None of these (*U.P. II Semester 2010*) **Ans.** (*ii*)

4. If the Fourier series of $f(x)$ has only cosine terms then $f(x)$ must be :
(*i*) odd function (*ii*) even function (*U.P. II Semester 2010*) **Ans.** (*ii*)

12

CHAPTER

UNIT – II

DIFFERENTIAL EQUATIONS OF FIRST ORDER

12.1 DEFINITION

An equation which involves differential co-efficient is called a differential equation.

For example,

1. $\frac{dy}{dx}=\frac{1+x^2}{1-y^2}$ **2.** $\frac{d^2y}{dx^2}+2\frac{dy}{dx}-8y=0$ **3.** $\left[1+\left(\frac{dy}{dx}\right)^2\right]^{3/2}=k\frac{d^2y}{dx^2}$

4. $x\frac{\partial u}{\partial x}+y\frac{\partial u}{\partial y}=nu,$ **5.** $\frac{\partial^2 z}{\partial x\partial y}=\frac{\partial z}{\partial y}$

There are two types of differential equations :

(1) Ordinary Differential Equation

A differential equation involving derivatives with respect to a single independent variable is called an ordinary differential equation.

(2) Partial Differential Equation

A differential equation involving partial derivatives with respect to more than one independent variable is called a partial differential equation.

12.2 ORDER AND DEGREE OF A DIFFERENTIAL EQUATION

The *order* of a differential equation is the order of the highest differential co-efficient present in the equation. Consider

1. $L\frac{d^2q}{dt^2}+R\frac{dq}{dt}+\frac{q}{c}=E\sin wt.$ **2.** $\cos x\frac{d^2y}{dx^2}+\sin x\left(\frac{dy}{dx}\right)^2+8y=\tan x$

3. $\left[1+\left(\frac{dy}{dx}\right)^2\right]^3=\left(\frac{d^2y}{dx^2}\right)^2$

The order of the above equations is 2.

The degree of a differential equation is the degree of the highest derivative after removing the radical sign and fraction.

The *degree* of the equation (1) and (2) is 1. The degree of the equation (3) is 2.

12.3 FORMATION OF DIFFERENTIAL EQUATIONS

The differential equations can be formed by differentiating the ordinary equation and eliminating the arbitrary constants.

Example 1. *Form the differential equation by eliminating arbitrary constants, in the following cases and also write down the order of the differential equations obtained.*

(*a*) $y = Ax + A^2$ (*b*) $y = A\cos x + B\sin x$ (*c*) $y^2 = Ax^2 + Bx + C$.

(*R.G.P.V. Bhopal, June 2008*)

Solution. (*a*) $y = Ax + A^2$ ... (1)

On differentiation $\dfrac{dy}{dx} = A$

Putting the value of A in (1), we get $y = x\dfrac{dy}{dx} + \left(\dfrac{dy}{dx}\right)^2$ **Ans.**

On eliminating one constant A we get the differential equation of order 1.

(*b*) $y = A\cos x + B\sin x$

On differentiation $\dfrac{dy}{dx} = -A\sin x + B\cos x$

Again differentiating

$$\frac{d^2y}{dx^2} = -A\cos x - B\sin x \quad \Rightarrow \quad \frac{d^2y}{dx^2} = -(A\cos x + B\sin x)$$

$$\Rightarrow \quad \frac{d^2y}{dx^2} = -y \Rightarrow \frac{d^2y}{dx^2} + y = 0 \qquad \textbf{Ans.}$$

This is differential equation of order 2 obtained by eliminating two constants A and B.

(*c*) $y^2 = Ax^2 + Bx + C$

On differentiation $2y\dfrac{dy}{dx} = 2Ax + B$

Again differentiating $2y\dfrac{d^2y}{dx^2} + 2\left(\dfrac{dy}{dx}\right)^2 = 2A$

On differentiating again $y\dfrac{d^3y}{dx^3} + \dfrac{dy}{dx}\dfrac{d^2y}{dx^2} + 2\dfrac{dy}{dx}\dfrac{d^2y}{dx^2} = 0 \Rightarrow y\dfrac{d^3y}{dx^3} + 3\dfrac{dy}{dx}\dfrac{d^2y}{dx^2} = 0$ **Ans.**

This is the differential equation of order 3, obtained by eliminating three constants *A, B, C.*

Example 2. *Determine the differential equation whose set of independent solution is* $\{e^x, xe^x, x^2e^x\}$

Solution. Here, we have

$$y = c_1 e^x + c_2 x e^x + c_3 x^2 e^x \qquad ...(1)$$

Differentiating both sides, we get

$$\begin{aligned} y' &= c_1 e^x + c_2 x e^x + c_2 e^x + c_3 x^2 e^x + c_3 2x e^x \\ &= (c_1 e^x + c_2 x e^x + c_3 x^2 e^x) + c_2 e^x + c_3 2x e^x \\ &= y + c_2 e^x + c_3 2x e^x \qquad \text{[Using (1)]} \end{aligned}$$

$$\Rightarrow \quad y' - y = c_2 e^x + c_3 2x e^x \qquad ...(2)$$

Again, differentiating both sides, we get

$$\Rightarrow \quad y'' - y' = c_2 e^x + 2c_3 e^x + 2x c_3 e^x \qquad \text{[Using (2)]}$$

$$\Rightarrow \quad y'' - y' = (c_2 e^x + 2x c_3 e^x) + 2c_3 e^x \qquad ...(3)$$

$\Rightarrow \quad y'' - y' = y' - y + 2c_3 e^x$

$\Rightarrow \quad y'' - 2y' + y = 2c_3 e^x$

Finally, on differentiating both sides, we get

$\Rightarrow \quad y''' - 2y'' + y' = 2c_3 e^x$

$\Rightarrow \quad y''' - 2y'' + y' = y'' - 2y' + y$ [Using (3)]

$\Rightarrow \quad y''' - 2y'' - y'' + y' + 2y' - y = 0$

$\Rightarrow \quad y''' - 3y'' + 3y' - y = 0$

$\Rightarrow \quad (D-1)^3 y = 0$ **Ans.**

Example 3. *By the elimination of the constants A and B obtain the differential equation of which $xy = Ae^x + Be^{-x} + x^2$ is the solution.* *(U.P.B. Pharma (C.O.) 2005)*

Solution. We have $xy = Ae^x + Be^{-x} + x^2$ .... (1)

On differentiating (1), we get $x\frac{dy}{dx} + y = Ae^x - Be^{-x} + 2x$ ... (2)

Again differentiating (2), we get $x\frac{d^2y}{dx^2} + \frac{dy}{dx} + \frac{dy}{dx} = Ae^x + Be^{-x} + 2$ ... (3)

From (1), we have $Ae^x + Be^{-x} = xy - x^2$ ... (4)

Putting the value of $Ae^x + Be^{-x}$ from (4) in (3), we have

$x\frac{d^2y}{dx^2} + 2\frac{dy}{dx} = xy - x^2 + 2, \quad x\frac{d^2y}{dx^2} + 2\frac{dy}{dx} - xy = 2 - x^2$ **Ans.**

EXERCISE 12.1

1. Write the order and the degree of the following differential equations.

(*i*) $\frac{d^2y}{dx^2} + a^2x = 0;$ (*ii*) $\left[1 + \left(\frac{dy}{dx}\right)^2\right]^{\frac{3}{2}} = \frac{d^2y}{dx^2};$ (*iii*) $x^2\left(\frac{d^2y}{dx^2}\right)^3 + y\left(\frac{dy}{dx}\right)^4 + y^4 = 0;$

(*iv*) $\left(\frac{d^3y}{dx^3}\right)^4 + \frac{d^2y}{dx^2} + y\frac{dy}{dx}.$ *(D.U. 2014)* **Ans.** (*i*) 2,1 (*ii*) 2,2 (*iii*) 2,3, (*iv*) 3,4

2. Give an example of each of the following type of differential equations.

(*i*) A linear-differential equation of second order and first degree **Ans.** *Q.* 1 (*i*)

(*ii*) A non-linear differential equation of second order and second degree **Ans.** *Q.* 1 (*ii*)

(*iii*) Second order and third degree. **Ans.** *Q.* 1 (*iii*)

3. Obtain the differential equation of which $y^2 = 4a(x + a)$ is a solution.

Ans. $y^2\left(\frac{dy}{dx}\right)^2 + 2xy\frac{dy}{dx} - y^2 = 0$

4. Obtain the differential equation associated with the primitive $Ax^2 + By^2 = 1$.

Ans. $xy\frac{d^2y}{dx^2} + x\left(\frac{dy}{dx}\right)^2 - y\frac{dy}{dx} = 0$

5. Find the differential equation corresponding to

$y = a\,e^{3x} + be^x.$ **Ans.** $\frac{d^2y}{dx^2} - 4\frac{dy}{dx} + 3y = 0$

6. By the elimination of constants A and B, find the differential equation of which $y = e^x\,(A\cos x + B\sin x)$ is a solution. **Ans.** $\frac{d^2y}{dx^2} - 2\frac{dy}{dx} + 2y = 0$

7. Find the differential equation whose solution is $y = a\cos(x + 3)$.

Ans. $\frac{dy}{dx} = -\tan(x+3)$

8. Show that set of function $\left\{x, \frac{1}{x}\right\}$ forms a basis of the differential equation $x^2y'' + xy' - y = 0$. Obtain a particular solution when $y(1) = 1$, $y(1) = 2$. **Ans.** $y = \frac{3x}{2} - \frac{1}{2x}$

12.4 SOLUTION OF A DIFFERENTIAL EQUATION

In the example 1(*b*), $y = A\cos x + B\sin x$, on eliminating A and B we get the differential equation

$$\frac{d^2y}{dx^2} + y = 0$$

$y = A\cos x + B\sin x$ is called the solution of the differential equation $\frac{d^2y}{dx^2} + y = 0$.

The order of the differential equation $\frac{d^2y}{dx^2} + y = 0$ is two and the solution $y = A\cos x + B\sin x$ contains two arbitrary constants. The number of arbitrary constants in the solution is equal to the order of the differential equation.

An equation containing dependent variable (y) and independent variable (x) and free from derivative, which satisfies the differential equation, is called the solution (primative) of the differential equation.

12.5 GEOMETRICAL MEANING OF THE DIFFERENTIAL EQUATION OF FIRST ORDER AND FIRST DEGREE

Let $$f\left(x,\ y,\ \frac{dy}{dx}\right) = 0 \qquad \text{... (1)}$$

be a differential equation of the first order and first degree.

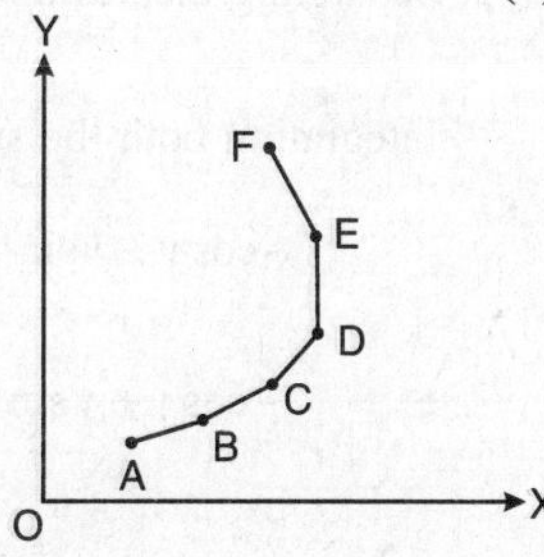

It is known that a direction of a curve at a particular point is given by the tangent line at that point and the slope of the tangent is $\frac{dy}{dx}$ at that point. Let $A(x_0, y_0)$ be any initial point. From (1), we can find $\frac{dy}{dx}$ at $A(x_0, y_0)$.

With the help of $\frac{dy}{dx}$ at $A(x_0, y_0)$ draw the tangent at the point A. On the tangent line take a neighbouring point $B(x_1, y_1)$.

Find $\frac{dy}{dx}$ at the point $B(x_1, y_1)$ from equation (1) and draw the tangent at B with the help of $\frac{dy}{dx}$ at (x_1, y_1).

Take a neighbouring point $C(x_2, y_2)$ on this tangent and in this way draw another tangent at the point C. Similarly draw, some more tangents by taking the neighbouring points on them. They form a smooth curve *i.e.* $y = f_1(x)$ which is the solution (1).

Again we take another starting point $A'x_0', y_0'$. We can draw another curve starting from A'. In this way we can draw a number of curves.

The given differential equation represents a family of curves.

12.6 DIFFERENTIAL EQUATIONS OF THE FIRST ORDER AND FIRST DEGREE

We will discuss the standard methods of solving the differential equations of the following types:

(*i*) Equations solvable by separation of the variables. (*ii*) Homogeneous equations.
(*iii*) Linear equations of the first order. (*iv*) Exact differential equations.

12.7 VARIABLES SEPARABLE

If a differential equation can be written in the form

$$f(y)dy = \phi(x)dx$$

We say that variables are separable, y on left hand side and x on right hand side.
We get the solution by integrating both sides.

Working Rule:

Step 1. Separate the variables as $f(y)dy = \phi(x)dx$

Step 2. Integrate both sides as $\int f(y)\,dy = \int \phi(x)\,dx$

Step 3. Add an arbitrary constant C on R.H.S.

Example 4. *Solve* : $\dfrac{dy}{dx} = \dfrac{x(2\log x+1)}{\sin y + y\cos y}$ *(U.P.B. Pharm (C.O.) 2005)*

Solution. We have, $\dfrac{dy}{dx} = \dfrac{x(2\log x+1)}{\sin y + y\cos y}$ *(U.P. II 2008)*

Separating the variables, we get

$$(\sin y + y \cos y)\, dy = \{x\,(2 \log x + 1)\}\, dx$$

Integrating both the sides, we get $\int (\sin y + y\cos y)dy = \int \{x(2\log x+1)\}dx$

$$-\cos y + y\sin y - \int (1)\cdot \sin y\,dy = 2\int \log x\cdot x\,dx + \int x\,dx + C$$

$$\Rightarrow \quad -\cos y + y\sin y + \cos y = 2\left[\log x\cdot\frac{x^2}{2} - \int \frac{1}{x}\cdot\frac{x^2}{2}dx\right] + \frac{x^2}{2}$$

$$\Rightarrow \quad y\sin y = 2\log x\cdot\frac{x^2}{2} - \int x\,dx + \frac{x^2}{2}$$

$$\Rightarrow \quad y\sin y = 2\log x\cdot\frac{x^2}{2} - \frac{x^2}{2} + \frac{x^2}{2}$$

$$\Rightarrow \quad y \sin y = x^2 \log x + C$$ **Ans.**

Example 5. *Solve the differential equation.*

$$x^4\frac{dy}{dx} + x^3 y = -\sec(x\,y).$$

Solution. $x^4\dfrac{dy}{dx} + x^3 y = -\sec(x\,y) \quad \Rightarrow \quad x^3\left(x\dfrac{dy}{dx} + y\right) = -\sec xy$

Put $v = xy, \quad \frac{dv}{dx} = x\frac{dy}{dx} + y \quad \Rightarrow \quad x^3\frac{dv}{dx} = -\sec v$

$$\Rightarrow \quad \frac{dv}{\sec v} = -\frac{dx}{x^3} \qquad \Rightarrow \quad \int \cos v \, dv = -\int \frac{dx}{x^3}.$$

$$\Rightarrow \sin v = \frac{1}{2x^2} + c \qquad \Rightarrow \sin xy = \frac{1}{2x^2} + c$$ **Ans.**

Example 6. *Solve :* $\cos(x + y)dy = dx$

Solution. $\cos(x + y)\, dy = dx \quad \Rightarrow \quad \frac{dy}{dx} = \sec(x + y)$

On putting $x + y = z$

So that $1 + \frac{dy}{dx} = \frac{dz}{dx} \quad \Rightarrow \quad \frac{dy}{dx} = \frac{dz}{dx} - 1$

$$\frac{dz}{dx} - 1 = \sec z \quad \Rightarrow \quad \frac{dz}{dx} = 1 + \sec z$$

Separating the variables, we get

$$\frac{dz}{1 + \sec z} = dx$$

On integrating,

$$\int \frac{\cos z}{\cos z + 1} dz = \int dx \quad \Rightarrow \quad \int \left[1 - \frac{1}{\cos z + 1}\right] dz = x + C$$

$$\int \left(1 - \frac{1}{2\cos^2 \frac{z}{2} - 1 + 1}\right) dz = x + C$$

$$\int \left(1 - \frac{1}{2}\sec^2 \frac{z}{2}\right) dz = x + C \quad \Rightarrow \quad z - \tan\frac{z}{2} = x + C$$

$$x + y - \tan\frac{x + y}{2} = x + C$$

$$y - \tan\frac{x + y}{2} = C$$ **Ans.**

Example 7. *Solve the equation.*

$(2x^2 + 3y^2 - 7)\, x\, dx - (3x^2 + 2y^2 - 8)\, y\, dy = 0$ *(U.P. II Semester, Summer 2005)*

Solution. We have

$$(2x^2 + 3y^2 - 7)\, x\, dx - (3x^2 + 2y^2 - 8)\, y\, dy = 0 \qquad ...(1)$$

Re-arranging (1), we get $\frac{x\,dx}{y\,dy} = \frac{3x^2 + 2y^2 - 8}{2x^2 + 3y^2 - 7}$

Applying componendo and dividendo rule, we get

$$\frac{x\,dx + y\,dy}{x\,dx - y\,dy} = \frac{5x^2 + 5y^2 - 15}{x^2 - y^2 - 1} \quad \Rightarrow \quad \frac{x\,dx + y\,dy}{x^2 + y^2 - 3} = 5\left(\frac{x\,dx - y\,dy}{x^2 - y^2 - 1}\right)$$

Multiplying by 2 both the sides, we get

$$\Rightarrow \quad \left(\frac{2x\,dx + 2y\,dy}{x^2 + y^2 - 3}\right) = 5\left(\frac{2x\,dx - 2y\,dy}{x^2 - y^2 - 1}\right)$$

Integrating both sides, we get

$$\log(x^2+y^2-3) = 5\log(x^2-y^2-1) + \log C$$

$$\Rightarrow \quad x^2+y^2-3 = C(x^2-y^2-1)^5 \qquad \textbf{Ans.}$$

where C is arbitrary constant of integration.

EXERCISE 12.2

Solve the following differential equations :

1. $\dfrac{dx}{x} = \tan y \cdot dy$ **Ans.** $x\cos y = C$
2. $\dfrac{dy}{dx} = \dfrac{\sqrt{1-y^2}}{\sqrt{1-x^2}}$ **Ans.** $\sin^{-1} y = \sin^{-1} x + C$
3. $y(1+x^2)^{1/2} dy + x\sqrt{1+y^2}\, dx = 0$ **Ans.** $\sqrt{1+y^2}+\sqrt{1+x^2} = C$
4. $\sec^2 x \tan y\, dx + \sec^2 y \tan x\, dy = 0$ **Ans.** $\tan x \tan y = C$
5. $(1+x^2)\, dy - x\, y\, dx = 0$ **Ans.** $y^2 = C(1+x^2)$
6. $(e^y+1)\cos x\, dx + e^y \sin x\, dy = 0$ **Ans.** $(e^y+1)\sin x = C$
7. $3\, e^x \tan y\, dx + (1-e^x)\sec^2 y\, dy = 0$ **Ans.** $(1-e^x)^3 = C\tan y$
8. $(e^y+2)\sin x\, dx - e^y \cos x\, dy = 0$ **Ans.** $(e^y+2)\cos x = C$
9. $\dfrac{dy}{dx} = e^{x-y} + x^2 e^{-y}$ **Ans.** $e^y = e^x + \dfrac{x^3}{3} + C$
10. $\dfrac{dy}{dx} = 1+\tan(y-x)$ [Put $y - x = z$] **Ans.** $\sin(y-x) = e^{x+c}$
11. $(4x+y)^2 \dfrac{dx}{dy} = 1$ **Ans.** $\tan^{-1}\dfrac{4x+y}{2} = 2x + C$
12. $\dfrac{dy}{dx} = (4x+y+1)^2$ [**Hint.** Put $4x + y + 1 = z$] **Ans.** $\tan^{-1}\dfrac{4x+y+1}{2} = 2x + C$
13. $\dfrac{dy}{dx} - y\tan 2x = 0,\ y(0) = 2$ *(DU, II Sem. 2012)* **Ans.** $y\sqrt{\cos 2x} = 2$

12.8 HOMOGENEOUS DIFFERENTIAL EQUATIONS

A differential equation of the form $\dfrac{dy}{dx} = \dfrac{f(x,y)}{\phi(x,y)}$

is called a homogeneous equation if each term of $f(x, y)$ and $\phi(x, y)$ is of the same degree *i.e.*,

$$\frac{dy}{dx} = \frac{3xy+y^2}{3x^2+xy}$$

In such case we put $\boxed{y = vx \text{ . and } \dfrac{dy}{dx} = v + x\dfrac{dv}{dx}}$

The reduced equation involves v and x only. This new differential equation can be solved by *variables separable* method.

Working Rule

Step 1. Put $y = vx$ so that $\dfrac{dy}{dx} = v + x\dfrac{dv}{dx}$ **Step 2.** Separate the variables.

Step 3. Integrate both the sides. **Step 4.** Put $v = \dfrac{y}{x}$ and simplify.

Example 8. *Solve the following differential equation*

$$(2xy + x^2)\, dy = (3y^2 + 2xy)\, dx$$ *(A.M.I.E.T.E. Dec. 2006)*

Solution. We have, $(2xy + x^2)\dfrac{dy}{dx} = 3y^2 + 2xy \Rightarrow \dfrac{dy}{dx} = \dfrac{3y^2 + 2xy}{2xy + x^2}$ ...(1)

Put $y = vx$ so that $\dfrac{dy}{dx} = v + x\dfrac{dv}{dx}$

On substituting, 1 the given equation becomes $v + x\dfrac{dv}{dx} = \dfrac{3v^2x^2 + 2vx^2}{2vx^2 + x^2} = \dfrac{3v^2 + 2v}{2v + 1}$

$$\Rightarrow \quad x\frac{dv}{dx} = \frac{3v^2 + 2v - 2v^2 - v}{2v + 1} \qquad \Rightarrow x\frac{dv}{dx} = \frac{v^2 + v}{2v + 1} \Rightarrow \left(\frac{2v+1}{v^2 + v}\right) dv = \frac{dx}{x}$$

$$\Rightarrow \quad \int\left(\frac{2v+1}{v^2 + v}\right) dv = \int\frac{dx}{x} \qquad \Rightarrow \log (v^2 + v) = \log x + \log c$$

$$\Rightarrow \quad v^2 + v = cx \qquad \Rightarrow \frac{y^2}{x^2} + \frac{y}{x} = cx$$

$$\Rightarrow \quad y^2 + xy = cx^3$$

Example 9. *Solve the equation* :

$$\frac{dy}{dx} = \frac{y}{x} + x\sin\frac{y}{x}$$

Solution.
$$\frac{dy}{dx} = \frac{y}{x} + x\sin\frac{y}{x} \qquad ...(1)$$

Put $y = vx$ in (1) so that $\dfrac{dy}{dx} = v + x\dfrac{dv}{dx}$

$$v + x\frac{dv}{dx} = v + x\sin v$$

$$\Rightarrow \quad x\frac{dv}{dx} = x\sin v \Rightarrow \frac{dv}{dx} = \sin v$$

Separating the variable, we get

$$\Rightarrow \quad \frac{dv}{\sin v} = dx \quad \Rightarrow \quad \int \operatorname{cosec} v\, dv = \int dx$$

$$\log\tan\frac{v}{2} = x + C \quad \Rightarrow \quad \log\tan\frac{y}{2x} = x + C$$ **Ans.**

EXERCISE 12.3

Solve the following differential equations:

1. $(y^2 - xy)\, dx + x^2\, dy = 0$ **Ans.** $\dfrac{x}{y} = \log x + C$
2. $(x^2 - y^2)\, dx + 2xy\, dy = 0$ *(AMIETE, June 2009)* **Ans.** $x^2 + y^2 = ax$
3. $x(y - x)\dfrac{dy}{dx} = y(y + x).$ **Ans.** $\dfrac{y}{x} - \log xy = a$
4. $x\,(x - y)\, dy + y^2\, dx = 0$ *(U.P. B. Pharm (C.O.) 2005)* **Ans.** $y = x \log C y$

5. $\frac{dy}{dx}+\frac{x-2y}{2x-y}=0$ **Ans.** $y-x=C(x+y)^3$

6. $\frac{dy}{dx}=\tan\frac{y}{x}+\frac{y}{x}$ **Ans.** $\sin\frac{y}{x}=Cx$

7. $\frac{dy}{dx}=\frac{3xy+y^2}{3x^2}$ **Ans.** $3x+y\log x+Cy=0$

8. $\frac{dy}{dx}=\frac{x^2-2y^2}{2xy}$ **Ans.** $4y^2-x^2=\frac{C}{x^2}$

9. $(x^2+y^2)\,dy=xy\,dx$ **Ans.** $-\frac{x^2}{2y^2}+\log y=C$

10. $x^2y\,dx-(x^3+y^3)\,dy=0$ **Ans.** $\frac{-x^3}{3y^3}+\log y=C$

11. $(y^2+2xy)\,dx+(2x^2+3xy)\,dy=0$ **Ans.** $xy^2(x+y)=C$

12. $(2xy^2-x^3)\,dy+(y^3-2yx^2)\,dx=0$ **Ans.** $y^2(y^2-x^2)=Cx^{-2}$

13. $(x^3-3xy^2)\,dx+(y^3-3x^2y)\,dy=0,\ y(0)=1$ **Ans.** $x^4-6x^2y^2+y^4=1$

14. $2xy^2\,dy-(x^3+2y^3)\,dx=0$ **Ans.** $2y^3=3x^3\log x+3x^3+C$

15. $x\sin\frac{y}{x}dy=\left(y\sin\frac{y}{x}-x\right)dx$ **Ans.** $\cos\frac{y}{x}=\log x+C$

16. $\left\{x\cos\frac{y}{x}+y\sin\frac{y}{x}\right\}y-\left\{y\sin\frac{y}{x}-x\cos\frac{y}{x}\right\}x\frac{dy}{dx}=0$ **Ans.** $xy\cos\frac{y}{x}=a$

17. $\frac{dy}{dx}=\frac{y+\sqrt{x^2+y^2}}{x}$ **Ans.** $y+\sqrt{x^2+y^2}=Cx^2$

18. $x\frac{dy}{dx}=y(\log y-\log x+1)$ **Ans.** $\log\frac{y}{x}=Cx$

19. $xy\log\frac{x}{y}dx+\left(y^2-x^2\log\frac{x}{y}\right)dy=0$ given that $y(1)=0$

Ans. $\frac{x^2}{2y^2}\log\frac{x}{y}-\frac{x^2}{4y^2}+\log y=1-\frac{3}{4e^2}$

20. $(1+e^{\frac{x}{y}})\,dx+e^{\frac{x}{y}}\left(1-\frac{x}{y}\right)dy=0$ *(AMIETE, June 2009)* **Ans.** $e^{\frac{x}{y}}+\frac{x}{y}=e^{-y}+C$

12.9 EQUATIONS REDUCIBLE TO HOMOGENEOUS FORM

Case I. $\frac{a}{A}+\frac{b}{B}$

The equations of the form

$$\boxed{\frac{dy}{dx}=\frac{ax+by+c}{Ax+By+C}}$$

can be reduced to the homogeneous form by the substitution if $\frac{a}{A}+\frac{b}{B}$

$$x=X+h,\ y=Y+k \qquad (h,k \text{ being constants})$$

$$\therefore \qquad \frac{dy}{dx}=\frac{dY}{dX}$$

The given differential equation reduces to

$$\frac{dY}{dX}=\frac{a(X+h)+b(Y+k)+c}{A(X+h)+B(Y+k)+C}=\frac{aX+bY+ah+bk+c}{AX+BY+Ah+Bk+C}$$

Choose h, k so that $\quad a\,h + b\,k + c = 0$

$$A\,h + B\,k + C = 0$$

Then the given equation becomes homogeneous $\dfrac{dY}{dX} = \dfrac{aX+bY}{AX+BY}$

Case II. If $\dfrac{a}{A} = \dfrac{b}{B}$ then the value of h, k will not be finite.

$$\frac{a}{A} = \frac{b}{B} = \frac{1}{m} \quad \text{(say)}$$

$$A = a\,m, \; B = b\,m$$

The given equation becomes $\dfrac{dy}{dx} = \dfrac{ax+by+c}{m(ax+by)+C}$

Now put $ax + by = z$ and apply the method of variables separable.

Example 10. *Solve* : $\dfrac{dy}{dx} = \dfrac{x+2y-3}{2x+y-3}$

Solution. Put $\quad x = X + h, \quad y = Y + k.$

The given equation reduces to

$$\therefore \quad \frac{dY}{dX} = \frac{(X+h)+2(Y+k)-3}{2(X+h)+(Y+k)-3} \qquad \left(\frac{1}{2} \neq \frac{2}{1}\right)$$

$$= \frac{X+2Y+(h+2k-3)}{2X+Y+(2h+k-3)} \qquad \text{.... (1)}$$

Now choose h and k so that $h + 2k - 3 = 0$ and $2h + k - 3 = 0$

Solving these equations, we get $h = k = 1$

$$\therefore \quad \frac{dY}{dX} = \frac{X+2Y}{2X+Y} \qquad \text{... (2)}$$

Put $Y = v\,X$, so that $\dfrac{dY}{dX} = v + X\dfrac{dv}{dX}$

The equation (2) is transformed as

$$v + X\frac{dv}{dX} = \frac{X+2vX}{2X+vX} = \frac{1+2v}{2+v}$$

$$X\frac{dv}{dX} = \frac{1+2v}{2+v} - v = \frac{1-v^2}{2+v} \quad \Rightarrow \quad \left(\frac{2+v}{1-v^2}\right)dv = \frac{dX}{X}$$

$$\Rightarrow \quad \frac{1}{2}\frac{1}{(1+v)}dv + \frac{3}{2}\frac{1}{1-v}dv = \frac{dX}{X} \qquad \text{(Partial fractions)}$$

On integrating, we have

$$\frac{1}{2}\log(1+v) - \frac{3}{2}\log(1-v) = \log X + \log C$$

$$\Rightarrow \quad \log\frac{1+v}{(1-v)^3} = \log C^2X^2 \quad \Rightarrow \quad \frac{1+v}{(1-v)^3} = C^2X^2$$

$$\frac{1+\dfrac{Y}{X}}{\left(1-\dfrac{Y}{X}\right)^3} = C^2X^2 \quad \Rightarrow \quad \frac{X+Y}{(X-Y)^3} = C^2 \quad \text{or} \quad X + Y = C^2\,(X - Y)^3$$

Put $X = x - 1$ and $Y = y - 1$ $\Rightarrow$ $x + y - 2 = a(x - y)^3$ **Ans.**

Example 11. *Solve* : $(x + 2y)(dx - dy) = dx + dy$

Solution. $(x + 2y)(dx - dy) = dx + dy \Rightarrow (x + 2y - 1)\,dx - (x + 2y + 1)\,dy = 0$

$$\Rightarrow \quad \frac{dy}{dx} = \frac{x+2y-1}{x+2y+1} \qquad ...(1)$$

Hence $\dfrac{a}{A} = \dfrac{b}{B}$ *i.e.,* $\left(\dfrac{1}{1} = \dfrac{2}{2}\right)$ (Case of failure)

Now put $x + 2y = z$ so that $1 + 2\dfrac{dy}{dx} = \dfrac{dz}{dx}$

Equation (1) becomes

$$\frac{1}{2}\frac{dz}{dx} - \frac{1}{2} = \frac{z-1}{z+1} \quad \Rightarrow \quad \frac{dz}{dx} = 2\frac{(z-1)}{z+1} + 1 = \frac{3z-1}{z+1}$$

$$\Rightarrow \quad \frac{z+1}{3z-1}dz = dx \quad \Rightarrow \quad \left(\frac{1}{3} + \frac{4}{3}\frac{1}{3z-1}\right)dz = dx$$

On integrating, $\dfrac{z}{3} + \dfrac{4}{9}\log(3z-1) = x + C$

$$3z + 4\log(3z - 1) = 9x + 9C \qquad [9C = a]$$

$$\Rightarrow \quad 3(x + 2y) + 4\log(3x + 6y - 1) = 9x + a$$

$$6x - 6y + a = 4\log(3x + 6y - 1)$$ **Ans.**

EXERCISE 12.4

Solve the following differential equations:

1. $\dfrac{dy}{dx} = \dfrac{2x+9y-20}{6x+2y-10}$ **Ans.** $(2x - y)^2 = C(x + 2y - 5)$

2. $\dfrac{dy}{dx} = \dfrac{y-x+1}{y+x+5}$ **Ans.** $\log[(y+3)^2 + (x+2)^2] + 2\tan^{-1}\dfrac{y+3}{x+2} = a$

3. $\dfrac{dy}{dx} = \dfrac{x-y-2}{x+y+6}$ **Ans.** $(y + 4)^2 + 2(x + 2)(y + 4) - (x + 2)^2 = a^2$

4. $\dfrac{dy}{dx} = \dfrac{y+x-2}{y-x-4}$ *(AMIETE, Dec. 2009)* **Ans.** $-(y - 3)^2 + 2(x + 1)(y - 3) + (x + 1)^2 = a$

5. $\dfrac{dy}{dx} = \dfrac{2x-5y+3}{2x+4y-6}$ **Ans.** $(x - 4y + 3)^2(2x + y - 3) = a$

6. $(2x + y + 1)\,dx + (4x + 2y - 1)\,dy = 0$ **Ans.** $2(2x + y) + \log(2x + y - 1) = 3x + C$

7. $(x - y - 2)\,dx - (2x - 2y - 3)\,dy = 0$ **Ans.** $\log(x - y - 1) = x - 2y + C$

(U.P. B. Pharm (C.O.) 2005)

8. $(6x - 4y + 1)\,dy - (3x - 2y + 1)\,dx = 0$ *(A.M.I.E.T. E., Dec. 2006)*

Ans. $4x - 8y - \log(12x - xy + 1) = c$

9. $\dfrac{dy}{dx} = -\dfrac{3y-2x+7}{7y-3x+3}$ **Ans.** $(x + y - 1)^5(x - y - 1)^2 = 1$

10. $\dfrac{dy}{dx} = \dfrac{2y-x-4}{y-3x+3}$ **Ans.** $X^2 - 5XY + Y^2 = C\left[\dfrac{2Y + (-5+\sqrt{21})X}{2Y - (5-\sqrt{21})X}\right]^{1/\sqrt{21}}$, $\begin{array}{l} X = x-2 \\ Y = y-3 \end{array}$

(AMIETE, Dec. 2010)

12.10 LINEAR DIFFERENTIAL EQUATIONS

A differential equation of the form

$$\boxed{\frac{dy}{dx}+Py=Q} \quad \text{... (1)}$$

is called a linear differential equation, where P and Q, are functions of x (but not of y) or constants.

In such case, multiply both sides of (1) by $e^{\int P dx}$

$$e^{\int P dx}\left(\frac{dy}{dx}+Py\right)=Qe^{\int P dx} \quad \text{... (2)}$$

The left hand side of (2) is

$$\frac{d}{dx}\left[y.e^{\int P dx}\right]$$

(2) becomes $$\frac{d}{dx}\left[y.e^{\int P dx}\right]=Q.e^{\int P dx}$$

Integrating both sides, we get

$$y.e^{\int P dx}=\int Q.e^{\int P dx}\,dx+C$$

This is the required solution.

Note. $e^{\int P dx}$ is called the integrating factor.

Solution is $$\boxed{y\times [I.F.]=\int Q\,[I.F.]\,dx+C}$$

Working Rule

Step 1. Convert the given equation to the standard form of linear differential equation

i.e. $$\frac{dy}{dx}+Py=Q$$

Step 2. Find the integrating factor i.e. I.F. = $e^{\int P dx}$

Step 3. Then the solution is $y(I.F.)=\int Q\ (I.F.)dx+C$

Example 12. *Solve*: $(x+1)\frac{dy}{dx}-y=e^x(x+1)^2$

Solution. $$\frac{dy}{dx}-\frac{y}{x+1}=e^x(x+1)$$

$$\text{Integrating factor } = e^{-\int\frac{dx}{x+1}}=e^{-\log(x+1)}=e^{\log(x+1)^{-1}}=\frac{1}{x+1}$$

The solution is $$y.\frac{1}{x+1}=\int e^x.(x+1).\frac{1}{x+1}dx=\int e^x dx$$

$$\frac{y}{x+1}=e^x+C$$ **Ans.**

Example 13. *Solve a differential equation*

$$(x^3-x)\frac{dy}{dx}-(3x^2-1)y=x^5-2x^3+x.$$ *(Nagpur University, Summer 2008)*

Solution. We have $(x^3-x)\dfrac{dy}{dx}-(3x^2-1)y = x^5-2x^3+x$

$$\Rightarrow \quad \frac{dy}{dx}-\frac{3x^2-1}{x^3-x}y = \frac{x^5-2x^3+x}{x^3-x} \quad \Rightarrow \quad \frac{dy}{dx}-\frac{3x^2-1}{x^3-x}y = x^2-1$$

$$\text{I.F.} = e^{\int -\frac{3x^2-1}{x^3-x}dx} = e^{-\log(x^3-x)} = e^{\log(x^3-x)^{-1}} = \frac{1}{x^3-x}$$

Its solution is

$$y(\text{I.F.}) = \int Q(I.F.)\,dx + C \quad \Rightarrow \quad y\left(\frac{1}{x^3-x}\right) = \int \frac{x^2-1}{x^3-x}dx + C$$

$$\Rightarrow \quad \frac{y}{x^3-x} = \int \frac{x^2-1}{x(x^2-1)}dx + C \quad \Rightarrow \quad \frac{y}{x^3-x} = \int \frac{1}{x}dx + C$$

$$\Rightarrow \quad \frac{y}{x^3-x} = \log x + C \qquad \Rightarrow \quad y = (x^3-x)\log x + (x^3-x)\,C$$ **Ans.**

Example 14. *Solve* $\sin x\dfrac{dy}{dx}+2y = \tan^3\left(\dfrac{x}{2}\right)$

Solution. Given equation: $\sin x\,\dfrac{dy}{dx}+2y = \tan^3\dfrac{x}{2} \quad \Rightarrow \quad \dfrac{dy}{dx}+\dfrac{2}{\sin x}y = \dfrac{\tan^3\frac{x}{2}}{\sin x}$

This is linear form of $\dfrac{dy}{dx}+Py = Q$

$$\therefore \qquad P = \frac{2}{\sin x} \quad \text{and} \quad Q = \frac{\tan^3\frac{x}{2}}{\sin x}$$

$$\therefore \qquad \text{I.F.} = e^{\int P dx} = e^{\int \frac{2}{\sin x}dx} = e^{2\int \operatorname{cosec} x\,dx} = e^{2\log\tan\frac{x}{2}} = \tan^2\frac{x}{2}$$

$\therefore$ Solution is $\quad y.(\text{I.F.}) = \int \text{I.F.}(Q\,dx) + C$

$$y\tan^2\frac{x}{2} = \int \tan^2\frac{x}{2}\cdot\frac{\tan^3\frac{x}{2}}{2\sin\frac{x}{2}\cdot\cos\frac{x}{2}}dx + C = \frac{1}{2}\int\frac{\tan^4\frac{x}{2}}{\cos^2\frac{x}{2}}dx + C$$

$$= \frac{1}{2}\int \tan^4\frac{x}{2}.\sec^2\frac{x}{2}dx + C \qquad ...(1)$$

Putting $\tan\dfrac{x}{2} = t$ so that $\dfrac{1}{2}\sec^2\dfrac{x}{2}dx = dt$ on R.H.S. (1), we get

$$y.\tan^2\frac{x}{2} = \frac{1}{2}\int t^4(2dt) + C \quad \Rightarrow \quad y\tan^2\frac{x}{2} = \frac{t^5}{5} + C$$

$$y\tan^2\frac{x}{2} = \frac{\tan^5\frac{x}{2}}{5} + C$$ **Ans.**

EXERCISE 12.5

Solve the following differential equations:

1. $\dfrac{dy}{dx}+\dfrac{1}{x}y=x^3-3$ **Ans.** $xy=\dfrac{x^5}{5}-\dfrac{3x^2}{2}+C$

2. $(2y-3x)\,dx+x\,dy=0$ **Ans.** $y\,x^2=x^3+C$

3. $\dfrac{dy}{dx}+y\cot x=\cos x$ **Ans.** $y\sin x=\dfrac{\sin^2 x}{2}+C$

4. $\dfrac{dy}{dx}+y\sec x=\tan x$ **Ans.** $y=\dfrac{C-x}{\sec x+\tan x}+1$

5. $\cos^3 x\dfrac{dy}{dx}+y\cos x=\sin x$ *(DU, II Sem. 2012)* **Ans.** $y=\tan x-1+Ce^{-\tan x}$

6. $(x+a)\dfrac{dy}{dx}-3y=(x+a)^5$ **Ans.** $2y=(x+a)^5+2C(x+a)^3$

7. $x\cos x\dfrac{dy}{dx}+y(x\sin x+\cos x)=1$ **Ans.** $x\,y=\sin x+C\cos x$

8. $x\log x\dfrac{dy}{dx}+y=2\log x$ **Ans.** $y\log x=(\log x)^2+C$

9. $x\dfrac{dy}{dx}+2y=x^2\log x$ **Ans.** $yx^2=\dfrac{x^4}{4}\log x-\dfrac{x^4}{16}+C$

10. $dr+(2r\cot\theta+\sin 2\theta)\,d\theta=0$ **Ans.** $r\sin^2\theta=\dfrac{-\sin^4\theta}{2}+C$

11. $\dfrac{dy}{dx}+y\cos x=\dfrac{1}{2}\sin 2x$ **Ans.** $y=\sin x-1+Ce^{-\sin x}$

12. $(1-x^2)\dfrac{dy}{dx}+2xy=x(1-x^2)^{1/2}$ **Ans.** $y=\sqrt{1-x^2}+C(1-x^2)$

13. $\sec x\dfrac{dy}{dx}=y+\sin x$ *(A.M.I.E.T.E., Dec 2005)* **Ans.** $y=-\sin x-1+ce^{\sin x}$

14. $y'+y\tan x=\cos x,\ y(0)=0$ *(A.M.I.E.T.E., June 2006)* **Ans.** $y=x\cos x$

15. Solve $(1+y^2)\,dx=(\tan^{-1}y-x)\,dy$ *(AMIETE, Dec. 2009)* **Ans.** $x=-\tan^{-1}y-1+ce^{\tan^{-1}y}$

16. Find the value of α so that $e^{\alpha x^2}$ is an integrating factor of differential equation $x(1-y)\,dx-dy=0$.

(A.M.I.E.T.E., Summer 2005) **Ans.** $\alpha=\dfrac{1}{2}$

17. Slove the differential equation $\cot 3x\dfrac{dy}{dx}-3y=\cos 3x+\sin 3x,\ 0<x<\dfrac{\pi}{2}$. *(AMIETE, Dec. 2009)*

Ans. $y\cos 3x=\dfrac{1}{12}[6x-\sin 6x-\cos 6x]$

18. The value of α so that $e^{\alpha y^2}$ is an integrating factor of the differential equation

$(e^{\frac{-y^2}{2}}-xy)\,dy-dx=0$ is *(A.M.I.E.T.E. Dec., 2005)*

(*a*) -1 (*b*) 1 (*c*) $\dfrac{1}{2}$ (*d*) $-\dfrac{1}{2}$ **Ans.** (*c*)

19. The solution of the differential equation $(y+x)^2\dfrac{dy}{dx}=a^2$ is given by

(*a*) $y+x=a\tan\left(\dfrac{y-c}{a}\right)$ (*b*) $y-x=\tan\left(\dfrac{y-c}{a}\right)$

(*c*) $y-x=a\tan(y-c)$ (*d*) $a(y-x)=\tan\left(y-\dfrac{c}{a}\right)$ **Ans.** (*a*)

(AMIETE, June 2010)

12.11 EQUATIONS REDUCIBLE TO THE LINEAR FORM (BERNOULLI EQUATION)

The equation of the form

$$\frac{dy}{dx} + Py = Qy^n \qquad ...(1)$$

where **P** and **Q** are constants or functions of x can be reduced to the linear form on dividing by y^n and substituting $\frac{1}{y^{n-1}} = z$

On dividing both sides of (1) by y^n, we get

$$\frac{1}{y^n}\frac{dy}{dx} + \frac{1}{y^{n-1}}P = Q \qquad ...(2)$$

Put $\frac{1}{y^{n-1}} = z$, so that $\frac{(1-n)}{y^n}\frac{dy}{dx} = \frac{dz}{dx} \Rightarrow \frac{1}{y^n}\frac{dy}{dx} = \frac{1}{1-n}\frac{dz}{dx}$

$\therefore$ (2) becomes $\frac{1}{1-n}\frac{dz}{dx} + Pz = Q$ or $\frac{dz}{dx} + P(1-n)z = Q(1-n)$

which is a linear equation and can be solved easily by the previous method discussed in article 12.10 on page 307.

Example 15. *Solve:* $\frac{dy}{dx} + xy = x^3 y^3$ (*Delhi University, April 2014*)

Solution. We have, $\frac{dy}{dx} + xy = x^3 y^3$

$$\Rightarrow \frac{1}{y^3}\frac{dy}{dx} + \frac{x}{y^2} = x^3 \qquad ...(1)$$

Putting $\frac{1}{y^2} = z \Rightarrow \frac{-2}{y^3}\frac{dy}{dx} = \frac{dz}{dx} \Rightarrow \frac{1}{y^3}\frac{dy}{dx} = \frac{-1}{2}\frac{dz}{dx}$, in (1), we get

$$\therefore \quad -\frac{1}{2}\frac{dz}{dx} + xz = x^3 \Rightarrow \frac{dz}{dx} - 2xz = -2x^3 \qquad ...(2)$$

$$\text{I.F.} = e^{-\int 2x\,dx} = e^{-x^2}$$

$$\therefore \quad ze^{-x^2} = -2\int x^3 e^{-x^2}\,dx$$

Let $-x^2 = t \Rightarrow -2x\,dx = dt$

$$ze^{-x^2} = \int te^t\,dt = te^t - e^t + c$$

Put $z = y^{-2}$ and $t = -x^2$

$$\therefore \quad \frac{e^{-x^2}}{y^2} = -x^2 e^{-x^2} - e^{-x^2} + c$$

$$\Rightarrow \quad \frac{1}{y^2} = -x^2 - 1 + Ce^{x^2}$$ **Ans.**

Example 16. *Solve* $x^2 dy + y(x + y)\,dx = 0$ (*U.P. II Semester Summer 2006*)

Solution. We have, $x^2\,dy + y\,(x + y)\,dx = 0$

$$\Rightarrow \frac{dy}{dx} + \frac{y}{x} = -\frac{y^2}{x^2} \Rightarrow \frac{1}{y^2}\frac{dy}{dx} + \frac{1}{xy} = -\frac{1}{x^2}$$

Put $$-\frac{1}{y}=z \text{ so that } \frac{1}{y^2}\frac{dy}{dx}=\frac{dz}{dx}$$

The given equation reduces to a linear differential equation in z.

$$\frac{dz}{dx}-\frac{z}{x}=-\frac{1}{x^2}$$

$$\text{I.F.}=e^{-\int\frac{1}{x}dx}=e^{-\log x}=e^{\log 1/x}=\frac{1}{x}.$$

Hence the solution is

$$z.\frac{1}{x}=\int-\frac{1}{x^2}.\frac{1}{x}\,dx+C \quad\Rightarrow\quad \frac{z}{x}=\int -x^{-3}\,dx+C$$

$$\Rightarrow\quad -\frac{1}{xy}=-\frac{x^{-2}}{-2}+C \quad\Rightarrow\quad \frac{1}{xy}=-\frac{1}{2x^2}-C \qquad \textbf{Ans.}$$

Example 17. *Solve*: $x\frac{dy}{dx}+y\log y=xy\ e^x$

Solution. $$x\frac{dy}{dx}+y\log y=xy e^x$$

Dividing by xy, we get

$$\frac{1}{y}\frac{dy}{dx}+\frac{1}{x}\log y=e^x \qquad ...(1)$$

Put $\log y=z$, so that $\frac{1}{y}\frac{dy}{dx}=\frac{dz}{dx}$

Equation (1) becomes, $\frac{dz}{dx}+\frac{z}{x}=e^x$

$$\text{I.F.}=e^{\int\frac{1}{x}dx}=e^{\log x}=x$$

Solution is $$zx=\int xe^x\,dx+C$$

$$zx=xe^x-e^x+C$$

$$\Rightarrow\quad x\log y=xe^x-e^x+C \qquad \textbf{Ans.}$$

Example 18. *Solve*: $\frac{dy}{dx}-\frac{\tan y}{1+x}=(1+x)e^x\sec y.$

Solution. $$\frac{dy}{dx}-\frac{\tan y}{1+x}=(1+x)e^x\sec y$$

$$\Rightarrow\quad \cos y\frac{dy}{dx}-\frac{\sin y}{1+x}=(1+x)e^x \qquad ...(1)$$

Put $\sin y=z$, so that $\cos y\frac{dy}{dx}=\frac{dz}{dx}$

(1) becomes $$\frac{dz}{dx}-\frac{z}{1+x}=(1+x)e^x$$

$$I.F.=e^{-\int\frac{1}{1+x}dx}=e^{-\log(1+x)}=e^{\log\frac{1}{1+x}}=\frac{1}{1+x}$$

Solution is $$z.\frac{1}{1+x}=\int(1+x)e^x.\frac{1}{1+x}dx=\int e^x dx$$

$$\frac{\sin y}{1+x}=e^x+C$$ **Ans.**

Example 19. *Solve*: $\tan y\frac{dy}{dx}+\tan x=\cos y\cos^2 x$

Solution. $$\tan y\frac{dy}{dx}+\tan x=\cos y\cos^2 x$$

$$\sec y\tan y\frac{dy}{dx}+\sec y\tan x=\cos^2 x$$

Writing $z=\sec y$, so that $\frac{dz}{dx}=\sec y\tan y\frac{dy}{dx}$

The equation becomes $\frac{dz}{dx}+z\tan x=\cos^2 x$

$$\text{I.F.}=e^{\int\tan x\,dx}=e^{\log\sec x}=\sec x$$

$\therefore$ The solution of the equation is

$$z\sec x=\int\cos^2 x\sec x\,dx+C$$

$$\sec y\,\sec x=\int\cos x\,dx+C=\sin x+C$$

$$\sec y=(\sin x+C)\cos x$$ **Ans.**

Example 20. *Solve differential equation*

$$\frac{dy}{dx}+\frac{1}{x}\tan y=\frac{1}{x^2}\tan y\sin y$$

Solution. We have, $\frac{dy}{dx}+\frac{1}{x}\tan y=\frac{1}{x^2}\tan y\sin y \Rightarrow \frac{1}{\tan y\sin y}\frac{dy}{dx}+\frac{1}{x}\frac{1}{\sin y}=\frac{1}{x^2}$

$$\Rightarrow \cot y\,\text{cosec}\,y\frac{dy}{dx}+\frac{1}{x}\text{cosec}\,y=\frac{1}{x^2} \quad \text{... (1)}$$

Putting $\text{cosec}\,y=z$, and $-\text{cosec}\,y\cot y\frac{dy}{dx}=\frac{dz}{dx}$ in (1), we get

$$-\frac{dz}{dx}+\frac{1}{x}z=\frac{1}{x^2}$$

$$\frac{dz}{dx}-\frac{1}{x}z=-\frac{1}{x^2}$$

$$\text{I.F.}=e^{-\int\frac{1}{x}dx}=e^{-\log x}=e^{\log x^{-1}}=x^{-1}=\frac{1}{x}$$

Its solution is $$z\,(\text{I.F.})=\int Q(\text{I.F.})\,dx+C$$

$$z\cdot\frac{1}{x}=\int\left(\frac{-1}{x^2}\right)\frac{1}{x}dx=-\int\frac{1}{x^3}dx=-\frac{x^{-2}}{-2}$$

$$z=\frac{1}{2x}+Cx \Rightarrow \text{cosec}\,y=\frac{1}{2x}+Cx$$ **Ans.**

Example 21. *Solve* $x\left[\dfrac{dy}{dx}+y\right]=1-y$

Solution. $x\left(\dfrac{dy}{dx}+y\right)=(1-y)$

$$\Rightarrow \quad \frac{dy}{dx}+y=\frac{1}{x}-\frac{y}{x} \quad \Rightarrow \quad \frac{dy}{dx}+\left(1+\frac{1}{x}\right)y=\frac{1}{x}$$

which is in linear form of $\dfrac{dy}{dx}+Py=Q.$

$$\therefore \quad P=\left(1+\frac{1}{x}\right), \quad Q=\frac{1}{x}$$

$$\text{I.F.}=e^{\int P dx}=e^{\int\left(1+\frac{1}{x}\right)dx}=e^{x+\log x}=e^x.e^{\log x}=e^x.x=xe^x$$

Its solution is

$$y(\text{I.F.})=\int \text{I.F.}(Q\,dx)+C$$

$$y(x.e^x)=\int (x.e^x)\times\frac{1}{x}dx+C \Rightarrow y(x.e^x)=\int e^x dx$$

$$y\,(x\,.\,e^x)=e^x+C$$

$$\therefore \quad y=\frac{1}{x}+\frac{C}{x}e^{-x} \qquad \textbf{Ans.}$$

Example 22. *Solve the differential equation.*

$y \log y\, dx+(x-\log y)\, dy=0$ *(Uttarakhand II Semester, June 2007)*

Solution. We have,

$$y \log y\, dx+(x-\log y)\, dy=0$$

$$\Rightarrow \quad \frac{dx}{dy}=\frac{-x+\log y}{y\log y} \quad \Rightarrow \quad \frac{dx}{dy}=\frac{-x}{y\log y}+\frac{\log y}{y\log y}$$

$$\Rightarrow \quad \frac{dx}{dy}+\frac{x}{y\log y}=\frac{1}{y}$$

$$\text{I.F.}=e^{\int\frac{1}{y\log y}dy}=e^{\log(\log y)}=\log y$$

Its solution is $\quad x.\log y=\int \dfrac{1}{y}(\log y)\,dy$

$$x.\log y=\frac{(\log y)^2}{2}+C \qquad \textbf{Ans.}$$

Example 23. *Solve* $\quad ye^y dx=(y^3+2xe^y)dy$

Solution. $\quad ye^y dx=(y^3+2xe^y)dy \Rightarrow \dfrac{dx}{dy}-\dfrac{2x}{y}=\dfrac{y^2}{e^y}$

which is linear in x

$$\therefore \quad P=\frac{-2}{y} \quad \text{and} \quad Q=\frac{y^2}{e^y}$$

$$\text{I.F.} = e^{\int P\,dy} = e^{\int \frac{-2}{y} dy} = e^{-2\log y} = e^{\log y^{-2}} = y^{-2} = \frac{1}{y^2}$$

Its solution is $x(\text{I.F.}) = \int Q.(\text{I.F.}\,dy) + C \quad \Rightarrow \quad x.\frac{1}{y^2} = \int \frac{y^2}{e^y} \times \frac{1}{y^2} dy + C$

$$\frac{x}{y^2} = \int e^{-y}.dy + C \quad \Rightarrow \quad \frac{x}{y^2} = -e^{-y} + C$$

$\therefore \qquad \frac{x}{y^2} + e^{-y} = C$ **Ans.**

Example 24. *Solve* $\frac{dy}{dx} = \frac{y}{2y\log y + y - x}$

Solution. $\frac{dx}{dy} = \frac{2y\log y + y - x}{y}$

$\Rightarrow \qquad \frac{dx}{dy} + \frac{1}{y}x = 1 + 2\log y$

Which is of the form $\frac{dx}{dy} + Px = Q$

Here $P = \frac{1}{y}$ and $Q = 1 + 2\log y$

$$\text{I.F.} = e^{\int P\,dy} = e^{\int \frac{1}{y} dy} = e^{\log y} = y$$

Its solution is

$$x.(I.F.) = \int Q(I.F.)dy + C \quad \Rightarrow \quad xy = \int (1 + 2\log y)\,y\,dy + C$$

$\Rightarrow \qquad xy = \int (y + 2y\log y)\,dy + C = \frac{y^2}{2} + 2\left[\log y.\frac{y^2}{2} - \int \frac{1}{y}.\frac{y^2}{2} dy\right] + C$

$$= \frac{y^2}{2} + 2\left[\frac{y^2}{2}\log y - \frac{1}{4}y^2\right] + C = y^2 \log y + C$$

$\Rightarrow \qquad x = y\log y + \frac{C}{y}$ **Ans.**

Example 25. *Solve*: $\frac{dy}{dx} = \frac{y+1}{(y+2)\,e^y - x}$

Solution. $\frac{dx}{dy} = \frac{(y+2)\,e^y}{y+1} - \frac{x}{y+1}$

$\Rightarrow \qquad \frac{dx}{dy} + \frac{x}{y+1} = \frac{(y+2)\,e^y}{y+1}$

which is linear in x

Here $P = \frac{1}{y+1}$ and $Q = \frac{y+2}{y+1}e^y$

$$\text{I.F.} = e^{\int P\,dy} = e^{\int \frac{1}{y+1}dy} = e^{\log(y+1)} = y+1$$

Its solution is

$$x(\text{I.F.}) = \int Q(\text{I.F.}dy)$$

$$\Rightarrow \quad x(y+1) = \int \frac{y+2}{y+1} e^y (y+1)\,dy$$

$$\Rightarrow \quad x(y+1) = \int (y+2)e^y\,dy = (y+2)e^y - \int \frac{d}{dy}(y+2).e^y\,dy$$

$$\Rightarrow \quad x\,(y+1) = (y+2)e^y - \int e^y\,dy$$

$$\Rightarrow \quad x\,(y+1) = (y+2)\,e^y - e^y + C$$

$$\Rightarrow \quad x\,(y+1) = (y+1)\,e^y + C \;\Rightarrow\; x = e^y + \frac{C}{y+1}$$ **Ans.**

Example 26. *Solve*: $(1 + y^2)\,dx = (\tan^{-1} y - x)\,dy$.

(AMIETE, June 2010, R.G.P.V., Bhopal, April 2010, June 2008, U.P. (B. Pharm) 2005)

Solution. $(1 + y^2)\,dx = (\tan^{-1} y - x)\,dy$

$$\frac{dx}{dy} = \frac{\tan^{-1} y - x}{1+y^2} \quad\Rightarrow\quad \frac{dx}{dy} + \frac{x}{1+y^2} = \frac{\tan^{-1} y}{1+y^2}$$

This is a linear differential equation.

$$\text{I.F.} = e^{\int \frac{1}{1+y^2}dy} = e^{\tan^{-1} y}$$

Its solution is $\quad x.e^{\tan^{-1} y} = \int e^{\tan^{-1} y} \frac{\tan^{-1} y}{1+y^2}\,dy + C$

Put $\quad \tan^{-1} y = t$ on R.H.S., so that $\frac{1}{1+y^2}dy = dt$

$$x.e^{\tan^{-1} y} = \int e^t.t\,dt + C = t.e^t - e^t + C = e^{\tan^{-1} y}\left(\tan^{-1} y - 1\right) + C$$

$$x = (\tan^{-1} y - 1) + Ce^{-\tan^{-1} y}$$ **Ans.**

Example 27. *Solve*: $r\sin\theta - \frac{dr}{d\theta}\cos\theta = r^2$ *(Nagpur University, Summer 2005)*

Solution. The given equation can be written as $-\frac{dr}{d\theta}\cos\theta + r\sin\theta = r^2$ ... (1)

Dividing (1) by $r^2 \cos\theta$ we get $-r^{-2}\frac{dr}{d\theta} + r^{-1}\tan\theta = \sec\theta$ ... (2)

Putting $\quad r^{-1} = v$ so that $-r^{-2}\frac{dr}{d\theta} = \frac{dv}{d\theta}$ in (2), we get

$$\frac{dv}{d\theta} + v\tan\theta = \sec\theta$$

$$\text{I.F.} = e^{\int \tan\theta\,d\theta} = e^{\log\sec\theta} = \sec\theta$$

Solution is $v \sec\theta = \int \sec\theta . \sec\theta \, d\theta + C \quad \Rightarrow \quad v\sec\theta = \int \sec^2\theta \, d\theta + C$

$$\frac{\sec\theta}{r} = \tan\theta + C \quad \Rightarrow \quad r^{-1} = (\sin\theta + C\cos\theta)$$

$\therefore \qquad r = \dfrac{1}{\sin\theta + C\cos\theta}$ **Ans.**

EXERCISE 12.6

Solve the following differential equations:

1. $\dfrac{1}{y^2}\dfrac{dy}{dx} - \dfrac{1}{y} = 2xe^{-x}$ **Ans.** $e^x + x^2y + Cy = 0$

2. $3\dfrac{dy}{dx} + 3\dfrac{y}{x} = 2x^4y^4$ **Ans.** $\dfrac{1}{y^3} = x^5 + Cx^3$

3. $\dfrac{dy}{dx} = y\tan x - y^2 \sec x$ *(DU, II Sem. 2012)* **Ans.** $\sec x = (\tan x + C)\, y$

4. $\dfrac{dy}{dx} = 2y\tan x + y^2 \tan^2 x$, if $y = 1$ at $x = 0$ **Ans.** $\dfrac{1}{y}\sec^2 x = -\dfrac{\tan^3 x}{3} + 1$

5. $\dfrac{dy}{dx} + \tan x \tan y = \cos x \sec y$ **Ans.** $\sin y \sec x = x + C$

6. $dy + y\tan x \,.\, dx = y^2 \sec x \,.\, dx$ **Ans.** $y\,(x + C) + \cos x = 0$

7. $(x^2y^2 + xy)\, y\, dx + (x^2y^2 - 1)\, x\, dy = 0$ **Ans.** $x\,y = \log C\,y$

8. $(x^2 + y^2 + x)\, dx + xy\, dy = 0$ **Ans.** $x^2y^2 = -\dfrac{x^4}{2} - \dfrac{2x^3}{3} + C$

9. $\dfrac{dy}{dx} + y = 3e^x y^3$ **Ans.** $\dfrac{1}{y^2} = 6e^x + Ce^{2x}$

10. $(x - y^2)\, dx + 2\,x\,y\, dy = 0$ **Ans.** $\dfrac{y^2}{x} + \log x = C$

11. $e^y\left(\dfrac{dy}{dx} + 1\right) = e^x$ **Ans.** $e^{x+y} = \dfrac{e^{2x}}{2} + C$

12. $x^2y - x^3\dfrac{dy}{dx} = y^4 \cos x$ **Ans.** $x^3 = y^3(3\sin x - C)$

13. $3\dfrac{dy}{dx} + \dfrac{2}{x+1} . y = \dfrac{x^2}{y^2}$ **Ans.** $y^3(x+1)^2 = \dfrac{x^5}{5} + \dfrac{x^4}{2} + \dfrac{x^3}{3} + C$

14. $\cos x\dfrac{dy}{dx} + 4y\sin x = 4\sqrt{y}\sec x$ **Ans.** $\sqrt{y}\sec^2 x = 2\left[\tan x + \dfrac{\tan^3 x}{3}\right] + C$

15. $\dfrac{dy}{dx} + x\sin 2y = x^3\cos^2 y$ **Ans.** $\tan y = \dfrac{1}{2}(x^2 - 1) + Ce^{-x^2}$

16. $\dfrac{1}{1+y^2}\dfrac{dy}{dx} + 2x\tan^{-1} y = x^3$ **Ans.** $\tan^{-1} y = \dfrac{1}{2}(x^2 - 1) + Ce^{-x^2}$

17. $e^{-y}\sec^2 y\, dy = dx + x\, dy$ **Ans.** $x\,e^y = \tan y + C$

18. $(x + y + 1)\dfrac{dy}{dx} = 1$ **Ans.** $x + y + 2 = C\,e^y$

19. $\frac{dy}{dx}=\frac{y^3}{e^{2x}+y^2}$ **Ans.** $e^{-2x}y^2+2\log y+C=0$

20. $dx - xy(1+xy^2)\,dy = 0$ **Ans.** $-\frac{1}{x}=y^2-2+Ce^{-y^2/2}$

21. $\frac{dy}{dx}+\frac{y}{x}\log y=\frac{y}{x^2}(\log y)^2$ **Ans.** $\frac{1}{x\log y}=\frac{1}{2x}+C$

22. $3\frac{dy}{dx}+xy=xy^{-2}$ (*A.M.I.E.T.E., June 2009*) **Ans.** $y^3=1+Ce^{-x^2/2}$

23. $x\frac{dy}{dx}+y=x^3y^6$ (*AMIETE, June 2010*) **Ans.** $\frac{1}{y^5x^5}=\frac{5}{2x^2}+C$

24. General solution of linear differential equation of first order $\frac{dx}{dy}+Px=Q$ (where P and Q are constants or functions of y) is

(*a*) $ye^{\int P.dx}=\int Qe^{\int P.dx}\,dx+c$ (*b*) $xe^{\int P.dy}=\int Qe^{\int P.dy}\,dy+c$

(*c*) $y=\int Qe^{\int P.dx}\,dx+c$ (*d*) $x=\int Qe^{\int P.dy}\,dy+c$ (*AMIETE, June, 2010*) **Ans.** (*b*)

12.12 EXACT DIFFERENTIAL EQUATION

An exact differential equation is formed by directly differentiating its primitive (solution) without any other process

$$\boldsymbol{Mdx + Ndy = 0}$$

is said to be an exact differential equation if it satisfies the following condition $\boxed{\frac{\partial M}{\partial y}=\frac{\partial N}{\partial x}}$

where $\frac{\partial M}{\partial y}$ denotes the differential co-efficient of M with respect to y keeping x constant and $\frac{\partial N}{\partial x}$, the differential co-efficient of N with respect to x, keeping y constant.

Method for Solving Exact Differential Equations

Step I. Integrate M w.r.t. x keeping y constant

Step II. Integrate w.r.t. y, only those terms of N which do not contain x.

Step III. Result of I + Result of II = Constant.

Example 28. *Solve* :

$$(5x^4+3x^2y^2-2xy^3)\,dx+(2x^3y-3x^2y^2-5y^4)\,dy=0$$

Solution. Here, $M=5x^4+3x^2y^2-2xy^3$, $N=2x^3y-3x^2y^2-5y^4$

$$\frac{\partial M}{\partial y}=6x^2y-6xy^2,\quad \frac{\partial N}{\partial x}=6x^2y-6xy^2$$

Since, $\frac{\partial M}{\partial y}=\frac{\partial N}{\partial x}$, the given equation is exact.

Now $\int M\,dx+\int(\text{terms of } N \text{ is not containing } x)\,dy=C$ (y constant)

$$\int(5x^4+3x^2y^2-2xy^3)\,dx+\int-5y^4dy=C$$

$\Rightarrow \qquad x^5 + x^3y^2 - x^2y^3 - y^5 = C$ **Ans.**

Example 29. *Solve*: $\left\{2xy\cos x^2 - 2xy + 1\right\}dx + \left\{\sin x^2 - x^2 + 3\right\}dy = 0$

Solution. Here we have

$$\left\{2xy\cos x^2 - 2xy + 1\right\}dx + \left\{\sin x^2 - x^2 + 3\right\}dy = 0 \qquad \text{... (1)}$$

$$M\,dx + N\,dy = 0 \qquad \text{... (2)}$$

Comparing (1) and (2), we get

$$M = 2xy\cos x^2 - 2xy + 1 \quad \Rightarrow \quad \frac{\partial M}{\partial y} = 2x\cos x^2 - 2x$$

$$N = \sin x^2 - x^2 + 3 \quad \Rightarrow \quad \frac{\partial N}{\partial x} = 2x\cos x^2 - 2x$$

Here, $\therefore \qquad \dfrac{\partial M}{\partial y} = \dfrac{\partial N}{\partial x}$

So the given differential equation is exact differential equation.

Hence solution is $\displaystyle\int_{y \text{ as const}} M\,dx + \int$ (terms of N not containing x) $dy = C$

$$\int (2xy\cos x^2 - 2xy + 1)\,dx + \int 3\,dy = C$$

$$\Rightarrow \quad \int[y(2x\cos x^2) - y(2x) + 1]\,dx + 3\int dy = C$$

$$\Rightarrow \quad y\int 2x\cos x^2 dx - y\int 2x\,dx + \int 1\,dx + 3\int dy = C$$

Put $\quad x^2 = t \quad$ so that $\quad 2x\,dx = dt$

$$y\int \cos t\,dt - 2y\frac{x^2}{2} + x + 3y = C$$

$$\Rightarrow \quad y\sin t - x^2 y + x + 3y = C$$

$$y\sin x^2 - yx^2 + x + 3y = C$$ **Ans.**

Example 30. *Solve* :

$$(1 + e^{x/y}) + e^{x/y}\left(1 - \frac{x}{y}\right)\frac{dy}{dx} = 0$$

(*Nagpur University, Summer 2008, A.M.I.E.T.E. June, 2009*)

Solution. We have,

$$\left(1 + e^{\frac{x}{y}}\right) + e^{\frac{x}{y}}\left(1 - \frac{x}{y}\right)\frac{dy}{dx} = 0 \quad \Rightarrow \quad \left(1 + e^{\frac{x}{y}}\right)dx + \left(e^{\frac{x}{y}} - e^{\frac{x}{y}}\frac{x}{y}\right)dy = 0$$

$$M = 1 + e^{\frac{x}{y}} \quad \Rightarrow \quad \frac{\partial M}{\partial y} = -\frac{x}{y^2}e^{\frac{x}{y}}$$

$$N = e^{\frac{x}{y}} - e^{\frac{x}{y}}\frac{x}{y} \quad \Rightarrow \quad \frac{\partial N}{\partial x} = \frac{1}{y}e^{\frac{x}{y}} - \frac{1}{y}e^{\frac{x}{y}} - \frac{x}{y^2}e^{\frac{x}{y}} = -\frac{x}{y^2}e^{\frac{x}{y}}$$

$$\Rightarrow \quad \frac{\partial M}{\partial y} = \frac{\partial N}{\partial x}$$

$\therefore$ Given equation is exact.

Its solution is $\int\left(1+e^{\frac{x}{y}}\right)dx+\int(\text{terms of } N \text{ not containing } x)\,dy=C$

$$\Rightarrow \quad \int\left(1+e^{\frac{x}{y}}\right)dx+\int 0\,dy=C \quad \Rightarrow \quad x+ye^{\frac{x}{y}}=C$$ **Ans.**

Example 31. *Solve*: $x\,dx+y\,dy=\dfrac{a^2(x\,dy-y\,dx)}{x^2+y^2}$ *(U.P. Second Semester Summer 2005)*

Solution. We have, $x\,dx+y\,dy=\dfrac{a^2(x\,dy-y\,dx)}{x^2+y^2}$

$$\Rightarrow \quad \left(x+\frac{a^2y}{x^2+y^2}\right)dx+\left(y-\frac{(a^2x)}{x^2+y^2}\right)dy=0$$

Here, $M=x+\dfrac{a^2y}{x^2+y^2}, \qquad N=y-\dfrac{a^2x}{x^2+y^2}$

Now, $\dfrac{\partial M}{\partial y}=\dfrac{a^2(x^2-y^2)}{(x^2+y^2)^2}, \qquad \dfrac{\partial N}{\partial x}=\dfrac{a^2(x^2-y^2)}{(x^2+y^2)^2}$

Since, $\dfrac{\partial M}{\partial y}=\dfrac{\partial N}{\partial x}$

Therefore equation is exact. Hence,

$$\int\left(x+\frac{a^2y}{x^2+y^2}\right)dx+\int y\,dy=C$$

$$\Rightarrow \quad \frac{x^2}{2}+a^2y.\frac{1}{y}\tan^{-1}\left(\frac{x}{y}\right)+\frac{y^2}{2}=C$$

$$\Rightarrow \quad \left(\frac{x^2+y^2}{2}\right)+a^2\tan^{-1}\left(\frac{x}{y}\right)=C$$ **Ans.**

Example 32. *Solve*: $[1+\log(x\,y)]dx+\left[1+\dfrac{x}{y}\right]dy=0$

Solution. $[1+\log x\,y]dx+\left[1+\dfrac{x}{y}\right]dy=0$

$$\therefore \quad [1+\log x+\log y]dx+\left[1+\frac{x}{y}\right]dy=0$$

which is in the form

$$M\,dx+N\,dy=0$$

$$M=[1+\log x+\log y] \quad \text{and} \quad N=1+\frac{x}{y}$$

$$\Rightarrow \quad \frac{\partial M}{\partial y}=\frac{1}{y} \quad \text{and} \quad \Rightarrow \quad \frac{\partial N}{\partial x}=\frac{1}{y}$$

$$\Rightarrow \qquad \frac{\partial M}{\partial y} = \frac{\partial N}{\partial x}$$

Hence the given differential equation is exact.

$\therefore$ Solution is $\displaystyle\int_{y \text{ constant}} M\,dx + \int N(\text{terms not containing } x)\,dy = C$

$$\therefore \qquad \int (1+\log x+\log y)\,dx + \int dy = C$$

$$\Rightarrow \qquad x + \int \log x\,dx + \int \log y\,dx + y = C \qquad \text{... (1)}$$

Now, $\displaystyle\int \log x\,dx = \int \log x.(1)\,dx = (\log x)x - \int\left[\frac{d}{dx}(\log x)x\right]dx = x\log x - \int \frac{1}{x}.x\,dx$

$$= x\log x - \int dx = x\log x - x = x[\log x - 1]$$

$\therefore$ Equation (1) becomes $\Rightarrow$ $x + x\log x - x + x\log y + y = C$

$x[\log x + \log y] + y = C \Rightarrow x\log xy + y = C$ **Ans.**

Example 33. *Find the value of λ, for the differential equation*

$$(xy^2 + \lambda x^2 y)dx + (x+y)x^2 dy = 0 \text{ is exact}$$

Solve the equation for this value of λ, (*Uttarakhand Summer 2010*)

Solution. Here $(xy^2 + \lambda x^2 y)\,dx + (x+y)x^2 dy = 0$ ... (1)

which is of the form $M\,dx + N\,dx = 0$

Where $M = xy^2 + \lambda x^2 y$ and $N = (x+y)\,x^2 = x^3 + x^2 y$

$$\frac{\partial M}{\partial y} = 2xy + \lambda x^2, \qquad \frac{\partial N}{\partial x} = 3x^2 + 2xy$$

Condition of to be exact is

$$\frac{\partial M}{\partial y} = \frac{\partial N}{\partial x}$$

$$2xy + \lambda x^2 = 3x^2 + 2xy$$

$$\Rightarrow \qquad \lambda x^2 = 3x^2 \quad \Rightarrow \quad \lambda = 3$$

If $\lambda = 3$ then (1) becomes an excat differential equation.

$$(xy^2 + 3x^2 y)\,dx + (x+y)\,x^2\,dy = 0$$

Its solution is given by

$$\int_{(y = \text{constant})} M\,dx + \int (\text{terms of } N \text{ not containing } x)\,dy = \text{constant}$$

$$\Rightarrow \qquad \int (xy^2 + 3x^2 y)dx + \int 0\,dy = C$$

$$\Rightarrow \qquad \frac{x^2y^2}{2} + \frac{3x^3y}{3} = C \qquad \Rightarrow \qquad \frac{x^2y^2}{2} + x^3 y = C$$

$x^2y^2 + 2x^3y = C_1$ **Ans.**

EXERCISE 12.7

Solve the following differential equation (1 – 11).

1. $(x + y - 10)\, dx + (x - y - 2)\, dy = 0$ **Ans.** $\dfrac{x^2}{2} + xy - 10x - \dfrac{y^2}{2} - 2y = C$

2. $(y^2 - x^2)\, dx + 2x\, y\, dy = 0$ **Ans.** $\dfrac{x^3}{3} = x\, y^2 + C$

3. $(1 + 3e^{x/y})dx + 3e^{x/y}\left(1 - \dfrac{x}{y}\right)dy = 0$ *(R.G.P.V. Bhopal, Winter 2010)* **Ans.** $x + 3y\, e^{x/y} = C$

4. $(2x - y)\, dx = (x - y)\, dy$ **Ans.** $xy = x^2 + \dfrac{y^2}{2} + C$

5. $(y \sec^2 x + \sec x \tan x)\, dx + (\tan x + 2y)\, dy = 0$ **Ans.** $y \tan x + \sec x + y^2 = C$

6. $(ax + hy + g)\, dx + (hx + by + f)\, dy = 0$ **Ans.** $ax^2 + 2h\, xy + by^2 + 2gx + 2fy + C = 0$

7. $(x^4 - 2xy^2 + y^4)\, dx - (2x^2y - 4xy^3 + \sin y)\, dy = 0$ **Ans.** $\dfrac{x^5}{5} - x^2y^2 + xy^4 + \cos y = C$

8. $(2xy + e^y)\, dx + (x^2 + xe^y)\, dy = 0$ **Ans.** $x^2y + xe^y = C$

9. $(x^2 + 2ye^{2x})\, dy + (2xy + 2y^2e^{2x})\, dx = 0$ **Ans.** $x^2y + y^2\, e^{2x} = C$

10. $\left[y\left(1 + \dfrac{1}{x}\right) + \cos y\right]dx + (x + \log x - x \sin y)dy = 0$ *(M.D.U., 2010)* **Ans.** $y\, (x + \log x) + x \cos y = C$

11. $(x^3 - 3xy^2)\, dx + (y^3 - 3x^3y)\, dy = 0,\ y(0) = 1$ **Ans.** $x^4 - 6x^2y^2 + y^4 = 1$

12. $\left(e^x + \dfrac{2x}{y}\right)dx - \dfrac{x^2}{y^2}dy = 0$ *(D.U. 2014)* **Ans.** $e^x + \dfrac{x^2}{y} = c$

13. The differential equation $M\, (x, y)\, dx + N\, (x, y)\, dy = 0$ is an exact differential equation if

(*a*) $\dfrac{\partial M}{\partial y} + \dfrac{\partial N}{\partial x} = 0$ (*b*) $\dfrac{\partial M}{\partial y} - \dfrac{\partial N}{\partial x} = 0$ (*c*) $\dfrac{\partial M}{\partial y} \times \dfrac{\partial N}{\partial x} = 1$ (*d*) None of the above

(A.M.I.E.T.E. Dec. 2010, Dec 2006) **Ans.** (*b*)

12.13 EQUATIONS REDUCIBLE TO THE EXACT EQUATIONS

Sometimes a differential equation which is not exact may become so, on multiplication by a suitable function known as the integrating factor.

Rule 1. If $\dfrac{\dfrac{\partial M}{\partial y} - \dfrac{\partial N}{\partial x}}{N}$ is a function of x alone, say $f(x)$, then $\text{I.F.} = e^{\int f(x)dx}$

Example 34. *Solve* $(2x \log x - xy)\, dy + 2y\, dx = 0$ ...(1)

Solution. $M = 2y, \quad N = 2x \log x - xy$

$$\frac{\partial M}{\partial y} = 2, \qquad \frac{\partial N}{\partial x} = 2(1 + \log x) - y$$

Here, $$\frac{\dfrac{\partial M}{\partial y} - \dfrac{\partial N}{\partial x}}{N} = \frac{2 - 2 - 2\log x + y}{2x \log x - xy} = \frac{-(2 \log x - y)}{x\,(2 \log x - y)} = -\frac{1}{x} = f(x)$$

$$\text{I.F.} = e^{\int f(x)dx} = e^{\int -\frac{1}{x}dx} = e^{-\log x} = e^{\log x^{-1}} = x^{-1} = \frac{1}{x}$$

On multiplying the given differential equation (1) by $\dfrac{1}{x}$, we get

$$\frac{2y}{x}dx + (2\log x - y)dy = 0 \quad \Rightarrow \quad \int \frac{2y}{x}dx + \int -y\, dy = c$$

$$\Rightarrow \quad 2y \log x - \frac{1}{2}y^2 = c$$ **Ans.**

EXERCISE 12.8

Solve the following differential equations:

1. $(y \log y)\, dx + (x - \log y)\, dy = 0$ **Ans.** $2x \log y = c + (\log y)^2$

2. $\left(y + \frac{1}{3}y^3 + \frac{1}{2}x^2\right)dx + \frac{1}{4}(1 + y^2)x\,dy = 0$ **Ans.** $\frac{yx^4}{4} + \frac{y^3x^4}{12} + \frac{x^6}{12} = c$

3. $(y - 2x^3)\, dx - x(1 - xy)\, dy = 0$ **Ans.** $-\frac{y}{x} - x^2 + \frac{y^2}{2} = c$

4. $(x \sec^2 y - x^2 \cos y)\, dy = (\tan y - 3x^4)\, dx$ **Ans.** $-\frac{1}{x}\tan y - x^3 + \sin y = c$

5. $(x - y^2)\, dx + 2xy\, dy = 0$ **Ans.** $y^2 = cx - x \log x$

Rule II. If $\dfrac{\frac{\partial N}{\partial x} - \frac{\partial M}{\partial y}}{M}$ is a function of y alone, say $f(y)$, then

$$\text{I.F.} = e^{\int f(y)dy}$$

Example 35. *Solve* $(y^4 + 2y)\, dx + (xy^3 + 2y^4 - 4x)\, dy = 0$

Solution. Here $M = y^4 + 2y;$ $N = xy^3 + 2y^4 - 4x$ ...(1)

$$\therefore \quad \frac{\partial M}{\partial y} = 4y^3 + 2; \qquad \frac{\partial N}{\partial x} = y^3 - 4$$

$$\therefore \quad \frac{\frac{\partial N}{\partial x} - \frac{\partial M}{\partial y}}{M} = \frac{(y^3 - 4) - (4y^3 + 2)}{y^4 + 2y} = \frac{-3(y^3 + 2)}{y(y^3 + 2)} = -\frac{3}{y} = f(y)$$

$$\text{I.F.} = e^{\int f(y)dy} = e^{\int -\frac{3}{y}dy} = e^{-3\log y} = e^{\log y^{-3}} = y^{-3} = \frac{1}{y^3}$$

On multiplying the given equation (1) by $\frac{1}{y^3}$ we get the exact differential equation.

$$\left(y + \frac{2}{y^2}\right)dx + \left(x + 2y - \frac{4x}{y^3}\right)dy = 0$$

$$\int\left(y + \frac{2}{y^2}\right)dx + \int 2y\,dy = c \quad \Rightarrow \quad x\left(y + \frac{2}{y^2}\right) + y^2 = c \qquad \textbf{Ans.}$$

EXERCISE 12.9

Solve the following differential equations:

1. $(3x^2y^4 + 2xy)\, dx + (2x^3y^3 - x^2)\, dy = 0$ **Ans.** $x^3y^2 + \frac{x^2}{y} = c$

2. $(xy^3 + y)\, dx + 2(x^2y^2 + x + y^4)\, dy = 0$ **Ans.** $\frac{x^2y^4}{2} + xy^2 + \frac{y^6}{3} = c$

3. $y(x^2y + e^x)dx - e^x dy = 0$ **Ans.** $\frac{x^3}{3} + \frac{e^x}{y} = c$

4. $(2x^4y^4e^y + 2xy^3 + y)\, dx + (x^2y^4e^y - x^2y^2 - 3x)\, dy = 0$ **Ans.** $x^2e^y + \frac{x^2}{y} + \frac{x}{y^3} = c$

Rule III. If M is of the form $M = y\, f_1(xy)$ and N is of the form $N = x\, f_2(xy)$

Then $$\text{I.F.} = \frac{1}{M.x - N.y}$$

Example 36. *Solve* $y\,(xy + 2x^2y^2)\,dx + x\,(xy - x^2y^2)\,dy = 0$

Solution. $$y\,(xy + 2x^2y^2)\,dx + x\,(xy - x^2y^2)\,dy = 0 \qquad \dots (1)$$

Dividing (1) by xy, we get

$$y\,(1 + 2xy)\,dx + x\,(1 - xy)\,dy = 0 \qquad \dots (2)$$

$$M = y\,f_1(xy), \quad N = x\,f_2(xy)$$

$$\text{I.F.} = \frac{1}{Mx - Ny} = \frac{1}{xy(1+2xy) - xy(1-xy)} = \frac{1}{3x^2y^2}$$

On multiplying (2) by $\frac{1}{3x^2y^2}$, we have an exact differential equation

$$\left(\frac{1}{3x^2y} + \frac{2}{3x}\right)dx + \left(\frac{1}{3xy^2} - \frac{1}{3y}\right)dy = 0 \quad \Rightarrow \quad \int\left(\frac{1}{3x^2y} + \frac{2}{3x}\right)dx + \int -\frac{1}{3y}\,dy = c$$

$$\Rightarrow \quad -\frac{1}{3xy} + \frac{2}{3}\log x - \frac{1}{3}\log y = c \quad \Rightarrow \quad -\frac{1}{xy} + 2\log x - \log y = b \qquad \textbf{Ans.}$$

EXERCISE 12.10

Solve the following differential equations:

1. $(y - xy^2)\,dx - (x + x^2y)\,dy = 0$ **Ans.** $\log\left(\frac{x}{y}\right) - xy = A$
2. $y\,(1 + xy)\,dx + x(1 - xy)\,dy = 0$ **Ans.** $xy\log\left(\frac{y}{x}\right) = cxy - 1$
3. $y\,(1 + xy)\,dx + x\,(1 + xy + x^2y^2)\,dy = 0$ **Ans.** $\frac{1}{2x^2y^2} + \frac{1}{xy} - \log y = c$
4. $(xy \sin xy + \cos xy)\,y\,dx + (xy \sin xy - \cos xy)\,x\,dy = 0$ **Ans.** $y \cos xy = cx$

Rule IV. For of this type of $x^m y^n(ay\,dx + bx\,dy) + x^{m'} y^{n'}(a'\,y\,dx + b'\,x\,dy) = 0,$ the integrating factor is $x^h\, y^k$.

where $$\frac{m+h+1}{a} = \frac{n+k+1}{b}, \quad \text{and} \quad \frac{m'+h+1}{a'} = \frac{n'+k+1}{b'}$$

Example 37. *Solve* $(y^3 - 2x^2y)\,dx + (2xy^2 - x^3)\,dy = 0$

Solution. $$(y^3 - 2x^2y)\,dx + (2xy^2 - x^3)\,dy = 0$$

$$y^2\,(ydx + 2xdy) + x^2\,(-2ydx - xdy) = 0$$

Here $m = 0$, $n = 2$, $a = 1$, $b = 2$, $\quad m' = 2$, $n' = 0$, $a' = -2$, $b' = -1$

$$\frac{0+h+1}{1} = \frac{2+k+1}{2} \quad \text{and} \quad \frac{2+h+1}{-2} = \frac{0+k+1}{-1}$$

$$\Rightarrow \quad 2h + 2 = 2 + k + 1 \text{ and } h + 3 = 2k + 2$$

$$\Rightarrow \quad 2h - k = 1 \text{ and } h - 2k = -1$$

On solving $h = k = 1$. Integrating Factor $= xy$

Multiplying the given equation by xy, we get

$$(xy^4 - 2x^3y^2)\,dx + (2x^2y^3 - x^4y)\,dy = 0$$

which is an exact differential equation.

$$\int (xy^4 - 2x^3y^2)dx = C \quad \Rightarrow \quad \frac{x^2y^4}{2} - \frac{2x^4y^2}{4} = C$$

$$\Rightarrow \quad x^2y^4 - x^4y^2 = C' \quad \Rightarrow \quad x^2y^2(y^2 - x^2) = C' \qquad \textbf{Ans.}$$

Example 38. Solve $(3y - 2xy^3)\,dx + (4x - 3x^2y^2)\,dy = 0$. (*U.P., II Semester, June 2007*)

Solution. $$(3y - 2xy^3)\,dx + (4x - 3x^2y^2)\,dy = 0$$

$$\Rightarrow \quad (3y\,dx + 4x\,dy) + xy^2(-2y\,dx - 3x\,dy) = 0 \qquad \text{...(1)}$$

Comparing the coefficients of (1) with

$x^m y^n (a\,y\,dx + b\,x\,dy) + x^{m'} y^{n'} (a'\,y\,dx + b'\,x\,dy) = 0$, we get

$m = 0,\ n = 0,\ a = 3,\ b = 4$

$m' = 1,\ n' = 2,\ a' = -2,\ b' = -3$

To find the integrating factor $x^h y^k$

$$\frac{m+h+1}{a} = \frac{n+k+1}{b} \quad \text{and} \quad \frac{m'+h+1}{a'} = \frac{n'+k+1}{b'}$$

$$\frac{0+h+1}{3} = \frac{0+k+1}{4} \quad \text{and} \quad \frac{1+h+1}{-2} = \frac{2+k+1}{-3}$$

$$\Rightarrow \quad \frac{h+1}{3} = \frac{k+1}{4} \quad \text{and} \quad \frac{h+2}{2} = \frac{k+3}{3} \quad \Rightarrow \quad 4h - 3k + 1 = 0 \qquad \text{... (2)}$$

and $$3h - 2k = 0 \quad \Rightarrow \quad h = \frac{2k}{3} \qquad \text{...(3)}$$

Putting the value of h from (3) in (2), we get

$$\frac{8k}{3} - 3k + 1 = 0 \quad \Rightarrow \quad -\frac{k}{3} + 1 = 0 \quad \Rightarrow \quad k = 3$$

Putting $k = 3$ in (2), we get $h = \frac{2k}{3} = \frac{2 \times 3}{3} = 2$

$$\text{I.F.} = x^h y^k = x^2y^3$$

On multiplying the given differential equation by x^2y^3, we get

$$x^2y^3\,(3y - 2xy^3)dx + x^2y^3(4x - 3x^2y^2)\,dy = 0$$

$$(3x^2y^4 - 2x^3y^6)\,dx + (4x^3y^3 - 3x^4y^5)\,dy = 0$$

This is the exact differential equation.

Its solution is $\int (3x^2y^4 - 2x^3y^6)dx = 0 \quad \Rightarrow \quad x^3y^4 - \frac{x^4}{2}y^6 = C$ **Ans.**

EXERCISE 12.11

Solve the following differential equations:

1. $(2y\,dx + 3x\,dy) + 2xy\,(3y\,dx + 4x\,dy) = 0$ **Ans.** $x^2y^3\,(1 + 2xy) = c$
2. $(y^2 + 2yx^2)\,dx + (2x^3 - xy)\,dy = 0$ **Ans.** $4(xy)^{1/2} - \frac{2}{3}\left(\frac{y}{x}\right)^{3/2} = c$
3. $(3x + 2y^2)y\,dx + 2x\,(2x + 3y^2)\,dy = 0$ **Ans.** $x^2y^4\,(x + y^2) = c$
4. $(2x^2y^2 + y)\,dx - (x^3y - 3x)\,dy = 0$ **Ans.** $\frac{7}{5}x^{10/7}y^{-5/7} - \frac{7}{4}x^{-4/7}y^{-12/7} = c$
5. $x\,(3y\,dx + 2x\,dy) + 8y^4\,(y\,dx + 3x\,dy) = 0$ **Ans.** $x^3y^2 + 4x^2y^6 = c$

Rule V.

If the given equation $M\,dx + N\,dy = 0$ is homogeneous equation and $Mx + Ny \neq 0$, then $\frac{1}{Mx+Ny}$ is an integrating factor.

Example 39. *Solve* $\qquad \frac{dy}{dx}=\frac{x^3+y^3}{xy^2}$

Solution. $\qquad (x^3 + y^3)\,dx - (xy^2)\,dy = 0 \qquad \ldots (1)$

Here $\qquad M = x^3 + y^3, \qquad N = -xy^2$

$$\text{I.F.} = \frac{1}{Mx+Ny}=\frac{1}{x(x^3+y^3)-xy^2(y)}=\frac{1}{x^4}$$

Multiplying (1) by $\frac{1}{x^4}$ we get $\frac{1}{x^4}(x^3+y^3)dx+\frac{1}{x^4}(-xy^2)dy=0$

$\Rightarrow \qquad \left(\frac{1}{x}+\frac{y^3}{x^4}\right)dx-\frac{y^2}{x^3}dy=0$, which is an exact differential equation.

$$\int\left(\frac{1}{x}+\frac{y^3}{x^4}\right)dx=c \quad \Rightarrow \quad \log x-\frac{y^3}{3x^3}=c \qquad \textbf{Ans.}$$

EXERCISE 12.12

Solve the following differential equations:

1. $x^2y\,dx - (x^3 + y^3)\,dy = 0$ **Ans.** $-\frac{x^3}{3y^3}+\log y=c$
2. $(y^3 - 3xy^2)\,dx + (2x^2y - xy^2)\,dy = 0$ **Ans.** $\frac{y}{x}+3\log x-2\log y=c$
3. $(x^2y - 2xy^2)\,dx - (x^3 - 3x^2y)\,dy = 0$ **Ans.** $\frac{x}{y}-2\log x+3\log y=c$
4. $(y^3 - 2yx^2)\,dx + (2xy^2 - x^3)\,dy = 0$ **Ans.** $x^2y^4 - x^4y^2 = c$

12.14 DIFFERENTIAL EQUATIONS REDUCIBLE TO EXACT FORM (BY INSPECTION)

The following differentials, which commonly occur, help in selecting the suitable integrating factor.

(*i*) $y\,dx + x\,dy = d\,[xy]$

(*ii*) $\frac{x\,dy-y\,dx}{x^2}=d\left[\frac{y}{x}\right]$

(*iii*) $\frac{y\,dx-x\,dy}{y^2}=d\left[\frac{x}{y}\right]$

(*iv*) $\frac{x\,dy-y\,dx}{x^2+y^2}=d\left[\tan^{-1}\left(\frac{y}{x}\right)\right]$

(*v*) $\frac{x\,dy-y\,dx}{xy}=d\left[\log\left(\frac{y}{x}\right)\right]$

(*vi*) $\frac{x\,dx-y\,dy}{x^2+y^2}=d\left[\frac{1}{2}\log\left(x^2+y^2\right)\right]$

(*vii*) $\frac{x\,dy-y\,dx}{x^2-y^2}=d\left[\frac{1}{2}\log\frac{x+y}{x-y}\right]$

(*viii*) $\frac{x\,dy+y\,dx}{x^2y^2}=d\left[-\frac{1}{xy}\right]$

12.15 EQUATIONS OF FIRST ORDER AND HIGHER DEGREE

The differential equations will involve $\frac{dy}{dx}$ in higher degree and $\frac{dy}{dx}$ will be denoted by p. The differential equation will be of the form $f(x, y, p) = 0$.

Case 1. Equations solvable for p. $\left[P = \frac{dy}{dx}\right]$

Example 40. *Solve* : $x^2 = 1 + p^2$

Solution. $x^2 = 1 + p^2 \quad \Rightarrow \quad p^2 = x^2 - 1$

$$\Rightarrow \quad p = \pm\sqrt{x^2-1} \quad \Rightarrow \quad \frac{dy}{dx} = \pm\sqrt{x^2-1} \quad \Rightarrow \quad dy = \pm\sqrt{x^2-1}\,dx$$

which gives on integration $y = \pm\frac{x}{2}\sqrt{x^2-1} \mp \frac{1}{2}\log\left(x+\sqrt{x^2-1}\right) + c$ **Ans.**

Case II. Equations solvable for y.

(*i*) Differentiate the given equation w.r.t. "x".

(*ii*) Eliminate p from the given equation and the equation obtained as above.

(*iii*) The eliminant is the required solution.

Example 41. *Solve*: $y = (x - a)\,p - p^2$.

Solution. $y = (x - a)\,p - p^2$ ... (1)

Differentiating (1) w.r.t. "x" we obtain

$$\frac{dy}{dx} = p + (x-a)\frac{dp}{dx} - 2p\frac{dp}{dx}$$

$$p = p + (x-a)\frac{dp}{dx} - 2p\frac{dp}{dx}$$

$$\Rightarrow \quad 0 = (x-a)\frac{dp}{dx} - 2p\frac{dp}{dx}$$

$$\Rightarrow \quad 0 = \frac{dp}{dx}[x-a-2p] \quad \Rightarrow \quad \frac{dp}{dx} = 0$$

On integration, we get $p = c$.

Putting the value of p in (1), we get

$$y = (x - a)\,c - c^2$$ **Ans.**

Case III. Equations solvable for x

(*i*) Differentiate the given equation w.r.t. "y".

(*ii*) Solve the equation obtained as in (1) for p.

(*iii*) Eliminate p, by putting the value of p in the given equation.

(*iv*) The eliminant is the required solution.

Example 42. *Solve*: $y = 2px + yp^2$

Solution. $y = 2px + yp^2$ ... (1)

$$\Rightarrow \quad 2px = y - yp^2 \quad \Rightarrow \quad 2x = \frac{y}{p} - yp \qquad \text{... (2)}$$

Differentiating (2) w.r.t. "y" we get

$$2\frac{dx}{dy} = \frac{1}{p} - \frac{y}{p^2}\frac{dp}{dy} - p - y\frac{dp}{dy}$$

$$\Rightarrow \quad \frac{2}{p} = \frac{1}{p} - \frac{y}{p^2}\frac{dp}{dy} - p - y\frac{dp}{dy} \quad \Rightarrow \quad \frac{1}{p} + p = -\frac{y}{p^2}\frac{dp}{dy} - y\frac{dp}{dy} \qquad \left[P = \frac{dy}{dx}\right]$$

$$\Rightarrow \quad \frac{1}{p}+p=-y\left(\frac{1}{p^2}+1\right)\frac{dp}{dy} \quad \Rightarrow \quad \frac{1+p^2}{p}=-y\frac{1+p^2}{p^2}\frac{dp}{dy}$$

$$\Rightarrow \quad 1=-\frac{y}{p}\frac{dp}{dy} \quad \Rightarrow \quad -\frac{dy}{y}=\frac{dp}{p} \quad \Rightarrow \quad -\log y=\log p+\log c'$$

$$\Rightarrow \quad \log p\,y=\log c \Rightarrow \quad p\,y=c \quad \Rightarrow \quad p=\frac{c}{y} \qquad [\log c'=\log c]$$

Putting the value of p in (1), we get

$$y=2\left(\frac{c}{y}\right)x+y\left(\frac{c^2}{y^2}\right) \Rightarrow \quad y^2=2\,cx+c^2$$

$$\Rightarrow \quad y^2=c(2x+c)$$ **Ans.**

Class IV. Clairaut's Equation.

The equation $y = px + f(p)$ is known as Clairaut's equation. ... (1)

Differentiating (1) w.r.t. "x", we get

$$\frac{dy}{dx}=p+x\frac{dp}{dx}+f'(p)\frac{dp}{dx}$$

$$\Rightarrow \quad p=p+x\frac{dp}{dx}+f'(p)\frac{dp}{dx} \quad \Rightarrow \quad 0=x\frac{dp}{dx}+f'(p)\frac{dp}{dx}$$

$$\Rightarrow \quad [x+f'(p)]\frac{dp}{dx}=0 \quad \Rightarrow \quad \frac{dp}{dx}=0 \quad \Rightarrow \quad p=a \qquad \text{(constant)}$$

Putting the value of p in (1), we have

$$y = ax + f(a)$$

which is the required solution.

Method. In the Clairaut's equation, on replacing p by a (constant), we get the solution of the equation.

Example 43. *Solve* : $p = \log(p\,x - y)$

Solution. $p = \log(p\,x - y)$ or $e^p = p\,x - y$ or $y = p\,x - e^p$

Which is Clairaut's equation.

Hence its solution is $y = a\,x - e^a$ **Ans.**

EXERCISE 12.13

Solve the following differential equations:

1. $xp^2 + x = 2yp$ **Ans.** $2cy = c^2x^2 + 1$
2. $x(1 + p^2) = 1$ **Ans.** $y-c=\sqrt{(x-x^2)}-\tan^{-1}\sqrt{\frac{1-x}{x}}$
3. $x^2p^2 + xyp - 6y^2 = 0$ **Ans.** $y=\frac{c}{x^3}, y=c_1x^2$
4. $\frac{dy}{dx}-\frac{dx}{dy}=\frac{x}{y}-\frac{y}{x}$ **Ans.** $xy = c,\ x^2 - y^2 = c$
5. $y = px + p^3$ **Ans.** $y = ax + a^3$
6. $x^2(y - px) = yp^2$ **Ans.** $y^2 = cx^2 + c^2$

12.16 ORTHOGONAL TRAJECTORIES

Two families of curves are such that every curve of either family cuts each curve of the other family at right angles. They are called orthogonal trajectories of each other.

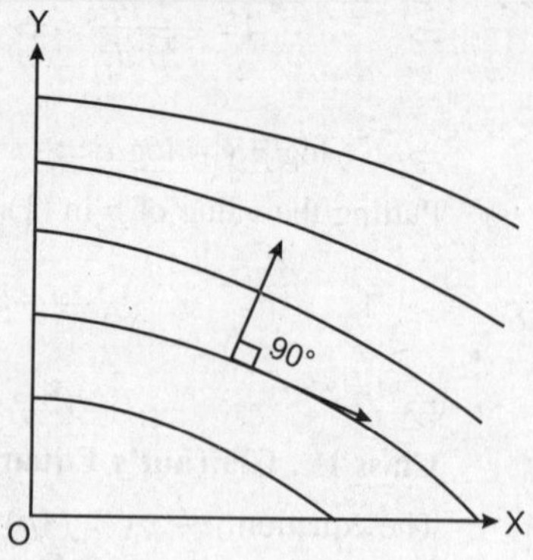

Orthogonal trajectories are very useful in engineering problems.

For example:

(*i*) The path of an electric field is perpendicular to equipotential curves.

(*ii*) In fluid flow, the stream lines and equipotential lines are orthogonal trajectories.

(*iii*) The lines of heat flow is perpendicular to isothermal curves.

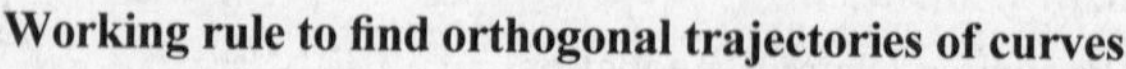

Working rule to find orthogonal trajectories of curves

Step 1. By differenciating the equaton of curves find the differential equations in the form $f\left(x, y, \dfrac{dy}{dx}\right) = 0$

Step 2. Replace $\dfrac{dy}{dx}$ by $-\dfrac{dx}{dy}$ $(M_1 . M_2 = -1)$

Step 3. Solve the differential equation of the orthogonal trajectories *i.e.*, $f\left(x, y - \dfrac{dx}{dy}\right) = 0$

Self-orthogonal. A given family of curves is said to be 'self-orthogonal' if the family of orthogonal trajectory is the same as the given family of curves.

Example 44. *Find the orthogonal trajectories of the family of curves $xy = c$.*

Solution. Here, we have

$$xy = c \quad \text{... (1)}$$

Differentiating (1), w.r.t., "x", we get

$$y + x\frac{dy}{dx} = 0 \quad \Rightarrow \quad \frac{dy}{dx} = -\frac{y}{x}$$

On replacing $\dfrac{dy}{dx}$ by $-\dfrac{dx}{dy}$, we get

$$\Rightarrow \quad -\frac{dx}{dy} = -\frac{y}{x} \quad \Rightarrow \quad \frac{dy}{dx} = \frac{x}{y}$$

$$y\,dy = x\,dx \quad \text{... (2)}$$

Integrating (2), we get $\dfrac{y^2}{2} = \dfrac{x^2}{2} + c$

$\Rightarrow \quad y^2 - x^2 = 2c$ **Ans.**

Example 45. *Find the orthogonal trajectories of $x^p + cy^p = 1$, p = constant.*

Solution. Here we have $x^p + cy^p = 1$ ... (1)

Differentiating (1), we get $px^{p-1} + pcy^{p-1}\dfrac{dy}{dx} = 0 \quad \Rightarrow \quad \dfrac{dy}{dx} = -\dfrac{x^{p-1}}{cy^{p-1}}$ ... (2)

Putting the value of $\dfrac{1}{c}$ in (2), we get $\dfrac{dy}{dx} = -\dfrac{x^{p-1}}{y^{p-1}}\dfrac{y^p}{1-x^p} \quad \Rightarrow \quad \dfrac{dy}{dx} = -\dfrac{x^{p-1}y}{1-x^p}$ ... (3)

Replacing $\frac{dy}{dx}$ by $-\frac{dx}{dy}$ for orthogonal trajectory in (3), we get

$$-\frac{dx}{dy}=-\frac{x^{p-1}y}{1-x^p} \quad \Rightarrow \quad \frac{dy}{dx}=\frac{1-x^p}{x^{p-1}y} \qquad \text{... (4)}$$

$$\Rightarrow \quad y\,dy=\frac{1-x^p}{x^{p-1}}dx \quad \Rightarrow \quad \int y\,dy=\int x^{1-p}dx-\int x\,dx$$

$$\frac{y^2}{2}=\frac{x^{1-p+1}}{1-p+1}-\frac{x^2}{2}+c \quad \Rightarrow \quad y^2=\frac{2x^{2-p}}{2-p}-x^2+2c \qquad \textbf{Ans.}$$

Putting $p = 2$ in (4), we get $\frac{dy}{dx}=\frac{1-x^2}{xy} \quad \Rightarrow \quad y\,dy=\frac{1-x^2}{x}dx$

$$y\,dy=\left(\frac{1}{x}-x\right)dx \quad \Rightarrow \quad \frac{y^2}{2}=\log x-\frac{x^2}{2}+\log c$$

$$\log x+\log c=\frac{x^2+y^2}{2} \quad \Rightarrow \quad 2\log x+2\log c=x^2+y^2$$

$$x^2c_2=e^{x^2+y^2}$$

$$c_1x^2=e^{x^2+y^2} \qquad \textbf{Ans.}$$

$$(c_1=c_2)$$

Example 46. *Show that the family of parabolas $y^2 = 2cx + c^2$ is "self-orthogonal."*

Solution. Here we have

$$y^2 = 2cx + c^2 \qquad \text{... (1)}$$

Differentiating (1), we get $2y\frac{dy}{dx}=2c \quad \Rightarrow \quad c=y\frac{dy}{dx}$

Putting the value of c in (1), we have $y^2=2\left(y\frac{dy}{dx}\right)x+\left(y\frac{dy}{dx}\right)^2$ ... (2)

Putting $\frac{dy}{dx}=p$ in (2), we get

$$y^2 = 2ypx + y^2p^2 \qquad \text{... (3)}$$

This is differential equation of give n family of parabolas.

For orthogonal trajectories we put $-\frac{1}{p}$ for p in (3)

$$y^2=2y\left(-\frac{1}{p}\right)x+y^2\left(-\frac{1}{p}\right)^2 x \quad \Rightarrow \quad y^2=-\frac{2yx}{p}+\frac{y^2}{p^2}$$

$$\Rightarrow \qquad y^2p^2 = -2pyx + y^2$$

Rewriting, we get

$$\Rightarrow \qquad y^2 = 2ypx + y^2p^2$$

Which is same as equation (3). Thus (2) is D.E. for the given family and its orthogonal trajectories.

Hence, the given family is self-orthogonal. **Proved.**

Example 47. *Show that the system of confocal conics*

$$\frac{x^2}{a^2+\lambda}+\frac{y^2}{b^2+\lambda}=1$$

Where λ is a parameter, is self orthogonal.

Solution. Here we have $\dfrac{x^2}{a^2+\lambda}+\dfrac{y^2}{b^2+\lambda}=1$ ... (1)

Differentiating (1), we get

$$\frac{2x}{a^2+\lambda}+\frac{2y}{b^2+\lambda}\frac{dy}{dx}=0 \qquad \left(\text{Put } \frac{dy}{dx}=p\right)$$

$$\Rightarrow \quad \frac{x}{a^2+\lambda}+\frac{yp}{b^2+\lambda}=0$$

$$\Rightarrow \quad x(b^2+\lambda)+py(a^2+\lambda)=0 \quad \Rightarrow \quad \lambda(x+py)=-b^2x-a^2yp$$

$$\Rightarrow \quad \lambda=\frac{-(b^2x+a^2yp)}{x+py}$$

Now $\quad a^2+\lambda=a^2-\dfrac{b^2x+a^2yp}{x+py}=\dfrac{a^2x+a^2py-b^2x-a^2yp}{x+py}=\dfrac{(a^2-b^2)x}{x+py}$

Again $\quad b^2+\lambda=b^2-\dfrac{b^2x+a^2yp}{x+py}=\dfrac{b^2x+b^2py-b^2x-a^2yp}{x+py}=\dfrac{-(a^2-b^2)yp}{x+py}$

Eliminating λ by putting the value of $a^2+\lambda$ and $b^2+\lambda$ in (1), we get

$$\frac{x^2(x+py)}{(a^2-b^2)x}+\frac{y^2(x+py)}{-(a^2-b^2)yp}=1 \quad \Rightarrow \quad \frac{x(x+py)}{(a^2-b^2)}-\frac{y(x+py)}{(a^2-b^2)p}=1$$

$$\frac{x+py}{a^2-b^2}\left[x-\frac{y}{p}\right]=1 \quad \Rightarrow \quad \frac{(x+py)\left(x-\dfrac{y}{p}\right)}{a^2-b^2}=1 \quad \Rightarrow \quad (x+py)\left(x-\frac{y}{p}\right)=a^2-b^2 \qquad \text{... (2)}$$

Equation (2) is the differential equation of (1),

To get the differential equation of orthogonal trajectory

Replace p by $-\dfrac{1}{p}$ in (2) $\left(x-\dfrac{1}{p}y\right)(x+py)=a^2-b^2$ ...(3)

Equation (3) is the same as eq. (2).

Thus the differential equation of the family of the orthogonal trajectory is the same as the differential equation of the family of the given curves.

Hence it is a self orthogonal family of curves. **Ans.**

EXERCISE 12.14

Find the orthogonal trajectories of the following family of curves:

1. $y^2=cx^3$ **Ans.** $(x+1)^2+y^2=a^2$
2. $x^2-y^2=cx$ **Ans.** $y(y^2+3x^2)=c$
3. $x^2-y^2=c$ **Ans.** $xy=c$
4. $(a+x)y^2=x^2(3a-x)$ **Ans.** $(x^2+y^2)^5=cy^3(5x^2+y^2)$
5. $y=ce^{-2x}+3x$, passing through the point (0, 3) **Ans.** $9x-3y+5=-4e^{6(3-y)}$

6. $16x^2 + y^2 = c$ **Ans.** $y^{16} = kx$

7. $y = \tan x + c$ **Ans.** $2x + 4y + \sin 2x = k$

8. $y = ax^2$ **Ans.** $x^2 + 2y^2 = c$

9. $x^2 + (y - c)^2 = c^2$ **Ans.** $x^2 + y^2 = cx$

10. $x^2 + y^2 + 2gx + 2fy + c = 0$ **Ans.** $x^2 + y^2 + 2fy - c = 0$

11. Family of parabolas through origin and focii on y-axis.

12. Show that the system of rectangular hyperbola $x^2 - y^2 = c^2$ and $xy = c^2$ are mutually orthogonal trajectories.

13. Show that the family of curves $y^2 = 4c(c + x)$ is self orthogonal.

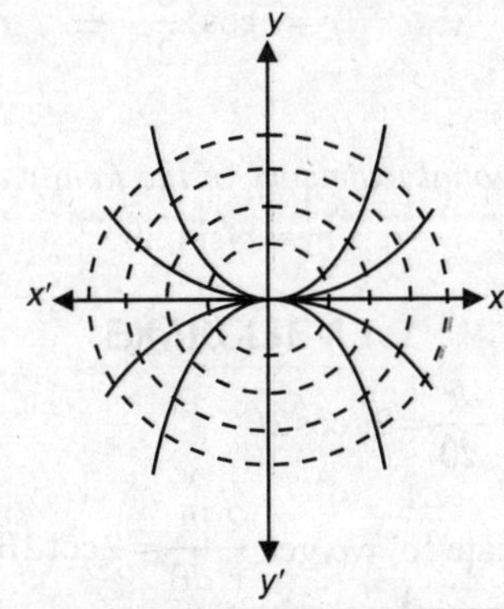

12.17 POLAR EQUATION OF THE FAMILY OF CURVES

Let the polar equation of the family of curves be $f(r, \theta, c) = 0$ ... (1)

Working Rule

Step 1. On differentiating and eliminating the arbitrary constant c between (1) and $f'(r, \theta, c) = 0$ we get the differential equation of (1) *i.e.,*

$$F\left(r, \theta, \frac{dr}{d\theta}\right) = 0 \quad \text{... (2)}$$

Step 2. Replace $\frac{dr}{d\theta}$ by $-r^2 \frac{d\theta}{dr}$ in (2). Here we will get the differential equation of orthogonal trajectory *i.e.,*

$$F\left(r, \theta - r^2 \frac{d\theta}{dr}\right) = 0 \quad \text{... (3)}$$

Step 3. Integrating (3) to get the equation of the orthogonal trajectory.

Example 48. *Find the orthogonal trajectory of the cardioids $r = a(1 - \cos\theta)$.*

Solution. We have, $r = a(1 - \cos\theta)$ ... (1)

Differentiating (1) w.r.t. θ, we get $\frac{dr}{d\theta} = a\sin\theta$ ... (2)

Dividing (2) by (1) to eliminate a, we get

$$\frac{1}{r}\frac{dr}{d\theta} = \frac{\sin\theta}{1-\cos\theta} = \frac{2\sin\frac{\theta}{2}\cos\frac{\theta}{2}}{1-1+2\sin^2\frac{\theta}{2}} = \cot\frac{\theta}{2} \quad \text{... (3)}$$

which is the differential equation of (1).

Replacing $\frac{dr}{d\theta}$ by $-r^2\frac{d\theta}{dr}$ in (3), we get $\frac{1}{r}\left(-r^2\frac{d\theta}{dr}\right)=\cot\frac{\theta}{2}$

$$r\frac{d\theta}{dr}=-\cot\frac{\theta}{2}$$

Separating the variables we get $\frac{dr}{r}=-\tan\frac{\theta}{2}d\theta$ ... (4)

Integrating (4), we get $\log r=2\log\cos\frac{\theta}{2}+\log c \;=\log c\cos^2\frac{\theta}{2}$

$\Rightarrow$ $$r=c\cos^2\frac{\theta}{2} \quad\Rightarrow\quad r=\frac{c}{2}(1+\cos\theta)$$

Which is the required trajectory. **Ans.**

Example 49. *Find the orthogonal trajectory of the family of curves*

$$r^2 = c\sin 2\theta$$

Solution. We have

$$r^2 = c\sin 2\theta \qquad \text{... (1)}$$

Differentiating (1), we get $2r\frac{dr}{d\theta}=2c\cos 2\theta$ ... (2)

Dividing (2) by (1), to eliminate 'c' we get $\frac{2}{r}\frac{dr}{d\theta}=2\cot 2\theta$ ... (3)

Replacing $\frac{dr}{d\theta}$ by $-r^2\frac{d\theta}{dr}$ in (3), we have $\frac{2}{r}\left(-r^2\frac{d\theta}{dr}\right)=2\cot 2\theta$

$$-2r\frac{d\theta}{dr}=2\cot 2\theta \qquad \text{... (4)}$$

Separating the variables of (4), we obtain $\frac{dr}{r}=-\tan 2\theta\, d\theta$ ... (5)

Integrating (5), we get $\log r=\frac{1}{2}\log\cos 2\theta+\log c$

$$2\log r = \log c\cos\theta$$

$$r^2 = c\cos 2\theta$$

which is the required trajectory **Ans.**

Example 50. *Find the orthogonal trajectory of the family of curves*

$$r = c(\sec\theta + \tan\theta)$$

Solution. We have $r = c(\sec\theta + \tan\theta)$ ... (1)

Differentiating (1) w.r.t. 'θ' we get $\frac{dr}{d\theta}=c(\sec\theta\tan\theta+\sec^2\theta)$ ...(2)

$$\frac{dr}{d\theta}=c\sec\theta\,(\tan\theta+\sec\theta)$$

Dividing (2) by (1), we get $\frac{1}{r}\frac{dr}{d\theta}=\sec\theta$ ... (3)

Separating the variables of (3), we have $\frac{1}{r}\frac{dr}{d\theta} = \sec\theta$ ... (4)

Replacing $\frac{dr}{d\theta}$ by $-r^2\frac{d\theta}{dr}$, we obtain

$$\frac{1}{r}\left(-r^2\frac{d\theta}{dr}\right) = \sec\theta \quad \Rightarrow \quad -r\frac{d\theta}{dr} = \sec\theta \qquad \text{... (5)}$$

Separating the variables of (5), we obtain $\frac{dr}{r} = -\cos\theta\, d\theta$ ... (6)

Integrating (6), we get $\log r = -\sin\theta + c \quad \Rightarrow \quad r = c'e^{-\sin\theta}$

which is the required orthogonal trajectory. **Ans.**

EXERCISE 12.15

Find the orthogonal trajectory of the following families of the curves:

1. $r = ce^{\theta}$ **Ans.** $r = ke^{-\theta}$

2. $r = c\theta^2$ **Ans.** $r = ke^{-\frac{\theta^2}{4}}$

3. $r = a(1 + \cos\theta)$ **Ans.** $r = a(1 - \cos\theta)$

4. $r^n \sin n\theta = a^n$ **Ans.** $r^n \sin n\theta = c^n$

5. $r = a\cos^2\theta$ **Ans.** $r^2 = a\sin\theta$

6. $r = 2a(\sin\theta + \cos\theta)$ **Ans.** $r = 2c(\sin\theta - \cos\theta)$

7. $r = c(1 + \sin^2\theta)$ **Ans.** $r^2 = k\cos\theta \cdot \cot\theta$

8. $r = \frac{a}{1+2\cos\theta}$ **Ans.** $r^2 \sin^3\theta = (1 + \cos\theta)$

LINEAR DIFFERENTIAL EQUATIONS OF SECOND ORDER

13.1 LINEAR DIFFERENTIAL EQUATIONS

If the degree of the dependent variable and all derivatives is one, such differential equations are called *linear differential equations e.g.*

(1) $\frac{d^2y}{dx^2}+5\frac{dy}{dx}+6y=x^2+x+1$ (2) $2\frac{d^2x}{dt^2}-\frac{dx}{dt}-3x=f(t)$

13.2 NON LINEAR DIFFERENTIAL EQUATIONS

If the degree of the dependent variable and / or its derivatives are of greater than 1 such differential equations are called non-linear differential equations.

(1) $\frac{d^2y}{dx^2}+\frac{dy}{dx}+y^2=\sin x$ (2) $\frac{d^2y}{dx^2}+2\left(\frac{dy}{dx}\right)^2+y^2=e^x$ (3) $\left(\frac{d^2x}{dt^2}\right)^2+\frac{dx}{dt}+x=f(t)$

The order of a differential equation is the highest order of the derivative involved. All the above differential equations are of second order.

Fourier and Laplace transforms are mathematical tools to solve the differential equations.

13.3 LINEAR DIFFERENTIAL EQUATIONS OF SECOND ORDER WITH CONSTANT COEFFICIENTS

The general form of the linear differential equation of second order is

$$\frac{d^2y}{dx^2}+P\frac{dy}{dx}+Qy=R$$

where P and Q are constants and R is a function of x or constant.

Differential operator. Symbol D stands for the operation of differential *i.e.*,

$$Dy=\frac{dy}{dx},\ D^2y=\frac{d^2y}{dx^2}$$

$\frac{1}{D}$ stands for the operation of integration.

$\frac{1}{D^2}$ stands for the operation of integration twice.

$$\frac{d^2y}{dx^2}+P\frac{dy}{dx}+Qy=R$$ can be written in the operator form.

$$D^2y + P\,Dy + Qy = R \qquad \Rightarrow \qquad (D^2 + PD + Q)\,y = R$$

13.4 DIMENSION OF SPACE OF SOLUTION

A differential equation is said to have dimension k if the differential equation has k linearly independent solutions, $y_1, y_2 \,......\, y_k$,

For example; $y''-5y'+6y=0$ has two dimensional solution e^{2x}, e^{3x}

$y'''-6y''+11y'+6y=0$ has three dimensional independent solution e^x, e^{2x}, e^{3x}.

13.5 NON-HOMOGENEOUS

Consider the differential equation

$$y''+P(x)y'+Q(x)y=F(x) \qquad ...(1)$$

(1) is said to be non-homogeneous if R.H.S. of (1) *i.e.*, $F(x) \neq 0$

13.6 HOMOGENEOUS

When $\qquad F(x) = 0$

Then (1) is said to be complementary homogeneous linear differential equation of (1).

13.7 SUPER POSITION OR LINEARITY PRINCIPLE

OR

FUNDAMENTAL THEOREM FOR HOMOGENEOUS DIFFERENTIAL EQUATION

Theorem. If $y_1(x)$ and $y_2(x)$ be two linearly independent solutions of the homogeneous differential equation

$$y'' + Py' + Qy = 0 \qquad ...(1)$$

then the general solution of (1) is given by

General solution $= y = A\,y_1 + B\,y_2$.

Where A and B are arbitrary constants.

Proof. Here we have $\qquad y = Ay_1 + By_2 \qquad ...(2)$

Differentiating (2), we get $\qquad y' = Ay_1' + By_2' \qquad ...(3)$

$$y'' = Ay_1''+ By_2'' \qquad ...(4)$$

On putting the values of y, y', y'' from (2), (3) and (4) in (1), we get

$$Ay_1''+ By_2''+ P(Ay_1' + By_2')+Q\,[Ay_1 + By_2] = 0$$

$$\Rightarrow \quad [Ay_1''+ PAy_1' + QAy_1]+[By_2''+ P.B.y_2' + Q.B.y_2] = 0$$

$$\Rightarrow \quad A\,[y_1''+ Py_1' + Q\,y_1]+ B[y_2''+ Py_2' + Qy_2] = 0 \qquad ...(5)$$

Since y_1 and y_2 are independent solution of (1), *i.e.*,

$$y_1''+ Py_1' + Qy_1 = 0 \qquad \text{[Form (3)]}$$

and $\qquad y_2''+ Py_2' + Qy_2 = 0 \qquad$ [From (4)]

Then (5) becomes $A(0) + B(0) = 0$

Thus (2) is the general solution of homogeneous differential equation (1).

13.8 LINEAR INDEPENDENCE AND DEPENDENCE

Two solutions $y_1(x)$ and $y_2(x)$ are said to be linearly independent if

$$Ay_1(x) + By_2(x) \neq 0$$

A and B are not equal to zero.

13.9 WRONSKIAN

We know that

$$W(y_1, y_2, x) = \begin{vmatrix} y_1(x) & y_2(x) \\ y_1'(x) & y_2'(x) \end{vmatrix} = y_1(x)y_2'(x) - y_1'(x)y_2(x)$$

(1) If $W(y_1, y_2, x) = 0$, then $y_1(x)$ and $y_2(x)$ are linearly dependent.

(2) If $W(y_1, y_2, x) \neq 0$, then $y_1(x)$, $y_2(x)$ are linearly independent.

13.10 EXISTENCE OF LINEARLY INDEPENDENCE

Consider the equation

$$y'' + P(x)y' + P(x)y = 0 \qquad \text{...(1)}$$

By the fundamental theorem we can say that (1) has two solutions $y_1(x)$ and $y_2(x)$

By Abel formula, these two solutions are independent.

The initial conditions are

$$y_1(x_0) = 1,\ y_1'(x_0) = 0$$

$$y_2(x_0) = 0,\ y_2'(x_0) = 1$$

$$W(y_1, y_2, x_0) = \begin{vmatrix} y_1(x_0) & y_2(x_0) \\ y_1'(x_0) & y_2'(x_0) \end{vmatrix} = \begin{vmatrix} 1 & 0 \\ 0 & 1 \end{vmatrix} = 1$$

Since $W(y_1, y_2, x_0) = 1$, the above pair is independent.

Example 1. *Check whether the following functions are linearly independent or not:* $e^x \cos x$, $e^x \sin x$. *(Delhi University, 2010)*

Solution. We have,

$$y_1 = e^x \cos x,\ y_2 = e^x \sin x$$

$$\Rightarrow \qquad y'_1 = e^x \cos x - e^x \sin x \text{ and } y'_2 = e^x \sin x + e^x \cos x$$

$$\text{W}(y_1, y_2, x) = \begin{vmatrix} y_1 & y_2 \\ y'_1 & y'_2 \end{vmatrix} = \begin{vmatrix} e^x \cos x & e^x \sin x \\ e^x \cos x - e^x \sin x & e^x \sin x + e^x \cos x \end{vmatrix}$$

$$= e^{2x} \begin{vmatrix} \cos x & \sin x \\ \cos x - \sin x & \sin x + \cos x \end{vmatrix}$$

$$= e^{2x} (\sin x \cdot \cos x + \cos^2 x - \sin x \cdot \cos x + \sin^2 x)$$

$$= e^{2x} \neq 0$$

Hence, $e^x \cos x$ and $e^x \sin x$ are linearly independent. **Ans.**

13.11 STRUCTURE THEOREM FOR

$$y'' + P(x)y' + Q(x)y = 0 \qquad ...(1)$$

We have seen that differential equation (1) has two solutions $y_1(x)$ and $y_2(x)$, secondly these solutions are linearly independent.

If third solution $y(x)$ exists then two unique constants A and B exist such that

$$y(x) = Ay_1(x) + By_2(x)$$

holds good.

For verification, we say that $y(x)$ be the solution of (1) and $c_0 = y(x_0)$ and $c_1 = y'(x_0)$

$$A_1y_1(x_0) + B_1y_2(x_0) = c_0 \qquad ...(2)$$

$$A_1y_1'(x_0) + B_1y_2'(x_0) = c_1 \qquad ...(3)$$

(2) and (3) is a pair of linear equation for A_1, B_1.

This pair of equations has unique solution.

Now $y(x) = A_1 y_1 + B_1 y_2(x)$ ...(4)

Equation (4) satisfy

$$A_1 \text{ and } B_1 \text{ satisfy (4)}$$

13.12 SUPERPOSITION PRINCIPLE DOES NOT HOLD GOOD FOR A NON HOMOGENEOUS DIFFERENTIAL EQUATION OR NONLINEAR EQUATION

Test for Dependence (WRONSKIAN TEST). If $y_1(x)$ and $y_2(x)$ are two linearly dependent solution, then

$$A\,y_1(x) + B\,y_2(x) = 0$$

$$y_1(x)y_2''(x) - y_1'(x)y_2(x) = 0$$

$$\Rightarrow \quad y_1(x)y_2''(x) - y_1'(x)y_2(x) = 0$$

$$\begin{vmatrix} y_1(x) & y_2(x) \\ y_1'(x) & y_2'(x) \end{vmatrix} = 0$$

$$W(y_1, y_2, x) = 0$$

Conversely. If $W(y_1, y_2, x) = 0$ then $y_1(x)$ and $y_2(x)$ are linearly dependent.

$$W(y_1, y_2, x) = 0$$

$$\Rightarrow \quad y_1(x)y_2'(x) - y_1'(x)\,y_2(x) = 0$$

$$\Rightarrow \quad A\,y_1(x) + B\,y_2(x) = 0$$

$$\Rightarrow \quad y_1(x) = C\,y_2(x)$$

$y_1(x)$ and $y_2(x)$ are linearly dependent.

ABELS–Liouville formula

If $y_1(x)$ and $y_2(x)$ are two linearly independent solutions of

$$y'' + P(x)\,y' + Q(x)y = 0, \text{ then} \qquad ...(1)$$

$$W(y_1, y_2, x) = W(y_1, y_2, x_0)\, e^{-\int_{x_1}^{x_2} a(x)\,dx} \qquad ...(2)$$

Now we can say that

$$W(y_1, y_2, x) = 0 \text{ if and only if } W(y_1, y_2, x_0) = 0$$

Thus linear dependence can be checked by finding the value of $W(y_1, y_2, x.)$

Verification.

We know that $W(y_1, y_2, x) = \begin{vmatrix} y_1(x) & y_2(x) \\ y_1'(x) & y_2'(x) \end{vmatrix}$

$$= y_1(x)\,y_2'(x) - y_2(x)\,y_1'(x) \qquad \text{...(3)}$$

Differentiating (3), we get

$$\frac{dW(y_1, y_2, x)}{dx} = y_1(x)y_2''(x) + y_1'(x)y_2'(x) - y_2(x)y_1''(x) - y_2'(x)y_1'(x)$$

$$= y_1(x)y_2''(x) - y_2(x)y_1''(x) = -a(x)\,(y_1y_2' - y_1'y_2)$$

$$\frac{dW(y_1, y_2, x)}{dx} = -a(x)\,W(y_1, y_2, x) \qquad \text{...(4)}$$

$$\Rightarrow \qquad \frac{d}{dx}W(y_1, y_2, x) + a(x)\,W(y_1, y_2, x) = 0$$

Equation (4) is a linear differential equation of first order

$$\text{Its I.F.} = e^{\int_{x_0}^{x} a(x).dx}$$

Hence, the solution of (4) is

$$W(y_1, y_2, x)\,e^{\int a\,x dx} = \text{constant}$$

$$W(y_1, y_2, x)\,e^{\int a(x)dx} = W[y_1, y_2, x_0]$$

$$[W(y_1, y_2, x_0) = 1]$$

$$W(y_1, y_2, x) = W(y_1, y_2, x_0)\; e^{-\int_{x_0}^{x} a(x)dx}$$ **Verified.**

Example 2. (*a*) *Find Wronskian determinant*

(*b*) *Verify that the solutions satisfy the differential equation*

(*c*) *Show by Wronskian test the solutions are independent*

if $y_1 = e^{2x}$, $y_2 = e^{3x}$ and $y'' - 5y' + 6y = 0$

Solution (*a*) $y_1 = e^{2x}$, $y_2 = e^{3x}$ and $y_1' = 2e^{2x}$, $y_2' = 3e^{3x}$

$$W(y_1, y_2, x) = \begin{vmatrix} y_1 & y_2 \\ y_1' & y_2' \end{vmatrix} = \begin{vmatrix} e^{2x} & e^{3x} \\ 2e^{2x} & 3e^{3x} \end{vmatrix} = 3e^{5x} - 2e^{5x} = e^{5x} \qquad \text{...(1)}$$

(*b*) Now $\quad y'' - 5y' + 6y = 0 \qquad$ ...(2)

If $y_1 = e^{2x}$, then (2) becomes $\quad 4e^{2x} - 5(2e^{2x}) + 6e^{2x} = 0$

$\Rightarrow \quad (4 - 10 + 6)\,e^{2x} = 0 \Rightarrow 0 = 0$

If $y_2 = e^{3x}$ then (2) becomes $9e^{3x} - 5(3e^{3x}) + 6e^{3x} = 0$

$\Rightarrow \quad (9 - 15 + 6)\,e^{3x} = 0 \Rightarrow 0 = 0$

Thus, y_1 and y_2 satisfy the differential equation (2)

$$W(y_1, y_2, x) = e^5 \neq 0$$

Hence y_1 and y_2 are linearly independent. **Ans.**

Example 3. *If* $y_1 = e^{-x}\cos x$, $y_2 = e^{-x}\sin x$ and $\frac{d^2y}{dx^2}+\frac{2dy}{dx}+2y=0$, *then*

(a) Calculate Wronskian determinant

(b) Verify that y_1 *and* y_2 *satisfy the given differential equation.*

(c) Apply wronskian test to check that y_1, y_2 *are linearly independent.*

Solution. (*a*) $y_1 = e^{-x}\cos x, \qquad y_2 = e^{-x}\sin x$

$$y_1' = -e^{-x}\cos x - e^{-x}\sin x, \qquad y_2' = -e^{-x}\sin x + e^{-x}\cos x$$

$$W(y_1, y_2, x) = \begin{vmatrix} y_1 & y_2 \\ y_1' & y_2' \end{vmatrix} = \begin{vmatrix} e^{-x}\cos x & e^{-x}\sin x \\ -e^{-x}\cos x - e^{-x}\sin x & -e^{-x}\sin x + e^{-x}\cos x \end{vmatrix}$$

$$= e^{-x}\cos x\,.(-e^{-x}\sin x + e^{-x}\cos x) - e^{-x}\sin x\,(-e^{-x}\cos x - e^{-x}\sin x)$$

$$= e^{-2x}(-\sin x\cos x + \cos^2 x) + e^{-2x}(\sin x\cos x + \sin^2 x)$$

$$= e^{-2x}(\sin^2 x + \cos^2 x) \qquad \text{...(1)}$$

$$= e^{-2x}$$

(*b*) $y''(x) + 2y'(x) + 2y = 0$ ...(2)

$$y_1 = e^{-x}\cos x$$

Patting the values y'', y' of and y in (2), we get

$$\text{L.H.S. of (2)} = e^{-x}(2\sin x) - 2e^{-x}(\cos x + \sin x) + 2e^{-x}\cos x$$

$$= e^{-x}(2\sin x - 2\cos x - 2\sin x + 2\cos x) = 0$$

Thus $\qquad y_1 = e^{-x}\cos x$ satisfies (2).

Similarly $\qquad y_2 = e^{-x}\sin x$ satisfies (2).

(*c*) $\qquad W(y_1, y_2, x) = e^{-x} \neq 0$

Hence y_1 and y_2 are linearly independent. **Ans.**

13.13 ABEL'S FORMULA

If $y_1, \ldots\ldots, y_n$ are n solutions of $Y' = AY$

then $W(y_1, y_2, \ldots\ldots, y_n, x) = W(y_1, y_2, \ldots\ldots, y_n, x_0)\, e^{\int_{x_0}^{x} \mathrm{tr}\,A(t)\,dt}$ for every $x \in I$, where

$$\mathrm{tr}\,A(t) = \sum_{i=1}^{n} a_{ii}(t).$$

Example 4. *Consider the linear system*

$$y_1' = -\frac{2}{x}y_1 + \frac{1}{x^2}y_2 \qquad \text{... (1)}$$

$$y_2' = -y_1 - \frac{3}{x}y_2, \qquad \text{... (2)}$$

over $I = \{x : x > 0\}$. It may be verified that

$$Y_1 = \begin{pmatrix} \frac{1}{x^3} \\ -\frac{1}{x^2} \end{pmatrix} \quad \text{and} \quad Y_2 = \begin{pmatrix} \frac{\log x}{x^3} \\ \frac{1-\log x}{x^2} \end{pmatrix}$$

are solutions of the above equation.

Verification for equation (1).

$$Y_1 = \begin{pmatrix} y_1 \\ y_2 \end{pmatrix} \text{ where } y_1 = \frac{1}{x^3} \text{ and } y_2 = -\frac{1}{x^2}$$

$$y_1' = -\frac{3}{x^4}$$

On putting the value of y_1, y_2 and y_1' in (1), we get

$$-\frac{3}{x^4} = -\frac{2}{x}\left(\frac{1}{x^3}\right) + \frac{1}{x^2}\left(-\frac{1}{x^2}\right)$$

$$\Rightarrow \quad -\frac{3}{x^4} = -\frac{2}{x^4} - \frac{1}{x^4} \Rightarrow \quad -\frac{3}{x^4} = -\frac{3}{x^4}, \text{ which is true.}$$

Verification for equation (2).

$$Y_2 = \begin{pmatrix} y_1 \\ y_2 \end{pmatrix} \quad \text{where, } y_1 = \frac{\log x}{x^3} \text{ and } y_2 = \frac{1-\log x}{x^2}$$

$$y_2' = -\frac{2}{x^3} + \frac{2\log x}{x^3} - \frac{1}{x^3} = -\frac{3}{x^3} + \frac{2\log x}{x^3}$$

On putting the value of y_1, y_2 and y_2' in (2), we get

$$-\frac{3}{x^3} + \frac{2\log x}{x^3} = -\frac{\log x}{x^3} - \frac{3}{x}\left(\frac{1-\log x}{x^2}\right)$$

$$\Rightarrow \quad -\frac{3}{x^3} + \frac{2\log x}{x^3} = -\frac{\log x}{x^3} - \frac{3}{x^3} + \frac{3\log x}{x^3}$$

$$\Rightarrow \quad -\frac{3}{x^3} + \frac{2\log x}{x^3} = -\frac{2\log x}{x^3} - \frac{3}{x^3}, \text{ which is true.}$$

Its Wronskian is $W(Y_1, Y_2) = \begin{vmatrix} \frac{1}{x^3} & \frac{\log x}{x^3} \\ -\frac{1}{x^2} & \frac{1-\log x}{x^2} \end{vmatrix} = \frac{1}{x^5} - \frac{\log x}{x^5} + \frac{\log x}{x^5} = \frac{1}{x^5} \neq 0,$

for any $x > 0$. Thus (Y_1, Y_2) is a fundamental matrix and hence a general solution of the system is given by

$$y_1 = \frac{c_1}{x^3} + \frac{c_2 \log x}{x^3}$$

$$y_2 = -\frac{c_1}{x^2} + \frac{c_2(1-\log x)}{x^2}$$ **Ans.**

Example 5. *Consider the system*

$$y_1' = -\frac{2}{x}y_1 + \frac{1}{x^2}y_2 + x$$

$$y_2' = -y_1 - \frac{3}{x}y_2 + x^2 \qquad \text{... (1)}$$

$$Y_1 = \begin{bmatrix} \frac{1}{x^3} \\ -\frac{1}{x^2} \end{bmatrix} \text{ and } Y_2 = \begin{bmatrix} \frac{\log x}{x^3} \\ \frac{1-\log x}{x^2} \end{bmatrix}$$

over $I=\{x \mid x>0\}$. As seen in Example 4, a fundamental matrix for the above system is given.

$$W(y_1, y_2, x) = \begin{pmatrix} \frac{1}{x^3} & \frac{\log x}{x^3} \\ -\frac{1}{x^2} & \frac{1-\log x}{x^2} \end{pmatrix} = \frac{1}{x^5}(1-\log x) + \frac{1}{x^5}\log(x) = \frac{1}{x^5}$$

and $|W(y_1, y_2, x)| = \frac{1}{x^5}$. A general solution of the corresponding homogeneous system is

$$Y = c_1 \begin{pmatrix} \frac{1}{x^3} \\ -\frac{1}{x^2} \end{pmatrix} + c_2 \begin{pmatrix} \frac{\log x}{x^3} \\ \frac{1-\log x}{x^2} \end{pmatrix},$$

so that a solution of (1) is of the form

$$Y = u_1(x) \begin{pmatrix} \frac{1}{x^3} \\ -\frac{1}{x^2} \end{pmatrix} + u_2(x) \begin{pmatrix} \frac{\log x}{x^3} \\ \frac{1-\log x}{x^2} \end{pmatrix},$$

Substitution in (1) gives

$$\frac{1}{x^3}u_1'(x) + \frac{\log x}{x^3}u_2'(x) = x$$

$$-\frac{1}{x^2}u_1'(x) + \frac{1-\ln x}{x^3}u_2'(x) = x^2.$$

Solving for $u_1'(x)$ and $u_2'(x)$, we get

$$u_1'(x) = \frac{\begin{vmatrix} x & \frac{\ln x}{x^3} \\ x^2 & \frac{1-\ln x}{x^2} \end{vmatrix}}{1/x^5} = x^4(1-2\ln x), \quad u_2'(x) = \frac{\begin{vmatrix} \frac{1}{x^3} & x \\ -\frac{1}{x^2} & x^2 \end{vmatrix}}{1/x^5} = 2x^4.$$

$$u_1(x) = \frac{x^5}{5}\left(\frac{7}{5} - 2\ln x\right) \text{ and } u_2(x) = \frac{2}{5}x^5.$$

Thus a general solution of (1) is given by

$$Y = c_1 \begin{pmatrix} \frac{1}{x^3} \\ -\frac{1}{x^2} \end{pmatrix} + c_2 \begin{pmatrix} \frac{\ln x}{x^3} \\ \frac{1-\ln x}{x^2} \end{pmatrix} + \frac{x^5}{5}\left(\frac{7}{5} - 2\ln x\right) \begin{pmatrix} \frac{1}{x^3} \\ -\frac{1}{x^2} \end{pmatrix} + \frac{2}{5}x^5 \begin{pmatrix} \frac{(\ln x)}{x^3} \\ \frac{1-\ln x}{x^2} \end{pmatrix}$$

$$= \begin{pmatrix} \frac{c_1}{x^3} + c_2\frac{\ln x}{x^3} + \frac{7}{25}x^2 \\ -\frac{c_1}{x^2} + c_2\frac{(1-\ln x)}{x^2} + \frac{3}{25}x^3 \end{pmatrix}$$

Ans.

EXERCISE 13.1

Show that the following functions are linearly independent with the help of Wronskian.

1. e^x, x^{ex}

2. $x^2, x^2 \log x$

3. x, $\sin x$ (*Gujarat, II Semester, June 2009*)

4. $x^a \cos(2 \log x), x^a \sin[2 \log x]$

5. e^{ax}, e^{bx}

6. $e^{ax}. \sin bx, e^{ax} \cos bx.$

Find a second order Homogeneous Linear differential equation whose solutions are given below. Find Wronskian to show that the following functions are linearly independent.

7. $\cosh 2x, \sinh 2x$ **Ans.** $y'' - 4y = 0, W = 2$

8. $\cos 2\pi x, \sin 2\pi x$ **Ans.** $y'' + 4\pi^2 y = 0, W = 2\pi$

9. $x^{\frac{3}{2}}, x^{-\frac{3}{2}},$ **Ans.** $4x^2 y'' + 4xy' - 9y = 0, = -\frac{3}{x}$

10. e^{3x}, xe^{3x} **Ans.** $y'' - 6y' + 9y = 0, W = e^{6x}$

11. $1, e^{-2x}$ **Ans.** $y'' + 2y' = 0, W = -2e^{-2x}$

12. $\cos(\log x), \sin(\log x)$ **Ans.** $x^2 y'' + xy' + y = 0, W = \frac{1}{2}$

From the following

(*a*) find Wronskian determinant

(*b*) verify that the solutions satisfy the differential equation by Wronskian test.

(*c*) show that the solutions are linearly independent.

13. $y_1(x) = e^{5x}, y_2(x) = e^{-6x}$ and $y'' - y' = 0$ **Ans.** $W(y_1, y_2, x) = -11\, e^{-x}$

14. $y_1(x) = e^{-x}, y_2(x) = e^{-6x}$ and $y'' - y' = 0$ **Ans.** $W(y_1, y_2, x) = -2$

15. $y_1(x) = \cosh(x - x_0), y_2(x) = \sinh(x - x_0), y'' - y' = 0$

Ans. $W(y_1, y_2, x) = \cosh^2(x - x_0) - \sinh^2(x - x_0)$

16. Evaluate $W(y_1, y_2, x)$ by using Abel's formula

$2y'' + 3y' + xy = 0, \quad y(0) = 1, y_2(0) = 2, y_2'(0) = 4$ **Ans.** $W(y_1, y_2, x) = -3e^{\frac{x^2}{4}}$

13.14 COMPLETE SOLUTION = COMPLEMENTARY FUNCTION + PARTICULAR INTEGRAL

Let us consider a linear differential equation of the first order

$$\frac{dy}{dx} + Py = Q \qquad ...(1)$$

Its solution is $ye^{\int P dx} = \int (Q\, e^{\int P dx})\, dx + c$

$$\Rightarrow \qquad y = ce^{-\int P dx} + e^{-\int P dx} \int (Qe^{\int P dx})\, dx$$

$$\Rightarrow \qquad y = cu + v \qquad \text{(say)} \qquad ...(2)$$

where $u = e^{-\int P dx}$ and $v = e^{-\int P dx} \int Q\, e^{\int P dx}\, dx$

(*i*) Now differentiating $u = e^{-\int P dx}$ w.r.t. x, we get $\frac{du}{dx} = -Pe^{-\int P dx} = -Pu$

$$\Rightarrow \quad \frac{du}{dx} + Pu = 0 \quad \Rightarrow \quad \frac{d(cu)}{dx} + P(cu) = 0$$

which shows that $y = c.u$ is the solution of $\frac{dy}{dx} + Py = 0$

(*ii*) Differentiating $v = e^{-\int P dx} \int (Qe^{\int P dx}\ dx$ with respect to x, we get

$$\frac{dv}{dx} = -Pe^{-\int P dx} \int (Qe^{\int P dx})\, dx + e^{-\int P dx} Qe^{\int P dx} \quad \Rightarrow \quad \frac{dv}{dx} = -Pv + Q$$

$\Rightarrow \quad \frac{dv}{dx} + Pv = Q$ which shows that $y = v$ is the solution of $\boxed{\frac{dy}{dx} + Py = Q}$

Solution of the differential equation (1) is (2) consisting of two parts *i.e.* *cu* and *v*. *cu* is the solution of the differential equation whose R.H.S. is zero. *cu* is known as *complementary function*. Second part of (2) is *v* free from any arbitrary constant and is known as *particular integral*.

Complete Solution = Complementary Function + Particular Integral.

$$\boxed{y = C.F. + P.I.}$$

13.15 METHOD FOR FINDING THE COMPLEMENTARY FUNCTION

Consider the differential equation

$$(D^n + a_1D^{n-1}) + a_2D^{n-2} + \ldots + a_nD^0)\, y = 0 \qquad \ldots(1)$$

Where $a_1, a_2, \ldots a_n$ are all constants.

Let $\qquad y = C_1e^{mx}$ be its solution

$$Dy = C_1me^{mx}, \quad D^2y = C_1m^2e^{mx} \ldots \quad D^ny = C_1m^ne^{mx} \ldots$$

Putting these value in (1) we get

$$C_1(m^n + a_1m^{n-1} + a_2m^{n-2} + \ldots + a_n)\, e^{mx} = 0 \qquad \ldots(2)$$

Its auxiliary equation is $m^n + a_1m^{n-1} + a_2m^{n-2} + \ldots + a_n = 0$

This is plynomial equation of degree n. So, it has n roots i.e.

$$m_1, \quad m_2, \quad m_3, \quad m_n$$

The complementary function of equation (1) depends upon the nature of the roots. The six nature of roots are as follows:

In brief Nature of roots and corresponding C.F.

S.l.	*Nature of Roots of A.E.*	*Roots*	*C.F.*
1.	Real (ational) and Distinct roots	m_1, m_2, m_3	$C_1e^{m_1x} + C_2e^{m_2x} + C_3e^{m_3x}$
2.	Repeated roots	$m_1 = m_2$ $m_1 = m_2 = m_3$	$(C_1 + C_2x)\ e^{m_1x}$ $(C_1 + C_2x + C_3x^2)\ e^{m_1x}$
3.	Complex roots	$m_1 = \alpha + i\beta$ $m_2 = \alpha - i\beta$	$e^{\alpha x}\,[C_1 \cos \beta x + C_2 \sin \beta x]$

4.	Repeated Complex roots	$m_1 = m_2 = \alpha + i\beta$ $m_3 = m_4 = \alpha - i\beta$	$e^{\alpha x}\,[(C_1 + C_2x)\cos\beta x + (C_3 + C_4x)\sin\beta x]$
5.	Irrational root	$m_1 = a + \sqrt{b}$ $m_2 = a - \sqrt{b}$	$e^{ax}\,[C_1\cosh\sqrt{b}\,x + C_2\sinh\sqrt{b}\,x]$
6.	Repeated irrational roots	$m_1 = m_2 = a + \sqrt{b}$ $m_3 = m_4 = a - \sqrt{b}$	$e^{ax}\,[(C_1 + C_2x)\cosh\sqrt{b}\,x + (C_3 + C_4x)\sinh\sqrt{b}\,x]$

Case I. When the roots of auxiliary equation are (rational) and distinct.

The given equation (1) is equivalent to

$$(D - m_1)(D - m_2)(D - m_3)\,...\,(D - m_n)\,y = 0$$

$$\Rightarrow \qquad (D - m_1)\,y = 0,\ (D - m_2)\,y = 0\ ...\ (D - m_3)\,y = 0$$

Let us discuss $(D - m_1)\,y = 0 \Rightarrow \dfrac{dy}{dx} - m_1 y = 0$

This is the linear differential equation.

$$\text{I.F} = e^{-\int m_1\,dx} = e^{-m_1x}$$

Nos its solution is $ye^{-m_1x} = \int 0.e^{-m_1x}\,dx$

$$\Rightarrow \qquad ye^{-m_1x} = 0 + C_1$$

$$\Rightarrow \qquad ye^{-m_1x} = C_1$$

$$\Rightarrow \qquad y = C_1e^{m_1x}$$

Similarly the solution of $(D - m_2)\,y = 0$ is $y = C_2\,e^{m_2x}$, the solution of $(D - m_n)\,y = 0$ is $y = C_n e^{m_nx}$

C.F. of equation (1) is given by

$$\text{C.F.} = C_1e^{m_1x} + C_2e^{m_2x} + ... + C_ne^{m_nx}$$

Example 1. *Solve* $\dfrac{d^2y}{dx^2} - 5\dfrac{dy}{dx} + 6y = 0$

Solution. $(D^2 - 5D + 6)\,y = 0$

A.E. is $m^2 - 5m + 6 = 0 \Rightarrow (m - 2)(m - 3) = 0 \Rightarrow m = 2, 3$

$$\text{C.F.} = C_1e^{2x} + C_2e^{3x} \text{ and P.I.} = 0$$

$$y = \text{C.F.} + \text{P.I.} = C_1e^{2x} + C_2e^{3x}$$

Ans.

Example 2. *Solve:* $\dfrac{d^3y}{dx^3} - 6\dfrac{d^2y}{dx^2} + 11\dfrac{dy}{dx} - 6y = 0$

Solution. The auxiliary equation is $m^3 - 6m^2 + 11\,m - 6 = 0$

$\Rightarrow \quad (m - 1)(m - 2)(m - 3) = 0 \quad \Rightarrow \quad m = 1, 2, 3$

The roots are real and distinct.

$\therefore$ Complementary function (C.F.) $= C_1e^x + C_2e^{2x} + C_3e^{3x}$

Particular Integral (P.I.) = 0

Hence, the complete solution is $y = \text{C.F.} + \text{P.I.} = C_1e^x + C_2e^{2x} + C_3e^{3x}$

where C_1, C_2 and C_3 are arbitrary constants of integration. **Ans.**

Case II. What the roots of auxiliary equation are equal.

The given equation (1) is equivalent to

$$(D - m_1)(D - m_2)(D - m_3) \ldots (D - m_n) y = 0 \qquad \ldots(1)$$

If $m_1 = m_2$ then $(D - m_1)(D - m_1)(D - m_3) \ldots (D - m_n) y = 0$

Part of C.F. is the solution of the equation $(D - m_1)(D - m_1) y = 0$ ...(2)

Replacing $(D - m_1) y = v$ in (2), we get

$$(D - m_1) v = 0 \qquad \ldots(3)$$

$$\frac{dv}{dx} - m_1 v = 0 \quad \Rightarrow \quad \frac{dv}{v} = m_1 dx$$

$$\Rightarrow \quad \log v = m_1 x + \log C_2 \Rightarrow \log v - \log C_2 = m_1 x \Rightarrow \log \frac{v}{C_2} = m_1 x \Rightarrow \frac{v}{C_2} = e^{m_1 x}$$

$$v = C_2 e^{m_1 x}$$

From (3) $\qquad (D - m_1) = C_2 e^{m_1 x}$

Linear Differential Equations of *n*th order

This is the linear differential equation.

$$\text{I.F.} = e^{\int -m_1 dx} = e^{-m_1 x}$$

Solution is $y . e^{-m_1 x} = \int (C_2 e^{m_1 x})(e^{-m_1 x})\, dx + C_1 = \int C_2\, dx + C_1 = C_2 x = C_1$

$$y = (C_2 x + C_1) e^{m_1 x} \qquad \ldots(4)$$

From (1) and (4) we have complete C.F. is

$$\text{C.F.} = (C_1 + C_2 x) e^{m_1 x} + C_3 e^{m_3 x} + C_4 e^{m_4 x} + \ldots + C_n e^{m_n x}$$

If $m_1 = m_2 = m_3$, then

$$\text{C.F.} = (C_1 + C_2 x + C_3 x^2) e^{m_1 x} + C_4 e^{m_4 x} + \ldots + C_n e^{m_n x}$$

Example 3. *Solve:* $\frac{d^2 y}{dx^2} - 6 \frac{dy}{dx} + 9y = 0$

Solution. Here, then

$$\frac{d^2 y}{dx^2} - 6 \frac{dy}{dx} + 9y = 0 \Rightarrow (D^2 - 6D + 9) y = 0, \text{ where } D \equiv \frac{d}{dx}$$

A.E. is $m^2 - 6m + 9 = 0 \Rightarrow (m - 3)^2 = 0 \Rightarrow m = 3, 3.$

$$\text{C.F.} = (C_1 + C_2 x) e^{3x} \text{ and P.I.} = 0$$

$$y = \text{C.F.} + \text{P.I.}$$

$$= (C_1 + C_2 x) e^{3x}$$

Ans.

Example 4. *Solve:* $\frac{d^4 y}{dx^4} - 7\frac{d^3 y}{dx^3} + 15\frac{d^2 y}{sx^2} - 13\frac{dy}{dx} + 4y = 0$

Solution. $(D^4 - 7D^3 + 15D^2 - 13D + 4) y = 0$, where $D = \frac{d}{dx}$

The auxiliary equation is $m^4 - 7m^3 + 15 m^2 - 13 m + 4 = 0$

$$(m - 1)^3 (m - 4) = 0 \Rightarrow m = 1, 1, 1, 4$$

$$\text{C.F.} = (C_1 + C_2 x + C_3 x^2) e^x + C_4 e^{4x} \text{ and P.I.} = 0$$

$$y = \text{C.F.} + \text{P.I.} = (C_1 + C_2 x + C_3 x^2) e^x + C_4 e^{4x}$$

Example 5. *Find the general solution of the differential equation.*

$$\frac{d^5y}{dx^5} - \frac{d^3y}{dx^3} = 0$$ *(U.P., II Semester 2009)*

Solution. Here, we have

$\Rightarrow \quad D^5 y - D^3 y = 0 \quad \Rightarrow \quad (D^5 - D^3)\, y = 0 \quad \Rightarrow \quad D^3 (D^2 - 1)\, y = 0$

A.E. is $m^3 (m^2 - 1) = 0 \Rightarrow m = 0, 0, 0, 1, -1$

Hence, its solution is $y\,(C_1 + C_2 + C_3x^2) + C_4e^x + C_5e^{-x}$ **Ans.**

Case III. When two roots of auxiliary equation are complex,

Let $\qquad m_1 = \alpha + i\beta$ and $m_2 = \alpha - i\beta$

Then C.F. $= C_1e^{(\alpha + i\beta)x} + C_2e^{(\alpha - i\beta)x} + C_3e^{m_3x} + ... + C_ne^{m_nn}$

$$= e^{\alpha x}(C_1e^{i\beta x} + C_2e^{-i\beta x}) + C_3e^{m_3x} + ... + C_ne^{m_nn}$$

$$= e^{\alpha x}[C_1(\cos\beta x + i\sin\beta x) + C_2(\cos\beta x - i\sin\beta x)] + C_3e^{m_3x} + ... + C_ne^{m_nx}$$

$$= e^{\alpha x}[(C_1 + C_2)\cos\beta x + i(C_1 - C_2)\sin\beta x] + C_3e^{m_3x} + ... + C_ne^{m_nx}$$

C.F. $= e^{\alpha x}(c_1\cos\beta x + c_2\sin\beta x) + C_3e^{m_3x} + ... + C_ne^{m_nx}$

where $c_1 = C_1 + C_2,\ c_2 = i\,(C_1 - C_2)$

This is the C.F. of differential equation,

Example 6. *Solve:* $\dfrac{d^2y}{dx^2} + 4\dfrac{dy}{dx} + 5y = 0,$

Subject to $y = 2$ *and* $\dfrac{dy}{dx} = \dfrac{d^2y}{dx^2}$ *when* $x = 0$

Solution. Here the auxiliary equation is

$$m^2 + 4m + 5 = 0 \Rightarrow \frac{-4 \pm \sqrt{16 - 20}}{2} = -2 \pm i$$

Its roots are $-2 \pm i$

The complementary function is

$$y = e^{-2x}(C_1\cos x + C_2\sin x) \qquad ...(1)$$

On putting $y = 2$ and $x = 0$, we have

$$2 = C_1$$

On putting $C_1 = 2$ in (1), we have

$$y = e^{-2x}[2\cos x + C_2\sin x] \qquad ...(2)$$

On differentiating (2), we have

$$\frac{dy}{dx} = e^{-2x}[-2\sin x + C_2\cos x] - 2e^{-2x}[2\cos x + C_2\sin x]$$

$$= e^{-2x}[(-2C_2 - 2)\sin x + (C_2 - 4)\cos x]$$

$$\frac{d^2y}{dx^2} = e^{-2x}[(-2C_2 - 2)\cos x - (C_2 - 4)\sin x]$$

$$-2e^{-2x}[(-2C_2 - 2)\sin x + (C_2 - 4)\cos x]$$

$$= e^{-2x}[(4C_2 + 4)\cos x + (-2C_2 + 8)\sin x]$$

But $\qquad \dfrac{dy}{dx} = \dfrac{d^2y}{dx^2}$ *(given)*

$$e^{-2x}[(-2C_2-2)\sin x+(C_2-4)\cos x]=e^{-2x}[(-4C_2+6)\cos x+(3C_2+8)\sin x]$$

On putting $x = 0$, we get

$$C_2-4=-4C_2+6 \Rightarrow C_2=2$$

(2) becomes, $y = e^{-2x}[(2\cos x + 2\sin x]$

$$y = 2e^{-2x}[\sin x + \cos x]$$ **Ans.**

Example 7. *Solve:* $\dfrac{d^4y}{dx^4}+81y=0$

Solution. $(D^4+81)y=0$ where $D=\dfrac{d}{dx}$

A.E. is $m^4+81=0$

$$m^4+18m^2+81=18m^2$$

$$(m^2+9)^2=18m^2$$

$$m^2+9=\pm 3\sqrt{2}\,m$$

Now, $m^2-3\sqrt{2}\,m+9=0$

$$m=\frac{3\sqrt{2}\pm\sqrt{18-36}}{2}=\frac{3\sqrt{2}\pm\sqrt{-18}}{2}=\frac{3\sqrt{2}\pm 3i\sqrt{2}}{2}$$

Again $m^2+3\sqrt{2}\,m+9=0$

$$\Rightarrow \quad m=\frac{-3\sqrt{2}\pm 3i\sqrt{2}}{2}$$

The complex roots are $\dfrac{3\sqrt{2}\pm 3i\sqrt{2}}{2}$ and $\dfrac{-3\sqrt{2}\pm 3i\sqrt{2}}{2}$

Hence,

$$\text{C.F.}=e^{\frac{3\sqrt{2}}{2}x}\left[C_1\cos\frac{3\sqrt{2}}{2}x+C_2\sin\frac{3\sqrt{2}}{2}x\right]+e^{\frac{-3\sqrt{2}x}{2}}\left[C_3\cos\frac{3\sqrt{2}}{2}x+C_4\sin\frac{3\sqrt{2}}{2}x\right]$$ **Ans.**

Case IV. When roots of auxiliary equation are repeated imaginary

Let $m_1=m_2=\alpha+i\beta,\ m_3=m_4=\alpha-i\beta$

$$\text{C.F.}=(C_1+C_2x)e^{(\alpha+i\beta)x}+(C_3+C_4x)e^{(\alpha-i\beta)x}+C_5e^{m_5x}+\ldots+C_ne^{m_nx}$$

$$=e^{\alpha x}[(C_1+C_2x)e^{i\beta x}+(C_3+C_4x)e^{-i\beta x}]+C_5e^{m_3x}+\ldots+C_ne^{m_nx}$$

$$=e^{\alpha x}[(C_1+C_2x)(\cos\beta x+i\sin\beta x)+(C_3+C_4x)(\cos\beta x-i\sin\beta)+C_5e^{m_5x}+\ldots+C_ne^{m_nx}$$

$$=e^{\alpha x}[\{(C_1+C_3x)\}\cos\beta x+\{(C_1+C_3)+(C_2-C_4x)\}i\sin\beta x]+C_5e^{m_5x}+\ldots+C_ne^{m_nx}$$

$$\text{C.F.}=e^{\alpha x}[(c_1+c_2x)\cos\beta+(c_3+c_4x)\sin\beta x]+C_5e^{m_3x}+\ldots+C_ne^{m_nx}$$

where $c_1+c_2=C_1+C_3x,\quad c_2=C_2+C_4,\quad c_4=C_2-C_4$

Example 8. *Solve* $(D^2-6D+25)^2y=0$

Solution. Auxiliary equation is

$$(m^2-6m+25)^2=0$$

$$\Rightarrow \quad m=\frac{6\pm\sqrt{36-100}}{2}\text{ and }\frac{6\pm\sqrt{36-100}}{2}=\frac{6\pm 8i}{2}\text{ and }\frac{6\pm 8i}{2}$$

$$=3\pm 4i\text{ and }3\pm 4i$$

The roots are repeated complex.

$\therefore$ C.F. $= e^{3x}\,[(C_1 + C_2x)\cos 4x + (C_3 + C_4x)\sin 4x]$ and P.I. $= 0$

The complete solution is

$$y = \text{C.F.} + \text{P.I.}$$

$$\Rightarrow \qquad y = e^{3x}\,[(C_1 + C_2x)\cos 4x + (C_3 + C_4x)\sin 4x]$$

where C_1, C_2, C_3 and C_4 are arbitrary constant of integration.

Example 9. *Solve:* $(D^2 + 1)^2 (D - 1)\, y = 0$

Solution. The auxiliary equation is

$$(m^2 + 1)^2 (m - 1) = 0$$

Either $\qquad m - 1 = 0 \Rightarrow m = 1$

$$(m^2 + 1)^2 = 0 \Rightarrow m = \pm i, \pm i$$

$$\text{C.F.} = C_1e^x + [(C_2 + C_3x)\cos x + (C_4 + C_5x)\sin x]\,e^{0x}$$

$$\text{P.I.} = 0$$

The complete solution is

$$y = \text{C.F.} + \text{P.I.}$$

$$= C_1e^x + [(C_2 + C_3x)\cos x + (C_4 + C_5x)\sin x]$$

where C_1, C_2, C_3, C_4 and C_5 are arbitrary constant of integration.

Example 10. *Solve the differential equation:*

$$(D^2 + 1)^3 (D^2 + D + 1)^2\, y = 0 \text{ where } D = \frac{d}{dx}.$$

Solution. Here, we have

$$(D^2 + 1)^3 (D^2 + D + 1)^2\, y = 0$$

Auxiliary equation is

$$(m^2 + 1)^3 (m^2 + m + 1)^2 = 0$$

$$\Rightarrow \qquad (m^2 + 1)^3 = 0 \text{ gives } m = \pm i, \pm i, \pm i.$$

and $\qquad (m^2 + m + 1)^2 = 0$

$$\Rightarrow \qquad m = \frac{-1 \pm \sqrt{1-4}}{2}, \frac{-1 \pm \sqrt{1-4}}{2}$$

$$\Rightarrow \qquad m = \frac{-1 \pm \sqrt{3}i}{2}, \frac{-1 \pm \sqrt{3}i}{2}$$

$$\text{C.F.} = [(C_1 + C_2x + C_3x^2)\cos x + (C_4 + C_5x + C_6x^2)\sin x]\,e^{0x}$$

$$+ \left[(C_7 + C_8x)\cos\frac{\sqrt{3}}{2}x + (C_9 + C_{10}x)\sin\frac{\sqrt{3}}{2}x\right]e^{-\frac{x}{2}}, \text{ P.I.} = 0$$

$$y = \text{C.F.} + \text{P.I.}$$

$$= (C_1 + C_2x + C_3x^2)\cos x + (C_4 + C_5x + C_6x^2)\sin x$$

$$\left[(C_7 + C_8x)\cos\frac{\sqrt{3}}{2}x + (C_9 + C_{10}x)\sin\frac{\sqrt{3}}{2}x\right]e^{-\frac{x}{2}}$$

where, C_1, C_2, C_3, C_4, C_5, C_6, C_7, C_8, C_9, C_{10}, are arbitrary constants.

Case V. When roots of auxiliary equation are irrational

Let $m_1 = a+\sqrt{b}, m_2 = a-\sqrt{b}$

$$\text{C.F.} = C_1 e^{(a+\sqrt{b})} + C_2 e^{(a-\sqrt{b})} + C_3 e^{m_3 x} + C_n e^{m_n x}$$

$$= e^{ax}\left[C_1 e^{\sqrt{6}x} + C_2 e^{-\sqrt{6}x}\right] + C_3 e^{m_3 x} + \ldots + C_n e^{m_n x}$$

$$= e^{ax}\left[C_1 (\cosh \sqrt{b}x + \sinh \sqrt{b}x) + C_2 (\cosh \sqrt{b}x - \sinh \sqrt{b}x)\right] + C_3 e^{m_3 x} + \ldots + C_n e^{m_n x}$$

$$= e^{ax}\left[(C_1 + C_2)\cosh \sqrt{b}x + (C_1 - C_2) + \sinh \sqrt{b}x\right] + C_3 e^{m_3 x} + \ldots + C_n e^{m_n x}$$

$$= e^{ax}\left[C_1 \cosh \sqrt{b}x + C_2 \sinh \sqrt{b}x\right] + C_3 e^{m_3 x} + \ldots + C_n e^{m_n x}$$

where $C_1 = C_1 + C_2$ and $C_2 = C_1 - C_2$.

Example 11. *Solve* $\frac{d^3 y}{dx^3} - 5\frac{d^2 y}{dx^2} + 5\frac{dy}{dx} - 1 = 0$

Solution. The given equation is

$$(D^3 - 5D^2 + 5D - 1)\, y = 0, \text{ where } D = \frac{d}{dx}$$

A.E. is $m^3 - 5m^2 + 5m - 1 = 0$

$(m - 1)(m^2 - 4m + 1) = 0$

Either $m - 1 = 0 \Rightarrow m = 1$,

or $m^2 - 4m + 1 = 0 \Rightarrow m = \frac{4 \pm \sqrt{16-4}}{2} = \frac{4 \pm 2\sqrt{3}}{2} = 2 \pm \sqrt{3}$

$$\text{C.F.} = C_1 e^x + e^{2x}\left[C_2 \cosh \sqrt{3}x + C_3 \sinh \sqrt{3}x\right] \text{ and P.I.} = 0$$

Complete solution = C.F. + P.I.

$$y = C_1 e^x + e^{2x}\left(C_1 \cosh \sqrt{3}x + C_3 \sinh \sqrt{3}x\right).$$ **Ans.**

Case VI. When roots of auxiliary equations are repeated irrational.

$$m_1 = m_2 = a + \sqrt{b} \text{ and } m_3 = m_4 = a - \sqrt{b}$$

$$\text{C.F.} = e^{ax}\left[(C_1 + C_2 x)\cosh \sqrt{b}x + (C_3 + C_4 x)\sinh \sqrt{b}x\right] + C_3 e^{m_3 x} + \ldots + C_n e^{m_n x}$$

Example 12. *Solve: $(D^2 - 4D + 1)^2\, y = 0$*

Solution. We have $(D^2 - 4D + 1)^2\, y = 0$

Its auxiliary equation is $(m^2 - 4m + 1)^2 = 0$

$$\Rightarrow \quad m = \frac{4 \pm \sqrt{16-4}}{2}, \frac{4 \pm \sqrt{16-4}}{2}$$

$$= \frac{4 \pm \sqrt{12}}{2}, \frac{4 \pm \sqrt{12}}{2}$$

$$= 2 \pm \sqrt{3}, 2 \pm \sqrt{3}$$

These roots are repeated and irrational

C.F. $= e^{2x}\left[(C_1 + C_2 x)\cos \sqrt{3}x + (C_3 + C_4 x)\sinh \sqrt{3}x\right]$, P.I. = 0

$$\Rightarrow \quad y = e^{2x}\left[(C_1 + C_2 x)\cos \sqrt{3}x + (C_3 + C_4 x)\sinh \sqrt{3}x\right]$$ **Ans.**

EXERCISE 13.2

Solve the following differential equations :

1. $\frac{d^2y}{dx^2}-3\frac{dy}{dx}+2y=0$ **Ans.** $y = C_1 e^x + C_2 e^{2x}$

2. $\frac{d^2y}{dx^2}+\frac{dy}{dx}-30y=0$. **Ans.** $y = C_1e^{5x} + C_2e^{-6x}$

3. $\frac{d^2y}{dx^2}-8\frac{dy}{dx}+16y=0$ **Ans.** $y = (C_1 + C_2x)\, e^{4x}$

4. $\frac{d^2y}{dx^2}+\mu^2y=0$ **Ans.** $y=C_1\cos\mu x+C_2\sin\mu x$

5. $(D^2+2D+2)\,y=0,\ y(0)=0,\ y'(0)=1$ **Ans.** $y = e^{-x}\sin x$ *(A.M.I.E.T.E., June 2006)*

6. $\frac{d^3y}{dx^3}-2\frac{d^2y}{dx^2}+4\frac{dy}{dx}-8y=0$ **Ans.** $y = C_1e^{2x} + [(C_2 + C_3x)\cos 2x + (C_4 + C_5)\sin 2x]$

7. $\frac{d^4y}{dx^4}-32\frac{d^2y}{dx^2}+256=0$

(A.M.I.E.T.E., Dec. 2004) **Ans.** $y = (C_1 + C_2x)\cos 4x + (C_3 + C_4x)\sin 4x$

8. $\frac{d^4y}{dx^4}+4\frac{d^3y}{dx^3}+8\frac{d^2y}{dx^2}-8\frac{dy}{dx}+4y=0$ **Ans.** $y = e^x\,[(C_1 + C_2x)\cos x + (C_3 + C_4\,x)\sin x]$

9. $\frac{d^4y}{dx^4}+\frac{d^2y}{dx^2}=0,\ y(0) = y'(0) = y''(0) = 0,\ y'''(0) = 1$ **Ans.** $y = x - \sin x$

10. $\frac{d^3y}{dx^3}+6\frac{d^2y}{dx^2}+12\frac{dy}{dx}+8y=0,\ y(0) = 0,$ and $y'(0)=0$ and $y''(0)=2$

(A.M.I.E.T.E. Dec. 2008) **Ans.** $y = x^2e^{-2x}$

11. $\frac{d^3y}{dx^3}+\frac{d^2y}{dx^2}+\frac{4dy}{dx}+4y=0,\ y(0) = 0,\ y'(0)=0, y''(0)=-5,$ **Ans.** $y=-e^x+\cos 2x-\frac{1}{2}\sin 2x$

12. $(D^8+6D^6-32D^2)y=0$ *(A.M.I.E.T.E., Summer 2005)*

Ans. $y = C_1 + C_2x + C_3e^{\sqrt{2}x} + C_4e^{-\sqrt{2}x} + (C_5+xC_6)\cos 2x + (C_7+xC_8)\sin 2x$

13. Show that non-trivial solutions of the boundary value problem $y^{(iv)}-w^4y=0,\ y(0)=0=y''(0),$ $y(L)=0, y''(L)=0$ are $y(x) = \sum_{n=1}^{\infty} D_n \sin\left(\frac{n\pi x}{L}\right)$ where D_n are constants.

(A.M.I.E.T.E. Dec. 2005)

14. Solve the initial value problem $y'''+6y''+11y'+6y=0,\ y(0) = 0, y'(0)=1, y''(0)=-1.$

(A.M.I.E.T.E., Dec. 2006) **Ans.** $y = 2e^{-x} - 3e^{-2x} + e^{-3x}$.

15. Let y_1, y_2 be two linearly independent solutions of the differential equation $yy''-(y')^2=0$. Then, $c_1y_1 + c_2y_2$, where c_1, c_2 are constants is a solution of this differential equation for

(*a*) $c_1 = c_2 = 0$ only. (*b*) $c_1 = 0$ or $c_2 = 0$ (*c*) no value of c_1, c_2. (*d*) all real c_1, c_2

16. The solution of the differential equation $\frac{d^2y}{dx^2}+y=0$ satisfying the initial conditions $y(0) = y\left(\frac{\pi}{2}\right)=2$ is (*a*) $y = 2\cos(x) + \sin(x)$ (*b*) $y = \cos(x) + 2\sin(x)$

(*c*) $y = \cos(x) + \sin(x)$ (*d*) $y = 2\cos(x) + 2\sin(x)$ (*AMIETE, Dec. 2009*) **Ans.** (*b*)

17. Find the complementary function of $(D-2)^2 = 8(e^{2x} + \sin 2x - x^2)$

(*a*) $(C_1 + C_2 e^{2x})\, x$ (*b*) $(C_1 + C_2 x)\, e^{2x}$

(*c*) $(C_1 x + C_2 x^2)\, e^{2x}$ (*d*) $(C_1 x + C_2 e x^2)\,.\, 2^{2x}$ (*AMIETE, Dec. 2010*) **Ans.** (*b*)

18. Solution of $\dfrac{d^3y}{dx^3} + 2\dfrac{d^2y}{dx^2} + \dfrac{dy}{dx} = 0$ is

(*a*) $y = C_1 e^x + C_2 e^{2x}$ (*b*) $y = C_1 + (C_2 + C_3 x)\, e^{-x}$

(*c*) $y = (C_1 + C_2 x + C_3 x^2)\, e^{-x}$ (*d*) $y = C_1 + C_2 e^{-x}$ (*AMIETE, June 2009*) **Ans.** (*b*)

13.16 RULES TO FIND PARTICULAR INTEGRAL

(*i*) $\dfrac{1}{f(D)} e^{ax} = \dfrac{1}{f(a)} e^{ax}$

If $f(a) = 0$ then $\dfrac{1}{f(D)} \cdot e^{ax} = x \cdot \dfrac{1}{f'(a)} \cdot e^{ax}$

If $f'(a) = 0$ then $\dfrac{1}{f(D)} \cdot e^{ax} = x^2 \dfrac{1}{f''(a)} \cdot e^{ax}$

(*ii*) $\dfrac{1}{f(D)} x^n = [f(D)]^{-1} x^n$ Expand $[f(D)]^{-1}$ and then operate.

(*iii*) $\dfrac{1}{f(D^2)} \sin ax = \dfrac{1}{f(-a^2)} \sin ax$ and $\dfrac{1}{f(D^2)} \cos ax = \dfrac{1}{f(-a^2)} \cos ax$

If $f(-a^2) = 0$ then $\dfrac{1}{f(D^2)} \sin ax = x \cdot \dfrac{1}{f'(-a^2)} \cdot \sin ax$

(*iv*) $\dfrac{1}{f(D)} e^{ax} \cdot \phi(x) = e^{ax} \cdot \dfrac{1}{f(D+a)} \phi(x)$

(*v*) $\dfrac{1}{D+a} \phi(x) = e^{-ax} \int e^{ax} \cdot \phi(x)\, dx$

13.17 $\boxed{\dfrac{1}{f(D)} e^{ax} = \dfrac{1}{f(a)} e^{ax}}$.

We know that, $D.e^{ax} = a.e^{ax}$, $D^2 e^{ax} = a^2.e^{ax}$, ..., $D^n e^{ax} = a^n e^{ax}$

Let $f(D)\, e^{ax} = (D^n + K_1 D^{n-1} + \ldots + K_n)\, e^{ax} = (a^n + K_1 a^{n-1} + \ldots + K_n) e^{ax} = f(a)\, e^{ax}$.

Operating both sides by $\dfrac{1}{f(D)}$

$$\frac{1}{f(D)} \cdot f(D) e^{ax} = \frac{1}{f(D)} \cdot f(a) e^{ax}$$

$$\Rightarrow \qquad e^{ax} = f(a) \frac{1}{f(D)} \cdot e^{ax} \quad \Rightarrow \quad \boxed{\frac{1}{f(D)} e^{ax} = \frac{1}{f(a)} e^{ax}}$$

If $f(a) = 0$, then the above rule fails.

Then $\frac{1}{f(D)}e^{ax} = x\cdot\frac{1}{f'(D)}e^{ax} = x\frac{1}{f'(a)}e^{ax} \Rightarrow$ $\boxed{\frac{1}{f(D)}e^{ax} = x\,.\frac{1}{f'(a)}e^{ax}}$

If $f(a) = 0$, then $\boxed{\frac{1}{f(D)}e^{ax} = x^2\frac{1}{f''(a)}e^{ax}}$

Example 13. *Solve the differential equaton* $\frac{d^2x}{dt^2}+\frac{g}{l}x=\frac{g}{l}L$

where g, l, L are constants subject to the conditions,

$$x = a,\ \frac{dx}{dt} = 0 \ \text{at } t = 0.$$

Solution. We have, $\frac{d^2x}{dt^2}+\frac{g}{l}x=\frac{g}{l}L \quad \Rightarrow \quad \left(D^2+\frac{g}{l}\right)x=\frac{g}{l}L$

A.E. is $m^2+\frac{g}{l}=0 \quad \Rightarrow \quad m=\pm\, i\sqrt{\frac{g}{l}}$

$$\text{C.F.} = C_1\cos\sqrt{\frac{g}{l}}\,t+C_2\sin\sqrt{\frac{g}{l}}\,t$$

$$\text{P.I.} = \frac{1}{D^2+\frac{g}{l}}\cdot\frac{g}{l}L=\frac{g}{l}L\frac{1}{D^2+\frac{g}{l}}e^{0t}=\frac{g}{l}L\frac{1}{0+\frac{g}{l}}=L \qquad [D=0]$$

$\therefore$ General solution is $x =$ C.F. + P.I.

$$x=C_1\cos\left(\sqrt{\frac{g}{l}}\right)t+C_2\sin\left(\sqrt{\frac{g}{l}}\right)t+L \qquad ...(1)$$

$$\frac{dx}{dt}=-C_1\sqrt{\frac{g}{l}}\sin\left(\sqrt{\frac{g}{l}}\right)t+C_2\sqrt{\frac{g}{l}}\cos\left(\sqrt{\frac{g}{l}}\right)t$$

Put $t = 0$ and $\frac{dx}{dt}=0$

$$0=C_2\sqrt{\frac{g}{l}} \qquad \therefore \quad C_2=0$$

(1) becomes $x=C_1\cos\sqrt{\frac{g}{l}}\,t+L$ ...(2)

Put $x = a$ and $t = 0$ in (2), we get

$$a=C_1+L \qquad \Rightarrow \quad C_1=a-L$$

On putting the value of C_1 in (2), we get $x=(a-L)\cos\left(\sqrt{\frac{g}{l}}\right)t+L$ **Ans.**

Example 14. *Solve* : $\frac{d^2y}{dx^2}+6\frac{dy}{dx}+9y=5e^{3x}$

Solution. $(D^2 + 6D + 9)y = 5e^{3x}$

Auxiliary equation is $m^2 + 6m + 9 = 0 \Rightarrow \quad (m + 3)^2 = 0 \Rightarrow \quad m = -3, -3,$

$$\text{C.F.} = (C_1 + C_2x)\,e^{-3x}$$

$$\text{P.I.} = \frac{1}{D^2+6D+9}.5.e^{3x} = 5\frac{e^{3x}}{(3)^2+6(3)+9} = \frac{5e^{3x}}{36}$$

The complete solution is $y = (C_1 + C_2x)e^{-3x} + \frac{5e^{3x}}{36}$ **Ans.**

Example 15. *Solve:* $\frac{d^2y}{dx^2} - 6\frac{dy}{dx} + 9y = 6e^{3x} + 7e^{-2x} - \log 2$

Solution. $(D^2 - 6D + 9)y = 6e^{3x} + 7e^{-2x} - \log 2$

A.E. is $(m^2 - 6m + 9) = 0 \Rightarrow (m - 3)^2 = 0, \Rightarrow m = 3, 3$

$$\text{C.F.} = (C_1 + C_2x)e^{3x}$$

$$\text{P.I.} = \frac{1}{D^2-6D+9}6e^{3x} + \frac{1}{D^2-6D+9}7e^{-2x} + \frac{1}{D^2-6D+9}(-\log 2)$$

$$= x\frac{1}{2D-6}6e^{3x} + \frac{1}{4+12+9}7e^{-2x} - \log 2\frac{1}{D^2-6D+9}e^{0x}$$

$$= x^2\frac{1}{2}\cdot 6\cdot e^{3x} + \frac{7}{25}e^{-2x} - \log 2\left(\frac{1}{9}\right) = 3x^2e^{3x} + \frac{7}{25}e^{-2x} - \frac{1}{9}\log 2$$

Complete solution is $y = (C_1 + C_2\,x)e^{3x} + 3x^2e^{3x} + \frac{7}{25}e^{-2x} - \frac{1}{9}\log 2$ **Ans.**

EXERCISE 13.3

Solve the following differential equations:

1. $[D^2 + 5D + 6]\,[y] = e^x$ **Ans.** $C_2\,e^{-2x} + C_2e^{-3x} + \frac{e^x}{12}$

2. $\frac{d^2y}{dx^2} - 3\frac{dy}{dx} + 2y = e^{3x}$ **Ans.** $C_1e^x + C_2e^{2x} + \frac{e^{3x}}{2}$ *(A.M.I.E.T.E. June 2010, 2007)*

3. $\frac{d^2y}{dx^2} + 2\frac{dy}{dx} + 2y = \sinh x$ **Ans.** $e^{-x}[C_1\cos x + C_2\sin x] + \frac{e^x}{10} - \frac{e^{-x}}{2}$

4. $\frac{d^2y}{dx^2} + 4\frac{dy}{dx} + 5y = -2\cosh x$ **Ans.** $e^{-2x}(C_1\cos x + C_2\sin x) - \frac{1}{10}e^x - \frac{e^{-x}}{2}$

5. $(D^3 - 2D^2 - 5D + 6)\,y = e^{3x}$ **Ans.** $C_1e^x + C_2e^{-2x} + C_3e^{3x} + \frac{x.e^{3x}}{10}$

6. $\frac{d^2y}{dx^2} - 6\frac{dy}{dx} + 9y = e^{3x}$ **Ans.** $(C_1 + C_2x)e^{3x} + \frac{x^2}{2}e^{3x}$

7. $\frac{d^3y}{dx^3} + 3\frac{d^2y}{dx^2} + 3\frac{dy}{dx} + y = e^{-x}$ **Ans.** $(C_1 + C_2x + C_3x^2)e^{-x} + \frac{x^3}{6}e^{-x}$

8. $\frac{d^2y}{dx^2} - \frac{dy}{dx} - 6y = e^x\cosh 2x$ **Ans.** $C_1e^{3x} + C_2e^{-2x} + \frac{1}{10}xe^{3x} - \frac{1}{8}e^{-x}$

9. $(D-2)(D+1)^2 y = e^{2x} + e^x$ **Ans.** $C_1 e^{2x} + (C_2 + C_3 x)e^{-x} + \frac{x}{9}e^{2x} - \frac{e^x}{4}$

10. $(D-1)^3 y = 16\, e^{3x}$ **Ans.** $(C_1 + C_2 x + C_3 x^2)\, e^x + 2e^{3x}$

11. The particular integral (PI) of differential equation $[D^2 + 5D + 6]\, y = e^x$ is

(*a*) $\frac{e^x + x}{12}$ (*b*) $\frac{e^x - x}{12}$ (*c*) $\frac{e^{-x}}{12}$ (*d*) $\frac{e^x}{12}$ (*AMIETE, June 2010*) **Ans.** (*d*)

13.18 $\frac{1}{f(D)} x^n = [f(D)]^{-1} x^n.$

Expand $[f(D)]^{-1}$ by the Binomial theorem in ascending powers of D as far as the result of operation on x^n is zero.

Example 16. *Solve the differential equation* $\frac{d^2 y}{dx^2} + a^2 y = \frac{a^2 R}{p}(l - x)$

where a, R, p and l are constants subject to the conditions $y = 0,\ \frac{dy}{dx} = 0$ *at* $x = 0$.

Solution. $\frac{d^2 y}{dx^2} + a^2 y = \frac{a^2}{p} R(l - x) \quad \Rightarrow \quad (D^2 + a^2) y = \frac{a^2}{p} R(l - x)$

A.E. is $m^2 + a^2 \Rightarrow m = \pm\, ia$

C.F. $= C_1 \cos ax + C_2 \sin ax$

$$\text{P.I.} = \frac{1}{D^2 + a^2} \frac{a^2}{p} R(l - x) = \frac{a^2 R}{p} \frac{1}{a^2}\left[\frac{1}{1 + \frac{D^2}{a^2}}\right](l - x) = \frac{R}{p}\left[1 + \frac{D^2}{a^2}\right]^{-1}(l - x)$$

$$= \frac{R}{p}\left[1 - \frac{D^2}{a^2}\right](l - x) = \frac{R}{p}(l - x)$$

$$y = C_1 \cos ax + C_2 \sin ax + \frac{R}{p}(l - x) \qquad \text{...(1)}$$

On putting $y = 0$, and $x = 0$ in (1), we get $0 = C_1 + \frac{R}{p} l \Rightarrow C_1 = -\frac{R\, l}{p}$

On differentiating (1), we get $\frac{dy}{dx} = -a\, C_1 \sin ax + a\, C_2 \cos ax - \frac{R}{p}$ ...(2)

On putting $\frac{dy}{dc} = 0$ and $x = 0$ in (2), we have

$$0 = a\, C_2 - \frac{R}{p} \quad \Rightarrow \quad C_2 = \frac{R}{a.p}$$

On putting the values of C_1 and C_2 in (1), we get

$$y = -\frac{R}{p} l \cos a.x + \frac{R}{a.p} \sin a.x + \frac{R}{p}(l - x) \Rightarrow y = \frac{R}{p}\left[\frac{\sin ax}{a} - l \cos a.x + l - x\right] \qquad \textbf{Ans.}$$

EXERCISE 13.4

Solve the following differential equations :

1. $(D^2 + 5D + 4)\,y = 3 - 2x$ **Ans.** $C_1e^{-x} + C_2e^{-4x} + \frac{1}{8}(11-4x)$

2. $\frac{d^2y}{dx^2} + 2\frac{dy}{dx} + y = x$ **Ans.** $(C_1 + C_2\,x)\,e^{-x} + x - 2$

3. $(2D^2 + 3D + 4)\,y = x^2 - 2x$ **Ans.** $e^{-\frac{3}{4}x}[A\cos\frac{\sqrt{23}}{4}x + B\sin\frac{\sqrt{23}}{4}x] + \frac{1}{32}[8x^2 - 28x + 13]$

4. $(D^2 - 4D + 3)\,y = x^3$ **Ans.** $C_1e^x + C_2e^{3x} + \frac{1}{27}(9x^3 + 36x^2 + 78x + 80)$.

(D.U. III Sem 2014)

5. $\frac{d^3y}{dx^3} - \frac{d^2y}{dx^2} - 6\frac{dy}{dx} = 1 + x^2$. **Ans.** $A + Be^{-2x} + Ce^{3x} - \frac{1}{36}\left(2x^3 - x^2 + \frac{25}{3}x\right)$

6. $\frac{d^4y}{dx^4} + 4y = x^4$ **Ans.** $e^x(C_1\cos x + C_2\sin x) + e^{-x}(C_3\cos x + C_4\sin x) + \frac{1}{4}(x^4 - 6)$

7. $\frac{d^2y}{dx^2} + 2p\frac{dy}{dx} + (p^2 + q^2)y = e^{cx} + p.q\,x^2$

Ans. $e^{-px}[C_1\cos qx + C_2\sin qx] + \frac{e^{Cx}}{(p+C)^2 + q^2} + \frac{pq}{p^2+q^2}\left[x^2 - \frac{4px}{p^2+q^2} + \frac{6p^2 - 2q^2}{(p^2+q^2)^2}\right]$

8. $D^2\,(D^2 + 4)\,y = 96\,x^2$ **Ans.** $C_1 + C_2x + C_3\cos 2x + C_4\sin 2x + 2x^2\,(x^2 - 3)$

13.19 $\frac{1}{f(D^2)}\sin ax = \frac{\sin ax}{f(-a^2)}$. $\frac{1}{f(D^2)}\cdot\cos ax = \frac{\cos ax}{f(-a^2)}$

$D\,(\sin ax) = a.\cos ax$, $D^2\,(\sin ax) = D\,(a\cos ax) = -a^2.\sin ax$

$D^4\,(\sin ax) = D^2.D^2\,(\sin ax) = D^2\,(-a^2\sin ax) = (-a^2)^2\sin ax$

$(D^2)^n\sin ax = (-a^2)^n\sin ax$

Hence, $f(D^2)\sin ax = f(-a^2)\sin ax$

$$\frac{1}{f(D^2)}\cdot f(D^2)\sin ax = \frac{1}{f(D^2)}.f(-a^2).\sin ax$$

$$\sin ax = f(-a^2)\frac{1}{f(D^2)}\sin ax \quad\Rightarrow\quad \frac{1}{f(D^2)}\cdot\sin ax = \frac{\sin ax}{f(-a^2)}$$

Similarly $$\frac{1}{f(D^2)}\cos ax = \frac{\cos ax}{f(-a^2)},$$

If $f(-a^2) = 0$ then above rule fails

$$\frac{1}{f(D^2)}\sin ax = x\frac{\sin ax}{f'(-a^2)}$$

If $f'(-a^2) = 0$ then, $\frac{1}{f(D^2)}\sin ax = x^2\frac{\sin ax}{f''(-a^2)}$

Example 17. *Solve* : $(D^2 + 4)\, y = \cos 2x$

(*R.G.P.V., Bhopal June, 2008, A.M.I.E.T.E. Dec 2008*)

Solution. $(D^2 + 4)\, y = \cos 2x$

Auxiliary equation is $m^2 + 4 = 0$

$m = \pm 2i,\qquad$ C.F. $= A \cos 2x + B \sin 2x$

$$\text{P.I. } \frac{1}{D^2+4}\cos 2x = x.\frac{1}{2D}\cos 2x = \frac{x}{2}\left(\frac{1}{2}\sin 2x\right) = \frac{x}{4}\sin 2x$$

Complete solution is $y = A\cos 2x + B\sin 2x + \frac{x}{4}\sin 2x$ **Ans.**

Example 18. *Solve:* $\frac{d^3y}{dx^3} - 3\frac{d^2y}{dx^2} + 4\frac{dy}{dx} - 2y = e^x + \cos x$

(*U.P., II Semester, Summer* 2006)

Solution. Given $(D^3 - 3D^2 + 4D - 2)\, y = e^x + \cos x$

A.E. is $m^3 - 3m^2 + 4m - 2 = 0$

$(m - 1)(m^2 - 2m + 2) = 0$, *i.e.*, $m = 1, 1 \pm i$

$\therefore\qquad$ C.F. $= C_1e^x + e^x (C_2 \cos x + C_3 \sin x)$

$$\text{P.I.} = \frac{1}{D^3-3D^2+4D-2}e^x + \frac{1}{D^3-3D^2+4D-2}\cos x$$

$$= x\frac{1}{3D^2-6D+4}e^x + \frac{1}{(-1)D-3(-1)+4D-2}\cos x$$

$$= x\frac{1}{3(1)^2-6(1)+4}e^x + \frac{1}{3D+1}\cos x = x\frac{1}{3-6+4}e^x + \frac{3D-1}{9D^2-1}\cos x$$

$$= e^x.x + \frac{(-3\sin x - \cos x)}{-9-1} = e^x.x + \frac{1}{10}(3\sin x + \cos x)$$

Hence, complete solution is

$$y = C_1e^x + e^x(C_2\cos x + C_3\sin x) + x\,e^x + \frac{1}{10}(3\sin x + \cos x)$$ **Ans.**

Example 19. *Solve* : $(D^4 - 3D^2 - 4)y = 5 \sin 2x - e^{-2x}$

Solution. Auxiliary equation is

$m^4 - 3m^2 - 4 = 0$

$(m^2 + 1)(m^2 - 4) = 0$

$\Rightarrow\qquad m^2 + 1 = 0 \;\Rightarrow\; m^2 - 4 = 0 \;\Rightarrow\; m = \pm i \;\Rightarrow\; m = \pm 2$

$$\text{C.F.} = C_1e^{2x} + C_2e^{-2x} + C_3\cos x + C_4\sin x$$

$$\text{P.I.} = \frac{1}{D^4-3D^2-4}(5\sin 2x - e^{-2x}) = 5.\frac{1}{D^4-3D^2-4}\sin 2x - \frac{1}{D^4-3D^2-4}e^{-2x}$$

$$= 5\frac{1}{(-2^2)^2-3(-2^2)-4}\sin 2x - \frac{1}{16-12-4}e^{-2x}$$ (Rule fails)

$$= \frac{5}{24}\sin 2x - x\frac{1}{4D^3-6D}e^{-2x}$$

$$= \frac{5}{24}\sin 2x - x\frac{1}{4(-2)^3-6(-2)}e^{-2x} = \frac{5}{24}\sin 2x + \frac{xe^{-2x}}{20}$$

Hence, the complete solution is

$$y = C_1e^{2x} + C_2e^{-2x} + C_3\cos x + C_4\sin x + \frac{5}{24}\sin 2x + \frac{xe^{-2x}}{20}$$ **Ans.**

Example 20. *Solve:* $(D^3+1)y = \cos^2\left(\frac{x}{2}\right) + e^{-x}$:

Solution. $(D^3+1)y = \cos^2\left(\frac{x}{2}\right) + e^{-x}$

A.E. is $m^3 + 1 = 0$

$(m+1)(m^2 - m + 1) = 0$ $\qquad m = -1$

or $m = \frac{-(-1)\pm\sqrt{1-4}}{2} = \frac{1\pm i\sqrt{3}}{2} \Rightarrow m = \frac{1}{2} \pm i\frac{\sqrt{3}}{2}$

$\therefore$ $$\text{C.F.} = C_1e^{-x} + e^{\frac{x}{2}}\left[C_2\cos\frac{\sqrt{3}}{2}x + C_3\sin\frac{\sqrt{3}}{2}x\right]$$

$$\text{P.I.} = \frac{1}{D^3+1}\left[\cos^2\left(\frac{x}{2}\right) + e^{-x}\right] = \frac{1}{D^3+1}\cos^2\left(\frac{x}{2}\right) + \frac{1}{D^3+1}e^{-x} \quad [\text{Put } D = -1]$$

$$= \frac{1}{D^3+1}\left(\frac{1+\cos x}{2}\right) + \frac{x.1}{3D^2}e^{-x}$$

$$= \frac{1}{2}\left(\frac{1}{D^3+1}e^{0x}\right) + \left(\frac{1}{2}\frac{1}{D^3+1}\cos x\right) + \frac{x.1}{3(-1)^2}e^{-x} = \frac{1}{2} + \frac{1}{2}\frac{1}{-D+1}\cos x + \frac{x}{3}e^{-x}$$

$$= \frac{1}{2} - \frac{1}{2}\frac{(D+1)\cos x}{(D-1)(D+1)} + \frac{x}{3}e^{-x} = \frac{1}{2} - \frac{1}{2}\frac{(-\sin x + \cos x)}{(D^2-1)} + \frac{x}{3}e^{-x}$$

$$= \frac{1}{2} + \frac{1}{2}\frac{\sin x}{(D^2-1)} - \frac{1}{2}\frac{1}{(D^2-1)}\cos x + \frac{x}{3}e^{-x} \qquad [\text{Put } D^2 = -1]$$

$$= \frac{1}{2} + \frac{1}{2}\frac{\sin x}{(-1-1)} - \frac{1}{2}\frac{1}{(-1-1)}\cos x + \frac{x}{3}e^{-x} = \frac{1}{2} - \frac{\sin x}{4} + \frac{\cos x}{4} + \frac{x}{3}e^{-x}$$

$$\text{P.I.} = \frac{1}{2} + \frac{1}{4}(\cos x - \sin x) + \frac{x}{3}e^{-x}$$

Hence, the complete solution is

$$y = C_1e^{-x} + e^{\frac{x}{2}}\left[C_2\cos\frac{\sqrt{3}}{2}x + C_3\sin\frac{\sqrt{3}}{2}x\right] + \frac{1}{2} + \frac{1}{4}(\cos x - \sin x) + \frac{x}{3}e^{-x}$$ **Ans.**

Example 21. *Solve the differential equation*

$$\frac{d^2y}{dx^2} - 2\frac{dy}{dx} + 2y = \sinh x + \sin\sqrt{2}\,x.$$

Solution. A.E. is $m^2 - 2m + 2 = 0$

$\therefore$ $$m = \frac{2\pm\sqrt{4-8}}{2} = \frac{2\pm 2i}{2} = 1\pm i$$

$$\text{C.F.} = e^x (C_1 \cos x + C_2 \sin x) \qquad \left(\sinh x = \frac{e^x - e^{-x}}{2}\right)$$

$$\text{P.I.} = \frac{1}{D^2 - 2D + 2} \sinh x + \frac{1}{D^2 - 2D - 2} \sin \sqrt{2} x$$

$$\text{P.I.} = \frac{1}{D^2 - 2D + 2}\left(\frac{e^x - e^{-x}}{2}\right) + \frac{1}{D^2 - 2D + 2} \sin \sqrt{2} x$$

$$= \frac{1}{2}\left[\frac{1}{1 - 2 + 2} e^x - \frac{1}{1 + 2 + 2} e^{-x}\right] + \frac{1}{-2 - 2D + 2} \sin \sqrt{2} x$$

$$= \frac{1}{2} e^x - \frac{1}{10} e^{-x} + \frac{1}{2\sqrt{2}} \cos \sqrt{2} x \qquad \left(\frac{1}{D} \sin \sqrt{2} x = \int \sin \sqrt{2}\, x\, dx\right)$$

Hence the solution is,

$$y = e^x (C_1 \cos x + C_2 \sin x) + \frac{1}{2} e^x - \frac{1}{10} e^{-x} + \frac{1}{2\sqrt{2}} \cos \sqrt{2} x. \qquad \textbf{Ans.}$$

Example 22. *Solve:* $\frac{d^2 y}{dx^2} + y = 2 \cos x \cos 3x$ *(Delhi University, April 2010)*

Solution. We have, $\frac{d^2 y}{dx^2} + y = 2 \cos x \cos 3x$

$$\Rightarrow \qquad (D^2 + 1) y = 2 \cos x \cos 3x$$

A.E. is $m^2 + 1 = 0 \Rightarrow m^2 = -1 \Rightarrow m = \pm i$

$$\therefore \qquad \text{C.F.} = A \cos x + B \sin x \qquad ...(1)$$

$$\text{P.I.} = \frac{1}{D^2 + 1}[2 \cos x \cos 3x] = \frac{1}{D^2 + 1}[\cos 4x + \cos 2x]$$

$$= \frac{1}{D^2 + 1} \cos 4x + \frac{1}{D^2 + 1} \cos 2x$$

$$= \frac{\cos 4x}{-4^2 + 1} + \frac{\cos 2x}{-2^2 + 1} = -\frac{\cos 4x}{15} - \frac{\cos 2x}{3} \qquad ...(2)$$

Hence the solution is y = C.F. + P.I.

$$y = A \cos x + B \sin x - \frac{\cos 4x}{15} - \frac{\cos 2x}{3} \qquad \textbf{Ans.}$$

EXERCISE 13.5

Solve the following differential equations :

1. $\frac{d^2 y}{dx^2} + 6y = \sin 4x$ **Ans.** $C_1 \cos \sqrt{6} x + C_2 \sin \sqrt{6} x - \frac{1}{10} \sin 4x$

2. $\frac{d^2 x}{dt^2} + 2\frac{dx}{dt} + 3x = \sin t$ **Ans.** $e^{-t}[A \cos \sqrt{2} t + B \sin \sqrt{2} t] - \frac{1}{4}(\cos t - \sin t)$

3. $\frac{d^2 x}{dt^2} + 2\frac{dx}{dt} + 5x = \sin 2t$, given that when $t = 0$, $x = 3$ and $\frac{dx}{dt} = 0$

Ans. $e^{-t}\left[\frac{55}{17}\cos 2t+\frac{53}{34}\sin 2t\right]-\frac{1}{17}(4\cos 2t-\sin 2t)$

4. $\frac{d^2y}{dx^2}-7\frac{dy}{dx}+6y=2\sin 3x$, given that $y=1$, $\frac{dy}{dx}=0$ when $x=0$.

Ans. $-\frac{13}{75}e^{6x}+\frac{27}{25}e^{x}+\frac{1}{75}(7\cos 3x-\sin 3x)$

5. $(D^3+1)\,y=2\cos^2 x$

Ans. $C_1e^{-x}+e^{\frac{1}{2}x}\left(C_2\cos\frac{\sqrt{3}}{2}x+C_3\sin\frac{\sqrt{3}}{2}x\right)+1+\frac{1}{65}(-8\sin 2x+\cos 2x)$

6. $(D^2+a^2)\,y=\sin ax$ *(A.M.I.E.T.E., June 2009)* **Ans.** $C_1\cos ax+C_2\sin ax-\frac{x}{2a}\cos ax$

7. $(D^4+2a^2D^2+a^4)\,y=8\cos ax$ **Ans.** $(C_1+C_2x)+(C_3\cos ax+C_4\sin ax)-\frac{x^2}{a^2}\cos ax$

8. $\frac{d^2y}{dx^2}+3\frac{dy}{dx}+2y=\sin 2x$

Ans. $C_1e^{-x}+C_2e^{-2x}-\frac{1}{20}(3\cos 2x+\sin 2x)$

9. $\frac{d^2y}{dx^2}+y=\sin 3x\cos 2x$ **Ans.** $C_1\cos x+C_2\sin x+\frac{1}{48}[-\sin 5x-12x\cos x]$

10. $\frac{d^2y}{dx^2}-2\frac{dy}{dx}-3y=2e^{2x}+10\sin 3x$ given that $y\,(0)=2$ and $y'(0)$ 4

Ans. $\frac{29}{12}e^{3x}-\frac{1}{12}e^{-x}-\frac{2}{3}e^{2x}+\frac{1}{3}[\cos 3x-2\sin 3x]$

11. $\frac{d^2y}{dx^2}+3\frac{dy}{dx}+2y=4\cos^2 x$ *(R.G.P.V., Bhopal, I Semester, June 2007)*

Ans. $C_1e^{-x}+C_2e^{-2x}-e^{2x}+\frac{1}{10}(3\sin 2x-\cos 2x)+1$

12. $\frac{d^2y}{dx^2}-2\frac{dy}{dx}+3y=\cos x+x^2$ **Ans.** $e^{x}[C_1\cos\sqrt{2}\,x+C_2\sin\sqrt{2}\,x]+\frac{1}{4}(\cos x-\sin x)+\frac{1}{3}(x^2+\frac{4}{3}x+\frac{2}{9})$

13. $(D^3-3D^2+4D-2)y=e^x+\cos x$ **Ans.** $(C_1+C_2\cos x+C_3\sin x)e^x+\frac{1}{10}(3\sin x+\cos x)+xe^x$

14. $(D^3-4D^2+13\,D)\,y=1+\cos 2x$

Ans. $C_1+e^{2x}(C_2\cos 3x+C_3\sin 3x)+\frac{1}{290}(9\sin 2x+8\cos 2x)+\frac{x}{13}$

15. $(D^2-4D+4)\,y=e^{2x}+x^3+\cos 2x$ **Ans.** $(C_1+C_2x)e^{2x}+\frac{1}{2}x^2e^{2x}+\frac{1}{8}(2x^3+6x^2+9x+6)-\frac{1}{8}\sin 2x$

(DU III Sem 2014)

16. $\frac{d^2y}{dx^2}+n^2y=h\sin px$

where h, p and n are constants satisfying the conditions

$y=a$, $\frac{dy}{dx}=b$ for $x=0$ **Ans.** $a\cos nx+\left(\frac{b}{n}-\frac{ph}{n(n^2-p^2)}\right)\sin nx+\frac{h\sin px}{(n^2-p^2)}$

17. $y''+y'-2y=-6\sin 2x-18\cos 2x$, $y\,(0)=2$, $y'(0)=2$ **Ans.** $-e^{-2x}+3\cos 2x$

13.20 $\boxed{\dfrac{1}{f(D)}\cdot e^{ax}\cdot\phi(x) = e^{ax}\cdot\dfrac{1}{f(D+a)}\cdot\phi(x)}$

$$D[e^{ax}\phi(x)] = e^{ax}D\phi(x) + ae^{ax}\phi(x) = e^{ax}(D+a)\phi(x)$$

$$D^2[e^{ax}\phi(x)] = D[e^{ax}(D+a)\phi(x)] = e^{ax}(D^2+aD)\phi(x) + ae^{ax}(D+a)\phi(x)$$

$$= e^{ax}(D^2+2aD+a^2)\phi(x) = e^{ax}(D+a)^2\phi(x)$$

Similarly, $\quad D^n[e^{ax}\phi(x)] = e^{ax}(D+a)^n\phi(x)$

$$f(D)[e^{ax}\phi(x)] = e^{ax}f(D+a)\phi(x)$$

$$e^{ax}\phi(x) = \frac{1}{f(D)}.[e^{ax}f(D+a)\phi(x)] \qquad ...(1)$$

Put $f(D+a)\phi(x) = X$, so that $\phi(x) = \dfrac{1}{f(D+a)}.X$

Substituting these values in (1), we get

$$e^{ax}\frac{1}{f(D+a)}X = \frac{1}{f(D)}[e^{ax}.X] \quad \Rightarrow \quad \frac{1}{f(D)}[e^{ax}.\phi(x)] = e^{ax}\frac{1}{f(D+a)}\phi(x)$$

Example 23. *Obtain the general solution of the differential equation*

$$y'' - 2y' + 2y = x + e^x\cos x.$$

Solution. $y'' - 2y' + 2y = x + e^x\cos x$

A.E. is $\quad m^2 - 2m + 2 = 0 \quad \Rightarrow \quad m = 1 + i$

C.F. $= e^x(A\cos x + B\sin x)$

$$\text{P.I.} = \frac{1}{D^2-2D+2}x + \frac{1}{D^2-2D+2}e^x\cos x$$

where $\quad I_1 = \dfrac{1}{D^2-2D+2}x = \dfrac{1}{2\left[1-D+\dfrac{D^2}{2}\right]}x = \dfrac{1}{2\left[1-\left(D-\dfrac{D^2}{2}\right)\right]}x$

$$= \frac{1}{2}\left[1-\left(D-\frac{D^2}{2}\right)\right]^{-1}x = \frac{1}{2}\left[1+\left(D-\frac{D^2}{2}\right)+\ldots\right]x = \frac{1}{2}\left[x+Dx-\frac{D^2}{2}x+\ldots\right] = \frac{1}{2}[x+1]$$

and, $\quad I_2 = \dfrac{1}{D^2-2D+2}e^x\cos x = e^x\dfrac{1}{(D+1)^2-2(D+1)+2}\cos x = e^x\dfrac{1}{D^2+1}\cos x = e^x.x\dfrac{1}{2D}\cos x$

$$\left[\text{If } f(-a^2)=0, \text{ then } \frac{1}{f(D^2)}\cos ax = x\frac{1}{f'(-a^2)}\cos ax\right]$$

$$= \frac{1}{2}xe^x\sin x$$

y = C.F. + P.I.

$$= e^x(A\cos x + B\sin x) + \frac{1}{2}(x+1) + \frac{1}{2}xe^x\sin x. \qquad \textbf{Ans.}$$

Example 24. *Solve* : $(D^2 - 4D + 4)\, y = x^3\, e^{2x}$

Solution. $(D^2 - 4D + 4)\, y = x^3\, e^{2x}$

A.E. is $m^2 - 4m + 4 = 0 \;\Rightarrow\; (m-2)^2 = 0 \;\Rightarrow\; m = 2, 2$

C.F. $= (C_1 + C_2\, x)\, e^{2x}$

$$\text{P.I.} = \frac{1}{D^2 - 4D + 4} x^3 \cdot e^{2x} = e^{2x} \frac{1}{(D+2)^2 - 4(D+2) + 4} x^3$$

$$= e^{2x} \frac{1}{D^2} x^3 = e^{2x} . \frac{1}{D}\left(\frac{x^4}{4}\right) = e^{2x} . \frac{x^5}{20}$$

The complete solution is $y = (C_1 + C_2 x) e^{2x} + e^{2x} . \dfrac{x^5}{20}$ **Ans.**

Example 25. *Solve* :

$$\frac{d^4 y}{dx^4} - y = \cos x . \cosh x$$

Solution. We have, $(D^4 - 1)\, y = \cos x \cosh x$

A.E. is $m^4 - 1 = 0 \;\Rightarrow\; (m^2 - 1)(m^2 + 1) = 0 \;\Rightarrow\; m = \pm 1, \pm i$

C.F. $= C_1\, e^x + C_2\, e^{-x} + C_3 \cos x + C_4 \sin x$

$$\text{P.I.} = \frac{1}{D^4 - 1} \cos x \; \cosh x = \frac{1}{D^4 - 1} \cos x \left(\frac{e^x + e^{-x}}{2}\right)$$

$$= \frac{1}{2}\left[\frac{1}{D^4 - 1} e^x \cos x + \frac{1}{D^4 - 1} e^{-x} \cos x\right] = \frac{1}{2}\left[e^x \frac{1}{(D+1)^4 - 1} \cos x + e^{-x} \frac{1}{(D-1)^4 - 1} \cos x\right]$$

$$= \frac{1}{2}\left[e^x \frac{1}{D^4 + 4D^3 + 6D^2 + 4D} \cos x + e^{-x} \frac{1}{D^4 - 4D^3 + 6D^2 - 4D} \cos x\right]$$

$$= \frac{1}{2}\left[e^x \frac{1}{(-1)^2 + 4D(-1) + 6(-1) + 4D} \cos x + e^{-x} \frac{1}{(-1)^2 - 4D(-1) + 6(-1) - 4D} \cos x\right]$$

$$= \frac{1}{2}\left[e^x \frac{1}{-5} \cos x + e^{-x} \frac{1}{-5} \cos x\right] = -\frac{1}{5}\left(\frac{e^x + e^{-x}}{2}\right) \cos x = -\frac{1}{5} \cosh x \cos x$$

Hence, the complete solution is

$$y = C_1 e^x + C_2 e^{-x} + C_3 \cos x + C_4 \sin x - \frac{1}{5} \cos x \cosh x$$ **Ans.**

Example 26. *Solve the differential equation:*

$$\frac{d^3 y}{dx^3} - 7\frac{d^2 y}{dx^2} + 10\frac{dy}{dx} = e^{2x} \sin x :$$ (*AMIETE, June 2010, Nagpur University, Summer* 2005)

Solution. $\dfrac{d^3 y}{dx^3} - 7\dfrac{d^2 y}{dx^2} + 10\dfrac{dy}{dx} = e^{2x} \sin x$

$\Rightarrow \quad D^3 y - 7D^2 y + 10\, Dy = e^{2x} \sin x$

A.E. is

$m^3 - 7m^2 + 10\, m = 0 \;\Rightarrow\; (m-2)(m^2 - 5m) = 0$

$\Rightarrow \; m(m-2)(m-5) = 0 \;\Rightarrow\; m = 0, 2, 5$

$$\text{C.F} = C_1e^{0x} + C_2e^{2x} + C_3e^{5x}$$

$$\text{P.I.} = \frac{1}{D^3 - 7D^2 + 10D} e^{2x} \sin x = e^{2x} \frac{1}{(D+2)^3 - 7(D+2)^2 + 10(D+2)} . \sin x$$

$$= e^{2x} \frac{1}{D^3 + 6D^2 + 12D + 8 - 7D^2 - 28D - 28 + 10D + 20} . \sin x$$

$$= e^{2x} \frac{1}{D^3 - D^2 - 6D} \sin x = e^{2x} \frac{1}{(-1^2) D - (-1^2) - 6D} \sin x$$

$$= e^{2x} \frac{1}{-D + 1 - 6D} \sin x = e^{2x} \frac{1}{1 - 7D} \sin x = e^{2x} \frac{1 + 7D}{1 - 49D^2} \sin x = e^{2x} \frac{1 + 7D}{1 - 49(-1^2)} \sin x$$

$$= \frac{e^x}{50} (\sin x + 7 \cos x)$$

Complete solution is

$$y = \text{C.F.} + \text{P.I.}$$

$$\Rightarrow \quad y = = C_1 + C_2 e^{2x} + C_3 e^{5x} + \frac{e^{2x}}{50} (\sin x + 7 \cos x)$$ **Ans.**

Example 27. *A body executes damped forced vibrations given by the equation*

$$\frac{d^2x}{dt^2} + 2k\frac{dx}{dt} + b^2x = e^{-kt} \sin \omega t.$$

Solve the differential equation for both the cases when $\omega^2 \neq b^2 - k^2$ *and when* $\omega^2 = b^2 - k^2$.

Solution. The given equation is $(D^2 + 2kD + b^2) x = e^{-kt} \sin \omega t$, ...(1)

which is a linear differential equation with constant coefficients.

A.E. is $m^2 + 2km + b^2 = 0$ or $m = \dfrac{-2k \pm \sqrt{(4k^2 - 4b^2)}}{2} = -k \pm \sqrt{(k^2 - b^2)}$

As the given problem is on vibration, we must have $k^2 < b^2$

$$m = -k \pm \sqrt{-(b^2 - k^2)} = -k \pm i\sqrt{(b^2 - k^2)}$$

$$\text{C.F.} = e^{-kt} \left\{ C_1 \cos \sqrt{(b^2 - k^2)}\, t + C_2 \sin \sqrt{(b^2 - k^2)}\, t \right\}$$

$$\text{P.I.} = \frac{1}{D^2 + 2kD + b^2} e^{-kt} \sin \omega t = e^{-kt} \frac{1}{(D-k)^2 + 2k(D-k) + b^2} \sin \omega t$$

$$e^{-kt} \frac{1}{D^2 + (b^2 - k^2)} \sin \omega t = e^{-kt} \frac{1}{-\omega^2 + (b^2 - k^2)} \sin \omega t, \text{ if } \omega^2 \neq b^2 - k^2 \quad ...(3)$$

If $\omega^2 = b^2 - k^2$, then $\text{P.I.} = e^{-kt} .t \dfrac{1}{2D} \sin \omega t = e^{-kt} \left(-\dfrac{t}{2\omega} \cos \omega t \right)$, ...(3)

Case. I. If $\omega^2 \neq b^2 - k^2$, the complete solution of (1) is

$$x = e^{-kt} \left\{ C_1 \cos \sqrt{(b^2 - k^2)}\, t + C_2 \sin \sqrt{(b^2 - k^2)}\, t \right\} + \frac{e^{-kt}}{(b^2 - k^2) - \omega^2} \sin \omega t \quad \text{[From (2)]}$$

Case II. If $\omega^2 = b^2 - k^2$, the complete solution of (1) is

$$x = e^{-kt} \{ C_1 \cos \omega t + C_2 \sin \omega t \} - \frac{e^{-kt} t \cos \omega t}{2\omega} \quad \text{[From (3)]}$$ **Ans.**

Example 28. *Solve* : $(D^2+6D+9)y=\dfrac{e^{-3x}}{x^3}$. *(A.M.I.E.T.E., June 2009)*

Solution A.E. is $m^2 + 6m + 9 = 0$

$(m + 3)^2 = 0 \qquad \therefore\ m = -3, -3$

C.F. $= (C_1 + C_2x)\, e^{-3x}$

$$\text{P.I.} = \frac{1}{D^2+6D+9}\frac{e^{-3x}}{x^3} = e^{-3x}\frac{1}{(D-3)^2+6(D-3)+9}\frac{1}{x^3}$$

$$= e^{-3x}\frac{1}{D^2-6D+9+6D-18+9}\frac{1}{x^3} = e^{-3x}\frac{1}{D^2}(x^{-3})$$

$$= e^{-3x}\frac{1}{D}\left(\frac{x^{-2}}{-2}\right) = e^{-3x}\frac{x^{-1}}{(-2)\,(-1)} = \frac{e^{-3x}x^{-1}}{2} = \frac{e^{-3x}}{2x}$$

Hence, the solution is $y = (C_1+C_2x)e^{-3x}+\dfrac{e^{-3x}}{2x}$ **Ans.**

Example 29. *Solve* $(D^2 - 4D + 3)\, y = 2x\, e^{3x} + 3e^x \cos 2x$

Solution. The auxiliary equation is

$m^2 - 4m + 3 = 0$ which gives $m = 1, 3$

C.F. $= C_1e^x + C_2e^{3x}$

$$\text{P.I.} = \frac{1}{D^2-4D+3}2xe^{3x}+\frac{1}{D^2-4D+3}3e^x\cos 2x$$

$$= 2e^{3x}.\frac{1}{(D+3)^2-4(D+3)+3}x+3e^x.\frac{1}{(D+1)^2-4(D+1)+3}\cos 2x$$

$$= 2e^{3x}.\frac{1}{D^2+2D}x+3e^x.\frac{1}{D^2-2D}\cos 2x = 2e^{3x}.\frac{1}{2D(1+D/2)}x+3e^x.\frac{1}{-4-2D}\cos 2x$$

$$= e^{3x}.\frac{1}{D}\left(1+\frac{D}{2}\right)^{-1}x-\frac{3e^x}{2}.\frac{1}{2+D}\cos 2x = e^{3x}.\frac{1}{D}\left(1-\frac{D}{2}+\frac{D^2}{4}...\right)x-\frac{3e^x}{2}.\frac{2-D}{4-D^2}\cos 2x$$

$$= e^{3x}.\left(\frac{1}{D}-\frac{1}{2}+\frac{D}{4}...\right)x-\frac{3e^x}{2}.\frac{2-D}{4+4}\cos 2x = e^{3x}.\left(\frac{x^2}{2}-\frac{x}{2}+\frac{1}{4}\right)-\frac{3e^x}{16}(2\cos 2x+2\sin 2x)$$

The complete solution is

$$y = C_1e^x+C_2e^{3x}+e^{3x}\left(\frac{x^2}{2}-\frac{x}{2}+\frac{1}{4}\right)-\frac{3e^x}{8}(\cos 2x+\sin 2x)$$

The term $\dfrac{e^{3x}}{4}$ has been omitted from the P.I., since C_2e^{3x} is present in the C.F. **Ans.**

Example 30. *Find the complete solution of* $\frac{d^2y}{dx^2} - 3\frac{dy}{dx} + 2y = xe^{3x} + \sin 2x$

Solution. The auxiliary equation is

$$m^2 - 3m + 2 = 0 \quad \Rightarrow \quad m^2 - 2m - m + 2 = 0$$

$$(m-2)(m-1) = 0 \quad \Rightarrow \quad m = 1, 2$$

$$\text{C.F.} = C_1e^x + C_2e^{2x}$$

$$\text{P.I.} = \frac{1}{D^2 - 3D + 2}(xe^{3x} + \sin 2x) = \frac{1}{D^2 - 3D + 2}xe^{3x} + \frac{1}{D^2 - 3D + 2}\sin 2x$$

$$= e^{3x}\frac{1}{(D+3)^2 - 3(D+3) + 2}x + \frac{1}{-4 - 3D + 2}\sin 2x$$

$$= e^{3x}\frac{1}{D^2 + 6D + 9 - 3D - 9 + 2}x + \frac{1}{-3D - 2}\sin 2x = e^{3x}\frac{1}{D^2 + 3D + 2}x - \frac{1}{3D + 2}\sin 2x$$

$$= \frac{e^{3x}}{2}\left[1 + \left(\frac{3D + D^2}{2}\right)\right]^{-1} x - \frac{(3D-2)}{9D^2 - 4}\sin 2x = \frac{e^{3x}}{2}\left[1 - \left(\frac{3D + D^2}{2}\right) + \ldots\right]x - \frac{(3D-2)}{9(-4) - 4}\sin 2x$$

$$= \frac{e^{3x}}{2}\left[x - \left(\frac{3D + D^2}{2}\right)x + \ldots\right] - \frac{(3D-2)}{-36-4}\sin 2x = \frac{e^{3x}}{2}\left[x - \frac{3}{2}\right] + \frac{3D - 2}{40}\sin 2x$$

$$\text{P.I.} = \frac{e^{3x}}{4}(2x - 3) + \frac{1}{40}(6\cos 2x - 2\sin 2x) = \frac{e^{3x}}{4}(2x - 3) + \frac{3}{20}\cos 2x - \frac{1}{20}\sin 2x$$

The complete solution is

$$y = \text{C.F.} + \text{P.I.}$$

$$y = c_1e^x + c_2e^{2x} + \frac{e^{3x}}{4}(2x - 3) + \frac{3}{20}\cos 2x - \frac{1}{20}\sin 2x \qquad \textbf{Ans.}$$

Example 31. *Solve* : $\frac{d^2y}{dx^2} - 2\frac{dy}{dx} + y = xe^x \sin x$ *(U.P., II Semester 2009, 2005)*

Solution. The given equation is $(D^2 - 2D + 1)\, y = xe^x \sin x$

A.E. is $m^2 - 2m + 1 = 0 \quad \therefore \quad m = 1, 1$

C.F. $= (C_1 + C_2x)\, e^x$

$$\text{P.I.} = \frac{1}{(D-1)^2}e^x.x\sin x = e^x\frac{1}{(D+1-1)^2}x\sin x = e^x\frac{1}{D^2}x\sin x = e^x.\frac{1}{D}\int x\sin x\, dx$$

Integrating by parts

$$= e^x\frac{1}{D}[x(-\cos x) - \int(-\cos x)\, dx] = e^x.\frac{1}{D}(-x\cos x + \sin x)$$

$$= e^x\int(-x\cos x + \sin x)\, dx = e^x\left\{-x\sin x + \int 1.\sin x\, dx - \cos x\right\}$$

$$= e^x\,[-x\sin x - \cos x - \cos x] \qquad = -e^x(x\sin x + 2\cos x)$$

Hence, the complete solution is $y = (C_1 + C_2x)\, e^x - e^x(x\sin x + 2\cos x)$. **Ans.**

Example 32. *Solve* $(D^2 + 5D + 6)\, y = e^{-2x} \sec^2 x\, (1 + 2 \tan x)$

Solution. $(D^2 + 5D + 6)\, y = e^{-2x} \sec^2 x\, (1 + 2 \tan x)$

Auxiliary Equation is $m^2 + 5m + 6 = 0$

$$\Rightarrow \quad (m + 2)(m + 3) = 0 \quad \Rightarrow \quad m = -2, \text{ and } m = -3$$

Hence, complementry function (C.F.) $= C_1 e^{-2x} + C_2 e^{-3x}$

$$\text{P.I.} = \frac{1}{D^2+5D+6} e^{-2x} \cdot \sec^2 x(1+2\tan x) = e^{-2x} \frac{1}{(D-2)^2+5(D-2)+6} \sec^2 x(1+2\tan x)$$

$$e^{-2x} \frac{1}{D^2-4D+4+5D-10+6} \sec^2 x(1+2\tan x)$$

$$= e^{-2x} \frac{1}{D^2+D} \sec^2 x(1+2\tan x) = e^{-2x} \left[\frac{\sec^2 x}{D^2+D} + \frac{2\tan x \sec^2 x}{D^2+D} \right]$$

$$= e^{-2x} \left[\frac{1}{D(D+1)} \sec^2 + \frac{1}{D(D+1)} 2\tan x \sec^2 x \right]$$

$$= e^{-2x} \left[\left(\frac{1}{D} - \frac{1}{D+1} \right) \sec^2 x + \left(\frac{1}{D} - \frac{1}{D+1} \right) 2\tan x \sec^2 x \right]$$

$$= e^{-2x} \left[\frac{1}{D} \sec^2 x - \frac{1}{D+1} \sec^2 x + \frac{1}{D} 2\tan x \sec^2 x - \frac{1}{D+1} 2\tan x \sec^2 x \right]$$

$$= e^{-2x} \left[\tan x - e^{-x} \int e^x . \sec^2 x\, dx + \tan^2 x - e^{-x} \int 2e^x \tan x \sec^2 x\, dx \right] \qquad ...(1)$$

$$\text{Now,} = \int e^x \sec^2 x\, dx = e^x \sec^2 x - \int e^x . 2\sec x \sec x \tan x\, dx$$

$$= e^x \sec^2 x - 2\int e^x \sec^2 x . \tan x\, dx \qquad ...(2)$$

Putting the value of $\int e^x \sec^2 x\, dx$ from (2) in (1), we get

$$\therefore \text{ P.I.} = e^{-2x} \left[\tan x - e^{-x} . e^x \sec^2 x + 2e^{-x} \int e^x \sec x \tan x\, dx + \tan^2 x - 2e^{-x} \int e^x \sec^2 x \tan x\, dx \right]$$

$$= e^{-2x} [\tan x - \sec^2 x + \tan^2 x] = e^{-2x} [\tan x - (\sec^2 x - \tan^2 x] = e^{-2x} (\tan x - 1)$$

$\therefore$ Complete solution is

$$\Rightarrow y = C.F. + P.I. = C_1 e^{-2x} + C_2 e^{-3x} + e^{-2x} (\tan x - 1) \qquad \textbf{Ans.}$$

Example 33. *Solve the differential equation* $(D^2 - 4D + 4)\, y = 8x^2 e^{2x} \sin 2x$

(U.P. II Semester, Summer 2008, Uttrakhand 2007, 2005, Nagpur University June 2008)

Solution. $(D^2 - 4D + 4)\, y = 8x^2 e^{2x} \sin 2x$

A.E. is $(m^2 - 4m + 4) = 0 \Rightarrow (m - 2)^2 = 0 \Rightarrow m = 2, 2$

C.F. $= (C_1 + C_2 x)\, e^{2x}$

$$\text{P.I.} = \frac{1}{D^2-4D+4} 8x^2 e^{2x} \sin 2x = 8 \frac{1}{(D-2)^2} x^2 e^{2x} \sin 2x$$

$$= 8e^{2x} \frac{1}{(D-2+2)^2} x^2 \sin 2x = 8e^{2x} \frac{1}{D^2} x^2 \sin 2x$$

$$= 8e^{2x}\frac{1}{D}\left[x^2\frac{(-\cos 2x)}{2} - 2x\left(-\frac{\sin 2x}{4}\right) + 2\frac{\cos 2x}{8}\right] = 8e^{2x}\frac{1}{D}\left[-\frac{x^2}{2}\cos 2x + \frac{x\sin 2x}{2} + \frac{\cos 2x}{4}\right]$$

$$= 8e^{2x}\left[\begin{array}{r}-\frac{x^2}{2}\left(\frac{\sin 2x}{2}\right) - \left(\frac{-2x}{2}\right)\left(-\frac{\cos 2x}{4}\right) + (-1)\left(-\frac{\sin 2x}{8}\right) + \frac{x}{2}\left(-\frac{\cos 2x}{2}\right) \\ -\left(\frac{1}{2}\right)\left(-\frac{\sin 2x}{4}\right) + \frac{\sin 2x}{8}\end{array}\right]$$

$= e^{2x}$ [– $2x^2$ sin 2*x* – 2*x* cos 2*x* + sin 2*x* – 2*x* cos 2*x* + sin 2*x* + sin 2*x*]

$= e^{2x}$ [– $2x^2$ sin 2*x* – 4*x* cos 2*x* + 3 sin 2*x*] = – e^{2x} [4*x* cos 2*x* + ($2x^2$ – 3) sin 2*x*]

Complete solution is, *y* = C.F. + P.I.

$$y = (C_1 + C_2x)\, e^{2x} - e^{2x}\, [4x \cos 2x + (2x^2 - 3) \sin 2x]$$ **Ans.**

EXERCISE 13.6

Solve the following differential equations :

1. $(D^2 - 5D + 6)\, y = e^x \sin x$ **Ans.** $y = C_1e^{2x} + C_2e^{3x} + \frac{e^x}{10}(3\cos x + \sin x)$

2. $\frac{d^2y}{dx^2} - 7\frac{dy}{dx} + 10y = e^{2x}\sin x$ **Ans.** $y = C_1e^{2x} + C_2e^{5x} + \frac{e^{2x}}{10}(3\cos x - \sin x)$

3. $\frac{d^3y}{dx^3} - 2\frac{dy}{dx} + 4y = e^x\cos x$ **Ans.** $y = C_1e^{-2x} + e^x(C_2\cos x + C_3\sin x) + \frac{xe^x}{20}(3\sin x - \cos x)$

4. $\frac{d^2y}{dx^2} + 2\frac{dy}{dx} + y = \frac{e^{-x}}{x^2}$ **Ans.** $y = (C_1 + C_2\, x)\, e^{-x} - e^{-x} \log x$

5. $(D^2 - 4)\, y = x^2\, e^{3x}$ **Ans.** $y = C_1e^{2x} + C_2e^{-2x} + \frac{e^{3x}}{5}\left[x^2 - \frac{12x}{5} + \frac{62}{25}\right]$

6. $(D^2 - 3D + 2)\, y = 2x^2\, e^{4x} + 5e^{3x}$ **Ans.** $y = C_1e^x + C_2e^{2x} + \frac{e^{4x}}{54}[18x^2 - 30x + 19] + \frac{5}{2}e^{3x}$

7. $\frac{d^2y}{dx^2} - 4y = x\sinh x$ **Ans.** $y = C_1e^{2x} + C_2e^{-2x} - \frac{x}{3}\sinh x - \frac{2}{9}\cosh x$

8. $\frac{d^2y}{dt^2} + 2h\frac{dy}{dt} + (h^2 + p^2)y = ke^{-ht}\cos pt$ **Ans.** $y = e^{-ht}[A\cos pt + B\sin pt] + \frac{k}{2p}te^{-ht}\sin pt$

13.21 TO FIND THE VALUE OF $\frac{1}{f(D)}x^n \sin ax$.

Now $$\frac{1}{f(D)}x^n(\cos ax + i\sin ax) = \frac{1}{f(D)}x^n e^{iax} = e^{iax}\frac{1}{f(D+ia)}x^n$$

$$\boxed{\frac{1}{f(D)}\cdot x^n \sin ax = \text{Imaginary part of } e^{iax}\frac{1}{f(D+ia)}\cdot x^n}$$

$$\boxed{\frac{1}{f(D)}\cdot x^n \cos ax = \text{Real part of } e^{iax}\frac{1}{f(D+ia)}\cdot x^n}$$

Example 34. *Solve* $\frac{d^2y}{dx^2}-2\frac{dy}{dx}+y = x\sin x$

Solution. Auxiliary equation is $m^2 - 2m + 1 = 0$ or $m = 1, 1$

C.F. $= (C_1 + C_2\, x)\, e^x$

$$\text{P.I.} = \frac{1}{D^2-2D+1}x\cdot\sin x \quad (e^{ix} = \cos x + i\sin x)$$

$$= \text{Imaginary part of } \frac{1}{D^2-2D+1}x(\cos x + i\sin x) = \text{Imaginary part of } \frac{1}{D^2-2D+1}x\cdot e^{ix}$$

$$= \text{Imaginary part of } e^{ix}\frac{1}{(D+i)^2-2(D+i)+1}\cdot x$$

$$= \text{Imaginary part of } e^{ix}\frac{1}{D^2-2(1-i)D-2i}\cdot x$$

$$= \text{Imaginary part of } e^{ix}\frac{1}{-2i}\left[1-(1+i)D-\frac{1}{2i}D^2\right]^{-1}\cdot x$$

$$= \text{Imaginary part of } (\cos x + i\sin x)\left(\frac{i}{2}\right)[1+(1+i)D]x$$

$$= \text{Imaginary part of } \frac{1}{2}(i\cos x - \sin x)\,[x+1+i]$$

$$\text{P.I.} = \frac{1}{2}x\cos x + \frac{1}{2}\cos x - \frac{1}{2}\sin x$$

Complete solution is $y = (C_1 + C_2x)e^x + \frac{1}{2}(x\cos x + \cos x - \sin x)$ **Ans.**

Example 35. *Solve* : $\frac{d^2y}{dx^2}-2\frac{dy}{dx}+y = xe^x\sin x$ (*U.P., II Semester 2009, 2005*)

Solution. The given equation is $(D^2 - 2D + 1)\, y = xe^x \sin x$

A.E. is $m^2 - 2m + 1 = 0$ $\therefore$ $m = 1, 1$

C.F. $= (C_1 + C_2x)\, e^x$

$$\text{P.I.} = \frac{1}{(D-1)^2}e^x.x\sin x = e^x\frac{1}{(D+1-1)^2}x\sin x = e^x\frac{1}{D^2}x\sin x = e^x.\frac{1}{D}\int x\sin x\, dx$$

Integrating by parts

$$= e^x\frac{1}{D}[x(-\cos x) - \int(-\cos x)\,dx] = e^x.\frac{1}{D}(-x\cos x + \sin x)$$

$$= e^x\int(-x\cos x + \sin x)\,dx = e^x\left\{-x\sin x + \int 1.\sin x\, dx - \cos x\right\}$$

$$= e^x\,[-x\sin x - \cos x - \cos x] \qquad = -e^x\,(x\sin x + 2\cos x)$$

Hence, the complete solution is $y = (C_1 + C_2x)\, e^x - e^x\,(x \sin x + 2 \cos x)$. **Ans.**

EXERCISE 13.7

Solve the following differential equations :

1. $(D^2+4)y = 3x \sin x$ *(DU, II Sem. 2012)* **Ans.** $C_1 \cos 2x + C_2 \sin 2x + x \sin x - \frac{2}{3} \cos x$

2. $\frac{d^2y}{dx^2} - y = x \sin 3x + \cos x$ **Ans.** $C_1e^x + C_2e^{-x} - \frac{1}{10}\left[\frac{3}{5}\cos 3x + x \sin 3x + 5\cos x\right]$

3. $\frac{d^2y}{dx^2} - y = x \sin x + e^x + x^2e^x$ **Ans.** $C_1e^x + C_2e^{-x} - \frac{1}{2}[x \sin x + \cos x] + \frac{x}{12}e^x(2x^2 - 3x + 9)$

4. $(D^4 + 2D^2 + 1)\, y = x^2 \cos x$

   Ans. $(C_1 + C_2x)\cos x + (C_3 + C_4x)\sin x + \frac{1}{12}x^3 \sin x - \frac{1}{48}(x^4 - 9x^2)\cos x$

5. $(D^2+4)y = x \sin 2x$ **Ans.** $(C_1 \cos 2x + C_2 \sin 2x) - \frac{1}{16}(2x^2 \cos 2x + x \sin 2x)$ *(DU II Sem 2014)*

13.22 GENERAL METHOD OF FINDING THE PARTICULAR INTEGRAL OF ANY FUNCTION ϕ (*x*)

$$\text{P.I.} = \frac{1}{D-a}\phi(x) = y$$

or
$$(D-a)\frac{1}{D-a}\cdot\phi(x) = (D-a)\cdot y$$

$$\phi(x) = (D-a)y \qquad \text{or} \qquad \phi(x) = Dy - ay$$

$$\frac{dy}{dx} - ay = \phi(x) \qquad \text{which is the linear differential equation.}$$

Its solution is $ye^{-\int a\,dx} = \int e^{-\int a\,dx}\cdot\phi(x)\,dx \qquad ye^{-ax} = \int e^{-ax}\cdot\phi(x)\,dx$

$$y = e^{ax}\int e^{-ax}\cdot\phi(x)\,dx \qquad \boxed{\frac{1}{D-a}\cdot\phi(x) = e^{ax}\int e^{-ax}\cdot\phi(x)\,dx}$$

Example 36. *Solve* $\frac{d^2y}{dx^2} + 9y = \sec 3x.$

Solution. Auxiliary equation is $m^2 + 9 = 0$ or $m = \pm 3i$,

$$\text{C.F.} = C_1 \cos 3x + C_2 \sin 3x$$

$$\text{P.I.} = \frac{1}{D^2+9}\cdot\sec 3x = \frac{1}{(D+3i)\,(D-3i)}\cdot\sec 3x = \frac{1}{6i}\left[\frac{1}{D-3i} - \frac{1}{D+3i}\right]\cdot\sec 3x$$

$$= \frac{1}{6i}\cdot\frac{1}{D-3i}\cdot\sec 3x - \frac{1}{6i}\frac{1}{D+3i}\cdot\sec 3x \qquad ...(1)$$

Now, $\frac{1}{D-3i}\sec 3x = e^{3ix}\int e^{-3ix}\sec 3x\,dx \qquad \left[\frac{1}{D-a}\phi(x) = e^{ax}\int e^{-ax}\phi(x)\,dx\right]$

$$= e^{3ix}\int\frac{\cos 3x - i\sin 3x}{\cos 3x}dx = e^{3ix}\int(1 - i\tan 3x)\,dx = e^{3ix}\,(x + \frac{i}{3}\log\cos 3x)$$

Changing i to $-i$, we have $\frac{1}{D+3i}\sec 3x = e^{-3ix}\,(x - \frac{i}{3}\log\cos 3x)$

Putting these values in (1), we get

$$\text{P.I.} = \frac{1}{6i}\left[e^{3ix}\left(x + \frac{i}{3}\log\cos 3x\right) - e^{-3ix}\left(x - \frac{i}{3}\log\cos 3x\right)\right]$$

$$= \frac{x}{6i}e^{3ix} + \frac{e^{3ix}\log\cos 3x}{18} - \frac{xe^{-3ix}}{6i} + \frac{e^{-3ix}}{18}\log\cos 3x$$

$$= \frac{x}{3}\frac{e^{3ix} - e^{-3ix}}{2i} + \frac{1}{9}.\frac{e^{3ix} + e^{-3ix}}{2}\log\cos 3x = \frac{x}{3}\sin 3x + \frac{1}{9}\cos 3x \cdot \log\cos 3x$$

Hence, complete solution is $y = C_1\cos 3x + C_2\sin 3x + \frac{x}{3}\sin 3x + \frac{1}{9}\cos 3x \cdot \log\cos 3x$ **Ans.**

Example 37. *Solve* $\frac{d^2y}{dx^2} + 2\frac{dy}{dx} + 2y = e^{-x}\sec^3 x.$

Solution. Here we have

$$\frac{d^2y}{dx^2} + 2\frac{dy}{dx} + 2y = e^{-x}\sec^3 x$$

$$(D^2 + 2D + 2)y = e^{-x}\sec^3 x$$

A.E. is $m^2 + 2m + 2 = 0 \quad \Rightarrow \quad m = \frac{-2 \pm \sqrt{4-8}}{2} = -1 + i$

$C.F. = e^{-x}(C_1\cos x + C_2\sin x)$

$$\text{P.I.} = \frac{1}{D^2 + 2D + 2}e^{-x}\sec^3 x = e^{-x}\frac{1}{(D-1)^2 + 2(D-1) + 2}\sec^3 x$$

$$e^{-x}\frac{1}{D^2+1}\sec^3 x = \frac{1}{(D+i)(D-i)}\sec^3 x = e^{-x}\frac{1}{2i}\left[\frac{1}{D-i} - \frac{1}{D+i}\right]\sec^3 x \qquad ...(1)$$

$$\text{Now } \frac{1}{D-i}\sec^3 x = e^{ix}\int e^{-ix}\sec^3 x\,dx \quad \left[\frac{1}{D-a}\phi(x) = e^{ax}\int e^{-ax}\phi(x)\,dx\right]$$

$$= e^{ix}\int(\cos x - i\sin x)\sec^3 x\,dx = e^{ix}\int[\sec^2 x - i\tan x\sec^2 x]dx = e^{ix}\left[\tan x - \frac{i\tan^2 x}{2}\right] \qquad ...(2)$$

Similarly $\frac{1}{D+i}\sec^3 x = e^{-ix}\left[\tan x + i\frac{\tan^2 x}{2}\right]$ ...(3) [changing i to $-i$]

Putting the values from (2) and (3) in (1), we get

$$P.I. = \frac{e^{-x}}{2i}\left[e^{ix}\left(\tan x - \frac{i\tan^2 x}{2}\right) - e^{-ix}\left(\tan x + i\frac{\tan^2 x}{2}\right)\right]$$

$$= e^{-x}\left[\tan x\frac{e^{ix} - e^{-ix}}{2i} - i\frac{\tan^2 x}{2}\left(\frac{e^{ix} + e^{-ix}}{2i}\right)\right] = e^{-x}\left[\tan x\sin x - \frac{\tan^2 x}{2}\cos x\right]$$

$$= e^{-x}\left[\tan x.\sin x - \frac{\tan x}{2}\frac{\sin x}{\cos x}\cos x\right] = e^{-x}\left[\tan x\sin x - \frac{\tan x}{2}\sin x\right] = e^{-x}\left(\frac{1}{2}\tan x\sin x\right)$$

Complete solution = C.F. + P.I.

$$e^{-x}(c_1\cos x + c_2\sin x) + \frac{e^{-x}}{2}\tan x\sin x = e^{-x}\left[C_1\cos x + C_2\sin x + \frac{\sin x\tan x}{2}\right] \qquad \textbf{Ans.}$$

Example 38. *Solve* . $\frac{d^2y}{dx^2} - 2\frac{dy}{dx} + 2y = e^x\tan x$.

Solution. Here we have $\frac{d^2y}{dx^2} - 2\frac{dy}{dx} + 2y = e^x\tan x$

$$(D^2 - 2D + 2)y = e^x \tan x$$

A.E. is $m^2 - 2m + 2 = 0 \quad \Rightarrow m = \dfrac{2 \pm \sqrt{4-8}}{2} = 1 \pm i$

$C.F. = e^x (C_1 \cos x + C_2 \sin x)$

$$P.I. = \frac{1}{D^2 - 2D + 2} e^x \tan x = e^x \frac{1}{(D+1)^2 - 2(D+1) + 2} \tan x$$

$$= e^x \frac{1}{D^2 + 1} \tan x = e^x \frac{1}{(D+i)(D-i)} \tan x = \frac{e^x}{2i}\left[\frac{1}{D-i} - \frac{1}{D+i}\right] \tan x \quad ...(1)$$

Now $\quad \dfrac{1}{D-i} \tan x = e^{ix} \int e^{-ix} \tan x \, dx$

$$= e^{ix} \int (\cos x - i \sin x) \tan x \, dx = e^{ix} \int \left(\sin x - i \frac{\sin^2 x}{\cos x} \right) dx$$

$$= e^{ix} \int \left[\sin x - \frac{i(1 - \cos^2 x)}{\cos x} \right] dx = e^{ix} \int (\sin x - i \sec x + i \cos x] dx$$

$$e^{ix} [-\cos x + i \sin x - i \log (\sec x + \tan x)] \quad ...(2)$$

Similarly $\quad \dfrac{1}{D+i} \tan x = e^{-ix} [-\cos x - i \sin x + i \log (\sec x + \tan x)] \quad ...(3)$

On putting the values from (2) and (3) in (1), we get

$$P.I. = \frac{e^x}{2i} [e^{ix} (-\cos x + i \sin x - i \log (\sec x + \tan x) - e^{-ix} (-\cos x - i \sin x + i \log (\sec x + \tan x)]$$

$$= e^x \left[-\cos x \frac{e^{ix} - e^{-ix}}{2i} + \sin x \frac{e^{ix} + e^{-ix}}{2} - \log (\sec x + \tan x) \frac{e^{ix} + e^{-ix}}{2} \right]$$

$$= e^x [-\cos x \sin x + \sin x \cos x - \cos x \log (\sec x + \tan x)]$$

$$= - e^x \cos x \log (\sec x + \tan x)$$

Complete solution $= C.F. + P.I.$

$= e^x (C_1 \cos x + C_2 \sin x) - e^x \cos x \log (\sec x + \tan x)$ **Ans.**

EXERCISE 13.8

Solve the following differential equations :

1. $\dfrac{d^2y}{dx^2} + a^2 y = \sec ax$ *(R.G.P.V., Bhopal April, 2010)*

Ans. $C_1 \cos ax + C_2 \sin ax + \dfrac{x}{a} \sin ax + \dfrac{1}{a^2} \cos ax \cdot \log \cos ax$

2. $\dfrac{d^2y}{dx^2} + y = \operatorname{cosec} x$ **Ans.** $C_1 \cos x + C_2 \sin x - x \cos x + \sin x \log \sin x$

3. $(D^2 + 4)\, y = \tan 2x$ **Ans.** $C_1 \cos 2x + C_2 \sin 2x - \dfrac{1}{4} \cos 2x \log (\sec 2x + \tan 2x)$

4. $\dfrac{d^2y}{dx^2} + y = (x - \cot x)$ **Ans.** $C_1 \cos x + C_2 \sin x - x \cos^2 x - \sin x \log (\operatorname{cosec} x - \cot x)$

Cauchy – Euler Equations, Method of Variation of Parameters

14.1 CAUCHY EULER HOMOGENEOUS LINEAR EQUATIONS

$$a_n x_n \frac{d^n y}{dx^n} + a_{n-1} x^{n-1} \frac{d^{n-1} y}{d x^{n-1}} + ... + a_0 y = \phi(x) \qquad ... (1)$$

where a_0, a_1, a_2, ... are constants, is called a homogeneous equation.

Put $\quad x = e^z, \quad z = \log_e x, \quad \frac{d}{dz} \equiv D$

$$\frac{dy}{dx} = \frac{dy}{dz} \cdot \frac{dz}{dx} = \frac{1}{x} \frac{dy}{dz} \Rightarrow x \frac{dy}{dx} = \frac{dy}{dz} \Rightarrow x \frac{dy}{dx} = Dy$$

Again, $$\frac{d^2 y}{dx^2} = \frac{d}{dx}\left(\frac{dy}{dx}\right) = \frac{d}{dx}\left(\frac{1}{x}\frac{dy}{dz}\right) = -\frac{1}{x^2}\frac{dy}{dz} + \frac{1}{x}\frac{d^2 y}{dz^2}\frac{dz}{dx}$$

$$= -\frac{1}{x^2}\frac{dy}{dz} + \frac{1}{x}\frac{d^2 y}{dz^2}\frac{1}{x} = \frac{1}{x^2}\left(\frac{d^2 y}{dz^2} - \frac{dy}{dz}\right) = \frac{1}{x^2}(D^2 - D)\, y$$

$$x^2 \frac{d^2 y}{dx^2} = (D^2 - D)\, y$$

or $$\boxed{x^2 \frac{d^2 y}{dx^2} = D(D-1)\, y}$$

Similarly. $$\boxed{x^3 \frac{d^3 y}{dx^3} = D(D-1)(D-2)\, y}$$

The substitution of these values in (1) reduces the given homogeneous equation to a differential equation with constant coefficients.

Example 1. *Solve*: $x^2 \frac{d^2 y}{dx^2} - 2x \frac{dy}{dx} - 4y = x^4$

Solution. We have, $$x^2 \frac{d^2 y}{dx^2} - 2x \frac{dy}{dx} - 4y = x^4 \qquad ... (1)$$

Putting $x = e^z$, $\quad D = \frac{d}{dz},\ x\frac{dy}{dx} = Dy,\ x^2 \frac{d^2 y}{dx^2} = D(D-1)\, y$ in (1), we get

$$D(D-1)\, y - 2\, Dy - 4y = e^{4z} \quad \text{or} \quad (D^2 - 3D - 4)\, y = e^{4z}$$

A.E. is $m^2 - 3m - 4 = 0 \Rightarrow (m-4)(m+1) = 0 \Rightarrow m = -1, 4$

$$\text{C.F.} = C_1 e^{-z} + C_2 e^{4z}$$

$$\text{P.I.} = \frac{1}{D^2 - 3D - 4} e^{4z} \qquad \text{[Rule Fails]}$$

$$= z\frac{1}{2D-3} e^{4z} = z\frac{1}{2(4)-3} e^{4z} = \frac{z e^{4z}}{5}$$

Thus, the complete solution is given by

$$y = C_1 e^{-z} + C_2 e^{4z} + \frac{z e^{4z}}{5} \Rightarrow y = \frac{C_1}{x} + C_2 x^4 + \frac{1}{5} x^4 \log x \qquad \textbf{Ans.}$$

Example 2. *Solve:* $x^2 \dfrac{d^2y}{dx^2} + 4x\dfrac{dy}{dx} + 2y = e^x$

(U.P., II Semester 2005;)

Solution. Given equation is $x^2 \dfrac{d^2y}{dx^2} + 4x\dfrac{dy}{dx} + 2y = e^x$ ...(1)

To solve (1), we put $x = e^z$ or $z = \log x$ and $D \equiv \dfrac{d}{dz}$

$$x\frac{dy}{dx} = Dy \quad \text{and} \quad x^2 \frac{d^2y}{dx^2} = D(D-1)\,y$$

Substituting these values in (1), it reduces

$$[D(D-1) + 4D + 2]\,y = e^{e^z} \Rightarrow (D^2 + 3D + 2)\,y = e^{e^z}$$

$\therefore$ It's *A.E.* is $m^2 + 3m + 2 = 0$

$\therefore$ $(m+1)(m+2) = 0 \quad \therefore \quad m = -1, -2$

$\therefore$ $\text{C.F.} = C_1 e^{-z} + C_2 e^{-2z} = \dfrac{C_1}{x} + \dfrac{C_2}{x^2}$

$$\text{P.I.} = \frac{1}{D^2 + 3D + 2} e^{e^z} = \frac{1}{(D+1)(D+2)} e^{e^z}$$

$$= \left(\frac{1}{D+1} - \frac{1}{D+2}\right) e^{e^z} = \frac{1}{D+1} e^{e^z} - \frac{1}{D+2} e^{e^z}$$

$$= e^{-z}\int e^{e^z}.e^z\,dz - e^{-2z}\int e^{e^z}.e^{2z}\,dz \qquad \left[\text{Since } \frac{1}{D-a}X = e^{ax}\int Xe^{-ax}\,dx\right]$$

Put $e^z = t$ so that $e^z\,dz = dt$

$$\text{P.I.} = e^{-z}\int e^t\,dt - e^{-2z}\int e^t.t\,dt = e^{-z}e^t - e^{-2z}(te^t - e^t)$$

$$= e^{-z}e^{e^z} - e^{-2z}(e^z e^{e^z} - e^{e^z})$$

$$= e^{-z}e^{e^z} - e^{-z}e^{e^z} + e^{-2z}e^{e^z} = e^{-2z}e^{e^z} = x^{-2}e^x = \frac{e^x}{x^2} \qquad (\because x = e^z)$$

Hence, the C.S. of (1) is $y = \text{C.F.} + \text{P.I.}$

$$y = \frac{C_1}{x} + \frac{C_2}{x^2} + \frac{e^x}{x^2} \qquad \textbf{Ans.}$$

Example 3. *Solve* $(x^3 D^3 + x^2 D^2 - 2)\, y = x - \frac{1}{x^3}$

Solution. Put $x = e^z$, $z = \log x$

Let $D_1 \equiv \frac{d}{dz}$

Then $x^2 \frac{d^2 y}{dx^2} = D_1 (D_1 - 1)\, y \quad \Rightarrow \quad x^3 \frac{d^3 y}{dx^3} = D_1 (D_1 - 1)(D_1 - 2)\, y$

Substituting in the given equation, we get

$$[\, D_1 (D_1 - 1)(D_1 - 2) + D_1 (D_1 - 1) - 2\,]\, y = e^z + e^{-3z}$$

$$\Rightarrow \quad [\, D_1^3 - 3D_1^2 + 2D_1 + D_1^2 - D_1 - 2\,]\, y = e^z + e^{-3z}$$

$$\Rightarrow \quad (\, D_1^3 - 2D_1^2 + D_1 - 2\,)\, y = e^z + e^{-3z}$$

A.E. is $m^3 - 2m^2 + m - 2 = 0$

i.e. $(m - 2)(m^2 + 1) = 0$ *i.e.* $m = 2, \quad \pm i.$

$$\text{C.F.} = C_1 e^{2z} + C_2 \cos z + C_3 \sin z$$

$$\text{P.I.} = \frac{1}{D_1^3 - 2D_1^2 + D_1 - 2} e^z + \frac{1}{D_1^3 - 2D_1^2 + D_1 - 2} e^{-3z}$$

$$= \frac{1}{1 - 2 + 1 - 2} e^z + \frac{1}{-27 - 18 - 3 - 2} e^{-3z} \qquad \begin{bmatrix} D_1 = 1 \\ D_1 = -3 \end{bmatrix}$$

$$= -\frac{e^z}{2} - \frac{1}{50} e^{-3z}$$

$$\therefore \quad y = C_1 e^{2z} + C_2 \cos z + C_3 \sin z - \frac{1}{2} e^z - \frac{1}{50} e^{-3z}$$

$$= C_1 x^2 + C_2 \cos(\log x) + C_3 \sin(\log x) - \frac{1}{2} x - \frac{1}{50} \cdot \frac{1}{x^3}$$ **Ans.**

Example 4. *Solve* $x^2 \frac{d^2 y}{dx^2} + x \frac{dy}{dx} + y = \sin(\log x^2)$ (*Nagpur University, Summer* 2005)

Solution. We have, $x^2 \frac{d^2 y}{dx^2} + x \frac{dy}{dx} + y = \sin(\log x^2)$ ... (1)

Let $x = e^z$, so that $z = \log x$, $D \equiv \frac{d}{dz}$

(1) becomes

$$D(D - 1)\, y + Dy + y = \sin(2z) \quad \Rightarrow \quad (D^2 + 1)\, y = \sin 2z$$

A.E. is $m^2 + 1 = 0$ or $m = \pm i$

$$C.F. = C_1 \cos z + C_2 \sin z$$

$$\text{P.I} = \frac{1}{D^2 + 1} \sin 2z = \frac{1}{-4 + 1} \sin 2z = -\frac{1}{3} \sin 2z$$

$$y = \text{C.F.} + \text{P.I.} = C_1 \cos z + C_2 \sin z - \frac{1}{3} \sin 2z$$

$$= C_1 \cos(\log x) + C_2 \sin(\log x) - \frac{1}{3} \sin(\log x^2)$$ **Ans.**

Example 5. *Solve* $x^3 \dfrac{d^3y}{dx^3} + 3x^2 . \dfrac{d^2y}{dx^2} + x\dfrac{dy}{dx} + y = x + \log x$

(*AMIETE, June 2010*)

Solution. Putting $x = e^z$ or $z = \log x$ and denoting $\dfrac{d}{dz}$ by D the equation becomes

$$[D(D-1)(D-2) + 3D(D-1) + D + 1]\, y = e^z + z$$

$$\Rightarrow \qquad [D^3 + 1]\, y = e^z + z$$

$\therefore$ A.E. is $\qquad m^3 + 1 = 0$

$$\Rightarrow \quad (m+1)(m^2 - m + 1) = 0 \Rightarrow \quad m = -1, \frac{1 \pm i\sqrt{3}}{2}$$

$$\text{C.F.} = C_1 e^{-z} + e^{\frac{1}{2}z}\left\{C_2 \cos\frac{\sqrt{3}}{2}z + C_3 \sin\frac{\sqrt{3}}{2}z\right\}$$

$$\text{P.I.} = \frac{1}{D^3+1}\{e^z + z\}$$

$$= \frac{1}{D^3+1}e^z + \frac{1}{D^3+1}z = \frac{e^z}{1+1} + (1+D^3)^{-1} z$$

$$= \frac{1}{2}e^z + (1 - D^3 + \ldots)\, z = \frac{1}{2}e^z + z.$$

$\therefore$ Complete solution is

$$y = C_1 e^{-z} + e^{z/2}\left\{C_2 \cos\frac{\sqrt{3}}{2}z + C_3 \sin\frac{\sqrt{3}}{2}z\right\} + \frac{1}{2}e^z + z$$

$$\Rightarrow \quad y = C_1 x^{-1} + \sqrt{x}\left\{C_2 \cos\left(\frac{\sqrt{3}}{2}\log x\right) + C_3 \sin\left(\frac{\sqrt{3}}{2}\log x\right)\right\} + \frac{1}{2}x + \log x \qquad \textbf{Ans.}$$

Example 6. *Solve the following:*

$$x^2\frac{d^2y}{dx^2} + x\frac{dy}{dx} + y = x\log x.$$

(*Delhi University, April 2010*)

Solution. We have $x^2\dfrac{d^2y}{dx^2} + x\dfrac{dy}{dx} + y = x\log x$ ...(1)

Let $x = e^z$. So that $z = \log x$, $D \equiv \dfrac{d}{dz}$

(1) becomes

$$D(D-1)\, y + Dy + y = ze^z$$

$$\Rightarrow (D^2 - D + D + 1)y = ze^z \Rightarrow (D^2+1)y = ze^z$$

A.E. is $\qquad m^2 + 1 = 0 \Rightarrow m = \pm i$

$\therefore \qquad$ C.F. $= C_1 \cos z + C_2 \sin z$

$$\text{P.I.} = \frac{1}{D^2+1}ze^z = e^z\frac{1}{(D+1)^2+1}z = e^z\frac{1}{D^2+2D+2}z$$

$$= \frac{e^z}{2}\frac{1}{1+\left(\dfrac{D^2}{2}+D\right)}z = \frac{e^z}{2}\left[1+\left(\frac{D^2}{2}+D\right)\right]^{-1} z$$

$$= \frac{e^z}{2}\left[1 - \frac{D^2}{2} - D\right] z = \frac{e^z}{2}[z - 0 - 1] = \frac{e^z}{2}[z - 1]$$

Hence, the solution is $\quad y = \text{C.F.} + \text{P.I.}$

$$\therefore \quad y = C_1 \cos z + C_2 \sin z + \frac{e^z}{2}[z - 1]$$

$$\Rightarrow \quad y = C_1 \cos(\log x) + C_2 \sin(\log x) + \frac{x}{2}[\log x - 1] \qquad \textbf{Ans.}$$

Example 7. *Solve:* $x^2 \dfrac{d^3y}{dx^3} + 3x \dfrac{d^2y}{dx^2} + \dfrac{dy}{dx} = x^2 \log x$

Solution. We have, $x^3 \dfrac{d^3y}{dx^3} + 3x^2 \dfrac{d^2y}{dx^2} + x\dfrac{dy}{dx} = x^3 \log x$

Let $x = e^z$ so that $z = \log x$, $D \equiv \dfrac{d}{dz}$

The equation becomes after substitution

$$[D(D-1)(D-2) + 3D(D-1) + D]\, y = z\, e^{3z} \quad \Rightarrow \quad D^3 y = z e^{3z}$$

Auxiliary equation is $m^3 = 0 \Rightarrow \quad m = 0, 0, 0.$

C.F. $= C_1 + C_2 z + C_3 z^2 = C_1 + C_2 \log x + C_3 (\log x)^2$

$$\text{P.I.} = \frac{1}{D^3}\,.\, z e^{3z} = e^{3z}\,.\, \frac{1}{(D+3)^3}\,.z$$

$$= e^{3z}\,\frac{1}{27}\left(1 + \frac{D}{3}\right)^{-3} z = \frac{e^{3z}}{27}(1 - D)\, z = \frac{e^{3z}}{27}(z - 1) = \frac{x^3}{27}(\log x - 1)$$

Complete solution is $\quad y = C_1 + C_2 \log x + C_3 (\log x)^2 + \dfrac{x^3}{27}(\log x - 1)$ **Ans.**

Example 8. *Solve the differential equation*

$$x^2 \frac{d^2y}{dx^2} - 3x \frac{dy}{dx} + 5y = x \log x.$$

Solution. Putting $x = e^z$ or $z = \log x$ and $D \equiv \dfrac{d}{dz}$

On putting $\quad x\dfrac{dy}{dx} = Dy \quad$ and $\quad x^2 \dfrac{d^2y}{dx^2} = D(D-1)\, y \;$ in the given equation, we get

$$[D(D-1) - 3D + 5]\, y = z e^z$$

i.e. $\quad (D^2 - 4D + 5)\, y = z e^z$

It's A.E. is $\quad m^2 - 4m + 5 = 0$

$$\therefore \quad m = \frac{4 \pm \sqrt{16 - 20}}{2} = \frac{4 \pm 2i}{2} = 2 \pm i$$

$$\text{C.F.} = e^{2z}(C_1 \cos z + C_2 \sin z)$$

$$\text{P.I.} = \frac{1}{D^2 - 4D + 5}\, z e^z$$

$$= e^z\,.\, \frac{1}{(D+1)^2 - 4(D+1) + 5}\, z \qquad \text{(by replacing } D \text{ by } D + 1)$$

$$= e^z . \frac{1}{D^2 - 2D + 2} z = \frac{e^z}{2} . \frac{1}{1 + \left(\frac{D^2 - 2D}{2}\right)} z$$

$$= \frac{e^z}{2}\left[1 + \left(\frac{D^2 - 2D}{2}\right)\right]^{-1} z = \frac{e^z}{2}\left[1 - \left(\frac{D^2 - 2D}{2}\right) + ...\right] z$$

$$= \frac{e^z}{2}[z + Dz] = \frac{e^z}{2}(z + 1).$$

Hence, the solution is

$$y = e^{2z}[C_1 \cos z + C_2 \sin z] + \frac{e^z}{2}(z + 1)$$

$$y = x^2 [C_1 \cos(\log x) + C_2 \sin(\log x)] + \frac{x}{2}(1 + \log x)$$ **Ans.**

Example 9. *Solve the homogeneous linear differential equation.*

$$x^2 \frac{d^2 y}{dx^2} + x \frac{dy}{dx} + y = (\log x) \sin(\log x)$$

Solution. Since given equation is homogeneous,

Put $x = e^z \Rightarrow \log x = z$

Also, $x \frac{dy}{dx} = Dy, \quad x^2 \frac{d^2 y}{dx^2} = D(D - 1) y, \quad D \equiv \frac{d}{dz}$

The transformed equation is

$$D(D - 1) y + Dy + y = z \sin z$$

$$(D^2 - D + D + 1) y = z \sin z$$

$$(D^2 + 1) y = z \sin z$$

A.E. is $m^2 + 1 = 0 \Rightarrow m = \pm i$

$$\text{C.F.} = C_1 \cos z + C_2 \sin z$$

$$P.I. = \frac{1}{D^2 + 1} z \sin z$$

$$= \text{Imaginary part of } \frac{1}{D^2 + 1} z(\cos z + i \sin z)$$

$$= \text{Imaginary part of } \frac{1}{D^2 + 1} ze^{iz} = \text{Imaginary part of } e^{iz} \frac{1}{(D + i)^2 + 1} z$$

$$= \text{Imaginary part of } e^{iz} \frac{1}{D^2 + 2iD - 1 + 1} z = \text{Imaginary part of } e^{iz} \frac{1}{D^2 + 2iD} z$$

$$= \text{Imaginary part of } e^{iz} \frac{1}{2iD} \frac{1}{\left(1 + \frac{D}{2i}\right)} z = \text{Imaginary part of } e^{iz} \frac{1}{2iD}\left(1 - \frac{D}{2i}\right) z$$

$$= \text{Imaginary part of } e^{iz} \frac{1}{2iD}\left(z - \frac{1}{2i}\right) = \text{Imaginary part of } e^{iz} \frac{1}{2i}\left(\frac{z^2}{2} - \frac{z}{2i}\right)$$

$= \text{Imaginary part of } \frac{1}{2i}\ (\cos z + i \sin z)\left(\frac{z^2}{2} - \frac{z}{2i}\right)$

$= \text{Imaginary part of } (\cos z + i \sin z)\left(\frac{z^2}{4i} + \frac{z}{4}\right)$

$= \text{Imaginary part of } (\cos z + i \sin z)\left(-i\frac{z^2}{4} + \frac{z}{4}\right) = -\frac{z^2}{4}\cos z + \frac{z}{4}\sin z$

Complete solution is, y = C.F. + P.I.

$$y = C_1 \cos z + C_2 \sin z - \frac{z^2}{4}\cos z + \frac{z}{4}\sin z$$

$$y = C_1 \cos(\log x) + C_2 \sin(\log x) - \frac{1}{4}(\log x)^2 \cos(\log x) + \frac{1}{4}(\log x)\sin(\log x) \qquad \textbf{Ans.}$$

Example 10. *The radial displacement in a rotating disc at a distance r from the axis is given by* $r^2\frac{d^2u}{dr^2} + r\frac{du}{dr} - u + kr^3 = 0$, *where k is a constant. Solve the equation under the conditions* $u = 0$ *when* $r = 0$, $u = 0$ *when* $r = a$. (*Nagpur University, Summer* 2008)

Solution. Here, we have

$$r^2\frac{d^2u}{dr^2} + r\frac{du}{dr} - u + kr^3 = 0 \qquad \text{... (1)}$$

On putting $r = e^z$, $\quad r\frac{du}{dr} = Dz$, $\quad r^2\frac{d^2u}{dr^2} = D(D-1)z \quad$ in (1), we get

$$D(D-1)\,u + Du - u = -ke^{3z} \qquad \left[D \equiv \frac{d}{dz}\right]$$

$$\Rightarrow \qquad (D^2 - D + D - 1)\,u = -ke^{3z} \quad \Rightarrow \quad (D^2 - 1)u = -k\,e^{3z}$$

A.E. is $\quad m^2 - 1 = 0 \quad \Rightarrow \quad m = \pm 1$

$$\text{C.F.} = C_1 e^z + C_2 e^{-z} = C_1 r + \frac{C_2}{r}$$

$$\text{P.I.} = \frac{1}{D^2 - 1}(-ke^{3z}) = -k\frac{1}{(3)^2 - 1}e^{3z} = -\frac{k}{8}e^{3z} = \frac{-k}{8}r^3$$

C.S. = C.F. + P.I.

$$u = C_1 r + \frac{C_2}{r} - \frac{k}{8}r^3 \qquad \text{... (2)}$$

Putting $\quad u = 0, \quad r = 0 \quad$ in (2), we get

$$0 = C_2$$

Putting $C_2 = 0$ in (2), we get

$$u = C_1 r - \frac{k}{8}r^3 \qquad \text{... (3)}$$

Putting $u = 0$, $r = a$ in (3), we get

$$0 = C_1 a - \frac{k}{8}a^3 \quad \Rightarrow \quad C_1 = \frac{k}{8}a^2$$

Putting $\qquad C_1 = \frac{k}{8}a^2$ in (3), we get

$$u = \frac{k}{8}a^2 r - \frac{k}{8}r^3$$

$$u = \frac{kr}{8}(a^2 - r^2)$$ **Ans.**

14.2 LEGENDRE'S HOMOGENEOUS DIFFERENTIAL EQUATIONS

A linear differential equation of the form

$$(a+bx)^n \frac{d^n y}{dx^n} + a_1 (a+bx)^{n-1} \frac{d^{n-1} y}{dx^{n-1}} + \ldots + a_n y = X \qquad \ldots (1)$$

where a, b, a_1, a_2, ... a_n are constants and X is a function of x, is called Legendre's linear equation.

Equation (1) can be reduced to linear differential equation with constant coefficients by the substitution.

$$a + bx = e^z \qquad \Rightarrow \qquad z = \log(a+bx)$$

so that $$\frac{dy}{dx} = \frac{dy}{dz}\cdot\frac{dz}{dx} = \frac{b}{a+bx}\cdot\frac{dy}{dz}$$

$$\Rightarrow \quad (a+bx)\frac{dy}{dx} = b\frac{dy}{dz} = b\,Dy, \quad D \equiv \frac{d}{dz} \quad \Rightarrow \quad (a+bx)\frac{dy}{dx} = b\,Dy$$

where

Again $$\frac{d^2y}{dx^2} = \frac{d}{dx}\left(\frac{dy}{dx}\right) = \frac{d}{dx}\left(\frac{b}{a+bx}\cdot\frac{dy}{dz}\right)$$

$$= -\frac{b^2}{(a+bx)^2}\frac{dy}{dz} + \frac{b}{(a+bx)}\cdot\frac{d^2y}{dz^2}\cdot\frac{dz}{dx}$$

$$= -\frac{b^2}{(a+bx)^2}\frac{dy}{dz} + \frac{b}{(a+bx)}\cdot\frac{d^2y}{dz^2}\cdot\frac{b}{(a+bx)}$$

$$\Rightarrow \quad (a+bx)^2\frac{d^2y}{dx^2} = -b^2\frac{dy}{dz} + b^2\frac{d^2y}{dz^2}$$

$$= b^2\left(\frac{d^2y}{dz^2} - \frac{dy}{dz}\right) = b^2(D^2 y - D y) = b^2 D(D-1)y$$

$$\Rightarrow \quad (a+bx)^2\frac{d^2y}{dx^2} = b^2 D(D-1)$$

Similarly, $(a+bx)^3\frac{d^3y}{dx^3} = b^3 D(D-1)(D-2)y$

....

$$(a+bx)^n\frac{d^ny}{dx^n} = b^n D(D-1)(D-2)\ldots\ldots(D-n+1)y$$

Substituting these values in equation (1), we get a linear differential equation with constant coefficients, which can be solved by the method given in the previous section.

Example 11. *Solve* $(1+x)^2 \dfrac{d^2 y}{dx^2} + (1+x)\dfrac{dy}{dx} + y = \sin 2\{\log(1+x)\}$ (*DU III Sem 2014*)

Solution. We have, $(1+x)^2 \dfrac{d^2 y}{dx^2} + (1+x)\dfrac{dy}{dx} + y = \sin 2\{\log(1+x)\}$

Put $1 + x = e^z$ or $\log(1 + x) = z$

$$(1+x)\frac{dy}{dx} = Dy \text{ and } (1+x)^2 \frac{d^2 y}{dx^2} = D(D-1)y, \text{ where } D \equiv \frac{d}{dz}$$

Putting these values in the given differential equation, we get

$$D(D-1)y + Dy + y = \sin 2z \quad \text{or} \quad (D^2 - D + D + 1)y = \sin 2z$$

$$(D^2 + 1)y = \sin 2z$$

A.E. is $m^2 + 1 = 0 \Rightarrow \quad m = \pm i$

$$C.F. = A\cos z + B\sin z$$

$$\text{P.I.} = \frac{1}{D^2+1}\sin 2z = \frac{1}{-4+1}\sin 2z = -\frac{1}{3}\sin 2z$$

Now, complete solution is $y = \text{C.F.} + \text{P.I.}$

$$\Rightarrow \quad y = A\cos z + B\sin z - \frac{1}{3}\sin 2z$$

$$\Rightarrow \quad y = A\cos\{\log(1+x)\} + B\sin\{\log(1+x)\} - \frac{1}{3}\sin 2\{\log(1+x)\}$$ **Ans.**

Example 12. *Solve:* $(3x+2)^2 \dfrac{d^2 y}{dx^2} + 3(3x+2)\dfrac{dy}{dx} - 36y = 3x^2 + 4x + 1.$

(*Uttarakhand II, Semester, June* 2007)

Solution. Let $3x + 2 = e^z \quad \Rightarrow \quad z = \log(3x + 2) \qquad \left[x = \dfrac{e^z - 2}{3}\right]$

So that $(3x+2)\dfrac{dy}{dx} = 3Dy$ and $(3x+2)^2 \dfrac{d^2 y}{dx^2} = 9D(D-1)y$ where $D \equiv \dfrac{d}{dz}$

Putting these values in the given differential equation, we get

$$9D(D-1)y + 9Dy - 36y = 3\left(\frac{e^z - 2}{3}\right)^2 + 4\left(\frac{e^z - 2}{3}\right) + 1$$

$$\therefore \quad (9D^2 - 36)y = \frac{1}{3}(e^{2z} - 4e^z + 4) + \frac{4}{3}e^z - \frac{8}{3} + 1 = \frac{e^{2z}}{3} - \frac{1}{3}$$

A.E. is $\quad 9m^2 - 36 = 0$

$$m^2 - 4 = 0 \qquad \therefore \quad m = \pm 2$$

$$\text{C.F.} = C_1 e^{2z} + C_2 e^{-2z}$$

$$\text{P.I.} = \frac{1}{9D^2 - 36}\left[\frac{e^{2z}}{3} - \frac{1}{3}\right] = \frac{1}{27}\frac{1}{D^2 - 4}e^{2z} - \frac{1}{3}\frac{1}{9D^2 - 36}e^{0z}$$

$$= \frac{1}{27}z\frac{1}{2D}e^{2z} - \frac{1}{3}\frac{1}{0-36} = \frac{1}{27}z\left(\frac{e^{2z}}{4}\right) + \frac{1}{108} = \frac{1}{108}[ze^{2z} + 1]$$

Complete solution is

$$y = C.F. + P.I. = C_1 e^{2z} + C_2 e^{-2z} + \frac{1}{108}(ze^{2z} + 1)$$

$$y = C_1(3x+2)^2 + \frac{C_2}{(3x+2)^2} + \frac{1}{108}[(3x+2)^2 \log(3x+2) + 1]$$ **Ans.**

EXERCISE 14.1

Solve the following differential equations:

1. $x^2 \frac{d^2y}{dx^2} - 4x\frac{dy}{dx} + 6y = \frac{42}{x^4}$ **Ans.** $C_1 x^2 + C_2 x^3 + \frac{1}{x^4}$

2. $(x^2 D^2 - 3xD + 4)\, y = 2x^2$ **Ans.** $(C_1 + C_2 \log x)\, x^2 + x^2 (\log x)^2$

3. $x^2 \frac{d^2y}{dx^2} - x\frac{dy}{dx} + y = \log x$ **Ans.** $(C_1 + C_2 \log x)\, x + \log x + 2$ *(AMIETE, June 2010)*

4. $\frac{d^2y}{dx^2} + \frac{1}{x}\frac{dy}{dx} = \frac{12 \log x}{x^2}$ **Ans.** $C_1 + C_2 \log x + 2(\log x)^3$

5. $(x^2 D^2 - xD - 3)\, y = x^2 \log x$ **Ans.** $\frac{C_1}{x} + C_2 x^3 - \frac{x^2}{3}\left(\log x + \frac{2}{3}\right)$

6. $x^2 \frac{d^2y}{dx^2} - 2x\frac{dy}{dx} + 2y = x^2 + \sin(5 \log x)$

Ans. $c_1 x + c_2 x^2 + x^2 \log x + \frac{1}{754}[15 \cos(5 \log x) - 23 \sin(5 \log x)]$

7. $x^2 \frac{d^2y}{dx^2} - 3x\frac{dy}{dx} + y = \log x \frac{\sin(\log x) + 1}{x}$ *(AMIETE, Dec. 2009)*

Ans. $y = C_1 x^{2+\sqrt{3}} + C_2 x^{2-\sqrt{3}} + \frac{1}{x}\left[\frac{382}{61} \cos \log x + \frac{54}{61} \sin(\log x) + 6 \log x \cos(\log x)\right.$ $\left. + 5 \log x \sin(\log x)\right] + \frac{1}{6x}$

8. $(1+x)^2 \frac{d^2y}{dx^2} + (1+x)\frac{dy}{dx} + y = 2 \sin \log(1+x)$

Ans. $y = C_1 \cos \log(1+x) + C_2 \sin \log(1+x) - \log(1+x) \cos \log(1+x)$

9. Which of the basis of solutions are for the differential equation $x^2 \frac{d^2y}{dx^2} - x\frac{dy}{dx} + y = 0$

(a) x, xlpx, *(b)* $\ln x, e^x$ *(c)* $\frac{1}{x}, \frac{1}{x^2}$, *(d)* $\frac{1}{x^2} e^x, x \ln x$

Ans. *(a)*

10. The general solution of $x^2 \frac{d^2y}{dx^2} - 5x\frac{dy}{dx} + 9y = 0$ is

(a) $(C_1 + C_n x)\, e^{3x}$ *(b)* $(C_1 + C_2 x)\, x^3$ *(c)* $(C_1 + C_2 x)\, x^3$ *(d)* $(C_1 + C_2 l_n x)\, e^{x^3}$

(AMIETE, Dec. 2009) **Ans.** *(b)*

11. To transform $x\frac{d^2y}{dx^2} + \frac{dy}{dx} = \frac{1}{x}$ into a linear differential equation with constant coefficients, the required substitution is (*a*) $x = \sin t$ (b) $x = t^2 + 1$ (*c*) $x = \log t$ (*d*) $x = e^t$

(*AMIETE, June 2010*) **Ans.** (*d*)

14.3 METHOD OF VARIATION OF PARAMETERS

To find particular integral of

$$\frac{d^2y}{dx^2} + b\frac{dy}{dx} + cy = X \qquad \ldots (1)$$

Let complementary function $= Ay_1 + By_2$, so that y_1 and y_2 satisfy

$$\frac{d^2y}{dx^2} + b\frac{dy}{dx} + cy = 0 \qquad \ldots (2)$$

Let us assume particular integral $y = uy_1 + vy_2$, ... (3)

where u and v are unknown function of x.

Differentiating (3) *w.r.t.* x, we have $y' = uy_1' + vy_2' + u'y_1 + v'y_2$ assuming that u,v satisfy the equation

$$u'y_1 + v'y_2 = 0 \qquad \ldots (4)$$

then $\qquad y' = uy_1' + vy_2'$ ... (5)

Differentiating (5) *w.r.t.x*, we have $\quad y'' = uy_1'' + u'y_1' + vy_2'' + v'y_2'$

Substituting the values of y, y' and y'' in (1), we get

$$a(uy_1'' + u'y_1' + vy_2'' + v'y_2') + b(uy_1' + vy_2') + c(uy_1 + vy_2) = X$$

$$\Rightarrow \quad u(y_1'' + by_1' + cy_1) + v(ay_2'' + by_2' + cy_2) + a(u'y_1' + v'y_2') = X \qquad \ldots (6)$$

y_1 and y_2 will satisfy equation (1)

$\therefore \qquad y_1'' + by_1' + cy_1 = 0$ ... (7)

and $\qquad y_2'' + by_2' + cy_2 = 0$ ... (8)

Putting the values of expressions from (7) and (8) in (6), we get

$$\Rightarrow \quad u'y_1' + v'y_2' = X \qquad \ldots (9)$$

Solving (4) and (9), we get

$$u' = \begin{vmatrix} 0 & y_2 \\ X & y_2' \end{vmatrix} \div \begin{vmatrix} y_1 & y_2 \\ y_1' & y_2' \end{vmatrix} = \frac{-y_2X}{y_1y_2' - y_1'y_2}$$

$$v' = \begin{vmatrix} y_1 & 0 \\ y_1' & X \end{vmatrix} \div \begin{vmatrix} y_1 & y_2 \\ y_1' & y_2' \end{vmatrix} = \frac{y_1X}{y_1y_2' - y_1'y_2}$$

$$\boxed{u = \int \frac{-y_2X}{y_1y_2' - y_1'y_2}\,dx}$$

$$\boxed{v = \int \frac{y_1X}{y_1y_2' - y_1'y_2}\,dx}$$

General solution = complementary function + particular integral.

Working Rule

Step 1. Find out the *C.F. i.e.*, $Ay_1 + By_2$

Step 2. Particular integral $= u\, y_1 + v\, y_2$

Step 3. Find u and v by the formulae

$$u = \int \frac{-y_2 X}{y_1 y_2' - y_1' y_2}\, dx, \qquad v = \int \frac{y_1 X}{y_1 y_2' - y_1' y_2}\, dx$$

Example 13. *Solve* $\dfrac{d^2 y}{dx^2} + y = \operatorname{cosec} x.$ (*Nagpur University, Summer* 2005)

Solution. $(D^2 + 1)\, y = \operatorname{cosec} x$

A.E. is $m^2 + 1 = 0 \quad \Rightarrow \quad m = \pm i$

$$\text{C.F.} = A \cos x + B \sin x$$

Here $y_1 = \cos x, \qquad y_2 = \sin x$

$$\text{P.I.} = y_1\, u + y_2\, v$$

where
$$u = \int \frac{-y_2 \,.\, \operatorname{cosec} x\, dx}{y_1 \,.\, y'_2 - y'_1 \,.\, y_2} = \int \frac{-\sin x \,.\, \operatorname{cosec} x\, dx}{\cos x\,(\cos x) - (-\sin x)\,(\sin x)}$$

$$= \int \frac{-\sin x \,.\, \dfrac{1}{\sin x}\, dx}{\cos^2 x + \sin^2 x} = -\int dx = -x$$

$$v = \int \frac{y_1 \,.\, X\, dx}{y_1 \,.\, y'_2 - y'_1\, y_2} = \int \frac{\cos x \,.\, \operatorname{cosec} x\, dx}{\cos x\,(\cos x) - (-\sin x)\,(\sin x)}$$

$$= \int \frac{\cos x \,.\, \dfrac{1}{\sin x}}{\cos^2 x + \sin^2 x}\, dx = \int \frac{\cot x\, dx}{1} = \log \sin x$$

$$P.I. = uy_1 + vy_2 = -x \cos x + \sin x\,(\log \sin x)$$

General solution = C.F. + P.I.

$$y = A \cos x + B \sin x - x \cos x + \sin x\,.(\log \sin x)$$ **Ans.**

Example 14. *Apply the method of variation of parameters to solve*

$$\frac{d^2 y}{dx^2} + y = \tan x$$ (*A.M.I.E.T.E., Dec. 2010,*)

Solution. We have, $\dfrac{d^2 y}{dx^2} + y = \tan x$

$$(D^2 + 1)y = \tan x$$

A.E. is $m^2 = -1$ or $m = \pm i$

C. F. $y = A \cos x + B \sin x$

Here, $y_1 = \cos x, \qquad y_2 = \sin x$

$$y_1 \,.\, y'_2 - y'_1 \,.\, y_2 = \cos x\,(\cos x) - (-\sin x)\sin x = \cos^2 x + \sin^2 x = 1$$

P. I. $= u.\, y_1 + v.\, y_2$ where

$$u = \int \frac{-y_2 \tan x}{y_1 \,.\, y'_2 - y'_1 \,.\, y_2}\, dx = -\int \frac{\sin x \tan x}{1}\, dx$$

$$= -\int \frac{\sin^2 x}{\cos x}\, dx = -\int \frac{1 - \cos^2 x}{\cos x}\, dx$$

$$= \int (\cos x - \sec x)\, dx = \sin x - \log(\sec x + \tan x)$$

$$v = \int \frac{y_1 \tan x}{y_1 \cdot y'_2 - y'_1 \cdot y_2}\, dx = \int \frac{\cos x \cdot \tan x}{1}\, dx = \int \sin x\, dx = -\cos x$$

$$\text{P. I.} = u.\, y_1 + v.\, y_2$$

$$= [\sin x - \log(\sec x + \tan x)]\cos x - \cos x \sin x$$

$$= -\cos x \log(\sec x + \tan x)$$

Complete solution is

$$y = A\cos x + B\sin x - \cos x \log(\sec x + \tan x) \qquad \textbf{Ans.}$$

Example 15. *Use variation of parameters method to solve* $y'' + y = \sec x$

(*AMIETE, June 2010,*)

Solution. We have, $\dfrac{d^2y}{dx^2} + y = \sec x$ ... (1)

A. E. is $(m^2 + 1) = 0 \Rightarrow m^2 = -1 \Rightarrow m = \pm i$

Complementary Function of (1) is

$$\text{C. F.} = A\cos x + B\sin x$$

Here $y_1 = \cos x, \quad y_2 = \sin x$

$$\text{P. I.} = uy_1 + vy_2 \qquad \text{... (2)}$$

where $u = \displaystyle\int \frac{-y_2 \sec x}{y_1 \cdot y'_2 - y'_1 \cdot y_2}\, dx$ $\left[\begin{aligned} y_1 y'_2 - y_1' y_2 &= \cos x \cos x - (-\sin x)\sin x \\ &= \cos^2 x + \sin^2 x = 1\end{aligned}\right]$

On putting the values of y_2 and $y_1 y_2' - y_1' y_2$, we get

$$u = \int \frac{-\sin x \sec x}{1}\, dx = -\int \tan x\, dx = \log \cos x$$

$$v = \int \frac{y_1 \sec x}{y_1 y'_2 - y'_1 y_2}\, dx$$

On putting the values of y_1 and $y_1 y_2' - y_1' y_2$, we get

$$v = \int \frac{\cos x \cdot \sec x}{1}\, dx = \int dx = x$$

Putting the values of u and v in (2), we get

$$P.\ I. = \cos x \,.\, \log \cos x + x \sin x$$

Complete solution is

$$y = C.\ F. + P.\ I.$$

$$\Rightarrow \quad y = A\cos x + B\sin x + \cos x \log \cos x + x \sin x \qquad \textbf{Ans.}$$

Example 16. *Obtain general solution of the differential equation* $x^2 y'' + xy' - y = x^3 e^x$.

Solution. The given differential equation is $x^2 y'' + xy' - y = x^3 e^x$

$$\Rightarrow \quad x^2 \frac{d^2y}{dx^2} + x\frac{dy}{dx} - y = x^3 e^x \qquad \text{...(1)}$$

Putting $x = e^z \Rightarrow \quad D = \frac{d}{dz}, x\frac{dy}{dx} = Dy, \qquad x^2\frac{d^2y}{dx^2} = D(D-1)\,y,$ in (1), we get

$$D(D-1)\,y + D\,y - y = e^{3z}e^{e^z}$$

$$\Rightarrow \qquad (D^2-1)y = e^{3z}e^{e^z}$$

A. E. is $m^2 - 1 = 0 \quad \Rightarrow \quad m = \pm 1$

$$\therefore \qquad \text{C. F.} = c_1 e^z + c_2 e^{-z}$$

$$\Rightarrow \qquad \text{C. F.} = c_1 x + c_2 x^{-1} = c_1 x + \frac{c_2}{x}$$

P. I. $= u\,y_1 + vy_2$, where $y_1 = e^z, y_2 = e^{-z}$, $\quad \left[y_1 = x, y_2 = \frac{1}{x}\right]$

Also,

$$u = -\int \frac{-y_2 f(z)}{y_1 y_2' - y_1' y_2}\,dz = -\int \frac{e^{-z}\,.\,e^{3z}\,.\,e^{e^z}}{e^z(-e^{-z}) - e^z(e^{-z})}\,dz = -\int \frac{e^{2z}\,e^{e^z}}{-1-1}$$

$$= \frac{1}{2}\int e^{2z}\,e^{e^z}\,dz = \frac{1}{2}\int x^2\,e^x\frac{dx}{x} = \frac{1}{2}\int x\,e^x\,dx \qquad \left[\begin{array}{l} x = e^z, dx = e^z\,dz \\ dz = \frac{dx}{e^z} = \frac{dx}{x}\end{array}\right]$$

$$= \frac{1}{2}[xe^x - (1)\,e^x] = \frac{1}{2}(xe^x - e^x)$$

and

$$v = \int \frac{y_1 f(z)}{y_1 y_2' - y_1' y_2}\,dz = \int \frac{e^z\,.\,e^{3z}\,.\,e^{e^z}}{e^z(-e^{-z}) - e^z(e^{-z})}\,dz$$

$$= \int \frac{e^{4z}\,e^{e^z}}{-1-1}\,dz = \int \frac{x^4\,e^x}{-2}\frac{dx}{x} = -\frac{1}{2}\int x^3 e^x\,dx$$

$$= -\frac{1}{2}[x^3e^x - 3x^2e^x + 6xe^x - 6e^x]$$

$$\text{P.I.} = uy_1 + vy_2 = \frac{1}{2}(xe^x - e^x)\,x - \frac{1}{2}(x^3e^x - 3x^2e^x + 6xe^x - 6e^x)\frac{1}{x}$$

$$= \frac{1}{2}\left[x^2 - x - x^2 + 3x - 6 + \frac{6}{x}\right]e^x = \frac{1}{2}\left(2x - 6 + \frac{6}{x}\right)e^x = \left(x - 3 + \frac{3}{x}\right)e^x$$

Complete solution = C.F. + P.I.

$$y = (c_1e^z + c_2e^{-z}) + \left(x - 3 + \frac{3}{x}\right)e^x$$

$$= c_1x + \frac{c_2}{x} + \left(x - 3 + \frac{3}{x}\right)e^x \qquad \textbf{Ans.}$$

Example 17. *Solve by method of variation of parameters:*

$$\frac{d^2y}{dx^2} - y = \frac{2}{1+e^x}$$

(Uttarakhand, II Semester, June 2007.)

Solution. $\qquad \frac{d^2y}{dx^2} - y = \frac{2}{1+e^x}$

A. E. is $\qquad (m^2 - 1) = 0$

$$m^2 = 1, \qquad m = \pm 1$$

$$C.F. = C_1 e^x + C_2 e^{-x}$$

$$\therefore \quad P.I. = uy_1 + vy_2$$

Here, $\quad y_1 = e^x, \quad y_2 = e^{-x}$

and $\quad y_1 \cdot y'_2 - y'_1 \cdot y_2 = -e^x \cdot e^{-x} - e^x \cdot e^{-x} = -2$

$$u = \int \frac{-y_2 X}{y_1 . y'_2 - y'_1 . y_2} dx = -\int \frac{e^{-x}}{-2} \times \frac{2}{1+e^x} dx$$

$$= \int \frac{e^{-x}}{1+e^x} dx = \int \frac{dx}{e^x (1+e^x)} = \int \left(\frac{1}{e^x} - \frac{1}{1+e^x} \right) dx$$

$$= \int e^{-x} dx - \int \frac{e^{-x}}{e^{-x}+1} dx = -e^{-x} + \log (e^{-x}+1)$$

$$v = \int \frac{y_1 X}{y_1 . y'_2 - y'_1 . y_2} dx = \int \frac{e^x}{-2} \frac{2}{1+e^x} dx = -\int \frac{e^x}{1+e^x} dx = -\log (1+e^x)$$

$$\text{P.I.} = u . y_1 + v . y_2 = [-e^{-x} + \log (e^{-x}+1)] e^x - e^{-x} \log (1+e^x)$$

$$= -1 + e^x \log (e^{-x}+1) - e^{-x} \log (e^x+1)$$

Complete solution = $y = C_1 e^x + C_2 e^{-x} - 1 + e^x \log (e^{-x}+1) - e^{-x} \log (e^x+1)$ **Ans.**

Example 18. *Apply the method of variation of parameters to solve*

$$\frac{d^2y}{dx^2} - 3\frac{dy}{dx} + 2y = \frac{e^x}{1+e^x}$$

(*U.P. II Semester Summer* 2005)

Solution. Given equation is $\dfrac{d^2y}{dx^2} - 3\dfrac{dy}{dx} + 2y = \dfrac{e^x}{1+e^x}$

Auxiliary equation is $\quad m^2 - 3m + 2 = 0 \Rightarrow (m-1)(m-2) = 0 \Rightarrow m = 1, 2$

$$\text{C. F.} = C_1 e^x + C_2 e^{2x}$$

$$= C_1 y_1 + C_2 y_2, \qquad \text{Here} \quad y_1 = e^x, \quad y_2 = e^{2x}$$

$$\text{P. I.} = uy_1 + vy_2,$$

$$u = \int \frac{-y_2 X dx}{y_1 y_2' - y_2 y_1'} = \int \frac{-e^{2x} \frac{e^x}{1+e^x}}{e^x (2e^{2x}) - e^{2x} (e^x)} dx = \int \frac{\frac{-e^{3x}}{1+e^x}}{2e^{3x} - e^{3x}} dx$$

$$= \int \frac{-e^{3x}}{e^{3x}(1+e^x)} dx = -\int \frac{1}{1+e^x} dx = -\int \frac{e^{-x}}{e^{-x}+1} \qquad \text{[Dividing by } e^x\text{]}$$

$$= \log (e^{-x}+1)$$

Now, $$v = \int \frac{y_1 X dx}{y_1 y_2' - y_2 y_1'} = \int \frac{e^x \left(\frac{e^x}{1+e^x} \right) dx}{e^x (2e^{2x}) - e^{2x} (e^x)} = \int \frac{\frac{e^{2x}}{1+e^x}}{2e^{3x} - e^{3x}} dx = \int \frac{e^{2x}}{e^{3x}(1+e^x)} dx$$

$$= \int \frac{1}{e^x (1+e^x)} dx = \int \left(\frac{1}{e^x} - \frac{1}{1+e^x} \right) dx \qquad \text{[By Partial fraction]}$$

$$= \int \left(e^{-x} - \frac{e^{-x}}{e^{-x}+1} \right) dx = -e^{-x} + \log(e^{-x}+1)$$

P.I. $= uy_1 + vy_2$

P.I. $= e^x \log(e^{-x}+1) + e^{2x} \{-e^{-x} + \log(e^{-x}+1)\}$

P.I. $= e^x \log(e^{-x}+1) - e^x + e^{2x} \log(e^{-x}+1)$

Complete solution is

$$y = \text{C. F.} + \text{P. I.}$$

$$= C_1 e^x + C_2 e^{2x} + e^x \log(e^{-x}+1) - e^x + e^{2x} \log(e^{-x}+1)$$ **Ans.**

Example 19. *Solve by method of variation of parameter.*

$$\frac{d^2y}{dx^2} - y = \left(1 + \frac{1}{e^x}\right)^{-2}$$

Solution. $\frac{d^2y}{dx^2} - y = \left(\frac{e^x+1}{e^x}\right)^{-2} = \frac{e^{2x}}{(e^x+1)^2}$

A. E. is $m^2 - 1 = 0 \qquad \therefore \quad m = \pm 1$

C. F. $= C_1 e^{-x} + C_2 e^x$

Let P. I. $= uy_1 + vy_2$, where $y_1 = e^{-x}$, $y_2 = e^x$

$$y_1 y_2' - y_1' y_2 = e^{-x} \cdot e^x + e^{-x} \cdot e^x = 2$$

$$u = \int \frac{-y_2 X}{y_1 y_2' - y_1' y_2} dx = \frac{1}{2} \int -e^x \cdot \frac{e^{2x}}{(1+e^x)^2} dx = -\frac{1}{2} \int \frac{e^{2x}}{(1+e^x)^2} e^x dx$$

Putting $t = 1 + e^x$, $dt = e^x dx$, we get

$$u = -\frac{1}{2} \int (t-1)^2 \frac{dt}{t^2}$$

$$u = -\frac{1}{2} \int \left(1 - \frac{2}{t} + t^{-2}\right) dt = -\frac{1}{2}(t - 2\log t - t^{-1})$$

$$u = -\frac{1}{2}(1+e^x) + \log(1+e^x) + \frac{1}{2}(1+e^x)^{-1}$$

Now, $v = \int \frac{y_1 X}{y_1 y_2' - y_1' y_2} dx = \frac{1}{2} \int \frac{e^{-x} \cdot e^{2x}}{(1+e^x)^2} dx = \frac{1}{2} \int \frac{e^x \, dx}{(1+e^x)^2}$

$$= \frac{1}{2} \int \frac{1}{t^2} dt = -\frac{1}{2} \frac{1}{t} = -\frac{1}{2}(1+e^x)^{-1}$$ where $t = 1 + e^x$

P.I. $= uy_1 + vy_2$

$$\text{P.I.} = e^{-x} \left[-\frac{1}{2}(1+e^x) + \log(1+e^x) + \frac{1}{2}(1+e^x)^{-1} \right] - \frac{1}{2} e^x (1+e^x)^{-1}$$

$$= -\frac{1}{2}(1+e^x)^{-1} \{e^x - e^{-x}\} + e^{-x} \left(-\frac{1}{2}\right) [(1+e^x) + e^{-x} \log(1+e^x)]$$

$$= -(1+e^x)^{-1} \sinh x - \frac{1}{2} e^{-x} (1+e^x) + e^{-x} \log(1+e^x)$$

Hence the solution is

$$y = C_1 e^{-x} + C_2 e^{x} - (1+e^{x})^{-1} \sinh x - \frac{1}{2} e^{-x} (1+e^{x}) + e^{-x} \log (1+e^{x}) \qquad \textbf{Ans.}$$

Example 20. *Solve by method of variation of parameters*

$$\frac{d^2 y}{dx^2} - 2\frac{dy}{dx} = e^{x} \sin x$$

Solution.

$$\frac{d^2 y}{dx^2} - 2\frac{dy}{dx} = e^{x} \sin x$$

$\Rightarrow \quad (D^2 - 2D)y = e^x \sin x$

A. E. is $m^2 - 2m = 0 \quad \Rightarrow \quad m(m-2) = 0 \Rightarrow m = 0, 2$

C. F. $= C_1 + C_2 e^{2x}$

P. I. $= uy_1 + vy_2$ where, $\quad y_1 = 1, \quad y_2 = e^{2x}$

$$\therefore \quad u = \int \frac{-y_2 X\,dx}{y_1 y_2' - y_1' y_2} = \int \frac{-e^{2x}.e^{x} \sin x}{1(2e^{2x}) - 0\,(e^{2x})} dx$$

$$= -\frac{1}{2}\int e^{x} \sin x\, dx \qquad \left[\int e^{ax} \sin bx\, dx = \frac{e^{ax}}{a^2+b^2}(a \sin bx - b \cos bx)\right]$$

$$= -\frac{1}{2}\frac{e^{x}}{1+1}\{\sin x - \cos x\} = -\frac{e^{x}}{4}(\sin x - \cos x)$$

$$v = \int \frac{y_1 X\,dx}{y_1 y_2' - y_1' y_2} = \int \frac{1.e^{x} \sin x\, dx}{1(2e^{2x}) - 0\,(e^{2x})} = \frac{1}{2}\int e^{-x} \sin x\, dx$$

$$= \frac{1}{2}\frac{e^{-x}}{(-1)^2 + (1)^2}[-\sin x - \cos x]$$

$$= -\frac{1}{2}\frac{e^{-x}}{2}(\sin x + \cos x) = -\frac{e^{-x}}{4}(\sin x + \cos x)$$

$$\text{P. I.} = uy_1 + vy_2 = -\frac{e^{x}}{4}(\sin x - \cos x)\,1 - \frac{e^{-x}}{4}(\sin x + \cos x)\, e^{2x}$$

$$= -\frac{e^{x}}{4}[\sin x - \cos x + \sin x + \cos x] = -\frac{e^{x}}{2} \sin x$$

The complete solution is

$y = C.\,F. + P.\,I.$

$$\Rightarrow \quad y = C_1 + C_2 e^{2x} - \frac{e^{x}}{2}\sin x \qquad \textbf{Ans.}$$

Example 21. *Solve :* $\dfrac{d^2 y}{dx^2} + 3\dfrac{dy}{dx} + 2y = \sin(e^{x})$

Solution. The given equation can be written as

$(D^2 + 3D + 2)y = \sin e^x$

A. E. is $\quad (m^2 + 3m + 2) = 0 \quad \Rightarrow \quad (m+1)(m+2) = 0$

$m = -1, -2,$

$\therefore$ C. F. $= C_1 e^{-x} + C_2 e^{-2x}$

P. I. $= uy_1 + vy_2$ ...(1)

where $y_1 = e^{-x}, \quad y_2 = e^{-2x}$

$y_1' = -e^{-x}, \quad y_2' = -2e^{-2x}$

$$y_1 y_2' - y_2 y_1' = e^{-x}(-2e^{-2x}) - e^{-2x}(-e^{-x}) = -2e^{-3x} + e^{-3x} = -e^{-3x}$$

$$u = \int \frac{-y_2 X}{y_1 y'_2 - y'_1 y_2} dx = \int \frac{-e^{-2x}}{-e^{-3x}} \sin(e^x)\, dx$$

$$= \int e^x \sin(e^x)\, dx = -\cos(e^x)$$

$$v = \int \frac{y_1 X}{y_1 y'_2 - y'_1 y_2} dx = \int \frac{e^{-x} \sin(e^x)\, dx}{-e^{-3x}}$$

$$= -\int e^{2x} \sin(e^x)\, dx = -\int t \sin t\, dt \qquad [t = e^x \text{ so that } dt = e^x\, dx]$$

$$= t\cos t - \sin t = e^x \cos(e^x) - \sin(e^x)$$

Putting the values of u, v, y_1 and y_2 in (1), we get

$$\text{P.I.} = e^{-x}[-\cos(e^x)] + e^{-2x}[e^x \cos(e^x) - \sin(e^x)] = -e^{-2x} \sin(e^x)$$

The solution is $y = C_1 e^{-x} + C_2 e^{-2x} - e^{-2x} \sin(e^x)$. **Ans.**

Example 22. *Solve by method of variation of parameters:*

$$\frac{d^2 y}{dx^2} + y = (x - \cot x)$$ *(DU, II Sem. 2012,)*

Solution. Here A. E. is $m^2 + 1 = 0 \qquad \therefore \qquad m = \pm i$

C. F. $= C_1 \cos x + C_2 \sin x = C_1 y_1 + C_2 y_2$

where $y_1 = \cos x, \qquad y_2 = \sin x$

$y'_1 = -\sin x, \qquad y'_2 = \cos x$

$y_1 y'_2 - y_2 y'_1 = \cos x \,.\cos x + \sin x.\, \sin x = 1$

Let P. I. $= uy_1 + vy_2$ ... (1)

Where $u = \int \frac{-y_2 X}{y_1 y_2' - y_2 y'_1} dx = \int \frac{-\sin x(x - \cot x)}{1} dx$

$$= \int \cos x\, dx - \int x \sin x\, dx = \sin x - \{x(-\cos x) + \sin x\} = x\cos x$$

$$v = \int \frac{y_1 X}{y_1 y'_2 - y_2 y'_1} dx = \int \cos x\,(x - \cot x)\, dx$$

$$= \int x\cos x\, dx - \int \frac{\cos^2 x}{\sin x} dx = \int x \cos x\, dx - \int \frac{1 - \sin^2 x}{\sin x} dx$$

$$= \int x\cos x\, dx - \int \operatorname{cosec} x\, dx + \int \sin x\, dx$$

$$= x \sin x + \cos x - \log(\operatorname{cosec} x - \cot x) - \cos x$$

$$= x\sin x - \log(\operatorname{cosec} x - \cot x)$$

Putting the values of u, v, y_1, y_2 in (1), we get

$$\text{P. I.} = \cos x.\, x\cos x + \sin x\,\{x\sin x - \log(\operatorname{cosec} x - \cot x)\}$$

$$= x\cos^2 x + x\sin^2 x - \sin x\log(\operatorname{cosec} x - \cot x)$$

$$= x - \sin x\log(\operatorname{cosec} x - \cot x)$$

Hence, complete solution is

$$y = C_1\cos x + C_2\sin x + x - \sin x\log(\operatorname{cosec} x - \cot x)$$ **Ans.**

Example 23. *Solve* $\dfrac{d^2y}{dx^2} + 2\dfrac{dy}{dx} + 2y = e^{-x}\sec^3 x.$

Solution. A. E. is $m^2 + 2m + 2 = 0$

$$m = \frac{-2+\sqrt{4-8}}{2} = \frac{-2\pm 2i}{2} = -1+i$$

C. F. $= e^{-x}(C_1\cos x + C_2\sin x) = C_1y_1 + C_2y_2$

where $y_1 = e^{-x}\cos x \quad \Rightarrow \quad y_1' = -e^{-x}\sin x - e^{-x}\cos x$

$y_2 = e^{-x}\sin x \quad \Rightarrow \quad y'_2 = -e^{-x}\sin x + e^{-x}\cos x$

$$y_1y'_2 - y'_1y_2 = e^{-x}\cos x(-e^{-x}\sin x + e^{-x}\cos x) - (-e^{-x}\cos x - e^{-x}\sin x)\,e^{-x}\sin x$$

$$= e^{-2x}(\sin^2 x + \cos^2 x - \sin x\cos x + \sin x\cos x)$$

$$= e^{-2x}(\sin^2 x + \cos^2 x) = e^{-2x}$$

Let P.I. $= uy_1 + vy_2$

where $$u = -\int\frac{y_2X}{y_1y'_2 - y'_1y_2}\,dx = -\int\frac{e^{-x}\sin x.\,e^{-x}\sec^3 x}{e^{-2x}}\,dx$$

$$= -\int\sin x.\sec^3 x\,dx = -\int\tan x\cdot\sec^2 x\,dx = -\frac{\tan^2 x}{2}$$

$$v = \int\frac{y_1X}{y_1y'_2 - y'_1y_2}\,dx = \int\frac{e^{-x}\cos x.\,e^{-x}\sec^3 x}{e^{-2x}}\,dx = \int\cos x.\sec^3 x\,dx = \int\sec^2 x\,dx = \tan x$$

$$\text{P. I.} = uy_1 + vy_2$$

$$= \frac{-\tan^2 x}{2}e^{-x}\cos x + \tan x.\,e^{-x}\sin x = e^{-x}\left[\frac{-\tan^2 x}{2}\cos x + \tan x.\sin x\right]$$

$$= e^{-x}\left[\frac{-\sin x.\tan x}{2} + \tan x.\sin x\right] = \frac{1}{2}e^{-x}\sin x\tan x$$

Complete solution is

$$y = C.\,F. + P.\,I.$$

$$= e^{-x}(C_1\cos x + C_2\sin x) + \frac{1}{2}e^{-x}\sin x\tan x$$ **Ans.**

Example 24. *Apply the method of variation of parameters to solve*

$$\frac{d^2y}{dx^2} - y = e^{-x}\sin(e^{-x}) + \cos(e^{-x})$$

Solution. The auxiliary equation is $m^2 - 1 = 0$

$$\therefore \quad m = \pm 1$$

$$\therefore \quad C.\,F. = C_1e^x + C_2e^{-x} = C_1y_1 + C_2y_2$$

where $y_1 = e^x$ and $y_2 = e^{-x}$

Let P. I. $= uy_1 + vy_2$

where $u = \int \dfrac{-y_2\, X}{y_1 y'_2 - y'_1 y_2}\, dx = \int -\dfrac{e^{-x}\{e^{-x}\sin(e^{-x}) + \cos(e^{-x})\}}{e^x(-e^{-x}) - e^x e^{-x}}\, dx$

i.e. $u = \dfrac{1}{2}\int [e^{-2x}\sin(e^{-x}) + e^{-x}\cos(e^{-x})]\, dx$ (Put $e^{-x} = t, -e^{-x}\, dx = dt$)

$$= -\frac{1}{2}\int (t\sin t + \cos t)\, dt = -\frac{1}{2}\{-t\cos t + \sin t + \sin t\}$$

$$= \frac{1}{2}[t\cos t - 2\sin t] = \frac{1}{2}[e^{-x}\cos(e^{-x}) - 2\sin(e^{-x})]$$

$$v = \int \frac{y_1\, X}{y_1 y'_2 - y'_1 y_2}\, dx = \int \frac{e^x\{e^{-x}\sin(e^{-x}) + \cos(e^{-x})\}}{e^x(-e^{-x}) - e^x e^{-x}}$$

$$= -\frac{1}{2}\int \{\sin(e^{-x}) + e^x\cos(e^{-x})\}dx$$

$$\left[e^{-x} = t \text{ and } -e^{-x}\, dx = dt \text{ i.e. } dx = -\frac{dt}{t}\right]$$

$$= \frac{1}{2}\int \left(\frac{\sin t}{t} + \frac{\cos t}{t^2}\right) dt = \frac{1}{2}\left\{\frac{-\cos t}{t} - \int \frac{\cos t}{t^2}\, dt + \int \frac{\cos t}{t^2}\, dt\right\}$$

$$= -\frac{1}{2}\frac{\cos t}{t} = -\frac{1}{2}e^x\cos(e^{-x})$$

$$P.\,I. = e^x\left\{\frac{1}{2}e^{-x}\cos(e^{-x}) - \sin(e^{-x})\right\} - \frac{1}{2}e^{-x}e^x\cos(e^{-x})$$

$$= -e^x\sin(e^{-x})$$

$\therefore$ Required solution is $y =$ C. F. + P. I.

i.e. $y = C_1 e^x + C_2 e^{-x} - e^x\sin(e^{-x})$ **Ans.**

Example 25. *Solve* $(D^2 + 2D + 1)\, y = 4e^{-x}\log x$ *by method of variation of parameters.*

Solution. $(D^2 + 2D + 1)\, y = 4e^{-x}\log x$

A. E. is $m^2 + 2m + 1 = 0$

$$(m+1)^2 = 0$$

$$m = -1, -1$$

$\therefore$ $C.\,F. = (C_1 + C_2 x)\, e^{-x} \Rightarrow C.\,F. = C_1 e^{-x} + C_2 x e^{-x} = C_1 y_1 + C_2 y_2$

$$y_1 = e^{-x} \quad \Rightarrow \quad y'_1 = -e^{-x}$$

$$y_2 = xe^{-x} \quad \Rightarrow \quad y'_2 = -xe^{-x} + e^{-x}$$

$$y_1 y'_2 - y'_1 y_2 = e^{-x}(-xe^{-x} + e^{-x}) + e^{-x}(x\, e^{-x})$$

$$= -xe^{-2x} + e^{-2x} + xe^{-2x} = e^{-2x}$$

Let P. I. $= uy_1 + vy_2$ ... (1)

where $u = -\int \dfrac{y_2\, X}{y_1 y'_2 - y'_1 y_2}\, dx$

$$u = -\int \frac{xe^{-x}\; 4e^{-x}\log x}{e^{-2x}}\, dx = -4\int x\log x\, dx = -4\left[\log x\frac{x^2}{2} - \int \frac{1}{x}\frac{x^2}{2}\, dx\right]$$

$$= -4\left[\frac{x^2}{2}\log x - \frac{1}{2}\frac{x^2}{2}\right] = -2x^2\log x + x^2$$

$$u = x^2(1 - 2\log x)$$

and $$v = \int \frac{y_1 X}{y_1 y'_2 - y'_1 y_2}\, dx$$

$$v = \int \frac{e^{-x} . 4e^{-x}\log x}{e^{-2x}}\, dx = 4\int \log x\, dx = 4\int 1\cdot \log x\, dx = 4\left[(\log x)\, x - \int \frac{1}{x}. x\, dx\right]$$

$$= 4\,[x \log x - x]$$

$$v = 4x(\log x - 1)$$

Putting the values of u, v, y_1 and y_2 in (1), we get

$$\text{P. I.} = x^2(1 - 2\log x)\, e^{-x} + 4x(\log x - 1) x e^{-x} = x^2 e^{-x}[1 - 2\log x + 4\log x - 4]$$

$$\text{P.I.} = x^2 e^{-x}\,[2\log x - 3]$$

$$\text{C. S.} = \text{C. F.} + \text{P. I.}$$

$$y = (C_1 + C_2 x)\, e^{-x} + x^2 e^{-x}\,[2\log x - 3]$$ **Ans.**

EXERCISE 14.2

Solve the following differential equations by variation of parameters method:

1. $\dfrac{d^2y}{dx^2} - 4y = e^{2x}$ **Ans.** $y = C_1 e^{2x} + C_2 e^{-2x} + \dfrac{x}{4}e^{2x} - \dfrac{e^{2x}}{16}$

2. $\dfrac{d^2y}{dx^2} + y = \sin x$ **Ans.** $y = C_1\cos x + C_2 \sin x - \dfrac{x}{2}\cos x + \dfrac{1}{4}\sin x$

3. $\dfrac{d^2y}{dx^2} - 3\dfrac{dy}{dx} + 2y = \sin x$ **Ans.** $y = C_1 e^{x} + C_2 e^{2x} + \dfrac{1}{10}(3\cos x + \sin x)$

4. $\dfrac{d^2y}{dx^2} + y = \sec x \tan x$ **Ans.** $y = C_1\cos x + C_2 \sin x + x\cos x + \sin x \log \sec x - \sin x$

5. $y'' - 6y' + 9y = \dfrac{e^{3x}}{x^2}$ *(AMIETE, June 2010, 2009)* **Ans.** $y = (c_1 + xc_2)\, e^{3x} - e^{3x}(\log x)$

14.4 METHOD OF UNDETERMINED COEFFICIENTS

To solve a differential equation with RHS X, we assume a trial solution containing unknown constants by trial solution. The trial solution to be assumed depending upon on the form of x.

(i) If $X = 2e^{3x}$, then trial solution is Ae^{3x}

(ii) If $X = 5\sin 3x$, then the trial solution is $A\sin 3x + B\cos 3x$

(iii) If $X = 3x^4$, then the trial solution is $Ax^4 + Bx^3 + Cx^2 + Dx + E$

(iv) If $X = \tan x$ or $\sec x$, then the method fails since number of terms after differentiating becomes infinite.

If any term of the trial solution appears in the C.F., we multiply the trial solution by the lowest positive integral power of x.

Example 26. *Solve using the method of undetermined coefficients the equation:*

$$\frac{d^2y}{dx^2}+10\frac{dy}{dx}+25y = 14e^{-5x}$$ *(Delhi University, April 2010)*

Solution. Here, we have

$$\frac{d^2y}{dx^2}+10\frac{dy}{dx}+25y = 14e^{-5x} \quad \text{... (1)}$$

$$\Rightarrow \quad (D^2 + 10D + 25)\,y = 14e^{-5x}$$

A.E. is $m^2 + 10\,m + 25 = 0$

$$\Rightarrow \quad (m+5)^2 = 0 \quad \Rightarrow \quad m = -5, -5$$

C.F. is $y = (c_1 + c_2 x)\,e^{-5x}$

Here $X = 14e^{-5x}$, since e^{-5x} is present in *C.F.*

So trial solution is $y = Axe^{-5x}$

$$\Rightarrow \quad \frac{dy}{dx} = -5\,A\,x\,e^{-5x} + Ae^{-5x}$$

$$\frac{d^2y}{dx^2} = 25\,Axe^{-5x} - 5A\,e^{-5x} - 5Ae^{-5x}$$

$$= 25\,A\,xe^{-5x} - 10\,Ae^{-5x}$$

Substituting the values of $\frac{d^2y}{dx^2}$ and $\frac{dy}{dx}$, y in (1), we get

$$(25Axe^{-5x} - 10Ae^{-5x}) + 10\,(-5Axe^{-5x} + Ae^{-5x}) + 25\,(Axe^{-5x}) = 14e^{-5x}$$

$$\Rightarrow \quad Ae^{-5x}\,(25x - 10 - 50x + 10 + 25x) = 14e^{-5x}$$

$$\Rightarrow \quad Ae^{-5x}\,(-50x) = 14e^{-5x}$$

$$\Rightarrow \quad A = -\frac{14}{50x} = -\frac{7}{25x}$$

Putting the value of A in the solution, we get

$$y = -\frac{7}{25x}\,.\,xe^{-5x} = -\frac{7}{25}\,e^{-5x}$$

Thus, the particular solution is $y = -\frac{7}{25}e^{-5x}$

Hence, the complete solution is

$$y = (c_1 + c_2x)\,e^{-5x} - \frac{7}{25}e^{-5x}$$ **Ans.**

Example 27. *Solve using the method of undetermined coefficient of the equation.*

$$x^2\frac{d^2y}{dx^2} - 2x\,(1+x)\frac{dy}{dx} + 2\,(1+x)\,y = x^3$$

Solution. Here, we have

$$x^2\frac{d^2y}{dx^2} - 2x\,(1+x)\frac{dy}{dx} + 2\,(1+x)\,y = x^3$$

Dividing the given equation by x^2, we get

$$\frac{d^2y}{dx^2}-\frac{2(1+x)}{x}\frac{dy}{dx}+\frac{2(1+x)y}{x^2}=x \qquad \text{.... (1)}$$

Now suppose that the complete solution of (1) is given by

$$y = Ax + B\,x\,e^{2x} \qquad \text{...(2)}$$

where A and B are functions of x.

$$\therefore \qquad \frac{dy}{dx}=A+Be^{2x}+2Bxe^{2x}+\frac{dA}{dx}.x+\frac{dB}{dx}xe^{2x}=A+Be^{2x}+2B\,xe^{2x} \qquad \text{...(3)}$$

where $\frac{dA}{dx}+\frac{dB}{dx}.e^{2x}=0$ ...(4)

Differentiating (3), we have

$$\frac{d^2y}{dx^2}=\frac{dA}{dx}+\left(\frac{dB}{dx}e^{2x}+2Be^{2x}\right)+\left(2\frac{dB}{dx}x\,e^{2x}+2B\,e^{2x}+4B\,xe^{2x}\right)$$

$$=\left(\frac{dA}{dx}+e^{2x}\frac{dB}{dx}\right)+\left(4\,Be^{2x}+4\,B\,xe^{2x}+2xe^{2x}\frac{dB}{dx}\right)$$

$$=0+4Be^{2x}+4B\,xe^{2x}+2xe^{2x}\frac{dB}{dx} \qquad \text{...(5)} \qquad \text{[Using (4)]}$$

Substituting the values of $\frac{d^2y}{dx^2}, \frac{dy}{dx}$ and y from (5), (3) and (2) respectively in (1), we get

$$\frac{dA}{dx}+e^{2x}(2x+1)\frac{dB}{dx}=x \qquad \text{...(6)}$$

Solving (4) and (6), we get

$$\frac{dA}{dx}=-\frac{1}{2}, \qquad \frac{dB}{dx}=\frac{1}{2}e^{-2x}$$

Integrating $\qquad A=-\frac{1}{2}x+C_1, \qquad B=-\frac{1}{4}e^{-2x}+C_2$

Substituting values of A and B in (2), the complete solution of (1) is given by

$$y=\left(-\frac{1}{2}x+c_1\right)x+\left(-\frac{1}{4}e^{-2x}+c_2\right)xe^{2x}=C_1x+C_2xe^{2x}-\frac{1}{2}x^2-\frac{1}{4}x. \qquad \textbf{Ans.}$$

Example 28. *Solve by using the method of undetermined coefficients:*

$$\frac{d^2y}{dx^2}+(1-\cot x)\frac{dy}{dx}-y\cot x=\sin^2 x$$

Solution. First we shall find the C.F. of the given equation *i.e.,* the solution of the equation.

$$\frac{d^2y}{dx^2}+(1-\cot x)\frac{dy}{dx}-y\cot x=0 \qquad \left[\frac{d^2y}{dx^2}+P\frac{dy}{dx}+Qy=0\right] \qquad \text{...(1)}$$

Here, $P = 1 - \cos x$, $Q = -\cot x$

Now, $\quad 1 - P + Q = 1 - (1 - \cot x) - \cot x = 0$, $\quad \therefore\ y = e^{-x}$ is a part of the C.F.

Putting $y = ve^{-x}$, $\qquad \frac{dy}{dx}=-ve^{-x}+e^{-x}\frac{dv}{dx}$

$$\frac{d^2y}{dx^2}=e^{-x}\frac{d^2v}{dx^2}-2e^{-x}\frac{dv}{dx}+ve^{-x}$$ the equation (1) reduces to

$$\frac{d^2v}{dx^2}-(1+\cot x)\frac{dv}{dx}=0$$

$$\Rightarrow \quad \frac{dp}{dx} - (1+\cot x)p = 0 \qquad \text{where } p = \frac{dv}{dx}$$

$$\Rightarrow \quad \frac{dp}{p} = (1 + \cot x)\, dx$$

Integrating, we get $\log p = x + \log \sin x + \log C_1$

$$\Rightarrow \quad \log p - \log \sin x - \log C_1 = x$$

$$\Rightarrow \quad \log \frac{p}{C_1 \sin x} = x$$

$$\Rightarrow \quad \frac{p}{C_1 \sin x} = e^x \Rightarrow p = C_1\, e^x \sin x$$

$$\therefore \quad p = \frac{dv}{dx} = C_1 e^x \sin x \;\Rightarrow\; v = C_1 \int e^x \sin x\, dx$$

$$\left[\int e^{ax} \sin bx\, dx = \frac{e^{ax}}{a^2+b^2}[a \sin bx - b \cos bx]\right]$$

$$\Rightarrow \quad v = C_1 . \frac{1}{2} e^x (\sin x - \cos x) + C_2 \;= \text{only one part of C.F. of equation (1),}$$

R.H.S. of the given differential equation (1) is zero.

$\therefore$ The solution of (1) *i.e.*, C.F. of the given equation is

$$\text{C.F.} = ve^{-x} = C_1 . \frac{1}{2} (\sin x - \cos x) + C_2\, e^{-x}$$

Let $y = A(\sin x - \cos x) + Be^{-x}$ be the complete solution of the given equation where A and B are function of x, so chosen that the given equation will be satisfied.

$$\therefore \quad \frac{dy}{dx} = A(\cos x + \sin x) - Be^{-x} + \frac{dA}{dx}(\sin x - \cos x) + \frac{dB}{dx} e^{-x}. \qquad ...(2)$$

Let us choose A and B such that

$$\frac{dA}{dx}(\sin x - \cos x) + \frac{dB}{dx} e^{-x} = 0. \qquad ...(3)$$

Now (2) becomes

$$\therefore \quad \frac{dy}{dx} = A(\cos x + \sin x) - Be^{-x}$$

$$\text{and} \quad \frac{d^2y}{dx^2} = \frac{dA}{dx}(\cos x + \sin x) - \frac{dB}{dx} e^{-x} + A(-\sin x + \cos x) + Be^{-x}$$

Putting these values of $\frac{d^2y}{dx^2}, \frac{dy}{dx}$ and y in the given equation, we get

$$\frac{dA}{dx}(\cos x + \sin x) - \frac{dB}{dx} e^{-x} + A(-\sin x + \cos x) + Be^{-x} + (1 - \cot x)$$

$$[A (\cos x + \sin x) - B\, e^{-x}] - [A(\sin x - \cos x) + Be^{-x}] \cot x = \sin^2 x$$

$$\Rightarrow \quad \frac{dA}{dx}(\cos x + \sin x) - \frac{dB}{dx} e^{-x} = \sin^2 x$$

Solving (3) and (4), we get

$$\frac{dA}{dx} = \frac{1}{2} \sin x \quad \text{and} \quad \frac{dB}{dx} = \frac{1}{2} e^x (\sin x \cos x - \sin^2 x)$$

$$= \frac{e^x}{4} (\sin 2x + \cos 2x - 1)$$

Integrating these, we get

$$A = -\frac{1}{2}\cos x + C_1$$

and $$B = \frac{1}{4}\int e^x(\sin 2x - 1 + \cos 2x)\,dx = \frac{1}{4}\int e^x \sin 2x\,dx - \frac{1}{4}\int e^x\,dx + \frac{1}{4}\int e^x \cos 2x\,dx$$

$$= \frac{1}{4}.\frac{e^x}{5}(\sin 2x - 2\cos 2x) - \frac{e^x}{4} + \frac{1}{4}.\frac{e^x}{5}(\cos 2x + 2\sin 2x) + C_3$$

$$= \frac{e^x}{20}(3\sin 2x - \cos 2x) - \frac{e^x}{4} + C_2$$

Putting the values of A and B in (2), the general solution of the given equation is

$$y = \left(-\frac{1}{2}\cos x + C_1\right)(\sin x - \cos x) + \left\{\frac{e^x}{20}(3\sin 2x - \cos 2x) - \frac{e^x}{4} + C_2\right\}e^{-x}$$

$$= -\frac{1}{2}\cos x\sin x + \frac{1}{2}\cos^2 x + C_1\sin x - C_1\cos x + \frac{3}{20}\sin 2x - \frac{1}{20}\cos x - \frac{1}{4} + C_2 e^{-x}$$

$$= -\frac{1}{4}(\sin 2x) + \frac{1}{4}(\cos 2x + 1) + C_1(\sin x - \cos x) + \frac{3}{20}\sin 2x - \frac{1}{20}\cos 2x - \frac{1}{4} + C_2 e^{-x}$$

$$y = C_1(\sin x - \cos x) + C_2 e^{-x} - \frac{1}{10}(\sin 2x - 2\cos 2x)$$ **Ans.**

EXERCISE 14.3

Solve the following differential equations by method of undetermined coefficients:

1. $\frac{d^2y}{dx^2} + 2\frac{dy}{dx} + 4y = 2x^2 + 3e^{-x}$ **Ans.** $y = e^{-x}(c_1 \cos\sqrt{3}\,x + c_2 \sin\sqrt{3}\,x) + \frac{1}{2}x^2 - \frac{1}{2}x + e^{-x}$

2. $\frac{d^2y}{dx^2} + y = \sin x$ **Ans.** $y = c_1 \cos x + c_2 \sin x - \frac{1}{2}x\sin x.$

3. $\frac{d^2y}{dx^2} - 2\frac{dy}{dx} = 3e^x \sin x$ **Ans.** $y = c_1 + c_2 e^{2x} - \frac{3}{2}e^x \sin x$

4. $\frac{d^2y}{dx^2} - 2\frac{dy}{dx} + 3y = x^3 + \cos x$

Ans. $y = e^x(c_1 \cos\sqrt{2}\,x + c_2 \sin\sqrt{2}\,x) + \frac{1}{27}(9x^3 + 18x^2 + 6x - 8) + \frac{1}{4}(\cos x - \sin x)$

5. $\frac{d^2y}{dx^2} + 4y = 4\sin 2x$ **Ans.** $y = c_1 \cos 2x + c_2 \sin 2x - x\sin 2x.$

DIFFERENTIAL EQUATION OF OTHER TYPES

15.1 INTRODUCTION

In this chapter we have to solve the following eight different types of equations:

1. $\frac{d^n y}{d x^n} = f(x)$. **2.** $\frac{d^n y}{d x^n} = f(y)$.

3. Equation which do not contain y directly.
4. Equations that do not contain x directly.
5. (*i*) Equations whose one part (u) of C.F. is given
(*ii*) Not a single part u or v of C.F. of the equation is given
6. Normal form (removal of first derivative).
7. By changing the independent variable of the different equation.
8. By variation of parameters.

15.2 EQUATION OF THE TYPE $\frac{d^n y}{dx^n} = f(x)$

This type of exact differential equations are solved by successive integration.

Example 1. *Solve* $\frac{d^2 y}{dx^2} = x^2 \sin x$.

Solution. We have $\frac{d^2 y}{dx^2} = x^2 \sin x$ ...(1)

Integrating the differential equation (1), we get

$$\frac{dy}{dx} = x^2(-\cos x) - (2x)(-\sin x) + (2)(\cos x) + c_1$$

$$\frac{dy}{dx} \equiv -x^2 \cos x + 2x \sin x + 2 \cos x + c_1$$

Integrating again, we have $y = [(-x^2)(\sin x) - (-2x)(-\cos x) + (-2)(-\sin x)]$
$+ [(2x)(-\cos x) - 2(-\sin x)] + 2 \sin x + c_1 x + c_2$
$= -x^2 \sin x - 4x \cos x + 6 \sin x + c_1 x + c_2$ **Ans.**

Example 2. *Solve* $\frac{d^3 y}{dx^3} = x + \log x$.

Solution. We have, $\dfrac{d^3 y}{dx^3} = x + \log x$ ...(1)

Integrating the differential equation (1), we get $\dfrac{d^2 y}{dx^2} = \dfrac{x^2}{2} + (\log x)(x) - \int \dfrac{1}{x} \cdot x\, dx + c_1$

$$\frac{d^2 y}{dx^2} = \frac{x^2}{2} + x \log x - x + c_1 \quad \text{...(2)}$$

Again integrating (2), we have,

$$\frac{dy}{dx} = \frac{x^3}{6} + (\log x)\left(\frac{x^2}{2}\right) - \int \frac{1}{x} \cdot \frac{x^2}{2} dx - \frac{x^2}{2} + c_1 x + c_2$$

$$\frac{dy}{dx} = \frac{x^3}{6} + \frac{x^2}{2} \log x - \frac{x^2}{4} - \frac{x^2}{2} + c_1 x + c_2 \quad \text{...(3)}$$

Again integrating (3), we obtain

$$y = \frac{x^4}{24} + (\log x)\frac{x^3}{6} - \int \frac{1}{x}\frac{x^3}{6} dx - \frac{x^3}{12} - \frac{x^3}{6} + c_1 \frac{x^2}{2} + c_2 x + c_3$$

$$\Rightarrow \quad y = \frac{x^4}{24} + \frac{x^3}{6} \log x - \frac{x^3}{18} - \frac{x^3}{12} - \frac{x^3}{6} + \frac{c_1 x^2}{2} + c_2 x + c_3$$

$$\Rightarrow \quad y = \frac{x^4}{24} + \frac{x^3}{6} \log x - \frac{11x^3}{36} + c_1 \frac{x^2}{2} + c_2 x + c_3$$ **Ans.**

EXERCISE 15.1

Solve the following differential equations:

1. $\dfrac{d^5 y}{dx^5} = x$ **Ans.** $y = \dfrac{x^6}{720} + \dfrac{c_1 x^4}{24} + \dfrac{c_2 x^3}{6} + \dfrac{c_3 x^2}{2} + c_4 x + c_5$

2. $\dfrac{d^2 y}{dx^2} = x e^x$ **Ans.** $y = (x - 2) e^x + c_1 x + c_2$

3. $\dfrac{d^4 y}{dx^4} = x + e^{-x} - \cos x$ **Ans.** $y = \dfrac{x^5}{120} + e^{-x} - \cos x + c_1 \dfrac{x^3}{6}$

4. $x^2 \dfrac{d^2 y}{dx^2} = \log x$ **Ans.** $y = -\dfrac{1}{2}(\log x)^2 + \log x - c_1 x + c_2$

5. $\dfrac{d^3 y}{dx^3} = \log x$ **Ans.** $y = \dfrac{1}{36}[6x^3 \log x - 11x^3 + c_1 x^2 + c_2 x + c_3]$

6. $\dfrac{d^3 y}{dx^3} = \sin^2 x$ **Ans.** $y = \dfrac{x^3}{12} + \dfrac{\sin 2x}{16} + \dfrac{c_1 x^2}{2} + c_2 x + c_3$

15.3 EQUATION OF THE TYPE $\dfrac{d^n y}{dx^n} = f(y)$

Multiplying by $2\dfrac{dy}{dx}$, we get $2\dfrac{dy}{dx}\dfrac{d^2 y}{dx^2} = 2 f(y) \dfrac{dy}{dx}$ ...(1)

Integrating (1), we have $\left(\dfrac{dy}{dx}\right)^2 = 2\int f(y)\, dy + c = \phi(y)$ (say)

$$\frac{dy}{dx} = \sqrt{\phi(y)} \Rightarrow \frac{dy}{\sqrt{\phi(y)}} = dx \Rightarrow \int \frac{dy}{\sqrt{\phi(y)}} = x + c$$

Example 3. *Solve* $\frac{d^2 y}{dx^2} = \sqrt{y}$, *under the condition* $y = 1$ *and* $\frac{dy}{dx} = \frac{2}{\sqrt{3}}$ *at* $x = 0$

Solution. We have $\frac{d^2 y}{dx^2} = \sqrt{y}$ ...(1)

Multiplying (1) by $2\frac{dy}{dx}$, we get $2\frac{dy}{dx}\frac{d^2 y}{dx^2} = 2\sqrt{y}\,\frac{dy}{dx}$ ...(2)

Integrating (2), we get $\left(\frac{dy}{dx}\right)^2 = \frac{4}{3}y^{3/2} + c_1$ ...(3)

On putting $y = 1$ and $\frac{dy}{dx} = \frac{2}{\sqrt{3}}$, we have $c_1 = 0$

Equation (3) becomes $\left(\frac{dy}{dx}\right)^2 = \frac{4}{3}y^{3/2}$ or $\frac{dy}{dx} = \frac{2}{\sqrt{3}}y^{3/4}$ or $y^{-3/4}\,dy = \frac{2}{\sqrt{3}}dx$

Again integrating $\frac{y^{1/4}}{\frac{1}{4}} = \frac{2}{\sqrt{3}}x + c_2 \quad \Rightarrow 4y^{\frac{1}{4}} = \frac{2}{\sqrt{3}}x + c_2$ ...(4)

On putting $x = 0$, $y = 1$, we get $c_2 = 4$

(4) becomes $4y^{1/4} = \frac{2}{\sqrt{3}}x + 4$ **Ans.**

Example 4. *Solve* $\frac{d^2 y}{dx^2} = \sec^2 y \tan y$ *under the condition* $y = 0$ *and* $\frac{dy}{dx} = 1$ *when* $x = 0$.

Solution. $\frac{d^2 y}{dx^2} = \sec^2 y \tan y \quad \Rightarrow \quad 2\frac{dy}{dx}\frac{d^2 y}{dx^2} = 2\sec^2 y \tan y\frac{dy}{dx}$

$$\int 2\frac{dy}{dx}\frac{d^2 y}{dx^2} = \int 2\sec^2 y \tan y\,\frac{dy}{dx}$$

$$\left(\frac{dy}{dx}\right)^2 = \tan^2 y + c_1 \quad \text{or} \quad \frac{dy}{dx} = \sqrt{\tan^2 y + c_1}$$

On putting $y = 0$, and $\frac{dy}{dx} = 1$, we get $c_1 = 1$

Now, $\frac{dy}{dx} = \sqrt{\tan^2 y + 1} = \sec y$

$\Rightarrow$ $\cos y\, dy = dx$

On integrating we get $\sin y = x + c$

On putting $y = 0$, $x = 0$, we have $c = 0$

$\sin y = x \Rightarrow y = \sin^{-1} x$ **Ans.**

Example 5. *Solve* $\frac{d^2 y}{dx^2} = 2(y^3 + y)$, *under the condition* $y = 0$, $\frac{dy}{dx} = 1$ *when* $x = 0$.

Solution. $\frac{d^2 y}{dx^2} = 2(y^3 + y)$ or $2\frac{dy}{dx}\frac{d^2 y}{dx^2} = 4(y^3 + y)\frac{dy}{dx}$

Integrating, we get

$$\left(\frac{dy}{dx}\right)^2 = 4\left(\frac{y^4}{4} + \frac{y^2}{2}\right) + c_1 = y^4 + 2y^2 + c_1 \qquad ...(1)$$

On putting $y = 0$ and $\frac{dy}{dx} = 1$ in (1), we get $1 = c_1$

Equation (1) becomes $\left(\frac{dy}{dx}\right)^2 = y^4 + 2y^2 + 1 = (y^2 + 1)^2$

$$\frac{dy}{dx} = y^2 + 1 \text{ or } \frac{dy}{1+y^2} = dx$$

Again integraiting, we get $\tan^{-1} y = x + c_2$ ...(2)

On putting $y = 0$ and $x = 0$ in (2), we have $0 = c_2$

Equation (2) is reduced to $\tan^{-1} y = x \Rightarrow y = \tan x$ **Ans.**

Example 6. *A motion is governed by* $\frac{d^2x}{dt^2} = 36x^{-2}$, *given that at t = 0, x = 8 and* $\frac{dx}{dt} = 0$, *find the displacement at any time t.*

Solution. We have $\frac{d^2x}{dt^2} = 36x^{-2} \Rightarrow 2\frac{d^2x}{dt^2}\frac{dx}{dt} = 2 \times 36x^{-2}\frac{dx}{dt}$ ...(1)

Integrating (1), we have $\left(\frac{dx}{dt}\right)^2 = -72x^{-1} + c_1$ ...(2)

Putting $x = 8$ and $\frac{dx}{dt} = 0$ in (2), we get $0 = -\frac{72}{8} + c_1$ or $c_1 = 9$

(2) becomes $\left(\frac{dx}{dt}\right)^2 = -\frac{72}{x} + 9$ or $\left(\frac{dx}{dt}\right)^2 = \frac{-72 + 9x}{x} \Rightarrow \frac{dx}{dt} = 3\sqrt{\frac{(x-8)}{x}}$

$$\Rightarrow \int \frac{\sqrt{x}\,dx}{\sqrt{x-8}} = 3\int dt + c_2 \Rightarrow \int \frac{x\,dx}{\sqrt{x^2 - 8x}} = 3t + c_2$$

$$\frac{1}{2}\int \frac{2x - 8 + 8}{\sqrt{x^2 - 8x}}\,dx = 3\text{t} + c_2$$

$$\frac{1}{2}\int \frac{2x-8}{\sqrt{x^2 - 8x}}\,dx + 4\int \frac{1}{\sqrt{(x-4)^2 - (4)^2}}\,dx = 3t + c_2$$

$$\sqrt{x^2 - 8x} + 4\cosh^{-1}\frac{x-4}{4} = 3t + c_2 \qquad ...(3)$$

On putting $x = 8$ and $t = 0$ in (3), we get $c_2 = 0$

(3) becomes $\sqrt{x^2 - 8x} + 4\cos h^{-1}\frac{x-4}{4} = 3t$ **Ans.**

EXERCISE 15.2

Solve the following differential equations (1 – 3):

1. $y^3 \frac{d^2y}{dx^2} = a$ **Ans.** $c_1 y^2 = (c_1 x + c_2)^2$

2. $e^{2y} \frac{d^2y}{dx^2} = 1$ **Ans.** $c_1 e^y = \cos h\,(c_1 x + c_2)$

3. $\sin^3 y \frac{d^2y}{dx^2} = \cos y$ **Ans.** $\sin[(x + c_2)\sqrt{(1 + c_1)}] + \sqrt{\left(\frac{1+c_1}{c_1}\right)} \cos y = 0$

4. A particle is acted upon by a force $\mu\left(x+\frac{a^4}{x^3}\right)$ per unit mass towards the origin where x is the distance from the origin at time t. If it starts that it will arrive at the origin in time $\frac{\pi}{4\sqrt{\mu}}$.

5. In the case of a stretched elastic string which has one end fixed and a particle of mass m attached to the other end, the equation of motion is

$$\frac{d^2s}{dt^2} = -\frac{mg}{e}(s-l)$$

where l is the natural length of the string and e its elongwith due to a weight mg. Find s and v determining the constants, so that $s = s_0$ at the time $t = 0$ and $v = 0$ when $t = 0$.

Ans. $v = -\sqrt{\left(\frac{g}{e}\right)}\,[(s_0-l)^2-(s-l)^2]^{1/2},\ s-l=(s_0-l)\cos\left[\sqrt{\left(\frac{g}{e}\right)}.t\right]$

15.4 EQUATIONS WHICH DO NOT CONTAIN 'y' DIRECTLY

The equation which do not contain y directly, can be written

$$f\left(\frac{d^ny}{dx^n}, \frac{d^{n-1}y}{dx^{n-1}}, \ldots\ldots \frac{dy}{dx}, x\right) = 0 \qquad \ldots(1)$$

On substituting $\frac{dy}{dx} = P$ *i.e.*, $\frac{d^2y}{dx^2} = \frac{dP}{dx}$ and $\frac{d^3y}{dx^3} = \frac{d^2P}{dx^2}$ etc. in (1), we get

$$f\left(\frac{d^{n-1}p}{dx^{n-1}}, \ldots\ldots P, x\right) = 0$$

Example 7. *Solve* $\frac{d^2y}{dx^2} + \frac{dy}{dx} + \left(\frac{dy}{dx}\right)^3 = 0 \qquad \ldots(1)$

Solution. On putting $\frac{dy}{dx} = P$ and $\frac{d^2y}{dx^2} = \frac{dP}{dx}$, equation (1) becomes

$$\frac{dP}{dx} + P + P^3 = 0 \text{ or } \frac{dP}{dx} + P(1+P^2) = 0$$

$$\frac{dp}{dx} = -P(1+P^2) \text{ or } \frac{dP}{P(1+P^2)} = -dx \;\Rightarrow\; \left(\frac{1}{P} - \frac{P}{1+P^2}\right)dP = -dx$$

On integrating, we have

$$\log P - \frac{1}{2}\log(1+P^2) = -x + c_1 \;\Rightarrow\; \log\frac{P}{\sqrt{1+P^2}} = -x + c_1$$

$$\frac{P}{\sqrt{1+P^2}} = e^{-x+c_1} \Rightarrow \frac{P^2}{1+P^2} = a^2e^{-2x} \;\Rightarrow\; P^2 = (1+P^2)\,a^2e^{-2x}$$

$$\Rightarrow\; P^2(1-a^2e^{-2x}) = a^2e^{-2x} \Rightarrow P = \frac{ae^{-x}}{\sqrt{1-a^2e^{-2x}}} \;\Rightarrow \frac{dy}{dx} = \frac{ae^{-x}}{\sqrt{1-a^2e^{-2x}}}$$

$$dy = \frac{ae^{-x}}{\sqrt{1-a^2e^{-2x}}}\,dx$$

On integration, we get $\quad y = -\sin^{-1}(ae^{-x}) + b$ **Ans.**

Example 8. *Solve* $\frac{d^2y}{dx^2} = \left[1-\left(\frac{dy}{dx}\right)^2\right]^{1/2}$

Solution. We have, $\dfrac{d^2y}{dx^2} = \left[1 - \left(\dfrac{dy}{dx}\right)^2\right]^{1/2}$ ... (1)

Putting $P = \dfrac{dy}{dx} \Rightarrow \dfrac{dP}{dx} = \dfrac{d^2y}{dx^2}$ in (1), we get $\dfrac{dP}{dx} = \sqrt{1-P^2} \Rightarrow \dfrac{dP}{\sqrt{1-P^2}} = dx$

On integrating, we have

$$\sin^{-1} P = x + c \quad \Rightarrow \quad P = \sin(x + c)$$

$$\frac{dy}{dx} = \sin(x + c)$$

On integrating, we have $\quad y = -\cos(x + c) + c_1$ **Ans.**

Example 9. *Solve* $x\dfrac{d^2y}{dx^2} + x\left(\dfrac{dy}{dx}\right)^2 - \dfrac{dy}{dx} = 0$ *(U.P. II Semester, 2010)*

Solution. On putting $\dfrac{dy}{dx} = P$ and $\dfrac{d^2y}{dx^2} = \dfrac{dP}{dx}$ in the given equation, we get

$$x\frac{dP}{dx} + x\,P^2 - P = 0 \Rightarrow \frac{1}{P^2}\frac{dP}{dx} - \frac{1}{P}\frac{1}{x} = -1 \qquad ...(1)$$

Again putting $\dfrac{1}{P} = z$ so that $-\dfrac{1}{P^2}\dfrac{dP}{dx} = \dfrac{dz}{dx}$

Equation (1) becomes $\dfrac{-dz}{dx} - \dfrac{z}{x} = -1 \Rightarrow \dfrac{dz}{dx} + \dfrac{z}{x} = 1$

$$\text{I.F.} = e^{\int \frac{1}{x} dx} = e^{\log x} = x$$

Hence, solution is $\quad z\,x = \int x\,dx + c_1$ or $z\,x = \dfrac{x^2}{2} + c_1$ or $\dfrac{1}{P}x = \dfrac{x^2}{2} + c_1$

$$\Rightarrow \quad \frac{x}{P} = \frac{x^2 + 2c_1}{2} \Rightarrow P = \frac{2x}{x^2 + 2c_1} \Rightarrow \frac{dy}{dx} = \frac{2x}{x^2 + 2c_1} \Rightarrow dy = \frac{2x}{x^2 + 2c_1}dx$$

On integrating, we have $\quad y = \log(x^2 + 2c_1) + c_2$ **Ans.**

EXERCISE 15.3

Solving the following differential equations:

1. $(1 + x^2)\dfrac{d^2y}{dx^2} + x\dfrac{dy}{dx} + ax = 0$ **Ans.** $y = c_2 - ax + c_1 \log[x + \sqrt{(1 + x^2)}]$

2. $(1 + x^2)\dfrac{d^2y}{dx^2} + 1 + \left(\dfrac{dy}{dx}\right)^2 = 0$ **Ans.** $y = -\dfrac{x}{k} + \dfrac{1+k^2}{k^2}\log(1 + kx) + a$

3. $\dfrac{d^4y}{dx^4} - \cot x\dfrac{d^3y}{dx^3} = 0$ **Ans.** $y = c_1 \cos x + c_2x^2 + c_3x + c_4$

4. $2x\dfrac{d^3y}{dx^3}\cdot\dfrac{d^2y}{dx^2} = \left[\dfrac{d^2y}{dx^2}\right]^2 - a^2$ **Ans.** $15\,c_1^2 y = 4(c_1x + a^2)^{5/2} + c_2x + c_3$

5. $e^{x^2/2}\left[x\dfrac{d^2y}{dx^2} - \dfrac{dy}{dx}\right] = x^3$ **Ans.** $y = e^{-x^2/2} + c_1\dfrac{x^2}{2} + c_2$

6. $\alpha\dfrac{d^2y}{dx^2} = \left[1 + \left(\dfrac{dy}{dx}\right)^2\right]^{1/2}$ *(D.U. April, 2010)* **Ans.** $y = -a\cosh\left(\dfrac{x + c_1}{a}\right) + c_2$

15.5 EQUATIONS THAT DO NOT CONTAIN 'x' DIRECTLY

The equations that do not contain x directly are of the form

$$f\left(\frac{d^n y}{dx^n}, \frac{d^{n-1}y}{dx^{n-1}}, \ldots\ldots \frac{dy}{dx}, y\right) = 0 \qquad \ldots(1)$$

On substituting $\frac{dy}{dx} = P, \frac{d^2y}{dx^2} = \frac{dP}{dx} = \frac{dP}{dy}\cdot\frac{dy}{dx} = \frac{dP}{dy}P$ in the equation (1), we get

$$\left[\frac{dP^{n-1}}{dy^{n-1}}, \ldots\ldots P, y\right] = 0 \qquad \ldots(2)$$

Equation (2) is solved for P.

Let $P = f_1(y) \Rightarrow \frac{dy}{dx} = f_1(y) \Rightarrow \frac{dy}{f_1(y)} = dx \Rightarrow \int \frac{dy}{f_1(y)} = x + c$

Example 10. *Solve* $y\frac{d^2y}{dx^2} + \left(\frac{dy}{dx}\right)^2 = \frac{dy}{dx}$ ...(1)

Solution. Put $\frac{dy}{dx} = P, \frac{d^2y}{dx^2} = \frac{dP}{dx} = \frac{dP}{dy}\frac{dy}{dx} = P\frac{dP}{dy}$ in equation (1)

$$yP\frac{dP}{dy} + P^2 = P \Rightarrow y\frac{dP}{dy} = 1 - P$$

$$\Rightarrow \frac{dP}{1-P} = \frac{dy}{y} \Rightarrow -\log(1-P) = \log y + \log c_1$$

$$\Rightarrow \frac{1}{1-P} = c_1 y \Rightarrow P = 1 - \frac{1}{c_1 y} \Rightarrow \frac{dy}{dx} = \frac{c_1 y - 1}{c_1 y}$$

$$\Rightarrow \frac{c_1 y}{c_1 y - 1}dy = dx \Rightarrow \left(1 + \frac{1}{c_1 y - 1}\right)dy = dx$$

$$y + \frac{1}{c_1}\log(c_1 y - 1) = x + c_2$$ **Ans.**

Example 11. *Solve* $y\frac{d^2y}{dx^2} + \left(\frac{dy}{dx}\right)^2 = y^2$ ...(1)

Solution. Put $\frac{dy}{dx} = P, \frac{d^2y}{dx^2} = \frac{dP}{dx} = \frac{dP}{dy}\frac{dy}{dx} = P\frac{dP}{dy}$ in (1)

$$yP\frac{dP}{dy} + P^2 = y^2 \text{ or } P\frac{dP}{dy} + \frac{P^2}{y} = y \qquad \ldots(2)$$

Put $P^2 = z$ or $2P\frac{dP}{dy} = \frac{dz}{dy}$ in (2), $\frac{1}{2}\frac{dz}{dy} + \frac{z}{y} = y$ or $\frac{dz}{dy} + \frac{2z}{y} = 2y$

$$\text{I.F.} = e^{\int \frac{2}{y}dy} = e^{2\log y} = e^{\log y^2} = y^2$$

Hence, the solution is $\quad z\,y^2 = \int 2y.(y^2)\,dy + c$

$\Rightarrow \quad P^2 y^2 = \frac{y^4}{2} + c \quad \Rightarrow \quad 2P^2 y^2 = y^4 + k$ or $\sqrt{2}\,y\,P = \sqrt{y^4 + k}$ [Put $2c = k$]

$$\Rightarrow \sqrt{2}\,y\frac{dy}{dx} = \sqrt{y^4 + k} \text{ or } \sqrt{2}\frac{y\,dy}{\sqrt{y^4+k}} = dx$$

$$\Rightarrow \quad \frac{1}{\sqrt{2}}\frac{dt}{\sqrt{t^2+k}} = dx \text{ [Put } y^2 = t,\ 2y\,dy = dt] \quad \Rightarrow \frac{1}{\sqrt{2}}\sin h^{-1}\frac{t}{\sqrt{k}} = x + c$$

$$\Rightarrow \quad \sin h^{-1}\frac{y^2}{\sqrt{k}} = \sqrt{2}\,x + c \text{ or } y^2 = \sqrt{k}\sin h(\sqrt{2}\,x + c) \qquad \textbf{Ans.}$$

EXERCISE 15.4

Solve the following differential equations:

1. $\frac{d^2y}{dx^2} + \left(\frac{dy}{dx}\right)^2 = 1$ **Ans.** $y^2 = x^2 + ax + b$ **2.** $y\frac{d^2y}{dx^2} - \left(\frac{dy}{dx}\right)^2 + 2\frac{dy}{dx} = 0$ **Ans.** $c^y + 2 = d\,e^{cx}$

3. $2y\frac{d^2y}{dx^2} - 3\left(\frac{dy}{dx}\right)^2 - 4y^2 = 0$ **Ans.** $y = a\sec^2(x+b)$ **4.** $y\frac{d^2y}{dx^2} + \frac{dy}{dx} + \left(\frac{dy}{dx}\right)^3 = 0$ **Ans.** $y = a - \sin^{-1}(b\,e^{-x})$

5. $y\frac{d^2y}{dx^2} = \left\{\frac{dy}{dx}\right\}^2\left[1 - \frac{dy}{dx}\cos y + y\frac{dy}{dx}\sin y\right]$ **Ans.** $x = c_1 + c_2\log y + \sin y$

6. $y\frac{d^2y}{dx^2} - \left(\frac{dy}{dx}\right)^2 = y^2\log y$ **Ans.** $\log y = b.\,e^x + a\,e^{-x}$

15.6 EQUATION WHOSE ONE SOLUTION IS KNOWN

If $y = u$ is given solution belonging to the complementary function of the differential equation. Let the other solution be $y = v$. Then $y = u.\,v$ is complete solution of the differential equation.

Let $\frac{d^2y}{dx^2} + P\frac{dy}{dx} + Qy = R$ ....(1), be the differential equation and u is the solution included in the complementary function of (1)

$$\therefore \quad \frac{d^2u}{dx^2} + P\frac{du}{dx} + Qu = 0 \qquad ...(2)$$

$$y = u.\,v$$

$$\frac{dy}{dx} = v\frac{du}{dx} + u\frac{dv}{dx}$$

$$\frac{d^2y}{dx^2} = v\frac{d^2u}{dx^2} + 2\frac{dv}{dx}\frac{du}{dx} + u\frac{d^2v}{dx^2}$$

Substituting the values of $y, \frac{dy}{dx}, \frac{d^2y}{dx^2}$ in (1), we get

$$v\frac{d^2u}{dx^2} + 2\frac{dv}{dx}\frac{du}{dx} + u\frac{d^2v}{dx^2} + P\left(v\frac{du}{dx} + u\frac{dv}{dx}\right) + Qu.v = R$$

On arranging

$$\Rightarrow v\left[\frac{d^2u}{dx^2} + P\frac{du}{dx} + Qu\right] + u\left[\frac{d^2v}{dx^2} + P\frac{dv}{dx}\right] + 2\frac{du}{dx}\cdot\frac{dv}{dx} = R$$

The first bracket is zero by virtue of relation (2), and the remaing is divided by u.

$$\frac{d^2v}{dx^2} + P\frac{dv}{dx} + \frac{2}{u}\frac{du}{dx}\frac{dv}{dx} = \frac{R}{u}$$

$$\Rightarrow \quad \frac{d^2v}{dx^2} + \left[P + \frac{2}{u}\frac{du}{dx}\right]\frac{dv}{dx} = \frac{R}{u} \qquad ...(3)$$

Let $\frac{dv}{dx} = z$, so that $\frac{d^2v}{dx^2} = \frac{dz}{dx}$

Equation (3) becomes

$$\boxed{\frac{dz}{dx} + \left[P + \frac{2}{u}\frac{du}{dx}\right] z = \frac{R}{u}}$$

This is the linear differential equation of first order and can be solved (z can be found), which will contain one constant.

On integration $z = \frac{dv}{dx}$, we can get v.

Having found v, the solution is $y = uv$.

Note: Rule to find out the integral belonging to the complementary function

Rule	*Condition*	u
1	$1 + P + Q = 0$	e^x
2	$1 - P + Q = 0$	e^{-x}
3	$1 + \frac{P}{a} + \frac{Q}{a^2} = 0$	e^{ax}
4	$P + Qx = 0$	x
5	$2 + 2Px + Qx^2 = 0$	x^2
6	$n(n-1) + Pnx + Qx^2 = 0$	x^n

Example 12. *Solve $y'' - 4xy' + (4x^2 - 2)y = 0$ given that $y = e^{x^2}$ is an integral included in the complementary function.*

Solution. $y'' - 4xy' + (4x^2 - 2)y = 0$ ...(1)

On putting $y = v.e^{x^2}$ in (1), the reduced equation as in the article 15.6.

$$\frac{d^2v}{dx^2} + \left[P + \frac{2}{u}\frac{du}{dx}\right]\frac{dv}{dx} = 0 \qquad [P = -4x,\ Q = 4x^2 - 2,\ R = 0]$$

$$\Rightarrow \frac{d^2v}{dx^2} + \left[-4x + \frac{2}{e^{x^2}}(2x\,e^{x^2})\right]\frac{dv}{dx} = 0$$

$$\Rightarrow \frac{d^2v}{dx^2} + [-4x + 4x]\frac{dv}{dx} = 0 \qquad \Rightarrow \frac{d^2v}{dx^2} = 0 \Rightarrow \frac{dv}{dx} = c_1 \Rightarrow v = c_1x + c_2$$

$$\therefore \qquad y = uv \qquad [u = e^{x^2}]$$

$$y = e^{x^2}(c_1x + c_2)$$ **Ans.**

Example 13. *Solve* $x\frac{d^2y}{dx^2} - (2x-1)\frac{dy}{dx} + (x-1)y = 0$

given that $y = e^x$ is an integral included in the complementary function.

Solution. $x\frac{d^2y}{dx^2} - (2x-1)\frac{dy}{dx} + (x-1)\,y = 0$

$$\Rightarrow \frac{d^2y}{dx^2} - \frac{2x-1}{x}\frac{dy}{dx} + \frac{x-1}{x}y = 0 \qquad [1 + P + Q = 0] \qquad ...(1)$$

By putting $y = ve^x$ in (1),we get the reduced equation as in the article 15.6.

$$\frac{d^2v}{dx^2}+\left[P+\frac{2}{u}\frac{du}{dx}\right]\frac{dv}{dx}=0 \qquad ...(2)$$

Putting $u = e^x$ and $\frac{dv}{dx}=z \quad \frac{d^2v}{dx^2}=\frac{dz}{dx}$ (2), we get $\frac{dz}{dx}+\left[-\frac{2x-1}{x}+\frac{2}{e^x}e^x\right]z=0$

$$\Rightarrow \quad \frac{dz}{dx}+\frac{-2x+1+2x}{x}z=0 \quad \Rightarrow \quad \frac{dz}{dx}+\frac{z}{x}=0$$

$$\Rightarrow \quad \frac{dz}{z}=-\frac{dx}{x}\Rightarrow \log z=-\log x+\log c_1$$

$$\Rightarrow \quad z=\frac{c_1}{x} \text{ or } \frac{dv}{dx}=\frac{c_1}{x} \text{ or } dv=c_1\frac{dx}{x}\Rightarrow v=c_1\log x+c_2$$

$$y=u.\,v=e^x\,(c_1\log x+c_2)$$ **Ans.**

Example 14. *Solve* $x^2\frac{d^2y}{dx^2}-2x\,[1+x]\frac{dy}{dx}+2(1+x)y=x^3$

Solution. $x^2\frac{d^2y}{dx^2}-2x(1+x)\frac{dy}{dx}+2(1+x)y=x^3$

$$\Rightarrow \quad \frac{d^2y}{dx^2}-\frac{2x(1+x)}{x^2}\frac{dy}{dx}+\frac{2(1+x)y}{x^2}=x \qquad ...(1)$$

Here $\qquad P+Qx=-\frac{2x(1+x)}{x^2}+\frac{2(1+x)}{x^2}x=0$

Hence $y = x$ is a solution of the C.F. and the other solution is v.

Putting $y = vx$ in (1), we get the reduced equation as in article 15.6

$$\frac{d^2v}{dx^2}+\left\{P+\frac{2}{u}\frac{du}{dx}\right\}\frac{dv}{dx}=\frac{x}{u}$$

$$\frac{d^2v}{dx^2}+\left[\frac{-2x\,(1+x)}{x^2}+\frac{2}{x}(1)\right]\frac{dv}{dx}=\frac{x}{x}$$

$$\Rightarrow \quad \frac{d^2v}{dx^2}-2\frac{dv}{dx}=1\Rightarrow\frac{dz}{dx}-2z=1 \qquad \left[\frac{dv}{dx}=z\right]$$

which is a linear differential equation of first order and $I.F. = e^{\int -2\,dx = e^{-2x}}$

Its solution is $z\,e^{-2x}=\int e^{-2x}\,dx+c_1$

$$z\,e^{-2x}=\frac{e^{-2x}}{-2}+c_1 \text{ or } z=\frac{-1}{2}+c_1e^{2x}$$

$$\Rightarrow \quad \frac{dv}{dx}=-\frac{1}{2}+c_1e^{2x} \text{ or } dv=\left(-\frac{1}{2}+c_1e^{2x}\right)dx \quad \Rightarrow v=\frac{-x}{2}+\frac{c_1}{2}e^{2x}+c_2$$

$$y=uv=x\left(\frac{-x}{2}+\frac{c_1}{2}e^{2x}+c_2\right)$$ **Ans.**

Example 15. *Verify that $y = e^{2x}$ is a solution of $(2x + 1) y'' - 4 (x + 1) y' + 4y = 0$. Hence find the general solution.*

Solution. We have

$$(2x+1)\frac{d^2y}{dx^2} - 4(x+1)\frac{dy}{dx} + 4y = 0 \qquad \text{... (1)}$$

Let $y = e^{2x}$, $y' = 2e^{2x}$, $y'' = 4e^{2x}$

Substituting the values of y, y' and y'' in (1), we get

$$(2x + 1)\, 4e^{2x} - 4\,(x + 1)\, 2e^{2x} + 4e^{2x} = 0$$

or $[8x + 4 - 8x - 8 + 4]\, e^{2x} = 0 \Rightarrow 0 = 0$

Thus $y_1 = e^{2x}$ is a solution

Equation (1) in the standard form is

$$y'' - \frac{4\,(x+1)}{(2x+1)} y' + \frac{4}{(2x+1)} y = 0$$

So $$P\,(x) = -\frac{4\,(x+1)}{(2x+1)}.$$

Then $$\omega\,(x) = \frac{1}{y_1^2} e^{-\int P\,dx} \qquad \text{... (2)}$$

Now $$\int -P\,dx = -\int -\frac{4\,(x+1)}{(2x+1)}\,dx = \int \left(\frac{4x+2}{2x+1} + \frac{2}{2x+1}\right) dx$$

$$= 2x + ln\,(2x + 1)$$

Then $$\omega = \frac{1}{(e^{2x})^2} e^{2x + ln\,(2x+1)} = \frac{e^{2x}}{(e^{2x})^2} . (2x+1) \qquad \text{[From (2)]}$$

$$\omega\,(x) = \frac{2x+1}{e^{2x}}$$

Now $$v\,(x) = \int \omega\,(x)\,dx = \int \frac{2x+1}{e^{2x}}\,dx$$

Integrating by parts

$$y\,(x) = (2x + 1)\frac{e^{-2x}}{-2} - 2 . \frac{e^{-2x}}{4}$$

The required second solution

$$y_2\,(x) = y_1\,(x)\, v\,(x) = e^{2x}\left[-\frac{2x+1}{2} . \frac{1}{e^{2x}} - \frac{1}{2}\frac{1}{e^{2x}}\right] = -x - 1 = -(x + 1)$$

Then the general solution is

$y\,(x) = c_1 y_1\,(x) + c_2 y_2\,(x) = c_1 e^{2x} - c_2\,(x + 1)$ **Ans.**

Example 16. *Solve $x^2 y'' - (x^2 + 2x)y' + (x + 2)\, y = x^3 e^x$ given that $y = x$ is a solution.*

Solution. $x^2 y'' - (x^2 + 2x)y' + (x + 2)\, y = x^3 e^x$

$$\Rightarrow \quad y'' - \frac{x^2 + 2x}{x^2} y' + \frac{x+2}{x^2} y = x e^x \qquad \text{...(1)}$$

On putting $y = vx$ in (1), the reduced equation as in the article 15.6.

$$\frac{d^2v}{dx^2}+\left\{P+\frac{2}{u}\frac{du}{dx}\right\}\frac{dv}{dx}=\frac{R}{u} \Rightarrow \frac{d^2v}{dx^2}+\left[-\frac{x^2+2x}{x^2}+\frac{2}{x}(1)\right]\frac{dv}{dx}=\frac{xe^x}{x}$$

$$\Rightarrow \qquad \frac{d^2v}{dx^2}-\frac{dv}{dx}=e^x \Rightarrow \frac{dz}{dx}-z=e^x \qquad \left(z=\frac{dv}{dx}\right)$$

which is a linear differential equation

$$I.F. = e^{-\int dx}=e^{-x} \quad \Rightarrow \quad z\,e^{-x} = \int e^x . e^{-x}\,dx + c$$

$$z\,e^{-x} = x + c \text{ or } z = e^x . x + c\,e^x \Rightarrow \frac{dv}{dx} = e^x . x + c\,e^x$$

$$v = x.e^x - e^x + c\,e^x + c_1 \;\Rightarrow\; v = (x-1)\,e^x + c\,e^x + c_1$$

$$y = vx = (x^2 - x + cx)\,e^x + c_1x$$

Ans.

Example 17. *Solve* $(x+2)\dfrac{d^2y}{dx^2}-(2x+5)\dfrac{dy}{dx}+2y=(x+1)e^x$

Solution. $\dfrac{d^2y}{dx^2}-\dfrac{2x+5}{x+2}\dfrac{dy}{dx}+\dfrac{2y}{x+2}=\dfrac{(x+1)\,e^x}{x+2}$ ...(1)

Here $\quad P=-\dfrac{2x+5}{x+2},\quad Q=\dfrac{2}{x+2},\quad R=\dfrac{(x+1)\,e^x}{x+2}$

$$1+\frac{P}{a}+\frac{Q}{a^2}=0, \text{ Choosing } a=2$$

$$1+\frac{P}{2}+\frac{Q}{4}=1-\frac{2x+5}{2x+4}+\frac{2}{4x+8}=0$$

Hence $y = e^{2x}$ is a part of C.F.

Putting $y = e^{2x}v$ in (1), the reduced equation as in the article 15.6.

$$\frac{d^2v}{dx^2}+\left[P+\frac{2}{u}\frac{du}{dx}\right]\frac{dv}{dx}=\frac{R}{u}$$

On putting the values of P, Q and R, we get

$$\Rightarrow \quad \frac{d^2v}{dx^2}+\left[-\frac{2x+5}{x+2}+\frac{2}{e^{2x}}2e^{2x}\right]\frac{dv}{dx}=\frac{(x+1)\,e^x}{e^{2x}(x+2)}$$

$$\Rightarrow \quad \frac{d^2v}{dx^2}+\left[-\frac{2x+5}{x+2}+4\right]\frac{dv}{dx}=\frac{x+1}{x+2}e^{-x}$$

$$\Rightarrow \quad \frac{dz}{dx}-\frac{2x+3}{x+2}z=\frac{x+1}{x+2}e^{-x} \qquad \left(\frac{dv}{dx}=z\right)$$

which is a linear differential equation, $I.F. = e^{\int\frac{2x+3}{x+2}dx} = e^{\int\left(2-\frac{1}{x+2}\right)dx} = e^{2x-\log(x+2)} = \dfrac{e^{2x}}{x+2}$

Its solution is $\qquad z.\dfrac{e^{2x}}{x+2}=\displaystyle\int\frac{e^{2x}}{x+2}\,\frac{x+1}{x+2}\,e^{-x}\,dx+c$

$$=\int\frac{e^x\,(x+1)}{(x+2)^2}dx+c=\int e^x\left[\frac{1}{x+2}-\frac{1}{(x+2)^2}\right]dx+c=\int\frac{e^x dx}{x+2}-\int\frac{e^x dx}{(x+2)^2}+c$$

$$= \frac{e^x}{x+2} + \int \frac{e^x dx}{(x+2)^2} - \int \frac{e^x\, dx}{(x+2)^2} + c = \frac{e^x}{x+2} + c$$

$$\Rightarrow \quad z = e^{-x} + c\,(x+2)e^{-2x} \Rightarrow \frac{dv}{dx} = e^{-x} + c(x+2)e^{-2x}$$

$$v = \int e^{-x}\, dx + c\int (x+2)e^{-2x} dx + c_1 = -e^{-x} + c\left[\frac{(x+2)\,e^{-2x}}{-2} - \frac{e^{-2x}}{4}\right] + c_1$$

$$= -e^{-x} - \frac{ce^{-2x}}{4}[2x+5] + c_1$$

$$y = u.v$$

$$y = e^{2x}\left[-e^{-x} - \frac{c\,e^{-2x}}{4}(2x+5) + c_1\right] \Rightarrow y = -e^x + \frac{c}{4}(2x+5) + c_1\, e^{2x} \qquad \textbf{Ans.}$$

EXERCISE 15.5

Solve the following differential equations:

1. $(3-x)\dfrac{d^2y}{dx^2} - (9-4x)\dfrac{dy}{dx} + (6-3x)\,y = 0$, given $y = e^x$ is a solution.

Ans. $y = \dfrac{c_1}{8}e^{3x}\,(4x^3 - 42x^2 + 150x - 183) + c_2e^x$

2. $x\dfrac{d^2y}{dx^2} - \dfrac{dy}{dx} + (1-x)y = x^2\, e^{-x}$ given $y = e^x$ is an integral included in C.F.

Ans. $y = c_2\, e^x + c_1\,(2x+1)\,e^{-x} - \dfrac{1}{4}(2x^2 + 2x + 1)e^{-x}$

3. $(1-x^2)\dfrac{d^2y}{dx^2} + x\dfrac{dy}{dx} - y = x(1-x^2)^{3/2}$, given $y = x$ is part of C.F.

Ans. $y = -\dfrac{x}{9}(1-x^2)^{3/2} - c_1\,[\sqrt{(1-x^2)} + x\sin^{-1} x] + c_2x.$

4. $\sin^2 x\dfrac{d^2y}{dx^2} = 2y$, given that $y = \cot x$ is a solution **Ans.** $cy = 1 + (c_1 - x)\cot x$

5. $\dfrac{d^2y}{dx^2} - x^2\dfrac{dy}{dx} + xy = x$, given $y = x$ is a part of C.F. **Ans.** $y = 1 + c_1x\displaystyle\int \frac{1}{x^2}e^{\frac{x^3}{3}}\, dx + c_2x$

6. $(x\sin x + \cos x)\dfrac{d^2y}{dx^2} - x\cos x\dfrac{dy}{dx} + y\cos x = 0$ given $y = x$ is solution. **Ans.** $y = c_2x - c_1\cos x$

7. $x^2\dfrac{d^2y}{dx^2} + x\dfrac{dy}{dx} - y = 0$, given that $y = x + \dfrac{1}{x}$ is one integral. **Ans.** $y = c_2\left(x + \dfrac{1}{x}\right) + \dfrac{c_1}{x}$

8. $x^2\dfrac{d^2y}{dx^2} + 2x\dfrac{dy}{dx} - 12y = x^3\log x$

[**Hint.** $(n\,(n-1) + pnx + Qx^2 = 0)$, $n = 3$, satisfies this equation. Put $y = vx^3, \dfrac{dv}{dx} = z$]

Ans. $y = \left(c_1x^3 + \dfrac{c_2}{x^4}\right) + \dfrac{x^3}{98}\log x\,(7\log x - 2)$

15.7 NORMAL FORM (REMOVAL OF FIRST DERIVATIVE)

Consider the differential equation $\frac{d^2y}{dx^2}+P\frac{dy}{dx}+Qy=R$ ...(1)

Put $y = uv$ where v is not an integral solution of C.F.

$$\frac{dy}{dx} = v\frac{du}{dx}+u\frac{dv}{dx}$$

$$\frac{d^2y}{dx^2} = u\frac{d^2v}{dx^2}+2\frac{du}{dx}\frac{dv}{dx}+v\frac{d^2u}{dx^2}$$

On putting the values of $y, \frac{dy}{dx}, \frac{d^2y}{dx^2}$ in (1) we get

$$\left(u\frac{d^2v}{dx^2}+2\frac{dv}{dx}\frac{du}{dx}+v\frac{d^2u}{dx^2}\right)+P\left(u\frac{dv}{dx}+v\frac{du}{dx}\right)+Q.uv=R$$

$$\Rightarrow \quad v\frac{d^2u}{dx^2}+\frac{du}{dx}\left(Pv+2\frac{dv}{dx}\right)+u\left(\frac{d^2v}{dx^2}+P\frac{dv}{dx}+Q.v\right)=R$$

$$\Rightarrow \quad \frac{d^2u}{dx^2}+\frac{du}{dx}\left(P+\frac{2}{v}\frac{dv}{dx}\right)+\frac{u}{v}\left(\frac{d^2v}{dx^2}+P\frac{dv}{dx}+Q.v\right)=\frac{R}{v} \quad ...(2)$$

Here in the last bracket on L.H.S. is not zero $y = v$ is not a part of C.F.

Here we shall remove the first derivative.

$$P+\frac{2}{v}\frac{dv}{dx}=0 \text{ or } \frac{dv}{v}=-\frac{1}{2}P\,dx \text{ or } \log v=\frac{-1}{2}\int P\,dx$$

$$v=e^{-\frac{1}{2}\int P\,dx}$$

In (2) we have to find out the value of the last bracket *i.e.*, $\frac{d^2v}{dx^2}+P\frac{dv}{dx}+Qv$

$$\frac{dv}{dx} = -\frac{P}{2}e^{-\frac{1}{2}\int P\,dx}=-\frac{1}{2}Pv \qquad \left[\because v=e^{-1/2\int P dx}\right]$$

$$\frac{d^2v}{dx^2} = -\frac{1}{2}\frac{dP}{dx}v-\frac{P}{2}\frac{dv}{dx}=-\frac{1}{2}\frac{dP}{dx}v-\frac{P}{2}\left(-\frac{1}{2}Pv\right)=-\frac{1}{2}\frac{dP}{dx}v+\frac{1}{4}P^2v$$

$$\therefore \frac{d^2v}{dx^2}+P\frac{dv}{dx}+Qv = -\frac{1}{2}\frac{dP}{dx}v+\frac{1}{4}P^2v+P\left(-\frac{1}{2}Pv\right)+Qv=v\left[Q-\frac{1}{2}\frac{dP}{dx}-\frac{1}{4}P^2\right]$$

Equation (1) is transformed as

$$\frac{d^2u}{dx^2}+\frac{u}{v}v\left\{Q-\frac{1}{2}\frac{dP}{dx}-\frac{P^2}{4}\right\}=\frac{R}{v}$$

$$\Rightarrow \quad \frac{d^2u}{dx^2}+u\left\{Q-\frac{1}{2}\frac{dP}{dx}-\frac{P^2}{4}\right\}=Re^{\frac{1}{2}\int P dx}$$

$$\frac{d^2u}{dx^2}+Q_1u = R_1 \qquad \text{where } Q_1=\left[Q-\frac{1}{2}\frac{dP}{dx}-\frac{P^2}{4}\right]$$

$$R_1 = Re^{\frac{1}{2}\int P\,dx} \text{ or } \frac{R}{v}$$

$$y=uv \quad \text{and} \quad v=e^{-\frac{1}{2}\int P\,dx}$$ **Ans.**

Example 18. *Solve* $\frac{d}{dx}\left[\cos^2 x\frac{dy}{dx}\right]+\cos^2 x.\,y=0$

Solution. We have, $\frac{d}{dx}\left(\cos^2 x\frac{dy}{dx}\right)+\cos^2 x.\,y=0$

$$\Rightarrow \quad \frac{d^2y}{dx^2}\cos^2 x-2\cos x\sin x\frac{dy}{dx}+(\cos^2 x)y=0 \Rightarrow \frac{d^2y}{dx^2}-2\tan x.\frac{dy}{dx}+y=0$$

Here, $\quad P=-2\tan x,\ Q=1,\ R=0$

$$Q_1=Q-\frac{1}{2}\frac{dP}{dx}-\frac{P^2}{4}=1-\frac{1}{2}(-2\sec^2 x)-\frac{4\tan^2 x}{4}$$

$$=1+\sec^2 x-\tan^2 x=1+1=2$$

$$R_1=Re^{\frac{1}{2}\int P\,dx}=0$$

$$v=e^{-\frac{1}{2}\int P dx}=e^{-\frac{1}{2}\int(-2\tan x)dx}=e^{\int\tan x\,dx}=e^{\log\sec x}=\sec x$$

Normal equation is

$$\frac{d^2u}{dx^2}+Q_1u=R_1$$

$$\frac{d^2u}{dx^2}+2u=0\Rightarrow(D^2+2)\,u=0$$

$$\Rightarrow \quad \text{A.E. is } m^2+2=0 \Rightarrow m=\pm i\sqrt{2}$$

$$u=c_1\cos\sqrt{2}\,x+c_2\sin\sqrt{2}\,x$$

$$y=u.v=[c_1\cos\sqrt{2}\,x+c_2\sin\sqrt{2}x]\sec x$$ **Ans.**

Example 19. *Solve* $x^2\frac{d^2y}{dx^2}-2(x^2+x)\frac{dy}{dx}+(x^2+2x+2)y=0$

Solution. We have, $\frac{d^2y}{dx^2}-\frac{2(x^2+x)}{x^2}\frac{dy}{dx}+\left(\frac{x^2+2x+2}{x^2}\right)y=0$ ...(1)

Here $p=-2\left(1+\frac{1}{x}\right),\ Q=\frac{x^2+2x+2}{x^2},\ \ R=0$

In order to remove the first derivative, we put $y=u.v$ in (1) to get the normal equation

$$\frac{d^2u}{dx^2}+Q_1u=R_1 \quad ...(2)$$

where $v=e^{-\frac{1}{2}\int pdx}=e^{-\frac{1}{2}\int-2\left(1+\frac{1}{x}\right)dx}=e^{\int\left(1+\frac{1}{x}\right)dx}=e^x.e^{\log x}=x\,e^x$

$$Q_1=Q-\frac{1}{2}\frac{dp}{dx}-\frac{p^2}{4}=\frac{x^2+2x+2}{x^2}-\frac{1}{2}\left(\frac{2}{x^2}\right)-\frac{4}{4}\left(1+\frac{1}{x}\right)^2$$

$$=1+\frac{2}{x}+\frac{2}{x^2}-\frac{1}{x^2}-1-\frac{1}{x^2}-\frac{2}{x}=0$$

$$R_1=Re^{\frac{1}{2}\int p\,dx}=0$$

On putting the values of Q_1 and R_1 in (2), we get

$$\frac{d^2u}{dx^2} + 0(u) = 0 \Rightarrow \frac{d^2u}{dx^2} = 0$$

$$\frac{du}{dx} = c_1 \Rightarrow u = c_1 x + c_2$$

$\therefore$ $$y = u.v = (c_1 x + c_2)\, x\, e^x$$ **Ans.**

Example 20. *Solve* $\dfrac{d^2y}{dx^2} - 4x\dfrac{dy}{dx} + (4x^2 - 1)y = -3e^{x^2} \sin 2x$

Solution. We have, $\dfrac{d^2y}{dx^2} - 4x\dfrac{dy}{dx} + (4x^2 - 1)\, y = -3e^{x^2} \sin 2x$ ...(1)

Here $p = -4x$, $Q = 4x^2 - 1$, $R = -3e^{x^2} \sin 2x$

In order to remove the first derivative, $u = e^{-\frac{1}{2}\int p dx} = e^{-\frac{1}{2}\int -4x\, dx} = e^{2\int x\, dx} = e^{x^2}$

On putting $y = uv$, the normal equation is $\dfrac{d^2u}{dx^2} + Q_1 u = R_1$ ...(2)

where $Q_1 = Q - \dfrac{1}{2}\dfrac{dp}{dx} - \dfrac{p^2}{4} = (4x^2 - 1) - \dfrac{1}{2}(-4) - \dfrac{16x^2}{4} = 4x^2 - 1 + 2 - 4x^2 = 1$

$$R_1 = \frac{R}{v} = \frac{-3e^{x^2} \sin 2x}{e^{x^2}} = -3 \sin 2x$$

Equation (2) becomes $\dfrac{d^2u}{dx^2} + u = -3 \sin 2x \quad \Rightarrow \quad (D^2 + 1)u = -3 \sin 2x$

A.E. is $m^2 + 1 = 0 \Rightarrow m = \pm i \Rightarrow C.F. = c_1 \cos x + c_2 \sin x$

$$P.I. = \frac{1}{D^2 + 1}(-3 \sin 2x) = \frac{-3 \sin 2x}{-4 + 1} = \sin 2x$$

$$u = c_1 \cos x + c_2 \sin x + \sin 2x$$

$$y = u.\, v = (c_1 \cos x + c_2 \sin x + \sin 2x)e^{x^2}$$ **Ans.**

EXERCISE 15.6

Solve the following differential equations:

1. $\dfrac{d^2y}{dx^2} - 2\tan x.\, y - 5y = 0$ **Ans.** $y = (a\, e^{2x} + e^{-3x})\sec x$

2. $\dfrac{d^2y}{dx^2} - 4x\dfrac{dy}{dx} + (4x^2 - 3)\, y = e^{x^2}$ **Ans.** $y = (c_1 e^x + c_2 e^{-x} - 1)\, e^{x^2}$

3. $\dfrac{d^2y}{dx^2} - 2x\dfrac{dy}{dx} + (x^2 + 2)y = e^{\frac{1}{2}(x^2 + 2x)}$ **Ans.** $y = (c_1 \cos\sqrt{3}x + c_2 \sin\sqrt{3}\, x)\, e^{\frac{x^2}{2}} + \frac{1}{4}e^x.e^{\frac{x^2}{2}}$

4. $\dfrac{d^2y}{dx^2} - 2x\dfrac{dy}{dx} + \left(n^2 + \dfrac{2}{x^2}\right)y = 0$ **Ans.** $y = (c_1 \cos nx + c_2 \sin nx)x$

5. $\dfrac{d^2y}{dx^2} + \dfrac{2}{x}\dfrac{dy}{dx} - n^2 y = 0$ **Ans.** $y = (c_1 e^{nx} + c_2 + e^{-nx})\dfrac{1}{x}$

6. $\dfrac{d^2y}{dx^2}+\dfrac{1}{x^{\frac{1}{3}}}\dfrac{dy}{dx}+\left(\dfrac{1}{4x^{2/3}}-\dfrac{1}{x^{4/3}}-\dfrac{6}{x^2}\right)y=0$ **Ans.** $y=(c_1x^3+c_2\,x^{-2})e^{-\frac{3}{4}x^{\frac{2}{3}}}$

7. $\dfrac{d^2y}{dx^2}-\dfrac{1}{\sqrt{x}}\dfrac{dy}{dx}+\dfrac{y}{4x^2}(-8+\sqrt{x}+x)=0$ **Ans.** $y=(c_1x^2+c_2x^{-1})\,e^{\sqrt{x}}$

15.8 METHOD OF SOLVING LINEAR DIFFERENTIAL EQUATIONS BY CHANGING THE INDEPENDENT VARIABLE

Consider, $$\frac{d^2y}{dx^2}+P\frac{dy}{dx}+Qy=R \qquad ...(1)$$

Let us change the independent variable x to z and $z=f(x)$

$$\frac{dy}{dx}=\frac{dy}{dz}\frac{dz}{dx} \Rightarrow \frac{d^2y}{dx^2}=\frac{d^2y}{dz^2}\left(\frac{dz}{dx}\right)^2+\frac{dy}{dz}\frac{d^2z}{dx^2}$$

Putting the values of $\dfrac{dy}{dx}$ and $\dfrac{d^2y}{dx^2}$ in (1), we get

$$\left(\frac{d^2y}{dz^2}\left(\frac{dz}{dx}\right)^2+\frac{dy}{dz}\frac{d^2z}{dx^2}\right)+P\left(\frac{dy}{dz}\frac{dz}{dx}\right)+Qy=R$$

$$\Rightarrow \frac{d^2y}{dz^2}\left(\frac{dz}{dx}\right)^2+\left(P\frac{dz}{dx}+\frac{d^2z}{dx^2}\right)\frac{dy}{dz}+.Qy=R$$

$$\Rightarrow \frac{d^2y}{dz^2}+\frac{P\left(\dfrac{dz}{dx}+\dfrac{d^2z}{dx^2}\right)}{\left(\dfrac{dz}{dx}\right)^2}\frac{dy}{dz}+\frac{Qy}{\left(\dfrac{dz}{dx}\right)^2}=\frac{R}{\left(\dfrac{dz}{dx}\right)^2} \Rightarrow \frac{d^2y}{dz^2}+P_1\frac{dy}{dz}+Q_1y=R_1 \qquad ...(2)$$

where $$P_1=\frac{P\left(\dfrac{dz}{dx}+\dfrac{d^2z}{dx^2}\right)}{\left(\dfrac{dz}{dx}\right)^2},\quad Q_1=\frac{Q}{\left(\dfrac{dz}{dx}\right)^2},\quad R_1=\frac{R}{\left(\dfrac{dz}{dx}\right)^2}$$

Equation (2) is solved either by taking $P_1=0$ or $Q_1=a$ constant.

Equation (2) can be solved by by two methods, by taking

First Method, $P_1=0$

Second Method, $Q_1=$ constant

Working Rule

Step 1. Coefficient of $\dfrac{d^2y}{dx^2}$ should be made as 1 if it is not so.

Step 2. To get P, Q and R, compare the given differential equation with the standard form $y''+Py'+Qy=R$.

Step 3. Find P_1, Q_1 and R_1 by the following formulae.

$$P_1 = \frac{\frac{d^2y}{dx^2} + p\frac{dz}{dx}}{\left(\frac{dz}{dx}\right)^2}, \quad R_1 = \frac{R}{\left(\frac{dz}{dx}\right)^2}$$

Step 4. Find out the value of z by taking

First Method, $P_1 = 0$ **Second Method.** Q_1 = constant

Step 5. We get a reduced equation $\frac{d^2y}{dz^2} + P_1\frac{dy}{dz} + Q_1 y = R_1$

On solving this equation we can find out the value of y in terms of z.
Then write down the solution in terms of x by replacing the value of z.

Example 21. *Solve* $\frac{d^2y}{dx^2} + \cot x \frac{dy}{dx} + 4\, y \operatorname{cosec}^2 x = 0$

Solution. We have, $\frac{d^2y}{dx^2} + \cot x \frac{dy}{dx} + 4\, y \operatorname{cosec}^2 x = 0$ ...(1)

Here, $P = \cot x$, $Q = 4 \operatorname{cosec}^2 x$ and $R = 0$

Changing the independent variable from x to z, the equation becomes

$$\frac{d^2y}{dx^2} + P_1\frac{dy}{dz} + Q_1 y = 0 \qquad ...(2)$$

where
$$P_1 = \frac{P\frac{dz}{dx} + \frac{d^2z}{dx^2}}{\left(\frac{dz}{dx}\right)^2}, \quad Q_1 = \frac{Q}{\left(\frac{dz}{dx}\right)^2}$$

Case I. Let us take $P_1 = 0$

$$\frac{p\frac{dz}{dx} + \frac{d^2z}{dx^2}}{\left(\frac{dz}{dx}\right)^2} = 0 \text{ or } P\frac{dz}{dx} + \frac{d^2z}{dx^2} = 0 \Rightarrow \frac{d^2z}{dx^2} + \cot x \frac{dz}{dx} = 0 \qquad ...(3)$$

Put
$$\frac{dz}{dx} = v, \quad \frac{d^2z}{dx^2} = \frac{dv}{dx}$$

(3) becomes
$$\frac{dv}{dx} + (\cot x)\, v = 0 \Rightarrow \frac{dv}{x} = -\cot x.\, dx$$

$$\Rightarrow \quad \log v = -\log \sin x + \log c = \log c \operatorname{cosec} x \Rightarrow v = c \operatorname{cosec} x$$

$$\frac{dz}{dx} = c \operatorname{cosec} x \text{ or } dz = (c \operatorname{cosec} x)\, dx \Rightarrow z = c \log \tan \frac{x}{2} + c_1$$

Case II.

Now,

$$Q_1 = \frac{Q}{\left(\frac{dz}{dx}\right)^2} = \frac{4 \operatorname{cosec}^2 x}{c^2 \operatorname{cosec}^2 x} = \frac{4}{c^2} \text{ which is constant}$$

Hence the equation (2) reduces to

$$\frac{d^2y}{dz^2}+0\frac{dy}{dz}+\frac{4}{c^2}y=0 \text{ or } \frac{d^2y}{dz^2}+\frac{4}{c^2}y=0 \qquad \left[\because P_1=0, Q_1=\frac{4}{c^2}\right]$$

$$\Rightarrow \qquad \left(D^2+\frac{4}{c^2}\right)y=0, \text{ A.E. is } m^2+\frac{4}{c^2}=0 \Rightarrow m=\pm i\frac{2}{c}$$

$$\text{C.F.}=c_1\cos\frac{2z}{c}+c_2\sin\frac{2z}{c} \qquad \left(z=c\log\tan\frac{x}{2}\right)$$

$$\Rightarrow \qquad y=c_1\cos\left(2\log\tan\frac{x}{2}\right)+c_2\sin\left(2\log\tan\frac{x}{2}\right) \qquad \textbf{Ans.}$$

Example 22. *Solve* $x^6\frac{d^2y}{dx^2}+3x^5\frac{dy}{dx}+a^2y=\frac{1}{x^2}$

Solution. We have, $\frac{d^2y}{dx^2}+\frac{3}{x}\frac{dy}{dx}+a^2\frac{y}{x^6}=\frac{1}{x^8}$ ...(1)

Here $P=\frac{3}{x}$ and $Q=\frac{a^2}{x^6}$

On changing the independent variable x to z, the equation (1) is reduced to

$$\frac{d^2y}{dz^2}+P_1\frac{dy}{dz}+Q_1y=R_1 \qquad ...(2)$$

Using Second Method

Let $Q_1=a^2$ (constant) $\qquad Q_1=\frac{Q}{\left(\frac{dz}{dx}\right)^2}=\frac{a^2}{x^6\left(\frac{dz}{dx}\right)^2}$ = constant = a^2 (say)

$$\therefore \quad x^6\left(\frac{dz}{dx}\right)^2=1 \Rightarrow x^3\left(\frac{dz}{dx}\right)=1 \Rightarrow \frac{dz}{dx}=\frac{1}{x^3} \Rightarrow dz=\frac{dx}{x^3} \Rightarrow z=\frac{x^{-2}}{-2}+c$$

On differentiating twice, we have $\frac{d^2z}{dx^2}=\frac{-3}{x^4}$

$$P_1=\frac{P\frac{dz}{dx}+\frac{d^2z}{dx^2}}{\left(\frac{dz}{dx}\right)^2}=\frac{\frac{3}{x}\cdot\frac{1}{x^3}+\left(\frac{-3}{x^4}\right)}{\left(\frac{1}{x^3}\right)^2}=0 \Rightarrow R_1=\frac{R}{\left(\frac{dz}{dx}\right)^2}=\frac{\frac{1}{x^8}}{\frac{1}{x^6}}=\frac{1}{x^2}=-2z$$

On putting the values of P_1, Q_1 and R_1 in (2), we get

$$\frac{d^2y}{dz^2}+a^2y=-2z \qquad \Rightarrow \qquad (D^2+a^2)y=-2z$$

A.E. is $m^2+a^2=0$, $m=\pm ia$, $\Rightarrow$ C.F. $=c_1\cos az+c_2\sin az$

$$\text{P.I.}=\frac{1}{D^2+a^2}(-2z)=\frac{1}{a^2}\left[1+\frac{D^2}{a^2}\right]^{-1}(-2z)=\frac{1}{a^2}\left[1-\frac{D^2}{a^2}\right](-2z)=\frac{-2z}{a^2}=\frac{1}{a^2x^2}$$

$$y=\text{C.F.}+\text{P.I.}$$

$$y=c_1\cos\frac{a}{2x^2}+c_2\sin\frac{a}{2x^2}+\frac{1}{a^2x^2} \qquad \textbf{Ans.}$$

Example 23. *Solve* $\frac{d^2y}{dx^2} - \frac{1}{x}\frac{dy}{dx} + 4x^2 y = x^4$

Solution. We have, $\frac{d^2y}{dx^2} - \frac{1}{x}\frac{dy}{dx} + 4x^2 y = x^4$ ...(1)

Hence $P = -\frac{1}{x},\ Q = 4x^2,\ R = x^4$

On changing the independent variable x to z, the equation (1) is transformed as

$$\frac{d^2y}{dz^2} + P_1\frac{dy}{dz} + Q_1 y = R_1 \quad \text{...(2)}$$

Using Second Method

Let $Q_1 = 1$ (constant)

but $Q_1 = \frac{Q}{\left(\frac{dz}{dx}\right)^2} = \frac{4x^2}{\left(\frac{dz}{dx}\right)^2} = \text{constant} = 1 \quad \text{(say)}$

$$\left(\frac{dz}{dx}\right)^2 = 4x^2 \Rightarrow \frac{dz}{dx} = 2x$$

$\Rightarrow$ $dz = 2x\,dx \Rightarrow z = x^2 + c \Rightarrow x^2 = z - c$ ...(3)

$$P_1 = \frac{P\left(\frac{dz}{dx}\right) + \frac{d^2z}{dx^2}}{\left(\frac{dz}{dx}\right)^2} = \frac{-\frac{1}{x}(2x) + 2}{(2x)^2} = 0$$

$$R_1 = \frac{R}{\left(\frac{dz}{dx}\right)^2} = \frac{x^4}{4x^2} = \frac{x^2}{4} = \frac{z-c}{4} \quad \text{[Using (3)]}$$

On putting the value of P_1, Q_1 and R_1 in (2), we get

$$\frac{d^2y}{dz^2} + (0)\frac{dy}{dz} + (1)y = \frac{z-c}{4} \Rightarrow \frac{d^2y}{dz^2} + y = \frac{z-c}{4} \Rightarrow (D^2+1)\,y = \frac{z-c}{4}$$

A.E. is $m^2 + 1 = 0$, $m = \pm i$

$$\text{C.F.} = c_1 \cos z + c_2 \sin z$$

$$\text{P.I.} = \frac{1}{D^2+1}\left(\frac{z-c}{4}\right) = (1+D^2)^{-1}\frac{z-c}{4} = (1-D^2)\frac{z-c}{4} = \frac{z-2c}{4}$$

Now complete solution = C.F. + P.I.

$\Rightarrow$ $y = c_1 \cos z + c_2 \sin z + \frac{z-2c}{4} \Rightarrow y = c_1 \cos x^2 + c_2 \sin x^2 + \frac{x^2}{4}$ **Ans.**

Example 24. *Solve the following differential equation by changing the independent variable*

$$x\frac{d^2y}{dx^2} + (4x^2-1)\frac{dy}{dx} + 4x^3 y = 2x^3$$ *(U.P., II Semester, Summer 2006)*

Solution. We have

$$x\frac{d^2y}{dx^2} + (4x^2-1)\frac{dy}{dx} + 4x^3 y = 2x^3$$

$$\Rightarrow \quad \frac{d^2y}{dx^2}+\left(\frac{4x^2-1}{x}\right)\frac{dy}{dx}+4x^2y=2x^2 \qquad \text{...(1)}$$

Here, $\quad P=\dfrac{4x^2-1}{x},\ Q=4x^2$ and $R=2x^2$

Changing the independent variable from x to z, the equation becomes

$$\frac{d^2y}{dz^2}+P_1\frac{dy}{dz}+Q_1y=R_1 \qquad \text{...(2)}$$

where $\quad P_1=\dfrac{P\dfrac{dz}{dx}+\dfrac{d^2z}{dx^2}}{\left(\dfrac{dz}{dx}\right)^2},\ Q_1=\dfrac{Q}{\left(\dfrac{dz}{dx}\right)^2},\ R_1=\dfrac{R}{\left(\dfrac{dz}{dx}\right)^2}$

Using Second Method

Choose $Q_1=1$ (constant) $\quad$ therefore $1=\dfrac{Q}{\left(\dfrac{dz}{dx}\right)^2}$

$$\Rightarrow \quad \left(\frac{dz}{dx}\right)^2=4x^2 \qquad [\because Q=4x^2]$$

$$\Rightarrow \quad \frac{dz}{dx}=2x \quad \Rightarrow dz=2x\,dx \quad \Rightarrow \quad z=x^2$$

$$\frac{d^2z}{dx^2}=2$$

and $\quad P_1=\dfrac{P\left(\dfrac{dz}{dx}\right)+\dfrac{d^2z}{dx^2}}{\left(\dfrac{dz}{dx}\right)^2}=\dfrac{\left(\dfrac{4x^2-1}{x}\right)(2x)+2}{(2x)^2} \quad \Rightarrow P_1=\dfrac{8x^2-2+2}{4x^2}=2$

$$R_1=\frac{R}{\left(\dfrac{dz}{dx}\right)^2}=\frac{2x^2}{4x^2}=\frac{1}{2}$$

The equation (2) is transformed to

$$\frac{d^2y}{dz^2}+2\frac{dy}{dz}+y=\frac{1}{2}$$

A.E. is $\quad m^2+2m+1=0$

$\Rightarrow \quad (m+1)^2=0 \quad \Rightarrow \quad m=-1,-1$

$$\text{C.F.}=(C_1+C_2z)\,e^{-z}$$

$$\text{P.I.}=\frac{1}{D^2+2D+1}\left(\frac{1}{2}\right)=\frac{1}{0+0+1}\left(\frac{1}{2}\right)=\frac{1}{2}$$

Complete solution is

$$y=C.F.+P.I.=(c_1+c_1z)e^{-z}+\frac{1}{2}$$

$$\Rightarrow \quad y=(c_1+c_2x^2)\,e^{-x^2}+\frac{1}{2} \qquad \textbf{Ans.}$$

EXERCISE 15.7

Solve the following differential equations:

1. $x^4 \dfrac{d^2y}{dx^2} + 2x^3 \dfrac{dy}{dx} + a^2 y = 0$ **Ans.** $y = c_1 \cos \dfrac{a}{x} + c_2 \sin \dfrac{a}{x}$

2. $\cos x \dfrac{d^2y}{dx^2} + \dfrac{dy}{dx} \sin x - 2y\cos^3 x = 2\cos^5 x$ **Ans.** $y = c_1 e^{\sqrt{2}\sin x} + c_2 e^{-\sqrt{2}\sin x} + \sin^2 x$

3. $\dfrac{d^2y}{dx^2} + \tan x \dfrac{dy}{dx} + y\cos^2 x = 0$ **Ans.** $y = c_1 \cos(\sin x) + c_2 \sin(\sin x)$

4. $x\dfrac{d^2y}{dx^2} - \dfrac{dy}{dx} - 4x^3 y = 8x^2 \sin x^2$ **Ans.** $y = c_1 e^{x^2} + c_2 e^{\cos x} + \dfrac{1}{6} e^{-\cos x}$

5. $\dfrac{d^2y}{dx^2} + (3\sin x - \cot x)\dfrac{dy}{dx} + 2y\sin^2 x = e^{-\cos x}\sin^2 x$ **Ans.** $y = c_1 e^{2\cos x} + c_2 e^{\cos x} + \dfrac{1}{6} e^{-\cos x}$

6. $\dfrac{d^2y}{dx^2} + (\tan x - 1)^2 \dfrac{dy}{dx} - n(n-1)y\sec^4 x = 0$ **Ans.** $y = C_1 e^{-n \tan x} + C_2 e^{(n-1)\tan x}$

7. $\dfrac{d^2y}{dx^2} - \cot x \dfrac{dy}{dx} - y\sin^2 x = \cos x - \cos^3 x$ **Ans.** $y = C_1 e^{-\cos x} + C_2 e^{\cos x} - \cos x$

8. $\dfrac{d^2y}{dx^2} + (3\sin x - \cot x)\dfrac{dy}{dx} + 2y\sin^2 x = e^{-\cos x}\sin^2 x$ **Ans.** $y = C_1 e^{\cos x} + C_2 e^{2\cos x} + \dfrac{1}{6} e^{-\cos x}$

15.9 METHOD OF VARIATION OF PARAMETERS

Here, the method to find out C.F. is different from the methods discussed earlier. Now, we will find C.F. by new methods and then will apply the method of variation of parameters as discussed in Article 14.3 on page 381.

Example 25. *By the method of variation of parameters, solve the differential equation.*

$$\frac{d^2y}{dx^2} + (1 - \cot x)\frac{dy}{dx} - y\cot x = \sin^2 x$$

Solution. First we shall find the C.F. of the given equation *i.e.*, the solution of the equation.

$$\frac{d^2y}{dx^2} + (1 - \cot x)\frac{dy}{dx} - y\cot x = 0 \qquad \left[\frac{d^2y}{dx^2} + P\frac{dy}{dx} + Qy = 0\right] \qquad ...(1)$$

Here, $P = 1 - \cot x,\ Q = -\cot x$

Now, $1 - P + Q = 1 - (1 - \cot x) - \cot x = 0$, $\therefore\ y = e^{-x}$ is a part of the C.F.

Putting $y = ve^{-x}$, $$\frac{dy}{dx} = -ve^{-x} + e^{-x}\frac{dv}{dx}$$

$$\frac{d^2y}{dx^2} = e^{-x}\frac{d^2v}{dx^2} - 2e^{-x}\frac{dv}{dx} + ve^{-x}$$ the equation (1) reduces to

$$\frac{d^2v}{dx^2} - (1 + \cot x)\frac{dv}{dx} = 0$$

$$\Rightarrow \quad \frac{dp}{dx} - (1 + \cot x)p = 0 \qquad \text{where } p = \frac{dv}{dx}$$

$$\Rightarrow \qquad \frac{dp}{p} = (1 + \cot x)\, dx$$

Integrating, we get $\qquad \log p = x + \log \sin x + \log C_1$

$\Rightarrow \quad \log p - \log \sin x - \log C_1 = x$

$$\Rightarrow \qquad \log \frac{p}{C_1 \sin x} = x$$

$$\Rightarrow \qquad \frac{p}{C_1 \sin x} = e^x \Rightarrow p = C_1\, e^x \sin x$$

$$\therefore \qquad p = \frac{dv}{dx} = C_1 e^x \sin x \;\Rightarrow\; v = C_1 \int e^x \sin x\, dx$$

$$\left[\int e^{ax} \sin bx\, dx = \frac{e^{ax}}{a^2 + b^2}[a \sin bx - b \cos bx]\right]$$

$\Rightarrow \quad v = C_1.\frac{1}{2} e^x (\sin x - \cos x) + C_2$ = only one part of C.F. of equation (1),

R.H.S. of the given differential equation (1) is zero.

$\therefore$ The solution of (1) i.e., C.F. of the given equation is

$$\text{C.F.} = v e^{-x} = C_1.\frac{1}{2} e^x (\sin x - \cos x)\, e^{-x} + C_2\, e^{-x} = \frac{C_1}{2} (\sin x - \cos x) + C_2 e^{-x}$$

$$y_1 = (\sin x - \cos x),\; y_2 = e^{-x}$$

$$\text{P. I.} = U y_1 + V y_2$$

$$P.I. = U(\sin x - \cos x) + V \,.\, e^{-x} \qquad \text{... (2)}$$

where
$$U = \int \frac{-y_2 X}{y_1 y_2' - y_1' y_2}\, dx, = \int \frac{-e^{-x} \,.\, \sin^2 x}{(\sin x - \cos x)(-e^{-x}) - (\cos x + \sin x) e^{-x}}\, dx$$

$$= \int \frac{-\sin^2 x}{-\sin x + \cos x - \cos x - \sin x} dx = \int -\frac{\sin^2 x}{-2 \sin x} dx$$

$$= \frac{1}{2} \int \sin x\, dx = -\frac{1}{2} \cos x$$

$$V = \int \frac{y_1 X}{y_1 y_2' - y_1' y_2}\, dx = \int \frac{(\sin x - \cos x).\sin^2 x}{(\sin x - \cos x)(-e^{-x}) - (\cos x + \sin x)\, e^{-x}}\, dx$$

$$= \int \frac{(\sin x - \cos x) \sin^2 x.\, e^x}{-\sin x + \cos x - \cos x - \sin x} dx = \int \frac{(\sin x - \cos x) \sin^2 x.\, e^x\, dx}{-2 \sin x}$$

$$= \frac{1}{2} \int (-\sin x + \cos x) \sin x\, e^x\, dx = \frac{1}{2} \int (-\sin^2 x\, e^x + \sin x \cos x\, e^x)\, dx$$

$$= \frac{1}{2} \int \frac{1}{2} (\cos 2x - 1)\, e^x\, dx + \frac{1}{4} \int \sin 2x.\, e^x dx$$

$$= \frac{1}{4} \int \cos 2x.\, e^x\, dx - \frac{1}{4} \int e^x dx + \frac{1}{4} \int \sin 2x\, e^x dx$$

$$= \frac{1}{4} \frac{e^x}{1+4} [\cos 2x + 2 \sin 2x] - \frac{1}{4} e^x + \frac{1}{4} \frac{e^x}{1+4} [\sin 2x - 2 \cos 2x]$$

$$= \frac{e^x}{20}[\cos 2x + 2\sin 2x + \sin 2x - 2\cos 2x] - \frac{e^x}{4}$$

$$= \frac{e^x}{20}[-\cos 2x + 3\sin 2x] - \frac{e^x}{4}$$

On putting the values of U and V in (2), we get

$$\text{P.I.} = Uy_1 + Vy_2$$

$$\text{P.I.} = -\frac{1}{2}\cos x(\sin x - \cos x) + \left[\frac{e^x}{20}(-\cos 2x + 3\sin 2x) - \frac{e^x}{4}\right]e^{-x}$$

$$= -\frac{1}{2}\sin x\cos x + \frac{1}{2}\cos^2 x + \frac{1}{20}(-\cos 2x + 3\sin 2x) - \frac{1}{4}$$

$$= -\frac{1}{4}\sin 2x + \frac{1}{2}\cos^2 x - \frac{1}{20}\cos 2x + \frac{3}{20}\sin 2x - \frac{1}{4}$$

$$= -\frac{1}{10}\sin 2x + \frac{1}{4}(\cos 2x + 1) - \frac{1}{20}\cos 2x - \frac{1}{4}$$

$$= -\frac{1}{10}\sin 2x + \frac{1}{5}\cos 2x$$

$$y = \text{C.F.} + \text{P.I.}$$

$$y = C_1(\sin x - \cos x) + C_2 e^{-x} - \frac{1}{10}\sin 2x + \frac{1}{5}\cos 2x$$

$$y = C_1(\sin x - \cos x) + C_2 e^{-x} - \frac{1}{10}(\sin 2x - 2\cos 2x)$$ **Ans.**

Example 26. *Solve by the method of variation of parameters:*

$$x^2\frac{d^2y}{dx^2} - 2x(1+x)\frac{dy}{dx} + 2(1+x)y = x^3.$$

Solution. Dividing the given equation by x^2, we get

$$\frac{d^2y}{dx^2} - \frac{2(1+x)}{x}\frac{dy}{dx} + \frac{2(1+x)}{x^2}y = x. \quad ...(1)$$

Firstly, we shall find the C.F. of (1) *i.e.,* the complete solution of the following differential equation

$$\frac{d^2y}{dx^2} - \frac{2(1+x)}{x}\frac{dy}{dx} + \frac{2(1+x)}{x^2}y = 0 \quad ...(2)$$

Here $\quad P = -\frac{2(1+x)}{x},\ Q = \frac{2(1+x)}{x^2},\ R = 0$

Clearly, $\quad P + Qx = -\frac{2(1+x)}{x} + \frac{2(1+x)}{x} = 0$

$\therefore \quad y = x$ is part of C.F

Hence, $\quad y = x$ here

Let $y = vx$ be complete primitive of (2). So putting $y = vx$ in (2), the reduced equation is

$$\frac{d^2v}{dx^2} + \left(\frac{2}{u}\frac{du}{dx} + P\right)\frac{dv}{dx} = \frac{R}{u}$$

$$i.e., \quad \frac{d^2u}{dx^2}+\left\{\frac{2}{x}.1-\frac{2(1+x)}{x}\right\}\frac{dv}{dx}=0$$

$$\Rightarrow \quad \frac{d^2v}{dx^2}-\frac{2\,dv}{dx}=0 \Rightarrow (D^2-2D)v=0 \qquad [D \equiv d/dx]$$

$$\Rightarrow \quad D(D-2)\,v=0$$

$$v = C_1 + C_2 e^{2x}$$

Thus from $y = vx$, the complete solution of (2) *i.e.*, the C.F. of equation (1) is given by

$$\text{C.F.} = C_1\,x + C_2\,xe^{2x} \Rightarrow C.F. = C_1\,y_1 + C_2\,y_2$$

$$\text{P.I.} = u\,y_1 + v\,y_2$$

$$\text{P.I.} = u\,x + v\,x\,e^{2x} \qquad ...(3)$$

where $$u = \int -\frac{y_2\,X}{y_1\,y_2' - y_1'\,y_2}\,dx \text{ and } v = \int \frac{y_1\,X}{y_1\,y_2' - y_1'\,y_2}\,dx$$

$$u = \int \frac{-x e^{2x}.x}{x(e^{2x}+2x\,e^{2x})-1.(x e^{2x})}\,dx$$

$$= \int \frac{-x^2}{x+2\,x^2-x}\,dx = -\int \frac{x^2}{2x^2}\,dx = -\frac{1}{2}\int dx = -\frac{x}{2}$$

$$v = \int \frac{x.x}{x(e^{2x}+2x\,e^{2x})-1.(x\,e^{2x})}\,dx$$

$$= \int \frac{x}{e^{2x}+2x\,e^{2x}-e^{2x}}\,dx = \frac{1}{2}\int \frac{1}{e^{2x}}\,dx = \frac{1}{2}\int e^{-2x}\,dx = -\frac{e^{-2x}}{4}$$

On putting the values of u and v in (3), we get

$$\text{P.I.} = -\frac{x}{2}x - \frac{e^{-2x}}{4}(x\,e^{2x}) = -\frac{x^2}{2} - \frac{x}{4}$$

$$y = \text{C.F.} + \text{P.I.} = C_1 x + C_2 x\,e^{2x} - \frac{x^2}{2} - \frac{x}{4} \qquad \textbf{Ans.}$$

EXERCISE 15.8

Solve the following differential equations by the method of variation of parameters:

1. $(1-x)\frac{d^2y}{dx^2} + x\frac{dy}{dx} - y = (1-x)^2.$ **Ans.** $y = C_1e^x + C_2x + x^2 + x + 1$

2. $(1-x^2)\frac{d^2y}{dx} - 4x\frac{dy}{dx} - (1+x^2)y = x$ **Ans.** $y = \frac{1}{1-x^2}(C_1\cos x + C_2 \sin x + x)$

3. $(1-x)\frac{d^2y}{dx^2} + x\frac{dy}{dx} - y = x(1-x^2)^{\frac{3}{2}}$ **Ans.** $y = C_1(\sqrt{1-x^2} + x\sin^{-1}x) + C_2x - \frac{x}{9}(1-x^2)^{3/2}$

4. $x\frac{d^2y}{dx^2} - (2x+1)\frac{dy}{dx} + (x+1)y = (x^2+x-1)\,e^{2x}$ **Ans.** $y = C_1x + C_2x^{-1}x + x\int e^x x^{-3}(x^2-2x+2)\,dx$

5. $x\frac{d^2y}{dx^2} - 2(x+1)\frac{dy}{dx} + (x+2)y = (x-2)\,e^x$ **Ans.** $y = \frac{C_1}{3}x^3e^x + C_2e^x + \left(x - \frac{x^2}{2}\right)e^x.$

6. $x^2\frac{d^2y}{dx^2} + \frac{dy}{dx} - y = x^2\,e^x$ **Ans.** $y = C_1x + \frac{C_2}{x} + e^x, (x>0)$

COUPLED DIFFERENTIAL EQUATIONS

16.1 COUPLED DIFFERENTIAL EQUATIONS

If two or more dependent variables are functions of a single independent variable, the equations involving their derivatives are called coupled differential equations, *e.g.*

$$\frac{dx}{dt} + 4y = t$$

$$\frac{dy}{dt} + 2x = e^t$$

The method of solving these equations is based on the process of elimination, as we solve algebraic coupled equations.

Example 1. *The equations of motions of a particle are given by*

$$\frac{dx}{dt} + \omega y = 0$$

$$\frac{dy}{dt} - \omega x = 0$$

Find the path of the particle and show that it is a circle.

(*R.G.P.V. Bhopal, Feb. 2006, U.P. II Semester summer* 2009)

Solution. On putting $\frac{d}{dt} \equiv D$ in the equations, we have

$$Dx + \omega y = 0 \qquad \ldots(1)$$

$$-\omega x + Dy = 0 \qquad \ldots(2)$$

On multiplying (1) by ω and (2) by D, we get

$$\omega Dx + \omega^2 y = 0 \qquad \ldots(3)$$

$$-\omega Dx + D^2 y = 0 \qquad \ldots(4)$$

On adding (3) and (4), we obtain

$$\omega^2 y + D^2 y = 0 \quad \Rightarrow \quad (D^2 + \omega^2)\, y = 0 \qquad \ldots(5)$$

Now, we have to solve (5) to get the value of y.

A.E. is $m^2 + \omega^2 = 0 \quad \Rightarrow \quad m^2 = -\omega^2 \quad \Rightarrow \quad m = \pm i\omega$

$$\therefore \qquad y = A \cos \omega t + B \sin \omega t \qquad \ldots(6)$$

$\Rightarrow Dy = -A\,\omega \sin \omega t + B\omega \cos \omega t$

On putting the value of Dy in (2), we get

$$-\omega x - A\omega \sin \omega t + B\omega \cos \omega t = 0$$

$$\Rightarrow \quad \omega x = -A\omega \sin \omega t + B\omega \cos \omega t$$

$$\Rightarrow \quad x = -A \sin \omega t + B \cos \omega t \quad ...(7)$$

On squaring (6) and (7) and adding , we get

$$x^2 + y^2 = A^2(\cos^2\omega t + \sin^2\omega t) + B^2 (\cos^2\omega t + \sin^2\omega t)$$

$$\Rightarrow \quad x^2 + y^2 = A^2 + B^2$$

This is the equation of circle. **Proved.**

Example 2. *Solve the following differential equation*

$$\frac{dx}{dt} = y + 1, \frac{dy}{dt} = x + 1$$

(*U.P. II Semester, 2009*)

Solution. Here, we have

$$Dx - y = 1 \quad ...(1)$$

$$-x + Dy = 1 \quad ...(2)$$

Multiplying (1) by D, we get

$$D^2x - Dy = D.1 \quad ...(3)$$

Adding (2) and (3), we get

$$(D^2 - 1)\, x = 1 + D.1$$

$$\Rightarrow \quad (D^2 - 1)\, x = 1 \text{ or } (D^2 - 1)x = e^0 \quad [D.(1) = 0]$$

A.E. is $\quad m^2 - 1 = 0 \Rightarrow m = \pm 1$

$$\therefore \quad \text{C.F.} = c_1 e^t + c_2\, e^{-t}$$

$$\text{P.I.} = \frac{1}{D^2 - 1} \cdot e^0 = \frac{1}{0 - 1} e^0 = -1$$

$$\therefore \quad x = \text{C.F.} + \text{P.I.} = c_1 e^t + c_2\, e^{-t} - 1$$

From (1), $y = \dfrac{dx}{dt} - 1$

$$\Rightarrow \quad y = \frac{d}{dt}(c_1 e^t + c_2 e^{-t} - 1) - 1$$

$$\Rightarrow \quad \left.\begin{aligned} y &= c_1 e^t - c_2 e^{-t} - 1 \\ \text{and} \quad x &= c_1 e^t + c_2 e^{-t} - 1 \end{aligned}\right\}$$

Ans.

Example 3. *Solve*

$$\frac{dx}{dt} + y = \sin t$$

$$\frac{dy}{dt} + x = \cos t \quad \text{where } y(0) = 0, \quad x(0) = 2$$

(*R.G.P.V Bhopal I Semester April 2010 June 2007*)

Solution. We have,

$$\frac{dx}{dt} + y = \sin t \quad \Rightarrow \quad Dx + y = \sin t \quad ...(1)$$

$$\frac{dy}{dt} + x = \cos t \quad \Rightarrow \quad Dy + x = \cos t \quad ...(2)$$

Multiplying (2) by D, we get

$$D^2y + Dx = D \cos t$$

$$D^2y + Dx = -\sin t \qquad \text{...(3)}$$

Subtracting (1) from (3), we have

$$D^2y - y = -2 \sin t$$

$$\Rightarrow \qquad (D^2 - 1)y = -2 \sin t$$

A.E. is $m^2 - 1 = 0 \Rightarrow m^2 = 1 \Rightarrow m = \pm 1$

$$\text{C.F.} = C_1e^t + C_2e^{-t}$$

$$\text{P.I.} = \frac{1}{D^2 - 1}(-2 \sin t)$$

$$\Rightarrow \qquad \text{P.I.} = \frac{1}{-1-1}(-2 \sin t) = \sin t$$

Complete solution = C.F. + P.I.

$$y = C_1e^t + C_2e^{-t} + \sin t \qquad \text{...(4)}$$

Putting $y = 0$ and $t = 0$ in (4), we get

$$0 = C_1 + C_2 \quad \text{or} \quad C_2 = -C_1$$

On putting $C_2 = -C_1$ in (4), we get

$$y = C_1e^t - C_1e^{-t} + \sin t$$

On putting the value of y in (2), we get

$$D(C_1e^t - C_1e^{-t} + \sin t) + x = \cos t$$

$$C_1e^t + C_1e^{-t} + \cos t + x = \cos t$$

$$x = -C_1e^t - C_1e^{-t} \qquad \text{...(5)}$$

On putting $x = 2$, $t = 0$ in (5), we get

$$2 = -C_1 - C_1 \quad \Rightarrow \quad C_1 = -1$$

Putting the value of C_1 in (5) and (4), we have

$$x = e^t + e^{-t}$$

$$y = -e^t + e^{-t} + \sin t$$

Which is the required solution. **Ans.**

Example 4. *Solve:* $\frac{dx}{dt} + 4x + 3y = t$

$$\frac{dy}{dt} + 2x + 5y = e^t$$

[U.P. II Semester, 2006]

Solution. Here, we have

$$(D + 4)x + 3y = t \qquad \text{...(1)}$$

$$2x + (D + 5)y = e^t \qquad \text{...(2)} \quad \left(D \equiv \frac{d}{dt}\right)$$

To eliminate y, operating (1) by $(D + 5)$ and multiplying (2) by 3 then subtracting, we get

$$(D + 5)(D + 4)x + 3(D + 5)y - 3(2x) - 3(D + 5)y = (D + 5)t - 3e^t$$

$$[(D + 4)(D + 5) - 6]x = (D + 5)t - 3e^t$$

$$(D^2 + 9D + 14)x = 1 + 5t - 3e^t$$

Auxiliary equation is

$$m^2 + 9m + 14 = 0 \Rightarrow m = -2, -7$$

$$\therefore \qquad \text{C.F.} = c_1e^{-2t} + c_2e^{-7t}$$

$$\text{P.I.} = \frac{1}{D^2 + 9D + 14}(1 + 5t - 3e^t)$$

$$= \frac{1}{D^2+9D+14}e^{0t} + 5\frac{1}{D^2+9D+14}t - 3\frac{1}{D^2+9D+14}e^t$$

$$= \frac{1}{0^2+9(0)+14}e^{0t} + 5.\frac{1}{14\left(1+\frac{9D}{14}+\frac{D^2}{14}\right)}t - 3\frac{1}{1^2+9(1)+14}e^t$$

$$= \frac{1}{14} + \frac{5}{14}\left[1+\left(\frac{9D}{14}+\frac{D^2}{14}\right)\right]^{-1} t - \frac{1}{8}e^t = \frac{1}{14} + \frac{5}{14}\left[1-\left(\frac{9D}{14}+\frac{D^2}{14}\right)+\cdots\right]t - \frac{1}{8}e^t \quad ...(5)$$

$$= \frac{1}{14} + \frac{5}{14}\left(t - \frac{9}{14}\right) - \frac{1}{8}e^t = \frac{5}{14}t - \frac{31}{196} - \frac{1}{8}e^t$$

$$x = c_1e^{-2t} + c_2e^{-7t} + \frac{5}{14}t - \frac{31}{196} - \frac{1}{8}e^t$$

$$3y = t - \frac{dx}{dt} - 4x \qquad \text{[From (1)]}$$

$$= t - \frac{d}{dt}\left[c_1e^{-2t} + c_2e^{-7t} + \frac{5}{14}t - \frac{31}{196} - \frac{1}{8}e^t\right] - 4\left[c_1e^{-2t} + c_2e^{-7t} + \frac{5}{14}t - \frac{31}{196} - \frac{e^t}{8}\right]$$

$$3y = t + 2c_1e^{-2t} + 7c_2e^{-7t} - \frac{5}{14} + \frac{1}{8}e^t - 4c_1e^{-2t} - 4c_2e^{-7t} - \frac{10}{7}t + \frac{31}{49} + \frac{1}{2}e^t$$

$$\therefore \quad y = \frac{1}{3}\left[-2c_1e^{-2t} + 3c_2e^{-7t} - \frac{3}{7}t + \frac{27}{98} + \frac{5}{8}e^t\right]$$

Hence,
$$x = c_1e^{-2t} + c_2e^{-7t} + \frac{5}{14}t - \frac{31}{196} - \frac{1}{8}e^t$$

$$y = -\frac{2}{3}c_1e^{-2t} + c_2e^{-7t} - \frac{1}{7}t + \frac{9}{98} + \frac{5}{24}e^t$$ **Ans.**

Example 5. *Solve:* $\frac{dx}{dt} + 5x + y = e^t,\ \frac{dy}{dt} + x + 5y = e^{5t}.$

Solution. Here, we have

$$(D + 5)\,x + y = e^t \quad ...(1)$$

$$x + (D + 5)y = e^{5t} \quad ...(2) \quad \left(D \equiv \frac{d}{dt}\right)$$

Multiplying (1) by $(D + 5)$, we get

$$(D + 5)^2x + (D + 5)y = (D + 5)\,e^t \quad ...(3)$$

Subtracting (3) form (2), we get

$$\left\{1-(D+5)^2\right\}x = e^{5t} - (D + 5)e^t$$

$$\Rightarrow \quad [\,1 - D^2 - 10D - 25]x = e^{5t} - e^t - 5e^t$$

$$\Rightarrow \quad (D^2 + 10D + 24)x = 6e^t - e^{5t}$$

Auxiliary equation is

$$m^2 + 10\,m + 24 = 0 \quad \Rightarrow \quad (m + 4)\,(m + 6) = 0$$

$$\Rightarrow \quad m = -4, -6$$

$$\therefore \quad \text{C.F.} = c_1e^{-4t} + c_2e^{-6t}$$

$$\text{P.I.} = \frac{1}{D^2+10D+24}(6e^t - e^{5t})$$

$$= \frac{6}{1^2+10(1)+24}e^t - \frac{1}{(5)^2+10(5)+24}e^{5t} = \frac{6e^t}{35} - \frac{e^{5t}}{99}$$

Thus $\quad x = \text{C.F.} + \text{P.I.} = c_1e^{-4t} + c_2e^{-6t} + \frac{6e^t}{35} - \frac{e^{5t}}{99}$

From (1),

$$y = e^t - (D+5)x = e^t - \frac{dx}{dt} - 5x$$

$$= e^t - \frac{d}{dt}\left(c_1e^{-4t} + c_2e^{-6t} + \frac{6e^t}{35} - \frac{e^{5t}}{99}\right) - 5\left(c_1e^{-4t} + c_2e^{-6t} + \frac{6e^t}{35} - \frac{e^{5t}}{99}\right)$$

$$= e^t + 4c_1e^{-4t} + 6c_2e^{-6t} - \frac{6e^t}{35} + \frac{5e^{5t}}{99} - 5c_1e^{-4t} - 5c_2e^{-6t} - \frac{30e^t}{35} + \frac{5e^{5t}}{99}$$

$$y = -\frac{1}{35}e^t - c_1e^{-4t} + \frac{10}{99}e^{5t} + c_2e^{-6t}$$

$$x = c_1e^{-6t} + c_2e^{-4t} + \frac{6e^t}{35} - \frac{e^{5t}}{99}$$ **Ans.**

Example 6. *Solve the following system of differential equations*

$$Dx + Dy + 3x = \sin t$$

and $\quad Dx + y - x = \cos t$

Solution. We have, $(D + 3)\,x + Dy = \sin t$ ...(1)

$$(D-1)\,x + y = \cos t \quad \text{...(2)}$$

Operating (2) by D, we get

$$D\,(D-1)\,x + Dy = -\sin t \quad \text{...(3)}$$

Subtracting (1) from (3), we get

$$\{D\,(D-1) - (D+3)\}\,x = -2\sin t$$

$$\Rightarrow \quad \{D^2 - D - D - 3\}\,x = -2\sin t$$

$$\Rightarrow \quad (D^2 - 2D - 3)\,x = -2\sin t$$

A.E. is $\quad m^2 - 2m - 3 = 0$

$$\Rightarrow \quad (m-3)(m+1) = 0 \quad \Rightarrow \quad m = 3, -1$$

$$\therefore \quad C.F. = c_1\,e^{3t} + c_2\,e^{-t}$$

$$P.I. = \frac{1}{D^2-2D-3}(-2\sin t) = -2\frac{1}{(-1)-2D-3}\sin t$$

$$= 2\cdot\frac{1}{2\,(D+2)}\sin t = \frac{(D-2)}{D^2-4}\sin t$$

$$= \frac{D-2}{-1-4}\sin t = \frac{\cos t - 2\sin t}{-5} = \frac{1}{5}(2\sin t - \cos t)$$

$$x = \text{C.F.} + \text{P.I.} = c_1e^{3t} + c_2\,e^{-t} + \frac{1}{5}(2\sin t - \cos t) \quad \text{...(4)}$$

From (2), we get

$$(D-1)\,x + y = \cos t$$

$$\Rightarrow \quad (D-1)\left\{c_1 e^{3t} + c_2 e^{-t} + \frac{1}{5}(2\sin t - \cos t)\right\} + y = \cos t$$

$$\Rightarrow \quad y = \cos t - D\left\{c_1 e^{3t} + c_2 e^{-t} + \frac{1}{5}(2\sin t - \cos t)\right\} + \left\{c_1 e^{3t} + c_2 e^{-t} + \frac{1}{5}(2\sin t - \cos t)\right\}$$

$$= \cos t - 3c_1 e^{3t} + c_2 e^{-t} - \frac{1}{5}(2\cos t + \sin t) + c_1 e^{3t} + c_2 e^{-t} + \frac{1}{5}(2\sin t - \cos t)$$

$$= \cos t - 2c_1 e^{3t} + 2c_2 e^{-t} - \frac{1}{5}[3\cos t - \sin t]$$

$$= \frac{2}{5}\cos t + \frac{1}{5}\sin t + 2\,c_2 e^{-t} - 2\,c_1 e^{3t}$$

Here, $y = \frac{1}{5}(2\cos t + \sin t) - 2c_1 e^{3t} + 2c_2 e^{-t}$ ...(5)

(4) and (5) are the required solutions **Ans.**

Example 7. *Solve the simultaneous equations*

$$\frac{dx}{dt} + \frac{dy}{dt} - 2y = 2\cos t - 7\sin t \qquad ...(1)$$

$$\frac{dx}{dt} - \frac{dy}{dt} + 2x = 4\cos t - 3\sin t \qquad ...(2)$$

(*U.P., B. Pharm* 2005, *II Semester*)

Solution. $Dx + (D-2)\,y = 2\cos t - 7\sin t$ ...(3)

$(D+2)\,x - Dy = 4\cos t - 3\sin t$ ...(4)

Operating (3) by D (4) by $(D-2)$, we get

$\Rightarrow \quad D^2x + D(D-2)\,y = -2\sin t - 7\cos t$ ...(5)

$(D^2-4)\,x - D(D-2)\,y = (D-2)\,4\cos t - (D-2)\,3\sin t$ ...(6)

On adding (5) and (6), we get

$$(D^2 + D^2 - 4)\,x = -2\sin t - 7\cos t - 4\sin t - 8\cos t - 3\cos t + 6\sin t$$

$$\Rightarrow \quad (2D^2 - 4)\,x = -18\cos t$$

$$\Rightarrow \quad (D^2 - 2)\,x = -9\cos t$$

A.E. is $m^2 - 2 = 0 \Rightarrow m = \pm\sqrt{2}$, C.F. $= c_1 e^{\sqrt{2}t} + c_2 e^{-\sqrt{2}t}$

$$P.I. = \frac{1}{D^2-2}(-9\cos t) = \frac{-9}{-1-2}\cos t = 3\cos t$$

$$x = c_1 e^{\sqrt{2}t} + c_2 e^{-\sqrt{2}t} + 3\cos t \qquad ...(7)$$

Putting the value of x in (2), we get

$$\frac{dy}{dt} = \sqrt{2}\,c_1 e^{\sqrt{2}t} - \sqrt{2}\,c_2 e^{-\sqrt{2}t} - 3\sin t + 2c_1 e^{\sqrt{2}t} + 2c_2 e^{-\sqrt{2}t} + 6\cos t - 4\cos t + 3\sin t$$

$$\frac{dy}{dt} = (2+\sqrt{2})\,c_1 e^{\sqrt{2}t} + (2-\sqrt{2})\,c_2 e^{-\sqrt{2}t} + 2\cos t$$

On integrating, we get

$$\Rightarrow y = (\sqrt{2}+1)\,c_1 e^{\sqrt{2}t} - (\sqrt{2}-1)\,c_2 e^{-\sqrt{2}t} + 2\sin t + c_3 \qquad ...(8)$$

Relations (7) and (8) are the required solutions **Ans.**

Example 8. *Solve* $\frac{dx}{dt}=2y,\ \frac{dy}{dt}=2z,\ \frac{dz}{dt}=2x$ *(Uttarakhand, II Semester, June 2007)*

Solution. Here, we have

$$\frac{dx}{dt}=2y \quad \Rightarrow \quad Dx = 2y \qquad ...(1)$$

$$\frac{dy}{dt}=2z \quad \Rightarrow \quad Dy = 2z \qquad ...(2)$$

$$\frac{dz}{dt}=2x \quad \Rightarrow \quad Dz = 2x \qquad ...(3)$$

From (1), we have $\frac{dx}{dt}=2y$

$$\Rightarrow \quad \frac{d^2x}{dt^2}=\frac{2\,dy}{dt}=2\,(2z)=4z \qquad \left[\text{Using (2), } \frac{dy}{dt}=2z\right]$$

$$\frac{d^3x}{dt^3}=4\frac{dz}{dt}=4\,(2x)=8x \qquad \left[\text{Using (3), } \frac{dz}{dt}=2x\right]$$

$$\Rightarrow \quad \frac{d^3x}{dt^3}-8x=0 \quad \Rightarrow \quad (D^3-8)\,x=0$$

A.E. is $m^3-8=0 \quad \Rightarrow \quad (m-2)\,(m^2+2m+4)=0$

$$\Rightarrow \quad m-2=0 \quad \Rightarrow \quad m=2$$

$$\Rightarrow \quad m^2+2m+4=0 \quad \Rightarrow \quad m=\frac{-2\pm\sqrt{4-16}}{2}=\frac{-2\pm i\sqrt{12}}{2}=-1\pm i\sqrt{3}$$

So the C.F. of x is

$$x = C_1\,e^{2t}+e^{-t}\,(A\cos\sqrt{3}\,t+B\sin\sqrt{3}\,t) \qquad ...(4)$$

$$[A=C_2\cos\alpha,\ B=C_2\sin\alpha] \qquad \left[\begin{array}{l}\tan\alpha=\frac{B}{A}\\ \alpha=\tan^{-1}\left(\frac{B}{A}\right)\end{array}\right]$$

$$x = C_1\,e^{2t}+e^{-t}\,[C_2\cos\alpha\cos\sqrt{3}\,t+C_2\sin\alpha\sin\sqrt{3}\,t]$$

$$x = C_1\,e^{2t}+e^{-t}\,C_2\cos(\sqrt{3}\,t-\alpha) = C_1\,e^{2t}+C_2\,e^{-t}\cos(\sqrt{3}\,t-\alpha)$$

From (3), we have $\frac{dz}{dt}=2x$

$$\Rightarrow \quad \frac{dz}{dt}=2C_1\,e^{2t}+2C_2\,e^{-t}\cos(\sqrt{3}\,t-\alpha) \qquad \text{[On putting the value of } x]$$

$$z = C_1\,e^{2t}+2C_2\,\frac{e^{-t}}{\sqrt{1+3}}\cos(\sqrt{3}\,t-\alpha-\beta) \qquad \left[\int e^{ax}\cos bx\,dx=\frac{e^{ax}}{\sqrt{a^2+b^2}}\cos(bx-\beta)\right]$$

$$\Rightarrow \quad z = C_1\,e^{2t}+2C_2\,\frac{e^{-t}}{\sqrt{1+3}}\cos\left[\sqrt{3}\,t-\alpha-\frac{2\pi}{3}\right] \qquad \left[\beta=\tan^{-1}\frac{\sqrt{3}}{-1}=\frac{2\pi}{3}\right]$$

$$\Rightarrow \quad z = C_1\,e^{2t}+C_2\,e^{-t}\cos\left(\sqrt{3}\,t-\alpha+\frac{4\pi}{3}\right) \qquad ...(5) \qquad \left[-\frac{2\pi}{3}=\frac{4\pi}{3}\right]$$

From (2), we have $\dfrac{dy}{dt} = 2z$

$$\Rightarrow \quad \frac{dy}{dt} = 2C_1 e^{2t} + 2C_2 e^{-t} \cos\left(\sqrt{3}t - \alpha + \frac{4\pi}{3}\right) \qquad \text{[On putting the value of } z]$$

$$\Rightarrow \quad y = \int 2C_1 e^{2t}\, dt + 2C_2 \int e^{-t} \cos\left(\sqrt{3}t - \alpha + \frac{4\pi}{3}\right) dt \qquad \left(\gamma = \tan^{-1}\frac{\sqrt{3}}{-1} = \frac{2\pi}{3}\right)$$

$$y = C_1 e^{2t} + 2C_2 \frac{e^{-t}}{\sqrt{1+3}} \cos\left(\sqrt{3}t - \alpha + \frac{4\pi}{3} - \gamma\right)$$

$$\Rightarrow \quad y = C_1 e^{2t} + 2C_2 \frac{e^{-t}}{\sqrt{1+3}} \cos\left(\sqrt{3}t - \alpha + \frac{4\pi}{3} - \frac{2\pi}{3}\right)$$

$$y = C_1 e^{2t} + C_2 e^{-t} \cos\left(\sqrt{3}t - \alpha + \frac{2\pi}{3}\right) \qquad ...(6)$$

Relations (4), (5) and (6) are the required solutions. **Ans.**

Example 9. *Solve the following simultaneous equations*:

$$\frac{d^2x}{dt^2} - 3x - 4y = 0, \quad \frac{d^2y}{dt^2} + x + y = 0 \qquad (\textit{U.P. II Semester, Summer } 2005)$$

Solution. We have, $\dfrac{d^2x}{dt^2} - 3x - 4y = 0$

$$\frac{d^2y}{dt^2} + x + y = 0$$

$$(D^2 - 3)x - 4y = 0 \qquad ...(1)$$

$$x + (D^2 + 1)\, y = 0 \qquad ...(2)$$

Operating equation (2) by $(D^2 - 3)$, we get

$$(D^2 - 3)x + (D^2 - 3)(D^2 + 1)\, y = 0 \qquad ...(3)$$

Subtracting (3) from (1), we get

$$-4y - (D^2 - 3)(D^2 + 1)\, y = 0 = \;\Rightarrow\; -4y - (D^4 - 2D^2 - 3)\, y = 0$$

$$\Rightarrow \quad (D^4 - 2D^2 - 3 + 4)\, y = 0 \quad \Rightarrow \quad (D^4 - 2D^2 + 1)\, y = 0$$

$$\Rightarrow \quad (D^2 - 1)^2 y = 0$$

A.E. is $\quad (m^2 - 1)^2 = 0 \quad \Rightarrow \quad (m^2 - 1) = 0 \Rightarrow \quad m = \pm 1$

$$y = (c_1 + c_2 t)e^t + (c_3 + c_4 t)e^{-t} \qquad ...(4)$$

From (2), we have

$$x = -(D^2 + 1)y$$

$$= -D^2 y - y$$

$$= -D^2 [(c_1 + c_2 t)\, e^t + (c_3 + c_4 t)e^{-t}] - [(c_1 + c_2 t)e^t + (c_3 + c_4 t)e^{-t}]$$

$$= -D\, [\{(c_1 + c_2 t\,)e^t + c_2 e^t\} + \{(c_3 + c_4 t)\,(-e^{-t}) + c_4 e^{-t}\}] - [(c_1 + c_2 t)e^t + (c_3 + c_4 t)e^{-t}]$$

$$= -[(c_1 + c_2 t)\, e^t + c_2 e^t + c_2 e^t + (c_3 + c_4 t)\,(e^{-t}) - c_4 e^{-t} - c_4 e^{-t}] - [(c_1 + c_2 t)\, e^t + (c_3 + c_4 t)e^{-t}]$$

$$= -[(c_1 + c_2 t + 2c_2 + c_1 + c_2 t)\, e^t + (c_3 + c_4 t - 2c_4 + c_3 + c_4\, t)\, e^{-t}]$$

$$= -[(2c_1 + 2c_2 + 2c_2 t)e^t + (2c_3 - 2c_4 + 2c_4 t)e^{-t}] \qquad ...(5)$$

Relations (4) and (5) are the required solutions. **Ans.**

Example 10. *A mechanical system with two degrees of freedom satisfies the equations:*

$$2\frac{d^2x}{dt^2}+3\frac{dy}{dt}=4$$

$$2\frac{d^2y}{dt^2}-3\frac{dx}{dt}=0$$

Obtain expressions for x and y in terms of t, given $x, y\frac{dx}{dt}, \frac{dy}{dt}$ *all vanish at t = 0*

Solution. $2D^2x + 3Dy = 4$

$-3Dx + 2D^2y = 0$

Multiplying (1) by 3 and (2) by $2D$ We get

$6D^2x + 9Dy = 12$

$-6D^2x + 4D^3y = 0$

Adding (3) and (4), we have

$4D^3y + 9Dy = 12$

$\Rightarrow$ $(4D^3 + 9D)y = 12$

A.E. is $4m^3 + 9m = 0$

$\Rightarrow$ $m(4m^2 + 9) = 0$

$$m = 0,\ m = \frac{3}{2}i; -\frac{3}{2}i$$

$$\therefore \quad y = c_1 + \left(c_2 \cos\frac{3}{2}t + c_3 \sin\frac{3}{2}t\right)$$

$$\text{P.I.} = \frac{12}{4D^3+9D}e^0 = t\cdot\frac{12}{12\,D^2+9}e^0 \quad\Rightarrow\quad \text{P.I.} = t\frac{12}{9}=\frac{12t}{9}=\frac{4t}{3}$$

$\therefore$ G.S = C.F. + P.I.

$$\Rightarrow \quad y = c_1 + c_2\cos\frac{3}{2}t + c_3\sin\frac{3}{2}t + \frac{4t}{3}$$

$$y = 0, t = 0 \Rightarrow c_1 + c_2 = 0 \Rightarrow c_2 = -c_1$$

$$\therefore \quad y = c_1 - c_1\cos\frac{3}{2}t + c_3\sin\frac{3}{2}t + \frac{4}{3}t$$

$$\frac{dy}{dt} = \frac{3}{2}c_1\sin\frac{3}{2}t + \frac{3}{2}c_3\cos\frac{3}{2}t + \frac{4}{3}$$

$$t = 0, \frac{dy}{dt} = 0 \Rightarrow 0 = \frac{3}{2}c_3 + \frac{4}{3} \Rightarrow c_3 = -\frac{8}{9}$$

Now, $y = c_1 - c_1\cos\frac{3}{2}t - \frac{8}{9}\sin\frac{3}{2}t + \frac{4}{3}t$

Putting the value of y in (1). we get

$$\frac{2d^2x}{dt^2} + \frac{3d}{dt}\left(c_1 - c_1\cos\frac{3}{2}t - \frac{8}{9}\sin\frac{3}{2}t + \frac{4t}{3}\right) = 4$$

$$\Rightarrow \quad 2\frac{d^2x}{dt^2} + 3\left[0 + c_1\frac{3}{2}\sin\frac{3}{2}t - \frac{3}{2}\times\frac{8}{9}\cos\frac{3}{2}t + \frac{4}{3}\right] = 4$$

$$\Rightarrow \qquad 2\frac{d^2x}{dt^2}+\frac{9}{2}c_1\sin\frac{3}{2}t-4\cos\frac{3}{2}t+4 = 4$$

$$\Rightarrow \qquad 2\frac{d^2x}{dt^2}=-\frac{9}{2}c_1\sin\frac{3}{2}t+4\cos\frac{3}{2}t$$

Integration, we get

$$\therefore \qquad 2\frac{dx}{dt}=\frac{9}{2}c_1\times\frac{2}{3}\cos\frac{3}{2}t+4\times\frac{2}{3}\sin\frac{3}{2}t+c_4$$

$$\therefore \qquad t=0, \frac{dx}{dt}=0 \quad\Rightarrow\quad 0=3c_1+c_4$$

$$\Rightarrow \qquad c_4=-3c_1$$

Now, $$2\frac{dx}{dt}=3c_1\cos\frac{3}{2}t+\frac{8}{3}\sin\frac{3}{2}t-3c_1$$

Again integrating we have

$$\therefore \qquad 2x = 3c_1\times\frac{2}{3}\sin\frac{3}{2}t-\frac{8}{3}\times\frac{2}{3}\cos\frac{3}{2}t-3c_1\,t+c_5$$

$$t=0, x=0 \quad\Rightarrow\quad 0=-\frac{16}{9}+c_5 \quad\Rightarrow\quad c_5=\frac{16}{9}$$

$$\therefore \qquad 2x = 2c_1\sin\frac{3}{2}t-\frac{16}{9}\cos\frac{3t}{2}-3c_1\,t+\frac{16}{9}$$

$$\Rightarrow \qquad x = c_1\sin\frac{3}{2}t-\frac{8}{9}\cos\frac{3}{2}t-\frac{3}{2}c_1t+\frac{8}{9}$$

Hence, $$x = c_1\sin\frac{3}{2}t-\frac{8}{9}\cos\frac{3}{2}t-\frac{3}{2}c_1t+\frac{8}{9}$$

$$y = c_1-c_1\cos\frac{3}{2}t-\frac{8}{9}\sin\frac{3}{2}t+\frac{4}{3}t$$ **Ans.**

Example 11. *Solve*: $\dfrac{d^2x}{dt^2}+\dfrac{dy}{dt}+3x = e^{-t},$

$$\frac{d^2y}{dt^2}-4\frac{dx}{dt}+3y = \sin 2t$$ (*U.P., II Semester, June* 2007)

Solution. We have, $D^2x+Dy+3x=e^{-t} \quad\Rightarrow\quad (D^2+3)\,x+D\,y=e^{-t}$ ...(1)

$D^2y-4Dx+3y=\sin 2t \quad\Rightarrow\quad -4Dx+(D^2+3)\,y=\sin 2t$ ...(2)

To eliminate y operating (1) by (D^2+3) and (2) by D, we get

$(D^2+3)^2\,x+D\,(D^2+3)\,y=(D^2+3)e^{-t}$

$-4D^2\,x+D(D^2+3)\,y \;=D\sin 2t$

$(D^4+6D^2+9)x+D\,(D^2+3)\,y=e^{-t}+3e^{-t}$ ...(3)

$-4D^2x+D(D^2+3)\,y=2\cos 2t$ ...(4)

Subtracting (4) from (3), we get

$(D^4+10\,D^2+9)x=4e^{-t}-2\cos 2t$

A.E. is $m^4+10m^2+9=0 \quad\Rightarrow (m^2+1)\,(m^2+9)=0 \quad\Rightarrow\quad m=\pm\, i,\ m=\pm 3i$

C.F. $= C_1\cos t+C_2\sin t+C_3\cos 3t+C_4\sin 3t$

$$\text{P.I.} = \frac{1}{D^4+10D^2+9}4e^{-t} - \frac{1}{D^4+10D^2+9}(2\cos 2t)$$

$$= \frac{4}{1+10+9}e^{-t} - \frac{1}{(-4)^2+10(-4)+9}(2\cos 2t) = \frac{e^{-t}}{5}+\frac{2}{15}\cos 2t$$

$$x = C_1 \cos t + C_2 \sin t + C_3 t \cos 3t + C_4 \sin 3t + \frac{e^{-t}}{5}+\frac{2}{15}\cos 2t$$

Putting the value of x in (2), we get

$$-4D\left[C_1 \cos t + C_2 \sin t + C_3 \cos 3t + C_4 \sin 3t + \frac{e^{-t}}{5}+\frac{2}{15}\cos 2t\right] + (D^2+3)\,y = \sin 2t$$

$$\Rightarrow \quad -4\left(-C_1 \sin t + C_2 \cos t - 3C_3 \sin 3t + 3C_4 \cos 3t - \frac{e^{-t}}{5} - \frac{4}{15}\sin 2t\right) + (D^2+3)\,y = \sin 2t$$

$$\Rightarrow \quad (D^2+3)\,y = \sin 2t - 4C_1 \sin t + 4C_1 \cos t - 12\,C_3 \sin 3t + 12C_4 \cos 3t - \frac{4}{5}e^{-t} - \frac{16}{15}\sin 2t$$

A.E. is $m^2 + 3 = 0 \Rightarrow m = \pm i\sqrt{3}$; C.F. $= C_1 \cos\sqrt{3}\,t + C_2 \sin\sqrt{3}\,t$

$$\text{P.I.} = \frac{1}{D^2+3}\left[\sin 2t - 4C_1 \sin t + 4C_2 \cos t - 12C_3 \sin 3t + 12C_4 \cos 3t - \frac{4}{5}e^{-t} - \frac{16}{15}\sin 2t\right]$$

$$= \frac{1}{D^2+3}\left(-\frac{1}{15}\sin 2t\right) + \frac{1}{D^2+3}(-4C_1 \sin t) + \frac{1}{D^2+3}4\,C_2 \cos t$$
$$+ \frac{1}{D^2+3}(-12C_3 \sin 3t) + \frac{1}{D^2+3}(12C_4 \cos 3t) + \frac{1}{D^2+3}\left(\frac{-4}{5}e^{-t}\right)$$

$$= \frac{1}{-4+3}\left(-\frac{1}{15}\sin 2t\right) + \frac{1}{-1+3}(-4C_1 \sin t) + \frac{1}{-1+3}4C_2 \cos t + \frac{1}{-9+3}(-12C_3 \sin 3t)$$
$$+ \frac{1}{-9+3}(12\,C_4 \cos 3t) + \frac{1}{1+3}\left(\frac{-4}{5}e^{-t}\right)$$

$$= \frac{1}{15}\sin 2t - 2C_1 \sin t + 2C_2 \cos t + 2C_3 \sin 3t - 2C_4 \cos 3t - \frac{1}{5}e^{-t}$$

$y =$ C.F. + P.I.

$$y = C_1 \cos\sqrt{3}\,t + C_2 \sin\sqrt{3}t + \frac{1}{15}\sin 2t - 2C_1 \sin t + 2C_2 \cos t + 2C_3 \sin 3t$$
$$- 2C_4 \cos 3t - \frac{1}{5}e^{-t}$$

$$x = C_1 \cos t + C_2 \sin t + C_3 \cos 3t + C_4 \sin 3t + \frac{e^{-t}}{5} + \frac{2}{15}\cos 2t$$ **Ans.**

Example 12. *Solve the simultaneous differential equations*

$$\frac{d^2x}{dt^2} - 4\frac{dx}{dt} + 4x = y,$$

$$\frac{d^2y}{dt^2} + 4\frac{dy}{dt} + 4y = 25x + 16e^t$$

Solution. Here, we have

$$(D^2 - 4D + 4)x - y = 0 \qquad ...(1)$$

$$-25x + (D^2 + 4D + 4)y = 16e^t \qquad ...(2) \quad \left(D \equiv \frac{d}{dt}\right)$$

Operating (1) by $D^2 + 4D + 4$ and adding to (2), we get

$$\Rightarrow \quad (D^2 + 4D + 4)(D^2 - 4D + 4)x - (D^2 + 4D + 4)y - 25x + (D^2 + 4D + 4)y = 16e^t$$

$$(D^2 - 4D + 4)(D^2 + 4D + 4)x - 25x = 16e^t \quad \Rightarrow \quad (D^4 - 8D^2 - 9)x = 16e^t$$

Auxiliary equation is

$$m^4 - 8m^2 - 9 = 0 \quad \Rightarrow (m^2 - 9)(m^2 + 1) = 0 \Rightarrow m = \pm i, \pm 3$$

$$\text{C.F.} = c_1e^{3t} + c_2e^{-3t} + c_3 \cos t + c_4 \sin t, \quad \text{P.I.} = \frac{1}{D^4 - 8D^2 - 9}(16e^t) = -e^t$$

$$x = \text{C.F.} + \text{P.I.}$$

$$x = c_1e^{3t} + c_2e^{-3t} + c_3 \cos t + c_4 \sin t - e^t$$

From (1), $y = \dfrac{d^2x}{dt^2} - 4\dfrac{dx}{dt} + 4x$

$$= \frac{d^2}{dt^2}\left(c_1e^{3t} + c_2e^{-3t} + c_3 \cos t + c_4 \sin t - e^t\right) - 4\frac{d}{dt}\left(c_1e^{3t} + c_2e^{-3t} + c_3 \cos t + c_4 \sin t - e^t\right)$$
$$+ 4\left(c_1e^{3t} + c_2e^{-3t} + c_3 \cos t + c_4 \sin t - e^t\right)$$

$$= \frac{d}{dt}\left[3c_1e^{3t} - 3c_2e^{-3t} + c_3(-\sin t) + c_4 \cos t - e^t\right]$$
$$- 4\left[3c_1e^{3t} - 3c_2e^{-3t} + c_3(-\sin t) + c_4 \cos t - e^t\right]$$
$$+ 4\left[c_1e^{3t} + c_2e^{-3t} + c_3 \cos t + c_4 \sin t - e^t\right]$$

$$= 9c_1e^{3t} + 9c_2e^{-3t} - c_3 \cos t - c_4 \sin t - e^t +$$
$$\left[-12c_1e^{3t} + 12c_2e^{-3t} + 4c_3 \sin t - 4c_4 \cos t + 4e^t\right]$$
$$+ \left[4c_1e^{3t} + 4c_2e^{-3t} + 4c_3 \cos t + 4c_4 \sin t - 4e^t\right]$$

$$y = c_1e^{3t} + 25c_2e^{-3t} + (3c_3 - 4c_4)\cos t + (4c_3 + 3c_4)\sin t - e^t$$

$$x = c_1e^{3t} + c_2e^{-3t} + c_3 \cos t + c_4 \sin t - e^t \qquad \text{[From (3)]} \qquad \textbf{Ans.}$$

EXERCISE 16.1

Solve the following coupled differential equations:

1. $\dfrac{dx}{dt} + 2x - 3y = 0, \quad \dfrac{dy}{dt} - 3x + 2y = 0$ **Ans.** $x = c_1 e^t - c_2 e^{-5t}, y = c_1 e^t + c_2 e^{-5t}$

2. $\dfrac{d^2y}{dt^2} = x, \quad \dfrac{d^2x}{dt^2} = y$ **Ans.** $x = c_1 e^t + c_2 e^{-t} + (c_3 \cos t + c_4 \sin t)$

$y = c_1 e^t + c_2 e^{-t} - (c_3 \cos t + c_4 \sin t)$

3. $\dfrac{dx}{dt} + 5x - 2y = t, \quad \dfrac{dy}{dt} + 2x + y = 0$ **Ans.** $x = -\dfrac{1}{27}(1 + 6t)e^{-3t} + \dfrac{1}{27}(1 + 3t)$

So that $x = y = 0$ when $t = 0$ (*AMIETE, June 2009, U.P., II Semester, June* 2008)

Ans. $y = -\dfrac{2}{27}(2 + 3t)e^{-3t} + \dfrac{2}{27}(2 - 3t)$

4. $\dfrac{dx}{dt} - y = t\,,\ \dfrac{dy}{dt} = t^2 - x$ **Ans.** $x = c_1 \cos t + c_2 \sin t + t^2 - 1\ ;\ y = -c_1 \sin t + c_2 \cos t + t$

5. $\dfrac{dx}{dt} + 2y + \sin t = 0$

$\dfrac{dy}{dt} - 2x - \cos t = 0$ **Ans.** $x = c_1 \cos 2t + c_2 \sin 2t - \cos t\ ;\ y = c_1 \sin 2t - c_2 \cos 2t - \sin t$

6. $4\dfrac{dx}{dt} - \dfrac{dy}{dt} + 3x = \sin t$

$\dfrac{dx}{dt} + y = \cos t$ **Ans.** $x = c_1 e^{-t} + c_2 e^{-3t},\ y = c_1 e^{-t} + 3 c_2 e^{-3t} + \cos t$

7. $\dfrac{dy}{dt} = x$ and $\dfrac{dx}{dt} = y + e^{2t}$ **Ans.** $x = C_1 e^t + C_2 e^{-t} + \dfrac{2}{3}e^{2t},\ y = C_1 e^t - C_2 e^{-t} + \dfrac{1}{3}e^{2t}$

8. $\dfrac{dx}{dt} = y + t,\ \dfrac{dy}{dt} = -2x + 3y + 1$ **Ans.** $x = c_1 e^t + \dfrac{1}{2}c_2 e^{2t} - \dfrac{3}{2}t - \dfrac{5}{4},\ y = c_1 e^t + c_2 e^{2t} - t - \dfrac{3}{2}$

9. $t\dfrac{dx}{dt} + y = 0,\ t\dfrac{dy}{dt} + x = 0$ **Ans.** $x = c_1 t + c_2 t^{-1},\ y = c_2 t^{-1} - c_1 t$

given $x(1) = 1$ and $y(-1) = 0$

10. $\dfrac{dx}{dt} + y = \sin t,\ \dfrac{dy}{dt} + x = \cos t,$ given that $x = 2,\ y = 0$ when $t = 0$

Ans. $x = e^t + e^{-t},\ y = \sin t - e^t + e^{-t}$

11. $(D - 1)\,x + Dy = 2t + 1$

$(2D + 1)\,x + 2\,Dy = t$ **Ans.** $x = -t - \dfrac{2}{3},\ y = \dfrac{t^2}{2} + \dfrac{4}{3}t + C$

12. $\dfrac{dx}{dt} + \dfrac{2}{t}(x - y) = 1,$ *(U.P., II Semester, Summer (C.O.) 2005)*

$\dfrac{dy}{dt} + \dfrac{1}{t}(x + 5y) = t$ **Ans.** $x = At^{-4} + Bt^{-3} + \dfrac{t^2}{15} + \dfrac{3y}{10},\ y = -At^{-4} - \dfrac{1}{2}Bt^{-3} + \dfrac{2t^2}{15} - \dfrac{t}{20}$

13. $(D^2 - 1)x + 8Dy = 16e^t$ and $Dx + 3(D^2 + 1)y = 0$

Ans. $y = c_1 \cos\dfrac{t}{\sqrt{3}} + c_2 \sin\dfrac{t}{\sqrt{3}} + c_3 \cosh\sqrt{3}\,t + c_4 \sinh\sqrt{3}\,t + 2e^t$

$x = \sqrt{3}\,c_1 \sin\dfrac{t}{\sqrt{3}} - \sqrt{3}\,c_2 \cos\dfrac{t}{\sqrt{3}} - 3\sqrt{3}\,c_3 \sinh\sqrt{3}\,t - 3\sqrt{3}\,c_4 \cosh\sqrt{3}\,t - 6e^t - 3t.$

14. $\dfrac{dx}{dt} + \dfrac{2}{t}(x - y) = 1,$ *(U.P. II Semester, 2005)*

$\dfrac{dy}{dt} + \dfrac{1}{t}(x + 5y) = t.$ **Ans.** $x = At^{-4} + Bt^{-3} + \dfrac{t^2}{15} + \dfrac{3t}{10},\ y = -At^{-4} - \dfrac{1}{2}Bt^{-3} + \dfrac{2t^2}{15} - \dfrac{t}{20}$

APPLICATIONS TO DIFFERENTIAL EQUATIONS

17.1 INTRODUCTION

In this chapter, we shall study the applications of differential equations to various physical problems.

17.2 ELECTRICAL CIRCUIT

We will consider circuits made up of

(*i*) Voltage source which may be a battery or a generator.

(*ii*) Resistance, inductance and capacitance.

Table of Elements, Symbols and Units

	Element	*Symbol*	*Unit*
1.	Charge	q	coulomb
2.	Current	i	ampere
3.	Resistance,	R	ohm
4.	Inductance,	L	henry
5.	Capacitance,	C	farad
6.	Electromotive force or voltage (constant)	constant V	volt
7.	Variable voltage	variable V	volt

The formation of differential equation for an electric circuit depends upon the following laws.

(*i*) $i = \dfrac{dq}{dt}$, (*ii*) Voltage drop across resistance $R = Ri$

(*iii*) Voltage drop across inductance $L = L.\dfrac{di}{dt}$

(*iv*) Voltage drop across capacitance $C = \dfrac{q}{C}$

Kirchhoff's laws

I. Voltage law. The algebraic sum of the voltage drop around any closed circuit is equal to the resultant electromotive force in the circuit.

II. Current law. At a junction or node, current coming is equal to current going.

(*i*) **L - R series circuit.** Let i be the current flowing in the circuit containing resistance R and inductance L in series, with voltage source E, at any time t.

By voltage law $\quad Ri + L\frac{di}{dt} = E \quad \Rightarrow \quad \frac{di}{dt} + \frac{R}{L}i = \frac{E}{L}$ ...(1) (*M.U. II Semester, 2009*)

This is the linear differential equation

$$\text{I.F.} = e^{\int \frac{R}{L}dt} = e^{\frac{R}{L}t}$$

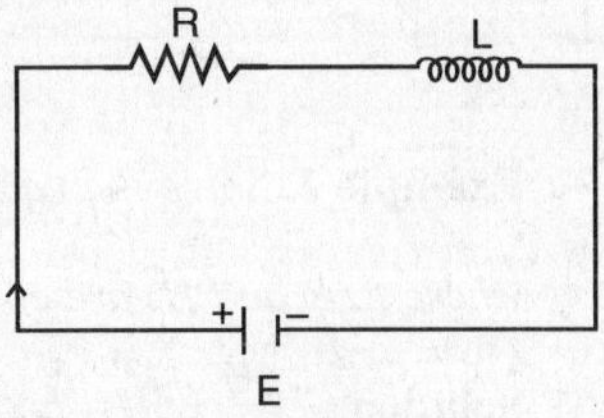

Its solution is $\quad i.e^{\frac{R}{L}t} = \int \frac{E}{L} e^{\frac{R}{L}t}\, dt + A$

$\Rightarrow \quad i.e^{\frac{R}{L}t} = \frac{E}{L} \times \frac{L}{R} e^{\frac{R}{L}t} + A$

$\Rightarrow \quad i = \frac{E}{R} + Ae^{-\frac{Rt}{L}}$ ...(2)

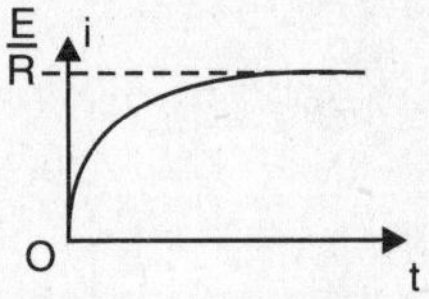

At $t = 0$, $\quad i = 0 \Rightarrow A = -\frac{E}{R}$

Thus, (2) becomes $i = \frac{E}{R}\left[1 - e^{-\frac{Rt}{L}}\right]$

(*ii*) **C-R series circuit.** Let i be current in the circuit containing resistance R, and capacitance C in series with voltage source E, at any time t.

By voltage law

$$Ri + \frac{q}{C} = E \qquad \left[i = \frac{dq}{dt}\right]$$

$\Rightarrow \quad R\frac{dq}{dt} + \frac{q}{C} = E$

Example 1. *An inductance of 2 henries and a resistance of 20 ohms are connected in series with an e.m.f. E volts. If the current is zero when t = 0, find the current at the end of 0.01 sec if E = 100 Volts.* (*U.P., II Semester, June 2008*)

Solution. Differential equation of the above circuit is as

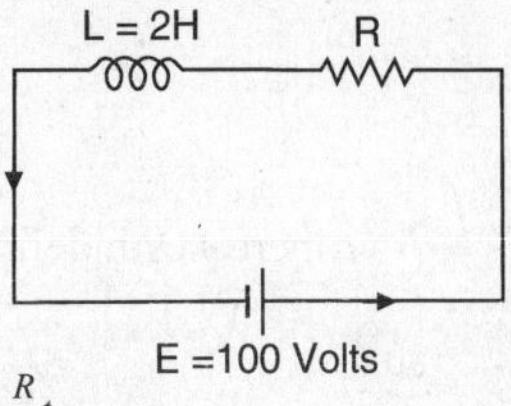

L = 2H
R
E =100 Volts

1st case: $\quad L\frac{di}{dt} + Ri = E \Rightarrow \frac{di}{dt} + \frac{R}{L}i = \frac{E}{L}$

$$\text{I.F.} = e^{\int \frac{R}{L}dt} = e^{\frac{Rt}{L}}$$

Its solution is $\quad ie^{\frac{Rt}{L}} = \frac{E}{L}\int e^{\frac{Rt}{L}}dt \quad \Rightarrow \quad ie^{\frac{Rt}{L}} = \frac{E}{L}\frac{L}{R}e^{\frac{R}{L}t} + A$

$\Rightarrow \quad ie^{\frac{Rt}{L}} = \frac{E}{R}e^{\frac{R}{L}t} + A$ ...(1)

Putting $t = 0$, $i = 0$; in (1), we get $0 = \frac{E}{R} + A \Rightarrow A = -\frac{E}{R}$

Putting the value of A in (1), we get

$$i e^{\frac{Rt}{L}} = \frac{E}{R} e^{\frac{Rt}{L}} - \frac{E}{R} \quad \Rightarrow \quad i = \frac{E}{R} - \frac{E}{R} e^{-\frac{R}{L}t}$$

$$i = \frac{E}{R}\left[1 - e^{-\frac{R}{L}t}\right] \qquad \text{...(2)}$$

On putting the values of E, R and L in (2), we get

$$i = \frac{100}{20}\left[1 - e^{-\frac{20}{2}t}\right] = 5[1 - e^{-10t}]$$

$$= 5\,[1 - e^{-10\times 0.01}] = 5[1 - e^{-0.1}] = 5\left[1 - \frac{1}{e^{0.1}}\right] \quad \text{at } [t = 0.01 \text{ sec}]$$

$$= 0.475 \text{ Approx.}$$ **Ans.**

Example 2. *Solve the equation* $L\dfrac{di}{dt} + Ri = E_0 \sin \omega t$

where L, R and E_0 are constants and discuss the case when t increases indefinitely.

Solution. $L\dfrac{di}{dt} + Ri = E_0 \sin \omega t$

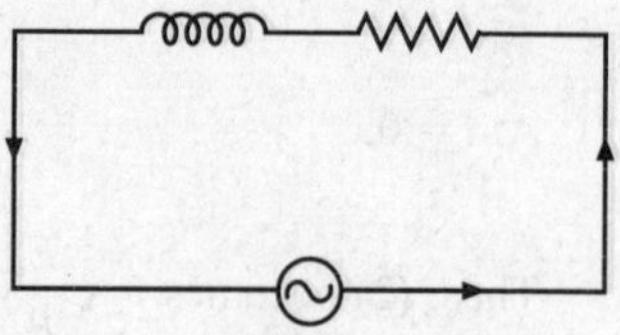

L R

E sin ωt

$$\Rightarrow \qquad \frac{di}{dt} + \frac{R}{L} i = \frac{E_0}{L} \sin \omega t$$

$$\text{I.F.} = e^{\int \frac{R}{L} dt} = e^{\frac{R}{L}t}$$

Solution is

$$i.e^{\frac{R}{L}t} = \frac{E_0}{L} \int e^{\frac{R}{L}t} \sin \omega t \, dt + A$$

$$\left[\int e^{ax} \sin bx \, dx = \frac{e^{ax}}{\sqrt{a^2 + b^2}} \sin\left(bx - \tan^{-1}\frac{b}{a}\right)\right]$$

$$\Rightarrow \qquad i.e^{\frac{R}{L}t} = \frac{E_0}{L} \frac{e^{\frac{R}{L}t}}{\sqrt{\frac{R^2}{L^2} + \omega^2}} \sin\left(\omega t - \tan^{-1}\frac{L\omega}{R}\right) + A$$

$$i = \frac{E_0}{\sqrt{R^2 + L^2\omega^2}} \sin\left(\omega t - \tan^{-1}\frac{L\omega}{R}\right) + Ae^{-\frac{R}{L}t}$$

As t increases indefinitely, then $Ae^{-\frac{Rt}{L}}$ tends to zero.

so

$$i = \frac{E_0}{\sqrt{R^2 + L^2\omega^2}} \sin\left(\omega t - \tan^{-1}\frac{L\omega}{R}\right)$$ **Ans.**

Example 3. *A condenser of capacity C farads with V_0 is discharged through a resistance R ohms. Show that if q coulomb is the charge on the condenser, i ampere the current and V the voltage at time t.*

$$q = CV,\ V = Ri \text{ and } i = \frac{dq}{dt}, \text{ hence show that } V = V_0 e^{-\frac{t}{RC}}$$

Solution. Voltage across $R = Ri$

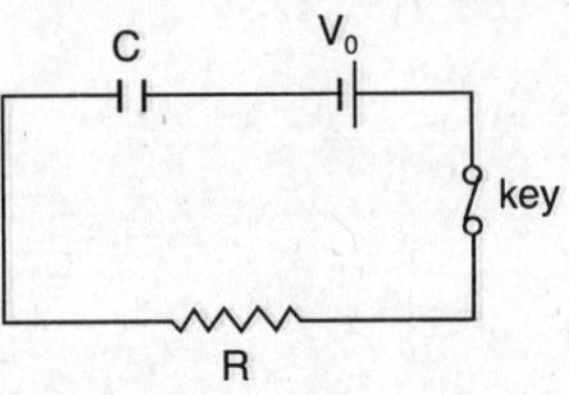

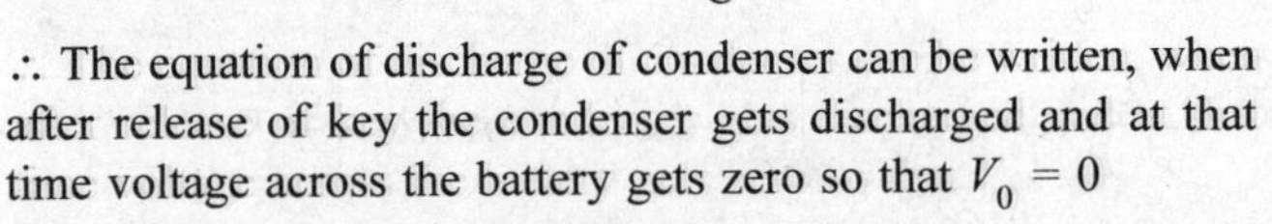

Voltage drop across capacitance $= \dfrac{q}{C}$

$\therefore$ The equation of discharge of condenser can be written, when after release of key the condenser gets discharged and at that time voltage across the battery gets zero so that $V_0 = 0$

The differential equation of the above circuit is

$$Ri + \frac{q}{C} = 0 \quad \Rightarrow \quad R\frac{dq}{dt} + \frac{q}{C} = 0 \qquad \left(\text{as } i = \frac{dq}{dt}\right)$$

$$\Rightarrow \quad \frac{dq}{dt} + \frac{q}{RC} = 0 \quad \Rightarrow \quad \frac{dq}{dt} = -\frac{q}{RC} \Rightarrow \frac{dq}{q} = -\frac{1}{RC}dt$$

Integrating both sides, we get

$$\int \frac{dq}{q} = -\frac{1}{RC}\int dt \quad \Rightarrow \quad \log q = -\frac{1}{RC}t + A \qquad ...(1)$$

But at $t = 0$, the charge at the condenser is q_0 such that

$$\log q_0 = -\frac{1}{RC}(0) + A \quad \Rightarrow \quad A = \log q_0 \qquad ...(2)$$

Putting the value of A from (2) in (1), we have

$$\log q = -\frac{1}{RC}t + \log q_0 \quad \Rightarrow \quad \log q - \log q_0 = -\frac{1}{RC}t$$

$$\Rightarrow \quad \log\frac{q}{q_0} = -\frac{1}{RC}t \quad \Rightarrow \quad \frac{q}{q_0} = e^{-\frac{t}{RC}}$$

$$\Rightarrow \quad q = q_0\, e^{-\frac{t}{RC}} \qquad ...(3)$$

Dividing both side of (3) by C, we get

$$\frac{q}{C} = \frac{q_0}{C}e^{-\frac{t}{RC}} \quad \Rightarrow \quad V = V_0 e^{-\frac{t}{RC}} \qquad \left[\text{as } \frac{q}{C} = V\right] \quad \textbf{Proved.}$$

Example 4. *The equations of electromotive force in terms of current i for an electrical circuit having resistance R and a condenser of capacity C, in series, is* $E = Ri + \int \frac{i}{C}dt$. *Find the current i at any time t, when* $E = E_0 \sin wt$. (U.P. II Semester, Summer 2006)

Solution. We have,

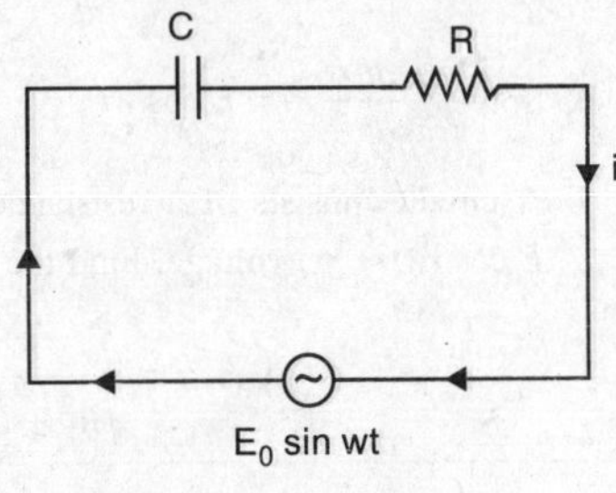

$$Ri + \int \frac{i}{C}dt = E_0 \sin wt$$

Differentiating both the sides, we get

$$\frac{R\,di}{dt} + \frac{i}{C} = E_0\, w \cos wt$$

$$\Rightarrow \quad \frac{di}{dt} + \frac{i}{RC} = \frac{E_0}{R} w \cos wt$$

$$I.F. = e^{\int \frac{1}{RC}dt} = e^{\frac{t}{RC}}$$

Its solution is $i.\,(I.F.) = \int \dfrac{E_0 w}{R} \cos wt\,(I.F.)\,dt \;\Rightarrow\; i.e^{\frac{t}{RC}} = \dfrac{E_0 w}{R}\int \cos wt.\, e^{\frac{t}{RC}}\,dt + A$

$$\left[\int e^{ax}\cos bx\,dx = \frac{e^{ax}}{a^2+b^2}[a\cos bx + b\sin bx]\right]$$

$$\Rightarrow \quad i.e^{\frac{t}{RC}} = \frac{E_0 w}{R}\frac{e^{\frac{t}{RC}}}{\frac{1}{R^2C^2}+w^2}\left[\frac{1}{RC}\cos\omega t + \omega\sin\omega t\right] + A$$

$$= \frac{E_0 w}{R}\frac{R^2C^2\, e^{\frac{t}{RC}}}{1+w^2R^2C^2}\left[\frac{1}{RC}\cos\omega t + \omega\sin\omega t\right] + A$$

$$i = E_0 w.\frac{RC^2}{1+w^2R^2C^2}\left[\frac{1}{RC}\cos wt + w\sin wt\right] + Ae^{-\frac{t}{RC}}$$

$$i = E_0 w.\frac{C}{1+w^2R^2C^2}[\cos wt + w\,RC\sin wt] + Ae^{-\frac{t}{RC}} \qquad \textbf{Ans.}$$

EXERCISE 17.1

1. A coil having a resistance of 15 ohms and an inductance of 10 henries is connected to 90 volts supply. Determine the value of current after 2 seconds. ($e^{-3} = 0.05$) **Ans.** 5.985 amp.
2. A resistance of 70 ohms, an inductance of 0.80 henry are connected in series with a battery of 10 volts. Determine the expression for current as a function of time after $t = 0$.

   Ans. $i = \frac{1}{7}\left(1 - e^{-\frac{175}{2}t}\right)$
3. A circuit consists of resistance R ohms and a condenser of C farads connected to a constant e.m.f. E; if $\frac{q}{C}$ is the voltage of the condenser at time t after closing the circuit Show that $\frac{q}{C} = E - Ri$ and hence show that the voltage at time t is $E\left(1 - e^{-\frac{t}{CR}}\right)$.
4. Show that the current $i = \frac{q}{CR}e^{-\frac{t}{R.C.}}$ during the discharge of a condenser of charge Q coulomb through a resistance R ohms.
5. A condenser of capacity C farads with voltage v_0 is discharged through a resistance R ohms. Show that if q coulomb is the charge on the condenser, i ampere the current and v the voltage at time t.

   $q = Cv$, $v = Ri$ and $i = -\frac{dq}{dt}$, hence show that $v = v_0\, e^{-\frac{1}{Rc}}$.
6. Solve $L\frac{di}{dt} + Ri = E\cos wt$ **Ans.** $i = \frac{E}{L^2w^2+R^2}\left(R\cos wt + Lw\sin wt - \text{Re}^{-\frac{Rt}{L}}\right)$
7. A circuit consists of a resistance R ohms and an inductance of L henry connected to a generator of $E\cos(wt + \alpha)$ voltage. Find the current in the circuit. ($i = 0$, when $t = 0$).

   Ans. $i = \frac{E}{\sqrt{R^2+L^2w^2}}\cos\left[wt + \alpha - \tan^{-1}\frac{Lw}{R}\right] - \frac{E}{\sqrt{R^2+L^2w^2}}.e^{-\frac{R}{L}t}\cos\left[\alpha - \tan^{-1}\frac{Lw}{R}\right]$

17.3 SECOND ORDER DIFFERENTIAL EQUATION

We have already discussed $R - L$ and $R - L - C$ electric circuits. Here we want to do circuit problems involving second order differential equations.

Example 5. *The damped LCR circuit is governed by the equation*

$$L\frac{d^2q}{dt^2}+R\frac{dq}{dt}+\left(\frac{1}{C}\right)q=0$$

where L, R, C are positive constants. Find the conditions under which the circuit is overdamped, underdamped and critically damped. Find also the critical resistance.

(*U.P. II Semester, Summer 2005*)

Solution. Given equation is

$$L\frac{d^2q}{dt^2}+R\frac{dq}{dt}+\left(\frac{1}{C}\right)q=0 \Rightarrow \frac{d^2q}{dt^2}+\frac{R}{L}\frac{dq}{dt}+\left(\frac{1}{LC}\right)q=0 \quad ...(1)$$

Let $\quad \frac{R}{L}=2p$ and $\frac{1}{LC}=w^2$

Thus equation (1) becomes

$$\frac{d^2q}{dt^2}+2p\frac{dq}{dt}+w^2q=0$$

Its auxiliary equation is

$$m^2+2pm+w^2=0$$

$$\Rightarrow \quad m=-p\pm\sqrt{p^2-w^2}$$

Case 1. When $p > w$, roots are real and distinct solution of equation (1) is

$$q=Ae^{(-p+\sqrt{p^2-w^2})t}+Be^{(-p-\sqrt{p^2-w^2})t}$$

In this case q is always positive, this is a condition of over damping.

Thus if $p > w$

$$\frac{R}{2L}>\frac{1}{\sqrt{LC}}$$

$$R>2\sqrt{\frac{L}{C}}$$

Case 2. When $p < w$, roots are imaginary

$$q=e^{-pt}\left(A\cos\sqrt{w^2-p^2}+B\sin\sqrt{w^2-p^2}\right)$$

period of oscillation decreases and this condition is of under damping.

Case 3. When $p = w$, roots are equal $q = (A + Bt)e^{-pt}$,

This is a condition of critically damped.

Critical resistance is given by

$$p=w$$

$$\Rightarrow \quad \frac{R}{2L}=\frac{1}{\sqrt{LC}}$$

$$R=2\sqrt{\frac{L}{C}} \quad \textbf{Ans.}$$

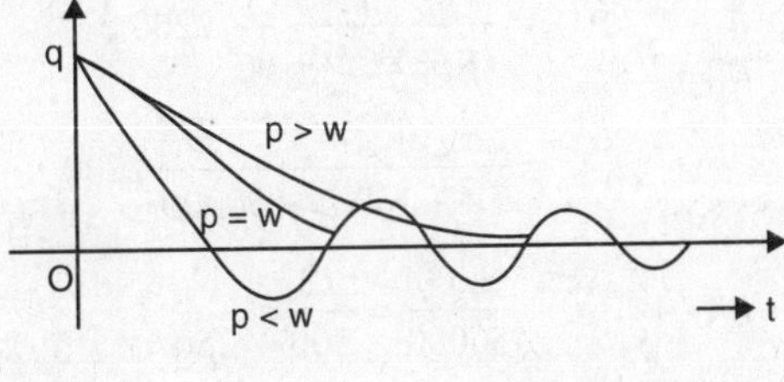

Example 6. *A circuit consists of resistance of 5ohms, inductance of 0.05 Henrys and capacitance of 4 × 10⁻⁴ farads. If q (0) = 0, i (0) = 0 find q (t) and i (t), when an emf of 110 volts is applied.*

(*M.D.U., 2010*)

Solution.

$$L\frac{d^2Q}{dt^2}+R\frac{dQ}{dt}+\left(\frac{1}{C}\right)Q=110 \qquad \text{... (1)}$$

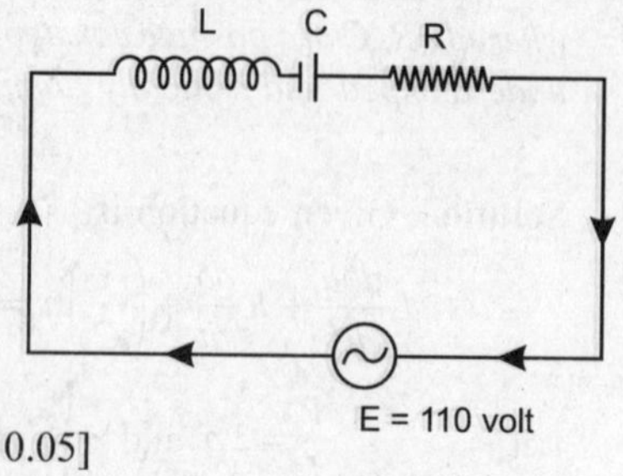

$$\Rightarrow \quad L\frac{d^2Q}{dt^2}+\frac{R}{L}\frac{dQ}{dt}+\frac{1}{LC}Q=\frac{110}{L}$$

Let $\frac{R}{L}=2p$ and $\frac{1}{LC}=w^2$

Thus equation is

$$\frac{d^2Q}{dt^2}+2p\frac{dQ}{dt}+w^2Q=\frac{110}{0.05} \qquad [\text{L} = 0.05]$$

$$\Rightarrow \quad \frac{d^2Q}{dt^2}+2p\frac{dQ}{dt}+w^2Q=2200$$

Its auxiliary equation is

$$m^2 + 2\,p\,m + w^2 = 0$$

$$m = -p \pm \sqrt{p^2-w^2} \qquad \text{... (2)}$$

Here, we have

$$R = 5 \text{ ohms}, \quad L = 0.05 \text{ Henrys}, \quad C = 4 \times 10^{-4} \text{ farads}$$

$$\therefore \quad 2p=\frac{R}{L}=\frac{5}{0.05}=100 \quad \Rightarrow \quad p = 50$$

$$w^2=\frac{1}{LC}=\frac{1}{0.05\times 4\times 10^{-4}}=50000$$

Putting the values of p and w in (2), we get

$$m = -50 \pm \sqrt{(50)^2-50000} = -50 \pm \sqrt{2500-50000}$$

$$\Rightarrow \quad m = -50 \pm \sqrt{-47500} = -50 \pm 50\sqrt{-19} = -50 \pm 50\sqrt{19}\, i$$

$$\text{C.F.} = e^{-50t}\left(A\cos 50\sqrt{19}\,t + B\sin 50\sqrt{19}\,t\right) \qquad \text{... (3)}$$

$$\text{P.I.} = \frac{1}{D^2+2PD+w^2}2200$$

$$= \frac{1}{D^2+100\,D+50000}2200 \qquad [D = 0]$$

$$= \frac{2200}{50000}=\frac{22}{500}=\frac{11}{250}$$

Complete solution = C.F. + P.I.

$$Q = e^{-50t}\left[A\cos 50\sqrt{19}\,t + B\sin 50\sqrt{19}\,t\right]+\frac{11}{250} \qquad \text{... (4)}$$

On putting $Q = 0$, $t = 0$ in (4), we get

$$Q = A + \frac{11}{250} \qquad \Rightarrow \qquad A = -\frac{11}{250}$$

On differentiating (4), we get

$$i = \frac{dQ}{dt} = -50\,e^{-50t} \quad [A \cos 50\sqrt{19}\,t + B \sin 50\sqrt{19}\,t] +$$

$$e^{-50t}\ [-50\sqrt{19}\ A \sin 50\sqrt{19}\,t + 50\sqrt{19}\ B \cos 50\sqrt{19}\,t] \qquad \text{... (5)}$$

On putting $i = 0$, $t = 0$ in (5), we get

$$0 = -50\,A + 50\sqrt{19}\ B \qquad \text{... (6)}$$

On putting $A = -\dfrac{11}{250}$ in (6), we get

$$0 = -50\left(-\frac{11}{250}\right) + 50\sqrt{19}\,B \qquad \Rightarrow \qquad B = -\frac{11}{5 \times 50\sqrt{19}}$$

$$B = -\frac{11}{250\sqrt{19}}$$

On putting the values of A and B in (4), we get

$$Q = e^{-50t}\left[-\frac{11}{250}\cos 50\sqrt{19}\,t - \frac{11}{250\sqrt{19}}\sin 50\sqrt{19}\,t\right] + \frac{11}{250}$$

On putting the values of A and B in (5), we get

$$i = -50\,e^{-50t}\left[\left(-\frac{11}{250}\right)\cos 50\sqrt{19}\,t - \frac{11}{250\sqrt{19}}\sin 50\sqrt{19}\,t\right] +$$

$$e^{-50t}\left[-50\sqrt{19}\cdot\left(\frac{-11}{250}\right)\sin 50\sqrt{19}\,t + 50\sqrt{19}\left(\frac{-11}{250\sqrt{19}}\right)\cos 50\sqrt{19}\,t\right]$$

$$= e^{-50t}\left[\left(\frac{11}{5} - \frac{11}{5}\right)\cos 50\sqrt{19}\,t + \left(\frac{11}{5\sqrt{19}} + \frac{11\sqrt{19}}{5}\right)\sin 50\sqrt{19}\,t\right]$$

$$\Rightarrow \qquad i = e^{-50t}\,\frac{11 + 11 \times 19}{5 \times 19}\sin 50\sqrt{19}\,t$$

$$\Rightarrow \qquad i = e^{-50t}\,\frac{44}{\sqrt{19}}\sin 50\sqrt{19}\,t = \frac{44}{\sqrt{19}}\,e^{-50t}\sin 50\sqrt{19}\,t \qquad \textbf{Ans.}$$

Example 7. *An electric circuit consists of an inductance 0.1 henry, a resistance of 20 ohms and a condenser of capacitance 25 microfarads. Find the charge q and the current i at time t, given the initial conditions q = 0.05 coulombs, i = 0 when t = 0*

Solution. The differential equation of the above given circuit can be written as

$$L\frac{di}{dt} + Ri + \frac{q}{C} = 0 \quad \Rightarrow \quad L\frac{d^2q}{dt^2} + R\frac{dq}{dt} + \frac{q}{C} = 0 \qquad \left[i = \frac{dq}{dt}\right]$$

$$\Rightarrow \qquad \frac{d^2q}{dt^2} + \frac{R}{L}\frac{dq}{dt} + \frac{q}{LC} = 0$$

L=0.1H R=20Ω C=25μF

First, we will solve the equation and then put the values of R, L and C. For convenience we put

$$\frac{R}{L} = 2b, b = \frac{R}{2L} = \frac{20}{2 \times 0.1} = 100$$

Let $\frac{1}{LC} = k^2$

$$\Rightarrow \quad k = \sqrt{\frac{1}{LC}} = \sqrt{\frac{1}{0.1 \times 25 \times 10^{-6}}} = \sqrt{\frac{10^7}{25}} = 632.5 \geq 100$$

Our equation reduces to

$$\frac{d^2q}{dt^2} + 2b\frac{dq}{dt} + k^2 q = 0$$

A.E. is $m^2 + 2b\,m + k^2 = 0$

So that $m = \dfrac{-2b \pm \sqrt{4b^2 - 4k^2}}{2} = -b \pm \sqrt{b^2 - k^2} = -b \pm i\sqrt{k^2 - b^2}$

C.F. is $q = e^{-bt}[A\cos\sqrt{k^2 - b^2}\,t + B\sin\sqrt{k^2 - b^2}\,t]$ ...(1)

On putting $q = 0.05$ and $t = 0$ in (1), we get $0.05 = A$

On differentiating (1), we get

$$\frac{dq}{dt} = -be^{-bt}[A\cos\sqrt{k^2 - b^2}\,t + B\sin\sqrt{k^2 - b^2}\,t]$$

$$+ e^{-bt}[-A\sqrt{k^2 - b^2}\sin\sqrt{k^2 - b^2}\,t + B\sqrt{k^2 - b^2}\cos\sqrt{k^2 - b^2}\,t] \quad ...(2)$$

On putting $\frac{dq}{dt} = 0$ and $t = 0$ in (2), we get

$$0 = -bA + B\sqrt{k^2 - b^2} \quad \Rightarrow \quad B = \frac{bA}{\sqrt{k^2 - b^2}} = \frac{0.05\,b}{\sqrt{k^2 - b^2}}$$

Substituting the values of A and B in (1), we have

$$q = e^{-bt}[0.05\cos\sqrt{k^2 - b^2}\,t + \frac{0.05\,b}{\sqrt{k^2 - b^2}}\sin\sqrt{k^2 - b^2}\,t] \quad ...(3)$$

Now, $\sqrt{k^2 - b^2} = \sqrt{\frac{10^7}{25} - (100)^2} = \sqrt{400000 - 10000} = \sqrt{390000} = 624.5$

On putting these values in (3), we have

$$q = e^{-100t}[0.05\cos 624.5t + \frac{0.05 \times 100}{624.5}\sin 624.5t]$$

$$\Rightarrow \quad q = e^{-100t}[0.05\cos 624.5t + 0.008\sin 624.5t] \quad ...(4)$$ **Ans.**

On differentiating (4), we have

$$\frac{dq}{dt} = -100\,e^{-100t}[0.05\cos 624.5t + 0.008\sin 624.5t]$$

$$+ e^{-100t}[-0.05 \times 624.5\sin 624.5t + 0.008 \times 624.5\cos 624.5t]$$

$$\Rightarrow \quad i = e^{-100t}[(-5 + 4.996)\cos 624.5t - (0.8 + 31.225)\sin 624.5t]$$

$$= e^{-100t}[-0.004\cos 624.5t - 32.025\sin 624.5t]$$

$$= -32\,e^{-100t}\sin 624.5t. \text{ approximately}$$ **Ans.**

Example 8. *An alternating e.m.f. E sin wt is applied to an inductance L and capacitance C in series. Show that the current in the circuit is* $\dfrac{Ew}{(n^2 - w^2)L}$ *(cos wt – cos nt), where* $n^2 = \dfrac{1}{LC}$.

(U.P. II Semester, June 2010, 2009)

Solution. The differential equation for the above circuit is

L C i i E sin wt

$$L\frac{d^2q}{dt^2} + \frac{q}{C} = E \sin wt$$

$$\Rightarrow \quad \frac{d^2q}{dt^2} + \frac{q}{LC} = \frac{E}{L} \sin wt$$

$$\Rightarrow \quad \left(D^2 + \frac{1}{LC}\right) q = \frac{E}{L} \sin wt$$

A.E. is $m^2 + \dfrac{1}{LC} = 0 \Rightarrow m^2 + n^2 = 0 \Rightarrow m = \pm i\, n$ $\qquad \left(\because \dfrac{1}{LC} = n^2\right)$

$$\text{C.F.} = A \cos nt + B \sin nt$$

$$\text{P.I.} = \frac{1}{D^2 + n^2} \frac{E}{L} \sin wt$$

$$\Rightarrow \quad \text{P.I.} = \frac{1}{D^2 + n^2} \frac{E}{L} \sin wt$$

Complete solution is $\quad q = A \cos n\,t + B \sin n\,t + \dfrac{E}{(n^2 - w^2)\,L} \sin wt \quad$ ...(1)

On putting $q = 0$, $t = 0$ in (1), we get

$$0 = A$$

On putting the value of A in (1), we get

$$q = B \sin nt + \frac{E}{(n^2 - w^2)L} \sin wt \qquad ...(2)$$

On differentiating (2) w.r.t., 't', we get

$$\frac{dq}{dt} = B\,n \cos nt + \frac{Ew}{(n^2 - w^2)\,L} \cos wt$$

$$\Rightarrow \quad i = B\,n \cos nt + \frac{Ew}{(n^2 - w^2)\,L} \cos wt \qquad ...(3)$$

On putting $i = 0$, $t = 0$ in (3), we get

$$0 = Bn + \frac{Ew}{(n^2 - w^2)L} \quad \Rightarrow \quad B = -\frac{Ew}{n(n^2 - w^2)L}$$

Putting the value of B in (3), we get

$$i = -\frac{Ewn}{n(n^2 - w^2)L} \cos nt + \frac{Ew}{(n^2 - w^2)L} \cos wt$$

$$\Rightarrow \quad i = \frac{Ew}{(n^2 - w^2)L} (\cos wt - \cos nt)$$ **Proved.**

Example 9. *For an electric circuit with circuit constants, L, R, C the charge q on a plate condenser is given by*

$$L\frac{d^2q}{dt^2} + R\frac{dq}{dt} + \frac{q}{C} = E \text{ and the current by } i = \frac{dq}{dt}$$

Let L = 1 henry, $C = 10^{-4}$ *farad, R = 100 ohms, E = 100 volts,*

Suppose that no charge present and no current is flowing at time t = 0, when the e.m.f. is applied. Determine q and i at any time t.

Solution. The differential equation is

$$L\frac{d^2q}{dt^2}+R\frac{dq}{dt}+\frac{q}{C}=E$$

$$\Rightarrow \quad \frac{d^2q}{dt^2}+\frac{R}{L}\frac{dq}{dt}+\frac{q}{LC}=\frac{E}{L}$$

L R C E

Putting $\frac{R}{L}=2b$ and $\frac{1}{LC}=k^2$ in (1), we have

$$\frac{d^2q}{dt^2}+2b\frac{dq}{dt}+k^2q=\frac{E}{L} \qquad ...(1)$$

This equation is exactly identical, we have

$$\Rightarrow \qquad q=\frac{E}{k^2L}+e^{-bt}[A\cos\sqrt{k^2-b^2}\,t+B\sin\sqrt{k^2-b^2}\,t] \qquad ...(2)$$

On putting $q=0$ and $t=0$ in (1), we get

$$0=\frac{E}{k^2L}+A \quad \Rightarrow \quad A=-\frac{E}{k^2L}$$

Differentiating (2), we have

$$\frac{dq}{dt}=-be^{-bt}[A\cos\sqrt{k^2-b^2}\,t+B\sin\sqrt{k^2-b^2}\,t]$$

$$+e^{-bt}[-A\sqrt{k^2-b^2}\sin\sqrt{k^2-b^2}\,t+B\sqrt{k^2-b^2}\cos\sqrt{k^2-b^2}\,t] \qquad ...(3)$$

On putting $\frac{dq}{dt}=0$ and $t=0$ in (3), we have

$$0=-bA+B\sqrt{k^2-b^2} \quad \Rightarrow \quad B=\frac{bA}{\sqrt{k^2-b^2}}=\frac{-\frac{bE}{k^2L}}{\sqrt{k^2-b^2}}=-\frac{bE}{k^2L\sqrt{k^2-b^2}}$$

Substituting the values of A and B in (2), we have

$$q=\frac{E}{k^2L}+e^{-bt}\left[-\frac{E}{k^2L}\cos\sqrt{k^2-b^2}\,t-\frac{bE}{k^2L\sqrt{k^2-b^2}}\sin\sqrt{k^2-b^2}\,t\right]$$

$$=\frac{E}{k^2L}\left[1-e^{-bt}\left(\cos\sqrt{k^2-b^2}\,t+\frac{b}{\sqrt{k^2-b^2}}\sin\sqrt{k^2-b^2}\,t\right)\right] \qquad ...(4)$$

Now,
$$\frac{E}{k^2L}=\frac{E}{\frac{1}{LC}.L}=EC=100\times10^{-4}=\frac{1}{100}$$

$$b=\frac{R}{2L}=\frac{100}{2\times1}=50$$

$$\sqrt{k^2-b^2}=\sqrt{\frac{1}{LC}-(50)^2}=\sqrt{\frac{1}{10^{-4}}-(50)^2}=\sqrt{10000-2500}=\sqrt{7500}=50\sqrt{3}$$

On putting these values in (4), we get

$$q=\frac{1}{100}\left[1-e^{-50t}(\cos 50\sqrt{3}\,t+\frac{1}{\sqrt{3}}\sin 50\sqrt{3}\,t)\right] \qquad \textbf{Ans.}$$

Example 10. *The voltage V and the current i at a distance x from the sending end of the transmission line satisfy the equations.*

$$-\frac{dV}{dx} = Ri, \quad -\frac{di}{dx} = GV$$

where R and G are constants. If $V = V_0$ *at the sending end* $(x = 0)$ *and* $V = 0$ *at receiving end* $(x = 1)$.

Show that $$V = V_0 \left\{\frac{\sinh n\,(l - x)}{\sinh nl}\right\}, \text{ when } n^2 = RG$$

Solution. We have, $$-\frac{dV}{dx} = Ri \qquad \text{...(1)}$$

$$-\frac{di}{dx} = GV \qquad \text{...(2)}$$

When $x = 0$, $V = V_0$; When $x = l$, $V = 0$

Putting the value of i from (1) in (2), we get

$$-\frac{d}{dx}\left(-\frac{dV}{dx}\frac{1}{R}\right) = GV \quad \Rightarrow \quad \frac{d^2 V}{dx^2} = RGV$$

$$\Rightarrow \quad \frac{d^2 V}{dx^2} - (RG)V = 0 \quad \Rightarrow \quad (D^2 - RG)V = 0 \qquad (RG = n^2)$$

A.E. is $\quad m^2 - n^2 = 0, \quad m = \pm\, n$

$$\therefore \quad V = A\, e^{nx} + B\, e^{-nx} \qquad \text{...(3)}$$

Now, we have to find out the values of A and B with the help of given conditions.

On putting $x = 0$ and $V = V_0$ in (3), we get

$$V_0 = A + B \qquad \text{...(4)}$$

On putting $x = 1$ and $V = 0$ in (3), we get

$$0 = A\, e^{nl} + B\, e^{-nl} \qquad \text{...(5)}$$

On solving (4) and (5), we have

$$A = \frac{V_0}{1 - e^{2nl}}, \quad B = \frac{-V_0\, e^{2nl}}{1 - e^{2nl}}$$

Substituting the values of A and B in (3), we have

$$V = \frac{V_0\, e^{nx}}{1 - e^{2nl}} - \frac{V_0\, e^{2nl}\, e^{-nx}}{1 - e^{2nl}} = \frac{V_0\,[e^{nx} - e^{2nl - nx}]}{1 - e^{2nl}}$$

$$= \frac{V_0\,[e^{(nl - nx)} - e^{-(nl - nx)}]}{e^{nl} - e^{-nl}} = V_0 \left\{\frac{\sinh n\,(l - x)}{\sinh nl}\right\} \qquad \textbf{Proved.}$$

EXERCISE 17.2

1. For an electric circuit with circuit constants L, R, C the charge q on the plate of the condensor is given by :

$$L\frac{d^2 q}{dt^2} + R\frac{dq}{dt} + \frac{q}{C} = 0$$

Find q at any time t.

Discuss the case when R is negligible and show that q is oscillatory. Calculate its period and frequency.

$$\textbf{Ans. } q = e^{-\frac{R}{2L}t}\left[A\cos\frac{\sqrt{4CL - R^2C^2}}{2LC}t + B\sin\frac{\sqrt{4LC - R^2C^2}}{2LC}t\right]$$

$$q = A\cos\frac{1}{\sqrt{LC}}t + B\sin\frac{1}{\sqrt{LC}}t, \text{ Period} = 2\pi\sqrt{LC}, \text{ frequency} = \frac{1}{2\pi\sqrt{LC}}$$

2. A condenser of capacity C is discharged through an iductance L and a resistance R in series and the charge q at any time t is given by

$$L\frac{d^2q}{dt^2} + R\frac{dq}{dt} + \frac{q}{C} = 0$$

If L = 10 milli henry, R = 200 ohms, C = 0.1 μ F and also when t = 0, charge q = 0.01 coulomb and current $\frac{dq}{dt} = 0$. Find the value of q at any time t and find the frequency of the circuit if the discharge is oscillatory. **Ans.** $q = e^{-10000\,t}\,[\,0.01 \cos 3 \times 10^4 t + 0.33 \times 10^{-2} \sin 3 \times 10^4 t]$

$$\text{frequency} = \frac{3\times 10^4}{2\pi}$$

3. A condenser of capacity C is discharged through L and a resistance R in series and the charge q at any time t is given by the equation

$$L\frac{d^2q}{dt^2} + R\frac{dq}{dt} + \frac{q}{C} = 0$$

If L = 0.5 henry, R = 300 ohms, $C = 2 \times 10^{-6}$ farad and also when t = 0, charge q = 0.01 and current $\frac{dq}{dt} = 0$, find the value of q in terms of t. **Ans.** $q = e^{-200\,t}\,[0.01 \cos 100\sqrt{91}\,t + 0.0031 \sin 100\sqrt{91}\,t]$

4. A 10^{-3} farad capacitor is connected in series with 0.05 henry inductor and 10 ohms resistor. Initially, the current in the circuit is zero and the charge on the capacitor is also zero. If the e.m.f. is 50 sin 200 t. Find the charge t seconds after the circuit is closed.

Ans. $q = 0.125\, e^{-100\,t}\,[\cos 100\,t + \sin 100\,t] - 25 \cos 200\,t + 0.125 \sin 200\,t$

5. For an electric circuit with circuit constants L, R, C, the charge q on a plate on the condenser is given by

$$L\frac{d^2q}{dt^2} + R\frac{dq}{dt} + \frac{q}{C} = E\sin wt$$

and the current $i = \frac{dq}{dt}$. The circuit is tuned to resonance so that $w^2 = \frac{1}{LC}$

If $R^2 = \frac{4L}{C}$ and $q = i = 0$ at $t = 0$, show that

$$q = \frac{E}{Rw}\left[-\cos wt + e^{-\frac{Rt}{2L}}\left(\cos pt + \frac{R}{2LP}\sin pt\right)\right] \qquad \text{where } p^2 = \frac{1}{LC} - \frac{R^2}{4L^2}$$

6. If the charge on one of the coatings of a layden jar be q when a force $E\cos pt$ acts in the circuit connecting the coatings and the circuit contains inductance L. Resistance R and capacitance C satisfying the differential equation:

$$\left(LD^2 + RD + \frac{1}{C}\right)q = E\cos pt$$

$D = \frac{d}{dt}$, find an expression for the charge given

$q = \frac{dq}{dt} = 0$ when $t = 0$.

7. An e.m.f. $E\sin pt$ is applied at $t = 0$ to a circuit containing a condenser C and inductance L in series. The current x satisfies the equation

$$L\frac{dx}{dt}+\frac{1}{C}\int x\,dt = E\sin pt$$

If $p^2 = \frac{1}{LC}$, and initially the current x and the charge q are zero, show that the current in the circuit at time t is given by

$$x = \frac{E}{2L}t\sin pt, \text{ where } x = -\frac{dq}{dt}.$$

8. An L-C-R circuit has $R = 180$ ohm, $C = \frac{1}{280}$ farad, $L = 20$ henries and an applied voltage $E(t) = 10\sin t$. Assuming that no charge is present but an initial current of i ampere is flowing at $t = 0$ when the voltage is first applied, find q and $i = \frac{dq}{dt}$ at any time t. q is given by the differential equation

$$L\frac{d^2q}{dt^2}+R\frac{dq}{dt}+\frac{q}{C} = E(t)$$

Ans. $q = \frac{11}{50}e^{-2t} - \frac{101}{500}e^{-7t} + \frac{1}{1000}[26\sin t - 18\cos t]$

17.4 MECHANICAL ENGINEERING PROBLEMS

Rectilinear Motion

When a body moves in a straight line the motion is called rectilinear motion. If x be the distance of the body at any time t from starting point then we have its velocity v given by

$$\text{velocity} = v = \frac{dx}{dt} \qquad \boxed{\boldsymbol{v = \frac{dx}{dt}}}$$

If the acceleration of the body be 'a' then

$$\text{Acceleration} = a = \frac{dv}{dt} = \frac{d^2x}{dt^2}$$

Since

$$a = \frac{dv}{dt} = \frac{dv}{dx}\cdot\frac{dx}{dt} = v\frac{dv}{dx}$$

$$\boxed{\boldsymbol{a = \frac{d^2x}{dt^2}}}$$

$$\boxed{\boldsymbol{a = v\frac{dv}{dx}}}$$

If mass of a body is m and the body is moving with acceleration a by a force F acting on it, then

$$\boxed{\boldsymbol{F = ma}}$$

$$\boxed{\boldsymbol{F = m\frac{d^2x}{dt^2}}}$$

$$\boxed{\boldsymbol{F = mv\frac{dv}{dx}}}$$

Example 11. *A moving body is opposed by a force per unit mass of value cx and resistance per unit mass of value bv^2 where x and v are the displacement and velocity of the particle at that instant. Find the velocity of the particle in terms of x, if it starts from rest.*

Solution. By Newton's second law of motion, the equation of motion of the body is

$$v\frac{dv}{dx} = -cx - bv^2 \quad\Rightarrow\quad v\frac{dv}{dx} + bv^2 = -cx \qquad ...(1)$$

Putting $v^2 = z,\ 2v\dfrac{dv}{dx} = \dfrac{dz}{dx}$, in (1), we get

$$\frac{1}{2}\frac{dz}{dx} + bz = -cx \quad \Rightarrow \quad \frac{dz}{dx} + 2bz = -2cx$$

$$\text{I.F.} = e^{\int 2b\,dx} = e^{2bx}$$

Its solution is
$$z.\,e^{2bx} = \int -2cx\,e^{2bx}\,dx + A$$

$$= -2c\left[\frac{x\,e^{2bx}}{2b} - \int \frac{e^{2bx}}{2b}\,dx\right] + A'$$

$$= -\frac{c}{b}\,x e^{2bx} + \frac{c}{b}\frac{e^{2bx}}{2b} + A'$$

$$\Rightarrow \qquad z = -\frac{c}{b}x + \frac{c}{2b^2} + A'\,e^{-2bx}$$

$$v^2 = -\frac{cx}{b} + \frac{c}{2b^2} + A'\,e^{-2bx} \qquad \text{...(2)}$$

Putting $v = 0$ and $x = 0$ in (2), we have

$$\therefore \qquad 0 = \frac{c}{2b^2} + A' \quad \Rightarrow \quad A' = -\frac{c}{2b^2}$$

$\therefore$ Equation (2) becomes $\quad v^2 = -\dfrac{cx}{b} + \dfrac{c}{2b^2} - \dfrac{c}{2b^2}e^{-2bx}$ **Ans.**

Example 12. *A particle of mass m moves in a straight line under the action of force mn^2x which is always directed towards a fixed point O on the line. Determine the displacement x(t) if the resistance to the motion is $2\lambda mnv$ given that initially $x = 0$, $\dfrac{dx}{dt} = 0$, $(0 < \lambda < 1)$.*

(U.P. II Sem 2010)

Solution. Equation of motion is

$$mv\frac{dv}{dx} = -mn^2x - 2\lambda mnv \qquad \text{...(1)}$$

$$\frac{dv}{dx} = -\frac{(n^2x + 2\lambda nv)}{v}$$

Let
$$v = Zx$$

$$\frac{dv}{dx} = Z + x\frac{dZ}{dx}$$

Putting the value of v, $\dfrac{dv}{dx}$ in (1), we get

$$Z + x\frac{dZ}{dx} = -\left(\frac{n^2x + 2\lambda nZx}{Zx}\right) = -\left(\frac{n^2 + 2\lambda nZ}{Z}\right)$$

$$x\frac{dZ}{dx} = \frac{-n^2 - 2\lambda nZ - Z^2}{Z}$$

$$\int \frac{Z\,dZ}{n^2 + 2\lambda nZ + Z^2} = -\int \frac{dx}{x}$$

$$\Rightarrow \quad \int \frac{Zd\,Z}{z^2 + 2\lambda n Z + n^2} = -\int \frac{dx}{x}$$

$$\Rightarrow \quad \frac{1}{2}\int \frac{2Z + 2n\lambda - 2n\lambda}{z^2 + 2\lambda nZ + n^2}\, dZ = -\int \frac{dx}{x}$$

$$\Rightarrow \quad \frac{1}{2}\int \frac{2Z + 2\lambda n}{Z^2 + 2\lambda nZ + n^2}\, dz - \int \frac{\lambda n}{Z^2 + 2\lambda nZ + n^2}\, dZ = -\int \frac{dx}{x}$$

$$\Rightarrow \quad \frac{1}{2}\log (Z^2 + 2\lambda nZ + n^2) - \int \frac{\lambda n}{Z^2 + 2\lambda nZ + \lambda^2 n^2 + n^2 - \lambda^2 n^2}\, dZ = -\log x + C$$

$$\Rightarrow \quad \frac{1}{2}\log (Z^2 + 2\lambda nZ + n^2) - \lambda n \int \frac{1}{(Z + \lambda n)^2 + n^2(1 - \lambda^2)}\, dZ = -\log x + C$$

$$\Rightarrow \quad \frac{1}{2}\log (Z^2 + 2\lambda nZ + n^2) - \lambda n \int \frac{1}{(Z + \lambda n)^2 + \left[n\sqrt{1 - \lambda^2}\right]^2}\, dZ = -\log x + C$$

$$\Rightarrow \quad \frac{1}{2}\log (Z^2 + 2\lambda nZ + n^2) + \log x - \frac{\lambda n}{n\sqrt{1 - \lambda^2}} \tan^{-1}\left(\frac{Z + \lambda n}{n\sqrt{1 - \lambda^2}}\right) = C$$

$$\Rightarrow \quad \log x\sqrt{Z^2 + 2\lambda nZ + n^2} - \frac{\lambda}{\sqrt{1 - \lambda^2}} \tan^{-1}\left(\frac{\frac{v}{x} + \lambda n}{n\sqrt{1 - \lambda^2}}\right) = C$$

$$\Rightarrow \quad \log x\sqrt{\frac{v^2}{x^2} + 2\lambda n\left(\frac{v}{x}\right) + n^2} - \frac{\lambda}{\sqrt{1 - \lambda^2}} \tan^{-1}\left(\frac{\frac{v}{x} + \lambda n}{n\sqrt{1 - \lambda^2}}\right) = C$$

$$\Rightarrow \quad \log \sqrt{v^2 + 2\lambda nvx + n^2x^2} - \frac{\lambda}{\sqrt{1 - \lambda^2}} \tan^{-1}\left(\frac{v + \lambda nx}{nx\sqrt{1 - \lambda^2}}\right) = C \qquad \text{....(2)}$$

On putting $v = 0$ and $x = 0$ in (2) we get

$$\log 0 - \frac{\lambda}{\sqrt{1 - \lambda^2}} \tan^{-1} 0 = C \quad \Rightarrow \quad C = 0$$

Putting $C = 0$ in (2), we get

$$\log \sqrt{v^2 + 2\lambda nvx + n^2x^2} - \frac{\lambda}{\sqrt{1 - \lambda^2}} \tan^{-1}\left(\frac{v + \lambda nx}{nx\sqrt{1 - \lambda^2}}\right) = 0 \qquad \textbf{Ans.}$$

17.5 VERTICAL MOTION

Example 13. *A body falling vertically under gravity encounters resistance of the atmosphere. If the resistance varies as the velocity, show that the equation of motion is given by*

$$\frac{du}{dt} = g - ku$$

where u is the velocity, k is a constant and g is the acceleration due to gravity. Show that as t increases, u approaches the value g/k. Also, if $u = \frac{dx}{dt}$ *where x is the distance fallen by the body from rest in time t, show that*

$$x = \frac{gt}{k} - \frac{g}{k^2}(1 - e^{-kt})$$

Solution. Let the mass of the falling body be unity.

$$\text{Acceleration} = \frac{du}{dt}$$

$$\text{Force acting downward} = 1.\frac{du}{dt} = \frac{du}{dt}$$

$$\text{Force of resistance} = ku$$

$$\text{Net force acting downward} = g - ku \quad \Rightarrow \quad \frac{du}{dt} = g - ku \qquad \text{...(1)} \quad \textbf{Proved.}$$

$$\Rightarrow \qquad \frac{du}{g - ku} = dt$$

Integrating, we get $\int \frac{du}{g - ku} = \int dt$

$$\Rightarrow \qquad t = -\frac{1}{k} \log (g - ku) + \log A = \log (g - ku)^{-1/k} A$$

$$A(g - ku)^{-1/k} = e^t \quad \Rightarrow \quad (g - ku) = A^k e^{-kt}$$

$$\Rightarrow \qquad u = \frac{g}{k} - \frac{A^k}{k} e^{-kt}$$

If t increases very large then $\frac{A^k}{k} e^{-kt} = 0$

$$\Rightarrow \qquad u = \frac{g}{k} \qquad \text{when } t \to \infty$$

Given $\quad u = \frac{dx}{dt} \quad \Rightarrow \quad \frac{du}{dt} = \frac{d^2x}{dt^2}$ **Proved.**

Putting the values of $\frac{du}{dt}$ and u in (1), we get

$$\frac{d^2x}{dt^2} + k\frac{dx}{dt} = g \quad \Rightarrow \quad (D^2 + kD)\, x = g$$

$$\text{A.E. is} \quad m(m + k) = 0 \quad \Rightarrow \quad m = 0,\ m = -k$$

$$\text{C.F.} = A_1 + A_2 e^{-kt}$$

$$\text{P.I.} = \frac{1}{D^2 + kD} g = t\frac{1}{2D + k} g$$

$$= \frac{t}{k} \frac{1}{\left(1 + \frac{2D}{k}\right)} g = \frac{t}{k}\left(1 + \frac{2D}{k}\right)^{-1} g = \frac{t}{k}\left(1 - \frac{2D}{k}\right) g = \frac{t\,g}{k}$$

$$x = A_1 + A_2 e^{-kt} + \frac{g\,t}{k} \qquad \text{...(2)}$$

Putting the values of $t = 0$ and $x = 0$ in (2), we get

$$0 = A_1 + A_2 \quad \Rightarrow \quad A_2 = -A_1$$

(2) becomes $\quad x = A_1 - A_1 e^{-kt} + \frac{gt}{k} \qquad \text{...(3)}$

On differentiating (3), we get $\frac{dx}{dt} = A_1 ke^{-kt} + \frac{g}{k} \qquad \text{...(4)}$

On putting $\frac{dx}{dt} = 0$, when $t = 0$ in (4), we get $0 = A_1 k + \frac{g}{k} \Rightarrow A_1 = -\frac{g}{k^2}$

Putting the value of A_1 in (3), we get

$$x = -\frac{g}{k^2} + \frac{g}{k^2} e^{-kt} + \frac{gt}{k} \quad \Rightarrow \quad x = \frac{gt}{k} - \frac{g}{k^2}(1 - e^{-kt}) \qquad \textbf{Proved.}$$

Example 14. *A particle falls under gravity in a resisting medium whose resistance varies with velocity. Find the relation between distance and velocity if initially the particle starts from rest.* (*U.P., II Semester, Summer 2003*)

Solution. By Newton's second law of motion, the equation of motion of the body is

$$m \frac{V\,dV}{dx} = m\,g - m\,k\,V$$

$$\Rightarrow \qquad \frac{V\,dV}{dx} = g - k\,V$$

$$\Rightarrow \qquad \frac{V\,dV}{g - k\,V} = dx \quad \Rightarrow \quad \frac{-dV}{k} + \frac{g}{k}\frac{dV}{g - kV} = d\,x$$

Integrating, we get

$$-\frac{V}{k} + \frac{g}{k}\left(-\frac{1}{k}\right) \log\,(g + k\,V) = x + A$$

$$-\frac{V}{k} - \frac{g}{k^2} \log\,(g - kV) = x + A \qquad ...(1)$$

Initially, $x = 0$, $V = 0$

$$-\frac{g}{k^2} \log g = A$$

(1) becomes

$$-\frac{V}{k} - \frac{g}{k^2} \log\,(g - k\,V) = x - \frac{g}{k^2} \log g$$

$$\Rightarrow \qquad -\frac{V}{k} - \frac{g}{k^2} \log \frac{g - k\,V}{g} = x \qquad \textbf{Ans.}$$

Example 15. *The acceleration and velocity of a body falling in the air approximately satisfy the equation :*
Acceleration = $g - kv^2$, where v is the velocity of the body at any time t, and g, k are constants. Find the distance traversed as a function of the time, if the body falls from rest.

Show that value of v will never exceed $\sqrt{\frac{g}{k}}$.

Solution Acceleration $= g - k\,v^2 \quad \Rightarrow \quad \frac{dv}{dt} = g - k\,v^2 \quad \Rightarrow \quad \frac{dv}{g - k\,v^2} = dt.$

$$\Rightarrow \qquad \frac{1}{2\sqrt{g}}\left[\frac{1}{\sqrt{g} + \sqrt{k}.\,v} + \frac{1}{\sqrt{g} - \sqrt{k}.v}\right] dv = dt$$

On integrating, we get

$$\frac{1}{2\sqrt{g}}\,\frac{1}{\sqrt{k}} \log\,(\sqrt{g} + \sqrt{k}.\,v) - \frac{1}{2\sqrt{gk}} \log\,(\sqrt{g} - \sqrt{k}.\,v) = t + A$$

$$\Rightarrow \qquad \frac{1}{2\sqrt{gk}} \log \frac{\sqrt{g} + \sqrt{k}.\,v}{\sqrt{g} - \sqrt{k}.\,v} = t + A \qquad ...(1)$$

On putting $t = 0$, $v = 0$ in (1), we get $\frac{1}{2\sqrt{gk}} \log 1 = 0 + A \Rightarrow A = 0$

Equation (1) becomes $\frac{1}{2\sqrt{gk}} \log \frac{\sqrt{g}+\sqrt{k}.v}{\sqrt{g}-\sqrt{k}.v} = t \Rightarrow \log \frac{\sqrt{g}+\sqrt{k}.v}{\sqrt{g}-\sqrt{k}.v} = 2\sqrt{gk}\, t$

$$\Rightarrow \qquad \frac{\sqrt{g}+\sqrt{k}.v}{\sqrt{g}-\sqrt{k}.v} = e^{2\sqrt{gk}\,t}$$

By componendo and dividendo, we have

$$\frac{\sqrt{k}.v}{\sqrt{g}} = \frac{e^{2\sqrt{gk}t}-1}{e^{2\sqrt{gk}t}+1} = \frac{e^{\sqrt{gk}\,t}-e^{-\sqrt{gk}\,t}}{e^{\sqrt{gk}\,t}+e^{-\sqrt{gk}\,t}} = \tanh\sqrt{gk}\,t$$

$$\Rightarrow \qquad v = \sqrt{\frac{g}{k}}\tanh\sqrt{gk}\,t$$

Whatever the value of t may be $\tanh\sqrt{gk}\,t \le 1$.

Hence the value of v will never exceed $\sqrt{\frac{g}{k}}$. **Proved.**

$$\frac{dx}{dt} = \sqrt{\frac{g}{k}}\tanh\sqrt{gk}\,t$$

Integrating again, we get $\quad x = \sqrt{\frac{g}{k}}\int\tanh\sqrt{gk}\,t\,dt$

$$= \frac{1}{k}\log\cosh\sqrt{gk}\,t + B$$

when $\quad t = 0, \quad x = 0$ then $\quad B = 0$

$$\therefore \qquad x = \frac{1}{k}\log\cosh\sqrt{gk}\,t$$ **Ans.**

EXERCISE 17.3

1. A moving body is opposed by a force proportional to the displacement and by a resistance proportional to the square of velocity. Prove that the velocity is given by

   $V^2 = ae - \frac{cx}{b} + \frac{c}{ab^2}$ **Hint.** Equation of motion is $m\frac{VdV}{dx} = -K_1x - K_2V^2$

2. A particle of mass m is projected vertically upward with an initial velocity v_0. The resisting force at any time is K times the velocity. Formulate the differential equation of motion and show that the distance s covered by the particle at any time t is given by

   $s = \left(\frac{g}{K^2}+\frac{v_0}{K}\right)(1-e^{-Kt}) - \frac{g}{K}t$

3. A particle falls in a vertical line under gravity (supposed constant) and the force of air resistance to its motion is proportional to its velocity. Show that its velocity cannot exceed a particular limit.

   Ans. $V = \frac{g}{K}$

4. A body falling from rest is subjected to a force of gravity and an air resistance of $\frac{n^2}{g}$ times the square of velocity. Show that the distance travelled by the body in t seconds in $\frac{g}{n^2}$ log cosh nt.

5. A body of mass m, falling from rest is subject to the force of gravity and an air resistance proportional to the square of the velocity Kv^2. If it falls through a distance x and possesses a velocity v, at the instant, prove that

$$\frac{2kx}{m} = \log\left(\frac{a^2}{a^2 - v^2}\right) \quad \text{where} \quad \frac{mg}{k} = a^2$$ *(A.M.I.E.T.E., June 2009)*

17.6 VERTICAL ELASTIC STRING

Let an elastic string OA of length l be attached to a fixed point O and a particle of mass m be attached at the other end A. When a mass m hangs freely then it pulls the string OA and it stays at B, in the position of equilibrium. The tension T in the string balances the mass mg hanging at A.

a
A
mg B
x
P
C

By Hooke's Law

$$\frac{\text{Stress } (T)}{\text{Strain}} = \text{Constant} = \text{Modulus of Elasticity } (E)$$

and $$\text{Strain} = \frac{\text{Extension in length}}{\text{Original length}} = \frac{a}{l}$$

$$T = \frac{Ea}{l} \quad \text{or} \quad mg = \frac{Ea}{l} \qquad ...(1)$$

The string is further pulled down to a point C and then released. Then the particle at the lower end C will make motion up and down between B and C. Let the particle be at P at any time t, where $BP = x$.

$$\text{The down ward force} = mg$$

$$\text{Upward force} = \text{Tension } T$$

$$T = E\left(\frac{a + x}{l}\right) \qquad ...(2) \quad (\text{Extension} = a + x)$$

$$\text{Resultant down force} = mg - T = mg - E\left(\frac{a+x}{l}\right)$$

$$= mg - \frac{Ea}{l} - \frac{Ex}{l} = mg - mg - \frac{Ex}{l} = \frac{-Ex}{l}$$

$$\text{Downward Acceleration} = -\frac{Ex}{ml}$$

Equation of motion is $$\frac{d^2x}{dt^2} = -\frac{Ex}{ml}, \text{ which is a S.H.M.}$$

Its time period $$T = \frac{2\pi}{\sqrt{\frac{E}{ml}}} = 2\pi\sqrt{\frac{ml}{E}}$$

17.7 HORIZONTAL ELASTIC STRING

If one end of the elastic string be fixed at O on a table. The other end A of the elastic string of length l is attached to a particle of mass m.

The string is stretched to a point B and then released. The particle comes into motion. Let the particle be at a distance x from A at any time t. The weight mg of the particle is acting downward and is balanced by the reaction R of the table.

The only force acting upon the particle is the tension of the string.

B′ A′ O A P B →X (l from O to A, x from A to P, a from A to B)

By Hook's Law $\dfrac{\text{stress}}{\text{strain}}$ = constant of elasticity

$$T = E\frac{x}{l} \qquad (x = \text{Extension of the length of the string})$$

$(E$ = Modulus of elasticity)

Equation of motion is $m\dfrac{d^2x}{dt^2} = -\dfrac{Ex}{l}$

$$\Rightarrow \qquad \frac{d^2x}{dt^2} = -\left(\frac{E}{ml}\right)x \qquad ...(1)$$

The motion of the equation is S.H.M.

On multiplying (1) by $\dfrac{2dx}{dt}$, we get

$$\frac{2d^2x}{dt^2}\frac{dx}{dt} = \frac{-2\,Ex}{ml}\frac{dx}{dt}$$

On integrating, we get $\left(\dfrac{dx}{dt}\right)^2 = \dfrac{-Ex^2}{ml} + A \Rightarrow v^2 = \dfrac{-Ex^2}{ml} + A$ ....(2)

If the velocity of the particle is zero at B, amplitude $AB = a$. The particle moves from A to B and back B to A will be a S.H.M. The particle moves towards O. Then it moves with uniform velocity upto A'.

On putting $v = 0$, $x = a$, we get $0 = \dfrac{-Ea^2}{ml} + A \Rightarrow A = \dfrac{Ea^2}{ml}$

$$v^2 = \frac{E}{ml}(a^2 - x^2)$$

At A, $x = 0$, $v = \sqrt{\left(\dfrac{E}{\lambda_1}\right)}\,a$. This is the maximum velocity. The particle moves from A to A' with this velocity. After that the string again stretches and motion becomes S.H.M.

Periodic Time of S.H.M. (from A to B, B to A, A' to B', B' to A') + time taken by the particle from A to A' and A' to A with constant velocity $\sqrt{\left(\dfrac{E}{l}\right)}\,a$.

$$= \frac{2\pi}{\sqrt{\left(\dfrac{E}{lm}\right)}} + \frac{4l}{\sqrt{\left(\dfrac{E}{lm}\right)}\,a} = 2\sqrt{\frac{lm}{E}}\left(\pi + \frac{2l}{a}\right)$$

Example 16. *A light elastic string of original length l is hung by one end to the other end are tied successively with particles of masses m, m′. If t_1 and t_2 be the periods of small oscillations corresponding to these weights and c_1, c_2 are the statical extensions, prove that*

$$g(t_1^2 - t_2^2) = 4\pi^2(c_1 - c_2)$$

Solution. $\qquad mg = T_1 \qquad ...(1)$

$$mg = E\frac{c_1}{l} \qquad ...(2)$$

$$m'g = E\frac{c_2}{l}E\frac{c_2}{l} \quad ...(3)$$

Equation of motion of first particle, $m\frac{d^2x}{dt^2} = mg - E\frac{(x+c_1)}{l}$

$$\Rightarrow \quad m\frac{d^2x}{dt^2} = mg - E\frac{c_1}{l} - \frac{Ex}{l} = -E\frac{x}{l} \quad \left[\text{From (2)}, mg = E\frac{c_1}{l}\right]$$

Motion of S.H.M. with $t_1 = \frac{2\pi}{\sqrt{\frac{E}{lm}}}$ similarly, $t_2 = \frac{2\pi}{\sqrt{\frac{E}{lm'}}}$

$$\therefore \quad t_1^2 - t_2^2 = 4\pi^2\frac{l}{E}(m - m') = 4\pi^2\left(\frac{c_1}{g} - \frac{c_2}{g}\right) \quad \text{[Using (2) and (3)]}$$

$$\Rightarrow \quad g(t_1^2 - t_2^2) = 4\pi^2(c_1 - c_2)$$

Ans.

17.8 SIMPLE HARMONIC MOTION

If a particle moves in a straight line such that its acceleration is always directed towards a fixed point in the line and is proportional to the distance of the particle from the fixed point.

Let O be the fixed point in the line $A'A$. Let the position of the particle be P at time t. Since the acceleration is always directed towards O, *i.e.*, the acceleration is in the direction opposite to that in which x increases, then

Fixed point ← $\mu^2 x$ = Force; A' O P A; x; a

$$\frac{d^2x}{dt^2} = -\mu^2 x \quad ...(1) \quad (\mu \text{ is constant, } OP = x)$$

$$D^2x = -\mu^2 x \Rightarrow (D^2 + \mu^2)x = 0$$

A.E. is $m^2 + \mu^2 = 0, \quad m = \pm i\mu$

$$x = A\cos\mu t + B\sin\mu t \quad ...(2)$$

The solution of (1) is

Velocity of particle at $P = \frac{dx}{dt} = -A\mu\sin\mu t + B\mu\cos\mu t \quad ...(3)$

At A, velocity $= \frac{dx}{dt} = 0, x = a, t = 0$

Putting $x = a,\ t = 0$ in (2), we get $a = A$

Putting $\frac{dx}{dt} = 0,\ t = 0$ in (3), we get $0 = 0 + B\mu \Rightarrow B = 0$

Equation (2) becomes $x = a\cos\mu t$

Equation (3) becomes $\frac{dx}{dt} = -a\mu\sin\mu t$

$$\text{velocity} = -a\mu\sqrt{1-\cos^2\mu t} = -a\mu\sqrt{1-\left(\frac{x}{a}\right)^2} \quad (x = a\cos\mu t)$$

$$= -\mu\sqrt{a^2 - x^2} \quad ...(4)$$

At A, $x = a$ and $v = 0$

At O, $x = 0$, and acceleration = 0, velocity is maximum.

O is called the centre of motion or the mean position.

Amplitude. In S.H.M. the distance from the centre to the position of maximum displacement is called the amplitude of the motion. OA is the maximum distance and is called the amplitude.

From (4), $$-\frac{dx}{\sqrt{a^2 - x^2}} = \mu\, dt$$

Integrating, we get $$\cos^{-1}\frac{x}{a} = \mu\, t + A \qquad \text{...(5)}$$

Putting $t = 0$, $x = a$ in (5), we get

$$0 = 0 + A \quad \Rightarrow \quad A = 0$$

On putting the value of A, (5) becomes, $\cos^{-1}\frac{x}{a} = \mu\, t \quad \Rightarrow \quad x = a \cos \mu\, t$

Particle will reach O in time t_1,

$$0 = a \cos \mu\, t_1 \quad \Rightarrow \quad 0 = \cos \mu\, t_1$$

$$\Rightarrow \quad \cos\frac{\pi}{2} = \cos \mu\, t_1, \quad \Rightarrow \quad \frac{\pi}{2} = \mu\, t_1 \quad \Rightarrow \quad t_1 = \frac{\pi}{2\mu}$$

Time period: In a S.H.M., the time taken to make a complete oscillation is called time period.

Frequency: The number of complete oscillations per second is called the frequency of motion. If η is the frequency and T is the time period.

$$\text{Time period} = T = 4\left(\frac{\pi}{2\mu}\right) = \frac{2\pi}{\mu}$$

$$\text{Frequency} = n = \frac{1}{T} = \frac{\mu}{2\pi}$$

(*i*) The equation of S.H.M. is $\frac{d^2x}{dt^2} = -\,\omega^2 x$

(*ii*) The velocity v at a distance x from the centre at time t is

$$v^2 = \omega^2\,(a^2 - x^2)$$

$x = a \cos \omega t$, where a is the amplitude and ω is the angular velocity.

(*iii*) Maximum acceleration = $\omega^2\, a$ (At the extreme point)

(*iv*) Maximum velocity = $\omega\, a$ (at the centre)

(*v*) Time period = $T = \frac{2\pi}{\omega}$

Example 17. *A particle moves with S.H.M. of period 12 secs, travels 8 cm from the position of rest in 2 secs. Find the amplitude, the maximum velocity and the velocity at the end of 2 secs.*

Solution. $$T = \frac{2\pi}{\sqrt{\mu}} = 12 \Rightarrow \sqrt{\mu} = \frac{\pi}{6}$$

Let a be the amplitude OA

$$AP = 8 \text{ cm}, \; OP = x = a - 8, \; t = 2 \text{ secs.}$$

We know that $$x = a \cos \sqrt{\mu}\, t$$

$$a - 8 = a \cos 2\sqrt{\mu} \qquad \left(\sqrt{\mu} = \frac{\pi}{6}\right)$$

$$= a\cos\frac{\pi}{3} = \frac{a}{2} \quad \Rightarrow \quad a = 16$$

Maximum velocity $= \sqrt{\mu}\, a = \frac{\pi}{6} \times 16 = \frac{8\pi}{3} = 4.619$ cm/sec.

Velocity v at the end of two seconds $= \sqrt{\mu}\,\sqrt{a^2 - x^2} = \frac{\pi}{6}\sqrt{256 - 64}$

$$= \frac{\pi}{6}\sqrt{192} = \frac{4\pi\sqrt{3}}{3} \text{ cm/sec.}$$ **Ans.**

Example 18. *A particle is performing a simple harmonic motion of period T about a centre O and it passes through a point P where OP = b with velocity v in the direction OP. Prove that the time which elapses before it returns to P is*

$$\frac{T}{\pi}\tan^{-1}\frac{vT}{2\pi b}.$$

A' —— O —— P —— A

Solution. We have to find time taken from P to A and then A to P

$$t = 2 \text{ (time from } A \text{ to } P)$$

$$= 2\int_0^t dt = 2\int_a^p \frac{dx}{\sqrt{\mu}\,\sqrt{a^2 - x^2}} \text{ (Ignoring } -ve \text{ sign)} \left(\frac{dx}{dt} = -\sqrt{\mu}\,\sqrt{a^2 - x^2}\right)$$

$$= \frac{2}{\sqrt{\mu}}\left[\cos^{-1}\frac{x}{a}\right]_a^b = \frac{2}{\sqrt{\mu}}\left[\cos^{-1}\frac{b}{a} - \cos^{-1}\frac{a}{a}\right] = \frac{2}{\sqrt{\mu}}\cos^{-1}\frac{b}{a}$$

$$\Rightarrow \quad t = \frac{2}{\sqrt{\mu}}\tan^{-1}\left(\frac{\sqrt{a^2 - b^2}}{b}\right)$$

(Right triangle: hypotenuse a, base b, perpendicular $\sqrt{a^2 - b^2}$, angle θ)

$$= \frac{2}{\sqrt{\mu}}\tan^{-1}\left(\frac{v}{b\sqrt{\mu}}\right)$$

$$\left[\begin{aligned} v^2 &= \mu(a^2 - b^2) \\ \Rightarrow \quad v &= \sqrt{\mu}\,\sqrt{(a^2 - b^2)} \\ \Rightarrow \quad \frac{v}{\sqrt{\mu}} &= \sqrt{a^2 - b^2} \end{aligned}\right]$$

$$= \frac{2}{\frac{2\pi}{T}}\tan^{-1}\left[\frac{v}{b\left(\frac{2\pi}{T}\right)}\right] \qquad \left[T = \frac{2\pi}{\sqrt{\mu}} \Rightarrow \sqrt{\mu} = \frac{2\pi}{T}\right]$$

$$= \frac{T}{\pi}\tan^{-1}\left[\frac{vT}{2\pi b}\right]$$ **Proved.**

Example 19. *At the end of three successive seconds, the distances of a point moving with S.H.M. from its mean position are x_1, x_2, x_3. Show that the time of complete oscillation is*

$$\frac{2\pi}{\cos^{-1}\left(\frac{x_1 + x_3}{2x_2}\right)}$$

Solution. Let x be the distance from the centre at any time t,

$$x = a\sin\sqrt{\mu}\, t$$

$$x_1 = a\sin\sqrt{\mu}\, t,\ x_2 = a\sin\sqrt{\mu}\,(t+1),\ x_3 = a\sin\sqrt{\mu}\,(t+2)$$

$$\Rightarrow \quad x_1 + x_3 = a[\sin\sqrt{\mu}\, t + \sin\sqrt{\mu}\,(t+2)]$$

$$= 2a \sin \sqrt{\mu}\,(t+1) \cos \sqrt{\mu} = 2x_2 . \cos \sqrt{\mu}$$

$$\Rightarrow \qquad \frac{x_1 + x_3}{2x_2} = \cos \sqrt{\mu} \quad \Rightarrow \quad \sqrt{\mu} = \cos^{-1} . \frac{x_1 + x_3}{2x_2}$$

$$\text{Time period} = \frac{2\pi}{\sqrt{\mu}} = \frac{2\pi}{\cos^{-1} \dfrac{x_1 + x_3}{2x_2}}$$ **Proved.**

Example 20. *A point moves in a straight line towards a centre of force* μ */(distance)*^{3}, *starting from rest at a distance 'a' from the centre of force; show that the time of reaching a point distant b from the centre of force is* $(a/\sqrt{\mu})\sqrt{(a^2 - b^2)}$, *and that its velocity is* $\dfrac{\sqrt{\mu}}{ab}\sqrt{(a^2 - b^2)}$.

Solution. Let a point move from A towards the centre of force O.

$$\therefore \qquad \frac{m\mu}{x^3} = -m\frac{d^2x}{dt^2} \Rightarrow \frac{d^2x}{dt^2} = -\frac{\mu}{x^3}$$

A → B, O, ← b →, x = a, a, x = 0, $\frac{dx}{dt} = 0$, $\frac{dx}{dt} = 0$

$$\Rightarrow \qquad 2\frac{dx}{dt} . \frac{d^2x}{dt^2} = -2\frac{\mu}{x^3} . \frac{dx}{dt}$$

Integrating, we get

$$\left(\frac{dx}{dt}\right)^2 = -2 . \mu \frac{x^{-2}}{-2} + C$$

$$\Rightarrow \qquad V^2 = \frac{\mu}{x^2} + C \qquad \text{...(1)}$$

At A, $V = 0$ and $x = a$, $\quad \therefore \; 0 = \dfrac{\mu}{a^2} + C \quad \Rightarrow \quad C = -\dfrac{\mu}{a^2}$

On putting the value of C, (1) becomes

$$V^2 = \frac{\mu}{x^2} - \frac{\mu}{a^2} = \mu\left(\frac{a^2 - x^2}{x^2 a^2}\right) \qquad \text{...(2)}$$

Therefore velocity when $x = b$ is given by

$$V^2 = \mu\left(\frac{a^2 - b^2}{a^2 b^2}\right) \quad \Rightarrow \quad V = \pm\sqrt{\mu}\,\frac{\sqrt{a^2 - b^2}}{ab}$$

$$\Rightarrow \qquad V = -\sqrt{\mu}\,\frac{\sqrt{a^2 - b^2}}{ab}$$

From (2), $\quad \left(\dfrac{dx}{dt}\right)^2 = \mu\dfrac{(a^2 - x^2)}{x^2 a^2} \quad \Rightarrow \quad \dfrac{dx}{dt} = -\sqrt{\mu}\,\dfrac{\sqrt{a^2 - x^2}}{xa}$

$$dt = -\frac{1}{\sqrt{\mu}}\frac{xa}{\sqrt{a^2 - x^2}}\,dx$$

Integrating, we get

$$t = -\frac{1}{\sqrt{\mu}}\int \frac{xa\,dx}{\sqrt{a^2 - x^2}}$$

Let $\quad a^2 - x^2 = z^2 \Rightarrow -2x\,dx = 2\,z\,dz$

$$t = \frac{a}{\sqrt{\mu}} \int \frac{z\,dz}{z} = \frac{a}{\sqrt{\mu}} \int dz = \frac{a}{\sqrt{\mu}} z + C_1 = \frac{a}{\sqrt{\mu}} \sqrt{a^2 - x^2} + C_1 \quad ...(3)$$

At A, $t = 0$, $x = a$,

On putting $t = 0$, $x = a$ in (3), we get $0 = 0 + C_1 \Rightarrow C_1 = 0$

On putting the value of C_1 in (3), we have

$$t = \frac{a}{\sqrt{\mu}} \sqrt{a^2 - x^2}$$

At B, $x = b$, $t = \frac{a}{\sqrt{\mu}} \sqrt{a^2 - b^2}$ **Proved.**

Example 21. *The radial displacement 'u' in a rotating disc at a distance r from the axis is given by*

$$\frac{d^2u}{dr^2} + \frac{1}{r}\frac{du}{dr} - \frac{u}{r^2} + kr = 0$$

Find the displacement if $u = 0$ at $r = 0$ and at $r = a$. *(M.U. II Semester, 2008)*

Solution. Here, we have

$$\frac{d^2u}{dr^2} + \frac{1}{r}\frac{du}{dr} - \frac{u}{r^2} + kr = 0$$

$$\Rightarrow \quad r^2 \frac{d^2u}{dr^2} + r\frac{du}{dr} - u = -k\,r^3 \quad ...(1)$$

This is the homogeneous equation.

On putting $z = \log r \Rightarrow r = e^z$ and $r^2 \frac{d^2u}{dr^2} = D(D-1)u$, $r\frac{du}{dr} = Du$ in (1), we get

$$D(D-1)u + Du - u = -k\,e^{3z}$$

$$\Rightarrow \quad (D^2 - D + D - 1)u = k\,e^{3z}$$

$$\Rightarrow \quad (D^2 - 1)u = -k\,e^{3z}$$

A.E. is $m^2 - 1 = 0 \Rightarrow m = 1, m = -1$

$$\text{C.F.} = A\,e^z + B\,e^{-z}$$

$$\text{P.I.} = \frac{1}{D^2 - 1}(-k\,e^{3z}) = -k\frac{1}{(3)^2 - 1}e^{3z} = -\frac{k}{8}e^{3z}$$

Complete solution is $u = \text{C.F.} + \text{P.I.}$

$$\Rightarrow \quad u = Ae^z + Be^{-z} - \frac{k}{8}e^{3z}$$

$$\Rightarrow \quad u = Ar + \frac{B}{r} - \frac{k}{8}r^3$$ **Ans.**

EXERCISE 17.4

1. At what distance from the centre the velocity in a S.H.M. will be one fourth of the maximum?

Ans. $x = \pm \frac{\sqrt{15}\,a}{4}$

2. A particle moving in a straight line with S.H.M. has velocities 3m/sec. and 2m/sec. respectively when it is at distances 1 metre and 1.3 metre from the centre of its path. Find its period and acceleration at the greatest distance from the centre of motion. **Ans.** Period = 2.33 sec; Acceleration = 4.0293 m/sec^2

3. A particle moves with S.H.M. If, when at a distance of 3 and 4 cm from the centre of the path, its velocities are 8 and 6 cm per sec. respectively. Find its period, maximum velocity and acceleration when at its greatest distance from the centre. **Ans.** π secs, 10 cm/sec, 20 cm/sec^2.
4. A point executes S.H.M. such that in two of its positions, the velocities are u, v and the corresponding accelerations α, β. Show that the distance between the positions is $\frac{v^2-u^2}{\alpha+\beta}$, and find the amplitude of the motion.
5. A particle of mass m is oscillating in a straight line about a centre of force, O towards which when at a distance r the force is $m\,n^2\,r$ and a is the amplitude of oscillation, when at a distance $\frac{a\sqrt{3}}{2}$ from O, the particle receives a blow in the direction of motion which generates a velocity na. If this velocity be away from O, show that the new amplitude is $a\sqrt{3}$.
6. The speed v of the point P which moves in a line is given by the relation $v^2 = a + 2bx - cx^2$ where x is the distance of the point P from a fixed point on the path, and a, b, c are constants the motion is simple harmonic if c is positive; determine the period and the amplitude of the motion.

   Ans. $T=\frac{2\pi}{\sqrt{c}}$, Amplitude $=\frac{\sqrt{b^2+ab}}{c}$
7. In the case of a stretched elastic string which has one end fixed and a particle of mass m attached to the other end, the equation of motion is

   $$\frac{d^2s}{dt^2}=-\frac{mg}{e}(s-1)$$

   where l is the natural length of the string and e its elongation due to a weight mg. Find s and v determining the constants, so that $s = s_0$ at the time $t = 0$ and $v = 0$ when $t = 0$.

   Ans. $v=-\sqrt{\left(\frac{g}{e}\right)}\,[(s_0-l)^2-(s-l)^2]^{1/2},\ s-l=(s_0-l)\cos\left[\sqrt{\left(\frac{g}{e}\right)}\,.t\right]$

17.9 THE SIMPLE PENDULUM

A particle of mass m suspended vertically by a light inextensible string oscillating under gravity constitutes a simple pendulum.

Let l be the length of the string, O be the fixed point. A be the initial position of the bob. In the displaced position P at any time t, then forces acting on the bob are

(*i*) weight mg acting vertically downward

(*ii*) the tension T in the string.

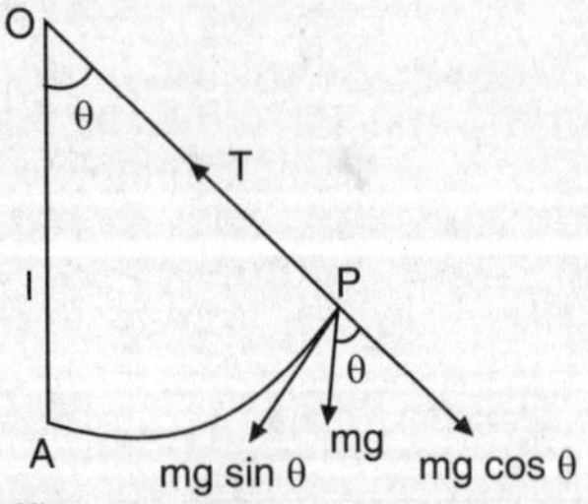

O θ T l P θ A mg mg sin θ mg cos θ

$$mg\cos\theta = T$$

$$\text{Restoring force} = m\frac{d^2x}{dt^2} = -\,mg\sin\theta \quad (x = AP)$$

$$\Rightarrow \quad \frac{d^2x}{dt^2} = -\,g\sin\theta$$

$$= -\,g\,\theta \quad (\sin\theta = \theta \text{ if } \theta \text{ is small})$$

$$= -\,g\frac{x}{l}$$

$$\Rightarrow \quad D^2x+\frac{g}{l}x = 0 \;\Rightarrow\; \left(D^2+\frac{g}{l}\right)x=0$$

A.E. is $m^2 + \frac{g}{l} = 0 \;\Rightarrow\; m^2 = \frac{-g}{l},\ m=\pm i\sqrt{\frac{g}{l}}$

$$\therefore \qquad x = c_1 \cos\sqrt{\frac{g}{l}}\,t + c_2 \sin\sqrt{\frac{g}{l}}\,t$$

Thus, the motion of the bob is simple harmonic and

$$\text{Period of Oscillation} = \frac{2\pi}{\sqrt{\dfrac{g}{l}}} = 2\pi\sqrt{\frac{l}{g}}$$

Note. The motion of the bob from one end to the other end is half oscillation and is called a **beat** or **swing.**

In a second pendulum, the time of one beat is one second. The number of beats in a day

$$= 24 \times 60 \times 60 = 86400.$$

17.10 OSCILLATIONS OF A SPRING

(*a*) **Free oscillations.** Let a spring be fixed at *O* and a mass *m* is suspended from the lower end A.

Let *S* (= *AB*) be the elongation produced by the mass *m* hanging. *B* is called the position of static equilibrium and *S* is called the static extension.

We choose the downward direction as the positive direction and regard forces which act down as positive and upward forces as negative.

If *k* be the restoring force per unit stretch of the spring due to elasticity.

Equilibrium at *B* $mg = ks$

Let the mass *m* be displaced through a further distance *x* (= *BP*) from the equilibrium position *B*.

Weight *mg* is acting downward

Restoring force $k(s + x)$ is acting upward.

Equation of motion is $m\dfrac{d^2x}{dt^2} = mg - K(s + x)$

$$= mg - ks - kx \qquad (mg = ks)$$

$$= mg - mg - kx = -kx$$

$$\Rightarrow \qquad \frac{d^2x}{dt^2} = -\frac{k}{m}x \qquad \Rightarrow \qquad \frac{d^2x}{dt^2} + \frac{k}{m}x = 0$$

$$\Rightarrow \qquad \frac{d^2x}{dt^2} + \omega^2 x = 0 \qquad \left(\omega^2 = \frac{k}{m}\right)$$

O A S B x P

Example 22. *A spring of negligible weight hangs vertically. A mass m is attached to the other end. If the mass is moving with velocity u when the spring is unstretched, find the velocity v as a function of the stretch x.*

Solution. If *x* is the increase in length of the spring when velocity of the mass *m* is *v*, then the equation of motion is

$$m.v\frac{dv}{dx} = mg - T$$

where $T = kx$, by Hooke's law, *k* being Young's modulus

$$\therefore \qquad mv\frac{dv}{dx} = mg - kx \qquad \Rightarrow \qquad mv\,dv = (mg - kx)\,dx$$

Integrating, we get $$m\frac{v^2}{2} = mgx - k\frac{x^2}{2} + c$$

Now, $v = u$ when $x = 0 \Rightarrow c = m\frac{u^2}{2}$

$$\therefore \quad m\frac{v^2}{2} = mgx - k\frac{x^2}{2} + m\frac{u^2}{2} \Rightarrow mv^2 = 2\,mgx - kx^2 + mu^2$$

$$\Rightarrow \quad v^2 = 2gx + u^2 - \frac{k}{m}x^2$$ **Ans.**

Example 23. *A spring for which stiffness k = 700 Newtons/m hangs in a vertical position with its upper end fixed. A mass of 7 kg is attached to the lower end. After coming to rest, the mass is pulled down 0.05 m and released. Discuss the resulting motion of the mass, neglecting air resistance.*

Solution. The equation of motion is $m\frac{d^2x}{dt^2} = -kx$

$$\Rightarrow \quad 7\frac{d^2x}{dt^2} = -700\,x \Rightarrow \frac{d^2x}{dt^2} + 100x = 0 \Rightarrow D^2x + 100\,x = 0$$

A.E. is $m^2 + 100 = 0 \Rightarrow m = \pm\, i\,10$

Its solution is $x = C_1 \cos 10\,t + C_2 \sin 10t$ ...(1)

Putting $t = 0$, $x = 0.05$ in (1), we get $0.05 = C_1$

Differentiating (1), we have $\frac{dx}{dt} = -10\,C_1 \sin 10\,t + 10\,C_2 \cos 10t$ ...(2)

On putting $v = \frac{dx}{dt} = 0$ and $t = 0$ in (2), $0 = 10\,C_2 \Rightarrow C_2 = 0$

On substituting the values of C_1 and C_2 in (1), we obtain

$$x = 0.05 \cos 10\,t$$

This is S.H.M. $\quad$ Period $= \frac{2\pi}{\omega} = \frac{2\pi}{10} = 0.628$ secs.

Frequency $= \frac{10}{2\pi} = 1.59$ cycle/sec

Amplitude $= 0.50\,m$ **Ans.**

Example 24. *A mass M suspended from the end of a helical spring is subjected to a periodic force f = F sin ω t in the direction of its length. The force f is measured positive vertically downwards and at zero time M is at rest. If the spring stiffness is S, prove that the displacement of M at time t from the commencement of motion is given by*

$$x = \frac{F}{M\,(P^2 - \omega^2)}\left[\sin \omega t - \frac{\omega}{P}\sin pt\right] \quad \textit{where } P^2 = \frac{S}{M}$$

and damping effects are neglected.

Solution. Let x be the displacement from the equilibrium position, the equation of motion is

$$M\frac{d^2x}{dt^2} = -S\,x + F \sin \omega\,t$$

$$\frac{d^2x}{dt^2} + \frac{S}{M}x = \frac{F}{M}\sin \omega\,t \Rightarrow \frac{d^2x}{dt^2} + P^2x = \frac{F}{M}\sin \omega\,t \qquad \left(\because \frac{S}{M} = P^2\right)$$

$$\Rightarrow \qquad (D^2 + P^2) = \frac{F}{M} \sin wt$$

A.E. is $m^2 + P^2 = 0$, $m = \pm iP$

$$\therefore \qquad \text{C.F.} = (c_1 \cos pt + c_2 \sin pt)$$

$$\text{P.I.} = \frac{1}{D^2 + P^2} \frac{F}{M} \sin \omega t = \frac{F}{M} \frac{1}{-\omega^2 + P^2} \sin \omega t$$

$$\therefore \qquad x = c_1 \cos pt + c_2 \sin pt + \frac{F}{M} \frac{1}{P^2 - \omega^2} \sin \omega t \qquad ...(1)$$

Putting $t = 0$ and $x = 0$ in (1), we get, $0 = c_1$

Equation (1) becomes $\qquad x = c_2 \sin pt + \frac{F}{M} \frac{1}{P^2 - \omega^2} \sin \omega t \qquad ...(2)$

Differentiating (2), we obtain $\frac{dx}{dt} = c_2\, p \cos pt + \frac{F}{M} \frac{\omega}{P^2 - \omega^2} \cos \omega t \qquad ...(3)$

Putting $\frac{dx}{dt} = 0$ and $t = 0$ in (3), we have

$$0 = c_2\, P + \frac{F}{M} \frac{\omega}{P^2 - \omega^2} \Rightarrow c_2 = -\frac{F\,\omega}{PM\,(P^2 - \omega^2)}$$

On substituting the value of c_2 in (2), we get

$$x = -\frac{\omega}{P} \frac{F}{M\,(P^2 - \omega^2)} \sin pt + \frac{F}{M} \frac{1}{P^2 - \omega^2} \sin \omega t$$

$$= \frac{F}{M(P^2 - \omega^2)} \left[\sin \omega t - \frac{\omega}{P} \sin pt \right] \qquad \textbf{Proved.}$$

***(b)* Damped Free Oscillations:** If the motion of the mass m be opposed by some resistance, proportional to the velocity $\left(v = K_1 \frac{dx}{dt} \right)$, the oscillations are said to be damped.

The equation of motion is $m \frac{d^2x}{dt^2} = mg - k(x + s) - k_1 \frac{dx}{dt} \qquad [mg = ks]$

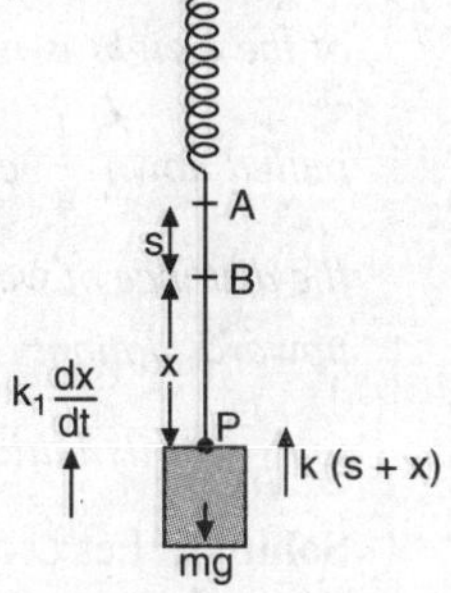

$$\Rightarrow m \frac{d^2x}{dt^2} = mg - mg - kx - k_1 \frac{dx}{dt} \Rightarrow m \frac{d^2x}{dt^2} = -kx - k_1 \frac{dx}{dt}$$

$$\Rightarrow \qquad \frac{d^2x}{dt^2} + \frac{k_1}{m} \frac{dx}{dt} + \frac{k}{m} x = 0$$

On putting $\frac{k_1}{m} = 2\lambda, \frac{k}{m} = \mu^2$, we get

$$\Rightarrow \qquad \frac{d^2x}{dt^2} + 2\lambda \frac{dx}{dt} + \mu^2 x = 0 \qquad \Rightarrow \qquad (D^2 + 2\lambda\, D + \mu^2)x = 0$$

A.E. is $\qquad m^2 + 2\,\lambda\, m + \mu^2 = 0 \qquad ...(1)$

$$\Rightarrow \qquad m = \frac{-2\,\lambda \pm \sqrt{4\lambda^2 - 4\,\mu^2}}{2} \qquad \Rightarrow \qquad m = -\lambda \pm \sqrt{\lambda^2 - \mu^2} \qquad ...(2)$$

Case I. If $\lambda > \mu$

Then m is real and distinct.

Hence, the solution of (1) is $x = c_1 e^{[-\lambda + \sqrt{\lambda^2 - \mu^2}]t} + c_2 e^{[-\lambda - \sqrt{\lambda^2 - \mu^2}]t}$ ...(3)

From (3), it is obvious that x is positive and decreases to zero as $t \Rightarrow \infty$.

The restoring force, in this case, so great that the motion is non-oscillatory and is known as **over damped** or **dead beat motion.**

Case II. If $\lambda = \mu$

From (2), $m = -\lambda \pm \sqrt{\mu^2 - \mu^2} \Rightarrow m = -\lambda$

The roots of (1) are real and equal.

$\therefore \quad x = (c_1 + c_2 t)e^{-\lambda t}$

Here also x is non-oscillatory and $x \Rightarrow 0$ as $t \Rightarrow \infty$. This is the critical damping. There is no oscillatory term in the solution hence motion becomes a periodic or non-oscillatory.

Case III. If $\lambda < \mu$

From (2),

$$m = -\lambda \pm \sqrt{\lambda^2 - \mu^2}$$

$$m = -\lambda \pm i\sqrt{\mu^2 - \lambda^2}$$

$$\therefore \quad x = e^{-\lambda t}[c_1 \cos \sqrt{\mu^2 - \lambda^2}\, t + c_2 \sin \sqrt{\mu^2 - \lambda^2}\, t]$$

$$x = e^{-\lambda t}\, r \sin (\sqrt{\mu^2 - \lambda^2}\, t + \alpha) \qquad [c_1 = r \sin \alpha,\ c_2 = r \cos \alpha]$$

Hence, the motion is oscillatory.

The periodic time of the oscillation $= T = \dfrac{2\pi}{\sqrt{\mu^2 - \lambda^2}}$ which is greater than the periodic time in case of free oscillations which is $\dfrac{2\pi}{\mu}$. Thus, the effect of damping is to increase the periodic time of oscillation.

Example 25. *A spring fixed at the upper end supports a weight of 980 gm at its lower end. The spring stretches $\frac{1}{2}$ cm under a load of 10 gm and the resistance (in gm ωt) to the motion of the weight is numerically equal to $\frac{1}{10}$ of the speed of the weight in cm/sec. The weight is pulled down $\frac{1}{4}$ cm below its equilibrium position and then released. Find the expression for the distance of weight from its equilibrium position at time t during its first upward motion. Also find the time t it takes the damping factor to drop to $\frac{1}{10}$ of its initial value.*

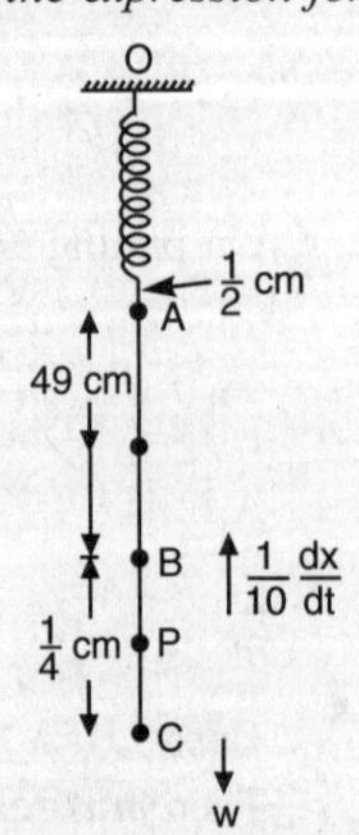

Solution. Let OA be a spring fixed at O and a load of 10 gm is attached at A. The spring is stretched by $\frac{1}{2}$ cm.

$$mg = T_0 \Rightarrow 10 = T_0 = k \cdot \frac{1}{2} \Rightarrow k = 20 \text{ gm/cm}$$

Let B be the equilibrium after attaching a weight 980 gm at A.

$$mg = kx \text{ where } x = AB$$

$$980 = 20x \quad \Rightarrow \quad x = AB = \frac{980}{20} = 49 \text{ cm}$$

After static equilibrium the weight is pulled down to C and released. $\left(BC = \frac{1}{4}\text{ cm}\right)$. After release, the weight be at P after time t.

$$BP = x$$

$$T = K \,.\, AP = 20\,(49 + x) = 980 + 20x$$

Equation of motion is $\quad m\dfrac{d^2x}{dt^2} = \omega - T - \dfrac{1}{10}\dfrac{dx}{dt} \quad \left(\text{Resistance} = \dfrac{1}{10}\dfrac{dx}{dt}\right)$

$$\frac{980}{g}\frac{d^2x}{dt^2} = 980 - (980 + 20x) - \frac{1}{10}\frac{dx}{dt} \qquad [g = 980 \text{ cm/sec}^2]$$

$$\Rightarrow \quad 10\frac{d^2x}{dt^2} + \frac{dx}{dt} + 200x = 0 \Rightarrow 10\,D^2x + Dx + 200\,x = 0$$

A.E. is $10\,m^2 + m + 200 = 0$

$$\Rightarrow \quad m = \frac{-1 \pm \sqrt{1 - 8000}}{20} = \frac{-1 \pm i\,89.4}{20}$$

$$\Rightarrow \quad m = -\,0.05 \pm i\,4.5$$

$$\therefore \quad \text{C.F.} \quad x = e^{-0.05\,t}\,[c_1 \cos 4.5\,t + c_2 \sin 4.5\,t] \qquad \ldots(1)$$

On putting $t = 0$ and $x = \dfrac{1}{4}$ in (1), we get $\dfrac{1}{4} = c_1$

On differentiating (1), we get

$$\frac{dx}{dt} = -\,0.05\,e^{-0.05\,t}[c_1 \cos 4.5\,t + c_2 \sin 4.5\,t] + e^{-0.05t}[-4.5\,c_1 \sin 4.5t + 4.5\,c_2 \cos 4.5\,t] \quad \ldots(2)$$

On putting $\dfrac{dx}{dt} = 0$ and $t = 0$ in (2), we have

$$0 = -\,0.05\,c_1 + 4.5\,c_2 \Rightarrow c_2 = \frac{0.05}{4.5}\,c_1 = \frac{0.05}{4.5}\left(\frac{1}{4}\right) = 0.0028$$

On substituting the values of c_1 and c_2 in (1), we obtain

$$x = e^{-0.05\,t}[0.25 \cos 4.5t + 0.0028 \sin 4.5\,t]$$

Damping factor $= b\,e^{-0.05t}$ $\qquad$ (b = constant of proportionality)

Initial value of damping factor $= b\,e^0 = b$

Let damping factor after time t be $\dfrac{b}{10}$.

$$\frac{b}{10} = b\,e^{-\,0.05t} \quad \Rightarrow \quad e^{\frac{t}{20}} = 10 \quad \Rightarrow \quad \frac{t}{20} = \log_e 10$$

$$\Rightarrow \quad t = 20\,\log_e 10 \Rightarrow t = 20\frac{\log_{10} 10}{\log_{10} e} = \frac{20}{\log_{10} e} = 20 \times 2.3 = 46 \text{ secs.} \qquad \textbf{Ans.}$$

(c) Forced Oscillations (without damping)

If an external force is applied on the point of support of the spring, it oscillates. The motion is called the forced oscillatory motion.

Let the external force be $q \cos nt$.

Equation of motion is $\quad m\dfrac{d^2x}{dt^2} = mg - ks - kx + q \cos nt \qquad (mg = ks)$

$$\Rightarrow \quad m\frac{d^2x}{dt^2} = -kx + q\cos nt \Rightarrow \frac{d^2x}{dt^2} = -\frac{k}{m}x + \frac{q}{m}\cos nt \qquad ...(1)$$

Let $\frac{k}{m} = \mu^2$ and $\frac{q}{m} = e,$ then (1) becomes

$$\frac{d^2x}{dt^2} = -\mu^2 x + e\cos nt$$

$$\Rightarrow \quad \frac{d^2x}{dt^2} + \mu^2 x = e\cos nt$$

$$\Rightarrow \quad (D^2 + \mu^2)x = e\cos nt \qquad ...(2)$$

A.E. is $m^2 + \mu^2 = 0 \;\Rightarrow\; m = \pm i\mu$

$$\text{C.F.} = c_1 \cos \mu t + c_2 \sin \mu t$$

$$\text{P.I.} = \frac{1}{D^2 + \mu^2} e\cos nt$$

Case (*a*). If $\mu \neq n$, $\quad \text{P.I.} = e\frac{1}{-n^2 + \mu^2}\cos nt$

Complete solution of (1) is $\quad x = c_1 \cos \mu t + c_2 \sin \mu t + \frac{e}{\mu^2 - n^2}\cos nt \qquad \begin{bmatrix} c_1 = A\cos\alpha \\ c_2 = A\sin\alpha \end{bmatrix}$

$$x = A\cos(\mu t + \alpha) + \frac{e}{\mu^2 - n^2}\cos nt \qquad ...(3)$$

Equation (2) shows that the motion is the resultant of two oscillatory motions, *i.e.*, the first due to $A\cos(\mu t + \alpha)$ gives free oscillation of period $\frac{2\pi}{\mu}$ and the second due to $\frac{e}{\mu^2 - n^2}\cos nt$ gives forced oscillations of period $\frac{2\pi}{n}$. If μ is large, then the frequency of free oscillations is very high, then the amplitude $\frac{e}{\mu^2 - n^2}$ of forced oscillations is small.

Case (*b*). If $\mu = n$ $\quad \text{P.I.} = \frac{1}{D^2 + \mu^2} e\cos nt = e \cdot \frac{1}{2D}\cos nt = \frac{e}{2}\int \cos nt\, dt$

$$\text{P.I.} = \frac{et}{2}\left(\frac{\sin nt}{n}\right)$$

$$x = c_1 \cos \mu t + c_2 \sin \mu t + \frac{et\sin nt}{2n}$$

$$= c_1 \cos \mu t + c_2 \sin \mu t + e\frac{t}{2\mu}\sin \mu t \qquad (n = \mu)$$

$$= c_1 \cos \mu t + \left(c_2 + \frac{et}{2\mu}\right)\sin \mu t$$

Let $c_1 = r\sin\phi$ and $\left(c_2 + \frac{et}{2\mu}\right) = r\cos\phi$ $\qquad \left[\begin{array}{l} r = \sqrt{c_1^2 + \left(c_2 + \frac{et}{2\mu}\right)^2} \\ \phi = \tan^{-1}\left(\frac{c_1}{c_2 + \frac{et}{2\mu}}\right) \end{array}\right]$

$$\Rightarrow \quad x = r\sin\phi\cos\mu t + r\cos\phi\sin\mu t$$

$$= r\sin(\mu t + \phi)$$

The period of oscillation $= \frac{2\pi}{\mu}$

Amplitude $= \sqrt{c_1^2 + \left(c_2 + \dfrac{et}{2\mu}\right)^2}$ and it increases as t increases.

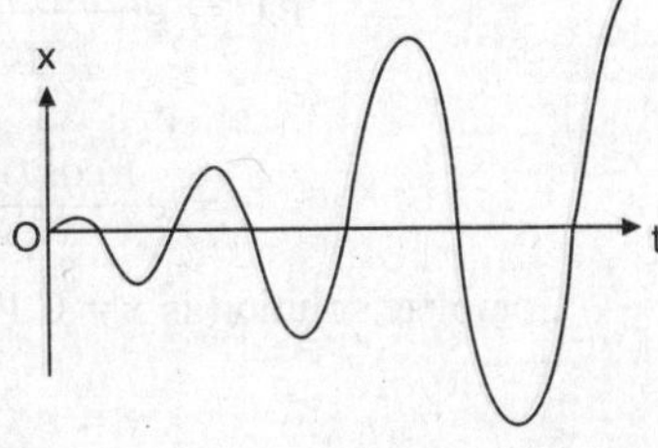

After long time, the amplitude of the oscillation may become abnormally large causing over strain and consequently break down the system. But it does not happen as there is always some resistance in the system.

Resonance. If the frequency due to external periodic force becomes equal to the natural frequency of the system, the phenomenon is known as *resonance.*

In designing a machine or structure, occurrence of the resonance is always to be avoided so that the system may not break down. While marching over a bridge, the soldiers avoid that their steps may not be in rythm with the natural frequency of the bridge. Resonance may cause the collapse the bridge.

(*d*) **Forced Oscillations (with damping)**

If there is an additional damping force, proportional to velocity, then equation of motion is

$$m\frac{d^2x}{dt^2} = mg - ks - kx - k_1\frac{dx}{dt} + q\cos nt \qquad (mg = ks)$$

$$\Rightarrow \quad m\frac{d^2x}{dt^2} = -kx - k_1\frac{dx}{dt} + q\cos nt$$

$$\Rightarrow \quad \frac{d^2x}{dt^2} = -\frac{k}{m}x - \frac{k_1}{m}\frac{dx}{dt} + \frac{q}{m}\cos nt \qquad \ldots(1)$$

On putting $\dfrac{k}{m} = \mu^2, \dfrac{k_1}{m} = 2\lambda, \dfrac{q}{m} = e$ in equation (1), we get

$$\Rightarrow \quad \frac{d^2x}{dt^2} = -\mu^2 x - 2\lambda\frac{dx}{dt} + e\cos nt$$

$$\Rightarrow \quad \frac{d^2x}{dt^2} + 2\lambda\frac{dx}{dt} + \mu^2 x = e\cos nt \qquad \ldots(2)$$

$$\Rightarrow \quad (D^2 + 2\lambda D + \mu^2)x = e\cos nt$$

A.E. is $\quad m^2 + 2\lambda m + \mu^2 = 0 \Rightarrow m = \dfrac{-2\lambda \pm \sqrt{4\lambda^2 - 4\mu^2}}{2}$

$$m = -\lambda \pm \sqrt{\lambda^2 - \mu^2}$$

$$\text{C.F.} = c_1 e^{[-\lambda + \sqrt{\lambda^2 - \mu^2}]t} + c_2 e^{[-\lambda - \sqrt{\lambda^2 - \mu^2}]t} = e^{-\lambda t}\left[c_1 e^{\sqrt{\lambda^2 - \mu^2}\,t} + c_2 e^{-\sqrt{\lambda^2 - \mu^2}\,t}\right]$$

$$\text{P.I.} = \frac{1}{D^2 + 2\lambda D + \mu^2} e\cos nt = e\frac{1}{-n^2 + 2\lambda D + \mu^2}\cos nt$$

$$= e\frac{1}{(\mu^2 - n^2) + 2\lambda D}\cos nt = e\frac{(\mu^2 - n^2) - 2\lambda D}{(\mu^2 - n^2)^2 - 4\lambda^2 D^2}\cos nt$$

$$= e\frac{(\mu^2 - n^2) - 2\lambda D}{(\mu^2 - n^2)^2 - 4\lambda^2(-n^2)}\cos nt = e\frac{(\mu^2 - n^2)\cos nt + 2n\lambda\sin nt}{(\mu^2 - n^2)^2 + 4\lambda^2 n^2}$$

Put $\mu^2 - n^2 = r\cos\phi$ and $2n\lambda = r\sin\phi$

$$\text{P.I.} = e\frac{r\cos\phi\cos nt + r\sin\phi\sin nt}{(\mu^2-n^2)^2+4\lambda^2 n^2} \qquad \left[\begin{array}{l} r = \sqrt{(\mu^2-n^2)^2+4n^2\lambda^2} \\ \phi = \tan^{-1}\dfrac{2n\lambda}{\mu^2-n^2}\end{array}\right]$$

$$= e\frac{r\cos(nt-\phi)}{r^2} = \frac{e}{r}\cos(nt-\phi)$$

Complete solution is x = C.F. + P.I.

$$x = e^{-\lambda t}[c_1 e^{\sqrt{\lambda^2-\mu^2}\,t} + c_2 e^{-\sqrt{\lambda^2-\mu^2}\,t}] + \frac{e}{r}\cos(nt-\phi) \qquad ...(3)$$

C.F. in (3) represents free oscillations of the system, which die out as $t \to \infty$, due to $e^{-\lambda t}$. $\frac{e}{r}\cos(nt-\phi)$ represents the forced oscillation.

Its constant amplitude $= \dfrac{e}{r} = \dfrac{e}{\sqrt{(\mu^2-n^2)^2+4n^2\lambda^2}}$

Its period $= \dfrac{2\pi}{n}$ which is the same as that of impressed force. As t increases, the free oscillations (given in the C.F.) die out while the forced oscillations persist giving the steady state of motion.

Example 26. *A body executes damped forced vibrations given the equation* $\dfrac{d^2x}{dt^2} + 2k\dfrac{dx}{dt} + b^2x = e^{-kt}\sin\omega t$. *Solve the equation for both the cases, when* $w^2 \neq b^2 - k^2$ *and when* $w^2 = b^2 - k^2$. *(Uttarakhand, II Semester, June 2007)*

Solution. We have,

$$\frac{d^2x}{dt^2} + 2k\frac{dx}{dt} + b^2x = e^{-kt}\sin\omega t$$

$$\Rightarrow \quad (D^2 + 2kD + b^2)x = e^{-kt}\sin\omega t \qquad ...(1)$$

Which is a linear differential equation with constant coefficients.

A.E. is $m^2 + 2\,km + b^2 = 0 \quad \Rightarrow \quad m = \dfrac{-2k \pm \sqrt{4k^2-4b^2}}{2} = -k \pm \sqrt{k^2-b^2}$

As the given problem is on vibrations, we must have $k^2 < b^2$

$$\therefore \quad m = -k \pm \sqrt{\{-(b^2-k^2)\}} = -k \pm i\sqrt{(b^2-k^2)}$$

$$\therefore \Rightarrow \quad \text{C.F.} = e^{-kt}[C_1\cos\sqrt{b^2-k^2}\,t + C_2\sin\sqrt{b^2-k^2}\,t]$$

$$= e^{-kt}A\cos\{\sqrt{b^2-k^2}\,t + \beta\}$$

where A and B are arbitrary constants.

$$\text{P.I.} = \frac{1}{D^2+2kD+b^2}e^{-kt}\sin\omega t = e^{-kt}\frac{1}{(D-k)^2+2k(D-k)+b^2}\sin\omega t$$

$$= e^{-kt}\frac{1}{D^2+(b^2-k^2)}\sin\omega t = e^{-kt}\frac{1}{-\omega^2+(b^2-k^2)}\sin\omega t \quad \text{if } \omega^2 \neq b^2-k^2$$

When $\omega^2 - (b^2 - k^2) = 0$

$$\text{P.I.} = e^{-kt}\,t\frac{1}{2D}\sin\omega t = e^{-kt}\left(-\frac{t}{2\omega}\cos\omega t\right),$$

Case I. If $w^2 \neq b^2 - k^2$, the complete solution of (1) is

$$x = Ae^{-kt}\cos\{\sqrt{(b^2-k^2)}\,t+\beta\} + \frac{e^{-kt}}{(b^2-k^2)-\omega^2}\sin\omega t$$

Case II. If $\omega^2 = b^2 - k^2$, the complete solution of (1) is

$$x = Ae^{-kt}\cos(\sqrt{b^2-k^2}\,t+\beta) + \frac{-e^{-kt}\,t\cos wt}{2\omega} \qquad (b^2-k^2=\omega^2)$$

$$x = Ae^{-kt}\cos(\omega t+\beta) - \frac{e^{-kt}\,t\cos\omega t}{2\omega}$$ **Ans.**

Example 27. *A spring of negligible weight which stretches 1 inch under tension of 2 lb is fixed at one end and is attached to a weight of w lb at the other. It is found that resonance occurs when an axial periodic force 2 cos 2t lb acts on the weight. Show that when the free vibrations have died out, the forced vibrations are given by x = ct sin 2t and find values of w and c.* *(Uttarakhand, II Semester, June 2007)*

Solution. When a weight of 2 lb is attached to A, spring stretches by $\frac{1}{12}$ *ft.*

Stress = k strain

$$\therefore \quad 2 = k.\frac{1}{12} \quad \Rightarrow \quad k = 24 \text{ lb/ft.}$$

Let B be the position of the weight ω attached to A then,

Stress = k strain

$$\omega = k \times AB \;\Rightarrow\; \omega = 24\,AB \;\Rightarrow\; AB = \frac{\omega}{24}\,ft.$$

At any time t, let the weight T_p be at P where $BP = x$.

Tension at P, $$T_P = k \times AP = 24\left(\frac{\omega}{24}+x\right) \qquad [AP = AB + BP]$$

$$T_P = \omega + 24x$$

Its equation of motion is

$$\frac{\omega}{g}\cdot\frac{d^2x}{dt^2} = -T + \omega + 2\cos 2t \qquad [\text{Mass} \times \text{acceleration} = \text{Force}]$$

$$= -\omega - 24x + \omega + 2\cos 2t \qquad [T = \omega + 24x]$$

$$\Rightarrow \quad \omega\frac{d^2x}{dt^2} + 24gx = 2g\cos 2t \qquad ...(1)$$

The phenomenon of resonance occurs when the period of free oscillations is equal to the period of forced oscillations,

From (1), $$\frac{d^2x}{dt^2} + \mu^2 x = \frac{2g}{\omega}\cos 2t \qquad ...(2) \left[\text{where } \mu^2 = \frac{24g}{\omega}\right]$$

The period of free oscillations is found as $\frac{2\pi}{\mu}$

and the period of the force $\left(\frac{2g}{\omega}\right)\cos 2t$ is π.

$$\therefore \quad \frac{2\pi}{\mu} = \pi \;\Rightarrow\; \mu = 2$$

On putting the value of μ in $\mu^2 = \frac{24g}{\omega}$, we get $4 = \frac{24g}{\omega} \;\Rightarrow\; \omega = 6g$

On putting the values of μ and w in (2), we get

$$\frac{d^2x}{dt^2}+4x=\frac{1}{3}\cos 2t \qquad ...(3)\ [\because\ \omega = 6g \text{ and } \mu = 2]$$

We know that the free oscillations are given by the C.F. and the forced oscillations are given by P.I. Thus, when the free oscillations have died out, the forced oscillations are given by the P.I. of (3).

$$\text{P.I.} = \frac{1}{3}\left(\frac{1}{D^2+4}\cos 2t\right)=\frac{1}{3}t.\frac{1}{2D}\cos 2t=\frac{t}{12}\sin 2t$$

Hence, $c=\dfrac{1}{12}$. **Ans.**

EXERCISE 17.5

1. A mass of 30 kg is attached to a spring for which k = 750 Newton/m and brought to rest. Find the position of the mass at time t if a force equal to 20 sin 2 t is applied to it.

   Ans. $x = -\,0.013 \sin 5t + 0.032 \sin 2\,t$

2. A body weighing 4.9 kg is hung from a spring. A pull of 10 kg will stretch the spring to 5 cm. The body is pulled down 6 cm below the static equilibrium position and then released. Find the displacement of the body from its equilibrium position at time t seconds, the maximum velocity and the period of oscillation. **Ans.** 0.06 cos 20 t m; 1.2 m/sec; 0.314 sec.

3. A spring is such that it would stretch 72 mm by a mass of 15 kg. A mass of 30 kg is attached and brought to rest. The resistance of the medium is numerically equal to $20\dfrac{dx}{dt}$ Newtons. Find the equation of the motion of the weight if it is pulled down 140 mm and given an upward velocity 3m/sec.

   Ans. $x = e^{-2t}\,(0.14 \cos 8t - 0.34 \sin 8t)$

4. A body weighing 10 kg is hung from a spring. A pull of 20 kg wt. will stretch the spring to 10 cm. The body is pulled down to 20 cm below the static equilibrium position and then released. Find the displacement of the body from its equilibrium position at time t sec, the maximum velocity and the period of oscillation. **Ans.** x = 0.2 cos 14 t, Max. Vel = 2.8 m/sec, period of oscillation = 0.45 sec

5. A particle of mass m is attached to one end of a light spring of modulus λ, the other end being fixed and the spring vertical. Prove that the velocity of the particle when it has traversed a distance 'a' is $\sqrt{2ag-\dfrac{\lambda}{m}a^2}$

17.11 BEAM

A bar whose length is much greater than its cross-section and its thickness is called a *beam.*

A B

Supported beam. If a beam may just rest on a support like a knife edge is called a *supported beam.*

Fixed beam. If one or both ends of a beam are firmly fixed then it is called *fixed beam.*

Cantilever. If one end of a beam is fixed and the other end is loaded, it is called a *cantilever.*

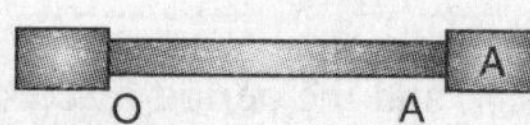

Bending of Beam. Let a beam be fixed at one end and the other end is loaded. Then the upper surface is elongated and therefore under tension and the lower surface is shortened so under compression.

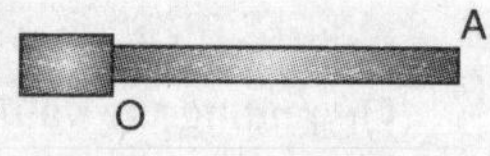

Neutral Surface. In between the lower and upper surface there is a surface which is neither stretched nor compressed. It is known as a *neutral surface.*

Bending Moment. Whenever a beam is loaded it deflects from its original position. If M is the bending moment of the forces acting on it, then

$$M = \frac{EI}{R} \qquad ...(1)$$

where E = Modulus of elasticity of the beam,

I = Moment of inertia of the cross-section of beam about neutral axis.

R = Radius of curvature of the curved beam

$$R = \frac{\left[1+\left(\frac{dy}{dx}\right)^2\right]^{3/2}}{\frac{d^2y}{dx^2}} = \frac{1}{\frac{d^2y}{dx^2}} \qquad \left[\text{neglecting } \frac{dy}{dx}\right]$$

Thus equation (1) becomes $M = EI\,\dfrac{d^2y}{dx^2}$

Boundary Conditions

(*i*) *End freely supported.* At the freely supported end there will be no deflection and no bending moment.

$$y = 0, \quad \frac{d^2y}{dx^2} = 0$$

(*ii*) *Fixed end horizontally.* Deflection and slope of the beam are zero.

$$y = 0, \quad \frac{dy}{dx} = 0$$

(*iii*) *Perfectly free end.* At the free end there is no bending moment or shear force.

$$\frac{d^2y}{dx^2} = 0, \quad \frac{d^3y}{dx^3} = 0$$

Convention of signs

The sign of the moment about NN' on the left NN' is positive if anticlockwise and negative if clockwise.

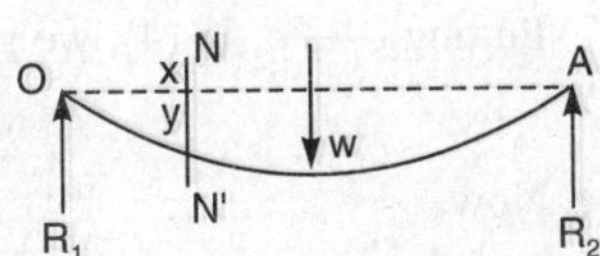

The downward deflection is positive and length x on right-side is also positive. Slope $\dfrac{dy}{dx}$ is positive if downward.

Example 28. *The differential equation satisfied by a beam uniformly loaded (W kg/metre), with one end fixed and the second end subjected to tensile force P, is given by*

$$E.I.\frac{d^2y}{dx^2} = Py - \frac{1}{2}Wx^2$$

Show that the elastic curve for the beam with conditions

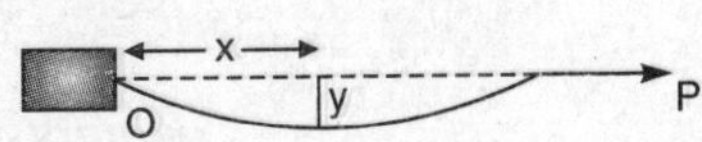

$$y = 0 = \frac{dy}{dx} \text{ at } x = 0, \text{ is given by}$$

$$y = \frac{W}{Pn^2}(1-\cosh nx) + \frac{Wx^2}{2P} \text{ where } n^2 = \frac{P}{EI}$$

Solution. We have, $E.I.\dfrac{d^2y}{dx^2} = Py - \dfrac{1}{2}W.x^2$ ...(1)

$$\Rightarrow \quad \frac{d^2y}{dx^2} - \frac{P}{E.I.}y = -\frac{W}{2E.I.}x^2 \quad \Rightarrow \quad \left(D^2 - \frac{P}{E.I.}\right)y = -\frac{W}{2E.I.}x^2$$

A.E. is $m^2 - \dfrac{P}{E.I.} = 0 \Rightarrow m^2 = \dfrac{P}{E.I.} = n^2 \Rightarrow m = \pm n$ $\left(n^2 = \dfrac{P}{EI}\right)$

C.F. $= c_1 e^{nx} + c_2 e^{-nx}$

$$\text{P.I.} = \frac{1}{D^2 - \frac{P}{E.I.}}\left(-\frac{W}{2E.I.}\right)x^2 = -\frac{W}{2E.I.}\frac{1}{D^2 - n^2}.x^2$$

$$= \frac{W}{2n^2.E.I.}\left(1 - \frac{D^2}{n^2}\right)^{-1}.x^2 = \frac{W}{2n^2\,E.I.}\left(1 + \frac{D^2}{n^2}\right).x^2 = \frac{W}{2n^2\,E.I.}\left(x^2 + \frac{2}{n^2}\right)$$

$$\therefore \quad y = c_1 e^{nx} + c_2 e^{-nx} + \frac{W}{2n^2\,E.I.}\left(x^2 + \frac{2}{n^2}\right) \quad \text{...(2)}$$

Differentiating (2) w.r.t. x, we get

$$\frac{dy}{dx} = nc_1 e^{nx} - nc_2 e^{-nx} + \frac{W}{2n^2\,E.I.}(2x) \quad \text{...(3)}$$

Putting $x = 0$, $\dfrac{dy}{dx} = 0$ in (3), we get

$$0 = n\,c_1 - n\,c_2 \quad \Rightarrow \quad c_1 = c_2$$

Putting $x = 0$, $y = 0$ in (2), we get

$$0 = c_1 + c_2 + \frac{W}{2n^2\,E.I.}\frac{2}{n^2} \quad \Rightarrow \quad 0 = c_1 + c_2 + \frac{W}{n^4\,E.I.} \quad \text{...(4)}$$

Putting $c_1 = c_2$ in (4), we get $0 = 2c_1 + \dfrac{W}{n^4\,E.I.} \Rightarrow c_1 = -\dfrac{W}{2n^4\,E.I.}$,

Now, $\quad n^2 = \dfrac{P}{E.I.} \Rightarrow n^2\,E.I. = P$

$$\Rightarrow \quad c_1 = c_2 = -\frac{W}{2n^2P}$$

Putting the values of c_1 and c_2 in (2), we get

$$y = \frac{-W}{2n^2\,P}(e^{nx} + e^{-nx}) + \frac{W}{2P}\left(x^2 + \frac{2}{n^2}\right)$$

$$y = \frac{-W}{n^2P}\cosh nx + \frac{W}{2P}x^2 + \frac{W}{Pn^2}$$

$$y = \frac{W}{P.n^2}(1 - \cosh nx) + \frac{W\,x^2}{2P}$$ **Ans.**

EXERCISE 17.6

1. A beam of length l and of uniform cross-section has the differential equation of its elastic curve as

$$E.I.\frac{d^2y}{dx^2}=\frac{w}{2}\left(\frac{l^2}{4}-x^2\right)$$

where E is the modulus of elasticity, I is the moment of inertia of the cross-section, w is weight per unit length and x is measured from the centre of span.

If at $x=0, \frac{dy}{dx}=0$. Prove that the equation of the elastic curve is

$$y=\frac{1}{2}\cdot\frac{2}{E.I.}\left(\frac{l^3.x^2}{8}-\frac{x^4}{12}\right)+\frac{5w.l^4}{384\,E.I.}$$

2. A horizontal tie rod of length l is freely pinned at each end. It carries a uniform load w kg per unit length and has a horizontal pull P. Find the central deflection and the maximum bending moment, taking the origin at one of its ends. **Ans.** $\frac{w}{a}\left(\sec h\frac{al}{2}-1\right)$ where $a^2=\frac{P}{EI}$

3. A light horizontal strut AB is freely pinned at A and B. It is under the action of equal and opposite compressive forces P at its ends and it carries a load W at its centre. Then for $0<x<\frac{l}{2}$,

$$EI\frac{d^2y}{dx^2}+Py+\frac{1}{2}Wx=0$$

Also $y=0$ at $x=0$ and $\frac{dy}{dx}=0$ at $x=\frac{l}{2}$. Prove that $y=\frac{W}{2P}\left(\frac{\sin ax}{a\cos\frac{al}{2}}-x\right)$, where $a^2=\frac{P}{EI}$

4. A horizontal tie-rod of length $2l$ with concentrated load W at its centre and ends freely hinged satisfies the differential equation $EI\frac{d^2y}{dx^2}=Py-\frac{W}{2}x$. With conditions $x=0, y=0$ and $x=l, \frac{dy}{dx}=0$. Prove that the deflection δ and bending moment M at the centre $(x=l)$ are given by $\delta=\frac{W}{2Pn}(nl-\tan nl)$ and $M=-\frac{W}{2n}\tan h\,nl$, where $n^2\,EI=P$.

17.12 PROJECTILE

Example 29. *A particle is projected with velocity u making an angle α with the horizontal. Neglecting air resistance, show that the equation of its path is the parabola.*

$$y=x\tan\alpha-\frac{gx^2}{2u^2\cos^2\alpha}.$$

Find the time of flight, the greatest height attained and the range on the horizontal plane.

Solution. If a particle of mass m is projected from 0 with velocity u in a direction making an angle α with the horizontal.

Let horizontal line and vertical line be taken as x-axis and y-axis. Let $P(x, y)$ be the position of the particle at time t.

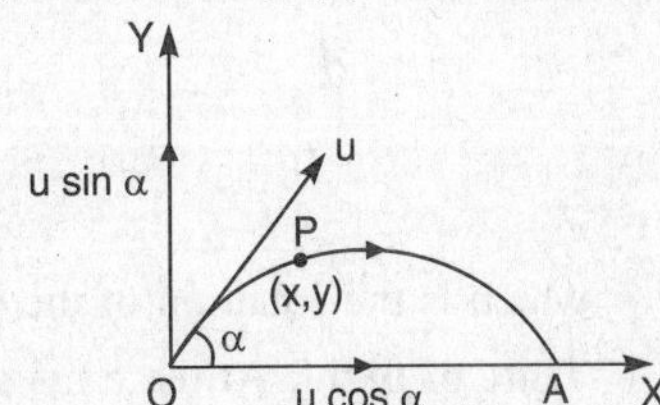

Horizontal component of $u = u\cos\alpha$

Vertical component of $u = u\sin\alpha$

The force acting on the particle = weight = mg

The equations of motion are

Horizontal components: $m\frac{d^2x}{dt^2}=0 \Rightarrow \frac{d^2x}{dt^2}=0$ ...(1)

Vertical components: $m\frac{d^2y}{dx^2}=-mg \Rightarrow \frac{d^2y}{dx^2}=-g$ ...(2)

Integrating (1), we have $\frac{dx}{dt}=c_1$ ...(3)

Initially $\frac{dx}{dt}=u\cos\alpha$, time $t=0$, putting in (3), we get

$$u\cos\alpha=c_1$$

On putting the value of c_1 in (3), we get

$$\frac{dx}{dt}=u\cos\alpha \quad ...(4)$$

Integrating (4), we have $x=(u\cos\alpha)\,t+c_2$ ...(5)

Putting (initial condition) $t=0$, $\alpha=0$, we get $c_2=0$

(5) becomes, $x=(u\cos\alpha)\,t$ ...(6)

Integrating (2), we have $\frac{dy}{dt}=-gt+c_3$ ...(7)

Putting (initial condition) $t=0$, $\frac{dy}{dt}=u\sin\alpha$ in (7), we get

$$u\sin\alpha=c_3$$

(7) becomes $\frac{dy}{dt}=-gt+u\sin\alpha$ ...(8)

Integrating (8), we get $y=-g\frac{t^2}{2}+(u\sin\alpha)\,t+c_4$ ...(9)

Putting (initial condition), $t=0$, $y=0$, we obtain $0=c_4$

(9) becomes, $y=-\frac{gt^2}{2}+(u\sin\alpha)\,t$ ...(10)

Equation (6) and (10) give the position of the particle at any time t. The equation of the path described by the particle is obtained by eliminating t, between the equations (6) and (10),

From (6), $t=\frac{x}{u\cos\alpha}$

Substituting this value of t in (10), we get

$$y=(u\sin\alpha)\frac{x}{u\cos\alpha}-\frac{g}{2}\left(\frac{x^2}{u^2\cos^2\alpha}\right)$$

$$y=x\tan\alpha-\frac{gx^2}{2u^2\cos^2\alpha}$$

which is the equation of the projectile, the path is parabola.

Time of flight. At the point A, $y=0$

On putting $y = 0$ in (10), we get

$$0 = -\frac{gt^2}{2} + (u \sin \alpha)t \quad \Rightarrow \quad t = \frac{2u \sin \alpha}{g}$$

Greatest Height

At the highest point, the vertical component becomes zero, *i.e.*, $\frac{dy}{dt} = 0$

Putting $\frac{dy}{dt} = 0$ in (8), we get $0 = -gt + u \sin \alpha \quad \Rightarrow \quad t = \frac{u \sin \alpha}{g}$

For the highest point, we substitute the value of t in (10), we get

$$y = -\frac{g}{2}\left(\frac{u \sin \alpha}{g}\right)^2 + u \sin \alpha\left(\frac{u \sin \alpha}{g}\right) = -\frac{u^2 \sin^2 \alpha}{2g} + \frac{u^2 \sin^2 \alpha}{g} = \frac{u^2 \sin^2 \alpha}{2g}$$

Range, *i.e., OA* is the horizontal distance covered during flight.

Putting the value of $t = \frac{2u \sin \alpha}{g}$ in (6), we get

$$\text{Range} = (u \cos \alpha)\left(\frac{2u \sin \alpha}{g}\right) = \frac{u^2 \sin 2\alpha}{g}$$

$$\text{Maximum range} = \frac{u^2}{g} \text{ if } \sin 2\alpha = 1 \quad \Rightarrow \quad \alpha = \frac{\pi}{4}$$ **Ans.**

Example 30. *The equation of motion under certain conditions are*

$$m\frac{d^2x}{dt^2} + eh\frac{dy}{dt} = eE \qquad \text{...(1)}$$

$$m\frac{d^2y}{dt^2} - eh\frac{dx}{dt} = 0 \qquad \text{..(2)}$$

With condition $x = \frac{dx}{dt} = y = \frac{dy}{dt} = 0$ *when* $t = 0$, *find the path of electron.*

Solution. Multiplying (2) by k and adding to (1), we get

$$m\frac{d^2x}{dt^2} + mk\frac{d^2y}{dt^2} + eh\frac{dy}{dt} - ehk\frac{dx}{dt} = eE$$

$$\Rightarrow \quad m\frac{d^2}{dt^2}(x + ky) - ehk\frac{d}{dt}\left(-\frac{y}{k} + x\right) = e.\, E \qquad \text{...(3)}$$

Let us choose k such that $x + ky = x - \frac{y}{k}$

$$\Rightarrow \qquad k = -\frac{1}{k} \quad \Rightarrow \quad k^2 = -1 \Rightarrow k = \pm i$$

Putting $x + ky = u$ in (3), we have

$$m\frac{d^2u}{dt^2} - ehk\frac{du}{dt} = eE \quad \Rightarrow \quad \frac{d^2u}{dt^2} - wk\frac{du}{dt} = \frac{eE}{m} \qquad \left(w = \frac{eh}{m}\right)$$

$$\Rightarrow \qquad D^2 u - wk\, Du = \frac{eE}{m}$$

A.E. is $m^2 - wkm = 0 \quad \Rightarrow \quad m(m - wk) = 0 \quad \Rightarrow \quad m = 0 \quad \text{or} \quad m = wk$

$$\text{C.F.} = c_1 + c_2\, e^{wkt}$$

$$\text{P.I.} = \frac{1}{D^2 - wkD}\frac{eE}{m} = \frac{eE}{m}\frac{1}{D^2 - wkD}e^{0t}$$

$$= \frac{eEt}{m}\frac{1}{2D - wk}e^{0t} = \frac{wEt}{h}\frac{1}{0 - wk} = -\frac{Et}{hk} \qquad \left(w = \frac{eh}{m}\right)$$

The complete solution is

$$u = c_1 + c_2\, e^{wkt} - \frac{Et}{h.k} \text{ or } x + ky = c_1 + c_2\, e^{wkt} - \frac{Et}{hk} \qquad ...(4)$$

Putting the value of *k, i.e., i*, – *i* in (4), we get

$$x + iy = c_1 + c_2\, e^{iwt} - \frac{Et}{ih} \qquad ...(5)$$

$$x - iy = c_3 + c_4\, e^{-iwt} + \frac{Et}{ih} \qquad ...(6)$$

Differentiating (5) and (6), we get

$$\frac{dx}{dt} + i\frac{dy}{dt} = c_2\, iwe^{iwt} + \frac{iE}{h} \qquad ...(7)$$

$$\frac{dx}{dt} - i\frac{dy}{dt} = -iwc_4\, e^{-iwt} - \frac{iE}{h} \qquad ...(8)$$

Initial conditions $\quad x = y = \frac{dx}{dt} = \frac{dy}{dt} = 0$ when $t = 0$

Putting these values in (5), (6), (7) and (8), we get

$$0 = c_1 + c_2 \quad \Rightarrow \quad c_2 = -c_1$$

$$0 = c_3 + c_4 \quad \Rightarrow \quad c_4 = -c_3$$

$$0 = iwc_2 + \frac{iE}{h} \quad \Rightarrow \quad c_2 = -\frac{E}{wh}$$

$$0 = -iwc_4 - \frac{iE}{h} \quad \Rightarrow \quad c_4 = -\frac{E}{wh}$$

On substituting the values of c_1, c_2, c_3 and c_4 in (5) and (6), we get

$$x + iy = \frac{E}{wh} - \frac{E}{hw}e^{iwt} + i\frac{Et}{h} \qquad ...(9)$$

$$x - iy = \frac{E}{wh} - \frac{E}{wh}e^{-iwt} - i\frac{Et}{h} \qquad ...(10)$$

On adding (9) and (10), we get

$$2x = \frac{2E}{hw} - \frac{E}{hw}(e^{iwt} + e^{-iwt})$$

$$x = \frac{E}{hw} - \frac{E}{hw}\cos wt$$

$$x = \frac{E}{hw}(1 - \cos wt)$$

Subtracting (10) from (9), we obtain

$$2iy = -\frac{E}{hw}(e^{iwt} - e^{-iwt}) + \frac{2iEt}{h}$$

$$\Rightarrow \qquad y = -\frac{E}{wh}\left(\frac{e^{iwt} - e^{-iwt}}{2i}\right) + \frac{Et}{h} = -\frac{E}{wh}\sin wt + \frac{Et}{h} \Rightarrow y = \frac{E}{hw}(wt - \sin wt) \quad \textbf{Ans.}$$

Example 31. *Assuming that a spherical rain drop evaporates at rate proportional to its surface area and if its radius originally is 3 mm and one hour later has been reduced to 2 mm, find an expression for the radius of the rain drop at any time.*

Solution. Evaporation $\propto$ surface area

$$\frac{dV}{dt} \propto S \quad \Rightarrow \quad \frac{dV}{dt} = kS \qquad ...(1)$$

$$V = \frac{4}{3}\pi r^3 \quad \Rightarrow \quad \frac{dV}{dt} = 4\pi r^2 \frac{dr}{dt} \qquad ...(2)$$

Putting the value of $\frac{dV}{dt}$ from (1) in (2), we have

$$4\pi r^2 \frac{dr}{dt} = kS \Rightarrow S\frac{dr}{dt} = kS \quad \Rightarrow \quad \frac{dr}{dt} = k$$

$$r = kt + c \qquad ...(3)\ [S = 4\pi r^2]$$

Putting $t = 0, r = 3$ in (3), we get $3 = c$

(3) becomes $\quad r = kt + 3 \quad$ ...(4)

Putting $t = 1$ and $r = 2$ in (4), we get $2 = k + 3 \;\Rightarrow\; k = -1$

(4) becomes $\quad r = -t + 3 \quad$ **Ans.**

EXERCISE 17.7

1. The current i_1 and i_2 in mesh are given by the differential equations

$$\frac{di_1}{dt} - wi_2 = a\cos pt, \frac{di_2}{dt} + wi_1 = a\sin pt$$

Find the currents i_1 and i_2 if $i_1 = i_2 = 0$ at $t = 0$. **Ans.** $i_1 = \frac{a}{P+w}\sin pt, i_2 = \frac{a}{P+w}\cos pt$

2. A particle moving in a plane is subjected to a force directed towards a fixed point O and proportional to the distance of the particle from *O*. Show that the differential equations of motion are of the form $\frac{d^2x}{dt^2} = -k^2x, \frac{d^2y}{dt^2} = -k^2y$. Find the cartesian equation of the path of the particle if $x = 1, y = 0, \frac{dx}{dt} = 0$ and $\frac{dy}{dt} = 2$ when $t = 0$. **Ans.** $4x^2 + k^2y^2 = 4$

3. A projectile of mass m is fired into the air with initial velocity v_0 at an angle θ with the ground. Neglecting all forces except gravity and the resistance of air, assumed proportional to velocity. Find the position of the projectile at time t.

$$x = \frac{1}{k}(v_0\cos\theta)(1 - e^{kt}), \quad y = \frac{1}{k}\left\{\frac{g}{k} + v_0\sin\theta\,(1 - e^{-kt}) - gt\right\}$$

4. A particle of unit mass is projected with velocity u at an inclination α above the horizon in a medium whose resistance is k times the velocity. Show that its direction will again make an angle α with the horizon after a time $\frac{1}{k}\log\left\{1 + \frac{2ku}{g}\sin\alpha\right\}$.

5. An inclined plane makes angle α with the horizontal. A projectile is launched from the bottom of the inclined plane with speed v in a direction making an angle β with the horizontal. Set up the differential equations and find

(*i*) the range on the incline (*ii*) the maximum range up the incline.

Ans. (*i*) $\frac{2u^2\sin(\alpha - \beta)\cos\alpha}{g\cos^2\beta}$ (*ii*) $u^2\, g(1 + \sin\beta)$

CALCULUS OF VARIATION

18.1 INTRODUCTION

The calculus of variations primarily deals with finding maximum or minimum value of a definite integral involving a certain function.

18.2 FUNCTIONALS

A simple example of functional is the shortest length of a curve through two points $A\ (x_1, y_1)$ and $B\ (x_2, y_2)$. In other words, the determination of the curve $y = y(x)$ for which $y = (x_1) = y_1$, $y\ (x_2) = y_2$ such that

$$\int_{x_1}^{x_2} \sqrt{1+\left(\frac{dy}{dx}\right)^2}\, dx \quad ...(1)$$

is a minimum.

An integral such as (1) is called a *Functional.*

In general, it is required to find the curve $y = y\ (x)$ where $y\ (x_1) = y_1$ and $y\ (x_2) = y_2$ such that for a given function $f\left(x, y, \frac{dy}{dx}\right)$,

$$\int_{x_1}^{x_2} f\left(x, y, \frac{dy}{dx}\right) dx \quad ...(2)$$

is maximum or minimum.

Integral (2) is known as the functional.

In differential calculus, we find the maximum or minimum value of functions. But the calculus of variations deals with the problems of maxima or minima of functionals.

A functional $I\ [y\ (x)]$ is said to be linear if it satisfies.

(*i*) $I\ [cy\ (x)] = c\ I\ [y\ (x)]$, where c is an arbitrary constant.

(*ii*) $I\ [y_1\ (x) + y_2\ (x)] = I\ [y_1\ (x)] + I\ [y_2\ (x)]$, where $y_1\ (x) \in M$ and $y_2\ (x) \in M$

18.3 DEFINITION

A functional $I\ [y\ (x)]$ is maximum on a curve $y = y\ (x)$, if the values of $I\ [y\ (x)]$ on any curve close to $y = y_1\ (x)$ do not exceed $I\ [y_1\ (x)]$. It means $\Delta I = I\ [y\ (x)] - I\ [y_1\ (x)] \leq 0$ and $\Delta I = 0$ *on* $y = y_1\ (x)$.

In case of minimum of $I[y(x)]$, $\Delta I = 0$.

Extremal: A function $y = y(x)$ which extremizes a functional is called extremal or extremizing function.

18.4 EULER'S EQUATION IS

$$\frac{\partial f}{\partial y} - \frac{d}{dx}\left(\frac{\partial f}{\partial y'}\right) = 0 \qquad \textit{(DU, II Sem. 2012)}$$

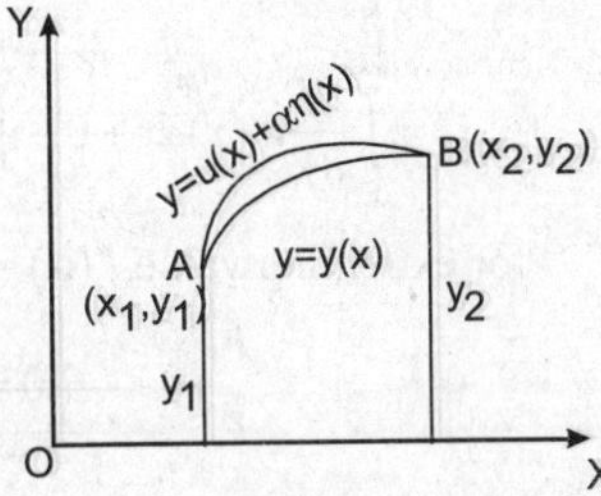

This is the necessary condition for $I = \int_{x_1}^{x_2} f(x, y, y')\,dx$ to be maximum or minimum.

Proof: Let $y = y(x)$ be the curve AB which makes the given function I an extremum.

Consider a family of neighbouring curves

$$Y = u(x) + \alpha\eta(x) \qquad \text{...(1)}$$

where α is a parameter, and $\eta(x)$ is an arbitrary differentiable function.

At the end points A and B,

$$\eta(x_1) = \eta(x_2) = 0$$

when $\alpha = 0$, neighbouring curves become $y = y(x)$, which is extremal.

The family of neighbouring curves is called the family of *comparison functions.*

If in the functional $\int_{x_1}^{x_2} f(x, y, y')\,dx$ We replace y by Y, we get

$$\int_{x_1}^{x^2} f(x, Y, Y')\,dx = \int_{x_1}^{x_2} f[x, y(x) + \alpha\eta(x), y'(x) + \alpha\eta'(x)]\,dx.$$

which is a function of α, say $I(\alpha)$.

$$\therefore \qquad I(\alpha) = \int_{x_1}^{x_2} f(x, Y, Y')\,dx$$

For $\alpha = 0$, the neighbouring curves become the extremal, an extremum for $\alpha = 0$.

The necessary condition for this is $I'(\alpha) = 0$ ...(2)

Differentiating I under the integral sign by Leibnitz's rule, we have

$$I'(\alpha) = \int_{x_1}^{x_2}\left(\frac{\partial f}{\partial x}\frac{\partial x}{\partial \alpha} + \frac{\partial f}{\partial Y}\frac{\partial Y}{\partial \alpha} + \frac{\partial f}{\partial Y'}\frac{\partial Y'}{\partial \alpha}\right)dx$$

$$I'(\alpha) = \int_{x_1}^{x_2}\left(\frac{\partial f}{\partial Y}\frac{\partial Y}{\partial \alpha} + \frac{\partial f}{\partial Y'}\frac{\partial Y'}{\partial \alpha}\right)dx \qquad \left(\frac{\partial x}{\partial \alpha} = 0 \text{ as } \alpha \text{ is indenpendent of } x\right) \qquad \text{...(3)}$$

On differentiating (1), *w. r. t.* 'x', we get, $Y' = y'(x) + \alpha\eta'(x)$

Again differentiating *w.r. t.* 'α', we get $\dfrac{\partial Y'}{\partial \alpha} = \eta'(x)$

Differentiating (1), *w. r. t.*, we get $\alpha\dfrac{\partial Y}{\partial \alpha} = \eta(x)$

Now (3) becomes $I'(\alpha) = \int_{x_1}^{x_2}\left[\dfrac{\partial f}{\partial Y}\eta(x) + \dfrac{\partial f}{\partial Y'}\eta'(x)\right]dx$

Integrating the second term on the right by parts, we get

$$= \int_{x_1}^{x_2} \frac{\partial f}{\partial Y}\eta(x)dx + \left[\left\{\frac{\partial f}{\partial Y'}\eta(x)\right\}_{x_1}^{x_2} - \int_{x_1}^{x_2} \frac{d}{dx}\left(\frac{\partial f}{\partial Y'}\right)\eta(x)dx\right]$$

$$= \int_{x_1}^{x_2} \frac{\partial f}{\partial Y}\eta(x)dx + \left[\frac{\partial f}{\partial Y'}\eta(x_2) - \frac{\partial f}{\partial Y'}\eta(x_1)\right] - \int_{x_1}^{x_2} \frac{d}{dx}\left[\frac{\partial f}{\partial Y'}\right]\eta(x)dx$$

$$= \int_{x_1}^{x_2} \frac{\partial f}{\partial Y}\eta(x)dx + 0 - \int_{x_1}^{x_2} \frac{d}{dx}\left(\frac{\partial f}{\partial Y'}\right)\eta(x)dx = \int_{x_1}^{x_2}\left[\frac{\partial f}{\partial Y} - \frac{d}{dx}\left(\frac{\partial f}{\partial Y'}\right)\right]\eta(x)dx \left[\eta(x_1) = \eta(x_2) = 0\right]$$

for extremum value, $I'(\alpha) = 0$

$$0 = \int_{x_1}^{x_2}\left[\frac{\partial f}{\partial Y} - \frac{d}{dx}\left(\frac{\partial f}{\partial Y'}\right)\right]\eta(x)dx$$

$\eta(x)$ is an arbitrary continuous function.

$\therefore$ $\quad \dfrac{\partial f}{\partial y} - \dfrac{d}{dx}\left(\dfrac{\partial f}{\partial y'}\right) = 0$ which is a required Euler's equation.

Note: Other Forms of Euler's equation

1. $\dfrac{d}{dx} f(x, y, y') = \dfrac{\partial f}{\partial x}\dfrac{dx}{dx} + \dfrac{\partial f}{\partial y}\dfrac{dy}{dx} + \dfrac{\partial f}{\partial y'}\dfrac{dy'}{dx}$

or
$$\frac{df}{dx} = \frac{\partial f}{\partial x} + \frac{\partial f}{\partial y}y' + \frac{\partial f}{\partial y'}y'' \qquad ...(4)$$

But
$$\frac{d}{dx}\left(y'\frac{\partial f}{\partial y'}\right) = y'\frac{d}{dx}\left(\frac{\partial f}{\partial y'}\right) + \frac{\partial f}{\partial y'}y'' \qquad ...(5)$$

On subtracting (5) from (4), we have

$$\frac{df}{dx} - \frac{d}{dx}\left(y'\frac{\partial f}{\partial y'}\right) = \frac{\partial f}{\partial x} + \frac{\partial f}{\partial y}y' - y'\frac{d}{dx}\left(\frac{\partial f}{\partial y'}\right)$$

$$\frac{d}{dx}\left[f - y'\frac{\partial f}{\partial y'}\right] - \frac{\partial f}{\partial x} = y'\left[\frac{\partial f}{\partial y} - \frac{d}{dx}\frac{\partial f}{\partial y'}\right] = (y')(0) = 0 \qquad [\textit{Euler's equation}]$$

Hence
$$\frac{d}{dx}\left[f - y'\frac{\partial f}{\partial y'}\right] - \frac{\partial f}{\partial x} = 0 \qquad ...(6)$$

Which is an another form of Euler's equation.

2. We know that $\dfrac{\partial f}{\partial y'}$ is also a function x, y, y' say $\phi(x, y, y')$.

$$\frac{d}{dx}\left(\frac{\partial f}{\partial y'}\right) = \frac{\partial \phi}{\partial x}\frac{dx}{dx} + \frac{\partial \phi}{\partial y}\frac{dy}{dx} + \frac{\partial \phi}{\partial y'}\frac{dy'}{dx} = \frac{\partial \phi}{\partial x} + \frac{\partial \phi}{\partial y}y' + \frac{\partial \phi}{\partial y'}y''$$

$$\frac{\partial}{\partial x}\left(\frac{\partial f}{\partial y'}\right) + \frac{\partial}{\partial y}\left(\frac{\partial f}{\partial y'}\right)y' + \frac{\partial}{\partial y'}\left(\frac{\partial f}{\partial y'}\right)y'' = \frac{\partial^2 f}{\partial x \partial y'} + y'\frac{\partial^2 f}{\partial y \partial y'} + y''\frac{\partial^2 f}{\partial y'^2}$$

Putting the value of $\dfrac{d}{dx}\left(\dfrac{\partial f}{\partial y'}\right)$ in Euler's equation, we get

$$\frac{\partial f}{\partial y} - \frac{\partial^2 f}{\partial x \partial y'} - y'\frac{\partial^2 f}{\partial y \partial y'} - y''\frac{\partial^2 f}{\partial y'^2} = 0 \qquad ...(7)$$

This is the third form of Euler's equation.

18.5 EXTREMAL

Any function which satisfies Euler's equation is known as Extremal. Extremal is obtained by solving the Euler's equation.

Case 1. If f is independent of $x, i.e., \frac{\partial f}{\partial x} = 0.$

On substituting the value of $\frac{\partial f}{\partial x}$ in (6), we have $\frac{d}{dx}\left[f - y'\frac{\partial f}{\partial y'}\right] = 0$

Integrating, we get $f - y'\frac{\partial f}{\partial y'}$ = constant

Case 2. When f is independent of y, $i.e., \frac{\partial f}{\partial y} = 0.$

Putting the value of $\frac{\partial f}{\partial y}$ in Euler's equation, we get

$$\frac{d}{dx}\left(\frac{\partial f}{\partial y'}\right) = 0, \text{ Integrating we get } \frac{\partial f}{\partial y'} = \textbf{constant}$$

Case 3. If f is an independent of $y', i.e., \frac{\partial f}{\partial y'} = 0.$ On substituting the value of $\frac{\partial f}{\partial y'}$ in the Euler's equation, we get $\boldsymbol{\frac{\partial f}{\partial y} = 0}$

This is the desired solution.

Case 4. If f is independent of x and y,

we have $\frac{\partial f}{\partial x} = 0$ and $\frac{\partial f}{\partial y} = 0$ or $\frac{\partial^2 f}{\partial x \partial y'} = 0$ and $y'\frac{\partial^2 f}{\partial y \partial y'} = 0$

Putting these value in Euler's equation (7), we have $-y''\frac{\partial^2 f}{\partial y'^2} = 0$

If $\frac{\partial^2 f}{\partial y'^2} \neq 0$ then $y'' = 0$ whose solution is $\boldsymbol{y = ax+b}$.

Example 1. *Write the Euler-Lagrange's equation and explain the terms involved.*

(D.U., II Sem. 2012, April 2010)

Solution. The Euler Lagrange's equation is $\frac{d}{dx}\left(\frac{\partial f}{\partial y'}\right) - \frac{\partial f}{\partial y} = 0$

Example 2. *Prove that if f does not depend on x explicity, then $f - f'\frac{\partial f}{\partial y'}$ = constant.*

(D.U., II Sem. 2012)

Solution. The Euler Lagrange's differential equation is

$$\frac{d}{dx}\left(\frac{\partial f}{\partial y'}\right) - \frac{\partial f}{\partial y} = 0$$

Multiplying above equation by y' and adding subtracting the expression $y''\frac{\partial f}{\partial y'}$ $\left(\text{where } y'' = \frac{\partial y'}{\partial x} \text{ and } y' = \frac{\partial y}{\partial x}\right)$, we get

$$y'\frac{d}{dx}\left(\frac{\partial f}{\partial y'}\right)-y'\frac{\partial f}{\partial y}+y''\frac{\partial f}{\partial y'}-y''\frac{\partial f}{\partial y'}=0 \quad\Rightarrow\quad y'\frac{d}{dx}\left(\frac{\partial f}{\partial y'}\right)+y''\frac{\partial f}{\partial y'}-y''\frac{\partial f}{\partial y'}-y'\frac{\partial f}{\partial y}=0$$

$$\Rightarrow\quad \frac{d}{dx}\left(y'\frac{\partial f}{\partial y'}\right)-y''\frac{\partial f}{\partial y'}-y'\frac{\partial f}{\partial y}-\frac{\partial f}{\partial x}+\frac{\partial f}{\partial x}=0 \qquad \left(\text{adding and subtracting } \frac{\partial f}{\partial x}\right)$$

$$\Rightarrow\quad \frac{d}{dx}\left(y'\frac{\partial f}{\partial y'}\right)-\left[y''\frac{\partial f}{\partial y'}+y'\frac{\partial f}{\partial y}+\frac{\partial f}{\partial x}\right]+\frac{\partial f}{\partial x}=0$$

$$\Rightarrow\quad \frac{d}{dx}\left(y'\frac{\partial f}{\partial y'}\right)-\frac{\partial f}{dx}+\frac{\partial f}{\partial x}=0 \qquad \{\because f=f(y,y',x)\}$$

$$\Rightarrow\quad \frac{d}{dx}\left[y'\frac{\partial f}{\partial y'}-f\right]+\frac{\partial f}{\partial x}=0 \qquad \ldots(1)$$

If f does not depend upon x explicitly, then $\frac{\partial f}{\partial x}=0$ and so we must have

$$\frac{d}{dx}\left[y'\frac{\partial f}{\partial y'}-f\right]=0 \quad\Rightarrow\quad y'\frac{\partial f}{\partial y'}-f=\text{ constant}$$

$$\Rightarrow\qquad f-y'\frac{\partial f}{\partial y'}=\text{ constant.} \qquad \ldots(2)$$ **Proved**

Example 3. *Test for an extremum the functional*

$$I[y(x)]=\int_0^1\left(xy+y^2-2y^2y'\right)dx,\ y(0)=1, y(1)=2$$

Solution. Euler's equation

$$\frac{\partial f}{\partial y}-\frac{d}{dx}\left(\frac{\partial f}{\partial y'}\right)=0 \qquad \ldots(1)$$

Here $f=xy+y^2-2y^2y'$

$$\frac{\partial f}{\partial y}=x+2y-4yy' \quad\text{and}\quad \frac{\partial f}{\partial y'}=-2y^2$$

$$\frac{d}{dx}\left(\frac{\partial f}{\partial y'}\right)=\frac{d}{dx}(-2y^2)=-4yy'$$

Putting these values in (1), we get $x+2y-4yy'-(-4yy')=0$

or $\quad x+2y=0$ or $y=-\frac{x}{2}$ At $x=0, y=0$; At $x=1, y=-\frac{1}{2}$.

This extremal does not satisfy the boundary conditions $y(0)=1$, $y(1)=2$.
Hence there is no extremal. **Ans.**

Example 4. *Prove that the shortest distance between two points is along a straight line.*

(*DU, II Sem. 2012*)

Solution. Let $A(x_1, y_1)$ and $B(x_2, y_2)$ be the two given points and s the length of the arc joining these points.

Then $$s=\int_{x_1}^{x_2}ds=\int_{x_1}^{x_2}\sqrt{1+\left(\frac{dy}{dx}\right)^2}\,dx=\int_{x_1}^{x_2}\sqrt{(1+y'^2)}\,dx \qquad \ldots(1)$$

$$y(x_1)=y_1,\quad y(x_2)=y_2$$

Y
B (x_2,y_2)
A (x_1,y_1)
0
X

If s satisfies the Euler's equation, then it will be minimum

$$\frac{\partial f}{\partial y}-\frac{d}{dx}\left(\frac{\partial f}{\partial y'}\right)=0 \qquad \text{(Euler's equation)}$$

Here in (1), $f=\sqrt{(1+y'^2)}$

f is independent of y, *i.e.*, $\frac{\partial f}{\partial y}=0$

$$\frac{d}{dx}\left(\frac{\partial f}{\partial y'}\right)=\frac{d}{dx}\left(\frac{\partial}{\partial y'}\sqrt{\left(1+y'^2\right)}\right)=\frac{d}{dx}\left[\frac{1}{2}\left(1+y'^2\right)^{-\frac{1}{2}}2y'\right]=\frac{d}{dx}\frac{y'}{\sqrt{\left(1+y'^2\right)}}$$

Putting these values in Euler's Equation, we have

$$0-\frac{d}{dx}\frac{y'}{\sqrt{\left(1+y'^2\right)}}=0 \quad \text{or} \quad \frac{d}{dx}\frac{y'}{\sqrt{(1+y'^2)}}=0$$

On integrating $\frac{y'}{\sqrt{(1+y'^2)}}$ = constant (c), *i.e.*, $(y')^2=c^2(1+y'^2)$

or $\quad y'^2(1-c^2)=c^2 \quad$ or $\quad y'^2=\frac{c^2}{1-c^2}=m^2 \quad$ or $\quad y'=m$ or $\frac{dy}{dx}=m$

Integrating $\quad y = mx + c$ ...(2)

which is a straight line. **Ans.**

Now $\quad y(x_1)=y_1 \quad$ and $\quad y(x_2)=y_2$

$mx_1+c=y_1$ and $mx_2+c=y_2$ ...(3)

on subtracting, we get

or $\quad y_2-y_1=m(x_2-x_1)$ or $m=\frac{y_2-y_1}{x_2-x_1}$

Subtracting (3) from (2), we get

$$y-y_1=m(x-x_1)$$

$$y-y_1=\frac{y_2-y_1}{x_2-x_1}(x-x_1)$$ **Proved.**

Example 5. *Find the curve connecting the points* (x_1, x_2) *and* (x_2, y_2) *which when rotated about the x-axis gives a minimum surface.*

Find the extremal of the functional.

$$\int_{x_1}^{x_2} 2\pi\, y\, ds \text{ or } 2\pi\int_{x_1}^{x_2} y\sqrt{(1+y'^2)}\,dx$$

Subject to $y(x_1)=y_1,\ y(x_2)=y_2$ *(D.U., April 2010)*

Solution. 2π is constant so we have to find the extremal of

$$\int_{x_1}^{x_2} y\sqrt{(1+y'^2)}\,dx$$

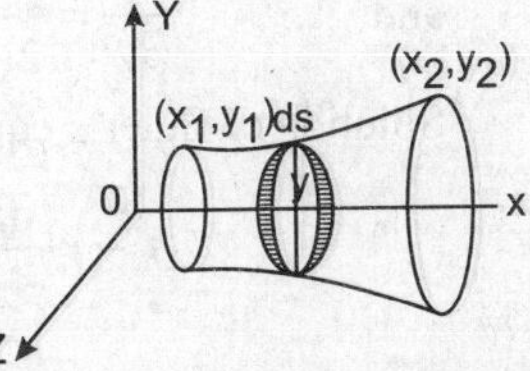

Here $f=y\sqrt{(1+y'^2)}$ which is independent of x. $\frac{\partial f}{\partial x}=0$

One form of Euler's equation is

$$\frac{d}{dx}\left[f-y'\frac{\partial f}{\partial y'}\right]-\frac{\partial f}{\partial x}=0 \quad \text{or} \quad \frac{d}{dx}\left[f-y'\frac{\partial f}{\partial y'}\right]=0 \qquad \left(\frac{\partial f}{\partial x}=0\right)$$

On integrating, we get, $f-y'\frac{\partial f}{\partial y'}$ = constant (c) ...(1)

$$f = y\sqrt{(1+y'^2)}, \frac{\partial f}{\partial y'} = \frac{2y'}{2\sqrt{1+y'^2}}y$$

Putting the value of f and $\frac{\partial f}{\partial y'}$ (1), we have

$$y\sqrt{(1+y'^2)} - y'\frac{2y'}{2\sqrt{(1+y'^2)}}y = c$$

$$\Rightarrow \quad y\sqrt{(1+y'^2)} - \frac{yy'^2}{\sqrt{(1+y'^2)}} = c \text{ or } y(1+y'^2) - yy'^2 = c\sqrt{(1+y'^2)}$$

$$y = c\sqrt{(1+y'^2)} \text{ or } y'^2 = c^2(1+y'^2)$$

$$\Rightarrow \quad y'^2 = \frac{y^2 - c^2}{c^2} \; or \; y' = \frac{\sqrt{y^2-c^2}}{c} \; or \frac{dy}{dx} = \frac{\sqrt{y^2-c^2}}{c}$$

$$\frac{dy}{\sqrt{y^2-c^2}} = \frac{dx}{c} \Rightarrow \int \frac{dy}{\sqrt{y^2-c^2}} = \int \frac{dx}{c} \Rightarrow \cosh^{-1}\frac{y}{c} = \frac{x}{c} + b$$

$y = c\cosh\left(\frac{x}{c} + b\right)$ which is the equation of catenary. This is the required extremal. **Ans.**

Example 6. *Find the curve connecting two points (not on a vertical line), such that a particle sliding down this curve under gravity (in absence of resistance) from one point to another reaches in the shortest time. (Brachistochrone problem).*

Solution. Let the particle slide on the curve OA from O with zero velocity. Let OP = s and time taken from 0 to P = t. By the law of conservation of energy, we have

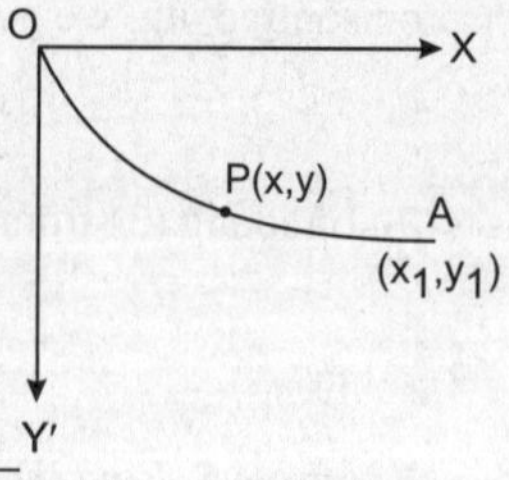

K.E. at P – K.E. at O = potential energy at P.

$$\frac{1}{2}mv^2 - 0 = mgh$$

$$\Rightarrow \quad \frac{1}{2}m\left(\frac{ds}{dt}\right)^2 = mgh \; or \; \frac{ds}{dt} = \sqrt{(2gy)}$$

Time taken by the particle to move from O to A

$$T = \int_0^T dt = \int_0^{x_1} \frac{ds}{\sqrt{(2gy)}} = \frac{1}{\sqrt{(2g)}}\int_0^{x_1}\frac{ds}{\sqrt{y}} = \frac{1}{\sqrt{(2g)}}\int_0^{x_1}\frac{\sqrt{(1+y'^2)}}{\sqrt{y}}dx$$

Here, $f = \frac{\sqrt{(1+y'^2)}}{\sqrt{y}}$ which is independent of x, i.e., $\frac{\partial f}{\partial x} = 0$.

and $$\frac{\partial f}{\partial y'} = \frac{1}{2\sqrt{y}}\frac{2y'}{\sqrt{(1+y'^2)}} = \frac{y'}{\sqrt{y}\sqrt{(1+y'^2)}}$$

Solution of Euler's equation is

$$f - y'\frac{\partial f}{\partial y'} = \text{constant } c$$

On substituting the values of f and $\frac{\partial f}{\partial y'}$, we get

$$\frac{\sqrt{(1+y'^2)}}{\sqrt{y}} - y'\frac{y'}{\sqrt{y}\sqrt{(1+y'^2)}} = c$$

$$\Rightarrow \quad \sqrt{1+y'^2} - \frac{y'^2}{\sqrt{(1+y'^2)}} = c\sqrt{y} \quad \text{or} \quad 1+y'^2-y'^2 = c\sqrt{(1+y'^2)}\sqrt{y}$$

$$\Rightarrow \quad 1 = c\sqrt{y(1+y'^2)} \quad \text{or} \quad 1+\left(\frac{dy}{dx}\right)^2 = \frac{1}{yc^2} \quad \text{or} \quad \frac{dy}{dx} = \frac{\sqrt{1-yc^2}}{yc^2}$$

$$\Rightarrow \quad \frac{dy}{dx} = \frac{\sqrt{1/c^2-y}}{y} = \frac{\sqrt{a-y}}{y} \qquad \left(\frac{1}{c^2}=a\right)$$

$$dx = \sqrt{\frac{y}{a-y}}\,dy$$

$$\int_0^x x\,dx = \int_0^y \sqrt{\left(\frac{y}{a-y}\right)}\,dy$$

Put $y = a\sin^2\theta$

$dy = 2a\sin\theta\cos\theta\,d\theta$

$$x = \int_0^\theta \sqrt{\left(\frac{a\sin^2\theta}{a-a\sin^2\theta}\right)}\,2a\sin\theta\cos\theta\,d\theta = \int_0^\theta \left(\frac{\sin\theta}{\cos\theta}\right)2a\sin\theta\cos\theta\,d\theta = \int_0^\theta 2a\sin^2\theta\,d\theta$$

$$= a\int_0^\theta (1-\cos 2\theta)\,d\theta = a\left(\theta - \frac{\sin 2\theta}{2}\right)_0^\theta$$

$$\Rightarrow \quad x = \frac{a}{2}(2\theta - \sin 2\theta) \quad \text{and} \quad y = a\sin^2\theta = \frac{a}{2}(1-\cos 2\theta)$$

On putting $\frac{a}{2} = A$ and $2\theta = \Theta$ $\left.\begin{aligned} x &= A(\Theta - \sin\Theta) \\ y &= A(1-\cos\Theta)\end{aligned}\right]$ which is a cycloid. **Ans.**

EXERCISE 18.1

1. Find the extremal of the functional

   $I[y(x)] = \int_{x_0}^{x_1} \frac{1+y^2}{y'^2}\,dy$ **Ans.** $y = \sinh(c_1x+c_2)$

2. Solve the Euler's equation for $\int_{x_0}^{x_1}(x+y')y'\,dx.$ **Ans.** $y = -\frac{x^2}{4} + c_1x + c_2$

3. Solve the Euler's equation for $\int_{x_0}^{x_1}(1+x^2y')y'\,dx$ **Ans.** $y = cx^{-1} + c_2$

Find the extremals of the functional and extremum value of the following:

4. $I[y(x)] = \int_{x_0}^{x_1} \frac{1+y^2}{y'^2}\,dx$ **Ans.** $y = \sinh(c_1x + c_2)$

5. $I[y(x)] = \int_{\frac{1}{2}}^{1} x^2y'^2\,dx$ subject to $y\left(\frac{1}{2}\right) = 1, y(1) = 2.$ **Ans.** $y = -\frac{c}{x} + d$, value = 1

6. $I[y(x)] = \int_0^2 (x-y')^2\,dx$ subject to $y(0) = 0, y(2) = 4.$ **Ans.** $y = \frac{x^2}{2} + cx + d$, value = 2

7. $\int_0^{\frac{\pi}{2}} (y'^2 - y^2)\,dx$ subject to $y(0) = 0$, $y\left(\frac{\pi}{2}\right) = 1$ **Ans.** $y = \sin x$, value = 0

8. $\int_0^1 (y'^2 + 12xy)\,dx$ subject to $y(0) = 0,\ y(1) = 1$ **Ans.** $y = x^3$, value$= \frac{21}{5}$

9. $\int_1^2 \frac{\sqrt{(1+y'^2)}}{x}\,dx$ subject to $y(1) = 0, y(2) = 1.$ **Ans.** $y = x^3$

18.6 ISOPERIMETRIC PROBLEMS

The determination of the shape of a closed curve of the given perimeter enclosing maximum area is the example of isoperimetric problem. In certain problems it is necessary to make a given integral.

$$I = \int_{x_1}^{x_2} f(x, y, y')\,dx \qquad ....(1)$$

maximum or minimum while keeping another integral

$$I = \int_{x_1}^{x_2} g(x, y, y')\,dx = K\,(\text{Constant}) \qquad ...\ (2)$$

Problems of this type are solved by Lagrange's multipliers method. We multiply (2) by λ and add to (1) to extremize (1)

$$I^* = \int_{x_1}^{x_2} f(x, y, y')\,dx + \lambda \int_{x_1}^{x_2} g(x, y, y')\,dx = \int_{x_1}^{x_2} F\,dx \text{ (say)}$$

Then by Euler's equation $\frac{\partial F}{\partial y} - \frac{d}{dx}\left(\frac{\partial F}{\partial y'}\right) I = 0$.

Note. Isoperimetric problem. To find out possible curves having the same perimeter, the one which encloses the maximum area.

Example 7. *Find the shape of the curve of the given perimeter enclosing maximum area.*

Solution. Let P be the perimeter of the closed curve,

Then $$P = \int_{x_1}^{x_2} \sqrt{1 + y'^2}\;dx \qquad ...(1)$$

The area enclosed by the curve, x-axis and two perpendicular lines is

$$A = \int_{x_1}^{x_2} y\,dx \qquad ....(2)$$

We have to find the maximum value of (2) under the condition (1).

By Lagrange's multiplier method.

$$f = y + \lambda\sqrt{1 + y'^2}$$

For maximum or minimum value of A, F must satisfy Euler's equation

$$\frac{\partial f}{\partial y} - \frac{d}{dx}\left(\frac{\partial F}{\partial y'}\right) = 0$$

$$1 - \lambda \frac{d}{dx}\left[\frac{1}{2}(1 + y'^2)^{-\frac{1}{2}}(2y')\right] = 0 \text{ or } 1 - \lambda \frac{d}{dx}\left(\frac{y'}{\sqrt{1 + y'^2}}\right) = 0$$

Integraing w.r.t. 'x', we get $x - \frac{\lambda y'}{\sqrt{(1 + y'^2)}} = a$

$$\Rightarrow \qquad \frac{\lambda y'}{\sqrt{(1+y'^2)}} = x - a \text{ or } \lambda^2 y'^2 = (1 + y'^2)(x-a)^2$$

$$[\lambda^2 - (x-a)^2]\, y'^2 = (x-a)^2$$

$$\Rightarrow \qquad y' = \frac{x-a}{\sqrt{[\lambda^2 - (x-a)^2]}} \quad \text{or} \quad \frac{dy}{dx} = \frac{x-a}{\sqrt{[\lambda^2-(x-a)^2]}}$$

Integrating w.r.t. (x), we obtain

$$y = -\sqrt{[\lambda^2 - (x-a)^2]} + b$$

$$\Rightarrow \qquad y - b = -\sqrt{[\lambda^2-(x-a)^2]} \;\Rightarrow\; (y-b)^2 = \lambda^2 - (x-a)^2 \;\Rightarrow\; (x-a)^2 + (y-b)^2 = \lambda^2$$

This is the equation of a circle whose centre is *(a, b)* and radius λ. **Ans.**

Example 8. *Find the extremal of the functional* $A = \int_{t_1}^{t_2} \frac{1}{2}(x\dot{y} - y\dot{x})dt$ *subject to the integral constraint* $\int_{t_1}^{t_2} \frac{1}{2}\sqrt{(\dot{x}^2 - \dot{y}^2)}\, dt = l.$

Solution. Here $f = \frac{1}{2}(x\dot{y} - \dot{x}y)$, $g = \sqrt{\dot{x}^2 - \dot{y}^2}$

$$F = f + \lambda g$$

$$F = \frac{1}{2}(x\dot{y} - y\dot{x}) + \lambda\sqrt{\dot{x}^2 + \dot{y}^2}$$

For A to have extremal F must satisfy the Euler's equation

$$\frac{\partial F}{\partial x} - \frac{d}{dx}\left[\frac{\partial F}{\partial \dot{x}}\right] = 0 \qquad \text{...(1)}$$

$$\frac{\partial F}{\partial y} - \frac{d}{dt}\left[\frac{\partial F}{\partial \dot{y}}\right] = 0 \qquad \text{...(2)}$$

From (1) $$\frac{1}{2}\dot{y} - \frac{d}{dt}\left(-\frac{y}{2} + \frac{\lambda 2\dot{x}}{2\sqrt{\dot{x}^2+\dot{y}^2}}\right) = 0$$

$$\frac{d}{dt}\left(y - \frac{\lambda\dot{x}}{2\sqrt{\dot{x}^2+\dot{y}^2}}\right) = 0 \qquad \text{...(3)}$$

From (2) $$-\frac{1}{2}\dot{x} - \frac{d}{dt}\left[\frac{x}{2} + \frac{\lambda\dot{y}}{\sqrt{\dot{x}^2+\dot{y}^2}}\right] = 0 \qquad \text{...(4)}$$

$$\frac{d}{dt}\left[x - \frac{\lambda\dot{y}}{\sqrt{\dot{x}^2+\dot{y}^2}}\right] = 0$$

Integrating (3) and (4), we have

$$y-\frac{\lambda\dot{x}}{\sqrt{\dot{x}^2+\dot{y}^2}}=c_1 \quad\Rightarrow\quad y-c_1=\frac{\lambda\dot{x}}{\sqrt{\dot{x}^2+\dot{y}^2}} \qquad ...(5)$$

$$x-\frac{\lambda\dot{y}}{\sqrt{\dot{x}^2+\dot{y}^2}}=c_2 \quad\Rightarrow\quad x-c_2=\frac{\lambda\dot{y}}{\sqrt{\dot{x}^2+\dot{y}^2}} \qquad ...(6)$$

Squaring (5) , (6) and adding, we get

$$(x-c_2)^2+(y-c_1)^2=\lambda^2\left(\frac{\dot{x}^2+\dot{y}^2}{\dot{x}^2+\dot{y}^2}\right)$$

$$(x-c_2)^2+(y-c_1)^2=\lambda^2$$

This is the equation of circle. **Ans.**

Example 9. *Find the solid of maximum volume formed by the revolution of a given surface area.*

Solution. Let the curve PA pass through origin and it is rotated about the x - axis.

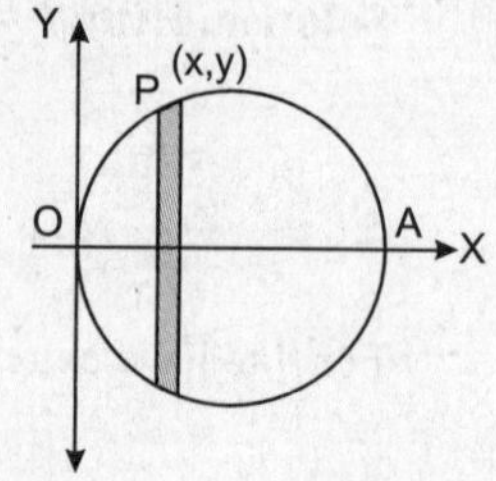

$$S=\int_0^a 2\pi y\,ds$$

$$S=\int_0^a 2\pi y\sqrt{(1+y'^2)}\,dx \qquad ...(1)$$

$$V=\int_0^a \pi y^2\,dx \qquad ...(2)$$

Here we have to extremize V with the given S.

Here $\qquad f=\pi y^2,\ g=2\pi y\sqrt{(1+y'^2)}$

$$F=f+\lambda g$$

$$F=\pi y^2+\lambda 2\pi y\sqrt{(1+y'^2)}$$

For maximum V, F must satisfy Euler's equation. But F does not contain x.

$$\therefore \qquad F-y'\frac{\partial F}{\partial y'}=C$$

$$\Rightarrow \qquad \pi y^2+\lambda 2\pi y\sqrt{(1+y'^2)}-y'\frac{1}{2}\frac{2\pi y\ \lambda y'}{\sqrt{(1+y'^2)}}=C$$

$$\Rightarrow \qquad \pi y^2+2\pi y\lambda\sqrt{(1+y'^2)}-\frac{2\pi\lambda y y'^2}{\sqrt{(1+y'^2)}}=C$$

$$\Rightarrow \qquad \pi y^2+\frac{2\pi y\lambda}{\sqrt{(1+y'^2)}}=C$$

As the curve passes through origin (0, 0), so $C = 0$.

$$\pi y^2+\frac{2\pi y\lambda}{\sqrt{(1+y'^2)}}=0$$

$$\Rightarrow \qquad y+\frac{2\lambda}{\sqrt{(1+y'^2)}}=0 \quad \Rightarrow \quad y\sqrt{(1+y'^2)}=-2\lambda$$

$$\Rightarrow \qquad 1+y'^2=\frac{4\lambda^2}{y^2} \Rightarrow \quad y'^2=\frac{4\lambda^2}{y^2}-1=\frac{4\lambda^2-y^2}{y^2}$$

$$\Rightarrow \qquad \frac{dy}{dx}=\frac{\sqrt{(4\lambda^2-y^2)}}{y}$$

$$\int\frac{ydy}{\sqrt{(4\lambda^2-y^2)}}=\int dx+C$$

$$-\sqrt{4\lambda^2-y^2}=x+C \qquad \text{...(1)}$$

$$\Rightarrow \qquad \sqrt{4\lambda^2-y^2}=-x-C$$

The curve xpasses through (0, 0). On putting $x = 0,\ y = 0$ in (1) we get

$$-C = 2\lambda$$

(1) becomes $\qquad \sqrt{4\lambda^2-y^2}=-x+2\lambda$

Squaring $\qquad 4\lambda^2-y^2=(x-2\lambda)^2$

$$\Rightarrow \qquad (x-2\lambda)^2+y^2=4\lambda^2$$

This is the equation of a circle.
Hence, on revolving the circle about x- axis, the solid formed is a sphere. **Ans.**

EXERCISE 18.2

1. Show that an isosceles triangle has the smallest perimeter for a given area and a given base.
2. Find the extremal in the isoperimetric problem of the extremum of

$$\int_0^1 (y'^2+z'^2-4xz'-4z)\,dx$$

subject to $\int_0^1 (y'^2+xy'-z'^2)\,dx=2,\ y(0)=0,\ z(0)=0,\ y(1)=1,\ z(1)=1.$ **Ans.** $y=\frac{-5x^2}{2}+\frac{7x}{2}, z=x.$

3. Find the surface with the smallest area which encloses a given volume. **Ans.** Sphere
4. Find the extremal of the functional $\int_{t_1}^{t_2}\sqrt{x^2+y^2+z^2}\,dt$ subject to $x^2+y^2+z^2=a^2$.

   Ans. Arc of a great circle of a sphere.
5. Find the extremals of the isoperimetric problem $\int_{x_0}^{x_1} y'^2 dx$ subject to $\int_{x_0}^{x_1} y\,dx=c.$ **Ans.** $y = x^2 + ax+b$

18.7 FUNCTIONALS OF SECOND ORDER DERIVATIVES

Let us consider the extremum of a functional.

$$\int_{x_1}^{x_2}[f(x,y,y',y'')]\ dx \qquad \text{...(1)}$$

The necessary condition for the above mentioned functional to be extremum is

$$\frac{\partial f}{\partial y}-\frac{d}{dx}\left(\frac{\partial f}{\partial y'}\right)+\frac{d^2}{dx^2}\left(\frac{\partial f}{\partial y''}\right)=0$$

Proof. Let the boundary conditions be

$$y(x_1) = y_1,\ y(x_2) = y_2,\ y'(x_1) = y'_1,\ y'(x_2) = y'_2$$

Let α be a parameter and $\eta(x)$ is a differentiable function.

At the end points $\eta(x_1) = \eta(x_2) = 0$ and $\eta'(x_1) = \eta'(x_2) = 0$

Putting $y + \alpha\eta(x)$for y in (1), we have

$$\int_{x_1}^{x_2} f[x, y+\alpha\eta(x), y'+\alpha\eta'(x), y''+\alpha\eta''(x)]dx$$

Writing $$\int_{x_1}^{x_2} f[x, y+\alpha\eta(x), y'+\alpha\eta'(x), y''+\alpha\eta''(x)]dx = \int_{x_1}^{x_2} F dx = 1$$

For extremum value of (1)

$$\frac{dI}{d\alpha} = 0 \qquad \frac{dI}{d\alpha} = \int_{x_1}^{x_2} \frac{\partial F}{\partial\alpha} dx$$

Differentiating under the sign of integral, we get

$$= \int_{x_1}^{x_2}\left(\frac{\partial F}{\partial y}\frac{\partial y}{\partial\alpha} + \frac{\partial F}{\partial y'}\frac{\partial y'}{\partial\alpha} + \frac{\partial F}{\partial y''}\frac{\partial y''}{\partial\alpha}\right)dx = \int_{x_1}^{x_2}\left(\frac{\partial F}{\partial y}\frac{\partial(\alpha n)}{\partial\alpha} + \frac{\partial F}{\partial y'}\frac{\partial(\alpha n')}{\partial\alpha} + \frac{\partial F}{\partial y''}\frac{\partial(\alpha n'')}{\partial\alpha}\right)dx$$

But $\frac{dI}{d\alpha} = 0$ when $\alpha = 0$

$$0 = \int_{x_1}^{x_2}\left[\frac{\partial f}{\partial y}\eta + \frac{\partial f}{\partial y}\eta' + \frac{\partial f}{\partial y}\eta''\right]dx \quad \text{or} \quad \int_{x_1}^{x_1}\frac{\partial f}{\partial y}\eta\, dx + \int_{x_1}^{x_2}\frac{\partial f}{\partial y'}\eta' dx + \int_{x_1}^{x_2}\frac{\partial f}{\partial y''}\eta'' dx = 0$$

Integrating by parts, w.r.t. 'x', we have

$$\int_{x_1}^{x_2}\frac{\partial f}{\partial y}\eta\, dx + \left[\frac{\partial f}{\partial y}\eta - \int_{x_1}^{x_2}\frac{\partial}{\partial x}\left(\frac{\partial f}{\partial y'}\right).\eta\, dx\right]_{x_1}^{x_2} + \left(\frac{\partial f}{\partial y''}\eta' - \frac{d}{dx}\left(\frac{\partial f}{\partial y''}\right).\eta + \frac{d^2}{dx^2}\left(\frac{\partial f}{\partial y''}\right)\int_{x_1}^{x_2}\eta\, dx\right)_{x_1}^{x_2} = 0$$

But $n\,(x_1) = n\,(x_2) = 0$ and $\eta'(x_1) = \eta'(x_2) = 0$

so $$\int_{x_1}^{x_2}\left[\frac{\partial f}{\partial y} - \frac{d}{dx}\left(\frac{\partial f}{\partial y'}\right) + \frac{d^2}{dx^2}\left(\frac{\partial f}{\partial y''}\right)\right]\eta(x)\,dx = 0 \Rightarrow \frac{\partial f}{\partial y} - \frac{d}{dx}\left(\frac{\partial f}{\partial y'}\right) + \frac{d^2}{dx^2}\left(\frac{\partial f}{\partial y''}\right) = 0$$ **Proved.**

EXERCISE 18.3

1. Find the extremal of $\int_{x_0}^{x_1}(16y^2 - y''^2 + x^2)\,dx$. **Ans.** $y = c_1e^{2x} + c_2e^{-2x} + c_3\cos 2x + c_4\sin 2x$

2. Find the extremal of $\int_{-c}^{c}(ay + \frac{1}{2}by''^2)\,dx$ subject to $y(-c) = 0, y'(-c) = 0$, $y(c) = 0, y'(c) = 0$. **Ans.** $y = -\frac{a}{24b}(x^2 - c^2)^2$

3. Find the extremal of $\int_0^\pi y''^2\,dx$ subject to $\int_0^\pi y^2\,dx = 1$, $y(0) = y(\pi) = 0$, $y''(0) = y''(\pi) = 0$. **Ans.** $y = a_1\sin x + a_2\sin 2x + ...$

4. Find the extremal of $\int_{x_0}^{x_1}(2xy + y'''^2)\,dx$. **Ans.** $y = \frac{x^7}{7!} + c_1x^5 + c_2x^4 + c_3x^3 + c_4x^2 + c_5x + c_6$

Maxima and Minima of Functions (Two Variables)

19.1 MAXIMUM VALUE

A function $f(x, y)$ is said to have a maximum value at $x = a$, $y = b$, if there exists a small neighbourhood of (a, b) such that,

$f(a, b) > f(a + h, b + k)$

Minimum Value. A function $f(x, y)$ is said to have a minimum value for $x = a$, $y = b$, if there exists a small neighbourhood of (a, b) such that.

$f(a, b) < f(a + h, b + k)$

The maximum and minimum values of a function are also called extreme or extremum values of the function.

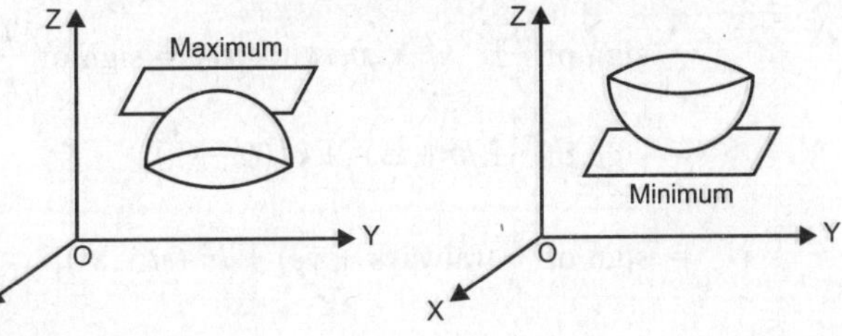

Saddle Point or Minimax

It is a point where function is neither maximum nor minimum.

Geometrically such a surface (looks like the leather seat on the back of a horse) forms a ridge rising in one direction and falling in another direction.

19.2 CONDITIONS FOR EXTREMUM VALUES

If $f(a + h, b + k) - f(a, b)$ remains of the same sign for all values (positive or negative) of h, k then $f(a, b)$ is said to be extremum value of $f(x, y)$ at (a, b)

(*i*) If $f(a + h, b + k) - f(a, b) < 0$, then $f(a, b)$ is maximum.

(*ii*) If $f(a + h, b + k) - f(a, b) > 0$, then $f(a, b)$ is minimum.

By Taylor's Theorem

$$f(a + h, b + k) = f(a, b) + \left(h\frac{\partial f}{\partial x} + k\frac{\partial f}{\partial y}\right)_{(a,b)} + \frac{1}{2!}\left[h^2\frac{\partial^2 f}{\partial x^2} + 2hk\frac{\partial^2 f}{\partial x \partial y} + k^2\frac{\partial^2 f}{\partial y^2}\right] + \ldots$$

$$\Rightarrow \quad f(a + h, b + k) - f(a, b) = \left(h\frac{\partial f}{\partial x} + k\frac{\partial f}{\partial y}\right)_{(a,b)} + \frac{1}{2!}\left[h^2\frac{\partial^2 f}{\partial x^2} + 2hk\frac{\partial^2 f}{\partial x \partial y} + k^2\frac{\partial^2 f}{\partial y^2}\right] + \ldots \quad \ldots(1)$$

$$\Rightarrow \quad f(a + h, b + k) - f(a, b) = \left(h\frac{\partial f}{\partial x} + k\frac{\partial f}{\partial y}\right)_{(a,b)} \quad \ldots(2)$$

For small values of h, k, the second and higher order terms are still smaller and hence may be neglected.

The sign of L.H.S. of (2) is governed by $h\frac{\partial f}{\partial x}+k\frac{\partial f}{\partial y}$ which may be positive or negative depending on h, k.

Hence, the necessary condition for (a, b) to be a maximum or minimum is that

$$\left(h\frac{\partial f}{\partial x}+k\frac{\partial f}{\partial y}\right)=0 \Rightarrow \frac{\partial f}{\partial x}=0, \quad \frac{\partial f}{\partial y}=0$$

By solving the equations, we get, point $x = a$, $y = b$ which may be maximum or minimum value.

Then from (1)

$$f(a + h, b + k) - f(a, b) = \frac{1}{2!}\left[h^2\frac{\partial^2 f}{\partial x^2}+2hk\frac{\partial^2 f}{\partial x\partial y}+k^2\frac{\partial^2 f}{\partial y^2}\right]$$

$$= \frac{1}{2!}[h^2r+2hks+k^2t] \qquad ...(3)$$

$$\text{where } r = \frac{\partial^2 f}{\partial x^2}, s =\frac{\partial^2 f}{\partial x\partial y}, t=\frac{\partial^2 f}{\partial y^2} \text{ at } (a, b)$$

Now the sign of L.H.S. of (3) is sign of $[rh^2 + 2hks + k^2 t]$

$$= \text{sign of } \frac{1}{r}[r^2h^2+2hkrs+k^2rt] = \text{sign of } \frac{1}{r}[(r^2h^2+2hkrs+k^2s^2)+(-k^2s^2+k^2rt)]$$

$$= \text{sign of } \frac{1}{r}[(hr+ks)^2+k^2(rt-s^2)]$$

$$= \text{sign of } \frac{1}{r}[(\text{always} + ve) + k^2 (rt - s^2)] \qquad [(hr + ks)^2 = + ve]$$

$$= \text{sign of } \frac{1}{r}[k^2 (rt - s^2)] = \text{sign of } r \text{ if } rt - s^2 > 0$$

Hence, if $rt - s^2 > 0$, then $f(x, y)$ has a maximum or minimum at (a, b) according as $r < 0$ or $r > 0$.

Note: (*i*) If $rt - s^2 < 0$, then L.H.S. will change with h and k hence there is no maximum or minimum at (a, b), *i.e.*, it is a saddle point.

(*ii*) If $rt - s^2 = 0$, then $rh^2 + 2shk + tk^2 = \frac{1}{r}[(rh + sk)^2 + k^2 (rt - s^2)]$

$$= \frac{1}{r}(rh + sk)^2 \text{ which is zero for values of } h, k, \text{ such that}$$

$$\Rightarrow \quad \frac{h}{k} = -\frac{s}{r}$$

This is, therefore, a doubtful case, further investigation is required.

19.3 WORKING RULE TO FIND EXTREMUM VALUES (*D.U., April 2010*)

(*i*) Differentiate $f(x, y)$ and find out

$$\frac{\partial f}{\partial x}, \frac{\partial f}{\partial y}, \frac{\partial^2 f}{\partial x^2}, \frac{\partial^2 f}{\partial x\partial y}, \frac{\partial^2 f}{\partial y^2}$$

(*ii*) Put $\frac{\partial f}{\partial x} = 0$ and $\frac{\partial f}{\partial y} = 0$ and solve these equations for x and y. Let (a, b) be the values of (x, y).

(*iii*) Evaluate $r = \dfrac{\partial^2 f}{\partial x^2}$, $s = \dfrac{\partial^2 f}{\partial x \partial y}$, $t = \dfrac{\partial^2 f}{\partial y^2}$ for these values (a, b).

(*iv*) If $rt - s^2 > 0$ and

(*a*) $r < 0$, then $f(x, y)$ has a maximum value.

(*b*) $r > 0$, then $f(x, y)$ has a minimum value.

(*v*) If $rt - s^2 < 0$, then $f(x, y)$ has no extremum value at the point (a, b).

(*iv*) If $rt - s^2 = 0$, then the case is doubtful and needs further investigation.

Note: The point (a, b), which are the roots of $\dfrac{\partial f}{\partial x} = 0$ and $\dfrac{\partial f}{\partial y} = 0$ are called stationary points.

Example 1. *Find the absolute maximum and minimum values of*

$$f(x, y) = 2 + 2x + 2y - x^2 - y^2$$

on triangular plane in the first quadrant, bounded by the lines $x = 0$, $y = 0$ and $y = 9 - x$. (*Gujarat, I Semester, Jan. 2009*)

Solution. We have, $f(x, y) = 2 + 2x + 2y - x^2 - y^2$

$$\frac{\partial f}{\partial x} = 2 - 2x, \qquad \frac{\partial f}{\partial y} = 2 - 2y$$

$$\frac{\partial^2 f}{\partial x^2} = -2, \qquad \frac{\partial^2 f}{\partial x \partial y} = 0, \qquad \frac{\partial^2 f}{\partial x^2} = -2$$

For maxima and minima,

$$\frac{\partial f}{\partial x} = 0 \quad \Rightarrow \quad 2 - 2x = 0 \quad \Rightarrow \quad x = 1$$

and $$\frac{\partial f}{\partial y} = 0 \quad \Rightarrow \quad 2 - 2y = 0 \quad \Rightarrow \quad y = 1$$

At (1, 1) $$rt - s^2 = (-2)(-2) - 0 = +4.$$

Here $$r = \frac{\partial^2 f}{\partial x^2} = -2 = -\text{ve}$$

Hence $f(x, y)$ is maximum at (1, 1).

Maximum value of $f(x, y) = 2 + 2 + 2 - 1 - 1 = 4$. **Ans.**

Example 2. *Examine the function $f(x, y) = y^2 + 4xy + 3x^2 + x^3$ for extreme values.* (*D.U., 2014 M.U. 2008*)

Solution. We have, $f(x, y) = y^2 + 4xy + 3x^2 + x^3$

$$p = \frac{\partial f}{\partial x} = 4y + 6x + 3x^2 \qquad q = \frac{\partial f}{\partial y} = 2y + 4x$$

$$r = \frac{\partial^2 f}{\partial x^2} = 6 + 6x \qquad s = \frac{\partial^2 f}{\partial x \partial y} = 4$$

$$t = \frac{\partial^2 f}{\partial y^2} = 2$$

For maxima or minima

$$\frac{\partial f}{\partial x} = 0 \qquad \frac{\partial f}{\partial y} = 0$$

$\Rightarrow$ $4y + 6x + 3x^2 = 0$ ...(1) $\qquad$ $2y + 4x = 0$

$y = -2x$ ...(2)

Putting the value of y from (2) in (1), we get

$$4(-2x) + 6x + 3x^2 = 0$$

$$\Rightarrow \quad 3x^2 + 6x - 8x = 0$$

$$\Rightarrow \quad 3x^2 - 2x = 0 \Rightarrow x(3x-2) = 0$$

$$\Rightarrow \quad x = 0 \text{ or } x = \frac{2}{3}$$

when $x = 0$ then $y = 0$

when $x = \frac{2}{3}$ then $y = -2\left(\frac{2}{3}\right) = -\frac{4}{3}$

Thus, the stationary points are (0, 0) and $\left(\frac{2}{3}, \frac{-4}{3}\right)$,

Stationary points	(0, 0)	$\left(\frac{2}{3}, \frac{-4}{3}\right)$
$r = 6 + 6x$	6	10
$s = 4$	4	4
$t = 2$	2	2
$rt - s^2$	-4	$+4$

At (0, 0) there is no extremum value, since $rt - s^2 < 0$;

At $\left(\frac{2}{3}, \frac{-4}{3}\right)$, $rt - s^2 > 0$, $r > 0$.

Therefore $\left(\frac{2}{3}, \frac{-4}{3}\right)$ is a point of minimum value.

The minimum value of $f\left(\frac{2}{3}, -\frac{4}{3}\right) = \left(\frac{-4}{3}\right)^2 + 4\left(\frac{2}{3}\right)\left(-\frac{4}{3}\right) + 3\left(\frac{2}{3}\right)^2 + \left(\frac{2}{3}\right)^3$

$$\Rightarrow \quad f\left(\frac{2}{3}, \frac{-4}{3}\right) = \frac{16}{9} - \frac{32}{9} + \frac{12}{9} + \frac{8}{27} = \frac{8}{27} - \frac{4}{9} = -\frac{4}{27}$$ **Ans.**

Example 3. *Show that the minimum value of* $u = xy + a^3\left(\frac{1}{x} + \frac{1}{y}\right)$ *is* $3a^2$.

Solution. We have,

$$f(x, y) = xy + a^3\left(\frac{1}{x} + \frac{1}{y}\right)$$

$$p = \frac{\partial f}{\partial x} = y - \frac{a^3}{x^2} \qquad q = \frac{\partial f}{\partial y} = x - \frac{a^3}{y^2}$$

$$r = \frac{\partial^2 f}{\partial x^2} = \frac{2a^3}{x^3} \qquad s = \frac{\partial^2 f}{\partial x \partial y} = 1$$

$$t = \frac{\partial^2 f}{\partial y^2} = \frac{2a^3}{y^3}$$

For maxima and minima

$$\frac{\partial f}{\partial x} = 0 \qquad\qquad \frac{\partial f}{\partial y} = 0$$

$$\Rightarrow \quad y - \frac{a^3}{x^2} = 0 \qquad\qquad \Rightarrow \quad x - \frac{a^3}{y^2} = 0$$

$$\Rightarrow \quad x^2 y = a^3 \qquad \text{...(1)}$$

$$\Rightarrow \quad x = \frac{a^3}{y^2} \qquad \text{...(2)}$$

Putting the value of x from (2) in (1), we get

$$\left(\frac{a^3}{y^2}\right)^2 y = a^3 \Rightarrow \frac{a^6}{y^3} = a^3 \Rightarrow y = a \qquad \text{...(3)}$$

On putting the value of y from (3) in (1), we get

$$x^2 a = a^3 \Rightarrow x^2 = a^2 \Rightarrow x = \pm a$$

Thus, the stationary pairs are (a, a) and $(-a, a)$

Stationary points	(a, a)	$(-a, a)$
$r = \dfrac{2a^3}{x^3}$	2	-2
$s = 1$	1	1
$t = \dfrac{2a^3}{y^3}$	2	2
$rt - s^2$	$+3$	-5

At $(-a, a)$, $rt - s^2 = -ve$

Hence, we reject this pair.

At (a, a), $r = +ve$, $rt - s^2 = +ve$

Hence, $f(x, y)$ is minimum at (a, a).

Minimum value $= a^2 + a^3\left(\dfrac{1}{a} + \dfrac{1}{a}\right) = 3a^2$ **Ans.**

Example 4. *Examine $f(x, y) = x^3 + y^3 - 3axy$ for maximum and minimum values.*

(GBTU, 2012, DU, II Sem. 2012,)

Solution. We have, $f(x, y) = x^3 + y^3 - 3axy$

$$p = \frac{\partial f}{\partial x} = 3x^2 - 3ay, \quad q = \frac{\partial f}{\partial y} = 3y^2 - 3ax$$

$$r = \frac{\partial^2 f}{\partial x^2} = 6x, \qquad s = \frac{\partial^2 f}{\partial x \partial y} = -3a, \qquad t = \frac{\partial^2 f}{\partial y^2} = 6y$$

For maxima and minima

$$\frac{\partial f}{\partial x} = 0 \text{ and } \qquad \frac{\partial f}{\partial y} = 0.$$

$$3x^2 - 3ay = 0 \qquad\qquad 3y^2 - 3ax = 0$$

$$\Rightarrow \quad x^2 = ay \Rightarrow y = \frac{x^2}{a} \quad \text{...(1)} \qquad\qquad \Rightarrow y^2 = ax \quad \text{...(2)}$$

Putting the value of y from (1) in (2), we get

$$x^4 = a^3 x \Rightarrow x(x^3 - a^3) = 0$$

$$\Rightarrow \quad x(x-a)(x^2 + ax + a^2) = 0$$

$$x = 0, a$$

Putting $x = 0$ in (1), we get $\quad y = 0$

Putting $x = a$ in (1), we get $\quad y = a,$

	(0, 0)	(a, a)
$r = 6x$	0	$6a$
$s = -3a$	$-3a$	$-3a$
$t = 6y$	0	$6a$
$rt - s^2$	$-9a^2 < 0$	$27a^2 > 0$

At (0, 0) there is no extremum value, since $rt - s^2 < 0$.

At (a, a), $\quad rt - s^2 > 0, \quad r > 0$

Therefore (a, a) is a point of minimum value.

The minimum value of $f(a, a) = a^3 + a^3 - 3a^3 = -a^3$ **Ans.**

Example 5. *Show that the function*

$$f(x, y) = x^3 + y^3 - 63(x + y) + 12xy$$

is maximum at $(-7, -7)$ *and minimum at* (3, 3.)

Solution. We have, $f(x, y) = x^3 + y^3 - 63(x + y) + 12xy$ ...(1)

$$\frac{\partial f}{\partial x} = 3x^2 - 63 + 12y, \qquad \frac{\partial f}{\partial y} = 3y^2 - 63 + 12x$$

$$\frac{\partial^2 f}{\partial x^2} = 6x, \quad \frac{\partial^2 f}{\partial x \partial y} = 12, \qquad \frac{\partial^2 f}{\partial y^2} = 6y$$

For extremum, we have

$$p = \frac{\partial f}{\partial x} = 3x^2 - 63 + 12y = 0 \qquad \Rightarrow x^2 + 4y - 21 = 0 \quad \text{...(2)}$$

$$q = \frac{\partial f}{\partial x} = 3y^2 - 63 + 12x = 0 \qquad \Rightarrow y^2 + 4x - 21 = 0 \quad \text{...(3)}$$

$$r = \frac{\partial^2 f}{\partial x^2} = 6x \quad , \quad s = \frac{\partial^2 f}{\partial x \partial y} = 12 \quad , \quad t = \frac{\partial^2 f}{\partial y^2} = 6y$$

We have to solve (2) and (3) for x, y.

On subtracting (3) from (2), we have

$$x^2 - y^2 - 4(x - y) = 0 \quad \Rightarrow \quad (x - y)(x + y - 4) = 0$$

$$x = y \text{ and } x + y = 4 \quad \text{...(4)}$$

If $x = y$ then (2) becomes, $x^2 + 4x - 21 = 0, \qquad (x + 7)(x - 3) = 0$

$$x = -7, \quad \text{and} \quad x = 3$$

$$y = -7, \quad \text{and} \quad y = 3$$

Two stationary points are $(-7, -7)$ and $(3, 3)$

On solving (2) and (4), we get

$$x^2 + 4(4 - x) - 21 = 0 \Rightarrow x^2 - 4x - 5 = 0$$

$$\Rightarrow \quad (x - 5)(x + 1) = 0$$

$$x = -1, x = 5$$

$$y = 5, y = -1$$

Two more stationary points are $(-1, 5)$ and $(5, -1)$

Hence four possible extremum points of $f(x, y)$ are $(-7, -7)$, $(3, 3)$, $(-1, 5)$ and $(5, -1)$ may be.

	$(-7, -7)$	$(3, 3)$	$(-1, 5)$	$(5, -1)$
$r = 6x$	-42	$+18$	-6	30
$s = 12$	12	12	12	12
$t = 6y$	-42	18	30	-6
$rt - s^2$	$+1620$	$+180$	-324	-324

At $(-7, -7)$

$$r = -\text{ve}, \text{ and } rt - s^2 = +\text{ve}$$

Hence, $f(x, y)$ is maximum at $(-7, -7)$.

At $(3, 3)$

$$r = +\text{ve}, \text{ and } rt - s^2 = +\text{ve}$$

Hence $f(x, y)$ is minimum at $(3, 3)$. **Proved.**

Example 6. *Find the points on the surface $z^2 = xy + 1$ nearest to the origin.*

(Nagpur University, Summer 2005,*)*

Solution. Let the point on the surface $z^2 = xy + 1$ be (x, y, z). ...(1)

$\therefore$ Its distance from origin is $r = \sqrt{x^2 + y^2 + z^2}$.

Let $$u = r^2 = x^2 + y^2 + z^2 \quad ...(2)$$

We have to find the values of x, y, z for which 'u' is minimum.

Put the value of z^2 from (1) in (2) to get the equation in two variables only.

$$u = x^2 + y^2 + xy + 1$$

For maximum or minimum, we have

$$\frac{\partial u}{\partial x} = 2x + y = 0 \quad ...(3) \qquad \text{and} \quad \frac{\partial u}{\partial y} = 2y + x = 0 \quad ...(4)$$

Solving equations (3) and (4), we get $x = 0, y = 0$

On putting $x = 0, y = 0$ in $z^2 = xy + 1$, we get

$$z^2 = 0 + 1 = 1 \qquad \therefore z = \pm 1$$

Now, $$\frac{\partial^2 u}{\partial x^2} = 2, \quad \frac{\partial^2 u}{\partial y^2} = 2, \quad \frac{\partial^2 u}{\partial x \partial y} = 1$$

$$r = 2, \quad t = 2, \quad s = 1$$

$$rt - s^2 = 2 \times 2 - (1)^2 = +3$$

But $r = +2$, so u in minimum for $(0, 0, \pm 1)$ **Ans.**

Example 7. *Divide* 120 *into three parts so that the sum of their products taken two at a time shall be maximum.*

Solution. Let x, y, z be the number whose sum is 120.

i.e., $x + y + z = 120 \Rightarrow z = 120 - x - y$ ...(1)

Let $$f = xy + yz + zx$$

$$\Rightarrow \quad f = xy + y(120 - x - y) + x(120 - x - y) \quad \text{[Using (1)]}$$

$$\Rightarrow \quad f = xy + 120y - xy - y^2 + 120x - x^2 - xy$$

$$\Rightarrow \quad f = 120x + 120y - xy - x^2 - y^2$$

$$p = \frac{\partial f}{\partial x} = 120 - y - 2x \qquad q = \frac{\partial f}{\partial y} = 120 - x - 2y$$

$$r = \frac{\partial^2 f}{\partial x^2} = -2 \qquad s = \frac{\partial^2 f}{\partial x \partial y} = -1$$

$$t = \frac{\partial^2 f}{\partial y^2} = -2$$

For maxima and minima

$$\frac{\partial f}{\partial x} = 0 \quad \text{and} \quad \frac{\partial f}{\partial y} = 0$$

$\Rightarrow \quad 120 - y - 2x = 0$ | $\Rightarrow 120 - x - 2y = 0 \quad ...(3)$

$\Rightarrow \quad y = 120 - 2x \quad ...(2)$

Putting the value of y from (2) in (3), we get

$$120 - x - 2(120 - 2x) = 0$$

$$\Rightarrow \quad 120 - x - 240 + 4x = 0 \quad \Rightarrow \quad 3x = 120 \quad \Rightarrow \quad x = 40$$

Putting the value of x in (1), we get

$$y = 120 - 2(40) = 120 - 80 = 40$$

Thus, the stationary pair is (40, 40).

	(40, 40)
$r = -2$	-2
$s = -1$	-1
$t = -2$	-2
$rt - s^2$	$+3$

At (40, 40), $r = -$ ve and $rt - s^2 = +$ ve

Hence, f is maximum at (40, 40).

Putting $x = 40$, $y = 40$ in (1), we get

$$40 + 40 + z = 120 \quad \Rightarrow \quad z = 40$$

Hence, f is maximum at $x = 40$, $y = 40$ and $z = 40$. **Ans.**

Example 8. *Divide 24 into three parts such that continued product of first, square of second and cube of third is a maximum.* *(Nagpur University, Summer 2005,)*

Solution. Let 24 be divided into x, y, z then $x + y + z = 24 \Rightarrow z = 24 - x - y$

and $f(x, y, z) = x^3 y^2 z = x^3 y^2 (24 - x - y)$

$$\frac{\partial f}{\partial x} = 72 x^2 y^2 - 4 x^3 y^2 - 3 x^2 y^3$$

$$\frac{\partial f}{\partial y} = 48 x^3 y - 2 x^4 y - 3 x^3 y^2$$

For a maximum value of f, $\frac{\partial f}{\partial x} = 0$ and $\frac{\partial f}{\partial y} = 0.$

Hence, we have $x^2y^2(72 - 4x - 3y) = 0$

$$\Rightarrow \quad 72 - 4x - 3y = 0 \quad ...(1)$$

and $x^3 y(48 - 2x - 3y) = 0$

$$\Rightarrow \quad 48 - 2x - 3y = 0 \quad ...(2)$$

By (2) × 2 – (1), we get $24 - 3y = 0 \quad \Rightarrow \quad y = 8$

Hence, $x = \frac{48 - 24}{2} = 12$

	$x = 12,\ y = 8$
$r = \dfrac{\partial^2 f}{\partial x^2} = 144\,xy^2 - 12x^2\,y^2 - 6xy^3$	$144.(12).(8)^2 - 12\,(12)^2\,8^2 - 6\,(12)\,8^3 = 12(64)(-48)$ $= -\,36864$
$t = \dfrac{\partial^2 f}{\partial y^2} = 48\,x^3 - 2x^4 - 6x^3\,y$	$48.(12)^3 - 2\,(12)^4 - 6(12)^3\,8 = (12)^3\,(-24) = -\,41472$
$s = \dfrac{\partial^2 f}{\partial x\,\partial y} = 144x^2\,y - 8x^3\,y - 9x^2\,y^2$	$144.(12)^2\,(8) - 8(12)^3\,8 - 9(12)^2\,(8)^2 = 144(8)(-24)$ $= -\,27648$
$rt - s^2$	$(-\,36864)\,(-\,41472) - (-\,27648)^2 = 764411904 > 0$

Since $rt - s^2 > 0$ and $r < 0$, so f is maximum at (12, 8)

Putting $x = 12$, $y = 8$ in $x + y + z = 24$, we get $z = 4$.

Hence, the division is 12, 8 and 4. **Ans.**

Example 9. *A rectangular box, open at the top, is to have a volume of 32 c.c. Find the dimensions of the box requiring least material for its construction.*

(*A.M.I.E.T.E., June 2009, M.U. 2009; U.P. I semester Dec. 2005,*)

Solution. Let l, b and h be the length, breadth, and height of the box respectively and S its surface area and V the volume.

$$V = 32 \text{ c.c.}$$

$$\Rightarrow \qquad l\,b\,h = 32 \text{ or } b = \frac{32}{lh}$$

$$S = 2\,(l + b)\,h + l\,b \qquad \text{...(1)}$$

Putting the value of b in (1), we get

$$S = 2\left(l + \frac{32}{l\,h}\right)h + l\left(\frac{32}{l\,h}\right)$$

$$S = 2\,l\,h + \frac{64}{l} + \frac{32}{h} \qquad \text{...(2)}$$

Differentiating (2) partially w.r.t. l, we get

$$\frac{\partial S}{\partial l} = 2\,h - \frac{64}{l^2} \qquad \text{...(3)}$$

Differentiating (2) partially w.r.t. h, we get

$$\frac{\partial S}{\partial h} = 2\,l - \frac{32}{h^2} \qquad \text{...(4)}$$

For maximum and minimum S, we get

$$\frac{\partial S}{\partial l} = 0 \;\Rightarrow\; 2\,h - \frac{64}{l^2} = 0 \;\Rightarrow\; h = \frac{32}{l^2} \qquad \text{...(5)}$$

$$\frac{\partial S}{\partial h} = 0 \;\Rightarrow\; 2\,l - \frac{32}{h^2} = 0 \;\Rightarrow\; l = \frac{16}{h^2} \qquad \text{...(6)}$$

From (5) and (6), $l = 4$, $h = 2$ and $b = 4$

$$\frac{\partial^2 S}{\partial l^2} = \frac{128}{l^3} = \frac{128}{64} = 2$$

$$\frac{\partial^2 S}{\partial l\,\partial h} = 2$$

$$\frac{\partial^2 S}{\partial h^2} = \frac{64}{h^3} = \frac{64}{8} = 8$$

$$\frac{\partial^2 S}{\partial l^2}\cdot\frac{\partial^2 S}{\partial h^2}-\left(\frac{\partial^2 S}{\partial l\,\partial h}\right)^2 = (2)\,(8) - (2)^2 = +\,12$$

$$\frac{\partial^2 S}{\partial l^2} = +\,2, \text{ so } S \text{ is minimum for } l = 4 \text{ cm}, b = 4 \text{ cm}, h = 2 \text{ cm}$$ **Ans.**

Example 10. *In a plane triangle ABC, find the maximum value of cos A cos B cos C.*

Solution.
$$\cos A \cos B \cos C = \cos A \cos B \cos [\pi - (A + B)]$$
$$= -\cos A \cos B \cos (A + B)$$

Let
$$f(A, B) = -\cos A \cos B \cos (A + B)$$
$$\frac{\partial f}{\partial A} = -\cos B\,[-\sin A \cos (A + B) - \cos A \sin (A + B)]$$
$$= \cos B \sin [A + (A + B)]$$
$$= \cos B \sin (2A + B)$$
$$\frac{\partial f}{\partial B} = -\cos A\,[-\sin B \cos (A + B) - \cos B \sin (A + B)]$$
$$= \cos A \sin [B + (A + B)] = \cos A \sin (A + 2B)$$
$$r = \frac{\partial^2 f}{\partial A^2} = 2 \cos B \cos (2A + B)$$
$$s = \frac{\partial^2 f}{\partial A \partial B} = -\sin B \sin (2A + B) + \cos B \cos (2A + B)$$
$$= \cos [B + (2A + B)] = \cos (2A + 2B)$$
$$t = \frac{\partial^2 f}{\partial B^2} = 2 \cos A \cos (A + 2B)$$

For maxima and minima
$$\frac{\partial f}{\partial A} = 0 \text{ and } \frac{\partial f}{\partial B} = 0$$

$$\cos B \sin (2A + B) = 0 \qquad ...(1)$$
$$\cos A \sin (A + 2B) = 0 \qquad ...(2)$$

From (1) if $\cos B = 0$, then $B = \pi/2$

$$\cos A \sin (A + 2B) = 0$$

From (2), $\cos A \sin (A + \pi) = 0 \quad \Rightarrow \quad \cos A\,(-\sin A) = 0$

$\Rightarrow$ either $\cos A = 0$, *i.e.* $A = \pi/2$ which is not possible.

$\sin A = 0$ *i.e.* $A = 0$ or π, which is not possible.

$$\left[\begin{array}{l}\because\ A + B + C = \pi \\ \dfrac{\pi}{2} + \dfrac{\pi}{2} + C = \pi \Rightarrow C = 0\end{array}\right]$$

From (1) if $\cos B \neq 0$, similarly $\cos A \neq 0$.

From (1), $\sin (2A + B) = 0 \quad \Rightarrow \quad 2A + B = \pi \qquad ...(3)$

From (2), $\sin (A + 2B) = 0 \quad \Rightarrow \quad A + 2B = \pi \qquad ...(4)$

Solving (3) and (4), we get $A = B = \pi/3$

$$r = 2 \cos \frac{\pi}{3} \cos \pi = 2 \times \frac{1}{2}(-1) = -1$$

$$s = \cos \frac{4\pi}{3} = -\frac{1}{2}$$

$$t = 2\cos\frac{\pi}{3}\cos\pi = 2\left(\frac{1}{2}\right)(-1) = -1$$

$$r\,t - s^2 = 1 - \frac{1}{4} = \frac{3}{4} = +\text{ ve}$$

Also, $r = -1 = -$ ve

$\Rightarrow$ $f(A, B)$ is maximum at $A = B = \dfrac{\pi}{3}$

Maximum value $= f(\pi/3, \pi/3) = \cos\dfrac{\pi}{3}\cos\dfrac{\pi}{3}\cos\dfrac{\pi}{3}$

$$= \left(\frac{1}{2}\right)\left(\frac{1}{2}\right)\left(\frac{1}{2}\right) = \frac{1}{8}.$$ **Ans.**

Example 11. *Prove that if the perimeter of a triangle is constant, its area is maximum when the triangle is equilateral.*

Solution. Let a, b, c, be the sides of a triangle whose perimeter $2s$ is constant.

Then $2s = a + b + c \Rightarrow c = 2s - a - b$ ...(1)

$$\Delta = \sqrt{s(s-a)(s-b)(s-c)} = \sqrt{s(s-a)(s-b)(a+b-s)}$$ [Using (1)]

Let $z = \Delta^2 = s(s-a)(s-b)(a+b-s) = f(a, b)$

$$\frac{\partial f}{\partial a} = s(s-b)\frac{\partial}{\partial a}[(s-a)(a+b-s)]$$

$$= s(s-b)[-(a+b-s) + (s-a)] = s(s-b)(2s-2a-b)$$

$$\frac{\partial f}{\partial b} = s(s-a)\frac{\partial}{\partial b}[(s-b)(a+b-s)]$$

$$= s(s-a)[-(a+b-s) + (s-b)] = s(s-a)(2s-a-2b)$$

Now, $\dfrac{\partial f}{\partial a} = 0$ and $\dfrac{\partial f}{\partial b} = 0$

$\Rightarrow$ $s(s-b)(2s-2a-b) = 0$ and $s(s-a)(2s-a-2b) = 0$

$\Rightarrow$ $(s-b)(2s-2a-b) = 0$ ...(2) $\Rightarrow$ $(s-a)(2s-a-2b) = 0$ ...(3) $[\because s \neq 0]$

From (2), $s = b$ or $2s = 2a + b$

When $s = b$, from (3), $(b-a)(-a) = 0 \Rightarrow b = a$ $[\because a \neq 0]$

When $2s = 2a + b$, from (3), $\dfrac{b}{2}(a-b) = 0$ or $a = b$ $[\because b \neq 0]$

If we express z as a function of b and c, we similarly get $b = c$

$\therefore$ $a = b = c = \dfrac{2s}{3}$

$$\frac{\partial^2 f}{\partial a^2} = -2s(s-b) \Rightarrow \frac{\partial^2 f}{\partial b^2} = -2s(s-a)$$

$$\frac{\partial^2 f}{\partial a\,\partial b} = s[-(2s-a-2b) - (s-a)] = s(2a+2b-3s)$$

$$\frac{\partial^2 f}{\partial a^2} = -2s\left(\frac{s}{3}\right) = -\frac{2s^2}{3} < 0$$

$$\frac{\partial^2 f}{\partial a\,\partial b} = s\left(\frac{4s}{3} + \frac{4s}{3} - 3s\right) = s\left(-\frac{s}{3}\right) = -\frac{s^2}{3}$$

$$\frac{\partial^2 f}{\partial b^2} = -2s\left(\frac{s}{3}\right) = -\frac{2s^2}{3}$$

$$\frac{\partial^2 f}{\partial a^2}\frac{\partial^2 f}{\partial b^2} - \left(\frac{\partial^2 f}{\partial a \partial b}\right)^2 = \frac{4s^4}{9} - \frac{s^4}{9} = \frac{s^4}{3} > 0. \text{ Also } \frac{\partial^2 f}{\partial a^2} < 0$$

Hence, Δ is maximum when $a = b = c = \frac{2s}{3}$ *i.e.* when the triangle is equilateral. **Proved.**

Example 12. *Show that the diameter of the right circular cylinder of greatest curved surface which can be inscribed in a given cone is equal to the radius of the cone.*

(*R.G.P.V. Bhopal, April 2010*)

Solution. Let R be the radius of the cone ABC and H be the height of the cone.

A cylinder PQRS of radius r and height h is inscribed in the cone.

In Δ APT and Δ PBQ

$$\frac{AT}{PQ} = \frac{PT}{BQ} \Rightarrow \frac{H-h}{h} = \frac{r}{R-r} \Rightarrow \left(\frac{H}{h} - 1\right) = \frac{r}{R-r} \Rightarrow \frac{H}{h} = \frac{r}{R-r} + 1$$

$$\frac{H}{h} = \frac{r+R-r}{R-r} \Rightarrow \frac{H}{h} = \frac{R}{R-r} \Rightarrow h = \frac{R-r}{R}H \quad \text{... (1)}$$

Let the curved surface of the cylinder be S.

$$S = 2\pi r h \quad \text{... (2)}$$

Putting the value of h from (1) in (2), we get

$$S = 2\pi r \frac{R-r}{R} H$$

$$= \frac{2\pi H}{R}(rR - r^2) \quad \text{... (3)}$$

Differentiating (3) w.r.t. 'r', we get

$$\frac{\partial S}{\partial r} = \frac{2\pi H}{R}(R - 2r)$$

$$\frac{\partial^2 S}{\partial r^2} = \frac{2\pi H}{R}(-2) = -\frac{4\pi H}{R} = -Ve$$

For maximum curved surface area $\frac{\partial S}{\partial r} = 0$

$$\Rightarrow \frac{2\pi H}{R}(R - 2r) = 0 \Rightarrow R - 2r = 0 \Rightarrow R = 2r \text{ and } \frac{\partial^2 S}{\partial r^2} = -Ve$$

Hence, for maximum 'S', Diameter of the cylinder = Radius of the cone. **Proved.**

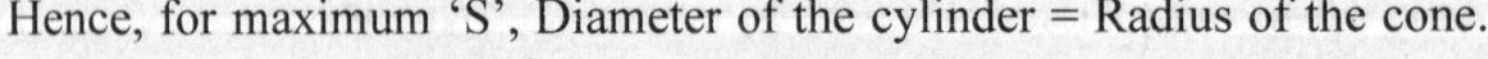

EXERCISE 19.1

Find the stationary points of the following functions

1. $f(x, y) = x^3 y^2 (1 - x - y)$ **Ans.** $\left(\frac{1}{2}, \frac{1}{3}\right)$, Maximum

2. $f(x, y) = x^3 + 3xy^2 - 15x^2 - 15y^3 + 72x.$ (*M.U. 2007, 2005,*) **Ans.** (6, 0), (4 0)

3. $f(x, y) = x^2 + 2xy + 2y^2 + 2x + 3y$ such that $x^2 - y = 1$. **Ans.** $\left(-\frac{3}{4}, -\frac{7}{16}\right), -\frac{155}{128}.$

4. $f(x, y) = xy\, e^{-(2x + 3y)}$ **Ans.** (3, 2)

5. Find the extreme value of the function $f(x, y) = x^2 + y^2 + xy + x - 4y + 5$.
State whether this value is a relative maximum or a relative minimum.
Ans. Minimum value of $f(x, y)$ at $(-2, 3) = -2$.

6. Find the values of x and y for which $x^2 + y^2 + 6x + 12$ has a minimum value and find this minimum value. **Ans.** $(-3, 0)$, 3.

7. In a plane triangle ABC, find the maximum value of $\cos A \cos B \cos C$.

8. Find a point within a triangle such that the sum of the square of its distances from the three angular points is a minimum.

9. Find the maximum and minimum values of $x^3 + 3xy^2 - 3x^2 - 3y^2 + 4$. *(M.U. 2006)*
Ans. Maximum at (0, 0) and mimum (2,0)

10. Test the function $f(x, y) = (x^2 + y^2)\, e^{-(x^2 + y^2)}$ for maxima and minima for limits not on the circle $x^2 + y^2 = 1$.

Tick (✓) the correct answer in the following :

12. One of the stationary values of the function $f(x, y) = x^4 + y^4 - 2x^2 + 4xy - 2y^2$ is

(*a*) $(\sqrt{2}, -\sqrt{2})$ (*b*) $(2, -2)$ (*c*) $(\sqrt{2}, \sqrt{2})$ (*d*) $(-2, 2)$

(AMIETE, June 2009) **Ans.** (*a*)

19.4 LAGRANGE METHOD OF UNDETERMINED MULTIPLIERS

Let $f(x, y, z)$ be a function of three variables x, y, z and the variables be connected by the relation.

$$\phi(x, y, z) = 0 \qquad \text{...(1)}$$

$\Rightarrow$ $f(x, y, z)$ to have stationary values,

$$\frac{\partial f}{\partial x} = 0, \quad \frac{\partial f}{\partial y} = 0, \quad \frac{\partial f}{\partial z} = 0$$

$$\Rightarrow \qquad \frac{\partial f}{\partial x}dx + \frac{\partial f}{\partial y}dy + \frac{\partial f}{\partial z}dz = 0 \qquad \text{...(2)}$$

By total differentiation of (1), we get

$$\frac{\partial \phi}{\partial x}dx + \frac{\partial \phi}{\partial y}dy + \frac{\partial \phi}{\partial z}dz = 0 \qquad \text{...(3)}$$

Multiplying (3) by λ and adding to (2), we get

$$\left(\frac{\partial f}{\partial x}dx + \lambda\frac{\partial \phi}{\partial x}dx\right) + \left(\frac{\partial f}{\partial y}dy + \lambda\frac{\partial \phi}{\partial y}dy\right) + \left(\frac{\partial f}{\partial z}dz + \lambda\frac{\partial \phi}{\partial z}dz\right) = 0$$

$$\left(\frac{\partial f}{\partial x} + \lambda\frac{\partial \phi}{\partial x}\right)dx + \left(\frac{\partial f}{\partial y} + \lambda\frac{\partial \phi}{\partial y}\right)dy + \left(\frac{\partial f}{\partial z} + \lambda\frac{\partial \phi}{\partial z}\right)dz = 0$$

This equation will hold good if

$$\frac{\partial f}{\partial x} + \lambda\frac{\partial \phi}{\partial x} = 0 \qquad \text{...(4)}$$

$$\frac{\partial f}{\partial y} + \lambda\frac{\partial \phi}{\partial y} = 0 \qquad \text{...(5)}$$

$$\frac{\partial f}{\partial z} + \lambda\frac{\partial \phi}{\partial z} = 0 \qquad \text{...(6)}$$

On solving (1), (4), (5), (6), we can find the values of x, y, z and λ for which $f(x, y, z)$ has stationary value.

Draw Back in Lagrange method is that the nature of stationary point cannot be determined.

Example 13. *The shape of a hole pored by a drill is a cone surmounted by cylinder. If the cylinder be of height h and radius r and the semi-vertical angle of the cone be α, where tan* $\alpha = \frac{h}{r}$ *show that for a total height H of the hole, the volume removed is maximum if* $h = H\,(\sqrt{7}+1)/6$.

Solution. Let *ABCD* be the given cylinder of height '*h*' and radius '*r*' and *DPC* be the cone of course, of radius *r*.

Now, since α is the semi-vertical angle of the cone.

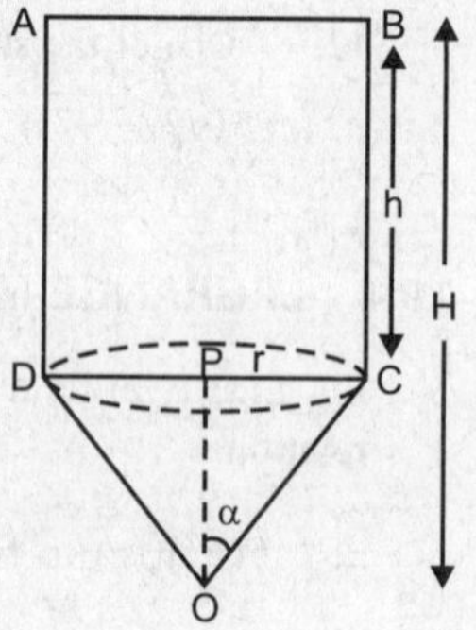

$$\therefore \quad \tan\alpha = \frac{PC}{OP} = \frac{r}{OP} \qquad ...(1)$$

but, given that $\tan\alpha = \frac{h}{r}$ ...(2)

From (1) and (2), we have $\frac{h}{r} = \frac{r}{OP}$

$$\Rightarrow \quad OP = \frac{r^2}{h} \qquad ...(3)$$

Total height of the hole = H

$$\Rightarrow \quad H = h + OP \Rightarrow OP = H - h \qquad ...(4)$$

From (3) and (4)

$$\frac{r^2}{h} = H - h \qquad ...(5)$$

Again, let $\phi = H - h - \frac{r^2}{h}$ ...(6)

In drilling a hole, the volume of the removed portion

$$\frac{\partial\phi}{\partial r} = -\frac{2r}{h}, \quad \frac{\partial\phi}{\partial h} = -1 + \frac{r^2}{h^2}$$

V = Volume of the cylinder + Volume of the cone.

$$= \pi r^2 h + \frac{1}{3}\pi r^2 (OP) = \pi r^2 h + \frac{1}{3}\pi r^2 . \frac{r^2}{h} \quad \text{[From (3)]}$$

$$V = \pi r^2 h + \frac{\pi r^4}{3h},$$

$$\frac{\partial V}{\partial r} = 2\pi rh + \frac{4\pi r^3}{3h}, \quad \frac{\partial V}{\partial h} = \pi r^2 - \frac{\pi r^4}{3h^2} \qquad ...(7)$$

By Legrange Method

$$\frac{\partial V}{\partial r} + \lambda \frac{\partial\phi}{\partial r} = 0 \quad \Rightarrow \quad 2\pi r h + \frac{4\pi r^3}{3h} + \lambda\left(\frac{-2r}{h}\right) = 0 \qquad ...(8)$$

$$\frac{\partial V}{\partial h} + \lambda \frac{\partial\phi}{\partial h} = 0 \quad \Rightarrow \pi r^2 - \frac{\pi r^4}{3h^2} + \lambda\left(-1 + \frac{r^2}{h^2}\right) = 0 \qquad ...(9)$$

Multiplying (8) by *r* and (9) by 2*h*, we get

$$2\pi r^2 h + \frac{4\pi r^4}{3h} + \lambda\left(\frac{-2r^2}{h}\right) = 0 \quad \Rightarrow \quad 2\pi r^2 h - \frac{2\pi r^4}{3h} + 2\lambda\left(-h + \frac{r^2}{h}\right) = 0$$

On subtracting, we get

$$\frac{6\pi r^4}{3h}+\lambda\left(\frac{-2r^2}{h}+2h-\frac{2r^2}{h}\right)=0 \Rightarrow \frac{6\pi r^4}{3h}+\lambda\left(\frac{-4r^2}{h}+2h\right)=0$$

$$2\pi r^4+\lambda(-4r^2+2h^2)=0 \Rightarrow \pi r^4+\lambda(-2r^2+h^2)=0$$

$$\Rightarrow \qquad \lambda=\frac{\pi r^4}{-h^2+2r^2}$$

Putting the value of λ in (8), we get

$$2\pi r h+\frac{4\pi r^3}{3h}+\left(\frac{\pi r^4}{-h^2+2r^2}\right)\left(-\frac{2r}{h}\right)=0$$

$$\Rightarrow \qquad h+\frac{2r^2}{3h}+\frac{r^4}{h(h^2-2r^2)}=0 \qquad \left[\frac{r^2}{h}=H-h\right]$$

$$h+\frac{2}{3}(H-h)+\frac{h^2(H-h)^2}{h[h^2-2h(H-h)]}=0$$

$$h+\frac{2}{3}(H-h)+\frac{(H^2+h^2-2hH)}{h-2H+2h}=0$$

$$h+\frac{2}{3}(H-h)+\frac{H^2+h^2-2hH}{3h-2H}=0$$

$$\Rightarrow \qquad 3h^2-2Hh+\frac{2}{3}(H-h)(3h-2H)+H^2+h^2-2hH=0$$

$$\Rightarrow 9h^2-6Hh+6Hh-4H^2-6h^2+4Hh+3H^2+3h^2-6hH=0$$

$$6h^2-2hH-H^2=0$$

$$h=\frac{2H\pm\sqrt{4H^2+24H^2}}{12}$$

$$h=\frac{H\pm H\sqrt{7}}{6}=H\frac{[\sqrt{7}+1]}{6} \text{ (–ve is not possible)}$$ **Proved.**

Example 14. *Find the minimum value of $x^2+y^2+z^2$ subject to the condition $xyz=a^3$.*

(A.M.I.T.E., June 2009)

Solution. Let
$f(x, y) = x^2 + y^2 + z^2$ and ...(1)
$\phi(x, y) = xyz - a^3$ ...(2)

By Lagrange's method

$$\frac{\partial f}{\partial x}+\lambda\frac{\partial \phi}{\partial x}=0 \Rightarrow 2x+\lambda(yz)=0 \qquad ...(3)$$

$$\frac{\partial f}{\partial y}+\lambda\frac{\partial \phi}{\partial y}=0 \Rightarrow 2y+\lambda(xz)=0 \qquad ...(4)$$

$$\frac{\partial f}{\partial z}+\lambda\frac{\partial \phi}{\partial z}=0 \Rightarrow 2z+\lambda(xy)=0 \qquad ...(5)$$

On multiplying (3) by x, (4) by y and (5) by z, we get

$2x^2+\lambda(xyz)=0$ ...(6)
$2y^2+\lambda(xyz)=0$ ...(7)
$2z^2+\lambda(xyz)=0$ ...(8)

On subtracting (7) from (6), we get

$$2x^2 - 2y^2 = 0 \quad \Rightarrow \quad x = y$$

On subtracting (8) from (7), we get

$$2y^2 - 2z^2 = 0 \quad \Rightarrow \quad y = z$$

so $x = y = z$

Now, putting the value of y and z in term of x in (2), we get

$$(x)\,(x)\,(x) = a^3 \quad \Rightarrow x^3 = a^3 \Rightarrow \quad x = a$$

$$\Rightarrow \qquad y = a,\; z = a$$

Putting the values of x, y, z in (1), we get

$$f(x, y) = a^2 + a^2 + a^2 = 3a^2$$

Hence, the minimum value of $f(x, y) = 3a^2$ **Ans.**

Example 15. *Find the point upon the plane $ax + by + cz = p$ at which the function*

$$f = x^2 + y^2 + z^2$$

has a minimum value and find this minimum f.

Solution. We have, $\quad f = x^2 + y^2 + z^2 \qquad$ ...(1)

$$ax + by + cz = p \quad \Rightarrow \quad \phi = ax + by + cz - p \qquad \text{...(2)}$$

By agrange's method.

$$\frac{\partial f}{\partial x} + \lambda \frac{\partial \varphi}{\partial x} = 0 \quad \Rightarrow \quad 2x + \lambda\, a = 0 \quad \Rightarrow \quad x = \frac{-\lambda a}{2}$$

$$\frac{\partial f}{\partial y} + \lambda \frac{\partial \varphi}{\partial y} = 0 \quad \Rightarrow \quad 2y + \lambda\, b = 0 \quad \Rightarrow \quad y = \frac{-\lambda b}{2}$$

$$\frac{\partial f}{\partial z} + \lambda \frac{\partial \varphi}{\partial z} = 0 \quad \Rightarrow \quad 2\,z + \lambda\, c = 0 \quad \Rightarrow \quad z = \frac{-\lambda c}{2}$$

Substituting the values of x, y, z in (2), we get

$$a\left(\frac{-\lambda a}{2}\right) + b\left(\frac{-\lambda b}{2}\right) + c\left(\frac{-\lambda c}{2}\right) = p$$

$$\lambda\,(a^2 + b^2 + c^2) = -2p \quad \Rightarrow \quad \lambda = \frac{-2p}{a^2 + b^2 + c^2}$$

$$\therefore \qquad x = \frac{a\,p}{a^2 + b^2 + c^2},\; y = \frac{b\,p}{a^2 + b^2 + c^2},\; z = \frac{c\,p}{a^2 + b^2 + c^2}$$

The minimum value of $\quad f = \dfrac{a^2 p^2}{(a^2 + b^2 + c^2)^2} + \dfrac{b^2 p^2}{(a^2 + b^2 + c^2)^2} + \dfrac{c^2 p^2}{(a^2 + b^2 + c^2)^2}$

$$= \frac{p^2 (a^2 + b^2 + c^2)}{(a^2 + b^2 + c^2)^2} = \frac{p^2}{a^2 + b^2 + c^2}$$ **Ans.**

Example 16. *Find the maximum value of $u = x^p\, y^q\, z^r$ when the variables x, y, z are subject to the condition $ax + by + cz = p + q + r$.*

Solution. Here, we have $\quad u = x^p\, y^q\, z^r \qquad$ ...(1)

If $\quad \log u = p \log x + q \log y + r \log z \qquad$ ...(2)

$$\frac{l}{u} \frac{\partial u}{\partial x} = \frac{p}{x} \quad \Rightarrow \quad \frac{\partial u}{\partial x} = \frac{pu}{x}$$

$$\frac{l}{u} \frac{\partial u}{\partial y} = \frac{q}{y} \quad \Rightarrow \quad \frac{\partial u}{\partial y} = \frac{qu}{y}$$

$$\frac{l}{u} \frac{\partial u}{\partial z} = \frac{r}{z} \quad \Rightarrow \quad \frac{\partial u}{\partial z} = \frac{ru}{z}$$

$$ax + by + cz = p + q + r$$

$$\phi(x, y, z) = ax + by + cz - p - q - r$$

$$\frac{\partial \phi}{\partial x} = a, \quad \frac{\partial \phi}{\partial y} = b, \quad \frac{\partial \phi}{\partial z} = c$$

Lagranges equations are

$$\frac{\partial u}{\partial x} + \lambda \frac{\partial \phi}{\partial x} = 0 \Rightarrow \frac{pu}{x} + \lambda a = 0 \Rightarrow x = -\frac{pu}{\lambda a}$$

$$\frac{\partial u}{\partial y} + \lambda \frac{\partial \phi}{\partial y} = 0 \Rightarrow \frac{qu}{y} + \lambda b = 0 \Rightarrow y = -\frac{qu}{\lambda b}$$

$$\frac{\partial u}{\partial z} + \lambda \frac{\partial \phi}{\partial z} = 0 \Rightarrow \frac{ru}{z} + \lambda c = 0 \Rightarrow z = -\frac{ru}{\lambda c}$$

Putting in $ax + by + cz = p + q + r$ we have

$$-\frac{pu}{\lambda} - \frac{qu}{\lambda} - \frac{ru}{\lambda} = p + q + r$$

$$-\frac{u}{\lambda}(p+q+r) = p + q + r, \Rightarrow -\frac{u}{\lambda} = 1, \Rightarrow \lambda = -u$$

$$x = -\frac{pu}{\lambda a} = \frac{-pu}{-ua} = \frac{p}{a} \qquad y = -\frac{qu}{\lambda b} = \frac{-qu}{-ub} = \frac{q}{b}$$

$$z = -\frac{ru}{\lambda c} = \frac{-ru}{-uc} = \frac{r}{c}$$

Putting in (1), we have

$$\text{Maximum value of} \quad u = \left(\frac{p}{a}\right)^p \left(\frac{q}{b}\right)^q \left(\frac{r}{c}\right)^r$$ **Ans.**

Example 17. *If $xyz = 8$, find the value of x, y, z for which $u = \dfrac{5xyz}{x + 2y + 4z}$ is maximum.*

(K. University Dec. 2008)

Solution. Here, we have

$$u = \frac{5xyz}{x + 2y + 4z} \qquad \text{... (1)}$$

$$xyz = 8 \Rightarrow xyz - 8 = 0 \quad \text{Let } \phi = xyz - 8 = 0 \qquad \text{... (2)}$$

$$\frac{\partial u}{\partial x} = \frac{(x + 2y + 4z)\, 5yz - 5xyz\,(1)}{(x+2y+4z)^2} = \frac{5yz\,(x+2y+4z-x)}{(x+2y+4z)^2}$$

$$= \frac{10y^2z + 20yz^2}{(x+2y+4z)^2}$$

Similarly $$\frac{\partial u}{\partial y} = \frac{5x^2z + 20xz^2}{(x+2y+4z)^2}, \quad \text{and} \quad \frac{\partial u}{\partial z} = \frac{5x^2y + 10xy^2}{(x+2y+4z)^2}$$

Lagranges equations are

$$\frac{\partial u}{\partial x} + \lambda \frac{\partial \phi}{\partial x} = 0 \Rightarrow \frac{10y^2z + 20yz^2}{(x+2y+4z)^2} + \lambda yz = 0 \qquad \text{... (3)}$$

$$\frac{\partial u}{\partial y}+\lambda\frac{\partial \phi}{\partial y}=0 \quad \Rightarrow \quad \frac{5x^2z+20\,xz^2}{(x+2y+4z)^2}+\lambda xz=0 \qquad \text{... (4)}$$

$$\frac{\partial u}{\partial z}+\lambda\frac{\partial \phi}{\partial z}=0 \quad \Rightarrow \quad \frac{5x^2y+10\,xy^2}{(x+2y+4z)^2}+\lambda\,xy=0 \qquad \text{... (5)}$$

On multiplying (3) by x and (4) by y and then subtracting, we get

$$\frac{10\,x\,y^2z+20\,x\,y\,z^2}{(x+2y+4z)^2}-\frac{5x^2\,y\,z+20\,x\,y\,z^2}{(x+2y+4z)^2}=0 \Rightarrow 10\,xy^2\,z+20xyz^2-5x^2yz-20xyz^2=0$$

$$\Rightarrow \quad 10\,x\,y^2\,z-5\,x^2\,y\,z=0, \quad \Rightarrow \quad 2y=x \qquad \text{... (6)}$$

On multiplying (4) by y and (5) by z and then subtracting, we get

$$\frac{5x^2yz+20\,xyz^2}{(x+2y+4z)^2}-\frac{5x^2\,y\,z+10x\,y^2z}{(x+2y+4z)^2}=0 \Rightarrow 5\,x^2y\,z+20\,x\,y\,z^2-5\,x^2y\,z-10\,x\,y^2z=0$$

$$\Rightarrow \quad 20\,x\,y\,z^2-10\,x\,y^2\,z=0 \quad \Rightarrow \quad 2z=y$$

But $x=2y=2\,(2\,z)=4z \quad \Rightarrow \quad x=4\,z$

On putting $x=4z,\ \ y=2z$ in (2), we get

$$(4\,z)\,(2\,z)\,z=8 \quad \Rightarrow \quad 8\,z^3=8 \quad \Rightarrow \quad z^3=1 \quad \Rightarrow \quad z=1$$

$$x=4, \quad y=2, \quad z=1$$

Ans.

Example 18. *Show that the stationary value of* $u=x^m\,y^n\,z^p$ *where* $x+y+z=a$ *is* $\dfrac{m^m n^n p^p\,a^{m+n+p}}{(m+n+p)^{m+n+p}}$

Solution. Here, we have $u=x^m\,y^n\,z^p$ *and* ...(1)

$$\phi=x+y+z-a \;... \qquad \text{...(2)}$$

$$\frac{\partial u}{\partial x}=mx^{m-1}\,y^n\,z^p, \qquad \frac{\partial \phi}{\partial x}=1$$

$$\frac{\partial u}{\partial y}=n\,x^m\,y^{n-1}\,z^p, \qquad \frac{\partial \phi}{\partial y}=1$$

$$\frac{\partial u}{\partial z}=p\,x^m\,y^n\,z^{p-1}, \qquad \frac{\partial \phi}{\partial z}=1$$

By Lagrange's Method

$$\frac{\partial u}{\partial x}+\lambda\frac{\partial \phi}{\partial x}=0 \quad \Rightarrow \quad mx^{m-1}\,y^nz^p+\lambda=0 \qquad \text{...(3)}$$

$$\frac{\partial u}{\partial y}+\lambda\frac{\partial \phi}{\partial y}=0 \quad \Rightarrow \quad n\,x^m\,y^{n-1}\,z^p+\lambda=0 \qquad \text{...(4)}$$

$$\frac{\partial u}{\partial z}+\lambda\frac{\partial \phi}{\partial z}=0 \quad \Rightarrow \quad p\,x^m\,y^n\,z^{p-1}+\lambda=0 \qquad \text{...(5)}$$

Multiplying (3), (4), (5) by x, y, z respectively and adding, we get

$$x^m\,y^n\,z^p\,(m+n+p)+\lambda\,(x+y+z)=0$$

$$\Rightarrow \quad x^m\,y^n\,z^p\,(m+n+p)+\lambda\,a=0 \qquad \text{[From (2)]}$$

$$\Rightarrow \quad \lambda=-\frac{x^m\,y^n\,z^p\,(m+n+p)}{a}$$

Putting the value of λ in (3), we get

$$m\,x^{m-1}\,y^n\,z^p - \frac{x^m y^n z^p (m+n+p)}{a} = 0 \Rightarrow m - x\,\frac{(m+n+p)}{a} = 0$$

$$\Rightarrow \quad \frac{m}{x} = \frac{m+n+p}{a} \quad \Rightarrow \quad x = \frac{am}{m+n+p}$$

Similarly, $\quad y = \dfrac{an}{m+n+p} \quad$ and $\quad z = \dfrac{ap}{m+n+p}$

Putting the values of x, y and z in (1), we get

$$u = \left(\frac{am}{m+n+p}\right)^m \cdot \left(\frac{an}{m+n+p}\right)^n \cdot \left(\frac{ap}{m+n+p}\right)^p$$

$$u = \frac{m^m n^n p^p a^{m+n+p}}{(m+n+p)^{m+n+p}}$$

Proved.

Example 19. *Show that the rectangular solid of maximum volume that can be inscribed in a sphere is a cube.*

Solution. Let $2x$, $2y$, $2z$ be the length, breadth and height of the rectangular solid.

Let R be the radius of the sphere.

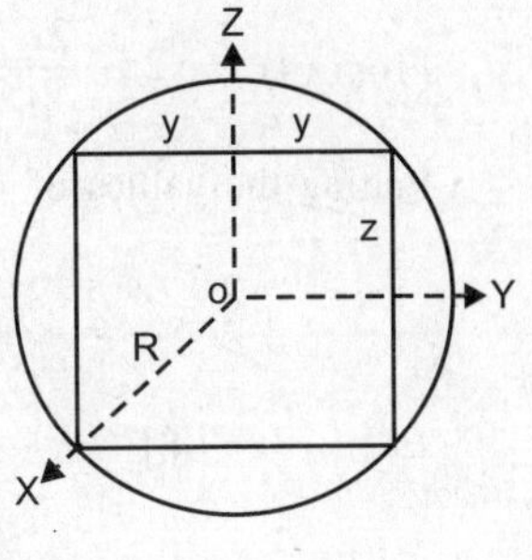

Volume of solid $V = 8\,x\,.\,y\,.\,z$ ...(1)

$$x^2 + y^2 + z^2 = R^2 \quad \text{....(2)}$$

$$\Rightarrow \quad \phi(x, y, z) = x^2 + y^2 + z^2 - R^2 = 0$$

$$\frac{\partial V}{\partial x} + \lambda \frac{\partial \phi}{\partial x} = 0 \Rightarrow 8yz + \lambda\,(2x) = 0 \quad \text{...(3)}$$

$$\frac{\partial V}{\partial y} + \lambda \frac{\partial \phi}{\partial y} = 0 \Rightarrow 8x\,z + \lambda\,(2\,y) = 0 \quad \text{...(4)}$$

$$\frac{\partial V}{\partial z} + \lambda \frac{\partial \phi}{\partial z} = 0 \Rightarrow 8xy + \lambda\,(2\,z) = 0 \quad \text{...(5)}$$

From (3) $\quad 2\,\lambda\,x = -\,8y\,z \quad \Rightarrow \quad 2\,\lambda\,x^2 = -\,8x\,y\,z$

From (4) $\quad 2\,\lambda\,y = -\,8x\,z \quad \Rightarrow \quad 2\,\lambda\,y^2 = -\,8x\,y\,z$

From (5) $\quad 2\,\lambda\,z = -\,8x\,y \quad \Rightarrow \quad 2\,\lambda\,z^2 = -\,8\,x\,y\,z$

$$2\,\lambda\,x^2 = 2\,\lambda\,y^2 = 2\,\lambda\,z^2$$

$$\Rightarrow \quad x^2 = y^2 = z^2 \quad \Rightarrow \quad x = y = z$$

Hence rectangular solid is a cube. **Proved.**

Example 20. *If x, y, z are the length of the perpendiculars dropped any point P to the three sides of a triangle of constant area A, find the minimum value of* $x^2 + y^2 + z^2$.

Solution. Let x, y, z are the length of the perpendiculars dropped from point P to the three sides a, b and c of a triangle.

Area ($\Delta\,ABC$) = Area ($\Delta\,PBC$) + Area ($\Delta\,PAC$) + Area ($\Delta\,PAB$)

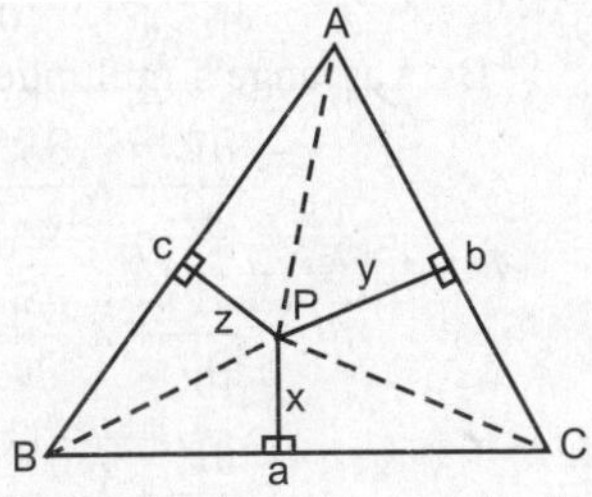

$$\therefore \quad \text{Area of triangle} = \frac{1}{2}xa + \frac{1}{2}yb + \frac{1}{2}zc$$

$$A = \frac{1}{2}xa + \frac{1}{2}yb + \frac{1}{2}zc \quad \text{...(1)}$$

$$\therefore \quad \phi(x, y, z) = \frac{1}{2}xa + \frac{1}{2}yb + \frac{1}{2}zc - A = 0$$

$$F(x, y, z) = x^2 + y^2 + z^2$$

$\therefore$ Lagranges multiplier method

$$\frac{\partial F}{\partial x}+\lambda\frac{\partial \phi}{\partial x} = 0 \quad \Rightarrow \quad 2x+\lambda\frac{1}{2}a = 0 \qquad ...(2)$$

$$\frac{\partial F}{\partial y}+\lambda\frac{\partial \phi}{\partial y} = 0 \quad \Rightarrow \quad 2y+\lambda\frac{1}{2}b = 0 \qquad ...(3)$$

$$\frac{\partial F}{\partial z}+\lambda\frac{\partial \phi}{\partial z} = 0 \quad \Rightarrow \quad 2z+\lambda\frac{1}{2}c = 0 \qquad ...(4)$$

Multiplying (2) by x, (3) by y and (4) by z and adding, we get

$$2(x^2+y^2+z^2)+\lambda\left[\frac{1}{2}xa+\frac{1}{2}yb+\frac{1}{2}zc\right] = 0$$

$$\therefore \qquad 2F+\lambda A = 0 \quad \Rightarrow \quad \lambda = \frac{-2F}{A}$$

From (2), $\quad 2x-\frac{2F}{A}.\frac{1}{2}a = 0 \quad \Rightarrow \quad x = \frac{aF}{2A}$

From (3), $\quad 2y-\frac{2F}{A}.\frac{1}{2}b = 0 \quad \Rightarrow \quad y = \frac{bF}{2A}$

From (4), $\quad 2z-\frac{2F}{A}.\frac{1}{2}c = 0 \quad \Rightarrow \quad z = \frac{cF}{2A}$

Putting the values of x, y, z in (1), we get

$$A = \frac{1}{2}\frac{aF}{2A}.a+\frac{1}{2}\frac{bF}{2A}.b+\frac{1}{2}\frac{cF}{2A}.c=\frac{F}{4A}[a^2+b^2+c^2]$$

$$\Rightarrow \quad 4A^2 = F(a^2+b^2+c^2) \quad \Rightarrow \quad F=\frac{4A^2}{a^2+b^2+c^2}$$

Hence, $x^2+y^2+z^2 = \dfrac{4A^2}{a^2+b^2+c^2}$ **Ans.**

Example 21. *Find the dimension of rectangular box of maximum capacity whose surface area is given when (a) box is open at the top (b) box is closed.* *(U.P., I Semester, Dec 2008)*

Solution. Let x, y, z be the length, breadth and height of rectangular box respectively. Then

volume, $\quad V = xyz \Rightarrow \frac{\partial V}{\partial x}=yz, \frac{\partial V}{\partial y}=xz, \frac{\partial V}{\partial z}=xy$ *(D.U. III Sem 2014)*

(*a*) **Surface of open rectangular box**

Surface area, $S_1 = xy + 2(yz+zx)$

$$\Rightarrow \quad \frac{\partial S_1}{\partial x}=y+2z, \frac{\partial S_1}{\partial y}=x+2z, \frac{\partial S_1}{\partial z}=2(x+y)$$

By Lagrange's multiplier method

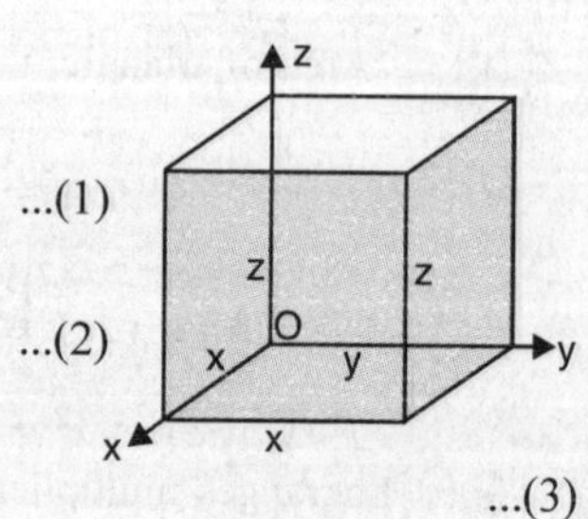

$$\frac{\partial V}{\partial x}+\lambda\frac{\partial S_1}{\partial x}=0 \Rightarrow yz+[y+2z]\lambda=0 \qquad ...(1)$$

$$\frac{\partial V}{\partial y}+\lambda\frac{\partial S_1}{\partial y}=0 \Rightarrow xz+[x+2z]\lambda=0 \qquad ...(2)$$

$$\frac{\partial V}{\partial z}+\lambda\frac{\partial S_1}{\partial z}=0 \Rightarrow xy+[2y+2x]\lambda=0 \qquad ...(3)$$

Multiplying (1) by x, (2) by y and subtracting, we get

$$2z\,(x - y)\,\lambda = 0 \Rightarrow x = y \qquad ....(4)$$

Multiplying (2) by y and (3) by z and then subtracting, we get

$$x\,(y - 2z)\,\lambda = 0 \Rightarrow y = 2z \qquad ....(5)$$

From (4) and (5), we get

$$x = y = 2z$$

$\Rightarrow$ Length = Breadth = 2 × Height.

(*b*) **Surface of closed rectangular box:**

Surface area $S_2 = 2\,(xy + yz + zx) \Rightarrow \dfrac{\partial S_2}{\partial x} = 2\,(y + z), \dfrac{\partial S_2}{\partial y} = 2\,(x + z), \dfrac{\partial S_2}{\partial z} = 2\,(x + y)$

By Lagrange's method, we have

$$\frac{\partial V}{\partial x} + \lambda \frac{\partial S_2}{\partial x} = 0 \Rightarrow yz + 2\,(y + z)\,\lambda = 0 \qquad ...(6)$$

$$\frac{\partial V}{\partial y} + \lambda \frac{\partial S_2}{\partial y} = 0 \Rightarrow xz + 2\,(x + z)\,\lambda = 0 \qquad ...(7)$$

$$\frac{\partial V}{\partial z} + \lambda \frac{\partial S_2}{\partial z} = 0 \Rightarrow xy + 2\,(y + x)\,\lambda = 0 \qquad ...(8)$$

Multiplying (6) by x and (7) by y and then subtracting, we get

$$2z\,(x - y)\,\lambda = 0 \Rightarrow x = y \qquad ...(9)$$

Multiplying (7) by y and (8) by z and then subtracting, we get

$$2x\,(y - z)\,\lambda = 0 \Rightarrow y = z \qquad ...(10)$$

From (9) and (10), we get

$$x = y = z$$

Thus, Length = Breadth = Height. **Ans.**

Example 22. *Find the area of a greatest rectangle that can be inscribed in an ellipse*

$$\frac{x^2}{a^2} + \frac{y^2}{b^2} = 1$$

Solution. Let, *ABCD* be the rectangle.

Let the co-ordinates of point *A* be (x, y).

$$AB = 2x,\ BC = 2y$$

$$\text{Area} = A = (2x)\,(2y) = 4\,xy, \quad \frac{\partial A}{\partial x} = 4y, \quad \frac{\partial A}{\partial y} = 4x \qquad ...(1)$$

$$\phi = \frac{x^2}{a^2} + \frac{y^2}{b^2} - 1 = 0, \quad \frac{\partial \phi}{\partial x} = \frac{2x}{a^2}, \quad \frac{\partial \phi}{\partial y} = \frac{2y}{b^2}$$

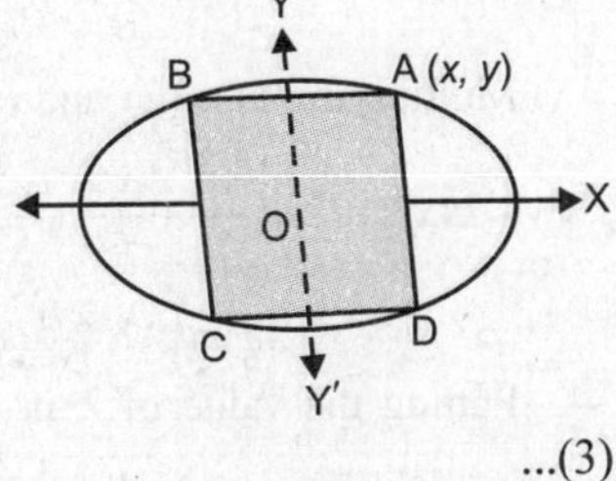

By Lagrange's multipler method

$$\frac{\partial A}{\partial x} + \lambda \frac{\partial \phi}{\partial x} = 0$$

$$\Rightarrow \quad 4y + \lambda\,\frac{2x}{a^2} = 0 \Rightarrow \lambda = -\frac{2ya^2}{x} \qquad ...(2)$$

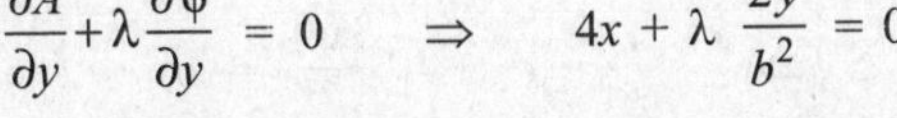

$$\frac{\partial A}{\partial y} + \lambda \frac{\partial \phi}{\partial y} = 0 \quad \Rightarrow \quad 4x + \lambda\,\frac{2y}{b^2} = 0 \qquad ...(3)$$

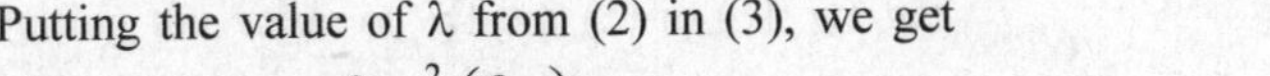

Putting the value of λ from (2) in (3), we get

$$4x - \frac{2ya^2}{x}\left(\frac{2y}{b^2}\right) = 0 \qquad \text{[From (2)]}$$

$$\Rightarrow \qquad x^2 - \frac{y^2a^2}{b^2} = 0 \qquad \Rightarrow \qquad \frac{x^2}{a^2} = \frac{y^2}{b^2} \qquad ...(4)$$

Putting the value of $\frac{y^2}{b^2}$ from (4) in $\frac{x^2}{a^2}+\frac{y^2}{b^2}=1$, we get

$$\frac{x^2}{a^2}+\frac{x^2}{a^2} = 1 \qquad \Rightarrow \frac{2x^2}{a^2}=1 \qquad \Rightarrow \qquad x = \pm\frac{a}{\sqrt{2}}$$

Similarly, $\qquad y = \pm\frac{b}{\sqrt{2}}$

$$\text{Area} = 4xy = 4\left(\frac{a}{\sqrt{2}}\right)\left(\frac{b}{\sqrt{2}}\right) = 2ab$$

Hence, area of greatest rectangle inscribed in the ellipse = $2ab$ **Ans.**

Example 23. *Use the method of the Lagrange's multipliers to find the volume of the largest rectangular parallelopiped that can be inscribed in the ellipsoid* $\frac{x^2}{a^2}+\frac{y^2}{b^2}+\frac{z^2}{c^2}=1$.

(DU, II Sem. 2012, Nagpur Univesity, Summer 2008,)

Solution. Here, we have $\frac{x^2}{a^2}+\frac{y^2}{b^2}+\frac{z^2}{c^2} = 1$

$$\Rightarrow \qquad \phi\,(x, y, z) = \frac{x^2}{a^2}+\frac{y^2}{b^2}+\frac{z^2}{c^2}-1=0 \qquad ...(1)$$

Let $2x$, $2y$, $2z$ be the length, breadth and height of the rectangular parallelopiped inscribed in the ellipsoid.

$$V = (2x)\,(2y)\,(2z) = 8\,xyz$$

$$\frac{\partial V}{\partial x} = 8\,yz; \quad \frac{\partial V}{\partial y}=8xz, \quad \frac{\partial V}{\partial z}=8xy$$

$$\frac{\partial \phi}{\partial x} = \frac{2x}{a^2}, \quad \frac{\partial \phi}{\partial y}=\frac{2y}{b^2}, \quad \frac{\partial \phi}{\partial z}=\frac{2z}{c^2}$$

Lagrange's equations are

$$\frac{\partial V}{\partial x}+\lambda\frac{\partial \phi}{\partial x} = 0 \qquad \Rightarrow \qquad 8\,yz+\lambda\,\frac{2x}{a^2} = 0 \qquad ...(1)$$

$$\frac{\partial V}{\partial y}+\lambda\frac{\partial \phi}{\partial y} = 0 \qquad \Rightarrow \qquad 8\,xz+\lambda\,\frac{2y}{b^2} = 0 \qquad ...(2)$$

$$\frac{\partial V}{\partial z}+\lambda\frac{\partial \phi}{\partial z} = 0 \qquad \Rightarrow \qquad 8\,xy+\lambda\,\frac{2z}{c^2} = 0 \qquad ...(3)$$

Multiplying (1), (2) and (3) by x, y, z respectively and adding, we get

$$24xyz+2\lambda\left[\frac{x^2}{a^2}+\frac{y^2}{b^2}+\frac{z^2}{c^2}\right] = 0 \qquad \left[\frac{x^2}{a^2}+\frac{y^2}{b^2}+\frac{z^2}{c^2}=1\right]$$

$$\Rightarrow \qquad 24\,xyz+2\,\lambda\,(1) = 0 \qquad \Rightarrow \qquad \lambda = -12\,xyz$$

Putting the value of λ in (1), we get

$$8\,yz+(-12\,xyz)\,\frac{2x}{a^2} = 0 \qquad \Rightarrow \qquad 1-\frac{3x^2}{a^2} = 0$$

$$\Rightarrow \qquad x = \frac{a}{\sqrt{3}}$$

Similarly from (2) and (3), we have

$$y = \frac{b}{\sqrt{3}}, \quad z = \frac{c}{\sqrt{3}}$$

Volume of the largest rectangular parallelopiped = 8 xyz

$$= 8\left(\frac{a}{\sqrt{3}}\right)\left(\frac{b}{\sqrt{3}}\right)\left(\frac{c}{\sqrt{3}}\right) = \frac{8abc}{3\sqrt{3}}$$ **Ans.**

Example 24. *The pressure P at any point (x, y, z) in space is* $P = 400\, xyz^2$. *Find the highest pressure at the surface of a unit sphere* $x^2 + y^2 + z^2 = 1$. *(Gujarat, I Semester, Jan. 2009)*

Solution. We have, $P = 400\, x\, y\, z^2$

$$x^2 + y^2 + z^2 = 1, \qquad \phi(x, y, z) = x^2 + y^2 + z^2 - 1$$

$$\frac{\partial P}{\partial x} + \lambda\frac{\partial \phi}{\partial x} = 0 \Rightarrow 400\, y\, z^2 + \lambda\,(2\,x) = 0 \quad ...(1)$$

$$\frac{\partial P}{\partial y} + \lambda\frac{\partial \phi}{\partial y} = 0 \Rightarrow 400\, x\, z^2 + \lambda\,(2\,y) = 0 \quad ...(2)$$

$$\frac{\partial P}{\partial z} + \lambda\frac{\partial \phi}{\partial z} = 0 \Rightarrow 800\, x\, y\, z + \lambda\,(2\,z) = 0 \quad ...(3)$$

Multiplying (1) by x, (2) by y and (3) by z and adding, we get

$$1600\, x\, y\, z^2 + 2\lambda\,(x^2 + y^2 + z^2) = 0$$

$$1600\, x\, y\, z^2 + 2\lambda\,(1) = 0 \qquad (x^2 + y^2 + z^2 = 1)$$

$$\Rightarrow \quad \lambda = -\,800\, x\, y\, z^2$$

Putting the value of λ in (l), we get

$$400\, y\, z^2 + 2\,x\,(-\,800\, x\, y\, z^2) = 0 \Rightarrow 1 - 4\,x^2 = 0 \Rightarrow x = \pm\frac{1}{2}$$

Similarly, $y = \pm\frac{1}{2}$

On putting the value of λ in (3), we get

$$800\, x\, y\, z - 1600\, x\, y\, z^3 = 0$$

$$1 - 2\,z^2 = 0 \Rightarrow z = \pm\frac{1}{\sqrt{2}}$$

On putting the values of x, y, z in P, we get

$$P = 400 \times \frac{1}{2} \times \frac{1}{2} \times \frac{1}{2} = 50$$ **Ans.**

Example 25. *A scope probe in the shape of ellipsoid* $4x^2 + y^2 + 4z^2 = 16$ *enters the earth atmosphere and its surface begins to heat. After one hour the temperature at the point (x, y, z) on the surface is* $T(x, y, z) = 8x^2 + 4yz - 16z + 600$. *Find the hottest point on the probe surface.*

Solution. Given temperature on the surface (x, y, z) is

$$T(x, y, z) = 8x^2 + 4yz - 16z + 600 \quad ...(1)$$

and ellipsoid is $4x^2 + y^2 + 4z^2 = 16$

Let $\phi = 4x^2 + y^2 + 4z^2 - 16 = 0$ ...(2)

$$\frac{\partial T}{\partial x} = 16x, \quad \text{and} \quad \frac{\partial \phi}{\partial x} = 8x$$

$$\frac{\partial T}{\partial y} = 4z, \quad \text{and} \quad \frac{\partial \phi}{\partial y} = 2y$$

$$\frac{\partial T}{\partial z} = 4y - 16, \quad \text{and} \quad \frac{\partial \phi}{\partial z} = 8z$$

By Lagrange's method

$$\frac{\partial T}{\partial x}+\lambda\frac{\partial \phi}{\partial x} = 0 \quad \Rightarrow \quad 16x + 8\lambda x = 0 \qquad ...(3)$$

$$\frac{\partial T}{\partial y}+\lambda\frac{\partial \phi}{\partial y} = 0 \quad \Rightarrow \quad 4z + 2\lambda y = 0 \qquad ...(4)$$

$$\frac{\partial T}{\partial z}+\lambda\frac{\partial \phi}{\partial z} = 0 \quad \Rightarrow \quad 4y - 16 + 8\lambda z = 0 \qquad ...(5)$$

From (3), $\lambda = -2$

Putting the value of λ in (4) and (5), we get

$$4z + 2(-2)y = 0 \quad \Rightarrow \quad z - y = 0 \qquad ...(6)$$

and $\quad 4y - 16 + 8(-2)z = 0 \quad \Rightarrow \quad y - 4z = 4 \qquad ...(7)$

Adding (6) and (7), we get

$$-3z = 4 \text{ and } z = -\frac{4}{3}$$

From (6), $y = z = -\dfrac{4}{3}$

On putting the values of y and z in (2), we get

$$4x^2 + \left(-\frac{4}{3}\right)^2 + 4\left(-\frac{4}{3}\right)^2 - 16 = 0 \quad \Rightarrow \quad x^2 = \frac{16}{9} \Rightarrow x = \pm\frac{4}{3}$$

Hence, the hottest points on the probe surface is $\left(\pm\frac{4}{3}, \frac{-4}{3}, \frac{-4}{3}\right)$ **Ans.**

Example 26. *A tent of a given volume has a square base of side 2a, has its four-side vertical of length b and is surmounted by a regular pyramid of height h. Find the values of a and b in terms of h such that the canvas required for its construction is minimum.*

Solution. Let V be the volume and S be the surface of the tent.

$$V = 4a^2b + \frac{1}{3}(4a^2)h \qquad \left[\text{Volume of pyramid} = \frac{1}{3}\text{ Area of the base} \times \text{height}\right]$$

$$S = 8ab + 4a\sqrt{a^2+h^2} \qquad \left[\text{Surface Area of pyramid} = \frac{1}{2}\text{ perimeter} \times \text{slant height}\right]$$

$$\frac{\partial S}{\partial a}+\lambda\frac{\partial V}{\partial a} = 0$$

$$\Rightarrow \quad 8b + 4\sqrt{a^2+h^2} + \frac{4a^2}{\sqrt{a^2+h^2}} + \lambda\left[8ab + \frac{8ah}{3}\right] = 0 \qquad ...(1)$$

$$\frac{\partial S}{\partial b}+\lambda\frac{\partial V}{\partial b} = 0 \quad \Rightarrow \quad 8a + 4\lambda a^2 = 0 \qquad ...(2)$$

$$\frac{\partial S}{\partial h}+\lambda\frac{\partial V}{\partial h} = 0 \quad \Rightarrow \quad \frac{4ah}{\sqrt{a^2+h^2}} + \frac{4}{3}\lambda a^2 = 0 \qquad ...(3)$$

From (2) $\quad \lambda a + 2 = 0 \quad \Rightarrow \quad \lambda a = -2 \qquad ...(4)$

From (3) $\quad 12ah + 4\lambda a^2\sqrt{a^2+h^2} = 0$

$$\Rightarrow \quad 3h + \lambda a\sqrt{a^2+h^2} = 0 \qquad ...(5)$$

Substituting the value of λa from (4) in (5), we get

$$3h+\lambda a\sqrt{a^2+h^2} = 0 \Rightarrow 9h^2 = 4a^2 + 4h^2 \Rightarrow 4a^2 = 5h^2$$

$$a = \frac{\sqrt{5}}{2}h$$

Substituting $\lambda a = -2$ and $a = \frac{\sqrt{5}}{2}h$ in (1) and simplifying, we get

$$8b+4\sqrt{\frac{5h^2}{4}+h^2}+\frac{5h^2}{\sqrt{\frac{5h^2}{4}+h^2}}-2\left[8b+\frac{8h}{3}\right] = 0$$

$$\Rightarrow \quad 8b + 6h + \frac{10h}{3} - 16b - \frac{16h}{3} = 0$$

$$\Rightarrow \quad -8b + 4h = 0 \Rightarrow b = \frac{h}{2}.$$

Thus, when $a = \frac{\sqrt{5}}{2}h$ and $b = \frac{h}{2}$ we get the stationary value of S. **Ans.**

Example 27. *Find the maximum and minimum distances of the point* (3, 4, 12) *from the sphere* $x^2 + y^2 + z^2 = 1$. *(Delhi University, April 2010, AMIETE, June 2010)*

Solution. Let the co-ordinates of the given point be (x, y, z), then its distance (D) from (3, 4, 12).

$$D = \sqrt{(x-3)^2+(y-4)^2+(z-12)^2}$$

$$\Rightarrow \quad F(x, y, z) = (x-3)^2 + (y-4)^2 + (z-12)^2$$

$$x^2 + y^2 + z^2 = 1$$

$$\phi(x, y, z) = x^2 + y^2 + z^2 - 1$$

$$\frac{\partial F}{\partial x}+\lambda\frac{\partial \phi}{\partial x} = 2(x-3) + 2\lambda x = 0 \quad \text{...(1)}$$

$$\frac{\partial F}{\partial y}+\lambda\frac{\partial \phi}{\partial y} = 2(y-4) + 2\lambda y = 0 \quad \text{...(2)}$$

$$\frac{\partial F}{\partial z}+\lambda\frac{\partial \phi}{\partial z} = 2(z-12) + 2\lambda z = 0 \quad \text{...(3)}$$

Multiplying (1) by x, (2) by y and (3) by z and adding, we get

$$(x^2 + y^2 + z^2) - 3x - 4y - 12z + \lambda(x^2 + y^2 + z^2) = 0$$

$$1 - 3x - 4y - 12z + \lambda = 0 \quad \text{...(4)}$$

From (1) $\quad x = \frac{3}{1+\lambda}$ ...(5)

From (2) $\quad y = \frac{4}{1+\lambda}$ ...(6)

From (3) $\quad z = \frac{12}{1+\lambda}$ ...(7)

Putting these values of x, y, z in (4),we have

$$1+\lambda-\frac{9}{1+\lambda}-\frac{16}{1+\lambda}-\frac{144}{1+\lambda} = 0 \Rightarrow (1+\lambda)^2 = 169$$

$$\Rightarrow \quad 1 + \lambda = \pm 13$$

Putting the value of $1 + \lambda$ in (5), (6) and (7) we have the points

$$\left(\frac{3}{13},\frac{4}{13},\frac{12}{13}\right) \text{ and } \left(\frac{-3}{13},\frac{-4}{13},\frac{-12}{13}\right)$$

The minimum distance $= \sqrt{\left(3-\frac{3}{13}\right)^2+\left(4-\frac{4}{13}\right)^2+\left(12-\frac{12}{13}\right)^2} = 12$

The maximum distance $= \sqrt{\left(3+\frac{3}{13}\right)^2+\left(4+\frac{4}{13}\right)^2+\left(12+\frac{12}{13}\right)^2} = 14$ **Ans.**

Example 28. *Use the method of Lagrange's multipliers to find the extreme values of $f(x, y, z) = 2x + 3y + z$ subject to $x^2 + y^2 = 5$ and $x + z = 1$.*

(A.M.I.E.T.E, June 2010, Dec. 2007, Uttarakhand, I Semester, Dec. 2006)

Solution. Let

$$f(x, y, z) = 2x + 3y + z \quad ...(1)$$
$$\phi(x, y) = x^2 + y^2 - 5 \quad ...(2)$$
$$\psi(x, z) = x + z - 1 \quad ...(3)$$

Lagranges Multipliers Equations are

$$\frac{\partial f}{\partial x}+\lambda\frac{\partial \phi}{\partial x}+\mu\frac{\partial \psi}{\partial x} = 0 \quad \Rightarrow \quad 2 + \lambda(2x) + \mu(1) = 0 \quad ...(4)$$

$$\frac{\partial f}{\partial y}+\lambda\frac{\partial \phi}{\partial y}+\mu\frac{\partial \psi}{\partial y} = 0 \quad \Rightarrow \quad 3 + \lambda(2y) + \mu(0) = 0 \quad ...(5)$$

$$\frac{\partial f}{\partial z}+\lambda\frac{\partial \phi}{\partial z}+\mu\frac{\partial \psi}{\partial z} = 0 \quad \Rightarrow 1 + \lambda(0) + \mu(1) = 0 \quad \Rightarrow \mu = -1 \quad ...(6)$$

Putting the value of μ in (4) and (5), we get

$$2 + 2\lambda x - 1 = 0 \quad \Rightarrow \quad 2\lambda x = -1, \quad \Rightarrow \quad x = -\frac{1}{2\lambda}$$

$$3 + 2\lambda y = 0 \quad \Rightarrow \quad 2\lambda y = -3, \quad \Rightarrow \quad y = -\frac{3}{2\lambda}$$

Putting the values of x, y in $x^2 + y^2 = 5$, we get

$$\frac{1}{4\lambda^2}+\frac{9}{4\lambda^2} = 5 \quad \Rightarrow \quad \frac{10}{4\lambda^2} = 5 \quad \Rightarrow \quad 2\lambda^2 = 1$$

$$\Rightarrow \quad \lambda^2 = \frac{1}{2} \quad \Rightarrow \quad \lambda = \pm\frac{1}{\sqrt{2}}$$

We know that $x = -\frac{1}{2\lambda} = \pm\frac{\sqrt{2}}{2} = \pm\frac{1}{\sqrt{2}}$

$$y = -\frac{3}{2\lambda} = \pm\frac{3\sqrt{2}}{2} = \pm\frac{3}{\sqrt{2}}$$

From (3), $x + z = 1 \quad \Rightarrow \quad z = 1 - x = 1 \mp \frac{1}{\sqrt{2}}$

Putting $x = \frac{1}{\sqrt{2}}$, $y = \frac{3}{\sqrt{2}}$ and $z = 1 - \frac{1}{\sqrt{2}}$ in (1), we get

$$f = \frac{2}{\sqrt{2}}+\frac{9}{\sqrt{2}}+1-\frac{1}{\sqrt{2}} = \frac{10}{\sqrt{2}}+1 = 5\sqrt{2}+1$$

Putting $x = -\frac{1}{\sqrt{2}}$, $y = -\frac{3}{\sqrt{2}}$ and $z = 1+\frac{1}{\sqrt{2}}$ in (1), we get

$$f = 2\left(-\frac{1}{\sqrt{2}}\right)+3\left(-\frac{3}{\sqrt{2}}\right)+\left(1+\frac{1}{\sqrt{2}}\right) = -\frac{2}{\sqrt{2}}-\frac{9}{\sqrt{2}}+1+\frac{1}{\sqrt{2}}$$

$$= 1-5\sqrt{2}$$

Ans.

Example 29. *A torpedo has the shape of a cylinder with conical ends. For given surface area, show that the dimensions which give maximum volume are,* $l = h = \frac{2}{\sqrt{5}}r$, *where l is the length of the cylinder, r its radius and h the altitude of the cone.*

Solution. Let V be the volume enclosed by the torpedo and S its surface. Then

$$V = \text{Volume of the cylinder} + \text{Volume of two cones}$$

$$= \pi r^2 l + 2.\frac{1}{3}\pi r^2 h$$

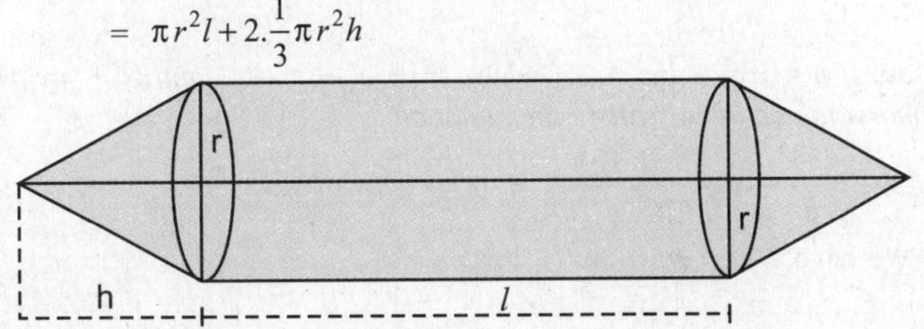

S = Surface of the cylinder + 2 (surface of the cone)

= (Circumference of the base × height) + 2 ($\frac{1}{2}$ × perimeter of the base × slant height)

$$\phi = S = 2\,\pi\, r\, l + 2\,\pi\, r\,\sqrt{(r^2 + h^2)} \qquad ...(1)$$

Let $$F\,(l, h, r) = V = \pi\, r^2\, l + \frac{2}{3}\pi r^2 h \qquad ...(2)$$

By Lagrange's method

$$\frac{\partial F}{\partial l} + \lambda\frac{\partial \phi}{\partial l} = 0 \quad \Rightarrow \quad \pi\, r^2 + 2\,\pi\, r\,\lambda = 0 \quad \Rightarrow \quad \lambda = -\frac{r}{2} \qquad ...(3)$$

$$\frac{\partial F}{\partial h} + \lambda\frac{\partial \phi}{\partial h} = 0 \quad \Rightarrow \quad \frac{2}{3}\pi r^2 + \frac{2\pi r h\lambda}{\sqrt{(r^2 + h^2)}} = 0 \qquad ...(4)$$

and $$\frac{\partial F}{\partial r} + \lambda\frac{\partial \phi}{\partial r} = 0 \Rightarrow 2\pi r l + \frac{4}{3}\pi r h + \lambda\left[2\pi l + 2\pi\sqrt{(r^2 + h^2)} + \frac{2\pi r^2}{\sqrt{(r^2 + h^2)}}\right] = 0 \qquad ...(5)$$

Putting the value of λ in (4), we get

$$\frac{2}{3}\pi r^2 + \frac{2\pi r h}{\sqrt{r^2 + h^2}}\left(-\frac{r}{2}\right) = 0 \Rightarrow \frac{2}{3} - \frac{h}{\sqrt{r^2 + h^2}} = 0$$

$$\Rightarrow \quad \frac{2}{3} = \frac{h}{\sqrt{r^2 + h^2}} \Rightarrow \frac{4}{9} = \frac{h^2}{r^2 + h^2} \Rightarrow 9h^2 = 4r^2 + 4h^2 \Rightarrow 5h^2 = 4r^2]$$

$$h = \frac{2}{\sqrt{5}}r$$

Putting $\lambda = -\frac{r}{2}$ and $h = \frac{2r}{\sqrt{5}}$ in (5), we get

$$2rl + \frac{4}{3}r\left(\frac{2}{\sqrt{5}}r\right) - \frac{r}{2}\left(2l + 2\sqrt{r^2 + \frac{4}{5}r^2} + \frac{2r^2}{\sqrt{r^2 + \frac{4r^2}{5}}}\right) = 0$$

$$2l+\frac{8r}{3\sqrt{5}}-l-\sqrt{\frac{9r^2}{5}-r^2}\sqrt{\frac{5}{9r^2}}=0 \quad \Rightarrow \quad l=\frac{-8r}{3\sqrt{5}}+\frac{3r}{\sqrt{5}}+\frac{\sqrt{5}}{3}r$$

$$l=\left(\frac{-8}{3\sqrt{5}}+\frac{3}{\sqrt{5}}+\frac{\sqrt{5}}{3}\right)r \quad \Rightarrow \quad l=\frac{2r}{\sqrt{5}}$$

Hence $\quad l = h = \dfrac{2r}{\sqrt{5}}$. **Proved.**

Example 30. *If $u = ax^2 + by^2 + cz^2$ where $x^2 + y^2 + z^2 = 1$ and $lx + my + nz = 0$ prove that stationary values of 'u' satisfy the equation*

$$\frac{l^2}{a-u}+\frac{m^2}{b-u}+\frac{n^2}{c-u} = 0$$

Solution. We have, $\quad u = ax^2 + by^2 + cz^2 \quad$ ...(1)

Let $\quad \phi = x^2 + y^2 + z^2 - 1 \quad$ ...(2)

$\quad \psi = lx + my + nz \quad$ ...(3)

$$\frac{\partial u}{\partial x} = 2\,a\,x, \qquad \frac{\partial u}{\partial y} = 2\,b\,y, \qquad \frac{\partial u}{\partial z} = 2\,c\,z$$

$$\frac{\partial \phi}{\partial x} = 2\,x, \qquad \frac{\partial \phi}{\partial y} = 2\,y, \qquad \frac{\partial \phi}{\partial z} = 2\,z$$

$$\frac{\partial \psi}{\partial x} = l, \qquad \frac{\partial \psi}{\partial y} = m, \qquad \frac{\partial \psi}{\partial z} = n$$

By Lagrange's method

$$\frac{\partial u}{\partial x}+\lambda_1\frac{\partial \phi}{\partial x}+\lambda_2\frac{\partial \psi}{\partial x} = 0, \qquad 2\,a\,x + 2\,x\,\lambda_1 + \lambda_2\,l = 0 \qquad \text{...(4)}$$

$$\frac{\partial u}{\partial y}+\lambda_1\frac{\partial \phi}{\partial y}+\lambda_2\frac{\partial \psi}{\partial y} = 0, \qquad 2\,b\,y + 2\,y\,\lambda_1 + \lambda_2\,m = 0 \qquad \text{...(5)}$$

$$\frac{\partial u}{\partial z}+\lambda_1\frac{\partial \phi}{\partial z}+\lambda_2\frac{\partial \psi}{\partial z} = 0, \qquad 2\,c\,z + 2\,z\,\lambda_1 + \lambda_2\,n = 0 \qquad \text{...(6)}$$

Multiplying (4), (5) and (6) by x, y and z respectively and adding, we get

$$(2\,ax^2 + 2\,by^2 + 2\,cz^2) + (2\,x^2 + 2\,y^2 + 2\,z^2)\,\lambda_1 + (lx + my + nz)\,\lambda_2 = 0$$

$$2u + 2\lambda_1 = 0, \qquad \lambda_1 = -\,u$$

Putting the value of λ_1 in (4), (5) and (6), we get

$$2\,a\,x - 2\,x\,u + \lambda_2\,l = 0, \qquad x = \frac{-\lambda_2 l}{2(a-u)}$$

$$2\,b\,y - 2\,y\,u + \lambda_2\,m = 0, \qquad y = \frac{-\lambda_2\,m}{2(b-u)}$$

$$2\,c\,z - 2\,z\,u + \lambda_2\,n = 0, \qquad z = \frac{-\lambda_2\,m}{2(b-u)}$$

Putting the values of x, y, z in (3), we get

$$\frac{-\lambda_2\,l^2}{2(a-u)}+\frac{-\lambda_2\,m^2}{2(b-u)}+\frac{-\lambda_2 n^2}{2(c-u)}a = 0$$

$$\frac{l^2}{a-u}+\frac{m^2}{b-u}+\frac{n^2}{c-u} = 0$$ **Proved.**

EXERCISE 19.2

1. Show that the greatest value of $x^m y^n$ where x and y are positive and $x + y = a$ is $\dfrac{m^m . n^n . a^{m+n}}{(m+n)^{m+n}}$, where a is constant.

2. Find the absolute maximum and minimum values of the function $f(x, y) = 3x^2 + y^2 - x$ over the region $2x^2 + y^2 \le 1$. *(A.M.I.E.T.E., Dec. 2008)*

3. Using Lagrange's method (of multipliers), find the critical (stationary values) of the function $f(x, y, z) = x^2 + y^2 + z^2$, given that $z^2 = xy + 1$. **Ans.** (0, 0, – 1), (0, 0, 1).

4. The sum, of three numbers is constant. Prove that their product is a maximum when they are equal.

5. Using the method of Lagrange's multipliers, find the largest product of the numbers x, y and z when $x^2 + y^2 + z^2 = 9$. **Ans.** $3\sqrt{3}$

6. Find a point in the plane $x + 2y + 3z = 13$ nearest to the point (1, 1, 1) using the method of Lagrange's multipliers. **Ans.** $\left(\frac{3}{2}, 2, \frac{5}{2}\right)$

7. Using the Lagrange's method (of multipliers), find the shortest distance from the point (1, 2, 2) to the sphere $x^2 + y^2 = 36$. **Ans.** 3

8. Find the shortest and the longest distances from the point (1, 2, – 1) to the $x^2 + y^2 + z^2 = 24$. *(U.P. I Semester, Dec 2009)* **Ans.** $\sqrt{6}, 3\sqrt{6}$

9. The sum of the surfaces of a sphere and a cube is given. Show that when the sum of the volumes is least, the diameter of the sphere is equal to the edge of the cube.

10. If $u = \dfrac{a^3}{x^2} + \dfrac{b^3}{y^2} + \dfrac{c^3}{z^2}$, where $x + y + z = 1$, prove that the stationary value of u is given by

$$x = \frac{a}{a+b+c},\ y = \frac{b}{a+b+c},\ z = \frac{c}{a+b+c}$$

11. If $u = a^3x^2 + b^3y^2 + c^2z^2$ where $x^{-1} + y^{-1} + z^{-1} = 1$, show that the stationary value of u is given by

$$x = \frac{\Sigma a}{a},\ y = \frac{\Sigma b}{b},\ z = \frac{\Sigma c}{c}$$ *(AMIETE, Dec. 2009)*

12. Find maximum value of the expression $\sum_{i=1}^{n} a_i x_i$ with $\sum_{i=1}^{n} x_i^2 = 1$, where $a_1, a_2, a_3 \ldots\ldots a_n$ are positive constants. **Ans.** $(a_1^2 + a_2^2 + \ldots\ldots + a_n^2)^{\frac{1}{2}}$

13. If r is the distance of a point on conic $ax^2 + by^2 + cz^2 = 1$, $lx + my + nz = 0$ from origin, then the stationary values of r are given by the equation.

$$\frac{l^2}{1-ar^2} + \frac{m^2}{1-br^2} + \frac{n^2}{1-cr^2} = 0$$

14. If x and y satisfy the relation $ax^2 + by^2 = ab$, prove that the extreme values of function $u = x^2 + xy + y^2$ are given by the roots of the equation $4(u - a)(u - b) = ab$

15. Use the Lagranges method of undetermined multipliers to find the minimum value of $x^2 + y^2 + z^2$ subject to the conditions $x + y + z = 1$, $xyz + 1 = 0$.

16. The temperature 'T' at any point (xyz) in space is $T(xyz) = kxyz^2$ where k is constant. Find the highest temperature on the surface of the sphere $x^2 + y^2 + z^2 = a^2$. **Ans.** $\dfrac{ka^4}{8}$

17. Prove that the stationary values of $u = \dfrac{x^2}{a^4} + \dfrac{y^2}{b^4} + \dfrac{z^2}{c^4}$, where $lx + my + nz = 0$ and $\dfrac{x^2}{a^2} + \dfrac{y^2}{b^2} + \dfrac{z^2}{c^2} = 1$

are the roots of the equation : $\dfrac{l^2 a^4}{l - a^2 u} + \dfrac{m^2 b^4}{1 - b^2 u} + \dfrac{n^2 c^4}{1 - c^2 u} = 0$

18. Using Lagrange method of undetermined multipliers. Find the maximum and minimum distance of the point (4.6) from $x^2 + y^2 = 4$. **Ans.** 5.21, 9.21

UNIT – III

Complex Numbers

20.1 INTRODUCTION

We have learnt the complex numbers in the previous class. Here we will review the complex number. In this chapter we will learn how to add, subtract, multiply and divide complex numbers.

20.2 COMPLEX NUMBERS

A number of the form $a + i\,b$ is called a complex number when a and b are real numbers and $i = \sqrt{-1}$. We call 'a' the real part and 'b' the imaginary part of the complex number $a + b$. If $a = 0$ the number $i\,b$ is said to be purely imaginary, if $b = 0$ the number a is real.

A complex number $x + iy$ is denoted by z.

20.3 GEOMETRICAL REPRESENTATION OF IMAGINARY NUMBERS

Let OA be positive numbers which is represented by x and OA' by $-x$.

And $\quad -x = (i)^2\, x = i\,(ix)$ is on OX'.

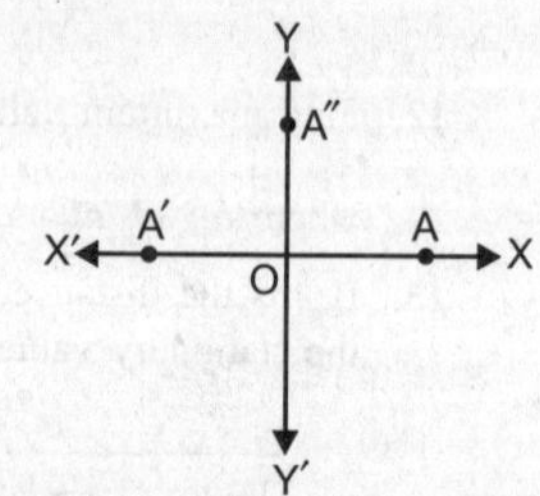

It means that the multiplication of the real number x by i twice amounts to the rotation of OA through two right angles to reach OA'.

Thus, it means that multiplication of x by i is equivalent to the rotation of x through one right angle to reach OA''.

Hence, y-axis is known as imaginary axis.

Multiplication by i rotates its direction through a right angle.

20.4 ARGAND DIAGRAM

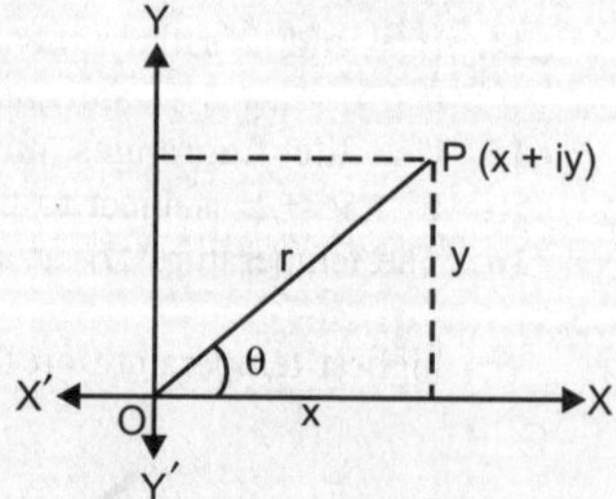

Mathematician (Argand) represented a complex number in a diagram known as Argand diagram. A complex number $x + iy$ can be represented by a point P whose co-ordinate are (x, y). The axis of x is called the real axis and the axis of y the imaginary axis. The distance OP is the **modulus of** $(x + iy)$ and the angle, OP makes with the x-axis, is the **argument** of $(x + iy)$.

20.5 EQUAL COMPLEX NUMBERS

If two complex numbers $a + ib$ and $c + id$ are equal, prove that

$$a = c \quad \text{and} \quad b = d$$

Solution. We have, $a + ib = c + id \Rightarrow a - c = i(d - b)$

$$(a - c)^2 = -(d - b)^2 \Rightarrow (a - c)^2 + (d - b)^2 = 0$$

Here sum of two positive numbers is zero. This is only possible if each number is zero.

i.e., $(a - c)^2 = 0 \Rightarrow a = c$ and $(d - b)^2 = 0 \Rightarrow b = d$ **Ans.**

20.6 ADDITION OF COMPLEX NUMBERS

Let $a + ib$ and $c + id$ be two complex numbers, then

$$(a + ib) + (c + id) = (a + c) + i(b + d)$$

Procedure. In addition of complex numbers we add real parts with real parts and imaginary parts with imaginary parts.

Example 1. *Add the following complex numbers:*

$$z_1 = 2+\frac{3}{2}i, \quad z_2 = -5+\frac{7}{4}i, \quad z_3 = \frac{5}{4}-\frac{8}{3}i, \quad z_4 = \frac{-11}{2}-i$$

Solution. $z_1 + z_2 + z_3 + z_4 = \left(2+\frac{3}{2}i\right)+\left(-5+\frac{7}{4}i\right)+\left(\frac{5}{4}-\frac{8}{3}i\right)+\left(-\frac{11}{2}-i\right)$

$$= \left(2-5+\frac{5}{4}-\frac{11}{2}\right)+\left(\frac{3}{2}+\frac{7}{4}-\frac{8}{3}-1\right)i$$

$$= \left(\frac{8-20+5-22}{4}\right)+\left(\frac{18+21-32-12}{12}\right)i$$

$$= -\frac{29}{4}-\frac{5}{12}i$$ **Ans.**

20.7 ADDITION OF COMPLEX NUMBERS BY GEOMETRY

Let $z_1 = x_1 + i\,y_1$ and $z_2 = x_2 + i\,y_2$ be two complex numbers represented by the points P and Q on the Argand diagram.

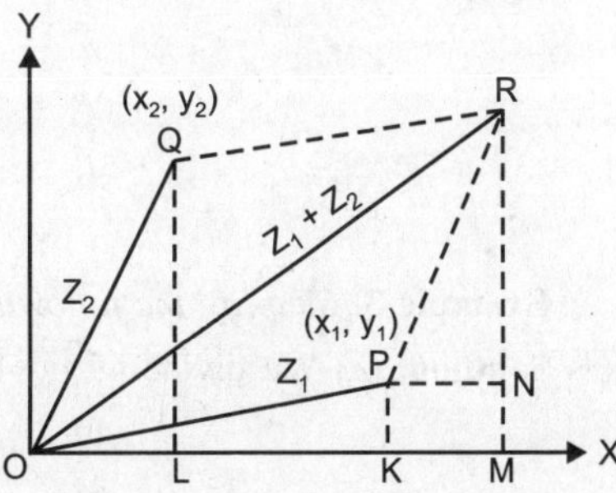

Complete the parallelogram $OPRQ$.

Draw PK, RM, QL, perpendiculars on OX.

Also draw $PN \perp$ to RM.

$$OM = OK + KM = OK + OL = x_1 + x_2$$

and $$RM = MN + NR = KP + LQ = y_1 + y_2$$

$\therefore$ The co-ordinates of R are $(x_1 + x_2, y_1 + y_2)$ and it represents the complex number.

$$(x_1 + x_2) + i(y_1 + y_2) = (x_1 + iy_1) + (x_2 + iy_2)$$

Thus the sum of two complex numbers is represented by the extremity of the diagonal of the parallelogram formed by OP (z_1) and OQ (z_2) as adjacent sides.

$$|z_1 + z_2| = OR \quad \text{and} \quad \text{amp}\,(z_1 + z_2) = \angle ROM.$$

20.8 SUBTRACTION

$$(a + ib) - (c + id) = (a - c) + i(b - d)$$

Procedure. In subtraction of complex numbers we subtract real parts from real parts and imaginary parts from imaginary parts.

Example 2. *Subtract* $z_1 = \frac{3}{4}-\frac{7}{3}i$ *from* $z_2 = \frac{-5}{3}+\frac{11}{5}i$.

Solution. $z_2 - z_1 = \left(\frac{-5}{3}+\frac{11}{5}i\right)-\left(\frac{3}{4}-\frac{7}{3}i\right) = \left(\frac{-5}{3}-\frac{3}{4}\right)+\left(\frac{11}{5}+\frac{7}{3}\right)i$

$= \left(\frac{-20-9}{12}\right)+\left(\frac{33+35}{15}\right)i = \frac{-29}{12}+\frac{68}{15}i$ **Ans.**

SUBTRACTION OF COMPLEX NUMBERS BY GEOMETRY.

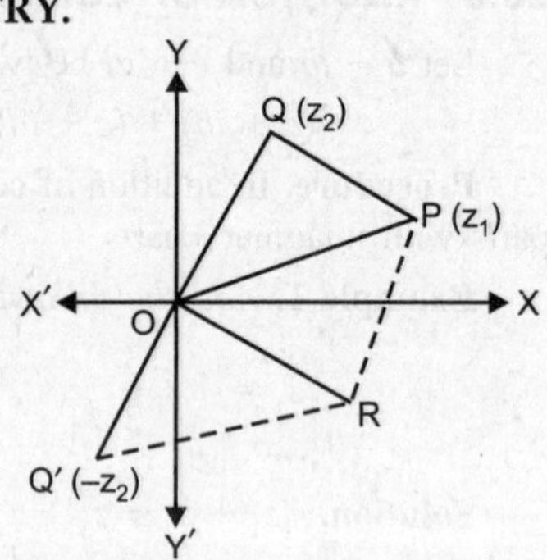

Let P and Q represent two complex numbers

$$z_1 = x_1 + iy_1 \text{ and } z_2 = x_2 + iy_2.$$

Then $z_1 - z_2 = z_1 + (-z_2)$

$z_1 - z_2$ means the addition of z_1 and $-z_2$.

$-z_2$ is represented by OQ' formed by producing OQ to OQ' such that $OQ = OQ'$.

Complete the parallelogram $OPRQ'$, then the sum of z_1 and $-z_2$ represented by OR.

$|z_2 - z_1|$ = OR, Argument of $(z_2 - z_1) = \angle ROX$

20.9 POWERS OF *i*

Some time we need various powers of i.

We know that $i = \sqrt{-1}$.

On squaring both sides, we get

$$i^2 = -1$$

Multiplying by i both sides, we get

$$i^3 = -i$$

Again,

$$i^4 = (i^3)(i) = (-i)(i) = -(i^2) = -(-1) = 1$$

$$i^5 = (i^4)(i) = (1)(i) = i$$

$$i^6 = (i^4)(i^2) = (1)(-1) = -1$$

$$i^7 = (i^4)(i^3) = 1(-i) = -i$$

$$i^8 = (i^4)(i^4) = (1)(1) = 1.$$

Example 3. *Simplify the following:* (*a*) i^{49}, (*b*) i^{103}.

Solution. (*a*) We divide 49 by 4 and we get

$$49 = 4 \times 12 + 1$$

$$i^{49} = i^{4 \times 12 + 1} = (i^4)^{12}(i^1) = (1)^{12}(i) = i$$

(*b*) We divide 103 by 4, we get

$$103 = 4 \times 25 + 3$$

$$i^{103} = i^{4 \times 25 + 3} = (i^4)^{25}(i^3) = (1)^{25}(-i) = -i$$ **Ans.**

20.10 MULTIPLICATION

$$(a + ib) \times (c + id) = ac - bd + i(ad + bc)$$

Proof. $(a + ib) \times (c + id) = ac + iad + ibc + i^2bd$

$= ac + i(ad + bc) + (-1)bd$ $[\because i^2 = -1]$

$$= (ac - bd) + (ad + bc)i$$

Example 4. *Multiply 3 + 4i by 7 – 3i.*

Solution. Let $z_1 = 3 + 4i$ and $z_2 = 7 - 3i$

$$z_1.z_2 = (3 + 4i).(7 - 3i)$$
$$= 21 - 9i + 28i - 12i^2$$
$$= 21 - 9i + 28i - 12(-1) \qquad [\because i^2 = -1]$$
$$= 21 - 9i + 28i + 12$$
$$= 33 + 19i$$

Ans.

Multiplication of complex numbers (Polar form) :

Let $z_1 = r_1(\cos\theta_1 + i\sin\theta_1)$ and $z_2 = r_2(\cos\theta_2 + i\sin\theta_2)$

$$x_1 = r_1\cos\theta_1, \quad y_1 = r_1\sin\theta_1$$
$$x_2 = r_2\cos\theta_2, \quad y_2 = r_2\sin\theta_2$$
$$z_1 = x_1 + iy_1 = r_1(\cos\theta_1 + i\sin\theta_1) \qquad |z_1| = r_1$$
$$z_2 = x_2 + iy_2 = r_2(\cos\theta_2 + i\sin\theta_2) \qquad |z_2| = r_2$$
$$z_1.z_2 = r_1r_2(\cos\theta_1 + i\sin\theta_1)(\cos\theta_2 + i\sin\theta_2)$$
$$= r_1r_2\,[\cos\theta_1\cos\theta_2 - \sin\theta_1\sin\theta_2 + i(\sin\theta_1\cos\theta_2 + \cos\theta_1\sin\theta_2)]$$
$$= r_1r_1[\cos(\theta_1 + \theta_2) + i\sin(\theta_1 + \theta_2)], \qquad |z_1z_2| = r_1r_2$$

The modulus of the product of two complex numbers is the product of their moduli and the argument of the product is the sum of their arguments.

Graphical method

Let P, Q represent the complex numbers.

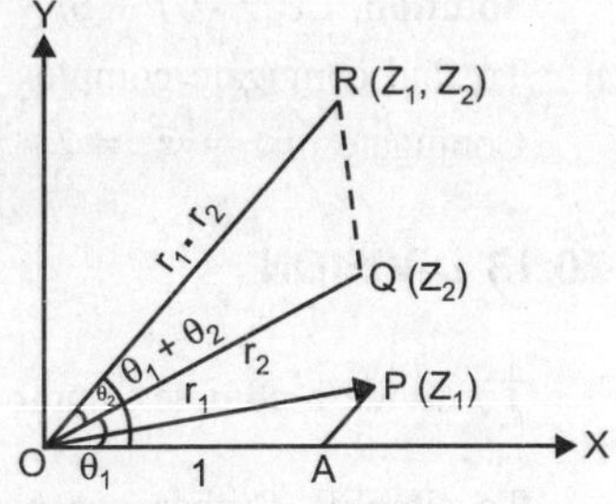

$$z_1 = x_1 + iy_1$$
$$= r_1(\cos\theta_1 + i\sin\theta_1)$$
$$z_2 = x_2 + iy_2$$
$$= r_2(\cos\theta_2 + i\sin\theta_2)$$

Cut off $OA = 1$ along x-axis. Construct $\Delta\ ORQ$ on OQ similar to $\Delta\ OAP$.

So that $\dfrac{OR}{OP} = \dfrac{OQ}{OA} \Rightarrow \dfrac{OR}{OP} = \dfrac{OQ}{1} \Rightarrow OR = OP.OQ = r_1r_2$

$$\angle XOR = \angle AOQ + \angle QOR = \theta_2 + \theta_1$$

Hence the product of two complex numbers z_1, z_2 is represented by the point R, such that

(*i*) $|z_1.z_2| = |z_1|.|z_2|$ (*ii*) Arg $(z_1.z_2)$ = Arg (z_1) + Arg (z_2)

20.11 *i* (IOTA) AS AN OPERATOR

Multipliction of a complex number by i.

Let $$z = x + iy = r(\cos\theta + i\sin\theta)$$

$$i = 0 + i\,.\,1 = \left[\cos\frac{\pi}{2} + i\sin\frac{\pi}{2}\right]$$

$$i\,.\,z = r(\cos\theta + i\sin\theta).\left[\cos\frac{\pi}{2} + i\sin\frac{\pi}{2}\right]$$

$$= r\left[\cos\left(\theta + \frac{\pi}{2}\right) + i\sin\left(\theta + \frac{\pi}{2}\right)\right]$$

Hence a complex number multiplied by i results :

The rotation of the complex number by $\frac{\pi}{2}$ in anticlockwise direction without change in magnitude.

20.12 CONJUGATE OF A COMPLEX NUMBER

Two complex numbers which differ only in the sign of imaginary parts are called conjugate of each other.

A pair of complex number $a + ib$ and $a - ib$ are said to be conjugate of each other.

Theorem. Show that the sum and product of a complex number and its conjugate complex are both real.

Proof. Let $x + iy$ be a complex number and $x - iy$ its conjugate complex.

$$\text{Sum} = (x + iy) + (x - iy) = 2x \qquad \text{(Real)}$$

$$\text{Product} = (x + iy).(x - iy) = x^2 + y^2. \qquad \text{(Real)}$$ **Proved.**

Note. Let a complex number be z. Then the conjugate complex number is denoted by $\overline{z}$.

Example 5. *Find out the conjugate of a complex number 7 + 6i.*

Solution. Let $z = 7 + 6i$

To find conjugate complex number of $7 + 6i$ we change the sign of imaginary number.

Conjugate of $z = \overline{z} = 7 - 6i$ **Ans.**

20.13 DIVISION

To divide a complex number $a + ib$ by $c + id$, we write it as $\frac{a+ib}{c+id}$.

To simplify further, we multiply the numerator and denominator by the conjugate of the denominator.

$$\frac{a+ib}{c+id} = \frac{(a+ib)}{(c+id)} \times \frac{(c-id)}{(c-id)} = \frac{ac - iad + ibc - i^2 bd}{(c)^2 - (id)^2}$$

$$= \frac{ac - i(ad - bc) + bd}{c^2 - d^2 i^2} \qquad [\because i^2 = -1]$$

$$= \frac{ac + bd}{c^2 + d^2} + \frac{bc - ad}{c^2 + d^2} i$$

Example 6. *Divide 1 + i by 3 + 4i.*

Solution.

$$\frac{1+i}{3+4i} = \frac{1+i}{3+4i} \times \frac{3-4i}{3-4i}$$

$$= \frac{3 - 4i + 3i - 4i^2}{9 - 16i^2}$$

$$= \frac{3 - i + 4}{9 + 16} = \frac{7}{25} - \frac{1}{25} i$$ **Ans.**

DIVISION (By Algebra)

Let $z_1 = r_1(\cos\theta_1 + i\sin\theta_1)$ and $z_2 = r_2(\cos\theta_2 + i\sin\theta_2)$

$$\frac{z_1}{z_2} = \frac{r_1(\cos\theta_1 + i\sin\theta_1)}{r_2(\cos\theta_2 + i\sin\theta_2)} = \frac{r_1(\cos\theta_1 + i\sin\theta_1)(\cos\theta_2 - i\sin\theta_2)}{r_2(\cos\theta_2 + i\sin\theta_2)(\cos\theta_2 - i\sin\theta_2)}$$

$$= \frac{r_1[(\cos\theta_1\cos\theta_2 + \sin\theta_1\sin\theta_2) + i(\sin\theta_1\cos\theta_2 - \sin\theta_2\cos\theta_1)]}{r_2(\cos^2\theta_2 + \sin^2\theta_2)}$$

$$= \frac{r_1}{r_2}[\cos(\theta_1 - \theta_2) + i\sin(\theta_1 - \theta_2)]$$

The modulus of the quotient of two complex numbers is the quotient of their moduli, and the argument of the quotient is the difference of their arguments.

20.14 DIVISION OF COMPLEX NUMBERS BY GEOMETRY

Let P and Q represent the complex numbers.

$$z_1 = x_1 + i\,y_1 = r_1(\cos\theta_1 + i\sin\theta_1)$$
$$z_2 = x_2 + i\,y_2 = r_2(\cos\theta_2 + i\sin\theta_2)$$

Cut off $OA = 1$, construct $\Delta\,OAR$ on OA similar to $\Delta\,OQP$.

So that $\dfrac{OR}{OA} = \dfrac{OP}{OQ} \Rightarrow \dfrac{OR}{1} = \dfrac{OP}{OQ}$

$$OR = \frac{OP}{OQ} = \frac{r_1}{r_2}$$

$$\angle AOR = \angle QOP = \angle AOP - \angle AOQ = \theta_1 - \theta_2$$

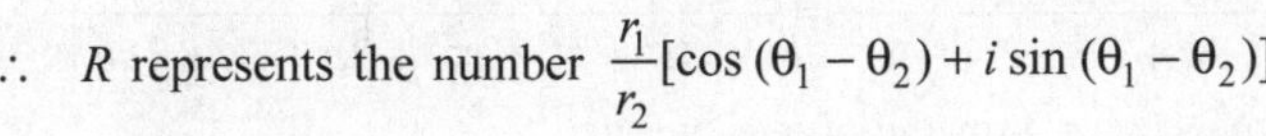

$\therefore$ R represents the number $\dfrac{r_1}{r_2}[\cos(\theta_1 - \theta_2) + i\sin(\theta_1 - \theta_2)]$

Hence the complex number $\dfrac{z_1}{z_2}$ is represented by the point R.

(*i*) $\left|\dfrac{z_1}{z_2}\right| = \dfrac{|z_1|}{|z_2|}$ (*ii*) $\text{Arg.}\left(\dfrac{z_1}{z_2}\right) = \text{Arg.}(z_1) - \text{Arg.}(z_2)$.

Example 7. *Express* $\dfrac{(6+i).(2-i)}{(4+3i).(1-2i)}$ *in the form of* $a + ib$.

Solution. $$\frac{(6+i).(2-i)}{(4+3i).(1-2i)} = \frac{12+1+i(2-6)}{4+6+i(3-8)} = \frac{13-4i}{10-5i}$$

$$= \frac{(13-4i)(10+5i)}{(10-5i)(10+5i)} = \frac{150+25i}{100+25} = \frac{6+i}{5} = \frac{6}{5} + \frac{1}{5}i.$$ **Ans.**

Example 8. *If* $a = \cos\theta + i\sin\theta$, *prove that* $1 + a + a^2 = (1 + 2\cos\theta)(\cos\theta + i\sin\theta)$.

Solution. Here we have $a = \cos\theta + i\sin\theta$

$$1 + a + a^2 = 1 + (\cos\theta + i\sin\theta) + (\cos\theta + i\sin\theta)^2$$
$$= 1 + \cos\theta + i\sin\theta + \cos^2\theta + 2i\sin\theta\cos\theta - \sin^2\theta$$
$$= (\cos\theta + i\sin\theta) + (1 - \sin^2\theta) + \cos^2\theta + 2i\sin\theta\cos\theta$$
$$= (\cos\theta + i\sin\theta) + \cos^2\theta + \cos^2\theta + 2i\sin\theta\cos\theta$$
$$= (\cos\theta + i\sin\theta) + 2\cos^2\theta + 2i\sin\theta\cos\theta$$

$$= (\cos\theta + i\sin\theta) + 2\cos\theta\,(\cos\theta + i\sin\theta)$$

$$= (\cos\theta + i\sin\theta)\,(1 + 2\cos\theta)$$ **Proved.**

Example 9. *Solve for θ such that the expression* $\dfrac{3+2i\sin\theta}{1-2i\sin\theta}$ *is imaginary.*

Solution. $$\frac{3+2i\sin\theta}{1-2i\sin\theta} = \frac{(3+2i\sin\theta)(1+2i\sin\theta)}{(1-2i\sin\theta)(1+2i\sin\theta)} = \frac{3-4\sin^2\theta+8i\sin\theta}{1+4\sin^2\theta}$$

If $3 - 4\sin^2\theta = 0$ then $\dfrac{3-4\sin^2\theta+8i\sin\theta}{1+4\sin^2\theta}$ = purely imaginary.

$$\sin^2\theta = \frac{3}{4} \quad\text{or}\quad \sin\theta = \frac{\sqrt{3}}{2} \quad\text{or}\quad \theta = \frac{\pi}{3}$$ **Ans.**

Example 10. *If* $a^2 + b^2 + c^2 = 1$ *and* $b + ic = (1 + a)z$, *prove that* $\dfrac{a+ib}{1+c} = \dfrac{1+iz}{1-iz}$.

Solution. Here, we have $b + ic = (1 + a)z \;\Rightarrow\; z = \dfrac{b+ic}{1+a}$

$$\frac{1+iz}{1-iz} = \frac{1+i\dfrac{b+ic}{1+a}}{1-i\dfrac{b+ic}{1+a}} = \frac{1+a+ib-c}{1+a-ib+c}$$

$$= \frac{[(1+a+ib)-c]}{(1+a+c-ib)} \times \frac{(1+a+ib+c)}{(1+a+c+ib)} = \frac{(1+a+ib)^2-c^2}{(1+a+c)^2+b^2}$$

$$= \frac{1+a^2-b^2+2a+2ib+2iab-c^2}{1+a^2+c^2+2a+2c+2ac+b^2} = \frac{1+a^2-b^2-c^2+2a+2ib+2iab}{1+(a^2+b^2+c^2)+2a+2c+2ac}$$

Putting the value of $a^2 + b^2 + c^2 = 1$ in the above, we get

$$= \frac{1+a^2-(1-a^2)+2a+2ib+2iab}{1+1+2a+2c+2ac} = \frac{2(a^2+a+ib+iab)}{2(1+a+c+ac)} = \frac{2(1+a)(a+ib)}{2(1+a)(1+c)} = \frac{a+ib}{1+c}$$

Proved.

Example 11. *If* $z = \cos\theta + i\sin\theta$, *prove that*

(*a*) $\dfrac{2}{1+z} = 1 - i\tan\dfrac{\theta}{2}$ (*b*) $\dfrac{1+z}{1-z} = i\cot\dfrac{\theta}{2}$

Solution. Here, we have $z = \cos\theta + i\sin\theta$

(*a*) $$\frac{2}{1+z} = \frac{2}{1+(\cos\theta+i\sin\theta)} = \frac{2}{(1+\cos\theta)+i\sin\theta} \times \frac{(1+\cos\theta)-i\sin\theta}{(1+\cos\theta)-i\sin\theta}$$

$$= \frac{2[(1+\cos\theta)-i\sin\theta]}{(1+\cos\theta)^2+\sin^2\theta}$$

$$= \frac{2[(1+\cos\theta)-i\sin\theta]}{2(1+\cos\theta)} = 1 - \frac{i\sin\theta}{1+\cos\theta}$$

$$= 1 - i\frac{2\sin\left(\frac{\theta}{2}\right)\cos\left(\frac{\theta}{2}\right)}{2\cos^2\left(\frac{\theta}{2}\right)} = 1 - i\tan\left(\frac{\theta}{2}\right)$$ **Proved**

$$\begin{aligned}&(1+\cos\theta)^2+\sin^2\theta\\ &= 1+\cos^2\theta+2\cos\theta+\sin^2\theta\\ &= 1+(\sin^2\theta+\cos^2\theta)+2\cos\theta\\ &= 1+1+2\cos\theta\\ &= 2+2\cos\theta\\ &= 2(1+\cos\theta)\end{aligned}$$

(*b*) $$\frac{1+z}{1-z} = \frac{(1+\cos\theta)+i\sin\theta}{(1-\cos\theta)-i\sin\theta} = \frac{2\cos^2\frac{\theta}{2}+2i\sin\frac{\theta}{2}\cos\frac{\theta}{2}}{2\sin^2\frac{\theta}{2}-2i\sin\frac{\theta}{2}\cos\frac{\theta}{2}}$$

$$= \frac{\cos\frac{\theta}{2}}{\sin\frac{\theta}{2}}\cdot\left(\frac{\cos\frac{\theta}{2}+i\sin\frac{\theta}{2}}{\sin\frac{\theta}{2}-i\cos\frac{\theta}{2}}\right) = \cot\frac{\theta}{2}\left(\frac{\cos\frac{\theta}{2}+i\sin\frac{\theta}{2}}{\sin\frac{\theta}{2}-i\cos\frac{\theta}{2}}\right) \quad ...(1)$$

Now, $$\frac{\cos\frac{\theta}{2}+i\sin\frac{\theta}{2}}{\sin\frac{\theta}{2}-i\cos\frac{\theta}{2}} = \left(\frac{\cos\frac{\theta}{2}+i\sin\frac{\theta}{2}}{\sin\frac{\theta}{2}-i\cos\frac{\theta}{2}}\right)\left(\frac{\sin\frac{\theta}{2}+i\cos\frac{\theta}{2}}{\sin\frac{\theta}{2}+i\cos\frac{\theta}{2}}\right)$$

$$= \frac{\cos\frac{\theta}{2}\sin\frac{\theta}{2}+i\cos^2\frac{\theta}{2}+i\sin^2\frac{\theta}{2}-\sin\frac{\theta}{2}\cos\frac{\theta}{2}}{\sin^2\frac{\theta}{2}+\cos^2\frac{\theta}{2}} = \frac{i\left(\cos^2\frac{\theta}{2}+\sin^2\frac{\theta}{2}\right)}{\left(\cos^2\frac{\theta}{2}+\sin^2\frac{\theta}{2}\right)} = i \quad ...(2)$$

From (1) and (2), we get

Thus, $$\frac{1+z}{1-z} = i\cot\frac{\theta}{2}$$ **Proved.**

Example 12. *If* $x = \cos\theta + i\sin\theta$, $y = \cos\phi + i\sin\phi$, *prove that*

$$\frac{x-y}{x+y} = i\tan\left(\frac{\theta-\phi}{2}\right)$$ (*M.U. 2008*)

Solution. We have,

$$\frac{x-y}{x+y} = \frac{(\cos\theta+i\sin\theta)-(\cos\phi+i\sin\phi)}{(\cos\theta+i\sin\theta)+(\cos\phi+i\sin\phi)}$$

$$= \frac{(\cos\theta-\cos\phi)+i(\sin\theta-\sin\phi)}{(\cos\theta+\cos\phi)+i(\sin\theta+\sin\phi)}$$

$$= \frac{\left[-2\sin\left(\frac{\theta+\phi}{2}\right)\sin\left(\frac{\theta-\phi}{2}\right)+2i\cos\left(\frac{\theta+\phi}{2}\right)\sin\left(\frac{\theta-\phi}{2}\right)\right]}{\left[2\cos\left(\frac{\theta+\phi}{2}\right)\cos\left(\frac{\theta-\phi}{2}\right)+2i\sin\left(\frac{\theta+\phi}{2}\right)\cos\left(\frac{\theta-\phi}{2}\right)\right]}$$

$$= \frac{2i\sin\left(\frac{\theta-\phi}{2}\right)\left[\cos\left(\frac{\theta+\phi}{2}\right)+i\sin\left(\frac{\theta+\phi}{2}\right)\right]}{2\cos\left(\frac{\theta-\phi}{2}\right)\left[\cos\left(\frac{\theta+\phi}{2}\right)+i\sin\left(\frac{\theta+\phi}{2}\right)\right]} = i\tan\left(\frac{\theta-\phi}{2}\right)$$ **Proved.**

EXERCISE 20.1

1. If $z = 1 + i$, find (*i*) z^2 (*ii*) $\frac{1}{z}$ and plot them on the Argand diagram. **Ans.** (*i*) $2i$, (*ii*) $\frac{1}{2}-\frac{i}{2}$

Express the following in the form *a* + *ib*, where *a* and *b* are real (2 to 4):

2. $\frac{2-3i}{4-i}$ **Ans.** $\frac{11}{17}-\frac{10}{17}i$

3. $\frac{(3+4i)(2+i)}{1+i}$ **Ans.** $\frac{13}{2}+\frac{9}{2}i$ **4.** $\frac{(1+2i)^3}{(1+i)(2-i)}$ **Ans.** $-\frac{7}{2}+\frac{1}{2}i$

5. The points A, B, C represent the complex numbers z_1, z_2, z_3 respectively, and G is the centroid of the triangle ABC, if $4z_1 + z_2 + z_3 = 0$, show that the origin is the mid-point of AG.
6. $ABCD$ is a parallelogram on the Argand plane. The affixes of A, B, C are $8 + 5i$, $-7 - 5i$, $-5 + 5i$, respectively. Find the affix of D. **Ans.** $10 + 15i$
7. If z_1, z_2, z_3 are three complex numbers and
$$a_1 = z_1 + z_2 + z_3$$
$$b_1 = z_1 + \omega\, z_2 + \omega^2 z_3$$
$$c_1 = z_1 + \omega^2 z_2 + \omega\, z_3$$
show that $|a_1|^2 + |b_1|^2 + |c_1|^2 = 3\{|z_1|^2 + |z_2|^2 + |z_3|^2\}$
where ω, ω^2 are cube roots of unity.
8. Find the complex conjugate of $\dfrac{2+3i}{1-i}$. **Ans.** $-\dfrac{1}{2}-\dfrac{5}{2}i$
9. If $x + iy = \dfrac{1}{a+ib}$, prove that $(x^2 + y^2)(a^2 + b^2) = 1$
10. Find the value of $x^2 - 6x + 13$, when $x = 3 + 2i$. **Ans.** 0
11. If $\alpha - i\beta = \dfrac{1}{a-ib}$, prove that $(\alpha^2 + \beta^2)(a^2 + b^2) = 1$. *(M.U. 2008)*
12. If $\dfrac{1}{\alpha+i\beta}+\dfrac{1}{a+ib} = 1$, where α, β, a, b are real, express b in terms of α, β.

Ans. $\dfrac{-\beta}{\alpha^2+\beta^2-2\alpha+1}$
13. If $(x + iy)^{1/3} = a + ib$, then show that $4(a^2 - b^2) = \dfrac{x}{a}+\dfrac{y}{b}$.
14. If $(x + iy)^3 = u + iv$, then show that $\dfrac{u}{x}+\dfrac{v}{y} = 4(x^2 - y^2)$.
15. Find the values of x and y, if $\dfrac{(1+i)x-2i}{3+i}+\dfrac{(2-3i)y+i}{3-i} = i$. **Ans.** $x = 3$ and $y = -1$
16. If $a + ib = \dfrac{(x+i)^2}{2x^2+1}$, prove that $a^2 + b^2 = \dfrac{(x^2+1)^2}{(2x^2+1)^2}$.

20.15 MODULUS AND ARGUMENT

Let $x + iy$ be a complex number.

Putting $x = r\cos\theta$ and $y = r\sin\theta$ so that $r = \sqrt{x^2+y^2}$

$$\cos\theta = \frac{x}{\sqrt{x^2+y^2}} \quad \text{and} \quad \sin\theta = \frac{y}{\sqrt{x^2+y^2}}$$

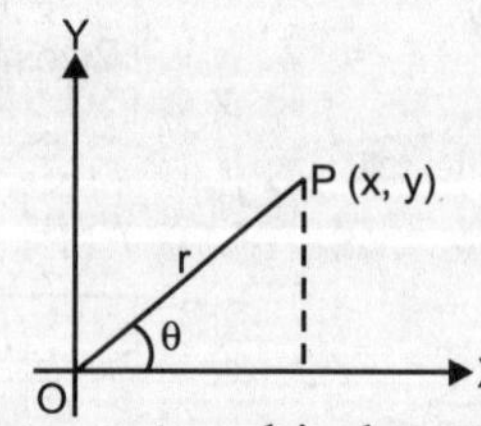

the positive value of the root being taken.

Then r called the *modulus* or absolute value of the complex number $x + iy$ and is denoted by $|x + iy|$.

The angle θ is called the *argument* or *amplitude* of the complex number $x + iy$ and is denoted by arg. $(x + iy)$.

It is clear that θ will have infinite number of values differing by multiples of 2π. The values of θ lying in the range $-\pi < \theta \le \pi$ $[(0 < \theta < \pi)$ or $(-\pi < \theta < 0)]$ is called the *principal value* of the argument.

The principal value of θ is written either between 0 and π or between 0 and $-\pi$.

A complex number $x + iy$ is denoted by a single letter z. The number $x - iy$ (conjugate) is denoted by $\overline{z}$. The complex number in polar form is $r(\cos\theta + i\sin\theta)$.

Modulus of z is denoted by $|z|$ and $|z|^2 = x^2 + y^2$.

Angle θ	**Principal value of θ**

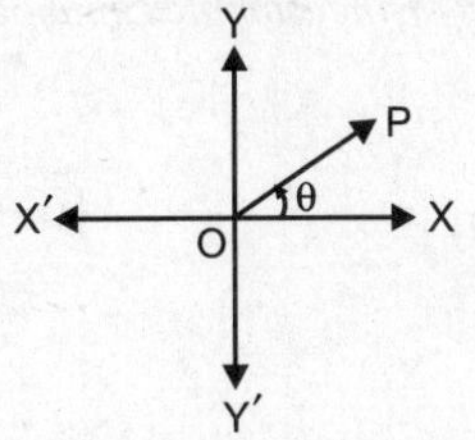

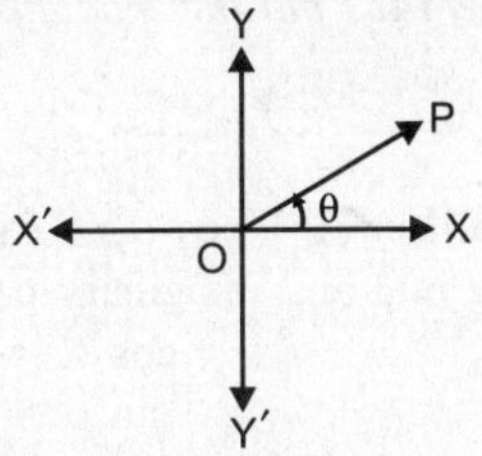

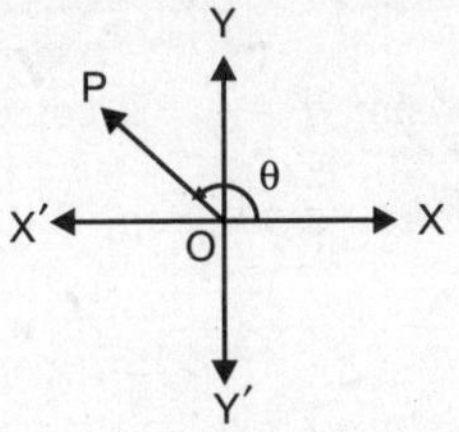

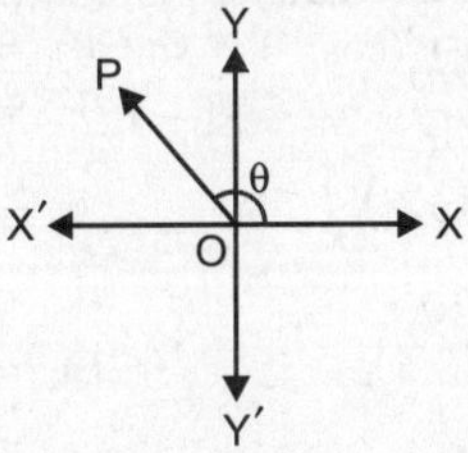

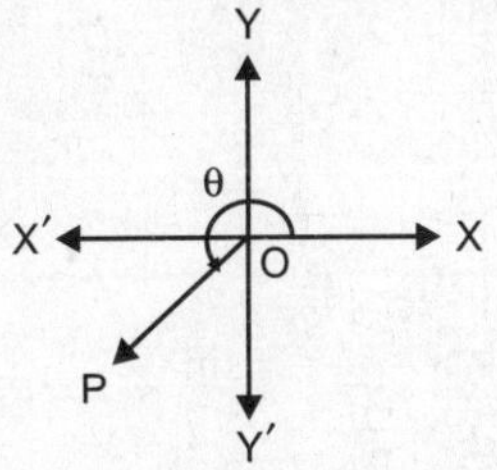

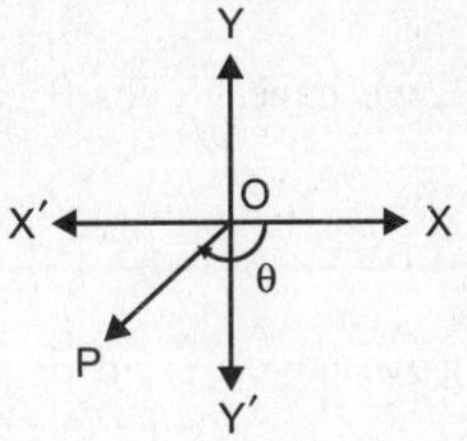

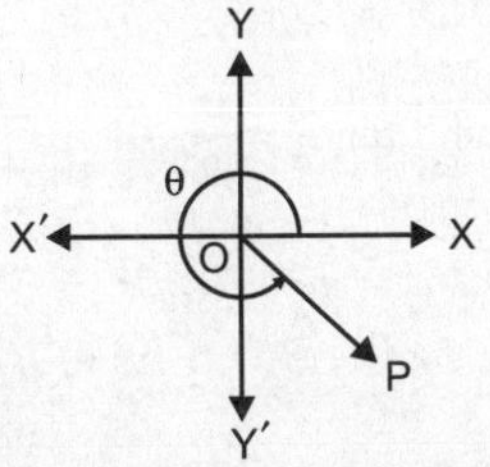

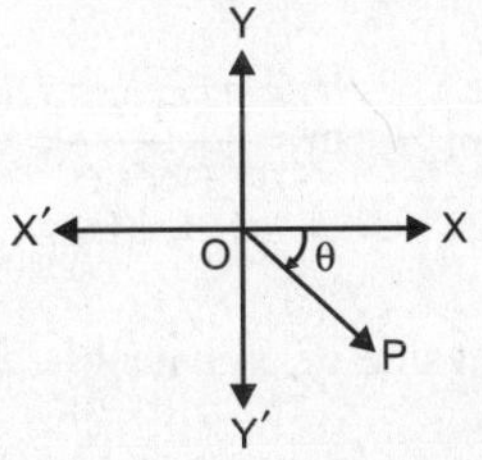

For example (*i*) the principal value of 240° is –120°.

(*ii*) the principal value of 330° is –30°.

Example 13. *Find the modulus and principal argument of the complex number*

$$\frac{1+2i}{1-(1-i)^2}.$$

Solution. $\dfrac{1+2i}{1-(1-i)^2} = \dfrac{1+2i}{1-(1-1-2i)} = \dfrac{1+2i}{1+2i} = 1 = 1 + 0i$

$\therefore \quad \left|\dfrac{1+2i}{1-(1-i)^2}\right| = |1+0i| = \sqrt{1^2} = 1$ **Ans.**

Principal argument of $\dfrac{1+2i}{1-(1-i)^2}$ = Principal argument of $1 + 0i$

$$= \tan^{-1}\frac{0}{1} = \tan^{-1} 0 = 0°.$$

Hence modulus = 1 and principal argument = 0°. **Ans.**

Example 14. *Find the modulus and principal argument of the complex number :*

$$1 + \cos\alpha + i\sin\alpha. \qquad \left(0 < \alpha < \frac{\pi}{2}\right)$$

Solution. Let $(1 + \cos\alpha) + i\sin\alpha = r(\cos\theta + i\sin\theta)$

Equating real and imaginary parts, we get

$$1 + \cos\alpha = r\cos\theta \qquad ...(1)$$

And

$$\sin\alpha = r\sin\theta \qquad ...(2)$$

Squaring and adding (1) and (2), we get

$$r^2(\cos^2\theta + \sin^2\theta) = (1 + \cos\alpha)^2 + (\sin\alpha)^2$$

$$\Rightarrow \quad r^2 = 1 + \cos^2\alpha + 2\cos\alpha + \sin^2\alpha = 1 + 2\cos\alpha + 1$$

$$= 2(1 + \cos\alpha) = 2\left(1 + 2\cos^2\frac{\alpha}{2} - 1\right) = 4\cos^2\frac{\alpha}{2}$$

$$\Rightarrow \quad r = 2\cos\frac{\alpha}{2}$$

From (1), we have,
$$\cos\theta = \frac{1+\cos\alpha}{r} = \frac{1 + 2\cos^2\frac{\alpha}{2} - 1}{2\cos\frac{\alpha}{2}} = \cos\frac{\alpha}{2} \qquad ...(3)$$

From (2), we have,
$$\sin\theta = \frac{\sin\alpha}{r} = \frac{\sin\alpha}{2\cos\frac{\alpha}{2}} = \frac{2\sin\frac{\alpha}{2}\cos\frac{\alpha}{2}}{2\cos\frac{\alpha}{2}} = \sin\frac{\alpha}{2} \qquad ...(4)$$

$$\text{Argument} = \tan^{-1}\frac{\sin\alpha}{1+\cos\alpha} = \tan^{-1}\frac{2\sin\frac{\alpha}{2}\cos\frac{\alpha}{2}}{1+2\cos^2\frac{\alpha}{2}-1} = \tan^{-1}\tan\frac{\alpha}{2} = \frac{\alpha}{2}$$

$$\text{General value of argument} = 2\pi k + \frac{\alpha}{2}$$

$\theta = \frac{\alpha}{2}$ satisfied both equations, (1) and (2),

$\text{Arg}(1 + \cos\alpha + i\sin\alpha) = \frac{\alpha}{2}$ and modulus of $(1 + \cos\alpha + i\sin\alpha) = r = 2\cos\frac{\alpha}{2}$ **Ans.**

EXERCISE 20.2

Find the modulus and principal argument of the following complex numbers:

1. $-\sqrt{3} - i$ **Ans.** $2, -\frac{5\pi}{6}$
2. $\frac{(1+i)^2}{1-i}$ **Ans.** $\sqrt{2}, \frac{3\pi}{4}$
3. $\sqrt{\left(\frac{1+i}{1-i}\right)}$ **Ans.** $1, \frac{\pi}{4}$
4. $\tan\alpha - i$ **Ans.** $\sec\alpha, -\left(\frac{\pi}{2} - \alpha\right)$
5. $1 - \cos\alpha + i\sin\alpha$ **Ans.** $2\sin\frac{\alpha}{2}, \frac{\pi - \alpha}{2}$

6. $(4+2i)(-3+\sqrt{2}i)$ **Ans.** $2\sqrt{55}, \tan^{-1}\left(\frac{3-2\sqrt{2}}{6+\sqrt{2}}\right)$

Find the modulus of the following complex numbers :

7. $\overline{(7+i^2)}+(6-i)-(4-3i^3)$ **Ans.** $4\sqrt{5}$

8. $\overline{(5-6i)}-(5+6i)+(8-i)$ **Ans.** $\sqrt{185}$

9. $(8-i^3)-(7i^2+5)+\overline{(9-i)}$ **Ans.** $\sqrt{365}$

10. $(5+6i^{11})+(8i^3+i^5)+(i^2-i^4)$ **Ans.** $\sqrt{178}$

11. If arg. $(z+2i) = \frac{\pi}{4}$ and arg. $(z-2i) = \frac{3\pi}{4}$, find z. **Ans.** $z = 2$

Example 15. *If $z_1 = \cos\alpha + i\sin\alpha$, $z_2 = \cos\beta + i\sin\beta$ show that*

$$\frac{1}{2i}\left(\frac{z_1}{z_2}-\frac{z_2}{z_1}\right) = \sin(\alpha-\beta)$$

Solution. We have

$$\frac{z_1}{z_2} = \frac{\cos\alpha+i\sin\alpha}{\cos\beta+i\sin\beta} = \frac{\cos\alpha+i\sin\alpha}{\cos\beta+i\sin\beta}\times\frac{\cos\beta-i\sin\beta}{\cos\beta-i\sin\beta}$$

$$= \frac{(\cos\alpha\cos\beta+\sin\alpha\sin\beta)+i(\sin\alpha\cos\beta-\cos\alpha\sin\beta)}{\cos^2\beta+\sin^2\beta}$$

$$= \cos(\alpha-\beta)+i\sin(\alpha-\beta) \quad ...(1)$$

$$\frac{z_2}{z_1} = \frac{1}{\cos(\alpha-\beta)+i\sin(\alpha-\beta)}\times\frac{\cos(\alpha-\beta)-i\sin(\alpha-\beta)}{\cos(\alpha-\beta)-i\sin(\alpha-\beta)}$$

$$= \frac{\cos(\alpha-\beta)-i\sin(\alpha-\beta)}{\cos^2(\alpha-\beta)+\sin^2(\alpha-\beta)} = \cos(\alpha-\beta)-i\sin(\alpha-\beta) \quad ...(2)$$

Subtracting (2) from (1), we get

$$\frac{z_1}{z_2}-\frac{z_2}{z_1} = 2i\sin(\alpha-\beta)$$ **Proved.**

Example 16. *If z_1 and z_2 are any two complex numbers, prove that*

$$|z_1+z_2|^2+|z_1-z_2|^2 = 2[|z_1|^2+|z_2|^2]$$

Solution. Let $z_1 = x_1+iy_1$

$z_2 = x_2+iy_2$

$$|z_1+z_2|^2 = |(x_1+iy_1)+(x_2+iy_2)|^2$$

$$= |(x_1+x_2)+i(y_1+y_2)|^2$$

$$= (x_1+x_2)^2+(y_1+y_2)^2 \quad ...(1)$$

Similarly $|z_1-z_2|^2 = (x_1-x_2)^2+(y_1-y_2)^2$ ...(2)

and $|z_1|^2 = x_1^2+y_1^2$ ...(3)

$|z_2|^2 = x_2^2+y_2^2$ ...(4)

L.H.S. $= |z_1+z_2|^2+|z_1-z_2|^2 = (x_1+x_2)^2+(y_1+y_2)^2+(x_1-x_2)^2+(y_1-y_2)^2$

[Using (1) and (2)]

$$= x_1^2+2x_1x_2+x_2^2+y_1^2+2y_1y_2+y_2^2+x_1^2-2x_1x_2+x_2^2+y_1^2-2y_1y_2+y_2^2$$

$$= 2[x_1^2 + x_2^2 + y_1^2 + y_2^2] = 2[(x_1^2 + y_1^2) + (x_2^2 + y_2^2)] \quad ...(5)$$

$$= 2[|z_1|^2 + |z_2|^2] = \text{R.H.S.}$$ **Proved.**

Example 17. *If z_1 and z_2 are two complex numbers such that $|z_1 + z_2| = |z_1 - z_2|$, prove that*

$$arg.\ z_1 - arg.\ z_2 = \frac{\pi}{2}$$ *(M.U. 2007,)*

Solution. Let

$$z_1 = x_1 + iy_1$$
$$z_2 = x_2 + iy_2$$

Given that $\quad |z_1 + z_2| = |z_1 - z_2|$

$\Rightarrow \quad |(x_1 + iy_1) + (x_2 + iy_2)| = |(x_1 + iy_1) - (x_2 + iy_2)|$

$\Rightarrow \quad |(x_1 + x_2) + i(y_1 + y_2)| = |(x_1 - x_2) + (y_1 - y_2)i|$

$\Rightarrow \quad (x_1 + x_2)^2 + (y_1 + y_2)^2 = (x_1 - x_2)^2 + (y_1 - y_2)^2$

$\Rightarrow \quad x_1^2 + x_2^2 + 2x_1x_2 + y_1^2 + y_2^2 + 2y_1y_2 = x_1^2 + x_2^2 - 2x_1x_2 + y_1^2 + y_2^2 - 2y_1y_2$

$\Rightarrow \quad 2x_1x_2 + 2y_1y_2 = -2x_1x_2 - 2y_1y_2$

$\Rightarrow \quad 4x_1x_2 + 4y_1y_2 = 0$

$\Rightarrow \quad x_1x_2 + y_1y_2 = 0 \quad ...(1)$

Now,

$$\text{arg. } z_1 - \text{arg. } z_2 = \tan^{-1}\left(\frac{y_1}{x_1}\right) - \tan^{-1}\left(\frac{y_2}{x_2}\right)$$

$$= \tan^{-1}\left[\frac{\frac{y_1}{x_1} - \frac{y_2}{x_2}}{1 + \left(\frac{y_1}{x_1}\right)\left(\frac{y_2}{x_2}\right)}\right] = \tan^{-1}\left(\frac{x_2y_1 - x_1y_2}{x_1x_2 + y_1y_2}\right)$$

$$= \tan^{-1}\left(\frac{x_2y_1 - x_1y_2}{0}\right) = \tan^{-1}\infty = \frac{\pi}{2} \quad \text{[Using (1)]}$$

$$\text{arg. } z_1 - \text{arg. } z_2 = \frac{\pi}{2}$$ **Proved.**

Example 18. *Find the complex number z if arg $(z + 1) = \frac{\pi}{6}$ and arg $(z - 1) = \frac{2\pi}{3}$.* *(M.U. 2009,)*

Solution. Let $\quad z = x + iy \quad ...(1)$

$\therefore \quad z + 1 = (x + 1) + iy$

We also given that

$$\text{Arg } (z + 1) = \tan^{-1}\left(\frac{y}{x+1}\right) = \frac{\pi}{6}$$

$$\therefore \quad \frac{y}{x+1} = \tan 30° = \frac{1}{\sqrt{3}}$$

$$\therefore \quad \sqrt{3}y = x + 1 \quad ...(2)$$

Now $\quad z - 1 = (x - 1) + iy \quad$ [From (1)]

and $\quad \tan^{-1}\left(\frac{y}{x-1}\right) = \frac{2\pi}{3} \Rightarrow \frac{y}{x-1} = \tan 120°$

$$\Rightarrow \qquad \frac{y}{x-1} = -\cot 30° = -\sqrt{3}$$

$$\therefore \qquad -y = \sqrt{3}x - \sqrt{3}$$

$$\Rightarrow \qquad -\sqrt{3}y = 3x - 3 \qquad ...(3)$$

Adding (2) and (3), we get

$$0 = 4x - 2 \quad \Rightarrow \quad 4x = 2 \Rightarrow x = \frac{1}{2}$$

Putting $x = \frac{1}{2}$ in (2), we get

$$\sqrt{3}y = \frac{1}{2} + 1 \Rightarrow \sqrt{3}y = \frac{3}{2} \Rightarrow y = \frac{\sqrt{3}}{2}$$

Putting the values of x and y in (1), we get

$$z = \frac{1}{2} + \frac{\sqrt{3}}{2}i$$ **Ans.**

Example 19. *Prove that*

(*i*) $|z_1 + z_2| \le |z_1| + |z_2|$ (*ii*) $|z_1 - z_2| \ge |z_1| - |z_2|$

Solution. (*a*) (By Geometry) Let $z_1 = x_1 + iy_1$ and $z_2 = x_2 + iy_2$ be the two complex numbers shown in the figure

$$|z_1| = OP, \quad |z_2| = OQ$$

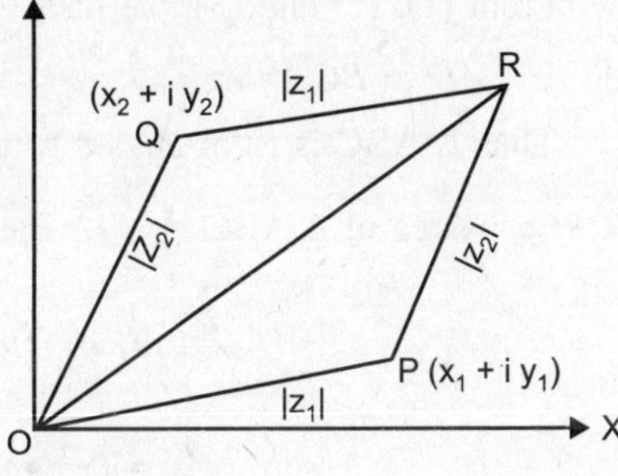

(*i*) Since in a triangle any side is less than the sum of the other two.

In Δ OPR, $OR < OP + PR$, $OR < OP + OQ$

$\Rightarrow \quad |z_1 + z_2| < |z_1| + |z_2|$

$OR = OP + PR$ if O, P, R are collinear.

or $\quad |z_1 + z_2| = |z_1| + |z_2|$

(*ii*) Again, any side of a triangle is greater than the difference between the other two, we have

In Δ OPR

$OR > OP - PR$, $\Rightarrow$ $OR > OP - OQ$

$|z_1 - z_2| > |z_1| - |z_2|$ **Proved.**

(*b*) By Algebra. $\quad z_1 + z_2 = x_1 + iy_1 + x_2 + iy_2 = (x_1 + x_2) + i(y_1 + y_2)$

(*i*) $\quad |z_1 + z_2|^2 = (x_1 + x_2)^2 + (y_1 + y_2)^2$

$$= x_1^2 + x_2^2 + 2x_1x_2 + y_1^2 + y_2^2 + 2y_1y_2 = (x_1^2 + y_1^2) + (x_2^2 + y_2^2) + 2(x_1x_2 + y_1y_2)$$

$$= (x_1^2 + y_1^2) + (x_2^2 + y_2^2) + 2\sqrt{(x_1x_2 + y_1y_2)^2}$$

$$= |z_1|^2 + |z_2|^2 + 2\sqrt{x_1^2x_2^2 + y_1^2y_2^2 + 2x_1x_2y_1y_2}$$

$$[\because (x_1y_2 - x_2y_1)^2 \ge 0 \text{ or } x_1^2 y_2^2 + x_2^2 y_1^2 \ge 2x_1x_2y_1y_2]$$

$$|z_1 + z_2|^2 \le |z_1|^2 + |z_2|^2 + 2\sqrt{x_1^2x_2^2 + y_1^2y_2^2 + x_1^2y_2^2 + x_2^2y_1^2}$$

$$\le |z_1|^2 + |z_2|^2 + 2\sqrt{(x_1^2 + y_1^2)(x_2^2 + y_2^2)}$$

$$\le |z_1|^2 + |z_2|^2 + 2|z_1|\,|z_2|$$

$$\le \{|z_1| + |z_2|\}^2$$

$$|z_1 + z_2| \le |z_1| + |z_2|$$

(*ii*) $$|z_1| = |(z_1 - z_2) + z_2| \le |z_1 - z_2| + |z_2|$$

$$|z_1| - |z_2| \le |z_1 - z_2|$$

$$|z_1 - z_2| \ge |z_1| - |z_2|$$ **Proved.**

Example 20. *Find the area of the triangle enclosed by the vectors z, z – iz and iz.*

(*Delhi University, April 2010*)

Solution. Let

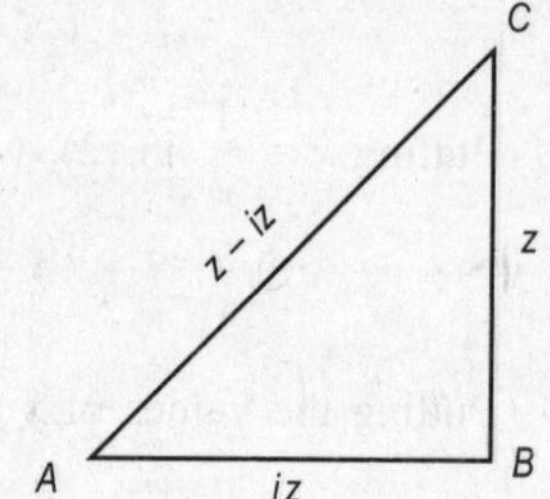

$AB = iz = ix - y$

$BC = z = x + iy$

$AC = z - iz$

$= (x + iy) - i(x + iy)$

$= (x + y) - (x - y)i$

$$|AB| = \sqrt{x^2 + y^2} \quad \text{... (1)}$$

$$|BC| = \sqrt{x^2 + y^2} \quad \text{... (2)}$$

$$|AC| = \sqrt{(x+y)^2 + (x-y)^2} = \sqrt{2}\sqrt{x^2 + y^2} \quad \text{... (3)}$$

From (1), (2) and (3), we have

$$AB^2 + BC^2 = AC^2$$

Thus Δ ABC is right angled triangle with $\angle B = 90°$

$$\Rightarrow \quad \text{Area of } \Delta \text{ ABC} = \frac{1}{2} \times \text{Base} \times \text{height}$$

$$= \frac{1}{2} \times |AB| \times |BC| = \frac{1}{2} \times \sqrt{x^2 + y^2} \times \sqrt{x^2 + y^2}$$

$$= \frac{1}{2}(x^2 + y^2)$$ **Ans.**

EXERCISE 20.3

1. If $z = x + iy$, prove that $\left(\dfrac{z}{\bar{z}} + \dfrac{\bar{z}}{z}\right) = 2\left(\dfrac{x^2 - y^2}{x^2 + y^2}\right)$.

2. If $z = a \cos\theta + ia \sin\theta$, prove that $\left(\dfrac{z}{\bar{z}} + \dfrac{\bar{z}}{z}\right) = 2\cos 2\theta$.

3. Prove that $\left|\dfrac{z - 1}{\bar{z} - 1}\right| = 1$.

4. Let $z_1 = 2 - i$, $z_2 = -2 + i$, find

(*i*) $\text{Re}\left[\dfrac{z_1 z_2}{\bar{z}_1}\right]$ (*ii*) $\text{Im}\left[\dfrac{1}{z_1 \bar{z}_2}\right]$ **Ans.** (*i*) $-\dfrac{2}{5}$, (*ii*) 0

5. If $|z| = 1$, prove that $\dfrac{z - 1}{z + 1}$ $(z \ne 1)$ is a pure imaginary number, what will you conclude, if $z = 1$?

Ans. If $z = 1$, $\dfrac{z - 1}{z + 1} = 0$, which is purely real.

20.16 POLAR FORM

Polar form of a complex number as we have discussed above

$$x = r\cos\theta \quad \text{and} \quad y = r\sin\theta$$

$$\Rightarrow \quad x + iy = r(\cos\theta + i\sin\theta)$$

$$= r\,e^{i\theta} \quad \text{(Exponential form)} \qquad (e^{i\theta} = \cos\theta + i\sin\theta)$$

Procedure. To convert $x + iy$ into polar.

We write $\quad x = r\cos\theta$

$$y = r\sin\theta$$

On solving these equations, we get the value of θ which satisfy both the equations and $r = \sqrt{x^2 + y^2}$.

20.17 TYPES OF COMPLEX NUMBERS

1. Cartesian form : $x + iy$
2. Polar form : $r(\cos\theta + i\sin\theta)$
3. Exponential form : $re^{i\theta}$

Example 21. *Express in polar form:* $1 - \sqrt{2} + i$

Solution. Let $\quad 1 - \sqrt{2} + i = r(\cos\theta + i\sin\theta)$

$$\therefore \quad 1 - \sqrt{2} = r\cos\theta \qquad ...(1)$$

$$1 = r\sin\theta \qquad ...(2)$$

Squaring and adding (1) and (2), we get

$$r^2(\cos^2\theta + \sin^2\theta) = (1 - \sqrt{2})^2 + 1^2$$

$$\Rightarrow \quad r^2 = 1 - 2\sqrt{2} + 2 + 1$$

$$\Rightarrow \quad r = \sqrt{4 - 2\sqrt{2}}$$

Putting the value of r in (1) and (2), we get

$$\cos\theta = \frac{1 - \sqrt{2}}{\sqrt{4 - 2\sqrt{2}}} \quad \text{and} \quad \sin\theta = \frac{1}{\sqrt{4 - 2\sqrt{2}}}$$

Hence, the polar form is $\sqrt{4 - 2\sqrt{2}}\left\{\dfrac{1 - \sqrt{2}}{\sqrt{4 - 2\sqrt{2}}} + i\dfrac{1}{\sqrt{4 - 2\sqrt{2}}}\right\}$ **Ans.**

Example 22. *Find the smallest positive integer n for which*

$$\left(\frac{1+i}{1-i}\right)^n = 1.$$

Solution.

$$\left[\frac{1+i}{1-i}\right]^n = 1$$

$$\left[\frac{1+i}{1-i} \times \frac{1+i}{1+i}\right]^n = 1 \quad \Rightarrow \quad \left(\frac{1-1+2i}{1+1}\right)^n = 1$$

$$(i)^n = 1 = (i)^4 \quad \Rightarrow \quad n = 4$$ **Ans.**

Example 23. *If* $i^{\alpha + i\beta} = \alpha + i\beta$, *prove that* $\alpha^2 + \beta^2 = e^{-(4n+1)\pi\beta}$

Solution. We have, $\alpha + i\beta = i^{(\alpha + i\beta)} = e^{\log i_{\alpha + i\beta}}$

$$\alpha + i\beta = e^{(\alpha + i\beta)\log i} = e^{(\alpha + i\beta)(\log i + 2n\pi i)}$$

$$= e^{(\alpha + i\beta)[\log(\cos \pi/2 + i \sin \pi/2) + 2n\pi i]}$$

$$\Rightarrow \quad \alpha + i\beta = e^{(\alpha + i\beta)[\log e^{i\pi/2} + 2n\pi i]} = e^{(\alpha + i\beta)[i\pi/2 + 2n\pi i]}$$

$$= e^{i\alpha(\pi/2 + 2n\pi) - \beta(\pi/2 + 2n\pi)} = e^{-\beta\pi(2n + 1/2)} \times e^{\pi\alpha(2n + 1/2)i}$$

$$\Rightarrow \quad \alpha + i\beta = e^{-\pi\beta(4n+1)/2}\left[\cos\left[\pi\alpha\frac{(4n+1)}{2}\right] + i\sin\left[\pi\alpha\frac{(4n+1)}{2}\right]\right]$$

Equating real and imaginary parts, we get

$$\alpha = e^{-\pi\beta(4n+1)/2}.\cos\left\{\frac{1}{2}\pi\alpha\,(4n+1)\right\} \qquad ...(1)$$

and

$$\beta = e^{-\pi\beta(4n+1)/2}.\sin\left\{\frac{1}{2}\pi\alpha\,(4n+1)\right\} \qquad ...(2)$$

Squaring and adding (1) and (2), we get

$$\alpha^2 + \beta^2 = e^{-\pi\beta(4n+1)}.\left[\cos^2\left\{\frac{1}{2}\pi\alpha\,(4n+1)\right\} + \sin^2\left\{\frac{1}{2}\pi\alpha\,(4n+1)\right\}\right]$$

$$\therefore \quad \alpha^2 + \beta^2 = e^{-\pi\beta(4n+1)}$$

Hence the result. **Proved.**

EXERCISE 20.4

Express the following complex numbers into polar form:

1. $\dfrac{1+i}{1-i}$ **Ans.** $\cos\dfrac{\pi}{2} + i\sin\dfrac{\pi}{2}$

2. $\dfrac{-35+5i}{4\sqrt{2}+3\sqrt{2}i}$ **Ans.** $5\left(\cos\dfrac{3\pi}{4} + i\sin\dfrac{3\pi}{4}\right)$

3. $\dfrac{3(-4-\sqrt{3}+4\sqrt{3}i-i)}{8+2i}$ **Ans.** $3\left(\cos\dfrac{2\pi}{3} + i\sin\dfrac{2\pi}{3}\right)$

4. $\dfrac{2+6\sqrt{3}i}{5+\sqrt{3}i}$ **Ans.** $2\left(\cos\dfrac{\pi}{3} + i\sin\dfrac{\pi}{3}\right)$

5. $\dfrac{2+3i}{3-7i}$ **Ans.** $r = \sqrt{754}$, $\theta = \tan^{-1}\left(-\dfrac{23}{15}\right)$

6. $\left(\dfrac{4-5i}{2+3i}\right)\cdot\left(\dfrac{3+2i}{7+i}\right)$ **Ans.** 0.905, $\theta = \tan^{-1}(-7.2)$

7. $\dfrac{(2+5i)(-3+i)}{(1-2i)^2}$ **Ans.** $\dfrac{\sqrt{290}}{5}$, $\tan^{-1}\left(-\dfrac{1}{17}\right)$

8. $\dfrac{1+7i}{(2-i)^2}$ **Ans.** $\sqrt{2}\left(\cos\dfrac{3\pi}{4} + i\sin\dfrac{3\pi}{4}\right)$

9. $\dfrac{1+3i}{1-2i}$ **Ans.** $\sqrt{2}\left(\cos\dfrac{3\pi}{4} + i\sin\dfrac{3\pi}{4}\right)$

10. $\dfrac{i-1}{\cos\dfrac{\pi}{3} + i\sin\dfrac{\pi}{3}}$ **Ans.** $\sqrt{2}\left(\cos\dfrac{5\pi}{12} + i\sin\dfrac{5\pi}{12}\right)$

20.18 SQUARE ROOT OF A COMPLEX NUMBER

Let $a + ib$ be a complex number and its square root is $x + iy$.

i.e.,
$$\sqrt{a+ib} = x + iy \qquad ...(1)$$

where x and $y \in R$.

Squaring both sides of (1), we get

$$a + ib = (x + iy)^2$$

$$\Rightarrow \quad a + ib = x^2 + i^2y^2 + i\,2xy$$

$$\Rightarrow \quad a + ib = (x^2 - y^2) + i\,2xy \qquad [\because i^2 = -1] \qquad ...(2)$$

Equating real and imaginary parts of (2), we get

$$x^2 - y^2 = a \qquad ...(3)$$

and

$$2xy = b \qquad ...(4)$$

Also, we know that

$$(x^2 + y^2)^2 = (x^2 - y^2)^2 + 4x^2y^2$$

$$\Rightarrow \quad (x^2 + y^2)^2 = a^2 + b^2 \qquad \text{[Using (3) and (4)]}$$

$$\Rightarrow \quad x^2 + y^2 = \sqrt{a^2 + b^2} \qquad ...(5)$$

Adding (3) and (5), we get

$$2x^2 = a + \sqrt{a^2 + b^2} \quad \Rightarrow \quad x = \sqrt{\frac{a + \sqrt{a^2 + b^2}}{2}}$$

Example 24. *Find the square root of the complex number 5 + 12i.*

Solution. Let $\sqrt{5 + 12i} = x + iy$ ...(1)

Squaring both sides of (1), we get $5 + 12i = (x + iy)^2 = (x^2 - y^2) + i\,2xy$ ...(2)

Equating real and imaginary parts of (2), we get

$$x^2 - y^2 = 5 \qquad ...(3)$$

and $$2xy = 12 \qquad ...(4)$$

Now, $$x^2 + y^2 = \sqrt{(x^2 - y^2)^2 + 4x^2y^2} = \sqrt{(5)^2 + (12)^2}$$

$$= \sqrt{25 + 144} = \sqrt{169} = 13$$

$$\Rightarrow \quad x^2 + y^2 = 13 \qquad ...(5)$$

Adding (3) and (5), we get $2x^2 = 5 + 13 = 18 \Rightarrow x = \sqrt{\frac{18}{2}} = \sqrt{9} = \pm 3$

Subtracting (3) from (5), we get $2y^2 = 13 - 5 = 8 \Rightarrow y = \sqrt{\frac{8}{2}} = \sqrt{4} = \pm 2$

Since, xy is positive, so x and y are of same sign. Hence, $x = \pm 3$, $y = \pm 2$

$\therefore \quad \sqrt{5 + 12i} = \pm 3 \pm 2i$ *i.e.* $(3 + 2i)$ or $-(3 + 2i)$ **Ans.**

Example 25. *Find the square root of –4 – 3i.*

Solution. Let $\sqrt{-4 - 3i} = x + iy$ ...(1)

Squaring both sides of (1), we get

$$-4 - 3i = (x + iy)^2 = (x^2 - y^2) + i\,2xy \qquad ...(2)$$

Equating real and imaginary parts of (2), we get

$$x^2 - y^2 = -4 \qquad ...(3)$$

And $$2xy = -3 \qquad ...(4)$$

Now, $$x^2 + y^2 = \sqrt{(x^2 - y^2)^2 + 4x^2y^2} = \sqrt{16 + 9} = \sqrt{25} = \pm 5$$

$$\Rightarrow \quad x^2 + y^2 = 5 \qquad ...(5) \quad (\because x^2 + y^2 \geq 0)$$

Adding (3) and (5), we get

$$2x^2 = 5 - 4 = 1 \Rightarrow x = \sqrt{\frac{1}{2}} = \pm\frac{1}{\sqrt{2}}$$

Subtracting (3) from (5), we get

$$2y^2 = 5 + 4 = 9 \Rightarrow y = \sqrt{\frac{9}{2}} = \pm\frac{3}{\sqrt{2}}$$

Since, xy is negative, so x and y will be of different signs. Hence, $x = \pm\frac{1}{\sqrt{2}}$, $y = \mp\frac{3}{\sqrt{2}}$

$$\therefore \quad \sqrt{-4 - 3i} = \pm\left(\frac{1}{\sqrt{2}} - \frac{3}{\sqrt{2}}i\right)$$ **Ans.**

Example 26. *Prove that if the sum and product of two complex numbers are real then the two numbers must be either real or conjugate.* *(M.U. 2008)*

Solution. Let z_1 and z_2 be the two complex numbers.

We are given that $z_1 + z_2 = a$ (real)

and $z_1.z_2 = b$ (real)

If sum and product of the roots of a quadratic equation are given. Then the equation becomes

$$x^2 - (\text{sum of the roots})\, x + \text{product of the roots} = 0$$

$$x^2 - ax + b = 0$$

$$\text{Root} = x = \frac{a \pm \sqrt{a^2 - 4b}}{2}$$

Case I. If $a^2 > 4b$ Then both the roots are real

Case II. If $a^2 < 4b$

Then
$$\text{one root} = \frac{a}{2} + i\frac{\sqrt{4b - a^2}}{2}$$

$$\text{Second root} = \frac{a}{2} - i\frac{\sqrt{4b - a^2}}{2}$$

These roots are conjugate to each other. **Proved**

EXERCISE 20.5

Find the square root of the following :

1. $1 + i$ **Ans.** $\left\{\pm\sqrt{\frac{\sqrt{2}+1}{2}} \pm \sqrt{\frac{\sqrt{2}-1}{2}}\,i\right\}$

2. $1 - i$ **Ans.** $\left\{\pm\sqrt{\frac{\sqrt{2}+1}{2}} \mp \sqrt{\frac{\sqrt{2}-1}{2}}\,i\right\}$

3. i **Ans.** $\left\{\pm\frac{1}{\sqrt{2}} \pm \frac{1}{\sqrt{2}}\,i\right\}$

4. $15 - 8i$ **Ans.** $1 - 4i, -1 + 4i$

5. $-2 + 2\sqrt{3}i$ **Ans.** $\pm(1 + \sqrt{3}i)$

6. $3 + 4\sqrt{7}i$ **Ans.** $\pm(\sqrt{7} + 2i)$

7. $\frac{2+3i}{5-4i} + \frac{2-3i}{5+4i}$ **Ans.** $\pm\frac{2}{\sqrt{41}}\,i$

8. $x^2 - 1 + i\,2x$ **Ans.** $\pm(x + i)$

9. $3 - 4i$ **Ans.** $\pm(2 - i)$

20.19 EXPONENTIAL AND CIRCULAR FUNCTIONS OF COMPLEX VARIABLES

Proof. $\boxed{\cos\theta + i\sin\theta = e^{i\theta}}$

$$e^z = 1 + z + \frac{z^2}{2!} + \frac{z^3}{3!} + \frac{z^4}{4!} + \ldots \qquad \ldots(1)$$

$$\sin z = z - \frac{z^3}{3!} + \frac{z^5}{5!} - \frac{z^7}{7!} + \ldots \qquad \ldots(2)$$

$$\cos z = 1 - \frac{z^2}{2!} + \frac{z^4}{4!} - \frac{z^6}{6!} + \ldots \qquad \ldots(3)$$

From (2) and (3), we have

$$\cos z + i\sin z = \left(1 - \frac{z^2}{2!} + \frac{z^4}{4!} - \frac{z^6}{6!} + \ldots\right) + i\left(z - \frac{z^3}{3!} + \frac{z^5}{5!} - \ldots\right)$$

$$= 1 + \frac{(iz)^1}{1!} + \frac{(iz)^2}{2!} + \frac{(iz)^3}{3!} + \ldots = e^{iz}$$

Therefore, $\cos z + i \sin z = e^{iz}$ ...(4)

Similarly, $\cos z - i \sin z = e^{-iz}$ ...(5)

From (4) and (5), we have

$$\cos z = \frac{e^{iz} + e^{-iz}}{2} \qquad \text{...(6)}$$

$$\sin z = \frac{e^{iz} - e^{-iz}}{2i} \qquad \text{...(7)}$$

20.20 DE MOIVRE'S THEOREM (By Exponential Function)

$$(\cos \theta + i \sin \theta)^n = \cos n\,\theta + i \sin n\,\theta$$

Proof. We know that $e^{i\theta} = \cos \theta + i \sin \theta$

$$(e^{i\theta})^n = (\cos \theta + i \sin \theta)^n$$

$$e^{in\theta} = (\cos \theta + i \sin \theta)^n$$

$$(\cos n\,\theta + i \sin n\,\theta) = (\cos \theta + i \sin \theta)^n$$ **Proved.**

If n is a fraction, then $\cos n\,\theta + i \sin n\,\theta$ is one of the values of $(\cos \theta + i \sin \theta)^n$

20.21 DE MOIVRE'S THEOREM (BY INDUCTION)

Statement: For any rational number n the value or one of the values of

$$(\cos \theta + i \sin \theta)^n = \cos n\,\theta + i \sin n\,\theta$$

Proof. Case I. Let n be a non-negative integer. By actual multiplication,

$$(\cos \theta_1 + i \sin \theta_1)(\cos \theta_2 + i \sin \theta_2) = (\cos \theta_1 \cos \theta_2 - \sin \theta_1 \sin \theta_2) + i(\cos \theta_1 \sin \theta_2 + \sin \theta_1 \cos \theta_2)$$

$$= \cos(\theta_1 + \theta_2) + i \sin(\theta_1 + \theta_2) \qquad \text{...(1)}$$

Similarly we can prove that

$$(\cos \theta_1 + i \sin \theta_1)(\cos \theta_2 + i \sin \theta_2)(\cos \theta_3 + i \sin \theta_3)$$
$$= \cos(\theta_1 + \theta_2 + \theta_3) + i \sin(\theta_1 + \theta_2 + \theta_3)$$

Continuing in this way, we can prove that

$$(\cos \theta_1 + i \sin \theta_1)(\cos \theta_2 + i \sin \theta_2) \ldots (\cos \theta_n + i \sin \theta_n)$$
$$= \cos(\theta_1 + \theta_2 \ldots + \theta_n) + i \sin(\theta_1 + \theta_2 + \ldots + \theta_n)$$

Putting $\theta_1 = \theta_2 = \theta_3 = \ldots \theta_n = \theta$, we get

$$(\cos \theta + i \sin \theta)^n = (\cos n\,\theta + i \sin n\,\theta)$$

Case II. Let n be a negative integer, say $n = -m$ where m is a positive integer. Then,

$$(\cos \theta + i \sin \theta)^n = (\cos \theta + i \sin \theta)^{-m}$$

$$= \frac{1}{(\cos \theta + i \sin \theta)^m} = \frac{1}{(\cos m\theta + i \sin m\theta)} \qquad \text{[By case I]}$$

$$= \frac{1}{(\cos m\theta + i \sin m\theta)} \cdot \frac{(\cos m\theta - i \sin m\theta)}{(\cos m\theta - i \sin m\theta)} = \frac{\cos m\theta - i \sin m\theta}{\cos^2 m\theta + \sin^2 m\theta}$$

$$= \cos m\,\theta - i \sin m\,\theta \qquad [\because \cos^2 m\,\theta + \sin^2 m\,\theta = 1]$$

$$= \cos(-m\,\theta) + i \sin(-m\,\theta) = \cos n\,\theta + i \sin n\,\theta$$

Hence, the theorem is true for negative integers also.

Case III. Let n be a proper fraction $\frac{p}{q}$ where p and q are integers. Without loss of generality we can select q to be positive integer, p may be a positive or negative integer.

Since q is a positive integer

Now, $\left(\cos\frac{\theta}{q}+i\sin\frac{\theta}{q}\right)^q = \cos q.\frac{\theta}{q}+i\sin q.\frac{\theta}{q}$ [By case I]

$$= \cos\theta + i\sin\theta$$

Taking the q th root of both sides, we get

$$(\cos\theta+i\sin\theta)^{\frac{1}{q}} = \cos\frac{\theta}{q}+i\sin\frac{\theta}{q}$$

Raising both sides to the power p,

$$(\cos\theta+i\sin\theta)^{\frac{p}{q}} = \left(\cos\frac{\theta}{q}+i\sin\frac{\theta}{q}\right)^p = \cos p.\frac{\theta}{q}+i\sin p.\frac{\theta}{q} \quad \text{[By case I and II]}$$

Hence, one of the values of $(\cos\theta + i\sin\theta)^n$ is $\cos n\theta + i\sin n\theta$ when n is a proper fraction. Thus, the theorem is true for all rational values of n.

Example 27. *Express* $\dfrac{(\cos\theta+i\sin\theta)^8}{(\sin\theta+i\cos\theta)^4}$ *in the form $(x + iy)$.*

Solution. $$\frac{(\cos\theta+i\sin\theta)^8}{(\sin\theta+i\cos\theta)^4} = \frac{(\cos\theta+i\sin\theta)^8}{(i)^4\left(\cos\theta+\frac{1}{i}\sin\theta\right)^4}$$

$$= \frac{(\cos\theta+i\sin\theta)^8}{(\cos\theta-i\sin\theta)^4} = \frac{(\cos\theta+i\sin\theta)^8}{[\cos(-\theta)+i\sin(-\theta)]^4}$$

$$= \frac{(\cos\theta+i\sin\theta)^8}{[(\cos\theta+i\sin\theta)^{-1}]^4} = \frac{(\cos\theta+i\sin\theta)^8}{(\cos\theta+i\sin\theta)^{-4}}$$

$$= (\cos\theta + i\sin\theta)^{12} = \cos 12\theta + i\sin 12\theta \quad \textbf{Ans.}$$

Example 28. *Prove that* $(1 + \cos\theta + i\sin\theta)^n + (1 + \cos\theta - i\sin\theta)^n = 2^{n+1}\cos^n\frac{\theta}{2}\cos\frac{n\theta}{2}$ *where n is an integer.*

Solution. L.H.S. $= (1 + \cos\theta + i\sin\theta)^n + (1 + \cos\theta - i\sin\theta)^n$

$$= \left[1+2\cos^2\frac{\theta}{2}-1+2i\sin\frac{\theta}{2}\cos\frac{\theta}{2}\right]^n + \left[1+2\cos^2\frac{\theta}{2}-1-i\,2\sin\frac{\theta}{2}\cos\frac{\theta}{2}\right]^n$$

$$= \left[2\cos^2\frac{\theta}{2}+2i\sin\frac{\theta}{2}\cos\frac{\theta}{2}\right]^n + \left[2\cos^2\frac{\theta}{2}-2i\sin\frac{\theta}{2}\cos\frac{\theta}{2}\right]^n$$

$$= \left(2\cos\frac{\theta}{2}\right)^n\left[\cos\frac{\theta}{2}+i\sin\frac{\theta}{2}\right]^n + \left(2\cos\frac{\theta}{2}\right)^n\left[\cos\frac{\theta}{2}-i\sin\frac{\theta}{2}\right]^n$$

$$= 2^n\cos^n\frac{\theta}{2}\left[\cos\frac{n\theta}{2}+i\sin\frac{n\theta}{2}\right] + 2^n\cos^n\frac{\theta}{2}\left[\cos\frac{n\theta}{2}-i\sin\frac{n\theta}{2}\right]$$

$$= 2^n\cos^n\frac{\theta}{2}\left[\cos\frac{n\theta}{2}+i\sin\frac{n\theta}{2}+\cos\frac{n\theta}{2}-i\sin\frac{n\theta}{2}\right]$$

$$= 2^n\cos^n\frac{\theta}{2}\left(2\cos\frac{n\theta}{2}\right) = 2^{n+1}\cos^n\frac{\theta}{2}\cos\frac{n\theta}{2} = \text{R.H.S.} \quad \textbf{Proved.}$$

Example 29. *Evaluate* $\left(\dfrac{1+\sin\alpha+i\cos\alpha}{1+\sin\alpha-i\cos\alpha}\right)^n$ *(M.U. 2005,)*

Solution. We know that,

$$1 = \sin^2\alpha + \cos^2\alpha$$

$$\Rightarrow \quad 1 = \sin^2\alpha - i^2\cos^2\alpha$$

$$\Rightarrow \quad 1 = (\sin\alpha + i\cos\alpha)(\sin\alpha - i\cos\alpha) \quad ...(1)$$

Adding $\sin\alpha + i\cos\alpha$ both sides of (1), we get

$$1 + \sin\alpha + i\cos\alpha = (\sin\alpha + i\cos\alpha)(\sin\alpha - i\cos\alpha) + (\sin\alpha + i\cos\alpha)$$

$$= (\sin\alpha + i\cos\alpha)(\sin\alpha - i\cos\alpha + 1)$$

$$\therefore \quad \frac{1+\sin\alpha+i\cos\alpha}{1+\sin\alpha-i\cos\alpha} = \sin\alpha + i\cos\alpha$$

$$= \cos\left(\frac{\pi}{2}-\alpha\right) + i\sin\left(\frac{\pi}{2}-\alpha\right) \quad ...(2)$$

$$\Rightarrow \quad \left(\frac{1+\sin\alpha+i\cos\alpha}{1+\sin\alpha-i\cos\alpha}\right)^n = \left\{\cos\left(\frac{\pi}{2}-\alpha\right) + i\sin\left(\frac{\pi}{2}-\alpha\right)\right\}^n$$

$$= \cos n\left(\frac{\pi}{2}-\alpha\right) + i\sin n\left(\frac{\pi}{2}-\alpha\right)$$ **Ans.**

Example 30. *If* $2\cos\theta = x+\frac{1}{x}$ *and* $2\cos\phi = y+\frac{1}{y}$, *then prove that*

$x^p . y^q + \dfrac{1}{x^p . y^q} = 2\cos(p\theta + q\phi)$.

Solution. We have,

$$x+\frac{1}{x} = 2\cos\theta \Rightarrow \quad x^2 - 2x\cos\theta + 1 = 0$$

$$\Rightarrow \quad x = \frac{2\cos\theta \pm \sqrt{4\cos^2\theta - 4}}{2} = \cos\theta \pm \sqrt{-\sin^2\theta} = 0$$

Putting i^2 for -1 and considering the positive sign, we get

$x = \cos\theta + i\sin\theta$ and similarly, $y = \cos\phi + i\sin\phi$

Now, $$x^p = (\cos\theta + i\sin\theta)^p = \cos p\theta + i\sin p\theta$$

and $$y^q = (\cos\phi + i\sin\phi)^q = \cos q\phi + i\sin q\phi$$

(By De-Moivre's theorem)

$$\therefore \quad x^p . y^q = (\cos p\theta + i\sin p\theta)(\cos q\phi + i\sin q\phi)$$

$$= \cos(p\theta + q\phi) + i\sin(p\theta + q\phi) \quad ...(1)$$

Also $$\frac{1}{x^p . y^q} = [\cos(p\theta + q\phi) + i\sin(p\theta + q\phi)]^{-1}$$

$$= \cos(p\theta + q\phi) - i\sin(p\theta + q\phi) \quad ...(2)$$

Adding (1) and (2), we get

$$\therefore \quad x^p . y^q + \frac{1}{x^p . y^q} = 2\cos(p\theta + q\phi)$$ **Proved**

Example 31. *Prove that the general value of θ which satisfies the equation (cos θ + i sin θ). (cos 2θ + i sin 2θ)...(cos n θ + i sin n θ) = 1 is* $\frac{4m\pi}{n(n+1)}$, *where m is any integer.*

Solution. $(\cos\theta + i\sin\theta)(\cos 2\theta + i\sin 2\theta)...(\cos n\theta + i\sin n\theta) = 1$

$$(\cos\theta + i\sin\theta)(\cos\theta + i\sin\theta)^2...(\cos\theta + i\sin\theta)^n = 1$$

$$(\cos\theta + i\sin\theta)^{1+2...+n} = 1$$

$$(\cos\theta + i\sin\theta)^{\frac{n(n+1)}{2}} = (\cos 2m\pi + i\sin 2m\pi)$$

$$\cos\frac{n(n+1)}{2}\theta + i\sin\frac{n(n+1)}{2}\theta = \cos 2m\pi + i\sin 2m\pi$$

$$\frac{n(n+1)}{2}\theta = 2m\pi \Rightarrow \theta = \frac{4m\pi}{n(n+1)}$$ **Proved**

Example 32. *If* $(a_1 + ib_1).(a_2 + ib_2)...(a_n + ib_n) = A + iB$, *then prove that*

(*i*) $\tan^{-1}\frac{b_1}{a_1} + \tan^{-1}\frac{b_2}{a_2} + ... + \tan^{-1}\frac{b_n}{a_n} = \tan^{-1}\frac{B}{A}$

(*ii*) $(a_1^2 + b_1^2)(a_2^2 + b_2^2) \;...\; (a_n^2 + b_n^2) = A^2 + B^2$

Solution. Let $a_1 = r_1\cos\alpha_1$, $b_1 = r_1\sin\alpha_1$

$a_2 = r_2\cos\alpha_2$, $b_2 = r_2\sin\alpha_2$

....

$a_n = r_n\cos\alpha_n$, $b_n = r_n\sin\alpha_n$

$A = R\cos\theta$, $B = R\sin\theta$,

$$(a_1 + ib_1).(a_2 + ib_2)...(a_n + ib_n) = A + iB \quad \text{(Given)}$$

$$r_1(\cos\alpha_1 + i\sin\alpha_1)\, r_2(\cos\alpha_2 + i\sin\alpha_2)...r_n(\cos\alpha_n + i\sin\alpha_n) = R(\cos\theta + i\sin\theta)$$

$$r_1 r_2 ... r_n[\cos(\alpha_1 + \alpha_2 + ...\alpha_n) + i\sin(\alpha_1 + \alpha_2 + ...\alpha_n)] = R(\cos\theta + i\sin\theta)$$

$\therefore \quad r_1 r_2 \,.................\, r_n = R$

$\Rightarrow \quad (a_1^2 + b_1^2)(a_2^2 + b_2^2)\;...\;(a_n^2 + b_n^2) = A^2 + B^2$

And $\quad \alpha_1 + \alpha_2 + ... + \alpha_n = \theta$

$\Rightarrow \quad \tan^{-1}\frac{b_1}{a_1} + \tan^{-1}\frac{b_2}{a_2} + ... + \tan^{-1}\frac{b_n}{a_n} = \tan^{-1}\frac{B}{A}$ **Proved**

Example 33. *If cos α + cos β + cos γ = 0 = sin α + sin β + sin γ, then prove that*

(*i*) $\sin^2\alpha + \sin^2\beta + \sin^2\gamma = \cos^2\alpha + \cos^2\beta + \cos^2\gamma = \frac{3}{2}$

(*ii*) $\cos 2\alpha + \cos 2\beta + \cos 2\gamma = 0$

(*iii*) $\cos(\alpha + \beta) + \cos(\beta + \gamma) + \cos(\gamma + \alpha) = 0$

(*iv*) $\sin(\alpha + \beta) + \sin(\beta + \gamma) + \sin(\gamma + \alpha) = 0$ (*M.U. 2009*)

Solution. Here, we have

$$(\cos\alpha + \cos\beta + \cos\gamma) + i(\sin\alpha + \sin\beta + \sin\gamma) = 0$$

$$(\cos\alpha + i\sin\alpha) + (\cos\beta + i\sin\beta) + (\cos\gamma + i\sin\gamma) = 0$$

$\therefore \quad a + b + c = 0$ say ...(1)

where, $a = \cos\alpha + i\sin\alpha$, $b = \cos\beta + i\sin\beta$ and $c = \cos\gamma + i\sin\gamma$

Also we can write

$$(\cos\alpha - i\sin\alpha) + (\cos\beta - i\sin\beta) + (\cos\gamma - i\sin\gamma) = 0$$

$\Rightarrow \quad (\cos\alpha + i\sin\alpha)^{-1} + (\cos\beta + i\sin\beta)^{-1} + (\cos\gamma + i\sin\gamma)^{-1} = 0$

$\Rightarrow \quad \frac{1}{a} + \frac{1}{b} + \frac{1}{c} = 0$

$\Rightarrow \quad \dfrac{bc + ca + ab}{abc} = 0 \Rightarrow ab + bc + ca = 0$ ...(2)

But $\quad (a + b + c)^2 = (a^2 + b^2 + c^2) + 2(ab + bc + ca)$

$0 = (a^2 + b^2 + c^2) + 0$ [From (1) and (2)]

$\Rightarrow \quad a^2 + b^2 + c^2 = 0$

$\Rightarrow \quad (\cos\alpha + i\sin\alpha)^2 + (\cos\beta + i\sin\beta)^2 + (\cos\gamma + i\sin\gamma)^2 = 0$

$\Rightarrow \quad (\cos 2\alpha + i\sin 2\alpha) + (\cos 2\beta + i\sin 2\beta) + (\cos 2\gamma + i\sin 2\gamma) = 0$

$\Rightarrow \quad (\cos 2\alpha + \cos 2\beta + \cos 2\gamma) + i(\sin 2\alpha + \sin 2\beta + \sin 2\gamma) = 0$

$\Rightarrow$ (*ii*) $\quad \cos 2\alpha + \cos 2\beta + \cos 2\gamma = 0.$ ...(3)

$\Rightarrow$ (*i*) $\quad 2\cos^2\alpha - 1 + 2\cos^2\beta - 1 + 2\cos^2\gamma - 1 = 0$

$\Rightarrow \quad \cos^2\alpha + \cos^2\beta + \cos^2\gamma = \dfrac{3}{2}$ ...(4)

Further $\quad 1 - \sin^2\alpha + 1 - \sin^2\beta + 1 - \sin^2\gamma = \dfrac{3}{2}$

$\Rightarrow \quad \sin^2\alpha + \sin^2\beta + \sin^2\gamma = \dfrac{3}{2}$ ...(5)

Again consider $\quad ab + bc + ca = 0$ [From (2)]

$\Rightarrow \quad (\cos\alpha + i\sin\alpha)(\cos\beta + i\sin\beta) + (\cos\beta + i\sin\beta)(\cos\gamma + i\sin\gamma)$
$+ (\cos\gamma + i\sin\gamma)(\cos\alpha + i\sin\alpha) = 0$

$\Rightarrow \quad [\cos(\alpha + \beta) + i\sin(\alpha + \beta)] + [\cos(\beta + \gamma) + i\sin(\beta + \gamma)]$
$+ [\cos(\gamma + \alpha) + i\sin(\gamma + \alpha)] = 0$

Equating real and imaginary parts, we get

(*iii*) $\quad \cos(\alpha + \beta) + \cos(\beta + \gamma) + \cos(\gamma + \alpha) = 0$

(*iv*) $\quad \sin(\alpha + \beta) + \sin(\beta + \gamma) + \sin(\gamma + \alpha) = 0$ **Proved.**

EXERCISE 20.6

1. If n is a positive integer show that $(a + ib)^n + (a - ib)^n = 2r^n \cos n\theta$ where $r^2 = a^2 + b^2$ and $\theta = \tan^{-1}\left(\dfrac{b}{a}\right)$. Hence deduce that $\left(1+i\sqrt{3}\right)^8+\left(1-i\sqrt{3}\right)^8 = -2^8.$

2. If n be a positive integer, prove that $(1 + i)^n + (1 - i)^n = 2^{\frac{n+2}{2}} \cos\dfrac{n\pi}{4}$

3. Show that $(a + ib)^{m/n} + (a - ib)^{m/n} = 2(a^2+b^2)^{\frac{m}{2n}} \cos\left(\dfrac{m}{n}\tan^{-1}\dfrac{b}{a}\right)$.

4. If $P = \cos\theta + i\sin\theta$, $q = \cos\phi + i\sin\phi$, show that

   (*i*) $\dfrac{P-q}{P+q} = i\tan\dfrac{\theta-\phi}{2}$ (*ii*) $\dfrac{(P+q)(Pq-1)}{(P-q)(Pq+1)} = \dfrac{\sin\theta+\sin\phi}{\sin\theta-\sin\phi}$

5. If $x = \cos\theta + i\sin\theta$, show that (*i*) $x^m + \dfrac{1}{x^m} = 2\cos m\theta$ (*ii*) $x^m - \dfrac{1}{x^m} = 2i\sin m\theta$.

6. Prove that $\tanh(\log\sqrt{3}) = \dfrac{1}{2}$

7. Prove that $[\sin(\alpha + \theta) - e^{i\alpha}\sin\theta]^n = \sin^n\alpha\, e^{-in\theta}$

8. If $x + \dfrac{1}{x} = 2\cos\theta$, $y + \dfrac{1}{y} = 2\cos\phi$, $z + \dfrac{1}{z} = 2\cos\psi$, show that

   $xyz + \dfrac{1}{xyz} = 2\cos(\theta + \phi + \psi)$

20.22 ROOTS OF A COMPLEX NUMBER

We know that $\cos\theta + i\sin\theta = \cos(2m\pi + \theta) + i\sin(2m\pi + \theta)$, $m \in I$

$$[\cos\theta + i\sin\theta]^{1/n} = [\cos(2m\pi + \theta) + i\sin(2m\pi + \theta)]^{1/n}$$

$$= \cos\frac{(2m\pi+\theta)}{n} + i\sin\frac{(2m\pi+\theta)}{n}$$

Giving m the values 0, 1, 2, 3, ... $n - 1$ successively, we get the following n values of $(\cos\theta + i\sin\theta)^{1/n}$.

when $m = 0$, $\quad \cos\frac{\theta}{n} + i\sin\frac{\theta}{n}$

When $m = 1$, $\quad \cos\left(\frac{2\pi+\theta}{n}\right) + i\sin\left(\frac{2\pi+\theta}{n}\right)$

When $m = 2$, $\quad \cos\left(\frac{4\pi+\theta}{n}\right) + i\sin\left(\frac{4\pi+\theta}{n}\right)$

When $m = n - 1$, $\quad \cos\left(\frac{2(n-1)\pi+\theta}{n}\right) + i\sin\left(\frac{2(n-1)\pi+\theta}{n}\right)$

When $m = n$, $\quad \cos\frac{2n\pi+\theta}{n} + i\sin\frac{2n\pi+\theta}{n} = \cos\left(2\pi+\frac{\theta}{n}\right) + i\sin\left(2\pi+\frac{\theta}{n}\right)$

$$= \cos\frac{\theta}{n} + i\sin\frac{\theta}{n}$$

which is the same as the value for $m = 0$. Thus, the values of $(\cos\theta + i\sin\theta)^{1/n}$ for $m = n, n + 1, n + 2$ etc., are the mere repetition of the first n values as obtained above.

Example 34. *Solve* $x^4 + i = 0$. *(M.U. 2008)*

Solution. Here, we have

$$x^4 = -i = \cos\frac{\pi}{2} - i\sin\frac{\pi}{2}$$

$$x^4 = \cos\left(2n\pi+\frac{\pi}{2}\right) - i\sin\left(2n\pi+\frac{\pi}{2}\right)$$

$$\Rightarrow \quad x = \left[\cos\left(2n\pi+\frac{\pi}{2}\right) - i\sin\left(2n\pi+\frac{\pi}{2}\right)\right]^{\frac{1}{4}}$$

$$= \cos(4n+1)\frac{\pi}{8} - i\sin(4n+1)\frac{\pi}{8}$$

Putting $n = 0, 1, 2, 3$ we get the roots as

$$x_1 = \cos\frac{\pi}{8} - i\sin\frac{\pi}{8}, \quad x_2 = \cos\frac{5\pi}{8} - i\sin\frac{5\pi}{8}$$

$$x_3 = \cos\frac{9\pi}{8} - i\sin\frac{9\pi}{8}, \quad x_4 = \cos\frac{13\pi}{8} - i\sin\frac{13\pi}{8}$$

Ans.

Example 35. *Solve* $x^5 = 1 + i$ *and find the continued product of the roots.* *(M.U. 2005,)*

Solution. $\quad x^5 = 1 + i$

$$= \sqrt{2}\left(\cos\frac{\pi}{4} + i\sin\frac{\pi}{4}\right) \Rightarrow \quad x = 2^{\frac{1}{10}}\left(\cos\frac{\pi}{4} + i\sin\frac{\pi}{4}\right)^{\frac{1}{5}}$$

$$\Rightarrow \quad x = 2^{\frac{1}{10}}\left[\cos\left(2k\pi+\frac{\pi}{4}\right).\frac{1}{5}+i\sin\left(2k\pi+\frac{\pi}{4}\right).\frac{1}{5}\right]$$

$$= 2^{\frac{1}{10}}\left[\cos(8k+1)\frac{\pi}{20}+i\sin(8k+1)\frac{\pi}{20}\right]$$

The roots are obtained by putting $k = 0, 1, 2, 3, 4...$

$$x_1 = 2^{\frac{1}{10}}\left[\cos\frac{\pi}{20}+i\sin\frac{\pi}{20}\right], \quad x_2 = 2^{\frac{1}{10}}\left[\cos\frac{9\pi}{20}+i\sin\frac{9\pi}{20}\right]$$

$$x_3 = 2^{\frac{1}{10}}\left[\cos\frac{17\pi}{20}+i\sin\frac{17\pi}{20}\right], \quad x_4 = 2^{\frac{1}{10}}\left[\cos\frac{5\pi}{4}+i\sin\frac{5\pi}{4}\right]$$

$$x_5 = 2^{\frac{1}{10}}\left[\cos\frac{33\pi}{20}+i\sin\frac{33\pi}{20}\right]$$

$$x_1 . x_2 . x_3 . x_4 . x_5 = \left(2^{\frac{1}{10}}\right)^5\left(\cos\frac{\pi}{20}+i\sin\frac{\pi}{20}\right)\left(\cos\frac{9\pi}{20}+i\sin\frac{9\pi}{20}\right)\left(\cos\frac{17\pi}{20}+i\sin\frac{17\pi}{20}\right)$$

$$\left(\cos\frac{5\pi}{4}+i\sin\frac{5\pi}{4}\right)\left(\cos\frac{33\pi}{20}+i\sin\frac{33\pi}{20}\right)$$

$$= 2^{\frac{1}{2}}\left[\cos\left(\frac{\pi}{20}+\frac{9\pi}{20}+\frac{17\pi}{20}+\frac{5\pi}{4}+\frac{33\pi}{20}\right)+i\sin\left(\frac{\pi}{20}+\frac{9\pi}{20}+\frac{17\pi}{20}+\frac{5\pi}{4}+\frac{33\pi}{20}\right)\right]$$

$$= \sqrt{2}\left[\cos\frac{17\pi}{4}+i\sin\frac{17\pi}{4}\right] = \sqrt{2}\left[\cos\left(4\pi+\frac{\pi}{4}\right)+i\sin\left(4\pi+\frac{\pi}{4}\right)\right]$$

$$= \sqrt{2}\left[\cos\frac{\pi}{4}+i\sin\frac{\pi}{4}\right] = \sqrt{2}\left[\frac{1}{\sqrt{2}}+i\frac{1}{\sqrt{2}}\right] = 1+i$$

Ans.

Example 36. *If $\alpha, \alpha^2, \alpha^3, \alpha^4$, are the roots of $x^5 - 1 = 0$ find them and show that $(1-\alpha)(1-\alpha^2)(1-\alpha^3)(1-\alpha^4) = 5$.* *(M.U. 2007)*

Solution. Here, we have

$$x^5 - 1 = 0$$

$$\Rightarrow \quad x^5 = 1 = \cos 0 + i\sin 0$$

$$\Rightarrow \quad x^5 = \cos(2k\pi) + i\sin(2k\pi)$$

$$\Rightarrow \quad x = (\cos 2k\pi + i\sin 2k\pi)^{1/5} = \cos\frac{2k\pi}{5}+i\sin\frac{2k\pi}{5}$$

Putting $k = 0, 1, 2, 3, 4$, we get the five roots as below

$$x_0 = \cos 0 + i\sin 0, \quad x_1 = \cos\frac{2\pi}{5}+i\sin\frac{2\pi}{5}$$

$$x_2 = \cos\frac{4\pi}{5}+i\sin\frac{4\pi}{5}, \quad x_3 = \cos\frac{6\pi}{5}+i\sin\frac{6\pi}{5}$$

$$x_4 = \cos\frac{8\pi}{5}+i\sin\frac{8\pi}{5}$$

Putting $x_1 = \cos\frac{2\pi}{5}+i\sin\frac{2\pi}{5} = \alpha$, we see that

$$x_2 = \left(\cos\frac{4\pi}{5}+i\sin\frac{4\pi}{5}\right) = \left(\cos\frac{2\pi}{5}+i\sin\frac{2\pi}{5}\right)^2 = \alpha^2$$

Similarly, $x_3 = \alpha^3$ and $x_4 = \alpha^4$

$\therefore$ The roots are $1, \alpha, \alpha^2, \alpha^3, \alpha^4$

Hence
$$x^5 - 1 = (x - 1)(x - \alpha)(x - \alpha^2)(x - \alpha^3)(x - \alpha^4)$$
$$\Rightarrow \quad \frac{x^5 - 1}{x - 1} = (x - \alpha)(x - \alpha^2)(x - \alpha^3)(x - \alpha^4)$$

On dividing $x^5 - 1$ by $x - 1$, we get
$$x^4 + x^3 + x^2 + x + 1 = (x - \alpha)(x - \alpha^2)(x - \alpha^3)(x - \alpha^4)$$
$$\therefore (x - \alpha)(x - \alpha^2)(x - \alpha^3)(x - \alpha^4) = x^4 + x^3 + x^2 + x + 1$$

Putting $x = 1$, we get
$$(1 - \alpha)(1 - \alpha^2)(1 - \alpha^3)(1 - \alpha^4) = 1 + 1 + 1 + 1 + 1 = 5.$$ **Proved.**

Example 37. *If ω is a cube root of unity, prove that*
$$(1 - \omega)^6 = -27$$

Solution. Let $x^3 = 1$
$$\Rightarrow \quad x = (1)^{1/3} = (\cos 0 + i \sin 0)^{1/3} = (\cos 2n\pi + i \sin 2n\pi)^{1/3}$$
$$= \cos\left(\frac{2n\pi}{3}\right) + i \sin\left(\frac{2n\pi}{3}\right)$$

Putting $n = 0, 1, 2$ the roots of unity are
$$x_0 = 1$$
$$x_1 = \cos\frac{2\pi}{3} + i \sin\frac{2\pi}{3} = \omega \text{ (say)}$$
$$x_2 = \cos\frac{4\pi}{3} + i \sin\frac{4\pi}{3} = \left[\cos\frac{2\pi}{3} + i \sin\frac{2\pi}{3}\right]^2 = \omega^2$$

Now,
$$1 + \omega + \omega^2 = 1 + \cos\frac{2\pi}{3} + i \sin\frac{2\pi}{3} + \cos\frac{4\pi}{3} + i \sin\frac{4\pi}{3}$$
$$= 1 + \cos\left(\pi - \frac{\pi}{3}\right) + i \sin\left(\pi - \frac{\pi}{3}\right) + \cos\left(\pi + \frac{\pi}{3}\right) + i \sin\left(\pi + \frac{\pi}{3}\right)$$
$$= 1 - \cos\frac{\pi}{3} + i \sin\frac{\pi}{3} - \cos\frac{\pi}{3} - i \sin\frac{\pi}{3}$$
$$= 1 - 2\cos\frac{\pi}{3} = 1 - 2\left(\frac{1}{2}\right) = 0$$
$$\Rightarrow \quad 1 + \omega + \omega^2 = 0$$
$$\Rightarrow \quad 1 + \omega^2 = -\omega \quad \text{...(1)}$$

Now,
$$(1 - \omega)^6 = [(1 - \omega)^2]^3 = [1 - 2\omega + \omega^2]^3 = [-\omega - 2\omega]^3$$
$$= (-3\omega)^3 = -27\omega^3 = -27$$ [Using (1)] **Proved.**

Example 38. *Use De Moivre's theorem to solve the equation $x^4 - x^3 + x^2 - x + 1 = 0$.*

Solution. $x^4 - x^3 + x^2 - x + 1 = 0$
$$(x + 1)(x^4 - x^3 + x^2 - x + 1) = 0$$
$$x^5 + 1 = 0$$
$$x^5 = -1 = (\cos \pi + i \sin \pi) = \cos(2n\pi + \pi) + i \sin(2n\pi + \pi)$$
$$x = [\cos(2n + 1)\pi + i \sin(2n + 1)\pi]^{1/5}$$
$$= \cos\frac{(2n+1)\pi}{5} + i \sin\frac{(2n+1)\pi}{5}$$

When $n = 0, 1, 2, 3, 4$, the values are

$$\cos\frac{\pi}{5}+i\sin\frac{\pi}{5},\ \cos\frac{3\pi}{5}+i\sin\frac{3\pi}{5},\ \cos\pi+i\sin\pi,\ \cos\frac{7\pi}{5}+i\sin\frac{7\pi}{5},\ \cos\frac{9\pi}{5}+i\sin\frac{9\pi}{5}.$$

$\cos\pi + i\sin\pi = -1$, which is rejected as it is corresponding to $x + 1 = 0$.
Hence, the required roots are

$$\cos\frac{\pi}{5}+i\sin\frac{\pi}{5},\ \cos\frac{3\pi}{5}+i\sin\frac{3\pi}{5},\ \cos\frac{7\pi}{5}+i\sin\frac{7\pi}{5},\ \cos\frac{9\pi}{5}+i\sin\frac{9\pi}{5}.$$ **Ans.**

EXERCISE 20.7

Find the values of:

1. $(1 + i)^{1/5}$. **Ans.** $2^{1/10}\left[\cos\frac{1}{5}\left(2n\pi+\frac{\pi}{4}\right)+i\sin\frac{1}{5}\left(2n\pi+\frac{\pi}{4}\right)\right]$, where $n = 0, 1, 2, 3, 4$

2. $(1+\sqrt{-3})^{3/4}$ **Ans.** $(2)^{3/4}\left[\cos\frac{3}{4}\left(2n\pi+\frac{\pi}{3}\right)+i\sin\frac{3}{4}\left(2n\pi+\frac{\pi}{3}\right)\right]$, where $n = 0, 1, 2, 3$.

3. $(-i)^{1/6}$ **Ans.** $\cos(4n+1)\frac{\pi}{12} - i\sin(4n+1)\frac{\pi}{12}$, where $n = 0, 1, 2, 3, 4, 5$.

4. $(1 + i)^{2/3}$ **Ans.** $2^{1/3}\left[\cos\left(\frac{4n\pi}{3}+\frac{\pi}{6}\right)+i\sin\left(\frac{4n\pi}{3}+\frac{\pi}{6}\right)\right]$, where $n = 0, 1, 2$

5. Solve the equation with the help of De Moivre's theorem $x^7 - 1 = 0$

Ans. $\cos\frac{2n\pi}{7}+i\sin\frac{2n\pi}{7}$, where $n = 0, 1, 2, 3, 4, 5, 6$.

6. Find the roots of the equation $x^3 + 8 = 0$.

Ans. $2\left[\cos\left(\frac{2n\pi+\pi}{3}\right)+i\sin\left(\frac{2n\pi+\pi}{3}\right)\right]$, where $n = 0, 1, 2$.

7. Use De-Moivre's theorem to solve $x^9 - x^5 + x^4 - 1 = 0$

Ans. $\left[\cos(2n+1)\frac{\pi}{5}+i\sin(2n+1)\frac{\pi}{5}\right]$, where $n = 0, 1, 2, 3, 4$,

and $\left[\cos\frac{n\pi}{2}+i\sin\frac{n\pi}{2}\right]$, where $n = 0, 1, 2, 3$.

8. Show that the roots of $(x + 1)^6 + (x - 1)^6 = 0$ are given by

$i\cot\frac{(2n+1)\pi}{12}$, $n = 0, 1, 2, 3, 4, 5$. Deduce $\tan^2\frac{\pi}{12}+\tan^2\frac{3\pi}{12}+\tan^2\frac{5\pi}{12} = 15$.

9. Show that all the roots of $(x + 1)^7 = (x - 1)^7$ are given by $\pm i\cot\left(\frac{n\pi}{7}\right)$, where $n = 1, 2, 3$, why $n \neq 0$.

20.23 CIRCULAR FUNCTIONS OF COMPLEX NUMBERS

We have already discussed circular functions in terms of exponential functions i.e, Euler's exponential form of circular functions:

$$\boldsymbol{\cos\theta = \frac{e^{i\theta}+e^{-i\theta}}{2}},\qquad \boldsymbol{\sin\theta = \frac{e^{i\theta}-e^{-i\theta}}{2i}}$$

If $\theta = z$, then $\boldsymbol{\cos z = \frac{e^{iz}-e^{-iz}}{2}}$ and $\boldsymbol{\sin z = \frac{e^{iz}-e^{-iz}}{2i}}$

20.24 HYPERBOLIC FUNCTIONS

(*i*) $\sinh x = \dfrac{e^x - e^{-x}}{2}$ (*ii*) $\cosh x = \dfrac{e^x + e^{-x}}{2}$ (*iii*) $\tanh x = \dfrac{e^x - e^{-x}}{e^x + e^{-x}}$

(*iv*) $\coth x = \dfrac{e^x + e^{-x}}{e^x - e^{-x}}$ (v) $\operatorname{sech} x = \dfrac{2}{e^x + e^{-x}}$ (*vi*) $\operatorname{cosech} x = \dfrac{2}{e^x - e^{-x}}$

(*vii*) $\cosh x + \sinh x = \dfrac{e^x + e^{-x}}{2} + \dfrac{e^x - e^{-x}}{2} = e^x$

(*viii*) $\cosh x - \sinh x = \dfrac{e^x + e^{-x}}{2} - \dfrac{e^x - e^{-x}}{2} = e^{-x}$

(*ix*) $(\cosh x + \sinh x)^n = \cosh nx + \sinh nx.$

20.25 RELATION BETWEEN CIRCULAR AND HYPERBOLIC FUNCTIONS

$\sin ix = i \sinh x$ $\qquad$ $\sinh ix = i \sin x$

$\cos ix = \cosh x$ $\qquad$ $\cosh ix = \cos x$

$\tan ix = i \tanh x$ $\qquad$ $\tanh ix = i \tan x$

20.26 FORMULAE OF HYPERBOLIC FUNCTIONS

A. (1) $\cosh^2 x - \sinh^2 x = 1,$ (2) $\operatorname{sech}^2 x = 1 - \tanh^2 x,$

(3) $\operatorname{cosech}^2 x = \coth^2 x - 1$

B. (1) $\sinh (x \pm y) = \sinh x \cosh y \pm \cosh x \sinh y$

(2) $\cosh (x \pm y) = \cosh x \cosh y \pm \sinh x \sinh y$

(3) $\tanh (x \pm y) = \dfrac{\tanh x \pm \tanh y}{1 \pm \tanh x \tanh y}$

C. (1) $\sinh 2x = 2 \sinh x \cosh x$ (2) $\cosh 2x = \cosh^2 x + \sinh^2 x$

(3) $\cosh 2x = 2 \cosh^2 x - 1$ (4) $\cosh 2x = 1 + 2 \sinh^2 x$

(5) $\tanh 2x = \dfrac{2 \tanh x}{1 + \tanh^2 x}$

D. (1) $\sinh x + \sinh y = 2 \sinh \dfrac{x+y}{2} \cosh \dfrac{x-y}{2}$

(2) $\sinh x - \sinh y = 2 \cosh \dfrac{x+y}{2} \sinh \dfrac{x-y}{2}$

(3) $\cosh x + \cosh y = 2 \cosh \dfrac{x+y}{2} \cosh \dfrac{x-y}{2}$

(4) $\cosh x - \cosh y = 2 \sinh \dfrac{x+y}{2} \sinh \dfrac{x-y}{2}$

Note: For proof, put $\sinh x = \dfrac{e^x - e^{-x}}{2}$ and $\cosh x = \dfrac{e^x + e^{-x}}{2}$.

Example 39. *Prove that*

$$(\cosh x - \sinh x)^n = \cosh nx - \sinh nx.$$

Solution. L.H.S. $= (\cosh x - \sinh x)^n$

$$= \left[\frac{e^x + e^{-x}}{2} - \frac{e^x - e^{-x}}{2}\right]^n = \left[\frac{2e^{-x}}{2}\right]^n = (e^{-x})^n = e^{-nx} \qquad \text{...(1)}$$

R.H.S. $= \cosh nx - \sinh nx$

$$= \left(\frac{e^{nx} + e^{-nx}}{2} - \frac{e^{nx} + e^{-nx}}{2}\right) = \frac{2e^{-nx}}{2} = e^{-nx} \qquad \text{...(2)}$$

From (1) and (2), we have

L.H.S. = R.H.S. **Proved.**

Example 40. *If $x = 2 \sin \alpha \cosh \beta$, $y = 2 \cosh \alpha \sinh \beta$, show that*

$$\text{cosec}\,(\alpha - i\beta) + \text{cosec}\,(\alpha + i\beta) = \frac{4x}{x^2 + y^2}$$

Solution. We know that $\text{cosec}\,(\alpha + i\beta) = \dfrac{1}{\sin(\alpha + i\beta)} = \dfrac{1}{\sin \alpha \cos i\beta + \cos \alpha \sin i\beta}$

$$= \frac{1}{\sin \alpha \cosh \beta + i \cos \alpha \sinh \beta} = \frac{1}{\frac{x}{2} + i\frac{y}{2}} = \frac{2}{x + iy} \qquad \text{...(1) (Given)}$$

$$\text{cosec}\,(\alpha - i\beta) = \frac{2}{x - iy} \qquad \text{... (2)}$$

Adding (1) and (2), we get

$$\text{cosec}\,(\alpha - i\beta) + \text{cosec}\,(\alpha + i\beta) = \frac{2}{x - iy} + \frac{2}{x + iy} = \frac{4x}{x^2 + y^2}$$ **Proved.**

Example 41. *If $\tan(x + iy) = i$, where x and y are real, prove that x is indeterminate and y is infinite.*

Solution. $\tan(x + iy) = i \Rightarrow \tan(x - iy) = -i$

$$\tan 2x = \tan(\overline{x + iy} + \overline{x - iy}) = \frac{\tan(x + iy) + \tan(x + iy)}{1 - \tan(x + iy)\tan(x - iy)}$$

$$= \frac{i - i}{1 - i(-i)} = \frac{i - i}{1 - 1} = \frac{0}{0}, \text{ which is indeterminate.}$$

Also $$\tan 2iy = \tan[(x + iy) - (x - iy)] = \frac{\tan(x + iy) - \tan(x - iy)}{1 + \tan(x + iy)\tan(x - iy)}$$

$$= \frac{i - (-i)}{1 + i(-i)} = \frac{2i}{1 + 1} = i$$

$$i \tanh 2y = i \quad \Rightarrow \quad \tanh 2y = 1 \quad \Rightarrow \quad 2y = \tanh^{-1}(1) = \frac{1}{2}\log\frac{1 + 1}{1 - 1} = \infty$$

$\therefore$ y is infinite. **Proved.**

Example 42. *If $\tan(\theta + i\phi) = \cos \alpha + i \sin \alpha$, prove that:*

$$\theta = \frac{n\pi}{2} + \frac{\pi}{4} \text{ and } \phi = \frac{1}{2}\log \tan\left(\frac{\pi}{4} + \frac{\alpha}{2}\right)$$

Solution. We have, $\tan(\theta + i\phi) = \cos \alpha + i \sin \alpha$

$\therefore$ $\tan(\theta - i\phi) = \cos \alpha - i \sin \alpha$

But $\tan 2\theta = \tan[(\theta + i\phi) + (\theta - i\phi)]$

$$= \frac{\tan(\theta + i\phi) + \tan(\theta - i\phi)}{1 - \tan(\theta + i\phi)\tan(\theta - i\phi)} = \frac{\cos\alpha + i\sin\alpha + \cos\alpha - i\sin\alpha}{1 - (\cos\alpha + i\sin\alpha)(\cos\alpha - i\sin\alpha)}$$

$$= \frac{2\cos\alpha}{1 - (\cos^2\alpha + \sin^2\alpha)} = \frac{2\cos\alpha}{1-1} = \infty = \tan\frac{\pi}{2}$$

$\therefore \quad 2\theta = \frac{\pi}{2}$ or for general values,

$$2\theta = n\pi + \frac{\pi}{2} \Rightarrow \theta = \frac{n\pi}{2} + \frac{\pi}{4}$$

Again, $\tan(2i\phi) = \tan[(\theta + i\phi) - (\theta - i\phi)] = \dfrac{\tan(\theta + i\phi) - \tan(\theta - i\phi)}{1 + \tan(\theta + i\phi)\tan(\theta - i\phi)}$

$$= \frac{\cos\alpha + i\sin\alpha - (\cos\alpha - i\sin\alpha)}{1 + (\cos\alpha + i\sin\alpha)(\cos\alpha - i\sin\alpha)}$$

$$= \frac{2i\sin\alpha}{1 + \cos^2\alpha + \sin^2\alpha} = \frac{2i\sin\alpha}{1+1} = \frac{2i\sin\alpha}{2} = i\sin\alpha$$

$\Rightarrow \quad i\tanh 2\phi = i\sin\alpha \qquad (\because \tan ix = i\tanh x)$

$\Rightarrow \quad \tanh 2\phi = \sin\alpha$

i.e., $\quad \dfrac{e^{2\phi} - e^{-2\phi}}{e^{2\phi} + e^{-2\phi}} = \dfrac{\sin\alpha}{1}$

$\therefore \dfrac{e^{2\phi} - e^{-2\phi} + e^{2\phi} + e^{-2\phi}}{(e^{2\phi} + e^{-2\phi}) - (e^{2\phi} - e^{-2\phi})} = \dfrac{1 + \sin\alpha}{1 - \sin\alpha} \qquad$ (Componendo and dividendo)

i.e. $\quad \dfrac{2e^{2\phi}}{2e^{-2\phi}} = \dfrac{1 - \cos\left(\frac{\pi}{2} + \alpha\right)}{1 + \cos\left(\frac{\pi}{2} + \alpha\right)} \qquad \because \cos\left(\frac{\pi}{2} + \theta\right) = -\sin\theta$

$\Rightarrow \quad e^{4\phi} = \dfrac{2\sin^2\left(\frac{\pi}{4} + \frac{\alpha}{2}\right)}{2\cos^2\left(\frac{\pi}{4} + \frac{\alpha}{2}\right)}$

$\Rightarrow \quad e^{4\phi} = \tan^2\left(\frac{\pi}{4} + \frac{\alpha}{2}\right) \Rightarrow e^{2\phi} = \tan\left(\frac{\pi}{4} + \frac{\alpha}{2}\right)$

Hence, $\quad 2\phi = \log_e \tan\left(\frac{\pi}{4} + \frac{\alpha}{2}\right) \Rightarrow \phi = \frac{1}{2}\log_e \tan\left(\frac{\pi}{4} + \frac{\alpha}{2}\right)$ **Proved.**

Example 43. *If* $u = \log_e \tan\left(\frac{\pi}{4} + \frac{\theta}{2}\right)$, *prove that*

$$\tanh\frac{u}{2} = \tan\frac{\theta}{2}$$

Solution. Here, we have,

$$u = \log_e \tan\left(\frac{\pi}{4} + \frac{\theta}{2}\right) \Rightarrow e^u = \tan\left(\frac{\pi}{4} + \frac{\theta}{2}\right)$$

$$\Rightarrow \quad e^u = \frac{\tan\frac{\pi}{4} + \tan\frac{\theta}{2}}{1 - \tan\frac{\pi}{4}.\tan\frac{\theta}{2}} = \frac{1 + \tan\frac{\theta}{2}}{1 - \tan\frac{\theta}{2}} \qquad ...(1)$$

$$\Rightarrow \qquad e^{-u} = \frac{1-\tan\frac{\theta}{2}}{1+\tan\frac{\theta}{2}} \qquad \text{... (2)}$$

By componendo and dividendo on (1), we have

$$\frac{e^u+1}{e^u-1} = \frac{2}{2\tan\frac{\theta}{2}} \quad \therefore \quad \frac{e^u-1}{e^u+1} = \tan\frac{\theta}{2} \qquad \text{...(3)}$$

Now, $$\tanh\frac{u}{2} = \frac{e^{\frac{u}{2}}-e^{-\frac{u}{2}}}{e^{\frac{u}{2}}+e^{-\frac{u}{2}}} = \frac{e^{\frac{u}{2}}-e^{-\frac{u}{2}}}{e^{\frac{u}{2}}+e^{-\frac{u}{2}}}\cdot\frac{e^{\frac{u}{2}}}{e^{\frac{u}{2}}}$$

$$\Rightarrow \qquad \tanh\frac{u}{2} = \frac{e^u-1}{e^u+1}$$

$$\Rightarrow \qquad \tanh\frac{u}{2} = \tan\frac{\theta}{2} \qquad \text{[Using (3) and (4)]}$$ **Proved**

Example 44. *If* $\cosh x = \sec\theta$, *prove that:*

(i) $$\theta = \frac{\pi}{2} - 2\tan^{-1}(e^{-x})$$

(ii) $$\tanh\frac{x}{2} = \tan\frac{\theta}{2}$$ (*M.U.* 2005)

Solution. (*i*) Let $\tan^{-1} e^{-x} = \alpha$

$$\Rightarrow \qquad e^{-x} = \tan\alpha \quad \text{and} \quad \alpha = \tan^{-1}(e^{-x}) \qquad \text{...(1)}$$

$$\Rightarrow \qquad e^{x} = \cot\alpha \qquad \text{...(2)}$$

Now, $$\sec\theta = \cosh x = \frac{e^x+e^{-x}}{2} \qquad \text{...(3)} \quad \text{(Given)}$$

Putting the values of e^{-x} and e^x from (1) and (2) in (3), we get

$$\sec\theta = \frac{\cot\alpha+\tan\alpha}{2}$$

$$\therefore \qquad 2\sec\theta = \cot\alpha+\tan\alpha = \frac{\cos\alpha}{\sin\alpha}+\frac{\sin\alpha}{\cos\alpha} = \frac{\cos^2\alpha+\sin^2\alpha}{\sin\alpha\cos\alpha}$$

$$= \frac{2}{2\sin\alpha\cos\alpha} \qquad [\because \cos^2\alpha+\sin^2\alpha = 1]$$

$$= \frac{2}{\sin 2\alpha}$$

$$\therefore \qquad \cos\theta = \sin 2\alpha$$

$$\Rightarrow \qquad \cos\theta = \cos\left(\frac{\pi}{2}-2\alpha\right)$$

$$\therefore \qquad \theta = \frac{\pi}{2}-2\alpha = \frac{\pi}{2}-2\tan^{-1}(e^{-x}) \qquad \text{[From (1)]}$$ **Proved**

(*ii*) We have, $$\cosh x = \sec\theta \qquad \text{(Given)}$$

$$\Rightarrow \qquad \frac{e^x+e^{-x}}{2} = \sec\theta \qquad \left[\because \cosh x = \frac{e^x+e^{-x}}{2}\right]$$

$$\therefore \qquad e^x - 2\sec\theta + e^{-x} = 0$$

$$\therefore \qquad (e^x)^2 - 2e^x\sec\theta + 1 = 0$$

Solving the quadratic equation in e^x.

$$e^x = \frac{2\sec\theta \pm \sqrt{4\sec^2\theta - 4}}{2}$$

$$\Rightarrow \quad e^x = \sec\theta \pm \sqrt{\sec^2\theta - 1}$$

$$\Rightarrow \quad e^x = \sec\theta \pm \tan\theta \qquad ...(4)$$

Now,
$$\tanh\frac{x}{2} = \frac{e^{\frac{x}{2}} - e^{-\frac{x}{2}}}{e^{\frac{x}{2}} + e^{-\frac{x}{2}}} = \frac{e^x - 1}{e^x + 1} \qquad ...(5)$$

Putting the value of e^x from (4) in (5), we get

$$\tanh\frac{x}{2} = \frac{\sec\theta + \tan\theta - 1}{\sec\theta + \tan\theta + 1} \qquad \text{[Using (1)]}$$

$$= \frac{1 + \sin\theta - \cos\theta}{1 + \sin\theta + \cos\theta} = \frac{(1 - \cos\theta) + \sin\theta}{(1 + \cos\theta) + \sin\theta}$$

$$= \frac{2\sin^2\frac{\theta}{2} + 2\sin\frac{\theta}{2}\cos\frac{\theta}{2}}{2\cos^2\frac{\theta}{2} + 2\sin\frac{\theta}{2}\cos\frac{\theta}{2}} = \frac{\sin\frac{\theta}{2}}{\cos\frac{\theta}{2}} = \tan\frac{\theta}{2}$$ **Proved**

EXERCISE 20.8

1. If $\tan\left(\frac{\pi}{8} + i\alpha\right) = x + iy$, prove that $x^2 + y^2 + 2x = 1$.

2. If $\cot\left(\frac{\pi}{8} + i\alpha\right) = x + iy$, prove that $x^2 + y^2 - 2x = 1$.

3. Prove that if $(1 + i\tan\alpha)^{1 + i\tan\beta}$ can have real values, one of them is $(\sec\alpha)^{\sec 2\beta}$.

4. If $\frac{(1+i)^{x+iy}}{(1-i)^{x-iy}} = \alpha + i\beta$, prove that the value of $\tan^{-1}\frac{\beta}{\alpha}$ *is* $\frac{\pi x}{2} + y\log 2$.

5. If $\tanh x = \frac{1}{2}$, find the value of $\sinh 2x$. **Ans.** $\frac{4}{3}$

6. If $\sin\alpha\cosh\beta = \frac{x}{2}$, $\cos\alpha\sinh\beta = \frac{y}{2}$, show that

(*i*) $\operatorname{cosec}(\alpha - i\beta) + \operatorname{cosec}(\alpha + i\beta) = \frac{4x}{x^2 + y^2}$

(*ii*) $\operatorname{cosec}(\alpha - i\beta) - \operatorname{cosec}(\alpha + i\beta) = \frac{4iy}{x^2 + y^2}$

7. Show that $\tan\left(\frac{u + iv}{2}\right) = \frac{\sin u + i\sinh v}{\cos u + \cosh v}$

8. If $\cot(\alpha + i\beta) = x + iy$, prove that

(*i*) $x^2 + y^2 - 2x\cot 2\alpha = 1$

(*ii*) $x^2 + y^2 + 2y\coth 2\beta + 1 = 0$]

9. If tan $\tan\frac{x}{2} = \tanh\frac{u}{2}$ prove that

(*i*) $\sinh u = \tan x$ (*ii*) $\cosh u = \sec x$ **Ans.** – log 5

10. Solve the following equation for real values of x.

$$17\cosh x + 18\sinh x = 1$$ **Ans.** – log 5

20.27 SEPARATION OF REAL AND IMAGINARY PARTS OF CIRCULAR FUNCTIONS

Example 45. *Separate the following into real and imaginary parts:*

(i) $\sin(x+iy)$ *(ii)* $\cos(x+iy)$ *(iii)* $\tan(x+iy)$

(iv) $\cot(x+iy)$ *(v)* $\sec(x+iy)$ *(vi)* $\operatorname{cosec}(x+iy)$.

Solution. *(i)* $\sin(x+iy) = \sin x\cos iy + \cos x\sin(iy) = \sin x\cosh y + i\cos x\sinh y.$

(ii) $\cos(x+iy) = \cos x\cos(iy) - \sin x\sin(iy) = \cos x\cosh y - i\sin x\sinh y.$

$$(iii)\ \tan(x+iy) = \frac{\sin(x+iy)}{\cos(x+iy)} = \frac{2\sin(x+iy)\cos(x-iy)}{2\cos(x+iy)\cos(x-iy)}$$

$$= \frac{\sin 2x + \sin(2iy)}{\cos 2x + \cos 2iy} = \frac{\sin 2x + i\sinh 2y}{\cos 2x + \cosh 2y}$$

$$\left\{\begin{array}{l}\because\ 2\sin A.\cos B = \sin(A+B) + \sin(A-B)\\ \text{and } 2\cos A.\cos B = \cos(A+B) + \cos(A-B)\end{array}\right\}$$

$$(iv)\ \cot(x+iy) = \frac{\cos(x+iy)}{\sin(x+iy)} = \frac{2\cos(x+iy)\sin(x-iy)}{2\sin(x+iy)\sin(x-iy)}$$

$$= \frac{\sin 2x - \sin(2iy)}{\cos(2iy) - \cos 2x} = \frac{\sin 2x - i\sinh 2y}{\cosh 2y - \cos 2x}$$

$$(v)\ \sec(x+iy) = \frac{1}{\cos(x+iy)} = \frac{2\cos(x-iy)}{2\cos(x+iy)\cos(x-iy)}$$

$$= \frac{2[\cos x\cos(iy) + \sin x\sin(iy)]}{\cos 2x + \cos(2iy)} = \frac{2[\cos x\cosh y + i\sin x\sinh y]}{\cos 2x + \cosh 2y}$$

$$(vi)\ \operatorname{cosec}(x+iy) = \frac{1}{\sin(x+iy)} = \frac{2\sin(x-iy)}{2\sin(x+iy)\sin(x-iy)} = \frac{2[\sin x\cos(iy) - \cos x\sin(iy)]}{\cos(2iy) - \cos 2x}$$

$$= \frac{2[\sin x\cosh y - i\cos x\sinh y]}{\cosh 2y - \cos 2x}$$ **Ans.**

Example 46. *If* $\tan(A+iB) = x+iy$, *prove that*

$$\tan 2A = \frac{2x}{1-x^2-y^2} \text{ and } \tanh 2B = \frac{2x}{1+x^2+y^2}$$

Solution. $\tan(A+iB) = x+iy$; $\tan(A-iB) = x-iy$

$$\tan 2A = \tan(A+iB+A-iB)$$

$$= \frac{\tan(A+iB) + \tan(A-iB)}{1-\tan(A+iB)\tan(A-iB)}$$

$$\tan 2A = \frac{(x+iy)+(x-iy)}{1-(x+iy)(x-iy)} = \frac{2x}{1-\left(x^2+y^2\right)} = \frac{2x}{1-x^2-y^2}$$

Again $$\tan 2iB = \tan(A+iB-A+iB) = \frac{\tan(A+iB)-\tan(A-iB)}{1+\tan(A+iB)\tan(A-iB)}$$

$$\tan 2iB = \frac{(x+iy)-(x-iy)}{1+(x+iy)(x-iy)} = \frac{(2y)i}{1+x^2+y^2}$$

$$\tanh 2B = \frac{2y}{1+x^2+y^2}$$ $\tan ix = i\tanh x$ **Proved.**

Example 47. *If sin (α + iβ) = x + iy, prove that*

$$(a)\ \frac{x^2}{\cosh^2\beta}+\frac{y^2}{\sinh^2\beta}=1 \qquad (b)\ \frac{x^2}{\sin^2\alpha}-\frac{y^2}{\cos^2\alpha}=1$$

Solution. (*a*) $x + iy = \sin(\alpha + i\beta) = \sin\alpha\cosh\beta + i\cos\alpha\sinh\beta$

Equating real and imaginary parts, we get

$$x = \sin\alpha\cosh\beta,\ y = \cos\alpha\sinh\beta$$

$$\sin\alpha = \frac{x}{\cosh\beta} \text{ and } \cos\alpha = \frac{y}{\sinh\beta}$$

Squaring and adding, $\sin^2\alpha + \cos^2\alpha = \frac{x^2}{\cosh^2\beta}+\frac{y^2}{\sinh^2\beta}$

$$\Rightarrow \quad 1 = \frac{x^2}{\cosh^2\beta}+\frac{y^2}{\sinh^2\beta}$$ **Proved.**

(*b*) Again $$\cosh\beta = \frac{x}{\sin\alpha} \text{ and } \sinh\beta = \frac{y}{\cos\alpha}$$

$$\cosh^2\beta - \sinh^2\beta = \frac{x^2}{\sin^2\alpha}-\frac{y^2}{\cos^2\alpha}$$

$$1 = \frac{x^2}{\sin^2\alpha}-\frac{y^2}{\cos^2\alpha}$$ **Proved.**

20.28 SEPARATION OF REAL AND IMAGINARY PARTS OF HYPERBOLIC FUNCTIONS

Example 48. *Separate the following into real and imaginary parts of hyperbolic functions.*

(a) sinh (x + iy) (b) cosh (x + iy) (c) tanh (x + iy)

Solution. (*a*) $\sinh(x+iy) = \sinh x\cosh(iy) + \cosh x\sinh(iy)$

$= \sinh x\cos y + i\sin y\cosh x.$ **Ans.**

(*b*) $\cosh(x+iy) = \cosh x\cosh(iy) - \sinh x\sinh iy = \cosh x\cos y - i\sinh x\sin y.$ **Ans.**

(*c*) $$\tanh(x+iy) = \frac{\sinh(x+iy)}{\cosh(x+iy)} = \frac{-i\sin i(x+iy)}{\cos i(x+iy)}$$

$$= \frac{-i\sin(ix-y)}{\cos(ix-y)} = \frac{-i\,2\sin(ix-y)\cos(ix+y)}{2\cos(ix-y)\cos(ix+y)}$$ (Note this step)

$$= -i\frac{\sin 2ix - \sin 2y}{\cos 2ix + \cos 2y} = -i\frac{i\sinh 2x - \sin 2y}{\cosh 2x + \cos 2y} = \frac{\sinh 2x + i\sin 2y}{\cosh 2x + \cos 2y}$$

$$= \frac{\sinh 2x}{\cosh 2x + \cos 2y} + i\frac{\sin 2y}{\cosh 2x + \cos 2y}$$ **Ans.**

Example 49. *If tan (x + iy) = sin (u + iv), prove that*

$$\frac{\sin 2x}{\sinh 2y} = \frac{\tan u}{\tanh v}$$

Solution. Now $\tan(x+iy) = \sin(u+iv)$ separating the real and imaginary parts of both sides, we have

$$\frac{\sin 2x}{\cos 2x + \cosh 2y} + \frac{i\sinh 2y}{\cos 2x + \cosh 2y} = \sin u\cosh v + i\cos u\sinh v$$

Equating real and imaginary parts, we get

$$\frac{\sin 2x}{\cos 2x + \cosh 2y} = \sin u\cosh v \qquad ...(1)$$

and $$\frac{\sinh 2y}{\cos 2x + \cosh 2y} = \cos u \sinh v \qquad ...(2)$$

Dividing (1) by (2), we obtain

$$\frac{\sin 2x}{\sinh 2y} = \frac{\sin u \cosh v}{\cos u \sinh v}$$

$$\Rightarrow \quad \frac{\sin 2x}{\sinh 2y} = \frac{\tan u}{\tanh v}$$ **Proved.**

Example 50. *If $\sin(\theta + i\phi) = \tan\alpha + i\sec\alpha$, show that $\cos 2\theta \cosh 2\phi = 3$*

Solution. $\sin(\theta + i\phi) = \tan\alpha + i\sec\alpha$

$\sin\theta\cosh\phi + i\cos\theta\sinh\phi = \tan\alpha + i\sec\alpha$

Equating real and imaginary parts, we get

$$\sin\theta\cosh\phi = \tan\alpha \qquad ...(1)$$

$$\cos\theta\sinh\phi = \sec\alpha \qquad ...(2)$$

We know that

$$\sec^2\alpha - \tan^2\alpha = 1$$

$\cos^2\theta\sinh^2\phi - \sin^2\theta\cosh^2\phi = 1$ [From (1) and (2)]

$$\left(\frac{1+\cos 2\theta}{2}\right)\left(\frac{\cosh 2\phi - 1}{2}\right) - \left(\frac{1-\cos 2\theta}{2}\right)\left(\frac{\cosh 2\phi + 1}{2}\right) = 1$$

$[-1 + \cosh 2\phi - \cos 2\theta + \cos 2\theta\cosh 2\phi] - [\cosh 2\phi + 1 - \cos 2\theta\cosh 2\phi - \cos 2\theta] = 4$

$\Rightarrow \quad -2 + 2\cos 2\theta\cosh 2\phi = 4$

$\Rightarrow \quad 2\cos 2\theta\cosh 2\phi = 6 \quad \Rightarrow \quad \cos 2\theta\cosh 2\phi = 3$ **Proved.**

Example 51. *If $\sinh(\theta + i\phi) = \cos\alpha + i\sin\alpha$, prove that $\sinh^4\theta = \cos^2\alpha = \cos^4\phi$.*

Solution. $\sinh(\theta + i\phi) = \cos\alpha + i\sin\alpha$

$\sinh\theta\cos\phi + i\sin\phi\cosh\theta = \cos\alpha + i\sin\alpha$

Equating real and imaginary parts, we have

$$\sinh\theta\cos\phi = \cos\alpha \text{ and} \qquad ...(1)$$

$$\sin\phi\cosh\theta = \sin\alpha \qquad ...(2)$$

Let us eliminate ϕ from (1) and (2).

$$\cos\phi = \frac{\cos\alpha}{\sinh\theta} \text{ and } \sin\phi = \frac{\sin\alpha}{\cosh\theta}$$

Squaring and adding, we get

$$1 = \frac{\cos^2\alpha}{\sinh^2\theta} + \frac{\sin^2\alpha}{\cosh^2\theta} \Rightarrow \frac{\cos^2\alpha}{\sinh^2\theta} = 1 - \frac{\sin^2\alpha}{\cosh^2\theta}$$

$$\Rightarrow \quad \frac{\cos^2\alpha}{\sinh^2\theta} = 1 - \frac{1-\cos^2\alpha}{1+\sinh^2\theta} = \frac{1+\sinh^2\theta - 1 + \cos^2\alpha}{1+\sinh^2\theta}$$

$$\frac{\cos^2\alpha}{\sinh^2\theta} = \frac{\sinh^2\theta + \cos^2\alpha}{1+\sinh^2\theta}$$

$\sinh^4\theta + \sinh^2\theta\cos^2\alpha = \cos^2\alpha + \cos^2\alpha\sinh^2\theta$

$\Rightarrow \quad \sinh^4\theta = \cos^2\alpha$ **Proved.**

For second result, let us eliminate θ.

$$\sinh\theta = \frac{\cos\alpha}{\cos\phi} \quad \text{and} \quad \cosh\theta = \frac{\sin\alpha}{\sin\phi}$$

$$\cosh^2\theta - \sinh^2\theta = \frac{\sin^2\alpha}{\sin^2\phi} - \frac{\cos^2\alpha}{\cos^2\phi} \quad \Rightarrow \quad 1 = \frac{1-\cos^2\alpha}{1-\cos^2\phi} - \frac{\cos^2\alpha}{\cos^2\phi}$$

$$\Rightarrow \quad \frac{\cos^2\alpha}{\cos^2\phi} = \frac{1-\cos^2\alpha - 1 + \cos^2\phi}{1-\cos^2\phi}$$

$$\frac{\cos^2\alpha}{\cos^2\phi} = \frac{\cos^2\phi - \cos^2\alpha}{1-\cos^2\phi}$$

$$\Rightarrow \quad \cos^4\phi - \cos^2\phi\cos^2\alpha = \cos^2\alpha - \cos^2\alpha\cos^2\phi$$

$$\Rightarrow \quad \cos^4\phi = \cos^2\alpha.$$ **Proved.**

Example 52. *If $e^z = \sin(u + iv)$ and $z = x + iy$, prove that*

$$2e^{2x} = \cosh 2v - \cos 2u$$ (*M.U.* 2006)

Solution. We have, $e^z = \sin(u + iv)$

$$\Rightarrow \quad e^{x+iy} = \sin(u+iv)$$

$$\Rightarrow \quad e^x . e^{iy} = \sin u \cos iv + \cos u \sin iv$$

$$\Rightarrow \quad e^x(\cos y + i\sin y) = \sin u \cosh v + i\cos u \sinh v$$

Equating real and imaginary parts, we get

$$e^x \cos y = \sin u \cosh v$$

and $$e^x \sin y = \cos u \sinh v$$

Squaring and adding, we get

$$e^{2x}(\cos^2 y + \sin^2 y) = \sin^2 u \cosh^2 v + \cos^2 u \sinh^2 v$$

$$\Rightarrow \quad e^{2x} = (1-\cos^2 u)\cosh^2 v + \cos^2 u(\cosh^2 v - 1)$$

$$\Rightarrow \quad e^{2x} = \cosh^2 v - \cos^2 u$$

$$\Rightarrow \quad e^{2x} = \frac{1}{2}(1+\cosh 2v) - \frac{1}{2}(1+\cos 2u)$$

$$\Rightarrow \quad e^{2x} = \frac{1}{2}(\cosh 2v - \cos 2u)$$

$$\Rightarrow \quad 2e^{2x} = \cosh 2v - \cos 2u$$ **Proved.**

Example 53. *If $\sin(\theta + i\phi) = \cos\alpha + i\sin\alpha$, prove that*

$$\cos^4\theta = \sin^2\alpha = \sinh^4\phi.$$

Solution. Here, we have

$$\sin(\theta + i\phi) = \cos\alpha + i\sin\alpha$$

$$\Rightarrow \quad \sin\theta\cosh\phi + i\cos\theta\sinh\phi = \cos\alpha + i\sin\alpha$$

Equating real and imaginary parts, we get

$$\sin\theta\cosh\phi = \cos\alpha \quad \Rightarrow \quad \cosh\phi = \frac{\cos\alpha}{\sin\theta} \qquad ...(1)$$

and $$\cos\theta\sinh\phi = \sin\alpha \quad \Rightarrow \quad \sinh\phi = \frac{\sin\alpha}{\cos\theta} \qquad ...(2)$$

But $$\cosh^2\phi - \sinh^2\phi = 1$$

$$\Rightarrow \quad \frac{\cos^2\alpha}{\sin^2\theta} - \frac{\sin^2\alpha}{\cos^2\theta} = 1$$ [Using (1) and (2)]

$$\Rightarrow \quad \cos^2\alpha . \cos^2\theta - \sin^2\alpha . \sin^2\theta = \sin^2\theta\cos^2\theta$$

$$\Rightarrow \quad (1-\sin^2\alpha)\cos^2\theta - \sin^2\alpha . \sin^2\theta = (1-\cos^2\theta)\cos^2\theta$$

$$\Rightarrow \quad \cos^2\theta - \sin^2\alpha(\cos^2\theta + \sin^2\theta) = \cos^2\theta - \cos^4\theta$$

$$\Rightarrow \quad \sin^2\alpha = \cos^4\theta \qquad ...(3)$$

Again $$\sin^2\theta + \cos^2\theta = 1$$

$$\therefore \quad \frac{\cos^2\alpha}{\cosh^2\phi} + \frac{\sin^2\alpha}{\sinh^2\phi} = 1$$ [Using (1) and (2)]

$$\Rightarrow \quad \cos^2\alpha . \sinh^2\phi + \sin^2\alpha\cosh^2\phi = \sinh^2\phi\cosh^2\phi$$

$$\Rightarrow \quad (1-\sin^2\alpha)\sinh^2\phi + \sin^2\alpha(1+\sinh^2\phi) = \sinh^2\phi(1+\sinh^2\phi)$$

$\Rightarrow \quad \sinh^2\phi - \sin^2\alpha \sinh^2\phi + \sin^2\alpha + \sin^2\alpha \sinh^2\phi = \sinh^2\phi + \sinh^4\phi$

$\Rightarrow \quad \sin^2\alpha = \sinh^4\phi.$ ...(4)

From (3) and (4), we have $\cos^4\theta = \sin^2\alpha = \sinh^4\phi$ **Proved.**

Example 54. *If cosec* $\left(\frac{\pi}{4} + ix\right) = u + iv$, *prove that*

$$(u^2 + v^2)^2 = 2(u^2 - v^2)$$ (*M.U. 2009*)

Solution. Here, we have

$$u + iv = \text{cosec}\left(\frac{\pi}{4} + ix\right)$$

$$= \frac{1}{\sin\left(\frac{\pi}{4} + ix\right)} = \frac{1}{\sin\frac{\pi}{4}\cos ix + \cos\frac{\pi}{4}\sin ix}$$

$$= \frac{1}{\frac{1}{\sqrt{2}}\cosh x + \frac{1}{\sqrt{2}} i \sinh x} = \frac{\sqrt{2}}{\cosh x + i\sinh x}$$

$$= \frac{\sqrt{2}\,(\cosh x - i\sinh x)}{\cosh^2 x + \sinh^2 x} = \frac{\sqrt{2}\,(\cosh x - i\sinh x)}{\cosh 2x}$$

Equating real and imaginary parts, we get $u = \frac{\sqrt{2}\cosh x}{\cos h\, 2x}, v = -\frac{\sqrt{2}\sinh x}{\cosh 2x}$

Squaring and adding, we get

$$u^2 + v^2 = \frac{2\,(\cosh^2 x + \sinh^2 x)}{\cosh^2 2x} = \frac{2\cosh 2x}{\cosh^2 2x}$$

$$\Rightarrow \quad (u^2 + v^2)^2 = \left(\frac{2}{\cosh 2x}\right)^2 = \frac{4}{\cosh^2 2x} \quad ...(1)$$

Also, $$u^2 - v^2 = \frac{2}{\cosh^2 2x}(\cosh^2 x - \sinh^2 x) = \frac{2}{\cosh^2 2x} \quad ...(2)$$

From (1) and (2), we have

$$(u^2 + v^2)^2 = 2\,(u^2 - v^2)$$ **Proved.**

Example 55. *Separate into real and imaginary parts* $\sqrt{i}^{\sqrt{i}}$. (*M.U. 2008*)

Solution. We have,

$$\sqrt{i} = i^{\frac{1}{2}} = \left(\cos\frac{\pi}{2} + i\sin\frac{\pi}{2}\right)^{\frac{1}{2}}$$

$$= \cos\frac{\pi}{4} + i\sin\frac{\pi}{4}$$

$$= \frac{1}{\sqrt{2}} + i\frac{1}{\sqrt{2}}$$

Also, $$\sqrt{i} = \left(\cos\frac{\pi}{2} + i\sin\frac{\pi}{2}\right)^{\frac{1}{2}} = \left(e^{i\frac{\pi}{2}}\right)^{\frac{1}{2}} = e^{i\frac{\pi}{4}}$$

$$\therefore \quad \left(\sqrt{i}\right)^{\sqrt{i}} = \left(e^{i\frac{\pi}{4}}\right)^{\left(\frac{1}{\sqrt{2}} + i.\frac{1}{\sqrt{2}}\right)} = e^{i\frac{\pi}{4\sqrt{2}} - \frac{\pi}{4\sqrt{2}}}$$

$$= e^{-\frac{\pi}{4\sqrt{2}}}.e^{i\frac{\pi}{4\sqrt{2}}} = e^{-\frac{\pi}{4\sqrt{2}}}\left(\cos\frac{\pi}{4\sqrt{2}} + i\sin\frac{\pi}{4\sqrt{2}}\right)$$

$\therefore$ Real part $= e^{-\frac{\pi}{4\sqrt{2}}} \cos\left(\dfrac{\pi}{4\sqrt{2}}\right)$

Imaginary part $= e^{-\frac{\pi}{4\sqrt{2}}} \sin\left(\dfrac{\pi}{4\sqrt{2}}\right)$ **Ans.**

EXERCISE 20.9

Separate into real and imaginary parts.

1. sech $(x + iy)$ **Ans.** $\dfrac{2\cosh x \cos y - 2i \sinh x \sin y}{\cosh 2x + \cos 2y}$

2. coth $i\,(x + iy)$ **Ans.** $\dfrac{-\sinh 2y - i \sin 2x}{\cosh 2x - \cos 2y}$

3. coth $(x + iy)$ **Ans.** $\dfrac{\sinh 2x - i \sin 2y}{\cosh 2x - \cos 2y}$

4. If $\sin(\theta + i\phi) = p(\cos\alpha + i\sin\alpha)$, prove that

$$p^2 = \frac{1}{2}[\cosh 2\phi - \cos 2\theta],\ \tan\alpha = \tanh\phi \cot\theta$$

5. If $\sin(\alpha + i\beta) = x + iy$, prove that $x^2 \operatorname{sech}^2\beta + y^2 \operatorname{cosech}^2\beta = 1$ and $x^2 \operatorname{cosec}^2\alpha - y^2 \sec^2\alpha = 1$

6. If $\cos(\theta + i\phi) = r(\cos\alpha + i\sin\alpha)$, prove that $\theta = \dfrac{1}{2}\log\left[\dfrac{\sin(\theta - \alpha)}{\sin(\theta + \alpha)}\right]$

7. If $\tan\left(\dfrac{\pi}{6} + i\alpha\right) = x + iy$, prove that $x^2 + y^2 + \dfrac{2x}{\sqrt{3}} = 1$

8. If $\tan(A + B) = \alpha + i\beta$, show that $\dfrac{1-(\alpha^2+\beta^2)}{1+(\alpha^2+\beta^2)} = \dfrac{\cos 2A}{\cosh 2B}$

9. If $\dfrac{x+iy-c}{x+iy+c} = e^{u+iv}$, prove that

$$x = -\frac{c\sinh u}{\cosh u - \cos v}, \qquad y = \frac{c\sinh v}{\cosh u - \cos v}$$

Further, if $v = (2n+1)\dfrac{\pi}{2}$, prove that $x^2 + y^2 = c^2$ where n is an integer.

10. If $\dfrac{u-1}{u+1} = \sin(x + iy)$, where $u = \alpha + i\beta$ show that the argument of u is $\theta + \phi$ where

$$\tan\theta = \frac{\cos x \sinh y}{1 + \sin x \cosh y} \text{ and },\quad \tan\phi = \frac{\cos x \sinh y}{1 - \sin x \sinh y}$$

11. If $A + iB = C\tan(x + iy)$, prove that $\tan 2x = \dfrac{2CA}{C^2 - A^2 - B^2}$

12. If $\cosh(\alpha + i\beta) = x + iy$, prove that

(*a*) $\dfrac{x^2}{\cosh^2\alpha} + \dfrac{y^2}{\sinh^2\alpha} = 1$ (*b*) $\dfrac{x^2}{\cos^2\beta} - \dfrac{y^2}{\sin^2\beta} = 1$

13. If $\cos(\theta + i\phi) = R(\cos\alpha + i\sin\alpha)$, prove that $\phi = \dfrac{1}{2}\log_e\left[\dfrac{\sin(\theta-\alpha)}{\sin(\theta+\alpha)}\right]$

14. If $\cos(\alpha + i\beta)\cos(\gamma + i\delta) = 1$, prove that $\tanh^2\delta \cosh^2\beta = \sin^2\alpha$

15. If $\dfrac{u-1}{u+1} = \sin(x + iy)$, find u. **Ans.** $\tan^{-1}\dfrac{2\cos x \sinh y}{\cos^2 x - \sinh^2 y}$

20.29 LOGARITHMIC FUNCTION OF A COMPLEX VARIABLE

Example 56. *Define logarithm of a complex number.*

Solution. If z and w are two complex numbers and $z = e^w$ then $w = \log z$, and if $w = \log z$; then $z = e^w$

Here $\log z$ is a many valued function. General value of $\log z$ is defined by Log z, where $\text{Log } z = \log z + 2\, n\, \pi\, i$.

Example 57. *Separate log $(x + iy)$ into its real and imaginary parts.*

Solution. Let $\qquad x = r\cos\theta \qquad$ ...(1)

and $\qquad y = r\sin\theta \qquad$ ...(2)

Squaring and adding (1) and (2) we have $x^2 + y^2 = r^2$

$$\therefore \quad r = \sqrt{x^2+y^2},$$

We have, $\quad \tan\theta = \dfrac{y}{x} \Rightarrow \theta = \tan^{-1}\left(\dfrac{y}{x}\right) \qquad$ [Dividing (2) by (1)]

$$\therefore \quad \log(x+iy) = \log[r(\cos\theta + i\sin\theta)]$$
$$= [\log r + \log(\cos\theta + i\sin\theta)]$$
$$\log(x+iy) = \log r + \log[\cos(2\,n\,\pi + \theta) + i\sin(2\,n\,\pi+\theta)]$$
$$= \log r + \log e^{i(2\,n\,\pi+\theta)} = \log r + i\,(2\,n\,\pi + \theta)$$
$$\text{Log}(x+iy) = \log\sqrt{x^2+y^2} + i\left(2n\pi + \tan^{-1}\frac{y}{x}\right)$$

and $\quad \log(x+iy) = \log\sqrt{x^2+y^2} + i\tan^{-1}\dfrac{y}{x} \qquad$ **Ans.**

Example 58. *Find the general value of Log $(1 + i)$ + Log $(1 - i)$.*

Solution. $\quad 1 + i = \sqrt{2}\left(\cos\dfrac{\pi}{4} + i\sin\dfrac{\pi}{4}\right) = \sqrt{2}\,e^{i\frac{\pi}{4}}$

$$\log(1+i) = \log\sqrt{2}.e^{i\frac{\pi}{4}} = \log\sqrt{2} + i\frac{\pi}{4}$$
$$\text{Log}(1+i) = \log\sqrt{2} + i\frac{\pi}{4} + 2\,n\pi i = \log\sqrt{2} + \left(2n\pi + \frac{\pi}{4}\right)i$$
$$\text{Log}(1-i) = \log\sqrt{2} + \left(2n\pi - \frac{\pi}{4}\right)i$$

Hence, $\quad \text{Log}(1+i) + \text{Log}(1-i) = \left[\log\sqrt{2} + \left(2n\pi + \dfrac{\pi}{4}\right)i\right] + \left[\log\sqrt{2} + \left(2n\pi - \dfrac{\pi}{4}\right)i\right]$

$$= 2\log\sqrt{2} + 4n\pi i = \log 2 + 4n\pi i$$ **Ans.**

Example 59. *Show that* $\log\dfrac{x+iy}{x-iy} = 2i\tan^{-1}\dfrac{y}{x}$.

Solution. Let $\quad \log(x+iy) = \log(r\cos\theta + ir\sin\theta) = \log r\,e^{i\theta}$

$$= \log r + i\,\theta \qquad \begin{bmatrix} x = r\cos\theta \\ y = r\sin\theta \end{bmatrix}$$

Similarly, $\quad \log(x - iy) = \log r - i\,\theta$

$$\log\frac{x+iy}{x-iy} = \log(x+iy) - \log(x-iy) = (\log r + i\theta) - (\log r - i\theta) = 2i\,\theta$$
$$= 2\,i\tan^{-1}\frac{y}{x}.$$ **Proved.**

Example 60. *Show that for real values of a and b*

$$e^{2ai\cot^{-1}b}\left[\frac{bi-1}{bi+1}\right]^{-a} = 1 \qquad \textit{(M.U. 2008)}$$

Solution. Consider $\dfrac{bi-1}{bi+1} = \dfrac{bi+i^2}{bi-i^2} = \dfrac{b+i}{b-i}$

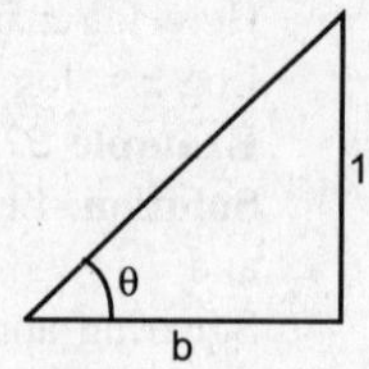

$$\Rightarrow \left(\frac{bi-1}{bi+1}\right)^{-a} = \left(\frac{b+i}{b-i}\right)^{-a}$$

$$\log\left[\frac{bi-1}{bi+1}\right]^{-a} = \log\left(\frac{b+i}{b-i}\right)^{-a} = -a\,[\log(b+i) - \log(b-i)]$$

$$= -a\left[\log\sqrt{b^2+1} + i\tan^{-1}\frac{1}{b} - \log\sqrt{b^2+1} + i\tan^{-1}\frac{1}{b}\right]$$

$$= -2ai\tan^{-1}\frac{1}{b}$$

$$\left(\frac{bi-1}{bi+1}\right)^{-a} = e^{-2ai\tan^{-1}\left(\frac{1}{b}\right)} \qquad \left[\begin{array}{l}\text{If } \cot\theta = b,\ \tan\theta = \dfrac{1}{b}\\ \text{Since } \cot^{-1}b = \tan^{-1}\left(\dfrac{1}{b}\right)\end{array}\right]$$

$$e^{2ai\cot^{-1}b}\left(\frac{bi-1}{bi+1}\right)^{-a} = \left[e^{2ai\tan^{-1}\left(\frac{1}{b}\right)}\right].\left[e^{-2ai\tan^{-1}\left(\frac{1}{b}\right)}\right] = 1$$

Proved.

EXERCISE 20.10

1. Find the general value of Log i. **Ans.** $(4n+1)\dfrac{\pi i}{2}$
2. Express Log (-5) in terms of $a + ib$. **Ans.** $\log 5 + i\,(2n+1)\,\pi$
3. Find the value of z if
   (*a*) $\cos z = 2$. **Ans.** $z = 2n\pi \pm i\log\left(2+\sqrt{3}\right)$
   (*b*) $\cosh z = -1$. **Ans.** $z = (2n+1)\,\pi\, i$
4. Find the general and principal values of i^i **Ans.** $e^{-\left(2n\pi+\frac{\pi}{2}\right)},\ e^{-\frac{\pi}{2}}$
5. If $i^{(\alpha+i\beta)} = x + iy$, prove that $x^2 + y^2 = e^{-(4m+1)\pi\theta}$.
6. Prove that $\log\dfrac{1}{1-e^{i\theta}} = \log\left(\dfrac{1}{2}\operatorname{cosec}\theta\right) + i\left(\dfrac{\pi}{2}-\dfrac{\theta}{2}\right)$
7. Show that $\log\sin(x+iy) = \dfrac{1}{2}\log\dfrac{\cosh 2y - \cos 2x}{2} + i\tan^{-1}(\cot x\tanh y)$.
8. Prove that $\tan\left[i\log\dfrac{a-ib}{a+ib}\right] = \dfrac{2ab}{a^2-b^2}$.
9. $\log\dfrac{\cos(x-iy)}{\cos(x+iy)} = 2\,i\tan^{-1}(\tan x\tanh y)$.
10. Separate $i^{(1+i)}$ into real and imaginary parts. **Ans.** $ie^{-\frac{\pi}{2}}$

20.30 INVERSE FUNCTIONS

If $\sin\theta = \frac{1}{2}$ then $\theta = \sin^{-1}\left(\frac{1}{2}\right)$, so here θ is called inverse sine of $\frac{1}{2}$.

Similarly, we can define inverse hyperbolic function sinh, cosh, tanh, etc. If $\cosh\theta = z$ then $\theta = \cosh^{-1} z$.

20.31 INVERSE HYPERBOLIC FUNCTIONS

Example 61. *Prove that* $\sinh^{-1} x = \log\,(x+\sqrt{x^2+1})$ *(M.U. 2009)*

Solution. Let $\sinh^{-1} x = y \quad\Rightarrow\quad x = \sinh y$

$$x = \frac{e^y - e^{-y}}{2}$$

$$\Rightarrow \quad e^y - e^{-y} = 2x$$

$$\Rightarrow \quad e^{2y} - 2x\,e^y - 1 = 0$$

This is quadratic in e^y.

$$e^y = \frac{2x \pm \sqrt{4x^2+4}}{2} = x \pm \sqrt{x^2+1}$$

$$y = \log\,(x+\sqrt{x^2+1}) \qquad \text{(Taking positive sign only)}$$

$$\sinh^{-1} x = \log\,(x+\sqrt{x^2+1})$$

Proved.

Example 62. *Prove that* $\cosh^{-1} x = \log\,(x+\sqrt{x^2-1})$. *(M.U. 2009)*

Solution. Let $y = \cosh^{-1} x \quad\Rightarrow\quad x = \cosh y$

$$x = \frac{e^y + e^{-y}}{2} \quad\Rightarrow\quad 2x = e^y + e^{-y}$$

$$\Rightarrow \quad e^{2y} - 2x\,e^y + 1 = 0 \qquad \text{(This is quadratic in } e^y\text{)}$$

$$\Rightarrow \quad e^y = \frac{2x \pm \sqrt{4x^2-4}}{2} = x \pm \sqrt{x^2-1}$$

$$\Rightarrow \quad y = \log\,(x+\sqrt{x^2-1}) \qquad \text{(Taking positive sign only)}$$

$$\Rightarrow \quad \cosh^{-1} x = \log\,(x+\sqrt{x^2-1})$$

Proved.

Example 63. *Prove that* $\tanh^{-1} x = \frac{1}{2}\log\frac{1+x}{1-x}$

Solution. Let $\tanh^{-1} x = y \quad\Rightarrow\quad x = \tanh y$

$$x = \frac{e^y - e^{-y}}{e^y + e^{-y}}$$

Applying componendo and dividendo, we obtain

$$\frac{1+x}{1-x} = \frac{e^y}{e^{-y}} = e^{2y}, \quad 2y = \log\frac{1+x}{1-x}$$

$$\Rightarrow \quad \tanh^{-1} x = \frac{1}{2}\log\frac{1+x}{1-x}.$$

Similarly, $$\coth^{-1} x = \frac{1}{2}\log\frac{x+1}{x-1}$$

Proved.

Example 64. *Prove that* $sech^{-1}\ x = \log \dfrac{1+\sqrt{1-x^2}}{x}$

Solution. Let
$$y = \text{sech}^{-1} x \Rightarrow x = \text{sech}\, y$$
$$x = \frac{2}{e^y + e^{-y}} \Rightarrow x = \frac{2e^y}{e^{2y}+1}$$
$$\Rightarrow \quad xe^{2y} - 2e^y + x = 0 \Rightarrow e^y = \frac{2 \pm \sqrt{4-4x^2}}{2x} = \frac{1 \pm \sqrt{1-x^2}}{x}$$

We take only positive sign
$$e^y = \frac{1+\sqrt{1-x^2}}{x} \Rightarrow y = \log \frac{1+\sqrt{1-x^2}}{x}$$
$$\text{sech}^{-1} x = \log \frac{1+\sqrt{1-x^2}}{x}$$

Similarly, $\text{cosech}^{-1} x = \log \dfrac{1+\sqrt{1+x^2}}{x}$ **Proved.**

Example 65. *If* $x + iy = \cos(\alpha + i\beta)$ *or if* $\cos^{-1}(x + iy) = \alpha + i\beta$ *express* x *and* y *in terms of* α *and* β. *Hence show that* $\cos^2 \alpha$ *and* $\cosh^2 \beta$ *are the roots of the equation* $\lambda^2 - (x^2 + y^2 + 1)\lambda + x^2 = 0$.

Solution. Here, we have
$$\cos(\alpha + i\beta) = x + iy$$
$$\Rightarrow \cos\alpha \cos i\beta - \sin\alpha \sin i\beta = x + iy$$
$$\Rightarrow \cos\alpha \cosh\beta - i \sin\alpha \sinh\beta = x + iy$$

Equating real and imaginary parts, we get
$$\cos\alpha \cosh\beta = x \text{ and } \sin\alpha \sinh\beta = -y$$

We want to find the equation whose roots are $\cos^2\alpha$ and $\cosh^2\beta$.

Now,
$$\begin{aligned} x^2 + y^2 + 1 &= \cos^2\alpha \cos h^2\beta + \sin^2\alpha \sinh^2\beta + 1 \\ &= \cos^2\alpha \cosh^2\beta + (1 - \cos^2\alpha)(\cosh^2\beta - 1) + 1 \\ &= \cos^2\alpha \cosh^2\beta + \cosh^2\beta - 1 - \cos^2\alpha \cosh^2\beta + \cos^2\alpha + 1 \\ &= \cos^2\alpha + \cosh^2\beta \end{aligned}$$
$$\begin{aligned} \text{Sum of the roots} &= \cos^2\alpha + \cosh^2\beta \\ &= x^2 + y^2 + 1 \end{aligned}$$
$$\begin{aligned} \text{And product of the roots} &= \cos^2\alpha \cosh^2\beta \\ &= x^2 \end{aligned}$$

Hence, the equation whose roots are $\cos^2\alpha$, $\cosh^2\beta$ is
$\lambda^2 - (x^2 + y^2 + 1)\lambda + x^2 = 0$ **Proved.**

Example 66. *Separate into real and imaginary parts* $\cos^{-1}\left(\dfrac{3i}{4}\right)$

Solution. Let $\cos^{-1}\left(\dfrac{3i}{4}\right) = x + iy$
$$\Rightarrow \quad \frac{3i}{4} = \cos(x + iy)$$
$$\Rightarrow \quad \frac{3i}{4} = \cos x \cosh y - i \sin x \sinh y$$

Equating real and imaginary parts, we get
$$\therefore \quad \cos x \cosh y = 0 \Rightarrow \cos x = 0 \Rightarrow x = \frac{\pi}{2}$$

and $\qquad - \sin x \sinh y = \dfrac{3}{4}$

$$-1 \sinh y = \frac{3}{4} \qquad\qquad \sin x = \sin\left(\frac{\pi}{2}\right) = 1$$

$$\therefore \qquad \sinh y = -\frac{3}{4}$$

$$\Rightarrow \quad y = \log\left(\frac{-3}{4} + \sqrt{1+\frac{9}{16}}\right) \quad \Rightarrow \quad y = \log\left(\frac{-3}{4}+\frac{5}{4}\right) = -\log 2 = \log\left(\frac{1}{2}\right)$$

$\therefore$ Real part $= \dfrac{\pi}{2}$ and imaginary Part $= -\log 2$ **Ans.**

20.32 SOME OTHER INVERSE FUNCTIONS

Example 67. *Separate* $\tan^{-1}(\cos\theta + i\sin\theta)$ *into real and imaginary parts.* (*M.U. 2009*)

Solution. Let $\tan^{-1}(\cos\theta + i\sin\theta) = x + iy$

$$\Rightarrow \qquad \cos\theta + i\sin\theta = \tan(x+iy)$$

Similarly, $\qquad \cos\theta - i\sin\theta = \tan(x-iy)$

$$\tan 2x = \tan[(x+iy)+(x-iy)] = \frac{\tan(x+iy)+\tan(x-iy)}{1-\tan(x+iy)\tan(x-iy)}$$

$$= \frac{(\cos\theta+i\sin\theta)+(\cos\theta-i\sin\theta)}{1-(\cos\theta+i\sin\theta)(\cos\theta-i\sin\theta)} = \frac{2\cos\theta}{1-(\cos^2\theta+\sin^2\theta)}$$

$$= \frac{2\cos\theta}{1-1} = \frac{2\cos\theta}{0} = \infty = \tan\frac{\pi}{2}$$

$$\tan 2x = \tan\left(n\pi+\frac{\pi}{2}\right) \quad \Rightarrow \quad 2x = n\pi + \frac{\pi}{2} \quad \Rightarrow \quad x = \frac{n\pi}{2}+\frac{\pi}{4}$$

$$\text{Now, } \tan 2iy = \tan[(x+iy)-(x-iy)] = \frac{\tan(x+iy)-\tan(x-iy)}{1+\tan(x+iy)\tan(x-iy)}$$

$$= \frac{(\cos\theta+i\sin\theta)-(\cos\theta-i\sin\theta)}{1+(\cos\theta+i\sin\theta)(\cos\theta-i\sin\theta)} = \frac{2i\sin\theta}{1+(\cos^2\theta+\sin^2\theta)} = \frac{2i\sin\theta}{1+1} = i\sin\theta$$

$$i\tanh 2y = i\sin\theta \quad \Rightarrow \quad \frac{e^{2y}-e^{-2y}}{e^{2y}+e^{-2y}} = \frac{\sin\theta}{1}$$

By componendo and dividendo, we have

$$\frac{2e^{2y}}{2e^{-2y}} = \frac{1+\sin\theta}{1-\sin\theta} \quad \Rightarrow \quad e^{4y} = \frac{1+\cos\left(\frac{\pi}{2}-\theta\right)}{1-\cos\left(\frac{\pi}{2}-\theta\right)} = \frac{1+2\cos^2\left(\frac{\pi}{4}-\frac{\theta}{2}\right)-1}{1-\left[1-2\sin^2\left(\frac{\pi}{4}-\frac{\theta}{2}\right)\right]}$$

$$= \frac{\cos^2\left(\frac{\pi}{4}-\frac{\theta}{2}\right)}{\sin^2\left(\frac{\pi}{4}-\frac{\theta}{2}\right)} = \cot^2\left(\frac{\pi}{4}-\frac{\theta}{2}\right) \Rightarrow e^{2y} = \cot\left(\frac{\pi}{4}-\frac{\theta}{2}\right)$$

$$\Rightarrow \qquad 2y = \log\cot\left(\frac{\pi}{4}-\frac{\theta}{2}\right) \quad \Rightarrow \quad y = \frac{1}{2}\log\cot\left(\frac{\pi}{4}-\frac{\theta}{2}\right)$$

$$\text{Imaginary part} = \frac{1}{2}\log\cot\left(\frac{\pi}{4}-\frac{\theta}{2}\right)$$

$$\text{Real part} = \frac{n\pi}{2}+\frac{\pi}{4}$$

$$\tan^{-1}(\cos\theta + i\sin\theta) = \frac{n\pi}{2}+\frac{\pi}{4}+\frac{i}{2}\log\cot\left(\frac{\pi}{4}-\frac{\theta}{2}\right) \qquad \textbf{Ans.}$$

Example 68. *Separate* $\sin^{-1}(\alpha + i\beta)$ *into real and imaginary parts.*

Solution. Let $\sin^{-1}(\alpha + i\beta) = x + iy$

$$\alpha + i\beta = \sin(x + iy)$$

$$\Rightarrow \quad \alpha + i\beta = \sin x \cos iy + \cos x \sin iy$$

$$= \sin x \cosh y + i \cos x \sinh y$$

Equating real and imaginary parts, we have

$$\alpha = \sin x \cosh y \qquad ...(1)$$

and $$\beta = \cos x \sinh y \qquad ....(2)$$

We know that $\cosh^2 y - \sinh^2 y = 1$

$$\left(\frac{\alpha}{\sin x}\right)^2-\left(\frac{\beta}{\cos x}\right)^2 = 1 \qquad \left[\cosh y=\frac{\alpha}{\sin x},\ \sinh y=\frac{\beta}{\cos x}\right]$$

$$\Rightarrow \quad \alpha^2\cos^2 x - \beta^2\sin^2 x = \sin^2 x\cos^2 x$$

$$\Rightarrow \quad \alpha^2(1-\sin^2 x) - \beta^2\sin^2 x = \sin^2 x(1-\sin^2 x)$$

$$\Rightarrow \quad \sin^4 x - (\alpha^2+\beta^2+1)\sin^2 x + \alpha^2 = 0$$

This is quadratic equation in $\sin^2 x$.

$$\sin^2 x = \frac{(\alpha^2+\beta^2+1)\pm\sqrt{(\alpha^2+\beta^2+1)^2-4\alpha^2}}{2}$$

$$\Rightarrow \quad \sin x = \sqrt{\frac{1}{2}\left[(\alpha^2+\beta^2+1)\pm\sqrt{(\alpha^2+\beta^2+1)^2-4\alpha^2}\right]}$$

$$\Rightarrow \quad x = \sin^{-1}\sqrt{\frac{1}{2}\left[(\alpha^2+\beta^2+1)\pm\sqrt{(\alpha^2+\beta^2+1)^2-4\alpha^2}\right]}$$

We know that $\sin^2 x + \cos^2 x = 1$

$$\Rightarrow \quad \left(\frac{\alpha}{\cosh y}\right)^2+\left(\frac{\beta}{\sinh y}\right)^2 = 1 \qquad \left[\text{From (1) and (2)}\ \sin x=\frac{\alpha}{\cosh y},\ \cos x=\frac{\beta}{\sinh y}\right]$$

$$\Rightarrow \quad \alpha^2\sinh^2 y + \beta^2\cosh^2 y = \sinh^2 y\cosh^2 y$$

$$\Rightarrow \quad \alpha^2\sinh^2 y + \beta^2(1+\sinh^2 y) = \sinh^2 y(1+\sinh^2 y)$$

$$\Rightarrow \quad \sinh^4 y - (\alpha^2+\beta^2-1)\sinh^2 y - \beta^2 = 0$$

This is quadratic equation in $\sinh^2 y$.

$$\sinh^2 y = \frac{(\alpha^2+\beta^2-1)\pm\sqrt{(\alpha^2+\beta^2-1)^2+4\beta^2}}{2}$$

$$\Rightarrow \quad \sinh y = \sqrt{\frac{1}{2}\left[(\alpha^2+\beta^2-1)\pm\sqrt{(\alpha^2+\beta^2-1)^2+4\beta^2}\right]}$$

$$\Rightarrow \qquad y = \sinh^{-1}\sqrt{\frac{1}{2}\left[(\alpha^2+\beta^2-1)\pm\sqrt{(\alpha^2+\beta^2-1)^2+4\beta^2}\right]}$$

$$\text{Real part} = \sin^{-1}\sqrt{\frac{1}{2}\left[(\alpha^2+\beta^2+1)\pm\sqrt{(\alpha^2+\beta^2+1)^2-4\alpha^2}\right]}$$

$$\text{Imaginary part} = \sinh^{-1}\sqrt{\frac{1}{2}\left[(\alpha^2+\beta^2-1)\pm\sqrt{(\alpha^2+\beta^2-1)^2+4\beta^2}\right]}$$ **Ans.**

Example 69. *Separate $\tan^{-1}(a + i\,b)$ into real and imaginary parts.*

(Nagpur University, Summer 2008)

Solution. Let $\tan^{-1}(a + i\,b) = x + i\,y$ ...(1)

$\therefore \qquad \tan(x + i\,y) = a + i\,b$

On both sides for i write $-i$ we get,

$\therefore \qquad \tan(x - i\,y) = a - i\,b$

Now, $\qquad \tan 2x = \tan[(x + i\,y) + (x - i\,y)]$

$$= \frac{\tan(x+i\,y)+\tan(x-i\,y)}{1-\tan(x+i\,y)\tan(x-i\,y)} = \frac{a+i\,b+a-i\,b}{1-(a+i\,b)(a-i\,b)} = \frac{2a}{1-a^2-b^2}$$

$$2x = \tan^{-1}\left[\frac{2a}{1-a^2-b^2}\right] \Rightarrow x = \frac{1}{2}\tan^{-1}\left[\frac{2a}{1-a^2-b^2}\right] \qquad ...(2)$$

and $\qquad \tan(2\,y\,i) = \tan[(x + i\,y) - (x - i\,y)]$

$$= \frac{\tan(x+i\,y)-\tan(x-i\,y)}{1+\tan(x+i\,y)\tan(x-i\,y)} = \frac{a+bi-a+bi}{1+(a+bi)(a-bi)}$$

$$i\tanh 2y = \frac{2bi}{1+a^2+b^2} \quad \text{so, } \tanh 2y = \frac{2b}{1+a^2+b^2}$$

$$2y = \tanh^{-1}\left[\frac{2b}{1+a^2+b^2}\right]$$

$$\text{so } y = \frac{1}{2}\tanh^{-1}\left[\frac{2b}{1+a^2+b^2}\right] \qquad ...(3)$$

From (1), (2) and (3), we have

$$\tan^{-1}(a+ib) = \frac{1}{2}\tan^{-1}\left[\frac{2a}{1-a^2-b^2}\right] + \frac{i}{2}\tanh^{-1}\left[\frac{2b}{1+a^2+b^2}\right]$$ **Ans.**

Example 70. *Show that* $\tan^{-1} i\left(\frac{x-a}{x+a}\right) = \frac{i}{2}\log\left(\frac{x}{a}\right)$. *(M.U. 2006)*

Solution. Let $\qquad \tan^{-1} i\left(\frac{x-a}{x+a}\right) = u + i\,v$ ...(1)

$$\Rightarrow \qquad \tan(u+iv) = i\left(\frac{x-a}{x+a}\right) \quad \text{and} \quad \tan(u-iv) = -i\left(\frac{x-a}{x+a}\right)$$

$$\tan 2u = \tan[(u+iv)+(u-iv)] = \frac{\tan(u+iv)+\tan(u-iv)}{1-\tan(u+iv)\tan(u-iv)}$$

$$= \frac{ix-ia-ix+ia}{x+a} = 0$$

$\therefore \qquad \tan 2u = 0 \Rightarrow 2u = 0 \Rightarrow u = 0$

Putting the value of u in (1), we get

$$\therefore \quad \tan^{-1} i\left(\frac{x-a}{x+a}\right) = i\,v \quad \therefore \quad i\left(\frac{x-a}{x+a}\right) = \tan i\,v = i\tanh v$$

$$\therefore \quad \frac{x-a}{x+a} = \tanh v = \frac{e^{v}-e^{-v}}{e^{v}+e^{-v}}$$

By Componendo and dividendo, we get

$$\frac{2x}{2a} = \frac{2e^{v}}{2e^{-v}} \Rightarrow \frac{x}{a} = e^{2v} \Rightarrow v = \frac{1}{2}\log\left(\frac{x}{a}\right)$$

$$\therefore \tan^{-1} i\left(\frac{x-a}{x+a}\right) = u + iv = 0 + \frac{i}{2}\log\frac{x}{a} = \frac{i}{2}\log\left(\frac{x}{a}\right)$$ **Proved**

Example 71. *Prove that*

(*i*) $\cosh^{-1}\sqrt{1+x^2} = \sinh^{-1} x$ (*M.U. 2007*)

(*ii*) $\cosh^{-1}\sqrt{1+x^2} = \tanh^{-1}\left(\frac{x}{\sqrt{1+x^2}}\right)$

Solution. (*i*) Let $\cosh^{-1}\sqrt{1+x^2} = y$ ...(1)

$$\Rightarrow \quad \sqrt{1+x^2} = \cosh y \quad \text{...(2)}$$

On squaring both sides, we get

$$1 + x^2 = \cosh^2 y$$

$$\therefore \quad x^2 = \cosh^2 y - 1$$

$$\Rightarrow \quad x^2 = \sinh^2 y$$

$$\Rightarrow \quad x = \sinh y \quad \text{...(3)}$$

$$\Rightarrow \quad y = \sinh^{-1} x$$

$$\Rightarrow \quad \cosh^{-1}\sqrt{1+x^2} = \sinh^{-1} x$$ [Using (1)] **Proved**

(*ii*) Dividing (3) by (2), we get

$$\frac{\sinh y}{\cosh y} = \frac{x}{\sqrt{1+x^2}}$$

$$\Rightarrow \quad \tanh y = \frac{x}{\sqrt{1+x^2}} \Rightarrow y = \tanh^{-1}\left(\frac{x}{\sqrt{1+x^2}}\right)$$

$$\Rightarrow \quad \cosh^{-1}\sqrt{1+x^2} = \tanh^{-1}\left(\frac{x}{\sqrt{1+x^2}}\right)$$ [Using (1)] **Proved**

EXERCISE 20.11

1. Prove that $\sin^{-1}(\operatorname{cosec}\theta) = \frac{\pi}{2} + i\log\cot\frac{\theta}{2}$.

2. If $\tan(\alpha + i\beta) = x + iy$, prove that

(*a*) $x^2 + y^2 + 2x\cot 2\alpha = 1$

(*b*) $x^2 + y^2 - 2y\coth 2\beta = -1$.

3. If $\tan(\theta + i\phi) = \sin(x + iy)$, then prove that $\coth y \sinh 2\phi = \cot x \sin 2\theta$.

4. If $\sin^{-1}(\cos\theta + i\sin\theta) = x + iy$, show that.

(*a*) $x = \cos^{-1}\sqrt{\sin\theta}$

(*b*) $y = \log[\sqrt{\sin\theta} + \sqrt{1+\sin\theta}]$.

5. Separate into real and imaginary parts $\sin^{-1}(e^{i\theta})$ **Ans.** $\cos^{-1}\sqrt{\sin\theta}+i\log[\sqrt{\sin\theta}+\sqrt{1+\sin\theta}]$

6. Prove that

$$\tan^{-1}\left(\frac{\tan 2\theta+\tan 2\phi}{\tan 2\theta-\tan 2\phi}\right)+\tan^{-1}\left(\frac{\tan\theta-\tan\phi}{\tan\theta+\tan\phi}\right)=\tan^{-1}(\cot\theta\coth\phi)$$

7. Prove that $\tanh^{-1} x = \sinh^{-1}\dfrac{x}{\sqrt{1-x^2}}$.

8. Prove that $\tanh^{-1}(\sin\theta) = \cosh^{-1}(\sec\theta)$

9. Prove that

$$\cosh^{-1}\left(\frac{b+a\cos x}{a+b\cos x}\right)=\log\left[\frac{\sqrt{b+a}+\sqrt{b-a}\tan\dfrac{x}{2}}{\sqrt{b+a}-\sqrt{b-a}\tan\dfrac{x}{a}}\right]$$

10. Prove that $\tan^{-1}(e^{i\theta}) = \dfrac{n\pi}{2}+\dfrac{\pi}{4} = \log\tan\left(\dfrac{\pi}{4}-\dfrac{\theta}{2}\right)$

11. If $\cosh^{-1}(x+iy)+\cosh^{-1}(x-iy)=\cosh^{-1} a$, prove that $2(a-1)x^2+2(a+1)y^2=a^2-1$.

12. Prove that : $\tanh^{-1}\cos\theta = \cosh^{-1}\operatorname{cosec}\theta$

13. Prove that : $\sinh^{-1}\tan\theta = \log(\sec\theta+\tan\theta)$

14. Prove that : $\sinh^{-1}\tan\theta = \log\tan\left(\dfrac{\theta}{2}+\dfrac{\pi}{4}\right)$

Separate into real and imaginary parts

15. $\cos^{-1} e^{i\theta}$ or $\cos^{-1}(\cos\theta+i\sin\theta)$ **Ans.** $\sin^{-1}\sqrt{\sin\theta}+i\log\left(\sqrt{1+\sin\theta}-\sqrt{\sin\theta}\right)$

16. If $\sinh^{-1}(x+iy)+\sinh^{-1}(x-iy)=\sinh^{-1} a$, prove that

$$2(x^2+y^2)\sqrt{a^2+1} = a^2-2x^2-2y^2.$$

CHAPTER

EXPANSION OF TRIGONOMETRIC FUNCTIONS

21.1 EXPANSION OF sin $n\theta$, cos $n\theta$ IN POWERS OF sin θ, cos θ

By De-Moivre's theorem, we know that

$$\cos n\,\theta + i\sin n\theta = (\cos\theta + i\sin\theta)^n. \qquad \text{(Binomial Theorem)}$$

$$\begin{aligned}
&= {}^nC_0(\cos\theta)^n + {}^nC_1(\cos\theta)^{n-1}(i\sin\theta) + {}^nC_2(\cos\theta)^{n-2}(i\sin\theta)^2\\
&\quad + {}^nC_3(\cos\theta)^{n-3}(i\sin\theta)^3 + {}^nC_4(\cos\theta)^{n-4}(i\sin\theta)^4\\
&\quad + {}^nC_5(\cos\theta)^{n-5}(i\sin\theta)^5 + {}^nC_6(\cos\theta)^{n-6}(i\sin\theta)^6\\
&\quad + {}^nC_7(\cos\theta)^{n-7}(i\sin\theta)^7 + \ldots + {}^nC_n(\cos\theta)^{n-n}(i\sin\theta)^n\\
&= \cos^n\theta + i\,{}^nC_1\cos^{n-1}\theta\sin\theta - {}^nC_2\cos^{n-2}\theta\sin^2\theta\\
&\quad - i\,{}^nC_3\cos^{n-3}\theta\sin^3\theta + {}^nC_4\cos^{n-4}\theta\sin^4\theta + i\,{}^nC_5\cos^{n-5}\theta\sin^5\theta\\
&\quad - {}^nC_6\cos^{n-6}\theta\sin^6\theta - i\,{}^nC_7\cos^{n-7}\theta\sin^7\theta + \ldots + (i\sin\theta)^n\\
&= \cos^n\theta - {}^nC_2\cos^{n-2}\theta\sin^2\theta + {}^nC_4\cos^{n-4}\theta\sin^4\theta\\
&\quad - {}^nC_6\cos^{n-6}\theta\sin^6\theta + \ldots + i\,[{}^nC_1\cos^{n-1}\theta\sin\theta\\
&\quad - {}^nC_3\cos^{n-3}\theta\sin^3\theta + {}^nC_5\cos^{n-5}\theta\sin^5\theta - {}^nC_7\cos^{n-7}\theta\sin^7\theta\ldots]
\end{aligned}$$

Equating real and imaginary parts, we get

$$\boxed{\cos n\,\theta = \cos^n\theta - {}^nC_2\cos^{n-2}\theta\sin^2\theta + {}^nC_4\cos^{n-4}\theta\sin^4\theta + \ldots} \qquad \ldots(1)$$

$$\boxed{\sin n\,\theta = {}^nC_1\cos^{n-1}\theta\sin\theta - {}^nC_3\cos^{n-3}\theta\sin^3\theta + {}^nC_5\cos^{n-5}\theta\sin^5\theta + \ldots} \qquad \ldots(2)$$

Replacing every $\sin^2\theta$ by $(1-\cos^2\theta)$ in (1) and every $\cos^2\theta$ by $(1-\sin^2\theta)$ in (2), we get the expansions of $\cos n\,\theta$ in powers of $\cos\theta$ and $\sin n\,\theta$ in powers of $\sin\theta$.

Dividing (2) by (1), we get

$$\frac{\sin n\,\theta}{\cos n\,\theta} = \frac{{}^nC_1\cos^{n-1}\theta\sin\theta - {}^nC_3\cos^{n-3}\theta\sin^3\theta + {}^nC_5\cos^{n-5}\theta\sin^5\theta\ldots}{\cos^n\theta - {}^nC_2\cos^{n-2}\theta\sin^2\theta + {}^nC_4\cos^{n-4}\theta\sin^4\theta - \ldots}$$

and dividing numerator and denominator by $\cos^n\theta$, we get

$$\boxed{\tan n\,\theta = \frac{{}^nC_1\tan\theta - {}^nC_3\tan^3\theta + {}^nC_5\tan^5\theta\ldots}{1 - {}^nC_2\tan^2\theta + {}^nC_4\tan^4\theta\ldots}}$$

Example 1. *Expand cos 6 θ and sin 6 θ in terms of cos θ and sin θ.*

Solution. $\cos 6\,\theta + i\sin 6\,\theta = (\cos\theta + i\sin\theta)^6$

Expansion by Binomial theorem

$$\cos 6\theta + i \sin 6\theta = \cos^6\theta + {}^6C_1 \cos^5\theta\,(i\sin\theta) + {}^6C_2 \cos^4\theta\,(i\sin\theta)^2$$
$$+ {}^6C_3 \cos^3\theta\,(i\sin\theta)^3 + {}^6C_4 \cos^2\theta\,(i\sin\theta)^4 + {}^6C_5 \cos\theta\,(i\sin\theta)^5 + {}^6C_6\,(i\sin\theta)^6$$
$$= \cos^6\theta + i\,6\cos^5\theta\sin\theta - 15\cos^4\theta\sin^2\theta - i.20\cos^3\theta\sin^3\theta + 15\cos^2\theta\sin^4\theta$$
$$+ i\,6\cos\theta\sin^5\theta - \sin^6\theta$$

Equating real and imaginary parts, we have

$$\cos 6\theta = \cos^6\theta - 15\cos^4\theta\sin^2\theta + 15\cos^2\theta\sin^4\theta - \sin^6\theta$$
$$\sin 6\theta = 6\cos^5\theta\sin\theta - 20\cos^3\theta\sin^3\theta + 6\cos\theta\sin^5\theta$$

Ans.

Example 2. *Prove that* $\sin 7\theta = 7\sin\theta - 56\sin^3\theta + 112\sin^5\theta - 64\sin^7\theta$.

Solution. $\cos 7\theta + i\sin 7\theta = (\cos\theta + i\sin\theta)^7$

$$= \cos^7\theta + {}^7C_1\cos^6\theta\,(i\sin\theta) + {}^7C_2\cos^5\theta\,(i\sin\theta)^2 + {}^7C_3\cos^4\theta\,(i\sin\theta)^3 +$$
$${}^7C_4\cos^3\theta\,(i\sin\theta)^4 + {}^7C_5\cos^2\theta\,(i\sin\theta)^5 + {}^7C_6\cos\theta\,(i\sin\theta)^6 + {}^7C_7\,(i\sin\theta)^7$$

Equating imaginary parts, we get

$$\sin 7\theta = 7\cos^6\theta\sin\theta - 35\cos^4\theta\sin^3\theta + 21\cos^2\theta\sin^5\theta - \sin^7\theta$$
$$= 7(1-\sin^2\theta)^3\sin\theta - 35(1-\sin^2\theta)^2\sin^3\theta + 21(1-\sin^2\theta)\sin^5\theta - \sin^7\theta$$
$$= 7(1 - 3\sin^2\theta + 3\sin^4\theta - \sin^6\theta)\sin\theta - 35(1 - 2\sin^2\theta + \sin^4\theta)\sin^3\theta$$
$$+ 21\sin^5\theta - 21\sin^7\theta - \sin^7\theta$$
$$= 7\sin\theta - 21\sin^3\theta + 21\sin^5\theta - 7\sin^7\theta - 35\sin^3\theta$$
$$+ 70\sin^5\theta - 35\sin^7\theta + 21\sin^5\theta - 22\sin^7\theta$$
$$= 7\sin\theta - 56\sin^3\theta + 112\sin^5\theta - 64\sin^7\theta.$$

Proved.

Example 3. *Expand* $\tan 9\theta$ *in powers of* $\tan\theta$.

Solution. We know,

$$\tan n\theta = \frac{{}^nC_1\tan\theta - {}^nC_3\tan^3\theta + {}^nC_5\tan^5\theta + \ldots}{1 - {}^nC_2\tan^2\theta + {}^nC_4\tan^4\theta - {}^nC_6\tan^6\theta + \ldots}$$

$$\tan 9\theta = \frac{{}^9C_1\tan\theta - {}^9C_3\tan^3\theta + {}^9C_5\tan^5\theta - {}^9C_7\tan^7\theta + \tan^9\theta}{1 - {}^9C_2\tan^2\theta + {}^9C_4\tan^4\theta - {}^9C_6\tan^6\theta + {}^9C_8\tan^8\theta}$$

$$= \frac{9\tan\theta - 84\tan^3\theta + 126\tan^5\theta - 36\tan^7\theta + \tan^9\theta}{1 - 36\tan^2\theta + 126\tan^4\theta - 84\tan^6\theta + 9\tan^8\theta}$$

Ans.

EXERCISE 21.1

Prove that:

1. $\cos 4\theta = \cos^4\theta - 6\cos^2\theta\sin^2\theta + \sin^4\theta$.
2. $\sin 4\theta = 4\cos^3\theta\sin\theta - 4\cos\theta\sin^3\theta$.
3. $\dfrac{\sin 6\theta}{\cos\theta} = 32\sin^5\theta - 32\sin^3\theta + 6\sin\theta$.
4. $\sin 7\theta = 7\cos^6\theta\sin\theta - 35\cos^4\theta\sin^3\theta + 21\cos^2\theta\sin^5\theta - \sin^7\theta$.
5. $\dfrac{\sin 7\theta}{\sin\theta} = 7 - 56\sin^2\theta + 112\sin^4\theta - 64\sin^6\theta$.
6. $\cos 8\theta = \cos^8\theta - 28\cos^6\theta\sin^2\theta + 70\cos^4\theta\sin^4\theta - 28\cos^2\theta\sin^6\theta + \sin^8\theta$.
7. $\sin 10\theta = 10\cos^9\theta\sin\theta - 120\cos^7\theta\sin^3\theta + 210\cos^5\theta\sin^5\theta - 120\cos^3\theta\sin^7\theta$
$+10\cos\theta\sin^9\theta$

8. $1 + \cos 10\theta = 2(16\cos^5\theta - 20\cos^3\theta + 5\cos\theta)^2$.

9. $1 - \cos 10\theta = 2(16\sin^5\theta - 20\sin^3\theta + 5\sin\theta)^2$

10. $\tan 5\theta = \dfrac{5\tan\theta - 10\tan^3\theta + \tan^5\theta}{1 - 10\tan^2\theta + 5\tan^4\theta}$

21.2 EXPANSION OF $\cos^n\theta \sin^n\theta$ IN TERMS OF SINES AND COSINES OF MULTIPLES OF θ

Method:

Let $x = \cos\theta + i\sin\theta$, then $\dfrac{1}{x} = \cos\theta - i\sin\theta$

$$x + \frac{1}{x} = 2\cos\theta \text{ and } x - \frac{1}{x} = 2i\sin\theta$$

Again $\quad x^n = \cos n\theta + i\sin n\theta$, and

$$\frac{1}{x^n} = \cos n\theta - i\sin n\theta$$

$$x^n + \frac{1}{x^n} = 2\cos n\theta \text{ and } x^n - \frac{1}{x^n} = 2i\sin n\theta$$

To expand $\cos^n\theta$: Start from $(2\cos\theta)^n = \left(x + \dfrac{1}{x}\right)^n$

Expand R.H.S. and substitute the values of $\left(x + \dfrac{1}{x}\right), \left(x^2 + \dfrac{1}{x^2}\right)$ etc.

To expand $\sin^n\theta$: Start from $(2i\sin\theta)^n = \left(x - \dfrac{1}{x}\right)^n$

Expand R.H.S. and substitute the values of $\left(x - \dfrac{1}{x}\right), \left(x^2 - \dfrac{1}{x^2}\right)$ etc.

Example 4. *Express $\sin^5\theta$ in terms of sines of multiples of θ.*

Solution. $(2i\sin\theta)^5 = \left(x - \dfrac{1}{x}\right)^5$ $\quad \left[\begin{array}{l} x = \cos\theta + i\sin\theta \\ \dfrac{1}{x} = \cos\theta - i\sin\theta \end{array}\right]$

$$i\,32\sin^5\theta = x^5 + 5x^4\left(-\frac{1}{x}\right) + 10x^3\left(-\frac{1}{x}\right)^2 + 10x^2\left(-\frac{1}{x}\right)^3 + 5x\left(-\frac{1}{x}\right)^4 + \left(-\frac{1}{x}\right)^5$$

$$= \left(x^5 - \frac{1}{x^5}\right) - 5\left(x^3 - \frac{1}{x^3}\right) + 10\left(x - \frac{1}{x}\right)$$

$$= 2i\sin 5\theta - 5(2i\sin 3\theta) + 10(2i\sin\theta)$$

$$16\sin^5\theta = \sin 5\theta - 5\sin 3\theta + 10\sin\theta$$

$$\sin^5\theta = \frac{1}{16}(\sin 5\theta - 5\sin 3\theta + 10\sin\theta)$$ **Ans.**

Example 5. *Prove that*

$$-2^{12}\cos^6\theta\sin^7\theta = \sin 13\theta - \sin 11\theta - 6\sin 9\theta + 6\sin 7\theta + 15\sin 5\theta - 15\sin 3\theta - 20\sin\theta$$

Solution. $x^n = (\cos\theta + i\sin\theta)^n = \cos n\theta + i\sin n\theta$ $\left[\begin{array}{l} x = \cos\theta + i\sin\theta \\ \dfrac{1}{x} = \cos\theta - i\sin\theta \end{array}\right]$

$$\frac{1}{x^n} = (\cos\theta + i\sin\theta)^{-n} = \cos n\theta - i\sin n\theta$$

$$x^n + \frac{1}{x^n} = 2\cos n\theta \text{ and } x^n - \frac{1}{x^n} = 2i\sin n\theta \qquad \left[\begin{array}{l} x+\dfrac{1}{x} = 2\cos\theta \\ x-\dfrac{1}{x} = 2i\sin\theta \end{array}\right]$$

$$(2\cos\theta)^6(2i\sin\theta)^7 = \left(x+\frac{1}{x}\right)^6\left(x-\frac{1}{x}\right)^7 = \left(x^2-\frac{1}{x^2}\right)^6\left(x-\frac{1}{x}\right)$$

$$= \left[x^{12} + 6x^{10}\left(-\frac{1}{x^2}\right) + 15x^8\left(-\frac{1}{x^2}\right)^2 + 20x^6\left(-\frac{1}{x^2}\right)^3 + 15x^4\left(-\frac{1}{x^2}\right)^4 + 6x^2\left(-\frac{1}{x^2}\right)^5 + \left(-\frac{1}{x^2}\right)^6\right]\left(x-\frac{1}{x}\right)$$

$$= \left[x^{12} - 6x^8 + 15x^4 - 20 + \frac{15}{x^4} - \frac{6}{x^8} + \frac{1}{x^{12}}\right]\left[x - \frac{1}{x}\right]$$

$$= x^{13} - 6x^9 + 15x^5 - 20x + \frac{15}{x^3} - \frac{6}{x^7} + \frac{1}{x^{11}} - x^{11} + 6x^7 - 15x^3 + \frac{20}{x} - \frac{15}{x^5} + \frac{6}{x^9} - \frac{1}{x^{13}}$$

$$= \left(x^{13} - \frac{1}{x^{13}}\right) - \left(x^{11} - \frac{1}{x^{11}}\right) - 6\left(x^9 - \frac{1}{x^9}\right) + 6\left(x^7 - \frac{1}{x^7}\right) + 15\left(x^5 - \frac{1}{x^5}\right) - 15\left(x^3 - \frac{1}{x^3}\right) - 20\left(x - \frac{1}{x}\right)$$

$$= 2i\sin 13\theta - 2i\sin 11\theta - 6(2i\sin 9\theta) + 6(2i\sin 7\theta) + 15(2i\sin 5\theta) - 15(2i\sin 3\theta) - 20(2i\sin\theta)$$

$$-2^{12}\cos^6\theta\sin^7\theta = \sin 13\theta - \sin 11\theta - 6\sin 9\theta + 6\sin 7\theta + 15\sin 5\theta - 15\sin 3\theta - 20\sin\theta$$ **Proved.**

EXERCISE 21.2

1. Express $\sin^7\theta$ as a sum of sines of multiples of θ.
   Ans. $-\dfrac{1}{64}[\sin 7\theta - 7\sin 5\theta + 12\sin 3\theta - 35\sin\theta]$
2. Express $\cos^8\theta$ as a sum of cosines of multiples of θ.
   Ans. $\dfrac{1}{128}[\cos 8\theta + 8\cos 6\theta + 28\cos 4\theta + 56\cos 2\theta + 35]$
3. Prove that $2^7\cos^3\theta\sin^5\theta = \sin 8\theta - 2\sin 6\theta - 2\sin 4\theta + 6\sin 2\theta$.
4. Prove that $32\cos^6\theta = \cos 6\theta + 6\cos 4\theta + 15\cos 2\theta + 10$
5. Prove that $\sin^8\theta = 2^{-7}(\cos 8\theta - 8\cos 6\theta + 28\cos 4\theta - 56\cos 2\theta + 35)$
6. $32\sin^4\theta\cos^2\theta = \cos 6\theta - 2\cos 4\theta - \cos 2\theta + 2$.
7. $\sin^5\theta\cos^2\theta = \dfrac{1}{64}(\sin 7\theta - 3\sin 5\theta + \sin 3\theta + 5\sin\theta)$
8. Expand $\cos^5\theta\sin^7\theta$ in a series of sines and of multiples of θ.
   Ans. $-2^{-11}(\sin 12\theta - 2\sin 10\theta - 4\sin 8\theta + 10\sin 6\theta + 5\sin 4\theta - 20\sin 2\theta)$

21.3 SUMMATION OF SINES AND COSINES SERIES: 'C + *i*S' METHOD

Sum of cosines series is denoted by C and sum of sines series is denoted by S.

(*i*) Suppose we have to find C, the sum of cosines series. Then write a similar series of sines, S.

(*ii*) Multiply the sines series by i and add to series of cosines, we will get $C + iS$ as an exponential series. The sum of exponential series is calculated by using any one of the following series.

(*a*) **Geometric Series :** $a + ar + ar^2 + ... + ar^{n-1} = \dfrac{a(1-r^n)}{1-r}$

$$a + ar + ar^2 + ... \infty = \frac{a}{1-r}$$

(*b*) **Binomial series :** $1 + nx + \dfrac{n(n-1)}{2!}x^2 + ... \infty = (1+x)^n$

(*c*) **Exponential series :** $1 + x + \dfrac{x^2}{2!} + \dfrac{x^3}{3!} + ... \infty = e^x$

(*d*) **sin, cos, sinh or cosh series :**

$$x - \frac{x^3}{3!} + \frac{x^5}{5!} - ... \infty = \sin x$$

and

$$1 - \frac{x^2}{2!} + \frac{x^4}{4!} + ... \infty = \cos x$$

$$x + \frac{x^3}{3!} + \frac{x^5}{5!} + ... \infty = \sinh x$$

and

$$1 + \frac{x^2}{2!} + \frac{x^4}{4!} + ... \infty = \cosh x$$

(*e*) **Logarithmic series :** $x - \dfrac{x^2}{2} + \dfrac{x^3}{3} - ... \infty = \log(1+x)$

$$-x - \frac{x^2}{2} - \frac{x^3}{3} ... \infty = \log(1-x)$$

(*f*) **Gregory's series :** $x - \dfrac{x^3}{3} + \dfrac{x^5}{5} - ... \infty = \tan^{-1} x$

$$x + \frac{x^3}{3} + \frac{x^5}{5} + ... \infty = \tanh^{-1} x = \frac{1}{2}\log\frac{1+x}{1-x}$$

Example 6. *Sum the series*

$\cos\alpha + x\cos(\alpha+\beta) + \dfrac{x^2}{2!}\cos(\alpha+2\beta) + ... \infty$ (*Nagpur University Summer 2005*)

Solution. Let

$$C = \cos\alpha + x\cos(\alpha+\beta) + \frac{x^2}{2!}\cos(\alpha+2\beta) + ... \infty$$

and

$$S = \sin\alpha + x\sin(\alpha+\beta) + \frac{x^2}{2!}\sin(\alpha+2\beta) + ... \infty$$

$\therefore$

$$C + iS = (\cos\alpha + i\sin\alpha) + x[\cos(\alpha+\beta) + i\sin(\alpha+\beta)] + \frac{x^2}{2!}[\cos(\alpha+2\beta) + i\sin(\alpha+2\beta)] + ... \infty$$

$$= e^{i\alpha} + xe^{i(\alpha+\beta)} + \frac{x^2}{2!}\, e^{i(\alpha+2\beta)} + \ldots \infty$$

$$= e^{i\alpha}\left[1 + x\,e^{i\beta} + \frac{x^2 e^{i2\beta}}{2!} + \ldots \infty\right]$$

This is an exponential series.

$$= e^{i\alpha}\left(e^{x.e^{i\beta}}\right) = e^{i\alpha}\,.\,e^{x(\cos\beta + i\sin\beta)} = e^{x\cos\beta}\,.\,e^{i(\alpha + x\sin\beta)}$$

$$= e^{x\cos\beta}\,[\cos(\alpha + x\sin\beta) + i\sin(\alpha + x\sin\beta)]$$

Equating real parts, we have

$$C = e^{x\cos\beta}\,.\,\cos(\alpha + x\sin\beta)$$ **Ans.**

Example 7. *Sum the series*

$$\sin\alpha + \sin(\alpha+\beta) + \sin(\alpha+2\beta) \ldots + \sin(\alpha + \overline{n-1}\,\beta)$$

Solution. Let $S = \sin\alpha + \sin(\alpha+\beta) + \sin(\alpha+2\beta) + \ldots + \sin(\alpha + \overline{n-1}\,\beta)$

and $C = \cos\alpha + \cos(\alpha+\beta) + \cos(\alpha+2\beta) + \ldots + \cos(\alpha + \overline{n-1}\,\beta)$

$$\therefore\quad C + iS = [\cos\alpha + i\sin\alpha] + [\cos(\alpha+\beta) + i\sin(\alpha+\beta)] + [\cos(\alpha+2\beta) + i\sin(\alpha+2\beta)] + \ldots + [\cos(\alpha + \overline{n-1}\,\beta) + i\sin(\alpha + \overline{n-1}\,\beta)]$$

$$= e^{i\alpha} + e^{i(\alpha+\beta)} + e^{i(\alpha+2\beta)} + \ldots + e^{i(\alpha + \overline{n-1}\beta)}$$

$$= e^{i\alpha}\,[1 + e^{i\beta} + e^{i2\beta} + \ldots + e^{i\overline{n-1}\beta}]$$

This is a geometric series.

$$= e^{i\alpha}\,\frac{1-(e^{i\beta})^n}{1-e^{i\beta}} = e^{i\alpha}.\frac{1-e^{in\beta}}{1-e^{i\beta}} = e^{i\alpha}\,\frac{1-\cos n\beta - i\sin n\beta}{1-\cos\beta - i\sin\beta}$$

$$= e^{i\alpha}\,\frac{2\sin^2\frac{n\beta}{2} - 2i\sin\frac{n\beta}{2}\cos\frac{n\beta}{2}}{2\sin^2\frac{\beta}{2} - 2i\sin\frac{\beta}{2}\cos\frac{\beta}{2}}$$

$$= e^{i\alpha}\,\frac{\sin\frac{n\beta}{2}}{\sin\frac{\beta}{2}} \times \frac{\sin\frac{n\beta}{2} - i\cos\frac{n\beta}{2}}{\sin\frac{\beta}{2} - i\cos\frac{\beta}{2}} = e^{i\alpha}\,\frac{\cos\frac{n\beta}{2} + i\sin\frac{n\beta}{2}}{\cos\frac{\beta}{2} + i\sin\frac{\beta}{2}} \times \frac{\sin\frac{n\beta}{2}}{\sin\frac{\beta}{2}}$$

$$= \frac{e^{i\alpha}\,.\,e^{i\frac{n\beta}{2}}}{e^{i\frac{\beta}{2}}}\sin\frac{n\beta}{2}\,\text{cosec}\,\frac{\beta}{2} = e^{i\left[\alpha + (n-1)\frac{\beta}{2}\right]}\sin\frac{n\beta}{2}\,\text{cosec}\,\frac{\beta}{2}$$

$$= \left[\cos\left\{\alpha + (n-1)\frac{\beta}{2}\right\} + i\sin\left\{\alpha + (n-1)\frac{\beta}{2}\right\}\right]\sin\frac{n\beta}{2}\,\text{cosec}\,\frac{\beta}{2}$$

Equating imaginary parts, we get $S = \sin\left\{\alpha + (n-1)\frac{\beta}{2}\right\}\sin\frac{n\beta}{2}\,\text{cosec}\,\frac{\beta}{2}$ **Ans.**

Example 8. *Sum the series*

$$n\sin\alpha + \frac{n(n+1)}{1.2}\sin 2\alpha + \frac{n(n+1)(n+2)}{1.2.3}\sin 3\alpha + \ldots \infty$$

Solution.

Let $$S = n\sin\alpha + \frac{n(n+1)}{1.2}\sin 2\alpha + \frac{n(n+1)(n+2)}{1.2.3}\sin 3\alpha + \dots\infty$$

and $$C = n\cos\alpha + \frac{n(n+1)}{1.2}\cos 2\alpha + \frac{n(n+1)(n+2)}{1.2.3}\cos 3\alpha + \dots\infty$$

$$\therefore \quad C + iS = n(\cos\alpha + i\sin\alpha) + \frac{n(n+1)}{1.2}(\cos 2\alpha + i\sin 2\alpha) + \frac{n(n+1)(n+2)}{1.2.3}(\cos 3\alpha + i\sin 3\alpha) + \dots\infty$$

$$= n\,e^{i\alpha} + \frac{n(n+1)}{1.2}e^{i2\alpha} + \frac{n(n+1)(n+2)}{1.2.3}e^{i3\alpha} + \dots\infty$$

$$= -1 + 1 + (-n)(-e^{i\alpha}) + \frac{-n(-n-1)}{1.2}(-e^{i\alpha})^2 + \frac{-n(-n-1)(-n-2)}{1.2.3}(-e^{i\alpha})^3 + \dots\infty$$

This is a binomial series.

$$= -1 + [1 - e^{i\alpha}]^{-n} = -1 + (1 - \cos\alpha - i\sin\alpha)^{-n}$$

$$= -1 + \left[2\sin^2\frac{\alpha}{2} - i2\sin\frac{\alpha}{2}\cos\frac{\alpha}{2}\right]^{-n} = -1 + \left(2\sin\frac{\alpha}{2}\right)^{-n}\left[\sin\frac{\alpha}{2} - i\cos\frac{\alpha}{2}\right]^{-n}$$

$$= -1 + \left(2\sin\frac{\alpha}{2}\right)^{-n}\left[\cos\left(\frac{\pi}{2} - \frac{\alpha}{2}\right) - i\sin\left(\frac{\pi}{2} - \frac{\alpha}{2}\right)\right]^{-n}$$

$$= -1 + \frac{\operatorname{cosec}^n \frac{\alpha}{2}}{2^n}\left[\cos n\left(\frac{\pi}{2} - \frac{\alpha}{2}\right) + i\sin n\left(\frac{\pi}{2} - \frac{\alpha}{2}\right)\right]$$

Equating imaginary parts, we get $S = \dfrac{\operatorname{cosec}^n \frac{\alpha}{2}}{2^n}\sin\dfrac{n}{2}(\pi - \alpha)$ **Ans.**

Example 9. *Sum the series*

$$\frac{\sin\theta}{1!} - \frac{\sin 2\theta}{2!} + \frac{\sin 3\theta}{3!} - \frac{\sin 4\theta}{4!} + \dots\infty$$

Solution. Let $$S = \frac{\sin\theta}{1!} - \frac{\sin 2\theta}{2!} + \frac{\sin 3\theta}{3!} - \frac{\sin 4\theta}{4!} + \dots\infty$$

and $$C = \frac{\cos\theta}{1!} - \frac{\cos 2\theta}{2!} + \frac{\cos 3\theta}{3!} - \frac{\cos 4\theta}{4!} + \dots\infty$$

$$\therefore \quad C + iS = \frac{1}{1!}(\cos\theta + i\sin\theta) - \frac{1}{2!}(\cos 2\theta + i\sin 2\theta) + \frac{1}{3!}(\cos 3\theta + i\sin 3\theta) - \dots\infty$$

$$= \frac{1}{1!}e^{i\theta} - \frac{1}{2!}e^{2i\theta} + \frac{1}{3!}e^{3i\theta} - \dots\infty = 1 - \left[1 - \frac{e^{i\theta}}{1!} + \frac{e^{2i\theta}}{2!} - \frac{e^{3i\theta}}{3!} + \dots\infty\right]$$

$$= 1 - e^{-e^{i\theta}} \qquad \left(\because 1 - \frac{x}{1!} + \frac{x^2}{2!} - \frac{x^3}{3!} + \dots\infty = e^{-x}\right)$$

$$= 1 - e^{-(\cos\theta + i\sin\theta)} = 1 - e^{-\cos\theta}.(e^{-i\sin\theta})$$

$$= 1 - e^{-\cos\theta}.[\cos(\sin\theta) - i\sin(\sin\theta)]$$
$$= [1 - e^{-\cos\theta}.\cos(\sin\theta)] + i[e^{-\cos\theta}.\sin(\sin\theta)]$$

Equating imaginary parts on both sides, we get

$$S = e^{-\cos\theta}.\sin(\sin\theta)$$ **Ans.**

Example 10. *Sum the series*

$$1+\frac{1}{2}\cos 2\theta+\frac{1.3}{2.4}\cos 4\theta+\frac{1.3.5}{2.4.6}\cos 6\theta+\ldots\infty$$

Solution. Let $C = 1+\frac{1}{2}\cos 2\theta+\frac{1.3}{2.4}\cos 4\theta+\frac{1.3.5}{2.4.6}\cos 6\theta+\ldots\infty$

and $S = \frac{1}{2}\sin 2\theta+\frac{1.3}{2.4}\sin 4\theta+\frac{1.3.5}{2.4.6}\sin 6\theta+\ldots\infty$

$$\therefore\quad C + iS = 1+\frac{1}{2}(\cos 2\theta + i\sin 2\theta)+\frac{1.3}{2.4}(\cos 4\theta + i\sin 4\theta)$$
$$+\frac{1.3.5}{2.4.6}(\cos 6\theta + i\sin 6\theta)+\ldots\infty$$

$$= 1+\frac{1}{2}e^{2i\theta}+\frac{1.3}{2.4}e^{4i\theta}+\frac{1.3.5}{2.4.6}e^{6i\theta}+\ldots\infty$$

$$= 1-\left(-\frac{1}{2}\right)e^{2i\theta}+\frac{-\frac{1}{2}\left(-\frac{1}{2}-1\right)}{1.2}(e^{2i\theta})^2-\frac{-\frac{1}{2}\left(-\frac{1}{2}-1\right)\left(-\frac{1}{2}-2\right)}{1.2.3}(e^{2i\theta})^3+\ldots\infty$$

(Binomial Theorem)

$$= (1 - e^{2i\theta})^{-1/2} = (1 - \cos 2\theta - i\sin 2\theta)^{-1/2}$$
$$= [2\sin^2\theta - i\,.\,2\sin\theta\cos\theta]^{-1/2} = (2\sin\theta)^{-1/2}(\sin\theta - i\cos\theta)^{-1/2}$$
$$= (2\sin\theta)^{-1/2}\left[\cos\left(\frac{\pi}{2}-\theta\right) - i\sin\left(\frac{\pi}{2}-\theta\right)\right]^{-1/2}$$
$$= (2\sin\theta)^{-1/2}\left[\cos\left(\frac{\pi}{4}-\frac{\theta}{2}\right) - i\sin\left(\frac{\pi}{4}-\frac{\theta}{2}\right)\right]$$

(De-Moivre's Theorem)

Equating real parts on both sides, we get

$$C = (2\sin\theta)^{-1/2}\cos\left(\frac{\pi}{4}-\frac{\theta}{2}\right)$$ **Ans.**

Example 11. *Sum the series*

$$\sin\alpha - \frac{\sin(\alpha+2\beta)}{2!}+\frac{\sin(\alpha+4\beta)}{4!}-\ldots\infty$$

Solution. Let $S = \sin\alpha - \frac{\sin(\alpha+2\beta)}{2!}+\frac{\sin(\alpha+4\beta)}{4!}-\ldots\infty$

and $C = \cos\alpha - \frac{\cos(\alpha+2\beta)}{2!}+\frac{\cos(\alpha+4\beta)}{4!}-\ldots\infty$

$$\therefore\quad C + iS = (\cos\alpha + i\sin\alpha) - \frac{1}{2!}[\cos(\alpha+2\beta) + i\sin(\alpha+2\beta)]$$
$$+\frac{1}{4!}[\cos(\alpha+4\beta) + i\sin(\alpha+4\beta)] - \ldots\infty$$

$$= e^{i\alpha} - \frac{e^{i(\alpha+2\beta)}}{2!} + \frac{e^{i(\alpha+4\beta)}}{4!} - \ldots \infty = e^{i\alpha}\left[1 - \frac{e^{i2\beta}}{2!} + \frac{e^{i4\beta}}{4!} - \ldots \infty\right]$$

This series of R.H.S. is a cosine series.

$$= e^{i\alpha} \cos (e^{i\beta}) = (\cos \alpha + i \sin \alpha) \cos (\cos \beta + i \sin \beta)$$

$$= (\cos \alpha + i \sin \alpha) [\cos (\cos \beta) \cos (i \sin \beta) - \sin (\cos \beta) \sin (i \sin \beta)]$$

$$= (\cos \alpha + i \sin \alpha)[\cos (\cos \beta) \cosh (\sin \beta) - i \sin (\cos \beta) \sinh (\sin \beta)]$$

Equating imaginary parts, we get

$$S = \sin \alpha \cos (\cos \beta) \cosh (\sin \beta) - \cos \alpha \sin (\cos \beta) \sinh (\sin \beta)$$ **Ans.**

Example 12. *Find Sum the series*

$$x \sin \theta - \frac{1}{2} . x^2 \sin 2\theta + \frac{1}{3} . x^3 \sin 3\theta - \ldots \infty$$

Solution. Let
$$S = x \sin \theta - \frac{1}{2} . x^2 \sin 2\theta + \frac{1}{3} . x^3 \sin 3\theta - \ldots \infty$$

and
$$C = x \cos \theta - \frac{1}{2} . x^2 \cos 2\theta + \frac{1}{3} . x^3 \cos 3\theta - \ldots \infty$$

$$\therefore \quad C + iS = x(\cos \theta + i \sin \theta) - \frac{x^2}{2}(\cos 2\theta + i \sin 2\theta) + \frac{x^3}{3}(\cos 3\theta + i \sin 3\theta) - \ldots \infty$$

$$= x e^{i\theta} - \frac{x^2}{2} e^{2i\theta} + \frac{x^3}{3} e^{i3\theta} - \ldots \infty$$

The series of R.H.S. is a logarithmic series.

$$= \log (1 + x e^{i\theta}) = \log (1 + x \cos \theta + i x \sin \theta)$$

Separating real and imaginary parts,

$$= \log \sqrt{(1 + x \cos \theta)^2 + (x \sin \theta)^2} + i \tan^{-1} \frac{x \sin \theta}{1 + x \cos \theta}$$

Equating imaginary parts, we have

$$S = \tan^{-1} \frac{x \sin \theta}{1 + x \cos \theta}$$ [except when $x \cos \theta = -1$] **Ans.**

Example 13. *Find sum of the series*

$$c \sin \alpha + c^2 \sin 2\alpha + c^3 \sin 3\alpha + \ldots$$

Solution. Let $S = c \sin \alpha + c^2 \sin 2\alpha + c^3 \sin 3\alpha + \ldots$

and $C = c \cos \alpha + c^2 \cos 2\alpha + c^3 \cos 3\alpha + \ldots$

$$C + iS = c(\cos \alpha + i \sin \alpha) + c^2(\cos 2\alpha + i \sin 2\alpha) + c^3(\cos 3\alpha + i \sin 3\alpha) + \ldots$$

$$C + iS = c\, e^{i\alpha} + c^2 e^{i2\alpha} + c^3 e^{i3\alpha} + \ldots$$

$$= \frac{ce^{i\alpha}}{1 - ce^{i\alpha}} = \frac{ce^{i\alpha}(1 - ce^{-i\alpha})}{(1 - ce^{i\alpha})(1 - ce^{-i\alpha})}$$ (*Infinite Geometric series*) $\left(S = \frac{a}{1-r}\right)$

$$= \frac{ce^{i\alpha} - c^2}{1 - c(e^{i\alpha} + e^{-i\alpha}) + c^2} = \frac{c(\cos \alpha + i \sin \alpha) - c^2}{1 - 2c \cos \alpha + c^2}$$

Equating imaginary part, we get

$$\therefore \quad S = \frac{c \sin \alpha}{1 + c^2 - 2c \cos \alpha}$$ **Ans.**

Example 14. *Find the sum of the series*

$$c \sin\theta - \frac{1}{2}c^2 \sin 2\theta + \frac{1}{3}c^3 \sin 3\theta - \ldots \infty$$

Solution. Let $\quad S = c\sin\theta - \frac{1}{2}c^2 \sin 2\theta + \frac{1}{3}c^3 \sin 3\theta - \ldots \infty$

and $\quad C = c\cos\theta - \frac{1}{2}c^2 \cos 2\theta + \frac{1}{3}c^3 \cos 3\theta - \ldots \infty$

$$\therefore \quad C + iS = c(\cos\theta + i\sin\theta) - \frac{1}{2}c^2(\cos 2\theta + i\sin 2\theta) + \frac{1}{3}c^3(\cos 3\theta + i\sin 3\theta) - \ldots \infty$$

$$= ce^{i\theta} - \frac{1}{2}c^2 e^{2i\theta} + \frac{1}{3}c^3 e^{3i\theta} - \ldots \text{ to } \infty$$

$$= ce^{i\theta} - \frac{1}{2}(ce^{i\theta})^2 + \frac{1}{3}(ce^{i\theta})^3 - \ldots \text{ to } \infty$$

$$= \log_e(1 + ce^{i\theta}) \qquad \left[\log(1+x) = x - \frac{x^2}{2} + \frac{x^3}{3} + \ldots\right]$$

$$= \log_e(1 + c\cos\theta + i\,c\sin\theta) \qquad \ldots(1)$$

Now, put $1 + c\cos\theta + i\,c\sin\theta = r(\cos\alpha + i\sin\alpha)$

$\therefore \quad 1 + c\cos\theta = r\cos\theta, \quad c\sin\theta = r\sin\theta$

$$r^2 = (1 + c\cos\theta)^2 + c^2\sin^2\theta = 1 + 2c\cos\theta + c^2\cos^2\theta + c^2\sin^2\theta$$

$$= 1 + 2c\cos\theta + c^2(\cos^2\theta + \sin^2\theta)$$

$$r^2 = 1 + 2c\cos\theta + c^2$$

$$\therefore \quad r^2 = \sqrt{1 + 2c\cos\theta + c^2} \quad \text{and} \quad \alpha = \tan^{-1}\frac{c\sin\theta}{1 + c\cos\theta}$$

From equation (1), we have

$$C + iS = \log_e\{r(\cos\alpha + i\sin\alpha)\} = \log_e re^{i\alpha} = \log_e r + \log_e e^{i\alpha}$$

$$\therefore \quad C + iS = \log_e r + i\alpha$$

Putting the values of r and α, we get

$$C + iS = \log_e\sqrt{1 + 2c\cos\theta + c^2} + i\tan^{-1}\frac{c\sin\theta}{1 + c\cos\theta}$$

Equating imaginary part, we get

$$S = \tan^{-1}\frac{c\sin\theta}{1 + c\cos\theta}, \qquad \textbf{Ans.}$$

Example 15. *Sum the series*

$$c\cos\alpha - \frac{c^2}{2}\cos 2\alpha + \frac{c^3}{3}\cos 3\alpha - \ldots \infty$$

where, $0 < c < 1$.

[**Hint.** From example 14, equating real parts, we get

$$c\cos\alpha - \frac{c^2}{2}\cos 2\alpha + \frac{c^3}{3}\cos 3\alpha \ldots = \log_e\sqrt{1 + 2c\cos\alpha + c^2}\Bigg] \qquad \textbf{Ans.}$$

Example 16. *Sum the series* $\cos\theta + \sin\theta\cos 2\theta + \frac{\sin^2\theta}{1.2}\cos 3\theta + \ldots \infty$

Solution. Let
$$C = \cos\theta + \sin\theta.\cos 2\theta + \frac{\sin^2\theta}{1.2}\cos 3\theta + \ldots\infty$$
and
$$S = \sin\theta + \sin\theta.\sin 2\theta + \frac{\sin^2\theta}{1.2}\sin 3\theta + \ldots\infty$$
$\therefore$
$$C + iS = (\cos\theta + i\sin\theta) + \sin\theta(\cos 2\theta + i\sin 2\theta) + \frac{\sin^2\theta}{2!}(\cos 3\theta + i\sin 3\theta) + \ldots\infty$$
$$= e^{i\theta} + \sin\theta.e^{2i\theta} + \frac{\sin^2\theta}{2!}e^{3i\theta} + \ldots\infty$$
$$= e^{i\theta}\left[1 + \frac{\sin\theta}{1!}e^{i\theta} + \frac{\sin^2\theta}{2!}e^{2i\theta} + \ldots\infty\right]$$
$$= e^{i\theta}.e^{\sin\theta.e^{i\theta}} \qquad \left(\because e^x = 1 + x + \frac{x^2}{2!} + \ldots\infty\right)$$
$$= e^{i\theta}.e^{\sin\theta(\cos\theta + i\sin\theta)} = e^{\sin\theta.\cos\theta}.e^{i(\theta + \sin^2\theta)}$$
$$= e^{\sin\theta\,.\,\cos\theta}\,.\,[\cos(\theta + \sin^2\theta) + i\sin(\theta + \sin^2\theta)]$$
Equating real parts on both sides, we get
$$C = e^{\sin\theta\,.\,\cos\theta}\,.\,\cos(\theta + \sin^2\theta)$$ **Ans.**

EXERCISE 21.3

Find the sum of the following series :

1. $\sin\alpha + x\sin(\alpha + \beta) + \frac{x^2}{2!}\sin(\alpha + 2\beta) + \ldots\infty$ **Ans.** $e^{x\cos\beta}\sin(\alpha + x\beta)$

2. $\cos\alpha + \cos(\alpha + \beta) + \cos(\alpha + 2\beta) + \ldots + \cos(\alpha + \overline{n-1}\beta)$
Ans. $\cos\left\{\alpha + (n-1)\frac{\beta}{2}\right\}\sin\frac{n\beta}{2}\operatorname{cosec}\frac{\beta}{2}$

3. $n\cos\alpha + \frac{n(n+1)}{1.2}\cos 2\alpha + \frac{n(n+1)(n+2)}{1.2.3}\cos 3\alpha + \ldots\infty$ **Ans.** $-1 + \frac{\operatorname{cosec}^n\frac{\alpha}{2}}{2^n}\cos\frac{n}{2}(\pi - \alpha)$

4. $e^{\alpha}\sin\beta - \frac{e^{3\alpha}}{3}\sin 3\beta + \frac{e^{5\alpha}}{5}\sin 5\beta + \ldots\infty$ **Ans.** $\frac{1}{2}\tanh^{-1}(\operatorname{sech}\alpha\sin\beta)$

5. $\sin\alpha.\cos\alpha - \frac{1}{2}\sin^2\alpha\cos 2\alpha + \frac{1}{3}\sin^3\alpha\cos 3\alpha + \ldots\infty$ **Ans.** $\log(1 + \sin\alpha)$

6. $-\frac{1}{2}\sin\alpha + \frac{1.3}{2.4}\sin 2\alpha - \frac{1.3.5}{2.4.6}\sin 3\alpha + \ldots\infty$ **Ans.** $-\frac{\sin\frac{\alpha}{4}}{\sqrt{2\cos\frac{\alpha}{2}}}$

7. $1 + \frac{1}{2}\cos 2\alpha - \frac{1.3}{2.4}\cos 4\alpha + \frac{1.3.5}{2.4.6}\cos 6\alpha + \ldots\infty$ **Ans.** $\sqrt{[\cos\alpha(1 + \cos\alpha)]}$

8. $\sin\alpha + \frac{1}{3}\sin 3\alpha + \frac{1}{3^2}\sin 5\alpha + \ldots\infty$ **Ans.** $\frac{6\sin\alpha}{5 - 3\cos 2\alpha}$

9. $1 + x\cos\theta + x^2\cos 2\theta + x^3\cos 3\alpha + \ldots$ **Ans.** $\dfrac{1 - x\cos\theta - x^n\cos n\theta + x^{n+1}\cos(n-1)\theta}{1 - 2x\cos\theta + x^2}$

10. $1 - \dfrac{1}{2}\cos\theta + \dfrac{1.3}{2.4}\cos 2\theta - \dfrac{1.3.5}{2.4.6}\cos 3\theta + \ldots$ **Ans.** $\left(2\cos\dfrac{\theta}{2}\right)^{-1/2}\cos\dfrac{\theta}{4}$

11. $\sin\theta - \dfrac{\sin 2\theta}{2!} + \dfrac{\sin 3\theta}{3!} - \ldots\infty$ **Ans.** $e^{-\cos\theta}\sin(\sin\theta)$

12. $1 + \dfrac{1}{3}\cos x + \dfrac{1}{9}\cos 2x + \dfrac{1}{27}\cos 3x + \ldots\infty$ **Ans.** $\dfrac{9 - 3\cos x}{10 - 6\cos x}$

13. $x\cos\theta - \dfrac{x^2}{2}\cos 2\theta + \dfrac{x^3}{3}\cos 3\theta - \ldots\infty$ **Ans.** $\dfrac{1}{2}\log(1 + 2x\cos\theta + x^2)$

21.4 APPROXIMATION

Example 17. *If* $\dfrac{\sin x}{x} = \dfrac{2399}{2400}$, *find an approximate value of x in radians.*

Solution. $\dfrac{\sin x}{x} = \dfrac{2399}{2400}$

We know that $\sin x = x - \dfrac{x^3}{3!} + \dfrac{x^5}{5!} - \ldots$ or $\dfrac{\sin x}{x} = 1 - \dfrac{x^2}{6} + \dfrac{x^4}{120}$

$$\frac{2399}{2400} = 1 - \frac{x^2}{6} \qquad \text{(Ignoring } x^4 \text{ and higher powers)}$$

$$\frac{x^2}{6} = 1 - \frac{2399}{2400} \Rightarrow \frac{x^2}{6} = \frac{1}{2400}$$

$$x^2 = \frac{1}{400} \quad \text{or} \quad x = \frac{1}{20} \text{ radian.}$$ **Ans.**

Example 18. *If* $\cos x = \dfrac{1681}{1682}$, *find x approximately.*

Solution. $\cos x = \dfrac{1681}{1682}$

$$1 - 2\sin^2\frac{x}{2} = \frac{1681}{1682} \quad \text{or} \quad 2\sin^2\frac{x}{2} = \frac{1}{1682} \quad \text{or} \quad \sin^2\frac{x}{2} = \frac{1}{3364}$$

$$\sin\frac{x}{2} = \frac{1}{58}$$

We know that $\lim\limits_{x\to 0}\dfrac{\sin\frac{x}{2}}{\frac{x}{2}} = 1 \Rightarrow \sin\dfrac{x}{2} = \dfrac{x}{2} \Rightarrow \dfrac{x}{2} = \dfrac{1}{58} \Rightarrow x = \dfrac{1}{29}$ radian

$$x = 1° \, 58.5'.$$ **Ans.**

Example 19. *Solve* $\cos\left(\dfrac{\pi}{3} + \theta\right) = 0.49.$

Solution. $\cos\left(\dfrac{\pi}{3} + \theta\right) = 0.49$

$$\cos\frac{\pi}{3}\cos\theta - \sin\frac{\pi}{3}\sin\theta = 0.49$$

$$\frac{1}{2}\cos\theta - \frac{\sqrt{3}}{2}\sin\theta = 0.49$$

$$\frac{1}{2}\left[1-\frac{\theta^2}{2!}+...\right]-\frac{\sqrt{3}}{2}\left[\theta-\frac{\theta^3}{3!}+...\right] = 0.49$$

$$\frac{1}{2}-\frac{\theta^2}{4}-\frac{\sqrt{3}}{2}\theta+\frac{\sqrt{3}}{12}\theta^3+... = 0.49$$

$$\frac{1}{2}-\frac{\sqrt{3}}{2}\theta+... = 0.49 \quad \text{(Ignoring higher powers of } \theta\text{)}$$

$$\Rightarrow \quad \frac{\sqrt{3}}{2}\theta = 0.5-0.49 \quad \Rightarrow \quad \theta = \frac{0.02}{\sqrt{3}} = 0.011547$$

$$\theta = 39.696'.$$

Ans.

Example 20. *Evaluate* $\lim\limits_{\theta\to 0}\dfrac{\tan\theta-\sin\theta}{\theta^3}$

Solution. $\lim\limits_{\theta\to 0}\dfrac{\tan\theta-\sin\theta}{\theta^3} = \lim\limits_{\theta\to 0}\dfrac{\left[\theta+\dfrac{\theta^3}{3}+\dfrac{2}{15}\theta^5+...\right]-\left[\theta-\dfrac{\theta^3}{3!}+\dfrac{\theta^5}{5!}-...\right]}{\theta^3}$

$$= \lim_{\theta\to 0}\frac{1+\dfrac{\theta^2}{3}+\dfrac{2}{15}\theta^4+...-1+\dfrac{\theta^2}{6}-\dfrac{\theta^4}{120}+...}{\theta^2} = \lim_{\theta\to 0}\frac{\dfrac{\theta^2}{2}+\dfrac{\theta^4}{8}+...}{\theta^2}$$

$$= \lim_{\theta\to 0}\left(\frac{1}{2}+\frac{\theta^2}{8}\right) = \frac{1}{2}.$$

Ans.

EXERCISE 21.4

1. If $\dfrac{\sin x}{x}=\dfrac{559}{600}$, find an approximate value of x. **Ans.** $x=\dfrac{1}{10}$ radian

2. If $\dfrac{\sin x}{x}=\dfrac{5045}{5046}$, find an approximate value of x. **Ans.** $x = 1°, 58'$

3. Solve $\sin\left(\dfrac{x}{6}+x\right) = 0.51$ approximately. **Ans.** $x = 39.7'$

4. Evaluate $\lim\limits_{x\to 0}\dfrac{\sin x - x\cos x}{x^3}$. **Ans.** $\dfrac{1}{3}$

FUNCTIONS OF COMPLEX VARIABLE, ANALYTIC FUNCTION

22.1 INTRODUCTION

The theory of functions of a complex variable is of atmost importance in solving a large number of problems in the field of engineering and science. Many complicated integrals of real functions are solved with the help of functions of a complex variable.

22.2 COMPLEX VARIABLE

$x + iy$ is a complex variable and it is denoted by z.

(1) $z = x + iy$. where $i = \sqrt{-1}$ (Cartesian form)

(2) $z = r(\cos\theta + i\sin\theta)$ (Polar form)

(3) $z = re^{i\theta}$ (Exponential form)

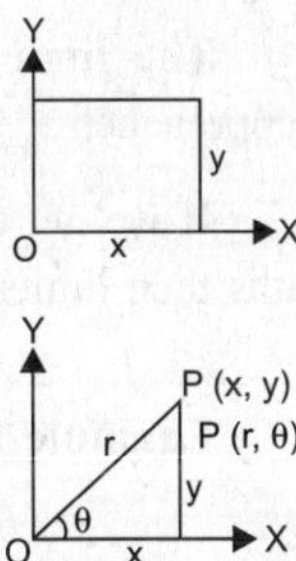

22.3 FUNCTIONS OF A COMPLEX VARIABLE

$f(z)$ is a function of a complex variable z and is denoted by w.

$$w = f(z)$$
$$w = u + iv$$

where u and v are the real and imaginary parts of $f(z)$.

22.4 NEIGHBOURHOOD OF Z_0

Let z_0 is a point in the complex plane and let z be any positive number, then the set of points z such that

$$|z - z_0| < \in$$

is called $\in$-negihbourhood of z_0.

Closed set

A set s is said to be closed if it contains all of its limits point.

Interior Point

A pint z_0 is called a interior point of a point set S if there exists a neighbourhood of z_0 lying whoolly in S.

22.5 LIMIT OF A FUNCTION OF A COMPLEX VARIABLE

Let $f(z)$ be a single valued function defined at all points in some neighbourhood of point z_0. Then $f(t)$ is said to have the limit l as z approaces z_0 along any path if given an arbitrary real number $\in > 0$, however small there exists a real number $\delta > 0$, such that

$$|f(z) - l| < \in \text{ whenever } 0 < |z - z_0| < \delta$$

i.e. for every $z \neq z_0$ in δ-disc (dotted) of z-plane, $f(z)$ has a value lying in the $\in$-disc of w-plane

In symbolic form, $\lim\limits_{z \to z_0} f(z) = l$

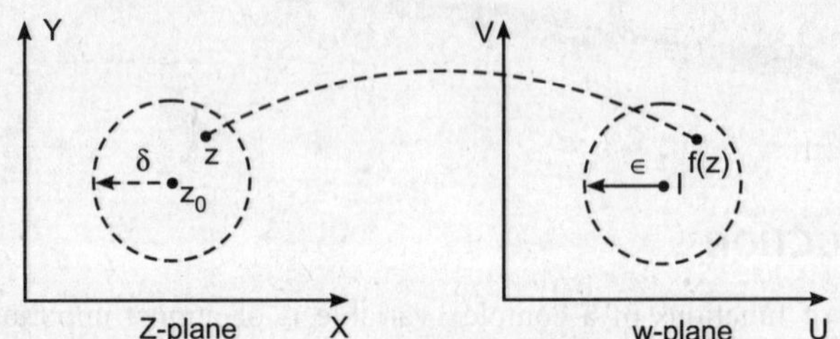

Note: (1) δ usually depends upon $\in$.

(II) $z \to z_0$ implies that z approaches z_0 along any path.

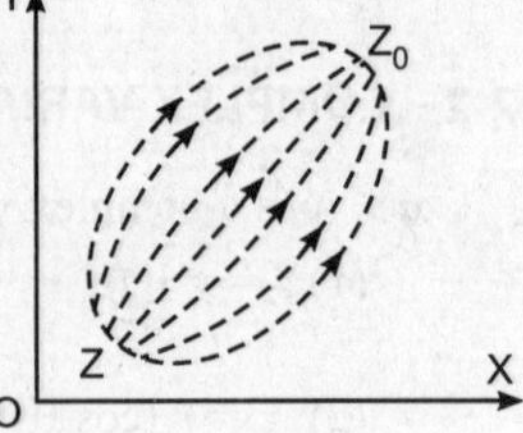

The limit must be independent of the manner in which z approaches z_0

If we get two different limits as $z \to z_0$ along two different paths then limits does not exist.

Example 1. *Prove that* $\lim\limits_{z \to 1-i} \dfrac{(z^2 + 4z + 3)}{z+1} = 4 - i$

Solution. $\lim\limits_{z \to 1-i} \dfrac{z^2 + 4z + 3}{z+1} = \lim\limits_{z \to 1-i} \dfrac{(z+1)(z+3)}{(z+1)} = \lim\limits_{z \to 1-i} (z+3) = (1-i) + 3 = 4 - i$

Example 2. *Show that* $\lim\limits_{z \to 0} \dfrac{z}{|z|}$ *does not exist.*

Solution. $\lim\limits_{z \to 0} \dfrac{z}{|z|} = \lim\limits_{\substack{x \to 0 \\ y \to 0}} \dfrac{x + iy}{\sqrt{x^2 + y^2}}$

Let $y = mx$,

$$= \lim_{x \to 0} \frac{x + imx}{\sqrt{x^2 + m^2 x}} = \lim_{x \to 0} \frac{1 + im}{\sqrt{1 + m^2}} = \frac{1 + mi}{\sqrt{1 + m^2}}$$

The value of $\dfrac{1 + mi}{\sqrt{1 + m^2}}$ are different for different values of m.

Hence, limit of the function does not exist.

Example 3. *Prove that* $\lim\limits_{z \to 0} \dfrac{z}{\bar{z}}$ *does not exist.*

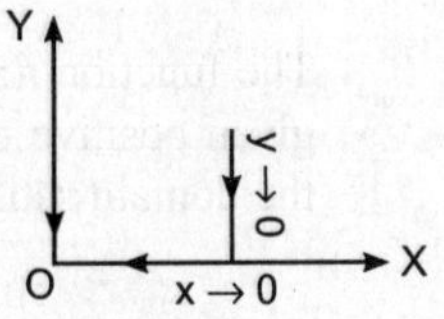

Solution. Case I. $\lim\limits_{z \to 0} \dfrac{z}{\bar{z}} = \lim\limits_{\substack{x \to 0 \\ y \to 0}} \dfrac{x+iy}{x-iy} = \lim\limits_{x \to 0} \left[\lim\limits_{y \to 0} \dfrac{x+iy}{x-iy} \right]$

$$= \lim_{x \to 0} \frac{x}{x} = 1$$

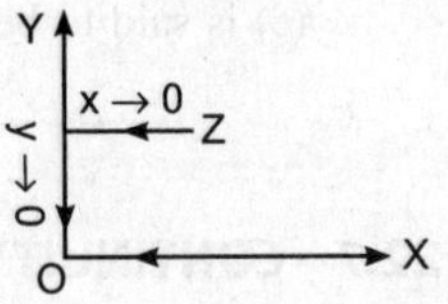

Here the path is $y \to 0$ and then $x \to 0$

Case II. Again $\lim\limits_{z \to 0} \dfrac{z}{\bar{z}} = \lim\limits_{y \to 0} \dfrac{z}{\bar{z}} = \left[\dfrac{x+iy}{x-iy} \right] = \lim\limits_{y \to 0} \dfrac{iy}{-iy} = -1$

In this case, we have a different path first $x \to 0$, then $y \to 0$

As $z \to 0$ along two different paths we get different limits.

Hence the limit does not exist. **Proved.**

Example 4. *Find the limit of the following* $\lim\limits_{z \to \infty} \dfrac{iz^3 + iz - 1}{(2z+3i)(z-i)^2}$

Solution. On dividing numerator and denominator by z^3 get

$$\lim_{z \to \infty} \frac{iz^3 + iz - 1}{(2z+3i)(z-i)^2} = \lim_{z \to \infty} \frac{\left(i + \frac{i}{z^2} - \frac{1}{z^3}\right)}{\left(2 + \frac{3i}{z}\right)\left(1 - \frac{i}{z}\right)^2} = \frac{i}{2}$$ **Ans.**

Example 5. *Find the limit of the following* $\lim\limits_{z \to 1+i} \dfrac{z^2 - z + 1 - i}{z^2 - 2z + 2}$

Solution.

$$\lim_{z \to 1+i} \frac{z^2 + 1 - z - i}{z^2 - 2z + 2} = \lim_{z \to 1+i} \frac{(z+i)(z-i) - (z+i)}{(z-1-i)(z+1+i)}$$

$$= \lim_{z \to 1+i} \frac{(z+i)(z-i-1)}{(z-1-i)(z-1+i)} = \lim_{z \to 1+i} \frac{z+i}{z-1+i}$$

$$= \frac{1+i+i}{1+i-1+i} = \frac{1+2i}{2i} = \frac{(1+2i)(-i)}{2(i)(-i)} = \frac{-i+2}{2(1)} = \frac{2-i}{2} = 1 - \frac{i}{2}$$ **Ans.**

EXERCISE 22.1

Show that the following limits do not exist:

1. $\lim\limits_{z \to 0} \dfrac{\operatorname{Im}(z^3)}{\operatorname{Re}(z^3)}$ 2. $\lim\limits_{z \to -i} \dfrac{z^2}{z+i}$ 3. $\lim\limits_{z \to -i} \dfrac{\operatorname{Re}(z)^2}{\operatorname{Im} z}$ 4. $\lim\limits_{z \to 0} \dfrac{z}{(\bar{z})^2}$

Find the Limits of the following:

5. $\lim\limits_{z \to 0} \dfrac{\operatorname{Re}(z^3)}{\operatorname{Re}(z^2)}$ **Ans.** 0 6. $\lim\limits_{z \to 1+i} \dfrac{2z^3}{\operatorname{Im}(z)^2}$ **Ans.** $2(-1+i)$ 7. $\lim\limits_{z \to 0} \dfrac{z^2 + 6z + 3}{z^2 + 2z + 2}$ **Ans.** $\dfrac{3}{2}$

22.6 CONTINUITY

The function $f(z)$ of a complex variable z is said to be continuous at the point z_0 if for any given positive number $\in$, we can find a number δ such that $| f(z_0) | < \in$ for all points z of the domain satisfying

$$| z - z_0 | < \delta$$

$f(z)$ is said to be continuous at $z = z_0 =$ if

$$\lim_{z \to 0} f(z) = f(z_0)$$

22.7 CONTINUITY IN TERMS OF REAL AND IMAGINARY PARTS

Let $w = f(z) = u(x, y) + iv(x, y)$ is continuous function at $z = z_0$ then $u(x, y)$ and $v(x, y)$ are separately continuous functions of x, y at (x_0, y_0) where $z_0 = z_0 + iy_0$.

Conversely, if $u(x, y)$ and $v(x, y)$ are continuous functions of x, y at (x_0, y_0) then $f(z)$ is continuous at $z = z_0$.

Example 6. *Examine the continuity of the following*

$$f(z) = \begin{cases} \dfrac{z^3 - iz^2 + z - i}{z - i}, & z \neq i \\ 0 \quad , & z = i \end{cases} \qquad \text{at } z = i$$

Solution. $\displaystyle \lim_{z \to i} \frac{z^3 - iz^2 + z - i}{z - i} = \lim_{z \to i} \frac{z^2(z - i) + 1(z - i)}{z - i}$

$$\lim_{z \to i} \frac{(z - i)(z^2 + 1)}{z - i} = \lim_{z \to i} (z^2 + 1) = -1 + 1 = 0$$

$$f(i) = 0$$

$$\boxed{\lim_{z \to i} f(z) = f(i)}$$

Hence $f(z)$ is continuous at $z = i$ **Ans.**

Example 7. *Show that the function $f(z)$ defined by*

$$f(z) = \begin{cases} \dfrac{\text{Re}(z)}{z}, & z \neq 0 \\ 0 \quad , & z = 0 \end{cases}$$

is not continuous at $z = 0$

Solution. Here $f(z) = \dfrac{\text{Re}(z)}{z}$ when $z \neq 0$

$$\lim_{z \to 0} \frac{\text{Re}(z)}{z} = \lim_{\substack{x \to 0 \\ y \to 0}} \frac{x}{x + iy} = \lim_{z \to 0} \left[\frac{x}{x + iy} \right] = \lim_{x \to 0} \frac{x}{x} = 1$$

Again $\displaystyle \lim_{z \to 0} \frac{\text{Re}(z)}{z} = \lim_{y \to 0} \left[\lim_{x \to 0} \frac{x}{x + iy} \right] = 0$

As $z \to 0$, for two different paths limit have two different values. So, limit does not exist.

Hence $f(z)$ is not continuous at $z = 0$ **Proved.**

EXERCISE 22.2

Examine the continuity of the following functions.

1. $f(z) = \begin{cases} \dfrac{\text{Im}(z)}{|z|}, & z \neq 0 \\ 0\ , & z \neq 0 \end{cases}$ at $z = 0$ **Ans.** Not Continuous

2. $f(z) = \dfrac{z^2 + 3z + 4}{z^2 + i}$ at $z = 1 - t$ **Ans.** Continuous

3. Show that the following functions are continuous for z

(*i*) $\cos z$ (*ii*) e^{2z}

22.8 DIFFERENTIABILITY

Let $f(z)$ be a single valued function of the variable z, then

$$f'(z) = \lim_{\delta z \to 0} \frac{f(z+\delta z) - f(z)}{\delta z}$$

provided that the limit exists and is independent of the path along which $\delta z \to 0$.

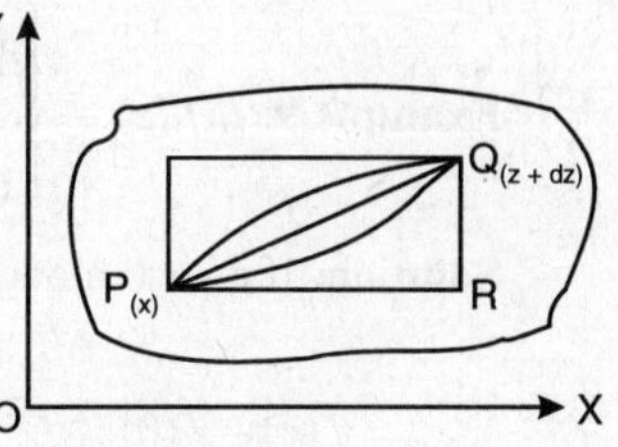

Let P be a fixed point and Q be a neighbouring point. The point Q may approach P along any straight line or curved path.

Example 8. *Consider the function*

$$f(z) = 4x + y + i(-x + 4y)$$

and discuss $\dfrac{df}{dz}$.

Solution. Here, $f(z) = 4x + y + i(-x + 4y)$

$$= u + iv$$

so $u = 4x + y$

and $v = -x + 4y$

$$f(z+\delta z) = 4(x+\delta x) + (y+\delta y) - i(x+\delta x) + 4i(y+\delta y)$$

$$f(z+\delta z) - f(z) = 4(x+\delta x) + (y+\delta y) - i(x+\delta x) + 4i(y+\delta y) - 4x - y + ix - 4iy$$

$$= 4\delta x + \delta y - i\delta x + 4i\delta y$$

$$\frac{f(z+\delta z) - f(z)}{\delta z} = \frac{4\delta x + \delta y - i\delta x + 4i\delta y}{\delta x + i\delta y}$$

$$\Rightarrow \quad \frac{\delta f}{\delta z} = \frac{4\delta x + \delta y - i\delta x + 4i\delta y}{\delta x + i\delta y} \quad \text{... (1)}$$

(*a*) **Along real axis:** If Q is taken on the horizontal line through $P(x, y)$ and Q then approaches P along this line, we shall have $\delta y = 0$ and $\delta y = \delta x$.

$$\frac{\delta f}{\delta z} = \frac{4\delta x - i\delta x}{\delta x} = 4 - i$$

(*b*) **Along imaginary axis:** If Q is taken on the vertical line through P and then Q approaches P along this line, we have

$$z = x + iy = 0 + iy,\ \delta z = i\delta y,\ \delta x = 0.$$

Putting these values in (1), we have

$$\frac{\delta f}{\delta z}=\frac{\delta y+4i\delta y}{i\delta y}=\frac{1}{i}(1+4i)=4-i$$

(*c*) **Along a line $y = x$:** If Q is taken on a line $y = x$.

$$z = x+iy = x+ix=(1+i)\,x$$

$$\delta z=(1+i)\,\delta x \quad \text{and} \quad \delta y=\delta x$$

On putting these values in (1), we have

$$\frac{\delta f}{\delta z}=\frac{4\delta x+\delta x-i\delta x+4i\delta x}{\delta x+i\delta x}=\frac{4+1-i+4i}{1+i}=\frac{5+3i}{1+i}=\frac{(5+3i)(1-i)}{(1+i)(1-i)}=4-i$$

In all the three different paths approaching Q from P, we get the same values of $\frac{\delta f}{\delta z}=4-i$.

In such a case, the function is said to be differentiable at the point z in the given region.

Example 9. *If* $f(z)=\begin{cases}\dfrac{x^3y(y-ix)}{x^6+y^2}, & z\neq 0,\\ 0, & z=0\end{cases}$ *then discuss* $\frac{df}{dz}$ *at* $z = 0$.

Solution. If $z \to 0$ along radius vector $y = mx$

$$\lim_{z\to 0}\frac{f(z)-f(0)}{z}=\lim_{z\to 0}\left[\frac{\dfrac{x^3y(y-ix)}{x^6+y^2}-0}{x+iy}\right]=\lim_{z\to 0}\left[\frac{-ix^3y(x+iy)}{(x^6+y^2)(x+iy)}\right]$$

$$=\lim_{z\to 0}\left[\frac{-ix^3y}{x^6+y^2}\right]=\lim_{z\to 0}\left[\frac{-ix^3(mx)}{x^6+m^2x^2}\right] \qquad [\because y=mx]$$

$$=\lim_{x\to 0}\left[\frac{-imx^2}{x^4+m^2}\right]=0$$

But along $\quad y = x^3$

$$\lim_{z\to 0}\frac{f(z)-f(0)}{z}=\lim_{z\to 0}\left[\frac{-ix^3y}{x^6+y^2}\right]=\lim_{x\to 0}\frac{-ix^3(x^3)}{x^6+(x^3)^2}=-\frac{i}{2}$$

In different paths we get different values of $\frac{df}{dz}$ *i.e.* 0 and $\frac{-i}{2}$. In such a case, the function is not differentiable at $z = 0$.

Theorem: Continuity is a necessary condition but not sufficient condition for the existence of a finite derivative.

Proof. We have, $f(z_0+\delta z)-f(z_0)=\delta z\left\{\dfrac{f(z_0+\delta z)-f(z_0)}{\delta z}\right\}$ ... (1)

Taking limit of both sides of (1), as $\delta z \to 0$, we get

$$\operatorname*{Lim}_{\delta z\to 0}\,[f(z_0+\delta z)-f(z_0)]=0.f'(z_0) \;\Rightarrow\; \operatorname*{Lim}_{\delta z\to 0}\,[f(z_0+\delta z)-f(z_0)]=0$$

$$\Rightarrow \quad \operatorname*{Lim}_{z\to z_0}\,[f(z)-f(z_0)]=0 \;\Rightarrow\; \operatorname*{Lim}_{z\to z_0} f(z)=f(z_0)$$

$\Rightarrow$ $\quad f(z)$ is continuous at $z = z_0$. **Proved.**

The converse of the above theorem is not true.

This can be shown by the following example.

Example 10. *Prove that the function $f(z) = |z|^2$ is continuous everywhere but no where differentiable except at the origin.*

Solution. Here, $f(z) = |z|^2$.

$\therefore$ But $|z| = \sqrt{(x^2 + y^2)} \Rightarrow |z|^2 = x^2 + y^2$

Since x^2 and y^2 are polynomial so $x^2 + y^2$ is continuous everywhere, therefore, $|z|^2$ is continuous everywhere.

Now, we have $f'(z) = \lim\limits_{\delta z \to 0} \dfrac{f(z+\delta z) - f(z)}{\delta z}$

$$f'(z) = \lim_{\delta z \to 0} \frac{|z + \delta z|^2 - |z|^2}{\delta z} \qquad (z\bar{z} = |z|^2)$$

$$= \lim_{\delta z \to 0} \frac{(z + \delta z)(\bar{z} + \delta \bar{z}) - z\bar{z}}{\delta z} = \lim_{\delta z \to 0} \frac{z\bar{z} + z\delta\bar{z} + \delta z.\bar{z} + \delta z.\delta\bar{z} - z\bar{z}}{\delta z}$$

$$= \lim_{\delta z \to 0} \frac{z.\delta\bar{z} + \delta z.\bar{z} + \delta z\,\delta\bar{z}}{\delta z} = \lim_{\delta z \to 0}\left\{\bar{z} + \delta\bar{z} + z\frac{\delta\bar{z}}{\delta z}\right\} = \lim_{\delta z \to 0}\left\{\bar{z} + z\frac{\delta\bar{z}}{\delta z}\right\} \quad ...(1)$$

[Since, $\delta z \to 0$ so $\delta\bar{z} \to 0$]

Let $\delta z = r(\cos\theta + i\sin\theta)$ and $\delta\bar{z} = r(\cos\theta - i\sin\theta)$

$$\Rightarrow \frac{\delta\bar{z}}{\delta z} = \frac{\cos\theta - i\sin\theta}{\cos\theta + i\sin\theta} \qquad \Rightarrow \frac{\delta\bar{z}}{\delta z} = (\cos\theta - i\sin\theta)(\cos\theta + i\sin\theta)^{-1}$$

$$\Rightarrow \frac{\delta\bar{z}}{\delta z} = (\cos\theta - i\sin\theta)(\cos\theta - i\sin\theta) \Rightarrow \frac{\delta\bar{z}}{\delta z} = (\cos\theta - i\sin\theta)^2$$

$$\Rightarrow \frac{\delta\bar{z}}{\delta z} = \cos 2\theta - i\sin 2\theta$$

Since $\dfrac{\delta\bar{z}}{\delta z}$ depends on θ. It means for different values of θ, $\dfrac{\delta\bar{z}}{\delta z}$ has different values.

It means $\dfrac{\delta\bar{z}}{\delta z}$ has different values for different z. $\qquad z = r(\cos\theta + i\sin\theta)$

Therefore $\lim\limits_{\delta z \to 0}\dfrac{\delta\bar{z}}{\delta z}$ does not tend to a unique limit when $z \neq 0$.

Thus, from (1), it follows that $f'(z)$ is not unique and hence $f(z)$ is not differentiable when $z \neq 0$.

But when $z = 0$ then $f'(z) = 0$ *i.e.*, $f'(0) = 0$ and is unique.

Hence, the function is differentiable at $z = 0$. **Ans.**

22.9 ANALYTIC FUNCTION

A function $f(z)$ is said to be **analytic** at a point z_0, if f is differentiable not only at z_0 but at every point of some neighbourhood of z_0.

A function $f(z)$ is analytic in a domain if it is **analytic** at every point of the domain.

The point at which the function is not differentiable is called a **singular point** of the function.

An analytic function is also known as "holomorphic", "regular", "monogenic."

Entire Function. A function which is analytic everywhere (for all z in the complex plane) is known as an entire function.

For Example 1. Polynomials rational functions are entire.

2. $|\bar{z}|^2$ is differentiable only at $z = 0$. So it is no where analytic.

Note: (*i*) An entire is always analytic, differentiable and continuous function. But converse is not true.

(*ii*) Analytic function is always differentiable and continuous. But converse is not true.

(*iii*) A differentiable function is always continuous. But converse is not true

22.10 THE NECESSARY CONDITION FOR *f* (*z*) TO BE ANALYTIC

(*Delhi University, April 2010*)

Theorem. *The necessary conditions for a function f (z) = u + iv to be analytic at all the points in a region R are*

(*i*) $\frac{\partial u}{\partial x} = \frac{\partial v}{\partial y}$ (*ii*) $\frac{\partial u}{\partial y} = -\frac{\partial v}{\partial x}$ provided $\frac{\partial u}{\partial x}, \frac{\partial u}{\partial y}, \frac{\partial v}{\partial x}, \frac{\partial v}{\partial y}$ exist.

Proof: Let $f(z)$ be an analytic function in a region R,

$$f(z) = u + iv,$$

where u and v are the functions of x and y.

Let δu and δv be the increments of u and v respectively corresponding to increments δx and δy of x and y.

$$\therefore \quad f(z+\delta z) = (u+\delta u) + i(v+\delta v)$$

$$\text{Now} \quad \frac{f(z+\delta z) - f(z)}{\delta z} = \frac{(u+\delta u) + i(v+\delta v) - (u+iv)}{\delta z} = \frac{\delta u + i\delta v}{\delta z} = \frac{\delta u}{\delta z} + i\frac{\delta v}{\delta z}$$

$$\lim_{\delta z \to 0} \frac{f(z+\delta z) - f(z)}{\delta z} = \lim_{\delta z \to 0}\left(\frac{\delta u}{\delta z} + i\frac{\delta v}{\delta z}\right) \text{ or } f'(z) = \lim_{\delta z \to 0}\left(\frac{\delta u}{\delta z} + i\frac{\delta v}{\delta z}\right) \quad \text{... (1)}$$

since $f(z)$ can approach zero along any path.

(*a*) **Along real axis (*x*-axis)**

$z = x + iy$ but on x-axis, $y = 0$

$\therefore$ $z = x,$ $\delta z = \delta x, \delta y = 0$

Y, (z = x), P, Q, O, X, δz = δx..

Putting these values in (1), we have

$$f'(z) = \lim_{\delta x \to 0}\left(\frac{\delta u}{\delta x} + i\frac{\delta v}{\delta x}\right) = \frac{\partial u}{\partial x} + i\frac{\partial v}{\partial x} \quad \text{... (2)}$$

(*b*) **Along imaginary axis (*y*-axis)**

$z = x + iy$ but on y-axis, $x = 0$

$z = 0 + iy$ $\delta x = 0, \delta z = i\delta y.$

Y, δz = i δy, Q, P(z = iy), O, X

Putting these values in (1), we get

$$f'(z) = \lim_{\delta y \to 0}\left(\frac{\delta u}{i\delta y} + \frac{i\delta v}{i\delta y}\right) = \lim_{\delta y \to 0}\left(-i\frac{\delta u}{\delta y} + \frac{\delta v}{\delta y}\right) = -i\frac{\partial u}{\partial y} + \frac{\partial v}{\partial y} \quad \text{... (3)}$$

If $f(z)$ is differentiable, then two values of $f'(z)$ must be the same.

Equating (2) and (3), we get $\frac{\partial u}{\partial x} + i\frac{\partial v}{\partial x} = -i\frac{\partial u}{\partial y} + \frac{\partial v}{\partial y}$

Equating real and imaginary parts, we have

$$\frac{\partial u}{\partial x} = \frac{\partial v}{\partial y}, \qquad \frac{\partial v}{\partial x} = -\frac{\partial u}{\partial y}$$

$$\boxed{\frac{\partial u}{\partial x}=\frac{\partial v}{\partial y}}$$

$$\boxed{\frac{\partial u}{\partial y}=-\frac{\partial v}{\partial x}}$$ **are known as Cauchy Riemann equations.**

22.11 SUFFICIENT CONDITION FOR *f* (*z*) TO BE ANALYTIC (*D.U., APRIL 2010*)

Theorem. *The sufficient condition for a function f (z) = u + iv to be analytic at all the points in a region R are*

(*i*) $\frac{\partial u}{\partial x}=\frac{\partial v}{\partial y}, \qquad \frac{\partial u}{\partial y}=-\frac{\partial v}{\partial x}$

(*ii*) $\frac{\partial u}{\partial x}, \frac{\partial u}{\partial y}, \frac{\partial v}{\partial x}, \frac{\partial v}{\partial y}$ *are continuous functions of x and y in region R.*

Proof. Let $f(z)$ be a single-valued function having

$$\frac{\partial u}{\partial x}, \frac{\partial u}{\partial y}, \frac{\partial v}{\partial x}, \frac{\partial v}{\partial y}$$

at each point in the region R. Then the $C-R$ equations are satisfied.

By Taylor's Theorem:

$$f(z+\delta z)=u(x+\delta x,\, y+\delta y)+iv(x+\delta x,\, y+\delta y)$$

$$= u(x,y)+\left(\frac{\partial u}{\partial x}\delta x+\frac{\partial u}{\partial y}\delta y\right)+\ldots+i\left[v(x,y)+\left(\frac{\partial v}{\partial x}\delta x+\frac{\partial v}{\partial y}\delta y\right)+\ldots\right]$$

$$= [u(x,y)+iv(x,y)]+\left[\frac{\partial u}{\partial x}\cdot\delta x+i\frac{\partial v}{\partial x}\cdot\delta x\right]+\left[\frac{\partial u}{\partial y}\delta y+i\frac{\partial v}{\partial y}\cdot\delta y\right]+\ldots$$

$$= f(z)+\left(\frac{\partial u}{\partial x}+i\frac{\partial v}{\partial x}\right)\delta x+\left(\frac{\partial u}{\partial y}+i\frac{\partial v}{\partial y}\right)\delta y+\ldots$$

(Ignoring the terms of second power and higher powers)

$$\Rightarrow \quad f(z+\delta z)-f(z)=\left(\frac{\partial u}{\partial x}+i\frac{\partial v}{\partial x}\right).\delta x+\left(\frac{\partial u}{\partial y}+i\frac{\partial v}{\partial y}\right)\delta y \qquad \ldots (1)$$

We know $C-R$ equations *i.e.*,

$$\frac{\partial u}{\partial x}=\frac{\partial v}{\partial y} \text{ and } \frac{\partial u}{\partial y}=-\frac{\partial v}{\partial x}$$

Replacing $\frac{\partial u}{\partial y}$ and $\frac{\partial v}{\partial y}$ by $-\frac{\partial v}{\partial x}$ and $\frac{\partial u}{\partial x}$ respectively in (1), we get

$$f(z+\delta z)-f(z)=\left(\frac{\partial u}{\partial x}+i\frac{\partial v}{\partial x}\right).\delta x+\left(-\frac{\partial v}{\partial x}+i\frac{\partial u}{\partial x}\right).\delta y \qquad \text{(taking } i \text{ common)}$$

$$=\left(\frac{\partial u}{\partial x}+i\frac{\partial v}{\partial x}\right).\delta x+\left(i\frac{\partial v}{\partial x}+\frac{\partial u}{\partial x}\right).i\delta y=\left(\frac{\partial u}{\partial x}+i\frac{\partial v}{\partial x}\right).(\delta x+i\delta y)=\left(\frac{\partial u}{\partial x}+i\frac{\partial v}{\partial x}\right).\delta z$$

$$\Rightarrow \quad \frac{f(z+\delta z)-f(z)}{\delta z}=\frac{\partial u}{\partial x}+i\frac{\partial v}{\partial x}$$

$$\Rightarrow \quad \lim_{\delta z\to 0}\frac{f(z+\delta z)-f(z)}{\delta z}=\frac{\partial u}{\partial x}+i\frac{\partial v}{\partial x}$$

$$\Rightarrow \quad \boxed{f'(z) = \frac{\partial u}{\partial x} + i\frac{\partial v}{\partial x}}$$

$$\Rightarrow \quad \boxed{f'(z) = \frac{\partial v}{\partial y} - i\frac{\partial u}{\partial y}}$$

Proved.

Remember: 1. If a function is analytic in a domain D, then u, v satisfy $C - R$ conditions at all points in D.

2. $C - R$ conditions are necessary but not sufficient for analytic function.

3. $C - R$ conditions are sufficient if the partial derivative are continuous.

Example 11. *Determine whether $\frac{1}{z}$ is analytic or not ?*

Solution. Let $w = f(z) = u + iv = \frac{1}{z} \quad \Rightarrow \quad u + iv = \frac{1}{x+iy} = \frac{x-iy}{x^2+y^2}$

Equating real and imaginary parts, we get

$$u = \frac{x}{x^2+y^2}, \qquad v = \frac{-y}{x^2+y^2}$$

$$\frac{\partial u}{\partial x} = \frac{(x^2+y^2).1 - x.2x}{(x^2+y^2)^2} = \frac{y^2-x^2}{(x^2+y^2)^2}, \qquad \frac{\partial u}{\partial y} = \frac{-2xy}{(x^2+y^2)^2}.$$

$$\frac{\partial v}{\partial x} = \frac{2xy}{(x^2+y^2)^2}, \qquad \frac{\partial v}{\partial y} = \frac{y^2-x^2}{(x^2+y^2)^2}$$

Thus, $\quad \frac{\partial u}{\partial x} = \frac{\partial v}{\partial y}$ and $\frac{\partial u}{\partial y} = -\frac{\partial v}{\partial x}$.

Thus C – R equations are satisfied. Also partial derivatives are continuous except at (0, 0).

Therefore $\frac{1}{z}$ is analytic everywhere except at $z = 0$.

Also $\quad \frac{dw}{dz} = -\frac{1}{z^2}$

This again shows that $\frac{dw}{dz}$ exists everywhere except at $z = 0$. Hence $\frac{1}{z}$ is analytic everywhere except at $z = 0$. **Ans.**

Example 12. *Show that the function e^x (cos y + i sin y) is an analytic function, find its derivative.* (*R.G.P.V., Bhopal, III Semester, June 2008*)

Solution. Let $e^x (\cos y + i \sin y) = u + iv$

So, $e^x \cos y = u \quad$ and $\quad e^x \sin y = v \quad$ then $\quad \frac{\partial u}{\partial x} = e^x \cos y, \qquad \frac{\partial v}{\partial y} = e^x \cos y$

$$\frac{\partial u}{\partial y} = -e^x \sin y, \qquad \frac{\partial v}{\partial x} = e^x \sin y$$

Here we see that $\quad \frac{\partial u}{\partial x} = \frac{\partial v}{\partial y}, \qquad \frac{\partial u}{\partial y} = -\frac{\partial v}{\partial x}$

These are $C - R$ equations and are satisfied and the partial derivatives are continuous. Hence, $e^x (\cos y + i \sin y)$ is analytic.

$$f(z) = u + iv = e^x(\cos y + i\sin y) \text{ and } \frac{\partial u}{\partial x} = e^x \cos y, \quad \frac{\partial v}{\partial x} = e^x \sin y$$

$$f'(z) = \frac{\partial u}{\partial x} + i\frac{\partial v}{\partial x} = e^x \cos y + ie^x \sin y = e^x(\cos y + i\sin y) = e^x.e^{iy} = e^{x+iy} = e^z.$$

Which is the required derivative. **Ans.**

Example 13. *Test the analyticity of the function w = sin z and hence derive that:*

$$\frac{d}{dz}(\sin z) = \cos z$$

Solution. $w = \sin z = \sin(x + iy) = \sin x \cos iy + \cos x \sin iy$

$= \sin x \cosh y + i \cos x \sinh y$

$u = \sin x \cosh y$, $v \cos x \sinh y$ $\left[\begin{array}{l} \cos iy = \cosh y \\ \sin iy = i\sinh y \end{array}\right]$

$$\frac{\partial u}{\partial x} = \cos x \cosh y, \quad \frac{\partial u}{\partial y} = \sin x \sinh y$$

$$\frac{\partial v}{\partial x} = -\sin x \sinh y, \quad \frac{\partial v}{\partial y} = \cos x \cosh y$$

Thus $\quad \dfrac{\partial u}{\partial x} = \dfrac{\partial v}{\partial y} \quad$ and $\quad \dfrac{\partial u}{\partial y} = -\dfrac{\partial v}{\partial x}$

So $C - R$ equations are satisfied and partial derivatives are continuous.
Hence, sin z is an analytic function.

$$\cosh x = \frac{e^x + e^{-x}}{2} \quad ...(1)$$

$$\sinh x = \frac{e^x - e^{-x}}{2} \quad ...(2)$$

$$\cos x = \frac{e^{ix} + e^{-ix}}{2} \quad ...(3)$$

$$\sin x = \frac{e^{ix} - e^{-ix}}{2i} \quad ...(4)$$

From (1) $\cosh ix = \dfrac{e^{ix} + e^{-ix}}{2} = \cos x$

From (3) $\cos ix = \dfrac{e^{i(ix)} + e^{-i(ix)}}{2}$

$= \dfrac{e^x + e^{-x}}{2} = \cosh x$

From (4) $\sin ix = \dfrac{e^{i(ix)} - e^{-i(ix)}}{2i}$

$= i\dfrac{e^x - e^{-x}}{2} = i \sinh x$

From (2) $\sinh ix = \dfrac{e^{ix} - e^{-ix}}{2} = i \sin x$

$$\frac{d}{dz}(\sin z) = \frac{d}{dz}[\sin x \cosh y + i \cos x \sinh y]$$

$$= \frac{\partial}{\partial x}(\sin x \cosh y + i \cos x \sinh y)$$

$= \cos x \cosh y - i \sin x \sinh y = \cos x \cos iy - \sin y$

$= \cos(x + iy) = \cos z$ **Ans.**

Example 14. *Show that the real and imaginary parts of the function w = log z satisfy the Cauchy-Riemann equations when z is not zero. Find its derivative.*

Solution. To separate the real and imaginary parts of log z, we put $x = r\cos\theta$; $y = r\sin\theta$

$$w = \log z = \log(x + iy)$$

$$\Rightarrow \quad u + iv = \log(r\cos\theta + ir\sin\theta) = \log r(\cos\theta + i\sin\theta) = \log_e r.e^{i\theta}$$

$$= \log_e r + \log_e e^{i\theta} = \log r + i\theta = \log\sqrt{x^2+y^2} + i\tan^{-1}\frac{y}{x} \qquad \left[\begin{array}{l} r = \sqrt{x^2+y^2} \\ \theta = \tan^{-1}\dfrac{y}{x} \end{array}\right]$$

So $\qquad u = \log\sqrt{x^2+y^2} = \frac{1}{2}\log(x^2+y^2),\ v = \tan^{-1}\frac{y}{x}$

On differentiating u, v, we get

$$\frac{\partial u}{\partial x} = \frac{1}{2}\frac{1}{x^2+y^2}.(2x) = \frac{x}{x^2+y^2} \qquad \text{... (1)}$$

$$\frac{\partial v}{\partial y} = \frac{1}{1+\dfrac{y^2}{x^2}}\left(\frac{1}{x}\right) = \frac{x}{x^2+y^2} \qquad \text{... (2)}$$

From (1) and (2), $\dfrac{\partial u}{\partial x} = \dfrac{\partial v}{\partial y}$ ... (A)

Again differentiating u, v, we have

$$\frac{\partial u}{\partial y} = \frac{1}{2}\frac{1}{x^2+y^2}(2y) = \frac{y}{x^2+y^2} \qquad \text{... (3)}$$

$$\frac{\partial v}{\partial x} = \frac{1}{1+\dfrac{y^2}{x^2}}\left(-\frac{y}{x^2}\right) = -\frac{y}{x^2+y^2} \qquad \text{... (4)}$$

From (3) and (4), we have

$$\frac{\partial u}{\partial y} = -\frac{\partial v}{\partial x} \qquad \text{... (B)}$$

Equations (A) and (B) are $C - R$ equations and partial derivatives are continuous.

Hence, $w = \log z$ is an analytic function except

when $\qquad x^2 + y^2 = 0 \Rightarrow x = y = 0 \Rightarrow x + iy = 0 \Rightarrow z = 0$

Now $\qquad w = u + iv$

$$\frac{dw}{dz} = \frac{\partial u}{\partial x} + i\frac{\partial v}{\partial x} = \frac{x}{x^2+y^2} - i\frac{y}{x^2+y^2} = \frac{x-iy}{x^2+y^2}$$

$$= \frac{x-iy}{(x+iy)(x-iy)} = \frac{1}{x+iy} = \frac{1}{z}$$

Which is the required derivative. **Ans.**

Example 15. *Discuss the analyticity of the function $f(z) = z\bar{z}$.*.

Solution. $f(z) = z\bar{z} = (z+iy)(z-iy) = x^2 - i^2y^2 = x^2 + y^2$

$f(z) = x^2 + y^2 = u + iv.$

$u = x^2 + y^2,\ v = 0$

At origin, $\dfrac{\partial u}{\partial x} = \lim\limits_{h\to 0}\dfrac{u(0+h,0)-u(0,0)}{h} = \lim\limits_{h\to 0}\dfrac{h^2}{h} = 0$

$$\frac{\partial u}{\partial y} = \lim_{h\to 0}\frac{u(0,0+k)-u(0,0)}{k} = \lim_{h\to 0}\frac{k^2}{k} = 0$$

Also, $\dfrac{\partial v}{\partial x} = \lim\limits_{h\to 0}\dfrac{v(0,0+k)-v(0,0)}{k} = 0$

$$\frac{\partial v}{\partial y} = \lim_{h\to 0}\frac{v(0,0+k)-v(0,0)}{k} = 0$$

Thus, $\dfrac{\partial u}{\partial x} = \dfrac{\partial v}{\partial y}$ and $\dfrac{\partial u}{\partial y} = -\dfrac{\partial v}{\partial x}$.

Hence, C – R equation are satisfied at the origin.

$$f'(0) = \lim_{z\to 0} \frac{f(z) - f(0)}{z} = \lim_{z\to 0} \frac{(x^2 + y^2)}{1 + im} = 0$$

Let $z \to 0$ along the line $y = mx$, then

$$f'(0) = \lim_{x\to 0} \frac{(x^2 + m^2x^2)}{(x + imx)} = \lim_{x\to 0} \frac{(1 + m^2)\,x}{1 + im} = 0$$

Therefore, $f'(0)$ is unique. Hence the function $f(z)$ s analytic at $z = 0$. **Ans.**

Example 16. *Show that the function $f(z) = u + iv$ where*

$$f(z) = \begin{cases} \dfrac{x^3(1+i) - y^3(1-i)}{x^2 + y^2}, & z \neq 0 \\ & z = 0 \end{cases}$$

satisfies the Cauchy-Riemann equations at $z = 0$. Is the function analytic at $z = 0$? Justify yuor answer. *(MDU Dec. 2009; AMIETE, Dec. 2015)*

Solution. $$f(z) = \frac{x^3(1+i) - y^3(1-i)}{x^2 + y^2} = u + iv$$

$$u = \frac{x^3 - y^3}{x^2 + y^2}, \qquad v = \frac{x^3 + y^3}{x^2 + y^2}$$

[By differentiation the value of $\dfrac{\partial u}{\partial x}, \dfrac{\partial u}{\partial y}, \dfrac{\partial v}{\partial x}, \dfrac{\partial v}{\partial y}$ at (0, 0) we get $\dfrac{0}{0}$, so we apply first principle method]

At the origin

$$\frac{\partial u}{\partial x} = \lim_{h\to 0} \frac{y(0 + h, 0) - u(0, 0)}{h} = \lim_{h\to 0} \frac{\frac{h^3}{h^2}}{h} = 1 \qquad \text{(Along } x\text{-axis)}$$

$$\frac{\partial u}{\partial y} = \lim_{k\to 0} \frac{u(0 + 0 + k) - u(0, 0)}{k} = \lim_{k\to 0} \frac{\frac{-k^3}{k^2}}{k} = -1 \qquad \text{(Along } y\text{-axis)}$$

$$\frac{\partial v}{\partial x} = \lim_{h\to 0} \frac{v(0 + h, 0) - v(0, 0)}{h} = \lim_{h\to 0} \frac{\frac{h^3}{h^2}}{h} = 1 \qquad \text{(Along } x\text{-axis)}$$

$$\frac{\partial v}{\partial y} = \lim_{k\to 0} \frac{v(0, 0 + h) - v(0, 0)}{k} = \lim_{k\to 0} \frac{\frac{k^3}{k^2}}{k} = 1 \qquad \text{(Along } y\text{-axis)}$$

Thus we see that

$$\frac{\partial u}{\partial x} = \frac{\partial v}{\partial y} \quad \text{and} \quad \frac{\partial u}{\partial y} = -\frac{\partial v}{\partial x}$$

Hence, Cauchy-Riemann equations are satisfied at $z = 0$.

Again $$f'(0) = \lim_{z \to 0} \frac{f(0+z) - f(0)}{z} = \lim_{z \to 0} \left[\frac{\dfrac{x^3 - y^3 + i(x^3 + y^3)}{x^2 + y^2} - (0)}{x + iy} \right]$$

$$= \lim_{z \to 0} \left[\frac{x^3 - y^3 + i(x^3 + y^3)}{x^2 + y^2} \cdot \frac{1}{x + iy} \right]$$

Now let $z \to 0$ along $y = x$, then

$$f'(0) = \lim_{x \to 0} \frac{x^3 - x^3 + i(x^3 + x^3)}{x^2 + x^2} \left(\frac{1}{x + ix} \right)$$

$$= \frac{2i}{2(1+i)} = \frac{i}{1+i} = \frac{i(1-i)}{(1+i)(1-i)} = \frac{i+1}{1+1} = \frac{1}{2}(1+i) \qquad \text{...(1)}$$

Again let $z \to 0$ along $y = 0$, then

$$f'(0) = \lim_{x \to 0} \frac{x^3 + ix^3}{x^2} \cdot \frac{1}{x} = (1+i) \qquad [\text{Increment} = z] \quad \text{...(2)}$$

From (1) and (2), we see that $f'(0)$ is not unique. Hence the function $f(z)$ is not analytic at $z = 0$. **Ans.**

Example 17. *Show that the function*

$$f(z) = e^{-z^{-4}}, \quad (z \neq 0) \quad \text{and}$$
$$f(0) = 0$$

is not analytic at $z = 0$.
although, Cauchy-Riemann euquations are satisfied at the point. How would you explain this.

Solution.

$$f(z) = u + iy = e^{-z^{-4}} = e^{-(x+iy)^{-4}} = e^{-\frac{1}{(x+iy)^4}}$$

$$\Rightarrow \qquad u + iv = e^{-\frac{(x-iy)^4}{(x^2+y^2)^4}} = e^{-\frac{1}{(x^2+y^2)^4}[(x^4+y^4-6x^2y^2) - i4xy(x^2-y^2)]}$$

$$\Rightarrow \qquad u + iv = e^{-\frac{x^4+y^4-6x^2y^2}{(x^2+y^2)^4}} \cdot e^{\frac{-i4xy(x^2-y^2)}{(x^2+y^2)^4}}$$

$$\Rightarrow \qquad u + iv = e^{-\frac{x^4+y^4-6x^2y^2}{(x^2+y^2)^4}}$$
$$\left[\cos \frac{4xy(x^2-y^2)}{(x^2+y^2)^4} - i \sin \frac{4xy(x^2-y^2)}{(x^2+y^2)^4} \right]$$

Equating real and imaginary parts, we get

$$u = e^{-\frac{x^4+y^4-6x^2y^2}{(x^2+y^2)^4}} \cos \frac{4xy(x^2-y^2)}{(x^2+y^2)^4}, \; v = e^{-\frac{x^4+y^4-6x^2y^2}{(x^2+y^2)^4}} \sin \frac{4xy(x^2-y^2)}{(x^2+y^2)^4}$$

$$\frac{\partial u}{\partial x} = \lim_{h \to 0} \frac{u(0+h,\, 0) - u(0,\, 0)}{h} = \lim_{h \to 0} \frac{e^{-h^{-4}}}{h} = \lim_{h \to 0} \frac{1}{h e^{\frac{1}{h^4}}}$$

$$= \lim_{h\to 0}\left[\frac{1}{h\left[1+\frac{1}{h^4}+\frac{1}{2!h^8}+\frac{1}{3!h^{12}}+.....\right]}\right], \quad \left(e^x = 1+x+\frac{x^2}{2!}+.....\right)$$

$$= \lim_{h\to 0}\left[\frac{1}{\left[h+\frac{1}{h^3}+\frac{1}{2\,h^7}+\frac{1}{6h^{11}}.....\right]}\right] = \frac{1}{0+\infty} = \frac{1}{\infty} = 0$$

$$\frac{\partial u}{\partial y} = \lim_{k\to 0}\frac{u(0,\,0+k)-u(0,\,0)}{k} = \lim_{k\to 0}\frac{e^{-k^{-4}}}{k} = \lim_{k\to 0}\frac{1}{k\,e^{\frac{1}{k^4}}} = 0$$

$$\frac{\partial v}{\partial x} = \lim_{h\to 0}\frac{v(0+h,\,0)-v(0,\,0)}{h} = \lim_{h\to 0}\frac{e^{-h^{-4}}}{h} = \lim_{h\to 0}\frac{1}{h.e^{\frac{1}{h^4}}} = 0$$

$$\frac{\partial v}{\partial y} = \lim_{k\to 0}\frac{v(0,\,0+k)-v(0,\,0)}{k} = \lim_{k\to 0}\frac{e^{-k^{-4}}}{k} = \lim_{k\to 0}\frac{1}{k.e^{\frac{1}{k^4}}} = 0$$

Hence $\dfrac{\partial u}{\partial x} = \dfrac{\partial v}{\partial y}$ and $\dfrac{\partial u}{\partial y} = -\dfrac{\partial v}{\partial x}$ ($C - R$ equations are satisfied at $z = 0$)

But $$f'(0) = \lim_{z\to 0}\frac{f(z)-f(0)}{z} = \lim_{z\to 0}\frac{e^{-z^{-4}}}{z}$$

Along $z = re^{i\frac{\pi}{4}}$ $$f'(0) = \lim_{r\to 0}\frac{e^{-r^{-4}}\cdot e^{-\left(e^{i\frac{\pi}{4}}\right)^{-4}}}{re^{i\frac{\pi}{4}}} = \lim_{r\to 0}\frac{e^{-r^{-4}}\cdot e^{-\left(\cos\frac{\pi}{4}+i\sin\frac{\pi}{4}\right)^{-4}}}{re^{i\frac{\pi}{4}}}$$

$$= \lim_{r\to 0}\frac{e^{-r^{-4}}\,e^{-\cos\pi}}{re^{i\frac{\pi}{4}}} = \lim_{r\to 0}\frac{e^{-r^{-4}}\cdot e}{re^{i\frac{\pi}{4}}} = \infty$$

Showing that $f'(z)$ does not exist at $z = 0$. Hence $f(z)$ is not analytic at $z = 0$. **Proved.**

Example 18. *Examine the nature of the function*

$$f(z) = \frac{x^2y^5(x+iy)}{x^4+y^{10}};\, z \neq 0$$

$$f(0) = 0$$

in the region including the origin

Solution. Here $f(z) = u + iv = \dfrac{x^2y^5(x+iy)}{x^4+y^{10}};\, z \neq 0$

Equating real and imaginary parts, we get

$$u = \frac{x^3y^5}{x^4+y^{10}}, \quad v = \frac{x^2y^6}{x^4+y^{10}}$$

$$\frac{\partial u}{\partial x}=\lim_{h\to 0}\frac{u(0+h,0)-u(0,0)}{h}=\lim_{h\to 0}\frac{\frac{0}{h^4}}{h}=\lim_{h\to 0}\frac{0}{h}=0$$

$$\frac{\partial u}{\partial y}=\lim_{k\to 0}\frac{u(0,0+k)-u(0,0)}{k}=\lim_{k\to 0}\frac{\frac{0}{k^{10}}}{k}=\lim_{k\to 0}\frac{0}{k}=0$$

$$\frac{\partial v}{\partial x}=\lim_{h\to 0}\frac{v(0+h,0)-v(0,0)}{h}=\lim_{h\to 0}\frac{\frac{0}{h^4}}{h}=\lim_{h\to 0}\frac{0}{h}=0$$

$$\frac{\partial v}{\partial y}=\lim_{k\to 0}\frac{v(0,0+k)-v(0,0)}{k}=\lim_{k\to 0}\frac{\frac{0}{k^{10}}}{k}=\lim_{k\to 0}\frac{0}{k}=0$$

From the above results, it is clear that

$$\frac{\partial u}{\partial x}=\frac{\partial v}{\partial y}\quad\text{and}\quad\frac{\partial u}{\partial y}=-\frac{\partial v}{\partial x}$$

Hence, C-R equations are satisfied at the origin.

But $$f'(0)=\lim_{z\to 0}\frac{f(0+z)-f(0)}{z}=\lim_{\substack{x\to 0\\ y\to 0}}\left[\frac{x^2y^5(x+iy)}{x^4+y^{10}}-0\right]\cdot\frac{1}{x+iy}\quad(\text{Increment}=z)$$

$$=\lim_{\substack{x\to 0\\ y\to 0}}\frac{x^2y^5}{x^4+y^{10}}$$

Let $z\to 0$ along the radius vector $y=mx$, then

$$f'(0)=\lim_{x\to 0}\frac{m^5x^7}{x^4+m^{10}x^{10}}=\lim_{x\to 0}\frac{m^5x^3}{1+m^{10}x^6}=\frac{0}{1}=0 \qquad ...(1)$$

Again let $z\to 0$ along the curve $y^5=x^2$

$$f'(0)=\lim_{x\to 0}\frac{x^4}{x^4+x^4}=\frac{1}{2} \qquad ...(2)$$

(1) and (2) shows that $f'(0)$ does not exist. Hence, $f(z)$ is not analytic at origin although Cauchy-Riemann equations are satisfied there. **Ans.**

Example 19. *Find the values of C_1 and C_2 such that the function*

$$f(z)=x^2+C_1y^2-2xy+i(C_2x^2-y^2+2xy)\ \textit{is analytic. Also find } f'(z).$$

Solution. Let $$f(z)=u+iv=x^2+C_1y^2-2xy+i(C_2x^2-y^2+2xy)$$

Equating real and imaginary parts, we get

$$u=x^2+C_1y^2-2xy\ \text{and}\ v=C_2x^2-y^2+2xy$$

$$\frac{\partial u}{\partial x}=2x-2y\ \text{and}\ \frac{\partial v}{\partial x}=2C_2x+2y$$

$$\frac{\partial u}{\partial y}=2C_1y-2x\ \text{and}\ \frac{\partial v}{\partial y}=-2y+2x$$

$C - R$ equations are

$$\left.\begin{aligned}\frac{\partial u}{\partial x} &= \frac{\partial v}{\partial y}\\ \frac{\partial u}{\partial y} &= -\frac{\partial v}{\partial x}\end{aligned}\right] \Rightarrow \begin{aligned}&2x-2y=-2y+2x & \text{...(1)}\\ &2C_1y-2x=-2C_2x-2y & \text{...(2)}\end{aligned}$$

From (2) equating the coefficient of x and y.

$$2C_1 = -2 \quad \Rightarrow \quad C_1 = -1$$

$$-2 = -2C_2 \Rightarrow \quad C_2 = 1$$

Hence, $\qquad C_1 = -1$ and $C_2 = 1$ **Ans.**

On putting the value of C_2, we get

$$\frac{\partial u}{\partial x} = 2x - 2y, \qquad \frac{\partial v}{\partial x} = 2x + 2y$$

$$f'(z) = \frac{\partial u}{\partial x} + i\frac{\partial v}{\partial x} = (2x-2y) + i(2x+2y) = 2[(x+ix)+(-y+iy)]$$

$$= 2[(1+i)x + i(1+i)y] = 2(1+i)(x+iy) = 2(1+i)z$$

This is the required derivative. **Ans.**

Example 20. *Show that the function defined by* $f(z) = \sqrt{|xy|}$

Satisfies Cauchy-Riemann equation at the origin but is not analytic at that point.

Solution. Let $\qquad f(z) = u + iv = \sqrt{|xy|}$

Equating real and imaginary parts, we get $u = \sqrt{|xy|}$, $v = 0$

At origin

$$\frac{\partial u}{\partial x} = \lim_{h\to 0}\frac{u(0+h,0)-u(0,0)}{h} = \lim_{h\to 0}\frac{0-0}{h} = 0$$

$$\frac{\partial u}{\partial y} = \lim_{k\to 0}\frac{u(0,0+k)-u(0,0)}{k} = \lim_{k\to 0}\frac{0-0}{k} = 0$$

Also

$$\frac{\partial v}{\partial x} = \lim_{h\to 0}\frac{v(0+h,0)-v(0,0)}{h} = \lim_{h\to 0}\frac{0-0}{h} = 0$$

$$\frac{\partial v}{\partial y} = \lim_{k\to 0}\frac{v(0,0+k)-v(0,0)}{k} = \lim_{k\to 0}\frac{0-0}{k} = 0$$

From the above results, it is clear that

$$\frac{\partial u}{\partial x} = \frac{\partial v}{\partial y} \text{ and } \frac{\partial u}{\partial y} = -\frac{\partial v}{\partial x}$$

Hence, $C - R$ equations are satisfied at the origin

$$f'(0) = \lim_{z\to 0}\frac{f(z)-f(0)}{z} = \lim_{z\to 0}\frac{\sqrt{|xy|}-0}{x+iy}$$

Let $z \to 0$ along the line $y = mx$, then

$$f'(0) = \lim_{x\to 0}\frac{\sqrt{|mx^2|}-0}{x(1+im)} = \lim_{x\to 0}\frac{\sqrt{|m|}}{1+im}$$

Thus, the limit on R.H.S. depends upon m and hence will have different values for different values of m.

Therefore, $f'(0)$ is not unique.

Hence the function $f(z)$ is not analytic at $z = 0$. **Ans.**

Example 20a. *Examine the nature of the function*

$$f(z) = \frac{x^2\, y^5\, (x+iy)}{x^4 + y^{10}}\,; z \neq 0$$

$$f(0) = 0$$

in the region including the origin.

Solution. Here $\quad f(z) = u + iv = \dfrac{x^2 y^5\,(x+iy)}{x^4 + y^{10}}\,;\ z \neq 0$

Equating real and imaginary parts, we get

$$u = \frac{x^3 y^5}{x^4 + y^{10}}, \quad v = \frac{x^2 y^6}{x^4 + y^{10}}$$

$$\frac{\partial u}{\partial x} = \lim_{h \to 0} \frac{u(0+h, 0) - u(0, 0)}{h} = \lim_{h \to 0} \frac{\frac{0}{h^4}}{h} = \lim_{h \to 0} \frac{0}{h} = 0$$

$$\frac{\partial u}{\partial y} = \lim_{k \to 0} \frac{u(0, 0+k) - u(0, 0)}{k} = \lim_{k \to 0} \frac{\frac{0}{k^{10}}}{k} = \lim_{k \to 0} \frac{0}{k} = 0$$

$$\frac{\partial v}{\partial x} = \lim_{h \to 0} \frac{v(0+h, 0) - v(0, 0)}{h} = \lim_{h \to 0} \frac{\frac{0}{h^4}}{h} = \lim_{h \to 0} \frac{0}{h} = 0$$

$$\frac{\partial v}{\partial y} = \lim_{k \to 0} \frac{v(0, 0+k) - v(0, 0)}{k} = \lim_{k \to 0} \frac{\frac{0}{k^{10}}}{k} = \lim_{k \to 0} \frac{0}{k} = 0$$

From the above results, it is clear that

$$\frac{\partial u}{\partial x} = \frac{\partial v}{\partial y} \quad \text{and} \quad \frac{\partial u}{\partial y} = -\frac{\partial v}{\partial x}$$

Hence, C-R equations are satisfied at the origin.

But $\quad f'(0) = \lim\limits_{z \to 0} \dfrac{f(0+z) - f(0)}{z} = \lim\limits_{\substack{x \to 0 \\ y \to 0}} \left[\dfrac{x^2 y^5\,(x+iy)}{x^4 + y^{10}} - 0\right] \cdot \dfrac{1}{x+iy} \quad$ (Increment $= z$)

$$= \lim_{\substack{x \to 0 \\ y \to 0}} \frac{x^2 y^5}{x^4 + y^{10}}$$

Let $z \to 0$ along the radius vector $y = mx$, then

$$f'(0) = \lim_{x \to 0} \frac{m^5 x^7}{x^4 + m^{10} x^{10}} = \lim_{x \to 0} \frac{m^5 x^3}{1 + m^{10} x^6} = \frac{0}{1} = 0 \qquad \text{... (1)}$$

Again let $z \to 0$ along the curve $y^5 = x^2$

$$f'(0) = \lim_{x \to 0} \frac{x^4}{x^4 + x^4} = \frac{1}{2} \qquad \text{... (2)}$$

(1) and (2) shows that $f'(0)$ does not exist. Hence, $f(z)$ is not analytic at origin although Cauchy-Riemann equations are satisfied there. **Ans.**

22.12 C–R EQUATIONS IN POLAR FORM

$$\frac{\partial u}{\partial r} = \frac{1}{r}\frac{\partial v}{\partial \theta}$$

$$\frac{\partial u}{\partial \theta} = -r\,\frac{\partial v}{\partial r}$$ *(DU, III Sem. 2012, RGPV., Bhopal, III Sem. Dec. 2007)*

Proof. We know $x = r\cos\theta$, and u is a function of x and y.

$$z = x+iy = r(\cos\theta + i\sin\theta) = r\,e^{i\theta}$$

$$u+iv = f(z) = f(r\,e^{i\theta}) \qquad \text{...(1)}$$

Differentiating (1) partially w.r.t., "r", we get

$$\frac{\partial u}{\partial r} + i\frac{\partial v}{\partial r} = f'(r\,e^{i\theta}).e^{i\theta} \qquad \text{...(2)}$$

Differentiating (1) w.r.t. "θ", we get

$$\frac{\partial u}{\partial \theta} + i\frac{\partial v}{\partial \theta} = f'(r\,e^{i\theta})\,r\,e^{i\theta}\,i \qquad \text{...(3)}$$

Substituting the value of $f'(r\,e^{i\theta})e^{i\theta}$ from (2) in (3), we obtain

$$\frac{\partial u}{\partial \theta} + i\frac{\partial v}{\partial \theta} = r\left(\frac{\partial u}{\partial r} + i\frac{\partial v}{\partial r}\right)i \quad \text{or} \quad \frac{\partial u}{\partial \theta} + i\frac{\partial v}{\partial \theta} = ir\,\frac{\partial u}{\partial r} - r\,\frac{\partial v}{\partial r}$$

Equating real and imaginary parts, we get

$$\boxed{\frac{\partial u}{\partial \theta} = -r\,\frac{\partial v}{\partial r}} \quad \Rightarrow \quad \frac{\partial v}{\partial r} = \frac{-1}{r}\frac{\partial u}{\partial \theta}$$

And $$\boxed{\frac{\partial u}{\partial r} = \frac{1}{r}\frac{\partial v}{\partial \theta}}$$ **Proved.**

22.13 DERIVATIVE OF *W* IN POLAR FORM

We know that $w = u + iv$, $\dfrac{\partial w}{\partial x} = \dfrac{\partial u}{\partial x} + i\dfrac{\partial v}{\partial x}$

But $$\frac{dw}{dz} = \frac{\partial w}{\partial r}\frac{\partial r}{\partial x} + \frac{\partial w}{\partial \theta}\frac{\partial \theta}{\partial x} = \frac{\partial w}{\partial r}\cos\theta - \left(\frac{\partial u}{\partial \theta} + i\frac{\partial v}{\partial \theta}\right)\frac{\sin\theta}{r}$$

$$= \frac{\partial w}{\partial r}\cos\theta - \left(-r\frac{\partial v}{\partial r} + i\cdot r\frac{\partial u}{\partial r}\right)\frac{\sin\theta}{r} \qquad \frac{\partial u}{\partial \theta} = -r\frac{\partial v}{\partial r}$$

$$= \frac{\partial w}{\partial r}\cos\theta - i\left(\frac{\partial u}{\partial r} + i\frac{\partial v}{\partial r}\right)\sin\theta \qquad \frac{\partial v}{\partial \theta} = r\frac{\partial u}{\partial r}$$

$$= \frac{\partial w}{\partial r}\cos\theta - i\frac{\partial}{\partial r}(u+iv)\sin\theta = \frac{\partial w}{\partial r}\cos\theta - i\,\frac{\partial w}{\partial r}\sin\theta \qquad [\ldots\ w = u + iv]$$

$$= (\cos\theta - i\sin\theta)\,\frac{\partial w}{\partial r} \qquad \text{...(1)}$$

Second form of $\dfrac{dw}{dz}$

$$\frac{dw}{dz} = \frac{\partial w}{\partial r}\frac{\partial r}{\partial x} + \frac{\partial w}{\partial \theta}\frac{\partial \theta}{\partial x} = \frac{\partial(u+iv)}{\partial r}\cos\theta - \frac{\partial w}{\partial \theta}\frac{\sin\theta}{r} \qquad [w = u + iv]$$

$$= \left(\frac{\partial u}{\partial r} + i\frac{\partial v}{\partial r}\right)\cos\theta - \frac{\partial w}{\partial \theta}\frac{\sin\theta}{r}$$

$$= \left(\frac{1}{r}\frac{\partial v}{\partial \theta} - i\frac{1}{r}\cdot\frac{\partial u}{\partial \theta}\right)\cos\theta - \frac{\partial w}{\partial \theta}\cdot\frac{\sin\theta}{r} \qquad \left[\begin{array}{l}\frac{\partial u}{\partial \theta} = -r\frac{\partial v}{\partial r}\\ \frac{\partial v}{\partial \theta} = r\frac{\partial u}{\partial r}\end{array}\right]$$

$$= -\frac{i}{r}\left(\frac{\partial u}{\partial \theta} + i\frac{\partial v}{\partial \theta}\right)\cos\theta - \frac{\partial w}{\partial \theta}\left(\frac{\sin\theta}{r}\right)$$

$$= -\frac{i}{r}\frac{\partial}{\partial \theta}(u+iv)\cos\theta - \frac{\partial w}{\partial \theta}\left(\frac{\sin\theta}{r}\right) \qquad [w = 0]$$

$$= -\frac{i}{r}\frac{\partial w}{\partial \theta}\cos\theta - \frac{\partial w}{\partial \theta}\frac{\sin\theta}{r} = -\frac{i}{r}(\cos\theta - i\sin\theta)\frac{\partial w}{\partial \theta} \qquad \text{... (2)}$$

$$\boxed{\frac{dw}{dz} = (\cos\theta - i\sin\theta)\frac{\partial w}{\partial r}} \qquad \left[-\frac{i}{r}\frac{\partial w}{\partial \theta} = \frac{\partial w}{\partial r}\right]$$

$$\boxed{\frac{dw}{dz} = -\frac{i}{r}(\cos\theta - i\sin\theta)\frac{\partial w}{\partial \theta}}$$

These are the two forms for $\frac{dw}{dz}$.

Example 21. *If n is real, show that r^n (cos nθ + i sin nθ) is analytic except possible when r = 0 and that its derivative is*

$$nr^{n-1}[\cos(n-1)\theta + i\sin(n-1)\theta].$$

Solution. Let $w = f(z) = u + iv = r^n(\cos n\theta + i\sin n\theta)$

Here, $u = r^n\cos n\theta, \quad v = r^n\sin n\theta$

then, $\frac{\partial u}{\partial r} = nr^{n-1}\cos n\theta, \quad \frac{\partial v}{\partial r} = nr^{n-1}\sin n\theta$

$$\frac{\partial u}{\partial \theta} = -nr^n\sin n\theta, \quad \frac{\partial v}{\partial \theta} = nr^n\cos n\theta$$

Here, $\frac{\partial u}{\partial r} = nr^{n-1}\cos n\theta = \frac{1}{r}(nr^n\cos n\theta)$

$$\frac{\partial u}{\partial r} = \frac{1}{r}\frac{\partial v}{\partial \theta} \qquad \text{...(1)}$$

and $\frac{\partial v}{\partial r} = nr^{n-1}\sin n\theta \quad \frac{\partial v}{\partial r} = -\frac{1}{r}(-nr^n\sin n\theta)$

$$\frac{\partial v}{\partial r} = -\frac{1}{r}\frac{\partial u}{\partial \theta} \qquad \text{...(2)}$$

Equations (1) and (2) satisfied C-R equations.

We have, $\frac{dw}{dz} = (\cos\theta - i\sin\theta)\frac{\partial w}{\partial r} = (\cos\theta - i\sin\theta)\left(\frac{\partial u}{\partial r} + i\frac{\partial v}{\partial r}\right)$

$$= (\cos\theta - i\sin\theta)(nr^{n-1}\cos n\theta + inr^{n-1}\sin n\theta)$$

$$= (\cos\theta - i\sin\theta)\, nr^{n-1} (\cos n\theta + i\sin n\theta)$$
$$= nr^{n-1}\{(\cos n\theta\cos\theta + \sin n\theta\sin\theta) + i(\sin n\theta\cos\theta - \cos n\theta\sin\theta)\}$$
$$= nr^{n-1}\{\cos(n-1)\theta + i\sin(n-1)\theta\}$$

This exists for all finite values of r including zero, except when $r = 0$ and $n \le 1$. **Proved.**

EXERCISE 22.3

Determine which of the following functions are analytic:

1. $x^2 + iy^2$ **Ans.** Analytic at all points $y = x$

2. $2xy + i(x^2 - y^2)$ **Ans.** Not analytic

3. $\dfrac{x - iy}{x^2 + y^2}$ **Ans.** Not analytic

4. $\dfrac{1}{(z-1)(z+1)}$ **Ans.** Analytic at all points, except $z = \pm 1$

5. $\dfrac{x - iy}{x - iy + a}$ **Ans.** Not analytic

6. $\sin x \cosh y + \cos x \sinh y$ **Ans.** Yes, analytic

7. $xy + iy^2$ **Ans.** Yes, analytic at origin

8. Discuss the analyticity of the function $f(z) = z\bar{z} + \bar{z}^2$ in the complex plane, where $\bar{z}$ is the complex conjugate of z. Also find the points where it is differentiable but not analytic.

Ans. Differentiable only at $z = 0$, No where analytic.

9. Show the function of $\bar{z}$ is not analytic any where.

10. If $\begin{cases} \dfrac{x^2 y(y - ix)}{x^4 + y^2}, & \text{when } z \ne 0 \\ 0, & \text{when } z = 0 \end{cases}$

prove that $\dfrac{f(z) - f(0)}{z} \to 0$, as $z \to 0$, along any radius vector but not as $z \to 0$ in any manner.

(AMIETE, Dec. 2010)

11. If $f(z)$ is an analytic function with constant modulus, show that $f(z)$ is constant. *(AMIETE, Dec. 2009)*

Choose the correct answer :

12. The Cauchy-Riemann equations for $f(z) = u(x, y) + iv(x, y)$ to be analytic are :

(*a*) $\dfrac{\partial^2 u}{\partial x^2} + \dfrac{\partial^2 u}{\partial y^2} = 0,\ \dfrac{\partial^2 v}{\partial x^2} + \dfrac{\partial^2 v}{\partial y^2} = 0$ (*b*) $\dfrac{\partial u}{\partial x} = -\dfrac{\partial v}{\partial y},\ \dfrac{\partial u}{\partial y} = -\dfrac{\partial v}{\partial x}$

(*c*) $\dfrac{\partial u}{\partial x} = \dfrac{\partial v}{\partial y},\ \dfrac{\partial u}{\partial y} = -\dfrac{\partial v}{\partial x}$ (*d*) $\dfrac{\partial u}{\partial x} = -\dfrac{\partial v}{\partial y},\ \dfrac{\partial u}{\partial y} = \dfrac{\partial v}{\partial x}$ **Ans.** (*c*)

(R.G.P.V., Bhopal, III Semester, Dec. 2006)

13. Polar form of *C-R* equations are :

(*a*) $\dfrac{\partial u}{\partial \theta} = \dfrac{1}{r}\dfrac{\partial v}{\partial r},\ \dfrac{\partial u}{\partial r} = r\dfrac{\partial v}{\partial \theta}$ (*b*) $\dfrac{\partial u}{\partial \theta} = r\dfrac{\partial v}{\partial r},\ \dfrac{\partial u}{\partial r} = \dfrac{1}{r}\dfrac{\partial v}{\partial \theta}$

(*c*) $\dfrac{\partial u}{\partial r} = \dfrac{1}{r}\dfrac{\partial v}{\partial \theta},\ \dfrac{\partial u}{\partial \theta} = -r\dfrac{\partial v}{\partial r}$ (*d*) $\dfrac{\partial u}{\partial r} = r\dfrac{\partial v}{\partial \theta},\ \dfrac{\partial u}{\partial \theta} = -r\dfrac{\partial v}{\partial r}$ **Ans.** (*c*)

(R.G.P.V., Bhopal, III Semester, June, 2007)

14. The curve $u(x, y) = C$ and $v(x, y) = C'$ are orthogonal if

(*a*) u and v are complex functions (*b*) $u + iv$ is an analytic function.

(*c*) $u - v$ is an analytic function. (*d*) $u + v$ is an analytic function **Ans.** (*b*)

15. If $f(z) = \dfrac{1}{2}\log_e(x^2 + y^2) + i\tan^{-1}\left(\dfrac{\alpha x}{y}\right)$ be an analytic function if α is equal to

(*a*) $+1$ (*b*) -1 (*c*) $+2$ (*d*) -2 *(AMIETE, Dec. 2009)* **Ans.** (*b*)

22.14 ORTHOGONAL CURVES

(U.P. III Semester, June 2009)

Two curves are said to be orthogonal to each other, when they intersect at right angle at each of their points of intersection.

The analytic function $f(z) = u + iv$ consists of two families of curves $u(x, y) = c_1$ and $v(x, y) = c_2$ which form an orthogonal system.

$$u(x, y) = c_1 \quad \text{...(1)}$$

$$v(x, y) = c_2 \quad \text{...(2)}$$

Differentiating (1), $\frac{\partial u}{\partial x}dx + \frac{\partial u}{\partial y}dy = 0$

$$\Rightarrow \quad \frac{dy}{dx} = -\frac{\frac{\partial u}{\partial x}}{\frac{\partial u}{\partial y}} = m_1 \text{ (say)}$$

Similarly from (2), $\frac{dy}{dx} = -\frac{\frac{\partial v}{\partial x}}{\frac{\partial v}{\partial y}} = m_2 \text{ (say)}$

Y

u (x, y) = c1

v (x, y) = c2

O X

The product of two slopes

$$m_1 m_2 = \left(-\frac{\frac{\partial u}{\partial x}}{\frac{\partial u}{\partial y}}\right)\left(-\frac{\frac{\partial v}{\partial x}}{\frac{\partial v}{\partial y}}\right) = \left(-\frac{\frac{\partial u}{\partial x}}{\frac{\partial u}{\partial y}}\right)\left(-\frac{-\frac{\partial u}{\partial y}}{\frac{\partial u}{\partial x}}\right) \qquad (C-R \text{ equations})$$

$$= -1$$

Since $m_1\, m_2 = -1$, two curves $u = c_1$ and $v = c_2$ are orthogonal, and c_1, c_2 are parameters, $u = c_1$ and $v = c_2$ form an orthogonal system.

22.15 HARMONIC FUNCTION

(U.P., III SEMESTER 2009-2010)

Any function which satisfies the Laplace's equation is known as a harmonic function.

Theorem. If $f(z) = u + iv$ is an analytic function, then u and v are both harmonic functions.

Proof. Let $f(z) = u + iv$, be an analytic function, then we have

$$\left.\begin{aligned} \frac{\partial u}{\partial x} &= \frac{\partial v}{\partial y} \quad &\text{...(1)} \\ \frac{\partial u}{\partial y} &= -\frac{\partial v}{\partial x} \quad &\text{...(2)} \end{aligned}\right] \; C-R \text{ equations.}$$

Differentiating (1) with respect to x, we get $\frac{\partial^2 u}{\partial x^2} = \frac{\partial^2 v}{\partial x \partial y}$ ... (3)

Differentiating (2) w.r.t. 'y' we have $\frac{\partial^2 u}{\partial y^2} = -\frac{\partial^2 v}{\partial y \partial x}$ ... (4)

Adding (3) and (4) we have $\frac{\partial^2 u}{\partial x^2} + \frac{\partial^2 u}{\partial y^2} = \frac{\partial^2 v}{\partial x \partial y} - \frac{\partial^2 v}{\partial y \partial x}$

$$\frac{\partial^2 u}{\partial x^2} + \frac{\partial^2 u}{\partial y^2} = 0 \qquad \left(\frac{\partial^2 v}{\partial x \partial y} = \frac{\partial^2 v}{\partial y \partial x}\right)$$

Similarly $$\frac{\partial^2 v}{\partial x^2}+\frac{\partial^2 v}{\partial y^2}=0$$

Therefore both u and v are harmonic functions.

Such functions u, v are called **Conjugate harmonic functions if $u + iv$ is also analytic function.**

Example 22. *Define a harmonic function and conjugate harmonic function. Find the harmonic conjugate function of the function $U(x, y) = 2x(1 - y)$.* (U.P., III Semester Dec. 2009)

Solution. See Art. 22.15

Here, we have $U(x, y) = 2x(1 - y)$ Let V be the harmonic conjugate of U.

By total differentiation

$$dV = \frac{\partial V}{\partial x}dx+\frac{\partial V}{\partial y}dy$$

$$= -\frac{\partial U}{\partial y}dx+\frac{\partial U}{\partial x}dy$$

$$= -(-2x)\,dx + (2 - 2y)\,dy$$

$$= 2x\,dx + (2\,dy - 2y\,dy)$$

$$V = x^2 + 2y - y^2 + C$$

Hence, the harmonic conjugate of U is $x^2 + 2y - y^2 + C$ **Ans.**

Example 23. *Show that the function $u=\frac{1}{2}\log(x^2+y^2)$ is harmonic. Find its harmonic conjugate.*

Solution. $$u=\frac{1}{2}\log(x^2+y^2)$$

$$\frac{\partial u}{\partial x}=\frac{1}{2}\frac{1}{x^2+y^2}.(2x)=\frac{x}{x^2+y^2} \qquad \frac{\partial u}{\partial y}=\frac{y}{x^2+y^2}$$

$$\frac{\partial^2 u}{\partial x^2}=\frac{(x^2+y^2).1-x(2x)}{(x^2+y^2)^2}=\frac{y^2-x^2}{(x^2+y^2)^2}$$

$$\frac{\partial^2 u}{\partial y^2}=\frac{(x^2+y^2).1-y(2y)}{(x^2+y^2)^2}=\frac{x^2-y^2}{(x^2+y^2)^2} \quad \Rightarrow \quad \frac{\partial^2 u}{\partial x^2}+\frac{\partial^2 u}{\partial y^2}=0$$

Hence u is a harmonic function.

Let v be the harmonic conjugate of u.

$$dv=\frac{\partial v}{\partial x}dx+\frac{\partial v}{\partial y}dy=-\frac{\partial u}{\partial y}dx+\frac{\partial u}{\partial x}dy \qquad \text{(By C – R equations)}$$

$$dv=-\frac{y}{x^2+y^2}dx+\frac{x}{x^2+y^2}dy$$

$$dv=\frac{xdy-ydx}{x^2+y^2}=d\left(\tan^{-1}\frac{y}{x}\right)$$

Integrating, we get $v=\tan^{-1}\frac{y}{x}+C.$, where C is a real constant.

This is the required harmonic conjugate. **Ans.**

Example 24. *Prove that* $u = x^2 - y^2$ *and* $v = \frac{y}{x^2 + y^2}$ *are harmonic functions of* (x, y), *but are not harmonic conjugates.*

Solution. We have, $u = x^2 - y^2$

$$\frac{\partial u}{\partial x} = 2x, \quad \frac{\partial^2 u}{\partial x^2} = 2, \quad \frac{\partial u}{\partial y} = -2y, \quad \frac{\partial^2 u}{\partial y^2} = -2$$

$$\frac{\partial^2 u}{\partial x^2} + \frac{\partial^2 u}{\partial y^2} = 2 - 2 = 0$$

$u(x, y)$ satisfies Laplace equation, hence $u(x, y)$ is harmonic

$$v = \frac{y}{x^2 + y^2}, \quad \frac{\partial v}{\partial x} = -\frac{2xy}{(x^2 + y^2)^2}$$

$$\frac{\partial^2 v}{\partial x^2} = \frac{(x^2 + y^2)^2(-2y) - (-2xy)2(x^2 + y^2)2x}{(x^2 + y^2)^4}$$

$$= \frac{(x^2 + y^2)(-2y) - (-2xy)4x}{(x^2 + y^2)^3} = \frac{6x^2y - 2y^3}{(x^2 + y^2)^3} \quad \text{... (1)}$$

$$\frac{\partial v}{\partial y} = \frac{(x^2 + y^2) \cdot 1 - y(2y)}{(x^2 + y^2)^2} = \frac{x^2 - y^2}{(x^2 + y^2)^2}$$

$$\frac{\partial^2 v}{\partial y^2} = \frac{(x^2 + y^2)^2(-2y) - (x^2 - y^2)2(x^2 + y^2)(2y)}{(x^2 + y^2)^4} = \frac{(x^2 + y^2)(-2y) - (x^2 - y^2)(4y)}{(x^2 + y^2)^3}$$

$$= \frac{-2x^2y - 2y^3 - 4x^2y + 4y^3}{(x^2 + y^2)^3} = \frac{-6x^2y + 2y^3}{(x^2 + y^2)^3} \quad \text{... (2)}$$

On adding (1) and (2), we get $\frac{\partial^2 v}{\partial x^2} + \frac{\partial^2 v}{\partial y^2} = 0$

$v(x, y)$ also satisfies Laplace equations, hence $v(x, y)$ is also harmonic function.

But $\quad \frac{\partial u}{\partial x} \neq \frac{\partial v}{\partial y} \quad$ and $\quad \frac{\partial u}{\partial y} \neq -\frac{\partial v}{\partial x}$

Therefore u and v are not harmonic conjugates. **Proved.**

Example 25. *Find A and B so that*

$f(x, y) = x^2 + A\,xy + By^2$ *is harmonic.* (*Delhi University, April 2010*)

Solution. Here, we have

$$f(x, y) = x^2 + Axy + By^2$$

$$\frac{\partial f}{\partial x} = 2x + Ay \Rightarrow \frac{\partial^2 f}{\partial x^2} = 2 \quad \text{... (1)}$$

$$\frac{\partial f}{\partial y} = Ax + 2By \Rightarrow \frac{\partial^2 f}{\partial y^2} = 2B \quad \text{... (2)}$$

Adding (1) (2), we get

$$\frac{\partial^2 f}{\partial x^2} + \frac{\partial^2 f}{\partial y^2} = 2 + 2B \quad \text{... (3)}$$

But for harmonic function

$$\frac{\partial^2 f}{\partial x^2}+\frac{\partial^2 f}{\partial y^2}=0 \qquad \dots (4)$$

From (3) and (4)

$$2 + 2B = 0 \quad \Rightarrow \quad B = -1$$ **Ans.**

Example 26. *Show that the function $x^2 - y^2 + 2y$ which is harmonic remains harmonic under the transformation $z = w^3$*

Solution. $u = x^2 - y^2 + 2y$

$$\frac{\partial u}{\partial x}=2x, \qquad \frac{\partial^2 u}{\partial x^2}=2 \qquad \dots (1)$$

$$\Rightarrow \qquad \frac{\partial u}{\partial y}=-2y+2 \qquad \frac{\partial^2 u}{\partial y^2}=-2 \qquad \dots (2)$$

Adding (1) and (2), we get

$$\frac{\partial^2 u}{\partial x^2}+\frac{\partial^2 u}{\partial y^2}=2-2=0 \Rightarrow \frac{\partial^2 u}{\partial x^2}+\frac{\partial^2 u}{\partial y^2}=0$$

Hence function is harmonic.

Transformation: $z=w^3$, $z=re^{i\theta}$ and $w=Re^{i\phi}$

$$\Rightarrow \qquad re^{i\theta}=(Re^{i\phi})^3 \quad \Rightarrow \quad re^{i\theta}=R^3e^{3i\phi}$$

By comparing both side $r = R^3$, $\theta = 3\phi$

Given function, $f(x, y) = x^2 - y^2 + 2y$ where $x = r\cos\theta$ and $y = r\sin\theta$

$$f(r\cos\theta,\ r\sin\theta)=(r\cos\theta)^2-(r\sin\theta)^2+2\times r\sin\theta$$

$$= r^2\cos^2\theta-r^2\sin^2\theta+2r\sin\theta$$

$$= r^2(\cos^2\theta-\sin^2\theta)+2r\sin\theta=r^2\cos 2\theta+2r\sin\theta$$

$$f(R^3\cos 3\phi,\ R^3\sin 3\phi)=R^6\cos 6\phi+2R^3\sin 3\phi$$

This is a function in cosine and sine. Hence it will be harmonic function. **Proved.**

Example 27. *If $u(x, y)$ and $v(x, y)$ are harmonic functions in a region R, prove that the function*

$$\left[\left(\frac{\partial u}{\partial y}-\frac{\partial v}{\partial x}\right)+i\left(\frac{\partial u}{\partial x}+\frac{\partial v}{\partial y}\right)\right]$$

is an analytic function of $z = x + iy$.

Solution. Since $u(x, y)$ and $v(x, y)$ are harmonic functions in a region R, therefore

$$\frac{\partial^2 u}{\partial x^2}+\frac{\partial^2 u}{\partial y^2}=0 \qquad \dots (1)$$

and

$$\frac{\partial^2 v}{\partial x^2}+\frac{\partial^2 v}{\partial y^2}=0 \qquad \dots (2)$$

Let

$$F(z)=R+iS=\left(\frac{\partial u}{\partial y}-\frac{\partial v}{\partial x}\right)+i\left(\frac{\partial u}{\partial x}+\frac{\partial v}{\partial y}\right)$$

Equating real and imaginary parts, we get

$$R = \frac{\partial u}{\partial y} - \frac{\partial v}{\partial x},$$

$$\frac{\partial R}{\partial x} = \frac{\partial^2 u}{\partial x \partial y} - \frac{\partial^2 v}{\partial x^2} \quad \text{... (3)}$$

$$\frac{\partial R}{\partial y} = \frac{\partial^2 u}{\partial y^2} - \frac{\partial^2 v}{\partial x \partial y} \quad \text{... (4)}$$

$$S = \frac{\partial u}{\partial x} + \frac{\partial v}{\partial y}$$

$$\frac{\partial S}{\partial x} = \frac{\partial^2 u}{\partial x^2} + \frac{\partial^2 v}{\partial x \partial y} \quad \text{... (5)}$$

$$\frac{\partial S}{\partial y} = \frac{\partial^2 u}{\partial x \partial y} + \frac{\partial^2 v}{\partial y^2} \quad \text{... (6)}$$

Putting the value of $\frac{\partial^2 u}{\partial x^2}$ from (1) in (5), we get

$$\frac{\partial S}{\partial x} = -\frac{\partial^2 u}{\partial y^2} + \frac{\partial^2 v}{\partial x \partial y} \quad \text{... (7)}$$

Putting the value of $\frac{\partial^2 v}{\partial y^2}$ from (2) in (6), we get

$$\frac{\partial S}{\partial y} = \frac{\partial^2 u}{\partial x \partial y} - \frac{\partial^2 v}{\partial x^2} \quad \text{... (8)}$$

From (3) and (8), $\frac{\partial R}{\partial x} = \frac{\partial S}{\partial y}$

From (4) and (7), $\frac{\partial R}{\partial y} = -\frac{\partial S}{\partial x}$

Therefore, C-R equations are satisfied and hence the given function is analytic. **Proved.**

22.16 APPLICATIONS TO FLOW PROBLEMS

As the real part u and imaginary part v of an analytic function $f(z)$ are the solution of Laplace equation.

$$\frac{\partial^2 u}{\partial x^2} + \frac{\partial^2 u}{\partial y^2} = 0$$

and

$$\frac{\partial^2 v}{\partial x^2} + \frac{\partial^2 v}{\partial y^2} = 0$$

So, we get a solution to a number of field and flow problems.

22.17 VELOCITY POTENTIAL

Consider the two dimensional irrotational motion of an incompressible fluid, in planes parallel to *xy*-plane.

Let $\vec{V}$ be the velocity of a fluid particle, then it can be expressed as

$$\vec{V} = v_x\,\hat{i} + v_y\,\hat{j} \qquad ...(1)$$

Since the motion is irrotational, there exists a scalar function $\phi(x, y)$, such that

$$\vec{V} = \nabla\phi(x,y) = \frac{\partial\phi}{\partial x}\hat{i} + \frac{\partial\phi}{\partial y}\hat{j} \qquad ...(2)$$

From (1) and (2), we have $v_x = \dfrac{\partial\phi}{\partial x}$ and $v_y = \dfrac{\partial\phi}{\partial y}$ ...(3)

The scalar function $\phi(x, y)$, which gives the velocity components, is called the velocity potential function or simply the velocity potential.

Also the fluid being incompressible, div $\vec{V} = 0$

$$\Rightarrow \left(\hat{i}\frac{\partial}{\partial x} + \hat{j}\frac{\partial}{\partial y}\right).(v_x\hat{i} + v_y\hat{j}) = 0$$

$$\Rightarrow \frac{\partial v_x}{\partial x} + \frac{\partial v_y}{\partial y} = 0 \qquad ...(4)$$

Substituting the values of v_x and v_y from (3) in (4), we get

$$\frac{\partial}{\partial x}\left(\frac{\partial\phi}{\partial x}\right) + \frac{\partial}{\partial y}\left(\frac{\partial\phi}{\partial y}\right) = 0 \;\Rightarrow\; \left(\frac{\partial^2\phi}{\partial x^2}\right) + \left(\frac{\partial^2\phi}{\partial y^2}\right) = 0$$

Thus the function ϕ is harmonic.

This is the real part of an analytic function.

$$f(z) = \phi\,(x, y) + i\,\psi(x, y)$$

$$\frac{df}{dz} = \frac{\partial\phi}{\partial x} + i\frac{\partial\psi}{\partial x} = \frac{\partial\phi}{\partial x} - i\frac{\partial\phi}{\partial y} \qquad \left(\frac{\partial\psi}{\partial x} = -\frac{\partial\phi}{\partial y}\right)$$

$$= v_x - iv_y$$

The magnitude of the resultant velocity

$$= \left|\frac{df}{dz}\right| = \sqrt{v_x^2 + v_y^2}.$$

$\phi(x, y) = C_1$ and $\psi(x, y) = C_2$ are called equipotential lines and lines of force respectively.

In heat flow problem the curves $\phi\,(x, y) = C_1$ and $\psi\,(x, y) = C_2$ are known as isothermals and heat flow lines respectively.

22.18 METHOD TO FIND THE CONJUGATE FUNCTION

Case I. Given. If $f(z) = u + iv$, and u is known.

To find. v, conjugate function.

Method. We know that $\quad dv = \dfrac{\partial v}{\partial x}\cdot dx + \dfrac{\partial v}{\partial y}\cdot dy$ ...(1)

Replacing $\frac{\partial v}{\partial x}$ *by* $-\frac{\partial u}{\partial y}$ and $\frac{\partial v}{\partial y}$ *by* $\frac{\partial u}{\partial x}$ in (1), we get [C-R equations]

$$dv = -\frac{\partial u}{\partial y}\cdot dx + \frac{\partial u}{\partial x}\cdot dy$$

$$v = -\int \frac{\partial u}{\partial y} dx + \int \frac{\partial u}{\partial x}\cdot dy$$

$\Rightarrow$
$$v = \int M\,dx + \int N\,dy \qquad ...(2)$$

where
$$M = -\frac{\partial u}{\partial y} \text{ and } N = \frac{\partial u}{\partial x}$$

so that
$$\frac{\partial M}{\partial y} = \frac{\partial}{\partial y}\left(-\frac{\partial u}{\partial y}\right) = -\frac{\partial^2 u}{\partial y^2} \text{ and } \frac{\partial N}{\partial x} = \frac{\partial}{\partial x}\left(\frac{\partial u}{\partial x}\right) = \frac{\partial^2 u}{\partial x^2}$$

since u is a conjugate function, so $\frac{\partial^2 u}{\partial x^2} + \frac{\partial^2 u}{\partial y^2} = 0$

$\Rightarrow$
$$-\frac{\partial^2 u}{\partial y^2} = \frac{\partial^2 u}{\partial x^2} \quad \Rightarrow \quad \frac{\partial M}{\partial y} = \frac{\partial N}{\partial x} \qquad ...(3)$$

Equation (3) satisfies the condition of an exact differential equation.
So equation (2) can be integrated and thus v is determined.

Case II. Similarly, if $v = v(x, y)$ is given
To find out u.

We know that
$$du = \frac{\partial u}{\partial x} dx + i\frac{\partial u}{\partial y} dy \qquad ...(4)$$

On substituting the values of $\frac{\partial u}{\partial x}$ and $\frac{\partial u}{\partial y}$ in (4) , we get

$$du = \frac{\partial v}{\partial y} dx - \frac{\partial v}{\partial x} dy$$

On integrating, we get

$$u = \int \frac{\partial v}{\partial y} dx - \int \frac{\partial v}{\partial x} dy \qquad ...(5)$$

(since v is already known so $\frac{\partial v}{\partial y}, \frac{\partial v}{\partial x}$ on R.H.S. are also known)

Equation (5) is an exact differential equation. On solving (5), u can be determined.
Consequently $f(z) = u + iv$ can also be determined.

Example 28. *Let $f(z) = u(x, y) + iv(x, y)$ be an analytic function. If $u = 3x - 2xy$, then find v and express $f(z)$ in terms of z.*

Solution. Here, we have $u = 3x - 2xy$

$$\frac{\partial u}{\partial x} = 3 - 2y, \qquad \frac{\partial u}{\partial y} = -2x$$

We know that
$$dv = \frac{\partial v}{\partial x} dx + \frac{\partial v}{\partial y} dy \qquad \text{(Total differentiation)}$$

$$= \left(-\frac{\partial u}{\partial y}\right) dx + \left(\frac{\partial u}{\partial x}\right) dy \qquad \left(\frac{\partial v}{\partial x} = -\frac{\partial u}{\partial y}, \frac{\partial v}{\partial y} = \frac{\partial u}{\partial x}\right)$$

$$= 2\,x\,dx + (3 - 2\,y)\,dy$$

$$v = \int 2x\,dx + \int (3 - 2y)\,dy = x^2 + 3y - y^2 + c$$

$$f(z) = u\,(x, y) + iv(x, y) = (3x - 2xy) + i(x^2 + 3y - y^2 + c)$$

$$= (ix^2 - iy^2 - 2xy) + (3x + 3yi) + ic = i(x^2 - y^2 + 2ixy) + 3(x + iy) + ic$$

$$= i(x + iy)^2 + 3(x + iy) + ic = iz^2 + 3z + ic \qquad \textbf{Ans.}$$

Which is the required expression of $f(z)$ in terms of z.

Example 29. *Prove that $u = x^2 - y^2 - 2xy - 2x + 3y$ is harmonic. Find a function v such that $f(z) = u + iv$ is analytic. Also express $f(z)$ in terms of z.*

(R.G.P.V., Bhopal, III Semester, June 2005)

Solution. We have,

$$u = x^2 - y^2 - 2xy - 2x + 3y$$

$$\frac{\partial u}{\partial x} = 2x - 2y - 2 \quad \Rightarrow \quad \frac{\partial^2 u}{\partial x^2} = 2$$

$$\frac{\partial u}{\partial y} = -2y - 2x + 3 \quad \Rightarrow \quad \frac{\partial^2 u}{\partial y^2} = -2$$

$$\frac{\partial^2 u}{\partial x^2} + \frac{\partial^2 u}{\partial y^2} = 2 - 2 \quad \Rightarrow \quad \frac{\partial^2 u}{\partial x^2} + \frac{\partial^2 u}{\partial y^2} = 0$$

Since Laplace equation is satisfied, therefore u is harmonic. **Proved.**

We know that $$dv = \frac{\partial v}{\partial x} dx + \frac{\partial v}{\partial y} dy$$

$$\Rightarrow \quad dv = -\frac{\partial u}{\partial y} dx + \frac{\partial u}{\partial x} dy \quad \ldots(1) \qquad \left[\because \frac{\partial v}{\partial x} = -\frac{\partial u}{\partial y} \text{ and } \frac{\partial v}{\partial y} = \frac{\partial u}{\partial x}\right]$$

Putting the values of $\frac{\partial u}{\partial y}$ and $\frac{\partial u}{\partial x}$ in (1), we get

$$dv = -(-2y - 2x + 3)\,dx + (2x - 2y - 2)dy$$

$$\Rightarrow \quad v = \int (2y + 2x - 3)dx + \int (-2y - 2)dy + C \qquad \text{(Ignoring } 2x)$$

Hence, $$v = 2xy + x^2 - 3x - y^2 - 2y + C$$ **Ans.**

Now, $$f(z) = u + iv$$

$$= (x^2 - y^2 - 2xy - 2x + 3y) + i\,(2xy + x^2 - 3x - y^2 - 2y) + iC$$

$$= (x^2 - y^2 + 2ixy) + (ix^2 - iy^2 - 2xy) - (2 + 3i)\,x - i\,(2 + 3i)\,y + iC$$

$$= (x^2 - y^2 + 2ixy) + i\,(x^2 - y^2 + 2ixy) - (2 + 3i)\,x - i\,(2 + 3i)y + iC$$

$$= (x + iy)^2 + i\,(x + iy)^2 - (2 + 3i)\,(x + iy) + iC$$

$$= z^2 + iz^2 - (2 + 3i)\,z + iC$$

$$= (1 + i)\,z^2 - (2 + 3i)\,z + iC$$

Which is the required expression of $f(z)$ in terms of z. **Ans.**

Example 30. *Define a harmonic function. Show that the function $u(x, y) = x^4 - 6x^2y^2 + y^2 + y^4$ is harmonic. Also find the analytic function $f(z) = u(x, y) + iv + (x, y)$.*

Solution. See Art. 22.15 on page 602 for definition of harmonic function.

We have,

$$u(x,y) = x^4 - 6x^2y^2 + y^4\ , \qquad \frac{\partial u}{\partial x} = 4x^3 - 12xy^2$$

$$\frac{\partial u}{\partial y} = -12x^2y + 4y^3\ , \qquad \frac{\partial^2 u}{\partial x^2} = 12x^2 - 12y^2 \qquad \text{... (1)}$$

$$\frac{\partial^2 u}{\partial y^2} = -12x^2 + 12y^2 \qquad \text{... (2)}$$

Adding (1), and (2), we get

$$\frac{\partial^2 u}{\partial x^2} + \frac{\partial^2 u}{\partial y^2} = 12x^2 - 12y^2 - 12x^2 + 12y^2$$

$$\Rightarrow \qquad \frac{\partial^2 u}{\partial x^2} + \frac{\partial^2 u}{\partial y^2} = 0$$

Hence, u is a harmonic function. **Proved.**

Let us find out v:

We know that $dv = \frac{\partial v}{\partial x} dx + \frac{\partial v}{\partial y} dy$

$$\Rightarrow \qquad dv = -\frac{\partial u}{\partial y} dx + \frac{\partial u}{\partial x} dy \qquad \left[\because \frac{\partial v}{\partial x} = -\frac{\partial u}{\partial y} \text{ and } \frac{\partial v}{\partial y} = \frac{\partial u}{\partial x}\right]$$

$$\Rightarrow \qquad dv = (12x^2y - 4y^3)\, dx + (4x^3 - 12xy^2)\, dy$$

$$v = \int (12x^2y - 4y^3)\, dx + \int (4x^3 - 12xy^2)\, dy$$

(y is constant) (Integrate only those terms which do not contain x)

$$v = 4x^3y - 4xy^3 + C$$

$$f(z) = u + iv = x^4 - 6x^2y^2 + y^4 + i4x^3y - 4ixy^3 + iC$$

$$f(z) = x^4 + 4x^3(iy) + 6x^2(iy)^2 + 4x(iy)^3 + (iy)^4 + iC$$

$$= (x + iy)^4 + iC$$

$$= z^4 + iC \qquad [\because z = x + iy]$$

This is the required analytic function. **Ans.**

Example 31. *If $w = \phi + i\psi$ represents the complex potential for an electric field and*

$$\psi = x^2 - y^2 + \frac{x}{x^2 + y^2},$$

determine the function ϕ.

Solution. $w = \phi + i\psi$ and $\psi = x^2 - y^2 + \frac{x}{x^2 + y^2}$

$$\frac{\partial \psi}{\partial x} = 2x + \frac{(x^2 + y^2)\cdot 1 - x\cdot 2x}{(x^2 + y^2)^2} = 2x + \frac{y^2 - x^2}{(x^2 + y^2)^2}$$

$$\frac{\partial \psi}{\partial y} = -2y - \frac{x(2y)}{(x^2+y^2)^2} = -2y - \frac{2xy}{(x^2+y^2)^2}$$

We know that, $d\phi = \frac{\partial \phi}{\partial x} dx + \frac{\partial \phi}{\partial y} dy = \frac{\partial \psi}{\partial y} dx - \frac{\partial \psi}{\partial x} dy$

$$= \left(-2y - \frac{2xy}{(x^2+y^2)^2}\right) dx - \left(2x + \frac{y^2 - x^2}{(x^2+y^2)^2}\right) dy$$

$$\phi = \int \left[-2y - \frac{2xy}{(x^2+y^2)^2}\right] dx + c$$

This is an exact differential equation.

$$\phi = -2xy + \frac{y}{x^2+y^2} + C$$ **Ans.**

Which is the required function.

Example 32. *An electrostatic field in the xy-plane is given by the potential function $\phi = 3x^2y - y^3$, find the stream function.* (*R.G.P.V., Bhopal, III Semester, Dec. 2001*)

Solution. Let $\psi(x, y)$ be a stream function

We know that $d\psi = \frac{\partial \psi}{\partial x} dx + \frac{\partial \psi}{\partial y} dy = \left(-\frac{\partial \phi}{\partial y}\right) dx + \left(\frac{\partial \phi}{\partial x}\right) dy$ [C-R equations]

$$= \{-(3x^2 - 3y^2)\}\, dx + 6xy\, dy$$

$$= -3x^2\, dx + (3y^2\, dx + 6xy\, dy)$$

$$= -d\,(x^3) + 3d\,(xy^2)$$

$$\psi = \int -d\,(x^3) + 3d\,(xy^2) + c$$

$$\psi = -x^3 + 3xy^2 + c$$

ψ is the required stream function. **Ans.**

Example 33. *Find the imaginary part of the analytic function whose real part is* $x^3 - 3xy^2 + 3x^2 - 3y^2$. (*R.G.P.V., Bhopal, III Semester, Dec. 2008, 2005*)

Solution.

Let $u(x, y) = x^3 - 3xy^2 + 3x^2 - 3y^2$

$$\frac{\partial u}{\partial x} = 3x^2 + 3y^2 + 6x$$

$$\frac{\partial u}{\partial y} = -6xy - 6y$$

We know that

$$dv = \frac{\partial v}{\partial x} dx + \frac{\partial v}{\partial y} dy \Rightarrow dv = -\frac{\partial u}{\partial y} dx + \frac{\partial u}{\partial x} dy$$

$\Rightarrow$ $dv = (6xy + 6y)\, dx + (3x^2 - 3y^2 + 6x)\, dy$

This is an exact differential equation.

$$v = \int (6xy + 6y)\, dx + \int -3y^2 dy + C$$

$$= 3x^2 y + 6xy - y^3 + C$$

Which is the required imaginary part. **Ans.**

Example 34. *If $u - v = (x - y)(x^2 + 4xy + y^2)$ and $f(z) = u + iv$ is an analytic function of $z = x + iy$, find $f(z)$ in terms of z.*

Solution. $u + iv = f(z) \Rightarrow iu - v = if(z)$

Adding these, $(u - v) + i(u + v) = (1 + i) f(z)$

Let $U + iV = (1 + i) f(z)$ where $U = u - v$ and $V = u + v$

$$F(z) = (1 + i) f(z)$$

$$U = u - v = (x - y)(x^2 + 4xy + y^2)$$

$$= x^3 + 3x^2y - 3xy^2 - y^3$$

$$\frac{\partial U}{\partial x} = 3x^2 + 6xy - 3y^2,$$

$$\frac{\partial U}{\partial y} = 3x^2 - 6xy - 3y^2$$

We know that $dV = \frac{\partial V}{\partial x} \cdot dx + \frac{\partial V}{\partial y} dy = -\frac{\partial U}{\partial y} \cdot dx + \frac{\partial U}{\partial x} \cdot dy$ [C-R equations]

On putting the values of $\frac{\partial U}{\partial x}$ and $\frac{\partial U}{\partial y}$, we get

$$= (-3x^2 + 6xy + 3y^2)\, dx + (3x^2 + 6xy - 3y^2) \,.\, dy$$

Integrating, we get

$$V = \int (-3x^2 + 6xy + 3y^2)\, dx + \int (-3y^2)\, dy$$

(y as constant) (Ignoring terms of x)

$$= -x^3 + 3x^2y + 3xy^2 - y^3 + c$$

$$F(z) = U + iV$$

$$= (x^3 + 3x^2y - 3xy^2 - y^3) + i(-x^3 + 3x^2y + 3xy^2 - y^3) + ic$$

$$= (1 - i)x^3 + (1 + i)3x^2y - (1 - i)3xy^2 - (1 + i)y^3 + ic$$

$$= (1 - i)x^3 + i(1 - i)3x^2y - (1 - i)3xy^2 - i(1 - i)y^3 + ic$$

$$= (1 - i)[x^3 + 3ix^2y - 3xy^2 - iy^3] + ic$$

$$= (1 - i)(x + iy)^3 + iC = (1 - i)z^3 + ic$$

$$(1 + i) f(z) = (1 - i) z^3 + ic, \qquad [F(z) = (1 + i) f(z)]$$

$$f(z) = \frac{1-i}{1+i} z^3 + \frac{ic}{1+i} = -\frac{i(1+i)}{(1+i)} z^3 + \frac{i(1-i)}{(1+i)(1-i)} c = -iz^3 + \frac{1+i}{2} c$$ **Ans.**

Example 35. *If $f(z) = u + iv$, is any analytic function of the complex variable z and $u - v = e^x(\cos y - \sin y)$, find $f(z)$ in terms of z*

Solution. $u + iv = f(z) \Rightarrow iu - v = i f(z)$

Adding, we have

$$u + iv + iu - v = f(z) + i f(z)$$

$$(u - v) + i(u + v) = (1 + i) f(z) = F(z) \text{ say}$$

Put $u - v = U$ and $u + v = V$, then $F(z) = U + iV$ is an analytic function.

Now $U = e^x(\cos y - \sin y)$

$\therefore$ $\frac{\partial U}{\partial x} = e^x(\cos y - \sin y)$ and $\frac{\partial U}{\partial y} = e^x(-\sin y - \cos y)$

We know that $dV = \frac{\partial V}{\partial x} dx + \frac{\partial V}{\partial y} dy = -\frac{\partial U}{\partial y} dx + \frac{\partial U}{\partial x} dy$

$$= e^x (\sin y + \cos y)\, dx + e^x (\cos y - \sin y) dy.$$

Integrating, we have

$$V = e^x (\sin y + \cos y) + c$$

$$F(z) = U + iV$$

$$= e^x (\cos y - \sin y) + ie^x (\sin y + \cos y) + ic$$

$$= e^x (\cos y + i \sin y) + ie^x (\cos y + i \sin y) + ic$$

$$= e^x . e^{iy} + ie^x e^{iy} + ic = e^{x+iy} + ie^{x+iy} + ic = e^z + ie^z + ic$$

$$(1 + i) f(z) = (1 + i) e^z + ic$$

$$f(z) = e^z + \frac{ic}{1+i}, \quad f(z) = e^z + c_1$$ **Ans.**

This is the required result.

Example 36. *If $f(z) = u + iv$ is an analytic function of $z = x + iy$ and*

$$u - v = e^{-x} [(x - y) \sin y - (x + y) \cos y]$$ *(U.P. III Semester, 2009-2010)*

Solution. We know that,

$$f(z) = u + iv \quad \text{... (1)}$$

$$i f(z) = i u - v \quad \text{... (2)}$$

$$F(z) = U + iV$$

$$U = u - v = e^{-x} [(x - y) \sin y - (x + y) \cos y]$$

$$\frac{\partial U}{\partial x} = - e^{-x} [(x - y) \sin y - (x + y) \cos y] + e^{-x} [\sin y - \cos y]$$

$$\frac{\partial U}{\partial y} = e^{-x} [(x - y) \cos y - \sin y - (x + y) (- \sin y) - \cos y)]$$

We know that,

$$dV = \frac{\partial V}{\partial x} dx + \frac{\partial V}{\partial y} dy = -\frac{\partial U}{\partial y} dx + \frac{\partial U}{\partial x} dy \quad \text{[C – R equations]}$$

$$= - e^{-x} [(x - y) \cos y - \sin y + (x + y) \sin y - \cos y]\, dx$$
$$- e^{-x} [(x - y) \sin y - (x + y) \cos y - \sin y + \cos y]\, dy$$

$$= - e^{-x} x \{(\cos y + \sin y)\, dx - e^{-x} (-y \cos y - \sin y + y \sin y - \cos y)\, dx$$
$$- e^{-x} [(x - y) \sin y - (x + y) \cos y - \sin y + \cos y]\, dy$$

$$V = (\cos y + \sin y) (x e^{-x} + e^{-x}) + e^{-x} (-y \cos y - \sin y + y \sin y - \cos y) + C$$

$$F(z) = U + iV$$

$$F(z) = e^{-x} [(x - y) \sin y - (x + y) \cos y] + i e^{-x} [x \cos y + \cos y + x \sin y + \sin y$$
$$- y \cos y - \sin y + y \sin y - \cos y] + iC$$

$$= e^{-x} [\{x \sin y - y \sin y - x \cos y - y \cos y\} + i \{x \cos y + x \sin y - y \cos y + y \sin y\}] + iC$$

$$= e^{-x} [(x + iy) \sin y - (x + i y) \cos y + (-y + i x) \sin y + (- y + i x) \cos y] + iC$$

$$= e^{-x} [(x + iy) \sin y - (x + i y) \cos y + i (x + i y) \sin y + i (x + i y) \cos y] + iC$$

$$= e^{-x} (x + i y) [\sin y - \cos y + i \sin y + i \cos y] + i C$$

$$= e^{-x} (x + i y) [(1 + i) \sin y + i (1 + i) \cos y] + i C$$

$$(1 + i) f(z) = e^{-x} (x + iy) (1 + i) (\sin y + i \cos y) + i C$$

$$f(z) = e^{-x} (x + i y) (\sin y + i \cos y) + \frac{iC}{1+i}$$

$$= i\,z\,e^{-x}(\cos y - i\sin y) + \frac{iC}{1+i}$$

$$= i\,z\,e^{-x}\,e^{-iy} = i\,z\,e^{-(x+iy)} = i\,z\,e^{-z} + \frac{iC}{1+i}$$ **Ans.**

Example 37. *Let $f(z) = u(r, \theta) + iv(r, \theta)$ be an analytic function and $u = -r^3 \sin 3\theta$, then construct the corresponding analytic function $f(z)$ in terms of z.*

Solution. $u = -r^3 \sin 3\theta$

$$\frac{\partial u}{\partial r} = -3r^2 \sin 3\theta, \qquad \frac{\partial u}{\partial \theta} = -3r^3 \cos 3\theta$$

We know that $$dv = \frac{\partial v}{\partial r} dr + \frac{\partial v}{\partial \theta} d\theta$$

$$= \left(-\frac{1}{r}\frac{\partial u}{\partial \theta}\right) dr + \left(r\frac{\partial u}{\partial r}\right) d\theta \qquad \left(\begin{array}{c} C-R \text{ equations} \\ \dfrac{\partial u}{\partial r} = \dfrac{1}{r}\dfrac{\partial v}{\partial \theta} \\ \dfrac{\partial u}{\partial \theta} = -r\dfrac{\partial v}{\partial r} \end{array}\right)$$

$$= -\frac{1}{r}(-3r^3 \cos 3\theta)\, dr + r(-3r^2 \sin 3\theta)\, d\theta$$

$$= 3r^2 \cos 3\theta \cdot dr - 3r^3 \sin 3\theta\, d\theta$$

$$v = \int (3r^2 \cos 3\theta) dr - c = r^3 \cos 3\theta + c$$

$$f(z) = u + iv = -r^3 \sin 3\theta + ir^3 \cos 3\theta + ic = ir^3(\cos 3\theta + i \sin 3\theta) + ic$$

$$= ir^3 e^{i3\theta} + ic = i(re^{i\theta})^3 + ic = iz^3 + ic$$ **Ans.**

This is the required analytic function.

Example 38. *Find analytic function $f(z)$ $u(r, \theta) + iv(r, \theta)$ such that*

$$v(r,\theta) = r^2 \cos 2\theta - r\cos\theta + 2.$$

Solution. We have, $$v = r^2 \cos 2\theta - r\cos\theta + 2 \qquad \text{... (1)}$$

Differentiating (1), we get

$$\frac{\partial v}{\partial \theta} = -2r^2 \sin 2\theta + r\sin\theta \qquad \text{... (2)}$$

$$\frac{\partial v}{\partial r} = 2r\cos 2\theta - \cos\theta \qquad \text{... (3)}$$

Using $C - R$ equations in polar coordinates, we get

$$r\frac{\partial u}{\partial r} = \frac{\partial v}{\partial \theta} = -2r^2 \sin 2\theta + r\sin\theta \qquad \text{[From (2)]}$$

$$\Rightarrow \quad \frac{\partial u}{\partial r} = -2r\sin 2\theta + \sin\theta \qquad \text{... (4)}$$

$$-\frac{1}{r}\frac{\partial u}{\partial \theta} = \frac{\partial v}{\partial r} = 2r\cos 2\theta - \cos\theta \qquad \text{[From (3)]}$$

$$\Rightarrow \qquad \frac{\partial u}{\partial \theta} = -2r^2 \cos 2\theta + r\cos\theta \qquad \text{... (5)}$$

By total differentiation formula

$$du = \frac{\partial u}{\partial r}dr + \frac{\partial u}{\partial \theta}d\theta = (-2r\sin 2\theta + \sin\theta)dr + (-2r^2\cos 2\theta + r\cos\theta)d\theta$$

$$= -[(2r\,dr)\sin 2\theta + r^2(2\cos 2\theta\, d\theta)] + [\sin\theta\cdot dr + r(\cos\theta\, d\theta)]$$

$$= -[(2r\,dr)\sin 2\theta - \sin\theta dr] + [-r^2 2\cos 2\theta\, d\theta + r\cos\theta\, d\theta]$$

$$= -d(r^2\sin 2\theta) + d(r\sin\theta) \qquad \text{(Exact differential equation)}$$

Integrating, we get

$$u = -r^2\sin 2\theta + r\sin\theta + c$$

Hence, $\quad f(z) = u + iv$

$$= (-r^2\sin 2\theta + r\sin\theta + c) + i(r^2\cos 2\theta - r\cos\theta + 2)$$

$$= ir^2(\cos 2\theta + i\sin 2\theta) - ir(\cos\theta + i\sin\theta) + 2i + c$$

$$= ir^2 e^{2i\theta} - ir\,e^{i\theta} + 2i + c = i(r^2 e^{2i\theta} - r\,e^{i\theta}) + 2i + c. \quad \textbf{Ans.}$$

This is the required analytic function.

Example 39. *Deduce the following with the polar form of Cauchy-Riemann equations:*

$$(a)\ \frac{\partial^2 u}{\partial r^2} + \frac{1}{r}\frac{\partial u}{\partial r} + \frac{1}{r^2}\frac{\partial^2 u}{\partial \theta^2} = 0 \qquad (b)\ f'(z) = \frac{r}{z}\left[\frac{\partial u}{\partial r} + i\frac{\partial v}{\partial r}\right]$$

Solution. We know that polar form of C-R equations are

$$\frac{\partial u}{\partial r} = \frac{1}{r}\frac{\partial v}{\partial \theta} \qquad \text{... (1)}$$

$$\frac{\partial u}{\partial \theta} = -r\frac{\partial v}{\partial r} \qquad \text{... (2)}$$

(*a*) Differentiating (1) partially w.r.t. r., we get

$$\frac{\partial^2 u}{\partial r^2} = -\frac{1}{r^2}\frac{\partial v}{\partial \theta} + \frac{1}{r}\frac{\partial^2 v}{\partial r\partial \theta} \qquad \text{... (3)}$$

Differentiating (2) partially w.r.t. θ, we have

$$\frac{\partial^2 u}{\partial \theta^2} = -r\frac{\partial^2 v}{\partial \theta\partial r} \qquad \text{...(4)}$$

Thus using (1), (3) and (4), we get

$$\frac{\partial^2 u}{\partial r^2} + \frac{1}{r}\frac{\partial u}{\partial r} + \frac{1}{r^2}\frac{\partial^2 u}{\partial \theta^2} = -\frac{1}{r^2}\frac{\partial v}{\partial \theta} + \frac{1}{r}\frac{\partial^2 v}{\partial r\partial \theta} + \frac{1}{r}\left(\frac{1}{r}\frac{\partial v}{\partial \theta}\right) + \frac{1}{r^2}\left(-r\frac{\partial^2 v}{\partial \theta\partial r}\right) = 0 \qquad \left[\frac{\partial^2 v}{\partial \theta dr} = \frac{\partial^2 v}{\partial r\partial \theta}\right]$$

Proved.

$$(b)\text{ Now,}\quad r\left(\frac{\partial u}{\partial r} + i\frac{\partial v}{\partial r}\right) = r\left[\left(\frac{\partial u}{\partial x}\frac{\partial x}{\partial r} + \frac{\partial u}{\partial y}\frac{\partial y}{\partial r}\right) + i\left(\frac{\partial v}{\partial x}\frac{\partial x}{\partial r} + \frac{\partial v}{\partial y}\frac{\partial y}{\partial r}\right)\right]$$

$$= r\left[\left(\frac{\partial u}{\partial x}\cos\theta+\frac{\partial u}{\partial y}\sin\theta\right)+i\left(\frac{\partial v}{\partial x}\cos\theta+\frac{\partial v}{\partial y}\sin\theta\right)\right]$$

$$= r\cos\theta\left(\frac{\partial u}{\partial x}+i\frac{\partial v}{\partial x}\right)+r\sin\theta\left(\frac{\partial u}{\partial y}+i\frac{\partial v}{\partial y}\right)$$

$$= x\left(\frac{\partial u}{\partial x}+i\frac{\partial v}{\partial x}\right)+iy\left(\frac{\partial v}{\partial y}-i\frac{\partial u}{\partial y}\right) \qquad \text{(By C-R equations)}$$

$$= x f'(z) + iy f'(z) = (x+iy) f'(z) = z f'(z).$$

$$\therefore \quad f'(z)=\frac{r}{z}\left(\frac{\partial u}{\partial r}+i\frac{\partial v}{\partial r}\right)$$

Proved.

22.19 MILNE THOMSON METHOD (To construct an Analytic function)

By this method $f(z)$ is directly constructed without finding v and the method is given below:

Since $\quad z = x + iy$ and $\bar{z} = x - iy$

$$\therefore \quad x=\frac{z+\bar{z}}{2},\; y=\frac{z-\bar{z}}{2i}$$

$$f(z)\equiv u(x,y)+iv(x,y) \qquad \text{... (1)}$$

$$f(z)\equiv u\left(\frac{z+\bar{z}}{2},\frac{z-\bar{z}}{2i}\right)+iv\left(\frac{z+\bar{z}}{2},\frac{z-\bar{z}}{2i}\right)$$

This relation can be regarded as a formal identity in two independent variables z and $\bar{z}$. Replacing $\bar{z}$ by z, we get

$$f(z)\equiv u(z,0)+iv(z,0)$$

Which can be obtained by replacing x by z and y by 0 in (1)

Case I. If *u* is given

We have $\quad f(z) = u + iv$

$$\therefore \quad f'(z)=\frac{\partial u}{\partial x}+i\frac{\partial v}{\partial x},\qquad f'(z)=\frac{\partial u}{\partial x}-i\frac{\partial u}{\partial y} \qquad (C-R \text{ equations})$$

If we write $\quad \frac{\partial u}{\partial x}=\phi_1(x,y),\qquad \frac{\partial u}{\partial y}=\phi_2(x,y)$

$$f'(z)=\phi_1(x,y)-i\phi_2(x,y) \text{ or } f'(z)=\phi_1(z,0)-i\phi_2(z,0)$$

On integrating $\quad f(z)=\int\phi_1(z,0)\,dz-i\int\phi_2(z,0)dz+c$

Case II. If *v* is given

$$f(z) = u + iv$$

$$f'(z)=\frac{\partial u}{\partial x}+i\frac{\partial v}{\partial x}=\frac{\partial v}{\partial y}+i\frac{\partial v}{\partial x}=\psi_1(x,y)+i\,\psi_2(x,y)$$

when $\quad \psi_1(x,y)=\frac{\partial v}{\partial y},\qquad \psi_2(x,y)=\frac{\partial v}{\partial x}.$

$$f(z)=\int\psi_1(z,0)\,dz+i\int\psi_2(z,0)\,dz+c$$

22.20 WORKING RULE: TO CONSTRUCT AN ANALYTIC FUNCTION BY MILNE THOMSON METHOD

Case I. When u is given

Step 1. Find $\frac{\partial u}{\partial x}$ and equate it to $\phi_1(x, y)$.

Step 2. Find $\frac{\partial u}{\partial y}$ and equate it to $\phi_2(x, y)$.

Step 3. Replace x by z and y by 0 in $\phi_1(x, y)$ to get $\phi_1(z, 0)$.

Step 4. Replace x by z and y by 0 in $\phi_2(x, y)$ to get $\phi_2(z, 0)$.

Step 5. Find $f(z)$ by the formula $f(z) = \int\{\phi_1(z,0) - i\phi_2(z,0)\}\,dz + c$

Case II. When v is given

Step 1. Find $\frac{\partial v}{\partial x}$ and equate it to $\psi_1(x, y)$.

Step 2. Find $\frac{\partial v}{\partial y}$ and equate it to $\psi_2(x, y)$.

Step 3. Replace x by z and y by 0 in $\psi_1(x, y)$ to get $\psi_1(z, 0)$.

Step 4. Replace x by z and y by 0 in $\psi_2(x, y)$ to get $\psi_2(z, 0)$.

Step 5. Find $f(z)$ by the formula

$$f(z) = \int\{\psi_1(z,0) + i\psi_2(z,0)\}\,dz + c$$

Case III. When $u - v$ *is given.*

We know that
$$f(z) = u + iv \qquad ...(1)$$
$$if(z) = iu - v \qquad ...(2) \text{ [Multiplying by } i]$$

Adding (1) and (2), we get
$$(1 + i)\,f(z) = (u - v) + i(u + v)$$
$$\Rightarrow \quad F(z) = U + iV$$
where
$$F(z) = (1 + i)\,f(z) \qquad ...(3) \quad \begin{bmatrix} U = u - v \\ V = u + v \end{bmatrix}$$

Here, $\quad U = (u - v)$ is given

Find out F(z) by the method described in case I, then substitute the value of F(z) in (3), we get
$$f(z) = \frac{F(z)}{1+i}$$

Case IV. When $u + v$ is given.

We know that
$$f(z) = u + iv \qquad ...(1)$$
$$if(z) = iu - v \qquad \text{[Multiplying by } i] \text{ ...(2)}$$

Adding (1) and (2), we get
$$(1 + i)\,f(z) = (u - v) + i(u + v)$$
$$\Rightarrow \quad F(z) = U + iV$$
where
$$F(z) = (1 + i)\,f(z) \qquad ...(3) \quad \begin{bmatrix} U = u - v \\ V = u + v \end{bmatrix}$$

Here, $\quad V = (u + v)$ is given

Find out F(z) by the method described in case II, then substitute the value of F(z) in (3), we get
$$f(z) = \frac{F(z)}{1+i}$$

Example 40. *If $u = x^2 - y^2$, find a corresponding analytic function.*

Solution. $$\frac{\partial u}{\partial x} = 2x, \quad \frac{\partial u}{\partial y} = -2y$$

On replacing x by z and y by 0, we have

$$u_x(z,0) = 2z \text{ and } u_y(z,0) = 0$$

$$f(z) = \int [u_x(z,0) - iu_y(z,0)]dz + C$$

$$= \int [2z - i(0)]dz + c = \int 2z\,dz + c = z^2 + C \qquad \textbf{Ans.}$$

This is the required analytic function.

Example 41. *Show that the function $u = e^{-2xy}\sin(x^2 - y^2)$ is harmonic. Find the conjugate function v and express u + iv as an analytic function of z.*

(R.G.P.V. Bhopal, III Semester, June, 2007, Dec. 2006)

Solution. We have,

$$u = e^{-2xy}\sin(x^2 - y^2) \qquad \text{... (1)}$$

Differentiating (1), w.r.t. x, we get

$$\frac{\partial u}{\partial x} = 2x\,e^{-2xy}\cos(x^2 - y^2) - 2ye^{-2xy}\sin(x^2 - y^2)$$

$$\Rightarrow \quad \frac{\partial u}{\partial x} = e^{-2xy}[2x\cos(x^2 - y^2) - 2y\sin(x^2 - y^2)] = u_x(x,y) \qquad \text{... (2)}$$

$$u_x(z, 0) = 2z\cos z^2$$

Differentiating (1), w.r.t. y, we get

$$\frac{\partial u}{\partial y} = -2ye^{-2xy}\cos(x^2 - y^2) - 2xe^{-2xy}\sin(x^2 - y^2)$$

$$\Rightarrow \quad \frac{\partial u}{\partial y} = e^{-2xy}[-2y\cos(x^2 - y^2) - 2x\sin(x^2 - y^2)] = u_y(x,y) \qquad \text{... (3)}$$

$$u_y(z, 0) = -2z\sin z^2$$

Differentiating (2), w.r.t, 'x', we get

$$\frac{\partial^2 u}{\partial x^2} = -2ye^{-2xy}[2x\cos(x^2 - y^2) - 2y\sin(x^2 - y^2)]$$

$$+ e^{-2xy}[2\cos(x^2 - y^2) + 2x(2x)\{-\sin(x^2 - y^2)\} - 2y(2x)\cos(x^2 - y^2)]$$

$$\Rightarrow \quad \frac{\partial^2 u}{\partial x^2} = e^{-2xy}[-4xy\cos(x^2 - y^2) + 4y^2\sin(x^2 - y^2) + 2\cos(x^2 - y^2)$$

$$-4x^2\sin(x^2 - y^2) - 4xy\cos(x^2 - y^2)]$$

$$= e^{-2xy}[-8xy\cos(x^2 - y^2) + 4y^2\sin(x^2 - y^2) + 2\cos(x^2 - y^2) - 4x^2\sin(x^2 - y^2)] \qquad \text{...(4)}$$

Differentiating (3), w.r.t. 'y', we get

$$\frac{\partial^2 u}{\partial y^2} = -2xe^{-2xy}[-2y\cos(x^2 - y^2) - 2x\sin(x^2 - y^2)]$$

$$+ e^{-2xy}[-2\cos(x^2 - y^2) + 2y(-2y)\sin(x^2 - y^2) - 2x(-2y)\cos(x^2 - y^2)]$$

$$\Rightarrow \quad \frac{\partial^2 u}{\partial y^2} = e^{-2xy}[4xy\cos(x^2-y^2)+4x^2\sin(x^2-y^2)-2\cos(x^2-y^2)$$
$$-4y^2\sin(x^2-y^2)+4xy\cos(x^2-y^2)]$$

$$\frac{\partial^2 u}{\partial y^2} = e^{-2xy}\,[8xy\cos(x^2-y^2)+4x^2\sin(x^2-y^2)-2\cos(x^2-y^2)-4y^2\sin(x^2-y^2)] \quad \ldots (5)$$

Adding (4) and (5), we get $\dfrac{\partial^2 u}{\partial x^2}+\dfrac{\partial^2 u}{\partial y^2}=0$

Which proves that u is harmonic.

Now we have to express $u + iv$ as a function of z

$$f(z) = \int [u_x(z, 0) - i\, u_y(z, 0)]\, dz = \int [2z\cos z^2 - i(-2z\sin z^2)]\, dz$$
$$= \sin z^2 - i\cos z^2 + C = -i(\cos z^2 + i\sin z^2) + C = -i\, e^{iz^2} + C \qquad \textbf{Ans.}$$

Example 42. *If* $u = \dfrac{\sin 2x}{\cosh 2y + \cos 2x}$*, find* $f(z)$. *(R.G.P.V., Bhopal, III Semester, Dec. 2003)*

Solution. $$\frac{\partial u}{\partial x} = \frac{(\cosh 2y+\cos 2x)2\cos 2x - \sin 2x(-2\sin 2x)}{(\cosh 2y+\cos 2x)^2} \quad \ldots(1)$$

$$= \frac{2\cosh 2y\cos 2x + 2(\cos^2 2x+\sin^2 2x)}{(\cosh 2y+\cos 2x)^2} = \frac{2\cosh 2y\cos 2x+2}{(\cosh 2y+\cos 2x)^2} = u_x(x,y)$$

On putting $x = z$ and $y = 0$ in (1), we get

$$u_x(z, 0) = \frac{2\cos 2z+2}{(1+\cos 2z)^2}$$

$$\frac{\partial u}{\partial y} = \frac{-\sin 2x(2\sinh 2y)}{(\cosh 2y+\cos 2x)^2} = \frac{-2\sin 2x\sinh 2y}{(\cosh 2y+\cos 2x)^2} = u_y(x,y) \quad \ldots(2)$$

On putting $x = z$ and $y = 0$ in (2), we get

$$u_y(z, 0) = 0$$

$$f(z) = \int [u_x(z,0) - i u_y(z,0)]\, dz + C = \int \frac{(2\cos 2z+2)}{(1+\cos 2z)^2} dz + C = 2\int \frac{1}{1+\cos 2z} dz + C$$

$$= 2\int \frac{1}{2\cos^2 z} dz + C = \int \sec^2 z\, dz + C = \tan z + C \qquad \textbf{Ans.}$$

which is the required function.

Example 43. *Find the analytic function* $f(z) = u + iv$*, given that*
$v = e^x(x\sin y + y\cos y)$.

Solution. $$\frac{\partial v}{\partial x} = e^x(x\sin y + y\cos y) + e^x\sin y = v_x(x,y) \quad \Rightarrow \quad v_2(z, 0) = 0$$

$$\frac{\partial v}{\partial y} = e^x(x\cos y + \cos y - y\sin y) = v_y(x,y) \Rightarrow v_y(z,0) = ze^z + e^z$$

$$f(z) = \int [v_y(z,0) + i v_x(z,0)] dz + C$$
$$= \int [e^z(z+1) + i(0)] dz + c = (z+1)e^z - \int e^z\, dz + C$$
$$= (z+1)\, e^z - e^z + c = z\, e^z + C \qquad \textbf{Ans.}$$

Which is the required function.

Example 44. *Show that e^x (x cos y − y sin y) is a harmonic function. Find the analytic function for which e^x (x cos y − y sin y) is imaginary part.*

(*U.P., III Semester, June 2009*)

Solution. Here $v = e^x (x \cos y - y \sin y)$

Differentiating partially w.r.t. x and y, we have

$$\frac{\partial v}{\partial x} = e^x (x\cos y - y\sin y) + e^x \cos y = v_x (x, y), \quad \text{(say)} \quad ...(1)$$

$$\frac{\partial v}{\partial y} = e^x (-x\sin y - y\cos y - \sin y) = v_y (x, y) \quad \text{(say)} \quad ...(2)$$

$$\frac{\partial^2 v}{\partial x^2} = e^x (x\cos y - y\sin y) + e^x \cos y + e^x \cos y$$

$$= e^x (x\cos y - y\sin y + 2\cos y) \quad ...(3)$$

and
$$\frac{\partial^2 v}{\partial y^2} = e^x (-x\cos y + y\sin y - 2\cos y) \quad ...(4)$$

Adding equations (3) and (4), we have

$$\frac{\partial^2 v}{\partial x^2} + \frac{\partial^2 v}{\partial y^2} = 0 \quad \Rightarrow v \text{ is a harmonic function.}$$

Now putting $x = z$, $y = 0$ in (1) and (2), we get

$$v_x = (z, 0) = ze^z + e^z, \qquad v_y = (z, 0) = 0$$

Hence by Milne-Thomson method, we have

$$f(z) = \int [v_y (z, 0) + i v_x (z, 0)] \, dz + C$$

$$= \int [0 + i(ze^z + e^z)] \, dz + C = i(ze^z - e^z + e^z) + C = ize^z + C.$$

This is the required analytic function. **Ans.**

Example 45. *If $u - v = (x - y)(x^2 + 4xy + y^2)$ and $f(z) = u + iv$ is an analytic function of $z = x + iy$, find $f(z)$ in terms of z by Milne Thomson method.*

Solution. We know that

$$f(z) = u + iv \quad ...(1)$$

$$i f(z) = i u - v \quad ...(2)$$

Adding (1) and (2), we get

$$(1 + i) f(z) = (u - v) + i (u + v)$$

Let
$$F(z) = U + i V$$

$$U = u - v = (x - y)(x^2 + 4xy + y^2)$$

$$\frac{\partial U}{\partial x} = (x^2 + 4xy + y^2) + (x - y)(2x + 4y)$$

$$= x^2 + 4xy + y^2 + 2x^2 + 4xy - 2xy - 4y^2 = 3x^2 + 6xy - 3y^2$$

$$u_x(x, y) = 3x^2 + 6xy - 3y^2$$

$$u_x (z, 0) = 3z^2$$

$$\frac{\partial U}{\partial y} = -(x^2 + 4xy + y^2) + (x - y)(4x + 2y)$$

$$= -x^2 - 4xy - y^2 + 4x^2 + 2xy - 4xy - 2y^2 = 3x^2 - 6xy - 3y^2$$

$$u_y (x, y) = 3x^2 - 6xy - 3y^2$$

$$u_y (z, 0) = 3z^2$$

$$F(z) = U + iV$$

$$F'(z) = \frac{\partial U}{\partial x} + i\frac{\partial V}{\partial x} = \frac{\partial U}{\partial x} - i\frac{\partial U}{\partial y} = u_x, (z, 0) - i\,u_y\,(z, 0) = 3z^2 - i\,3z^2$$

$$= 3\,(1 - i)\,z^2$$

$$F(z) = (1 - i)\,z^3 + C$$

$$(1 + i)\,f(z) = (1 - i)\,z^3 + C$$

$$f(z) = \frac{1-i}{1+i}z^3 + \frac{C}{1+i} = \frac{(1-i)(1-i)}{(1+i)(1-i)}z^3 + C_1$$

$$= \frac{1-2i+(-i)^2}{1+1}z^3 + C_1 = \frac{1-2i-1}{2}z^3 + C_1 = -\,i\,z^3 + C_1 \qquad \textbf{Ans.}$$

Note: This example has already been solved on page 612 as Example 34.

Example 46. *If $f(z) = u + iv$ is an analytic function of z and* $u - v = \dfrac{\cos x + \sin x - e^{-y}}{2\cos x - 2\cosh y}$, *prove that*

$$f(z) = \frac{1}{2}\left[1 - \cot\frac{z}{2}\right] \text{ when } f\left(\frac{\pi}{2}\right) = 0. \quad \textit{(R.G.P.V. Bhopal, III Semester, Dec. 2007)}$$

Solution. We know that $f(z) = u + iv$

$\therefore$ $if(z) = iu - v$ [Multiplying by i]

On adding, we get $(1 + i)\,f(z) = (u - v) + i(u + v)$

$\Rightarrow$ $F(z) = U + iV$

$$\begin{bmatrix} U = u - v \\ V = u + v \end{bmatrix}$$

$\Rightarrow$ $(1 + i)\,f(z) = F(z)$

We have, $U = u - v = \dfrac{\cos x + \sin x - e^{-y}}{2\cos x - 2\cosh y}$

$$\Rightarrow \quad U = \frac{\cos x + \sin x - \cosh y + \sinh y}{2\cos x - 2\cosh y} \qquad [\because e^{-y} = \cosh y - \sinh y]$$

$$= \frac{\cos x - \cosh y}{2(\cos x - \cosh y)} + \frac{\sin x + \sinh y}{2(\cos x - \cosh y)} = \frac{1}{2} + \frac{\sin x + \sinh y}{2(\cos x - \cosh y)} \qquad ...(1)$$

Differentiating (1) w.r.t. x partially, we get

$$\frac{\partial U}{\partial x} = \frac{1}{2}\left[\frac{(\cos x - \cosh y)\cos x - (\sin x + \sinh y)(-\sin x)}{(\cos x - \cosh y)^2}\right]$$

$$= \frac{1}{2}\left[\frac{(\cos^2 x + \sin^2 x - \cosh y \cos x + \sinh y \sin x)}{(\cos x - \cosh y)^2}\right]$$

$$u_x(x, y) = \frac{1}{2}\left[\frac{1 - \cosh y\,\cos x + \sinh y\,\sin x}{(\cos x - \cosh y)^2}\right] = \frac{1}{2}\left[\frac{1 - \cos iy\,\cos x - i\sin iy\,\sin x}{(\cos x - \cosh y)^2}\right] \quad ...\,(2)$$

Replacing x by z and y by 0 in (2), we get

$$u_x(z, 0) = \frac{1}{2}\left[\frac{1-\cos z}{(\cos z-1)^2}\right] = \frac{1}{2}\frac{-(1-\cos z)}{(1-\cos z)^2} = \frac{1}{2(1-\cos z)}$$

Differentiating (1) partially w.r.t. y, we get

$$\frac{\partial U}{\partial y} = \frac{1}{2}\left[\frac{(\cos x-\cosh y).\cosh y-(\sin x+\sinh y)(-\sinh y)}{(\cos x-\cosh y)^2}\right]$$

$$= \frac{1}{2}\left[\frac{(\cos x\cosh y)+\sin x\sinh y-(\cosh^2 y-\sinh^2 y)}{(\cos x-\cosh y)^2}\right]$$

$$u_y(x, y) = \frac{1}{2}\left[\frac{\cos x\cosh y+\sin x\sinh y-1}{(\cos x-\cosh y)^2}\right] \quad ...(3)$$

Replacing x by z and y by 0 in (3), we have

$$u_x(z, 0) = \frac{1}{2}\left[\frac{\cos z-1}{(\cos z-1)^2}\right] = \frac{1}{2}.\left(\frac{-1}{1-\cos z}\right)$$

$$F'(z) = \frac{\partial U}{\partial x}+i\frac{\partial V}{\partial x} = \frac{\partial U}{\partial x}-i\frac{\partial U}{\partial y} \quad \text{[C–R equations]}$$

$$= u_x(z, 0) - i\, u_y(z, 0)$$

By Milne Thomson Method,

$$F(z) = \int\left[u_x(z,0)-iu_y(z,0)\right]dz$$

$$= \int\left[\frac{1}{2}.\frac{1}{(1-\cos z)}+\frac{i}{2}.\frac{1}{1-\cos z}\right]dz$$

$$= \frac{1+i}{2}\int\frac{1}{2\sin^2 z/2}dz = \frac{1+i}{4}\int \text{cosec}^2(z/2)\,dz$$

$$= \left(\frac{1+i}{4}\right).\frac{(-\cot z/2)}{\left(\frac{1}{2}\right)}+C = -\left(\frac{1+i}{2}\right)\cot\frac{z}{2}+C \quad [F(z) = (1+i)f(z)]$$

$$\Rightarrow \quad (1+i)f(z) = -\left(\frac{1+i}{2}\right)\cot\frac{z}{2}+C \quad \Rightarrow f(z) = -\frac{1}{2}\cot\frac{z}{2}+\frac{C}{1+i} \quad ...(4)$$

On putting $z = \frac{\pi}{2}$ in (4), we get

$$f\left(\frac{\pi}{2}\right) = -\frac{1}{2}\cot\frac{\pi}{4}+\frac{C}{1+i}$$

$$0 = -\frac{1}{2}+\frac{C}{1+i} \Rightarrow \frac{C}{1+i} = \frac{1}{2} \quad \left[f\left(\frac{\pi}{2}\right) = 0, \text{given}\right]$$

On putting the value of $\frac{C}{1+i}$ in (4), we get

$$f(z) = -\frac{1}{2}\cot\frac{z}{2}+\frac{1}{2}$$

Hence, $f(z) = \frac{1}{2}\left(1-\cot\frac{z}{2}\right)$, when $f\left(\frac{\pi}{2}\right) = 0$. **Proved.**

EXERCISE 22.4

Show that the following functions are harmonic and determine the conjugate functions.

1. $u = 2x(1-y)$ **Ans.** $v = x^2 - y^2 + 2y + C$ 2. $u = 2x - x^3 + 3xy + 3xy^2$

Ans. $v = -\frac{3}{2}(x^2 - y^2) + \frac{3}{2}y^2 + y^3 + 2y + C$

Determine the analytic function, whose real part is

3. $\log\sqrt{x^2+y^2}$ **Ans.** $\log z + C$ 4. $\cos x \cosh y$ **Ans.** $\cos z + c$

5. $e^{-x}(\cos y + \sin y)$ *(AMIETE, June 2010)* **Ans.** $e^{-z}(1+i) + C$

6. $e^{2x}(x\cos 2y - y\sin 2y)$ **Ans.** $ze^{2z} + iC$ 7. $e^{-x}(x\cos y + y\sin y)$ and $f(0) = i$. **Ans.** $ze^{-z} + i$

Determine the analytic function, whose imaginary part is

8. $v = \log(x^2+y^2) + x - 2y$ **Ans.** $2i\log z - (2-i)z + C$ *(G.B.T.U. 2012)*

9. $v = \sinh x \cos y$ **Ans.** $\sin iz + C$

10. $v = \left(r - \frac{1}{r}\right)\sin\theta$ **Ans.** $z + \frac{1}{z} + C$

11. If $f(z) = u + iv$ is an analytic function of $z = x + iy$ and $u - v = \dfrac{e^y - \cos x + \sin x}{\cosh y - \cos x}$, find $f(z)$ subject to the condition that $f\left(\frac{\pi}{2}\right) = \frac{3-i}{2}$. **Ans.** $f(z) = \cot\frac{z}{2} + \frac{1-i}{2}$

12. Find an analytic function $f(z) = u(r, \theta) + iv(r, \theta)$ such that $V(r, \theta) = r^2\cos 2\theta - r\cos\theta + 2$.
Ans. $i[z^2 - z + 2]$

13. Show that the function $u = x^2 - y^2 - 2xy - 2x - y - 1$ is harmonic. Find the conjugate harmonic function v and express $u + iv$ as a function of z where $z = x + iy$. **Ans.** $(1+i)z^2 + (-2+i)z - 1$

14. Construct an analytic function of the form $f(z) = u + iv$, where v is $\tan^{-1}(y/x)$, $x \neq 0$, $y \neq 0$.
Ans. $\log cz$

15. Show that the function $v(x, y) = e^x \sin y$ is harmonic. Find its conjugate harmonic function $u(x, y)$ and the corresponding analytic function $f(z)$. *(AMIETE, June 2009)* **Ans.** $u = e^x\cos y$, $f(z) = e^z$

Choose the correct answer:

16. The harmonic conjugate of $u = x^3 - 3xy^2$ is
(a) $y^3 - 3xy^2$ (b) $3x^2y - y^3$ (c) $3xy^2 - y^3$ (d) $3xy^2 - x^3$ *(AMIETE, June 2010)*
Ans. (b)

22.21 PARTIAL DIFFERENTIATION OF FUNCTION OF COMPLEX VARIABLE

Example 47. *Prove that*

$$4\frac{\partial^2}{\partial z\,\partial \bar{z}} = \left[\frac{\partial^2}{\partial x^2} + \frac{\partial^2}{\partial y^2}\right]$$

Solution. We know that

$$x + iy = z \quad \text{... (1)}, \qquad x - iy = \bar{z} \quad \text{... (2)}$$

From (1) and (2), we get

$$x = \frac{1}{2}(z + \bar{z}), \qquad y = \frac{-i}{2}(z - \bar{z})$$

$$\Rightarrow \qquad \frac{\partial x}{\partial z}=\frac{1}{2}, \qquad \frac{\partial y}{\partial z}=-\frac{i}{2}$$

and
$$\frac{\partial x}{\partial \bar{z}}=\frac{1}{2}, \qquad \frac{\partial y}{\partial \bar{z}}=\frac{i}{2}$$

We know that,

$$\frac{\partial}{\partial z}=\frac{\partial}{\partial x}\left(\frac{\partial x}{\partial z}\right)+\frac{\partial}{\partial y}\left(\frac{\partial y}{\partial z}\right)=\frac{\partial}{\partial x}\left(\frac{1}{2}\right)+\frac{\partial}{\partial y}\left(\frac{-i}{2}\right)=\frac{1}{2}\left(\frac{\partial}{\partial x}-i\frac{\partial}{\partial y}\right) \quad ...(3)$$

$$\frac{\partial}{\partial \bar{z}}=\frac{\partial}{\partial x}\left(\frac{\partial x}{\partial \bar{z}}\right)+\frac{\partial}{\partial y}\left(\frac{\partial y}{\partial \bar{z}}\right)=\frac{\partial}{\partial x}\left(\frac{1}{2}\right)+\frac{\partial}{\partial y}\left(\frac{i}{2}\right)=\frac{1}{2}\left(\frac{\partial}{\partial x}+\frac{i\partial}{\partial y}\right) \quad ...(4)$$

From (3) and (4), we get

$$\frac{\partial^2}{\partial z\partial \bar{z}}=\frac{\partial}{\partial z}\left(\frac{\partial}{\partial \bar{z}}\right)=\frac{1}{4}\left(\frac{\partial}{\partial x}-i\frac{\partial}{\partial y}\right)\left(\frac{\partial}{\partial x}+i\frac{\partial}{\partial y}\right)$$

$$=\frac{1}{4}\left(\frac{\partial^2}{\partial x^2}+\frac{\partial^2}{\partial y^2}+i\frac{\partial^2}{\partial x\partial y}-i\frac{\partial^2}{\partial x\partial y}\right)=\frac{1}{4}\left(\frac{\partial^2}{\partial x^2}+\frac{\partial^2}{\partial y^2}\right)$$

$$\Rightarrow \qquad 4\frac{\partial^2}{\partial z\,\partial \bar{z}}=\left[\frac{\partial^2}{\partial x^2}+\frac{\partial^2}{\partial y^2}\right]$$

Proved.

Example 48. *If f (z) is a harmonic function of z, show that*

$$\left\{\frac{\partial}{\partial x}|f(z)|\right\}^2+\left\{\frac{\partial}{\partial y}|f(z)|\right\}^2=|f'(z)|^2$$

(*U.P. III Semester, June 2009*)

Solution. Since $f(z) = u(x, y) + iv(x, y)$

so
$$|f(z)| = \sqrt{u^2+v^2} \quad ...(1)$$

Differentiating (1) partially w.r.t. 'x', we get

$$\frac{\partial}{\partial x}|f(z)| = \frac{\partial}{\partial x}(\sqrt{u^2+v^2})$$

$$=\frac{1}{2}(u^2+v^2)^{\frac{-1}{2}}\left(2u\frac{\partial u}{\partial x}+2v\frac{\partial v}{\partial x}\right)=\frac{u\frac{\partial u}{\partial x}+v\frac{\partial v}{\partial x}}{\sqrt{u^2+v^2}}=\frac{u\frac{\partial u}{\partial x}+v\frac{\partial v}{\partial x}}{|f(z)|} \quad ...(2)$$

Similarly
$$\frac{\partial}{\partial y}|f(z)| = \frac{u\frac{\partial u}{\partial y}+v\frac{\partial v}{\partial y}}{|f(z)|} \quad ...(3)$$

Squaring (2) and (3) adding, we get

$$\left[\frac{\partial}{\partial x}|f(z)|\right]^2+\left[\frac{\partial}{\partial y}|f(z)|\right]^2=\frac{\left(u\frac{\partial u}{\partial x}+v\frac{\partial v}{\partial x}\right)^2+\left(u\frac{\partial u}{\partial y}+v\frac{\partial v}{\partial y}\right)^2}{|f(z)|^2}$$

$$=\frac{\left(u\frac{\partial u}{\partial x}\right)^2+2uv\frac{\partial u}{\partial x}\frac{\partial v}{\partial x}+\left(v\frac{\partial v}{\partial x}\right)^2+\left(u\frac{\partial u}{\partial y}\right)^2+2u\frac{\partial u}{\partial y}.v\frac{\partial v}{\partial y}+\left(v\frac{\partial v}{\partial y}\right)^2}{|f(z)|^2}.$$

By C-R equation $\quad \frac{\partial u}{\partial x}=\frac{\partial v}{\partial y}$ and $\frac{\partial u}{\partial y}=-\frac{\partial v}{\partial x}$

Now, $$2uv\frac{\partial u}{\partial x}\frac{\partial v}{\partial x} = 2uv\left(\frac{\partial v}{\partial y}\right)\left(-\frac{\partial u}{\partial y}\right)$$

Putting the value of $2uv\frac{\partial u}{\partial x}.\frac{\partial v}{\partial x} = -2uv\frac{\partial v}{\partial y}.\frac{\partial u}{\partial y}$ in (4), we get

$$\left[\frac{\partial}{\partial x}|f(z)|\right]^2+\left[\frac{\partial}{\partial y}|f(z)|\right]^2=\frac{\left(u\frac{\partial u}{\partial x}\right)^2-2uv\frac{\partial v}{\partial y}.\frac{\partial u}{\partial y}+\left(v\frac{\partial v}{\partial x}\right)^2+\left(u\frac{\partial u}{\partial y}\right)^2+2uv\frac{\partial u}{\partial y}.\frac{\partial v}{\partial y}+\left(v\frac{\partial v}{\partial y}\right)^2}{|f(z)|^2}$$

$$=\frac{u^2\left(\frac{\partial u}{\partial x}\right)^2+u^2\left(\frac{\partial u}{\partial y}\right)^2+v^2\left(\frac{\partial v}{\partial x}\right)^2+v^2\left(\frac{\partial v}{\partial y}\right)^2}{|f(z)|^2}$$

$$=\frac{u^2\left[\left(\frac{\partial u}{\partial x}\right)^2+\left(\frac{\partial u}{\partial y}\right)^2\right]+v^2\left[\left(\frac{\partial v}{\partial x}\right)^2+\left(\frac{\partial v}{\partial y}\right)^2\right]}{|f(z)|^2}$$

$$=\frac{u^2\left[\left(\frac{\partial u}{\partial x}\right)^2+\left(\frac{-\partial v}{\partial x}\right)^2\right]+v^2\left[\left(\frac{\partial v}{\partial x}\right)^2+\left(\frac{\partial u}{\partial x}\right)^2\right]}{|f(z)|^2} \qquad \text{[C – R equations]}$$

$$=\frac{(u^2+v^2)\left[\left(\frac{\partial u}{\partial x}\right)^2+\left(\frac{\partial v}{\partial x}\right)^2\right]}{|f(z)|^2}=\frac{|f(z)|^2\left[\left(\frac{\partial u}{\partial x}\right)^2+\left(\frac{\partial v}{\partial x}\right)^2\right]}{|f(z)|^2}=\left(\frac{\partial u}{\partial x}\right)^2+\left(\frac{\partial v}{\partial x}\right)^2$$

$[\,|f(z)|^2 = u^2+v^2\,]$

$$=|f'(z)|^2$$ **Proved.**

Example 49. *Prove that* $\left(\frac{\partial^2}{\partial x^2}+\frac{\partial^2}{\partial y^2}\right)|u|^P = P(P-1)|u|^{P-2}|f'(z)|^2$

Solution. We know that $\frac{\partial^2}{\partial x^2}+\frac{\partial^2}{\partial y^2}=4\frac{\partial^2}{\partial z\,\partial \bar{z}}$ [Example 47, page 623]

$$\left(\frac{\partial^2}{\partial x^2}+\frac{\partial^2}{\partial y^2}\right)|u|^P=\frac{1}{2^P}\frac{4\partial^2}{\partial z\partial \bar{z}}[f(z)+f(\bar{z})]^P \qquad \left[\because\ u=\frac{1}{2}[f(z)+f(\bar{z})]\right]$$

$$=\frac{4}{2^P}\frac{\partial}{\partial \bar{z}}P[f(z)+f(\bar{z})]^{P-1}f'(z) = \frac{1}{2^{P-2}}P(P-1)[f(z)+f(\bar{z})]^{P-2}f'(z)f'(\bar{z})$$

$$=P(P-1)\left[\frac{1}{2}\{f(z)+f(\bar{z})\}\right]^{P-2}[f'(z)f'(\bar{z})]$$

$$=P(P-1)|u|^{P-2}|f'(z)|^2$$ **Proved.**

Example 50. *Prove that* $\left(\frac{\partial^2}{\partial x^2}+\frac{\partial^2}{\partial y^2}\right)\log|f'(z)|=0$

Solution. We have, $\left(\frac{\partial^2}{\partial x^2}+\frac{\partial^2}{\partial y^2}\right)=4\frac{\partial^2}{\partial z\partial \bar{z}}$ (Example 47 on page 623)

Hence $\left(\frac{\partial^2}{\partial x^2}+\frac{\partial^2}{\partial y^2}\right)\{\log|f'(z)|\} = 4\frac{\partial^2}{\partial z\,\partial \bar{z}}\{\log|f'(z)|\} = 4\frac{\partial^2}{\partial z\partial \bar{z}}\frac{1}{2}\log|f'(z)|^2$

$$= 2\frac{\partial^2}{\partial z\partial \bar{z}}\log\{f'(z)\overline{f'(z)}\} = 2\frac{\partial^2}{\partial z\partial \bar{z}}[\log f'(z)+\log f'(\bar{z})] = 2\frac{\partial}{\partial z}\left(0+\frac{1}{f'(\bar{z})}f''(\bar{z})\right)$$

$$= 2\frac{\partial}{\partial z}\frac{f''(\bar{z})}{f'(\bar{z})}$$ [$\bar{z}$ is constant in regards to differentiation *w.r.t.* z]

$= 2 \times 0$

$= 0$

Proved.

Example 51. *Prove that* $\left(\frac{\partial^2}{\partial x^2}+\frac{\partial^2}{\partial y^2}\right)|Rf(z)|^2 = 2|f'(z)|^2$ [*G.B.T.U., 2012*]

Solution. $f(z) = u + iv$ or $R\,f(z) = u \Rightarrow$ Real part of $f(z) = u$

$$\frac{\partial}{\partial x}u^2 = 2u\frac{\partial u}{\partial x}$$

$$\frac{\partial^2}{\partial x^2}u^2 = 2\left(\frac{\partial u}{\partial x}\right)^2 + 2u\frac{\partial^2 u}{\partial x^2} \quad \text{... (1)}$$

Similarly,
$$\frac{\partial^2}{\partial y^2}u^2 = 2\left(\frac{\partial u}{\partial y}\right)^2 + 2u\frac{\partial^2 u}{\partial y^2} \quad \text{... (2)}$$

Adding (1) and (2), we get

$$\left(\frac{\partial^2}{\partial x^2}+\frac{\partial^2}{\partial y^2}\right)u^2 = 2\left[\left(\frac{\partial u}{\partial x}\right)^2+\left(\frac{\partial u}{\partial y}\right)^2\right]+2u\left(\frac{\partial^2 u}{\partial x^2}+\frac{\partial^2 u}{\partial y^2}\right)$$

$$= 2\left[\left(\frac{\partial u}{\partial x}\right)^2+\left(\frac{\partial u}{\partial y}\right)^2\right]+0 = 2\left[\left(\frac{\partial u}{\partial x}\right)^2+\left(-\frac{\partial v}{\partial x}\right)^2\right] = 2|f'(z)|^2 \qquad \left(\frac{\partial u}{\partial y} = -\frac{\partial v}{\partial x}\right)$$

$$\Rightarrow \quad \left(\frac{\partial^2}{\partial x^2}+\frac{\partial^2}{\partial y^2}\right)|R\,f(z)|^2 = 2\,|f'(z)|^2$$ **Proved.**

Example 52. *If $f(z)$ is regular function of z, show that*

$$\left(\frac{\partial^2}{\partial x^2}+\frac{\partial^2}{\partial y^2}\right)|f(z)|^2 = 4\,|f'(z)|^2$$

Solution. $f(z) = u + iv$

$$|f(z)|^2 = u^2 + v^2 \quad \text{... (1)}$$

Let $\phi = u^2 + v^2$

Differentiating (1) *w.r.t.* x, we get

$$\frac{\partial \phi}{\partial x} = 2u\frac{\partial u}{\partial x} + 2v\frac{\partial v}{\partial x}$$

$$\frac{\partial^2 \phi}{\partial x^2} = 2\left[u\frac{\partial^2 u}{\partial x^2}+\left(\frac{\partial u}{\partial x}\right)^2 + v\frac{\partial^2 v}{\partial x^2}+\left(\frac{\partial v}{\partial x}\right)^2\right] \quad \text{...(2)}$$

Similarly, $$\frac{\partial^2\phi}{\partial y^2}=2\left[u\frac{\partial^2 u}{\partial y^2}+\left(\frac{\partial u}{\partial y}\right)^2+v\,\frac{\partial^2 v}{\partial y^2}+\left(\frac{\partial v}{\partial y}\right)^2\right] \qquad \text{... (3)}$$

Adding (2) and (3), we get

$$\frac{\partial^2\phi}{\partial x^2}+\frac{\partial^2\phi}{\partial y^2}=2\left[u\left(\frac{\partial^2 u}{\partial x^2}+\frac{\partial^2 u}{\partial y^2}\right)+v\left(\frac{\partial^2 v}{\partial x^2}+\frac{\partial^2 v}{\partial y^2}\right)+\left\{\left(\frac{\partial u}{\partial x}\right)^2+\left(\frac{\partial u}{\partial y}\right)^2+\left(\frac{\partial v}{\partial x}\right)^2+\left(\frac{\partial v}{\partial y}\right)^2\right\}\right] \qquad \text{... (4)}$$

By C – R equations $$\left(\frac{\partial u}{\partial x}\right)^2=\left(\frac{\partial v}{\partial y}\right)^2$$

$$\left(\frac{\partial u}{\partial y}\right)^2=\left(-\frac{\partial v}{\partial x}\right)^2=\left(\frac{\partial v}{\partial x}\right)^2 \qquad \text{... (5)}$$

By Laplace equations $$\frac{\partial^2 u}{\partial x^2}+\frac{\partial^2 u}{\partial y^2}=0$$

$$\frac{\partial^2 v}{\partial x^2}+\frac{\partial^2 v}{\partial y^2}=0$$

On putting the values of $\left(\frac{\partial^2 u}{\partial x^2}+\frac{\partial^2 u}{\partial y^2}\right),\left(\frac{\partial^2 v}{\partial x^2}+\frac{\partial^2 v}{\partial y^2}\right),\frac{\partial v}{\partial x},\frac{\partial v}{\partial y}$ from (5) in (4), we get

$$\frac{\partial^2\phi}{\partial x^2}+\frac{\partial^2\phi}{\partial y^2}=4\left[\left(\frac{\partial u}{\partial x}\right)^2+\left(\frac{\partial v}{\partial y}\right)^2\right],\quad\left(\frac{\partial^2}{\partial x^2}+\frac{\partial^2}{\partial y^2}\right)\phi=4\left|\frac{\partial u}{\partial x}+i\,\frac{\partial v}{\partial x}\right|^2$$

$$\left(\frac{\partial^2}{\partial x^2}+\frac{\partial^2}{\partial y^2}\right)|f(z)|^2=4\,|f'(z)|^2$$ **Proved.**

Example 53. *If $|f(z)|$ is constant, prove that $f(z)$ is also constant.*

Solution.

$$f(z)=u+iv$$

$$|f(z)|^2=u^2+v^2$$

$$|f(z)|=\text{constant}=c\ \text{(given)}$$

$$u^2+v^2=c^2 \qquad \text{... (1)}$$

Differentiating (1) w.r.t. x, we get $2u\frac{\partial u}{\partial x}+2v\frac{\partial v}{\partial x}=0 \quad\Rightarrow\quad u\frac{\partial u}{\partial x}+v\frac{\partial v}{\partial x}=0$ ... (2)

Differentiating (1) w.r.t. 'y', we get $2u\frac{\partial u}{\partial y}+2v\frac{\partial v}{\partial y}=0$

$$-u\frac{\partial v}{\partial x}+v\frac{\partial u}{\partial x}=0 \qquad \text{... (3)}$$

Squaring (2) and (3) and then adding, we get

$$u^2\left(\frac{\partial u}{\partial x}\right)^2+v^2\left(\frac{\partial v}{\partial x}\right)^2+u^2\left(\frac{\partial v}{\partial x}\right)^2+v^2\left(\frac{\partial u}{\partial x}\right)^2=0$$

$$(u^2+v^2)\left[\left(\frac{\partial u}{\partial x}\right)^2+\left(\frac{\partial v}{\partial x}\right)^2\right]=0$$

$$\left(\frac{\partial u}{\partial x}\right)^2+\left(\frac{\partial v}{\partial x}\right)^2=0$$

As $$f(z)=u+iv \Rightarrow f'(z)=\frac{\partial u}{\partial x}+i\frac{\partial v}{\partial x}$$

$$f'(\bar{z})=\frac{\partial u}{\partial x}-i\frac{\partial v}{\partial x}$$

$$|f'(z)|^2=\left(\frac{\partial u}{\partial x}\right)^2+\left(\frac{\partial v}{\partial x}\right)^2=0$$

$|f'(z)|^2=0 \Rightarrow f(z)$ is constant. **Proved.**

EXERCISE 22.5

1. If $f(z)=u+iv$ is an analytic function of $z=x+iy$, and ψ is any function of x and y with differential coefficients of the first two orders, then show that

$$\left(\frac{\partial \psi}{\partial x}\right)^2+\left(\frac{\partial \psi}{\partial y}\right)^2 = \left[\left(\frac{\partial \psi}{\partial u}\right)^2+\left(\frac{\partial \psi}{\partial v}\right)^2\right]|f'(z)|^2$$

and $$\frac{\partial^2 \psi}{\partial x^2}+\frac{\partial^2 \psi}{\partial y^2} = \left(\frac{\partial^2 \psi}{\partial u^2}+\frac{\partial^2 \psi}{\partial v^2}\right)|f'(z)|^2.$$

2. If $|f'(z)|$ is the product of a function of x and a function of y, show that

$$f'(z) = \exp(\alpha z^2+\beta z+\gamma)$$

where α is a real and β and γ are complex constants.

Choose the correct alternative:

3. If $|f(z)|$ is constant then $f(z)$ is

(*a*) Variable (*b*) Partially variable and constant (*c*) Constant (*d*) None of these **Ans.** (*c*)

4. If $f(z)=u+iv$ then $|f(z)|$ is equal to

(*a*) $\sqrt{u^2+v^2}$ (*b*) u^2+v^2 (*c*) $u+v$ (*d*) $\sqrt{u^2-v^2}$ **Ans.** (*a*)

5. If $z=r(\cos\theta+i\sin\theta)$ then $|z|^3$ is equal to

(*a*) $(\cos\theta+i\sin\theta)^3$ (*b*) $r^3(\cos\theta+i\sin\theta)^3$ (*c*) $r^3/2$ (*d*) r^3 **Ans.** (*d*)

CHAPTER 23 CONFORMAL TRANSFORMATION

23.1 GEOMETRICAL REPRESENTATION

To draw a curve of complex variable (x, y) on z-plane we take two axes *i.e.,* one real axis and the other imaginary axis. A number of points (x, y) are plotted on z-plane, by taking different value of z (different values of x and y). The curve C is drawn by joining the plotted points. The diagram obtained is called *Argand diagram* in z-plane.

But a complex function $w = f(z)$ *i.e.,* $(u + iv) = f(x + iy)$ involves four variables x, y and u, v.

A figure of only three dimensions (x, y, z) is possible in a plane. A figure of four dimensional region for x, y, u, v is not possible.

So, we choose two complex planes z-plane and w-plane. In the w-plane we plot the corresponding points $w = u + iv$. By joining these points we have a corresponding curve C' in w-plane.

23.2 TRANSFORMATION

For every point (x, y) in the z-plane, the relation $w = f(z)$ defines a corresponding point (u, v) in the w-plane. We call this "transformation or mapping of z-plane into w-plane." If a point z_0 maps into the point w_0, w_0 is also known as the image of z_0.

If the point $P(x, y)$ moves along a curve C in z-plane, the point $P'(u, v)$ will move along a corresponding curve C' in w-plane, then we say that a curve C in the z-plane is mapped into the corresponding curve C' in the w-plane by the relation $w = f(z)$.

Example 1. *Transform the rectangular region ABCD in z-plane bounded by x = 1, x = 3; y = 0 and y = 3. Under the transformation w = z + (2 + i).*

Solution. Here,

$$w = z + (2 + i)$$

$$\Rightarrow \quad u + iv = x + iy + (2 + i)$$

$$= (x + 2) + i(y + 1)$$

By equating real and imaginary quantities, we have $u = x + 2$ and $v = y + 1$.

z-plane	w-plane	z-plane	w-plane
x	$u = x + 2$	y	$v = y + 1$
1	$= 1 + 2 = 3$	0	$= 0 + 1 = 1$
3	$= 3 + 2 = 5$	3	$= 3 + 1 = 4$

Here the lines $x = 1$, $x = 3$; $y = 0$ and $y = 1$ in the z-plane are transformed onto the line $u = 3$, $u = 5$; $v = 1$ and $v = 4$ in the w-plane. The region $ABCD$ in z-plane is transformed into the region $EFGH$ in w-plane.

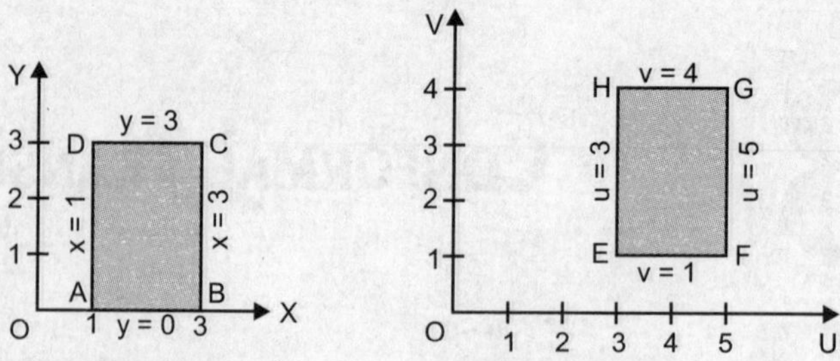

Example 2. *Transform the curve $x^2 - y^2 = 4$ under the mapping $w = z^2$.*

Solution. $w = z^2 = (x + iy)^2, \; u + iv = x^2 - y^2 + 2ixy$

This gives $u = x^2 - y^2$ and $v = 2xy$

Table of (*x*, *y*) and (*u*, *v*)

x	2	2.5	3	3.5	4	4.5	5
$y = \pm\sqrt{x^2 - 4}$	0	± 1.5	± 2.2	± 2.9	± 3.5	± 4.1	± 4.6
$u = x^2 - y^2$	4	4	4	4	4	4	4
$v = 2xy$	0	± 7.5	± 13.2	± 20.3	± 28	± 36.0	± 46

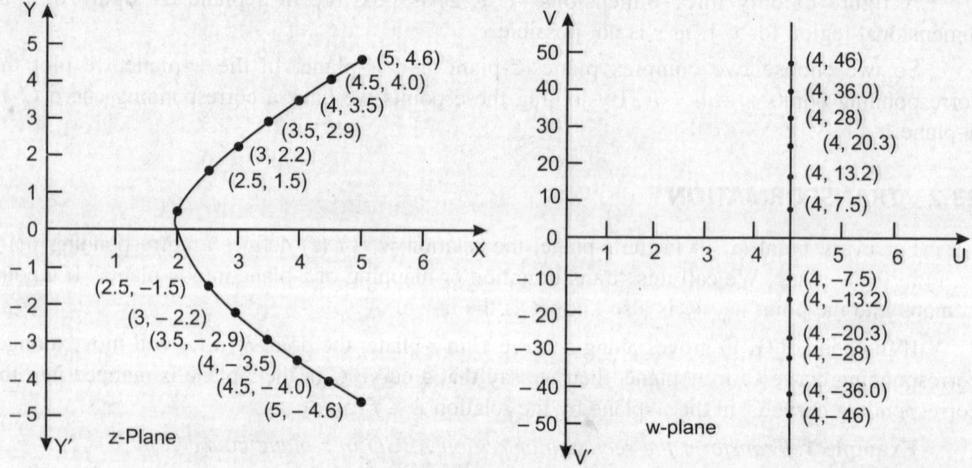

Image of the curve $x^2 - y^2 = 4$ is a straight line, $u = 4$ parallel to the v-axis in w-plane. **Ans.**

23.3 CONFORMAL TRANSFORMATION (*U.P. III Semester Dec., 2006, 2005*)

Let two curves C, C_1 in the z-plane intersect at the point P and the corresponding curve C', C_1' in the w-plane intersect at P'. **If the angle of intersection of the curves at P in z-plane is the same as the angle of intersection of the curves of w-plane at P' in magnitude and sense, then the transformation is called conformal:**

conditions: (*i*) $f(z)$ is analytic. (*ii*) $f'(z) \neq 0$ Or

If the sense of the rotation as well as the magnitude of the angle is preserved, the transformation is said to be **conformal**.

If only the magnitude of the angle is preserved, transformation is **Isogonal**.

23.4 THEOREM. If $f(z)$ is analytic, mapping is conformal

(U.P. III Semester Dec. 2005)

Proof. Let C_1 and C_2 be the two curves in the z-plane intersecting at the point z_0 and let the tangents at this point make angles α_1 and α_2 with the real axis. Let z_1 and z_2 be the points on the curves C_1 and C_2 near to z_0 at the same distance r from z_0, so that we have

$$z_1 - z_0 = r e^{i\theta_1}, \; z_2 - z_0 = r e^{i\theta_2}$$

As $\quad r \to 0, \theta_1 \to \alpha_1$ and $\theta_2 \to \alpha_2$.

Let the image of the curves C_1, C_2 be C_1' and C_2' in w-plane and images of z_0, z_1 and z_2 be w_0, w_1 and w_2

Let $\quad w_1 - w_0 = r.e^{i\phi_1}, \; w_2 - w_0 = r.e^{i\phi_2}$

$$f'(z_0) = \lim_{z_1 \to z_0} \frac{w_1 - w_0}{z_1 - z_0}$$

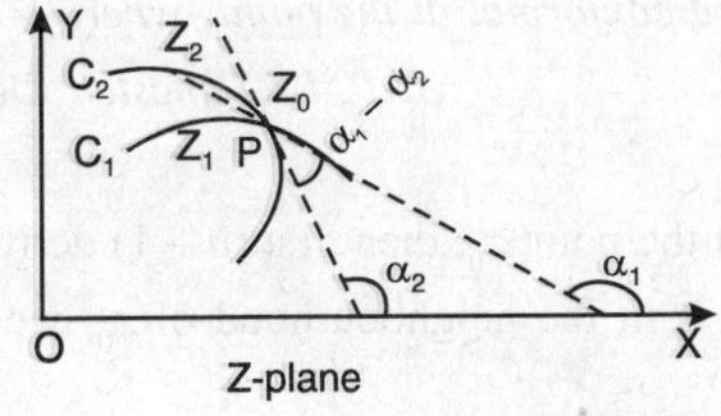

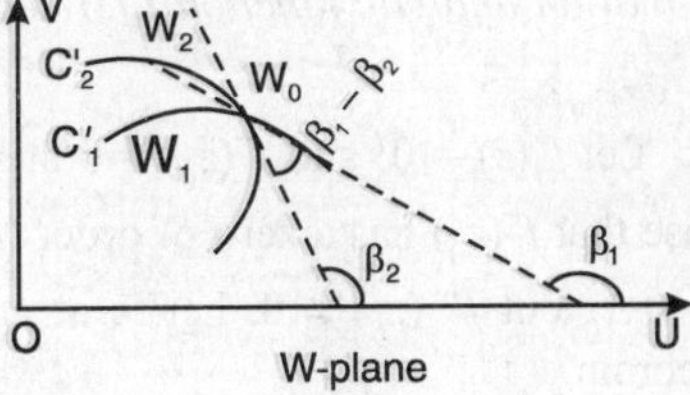

$$\mathrm{Re}^{i\lambda} = \lim_{r \to 0} \frac{r_1 e^{i\phi_1}}{r e^{i\theta_1}} \qquad \text{(since } f'(z_0) = \mathrm{Re}^{i\lambda}\text{)}$$

$$\mathrm{Re}^{i\lambda} = \frac{r_1}{r} e^{i\phi_1 - i\theta_1} = \frac{r_1}{r} e^{i(\phi_1 - \theta_1)}$$

Hence $\quad \lim_{r \to 0} \left[\frac{r_1}{r}\right] = R = |f'(z_0)|$ and $\lim (\phi_1 - \theta_1) = \lambda$

$\Rightarrow \quad \lim \phi_1 - \lim \theta_1 = \lambda$ or $\beta_1 - \alpha_1 = \lambda$ *i.e.*, $\beta_1 = \alpha_1 + \lambda$

Similarly it can be proved $\beta_2 = \alpha_2 + \lambda$ curve C'_1 has a definite tangent at w_0 making angles $\alpha_2 + \lambda$ and $\alpha_2 + \lambda$ so curve C'_2.

Angle between two curves C'_1 and C'_2

$$= \beta_1 - \beta_2 = (\alpha_1 + \lambda) - (\alpha_2 + \lambda) = (\alpha_1 - \alpha_2)$$

so the transformation is conformal at each point where $f'(z) = 0$

Note 1. The point at which $f'(z) = 0$ is called a **critical point** of the transformation. Also the points where $\frac{dw}{dz} \neq 0$ are called **ordinary points**.

2. Let $\phi = \alpha_1 - \alpha_2$ or $\alpha_1 = \alpha_2 + \phi$ shows that the tangent at P to the curve is rotated through an $\angle\phi = \text{amp}\ \{f'(z)\}$ under the given transformation.

Angle of rotation $= \tan^{-1}\frac{v}{u}$.

3. In formal transformation, element of arc passing through P is magnified by the factor $|f'(z)|$. The area element is also magnified by the factor $|f'(z)|$ or $J = \frac{\partial(u,v)}{\partial(x,y)}$ in a conformal transformation.

$$J = \frac{\partial(u,v)}{\partial(x,y)} = \begin{vmatrix} \frac{\partial u}{\partial x} & \frac{\partial u}{\partial y} \\ \frac{\partial v}{\partial x} & \frac{\partial v}{\partial y} \end{vmatrix} = \begin{vmatrix} \frac{\partial u}{\partial x} & -\frac{\partial v}{\partial x} \\ \frac{\partial v}{\partial x} & \frac{\partial u}{\partial x} \end{vmatrix} = \left(\frac{\partial u}{\partial x}\right)^2 + \left(\frac{\partial v}{\partial x}\right)^2$$

$$= \left|\frac{\partial u}{\partial x} + i\frac{\partial v}{\partial x}\right|^2 = |f'(z)|^2 = |f'(x+iy)|^2$$

$|f'(z)|$ is called the **coefficient of magnification.**

4. Conjugate functions remain conjugate functions after conformal transformation. A function which is the solution of Laplace's equation, its transformed function again remains the solution of Laplace's equation after conformal transformation.

23.5 THEOREM

Prove that an analytic function $f(z)$ ceases to be conformal at the points where $f'(z) = 0$.

(*U.P. III Semester, Dec. 2006*)

Proof. Let $f'(z) = 0$ and $f'(z_0) = 0$ at $z = z_0$

Suppose that $f'(z_0)$ has a zero of order $(n-1)$ at the point z_0, then first $(n-1)$ derivatives of $f(z)$ vanish at z_0 but $f^n(z_0) \neq 0$, hence at any point z in the neighbourhood of z_0, we have by Taylor's Theorem.

$$f(z) = f(z_0) + a_n(z-z_0)^n + \dots\dots$$

where $a_n = \dfrac{f^n(z_0)}{n!}$, so that $a_n \neq 0$.

Hence, $f(z_1) - f(z_0) = a_n(z_1 - z_0)^n + \dots\dots$

i.e. $w_1 - w_0 = a_n(z_1 - z_0)^n + \dots$

or $\rho_1 e^{i\phi_1} = |a_n| . r^n e^{i(n\theta_1 + \lambda)} + \dots\dots$ where $\lambda = \text{amp } a_n$

Hence, $\text{Lim } \phi_1 = \text{Lim } (n\theta_1 + \lambda) = n\alpha_1 + \lambda$

Similary, $\text{Lim } \phi_2 = n\alpha_2 + \lambda.$

Thus the curves γ_1 and γ_2 still have definite tangents at w_0.

But the angle between the tangents

$$= \text{Lim}\,\phi_2 - \text{Lim}\,\phi_1 = n(\alpha_2 - \alpha_1).$$

So magnitude of the angle is not preserved.

Also the linear magnification $R = \text{Lim}(\rho_1 / r) = 0.$

Hence, the conformal property does not hold good at a point where $f'(z) = 0$.

Example 3. *If $u = 2x^2 + y^2$ and $v = \dfrac{y^2}{x}$, show that the curves u = constant and v= constant cut orthogonally at all intersections but that the transformation $w = u + iv$ is not conformal.*

Solution. For the curve, $2x^2 + y^2 = u$

$$2x^2 + y^2 = \text{constant} = k_1 \text{ (say)} \quad \dots(1)$$

Differentiating (1), we get

$$4x + 2y\frac{dy}{dx} = 0 \quad \Rightarrow \quad \frac{dy}{dx} = -\frac{2x}{y} \qquad ...(2)$$

$$\frac{y^2}{x} = v$$

For the curve, $\frac{y^2}{x} =$ constant $= k_2$ (say),

$\Rightarrow$ $$y^2 = k_2 x. \qquad ...(3)$$

Differentiating (3), we get

$$2y\frac{dy}{dx} = k_2 \quad \Rightarrow \quad \frac{dy}{dx} = \frac{k_2}{2y} = \frac{y^2}{x} \times \frac{1}{2y} = \frac{y}{2x} \qquad ...(4)$$

From (2) and (4), we see that

$$m_1 m_2 = \left(\frac{-2x}{y}\right)\left(\frac{y}{2x}\right) = -1$$

Hence, two curves cut orthogonally.

However, since $\frac{\partial u}{\partial x} = 4x, \quad \frac{\partial u}{\partial y} = 2y$

$$\frac{\partial v}{\partial x} = -\frac{y^2}{x^2}, \quad \frac{\partial v}{\partial y} = \frac{2y}{x}$$

The Cauchy-Riemann equations are not satisfied by u and v.

Hence, the function $u + iv$ is not analytic. So, the transformation is not conformal. **Proved**

Example 4. (*i*) *For the conformal transformation $w = z^2$, show that*

(*a*) *The coefficient of magnification at $z = 2 + i$ is $2\sqrt{5}$.*

(*b*) *The angle of rotation at $z = 2 + i$ is $\tan^{-1} 0.5$.*

(*ii*) *For the conformal transformation $w = z^2$, show that*

(*a*) *The co-efficient of magnification at $z = 1 + i$ is $2\sqrt{2}$.*

(*b*) *The angle of rotation at $z = 1 + i$ is $\frac{\pi}{4}$.*

Solution. (*i*) Let $w = f(z) = z^2$

$\therefore$ $f'(z) = 2z$

$f'(2 + i) = 2(2 + i) = 4 + 2i.$

(*a*) Coefficient of magnification at $z = 2 + i$ is $= |f'(2 + i)| = |4 + 2i| = \sqrt{16 + 4} = 2\sqrt{5}.$

(*b*) Angle of rotation at $z = 2 + i$ is $=$ amp. $f'(2 + i) =$ amp. $(4 + 2i) = \tan^{-1}\left(\frac{2}{4}\right) = \tan^{-1}(0.5).$

(*ii*) Here $f(z) = w = z^2$

$\therefore$ $f'(z) = 2z$

and $f'(1 + i) = 2(1 + i) = 2 + 2i$

$\therefore$ (*a*) The co-efficient of magnification at $z = 1 + i$ is $|f'(1+i)| = |2 + 2i| = \sqrt{4+4} = 2\sqrt{2}$

(*b*) The angle of rotation at $z = 1 + i$ is amp. $|f'(1+i)| =$ amp. $2(1 + i) = 2 + 2i = \tan^{-1}\frac{2}{2} = \frac{\pi}{4}$

Ans.

Standard transformations

23.6 TRANSLATION $w = z + C$,

where $$C = a + ib$$
$$u + iv = x + iy + a + ib$$
$$u = x + a \text{ and } v = y + b$$
$$x = u - a \text{ and } y = v - b$$

On substituting the values of x and y in the equation of the curve to be transformed, we get the equation of the image in the w-plane.

The point $P(x, y)$ in the z-plane is mapped onto the point $P'(x + a, y + b)$ in the w-plane. Similarly other points of z-plane are mapped onto w-plane. Thus if w-plane is superimposed on the z-plane, the figure of w-plane is shifted through a vector C.

In other words the transformation is mere translation of the axes.

23.7 ROTATION $w = ze^{i\theta}$

The figure in z-plane rotates through an angle θ in anticlockwise in w-plane.

Example 5. *Consider the transformation $w = ze^{i\pi/4}$ and determine the region R' in w-plane corresponding to the triangular region R bounded by the lines $x = 0$, $y = 0$ and $x + y = 1$ in z-plane.*

Solution. $$w = ze^{i\pi/4}$$
$$w = (x + iy)\left(\cos\frac{\pi}{4} + i\sin\frac{\pi}{4}\right)$$
$$\Rightarrow \quad u + iv = (x + iy)\left(\frac{1+i}{\sqrt{2}}\right) = \frac{1}{\sqrt{2}}[x - y + i(x + y)]$$

Equating real and imaginay parts, we get

$$\Rightarrow \quad u = \frac{1}{\sqrt{2}}(x - y), \quad v = \frac{1}{\sqrt{2}}(x + y)$$

(*i*) Put $x = 0$, $\quad u = -\frac{1}{\sqrt{2}}y, \quad v = \frac{1}{\sqrt{2}}y$ or $v = -u$

(*ii*) Put $y = 0$, $\quad u = \frac{1}{\sqrt{2}}x, \quad v = \frac{1}{\sqrt{2}}x$ or $v = u$

(*iii*) Putting $x + y = 1$ in (1), we get $v = \frac{1}{\sqrt{2}}$

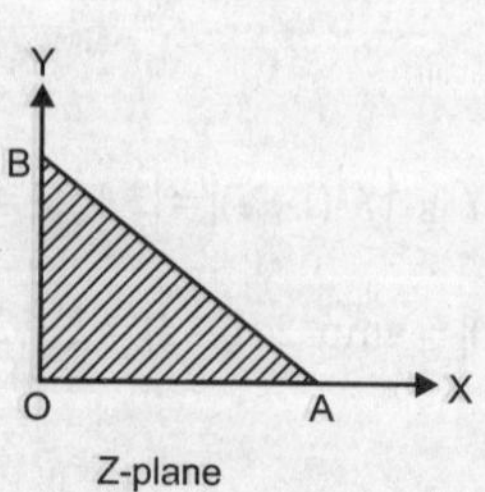

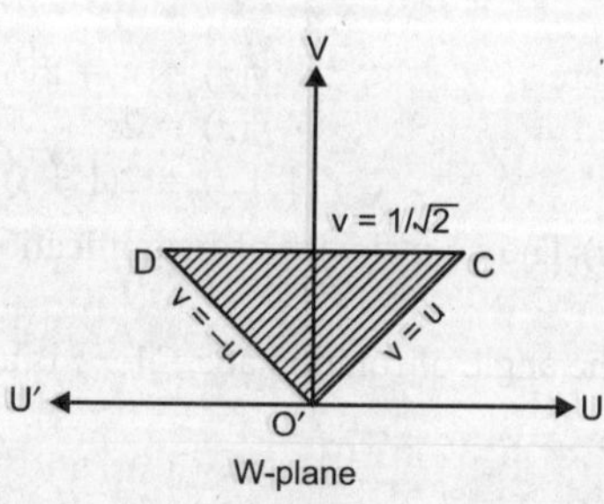

Hence the triangular region OAB in z-plane is mapped on a triangular region O′CD of w-plane bounded by the lines $v = u,\ v = -u,\ v = \frac{1}{\sqrt{2}}$.

$$f'(z) = \frac{1}{\sqrt{2}}(1 + i)$$

$$f(z) = \frac{1}{\sqrt{2}}[(x - y) + i(x + y)]$$

Amp. $f'(z) = \tan^{-1}(1) = \frac{\pi}{4}$

The mapping $w = ze^{i\pi/4}$ performs a rotation of R through an angle $\pi/4$. **Ans.**

23.8 MAGNIFICATION

$$\boxed{w = cz}$$

where c is a real quantity.

(*i*) The figure in w-plane is magnified c-times the size of the figure in z-plane.

(*ii*) Both figures in z-plane and w-plane are singular.

Example 6. *Determine the region in w-plane on the transformation of rectangular region enclosed by $x = 1,\ y = 1,\ x = 2$ and $y = 2$ in the z-plane. The transformation is $w = 3z$.*

Solution. We have, $\quad w = 3z$

$$u + iv = 3(x + iy)$$

Equating the real and imaginary parts, we get

$$u = 3x \quad \text{and} \quad v = 3y$$

z-plane		*w*-plane	
x	y	$u = 3x$	$v = 3y$
1	1	3	3
2	2	6	6

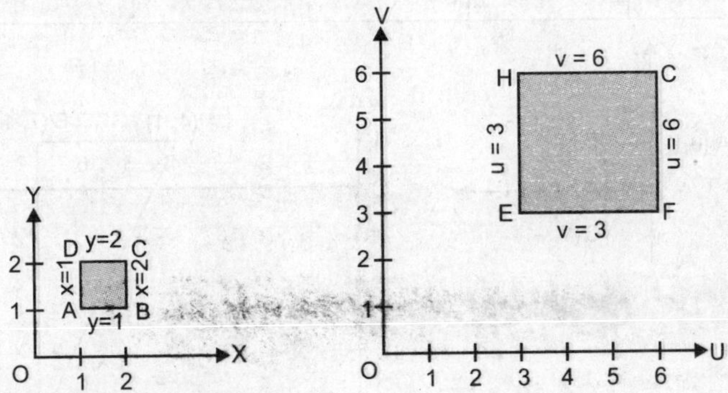

23.9 MAGNIFICATION AND ROTATION $w = cz$ ... (1)

where c, z, w all are complex numbers.

$$c = ae^{i\alpha}, \quad z = re^{i\theta}, \quad w = Re^{i\phi}$$

Putting these values in (1), we have

$$Re^{i\phi} = (ae^{i\alpha})(re^{i\theta}) = are^{i(\theta+\alpha)}$$

i.e. $R = ar$ and $\phi = \theta + \alpha$

Thus we see that the transform $w = cz$ corresponding to a rotation, together with magnification.

Algebraically $w = c\,z$ or $u + iv = (a + ib)(x + iy)$

$\Rightarrow$ $u + iv = ax - by + i(ay + bx)$

$$u = ax - by \text{ and } v = ay + bx$$

On solving these equations, we can get the values of x and y.

$$x = \frac{au + bv}{a^2 + b^2},\ y = \frac{-bu + av}{a^2 + b^2}$$

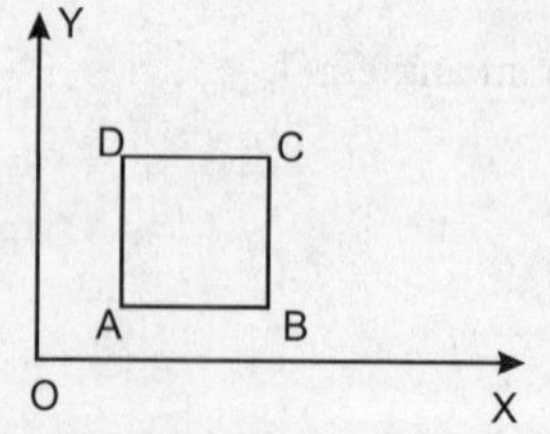

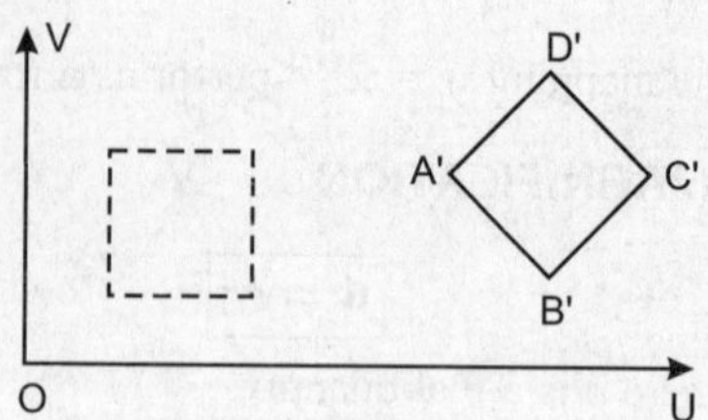

On putting values of x and y in the equation of the curve to be transformed we get the equation of the image.

Example 7. *Find the image of the triangle with vertices at i, 1 + i, 1 – i in the z-plane, under the transformation*

(*i*) $w = 3z + 4 - 2i$, (*ii*) $w = e^{\frac{5\pi i}{3}}.z - 2 + 4i$

Solution. (*i*) $w = 3z + 4 - 2i$,

$\Rightarrow$ $u + iv = 3(x + iy) + 4 - 2i \Rightarrow u = 3x + 4,\ v = 3y - 2$

	z-plane				*w-plane*	
S. No.	x	y	*points*	$u = 3x + 4$	*points*	$v = 3y - 2$
1.	0	1	A	4	A'	1
2.	1	1	B	7	B'	1
3.	1	–1	C	7	C'	–5

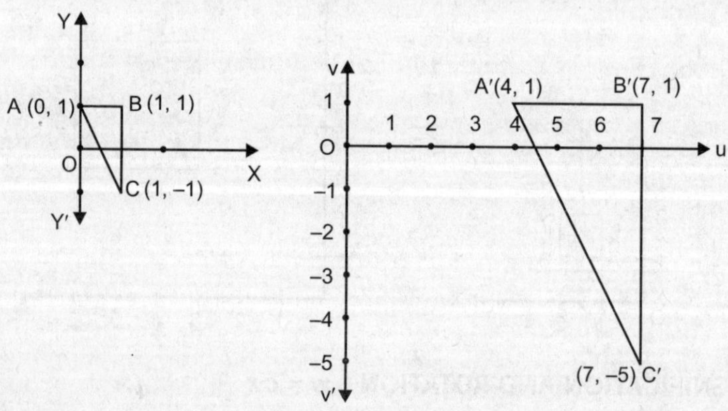

(*ii*) $w = e^{\frac{5\pi i}{3}}.z - 2 + 4i$

$\Rightarrow$ $u + iv = \left(\cos\frac{5\pi}{3} + i\sin\frac{5\pi}{3}\right)(x + iy) - 2 + 4i$

$$= \left(\frac{1}{2} - \frac{\sqrt{3}}{2} i\right)(x + iy) - 2 + 4i$$

$$= \frac{x}{2} - 2 + \frac{\sqrt{3}}{2} y + i\left(-\frac{\sqrt{3}}{2} x + \frac{y}{2} + 4\right)$$

$$\Rightarrow \qquad u = \frac{x}{2} - 2 + \frac{\sqrt{3}}{2} y \quad \text{and} \quad v = -\frac{\sqrt{3}}{2} x + \frac{y}{2} + 4$$

S.No.	z-Plane			w-plane		
	x	y	Points	u	Points	v
				$u = \frac{x}{2} - 2 + \frac{\sqrt{3}}{2} y$		$v = -\frac{\sqrt{3}}{2} x + \frac{y}{2} + 4$
1.	0	1	A	$-2 + \frac{\sqrt{3}}{2} = -1.1$	A′	$\frac{9}{2} = 4.5$
2.	1	1	B	$-\frac{3}{2} + \frac{\sqrt{3}}{2} = -0.6$	B′	$-\frac{\sqrt{3}}{2} + \frac{9}{2} = 3.6$
3.	1	−1	C	$-\frac{3}{2} - \frac{\sqrt{3}}{2} = -2.3$	C′	$-\frac{\sqrt{3}}{2} + \frac{7}{2} = 2.6$

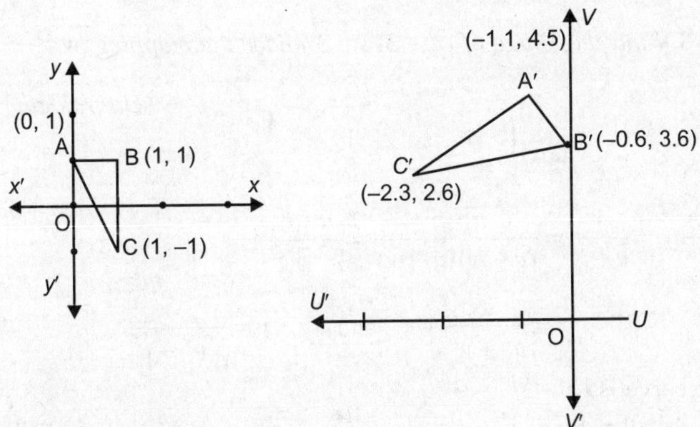

23.10 INVERSION AND REFLECTION

$$w = \frac{1}{z} \qquad \text{... (1)}$$

If $\qquad z = r e^{i\theta}$ and $w = R e^{i\phi}$

Putting these values in (1), we get

$$\text{R}e^{i\phi} = \frac{1}{r e^{i\theta}} = \frac{1}{r} e^{-i\theta}$$

On equating, $\qquad R = \frac{1}{r}$ and $\phi = -\theta$

Thus the point $P(r, \theta)$ in the z-plane is mapped into the point $P'\left(\frac{1}{r}, -\theta\right)$ in the w-plane.

Hence the transformation is an inversion of z and followed by reflection into the real axis. The points inside the unit circle ($|z| = 1$) map into points outside it, and points outside the unit circle into points inside it.

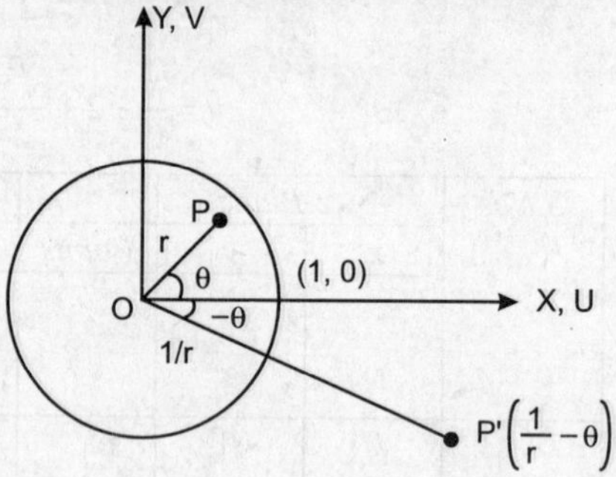

Algebraically $\quad w = \frac{1}{z}$ or $z = \frac{1}{w}$

$$x + iy = \frac{1}{u + iv}$$

$$\Rightarrow \quad x + iy = \frac{u - iv}{(u + iv)(u - iv)} = \frac{u - iv}{u^2 + v^2}$$

$$x = \frac{u}{u^2 + v^2}, \; y = -\frac{v}{u^2 + v^2}$$

Let the circle $x^2 + y^2 + 2gx + 2fy + c = 0$ ... (1) be in z-plane.

On substituting the values of x and y in (1), we get

$$\frac{u^2}{(u^2 + v^2)^2} + \frac{v^2}{(u^2 + v^2)^2} + 2g\frac{u}{u^2 + v^2} + 2f\frac{(-v)}{u^2 + v^2} + c = 0$$

This is the equation of circle in w-plane. This shows that a circle in z-plane transforms to another circle in w-plane.
But a circle through origin transforms into a straight line.

Example 8. *Find the image of* $|z - 3i| = 3$ *under the mapping* $w = \frac{1}{z}$.

(*Uttarakhand, III Semester 2008*)

Solution. $\quad w = \frac{1}{z} \Rightarrow z = \frac{1}{w}$

$$\Rightarrow \quad x + iy = \frac{1}{u + iv} = \frac{u - iv}{(u + iv)(u - iv)} = \frac{u - iv}{u^2 + v^2}$$

$$\Rightarrow \quad x = \frac{u}{u^2 + v^2} \quad \text{and} \quad y = -\frac{v}{u^2 + v^2} \qquad \text{... (1)}$$

The given curve is $|z - 3i| = 3$

$\Rightarrow \quad |x + iy - 3i| = 3 \quad \Rightarrow \quad x^2 + (y - 3)^2 = 9 \qquad$ ... (2)

On substituting the values of x and y from (1) into (2), we get

$$\frac{u^2}{(u^2 + v^2)^2} + \left(-\frac{v}{u^2 + v^2} - 3\right)^2 = 9$$

$$\frac{u^2}{(u^2 + v^2)^2} + \frac{(-v - 3u^2 - 3v^2)^2}{(u^2 + v^2)^2} = 9$$

$$\Rightarrow \quad u^2 + (-v - 3u^2 - 3v^2)^2 = 9(u^2 + v^2)^2$$

$$\Rightarrow \quad u^2 + v^2 + 9u^4 + 9v^4 + 6u^2v + 6v^3 + 18u^2v^2 = 9u^4 + 18u^2v^2 + 9v^4$$

$$\Rightarrow \quad u^2 + v^2 + 6u^2v + 6v^3 = 0$$

$$\Rightarrow \quad u^2 + v^2 + 6v(u^2 + v^2) = 0$$

$\Rightarrow \quad (u^2+v^2)(6v+1)=0$

$\Rightarrow \quad 6v + 1 = 0$ is the equation of the image. **Ans.**

Aliter. $|z-3i|=3, \quad z=\dfrac{1}{w}$

$$\left|\frac{1}{w}-3i\right|=3 \quad \Rightarrow |1-3iw|=3|w|$$

$\Rightarrow \quad |1-3i(u+iv)|=3|u+iv| \quad \Rightarrow \quad |1+3v-3iu|=3|u+iv|$

$\Rightarrow \quad (1+3v)^2+9u^2=9(u^2+v^2) \quad \Rightarrow \quad 1+6v+9v^2+9u^2=9(u^2+v^2)$

$\Rightarrow \quad 1+6v=0$ **Ans.**

Aliter. $|z-3i|=3 \Rightarrow z-3i=3e^{i\theta} \Rightarrow z=3i+3e^{i\theta}$

$$w=\frac{1}{z}=\frac{1}{3i+3e^{i\theta}} \quad \Rightarrow \quad 3w=\frac{1}{i+e^{i\theta}}$$

$$\Rightarrow \quad 3(u+iv)=\frac{1}{i+\cos+i\sin\theta}$$

$$(3u+3iv)=\frac{\cos\theta-i(1+\sin\theta)}{\cos^2\theta+(1+\sin\theta)^2} \quad \Rightarrow \quad 3v=-\frac{1+\sin\theta}{2+2\sin\theta}=-\frac{1}{2}$$

$6v+1=0$ **Ans.**

Example 9. *Image of* $|z+1|=1$ *under the mapping* $\omega=\dfrac{1}{z}$ is

(*a*) $2v+1=1$ (*b*) $2v-1=0$ (*c*) $2u+1=0$ (*d*) $2u-1=0$

(*AMIETE, June 2009*)

Solution. $\omega=\dfrac{1}{z} \quad \Rightarrow u+iv=\dfrac{1}{x+iy}=\dfrac{x-iy}{x^2+y^2}$

$$\Rightarrow \quad u=\frac{x}{x^2+y^2}, \quad v=-\frac{y}{x^2+y^2}$$

Given $|z+1|=1 \quad \Rightarrow |x+iy+1|=1 \Rightarrow (x+1)^2+y^2=1$

$\Rightarrow \quad x^2+y^2+2x=0 \quad \Rightarrow x^2+y^2=-2x \quad \Rightarrow \quad \dfrac{1}{2}=\dfrac{-x}{x^2+y^2}=-u$

$\Rightarrow \quad \dfrac{1}{2}=-u \quad \Rightarrow 2u+1=0$

Hence (c) is correct answer. **Ans.**

Example 10. *Show that under the transformation* $w=\dfrac{1}{z}$, *the image of the hyperbola* $x^2-y^2=1$ *is the lemniscate* $R=\cos 2\phi$.

Solution. $x^2-y^2=1$

Putting $x=r\cos\theta$

and $y=r\sin\theta$

$\Rightarrow \quad r^2\cos^2\theta-r^2\sin^2\theta=1 \quad \Rightarrow \quad r^2(\cos^2\theta-\sin^2\theta)=1$

$\Rightarrow \quad r^2\cos 2\theta=1$ ... (1)

And $\quad w=\dfrac{1}{z} \quad \Rightarrow \quad z=\dfrac{1}{w} \quad \Rightarrow \quad re^{i\theta}=\dfrac{1}{Re^{i\phi}} \quad \Rightarrow \quad re^{i\theta}=\dfrac{1}{R}e^{-i\phi}$

$$\therefore \qquad r = \frac{1}{R} \quad \text{and} \quad \theta = \phi$$

Putting the values of r and θ in (1), we get

$$\frac{1}{R^2}\cos 2(-\phi) = 1 \quad \Rightarrow \quad R^2 = \cos 2\phi$$ **Proved.**

EXERCISE 23.1

1. Find the image of the semi infinite, strip $x > 0$, $0 < y < 2$ under the transformation $w = iz + 1$.

   Ans. Strip $-1 < u < 1$, $v > 0$

2. Determine the region in the w-plane in which the rectangle bounded by the lines $x = 0$, $y = 0$, $x = 2$ and $y = 1$ is mapped under the transformation $w = \sqrt{2}\, e^{i\pi/4}\, z$.

   Ans. Region bounded by the lines $v = -u$, $v = u$, $u + v = 4$ and $v - u = 2$.

3. Show that the condition for transformation $w = a^2 + blcz + d$ to make the circle $|w| = 1$ correspond to a straight line in the z-plane is $(a) = (c)$.

4. What is the region of the w-plane in two ways the rectangular region in the z-plane bounded by the lines $x = 0$, $y = 0$, $x = 1$ and $y = 2$ is mapped under the transformation $w = z + (2 - i)$?

   Ans. Region bounded by $u = 2$, $v = -1$, $u = 3$ and $v = 1$.

5. For the mapping $w(z) = 1/z$, find the image of the family of circles $x^2 + y^2 = ax$, where a is real.

   Ans. $u = \frac{1}{a}$ is a straight line || to v-axis.

6. Show that the function $w = \frac{4}{z}$ transforms the straight line $x = c$ in the z-plane into a circle in the w-plane.

7. If $(w+1)^2 = \frac{4}{z}$, then prove that the unit circle in the w-plane corresponds to a parabola in the z-plane, and the inside of the circle to the outside of the parabola.

8. Find the image of $|z - 2i| = 2$ under the mapping $w = \frac{1}{z}$ **Ans.** $4v + 1 = 0$

9. Determine the region in the z plane by $4 < |z + i| < 8$ *(DU. 2014)* **Ans.**

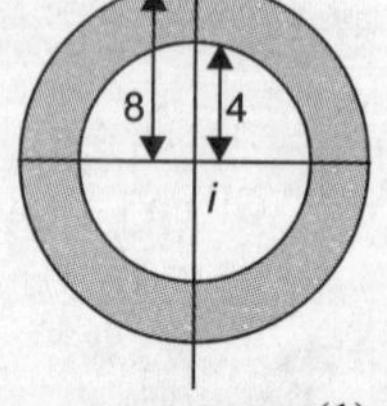

23.11 BILINEAR TRANSFORMATION (Mobius Transformation)

$$\boxed{w = \frac{az+b}{cz+d}} \qquad \text{...(1)}$$

(1) is known as bilinear transformation.

If then $\frac{dw}{dz} \neq 0$ *i.e.* transformation is conformal.

From (1), $$z = \frac{-dw+b}{cw-a} \qquad \text{...(2)}$$

This is also bilinear except $w = \frac{a}{c}$.

Note. From (1), every point of z-plane is mapped into unique point in w-plane except $z = -\frac{d}{c}$.

From (2), every point of w-plane is mapped into unique point in z-plane except $w = \frac{a}{c}$.

23.12 INVARIANT POINTS OF BILINEAR TRANSFORMATION.

We know that $$w = \frac{az+b}{cz+d} \qquad \text{... (1)}$$

If z maps into itself, then $w = z$

(1) becomes $$z = \frac{az+b}{cz+d} \qquad \text{... (2)}$$

Roots of (2) are the invariants or fixed points of the bilinear transformation.

If the roots are equal, the bilinear transformation is said to be parabolic.

23.13 CROSS-RATIO

If there are four points z_1, z_2, z_3, z_4 taken in order, then the ratio $\frac{(z_1 - z_2)(z_3 - z_4)}{(z_2 - z_3)(z_4 - z_1)}$ is called the cross-ratio of z_1, z_2, z_3, z_4.

23.14 THEOREM

A bilinear transformation preserves cross-ratio of four points

Proof. We know that $$w = \frac{az+b}{cz+d}.$$

As w_1, w_2, w_3, w_4 are images of z_1, z_2, z_3, z_4 respectively, so

$$w_1 = \frac{az_1+b}{cz_1+d}, \quad w_2 = \frac{az_2+b}{cz_2+d}$$

$$\therefore \quad w_1 - w_2 = \frac{(ad-bc)}{(cz_1+d)(cz_2+d)}(z_1 - z_2) \qquad \text{...(1)}$$

Similarly $$w_2 - w_3 = \frac{ad-bc}{(cz_2+d)(cz_3+d)}(z_2 - z_3) \qquad \text{...(2)}$$

$$w_3 - w_4 = \frac{ad-bc}{(cz_3+d)(cz_4+d)}(z_3 - z_4) \qquad \text{...(3)}$$

$$w_4 - w_1 = \frac{ad-bc}{(cz_4+d)(cz_1+d)}(z_4 - z_1) \qquad \text{...(4)}$$

From (1), (2), (3) and (4), we have

$$\frac{(w_1 - w_2)(w_3 - w_4)}{(w_1 - w_4)(w_3 - w_2)} = \frac{(z_1 - z_2)(z_3 - z_4)}{(z_1 - z_4)(z_3 - z_2)}$$

$$\Rightarrow \quad (w_1, w_2, w_3, w_4) = (z_1, z_2, z_3, z_4).$$

23.15 PROPERTIES OF BILINEAR TRANSFORMATION

1. A bilinear transformation maps circles into circles.

2. A bilinear transformation preserves cross ratio of four points.

If four points z_1, z_2, z_3, z_4 of the z-plane map onto the points w_1, w_2, w_3, w_4 of the w-plane respectively.

$$\Rightarrow \quad \frac{(w_1 - w_2)(w_3 - w_4)}{(w_1 - w_4)(w_3 - w_2)} = \frac{(z_1 - z_2)(z_3 - z_4)}{(z_1 - z_4)(z_3 - z_2)}$$

Hence, under the **bilinear** transform of four points cross-ratio is preserved.

23.16 METHODS TO FIND BILINEAR TRANSFORMATION

1. By finding a, b, c, d for $w=\dfrac{az+b}{cz+d}$ with the given conditions.

2. With the help of cross-ratio

$$\boxed{\frac{(w-w_1)(w_2-w_3)}{(w-w_3)(w_2-w_1)}=\frac{(z-z_1)(z_2-z_3)}{(z-z_3)(z_2-z_1)}}$$

Example 11. *Find the bilinear transformation which maps the points* $z = 1, i, -1$ *into the points* $w = i, 0, -i$.
Hence find the image of $|z| < 1$.

(*U.P., III Semester, 2008,*)

Solution. Let the required transformation be $w=\dfrac{az+b}{cz+d}$

or $w=\dfrac{\frac{a}{d}z+\frac{b}{d}}{\frac{c}{d}z+1}=\dfrac{pz+q}{rz+1}$ ...(1) $\left[p=\dfrac{a}{d}, q=\dfrac{b}{d}, r=\dfrac{c}{d}\right]$

On substituting the values of z and corresponding values of w in (1), we get

z	w
1	i
i	0
-1	$-i$

$$i=\frac{p+q}{r+1} \Rightarrow p+q=ir+i \quad ...(2)$$

$$0=\frac{pi+q}{ri+1} \qquad pi+q=0 \quad ...(3)$$

$$-i=\frac{-p+q}{-r+1} \qquad -p+q=ir-i \quad ...(4)$$

On subtracting (4) from (2), we get $2p = 2i$ or $p = i$
On putting the value of p in (3), we have $i\,(i) + q = 0$ or $q = 1$
On substituting the values of p and q in (2), we obtain

$$i+1=ir+i \quad \text{or} \quad 1=ir \quad \text{or} \quad r=-i$$

By using the values of p, q, r and (1), we have

$$w=\frac{iz+1}{-iz+1}$$

$$u+iv=\frac{i(x+iy)+1}{-i(x+iy)+1}=\frac{(ix-y+1)(ix+y+1)}{(-ix+y+1)(ix+y+1)}=\frac{-x^2-y^2+1+2ix}{x^2+(y+1)^2}$$

Equating real and imaginary parts, we get

$$u=\frac{-x^2-y^2+1}{x^2+(y+1)^2} \quad ...(5)$$

But $\quad |z|<1 \Rightarrow x^2+y^2<1 \quad \Rightarrow \quad 1-x^2-y^2>0$

From (5) $\quad u>0$ As denominator is positive **Ans.**

Example 12. *Find a bilinear transformation which maps the points* $i, -i, 1$ *of the z-plane into* $0, 1, \infty$ *of the w-plane respectively.*

Solution. By Cross ratio

$$\frac{(w-w_1)(w_2-w_3)}{(w-w_3)(w_2-w_1)}=\frac{(z-z_1)(z_2-z_3)}{(z-z_3)(z_2-z_1)} \quad ...(1)$$

z	w
i	0
$-i$	1
1	∞

Here $w_1 = 0, w_2 = 1, w_3 = \infty$ and $z_1 = i, z_2 = -i, z_3 = 1$

From (1),

$$\frac{(w-w_1)\left(\dfrac{w_2}{w_3}-1\right)}{\left(\dfrac{w}{w_3}-1\right)(w_2-w_1)} = \frac{(z-z_1)(z_2-z_3)}{(z-z_3)(z_2-z_1)}$$

$$\Rightarrow \quad \frac{(w-0)\left(\dfrac{1}{\infty}-1\right)}{\left(\dfrac{w}{\infty}-1\right)(1-0)} = \frac{(z-i)(-i-1)}{(z-1)(-i-i)}$$

$$\Rightarrow \quad w = \frac{(z-i)(i+1)}{(z-1)(2i)} = \frac{(z-i)(-1+i)}{(z-1)(-2)} = \frac{(i-1)z+(i+1)}{-2z+2} \qquad \textbf{Ans.}$$

Example 13. *Find the bilinear transformation which maps the points $z = 1, -i, -1$ to the points $w = i, 0, -i$ respectively. Show also that transformation maps the region outside the circle $|z| = 1$ into the half-plane $R(w) \geq 0$.*

Solution. On putting $z = 1, -i, -1$ and $w = i, 0, -i$ in

$$\frac{(w-w_1)(w_2-w_3)}{(w-w_3)(w_2-w_1)} = \frac{(z-z_1)(z_2-z_3)}{(z-z_3)(z_2-z_1)}$$

z	w
1	i
$-i$	0
-1	$-i$

$$\Rightarrow \quad \frac{(w-i)(0+i)}{(w+i)(-i)} = \frac{(z-1)(-i+1)}{(z+1)(-i-1)} \quad \frac{i-w}{i+w} = \frac{(z-1)(i-1)}{(z+1)(i+1)}$$

$$\frac{i-w}{i+w} = \frac{(i-1)z+1-i}{(i+1)z+1+i}$$

$$\Rightarrow \quad \frac{2i}{-2w} = \frac{2iz+2}{-2z-2i} \qquad \text{(Applying Componendo and Dividendo)}$$

$$\Rightarrow \quad \frac{i}{-w} = \frac{iz+1}{-(z+i)} \quad \Rightarrow \quad w = \frac{i(z+i)}{iz+1}$$

$$\Rightarrow \quad w = \frac{iz-1}{iz+1} \qquad \text{... (1)} \qquad \textbf{Ans.}$$

From (1), $\quad z = i\left(\dfrac{w+1}{w-1}\right)$

$|z| \geq 1$ is transformed into $\left|\dfrac{w+1}{w-1}\right||i| \geq 1$

$\Rightarrow \quad |w+1|^2 \geq |w-1|^2 \qquad \Rightarrow \quad |u+iv+1|^2 \geq |u+iv-1|^2$

$\Rightarrow \quad (u+1)^2 + v^2 \geq (u-1)^2 + v^2 \qquad \Rightarrow \quad u \geq 0$

$\Rightarrow \quad R(w) \geq 0.$

Thus exterior of the circle $|z| = 1$ is transformed into the half-plane $R(w) \geq 0$. **Proved.**

Example 14. *Find the bilinear transformation which maps the points $z = 0, -1, i$ onto $w = i, 0, \infty$. Also find the image of the unit circle $|z| = 1$.*

[*Uttarakhand, III Semester 2008,*]

Solution. On putting $z = 0, -1, i$ into $w = i, 0, \infty$ respectively in

$$\frac{(w-w_1)(w_2-w_3)}{(w-w_3)(w_2-w_1)} = \frac{(z-z_1)(z_2-z_3)}{(z-z_3)(z_2-z_1)} \quad ...(1)$$

z	w
0	i
-1	0
i	∞

$$\Rightarrow \frac{(w-w_1)\left(\frac{w_2}{w_3}-1\right)}{\left(\frac{w}{w_3}-1\right)(w_2-w_1)} = \frac{(z-z_1)(z_2-z_3)}{(z-z_3)(z_2-z_1)}$$

$$\Rightarrow \frac{(w-i)(-1)}{(-1)(0-i)} = \frac{(z-0)(-1-i)}{(z-i)(-1-0)} \Rightarrow \left(\frac{w-i}{-i}\right) = \frac{z(1+i)}{z-i}$$

$$\Rightarrow w - i = \frac{(-i+1)z}{z-i} \Rightarrow w = \frac{(1-i)z}{z-i} + i = \frac{(1-i)z + iz + 1}{z-i}$$

$$\Rightarrow w = \frac{z+1}{z-i} \quad ...(2) \textbf{ Ans.}$$

From (2) $\quad z = \dfrac{iw+1}{w-1} \quad ...(3) \quad \left[\text{Inverse transformation is } z = \dfrac{-dw+b}{cw-a}\right]$

And $\quad |z| = 1$

$$\Rightarrow \left|\frac{iw+1}{w-1}\right| = 1 \Rightarrow |1+iw| = |w-1|$$

$$\Rightarrow |1+i(u+iv)| = |u+iv-1| \Rightarrow |1-v+iu| = |u-1+iv|$$

$$\Rightarrow (1-v)^2 + u^2 = (u-1)^2 + v^2 \Rightarrow 1 + v^2 - 2v + u^2 = u^2 + 1 - 2u + v^2$$

$$\Rightarrow u - v = 0 \Rightarrow v = u$$

Ans.

Example 15. *Find the fixed points and the normal form of the following bilinear transformations.*

(*a*) $w = \dfrac{3z-4}{z-1}$ and (*b*) $w = \dfrac{z-1}{z+1}$

Discuss the nature of these transformations.

Solution. (*a*) The fixed points are obtained by

$$z = \frac{3z-4}{z-1} \quad \text{or} \quad z^2 - 4z + 4 = 0 \quad \text{or} \quad (z-2)^2 = 0 \Rightarrow z = 2$$

$z = 2$ in the only fixed point. This transformation is parabolic.

Normal Form

$$w = \frac{3z-4}{z-1} \Rightarrow \frac{1}{w-2} = \frac{1}{\frac{3z-4}{z-1} - 2} = \frac{z-1}{3z-4-2z+2} = \frac{z-1}{z-2}$$

$$\Rightarrow \frac{1}{w-2} = \frac{1}{z-2} + 1$$

(*b*) The fixed points are obtained by

$$z = \frac{z-1}{z+1} \quad \Rightarrow \quad z^2 + z = z - 1 \;\Rightarrow\; z^2 = -1 \;\Rightarrow\; z = \pm i$$

Hence $\pm\, i$ are the two fixed points.

Normal Form

$$w = \frac{z-1}{z+1}$$

$$w - i = \frac{z-1}{z+1} - i = \frac{z-1-i(z+1)}{z+1} \qquad \text{... (1)}$$

and $$w + i = \frac{z-1}{z+1} + i = \frac{z-1+i\,(z+1)}{z+1} \qquad \text{...(2)}$$

On dividing (1) by (2), we get

$$\frac{w-i}{w+i} = \frac{z-1-iz-i}{z-1+iz+i} = \frac{(1-i)(z-i)}{(1+i)(z+i)} = \frac{(-i^2-i)(z-i)}{(1+i)(z+i)}$$

$$\frac{w-1}{w+1} = -i\left(\frac{z-i}{z+i}\right) = k\left(\frac{z-i}{z+i}\right), \text{ where } k = -\,i$$

The transformation is elliptic. **Ans.**

Example 16. *The fixed points of the transformation* $w = \dfrac{2z-5}{z+4}$ *are given by:*

(*a*) $\left(\dfrac{5}{2}, 0\right)$ (*b*) (–4, 0) (*c*) (–1 + 2*i*, – 1 – 2*i*) (*d*) $(-1+\sqrt{6}, -1-\sqrt{6})$

(*AMIETE, Dec. 2010*)

Solution. Here $f(z) = \dfrac{2z-5}{z+4}$

In the case of fixed point $z = \dfrac{2z-5}{z+4}$

$\Rightarrow z^2 + 4z = 2z - 5 \quad \Rightarrow \quad z^2 + 2z + 5 = 0$

$\Rightarrow z = \dfrac{-2 \pm \sqrt{4-20}}{2} = \dfrac{-2 \pm 4i}{2} = -\,1 \pm 2i$

$z = -1 \pm 2i$ are the only fixed points

Hence (c) is correct answer. **Ans.**

Example 17. *Show that the transformation* $w = i\dfrac{1-z}{1+z}$ *transforms the circle* $|z| = 1$ *into the real axis of the w-plane and the interior of the circle into the upper half of the w-plane.*

Solution. $w = i\left(\dfrac{1-z}{1+z}\right)$

$$(u+iv) = i\left(\frac{1-(x+iy)}{1+(x+iy)}\right) = \frac{(i-ix+y)}{[1+(x+iy)]}\frac{[(1+x)-iy]}{[(1+x)-iy]}$$

$$= \frac{i+ix+y-ix-ix^2-xy+y+xy-iy^2}{(1+x)^2+y^2} = \frac{y-xy+y+xy+i+ix-ix-ix^2-iy^2}{(1+x)^2+y^2}$$

$$= \frac{2y+i(1-x^2-y^2)}{(1+x)^2+y^2}$$

Equating the real and imaginary parts, we get

$$u = \frac{2y}{(1+x)^2 + y} \quad \text{... (1)}$$

and $$v = \frac{1-(x^2+y^2)}{(1+x)^2 + y^2} \quad \text{... (2)}$$

when $x^2 + y^2 = 1$ then $v = \dfrac{1-1}{(1+x)^2+y^2} = 0$

$v = 0$ is the equation of the real axis in the w – plane. **Proved.**

(*b*) Now the equation of the interior of the circle is $x^2 + y^2 < 1$.

Dividing (1) by (2), we get

$$1 - (x^2 + y^2) > 0$$

$$\Rightarrow \quad \frac{1-(x^2+y^2)}{(1+x)^2+y^2} > 0$$

$$\Rightarrow \quad v > 0$$

$v > 0$ is the equation of the upper half of w-plane. **Proved.**

Example 18. *Show that* $\omega = \dfrac{i-z}{i+z}$ *maps the real axis of the z-plane into the circle* $| w | = 1$ *and (ii) the half-plane* $y > 0$ *into the interior of the unit circle* $| \omega | < 1$ *in the w-plane.*

(*U.P., III Semester, Dec. 2005,*)

Solution. We have $\omega = \dfrac{i-z}{i+z}$

$$|\omega| = \left|\frac{i-z}{i+z}\right| = \frac{|i-z|}{|i+z|} = \frac{|i-x-iy|}{|i+x+iy|}$$

$$|\omega| = \left|\frac{-x+i(1-y)}{x+i(1+y)}\right|, \quad |\omega| = \frac{\sqrt{x^2+(1-y)^2}}{\sqrt{x^2+(1+y)^2}}$$

Now the real axis in z-plane *i.e.*, $y = 0$, transform into

$$|\omega| = \frac{\sqrt{x^2+1}}{\sqrt{x^2+1}} = 1, \; |\omega| = 1$$

Hence the real axis in the z-plane is mapped into the circle $| \omega | = 1$.

(*ii*) The interior of the circle *i.e.*, $| w | < 1$ gives.

$$\frac{\sqrt{x^2+(1-y)^2}}{\sqrt{x^2+(1+y)^2}} < 1$$

$$\Rightarrow \quad \frac{x^2+(1-y)^2}{x^2+(1+y)^2} < 1 \quad \Rightarrow \quad x^2+(1-y)^2 < x^2+(1+y)^2$$

$$\Rightarrow \quad 1+y^2-2y < 1+y^2+2y \quad \Rightarrow \quad -4y < 0, \quad \Rightarrow \quad y > 0.$$

Thus the upper half of the z-plane corresponds to the interior of the circle $| w | = 1$ **Proved.**

Example 19. *Show that the transformation* $w=\dfrac{3-z}{z-2}$ *transforms the circle with centre* $\left(\dfrac{5}{2},0\right)$ *and radius* $\dfrac{1}{2}$ *in the z-plane into the imaginary axis in the w-plane and the interior of the circle into the right half of the plane.*

Solution. $w=\dfrac{3-z}{z-2} \quad \Rightarrow \quad u+iv=\dfrac{3-x-iy}{x+iy-2} \quad \Rightarrow \quad (u+iv)(x+iy-2)=3-x-iy$

$\Rightarrow \quad ux+iuy-2u+ivx-vy-2iv=3-x-iy$

$\Rightarrow \quad ux-2u-vy+i(uy+vx-2v)=3-x-iy$

Equating real and imaginary quantities, we have

$$ux-vy-2u=3-x \text{ and } vx-2v+uy=-y$$

$$\Rightarrow \quad (u+1)x-vy=2u+3 \text{ and } vx+(u+1)y=2v$$

On solving the equations for x and y, we have

$$x=\frac{2u^2+2v^2+5u+3}{u^2+v^2+2u+1}, \quad y=\frac{-v}{u^2+v^2+2u+1}$$

Here, the equation of the given circle is $\left(x-\dfrac{5}{2}\right)^2+y^2=\dfrac{1}{4}$ ... (1)

Putting the values of x and y in (1), we have

$$\left(\frac{2u^2+2v^2+5u+3}{u^2+v^2+2u+1}-\frac{5}{2}\right)^2+\left(\frac{-v}{u^2+v^2+2u+1}\right)^2=\frac{1}{4}$$

$$\Rightarrow \quad \left(\frac{-u^2-v^2+1}{2(u^2+v^2+2u+1)}\right)^2+\left(\frac{-v}{u^2+v^2+2u+1}\right)^2=\frac{1}{4}$$

$$\Rightarrow \quad (-u^2-v^2+1)^2+4v^2=(u^2+v^2+2u+1)^2$$

$$\Rightarrow \quad (u^2+v^2-1)^2+4v^2=[(u^2+v^2-1)+(2u+2)]^2$$

$$\Rightarrow \quad (u^2+v^2-1)^2+4v^2=(u^2+v^2-1)^2+(2u+2)^2+2(u^2+v^2-1)(2u+2)$$

$$\Rightarrow \quad v^2=(u+1)^2+(u^2+v^2-1)(u+1)$$

$$\Rightarrow \quad v^2=u^2+2u+1+u^3+uv^2-u+u^2+v^2-1$$

$$\Rightarrow \quad 0=u^3+2u^2+u+uv^2$$

$\Rightarrow \quad u(u^2+2u+1+v^2)=0 \quad \Rightarrow \quad u=0$ *i.e.*, equation of imaginary axis.

Equation of the interior of the circle is $\left(x-\dfrac{5}{2}\right)^2+y^2<\dfrac{1}{4}$.

Then corresponding equation in u, v is

$$u(u^2+2u+1+v^2)>0 \text{ or } u[(u+1)^2+v^2]>0$$

As $(u+1)^2+v^2>0$ so $u>0$ *i.e.*, equation of the right half plane. **Ans.**

23.17 INVERSE POINT WITH RESPECT TO A CIRCLE

Two points P and Q are said to be the inverse points with respect a circle S if they are collinear with the centre C on the same side of it, and if the product of their distances from the centre is equal to r^2 where r is the radius of the circle.

Thus when P and Q are the inverse points of the circle, then the three points C, P, Q are collinear, and also $CP.CQ=r^2$

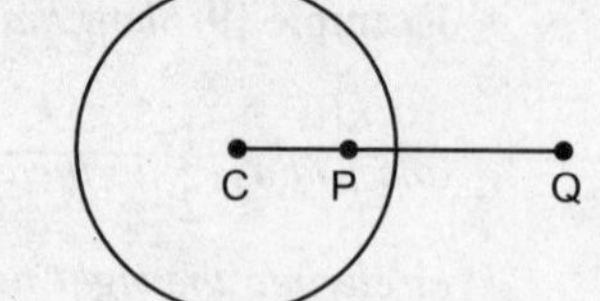

Example 20. *Show that the inverse of a point a, with respect to the circle* $|z - c| = R$ *is the point* $c + \dfrac{R^2}{\bar{a} - \bar{c}}$

Solution. Let b be the inverse point of the point a' with respect to the circle $|z - c| = R$.

Condition I. The points a, b, c are collinear. Hence

$$arg\ (b - c) = arg\ (a - c) = -arg\ (\bar{a} - \bar{c}) \qquad \text{(since arg } z = -\text{ arg } \bar{z}\text{)}$$

$$\Rightarrow \quad arg\ (\bar{b} - \bar{c}) + arg\ (\bar{a} - \bar{c}) = 0 \quad \text{or} \quad arg\ (\bar{b} - \bar{c})(\bar{a} - \bar{c}) = 0$$

$$\therefore \quad (\bar{b} - \bar{c})(\bar{a} - \bar{c}) \text{ is real, so that}$$

$$(\bar{b} - \bar{c})(\bar{a} - \bar{c}) = |(\bar{b} - \bar{c})(\bar{a} - \bar{c})|$$

Condition II. $|b - c||a - c| = R^2 \Rightarrow |\bar{b} - \bar{c}||\bar{a} - \bar{c}| = R^2$ $\qquad \{|z| = |\bar{z}|\}$

$$|(\bar{b} - \bar{c})(\bar{a} - \bar{c})| = R^2 \Rightarrow (\bar{b} - \bar{c})(\bar{a} - \bar{c}) = R^2 \Rightarrow \bar{b} - \bar{c} = \frac{R^2}{\bar{a} - \bar{c}}$$

$$\Rightarrow \qquad b = c + \frac{R^2}{\bar{a} - \bar{c}}.$$

Proved.

Example 21. *Find a Mobius transformation which maps the circle* $|w| \leq 1$ *into the circle* $|z - 1| < 1$ *and maps* $w = 0$, $w = 1$ *respectively into* $z = \frac{1}{2}$, $z = 0$.

Solution. Let the transformation be,

$$w = \frac{az + b}{cz + d} \qquad \text{... (1)}$$

z	w
$\frac{1}{2}$	0
0	1

Since, $w = 0$ maps into $z = \frac{1}{2}$,

From (1), we get

$$0 = \frac{\frac{a}{2} + b}{\frac{c}{2} + d} \quad \Rightarrow \quad \frac{a}{2} + b = 0 \qquad \text{... (2)}$$

Since $w = 1$ maps into $z = 0$, from (1), we get

$$1 = \frac{0 + b}{0 + d} \quad \Rightarrow \quad b = d \qquad \text{... (3)}$$

Here $\qquad |w| = 1$ corresponding to $|z - 1| = 1$

Therefore points $w, \frac{1}{\bar{w}}$ inverse with respect to the circle $|w| = 1$ correspond to the points $z, 1 + \frac{1}{\bar{z} - 1}$ inverse with respect to the circle $|z - 1| = 1$

[z and $a + \frac{R^2}{\bar{z} - \bar{a}}$ are inverse points on the circle $|z - a| = R$]

Particular $w = 0$ and ∞ correspond to

$$z = \frac{1}{2}, 1 + \frac{1}{\frac{1}{2} - 1} \Rightarrow z = \frac{1}{2}, -1$$

Since $w = 0$ maps into $z = -1$, from (1), we get

$$\infty = \frac{-a+b}{-c+d} \Rightarrow -c+d = 0 \Rightarrow c = d \qquad \text{... (4)}$$

From (2), (3) and (4), $b = -\dfrac{a}{2}$, $b = c = d$

From (1) $w = \dfrac{az+b}{cz+d} = \dfrac{-2bz+b}{bz+b} = \dfrac{-2z+1}{z+1}$ **Ans.**

Example 22. *Find two bilinear transformations whose fixed points are 1 and 2.*

Solution. We have, $w = \dfrac{az+b}{cz+d}$ ... (1)

Fixed points are given by

$$z = \frac{az+b}{cz+d}$$

$$\Rightarrow \quad cz^2 - (a-d)\,z - b = 0 \quad \Rightarrow \quad z^2 - \frac{(a-d)}{c} z - \frac{b}{c} = 0 \qquad \text{... (2)}$$

Fixed points are 1 and 2, so

$$(z-1)(z-2) = 0$$

$$\Rightarrow \quad z^2 - 3z + 2 = 0$$

Equating the coefficeints of z and constants in (2) and (3), we get

$$\therefore \quad \frac{a-d}{c} = 3 \quad \text{and} \quad -\frac{b}{c} = 2$$

$$\Rightarrow \quad b = -2c \quad \text{and} \quad d = a - 3c$$

Putting the values of b and d in (1), we get

$w = \dfrac{az-2c}{cz+a-3c}$ has its fixed points at $z = 1$ and $z = 2$.

Taking $a = 1$, $c = -1$ and $a = 2$, $c = -1$, we have

$$w = \frac{z+2}{4-z} \quad \text{and} \quad w = \frac{2(z+1)}{5-z}$$ **Ans.**

Example 23. *Show that the transformation* $w = \dfrac{2z+3}{z-4}$ *maps the circle* $x^2 + y^2 - 4x = 0$ *onto the straight line* $4u + 3 = 0$.

Solution. We have, $w = \dfrac{2z+3}{z-4}$

The inverse transformation is $z = \dfrac{4w+3}{w-2}$ ... (1)

Now the circle $x^2 + y^2 - 4x = 0$ can be written as $z\bar{z} - 2(z + \bar{z}) = 0$ $\left[\begin{array}{l} z = x+iy \\ \bar{z} = x - iy \end{array}\right]$

Substituting for z and $\bar{z}$ from (1), we get

$$\frac{4w+3}{w-2} \cdot \frac{4\bar{w}+3}{\bar{w}-2} - 2\left(\frac{4w+3}{w-2} + \frac{4\bar{w}+3}{\bar{w}-2}\right) = 0$$

$$\Rightarrow 16w\bar{w} + 12w + 12\bar{w} + 9 - 2(4w\bar{w} + 3\bar{w} - 8w - 6 + 4w\bar{w} + 3w - 8\bar{w} - 6) = 0$$

$$\Rightarrow \quad 22(w+\overline{w})+33 = 0 \Rightarrow 22(2u)+33=0 \Rightarrow 4u+3=0 \qquad \begin{bmatrix} w=u+iv \\ \overline{w}=u-iv \end{bmatrix}$$

Thus, circle is transformed into a straight line. **Ans.**

Example 24. *Show that the transformation* $w=\dfrac{5-4z}{4z-2}$ *transform the circle* $|z|=1$ *into a circle of radius unity in w-plane and find the centre of the circle.*

Solution. Here, $w = \dfrac{5-4z}{4z-2}$

$$\Rightarrow \quad z = \frac{2w+5}{4w+4} \quad \Rightarrow \quad |z| = \left|\frac{2w+5}{4w+4}\right|$$

$$|z| = 1 \quad \Rightarrow \quad \left|\frac{2w+5}{4w+4}\right| = 1$$

$$\Rightarrow \quad |2w+5| = |4w+4|$$

$$\Rightarrow \quad |2u+5+2iv| = |4u+4+4iv| \qquad [\because w = u+iv]$$

$$\Rightarrow \quad (2u+5)^2+4v^2 = (4u+4)^2+(4v)^2 \qquad ...(1)$$

$$\Rightarrow \quad 4u^2+25+20u+4v^2 = 16u^2+16+32u+16v^2$$

$$\Rightarrow \quad 12u^2+12v^2+12u-9=0$$

$$\Rightarrow \quad u^2+v^2+u-\frac{3}{4} = 0.$$

This is the equation of circle in w plane. ...(2)

Now we have to find its centre.

$$u^2+v^2+2gu+2fv+c=0 \qquad ...(3)$$

From (2) and (3), $\quad g = \dfrac{1}{2},\ f=0,\ c=-\dfrac{3}{4}$

Centre is $(-g, -f)$ *i.e.*, $\left(-\dfrac{1}{2}, 0\right)$

$$\text{Radius} = \sqrt{g^2+f^2-c} = \sqrt{\frac{1}{4}+0+\frac{3}{4}} = 1$$

Thus (2) is circle with its centre at $\left(-\dfrac{1}{2}, 0\right)$ and of radius unity in w-plane. **Proved.**

Example 25. *Find the image of* $x^2+y^2-4y+2=0$ *under the mapping* $w=\dfrac{z-i}{iz-1}$.

Solution. $\quad w = \dfrac{z-i}{iz-1} \quad \Rightarrow \quad w(iz-1) = z-i$

$$x^2+y^2-4y+2=0 \qquad ...(1)$$

$$\Rightarrow \quad z = \frac{w-i}{iw-1}$$

$$\Rightarrow \quad x+iy = \frac{u+i(v-1)}{iu-(v+1)} \qquad ...(2)$$

$$\therefore \Rightarrow \quad x-iy = \frac{u-i(v-1)}{-iu-(v+1)} \qquad ..(3)$$

Multiplying (2) and (3) we get

$$\Rightarrow \qquad x^2 + y^2 = \frac{u^2 + (v-1)^2}{u^2 + (v+1)^2} \qquad \text{... (4)}$$

Subtracting (3) from (2), we get

$$2iy = \frac{-2iu^2 - 2i(v^2 - 1)}{u^2 + (v+1)^2} \qquad \text{... (5)}$$

Putting the values of $(x^2 + y^2)$ and y in (1), we get

$$\frac{u^2 + (v-1)^2}{u^2 + (v+1)^2} + 4 . \frac{u^2 + (v^2 - 1)}{u^2 + (v+1)^2} + 2 = 0$$

$$\Rightarrow \qquad u^2 + (v-1)^2 + 4[u^2 + (v^2 - 1)] + 2[u^2 + (v+1)^2] = 0$$

$$\Rightarrow \qquad 7(u^2 + v^2) + 2v - 1 = 0$$

This is the image. **Ans.**

EXERCISE 23.2

1. Find the bilinear transformation that maps the points $z_1 = 2$, $z_2 = i$, $z_3 = -2$ into the points $w_1 = 1$, $w_2 = i$ and $w_3 = -1$ respectively. **Ans.** $w = \frac{3z + 2i}{iz + 6}$

2. Determine the bilinear transformation which maps $z_1 = 0$, $z_2 = 1$, $z_3 = \infty$ onto $w_1 = i$, $w_2 = -1$, $w_3 = -$ i respectively. **Ans.** $w = \frac{z - i}{iz - 1}$

3. Verify that the equation $w = \frac{1 + iz}{1 + z}$ maps the exterior of the circle $|z| = 1$ into the upper half plane $v > 0$.

4. Find the bilinear transformation which maps $1, i, -1$ to $2, i, -2$ respectively. Find the fixed and critical points of the transformation. **Ans.** i, $2i$

5. Show that the transformation $w = \frac{i(1 - z)}{1 + z}$ maps the circle $|z| = 1$ into the real axis of the w-plane and the interior of the circle $|z| < 1$ into the upper half of the w-plane.

6. Show that the transformation $w = \frac{iz + 2}{4z + i}$ transforms the real axis in the z-plane into circle in the w-plane. Find the centre and the radius of this circle. **Ans.** $\left(0, \frac{7}{8}\right), \frac{9}{8}$

7. If z_0 is the upper half of the z-plane show that the bilinear transformation

$$w = e^{i\alpha}\left(\frac{z - z_0}{z - \bar{z}_0}\right)$$

maps the upper half of the z-plane into the interior of the unit circle at the origin in the w-plane.

8. Find the condition that the transformation $w = \frac{az + b}{cz + d}$ transforms the unit circle in the w-plane into straight line in the z-plane. **Ans.** If $\left|\frac{c}{a}\right| = 1$ or $|a| = |c|$

9. Prove that $w = \frac{z}{1 - z}$ maps the upper half of the z-plane into the upper half of the w-plane. What is the image of the circle $|z| = 1$ under this transformation ? **Ans.** Straight line $2u + 1 = 0$

10. Show that the map of the real axis of the z-plane on the w-plane by the transformation $w = \frac{1}{z + i}$ is a circle and find its centre and radius. **Ans.** Centre $\left(0, -\frac{1}{2}\right)$, Radius $= \frac{1}{2}$

11. Find the invariant points of the transformation $w = -\left(\dfrac{2z+4i}{iz+1}\right)$. Prove also that these two points together with any point z and its image w, form a set of four points having a constant cross ratio.

Ans. $4i$ and $-i$

12. Show that under the transformation $w = \dfrac{z-i}{z+i}$, the real axis in z-plane is mapped into the circle $|w| = 1$. What portion of the z-plane corresponds to the interior of the circle ?

Ans. The half z-plane above the real axis corresponds to the interior of the circle $|w| = 1$.

13. Discuss the application of the transformation $w = \dfrac{iz+1}{z+i}$ to the areas in the z-plane which are respectively inside and outside the unit circle with its centre at the origin.

14. What is the form of a bilinear transformation which has one fixed point α and the other fixed point ∞?

Choose the correct alternative:

15. The fixed points of the mapping w = (5z + 4)/(z + 5) are

(*i*) – 4/5, –5 (*ii*) 2, 2 (*iii*) – 2, – 2 (*iv*) 2, –2 **Ans.** (*iv*)

16. The invariant points of the bilinear transformation are

(*i*) $1 \pm 2i$ (*ii*) $-1 \pm 2i$ (*iii*) $\pm 2i$ (*iv*) invariant point does not exit

(*AMIETE, June 2010*) **Ans.** (*iv*)

23.18 TRANSFORMATION: $w = z^2$

Solution.
$$w = z^2$$
$$u + iv = (x+iy)^2 = x^2 - y^2 + 2ixy$$

Equating real and imaginary parts, we get $u = x^2 - y^2, \ v = 2xy$

(*i*) (*a*) Any line parallel to x-axis, *i.e.*, $y = c$, maps into
$$u = x^2 - c^2, \quad v = 2cx$$

Eliminating x, we get $v^2 = 4c^2(u + c^2)$ ... (1) which is a parabola.

(*b*) Any line parallel to y-axis, *i.e.*, $x = b$, maps into a curve
$$u = b^2 - y^2, \ v = 2by$$

Eliminating y, we get $v^2 = -4b^2(u - b^2)$, ... (2) which is a parabola.

(*c*) The rectangular region bounded by the lines $x = 1$, $x = 2$, and $y = 1$, $y = 2$ maps into the region bounded by the parabolas.

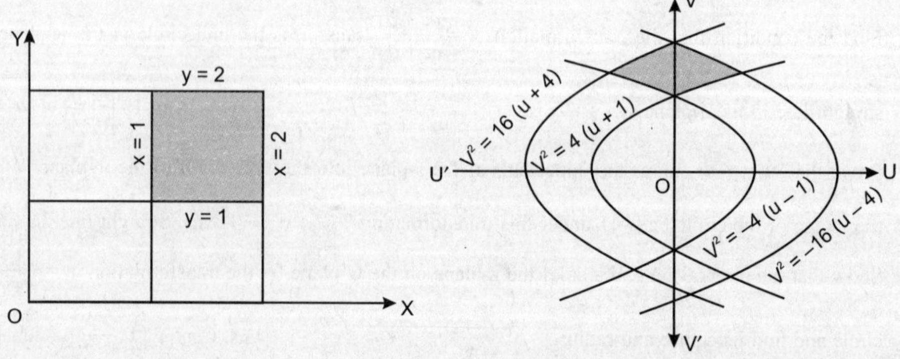

By putting $x = 1 = b$ in (2) we get $v^2 = -4(u - 1)$,

By putting $x = 2 = b$ in (2) we get $v^2 = -16(u - 4)$

By putting $y = 1 = c$ in (1) we get $v^2 = 4(u + 1)$,

By putting $y = 2 = c$ in (1) we get $v^2 = 16(u + 4)$

(*ii*) (*a*) In polar co-ordinates: $z = re^{i\theta}$, $w = Re^{i\phi}$

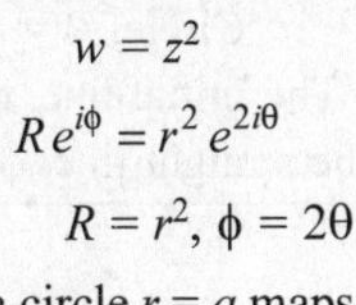

$$w = z^2$$

$$Re^{i\phi} = r^2 e^{2i\theta}$$

Then $\qquad R = r^2,\ \phi = 2\theta$

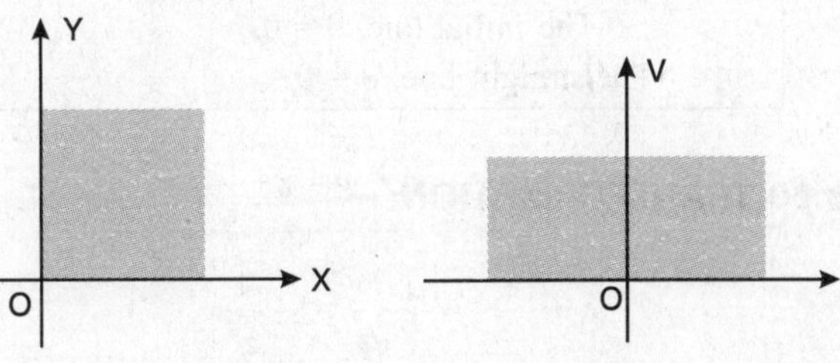

In *z*-plane, a circle $r = a$ maps into $R = a^2$ in *w*-plane.

Thus, circles with centre at the origin map into circles with centre at the origin.

(*b*) If $\theta = 0$, $\phi = 0$, *i.e.,* real axis in *z*-plane maps into real axis in *w*-plane

If $\theta = \frac{\pi}{2}$, $\phi = \pi$, *i.e.*, the positive imaginary axis in *z*-plane maps into negative real axis in *w*-plane.

Thus, the first quadrant in *z*-plane $0 \le \theta \le \frac{\pi}{2}$, maps into upper half of *w*-plane $0 \le \phi \le \pi$.

The angles in *z*-plane at origin maps into double angle in *w*-plane at origin.

Hence, the mapping $w = z^2$ is not conformal at the origin.

It is conformal in the entire *z*-plane except origin. Since $\frac{dw}{dz} = 2z = 0$ for $z = 0$, therefore, it is critical point of mapping.

Example 27. *For the conformal transformation $w = z^2$, show that the circle $|z - 1| = 1$ transforms into the cardioid $R = 2(1 + \cos\phi)$ where $w = Re^{i\phi}$ in the w-plane.*

Solution. $\qquad |z - 1| = 1 \qquad$ ... (1)

Equation (1) represents a circle with centre at (1, 0) and radius 1.

Shifting the pole to the point (1, 0), any point on (1) is $1 + e^{i\theta}$

Transformation is under $w = z^2$.

$$Re^{i\phi} = (1 + e^{i\theta})^2$$

$$= e^{i\theta}\left(e^{\frac{i\theta}{2}} + e^{-\frac{i\theta}{2}}\right)^2$$

$$= e^{i\theta}\left(2\cos\frac{\theta}{2}\right)^2 = 4e^{i\theta}\cos^2\frac{\theta}{2}$$

This gives $\qquad R = 4\cos^2\frac{\theta}{2}$,

$\Rightarrow \qquad R = 2\left(2\cos^2\frac{\phi}{2}\right) \qquad [\phi = \theta]$

$\Rightarrow \qquad R = 2(\cos\phi + 1) \qquad$ **Proved**

23.19 TRANSFORMATION: $W = Z^n \ (n \in N)$

$$Re^{i\phi} = (re^{i\theta})^n = r^n e^{in\theta}$$

Hence, $R = r^n,\ \phi = n\theta$

Mapping of simple figures

z-plane	w-plane
Circle, $r = a$	Circle, $R = a^n$
The initial line, $\theta = 0$	The initial line, $\phi = 0$
The straight line, $\theta = \theta_0$	The straight line, $\phi = n\theta_0$

23.20 TRANSFORMATION: $w = z + \dfrac{1}{z}$

$$\frac{dw}{dz} = 1 - \frac{1}{z^2}$$

At $z = \pm 1$, $\dfrac{dw}{dz} = 0$, so transformation is not conformal at $z = \pm 1$.

$$w = z + \frac{1}{z} = r(\cos\theta + i\sin\theta) + \frac{1}{r(\cos\theta + i\sin\theta)}$$

$$= r(\cos\theta + i\sin\theta) + \frac{1}{r}(\cos\theta - i\sin\theta)$$

$$u + iv = \left(r + \frac{1}{r}\right)\cos\theta + i\left(r - \frac{1}{r}\right)\sin\theta$$

$$u = \left(r + \frac{1}{r}\right)\cos\theta \quad \text{and} \quad v = \left(r - \frac{1}{r}\right)\sin\theta$$

$$\frac{u}{r + \frac{1}{r}} = \cos\theta \quad \text{and} \quad \frac{v}{r - \frac{1}{r}} = \sin\theta$$

$$\sin^2\theta + \cos^2\theta = \frac{u^2}{\left(r + \frac{1}{r}\right)^2} + \frac{v^2}{\left(r - \frac{1}{r}\right)^2} \Rightarrow 1 = \frac{u^2}{\left(r + \frac{1}{r}\right)^2} + \frac{v^2}{\left(r - \frac{1}{r}\right)^2}$$

z-plane	w-plane
Circle, $r = r$	Ellipses
Circle, $r = 1$	Lines $u = 2\cos\theta$, $v = 0$
Lines, $\theta = \theta_0$	Hyperbola: $\dfrac{u^2}{4\cos^2\theta} - \dfrac{v^2}{4\sin^2\theta} = 1$

23.21 TRANSFORMATION: $w = e^z$

$$u + iv = e^{x+iy} = e^x(\cos y + i\sin y)$$

Equating real and imaginary parts, we have

$$u = e^x \cos y, \qquad v = e^x \sin y$$

Again $\quad w = e^z$

$$Re^{i\phi} = e^{x+iy} = e^x . e^{iy}$$

Hence $\quad R = e^x$ or $x = \log_e R$ and $y = \phi$

Mapping of simple figures

z-plane	w-plane
The straight line $x = c$	Circle $R = e^c$
y-axis ($x = 0$)	Unit Circle $R = e^0 = 1$
Region between $y = 0, y = \pi$	Upper half plane
Region between $y = 0, y = y = -\pi$	Lower half plane
Region between the lines $y = c$ and $y = c + 2\pi$	Whole plane

Example 28. *Find the image and draw a rough sketch of the mapping of the region* $1 \le x \le 2$ and $2 \le y \le 3$ *under the mapping* $w = r^z$.

Solution. $z = x + iy$

Let $w = Re^{i\phi}$ ...(1)

But $w = e^z = e^{x+iy}$ ... (2)

From (1) and (2); $Re^{i\phi} = e^{x+iy} = e^x . e^{iy}$

Equating real and imaginary parts, we get $R = e^x$ ... (3) and $\phi = y$

(*i*) Here $1 \le x$, then $R = e^1$ is circle of radius $e^1 = 2.7$

$x = 2$, then $R = e^2$ represents a circle of radius $e^2 = 7.4$

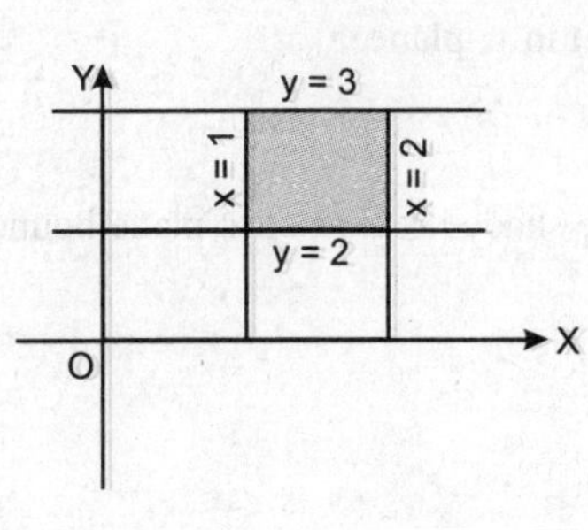

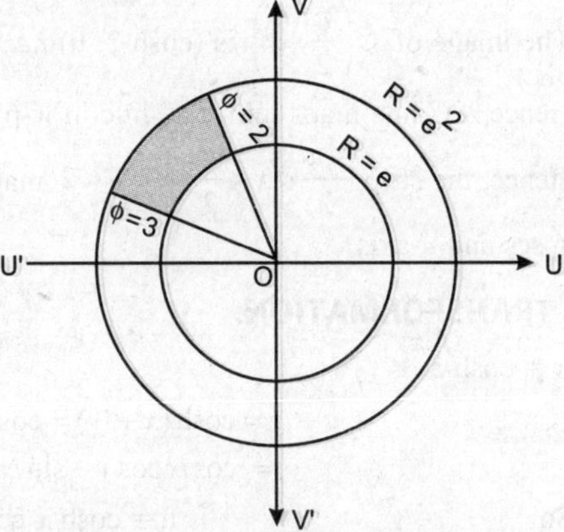

(*ii*) $y = 2$, then $\phi = 2$ represents radial line making an angle of 2 radians with the x-axis. $y = 3$, then $\phi = 3$ represents radial line making an angle 3 radians with x-axis.

Hence, the mapping of the region $1 \le x \le 2$ and $2 \le y \le 3$ maps the shaded sectors in the figure. **Ans.**

Example 29. *Find the image of the strip* $-\frac{\pi}{2} < x < \frac{\pi}{2}, 1 < y < 2$ *under the mapping* $w(z) = \sin z$.

Solution. $w(z) = \sin z = \sin (x + iy)$

$= \sin x \cos iy + \cos x \sin iy$

$u + iv = \sin x \cosh y + i \cos x \sinh y$

$u = \sin x \cosh y \quad \Rightarrow \quad \sin x = \frac{u}{\cosh y}$ ...(1)

$v = \cos x \sinh y \quad \Rightarrow \quad \cos x = \frac{v}{\sinh y}$ ...(2)

Eliminating x from (1) and (2), we get

$$\sin^2 x+\cos^2 x=\frac{u^2}{\cosh^2 y}+\frac{v^2}{\sinh^2 y}\Rightarrow 1=\frac{u^2}{\cosh^2 y}+\frac{v^2}{\sinh^2 y}$$

Hence $y = 2$, maps into the ellipse

$$\frac{u^2}{\cosh^2 2}+\frac{v^2}{\sinh^2 2}=1 \Rightarrow \frac{u^2}{14.15}+\frac{v^2}{13.15}=1$$

Also $y = 1$, maps into the ellipse.

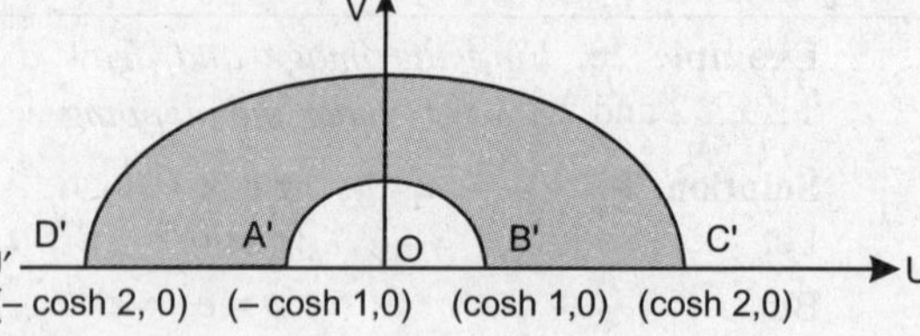

The image of $A\left(\frac{-\pi}{2}, 1\right)$ in z-plane is $(-\cosh 1, 0)$ *i.e.* $(-1.543, 0)$ in w-plane

The image of the point $D\left(-\frac{\pi}{2}, 2\right)$ in z-plane is $(-\cosh 2, 0)$ *i.e.*, $(-7.524, 0)$.

Hence, AD line in z-plane maps into $A'D'$ line in w-plane.

The image of $B\left(\frac{\pi}{2}, 1\right)$ is $(\cosh 1, 0)$ *i.e.*, $(1.543, 0)$ in w-plane.

The image of $C\left(\frac{\pi}{2}, 2\right)$ is $(\cosh 2, 0)$ *i.e.*, $(7.524, 0)$ in w-plane.

Hence, BC line maps into $B'C'$ line in w-plane.

Hence, the strip $\frac{-\pi}{2}<x<\frac{\pi}{2},\ 1<y<2$ maps into the shaded region of w-plane bounded by the ellipses and u-axis. **Ans.**

23.22 TRANSFORMATION:

$w = \cosh z$

$$u+iv=\cosh(x+iy)=\cos i(x+iy)=\cos(ix-y)$$
$$=\cos ix\cos y+\sin ix\sin y=\cosh x\cos y+i\sinh x\sin y$$

So $\quad u = \cosh x \cos y \quad v = \sinh x \sin y$

On eliminating x, we get $(\cosh^2 x - \sinh^2 x = 1)$ $\frac{u^2}{\cos^2 y}-\frac{v^2}{\sin^2 y}=1$ ... (1)

If y is eliminated $(\cos^2 y + \sin^2 y = 1)$ $\frac{u^2}{\cosh^2 x}+\frac{v^2}{\sinh^2 x}=1$ ... (2)

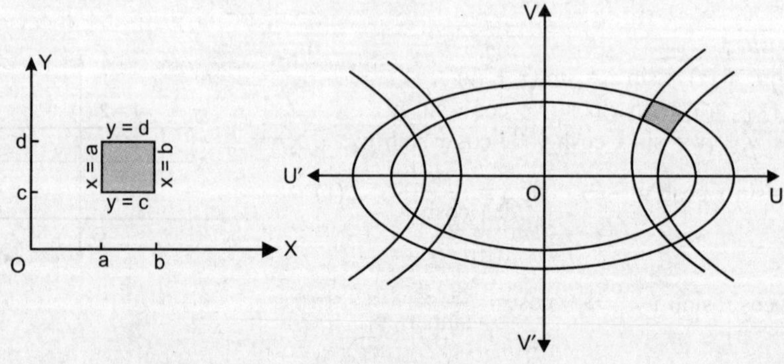

(*a*) On putting $y = a$ (constant) in (1), we get

$$\frac{u^2}{\cos^2 a} - \frac{v^2}{\sin^2 a} = 1 \quad i.e., \text{ Hyperbola.}$$

It shows that the lines parallel to x-axis in the z-plane map into hyperbola in the w-plane.

(*b*) On substituting $x = b$ (constant) in (2), we obtain

$$\frac{u^2}{\cosh^2 b} + \frac{v^2}{\sinh^2 b} = 1$$

It means that lines parallel to y-axis in the z-plane map into ellipses in w-plane.

(*c*) The rectangular region $a \le x \le b,\ c \le y \le d$ in the z-plane transforms into the shaded portion in the w-plane.

EXERCISE 23.3

1. Determine the region of the w-plane into which the region bounded by $x = 1, y = 1, x + y = 1$ is mapped by the transformation $w = z^2$ **Ans.** $4u + v^2 = 4,\ 4u - v^2 = -4,\ u^2 - 2v^2 = 1$
2. By the transformation $w = z^2$, show that the circle $|z - a| = c$ in the z-plane correspond to the limacon in the w-plane. **Ans.** $R = 2c(a + c \cos \phi)$
3. Under the mapping $w = z^2$, show that the family of circles $|w - 1| = c$ is transformed into the family of leminscates $|z - 1||z + 1| = c$, where c is the parameter.
4. Discuss the conformal transformation $w = \sqrt{z}$.
5. Show that the transformation $w(z + i)^2 = 1$ maps the inside of the circle $|z| = 1$ in the z-plane on the exterior of the parabola.
6. Show that the transformation $w = z + \frac{1}{z}$ maps the circle $|r| = c$ into the ellipse in w-plane. Discuss the case $c = 1$. $u = \left(c + \frac{1}{c}\right)\cos\theta,\quad v = \left(c - \frac{1}{c}\right)\sin\theta$

   Discuss the case when $c = 1$.
7. Show that the transformation $w = z + 1/z$ converts the straight line $\arg z = \alpha \ (|\alpha| < \pi/2)$ into a branch of hyperbola of eccentricity sec α.
8. If $w = z + \frac{a^2}{z}$, prove that when z describes the circle $x^2 + y^2 = a^2$, w describes a st. line of length $4a$.

   Also prove that if z describes the circle $x^2 + y^2 = b^2$, $(b > a)$ w describes an ellipse whose foci are the extermities of the above line.
9. Find the region of the z-plane which corresponds to the interior of the circle $|w| = 1$ by means of the transformation $(w + 1)^2 z = 4$ **Ans.** $y^2 + 4x - 4 > 0$ which is the exterior of the parabola.
10. Show that the transformation $\omega = \sin z$ maps the families of lines x = constant and y = cos into confocal hyperbolas and confocal ellipse respectively. (*AMIETE, Dec. 2009*)
11. Let $z = re^{i\theta}$. Then the image of θ = constant, under the mapping

    $w = Re^{i\phi} = iz^2$ is (*AMIETE, Dec. 2009*)

23.23 SCHWARZ-CHRISTOFFEL TRANSFORMATION

The interior of a polygon of w-plane is mapped into the upper half of the z-plane and the sides of the polygon into the real axis. This transformation is called Schwarz-Christoffel transformation.

The formula of the mapping is

$$\frac{dw}{dz} = A(z-x_1)^{\alpha_1/\pi-1}(z-x_2)^{\alpha_2/\pi-1}....(z-x_n)^{\alpha_n/\pi-1} \quad ...(1)$$

where A is a complex constant.

$\alpha_1, \alpha_2, ... \alpha_n$ are the interior angles of the polygon of the w-plane with vertices $w_1, w_2, ... w_n$ are mapped into the points $x_1, x_2, ... x_n$ on the real axis of the z-plane.

$$w = A\int (z-x_1)^{\alpha_1/\pi-1}(z-x_2)^{\alpha_2/\pi-1}....(z-x_n)^{\alpha_n/\pi-1}\,dz + B \quad ...(2)$$

Proof. From (1), we have

$$\text{Arg. } dw = \arg dz + \arg A + \left(\frac{\alpha_1}{\pi}-1\right)\arg(z-x_1) + ... + \left(\frac{\alpha_n}{\pi}-1\right)\arg(z-x_n) \quad ...(3)$$

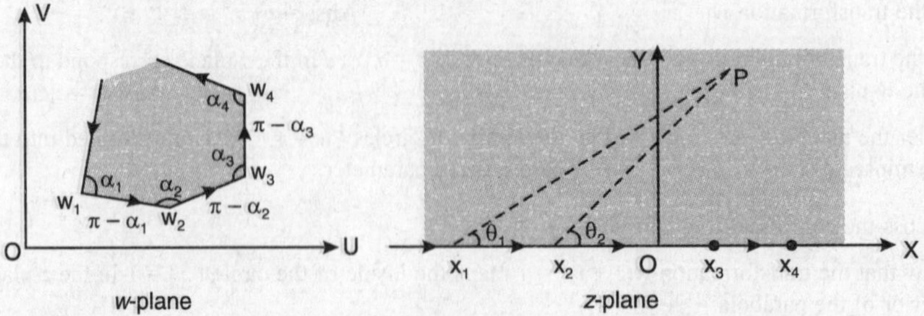

As z moves from left towards x_1, let w move along a side of a polygon towards the vertex w_1. So long as z remains to the left of x_1, the arg of w remains unchanged.

When z crosses x_1 from left to right, $\theta_1 = \arg(z-x_1)$ changes from π to zero, while all other terms in (3) remain uchanged. Hence, the arg dw decreases by $\left(\frac{\alpha_1}{\pi}-1\right)\pi = \alpha_1 - \pi$, *i.e.*, increases by $\pi - \alpha$, in the anticlockwise direction. It means amp $\frac{dw}{dz}$ increases by $\pi - \alpha_1$. Thus, the direction of w_1 turns through the angle $\pi - \alpha_1$ and w now moves along the side w_1w_2 of the polygon.

Similarly, as the point z moves from left to right of x_2 the point w moves along the side w_1w_2 of the polygon. Now as the point z crosses x_2 from left to right, the amptitude $(z-x_2)$ changes from π to 0, while the amplitude of all the other terms unaltered. Thus, amp dw increasingly $\left(\frac{\alpha_2}{\pi}-1\right)(-\pi)$ by $\pi-\alpha_2$. Hence, the direction at the point w_2 rotates through an angle $\pi-\alpha_2$. Consequently w begins moving along the third side of the polygon towards w_3.

By continuing the process we find that as the point w moves along the sides of the polygon in the w-plane, the point z moves along the x-axis in the z-plane.

It can also be proved that interior of polygon (if it is closed) maps into upper half of the z-plane.

Example 30. *Find the transformation which transforms, the semi infinite strip bounded by $v = 0$, $v = \pi$ and $u = 0$ onto the upper half z-plane.*

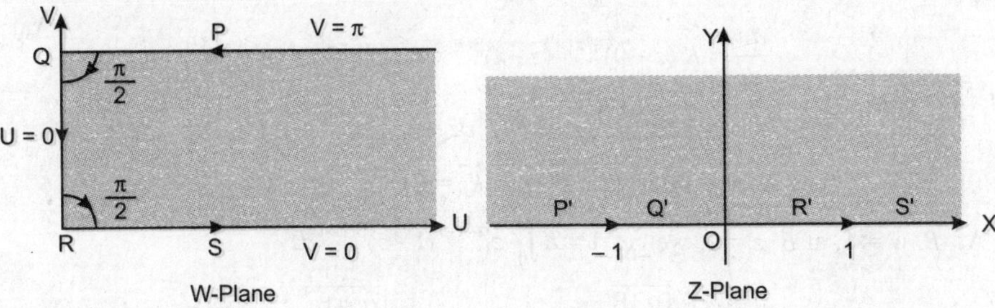

Solution.

Let the points, P, Q, R, S map into P', Q', R', S' respectively. Let us consider $PQRS$ as a limiting case of polygon with vertices Q and R and the third vertex P or S at infinity.

By Schwarz-Christoffel transformation

$$\frac{dw}{dz} = A(z+1)^{\frac{\pi/2}{\pi}-1}(z-1)^{\frac{\pi/2}{\pi}-1} \qquad \left\{\angle Q = \frac{\pi}{2}, \angle R = \frac{\pi}{2}\right\}$$

$$= A(z+1)^{-1/2}(z-1)^{-1/2} = \frac{A}{\sqrt{z^2-1}}$$

$$w = A\int \frac{1}{\sqrt{z^2-1}}\,dz + B$$

$$w = A\cosh^{-1} z + B \qquad \ldots (1)$$

when $\quad w = 0, z = 1$ (At R, $w = 0$ and $R', z = 1$)

$$0 = A\cosh^{-1}(1) + B \qquad [\cosh^{-1} x = \log\,(x + \sqrt{x^2-1})]$$

$$0 = A\log\,(1+\sqrt{1-1}) + B$$

$$0 = 0 + B \text{ or } B = 0$$

Equation (1) is reduced to $w = A\cosh^{-1} z$ ... (2)

At Q, $w = \pi i$; at Q', $z = -1$,

Putting these values of w, z in (2), we get $\pi i = A\cosh^{-1}(-1) = A(\pi i) \Rightarrow A = 1$

Equation (2) is reduced to $w = \cosh^{-1} z$ or $\cosh z = w$ is the required transformation. **Ans.**

Example 31. *Find a transformation which maps the interior of a triangle ABC in w-plane into the upper half of the z-plane.*

Solution. Let us consider a triangle ABC in w-plane. Suppose vertices A, B, of the triangle ABC map into A', B' and C be mapped into C' at infinity.

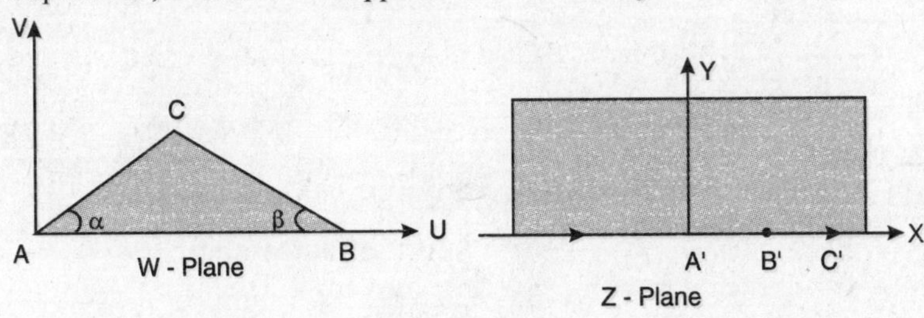

By Schwarz-Christoffel transformation

$$[\angle A = \alpha,\ \angle B = \beta;\ \text{At } A', z = 0;\ \text{At } B',\ z = 1]$$

$$\frac{dw}{dz} = K(z-0)^{\theta/\pi - 1}(z-1)^{\beta/\pi - 1} = Kz^{\alpha/\pi - 1}(z-1)^{\beta/\pi - 1}$$

$$w = \int Kz^{\alpha/\pi - 1}(z-1)^{\beta/\pi - 1}\,dz + k_1 \qquad \text{... (1)}$$

(At A, $w = 0$; at A', $z = 0$, so $k_1 = 0$)

At B, $w = 1$, at B' $z = 1$, we get $1 = K\int_0^1 z^{\alpha/\pi - 1}(1-z)^{\beta/\pi - 1}\,dz$

$$1 = K\frac{\left|\frac{\alpha}{\pi}\right.\left|\frac{\beta}{\pi}\right.}{\left|\frac{\alpha+\beta}{\pi}\right.} \quad \Rightarrow \quad K = \frac{\left|\frac{\alpha+\beta}{\pi}\right.}{\left|\frac{\alpha}{\pi}\right.\left|\frac{\beta}{\pi}\right.}$$

The required transformation, from (1)

$$w = \frac{\left|\frac{\alpha+\beta}{\pi}\right.}{\left|\frac{\alpha}{\pi}\right.\left|\frac{\beta}{\pi}\right.}\int_0^1 z^{\alpha/\pi - 1}(1-z)^{\beta/\pi - 1}\,dz \qquad \textbf{Ans.}$$

EXERCISE 23.4

1. Find the transformation which will map the interior of the infinite strip bounded by the lines $v = 0$, $v = \pi$ into the upper half of the z-plane. **Ans.** $w = \log z$
2. Determine the function which maps the semi-infinite strip bounded by $v = -b$, $u = 0$ and $v = b$ into upper half of the z-plane. **Ans.** $w = \frac{2b}{\pi}\cosh^{-1} z - bi$
3. Find the transformation that maps the semi-infinite strip $u = b$, $u = -b$, $v = 0$ in w-plane into the upper half of z-plane. **Ans.** $z = \sin\frac{\pi w}{2b}$

COMPLEX INTEGRATION
(CAUCHY'S INTEGRAL THEOREM, CAUCHY'S INTEGRAL FORMULA FOR DERIVATIVES OF ANALYTIC FUNCTION)

24.1 INTRODUCTION (LINE INTEGRAL)

In case of real variable, the path of integration of $\int_a^b f(x)\,dx$ is always along the x-axis from $x = a$ to $x = b$. But in case of a complex function $f(z)$ the path of the definite integral $\int_a^b f(z)\,dz$ can be along any curve from $z = a$ to $z = b$.

$z = x + iy \quad \Rightarrow dz = dx + idy \quad ... (1) \quad dz = dx$ if $y = 0 \quad ... (2)\ dz = idy$ if $x = 0 \quad ...(3)$

In (1), (2), (3) the direction of dz are different. Its value depends upon the path (curve) of integration. But the value of integral from a to b remains the same along any regular curve from a to b.

In case the initial point and final point coincide so that c is a closed curve, then this integral is called *contour integral* and is denoted by $\oint_C f(z)\,dz$.

If $f(z) = u(x, y) + iv(x, y)$, then since $dz = dx + idy$, we have

$$\oint_C f(z)\,dz = \int_C (u + iv)(dx + idy) = \int_C (u\,dx - v\,dy) + i\int_C (v\,dx + u\,dy)$$

which shows that the evaluation of the line integral of a complex function can be reduced to the evaluation of two line integrals of real functions.

Let us consider a few examples:

Real integral

Example 1. *Find the value of the integral* $\int_C (x + y)\,dx + x^2 y\,dy$

(*a*) *along* $y = x^2$, *having* (0, 0), (3, 9) *end points.*

(*b*) *along* $y = 3x$ *between the same points.*

Do the values depend upon path.

Solution. $\int_C (x + y)\,dx + x^2 y\,dy \quad ... (1)$

Let $\quad P = x + y,\ Q = x^2 y$

$$\frac{\partial P}{\partial y} = 1, \quad \frac{\partial Q}{\partial x} = 2xy \quad \text{or} \quad \frac{\partial P}{\partial y} \neq \frac{\partial Q}{\partial x}$$

The integrals are not independent of path.

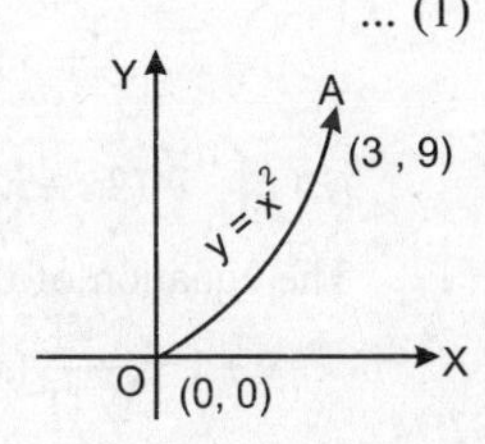

(*a*) Along $y = x^2 \Rightarrow dy = 2x dx$

Putting the values of y and dy in (1), we get

$$\int_0^3 (x+x^2)\,dx + x^2x^2(2x\,dx) = \int_0^3 (x+x^2+2x^5)\,dx = \left[\frac{x^2}{2}+\frac{x^3}{3}+\frac{x^6}{3}\right]_0^3$$

$$= \frac{9}{2}+9+243 = 256\frac{1}{2}$$

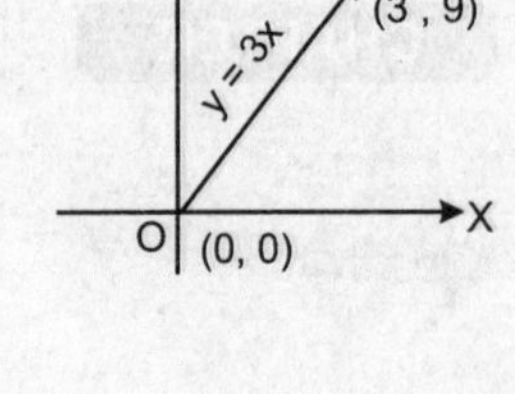

Which is the required value of given integral.

(*b*) Along $y = 3x$, $dy = 3dx$

Substituting these values in (1), we get

$$\int_0^3 (x+3x)\,dx + x^2(3x)(3dx) = \int_0^3 (4x+9x^3)\,dx = \left[2x^2+\frac{9x^4}{4}\right]_0^3$$

$$= 18+\frac{729}{4} = 200\frac{1}{4}$$

Which is the required value of given integral. **Ans.**

Example 2. *Evaluate* $\int_{1-i}^{2+i} (2x+iy+1)\,dz$ *along the two paths:*

(*i*) $x = t + 1,\; y = 2t^2 - 1$

(*ii*) *The straight line joining* $1 - i$ *and* $2 + i$. (*RGPV, Bhopal, Dec. 2008*)

Solution. (i) Here we have

$$x = t + 1 \Rightarrow dx = dt$$
$$y = 2t^2 - 1 \Rightarrow dy = 4t\,dt$$
$$dz = dx + i dy$$
$$= dt + i\,4t\,dt = (1 + i4t)\,dt$$

The limits of the integral are (1, – 1) and (2 , 1).

Corresponding to these points, the limits of t are 0 and 1. $\left[\begin{array}{lll}\text{If } t=0, & x=1, & y=-1\\ \text{If } t=1, & x=2, & y=1\end{array}\right]$

$$\int_{1-i}^{2+i} (2x+iy+1)\,dz$$

$$= \int_0^1 [2(t+1) + i(2t^2-1) + 1]\,(1+4it)\,dt$$

$$= \int_0^1 [\{2(t+1) + i(2t^2-1) + 1\} + \{8it(t+1) - 4t(2t^2-1) + 4it\}]\,dt$$

$$= \int_0^1 [2(t+1) + 1 - 4t(2t^2-1) + i\{2t^2 - 1 + 8t(t+1) + 4t\}]\,dt$$

$$= \int_0^1 \{(2t + 2 + 1 - 8t^3 + 4t) + i(2t^2 - 1 + 8t^2 + 8t + 4t)\}\,dt$$

$$= \int_0^1 \{(-8t^3 + 6t + 3) + i(10t^2 + 12t - 1)\}\,dt$$

$$= \left[-2t^4+3t^2+3t+i\left(\frac{10}{3}t^3+6t^2-t\right)\right]_0^1$$

$$= -2 + 3 + 3 + i\left(\frac{10}{3}+6-1\right) - 0 = 4 + \frac{25}{3}i$$ **Ans.**

(*ii*) $\int_{1-i}^{2+i} (2x+iy+1)\,dz$

The equation of the straight line joining (1, –1) and (2, 1) is

$$y+1 = \frac{1+1}{2-1}(x-1) \Rightarrow y + 1 = 2x - 2 \Rightarrow y = 2x - 3 \Rightarrow dy = 2dx$$

$$\int_{(1,-1)}^{(2,1)} (2x+iy+1)(dx+i\,dy)$$

$$= \int_1^2 (2x + 2ix - 3i + 1)(dx + 2idx)$$

$$= \int_1^2 (2x + 2ix - 3i + 1)(1 + 2i)\,dx$$

$$= \int_1^2 (2x + 2ix - 3i + 1 + 4ix - 4x + 6 + 2i)\,dx$$

$$= \int_1^2 (-2x + 6ix - i + 7)\,dx = (-x^2 + 3ix^2 - ix + 7x)_1^2$$

$$= [-4 + 12i - 2i + 14 + 1 - 3i + i - 7]$$

$$= 4 + 8i$$ **Ans.**

Example 3. *Evaluate* $\int_0^{2+i} (\bar{z})^2 dz$ *along the real axis from* $z = 0$ *to* $z = 2$ *and then along a line parallel to y-axis from* $z = 2$ *to* $z = 2 + i$.

(R.G.P.V., Bhopal, III Semester, June 2005)

Solution. $\int_0^{2+i} (\bar{z})^2 dz = \int_0^{2+i} (x-iy)^2 (dx+idy)$

$$= \int_{OA} (x^2)\,dx + \int_{AB} (2-iy)^2\, i\,dy$$

[Along OA, $y = 0$, $dy = 0$, x varies 0 to 2.
Along AB, $x = 2$, $dx = 0$ and y varies 0 to 1

$$= \int_0^2 x^2 dx + \int_0^1 (2-iy)^2\, i\,dy$$

$$= \left[\frac{x^3}{3}\right]_0^2 + i\int_0^1 (4-4iy-y^2)\,dy = \frac{8}{3} + i\left[4y - 2iy^2 - \frac{y^3}{3}\right]_0^1$$

$$= \frac{8}{3} + i\left[4 - 2i - \frac{1}{3}\right] = \frac{8}{3} + \frac{i}{3}(11-6i) = \frac{1}{3}(8+11i+6) = \frac{1}{3}(14+11i)$$

Which is the required value of the given integral. **Ans.**

Example 4. *Find the value of the integral*

$$\int_0^{1+i} (x-y+ix^2)\,dz$$

(*a*) *Along the straight line from* $z = 0$ *to* $z = 1 + i$;
(*b*) *Along real axis from* $z = 0$ *to* $z = 1$ *and then along a line parallel to the imaginary axis from* $z = 1$ *to* $z = 1 + i$.

Solution. (*a*) Along OA line: Equation of a straight line OA passing through (0, 0) and (1, 1) is

$$y - 0 = \frac{1-0}{1-0}(x-0) \quad \Rightarrow \quad y = x.$$

$$z = x + iy = x + ix = (1+i)x$$

$$dz = (1+i)\,dx$$

$$\int_{OA} (x-y+ix^2)\,dz = \int_0^1 (x-x+ix^2)(1+i)dx$$

$$= (1+i)\int_0^1 ix^2 dx = (1+i)i\left(\frac{x^3}{3}\right)_0^1 = (1+i)\frac{i}{3} = \frac{1}{3}(i-1)$$ **Ans.**

(*b*) Along OB and then along BA. Along OB, from $z = 0$ to $z = 1$. Along BA, from $z = 1$ to $z = 1 + i$.

Required integral $= \int_{OB} (x - y + ix^2)\,dz + \int_{BA} (x - y + ix^2)\,dz$ ... (1)

Now first integral $= \int_{OB} (x - y + ix^2)\,dz = \int_0^1 (x - 0 + ix^2)\,dx$

(Along OB, $y = 0$, $dy = 0$, x varies 0 to 1)

$$= \left(\frac{x^2}{2} + i\frac{x^3}{3}\right)_0^1 = \frac{1}{2} + \frac{i}{3} \qquad ...(2)$$

Y, A (1 + i), dz = idy, O, B, X, x = 1

Now second integral $= \int_{BA} (x - y + ix^2)\,dz$

$$= \int_0^1 (1 - y + i)\,i\,dy \qquad [z = x + iy = 1 + iy \Rightarrow dz = idy]$$

(Along BA, $x = 1$, $dx = 0$, and y varies 0 to 1.)

$$= i\left(y - \frac{y^2}{2} + iy\right)_0^1 = i\left(1 - \frac{1}{2} + i\right) = \frac{i}{2} - 1 \qquad ...(3)$$

Substituting the values of the first and second integral from (2) and (3) in (1), we get

Required integral $= \left(\frac{1}{2} + \frac{i}{3}\right) + \left(\frac{i}{2} - 1\right) = -\frac{1}{2} + \frac{5}{6}i$ **Ans.**

Example 5. *Evaluate* $\int_0^{1+i} (x^2 - iy)\,dz$, *along the path*

(*a*) $y = x$ *(R.G.P.V., Bhopal, III Semester, Dec. 2007)* (*b*) $y = x^2$.

Solution. (*a*) Along the line $y = x$,

$dy = dx$ so that $dz = dx + i\,dy \Rightarrow dz = dx + i\,dx = (1 + i)\,dx$

$\therefore \quad \int_0^{1+i} (x^2 - iy)\,dz$

[On putting $y = x$ and $dz = (1 + i)dx$]

$$= \int_0^1 (x^2 - ix)(1 + i)\,dx$$

$$= (1 + i)\left[\frac{x^3}{3} - i\frac{x^2}{2}\right]_0^1 = (1 + i)\left(\frac{1}{3} - \frac{1}{2}i\right)$$

$$= \frac{(1 + i)(2 - 3i)}{6} = \frac{1}{6}(2 - 3i + 2i + 3) = \frac{5}{6} - \frac{1}{6}i.$$

Y, P(1, 1), y = x, y = x², O, X

Which is the required value of the given integral. **Ans.**

(*b*) Along the parabola $y = x^2$, $dy = 2x\,dx$ so that

$dz = dx + i\,dy \Rightarrow dz = dx + 2ix\,dx = (1 + 2ix)\,dx$

and x varies from 0 to 1.

$$\therefore \int_0^{1+i} (x^2 - iy)\,dx = \int_0^1 (x^2 - ix^2)(1 + 2ix)dx = \int_0^1 x^2\,(1 - i)(1 + 2ix)\,dx$$

$$= (1 - i)\int_0^1 x^2\,(1 + 2ix)\,dx = (1 - i)\left[\frac{x^3}{3} + i\frac{x^4}{2}\right]_0^1$$

$$= (1 - i)\left[\frac{1}{3} + \frac{1}{2}i\right] = \frac{(1 - i)(2 + 3i)}{6} = \frac{1}{6}(2 + 3i - 2i + 3) = \frac{5}{6} + \frac{1}{6}i$$

Which is the required value of the given integral. **Ans.**

Example 6. *Integrate z^2 along the straight line OM and also along the path OLM consisting of two line segments OL and OM where O is the origin, L is the point $z = 3$ and M the point $z = 3 + i$.*

Solution. We have, $z = x + iy$, $dz = dx + idy$. *curve C is line OM.*

$$\int_C z^2\, dz = \int_C (x+iy)^2\ (dx+idy) = \int_C (x^2-y^2+2ixy)\ (dx+idy) \qquad \text{... (1)}$$

The point M is $z = 3 + i$, *i.e*, M is (3, 1).

The equation of the line OM is $y-0 = \dfrac{1-0}{3-0}(x-0)$ *i.e.*, $x = 3y$

Now on the line OM, $x = 3y$, $\therefore\ dx = 3dy$ and y varies from 0 to 1. Therefore, from (1), we have

$$\int_{OM} z^2\, dz = \int_0^1 (9y^2-y^2+2i3y.y).(3dy+idy) = \int_0^1 (8+6i)\ (3+i)\, y^2\, dy$$

$$= (18+26i)\left[\frac{y^3}{3}\right]_0^1 = \frac{1}{3}(18+26i)$$

$$= 6+i\,\frac{26}{3} \qquad \text{... (2)}$$

Now, we have to integrate along OL and LM.

$$\text{Again, } \int_{OLM} z^2 dz = \int_{OL+LM} z^2\, dz = \int_{OL} z^2\, dz + \int_{LM} z^2 dz$$

$$= \int_{OL} (x^2-y^2+2ixy)\ (dx+idy) + \int_{LM} (x^2-y^2+2ixy)\ (dx+idy) \qquad \text{... (3)}$$

On the line OL, $y = 0$ $\quad\therefore\ dy = 0$ and x varies from 0 to 3.
On the line LM, $x = 3$ $\quad\therefore\ dx = 0$ and y varies from 0 to 1.
$\therefore$ From (3), we obtain

$$\int_{OLM} z^2 dz = \int_0^3 x^2\, dx + \int_0^1 (9-y^2+6iy)\, idy = \left[\frac{x^3}{3}\right]_0^3 + i\left[9y-\frac{y^3}{3}+3iy^2\right]_0^1$$

$$= 9+i\left[9-\frac{1}{3}+3i\right] = 6+i\,\frac{26}{3} \qquad \text{... (4)}$$

Which is the required result. **Ans.**

Example 7. *Evaluate $\int_C (12z^2-4iz)$ dz along the curve C joining the points (1, 1) and (2, 3)*

(U.P., III Semester, Dec. 2009)

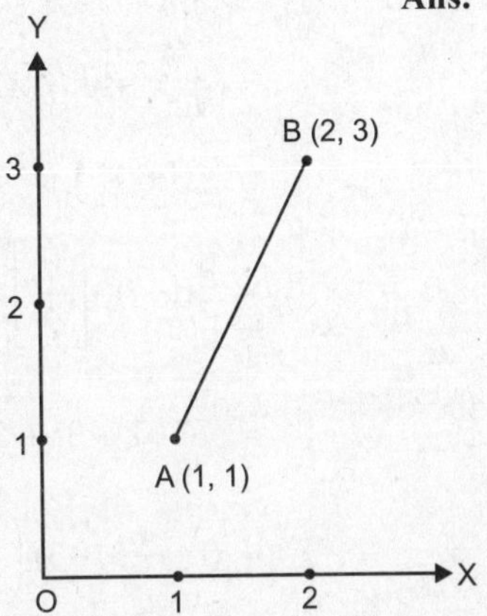

Solution. Here, we have

$$\int_C (12z^2-4iz)\, dz$$

$$= \int_C [12\,(x+iy)^2 - 4i\,(x+iy)]\ (dx+i\,dy)$$

$$= \int_C\ [12\,(x^2-y^2+2i\,xy) - 4ix + 4y]\ (dx + i\,dy)$$

$$= \int_C\ (12\,x^2 - 12y^2 + 24\,ixy - 4ix + 4y)\ (dx + i\,dy) \qquad \text{... (1)}$$

Equation of the line AB passing through (1, 1) and (2, 3) is

$$y-1 = \frac{3-1}{2-1}(x-1)$$

$$y-1 = 2(x-1) \Rightarrow y = 2x-1 \Rightarrow dy = 2\,dx$$

Putting the values of y and dy in (1), we get

$$= \int_1^2 (12x^2 - 12(2x-1)^2 + 24\,ix(2x-1) - 4ix + 4(2x-1)]\,(dx + i2\,dx)$$

$$= \int_1^2 [12x^2 - 48x^2 + 48x - 12 + 48\,ix^2 - 24\,ix - 4ix + 8x - 4]\,(1+2i)\,dx$$

$$= (1+2i)\int_1^2 [(-36+48i)\,x^2 + (56-28i)\,x - 16]\,dx$$

$$= (1+2i)\left[(-36+48i)\frac{x^3}{3} + (56-28i)\frac{x^2}{2} - 16x\right]_1^2$$

$$= (1+2i)\left[(-36+48i)\frac{8}{3} + (56-28i)\,2 - 16\times 2 - (36+48i)\frac{1}{3} - (56-28i)\frac{1}{2} + 16\right]$$

$$= (1+2i)(-96 + 128i + 112 - 56i - 32 + 12 - 16i - 28 + 14i + 16)$$

$$= (1+2i)(-16+70i) = -16 + 70i - 32i - 140 = -156 + 38i$$ **Ans.**

Example 8. *Evaluate the line integral* $\int_C z^2\,dz$ *where C is the boundary of a triangle with vertices 0, 1 + i, – 1 + i clockwise.*

Solution. Here, $I = \int_c z^2\,dz$

where C is the boundary of a triangle with vertices 0, $1+i$, $-1+i$ clockwise.

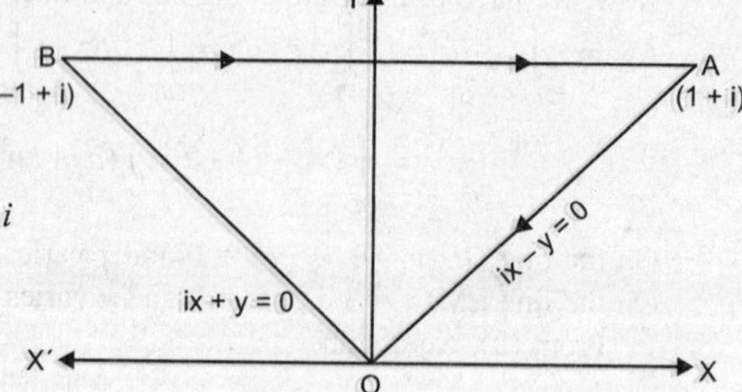

$$I = \int_{AO}(x+iy)^2\,(dx+i\,dy) + \int_{OB}(x+iy)^2\,(dx+i\,dy) + \int_{BA}(x+iy)^2\,(dx+i\,dy)$$

$(x = y)$, $dx = dy$ $\qquad (x = -y)$, $dx = -dy$ $\qquad (y = 1)$, $dy = 0$

$$= \int_1^0 (x+ix)^2\,(dx+i\,dx) + \int_0^{-1}(x-ix)^2\,(dx-i\,dx) + \int_{-1}^1 (x+i)^2\,(dx+0)$$

$$= \int_1^0 x^2(1+i)^2\,(1+i)\,dx + \int_0^{-1} x^2(1-i)^2\,(1-i)\,dx + \int_{-1}^1 (x+i)^2\,dx$$

$$= (1+i)^3\int_1^0 x^2\,dx + (1-i)^3\int_0^{-1} x^2\,dx + \int_{-1}^1 (x+i)^2\,dx$$

$$= (1+i)^3\left[\frac{x^3}{3}\right]_1^0 + (1-i)^3\left[\frac{x^3}{3}\right]_0^{-1} + \left[\frac{(x+i)^3}{3}\right]_{-1}^1$$

$$= (1+i^3+3i+3i^2)\left[0-\frac{1}{3}\right] + (1-i^3-3i+3i^2)\left[\frac{(-1)^3}{3}-0\right] + \left[\frac{(1+i)^3}{3} - \frac{(-1+i)^3}{3}\right]$$

$$= (1-i+3i-3)\left(-\frac{1}{3}\right) + (1+i-3i-3)\left(-\frac{1}{3}\right) + \frac{1}{3}(1+i^3+3i+3i^2+1-i^3-3i+3i^2)$$

$$= (2i-2)\left(-\frac{1}{3}\right)+(-2-2i)\left(-\frac{1}{3}\right)+\frac{1}{3}[1-i+3i-3+1+i-3i-3]$$

$$= \frac{1}{3}[2-2i+2i+2-4] = 0$$

Which is the required value of the given integral. **Ans.**

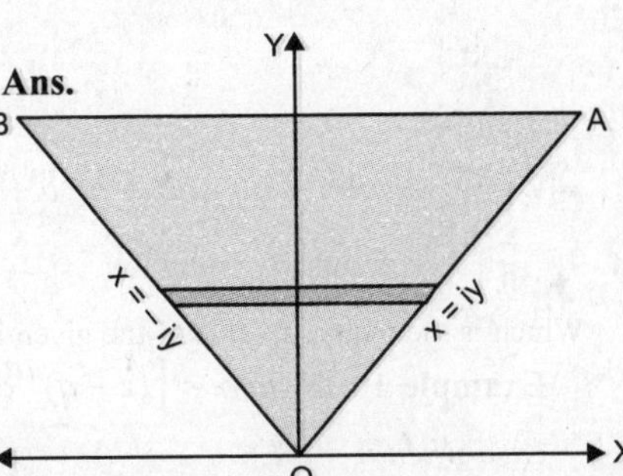

By Green's Theorem

$$I = \int (x+iy)^2\,dx + (x+iy)^2\,i\,dy$$

$$= \iint [2i(x+iy) - 2(x+iy)i]\,dx\,dy$$

$$= 0 \qquad \textbf{Ans.}$$

Which is the required value of the given integral.

Example 9. *Evaluate* $\int_C (z+1)^2\,dz$ *where C is the boundary of the rectangle with vertices at the points a + ib, – a + ib, – a – ib, a – ib.*

Solution. Here, $I = \int_C (z+1)^2\,dz$

Where C is the boundary of the rectangle with vertices at the points $a+ib$, $-a+ib$, $-a-ib$ and $a-ib$.

$$I = \int_{ABCD} (x+iy+1)^2\,(dx+i\,dy)$$

$$= \int_{ABCD} (x^2-y^2+1+2ixy+2x+2iy)\,(dx+i\,dy)$$

$$= \int_{AB} (x^2+b^2+1-2bx+2x-2b)\,dx + \int_{BC} (a^2-y^2+1-2iay-2a+2iy)\,(i\,dy)$$
$$+ \int_{CD} [x^2+b^2+1+2ix(-ib)+2x+2i(-ib)]\,dx + \int_{DA} (a^2-y^2+1+2iay+2a+2iy)\,(i\,dy)$$

Along AB	$y = ib$	$dy = 0$
Along BC	$x = -a$	$dx = 0$
Along CD	$y = -ib$	$dy = 0$
Along DA	$x = a$	$dx = 0$

$$= \int_a^{-a} (x^2+b^2+1-2bx+2x-2b)\,dx + \int_{ib}^{-ib} (ia^2-iy^2+i+2ay-2ia-2y)\,dy$$
$$+ \int_{-a}^{a} (x^2+b^2+1+2bx+2x+2b)\,dx + \int_{-ib}^{ib} (a^2i-iy^2+i-2ay+2ia-2y)\,dy$$

$$\left[\int_{-a}^{a} f(x)\,dx = 2\int_0^a f(x)\,dx, \text{ if } f(x) \text{ is even.} \quad = 0, \text{ if } f(x) \text{ is odd}\right]$$

$$= -2\int_0^a (x^2+b^2+1-2b)\,dx - 2\int_0^{ib} (ia^2-iy^2+i-2ia)\,dy + 2\int_0^a (x^2+b^2+1+2b)\,dx + 2\int_0^{ib} (a^2i-iy^2+i+2ia)\,dy$$

$$= -2\left[\frac{x^3}{3}+b^2x+x-2bx\right]_0^a - 2\left[ia^2y - i\frac{y^3}{3}+iy-2iay\right]_0^{ib} + 2\left[\frac{x^3}{3}+b^2x+x+2bx\right]_0^a + 2\left[a^2iy-\frac{iy^3}{3}+iy+2iay\right]_0^{ib}$$

$$= -2\left[\frac{a^3}{3}+ab^2+a-2ab\right]-2\left[-a^2b-\frac{b^3}{3}-b+2ab\right]+2\left[\frac{a^3}{3}+ab^2+a+2ab\right]$$

$$+2\left[-a^2b-\frac{b^3}{3}-b-2ab\right]$$

$$=2\left[-\frac{a^3}{3}-ab^2-a+2ab+a^2b+\frac{b^3}{3}+b-2ab+\frac{a^3}{3}+ab^2+a+2ab-a^2b-\frac{b^3}{3}-b-2ab\right]$$

$$= 0$$

Which is the required value of the given integral. **Ans.**

Example 10. *Evaluate* $\int (z-a)^n\, dz$ *where c is the circle with centre a and r. Discuss the case when* $n = -1$.

Solution. The equation of circle C is $|z-a| = r$ or $z-a = re^{i\theta}$ where θ varies from 0 to 2π

$$dz = ire^{i\theta}\, d\theta$$

$$\oint_C (z-a)^n\, dz = \int_0^{2\pi} r^n e^{in\theta}\,.ire^{i\theta}\, d\theta$$

$$= ir^{n+1}\int_0^{2\pi} e^{i(n+1)\theta}\, d\theta = ir^{n+1}\left[\frac{e^{i(n+1)\theta}}{i(n+1)}\right]_0^{2\pi} \qquad [\because n \neq -1]$$

$$= \frac{r^{n+1}}{n+1}[e^{i2(n+1)\pi}-1]$$

$$= \frac{r^{n+1}}{n+1}[\cos 2(n+1)\pi + i\sin 2(n+1)\pi - 1] = \frac{r^{n+1}}{n+1}[1+0i-1]$$

$$= 0. \qquad [\text{When } n \neq -1]$$

Which is the required value of the given integral.

When $n = -1$

$$\oint_C \frac{dz}{z-a} = \int_0^{2\pi} \frac{ire^{i\theta}\, d\theta}{re^{i\theta}} = i\int_0^{2\pi} d\theta = 2\pi i.$$ **Ans.**

Example 11. *Evaluate the integral* $\int_c |z|\, dz$, *where c is the straight line from* $z = -i$ *to* $z = i$.

Solution. Equation of the straight line AB from $z = -i$ to $z = i$ is $x = 0$

$\Rightarrow$ $z = x + iy = 0 + iy = iy$ so that $dz = idy$

$$\therefore \quad \int_c |z|\, dz = \int_{-1}^{1} |iy|\, i\,dy = i\int_{-1}^{0}(-y)\, dy + i\int_0^1 y\, dy$$

$$= -i\left(\frac{y^2}{2}\right)_{-1}^{0} + i\left(\frac{y^2}{2}\right)_0^1 = -i\left(-\frac{1}{2}\right) + i\left(\frac{1}{2}\right) = i$$

Which is the required value of the given integral. **Ans.**

Example 12. *Evaluate the integral* $\int_c |z|\, dz$, *where c is the left half of the unit circle* $|z| = 1$ *from* $z = -i$ *to* $z = i$.

Solution. For a point on the unit circle $|z| = 1$,

$$z = e^{i\theta}$$

$$\therefore \quad dz = ie^{i\theta}\, d\theta.$$

The points $z = -i$ and i correspond to $\theta = \frac{3\pi}{2}$ and $\theta = \frac{\pi}{2}$ respectively.

i, $\theta = \frac{\pi}{2}$, X′, X, O, $-I$, $\theta = \frac{3\pi}{2}$, Y′

$$\therefore \quad \int_c |z|\, dz = \int_{3\pi/2}^{\pi/2} 1 \cdot e^{i\theta}\, i d\theta$$

$$= \left(e^{i\theta}\right)_{3\pi/2}^{\pi/2} = e^{i\pi/2} - e^{3i\pi/2}$$

$$= \cos\frac{\pi}{2} + i\sin\frac{\pi}{2} - \cos\frac{3\pi}{2} - i\sin\frac{3\pi}{2}$$

$$= 0 + i - 0 - i(-1) = 2i.$$

Which is the required value of the given integral. **Ans.**

Example 13. *Evaluate the integral* $\int_c \log z\, dz$, *where c is the unit circle* $|z|=1$.

Solution. Here, $c \equiv |z| = 1$ ...(1)

$$\int_c \log z\, dz = \int_c \log(x+iy)\, dz = \int_c \left[\frac{1}{2}\log(x^2+y^2) + i\tan^{-1}\frac{y}{x}\right] dz$$

$$= \int_c \left[\frac{1}{2}\log 1 + i\tan^{-1}\frac{y}{x}\right] dz \qquad [\because x^2+y^2=1]$$

$$= \int_c \left[0 + i\tan^{-1}\frac{y}{x}\right] dz = i\int_c \tan^{-1}\left(\frac{y}{x}\right) dz \qquad ...(2)$$

On the unit circle, $z = e^{i\theta}$

$\therefore$ $dz = ie^{i\theta}\, d\theta$

Now (2) becomes,

$$\int_c \log z\, dz = i\int_0^{2\pi} \tan^{-1}(\tan\theta)\, ie^{i\theta}\, d\theta = -\int_0^{2\pi} \theta e^{i\theta}\, d\theta$$

$$= -\left[\left(\theta\frac{e^{i\theta}}{i}\right)_0^{2\pi} - \int_0^{2\pi} 1 \cdot \frac{e^{i\theta}}{i}\, d\theta\right] = -\left[\frac{2\pi}{i}e^{2\pi i} - \frac{1}{i}\left(\frac{e^{i\theta}}{i}\right)_0^{2\pi}\right]$$

$$= -\left[\frac{2\pi}{i}e^{2\pi i} + e^{2\pi i} - 1\right] = 2\pi i e^{2\pi i} + 1 - e^{2\pi i}$$

$$= (2\pi i - 1)\, e^{2\pi i} + 1 = (2\pi i - 1)(\cos 2\pi + i\sin 2\pi) + 1$$

$$= 2\pi i - 1 + 1 = 2\pi i$$

Which is the required value of the given integral. **Ans.**

EXERCISE 24.1

1. Integrate $f(z) = x^2 + ixy$ from A (1, 1) to B (2, 8) along

   (*i*) the straight line AB; (*ii*) the curve C, $x = t$, $y = t^3$. **Ans.** (*i*) $-\frac{1}{3}(147-71)i$ (*ii*) $-\left(\frac{1094}{21} - \frac{124i}{5}\right)$

2. Evaluate $\int_{1-i}^{2+i} (2x+iy+1)\, dz$ along

   (*i*) $x = t+1$, $y = 2t^2 - 1$; (*ii*) the straight line joining $1-i$ and $2+i$. **Ans.** (*i*) $4 + \frac{25}{3}i$ (*ii*) $4+8i$

   (*R.G.P.V., Bhopal, Dec. 2008*)

3. Evaluate the line integral $\int_c (3y^2 dx + 2y\, dy)$, where c is the circle $x^2+y^2 = 1$, counter clockwise from (1, 0) to (0, 1). **Ans.** – 1.

4. Integrate xz along the straight line from A (1, 1) to B (2, 4) in the complex plane. Is the value the same if the path of integration from A to B is along the curve $x = t, y = t^2$? **Ans.** $-\frac{151}{15}+\frac{45i}{4}$

5. Evaluate $\int_0^{2+i} (\bar{z})^2 dz$, along the line $y = x/2$. *(U.P., III Semester, June 2009)* **Ans.** $\frac{5}{3}(2-i)$

Choose the correct answer:

6. The value of the integral $\int_C (4x^3 dx + 3y^2z^2 dy + 2y^3 z\, dz)$ where C is any path joining A (– 1, 1, 0) to B (1, 2, 1) is
   (*i*) 1 (*ii*) 0 (*iii*) – 8 (*iv*) 8 **Ans.** (*iv*)

7. The value of $\int_C \frac{4z^2+z+5}{z-4} dz$, where C : $9x^2 + 4y^2 = 36$
   (*i*) – 1 (*ii*) 1 (*iii*) 2 (*iv*) 0 *(AMIETE, June 2009)* **Ans.** (*iv*)

24.2 IMPORTANT DEFINITIONS

(*i*) **Simply connected Region.** A connected region is said to be a simply connected if all the interior points of a closed curve C_1 drawn in the region D are the points of the region D.

(*ii*) **Multi-Connected Region.** Multi-connected region is bounded by more than one curve. We can convert a multi-connected region into a simply connected one, by giving it one or more cuts.

Note. A function $f(z)$ is said to be **meromorphic** in a region R if it is analytic in the region R except at a finite number of poles.

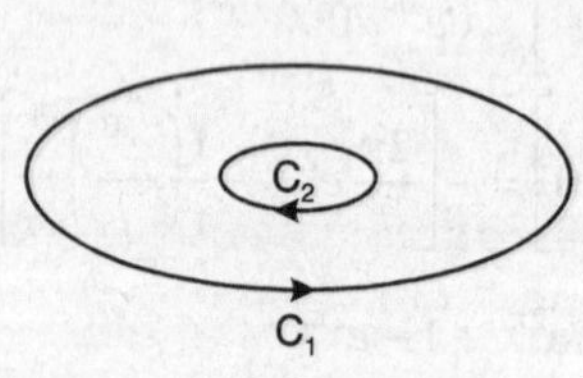

Multi-Connected Region

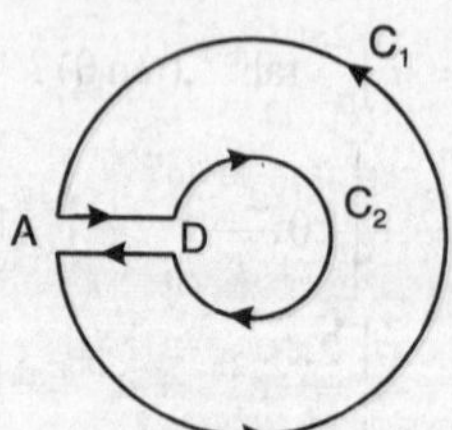

Simply Connected Region

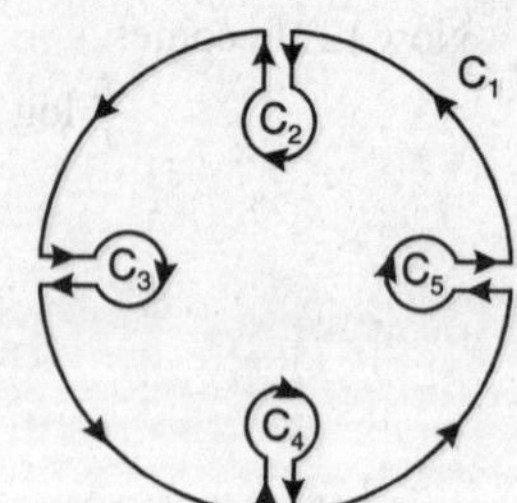

Multi Connected Region

(*iii*) **Single-valued and Multi-valued function**

If a function has only one value for a given value of z, then it is a single valued function.

For example $f(z) = z^2$

If a function has more than one value, it is known as multi-valued function,

For example $f(z) = z^{\frac{1}{2}}$

(*iv*) **Limit of a function**

A function $f(z)$ is said to have a limit l at a point $z = z_0$, if for a given an arbitrary chosen positive number ε, there exists a positive number δ, such that

$$|f(z)-l| < \varepsilon \text{ for } |z-z_0| < \delta$$

It may be written as $\lim_{z \to z_0} f(z) = l$

A
B
D

(*v*) **Continuity**

A function $f(z)$ is said to be continuous at a point $z = z_0$ if for a given an arbitrary positive number ε, there exists a positive number δ, such that

$$|f(z)-f(z_0)|<\varepsilon \text{ for } |z-z_0|<\delta$$

In other words, a function $f(z)$ is continuous at a point $z = z_0$ if

(*a*) $f(z_0)$ exists (*b*) $\lim_{z \to z_0} f(z) = f(z) z = z_0$

(*vi*) **Multiple point.** If an equation is satisfied by more than one value of the variable in the given range, then the point is called a multiple point of the arc.

(*vii*) **Jordan arc.** A continuous arc without multiple points is called a Jordan arc.

Regular arc. If the derivatives of the given function are also continuous in the given range, then the arc is called a regular arc.

(*viii*) **Contour.** A contour is a Jordan curve consisting of continuous chain of a finite number of regular arcs.

The contour is said to be closed if the starting point A of the arc coincides with the end point B of the last arc.

(*ix*) **Zeros of an Analytic function.**

The value of z for which the analytic function $f(z)$ becomes zero is said to be the zero of $f(z)$. **For example,** (1) Zeros of $z^2 - 3z + 2$ are $z = 1$ and $z = 2$.

(2) Zeros of $\cos z$ is $\pm(2n-1)\frac{\pi}{2}$, where $n = 1, 2, 3.........$

24.3 CAUCHY'S INTEGRAL THEOREM

(*AMIETE, Dec. 2009, U.P. III Semester, 2009-2010*)

If a function $f(z)$ is analytic and its derivative $f'(z)$ ontinuous at all points inside and on a simple closed curve c, then $\int_c f(z)\,dz = 0$.

Proof. Let the region enclosed by the curve c be R and let

$$f(z) = u + iv, \quad z = x + iy, \quad dz = dx + i dy$$

$$\int_c f(z)\,dz = \int_c (u+iv)(dx + i dy) = \int_c (u\,dx - v\,dy) + i\int_c (v\,dx + u\,dy)$$

$$= \iint_R \left(-\frac{\partial v}{\partial x} - \frac{\partial u}{\partial y}\right) dx\,dy + i\iint_c \left(\frac{\partial u}{\partial x} - \frac{\partial v}{\partial y}\right) dx\,dy \qquad \text{(By Green's theorem)}$$

Replacing $-\frac{\partial v}{\partial x}$ by $\frac{\partial u}{\partial y}$ and $\frac{\partial v}{\partial y}$ by $\frac{\partial u}{\partial x}$, we get

$$\int_c f(z)\,dz = \iint_R \left(\frac{\partial u}{\partial y} - \frac{\partial u}{\partial y}\right) dx\,dy + i\iint_c \left(\frac{\partial u}{\partial x} - \frac{\partial u}{\partial x}\right) dx\,dy = 0 + i0 = 0$$

$$\Rightarrow \quad \int_C f(z)\,dz = 0$$

Proved.

Note. If there is no pole inside and on the contour then the value of the integral of the function is zero.

Example 14. *Find the integral* $\int_c \frac{3z^2 + 7z + 1}{z+1}\,dz$, *where C is the circle* $|z| = \frac{1}{2}$.

Solution. Poles of the integrand are given by putting the denominator equal to zero.

$z + 1 = 0 \quad \Rightarrow \quad z = -1$

The given circle $|z| = \frac{1}{2}$ with centre at $z = 0$ *and radius* $\frac{1}{2}$ does not enclose any singularity of the given function.

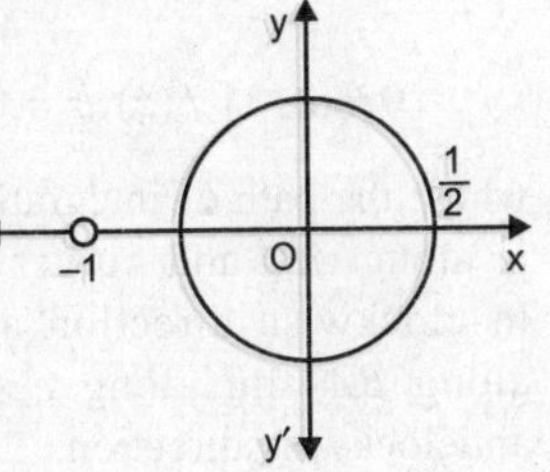

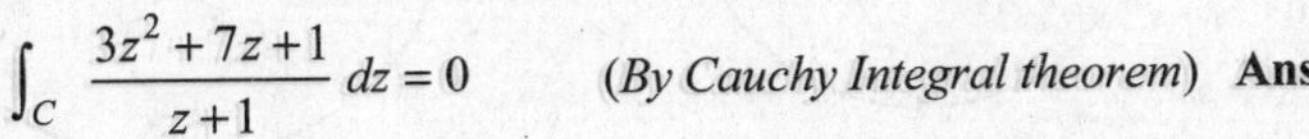

$$\int_C \frac{3z^2 + 7z + 1}{z+1}\,dz = 0 \qquad \text{(By Cauchy Integral theorem)} \quad \textbf{Ans.}$$

Example 15. *Find the value of* $\int_c \frac{z+4}{z^2+2z+5}\,dz$, *if C is the circle* $|z+1| = 1$.

Solution. Poles of integrand are given by putting the denominator equal to zero.

$$z^2+2z+5=0$$

$$z=\frac{-2\pm\sqrt{4-20}}{2}=\frac{-2\pm 4i}{2}=-1\pm 2i$$

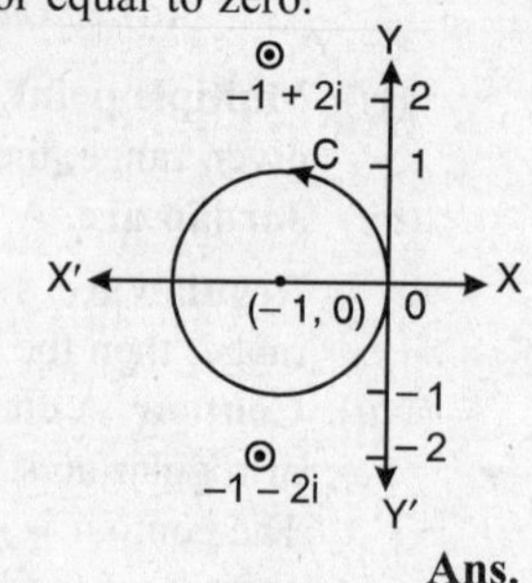

The given circle $|z+1| = 1$ with centre at $z = -1$ and radius unity does not enclose any singularity of the function $\frac{z+4}{z^2+2z+5}$.

$$\therefore \quad \int_c \frac{z+4}{z^2+2z+5}dz = 0 \text{ (By Cauchy Integral Theorem)}$$ **Ans.**

Example 16. *Evaluate* $\oint_c \frac{e^{-z}}{z+1}dz$, *where c is the circle* $|z|=\frac{1}{2}$.

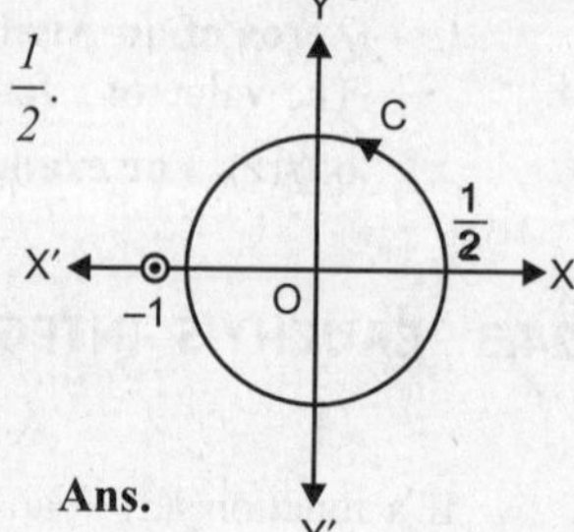

Solution. The point $z = -1$ lies outside the circle $|z| = \frac{1}{2}$.

$\therefore$ The function $\frac{e^{-z}}{z+1}$ is analytic within and on C.

By Cauchy's integral theorem, we have $\oint_c \frac{e^{-z}}{z+1}dz = 0.$ **Ans.**

Example 17. *Evaluate:* $\oint_C \frac{2z^2+5}{(z+2)^3\,(z^2+4)}dz.$

where C is the square with the vertices at $1+i,\ 2+i,\ 2+2i,\ 1+2i.$

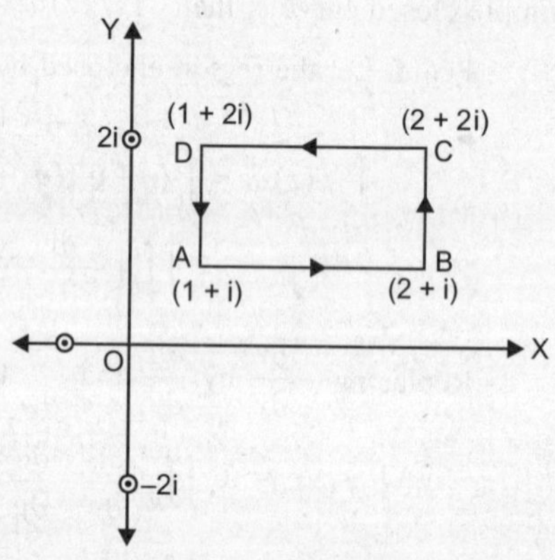

Solution. Here, $f(z) = \frac{2z^2+5}{(z+2)^3\,(z^2+4)}$

Poles are given by

$z = -2$ (pole of order 3)

$z = \pm 2i$ (simple poles).

Since the obtained poles do not lie inside the contour C with vertices at $1+i$, $2+i$, $2+2i$ and $1+2i$, hence by Cauchy Integral theorem.

$$\oint_C \frac{2z^2+5}{(z+2)^3\,(z^2+4)}dz = 0.$$ **Ans.**

24.4 EXTENSION OF CAUCHY'S THEOREM TO MULTIPLE CONNECTED REGION

If $f(z)$ is analytic in the region R between two simple closed curves c_1 and c_2 then

$$\int_{c_1} f(z)dz = \int_{c_2} f(z)dz$$

Proof. $\int f(z)\,dz = 0$

where the path of integration is along AB, and curves c_2 in clockwise direction and along BA and along c_1 in anticlockwise direction.

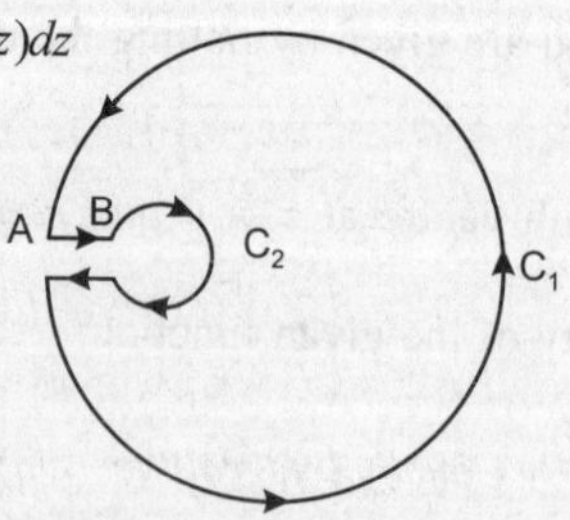

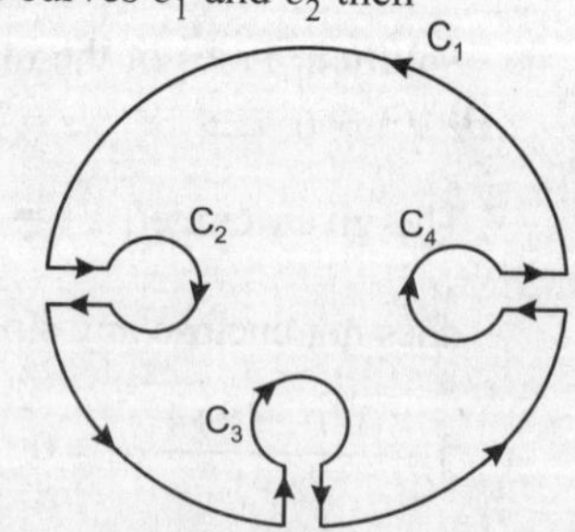

$$\int_{AB} f(z)\,dz - \int_{c2} f(z)\,dz + \int_{BA} f(z)\,dz + \int_{c1} f(z)\,dz = 0$$

$$\Rightarrow \quad -\int_{c2} f(z)\,dz + \int_{c1} f(z)\,dz = 0 \qquad \text{as } \int_{AB} f(z)\,dz = -\int_{BA} f(z)\,dz$$

$$\int_{c1} f(z)\,dz = \int_{c2} f(z)\,dz$$

Proved.

Corollary. $\int_{c1} f(z)\,dz = \int_{c2} f(z)\,dz + \int_{c3} f(z)\,dz + \int_{c4} f(z)\,dz$

24.5 CAUCHY INTEGRAL FORMULA

If $f(z)$ is analytic within and on a closed curve C, and if a is any point within C, then

$$f(a) = \frac{1}{2\pi i}\int_z \frac{f(z)}{z-a}\,dz$$

(*AMIETE June 2010, U.P., III Semester Dec. 2009 R.G.P.V., Bhopal, III Semester, June 2008*)

Proof. Consider the function $\frac{f(z)}{z-a}$, which is analytic at all points within C, except $z = a$. With the point a as centre and radius r, draw a small circle C_1 lying entirely within C.

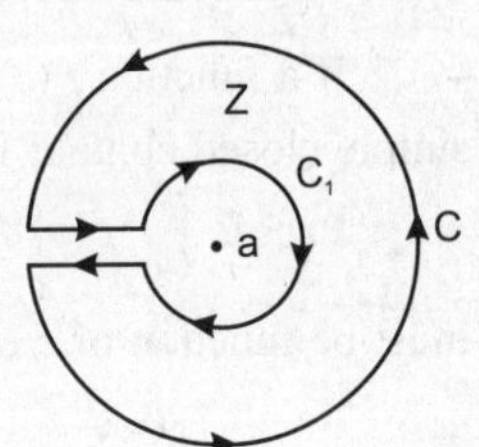

Now $\frac{f(z)}{z-a}$ is analytic in the region between C and C_1; hence by Cauchy's Integral Theorem for multiple connected region, we have

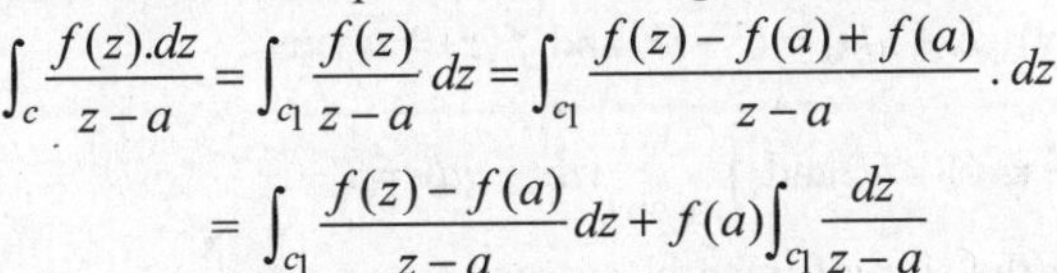

$$\int_c \frac{f(z).dz}{z-a} = \int_{c1} \frac{f(z)}{z-a}\,dz = \int_{c1} \frac{f(z)-f(a)+f(a)}{z-a}\,.\,dz$$

$$= \int_{c1} \frac{f(z)-f(a)}{z-a}\,dz + f(a)\int_{c1} \frac{dz}{z-a} \qquad \dots(1)$$

For any point on C_1

Now, $\int_{c1} \frac{f(z)-f(a)}{z-a}\,dz = \int_0^{2\pi} \frac{f(a+re^{i\theta})-f(a)}{re^{i\theta}}\, ire^{i\theta}\,d\theta$, $\quad [z-a = re^{i\theta} \text{ and } dz = ire^{i\theta}\,d\theta]$

$$= \int_0^{2\pi} [f(a+re^{i\theta}) - f(a)]\, i\,d\theta = 0 \qquad \text{(where } r \text{ tends to zero).}$$

$$\int_{c1} \frac{dz}{z-a} = \int_0^{2\pi} \frac{ire^{i\theta}\,d\theta}{re^{i\theta}} = \int_0^{2\pi} i\,d\theta = i[\theta]_0^{2\pi} = 2\pi i$$

Putting the values of the integrals in R.H.S. of (1), we have

$$\int_c \frac{f(z)\,dz}{z-a} = 0 + f(a)\,(2\pi i) \Rightarrow f(a) = \frac{1}{2\pi i}\int_c \frac{f(z)}{z-a}\,dz$$

Proved.

24.6 CAUCHY INTEGRAL FORMULA FOR THE DERIVATIVE OF AN ANALYTIC FUNCTION

(*R.G.P.V., Bhopal, III Semester, Dec. 2007 DU, 2014*)

If a function $f(z)$ is analytic in a region R, then its derivative at any point $z = a$ of R is also analytic in R, and is given by

$$f'(a) = \frac{1}{2\pi i}\int_c \frac{f(z)}{(z-a)^2}\,dz$$

where c is any closed curve in R surrounding the point $z = a$.

Proof. We know Cauchy's Integral formula

$$f(a)=\frac{1}{2\pi i}\int_c \frac{f(z)}{(z-a)}dz \qquad \text{... (1)}$$

Differentiating (1) w.r.t. 'a', we get

$$f'(a)=\frac{1}{2\pi i}\int_c f(z)\frac{\partial}{\partial a}\left(\frac{1}{z-a}\right)dz$$

$$\boxed{f'(a)=\frac{1}{2\pi i}\int_c \frac{f(z)}{(z-a)^2}dz}$$

Similarly,

$$\boxed{f''(a)=\frac{2!}{2\pi i}\int_c \frac{f(z)\,dz}{(z-a)^3}}$$

$$\boxed{f^n(a)=\frac{n!}{2\pi i}\int_c \frac{f(z)\,dz}{(z-a)^{n+1}}}$$

24.7 MORERA THEOREM (Converse of Cauchy's Theorem)

If a function $f(z)$ is continuous in region D and if the integral $\int f(z)\,dz$, taken around any simple closed contour in D, is zero then $f(z)$ is an analytic function inside D.

Proof. $\int_{z_0}^{z} f(z)\,dz$ is independent of path from z_0 fixed point to a variable point z and hence must be function of z only. Thus $\int_{z_0}^{z} f(z)\,dz=\phi(z)$

$$\int_{(x_0,y_0)}^{(x,y)} (u+iv)(dx+idy)=U+iV \text{ and } f(z)=u+iv$$

$$\Rightarrow \quad \int_{(x_0,y_0)}^{(x,y)} (udx-vdy)=U \text{ and } \int_{(x_0,y_0)}^{(x,y)} vdx+udy=V$$

Differentiating under the sign of integral, we get

$$\frac{\partial U}{\partial x}=u,\quad \frac{\partial V}{\partial x}=v,\qquad \frac{\partial U}{\partial y}=-v,\quad \frac{\partial V}{\partial y}=u$$

$$\therefore \quad \frac{\partial U}{\partial x}=\frac{\partial V}{\partial y} \text{ and } \frac{\partial U}{\partial y}=\frac{-\partial V}{\partial x}$$

Thus, U and V satisfy $C-R$ equations.

$$\therefore \quad \phi(z)=U+iV \text{ is an analytic function.}$$

$$\phi'(z)=\frac{\partial U}{\partial x}+i\frac{\partial V}{\partial x}=u+iv=f(z)$$

$f(z)$ is the derivative of an analytic function $\phi(z)$. **Proved.**

24.8 CAUCHY'S INEQUALITY

If $f(z)$ is analytic within a circle C *i.e.*, $|z-a|=R$ and if $|f(z)|\le M$ on C, then

$$|f^n(a)| \le \frac{Mn!}{R^n}$$

Proof. We know that $f^n(a)=\dfrac{n!}{2\pi i}\displaystyle\int_c \frac{f(z)dz}{(z-a)^{n+1}} \le \frac{n!}{|2\pi i|}\int_c \frac{|f(z)||dz|}{|z-a|^{n+1}}$

$$\le \frac{n!}{2\pi}\frac{M}{R^{n+1}}\int_0^{2\pi} Rd\theta \qquad [\text{since } z=Re^{i\theta}, |dz|=|iRe^{i\theta}d\theta|=Rd\theta]$$

$$\le \frac{n!}{2\pi}\frac{M}{R^{n+1}}2\pi R \le \frac{Mn!}{R^n}$$

Proved.

Second Proof. Let a and b be any two points of z plane. Draw a large circle c_n with centre at the origin, of radius R, enclosing the points a and b. So that

$R > |a|$, and also $R > |b|$.

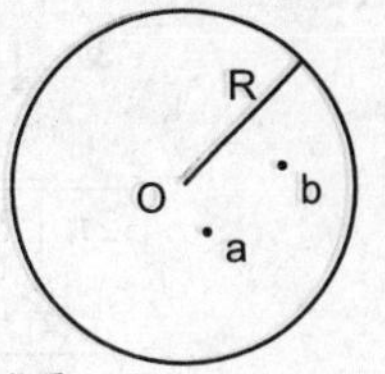

By Cauchy's integral formula

$$\int_c \frac{f(z)\,dz}{z-a} = 2\pi i f(a) \text{ and } \int_c \frac{f(z)}{z-b} dz = 2\pi i f(b)$$

$$2\pi i[f(a) - f(b)] = \int_c \frac{f(z)dz}{z-a} - \int_c \frac{f(z)}{z-b} dz = \int_c \frac{a-b}{(z-a)(z-b)} f(z)\,dz$$

$$|2\pi i\,[f(a) - f(b)]| = \left|\int_c \frac{a-b}{(z-a)(z-b)} f(z)\,dz\right|$$

$$|f(a) - f(b)| \le \left|\frac{1}{2\pi i}\right| \int_c \frac{|a-b||f(z)||dz|}{|z-a||z-b|}$$

$$\le \frac{1}{2\pi}|a-b|M \int_c \frac{|dz|}{(|z|-|a|)(|z|-|b|)} \quad \text{[Since } f(z) \le M]$$

$$= \frac{1}{2\pi} \frac{|a-b|M}{(R-|a|)(R-|b|)} \int_c |dz| \quad \text{[Since } |z| = R]$$

$$= \frac{1}{2\pi} \frac{(a-b)M}{(R-|a|)(R-|b|)} \int_0^{2\pi} R\,d\theta \quad [z = Re^{i\theta},\ |dz| = R\,d\theta]$$

$$= \frac{|a-b|M 2\pi R}{2\pi(R-|a|)(R-|b|)} = \frac{|a-b|M}{R\left(1-\frac{a}{R}\right)\left(1-\frac{b}{R}\right)}$$

$$= 0 \text{ as } R \to \infty$$

$\therefore$ $f(a) = f(b)$. Since this holds for all values of a and b, therefore $f(x)$ is constant.

24.9 LIOUVILLE THEOREM

(U.P., III Semester, June 2009)

If a function $f(z)$ is analytic for all finite values of z, and is bounded, is a constant.

Proof. Since $f(z)$ is bounded so $|f(z)| \le M$, where M is positive constant.

Let z_1, z_2 be any two points of the z-plane.

Let us draw a circle with centre at origin and large radius R enclosing the points z_1 and z_2. So that $R > |z_1|$ and $R > |z_2|$

$$f(z_1) - f_2(z_2) = \frac{1}{2\pi i}\int_c \frac{f(z)}{z-z_1} dz - \frac{1}{2\pi i}\int_c \frac{f(z)}{z-z_2} dz$$

(Cauchy's Integral formula)

$$= \frac{1}{2\pi i}\int_c \frac{z_1 - z_2}{(z-z_1)(z-z_2)} \cdot f(z)\,dz$$

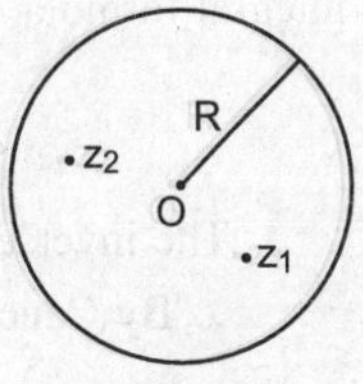

$$|f(z_1) - f(z_2)| = \left|\frac{1}{2\pi i}\int_c \frac{(z_1-z_2)f(z)}{(z-z_1)(z-z_2)} dz\right|$$

$$\le \left|\frac{1}{2\pi i}\right| \int_c \frac{|(z_1-z_2)||f(z)|}{|z-z_1||z-z_2|}|dz|$$

$$\le \frac{1}{2\pi}|z_1 - z_2| M . \int_c \frac{|dz|}{[|z|-|z_1|][|z|-|z_2|]} \quad [|f(z)| < M]$$

$$\leq \frac{1}{2\pi} \cdot \frac{|z_1 - z_2| M}{[R - |z_1|][R - |z_2|]} \int_c |dz| \qquad \text{(Since } |z| = R)$$

$$= \frac{1}{2\pi} \frac{|z_1 - z_2| M}{[R - |z_1|][R - |z_2|]} \int_0^{2\pi} R\, d\theta \qquad (\because z = Re^{i\theta}) \quad |dz| = R\, d\theta$$

$$= \frac{(z_1 - z_2) M . 2\pi R}{2\pi [R - |z_1|] [R - |z_2|]} = 0 \quad \text{Since } R \to \infty$$

Hence, $f(z_1) = f(z_2)$

$f(z)$ is constant. **Proved.**

Alternative. On putting $n = 1$ in Cauchy's inequality

$$|f'(z)| \leq \frac{M}{R}$$

As $R \to \infty$, $f'(z) = 0$, *i.e.*, $f(z)$ is constant for all finite values of z. **Proved.**

24.10 FUNDAMENTAL THEOREM OF ALGEBRA

Every polynomial of degree ≥1 has alteast one zero (root) in C.

Proof. Let $f(z)$ be a polynomial of degree ≥1. Suppose, $f(z)$ has no zero in C, then $f(z) \neq 0$ for all z.

Further $f(z)$ is an entire function in the complex plane.

$\therefore \quad \dfrac{1}{f(z)}$ is also an entire function. Also as $z \to \infty$, $f(z) \to \infty$

$\therefore \quad \dfrac{1}{f(z)} \to 0$ as $z \to \infty$ $\qquad \therefore \quad \dfrac{1}{f(z)}$ is a bounded function.

Thus, by Liouville's theorem $\dfrac{1}{f(z)}$ is a constant function.

$\therefore f(z)$ is a constant function and hence it is a polynomial of degree zero which is a contradiction. Hence, $f(z)$ has at least one root in C. **Proved.**

24.11 POISSON INTEGRAL FORMULA FOR A CIRCLE

If $f(z)$ is analytic within and on the circle C given by $|z| = R$ and $z = re^{i\theta}$ is any point within C, then

$$f(re^{i\theta}) = \frac{1}{2\pi} \int_0^{2\pi} \frac{(R^2 - r^2) f(Re^{i\phi})}{R^2 - 2rR\cos(\theta - \phi) + r^2} d\phi$$

y
$\left|\frac{R^2}{z}\right|$
C
R
|Z|
O
X

Proof. Since $z = re^{i\theta}$ is any point within C, by Cauchy's integral formula, we have

$$f(z) = \frac{1}{2\pi i} \oint_c \frac{f(w)}{w - z} dw \qquad \text{... (1)}$$

The inverse of the point z w.r.t. C is given by $\dfrac{R^2}{\bar{z}}$ and lies outside C.

$\therefore$ By Cauchy's theorem, we have

$$0 = \oint_c \frac{f(w)}{w - \frac{R^2}{\bar{z}}} dw \quad \text{or} \quad 0 = \frac{1}{2\pi i} \oint_c \frac{f(w)}{w - \frac{R^2}{\bar{z}}} dw \qquad \text{...(2)}$$

Subtracting (2) from (1), we get

$$f(z)=\frac{1}{2\pi i}\oint_C\left(\frac{1}{w-z}-\frac{1}{w-\frac{R^2}{\bar{z}}}\right)f(w)dw=\frac{1}{2\pi i}\oint_C\frac{z\bar{z}-R^2}{\bar{z}w^2-(z\bar{z}+R^2)w+R^2z}f(w)\,dw$$

$$=\frac{1}{2\pi i}\oint_C\frac{r^2-R^2}{\bar{z}w^2-(r^2+R^2)w+R^2z}f(w)dw \quad \text{... (3)} \quad [\text{since } z\bar{z}=|z|^2=r^2]$$

Taking $w=\mathrm{Re}^{i\phi}$, we have from (3),

$$f(re^{i\theta})=\frac{1}{2\pi i}\int_0^{2\pi}\frac{(r^2-R^2)f(Re^{i\phi}).Rie^{i\phi}d\phi}{re^{-i\theta}.R^2e^{2i\phi}-(r^2+R^2)Re^{i\phi}+R^2re^{i\theta}}$$

$$=\frac{1}{2\pi}\int_0^{2\pi}\frac{(r^2-R^2)f(Re^{i\phi}).e^{i\phi}d\phi}{rRe^{i(2\phi-\theta)}-(r^2+R^2)e^{i\phi}+Rre^{i\theta}}$$

Dividing the numerator and denominator by $e^{i\phi}$, we get

$$=\frac{1}{2\pi}\int_0^{2\pi}\frac{(r^2-R^2)f(Re^{i\phi})d\phi}{rRe^{i(\phi-\theta)}-(r^2+R^2)+Rre^{i(\theta-\phi)}}=\frac{1}{2\pi}\int_0^{2\pi}\frac{(R^2-r^2)f(Re^{i\phi})}{r^2+R^2-rR[e^{i(\theta-\phi)}+e^{-i(\theta-\phi)}]}d\phi$$

$$=\frac{1}{2\pi}\int_0^{2\pi}\frac{(R^2-r^2)f(Re^{i\phi})}{R^2-2rR\cos(\theta-\phi)+r^2}d\phi \qquad [\text{Since } e^{ix}+e^{-ix}=2\cos x.]$$

This is called *Poisson's integral formula for a circle*. It expresses the values of a harmonic function within a circle in terms of its values on the boundary.

If $\quad f(re^{i\theta})=u(r,\theta)+iv(r,\theta)$ and $f(Re^{i\phi})=u(R,\phi)+iv(R,\phi)$, then we have

$$u(r,\theta)+iv(r,\theta)=\frac{1}{2\pi}\int_0^{2\pi}\frac{(R^2-r^2)[u(R,\phi)+iv(R,\phi)]}{R^2-2rR\cos(\theta-\phi)+r^2}d\phi$$

Equating real and imaginary parts, we get

$$u(r,\theta)=\frac{1}{2\pi}\int_0^{2\pi}\frac{(R^2-r^2)u(R,\phi)}{R^2-2rR\cos(\theta-\phi)+r^2}d\phi$$

and

$$v(r,\theta)=\frac{1}{2\pi}\int_0^{2\pi}\frac{(R^2-r^2)v(R,\phi)}{R^2-2rR\cos(\theta-\phi)+r^2}d\phi$$

Example 18. *Using Poisson's integral formula for a circle, show that*

$$\int_0^{2\pi}\frac{e^{\cos\phi}\cos(\sin\phi)}{5-4\cos(\theta-\phi)}d\phi=\frac{2\pi}{3}e^{\cos\theta}\cos(\sin\theta).$$

Solution. Poisson's integral formula for the circle $|z|=R$ is

$$f(re^{i\theta})=\frac{1}{2\pi}\int_0^{2\pi}\frac{(R^2-r^2)f(Re^{i\phi})}{R^2-2rR\cos(\theta-\phi)+r^2}d\phi \qquad \text{...(1)}$$

Comparing the given integral with the integral on R.H.S. of (1), we get

$$R^2=5,\ rR=2,\ f(Re^{i\phi})=e^{\cos\phi}\cos(\sin\phi)$$

when $R=2, r=1, f(re^{i\theta})=e^{\cos\theta}\cos(\sin\theta)$

Substituting these values in (1), we get

$$e^{\cos\theta}\cos(\sin\theta) = \frac{1}{2\pi}\int_0^{2\pi}\frac{(2^2-1^2)\,e^{\cos\phi}\cos(\sin\phi)}{5-4\cos(\theta-\phi)}d\phi$$

$$\therefore \int_0^{2\pi}\frac{e^{\cos\phi}\cos(\sin\phi)}{5-4\cos(\theta-\phi)}d\phi = \frac{2\pi}{3}e^{\cos\theta}\cos(\sin\theta).$$ **Proved.**

24.12 POISSON INTEGRAL FORMULA FOR A HALF PLANE

If $f(z)$ is an analytic function in the upper half of the z-plane and $a = \alpha + i\beta$ is any point in this upper half plane, then

$$f(a) = \frac{1}{\pi}\int_{-\infty}^{\infty}\frac{\beta f(x)}{(x-\alpha)^2+\beta^2}dx \qquad \text{... (1)}$$

Consider a semi-circle of radius R with centre at the origin. Let C denote the boundary of the semi-circle along with its bounding diameter. Since $a = \alpha + i\beta$ is an interior point, $\bar{a} = \alpha - i\beta$ lies outside C.

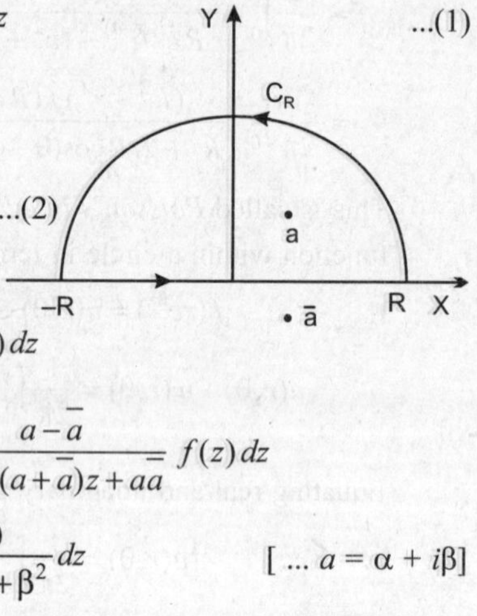

$\therefore$ By Cauchy integral formula, $f(a) = \dfrac{1}{2\pi i}\oint_C \dfrac{f(z)}{z-a}dz$ ...(1)

Also, by Cauchy theorem, we have $\oint_C \dfrac{f(z)}{z-\bar{a}}dz = 0$

$$\Rightarrow \quad 0 = \frac{1}{2\pi i}\oint_C \frac{f(z)}{z-\bar{a}}dz \qquad \text{...(2)}$$

Subtracting (2) from (1), we get

$$f(a) = \frac{1}{2\pi i}\oint_C\left(\frac{1}{z-a}-\frac{1}{z-\bar{a}}\right)f(z)\,dz$$

$$= \frac{1}{2\pi i}\oint_C \frac{a-\bar{a}}{(z-a)(z-\bar{a})}f(z)\,dz = \frac{1}{2\pi i}\oint_C \frac{a-\bar{a}}{z^2-(a+\bar{a})z+a\bar{a}}f(z)\,dz$$

$$= \frac{1}{2\pi i}\oint_C \frac{i\,2\beta f(z)}{z^2-2\alpha z+\alpha^2+\beta^2}dz = \frac{1}{\pi}\oint_C \frac{\beta f(z)}{(z-\alpha)^2+\beta^2}dz \qquad [\because a = \alpha + i\beta]$$

$$= \frac{1}{\pi}\int_{-R}^{R}\frac{\beta f(x)}{(x-\alpha)^2+\beta^2}dx + \frac{1}{\pi}\oint_{C_R}\frac{\beta f(z)}{(z-\alpha)^2+\beta^2}dz$$

where C_R is the semi-circular arc of C.

As $R \to \infty$, the second integral becomes zero.

$$\therefore \quad f(a) = \frac{1}{\pi}\int_{-\infty}^{\infty}\frac{\beta f(x)}{(x-\alpha)^2+\beta^2}dx \qquad \text{... (3)}$$

This is called *Poisson integral formula for a half-plane*. It expresses the values of a harmonic function in the upper half-plane in terms of the values on the real axis.

If $\qquad f(a) = f(\alpha+i\beta) = u(\alpha,\beta)+iv(\alpha,\beta)$

and $\qquad f(x) = u(x,0)+iv(x,0)$, then we have

$$u(\alpha,\beta)+iv(\alpha,\beta) = \frac{1}{\pi}\int_{-\infty}^{\infty}\frac{\beta[u(x,0)+iv(x,0)]}{(x-\alpha)^2+\beta^2}dx$$

Equating real and imaginary parts, we get

$$u(\alpha,\beta)=\frac{1}{\pi}\int_{-\infty}^{\infty}\frac{\beta u(x,0)}{(x-\alpha)^2+\beta^2}dx \text{ and } v(\alpha,\beta)=\frac{1}{\pi}\int_{-\infty}^{\infty}\frac{\beta v(x,0)}{(x-\alpha)^2+\beta^2}dx.$$

Example 19. *Find a function harmonic in the upper half of the z-plane which takes the following values on the x-axis.*

$$G(x)=\begin{cases}1, & x<-1\\ 0, & -1<x<1\\ -1, & x>1\end{cases}$$

Solution. Let $\phi(x,y)$ be the function harmonic in the upper half of the z-plane, then

$$\phi(x,y)=\frac{1}{\pi}\int_{-\infty}^{\infty}\frac{yG(w)}{(w-x)^2+y^2}dw \qquad \text{(Poisson's Integral Formula)}$$

[Obtained from (Article 24.12) on page 678, on replacing α, β by x, y and the variable of integration x by w]

$$=\frac{1}{\pi}\left[\int_{-\infty}^{-1}\frac{y.1}{(w-x)^2+y^2}dw+\int_{-1}^{1}\frac{y.0}{(w-x)^2+y^2}dw+\int_{1}^{\infty}\frac{y.(-1)}{(w-x)^2+y^2}dw\right]$$

$$=\frac{1}{\pi}\left[\left\{\tan^{-1}\frac{w-x}{y}\right\}_{-\infty}^{-1}+0-\left\{\tan^{-1}\frac{w-x}{y}\right\}_{1}^{\infty}\right]$$

$$=\frac{1}{\pi}\left[-\tan^{-1}\frac{1+x}{y}+\frac{\pi}{2}-\left\{\frac{\pi}{2}-\tan^{-1}\frac{1-x}{y}\right\}\right]=\frac{1}{\pi}\left[\tan^{-1}\frac{1-x}{y}-\tan^{-1}\frac{1+x}{y}\right]$$

$$=\frac{1}{\pi}\tan^{-1}\frac{\frac{1-x}{y}-\frac{1+x}{y}}{1+\frac{1-x}{y}.\frac{1+x}{y}}=\frac{1}{\pi}\tan^{-1}\frac{2xy}{x^2-y^2-1}.$$

Which is the required harmonic function. **Ans.**

Example 20. *Prove that* $\int_C \frac{dz}{z-a}=2\pi i$, *where C is the circle* $|z-a|=r$

(R.G.P.V., Bhopal, III Semester, Dec. 2006)

Solution. We have,

$\int_C \frac{dz}{z-a}$, where C is the circle with centre $(a, 0)$ and radius r.

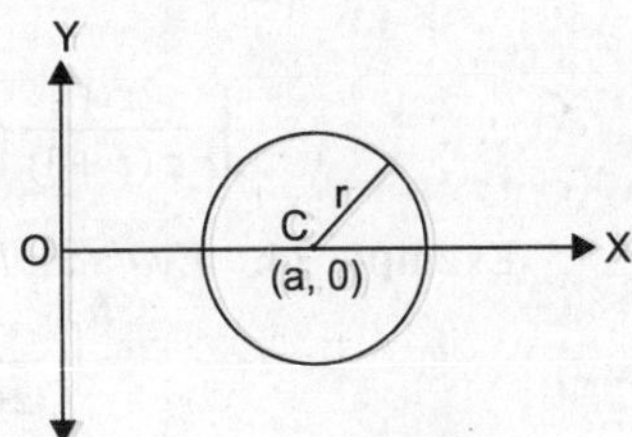

By Cauchy Integral Formula

$$\left[\int_C \frac{f(z)}{z-a}dz=2\pi i\, f(a)\right]$$

$$\int_C \frac{dz}{z-a}=2\pi i \text{ (1)}$$

$$\Rightarrow \qquad \int_C \frac{dz}{z-a}=2\pi i \qquad \textbf{Proved.}$$

Example 21. *Evaluate the following integral*

$$\int_c \frac{1}{z}\cos z dz.$$

where C is the ellipse $9x^2+4y^2=1$.

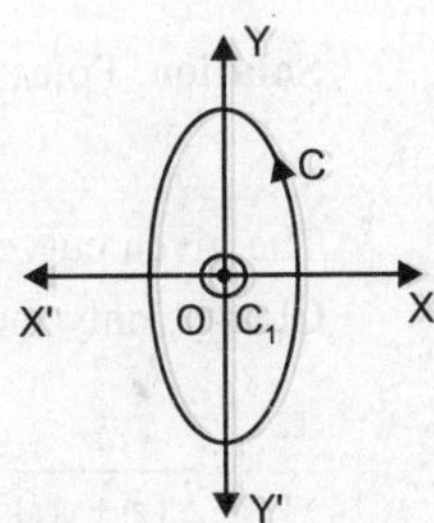

Solution. The given ellipse $9x^2+4y^2=1$ encloses a simple pole $z=0$.

By Cauchy Integral formula

$$\int_c \frac{\cos z}{z} dz = 2\pi i (\cos z)_{z=0} = 2\pi i.$$

Which is the required value of the given integral. **Ans.**

Example 22. *Use Cauchy's integral formula to evaluate* $\int_c \frac{z}{(z^2 - 3z + 2)} dz$

where c is the circle $|z - 2| = \frac{1}{2}$ *(U.P. III Semester, June 2009)*

Solution. Here, we have

$$\int_c \frac{z}{(z^2 - 3z + 2)} dz$$

The poles are determined by putting the denominator equal to zero

i.e.; $z^2 - 3z + 2 = 0 \Rightarrow (z-1)\ (z-2) = 0$

$\Rightarrow z = 1, 2$

So, there are two poles $z = 1$ and $z = 2$.

There is only one pole at $z = 2$ inside the given circle.

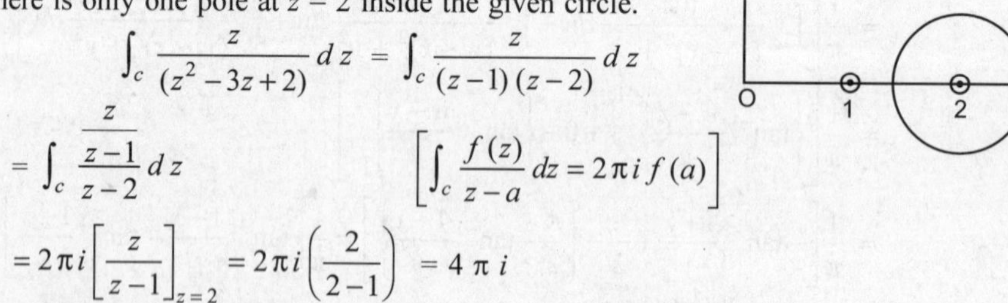

$$\int_c \frac{z}{(z^2 - 3z + 2)} dz = \int_c \frac{z}{(z-1)(z-2)} dz$$

$$= \int_c \frac{\frac{z}{z-1}}{z-2} dz \qquad \left[\int_c \frac{f(z)}{z-a} dz = 2\pi i f(a)\right]$$

$$= 2\pi i \left[\frac{z}{z-1}\right]_{z=2} = 2\pi i \left(\frac{2}{2-1}\right) = 4\pi i$$ **Ans.**

Example 23. *Use Cauchy's integral formula to calculate*

$\int_C \frac{2z+1}{z^2+z} dz$ *where C is* $|z| = \frac{1}{2}$. *(AMIETE, Dec. 2009)*

Solution. Poles are given by

$$z^2 + z = 0$$

$\Rightarrow z(z+1) = 0 \Rightarrow z = 0, -1$

$|z| = \frac{1}{2}$ is a circle with centre at origin and radius $\frac{1}{2}$.

Therefore it encloses only one pole $z = 0$.

$$\therefore \int_C \frac{2z+1}{z(z+1)} dz = \int_C \frac{\frac{2z+1}{z+1}}{z} dz = 2\pi i \left[\frac{2z+1}{z+1}\right]_{z=0} = 2\pi i$$ **Ans.**

Example 24. *Evaluate by Cauchy's integral formula*

$$\int_C \frac{dz}{z(z+\pi i)}, \text{ where C is } |z + 3i| = 1$$

Solution. Poles of the integrand are

$$z = 0, -\pi i \qquad \text{(simple poles)}$$

The given curve C is a circle with centre at $z = -3i$ *i.e.* at $(0, -3)$ and radius 1.

Clearly, only the pole $z = -\pi i$ lies inside the circle.

$$\therefore \int_C \frac{dz}{z(z+\pi i)} = \int_C \frac{\frac{1}{z}}{z+\pi i} dz$$

$$= 2\pi i \left(\frac{1}{z}\right)_{z=-\pi i}$$

[By Cauchy's Integral formula]

$$= \frac{2\pi i}{-\pi i} = -2$$

Which is the required value of the given integral. **Ans.**

Example 25. *Evaluate the complex integral*

$$\int_c \tan z.d z \text{ where } c \text{ is } |z| = 2.$$

Solution. $\int_C \tan z.dz = \int_C \frac{\sin z}{\cos z}.dz$

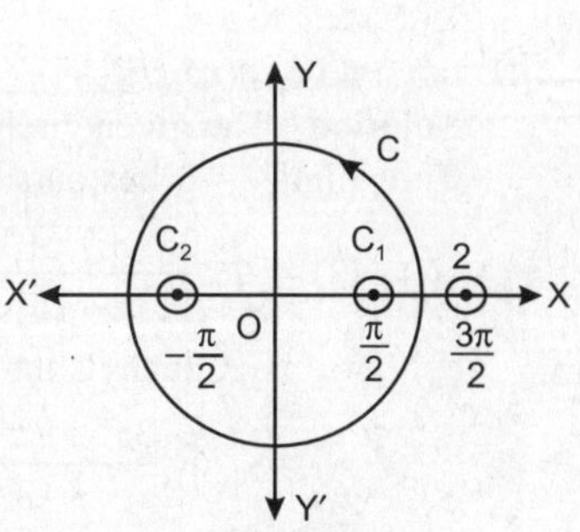

$|z| = 2$, is a circle with centre at origin and radius = 2.
Poles are given by putting the denominator equal to zero.

$$\cos z = 0,\ z = -\frac{\pi}{2}, \frac{\pi}{2}, \frac{3\pi}{2}, \ldots$$

The integrand has two poles at $z = \frac{\pi}{2}$ and $z = -\frac{\pi}{2}$ inside the given circle $|z| = 2$.
On applying Cauchy integral formula

$$\int_C \frac{\sin z}{\cos z} dz = \int_{c_1} \frac{\sin z}{\cos z} dz + \int_{c_2} \frac{\sin z}{\cos z} dz = 2\pi i\ [\sin z]_{z=\frac{\pi}{2}} + 2\pi i\ [\sin z]_{z=\frac{-\pi}{2}}$$

$$= 2\pi i(1) + 2\pi i(-1) = 0$$

Which is the required value of the given integral. **Ans.**

Example 26. *Evaluate* $\oint_C \frac{e^{-z}}{z+1} dz$, *where C is the circle* $|z| = 2$

Solution. $f(z) = e^{-z}$ is an analytic function
The point $z = -1$ lies inside the circle $|z| = 2$.
$\therefore$ By Cauchy's integral formula,

$$\oint_C \frac{e^{-z}}{z+1} dz = 2\pi i(e^{-z})_{z=-1} = 2\pi i e.$$ **Ans.**

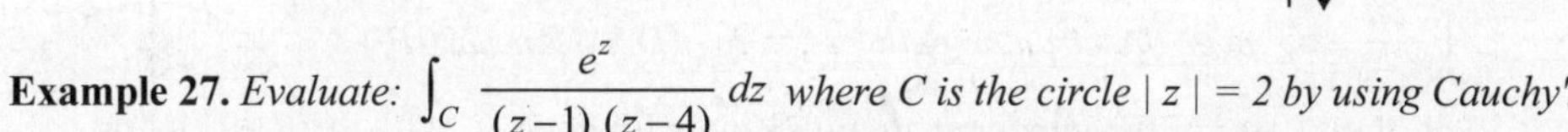

Example 27. *Evaluate:* $\int_C \frac{e^z}{(z-1)(z-4)} dz$ *where C is the circle* $|z| = 2$ *by using Cauchy's Integral Formula.* (*R.G.P.V., Bhopal, III Semester, June 2006*)

Solution. We have,

$$\int_C \frac{e^z}{(z-1)(z-4)} dz$$ where C is the circle with centre at origin and radius 2.

Poles are given by putting the denominator equal to zero.

$$(z-1)(z-4) = 0$$

$$\Rightarrow \quad z = 1, 4$$

Here there are two simple poles at $z = 1$ and $z = 4$.

There is only one pole at $z = 1$ inside the contour. Therefore

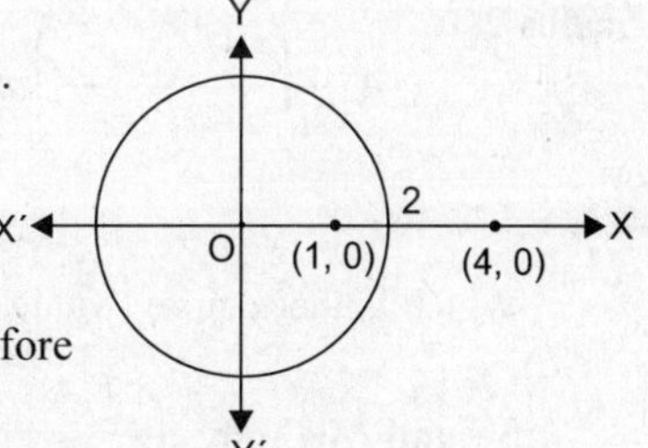

$$\int_C \frac{e^z}{(z-1)(z-4)} dz = \int \frac{\frac{e^z}{(z-4)}}{(z-1)} dz = 2\pi i \left[\frac{e^z}{z-4}\right]_{z=1} \quad \text{(By Cauchy Integral Theorem)}$$

$$= 2\pi i \left(\frac{e}{1-4}\right) = -\frac{2\pi i e}{3}$$

Which is the required value of the given integral. **Ans.**

Example 28. *If* $f(z_1) = \int_C \frac{3z^2+7z+1}{z-z_1} dz$, *where C is the circle* $x^2 + y^2 = 4$, *find the values of (i)* $f(3)$, *(ii)* $f'(1-i)$, *(iii)* $f''(1-i)$.

Solution. The given circle C is $x^2 + y^2 = 4$ or $|z| = 2$.

The point $z_1 = 3$ lies outside the circle $|z| = 2$.

(i) $f(3) = \oint_C \frac{3z^2+7z+1}{z-3} dz$ and $\frac{3z^2+7z+1}{z-3}$ is analytic within and on C.

$\therefore$ By Cauchy's integral theorem, we have

$$\oint_C \frac{3z^2+7z+1}{z-3} dz = 0 \Rightarrow f(3) = 0. \quad \textbf{Ans.}$$

(ii) $z_1 = 1 - i$ lies inside the circle *C*.

By Cauchy's Integral formula, we have

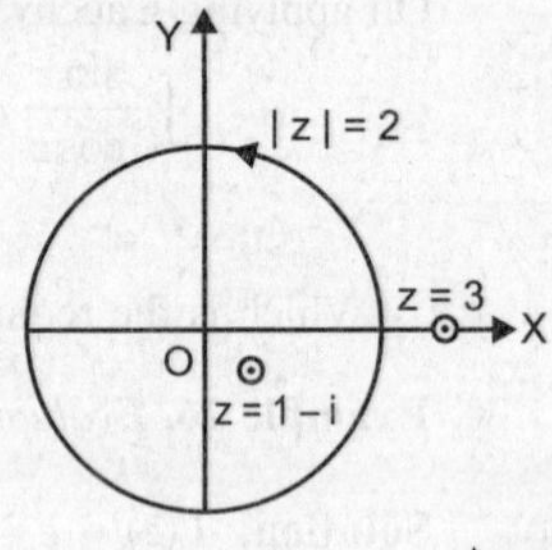

$$\int_C \frac{3z^2+7z+1}{z-z_1} dz = 2\pi i\ (3z^2+7z+1)_{z=z_1}$$

$$f(z) = 2\pi i\ (3z^2 + 7z + 1)$$

$$f'(z) = 2\pi i\ (6z + 7)$$

$$f'(1-i) = 2\pi i\ [6(1-i) + 7]$$

$$\Rightarrow \quad f'(1-i) = 2\pi i\ [13 - 6i]$$

$$\Rightarrow \quad f'(1-i) = 2\pi\ [6 + 13i] \quad \textbf{Ans.}$$

(iii)

$$f''(z) = 2\pi i\ 6$$

$$f''(1-i) = 12\pi i \quad \textbf{Ans.}$$

Example 29. *Evaluate*

$\int_c \frac{e^z}{z^2+1} dz$ *over the circular path* $|z| = 2$. *(D.U., April 2010)*

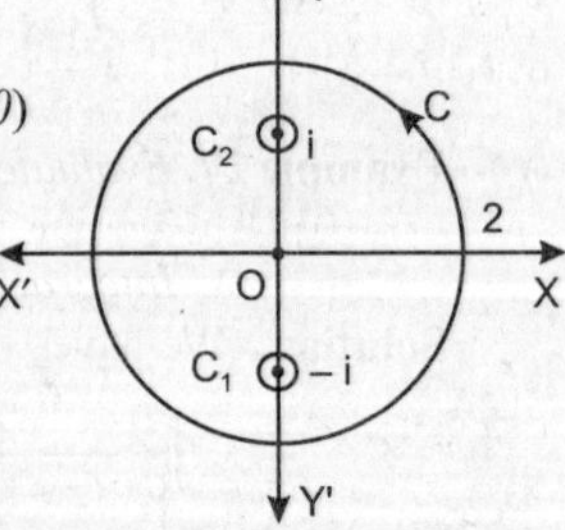

Solution. Poles of the integrand are given by putting the denominator equal to zero.

$$z^2 + 1 = 0 \Rightarrow z^2 = -1 \Rightarrow z = \pm i$$

The integrand has two simple poles at $z = i$ and $z = -i$.

Both poles are inside the given circle with centre at origin and radius 2.

$$\int_c \frac{1}{2i}\left(\frac{e^z}{z-i} - \frac{e^z}{z+i}\right) dz = \int_c \frac{1}{2i}\frac{e^z}{z-i} dz - \frac{1}{2i}\int_c \frac{e^z}{z+i} dz = \frac{1}{2i}\left[2\pi i(e^z)_{z=i} - 2\pi i(e^z)_{z=-i}\right]$$

$$= \frac{2\pi i}{2i}[e^i - e^{-i}] = \pi\ 2\ i \sin 1 = 2\pi i \sin 1$$

Which is the required value of the given integral. **Ans.**

Second Method. $\int_C \frac{e^z}{z^2+1} dz = \int_C \frac{e^z dz}{(z+i)(z-i)} = \int_{C_1} \frac{\frac{e^z}{z-i}}{z+i} dz + \int_{C_2} \frac{\frac{e^z}{z+i}}{z-i} dz$

$$= 2\pi i\left(\frac{e^z}{z-i}\right)_{z=-i} + 2\pi i\left(\frac{e^z}{z+i}\right)_{z=i} = \left[2\pi i\frac{e^{-i}}{-i-i} + 2\pi i\frac{e^i}{i+i}\right] = \pi[-e^{-i} + e^i]$$

$$= \pi(2i\sin 1) = 2\pi i\sin 1$$

Which is the required value of the given integral. **Ans.**

Example 30. *State the Cauchy's integral formula. Show that* $\int_C \frac{e^{zt}}{z^2+1}\,dz = 2\pi i \sin t$

if $t > 0$ *and C is the circle* $|z| = 3$ *(U.P., III Semester, Dec. 2009)*

Solution. See Art. 24.5 on page 673.

Here, we have $\int_C \frac{e^{zt}}{z^2+1}\,dz$

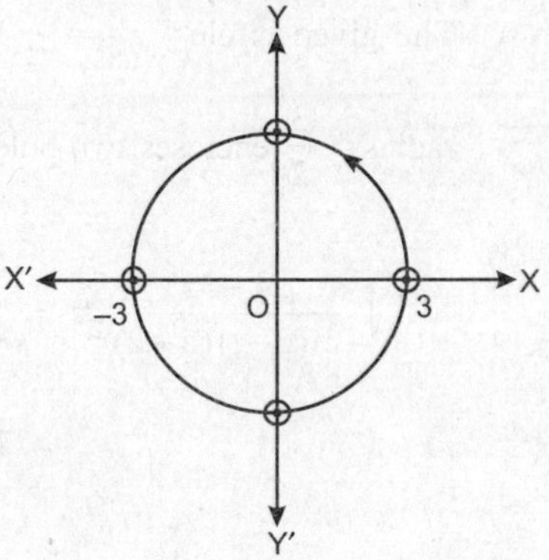

The poles are determined by putting the denominator equal to zero.

i.e., $z^2 + 1 = 0$

$\Rightarrow \quad z^2 = -1$

$\Rightarrow \quad z = \pm\sqrt{-1} = \pm i$

$\Rightarrow \quad z = i, -i$

The integrand has two simple poles at $z = i$ and at $z = -i$. Both poles are inside the given circle with centre at origin and radius 3.

Now, $\int_C \frac{e^{zt}}{z^2+1}\,dz = \frac{1}{2i}\int_C\left(\frac{e^{zt}}{z-i} - \frac{e^{zt}}{z+i}\right)dz$ [By partial fraction]

$$= \frac{1}{2i}\left[\int_{C_1}\frac{e^{zt}}{z-i}\,dz - \int_{C_2}\frac{e^{zt}}{z+i}\,dz\right] = \frac{1}{2i}\left[2\pi i\,(e^{zt})_{z=i} - 2\pi i\,(e^{zt})_{z=-i}\right]$$

$$= \frac{2\pi i}{2i}\left[e^{ti} - e^{-ti}\right] = \pi.\,2i\sin t = 2\pi i\sin t$$

Example 31. *Evaluate* $\int_C \frac{dz}{z^2-1}$, *where C is the circle* $x^2 + y^2 = 4$.

Solution. Poles are given by putting the denominator equal to zero.

$z^2 - 1 = 0, z^2 = 1, z = \pm 1$

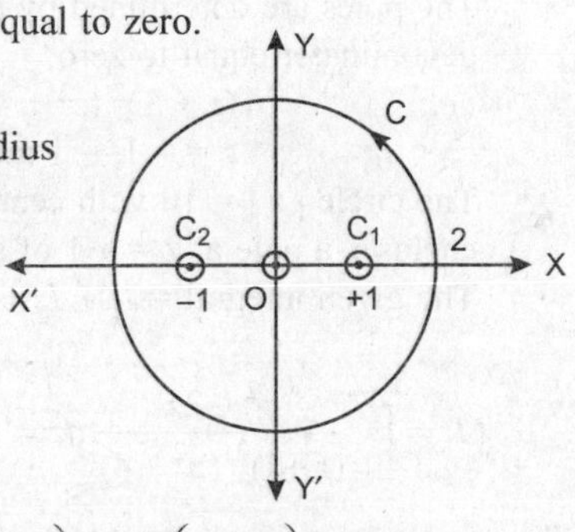

The given circle $x^2 + y^2 = 4$ with centre at $z = 0$ and radius 2 encloses two simple poles at $z = 1$ and $z = -1$.

$$\therefore \int_C \frac{dz}{z^2-1} = \int_{C_1}\frac{dz}{z^2-1} + \int_{C_2}\frac{dz}{z^2-1}$$

$$= \int_{C_1}\frac{\frac{1}{z+1}}{z-1}\,dz + \int_{C_2}\frac{\frac{1}{z-1}}{z+1}\,dz$$

$$= 2\pi i\left(\frac{1}{z+1}\right)_{z=1} + 2\pi i\left(\frac{1}{z-1}\right)_{z=-1} = 2\pi i\left(\frac{1}{1+1}\right) + 2\pi i\left(\frac{1}{-1-1}\right)$$

$$= \pi i - \pi i = 0$$

Which is the required value of the given integral. **Ans.**

Example 32. *Evaluate the following integral using Cauchy integral formula*

$$\int_c \frac{4-3z}{z(z-1)(z-2)}\,dz \quad \text{where } c \text{ is the circle } |z| = \frac{3}{2}.$$

(AMIETE, Dec. 2009, R.G.P.V., Bhopal, III Semester, June 2008)

Solution. Poles of the integrand are given by putting the denominator equal to zero.

$$z(z-1)(z-2) = 0 \text{ or } z = 0, 1, 2$$

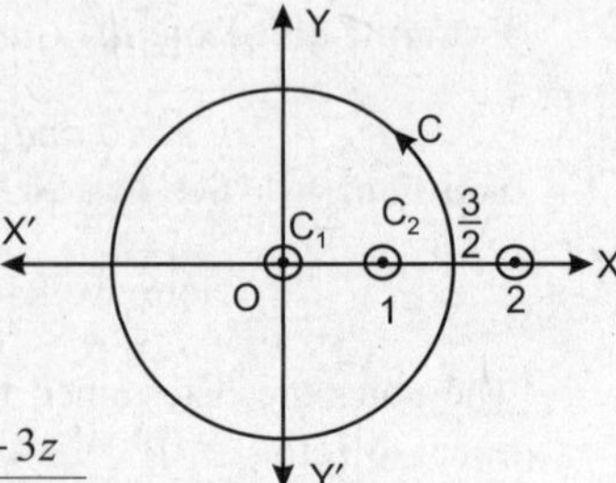

The integrand has three simple poles at $z = 0, 1, 2$.

The given circle $|z| = \frac{3}{2}$ with centre at $z = 0$ and radius $= \frac{3}{2}$ encloses two poles $z = 0$, and $z = 1$.

$$\int_C \frac{4-3z}{z(z-1)(z-2)}\,dz = \int_{c1} \frac{\frac{4-3z}{(z-1)(z-2)}}{z}\,dz + \int_{c2} \frac{\frac{4-3z}{z(z-2)}}{z-1}\,dz$$

$$= 2\pi i \left[\frac{4-3z}{(z-1)(z-2)}\right]_{z=0} + 2\pi i \left[\frac{4-3z}{z(z-2)}\right]_{z=1}$$

$$= 2\pi i.\frac{4}{(-1)(-2)} + 2\pi i \frac{4-3}{1(1-2)} = 2\pi i(2-1)$$

$$= 2\pi i$$

Which is the required value of the given integral. **Ans.**

Example 33. *Evaluate* $\int_c \frac{z^2-2z}{(z+1)^2\,(z^2+4)}\,dz$

where c is the circle $|z| = 10$. *(U.P. III Semester, June 2009)*

Solution. Here, we have

$$\int_c \frac{z^2-2z}{(z+1)^2\,(z^2+4)}\,dz$$

The poles are determined by putting the denominator equal to zero.

i.e.; $(z+1)^2\,(z^2+4) = 0$

$\Rightarrow$ $z = -1, -1$ and $z = \pm 2i$

The circle $|z| = 10$ with centre at origin and radius = 10.

encloses a pole at $z = -1$ of second order and simple poles $z = \pm 2i$

The given integral $= I_1 + I_2 + I_3$

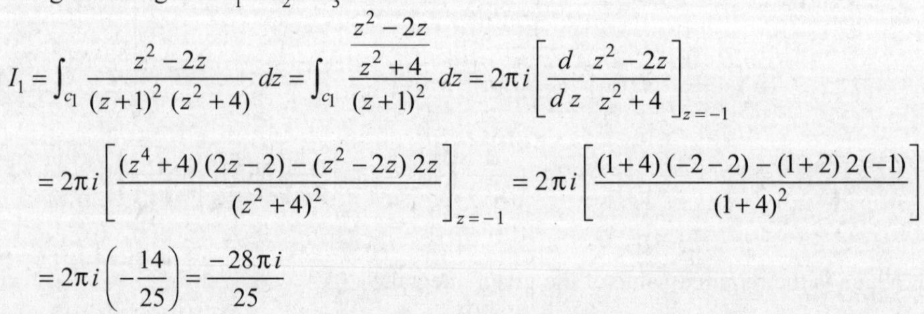

$$I_1 = \int_{c1} \frac{z^2-2z}{(z+1)^2\,(z^2+4)}\,dz = \int_{c1} \frac{\frac{z^2-2z}{z^2+4}}{(z+1)^2}\,dz = 2\pi i \left[\frac{d}{dz}\frac{z^2-2z}{z^2+4}\right]_{z=-1}$$

$$= 2\pi i \left[\frac{(z^4+4)\,(2z-2) - (z^2-2z)\,2z}{(z^2+4)^2}\right]_{z=-1} = 2\pi i \left[\frac{(1+4)\,(-2-2) - (1+2)\,2(-1)}{(1+4)^2}\right]$$

$$= 2\pi i\left(-\frac{14}{25}\right) = \frac{-28\pi i}{25}$$

$$I_2 = \int_{c2} \frac{\dfrac{z^2-2z}{(z+1)^2\,(z+2i)}}{(z-2i)}$$

$$dz = 2\pi i\left[\frac{z^2-2z}{(z+1)^2\,(z+2i)}\right]_{z=2i} = 2\pi i\left[\frac{-4-4i}{(2i+1)^2\,(2i+2i)}\right] = 2\pi i\,\frac{(1+i)}{4+3i}$$

$$I_3 = \int_{c3} \frac{\dfrac{z^2-2z}{(z+1)^2\,(z-2i)}}{(z+2i)}\,dz \quad = 2\pi i\left[\frac{z^2-2z}{(z+1)^2\,(z-2i)}\right]_{z=-2i}$$

$$= 2\pi i\left[\frac{-4+4i}{(-2i+1)^2\,(-2i-2i)}\right] = 2\pi i\,\frac{(i-1)}{(3i-4)}$$

$$\int_c \frac{z^2-2z}{(z+1)^2(z^2+4)}\,dz \quad = I_1 + I_2 + I_3$$

$$= \frac{-28\pi i}{25} + 2\pi i\left(\frac{1+i}{4+3i}\right) + 2\pi i\left(\frac{i-1}{3i-4}\right)$$

$$= 2\pi i\left[\frac{-14}{25} + \frac{1+i}{(4+3i)} + \frac{(i-1)}{(3i-4)}\right]$$

$$= 2\pi i\left[\frac{-14}{25} + \frac{(1+i)\,(3i-4)+(i-1)\,(4+3i)}{(-9-16)}\right]$$

$$= \frac{2\pi i}{-25}\;[14 + (3i - 4 - 3 - 4i) + (4i - 3 - 4 - 3i)]$$

$$= 0$$

Ans.

Example 34. *Integrate* $\dfrac{1}{(z^3-1)^2}$ *the counter clock-wise sense around the circle* $|z - 1| = 1$.

Solution. Poles of the given function are found by putting denominator equal to zero.

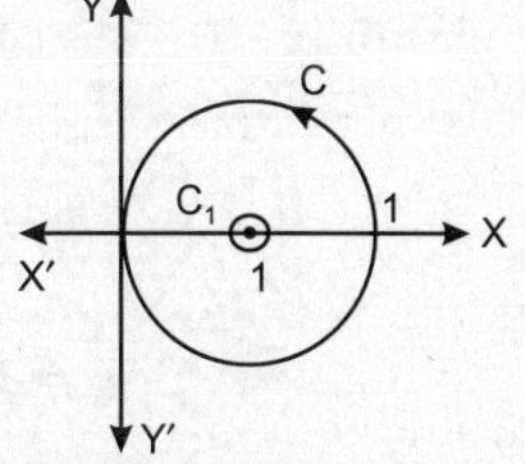

$$(z^3-1)^2 = 0,$$

$$(z-1)^2(z^2+z+1)^2 = 0$$

$$z = 1, 1, \qquad z = \frac{-1\pm\sqrt{1-4}}{2} = \frac{-1\pm\sqrt{-3}}{2} = -\frac{1}{2}\pm\frac{\sqrt{3}}{2}i$$

The circle $|z - 1| = 1$ with centre at $z = 1$ and unit radius encloses a pole of order two at $z = 1$. By Cauchy Integral formula

$$\int_C \frac{1}{(z^3-1)^2}dz = \int_{C_1} \frac{1}{(z-1)^2(z^2+z+1)^2}\,dz = \int_{C_1} \frac{\dfrac{1}{(z^2+z+1)^2}}{(z-1)^2}\,dz$$

$$= 2\pi i\left[\frac{d}{dz}\frac{1}{(z^2+z+1)^2}\right]_{z=1} = 2\pi i\left[\frac{-2(2z+1)}{(z^2+z+1)^3}\right]_{z=1} = 2\pi i\left[\frac{-2(2+1)}{(1+1+1)^3}\right] = -\frac{4\pi i}{9}$$

Ans.

Example 35. *Find the value of* $\displaystyle\int_C \frac{3z^2+z}{z^2-1}\,dz$.

If C is circle $|\,z - 1\,| = 1$ (*R.G.P.V., Bhopal, III Semester, June 2007*)

Solution. Poles of the integrand are given by putting the denominator equal to zero.

$$z^2-1 = 0,\; z^2 = 1,\; z = \pm 1$$

The circle with centre $z = 1$ and radius unity encloses a simple pole at $z = 1$.

By Cauchy Integral formula

$$\int_C \frac{3z^2+z}{z^2-1}\,dz = \int_C \frac{\frac{3z^2+z}{z+1}}{z-1}\,dz$$

$$= 2\pi i\left[\frac{3z^2+z}{z+1}\right]_{z=1} = 2\pi i\left(\frac{3+1}{1+1}\right) = 4\pi i$$

Which is the required value of the given integral. **Ans.**

Example 36. *Evaluate* $\oint_C \frac{z^2+1}{z^2-1}\,dz$ *where C is circle,*

(*i*) $|z| = \frac{3}{2}$ (*ii*) $|z-1| = 1,$ (*iii*) $|z| = \frac{1}{2}.$

Solution. Poles of the integrand are given by putting the denominator equal to zero.

i.e.; $z^2 - 1 = 0 \Rightarrow z = 1 - 1$

(*i*) $|z| = \frac{3}{2}$ is equation of circle C with centre O and radius $\frac{3}{2}$.

Both poles $z = 1, -1$ lie inside C.

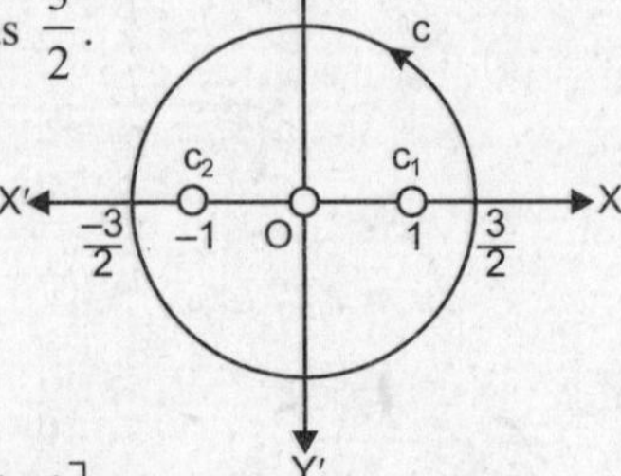

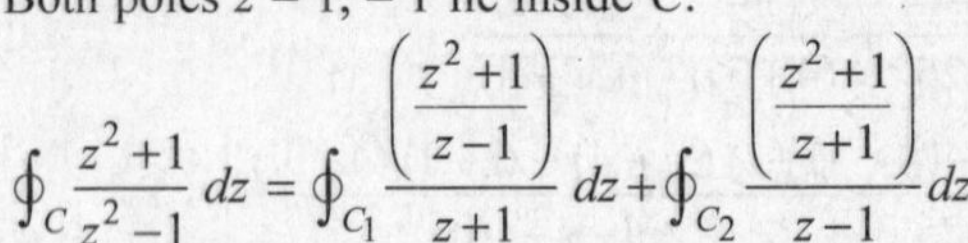

$$\oint_C \frac{z^2+1}{z^2-1}\,dz = \oint_{C_1} \frac{\left(\frac{z^2+1}{z-1}\right)}{z+1}\,dz + \oint_{C_2} \frac{\left(\frac{z^2+1}{z+1}\right)}{z-1}\,dz$$

$$= 2\pi i\left[\frac{z^2+1}{z-1}\right]_{z=-1} + 2\pi i\left[\frac{z^2+1}{z+1}\right]_{z=1} = 2\pi i\left[\frac{1+1}{-1-1}\right] + 2\pi i\left[\frac{1+1}{1+1}\right] = -2\pi i + 2\pi i = 0.$$

Which is the required value of the given integral. **Ans.**

(*ii*) $|z-1| = 1$ is equation of circle C with centre 1 and radius 1 encloses only pole $z = 1$.

$$\oint_C \frac{z^2+1}{z^2-1}\,dz = \oint_{C_1} \frac{\left(\frac{z^2+1}{z+1}\right)}{z-1}\,dz.$$

$$= 2\pi i\left[\frac{z^2+1}{z+1}\right]_{z=1}$$

$$= 2\pi i\left[\frac{1+1}{1+1}\right] = 2\pi i$$

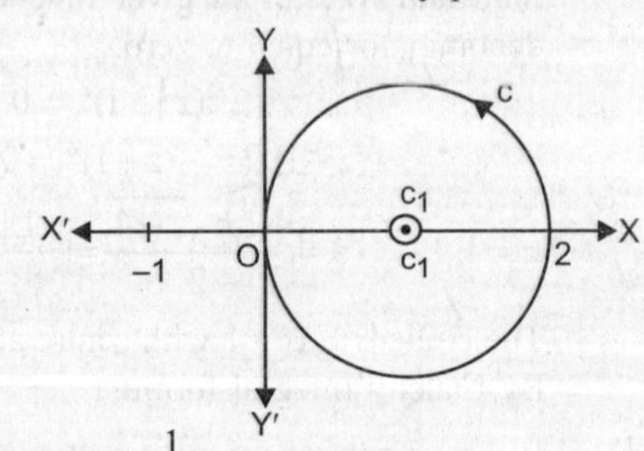

(*iii*) $|z| = \frac{1}{2}$ is equation of circle C with centre O and radius $\frac{1}{2}$.

There is no pole inside C.

Hence, $\oint_C \frac{z^2+1}{z^2-1}\,dz = 0.$ **Ans.**

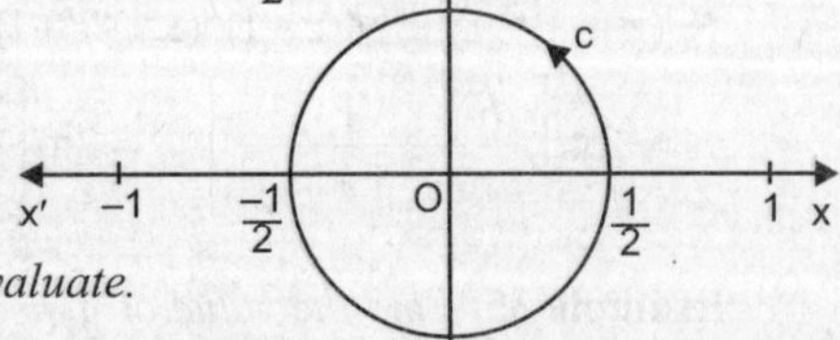

Example 37. *Use Cauchy integral formula to evaluate.*

$$\int_C \frac{\sin \pi z^2 + \cos \pi z^2}{(z-1)(z-2)}\,dz$$

where C is the circle $|z| = 3$.

(*AMIETE, Dec. 2010, R.G.P.V.,*)

Solution. $\int_C \frac{\sin \pi z^2 + \cos \pi z^2}{(z-1)(z-2)} dz$

Poles of the integrand are given by putting the denominator equal to zero.

$$(z-1)(z-2) = 0 \Rightarrow z = 1,\ 2$$

The integrand has two poles at $z = 1, 2$.

The given circle $|z| = 3$ with centre at $z = 0$ and radius 3 encloses both the poles $z = 1$, and $z = 2$.

$$\oint_C \frac{\sin \pi z^2 + \cos \pi z^2}{(z-1)(z-2)} dz = \int_{C_1} \frac{\frac{\sin \pi z^2 + \cos \pi z^2}{(z-2)}}{(z-1)} dz + \int_{C_2} \frac{\frac{\sin \pi z^2 + \cos \pi z^2}{(z-1)}}{(z-2)} dz$$

$$= 2\pi i \left[\frac{\sin \pi z^2 + \cos \pi z^2}{z-2}\right]_{z=1} + 2\pi i \left[\frac{\sin \pi z^2 + \cos \pi z^2}{z-1}\right]_{z=2}$$

$$= 2\pi i \left(\frac{\sin \pi + \cos \pi}{1-2}\right) + 2\pi i \left(\frac{\sin 4\pi + \cos 4\pi}{2-1}\right)$$

$$= 2\pi i \left(\frac{-1}{-1}\right) + 2\pi i \left(\frac{1}{1}\right) = 4\pi i$$

Which is the required value of the given integral. **Ans.**

Example 38. *Evaluate the following complex integration using Cauchy's integral formula*

$$\int_C \frac{3z^2 + z + 1}{(z^2-1)(z+3)} dz$$

Where C is the circle $|z| = 2$.

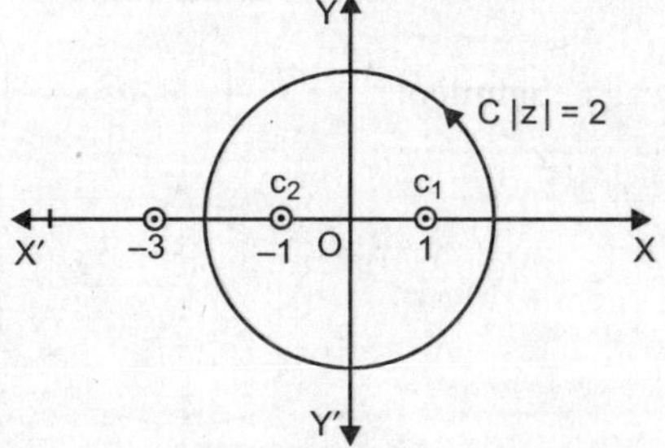

Solution. Poles of the integrand are given by putting the denominator equal to zero.

i.e., $(z^2 - 1)(z + 3) = 0$

$\Rightarrow \qquad z = 1, -1, -3 \qquad$ (Simple poles)

The circle $|z| = 2$ has centre at $z = 0$ and radius 2. Clearly the poles $z = 1$ and $z = -1$ lie inside the given circle while the pole $z = -3$ lies outside it.

$$\therefore \quad \int_C \frac{3z^2 + z + 1}{(z^2-1)(z+3)} dz = \int_{C_1} \frac{\frac{3z^2 + z + 1}{(z+1)(z+3)}}{z-1} dz + \int_{C_2} \frac{\frac{3z^2 + z + 1}{(z-1)(z+3)}}{z+1} dz$$

$$= 2\pi i \left[\frac{3z^2 + z + 1}{(z+1)(z+3)}\right]_{z=1} + 2\pi i \left[\frac{3z^2 + z + 1}{(z-1)(z+3)}\right]_{z=-1} \quad \text{(Using Cauchy's Integral formula)}$$

$$= 2\pi i \left(\frac{5}{8}\right) + 2\pi i \left(-\frac{3}{4}\right) = 2\pi i \left(\frac{-1}{8}\right) = -\frac{\pi i}{4}$$

Which is the required value of the given integral. **Ans.**

Example 39. *Using Cauchy's integral formula, evaluate* $\frac{1}{2\pi i} \int_C \frac{ze^z}{(z-a)^3} dz$, *where the point a lies within the closed curve C.*

Solution. $\int_C \frac{ze^z}{(z-a)^3} dz = \frac{2\pi i}{2!} \left[\frac{d^2}{dz^2}(ze^z)\right]_{z=a} = \frac{2\pi i}{2} \left[\frac{d}{dz}\{(z+1)e^z\}\right]_{\text{at } z=a}$

$$= \frac{2\pi i}{2}\Big[(z+1)e^z + e^z \cdot 1\Big]_{\text{at } z=a} = \frac{2\pi i}{2}\Big[(z+2)e^z\Big]_{\text{at } z=a} = 2\pi i\,\frac{(a+2)e^a}{2} = \pi i(a+2)e^a$$

$$\frac{1}{2\pi i}\int_C \frac{ze^z}{(z-a)^3}\,dz = \frac{\pi i}{2\pi i}(a+2)e^a = \frac{1}{2}(a+2)e^a$$

Which is the required value of the given integral. **Ans.**

Example 40. *Derive Cauchy's Integral Formula.*

Evaluate $\int_C \frac{e^{3iz}}{(z+\pi)^3}\,dz$

where C is the circle $|z-\pi| = 3.2$

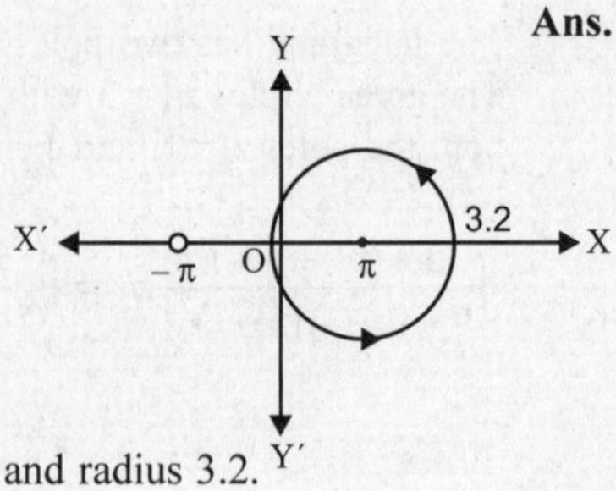

Y, X′, X, Y′, O, −π, π, 3.2

Solution. Here, $I = \int_C \frac{e^{3iz}}{(z+\pi)^3}\,dz$

Where C is a circle $\{|z-\pi| = 3.2\}$ with centre (π, 0) and radius 3.2.

Poles are determined by putting the denominator equal to zero.

$$(z+\pi)^3 = 0 \quad \Rightarrow \quad z = -\pi, -\pi, -\pi$$

There is a pole at $z = -\pi$ of order 3. But there is no pole within C.

By Cauchy Integral Formula $\int_C \frac{e^{3iz}}{(z+\pi)^3}\,dz = 0$ **Ans.**

Example 41. *Verify, Cauchy theorem by integrating e^{iz} along the boundary of the triangle with the vertices at the points $1+i$, $-1+i$ and $-1-i$.*

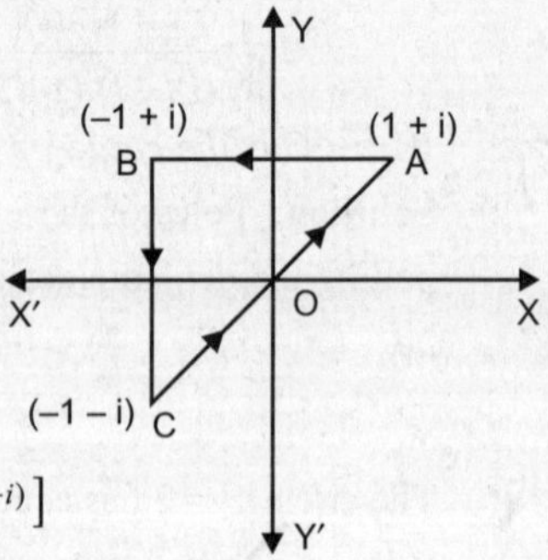

Y, (−1 + i), (1 + i), B, A, X′, O, X, (−1 − i) C, Y′

Solution.
$$\int_{AB} e^{iz}\,dz = \left(\frac{e^{iz}}{i}\right)_{1+i}^{-1+i}$$

$$= \frac{1}{i}\left[e^{i(-1+i)} - e^{i(1+i)}\right]$$

$$= \frac{1}{i}\left[e^{-i-1} - e^{i-1}\right] \qquad \text{...(1)}$$

$$\int_{BC} e^{iz}\,dz = \left[\frac{e^{iz}}{i}\right]_{-1+i}^{-1-i} = \frac{1}{i}\left[e^{i(-1-i)} - e^{i(-1+i)}\right]$$

$$= \frac{1}{i}\left[e^{-i+1} - e^{-i-1}\right] \qquad \text{...(2)}$$

$$\int_{CA} e^{iz}\,dz = \left[\frac{e^{iz}}{i}\right]_{-1-i}^{1+i} = \frac{1}{i}\left[e^{i(1+i)} - e^{i(-1-i)}\right] = \frac{1}{i}\left[e^{i-1} - e^{-i+1}\right] \qquad \text{...(3)}$$

On adding (1), (2) and (3), we get

$$\int_{AB} e^{iz}\,dz + \int_{BC} e^{iz}\,dz + \int_{CA} e^{iz}\,dz = \frac{1}{i}\left[\left(e^{-i-1} - e^{i-1}\right) + \left(e^{-i+1} - e^{-i-1}\right) + \left(e^{i-1} - e^{-i+1}\right)\right]$$

$$\Rightarrow \qquad \int_{\Delta ABC} e^{iz}\,dz = 0 \qquad \text{...(4)}$$

The given function has no pole. So there is no pole in Δ ABC.

The given function e^{iz} is analytic inside and on the triangle ABC.

By Cauchy Theorem, we have $\int_C e^{iz}\,dz = 0$ ...(5)

From (4) and (5) theorem is verified.

EXERCISE 24.2

Evaluate the following

1. $\int_C \frac{1}{z-a}\,dz$, where c is a simple closed curve and the point $z = a$ is

   (*i*) outside c; (*ii*) inside c. **Ans.** (*i*) 0 (*ii*) $2\pi i$

2. $\int_c \frac{e^z}{z-1}\,dz$, where c is the circle $|z| = 2$. **Ans.** $2\pi ie$

3. $\int_C \frac{\cos \pi z}{z-1}\,dz$, where c is the circle $|z| = 3$. **Ans.** $-2\pi i$

4. $\int_C \frac{\cos \pi z^2}{(z-1)(z-2)}\,dz$, where c is the circle $|z| = 3$. **Ans.** $4\pi i$

5. $\int_C \frac{e^{-z}}{(z+2)^5}\,dz$, where c is the circle $|z| = 3$. **Ans.** $\frac{\pi i e^2}{12}$

6. $\int_C \frac{e^{2z}}{(z+1)^4}\,dz$, where c is the circle $|z| = 2$. **Ans.** $\frac{8\pi}{3} ie^{-2}$

7. $\int_C \frac{2z^2+z}{z^2-1}\,dz$ where c is the circle $|z-1| = 1$ **Ans.** $3\pi i$

8. $\int_C \frac{e^z}{z^2(z+1)^3}\,dz$, C : $|z| = 2$. (*AMIETE, June 2009*) **Ans.** $2\pi i\left(\frac{11}{2e} - 2\right)$

Choose the correct alternative:

9. The value of the integral $\int_C \frac{z^2+1}{(z+1)(z+2)}\,dz$, where C is $|z| = \frac{3}{2}$ is

   (*i*) πi (*ii*) 0 (*iii*) $2\pi i$ (*iv*) $4\pi i$ **Ans.** (*iv*)

   (*AMIETE, June 2010*)

10. Cauchy's Integral formula states that if $f(z)$ is analytic within a and on a closed curve C and if a is any point within C then $f(a) =$: (*R.G.P.V., Bhopal, III Semester, June 2007*)

    (*i*) $\frac{1}{2\pi i}\oint \frac{f(z)\,dz}{z-a}$ (*ii*) $\frac{1}{2\pi i}\oint f(z)\,dz$ (*iii*) $\frac{1}{2\pi i}\oint \frac{dz}{z-a}$ (*iv*) none of these. **Ans.** (*i*)

11. The value of $\int_C \frac{z^2-z+1}{z-1}\,dz$, C being $|z| = \frac{1}{2}$ is :

    (*i*) $2\pi i$ (*ii*) $\frac{1}{2\pi i}$ (*iii*) 0 (*iv*) πi (*R.G.P.V., Bhopal, III Sem., Dec. 2006*) **Ans.** (*iii*)

12. If $f(z) = \frac{z^2}{(z-1)^2(z+2)}$, then Res. $f(-2)$ is :

    (*i*) $\frac{5}{9}$ (*ii*) $\frac{4}{9}$ (*iii*) $\frac{1}{9}$ (*iv*) $\frac{3}{9}$ (*RGPV, Bhopal, III Sems, Dec. 2006*) **Ans.** (*ii*)

13. Let $f(z) = \frac{1}{(z-2)^4 (z+3)^6}$, then $z = 2$ and $z = -3$ are the poles of order :

    (*i*) 6 and 4 (*ii*) 2 and 3 (*iii*) 3 and 4 (*iv*) 4 and 6 (*RGPV, Bhopal, III Sem., June 2006*) **Ans.** (*iv*)

14. The value of the integral $\int_C \frac{z+1}{z^3-2z^2}\,dz$, where C is the circle $|z| = 1$ is equal to.

    (*i*) $2\pi i$ (*ii*) $-\frac{2}{3}\pi i$ (*iii*) zero (*iv*) $-\frac{3}{2}\pi i$ (*AMIETE, Dec. 2010*) **Ans.** (*iv*)

TAYLOR'S AND LAURENT'S SERIES

25.1 INTRODUCTION

An analytic function within a circle can be expanded by Taylor's series.

If a function which is not analytic within a circle is expanded by Laurent's series.

25.2 CONVERGENCE OF A SERIES OF COMPLEX TERMS

Let $$(u_1+iv_1)+(u_2+iv_2)+(u_3+iv_3)+...+(u_n+iv_n)+... \qquad ...(1)$$

be an infinite series of complex terms: $u_1, u_2,... v_1, v_2...$ being real numbers.

(*a*) If the series Σu_n and Σv_n converge to the sums U and V then series (1) is said to converge to the sum $U + iV$.

(*b*) If (1) is a convergent series, then

$$\lim_{n\to\infty}(u_n+iv_n)=0$$

(*c*) The series (1) is said to be **absolutely convergent** if the series

$$|u_1+iv_1|+|u_2+iv_2|+|u_3+iv_3|+...+|u_n+iv_n|+...$$

is convergent. Since $|u_n|$ and $|v_n|$ are both less than $|u_n + iv_n|$.

(*d*) Let the series

$$a_1(z)+a_2(z)+a_3(z)+...+a_n(z)+... \qquad ...(2)$$

converge to the sum $S(z)$ and $S_n(z)$ be the sum of its first n terms.

The series (2) is said to be absolutely convergent in region R, if corresponding to any positive number ε, there exists a positive number N.

$|S(z) - S_n(z)| < \varepsilon$ for $n > N$

(*e*) Weirstras's, M-test holds good for series of complex terms also.

Series (2) is uniformly convergent in a region R if there is a convergent series ΣM_n. Such that $|a_n(z)| \le M_n$

A uniformly convergent series can be integrated term by term.

25.3 POWER SERIES

A series in powers of $(z - z_0)$ is called power series.

$$\sum_{n=0}^{\infty} a_n\,(z-z_0)^n = a_0 + a_1(z-z_0) + a_2\,(z-z_0)^2 + ...+ a_n\,(z-z_0)^n + ... \qquad ...(1)$$

Here a_0, a_1, a_2, ..., a_n, ... are known as the coefficient of the series.

Here z is a complex variable and z_0 is called the centre of the series.

(1) is also called the power series about the point z_0

$$\sum_{n=0}^{\infty} a_n z^n = a_0 + a_1 z + a_2 z^2 + \ldots + a_n z^n + \ldots$$

Here the centre of the series is zero.

25.4 REGION OF CONVERGENCE

The region of convergence is the set of all points z for which the series converges.

There are three distinct possibilities for a convergent series.

(1) The series converges only at the point $z = z_0$.

(2) The series converges for all the points in the whole plane.

(3) The series converges everywhere inside a circular plane $|z - z_0| < R$, where R is the radius of convergence and diverges everywhere outside the circle/circular ring.

25.5 RADIUS OF CONVERGENCE OF POWER SERIES

Consider the power series $\Sigma a_n z^n$.

By Cauchy theorem on limits, radius of convergence R is given by

(i) $\dfrac{1}{R} = \lim\limits_{n\to\infty} |a_n|^{1/n}$ (ii) $\dfrac{1}{R} = \lim\limits_{n\to\infty} \left|\dfrac{a_{n+1}}{a_n}\right|$

Example 1. *Find the radius of convergence of the power series*

$$f(z) = \sum_{n=0}^{\infty} \frac{z^n}{n!}$$

Solution. Here, $a_n = \dfrac{1}{n!} \Rightarrow a_{n+1} = \dfrac{1}{(n+1)!}$

Radius of convergence is given by

$$\frac{1}{R} = \lim_{n\to\infty} \left|\frac{a_{n+1}}{a_n}\right| = \lim_{n\to\infty} \frac{1}{n+1} = 0$$

$$R = \infty$$

Hence, the radius of convergence of the given power series is ∞. **Ans.**

Example 2. *Find the radius of convergence of the power series*

$$f(z) = \sum_{n=0}^{\infty} \frac{z^n}{2^n + 3}$$

Solution. Here,

$$a_n = \frac{1}{2^n + 3} \Rightarrow a_{n+1} = \frac{1}{2^{n+1} + 3}$$

Radius of convergence is given by

$$\frac{1}{R} = \lim_{n\to\infty} \left|\frac{a_{n+1}}{a_n}\right| = \lim_{n\to\infty} \frac{2^n + 3}{2^{n+1} + 3} = \lim_{n\to\infty} \frac{1 + \frac{3}{2^n}}{2 + \frac{3}{2^n}} = \frac{1}{2}$$

$$R = 2$$

Hence, the radius of convergence of the given power series is 2. **Ans.**

Example 3. *Find the radius of convergence of the power series:*

$$f(z) = \sum_{n=0}^{\infty} \frac{z^n}{n^n}$$

Solution. Here,

$$a_n = \frac{1}{n^n}, \qquad a_{n+1} = \frac{1}{(n+1)^{n+1}}$$

Radius of convergence is given by

$$\frac{1}{R} = \lim_{n\to\infty}\left|\frac{a_{n+1}}{a_n}\right| = \lim_{n\to\infty}\frac{n^n}{(n+1)^{n+1}}$$

$$= \lim_{n\to\infty}\frac{n^n}{(n+1)^n\,(n+1)} = \lim_{n\to\infty}\frac{1}{\left(1+\frac{1}{n}\right)^n(n+1)} = 0$$

$\Rightarrow$ $\qquad R = \infty$

Hence, the radius of convergence of the given power series is ∞. **Ans.**

Example 4. *Find the radius of convergence of the power series:*

$$f(z) = \sum_{n=0}^{\infty}\frac{n!}{n^n}z^n$$

Solution. Here, $\quad a_n = \dfrac{n!}{n^n}$

$\therefore \qquad a_{n+1} = \dfrac{(n+1)!}{(n+1)^{n+1}}$

Now,

$$\frac{a_{n+1}}{a_n} = \frac{(n+1)!}{(n+1)^{n+1}}\cdot\frac{n^n}{n!} = \frac{n!}{(n+1)^n}\cdot\frac{n^n}{n!} = \frac{n^n}{(n+1)^n} = \left(\frac{n}{n+1}\right)^n = \frac{1}{\left(1+\frac{1}{n}\right)^n}$$

Radius of convergence is given by

$$\frac{1}{R} = \lim_{n\to\infty}\left|\frac{a_{n+1}}{a_n}\right| = \lim_{n\to\infty}\frac{1}{\left(1+\frac{1}{n}\right)^n} = \frac{1}{e}$$

$\Rightarrow$ $\qquad R = e.$

Hence, the radius of convergence of the given power series is e. **Ans.**

EXERCISE 25.1

Find the radius of convergence of following power series:

1. $\displaystyle\sum_{n=1}^{\infty}\frac{z^n}{n^p}$ **Ans.** 1
2. $\displaystyle\sum_{n=0}^{\infty}\frac{z^n}{3^{n+1}}$ **Ans.** 3
3. $\displaystyle\sum_{n=0}^{\infty}\left(1+\frac{1}{n}\right)^{n^2}z^n$ **Ans.** $\dfrac{1}{e}$
4. $\displaystyle\sum_{n=0}^{\infty}(5+12i)^n z^n$ **Ans.** 13
5. $\displaystyle\sum_{n=0}^{\infty}\frac{2n+3}{(2n+5)(n+5)}z^n$ **Ans.** 1

25.6 METHOD OF EXPANSION OF A FUNCTION

(1) Taylor's series (2) Binomial series (3) Exponential series

25.7 TAYLOR'S TVHEOREM

(Delhi University, April 2010)

If a function $f(z)$ is analytic at all points inside a circle C, with its centre at the point a and radius R, then at each point z inside C.

$$f(z) = f(a) + f'(a)(z-a) + \frac{f''(a)}{2!}(z-a)^2 + \ldots + \frac{f^n(a)}{n!}(z-a)^n + \ldots$$

Proof. Take any point z inside C. Draw a circle C_1 with centre a, enclosing the point z. Let w be a point on circle C_1.

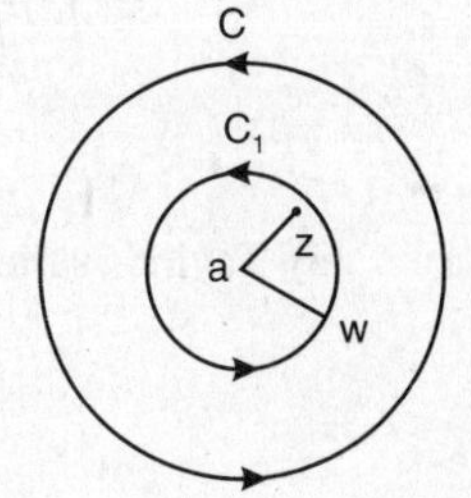

$$\frac{1}{w-z}=\frac{1}{w-a+a-z}=\frac{1}{w-a-(z-a)}$$

$$=\frac{1}{(w-a)\left(1-\frac{z-a}{w-a}\right)}=\frac{1}{w-a}\left(1-\frac{z-a}{w-a}\right)^{-1}$$

Apply Binomial theorem

$$\frac{1}{w-z}=\frac{1}{w-a}\left[1+\frac{z-a}{w-a}+\left(\frac{z-a}{w-a}\right)^2+...+\left(\frac{z-a}{w-a}\right)^n+...\right]$$

$$\Rightarrow \qquad \frac{1}{w-z}=\frac{1}{w-a}+\frac{z-a}{(w-a)^2}+\frac{(z-a)^2}{(w-a)^3}+...+\frac{(z-a)^n}{(w-a)^{n+1}}+... \qquad ...(1)$$

As $\quad |z-a|<|w-a| \quad \Rightarrow \quad \frac{|z-a|}{|w-a|}<1,$

So the series converges uniformly. Hence the series is integrable.

Multiply (1) by $f(w)$.

$$\frac{f(w)}{w-z}=\frac{f(w)}{w-a}+(z-a)\frac{f(w)}{(w-a)^2}+(z-a)^2\frac{f(w)}{(w-a)^3}+...+(z-a)^n\frac{f(w)}{(w-a)^{n+1}}+,...$$

On integrating w.r.t. 'w', we get

$$\int_{C_1}\frac{f(w)}{w-z}dw=\int_{C_1}\frac{f(w)}{w-a}dw+(z-a)\int_{C_1}\frac{f(w)}{(w-a)^2}dw+(z-a)^2\int_{C_1}\frac{f(w)}{(w-a)^3}dw+$$

$$...+(z-a)^n\int_{C_1}\frac{f(w)}{(w-a)^{n+1}}dw+... \qquad ...(2)$$

We know that

$$\int_{C_1}\frac{f(w)}{w-z}dz=2\pi i\,f(z) \text{ and } \int_{C_1}\frac{f(w)}{w-a}dw=2\pi i\,f(a)$$

$$\int_{C_1}\frac{f(w)}{(w-a)^2}dw=2\pi i\,f'(a) \text{ and so on.}$$

Substituting these values in (2), we get

Taylor's series as given below

$$\boxed{f(z)=f(a)+f'(a)(z-a)+\frac{f''(a)}{2!}(z-a)^2+...+\frac{f^n(a)(z-a)^n}{n!}+...} \qquad ...(3) \qquad \textbf{Proved.}$$

(Delhi University, April 2010)

Corollary 1. Putting $z = a + h$ in (3), we get

$$f(a+h)=f(a)+hf(a)+\frac{h^2}{2!}f''(a)+......+\frac{h^n}{n!}f^n(a)+...$$

Corollary 2. If $a = 0$, the series (3) becomes

$$f(z)=f(0)+z\,f'(0)+\frac{z^2}{2!}f''(0)+......+\frac{z^n}{n!}f^n(0)+...$$

This series is called **Maclaurin series.**

Example 5. *Expand e^z about a.*

Solution. Here, we have

$$f(z)=e^z \qquad \Rightarrow \qquad f(a)=e^a$$

$$\Rightarrow \quad f'(z) = e^z \quad \Rightarrow \quad f'(a) = e^a$$
$$f''(z) = e^z \quad \Rightarrow \quad f''(a) = e^a$$
$$\cdots\cdots\cdots\cdots\cdots\cdots\cdots\cdots$$
$$\cdots\cdots\cdots\cdots\cdots\cdots\cdots\cdots$$
$$f^n(z) = e^z \quad \Rightarrow \quad f^n(a) = e^a$$

By Taylor's series of $f(z)$ is

$$f(z) = f(a) + (z-a) f'(a) + \frac{(z-a)^2}{2!} f''(a) + \ldots\ldots\ldots$$

$$e^z = e^a + \frac{(z-a)}{1!} e^a + \frac{(z-a)^2}{2!} e^a + \ldots\ldots\ldots$$ **Ans.**

Example 6. *Expand the function*

$$f(z) = \frac{1}{z}$$

about $z = 2$ *in Taylor's series. Obtain its radius of convergence.*

Solution. Here, we have,

$$f(z) = \frac{1}{z}$$

$$\Rightarrow \quad f'(z) = -\frac{1}{z^2}$$

$$\Rightarrow \quad f''(z) = \frac{2}{z^3}$$

$$\cdots\cdots\cdots\cdots\cdots\cdots\cdots\cdots$$

$$f^n(z) = (-1)^n \frac{n!}{z^{n+1}}$$

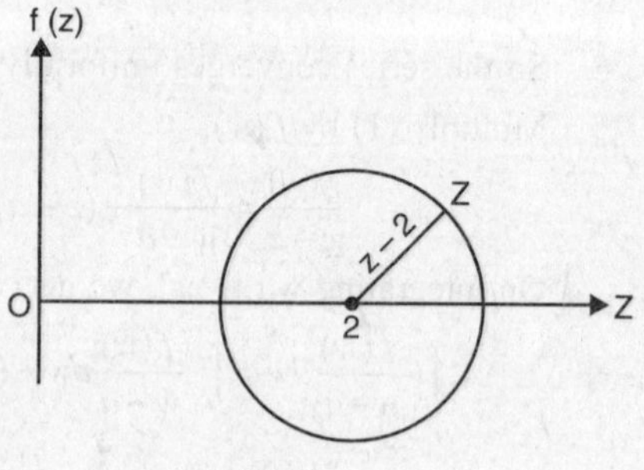

By Taylor's series

$$f(z) = f(2) + (z-2)\, f'(2) + \frac{(z-2)^2}{2!} f''(2) + \ldots\ldots\ldots + \frac{(z-2)^n}{n!} f^n(2) + \ldots\ldots$$

$$= \frac{1}{2} + (z-2)\left(-\frac{1}{2^2}\right) + \frac{(z-2)^2}{2!}\left(\frac{2}{2^3}\right) + \ldots\ldots \frac{(z-2)^n}{n!}(-1)^n \frac{n!}{2^{n+1}} + \ldots\ldots$$

$$= \frac{1}{2} - \frac{1}{4}(z-2) + \frac{1}{8}(z-2)^2 - \ldots\ldots\ldots + \frac{(-1)^n}{2^{n+1}}(z-2)^n + \ldots\ldots$$ **Ans.**

Alternative.

We can expand the given function by Binomial expansion.

$$\frac{1}{z} = \frac{1}{2+z-2} = \frac{1}{2}\left[\frac{1}{1+\frac{z-2}{2}}\right] = \frac{1}{2}\left[\left(1+\frac{z-2}{2}\right)^{-1}\right] \qquad \left|\frac{z-2}{2}\right| < 1$$

$$= \frac{1}{2}\left[1 - \frac{z-2}{2} + \frac{(-1)(-2)}{2!}\left(\frac{z-2}{2}\right)^2 + \frac{(-1)(-2)(-3)}{3!}\left(\frac{z-2}{2}\right)^3 + \ldots\ldots\right]$$

$$= \frac{1}{2} - \frac{z-2}{4} + \frac{1}{8}(z-2)^2 - \frac{1}{16}(z-2)^3 + \ldots\ldots$$ **Ans.**

Radius of convergence $\frac{1}{R} = \lim_{n\to\infty}\left(\frac{a_{n+1}}{a_n}\right) = \lim_{n\to\infty}\left|\frac{(-1)^{n+1}}{2^{n+2}} \cdot \frac{2^{n+1}}{(-1)^n}\right| = \frac{1}{2}$

$$\Rightarrow \quad R = 2$$ **Ans.**

Example 7. *Expand* $f(z) = \cosh z$ *about* πi.

Solution. Here, we have

$$f(z) = \cosh z = \frac{e^z + e^{-z}}{2} \Rightarrow f(\pi i) = \cosh(\pi i)$$

$$f'(z) = \sinh z \quad \Rightarrow \quad f'(\pi i) = \sinh(\pi i)$$

$$f''(z) = \cosh z \quad \Rightarrow \quad f''(\pi i) = \cosh(\pi i)$$

$$f'''(z) = \sinh z \quad \Rightarrow \quad f'''(\pi i) = \sinh(\pi i)$$

By Taylor's series

$$f(z) = f(\pi i) + (z - \pi i) f'(\pi i) + \frac{(z-\pi i)^2}{2!} f''(\pi i) + \frac{(z-\pi i)^3}{3!} f'''(\pi i) + \ldots..$$

$$= \cosh \pi i + (z - \pi i) \sinh \pi i + + \frac{(z-\pi i)^2}{2!} \cosh(\pi i)$$

$$+ \frac{(z-\pi i)^3}{3!} \sinh(\pi i) + \ldots\ldots\ldots.$$ **Ans.**

Example 8. *Expand* $f(z) = \dfrac{a}{bz+c}$ *about* z_0.

Solution. Here, we have

$$f(z) = \frac{a}{bz+c} = \frac{a}{bz - bz_0 + bz_0 + c} = \frac{a}{b(z-z_0) + bz_0 + c}$$

$$= \frac{a}{(bz_0 + c)\left[1 + \left(\dfrac{b(z-z_0)}{bz_0+c}\right)\right]} = \frac{a}{bz_0+c}\left[1 + \frac{b(z-z_0)}{bz_0+c}\right]^{-1} \quad \text{(Binomial series)}$$

$$= \frac{a}{bz_0+c}\left[1 - \frac{b(z-z_0)}{bz_0+c} + \frac{(-1)(-2)}{2!}\left(\frac{b(z-z_0)}{bz_0+c}\right)^2 + \frac{(-1)(-2)(-3)}{3!}\left(\frac{b(z-z_0)}{bz_0+c}\right)^3 + \ldots\ldots.\right]$$

$$= \frac{a}{bz_0+c}\left[1 - \frac{b(z-z_0)}{bz_0+c} + \left(\frac{b(z-z_0)}{bz_0+c}\right)^2 - \left(\frac{b(z-z_0)}{bz_0+c}\right)^3 + \ldots\ldots\right]$$

$$= \frac{a}{bz_0+c}\left[1 - \frac{b}{bz_0+c}(z-z_0) + \left(\frac{b}{bz_0+c}\right)^2 (z-z_0)^2 - \left(\frac{b}{bz_0+c}\right)^3 (z-z_0)^3 + \ldots\right]$$

Radius of curvature

$$\frac{1}{R} = \lim_{n\to\infty}\left(\frac{a_{n+1}}{a_n}\right) = \lim_{n\to\infty} \frac{\left(\dfrac{b}{bz_0+c}\right)^{n+1}}{\left(\dfrac{b}{bz_0+c}\right)^n} = \lim_{n\to\infty} \frac{b}{bz_0+c} = \frac{b}{bz_0+c}$$

$$\Rightarrow \quad R = \frac{bz_0+c}{b} = z_0 + \frac{c}{b}$$ **Ans.**

Example 9. *Show that :*

$$\log z = (z-1) - \frac{(z-1)^2}{2} + \frac{(z-1)^3}{3} + \ldots\ldots$$

Solution. Let $f(z) = \log(z) \quad \Rightarrow \quad f(1) = \log 1 = 0$

$$f'(z)=\frac{1}{z}, \qquad f'(1)=\frac{1}{1}=1$$

$$f''(z)=-\frac{1}{z^2}, \qquad f''(1)=-1$$

$$f'''(z)=\frac{2\times 1}{z^3}, \qquad f'''(1)=2$$

$$f^{iv}(z)=-3\times 2\times 1\times\frac{1}{z^4}, \quad f^{iv}(1)=-3!$$

By Taylor series

$$f(z)=f(a)+f'(a).(z-a)+\frac{f''(a)(z-a)^2}{2!}+\frac{f'''(a)}{3!}(z-a)^3+....$$

$$f(z) = \log z=\log\ (1+\overline{z-1})$$

On substituting the values of $f(1), f'(1), f''(1)$ etc., we get

$$\log z=0+1(z-1)-\frac{1}{2!}(z-1)^2+\frac{2}{3!}(z-1)^3-\frac{3!}{4!}(z-1)^4+.....$$

$$\Rightarrow \qquad \log z=(z-1)-\frac{1}{2}(z-1)^2+\frac{1}{3}(z-1)^3-\frac{1}{4}(z-1)^4+........$$ **Proved.**

Example 10. *Expand* $\frac{1}{z^2-3z+2}$ *in the region*

(*a*) $|z|<1$ (*b*) $|z|>2$. (*R.G.P.V., Bhopal, III Semester, Dec. 2005*)

Solution. Here, $f(z)=\frac{1}{z^2-3z+2}=\frac{1}{(z-1)(z-2)}=\frac{1}{z-2}-\frac{1}{z-1}$

(*a*) If $|z|<1$

Taking common, bigger term out of $|z|$ and 2, here 2 is bigger than $|z|$. So we take 2 common.

$$f(z)=\frac{1}{-2\left(1-\frac{z}{2}\right)}+\frac{1}{1-z}$$

$$=-\frac{1}{2}\left(1-\frac{z}{2}\right)^{-1}+(1-z)^{-1}$$

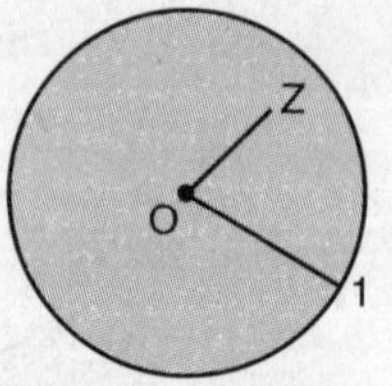

$$=-\frac{1}{2}\left(1+\frac{z}{2}+\frac{z^2}{4}+\frac{z^3}{8}+......\right)+(1+z+z^2+z^3+......)$$

[By Binomial theorem]

$$=\frac{1}{2}+\frac{3z}{4}+\frac{7}{8}z^2+\frac{15}{16}z^3+......$$

Which is the required expansion.

(*b*) If $|z|>2$

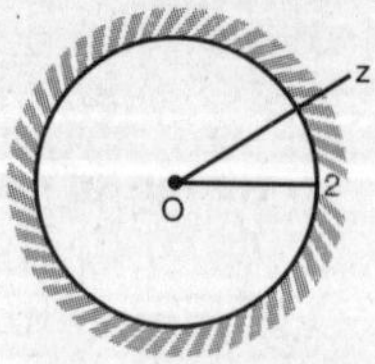

We have, $f(z)=\frac{1}{z-2}-\frac{1}{z-1}$

Taking common, bigger term out of $|z|$ and 2, here z is bigger than 2. So we take $|z|$ common.

$$f(z) = \frac{1}{z\left(1-\frac{2}{z}\right)}-\frac{1}{z\left(1-\frac{1}{z}\right)}=\frac{1}{z}\left(1-\frac{2}{z}\right)^{-1}-\frac{1}{z}\left(1-\frac{1}{z}\right)^{-1}$$

$$= \frac{1}{z}\left(1+\frac{2}{z}+\frac{4}{z^2}+\frac{8}{z^3}+......\right)-\frac{1}{z}\left(1+\frac{1}{z}+\frac{1}{z^2}+\frac{1}{z^3}+......\right)$$ [By Binomial theorem]

$$= z^{-2}+3z^{-3}+7z^{-4}+......$$

Which is the required expansion. **Ans.**

Example 11. *Show that when* $|z+1|<1$,

$$z^{-2} = 1 + \sum_{n=1}^{\infty} (n+1)(z+1)^n.$$

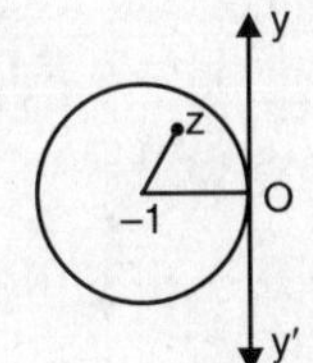

Solution. $f(z) = z^{-2} = \dfrac{1}{z^2} = \dfrac{1}{[(z+1)-1]^2}$

Taking common, bigger term out of 1 and $|z+1|$, here, $1 > |z+1|$.
So, we take 1 common.

$$f(z) = \frac{1}{[1-(z+1)]^2} = [1-(z+1)]^{-2} = 1 + 2(z+1) + 3(z+1)^2 + 4(z+1)^3 + \ldots\ldots$$

$$= 1 + \sum_{n=1}^{\infty} (n+1)(z+1)^n$$ **Proved.**

Example 12. *Find the first four terms of the Taylor's series expansion of the complex variable function*

$$f(z) = \frac{z+1}{(z-3)(z-4)}$$

about z = 2. Find the region of convergence.

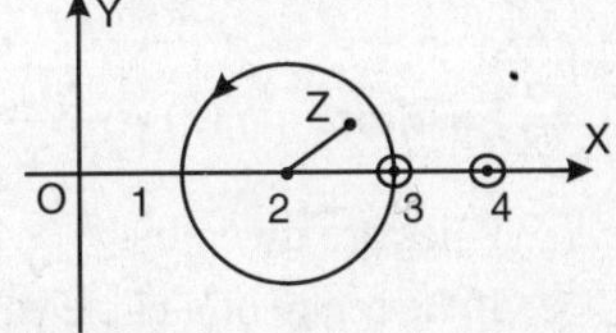

Solution. $f(z) = \dfrac{z+1}{(z-3)(z-4)}$

If centre of a circle is at $z = 2$, then the distances of the singularities $z = 3$ and $z = 4$ from the centre are 1 and 2.

Hence if a circle is drawn with centre $z = 2$ and radius 1, then within the circle $|z-2| = 1$, the given function $f(z)$ is analytic, hence it can be expanded in a Taylor's series within the circle $|z-2| = 1$, which is therefore the circle of convergence.

$$f(z) = \frac{z+1}{(z-3)(z-4)} = \frac{-4}{z-3} + \frac{5}{z-4}$$ (By Partial fraction method)

$$= \frac{-4}{(z-2)-1} + \frac{5}{(z-2)-2} = 4[1-(z-2)]^{-1} - \frac{5}{2}\left[1 - \frac{z-2}{2}\right]^{-1}$$ $[|z-2|<1]$

$$= 4[1+(z-2)+(z-2)^2+(z-2)^3+\ldots] - \frac{5}{2}\left[1 + \frac{z-2}{2} + \frac{(z-2)^2}{4} + \frac{(z-2)^3}{8} + \ldots\right]$$

$$= \left(4 - \frac{5}{2}\right) + \left(4 - \frac{5}{4}\right)(z-2) + \left(4 - \frac{5}{8}\right)(z-2)^2 + \left(4 - \frac{5}{16}\right)(z-2)^3 \ldots$$

$$= \frac{3}{2} + \frac{11}{4}(z-2) + \frac{27}{8}(z-2)^2 + \frac{59}{16}(z-2)^3 + \ldots$$ **Ans.**

Alternative method. In obtaining the Taylor series we evaluate the coefficients by contour integration.

$$f(z) = \frac{z+1}{(z-3)(z-4)}, \qquad f(2) = \frac{2+1}{(2-3)(2-4)} = \frac{3}{2}.$$

To make the differentiation easier let us convert the given fraction into partial fractions

$$f(z) = \frac{-4}{z-3} + \frac{5}{z-4}$$

$$f'(z) = \frac{4}{(z-3)^2} - \frac{5}{(z-4)^2}, \qquad f'(2) = \frac{4}{(2-3)^2} - \frac{5}{(2-4)^2} = \frac{11}{4}$$

$$f''(z)=\frac{-8}{(z-3)^3}+\frac{10}{(z-4)^3},\qquad f''(2)=\frac{-8}{(2-3)^3}+\frac{10}{(2-4)^3}=\frac{27}{4}$$

$$f'''(z)=\frac{24}{(z-3)^4}-\frac{30}{(z-4)^4}.\qquad f'''(2)=\frac{24}{(2-3)^4}-\frac{30}{(2-4)^4}=\frac{177}{8}$$

Taylor series is $f(z)=f(a)+(z-a)f'(a)+\frac{(z-a)^2}{2!}f''(a)+\frac{(z-a)^3}{3!}f'''(a)+...$

$$\frac{z+1}{(z-3)(z-4)}=\frac{3}{2}+(z-2)\frac{11}{4}+\frac{(z-2)^2}{2!}\left(\frac{27}{4}\right)+\frac{(z-2)^3}{3!}\frac{177}{8}+...$$

$$=\frac{3}{2}+(z-2)\frac{11}{4}+(z-2)^2\cdot\frac{27}{8}+(z-2)^3\cdot\frac{59}{16}+...$$ **Ans.**

Example 13. *Find the first three terms of the Taylor series expansion of* $f(z)=\frac{1}{z^2+4}$ *about* $z=-i$. *Find the region of convergence.*

Solution. $f(z)=\frac{1}{z^2+4}$

Poles are given by $z^2+4=0$, $\Rightarrow$ $z^2=-4$, $\Rightarrow$ $z=\pm 2i$

If the centre of a circle is $z=-i$, then the distances of the singularities $z=2i$ and $z=-2i$ from the centre are 3 and 1. Hence if a circle of radius 1 is drawn with centre at $z=-i$, then within the circle $|z+i|=1$, the given function $f(z)$ is analytic. Thus the function can be expanded in Taylor series within the circle $|z+i|=1$, which is therefore the circle of convergence.

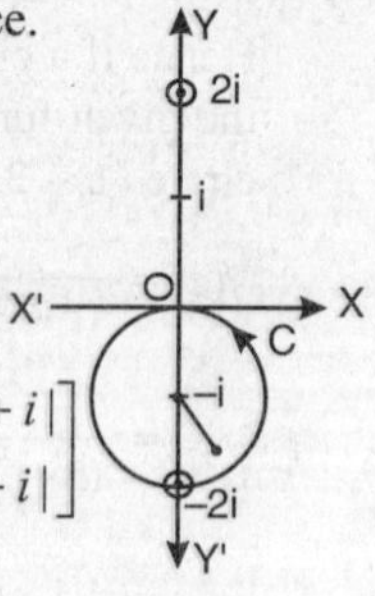

$$f(z)=\frac{1}{z^2+4}=\frac{1}{(z+2i)(z-2i)}=\frac{1}{4i}\left[\frac{1}{z-2i}-\frac{1}{z+2i}\right]$$

$$=\frac{1}{4i}\left[\frac{1}{(z+i)-3i}-\frac{1}{(z+i)+i}\right]$$

$$=\frac{1}{4i}\left[-\frac{1}{3i}\frac{1}{1-\frac{z+i}{3i}}-\frac{1}{i}\frac{1}{1+\frac{z+i}{i}}\right]=-\frac{1}{4}\left[\frac{1}{3}\frac{1}{1-\frac{z+i}{3i}}+\frac{1}{1+\frac{z+i}{i}}\right],\quad\left[\begin{matrix}3>|z+i|\\1>|z+i|\end{matrix}\right]$$

$$=-\frac{1}{4}\left[\frac{1}{3}\frac{1}{1+\frac{i}{3}(z+i)}+\frac{1}{1-i(z+i)}\right]=-\frac{1}{4}\left[\frac{1}{3}\left\{1+\frac{i}{3}(z+i)\right\}^{-1}+\{1-i(z+i)\}^{-1}\right]$$

$$=-\frac{1}{4}\left[\frac{1}{3}\left\{1-\frac{i}{3}(z+i)+\frac{(-1)(-2)}{2!}\left(\frac{i}{3}\right)^2(z+i)^2+\frac{(-1)(-2)(-3)}{3!}\left(\frac{i}{3}\right)^3(z+i)^3+...\right\}\right.$$
$$\left.+\left\{1+i(z+i)+\frac{(-1)(-2)}{2!}\cdot(-i)^2(z+i)^2+\frac{(-1)(-2)(-3)}{3!}(-i)^3(z+i)^3+...\right\}\right]$$

$$=-\frac{1}{4}\left[\frac{1}{3}-\frac{i}{9}(z+i)-\frac{1}{27}(z+i)^2+\frac{i}{81}(z+i)^3+...+1+i(z+i)-(z+i)^2-i(z+i)^3+...\right]$$

$$=-\frac{1}{4}\left[\frac{4}{3}+\frac{8i}{9}(z+i)-\frac{28}{27}(z+i)^2-\frac{80}{81}i(z+i)^3+...\right]$$

$$=-\frac{1}{3}+\frac{2i}{9}(z+i)-\frac{7}{27}(z+i)^2-\frac{20}{81}i(z+i)^3+...$$ **Ans.**

Alternative method

$$f(z)=\frac{1}{z^2+4}$$

By Taylor expansion $f(z)=f(a)+(z-a)f'(a)+\frac{(z-a)^2}{2!}f''(a)+\frac{(z-a)^3}{3!}f'''(a)+...$

Putting $a=-i$ in above, we get

$$f(z)=f(-i)+(z+i)f'(-i)+\frac{(z+i)^2}{2!}f''(-i)+...$$

$$f(z)=\frac{1}{z^2+4} \quad\Rightarrow\quad f(-i)=\frac{1}{(-i)^2+4}=\frac{1}{-1+4}=\frac{1}{3}$$

$$f'(z)=\frac{-2z}{(z^2+4)^2} \quad\Rightarrow\quad f'(-i)=\frac{2i}{(-1+4)^2}=\frac{2i}{9}$$

$$f''(z)=-\frac{(z^2+4)^2\,(2)-2z\cdot 2(z^2+4)2z}{(z^2+4)^4}$$

$$f''(z)=-\frac{(z^2+4)2-8z^2}{(z^2+4)^3} \quad\Rightarrow\quad f''(-i)=-\frac{(-1+4)2-8(-1)}{(-1+4)^3}=-\frac{14}{27}$$

On substituting the value of $f(-i), f'(-i), f''(-i)$, we get

$$f(z)=\frac{1}{3}+(z+i)\left(\frac{2i}{9}\right)+\frac{(z+i)^2}{2!}\left(-\frac{14}{27}\right)+...$$

$$\Rightarrow \qquad f(z)=\frac{1}{3}+\frac{2i}{9}(z+i)-\frac{7}{27}(z+i)^2+...$$

Region of convergence is $|z+i|<1$. **Ans.**

Example 14. *For the function* $f(z)=\frac{4z-1}{z^4-1}$, *find all Taylor series about the centre zero.* (*U.P., III Semester, Dec. 2006*)

Solution. $f(z)=\frac{4z-1}{z^4-1}$,

Poles are determined by $z^4-1=0$

$\Rightarrow \qquad (z-1)(z+1)(z^2+1)=0$

$\Rightarrow \qquad z=1,-1,\pm i$

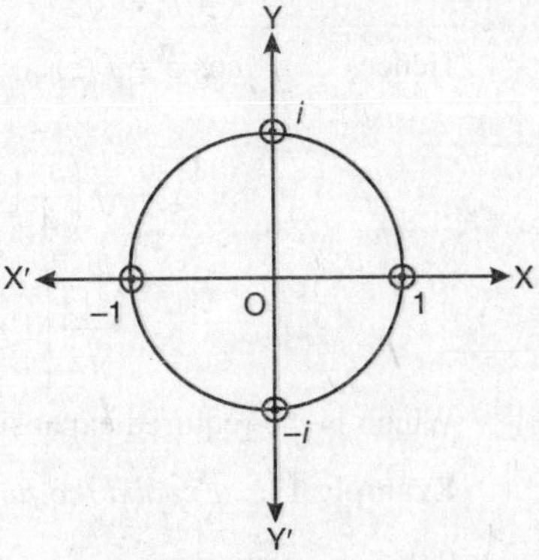

By Partial fractions

$$f(z)=\frac{\frac{3}{4}}{z-1}+\frac{\frac{5}{4}}{z+1}+\frac{-2z+\frac{1}{2}}{z^2+1}$$

$$=-\frac{3}{4}\frac{1}{1-z}+\frac{5}{4}\frac{1}{1+z}+\left(-2z+\frac{1}{2}\right)\frac{1}{1+z^2}$$

$$=-\frac{3}{4}(1-z)^{-1}+\frac{5}{4}(1+z)^{-1}+\left(-2z+\frac{1}{2}\right)(1+z^2)^{-1}$$

$$=-\frac{3}{4}[1+z+z^2+z^3+z^4+...]+\frac{5}{4}[1-z+z^2-z^3+z^4...]+\left(-2z+\frac{1}{2}\right)[1-z^2+z^4+...]$$

$$=-\frac{3}{4}-\frac{3}{4}z-\frac{3}{4}z^2-\frac{3}{4}z^3-\frac{3}{4}z^4+.........+\frac{5}{4}-\frac{5}{4}z+\frac{5}{4}z^2-\frac{5}{4}z^3+\frac{5}{4}z^4+...$$

$$-2z+2z^3-2z^5+....+\frac{1}{2}-\frac{z^2}{2}+\frac{z^4}{2}+...$$

$$=1-4z+z^4+......$$

Which is the required series. **Ans.**

Example 15. *Find Taylor expansion of* $f(z) = \dfrac{2z^3+1}{z^2+z}$ *about the point* $z = 1$.

Solution. $f(z) = \dfrac{2z^3+1}{z(z+1)}$, singularities are given by $z\,.\,(z+1) = 0 \Rightarrow z = 0, z = -1$

If centre of the circle is at $z = 1$, then the distance of the singularities $z = 0$ and $z = -1$ from the centre are 1 and 2. Hence, if a circle is drawn with centre $z = 1$ and radius 1, then within the circle $|z-1| = 1$, the given function $f(z)$ is analytic & therefore, it can be expanded in a Taylor series within the circle $|z-1| = 1$ which is thus the circle of convergence.

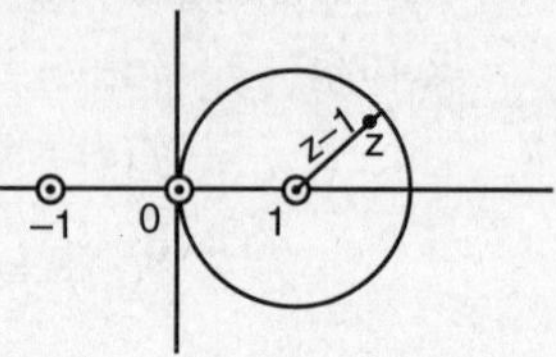

$$\frac{2z^3+1}{z(z+1)} = 2z-2+\frac{1}{z+1}+\frac{1}{z} = 2z-2+\frac{1}{z-1+2}+\frac{1}{z-1+1}, \quad [|z-1|<1]$$

$$= 2z-2+\frac{1}{2}\left(1+\frac{z-1}{2}\right)^{-1}+[1+(z-1)]^{-1}$$

$$= 2z-2+\frac{1}{2}\left[1-\left(\frac{z-1}{2}\right)+\left(\frac{z-1}{2}\right)^2-\left(\frac{z-1}{2}\right)^3+\ldots.\right]+-[1-(z-1)+(z-1)^2-(z-1)^3+\ldots]$$

$$= 2z-2+\frac{3}{2}-\frac{3}{2}\left(\frac{z-1}{2}\right)+\frac{9}{8}(z-1)^2-\frac{17}{16}(z-1)^3+\ldots.$$

Which is the required expansion. **Ans.**

Example 16. *Expand cos z in a Taylor series about* $z = \dfrac{\pi}{4}$.

Solution. Here $f(z) = \cos z$, $f'(z) = -\sin z$, $f''(z) = -\cos z$, $f'''(z) = \sin z$, ...

$$\therefore \quad f\left(\frac{\pi}{4}\right) = \frac{1}{\sqrt{2}}, f'\left(\frac{\pi}{4}\right) = -\frac{1}{\sqrt{2}}, f''\left(\frac{\pi}{4}\right) = -\frac{1}{\sqrt{2}}, f'''\left(\frac{\pi}{4}\right) = \frac{1}{\sqrt{2}}, \ldots\ldots$$

Hence $\cos z = f(z)$

$$= f\left(\frac{\pi}{4}\right)+\left(z-\frac{\pi}{4}\right)f'\left(\frac{\pi}{4}\right)+\frac{\left(z-\frac{\pi}{4}\right)^2}{2!}f''\left(\frac{\pi}{4}\right)+\frac{\left(z-\frac{\pi}{4}\right)^3}{3!}f'''\left(\frac{\pi}{4}\right)+\ldots.$$

$$= \frac{1}{\sqrt{2}}\left[1-\left(z-\frac{\pi}{4}\right)-\frac{1}{2!}\left(z-\frac{\pi}{4}\right)^2+\frac{1}{3!}\left(z-\frac{\pi}{4}\right)^3+\ldots\ldots\right]$$ **Ans.**

Which is the required expansion.

Example 17. *Expand the function* $\dfrac{\sin z}{z-\pi}$ *about* $z = \pi$.

Solution. Putting $z - \pi = t$, we have

$$\frac{\sin z}{z-\pi} = \frac{\sin(\pi+t)}{t} = \frac{-\sin t}{t}$$

$$= -\frac{1}{t}\left(t-\frac{t^3}{3!}+\frac{t^5}{5!}\ldots..\right) = -1+\frac{t^2}{3!}-\frac{t^4}{5!}+\ldots\ldots = -1+\frac{(z-\pi)^2}{3!}-\frac{(z-\pi)^4}{5!}+\ldots\ldots$$

Which is the required expansion. **Ans.**

Example 18. *Expand the function* $\sin^{-1} z$ *in powers of z.* (*U.P. III Semester, Dec. 2006*)

Solution. Let $w = \sin^{-1} z$

$$\frac{dw}{dz} = \frac{1}{\sqrt{1-z^2}} = (1-z^2)^{-\frac{1}{2}} \qquad \text{... (1)}$$

On expanding the R.H.S. of (1) by Binomial Theorem, we get

$$\frac{dw}{dz} = 1 - \frac{1}{2}(-z^2) + \frac{\frac{-1}{2}\left(-\frac{3}{2}\right)}{2!}(-z^2)^2 + \frac{\frac{-1}{2}\left(-\frac{3}{2}\right)\left(\frac{-5}{2}\right)}{3!}(-z^2)^3 + \ldots$$

$$\frac{dw}{dz} = 1 + \frac{z^2}{2} + \frac{3}{8}z^4 + \frac{5}{16}z^6 + \ldots.$$

On integrating, we get

$$w = z + \frac{z^3}{6} + \frac{3z^5}{40} + \frac{5z^7}{112} + \ldots..+ C \qquad \text{... (2)}$$

On putting $z =$ and $w = \sin^{-1} z = 0$ in (2), we get

$$0 = 0 + \mathrm{C} \quad \Rightarrow \quad C = 0$$

On putting the value of C in (2), we get

$$\sin^{-1} z = z + \frac{z^3}{6} + \frac{3z^5}{40} + \frac{5z^7}{112} + \ldots\ldots\ldots$$

Which is the required expansion. **Ans.**

Example 19. *Expand the function $f(z) = tan^{-1} z$ in powers of z.* *(U.P. III Sem. 2009-2010)*

Solution. We have, $f(z) = \tan^{-1} z$

$$\Rightarrow \quad \frac{d\,f(z)}{d\,z} = \frac{1}{1+z^2} = (1+z^2)^{-1} \qquad \text{... (1)}$$

On expanding the R.H.S of (1) by Binomial Theorem, we get

$$\frac{d\,f(z)}{d\,z} = 1 - z^2 + \frac{-1(-1-1)}{2!}(z^2)^2 + \frac{-1(-1-1)(-1-2)}{3!}(z^2)^3 + \ldots..$$

$$\Rightarrow \quad \frac{d\,f(z)}{d\,z} = 1 - z^2 + z^4 - z^6 + \ldots\ldots$$

On integration, we get

$$f(z) = z - \frac{z^3}{3} + \frac{z^5}{5} - \frac{z^7}{7} + \ldots\ldots\ldots\ldots + C \qquad \text{... (2)}$$

On putting $z = 0$ and $f(z) = \tan^{-1} z = 0$ in (2), we get

$$0 = 0 + C \quad \Rightarrow \quad C = 0$$

On putting the value of C in (2), we get

$$\tan^{-1} z = z - \frac{z^3}{3} + \frac{z^5}{5} - \frac{z^7}{7} + \ldots\ldots\ldots \qquad \textbf{Ans.}$$

EXERCISE 25.2

Expand the following functions in Taylor's series

1. $\frac{1}{z+1}$, about $z = 1$. **Ans.** $-\frac{1}{2}\left(1 - \frac{z-1}{2} + \frac{(z-1)^2}{4} + \ldots\right)$

2. $\frac{z-1}{z+1}$ (*a*) about $z = 0$ (*b*) about $z = 1$.

Ans. (*a*) $-1 + 2(z - z^2 + z^3 - z^4 + \ldots)$ (*b*) $\frac{1}{2}(z-1) - \frac{1}{2^2}(z-1)^2 + \frac{1}{2^3}(z-1)^3 - \ldots.$

3. $\dfrac{1}{4-3z}$ about $(1+i)$ **Ans.** $\dfrac{1}{1-3i}\sum_{n=0}^{\infty}(-1)^n\left(\dfrac{-3}{1-3i}\right)^n(z-(1+i))^n$

Region of convergence $|z-(1+i)| < \sqrt{\dfrac{10}{3}}$

4. $\dfrac{1}{z^2-z-6}$ about (*a*) $z = -1$, (*b*) $z = 1$ **Ans.** (*a*) $\dfrac{1}{20}\sum_{n=0}^{\infty}\dfrac{(-4)^{n+1}-1}{4^n}(z+1)^n$

Region of convergence $|z+1| \leq 1$.

Ans. (*b*) $-\dfrac{1}{10}\sum_{n=0}^{\infty}\left(\dfrac{z-1}{2}\right)^n - \dfrac{1}{15}\sum_{n=0}^{\infty}(-1)^n\left(\dfrac{z-1}{3}\right)^n$

Region of convergence $\left|\dfrac{z-1}{2}\right| < 1$ and $\left|\dfrac{z-1}{3}\right| < 1$ common region $|z-1| < 2$.

5. $\dfrac{2z^2+9z+5}{z^3+z^2-8z-12}$ about $z = 1$ **Ans.** $\sum_{n=0}^{\infty}\left[(-1)^n\dfrac{n+1}{3^{n+2}}-\dfrac{1}{2^n}\right](z-1)^n$

Region of convergence $|z-1| < 2$.

6. $\log\left(\dfrac{1+z}{1-z}\right)$ about $z = 0$ **Ans.** $\sum_{n=0}^{\infty}\dfrac{2z^{2n+1}}{2n+1}$, $|z| < 1$

7. $\dfrac{1}{z^2+(1+2i)z+2i}$ about $z = 0$. **Ans.** $\dfrac{1}{(1-2i)}\left[\sum_{r=0}^{\infty}\left\{\left(\dfrac{1}{2i}\right)^{n+1}-1\right\}(-1)^n z^n\right]$

8. $(z+i)^{\frac{1}{2}}$, about $z = 0$ **Ans.** $\dfrac{1+i}{\sqrt{2}}\left[\sum_{r=0}^{\infty}{}^{\frac{1}{2}}C_r\left(\dfrac{z}{i}\right)^r\right]$, $R = 1$

9. $\dfrac{1}{(z^2-1)(z^2-2)}$ about $z = 0$. **Ans.** $\sum_{n=0}^{\infty}\left[1-\dfrac{1}{2^{n+1}}\right]z^{2n}$, $R = 1$.

10. $\tan z$ about $z = 0$ **Ans.** $z-\dfrac{z^3}{3}+\dfrac{2z^5}{15}+\ldots\ldots$

11. $z\cot z$ about $z = 0$ **Ans.** $1-\dfrac{z^2}{3}-\dfrac{z^4}{45}+\ldots\ldots$ **12.** $\dfrac{e^z}{1+e^z}$ about $z = 0$ **Ans.** $\dfrac{1}{2}+\dfrac{z}{4}-\dfrac{z^3}{48}$

25.8 LAURENT'S THEOREM

(U.P., III Semester, Dec. 2009)

If we are required to expand $f(z)$ about a point where $f(z)$ is not analytic, then it is expanded by Laurent's Series and not by Taylor's Series.

Statement. If $f(z)$ is analytic on c_1 and c_2, and the annular region R bounded by the two concentric circles c_1 and c_2 of radii r_1 and r_2 $(r_2 < r_1)$ and with centre at a, then for all z in R

$$f(z) = a_0 + a_1(z-a) + a_2(z-a)^2 + \ldots + \frac{b_1}{z-a} + \frac{b_2}{(z-a)^2} + \ldots$$

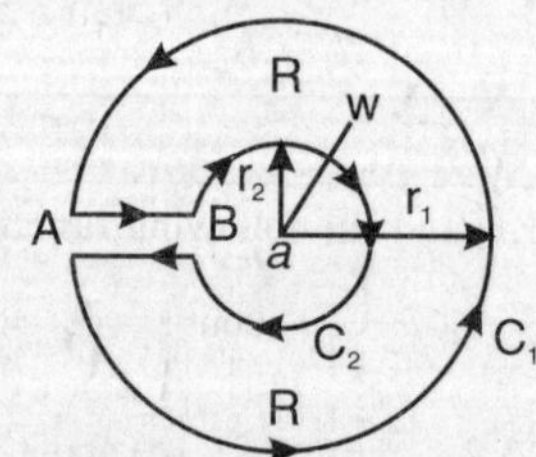

where $$a_n = \frac{1}{2\pi i}\int_{c_1}\frac{f(w)}{(w-a)^{n+1}}dw,$$

$$b_n = \frac{1}{2\pi i}\int_{c_2}\frac{f(w)}{(w-a)^{-n+1}}dw$$

Proof. By introducing a cross cut AB, multi-connected region R is converted to a simply connected region. Now $f(z)$ is analytic in this region.

Now by Cauchy's integral formula

$$f(z)=\frac{1}{2\pi i}\int_{c_1}\frac{f(w)\,dw}{w-z}+\frac{1}{2\pi i}\int_{AB}\frac{f(w)}{w-z}dw-\frac{1}{2\pi i}\int_{c_2}\frac{f(w)\,dw}{w-z}+\frac{1}{2\pi i}\int_{BA}\frac{f(w)}{w-z}dw$$

Integral along c_2 is clockwise, so it is negative. Integrals along AB and BA cancel.

$$\therefore \qquad f(z)=\frac{1}{2\pi i}\int_{c_1}\frac{f(w)\,dw}{w-z}-\frac{1}{2\pi i}\int_{c_2}\frac{f(w)\,dw}{w-z} \qquad \text{... (1)}$$

For the first integral, $\frac{f(w)}{w-z}$ can be expanded exactly as in Taylor's series as z lies on c_1.

$$\left(\frac{|z-a|}{|w-a|}\le 1 \qquad \because w \text{ lies on } c_1\right)$$

$$\frac{1}{2\pi i}\int_{c_1}\frac{f(w)\,dw}{w-z}=\frac{1}{2\pi i}\int_{c_1}\frac{f(w)}{w-a}dw+\frac{(z-a)}{2\pi i}\int_{c_1}\frac{f(w)}{(w-a)^2}dw+\frac{(z-a)^2}{2\pi i}\int_{c_1}\frac{f(w)}{(w-a)^3}dw+\ldots$$

$$= a_0 + a_1(z-a) + a_2(z-a)^2 + \quad \text{... (2)} \qquad \left[a_n=\frac{1}{2\pi i}\int_{c_1}\frac{f(w)}{(w-a)^{n+1}}dw\right]$$

In the second integral, z lies on c_2. Therefore

$$|w-a| < |z-a| \quad \text{or} \quad \frac{|w-a|}{|z-a|}<1$$

so here
$$\frac{1}{w-z}=\frac{1}{w-a+a-z}=\frac{1}{(w-a)-(z-a)}$$

$$=-\frac{1}{(z-a)}\frac{1}{\left(1-\frac{w-a}{z-a}\right)}=-\frac{1}{z-a}\left(1-\frac{w-a}{z-a}\right)^{-1}$$

$$=-\frac{1}{z-a}\left[1+\frac{w-a}{z-a}+\left(\frac{w-a}{z-a}\right)^2+\ldots+\left(\frac{w-a}{z-a}\right)^{n+1}+\ldots\right]$$

Multiplying by $-\frac{f(w)}{2\pi i}$, we get

$$-\frac{1}{2\pi i}\frac{f(w)}{w-z}=\frac{1}{2\pi i}\frac{f(w)}{z-a}+\frac{1}{2\pi i}\frac{(w-a)}{(z-a)^2}f(w)+\frac{1}{2\pi i}\frac{(w-a)^2}{(z-a)^3}f(w)+\ldots$$

$$=\left(\frac{1}{z-a}\right)\frac{1}{2\pi i}f(w)+\frac{1}{(z-a)^2}\frac{1}{2\pi i}\frac{f(w)}{(w-a)^{-1}}+\frac{1}{(z-a)^3}\frac{1}{2\pi i}\frac{f(w)}{(w-a)^{-2}}+\ldots$$

Integrating, we have

$$-\frac{1}{2\pi i}\int_{c_2}\frac{f(w)}{w-z}dw=\left(\frac{1}{z-a}\right)\frac{1}{2\pi i}\int_{c_2}f(w)\,dw+\frac{1}{(z-a)^2}\frac{1}{2\pi i}\int_{c_2}\frac{f(w)\,dw}{(w-a)^{-1}}$$

$$+\frac{1}{(z-a)^3}\frac{1}{2\pi i}\int_{c_2}\frac{f(w)\,dw}{(w-a)^{-2}}+\ldots,$$

$$=\frac{b_1}{z-a}+\frac{b_2}{(z-a)^2}+\frac{b_3}{(z-a)^3}+ \quad \text{... (3)} \qquad \left[b_n=\frac{1}{2\pi i}\int_{c_2}\frac{f(w)\,dw}{(w-a)^{-n+1}}\right]$$

Substituting the values of both integrals from (2) and (3) in (1), we get

$$f(z)=a_0+a_1(z-a)+a_2(z-a)^2+b_1(z-a)^{-1}+b_2(z-a)^{-2}+\ldots$$

$$\Rightarrow \qquad f(z)=\sum_{n=0}^{n=\infty}a_n(z-a)^n+\sum_{n=1}^{n=\infty}\frac{b_n}{(z-a)^n}$$ **Proved.**

Note. To expand a function by Laurent Theorem is cumbersome. By Binomial theorem, the expansion of a function can be done easily.

Example 20. *Expand* $f(z) = \dfrac{1}{(z-1)(z-2)}$ *for* $1 < |z| < 2$ [By partial Fraction]

Solution. $f(z) = \dfrac{1}{(z-1)(z-2)} = \dfrac{1}{z-2} - \dfrac{1}{z-1}$

In first bracket $|z| < 2$, we take out 2 as common and from second bracket z is taken out common as $1 < |z|$.

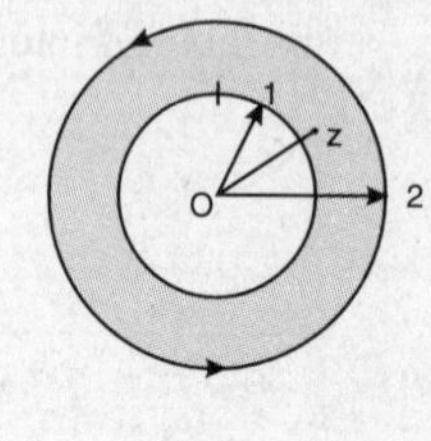

$$f(z) = -\frac{1}{2}\left(\frac{1}{1-\frac{z}{2}}\right) - \frac{1}{z}\left(\frac{1}{1-\frac{1}{z}}\right) = -\frac{1}{2}\left(1-\frac{z}{2}\right)^{-1} - \frac{1}{z}\left(1-\frac{1}{z}\right)^{-1}$$

$$= -\frac{1}{2}\left[1+\frac{z}{2}+\frac{z^2}{4}+\frac{z^3}{8}+...\right] - \frac{1}{z}\left[1+\frac{1}{z}+\frac{1}{z^2}+\frac{1}{z^3}+...\right]$$

$$= -\frac{1}{2}-\frac{z}{4}-\frac{z^2}{8}-\frac{z^3}{16}...-\frac{1}{z}-\frac{1}{z^2}-\frac{1}{z^3}-\frac{1}{z^4}...$$

Which is the required expansion. **Ans.**

Example 21. *Find the Laurent Series expansion of*

$$f(z) = \frac{z}{(z-1)(z-2)} \text{ valid for } |z-1| > 1. \qquad (D.U.\ April\ 2010)$$

Solution.

$$f(z) = \frac{z}{(z-1)(z-2)} = \frac{-1}{z-1} + \frac{2}{z-2} = \frac{-1}{z-1} + \frac{2}{z-1-1}$$

$$= -\frac{1}{z-1} + \frac{2}{z-1}\,\frac{1}{1-\frac{1}{z-1}} = -\frac{1}{z-1} + \frac{2}{z-1}\left(1-\frac{1}{z-1}\right)^{-1}$$

$$= -\frac{1}{z-1} + \frac{2}{z-1}\left(1+\frac{1}{z-1}+\frac{1}{(z-1)^2}+\frac{1}{(z-1)^3}+...\right)$$

$$= -\frac{1}{z-1} + \frac{2}{z-1} + \frac{2}{(z-1)^2} + \frac{2}{(z-1)^3} + \frac{2}{(z-1)^4} + ...$$

$$= \frac{1}{z-1} + \frac{2}{(z-1)^2} + \frac{2}{(z-1)^3} + \frac{2}{(z-1)^4} + ...$$ **Ans.**

Example 22. *Obtain the Taylor or Laurent series which represents the function*

$$f(z) = \frac{1}{(1+z^2)(z+2)} \text{ when}$$

(*i*) $1 < |z| < 2$; (*ii*) $|z| > 2$

Solution. $f(z) = \dfrac{1}{(1+z^2)(z+2)} = \dfrac{-\frac{z}{5}+\frac{2}{5}}{1+z^2} + \dfrac{\frac{1}{5}}{z+2} = -\dfrac{1}{5}\dfrac{z-2}{1+z^2} + \dfrac{1}{5}\dfrac{1}{z+2}$, $\quad 1 < |z| < 2$

(*i*) In first expression $1 < |z|$ and in second expression $|z| < 2$

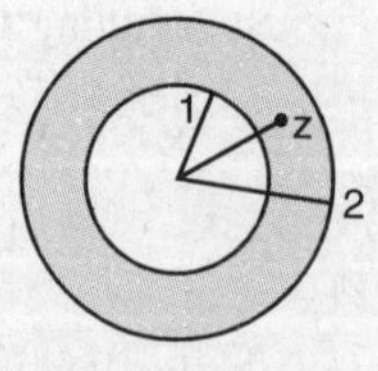

$$f(z) = -\frac{1}{5}\frac{1}{z^2}\frac{z-2}{1+\frac{1}{z^2}} + \frac{1}{5}\frac{1}{2}\frac{1}{1+\frac{z}{2}}$$

$$= -\frac{1}{5z^2}(z-2)\left(1+\frac{1}{z^2}\right)^{-1} + \frac{1}{10}\left(1+\frac{z}{2}\right)^{-1}$$

$$= -\frac{1}{5}\left(\frac{1}{z}-\frac{2}{z^2}\right)\left(1-\frac{1}{z^2}+\frac{1}{z^4}-\frac{1}{z^6}+\ldots\right)+\frac{1}{10}\left[1-\frac{z}{2}+\frac{z^2}{4}-\frac{z^3}{8}+\ldots\right]$$

$$=\frac{1}{5}\left[-\frac{1}{z}+\frac{1}{z^3}-\frac{1}{z^5}+\frac{1}{z^7}+\ldots+\frac{2}{z^2}-\frac{2}{z^4}+\frac{2}{z^6}-\frac{2}{z^8}+\ldots+\frac{1}{2}-\frac{z}{4}+\frac{z^2}{8}-\frac{z^3}{16}\ldots\right]$$

$$=\frac{1}{5}\left[\ldots-2z^{-8}+z^{-7}+2z^{-6}-z^{-5}-2z^{-4}+z^{-3}+2z^{-2}-z^{-1}+\frac{1}{2}-\frac{z}{4}+\frac{z^2}{8}-\frac{z^3}{16}\ldots\right]$$

Which is the required series. **Ans.**

(*ii*) Here $|z| > 2$

$$f(z)=-\frac{1}{5}\frac{z-2}{1+z^2}+\frac{1}{5}\frac{1}{z+2}=-\frac{1}{5}\frac{z-2}{z^2}\frac{1}{1+\frac{1}{z^2}}+\frac{1}{5}\frac{1}{z}\frac{1}{1+\frac{2}{z}}$$

$$=-\frac{1}{5z^2}(z-2)\left[1+\frac{1}{z^2}\right]^{-1}+\frac{1}{5z}\left[1+\frac{2}{z}\right]^{-1}$$

$$=\frac{1}{5}\left[-\frac{1}{z}+\frac{2}{z^2}\right]\left[1-\frac{1}{z^2}+\frac{1}{z^4}-\frac{1}{z^6}+\ldots\right]+\frac{1}{5z}\left[1-\frac{2}{z}+\frac{4}{z^2}-\frac{8}{z^3}+\frac{16}{z^4}+\ldots\right]$$

$$=\frac{1}{5}\left[-\frac{1}{z}+\frac{1}{z^3}-\frac{1}{z^5}+\frac{1}{z^7}+\ldots+\frac{2}{z^2}-\frac{2}{z^4}+\frac{2}{z^6}-\frac{2}{z^8}+\ldots+\frac{1}{z}-\frac{2}{z^2}+\frac{4}{z^3}-\frac{8}{z^4}+\ldots\right]$$

Which is the required series. **Ans.**

Example 23. *Expand* $f(z)=\dfrac{1}{(z+1)(z+3)}$ *in Laurent series valid for*

(*i*) $1 < |z| < 3$ (*ii*) $|z| > 3$

(*iii*) $0 < |z+1| < 2$ (*iv*) $|z| < 1$

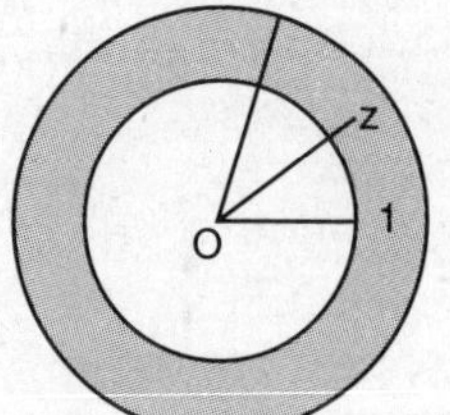

Solution. $f(z)=\dfrac{1}{(z+1)(z+3)}=\dfrac{1}{2}\left(\dfrac{1}{z+1}-\dfrac{1}{z+3}\right)$

(*i*) $1 < |z| < 3. \quad \Rightarrow \quad \frac{1}{|z|}<1$ and $\frac{|z|}{3}<1.$

$\Rightarrow \quad 1 < |z|$ and $|z| < 3.$

$$f(z)=\frac{1}{2}\left[\frac{1}{z\left(1+\frac{1}{z}\right)}-\frac{1}{3\left(1+\frac{z}{3}\right)}\right]=\frac{1}{2}\left[\frac{1}{z}\left(1+\frac{1}{z}\right)^{-1}-\frac{1}{3}\left(1+\frac{z}{3}\right)^{-1}\right]$$

$$=\frac{1}{2}\left[\frac{1}{z}\left(1-\frac{1}{z}+\frac{1}{z^2}-\frac{1}{z^3}\ldots\right)-\frac{1}{3}\left(1-\frac{z}{3}+\frac{z^2}{3^2}-\frac{z^3}{3^3}\ldots\right)\right]$$

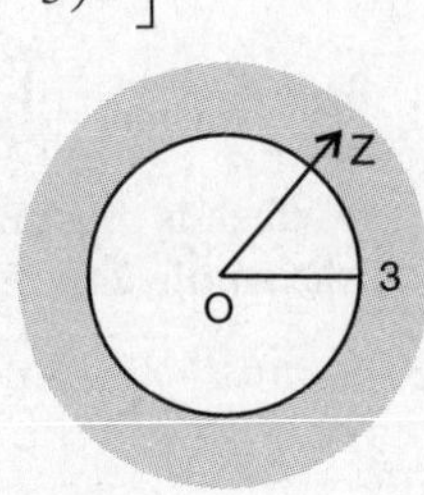

$$=\left(\frac{1}{2z}-\frac{1}{2z^2}+\frac{1}{2z^3}-\frac{1}{2z^4}\ldots\right)-\frac{1}{6}+\frac{z}{18}-\frac{z^2}{54}+\frac{z^3}{162}\ldots$$

$$=-\frac{1}{2z^4}+\frac{1}{2z^3}-\frac{1}{2z^2}+\frac{1}{2z}-\frac{1}{6}+\frac{z}{18}-\frac{z^2}{54}+\frac{z^3}{162}\ldots$$

Which is the required series. **Ans.**

(*ii*) $|z|>3 \quad \Rightarrow \quad \frac{3}{|z|}<1$

$$f(z)=\frac{1}{2}\left[\frac{1}{z}\left(1+\frac{1}{z}\right)^{-1}-\frac{1}{z}\left(1+\frac{3}{z}\right)^{-1}\right]$$

$$=\frac{1}{2z}\left(1-\frac{1}{z}+\frac{1}{z^2}-\frac{1}{z^3}\ldots\right)-\frac{1}{2z}\left(1-\frac{3}{z}+\frac{3^2}{z^2}\ldots\right)$$

$$= \left(\frac{1}{2z}-\frac{1}{2z}\right)+\left(\frac{-1}{2z^2}+\frac{3}{2z^2}\right)+\left(\frac{1}{2z^3}-\frac{9}{2z^3}\right)+\left(-\frac{1}{2z^4}+\frac{27}{2z^4}\right)$$

$$= \frac{1}{z^2}-\frac{4}{z^3}+\frac{13}{z^4}-\frac{50}{z^5}+....$$

Which is required expansion. **Ans.**

(*iii*) $0 < |z+1| < 2$.

$|z+1| > 0$ and $|z+1| < 2$.

$\frac{|z+1|}{2} < 1$.

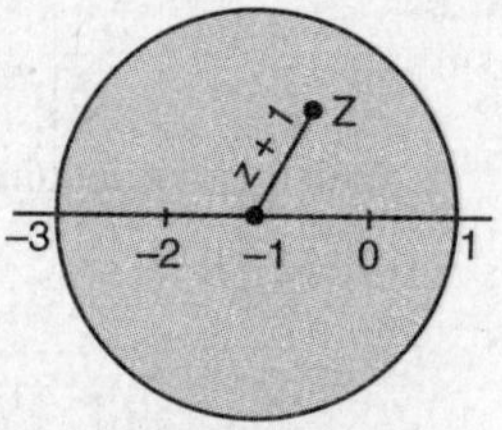

$$f(z) = \frac{1}{2}\left[\frac{1}{1+z}-\frac{1}{z+1+2}\right] = \frac{1}{2}\left[\frac{1}{1+z}-\left(2+\overline{z+1}\right)^{-1}\right]$$

$$= \frac{1}{2(1+z)}-\frac{1}{2.2}\left(1+\frac{z+1}{2}\right)^{-1}$$

$$= \frac{1}{2(1+z)}-\frac{1}{4}\left[1-\frac{(z+1)}{2}+\frac{(z+1)^2}{4}-...\right]$$

$$= \frac{1}{2(1+z)}-\frac{1}{4}+\frac{(z+1)}{8}-\frac{(z+1)^2}{16}+....$$

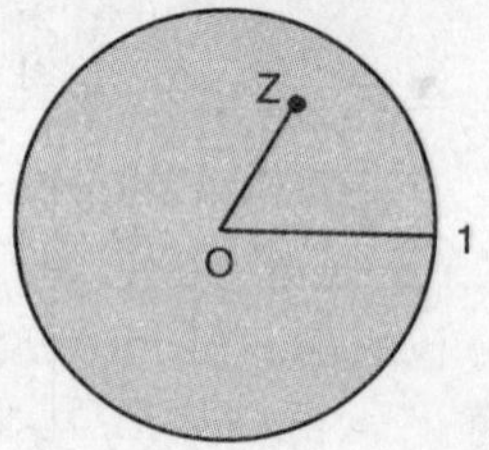

Which is required expansion. **Ans.**

(*iv*) $|z| < 1$

$$f(z) = \frac{1}{2}\left[\frac{1}{(1+z)}-\frac{1}{3+z}\right] = \frac{1}{2}(1+z)^{-1}-\frac{1}{2}(3+z)^{-1}$$

$$= \frac{1}{2}(1+z)^{-1}-\frac{1}{6}\left(1+\frac{z}{3}\right)^{-1} \qquad \left[|z|<1 \text{ and } \frac{|z|}{3}<1.\right]$$

$$= \frac{1}{2}(1-z+z^2-z^3+....)-\frac{1}{6}\left(1-\frac{z}{3}+\frac{z^2}{9}-\frac{z^3}{27}+...\right)$$

$$= \left(\frac{1}{2}-\frac{1}{6}\right)+\left(\frac{-z}{2}+\frac{z}{18}\right)+\left(\frac{z^2}{2}-\frac{z^2}{54}\right)+\left(-\frac{z^3}{2}+\frac{z^3}{162}\right)+...$$

$$= \frac{1}{3}-\frac{4}{9}z+\frac{13}{27}z^2-\frac{40}{81}z^3+...$$

Which is required expansion. **Ans.**

Example 24. *Expand the following function in a Laurent series about the point z = 0.*

$$f(z) = \frac{1-\cos z}{z^3}$$

Solution. $$f(z) = \frac{1-\cos z}{z^3} = \frac{1}{z^3}\left[1-\left\{1-\frac{z^2}{2!}+\frac{z^4}{4!}-......\right\}\right]$$

$$= \frac{1}{z^3}\left(\frac{z^2}{2!}-\frac{z^4}{4!}+\frac{z^6}{6!}-......\right) = \frac{1}{2!z}-\frac{1}{4!}z+\frac{1}{6!}z^3-......$$

$$= \sum_{n=2}^{\infty}\frac{(-1)^n z^{2n-5}}{(2n-2)!}$$

which is a Laurent's series. **Ans.**

Example 25. *Find the terms in the Laurent expansion of* $\dfrac{1}{z(e^z-1)}$ *for the region* $0<|z|<2\pi$.

(AMIETE, June 2010, 2009)

Solution. $f(z) = \dfrac{1}{z(e^z-1)} = \dfrac{1}{z\left[\left(1+z+\dfrac{z^2}{2!}+\dfrac{z^3}{3!}+\ldots\ldots\right)-1\right]}$

$$= z^{-1}\left(z+\frac{z^2}{2!}+\frac{z^3}{3!}+\ldots\ldots\right)^{-1} = z^{-2}\left(1+\frac{z}{2!}+\frac{z^2}{3!}+\ldots\ldots\right)^{-1}$$

$$= z^{-2}\left[1-\left(\frac{z}{2}+\frac{z^2}{6}+\frac{z^3}{24}+\ldots\right)+\left(\frac{z}{2}+\frac{z^2}{6}+\frac{z^3}{24}+\ldots\right)^2-\left(\frac{z}{2}+\frac{z^2}{6}+\frac{z^3}{24}+\ldots\right)^3+\ldots\right]$$

$$= z^{-2}\left[1-\left(\frac{z}{2}+\frac{z^2}{6}+\frac{z^3}{24}+\frac{z^4}{120}+\ldots\ldots\right)+\frac{1}{4}z^2\left(1+\frac{z}{3}+\frac{z^2}{12}+\ldots\ldots\right)^2\right.$$
$$\left.-\frac{1}{8}z^3\left(1+\frac{z}{3}+\ldots\ldots\right)^3+\frac{z^4}{16}(1+\ldots\ldots)^4+\ldots\ldots\right]$$

$$= z^{-2}\left[1-\frac{z}{2}+z^2\left(\frac{1}{4}-\frac{1}{6}\right)-z^3\left(\frac{1}{8}-\frac{1}{6}+\frac{1}{24}\right)+z^4\left(\frac{1}{16}-\frac{1}{8}+\frac{1}{24}+\frac{1}{36}-\frac{1}{120}\right)+\ldots\ldots\right]$$

$$= z^{-2}\left[1-\frac{z}{2}+\frac{z^2}{12}-\frac{z^4}{720}+\ldots\ldots\right] = z^{-2}-\frac{1}{2}z^{-1}+\frac{1}{12}-\frac{z^2}{720}+\ldots\ldots \qquad \textbf{Ans.}$$

25.9 IF *f* (*z*) HAS A POLE AT *z* = *a* THEN | *f* (*z*) | → ∞ AS *z* → *a*

Solution. Suppose the pole is of order *m*, then $f(z)=\sum\limits_{n=1}^{\infty} a_n(z-a)^n+\sum\limits_{n=0}^{m} b_n(z-a)^{-n}$

Its principal part is $\sum\limits_{n=0}^{m} b_n(z-a)^{-n}$

$$\sum_{n=1}^{m} b_n(z-a)^{-n} = \frac{b_1}{z-a}+\frac{b_2}{(z-a)^2}+\frac{b_3}{(z-a)^3}+\ldots+\frac{b_m}{(z-a)^m}$$
$$= \frac{1}{(z-a)^m}[b_m+b_{m-1}(z-a)+\ldots b_1(z-a)^{m-1}]$$
$$= \frac{1}{(z-a)^m}\left[b_m+\sum_{n=1}^{m-1} b_n(z-a)^{m-n}\right]$$

$$\left|\sum_{n=1}^{m} b_n(z-a)^{-n}\right| = \left|\frac{1}{(z-a)^m}\left[b_m+\sum_{n=1}^{m-1} b_n(z-a)^{m-n}\right]\right|$$
$$\geq \left|\frac{1}{(z-a)^m}\right|\left\{|b_m|-\sum_{n=1}^{m-1}|b_n||z-a|^{m-n}\right\}$$

This tends to b_m $\quad |a_1+a_2|\geq |a_1|-|a_2|$.

As $\quad z\to a \quad$ R.H.S $=\infty$.

Example 26. *Write all possible Laurent series for the function*

$$f(z)=\frac{1}{z(z+2)^3}$$

about the pole $z=-2$. *Using appropriate Laurent series*

Solution. To expand $\dfrac{1}{z(z+2)^3}$ about $z = -2$, *i.e.*, in powers of $(z+2)$, we put $z + 2 = t$.

Then $$f(z) = \frac{1}{z(z+2)^3} = \frac{1}{(t-2)t^3} = \frac{1}{t^3}\cdot\frac{1}{t-2}$$

$$= \frac{1}{t^3}\cdot\frac{1}{-2}\cdot\left(\frac{1}{1-\frac{t}{2}}\right) = -\frac{1}{2t^3}\left(1-\frac{t}{2}\right)^{-1}$$

$$[0 < | z - 2 | < 1 \text{ or } 0 < | t | < 1]$$

X' −2 −1 O X Y

$$= -\frac{1}{2t^3}\left[1+\frac{t}{2}+\frac{t^2}{4}+\frac{t^3}{8}+\frac{t^4}{16}+\frac{t^5}{32}+...\right]$$

$$= -\frac{1}{2t^3}-\frac{1}{4t^2}-\frac{1}{8t}-\frac{1}{16}-\frac{t}{32}-\frac{t^2}{64}+....$$

$$f(z) = -\frac{1}{2(z+2)^3}-\frac{1}{4(z+2)^2}-\frac{1}{8(z+2)}-\frac{1}{16}-\frac{z+2}{32}-\frac{(z+2)^2}{64}+...$$ **Ans.**

Example 27. *Expand* $\dfrac{e^z}{(z-1)^2}$ *about* $z = 1$

Solution. Let. $f(z) = \dfrac{e^z}{(z-1)^2} = \dfrac{e^{1+t}}{t^2}$ [Put $z - 1 = t \Rightarrow z = 1 + t$]

$$= \frac{e\cdot e^t}{t^2} = \frac{e}{t^2}\left[1+t+\frac{t^2}{2!}+\frac{t^3}{3!}+...\right] = e\left[\frac{1}{t^2}+\frac{1}{t}+\frac{1}{2!}+\frac{t}{3!}...\right]$$

$$= e\left[\frac{1}{(z-1)^2}+\frac{1}{z-1}+\frac{1}{2!}+\frac{z-1}{3!}+...\right]$$

Which is required expansion. **Ans.**

Example 28. *Expand* $f(z) = \sin\left\{c\left(z+\dfrac{1}{z}\right)\right\}$

Solution. $f(z)$ is not analytic at $z = 0$.
Therefore $f(z)$ can be expanded by Laurent theorem.

$$\therefore \quad f(z) = \sum_{n=0}^{\infty} a_n z^n + \sum_{n=1}^{\infty}\frac{b_n}{z^n}$$

where $$a_n = \frac{1}{2\pi i}\int_c \frac{f(z)dz}{(z-0)^{n+1}} \text{ and } b_n = \frac{1}{2\pi i}\int_c f(z)z^{n-1}dz$$

Now, $$a_n = \frac{1}{2\pi i}\int_c \frac{\sin c\left(z+\frac{1}{z}\right)dz}{z^{n+1}}$$ $\left[\begin{array}{l} z = e^{i\theta} \\ dz = ie^{i\theta}d\theta \end{array}\right]$

$$= \frac{1}{2\pi i}\int_0^{2\pi}\frac{\sin[c(2\cos\theta)]ie^{i\theta}d\theta}{e^{(n+1)i\theta}} = \frac{1}{2\pi}\int_0^{2\pi}\sin(2c\cos\theta)e^{-ni\theta}d\theta$$

$$= \frac{1}{2\pi}\int_0^{2\pi}\sin(2c\cos\theta)\cos n\theta\, d\theta - \frac{i}{2\pi}\int_0^{2\pi}\sin(2c\cos\theta)\sin n\theta d\theta$$

$$= \frac{1}{2\pi}\int_0^{2\pi} \sin(2c\cos\theta)\cos n\theta\, d\theta + 0 \qquad \left[\text{If } f(2a-x) = -f(x) \text{ then } \int_0^{2a} f(x)dx = 0\right]$$

Similarly, $$b_n = \frac{1}{2\pi}\int_0^{2\pi} \sin(2\,c\cos\theta)\cos n\theta d\theta$$

Since the function remains unaltered by putting z for $\frac{1}{z}$.

$$f(z) = a_0 + \sum_{n=1}^{\infty} a_n z^n + \sum_{n=1}^{\infty} \frac{b_n}{z^n} \qquad [a_n = b_n]$$

$$= a_0 + \Sigma a_n\left(z^n + \frac{1}{z^n}\right) \qquad \textbf{Ans.}$$

Example 29. *Expand* $f(z) = e^{\frac{c}{2}\left(z - \frac{1}{z}\right)} = \sum_{n=0}^{\infty} a_n z^n$

Solution. $f(z)$ is also analytic function at $z = 0$ so $f(z)$ can be expanded by Taylor's theorem.

$$f(z) = \sum_{n=0}^{\infty} a_n z^n$$

where $$a_n = \frac{1}{2\pi i}\int_c \frac{f(z)dz}{z^{n+1}} = \frac{1}{2\pi i}\int_c \frac{e^{\frac{c}{2}\left(z-\frac{1}{z}\right)}dz}{z^{n+1}} \qquad [z = e^{i\theta},, \ dz = i\, e^{i\theta}\, d\theta]$$

$$= \frac{\frac{1}{2\pi i}\int_0^{2\pi} e^{\frac{c}{2}(2i\sin\theta)}\, ie^{i\theta} d\theta}{e^{(n+1)i\theta}} = \frac{1}{2\pi}\int_0^{2\pi} e^{ci\sin\theta} e^{-ni\theta} d\theta$$

$$= \frac{1}{2\pi}\int_0^{2\pi} e^{i(c\sin\theta - n\theta)}\, d\theta$$

$$= \frac{1}{2\pi}\int_0^{2\pi} [\cos(c\sin\theta - n\theta) - i\sin(c\sin\theta - n\theta)]d\theta$$

$$= \frac{1}{2\pi}\int_0^{2\pi} \cos(c\sin\theta - n\theta)\, d\theta \qquad \left\{\begin{array}{l}\text{If } f(2a-x) = -f(x) \\ \text{then } \int_0^{2a} f(x)dx = 0\end{array}\right.$$

$b_n = (-1)^n a_n$ since $f(z)$ remains unaltered if $\frac{-1}{z}$ is written for z

So, $$f(z) = \Sigma a_n z^n + \Sigma \frac{b_n}{z^n}$$

$$= \sum_{n=0}^{\infty} a_n(z^n) + (-1)^n \sum_{n=1}^{n} \frac{a_n}{z^n} = \sum_{-\infty}^{\infty} a_n z^n \qquad \textbf{Ans.}$$

Example 30. *Find the Laurent expansion for* $f(z) = \dfrac{7z-2}{z^3 - z^2 - 2z}$ *in the regions given by*

(i) $0 < |z+1| < 1$ *(ii)* $1 < |z+1| < 3$ *(AMIETE, June 2010) (iii)* $|z+1| > 3$. *(G.B.T.U. 2012)*

Solution. We have,

$$f(z) = \frac{7z-2}{z^3 - z^2 - 2z} = \frac{1}{z} - \frac{3}{z+1} + \frac{2}{z-2} \qquad \text{[By partial fractions]}$$

$$= \frac{1}{(z+1)-1} - \frac{3}{z+1} + \frac{2}{(z+1)-3}$$

(*i*) $0 < |z+1| < 1$

$$f(z) = -\{1-(z+1)\}^{-1} - \frac{3}{z+1} - \frac{2}{3}\left\{1-\left(\frac{z+1}{3}\right)\right\}^{-1}$$

$$= -\left[1+(z-1)+(z-1)^2+(z-1)^3+.....\right] - \frac{3}{z+1}$$

$$-\frac{2}{3}\left[1+\frac{(z+1)}{3}+\frac{(z+1)^2}{9}+\frac{(z+1)^3}{27}+.....\right]$$

Which is the required Laurent expansion.

(*ii*) $1 < |z+1| < 3$

$$f(z) = \frac{1}{(z+1)-1} - \frac{3}{z+1} + \frac{2}{(z+1)-3}$$

Taking common, bigger term out of $|z+1|$ and 1in first fraction, here $|z+1|$ is bigger than 1, so we take $|z+1|$ common from the first fraction.

Similarly, we take 3 common from the third fraction as 3 is bigger than $|z+3|$.

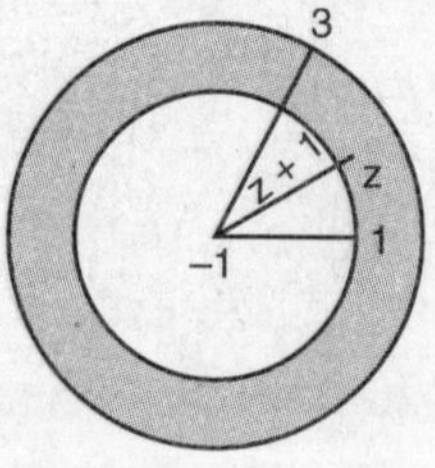

$$\Rightarrow \qquad f(z) = \frac{1}{z+1}\left[\frac{1}{1-\left(\frac{1}{z+1}\right)}\right] - \frac{3}{z+1} - \frac{2}{3}\left[\frac{1}{1-\left(\frac{z+1}{3}\right)}\right]$$

$$= \frac{1}{(z+1)}\left(1-\frac{1}{z+1}\right)^{-1} - \frac{3}{z+1} - \frac{2}{3}\left\{1-\left(\frac{z+1}{3}\right)\right\}^{-1}$$

$$= \frac{1}{z+1}\left[1+\frac{1}{z+1}+\frac{1}{(z+1)^2}+\frac{1}{(z+1)^3}+...\right]$$

$$-\frac{3}{z+1} - \frac{2}{3}\left[1+\frac{z+1}{3}+\frac{(z+1)^2}{9}+\frac{(z+1)^3}{27}+.....\right]$$

Which is the required Laurent expansion.

(*iii*) $|z+1| > 3$.

$$f(z) = \frac{1}{(z+1)-1} - \frac{3}{(z+1)} + \frac{2}{(z+1)-3}$$

Taking common bigger term out of $|z+1|$ and 3 here $|z+1|$ is greater than 3, so we take $|z+1|$ common from third fraction.

Similarly, $|z+1|$ is also greater than 1, so we take $|z+1|$ common from the first fraction.

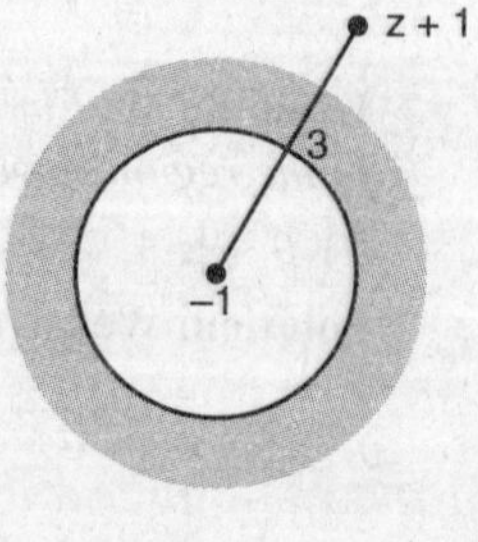

$$f(z) = \frac{1}{z+1}\left(\frac{1}{1-\frac{1}{z+1}}\right) - \frac{3}{z+1} + \frac{2}{z+1}\left(\frac{1}{1-\frac{3}{z+1}}\right)$$

$$\Rightarrow \qquad f(z) = \frac{1}{z+1}\left(1-\frac{1}{z+1}\right)^{-1} - \frac{3}{z+1} + \frac{2}{z+1}\left(1-\frac{3}{z+1}\right)^{-1}$$

$$= \frac{1}{z+1}\left[1+\frac{1}{z+1}+\frac{1}{(z+1)^2}+\frac{1}{(z+1)^3}+.....\right]$$

$$-\frac{3}{z+1}+\frac{2}{z+1}\left[1+\frac{3}{z+1}+\frac{9}{(z+1)^2}+\frac{27}{(z+1)^3}+......\right]$$

Which is the required Laurent expansion. **Ans.**

EXERCISE 25.3

1. Find the Taylor's and Laurent series which represents the function $\frac{z^2-1}{(z+2)(z+3)}$ when

   (*i*) $|z|<2$ (*ii*) $2<|z|<3$

   Ans. (*i*) $-\frac{1}{6}+\frac{5}{36}z+\frac{17}{216}z^2-\frac{115}{1296}z^3+...$ (*ii*) $\frac{5}{3}+\frac{3}{z}-\frac{6}{z^2}+\frac{12}{z^3}-\frac{24}{z^4}+...\frac{8z}{9}-\frac{8z^2}{27}+\frac{8z^3}{81}+...$

2. Find four terms of the Laurent series expansion valid in the region $0<|z-1|<1$ for the function $f(z)=\frac{2z+1}{z^3+z^2-2z}$

   Ans. $-\frac{1}{2}[1-(z-1)+(z-1)^2-(z-1)^3+...]-\frac{1}{6}\left[1-\frac{z-1}{3}+\frac{(z-1)^2}{9}-\frac{(z-1)^3}{27}+...\right]+\frac{1}{z-1}$

3. Expand $\frac{z}{(z^2-1)(z^2+4)}$ in $1<|z|<2$ **Ans.** $\frac{1}{10}\left[\left(\frac{2}{z}+\frac{2}{z^3}+\frac{2}{z^5}+...\right)-\left(\frac{z}{2}+\frac{z^3}{8}+...\right)\right]$

4. Represent the function $f(z)=\frac{4z+3}{z(z-3)(z+2)}$ in Laurent series

   (*i*) within $|z|=2$ (*ii*) in the annular region between $|z|=2$ and $|z|=3$.

   Ans. (*i*) $-\frac{1}{2z}+\sum_{n=0}^{\infty}\left[(-1)^{n+1}\frac{1}{2^{n+2}}-\frac{1}{3^{n+1}}\right]z^n$. (*ii*) $-\frac{1}{2z}-\frac{1}{2z}\sum_{n=0}^{\infty}(-1)^n\left(\frac{2}{z}\right)^n-\frac{1}{3}\sum_{n=0}^{\infty}\left(\frac{z}{3}\right)^n$.

5. Write all possible Laurent Series for the function $f(z)=\frac{z^2}{(z-1)^2(z+3)}$ about the singularity $z=1$, stating the region of convergence in each case.

   Ans. When $|z-1|>4$, $\frac{1}{z-1}-\frac{2}{(z-1)^2}+\frac{9}{(z-1)^3}-\frac{36}{(z-1)^4}+....$

   When $0<|z-1|<4$, $\frac{1}{4}\left[1-\frac{1}{(z-1)^2}+\frac{7}{4}\frac{1}{z-1}+\frac{9}{16}-\frac{9}{64}(z-1)-....\right]$

6. Obtain the expansion

$$f(z)=f(a)+2\left\{\frac{z-a}{2}f'\left(\frac{z+a}{2}\right)+\frac{(z-a)^3}{2^3.3!}f'''\left(\frac{z+a}{2}\right)+\frac{(z-a)^5}{2^5.5!}f^{v}\left(\frac{z+a}{2}\right)+...\right\}$$

7. Expand $\frac{z^2-6z-1}{(z-1)(z+2)(z-3)}$ in $3<|z+2|<5$.

   Ans. $\frac{2}{z+2}+\frac{3}{(z+2)^2}+\frac{3^2}{(z+2)^3}+...+\frac{1}{5}\left[1+\frac{z+2}{5}+\frac{(z+2)^2}{5}+\frac{(z+2)^3}{5^3}+...\right]$

9. Find Taylor expansion of $f(z)=\frac{2z^3+1}{z^2+z}$ about the point $z=i$. $\left[\textbf{Hint. } f(z)=2z-2+\frac{1}{z}+\frac{1}{z+1}\right]$

   Ans. $\left(\frac{i}{2}-\frac{3}{2}\right)+\left(3+\frac{i}{2}\right)(z-i)+\sum_{n=2}^{\infty}(-1)^n\left\{\frac{1}{(1+i)^{n+1}}+\frac{1}{(i)^{n+1}}\right\}(z-i)^n$

10. Find the Laurent's series of $f(z)=\frac{1}{z^2(1-z^2)}$ and determine the precise region of its convergence. (*AMIETE, Dec. 2010*)

THE CALCULUS OF RESIDUES (INTEGRATION)

26.1 ZERO OF ANALYTIC FUNCTION

A zero of analytic function $f(z)$ is the value of z for which $f(z) = 0$.

Example 1. *Find out the zeros and discuss the nature of the singularities of*

$$f(z) = \frac{(z-2)}{z^2} \sin\left(\frac{1}{z-1}\right)$$

Solution. Poles of $f(z)$ are given by equating to zero the denominator of $f(z)$ i.e. $z = 0$ is a pole of order two.

zeros of $f(z)$ are given by equating to zero the numerator of $f(z)$ *i.e.*, $(z-2)\sin\left(\frac{1}{z-1}\right) = 0$

$$\Rightarrow \quad \text{Either } z - 2 = 0 \quad \text{or} \quad \sin\left(\frac{1}{z-1}\right) = 0$$

$$\Rightarrow \quad z = 2 \quad \text{and} \quad \frac{1}{z-1} = n\pi$$

$$\Rightarrow \quad z = 2, \quad z = \frac{1}{n\pi} + 1, \quad n = \pm 1, \pm 2, \ldots\ldots\ldots$$

Thus, $z = 2$ is a simple zero. The limit point of the zeros are given by

$$z = \frac{1}{n\pi} + 1 \ (n = \pm 1, \pm 2, \ldots\ldots\ldots\ldots) \text{ is } z = 1.$$

Hence $z = 1$ is an isolated essential singularity. **Ans.**

26.2 SINGULAR POINT

A point at which a function $f(z)$ is not analytic is known as a singular point or **singularity** of the function.

For example, the function $\frac{1}{z-2}$ has a singular point at $z - 2 = 0$ or $z = 2$.

Isolated singular point. If $z = a$ is a singularity of $f(z)$ and if there is no other singularity within a small circle surrounding the point $z = a$, then $z = a$ is said to be an isolated singularity of the function $f(z)$; otherwise it is called non-isolated.

For example, the function $\frac{1}{(z-1)(z-3)}$ has two isolated singular points, namely $z = 1$ and $z = 3$. $[(z-1)(z-3) = 0 \text{ or } z = 1, 3]$.

Example of non-isolated singularity. Function $\dfrac{1}{\sin\frac{\pi}{z}}$ is not analytic at the points where $\sin\frac{\pi}{z} = 0$, *i.e.*, at the points $\frac{\pi}{z} = n\pi$ *i.e.*, the points $z = \frac{1}{n}\,(n = 1, 2, 3, ...)$. Thus $z = 1, \frac{1}{2}, \frac{1}{3}, ...,$ $z = 0$ are the points of singularity. $z = 0$ is the **non-isolated singularity** of the function $\dfrac{1}{\sin\frac{\pi}{z}}$ because in the neighbourhood of $z = 0$, there are infinite number of other singularities $z = \frac{1}{n}$, where n is very large.

Pole of order *m*. Let a function $f(z)$ have an isolated singular point $z = a$, $f(z)$ can be expanded in a Laurent's series around $z = a$, giving

$$f(z) = a_0 + a_1(z-a) + a_2(z-a)^2 + ...$$
$$+ \frac{b_1}{z-a} + \frac{b_2}{(z-a)^2} + ... + \frac{b_m}{(z-a)^m} + \frac{b_{m+1}}{(z-a)^{m+1}} + \frac{b_{m+2}}{(z-a)^{m+2}} + ... \quad ... (1)$$

In some cases it may happen that the coefficients $b_{m+1} = b_{m+2} = b_{m+3} = 0$, then (1) reduces to

$$f(z) = a_0 + a_1(z-a) + a_2(z-a)^2 + ... + \frac{b_1}{(z-a)} + \frac{b_2}{(z-a)^2} + ... + \frac{b_m}{(z-a)^m}$$

$$f(z) = a_0 + a_1(z-a) + a_2(z-a)^2 + ... + \frac{1}{(z-a)^m}\{b_1(z-a)^{m-1} + b_2(z-a)^{m-2}$$
$$+ b_3(z-a)^{m-3} + ... + b_m\}$$

then $z = a$ is said to be a **pole of order *m*** of the function $f(z)$, when $m = 1$, the pole is said to be **simple pole**. In this case

$$f(z) = a_0 + a_1(z-a) + a_2(z-a)^2 + ... + \frac{b_1}{z-a}$$

If the number of the terms of negative powers in expansion (1) is infinite, then $z = a$ is called an essential singular point of $f(z)$.

Example 2. *Define the singularity of a function. Find the singularity (ties) of the functions*

(*i*) $f(z) = \sin\frac{1}{z}$ (*ii*) $g(z) = \dfrac{e^{\frac{1}{z}}}{z^2}$ *(U.P. III Semester, 2009-2010)*

Solution. See Art. 26.2 on page 712 for definition.

(*i*) We know that

$$\sin\frac{1}{z} = \frac{1}{z} - \frac{1}{3!\,z^3} + \frac{1}{5!\,z^5} + ... + (-1)^n \frac{1}{(2n+1)!\,z^{2n+1}}$$

Obviously, there is a number of singularity.

$\sin\frac{1}{z}$ is not analytic at $z = 0$. $\left(\frac{1}{z} = \infty \text{ at } z = 0\right)$

Hence, $\sin\frac{1}{z}$ has a singularity at $z = 0$.

(ii) Here, we have $g(z) = \dfrac{e^{\frac{1}{z}}}{z^2}$

We know that, $\left(\frac{1}{z^2}\right)\left(e^{\frac{1}{z}}\right) = \frac{1}{z^2}\left(1 + \frac{1}{z} + \frac{1}{2!\,z^2} + \frac{1}{3!\,z^3} + ... + \frac{1}{n!\,z^n} .... + \right)$

$$= \frac{1}{z^2} + \frac{1}{z^3} + \frac{1}{2!\,z^4} + \frac{1}{3!\,z^5} + \ldots\ldots \frac{1}{n!\,z^{n+2}} + \ldots\ldots$$

Here, $f(z)$ has infinite number of terms in negative powers of z.
Hence, $f(z)$ has essential singularity at $z = 0$. **Ans.**

Example 3. *Find the pole of the function* $\dfrac{e^{z-a}}{(z-a)^2}$

Solution. $$\frac{e^{z-a}}{(z-a)^2} = \frac{1}{(z-a)^2}\left[1 + (z-a) + \frac{(z-a)^2}{2!} + \ldots\right]$$

The given function has negative power 2 of $(z - a)$.
So, the given function has a pole at $z = a$ of order 2. **Ans.**

Example 4. *Find the poles of* $f(z) = \sin\left(\dfrac{1}{z-a}\right)$

Solution. $$\sin\left(\frac{1}{z-a}\right) = \frac{1}{z-a} - \frac{1}{3!}\frac{1}{(z-a)^3} + \frac{1}{5!}\frac{1}{(z-a)^5} - \ldots\ldots$$

The given function $f(z)$ has infinite number of terms in the negative powers of $z - a$.
So, $f(z)$ has essential singularity at $z = a$. **Ans.**

Example 5. *Find the pole of* $f(z) = \dfrac{\sin(z-a)}{(z-a)^4}$

Solution. $$\frac{\sin(z-a)}{(z-a)^4} = \frac{1}{(z-a)^4}\left[(z-a) - \frac{(z-a)^3}{3!} + \frac{(z-a)^5}{5!} - \frac{(z-a)^7}{7!} + \ldots\ldots\right]$$

$$= \frac{1}{(z-a)^3}\left[1 - \frac{(z-a)^2}{3!} + \frac{(z-a)^4}{5!} - \frac{(z-a)^6}{7!} + \ldots\right]$$

The given function has a negative power 3 of $(z - a)$.
So, $f(z)$ has a pole at $z = a$ of order 3. **Ans.**

Example 6. *Prove that* $f(z) = \lim\limits_{z \to a} e^{\frac{1}{z-a}}$ *does not exist.*

Solution. $$\lim_{z \to a} e^{\frac{1}{z-a}} = \lim_{z \to a}\left(1 + \frac{1}{z-a} + \frac{1}{2!(z-a)^2} + \frac{1}{3!(z-a)^3} + \ldots + \frac{1}{n!(z-a)^n} + \ldots \infty\right)$$

Here $n \to \infty$, $f(z)$ has infinite number of terms in negative power of $(z - a)$.
Thus, $f(z)$ has essential singularity at $z = a$.

Hence, $f(z) = \lim\limits_{z \to a} e^{\frac{1}{z-a}}$ does not exist. **Ans.**

Example 7. *Discuss singularity of* $\dfrac{1}{1 - e^z}$ *at* $z = 2\pi\, i$.

Solution. We have, $$f(z) = \frac{1}{1 - e^z}$$

The poles are determined by putting the denominator equal to zero.

i.e., $1 - e^z = 0$

$\Rightarrow$ $e^z = 1 = (\cos 2n\pi + i \sin 2n\pi) = e^{2n\pi i}$

$\Rightarrow$ $z = 2n\pi i$ $\qquad (n = 0, \pm 1, \pm 2, \ldots\ldots)$

Clearly $z = 2\pi i$ is a simple pole. **Ans.**

Example 8. *Discuss singularity of* $\frac{\cot \pi z}{(z-a)^2}$ *at* $z = a$ *and* $z = \infty$.

Solution. Let $$f(z) = \frac{\cot \pi z}{(z-a)^2} = \frac{\cos \pi z}{\sin \pi z\,(z-a)^2}$$

The poles are given by putting the denominator equal to zero.

i.e., $\sin \pi z\,(z-a)^2 = 0 \;\Rightarrow\; (z-a)^2 = 0$ or $\sin \pi z = 0 = \sin n\pi$

$\Rightarrow \qquad z = a,\ \pi z = n\pi, \qquad (n \in I)$

$\Rightarrow \qquad z = a, n$

$f(z)$ has essential singularity at $z = \infty$.

Also, $z = a$ being repeated twice gives the double pole. **Ans.**

Example 9. *Show that* $e^{-\left(\frac{1}{z^2}\right)}$ *has no singularities.*

Solution. $$f(z) = e^{-\left(\frac{1}{z^2}\right)} = \frac{1}{e^{(1/z^2)}}$$

The poles are determined by putting the denominator

$$e^{\left(\frac{1}{z^2}\right)} = 0 \qquad \text{...(1)}$$

It is not possible to find the value of z which can satisfy equation (1).

Hence, there is no pole or singularity of the given function. **Proved.**

Example 10. *Find the nature of singularities of*

$$f(z) = \frac{z - \sin z}{z^3} \quad at \ z = 0.$$

Solution. $$f(z) = \frac{1}{z^3}(z - \sin z) = \frac{1}{z^3}\left[z - \left(z - \frac{z^3}{3!} + \frac{z^5}{5!} - \frac{z^7}{7!} + \ldots\right)\right]$$

$$= \frac{1}{z^3}\left(\frac{z^3}{3!} - \frac{z^5}{5!} + \frac{z^7}{7!} - \ldots\right) = \frac{1}{3!} - \frac{z^2}{5!} + \frac{z^4}{7!} - \ldots$$

There is no negative power of z.

Hence, there is no pole. **Ans.**

Example 11. *Determine the poles of the function z*

$$f(z) = \frac{1}{z^4 + 1}$$

Solution. $$f(z) = \frac{1}{z^4 + 1}$$

The poles of $f(z)$ are determined by putting the denominator equal to zero.

i.e., $$z^4 + 1 = 0 \quad \Rightarrow \quad z^4 = -1$$

$$z = (-1)^{\frac{1}{4}} = (\cos \pi + i \sin \pi)^{\frac{1}{4}}$$

$$= [\cos(2n+1)\pi + i \sin(2n+1)\pi]^{\frac{1}{4}} \quad \text{[By De Moiver's theorem]}$$

$$= \left[\cos \frac{(2n+1)\pi}{4} + i \sin \frac{(2n+1)\pi}{4}\right]$$

If $n = 0$, Pole at $z = \left[\cos\frac{\pi}{4} + i\sin\frac{\pi}{4}\right] = \left(\frac{1}{\sqrt{2}} + i\frac{1}{\sqrt{2}}\right)$

If $n = 1$, Pole at $z = \left[\cos\frac{3\pi}{4} + i\sin\frac{3\pi}{4}\right] = \left(-\frac{1}{\sqrt{2}} + i\frac{1}{\sqrt{2}}\right)$

If $n = 2$, Pole at $z = \left[\cos\frac{5\pi}{4} + i\sin\frac{5\pi}{4}\right] = \left(-\frac{1}{\sqrt{2}} - i\frac{1}{\sqrt{2}}\right)$

If $n = 3$, Pole at $z = \left[\cos\frac{7\pi}{4} + i\sin\frac{7\pi}{4}\right] = \left(\frac{1}{\sqrt{2}} - i\frac{1}{\sqrt{2}}\right)$ **Ans.**

Example 12. *Show that the function e^z has an isolated essential singularity at $z = \infty$.*

Solution. Let $f(z) = e^z$

Putting $z = \frac{1}{t}$, we get $f\left(\frac{1}{t}\right) = e^{\frac{1}{t}} = 1 + \frac{1}{t} + \frac{1}{2!.t^2} + \frac{1}{3!\,t^3} + \ldots..$

Here, the principal part of $f\left(\frac{1}{t}\right)$;

$$\frac{1}{t} + \frac{1}{2\,!\,t^2} + \frac{1}{3\,!\,t^3} + \ldots$$

Contains infinite number of terms.

Hence $t = 0$ is an isolated essential singularity of $\frac{1}{e^t}$ and $z = \infty$ is an isolated essential singularity of e^z. **Ans.**

EXERCISE 26.1

Find the poles or singularity of the following functions:

1. $\frac{1}{(z-2)(z-3)}$ **Ans.** 2 simple poles at $z = 2$ and $z = 3$.

2. $\frac{e^z}{(z-2)^3}$ **Ans.** Pole at $z = 2$ of order 3.

3. $\frac{1}{\sin z - \cos z}$ **Ans.** Simple pole at $z = \frac{\pi}{4}$

4. $\cot\frac{1}{z}$ **Ans.** Essential singularity at $z = 0$

5. z cosec z **Ans.** Non-isolated essential singularity

6. $\sin\frac{1}{z}$ **Ans.** Essential singularity

26.3 THEOREM

If $f(z)$ has a pole at $z = a$, then $|f(z)| \to \infty$ as $z \to a$.

Proof. Let $z = a$ be a pole of order m of $f(z)$. Then by Laurent's theorem

$$f(z) = \sum_0^{\infty} a_n (z-a)^n + \sum_1^{m} b_n (z-a)^{-n}$$

$$= \sum_0^{\infty} a_n (z-a)^n + \frac{b_1}{z-a} + \frac{b_2}{(z-a)^2} + \ldots.. + \frac{b_m}{(z-a)^m}$$

$$= \sum_0^{\infty} a_n (z-a)^n + \frac{1}{(z-a)^m}[b_1(z-a)^{m-1} + b_2(z-a)^{m-2} + ... + b_{m-1}(z-a) + b_m]$$

$$= \sum_0^{\infty} a_n (z-a)^n + \frac{\varphi(z)}{(z-a)^m}$$

Now $\varphi(z) \to b_m$ as $z \to a$.

Hence $|f(z)| \to \infty$ as $z \to a$. **Proved.**

Example 13. *If an analytic function f(z) has a pole of order m at z = a, then* $\frac{1}{f(z)}$ *has a zero of order m at z = a.*

Solution. If $f(z)$ has a pole of order m at $z = a$, then

$$f(z) = \frac{\varphi(z)}{(z-a)^m} \qquad \text{where } \varphi(z) \text{ is analytic and non-zero at } z = a.$$

$$\therefore \qquad \frac{1}{f(z)} = \frac{(z-a)^m}{\varphi(z)}$$

Clearly, $\frac{1}{f(z)}$ has a zero of order m at $z = a$, since $\phi(a) \neq 0$.

26.4 DEFINITION OF THE RESIDUE AT A POLE

Let $z = a$ be a pole of order m of a function $f(z)$ and C_1 circle of radius r with centre at $z = a$ which does not contain any other singularities except at $z = a$ then $f(z)$ is analytic within the annulus $r < |z - a| < R$ can be expanded within the annulus. Laurent's series:

$$f(z) = \sum_{n=0}^{\infty} a_n (z-a)^n + \sum_{n=1}^{\infty} b_n (z-a)^{-n} \qquad ...(1)$$

where $$a_n = \frac{1}{2\pi i}\int_C \frac{f(z)\,dz}{(z-a)^{n+1}} \qquad ...(2)$$

and $$b_n = \frac{1}{2\pi i}\int_{C_1} \frac{f(z)}{(z-a)^{-n+1}}\,dz \qquad ...(3)$$

$|z - a| = r$ being the circle C_1.

Particularly, $$b_1 = \frac{1}{2\pi i}\int_{C_1} f(z)\,dz$$

The coefficient b_1 is called residue of $f(z)$ at the pole $z = a$. It is denoted by symbol Res. $(z = a) = b_1$.

26.5 RESIDUE AT INFINITY

Residue of $f(z)$ at $z = \infty$ is defined as $-\frac{1}{2\pi i}\int_C f(z)\,dz$ where the integration is taken round C in anti-clockwise direction.

where C is a large circle containing all finite singularities of $f(z)$.

26.6 METHOD OF FINDING RESIDUES

(*a*) **Residue at simple pole**

(*i*) If $f(z)$ has a simple pole at $z = a$, then

$$\text{Res } f(a) = \lim_{z\to a}(z-a)f(z)$$

Proof. $$f(z) = a_0 + a_1(z-a) + a_2(z-a)^2 + ... + \frac{b_1}{z-a}$$

$$\Rightarrow \qquad (z-a)f(z) = a_0(z-a) + a_1(z-a)^2 + a_2(z-a)^3 + ... + b_1$$

$$\Rightarrow \qquad b_1 = (z-a)f(z) - [a_0(z-a) + a_1(z-a)^2 + a_2(z-a)^3 + \ldots]$$

Taking limit as $z \to a$, we have $b_1 = \lim_{z \to a}(z-a)f(z)$

$$\textbf{Res } (\textbf{\textit{at} } z = a) = \lim_{z \to a} (z-a)\, f(z)$$ **Proved.**

(*ii*) If *f*(*z*) is of the form $f(z) = \dfrac{\phi(z)}{\psi(z)}$ where $\psi(a) = 0$, but $\phi(a) \neq 0$

$$\textbf{Res } (\textbf{\textit{at} } z = a) = \frac{\phi(a)}{\psi'(a)}$$

Proof . $\quad f(z) = \dfrac{\phi(z)}{\psi(z)}$

$$\text{Res (at } z = a) = \lim_{z \to a}(z-a)f(z) = \lim_{z \to a}(z-a)\frac{\phi(z)}{\psi(z)}$$

$$= \lim_{z \to a} \frac{(z-a)[\phi(a) + (z-a)\,\phi'(a) + \ldots]}{\psi(a) + (z-a)\psi'(a) + \dfrac{(z-a)^2}{2!}\psi''(a) + \ldots} \quad \text{(By Taylor's Theorem)}$$

$$= \lim_{z \to a} \frac{(z-a)\,[\phi(a) + (z-a)\phi'(a) + \ldots]}{(z-a)\psi'(a) + \dfrac{(z-a)^2}{2!}\psi''\,(a) + \ldots} \quad [\text{since } \psi(a) = 0]$$

$$= \lim_{z \to a} \frac{\phi(a) + (z-a)\,\phi'(a) + \ldots}{\psi'(a) + \dfrac{z-a}{2!}\psi''(a) + \ldots}$$

$$\textbf{Res (at } z = a) = \frac{\phi(a)}{\psi'(a)}$$ **Proved.**

(*b*) **Residue at a pole of order *n*.** If $f(z)$ has a pole of order n at $z = a$, then

$$\textbf{Res (at } z = a) = \frac{1}{(n-1)!}\left\{\frac{d^{n-1}}{dz^{n-1}}[(z-a)^n f(z)]\right\}_{z=a}$$

Proof. If $z = a$ is a pole of order n of function $f(z)$, then by Laurent's theorem

$$f(z) = a_0 + a_1(z-a) + a_2(z-a)^2 + \ldots + \frac{b_1}{z-a} + \frac{b_2}{(z-a)^2} + \ldots + \frac{b_n}{(z-a)^n}$$

Multiplying by $(z-a)^n$, we get

$$(z-a)^n f(z) = a_0(z-a)^n + a_1(z-a)^{n+1} + a_2(z-a)^{n+2} + \ldots$$
$$+\; b_1(z-a)^{n-1} + b_2(z-a)^{n-2} + b_3(z-a)^{n-3} + \ldots + b_n$$

Differentiating both sides w.r.t. 'z' $n-1$ times and putting $z = a$, we get

$$\left\{\frac{d^{n-1}}{dz^{n-1}}[(z-a)^n f(z)]\right\}_{z=a} = (n-1)!\,b_1$$

$$\Rightarrow \qquad b_1 = \frac{1}{(n-1)!}\left\{\frac{d^{n-1}}{dz^{n-1}}[(z-a)^n f(z)]\right\}_{z=a}$$

$$\textbf{Residue } f(a) = \frac{1}{(n-1)!}\left\{\frac{d^{n-1}}{dz^{n-1}}[(z-a)^n f(z)]\right\}_{z=a}$$

(*c*) **Residue at a pole *z* = *a* of any order (simple or of order *m*)**

$$\textbf{Res } f(a) = \textbf{coefficient of } \frac{1}{t}$$

Proof. If $f(z)$ has a pole of order m, then by Laurent's theorem

$$f(z)=a_0+a_1(z-a)+a_2(z-a)^2...+\frac{b_1}{z-a}+\frac{b_2}{(z-a)^2}+...+\frac{b_m}{(z-a)^m}$$

If we put $z-a=t$ or $z=a+t$, then

$$f(a+t)=a_0+a_1t+a_2t^2+...+\frac{b_1}{t}+\frac{b_2}{t^2}+...+\frac{b_m}{t^m}$$

Res $f(a)=b_1$, **Res $f(a)$ = coefficient of $\frac{1}{t}$**

Rule. Put $z=a+t$ in the function $f(z)$, expand it in powers of t. Coefficient of $\frac{1}{t}$ is the residue of $f(z)$ at $z=a$.

(*d*) **Residue of $f(z)$ at $z=\infty$** $=\lim_{z\to\infty}\{-z\,f(z)\}$

or The residue of $f(z)$ at infinity $=-\frac{1}{2\pi i}\int_c f(z)\,dz$

26.7 RESIDUE BY DEFINITION

Example 14. *Find the residue at $z=0$ of $z\cos\frac{1}{z}$.*

Solution. Expanding the function in powers of $\frac{1}{z}$, we have

$$z\cos\frac{1}{z}=z\left[1-\frac{1}{2z^2}+\frac{1}{4!\,z^4}-.....\right]=z-\frac{1}{2z}+\frac{1}{24z^3}-......$$

This is the Laurent's expansion about $z=0$.

The coefficient of $\frac{1}{z}$ in it is $-\frac{1}{2}$. So the residue of $z\cos\frac{1}{z}$ at $z=0$ is $-\frac{1}{2}$. **Ans.**

Example 15. *Find the residue of $f(z)=\frac{z^3}{z^2-1}$ at $z=\infty$.*

Solution. We have, $f(z)=\frac{z^3}{z^2-1}$

$$f(z)=\frac{z^3}{z^2\left(1-\frac{1}{z^2}\right)}=z\left(1-\frac{1}{z^2}\right)^{-1}=z\left(1+\frac{1}{z^2}+\frac{1}{z^4}+.....\right)=z+\frac{1}{z}+\frac{1}{z^3}+.....$$

Residue at infinity $=-\left(\text{coeff. of }\frac{1}{z}\right)=-1.$ **Ans.**

26.8 FORMULA: RESIDUE = $\lim_{z\to a}(z-a)\,f(z)$

Example 16. *Determine the pole and residue at the pole of the function $f(z)=\frac{z}{z-1}$*

Solution. The poles of $f(z)$ are given by putting the denominator equal to zero.

$\therefore$ $z-1=0 \Rightarrow z=1$

The function $f(z)$ has a simple pole at $z=1$.

Residue is calculated by the formula

$$\text{Residue}=\lim_{z\to a}(z-a)\,f(z)$$

Residue of $f(z)$ at (at $z=1$) $=\lim_{z\to 1}(z-1)\left(\frac{z}{z-1}\right)=\lim_{z\to 1}(z)=1$

Hence, $f(z)$ has a simple pole at $z=1$ and residue at the pole is 1. **Ans.**

Example 17. *Determine the poles and the residue at simple pole of the function*

$$f(z)=\frac{z^2}{(z-1)^2(z+2)}$$

Solution. The pole of $f(z)$ are given by putting the denominator equal to zero.

$(z-1)^2(z+2)=0 \quad \Rightarrow \quad z=1,1,-2$

The function $f(z)$ has simple pole at $z=-2$ and at $z=1$ pole of second order.

Residue of $f(z)$ at $z=-2$ is $=\lim_{z\to-2}(z+2)f(z)$ [Residue $=\lim_{z\to a}(z-a)f(z)$]

$$=\lim_{z\to-2}(z+2)\frac{z^2}{(z-1)^2(z+2)}=\lim_{z\to-2}\frac{z^2}{(z-1)^2}=\frac{(-2)^2}{(-2-1)^2}=\frac{4}{9}$$

Hence, residue at simple pole is $\frac{4}{9}$. **Ans.**

Example 18. *Find the order of each pole and residue of* $\frac{1-2z}{z(z-1)(z-2)}$.

Solution. Let $f(z)=\frac{1-2z}{z(z-1)(z-2)}$

The poles of $f(z)$ are given by $z(z-1)(z-2)=0$

$\Rightarrow$ $z=0,1,2$ all are simple poles.

Residue of $f(z)$ at $(z=0)$ $=\lim_{z\to0}(z-0)f(z)=\lim_{z\to0}\frac{z(1-2z)}{z(z-1)(z-2)}$

$$=\lim_{z\to0}\frac{1-2z}{(z-1)(z-2)}=\frac{1}{2}$$

Residue of $f(z)$ at $(z=1)$ $=\lim_{z\to1}(z-1)f(z)=\lim_{z\to1}\frac{(z-1)(1-2z)}{z(z-1)(z-2)}=\lim_{z\to1}\frac{1-2z}{z(z-2)}=1$

Residue of $f(z)$ at $(z=2)=\lim_{z\to2}(z-2)f(z)=\lim_{z\to2}\frac{(z-2)(1-2z)}{z(z-1)(z-2)}$

$$=\lim_{z\to2}\frac{1-2z}{z(z-1)}=-\frac{3}{2}$$

Hence, the residue of $f(z)$ at $z=0$, $z=1$ and $z=2$ are $\frac{1}{2}$, 1 and $-\frac{3}{2}$ respectively. **Ans.**

Example 19. *Determine the residue of* $f(z)=\frac{z^3}{(z-1)^4(z-2)(z-3)}$ *at its simple poles.*

Solution. The poles of $f(z)$ are determined by putting the denominator equal to zero.

i.e. $(z-1)^4(z-2)(z-3)=0$

$\Rightarrow$ $z=1,1,1,1$ and $z=2$ and $z=3$

The simple poles of the function $f(z)$ are at $z=2$ and $z=3$.

Pole at $z=2$

Residue, $\quad R(2)=\lim_{z\to2}(z-2)\frac{z^3}{(z-1)^4(z-2)(z-3)}$

[Residue $R(2)=\lim_{z\to2}[(z-2)f(z)]$]

$$=\lim_{z\to2}\frac{z^3}{(z-1)^4(z-3)}=\frac{(2)^3}{(1)^4(-1)}=-8$$

Pole at $z=3$

Residue, $R(3) = \lim_{z \to 3} (z-3) \dfrac{z^3}{(z-1)^4 (z-2)(z-3)}$

$$= \lim_{z \to 3} \frac{z^3}{(z-1)^4 (z-2)} = \frac{(3)^3}{(3-1)^4 (3-2)} = \frac{27}{16}$$

Hence, residue at $z = 2$ and $z = 3$ are -8 and $\dfrac{27}{16}$ respectively. **Ans.**

Example 20. *Evaluate the residues of* $\dfrac{z^2}{(z-1)(z-2)(z-3)}$ *at* $z = 1, 2, 3$ *and infinity and show that their sum is zero.*

Solution. Let $f(z) = \dfrac{z^2}{(z-1)(z-2)(z-3)}$

The poles of $f(z)$ are determined by putting the denominator equal to zero.

$$\therefore \; (z-1)(z-2)(z-3) = 0 \; \Rightarrow \; z = 1, \; 2, \; 3$$

Residue of $f(z)$ at $(z = 1)$ $= \lim_{z \to 1} (z-1) f(z) = \lim_{z \to 1} (z-1) . \dfrac{z^2}{(z-1)(z-2)(z-3)}$

$$= \lim_{z \to 1} \frac{z^2}{(z-2)(z-3)} = \frac{1}{2}$$

Residue of $f(z)$ at $(z = 2)$ $= \lim_{z \to 2} (z-2) f(z) = \lim_{z \to 2} (z-2) \dfrac{z^2}{(z-1)(z-2)(z-3)}$

$$= \lim_{z \to 2} \frac{z^2}{(z-1)(z-3)} = \frac{4}{(1)(-1)} = -4$$

Residue of $f(z)$ at $(z = 3)$ $= \lim_{z \to 3} (z-3) f(z)$

$$= \lim_{z \to 3} (z-3) \frac{z^2}{(z-1)(z-2)(z-3)} = \lim_{z \to 3} \frac{z^2}{(z-1)(z-2)} = \frac{9}{2}$$

Residue of $f(z)$ at $(z = \infty)$ $= \lim_{z \to \infty} -z f(z) = \dfrac{-z(z^2)}{(z-1)(z-2)(z-3)}$

$$= \lim_{z \to \infty} \frac{-1}{\left(1 - \frac{1}{z}\right)\left(1 - \frac{2}{z}\right)\left(1 - \frac{3}{z}\right)} = -1$$

Sum of the residues at all the poles of $f(z) = \dfrac{1}{2} - 4 + \dfrac{9}{2} - 1 = 0$

Hence, the sum of the residues is zero. **Proved.**

26.9 FORMULA: RESIDUE OF $f(a) = \dfrac{1}{(n-1)!} \left\{ \dfrac{d^{n-1}}{dz^{n-1}} [(z-a)^n f(z)] \right\}_{z=a}$

Example 21. *Find the residue of a function*

$$f(z) = \frac{z^2}{(z+1)^2 (z-2)} \text{ at its double pole.}$$

Solution. We have, $f(z) = \dfrac{z^2}{(z+1)^2 (z-2)}$

Poles are determined by putting denominator equal to zero.

i.e.; $(z+1)^2 (z-2) = 0$

$\Rightarrow$ $z = -1, -1$ and $z = 2$

The function has a double pole at $z = -1$

$$\text{Residue at } (z=-1) = \lim_{z\to -1}\frac{1}{(2-1)!}\left[\frac{d}{dz}\left\{(z+1)^2\frac{z^2}{(z+1)^2(z-2)}\right\}\right]$$

$$=\left[\frac{d}{dz}\left(\frac{z^2}{z-2}\right)\right]_{z=-1} = \left(\frac{(z-2)\,2z - z^2\cdot 1}{(z-2)^2}\right)_{z=-1} = \left[\frac{z^2-4z}{(z-2)^2}\right]_{z=-1} = \frac{(-1)^2-4(-1)}{(-1-2)^2}$$

$$\text{Residue at } (z=-1) = \frac{1+4}{9} = \frac{5}{9}$$ **Ans.**

Example 22. *Find the residue of* $\dfrac{1}{(z^2+1)^3}$ *at* $z = i$.

Solution. Let $$f(z) = \frac{1}{(z^2+1)^3}$$

The poles of $f(z)$ are determined by putting denominator equal to zero.

i.e.; $(z^2+1)^3 = 0$

$\Rightarrow$ $(z+i)^3(z-i)^3 = 0$

$\Rightarrow$ $z = \pm i$

Here, $z = i$ is a pole of order 3 of $f(z)$.

Residue at $z = i$:

$$=\lim_{z\to i}\frac{1}{(3-1)!}\left\{\frac{d^{3-1}}{dz^{3-1}}\left[(z-i)^3\frac{1}{(z^2+1)^3}\right]\right\} = \lim_{z\to i}\frac{1}{2!}\left\{\frac{d^2}{dz^2}\left(\frac{1}{(z+i)^3}\right)\right\}$$

$$=\lim_{z\to i}\frac{1}{2}\left(\frac{3\times 4}{(z+i)^5}\right) = \frac{1}{2}\times\frac{12}{(i+i)^5} = \frac{6}{32i} = \frac{3}{16i} = -\frac{3i}{16}$$

Hence, the residue of the given function at $z = i$ is $-\dfrac{3i}{16}$. **Ans.**

26.10 FORMULA: RES. (AT $z = a$) = $\dfrac{\phi(a)}{\psi'(a)}$

Example 23. *Determine the poles and residue at each pole of the function* $f(z) = \cot z$.

Solution. $$f(z) = \cot z = \frac{\cos z}{\sin z}$$

The poles of the function $f(z)$ are given by

$$\sin z = 0 = \sin n\pi,\quad z = n\pi,\quad \text{where } n = 0, \pm 1, \pm 2, \pm 3 \ldots$$

$$\text{Residue of } f(z) \text{ at } z = n\pi \text{ is } = \frac{\cos z}{\frac{d}{dz}(\sin z)} = \frac{\cos z}{\cos z} = 1 \quad \left[\text{Res. at } (z=a) = \frac{\phi(a)}{\psi'(a)}\right].$$ **Ans.**

Example 24. *Determine the poles of the function and residue at the poles.*

$$f(z) = \frac{z}{\sin z}$$

Solution. $$f(z) = \frac{z}{\sin z}$$

Poles are determined by putting $\sin z = 0 = \sin n\pi \Rightarrow z = n\pi$

$$\text{Residue} = \left(\frac{z}{\cos z}\right)_{z=n\pi} \qquad \left[\text{Residue} = \frac{\phi(a)}{\psi'(a)}\right]$$

$$=\frac{n\pi}{\cos n\pi}=\frac{n\pi}{(-1)^n}$$

Hence, the residue of the given function at pole $z = n\pi$ is $\frac{n\pi}{(-1)^n}$. **Ans.**

26.11 FORMULA: RESIDUE = COEFFICIENT OF $\frac{1}{t}$ where $z = \frac{1}{t}$

Example 25. *Find the residue of* $\frac{z^3}{(z-1)^4(z-2)(z-3)}$ *at a pole of order 4.*

Solution. The poles of $f(z)$ are determined by $(z-1)^4(z-2)(z-3)=0 \Rightarrow z=1, 2, 3$

Here $z = 1$ is a pole of order 4.

$$f(z)=\frac{z^3}{(z-1)^4(z-2)(z-3)} \quad \text{...(1)}$$

Putting $z-1=t$ or $z=1+t$ in (1), we get

$$f(1+t)=\frac{(1+t)^3}{t^4(t-1)(t-2)}=\frac{1}{t^4}(t^3+3t^2+3t+1)(1-t)^{-1}\frac{1}{2}\left(1-\frac{t}{2}\right)^{-1}$$

$$=\frac{1}{2}\left(\frac{1}{t}+\frac{3}{t^2}+\frac{3}{t^3}+\frac{1}{t^4}\right)(1+t+t^2+t^3+\ldots)\times\left(1+\frac{t}{2}+\frac{t^2}{4}+\frac{t^3}{8}\ldots\right)$$

$$=\frac{1}{2}\left(\frac{1}{t}+\frac{3}{t^2}+\frac{3}{t^3}+\frac{1}{t^4}\right)\left(1+\frac{3}{2}t+\frac{7}{4}t^2+\frac{15}{8}t^3+\ldots\right)=\frac{1}{2}\left(\frac{1}{t}+\frac{9}{2t}+\frac{21}{4}\frac{1}{t}+\frac{15}{8}\frac{1}{t}\right)+\ldots$$

$$=\frac{1}{2}\left(1+\frac{9}{2}+\frac{21}{4}+\frac{15}{8}\right)\frac{1}{t} \qquad \left[\text{Res } f(a)=\text{coeffi. of }\frac{1}{t}\right]$$

Coefficient of $\frac{1}{t}=\frac{1}{2}\left(1+\frac{9}{2}+\frac{21}{4}+\frac{15}{8}\right)=\frac{101}{16}$,

Hence, the residue of the given function at a pole of order 4 is $\frac{101}{16}$. **Ans.**

Example 26. *Find the residue of* $f(z)=\frac{ze^z}{(z-a)^3}$ *at its pole.*

Solution. The pole of $f(z)$ is given by $(z-a)^3$ *i.e.,* $z = a$

Here $z = a$ is a pole of order 3.

Putting $z - a = t$ where t is small.

$$f(z)=\frac{ze^z}{(z-a)^3} \Rightarrow f(z)=\frac{(a+t)e^{a+t}}{t^3}=\left(\frac{a}{t^3}+\frac{1}{t^2}\right)e^{a+t}=e^a\left(\frac{a}{t^3}+\frac{1}{t^2}\right)e^t \qquad (z=a+t)$$

$$=e^a\left(\frac{a}{t^3}+\frac{1}{t^2}\right)\left(1+\frac{t}{1!}+\frac{t^2}{2!}+\ldots\right)=e^a\left[\frac{a}{t^3}+\frac{a}{t^2}+\frac{a}{2t}+\frac{1}{t^2}+\frac{1}{t}+\frac{1}{2}+\ldots\right]$$

$$=e^a\left[\frac{1}{2}+\left(\frac{a}{2}+1\right)\frac{1}{t}+(a+1)\frac{1}{t^2}+(a)\frac{1}{t^3}+\ldots\right]$$

Coefficient of $\frac{1}{t}=e^a\left(\frac{a}{2}+1\right)$

Hence the residue at $z=a$ is $e^a\left(\frac{a}{2}+1\right)$. **Ans.**

Example 27. *Find the sum of the residues of the function* $f(z) = \dfrac{\sin z}{z \cos z}$ *at its poles inside the circle* $|z| = 2$.

Solution. We have,

$$f(z) = \frac{\sin z}{z \cos z}$$

The pole can be determined by putting denominator

$$z \cos z = 0$$

$$\Rightarrow \quad z = 0, \pm\frac{\pi}{2}, \pm\frac{3\pi}{2}, \ldots\ldots$$

Of these poles only $z = 0$, $z = \pm\dfrac{\pi}{2}$ lie inside a circle $|z| = 2$.

Residue of $f(z)$ at $z = 0$ is $\lim\limits_{z\to 0} |z \cdot f(z)| = \lim\limits_{z\to 0} \dfrac{\sin z}{\cos z} = 0$ ...(1)

Residue of $f(z)$ at $z = \dfrac{\pi}{2}$ is

$$\lim_{z\to\frac{\pi}{2}} \left(z - \frac{\pi}{2}\right) f(z) = \lim_{z\to\frac{\pi}{2}} \frac{\left(z - \frac{\pi}{2}\right)\sin z}{z \cos z} \qquad \left[\text{From } \frac{0}{0}\right]$$

$$= \lim_{z\to\frac{\pi}{2}} \frac{\left(z - \frac{\pi}{2}\right)\cos z + \sin z}{\cos z - z \sin z} \qquad \text{[By L' Hopital's Rule]}$$

$$= \frac{1}{-\frac{\pi}{2}} = -\frac{2}{\pi} \qquad \ldots(2)$$

Similarly, residue of $f(z)$ at $z = -\dfrac{\pi}{2}$ is $\dfrac{2}{\pi}$ ...(3)

$\therefore$ Sum of the residues $= 0 - \dfrac{2}{\pi} + \dfrac{2}{\pi} = 0.$ **Ans.**

EXERCISE 26.2

1. Determine the poles of the following functions. Find the order of each pole.

(i) $\dfrac{z^2}{(z-a)(z-b)(z-c)}$ **Ans.** Simple poles at $z = a, z = b, z = c$

(ii) $\dfrac{z-3}{(z-2)^2(z+1)}$ **Ans.** Pole at $z = 2$ of second order and $z = -1$of first order.

(iii) $\dfrac{ze^{iz}}{z^2+a^2}$ **Ans.** Poles at $z = \pm ia$, order 1.

(iv) $\dfrac{1}{(z-1)(z-2)}$ **Ans.** $z = 2, z = 1$

Find the residue of

2. $\dfrac{z^3}{(z-2)(z-3)}$ at its poles. **Ans.** 27, –8

3. $\dfrac{z^2}{z^2+a^2}$ at $z = ia$. **Ans.** $\dfrac{1}{2}ia$

4. $\dfrac{1}{(z^2+a^2)^2}$ at $z = ia$ **Ans.** $-\dfrac{i}{4a^3}$

5. $\tan z$ at its pole. **Ans.** – 1 at pole $z = \left(n+\dfrac{1}{2}\right)\pi$

6. $f(z) = \dfrac{z^3}{z(z^2+4)}$ at $z = 0$, $z = 2i$. **Ans.** 0, – 2

7. $z^2 e^{1/z}$ at the point $z = 0$. **Ans.** $\dfrac{1}{6}$

8. $z^2 \sin\left(\dfrac{1}{z}\right)$ at $z = 0$ **Ans.** $-\dfrac{1}{6}$

9. $\dfrac{1}{z^2(z-i)}$ at $z = i$ **Ans.** –1

10. $\dfrac{e^{2z}}{1-e^z}$ at its pole **Ans.** – 1

11. $\dfrac{1+e^z}{\sin z + z\cos z}$ at $z = 0$ **Ans.** 1

12. $\dfrac{1}{z(e^z-1)}$ at its poles **Ans.** $-\dfrac{1}{2}$

26.12 RESIDUE THEOREM

If $f(z)$ is analytic in a closed curve C, except at a finite number of poles within C, then $\int_c f(z)\,dz = 2\pi i$ **(sum of residues at the poles within *C*).**

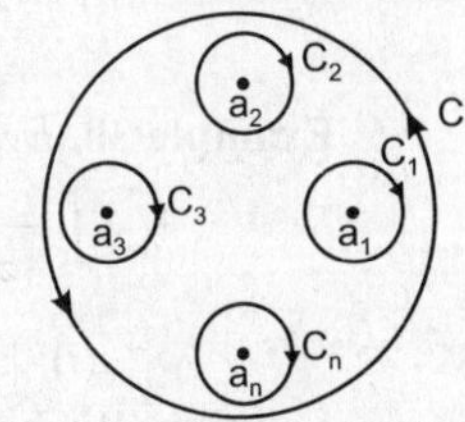

Proof. Let $C_1, C_2, C_3, ..., C_n$ be the non-intersecting circles with centres at $a_1, a_2, a_3, ..., a_n$ respectively, and radii so small that they lie entirely within the closed curve C. Then $f(z)$ is analytic in the multiple connected region lying between the curves C and $C_1, C_2, ..., C_n$.

Applying Cauchy's theorem

$$\int_c f(z)\,dz = \int_{c1} f(z)\,dz + \int_{c2} f(z)\,dz + \int_{c3} f(z)\,dz + ... + \int_{cn} f(z)\,dz.$$

$$= 2\pi i\,[\text{Res } f(a_1) + \text{Res } f(a_2) + \text{Res } f(a_3) + ... + \text{Res } f(a_n)]$$ **Proved.**

Example 28. *Evaluate the following integral using residue theorem*

$$\int_c \frac{1+z}{z\,(2-z)}\,dz$$

where c is the circle $|z| = 1$.

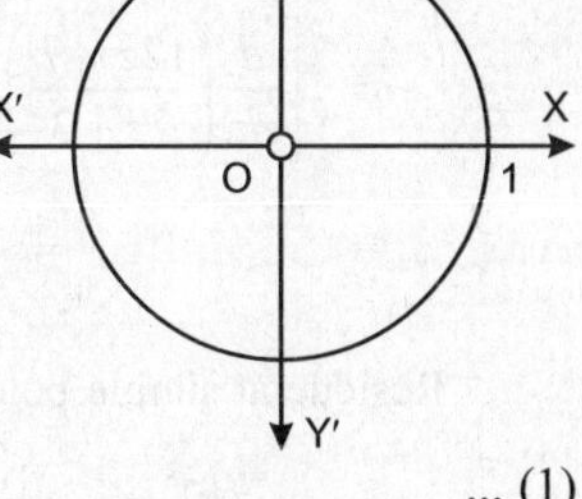

Solution. The poles of the integrand are given by putting the denominator equal to zero.

$$z(2-z) = 0 \text{ or } z = 0, 2$$

The integrand is analytic on $|z| = 1$ and all points inside except $z = 0$, as a pole at $z = 0$ is inside the circle $|z| = 1$.

Hence by residue theorem

$$\int_c \frac{1+z}{z(2-z)}\,dz = 2\pi i\ [\text{Res } f(0)] \qquad ...(1)$$

$$\text{Residue } f(0) = \lim_{z\to 0} z.\frac{1+z}{z\,(2-z)} = \lim_{z\to 0}\frac{1+z}{2-z} = \frac{1}{2}$$

Putting the value of Residue $f(0)$ in (1), we get

$$\int_c \frac{1+z}{z(2-z)}\,dz = 2\pi i\left(\frac{1}{2}\right) = \pi i$$ **Ans.**

Example 29. *Evaluate the following integral using residue theorem*

$$\int_c \frac{4-3z}{z(z-1)(z-2)}\,dz$$

where c is the circle $|z| = \dfrac{3}{2}$.

Solution. The poles of the function $f(z)$ are given by equating the denominator to zero.

$$z(z-1)(z-2) = 0, \quad z = 0, 1, 2$$

The function has poles at $z = 0$, $z = 1$ and $z = 2$ of which the given circle encloses the pole at $z = 0$ and $z = 1$.

Residue of $f(z)$ at the simple pole $z = 0$ is

$$= \lim_{z\to 0} z\frac{4-3z}{z(z-1)(z-2)} = \lim_{z\to 0}\frac{4-3z}{(z-1)(z-2)}$$

$$= \frac{4-0}{(0-1)(0-2)} = 2$$

Residue of $f(z)$ at the simple pole $z = 1$ is

$$= \lim_{z\to 1}(z-1)\frac{4-3z}{z(z-1)(z-2)} = \lim_{z\to 1}\frac{4-3z}{z(z-2)} = \frac{4-3}{1(1-2)} = -1$$

By Cauchy's integral formula

$$\int_C f(z)dz = 2\pi i\times \text{ sum of the residue within } c$$

$$= 2\pi i\times(2-1) = 2\pi i$$

Ans.

Example 30. *Evaluate*

$$\int_C \frac{12z-7}{(z-1)^2\,(2z+3)}\,dz, \text{ where } C \text{ is the circle}$$

(i) $|z| = 2$ (ii) $|z+i| = \sqrt{3}$

Solution. We have, $f(z) = \dfrac{12z-7}{(z-1)^2(2z+3)}$

Poles are given by

$$z = 1 \text{ (double pole) and } z = -\frac{3}{2} \text{ (simple pole)}$$

Residue at $(z = 1)$ is

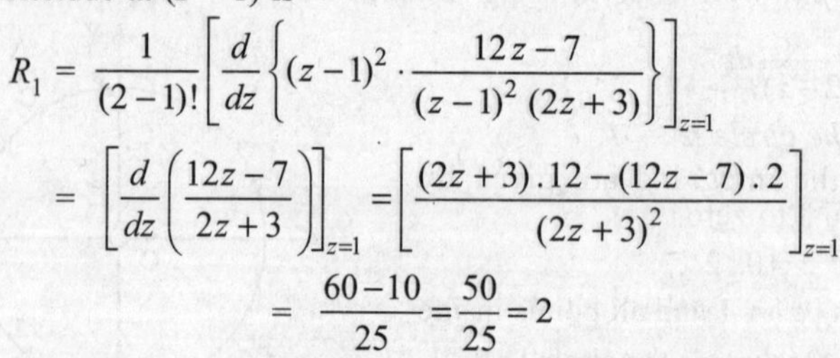

$$R_1 = \frac{1}{(2-1)!}\left[\frac{d}{dz}\left\{(z-1)^2\cdot\frac{12z-7}{(z-1)^2\,(2z+3)}\right\}\right]_{z=1}$$

$$= \left[\frac{d}{dz}\left(\frac{12z-7}{2z+3}\right)\right]_{z=1} = \left[\frac{(2z+3).12-(12z-7).2}{(2z+3)^2}\right]_{z=1}$$

$$= \frac{60-10}{25} = \frac{50}{25} = 2$$

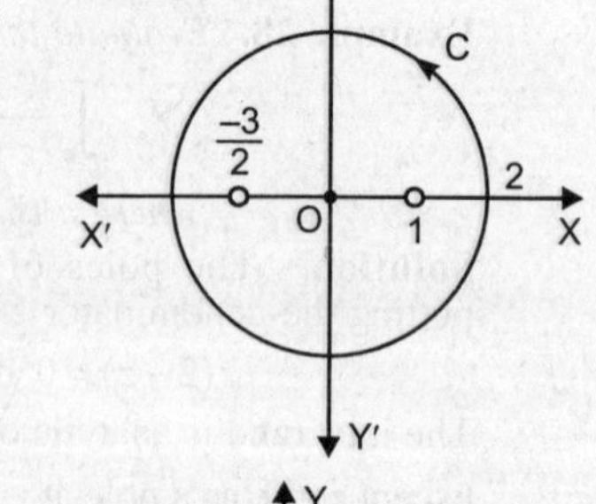

Residue at simple pole $\left(z = -\dfrac{3}{2}\right)$ is

$$R_2 = \lim_{z\to -3/2}\left(z+\frac{3}{2}\right).\frac{12z-7}{(z-1)^2\,(2z+3)}$$

$$= \lim_{z\to -3/2}\frac{1}{2}.\frac{(12z-7)}{(z-1)^2} = -2.$$

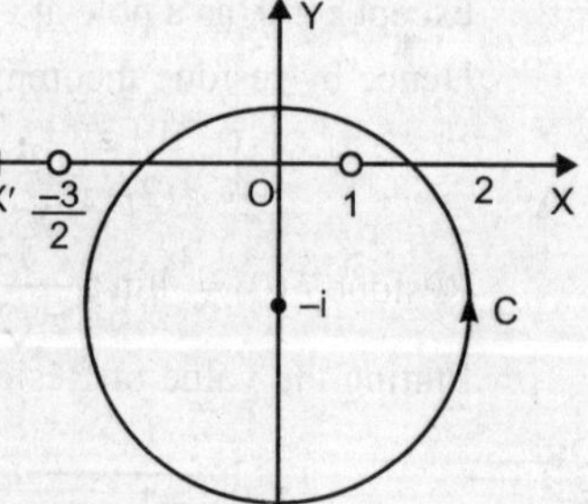

(i) The contour $|z| = 2$ encloses both the poles 1 and $-\dfrac{3}{2}$.

$\therefore$ The given integral $= 2\pi i\,(R_1 + R_2) = 2\pi i\,(2 - 2) = 0$.

(ii) The contour $|z+i| = \sqrt{3}$ is a circle of radius $\sqrt{3}$ and centre at $z = -i$.

The distances of the centre from $z = 1$ and $-\dfrac{3}{2}$ are respectively $\sqrt{2}$ and $\sqrt{\dfrac{13}{4}}$. and $\sqrt{\dfrac{13}{4}}$

The first of these is $< \sqrt{3}$ and the second is $> \sqrt{3}$.

$\therefore$ The second contour includes only the first singularity $z = 1$.

Hence, the given integral $= 2\pi i\,(R_1) = 2\pi i\,(2) = 4\pi i$. **Ans.**

Example 31. *Determine the poles of the following function and residue at each pole:*

$$f(z) = \frac{z^2}{(z-1)^2(z+2)} \quad \textit{and hence evaluate}$$

$$\int_c \frac{z^2 dz}{(z-1)^2(z+2)} \quad \textit{where } c: |z| = 3. \qquad \textit{(R.G.P.V. Bhopal, III Sem. Dec. 2007)}$$

Solution. $$f(z) = \frac{z^2}{(z-1)^2(z+2)}$$

Poles of $f(z)$ are given by $(z-1)^2(z+2) = 0$ *i.e.* $z = 1, 1, -2$

The pole at $z = 1$ is of second order and the pole at $z = -2$ is simple.

$$\text{Residue of } f(z) \text{ (at } z = 1) = \lim_{z\to 1} \frac{1}{(2-1)!} \frac{d}{dz} \frac{(z-1)^2 z^2}{(z-1)^2(z+2)}$$

$$= \lim_{z\to 1} \frac{d}{dz} \frac{z^2}{z+2} = \lim_{z\to 1} \frac{(z+2)2z - 1.z^2}{(z+2)^2}$$

$$= \lim_{z\to 1} \frac{z^2+4z}{(z+2)^2} = \frac{1+4}{(1+2)^2} = \frac{5}{9}$$

$$\text{Residue of } f(z) \text{ (at } z = -2) = \lim_{z\to -2} \frac{(z+2)z^2}{(z-1)^2(z+2)} = \lim_{z\to -2} \frac{z^2}{(z-1)^2} = \frac{4}{(-2-1)^2} = \frac{4}{9} \quad \textbf{Ans.}$$

$$\int_c \frac{z^2 dz}{(z-1)^2(z+2)} = 2\pi i \left(\frac{5}{9} + \frac{4}{9}\right) = 2\pi i \quad \textbf{Ans.}$$

Example 32. *Using Residue theorem, evaluate* $\frac{1}{2\pi i}\int_C \frac{e^{zt} dz}{z^2(z^2+2z+2)}$

where C is the cirlce $|z| = 3$. *(U.P., III Semester, Dec. 2009)*

Solution. Here, we have

$$\frac{1}{2\pi i}\int_C \frac{e^{zt} dz}{z^2(z^2+2z+2)}$$

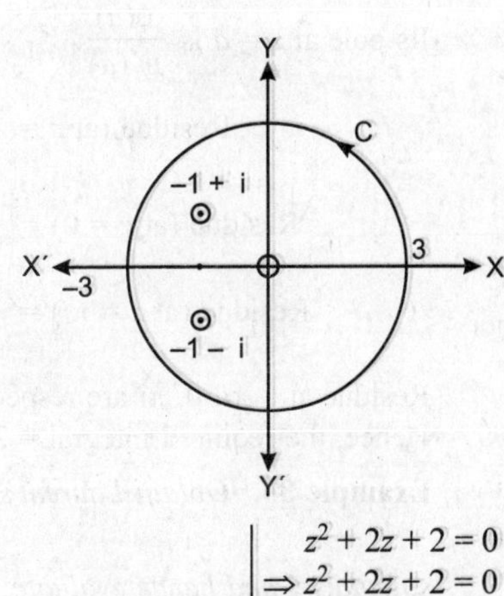

Poles are given by

$z = 0$ (double pole)

$z = -1 \pm i$ (simple poles)

All the four poles are inside the given circle.

$|z| = 3$

$$\text{Residue at } z = 0 \text{ is } \lim_{z\to 0} \frac{d}{dz} z^2 \frac{e^{zt}}{z^2(z^2+2z+2)}$$

$$= \lim_{z\to 0} \frac{d}{dz} \frac{e^{zt}}{z^2+2z+2}$$

$$= \lim_{z\to 0} \frac{(z^2+2z+2)\, t e^{zt} - (2z+2)e^{zt}}{(z^2+2z+2)^2}$$

$$= \frac{2te^{\circ} - 2e^{\circ}}{4} = \frac{(t-1)}{2}$$

$z^2 + 2z + 2 = 0$
$\Rightarrow z^2 + 2z + 2 = 0$
$\Rightarrow (z+1)^2 = -1$
$\Rightarrow z + 1 = \pm i$
$\Rightarrow z = -1 \pm i$

Residue at $z = -1 + i$

$$= \lim_{z \to -1+i} \frac{(z+1-i)\, e^{zt}}{z^2\,(z+1-i)\,(z+1+i)} = \lim_{z \to -1+i} \frac{e^{zt}}{z^2\,(z+1+i)}$$

$$= \frac{e^{(-1+i)t}}{(-1+i)^2\,(-1+i+1+i)} = \frac{e^{(-1+i)\,t}}{(1-2i-1)\,(2i)} = \frac{e^{(-1+i)\,t}}{4}$$

Similary Residue (at $z = -1 - i$) $= \dfrac{e^{(-1-i)\,t}}{4}$

$$\int \frac{e^{zt}}{z^2\,(z^2+2z+2)}\, dz = 2\pi i \text{ Sum of the Residues)}$$

$$\Rightarrow \quad \frac{1}{2\pi i} \int \frac{e^{zt}}{z^2\,(z^2+2z+2)}\, dz = \frac{t-1}{2} + \frac{e^{(-1+i)\,t}}{4} + \frac{e^{(-1-i)\,t}}{4}$$

$$= \frac{t-1}{2} + \frac{e^{-t}}{4}\,(e^{it} + e^{-it}) = \frac{t-1}{2} + \frac{e^{-t}}{4}\,(2\cos t)$$

$$= \frac{t-1}{2} + \frac{e^{-t}}{2}\cos t$$

Ans.

Example 33. *Evaluate* $\oint_C \dfrac{1}{\sinh z}\, dz$, *where C is the circle* $|z| = 4$.

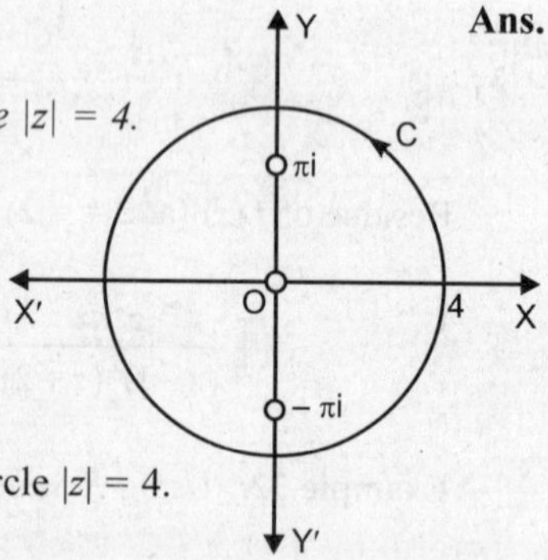

Solution. Here, $f(z) = \dfrac{1}{\sinh z}$.

Poles are given by

$$\sinh z = 0$$

$$\Rightarrow \quad \sin iz = 0$$

$$\Rightarrow \quad z = n\pi i \text{ where } n \text{ is an integer.}$$

Out of these, the poles $z = -\pi i$, 0 and πi lie inside the circle $|z| = 4$.

Y
C
πi
X′
O
4
X
– πi
Y′

The given function $\dfrac{1}{\sinh z}$ is of the form $\dfrac{\phi(z)}{\psi(z)}$

Its pole at $z = a$ is $\dfrac{\phi(a)}{\psi'(a)}$.

$$\text{Residue (at } z = -\pi i) = \frac{1}{\cosh(-\pi i)} = \frac{1}{\cos i(-\pi i)} = \frac{1}{\cos \pi} = \frac{1}{-1} = -1$$

$$\text{Residue (at } z = 0) = \frac{1}{\cosh 0} = \frac{1}{1} = 1$$

$$\text{Residue (at } z = \pi i) = \frac{1}{\cosh(\pi i)} = \frac{1}{\cos i\,(\pi i)} = \frac{1}{\cos(-\pi)} = \frac{1}{\cos \pi} = \frac{1}{-1} = -1$$

Residue at $-\pi i$, 0, πi are respectively -1, 1 and -1.

Hence, the required integral $= 2\pi i\,(-1 + 1 - 1) = -2\pi i$. **Ans.**

Example 34. *Obtain Laurent's expansion for the function* $f(z) = \dfrac{1}{z^2 \sinh z}$ *at the isolated singularity and hence evaluate* $\oint_C \dfrac{1}{z^2 \sinh z}\, dz$, *where C is the circle* $|z - 1| = 2$.

Solution. Here,

$$f(z) = \frac{1}{z^2 \sinh z} = \frac{2}{z^2\,(e^z - e^{-z})} = \frac{2}{z^2\left[\left(1 + z + \frac{z^2}{2!} + \frac{z^3}{3!} + \ldots\right) - \left(1 - z + \frac{z^2}{2!} - \frac{z^3}{3!} + \ldots\right)\right]}$$

$$= \frac{2}{z^2\left(2z+\frac{2z^3}{3!}+\frac{2z^5}{5!}+...\right)} = \frac{1}{z^3\left(1+\frac{z^2}{3!}+\frac{z^4}{5!}+...\right)}$$

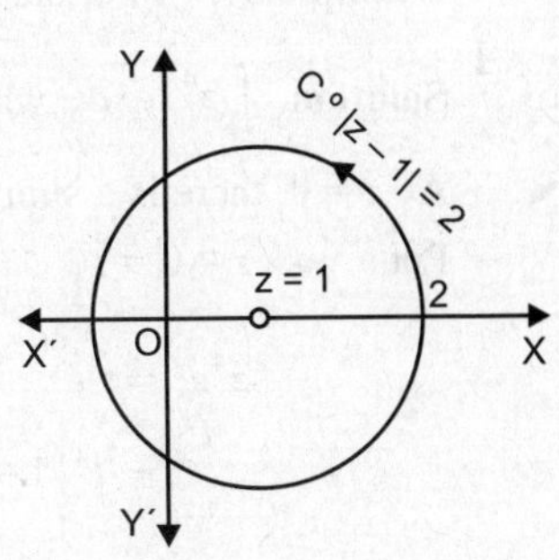

$$= z^{-3}\left[1+\left(\frac{z^2}{6}+\frac{z^4}{120}\right)+...\right]^{-1}$$

$$= z^{-3}\left(1-\frac{z^2}{6}-\frac{z^4}{120}+...\right)$$

$$= z^{-3}-\frac{z^{-1}}{6}-\frac{z}{120}+... = \frac{1}{z^3}-\frac{1}{6z}-\frac{z}{120}+...$$

which is the required Laurent's expansion.

Only pole $z = 0$ of order three lies inside the circle $C \equiv |z-1| = 2$.

Residue of $f(z)$ at $(z = 0)$ is

$$= \text{coeff. of } \frac{1}{z} \text{ in the Laurent's expansion of } f(z) = -\frac{1}{6}. \quad \textbf{Ans.}$$

Example 35. *Evaluate* $\int_c \frac{dz}{z\sin z}$ *: c is the unit circle about origin.*

Solution. $\frac{1}{z\sin z} = \frac{1}{z\left[z-\frac{z^3}{3!}+\frac{z^5}{5!}-...\right]} = \frac{1}{z^2\left[1-\frac{z^2}{3!}+\frac{z^4}{5!}-...\right]}$

$$= \frac{1}{z^2}\left[1-\left(\frac{z^2}{6}-\frac{z^4}{120}...\right)\right]^{-1} = \frac{1}{z^2}\left[1+\left(\frac{z^2}{6}-\frac{z^4}{120}+...\right)+\left(\frac{z^2}{6}-\frac{z^4}{120}...\right)^2...\right]$$

$$= \frac{1}{z^2}\left[1+\frac{z^2}{6}-\frac{z^4}{120}+\frac{z^4}{36}+...\right] = \frac{1}{z^2}+\frac{1}{6}-\frac{z^2}{120}+\frac{z^2}{36}...$$

$$= \frac{1}{z^2}+\frac{1}{6}+\frac{7}{360}z^2...$$

This shows that $z = 0$ is a pole of order 2 for the function $\frac{1}{z\sin z}$ and the residue at the pole is zero, (coefficient of $\frac{1}{z}$).

Now the pole at $z = 0$ lies within C.

$$\therefore \quad \int\frac{1}{z\sin z}dz = 2\pi i \text{ (Sum of Residues)} = 0 \quad \textbf{Ans.}$$

Example 36. *Find the value of* $\oint_C z\,e^{1/z}\,dz$*, around the unit circle.*

Solution. The only singularity of $z\,e^{1/z}$ is at the origin. Expanding $e^{1/z}$, we have

$$z\,e^{1/z} = z\left[1+\frac{1}{z}+\frac{1}{2z^2}+\frac{1}{6z^3}+...\right]$$

$$= z+1+\frac{1}{2z}+\frac{1}{6z^2}+...$$

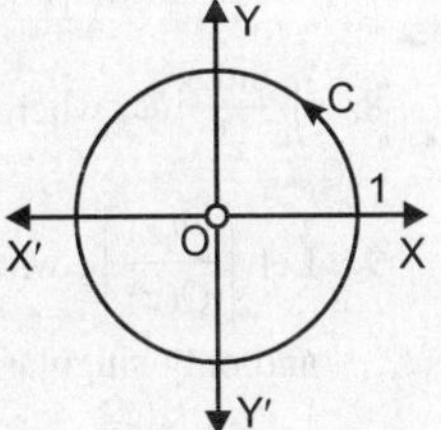

Residue at origin $= \text{coeff. of } \frac{1}{z} = \frac{1}{2}$.

Hence, the required integral $= 2\pi i\left(\frac{1}{2}\right) = \pi i.$ **Ans.**

Example 37. *Find the value of the complex integral* $\int_c z^4 e^{1/z} dz$, *where c is* $|z| = 1$.

Solution. $\int_c z^4 \cdot e^{\frac{1}{z}} dz$ where c is $|z| = 1$

At $z = 0$, there is a simple pole. Now we have to find out the residue at $z = 0$.

Put $z - 0 = t$

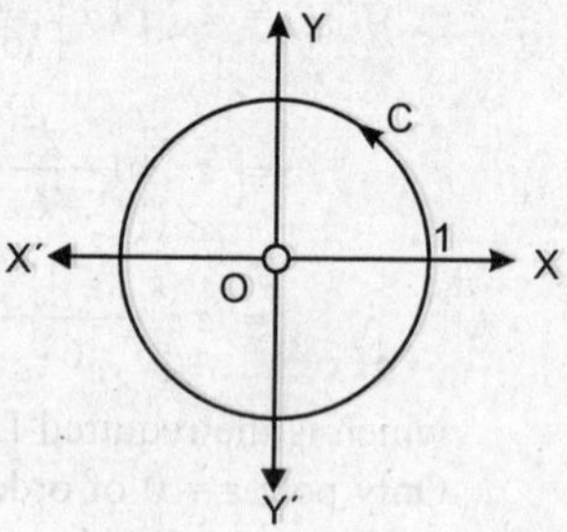

$$z^4 e^{\frac{1}{z}} = t^4 e^{\frac{1}{t}}$$

$$= t^4 \left[1 + \frac{1}{t} + \frac{1}{2!t^2} + \frac{1}{3!}\frac{1}{t^3} + \frac{1}{4!}\frac{1}{t^4} + \frac{1}{5!}\frac{1}{t^5} + \ldots\right]$$

$$= t^4 + t^3 + \frac{t^2}{2!} + \frac{t}{3!} + \frac{1}{4!} + \frac{1}{5!}\frac{1}{t} + \ldots$$

Coefficient of $\frac{1}{t} = \frac{1}{5!} = \frac{1}{120}$

$$\int_c z^4 e^{\frac{1}{z}} dz = 2\pi i \text{ Residue}$$

$$\Rightarrow \quad \int_C z^4 e^{\frac{1}{z}} dz = (2\pi i)\left(\frac{1}{120}\right) = \frac{\pi i}{60}$$ **Ans.**

EXERCISE 26.3

Evaluate the following complex integrals:

1. $\int_c \frac{1-2z}{z(z-1)(z-2)} dz$, where c is the circle $|z| = 1.5$ **Ans.** $2\pi i$

2. $\int_c \frac{z^2 e^{zt}}{z^2+1} dz$, where c is the circle $|z| = 2$ **Ans.** $-2\pi i \sin t$

3. $\int_c \frac{z-1}{(z+1)^2(z-2)} dz$, where c is the circle $|z - i| = 2$. **Ans.** $-\frac{2\pi i}{9}$

4. $\int_c \frac{2z^2+z}{z^2-1} dz$, where c is the circle $|z - 1| = 1$. **Ans.** $3\pi i$

5. $\int_c \frac{e^{2z}+z^2}{(z-1)^5} dz$, where c is the circle $|z| = 2$ **Ans.** $\frac{4\pi e^2 i}{3}$

6. $\int_c \frac{dz}{(z^2+1)(z^2-4)}$, where c is the circle $|z| = 1.5$ **Ans.** 0

7. $\int_c \frac{4z^2-4z+1}{(z-2)(z^2+4)} dz$, where c is the circle $|z| = 1$ **Ans.** 0

8. $\int_c \frac{\sin z}{z^6} dz$, where c is the circle $|z| = 2$ **Ans.** $\frac{\pi i}{60}$

9. Let $\left[\frac{P(z)}{Q(z)}\right]$, where both $P(z)$ and $Q(z)$ are complex polynomial of degree two. If $f(0) = f(-1) = 0$ and only singularity of $f(z)$ is of order 2 at $z = 1$ with residue -1, then find $f(z)$.

Ans. $f(z) = -\frac{1}{3}\frac{z(z+1)}{(z-1)^2}$

26.13 EVALUATION OF REAL DEFINITE INTEGRALS BY CONTOUR INTEGRATION

A large number of real definite integrals, whose evaluation by usual methods become sometimes very tedious, can be easily evaluated by using Cauchy's theorem of residues. For finding the integrals we take a closed curve C, find the poles of the function $f(z)$ and calculate residues at those poles only which lie within the curve C.

$$\int_C f(z)\,dz = 2\pi i \text{ (sum of the residues of } f(z) \text{ at the poles within } C)$$

We call the curve, a contour and the process of integration along a contour is called contour integration.

26.14 INTEGRATION ROUND UNIT CIRCLE OF THE TYPE

$$\int_0^{2\pi} f(\cos\theta, \sin\theta)\,d\theta$$

where $f(\cos\theta, \sin\theta)$ is a rational function of $\cos\theta$ and $\sin\theta$.

Convert $\sin\theta$, $\cos\theta$ into z.

Consider a circle of unit radius with centre at origin, as contour.

$$\sin\theta = \frac{e^{i\theta}-e^{-i\theta}}{2i} = \frac{1}{2i}\left[z-\frac{1}{z}\right], \qquad z = re^{i\theta} = 1.\,e^{i\theta} = e^{i\theta}$$

$$\cos\theta = \frac{e^{i\theta}+e^{-i\theta}}{2} = \frac{1}{2}\left[z+\frac{1}{z}\right]$$

As we know

$$z = e^{i\theta} \Rightarrow dz = e^{i\theta}\,i\,d\theta = z\,i\,d\theta \text{ or } d\theta = \frac{dz}{iz}$$

The integrand is converted into a function of z.

Then apply Cauchy's residue theorem to evaluate the integral.

Some examples of these are illustrated below.

Example 38. *Evaluate the integral:*

$$\int_0^{2\pi} \frac{d\theta}{5-3\cos\theta}$$

(R.G.P.V., Bhopal, III Semester, June 2007)

Solution.
$$\int_0^{2\pi} \frac{d\theta}{5-3\cos\theta} = \int_0^{2\pi} \frac{d\theta}{5-3\left(\frac{e^{i\theta}+e^{-i\theta}}{2}\right)} = \int_0^{2\pi} \frac{2d\theta}{10-3e^{i\theta}-3e^{-i\theta}}$$

$$= \int_C \frac{2}{10-3z-\frac{3}{z}}\,\frac{dz}{iz} = \frac{1}{i}\int_C \frac{2dz}{10z-3z^2-3} \qquad \left[\begin{array}{l} e^{i\theta} = z \Rightarrow i.e.^{i\theta}\,d\theta = dz \\ d\theta = \dfrac{dz}{iz} \end{array}\right]$$

[C is the unit circle $|z| = 1$]

$$= -\frac{2}{i}\int_C \frac{dz}{3z^2-10z+3}$$

$$= -\frac{2}{i}\int_C \frac{dz}{(3z-1)(z-3)} = 2i\int_C \frac{dz}{(3z-1)(z-3)}$$

Let $\quad I = 2i\int_C \frac{dz}{(3z-1)(z-3)}]$

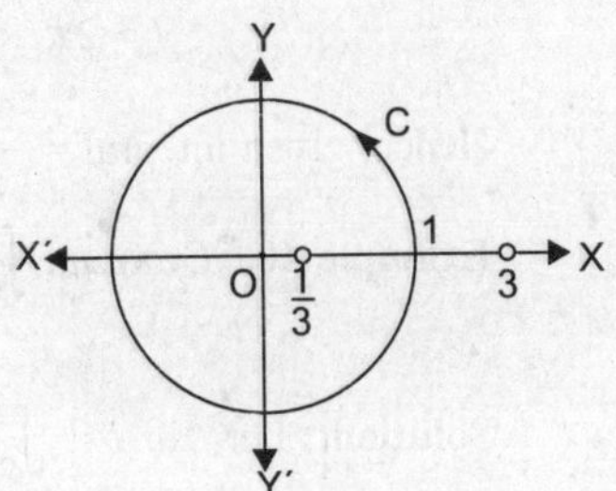

Poles of the integrand are given by

$$(3z-1)(z-3) = 0 \quad \Rightarrow \quad z = \frac{1}{3}, 3$$

There is only one pole at $z=\frac{1}{3}$ inside the unit circle C.

Residue at $z=\frac{1}{3}$ $=\lim\limits_{z\to\frac{1}{3}}\left(z-\frac{1}{3}\right)f(z)=\lim\limits_{z\to\frac{1}{3}}\frac{2i\left(z-\frac{1}{3}\right)}{(3z-1)(z-3)}=\lim\limits_{z\to\frac{1}{3}}\frac{2i}{3(z-3)}$

$$=\frac{2i}{3\left(\frac{1}{3}-3\right)}=-\frac{i}{4}$$

Hence, by Cauchy's Residue Theorem

$$I=2\pi i\text{ (Sum of the residues within Contour)}=2\pi i\left(-\frac{i}{4}\right)=\frac{\pi}{2}$$

$$\int_0^{2\pi}\frac{d\theta}{5-3\cos\theta}=\frac{\pi}{2}$$ **Ans.**

Example 39. *Use residue calculus to evaluate the following integral*

$$\int_0^{2\pi}\frac{1}{5-4\sin\theta}d\theta$$

Solution. Let $\text{I}=\int_0^{2\pi}\frac{1}{5-4\sin\theta}d\theta=\int_0^{2\pi}\frac{1}{5-4\left(\frac{e^{i\theta}-e^{-i\theta}}{2i}\right)}d\theta$

$$=\int_0^{2\pi}\frac{d\theta}{5+2ie^{i\theta}-2ie^{-i\theta}}$$ [writing $e^{i\theta}=z,\ d\theta=\frac{dz}{iz}$]

$$=\int_c\frac{1}{5+2iz-\frac{2i}{z}}\frac{dz}{iz}$$ where c is the unit circle $|z|=1$.

$$=\int_c\frac{dz}{5iz-2z^2+2}$$

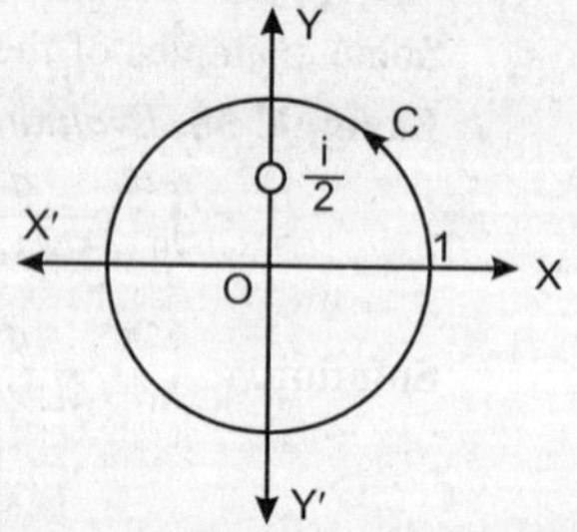

Poles of integrand are given by

$$-2z^2+5iz+2=0\Rightarrow z=\frac{-5i\pm\sqrt{-25+16}}{-4}=\frac{-5i\pm3i}{-4}=2i,\frac{i}{2}$$

Only $z=\frac{i}{2}$ lies inside c.

Residue at the simple pole at $z=\frac{i}{2}$ is

$$\lim_{z\to\frac{i}{2}}\left(z-\frac{i}{2}\right)\times\left[\frac{1}{(2z-i)(-z+2i)}\right]=\lim_{z\to\frac{i}{2}}\frac{1}{2(-z+2i)}=\frac{1}{2\left(-\frac{i}{2}+2i\right)}=\frac{1}{3i}$$

Hence, by Cauchy's residue theorem

$$I=2\pi i\times\text{ Sum of residues within the contour }=2\pi i\times\frac{1}{3i}=\frac{2\pi}{3}$$

Hence, given integral $=\frac{2\pi}{3}$ **Ans.**

Example 40. *Evaluate* $\int_0^{2\pi}\frac{d\theta}{a+b\sin\theta}$ if $a>|b|$

(U.P. III Semester 2009-2010; G.B.T.U., 2012)

Solution. Let $I=\int_0^{2\pi}\frac{d\theta}{a+b\sin\theta}$

$$= \int_0^{2\pi} \frac{1}{a + b\dfrac{e^{i\theta} - e^{-i\theta}}{2i}} d\theta \qquad \left[\text{Writing } e^{i\theta} = z, d\theta = \frac{dz}{iz}\right]$$

$$= \int_C \frac{1}{a + \dfrac{b}{2i}\left(z - \dfrac{1}{z}\right)} \frac{dz}{iz} \quad \text{(where } C \text{ is the unit circle } |z| = 1$$

$$= \int_c \frac{2}{2iaz + bz^2 - b} dz$$

$$= \int_c \frac{2}{bz^2 + 2aiz - b} dz = \frac{1}{b} \int \frac{2}{z^2 + \dfrac{2aiz}{b} - 1} dz$$

$$= \frac{1}{b} \int_c \frac{2}{(z-\alpha)(z-\beta)} dz \qquad \left[bz^2 + 2aiz - b = b\left\{z^2 + \frac{2aiz}{b} - 1\right\}\right]$$

Where $\alpha + \beta = -\dfrac{2ai}{b}$

$\alpha\beta = -1 \quad \Rightarrow |\alpha\beta| = 1$

$|\alpha| < 1$ then $|\beta| > 1$

$$\left[\begin{aligned}(\alpha-\beta)^2 &= (\alpha+\beta)^2 - 4\alpha\beta \\ &= -\frac{4a^2}{b^2} + 4 \\ \alpha - \beta &= 2\frac{\sqrt{b^2 - a^2}}{b}\end{aligned}\right]$$

i.e.; Pole lies at $z = \alpha$ in the unit circle.

Residue at $z = a = \lim_{z \to \alpha} (z-\alpha) \dfrac{2}{(z-\alpha)(z-\beta)} = \dfrac{2}{\alpha - \beta} = \dfrac{b}{\sqrt{b^2 - a^2}} = \dfrac{b}{i\sqrt{a^2 - b^2}}$

$$\int_0^{2\pi} \frac{1}{a + b\sin\theta} d\theta = \frac{1}{b} \int_c \frac{2}{z^2 + 2\dfrac{aiz}{b} - 1} dz = 2\pi i \frac{b}{bi\sqrt{a^2 - b^2}} = \frac{2\pi}{\sqrt{a^2 - b^2}} \qquad \textbf{Ans.}$$

Example 41. *Use the complex variable technique to find the value of the integral:*
$$\int_0^{2\pi} \frac{d\theta}{2 + \cos\theta}.$$

Solution. Let $I = \displaystyle\int_0^{2\pi} \frac{d\theta}{2 + \cos\theta} = \int_0^{2\pi} \frac{d\theta}{2 + \dfrac{e^{i\theta} + e^{-i\theta}}{2}} = \int_0^{2\pi} \frac{2d\theta}{4 + e^{i\theta} + e^{-i\theta}}$

Put $e^{i\theta} = z$ so that $e^{i\theta}(i\,d\theta) = dz \Rightarrow iz\,d\theta = dz \Rightarrow d\theta = \dfrac{dz}{iz}$

$$I = \int_C \frac{2\dfrac{dz}{iz}}{4 + z + \dfrac{1}{z}} \qquad \text{where } c \text{ denotes the unit circle } |z| = 1.$$

$$= \frac{1}{i} \int_c \frac{2\,dz}{z^2 + 4z + 1}$$

The poles are given by putting the denominator equal to zero.

$$z^2 + 4z + 1 = 0 \text{ or } z = \frac{-4 \pm \sqrt{16 - 4}}{2} = \frac{-4 \pm 2\sqrt{3}}{2} = -2 \pm \sqrt{3}$$

The pole within the unit circle C is a simple pole at $z = -2 + \sqrt{3}$. Now we calculate the residue at this pole.

$$\text{Residue at } (z=-2+\sqrt{3}) = \lim_{z\to(-2+\sqrt{3})} \frac{1}{i}\frac{(z+2-\sqrt{3})2}{(z+2-\sqrt{3})(z+2+\sqrt{3})}$$

$$= \lim_{z\to(-2+\sqrt{3})} \frac{2}{i(z+2+\sqrt{3})} = \frac{2}{i(-2+\sqrt{3}+2+\sqrt{3})} = \frac{1}{\sqrt{3}i}$$

Hence by Cauchy's Residue Theorem, we have

$$\int_0^{2\pi} \frac{d\theta}{2+\cos\theta} = 2\pi i \text{ (sum of the residues within the contour)}$$

$$= 2\pi i \frac{1}{i\sqrt{3}} = \frac{2\pi}{\sqrt{3}}$$ **Ans.**

Example 42. *Using complex variable techniques evaluate the real integral*

$$\int_0^{2\pi} \frac{\sin^2\theta}{5-4\cos\theta}\, d\theta$$ *(DU, III Sem. 2012)*

Solution. If we write $z = e^{i\theta}$

$$\cos\theta = \frac{1}{2}\left(z+\frac{1}{z}\right), \quad d\theta = \frac{dz}{iz}$$

and so $$I = \int_0^{2\pi} \frac{\sin^2\theta}{5-4\cos\theta}\, d\theta = \frac{1}{2}\int_0^{2\pi} \frac{1-\cos 2\theta}{5-4\cos\theta}\, d\theta$$

$$\text{I} = \text{Real part of } \frac{1}{2}\int_0^{2\pi} \frac{1-\cos 2\theta - i\sin 2\theta}{5-4\cos\theta}\, d\theta$$ [c is the unit $|z| = 1$]

$$= \text{Real part of } \frac{1}{2}\int_0^{2\pi} \frac{1-e^{2i\theta}}{5-4\cos\theta}\, d\theta$$

$$= \text{Real part of } \frac{1}{2}\int_c \frac{1-z^2}{5-2(z+\frac{1}{z})}\left(\frac{dz}{iz}\right) = \text{Real part of } \frac{1}{2i}\int_c \frac{1-z^2}{5z-2z^2-2}\, dz$$

$$= \text{Real part of } \frac{1}{2i}\int_c \frac{z^2-1}{2z^2-5z+2}\, dz$$

Poles are determined by $2z^2-5z+2=0$ or $(2z-1)(z-2) = 0$ or $z = \frac{1}{2}, 2$

So inside the contour C there is a simple pole at $z = \frac{1}{2}$

$$\text{Residue at the simple pole } \left(z=\frac{1}{2}\right) = \lim_{z\to\frac{1}{2}} \left(z-\frac{1}{2}\right)\frac{z^2-1}{(2z-1)(z-2)}$$

$$= \lim_{z\to\frac{1}{2}} \frac{z^2-1}{2(z-2)} = \frac{\frac{1}{4}-1}{2\left(\frac{1}{2}-2\right)} = \frac{1}{4}$$

$$I = \text{Real part of } \frac{1}{2i}\int_c \frac{(z^2-1)}{2z^2-5z+2}\, dz = \frac{1}{2i} 2\pi i \text{ (sum of the residues)}$$

$$\Rightarrow \quad \int_0^{2\pi} \frac{\sin^2\theta}{5-4\cos\theta}\, d\theta = \pi\left(\frac{1}{4}\right) = \frac{\pi}{4}$$ **Ans.**

Example 43. *Using contour integration, evaluate the real integral*

$$\int_0^{\pi} \frac{1+2\cos\theta}{5+4\cos\theta}\, d\theta$$

Solution. Let $I = \int_0^{\pi} \frac{1+2\cos\theta}{5+4\cos\theta}\, d\theta = \frac{1}{2}\int_0^{2\pi} \frac{1+2\cos\theta}{5+4\cos\theta}\, d\theta$ [Even function]

$$= \text{Real part of } \frac{1}{2}\int_0^{2\pi} \frac{1+2e^{i\theta}}{5+4\cos\theta}\, d\theta$$

$$= \text{Real part of } \frac{1}{2}\int_0^{2\pi} \frac{1+2e^{i\theta}}{5+2(e^{i\theta}+e^{-i\theta})}\, d\theta$$

writing $e^{i\theta} = z,\ d\theta = \frac{dz}{iz}$ where C is the unit circle $|z| = 1$.

$$= \text{Real part of } \frac{1}{2}\int_C \frac{1+2z}{5+2\left(z+\frac{1}{z}\right)} \frac{dz}{iz}, = \text{Real part of } \frac{1}{2}\int_C \frac{-i(1+2z)}{2z^2+5z+2}\, dz$$

$$= \text{Real part of } \frac{1}{2}\int_C \frac{-i(2z+1)}{(2z+1)(z+2)}\, dz = \text{Real part of } -\frac{i}{2}\int_C \frac{1}{z+2}\, dz$$

Pole is given by $z + 2 = 0$ *i.e.* $z = -2$.

Thus there is no pole of $f(z)$ inside the unit circle C. Hence $f(z)$ is analytic in C.

By Cauchy's Theorem $\int_C f(z)\, dz = 0$ if $f(z)$ is analytic in C.

$\therefore \quad I =$ Real part of zero $= 0$

Hence, the given integral $= 0$ **Ans.**

Example 44. *Using complex variables, evaluate the real integral*

$$\int_0^{2\pi} \frac{d\theta}{1-2p\sin\theta+p^2}, \text{ where } P^2 < 1.$$

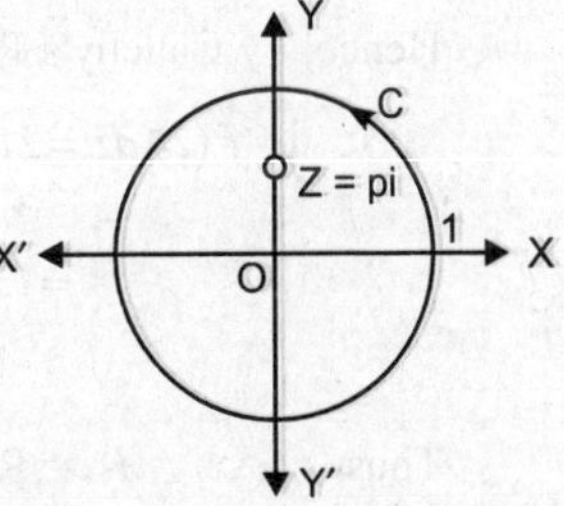

Solution. $\int_0^{2\pi} \frac{d\theta}{1-2p\sin\theta+p^2} = \int_0^{2\pi} \frac{d\theta}{1-2p\frac{(e^{i\theta}-e^{-i\theta})}{2i}+p^2}$

Let $I = \int_0^{2\pi} \frac{d\theta}{1+ip(e^{i\theta}-e^{-i\theta})+p^2}$

Writing $z = e^{i\theta},\ dz = ie^{i\theta}d\theta = iz\,d\theta,\ d\theta = \frac{dz}{zi}$

$$I = \int_C \frac{1}{1+ip\left(z-\frac{1}{z}\right)+p^2} \frac{dz}{zi}$$ where c is the unit circle $|z| = 1$.

$$= \int_C \frac{dz}{zi - pz^2 + p + p^2 zi} = \int_C \frac{dz}{-pz^2+ip^2z+zi+p} = \int_C \frac{dz}{(iz+p)(izp+1)}$$

Poles are given by $(iz+p)(ipz+1) = 0$

$\Rightarrow \quad z = -\frac{p}{i} = ip$ and $z = -\frac{1}{pi} = \frac{i}{p}$ $\qquad |ip| < 1$ and $\left|\frac{i}{p}\right| > 1$ as $p^2 < 1$

pi is the only pole inside the unit circle.

$$\text{Residue } (z = pi) = \lim_{z\to pi} \frac{(z-pi)}{(iz+p)(izp+1)} = \lim_{z\to pi}\left[\frac{1}{i(izp+1)}\right] = \frac{1}{i}\frac{1}{(-p^2+1)}$$

Hence by Cauchy's residue theorem

$$\int_0^{2\pi} \frac{d\theta}{1-2p\sin\theta+p^2} = 2\pi i\left(\frac{1}{i}\frac{1}{1-p^2}\right) = \frac{2\pi}{1-p^2}$$ **Ans.**

Example 45. *Apply calculus of residue to prove that:*

$$\int_0^{2\pi} \frac{\cos 2\theta\, d\theta}{1-2a\cos\theta+a^2} = \frac{2\pi a^2}{1-a^2}, \qquad (a^2 < 1)$$

Solution. Let $I = \int_0^{2\pi} \frac{\cos 2\theta\, d\theta}{1-2a\cos\theta+a^2} = \int_0^{2\pi} \frac{\cos 2\theta\, d\theta}{1-a\,(e^{i\theta}+e^{-i\theta})+a^2}$

$$= \text{Real part of } \int_0^{2\pi} \frac{e^{2i\theta}}{(1-ae^{i\theta})\,(1-ae^{-i\theta})}\, d\theta$$

$$= \text{Real part of } \oint_C \frac{z^2}{(1-az)\left(1-\frac{a}{z}\right)} \frac{dz}{iz} \qquad \left[\text{Put } e^{i\theta} = z \text{ so that } d\theta = \frac{dz}{iz}\right]$$

$$= \text{Real part of } \oint_C \frac{-iz^2}{(1-az)\,(z-a)}\, dz \qquad [C \text{ is the unit circle } |z| = 1]$$

Poles of $\frac{-iz^2}{(1-az)\,(z-a)}$ are given by

$$(1-az)\,(z-a) = 0$$

Thus, $z = \frac{1}{a}$ and $z = a$ are the simple poles. Only $z = a$ lies within the unit circle C as $a < 1$.

The residue of $f(z)$ at $(z = a)$ $= \lim_{z\to a} (z-a)\, \frac{-iz^2}{(1-az)\,(z-a)} = \lim_{z\to a} \frac{-iz^2}{(1-az)} = -\frac{ia^2}{1-a^2}$

Hence, by Cauchy's Residue Theorem, we have

$\oint_C f(z)\, dz = 2\pi i$ [Sum of residues within the contour]

$$= 2\pi i\left(-\frac{ia^2}{1-a^2}\right) = \frac{2\pi a^2}{1-a^2} \quad \text{which is purely real.}$$

Thus, $I = \text{Real part of } \oint_C f(z)\, dz = \frac{2\pi a^2}{1-a^2}$

Hence, $\int_0^{2\pi} \frac{\cos 2\theta}{1-2a\cos\theta+a^2} = \frac{2\pi a^2}{1-a^2}.$ **Proved.**

Example 46. *Using complex variable techniques, evaluate the integral*

$$\int_0^{2\pi} \frac{\sin^2\theta - 2\cos\theta}{2+\cos\theta}\, d\theta$$

Solution. $\int_0^{2\pi} \frac{\sin^2\theta - 2\cos\theta}{2+\cos\theta}\, d\theta = \int_0^{2\pi} \frac{\frac{1}{2}-\frac{1}{2}\cos 2\theta - 2\cos\theta}{2+\cos\theta}\, d\theta$

$$= \frac{1}{2}\int_0^{2\pi} \frac{1-\cos 2\theta - 4\cos\theta}{2+\cos\theta}\, d\theta = \text{Real part of } \frac{1}{2}\int_0^{2\pi} \frac{1-e^{2i\theta}-4e^{i\theta}}{2+\cos\theta}\, d\theta$$

write $e^{i\theta} = z$ so that $i e^{i\theta} d\theta = dz$ or $i z d\theta = dz$ or $d\theta = \frac{dz}{iz}$

$$= \text{Real part of } \frac{1}{2}\int_c \frac{1 - z^2 - 4z}{2 + \frac{1}{2}\left(z + \frac{1}{z}\right)} \frac{dz}{iz} = \text{Real part of } \frac{1}{i}\int_c \frac{(1 - z^2 - 4z)dz}{4z + z^2 + 1}$$

The poles are given by $z^2 + 4z + 1 = 0$

$$z = \frac{-4 \pm \sqrt{16 - 4}}{2} = -2 \pm \sqrt{3}$$

The pole within the unit circle C is $-2+\sqrt{3}$

Residue at the simple pole $z = -2+\sqrt{3}$

$$= \lim_{z \to -2+\sqrt{3}} (z + 2 - \sqrt{3}) \frac{1 - z^2 - 4z}{(z + 2 - \sqrt{3})(z + 2 + \sqrt{3})} = \lim_{z \to -2+\sqrt{3}} \frac{1 - z^2 - 4z}{z + 2 + \sqrt{3}}$$

$$= \frac{1 - (-2+\sqrt{3})^2 - 4(-2+\sqrt{3})}{(-2+\sqrt{3}) + 2 + \sqrt{3}} = \frac{1}{\sqrt{3}}$$

$$\text{Real part of } \frac{1}{i}\int_c \frac{(1 - z^2 - 4z)dz}{4z + z^2 + 1} = \text{Real part of } \left(\frac{1}{i}\right) 2\pi i \text{ (Residue)}$$

$$= \text{Real part of } 2\pi \left(\frac{1}{\sqrt{3}}\right) \text{ or } I = \frac{2\pi}{\sqrt{3}}$$

Hence, the given integral $= \frac{2\pi}{\sqrt{3}}$. **Ans.**

Example 47. *Evaluate:* $\int_0^{2\pi} \frac{\cos 2\theta}{5 + 4\cos\theta} d\theta$ *by using contour integration.*

(R.G.P.V., Bhopal, III Semester, June 2007)

Solution.

$$\left[\begin{array}{l} e^{i\theta} = z \\ \Rightarrow \quad i.e.^{i\theta} d\theta = dz \\ \Rightarrow d\theta = \frac{dz}{i.e^{i\theta}} = \frac{dz}{iz} \end{array}\right]$$

Let $I = \int_0^{2\pi} \frac{\cos 2\theta}{5 + 4\cos\theta} d\theta$

$$= \text{Real part of } \int_0^{2\pi} \frac{\cos 2\theta + i \sin 2\theta}{5 + 4\cos\theta} d\theta$$

$$= \text{Real part of } \int_0^{2\pi} \frac{e^{2i\theta}}{5 + 2(e^{i\theta} + e^{-i\theta})} d\theta$$

$$= \text{Real part of } \oint_C \frac{z^2}{5 + 2\left(z + \frac{1}{z}\right)} \frac{dz}{iz}$$ [C is the unit circle $|z| = 1$]

$$= \text{Real part of } \oint_C \frac{z^2}{5z + 2z^2 + 2} \frac{dz}{i}$$

$$= \text{Real part of } \oint_C \frac{-iz^2}{2z^2 + 5z + 2} dz$$

$$= \text{Real part of } \oint_C \frac{-iz^2}{(2z+1)(z+2)} dz$$

Poles are determined by putting denominator equal to zero.

$$(2z+1)(z+2) = 0 \quad \Rightarrow z = -\frac{1}{2}, -2$$

The only simple pole at $z = -\frac{1}{2}$ is inside the contour.

$$\text{Residue at } \left(z = -\frac{1}{2}\right) = \lim_{z \to -\frac{1}{2}} \left(z + \frac{1}{2}\right) f(z) = \lim_{z \to -\frac{1}{2}} \left(z + \frac{1}{2}\right) \frac{-iz^2}{(2z+1)(z+2)}$$

$$= \lim_{z \to -\frac{1}{2}} \frac{-iz^2}{2(z+2)} = \frac{-i\left(-\frac{1}{2}\right)^2}{2\left(-\frac{1}{2}+2\right)} = \frac{-i}{12}$$

By Cauchy's Integral Theorem

$$\int_C f(z)\, dz = 2\pi i \quad \text{(Sum of the residues within } C\text{)}$$

$$= 2\pi i \left(\frac{-i}{12}\right) = \frac{\pi}{6}, \text{ which is real}$$

$$\int_0^{2\pi} \frac{\cos 2\theta}{5+4\cos\theta} d\theta = \frac{\pi}{6}$$ **Ans.**

Example 48. *Evaluate contour integration of the real integral* $\int_0^{2\pi} \frac{\cos 3\theta}{5-4\cos\theta} d\theta$.

(*D.U., April 2010, U.P., III Sem., 2009, R.G.P.V., Bhopal, III Semester, Dec. 2007*)

Solution. $\int_0^{2\pi} \frac{\cos 3\theta}{5-4\cos\theta} d\theta = \text{Real part of } \int_0^{2\pi} \frac{e^{3i\theta}}{5-4\cos\theta} d\theta$

$$= \text{Real part of } \int_0^{2\pi} \frac{e^{3i\theta}}{5-2(e^{i\theta}+e^{-i\theta})} d\theta \qquad \text{On writing } z = e^{i\theta} \text{ and } d\theta = \frac{dz}{iz}$$

$$= \text{Real part of } \int_c \frac{z^3}{5-2\left(z+\frac{1}{z}\right)} \frac{dz}{iz} \qquad c \text{ is the unit circle.}$$

$$= \text{Real part of } \frac{1}{i}\int_c \frac{z^3}{5z-2z^2-2} dz = \text{Real part of } -\frac{1}{i}\int \frac{z^3}{2z^2-5z+2} dz$$

$$= \text{Real part of } i\int \frac{z^3}{(2z-1)(z-2)} dz$$

Poles are given by $(2z-1)(z-2) = 0$ *i.e.* $z = \frac{1}{2}, z = 2$

$z = \frac{1}{2}$ is the only pole inside the unit circle.

$$\text{Residue } \left(\text{at } z = \frac{1}{2}\right) = \lim_{z \to \frac{1}{2}} \frac{i\left(z-\frac{1}{2}\right)z^3}{(2z-1)(z-2)}$$

$$= \lim_{z \to \frac{1}{2}} \frac{iz^3}{2(z-2)} = \frac{i\frac{1}{8}}{2\left(\frac{1}{2}-2\right)} = -\frac{i}{24}$$

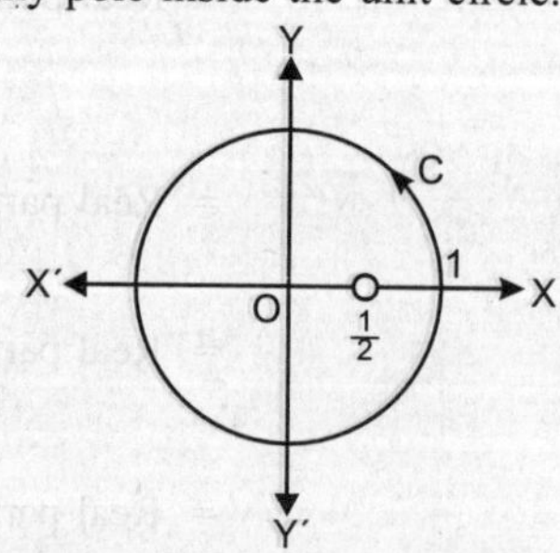

$$\int_0^{2\pi} \frac{\cos 3\theta}{5-4\cos\theta}\, d\theta = \text{Real part of } 2\pi i\left(-\frac{i}{24}\right) = \frac{\pi}{12}$$ **Ans.**

Question. *Evaluate* : $\int_0^{\infty} \frac{\cos 3\theta}{5+4\cos\theta}\, d\theta$ *(U.P. III Semester, Dec. 2008, 2006)*

Example 49. *Use the residue theorem to show that*

$$\int_0^{2\pi} \frac{d\theta}{(a+b\cos\theta)^2} = \frac{2\pi a}{(a^2-b^2)^{3/2}}$$ *where* $a>0, b>0, a>b$.

Solution. $\int_0^{2\pi} \frac{d\theta}{(a+b\cos\theta)^2} = \int_0^{2\pi} \frac{d\theta}{\left(a+b\cdot\frac{e^{i\theta}+e^{-i\theta}}{2}\right)^2}$

Put $e^{i\theta} = z$ so that $e^{i\theta}(i d\theta) = dz \Rightarrow izd\theta = dz \Rightarrow d\theta = \frac{dz}{iz}$

$$= \int_c \frac{1}{\left\{a+\frac{b}{2}\left(z+\frac{1}{z}\right)\right\}^2}\frac{dz}{iz}$$ where c is the unit circle $|z| = 1$.

$$\int_c \frac{1}{\left(a+\frac{bz}{2}+\frac{b}{2z}\right)^2}\frac{dz}{iz} = \int_c \frac{-4iz}{\left(a+\frac{bz}{2}+\frac{b}{2z}\right)^2}\frac{dz}{(2z)^2}$$

$$= \int_c \frac{-4izdz}{(bz^2+2az+b)^2} = \frac{-4i}{b^2}\int_c \frac{z\,dz}{\left(z^2+\frac{2az}{b}+1\right)^2}$$

The poles are given by putting the denominator equal to zero.

i.e., $\left(z^2+\frac{2a}{b}z+1\right)^2 = 0$

$\Rightarrow$ $(z-\alpha)^2(z-\beta)^2 = 0$

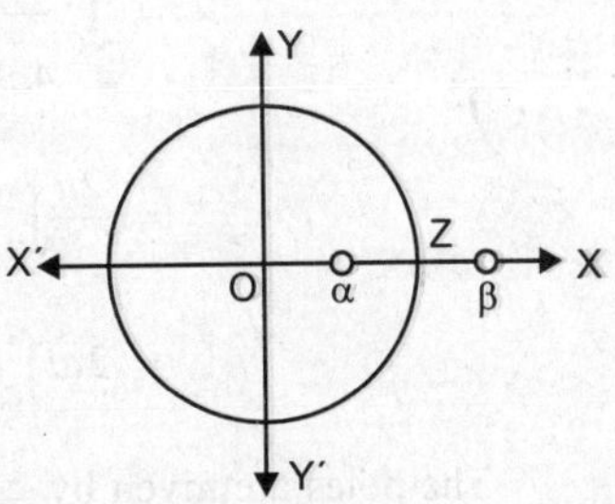

where $\alpha = \dfrac{-\frac{2a}{b}+\sqrt{\frac{4a^2}{b^2}-4}}{2} = \dfrac{-a+\sqrt{a^2-b^2}}{b}$

$$\beta = \frac{-\frac{2a}{b}-\sqrt{\frac{4a^2}{b^2}-4}}{2} = \frac{-a-\sqrt{a^2-b^2}}{b}$$

There are two poles, at $z = \alpha$ and at $z = \beta$, each of order 2.

Since $|\alpha\beta| = 1$ or $|\alpha||\beta| = 1$ if $|\alpha| < 1$ then $|\beta| > 1$.

There is only one pole [$|\alpha| < 1$] of order 2 within the unit circle c.

Residue (at the double pole $z = \alpha$) $= \lim_{z\to\alpha} \frac{d}{dz}(z-\alpha)^2 \frac{(-4iz)}{b^2(z-\alpha)^2(z-\beta)^2}$

$$= \lim_{z\to\alpha} \frac{d}{dz} \frac{-4iz}{b^2(z-\beta)^2}$$

$$= -\frac{4i}{b^2}\lim_{z\to\alpha}\frac{(z-\beta)^2.1-2(z-\beta)z}{(z-\beta)^4} = \frac{-4i}{b^2}\lim_{z\to\alpha}\frac{z-\beta-2z}{(z-\beta)^3} = \frac{-4i}{b^2}\lim_{z\to\alpha}\frac{-(z+\beta)}{(z-\beta)^3}$$

$$= \frac{4i}{b^2}\frac{(\alpha+\beta)}{(\alpha-\beta)^3} = \frac{4i}{b^2}\frac{\alpha+\beta}{[(\alpha+\beta)^2-4\alpha\beta]^{\frac{3}{2}}} = \frac{4i}{b^2}\frac{\frac{-2a}{b}}{\left[\left(-\frac{2a}{b}\right)^2-4(1)\right]^{\frac{3}{2}}}$$

$$= \frac{-8ai}{(4a^2-4b^2)^{\frac{3}{2}}} = -\frac{ai}{(a^2-b^2)^{\frac{3}{2}}}$$

Hence, $\int_0^{2\pi}\frac{d\theta}{(a+b\cos\theta)^2} = 2\pi i\times\frac{-ai}{(a^2-b^2)^{3/2}} = \frac{2\pi a}{(a^2-b^2)^{3/2}}$ **Proved.**

Example 50. *Show by the method of residues, that*

$$\int_0^{\pi}\frac{a\,d\theta}{a^2+\sin^2\theta} = \frac{\pi}{\sqrt{1+a^2}}$$

Solution. Let $I = \int_0^{\pi}\frac{a\,d\theta}{a^2+\sin^2\theta} = \int_0^{\pi}\frac{2a\,d\theta}{2a^2+2\sin^2\theta}$ $(\cos 2\theta = 1-2\sin^2\theta)$

$$= \int_0^{\pi}\frac{2a\,d\theta}{2a^2+1-\cos 2\theta} = \int_0^{2\pi}\frac{a\,d\phi}{2a^2+1-\cos\phi}$$ [Putting $2\theta=\phi$, $2d\theta = d\phi$]

$$= \int_0^{2\pi}\frac{a\,d\phi}{2a^2+1-\frac{1}{2}(e^{i\phi}+e^{-i\phi})} = \int_0^{2\pi}\frac{2a\,d\phi}{4a^2+2-(e^{i\phi}+e^{-i\phi})}$$

Writing $e^{i\phi}=z$, $e^{i\phi}(id\phi)=dz$ or $z(id\phi)=dz$, $d\phi=\frac{dz}{iz}$

$$= \int_C\frac{2a}{4a^2+2-\left(z+\frac{1}{z}\right)}.\frac{dz}{iz}$$ where C is unit circle $|z|=1$

$$= \frac{2a}{i}\int_C\frac{dz}{(4a^2+2)z-z^2-1} = \frac{2a}{-i}\int_c\frac{dz}{z^2-(4a^2+2)z+1}$$

$$= 2ai\int_C\frac{dz}{z^2-(4a^2+2)z+1}$$

The poles are given by $z^2-(4a^2+2)z+1=0$

$$\Rightarrow \quad z = \frac{(4a^2+2)\pm\sqrt{(4a^2+2)^2-4}}{2} = \frac{(4a^2+2)\pm\sqrt{16a^4+16a^2}}{2}$$

$$= 2a^2+1\pm 2a\sqrt{a^2+1}$$

Let $\alpha = 2a^2+1+2a\sqrt{a^2+1}$

$\beta = 2a^2+1-2a\sqrt{a^2+1}$

$$z^2-(4a^2+2)z+1 = (z-\alpha)(z-\beta)$$

$$\text{I} = 2ai\int\frac{dz}{(z-\alpha)(z-\beta)}$$

Product of the roots $= \alpha\beta = 1$ or $|\alpha\beta| = 1$

But $\quad |\alpha| > 1 \therefore |\beta| < 1$

Only β lies inside the circle c.

Now we calculate the residue at $z = \beta$.

Residue (at $z = \beta$) is $= \lim_{z\to\beta}(z-\beta)\dfrac{2ai}{(z-\alpha)(z-\beta)} = \lim_{z\to\beta}\dfrac{2ai}{(z-\alpha)}$

$$= \frac{2ai}{(\beta-\alpha)} = \frac{2ai}{(2a^2+1-2a\sqrt{a^2+1})-(2a^2+1+2a\sqrt{a^2+1})}$$

$$= \frac{2ai}{-4a\sqrt{a^2+1}} = -\frac{i}{2\sqrt{a^2+1}}$$

Hence by Cauchy's residue theorem

$$I = 2\pi i \text{ (sum of the residues within the contour } c)$$

$$= 2\pi i \frac{-i}{2\sqrt{a^2+1}} = \frac{\pi}{\sqrt{a^2+1}}.$$

Hence, $\displaystyle\int_0^{\pi} \frac{a\,d\theta}{a^2+\sin^2\theta} = \frac{\pi}{\sqrt{1+a^2}}$ **Proved.**

Example 51. *Evaluate by Contour integration:*

$$\int_0^{2\pi} e^{\cos\theta} \cos(\sin\theta - n\theta)\, d\theta.$$

Solution. Let $\quad I = \displaystyle\int_0^{2\pi} e^{\cos\theta}[\cos(\sin\theta - n\theta) + i\sin(\sin\theta - n\theta)]\, d\theta$

$$= \int_0^{2\pi} e^{\cos\theta} e^{i(\sin\theta - n\theta)}\, d\theta = \int_0^{2\pi} e^{\cos\theta + i\sin\theta}\cdot e^{-ni\theta}\, d\theta$$

$$= \int_0^{2\pi} e^{i\theta}\cdot e^{-in\theta}\, d\theta \qquad \text{...(1)}$$

Put $e^{i\theta} = z$ so that $d\theta = \dfrac{dz}{iz}$ then,

$$\text{I} = \int_C e^z \cdot \frac{1}{z^n}\cdot\frac{dz}{iz} = -i\int_C \frac{e^z}{z^{n+1}}\, dz$$

Pole is at $z = 0$ of order $(n+1)$.

It lies inside the unit circle.

Residue of $f(z)$ at $z = 0$ is

$$= \frac{1}{(n+1-1)!}\left[\frac{d^n}{dz^n}\left\{z^{n+1}\cdot\frac{-ie^z}{z^{n+1}}\right\}\right]_{z=0} = \frac{-i}{n!}\left[\frac{d^n}{dz^n}(e^z)\right]_{z=0} = \frac{-i}{n!}\left(e^z\right)_{z=0} = \frac{-i}{n!}$$

$\therefore$ By Cauchy's Residue theorem,

$$\text{I} = 2\pi i\left(\frac{-i}{n!}\right) = \frac{2\pi}{n!}$$

Comparing real part of $\displaystyle\int_0^{2\pi} e^{\cos\theta}[\cos(\sin\theta - n\theta) + i\sin(\sin\theta - n\theta)]\, d\theta = \frac{2\pi}{n!}$,

we have

$$\int_0^{2\pi} e^{\cos\theta}\cos(\sin\theta - n\theta)\, d\theta = \frac{2\pi}{n!}$$ **Ans.**

EXERCISE 26.4

Evaluate the following integrals:

1. $\int_0^{2\pi} \frac{\sin^2\theta}{a+b\cos\theta}\,d\theta$ (*R.G.P.V., Bhopal, III Semester, June 2008*) **Ans.** $\frac{2\pi}{b^2}\{a-\sqrt{a^2-b^2}\},\ a>b>0$

2. $\int_0^{2\pi} \frac{(1+2\cos\theta)^n \cos n\theta}{3+2\cos\theta}\,d\theta$ **Ans.** $\frac{2\pi}{\sqrt{5}}(3-\sqrt{5})^n,\ n>0$

3. $\int_0^{2\pi} \frac{4}{5+4\sin\theta}\,d\theta$ **Ans.** $\frac{8\pi}{3}$

4. $\int_0^{\pi} \frac{d\theta}{17-8\cos\theta}$ **Ans.** $\frac{\pi}{15}$

5. $\int_0^{\pi} \frac{d\theta}{a+b\cos\theta}$, *where* $a>|b|$. *Hence or otherwise evaluate* $\int_0^{2\pi} \frac{d\theta}{\sqrt{2}-\cos\theta}$. **Ans.** $\frac{\pi}{\sqrt{a^2-b^2}};\pi$

26.15 EVALUATION OF $\int_{-\infty}^{\infty} \frac{f_1(x)}{f_2(x)}\,dx$ WHERE $f_1(X)$ and $f_2(X)$ are polynomials in x.

Such integrals can be reduced to contour integrals, if

(*i*) $f_2(x)$ has no real roots.

(*ii*) the degree of $f_2(x)$ is greater than that of $f_1(x)$ by at least two.

Procedure: Let $f(x)=\frac{f_1(x)}{f_2(x)}$

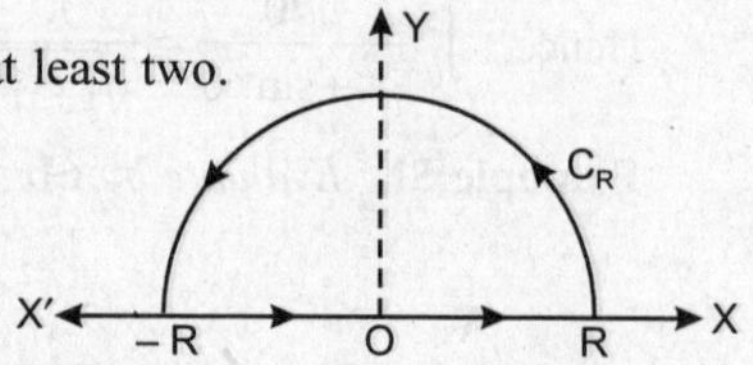

Consider $\int_C f(z)\,dz$

where C is a curve, consisting of the upper half C_R of the circle $|z|=R$, and part of the real axis from $-R$ to R.

If there are no poles of $f(z)$ on the real axis, the circle $|z|=R$ which is arbitrary can be taken such that there is no singularity on its circumference C_R in the upper half of the plane, but possibly some poles inside the contour C specified above.

Using Cauchy's theorem of residues, we have

$$\int_C f(z)\,dz = 2\pi i \times \text{(sum of the residues of } f(z) \text{ at the poles within } C)$$

i.e. $$\int_{-R}^{R} f(x)\,dx + \int_{CR} f(z)\,dz = 2\pi i \text{ (sum of residues within } C)$$

$$\Rightarrow \int_{-R}^{R} f(x)\,dx = -\int_{CR} f(z)\,dz + 2\pi i \text{ (sum of residues within } C)$$

$$\therefore \lim_{R\to\infty}\int_{-R}^{R} f(x)\,dx = -\lim_{R\to\infty}\int_{CR} f(z)\,dz + 2\pi i \text{ (sum of residues within } C) \quad ...(1)$$

Now, $$\lim_{R\to\infty}\int_{CR} f(z)\,dz = \int_0^{\pi} f(Re^{i\theta})\,Ri e^{i\theta}\,d\theta$$
$$= 0 \quad \text{when } R\to\infty$$

(1) reduces $$\int_{-\infty}^{\infty} f(x)\,dx = 2\pi i \text{ (sum of residues within } C)$$

Example 52. *Evaluate* $\int_0^{\infty} \frac{\cos mx}{(x^2+1)}\,dx$.

(*DU, III Sem. 2012, R.G.P.V. Bhopal, III Semester, Dec. 2006*)

Solution. $\int_0^{\infty} \frac{\cos mx}{x^2+1}\,dx$

Consider the integral $\int_C f(z)\,dz$, where

$f(z)=\dfrac{e^{imz}}{z^2+1}$, taken round the closed contour c consisting of the upper half of a large circle $|z|=R$ and the real axis from $-R$ to R.

Poles of $f(z)$ are given by

$$z^2+1=0 \text{ i.e. } z^2=-1 \text{ i.e. } z=\pm i$$

The only pole which lies within the contour is at $z=i$.

The residue of $f(z)$ at $z=i$

$$=\lim_{z\to i}\frac{(z-i)e^{imz}}{(z^2+1)}=\lim_{z\to i}\frac{e^{imz}}{z+i}=\frac{e^{-m}}{2i}$$

Hence by Cauchy's residue theorem, we have

$$\int_C f(z)\,dz=2\pi i\times \text{ sum of the residues}$$

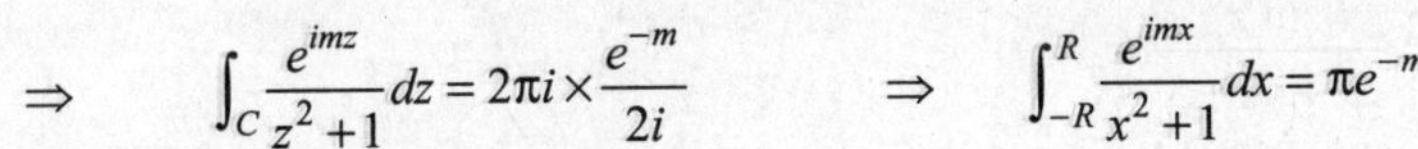

$$\Rightarrow \quad \int_C\frac{e^{imz}}{z^2+1}dz=2\pi i\times\frac{e^{-m}}{2i} \quad \Rightarrow \quad \int_{-R}^{R}\frac{e^{imx}}{x^2+1}dx=\pi e^{-m}$$

Equating real parts, we have

$$\int_{-\infty}^{\infty}\frac{\cos mx}{x^2+1}dx=\pi e^{-m} \quad \Rightarrow \quad \int_0^{\infty}\frac{\cos mx}{x^2+1}dx=\frac{\pi e^{-m}}{2}$$ **Ans.**

Example 53. *Evaluate* $\displaystyle\int_{-\infty}^{\infty}\frac{x\sin\pi x}{x^2+2x+5}dx$ *(U.P. III Semester 2009-2010)*

Solution. Here, we have $\displaystyle\int_{-\infty}^{\infty}\frac{x\sin\pi x}{x^2+2x+5}dx$

Let us consider $\displaystyle\int_c\frac{z\sin\pi z}{z^2+2z+5}dz$

The pole can be determined by putting the denominator equal to zero.

$$z^2+2z+5=0 \quad\Rightarrow\quad z=\frac{-2\pm\sqrt{4-20}}{2} \quad\Rightarrow\quad z=-1\pm 2i$$

Out of two poles, only $z=-1+2i$ is inside the contour.

Residue at $z=-1+2i$

$$=\lim_{z\to-1+2i}(z+1-2i)\frac{z\sin\pi z}{z^2+2z+5}=\lim_{z\to-1+2i}(z+1-2i)\frac{z\sin\pi z}{(z+1-2i)(z+1+2i)}$$

$$=\lim_{z\to-1+2i}\frac{z\sin\pi z}{(z+1+2i)}=\frac{(-1+2i)\sin\pi(-1+2i)}{(-1+2i+1+2i)}$$

$$=\frac{(-1+2i)\sin\pi(-1+2i)}{4i}$$

$$\int_{-R}^{R}\frac{z\sin\pi z}{z^2+2z+5}dz=2\pi i\,(\text{Residue})$$

$$=2\pi i\frac{(-1+2i)\sin\pi(-1+2i)}{4i}=\frac{\pi}{2}(2i-1)\sin(-\pi+2\pi i)$$

$$=\frac{\pi}{2}(2i-1)(-\sin 2\pi i) \qquad \left[\begin{aligned}\sin(-\pi+\theta)&=-\sin(\pi-\theta)\\&=-\sin\theta\end{aligned}\right]$$

$$= \frac{\pi}{2}(1-2i)\sin 2\pi i \quad = \frac{\pi}{2}(1-2i)\, i \sinh 2\pi$$

$$= \frac{\pi}{2}(i+2)\sinh 2\pi \qquad \text{(Taking real parts)}$$

Hence $\int_{-\infty}^{\infty} \frac{x \sin \pi x}{x^2+2x+5} = \pi \sinh 2\pi$ **Ans.**

Example 54. *Using contour integration, evaluate*

$$\int_{-\infty}^{\infty} \frac{\sin x}{x^2+4x+5}\, dx$$ (*D.U., April 2010*)

Solution. Here, we have

$$\int_{-\infty}^{\infty} \frac{\sin x}{x^2+4x+5}\, dx$$

Let us consider $\int \frac{\sin z}{z^2+4z+5}\, dz$

The poles can be determined by putting the denominator equal to zero.

$$z^2 + 4z + 5 = 0$$

$$z = \frac{-4 \pm \sqrt{16-20}}{2} = \frac{-4 \pm 2i}{2} = -2 \pm i$$

Out of the two poles only $z = -2 + i$ lies inside the contour.

$$\text{Residue at } (z = -2+i) = \lim_{z \to -2+i} (z+2-i)\left(\frac{\sin z}{z^2+4z+5}\right)$$

$$= \lim_{z \to -2+i} (z+2-i)\left(\frac{\sin z}{(z+2-i)(z+2+i)}\right)$$

$$= \lim_{z \to -2+i} \frac{\sin z}{z+2+i} = \frac{\sin(-2+i)}{-2+i+2+i} = \frac{\sin(-2+i)}{2i}$$

$$\int_{-R}^{R} \frac{\sin z}{z^2+4z+5}\, dz = 2\pi i \text{ (Residue)} = 2\pi i \frac{\sin(-2+i)}{2i}$$

$$= \pi \sin(-2+i)$$ **Ans.**

$$= \pi [\sin(-2)\cos i + \cos(-2)\sin i]$$

$$= \pi [-2 \sin 2 \cosh 1 + i \cos 2 \sinh 1]$$

Taking real parts, we get

$$\int_{-\infty}^{\infty} \frac{\sin x}{x^2+4x+5}\, dx = -2\pi \sin 2 \cosh 1$$ **Ans.**

Example 55. *Use contour integration to evaluate the real integral* $\int_0^{\infty} \frac{dx}{(1+x^2)^3}$.

Solution. Consider $\int_C f(z)dz$, where $f(z) = \frac{1}{(z^2+1)^3}$ taken round the closed contour C consisting of real axis and upper half C_R of a large semi-circle $|z| = R$.

Poles of $f(z)$ are given by

$$(z^{2+} + 1)^3 = 0 \text{ i.e. } (z-i)^3 (z+i)^3 = 0$$

i.e. $z = \pm i$ are the poles each of order 3.

The only pole which lies within C is $z = i$ of order 3.

$\therefore$ Residue of $\frac{1}{(z-i)^3} \cdot \frac{1}{(z+i)^3}$ (at $z = i$)

$$= \frac{1}{2}\left[\frac{d^2}{dz^2}(z-i)^3 \cdot \frac{1}{(z-i)^3(z+i)^3}\right]_{z=i} = \frac{1}{2}\left[\frac{d^2}{dz^2}\frac{1}{(z+i)^3}\right]_{z=i} = \frac{1}{2}\left[\frac{(-3)(-4)}{(z+i)^5}\right]_{z=i} = \frac{3}{16i}$$

Hence by Cauchy's residue theorem, we have

$\int f(z)dz = 2\pi i \times$ sum of residues within c.

$$\int_{-R}^{R} f(x)\,dx + \int_{CR} f(z)dz = 2\pi i \times \frac{3}{16i} \Rightarrow \int_{-R}^{R} \frac{1}{(x^2+1)^3}dx + \int_{CR} \frac{1}{(z^2+1)^3}dz = \frac{3\pi}{8} \quad \text{... (1)}$$

Now, $$\left|\int_{CR} \frac{1}{(z^2+1)^3}dz\right| \le \int_{CR} \left|\frac{1}{(z^2+1)^3}\right| |dz| \le \int_{CR} \frac{|dz|}{(|z|^2-1)^3}$$

$$= \int_0^{\pi} \frac{R d\theta}{(R^2-1)^3} \qquad [\text{since } z = Re^{i\theta}, \ |dz| = R d\theta]$$

$$= \frac{\pi R}{(R^2-1)^3}, \text{ which} \to 0 \text{ as } R \to \infty$$

Hence making $R \to \infty$, relation (1) becomes

$$\int_{-\infty}^{\infty} \frac{1}{(x^2+1)^3}dx = \frac{3\pi}{8} \quad \text{or} \quad \int_0^{\infty} \frac{1}{(x^2+1)^3}dx = \frac{3\pi}{16} \qquad \textbf{Ans.}$$

Example 56. *Evaluate by the method of complex variables, the integral*

$$\int_{-\infty}^{\infty} \frac{x^2}{(1+x^2)^3}dx$$

Solution. Consider $\int_C \frac{z^2}{(1+z^2)^3}dz$ where c is a closed contour consisting of the upper half C_R of a large circle $|z| = R$ and the real axis from $-R$ to $+R$.

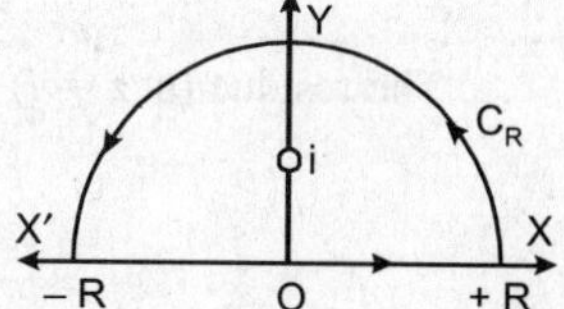

Poles of $\frac{z^2}{(1+z^2)^3}$ are given by

$$(z^2+1)^3 = 0 \quad \text{or} \quad z^2 = -1 \Rightarrow z = \pm i$$

$\therefore$ $z = i$ and $z = -i$ are the two poles each of order 3. But only $z = i$ lies within the contour.

To get residue at $z = i$, put $z = i + t$, then

$$\frac{z^2}{(1+z^2)^3} = \frac{(i+t)^2}{[1+(i+t)^2]^3} = \frac{-1+2it+t^2}{[1-1+2it+t^2]^3}$$

$$= \frac{(-1+2it+t^2)}{(2it)^3\left(1+\frac{1}{2i}t\right)^3} = \frac{(-1+2it+t^2)}{-8it^3}\left(1+\frac{t}{2i}\right)^{-3} = -\frac{1}{8i}\left(-\frac{1}{t^3}+\frac{2i}{t^2}+\frac{1}{t}\right)\left(1-\frac{3t}{2i}+\frac{(-3)(-4)}{2}\frac{t^2}{-4}+...\right)$$

$$= -\frac{1}{8i}\left[-\frac{1}{t^3}+\frac{2i}{t^2}+\frac{1}{t}\right]\left[1-\frac{3}{2i}t-\frac{3}{2}t^2+...\right].$$

Here coefficient of $\frac{1}{t}$ is $\frac{-1}{8i}\left(\frac{3}{2}-3+1\right)$ or $\frac{i}{8}\left(-\frac{1}{2}\right)$ or $\frac{-i}{16}$ which is therefore the residue at $z = i$.

Hence by Cauchy's residue theorem, we have

$$\int f(z)\,dz = 2\pi i \times \text{ sum of the residues within } c$$

i.e. $$\int_{-R}^{R} f(x)\,dx + \int_{C_R} f(z)\,dz = 2\pi i\left(-\frac{i}{16}\right)$$

$$\int_{-R}^{R} \frac{x^2}{(1+x^2)^3}dx + \int_{C_R} \frac{z^2}{(1+z^2)^3}dz = \frac{\pi}{8} \qquad \text{... (1)}$$

Now $$\left|\int_{C_R} \frac{z^2}{(1+z^2)^3}dz\right| \le \int_{C_R} \frac{|z|^2 |dz|}{|1+z^2|^3} \le \frac{R^2}{(R^2-1)^3}\int_0^{\pi} R\,d\theta$$

since $$z = Re^{\theta i}, \quad |dz| = R\,d\theta = \frac{R^3\pi}{(R^2-1)^3}, \text{ which} \to 0 \text{ as } R \to \infty$$

Hence by making $R \to \infty$, equation (1) becomes $\int_{-\infty}^{\infty} \frac{x^2}{(1+x^2)^3}dx = \frac{\pi}{8}$ **Ans.**

Example 57. *Evaluate* $\int_{-\infty}^{\infty} \frac{x^2 dx}{(x^2+1)(x^2+4)}$.

Solution. We consider $\int_C \frac{z^2 dz}{(z^2+1)(z^2+4)} = \int_C f(z)\,dz$

where C is the contour consisting of the semi-circle C_R of radius R together with the part of the real axis from $-R$ to $+R$.

The integral has simple poles at

$$z = \pm i, z = \pm 2i$$

of which $z = i$, $2i$ only lie inside C.

Y
z = 2i
C_R
z = i
X′
−R
O
+R
X

The residue (at $z = i$) $$= \lim_{z\to i}(z-i)\frac{z^2}{(z+i)(z-i)(z^2+4)}$$

$$= \lim_{z\to i}\frac{z^2}{(z+i)(z^2+4)} = \frac{-1}{2i(-1+4)} = \frac{-1}{6i}$$

The residue (at $z = 2i$) $$= \lim_{z\to 2i}(z-2i).\frac{z^2}{(z^2+1)(z+2i)(z-2i)}$$

$$= \lim_{z\to 2i}\frac{z^2}{(z^2+1)(z+2i)} = \frac{(2i)^2}{(-4+1)(2i+2i)} = \frac{1}{3i}$$

By theorem of residue;

$$\int_C f(z)\,dz = 2\pi i[\text{Res } f(i) + \text{Res } f(2i)] = 2\pi i\left(-\frac{1}{6i}+\frac{1}{3i}\right) = \frac{\pi}{3}$$

i.e. $$\int_{-R}^{R} f(x)dx + \int_{C_R} f(z)\,dz = \frac{\pi}{3} \qquad \text{... (1)}$$

Hence by making $R \to \infty$, relation (1) becomes

$$\int_{-\infty}^{\infty} f(x)dx + \lim_{z\to\infty}\int_{C_R} f(z)\,dz = \frac{\pi}{3}$$

Now $\quad R \to \infty, \int_{C_R} f(z)\,dz$ vanishes.

For any point on C_R as $|z| \to \infty$ $f(z) = 0$

$$\lim_{|z|\to\infty} \int_{C_R} f(z)\,dz = 0, \int_{-\infty}^{\infty} f(x)\,dx = \frac{\pi}{3}$$

$$\Rightarrow \quad \int_{-\infty}^{\infty} \frac{x^2 dx}{(x^2+1)(x^2+4)} = \frac{\pi}{3}$$ **Ans.**

Example 58. *Using the complex variable techniques, evaluate the integral*

$\int_{-\infty}^{\infty} \frac{1}{x^4+1}\,dx$ *(DU, III Sem. 2012, AMIETE, June 2010, U.P. III Semester, Dec. 2006)*

Solution. $\int_{-\infty}^{\infty} \frac{1}{x^4+1}\,dx$

Consider $\int_C f(z)\,dz$, $\quad$ where $f(z) = \frac{1}{z^4+1}$

taken around the closed contour consisting of real axis and upper half C_R, *i.e.* $|z| = R$.

Poles of $f(z)$ are given by

$$z^4 + 1 = 0 \text{ i.e. } z^4 = -1 = (\cos\pi + i\sin\pi)$$

$$\Rightarrow \quad z^4 = [\cos(2n+1)\pi + i\sin(2n+1)\pi]$$

$$z = [\cos(2n+1)\pi + i\sin(2n+1)\pi]^{\frac{1}{4}} = \left[\cos(2n+1)\frac{\pi}{4} + i\sin(2n+1)\frac{\pi}{4}\right]$$

If $\quad n = 0, \quad z_1 = \left(\cos\frac{\pi}{4} + i\sin\frac{\pi}{4}\right) = \left(\frac{1}{\sqrt{2}} + i\frac{1}{\sqrt{2}}\right) = e^{i\frac{\pi}{4}}$

$n = 1, \quad z_2 = \left(\cos\frac{3\pi}{4} + i\sin\frac{3\pi}{4}\right) = \left(-\frac{1}{\sqrt{2}} + i\frac{1}{\sqrt{2}}\right) = e^{i\frac{3\pi}{4}}$

$n = 2, \quad z_3 = \left(\cos\frac{5\pi}{4} + i\sin\frac{5\pi}{4}\right) = \left(-\frac{1}{\sqrt{2}} - i\frac{1}{\sqrt{2}}\right)$

$n = 3, \quad z_4 = \left(\cos\frac{7\pi}{4} + i\sin\frac{7\pi}{4}\right) = \left(\frac{1}{\sqrt{2}} - i\frac{1}{\sqrt{2}}\right)$

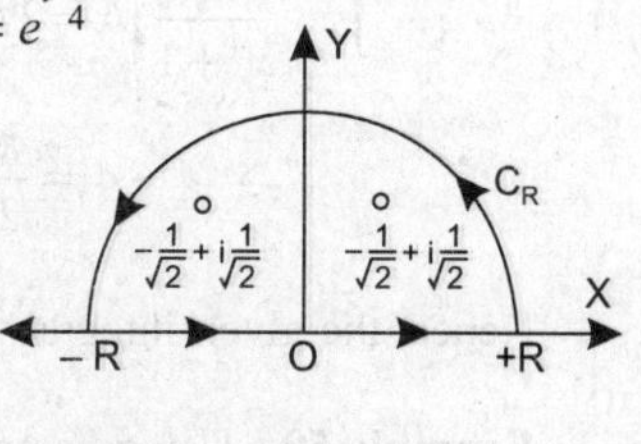

There are four poles, but only two poles at z_1 and z_2 lie within the contour.

$$\text{Residue}\left(\text{at } z = e^{\frac{i\pi}{4}}\right) = \left[\frac{1}{\frac{d}{dz}(z^4+1)}\right]_{z=e^{\frac{i\pi}{4}}} = \left[\frac{1}{4z^3}\right]_{z=e^{\frac{i\pi}{4}}} = \frac{1}{4\left(e^{i\frac{\pi}{4}}\right)^3} = \frac{1}{4e^{i\frac{3\pi}{4}}}$$

$$= \frac{1}{4}e^{-i\frac{3\pi}{4}} = \frac{1}{4}\left[\cos\frac{3\pi}{4} - i\sin\frac{3\pi}{4}\right] = \frac{1}{4}\left[-\frac{1}{\sqrt{2}} - i\frac{1}{\sqrt{2}}\right]$$

Residue $\left(\text{at } z = e^{\frac{3i\pi}{4}}\right) = \left[\dfrac{1}{\dfrac{d}{dz}(z^4+1)}\right]_{z=e^{\frac{3i\pi}{4}}} = \dfrac{1}{[4z^3]_{z=\frac{3i\pi}{4}}} = \dfrac{1}{4\left(e^{i\frac{3\pi}{4}}\right)^3} = \dfrac{1}{4e^{i\frac{9\pi}{4}}}$

$$= = \frac{1}{4}e^{-i\frac{9\pi}{4}} = \frac{1}{4}\left(\cos\frac{9\pi}{4} - i\sin\frac{9\pi}{4}\right) = \frac{1}{4}\left(\frac{1}{\sqrt{2}} - i\frac{1}{\sqrt{2}}\right)$$

$\int_C f(z)\,dz = 2\pi i$ (sum of residues at poles within c)

$\int_{-R}^{R} f(z)\,dz + \int_{C_R} f(z)\,dz = 2\pi i$ (sum of the residues)

$\int_{-R}^{R} \frac{1}{x^4+1}\,dx + \int_{C_R} \frac{1}{z^4+1}\,dz = 2\pi i$ (sum of the residues)

Now, $\left|\int_{C_R} \frac{1}{z^4+1}\,dz\right| \le \int_{C_R} \frac{1}{|z^4+1|}|dz|$

$\le \int_{C_R} \frac{1}{(|z^4|-1)}|dz|$ [Since $z = Re^{i\theta}, |dz| = |Re^{i\theta}\, i\,d\theta| = R\,d\theta$]

$\le \int_0^{\pi} \frac{1}{R^4-1} R\,d\theta \le \frac{R}{R^4-1}\int_0^{\pi} d\theta$

$\le \frac{R\pi}{R^4-1} = \frac{\pi/R^3}{1-1/R^4}$ which $\to 0$ as $R \to \infty$.

Hence, $\int_{-R}^{R} \frac{1}{x^4+1}\,dx = 2\pi i$ (Sum of the residues within contour)

As $R \to \infty$

Hence, $\int_{-\infty}^{\infty} \frac{1}{x^4+1}\,dx = 2\pi i$ (Sum of the residues within contour) ... (1)

$$\int_{-\infty}^{\infty} \frac{1}{x^4+1}\,dx = 2\pi i\left[\frac{1}{4}\left(-\frac{1}{\sqrt{2}} - i\frac{1}{\sqrt{2}}\right) + \frac{1}{4}\left(\frac{1}{\sqrt{2}} - i\frac{1}{\sqrt{2}}\right)\right]$$

$$= \frac{\pi}{2}i\left(-\frac{1}{\sqrt{2}} - i\frac{1}{\sqrt{2}} + \frac{1}{\sqrt{2}} - i\frac{1}{\sqrt{2}}\right) = \frac{\pi i}{2}\left(-i\frac{2}{\sqrt{2}}\right) = \frac{\pi}{\sqrt{2}}$$

Hence, the given integral $= \dfrac{\pi}{\sqrt{2}}$ **Ans.**

Example 59. *Using complex variable techniques, evaluate the real integral*

$$\int_0^{\infty} \frac{dx}{1+x^6}$$

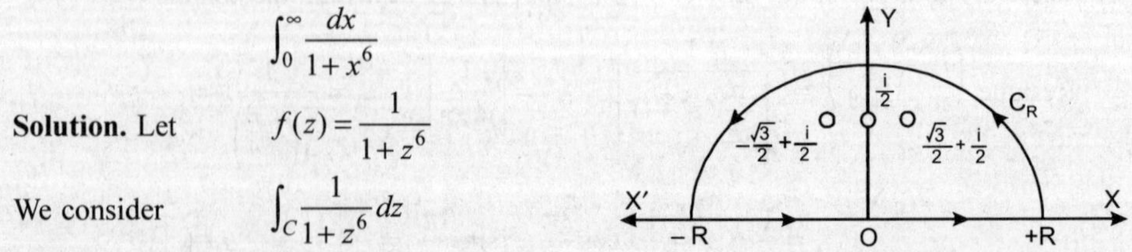

Solution. Let $f(z) = \dfrac{1}{1+z^6}$

We consider $\int_C \frac{1}{1+z^6}\,dz$

where C is the contour consisting of the semi-circle C_R of radius R together with the part of real axis from $-R$ to R.

Poles are given by $1 + z^6 = 0$.

$$z^6 = -1 = \cos\pi + i\sin\pi = \cos(2n\pi+\pi) + i\sin(2n\pi+\pi)$$
$$= e^{(2n+1)\pi i}$$
$$z = e^{\frac{2n+1}{6}\pi i} = \left[\cos\frac{2n\pi+\pi}{6} + i\sin\frac{2n\pi+\pi}{6}\right] \quad \text{where } n = 0, 1, 2, 3, 4, 5$$

If $n = 0$, $\quad z = e^{\frac{\pi i}{6}} = \cos\frac{\pi}{6} + i\sin\frac{\pi}{6} = \frac{\sqrt{3}}{2} + \frac{i}{2}$

If $n = 1$, $\quad z = e^{\frac{i\pi}{2}} = \cos\frac{\pi}{2} + i\sin\frac{\pi}{2} = i$

If $n = 2$ $\quad z = e^{i\frac{5\pi}{6}} = \cos\frac{5\pi}{6} + i\sin\frac{5\pi}{6} = \frac{-\sqrt{3}}{2} + \frac{i}{2}$

If $n = 3$, $\quad z = e^{i\frac{7\pi}{6}} = \cos\frac{7\pi}{6} + i\sin\frac{7\pi}{6} = \frac{-\sqrt{3}}{2} - \frac{i}{2}$

If $n = 4$, $\quad z = e^{i\frac{3\pi}{2}} = \cos\frac{3\pi}{2} + i\sin\frac{3\pi}{2} = -i$

If $n = 5$, $\quad z = e^{i\frac{11\pi}{6}} = \cos\frac{11\pi}{6} + i\sin\frac{11\pi}{6} = \frac{\sqrt{3}}{2} - \frac{i}{2}$

Only first three poles *i.e.*, $e^{i\frac{\pi}{6}}, e^{i\frac{\pi}{2}}, e^{i\frac{5\pi}{6}}$ are inside the contour.

Residue $\left(\text{at } z = e^{\frac{i\pi}{6}}\right) = \lim\limits_{z \to e^{\frac{i\pi}{6}}} \dfrac{1}{\frac{d}{dz}(1+z^6)} = \lim\limits_{z \to e^{\frac{i\pi}{6}}} \dfrac{1}{6z^5} = \dfrac{1}{6}e^{\frac{-i5\pi}{6}}$

Residue $\left(\text{at } z = e^{\frac{i\pi}{2}}\right) = \lim\limits_{z \to e^{\frac{i\pi}{2}}} \dfrac{1}{\frac{d}{dz}(1+z^6)} = \lim\limits_{z \to e^{\frac{i\pi}{2}}} \dfrac{1}{6z^5} = \dfrac{1}{6}e^{\frac{-i5\pi}{2}}$

Residue $\left(\text{at } z = e^{\frac{i5\pi}{6}}\right) = \lim\limits_{z \to e^{\frac{i5\pi}{6}}} \dfrac{1}{\frac{d}{dz}(1+z^6)} = \lim\limits_{z \to e^{\frac{i5\pi}{6}}} \dfrac{1}{6z^5} = \dfrac{1}{6}e^{\frac{-i25\pi}{6}}$

Sum of the residues $= \dfrac{1}{6}\left[e^{\frac{-5i\pi}{6}} + e^{-\frac{i5\pi}{2}} + e^{-\frac{i25\pi}{6}}\right] = \dfrac{1}{6}\left(-\dfrac{\sqrt{3}}{2} - \dfrac{i}{2} + 0 - i + \dfrac{\sqrt{3}}{2} - \dfrac{i}{2}\right) = \dfrac{1}{6}(-2i) = -\dfrac{i}{3}$

$\Rightarrow \quad \displaystyle\int_C \frac{dz}{1+z^6} = 2\pi i \text{ (Residue)} = 2\pi i\left(-\frac{i}{3}\right) = \frac{2\pi}{3}$

$\Rightarrow \quad \displaystyle\int_{-\infty}^{\infty} \frac{dz}{1+z^6} = \frac{2\pi}{3}$

$\Rightarrow \quad \displaystyle\int_0^{\infty} \frac{dx}{1+x^6} = \frac{\pi}{3}$ **Ans.**

Example 60. *Using complex variables, evaluate the real integral*

$$\int_0^{\infty} \frac{\cos 3x\, dx}{(x^2+1)(x^2+4)}$$

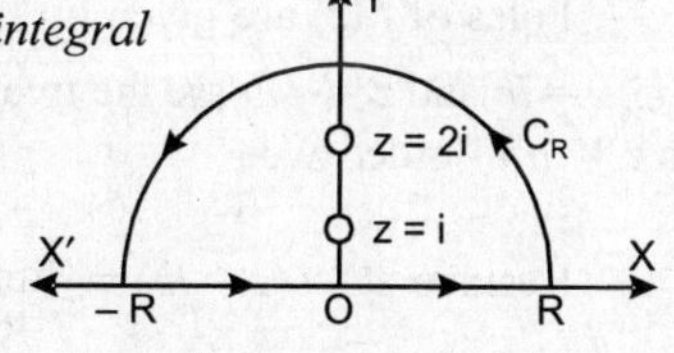

Solution. Let $\quad f(z) = \dfrac{e^{3iz}}{(z^2+1)(z^2+4)}$

Poles are given by

$$(z^2+1)(z^2+4)=0$$

i.e., $z^2 + 1 = 0$ or $z = \pm\, i$

$$z^2 + 1 = 0 \text{ or } z = \pm\, i$$

Let C be a closed contour consisting of the upper half C_R of a large circle $|\, z\,| = R$ and the real axis from $-R$ to $+R$. The poles at $z = i$ and $z = 2i$ lie within the contour.

$$\text{Residue (at } z = i) = \lim_{z\to i}\frac{(z-i)e^{3iz}}{(z^2+1)(z^2+4)} = \lim_{z\to i}\frac{e^{3iz}}{(z+i)(z^2+4)} = \frac{e^{-3}}{6i}$$

$$\text{Residue (at } z = 2i) = \lim_{z\to 2i}\frac{(z-2i)e^{3iz}}{(z^2+1)(z^2+4)} = \lim_{z\to 2i}\frac{e^{3iz}}{(z^2+1)(z+2i)} = \frac{e^{-6}}{-12i}$$

By theorem of Residue $\int_C f(z)dz = 2\pi i$ [Sum of Residues]

$$\int_{-R}^{R}\frac{e^{3iz}\,dz}{(z^2+1)(z^2+4)} + \int_{C_R}\frac{e^{3iz}\,dz}{(z^2+1)(z^2+4)} = 2\pi i\left[\frac{e^{-3}}{6i}+\frac{e^{-6}}{-12i}\right]$$

$$\left[\int_{C_R}\frac{e^{3iz}dz}{(z^2+1)(z^2+4)} = 0 \text{ as } z = Re^{i\theta} \text{ and } R\to\infty\right]$$

$$\int_{-R}^{R}\frac{e^{3ix}}{(x^2+1)(x^2+4)}dx = \pi\left[\frac{e^{-3}}{3}-\frac{e^{-6}}{6}\right]$$

$$\int_0^{\infty}\frac{\cos 3x\,dx}{(x^2+1)(x^2+4)} = \text{Real part of } \frac{1}{2}\int_{-\infty}^{\infty}\frac{e^{3ix}\,dx}{(x^2+1)(x^2+4)}$$

$$= \text{Real part of } \frac{\pi}{2}\left(\frac{e^{-3}}{3}-\frac{e^{-6}}{6}\right)$$

Hence, $\quad$ given integral $= \dfrac{\pi}{2}\left[\dfrac{e^{-3}}{3}-\dfrac{e^{-6}}{6}\right]$ **Ans.**

Example 61. *Using the calculus of residues, evaluate the integral given by the following:*

$$\int_0^{\infty}\frac{\cos ax}{(x^2+b^2)^2}dx,\ a>0,\ b>0$$

Solution. Consider the integral $\int_C f(z)\,dz$

where $\quad f(z) = \dfrac{e^{iaz}}{(z^2+b^2)^2}$

taken around the closed contour C consisting of the upper half of a large circle $|\, z\,| = R$ and the real axis from $-R$ to R.

Poles of $f(z)$ are given by $(z^2 + b^2)^2 = 0$

i.e., $z = ib$ and $z = -ib$ are the two poles of order two. The only pole which lies within the contour is $z = ib$ of order two.

$$\text{Residue at } (z = ib) = \lim_{z\to ib}\frac{d}{dz}(z-ib)^2\frac{e^{iaz}}{(z^2+b^2)^2} = \lim_{z\to ib}\frac{d}{dz}\frac{e^{iaz}}{(z+ib)^2}$$

$$= \lim_{z\to ib}\frac{(z+ib)^2 iae^{iaz} - e^{iaz}2(z+ib)}{(z+ib)^4} = \lim_{z\to ib}\frac{[(z+ib)ia-2]e^{iaz}}{(z+ib)^3}$$

$$= \frac{[(2ib)ia-2]e^{ia(ib)}}{(2ib)^3} = \frac{(-2ab-2)e^{-ab}}{-8ib^3} = \frac{(ab+1)e^{-ab}}{4ib^3}$$

Hence, by Cauchy's residue theorem, we have

$$\int_C f(z)dz = 2\pi i \times \text{ Sum of the residues within } C$$

$$\int_{-R}^{R} f(x)dx + \int_{C_R} f(z)dz = 2\pi i\frac{(ab+1)e^{-ab}}{4ib^3}$$

$$\int_{-R}^{R}\frac{e^{iax}}{(x^2+b^2)^2}dx + \int_{C_R}\frac{e^{iaz}}{(z^2+b^2)^2}dz = \frac{\pi(ab+1)e^{-ab}}{2b^3} \qquad \text{... (1)}$$

Now $\left|\int_{C_R}\frac{e^{iaz}dz}{(z^2+b^2)^2}\right| \le \int_{C_R}\frac{|e^{iaz}||dz|}{(z^2+b^2)^2} \le \int_{C_R}\frac{|e^{iaz}||dz|}{[|z|^2-b^2]^2}$

$$\le \int_0^{\pi}\frac{e^{-aR\sin\theta}Rd\theta}{(R^2-b^2)^2} \le \frac{R}{(R^2-b^2)^2}\int_0^{\pi} e^{-aR\sin\theta}d\theta \le \frac{R}{(R^2-b^2)^2}2\int_0^{\frac{\pi}{2}} e^{-aR\frac{2\theta}{\pi}}d\theta$$

$$\le \frac{R}{a(R^2-b^2)^2}(1-e^{-aR}) \qquad \text{which} \to 0, \text{ as } R\to\infty$$

Hence by making $R\to\infty$, (1) becomes $\int_{-\infty}^{\infty}\frac{e^{iax}dx}{(x^2+b^2)^2} = \frac{\pi(ab+1)e^{-ab}}{2b^3}$

Equating real parts, we have $\int_{-\infty}^{\infty}\frac{\cos ax}{(x^2+b^2)^2}dx = \frac{\pi(ab+1)e^{-ab}}{2b^3}$

$\Rightarrow \qquad \int_0^{\infty}\frac{\cos ax dx}{(x^2+b^2)^2} = \frac{\pi(ab+1)e^{-ab}}{4b^3}$ **Ans.**

Example 62. *Using complex variable techniques, evaluate the real integral*

$$\int_0^{\infty}\frac{\cos 2x}{(x^2+9)^2(x^2+16)}dx$$

Solution. Consider the integral $\int_C f(z)\,dz$,

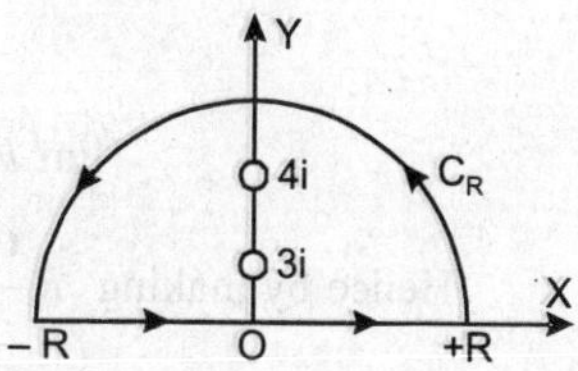

where $\qquad f(z) = \frac{e^{2iz}}{(z^2+9)^2(z^2+16)},$

taken around the closed contour C consisting of the upper half of a large circle $|z| = R$ and the real axis from $-R$ to R.

Poles of $f(z)$ are given by

$$(z^2+9)^2(z^2+16) = 0$$

i.e. $\qquad (z+3i)^2(z-3i)^2(z+4i)(z-4i) = 0$

i.e. $\qquad z = 3i, -3i, 4i, -4i$

The poles which lie within the contour are $z = 3i$ of the second order and $z = 4i$ simple pole.

Residue of $f(z)$ at $z = 3i$

$$= \frac{1}{1!}\left[\frac{d}{dz}\left\{(z-3i)^2 \frac{e^{2iz}}{(z-3i)^2(z+3i)^2(z^2+16)}\right\}\right]_{z=3i} = \left[\frac{d}{dz}\left\{\frac{e^{2iz}}{(z+3i)^2(z^2+16)}\right\}\right]_{z=3i}$$

$$= \left[\frac{(z+3i)^2(z^2+16)2ie^{2iz} - e^{2iz}[2(z+3i)(z^2+16)+2z(z+3i)^2]}{(z+3i)^4(z^2+16)^2}\right]_{z=3i}$$

$$= \left[\frac{(z+3i)(z^2+16)2ie^{2iz} - e^{2iz}[2(z^2+16)+2z(z+3i)]}{(z+3i)^3(z^2+16)^2}\right]_{z=3i}$$

$$= \frac{6i\times 7\times 2ie^{-6} - e^{-6}(2\times 7+6i\times 6i)}{(6i)^3(7)^2} = \frac{e^{-6}[-84+22]i}{216\times 49} = \frac{e^{-6}(-62)i}{216\times 49} = -\frac{i31e^{-6}}{108\times 49}$$

$$\text{Residue of } f(z) \text{ at } (z=4i) = \lim_{z\to 4i}(z-4i)\frac{e^{2iz}}{(z^2+9)^2(z-4i)(z+4i)}$$

$$= \frac{e^{-8}}{(-16+9)^2(4i+4i)} = \frac{e^{-8}}{49\times 8i} = \frac{-ie^{-8}}{392}$$

$$\text{Sum of the residues} = -\frac{i31e^{-6}}{108\times 49} - \frac{ie^{-8}}{392}$$

Hence by Cauchy's Residue Theorem, we have

$$\int_C f(z)dz = 2\pi i\times \text{ Sum of the residues within } C$$

i.e.
$$\int_{-R}^{R} f(x)dx + \int_{C_R} f(z)dz = 2\pi i\times \text{ sum of residues}$$

or
$$\int_{-R}^{R}\frac{e^{2ix}}{(x^2+9)^2(x^2+16)}dx + \int_{C_R}\frac{e^{2iz}}{(z^2+9)^2(z^2+16)}dz = 2\pi i\times \text{ Sum of residues} \quad \text{... (1)}$$

Now let $R\to\infty$, so as to show that the second integral in above relation vanishes. For any point on C_R, as $|z|\to\infty$

Let
$$F(z) = \frac{1}{z^6}\frac{e^{2iz}}{\left(1+\frac{9}{z^2}\right)^2\left(1+\frac{16}{z^2}\right)}$$

$$\lim_{|z|\to\infty} F(z) = 0 \quad \text{or} \quad \int_{C_R}\frac{e^{2iz}}{(z^2+9)^2(z^2+16)}dz = 0 \text{ as } z\to\infty$$

Hence by making $R\to\infty$, relation (1) becomes

$$\therefore \quad \int_{-\infty}^{\infty}\frac{e^{2ix}}{(x^2+9)^2(x^2+16)}dx = 2\pi i\left[\frac{-i31e^{-6}}{108\times 49} - i\frac{e^{-8}}{392}\right] = \frac{2\pi}{196}\left[\frac{31e^{-6}}{27}+\frac{e^{-8}}{2}\right]$$

Equating real parts, we have

$$\int_{-\infty}^{\infty}\frac{\cos 2x dx}{(x^2+9)^2(x^2+16)} = \frac{\pi}{98}\left(\frac{31e^{-6}}{27}+\frac{e^{-8}}{2}\right)$$

$$\int_0^{\infty}\frac{\cos 2x}{(x^2+9)^2(x^2+16)}dx = \frac{\pi}{196}\left(\frac{31e^{-6}}{27}+\frac{e^{-8}}{2}\right) \quad \left[\because \int_{-\infty}^{\infty} f(x)dx = 2\int_0^{\infty} f(x)dx \text{ If } f(x) \text{ is even function.}\right]$$

Ans.

EXERCISE 26.5

Evaluate the following :

1. $\int_0^{\infty} \frac{1}{1+x^2}\,dx$ **Ans.** $\frac{\pi}{2}$ **2.** $\int_{-\infty}^{\infty} \frac{1}{(x^2+1)^2}\,dx$ **Ans.** $\frac{\pi}{2}$

3. $\int_0^{\infty} \frac{x^3 \sin x}{(x^2+a^2)(x^2+b^2)}\,dx$ **Ans.** $\frac{\pi}{2(a^2-b^2)}[a^2e^{-a}-b^2e^{-b}]$

4. $\int_{-\infty}^{\infty} \frac{\cos x}{(x^2+a^2)(x^2+b^2)}\,dx,\quad a>b>0$ **Ans.** $\frac{\pi}{a^2-b^2}\left(\frac{e^{-b}}{b}-\frac{e^{-a}}{a}\right)$

5. Show that $\int_0^{\infty} \frac{\cos x}{x^2+a^2}\,dx = \frac{\pi e^{-a}}{2a}$

6. Show that $\int_0^{\infty} \frac{x^3 \sin x}{(x^2+a^2)}\,dx = -\frac{\pi}{4}(a-2)\,a^{-a},\ a>0$

Evaluate the following :

7. $\int_{-\infty}^{\infty} \frac{\sin mx}{x(x^2+a^2)}\,dx,\quad m>0, a>0$ **Ans.** $\frac{\pi}{a^2}(2-e^{-ma})$

8. $\int_0^{\infty} \frac{x^2}{x^6+1}\,dx$ **Ans.** $\frac{\pi}{6}$ **9.** $\int_0^{\infty} \frac{x\sin ax}{x^4+a^4}\,dx$ **Ans.** $\frac{\pi}{2a^2}e^{-\frac{a^2}{\sqrt{2}}}\sin\frac{a^2}{\sqrt{2}}$

10. $\int_0^{\infty} \frac{x^6}{(a^4+x^4)^2}\,dx$ **Ans.** $\frac{3\pi\sqrt{2}}{16a},\ a>0$ **11.** $\int_0^{\infty} \frac{\cos x^2+\sin x^2-1}{x^2}\,dx$ **Ans.** 0

12. $\int_0^{\infty} \frac{\cos mx}{x^4+x^2+1}\,dx$ **Ans.** $\frac{\pi}{\sqrt{3}}\sin\frac{1}{2}\left(m+\frac{\pi}{3}\right)e^{-\frac{1}{2}m\sqrt{3}}$ **13.** $\int_0^{\infty} \frac{\log(1+x^2)}{1+x^2}\,dx$ **Ans.** $\pi \log 2$

26.16 RECTANGULAR CONTOUR

Example 63. *Evaluate* $\int_{-\infty}^{+\infty} \frac{e^{ax}}{e^x+1}\,dx$

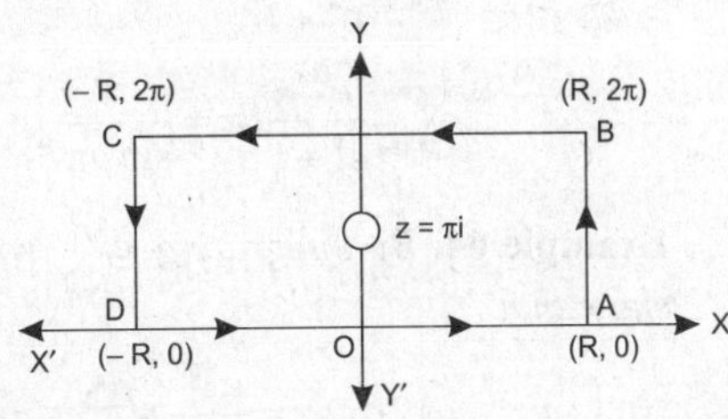

Solution. We consider $\int_C \frac{e^{az}}{e^z+1}\,dz = \int_C f(z)\,dz$

where C is the rectangle $ABCD$ with vertices at $(R, 0)$, $(R, 2\pi)$, $(-R, 2\pi)$ and $(-R, 0)$.

$f(z)$ has simple poles, $e^z = -1$

$$= \cos(2n+1)\pi + i\sin(2n+1)\pi = e^{i(2n+1)\pi}$$

$\Rightarrow$ $z = (2n+1)\pi i,$ where $n = 0, \pm 1, \pm 2....$

The only pole inside the rectangle is $z = \pi i$.

$\therefore$ By Residue theorem

$$\int_C f(z)\,dz = 2\pi i \text{ Residue } f(\pi i) = 2\pi i\left[\frac{e^{az}}{\frac{d}{dz}(e^z+1)}\right]_{z=\pi i} \qquad \left[R(a)=\frac{\phi(a)}{\psi'(a)}\right]$$

$$= 2\pi i\left[\frac{e^{az}}{e^z}\right]_{z=\pi i} = 2\pi i\frac{e^{a\pi i}}{e^{\pi i}} = -2\pi i\,e^{a\pi i} \qquad \left[\begin{aligned}\because e^{\pi i} &= \cos\pi + i\sin\pi\\ &= -1+0\\ &= -1\end{aligned}\right]$$

Also $\int_C f(z)\,dz = \int_{AB} f(z)\,dz + \int_{BC} f(z)\,dz + \int_{CD} f(z)\,dz + \int_{DA} f(z)\,dz$ ... (1)

$$= \int_0^{2\pi} f(R+iy)\,i\,dy + \int_R^{-R} f(x+2\pi i)\,dx + \int_{2\pi}^{0} (-R+iy)\,i\,dy + \int_{-R}^{R} f(x)\,dx \quad \text{... (2)}$$

[$\because$ $z = R + iy$ along AB, $z = x + 2\pi i$ along BC, $z = -R + iy$ along CD and $z = x$ along DA].

$$\int_C f(z)dz = i\int_0^{2\pi} \frac{e^{a(R+iy)}}{e^{R+iy}+1}dy - \int_{-R}^{+R} \frac{e^{a(x+2\pi i)}}{e^{x+2\pi i}+1}dx - i\int_0^{2\pi} \frac{e^{a(-R+iy)}}{e^{-R+iy}+1}dy + \int_{-R}^{R} \frac{e^{ax}}{e^x+1}dx$$

Now for any two complex numbers z_1, z_2, $|z_1| \geq |z_2|$

we have $|z_1 + z_2| \geq |z_1| - |z_2|$.

So that $|e^{R+iy}+1| \geq e^R - 1$. Also $|e^{a(R+iy)}| = e^{aR}$.

$\therefore$ For the integrand of first integral in (2), we have

$$\left|\frac{e^{a(R+iy)}}{e^{R+iy}+1}\right| \leq \frac{e^{aR}}{e^R - 1} \text{ which} \to 0, \text{ as } R \to \infty \; [\because a < 1]$$

Similarly, for the integrand of the third integral in (2), we get

$$\left|\frac{e^{a(-R+iy)}}{e^{-R+iy}+1}\right| \leq \frac{e^{-aR}}{1-e^{-r}} \text{ which also} \to 0, \text{ as } R \to \infty \; [\because a < 0]$$

Hence as $R \to \infty$, since the first and third integrals in (2) approach zero, we get

$$\int_C f(z)\,dz = -e^{2a\pi i}\int_{-\infty}^{\infty} \frac{e^{ax}}{e^x+1}dx + \int_{-\infty}^{\infty} \frac{e^{ax}}{e^x+1}dx$$

$$= (1-e^{2a\pi i})\int_{-\infty}^{\infty} \frac{e^{ax}}{e^x+1}dx. \quad \text{... (3)}$$

Thus from (1) and (3), we obtain

$$(1-e^{2a\pi i})\int_{-\infty}^{\infty} \frac{e^{ax}}{e^x+1}dx = -2\pi i e^{a\pi i} \quad \text{or} \quad \int_{-\infty}^{\infty} \frac{e^{ax}}{e^x+1}dx = \frac{2\pi i}{e^{a\pi i} - e^{-a\pi i}}$$

$$\int_{-\infty}^{\infty} \frac{e^{ax}}{e^x+1}dx = \frac{\pi}{\sin a\pi}$$ **Ans.**

Example 64. *By integrating* e^{-z^2} *round the rectangle whose vertices are 0, R, R + ia, ia, show that*

(i) $\int_0^\infty e^{-x^2} \cos 2ax\,dx = \frac{e^{-a^2}}{2}\sqrt{\pi}$ *and* *(ii)* $\int_0^\infty e^{-x^2} \sin 2ax\,dx = e^{-a^2}\int_0^a e^{y^2}\,dy$.

Solution. Let $f(z) = e^{-z^2}$

$\therefore$ $\int_C f(z)dz = \int_C e^{-z^2}\,dz.$

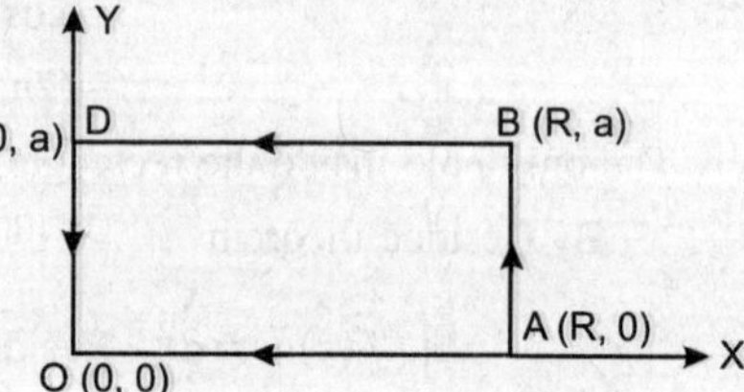

where C is the closed contour, a rectangle $OABD$.

Since $f(z)$ is analytic within and on the contour C (There is no pole within rectangle $OABD$). Hence by Cauchy's residue theorem, we have

$$\int_{OABD} e^{-z^2}\,dz = 0$$

i.e. $\int_{OA} e^{-z^2}\,dz + \int_{AB} e^{-z^2}\,dz + \int_{BD} e^{-z^2}\,dz + \int_{DO} e^{-z^2}\,dz = 0$ ... (1)

Since on $OA, z = x, dz = dx.$ On $AB, z = R + iy, dz = idy$

On $BD, z = x + ia, dz = dx.$ On $DO, z = iy, dz = idy$

Hence (1) becomes

$$\int_0^R e^{-x^2} dx + \int_0^a e^{-(R+iy)^2} .idy + \int_R^0 e^{-(x+ia)^2} .dx + \int_a^0 e^{-(iy)^2} .idy = 0 \quad \text{... (2)}$$

Now $$\left|\int_0^a e^{-(R+iy)^2} .idy\right| \le \int_0^a \left|e^{-(R+iy)^2}\right| |idy| \le \int_0^a e^{-R^2+y^2} dy$$

$$\le \int_0^a e^{-R^2+a^2} .dy \quad [\text{since } y \le a \text{ on } AB]$$

$$\le e^{-R^2+a^2} .a = 0 \text{ as } R \to \infty$$

Hence by making $R \to \infty$, equation (2) becomes

$$\int_0^\infty e^{-(x+ia)^2} dx = \int_0^\infty e^{-x^2} dx - i\int_0^a e^{y^2} dy$$

$$\Rightarrow \int_0^\infty e^{(-x^2+a^2-2aix)} dx = \frac{\sqrt{\pi}}{2} - i\int_0^a e^{y^2} dy \quad \left[\text{since } \int_0^\infty e^{-x^2} dx = \frac{\sqrt{\pi}}{2}\right]$$

$$\int_0^\infty e^{-x^2+a^2} .e^{-2aix} dx = \frac{\sqrt{\pi}}{2} - i\int_0^a e^{y^2} dy$$

$$\int_0^\infty e^{-x^2+a^2} (\cos 2ax - i\sin 2ax) dx = \frac{\sqrt{\pi}}{2} - i\int_0^a e^{y^2} dy$$

$$\int_0^\infty e^{-x^2} (\cos 2ax - i\sin 2ax) dx = \frac{\sqrt{\pi}}{2} e^{-a^2} - ie^{-a^2} \int_0^a e^{y^2} dy$$

Equating real and imaginary parts, we have

$$\int_0^\infty e^{-x^2} \cos 2ax\, dx = \frac{e^{-a^2}}{2} \sqrt{\pi}$$

$$\int_0^\infty e^{-x^2} \sin 2ax\, dx = e^{-a^2} \int_0^a e^{y^2} dy$$ **Proved.**

EXERCISE 26.6

1. Using contour integration, show that

$$\int_0^\infty \frac{x^6 dx}{(a^4 + x^4)^2} = \frac{3\sqrt{2}\pi}{16a}, \quad (a > 0)$$

2. Using method of contour integration, evaluate

$$\int_0^\infty \frac{x \sin ax\, dx}{x^4 + 4}$$ **Ans.** $\frac{\pi}{8} e^{-a} \sin a$

3. Integrating $\frac{e^{iz}}{z+a}$ along the boundary of the square defined by

$x = 0, x = R, y = 0, y = R.$

Prove that (*i*) $\int_0^\infty \frac{\cos x}{x+a} dx = \int_0^\infty \frac{xe^{-ax}}{1+x^2} dx$ (*ii*) $\int_0^\infty \frac{\sin x}{x+a} dx = \int_0^\infty \frac{e^{-ax}}{1+x^2} dx$

4. Evaluate, using Cauchy's integral formula

$\oint_c \frac{\cos \pi z\, dz}{z^2 - 1}$ around a rectangle

(*i*) $2 \pm i, -2 \pm i$ **Ans.** 0 (*ii*) $-i, 2 - i, 2 + i$ and i **Ans.** $-\pi i$

5. By integrating $\dfrac{e^{iaz^2}}{\sinh \pi z}$ round the rectangle with vertices $\pm R \pm \dfrac{i}{2}$, show that

$$\int_0^{\infty} \frac{\cos(ax^2)\cosh(ax)}{\cosh \pi x}\, dx = \frac{1}{2}\cos\left(\frac{a}{4}\right) \text{ and } \int_0^{\infty} \frac{\sin(ax^2)\cosh(ax)}{\cosh \pi x}\, dx = \frac{1}{2}\sin\left(\frac{a}{4}\right) \quad (0 < a \le \pi)$$

26.17 INDENTED SEMI-CIRCULAR CONTOUR

When the integrand has a simple pole on real axis, it is deleted from the region by indenting the contour (a small semi-circle having pole is drawn)

Example 65. *By contour integration, prove that*

$$\int_0^{\infty} \frac{\sin mx}{x}\, dx = \frac{\pi}{2}.$$

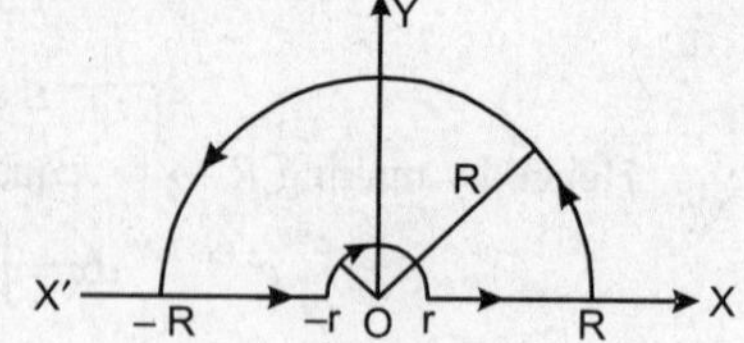

Solution. Consider the integral $\displaystyle\int_C \frac{e^{miz}}{z} dz$

where C is a large semi-circle $|z| = R$ indented at $z = 0$ (pole), let r be the radius of indentation. There is no singularity within the given contour.

Hence by Cauchy Theorem.

$$\int_C \frac{e^{miz}}{z} dz = 0$$

i.e.,
$$\int_{-R}^{-r} \frac{e^{mix}}{x} dx + \int_{C_r} \frac{e^{miz}}{z} dz + \int_r^R \frac{e^{mix}}{x} dx + \int_{C_R} \frac{e^{miz}}{z} dz = 0 \quad \ldots (1)$$

Substituting $-x$ for x in the first integral and combining it with the third integral, we get

$$\int_r^R \frac{e^{mix} - e^{-imx}}{x} dx + \int_{C_2} \frac{e^{miz}}{z} dz + \int_{C_1} \frac{e^{miz}}{z} dz = 0 \quad \left[z = Re^{i\theta} \Rightarrow dz = Rie^{i\alpha} d\theta\right]$$

$$\Rightarrow \qquad 2i\int_r^R \frac{\sin mx}{x} dx + \int_{C_2} \frac{e^{imz}}{z} dz + \int_{C_1} \frac{e^{imz}}{z} dz = 0 \quad \ldots (2)$$

Now
$$\int_{C_2} \frac{e^{imz}}{z} dz = \int_{C_2} \frac{1}{z} dz + \int_{C_2} \frac{e^{imz} - 1}{z} dz \quad \ldots (3)$$

On $\qquad C_2\ z = re^{i\theta}$

$$\therefore \qquad \int_{C_2} \frac{1}{z} dz = \int_{\pi}^{0} \frac{re^{i\theta} i\, d\theta}{re^{i\theta}} = -\int_0^{\pi} i\, d\theta = -i\pi$$

Also
$$\left| \int_{C_2} \frac{e^{imx} - 1}{z} dz \right| \le M \int_{C_2} \frac{|dz|}{|z|} = \pi M$$

where M is the maximum value on C_2 of $|e^{imx} - 1| = |e^{imr(\cos\theta + i\sin\theta)} - 1|$

Clearly, $M \to 0$ as $r \to 0$

$$\therefore \text{ From (3),} \qquad \int_{C_2} \frac{e^{imz}}{z} dz = -i\pi \quad \ldots (4)$$

Putting $\qquad z = Re^{i\theta}$ in the integral over C_1, we get

$$\int_{C_1} \frac{e^{imz}}{z} dz = \int_0^{\pi} \frac{e^{imR(\cos\theta + i\sin\theta)}}{Re^{i\theta}} Re^{i\theta} i\, d\theta = i\int_0^{\pi} e^{imR\cos\theta} . e^{-mR\sin\theta} d\theta$$

Since $|e^{imR\cos\theta}| \le 1$

$$\therefore \quad \left|\int_{C_1} \frac{e^{imz}}{z} dz\right| \le \int_0^{\pi} e^{-mR\sin\theta} d\theta = 2\int_0^{\pi/2} e^{-mR\sin\theta} d\theta$$

Also $\frac{\sin\theta}{\theta}$ continuously decreases from 1 to $\frac{2}{\pi}$ as θ increases from 0 to $\frac{\pi}{2}$.

$$\therefore \quad \text{For } 0 \le \theta \le \frac{\pi}{2}, \quad \frac{\sin\theta}{\theta} \ge \frac{2}{\pi} \text{ or } \sin\theta \ge \frac{2\theta}{\pi}$$

$$\therefore \quad \left|\int \frac{e^{imz}}{z} dz\right| \le 2\int_0^{\pi/2} e^{-2mR\theta/\pi} d\theta = \left[-\frac{\pi}{mR} e^{-2mR\theta/\pi}\right]_0^{\pi/2} = \frac{\pi}{mR}(1 - e^{-mR})$$

As $\quad R \to \infty, \frac{\pi}{mR}(1 - e^{-mR}) \to 0$

$$\therefore \quad \int_{C_1} \frac{e^{imz}}{z} dz = 0$$

Hence from (2), on taking the limit as $r \to 0$ and $R \to \infty$, we get

$$2i\int_0^{\infty} \frac{\sin mx}{x} dx - i\pi = 0 \text{ or } \int_0^{\infty} \frac{\sin mx}{x} dx = \frac{\pi}{2}.$$ **Ans.**

Example 66. *Show that, if $a \ge b \ge 0$, then* $\int_0^{\infty} \frac{\cos 2ax - \cos 2bx}{x^2} dx = \pi(b - a)$

Solution. Consider the integral $\int_C f(z)dz$

where $\quad f(z) = \frac{e^{i2az} - e^{i2bz}}{z^2}$

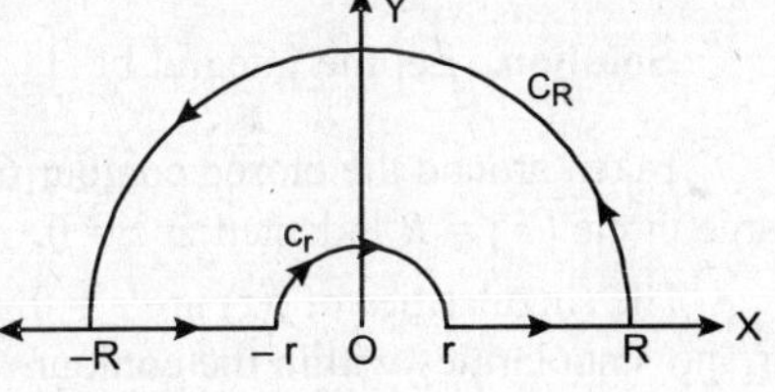

and C is a large semi-circle $|z| = R$ indented at $z = 0$ (pole), let r be the radius of indentation. Now there is no singularity within the given contour.

$\int_C f(z)dz = 0$ (By Cauchy Integral Theorem)

$$\Rightarrow \quad \int_{-R}^{-r} f(x)\, dx + \int_{C_R} f(z)\, dz + \int_r^R f(x)\, dx + \int_{C_R} f(z)\, dz = 0 \quad \text{... (1)}$$

Now $$\left|\int_{C_R} f(z)\, dz\right| \le \int_{C_R} \frac{|e^{2iaz} - e^{2ibz}|}{|z|^2} |dz|$$

$$\le \int_{C_R} \frac{|e^{2iaz}| + |e^{2ibz}|}{|z|^2} |dz|$$

$$= \int_0^{\pi} \frac{e^{-2aR\sin\theta} + e^{-2bR\sin\theta}}{R^2} R d\theta \quad (z\ Re^{i\theta})$$

$$\le \frac{2}{R}\int_0^{\frac{\pi}{2}} \left[e^{\frac{-4aR\theta}{\pi}} + e^{\frac{-4bR\theta}{\pi}}\right] d\theta \quad \text{[By Jordan's inequality]}$$

$$= \frac{2}{R}\left[\frac{\pi}{4aR}(1 - e^{-2aR}) + \frac{\pi}{4aR}(1 - e^{-2bR})\right] = 0 \text{ as } R \to \infty$$

We have, $\lim_{z\to 0}\{z(f(z))\} = \lim_{z\to 0}\left\{z\dfrac{e^{2iaz}-e^{2ibz}}{z^2}\right\} = \lim_{z\to 0}\{2i(a-b)-2(a^2-b^2)z^2....\} = 2i(a-b)$

$$\lim_{r\to 0}\int_{C_r} f(z)dz = -i(\pi-0)\times 2i(a-b) = -2\pi(b-a)$$

Hence, by making $R\to\infty$ and $r\to 0$, equation (1) reduces to

$$\int_{-\infty}^{0} f(x)\,dx - 2\pi(b-a) + \int_0^{\infty} f(x)\,dx + 0 = 0 \Rightarrow \int_{-\infty}^{\infty} f(x)dx = 2\pi(b-a)$$

$$\int_{-\infty}^{\infty}\frac{e^{2iax}-e^{i2bx}}{x^2}\,dx = 2\pi(b-a)$$

$$\int_{-\infty}^{\infty}\frac{(\cos 2ax + i\sin 2ax)-(\cos 2bx + i\sin 2bx)}{x^2}dx = 2\pi(b-a)$$

Equating real parts, we get

$$\int_{-\infty}^{\infty}\frac{\cos 2ax - \cos 2bx}{x^2}dx = 2\pi(b-a) \qquad \left[\begin{array}{l}\int_{-a}^{a} f(x)dx = 2\int_0^a f(x)dx \\ \text{If } f(x) \text{ is even function.}\end{array}\right]$$

Hence, $\displaystyle\int_0^{\infty}\frac{\cos 2ax - \cos 2bx}{x^2}dx = \pi(b-a)$ **Proved.**

Example 67. *Using contour integration method, prove the integral*

(*i*) $\displaystyle\int_0^{\infty}\frac{x^{a-1}}{1+x}dx = \frac{\pi}{\sin\pi a}$. $(0<a<1)$ (*ii*) $\displaystyle\int_0^{\infty}\frac{x^{a-1}}{1-x}dx = \pi\cot\pi a.$

Solution. Let the integral be $\int_C f(z)\,dz$, where $f(z) = \dfrac{z^{a-1}}{1-z}$

Taken around the closed contour C consisting of real axis from $-R$ to R, and upper half of a large circle $|z| = R$ indented at $z = 0$, $z = 1$, the radii of indentations being r and r' respectively.

The singularities of $f(z)$ are $z = 0$, $z = 1$ which have been avoided by the identation, so there are no singularities within the contour.

Hence, by Cauchy's residue theorem, we have

$$\int_{-R}^{-r} f(x)\,dx + \int_{Cr} f(z)dz + \int_r^{1-r'} f(x)dx + \int_{Cr'} f(z)dz + \int_{1+r'}^{R} f(x)dx + \int_{CR} f(z)dz = 0 \quad ...(1)$$

Since $\lim_{z\to\infty} z\,f(z) = \lim_{z\to\infty} z\dfrac{z^{a-1}}{1-z} = \lim_{z\to\infty}\dfrac{z^a}{1-z} = 0,\ 0<a<1.$

$$\therefore \operatorname{Lim}_{R\to\infty}\int_{C_R} f(z)dz = i(\pi-0)\,0 = 0$$

Again, $\operatorname{Lim}_{z\to 0}\{z\,f(z)\} = \operatorname{Lim}_{z\to 0}\left\{\dfrac{z.z^{a-1}}{1-z}\right\} = \lim_{z\to 0}\left(\dfrac{z^a}{1-z}\right) = 0,\ a>0.$

$$\therefore \operatorname{Lim}_{r\to 0}\int_{Cr} f(z)dz = -i(\pi-0).0 = 0$$

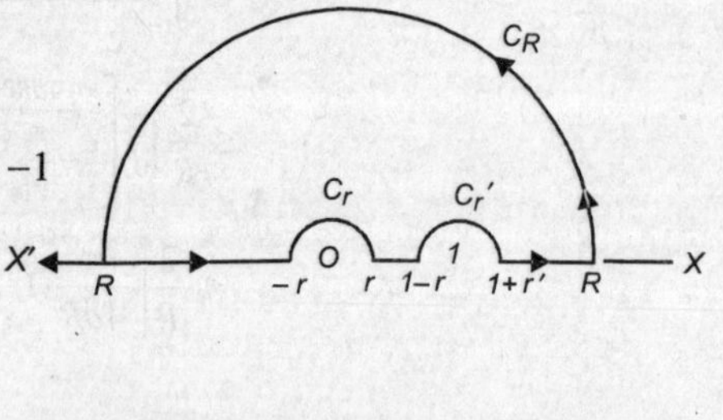

Also, $\operatorname{Lim}_{r\to 1}\{(z-1)f(z)\} = \operatorname{Lim}_{z\to 1}\left\{(z-1)\dfrac{z^{a-1}}{1-z}\right\} = -1$

$$\therefore \operatorname{Lim}_{r\to 0}\int_{C_{r'}} f(z)dz = -i(\pi-0)\,(-1) = i\pi$$

Hence making $R \to \infty$, $r \to 0$, $r' \to 0$, we have from (1)

$$\int_{-\infty}^{\infty} f(x)\,dx+\int_0^1 f(x)\,dx+\pi i+\int_1^{\infty} f(x)\,dx=0$$

$$\Rightarrow \quad \int_{-\infty}^{\infty} f(x)\,dx+\pi i = 0 \qquad \Rightarrow \quad \int_{-\infty}^{\infty} \frac{x^{a-1}}{1-x}dx+\pi i = 0$$

$$\Rightarrow \quad \int_{-\infty}^{0} \frac{x^{a-1}}{1-x}dx+\int_0^{\infty} \frac{x^{a-1}}{1-x}\,dx = -\pi i \qquad \Rightarrow \quad -\int_0^{-\infty} \frac{x^{a-1}}{1-x}\,dx + \int_0^{\infty} \frac{x^{a-1}}{1-x}\,dx = -\pi i$$

Putting $-x$ for x in the first integral, we have

$$\int_0^{\infty} \frac{(-1)^{a-1}x^{a-1}}{1+x}\,dx+\int_0^{\infty} \frac{x^{a-1}}{1-x}dx = -\pi i$$

$$\Rightarrow \quad \int_0^{\infty} \frac{(e^{i\pi})^{a-1}x^{a-1}}{1+x}\,dx+\int_0^{\infty} \frac{x^{a-1}}{1-x}\,dx = -\pi i$$

$$\Rightarrow \quad \int_0^{\infty} \frac{e^{-i\pi}e^{ia\pi}x^{a-1}}{1+x}\,dx + \int_0^{\infty} \frac{x^{a-1}}{1-x}dx = -\pi i$$

$$\therefore \quad -e^{ia\pi}\int_0^{\infty} \frac{x^{a-1}}{1+x}\,dx+\int_0^{\infty} \frac{x^{a-1}}{1-x}dx = -\pi i \qquad [\text{since } e^{-1\pi} = -1]$$

$$-(\cos a\pi + i\sin a\pi)\int_0^{\infty} \frac{x^{a-1}}{1+x}\,dx + \int_0^{\infty} \frac{x^{a-1}}{1-x}\,dx = -\pi i$$

Equating imaginary and real parts, we have

$$-\sin a\pi \int_0^{\infty} \frac{x^{a-1}}{1+x}dx = -\pi \Rightarrow \int_0^{\infty} \frac{x^{a-1}}{1+x}\,dx = \frac{\pi}{\sin a\pi} \qquad ...(1)$$

$$\text{and } -\cos a\pi \int_0^{\infty} \frac{x^{a-1}}{1+x}\,dx + \int_0^{\infty} \frac{x^{a-1}}{1-x}\,dx = 0$$

$$\Rightarrow \quad -\cos a\pi \int_0^{\infty} \frac{x^{a-1}}{1+x}\,dx = -\int_0^{\infty} \frac{x^{a-1}}{1-x}\,dx$$

$$\Rightarrow \quad \cos a\pi \times \frac{\pi}{\sin a\pi} = \int_0^{\infty} \frac{x^{a-1}}{1-x}\,dx \qquad [\text{From (1)}]$$

Thus
$$\int_0^{\infty} \frac{x^{a-1}}{1-x}\,dx = \pi \cot a\pi$$
Ans.

EXERCISE 26.7

Using the method of contour integration, evaluate the following :

1. $\int_0^{\infty} \frac{\cos x}{x}dx$ **Ans.** 0

2. $\int_0^{\infty} \frac{\log(1+x^2)}{x^{1+\alpha}}dx,\ 0<\alpha<1$ **Ans.** 0

3. $\int_0^{\infty} \frac{\log x}{(1+x)^3}\,dx$ **Ans.** $-\frac{1}{2}$

4. $\int_{-\infty}^{\infty} \frac{1}{x^3+1}\,dx$ **Ans.** $\frac{\pi}{\sqrt{3}}$

5. $\int_{-\infty}^{\infty} \frac{x^2}{(x^2+1)^2}dx$ **Ans.** $\frac{\pi}{2}$

6. $\int_0^{\infty} \frac{1}{x+1}\,dx$ **Ans.** $\frac{\pi}{2}$

7. $\int_0^{\infty} \frac{\log x}{1+x^2}\,dx$ **Ans.** $\frac{\pi^3}{8}$

7. $\int_0^{\infty} \sin x^2 dx$ *(D.U., April 2010)* **Ans.** $\frac{1}{2}\sqrt{\frac{\pi}{2}}$

SERIES SOLUTIONS OF SECOND ORDER DIFFERENTIAL EQUATIONS

27.1 INTRODUCTION

We have already studied how to find the solution of a differential equation with constant coefficients.

There are many differential equations with variable coefficients like Bessel's equation, Legender's equation, Hermite's equations, Laguerrer's differential equation whose solution are not the combination of elementary functions.

The solutions are infinite series. In this chapter we will solve the differential equations by power series method and Frobenius method (extended power series method).

27.2 POWER SERIES SOLUTIONS OF DIFFERENTIAL EQUATIONS

We know that the solution of the differential equation

$$\frac{d^2 y}{dx^2} - y = 0$$

are $\quad y = e^x = 1 + \frac{x}{1!} + \frac{x^2}{2!} + \frac{x^3}{3!} + \dots \quad$ and $\quad y = e^{-x} = 1 - \frac{x}{1!} + \frac{x^2}{2!} - \frac{x^3}{3!} + \dots$

These are power series solution of the given differential equation.

Another example of the differential equation $\frac{d^2 y}{dx^2} + y = 0$

is satisfied by the power series

$$y = \sin x = x - \frac{x^3}{3!} + \frac{x^5}{5!} - \frac{x^7}{7!} + \dots$$

This idea leads to the methods of obtaining the solution of a linear differential equation of second order in series form.

The solution of the differential equation will be a series of ascending powers of x, the infinite series solution obtained will have its own region of convergence or validity.

27.3 ANALYTIC FUNCTION

A function $f(x)$ which can be expended in Taylor's series on interval containing the point x_0. The series converges to $f(x)$ for all x in the interval of convergence.

27.4 ORDINARY POINT

Consider the equation

$$\frac{d^2y}{dx^2}+P\frac{dy}{dx}+Qy=0$$

where P, Q are polynomials in x.

$x = a$ is an ordinary point of the above equation if the denominators of P and Q do not vanish for $x = a$. *i.e.* $(P \neq \infty, Q \neq \infty)$

For example :

(*i*) $$(1+x^2)\frac{d^2y}{dx^2}+x\frac{dy}{dx}-y=0 \quad\Rightarrow\quad \frac{d^2y}{dx^2}+\frac{x}{1+x^2}\frac{dy}{dx}-\frac{1}{1+x^2}y=0$$

Here, denominator of P and Q *i.e.*, $1 + x^2$ is not equal to zero at $x = 0$. Therefore $x = 0$ is an ordinary point for this differential equation.

(*ii*) $$x^2\frac{d^2y}{dx^2}+(2x^2-x)\frac{dy}{dx}+y=0 \quad\Rightarrow\quad \frac{d^2y}{dx^2}+\frac{2x^2-x}{x^2}\frac{dy}{dx}+\frac{1}{x^2}y=0$$

Here, the denominator of P and Q *i.e.*, x^2 is equal to zero for $x = 0$ $(P = \infty, Q = \infty)$. So, $x = 0$ is not ordinary point for this equation.

Note. In this section, we have to solve those differential equation whose ordinary point is at $x = 0$.

27.5 SOLUTION OF THE DIFFERENTIAL EQUATION WHEN *X* = 0 IS AN ORDINARY POINT *i.e.* WHEN *P* DOES NOT VANISH FOR *X* = 0.

(*i*) Let $y = a_0 + a_1x + a_2x^2 + a_3x^3 + ... + a_k\,x^k + ... = \sum_{k=0}^{\infty} a_kx^k$ be the solution of the given differential equation.

(*ii*) Find $\frac{dy}{dx}, \frac{d^2y}{dx^2}$ etc.

$$\frac{dy}{dx} = a_1 + 2a_2x + 3a_3x^2 + ... + ka_kx^{k-1} + ... + ... = \sum_{k=1}^{\infty} ka_kx^{k-1}$$

$$\frac{d^2y}{dx^2} = 2a_2 + 2.3\,a_3x + ... + a_k\,k(k-1)\,x^{k-2} + ... = \sum_{k=2}^{\infty} a_k\,.\,k\,(k-1)\,.\,x^{k-2}$$

(*iii*) Substitute the values of $y, \frac{dy}{dx}, \frac{d^2y}{dx^2}$ etc. in the given differential equation.

(*iv*) Calculate a_0, a_1, a_2 .... coefficients of various powers of x by equating the coefficients to zero.

(*v*) Substitute the values of a_0, a_1, a_2 ... in the differential equation to get the required series solution.

27.6 WORKING RULE TO SOLVE A DIFFERENTIAL EQUATION IF *X* = 0 IS AN ORDINARY POINT OF THE EQUATION

Step 1. Assume the solution $y=\Sigma a_kx^k$ *i.e.*,

$$y=a_0+a_1x+a_2x^2+a_3x^3+a_4x^4+......$$

Step 2. Find $\frac{dy}{dx}$ and $\frac{d^2y}{dx^2}$ of step (1).

Step 3. Substitute the values of $y, \frac{dy}{dx}$ and $\frac{d^2y}{dx^2}$ in the given differential equation.

Step 4. Equate to zero the coefficients of various power of x, to find out a_2, a_3 ..... in terms of a_0 and a_1.

Step 5. Equate to zero the coefficient of x^k to get a recurrence relation of a's.

Step 6. Substitute $k = 0, 1, 2, 3,$ ..... in the recurrence relations to get the values of $a_2, a_3, ...$ $a_n, ...$.

Step 7. Substitute the values of $a_2, a_3,$ ...... etc (obtained in step 6) in the solution (1)

When $x = 0$ is the ordinary point.

Example 1. *Solve in series the equation* $\frac{d^2y}{dx^2} + x^2 y = 0.$ *(DU, II Sem 2014)*

Solution. We have, $\frac{d^2y}{dx^2} + x^2 y = 0$ ... (1)

The denominator of P and Q is not zero so $x = 0$ is the ordinary point.

Let $y = a_0 + a_1x + a_2x^2 + a_3x^3 + a_4x^4 + a_5x^5 + a_6x^6 + a_7x^7 + a_8a^8 + ... + a_nx^n + ...$ ... (2)

Here $P \neq \infty$, and $Q \neq \infty$ for $x = 0$. So, $x = 0$ is the ordinary point of the equation (1).

Then $\frac{dy}{dx} = a_1 + 2a_2 x + 3\, a_3\, x^2 + 4\, a_4\, x^3 + 5\, a_5\, x^4 + 6\, a_6\, x^5 + 7\, a_7\, x^6 + 8\, a_8\, x^7 + ...$ $\frac{d^2y}{dx^2}$

$= 2a_2 + 2\,.\,3\, a_3\, x + 3\,.\,4\, a_4\, x^2 + 4\,.\,5\, a_5\, x^3 + 5\,.\,6\, a_6\, x^4 + 6\,.\,7\, a_7\, x^5 + 7\,.\,8\, a_8\, x^6 + ...$

Substituting the values of $\frac{d^2y}{dx^2}$ and y in (1), we get

$$2\, a_2 + 2.3\, a_3\, x + 3.4\, a_4\, x^2 + 4.5\, a_5\, x^3 + 5.6\, a_6\, x^4 + 6.7\, a_7\, x^5 + 7.8\, a_8\, x^6 + ...$$
$$+ x^2\, (a_0 + a_1\, x + a_2\, x^2 + a_3\, x^3 + a_4\, x^4 + a_5\, x^5 + a_6\, x^6 + a_7\, x^7 + ...) = 0$$
$$2\, a_2 + 6\, a_3\, x + (a_0 + 12\, a_4)\, x^2 + (a_1 + 20a_5)\, x^3 + (a_2 + 30\, a_6)\, x^4 + ...$$
$$+ [a_{n-2} + (n+2)\,(n+1)\, a_{n+2}]\, x^n + ... = 0$$

Equating to zero the coefficients of the various powers of x, we obtain

$$a_2 = 0, \;\; a_3 = 0$$
$$a_0 + 12a_4 = 0 \;\; i.e.\; a_4 = -\frac{1}{12} a_0$$
$$a_1 + 20\, a_5 = 0 \;\; i.e.\; a_5 = -\frac{1}{20} a_1$$
$$a_2 + 30\, a_6 = 0 \;\; i.e.\; a_6 = -\frac{1}{30}\, a_2 = 0 \qquad (a_2 = 0)$$

and so on. In general

$$a_{n-2} + (n+2)\,(n+1)\, a_{n+2} = 0 \;\; \Rightarrow \;\; \boxed{\boldsymbol{a_{n+2} = -\frac{a_{n-2}}{(n+1)(n+2)}}}$$

Putting $n = 5$, $a_7 = -\frac{a_3}{6 \times 7} = 0$ $(a_3 = 0)$

Putting $n = 6$, $a_8 = -\frac{a_4}{7 \times 8} = \frac{a_0}{12 \times 7 \times 8}$

Putting $n = 7$, $a_9 = -\frac{a_5}{8 \times 9} = \frac{a_1}{20 \times 8 \times 9}$

Putting $n = 8$, $a_{10} = -\frac{a_6}{9 \times 10} = 0,$ $(a_6 = 0)$

Putting $n = 9$, $a_{11} = -\dfrac{a_7}{11\times 10} = 0$, $(a_7 = 0)$

Putting $n = 10$, $a_{12} = -\dfrac{a_8}{12\times 11} = -\dfrac{a_0}{12\times 8\times 7\times 11\times 12}$

Substituting these values in (2), we get

$$y = a_0 + a_1 x - \frac{1}{12}a_0 x^4 - \frac{a_1}{20}x^5 + \frac{a_0}{12\times 7\times 8}x^8 + \frac{a_1}{20\times 8\times 9}x^9 - \frac{a_0}{12\times 8\times 7\times 11\times 12}x^{12} + \ldots$$

$$y = a_0\left(1 - \frac{1}{12}x^4 + \frac{x^8}{12\times 7\times 8} - \frac{x^{12}}{12\times 8\times 7\times 11\times 12} + \ldots\right) + a_1\left(x - \frac{x^5}{20} + \frac{x^9}{20\times 8\times 9} - \ldots\right) \quad \textbf{Ans.}$$

Example 2. *Solve the following differential equation in series*

$$(1-x^2)\frac{d^2y}{dx^2} - x\frac{dy}{dx} + 4y = 0. \qquad \textit{(U.P. II Semester summer 2006)}$$

Solution. We have,

$$(1-x^2)\frac{d^2y}{dx^2} - x\frac{dy}{dx} + 4y = 0 \;\Rightarrow\; \frac{d^2y}{dx^2} - \frac{x}{1-x^2}\frac{dy}{dx} + \frac{4}{1-x^2}y = 0 \qquad \ldots(1)$$

Here, $P \neq \infty$ and $Q \neq \infty$ for $x = 0$. So, $x = 0$ is an ordinary point of the given equation.
Assume the solution

$$y = \Sigma a_k x^k$$

$$\Rightarrow \quad y = a_0 + a_1 x + a_2 x^2 + a_3 x^3 + \ldots + a_k x^k + \ldots\ldots \qquad \ldots(2)$$

$$y' = \Sigma a_k (k) x^{k-1}$$

$$y'' = \Sigma a_k (k)(k-1) x^{k-2}$$

Putting these values in the given equation, we get

$$(1-x^2)\Sigma a_k (k)(k-1)x^{k-2} - x\Sigma a_k (k) x^{k-1} + 4\Sigma a_k x^k = 0$$

$$\Rightarrow \quad \Sigma a_k (k)(k-1)x^{k-2} - \Sigma a_k (k)(k-1)x^k - \Sigma a_k (k) x^k + 4\Sigma a_k x^k = 0$$

$$\Rightarrow \quad \Sigma a_k (k)(k-1)x^{k-2} - \Sigma a_k x^k \{k(k-1) + k - 4\} = 0$$

$$\Rightarrow \quad \Sigma a_k k(k-1)x^{k-2} - \Sigma a_k x^k (k^2 - 4) = 0$$

Now, equating the coefficient of x^k equal to zero. By putting $k + 2$ for k in first summation and $k = k$ in second summation, we have

$$a_{k+2}(k+2)(k+1) - a_k(k^2 - 4) = 0$$

$$a_{k+2}(k+2)(k+1) = a_k(k^2 - 4)$$

$$a_{k+2} = \frac{k^2 - 4}{(k+2)(k+1)}a_k = \frac{k-2}{k+1}a_k \;\Rightarrow\; \boxed{\boldsymbol{a_{k+2} = \frac{k-2}{k+1}a_k}}$$

If $k = 0$, $a_2 = -2a_0$

If $k = 1$, $a_3 = -\dfrac{1}{2}a_1$

If $k = 2$, $a_4 = \dfrac{0}{3}a_2 = 0$

If $k = 3$, $a_5 = \dfrac{1}{4}a_3 = \dfrac{1}{4}\left(-\dfrac{1}{2}\right)a_1 = -\dfrac{a_1}{8}$

If $k = 4$, $a_6 = \dfrac{2}{5}a_4 = \dfrac{2}{5}\times 0 = 0$

and so on.

Substituting these values is (2), we get

$$y = a_0 + a_1x + (-2a_0)x^2 + \left(-\frac{1}{2}a_1\right)x^3 + 0x^4 + \left(\frac{-a_1}{8}\right)x^5 + 0.x^6 + ...$$

$$= a_0 + a_1x - 2a_0x^2 - \frac{1}{2}a_1x^3 - \frac{a_1}{8}x^5 + ...$$

$$= a_0(1 - 2x^2) + a_1x\left(1 - \frac{x^2}{2} - \frac{x^4}{8} + ...\right)$$ **Ans.**

Example 3. *Find the power series solution of $(1 - x^2)\, y'' - 2\, y' + 2\, y = 0$ about $x = 0$.*

Solution. Here, we have

$$(1 - x^2)\ \frac{d^2y}{dx^2} - 2x\frac{dy}{dx} + 2y = 0 \Rightarrow \frac{d^2y}{dx^2} - \frac{2x}{1-x^2}\frac{dy}{dx} + \frac{2}{1-x^2}y = 0$$

Here $P \neq \infty$ and $Q \neq \infty$ for $x = 0$. So $x = 0$ is an ordinary point.

Let $y = a_0 + a_1\, x + a_2\, x^2 + a_3\, x^3 + ... + a_n x^n + ...$ be the required solution.

$$y = \sum_{k=0}^{\infty} a_k x^k$$

Then $$\frac{dy}{dx} = \sum_{k=1}^{\infty} a_k \cdot k\, x^{k-1}, \qquad \frac{d^2y}{dx^2} = \sum_{k=2}^{\infty} a_k \cdot k\,(k-1)\, x^{k-2}$$

Substituting the values of $y, \dfrac{dy}{dx}$ and $\dfrac{d^2y}{dx^2}$ in the given equation, we get

$$(1 - x^2)\Sigma a_k\, k \cdot (k-1)\, x^{k-2} - 2x\Sigma a_k \cdot k\, x^{k-1} + 2\Sigma a_k\, x^k = 0$$

$$\Rightarrow \quad \Sigma a_k \cdot k \cdot (k-1)x^{k-2} - \Sigma a_k k(k-1)x^k - 2\,\Sigma a_k . k\, x^k + 2\Sigma a_k x^k = 0$$

$$\Rightarrow \quad \Sigma a_k \cdot k \cdot (k-1)x^{k-2} - \Sigma[k(k-1) + 2k - 2]a_k x^k = 0$$

$$\Rightarrow \quad \Sigma a_k \cdot k \cdot (k-1)x^{k-2} - \Sigma(k^2 + k - 2)\, a_k x^k = 0$$

where the first summation extends over all values of k from 2 to ∞ and the second from k=0 to ∞.

Now equating the coefficient of x^k equal to zero, we have

$$\Rightarrow \quad (k+2)\,(k+1)\,a_{k+2} - (k^2 + k - 2) = 0$$

$$\Rightarrow \quad a_{k+2} = \frac{k^2 + k - 2}{(k+2)(k+1)}a_k = \frac{(k+2)(k-1)}{(k+2)(k+1)}a_k$$

$$\Rightarrow \quad \boxed{a_{k+2} = \frac{k-1}{k+1}a_k}$$

For $k = 0$ $$a_2 = -a_0,\ a_3 = 0,\ a_4 = \frac{a_2}{3} = -\frac{a_0}{3},\ a_5 = \frac{2}{4}a_3 = 0$$

For $k = 4$ $$a_6 = \frac{3}{5}a_4 = \frac{3}{5}\left(-\frac{a_0}{3}\right) = -\frac{a_0}{5},\ a_7 = \frac{4}{6}a_5 = 0, \text{ etc.}$$

$$y = a_0 + a_1x + a_2x^2 + a_3x^3 + a_4x^4 + a_5x^5 + a_6x^6 + a_7x^7 + ...$$

$$y = a_0 + a_1x - a_0x^2 + 0 - \frac{a_0}{3}x^4 + 0 - \frac{a_0}{5}x^6 + 0 + ...$$

$$\Rightarrow \qquad y = a_0\left[1 - x^2 - \frac{x^4}{3} - \frac{x^6}{5} + \ldots\right] + a_1 x$$ **Ans.**

27.7 SINGULAR POINTS ABOUT x = a

(DU. II Semester 2014)

Definition. Consider the equation

$$y'' + P(x)\,y' + Q(x)\,y = 0 \qquad \ldots (1)$$

and assume that functions P and Q are not analytic ($P = \infty$ or $Q = \infty$) at $x = a$, so that $x = a$ is not *ordinary point* but a *singular point of* (1).

There are two types of singular points. (1) Regular singular point, (2) Irregular singular points.

1. Regular Singular Point:

If $(x - a)\,P$ and $(x - a)^2\,Q$ are not infinite at $x = a$, then $x = a$ is a regular singular point.

2. Irregular Singular Point:

If $(x - a)\,P$ and $(x - a)^2\,Q$ are infinite at $x = a$, then $x = a$ is an irregular singular point.

Example 4. *Solve the differential equation $y'' + (x - 1)^2\,y' - 4\,(x - 1)\,y = 0$ in series about the ordinary point $x = 1$.*

Solution. Put $\quad x = t + 1 \qquad$ (or $x - 1 = t$)

$$\frac{dy}{dx} = \frac{dy}{dt}\bullet\frac{dt}{dx} = \frac{dy}{dt} \qquad \left(\therefore \frac{dt}{dx} = 1\right)$$

$$\Rightarrow \qquad \frac{d}{dx} \equiv \frac{d}{dt}$$

$$\frac{d^2y}{dx^2} = \frac{d}{dx}\left(\frac{dy}{dx}\right) = \frac{d}{dt}\left(\frac{dy}{dt}\right) = \frac{d^2y}{dt^2}$$

$\therefore$ The given equation becomes,

$$\frac{d^2y}{dt^2} + t^2 y' - 4t\,y = 0 \qquad \ldots(1)$$

Now, $t = 0$ is an ordinary point. [given]

Assume the solution to be

$$y = a_0 + a_1 t + a_2 t^2 + a_3 t^3 + \ldots + a_n t^n + \ldots \qquad \ldots(2)$$

then $\quad y' = a_1 + 2a_2 t + 3a_3 t^2 + \ldots + na_n\,t^{n-1} + \ldots$

and $\quad y'' = 2a_2 + 3.2.\,a_3 t + \ldots + n\,(n - 1)\,a_n t^{n-2} + \ldots$

Substituting these values is equation (1), we get

$$[2a_2 + 3.2\,.\,a_3 t + 4.3\,.\,a_4 t^2 + \ldots + n\,(n-1)\,a_n t^{n-2} + \ldots]$$
$$+\, t^2\,[a_1 + 2a_2 t + 3a_3\,t^2 + 4a_4 t^3 + \ldots + n\,a_n t^{n-1} + \ldots\,]$$
$$-\, 4t\,[a_0 + a_1 t + a_2 t^2 + a_3 t^3 + \ldots + a_n t^n + \ldots] = 0$$

Coeffieient of $t^0 = 0$

$$\Rightarrow \qquad 2a_2 = 0 \qquad \Rightarrow \qquad \boxed{a_2 = 0}$$

Coefficient of $t = 0$

$$\Rightarrow \qquad 3.2.\,a_3 - 4a_0 = 0 \qquad \Rightarrow \qquad \boxed{a_3 = \frac{2a_0}{3}}$$

Coefficient of $t^2 = 0$

$$\Rightarrow \qquad 4.3.\,a_4 + a_1 - 4a_1 = 0$$

$$\Rightarrow \qquad 12a_4 = 3a_1 \qquad \Rightarrow \qquad \boxed{a_4 = \frac{a_1}{4}}$$

Coefficient of $t^3 = 0$

$$\Rightarrow \quad 5.4.\, a_5 + 2a_2 - 4a_2 = 0 \Rightarrow a_5 = \frac{1}{10}a_2 \quad \Rightarrow \quad \boxed{a_5 = 0}$$

Coefficient of $t^4 = 0$

$$\Rightarrow \quad 6.5.\, a_6 + 3a_3 - 4a_3 = 0$$

$$a_6 = \frac{a_3}{6.5} = \frac{2a_0}{6.5.3} \quad \Rightarrow \quad \boxed{a_6 = \frac{a_0}{45}}$$

Now, coefficient of $t^n = 0$

$$\Rightarrow (n+2)(n+1)\, a_{n+2} + (n-1)\, a_{n-1} - 4a_{n-1} = 0$$

$$\Rightarrow \quad a_{n+2} = -\frac{(n-5)}{(n+2)(n+1)} a_{n-1}$$

Putting $n = 5, 6, 7, 8, \ldots$, we get

$$a_7 = 0$$

$$a_8 = \frac{-1}{8.7} a_5 = 0$$

$$a_9 = \frac{-2}{9.8} a_6 = \frac{-2}{9.8}\frac{a_0}{45} = -\frac{a_0}{1620}$$

and so on.

Substituting these values in (2), we get

$$y = a_0 + a_1 t + \frac{2}{3} a_0 t^3 + \frac{a_1}{4} t^4 + \frac{a_0}{45} t^6 - \frac{a_0}{1620} t^9 + \ldots$$

$$= a_0 \left(1 + \frac{2}{3} t^3 + \frac{1}{45} t^6 - \frac{1}{1620} t^9 + \ldots\right) + a_1 \left(t + \frac{t^4}{4} + \ldots\right)$$

$$\Rightarrow \quad y = a_0 \left[1 + \frac{2}{3}(x-1)^3 + \frac{1}{45}(x-1)^6 - \frac{1}{1620}(x-1)^9 + \ldots\right] + a_1 \left[(x-1) + \frac{(x-1)^4}{4} + \ldots\right]$$

Where a_0 and a_1 are constants. **Ans.**

Note. Example 4 is solved about the regular singular point $x = 1$. Now we will solve the problems about the regular singular point $x = 0$.

We can also find the solution about a point other than $x = 0$, say about $x = a$. In this case we have to find out the series solution of powers of $(x - a)$, and the series is valid (convergent) around the point $x = a$.

In this method first we shift the origin to the point $x = c$, by putting $x = t + c$. The differential equation so obtained is solved by the method already discussed.

EXERCISE 27.1

Solve the following differential equation by power series method :

1. $\dfrac{d^2y}{dx^2} + xy = 0$ **Ans.** $y = a_0\left(1 - \dfrac{x^3}{3!} + \dfrac{4x^6}{6!} - \dfrac{28x^9}{9!} + \ldots\right) + a_1\left(x - \dfrac{2x^4}{4!} + \dfrac{10x^7}{7!} + \ldots\right)$

(AMIETE, June 2010)

2. $y'' - xy' + x^2 y = 0$ **Ans.** $y = a_0\left(1 - \dfrac{1}{12}x^4 - \ldots\right) + a_1\left(x + \dfrac{1}{6}x^3 - \dfrac{1}{40}x^5 \ldots\right)$

3. $(x^2+1)y'' + xy' - xy = 0$ (U.P. (C.O.) 2008)

Ans. $y = a_0\left(1 + \dfrac{x^3}{6} - \dfrac{3}{40}x^5 + \ldots\right) + a_1\left(x - \dfrac{x^3}{6} + \dfrac{x^4}{12} + \dfrac{3}{40}x^5 \ldots\right)$

4. $y'' - 2x^2 y' + 4xy = x^2 + 2x + 4$

Ans. $y = a_0\left(1 - \frac{2}{3}x^2 - \frac{2}{45}x^6 - \frac{2}{405}x^9 ....\right) + a_1\left(x - \frac{1}{6}x^4 - \frac{1}{63}x^7 - \frac{1}{567}x^{10} ...\right)$

$+ 2x^2 + \frac{1}{3}x^3 + \frac{1}{12}x^4 + \frac{1}{45}x^6 + \frac{1}{126}x^7 + \frac{1}{405}x^9 + \frac{1}{1134}x^{10} + ...$

5. $(x^2+2)y'' + xy' - (1+xy) = 0$ **Ans.** $y = a_0\left(1 + \frac{1}{4}x^2 + \frac{1}{12}x^3 - \frac{1}{32}x^4 ...\right) + a_1\left(x + \frac{1}{24}x^4 + ...\right)$

27.8 SINGULAR POINT ABOUT $x = 0$.

Consider the differential equation

$$\frac{d^2y}{dx^2} + P\frac{dy}{dx} + Qy = 0$$

If $(x - 0)\,P$ and $(x - 0)^2\,Q$ are not infinite at $x = 0$, then $x = 0$ is a regular singular point. Otherwise it is an irregular singular point.

Note: In this section we will solve those differential equation where x_0 is a regular singular point.

Example 5. *Find regular singular points of the differential equation:*

$$2x^2\frac{d^2y}{dx^2} + 3x\frac{dy}{dx} + (x^2 - 4)\,y = 0 \qquad ...(1)$$

Solution. We have, $\frac{d^2y}{dx^2} + \frac{3}{2x}\frac{dy}{dx} + \frac{x^2-4}{2x^2}\,y = 0$

$$P = \frac{3}{2x} \text{ and } Q = \frac{x^2-4}{2x^2}$$

P and Q are not analytic (infinity) at $x = 0$. So, $x = 0$ is not ordinary point but as $(x - 0)\,P$ and $(x - 0)^2\,Q$ are analytic (not infinite) so $x = 0$ is a regular singular point.

$$x \cdot P = x\left(\frac{3}{2x}\right) = \frac{3}{2} \neq \infty \text{ at } x = 0.$$

$$x^2 Q = x^2 \cdot \frac{x^2-4}{2x^2} = \frac{1}{2}(x^2 - 4) \neq \infty \text{ at } x = 0 \qquad \textbf{Ans.}$$

Example 6. *Find regular singular points of the differential equation:*

$$x(x-2)^2 y'' + 2(x-2)y' + (x+3)y = 0 \qquad ...(1)$$

Solution. Here, we have $P = \frac{2(x-2)}{x\,(x-2)^2} = \frac{2}{x\,(x-2)}$ and $Q = \frac{x+3}{x\,(x-2)^2}$

P and Q are not analytic ($P = \infty$, $Q = \infty$) at $x = 0$ and $x = 2$.

Hence both these points are singular points of (1).

(*i*) At $x = 0$

$$xP = x \cdot \frac{2}{x\,(x-2)} = \frac{2}{x-2} \neq \infty \quad \text{at } x = 0$$

$$x^2 Q = x^2 \cdot \frac{x+3}{x\,(x-2)^2} = \frac{x\,(x+3)}{(x-2)^2} \neq \infty \quad \text{at } x = 0$$

Hence, $x\,P$ and $x^2\,Q$ are analytic ($xP \neq \infty$, $x^2Q \neq \infty$) at $x = 0$. So $x = 0$ is a regular singular point.

(*ii*) At $x = 2$

$$(x-2)P=(x-2)\cdot\frac{2}{x(x-2)}=\frac{2}{x}\neq\infty \text{ at } x=2$$

$$(x-2)^2Q=(x-2)^2\frac{(x+3)}{x(x-2)^2}=\frac{x+3}{x}\neq\infty \text{ at } x=2.$$

Since both $(x-2)P$ and $(x-2)^2Q$ are analytic $((x-2)P\neq\infty, (x-2)^2Q\neq 0)$ at $x=2$, so $x=2$ is a regular singular point.

The solution of a differential equation about a regular singular point can be obtained.

The cases of irregular singular points are beyond the scope of this book. **Ans.**

27.9 FROBENIUS METHOD

If $x=0$ is a regular singularity of the equation:

$$\frac{d^2y}{dx^2}+P(x)\frac{dy}{dx}+Q(x)y=0 \qquad [P_0(0)=0] \quad ...(1)$$

Then the series solution is

$$\boxed{y=x^m(a_0+a_1x+a_2x^2+a_3x^3+...)=\sum_{k=0}^{\infty}a_kx^{m+k}}$$

The value of m will be determined by substituting the expressions for $y, \frac{dy}{dx}, \frac{d^2y}{dx^2}$ in (1), we get the identity.

On equating the coefficient of lowest power of x in the identity to zero, a quadratic equation in m (**indicial equation**) is obtained.

Thus, we will get two values of m. The series solution of (1) will depend on the nature of the roots of the indicial equation.

(*i*) **Case 1 : When roots m_1, m_2 are distinct and not differing by an integer i.e.** $m_1-m_2\neq 0$ or a positive integer. *e.g.*, $m_1=\frac{1}{2}$, $m_2=2$.

The complete solution is $\boxed{y=c_1(y)_{m_1}+c_2(y)_{m_2}}$

(*ii*) **Case 2 : When roots m_1, m_2 *are equal i.e.* $m_1=m_2$**

$$\boxed{y=c_1(y)_{m_1}+c_2\left(\frac{\partial y}{\partial m}\right)_{m_1}}$$

(*iii*) **Case 3 : When roots m_1, m_2 are distinct and differ by an integer** $(m_1<m_2)$ *e.g.*, $m_1=\frac{3}{2}$, $m_2=\frac{5}{2}$ or $m_1=2$, $m_2=4$.

If some of the coefficients of y series become infinite when $m=m_1$, to overcome this difficulty, replace a_0 by $b_0(m-m_1)$. We get a solution which is only a constant multiple of the first solution.

$$\boxed{a_0=b_0(m-m_1)}$$

Complete solution is $\boxed{y=c_1(y)_{m_1}+c_2\left(\frac{\partial y}{\partial m}\right)_{m_1}}$

(*iv*) **Case 4 : Roots are distinct and differing by an integer, making some coefficient indeterminate**

Complete solution is $\quad y = c_1(y)_{m_1} + c_2(y)_{m_2}$

if the coefficients do not become infinite when $m = m_1$, $m = m_2$

Case I : When the roots are distinct and not differing by an integer.

Example 7. *Find solution in generalized series form about x = 0 of the differential equation*

$$3x\frac{d^2y}{dx^2} + 2\frac{dy}{dx} + y = 0$$

Solution. We have,
$$3x\frac{d^2y}{dx^2} + 2\frac{dy}{dx} + y = 0 \qquad \text{... (1)}$$

Here, $xP(x)$ and $x^2Q(x)$ are analytic (not infinite). So, $x = 0$ is a regular singular point, we assume the solution in the form

$$y = \sum_{k=0}^{\infty} a_k x^{m+k}$$

Such that
$$\frac{dy}{dx} = \sum_{k=0}^{\infty} a_k (m+k) x^{m+k-1}$$

$$\frac{d^2y}{dx^2} = \sum_{k=0}^{\infty} a_k (m+k)(m+k-1) x^{m+k-2}$$

Substituting for y, $\frac{dy}{dx}$ and $\frac{d^2y}{dx^2}$ in the given equation (1), we get

$$3\Sigma a_k (m+k)(m+k-1) x^{m+k-1} + 2\Sigma a_k (m+k) x^{m+k-1} + \Sigma a_k x^{m+k} = 0$$

$$\Rightarrow \quad \Sigma a_k [3(m+k)(m+k-1) + 2(m+k)] x^{m+k-1} + \Sigma a_k x^{m+k} = 0 \qquad \text{... (2)}$$

The coefficient of the lowest degree term x^{m-1} in the identity (2) is obtained by putting $k = 0$ in first summation only and equating it to zero. Then the **indicial equation** is

$$a_0 [3m(m-1) + 2m] = 0 \;\Rightarrow\; a_0 [3m^2 - m] = 0 \;\Rightarrow\; \boxed{a_0 m(3m-1) = 0}$$

Since $\quad a_0 \neq 0,\ m = 0, \text{ or } 1/3$

The coefficient of next lowest degree term x^m in the identity (2) is obtained by putting $k = 1$ in first summation and $k = 0$ in the second summation and equating it to zero.

$$a_1 [3(m+1)m + 2(m+1)] + a_0 = 0$$

$$\Rightarrow \quad a_1 [3m^2 + 5m + 2] + a_0 = 0 \Rightarrow a_1 (3m+2)(m+1) + a_0 = 0$$

$$\Rightarrow \quad a_1 = -\frac{1}{(3m+2)(m+1)} a_0$$

Equating to zero the coefficient of x^{m+k}, the recurrence relation is given by

$$a_{k+1} [3(m+k+1)(m+k) + 2(m+k+1)] + a_k = 0.$$

$$\Rightarrow \quad a_{k+1} (m+k+1)(3m+3k+2) + a_k = 0 \Rightarrow \boxed{a_{k+1} = \frac{-1}{(m+k+1)(3m+3k+2)} a_k}$$

This gives

For $k = 0$, $\quad a_1 = \frac{-1}{(m+1)(3m+2)} a_0$

For $k = 1$, $\quad a_2 = \frac{-1}{(m+2)(3m+5)} a_1 = \frac{1}{(m+1)(m+2)(3m+2)(3m+5)} a_0$

For $k = 2$, $$a_3 = \frac{-1}{(m+3)(3m+8)} a_2$$

$$= \frac{-1}{(m+1)(m+2)(m+3)(3m+2)(3m+5)(3m+8)} a_0$$

For $m = 0$

$$a_1 = -\frac{1}{2} a_0, \qquad a_2 = \frac{1}{20} a_0, \qquad a_3 = -\frac{1}{480} a_0$$

Hence, for $m = 0$, $y_1 = a_0\left(1 - \frac{1}{2}x + \frac{1}{20}x^2 - \frac{1}{480}x^3 + ...\right)$

For $m = \frac{1}{3}$

$$a_1 = -\frac{1}{4} a_0, \qquad a_2 = \frac{1}{56} a_0, \qquad a_3 = \frac{-1}{1680} a_0$$

Hence for $m = \frac{1}{3}$, the second solution is

$$y_2 = a_0\left(x^{\frac{1}{3}} - \frac{1}{4}x^{\frac{4}{3}} + \frac{1}{56}x^{\frac{7}{3}} - \frac{1}{1680}x^{\frac{10}{3}} + ...\right)$$

Thus the complete solution is

$$y = Ay_1 + By_2$$

$$y = a_0\left(1 - \frac{x}{2} + \frac{x^2}{20} - \frac{x^3}{480} + ...\right) + b_0 x^{1/3}\left(1 - \frac{x}{4} + \frac{x^2}{56} - \frac{x^3}{1680} + ...\right)$$ **Ans.**

Example 8. *Solve* $9x(1-x)y'' - 12y' + 4y = 0$ ...(1)

Solution. Here, $xP(x)$ and $x^2Q(x)$ are analytic (not infinite) at $x = 0$. So, it is a regular singular point.

Let $y = \sum_{k=0}^{\infty} a_k x^{m+k}$ be the solution of (1)

Differentiating twice in succession, we get

$$y' = \sum_{k=0}^{\infty} a_k (m+k) x^{m+k-1}, \qquad y'' = \sum_{k=0}^{\infty} a_k (m+k)(m+k-1) x^{m+k-2}$$

Substituting the values of y' and y'' in (1), we have

$$9x(1-x)\sum_{k=0}^{\infty} a_k (m+k)(m+k-1) x^{m+k-2} - 12\sum_{k=0}^{\infty} a_k (m+k) x^{m+k-1} + 4\sum_{k=0}^{\infty} a_k x^{m+k} = 0$$

$$\Rightarrow \quad 9\sum_{k=0}^{\infty} a_k (m+k)(m+k-1) x^{m+k-1} - 9\sum_{k=0}^{\infty} a_k (m+k)(m+k-1) x^{m+k}$$

$$-12\sum_{k=0}^{\infty} a_k (m+k) x^{m+k-1} + 4\sum_{k=0}^{\infty} a_k x^{m+k} = 0$$

$$\Rightarrow \quad 3\sum_{k=0}^{\infty} a_k (m+k)(3m+3k-7) x^{m+k-1} - \sum_{k=0}^{\infty} a_k [9(m+k)^2 + (9m+9k+4)] x^{m+k} = 0$$

$$\Rightarrow \quad 3\sum_{k=0}^{\infty} a_k (m+k)(3m+3k-7) x^{m+k-1} - \sum_{k=0}^{\infty} a_k (3m+3k-4)(3m+3k+1) x^{m+k} = 0 \quad ...(2)$$

The coefficient of the lowest degree term x^{m-1} in the identity (2) is obtained by putting $k = 0$ in first summation only and equating it to zero. Then the **indicial equation** is

$$3a_0 m(3m-7) = 0 \Rightarrow m = 0,\ m = \frac{7}{3}$$

The coefficient of next lowest degree term x^m in the identity (2) is obtained by putting $k = 1$ in first summation and $k = 0$ in the second summation and equating it to zero.

$$3a_1(m+1)(3m+3-7) - a_0(3m-4)(3m+1) = 0$$

$$a_1 = \frac{(3m-4)(3m+1)}{3(m+1)(3m-4)}a_0 = \frac{3m+1}{3m+3}a_0$$

Equating to zero the coefficient of x^{m+k}, the recurrence relation is given by

$$3a_{k+1}(m+k+1)(3m+3k-4) - a_k(3m+3k-4)(3m+3k+1) = 0$$

$$a_{k+1} = \frac{(m+3k-4)(3m+3k+1)}{3(m+k+1)(3m+3k-4)}a_k$$

$$\boxed{\boldsymbol{a_{k+1} = \frac{3m+3k+1}{3m+3k+3}a_k}}$$

$m = 0$

$$a_{k+1} = \frac{3k+1}{3k+3}a_k$$

$k = 0,\quad a_1 = \frac{1}{3}a_0$

$k = 1,\quad a_2 = \frac{2}{3}a_1 = \frac{2}{9}a_0$

$k = 2,\quad a_3 = \frac{7}{9}a_2 = \frac{7}{9}\cdot\frac{2}{9}a_0 = \frac{14}{81}a_0$

and so on

$m = \frac{7}{3}$

$$a_{k+1} = \frac{3k+8}{3k+10}a_k$$

$k = 0,\quad a_1 = \frac{8}{10}a_0 = \frac{4}{5}a_0$

$k = 1,\quad a_2 = \frac{11}{13}a_1 = \frac{11}{13}\times\frac{4}{5}a_0 = \frac{44}{65}a_0$

$k = 2,\quad a_3 = \frac{14}{16}a_2 = \frac{7}{8}\times\frac{44}{65}a_0 = \frac{77}{130}a_0$

and so on

$$y_1 = a_0\left(1+\frac{1}{3}x+\frac{2}{9}x^2+\frac{14}{81}x^3+....\right),\quad y_2 = a_0 x^{\frac{7}{3}}\left(1+\frac{4}{5}x+\frac{44}{65}x^2+\frac{77}{130}x^3+...\right)$$

Since two solutions are linearly independent, the general solution of (1) may be written as

$$y = Ay_1 + By_2$$

Ans.

Example 9. *Solve the following equation in power series about x = 0.*

$$2x^2\frac{d^2y}{dx^2}+x\frac{dy}{dx}-(x+1)y = 0$$

(U.P. II Semester Summer 2011, 2005)

Solution. Given equation is $2x^2\frac{d^2y}{dx^2}+x\frac{dy}{dx}-(x+1)y = 0$ ... (1)

Here, $x\,P(x)$ and $x^2Q(x)$ are analytic (not infinite) at $x = 0$. So, $x = 0$ is a regular singular point, we assume the solution in the form

$$y = \sum_{k=0}^{\infty} a_k x^{m+k}$$

$$\frac{dy}{dx} = \Sigma a_k(m+k)x^{m+k-1}$$

$$\frac{d^2y}{dx^2} = \Sigma a_k(m+k)(m+k-1)x^{m+k-2}$$

Putting the values of y, $\frac{dy}{dx}$ and $\frac{d^2y}{dx^2}$ in equation (1), we get

$$2x^2\Sigma a_k(m+k)(m+k-1)x^{m+k-2}+x\Sigma a_k(m+k)x^{m+k-1}-(x+1)\Sigma a_k x^{m+k}=0$$

$$\Rightarrow \quad 2\Sigma a_k(m+k)(m+k-1)x^{m+k}+\Sigma a_k(m+k)x^{m+k}-\Sigma a_k x^{m+k+1}-\Sigma a_k x^{m+k}=0$$

$$\Rightarrow \quad \Sigma a_k\left[2(m+k)(m+k-1)+(m+k)-1\right]x^{m+k}-\Sigma a_k x^{m+k+1}=0 \qquad \text{... (2)}$$

Equating the coefficient of lowest degree term x^m to zero by putting $k = 0$ in first summation of (2)

$$\boxed{\boldsymbol{a_0[2m(m-1)+m-1]=0}} \qquad \text{[Indicial equation]}$$

Let $a_0 \neq 0$

$$2m(m-1)+m-1=0 \quad \Rightarrow \quad (m-1)(2m+1)=0 \quad \Rightarrow \quad m=1,-\frac{1}{2}$$

Now equating the coefficients of next lowest degree terms to zero by putting $k = 1$ and $k = 0$ in first and second summation of (2) respectively, we get

$$a_1\left[2(m+1)m+m+1-1\right]-a_0=0 \quad \Rightarrow \quad a_1(2m^2+2m+m)=a_0$$

$$\Rightarrow \quad a_1=\frac{a_0}{m(2m+3)}$$

Now equating to zero the coefficient of x^{m+k+1} by putting $k = k + 1$ in first summation and $k = k$ in second summation, we get

$$a_{k+1}\,[2(m+k+1)(m+k)+(m+k+1)-1]-a_k=0$$

$$\Rightarrow \quad a_{k+1}[(m+k+1)\{2(m+k)+1\}-1]=a_k$$

$$\Rightarrow \quad a_{k+1}[(m+k+1)(2m+2k+1)-1]=a_k$$

$$\Rightarrow \quad \boxed{\boldsymbol{a_{k+1}=\frac{a_k}{(m+k+1)(2m+2k+1)-1}}}$$

If $k = 1$, $\quad a_2=\dfrac{a_1}{(m+2)(2m+3)-1}$,

If $k = 2$, $\quad a_3=\dfrac{a_2}{(m+3)(2m+5)-1}$

If $k = 3$, $\quad a_4=\dfrac{a_3}{(m+4)(2m+7)-1}$

$m = 1$	$m=-\frac{1}{2}$
$a_1=\frac{a_0}{5}$,	$a_1=a_0$,
$a_2=\frac{a_1}{14}=\frac{a_0}{70}$,	$a_2=-\frac{a_0}{2}$,
$a_3=\frac{a_2}{27}=\frac{a_0}{1890}$,	$a_3=\frac{a_2}{9}=\frac{a_0}{9(-2)}=-\frac{a_0}{18}$,
$a_4=\frac{a_3}{44}=\frac{a_0}{44\times 1890}$	$a_4=\frac{a_3}{20}=-\frac{a_0}{18\times 20}=-\frac{a_0}{360}$

We have, $y = x^m (a_0 + a_1x + a_2x^2 + a_3x^3 + a_4x^4 + \ldots\ldots)$

For $m = 1$

$$(y)_{m=1} = x\left(a_0 + \frac{a_0}{5}x + \frac{a_0}{70}x^2 + \frac{a_0}{1890}x^3 + \frac{a_0}{44(1890)}x^4 + \ldots\right)$$

$$\Rightarrow \quad (y)_{m=1} = a_0x\left(1 + \frac{x}{5} + \frac{x^2}{70} + \frac{x^3}{1890} + \frac{x^4}{44(1890)} + \ldots\right)$$

For $m = -\frac{1}{2}$

$$(y)_{m=-\frac{1}{2}} = x^{-\frac{1}{2}}\left(a_0 - a_0x - \frac{a_0}{2}x^2 - \frac{a_0}{18}x^3 - \frac{a_0}{360}x^4 + \ldots\right)$$

$$\Rightarrow \quad (y)_{m=-\frac{1}{2}} = a_0x^{-\frac{1}{2}}\left(1 - x - \frac{1}{2}x^2 - \frac{1}{18}x^3 - \frac{1}{360}x^4 + \ldots\right)$$

Thus roots of indicial equation are distinct and not differing by an integer. Its solution is given by

$$y = c_1(y)_{m=1} + c_2(y)_{m=-\frac{1}{2}}$$

Thus, the required solution is

$$y = c_1a_0x\left(1 + \frac{x}{5} + \frac{x^2}{70} + \frac{x^3}{1890} + \frac{1}{44(1890)}x^4 + \ldots\right) + c_2a_0x^{-1/2}\left(1 - x - \frac{x^2}{2} - \frac{x^3}{18} - \frac{x^4}{360} - \ldots\right)$$

$$\Rightarrow \quad y = Ax\left(1 + \frac{x}{5} + \frac{x^2}{70} + \frac{x^3}{1890} + \frac{x^4}{44(1890)} + \ldots\ldots\right) + Bx^{-1/2}\left(1 - x - \frac{x^2}{2} - \frac{x^3}{18} - \frac{x^4}{360} \ldots\right)$$ **Ans.**

Example 10. *Using Frobenius method, obtain a series soultion in powers of x for differential equation:* $2x(1 - x)\frac{d^2y}{dx^2} + (1 - x)\frac{dy}{dx} + 3y = 0.$ *(U.P. II. Sem 2010)*

Solution. Here, we have

$$2x(1 - x)\frac{d^2y}{dx^2} + (1 - x)\frac{dy}{dx} + 3y = 0 \qquad \ldots(1)$$

Here $xP_1(x)$ and $x^2P_2(x)$ are analytic (Not infinite) at $x = 0$. So, $x = 0$ is a regular singular point, we assume the solution in the form

$$y = \sum_{k=0}^{\infty} a_k\, x^{m+k}$$

such that

$$\frac{dy}{dx} = \sum_{k=0}^{\infty} a_k\,(m + k)\,x^{m+k-1}$$

$$\frac{d^2y}{dx^2} = \sum_{k=0}^{\infty} a_k\,(m + k)(m + k - 1)\,x^{m+k-2}$$

Substituting the expressions for $y, \frac{dy}{dx}, \frac{d^2y}{dx^2}$ in (1), we have

$$2x(1-x)\sum_{k=0}^{\infty} a_k (m+k)(m+k-1)\,x^{m+k-2} + (1-x)\sum_{k=0}^{\infty} a_k (m+k)\,x^{m+k-1} + 3\sum_{k=0}^{\infty} a_k\, x^{m+k} = 0$$

$$\Rightarrow \sum_{k=0}^{\infty} 2a_k (m+k)(m+k-1)\,x^{m+k-1} - 2\sum_{k=0}^{\infty} a_k (m+k)(m+k-1)\,x^{m+k}$$

$$+\sum_{k=0}^{\infty} a_k (m+k)\,x^{m+k-1} - \sum_{k=0}^{\infty} a_k (m+k)\,x^{m+k} + 3\sum_{x=0}^{\infty} a_k\, x^{m+k} = 0$$

Collecting the coefficients of like powers of x, we get

$$\sum_{k=0}^{\infty} [-2a_k (m+k)(m+k-1) - a_k (m+k) + 3a_k]\,x^{m+k} +$$

$$\sum_{k=0}^{\infty} [2a_k (m+k)(m+k-1) + a_k (m+k)]\,x^{m+k-1} = 0$$

$$\Rightarrow \sum_{k=0}^{\infty} a_k [-2m^2 - 2mk + 2m - 2mk - 2k^2 + 2k - m - k + 3]\,x^{m+k} +$$

$$\sum_{k=0}^{\infty} a_k [2m^2 + 2mk - 2m + 2mk + 2k^2 - 2k + m + k]\,x^{m+k-1} = 0$$

$$\Rightarrow \sum_{k=0}^{\infty} a_k [-2m^2 - 4mk + m - 2k^2 + k + 3]\,x^{m+k}$$

$$+\sum_{k=0}^{\infty} a_k [2m^2 + (4k-1)\,m + 2k^2 - k)\,x^{m+k-1} = 0$$

$$\Rightarrow \sum_{k=0}^{\infty} a_k (m+k+1)(-2m-2k+3)\,x^{m+k} + \sum_{k=0}^{\infty} a_k (m+k)(2m+2k-1)\,x^{m+k-1} = 0 \quad ...(2)$$

The coefficient of the lowest degree term x^{m-1} in (2) is obtained by putting $k = 0$ in second summation of (2) only and equating it to zero.

Then the *indicial equation* is

$$a_0 m(2m-1) = 0 \Rightarrow m = 0,\ m = \frac{1}{2}, a_0 \neq 0$$

The coefficient of the next lowest degree term x^m in (2) is obtained by putting $k = 0$ in first summation and $k = 1$ in the second summation, we get

$a_0 (m+1)(-2m+3) + a_1 (m+1)(2m+1) = 0$

$$a_1 = -\frac{(m+1)(-2m+3)}{(m+1)(2m+1)} \quad \Rightarrow \quad a_1 = \frac{2m-3}{2m+1}$$

Equating the coefficient of x^2.

On putting $k \to k+1$ in second summation of (2), we get the coefficient of x^m and equation to zero, we get

$a_k (m+k+1)(-2m-2k+3) + a_{k+1} (m+k+1)(2m+2k+1) = 0$

$$\Rightarrow \quad a_{k+1} = -\frac{(m+k+1)(-2m-2k+3)}{(m+k+1)(2m+2k+1)}\, a_k$$

$$\Rightarrow \quad a_{k+1} = -\frac{-2m-2k+3}{2m+2k+1} a_k = \frac{2m+2k-3}{2m+2k+1} a_k$$

If $k = 0$, $\quad a_1 = \frac{2m-3}{2m+1} a_0$

If $k = 1$, $\quad a_2 = \frac{2m-1}{2m+3} a_1$

If $k = 2$, $\quad a_3 = \frac{2m+1}{2m+5} a_2$

If $k = 3$, $\quad a_4 = \frac{2m+3}{2m+7} a_3$

If $k = 4$, $\quad a_5 = \frac{2m+5}{2m+9} a_4$

$m = 0$	$m = \frac{1}{2}$
$a_1 = -3a_0$	$a_1 = \frac{2\left(\frac{1}{2}\right)-3}{2\left(\frac{1}{2}\right)+1} a_0 = -a_0$
$a_2 = -\frac{1}{3} a_1 = -\frac{1}{3}(-3a_0) = a_0$	$a_2 = \frac{2\left(\frac{1}{2}\right)-1}{2\left(\frac{1}{2}\right)+3} a_1 = 0$
$a_3 = \frac{1}{5} a_2 = \frac{1}{5}(a_0) = \frac{1}{5} a_0$	
$a_4 = \frac{3}{7} a_3 = \frac{3}{7}\left(-\frac{1}{5}\right) a_0 = \frac{3}{35} a_0$	$a_3 = a_4 = a_5 = ... = 0$
$a_5 = \frac{5}{9} a_4 = \frac{5}{9}\left(\frac{3}{35}\right) a_0 = \frac{1}{21} a_0$	
$y_1 = a_0\left(1 - 3x + x^2 + \frac{1}{5}x^3 + \frac{3}{35}x^4 + \frac{1}{21}x^5 + ...\right)$	$y_2 = a_0 x^{\frac{1}{2}} - a_0 x^{\frac{3}{2}}$
$y_1 = a_0\left(1 - 3x + \frac{3x^2}{1.3} + \frac{3x^3}{3.5} + \frac{3x^4}{5.7} + \frac{3x^5}{7.9} .....\right)$	$= a_0\sqrt{x}\,(1-x)$

General Solution is

$$y = Ay_1 + By_2$$

$$y = A\left(1 - 3x + \frac{3x^2}{1.3} + \frac{3x^3}{3.5} + \frac{3}{5.7}x^4 + \frac{3}{7.9}x^5 + ...\right) + B\sqrt{x}\,(1-x)$$ **Ans.**

EXERCISE 27.2

Solve in series the following differential equation:

1. $2x^2\dfrac{d^2y}{dx^2}-x\dfrac{dy}{dx}+(1-x^2)y=x^2$

Ans. $y=ax\left(1+\dfrac{x^2}{2.5}+\dfrac{x^4}{2.4.5.9}+....\right)+bx^{\frac{1}{2}}\left(1+\dfrac{x^2}{2.3}+\dfrac{x^4}{2.4.3.7}+....\right)$

2. $2x\,(1-x)\dfrac{d^2y}{dx^2}+(5-7x)\dfrac{dy}{dx}-3y=0.$ **Ans.** $y=a\left(1+\dfrac{3}{5}x+\dfrac{3}{7}x^2+\dfrac{3}{9}x^3+...\right)+bx^{-\frac{3}{2}}$

3. $2x^2\dfrac{d^2y}{dx^2}+(2x^2-x)\dfrac{dy}{dx}+y=0$

Ans. $ax\left(1-\dfrac{2}{3}x+\dfrac{2^2}{3.5}x^2-\dfrac{2^3}{3.5.7}x^3+....\right)+b\sqrt{x}\left(1-x+\dfrac{1}{2}x^2-\dfrac{1}{6}x^3+....\right)$

4. $x\,(2+x^2)\dfrac{d^2y}{dx^2}-\dfrac{dy}{dx}-6xy=0$

Ans. $y=a\left(1+3x^2+\dfrac{3}{5}x^4-\dfrac{1}{15}x^6+....\right)+bx^{\frac{3}{2}}\left(1+\dfrac{3}{8}x^2-\dfrac{3.1}{8.16}x^4+\dfrac{5.3.1}{8.16.24}x^6....\right)$

Case II. When the roots of indicial equation are equal.

Example 11. *Solve* $x\,(x-1)\,y''+(3x-1)\,y'+y=0$

Solution. $x(x-1)y''+(3x-1)\,y'+y=0$ ... (1)

Here $xP\,(x)$ and $x^2Q\,(x)$ are analytic (Not infinite) at $x = 0$. So, $x = 0$ is a regular singular point, we assume the solution in the form

$$y=\sum_{k=0}^{\infty}a_k\,x^{m+k}$$

such that
$$\frac{dy}{dx}=\sum_{k=0}^{\infty}a_k\,(m+k)\,x^{m+k-1}$$

$$\frac{d^2y}{dx^2}=\sum_{k=0}^{\infty}a_k\,(m+k)(m+k-1)\,x^{m+k-2}$$

Substituting the expressions for $y, \dfrac{dy}{dx}, \dfrac{d^2y}{dx^2}$ in (1), we have

$$x\,(x-1)\,\Sigma\,a_k\,(m+k)\,(m+k-1)\,x^{m+k-2} + (3x-1)\,\Sigma a_k\,(m+k)\,x^{m+k-1}+\Sigma a_k\,x^{m+k}=0$$

$$\Rightarrow\quad \Sigma a_k\,(m+k)\,(m+k-1)\,x^{m+k}-\Sigma a_k\,(m+k)\,(m+k-1)\,x^{m+k-1} + 3\,\Sigma a_k\,(m+k)\,x^{m+k}-\Sigma a_k\,(m+k)\,x^{m+k-1}+\Sigma a_k\,x^{m+k}=0$$

$$\Rightarrow\quad \Sigma a_k\,[(m+k)\,(m+k-1)+3(m+k)+1]\,x^{m+k} - \Sigma a_k\,[(m+k)\,(m+k-1)+(m+k)]\,x^{m+k-1}=0$$

$$\Rightarrow\quad \Sigma a_k\,[(m+k)\,(m+k+2)+1]\,x^{m+k}-\Sigma a_k\,(m+k)^2\,x^{m+k-1}=0 \qquad ...\,(2)$$

The coefficient of lowest degree term x^{m-1} in (2) is obtained by putting $k = 0$ in the second summation only of (2) and equating it to zero. Then the **indicial equation** is

$$\boxed{a_0(m+0)^2 = 0} \quad \Rightarrow \quad m = 0,\ 0 \text{ as } a_0 \neq 0$$

The coefficient of the next lowest degree term x^m in (2) is obtained by putting $k = 0$ in the first summation and $k = 1$ in the second summation only of (2) and equating it to zero, we get

$$a_0\,[(m+0)(m+2)+1] - a_1\,(m+1)^2 = 0$$

$$\Rightarrow \quad a_0(m^2+2m+1) - a_1(m^2+2m+1) = 0$$

$$a_1 - a_0 = 0 \quad \Rightarrow \quad a_1 = a_0 \quad (\text{as } m = 0)$$

Equating the coefficient of x^{m+k} to zero, the recurrence relation is given by

$$a_k\,[(m+k)(m+k+2)+1] - a_{k+1}\,(m+k+1)^2 = 0$$

$$a_k\,(m+k+1)^2 - a_{k+1}\,(m+k+1)^2 = 0$$

Hence, $\quad a_{k+1} = a_k$

$$y = x^m\,[a_0 + a_1 x + a_2 x^2 + \ldots\ldots..]$$

$$y = a_0\,x^m\,[1 + x + x^2 + x^3 + \ldots\ldots..]$$

For $m = 0$

when $m = 0, 0$, this gives only one solution instead of two.

Second solution is given by

$$\left(\frac{\partial y}{\partial m}\right)_{m=0} \quad \text{and} \quad y_1 = a_0\,(1 + x + x^2 + x^3)$$

$$\frac{\partial y}{\partial m} = a_0\,x^m \log x\,[1 + x + x^2 + x^3 + \ldots..]$$

$$y_2 = a_0 \log x\,[1 + x + x^2 + x^3 + \ldots\ldots.] \qquad m = 0$$

$$y_1 = a_0\,[1 + x + x^2 + x^3 + \ldots\ldots.] \qquad m = 0$$

$$y = A\,y_1 + B\,y_2$$

$$y = A\,[1 + x + x^2 + x^3 + \ldots\ldots] + B \log x\,(1 + x + x^2 + x^3 + \ldots\ldots.) \qquad \textbf{Ans.}$$

Example 12. *Using extended power series method find one solution of the differential equation $xy'' + y' + x^2 y = 0$. Indicate the form of a second solution which is linearly independent of the first obtained above.* (*U.P. II Semester, June 2007*)

Solution. Here $xP(x)$ and $x^2Q(x)$ are analytic (not infinity). So, $x = 0$ is a regular singular point of the equation. $x\dfrac{d^2y}{dx^2} + \dfrac{dy}{dx} + x^2 y = 0$

Let $\quad y = \Sigma a_k\,x^{m+k} \quad \ldots (1)$

$$\frac{dy}{dx} = \Sigma a_k (m+k) x^{m+k-1}$$

$$\frac{d^2y}{dx^2} = \Sigma a_k (m+k)(m+k-1) x^{m+k-2}$$

Substituting the values of y, $\dfrac{dy}{dx}$ and $\dfrac{d^2y}{dx^2}$ in (1), we get

$$x\,\Sigma a_k\,(m+k)(m+k-1)\,x^{m+k-2} + \Sigma a_k\,(m+k)\,x^{m+k-1} + x^2\,\Sigma\,a_k\,x^{m+k} = 0$$

$$\Rightarrow \quad \Sigma a_k\,(m+k)(m+k-1)\,x^{m+k-1} + \Sigma a_k\,(m+k)\,x^{m+k-1} + \Sigma a_k\,x^{m+k+2} = 0$$

$$\Rightarrow \quad \Sigma a_k\,[(m+k)(m+k-1) + (m+k)]\,x^{m+k-1} + \Sigma a_k\,x^{m+k+2} = 0$$

$$\Rightarrow \quad \Sigma\,a_k\,(m+k)^2\,x^{m+k-1} + \Sigma\,a_k\,x^{m+k+2} = 0 \qquad \ldots (2)$$

The coefficient of lowest degree term x^{m-1} in (2) is obtained by putting $k = 0$ in first summation of (2) only and equating it to zero. Then the **indicial equation** is

$$\boxed{a_0 m^2 = 0} \Rightarrow m^2 = 0 \text{ or } m = 0, 0$$

The coefficient of the next lowest degree term x^m in (2) is obtained by putting $k = 1$ in first summation only and equating it to zero.

$$a_1(m+1)^2 = 0 \quad \Rightarrow \quad a_1 = 0$$

Equating the coefficient of x^{m+1} for $k = 2$, we get $a_2(m+2)^2 = 0 \Rightarrow a_2 = 0$

Equating the coefficient of x^{m+k+2} to zero, we have $a_{k+3}(m+k+3)^2 + a_k = 0$

$$\boxed{\boldsymbol{a_{k+3} = -\frac{a_k}{(m+k+3)^2}}}$$

If $k = 0$, $\quad a_3 = -\dfrac{1}{(m+3)^2}a_0$

If $k = 1$, $\quad a_4 = -\dfrac{1}{(m+4)^2}a_1 = 0,\ a_7 = 0,\ a_{10} = 0$

If $k = 2$, $\quad a_5 = -\dfrac{1}{(m+5)^2}a_2 = 0,\ a_8 = 0,\ a_{11} = 0$

If $k = 3$, $\quad a_6 = -\dfrac{1}{(m+6)^2}a_3 = \dfrac{1}{(m+3)^2(m+6)^2}a_0$

$$a_9 = -\frac{1}{(m+9)^2}a_6 = -\frac{1}{(m+3)^2(m+6)^2(m+9)^2}a_0$$

$$y = x^m a_0\left[1 - \frac{x^3}{(m+3)^2} + \frac{x^6}{(m+3)^2(m+6)^2} - \frac{x^9}{(m+3)^2(m+6)^2(m+9)^2} + \ldots\right] \quad \ldots (3)$$

For $m = 0$

To get the first solution, put $m = 0$ in (3), then

$$y_1 = a_0\left[1 - \frac{x^3}{3^2} + \frac{x^6}{3^2\times 6^2} - \frac{x^9}{3^2\times 6^2\times 9^2} + \ldots\right] \quad \ldots (4)$$

To get the second independent solution, differentiate (3) *w.r.t.* m. Then

$$\frac{\partial y}{\partial m} = (x^m \log x)\, a_0\left[1 - \frac{x^3}{(m+3)^2} + \frac{x^6}{(m+3)^2(m+6)^2} - \frac{x^9}{(m+3)^2(m+6)^2(m+9)^2} + \ldots\right]$$

$$+ x^m a_0\left[\frac{2x^3}{(m+3)^3} - \frac{2x^6}{(m+3)^3(m+6)^2} - \frac{2x^6}{(m+3)^2(m+6)^3}\right.$$

$$+ \frac{2x^9}{(m+3)^3(m+6)^2(m+9)^2} + \frac{2x^9}{(m+3)^2(m+6)^3(m+9)^2}$$

$$\left. + \frac{2x^9}{(m+3)^2(m+6)^2(m+9)^3} + \ldots\right] \quad \ldots. (5)$$

Putting $m = 0$ in (5), we get

$$y_2 = (\log x)a_0\left[1 - \frac{x^3}{3^2} + \frac{x^6}{3^2\times 6^2} - \frac{x^9}{3^2\times 6^2\times 9^2} + \ldots\right]$$

$$+ a_0\left[\frac{2x^3}{3^3} - \frac{2x^6}{3^3\times 6^2} - \frac{2x^6}{3^2\times 6^3} + \frac{2x^9}{3^3\times 6^2\times 9^2} + \frac{2x^9}{3^2\times 6^3\times 9^2} + \frac{2x^9}{3^2\times 6^2\times 9^3} + \ldots\right] \quad \ldots (6)$$

Hence, the general solution is given by (4) and (6)

$$y = c_1 y_1 + c_2 y_2$$

$$y = c_1 a_0\left[1 - \frac{x^3}{3^2} + \frac{x^6}{3^2 \times 6^2} - \frac{x^9}{3^2 \times 6^2 \times 9^2} + \ldots\right]$$

$$+ c_2 (\log x) a_0 \left[1 - \frac{x^3}{3^2} + \frac{x^6}{3^2 \times 6^2} - \frac{x^9}{3^2 \times 6^2 \times 9^2} + \ldots\right]$$

$$+ c_2 a_0 \left[\frac{2x^3}{3^3} - \frac{2x^6}{3^2 \times 6^6}\left(\frac{1}{3} + \frac{1}{6}\right) + \frac{2\,x^9}{3^2 \times 6^2 \times 9^2}\left(\frac{1}{3} + \frac{1}{6} + \frac{1}{9}\right) + \ldots\right]$$

$\Rightarrow$

$$y = (c_1 + c_2 \log x)\, a_0 \left[1 - \frac{x^3}{3^2} + \frac{x^6}{3^2 \times 6^2} - \frac{x^9}{3^2 \times 6^2 \times 9^2} + \ldots\right]$$

$$+ 2 \cdot c_2 a_0 \left[\frac{x^3}{3^3} - \frac{x^6}{3^5 \times 2^2}\left(1 + \frac{1}{2}\right) + \frac{2x^9}{3^9 \times 2^2}\left(1 + \frac{1}{2} + \frac{1}{3}\right) + \ldots\right] \quad \textbf{Ans.}$$

Example 13. *Solve in series the differential equation:*

$$x\frac{d^2y}{dx^2} + \frac{dy}{dx} - y = 0$$

Solution. Comparing with the equation

$$\frac{d^2y}{dx^2} + P(x)\frac{dy}{dx} + Q(x)\,y = 0, \text{we get}$$

$$P(x) = \frac{1}{x} \quad \text{and} \quad Q(x) = -\frac{1}{x}$$

Since at $x = 0$, both $P(x)$ and $Q(x)$ are not analytic $\therefore$ $x = 0$ is a singular point.

Also, $x\,P(x) = 1$ and $x^2\,Q(x) = -x$

Both $x\,P(x)$ and $x^2\,Q(x)$ are analytic at $x = 0$ $\therefore$ $x = 0$ is a regular singular point.

Let us assume

$$y = a_0 x^m + a_1 x^{m+1} + a_2 x^{m+2} + a_3 x^{m+3} + \ldots \qquad \ldots(1)$$

Then, $y' = m a_0 x^{m-1} + (m+1)\,a_1\,x^m + (m+2)\,a_2 x^{m+1} + (m+3)\,a_3 x^{m+2} + \ldots$

and $y'' = m\,(m-1)\,a_0 x^{m-2} + (m+1)\,m a_1 x^{m-1}$
$+ (m+2)\,(m+1)\,a_2 x^m + (m+3)\,(m+2)\,a_3 x^{m+1} + \ldots$

Substituting these values in the given equation, we get

$$x\,[m\,(m-1)\,a_0 x^{m-2} + (m+1)m\,a_1 x^{m-1} + (m+2)\,(m+1)\,a_2 x^m + (m+3)\,(m+2)\,a_3 x^{m+1} + \ldots]$$
$$+ [m\,a_0 x^{m-1} + (m+1)\,a_1 x^m + (m+2)\,a_2 x^{m+1} + (m+3)\,a_3 x^{m+2} + \ldots]$$
$$- [a_0 x^m + a_1 x^{m+1} + a_2 x^{m+2} + a_3 x^{m+3} + \ldots] = 0$$

Now, coefficient of $x^{m-1} = 0$

$\Rightarrow$ $m(m-1)\,a_0 + m a_0 = 0$

$\Rightarrow$ $m^2 a_0 = 0$ $\Rightarrow$ $m^2 = 0$ $(\because a_0 \neq 0)$

Which is Indicial equation.

It roots are $\boxed{m = 0, 0}$ *which are equal.*

Coefficient of $x^m = 0$

$\Rightarrow \quad (m+1)\,ma_1 + (m+1)a_1 - a_0 = 0 \qquad \Rightarrow \qquad (m+1)^2 a_1 = a_0$

$$\Rightarrow \quad \boxed{a_1 = \frac{a_0}{(m+1)^2}}$$

Coefficient of $x^{m+1} = 0$

$\Rightarrow \quad (m+2)(m+1)\,a_2 + (m+2)\,a_2 - a_1 = 0 \Rightarrow (m+2)^2 a_2 = a_1$

$$\Rightarrow \quad a_2 = \frac{a_1}{(m+2)^2} \quad \Rightarrow \quad \boxed{a_2 = \frac{a_0}{(m+1)^2\,(m+2)^2}}$$

Similarly, $a_3 = \dfrac{a_0}{(m+1)^2\,(m+2)^2\,(m+3)^2}$ and so on.

From (1), $y = a_0 x^m \left[1 + \dfrac{x}{(m+1)^2} + \dfrac{x^2}{(m+1)^2\,(m+2)^2} + \dfrac{x^3}{(m+1)^2\,(m+2)^2\,(m+3)^2} + \ldots\right]$ ...(2)

Now, $\quad y_1 = (y)_{m=0} = a_0\left[1 + x + \dfrac{x^2}{(2!)^2} + \dfrac{x^3}{(3!)^2} + \ldots\right] d$ ...(3)

To get the second independent solution, differentiate (1) partially w.r.t. m, we get

$$\frac{\partial y}{\partial m} = a_0 x^m \log x\left[1 + \frac{x}{(m+1)^2} + \frac{x^2}{(m+1)^2\,(m+2)^2} + \frac{x^3}{(m+1)^2\,(m+2)^2\,(m+3)^2} + \ldots\right]$$

$$+ a_0 x^m\left[-\frac{2x}{(m+1)^3} - \frac{2}{(m+1)^2\,(m+2)^2}\left\{\frac{1}{m+1} + \frac{1}{m+2}\right\}x^2\right.$$

$$\left. - \frac{2}{(m+1)^2\,(m+2)^2\,(m+3)^2}\left\{\frac{1}{m+1} + \frac{1}{m+2} + \frac{1}{m+3}\right\}x^3 - \ldots\right]$$

The second solution is $y_2 = \left(\dfrac{\partial y}{\partial m}\right)_{m=0} = a_0 \log x\left[1 + x + \dfrac{x^2}{(2!)^2} + \dfrac{x^3}{(3!)^2} + \ldots\right]$

$$- 2a_0\left[x + \frac{1}{(2!)^2}\left(1 + \frac{1}{2}\right)x^2 \frac{1}{(3!)^2}\left(1 + \frac{1}{2} + \frac{1}{3}\right)x^3 + \ldots\right]$$

$$= y_1 \log x - 2a_0\left[x + \frac{1}{(2!)^2} + \left(1 + \frac{1}{2}\right)x^2 + \frac{1}{(3!)^2}\left(1 + \frac{1}{2} + \frac{1}{3}\right)x^3 + \ldots\right]$$

Hence the complete solution is

$$y = c_1 y_1 + c_2 y_2 = (c_1 a_0 + c_2 a_0 \log x)\left[1 + x + \frac{x^2}{(2!)^2} + \frac{x^3}{(3!)^2} + \ldots\right]$$

$$- 2c_2 a_0\left[x + \frac{1}{(2!)^2}\left(1 + \frac{1}{2}\right)x^2 + \frac{1}{(3!)^2}\left(1 + \frac{1}{2} + \frac{1}{3}\right)x^3 + \ldots\right]$$

$$\Rightarrow \quad y = (A + B\log x)\left[1 + x + \frac{x^2}{(2!)^2} + \frac{x^3}{(3!)^2} + \ldots\right]$$

$$-2B\left[x + \frac{1}{(2!)^2}\left(1 + \frac{1}{2}\right)x^2 + \frac{1}{(3!)^2}\left(1 + \frac{1}{2} + \frac{1}{3}\right)x^3 + \ldots\right]$$

Where $c_1 a_0 = A$, $c_2 a_0 = B$. **Ans.**

EXERCISE 27.3

Solve in series the following differential equations : (First part of the solution is denoted by y_1).

1. $x\dfrac{d^2y}{dx^2}+\dfrac{dy}{dx}+xy=0$ *(Delhi University, April 2010)*

Ans. $y=a\left(1-\dfrac{x^2}{2^2}+\dfrac{x^4}{2^2.4^2}-\dfrac{x^6}{2^2.4^2.6^2}+....\right)+b\left[y_1\log x+a_0\left\{\dfrac{x^2}{2^2}-\dfrac{1}{2^2.4^2}\left(1+\dfrac{1}{2}\right)x^4+...\right\}\right]$

2. $(x-x^2)\dfrac{d^2y}{dx^2}+(1-5x)\dfrac{dy}{dx}-4y=0$

Ans. $y=a\,(1^2+2^2x+3^2x^2+4^2x^3+...)+b(y_1\log x-2a_0(1.2x+2.3x^2+3.4x^3+...)$

3. $x\dfrac{d^2y}{dx^2}+(1+x)\dfrac{dy}{dx}+2y=0$

Ans. $y=a\left(1-2x+\dfrac{3}{2!}x^2-\dfrac{4}{3!}x^3+...\right)+b\left[y_1\log x+a_0\left(3x-\dfrac{13}{4}x^2+....\right)\right]$

4. $x^2\dfrac{d^2y}{dx^2}+3x\dfrac{dy}{dx}+(1-2x)\,y=0$ **Ans.** $y_1=\dfrac{1}{x}+\sum\limits_{n=1}^{\infty}\dfrac{2^n\,x^{n-1}}{(n!)^2}\quad y_2=y_1\,In\,x-\sum\limits_{n=1}^{\infty}\dfrac{2^{n+1}\,H_n\,x^{n-1}}{(n\,!)^2}$

5. $x^2\dfrac{d^2y}{dx^2}-x(1+x)\dfrac{dy}{dx}+y=0$

Ans. $y=ax\left(1+x+\dfrac{1}{2}x^2+\dfrac{1}{2.3}x^3+....\right)+b\left[y_1\log x+a_0x^2\left(-1-\dfrac{3}{4}x+...\right)\right]$

Case III : When m_1 and m_2 are distinct and differing by an integer, then

$$y=c_1(y)_{m_1}+c_2\left(\frac{\partial y}{\partial m}\right)_{m_2}\qquad\left[\begin{array}{r}\text{If coefficient}=\infty\\ \text{when } m=m_2\end{array}\right]$$

Example 14. *Solve*

$$x^2\,\frac{d^2y}{dx^2}+x\frac{dy}{dx}+(x^2-4)\,y=0 \qquad ...\ (1)$$

Solution. Here $xP\,(x)$ and $x^2Q\,(x)$ are analytic (not infinite) at $x = 0$. So, $x = 0$ is regular singular point of this equation.

Let $y=\Sigma a_k\,x^{m+k}$

$$\frac{dy}{dx}=\Sigma a_k(m+k)x^{m+k-1}$$

$$\frac{d^2y}{dx^2}=\Sigma a_k(m+k)(m+k-1)x^{m+k-2}$$

Substituting the values of $\dfrac{d^2y}{dx^2},\ \dfrac{dy}{dx}$ and y in (1), we get

$$x^2\,\Sigma a_k(m+k)(m+k-1)x^{m+k-2}+x\Sigma a_k(m+k)x^{m+k-1}+(x^2-4)\,\Sigma a_kx^{m+k}=0$$

$$\Rightarrow\quad \Sigma a_k[(m+k)(m+k-1)+(m+k)-4]x^{m+k}+\Sigma a_kx^{m+k+2}=0$$

$$\Rightarrow\quad \Sigma a_k(m+k+2)(m+k-2)\,x^{m+k}+\Sigma\,a_k\,x^{m+k+2}=0 \qquad ...\ (2)$$

The coefficient of lowest degree term x^m in (2) is obtained by putting $k = 0$ in first summation only and equating it to zero. Then the indicial equation is

$$\boxed{a_0(m+2)(m-2)=0} \Rightarrow m = 2, -2$$

The coefficient of next lowest term x^{m+1} in (2) is obtained by putting $k = 1$ in first summation only and equating it to zero.

$$a_1(m+3)(m-1)=0 \Rightarrow a_1 = 0$$

Equating to zero the coefficient of x^{m+k+2}, we get

$$a_{k+2}(m+k+4)(m+k)+a_k = 0 \quad \Rightarrow \quad \boxed{a_{k+2} = -\frac{a_k}{(m+k+4)(m+k)}}$$

$$a_1 = a_3 = a_5 = \ldots = 0$$

$$a_2 = -\frac{a_0}{m(m+4)}$$

$$a_4 = -\frac{a_2}{(m+2)(m+6)} = \frac{a_0}{m(m+2)(m+4)(m+6)}$$

$$a_6 = -\frac{a_4}{(m+4)(m+8)} = -\frac{a_0}{m(m+2)(m+4)^2(m+6)(m+8)}$$

Hence
$$y = a_0 x^m \left[1 - \frac{x^2}{m(m+4)} + \frac{x^4}{m(m+2)(m+4)(m+6)} - \frac{x^6}{m(m+2)(m+4)^2(m+6)(m+8)} + \ldots\right] \quad \ldots (3)$$

Putting $m = 2$ in (3), we get

$$y_1 = a_0 x^2\left[1 - \frac{x^2}{2\times 6} + \frac{x^4}{2\times 4\times 6\times 8} - \frac{x^6}{2\times 4\times 6^2\times 8\times 10} + \ldots\right] \quad \ldots (4)$$

For $m = -2$

Coefficient of x^4, x^6 etc. in (3) becomes infinite on putting $m = -2$. To overcome this difficulty, we put

$$a_0 = b_0(m+2) \text{ in (1) and we get}$$

$$y = b_0 x^m\left[(m+2) - \frac{(m+2)x^2}{m(m+4)} + \frac{x^4}{m(m+4)(m+6)} - \frac{x^6}{m(m+4)^2(m+6)(m+8)} + \ldots\right] \quad \ldots (5)$$

On differentiating (5) w.r.t. 'm', we get

$$\frac{\partial y}{\partial m} = b_0(x^m \cdot \log x)\left[(m+2) - \frac{(m+2)x^2}{m(m+4)} + \frac{x^4}{m(m+4)(m+6)} - \frac{x^6}{m(m+4)^2(m+6)(m+8)} + \ldots\right]$$

$$+ b_0 x^m\left[1 - \frac{(m+2)x^2}{m(m+4)}\left(\frac{1}{m+2} - \frac{1}{m} - \frac{1}{m+4}\right) + \frac{x^4}{m(m+4)(m+6)}\left(-\frac{1}{m} - \frac{1}{m+4} - \frac{1}{m+6}\right) + \ldots\right]$$

On replacing m by -2, we get

$$\left(\frac{\partial y}{\partial m}\right)_{m=-2} = (b_0 x^{-2}\log x)\left[0 - 0 + \frac{x^4}{(-2)(2)(4)} - \frac{x^6}{(-2)(2)^2(4)(6)} + \ldots\right]$$

$$+ b_0 x^{-2}\left[1 + \frac{x^2}{2^2} + \frac{x^4}{2^2\times 4}\left(\frac{1}{4}\right) + \ldots\right]$$

$$\Rightarrow \quad y_2 = b_0 x^2 \log x\left(-\frac{1}{2^2\times 4}+\frac{x^2}{2^3\times 4\times 6}-\frac{x^4}{2^3\times 4^2\times 6\times 8}+\ldots\right)+b_0 x^{-2}\left(1+\frac{x^2}{2^2}+\frac{x^4}{2^2\times 4^2}+\ldots\right)$$

General solution is $y = c_1 y_1 + c_2 y_2$

$$y = c_1 x^2\left(1-\frac{x^2}{2\times 6}+\frac{x^4}{2\times 4\times 6\times 8}-\frac{x^6}{2\times 4\times 6^2\times 8\times 10}+\ldots\right)$$

$$+c_2\left[x^2 \log x\left(-\frac{1}{2^2\times 4}+\frac{x^2}{2^3\times 4\times 6}-\frac{x^4}{2^3\times 4^2\times 6\times 8}+\ldots\right)+x^{-2}\left(1+\frac{x^2}{2^2}+\frac{x^4}{2^2\times 4^2}+\ldots\right)\right] \quad \textbf{Ans.}$$

Example 15. *Solve:* $x(1-x)\dfrac{d^2y}{dx^2}-(1+3x)\dfrac{dy}{dx}-y=0$ ...(1)

Solution. Here, $xP(x)$ and $x^2Q(x)$ are analytic (not infinite) at $x = 0$. So, $x = 0$ is regular singular point of the equation (1).

Let $\quad y = \Sigma a_k x^{m+k}$

$$\frac{dy}{dx} = \Sigma a_k (m+k) x^{m+k-1}$$

$$\frac{d^2y}{dx^2} = \Sigma a_k (m+k)(m+k-1)\, x^{m+k-2}$$

Substituting the values of $\dfrac{d^2y}{dx^2}, \dfrac{dy}{dx}$ and y in (1), we get

$$\sum_{k=0}^{\infty} a_k[(x-x^2)(m+k)(m+k-1)x^{m+k-2}-\sum_{k=0}^{\infty} a_k(1+3x)(m+k)x^{m+k-1}-\sum_{k=0}^{\infty} a_k x^{m+k}=0$$

$$\Rightarrow \sum_{k=0}^{\infty} a_k[(m+k)(m+k-1)x^{m+k-1}-\sum_{k=0}^{\infty} a_k(m+k)(m+k-1)x^{m+k}-\sum_{k=0}^{\infty} a_k(m+k)x^{m+k-1}$$

$$-3\sum_{k=0}^{\infty} a_k(m+k)x^{m+k}-\sum_{k=0}^{\infty} a_k x^{m+k}=0$$

$$-\sum_{k=0}^{\infty} a_k[(m+k)(m+k-1)+3(m+k)+1]\,x^{m+k}+\sum_{k=0}^{\infty} a_k[(m+k)(m+k-1)-(m+k)]\,x^{m+k-1}=0$$

$$\Rightarrow \quad -\sum_{k=0}^{\infty} a_k[(m+k)^2+2(m+k)+1]x^{m+k}+\sum_{k=0}^{\infty} a_k[(m+k)^2-2(m+k)]\,x^{m+k-1}=0$$

$$\Rightarrow \quad -\sum_{k=0}^{\infty} a_k(m+k+1)^2 x^{m+k}+\sum_{k=0}^{\infty} a_k(m+k)(m+k-2)x^{m+k-1}=0 \quad \text{...(2)}$$

Now (2), being an identity, we will equate the coefficients of various powers of x to zero.

The coefficient of lowest degree term x^{m-1} is obtained by putting $k = 0$ in the second summation of (2) and equating it to zero. Then the **indicial equation** is

$$\boxed{a_0 m(m-2)=0} \quad \Rightarrow \quad m = 0, 2$$

The coefficient of x^{m+k-1} is obtained by putting $k = k - 1$ in the first summation and $k = k$ in the second summation of (2), and equating it to zero.

$$-a_{k-1}(m+k)^2 + a_k(m+k)(m+k-2) = 0$$

$$\Rightarrow \quad -a_{k-1}(m+k) + a_k(m+k-2) = 0$$

$$\Rightarrow \quad a_k = \frac{(m+k)}{(m+k-2)} a_{k-1} \qquad ...(3)$$

If $k = 1$, $\quad a_1 = \frac{m+1}{m-1} a_0$

If $k = 2$, $\quad a_2 = \frac{m+2}{m} a_1 = \left(\frac{m+2}{m}\right)\left(\frac{m+1}{m-1}\right) a_0$

If $k = 3$, $\quad a_3 = \left(\frac{m+3}{m+1}\right) a_2 = \left(\frac{m+3}{m+1}\right)\left(\frac{m+2}{m}\right)\left(\frac{m+1}{m-1}\right) a_0$

We know that $\quad y = \sum_{k=0}^{\infty} a_k x^{m+k} = a_0 x^m + a_1 x^{m+1} + a_2 x^{m+2} + ...$

$$y = a_0 x^m + \left(\frac{m+1}{m-1}\right) a_0 x^{m+1} + \frac{(m+1)(m+2)}{(m-1)m} a_0 x^{m+2} + \frac{(m+1)(m+2)(m+3)}{(m-1)m(m+1)} a_0 x^{m+3} + ...$$

$$\Rightarrow \quad y = a_0 x^m \left[1 + \left(\frac{m+1}{m-1}\right) x + \frac{(m+1)(m+2)}{(m-1)m} x^2 + \frac{(m+1)(m+2)(m+3)}{(m-1)m(m+1)} x^3 + ...\right] \qquad ....(4)$$

For $m = 2$

$$y = a_0 x^2 \left[1 + 3x + 6x^2 + 10x^3 + ...\right] = a_0\, u \qquad ... (5)$$

For $m = 0$

If we put $m = 0$ in the above series (4), the coefficients become infinite. To remove this difficulty, we modify the form (4) of y by putting $a_0 = mb$, $b \neq 0$

$$y = bx^m \left[m + \frac{m(m+1)}{(m-1)} x + \frac{m(m+1)(m+2)}{(m-1)m} x^2 + \frac{m(m+1)(m+2)(m+3)}{(m-1)m(m+1)} x^3 + ...\right] \qquad ... (6)$$

Now equation (6), gives only one solution instead of two solutions. The second solution is given by $\frac{\partial y}{\partial m}$.

$$\frac{\partial y}{\partial m} = bx^m \log x \left[m + \frac{m(m+1)}{m-1} x + \frac{(m+1)(m+2)}{m-1} x^2 + ...\right]$$

$$+ bx^m \left[1 + \frac{m^2 - 2m - 1}{(m-1)^2} x + \frac{m^2 - m - 5}{(m-1)^2} x^2 + \frac{m^2 - 2m - 11}{(m-1)^2} x^3 + ...\right]$$

Putting $\boldsymbol{m = 0}$, we have

$$\left(\frac{\partial y}{\partial m}\right)_{m=0} = b \log x \left[\frac{2}{-1} x^2 + ...\right] + b\left[1 + \frac{(-1)}{1} x + (-5)x^2 + (-11)x^3 + ...\right]$$

$$= -b \log x\, [2x^2 + ...] + b\, [1 - x - 5x^2 - 11x^3 + ...]$$

$$= b\, v$$

The complete solution of (1) is given by $y = a_0 u + bv$ **Ans.**

Example 16. *Solve in series the differential equation*

$$x^2 \frac{d^2 y}{dx^2} + 5x \frac{dy}{dx} + x^2 y = 0$$

Solution. $\frac{d^2y}{dx^2}+\frac{5}{x}\frac{dy}{dx}y=0$ Comparing the given equation with the form

$$\frac{d^2y}{dx^2}+P(x)\frac{dy}{dx}+Q(x)\,y=0, \text{ we get}$$

$$P(x)=\frac{5}{x}, \qquad Q(x)=1$$

At $x = 0$, since $P(x)$ is not analytic therefore $x = 0$ is a singular point.

Also, $x\,P(x) = 5$

$x^2\,Q(x) = x^2$

Since both $x\,P(x)$ and $x^2Q(x)$ are analytic at $x = 0$ therefore $x = 0$ is a regular singular point.

Let us assume

$$y = a_0x^m + a_1x^{m+1} + a_2x^{m+2} + a_3x^{m+3} + \ldots \quad \ldots(1)$$

$$\therefore \quad \frac{dy}{dx} = ma_0x^{m-1} + (m+1)\,a_1x^m + (m+2)\,a_2x^{m+1} + \ldots\ldots \quad \ldots(2)$$

and

$$\frac{d^2y}{dx^2} = m(m-1)\,a_0x^{m-2} + (m+1)\,ma_1x^{m-1} + (m+2)(m+1)\,a_2x^m + \ldots\ldots \quad \ldots(3)$$

Substituting the above values in given equation, we get

$$x^2\,[m(m-1)\,a_0x^{m-2} + (m+1)\,ma_1\,x^{m-1} + (m+2)(m+1)\,a_2x^m + \ldots..]$$
$$+ 5x\,[ma_0\,x^{m-1} + (m+1)\,a_1x^m + (m+2)\,a_2\,x^{m+1} + \ldots..]$$
$$+ x^2\,[a_0x^m + a_1\,x^{m+1} + a_2x^{m+2} + a_3x^{m+3} + \ldots] = 0 \quad \ldots(4)$$

Equating the coefficient of lowest power of x to zero, we get

$$m(m-1)\,a_0 + 5ma_0 = 0 \qquad [\text{coeff. of } x^m = 0]$$

$$\Rightarrow \quad (m^2 + 4m)\,a_0 = 0$$

$$\Rightarrow \quad m(m+4) = 0 \qquad \text{(Indicial equation)} \qquad (\because a_0 \neq 0)$$

$$\Rightarrow \quad \boxed{m = 0, -4}$$

Hence the roots are distinct and differing by an integer. Equating to zero, the coefficients of successive powers of x, we get

$$\text{Coefficient of } x^{m+1} = 0$$

$$(m+1)\,ma_1 + 5(m+1)\,a_1 = 0$$

$$\Rightarrow \quad (m+5)(m+1)\,a_1 = 0 \quad \Rightarrow \quad \boxed{a_1 = 0} \quad \ldots(5) \qquad [\because m \neq -5, -1]$$

$$\text{Coefficient of } x^{m+2} = 0$$

$$(m+2)(m+1)\,a_2 + 5(m+2)\,a_2 + a_0 = 0$$

$$(m+2)(m+6)\,a_2 + a_0 = 0$$

$$\boxed{a_2 = \frac{-a_0}{(m+2)(m+6)}} \quad \ldots(6)$$

Again, $\quad$ Coefficient of $x^{m+3} = 0$

$$(m+3)(m+2)\,a_3 + 5(m+3)\,a_3 + a_1 = 0$$

$$(m+3)(m+7)\,a_3 + a_1 = 0$$

$$\Rightarrow \quad a_3 = \frac{-a_1}{(m+3)(m+7)}$$

$$\Rightarrow \quad \boxed{a_3 = 0} \quad \ldots(7)$$

Similarly, $\quad a_5 = a_7 = a_9 = \ldots\ldots = 0$

Now, $\quad$ coefficient of $x^{m+4} = 0$

$$(m+4)(m+3)\,a_4 + 5(m+4)\,a_4 + a_2 = 0$$

$$\Rightarrow \qquad (m+4)(m+8)\,a_4 = -a_2$$

$$a_4 = \frac{-a_2}{(m+4)(m+8)} = \frac{a_0}{(m+2)(m+4)(m+6)(m+8)} \text{ etc.} \qquad ...(8)$$

These give $\quad y = a_0 x^m \left[1 - \frac{x^2}{(m+2)(m+6)} + \frac{x^4}{(m+2)(m+4)(m+6)(m+8)} - ......\right] \quad ...(9)$

Putting $\quad m = 0$ in (9), we get

$$y_1 = (y)_{m=0} = a_0 \left[1 - \frac{x^2}{2.6} + \frac{x^4}{2.4.6.8} - ...\right] \qquad ...(10)$$

If we put $m = -4$ in the series given by equation (9), the coefficients become infinite. To avoid this difficulty, we put $a_0 = b_0 (m+4)$, so that

$$y = b_0 x^m \left[(m+4) - \frac{(m+4)x^2}{(m+2)(m+6)} + \frac{x^4}{(m+2)(m+6)(m+8)} - ...\right] \qquad ...(11)$$

Now, $\dfrac{\partial y}{\partial m} = \log x\, b_0 x^m \left[1 + \dfrac{m^2+8m+20}{(m^2+8m+12)^2} x^2 - \dfrac{(3m^2+32m+76)}{(m^3+16m^2+76m+96)^2} x^4 + ...\right]$

Second solution is given by

$$y_2 = \left(\frac{\partial y}{\partial m}\right)_{m=-4} = \log x\, b_0 x^{-4} \left(1 + \frac{x^2}{4} - \frac{x^4}{4} + ...\right)$$

$$= b_0\, x^{-4} \log x \left[0 - 0 + \frac{x^4}{(-2)(2)(4)} - \frac{x^6}{16} + ...\right] + b_0\, x^{-4} \left(1 + \frac{x^2}{4} - \frac{x^4}{4} + ...\right)$$

$$= b_0\, x^{-4} \log x \left(\frac{-x^4}{16} - \frac{x^6}{16} - ...\right) + b_0\, x^{-4} \left(1 + \frac{x^2}{4} - \frac{x^4}{4} + ...\right)$$

Hence the complete solution is given by

$y = c_1 y_1 + c_2 y_2$

$$= c_1 a_0 \left(1 - \frac{x^2}{12} + \frac{x^4}{384} - ...\right) + c_2 b_0\, x^{-4} \log x \left(-\frac{x^4}{16} - \frac{x^6}{16} - ...\right) + c_2 b_0\, x^{-4} \left(1 + \frac{x^2}{4} - \frac{x^4}{4} + ...\right)$$

$$\therefore \quad y = A\left(1 - \frac{x^2}{12} + \frac{x^4}{384} - ...\right) + Bx^{-4}\left(1 + \frac{x^2}{4} - \frac{x^4}{4} + ...\right) - B \log x \left(\frac{1}{16} + \frac{x^2}{16} + ...\right)$$

where $A = c_1 a_0$ and $B = c_2 b_0$. **Ans.**

EXERCISE 27.4

Solve in series the following differential equation

1. $x^2 \dfrac{d^2y}{dx^2} + x\dfrac{dy}{dx} + (x^2 - 1)y = 0$

Ans. $y = a_0 x \left(1 - \dfrac{x^2}{2.4} + \dfrac{x^4}{2.4^2.6} - ....\right) + b_0 x^{-1} \log x \left[-\dfrac{1}{2}x^2 + \dfrac{1}{2^2.4}x^4 - \dfrac{1}{2^2.4^2.6}x^6 + ...\right]$

$+ b_0 x^1 \left[1 + \dfrac{x^2}{2^2} - \dfrac{3}{2^2.2^3} x^4 + ...\right]$

2 $x(1-x)\frac{d^2y}{dx^2}-3x\frac{dy}{dx}-y=0$

Ans. $y=a_0x(1+2x+3x^2+.....)+b_0\log x[x+2x^2+3x^3+....]+b_0(1+x+x^2+....)$

Case IV. If the roots differ by an integer such that one or more coefficients are indeterminate.

Example 17. *Find the extended power series solution of the differential equation*

$$x^2y''+4xy'+(x^2+2)y=0 \qquad ...(1)$$

Solution. Here $xP(x)$ and $x^2Q(x)$ are analytic (not infinity) at $x = 0$. So, $x = 0$ is a regular singular point of this equation

Let $y=\Sigma a_k x^{m+k}$ be the required solution of the given equation.

Then $\quad \frac{dy}{dx}=\Sigma a_k(m+k)x^{m+k-1}, \quad \frac{d^2y}{dx^2}=\Sigma a_k(m+k)(m+k-1)x^{m+k-2}$

Substituting the values of y, $\frac{dy}{dx}$ and $\frac{d^2y}{dx^2}$ in the given equation, we get

$$x^2\,\Sigma a_k(m+k)(m+k-1)x^{m+k-2}+4x\,\Sigma a_k(m+k)x^{m+k-1}+(x^2+2)\Sigma a_k x^{m+k}=0$$

$$\Sigma a_k(m+k)(m+k-1)x^{m+k}+4\Sigma a_k(m+k)x^{m+k}+\Sigma a_k x^{m+k+2}+\Sigma 2a_k x^{m+k}=0$$

$$\Sigma a_k[(m+k)(m+k-1)+4(m+k)+2]x^{m+k}+\Sigma a_k x^{m+k+2}=0$$

$$\Sigma a_k[(m+k)^2+3(m+k)+2]x^{m+k}+\Sigma a_k x^{m+k+2}=0 \qquad ...(2)$$

The coefficient of lowest degree term x^m in (2) is obtained by putting $k = 0$ in first summation only and equating it to zero. Then the indicial equation is

$$\boldsymbol{a_0(m^2+3m+2)=0}$$

$$a_0\neq 0,\ m^2+3m+2=0 \quad\Rightarrow\quad (m+1)(m+2)=0 \;\Rightarrow\; m=-1,-2$$

The coefficient of next lowest degree term x^{m+1} in (2) is obtained by putting $k = 1$ in first summation only and equating it to zero.

$$a_1[m^2+5m+6]=0 \text{ or } a_1(m+2)(m+3)=0 \Rightarrow a_1=\frac{0}{(m+2)\,(m+3)}$$

when $m = -2$, a_1 becomes indeterminate $\left(\frac{0}{0}\right)$. When $m = -2$ we get the identity $a_1\,(0) = 0$ which is satisfied by every value of a_1. Therefore in this case we can take a_1 as arbitrary constant.

Equating to zero the coefficient of x^{m+k+2}

$$a_{k+2}\,[(m+2+k)^2+3\,(m+2+k)+2]+a_k=0$$

$$\Rightarrow \quad a_{k+2}[m^2+(2k+4+3)m+(k+2)^2+3(k+2)+2]+a_k=0$$

$$\Rightarrow \quad a_{k+2}[m^2+(2k+7)m+k^2+7k+12]+a_k=0$$

$$\Rightarrow \quad \boldsymbol{a_{k+2}=-\frac{1}{m^2+(2k+7)m+k^2+7k+12}a_k}$$

For $k = 0$, $\quad a_2=-\frac{1}{m^2+7m+12}a_0=-\frac{1}{(m+3)(m+4)}a_0$

$k = 1$ $\quad a_3=-\frac{1}{m^2+9m+20}a_1=-\frac{1}{(m+4)(m+5)}a_1$

$k = 2$ $\qquad a_4 = -\dfrac{1}{m^2+11m+30}a_2 = \dfrac{1}{(m+3)(m+4)(m+5)(m+6)}a_0$

$k = 3$ $\qquad a_5 = -\dfrac{1}{m^2+13m+42}a_3 = \left\{\dfrac{1}{(m+4)(m+5)(m+6)(m+7)}a_1\right\}$

For $m = -1$

$$a_2 = -\frac{1}{6}a_0, \quad a_3 = \frac{1}{12}a_1, \quad a_4 = \frac{1}{120}a_0, \quad a_5 = \frac{1}{360}a_1$$

First solution is $\quad y_1 = x^{-1}\left[1 - \dfrac{1}{6}x^2 + \dfrac{1}{120}x^4 + \ldots\right]a_0 + \left[1 - \dfrac{1}{12}x^2 + \dfrac{x^4}{360} + \ldots\right]a_1$

For $m = -2$

$$a_2 = -\frac{1}{2}a_0, \quad a_3 = -\frac{1}{6}a_1, \quad a_4 = \frac{1}{24}a_0, \quad a_5 = \frac{1}{120}a_1$$

Second solution is $\quad y_2 = x^{-2}\left[1 - \dfrac{x^2}{2} + \dfrac{x^4}{24} + \ldots\right]a_0 + \left[\dfrac{1}{x} - \dfrac{x}{6} + \dfrac{x^3}{120} + \ldots\right]a_1$

$$y_2 = x^{-2}\left[\left\{1 - \frac{x^2}{2} + \frac{x^4}{24} + \ldots\right\}a_0 + \left\{x - \frac{x^3}{6} + \frac{x^5}{120} + \ldots\right\}a_1\right]$$

$$y_2 = x^{-2}\left[a_0 \cos x + a_1 \sin x\right]$$

Thus the complete solution is $y = Ay_1 + By_2$ **Ans.**

Example 18. *Solve*

$$x^2 y'' + (x^2 + x)y' + (x - 9)y = 0 \qquad \ldots (1)$$

Solution. Here, $xP(x)$ and $x^2Q(x)$ are analytic (not infinity) at $x = 0$. So, $x = 0$ is a regular singular point of (1).

Let $\qquad y = \sum_{k=0}^{\infty} a_k x^{m+k}$

Differentiating it, we get

$$y' = \sum_{k=0}^{\infty}(m+k)a_k x^{m+k-1} \quad \text{and} \quad y'' = \sum_{k=0}^{\infty}(m+k)(m+k-1)a_k x^{m+k-2}$$

Substituting the values of y, y' and y'' in (1), we get

$$x^2\sum_{k=0}^{\infty}(m+k)(m+k-1)a_k x^{m+k-2} + (x^2+x)\sum_{k=0}^{\infty}(m+k)a_k x^{m+k-1} + (x-9)\sum_{k=0}^{\infty}a_k x^{m+k} = 0$$

$$\Rightarrow \sum_{k=0}^{\infty}(m+k)(m+k-1)a_k x^{m+k} + \sum_{k=0}^{\infty}(m+k)a_k x^{m+k+1} + \sum_{k=0}^{\infty}(m+k)a_k x^{m+k} + \sum_{k=0}^{\infty}a_k x^{m+k+1} - \sum_{k=0}^{\infty}9a_k x^{m+k} = 0$$

$$\Rightarrow \sum_{k=0}^{\infty}(m+k)^2 a_k x^{m+k} + \sum_{k=0}^{\infty}(m+k+1)a_k x^{m+k+1} - 9\sum_{k=0}^{\infty}a_k x^{m+k} = 0$$

$$\Rightarrow \sum_{k=0}^{\infty}[(m+k)^2 - 9]a_k x^{m+k} + \sum_{k=0}^{\infty}(m+k+1)a_k x^{m+k+1} = 0$$

$$\Rightarrow \sum_{k=0}^{\infty}[(m+k)^2 - 9]a_k x^{m+k} + \sum_{k=1}^{\infty}(m+k)a_{k-1} x^{m+k} = 0 \qquad \ldots (2)$$

The coefficient of lowest degree term x^m in (2) is obtained by putting $k = 0$ in first summation only and equating it to zero.

$$\boxed{(m^2-9)a_0=0} \quad \Rightarrow \; m^2-9=0 \; \Rightarrow \; m=\pm 3$$

The coefficient of next lowest degree x^{m+1} in (2) is obtained by putting $k = 1$ in first summation and second summation, we get

$$[(m+1)^2-9]a_1+(m+1)a_0=0$$

$$a_1=\frac{(m+1)a_0}{9-(m+1)^2}$$

Equating to zero the coefficient of x^{m+k}, we get $[(m+k)^2-9]a_k+(m+k)a_{k-1}=0$

$$\boxed{a_k=\frac{(m+k)\,a_{k-1}}{9-(m+k)^2}} \qquad k \geq 1$$

For $k = 2$, $$a_2=\frac{(m+2)a_1}{9-(m+2)^2}=\frac{(m+1)(m+2)a_0}{[9-(m+1)^2][9-(m+2)^2]}$$

For $k = 3$, $$a_3=\frac{(m+3)a_2}{9-(m+3)^2}=\frac{(m+1)(m+2)(m+3)a_0}{[9-(m+1)^2][9-(m+2)^2][9-(m+3)^2]}$$

and so on.

$$y=a_0x^m+a_1x^{m+1}+a_2x^{m+2}+a_3x^{m+3}+\ldots.$$

$$y=a_0x^m\left[1+\frac{(m+1)}{9-(m+1)^2}x+\frac{(m+1)(m+2)}{[9-(m+1)^2][9-(m+2)^2]}x^2+\frac{(m+1)(m+2)(m+3)}{[9-(m+1)^2][9-(m+2)^2][9-(m+3)^2]}x^3+\ldots\right]$$

For $m = 3$

$$y=a_0x^3\left[1+\frac{4}{-7}x+\frac{4\times5}{(-7)(-16)}x^2+\frac{4\times5\times6}{(-7)(-16)(-27)}x^3+\ldots\right] \quad \ldots (3)$$

For $m = -3$

Coefficients of x^3, x^4, x^5 become zero and the coefficients of remaining terms *indeterminate*. Thus for $m = -3$

$$y=a_0x^{-3}\left[1+\frac{-2}{5}x+\frac{(-2)(-1)}{(5)(8)}x^2\right]+a_6x^3\left[1+\frac{4}{-7}x+\frac{(4)(5)}{(-7)(-16)}x^2+\frac{(4)(5)(6)}{(-7)(-16)(-27)}x^3\ldots\right] \quad \ldots (4)$$

Series (3) is a constant multiple of second series in (4). Solution (4) contains two arbitrary constants, so it may be taken as the required solution.

Note. In general, if $m_1 - m_2 = a$ positive integer and some coefficients become indeterminate when $m = m_2$, the complete solution is given by putting $m = m_2$ is y which then contains two arbitrary constants. The result by putting $m = m_1$ in y nearly gives a numerical multiple of one of the series contained in the first solution.

Example 19. *Solve in series the differential equation:*

$$xy'' + 2y' + xy = 0.$$

Solution. Comparing the given equation with the form

$$\frac{d^2y}{dx^2}+P(x)\frac{dy}{dx}+Q(x)\,y = 0, \text{ we get}$$

$$P(x) = \frac{2}{x} \quad \text{and} \quad Q(x) = 1$$

At $x = 0$, $P(x)$ is not analytic $\therefore$ $x = 0$ is a *singular point.*

Also, $x\,P(x) = 2$ and $x^2\,Q(x) = x^2$

At $x = 0$, since $x\,P(x)$ and $x^2\,Q(x)$ are analytic $\therefore$ $x = 0$ is a *regular singular point.*

Let us assume

$$y = a_0 x^m + a_1 x^{m+1} + a_2 x^{m+2} + a_3 x^{m+3} + \ldots \qquad \ldots(1)$$

Then, $\dfrac{dy}{dx} = ma_0 x^{m+1} + (m+1)\,a_1 x^m + (m+2)\,a_2 x^{m+1} + (m+3)\,a_3 x^{m+2} + \ldots\ldots$

and $\dfrac{d^2y}{dx^2} = m\,(m-1)\,a_0 x^{m-2} + (m+1)\,ma_1 x^{m-1} + (m+2)\,(m+1)\,a_2 x^m$

$$+ (m+3)\,(m+2)\,a_3 x^{m+1} + \ldots$$

Substituting these values in the given equation, we get

$$x\,[m\,(m-1)\,a_0 x^{m-2} + (m+1)\,m\,a_1 x^{m-1} + (m+2)\,(m+1)\,a_2 x^m + (m+3)\,(m+2)\,a_3 x^{m+1} + \ldots]$$

$$+ 2\,[m\,a_0 x^{m-1} + (m+1)\,a_1 x^m + (m+2)\,a_2 x^{m+1} + \ldots]$$

$$+ x\,[a_0 x^m + a_1 x^{m+1} + a_2 x^{m+2} + a_3 x^{m+3} + \ldots] = 0$$

Now, Coefficient of $x^{m-1} = 0$

$\Rightarrow$ $m\,(m-1)\,a_0 + 2m\,a_0 = 0$

$(m^2 + m)\,a_0 = 0$

$\Rightarrow$ $m^2 + m = 0$ (Indicial equation) $[\because a_0 \neq 0]$

$\Rightarrow$ $\boxed{m = 0, -1}$

Hence, roots are distinct and differ by an integer.

Coefficient of $x^m = 0$

$\Rightarrow$ $(m+1)\,m\,a_1 + 2(m+1)a_1 = 0$

$\Rightarrow$ $(m+1)\,(m+2)\,a_1 = 0$

$\Rightarrow$ $(m+1)\,a_1 = 0$ $[\because m + 2 \neq 0]$

Since $m + 1$ may be zero, hence a_1 is arbitrary (or takes the form $\frac{0}{0}$). In other words, a_1 becomes indeterminate.

Hence the solution will contain a_0 and a_1 as arbitrary constants. The complete solution will be given by putting $m = -1$ in y.

Now, Coefficient of $x^{m+1} = 0$

$\Rightarrow$ $(m+2)\,(m+1)\,a_2 + 2\,(m+2)\,a_2 + a_0 = 0$

$\Rightarrow$ $(m+2)\,(m+3)\,a_2 + a_0 = 0$

$$\boxed{a_2 = \frac{-a_0}{(m+2)\,(m+3)}}$$

Coefficient of $x^{m+2} = 0$

$\Rightarrow \quad (m+3)(m+2)a_3 + 2(m+3)a_3 + a_1 = 0$

$(m+3)(m+4)a_3 + a_1 = 0$

$$\boxed{a_3 = \frac{-a_1}{(m+3)(m+4)}}$$

Coefficient of $x^{m+3} = 0$

$\Rightarrow \quad (m+4)(m+3)a_4 + 2(m+4)a_4 + a_2 = 0$

$\Rightarrow \quad (m+4)(m+5)a_4 = -a_2$

$\Rightarrow \quad a_4 = \dfrac{-a_2}{(m+4)(m+5)}$

$$\Rightarrow \quad \boxed{a_4 = \frac{a_0}{(m+2)(m+3)(m+4)(m+5)}}$$

Coefficient of $x^{m+4} = 0$

$(m+5)(m+4)a_5 + 2(m+5)a_5 + a_3 = 0$

$(m+5)(m+6)a_5 = -a_3$

$$\boxed{a_5 = \frac{a_1}{(m+3)(m+4)(m+5)(m+6)}}$$

and so on.

Substituting these values in equation (1), we get

$$y = x^m\left[a_0 + a_1 x - \frac{a_0}{(m+2)(m+3)}x^2 - \frac{a_1}{(m+3)(m+4)}x^3 + \frac{a_0}{(m+2)(m+3)(m+4)(m+5)}x^4 + \frac{a_1}{(m+3)(m+4)(m+5)(m+6)}x^5 + \ldots\right]$$

$$y = x^m\left[a_0\left\{1 - \frac{x^2}{(m+2)(m+3)} + \frac{x^4}{(m+2)(m+3)(m+4)(m+5)} - \ldots\right\} + a_1\left\{x - \frac{x^3}{(m+3)(m+4)} + \frac{x^5}{(m+3)(m+4)(m+5)(m+6)} - \ldots\right\}\right]$$

Now, $\quad (y)_{m=-1} = x^{-1}\left[a_0\left(1 - \dfrac{x^2}{1.2} + \dfrac{x^4}{1.2.3.4} - \ldots\right) + a_1\left(x - \dfrac{x^3}{2.3} + \dfrac{x^5}{2.3.4.5} - \ldots\right)\right]$

$= x^{-1}[a_0 \cos x + a_1 \sin x]$

Hence complete solution is given by

$y = (y)_{m=-1} \quad \Rightarrow \quad y = \dfrac{1}{x}(a_0 \cos x + a_1 \sin x).$ **Ans.**

Note. All those problems, in which $x = 0$, was an ordinary point of $y'' + P(x)y' + Q(x)y = 0$, can also be solved by Frobenius method as given in Art. 27.9 and explained in above illustrative example.

EXERCISE 27.5

Solve in series the following differential equations:

1. $(1-x^2)\dfrac{d^2y}{dx^2} + 2x\dfrac{dy}{dx} + y = 0$ **Ans.** $y = a_0\left(1 - \dfrac{1}{2}x^2 + \dfrac{1}{8}x^4 + \ldots\ldots\right) + b\left(x - \dfrac{1}{2}x^3 + \dfrac{1}{40}x^5 + \ldots\ldots\right)$

(AMIETE, June 2010)

2. $(2+x^2)\dfrac{d^2y}{dx^2} + x\dfrac{dy}{dx} + (1+x)y = 0$

Ans. $y = a_0\left(1 - \dfrac{1}{4}x^2 - \dfrac{1}{12}x^3 + \dfrac{5}{56}x^4 + \ldots\right) + b\left(x - \dfrac{1}{6}x^3 - \dfrac{1}{24}x^4 + \ldots\ldots\right)$

3. Find the power series solution about the point $x_0 = 2$ of the equation

$y'' + (x-1)y' + y = 0$ *(AMIETE, Dec. 2009)*

Ans. $y = a_0\,[1 - 1/2\,(x-2)^2 + 1/6\,(x-2)^3 + \ldots\ldots] + b\,[(x-2) - 1/2\,(x-2)^2 - 1/6\,(x-2)^3 + \ldots]$

CHAPTER

LEGENDRE'S FUNCTIONS

28.1 LEGENDRE'S EQUATION

The differential equation $$(1-x^2)\frac{d^2y}{dx^2}-2x\frac{dy}{dx}+n(n+1)\,y=0 \qquad \text{... (1)}$$

is known as Legendre's equation. The above equation can also be written as

$$\frac{d}{dx}\left\{(1-x^2)\frac{dy}{dx}\right\}+n(n+1)\,y=0 \qquad n \in I$$

This equation can be integrated in series of ascending or descending powers of x. *i.e.*, series in ascending or descending powers of x can be found which satisfy the equation (1).

Let the series in descending powers of x be

$$y = x^m(a_0+a_1x^{-1}+a_2x^{-2}+...) \qquad \text{... (2)}$$

$$\Rightarrow \qquad y=\sum_{r=0}^{\infty} a_r\, x^{m-r}$$

so that $$\frac{dy}{dx}=\sum_{r=0}^{\infty} a_r(m-r)^{m-r-1}$$

and $$\frac{d^2y}{dx^2}=\sum_{r=0}^{\infty} a_r(m-r)(m-r-1)\,x^{m-r-2}$$

Substituting these values in (1), we have

$$(1-x^2)\sum_{r=0}^{\infty} a_r(m-r)(m-r-1)\,x^{m-r-2}-2x\sum_{r=0}^{\infty} a_r(m-r)\,x^{m-r-1}+n(n+1)\sum_{r=0}^{\infty} a_r\,x^{m-r}=0$$

$$\Rightarrow \qquad \sum_{r=0}^{\infty}\left[(m-r)(m-r-1)\,x^{m-r-2}+\{n(n+1)-2(m-r)-(m-r)(m-r-1)\}\,x^{m-r}\right]a_r=0$$

$$\Rightarrow \qquad \sum_{r=0}^{\infty}[(m-r)(m-r-1)\,x^{m-r-2}+\{n(n+1)-(m-r)(m-r+1)\}\,x^{m-r}]\,a_r=0 \qquad \text{... (3)}$$

The equation (3) is an identity and therefore coefficients of various powers of x must vanish. Now equating to zero the coefficients of x^m *i.e.* by substituting $r=0$ in the second summation, we get,

$$a_0\,\{n(n+1)-m(m+1)\}=0$$

But $a_0 \neq 0$, as it is the coefficient of the very first term in the series

Hence $$n(n+1) - m(m+1) = 0 \qquad \text{... (4)}$$

i.e., $$n^2 + n - m^2 - m = 0 \quad \Rightarrow \quad (n^2 - m^2) + (n - m) = 0$$

$\Rightarrow$ $(n - m)(n + m + 1) = 0$, This is the indicial equation.

which gives $$m = n \quad \text{or} \quad m = -n - 1 \qquad \text{... (5)}$$

Next equating to zero the coefficient of x^{m-1} by putting $r = 1$, in the second summation

$$a_1[n(n+1) - (m-1)m] = 0$$

$$\Rightarrow \quad a_1(n^2 + n - m^2 + m) = 0 \quad \Rightarrow \quad a_1[(n^2 - m^2) + n + m] = 0$$

$$\Rightarrow \quad a_1[(m+n)(n - m + 1)] = 0$$

which gives $$a_1 = 0 \qquad \text{... (6)}$$

Since $(m + n)(m - n - 1) \neq 0$, by (5)

Again to find a relation in successive coefficients a_r, etc., equating the coefficient of x^{m-r-2} to zero, we get

$$(m - r)(m - r - 1)a_r + [n(n+1) - (m - r - 2)(m - r - 1)]a_{r+2} = 0 \qquad \text{... (7)}$$

Now
$$\left\{\begin{aligned} n(n+1) - (m-r-2)(m-r-1) &= n^2 + n - (m - r - 1 - 1)(m - r - 1) \\ &= -[(m - r - 1)^2 - (m - r - 1) - n^2 - n] \\ &= -[(m - r - 1 + n)(m - r - 1 - n) - (m - r - 1 + n)] \\ &= -[(m - r - 1 + n)(m - r - 1 - n - 1)] \\ &= -(m - r + n - 1)(m - r - n - 2) \end{aligned}\right\}$$

On simplification (7) becomes

$$\Rightarrow \quad (m - r)(m - r - 1)a_r - (m - r + n - 1)(m - r - n - 2)a_{r+2} = 0$$

$$\Rightarrow \quad a_{r+2} = \frac{(m-r)(m-r-1)}{(m-r+n-1)(m-r-n-2)}a_r \qquad \text{... (8)}$$

Now since $a_1 = a_3 = a_5 = a_7 = \ldots = 0$

For the two values given by (5) there arises following two cases.

Case I : When $m = n$

$$a_{r+2} = -\frac{(n-r)(n-r-1)}{(2n-r-1)(r+2)}a_r \qquad \text{[From (8)]}$$

If $r = 0$ $$a_2 = -\frac{n(n-1)}{(2n-1)2}a_0$$

If $r = 2$, $$a_4 = -\frac{(n-2)(n-3)}{(2n-3)\times 4}a_2 = \frac{n(n-1)(n-2)(n-3)}{(2n-1)(2n-3)2\cdot 4}a_0$$

and so on and $$a_1 = a_3 = a_5 = \ldots = 0$$

Hence the series (2) becomes

$$y = a_0\left[x^n - \frac{n(n-1)}{(2n-1).2}x^{n-2} + \frac{n(n-1)(n-2)(n-3)}{(2n-1)(2n-3)\,2.4}.x^{n-4} - \ldots\right]$$

Which is a solution of (1).

Case II : When $m = -(n + 1)$, we have

$$a_{r+2} = \frac{(n+r+1)(n+r+2)}{(r+2)(2n+r+3)}a_r \qquad \text{[From (8)]}$$

If $r = 0$, $$a_2 = \frac{(n+1)(n+2)}{2(2n+3)}a_0;$$

If $r = 2$, $$a_4 = \frac{(n+3)(n+4)}{4.(2n+5)} a_2 = \frac{(n+1)(n+2)(n+3)(n+4)}{2.4(2n+3)(2n+5)} a_0$$ and so on.

Hence the series (2) in this case becomes

$$y = a_0 \left[x^{-n-1} + \frac{(n+1)(n+2)}{2.(2n+3)} x^{-n-3} + \frac{(n+1)(n+2)(n+3)(n+4)}{2 \cdot 4(2n+3)(2n+5)} \cdot x^{-n-5} + \ldots \right] \quad \ldots (9)$$

This gives another solution of (1) in a series of descending powers of x.

Note. If we want to integrate the Legendre's equation in a series of ascending powers of x. We may proceed by taking

$$y = a_0 x^k + a_1 x^{k+1} + a_2 x^{k+2} + \ldots = \sum_0^{\infty} a_r x^{k+r}$$

But integration in descending powers of x is more important than that in ascending powers of x.

28.2 LEGENDRE'S POLYNOMIAL P_n (x).

Definition:

The Legendre's Equation is

$$(1-x^2)\frac{d^2 y}{dx^2} - 2x\frac{dy}{dx} + n(n+1)\, y = 0 \quad \ldots (1)$$

The solution of the above equation in the series of descending powers of x is

$$y = a_0 \left[x^n - \frac{n(n-1)}{(2n-1)2} x^{n-2} + \frac{n(n-1)(n-2)(n-3)}{(2n-1)(2n-3)2 \cdot 4} x^{n-4} \ldots \right]$$

where a_0 is an arbitrary constant.

Now if n is a positive integer and $a_0 = \dfrac{1.3.5\ldots(2n-1)}{n!}$ the above solution is $P_n(x)$, so that

$$P_n(x) = \frac{1.3.5\ldots(2n-1)}{n!}\left[x^n - \frac{n(n-1)}{(2n-1)\cdot 2} x^{n-2} + \ldots \right]$$

Note 1. This is a terminating series.

When n is even, it contains $\frac{1}{2}n+1$ terms, the last term being

$$(-1)^{\frac{1}{2}} \frac{n(n-1)(n-2)\ldots 1}{(2n-1)(2n-3)\ldots(n+1)\, 2.4.6\ldots(n-1)}$$

And when n is odd it contains $\frac{1}{2}(n+1)$ terms and the last term in this case is

$$(-1)^{\frac{1}{2}(n-1)} \frac{n(n-1)(n-2)\ldots 3.2}{(2n-1)(2n-3)\ldots(n+2)\, 2.4\ldots(n-1)} x$$

$P_n(x)$ is called the *Legendre's function of the first kind.*

Note. $P_n(x)$ is that solution of Legendre's equation (1) which is equal to unity when $x = 1$.

28.3 LEGENDRE'S FUNCTION OF THE SECOND KIND *i.e.* $Q_n(x)$

Another solution of Legendre's equation

$$(1-x^2)\frac{d^2 y}{dx^2} - 2x\frac{dy}{dx} + n(n+1)\, y = 0$$

when n is a positive integer

$$y = a_0 \left[x^{-n-1} + \frac{(n+1)(n+2)}{2\cdot(2n+3)} x^{-n-3} + \ldots \right]$$

If we take $$a_0 = \frac{n!}{1.3.5 \ldots (2n+1)}$$

the above solution is called $Q_n(x)$, so that

$$Q_n(x) = \frac{n!}{1.3.5 \ldots (2n+1)} \left[x^{-n-1} + \frac{(n+1)(n+2)}{2 \times (2n+3)} x^{-n-3} + \ldots \right]$$

The series for $Q_n(x)$ is a non-terminating series.

28.4 GENERAL SOLUTION OF LEGENDRE'S EQUATION

Since $P_n(x)$ and $Q_n(x)$ are two independent solutions of Legendre's equation, therefore the most general solution of Legendre's equation is

$$y = AP_n(x) + B\,Q_n(x)$$

Where A and B are two arbitrary constants.

28.5 RODRIGUE'S FORMULA

(AMIETE, June 2010, 2009, U.P., II Semester, 2010, 2007)

$$P_n(x) = \frac{1}{2^n \cdot n!} \frac{d^n}{dx^n}(x^2-1)^n$$

Proof. Let $$v = (x^2-1)^n \qquad \ldots (1)$$

Then $$\frac{dv}{dx} = n(x^2-1)^{n-1}(2x)$$

Multiplying both sides by (x^2-1), we get

$$(x^2-1)\frac{dv}{dx} = 2n(x^2-1)^n x.$$

$$\Rightarrow \quad (x^2-1)\frac{dv}{dx} = 2nvx \qquad \text{[Using (1)]} \qquad \ldots (2)$$

Now differentiating (2), $(n+1)$ times by Leibnitz's theorem, we have

$$(x^2-1)\frac{d^{n+2}v}{dx^{n+2}} + {}^{(n+1)}C_1\,(2x)\frac{d^{n+1}v}{dx^{n+1}} + {}^{(n+1)}C_2(2)\frac{d^n v}{dx^n} = 2n\left[x\frac{d^{n+1}v}{dx^{n+1}} + {}^{(n+1)}C_1(1)\frac{d^n v}{dx^n}\right]$$

$$\Rightarrow \quad (x^2-1)\frac{d^{n+2}v}{dx^{n+2}} + 2x\left[{}^{(n+1)}C_1 - n\right]\frac{d^{n+1}v}{dx^{n+1}} + 2\left[{}^{n+i}C_2 - n.\,{}^{(n+1)}C_1\right]\frac{d^n v}{dx^n} = 0$$

$$\Rightarrow \quad (x^2-1)\frac{d^{n+2}v}{dx^{n+2}} + 2x\frac{d^{n+1}v}{dx^{n+1}} - n(n+1)\frac{d^n v}{dx^n} = 0 \qquad \ldots (3)$$

If we put $\frac{d^n v}{dx^n} = y$, (3) becomes

$$(x^2-1)\frac{d^2y}{dx^2} + 2x\frac{dy}{dx} - n(n+1)\,y = 0$$

$$\Rightarrow \quad (1-x^2)\frac{d^2y}{dx^2} - 2x\frac{dy}{dx} + n(n+1)\,y = 0$$

This shows that $y = \frac{d^n v}{dx^n}$ is a solution of Legendre's equation.

$$\therefore \quad C\frac{d^n v}{dx^n} = P_n(x) \qquad \ldots (4)$$

Where C is a constant.

But $$v = (x^2 - 1)^n = (x + 1)^n (x - 1)^n$$

so that $$\frac{d^n v}{dx^n} = (x+1)^n \frac{d^n}{dx^n}(x-1)^n + {}^nC_1 \cdot n(x+1)^{n-1} \cdot \frac{d^{n-1}}{dx^{n-1}}(x-1)^n + ... + (x-1)^n \frac{d^n}{dx^n}(x+1)^n = 0$$

when $$x = 1, \quad \text{then} \quad \frac{d^n v}{dx^n} = 2^n \cdot n!$$

All the other terms disappear as $(x - 1)$ is a factor in every term except first.

Therefore when $x = 1$, (4) gives

$$C \cdot 2^n \cdot n! = P_n(1) = 1 \qquad [P_n(1) = 1]$$

$$C = \frac{1}{2^n \cdot n!}. \qquad ...(5)$$

Substituting the value of C from (5) in (4), we have

$$P_n(x) = \frac{1}{2^n \cdot n!} \frac{d^n v}{dx^n}$$

$$P_n(x) = \frac{1}{2^n \cdot n!} \frac{d^n}{dx^n}(x^2 - 1)^n \qquad [\because v = (x^2 - 1)^n]$$

Example 1. *Show that* $\int_{-1}^{+1} P_n(x)\,dx = 0, \quad n \neq 0$

and $\int_{-1}^{+1} P_n(x)\,dx = 2, \quad n = 0$

Solution.

(*i*) We know that $P_n(x) = \frac{1}{2^n \cdot n!} \frac{d^n}{dx^n}(x^2 - 1)^n$

Integrating, we get $$\int_{-1}^{+1} P_n(x)\,dx = \frac{1}{2^n\, n!} \int_{-1}^{+1} \frac{d^n}{dx^n}(x^2 - 1)^n\,dx = \frac{1}{2^n\, n!}\left[\frac{d^{n-1}}{dx^{n-1}}(x^2 - 1)^n\right]_{-1}^{+1}$$

$$= \frac{1}{2^n\, n!}[0 - 0] = 0$$

(*ii*) When $n = 0$

$$\int_{-1}^{1} P_0(x)\,dx = \int_{-1}^{+1} 1 \cdot dx = [x]_{-1}^{+1} = 2$$ **Proved.**

Example 2. *Let* $P_n(x)$ *be the Legendre polynomial of degree n. Show that for any function,* $f(x)$*, for which the nth derivative is continuous,*

$$\int_{-1}^{1} f(x) P_n(x)\,dx = \frac{(-1)^n}{2^n\, n!} \int_{-1}^{1} (x^2 - 1)^n f^n(x)\,dx.$$

Solution. $$\int_{-1}^{1} f(x) P_n(x)\,dx = \int_{-1}^{1} f(x) \cdot \frac{1}{2^n\, n!} \frac{d^n}{dx^n}(x^2 - 1)^n\,dx. \qquad \left[P_n(x) = \frac{1}{2^n\, n!} \frac{d^n}{dx^n}(x^2 - 1)^n\right]$$

$$= \frac{1}{2^n\, n!} \int_{-1}^{1} f(x) \cdot \frac{d^n}{dx^n}(x^2 - 1)^n\,dx$$

Integrating by parts, we get

$$= \frac{1}{2^n\, n!}\left[f(x) \cdot \frac{d^{n-1}}{dx^{n-1}}(x^2 - 1)^n - \int f'(x) \cdot \frac{d^{n-1}}{dx^{n-1}}(x^2 - 1)^n\,dx\right]_{-1}^{1}$$

$$= \frac{1}{2^n\, n!}\left[0 - \int_{-1}^{1} f'(x)\cdot\frac{d^{n-1}}{dx^{n-1}}(x^2-1)^n\, dx\right] = \frac{(-1)}{2^n\, n!}\int_{-1}^{1} f'(x)\cdot\frac{d^{n-1}}{dx^{n-1}}(x^2-1)^n\, dx$$

Again integrating by parts, we have

$$= \frac{(-1)}{2^n\, n!}\left[f'(x)\cdot\frac{d^{n-2}}{dx^{n-2}}(x^2-1)^n - \int f''(x)\cdot\frac{d^{n-2}}{dx^{n-2}}(x^2-1)^n\, dx\right]_{-1}^{1}$$

$$= \frac{(-1)^2}{2^n\, n!}\int_{-1}^{1} f''(x)\cdot\frac{d^{n-2}}{dx^{n-2}}(x^2-1)^n\, dx$$

Similarly, integrating $(n-2)$ times by parts, we get

$$\int_{-1}^{1} f(x)\, P_n(x)\, dx = \frac{(-1)^{n-2}}{2^n\, n!}\int_{-1}^{1} f^n(x)\,(x^2-1)^n\, dx = \frac{(-1)^n}{2^n\, n!}\int_{-1}^{+1} f^n(x)\,(x^2-1)^n\, dx$$

Proved.

28.6 LEGENDRE'S POLYNOMIALS

$$P_n(x) = \frac{1}{2^n\cdot n!}\cdot\frac{d^n}{dx^n}(x^2-1)^n \qquad \text{(Rodrigue's formula)}$$

If $n = 0$, $\quad P_0(x) = \dfrac{1}{2^0\cdot 0!} = 1$

If $n = 1$, $\quad P_1(x) = \dfrac{1}{2^1\cdot 1!}\dfrac{d}{dx}(x^2-1) = \dfrac{1}{2}(2x) = x$

If $n = 2$, $\quad p_2(x) = \dfrac{1}{2^2\cdot 2!}\dfrac{d^2}{dx^2}(x^2-1)^2 = \dfrac{1}{8}\dfrac{d}{dx}[2(x^2-1)(2x)]$

$$= \frac{1}{2}[(x^2-1)\cdot 1 + 2x\cdot x] = \frac{1}{2}(3x^2-1)$$

Similarly $\quad P_3(x) = \dfrac{1}{2}(5x^3-3x)$

$$P_4(x) = \frac{1}{8}(35x^4 - 30x^2 + 3)$$

$$P_5(x) = \frac{1}{8}(63x^5 - 70x^3 + 15x)$$

$$P_6(x) = \frac{1}{16}(231x^6 - 315x^4 + 105x^2 - 5)$$

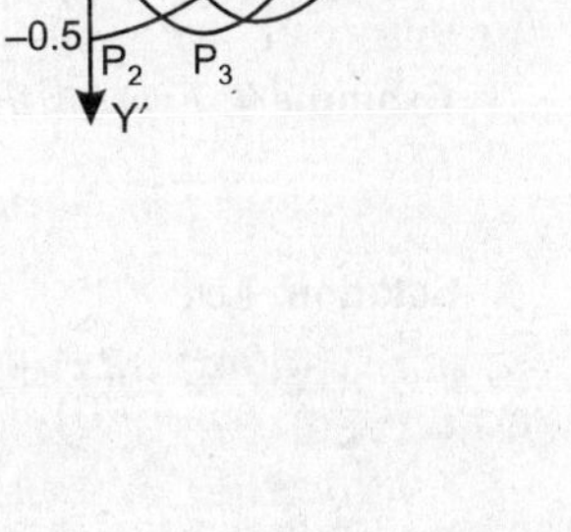

....

$$P_n(x) = \sum_{r=0}^{n}\frac{(-1)^r\,(2n-2r)!}{2^n\cdot r!\,(n-r)!\,(n-2r)!}x^{n-2r}$$

where $\quad N = \dfrac{n}{2}$ if n is even.

$$N = \frac{1}{2}(n-1) \text{ if } n \text{ is odd.}$$

Note. We can evaluate $P_n(x)$ by differentiating $(x^2-1)^n$, n times.

$$\frac{d^n}{dx^n}(x^2-1)^n = \sum_{r=0}^{r=n} {}^nC_r (x^2)^{n-r}(-1)^r = \sum_{r=0}^{r=n}(-1)^r \frac{n!}{r!(n-r)!}x^{2n-2r}$$

$$P_n(x) = \frac{1}{2^n \cdot n!}\frac{d^n}{dx^n}(x^2-1)^n = \frac{1}{2^n \cdot n!}\sum_{r=0}^{r=n}(-1)^r \frac{n!}{r!(n-r)!}(x^{2n-2r})$$

$$= \sum_{r=0}^{N} \frac{(-1)^r\,(2n-2r)!}{2^n \cdot r!(n-r)!(n-2r)!}x^{n-2r}$$

Either x^0 or x^1 is in the last term.

$$\therefore \quad n-2r=0 \quad \text{or} \quad r=\frac{n}{2} \qquad (n \text{ is even})$$

$$n-2r=1 \quad \text{or} \quad r=\frac{1}{2}(n-1) \qquad (n \text{ is odd})$$

Example 3. *Express $f(x) = 4x^3 + 6x^2 + 7x + 2$ in terms of Legendre Polynomials.*

Solution. Let $4x^3 + 6x^2 + 7x + 2 \equiv a\,P_3(x) + b\,P_2(x) + c\,P_1(x) + d\,P_0(x)$ ... (1)

$$\equiv a\left(\frac{5x^3}{2}-\frac{3x}{2}\right)+b\left(\frac{3x^2}{2}-\frac{1}{2}\right)+c(x)+d(1)$$

$$\equiv \frac{5ax^3}{2}-\frac{3ax}{2}+\frac{3bx^2}{2}-\frac{b}{2}+cx+d$$

$$4x^3 + 6x + 7x + 2 \equiv \frac{5ax^3}{2}+\frac{3bx^2}{2}+\left(\frac{-3a}{2}+c\right)x-\frac{b}{2}+d.$$

Equating the coefficients of like powers of x, we have

$$4=\frac{5a}{2}, \quad \Rightarrow \quad a=\frac{8}{5}$$

$$6=\frac{3b}{2} \quad \Rightarrow \quad b=4$$

$$7=\frac{-3a}{2}+c \quad \Rightarrow \quad 7=\frac{-3}{2}\left(\frac{8}{5}\right)+c \quad \Rightarrow \quad c=\frac{47}{5}$$

$$2=\frac{b}{-2}+d \quad \Rightarrow \quad 2=\frac{4}{-2}+d \quad \Rightarrow \quad d=4$$

Putting the values of a, b, c, d in (1), we get

$$4x^3 + 6x^2 + 7x + 2 = \frac{8}{5}P_3(x)+4P_2(x)+\frac{47}{5}P_1(x)+4P_0(x)$$ **Ans.**

Example 4. *Express the polynomial*

$f(x) = 4x^3 - 2x^2 - 3x + 8$ in terms of Legendre Polynomials.

(*U.P., II Semester, 2009*)

Solution. Let

$$4x^3 - 2x^2 - 3x + 8 \equiv a\,P_3(x) + bP_2(x) + cP_1(x) + dP_0(x) \qquad ...(1)$$

$$\equiv a\left(\frac{5x^3}{2}-\frac{3x}{2}\right)+b\left(\frac{3x^2}{2}-\frac{1}{2}\right)+c\,(x)+d\,(1)$$

$$\equiv \frac{5ax^3}{2}-\frac{3ax}{2}+\frac{3bx^2}{2}-\frac{b}{2}+cx+d$$

$$\Rightarrow \qquad 4x^3 - 2x^2 - 3x + 8 \equiv \frac{5ax^3}{2} + \frac{3bx^2}{2} + \left(\frac{-3a}{2} + c\right)x - \frac{b}{2} + d$$

Equating the coefficients of like powers of x, we have

$$4 = \frac{5a}{2} \qquad \Rightarrow \qquad a = \frac{8}{5}$$

$$-2 = \frac{3b}{2} \qquad \Rightarrow \qquad b = -\frac{4}{3}$$

$$-3 = -\frac{3a}{2} + c \qquad \Rightarrow \qquad c = -3 + \frac{3}{2}\left(\frac{8}{5}\right) = -3 + \frac{12}{5} = -\frac{3}{5}$$

$$8 = -\frac{b}{2} + d \qquad \Rightarrow \qquad d = 8 + \frac{b}{2} = 8 + \frac{1}{2}\left(-\frac{4}{3}\right) = 8 - \frac{2}{3} = \frac{22}{3}$$

Putting these values in (1), we get

$$f(x) = \frac{8}{5} P_3(x) - \frac{4}{3} P_2(x) - \frac{3}{5} P_1(x) + \frac{22}{3} P_0(x)$$ **Ans.**

Example 5. *Show that* $\quad x^4 = \frac{1}{35}[8P_4(x) + 20\,P_2(x) + 7P_0(x)]$

Solution. We know that $P_4(x) = \frac{1}{8}(35x^4 - 30x^2 + 3)$

$$P_2(x) = \frac{1}{2}(3x^2 - 1), \; P_0(x) = 1$$

$$\text{R.H.S.} = \frac{1}{35}[8P_4(x) + 20P_2(x) + 7P_0(x)]$$

$$= \frac{1}{35}[(35x^4 - 30x^2 + 3) + 10(3x^2 - 1) + 7]$$

$$= x^4 = \text{L.H.S.}$$ **Proved.**

28.7 A GENERATING FUNCTION OF LEGENDRE'S POLYNOMIAL

Prove that $P_n(x)$ is the coefficient of z^n in the expansion of $(1 - 2xz + z^2)^{-1/2}$ in ascending powers of x. *(U.P. II Semester summer 2005)*

Proof. $\quad (1 - 2\,xz + z^2)^{-1/2} = [1 - z\,(2x - z)]^{-1/2}$

Expanding R.H.S. by Binomial Theorem, we have

$$(1 - 2xz + z^2)^{\frac{-1}{2}} = 1 + \frac{1}{2} z\,(2x - z) + \frac{-\frac{1}{2}\left(-\frac{3}{2}\right)}{2\,!} z^2 (2x - z)^2 + \ldots$$

$$+ \frac{-\frac{1}{2}\left(-\frac{3}{2}\right)\left(-\frac{5}{2}\right)\ldots\left(-\frac{1}{2} - n + 1\right)}{n\,!}(-z)^n (2x - z)^n + \ldots \quad \ldots (1)$$

Now coefficient of z^n in $(n+1)$ th term *i.e.* $\frac{\left(-\frac{1}{2}\right)\left(-\frac{3}{2}\right)\left(-\frac{5}{2}\right)\ldots\left(-\frac{1}{2} - n + 1\right)}{n\,!}(-z)^n (2x - z)^n$

$$= \frac{\left(-\frac{1}{2}\right)\left(-\frac{3}{2}\right)\left(-\frac{5}{2}\right)\ldots\left(-\frac{1}{2} - n + 1\right)}{n\,!}(-1)^n (2x)^n$$

$$= \frac{1.3.5...(2n-1)}{2^n \cdot n!}(2)^n \cdot x^n = \frac{1.3.5...(2n-1)}{n!}x^n \qquad ...(2)$$

Coefficient of z^n in nth term *i.e.* $\dfrac{\left(-\frac{1}{2}\right)\left(-\frac{3}{2}\right)\left(-\frac{5}{2}\right)...\left(-\frac{1}{2}-n+2\right)}{(n-1)!}(-z)^{n-1}(2x-z)^{n-1}$

$$= \frac{\left(-\frac{1}{2}\right)\left(-\frac{3}{2}\right)\left(-\frac{5}{2}\right)...\left(-\frac{1}{2}-n+2\right)}{(n-1)!}(-1)^{n-1}[-(n-1)(2x)^{n-2}]$$

$$= \frac{1.3.5...(2n-3)}{2^{n-1}\cdot(n-1)!}(2)^{n-2}(n-1)x^{n-2} = \frac{1.3.5...(2n-3)}{2\cdot(n-1)!}(n-1)x^{n-2}$$

$$= \frac{1.3.5...(2n-3)}{2\cdot(n-1)!}\times\frac{(2n-1)}{(2n-1)}(n-1)x^{n-2} = \frac{1.3.5...(2n-3)(2n-1)}{n!}\times\frac{n(n-1)}{2(2n-1)}x^{n-2} \qquad ...(3)$$

Coefficient of x^n in

$$\frac{\left(-\frac{1}{2}\right)\left(-\frac{3}{2}\right)\left(-\frac{5}{2}\right)...\left(-\frac{1}{2}-n+3\right)}{(n-2)!}z^{n-2}(2x-z)^{(n-2)}$$

$$= \frac{\left(-\frac{1}{2}\right)\left(-\frac{3}{2}\right)\left(-\frac{5}{2}\right)...\left(-\frac{1}{2}-n+3\right)}{(n-2)!}\times(-1)^{n-2}\times\frac{(n-2)(n-3)}{2!}(2x)^{n-4}$$

$$= \frac{1.3.5...(2n-5)}{2^{n-2}(n-2)!}\times\frac{(n-2)(n-3)}{2!}(2x)^{n-4}$$

$$= \frac{1.3.5...(2n-5)(2n-3)(2n-1)}{4(n-2)!}\times\frac{(n-2)(n-3)}{2(2n-3)(2n-1)}x^{n-4}$$

$$= \frac{1.3.5...(2n-1)}{4n(n-1)(n-2)!}\times\frac{n(n-1)(n-2)(n-3)}{2(2n-3)(2n-1)}x^{n-4}$$

$$= \frac{1.3.5...(2n-1)}{n!}\times\frac{n(n-1)(n-2)(n-3)}{2.4(2n-1)(2n-3)}x^{n-4} \qquad ...(4)$$

and so on.

Thus coefficient of z^n in the expansion of (1) is sum of (2), (3) and (4) etc.

$$= \frac{1.3.5...(2n-1)}{n!}\left[x^n - \frac{n(n-1)}{2(2n-1)}\cdot x^{n-2} + \frac{n(n-1)(n-2)(n-3)}{2.4(2n-1)(2n-3)}x^{n-4} - ...\right] = P_n(x)$$

Thus coefficients of z, z^2, z^3 ... etc. in (1) are $P_1(x), P_2(x), P_3(x)$ ...

Hence

$$(1-2xz+z^2)^{-1/2} = P_0(x) + zP_1(x) + z^2P_2(x) + z^3P_3(x) + ... + z^n P_n(x) + ..$$

i.e.,

$$(1-2xz+z^2)^{-1/2} = \sum_{n=0}^{n=\infty} P_n(x)\cdot z^n.$$

Proved.

Example 6. *Prove that $P_n(1) = 1$.*

Solution. We know that

$$(1-2xz+z^2)^{-1/2} = 1 + zP_1(x) + z^2 P_2(x) + z^3 P_3(x) + \ldots + z^n P_n(x) + \ldots$$

Substituing 1 for x in the above equation, we get

$$(1-2z+z^2)^{-1/2} = 1 + z P_1(1) + z^2 P_2(1) + z^3 P_3(1) + \ldots + z^n P_n(1) + \ldots$$

$$[(1-z)^2]^{-1/2} = \sum_{n=0}^{\infty} z^n P_n(1) \quad \Rightarrow \quad (1-z)^{-1} = \sum z^n P_n(1)$$

$$\Rightarrow \quad \sum z^n P_n(1) = (1-z)^{-1} = 1 + z + z^2 + z^3 + \ldots + z^n + \ldots$$

Equating the coefficients of z^n on both sides, we get

$$P_n(1) = 1$$

Proved.

Example 7. *Prove that* $\sum_{n=0}^{\infty} P_n(x) = \frac{1}{\sqrt{2-2x}}$.

Solution. We know that $\quad (1-2xz+z^2)^{-\frac{1}{2}} = \sum_{n=0}^{\infty} z^n P_n(x) \quad \ldots (1)$

Putting $z = 1$ in (1), we get

$$(1-2x+1)^{-\frac{1}{2}} = \sum_{n=0}^{\infty} P_n(x)$$

$$\frac{1}{\sqrt{2-2x}} = \sum_{n=0}^{\infty} P_n(x)$$

Proved.

Example 8. *Prove that :* $\sum_{n=0}^{\infty} \frac{x^{n+1}}{n+1} P_n(x) = \frac{1}{2} \log\left(\frac{1+x}{1-x}\right)$

Solution. We know that $\quad \sum_{n=0}^{\infty} h^n P_n(x) = (1-2xh+h^2)^{-1/2}$

Intergrating both sides w.r.t. h from 0 to h, we get

$$\sum_{n=0}^{\infty} \frac{h^{n+1}}{n+1} P_n(x) = \int_0^h \frac{dh}{\sqrt{1-2hx+h^2}} = \int_0^h \frac{dh}{\sqrt{(h-x)^2+(1-x^2)}}; \; if\,|x|<1 \quad \left[\begin{array}{l}\text{Here } x \text{ is constant} \\ h \text{ is variable.}\end{array}\right]$$

$$= \log \frac{(h-x)+\sqrt{h^2-2hx+1}}{1-x} \qquad \left[\int \frac{dh}{\sqrt{h^2+a^2}} = \log \frac{h+\sqrt{h^2+a^2}}{a}\right]$$

Putting $h = x$ in the expression, we get

$$\sum_{n=0}^{\infty} \frac{x^{n+1}}{n+1} P_n(x) = \log\left(\frac{\sqrt{1-x^2}}{1-x}\right) = \frac{1}{2}\log\left(\frac{1+x}{1-x}\right)$$

Proved.

Example 9. *Show that*

$P_n(-x) = (-1)^n P_n(x)$ and $P_n(-1) = (-1)^n$. *(AMIETE, June 2010)*

Solution. We know that

$$(1-2xz+z^2)^{-\frac{1}{2}} = \sum_{n=0}^{\infty} z^n P_n(x) \quad \ldots (1)$$

Putting $-x$ for x in both sides of (1), we get

$$(1+2xz+z^2)^{-\frac{1}{2}} = \sum_{n=0}^{\infty} z^n P_n(-x) \quad \ldots (2)$$

Again putting $-z$ for z in (1), we obtain

$$(1+2xz+z^2)^{-\frac{1}{2}} = \sum_{n=0}^{\infty} (-1)^n z^n P_n(x) \qquad \text{... (3)}$$

Form (2) and (3), we have $\sum_{n=0}^{\infty} z^n P_n(-x) = \sum_{n=0}^{\infty} (-1)^n z^n P_n(x)$ ... (4)

Comparing the coefficients of z^n from both sides of (4), we obtain

$$P_n(-x) = (-1)^n P_n(x) \qquad \text{.... (5)}$$

Putting $x = 1$ in (5), we get $\quad P_n(-1) = (-1)^n P_n(1) = (1)(-1)^n \qquad [P_n(1) = 1]$

(1) If n is even, then from (5)

$$P_n(-x) = P_n(x),$$

So $P_n(x)$, is even function of x.

(*ii*) If n is odd, then from (5)

$P_n(-x) = -P_n(x)$, so $P_n(x)$ is an odd function. **Proved.**

Example 10. *Prove that* (*i*) $P'_n(1) = \frac{1}{2}n(n+1)$

(*ii*) $P'_n(-1) = (-1)^{n-1}\frac{n}{2}(n+1)$

Solution. Legendre's equation is

$$(1-x^2)\frac{d^2y}{dx^2} - 2x\frac{dy}{dx} + n(n+1)y = 0 \qquad \text{... (1)}$$

$$(1-x^2)P''_n(x) - 2xP'_n(x) + n(n+1)P_n(x) = 0 \qquad \text{... (2)}$$

[$P_n(x)$ is the solution of (1)]

(*i*) Putting $x = 1$ in (2), we get

$$0 - 2P'_n(1) + n(n+1)P_n(1) = 0 \Rightarrow 2P'_n(1) = n(n+1)(1) \qquad [P_n(1) = 1]$$

$$\Rightarrow \quad P'_n(1) = \frac{1}{2}n(n+1)$$ **Proved.**

(*ii*) On putting $x = -1$ in (2), we get

$$0 + 2P'_n(-1) + n(n+1)P_n(-1) = 0$$

$$\Rightarrow \quad 2P'_n(-1) + n(n+1)(-1)^n P_n(1) = 0$$

$$\Rightarrow \quad 2P'_n(-1) + n(n+1)(-1)^n(1) = 0 \quad \Rightarrow \quad 2P'_n(-1) - n(n+1)(-1)^{n-1} = 0$$

$$\Rightarrow \quad P'_n(-1) = (-1)^{n-1}\frac{1}{2}n(n+1)$$ **Proved.**

Example 11. *Show that*

(*i*) $P_{2n}(0) = (-1)^n \dfrac{1.3.5...(2n-1)}{2.4.6...2n}$ (*ii*) $P_{2n+1}(0) = 0$.

(U.P., II Semester, Summer 2008, 2005)

Solution. We know that $\Sigma z^{2n} P_{2n}(x) = (1-2xz+z^2)^{-1/2}$

On putting $x = 0$

$$\Sigma z^{2n} P_{2n}(0) = (1+z^2)^{-1/2}$$

$$= 1 + \left(-\frac{1}{2}\right)z^2 + \frac{\left(-\frac{1}{2}\right)\left(-\frac{3}{2}\right)}{2!}(z^2)^2 + \frac{\left(-\frac{1}{2}\right)\left(-\frac{3}{2}\right)\left(-\frac{5}{2}\right)}{3!}(z^2)^3$$

$$+ ... + \frac{\left(-\frac{1}{2}\right)\left(-\frac{3}{2}\right)\left(-\frac{5}{2}\right)...\left(-\frac{1}{2}-n+1\right)}{n!}(z^2)^n + ...$$

Equating the coefficient of z^{2n} both sides, we get

$$P_{2n}(0)=\frac{\left(-\frac{1}{2}\right)\left(-\frac{3}{2}\right)\left(-\frac{5}{2}\right)\cdots\left(-\frac{1}{2}-n+1\right)}{n!}=(-1)^n\frac{1.3.5...(2n-1)}{2^n.n!}$$

$$=(-1)^n\frac{1\cdot3\cdot5...(2n-1)}{2\cdot4\cdot6\cdot8...2n}$$ **Proved.**

Equating the coefficient of z^{2n+1} of both sides, we get $P_{2n+1}(0)=0$

Example 12. *Show that*

$$\frac{1-z^2}{(1-2xz+z^2)^{3/2}}=\sum_{n=0}^{\infty}(2n+1)\,P_n\cdot z^n$$

Solution. We know that

$$(1-2xz+z^2)^{-\frac{1}{2}}=\sum_{n=0}^{\infty}z^n\,P_n(x) \qquad \text{... (1)}$$

Differentiating both sides of (1) with respect to z, we get

$$-\frac{1}{2}(1-2xz+z^2)^{-3/2}(-2x+2z)=\sum_{n=0}^{\infty}n\,z^{n-1}\cdot P_n(x)$$

$$\Rightarrow \qquad \frac{x-z}{(1-2xz+z^2)^{3/2}}=\sum_{n=0}^{\infty}n\,z^{n-1}\,P_n(x) \qquad \text{... (2)}$$

Multiplying both sides of (2) by $2z$, we get

$$\frac{2xz-2z^2}{(1-2xz+z^2)^{3/2}}=\sum_{n=0}^{\infty}2n\,z^n\,P_n(x) \qquad \text{...(3)}$$

On adding (1) and (3), we get

$$(1-2xz+z^2)^{-\frac{1}{2}}+\frac{2xz-2z^2}{(1-2xz+z^2)^{3/2}}=\sum_{n=0}^{\infty}z^nP_n(x)+\sum_{n=0}^{\infty}2nz^n\,P_n(x)$$

$$\Rightarrow \qquad \frac{1-2xz+z^2+2xz-2z^2}{(1-2xz+z^2)^{3/2}}=\sum_{n=0}^{\infty}(2n+1)\,z^n\,P_n(x)$$

$$\Rightarrow \qquad \frac{1-z^2}{(1-2xz+z^2)^{3/2}}=\sum_{n=0}^{\infty}(2n+1)\,z^nP_n$$ **Proved.**

Example 13. *Prove that* $\dfrac{1+z}{z\sqrt{(1-2xz+z^2)}}-\dfrac{1}{z}=\sum_{n=0}^{\infty}(P_n+P_{n+1})\,z^n$

Solution.

$$\text{R.H.S.}=\sum_{n=0}^{\infty}(P_n+P_{n+1})\,z^n=\sum_{n=0}^{\infty}z^nP_n+\sum_{n=0}^{\infty}z^nP_{n+1}$$

$$=\sum_{n=0}^{\infty}z^nP_n+\frac{1}{z}\sum_{n=0}^{\infty}z^{n+1}P_{n+1} \qquad \text{... (1)}$$

But $$\sum_{n=0}^{\infty}z^nP_n=P_0+zP_1+z^2P_2+z^3P_3+\ldots$$

And $$\sum_{n=0}^{\infty}z^{n+1}\,P_{n+1}=zP_1+z^2\,P_2+z^3\,P_3+\ldots.$$

$$=-P_0+P_0+zP_1+z^2+z^3P_3+\ldots.=-P_0+\Sigma\,z^n\,P_n$$

Putting the value of $\sum_{n=0}^{\infty} z^{n+1}P_{n+1}$ in (1), we get

$$\text{R.H.S.} = \sum_{n=0}^{\infty} z^n P_n + \frac{1}{z}\left[\Sigma z^n P_n - P_0\right] = \left(1+\frac{1}{z}\right)\sum_{n=0}^{\infty} z^n P_n - \frac{P_0}{z}$$

$$= \left(1+\frac{1}{z}\right)(1-2xz+z^2)^{-\frac{1}{2}} - \frac{1}{z} \qquad (P_0 = 1)$$

$$= \frac{1+z}{z\sqrt{(1-2xz+z^2)}} - \frac{1}{z} = \text{L.H.S.}$$ **Proved.**

Example 14. *Prove that*

$$P_n\left(-\frac{1}{2}\right) = P_0\left(-\frac{1}{2}\right)P_{2n}\left(\frac{1}{2}\right) + P_1\left(-\frac{1}{2}\right)P_{2n-1}\left(\frac{1}{2}\right) + \ldots\ldots + P_{2n}\left(-\frac{1}{2}\right)P_0\left(\frac{1}{2}\right)$$

Solution. We know that

$$(1-2xz+z^2)^{-\frac{1}{2}} = \sum_{n=0}^{\infty} z^n\, P_n(x) \qquad \ldots (1)$$

Substituting $\frac{1}{2}$ for x in (1), we have

$$(1-z+z^2)^{-\frac{1}{2}} = \sum_{n=0}^{\infty} z^n\, P_n\left(\frac{1}{2}\right) \qquad \ldots (2)$$

Again putting $-\frac{1}{2}$ for x in (1), we have

$$(1+z+z^2)^{-\frac{1}{2}} = \sum_{n=0}^{\infty} z^n\, P_n\left(-\frac{1}{2}\right) \qquad \ldots (3)$$

Replacing z by z^2 in (3), we obtain

$$(1+z^2+z^4)^{-\frac{1}{2}} = \sum_{n=0}^{\infty} z^{2n}\, P_n\left(-\frac{1}{2}\right) \qquad \ldots (4)$$

But $$(1+z^2+z^4)^{-\frac{1}{2}} = [(1+z^2)^2 - z^2]^{-\frac{1}{2}}$$

$$= [(1+z^2+z)\,(1+z^2-z)]^{-\frac{1}{2}}$$

$$= (1+z+z^2)^{-\frac{1}{2}}(1-z+z^2)^{-\frac{1}{2}} \qquad \ldots (5)$$

Putting the values of $(1+z+z^2)^{-\frac{1}{2}}$ and $(1-z+z^2)^{-\frac{1}{2}}$ from (3) and (2) in (5), we get

$$(1+z^2+z^4)^{-\frac{1}{2}} = \sum_{n=0}^{\infty} z^n P_n\left(-\frac{1}{2}\right)\cdot\sum_{n=0}^{\infty} z^n P_n\left(\frac{1}{2}\right) \qquad \ldots (6)$$

Now substituting the value of $(1+z^2+z^4)^{-\frac{1}{2}}$ from (4) in (6), we get

$$\sum_{n=0}^{\infty} z^{2n} P_n\left(-\frac{1}{2}\right) = \sum_{n=0}^{\infty} z^n P_n\left(-\frac{1}{2}\right) \cdot \sum_{n=0}^{\infty} z^n P_n\left(\frac{1}{2}\right)$$

$$= \left[P_0\left(-\frac{1}{2}\right) + z P_1\left(-\frac{1}{2}\right) + \ldots + z^{2n-1} P_{2n-1}\left(-\frac{1}{2}\right) + z^{2n} P_{2n}\left(-\frac{1}{2}\right) + \ldots\right]$$

$$\left[P_0\left(\frac{1}{2}\right) + z P_1\left(\frac{1}{2}\right) + \ldots\ldots + z^{2n-1} P_{2n-1}\left(\frac{1}{2}\right) + z^{2n} P_{2n}\left(\frac{1}{2}\right) + \ldots\right]$$

On equating the coefficients of z^{2n} on both sides, we have

$$P_n\left(-\frac{1}{2}\right) = P_0\left(-\frac{1}{2}\right) P_{2n}\left(\frac{1}{2}\right) + P_1\left(-\frac{1}{2}\right) P_{2n-1}\left(\frac{1}{2}\right)$$

$$+ \ldots + P_{2n-1}\left(-\frac{1}{2}\right) P_1\left(\frac{1}{2}\right) + P_{2n}\left(-\frac{1}{2}\right) P_0\left(\frac{1}{2}\right) \quad \textbf{Proved.}$$

Example 15. *Prove that*

$$1 + \frac{1}{2} P_1(\cos\theta) + \frac{1}{3} P_2(\cos\theta) + \frac{1}{4} P_3(\cos\theta) + \ldots\ldots = \log \frac{1 + \sin\frac{\theta}{2}}{\sin\frac{\theta}{2}}$$

Solution. We know that

$$\sum_{n=0}^{\infty} z^n P_n(x) = (1 - 2xz + z^2)^{-\frac{1}{2}} = [z^2 - 2xz + x^2 + 1 - x^2]^{-\frac{1}{2}}$$

$$= \left[(z-x)^2 + (1-x^2)\right]^{-\frac{1}{2}} = \frac{1}{\sqrt{(z-x)^2 + (\sqrt{1-x^2})^2}} \qquad \ldots (1)$$

On integrating both sides of (1) w.r.t., 'z' between the limits 0 to 1, we have

$$\sum_{n=0}^{\infty} P_n(x) \int_0^1 z^n\, dz = \int_0^1 \frac{dz}{\sqrt{(z-x)^2 + (\sqrt{1-x^2})^2}}$$

$$\sum_{n=0}^{\infty} P_n(x) \cdot \left[\frac{z^{n+1}}{n+1}\right]_0^1 = \log\left[(z-x) + \sqrt{(z-x)^2 + (1-x^2)}\right]_0^1$$

$$\left[\int \frac{dx}{\sqrt{(x^2+a^2)}} = \log\left\{x + \sqrt{(x^2+a^2)}\right\}\right]$$

$$= \log\left[(1-x) + \sqrt{(1 - 2xz + z^2)}\right]_0^1$$

$$\sum_{n=0}^{\infty} P_n(x) \left(\frac{1}{n+1}\right) = \log\,[(1-x) + \sqrt{(1 - 2x + 1)}] - \log(-x+1)$$

$$= \log\left[1 - x + \sqrt{2 - 2x}\right] - \log(-x+1)$$

Replacing x by $\cos\theta$, we get

$$\sum_{n=0}^{\infty} \frac{P_n(\cos\theta)}{n+1} = \log[1 - \cos\theta + \sqrt{2 - 2\cos\theta}] - \log(-\cos\theta + 1) \qquad \left[\cos\theta = 1 - 2\sin^2\frac{\theta}{2}\right]$$

$$= \log\left[1-\left(1-2\sin^2\frac{\theta}{2}\right)+\sqrt{2-2\left(1-2\sin^2\frac{\theta}{2}\right)}\right]-\log\left\{1-\left(1-2\sin^2\frac{\theta}{2}\right)\right\}$$

$$= = \log\left\{2\sin^2\frac{\theta}{2}+2\sin\frac{\theta}{2}\right\}-\log 2\sin^2\frac{\theta}{2}=\log\frac{\left\{2\sin^2\frac{\theta}{2}+2\sin\frac{\theta}{2}\right\}}{2\sin^2\frac{\theta}{2}}$$

$$\Rightarrow \quad \sum_{n=0}^{\infty}\frac{P_n(\cos\theta)}{n+1} = \log\frac{\sin\frac{\theta}{2}+1}{\sin\frac{\theta}{2}}$$

$$\Rightarrow \quad \log\frac{1+\sin\frac{\theta}{2}}{\sin\frac{\theta}{2}} = P_0(\cos\theta)+\frac{1}{2}P_1(\cos\theta)+\frac{1}{3}P_2(\cos\theta)+\frac{1}{4}P_3(\cos\theta)+\ldots.$$

$$\Rightarrow \quad \log\frac{1+\sin\frac{\theta}{2}}{\sin\frac{\theta}{2}} = 1+\frac{1}{2}P_1(\cos\theta)+\frac{1}{3}P_2(\cos\theta)+\frac{1}{4}P_3(\cos\theta)+\ldots.$$

Proved.

28.8 ORTHOGONALITY OF LEGENDRE POLYNOMIALS

(DU, III Sem. 2012, AMIETE, June 2010)

$$\int_{-1}^{+1} P_m(x)\cdot P_n(x)\,dx = 0 \qquad n \neq m$$

Proof. $P_n(x)$ is a solution of

$$(1-x^2)\frac{d^2y}{dx^2}-2x\frac{dy}{dx}+n(n+1)\,y=0 \qquad \ldots (1)$$

$P_m(x)$ is the solution of

$$(1-x^2)\frac{d^2z}{dx^2}-2x\frac{dz}{dx}+m(m+1)\,z=0 \qquad \ldots (2)$$

Multiplying (1) by z and (2) by y and subtracting, we get

$$(1-x^2)\left[z\frac{d^2y}{dx^2}-y\frac{d^2z}{dx^2}\right]-2x\left[z\frac{dy}{dx}-y\frac{dz}{dx}\right]+[n(n+1)-m(m+1)]\,yz=0$$

$$(1-x^2)\left[\left\{z\frac{d^2y}{dx^2}+\frac{dz}{dx}\frac{dy}{dx}\right\}-\left\{\frac{dy}{dx}\frac{dz}{dx}+y\frac{d^2z}{dx^2}\right\}\right]-2x\left\{\frac{zdy}{dx}-\frac{ydz}{dx}\right\}+(n-m)(n+m+1)\,yz=0$$

$$\Rightarrow \quad \frac{d}{dx}\left[(1-x^2)\left(z\frac{dy}{dx}-y\frac{dz}{dx}\right)\right]+(n-m)(n+m+1)\,yz=0$$

Now integrating from -1 to 1, we get

$$\left[(1-x^2)\left(z\frac{dy}{dx}-y\frac{dz}{dx}\right)\right]_{-1}^{+1}+(n-m)(n+m+1)\int_{-1}^{+1} y.z\,dx=0.$$

$$\Rightarrow \quad 0+(n-m)(n+m+1)\int_{-1}^{+1} y\cdot z\,dx=0 \qquad \{y=P_n(x),\ z=P_m(x)\}$$

$$\int_{-1}^{+1} P_n(x)\cdot P_m(x)\,dx=0 \quad \text{if } n\neq m.$$

Proved.

Example 16. *Prove that*

$$\int_{-1}^{+1} [P_n(x)]^2 \, dx = \frac{2}{2n+1} \qquad \textit{(U.P., II Semester, June 2008, 2004, 2002)}$$

and hence show that $\int_{-1}^{+1} P_3^2(x) \, dx = \frac{2}{7}$ *(AMIETE, June 2010)*

Solution. We know that $(1 - 2xz + z^2)^{-1/2} = \Sigma z^n P_n(x)$

Squaring both sides, we get

$$(1 - 2xz + z^2)^{-1} = \Sigma z^{2n} P_n^2(x) + 2\Sigma z^{m+n} P_m(x) \cdot P_n(x)$$

Integrating both sides between -1 and $+1$, we have

$$\int_{-1}^{+1} \Sigma z^{2n} \cdot P_n^2(x) \, dx + \int_{-1}^{+1} 2\Sigma z^{m+n} \cdot P_m(x) \cdot P_n(x) \, dx = \int_{-1}^{+1} (1 - 2xz + z^2)^{-1} dx$$

$$\int_{-1}^{+1} \Sigma z^{2n} P_n^2(x) \, dx + 0 = \int_{-1}^{+1} \frac{1}{1 - 2xz + z^2} dx$$

$$\Rightarrow \qquad \Sigma z^{2n} \int_{-1}^{+1} P_n^2(x) \, dx = -\frac{1}{2z} \{\log(1 - 2xz + z^2)\}_{-1}^{+1}$$

$$= -\frac{1}{2z} \log \frac{1 - 2z + z^2}{1 + 2z + z^2} = -\frac{1}{2z} \log \left(\frac{1-z}{1+z}\right)^2$$

$$= \frac{1}{z} \log \frac{1+z}{1-z} = \frac{1}{z} [\log(1+z) - \log(1-z)]$$

$$= \frac{1}{z} \left[\left(z - \frac{z^2}{2} + \frac{z^3}{3} - \frac{z^4}{4} + \ldots + \frac{z^{2n+1}}{2n+1} + \ldots \right) - \left(-z - \frac{z^2}{2} - \frac{z^3}{3} - \frac{z^4}{4} - \ldots - \frac{z^{2n+1}}{2n+1} - \ldots \right) \right]$$

$$\Sigma z^{2n} \int_{-1}^{+1} P_n^2(x) \, dx = \frac{2}{z} \left[z + \frac{z^3}{3} + \frac{z^5}{5} + \ldots + \frac{z^{2n+1}}{2n+1} + \ldots \right] = 2 \left[1 + \frac{z^2}{3} + \frac{z^4}{5} + \ldots + \frac{z^{2n}}{2n+1} + \ldots \right]$$

Equating the coefficient of z^{2n} on both sides, we have

$$\int_{-1}^{+1} [P_n(x)]^2 \, dx = \frac{2}{2n+1}.$$ **Proved.**

Hence $$\int_{-1}^{+1} P_3^2(x) \, dx = \frac{2}{2 \times 3 + 1} = \frac{2}{7}.$$ **Proved.**

Example 17. *Prove that*

$$\int_{-1}^{1} P_n(x) (1 - 2xt + t^2)^{-1/2} \, dx = \frac{2t^n}{2n+1}$$

where n is a positive integer.

Solution. L.H.S. $= \int_{-1}^{1} P_n(x) (1 - 2xt + t^2)^{-1/2} \, dx = \int_{-1}^{1} P_n(x) \left\{ \sum t^n P_n(x) \right\} dx$

$$= \int_{-1}^{1} P_n(x) [P_0(x) + t P_1(x) + t^2 P_2(x) + t^3 P_3(x) + \cdots + t^n P_n(x) + \cdots]$$

$$= t^n \int_{-1}^{1} P_n^2(x) \, dx \qquad \left[\begin{array}{l} \text{All other terms vanish since} \\ \int_{-1}^{1} P_m(x) P_n(x) \, dx = 0; m \neq n \end{array} \right]$$

$= t^n \cdot \dfrac{2}{2n+1}$ = R.H.S. [By II orthogonal properety] **Proved.**

Example 18. *Show that*

$$\int_{-1}^{1}(1-x^2)\,P_m{}'P_n{}'\,dx = \begin{cases} 0, & \text{when } m \neq n \\ \dfrac{2n(n+1)}{2n+1}, & \text{when } m = n \end{cases}$$

where dashes denote differentiation w.r.t. x. (*U.P. II Semester summer 2006*)

Solution. We have, $\int_{-1}^{1}(1-x^2)\,P_m{}'\,P_n{}'\,dx$

$$= \left[(1-x^2)\,P_m{}'\,P_n\right]_{-1}^{1} - \int_{-1}^{1}P_n\left[\frac{d}{dx}(1-x^2)\,P_m{}'\right]dx \qquad \text{[Integrating by parts]}$$

$$= 0 - \int_{-1}^{1}P_n\,\frac{d}{dx}\{(1-x^2)\,P_m{}'\}\,dx$$

$$= -\int_{-1}^{1}P_n\,\{-m\,(m+1)P_m\}\,dx \qquad \left[\begin{array}{c}\text{From Legendre's differential equation} \\ \dfrac{d}{dx}\{(1-x^2)\,P_m{}'\} + m(m+1)\,P_m = 0 \\ \text{(Orthogonality)}\end{array}\right]$$

$$= m(m+1)\int_{-1}^{1}P_n\,P_m\,dx = m(m+1)\cdot 0 = 0 \qquad \left[\int_{-1}^{+1}P_n\,P_m dx = 0, \text{ when } m \neq n\right]$$

By putting $m = n$ in example 17, we can also prove that

$$\int_{-1}^{1}(1-x^2)\,(P_n{}')^2\,dx = \cdot\,\frac{2n(n+1)}{2n+1}$$

(Orthogonality Property) **Proved.**

Example 19. *Assuming that a polynomial f(x) of degree n can be written as*

$$f(x) = \sum_{0}^{\infty} C_m P_m(x),$$

show that

$$C_m = \frac{2m+1}{2}\int_{-1}^{1} f(x)P_m(x)dx$$

Solution. We have, $f(x) = \sum_{0}^{\infty} C_m P_m(x)$

$$= C_0P_0(x) + C_1P_1(x) + C_2P_2(x) + C_3P_3(x) + C_4P_4(x) + \ldots + C_mP_m(x) + \ldots.$$

Multiplying both sides by $P_m(x)$, we get

$$P_m(x)\,f(x) = C_0P_0(x)\,P_m(x) + C_1P_1(x)\,P_m(x) + C_2P_2(x)\,P_m(x) + \ldots + C_mP_m^2(x) + \ldots$$

$$\int_{-1}^{+1} f(x)P_m(x)\,dx = \int_{-1}^{+1}[C_0P_0(x)P_m(x) + C_1P_1(x)\,P_m(x) + C_2P_2(x)\,P_m(x) + \ldots + C_mP_m^2(x) + \ldots]\,dx$$

$$= \left[0 + 0 + \ldots + C_m\frac{2}{2m+1} + \ldots\right] = \frac{2C_m}{2m+1}$$

$$C_m = \frac{2m+1}{2}\int_{-1}^{+1} f(x)P_m(x)\,dx$$ **Proved.**

Example 20. *Using the Rodrigue's formula for Legendre function, prove that*

$\int_{-1}^{+1} x^m P_n(x)\,dx = 0,$ *where m, n are positive integers and* $m < n$.

Solution. $\int_{-1}^{+1} x^m P_n(x)\,dx = \int_{-1}^{+1} x^m \frac{1}{2^n\, n!}\frac{d^n}{dx^n}(x^2-1)^n\,dx = \frac{1}{2^n\, n!}\int_{-1}^{+1} x^m \frac{d^n}{dx^n}(x^2-1)^n\,dx$

On integrating by parts, we get

$$= \frac{1}{2^n\, n!}\left[\left\{x^m \frac{d^{n-1}}{dx^{n-1}}(x^2-1)^n\right\}_{-1}^{+1} - \int_{-1}^{+1} m\,x^{m-1}\frac{d^{n-1}}{dx^{n-1}}(x^2-1)^n\,dx\right]$$

$$= 0 - \frac{m}{2^n\, n!}\int_{-1}^{+1} x^{m-1}\frac{d^{n-1}}{dx^{n-1}}(x^2-1)^n\,dx$$

Similarly, $\frac{m}{2^n\, n!}\int_{-1}^{+} x^{m-1}\frac{d^{n-1}}{dx^{n-1}}(x^2-1)\,dx = -(-1)^2\frac{m(m-1)}{2^2\; n!}\int_{-1}^{+1} x^{m-2}\frac{d^{n-2}}{dx^{n-2}}(x^2-1)^n\,dx$

Integrating R.H.S., $m-2$ times, we get

$$\int_{-1}^{+1} x^m P_n(x)\,dx = (-1)^m\frac{m(m-1)\ldots 1}{2^n\; n!}\int_{-1}^{+1}\frac{d^{n-m}}{dx^{n-m}}(x^2-1)^n\,dx$$

$$= \frac{(-1)^m\; m!}{2^n\; n!}\int_{-1}^{+1}\frac{d^{n-m}}{dx^{n-m}}(x^2-1)^n\,dx$$

$$= \frac{(-1)^m\; m!}{2^n\; n!}\left[\frac{d^{n-m-1}}{dx^{n-m-1}}(x^2-1)^n\right]_{-1}^{+1} = 0$$ **Proved.**

28.9 RECURRENCE FORMULAE

The following recurrence formulae are derived from the generating function. These formulae are very useful in solving the questions. So they are to be committed to memory.

1.	$(n+1)P_{n+1}(x) = (2n+1)\,x\,P_n(x) - n\,P_{n-1}(x)$
2.	$nP_n(x) = x\,P_n'(x) - P_{n-1}'(x)$
3.	$(2n+1)\,P_n(x) = P_{n+1}'(x) - P_{n-1}'(x)$
4.	$P_n'(x) = x\,P_{n-1}' + nP_{n-1}'(x)$
5.	$(x^2-1)\,P_n'(x) = n[P_n - P_{n-1}]$
6.	$(x^2-1)\,P_n'(x) = (n+1)\,[P_{n+1}(x) - xP_n(x)]$

Formula I. $\boxed{(n+1)P_{n+1} = (2n+1)\,x\,P_n - n\,P_{n-1}}$

Solution. We know that, $(1-2xz+z^2)^{-1/2} = \Sigma z^n P_n(x)$

Differentiating w.r.t, 'z', we get

$$-\frac{1}{2}(1-2xz+z^2)^{-3/2}(-2x+2z) = \Sigma nz^{n-1}P_n(x)$$

Multiplying both sides by $(1-2xz+z^2)$, we get

$$(1-2xz+z^2)^{-1/2}(x-z) = (1-2xz+z^2)\Sigma nz^{n-1}P_n(x)$$

$$(x-z)\Sigma z^n P_n(x) = (1-2xz+z^2)\Sigma nz^{n-1}P_n(x) \qquad \ldots (1)$$

Equating the coefficients of z^n on both sides of (1), we get

$$xP_n(x) - P_{n-1}(x) = (n+1)P_{n+1}(x) - 2x\,(n)\,P_n(x) + (n-1)\,P_{n-1}(x)$$

$$(n+1)\,P_{n+1}(x) = (2n+1)\,x\,P_n - nP_{n-1}$$ **Proved.**

Formula II. $\boxed{nP_n = xP'_n - P'_{n-1}}$ (*DU, III Sem. 2012, U.P., II Sem., summer, 2009, 2006*)

Solution. We know that, $(1-2xz+z^2)^{-1/2} = \Sigma z^n P_n(x)$ ... (1)

Differentiating (1) with respect to z, we get

$$-\frac{1}{2}(1-2xz+z^2)^{-3/2}(-2x+2z) = \Sigma nz^{n-1} P_n(x)$$

$$\Rightarrow \quad (x-z)(1-2xz+z^2)^{-3/2} = \Sigma nz^{n-1} P_n(x) \quad ...(2)$$

Differentiating (1) with respect to x, we get

$$-\frac{1}{2}(1-2xz+z^2)^{-3/2}(-2z) = \Sigma z^n P'_n(x)$$

$$\Rightarrow \quad z(1-2xz+z^2)^{-3/2} = \Sigma z^n P'_n(x) \quad ...(3)$$

Dividing (2) by (3), we get

$$\frac{x-z}{z} = \frac{\Sigma nz^{n-1}P_n(x)}{\Sigma z^n P'_n(x)}$$

$$\Rightarrow \quad (x-z)\Sigma z^n P'_n(x) = \Sigma nz^n P_n(x)$$

Equating coefficients of z^n on both sides, we get

$$x P'_n(x) - P'_{n-1}(x) = n P_n(x)$$ **Proved.**

Formula III. $\boxed{P'_{n+1} - P'_{n-1} = (2n+1)P_n}$ (*Delhi University, April 2010*)

Solution. We know that, $(n+1)P_{n+1} = (2n+1)xP_n - nP_{n-1}$ (Recurrence Formula *I*) ... (1)

Differentiating (1) w.r.t. 'x', we get

$$(n+1)P'_{n+1} = (2n+1)P_n + (2n+1)xP'_n - nP'_{n-1} \quad ...(2)$$

$$xP'_n - P'_{n-1} = nP_n \quad \text{(Recurrence formula II) ... (3)}$$

Substituting the value of xP'_n from (3) into (2), we get

$$(n+1)P'_{n+1} = (2n+1)P_n + (2n+1)[nP_n + P'_{n-1}] - nP'_{n-1}$$

$$\Rightarrow \quad (n+1)P'_{n+1} - (n+1)P'_{n-1} = (2n+1)(1+n)P_n$$

$$\Rightarrow \quad (2n+1)P_n = P'_{n+1} - P'_{n-1}$$ **Proved.**

Formula IV. $\boxed{P'_n - xP'_{n-1} = nP_{n-1}}$ (*U.P. II Semester, 2010*)

Solution. We know that, $(n+1)P_{n+1} = (2n+1)xP_n - nP_{n-1}$ (Formula *I*)

Replacing n by $n-1$, we get

$$nP_n = (2n-1)xP_{n-1} - (n-1)P_{n-2}$$

Differentiating the above formula w.r.t. 'x', we get

$$nP'_n = (2n-1)P_{n-1} + (2n-1)xP'_{n-1} - (n-1)P'_{n-2}$$

$$\Rightarrow \quad n[P'_n - xP'_{n-1}] - (n-1)[xP'_{n-1} - P'_{n-2}] = (2n-1)P_{n-1}$$

$$n[P'_n - xP'_{n-1}] - (n-1)[(n-1)P_{n-1}] = (2n-1)P_{n-1} \quad \text{(Form formula II)}$$

$$\Rightarrow \quad n[P'_n - xP'_{n-1}] = [(n-1)^2 + (2n-1)]P_{n-1} = n^2 P_{n-1}$$

$$\Rightarrow \quad P'_n - xP'_{n-1} = nP_{n-1}.$$ **Proved.**

Formula V. $\boxed{(x^2-1)P'_n = n[xP_n - P_{n-1}]}$

Solution. We know that, $P'_n - xP'_{n-1} = nP_{n-1}$ ... (1) (Recurrence Formula *III*)

$xP'_n - P'_{n-1} = nP_n$ ... (2) (Recurrence Formula *II*)

Multiplying (2) by x and subtracting from (1), we get

$$(1-x^2)P'_n = n(P_{n-1} - xP_n).$$ **Proved.**

Formula VI. $\boxed{(x^2-1)P'_n = n[xP_n - P_{n-1}]}$ (*U.P. II Semester, June 2007*)

Solution. We know that, $(n+1)P_{n+1} = (2n+1)xP_n - nP_{n-1}$ (Recurrence formula *I*)

Which can be written as

$$(n+1)(P_{n+1} - xP_n) = n(xP_n - P_{n-1}) \quad \text{... (1)}$$

But $(x^2-1)P'_n = n(xP_n - P_{n-1})$ ... (2) (Recurrence formula *V*)

From (1) and (2), we get

$\Rightarrow$ $$(x^2-1)P'_n = (n+1)(P_{n+1} - xP_n).$$ **Proved.**

Example 21. *If $P_n(x)$ is a Legendre polynomial of degree n and α is such that $P_n(\alpha) = 0$. Show that $P_{n-1}(\alpha)$ and $P_{n+1}(\alpha)$ are of opposite signs.*

Solution. From Recurrence relation (1), we have

$$(2n+1)xP_n(x) = (n+1)P_{n+1}(x) + nP_{n-1}(x) \quad \text{...(1)}$$

Putting $x = \alpha$ in (1), we get

$$(2n+1)\alpha \,.\, P_n(\alpha) = (n+1)P_{n+1}(\alpha) + nP_{n-1}(\alpha) \quad \text{...(2)}$$

Putting $P_n(\alpha) = 0$ (given) in (2), we get

$$(2n+1)\alpha \,.\, 0 = (n+1)P_{n+1}(\alpha) + nP_{n-1}(\alpha)$$

$$\Rightarrow \quad \frac{P_{n+1}(a)}{p_{n-1}(a)} = \frac{n}{n+1} \quad \text{...(3)}$$

As n is a positive integer so R.H.S. of (3) is negative. Hence (3) shows that $P_{n+1}(\alpha)$ and $P_{n-1}(\alpha)$ are of opposite signs. **Proved.**

Example 22. *Prove that* $\int P_n dx = \frac{1}{2n+1}[P_{n+1} - P_{n-1}] + C$

Solution. We know that $(2n+1)P_n = P'_{n+1} - P'_{n-1}$ (Recurrence relation *III*)

Integrating $(2n+1)\int P_n\,dx = P_{n+1} - P_{n-1} + A$

$$\int P_n dx = \frac{P_{n+1} - P_{n-1}}{2n+1} + C$$ **Proved.**

Example 23. *Show that* $P'_{n+1} + P'_n = P_0 + 3P_1 + 5P_2 + \ldots + (2n+1)P_n$

Solution. We know that

$(2n+1)P_n = P'_{n+1} - P'_{n-1}$ (Recurrence formula III)

Putting $n = 1, 2, 3, \ldots n$ in succession, we get

$$3P_1 = P'_2 - P'_0 \quad \text{... (1)}$$

$$5P_2 = P'_3 - P'_1 \quad \text{... (2)}$$

$$7P_3 = P'_4 - P'_2 \quad \text{... (3)}$$

....

....

$$(2n-1)\, P_{n-1} = P'_n - P'_{n-2}$$

$$(2n+1)\, P_n = P'_{n+1} - P'_{n-1}$$

Adding (1), (2), (3) etc., we get

$$3P_1 + 5P_2 + 7P_3 + ... + (2n-1)\, P_{n-1} + (2n+1)\, P_n = P'_{n+1} + P'_n - P'_0 - P'_1 = P'_{n+1} + P'_n - P_0$$

$$\Rightarrow P'_{n+1} + P'_n = P_0 + 3P_1 + 5P_2 + 7P_3 + ... + (2n+1)P_n \qquad [P'_1 = 1 = P_0 \text{ and } P'_0 = 0]$$ **Proved.**

Example 24. *Prove that*

$$(2n+1)(x^2-1)\, P'_n(x) = n\,(n+1)\{p_{n+1}(x) - P_{n-1}(x)\}$$

Solution. We know that

$$(x^2-1)\, P'_n(x) = (n+1)\, [P_{n+1}(x) - xP_n(x)] \qquad ... (1)$$

(Recurrence Formula *VI*)

$$(x^2-1)\, P'_n(x) = n[x\, P_n(x) - P_{n-1}(x)] \qquad ... (2)$$

(Recurrence Formula *V*)

$P_n(x)$ does not occur in the required result so we have to eliminate $P_n(x)$.

Multiplying (1) by n and (2) by $(n+1)$ and then adding, we get

$$[n + (n+1)]\,(x^2-1)\, P'_n(x) = n(n+1)P_{n+1}(x) - n(n+1)\, P_{n-1}(x)$$

$$\Rightarrow \qquad (2n+1)(x^2-1)P'_n(x) = n(n+1)\,[P_{n+1}(x) - P_{n-1}(x)]$$ **Proved.**

Example 25. *Prove that:*

$$\int_{-1}^{1} (x^2-1)\, P_{n+1}\, P_n{}'\, dx = \frac{2n\,(n+1)}{(2n+1)\,(2n+3)}$$

Solution. $$\text{L.H.S.} = \int_{-1}^{1} (x^2-1)\, P_{n+1}\, P_n{}'\, dx \qquad ...(1)$$

From Recurrence relation (5),

$$n\,(P_{n-1} - xP_n) = (1-x^2)P_n{}'$$

$$\Rightarrow \qquad (x^2-1)\, P_n{}' = n(xP_n - P_{n-1}) \qquad ...(2)$$

Now, $$\int_{-1}^{1} (x^2-1)\, P_{n+1}\, P_n{}'\, dx = \int_{-1}^{1} \left\{(x^2-1)\, P_n{}'\right\} P_{n+1}\, dx$$

Putting the value of $(x^2-1)\, P_n$ from (2) in (1), we get

$$= \int_{-1}^{1} n\left(xP_n - P_{n-1}\right) P_{n+1}\, dx$$

$$= n\int_{-1}^{1} xP_n\, P_{n+1}\, dx - n\int_{-1}^{1} P_{n-1}\, P_{n+1}\, dx$$

$$= n\int_{-1}^{1} xP_n\, P_{n+1}\, dx + 0 \qquad ... (3) \qquad \left[\because \int_{-1}^{1} P_{n-1}\, P_{n+1}\, dx = 0\right]$$

From Recurrence relation (1), we have

$$(2n+1)\, xP_n = (n+1)\, P_{n+1} + nP_{n-1}$$

$$xP_n = \frac{(n+1)\, P_{n+1} + nP_{n-1}}{(2n+1)} \qquad ...(4)$$

Putting the value of xP_n from (4) in (3), we get

$$\int_{-1}^{1}(x^2-1)\,P_{n+1}\,P_n'\,dx = n\int_{-1}^{1}\left[\frac{(n+1)\,P_{n+1}+n\,P_{n-1}}{2n+1}\right].P_{n+1}\,dx \qquad \text{[From (2)]}$$

$$= \frac{n\,(n+1)}{2n+1}\int_{-1}^{1}P'_{n+1}dx+\frac{n^2}{2n+1}\int_{-1}^{1}P_{n-1}\,P_{n+1}\,dx$$

$$= \frac{n\,(n+1)}{2n+1}\bullet\frac{2}{2\,(n+1)+1}+0=\frac{2n\,(n+1)}{(2n+1)\,(2n+3)} \qquad \textbf{Proved.}$$

28.10 LAPLACE'S FIRST DEFINITE INTEGRAL FOR $P_n(x)$

$$P_n(x)=\frac{1}{\pi}\int_0^{\pi}[x\pm\sqrt{(x^2-1)}\cos\phi]^n\,d\phi$$

where n is a positive integer.

Solution. We know that

$$\int_0^{\pi}\frac{d\phi}{a\pm b\cos\phi}=\frac{\pi}{\sqrt{(a^2-b^2)}} \qquad \text{(If } a>b) \qquad \dots (1)$$

Put $\quad a=1-xz$ and $b=z^2\sqrt{(x-1)}$, so that

$$a^2-b^2=(1-xz)^2-z^2(x^2-1)=1-2xz+z^2$$

Substituting the values of a, b and a^2-b^2 in (1), we get

$$\int_0^{\pi}\frac{d\phi}{(1-xz)\pm z\sqrt{(x^2-1)}\,\cos\phi}=\frac{\pi}{\sqrt{(1-2xz+z^2)}}$$

$$\Rightarrow \qquad \pi(1-2xz+z^2)^{-\frac{1}{2}}=\int_0^{\pi}\frac{d\phi}{1-z\{x\pm\sqrt{(x^2-1)}\cos\phi\}}$$

$$=\int_0^{\pi}[1-z\{x\pm\sqrt{(x^2-1)}\cos\phi\}]^{-1}d\phi$$

$$\pi\sum_{n=0}^{\infty}z^nP_n(x)=\int_0^{\pi}[1-t]^{-1}\,d\phi$$

where $t=z\,\{x\mp\sqrt{(x^2-1)}\cos\phi\}$ and $z\,\{x\pm(\sqrt{x^2-1})\cos\phi\}<1$ and z is small quantity.

$$=\int_0^{\pi}[1+t+t^2+....]\,d\phi=\int_0^{\pi}\sum_{n=0}^{\infty}t^n\,d\phi$$

$$=\int_0^{\pi}\sum_{n=0}^{\infty}z^n\{x\pm\sqrt{x^2-1\cos\phi}\}^n d\phi$$

$$\pi\sum_{n=0}^{\infty}z^nP_n(x)=\int_{\phi=0}^{\pi}z^n\{x\pm\sqrt{(x^2-1)}\cos\phi\}^n\cdot d\phi$$

Equating the coefficients of z^n from both the sides, we obtain

$$\pi\,P_n(x)=\int_0^{\pi}\left\{x\mp\sqrt{(x^2-1)}\cos\phi\right\}^n\,d\phi$$

$$\Rightarrow \qquad P_n(x)=\frac{1}{\pi}\int_0^{\pi}\left\{x\mp\sqrt{(x^2-1)}\,\cos\phi\right\}^n\,d\phi \qquad \textbf{Proved.}$$

28.11 LAPLACE'S SECOND DEFINITE INTEGRAL FOR $P_n(x)$

$$P_n(x)=\frac{1}{\pi}\int_0^{\pi}\frac{d\phi}{\left\{x\pm\sqrt{(x^2-1)}\cos\phi\right\}^{n+1}}$$

where n is a positive integer.

Solution. We know that

$$\int_0^{\pi}\frac{d\phi}{a\pm b\cos\phi}=\frac{\pi}{\sqrt{(a^2-b^2)}}\qquad \text{(If } a>b\text{)} \quad \ldots (1)$$

Putting $a = xz - 1$ and $b = z\sqrt{(x^2-1)}$, we get

$$a^2-b^2=(x\,z-1)^2-z^2\,(x^2-1)=x^2z^2-2\,x\,z+1-z^2x^2+z^2=1-2\,x\,z+z^2$$

On substituting the values of a, b and $a^2 - b^2$ in (1), we get

$$\int_0^{\pi}\frac{d\phi}{z\{x\pm\sqrt{(x^2-1)}\cos\phi\}-1}=\pi(1-2xz+z^2)^{-\frac{1}{2}}$$

Put $z\{x\pm\sqrt{(x^2-1)}\cos\phi\}=t$ and $|t|>1$

$$\int_0^{\pi}\frac{d\phi}{t-1}=\frac{\pi}{z}\left\{1-\frac{2x}{z}+\frac{1}{z^2}\right\}^{-\frac{1}{2}}$$

$$\Rightarrow\qquad \int_0^{\pi}\frac{1}{t}\frac{d\phi}{1-\frac{1}{t}}=\frac{\pi}{z}\sum_{n=0}^{\infty}\frac{1}{z^n}P_n(x)$$

$$\Rightarrow\qquad \int_0^{\pi}\frac{1}{t}\left(1+\frac{1}{t}+\frac{1}{t^2}+\ldots\ldots\right)d\phi=\pi\sum_{n=0}^{\infty}\frac{1}{z^{n+1}}P_n(x)$$

$$\Rightarrow\qquad \int_0^{\pi}\sum_{n=0}^{\infty}\frac{1}{t^{n+1}}d\phi=\pi\sum_{n=0}^{\infty}\frac{1}{z^{n+1}}P_n(x)$$

On putting the value of t, we get

$$\Rightarrow\int_0^{\pi}\sum_{n=0}^{\infty}\frac{d\phi}{z^{n+1}\{x\pm\sqrt{(x^2-1)}\cos\phi\}^{n+1}}=\pi\sum_{n=0}^{\infty}\frac{1}{z^{n+1}}P_n(x)\qquad \ldots (2)$$

Equating the coefficient of $\frac{1}{z^{n+1}}$ on both sides of (2), we get

$$\int_0^{\pi}\frac{d\phi}{[x\pm\sqrt{(x^2-1)}\cos\phi]^{n+1}}=\pi\,P_n(x)$$

$$\Rightarrow\qquad P_n(x)=\frac{1}{\pi}\int_0^{\pi}\frac{d\phi}{[x\pm\sqrt{(x^2-1)}\cos\phi]^{n+1}}\qquad \textbf{Proved.}$$

Example 26. *Prove that*

$$\int_{-1}^{+1}x^2\,P_{n+1}(x)\cdot P_{n-1}(x)\,dx=\frac{2n(n+1)}{(2n-1)(2n+1)(2n+3)}$$

Deduce the value of $\int_0^1 x^2\,P_{n+1}\cdot P_{n-1}\,dx.$

Solution. The recurrence formula *1* is

$$(n+1)\,P_{n+1} = (2n+1)\,x\,P_n - n\,P_{n-1}$$

$$\Rightarrow \quad (2n+1)\,x\,P_n = (n+1)\,P_{n+1} + nP_{n-1}$$

Replacing n by $(n+1)$ and $(n-1)$, we have

$$(2n+3)\,xP_{n+1} = (n+2)\,P_{n+2} + (n+1)\,P_n \qquad \text{... (1)}$$

$$(2n-1)\,x\,P_{n-1} = nP_n + (n-1)\,P_{n-2} \qquad \text{[Replacing } n \text{ by } (n-2)\text{]. ... (2)}$$

Multiplying (1) and (2) and integrating in the limits – 1 to +1, we have

$$(2n+3)(2n-1)\int_{-1}^{+1} x^2\,P_{n+1}\cdot P_{n-1}dx = n(n+1)\int_{-1}^{1} P_n^2\,dx + n(n+2)\int_{-1}^{+1} P_n\cdot P_{n+2}\,dx$$

$$+(n^2-1)\int_{-1}^{+1} P_nP_{n-2}dx + (n-1)(n+2)\int_{-1}^{+1} P_{n+2}\cdot P_{n-2}\,dx$$

$$= n(n+1)\int_{-1}^{1} P_n^2 dx + 0 + 0 + 0 \qquad \text{(Orthogonality property)}$$

$$= n(n+1)\cdot\frac{2}{(2n+1)}$$

$$\int_{-1}^{+1} x^2\cdot P_{n+1}\cdot P_{n-1}dx = \frac{2n(n+1)}{(2n-1)(2n+1)(2n+3)} \qquad \text{as R.H.S. is even function.}$$

and $$\int_0^1 x^2\cdot P_{n+1}\cdot P_{n-1}\,dx = \frac{n(n+1)}{(2n-1)(2n+1)(2n+3)}$$ **Proved.**

Example 27. *Prove that*

$$P_{-(n+1)}(x) = P_n(x)$$

Solution. From Laplace's first integral

$$P_n(x) = \frac{1}{\pi}\int_0^{\pi} [x \pm \sqrt{(x^2-1)}\ \cos\phi]^n\ d\phi \qquad \text{... (1)}$$

On putting – $(n+1)$ for n in (1), we get

$$P_{-(n+1)}(x) = \frac{1}{\pi}\int_0^{\pi} \frac{d\phi}{[x \pm \sqrt{(x^2-1)}\ \cos\phi]^{n+1}}$$

$$= P_n(x) \text{ (from Laplace's second integral)}$$ **Proved.**

EXERCISE 28.1

1. Express in terms of Legendre Polynomials.

(*a*) $1 + x - x^2$ **Ans.** $-\frac{2}{3}P_2(x) + P_1(x) + \frac{2}{3}P_0(x)$

(*b*) $x^3 + 1$ **Ans.** $\frac{2}{5}P_3(x) + \frac{3}{5}P_1(x) + P_0(x)$

(*c*) $1 + 2x - 3x^2 + 4x^3$ **Ans.** $\frac{8}{5}P_3(x) - 2P_2(x) + \frac{22}{5}P_1(x)$

(*d*) $x^4 + 2x^3 - 6x^2 + 5x - 3$ (*AMIETE, Dec. 2009*)

Ans. $\frac{8}{35}P_4(x) + \frac{4}{5}P_3(x) - \frac{24}{7}P_2(x) + \frac{31}{5}P_1(x) - \frac{24}{5}P_0(x)$

(*e*) $x^4 + 3x^3 - x^2 + 5x - 2$ (*AMIETE, June 2010*)

Ans. $\frac{8}{35}P_4(x) + \frac{6}{5}P_3(x) - \frac{2}{21}P_2(x) + \frac{34}{5}P_1(x) - \frac{32}{15}$

2. Show that

(*a*) $x^5 = \frac{8}{63}\left[P_5(x)+\frac{7}{2}P_3(x)+\frac{27}{8}P_1(x)\right]$ (*b*) $x^3 = \frac{2}{5}P_3(x)+\frac{3}{5}P_1(x)$

Evaluate the following :

3. $\int_{-1}^{+1} x^3 P_4(x)dx$ **Ans.** 0

4. $\int_{-1}^{+1} x^3 P_3(x)dx$ **Ans.** $\frac{4}{35}$

Prove that

5. $\int_{-1}^{+1} x^n P_n(x)\,dx = \dfrac{\frac{1}{2}(n+1)}{2^n\left(\frac{2n+3}{2}\right)} - \dfrac{2^{n+1}(n!)^2}{(2n+1)}$ **6.** $\int_{-1}^{+1} xP_n(x)P_{n-1}(x)\,dx = \dfrac{2n}{4n^2-1}$

7. $\int_0^1 P_{2n}(x)\cdot P_{2n+1}(x)dx = \int_0^1 P_{2n}(x)\cdot P_{2n-1}(x)dx$ **8.** $\int_{\infty}^1 P_n(x)dx = \frac{1}{2n+1}\left[P_{n-1}(x) - P_{n+1}(x)\right]$

9. $P_n(x) = P'_{n+1}(x)$

10. $\frac{d}{dx}\left[(1-x^2)\frac{d}{dx}P_n(x)\right] + n(n+1)\,P_n(x) = 0$

11. $\int_{-1}^{+1} x\cdot P_n\cdot\frac{d}{dx}P_m dx$ is either 0, 2 or $\frac{2n}{2n+1}$

12. $P_n(x) = P'_{n+1}(x) - 2xP'_n(x) + P'_{n-1}(x)$

13. $P'_{2n+1} = (2n+1)\,P_{2n} + 2nx\,P_{2n-1} + (2n-1)\,x^2 P_{n-2} + ... + 2x^{2n-1}P_1 + x^{2n}P_0$

14. $(n+1)\{P_n P'_{n+1} - P_{n+1}P'_n\} = P_{.0}^2 + 3P_1^2 + 5P_2^2 + ... + (2n+1)P_n^2$

15. $(n+1)^2 P_n^2 - (x^2-1)P_n^{-2} = P_0^2 + 3P_1^2 + 5P_2^2 + ... + (2n+1)P_n^2$

16. $\int_0^{\pi} P_n(\cos\theta)(\cos n\theta\; d\theta) = \beta\left(n+\frac{1}{2}, \frac{1}{2}\right)$ if n is a positive integer.

17. $u = (1-2xz+z^2)^{-\frac{1}{2}}$ is a solution of the equation $z\frac{\partial^2 v}{\partial x^2}\left\{(1-x^2)\frac{\partial\theta}{\partial x}\right\} = 0$?

28.12 FOURIER-LEGENDRE EXPANSION

Let $f(x)$ be a function defined from $x = -1$ to $x = 1$.

The Fourier-Legendre expansion of $f(x)$

$$f(x) = C_1 P_1(x) + C_2 P_2(x) + C_3 P_3(x) + ... \quad ... (1)$$

$$\Rightarrow \quad C_n = \frac{2n+1}{2}\int_{-1}^{+1} f(x)P_n(x)\,dx$$

Multiplying both sides of (1) by $P_n(x)$, we have

$$f(x)\cdot P_n(x) = C_1P_1(x)\cdot P_n(x) + C_2P_2(x)P_n(x) + ... + C_nP_n^2(x) + ... \quad ... (2)$$

Integrating both sides of (2), we get

$$\int_{-1}^{+1} f(x)\cdot P_n(x)\,dx = C_1\int_{-1}^{+1} P_1(x)\cdot P_n(x)dx + C_2\int_{-1}^{+1} P_2(x)\cdot P_n(x)\,dx + ... + C_n\int_{-1}^{+1} P_n^2(x)dx + ...$$

$$\int_{-1}^{+1} f(x)\cdot P_n(x)\,dx = C_n\int_{-1}^{+1} P_n^2(x)dx \qquad \text{(Other integrals are equal to zero)}$$

$$= C_n \frac{2}{2n+1}$$

$$\Rightarrow \quad C_n = \frac{2n+1}{2}\int_{-1}^{+1} f(x)\cdot P_n(x)dx \quad \text{or} \quad C_n = \left(n+\frac{1}{2}\right)\int_{-1}^{+1} f(x)\cdot P_n(x)dx.$$

Example 28. *Evaluate*

$$\int_{-1}^{+1} x^2\, P_n^2(x)\,dx$$ (*U.P. II Semester 2010*)

Solution. We know that

$$(n + 1)\, P_{n+1} = (2n + 1)\, x\, P_n(x) - n\, P_{n-1}(x) \quad \text{(Recurrence formula I)}$$

$$(2n + 1)\, x\, P_n(x) = (n + 1)\, P_{n+1}(x) + n P_{n-1}(x)$$

Squaring both sides, we get

$$(2n+1)^2 x^2\, P_n^2(x) = (n+1)^2\, P_{n+1}^2(x) + n^2 P_{n-1}^2(x) + 2n(n+1) P_{n+1}(x) P_{n-1}(x)$$

Integrating both sides w.r.t. x between -1 and $+1$, we get

$$(2n+1)^2 \int_{-1}^{+1} x^2\, P_n^2(x)\,dx = (n+1)^2 \int_{-1}^{+1} P_{n+1}^2(x)\,dx + n^2 \int_{-1}^{+1} P_{n-1}^2(x)\,dx + 2n(n+1)\int_{-1}^{+1} P_{n+1}(x) P_{n-1}(x)\,dx$$

$$= (n+1)^2 \frac{2}{2(n+1)+1} + n^2 \frac{2}{2(n-1)+1} + 0 = \frac{2(n+1)^2}{2n+3} + \frac{2n^2}{2n-1}$$

$$\int_{-1}^{+1} x^2 P_n^2(x)\,dx = \frac{2(n+1)^2}{(2n+1)^2(2n+3)} + \frac{2n^2}{(2n+1)^2(2n-1)}$$ **Ans.**

Example 29. *Find the sum of the first (n + 1) terms of the series* $\sum_{m=0}^{n} (2m+1)\, P_m(x)\, P_m(y)$

[*Christoffel's Summation Formula*]

Solution. We know that

$$(2m + 1)\, x\, P_m(x) = (m + 1)\, P_{m+1}(x) + m\, P_{m-1}(x) \quad \text{... (1)}$$

and $$(2m + 1)\, y\, P_m(y) = (m + 1)\, P_{m+1}(y) + m\, P_{m-1}(y) \quad \text{... (2)}$$

Multiplying (1) by $P_m(y)$ and (2) by $P_m(x)$ and subtracting, we get

$$(2m + 1)(x - y)\, P_m(x)\, P_m(y) = (m + 1)\{P_{m+1}(x)\, P_m(y) - P_m(x)\, P_{m+1}(y)\}$$
$$- m\,\{P_m(x)\, P_{m-1}(y) - P_{m-1}(x)\, P_m(y)\}$$

Now putting $m = 1, 2, 3, \ldots\ldots n$ in succession, we have

$$3(x - y)\, P_1(x)\, P_1(y) = 2\{P_2(x)\, P_1(y) - P_1(x)\, P_2(y)\} - 1\{P_1(x)\, P_0(y) - P_0(x)\, P_1(y)\}$$
$$5(x - y)\, P_2(x)\, P_2(y) = \{P_3(x)\, P_2(y) - P_2(x)\, P_3(y)\} - 2\{P_2(x)\, P_1(y) - P_1(x)\, P_2(y)\}$$

. . . .

And again

$$(2n + 1)(x - y)\, P_n(x)\, P_n(y) = (n + 1)\{P_{n+1}(x)\, P_n(y) - P_n(x) + P_{n+1}(y)\}$$
$$- n\left\{P_n(x) P_{n-1}(y) - P_{n-1}(x)\, P_n(y)\right\}$$

Adding these equations, we get

$$(x - y)\,[3P_1(x)\, P_1(y) + 5P_2(x) P_2(y) + \ldots\ldots + (2n+1) P_n(x)\cdot P_n(y)]$$
$$= (n+1)[P_{n+1}(x) P_n(y) - P_n(x) P_{n+1}(y)] - [P_1(x) P_0(y) - P_0(x) P_1(y)]$$

But $P_0(x) = P_0(y) = 1$ and $P_1(x) = x,\ P_1(y) = y$

Therefore $P_1(x)\, P_2(y) - P_0(x) P_1(y) = x - y = (x - y)\, P_0(x)\, P_0(y)$

$$(x-y)\,[P_0(x)P_0(y)+3P_1(x),\,P_1(y)+5P_2(x)P_2(y)+....+(2n+1)\,P_n(x)P_n(y)]$$
$$=(n+1)[P_{n+1}(x)\,P_n(y)-P_n(x)P_{n+1}(y)]$$

$$\sum_{m=0}^{n}(2m+1)P_m(x)\cdot P_m(y)=(n+1)\frac{P_{n+1}(x)\cdot P_n(y)-P_n(x)P_{n+1}(y)}{x-y}$$ **Ans.**

Example 30. *Prove that*

$$\int_{-1}^{+1}\frac{d}{dx}((P_n(x))dx=n(n+1)$$

Solution. We know that *(Christoffel's series)*

$$\frac{d}{dx}P_n(x)=(2n-1)\,P_{n-1}(x)+(2n-5)\,P_{n-3}(x)+(2n-9)\,P_{n-5}(x)+....$$

the last term being $3P_1$ if n is even and P_0 if n is odd.

Squaring both sides and then integrating , we have

$$\int_{-1}^{+1}\frac{d}{dx}(P_n(x))dx = (2n-1)^2\int_{-1}^{+1}P_{n-1}^2(x)\,dx+(2n-5)^2\int_{-1}^{+1}P_{n-3}^2(x)\,dx+.....$$
$$+2(2n-1)(2n-5)\int_{-1}^{+1}P_{n-1}(x)\cdot P_{n-3}(x)\,dx+....$$
$$=(2n-1)^2\frac{2}{2(n-1)+1}+(2n-5)^2\frac{2}{2(n-3)+1}+....$$
$$=\frac{2(2n-1)^2}{(2n-1)}+\frac{2(2n-5)^2}{2n-5}+.....$$
$$=2\,(2n-1)+2(2n-5)+....$$
$$=2[(2n-1)+(2n-5)+......] \qquad ...(1)$$

(*i*) If n is even, then the last term is $3P_1$

$$(3)^2\int P_1^2(x)dx=9\frac{2}{2\cdot1+1}=2\times3=6$$

Equation (1) becomes

$$\int_{-1}^{+1}\frac{d}{dx}(P_n(x))dx=2[(2n-1)+(2n-5)+......+3]$$ (This is an *A. P.*, Number of terms = $\frac{n}{2}$)

$$=2\frac{n}{2\times2}\left[2\times3+\left(\frac{n}{2}-1\right)4\right]=\frac{n}{2}(6+2n-4)=n(n+1)$$

(*ii*) If n is odd, then the last term is $P_0\,(x)$.

$$\int_{-1}^{+1}P_0^2(x)\,dx=\int_{-1}^{+1}dx=2$$

$$\int_{-1}^{+1}\frac{d}{dx}(P_n(x))dx = 2[(2n-1)+(2n-5)+......+1]$$

[This is an A.P. Here number of terms = $\frac{n+1}{2}$]

$$=2\frac{n+1}{2\times2}\left[2\times1+\left(\frac{n+1}{2}-1\right)4\right]==\frac{n+1}{2}(2+2n+2-4)=n(n+1)$$ **Ans.**

Example 31. *Show that*

$$x P_n'(x) = n P_n(x) + (2n-3) P_{n-2}(x) + (2n-7) P_{n-4}(x) + \ldots.$$

Solution. We know that

$$n P_n(x) = x P_n'(x) - P_{n-1}'(x) \qquad \text{(Recurrence formula II)}$$

$$\Rightarrow \qquad x P_n'(x) = n P_n(x) + P_{n-1}'(x) \qquad \ldots (1)$$

$$P_{n+1}'(x) = (2n+1)\, P_n(x) + P_{n-1}'(x) \qquad \text{(Recurrence formula III)} \ldots (2)$$

Putting $n-2$ for n in (2), we get

$$P_{n-1}'(x) = (2n-3)\, P_{n-2}(x) + P_{n-3}'(x) \qquad \ldots (3)$$

Putting $n-4$ for n in (2), we get

$$P_{n-3}'(x) = (2n-7)\, P_{n-4}(x) + P_{n-5}'(x) \qquad \ldots (4)$$

Putting $n-6$ for n in (2), we have

$$P_{n-5}(x) = (2n-11)\, P_{n-6}(x) + P_{n-7}'(x) \qquad \ldots (5)$$

and so on.

Adding (1), (3), (4), (5) etc., we get

$$x P_n'(x) = nP_n(x) + (2n-3)\, P_{n-2}(x) + (2n-7) P_{n-4}(x) + (2n-11)\, P_{n-6}(x) + \ldots \qquad \textbf{Proved.}$$

Example 32. *Prove that* $\int_{-1}^{+1} x P_n\, P_n'\, dx = \dfrac{2n}{2n+1}$

Solution. In the last example we have proved that

$$x P_n'(x) = n P_n(x) + (2n-3)\, P_{n-2}(x) + (2n-7)\, P_{n-4}(x) + (2n-11) P_{n-6}(x) + \ldots.$$

Multiplying the above expansion by P_n and integrating between -1 and $+1$, we get

$$\int_{-1}^{+1} x \cdot P_n(x) P_n'(x) dx = n \int_{-1}^{+1} P_n^2(x)\, dx + (2n-3) \int_{-1}^{+1} P_n(x) \cdot P_{n-2}(x) dx + (2n-7)$$

$$\int_{-1}^{+1} P_n(x) \cdot P_{n-4}(x) dx$$

$$+ (2n-11) \int_{-1}^{+1} P_n(x) \cdot P_{n-6}(x) dx + \ldots..$$

$$= n \cdot \frac{2}{2n+1} + 0 + 0 + \ldots.. = \frac{2\,n}{2n+1} \qquad \textbf{Proved.}$$

Example 33. *Expand the function*

$$f(x) = \begin{cases} 0, & -1 < x < 0 \\ 1, & 0 < x < 1 \end{cases}$$

in terms of Legendre polynomials. *(GBTU, 2011)*

Solution. Let $\quad f(x) = \sum_{n=0}^{\infty} C_n\, P_n(x)$ then

$$C_n = \left(n + \frac{1}{2}\right) \int_{-1}^{+1} f(x) P_n(x)\, dx$$

$$= \left(n + \frac{1}{2}\right) \left[\int_{-1}^{0} 0 P_n(x)\, dx + \int_0^1 1 P_n(x)\, dx \right] = \left(n + \frac{1}{2}\right) \int_0^1 P_n(x) dx$$

$$C_0 = \frac{1}{2}\int_0^1 P_0(x)\,dx = \frac{1}{2}\int_0^1 1.\,dx = \frac{1}{2}[x]_0^1 = \frac{1}{2}$$

$$C_1 = \frac{3}{2}\int_0^1 P_1(x)dx = \frac{3}{2}\int_0^1 x\,dx = \frac{3}{2}\left[\frac{x^2}{2}\right]_0^1 = \frac{3}{4}$$

$$C_2 = \frac{5}{2}\int_0^1 P_2(x)dx = \frac{5}{2}\int_0^1 \frac{1}{2}(3x^2-1)dx = \frac{5}{4}(x^3-x)_0^1 = 0$$

$$C_3 = \frac{7}{2}\int_0^1 P_3(x)dx = \frac{7}{2}\int_0^1 \frac{1}{2}(5x^3-3x)dx = \frac{7}{4}\left[\frac{5x^4}{4}-\frac{3x^2}{2}\right]_0^1 = \frac{-7}{16}$$

$$C_4 = \frac{9}{2}\int_0^1 P_4(x)\,dx = \frac{9}{2}\int_0^1 \frac{1}{8}(35x^4-30x^2+3)dx$$

$$= \frac{9}{16}\left[7x^5-10x^3+3x\right]_0^1 = 0$$

$$C_5 = \frac{11}{2}\int_0^1 P_5(x)\,dx = \frac{11}{2}\int_0^1 \frac{1}{8}(63x^5-70x^3+15x)\,dx$$

$$= \frac{11}{16}\left[\frac{21}{2}x^6-\frac{35}{2}x^4+\frac{15}{2}x^2\right]_0^1 = \frac{11}{32}$$

Hence $$f(x) = \frac{1}{2}P_0(x)+\frac{3}{4}P_1(x)-\frac{7}{16}P_3(x)+\frac{11}{32}P_5(x)+\ldots$$ **Ans.**

Example 34. *Express the function*

$$f(x) = \begin{cases} 0, & -1 < x \le 0 \\ x, & 0 < x < 1 \end{cases}$$

in Fourier-Legendre expansion.

Solution. Let $f(x) = \sum_{n=0}^{\infty} C_n P_n(x)$, then

$$C_n = \left(n+\frac{1}{2}\right)\int_{-1}^{+1} f(x)\,P_n(x)\,dx$$

$$C_n = \left(n+\frac{1}{2}\right)\left[\int_{-1}^{0} 0\cdot P_n(x)\,dx + \left(n+\frac{1}{2}\right)\int_0^1 x\cdot P_n(x)\,dx\right] = \left(n+\frac{1}{2}\right)\int_0^1 x\cdot P_n(x)\,dx$$

$$C_0 = \frac{1}{2}\int_0^1 x\cdot P_0(x)\,dx = \frac{1}{2}\int_0^1 x\cdot 1\,dx = \frac{1}{2}\int_0^1 x\,dx = \frac{1}{2}\left[\frac{x^2}{2}\right]_0^1 = \frac{1}{4}$$

$$C_1 = \frac{3}{2}\int_0^1 x\cdot P_1(x)\,dx = \frac{3}{2}\int_0^1 x\cdot x\,dx = \frac{3}{2}\int_0^1 x^2\,dx = \frac{3}{2}\left[\frac{x^3}{3}\right]_0^1 = \frac{1}{2}$$

$$C_2 = \frac{5}{2}\int_0^1 x\cdot P_2(x)\,dx = \frac{5}{2}\int_0^1 x\cdot\frac{3x^2-1}{2}\,dx = \frac{5}{4}\int_0^1 (3x^3-x)dx = \frac{5}{4}\left(\frac{3x^4}{4}-\frac{x^2}{2}\right)_0^1 = \frac{5}{16}$$

$$C_3 = \frac{7}{2}\int_0^1 x \cdot P_3(x)\,dx = \frac{7}{2}\int_0^1 x \cdot \frac{5x^3 - 3x}{2}\,dx = \frac{7}{4}\int_0^1 (5x^4 - 3x^2)\,dx = \frac{7}{4}(x^5 - x^3)_0^1 = 0$$

$$C_4 = \frac{9}{2}\int_0^1 x \cdot P_4(x)\,dx = \frac{9}{2}\int_0^1 x \cdot \frac{35x^4 - 30x^2 + 3}{8}\,dx$$

$$= \frac{9}{16}\int_0^1 (35x^5 - 30x^3 + 3x)\,dx = \frac{9}{16}\left[\frac{35x^6}{6} - \frac{15}{2}x^4 + \frac{3x^2}{2}\right]_0^1 = \frac{-3}{32}$$

$$C_5 = \frac{11}{2}\int_0^1 x \cdot P_5(x)\,dx = \frac{11}{2}\int_0^1 x \cdot \frac{63x^5 - 70x^3 + 15x}{8}\,dx$$

$$= \frac{11}{16}\int_0^1 (63x^6 - 70x^4 + 15x^2)\,dx = \frac{11}{16}\left[9x^7 - 14x^5 + 5x^3\right]_0^1 = 0$$

Hence $f(x) = \frac{1}{4}P_0(x) + \frac{1}{2}P_1(x) + \frac{5}{16}P_2(x) - \frac{3}{32}P_4(x) + ...$ **Ans.**

Example 35. *Expand the function*

$$f(x) = \begin{cases} 0\ , & -1 < x \le 0 \\ x^2, & 0 < x < 1 \end{cases}$$

in terms of Legendre Polynomials.

Ans. $f(x) = \frac{1}{6}P_0(x) + \frac{3}{8}P_1(x) + \frac{1}{3}P_2(x) + \frac{7}{48}P_3(x) - \frac{11}{384}P_5(x) + ...$

28.13 STRUM-LIOUVILLE EQUATION

$$\frac{d}{dx}\left[P(x)\cdot\frac{dy}{dx}\right] + [\lambda q(x) + r(x)]y = 0$$

Solution. We know that Bessel's equation is

$$X^2\frac{d^2y}{dX^2} + X\frac{dy}{dX} + (X^2 - n^2)\,y = 0 \qquad ... (1)$$

Substituting $X = kx$ in (1), we get

$$\frac{dy}{dX} = \frac{dy}{dx}\frac{dx}{dX} = \frac{dy}{dx}\frac{1}{k}$$

$$k^2x^2\left(\frac{d^2y}{dx^2}\frac{1}{k^2}\right) + (kx)\left(\frac{dy}{dx}\frac{1}{k}\right) + (k^2x^2 - n^2)y = 0$$

$$\Rightarrow \qquad x^2\frac{d^2y}{dx^2} + x\frac{dy}{dx} + (k^2x^2 - n^2)\,y = 0$$

$$x\frac{d^2y}{dx^2} + \frac{dy}{dx} + \left(k^2x - \frac{n^2}{x}\right) = 0$$

$$\left(x\frac{d^2y}{dx^2} + \frac{dy}{dx}\right) + \left(\lambda x - \frac{n^2}{x}\right)y = 0 \qquad \text{(Put } k^2 = \lambda\text{)}$$

$$\frac{d}{dx}\left(x\frac{dy}{dx}\right) + \left(\lambda x - \frac{n^2}{x}\right)y = 0 \qquad ... (2)$$

Equations (1) and (2) are of the form.

$$\frac{d}{dx}\left[P(x)\cdot\frac{dy}{dx}\right]+[\lambda q(x)+r(x)]y=0 \quad \dots (3)$$

Equation (3) is known as the Strum-Liouville equation.

Equation (3) with the following conditions is known as Strum-Liouville problem.

$$\alpha_1 y(a) + \alpha_2 y'(a) = 0$$
$$\beta_1 y(b) + \beta_2 y'(b) = 0$$

Solution of Strum-Liouville problem is called an eigen function where λ is an eigen value.

Particular Case. Putting $p = 1$, $q = 1$, $r = 0$ in (3), we have

$$\frac{d^2y}{dx^2}+\lambda y=0$$

Now taking conditions as $\alpha_1 = \beta_1 = 1$ and $\alpha_2 = \beta_2 = 0$

$$y(a) = 0 \text{ and } y(b) = 0$$

Hence $\left.\begin{matrix} y''+\lambda y=0 \\ y(a)=0,\ y(b)=0 \end{matrix}\right]$ simplest form of Strum-Liouville problem.

28.14 ORTHOGONALITY

$$\int_a^b P(x)\,y_m(x)\cdot y_n(x)\,dx=0,\quad m\neq n$$

$$\int_a^b P(x)[y_m(x)]^2\,dx=\|y_m\|,\quad m=n$$

where $\| y_m \|$ is the norm of y_m.

Orthonormal : If the function is orthogonal and have norm equal to 1, then the function is known as orthonormal.

28.15 ORTHOGONALITY OF EIGEN FUNCTIONS

If P, q, r and r' are the functions in Strum-Liouville equation and $\lambda_m(x)$, $\lambda_n(x)$ be the eigen functions of Strum-Liouville problem, then

$$(\lambda_m-\lambda_n)\int_a^b q\,y_m y_n dx = y_m(P\,y'n)-y_n(P\,y'm)$$

$$=\frac{d}{dx}[(P\,y_n')\,y_m-(P\,y_m')\,y_n]_a^b$$

$$=P(b)\left[y_n'(b)\,y_m(b)-y_m'(b)\,y_n(b)\right]-P(a)\left[y_n'(a)\,y_m(a)-y_n'(a)\,y_n(a)\right] \quad \dots (1)$$

$$= 0 \text{ if}$$

(i) $y(a) = y(b)$ *(ii)* $y'(a) = y'(b)$

(iii) $\alpha_1 y(a) + \alpha_2 y'(a) = 0$ *(iv)* $\beta_1 y(b) + \beta_2 y'(b) = 0$

Equation (1) becomes $\int_a^b q\,y_m y_n dx=0 \quad (m\neq n)$

It means that eigen functions y_m, y_n are orthogonal with the weight $q(x)$.

OBJECTIVE TYPE QUESTIONS

Choose the correct or the best of the answer given in the following parts :

1. Let $P_n(x)$ be Legendre polynomial of degree $n > 1$, then

$\int_{-1}^{+1} (1+x)P_n(x)dx$ is equal to

(*i*) 0. (*ii*) $1/(2n+1)$. (*iii*) $2/(2n+1)$. (*iv*) $n/(2n+1)$

Ans. (*i*)

2. The Rodrigue formula for Legendre Polynomial $P_n(x)$ is given by:

(*i*) $P_n(x) = \frac{1}{n!2^n}\frac{d^n}{dx^n}(x^2-1)^n$ (*ii*) $P_n(x) = \frac{n!}{2^n}\frac{d^n}{dx^n}(x^2-1)^{n-1}$

(*iii*) $P_n(x) = \frac{n!}{2^{n-1}}\frac{d^n}{dx^n}(x^2-1)^{n-1}$ (*iv*) $P_n(x) = \frac{1}{n!2^n}(x^2-1)^n$ **Ans.** (*i*)

(*U.P., II Semester, 2009*)

3. The integral $\int_0^{\pi} P_n(\cos\theta)\sin 2\theta\, d\theta, n>1$, where $P_n(x)$ is the Legendre polynomial of degree n equals to

(*i*) 1 (*ii*) $\frac{1}{2}$ (*iii*) 0 (*iv*) 2 **Ans.** (*iii*)

BESSEL'S FUNCTIONS

29.1 BESSEL'S EQUATION

The differential equation

$$x^2\frac{d^2y}{dx^2}+x\frac{dy}{dx}+(x^2-n^2)y=0$$

is called the Bessel's differential equation, and particular solutions of this equation are called Bessel's functions of order n.

We find the Bessel's equation while solving Laplace equation in polar coordinates by the method of separation of variables. This equation has a number of applications in engineering.

Bessel's functions are involved in

(*i*) The Oscillatory motion of a hanging chain

(*ii*) Euler's theory of a circular membrane

(*iii*) The studies of planetary motion

(*iv*) The propagation of waves

(*v*) The Elasticity

(*vi*) The fluid motion

(*vii*) The potential theory

(*viii*) Cylindrical and spherical waves

(*ix*) Theory of plane waves

Bessel's functions are also known as cylindrical and spherical function.

29.2 SOLUTION OF BESSEL'S EQUATION

$$x^2\frac{d^2y}{dx^2}+x\frac{dy}{dx}+(x^2-n^2)y=0. \qquad \text{... (1)}$$

Let $$y=\sum_{r=0}^{\infty}a_r x^{m+r} \quad \text{or} \quad y=a_0x^m+a_1x^{m+1}+a_2x^{m+2}+\ldots \qquad \text{... (2)}$$

So that $$\frac{dy}{dx}=\sum_{r=0}^{\infty}a_r\,(m+r)x^{m+r-1}$$

and $$\frac{d^2y}{dx^2}=\sum_{r=0}^{\infty}a_r(m+r)(m+r-1)\,x^{m+r-2}$$

Substituting these values in (1), we get

$$x^2\sum_{r=0}^{\infty}a_r\,(m+r)(m+r-1)x^{m+r-2}+x\sum_{r=0}^{\infty}a_r\,(m+r)x^{m+r-1}+(x^2-n^2)\sum_{r=0}^{\infty}a_rx^{m+r}=0$$

$$\Rightarrow\sum_{r=0}^{\infty}a_r\,(m+r)(m+r-1)x^{m+r}+\sum_{r=0}^{\infty}a_r\,(m+r)x^{m+r}+\sum_{r=0}^{\infty}a_rx^{m+r+2}-n^2\sum_{r=0}^{\infty}a_rx^{m+r}=0$$

$$\Rightarrow\sum_{r=0}^{\infty}a_r\,[(m+r)(m+r-1)+(m+r)-n^2]x^{m+r}+\sum_{r=0}^{\infty}a_rx^{m+r+2}=0$$

$$\Rightarrow\sum_{r=0}^{\infty}a_r[(m+r)^2-n^2]x^{m+r}+\sum_{r=0}^{\infty}a_rx^{m+r+2}=0. \qquad \text{... (3)}$$

Equating the coefficient of lowest degree term of x^m in the identity (3) to zero, by putting $r = 0$ in the first summation we get the indicial equation.

$$a_0[(m+0)^2-n^2]=0. \qquad (r=0)$$

$$\Rightarrow \quad m^2 = n^2 \text{ i.e. } m = n,\ m = -n \qquad a_0 \neq 0$$

Equating the coefficient of the next lowest degree term x^{m+1} in the identity (3), we put $r = 1$ in the first summation

$$a_1\,[(m+1)^2-n^2]=0 \text{ i.e. } a_1=0, \text{ since } (m+1)^2-n^2\neq 0$$

Equating the coefficient of x^{m+r+2} in (3) to zero, to find relation in successive coefficients, we get

$$a_{r+2}[(m+r+2)^2-n^2]+a_r=0$$

$$\Rightarrow \quad a_{r+2}=-\frac{1}{(m+r+2)^2-n^2}.a_r$$

Therefore, $a_3 = a_5 = a_7 = ... = 0$, since $a_1 = 0$

If $r = 0$, $\quad a_2=-\dfrac{1}{(m+2)^2-n^2}a_0$

If $r = 2$, $\quad a_4=-\dfrac{1}{(m+4)^2-n^2}a_2=\dfrac{1}{[(m+2)^2-n^2][(m+4)^2-n^2]}a_0$ and so on.

On substituting the values of the coefficients a_1, a_2, a_3, a_4 .... in (2), we have

$$y=a_0x^m-\frac{a_0}{(m+2)^2-n^2}x^{m+2}+\frac{a_0}{[(m+2)^2-n^2][(m+4)^2-n^2]}x^{m+4}+...$$

$$y=a_0x^m\left[1-\frac{1}{(m+2)^2-n^2}x^2+\frac{1}{[(m+2)^2-n^2][(m+4)^2-n^2]}x^4-...\right]$$

For $m = n$

$$y=a_0x^n\left[1-\frac{1}{4(n+1)}x^2+\frac{1}{4^2.2!(n+1)(n+2)}x^4-...\right]$$

where a_0 is an arbitrary constant.

For $m = -n$

$$y=a_0x^{-n}\left[1-\frac{1}{4(-n+1)}x^2+\frac{1}{4^2.2!(-n+1)(-n+2)}x^4-...\right]$$

29.3 BESSEL'S FUNCTIONS, J_n (X)

The Bessel's equation is $x^2\frac{d^2y}{dx^2}+x\frac{dy}{dx}+(x^2-n^2)y=0$ ...(1)

Solution of (1) is

$$y=a_0x^n\left[1-\frac{x^2}{2.2(n+1)}+\frac{x^4}{2.4.2^2(n+1)(n+2)}-...+(-1)^r\frac{x^{2r}}{(2^r r!).2^r(n+1)(n+2)...(n+r)}+...\right]$$

$$=a_0x^n\sum_{r=0}^{\infty}(-1)^r\frac{x^{2r}}{2^{2r}.r!(n+1)(n+2)...(n+r)}$$

where a_0 is an arbitrary constant.

If $$a_0=\frac{1}{2^n\,\overline{|(n+1)}}$$

The above solution is called Bessel's, function denoted by J_n (x).

Thus $$J_n(x)=\frac{1}{2^n\,\overline{|(n+1)}}\Sigma(-1)^r\frac{x^{n+2r}}{2^{2r}.r!(n+1)(n+2)...(n+r)}\qquad(\overline{|n+1}=n!)$$

$$\Rightarrow\quad J_n(x)=\left(\frac{x}{2}\right)^n\left\{\frac{1}{\overline{|(n+1)}}-\frac{1}{1!\,\overline{|(n+2)}}\left(\frac{x}{2}\right)^2+\frac{1}{2!\,\overline{|(n+3)}}\left(\frac{x}{2}\right)^4-\frac{1}{3!\,\overline{|(n+4)}}\left(\frac{x}{2}\right)^6+...\right\}$$

$$\Rightarrow\quad J_n(x)=\frac{x^n}{2^n\,\overline{|n+1}}\left[1-\frac{x^2}{2.(2n+2)}+\frac{x^4}{2.4.(2n+2)(2n+4)}+\cdots\right]\quad ...(2)$$

$$\Rightarrow\quad J_n(x)=\sum_{r=0}^{\infty}\frac{(-1)^r}{r!\,\overline{|(n+r+1)}}\left(\frac{x}{2}\right)^{n+2r}\quad\Rightarrow\quad J_n(x)=\sum_{r=0}^{\infty}\frac{(-1)^r}{r!(n+r)!}\left(\frac{x}{2}\right)^{n+2r}$$

If $n=0$, $J_0(x)=\Sigma\frac{(-1)^r}{(r!)^2}\left(\frac{x}{2}\right)^{2r}\quad\Rightarrow\quad J_0(x)=1-\frac{x^2}{2^2}+\frac{x^4}{2^2.4^2}-\frac{x^6}{2^2.4^2.6^2}+...$

If $n=1$, $J_1(x)=\frac{x}{2}-\frac{x^3}{2^2.4}+\frac{x^5}{2^2.4^2.6}-...$

We draw the graph of these two functions. Both the functions are oscillatory with a varying period and a decreasing amplitude.

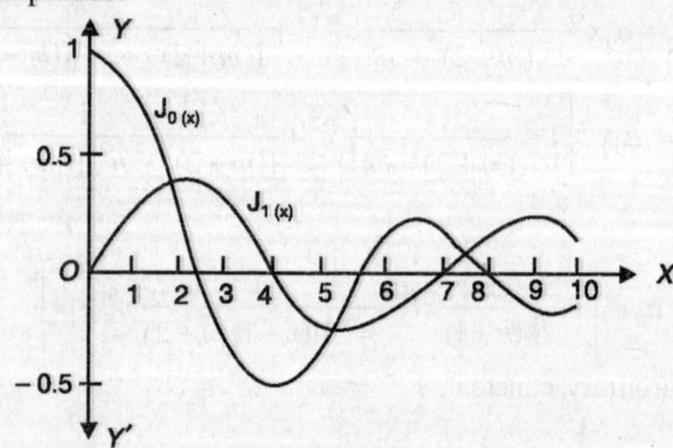

Replacing n by $-n$ in (2), we get $J_{-n}(x)=\sum_{r=0}^{\infty}\frac{(-1)^r}{r!\,\overline{|(-n+r+1)}}\left(\frac{x}{2}\right)^{-n+2r}$

Case I. If n is not integer or zero, then the complete solution of (1) is

$$y = AJ_n(x) + BJ_{-n}(x)$$

Case II. If $n = 0$, then $y_1 = y_2$ and complete solution of (1) is the Bessel's function of order zero.

Case III. If n is positive integer, then y_2 is not the solution of (1). And y_1 fails to give a solution for negative values of n. Let us find out the general solution when n is an integer.

Example 1. *Show that Bessel's function $J_n(x)$ is an even function when n is even and is odd function when n is odd.* (*U.P., II Semester, 2009*)

Solution. We know that

$$J_n(x) = \sum_{r=0}^{\infty} \frac{(-1)^r}{r!\,\overline{|n+r+1}} \left(\frac{x}{2}\right)^{n+2r} \quad \text{...(1)}$$

Replacing x by $-x$ in (1), we get

$$J_n(-x) = \sum_{r=0}^{\infty} \frac{(-1)^r}{r!\,\overline{|n+r+1}} \left(\frac{-x}{2}\right)^{n+2r} \quad \text{...(2)}$$

Case I. If n is even, then $n + 2r$ is even $\Rightarrow \left(\frac{-x}{2}\right)^{n+2r} = \left(\frac{x}{2}\right)^{n+2r}$

Thus (2), becomes

$$J_n(-x) = \sum_{r=0}^{\infty} \frac{(-1)^r}{r!\,\overline{|n+r+1}} \left(\frac{x}{2}\right)^{n+2r}$$

$$= J_n(x) \qquad \left[\begin{array}{c}\text{For even function} \\ f(-x) = f(x)\end{array}\right]$$

Hence, $J_n(x)$ is even function.

Case II. If n is odd, then $n + 2r$ is odd $\Rightarrow \left(\frac{-x}{2}\right)^{n+2r} = -\left(\frac{x}{2}\right)^{n+2r}$

Thus (2), becomes

$$J_n(-x) = -\sum_{r=0}^{\infty} \frac{(-1)^r}{r!\,\overline{|n+r+1}} \left(\frac{x}{2}\right)^{n+2r}$$

$$= -J_n(x) \qquad \left[\begin{array}{c}\text{For odd function} \\ f(-x) = -f(x)\end{array}\right]$$

Hence, $J_n(x)$ is odd function. **Proved.**

Example 2. *Prove that:*

$$\lim_{x \to 0} \frac{J_n(x)}{x_n} = \frac{1}{2^n \overline{|n+1}}; (n > -1).$$

Solution. From the equation (2) of Article 29.3 on page 798, we know that

$$J_n(x) = \frac{x^n}{2^n \overline{|n+1}} \left[1 - \frac{x^2}{2.(2n+2)} + \frac{x^4}{2.4.(2n+2)(2n+4)} - \cdots\right]$$

On taking limit on both sides when $x \to 0$, we get

$$\lim_{x \to 0} \frac{J_n(x)}{x^n} = \lim_{x \to 0} \frac{1}{2^n \overline{|n+1}} \left[1 - \frac{x^2}{2.(2n+2)} + \frac{x^4}{2.4.(2n+2)(2n+4)} - \cdots\right]$$

$$= \frac{1}{2^n \overline{|n+1}}$$

Proved.

29.4. BESSEL'S FUNCTION OF THE SECOND KIND OF ORDER *n*.

$$x^2\frac{d^2y}{dx^2}+x\frac{dy}{dx}+(x^2-n^2)y=0 \qquad ...(1)$$

Let $y = u(x)J_n(x)$ be the second solution of the Bessel's equation when n is integer.

$$\frac{dy}{dx}=u'J_n+uJ'_n$$

$$\frac{d^2y}{dx^2}=u''J_n+2u'J'_n+uJ_n''$$

Substituting these values of y, y', y'' in (1), we get

$$x^2(u''J_n+2u'J_n'+uJ_n'')+x(u'J_n+uJ_n')+(x^2-n^2)uJ_n=0$$

$$\Rightarrow \quad u[x^2J''_n+xJ'_n+(x^2-n^2)J_n]+x^2u''J_n+2x^2u'J'_n+xu'J_n=0 \qquad ...(2)$$

$$\Rightarrow \quad x^2J''_n+xJ'_n+(x^2-n^2)J_n=0 \quad [\text{Since } J_n \text{ is a solution of (1)}]$$

(2) becomes $x^2u''J_n+2x^2u'J_n'+xu'J_n=0$ ...(3)

Dividing (3) by $x^2u'J_n$, we have

$$\frac{u''}{u'}+2\frac{J'_n}{J_n}+\frac{1}{x}=0 \qquad ...(4)$$

(4) Can also be written as

$$\frac{d}{dx}[\log u']+2\frac{d}{dx}[\log J_n]+\frac{d}{dx}(\log x)=0$$

$$\Rightarrow \quad \frac{d}{dx}[\log u'+2\log J_n+\log x]=0$$

$$\Rightarrow \quad \frac{d}{dx}[\log(u'.J_n^2.x)=0 \qquad ...(5)$$

Integrating (5), we get

$$\log u'.J_n^2.x=\log C_1$$

$$\Rightarrow \quad u'.J_n^2.x=C_1 \quad \Rightarrow \quad u'=\frac{C_1}{J_n^2.x} \qquad ...(6)$$

On integrating (6), we obtain

$$u=\int\frac{C_1}{J_n^2.x}dx+C_2$$

Putting the value of u in the assumed solution $y = u(x).J_n(x)$, we get

$$y=\left[\int\frac{C_1dx}{J_n^2(x).x}+C_2\right]J_n(x)$$

$$y=C_2J_n(x)+C_1J_n(x)\int\frac{dx}{xJ_n^2(x)} \quad \Rightarrow \quad y=C_2J_n(x)+C_1y_n(x)$$

where, $y_n(x)=J_n(x)\int\frac{dx}{xJ_n^2(x)}$

The function $y_n(x)$ is known as Bessel's function of second kind of order n. It is also called **Neumann function**.

When n is not an integer.

$$y_n(x) = \frac{1}{\sin n\pi}[J_n(x)\cos n\pi - J_{-n}(x)]$$

When n is an integer

$$y_n(x) = \lim_{m\to n}\left[\frac{1}{\sin m\pi}\{J_m(x)\cos mx - J_{-m}(x)\}\right]$$

General solution of Bessel's Equation is

$$y = AJ_n(x) + BJ_{-n}(x)$$

Example 3. *Prove that,* $J_{-n}(x) = (-1)^n J_n(x)$ *where n is a positive integer.*

Solution. $J_{-n}(x) = \sum_{r=0}^{\infty}(-1)^r \frac{1}{r!\,\overline{|(r-n+1)}}\left(\frac{x}{2}\right)^{-n+2r}$

$$\Rightarrow \quad = \sum_{r=0}^{r=n-1}\frac{(-1)^r\left(\frac{x}{2}\right)^{-n+2r}}{r!\,\overline{|(-n+r+1)}} + \sum_{r=n}^{\infty}\frac{(-1)^r\left(\frac{x}{2}\right)^{-n+2r}}{r!\,\overline{|(-n+r+1)}}$$

$$\Rightarrow \quad = 0 + \sum_{r=n}^{\infty}\frac{(-1)^r\left(\frac{x}{2}\right)^{-n+2r}}{r!\,\overline{|(-n+r+1)}} \qquad \text{(since } \overline{|-\text{ve integer}} = \infty)$$

On putting $r = n + k$, we get

$$J_{-n}(x) = \sum_{k=0}^{\infty}\frac{(-1)^{n+k}\left(\frac{x}{2}\right)^{n+2k}}{(n+k)!\,\overline{|(k+1)}} = (-1)^n\sum_{k=0}^{\infty}\frac{(-1)^k\left(\frac{x}{2}\right)^{n+2k}}{(n+k)!\,k!}$$

$$= (-1)^n J_n(x)$$

Proved.

Example 4. *Prove that*

(*a*) $J_{1/2}(x) = \sqrt{\left(\frac{2}{\pi x}\right)}\sin x$ (*AMIETE, June 2010, U.P., II Semester, 2009*)

(*b*) $J_{-1/2}(x) = \sqrt{\left(\frac{2}{\pi x}\right)}\cos x$ (*AMIETE, June 2009*)

Solution. We know that,

$$J_n(x) = \frac{x^n}{2^n\,\overline{|n+1}}\left[1 - \frac{x^2}{2.2(n+1)} + \frac{x^4}{2.4.2^2(n+1)(n+2)} - \dots\dots\right] \quad \dots (1)$$

(*a*) Substituting $n = \frac{1}{2}$ in (1), we obtain

$$J_{1/2}(x) = \frac{x^{1/2}}{2^{1/2}\,\overline{|\frac{1}{2}+1}}\left[1 - \frac{x^2}{2.2.\left(\frac{1}{2}+1\right)} + \frac{x^4}{2.4.2^2\left(\frac{1}{2}+1\right)\left(\frac{1}{2}+2\right)} - \dots\dots\right]$$

$$= \frac{\sqrt{x}}{\sqrt{2}\,\overline{|3/2}}\left[1 - \frac{x^2}{2.3} + \frac{x^4}{2.3.4.5}\dots\right] = \frac{\sqrt{x}}{\sqrt{2}.\frac{1}{2}\overline{|\frac{1}{2}}}\ \frac{1}{x}\left[x - \frac{x^3}{3!} + \frac{x^5}{5!} - \dots\right]$$

$$\left(\text{since } \sqrt{\tfrac{1}{2}} = \sqrt{\pi}\right)$$

$$= \frac{1}{\sqrt{2x}.\ \frac{1}{2}\sqrt{\pi}} \sin x = \sqrt{\left(\frac{2}{\pi x}\right)} \sin x$$ **Proved.**

(*b*) Again substituting $n = -\frac{1}{2}$ in (1), we have

$$J_{-1/2}(x) = \frac{x^{-1/2}}{2^{-1/2}\sqrt{-\frac{1}{2}+1}}\left[1 - \frac{x^2}{2.2.\left(-\frac{1}{2}+1\right)} + \frac{x^4}{2.4.2^2\left(-\frac{1}{2}+1\right)\left(-\frac{1}{2}+2\right)} - ...\right]$$

$$= \frac{\sqrt{2}}{\sqrt{x}\sqrt{\frac{1}{2}}}\left[1 - \frac{x^2}{2!} + \frac{x^4}{4!} - ...\right] = \sqrt{\left(\frac{2}{\pi x}\right)} \cos x \quad \left(\text{since } \sqrt{\tfrac{1}{2}} = \sqrt{\pi}\right)$$ **Proved.**

Example 5. *Show that*

$$\sqrt{\left(\frac{1}{2}\pi x\right)}\ J_{\frac{3}{2}}(x) = \frac{\sin x}{x} - \cos x$$

Solution. We know that

$$J_n(x) = \frac{x^n}{2^n \sqrt{(n+1)}}\left[1 - \frac{x^2}{2.2.(n+1)} + \frac{x^4}{2.4.2^2(n+1)(n+2)} - ...\right] \quad ...(1)$$

Putting $n = \frac{3}{2}$ in (1), we get

$$J_{\frac{3}{2}}(x) = \frac{x^{3/2}}{2^{3/2}\sqrt{\frac{3}{2}+1}}\left[1 - \frac{x^2}{2.2.\left(\frac{3}{2}+1\right)} + \frac{x^4}{2.4.2^2.\left(\frac{3}{2}+1\right)\left(\frac{3}{2}+2\right)} - ...\right]$$

$$= \frac{x^{\frac{1}{2}}}{2\sqrt{2}.\frac{3}{2}.\frac{1}{2}\sqrt{\pi}}\left[x^2 - \frac{x^4}{2.5} + \frac{x^6}{2.4.5.7} - ...\right]$$

$$= \sqrt{\left(\frac{2}{\pi x}\right)}\left[\frac{x^2}{2!} - \frac{x^2}{3!} + \frac{x^4}{5!} - \frac{x^4}{4!} + \frac{x^6}{6!} - \frac{x^6}{7!} + ...\right]$$

$$= \sqrt{\frac{2}{\pi x}}\left[\frac{1}{x}\left\{x - \frac{x^3}{3!} + \frac{x^5}{5!} - ...\right\} - \left\{1 - \frac{x^2}{2!} + \frac{x^4}{4!} - ....\right\}\right] = \sqrt{\frac{2}{\pi x}}\left[\frac{\sin x}{x} - \cos x\right]$$

$$\Rightarrow \quad \sqrt{\frac{\pi x}{2}}\ J_{\frac{3}{2}}(x) = \frac{\sin x}{x} - \cos x$$ **Proved.**

Example 6. *Show that* $\sqrt{\left(\frac{1}{2}\pi x\right)}\ J_{-\frac{3}{2}}(x) = -\sin x - \frac{\cos x}{x}$

Solution. We know that

$$J_n(x) = \frac{x^n}{2^n \sqrt{n+1}}\left[1 - \frac{x^2}{2.2.(n+1)} + \frac{x^4}{2.4.2^2(n+1)(n+2)} - ...\right]$$

Multiplying numerator and denominator by $(n + 1)$, we get

$$J_n(x) = \frac{x^n (n+1)}{2^n \sqrt{n+2}} \left[1 - \frac{x^2}{2.2.(n+1)} + \frac{x^4}{2.4.2^2(n+1)(n+2)} - ...\right]$$

Putting $n = -\frac{3}{2}$, we get

$$J_{\frac{3}{2}}(x) = \frac{x^{-3/2}\left(-\frac{3}{2}+1\right)}{2^{-3/2}\sqrt{\frac{-3}{2}+2}}\left[1+\frac{x^2}{2}-\frac{x^4}{2.4}-...\right] = \frac{x^{-\frac{3}{2}}\left(-\frac{1}{2}\right)}{2^{-3/2}\sqrt{\frac{1}{2}}}\left[1+\frac{x^2}{2}-\frac{x^4}{2.4}...\right]$$

$$= -\sqrt{\left(\frac{2}{\pi x}\right)}\frac{1}{x}\left[1+\frac{x^2}{2}-\frac{x^4}{2.4}...\right] = \sqrt{\left(\frac{2}{\pi x}\right)}\left[-\frac{1}{x}\left\{1+x^2\left(1-\frac{1}{2}\right)-x^4\left(\frac{1}{6}-\frac{1}{2.4}\right)...\right\}\right]$$

$$= \sqrt{\left(\frac{2}{\pi x}\right)}\left[-\frac{1}{x}\left(1-\frac{x^2}{2!}+\frac{x^4}{4!}-...\right)-\left(x-\frac{x^3}{3!}+...\right)\right] = \sqrt{\left(\frac{2}{\pi x}\right)}\left[-\frac{1}{x}\cos x - \sin x\right]$$

Hence, $\sqrt{\left(\frac{\pi x}{2}\right)}\ J_{-\frac{3}{2}}(x) = -\frac{1}{x}\cos x - \sin x$ **Proved.**

29.5 RECURRENCE FORMULAE

These formulae are very useful in solving the questions. So, they are to be committed to memory.

1.	$x J_n' = n J_n - x J_{n+1}$
2.	$x J_n' = -n J_n + x J_{n-1}$
3.	$2J_n' = J_{n-1} - J_{n+1}$
4.	$2nJ_n = x(J_{n-1} - J_{n+1})$
5.	$\frac{d}{dx}(x^{-n} J_n) = -x^{-n} J_{n+1}$
6.	$\frac{d}{dx}(x^n J_n) = x^n J_{n-1}$

(*AMIETE, June 2010, DU, 2014*) [beside 4]

(*AMIETE, June 2010*) [beside 6]

Formula I. $xJ_n' = nJ_{n-1} - xJ_{n+1}$ (*DU, II Semester 2014*)

Proof. We know that

$$J_n = \sum_{r=0}^{\infty} \frac{(-1)^r}{r!\ \overline{|(n+r+1)}} \left(\frac{x}{2}\right)^{n+2r}$$

Differentiating with respect to x, we get

$$J_n' = \sum \frac{(-1)^r (n+2r)}{r!\ \overline{|(n+r+1)}} \left(\frac{x}{2}\right)^{n+2r-1} \frac{1}{2}$$

$$\Rightarrow \quad xJ_n' = n\sum \frac{(-1)^r}{r!\ \overline{|(n+r+1)}}\left(\frac{x}{2}\right)^{n+2r} + x\sum \frac{(-1)^r .2r}{2.r!\ \overline{|(n+r+1)}}\left(\frac{x}{2}\right)^{n+2r-1}$$

$$= nJ_n + x\sum_{r=1}^{\infty}\frac{(-1)^r}{(r-1)!\,\lceil(n+r+1)}\left(\frac{x}{2}\right)^{n+2r-1}$$

$$= nJ_n + x\sum_{s=0}^{\infty}\frac{(-1)^{s+1}}{s!\,\lceil(n+s+2)}\left(\frac{x}{2}\right)^{n+2s+1} \qquad \text{[Putting } r-1=s\text{]}$$

$$= nJ_n - x\sum_{s=0}^{\infty}\frac{(-1)^{s}}{s!\,\lceil[(n+1)+s+1]}\left(\frac{x}{2}\right)^{(n+1)+2s}$$

$$\boxed{xJ_n' = nJ_n - xJ_{n+1}}$$ **Proved.**

Formula II. $xJ_n' = -nJ_n + xJ_{n-1}$ (*U.P., II Semester, summer 2006*)

Proof. We know that $J_n = \sum_{r=0}^{\infty}\frac{(-1)^r}{r!\,\lceil(n+r+1)}\left(\frac{x}{2}\right)^{n+2r}$

Differentiating w.r.t. 'x', we get $J_n' = \sum_{r=0}^{\infty}\frac{(-1)^r(n+2r)}{r!\,\lceil(n+r+1)}\left(\frac{x}{2}\right)^{n+2r-1}\frac{1}{2}$

$$xJ_n' = \sum_{r=0}^{\infty}\frac{(-1)^r(n+2r)}{r!\,\lceil(n+r+1)}\left(\frac{x}{2}\right)^{n+2r} = \sum_{r=0}^{\infty}\frac{(-1)^r[(2n+2r)-n]}{r!\,\lceil(n+r+1)}\left(\frac{x}{2}\right)^{n+2r}$$

$$= \sum_{r=0}^{\infty}\frac{(-1)^r(2n+2r)}{r!\,\lceil(n+r+1)}\left(\frac{x}{2}\right)^{n+2r} - n\sum_{r=0}^{\infty}\frac{(-1)^r}{r!\,\lceil(n+r+1)}\left(\frac{x}{2}\right)^{n+2r}$$

$$= \sum_{r=0}^{\infty}\frac{(-1)^r 2}{r!\,\lceil(n+r)}\left(\frac{x}{2}\right)^{n+2r} - nJ_n = x\sum_{r=0}^{\infty}\frac{(-1)^r}{r!\,\lceil[(n-1)+r+1]}\left(\frac{x}{2}\right)^{(n-1+2r)} - nJ_n$$

$\Rightarrow$ $$\boxed{\boldsymbol{xJ_n' = xJ_{n-1} - nJ_n}}$$ **Proved.**

Formula III. $\boldsymbol{2J'_n = J_{n-1} - J_{n+1}}$

Proof. We know that $xJ_n' = nJ_n - xJ_{n+1}$ ... (1) (Recurrence formula *I*)

$xJ_n' = -nJ_n + xJ_{n-1}$ ... (2) (Recurrence formula *II*)

Adding (1) and (2), we get

$$2xJ_n' = -xJ_{n+1} + xJ_{n-1} \qquad \Rightarrow \qquad \boxed{\boldsymbol{2J'_n = J_{n-1} - J_{n+1}}}$$ **Proved.**

Formula IV. $\boldsymbol{2nJ_n = x(J_{n-1} + J_{n+1})}$ (*U.P. II Semester, June 2007*)

Proof. We know that

$xJ_n' = nJ_n - xJ_{n+1}$ ...(1) (Recurrence formula *I*)

$xJ_n' = -nJ_n + xJ_{n-1}$ ...(2) (Recurrence formula *II*)

Subtracting (2) from (1), we get

$$0 = 2nJ_n - xJ_{n+1} - xJ_{n-1} \Rightarrow \boxed{\boldsymbol{2nJ_n = x(J_{n-1} + J_{n+1})}} \; ...(3)$$ **Proved.**

The following examples are solved by using Recurrence formula IV.

Example 7. *Find the value of* $J_{-1}(x) + J_1(x)$. (*Delhi University, April 2010*)

Solution. By using Recurrence relation IV for $J_n(x)$ is

$$2nJ_n = x(J_{n-1} + J_{n+1})$$

$$J_{n-1}(x) + J_{n+1}(x) = \frac{2n}{x} J_n(x)$$

Put $n = 0$

$$J_{-1}(x) + J_1(x) = 0$$

Ans.

Example 8. *Prove that*

$$J_{5/2}(x) = \sqrt{\frac{2}{\pi x}}\left[\left(\frac{3 - x^2}{x^2}\right)\sin x - \frac{3\cos x}{x}\right]$$

(AMIETE, June 2010)

Solution. From Recurrence relation (IV), we have

$2n\, J_n(x) = x\,[J_{n-1}(x) + J_{n+1}(x)]$

$$J_{n+1}(x) = \frac{2n}{x} J_n(x) - J_{n-1}(x) \quad ...(1)$$

Putting $n = 1/2$ in (1), we get $J_{3/2}(x) = \frac{1}{x} J_{1/2}(x) - J_{-1/2}(x)$ ...(2)

Again putting $n = \frac{3}{2}$ in (1), we get $J_{5/2}(x) = \frac{3}{2} J_{3/2}(x) - J_{1/2}(x)$ ...(3)

Putting the value of $J_{\frac{3}{2}}(x)$ from (2) in (3), we get

$$J_{5/2}(x) = \frac{3}{x}\left[\frac{1}{x} J_{1/2}(x) - J_{-1/2}(x)\right] - J_{1/2}(x) \quad ...(4)$$

$$= \left(\frac{3}{x^2} - 1\right) J_{1/2}(x) - \frac{3}{x} J_{-1/2}(x)$$

Putting the values of $J_{\frac{1}{2}}(x)$ and $J_{-\frac{1}{2}}(x)$ in (4), we get

$$J_{5/2}(x) = \left(\frac{3 - x^2}{x^2}\right)\sqrt{\frac{2}{\pi x}}\sin x - \frac{3}{x}\cdot\sqrt{\frac{2}{\pi x}}\cos x = \sqrt{\frac{2}{\pi x}}\left[\left(\frac{3 - x^2}{x^2}\right)\sin x - \frac{3}{x}\cos x\right]$$ **Proved.**

Example 9. *Express $J_6(x)$ in terms of $J_0(x)$ and $J_1(x)$.* *(U.P., II Semester, 2009)*

Solution. We know that

$$J_{n+1} = \frac{2n}{x} J_n - J_{n-1} \quad ...(1) \left[\text{From Recurrence relation (IV)}\right]$$

Putting $n = 5$ in (1), we get

$$J_6 = \frac{10}{x} J_5 - J_4 \quad ...(2)$$

Putting $n = 4$ in (1), we get

$$J_5 = \frac{8}{x} J_4 - J_3$$

Putting the value of J_5 in (2), we get

$$J_6 = \frac{10}{x}\left[\frac{8}{x} J_4 - J_3\right] - J_4 = \left[\left(\frac{10}{x}\right)\left(\frac{8}{x}\right) - 1\right] J_4 - \frac{10}{x} J_3 \quad ...(3)$$

On putting the value of J_4 in (3), we get

$$J_6 = \left[\left(\frac{10}{x}\right)\left(\frac{8}{x}\right) - 1\right]\left(\frac{6}{x} J_3 - J_2\right) - \frac{10}{x} J_3, \left[J_4 = \frac{6}{x} J_3 - J_2\right]$$

$$= \left[\left(\frac{10}{x}\right)\left(\frac{8}{x}\right)\left(\frac{6}{x}\right)-\frac{16}{x}\right]J_3-\left[\left(\frac{10}{x}\right)\left(\frac{8}{x}\right)-1\right]J_2 \quad ...(4)$$

$$\left[\begin{array}{l}\text{If } n=2 \text{ in (1), then}\\ J_3=\frac{4}{x}J_2-J_1\end{array}\right]$$

On putting the values of J_3 in (4), we get

$$J_6 = \left[\left(\frac{10}{x}\right)\left(\frac{8}{x}\right)\left(\frac{6}{x}\right)-\frac{16}{x}\right]\left[\frac{4}{x}J_2-J_1\right]-\left[\left(\frac{10}{x}\right)\left(\frac{8}{x}\right)-1\right]J_2$$

$$= \left[\left(\frac{10}{x}\right)\left(\frac{8}{x}\right)\left(\frac{6}{x}\right)\left(\frac{4}{x}\right)-\left(\frac{16}{x}\right)\left(\frac{4}{x}\right)-\left(\frac{10}{x}\right)\left(\frac{8}{x}\right)+1\right]J_2$$

$$-\left[\left(\frac{10}{x}\right)\left(\frac{8}{x}\right)\left(\frac{6}{x}\right)-\frac{16}{x}\right]J_1 \quad ...(5) \quad \left[\begin{array}{l}\text{If } n=1 \text{ in (1) then}\\ J_2=\frac{2}{x}J_1-J_0\end{array}\right]$$

On putting the value of J_2 in (5), we get

$$J_6 = \left[\left(\frac{10}{x}\right)\left(\frac{8}{x}\right)\left(\frac{6}{x}\right)\left(\frac{4}{x}\right)-\left(\frac{16}{x}\right)\left(\frac{4}{x}\right)-\left(\frac{10}{x}\right)\left(\frac{8}{x}\right)+1\right]\left[\frac{2}{x}J_1-J_0\right]$$

$$-\left[\left(\frac{10}{x}\right)\left(\frac{8}{x}\right)\left(\frac{6}{x}\right)-\frac{16}{x}\right]J_1$$

$$\Rightarrow \quad J_6 = \left[\left(\frac{10}{x}\right)\left(\frac{8}{x}\right)\left(\frac{6}{x}\right)\left(\frac{4}{x}\right)\left(\frac{2}{x}\right)-\left(\frac{16}{x}\right)\left(\frac{4}{x}\right)\left(\frac{2}{x}\right)-\left(\frac{10}{x}\right)\left(\frac{8}{x}\right)\left(\frac{2}{x}\right)+\frac{2}{x}\right.$$

$$\left.-\left(\frac{10}{x}\right)\left(\frac{8}{x}\right)\left(\frac{6}{x}\right)+\frac{16}{x}\right]J_1$$

$$-\left[\left(\frac{10}{x}\right)\left(\frac{8}{x}\right)\left(\frac{6}{x}\right)\left(\frac{4}{x}\right)-\left(\frac{16}{x}\right)\left(\frac{4}{x}\right)-\left(\frac{10}{x}\right)\left(\frac{8}{x}\right)+1\right]J_0$$

$$\Rightarrow \quad J_6 = \left[\frac{3840}{x^5}-\frac{128}{x^3}-\frac{160}{x^3}+\frac{2}{x}-\frac{480}{x^3}+\frac{16}{x}\right]J_1$$

$$-\left[\frac{1920}{x^4}-\frac{64}{x^2}-\frac{80}{x^2}+1\right]J_0$$

$$= \left[\frac{3840}{x^5}-\frac{768}{x^3}+\frac{18}{x}\right]J_1-\left[\frac{1920}{x^4}-\frac{144}{x^2}+1\right]J_0 \qquad \textbf{Ans.}$$

Example 10. *Express J_5 in terms of J_1 and J_2.* *(AMIETE, Dec. 2009)*

Solution. Putting $n = 4$ in recurrence formula (IV), we get

$$n = 4,\ 8J_4 = x(J_3 + J_5) \quad \Rightarrow \quad J_5 = \frac{8}{x}J_4 - J_3$$

$$n = 3 \quad J_4 = \frac{6}{x}J_3 - J_2, \qquad J_5 = \frac{8}{x}\left(\frac{6}{x}J_3 - J_2\right) - J_3 = \frac{48}{x^2}J_3 - \frac{8}{x}J_2 - J_3$$

$$J_5 = \frac{48}{x^2}\left(\frac{4}{x}J_2 - J_1\right) - \frac{8}{x}J_2 - \left(\frac{4}{x}J_2 - J_1\right)\left(J_3 = \frac{4}{x}J_2' - J_1\right)$$

$$J_5 = \frac{192}{x^3} J_2 - \frac{48}{x^2} J_1 - \frac{8}{x} J_2 - \frac{4}{x} J_2 + J_1$$

$$\Rightarrow \qquad = \left(\frac{192}{x^3} - \frac{12}{x}\right) J_2 + \left(1 - \frac{48}{x^2}\right) J_1$$ **Ans.**

Formula V. $\frac{d}{dx}(x^{-n} \cdot J_n) = -x^{-n} J_{n+1}$

Proof. We know that $x J_n' = n J_n - x J_{n+1}$ (Recurrence formula *I*)

Multiplying by x^{-n-1}, we obtain $x^{-n} J_n' = n x^{-n-1} J_n - x^{-n} J_{n+1}$

i.e., $x^{-n} J_n' - n x^{-n-1} J_n = -x^{-n} J_{n+1}$

$$\Rightarrow \quad \boxed{\frac{d}{dx}(x^{-n} J_n) = -x^{-n} J_{n+1}}$$ **Proved.**

Example 11. *If n > –1, show that:*

$$\int_0^x x^{-n} J_{n+1}(x)\, dx = \frac{1}{2^n \lfloor n+1} - x^{-n} J_n(x)$$

Solution. From relation (5), we know that

$$\frac{d}{dx}[x^{-n} J_n(x)] = -x^{-n} J_{n+1}(x)$$

Integrating it between 0 and x, we get

$$\int_0^x x^{-n} J_{n+1}(x)\, dx = -\left[x^{-n} J_n(x)\right]_0^x$$

$$= -x^{-n} J_n(x) + \underset{x \to 0}{Lt} \left[\frac{J_n(x)}{x^n}\right]$$

$$= -x^{-n} J_n(x) + \frac{1}{2^n \lfloor n+1}$$

$$= \frac{1}{2^n \lfloor n+1} - x^{-n} J_n(x)$$ **Proved.**

Formula VI. $\frac{d}{dx}(x^n J_n) = x^n J_{n-1}$ (*U.P., II Semester, 2005*)

Proof. We know that

$$x J_n' = -n J_n + x J_{n-1}$$ (Recurrence formula *II*)

Multiplying by x^{n-1}, we have

$$x^n J_n' = -n x^{n-1} J_n + x^n J_{n-1} \qquad i.e., \qquad x^n J_n' + n x^{n-1} J_n = x^n J_{n-1}$$

$$\Rightarrow \quad \boxed{\frac{d}{dx}(x^n J_n) = x^n J_{n-1}}$$ **Proved.**

Example 12. *Prove that* $\frac{d}{dx}[J_0(x)] = -J_1(x)$

Solution. We know that $\frac{d}{dx}\left[x^{-n} J_n(x)\right] = -x^{-n} J_{n+1}(x)$ (Recurrence relation *VI)*

On putting $n = 0$ in the formula VI, we get $\frac{d}{dx}[x^0 J_0(x)] = -x^0 J_1(x)$

$$\frac{d}{dx}[J_0(x)] = -J_1(x)$$ **Proved.**

Example 13. *Prove that*

$$\int xJ_0^2 (x)\, dx = \frac{1}{2} x^2 [J_0^2 (x) + J_1^2 (x).] + c$$

Solution. $\quad$ L.H.S $= \int xJ_0^2 (x)\, dx$

On integrating by parts, we get

$$= J_0^2 (x).\frac{x^2}{2} - \int 2J_0 (x)\, J_0{}' (x).\frac{x^2}{2}\, dx + c$$

Putting the value of $J_0{}' (x) \; = - J_1 (x)$, we get

$$= \frac{x^2}{2} J_0^2 (x) - \int x^2 J_0 (x) \{- J_1 (x)\}\, dx + c$$

$$= \frac{x^2}{2} J_0^2 (x) + \int xJ_1 (x) \,.\, xJ_0 (x)\, dx + c$$

Putting the value of $xJ_0 (x)$ from recurrence relation VI, we get

$$= \frac{x^2}{2} J_0^2 (x) - \int xJ_1 (x) \,.\frac{d}{dx}[xJ_1 (x)]\, dx + c$$

$$= \frac{x^2}{2} J_0^2 (x) + \frac{[xJ_1 (x)]^2}{2} + c \qquad \left[\because \int t dt = \frac{t^2}{2}, \text{ put } x\, J_1 (x) = t\right]$$

$$= \frac{x^2}{2}[J_0^2 (x) + J_1^2 (x)] + c$$ **Proved.**

Example 14. *Show that* $\quad (a)\; J_{n+3} + J_{n+5} = \frac{2}{x}(n+4)\, J_{n+4}$

(b) Express $J_4(x)$ in terms of $J_0(x)$ and $J_1(x)$ (*A.M.I.E.T.E., Summer, 2007, DU. 2014*)

Solution. (*a*) Recurrence relation IV is $\quad 2n J_n = x\,(J_{n-1} + J_{n+1}) \quad \Rightarrow \quad J_{n-1} + J_{n+1} = \frac{2n}{x} J_n$

Putting $n + 4$ for n, we have $\quad J_{n+3} + J_{n+5} = \frac{2}{x}(n+4) J_{n+4}$ **Proved.**

(*b*) We know that $\quad J_{n+1} = \frac{2n}{x} J_n - J_{n-1}$ [From recurrence formula IV.]

If $n = 1$, $\quad J_2 = \frac{2}{x} J_1 - J_0$,

If $n = 2$, $\quad J_3 = \frac{4}{x} J_2 - J_1$,

If $n = 3$, $\quad J_4 = \frac{6}{x} J_3 - J_2, = \frac{6}{x}\left(\frac{4}{x} J_2 - J_1\right) - J_2$

$$J_4 = \frac{24}{x^2} J_2 - \frac{6}{x} J_1 - J_2 = \frac{24}{x^2}\left(\frac{2}{x} J_1 - J_0\right) - \frac{6}{x} J_1 - \left(\frac{2}{x} J_1 - J_0\right)$$

$$= \frac{48}{x^3} J_1 - \frac{24}{x^2} J_0 - \frac{6}{x} J_1 - \frac{2}{x} J_1 + J_0 = \left(\frac{48}{x^3} - \frac{8}{x}\right) J_1 + \left(1 - \frac{24}{x^2}\right) J_0 \quad \textbf{Ans.}$$

Example 15. *Prove that* $J_2'(x) = \left(1 - \frac{4}{x^2}\right) J_1(x) + \frac{2}{x} J_0(x)$ *where* $J_n(x)$ *is the Bessel's function of first kind.* *(AMIETE, Dec. 2010, U.P., III Semester, June 2008)*

Solution. $x J_n' = -n J_n + x J_{n-1}$ (Recurrence formula *II*) ... (1)

On putting $n = 2$ in (1), we have $x J_2' = -2 J_2 + x J_1$

$$\Rightarrow \quad J_2' = -\frac{2}{x} J_2 + J_1 \quad ...(2)$$

$$x J_n' = n J_n - x J_{n+1} \quad \text{(Recurrence formula } I) \text{ ... (3)}$$

From (1) and (3), we have $-n J_n + x J_{n-1} = n J_n - x J_{n+1}$

On putting $n = 1$, $-J_1 + x J_0 = J_1 - x J_2$

$$\Rightarrow \quad -\frac{1}{x} J_1 + J_0 = \frac{1}{x} J_1 - J_2 \quad \Rightarrow \quad J_2 = \frac{2}{x} J_1 - J_0 \quad ... (4)$$

Putting the value of J_2 from (4) in (2), we get

$$J_2' = -\frac{2}{x}\left(\frac{2}{x} J_1 - J_0\right) + J_1 = -\frac{4}{x^2} J_1 + \frac{2}{x} J_0 + J_1 = \left(1 - \frac{4}{x^2}\right) J_1 + \frac{2}{x} J_0 \quad \textbf{Proved.}$$

Example 16. *Using the recurrence relation, show that*

$$4 J_n''(x) = J_{n-2}(x) - 2 J_n(x) + J_{n+2}(x).$$

Solution. $2 J_n' = J_{n-1} - J_{n+1}$ (Recurrence formula *III*) ... (1)

On differentiating (1), we have

$$2 J_n'' = J_{n-1}' - J_{n+1}' \quad ... (2)$$

Replacing n by $n - 1$ and n by $n + 1$ in (1), we have

$$2 J_{n-1}' = J_{n-2} - J_n \quad \Rightarrow \quad J_{n-1}' = \frac{1}{2} J_{n-2} - \frac{1}{2} J_n \quad ... (3)$$

and

$$2 J_{n+1}' = J_n - J_{n+2} \quad \Rightarrow \quad J_{n+1}' = \frac{1}{2} J_n - \frac{1}{2} J_{n+2} \quad ....(4)$$

Putting the values of J'_{n-1} and J'_{n-1} from (3) and (4) in (2), we get

$$2 J_n'' = \frac{1}{2}[J_{n-2} - J_n] - \frac{1}{2}[J_n - J_{n+2}]$$

$$\Rightarrow \quad 4 J_n'' = J_{n-2} - J_n - J_n + J_{n+2}$$

$$\Rightarrow \quad 4 J_n'' = J_{n-2} - 2 J_n + J_{n+2} \quad \textbf{Proved.}$$

Example 17. *Show that*

$$\frac{d}{dx}(J_n^2 + J_{n+1}^2) = 2\left(\frac{n}{x} J_n^2 - \frac{n+1}{x} J_{n+1}^2\right)$$

(U.P. II Semester Summer 2005)

Solution. $x J_n' = n J_n - x J_{n+1}$ (Recurrence formula *I*) ... (1)

and $x J_n' = -n J_n + x J_{n-1}$ (Recurrence formula *II*) ... (2)

Putting $(n + 1)$ for n in (2), we get

$$x J_{n+1}' = -(n+1) J_{n+1} + x J_n \quad ... (3)$$

Now $\frac{d}{dx}(J_n^2 + J_{n+1}^2) = 2 J_n . J_n' + 2 J_{n+1} . J_{n+1}'$

$$= 2J_n \cdot \frac{1}{x}(nJ_n - xJ_{n+1}) + 2J_{n+1} \cdot \frac{1}{x}[-(n+1)J_{n+1} + xJ_n)]$$

[From (1) & (3)]

$$= 2\left[\frac{n}{x}J_n^2 - \frac{n+1}{x}J_{n+1}^2\right]$$

Proved.

Example 18. *Prove that following relation:*

$$x^2 J_n''(x) = (n^2 - n - x^2)\, J_n(x) + x J_{n+1}(x)$$

(*U.P. II Semester Summer, 2007, 2006*)

Solution. $x^2 \frac{d^2y}{dx^2} + x\frac{dy}{dx} + (x^2 - n^2)\, y = 0$ (Bessel's equation) ...(1)

$J_n(x)$ is the solution of (1)

So $\quad x^2 J_n'' + xJ_n' + (x^2 - n^2)\, J_n = 0$ ...(2)

We know that

$$xJ_n' = nJ_n - xJ_{n+1}$$ (Recurrence relation *I*) ...(3)

Putting the value of xJ_n' from (3) in (2), we get

$$x^2 J_n'' + nJ_n - xJ_{n+1} + (x^2 - n^2)\, J_n = 0$$

$$x^2 J'' = -nJ_n + xJ_{n+1} + (n^2 - x^2)\, J_n$$

$$\Rightarrow \quad x^2 J'' = (n^2 - n - x^2)\, J_n + xJ_{n+1}$$

Proved.

Example 19. *Show that* $J_1''(x) = -J_1(x) + \frac{1}{x}J_2(x)$ (*U.P. II Semester 2010*)

Solution. From example 18, we know that:

$$x^2 J_n''(x) = (n^2 - n - x^2)\, J_n(x) + x\, J_{n+1}(x), \qquad ...(1)$$

Putting $n = 1$, we get

$$x^2 J_1''(x) = (1^2 - 1 - x^2)\, J_1(x) + x\, J_{1+1}(x)$$

$$\Rightarrow \quad x^2 J_1''(x) = -x^2 J_1(x) + xJ_2(x)$$

$$\Rightarrow \quad J_1'' = -J_1(x) + \frac{1}{x}J_2(x)$$

Proved.

Example 20. *Prove that*

$$J_3(x) + 3J_0'(x) + 4J_0'''(x) = 0$$

Solution. We know that $2J_n' = J_{n-1} - J_{n+1}$ (Recurrence relation *II*)

Differentiating and multiplying by 2, we get

$$2^2 J_n'' = 2J_{n-1}' - 2J_{n+1}' = (J_{n-2} - J_n) - (J_n - J_{n+2})$$

$$= J_{n-2} - 2J_n + J_{n+2}$$

Differentiating again and multiplying by 2, we get

$$2^3 J_n''' = 2J_{n-2}' - 4J_n' + 2J_{n+2}'$$

$$= (J_{n-3} - J_{n-1}) - 2\,(J_{n-1} - J_{n+1}) + (J_{n+1} - J_{n+3})$$

$$= J_{n-3} - 3\,J_{n-1} + 3\,J_{n+1} - J_{n+3}$$

Putting $n = 0$, we get

$$2^3 . J_0''' = J_{-3} - 3J_{-1} + 3J_1 - J_3 = (-1)^3 J_3 - 3(-1)J_1 + 3\,J_1 - J_3$$

$$= -2\,J_3 + 6\,J_1$$

$$4J_0''' = -J_3 + 3\,J_1$$

$$= -J_3 + 3(-J_0') \qquad \text{[From example 11, } J_1 = -j_0'\text{]}$$

$$J_3 + 3J_0' + 4J_0''' = 0$$

$$\Rightarrow J_3(x) + 3J_0'(x) + 4J_0'''(x) = 0$$

Proved.

Example 21. *Prove that* $\dfrac{d}{dx}(x\,J_n J_{n+1}) = x(J_n^2 - J_{n+1}^2)$

Solution. $\dfrac{d}{dx}(x\,J_n J_{n+1}) = J_n J_{n+1} + x\dfrac{d}{dx}(J_n J_{n+1})$

$$= J_n J_{n+1} + x(J_n J_{n+1}' + J_n' J_{n+1})$$

$$= J_n J_{n+1} + (xJ_n')\,J_{n+1} + J_n(x\,J_{n+1}') \qquad \text{... (1)}$$

Recurrence formula *I*, $x\,J_n' = n\,J_n - x\,J_{n+1}$ ...(2)

Recurrence formula *II*, $x\,J_n' = -n\,J_n + x\,J_{n-1}$

Putting $n + 1$ for n, $x\,J_{n+1}' = -(n+1)\,J_{n+1} + x\,J_n$ ...(3)

Putting the values of xJ_n' and xJ_{n+1}' from (2) and (3) in (1), we obtain

$$\frac{d}{dx}(x\,J_n J_{n+1}) = J_n\,J_{n+1} + (n\,J_n - x\,J_{n+1})\,J_{n+1} + J_n[-(n+1)\,J_{n+1} + x\,J_n]$$

$$= (1 + n - n - 1)\,J_n \cdot J_{n+1} + x(J_n^2 - J_{n+1}^2)$$

$$= x(J_n^2 - J_{n+1}^2)$$

Proved.

Example 22. *Prove that*

$$\int J_3(x)\,dx + J_2(x) + \frac{2}{x}J_1(x) = 0$$

Solution. We know that

$$\frac{d}{dx}[x^{-n}J_n(x)] = -x^{-n}J_{n+1}(x) \qquad \text{(Recurrence Relation } V\text{)}$$

Integrating above relation, we get

$$x^{-n}J_n(x) = -\int x^{-n}J_{n+1}(x)\,dx \qquad \text{...(1)}$$

On taking $n = 2$ in (1), we have

$$\int x^{-2}J_3(x)\,dx = -x^{-2}J_2(x) \qquad \text{... (2)}$$

Again $\displaystyle\int J_3(x)\,dx = \int x^2(x^{-2})J_3(x)dx$

$$= x^2\int (x^{-2})J_3(x)dx - \int \left(2x\int (x^{-2}J_3(x)dx\right)dx \qquad \text{...(3)}$$

Putting the value of $\int x^{-2}J_3(x)dx$ from (2) in (3), we get

$$\int J_3(x)dx = x^2(-x^{-2}J_2(x)) - \int 2x(-x^{-2}J_2(x))dx$$

$$= -J_2(x) + 2\int x^{-1}J_2(x)\,dx = -J_2(x) + 2(-x^{-1}J_1(x))$$

On using (1), again, when $n = 1$

Hence, $\displaystyle\int J_3(x)dx + J_2(x) + \frac{2}{x}J_1(x) = 0$

Proved.

Example 23. *Show that* $\int_0^x x^{n+1} J_n(x)\,dx = x^{n+1} J_{n+1}(x)$ $(n > -1)$

Solution. Recurrence relation *VI* is $\frac{d}{dx}[x^n J_n(x)] = x^n J_{n-1}(x)$

Putting $n+1$ for n, we get $\frac{d}{dx}[x^{n+1} J_{n+1}(x)] = x^{n+1} J_n(x)$

Integrating both sides w.r.t. x between 0 and x, we get

$$x^{n+1} J_{n+1}(x) = \int_0^x x^{n+1} J_n(x)\,dx .$$ **Proved.**

Example 24. *Prove that*

$$2^r \cdot J_n^r = J_{n-r} - r J_{n-r+2} + \frac{r(r-1)}{2!} J_{n-r+4} + \ldots + (-1)^r J_{n+r}$$

Solution. We know that

$$2J'_n = J_{n-1} - J_{n+1} \qquad \text{(Recurrence formula } III)$$

Differentiating, we get

$$2J''_n = J'_{n-1} - J'_{n+1} \quad \Rightarrow \quad 4J''_n = 2J'_{n-1} - 2J'_{n+1}$$

$$\Rightarrow \quad 2^2 J''_n = (J_{n-2} - J_n) - (J_n - J_{n+2})$$

$$2^2 \cdot J''_n = J_{n-2} - 2J_n + J_{n+2}$$

Again differentiating and multiplying by 2, we get

$$2^3\ J'''_n = 2J'_{n-2} - 2^2 J'_n + 2J'_{n+2}$$

$$= [J_{n-3} - J_{n-1}] - 2(J_{n-1} - \ldots J_{n+1}) + (J_{n+1} - J_{n+3})$$

$$= J_{n-3} - 3J_{n-1} + 3J_{n+1} - \ldots J_{n+3}$$

And so on

$$2^r J_n^r = J_{n-r} - r J_{n-r+2} + \frac{2(2-1)}{21} J_{n-r+4} \cdots + (-1)^r J_{n+r}$$

Example 25. *Show that*

$$\frac{x}{2} J_{n-1} = n J_n - (n+2) J_{n+2} + (n+4) J_{n+4} + \ldots$$

Solution. We know that

$$2nJ_n = x[J_{n+1} + J_{n-1}] \qquad \text{(Recurrence Relation } IV)$$

$$\Rightarrow \quad nJ_n = \frac{x}{2}[J_{n+1} + J_{n-1}]$$

$$\Rightarrow \quad \frac{x}{2} J_{n-1} = nJ_n - \frac{x}{2} J_{n+1} \qquad \ldots(1)$$

Replacing n by $n+2$, we get $\frac{x}{2} J_{n+1} = (n+2)\, J_{n+2} - \frac{x}{2} J_{n+3}$ ...(2)

On putting the value of $\frac{x}{2} J_{n+1}$ from (2) in (1), we get

$$\frac{x}{2} J_{n-1} = n\,J_n - \left[(n+2) J_{n+2} - \frac{x}{2} J_{n+3}\right] = nJ_n - (n+2)\, J_{n+2} + \frac{x}{2} J_{n+3}$$

On putting the value of $\frac{x}{2} J_{n+3}$, we get

$$= nJ_n-(n+2)J_{n+2}+(n+4)J_{n-4}-\frac{x}{2}J_{n+5} \quad \text{and so on}$$

$$= nJ_n-(n+2)J_{n+2}+(n+4)J_{n+4}-(n+6)J_{n+6}+ \ldots..$$

Example 26. *Prove that*

$$J_n' = \frac{2}{x}\left[\frac{n}{2}J_n-(n+2)J_{n+2}+(n+4)J_{n+1}-\cdots\right]$$

Solution. From Recurrence formula (2), we have

$$J_n' = -\frac{n}{x}J_n+J_{n-1}$$

From example 25, Page 840, putting value J_{n-1}, we get

$$= -\frac{n}{x}J_n+\frac{2}{x}[nJ_n-(n+2)J_{n+2}+\cdots]$$

$$= \frac{2}{x}\left[\frac{n}{2}J_n-(n+2)J_{n+2}+\ldots\right]$$ **Proved.**

Example 27. *Prove that*

$$\frac{d}{dx}\left[\frac{J_{-n}(x)}{J_n(x)}\right]=\frac{-2\sin n\pi}{\pi x J_n^2}$$

Solution. $x^2\frac{d^2y}{dx^2}+x\frac{dy}{dx}+(x^2-n^2)y=0$ (Bessel's Equation)

$$\frac{d^2y}{dx^2}+\frac{1}{x}\frac{dy}{dx}+\left(1-\frac{n^2}{x^2}\right)y=0 \qquad \text{(Dividing by } x^2\text{)}$$

As J_n and J_{-n} are the solutions of the above equation, so

$$\Rightarrow \quad J_n''+\frac{1}{x}J_n'+\left(1-\frac{n^2}{x^2}\right)J_n=0 \qquad \ldots(1)$$

$$\Rightarrow \quad J_{-n}''+\frac{1}{x}J'_{-n}+\left(1-\frac{n^2}{x^2}\right)J_{-n}=0 \qquad \ldots(2)$$

On multiplying (1) by J_{-n} and (2) by J_n, we get

$$J_n''J_{-n}+\frac{1}{x}J_n'J_{-n}+\left(1-\frac{n^2}{x^2}\right)J_n.J_{-n}=0 \qquad \ldots(3)$$

$$J_{-n}''.J_n+\frac{1}{x}J'_{-n}.J_n+\left(1-\frac{n^2}{x^2}\right)J_{-n}.J_n=0 \qquad (n\to -n) \qquad \ldots(4)$$

On subtracting (4) from (3), we get

$$J_n''.J_{-n}-J_{-n}''.J_n+\frac{1}{x}\left(J'_n.J_{-n}-J'_{-n}.J_n\right)=0$$

$$\Rightarrow \quad \frac{J_n''J_{-n}-J_{-n}''J_n}{J'_nJ_{-n}-J'_{-n}J_n}=-\frac{1}{x}$$

On integrating, we get

$$\log\,(J_n'.J_{-n}-J_{-n}'.J_n) = -\log x+\log C=\log\frac{C}{x}$$

Therefore, $J_{-n} J'_n - J_n J'_{-n} = \dfrac{C}{x}$ ...(5)

Using definition of J_n and J_{-n}, (5) becomes

$$\frac{1}{2^{-n}\overline{|(-n+1)}}\left[x^{-n} - \frac{x^{-n+2}}{2\cdot(-2n+2)} + \frac{x^{-n+4}}{2\cdot 4\cdot(-2n+2)(-2n+4)} - \dots\right]$$

$$\times\frac{1}{2^{n}\overline{|(n+1)}}\left[nx^{n-1} - \frac{(n+2)x^{n+1}}{2\cdot(2n+2)} + \frac{(n+4)x^{n+3}}{2\cdot 4\cdot(2n+2)(2n+4)} - \dots\right]$$

$$-\frac{1}{2^{n}\overline{|(n+1)}}\left[x^{n} - \frac{x^{n+2}}{2\cdot(2n+2)} + \frac{x^{n+4}}{2\cdot 4\cdot(2n+2)(2n+4)} - \dots\right]$$

$$\times\frac{1}{2^{-n}\overline{|(-n+1)}}\left[-nx^{-n-1} - \frac{(-n+2)x^{-n+1}}{2\cdot(-2n+2)} + \frac{(-n+4)x^{-n+3}}{2\cdot 4(-2n+2)(-2n+4)}\right] = \frac{C}{x}$$

Now comparing the coefficients of $1/x$ on both sides, we get

$$\frac{n}{\overline{|n+1}\,\overline{|-n+1}} - \frac{-n}{\overline{|n+1}\,\overline{|-n+1}} = C$$

$$\frac{1}{\overline{|(-n+1)}\,\overline{|(n+1)}}[n-(-n)] = C$$

$$\Rightarrow \qquad C = \frac{2\sin n\pi}{\pi} \qquad \left[\text{Since } \overline{|(n)}\ \overline{|(1-n)} = \frac{\pi}{\sin n\pi}\right]$$

Substituting the values of C in (5), we get

$$J'_n \cdot J_{-n} - J'_{-n} J_n = \frac{2\sin n\pi}{\pi x}$$

$$\Rightarrow \qquad \frac{J'_n J_{-n} - J'_{-n} J_n}{J_n^2} = \frac{2\sin n\pi}{\pi x J_n^2}$$

$$\Rightarrow \qquad \frac{d}{dx}\left(\frac{J_{-n}(x)}{J_n(x)}\right) = \frac{-2\sin n\pi}{\pi x J_n^2}$$

Proved.

29.6 EQUATIONS REDUCIBLE TO BESSEL'S EQUATION

There are some differential equations which can be reduced to Bessel's equation and therefore, can be solved.

(*a*) We shall reduce the following differential equation to Bessel's equation.

$$x^2\frac{d^2y}{dx^2} + x\frac{dy}{dx} + (k^2x^2 - n^2)y = 0 \qquad \dots (1)$$

Put $t = kx$ $\quad \dfrac{dt}{dx} = k,\ \dfrac{dy}{dx} = \dfrac{dy}{dt}\dfrac{dt}{dx} = k\dfrac{dy}{dt} \quad \dfrac{d^2y}{dx^2} = \dfrac{d}{dx}\left(k\dfrac{dy}{dt}\right) = \dfrac{d}{dt}\left(k\dfrac{dy}{dt}\right)\dfrac{dt}{dx} = k^2\dfrac{d^2y}{dt^2}$

Thus (1) becomes

$$\left(\frac{t^2}{k^2}\right)\left(k^2\frac{d^2y}{dt^2}\right) + \left(\frac{t}{k}\right)\left(k\frac{dy}{dt}\right) + (t^2 - n^2)y = 0 \Rightarrow t^2\frac{d^2y}{dt^2} + t\frac{dy}{dt} + (t^2 - n^2)y = 0$$

Its solution is $y = c_1 J_n(t) + c_2 J_{-n}(t)$,

Hence solution of (1) is $y = c_1 J_n(kx) + c_2 J_{-n}(kx)$. **Ans.**

(*b*) Let us reduce the following differential equation to Bessel's equation.

$$x\frac{d^2y}{dx^2} + a\frac{dy}{dx} + k^2xy = 0 \qquad \text{... (2)}$$

Put $y = x^n z$, $\qquad \dfrac{dy}{dx} = x^n\dfrac{dz}{dx} + nx^{n-1}.z$

$$\frac{d^2y}{dx^2} = x^n\frac{d^2z}{dx^2} + nx^{n-1}\frac{dz}{dx} + nx^{n-1}\frac{dz}{dx} + n(n-1)x^{n-2}.z$$

$$= x^n\frac{d^2z}{dx^2} + 2nx^{n-1}\frac{dz}{dx} + n(n-1)x^{n-2}.z.$$

Then (2) becomes

$$x\left[x^n\frac{d^2z}{dx^2} + 2nx^{n-1}\frac{dz}{dx} + n(n-1)x^{n-2}z\right] + a\left[x^n\frac{dz}{dx} + nx^{n-1}z\right] + k^2.x.x^n z = 0$$

$$\Rightarrow \qquad x^{n+1}\frac{d^2z}{dx^2} + (2n+a)x^n\frac{dz}{dx} + [k^2x^2 + n^2 + (a-1)n]x^{n-1}z = 0 \qquad \text{...(3)}$$

Dividing (3) by x^{n-1}, we get

$$x^2\frac{d^2z}{dx^2} + (2n+a)x\frac{dz}{dx} + [k^2x^2 + n^2 + (a-1)n]z = 0$$

Let us put $2n + a = 1$, then $\qquad x^2\dfrac{d^2z}{dx^2} + x\dfrac{dz}{dx} + (k^2x^2 - n^2)z = 0$

Its solution is $\quad z = c_1 J_n(kx) + c_2 J_{-n}(kx)$

Hence, the solution of (2) is $y = x^n[c_1 J_n(kx) + c_2 J_{-n}(kx)]$, $n \notin 1$ **Ans.**

where $\qquad n = \dfrac{1-a}{2}$

(c) To reduce the following differential equation to Bessel's equation.

$$x\frac{d^2y}{dx^2} + c\frac{dy}{dx} + k^2x^r y = 0 \qquad \text{... (4)}$$

Put $\qquad x = t^m,\ t = x^{\frac{1}{m}}$

So that $\qquad \dfrac{dy}{dx} = \dfrac{dy}{dt}\dfrac{dt}{dx} = \dfrac{dy}{dt}\left(\dfrac{1}{m}x^{\frac{1}{m}-1}\right) = \dfrac{dy}{dt}\left(\dfrac{1}{m}t^{1-m}\right)$

$$\frac{d^2y}{dx^2} = \frac{d}{dx}\left(\frac{1}{m}t^{1-m}\frac{dy}{dt}\right) = \frac{d}{dt}\left(\frac{1}{m}t^{1-m}\frac{dy}{dt}\right)\frac{dt}{dx}$$

$$= \left(\frac{1}{m}t^{1-m}\frac{d^2y}{dt^2} + \frac{1}{m}(1-m)t^{-m}\frac{dy}{dt}\right)\frac{1}{m}t^{\frac{1}{m}-1} = \frac{1}{m^2}t^{2-2m}\frac{d^2y}{dt^2} + \frac{1-m}{m^2}t^{1-2m}\frac{dy}{dt}$$

Now, (4) becomes.

$$t^m\left[\frac{1}{m^2}t^{2-2m}\frac{d^2y}{dt^2} + \frac{1-m}{m^2}t^{1-2m}\frac{dy}{dt}\right] + c\frac{1}{m}t^{1-m}\frac{dy}{dt} + k^2t^{mr}y = 0$$

$$\Rightarrow \qquad \frac{1}{m^2} t^{2-m} \frac{d^2 y}{dt^2} + \frac{1-m+cm}{m^2} t^{1-m} \frac{dy}{dt} + k^2 t^{mr} y = 0$$

On multiplying by $\frac{m^2}{t^{1-m}}$, we get

$$t\frac{d^2 y}{dt^2} + (1-m+cm)\frac{dy}{dt} + (km)^2 t^{mr+m-1} y = 0 \qquad \text{... (5)}$$

Let us put $a = 1 - m + cm$ and $m = \frac{1}{r+1} \quad \Rightarrow \quad mr + m - 1 = \frac{r}{r+1} + \frac{1}{r+1} - 1 = 0$

Thus (5) becomes $t\frac{d^2 y}{dt^2} + a\frac{dy}{dt} + (km)^2 y = 0$

Its solution is $y = t^n [c_1 J_n(kmt) + c_2 J_{-n}(kmt)]$

Solution of (4) is $y = x^{n/m} [c_1 J_n(km\, x^{1/m}) + c_2 J_{-n}(km\, x^{1/m})]$ **Ans.**

29.7 ORTHOGONALITY OF BESSEL FUNCTION (*D.U., April 2010, AMIETE, June 2009*)

$$\int_0^1 x\, J_n(\alpha x). J_n(\beta x)\, dx = 0$$

where α and β are the roots of $j_n(x) = 0$

Proof. We know that

$$x^2 \frac{d^2 y}{dx^2} + x\frac{dy}{dx} + (\alpha^2 x^2 - n^2) y = 0 \qquad \text{... (1)}$$

$$\Rightarrow \qquad x^2 \frac{d^2 z}{dx^2} + x\frac{dz}{dx} + (\beta^2 x^2 - n^2) z = 0 \qquad \text{... (2)}$$

Solution of (1) and (2) are $y = J_n(\alpha x)$, $z = J_n(\beta x)$ respectively.

Multiplying (1) by $\frac{z}{x}$ and (2) by $-\frac{y}{x}$ and adding, we get

$$x\left(z\frac{d^2 y}{dx^2} - y\frac{d^2 z}{dx^2}\right) + \left(z\frac{dy}{dx} - y\frac{dz}{dx}\right) + (\alpha^2 - \beta^2) xyz = 0.$$

$$\Rightarrow \qquad \frac{d}{dx}\left[x\left(z\frac{dy}{dx} - y\frac{dz}{dx}\right)\right] + (\alpha^2 - \beta^2) xyz = 0$$

$$\frac{d}{dx}\left[x\left(z\frac{dy}{dx} - y\frac{dz}{dx}\right)\right] + (\alpha^2 - \beta^2) xyz = 0 \qquad \text{... (3)}$$

Integrating (3) w.r.t. 'x' between the limits 0 and 1, we get

$$\left[x\left(z\frac{dy}{dx} - y\frac{dz}{dx}\right)\right]_0^1 + (\alpha^2 - \beta^2)\int_0^1 x\, y\, z\, d\, x = 0$$

$$(\beta^2 - \alpha^2)\int_0^1 x\, y\, z dx = \left[x\left(z\frac{dy}{dx} - y\frac{dz}{dx}\right)\right]_0^1 = \left[z\frac{dy}{dx} - y\frac{dz}{dx}\right]_{x=1} \qquad \text{... (4)}$$

Putting the values of $y = J_n(\alpha x)$, $\frac{dy}{dx} = \alpha J_n'(\alpha x)$, $z = J_n(\beta x)$, $\frac{dz}{dx} = \beta, J_n'(\beta x)$ in (4), we get

$$(\beta^2-\alpha^2)\int_0^1 x J_n(\alpha x).J_n(\beta x)\,dx = \left[\alpha J_n'(\alpha x)J_n(\beta x)-\beta J_n'(\beta x)J_n(\alpha x)\right]_{x=1}$$

$$= \alpha J_n'(\alpha)J_n(\beta)-\beta J_n'(\beta)J_n(\alpha) \qquad \text{... (5)}$$

Since α, β are the roots of $J_n(x)=0$, so $J_n(\alpha)=J_n(\beta)=0$.
Putting the values of $J_n(\alpha)=J_n(\beta)=0$ in (5), we get

$$(\beta^2-\alpha^2)\int_0^1 x J_n(\alpha x).J_n(\beta x)\,dx = 0$$

$$\Rightarrow \qquad \boxed{\int_0^1 x J_n(\alpha x).J_n(\beta x)\,dx = 0} \qquad \textbf{Proved.}$$

Example 28. *Prove that*

$$\int_0^1 x[J_n(\alpha x)]^2\,dx = \frac{1}{2}[J_n'(\alpha)]^2.$$

Solution. From (5) of article 29.7, we know that

$$(\beta^2-\alpha^2)\int_0^1 x J_n(\alpha x).J_n(\beta x)\,dx = \alpha J_n'(\alpha).J_n(\beta)-\beta J_n'(\beta).J_n(\alpha)$$

when $\quad \beta=\alpha$

We also know that $J_n(\alpha)=0$. Let β be a neighbouring value of α, which tends to α. Then

$$\lim_{\beta\to\alpha}\int_0^1 xJ_n(\alpha x)\,J_n(\beta x)\,dx = \lim_{\beta\to\alpha}\frac{0+\alpha J_n'(\alpha).J_n(\beta)}{\beta^2-\alpha^2}$$

As the limit is of the form $\frac{0}{0}$, we apply L' Hopital's rule

$$\int_0^1 xJ_n^2(\alpha x)\,dx = \lim_{\beta\to\alpha}\frac{0+\alpha J_n'(\alpha).J_n'(\beta)}{2\beta} = \frac{1}{2}[J_n'(\alpha)]^2 \qquad [\because \alpha=\beta] \qquad \textbf{Proved.}$$

29.8 A GENERATING FUNCTION FOR J_n (*X*)

Prove that $J_n(x)$ is the coefficient of z^n in the expansion of $e^{\frac{x}{2}\left(z-\frac{1}{z}\right)}$.

Proof. We know that $e^t = 1+t+\frac{t^2}{2!}+\frac{t^3}{3!}+...$

$$e^{\frac{xz}{2}} = 1+\left(\frac{xz}{2}\right)+\frac{1}{2!}\left(\frac{xz}{2}\right)^2+\frac{1}{3!}\left(\frac{xz}{2}\right)^3+... \qquad \text{... (1)}$$

$$e^{-\frac{x}{2z}} = 1-\left(\frac{x}{2z}\right)+\frac{1}{2!}\left(\frac{x}{2z}\right)^2-\frac{1}{3!}\left(\frac{x}{2z}\right)^3+... \qquad \text{... (2)}$$

On multiplying (1) and (2), we get

$$e^{\frac{x}{2}\left(z-\frac{1}{z}\right)} = \left[1+\left(\frac{xz}{2}\right)+\frac{1}{2!}\left(\frac{xz}{2}\right)^2+\frac{1}{3!}\left(\frac{xz}{2}\right)^3+...\right]\times\left[1-\frac{x}{2z}+\frac{1}{2!}\left(\frac{x}{2z}\right)^2-\frac{1}{3!}\left(\frac{x}{2z}\right)^3+...\right] \qquad \text{... (3)}$$

The coefficient of z^n in the product of (3), we get

$$= \frac{1}{n!}\left(\frac{x}{2}\right)^n-\frac{1}{(n+1)!}\left(\frac{x}{2}\right)^{n+2}+\frac{1}{2!(n+2)!}\left(\frac{x}{2}\right)^{n+4}-... = J_n(x)$$

Similarly, coefficient of z^{-n} in the product (3) = $J_{-n}(x)$

$$\therefore \quad e^{\frac{x}{2}\left(z-\frac{1}{z}\right)} = J_0 + zJ_1 + z^2J_2 + z^3J_3 + \ldots + z^{-1}J_{-1} + z^{-2}J_{-2} + z^{-3}J_{-3} + \ldots$$

$$\boxed{e^{\frac{x}{2}\left(z-\frac{1}{z}\right)} = \sum_{n=-\infty}^{\infty} z^n J_n(x)}$$

For this reason $e^{\frac{x}{2}\left(z-\frac{1}{z}\right)}$ is known as the generating function of Bessel's functions. **Proved.**

Cor. In the expansion of (3), coefficient of z^0

$$= 1 - \frac{x^2}{2^2} + \frac{x^4}{2^2.4^2} - \ldots \quad \text{or} \quad J_0 = 1 - \frac{x^2}{2^2} + \frac{x^4}{2^2.4^2} - \ldots$$

29.9 TRIGONOMETRIC EXPANSION INVOLVING BESSEL FUNCTIONS

We know that

$$e^{\frac{x}{2}\left(z-\frac{1}{z}\right)} = J_0 + zJ_1 + z^2J_2 + z^3J_3 + \ldots + z^{-1}J_{-1} + z^{-2}J_{-2} + z^{-3}J_{-3} + \ldots \quad \ldots(1)$$

Putting $z = e^{i\theta}$ in (1), we get

$$e^{\frac{x}{2}(e^{i\theta}-e^{-i\theta})} = J_0 + J_1 e^{i\theta} + J_2 e^{2i\theta} + J_3 e^{3i\theta} + \ldots + J_{-1} e^{-i\theta} + J_{-2} e^{-2i\theta} + J_{-3} e^{-3i\theta} + \ldots$$

$$\left(\text{since } \frac{e^{i\theta} - e^{-i\theta}}{2i} = \sin\theta\right)$$

$$e^{ix\sin\theta} = J_0 + J_1 e^{i\theta} + J_2 e^{2i\theta} + J_3 e^{3i\theta} + \ldots - J_1 e^{-i\theta} + J_2 e^{-2i\theta} - J_3 e^{-3i\theta} - \ldots \quad (\text{since } J_{-n} = (-1)^n J_n)$$

$$\Rightarrow \cos(x\sin\theta) + i\sin(x\sin\theta) = J_0 + J_1(e^{i\theta} - e^{-i\theta}) + J_2(e^{2i\theta} + e^{-2i\theta}) + J_3(e^{3i\theta} - e^{-3i\theta}) + \ldots$$

$$\Rightarrow \cos(x\sin\theta) + i\sin(x\sin\theta) = J_0 + J_1(2i\sin\theta) + J_2(2\cos 2\theta) + J_3(2i\sin 3\theta) + \ldots$$

Now equating real and imaginary parts, we get

$$\cos(x\sin\theta) = J_0 + 2J_2\cos 2\theta + 2J_4\cos 4\theta + \ldots \quad \ldots(2)$$

$$\sin(x\sin\theta) = 2J_1\sin\theta + 2J_3\sin 3\theta + 2J_5\sin 5\theta + \ldots \quad \ldots(3)$$

On putting $\theta = \frac{\pi}{2} - \alpha$ in (2) and (3), we get

$$\cos(x\cos\alpha) = J_0 - 2J_2\cos 2\alpha + 2J_4\cos 4\alpha - \ldots$$

$$\sin(x\cos\alpha) = 2J_1\cos\alpha + 2J_3\cos 3\alpha + 2J_5\cos 5\alpha - \ldots$$

Example 29. *Prove that*

$$\cos x = J_0 - 2J_2 + 2J_4 - \ldots$$

$$\sin x = 2J_1 - 2J_3 + 2J_5 - \ldots$$

Solution. We know that

$$\cos(x\sin\theta) = J_0 + 2J_2\cos 2\theta + 2J_4\cos 4\theta + \ldots \quad \ldots(1)$$

$$\sin(x\sin\theta) = 2J_1\sin\theta + 2J_3\sin 3\theta + 2J_5\sin 5\theta + \ldots \quad \ldots(2)$$

Putting $\theta = \dfrac{\pi}{2}$ in (1) and (2), we get

$$\cos x = J_0 - 2J_2 + 2J_4 - \ldots$$

and $\sin x = 2J_1 - 2J_3 + 2J_5 - \ldots$ **Proved.**

Example 30. *Prove that*

$$x \sin x = 2[2^2 J_2 - 4^2 J_4 + 6^2 J_6 - \ldots]$$

$$x \cos x = 2[1^2 J_1 - 3^2 J_3 + 5^2 J_5 + \ldots]$$

Solution. We know that

$$\cos(x \sin\theta) = J_0 + 2J_2 \cos 2\theta + 2J_4 \cos 4\theta + \ldots \qquad \ldots(1)$$

and $\sin(x \sin\theta) = 2J_1 \sin\theta + 2J_3 \sin 3\theta + 2J_5 \sin 5\theta + \ldots$ ...(2)

Differentiating (1) w.r.t. "θ" we get

$$[-\sin(x \sin\theta)]x \cos\theta = 0 - 4J_2 \sin 2\theta - 8J_4 \sin 4\theta + \ldots \qquad \ldots(3)$$

Again differentiating (3) w.r.t., "θ", we get

$$[-\sin(x\sin\theta)](-x\sin\theta) + [-\cos(x\sin\theta)(x\cos\theta)]x\cos\theta$$

$$= -8J_2 \cos 2\theta - 32J_4 \cos 4\theta + \ldots \qquad \ldots (4)$$

Now putting $\theta = \dfrac{\pi}{2}$ in (4), we get

$$x \sin x = 8J_2 - 32J_4 + \ldots = 2[2^2 J_2 - 4^2 J_4 + 6^2 J_6 - \ldots]$$

Similarly differentiating (2) twice and putting $\theta = \dfrac{\pi}{2}$, we have

$$x \cos x = 2[1^2 J_1 - 3^2 J_3 + 5^2 J_5 - \ldots]$$ **Proved.**

Example 31. *Prove that* $J_0^2 + 2J_1^2 + 2J_2^2 + \ldots = 1$

Solution. $(J_0 + 2\,J_2 \cos 2\theta + 2\,J_4 \cos 4\theta + \ldots$

$= \cos(x \sin\theta)$ [From (29.9)] ... (1)

$2J_1 \sin\theta + 2J_3 \sin 3\theta + 2J_5 \sin 5\theta + \ldots = \sin(x \sin\theta)$ ... (2)

Now squaring (1) and integrating w.r.t. 'θ' between the limits 0 and π, we get

$$J_0^2 \pi + 2J_2^2 \pi + 2J_4^2 \pi + \ldots = \int_0^{\pi} \cos^2(x \sin\theta)\, d\theta \qquad \ldots (3)$$

$$\int_0^{\pi} 2\sin^2 n\theta\, d\theta = \pi$$

$$\int_0^{\pi} 2\cos^2 n\theta d\theta = \pi$$

$$\int_0^{\pi} 2\sin n\theta \sin m\theta\, d\theta = 0$$

$$\int_0^{\pi} 2\sin n\theta \sin m\theta\, d\theta = 0$$

Also squaring (2) and integrating w.r.t. "θ" between the limits 0 and π, we get

$$2J_1^2 \pi + 2J_3^2 \pi + 2J_5^2 \pi + \ldots = \int_0^{\pi} \sin^2(x \sin\theta)\, d\theta \qquad \ldots (4)$$

Adding (3) and (4), we get

$$\pi[J_0^2 + 2J_1^2 + 2J_2^2 + 2J_3^2 + \ldots] = \int_0^{\pi} \cos^2(x \sin\theta)\, d\theta + \int_0^{\pi} \sin^2(x\sin\theta)\, d\theta = \int_0^{\pi} d\theta = \pi$$

$\Rightarrow J_0^2 + 2J_1^2 + 2J_2^2 + 2J_3^2 + \ldots = 1$ **Proved.**

29.10 BESSEL'S INTEGRAL

To prove that

(a) $J_0(x) = \dfrac{1}{\pi}\displaystyle\int_0^{\pi} \cos(x \sin\theta)\, d\theta$ *(b)* $J_n(x) = \dfrac{1}{\pi}\displaystyle\int_0^{\pi} \cos(n\theta - x\sin\theta)\, d\theta$ [*G.B.T.U. 2011*]

Proof. We know that

$$\cos(x\sin\theta) = J_0 + 2J_2\cos 2\theta + 2J_4\cos 4\theta + ... \quad ...(1)$$

$$\sin(x\sin\theta) = 2J_1\sin\theta + 2J_3\sin 3\theta + 2J_5\sin 5\theta + ... \quad ...(2)$$

(*a*) Integrating (1) between the limits 0 and π, we have

$$\int_0^\pi \cos(x\sin\theta)\,d\theta = \int_0^\pi (J_0 + 2J_2\cos 2\theta + 2J_4\cos 4\theta + ...)\,d\theta$$

$$= J_0\int_0^\pi d\theta + 2J_2\int_0^\pi \cos 2\theta\,d\theta + 2J_4\int_0^\pi \cos 4\theta\,d\theta + ...$$

$$= J_0\pi + 0 + 0$$

i.e. $$J_0 = \frac{1}{\pi}\int_0^\pi \cos(x\sin\theta)\,d\theta$$ **Proved.**

(*b*) Multiplying (1) by $\cos n\theta$ and integrating between the limits 0 and π, we have

$$\int_0^\pi \cos(x\sin\theta)\cos n\theta\,d\theta = \int_0^\pi [J_0\cos n\theta + 2J_2\cos 2\theta\cos n\theta + 2J_4\cos 4\theta\cos n\theta + ...]\,d\theta$$

$$= J_0\int_0^\pi \cos n\theta\,d\theta + 2J_2\int_0^\pi \cos 2\theta\cos n\theta\,d\theta + ...$$

$$= 0, \quad \text{if } n \text{ is odd} \quad ...(3)$$

$$= \pi J_n, \quad \text{if } n \text{ is even} \quad ...(4)$$

Again multiplying (2) by $\sin n\theta$ and integrating between the limits 0 and π, we have

$$\int_0^\pi \sin(x\sin\theta)\sin n\theta\,d\theta = \int_0^\pi (2J_1\sin\theta\sin n\theta + 2J_3\sin 3\theta\sin n\theta + ...)\,d\theta$$

$$= 2J_1\int_0^\pi \sin\theta\sin n\theta\,d\theta + 2J_3\int_0^\pi \sin 3\theta\sin n\theta\,d\theta + ...$$

$$= 0 \quad \text{if } n \text{ is even} \quad ...(5)$$

$$= \pi J_n \quad \text{if } n \text{ is odd} \quad ...(6)$$

Adding (3) and (6) or (4) and (5), we get

$$\int_0^\pi [\cos(x\sin\theta)\cos n\theta + \sin(x\sin\theta)\sin n\theta]\,d\theta = \pi J_n$$

$$\Rightarrow \quad \int_0^\pi \cos(n\theta - x\sin\theta)\,d\theta = \pi J_n \text{ or } J_n = \frac{1}{\pi}\int_0^\pi \cos(n\theta - x\sin\theta)\,d\theta$$ **Proved.**

EXERCISE 29.1

Prove that

1. $J_1(x) = \frac{x}{2} - \frac{x^3}{2^3.1!.2!} + \frac{x^5}{2^5.2!.3!} - \frac{x^7}{2^7.3!.4!} + ...$

2. (*a*) $J_0(2) = 0.224$ (*b*) $J_1(2) = 0.44$

3. $J_2 = J_0'' - x^{-1}J_0'$

4. $\frac{d}{dx}[xJ_1(x)] = xJ_0(x)$

5. $\frac{d}{dx}[J_n^2(x) + J_{n+1}^2(x)] = 2\left(\frac{n}{x}J_n^2(x) - \frac{n+1}{x}J_{n+1}^2(x)\right)$

6. $\int_0^\pi x^{n+1}J_n(x)\,dx = x^{n+1}(x),$ $n > -1$

7. $x^2J_n''(x) = (n^2 - n - x^2)J_n(x) + xJ_{n+1}(x)$

8. $\int x^2 J_0 J_1 dx = \frac{1}{2} x^2 J_0' + c$

9. $J_{3/2}(x)\sin x - J_{-3/2}(x)\cos(x) = \frac{\sqrt{2\pi}}{x^3}$

10. $J_1'' = \left(\frac{2}{x^2} - 1\right) J_1(x) - \frac{1}{x} J_0(x)$ **11.** $J_0^2 + 2(J_1^2 + J_2^2 + J_3^2 + \ldots) = 1$

12. If J_0 (2) = a , J_1 (2) = b find J_2 (2), J_1' (2), J'_2 (2)
in terms of a and b where J_n (x) is the Bessel function of first kind.

Ans. J_2 (2) = $b - a$, J'_1 (2) = $a - \frac{b}{2}$, J_2' (2) $= a$

13. Prove that $J_n(x) = 0$ has no repeated root except $x = 0$.

14. Integrate $\int x^3 J_0(x) dx$, where J_n (x) is the Bessel's function of first kind, in terms of J_0 (x), J_1 (x) and J_2 (x).

29.11 FOURIER-BESSEL EXPANSION

If a function f (x) is continuous and has a finite number of oscillations in the interval $0 \le x \le a$, dthen f (x) can be expanded in a series.

$$f(x) = C_1 J_n(\alpha_1 x) + C_2 J_n(\alpha_2 x) + C_3 J_n(\alpha_3 x) + \ldots + C_n J_n(\alpha_n x) + \ldots\ldots$$

$$\Rightarrow \qquad f(x) = \sum_{i=1}^{\infty} C_i J_n(\alpha_i x).$$

where α_1, α_2, α_3 are the roots of the equation $J_n(x) = 0$.

[The orthogonal property of Bessel's functions enables us to expand a function in terms of Bessel's function].

Let $$f(x) = \sum_{i=1}^{\infty} C_i J_n(\alpha_i x) \qquad \ldots (1)$$

Multiplying both sides of (1) by $xJ_n(\alpha_j x)$, we get

$$x f(x) J_n(\alpha_j x) = \sum_{i=1}^{\infty} C_i x J_n(\alpha_j x).J_n(\alpha_i x) \qquad \ldots (2)$$

Integrating both sides of (2) from $x = 0$ to $x = a$, we have

$$\int_0^a x f(x).J_n(\alpha_j x) dx = \sum_{i=1}^{\infty} C_i \int_0^a x J_n(\alpha_j x).J_n(\alpha_i x) dx \qquad \ldots (3)$$

By orthogonal property of Bessel's functions, we know that

$$\int_0^a x J_n(\alpha_i x).J_n(\alpha_j x) dx = \begin{cases} 0 & \text{if } i \ne j \\ \frac{a^2}{2} J_{(n+1)}^2(\alpha_i a) \cdot & \text{if } i = j \end{cases}$$

On applying this property on the right-hand side of (3), it reduces to

$$\int_0^a x f(x) J_n(\alpha_i x) dx = C_i . \frac{a^2}{2} J_{n+1}^2(\alpha_i a)$$

$$\Rightarrow \qquad C_i = \frac{2\int_0^a x f(x) J_n(\alpha_i x) dx}{a^2 . J_{n+1}^2(\alpha_i a)}$$

By putting the values of the coefficient *C's* in (1), we get the Fourier-Bessel Expansions.

Ans.

Example 32. *Show that* $\sum_{i=1}^{\infty} \frac{2J_0(a_i x)}{a_i J_1(a_i)} = 1$, *where* a_1, a_2, a_3 .... *are the roots of* $J_0(x)$.

Solution. Let $$f(x) = \sum_{i=1}^{\infty} C_i J_n(a_i x), \quad \text{... (1)}$$

then $$C_n = \frac{2}{J_{n+1}^2(a_i)} \int_0^1 x J_n(a_i x) f(x)\, dx \quad \text{... (2)}$$

Putting $f(x) = 1$ and $n = 0$ in (1), we get

$$1 = \sum_{i=1}^{\infty} C_i J_0(a_i x) \quad \text{...(3)}$$

$$C_i = \frac{2}{J_1^2(a_i)} \int_0^1 x J_0(a_i x)\, dx = \frac{2}{J_1^2(a_i)} \left[\frac{J_1(a_i)}{a_i}\right] = \frac{2}{a_i J_1(a_i)}$$

Substituting the values of $C_i f(x)$ and n in (3), we obtain

$$1 = \sum_{i=1}^{\infty} \frac{2}{a_i J_1(a_i)} J_0(a_i x)$$

$$\Rightarrow \quad \sum \frac{2J_0(a_i x)}{a_i J_1(a_i)} = 1$$

Proved.

Example 33. *Expand* $f(x) = x^2$ *in the interval* $0 < x < 2$ *in terms of* $J_2(\alpha_n x)$ *where* α_n *are the roots of* $J2(2\alpha_n) = 0$.

Solution. $f(x) = x^2$

$$x^2 = \sum_{i=1}^{\infty} C_i J_2(\alpha_i x) \quad \text{...(1)}$$

Multiplying both sides of (1) by $xJ_2(\alpha_j x)$ we get

$$x^3 J_2(\alpha_j .x) = \sum_{i=1}^{\infty} C_i x J_2(\alpha_i x).J_2(\alpha_j .x) \quad \text{... (2)}$$

Integrating (2) w.r.t. x from $x = 0$ to $x = 2$, we get

$$\int_0^2 x^3 .J_2(\alpha_j\, x)\, dx = \sum_{i=1}^{\infty} C_i \int_0^2 x J_2(\alpha_i x).J_2(\alpha_j .x)\, dx$$

$$\left[\frac{x^3 .J_3(\alpha_i x)}{\alpha_i}\right]_0^2 = C_i \int_0^2 x J_2^2(\alpha_i x)\, dx \quad (i = j) \text{ (other integrals are zero)}$$

$$\frac{8J_3(2\alpha_i)}{\alpha_i} = C_i \frac{2^2}{2} J_3^2(2\alpha_i)$$

$$C_i = \frac{8J_3(2\alpha_i)}{\alpha_i} \frac{2}{4J_3^2(2\alpha_i)} = \frac{4}{\alpha_i J_3(2\alpha_i)}$$

On putting the values of coefficients C_i in (1), we get

$$x^2 = \sum_{i=1}^{\infty} \frac{4J_2(\alpha_i x)}{\alpha_i J_3(2\alpha_i)}$$

Ans.

EXERCISE 29.2

1. Expand $f(x) = x^3$ in the interval $0 < x < 3$ in terms of Bessel's functions $J_1(\alpha_n x)$ where α_n are the roots of $J_1(3\alpha) = 0$. **Ans.** $x^3 = \sum_{i=1}^{\infty} \frac{6}{\alpha_i^2} \frac{1}{J_2^2(3\alpha_i)} \{3\alpha_i J_2(3\alpha_i) - 2J_3(3\alpha_i)\}$

2. Show that the Fourier-Bessel series in $J_2(\alpha_i x)$ for $f(x) = x^2$, $(0 < x < \alpha)$ where α_i are positive roots of $J_2(x) = 0$ is

$$x^2 = 2a^2 \sum_{i=1}^{\infty} \frac{J_2(\alpha_i x)}{a\alpha_i J_3(\alpha_i a)}$$

29.12 BER AND BEI FUNCTIONS

The following differential equation is useful in certain problems in electrical engineering.

$$x\frac{d^2y}{dx^2} + \frac{dy}{dx} - ixy = 0 \qquad ...(1)$$

This equation (1) is a particular case of the differential equation

$$x^2\frac{d^2y}{dx^2} + x\frac{dy}{dx} + (k^2x^2 - n^2)y = 0 \qquad ...(2)$$

On putting $n = 0$, $k^2 = -i$ in equation (2), we get

$$x\frac{d^2y}{dx^2} + \frac{dy}{dx} - ixy = 0$$

Its solution is $y = J_0(kx) = J_0[(-i)^{1/2}x] = J_0(i^{3/2}x)$

$$y = J_0(i^{3/2}x) = 1 - \frac{i^3x^2}{2^2} + \frac{i^6x^4}{(2!)^2.2^4} - \frac{i^9x^6}{(3!)^2.2^6} + \frac{i^{12}x^8}{(4!)^2.2^8}....$$

$$= 1 + \sum_{k=1}^{\infty} (-1)^k \frac{x^{4k}}{2^2\cdot4^2\cdot6^2......(4k)^2} + i\left[-\sum_{k=1}^{\infty} \frac{(-1)^k x^{4k-2}}{2^2\cdot4^2\cdot6^2\cdot8^2...(4k-2)^2}\right]$$

$J_0(i^{3/2}x)$ = Ber (Bessel real) + Bei (Bessel imaginary)

$$\text{Ber } x = 1 + \sum_{k=1}^{\infty} (-1)^k \frac{x^{4k}}{2^2\cdot4^2\cdot6^2..........(4k)^2}$$

$$\text{Bei } x = -\sum_{k=1}^{\infty} \frac{(-1)^k x^{4k-2}}{2^2\cdot4^2\cdot6^2.8^2........(4k-2)^2}$$

Example 34. *Show that* $\frac{d}{dx}$ *(x Ber′ x) = − x Bei x.*

Solution. We know that

$$\text{Ber } x = 1 - \frac{x^4}{2^2.4^2} + \frac{x^8}{2^2\cdot4^2\cdot6^2\cdot8^2} + ...\infty$$

On differentiating, $$\text{Ber}' x = -\frac{4x^3}{2^2.4^2} + \frac{8x^7}{2^2\cdot4^2\cdot6^2\cdot8^2} + \cdots\infty$$

$$(x\ \text{Ber}'\ x) = -\frac{x^4}{2^2\cdot 4} + \frac{x^8}{2^2\cdot 4^2\cdot 6^2\cdot 8} + \ldots$$

$$\frac{d}{dx}(x\ \text{Ber}'\ x) = -\frac{x^3}{2^2} + \frac{x^7}{2^2\cdot 4^2\cdot 6} + \ldots$$

$$= -x\left[\frac{x^2}{2^2} - \frac{x^6}{2^2\cdot 4^2\cdot 6^2} + \ldots\ldots\infty\right] = -x\ \text{Bei}\ x \qquad \textbf{Proved.}$$

Example 35. *Show that* $\frac{d}{dx}(x\ \text{Bei}'\ x) = x\ Ber\ x$.

Solution. We know that $\text{Bei}\ x = \frac{x^2}{2^2} - \frac{x^6}{2^2\cdot 4^4\cdot 6^2} + \frac{x^{10}}{2^2\cdot 4^2\cdot 6^2\cdot 8^2\cdot 10^2}\ldots$ ... (1)

On differentiating (1) w.r.t. 'x', we get

$$\text{Bei}'x = \frac{x}{2} - \frac{x^5}{2^2\cdot 4^2\cdot 6} + \frac{x^9}{2^2\cdot 4^2\cdot 6^2\cdot 8^2\cdot 10}\ldots\ldots\infty$$

$$(x\ \text{Bei}'\ x) = \frac{x^2}{2} - \frac{x^6}{2^2\cdot 4^2\cdot 6} + \frac{x^{10}}{2^2\cdot 6^2\cdot 8^2\cdot 10}\ldots\ldots\infty$$

$$\frac{d}{dx}(x\,\text{Bei}'\ x) = x - \frac{x^5}{2^2\cdot 4^2} + \frac{x^9}{2^2\cdot 4^2\cdot 6^2\cdot 8^2}\ldots\ldots\ldots\infty$$

$$= x\left[1 - \frac{x^4}{2^2\cdot 4^2} + \frac{x^8}{2^2\cdot 4^2\cdot 6^2\cdot 8^2}\ldots\ldots\infty\right] = x\ \text{Ber}\ x \qquad \textbf{Proved.}$$

Example 36. *Show that*

$$\int_0^a x[Ber^2 x + Bei^2 x]\,dx = a[Ber\ \ a\ \ Bei'\ \ a - Bei\ \ a\ \ Ber'\,a]$$

Solution. $\frac{d}{dx}(x\ \text{Bei}'\ x) = x\ \text{Ber}\ x$ ... (1) (See example 34)

and $\frac{d}{dx}(x\ \text{Ber}'\ x) = -x\,\text{Bei}\,x$ ... (2) (See example 33)

Multiplying (1) by Ber x and (2) by Bei x and subtracting, we get

$$Ber\,x\frac{d}{dx}(x\ \text{Bei}'\ x) - Bei\ x\ \frac{d}{dx}(x\ Ber'\ x) = x\ \text{Ber}^2 x + x\,\text{Bei}^2 x \qquad \ldots (3)$$

Integrating both sides of (3) from 0 to a, we get

$$\int_0^a \left[\text{Ber}\ x\frac{d}{dx}(x\,\text{Bei}'\,x) - \text{Bei}\,x\frac{d}{dx}(x\,\text{Ber}'\,x)\right]dx = \int_0^a (x\,Ber^2 x + x\,\text{Bei}^2 x)dx$$

$$\Rightarrow \quad \int_0^a x(\text{Ber}^2 x + \text{Bei}^2 x)\,dx = \int_0^a \left[\text{Ber}\,x\frac{d}{dx}(x\,\text{Bei}'x) - \text{Bei}\ x\frac{d}{dx}(x\ \text{Ber}'\,x)\right]dx$$

On adding and subtracting $\text{Bei}'\ x\frac{d}{dx}(x\ \text{Ber}\ x)$ on R.H.S.

$$= \int_0^a \left[\text{Ber}\, x \frac{d}{dx}(x\, \text{Ber}'\, x) + \text{Bei}'\, x \frac{d}{dx}(x\, \text{Bei}\, x) - \text{Bei}'\, x \frac{d}{dx}(x\, \text{Ber}'\, x) - \text{Ber}'\, x \frac{d}{dx}(x\, \text{Ber}\, x) \right] dx$$

$$= \int_0^a \frac{d}{dx}[x(\text{Ber}\, x\, \text{Bei}'\, x - \text{Bei}\, x\, \text{Ber}'\, x)]\, dx$$

$$= [x\,(\text{Ber}\, x\, \text{Bei}'\, x - \text{Bei}\, x\, \text{Ber}'\, x)]_0^a$$

$$= a\,[\text{Ber}\, a\, \text{Bei}'\, a - \text{Bei}\, a\, \text{Ber}'\, a]$$ **Proved.**

OBJECTIVE TYPE QUESTIONS

Choose the correct or the best of the answers/statements given in the following parts:

1. $J_{\frac{1}{2}}$ is given by

(*i*) $\sqrt{\frac{2\pi}{x}}\sin x$ (*ii*) $\sqrt{\frac{2\pi}{x}}\cos x$ (*iii*) $\sqrt{\frac{\pi}{2x}}\cos x$ (*iv*) $\sqrt{\frac{2}{\pi x}}\sin x$

(UP., II Semester, 2010) **Ans.** (*iv*)

2. If $J_{n+1}(x) = \frac{2}{x} J_n(x) - J_0(x)$, then n is

(*i*) 0 (*ii*) 2 (*iii*) – 1 (*iv*) None of these **Ans.** (*iv*)

(R.G.P.V. Bhopal, II Semester, Feb 2006)

3. $J_0' =$

(*i*) J_1 (*ii*) $-J_1$ (*iii*) $J_2 - J_0$ (*iv*) J_2 **Ans.** (*ii*)

(R.G.P.V. Bhopal, II Semester, June 2006)

4. $J_{-\frac{1}{2}}(x) =$

(*i*) $\sqrt{\frac{2}{\pi x}}\sin x$ (*ii*) $\sqrt{\frac{2}{\pi}}\sin x$ (*iii*) $\sqrt{\frac{2}{\pi x}}\cos x$ (*iv*) $\sqrt{\frac{2\pi}{x}}\cos x$ **Ans.** (*iii*)

(R.G.P.V. Bhopal, II Semester, Feb. 2006)

5. Bessel's equation of order zero is:

(*i*) $xy + y = 0$ (*ii*) $xy_2 + xy_1 + y = 0$ (*iii*) $xy_2 + y_1 + xy = 0$ (*iv*) $xy_2 + y_1 + xy = 0$ **Ans.** (*iv*)

(R.G.P.V. Bhopal, II Semester, Dec. 2007, Feb. 2006)

6. The value of the integral $\int x^2 J_1(x)\, dx$ is

(*i*) $x^2 J_1(x) + c$ (*ii*) $x^2 J_{-1}(x) + c$ (*iii*) $x^2 J_2(x) + c$ (*iv*) $x^2 J_{-2}(x) + c$ **Ans.** (*iii*)

(AMIETE, June 2010)

CHAPTER

HERMITE FUNCTION

30.1 INTRODUCTION

In this chapter we will learn Hermite function in detail.

30.2 HERMITE'S EQUATION

$$\frac{d^2y}{dx^2}-2x\frac{dy}{dx}+2ny=0 \qquad ...(1)$$

The solution of (1) is known as Hermite's polynomial.

Solution Here, we have

$$\frac{d^2y}{dx^2}-2x\frac{dy}{dx}+2ny=0 \qquad ...(1)$$

Suppose its series solution is

$$y = a_0 x^m + a_1 x^{m+1} + a_2 x^{m+2} + a_3 x^{m+3} + ... + a_k x^{m+k}$$

or

$$y=\sum_{k=0}^{\infty} a_k x^{m+k} \qquad ...(2)$$

$$\frac{dy}{dx}=\sum a_k (m+k) x^{m+k-1}$$

$$\frac{d^2y}{dx^2}=\sum_{k=0}^{\infty} a_k (m+k)(m+k-1)\, x^{m+k-2}$$

Putting the values of y, $\frac{dy}{dx}$ and $\frac{d^2y}{dx^2}$ in (1), we get

$$\Rightarrow \quad \sum a_k (m+k)(m+k-1) x^{m+k-2} - 2x\sum a_k (m+k) x^{m+k-1} + 2n\sum a_k x^{m+k} = 0$$

$$\Rightarrow \quad \sum a_k (m+k)(m+k-1) x^{m+k-2} - 2\sum a_k (m+k) x^{m+k} + 2n\sum a_k\, x^{m+k} = 0$$

$$\Rightarrow \quad \sum a_k (m+k)(m+k-1) x^{m+k-2} - 2\sum a_k \left[(m+k)-n\right] x^{m+k} = 0 \qquad ...(3)$$

This equation holds good for $k = 0$ and all positive integer. By our assumption k cannot be negative.

To get the lowest degree term x^{m-2}, we put $k = 0$ in the first summation of (3) and we cannot have x^{m-2} from the second summation. Since $k \neq -2$.

The coefficient of x^{m-2} is

$a_0\, m(m-1) = 0 \;\Rightarrow\; m = 0,\ m = 1$, since $a_0 \neq 0$ ...(4)

This is the **indicial equation.**

Now equating the coefficient of next lowest degree term x^{m-1} to zero in (3), we get (by putting $k = 1$ in the first summation and we cannot have x^{m-1} from the second summation since $k \neq -1$.)

$$a_1 m(m+1) = 0 \Rightarrow \begin{bmatrix} a_1 \text{ may or may not be zero when } m = 0 \\ a_1 = 0, \text{ when } m = 1 \end{bmatrix} \quad \begin{pmatrix} m+1 \neq 0 \text{ as m is} \\ \text{already equal to zero.} \end{pmatrix}$$

Again equating the coefficient of the general term x^{m+k} to zero, we get

$a_{k+2}\,(m+k+2)(m+k+1) - 2a_k\,(m+k-n) = 0$

$$a_{k+2} = \frac{2(m+k-n)}{(m+k+2)(m+k+1)} a_k \quad \text{...(5)}$$

If $m = 0$, then,
$$a_{k+2} = \frac{2(k-n)}{(k+2)(k+1)} a_k \quad \text{...(6)}$$

If $m = 1$, then,
$$a_{k+2} = \frac{2(k+1-n)}{(k+3)(k+2)} a_k \quad \text{...(7)}$$

Case I. When $m = 0$, $a_{k+2} = \dfrac{2(k-n)}{(k+2)(k+1)} a_k$

If $k = 0$, then,
$$a_2 = \frac{-2n}{2} a_0 = -na_0$$

If $k = 1$, then,
$$a_3 = \frac{2(1-n)}{6} a_1 = -2\frac{(n-1)}{3!} a_1$$

If $k = 2$, then,
$$a_4 = \frac{2(2-n)}{12} a_2 = 2\frac{(2-n)}{12}(-na_0) = (2)^2 \frac{n(n-2)}{4!} a_0$$

If $k = 3$, then,
$$a_5 = \frac{2(3-n)}{20} a_3 = \frac{2(3-n)}{20}\left(-\frac{2(n-1)}{3!} a_1\right) = (2)^2 \frac{(n-1)(n-3)}{5!} a_1$$

$$a_{2r} = \frac{(-2)^r\, n(n-2)(n-4)........(n-2r+2)}{(2r)!} a_0$$

$$a_{2r+1} = \frac{(-2)^r\,(n-1)(n-3).......(n-2r+1)}{(2r+1)!} a_1 = 0$$

When $m = 0$, then there are two possibilities

Possibility I. When, $a_1 = 0$, then $a_3 = a_5 = a_7 = a_{2r+1} = ... = 0$.

Possibility II. When $a_1 \neq 0$

$$y = \sum_{k=0}^{\infty} a_k x^k$$

i.e. $y = a_0 + a_1x + a_2x^2 + a_3x^3 + a_4x^4 + a_5x^5 + ...$

$= a_0 + a_2x^2 + a_4x^4 + ... + a_1x + a_3x^3 + a_5x^5$ ...(8)

Putting the values of a_0, a_1, a_2, a_3, a_4 and a_5 in (8), we get

$$= a_0\left[1-\frac{2n}{2!}x^2+\frac{2^2 n(n-2)}{4!}x^4-...+(-1)^r\frac{2}{(2r)!}n(n-2)...(n-2r+2)x^{2r}+...\right]$$

$$+a_1 x\left[1-\frac{2(n-1)}{3!}x^2+\frac{2^2(n-1)(n-3)}{5!}-....\right.$$

$$\left.+(-1)^r\frac{2^r}{(2r+1)!}(n-1)(x-3)...(n-2r+1)x^{2r}+...\right] \quad ...(9)$$

$$= a_0\left[1+\sum_{r=1}^{\infty}\frac{(-1)^r 2^r}{(2r)!}n(n-2)...(n-2r+2)x^{2r}\right]$$

$$+a_0\left[x+\sum_{r=1}^{\infty}\frac{(-1)^r 2^r}{(2r+1)}(n-1)(n-3)...(n-2r+2)x^{2r+1}\right] \quad (\text{If } a_1 = a_0) \quad ...(10)$$

Case II. When $m = 1$, then $a_1 = 0$ and so by putting $k = 0, 1, 2, 3,.......$ in (7), we get

$$a_{k+2} = \frac{2(k+1-n)}{(k+3)(k+2)}a_k$$

$$a_2 = -\frac{2(n-1)}{3!}a_0$$

$$a_4 = \frac{2^2(n-1)(n-3)}{5!}a_0$$

....

$$a_{2r} = (-1)^r\frac{2^r(n-1)(n-3)...(n-2r+1)}{(2r+1)!}a_0$$

Hence, the solution is

$$= a_0 x\left[1-\frac{2(n-1)}{3!}x^2+\frac{2^2(n-1)(n-3)}{5!}x^4...+\frac{(-1)^r 2^r(n-1)(n-3)...(n-2r+1)}{(2r+1)!}x^{2r}+...\right] \quad ...(11)$$

It is clear that the solution (11) is included in the second part of (9) except that a_0 is replaced by a_1 and hence in order that the Hermite equation may have two independent solutions, a_1 must be zero, even if $m = 0$ and then (9) reduces to

$$y = a_0\left[1-\frac{2n}{2!}x^2+\frac{2^2 n(n-2)}{4!}x^4-...+(-1)^r\frac{2^r}{(2r)!}n(n-2)...(n-2r+2)x^{2r}+...\right] \quad ...(12)$$

The complete integral of (1) is then given by

$$y = A\left[1-\frac{2n}{2!}x^2+\frac{2^2 n(n-2)}{4!}x^4-...\right]+B\left[1-\frac{2(n-1)}{3!}x^2+\frac{2^2(n-1)(n-3)}{5!}x^4...\right] \quad ...(13)$$

where A and B are arbitrary constants.

30.3 GENERATING FUNCTION OF HERMITE POLYNOMIALS (RODRIGUE FORMULA)

We know that

$$e^{x^2}\frac{\partial^n}{\partial t^n}e^{\left\{-(t-x)^2\right\}} = H_n(x)+H_{n+1}(x)t+H_{n+2}(x).\,t^2+\ldots\ldots \qquad \ldots(1)$$

Now differentiating $e^{\left\{-(t-x)^2\right\}}$ w.r.t., t, we get

$$\frac{\partial}{\partial t}e^{\left\{-(t-x)^2\right\}} = -2(t-x)e^{\left\{-(t-x)^2\right\}}$$

Taking limit when $t\to 0$, we get

$$\lim_{t\to 0}\frac{\partial}{\partial t}e^{\left\{-(t-x)^2\right\}} = 2xe^{-x^2} \qquad \ldots(2)$$

Again differentiating $e^{\left\{-(t-x)^2\right\}}$ w.r.t. 'x', we get

$$\frac{\partial}{\partial x}e^{\left\{-(t-x)^2\right\}} = (-1)^2\,2(t-x)\,e^{\left\{-(t-x)^2\right\}}$$

Taking limit when $t\to 0$, we get

$$\lim_{t\to 0}\frac{\partial}{\partial x}e^{\left\{-(t-x)^2\right\}} = -2xe^{-x^2} \qquad \ldots(3)$$

From (2) and (3), we have

$$\lim_{t\to 0}\frac{\partial}{\partial t}e^{\left\{-(t-x)^2\right\}} = (-1)^1\lim_{t\to 0}\frac{\partial}{\partial x}e^{\left\{-(t-x)^2\right\}}$$

Similarly,

$$\lim_{t\to 0}\frac{\partial^2}{\partial t^2}e^{\left\{-(t-x)^2\right\}} = (-1)^2\lim_{t\to 0}\frac{\partial^2}{\partial x^2}e^{\left\{-(t-x)^2\right\}}$$

....

....

$$\lim_{t\to 0}\frac{\partial^n}{\partial t^n}e^{\left\{-(t-x)^2\right\}} = (-1)^n\lim_{t\to 0}\frac{\partial^n}{\partial x^n}e^{\left\{-(t-x)^2\right\}} = (-1)^n\frac{d^n}{dx^n}e^{-x^2}$$

[differentiating n times] ...(4)

Putting $t = 0$ in (1), we get

$$\lim_{t\to 0}e^{x^2}\frac{\partial^n}{\partial t^n}e^{\left\{-(t-x)^2\right\}} = H_n(x) \qquad \ldots(5)$$

Putting the value of $\lim_{t\to 0}\frac{\partial^n}{\partial t^n}e^{\left\{-(t-x)^2\right\}}$ from (4) in (5), we get

$$(-1)^n e^{x^2}\frac{d^n}{dx^n}e^{-x^2} = H_n(x)$$

$$H_n(x) = (-1)^n e^{x^2}\frac{d^n}{dx^n}e^{-x^2} \qquad \ldots(6)$$

$\boldsymbol{n = 0}$

On putting $n = 0$ in (6), we get

$$H_0(x) = (-1)^0 e^{x^2}e^{-x^2} = 1$$

$$H_0(x) = 1$$

***n* = 1**

On putting $n = 1$ in (6), we get

$$H_1(x) = (-1)^1 e^{x^2} \frac{d}{dx} e^{-x^2} = -e^{x^2}(-2x)e^{-x^2} = 2x$$

$$H_1(x) = 2x$$

***n* = 2**

On putting $n = 2$ in (6), we get

$$H_2(x) = (-1)^2 e^{x^2} \frac{d^2}{dx^2} e^{-x^2} = e^{x^2} \frac{d}{dx}\left(-2xe^{-x^2}\right)$$

$$= e^{x^2}\left[-2e^{x^2} - 2x(-2x)e^{-x^2}\right]$$

$$= -2 + 4x^2$$

$$H_2(x) = 4x^2 - 2$$

***n* = 3**

On putting $n = 3$ in (6), we get

$$H_3(x) = (-1)^3 e^{x^2} \frac{d^3}{dx^3}\left(e^{-x^2}\right) = -e^{x^2} \frac{d^2}{dx^2}\left(-2xe^{-x^2}\right)$$

$$= -e^{x^2} \frac{d}{dx}\left(-2e^{-x^2} + (-2x)(-2x)e^{-x^2}\right)$$

$$= -e^{x^2} \frac{d}{dx}\left(-2 + 4x^2\right)e^{-x^2} = -e^{x^2}\left[8xe^{-x^2} + \left(4x^2 - 2\right)(-2x)e^{-x^2}\right]$$

$$= -\left[8x + \left(4x^2 - 2\right)(-2x)\right] = -8x + 8x^3 - 4x = 8x^3 - 12x$$

$$H_3(x) = 8x^3 - 12x$$

Similarly $\quad H_4(x) = 16x^4 - 48x^2 + 12$

$$H_5(x) = 32x^5 - 160x^3 + 120x$$

$$H_6(x) = 64x^6 - 480x^4 + 720x^2 - 120$$

$$H_7(x) = 128x^7 - 1344x^5 + 3360x^3 - 1680x$$

Example 1. *Convert Hermite polynomial*

$$2H_4(x) + 3H_3(x) - H_2(x) + 5H_1(x) + 6H_0$$

into ordinary polynomial.

Solution. Here, we have

$$2H_4(x) + 3H_3(x) - H_2(x) + 5H_1(x) + 6H_0$$

$$= 2\left[16x^4 - 48x^2 + 12\right] + 3\left\{8x^3 - 12x\right\} - \left(4x^2 - 2\right) + 5(2x) + 6(1)$$

$$= 32x^4 - 96x^2 + 24 + 24x^3 - 36x - 4x^2 + 2 + 10x + 6$$

$$= 32x^4 + 24x^3 - 100x^2 - 26x + 32$$

Ans.

Example 2. *Convert ordinary polynomial*

$$64x^4 + 8x^3 - 32x^2 + 40x + 10$$

into Hermite polynomial.

Solution. Here, we have

Let $64x^4 + 8x^3 - 32x^2 + 40x + 10 = AH_4(x) + BH_3(x) + CH_2(x) + DH_1(x) + EH_0(x)$

$= A(16x^4 - 48x^2 + 12) + B(8x^3 - 12x) + C(4x^2 - 2) + D(2x) + E(1)$

$= 16Ax^4 + 8Bx^3 + (-48A + 4C)x^2 + (-12B + 2D)x + 12A - 2C + E$

Equating the coefficients of like powers of x, we get

$$16A = 64 \Rightarrow A = 4$$
$$8B = 8 \Rightarrow B = 1$$
$$-48A + 4C = -32 \Rightarrow 4C = -32 + 192 \Rightarrow C = 40$$
$$-12B + 2D = 40 \Rightarrow -12 + 2D = 40 \Rightarrow 2D = 52 \Rightarrow D = 26$$
$$12A - 2C + E = 10 \Rightarrow 12 \times 4 - 2(40) + E = 10 \Rightarrow E = 42$$

The required Hermite polynomial is

$$4H_4(x) + H_3(x) + 40H_2(x) + 26H_1(x) + 42H_0(x)$$ **Ans.**

30.4 ORTHOGONAL PROPERTY

The *orthogonal property* of Hermites polynomials is

$$\int_{-\infty}^{\infty} e^{-x^2} H_m(x) H_n(x)\,dx = \begin{cases} 0, & m \neq n \\ 2^n n! \sqrt{\pi}, & m = n \end{cases}$$

Solution. We know that,

$$e^{\{x^2-(t_1-x)^2\}} = \sum \frac{H_n(x)}{n!} t_1^n \qquad \text{(generating function)} \qquad ...(1)$$

and $$e^{\{x^2-(t_2-x)^2\}} = \sum \frac{H_m(x)}{m!} t_2^m \qquad ...(2)$$

Multiplying (1) and (2), we get

$$e^{\{x^2-(t_1-x)^2\}} . e^{\{x^2-(t_2-x)^2\}} = \left[\sum_{n=0}^{\infty} \frac{H_n(x)}{n!} t_1^n\right]\left[\sum_{m=0}^{\infty} \frac{H_m(x)}{m!} t_2^m\right]$$

$$= \sum_{\substack{n=0\\m=0}}^{\infty} \left[H_n(x) | H_m(x)\right] \frac{t_1^n . t_2^m}{n! m!}$$

Multiplying both the sides of this equation by e^{-x^2} and then integrating with the limits from $-\infty$ to ∞, we have

$$\sum_{nm} \left[\int_{-\infty}^{\infty} e^{-x^2} H_n(x) H_m(x)\,dx\right] \frac{t_1^n . t_2^m}{n! m!} = e^{-x^2} \int_{-\infty}^{\infty} e^{\{x^2-(t_1-x)^2\}} . e^{\{x^2-(t_2-x)^2\}} dx$$

$$= \int_{-\infty}^{\infty} e^{\{x^2-(t_1-x)^2\}-(t_2-x)^2} dx$$

$$= e^{\{-(t_1^2+t_2^2)\}} \int_{-\infty}^{\infty} e^{\{-x^2+2x(t_1+t_2)\}} dx \qquad ...(3)$$

We have already learnt that $\int_{-\infty}^{\infty} e^{\{-ax^2+2bx\}} dx = \sqrt{\frac{\pi}{2}} e^{\frac{b^2}{a}}$ [standard formula] ...(4)

Replacing by $(t_1 + t_2)$ and a by 1 in (4), we get

$$\int_{-\infty}^{\infty} e^{\{-x^2+2x(t_1+t_2)\}} dx = \sqrt{\pi}\, e^{(t_1+t_2)^2} \qquad ...(5)$$

Putting the value of $\int_{-\infty}^{\infty} e^{\{-x^2+2x(t_1+t_2)\}} dx$ from (5) in R.H.S. of (3), we get

$$e^{\left\{-\left(t_1^2+t_2^2\right)^2\right\}}.\sqrt{\pi}e^{(t_1+t_2)^2} = \sqrt{\pi}\,e^{-t_1^2-t_2^2+t_1^2+t_2^2+2t_1t_2} = \sqrt{\pi}\,e^{2t_1t_2}$$

$$= \sqrt{\pi}\left[1+2t_1t_2+\frac{(2t_1t_2)^2}{2!}+\frac{(2t_1t_2)^3}{3!}+.......\right] = \sqrt{\pi}\sum\frac{(2t_1t_2)^n}{n!}$$

$$= \sqrt{\pi}\sum\frac{2^n t_1^n t_2^n}{n!} = \sqrt{\pi}\sum_{\substack{m=0\\n=0}}^{\infty} 2^n t_1^n t_2^m \delta_{m,n} \qquad \left[t_2^n = t_2^m\,\delta_{n,m}\right]$$

From (3) , we have

$$\sum_{nm}\left[\int_{-\infty}^{\infty} e^{-x^2} H_n(x)H_m(x)\,dx\right]\frac{t_1^n t_2^m}{n!m!} = \sqrt{\pi}\sum_{nm}\frac{2^n}{n!}t_1^n t_2^m \delta_{n,m}$$

On equating the coefficients of t_1^n, t_2^m on both sides., we get

$$\int_{-\infty}^{\infty} e^{-x2}\frac{H_n(x)H_m(x)}{n!m!}dx = \frac{\sqrt{\pi}2^n}{n!}\delta_{n,m}$$

$$\Rightarrow \qquad \int_{-\infty}^{\infty} e^{-x2}H_n(x)H_m(x)\,dx = \sqrt{\pi}2^n m!\delta_{n,m}$$

$$\Rightarrow \qquad \int_{-\infty}^{\infty} e^{-x^2}H_m(x)H_n(x)\,dx = \begin{cases}0 & m \neq n\\ 2^n n!\sqrt{\pi}, & m = n\end{cases}\begin{bmatrix}\delta_{n,m} = 0, \text{if } m \neq n\\ = 1, \text{if } m = n\end{bmatrix}$$ **Proved.**

Example 3. *Find the value of* $\int_{-\infty}^{\infty} e^{-x^2} H_2(x)H_3(x)\,dx$.

Solution. We know that

$$\int_{-\infty}^{\infty} e^{-x^2} H_m(x)H_n(x)\,dx = 0 \text{ if } m \neq n$$

Here $m = 2$ and $n = 3$, $m \neq n$

Hence, $\int_{-\infty}^{\infty} e^{-x^2} H_2(x)H_3(x)\,dx = 0$ **Ans.**

Example 4. *Find the value of* $\int_{-\infty}^{\infty} e^{-x^2}\left[H_2(x)\right]^2 dx$

Solution. We know that

$$\int_{-\infty}^{\infty} e^{-x^2}\left[H_n(x)\right]^2 dx = 2^n (n)!\sqrt{\pi}$$

$$\int_{-\infty}^{\infty} e^{-x^2}\left[H_2(x)\right]^2 dx = 2^2(2!)\sqrt{\pi} = 8\sqrt{\pi}$$ **Ans.**

30.5 RECURRENCE FORMULAE FOR H_n (X) OF HERMITE EQUATION

Four recurrence Relations

1. $2n\,H_{n-1}(x) = H'_n(x)$
2. $2x\,H_n(x) = 2nH_{n-1}(x) + H_{n+1}(x)$
3. $H'_n(x) = 2x\,H_n(x) - H_{n+1}(x)$
4. $H'_n(x) = x\,H'_n(x) + 2n\,H_n(x) = 0$

Recurrence Relation I

Hermite equation is $y'' - 2xy' + 2ny = 0$ for integral values taking $v = n$.

Also, $e^{2tx-t^2}=\sum_{n=0}^{\infty}\frac{H_n(x)t^n}{n.!}$ ...(1)

I. Differentiating partially *w.r.t.* x, we have (*DU II Sem. 2014*)

$$2t\,e^{2tx-t^2}=\sum_{n=0}^{\infty}H_n^{'}(x)\frac{t^n}{n!} \quad ....(2)$$

Substituting the value of $2te^{2tx-t^2}$ from (1) in (2), we get

i.e $$2t\sum_{n=0}^{\infty}\frac{H_n(x)t^n}{n.!}=\sum_{n=0}^{\infty}H_n^{'}(x)\frac{t^n}{n!}$$

$\Rightarrow$ $$2\sum_{n=0}^{\infty}\frac{H_n(x)t^{n+1}}{n!}=\sum_{n=0}^{\infty}H_n^{'}(x)\frac{t^n}{n!}$$

On replacing n by $n-1$ on L.H.S, we get

$$2\frac{H_{n-1}(x)t^n}{(n-1)!}=H_n^{'}(x)\frac{t^n}{n!}$$

$\Rightarrow$ $$\frac{2nH_{n-1}(x)t^n}{n!}=H_n^{'}(x)\frac{t^n}{n!}$$

On equating the coefficients of $\frac{t^n}{n!}$ on both sides , we get

$$2\frac{n!}{(n-1)!}H_{n-1}(x)=H'_n(x)$$

i.e $$2nH_{n-1}(x)=H_n^{'}(x) \quad ...(3)$$

$$\boldsymbol{2nH_{n-1}(x)=H_n^{'}(x)}$$

II. Differentiating partially *w.r.t.* 't', both sides of (1), we get

$$2(x-t)e^{2tx-t^2}=\sum_{n=0}^{\infty}H_n(x)\frac{nt^{n-1}}{(n-1)!}$$

$$2(x-t)e^{2tx-t^2}=\sum_{n=1}^{\infty}H_n(x)\frac{t^{n-1}}{(n-1)!} \quad (n=0 \text{ vanishes on R.H.S}) \quad ...(4)$$

On putting $n = 0$ R.H.S becomes zero.

Putting the value of e^{2tx-t^2} from (1) in (4) , we get

$$2(x-t)\sum_{n=0}^{\infty}H_n(x)\frac{t^n}{n!}=\sum_{n=1}^{\infty}H_n(x)\frac{t^{n-1}}{(n-1)!}$$

or $$2x\sum_{n=0}^{\infty}H_n(x)\frac{t^n}{n!}-2\sum_{n=0}^{\infty}H_n(x)\frac{t^{n+1}}{n!}=\sum_{n=1}^{\infty}H_n(x)\frac{t^{n-1}}{(n-1)!}$$

Equating the coefficients of t^n on either side, we get

$$2x\frac{H_n(x)}{n!}-2\frac{H_{n-1}(x)}{(n-1)!}=\frac{H_{n+1}(x)}{n!} \quad \text{(Replacing } n \text{ by } n+1 \text{ on R.H.S)}$$

i.e. $2xH_n(x) = 2nH_{n-1}(x) + H_{n+1}(x)$

$$\mathbf{2xH_n(x) = 2nH_{n-1}(x) + H_{n+1}(x)} \quad ...(5)$$

III. Eliminating $H_{n-1}(x)$ from recurrence relation (1) and (2), we get

$$2xH_n(x) = H'_n(x) + H_{n+1}(x)$$

$$H'_n(x) = 2xH_n(x) - H_{n+1}(x)$$

$$\mathbf{H'_n(x) = 2xH_n(x) - H_{n+1}(x)}$$

IV. Differentiating recurrence relation (3), *w.r.t. x,* we get

$$H''_n(x) = 2xH'_n(x) + 2H_n(x) - H'_{n+1}(x)$$

Putting $H'_{n+1}(x) = 2(n+1)H_n(x)$ obtained from recurrence relation (1) on replacing n by $n+1$, we get

$$H''_n(x) = 2xH'_n(x) + 2H_n(x) - 2(n+1)H_n(x)$$

$$\Rightarrow \quad H''_n(x) - 2x\,H'_n(x) + 2n\,H_n(x) = 0$$

which clearly follows that $y = H_n(x)$ is a solution of Hermite eqaution.

$$H''_n(x) - 2x\,H'_n(x) + 2n\,H_n(x) = 0$$

Example 5. *Prove that*

$$H_{2n}(0) = (-1)^n \cdot 2^{2n}\left(\frac{1}{2}\right)^n$$

Solution. We know that

$$H_{2n}(x) = \sum \frac{(-1)^n (2n)!(2x)^{2n+2x}}{x!(2n-2x)!} \quad ...(1) \quad \left[\begin{matrix}\text{Even Hermite}\\ \text{polynomial}\end{matrix}\right]$$

Putting $x = 0$ in (1), we get

$$H_{2n}(0) = \frac{(-1)^n (2n)!}{(n)!} = (-1)^n \frac{(2n)(2n-1)(2n-2).....1}{n(n-1)(n-2)........1}$$

$$= (-1)^n \frac{2(2n-1)2(2n-3)2(2n-5)2.....2.1}{n!} n!$$

$$= (-1)^n\, 2^n.2^n \frac{(2n-1)}{2}\frac{(2n-3)}{2}\frac{(2n-5)}{2}............\frac{3}{2}.\frac{1}{2}$$

$$= (-1)^n\, 2^{2n}\left(\frac{1}{2}\right)\left(\frac{3}{2}\right)\left(\frac{5}{2}\right)\left(\frac{7}{2}\right).........\left(\frac{2n-3}{2}\right)\left(\frac{2n-1}{2}\right)$$

$$= (-1)^n\, 2^{2n}\left(\frac{1}{2}\right)^n$$

Proved.

Exmaple 6. *Prove that*

$$H'_{2n+1}(0) = (-1)^n\, 2^{2n+1}\left(\frac{3}{2}\right)^n$$

Solution. We know that by Recurrence Relation I

$$H'_n(x) = 2n\, H_{n-1}(x) \qquad ...(1)$$

Replacing n by $2n+1$ in (1), we get

$$H'_{2n+1}(x) = 2(2n+1)\, H_{2n}(x) \qquad ...(2)$$

Putting $x = 0$ in (2), we get

$$H'_{2n+1}(0) = 2(2n+1)\, H_{2n}(0)$$

$$= 2(2n+1)(-1)^n 2^{2n}\left(\frac{1}{2}\right)^n$$

$$= (2n+1)(-1)^n 2^{2n+1}\left[\frac{(2n-1)(2n-3).......3.1}{2^n}\right]$$

[Using previous exp]

$$= (-1)^n 2^{2n+1}\left[\frac{3}{2}\left(\frac{3}{2}+1\right).......\left(\frac{3}{2}+n-1\right)\right]$$

$$= (-1)^n \cdot 2^{2n+1}\left(\frac{3}{2}\right)^n$$

Proved.

Example 7. *Prove that* (*DU, III Sem 2012*)

$$H_{2n+1}(0) = 0$$

Solution. We know that

$$H_{2n+1}(x) = \sum_{k=0}^{2n+1/2} \frac{(-1)^k (2n+1)!(2x)^{2n+1-2k}}{k!(2n+1-2k)} \qquad \left[\begin{matrix}\text{Odd Hermite}\\ \text{Polynomial}\end{matrix}\right]$$

Putting $x = 0$ in above, we get

$\therefore$ $$H_{2n+1}(0) = 0$$ **Proved.**

Example 8. *Prove that*

$$H'_{2n}(0) = 0$$

Solution. From recurrence relation I, we know that

$$H'_n(x) = 2n H_{n-1}(x)$$

Replacing n by $2n$, we get

$$H'_{2n}(x) = 2(2n) H_{2n-1}(x) \qquad ...(1)$$

Putting $x = 0$ in (1), we get

$$H'_{2n}(0) = 4n H_{2n-1}(0)$$

$$= 4n\,(0) \qquad \left[\begin{matrix}\text{From example 7}\\ H_{2n-1}(0) = 0\end{matrix}\right]$$

$$= 0$$

Example 9. *Prove that*

$$\frac{d^m}{dx^m}\{H_n(x)\} = \frac{2^n (n)!}{(n-m)!} H_{n-m} \qquad m < n$$

Solution. From recurrence relation I, we know that

$$H'_n(x) = 2n\, H_{n-1}(x) \qquad ...(1)$$

$$\Rightarrow \qquad \frac{d}{dx}\{H_n(x)\} = 2nH_{n-1}(x)$$

$$\Rightarrow \qquad \frac{d^2}{dx^2}\{H_n(x)\} = 2n\frac{d}{dx}\left[H_{n-1}(x)\right]$$

$$= 2n\, H'_{n-1}(x)$$

$$= 2n\,[2\,(n-1)\, H_{n-2}(x)] \qquad \text{[From (1)]}$$

$$= 2^2 n\,(n-1)\, H_{n-2}(x)$$

Similarly $\frac{d^3}{dx^3}\{H_n(x)\} = 2^3 n(n-1)(n-2)H_{n-3}(x)$

Proceeding similarly m times, we get

$$\frac{d^m}{dx^m}\{H_n(x)\} = 2^m\, n(n-1)......(n-m+1)H_{n-m}(x), \cdot\ m < n$$

$$= \frac{2^m}{(n-m)!}H_{n-m}(x) \qquad \textbf{Proved.}$$

Example 10. *Prove that* $H_n(-x) = (-1)^n H_n(x)$.

Solution. Here, we have $\sum_{n=0}^{\infty}\frac{H_n(x)t^n}{n!} = e^{2tx-t^2} = e^{2tx}e^{-t^2} = \sum_{n=0}^{\infty}\frac{(2x)^n t^n}{n!} \times \sum_{n=0}^{\infty}\frac{(-1)t^{2n}}{n!}$

$$= \sum_{n=0}^{\infty}\sum_{k=0}^{n/2}\frac{(-1)^k (2x)^{n-2k}}{k!(n-2k)!}$$

Equating coefficient of $\frac{t^n}{n!}$ on either side, we get

$$H_n(x) = \sum_{k=0}^{n/2}\frac{(-1)^k\, n!(2x)^{n-2k}}{k!(n-2k)!}$$

Replacing x by $-x$, we get

$$H_n(-x) = \sum_{k=0}^{n/2}\frac{(-1)^k\, n!(-2x)^{n-2k}}{k!(n-2k)!}$$

$$= \sum_{k=0}^{n/2}\frac{(-1)^k (-1)^{n-2k}\, n!(2x)^{n-2k}}{k!(n-2k)!}$$

$$= (-1)^n \sum_{k=0}^{n/2}\frac{(-1)^k\, n!(2x)^{n-2k}}{k!(n-2k)!} = (-1)^n H_n(x)$$

Example 11. *Prove that*

$$\int_{-\infty}^{\infty} xe^{-x^2} H_n(x)H_m(x)\,dx = \sqrt{\pi}\left[2^{n-1} n!\delta_{m,n-1} + 2^n (n+1)\delta_{n+1,m}\right].$$

Solution. Integrating by parts, we have

$$\int_{-\infty}^{\infty} xe^{-x^2} H_n(x) H_m(x) dx = \left[-\frac{1}{2} e^{-x^2} H_n(x) H_m(x) dx \right]_{-\infty}^{\infty}$$

$$+ \frac{1}{2} \int_{-\infty}^{\infty} e^{-x^2} \frac{d}{dx} \{H_n(x) H_m(x)\} dx$$

$$= 0 + \frac{1}{2} \int_{-\infty}^{\infty} e^{-x^2} \frac{d}{dx} \{H_n(x) H_m(x)\} dx \text{ (Orthogonality property)}$$

Differentiating by product rule *w.r.t.* 'x' under the sign of integration, we get

$$= \frac{1}{2} \int_{-\infty}^{\infty} e^{-x^2} \{H'_n(x) H_m(x) + H_n(x) H'_m(x)\} dx$$

By putting the values of $H'_n(x) = 2nH_{n-1}(x)$ and $H'_m(x) = 2mH_{m-1}(x)$ by recurrence relation (1), we get

$$= \frac{1}{2} \int_{-\infty}^{\infty} e^{-x^2} [2nH_{n-1}(x) H_m(x) + 2mH_n(x) H_{m-1}(x)] dx$$

$$= n \int_{-\infty}^{\infty} e^{-x^2} H_{n-1}(x) H_m(x) dx + m \int_{-\infty}^{\infty} e^{-x^2} H_n(x) H_{m-1}(x) dx$$

$$= n\sqrt{\pi} 2^{n-1} (n-1)! \delta_{m,n-1} + m\sqrt{\pi} 2^n n! \delta_{n,m-1} \quad \text{(by orthogonal properties)}$$

By putting the value of $\delta_{n,m-1} = \delta_{n+1,m}$, we get

$$= \sqrt{\pi} \left[2^{n-1} n! \delta_{m,n-1} + 2^n (n+1)! \delta_{n+1,m} \right]$$

Proved.

EXERCISE 30.1

1. Find the value of $\int_{-\infty}^{\infty} e^{-x^2} H_{20}(x) H_{10}(x) dx$ **Ans. 0**

2. Find the value of $\int_{-\infty}^{\infty} e^{-x^2} [H_9(x)]^2 dx$ **Ans.** $2^9 (9)! \sqrt{\pi}$

3. Convert the Hermite polynomial $2H_3 - 4H_2 + H_1 + H_0$ into ordinary polynomial.
Ans. $16x^3 - 16x^2 - 22x + 9$

4. Convert $8x^3 + 8x^2 - 6x + 2$ into Hermite polynomial
Ans. $H_3 + 2H_2 + 3H_1 + 6H_0$

5. Convert ordinary polynomial $16x^4 + 4x^3 - 8x^2 + 20x + 8$ into Hermite polynomial.
Ans. $H_4(x) + \frac{1}{2} H_3(x) + 10H_2(x) + 13H_1(x) + 16H_0(x)$

6. Prove that $2n H_{n-1}(x) = H'_n(x)$

7. Verify, $P_n(x) = \frac{2}{\sqrt{\pi} n!} \int_0^{\infty} t^n e^{-t^2} H_n(xt) dt$.

8. Show that if m is an integer, $\int_{-\infty}^{\infty} x^m e^{-x^2} H_n(x) dx = 0$.

CHAPTER

LAGUERRES FUNCTIONS

31.1 LAGUERRES FUNCTION

The Laguerres differential equation is $x\frac{d^2y}{dx^2}+(1-x)\frac{dy}{dx}+ny=0$

Solution. Here, we have

$$x\frac{d^2y}{dx^2}+(1-x)\frac{dy}{dx}+ny=0 \quad \text{...(1)}$$

$$\Rightarrow \quad \frac{d^2y}{dx^2}+\left(\frac{1-x}{x}\right)\frac{dy}{dx}+\frac{n}{x}y=0$$

Here $x = 0$ is a regular singularity of (1). So we will solve it by series solution method. :

Let $\quad y=\sum_{k=0}^{\infty} a_k x^{m+k}$

$$\left[y=a_0x^m+a_1x^{m+1}+a_2x^{m+2}+a_3x^{m+3}+\text{.......}+a_kx^{m+k}+\text{.......}\right] \quad \text{...(2)}$$

$$\Rightarrow \quad \frac{dy}{dx}=\sum_{k=0}^{\infty}(m+k)a_kx^{m+k-1}$$

$$\Rightarrow \quad \frac{d^2y}{dx^2}=\sum_{k=0}^{\infty}(m+k)(m+k-1)a_kx^{m+k-2}$$

Putting the values of y, $\frac{dy}{dx}$ and $\frac{d^2y}{dx^2}$ in (1), we get

$$x\sum_{k=0}^{\infty}(m+k)(m+k-1)a_kx^{m+k-2}+(1-x)\sum_{k=0}^{\infty}(m+k)a_kx^{m+k-1}+n\sum_{k=0}^{\infty}a_kx^{m+k}=0$$

$$\Rightarrow \quad \sum_{k=0}^{\infty}(m+k)(m+k-1)a_kx^{m+k-1}+\sum_{k=0}^{\infty}(m+k)a_kx^{m+k-1}-\sum_{k=0}^{\infty}(m+k)a_kx^{m+k}+n\sum_{k=0}^{\infty}a_kx^{m+k}=0$$

$$\Rightarrow \quad \sum_{k=0}^{\infty}a_k\{(m+k)(m+k-1)+(m+k)\}x^{m+k-1}-\sum_{k=0}^{\infty}a_k\{m+k-n\}x^{m+k}=0$$

$$\Rightarrow \qquad \sum_{k=0}^{\infty} a_k \left\{(m+k)^2-(m+k)+(m+k)\right\} x^{m+k-1} - \sum_{k=0}^{\infty} a_k \left\{m+k-n\right\} x^{m+k} = 0$$

$$\Rightarrow \qquad \sum_{k=0}^{\infty} a_k (m+k)^2 x^{m+k-1} - \sum_{k=0}^{\infty} a_k (m+k-n) x^{m+k} = 0 \qquad ...(3)$$

Equating to zero the coefficient of lowest degree term x^{m-1}

($x^{m-1}$is obtained by putting $k = 0$ in the first summation of (3), but we cannot put $k = -1$ in second summation to get x^{m-1} since k is always positive.)

$$a_0 m^2 = 0 \qquad \text{(Indicial equation)}$$

$$\Rightarrow m = 0, \; m = 0, a_0 \neq 0$$

Again equating to zero, the coefficient of x^{m+k} in (2), we get

(To get x^{m+k} we put $k \Rightarrow k+1$ in first summation and $k \Rightarrow k$ in second summation of (2).

$$(m+k+1)^2 a_{k+1} - (m+k-n)\, a_k = 0$$

$$\Rightarrow \qquad a_{k+1} = \frac{(m+k-n)}{(m+k+1)^2} a_k \qquad ...(4)$$

For $m = 0$ in (4), we have

$$a_{k+1} = \frac{k-n}{(k+1)^2} a_k$$

If $k = 0$, then $\qquad a_1 = -na_0$

If $k = 1$, then $\qquad a_2 = \dfrac{1-n}{4} a_1 = \dfrac{(n-1)}{4} n a_0$

If $k = 2$, then $\qquad a_3 = \dfrac{2-n}{9} a_2 = \left(\dfrac{(2-n)}{9}\right)\left(\dfrac{(n-1)n}{4}\right) a_0 = (-1)^3 \dfrac{n(n-1)(n-2)}{(3!)^2} a_0$

$$a_k = (-1)^k \frac{n(n-1)(n-2)---(n-k+1)}{(k!)^2} a_0$$

On putting these values of the coefficients a_1, a_2, a_3 .... a_k and $m = 0$ in (2), we get

$y = a_0 + a_1 x + a_2 x^2 + a_3 x^3$ .... $+ a_k x^k +$ ....

$$y = a_0 - n a_0 x + \frac{n(n-1)}{(2!)^2} a_0 x^2 - \frac{n(n-1)(n-2)}{(3!)^2} a_0 x^3 + ...$$

$$+(-1)^k \frac{n(n-1)(n-2).........(n-k+1)}{(k!)^2} a_0 x^k + ...$$

$$= a_0 \left[1 - nx + \frac{n(n-1)}{(2!)^2} x^2 - \frac{n(n-1)(n-2)}{(3!)^2} x^3 + ... + (-1)^k \frac{n(n-1)(n-2)---(n-k+1)}{(k!)^2} x^k + ---\right]$$

$$= a_0 \sum_{k=0}^{\infty} (-1)^k \frac{(n!)}{(k!)^2 (n-k)!} x^k \text{ ,where } n \text{ is positive.}$$

If we take $a_0 = n!$, then solution of (1) becomes Laguerres polynomial

$$y = n!\left[1 - nx + \frac{n(n-1)}{(2!)^2}x^2 - \frac{n(n-1)(n-2)}{(3!)^2}x^3 + \dots\right.$$

$$\left. + \frac{(-1)^k n(n-1)(x-2)......(n-k+1)}{(k!)^2}x^k + \dots\right]$$

$$\Rightarrow \qquad L_n(x) = (-1)^n\left[x^n - \frac{n^2}{1!}x^{n-1} + \frac{n^2(n-1)^2}{2!}x^{n-2} + ... + (-1)^n n!\right]$$

This is the expression for Laguerre's polynomial.

31.2 LAGUERRES FUNCTION FOR DIFFERENT VALUES OF *n*.

$L_n(0) = n!$ (*DU, II Sem. 2014*)

$L_0(x) = 1$

$L_1(x) = 1 - x$

$L_2(x) = x^2 - 4x + 2$

$L_3(x) = -x^3 + 9x^2 - 18x + 6$

$L_4(x) = x^4 - 16x^3 + 72x^2 - 96x + 48$

and so on.

31.3 GENERATING FUNCTION OF LAGUERRE POLYNOMIAL

$$(1-t)\sum_{n=0}^{\infty}\frac{L_n(x)}{n!}t^n = e^{\frac{-xt}{1-t}}$$

Solution : Here, we have $(1-t)\sum_{n=0}^{\infty}\frac{L_n(x)}{n!}t^n = e^{\frac{-xt}{1-t}}$

$$\sum_{n=0}^{\infty}\frac{L_n(x)t^n}{n!} = \frac{1}{1-t}e^{\frac{-xt}{1-t}} = \frac{1}{1-t}\left[1 - \frac{xt}{(1-t)} + \frac{x^2t^2}{2!(1-t)^2} - \dots + \frac{(-1)^k x^k t^k}{k!(1-t)^k} + \dots\right]$$

$$= \sum_{k=0}^{\infty}-\frac{(-1)^k x^k t^k}{k!(1-t)^{k+1}} = \sum_{k=0}^{\infty}-\frac{(-1)^k x^k t^k}{k!}(1-t)^{-(k+1)}$$

$$= \sum_{k=0}^{\infty}-\frac{(-1)^k x^k}{k!}t^k\left[1 + (k+1)t + \frac{(k+1)(k+2)}{2!}t^2 + \dots + \frac{(k+1)(k+2)\dots(k+1)}{l!}t^l + \dots\right]$$

$$= \sum_{k=0}^{\infty}\sum_{l=0}^{\infty}\frac{(-1)^k(k+1)l}{k!l!}x^k t^{k+1}, \text{where } (k+1)_l = \frac{\Gamma(k+1+l)}{\Gamma(k+1)}$$

Equating the coefficients of t^n on both sides, we get

On putting $l = n - k$, we get the coefficient of x^k.

$$\frac{L_n(x)}{n!} = \sum_{k=0}^{\infty}\frac{(-1)^k(k+1)_{n-k}}{r!(n-k)!}x^k$$

Here $\qquad (k+1)_{n-k} = \dfrac{\Gamma(k+1+n-k)}{\Gamma(k+1)} + \dfrac{\Gamma(n+1)}{\Gamma(k+1)} = \dfrac{n!}{k!}$

$$\frac{(-1)^k}{(n-k)!}=\frac{(-1)^k n(n-1)...(n-k+1)}{n!}$$

$$=\frac{(-n)(-n+1)(-n+2)...(-n+k-1)}{n!}=\frac{(-n)_k}{n!}$$

$$L_n(-n)=n!\sum_{k=0}^{\infty}\frac{(-n)_k}{n!}.\frac{n!}{(k!)^2}x^k=n!\sum_{k=0}^{\infty}\frac{(-n)_k}{(k!)^2}x^k$$

$$=n!\left[1+\frac{(-n)}{1!1!}x+\frac{(-n)(-n+1)}{2!\,2!}x^2+\frac{(-n)(-n+1)(-n+2)}{3!\,3!}x^3+...\right]$$

$$=n!F(-n,1;x)$$

From which it follows that $L_n\ (x)$ is a polynomial of degree n in x and that the coefficient of x^n is $(-1)^n$.

31.4 RECURRENCE RELATION

Relation I : We know that

$$e^{\frac{-xt}{1-t}}=(1-t)\sum_{n=0}^{\infty}\frac{L_n(x)}{n!}t^n \qquad ...(1)$$

Differentiating (1) *w.r.t* *'t'*,we get

$$-\frac{x}{(1-t)^2}e^{\frac{-xt}{1-t}}=(1-t)\sum_{n=0}^{\infty}\frac{L_n(x)t^{n-1}}{(n-1)!}-\sum_{n=0}^{\infty}\frac{L_n(x)t^n}{n!}$$

$$\Rightarrow \qquad -\frac{x}{(1-t)^2}(1-t)\sum_{n=0}^{\infty}\frac{L_n(x)t^n}{n!}=(1-t)\sum_{n=0}^{\infty}\frac{L_n(x)t^{n-1}}{(n-1)!}-\sum_{n=0}^{\infty}\frac{L_n(x)t^n}{n!}$$

$$\Rightarrow \qquad x\sum_{n=0}^{\infty}\frac{L_n(x)t^n}{n!}+(1-t)^2\sum_{n=0}^{\infty}\frac{L_n(x)t^{n-1}}{(n-1)!}-(1-t)\sum_{n=0}^{\infty}\frac{L_n(x)t^n}{n!}=0$$

Equating to zero the coefficient of t^n, we get

$$x\frac{L_n(x)}{n!}+\frac{L_{n+1}(x)}{n!}-2\frac{L_n(x)}{(n-1)!}+\frac{L_{n-1}(x)}{(n-2)!}-\frac{L_n(x)}{n!}+\frac{L_{n-1}(x)}{(n-1)!}=0$$

$\Rightarrow \quad L_{n-1}(x) + (x-2n-1)\, L_n\ (x) + n^2\, L_{n-1}\ (x) = 0$

Relation II : We know that

$$e^{\frac{-xt}{1-t}}=(1-t)\sum_{n=0}^{\infty}\frac{L_n(x)}{n!}t^n \qquad ...(1)$$

On differentiating (1) w.r.t. x, we get

$$-\left(\frac{t}{1-t}\right)e^{\frac{-xt}{1-t}}=(1-t)\sum_{n=0}^{\infty}\frac{L'_n(x)t^n}{n!}$$

$$\Rightarrow \qquad -\left(\frac{t}{1-t}\right)(1-t)\sum_{n=0}^{\infty}\frac{L_n(x)}{n!}t^n=(1-t)\sum_{n=0}^{\infty}\frac{L'_n(x)}{n!}t^n$$

$$\Rightarrow \qquad t\sum_{n=0}^{\infty}\frac{L_n(x)t^n}{n!}+(1-t)\sum_{n=0}^{\infty}\frac{L'_n(x)t^n}{n!}=0$$

Equating to zero the coefficient of t^n, we get

$$\frac{L'_n(x)}{n!}-\frac{L'_{n-1}(x)}{(n-1)!}+\frac{L_n(x)}{(n-1)!}=0$$

i.e. $$L'_n(x) - n\,L'_{n-1}(x) + n\,L_{n-1}(x) = 0$$

Relation III : By recurrence relation (1), we have (*DU, II Sem. 2014*)

$$L_{n+1}(x) + (x - 2n - 1)\,L_n(x) + x^2\,L_{n-1}(x) = 0 \qquad \text{...(1)}$$

On differentiating (1) w.r.t 'x', we get

$$L'_{n+1}(x) - (x - 2n - 1)\,L'_n(x) + L_n(x) + n^2\,L'_{n-1}(x) = 0. \qquad \text{...(2)}$$

Differentiating (2) w.r.t 'x', we get

$$L''_{n+1}(x) + (x - 2n - 1)\,L''_n(x) + 2L'_n(x) + n^2\,L''_{n-1}(x) = 0.$$

Let us replace n by $(n + 1)$, we get

$$L''_{n+2}(x) + (x - 2n - 3)\,L''_{n-1}(x) + (n+1)^2\,L_n''(x) + 2L'_{n+1}(x) = 0. \qquad \text{...(3)}$$

From recurrence relation (2)

$$L_n'(x) = n\,\{L'_{n-1}(x) - L_{n-1}(x)\}$$

On replacing n by $(n+1)$, we get

$$L'_{n+1}(x) + (n+1)\,\{L'_n(x) - L_n(x)\} \qquad \text{...(4)}$$

On differentiating (4), we get

$$L''_{n+1}(x) = (n+1)\,\{L_n''(x) - L_n'(x)\}. \qquad \text{...(5)}$$

Again replacing n by $(n + 1)$, we get

$$L'_{n+2}(x) = (n+2)\,\{L''_{n+1}(x) - L'_{n+1}(x)\} \qquad \text{...(6)}$$

From (3),

$$(n+2)\,\{L''_{n+1}(x) - L'_{n+1}(x)\} + (x - 2n - 3)L''_{n+1}(x) + (n+1)^2\,L_n''(x) + 2L'_{n+1}(x) = 0$$

$$(x - n - 1)\,L''_{n+1}(x) - n\,L'_{n+1}(x) + (n+1)^2\,L_n''(x) = 0$$

Eliminating $L''_{n+1}(x)$ and $L'_{n+1}(x)$ by (4) and (5), we get

$$x\,L''_n(x) + (1-x)\,L'_n(x) + n\,L_n(x) = 0$$

So, $t\,y = A\,L_n(x)$ is a solution of Laguerre equation.

Relation IV : We know that

$$(1-t)^{-1}e^{\left(1-\frac{1}{1-t}\right)x}=\sum_{n=0}^{\infty}\frac{L_n(x)t^n}{n!} \qquad \text{...(1)}$$

Differentiating (1) w.r.t 't' n times by Leibnitz theorem, we have

$$e^x\frac{d^n}{dt^n}\left[(1-t)^{-1}e^{-\frac{x}{1-t}}\right]=L_n(x)+L_{n+1}(x)+\ldots\ldots$$

Now, $$\frac{d}{dt}\left[(1-t)^{-1}e^{-\frac{x}{1-t}}\right]=\frac{1-x-t}{(1-t)^3}e^{-\frac{x}{1-t}}$$

Taking limit when $t \to 0$, we get

$$\lim_{t\to 0}\frac{d}{dt}\left[(1-t)^{-1}e^{-\frac{x}{1-t}}\right]=(1-x)e^{-x}\frac{d}{dx}(xe^{-x})$$

Similarly

$$\lim_{t\to 0}\frac{d^2}{dt^2}\left[(1-t)^{-1}e^{-\frac{x}{1-t}}\right]=\frac{d^2}{dx^2}(x^2e^{-x})$$

$$\Rightarrow \qquad \lim_{t\to 0}\frac{d^n}{dt^n}\left[(1-t)^{-1}e^{-\frac{x}{1-t}}\right]=\frac{d^n}{dx^n}(x^ne^{-x})$$

Hence, proceeding to the limit as $t \to 0$, we get

$$e^x\frac{d^n}{dx^n}(x^ne^{-x})=L_n(x)$$

31.5 ORTHOGONAL PROPERTY

Let $\qquad f_n(x)=\frac{1}{n!}e^{-x/2}L_n(x)$ ...(1)

$$\int_0^\infty f_m(x)f_n(x)dx=\int_0^\infty e^{-x}\frac{L_m(x)}{m!}\frac{L_n(x)}{n!}dx=\delta_{m,n}$$

Over the interval $0\le x\le \infty 1$ when $\delta_{m,n}=\begin{cases}0 \text{ for } m\ne n\\ 1 \text{ for } m=n\end{cases}$

From recurrence relation (4), we know that

$$L_n(x)=e^x\frac{d^n}{dx^n}(x^ne^{-x})$$

So, we get

$$\int_0^\infty e^{-x}x^mL_n(x)dx=\int_0^\infty e^{-x}x^me^x\frac{d^n}{dx^n}(x^ne^{-x})dx$$

$$=\int_0^\infty x^m\frac{d^n}{dx^n}(x^ne^{-x})dx$$

Integrating the R.H.S by parts, we get

$$\int_0^\infty e^{-x}x^mL_n(x)dx=\left[x^m\frac{d^{n-1}}{dx^{n-1}}x^ne^{-x}\right]_0^\infty-\int_0^\infty mx^{m-1}\frac{d^{n-1}}{dx^{n-1}}(x^ne^{-x})dx$$

$$=(-1)m\int_0^\infty e^{m-1}\frac{d^{n-1}}{dx^{n-1}}(x^ne^{-x})dx$$

$$=(-1)^2m(m-1)\int_0^\infty e^{m-2}\frac{d^{n-2}}{dx^{n-2}}(x^ne^{-x})dx$$

$$= \ldots\ldots\ldots\ldots\ldots\ldots\ldots\ldots$$

$$=(-1)^nm!\int_0^\infty\frac{d^{n-m}}{dx^{n-m}}(x^ne^{-x})dx$$

$$=0 \text{ if } n>m.$$

Replacing n by m, we get $(m < n)$

$$\int_0^\infty e^{-x} x^n L_m dx = 0 \text{ for } m < n$$

$$\Rightarrow \int_0^\infty e^{-x} L_m(x) L_n(x) dx = 0, \text{ if } m \neq n$$

$$\Rightarrow \int_0^\infty e^{-x} \frac{L_m(x)}{m!} \frac{L_n(x)}{n!} dx = 0, \quad \text{if } m \neq n$$

Taking $m = n$, then

$$L_n(x) \text{ is } (-1)^n x^n,$$

$$\therefore \int_0^\infty e^{-x} \{L_n(x)\}^2 dx = (-1)^n \int_0^\infty e^{-x} x^n L_n(x) dx$$

$$= (-1)^n \int_0^\infty e^{-x} x^n e^x \frac{d^n}{dx^n}(x^n e^{-x}) dx$$

$$= (-1)^n \int_0^\infty x^n \frac{d^n}{dx^n}(x^n e^{-x}) dx$$

$$= (-1)^{2n} n! \int_0^\infty x^n e^{-x} dx$$

$$= (n!)^2$$

$$\Rightarrow \int_0^\infty e^{-x/2} \frac{L_n(x)}{n!} e^{-x/2} \frac{L_n(x)}{n!} dx = 1 \qquad ...(3)$$

Combining (2) and (3), we get

$$\int_0^\infty f_m(x) f_n(x) dx = \int_0^\infty e^{-x/2} \frac{L_m(x)}{m!} e^{-x/2} \frac{L_n(x)}{n!} dx = \delta_{m,n}$$

EXERCISE 31.1

Prove the following :

1. $L_3^1(x) = -18 + 18x - 3x^2$
2. $L_4^2(x) = 144 - 96x + 12x$
3. $L_4^4(x) = 24.$
4. Find a series solution of
   $xy'' + (1 + x) y' + y = 0.$ **Ans.** $y = e^{-x} L_0(x) = e^{-x}$

CHAPTER 32

UNIT – IV

ABSTRACT VECTOR SPACES

32.1 INTRODUCTION

The theory of vector spaces is of much use in solving systems of linear equations. Its origin can be traced back to some topics in geometry and physics though it is now used in several different contexts.

32.2 SOME USEFUL DEFINITIONS

Set: A well defined collection of distinct objects is called a set.

32.3 TYPES OF SETS

1. Empty set: A set, which does not contain any element, is called an empty set or null set or void set. It is denoted by ϕ or { }.

2. Singleton set : A set, consisting of a single element, is called a singleton set.

3. Finite set: A set, which is empty or consists of a definite number of elements, is called a finite set.

4. Infinite set : A set, which is not finite, is called an infinite set.

5. Subset : If every element of A is an element of B, then A is called a subset of B. (*i.e.* $A \subseteq B$)

6. Proper subset: If $A \subseteq B$ and $A \neq B$, then A is called a proper subset of B, written as $A \subset B$.

7. Power set: The collection of all subsets of a set A is called the power set and it is denoted by $P(A)$. The power of a set A having n elements is the number of subsets of A and equal to 2^n.

8. Union of sets: The union of two sets A and B is the set containing all the elements of A and B, and it is written as $A \cup B$.

9. Intersection of sets: The intersection of two sets A and B is the set of common elements of A and B, denoted by $A \cap B$.

We shall denote several sets of numbers by the following symbols:

1. N : the set of natural numbers.
2. W : the set of whole numbers.
3. Z : the set of integers.

4. Q : the set of rational numbers.
5. R : the set of real numbers.
6. Z^+: the set of positive integers.
7. Q^+: the set of positive rational numbers.
8. R^+: the set of positive real numbers.

32.4 BINARY OPERATION

Let A be a non-empty set. A function $f: A * A \rightarrow A$ is called a binary operation * on set A, and is denoted by $(A, *)$.

Thus, a binary operation * on a non-empty set A is a function which associates $(a, b) \in A$ to a uniquely defined element $a * b \in A$. *This is a closure property.*

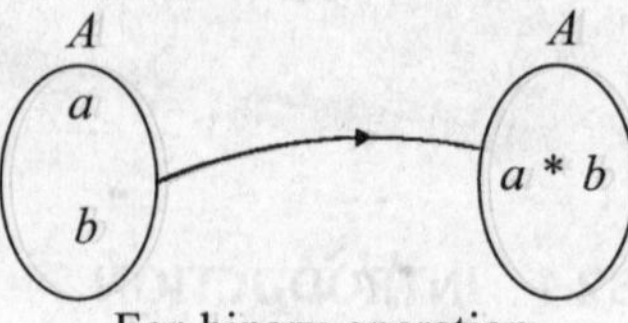

For binary operation

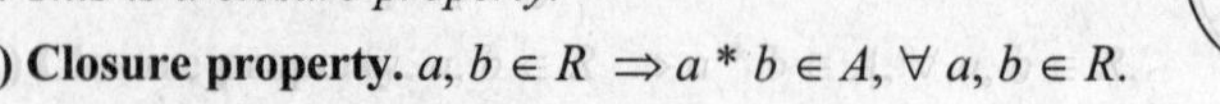

(1) Closure property. $a, b \in R \Rightarrow a * b \in A, \forall\ a, b \in R$.

> An operation * will be binary on set A,
>
> If $a, b \in A \Rightarrow a * b \in A$.

Notes 1. Instead of writing $*(a, b)$, we write $a * b$ for $a, b \in A$.

2. Binary operations are also denoted by 'O', '·', '+', '×'. We read *aob* as "*a* circle *b*", $a\,.\,b$ as "*a* dot *b*', $a + b$ as "*a* plus *b*", $a \times b$ as "*a* cross *b*".

1. Illustration. Addition (+) and multiplication (·) are binary operations on the set of even natural numbers, because the sum and product of two even natural numbers are also even natural numbers.

2. Illustration. Addition (+) is not binary operation on a set of odd natural numbers because sum of two odd numbers in not odd. Where as multiplication of odd natural numbers is binary operation because product of any two odd natural numbers is odd.

3. Illustration. Addition and multiplication are binary operations on Z (the set of integers), on Q (the set of rational numbers), on R (the set of real numbers) and on C (the set of complex numbers).

4. Illustration. Subtraction (–) is not a binary operation on the set of natural numbers N, because the subtraction of two natural numbers is not always natural number. **For example,** $4 \in N$ and $3 \in N$, but $3–4 \notin N$. Whereas, subtraction is a binary operation on Z, Q, R and C.

5. Illustration. Division (÷) is not binary operation on Z, because division of two integers need not be an integer.

Also, division is not binary operation on Q, R and C, since division by zero is not defined. But division is a binary operation on the set of all non-zero rational or real or complex numbers.

32.5 SEMI-GROUP

Algebraic Operation: A set with one or more binary compositions is called an *algebraic system.*

The simplest type of algebraic system is *semi-group.*

Definition. A non empty set G, with a binary composition * in G is said to be a semi-group, if * is associative *i.e.* if

$$a * (b * c) = (a * b) * c$$

If G is a semi-group with respect to *, then we say that $(G, *)$ is a semi-group. Hence a set G with * is a semi-group if it has the following properties.

1. Closure property. If $a \in G, b \in G \Rightarrow a * b \in G, \forall a, b \in G$

2. Associative property.

$a * (b * c) = (a * b) * c, \forall a, b, c \in G$

Illustration

1. The set N of natural numbers is a semi-group with respect to addition natural number, because sum of two natural number is again natural number. Secondly addition of natural numbers is associative. Thus, $(N, *)$ is a semi-group.

2. The set N, of natural numbers is not a semi- group with respect to subtraction in N, since 12–16 = – 4, which is not a natural number.

Secondly here associativity also fails.

32.6 COMMUTATIVE SEMI-GROUP

If a semi-group $(G, *)$, is commutative in G, then $(G, *)$ is known as *commutative semigroup*. Hence commutative semi-group has the following properties:

1. Closure Property.

If $a \in G, b \in G \Rightarrow a * b \in G, \forall a, b \in G$

2. Associative property.

$a * (b * c) = (a * b) * c, \forall a, b, c \in G$

3. Commutative property.

$a * b = b * a, \forall a, b, \in G$

EXERCISE 32.1

Examine which of the following structure are semi-groups.

1. $(R, +)$ **Ans.** Yes
2. $(C, \bullet)$ **Ans.** Yes
3. $(Z, \div)$, **Ans.** No
4. $(Q, -)$ **Ans.** No
5. $(N, *)$ where $m * n = l.c.m.$ of m and n for all $m, n \in N$. **Ans.** Yes
6. $(N, *)$ where $m * n = H.C.F.$ of m and n for all $m, n \in N$. **Ans.** Yes

32.7 GROUP

A non-empty set G with a binary composition * is said to be a group *w.r.t* * if the following conditions hold good.

1. Closure property. If $a \in G, b \in G \Rightarrow a * b \in G, \forall a, b \in G$

2. Associativity. $\forall a, b, c$ in G.

$$a * (b * c) = (a * b) * c$$

3. Existence of identity element.

If an identity element e exists in G such that

$a * e = e * a = a, \qquad \forall a, \in G.$

Where e is identity element.

0 is the identity element for addition.

1 is the identity element for multiplication.

4. Existence of inverses

If an element b exists in G such that

$a * b = b * a = e$ (b is an inverse element)

The inverse of a is denoted by a^{-1} for multiplication. $-a$ is the inverse element of a for addition.

Note. Parentheses can be dropped in the product of more than two elements of a group. In stead of writing

$a * (b * c)$ *or* $(a * b) * c$

we can simply write $a * b * c$.

32.8 COMMUTATIVE GROUP (ABELIAN GROUP)

A group $(G, *)$ is said to be commutative if $a * b = b * a$.

If a group is not commutative then the group is said to be non-abelian or non-commutative.

1. Closure property. If $a \in G, b \in G \Rightarrow a * b \in G, \forall a, b \in G$

2. Associative Property.

$a * (b * c) = (a * b) * c, \forall a, b, c \in G.$

3. Existence of identity element.

If an identity element e exists in G such that $a * e = e * a = a$ for all $a \in G$.
0 is the identity element for addition.
1 is the identity element for multiplication. (e is identity element)

4. Existence of inverse.

If an element b exists in G such that $a * b = b * a = e$ $\forall a, b \in G$, (b is inverse element)

The inverse of a is denoted by a^{-1} for multiplication, $-a$ is the inverse element of a for addition.

5. Commutative property.

$a * b = b * a \quad \forall a, b \in G.$

Example 1. *$(Z, +)$ is a commutative group as*

(i) it is associative

(ii) 0 belonging to Z is an identity element

(iii) each $a \in Z$ has $-a \in Z$.

(iv) Addition in Z is commutative

Similarly we can say that $(Q, +)$, $(R, +)$, $(C, +)$ are all commutative groups.

Example 2. *Let Q_1 denote the set of all non-zero-rational numbers and multiplication by '$\cdot$'. $(Q_1, \cdot)$ is a commutative group. 1 is the identity element of this group and the inverse of $a \in Q_1$ is $\frac{1}{a}$.*

Similarly, $(R_1, \cdot)$, $(C_1, \cdot)$, are also commutative group, where $R_1 = R - (0)$ and $C_1 = C - (0)$.

EXERCISE 32.2

Prove that the following (1 to 5) *w.r.t.* multiplication are groups.

1. Set of positive rational numbers.
2. Set of all positive real numbers.
3. $\{1, -1, i, -i\}$.
4. $\{ Z \in C \,;\, \overline{Z} = 1\}$
5. $\left\{(a+b\sqrt{2}) \;\; a, b \in Q,\; a^2+b^2 \neq 0\right\}$

Which of the following are groups ?

6. $x * y = x + y\ \sqrt{3}$ **Ans.** Group
7. $a * b = a + b - ab.$ **Ans.** Not a group
8. $a * b = a + b + 1$ **Ans.** Group

32.9 RING

A non empty set *R* with two binary compositions * and 0 is said to be a ring *w.r.t.* these compositions if the set with two compositions has the following properties.

[In place of * and ·, we use + and ·]

1. Closure property for addition.

$a, b \in R \quad \Rightarrow a + b \in R \qquad \forall\, a, b \in R.$

2. Associative property for addition.

$a + (b + c) = (a + b) + c \qquad \forall a, b, c \text{ in } R.$

3. Addition Identity element.

There exists an element $0 \in R$, such that

$a + 0 = 0 + a = a, \forall \;\; a \in R$

4. Additive Inverse element.

$\forall \;\; a \in R$ there exists $a^{-1} \in R$. Such that

$a + a^{-1} = a^{-1} + a = 0$

5. Commutative property for addition.

$a + b = b + a \qquad \forall\, a, b \in R.$

6. Closure property for multiplication.

$\forall \;\; a, b \in R \quad \Rightarrow \quad a \cdot b \in R \qquad \forall a, b \in R.$

7. Associative property for multiplication.

$(a \cdot b) \cdot c = a \cdot (b \cdot c) \qquad \forall\, a, b, c \in R.$

8. Distributive property.

(*a*) *Left distributive property*

$a \cdot (b + c) = a \cdot b + a \cdot c \qquad \forall\, a, b, c \in R.$

(*b*) *Right distributive property*

$(b + c) \cdot a = b \cdot a + c \cdot a \qquad \forall\, a, b, c \in R.$

In brief

$(R, + , \cdot)$ is said to be a ring if

1. $(R, +)$ is a commutative group
2. $(R, \cdot)$ is a semi-group and
3. Multiplication is left as well as right distributive *w.r.t.* to addition.

32.10 TYPES OF RINGS

By imposing some extra conditions on the composition in a ring we set several types of ring as mention below:

1. Null Ring.

The set $\{0\}$ with two binary operations $+, \cdot$

$$0 + 0 = 0$$

$$0 \cdot 0 = 0$$

is called a zero ring or null ring.

2. Ring with Unity.

If 1 exists in ring R such that

$$1 \cdot a = a \cdot 1 = a, \qquad \forall\, a, \in R$$

Then the ring is called ring with unity and the element 1 is called unit element or unity or identity of R.

3. Commutative Ring.

If the multiplication in a ring is also commutative then the ring is called commutative ring.

Note. If R is ring with unity then the element $a, \in R$ is called invertible if there exists an element $a^{-1} \in R$ such that

$$a \cdot a^{-1} = 1 = a^{-1} \,.\, a$$

4. Ring of Integers.

The set Z of all integers with two binary operations addition (+) and multiplication (·) of integers is known as the ring of integers.

32.11 SPECIAL TYPES OF RING

1. Zero Divisors

A ring $(R, +, \cdot)$ is said to be without zero divisors if the following property holds.

$$a \cdot b = 0 \quad \Rightarrow \quad a = 0 \text{ or } b = 0, \qquad \forall\, a, b \in R$$

2. Ring with or without zero Divisor.

A non zero element a in a ring R is called a zero divisor or divisor zero if there exists a non zero element b in R such that $a \cdot b = 0$ or $b \cdot a = 0$. Then R is said to be ring with zero divisor.

On the other hand, if in a ring R, $a \cdot b = 0 \quad \Rightarrow \quad a = 0$ or $b = 0$.

Then R is said to be a ring without zero divisor.

Example (*i*) *The ring* $(Z_6, +_6, \times_6)$ *is a ring with zero divisor where* $Z_6 = \{\overline{0}, \overline{1}, \overline{2}, \overline{3}, \overline{4}, \overline{5}\}$.

We have $\overline{2} \times_6 \overline{3} = 0, \overline{3} \times_6 \overline{4} = \overline{0}$ *i.e. the product of two non-zero elements is equal to the zero elements of the ring.*

(*ii*) *The ring of integers* $(Z, +, \cdot)$ *is a ring without zero divisors, as the product of two non-zero integers cannot be equal to the zero integer.*

Example. (*i*) *Let M be the ring of all 2 × 2 matrices whose elements are integers, the addition and multiplication of matrices being the two ring compositions. Then M is a ring without zero divisors.*

The null matrix $O = \begin{bmatrix} 0 & 0 \\ 0 & 0 \end{bmatrix}$ *is the zero element of this ring.*

(*ii*) *Now,* $A = \begin{bmatrix} 1 & 0 \\ 0 & 0 \end{bmatrix}, B = \begin{bmatrix} 0 & 0 \\ 1 & 0 \end{bmatrix}$ *are two non-zero elements of this ring,*

i.e. $A \neq 0, \ B \neq 0$, *we have*

$$\begin{bmatrix} 1 & 0 \\ 0 & 0 \end{bmatrix} \begin{bmatrix} 0 & 0 \\ 1 & 0 \end{bmatrix} = \begin{bmatrix} 0 & 0 \\ 0 & 0 \end{bmatrix} = 0.$$

Here, the product of two non-zero elements of the ring is equal to the zero element of ring.

Thus M is a ring with zero divisor.

Here, $BA = \begin{bmatrix} 0 & 0 \\ 1 & 0 \end{bmatrix} \begin{bmatrix} 1 & 0 \\ 0 & 0 \end{bmatrix} = \begin{bmatrix} 0 & 0 \\ 1 & 0 \end{bmatrix} \neq 0$

Thus in a ring R it is possible that $ab = 0$, but $ba \neq 0$.

32.12 INTEGRAL DOMAIN

A ring $(R, +, \cdot)$ is called an integral domain if it

1. is commutative

2. has unit element.

3. is without zero divisor.

32.13 FIELD

A ring with atleast two elements is called a field if

1. It is commutative

2. It has unit element

3. It is such that non-zero element posseses multiplicative inverse.

Thus a system $(F, +, \cdot)$ is a field if the following properties are satisfied.

$(F, +)$ is an obelian group.

1. Closure Property.

$a \in F, b \in F, \qquad \Rightarrow \quad a + b \in F, \qquad \forall a, b, \in F$

2. Addition is associative.

$(a + b) + c = a + (b + c), \ \forall \ a, b, c \in F$

3. Additive identity.

$a + 0 = a, \quad \forall \ a \in F$

4. Additive inverse.

$a + (-a) = (-a) + a = 0, \qquad \forall a \in F$

where $-a$ is the additive inverse of a.

5. Addition is commutative.

$a + b = b + a, \ \forall \ a, b \in F$

$(F', \cdot)$ is commutative group, where F' is the set of all non-zero elements of F.

1. Closure property (Multiplication).

$a \cdot b \in F, \quad \forall \ a \in F, \ \ b \in F$

2. Associative property (Multiplication).

$(a \cdot b) \cdot c = a \cdot (b \cdot c), \quad \forall \ a, b, c \in F$

3. Multiplicative identity.

$a \cdot 1 = a$ for $a \in F$

where $1 \in F$ is called unity or identity element.

4. Multiplicative inverse.

$a^{-1} \in F$ such that

$a \,.\, a^{-1} = 1$

where $a^{-1} \in F$ is called the multiplicative inverse of $a \in F$.

5. Multiplication is commutative.

$a \cdot b = b \cdot a,$ $\forall \ a, b, c \in F$

6. Distributive laws.

(*a*) $a \cdot (b + c) = a \cdot b + a \,.\, c$ (Left distributive law)

(*b*) $(b + c) \cdot \ a = b \cdot a + c \cdot a$ (Right distributive law)

Example 1. *The set of rational numbers with operation addition and multiplication is a field.*

2. *The set of real numbers with operation addition and multiplication is a field.*

32.14 VECTOR SPACE

(*R.G.P.V., Bhopal, III Semester, Dec. 2007*)

Let V be a non-empty set and F be the field of real numbers. Let we have two compositions one is plus (+) between two members of V and other is dot $(\cdot)$ between a member of V and a member of F. V is said to be vector space if the following properties hold good.

1. Closure property.

$\forall \ a, b \in R \quad \Rightarrow \quad a + b \in R.$

2. Associativity of addition.

For all $\alpha, \beta, \gamma \in V$.

$\alpha + (\beta + \gamma) = (\alpha + \beta) + \gamma.$

3. Existence of the neutral element.

There exists an element $0 \in V$, such that

$\alpha + 0 = 0 + \alpha = \alpha,$ for all $\alpha \in V$.

4. Existence of additive inverse.

For each $\alpha \in V$, there exists $\beta \in V$, such that

$\alpha + \beta = \beta + \alpha = 0.$

5. Commutativity of addition.

For all $\alpha, \beta \in V$, $\alpha + \beta = \beta + \alpha$

6. Closure property.

$\forall\, \alpha, \beta \in R \Rightarrow \alpha \,.\, \beta \subset R$

7. Associativity of scalar multiplication.

For all $x, y \in F$ and $\alpha \in V$

$x\,(y\,\alpha) = (x\,y)\,\alpha.$

8. Distributivity of scalar multiplication over addition.

For all $x \in F, \alpha, \beta \in V$,

$x\,(\alpha + \beta) = x\,\alpha + x\,\beta.$

9. Distributivity of scalar multiplication over addition in F.

For all $x, y \in F, \alpha \in V$.

$(x + y)\,\alpha = x\,\alpha + y\,\alpha.$

10. Property of unity.

If 1 be the identity in F, then for all $\alpha \in V$.

$1 \cdot \alpha = \alpha.$

Note. Vectors will be denoted by α, β, γ while scalars will be denoted by a, b, c, d or x, y, z.

Theorem 1. *Let $V\,(F)$ be a vector space, and*

(*i*) *If α is a non-zero element of V and $a, b \in F$, then*

$$a\alpha = b\alpha \;\Rightarrow\; a = b$$

(*ii*) *If a is a non-zero element of F and $\alpha, \beta \in V$, then*

$$a\alpha = a\,\beta \;\Rightarrow\; \alpha = \beta.$$

Proof. (*i*) We have, $a\alpha = b\alpha \Rightarrow a\alpha - b\alpha = \bar{0} \quad \Rightarrow (a - b)\,\alpha = \bar{0}.$

$\Rightarrow a - b = 0 \quad [\because \bar{\alpha} \neq 0] \quad \Rightarrow a = b.$

Hence, $a\alpha = b\alpha \quad \Rightarrow a = b \quad [\alpha \neq \bar{0}]$

(*ii*) We have, $a\alpha = a\beta \Rightarrow a\alpha - a\beta = 0 \quad \Rightarrow a\,(\alpha - \beta) = \bar{0}.$

$\Rightarrow \alpha - \beta = \bar{0} \quad [\because a \neq 0] \quad \Rightarrow \alpha = \beta.$

Hence, $a\alpha = a\beta \quad \Rightarrow \alpha = \beta \quad [a \neq 0]$

Example 1. *Let F be an arbitrary field. Prove that, the set of all ordered n–tuples of the elements of F with vector addition and scalar multiplication defined by*

$$(a_1, a_2, a_3 \ldots\ldots\ldots a_n) + (b_1, b_2 \ldots\ldots\ldots b_n) = (a_1 + b_1, a_2 + b_2, a_3 + b_3, \ldots\ldots\; a_n + b_n)$$

and $k\,(a_1, a_2, \ldots\ldots\ldots\ldots a_n) = (ka_1, ka_2, \ldots\ldots\ldots\; ka_n)$

where $a_i, b_i, k \in F$, is a vector space over F.

Solution. Let V be the set of all ordered n-tuples over F, *i.e.* let

$V = \{(a_1, a_2 \ldots\ldots\ldots\ldots\ldots a_n) \;:\; a_1, a_2 \ldots\ldots\ldots\ldots\ldots a_n \in F\}$

We have to prove that V_α is a vector space.

V_1. First of all we have to prove that (V, +) is always an abelian group.

1. Closure property.

Let $\alpha = (a_1, a_2 \ldots\ldots\ldots\ldots\ldots a_n)$ and

$\beta = (b_1, b_2 \ldots\ldots\ldots\ldots\ldots b_n)$

be two elements of V.

$\alpha + \beta = (a_1 + b_1, a_2 + b_2, \ldots\ldots\ldots\ldots\ldots, a_n + b_n)$

Since $a_1 + b_1, a_2 + b_2, \ldots\ldots\ldots\ldots\ldots, a_n + b_n$ are all elements of F.

Hence, $\alpha + \beta \in V, \forall \alpha, \beta \in V$ and thus V is closed with respect to addition of n-tuples.

2. Associative Property.

$$\begin{aligned}\alpha + (\beta + \gamma) &= (a_1, a_2, \ldots\ldots\ldots\ldots, a_n) + \{(b_1, b_2, \ldots\ldots\ldots b_n) + (c_1, c_2, \ldots\ldots\ldots c_n)\} \\ &= (a_1, a_2, \ldots\ldots\ldots\ldots, a_n) + (b_1 + c_1, b_2 + c_2, \ldots\ldots\ldots, b_n + c_n) \\ &= [a_1 + (b_1 + c_1), a_2 + (b_2 + c_2), \ldots\ldots\ldots, a_n + (b_n + c_n)] \\ &= [(a_1 + b_1) + c_1, (a_2 + b_2) + c_2, \ldots\ldots\ldots\ldots, (a_n + b_n) + c_n]\end{aligned}$$

[By associativity in F for +]

$$\begin{aligned}&= (a_1 + b_1, a_2 + b_2, \ldots\ldots\ldots\ldots, a_n + b_n) + (c_1, c_2, \ldots\ldots\ldots c_n) \\ &= [(a_1, a_2, \ldots\ldots\ldots a_n) + (b_1, b_2, \ldots\ldots\ldots\ldots, b_n)] + (c_1, c_2, \ldots\ldots\ldots c_n) \\ &= (\alpha + \beta) + \gamma\end{aligned}$$

3. Commutative property.

Here,

$$\begin{aligned}\alpha + \beta &= (a_1, a_2, \ldots\ldots\ldots\ldots a_n) + (b_1, b_2, \ldots\ldots\ldots\ldots b_n) \\ &= (a_1 + b_1, a_2 + b_2, \ldots\ldots\ldots\ldots\ldots, a_n + b_n) \\ &= (b_1 + a_1, b_2 + a_2, \ldots\ldots\ldots, b_n + a_n) \\ &= \beta + \alpha \qquad \text{[Commutativity for + in } F]\end{aligned}$$

4. Existence of additive identity or zero vector.

Here,

$\bar{0} = (0, 0, \ldots\ldots\ldots\ldots 0) \in V$ since $0 \in F$

If $\alpha = (a_1, a_2, \ldots\ldots\ldots\ldots, a_n) \in V$ is any element then

$$\begin{aligned}\alpha + \bar{0} &= (a_1, a_2, \ldots\ldots\ldots a_n) + (0, 0, \ldots\ldots\ldots\ldots, 0) \\ &= (a_1 + 0, a_2 + 0, \ldots\ldots\ldots\ldots a_n + 0) \\ &= (a_1, a_2 \ldots\ldots\ldots a_n) = \alpha\end{aligned}$$

Similarly, $\bar{0} + \alpha = \alpha$

Hence, $\bar{0}$ *i.e.* $(0, 0, 0, \ldots\ldots\ldots\ldots 0)$ is the additive identity *i.e.*, zero vector in V.

5. Existence of additive inverse. *i.e.* –ve vector in V.

If $\alpha = (a_1, a_2, \ldots\ldots\ldots\ldots a_n) \in V$ then $-\alpha = (-a_1, -a_2 \ldots\ldots\ldots - a_n) \in V$ is the additive inverse of α, since

$$\begin{aligned}\alpha + (-\alpha) &= (a_1, a_2, \ldots\ldots\ldots a_n) + (-a_1, -a_2, \ldots\ldots\ldots, -a_n) \\ &= (a_1 - a_1, a_2 - a_2, \ldots\ldots\ldots a_n - a_n) \\ &= (0, 0, \ldots\ldots\ldots\ldots 0) = \bar{0}\end{aligned}$$

Similarly, $(-\alpha) + \alpha = 0$

Hence V, the set of all ordered n-tuples is an additive abelian group.

V_2. Scalar multiplication operation in V over F.

It is given that $k\alpha = k(a_1, a_2, \ldots\ldots a_n) = (ka_1, ka_2, \ldots\ldots\ldots, ka_n), \forall k \in F, \forall \alpha \in V$

Since $ka_1, ka_2, \ldots\ldots\ldots\ldots, ka_n \in F$, and so $k\alpha \in V$. Hence V is closed with respect to the scalar multiplication.

V_3. Distributive Law.

(*i*) Left Distributive Law.

If $a \in F$ and $\alpha = (a_1, a_2, \ldots\ldots, a_n) \in V$, $\beta = (b_1, b_2, \ldots\ldots, b_n) \in V$, then

$$a(\alpha + \beta) = a[(a_1, a_2 \ldots\ldots, a_n) + (b_1, b_2, \ldots\ldots, b_n)]$$

$$= a(a_1 + b_1, a_2 + b_2, \ldots\ldots\ldots\ldots, a_n + b_n) = [a(a_1 + b_1), a(a_2 + b_2), \ldots\ldots\ldots\ldots, a(a_n + b_n)]$$

$$= (aa_1 + ab_1, aa_2 + ab_2, \ldots\ldots, aa_n + ab_n) \quad \text{[By distributive law in } F\text{]}$$

$$= (aa_1, aa_2, \ldots\ldots, aa_n) + (ab_1, ab_2, \ldots\ldots\ldots\ldots, ab_n)$$

$$= a(a_1, a_2, \ldots\ldots, a_n) + a(b_1, b_2, \ldots\ldots\ldots\ldots, b_n) = a\alpha + a\beta$$

(ii) Right Distributive Law. If $a, b \in F$ and $\alpha = (a_1, a_2 \ldots\ldots, a_n) \in V$, then

$$(a + b)\alpha = (a + b)(a_1, a_2, \ldots\ldots, a_n)$$

$$= [(a+b)a_1, (a+b)a_2, \ldots\ldots\ldots, (a+b)a_n] = (aa_1 + ba_1, aa_2 + ba_2, \ldots\ldots aa_n + ba_n)$$

$$= (aa_1, aa_2, \ldots\ldots, aa_n) + (ba_1, ba_2, \ldots\ldots, ba_n) = a(a_1, a_2, \ldots\ldots, a_n) + b(a_1, a_2, \ldots\ldots, a_n)$$

$$= a\alpha + b\alpha$$

(iii) Associative Property. If $a, b \in F$ and $\alpha = (a_1, a_2, \ldots\ldots\ldots, a_n) \in V$, then

$$(ab)\alpha = ab(a_1, a_2, \ldots\ldots, a_n) = [(ab)a_1, (ab)a_2, \ldots\ldots, (ab)a_n] = [a(ba_1), a(ba_2), \ldots\ldots, a(ba_n)]$$

$$= a(ba_1, ba_2, \ldots\ldots ba_n) = a(b\alpha)$$

(iv) Multiplicative Identity Element. If $1 \in F$ and $\alpha = (a_1, a_2, \ldots\ldots, a_n)$ then

$$1\alpha = 1(a_1, a_2, \ldots\ldots, a_n) = (1a_1, 1a_2, \ldots\ldots 1a_n) = (a_1, a_2, \ldots\ldots, a_n) = \alpha$$

Hence, V is a vector space (or linear space) over F. This vector space is denoted by

$V_n(V)$ or by $F^{(n)}$ or by F^n.

Note 1. Let $(F, +, \cdot)$ be any field. The above example 1 can be generalised by taking infinite elements of F. Let $V_\infty(F)$ denote the totality of ordered infinite-tuples (or infinite sequences) whose elements are those of F. The elements of $V_n(F)$ can be conveniently represented by row vectors, as

$$(a_1, a_2, a_3, \ldots\ldots\ldots, a_k) \in F,$$

which has infinite components. In this case $(a_1, a_2, \ldots\ldots) = (b_1, b_2, \ldots\ldots)$ if and only if $a_k = b_k, \forall k \in I_+$.

Vector addition is defined by

$$(a_1, a_2, \ldots\ldots) + (b_1, b_2, \ldots\ldots\ldots) = (a_1 + b_1, a_2 + b_2, \ldots\ldots)$$

Let $a \in F$, then scalar multiplication is defined by

$$a(a_1, a_2, \ldots) = (aa_1, aa_2, \ldots)$$

with respect to these compositions $V_\infty(F)$ is a vector space.

Example 2. *Let C be the field of complex numbers and R be the field of real numbers then*

(i) R is a vector space over R.

(ii) C is a vector space over C.

Solution. (*i*) V_1 **: Abelian Group :** $(R, +)$ is an abelian group as $(R, +)$ is a field.

V_2 **: Left distributive law:** $\alpha(a + b) = a\alpha + \alpha b$, $\forall \alpha \in R$ and $\forall a, b \in R$

V_3 **: Right distributive law:** $(\alpha + \beta)a = \alpha a + \beta a$, $\forall \alpha, \beta \in R$ an $\forall a \in R$

V_4 **: Associative law for multiplication.**

$\alpha(\beta\alpha) = (\alpha\beta)a$, $\forall \alpha, \beta \in R$ and $\forall a \in R$

V_5 **: Multiplicative identity.**

$1.\, a = a.1 = a$, $1 \in R$ and $\forall a \in R$

Hence, R is a vector space over R.

(ii) V_1 **: Abelian group.**

$(C, +)$ is an abelian group because C is a field.

V_2**: Left distributive law of multiplication.**

$\alpha\,(u + v) = \alpha u + \alpha v$, $\forall \alpha \in C$ and $\forall\, u, v \in C$.

V_3**: Right Distributive law.**

$(\alpha + \beta)u = \alpha u + \beta u$, $\forall \alpha, \beta \in C$ and $\forall u \in C$.

V_4**: Associative law of multiplication.**

$\alpha(\beta u) = (\alpha\beta)u$, $\forall \alpha, \beta \in C$ and $\forall u \in C$.

V_5**: Multiplicative identity.**

$1.u = u.1 = u$ $1 \in C$ and $\forall u \in C$.

Hence, C is a vector space over the field C.

Example 3. *If $V = \{(a, b) : a, b \in R\}$ and R is a field show that V is not a vector space over R under the addition and scalar multiplication defined by*

$(a, b) + (c, d) = (0, b + d)$

$\alpha\,(a, b) = (\alpha\, a, \alpha b)$ (*R.G.P.V., Bhopal, III Semester, June 2005, Dec. 2004*)

Solution. V_1 **: Closure property**

V is closed with respect to addition.

V_2 **: Associative law**

Associative law holds in V.

V_3 **: Identity element**

Identity element in V does not exist as there is no ordered pair (e, f) such that

$$(a, b) + (e, f) = (a, b) \qquad \forall\,(a, b) \in V$$

because by definition

$(a, b) + (e, f) = (0, b + f)$

Thus, $(V, +)$ is not an abelian group.

Hence, V is not a vector space over R. **Ans.**

Example 4. *Let V be the set of all pairs (x, y) of real numbers and let R be the field of real numbers. In each of the following, examine whether V is a vector space over the field of real numbers or not.*

(*i*) $(x, y) + (x_1, y_1) = (x + x_1, y + y_1)$
$c\,(x, y) = (|\,c\,|\,x, |\,c\,|\,y)$

(*ii*) $(x, y) + (x_1, y_1) = (x + x_1, y + y_1)$
$c\,(x, y) = (0, cy)$

(*iii*) $(x, y) + (x_1, y_1) = (x + x_1, y + y_1)$
$c\,(x, y) = (c^2\, x, c^2\, y)$

Solution. (*i*) Now we shall show that the distributive property $(a + b)\,\alpha = a\alpha + b\alpha$, $\forall a, b \in R$ and $\alpha \in V$ is not true.

Let $\alpha = (x, y)$ where $x, y \in R$ and $a, b \in R$ then

$(a + b)\,\alpha = (a + b)\,(x, y) = (|\,a + b\,|\,x, |\,a + b\,|\,y)$ (By definition) ... (1)

Again $a\,\alpha + b\alpha = a\,(x, y) + b\,(x, y)$

$= (\,|\,a\,|\,x, |\,a\,|\,y) + (|\,b\,|\,x, |\,b\,|\,y)$ (by definition)

$= (|\,a\,|\,x + |\,b\,|\,x, |\,a\,|\,y + |\,b\,|\,y)$ (by vector addition)

$= (\{|\,a\,| + |\,b\,|\}\,x, \{|\,a\,| + |\,b\,|\}\,y).$ ... (2)

But $|\,a + b\,| \le |\,a\,| + |\,b\,|$.

Thus, in general, it follows from (1) and (2), that

$(a + b)\,\alpha \ne a\,\alpha + b\,\alpha$ $\forall a, b \in R, \forall \alpha \in V.$

Hence, $V(R)$ is not a vector space.

(*ii*) Now we shall show that the multiplicative identity $1\alpha = \alpha$, $\forall \alpha \in V$ is not satisfied.

Let $\alpha = (x, y)$, where $x, y \in R$ then by scalar multiplication,

$$1\alpha = 1(x, y) = (0, 1\,y) = (0, y) \ne (x, y)$$

Thus, it follows that $1\alpha \ne \alpha$, $\forall \alpha \in V$

Hence, $V(R)$ is not a vector space.

(*iii*) Let $\alpha = (x, y)$ where $x, y \in R$ and $a, b \in R$ then

$(a + b)\,\alpha = (a + b)\,(x, y)$

$= [(a + b)^2\,x, (a + b)^2\,y]$ (by scalar multiplication) ...(1)

Also, $a\,\alpha + b\,\alpha = a\,(x, y) + b\,(x, y)$

$= (a^2\,x, a^2 y) + (b^2\,x, b^2\,y)$

$= (a^2\,x + b^2 x, a^2\,y + b^2\,y)$ [By vector addition]

$= [\{a^2 + b^2\}\,x, \{a^2 + b^2\}y]$ ... (2)

Thus, (1) and (2) $\Rightarrow (a + b)\,\alpha \ne a\alpha + b\alpha$

Hence, $V(R)$ is not a vector space. **Ans.**

Example 5. *Show that $V(F) = \{(a_1, a_2) : a_1, a_2 \in F\}$ is the vector space of all ordered pairs over F.*

Solution. All the above properties are satisfied by $V(F)$.

Hence, $V(F)$ is vector space.

Example 6. *Show that $V_3(F) = \{(a_1, a_2, a_3) : a_1, a_2, a_3 \in F\}$ is the vector space of all ordered triads over F.*

Solution. Proceed as in Example 5, above.

Example 7. *V is the set of 2×3 matrices with their elements as rational numbers and F is the field of real numbers. Then show that $V(F)$ is not a vector space.*

Solution. The members of V are rational numbers. Let $a \in V$.

The members of F are real numbers (including $\sqrt{2}$) *i.e.* $\sqrt{2} \in F$

The element of matrix $\sqrt{2}\,\alpha$ are not the rational numbers.

Since $\sqrt{2}\,\alpha$ is not rational number *i.e.* closure property is not satisfied by $V(F)$. Hence, $V(f)$ is not a vector space. **Ans.**

Example 8. *Determine whether the given set is a subspace of the given vector space*

(a) Let W be the set of matrices of the form $\begin{bmatrix} 5 & x_1 \\ 0 & x_2 \end{bmatrix}$. *Is this a subspace of 2 × 2 matrix.*

(b) Let W be the set of matrices of the form $\begin{bmatrix} 0 & x_1 \\ x_2 & x_3 \\ x_4 & x_5 \end{bmatrix}$. *Is this a subspace of 3 × 2 matrix.*

Solution. (*a*) Let $X, Y \in W$

$$X = \begin{bmatrix} 5 & x_1 \\ 0 & x_2 \end{bmatrix} \text{ and } Y = \begin{bmatrix} 5 & y_1 \\ 0 & y_2 \end{bmatrix}$$

$$\text{Now, } X + Y = \begin{bmatrix} 5 & x_1 \\ 0 & x_2 \end{bmatrix} + \begin{bmatrix} 5 & y_1 \\ 0 & y_2 \end{bmatrix} = \begin{bmatrix} 10 & x_1 + y_1 \\ 0 & x_2 + y_2 \end{bmatrix} \notin W$$

We know the form of X and Y where the element in the first column and first row is 5.
But in $(X + Y)$ the element in the first row and first column is 10.
Since it is not in the given form of matrices of W.
Thus, for every $X, Y \in W$, W is not a sub-space of 2 × 2 matrix.
(*b*) Let $X, Y \in W$, then,

$$X \text{ and } Y \text{ is of the form } \begin{bmatrix} 0 & x_1 \\ x_2 & x_3 \\ x_4 & x_5 \end{bmatrix} \text{ and } \begin{bmatrix} 0 & y_1 \\ y_2 & y_3 \\ y_4 & y_5 \end{bmatrix}$$

$$\text{Now, } X + Y = \begin{bmatrix} 0 & x_1 \\ x_2 & x_3 \\ x_4 & x_5 \end{bmatrix} + \begin{bmatrix} 0 & y_1 \\ y_2 & y_3 \\ y_4 & y_5 \end{bmatrix} = \begin{bmatrix} 0 & x_1 + y_1 \\ x_2 + y_2 & x_3 + y_3 \\ x_4 + y_4 & x_5 + y_5 \end{bmatrix} \in W$$

Let $X \in W$ and $k, \in W$

$$\text{Then} \qquad kX = k \begin{bmatrix} 0 & x_1 \\ x_2 & x_3 \\ x_4 & x_5 \end{bmatrix} = \begin{bmatrix} 0 & kx_1 \\ kx_2 & kx_3 \\ kx_4 & kx_5 \end{bmatrix} \in W$$

Since, it is of the given form.
Thus, for every $X, Y \in W$, we have $X, + Y \in W$.
and, for every scalar k and $X, \in W$, we have $kX, \in W$.
Hence, W is a vector space. **Proved.**

Example 9. *The field of all polynomials F [x] over a field F is a vector space over F.*

Solution. Let F be any field and let $V = F[x]$, the set of polynomials in x over F.
The addition of two polynomials in V is considered as addition of vectors.
A polynomial can be multiplied by an element of F.
This operation is considered as scalar multiplication.
Hence, V is a vector space over F. **Ans.**

32.15 EXTERNAL COMPOSITION OF THE VECTOR SPACE *V*

The scalar multiplication which may be regarded as a mapping from $F \times V \to V$ is known as the external composition of the vector space V.

32.16 INTERNAL COMPOSITION OF THE VECTOR SPACE V

Let F be a field, let V be a non-empty set, let $V \times V \rightarrow V$ be an internal composition in V called vector addition and let $F \times V \rightarrow V$ be an external composition in V called scalar multiplication; V is said to be a vector space over F if the following conditions holds :

1. $(V, +)$ is an abelian group.

2. $x(y\alpha) = (xy)\alpha$, for $x, y \in F$ and $\alpha \in V$.

3. (*i*) $x(\alpha + \beta) = x\alpha + x\beta$, for all $x \in F$, $\alpha, \beta \in V$.

(*ii*) $(x + y)\alpha = x\alpha + y\alpha$, for all $x, y \in F$, $\alpha \in V$.

4. $1.\alpha = \alpha$, for all $\alpha \in V$, where 1 is the unity of F.

EXERCISE 32.3

1. Show that the complex field C is a vector space over the real field R.

2. Let $R^2 = \{(a_1, a_2) : a_1, a_2 \in R\}$.
Show that R^2 is a vector space over the real field R with respect to the addition and scalar multiplication, defined by $(a_1, a_2) + (b_1, b_2) = (a_1 + b_1, a_2 + b_2)$
and $a(a_1, a_2) = (aa_1, aa_2)$.

3. Let V be the set of all ordered pairs (x, y) of real numbers, Define
(*i*) $(x_1, y_1) + (x_2, y_2) = (x_1 + x_2, y_1 + y_2)$.
and $\alpha(x, y) = (\alpha x, y)$.
(*ii*) $(x_1, y_1, z_1) + (x_2, y_2, z_2) = (z_1 + z_2, y_1 + y_2, x_1 + x_2)$
Show that with these operations, V is not a vector space over the real field R.

(*Gujarat, II Semester, June 2009*)

4. Let R be the field of real numbers and let P_n be the set of all polynomials of degree at most n, over the field R. Show that $P_n(R)$ is a vector space.

5. Prove that the set of all vectors in a plane over the field of real numbers is a vector space with respect to vector addition and scalar multiplication compositions.

6. Let $V = \{(x, y) : x, y \in R\}$
Show that V is not a vector space over the real field R, with respect to addition and scalar multiplication compositions defined in each of the following cases :
(*i*) $(x, y) + (x_1, y_1) = (x + x_1, y + y_1)$ & $c(x, y) = (0, cy)$;
(*ii*) $(x, y) + (x_1, y_1) = (x + y_1, y + x_1)$ & $c(x, y) = (cx, cy)$;
[Hint. $(c_1 + c_2)\alpha \neq c_1\alpha + c_2\alpha$]]
(*iii*) $(x, y) + (x_1, y_1) = (3y + 3y_1, -x - x_1)$ & $c(x, y) = (3cy, -cx)$

7. Prove that each of the following sets of matrices is a vector space over C with respect to matrix addition and multiplication of a matrix by a scalar :

(*a*) The set of all matrices of the form

$$\begin{pmatrix} x & y \\ z & 0 \end{pmatrix}, \text{ where } x, y, z \in C.$$

(*b*) The set of all matrices of the form

$$\begin{pmatrix} x & 0 \\ 0 & y \end{pmatrix}, \text{ where } x, y \in C.$$

(*c*) The set of all matrices of the form

$$\begin{pmatrix} x & 0 \\ 0 & 0 \end{pmatrix}, \text{ where } x \in C.$$

8. Prove that the set of all matrices of the form

$$\begin{pmatrix} a & b \\ -b & a \end{pmatrix} \text{ where } a, b \in C.$$

is a vector space over R with respect to matrix addition and multiplication of a matrix by a scalar.

CHAPTER 33

VECTORS IN R^n

33.1 INTRODUCTION

We have already done vectors in 2-space, and 3-space.

Here, we will study vectors in n-space; *i.e*, real Euclidean n-space (vectors in R^n). Vectors in R^n is the extension of vectors in 2-space and in 3-space.

33.2 ORDERED n-TUPLES OF REAL NUMBERS

A set $(x_1, x_2, \text{-----} x_n)$ of n real numbers arranged in the order is called an ordered n-tuple of reals.

For example: (1) In two dimensional coordinate geometry any point (3, 7) is nothing but an ordered 2- tuple.

(2) (– 2, 0,4) is 3- tuple (3) (4, 6, 8, 9) is 4- tuple.

(4) Each row of the matrix, $A = \begin{bmatrix} -1 & 2 & 5 & 0 \\ 1 & 2 & 3 & 4 \end{bmatrix}$ is an ordered 4-tuple. Each column of the matrix A is an ordered 2- tuple.

33.3 VECTORS IN R^n

Vectors in R^n can be represented horizontally (row vector) and vertically (column vector). It is denoted as

$$X = (x_1, x_2, \ldots x_n) \quad \text{or} \quad X = [x_1, x_2, \ldots x_n]$$

or $$X = \begin{bmatrix} x_1 \\ x_2 \\ \vdots \\ x_n \end{bmatrix}.$$

Definitions of vectors in R^n

Vectors in R^n is the collection of n-tuples of the form $(x_1, x_2, \ldots x_n)$ *i.e*; a matrix $1 \times n$. This can also be of the form a matrix $n \times 1$; $\begin{bmatrix} x_1 \\ x_2 \\ \vdots \\ x_n \end{bmatrix}$.

33.4 EQUALITY OF TWO n–TUPLES

If two n- tuples $X = (x_1, x_2, \ldots x_n)$ and $Y = (y_1, y_2, \ldots y_n)$are equal, then

$$x_1 = y_1, x_2 = y_2, \cdots, x_n = y_n.$$

For example: If $(x, z, 4 + y) = (10, 11, -20)$, then $x = 10, z = 11, 4 + y = -20 \Rightarrow y = -24$

33.5 ADDITION OF TWO *n*-TUPLES

Let $X = (x_1, x_2, ... x_n)$ and $y = (y_1, y_2, ... y_n)$
be two n–tuples, then
$X + Y = (x_1 + y_1, x_2 + y_2, ... x_n + y_n)$
For example: (0, 1, 6, 8) + (2, –2, 5, 0) = (0 + 2, 1 – 2, 6 + 5, 8 + 0) = (2, –1, 11, 8)

33.6 SUBTRACTION OF TWO *n*-TUPLES

Let two n -tuples be $X = (x_1, x_2, ... x_n)$ and $Y = (y_1, y_2, ... y_n)$, then
$X - Y = (x_1 - y_1, x_2 - y_2, ... x_n - y_n)$
For example: (0, 1, 6, 8) – (2, –2, 5, 0) = (0 – 2, 1 + 2, 6 – 5, 8 – 0) = (–2, 3, 1, 8)

33.7 MULTIPLICATION OF *n*-TUPLES BY REAL NUMBER

Let $X = (x_1, x_2, \cdots, x_n)$ be a n–tuples and k be a real number then the product
$k.X = (kx_1, kx_2, \cdots, kx_n)$

For example, $6 \cdot (3, 5, -10, 12) = (18, 30, -60, 72)$
$0 \cdot (-5, 0, 6, 15) = (0, 0, 0, 0)$

Example 1. *If* $X = (4, 5, 3, -1), Y = (0, 3, 10, 15)$ *then find* $4X - Y$.

Solution. $4X - Y = 4(4, 5, 3, -1) - (0, 3, 10, 15) = (16, 20, 12, -4) - (0, 3, 10, 15)$
$= (16 - 0, 20 - 3, 12 - 10, -4 - 15) = (16, 17, 2, -19)$ **Ans.**

33.8 PROPERTIES OF VECTOR IN R^n

Let $X = (x_1, x_2, \cdots, x_n)$
$Y = (y_1, y_2, \cdots, y_n)$
and $Z = (z_1, z_2, \cdots, z_n)$
be 3 vectors in R^n and a, b are scalars.

1. Commutative Property over addition

$$X + Y = Y + X$$

2. Associative Property over addition

$$X + (Y + Z) = (X + Y) + Z$$

3. Additive identity

$$X + 0 = 0 + X = X$$

4. Additive Inverse

$$X + (-X) = 0$$

5. Associative Property over multiplication.

$(ab)\,X = a(bX)$

6. Multiplicative Identity

$1 \cdot X = X$

7. Distribution of scalar multiplication over addition.

$a\,(X + Y) = aX + aY.$

Proof. $a(X+Y) = a\{(x_1, x_2, \cdots, x_n) + (y_1, y_2, \cdots, y_n)\}$

$= a\{(x_1 + y_1, x_2 + y_2, \cdots, x_n + y_n)\}$

$= [a(x_1 + y_1), a(x_2 + y_2), \cdots, a(x_n + y_n)]$

$= (ax_1 + ay_1, ax_2 + ay_2, \cdots, ax_n + ay_n)$

$= (ax_1, ax_2, \cdots, ax_n) + (ay_1, ay_2 \cdots, ay_n)$

$= a\,X + a\,Y.$

Proved.

8. Distribution of scalar addition over scalar multiplication.

$(a + b)\,X = a\,X + b\,X.$

Proof. $(a + b)\,X = (a + b)\,(x_1, x_2, \cdots, x_n)$

$= \{(a+b)x_1, (a+b)x_2, \cdots, (a+b)\,x_n\}$

$= (ax_1 + bx_1, ax_2 + bx_2, \cdots, ax_n + bx_n)$

$= (ax_1, ax_2, \cdots, ax_n) + (bx_1, bx_2, \cdots, bx_n)$

$= a(x_1, x_2, \cdots, x_n) + b(x_1, x_2, \cdots, x_n)$

$= aX + bX.$

Proved.

33.9 EUCLIDEAN NORM (OR MAGNITUDE)

Let $X = (x_1, x_2, \cdots x_n)$ be a vector in R^n, then the Euclidean norm is

$$\| X \| = (X.X)^{\frac{1}{2}} = \sqrt{x_1^2 + x_2^2 + \cdots x_n^2}$$

33.10 EUCLIDEAN DISTANCE

If $X = (x_1, x_2, \cdots, x_n)$ and $Y = (y_1, y_2, \cdots, y_n)$

be two points in R_n, then the Euclidean distance between them is defined as

$$d\,(X, Y) = \| X - Y \| = \sqrt{(x_1 - y_1)^2 + (x_2 - y_2)^2 + \cdots (x_n - y_n)^2}$$

33.11 UNIT VECTOR

A vector with norm 1 is called a unit vector.

33.12 THEOREM 1

If X and Y are two vectors in R^n and θ is the angle between them. Then,

$$X.Y = \| X \| \; \| Y \| \cos\theta$$

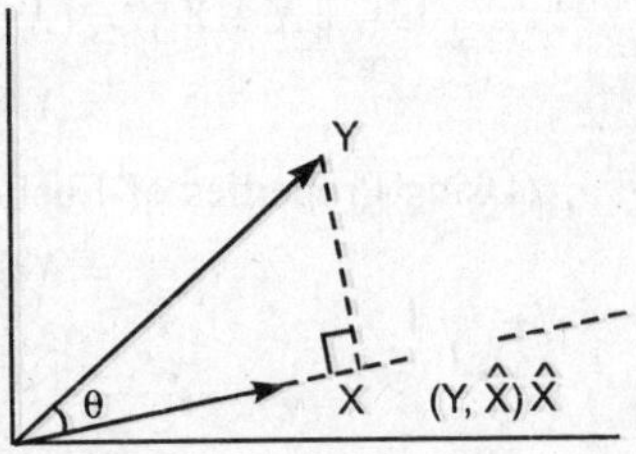

The inner product of X and Y is equal to product of norm of X, norm of Y and cos θ ,where θ is the angle between X and Y.

Inner product of vector X and unit vector Y is the projection of vector X in the direction of Y.

$$X.\hat{Y} = \| X \| (1) \cos\theta$$

$$= \text{Projection of } X \text{ in the direction of } Y.$$

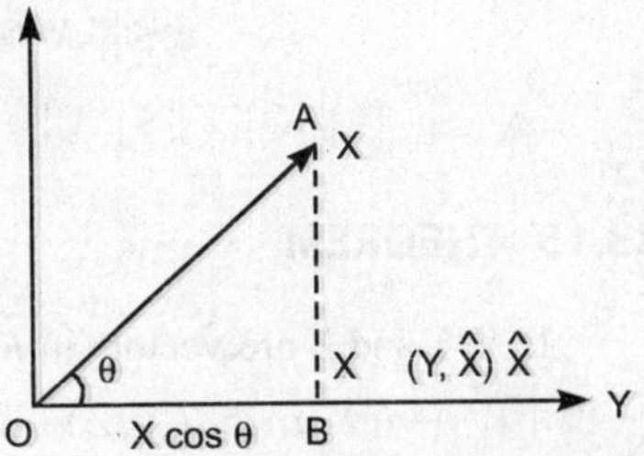

If two vectors X and Y are perpendicular to each other, then

$$X.Y = \| X \| \| Y \| \cos 90°$$

$$\Rightarrow \quad X.Y = \| X \| \| Y \| (0)$$

$$\Rightarrow \quad X \,.\, Y = 0$$

Inner product of two perpendicular vectors is always zero.

If $X.\ Y = 0$, then X and Y are known as **orthogonal vectors**.

33.13 CAUCHY- SCHWARZ INEQUALITY

If X and Y are two vectors in R^n,then $\| X.Y \| < \| X \| \; \| Y \|$

Proof. From theorem 1, we know that

$$X.Y = \| X \| \| Y \| \cos\theta$$

$$\| X.Y \| = \| X \| \| Y \| | \cos\theta |$$

$$\leq \| X \| \| Y \| (1) \qquad [\because | \cos\theta \leq 1]$$

$$\leq \| X \| \| Y \|$$

Proved.

33.14 THEOREM (PROPERTIES OF THE EUCLIDEAN NORM)

If X and Y are vectors in R^n and k is a scalar, then

(i) $\| X \| > 0$

(*ii*) $\| X \| = 0$ if and only if $X = 0$

(*iii*) (Scaling identity) $\| k X \| = | k | \, \| X \|$

(*iv*) (The triangle inequality) $\| X + Y \| \leq \| X \| + \| Y \|$

Proof : The proof of the first two parts is a direct consequence of the definition of the Euclidean norm and so we will not discuss here.

(*iii*) From the definition of the Euclidean norm, we have

$$\| k X \| = \sqrt{(kx_1)^2 + (kx_2)^2 + \cdots + (kx_n)^2}$$

$$= \sqrt{k^2 (x_1^2 + x_2^2 + \cdots + x_n^2)}$$

$$= | k | \, \| X \|.$$

(*iv*) From the definition of the Euclidean norm, we have

$$\| X+Y \|^2 = (X+Y).(X+Y)$$
$$= X\cdot(X+Y)+Y\cdot(X+Y)$$

(Using Properties of Euclidean inner product)

$$= X\cdot X + X\cdot Y + Y\cdot X + Y\cdot Y$$
$$= \| X \|^2 + 2X\cdot Y + \| Y \|^2$$
$$\le \| X \|^2 + 2\| X \| \| Y \| + \| Y \|^2 \qquad [X.Y \le \| X \| \ \| Y \|]$$
$$\le [\| X \| + \| Y \|]^2$$
$$\Rightarrow \qquad \| X+Y \| \le \| X \| + \| Y \|$$

33.15 THEOREM

If *X*, *Y* and *Z* are vectors in R^n, then

$$d(X,Y) \le d(X,Z) + d(Z,Y)$$

Proof. $d(X,Y) = \| X-Y \|$

$$= \| X-Z+Z-Y \| \qquad \text{[By adding and subtracting } z]$$
$$= \| (X-Z)+(Z-Y) \|$$
$$\le \| X-Z \| + \| Z-Y \| \qquad \text{(Using Triangular inequality for norms)}$$
$$\le d(X,Z) + d(Z,Y) \qquad \text{(Using definition of distance)}$$

Hence, $d(X,Y) \le d(X,Z) + d(Z,Y)$ **Proved.**

33.16 THEOREM (Pythagoras Theorem in R^n)

If *X* and *Y* are two orthogonal vectors in R^n then $\| X+Y \|^2 = \| X \|^2 + \| Y \|^2$

Proof: We know that

$$\| X+Y \|^2 = \| X \|^2 + 2(X.Y) + \| Y \|^2 \qquad \text{[From theorem]}$$
$$= \| X \|^2 + 0 + \| Y \|^2 \qquad [X, Y = 0]$$
$$= \| X \|^2 + \| Y \|^2$$

Proved.

33.17 THEOREM (Relationship between Euclidean inner Product and the Norm)

If *X* and *Y* are two vectors in R^n, then

$$X.Y = \frac{1}{4}\| X+Y \|^2 - \frac{1}{4}\| X-Y \|^2$$

Proof: We know that

$$\| X+Y \|^2 = \| X \|^2 + 2(X.Y) + \| Y \|^2 \qquad ...(1)$$
$$\| X-Y \|^2 = \| X \|^2 - 2(X.Y) + \| Y \|^2 \qquad ...(2)$$

Subtracting (2) from (1), we get

$$\|X+Y\|^2 - \|X-Y\|^2 = 4(X.Y)$$

$$\Rightarrow X.Y = \frac{1}{4}\|X+Y\|^2 - \frac{1}{4}\|X-Y\|^2$$ **Proved.**

Example 2. *Let* $X = (2,3,4)$, $Y = (3, 0, 5)$, $U = (2, 3, 4, 5)$, $V = (4, 6, 9, 2)$. *Find :*

(*i*) $X + Y$ (*ii*) $3V + 4U$ (*iii*) $3U - 4V$

Solution.

(*i*) $X + Y = (2, 3, 4) + (3, 0, 5)$
$= (2 + 3, 3 + 0, 4 + 5)$
$= (5, 3, 9)$ **Ans.**

(*ii*) $3V + 4U = 3(4, 6, 9, 2) + 4(2, 3, 4, 5)$
$= (12, 18, 27, 6) + (8, 12, 16, 20)$
$= (12 + 8, 18 + 12, 27 + 16, 6 + 20)$
$= (20, 30, 43, 26)$ **Ans.**

(*iii*) $3U - 4V = 3(2, 3, 4, 5) - 4(4, 6, 9, 2)$
$= (6, 9, 12, 15) - (16, 24, 36, 8)$
$= (6 - 16, 9 - 24, 12 - 36, 15 - 8)$
$= (-10, -15, -24, 7)$ **Ans.**

Example 3. *Given* $X = (9, 3, -4, 0, 1)$ *and* $Y = (0, -3, 2, -1, 7)$ *compute:*

(*a*) $X - 4Y$

(*b*) $Y. X$

(*c*) $X. X$

(*d*) $|X|$

(*e*) $d(X.Y)$

Solution. We have

$X = (9, 3, -4, 0, 1)$ and $Y = (0, -3, 2, -1, 7)$

(*a*) $X - 4Y = (9, 3, -4, 0, 1) - 4(0, -3, 2, -1, 7)$
$= (9, 3, -4, 0, 1) + (0, 12, -8, 4, -28)$
$= (9 + 0, 3 + 12, -4 - 8, 0 + 4, 1 - 28)$
$= (9, 15, -12, 4, -27)$ **Ans.**

(*b*) $Y. X = (0, -3, 2, -1, 7). (9, 3, -4, 0, 1)$
$= (0.9, -3.3, 2(-4), -1(0), 7.1)$
$= (0, -9, -8, 0, 7)$ **Ans.**

(*c*) $X. X = (9, 3, -4, 0, 1,).(9, 3, -4, 0, 1)$
$= (9.9, 3.3, (-4)(-4), 0.0, 1.1)$
$= (81, 9, 16, 0, 1)$ **Ans.**

(*d*) $|X| = |(9, 3, -4, 0, 1)| = \sqrt{9^2 + 3^2 + (-4)^2 + 0 + 1^2}$
$= \sqrt{81 + 9 + 16 + 1} = \sqrt{107}$

(*e*) $d(X,Y) = \| X - Y \|$

$$= \| (9, 3, -4, 0, 1) - (0, -3, 2, -1, 7) \|$$
$$= \| (9, 6, -6, 1, -6) \|$$
$$= \sqrt{9^2 + 6^2 + (-6)^2 + 1^2 + (-6)^2}$$
$$= \sqrt{81 + 36 + 36 + 1 + 36}$$
$$= \sqrt{190}$$

Ans.

EXERCISE 33.1

1. If $X = (3, 2, 0)$ and $Y = (-1, 6, 4)$. Find $X - 3Y$. **Ans.** $(6, -16, -12)$
2. If $X = (2, 3, 0, 5)$ and $Y = (0, 6, -1, 9)$ and $Z = (1, 1, 1, 0)$. Find $X + Y$, $Y - 3Z$. **Ans.** $(2, 9, -1, 14)$, $(-3, 3, -4, 9)$
3. Show that $X = (3, 0, 1, 0, 4, -1)$ and $Y = (-2, 5, 0, 2, -3, -18)$ are orthogonal and verify the Pythagorean theorem.
4. Prove that vectors X and Y are orthogonal if $X = (5, 1, -2, -2, 8)$ and $Y = (2, 2, 6, 4, 1)$.
5. If $X = \left(\frac{5}{2}, 4, 3\right)$ and $Y = \left(\frac{3}{2}, 2, -5\right)$, show that $X + Y$ and $X - Y$ are orthogonal.
6. If $X = (-2, 5, \sqrt{10}, 3, 4), Y = (1, -2, \sqrt{10}, -4, 3)$, then find

   (*a*) $\| X \|$ (*b*) $\| Y \|$ (*c*) $d(X,Y)$ **Ans.** (*a*) 8 (*b*) $2\sqrt{10}$ (*c*) $6\sqrt{3}$
7. Find a unit vector $\hat{X}$ orthogonal to (1, 1, 1) and (1, 0, 0).

   Ans. $X = \left(0, -\frac{1}{\sqrt{2}}, \frac{1}{\sqrt{2}}\right)$
8. Show that vectors $X - Y$ and $X + Y$ are othogonal if and only if $\| X \| = \| Y \|$.

Choose the correct answer

9. If vectors $(p+q, p-q) = (5, 3)$, then

   (*a*) $p = 1,\ q = 4$

   (*b*) $p = \frac{5}{2},\ q = \frac{5}{2}$

   (*c*) $p = 4,\ q = 1$

   (*d*) $p = 6,\ q = 3$ **Ans.** (*c*)
10. If vector $X = (12, 3, -4)$, then unit vector $\hat{X}$ =

    (*a*) $\left(\frac{12}{13}, \frac{3}{13}, -\frac{4}{13}\right)$ (*b*) $\left(\frac{12}{13}, \frac{-3}{13}, \frac{4}{13}\right)$

    (*c*) $\left(-\frac{12}{13}, \frac{3}{13}, \frac{4}{13}\right)$ (*d*) $\left(\frac{-12}{13}, \frac{-3}{13}, \frac{-4}{13}\right)$ **Ans** (*a*)
11. If $X = (3, 4, 12, 13)$, then $\| X \| =$

    (*a*) $2\sqrt{13}$ (*b*) $13\sqrt{2}$ (*c*) 13 (*d*) $\sqrt{32}$ **Ans.** (*b*)

12. If $X = (0, -5, 6)$ and $Y = (4, 7, 3)$, then $d(X, Y) =$

(*a*) –13 (*b*) 169 (*c*) 13 (*d*) –17 **Ans.** (*c*)

33.18 LINEAR COMBINATION OF VECTORS IN R^n

(1) **Parallel vectors**

Let two vectors X and Y in R^n are parallel and

$Y = aX$, where $a > 0$. Then X and Y have the same direction.

(2) If $Y = aX$, where $a < 0$, then X and Y are opposite in direction.

Let V be a vector space over a field F and if $v_1, v_2, \ldots\ldots\ldots v_n \in V$ then any element of the form $\alpha_1 v_1 + \alpha_2 v_2 + \alpha_3 v_3, + \cdots \alpha_n v_n$. where $\alpha_i \in F$, is called a linear combination of $v_1, v_2, \ldots\ldots. v_n$ over F.

33.19 LINEAR DEPENDENCE AND INDEPENDENCE OF VECTORS

Vectors (matrices) $X_1, X_2, \ldots\ldots\ldots X_n$ are said to be dependent if

(1) all the vectors (row or column matrices) are of the same order.

(2) n scalars $\lambda_1, \lambda_2 \ldots \lambda_n$ (not all zero) exist such that

$$\lambda_1 X_1 + \lambda_2 X_2 + \lambda_3 X_3 + \ldots\ldots\ldots + \lambda_n X_n = 0$$

Otherwise they are linearly independent.

Example 4. *Show that the vectors* $X_1 = (1, 2, 3)$, $X_2 = (3, -1, 4)$ *and* $X_3 = (4, 1, 7)$ *are linearly dependent.*

Solution. $X_1 = (1, 2, 3)$

$X_2 = (3, -1, 4)$

$X_3 = (4, 1, 7)$

If $\lambda_1 X_1 + \lambda_2 X_2 + \lambda_3 X_3 = 0$

$\Rightarrow \quad \lambda_1(1, 2, 3) + \lambda_2(3, -1, 4) + \lambda_3(4, 1, 7) = 0$

$\Rightarrow \quad [(\lambda_1 + 3\lambda_2 + 4\lambda_3), (2\lambda_1 - \lambda_2 + \lambda_3), (3\lambda_1 + 4\lambda_2 + 7\lambda_3)] = (0, 0, 0)$

$\Rightarrow \quad \lambda_1 + 3\lambda_2 + 4\lambda_3 = 0$ ... (1)

$2\lambda_1 - \lambda_2 + \lambda_3 = 0$ ... (2)

$3\lambda_1 + 4\lambda_2 + 7\lambda_3 = 0$ ... (3)

Solving (1), (2) and (3), we get

$$\lambda_1 = 1, \lambda_2 = 1 \text{ and } \lambda_3 = -1$$

So, the vectors are linearly dependent. **Proved.**

Example 5. *Are the vectors*

$X_1 = (1, 0, 0)$, $X_2 = (0, 1, 0)$ *and* $X_3 = (0, 0, 1)$

linearly dependent?

Solution. Consider the matrix equation $\lambda_1 X_1 + \lambda_2 X_2 + \lambda_3 X_3 = 0$

$\Rightarrow \lambda_1(1, 0, 0) + \lambda_2(0, 1, 0) + \lambda_3(0, 0, 1) = 0$

$\Rightarrow (\lambda_1 + 0.\lambda_2 + 0.\lambda_3, 0\lambda_1 + \lambda_2 + 0\lambda_3, 0\lambda_1 + 0\lambda_2 + \lambda_3) = (0, 0, 0)$

$\Rightarrow (\lambda_1, \lambda_2, \lambda_3) = (0, 0, 0)$

$\Rightarrow \lambda_1 = 0, \lambda_2 = 0, \lambda_3 = 0$

As $\lambda_1, \lambda_2, \lambda_3$ all are zero, therefore X_1, X_2, X_3 are linearly independent vectors. **Ans.**

Example 6. *Examine the following vectors for linear dependence and find the relation if it exists*

$$X_1 = (1, 2, 4), \quad X_2 = (2, -1, 3), \quad X_3 = (0, 1, 2), \quad X_4 = (-3, 7, 2)$$

Solution. Consider the matrix equation

$$\lambda_1 X_1 + \lambda_2 X_2 + \lambda_3 X_3 + \lambda_4 X_4 = 0 \qquad \ldots(1)$$

$$\Rightarrow \lambda_1(1,2,4) + \lambda_2(2,-1,3) + \lambda_3(0,1,2) + \lambda_4(-3,7,2) = 0$$

$$\lambda_1 + 2\lambda_2 + 0\lambda_3 - 3\lambda_4 = 0$$

$$2\lambda_1 - \lambda_2 + \lambda_3 + 7\lambda_4 = 0$$

$$4\lambda_1 + 3\lambda_2 + 2\lambda_3 + 2\lambda_4 = 0$$

This is the homogeneous system

$$\begin{bmatrix} 1 & 2 & 0 & -3 \\ 2 & -1 & 1 & 7 \\ 4 & 3 & 2 & 2 \end{bmatrix} \begin{bmatrix} \lambda_1 \\ \lambda_2 \\ \lambda_3 \\ \lambda_4 \end{bmatrix} = \begin{bmatrix} 0 \\ 0 \\ 0 \end{bmatrix} \quad \text{or} \quad A\lambda = 0$$

$R_2 \to R_2 - 2R_1, \; R_3 \to R_3 - 4R_1$

$$\begin{bmatrix} 1 & 2 & 0 & -3 \\ 0 & -5 & 1 & 13 \\ 0 & -5 & 2 & 14 \end{bmatrix} \begin{bmatrix} \lambda_1 \\ \lambda_2 \\ \lambda_3 \\ \lambda_4 \end{bmatrix} = \begin{bmatrix} 0 \\ 0 \\ 0 \end{bmatrix}$$

$R_3 \to R_3 - R_2$

$$\begin{bmatrix} 1 & 2 & 0 & -3 \\ 0 & -5 & 1 & 13 \\ 0 & 0 & 1 & 1 \end{bmatrix} \begin{bmatrix} \lambda_1 \\ \lambda_2 \\ \lambda_3 \\ \lambda_4 \end{bmatrix} = \begin{bmatrix} 0 \\ 0 \\ 0 \end{bmatrix}$$

$$\lambda_1 + 2\lambda_2 - 3\lambda_4 = 0$$

$$-5\lambda_2 + \lambda_3 + 13\lambda_4 = 0$$

$$\lambda_3 + \lambda_4 = 0$$

Let $\lambda_4 = t$, $\qquad \lambda_3 + t = 0 \Rightarrow \lambda_3 = -t$

$$-5\lambda_2 - t + 13t = 0 \Rightarrow \lambda_2 = \frac{12t}{5}$$

$$\lambda_1 + \frac{24t}{5} - 3t = 0 \quad \Rightarrow \quad \lambda_1 = \frac{-9t}{5}$$

Hence, the given vectors are linearly dependent.

Substituting the values of sum in (1), we get

$$-\frac{9tX_1}{5}+\frac{12t}{5}X_2-tX_3+tX_4=0 \quad \Rightarrow \quad -\frac{9X_1}{5}+\frac{12X_2}{5}-X_3+X_4=0$$

$$\Rightarrow \quad 9X_1-12X_2+5X_3-5X_4=0$$ **Ans.**

Example 7. *Examine for linear dependence* [1, 0, 2, 1], [3, 1, 2, 1], [4, 6, 2, – 4], [– 6, 0, – 3, – 4] *and find the relation between them if possible.*

Solution. Consider the matrix equation

$$\lambda_1X_1+\lambda_2X_2+\lambda_3X_3+\lambda_4X_4=0 \quad \ldots(1)$$

$$\lambda_1(1,0,2,1)+\lambda_2(3,1,2,1)+\lambda_3(4,6,2,-4)+\lambda_4(-6,0,-3,-4)=0$$

$$\lambda_1+3\lambda_2+4\lambda_3-6\lambda_4=0$$

$$0\lambda_1+\lambda_2+6\lambda_3+0\lambda_4=0$$

$$2\lambda_1+2\lambda_2+2\lambda_3-3\lambda_4=0$$

$$\lambda_1+\lambda_2-4\lambda_3-4\lambda_4=0$$

$$\begin{bmatrix}1&3&4&-6\\0&1&6&0\\2&2&2&-3\\1&1&-4&-4\end{bmatrix}\begin{bmatrix}\lambda_1\\\lambda_2\\\lambda_3\\\lambda_4\end{bmatrix}=\begin{bmatrix}0\\0\\0\\0\end{bmatrix}$$

$$R_3\to R_3-2R_1,\ R_4\to R_4-R_1$$

$$\begin{bmatrix}1&3&4&-6\\0&1&6&0\\0&-4&-6&9\\0&-2&-8&2\end{bmatrix}\begin{bmatrix}\lambda_1\\\lambda_2\\\lambda_3\\\lambda_4\end{bmatrix}=\begin{bmatrix}0\\0\\0\\0\end{bmatrix}$$

$$R_3\to R_3+4R_2,\ R_4\to R_4+2R_2$$

$$\begin{bmatrix}1&3&4&-6\\0&1&6&0\\0&0&18&9\\0&0&4&2\end{bmatrix}\begin{bmatrix}\lambda_1\\\lambda_2\\\lambda_3\\\lambda_4\end{bmatrix}=\begin{bmatrix}0\\0\\0\\0\end{bmatrix}$$

$$\begin{bmatrix}1&3&4&-6\\0&1&6&0\\0&0&18&9\\0&0&0&0\end{bmatrix}\begin{bmatrix}\lambda_1\\\lambda_2\\\lambda_3\\\lambda_4\end{bmatrix}=\begin{bmatrix}0\\0\\0\\0\end{bmatrix} \quad R_4\to R_4-\frac{2}{9}R_3$$

$$\lambda_1+3\lambda_2+4\lambda_3-6\lambda_4=0$$

$$\lambda_2+6\lambda_3=0$$

$$18\lambda_3+9\lambda_4=0$$

Let $\lambda_4=t$, $\quad 18\lambda_3+9t=0 \quad \Rightarrow \quad \lambda_3=\frac{-t}{2}$

$$\lambda_2-3t=0 \quad \Rightarrow \quad \lambda_2=3t$$

$$\lambda_1 + 9t - 2t - 6t = 0 \qquad \Rightarrow \qquad \lambda_1 = -t$$

Substituting the values of λ in (1), we get

$$-tX_1 + 3t\,X_2 - \frac{t}{2}X_3 + t\,X_4 = 0 \qquad \Rightarrow \quad 2X_1 - 6X_2 + X_3 - 2X_4 = 0$$ **Ans.**

Example 8. *Is the system of vectors*

$$X_1 = (2, 2, 1)^T, \quad X_2 = (1, 3, 1)^T, \quad X_3 = (1, 2, 2)^T$$

linearly dependent.

Solution. Here $X_1 = \begin{bmatrix} 2 \\ 2 \\ 1 \end{bmatrix}, X_2 = \begin{bmatrix} 1 \\ 3 \\ 1 \end{bmatrix}, \; X_3 = \begin{bmatrix} 1 \\ 2 \\ 2 \end{bmatrix}$ (T stands for transposition)

Consider the matrix equation

$$\lambda_1 X_1 + \lambda_2 X_2 + \lambda_3 X_3 = 0 \qquad \dots (1)$$

$$\lambda_1 \begin{bmatrix} 2 \\ 2 \\ 1 \end{bmatrix} + \lambda_2 \begin{bmatrix} 1 \\ 3 \\ 1 \end{bmatrix} + \lambda_3 \begin{bmatrix} 1 \\ 2 \\ 2 \end{bmatrix} = \begin{bmatrix} 0 \\ 0 \\ 0 \end{bmatrix}$$

$$2\lambda_1 + \lambda_2 + \lambda_3 = 0$$

$$2\lambda_1 + 3\lambda_2 + 2\lambda_3 = 0$$

$$\lambda_1 + \lambda_2 + 2\lambda_3 = 0$$

Which is the homogeneous equation.

$$\begin{bmatrix} 2 & 1 & 1 \\ 2 & 3 & 2 \\ 1 & 1 & 2 \end{bmatrix} \begin{bmatrix} \lambda_1 \\ \lambda_2 \\ \lambda_3 \end{bmatrix} = \begin{bmatrix} 0 \\ 0 \\ 0 \end{bmatrix} \qquad \Rightarrow \qquad \begin{bmatrix} 1 & 1 & 2 \\ 2 & 3 & 2 \\ 2 & 1 & 1 \end{bmatrix} \begin{bmatrix} \lambda_1 \\ \lambda_2 \\ \lambda_3 \end{bmatrix} = \begin{bmatrix} 0 \\ 0 \\ 0 \end{bmatrix} \; R_1 \leftrightarrow R_3$$

$$\begin{bmatrix} 1 & 1 & 2 \\ 0 & 1 & -2 \\ 0 & -1 & -3 \end{bmatrix} \begin{bmatrix} \lambda_1 \\ \lambda_2 \\ \lambda_3 \end{bmatrix} = \begin{bmatrix} 0 \\ 0 \\ 0 \end{bmatrix} \begin{matrix} \\ R_2 \to R_2 - 2R_1 \\ R_3 \to R_3 - 2R_1 \end{matrix} \qquad \Rightarrow \qquad \begin{bmatrix} 1 & 1 & 2 \\ 0 & 1 & -2 \\ 0 & 0 & -5 \end{bmatrix} \begin{bmatrix} \lambda_1 \\ \lambda_2 \\ \lambda_3 \end{bmatrix} = \begin{bmatrix} 0 \\ 0 \\ 0 \end{bmatrix} \; R_3 \to R_3 + R_2$$

$$\lambda_1 + \lambda_2 + 2\lambda_3 = 0$$

$$\lambda_2 - 2\lambda_3 = 0$$

$$-5\lambda_3 = 0 \Rightarrow \lambda_3 = 0$$

$$\therefore \qquad \lambda_2 = 0 \text{ and } \lambda_1 = 0$$

Thus non-zero values of $\lambda_1, \lambda_2, \lambda_3$ do not exist which can satisfy (1). Hence by definition, the given system of vectors is not linearly dependent. **Ans.**

Example 9. *Prove that the n-tuple vectors*

$$X_1 = (1, 0, 0, \cdots, 0)$$

$$X_2 = (0, 1, 0, \cdots, 0)$$

$$X_3 = (0, 0, 0, \cdots, 1)$$

are linearly independent.

Solution. Consider the matrix equation

$$\lambda_1 X_1 + \lambda_2 X_2 + \cdots + \lambda_n X_n = (0, 0, \cdots, 0)$$

$$\Rightarrow \quad \lambda_1(1, 0, 0 \cdots 0) + \lambda_2(0, 1, 0 \cdots 0) + \cdots + \lambda_n (0, 0, 0 \cdots 1) = (0, 0, \cdots, 0)$$

$$\Rightarrow \quad (\lambda_1, \lambda_2, \cdots, \lambda_n) = (0, 0, \cdots, 0)$$

$$\Rightarrow \quad \lambda_1 = 0, \lambda_2 = 0, \cdots, \lambda_n = 0$$

As all scalars $\lambda_1, \lambda_2, \cdots, \lambda_n$ are zero, therefore $X_1, X_2 \cdots X_n$ linearly independent. **Proved.**

Example 10. *Find whether the set of vectors $v_1 = (1, 2, 1)$, $v_2 = (3, 1, 5)$, $v_3 = (3, -4, 7)$ is linearly independent or dependent.*

Solution. If a_1, a_2, a_3 are three scalars, then $a_1 v_1 + a_2 v_2 + a_3 v_3 = 0$

$$\Rightarrow a_1 (1, 2, 1) + a_2 (3, 1, 5) + a_3 (3, -4, 7) = (0, 0, 0)$$

$$\Rightarrow (a_1 + 3a_2 + 3a_3, 2a_1 + a_2 - 4a_3, a_1 + 5a_2 + 7a_3) = (0,0,0)$$

$$\Rightarrow \left.\begin{aligned} a_1 + 3a_2 + 3a_3 &= 0 \\ 2a_1 + a_2 - 4a_3 &= 0 \\ a_1 + 5a_2 + 7a_3 &= 0 \end{aligned}\right\} \qquad \text{... (1)}$$

The coefficient matrix A of the system of equations (1) is

$$A = \begin{bmatrix} 1 & 3 & 3 \\ 2 & 1 & -4 \\ 1 & 5 & 7 \end{bmatrix}$$

$$= \begin{bmatrix} 1 & 3 & 3 \\ 0 & -5 & -10 \\ 0 & 2 & 4 \end{bmatrix} \begin{matrix} \\ R_2 \to R_2 - 2R_1 \\ R_3 \to R_3 - R_1 \end{matrix}$$

$$= \begin{bmatrix} 1 & 3 & 2 \\ 0 & 1 & 2 \\ 0 & 1 & 2 \end{bmatrix} \begin{matrix} R_2 \to \left(-\frac{1}{5}\right) R_2 \\ \\ R_3 \to \left(\frac{1}{2}\right) R_3 \end{matrix}$$

$$= \begin{bmatrix} 1 & 3 & 3 \\ 0 & 1 & 2 \\ 0 & 0 & 0 \end{bmatrix} \quad R_3 \to R_3 - R_2$$

$\therefore$ Rank $(A) = 2 <$ the number of unknowns.

The system (1) of equations has $3 - 2 = 1$ non-zero solution. Hence there exist scalars a_1, a_2, a_3 not all of them zero. Hence the given set of vectors is linearly dependent. **Ans.**

Example 11. *Prove that in R [x], the vector space of polynomials in x over R, the three polynomials.*

$$p(x) = 1 + x + 2x^2,\ q(x) = 2 - x + x^2,\ r(x) = -4 + 5x + x^2$$

Are linearly dependent.

Solution. We have,

$$2p(x) + (-3)q(x) + (-1)r(x) = 2(1 + x + 2x^2) + (-3)(2 - x + x^2) + (-1)(-4 + 5x + x^2)$$

$$= 0 + 0x + 0x^2 = 0, \text{ zero polynomial.}$$

Since scalar coefficients 2, – 3, –1 in the above relation are all non-zero and hence the given system is linearly dependent. **Proved.**

Example 12. *Prove that the four vectors $\alpha_1 = (1, 2, 3)$, $\alpha_2 = (1, 0, 0)$, $\alpha_3 = (0, 1, 0)$ and $\alpha_4 = (0, 0, 1)$ in V_3 (R) form a linearly dependent set.*

Solution. We have,

$$1.\alpha_1 + (-1)\alpha_2 + (-2)\alpha_3 + (-3)\alpha_4 = 1(1, 2, 3) + (-1)(1, 0, 0) + (-2)(0, 1, 0) + (-3)(0, 0, 1)$$

$$= (1, 2, 3) + (-1, 0, 0) + (0, -2, 0) + (0, 0, -3) = (0, 0, 0) = 0$$

i.e. $a_1\alpha_1 + a_2\alpha_2 + a_3\alpha_3 + a_4\alpha_4 = 0$,

where $a_1 = 1 \neq 0,\ a_2 = -1 \neq 0,\ a_3 = -2 \neq 0, a_4 = -3 \neq 0.$

Hence, $\{\alpha_1, \alpha_2, \alpha_3, \alpha_4\}$ is a linearly dependent (*L.D.*) subset of V_3 (*R*). **Proved.**

Example 13. *Are the vectors (2, 2, 2, 4), (2, – 2, – 4, 0), (4, –2, –5, 2), (4, 2, 1, 6) linearly independent ?* *(June 2007)*

Solution. Let $\alpha_1 = (2, 2, 2, 4)$, $\alpha_2 = (2, -2, -4, 0)$; $\alpha_3 = (4, -2, -5, 2)$; $\alpha_4 = (4, 2, 1, 6)$.

If a_1, a_2, a_3, a_4 be four scalars, then $a_1\alpha_1 + a_2\alpha_2 + a_3\alpha_3 + a_4\alpha_4 = 0$

$\Rightarrow a_1(2, 2, 2, 4) + a_2(2, -2, -4, 0) + a_3(4, -2, -5, 2) + a_4(4, 2, 1, 6) = (0, 0, 0, 0)$

$$\Rightarrow \left.\begin{aligned} 2a_1 + 2a_2 + 4a_3 + 4a_4 &= 0 \\ 2a_1 - 2a_2 - 2a_3 + 2a_4 &= 0 \\ 2a_1 - 4a_2 - 5a_3 + a_4 &= 0 \\ 4a_1 + 0a_2 + 2a_3 + 6a_4 &= 0 \end{aligned}\right\} \quad \text{... (1)}$$

The coefficient matrix A of the system of equations (1) is $A = \begin{bmatrix} 2 & 2 & 4 & 4 \\ 2 & -2 & -2 & 2 \\ 2 & -4 & -5 & 1 \\ 4 & 0 & 2 & 6 \end{bmatrix}$

$$= \begin{bmatrix} 2 & 2 & 4 & 4 \\ 0 & -4 & -6 & -2 \\ 0 & 12 & 18 & 6 \\ 0 & 0 & 0 & 0 \end{bmatrix} \begin{matrix} \\ \\ R_3 \to (-2)R_3 \\ R_4 \to R_4 - R_2 \end{matrix}$$

$$= \begin{bmatrix} 2 & 2 & 4 & 4 \\ 0 & -4 & -6 & -2 \\ 0 & -6 & -9 & -3 \\ 0 & -4 & -6 & -2 \end{bmatrix} \begin{matrix} \\ R_2 \to R_2 - R_1 \\ R_3 \to R_3 - R_1 \\ R_4 \to R_4 - 2R_1 \end{matrix}$$

$$= \begin{bmatrix} 2 & 2 & 4 & 4 \\ 0 & -4 & -6 & -2 \\ 0 & 0 & 0 & 0 \\ 0 & 0 & 0 & 0 \end{bmatrix} \quad R_3 \to R_3 + 3R_2$$

$\therefore$ Rank $(A) = 2 <$ the number of unknowns.

$\therefore$ The system (1) of equations has $4 - 2 = 2$ non-zero solutions. Hence the given set of vectors is linearly dependent. **Ans.**

Example 14. *Prove that the set (1, x, $1 + x + x^2$) is linearly independent set of vectors in the vector space of all polynomials over the real number field.*

Solution. Let $a_1, a_2, a_3 \in R$ be such that $a_1(1) + a_2(x) + a_3(1 + x + x^2) = 0$

$$\Rightarrow \quad (a_1 + a_3) + (a_2 + a_3)x + a_3 x^2 = 0 + 0x + 0x^2$$

$$\Rightarrow \quad a_1 + a_3 = 0,\ a_2 + a_3 = 0,\ a_3 = 0$$

$$\Rightarrow \quad a_1 = 0,\ a_2 = 0,\ a_3 = 0$$

Hence, the given set is linearly independent. **Proved.**

EXERCISE 33.2

Examine the following system of vectors for linear dependence. If dependent, find the relation between them.

1. $X_1 = (1, -1, 1), X_2 = (2, 1, 1), X_3 = (3, 0, 2).$ **Ans.** Dependent, $X_1 + X_2 - X_3 = 0$
2. $X_1 = (1, 2, 3), X_2 = (2, -2, 6).$ **Ans.** Independent
3. $X_1 = (3, 1, -4), X_2 = (2, 2, -3), X_3 = (0, -4, 1)$ **Ans.** Dependent, $2X_1 - 3X_2 - X_3 = 0$
4. $X_1 = (1, 1, 1, 3), X_2 = (1, 2, 3, 4), X_3 = (2, 3, 4, 7).$ **Ans.** Dependent, $X_1 + X_2 - X_3 = 0$
5. $X_1 = (1, 1, -1, 1), X_2 = (1, -1, 2, -1), X_3 = (3, 1, 0, 1).$

   Ans. Dependent, $2X_1 + X_2 - X_3 = 0$
6. $X_1 = (1, -1, 2, 0), X_2 = (2, 1, 1, 1), X_3 = (3, -1, 2, -1), X_4 = (3, 0, 3, 1).$

   Ans. Dependent, $X_1 + X_2 - X_4 = 0$

Find whether the following vectors are linearly independent or dependent?

7. (1, 2, 3, 0), (2, 3, 0, 1), (3, 0, 1, 2) **Ans.** Independent
8. (2, 6, – 1, 8), (0, 10, 4, 3), (10, 0, –1, 4) **Ans.** Independent
9. $(2, -6, 2, 8), \left(3, -9, \frac{3}{2}, 12\right), (0, 1, 5, 0), (7, 0, 0, -1)$ **Ans.** Independent
10. Find whether or not the following set of vectors are linearly dependent or independent:

    (*i*) (1, –2), (2, 1), (3, 2) (*ii*) (1, 1, 1, 1), (0, 1, 1, 1), (0, 0, 1, 1), (0, 0, 0, 1).
11. Express the vector (2, 5) as a linear combination of (1, 1) and (1, 0). **Ans.** 5(1,1) – 3 (1, 0)
12. Express the vector (15, 16, 21) as linear combination of (2, 3, 5), (4, 7, 3), (3, 0, 2) and (1, 2, 3).

Ans. $(15,16,21) = \frac{51}{4}(2,3,5) + \frac{5}{4}(4,7,3) + 0\,(3,0,2) - \frac{31}{2}(1,2,3)$

13. Show that vectors $X_1 = (2,3,1,-1), X_2 = (2,3,1,-2), X_3 = (4,6,2,1)$ are linearly dependent. Express one of the vectors as linear combination of the others.

14. Show that the vectors $X_1 = (a_1, b_1), X_2 = (a_2, b_2)$ are linearly dependent if $a_1b_2 - a_2b_1 = 0$.

15. Let $V_1 = (1,-1,0), V_2 = (0,1,-1), V_3 = (0,0,1)$ be elements of R^3. The set of vectors $\{V_1, V_2, V_3\}$ is

(*a*) Linearly independent (*b*) linearly dependent (*c*) null (*d*) none of these

Ans. (*b*) Linearly dependent $V_1 + V_2 - V_3 = 0$

16. Find whether the vectors $V_1 = (1,1,0,1), V_2 = (1,1,1,1), V_3 = (4,4,1,1)$ and $V_4 = (1,0,0,1)$ are linearly independent or dependent. **Ans.** Linearly independent

17. Find the value of α so that the vectors (1, 2, 9, 8), (2, 0, 0, α), (α, 0, 0, 8), (0, 0, 1, 0) are linearly independent. **Ans.** α = ± 4.

18. Express $\begin{bmatrix} 3 & 1 \\ 1 & -1 \end{bmatrix}$ as a linear combination of

$$A = \begin{bmatrix} 1 & 1 \\ 1 & 0 \end{bmatrix},\quad B = \begin{bmatrix} 0 & 0 \\ 1 & 1 \end{bmatrix},\quad C = \begin{bmatrix} 0 & 2 \\ 0 & -1 \end{bmatrix}$$

Ans. $3A - 2B - C$

19. Express $\begin{bmatrix} 3 & -1 \\ 1 & -2 \end{bmatrix}$ as a linear combination of

$$X = \begin{bmatrix} 1 & 1 \\ 0 & -1 \end{bmatrix}, Y = \begin{bmatrix} 1 & 1 \\ -1 & 0 \end{bmatrix}, Z = \begin{bmatrix} 1 & -1 \\ 0 & 0 \end{bmatrix}$$

Ans. $2X - Y + 2Z$

20. Prove that the vectors $\begin{bmatrix} 2 & -4 & 8 \\ 6 & 0 & -2 \end{bmatrix}$ and $\begin{bmatrix} 1 & -2 & 4 \\ 3 & 0 & -1 \end{bmatrix}$ are linearly independent.

21. Show that column vectors of following matrix A are linearly independent:

$$A = \begin{bmatrix} 1 & 0 & 0 \\ 6 & 2 & 1 \\ 4 & 3 & 2 \end{bmatrix}$$

Write down true or false against the following statements:

22. The vectors $(1,5), \left(\frac{1}{2}, \frac{5}{2}\right)$ and $\left(\frac{3}{2}, \frac{15}{2}\right)$ are linearly dependent. **(True)**

23. The vector (1, 1, 2) can be expressed as linear combination of (0, 2, 1) and (2, 2, 4). **(True)**

24. The vector (2, 2, 4) is a linear combination of the vectors (1, 1, 2) and (0, 2, 1). **(True)**

25. The vector (0, 2, 1) is a linear combination of the vectors (2, 2, 4) and (1, 1, 2). **(False)**

26. The vectors (1, 0, 0, 0), (0, 2, 0, 0), (0, 0, 3, 0) and (0, 0, 0, 4) are linearly dependent. **(False)**

33.20 LINEAR SPAN OF A SET

Set S be a non-empty subset of a vector space $V(F)$. The set of all linear combinations of finite sets of elements of S, is called the linear span of S and is denoted by $L(S)$. The set S is called a set of generators of $L(S)$.

Mathematically

$L(S) = \{a_1\alpha_1 + a_2\alpha_2 + ...a_n\alpha_n : \alpha_i \in S \text{ and } a_i \in F,\ i = 1, 2, ...n\}$

If S is empty, then we define : $L(S) = \{\bar{0}\}$

If $S = \{\alpha\}$, where α is a non-zero vector in R^2 or R^3, then $L(S)$, which is the set of all scalar products, $a\,\alpha$, represents a straight line defined by α.

If α_1 and α_2 are non-collinear vectos in R^3 with origin as initial points, then for $S = \{\alpha_1, \alpha_2\}$, $L(S)$ contains all linear combinations of the form $a\alpha_1 + b\alpha_2$, which is a plane represented by α_1 and α_2.

Example 15. *Solve if the vector (2, –5, 3) in the subspace of R^3 spanned by the vectors (1, –3, 2), (2, –4, 1), (1, –5, 7) ?* (*R.G.P.V., Bhopal, III Semester, June 2006*)

Solution. Let $\alpha = (2, -5, 3)$, $\alpha_1 = (1, -3, 2)$, $\alpha_2 = (2, -4, 1)$, $\alpha_3 = (1, -5, 7)$

We express $\alpha = a_1\alpha_1 + a_2\alpha_2 + a_3\alpha_3$

$$(2, -5, 3) = a_1(1, -3, 2) + a_2(2, -4, 1) + a_3(1, -5, 7) \quad ...(1)$$

$$\Rightarrow \quad (2, -5, 3) = (a_1 + 2a_2 + a_3, -3a_1 - 4a_2 - 5a_3, 2a_1 + a_2 + 7a_3)$$

$$\Rightarrow \quad a_1 + 2a_2 + a_3 = 2$$

$$-3a_1 - 4a_2 - 5a_3 = -5$$

$$2a_1 + a_2 + 7a_3 = 3$$

$$\Rightarrow \quad \begin{bmatrix} 1 & 2 & 1 \\ -3 & -4 & -5 \\ 2 & 1 & 7 \end{bmatrix}\begin{bmatrix} a_1 \\ a_2 \\ a_3 \end{bmatrix} = \begin{bmatrix} 2 \\ -5 \\ 3 \end{bmatrix}$$

$$\Rightarrow \quad \begin{bmatrix} 1 & 2 & 1 \\ 0 & 2 & -2 \\ 0 & -3 & 5 \end{bmatrix}\begin{bmatrix} a_1 \\ a_2 \\ a_3 \end{bmatrix} = \begin{bmatrix} 2 \\ 1 \\ -1 \end{bmatrix} \quad \begin{matrix} R_2 \to R_2 + 3R_1 \\ R_3 \to R_3 - 2R_1 \end{matrix}$$

$$\Rightarrow \quad \begin{bmatrix} 1 & 2 & 1 \\ 0 & 2 & -2 \\ 0 & 0 & 2 \end{bmatrix}\begin{bmatrix} a_1 \\ a_2 \\ a_3 \end{bmatrix} = \begin{bmatrix} 2 \\ 1 \\ \frac{1}{2} \end{bmatrix} \quad R_3 \to R_3 + \frac{3}{2}R_2$$

$$\Rightarrow \quad a_1 + 2a_2 + a_3 = 2 \quad ...(2)$$

$$2a_2 - 2a_3 = 1 \quad ...(3)$$

$$2a_3 = \frac{1}{2} \quad \Rightarrow \quad a_3 = \frac{1}{4} \quad ...(4)$$

Putting the value of a_3 from (4) in (3), we get

$$2a_2 - 2\left(\frac{1}{4}\right) = 1 \quad \Rightarrow 2a_2 = \frac{3}{2} \quad \Rightarrow a_2 = \frac{3}{4} \quad ...(5)$$

Putting the values of a_2 and a_3 from (5) and (4) in (2), we get

$$a_1 + 2\left(\frac{3}{4}\right) + \frac{1}{4} = 2 \quad \Rightarrow \quad a_1 = 2 - \frac{1}{4} - \frac{3}{2} = \frac{1}{4}$$

Putting the values of a_1, a_2, a_3 in (1), we get

$$(2, -5, 3) = \frac{1}{4}(1, -3, 2) + \frac{3}{4}(2, -4, 1) + \frac{1}{4}(1, -5, 7)$$ **Ans.**

33.21 VECTOR SUB SPACES

Let V be a vector space over a field F, then, a non-empty subset W of V is called a vector subspace of V, if W is a vector space in its own right with respect to the addition and scalar multiplication compositions on V, restricted only on points of W.

Remark. In an arbitrary vector space V, the sets $\{0\}$ and V are clearly subspaces of V and are known as trivial sub-spaces. However, our interest lies in non-trivial subspaces.

33.22 CRITERION FOR IDENTIFYING SUBSPACES

Theorem 2. Let V be a vector space over a field F. Then, the necessary and sufficient conditions for a non-empty subset W of V to be a subspace are that

$$\alpha \in W,\ \beta \in W \Rightarrow (\alpha + \beta) \in W.$$

and $$a \in F, \alpha \in W \Rightarrow a\,\alpha \in W.$$

Proof. First suppose that W is a subspace of V. Then, W is a vector space in its own right.

Consequently, W must be closed under addition, and the scalar multiplication on W over F must be well defined.

Hence, $\alpha \in W,\ \beta \in W \Rightarrow (\alpha + \beta) \in W.$

and $a \in F,\ \alpha \in W \Rightarrow a\alpha \in W.$

Conversely, let the given conditions be satisfied.

Now, if α be an arbitrary element of W and 1 be the unity of F, then $-1 \in F$ and therefore, according to the given conditions, we have :-

$$-1 \in F, \alpha \in W \Rightarrow (-1)\,\alpha \in W \Rightarrow -\alpha \in W,$$

Thus, every element in W has its additive inverse in W.

Consequently, $\alpha \in W,\ \beta \in W \Rightarrow \alpha \in W, -\beta \in W \Rightarrow [\alpha + (-\beta)] \in W$

$$\Rightarrow (\alpha - \beta) \in W.$$

This shows, that $(W, +)$ is a sub-group of $(V, +)$.

More over, all the elements of W being the elements of V, and the addition of vectors being commutative in V, so it is in W.

Thus, $(W, +)$ is an abelian group.

Also, it is being given that the scalar multiplication composition is well defined on W.

Further, all elements in W being elements of V, all the remaining four properties of a vector space are satisfied by elements of W.

Hence, W is a subspace of V.

Cor. 1. The necessary and sufficient conditions for a non-empty subset W of a vector space $V(F)$ to be a subspace are that

$$\alpha \in W, \beta \in W \Rightarrow \alpha - \beta \in W.$$

and $$a \in F,\ \alpha \in W \Rightarrow a\alpha \in W.$$

Cor. 2. The necessary and sufficient conditions for a non-empty subset W of a vector space $V(F)$ to be a subspace are that

$$\alpha \in W, \beta \in W \Rightarrow a\,\alpha + b\,\beta \in W\ \forall\ a, b \in F.$$

Example 16. *Consider the following subsets of the 3-tuple space R^3 (R)*

$W_1 = \{(\alpha_1, \alpha_2, 0) : \alpha_1, \alpha_2 \in R\}$; $W_2 = \{(0, \alpha_2, \alpha_3) : \alpha_2, \alpha_3 \in R\}$;

$W_3 = \{(\alpha_1, 0, \alpha_3) : \alpha_1, \alpha_3 \in R\}$; $W_4 = \{(\alpha_1, 0, 0): \alpha_1 \in R\}$;

$W_5 = \{(0, \alpha_2, 0) : \alpha_2 \in R\}$ & $W_6 = \{(0, 0, \alpha_3): \alpha_3 \in R\}$.

Prove that W_1, W_2, W_3, W_4, W_5 and W_6 are the sub space of R^3 (R).

Solution. It is easy to verify that each one of the above sets is a subspace of R^3 (*R*).

Let $\alpha = (\alpha_1, \alpha_2, 0) \in W_1$; $\beta = (\beta_1, \beta_2, 0) \in W_1$ and $a, b \in F$.

Then, $a\alpha + b\beta = a(\alpha_1, \alpha_2, 0) + b(\beta_1, \beta_2, 0)$

$= (a\alpha_1, a\alpha_2, 0) + (b\beta_1, b\beta_2, 0)$

$= (a\alpha_1 + b\beta_1, a\alpha_2 + b\beta_2, 0) \in W_1$

Hence, W_1 is a subspace of R^3 (*R*)

Similarly, we can show that w_2, w_3, w_4, w_5 and w_6 are the subspaces of R^3 (*R*). **Proved.**

Example 17. *In the vector space V (R) of all real valued functions, let*

$V_e = \{f \in V : f(-x) = f(x)\ \forall\, x \in R\}$

and $V_0 = \{f \in V : f(-x) = -f(x)\ \forall\, x \in R\}$

be the sets consisting of even and odd functions respectively.

Solution. Now, if $f, g \in V_e$, then for any scalars $a, b \in R$ and $\forall\, x \in R$, we have:

$$(af + bg)(-x) = (af)(-x) + (bg)(-x)$$
$$= af(-x) + bg(-x)$$
$$= af(x) + b.g(x) \qquad [\because f \,\&\, g \text{ are even}]$$
$$= (af)(x) + (bg)(x)$$
$$= (af + bg)(x)$$

This shows that whenever *f* and *g* are even functions, then $(af + bg)$ is also even.

Thus, $f \in V_e, g \in V_e \Rightarrow (af + bg) \in V_e\ \forall\, a, b \in F$.

Hence, V_e is a subspace of *V*.

Similarly, V_0 can be shown to be a subspace of *V*. **Ans.**

Example 18. *Let V (F) be the vector space of all n × n matrices over a field F and let W be the set of all symmetric matrices in V i.e.*

$W = \{A \in V : A^t = A\}$, where A^t denotes the transpose of A.

Solution. $\forall\, a, b \in F$ and $A, B \in V$, we have

$(aA + bB)^t = aA^t + bB^t = (aA + bB)$

Thus, $A \in W, B \in W \Rightarrow (aA + bB) \in W\ \forall\, a, b \in F$.

Hence, *W* is a subspace of *V*. **Ans.**

Example 19. *Let W be the set of all polynomials of degree not exceeding n in an indeterminate x, over the field F. Then, W is a subspace of the vector space F [x] of all polynomials over F.*

Solution. Let $f(x)$ and $g(x)$ be any two polynomials in *W* and let $a, b \in F$. Then, clearly $a.f(x)$ and $b.g(x)$ are polynomials in *W*. Also, by closure property of addition, $[a.f(x) + b.g(x)]$ is a polynomial in $F[x]$.

More-over, deg $[a.f(x) + b.g(x)]$

= max. {deg $[a.f(x)]$: deg. $[b.g(x)]$}

$\leq n$ $\left[\begin{array}{l}\because \deg[a.f(x)] = \deg f(x) \leq n \\ \& \deg[b.g(x)] = \deg g(x) \leq n\end{array}\right]$

Thus, $[a \,.\, f(x) + b \,.\, g(x)]$ is a polynomial of degree not exceeding n and so $[a \,.\, f(x) + b \,.\, g(x)] \in W$.

$\therefore\ f(x) \in W, g(x) \in W \Rightarrow [a \,.\, f(x) + b \,.\, g(x)] \in W, \forall\ a, b \in F$.

Hence, W is a subspace of $F(x)$. **Ans.**

Example 20. *Let R be the field of real numbers. Which of the following are subspaces of V_3 (R)?*

(i) $W_1 = \{(x, 2y, 3z) : x, y, z \in R\}$ *(ii)* $W_2 = \{(x, x, x) : x \in R\}$

(iii) $W_3 = \{(x, y, z) : x, y, z$ *are rational numbers*$\}$

Solution. (*i*) Here $W_1 = \{(x, 2y, 3z) : x, y, z \in R\}$.

Let $\alpha = (x_1, 2y_1, 3z_1)$ and $\beta = (x_2, 2y_2, 3z_2)$ be any two arbitrary elements of W_1, then $x_1, y_1, z_1, x_2, y_2, z_2 \in R$. If $a, b \in R$ be any two real numbers, then

$$a\alpha + b\beta = a(x_1, 2y_1, 3z_1) + b(x_2, 2y_2, 3z_2) = (ax_1 + bx_2, 2ay_1 + 2by_2, 3az_1 + 3bz_2)$$

$$= [ax_1 + bx_2, 2(ay_1 + by_2), 3(az_1 + bz_2)] \in W_1 \qquad [\because ax_1 + bx_2 \text{ etc. } \in R.]$$

$\therefore\ a, b \in R$ and $\alpha, \beta \in W_1 \Rightarrow a\alpha + b\beta \in W_1$.

Hence, W_1 is a subspace of V_3 (*R*).

(*ii*) Here $W_2 = \{(x, x, x) : x \in R\}$. Let $\alpha = (x_1, x_1, x_1)$ and $\beta = (x_2, x_2, x_2)$ be any two elements of W_2, then $x_1, x_2 \in R$. If $a, b \in R$ be any two real numbers, then we have

$a\alpha + b\beta = a(x_1, x_1, x_1) + b(x_2, x_2, x_2) = (ax_1 + bx_2, ax_1 + bx_2, ax_1 + bx_2) \in W_2$

$[\because ax_1 + bx_2 \in W_2]$

Hence W_2 is a subspace of V_3 (*R*).

(*iii*) Here $W_3 = \{(x, y, z) : x, y, z$ are rational numbers$\}$.

Let $\alpha = (4, 5, 7)$ be any element of W_3. If $a = \sqrt{6}$ is an element of R, then

$a\alpha = \sqrt{6}(4,5,7) = (4\sqrt{6},\ 5\sqrt{6},\ 7\sqrt{6}) \notin W_3$. Since $4\sqrt{6}, 5\sqrt{6}, 7\sqrt{6}$ are not rational numbers.

Consequently, W_3 is not closed with respect to scalar multiplication. Hence, W_3 is not a subspace of V_3 (*R*). **Ans.**

Example 21. *If V is a set of all (n × n) matrices over any field F, then a set W of all (n × n) symmetric matrices forms a vector subspace of V (F).*

(*R.G.P.V., Bhopal, III Semester, Dec. 2006*)

Solution. Let $A = \left[a_{ij}\right]_{n \times n}, B = \left[b_{ij}\right]_{n \times n} \in W$.

Since A and B are symmetric matrices,

$a_{ij} = a_{ji}$ and $b_{ij} = b_{ji}$.

If a , $b \in F$ be any two scalars, then

$aA + bB = a\,[a_{ij}]_{n \times n} + b\,[b_{ij}]_{n \times n} = [aa_{ij}]_{n \times n} + [bb_{ij}]_{n \times n} = [aa_{ij} + bb_{ij}]_{n \times n} = [aa_{ji} + bb_{ji}]_{n \times n}$.

Hence, $aA + bB$ is also a symmetric matrix of the same order $n \times n$.

$\therefore\ a, b \in F$ and $\alpha, \beta \in W \Rightarrow a\alpha + b\beta \in W$.

Hence, W is a subspace of V (*F*). **Proved.**

Example 22. *Let V be the vector space of all real valued continuous functions over R. Then show that the solution set W of the differential equation*

$$2\frac{d^2y}{dx^2} - 9\frac{dy}{dx} + 2y = 0.$$

where $y = f(x)$ is a subspace of V.

Solution. Here $$W = \left\{ y : 2\frac{d^2y}{dx^2} - 9\frac{dy}{dx} + 2y = 0 \right\}$$

where $y = f(x)$. Clearly $0 \in W$ since $y = 0$ satisfies the given differential equations.

Now, let $y_1, y_2 \in W$ be any two elements, then

$$2\frac{d^2y_1}{dx^2} - 9\frac{dy_1}{dx} + 2y_1 = 0 \quad \ldots (1)$$

and $$2\frac{d^2y_2}{dx^2} - 9\frac{dy_2}{dx} + 2y_2 = 0 \quad \ldots (2)$$

Now, if $a, b \in R$, then

$$2\frac{d^2}{dx^2}(ay_1 + by_2) - 9\frac{d}{dx}(ay_1 + by_2) + 2(ay_1 + by_2) = 0$$

$$\Rightarrow \quad a\left(2\frac{d^2y_1}{dx^2} - 9\frac{dy_1}{dx} + 2y_1\right) + b\left(2\frac{d^2y_2}{dx^2} - 9\frac{dy_2}{dx} + 2y_2\right) = 0$$

$\Rightarrow$ $a.0 + b.0 = 0$ [Using (1) and (2)]

Thus, $ay_1 + by_2$ is also a solution of the given differential equation, so $ay_1 + by_2 \in W$.

$\therefore$ $a, b \in R$ and $y_1, y_2 \in W \Rightarrow ay_1 + by_2 \in W$.

Hence, W is a subspace of V. **Proved.**

33.23 ALGEBRA OF VECTOR SUBSPACES

Theorem 3. *The intersection of any two subspaces of a vector space is also a subspace of the same.*

Proof. Let W_1 and W_2 be two subspaces of a vector space V over field F. Then, clearly $0 \in W_1$ and $0 \in W_2$ and therefore, $0 \in W_1 \cap W_2$. This shows that $W_1 \cap W_2 \neq \phi$.

Now, let $\alpha, \beta \in W_1 \cap W_2$ and $a, b \in F$.

Then, clearly $\alpha, \beta \in W_1$ and $\alpha, \beta \in W_2$

Since W_1 as well as W_2 is a subspace of V,

$$\left.\begin{array}{l} \alpha \in W_1, \beta \in W_1 \Rightarrow a\alpha + b\beta \in W_1 \\ \alpha \in W_2, \beta \in W_2 \Rightarrow a\alpha + b\beta \in W_2 \end{array}\right\} \Rightarrow a\alpha + b\beta \in W_1 \cap W_2.$$

Thus, $\alpha \in W_1 \cap W_2;\ \beta \in W_1 \cap W_2 \Rightarrow a\alpha + b\beta \in W_1 \cap W_2,\ \forall\, a, b \in F.$

Hence, $W_1 \cap W_2$ is also a subspace of V.

Remark. The union of two subspaces of a vector space $V(F)$ may not be a subspace of $V(F)$.

For example; Let $S_1 = \{(0, x) : x \text{ is real}\}$

and $S_2 = \{(x, 0) : x \text{ is real}\}$

are two subspaces of the vector space V_2.

We observe that $\alpha = (0, 4) \in S_1 \cup S_2$ $(\because \alpha \in S_1)$

and $\beta = (4, 0) \in S_1 \cup S_2$ $(\because \beta \in S_2)$

But $\alpha + \beta = (0, 4) + (4, 0) = (4, 4) \notin S_1 \cup S_2$

Because it neither belongs to S_1 nor to S_2.

$\therefore$ Closure property does not hold in $S_1 \cup S_2$.

So, $S_1 \cup S_2$ is not a subspace.

Example. $W_1 = \{(a_1, a_2, 0) : a_1, a_2 \in R\}$ & $W_2 = \{(0, a_2, a_3) : a_2, a_3 \in R\}$ are two subspaces of $V_3 (R)$.

Now, if we consider the elements $\alpha = (1, 2, 0)$ and $\beta = (0, 3, 4)$ of $W_1 \cup W_2$, then $\alpha + \beta = (1, 5, 4) \notin W_1 \cup W_2$.

Thus, $\alpha \in W_1 \cup W_2$; $\beta \in W_1 \cup W_2$ but $\alpha + \beta \notin W_1 \cup W_2$.

Hence, $W_1 \cup W_2$ is not a subspace of $V_3 (R)$.

Theorem 4. *The intersection of an arbitrary collection of subspaces of a vector space is a subspace of the same.*

Proof. Let $\{W_k : k \in A\}$ be an arbitrary collection of subspaces of a vector space $V (F)$. Then, $0 \in$ each W_k and therefore, $0 \in \cap \{W_k : k \in A\}$. Consequently, $\cap \{W_k : k \in A\} \neq \phi$.

Now, let $\alpha, \beta \in \cap \{W_k : k \in A\}$ and let $a, b \in F$.

then, $\alpha, \beta \in \cap \{W_k : k \in A\}$

$\Rightarrow$ $\alpha, \beta \in$ each W_k

$\Rightarrow$ $(a\,\alpha + b\,\beta) \in$ each W_k [$\because$ each W_k is a subspace]

$\Rightarrow$ $(a\alpha + b\beta) \in \cap \{W_k : k \in A\}$.

Hence, $\cap \{W_k : k \in A\}$ is a subspace of V.

33.24 SMALLEST SUBSPACE CONTAINING A SUBSET

Let S be a subset of a vector space $V (F)$. Then, a subset U of V is called the smallest subspace containing S, if U is a subspace of V containing S and is itself contained in every subspace of V containing S. Such a subspace is also called a subspace generated or spanned by S and we shall denote it by $\{S\}$.

Clearly, $\{S\}$ is the intersection of all subspaces of V, each containing S.

Theorem 5. *The union of two subspaces of a vector space is its subspace iff one is contained in the other.*

Proof. Let W_1 and W_2 be the two subspaces of a vector space $V (F)$.

Let $W_1 \subseteq W_2$ or $W_2 \subseteq W_1$, then

either $W_1 \cup W_2 = W_2$ or $W_1 \cup W_2 = W_1$.

But, W_1 and W_2 being the subspaces of V and $W_1 \cup W_2$ being either equal to W_2 or equal to W_1, so in each case $W_1 \cup W_2$ is a sub-space of V.

Conversely. Let W_1 and W_2 be two subspaces of V given in such a way that $W_1 \cup W_2$ is also a subspace.

If possible, let $W_1 \not\subseteq W_2$ and $W_2 \not\subseteq W_1$.

Then, $\exists$ elements α and β such that

$\alpha \in W_1$ but $\alpha \notin W_2$ and $\beta \in W_2$ but $\beta \in W_1$.

Now, $\alpha \in W_1 \Rightarrow \alpha \in W_1 \cup W_2$ & $\beta \in W_2 \Rightarrow \beta \in W_1 \cup W_2$.

But, $W_1 \cup W_2$ being a subspace $\alpha \in W_1 \cup W_2$, $\beta \in W_1 \cup W_2 \Rightarrow \alpha + \beta \in W_1 \cup W_2$

$\Rightarrow \alpha + \beta \in W_1$ or $\alpha + \beta \in W_2$.

Now, if $\alpha + \beta \in W_1$, then W_1 being a subspace.

$\alpha + \beta \in W_1, \alpha \in W_1 \Rightarrow (\alpha + \beta) - \alpha \in W_1 \Rightarrow \beta \in W_1$, which is a contradiction.

Again, if $\alpha + \beta \in W_2$, then W_2 being a subspace.

$\alpha + \beta \in W_2, \beta \in W_2 \Rightarrow (\alpha + \beta) - \beta \in W_2 \Rightarrow \alpha \in W_2$, which is again a contradiction.

Since the contradiction arises by assuming that $W_1 \not\subseteq W_2$ and $W_2 \not\subseteq W_1$, hence either $W_1 \subseteq W_2$ or $W_2 \subseteq W_1$. **Proved.**

33.25 LINEAR SUM OF TWO SUB SPACES

The linear sum of two subspaces W_1 and W_2 of a vector space $V(F)$, written as $W_1 + W_2$ is defined as

$W_1 + W_2 = \{\alpha_1 + \alpha_2 : \alpha_1 \in W_1 \text{ and } \alpha_2 \in W_2\}$.

Clearly, each element of $W_1 + W_2$ is expressible as sum of an element of W_1 and an element of W_2.

It is evident that $W_1 \subseteq W_1 + W_2$, for it if $\alpha_1 \in W_1$, then

$\alpha_1 \in W_1 \Rightarrow \alpha_1 = \alpha_1 + 0$ where $\alpha_1 \in W_1$ & $0 \in W_2 \Rightarrow \alpha_1 \in W_1 + W_2$.

Similarly, $W_2 \subseteq W_1 + W_2$.

Theorem 6. *The linear sum of two subspaces of a vector space is also a sub-space of the same.*

Proof. Let W_1 and W_2 be two subspaces of a vector space $V(F)$.

Then, each one of W_1 and W_2 is non-empty and therefore, $W_1 + W_2 \neq \phi$.

Let α and β be two arbitrary elements of $W_1 + W_2$.

Then, $\alpha = \alpha_1 + \alpha_2$ for some $\alpha_1 \in W_1$ & $\alpha_2 \in W_2$.

and $\beta = \beta_1 + \beta_2$ for some $\beta_1 \in W_1$ and $\beta_2 \in W_2$.

Now, for any scalars a and b, we have :-

$$a\alpha + b\beta = a(\alpha_1 + \alpha_2) + b(\beta_1 + \beta_2)$$
$$= (a\alpha_1 + b\beta_1) + (a\alpha_2 + b\beta_2) \in W_1 + W_2.$$

$$\left[\begin{array}{l} \because W_1 \text{ being a subspace, } \alpha_1 \in W_1, \beta_1 \in W_1 \Rightarrow a\alpha_1 + b\beta_1 \in W_1 \\ \quad \text{and similarly, } \alpha_2 \in W_2, \beta_2 \in W_2 \Rightarrow a\alpha_2 + b\beta_2 \in W_2 \end{array}\right]$$

Thus, $\alpha \in W_1 + W_2; \beta \in W_1 + W_2 \Rightarrow a\alpha + b\beta \in W_1 + W_2, \forall a, b \in F$.

Hence, $W_1 + W_2$ is a subspace of $V(F)$.

33.26 DIRECT SUM OF VECTOR SUB-SPACES

A vector space V is said to be the direct sum of two of its subspaces W_1 and W_2, written as $V = W_1 \oplus W_2$, if each element of V is uniquely expressible as sum of an element of W_1 and an element of W_2.

In this case, W_1 and W_2 are said to be complementary sub-spaces.

The above definition can be extended for more than two subspaces. Thus, $V = W_1 \oplus W_2 \oplus \ldots \oplus W_n$ each $\alpha \in V$ is uniquely expressible as

$\alpha = \alpha_1 + \alpha_2 + ...\alpha_n$, where $\alpha_i \in W_i$ for each $i = 1, 2, ..., n$.

The following theorem gives a criteria for a vector space to be the direct sum of two of its subspaces.

Theorem 7. The necessary and sufficient conditions for a vector space $V(F)$ to be the direct sum of its subspaces W_1 and W_2 are

(*i*) $V = W_1 + W_2$ and (ii) $W_1 \cap W_2 = \{0\}$

Proof. First suppose that $V = W_1 \oplus W_2$. Then, each element of V is expressible uniquely as sum of an element of W_1 and an element of W_2.

In particular, each element of V is expressible as sum of an element of W_1 and an element of W_2.

Consequently, $V = W_1 + W_2$ and therefore, (*i*) is satisfied.

Now, to verify the validity of (ii), if possible, let

$0 \neq \alpha \in W_1 \cap W_2$.

Then, α is a non-zero vector common to both W_1 and W_2.

Now, we may write:-

$\alpha = \alpha + 0$, where $\alpha \in W_1$, & $0 \in W_2$.

and $\alpha = 0 + \alpha$, where $0 \in W_1$ & $\alpha \in W_2$.

This shows that a non-zero element $\alpha \in V$ is expressible in at least two ways as sum of an element of W_1 and an element of W_2.

This contradicts the hypothesis that $V = W_1 \oplus W_2$.

Hence, 0 is the only vector common to both W_1 and W_2. *i.e.* $W_1 \cap W_2 = \{0\}$.

$\therefore V = W_1 \oplus W_2 \Rightarrow V = W_1 + W_2$ and $W_1 \cap W_2 = \{0\}$

Conversely, suppose that the conditions

(*i*) $V = W_1 + W_2$ and (ii) $W_1 \cap W_2 = \{0\}$ hold and we must show that $V = W_1 \oplus W_2$.

This is equivalent to show that each element of V is uniquely expressible as sum of an element of W_1 and an element of W_2.

Now, the condition $V = W_1 + W_2$ reveals that each element of V is expressible as sum of an element of W_1 and an element of W_2.

Now, we have to show that this representation is unique.

Let, if possible, $0 \neq \alpha \in V$ be expressible as

$\alpha = \alpha_1 + \alpha_2$, where $\alpha_1 \in W_1$ and $\alpha_2 \in W_2$

and $\alpha = \beta_1 + \beta_2$, where $\beta_1 \in W_1$ and $\beta_2 \in W_2$.

Then, $\alpha_1 + \alpha_2 = \beta_1 + \beta_2$

or $\alpha_1 - \beta_1 = \beta_2 - \alpha_2 \in W_1 \cap W_2$ $[\because \alpha_1 - \beta_1 \in W_1$ and $\beta_2 - \alpha_2 \in W_2]$

But, it being given that $W_1 \cap W_2 = \{0\}$.

$\therefore \alpha_1 - \beta_1 = 0$ and $\beta_2 - \alpha_2 = 0$

and so, $\alpha_1 = \beta_1$ and $\alpha_2 = \beta_2$.

Thus, each element of V is uniquely expressible as sum of an element of W_1 and an element of W_2.

Hence, $V = W_1 \oplus W_2$.

Example 23. *In the 3-tuple space V_3 (R), consider the subspaces:-*

$W_1 = \{(\alpha_1, 0, 0) : \alpha_1 \in R\}$; $W_2 = \{(0, \alpha_2, 0) : \alpha_2 \in R\}$;

$W_3 = \{(0, 0, \alpha_3) : \alpha_3 \in R\}$; $W_4 = \{(\alpha_1, \alpha_2, 0) : \alpha_1, \alpha_2 \in R\}$

$W_5 = \{(0, \alpha_2, \alpha_3) : \alpha_2, \alpha_3 \in R\}$ & $W_6 = \{(\alpha_1, 0, \alpha_3) : \alpha_1, \alpha_3 \in R\}$.

Then prove that, $V_3(R) = W_1 \oplus W_5 = W_2 \oplus W_6 = W_3 \oplus W_4 = W_1 \oplus W_2 \oplus W_3$

Solution. For, if $(\alpha_1,\alpha_2,\alpha_3) \in V_3(R)$, then

$(\alpha_1,\alpha_2,\alpha_3) = (\alpha_1,0,0) + (0, \alpha_2, \alpha_3) \in W_1 + W_5$ and it is also clear that there is no non-zero element common to both W_1 and W_5 *i.e.* $W_1 \cap W_5 = \{0\}$.

$\Rightarrow \quad V_3(R) = W_1 \oplus W_5$

Similarly, it may be proved that

$V_3(R) = W_2 \oplus W_6 = W_3 \oplus W_4 = W_1 \oplus W_2 \oplus W_3$. **Proved.**

Example 24. *In the vector space V of all real valued continuous functions, defined on the set R of all real numbers, let V_e and V_0 denote the sets of even and odd functions respectively.*

Then, V_e and V_0 are subspaces of V such that $V = V_e \oplus V_0$.

Solution. It has already been shown that V_e as well as V_0 is a subspace of V. Now, in order to show that $V = V_e \oplus V_0$, we must prove that $V = V_e + V_0$ and $V_e \cap V_0 = \{\hat{0}\}$.

Now, let f be an arbitrary element of V. Then, $\forall x \in R$, we have

$$f(x) = \frac{1}{2}[f(x)+f(-x)] + \frac{1}{2}[f(x) - f(-x)]$$

$$= f_e(x) + f_0(x), \qquad \text{where } f_e(x) = \frac{1}{2}[f(x)+f(-x)]$$

$$= (f_e + f_0)(x). \qquad \text{and } f_0(x) = \frac{1}{2}[f(x) - f(-x)]$$

Now, $f_e(-x) = \frac{1}{2}[f(-x) + f(x)] = f_e(x)$

and $f_0(-x) = \frac{1}{2}[f(-x) - f(x)] = -\frac{1}{2}[f(x) - f(-x)] = -f_0(x)$.

Thus, $f_e \in V_e$ and $f_0 \in V_0$.

$\therefore f = f_e + f_0$, where $f_e \in V_e$ and $f_0 \in V_0$.

It is clear that every element of V is expressible as sum of an element of V_e and an element of V_0.

So, $V = V_e + V_0$.

Also, the condition $V_e \cap V_0 = \{\hat{0}\}$ follows from the fact that the zero function is the only real valued function on R, which is both even and odd.

Thus, $V = V_e + V_0$ and $V_e \cap V_0 = \{\hat{0}\}$

Hence, $V = V_e \oplus V_0$. **Proved.**

33.27 QUOTIENT SPACE

Let W be a subspace of a vector space $V(F)$. Then, clearly $(W, +)$ is a subgroup of $(V, +)$. Also $(V,+)$ being an abelian group, every right coset $W + \alpha$ is equal to the corresponding left coset $\alpha + W$.

More-over, $W + \alpha = W + \beta \leftrightarrow \alpha - \beta \in W$.

Theorem 8. *Let W be a subspace of a vector space V (F). Then, the set (V/W) of all cosets $(W + \alpha)$, where $\alpha \in V$, is a vector space over the field F with respect to addition and scalar multiplication compositions, defined by*

$$(W + \alpha) + (W + \beta) = W + (\alpha + \beta)$$

and $$a.(W + \alpha) = (W + a\alpha) \ \forall \ \alpha, \beta \in V \text{ and } a \in F.$$

Proof. First we show that the above compositions are well defined.

Let $W + \alpha = W + \alpha'$ and $W + \beta = W + \beta'$. Then,

$$W + \alpha = W + \alpha' \text{ and } W + \beta = W + \beta'$$
$$\Rightarrow \alpha - \alpha' \in W \text{ and } \beta - \beta' \in W$$
$$\Rightarrow (\alpha - \alpha') + (\beta - \beta') \in W$$
$$\Rightarrow (\alpha + \beta) - (\alpha' + \beta') \in W$$
$$\Rightarrow W + (\alpha + \beta) = W + (\alpha' + \beta')$$
$$\Rightarrow (W + \alpha) + (W + \beta) = (W + \alpha') + (W + \beta').$$

We have proved that the addition composition on (V/W) is well defined.

Again, $W + \alpha = W + \alpha'$

$$\Rightarrow \alpha - \alpha' \in W$$
$$\Rightarrow a(\alpha - \alpha') \in W$$
$$\Rightarrow a\alpha - a\alpha' \in W$$
$$\Rightarrow W + a\alpha = W + a\alpha'$$
$$\Rightarrow a(W + \alpha) = a(W + \alpha').$$

So, the scalar multiplication is well defined.

Now, it is easy to show that $(V/W, +)$ is an abelian group.

In fact, the coset $W + 0 = W$ is the additive identity and every coset $W + \alpha$ in (V/W) has its additive inverse $W + (-\alpha)$ in (V/W).

More-over, $\forall \ (W + \alpha), (W + \beta) \in V/W$ and $a, b \in F$,

We have:-

(*i*) $$a[(W + \alpha) + (W + \beta)] = a[W + (\alpha + \beta)] = W + a(\alpha + \beta) = W + (a\alpha + a\beta) = (W + a\alpha) + (W + a\beta) = a(W + \alpha) + a(W + \beta);$$

(*ii*) $$(a + b)(W + \alpha) = W + (a + b)\alpha = W + (a\alpha + b\alpha) = (W + a\alpha) + (W + b\alpha) = a(W + \alpha) + b(W + \alpha)$$

(iii) $(ab)(W + \alpha) = W + (ab)\alpha$
$= W + a(b\alpha)$
$= a(W + b\alpha)$
$= a[b(W + \alpha)];$

and (iv) $1.(W + \alpha) = W + (1.\alpha) = (W + \alpha)$

Hence, (V/W) is a vector space.

Remark. The vector space (V/W) is called a quotient space of V of W.

EXERCISE 33.3

1. Show that the set W of the elements of the vector space $V_3(R)$ of the form
$(x + 2y, y, -x+3y)$, where $x, y \in R$ is a subspace of $V_3(R)$.

2. Let R be the field of real numbers and W be a subset of $V_3(R)$, defined by
$W = \{(x, 2y, 3z) : x, y, z \in R\}$.
Show that W is a subspace of $V_3(R)$.

3. Show that the set $W = \{(a, b, 0) : a, b \in F\}$ is a subspace of $V_3(F)$.
(R.G.P.V., Bhopal, III Semester, June 2007)

4. In the 3-tuple space $V_3(R)$, show that the set $W = \{(x, 2x, 2x + 1) : x \in R\}$ is not a subspace of $V_3(R)$.

5. In the 3-tuple space $V_3(R)$, show that the set $W = \{(x, y, z) : x, y, z \in Q\}$ is not a subspace of $V_3(R)$.
[**Hint.** Take $a = (2, 3, 4) \in W$ & $a = \sqrt{3} \in R$. Clearly, $a\,\alpha \notin W$.]

6. Find whether the set $W = \{(x_1, x_2) : x_1, x_2$ are reals$\}$ is a vector space where the compositions are defined as
(i) $(x_1, y_1) + (x_2, y_2) = (3x_2 + 3y_2, -x_1 - x_2)$ and $c.(x, y) = (3cy, -cx)$
(ii) $(x_1, y_1) + (x_2, y_2) = (x_1 + x_2, y_1 + y_2)$, $c.(x_1, y_1) = (0, cy_1)$ **Ans.** (i) Not, (ii) Not

7. Find whether the set of all ordered 2-tuple is a vector space with respect to the following operations of addition and scalar multiplication:
(i) $(x_1, y_1) + (x_2, y_2) = (x_1 + x_2, y_1 + y_2)$, $k.(x_1, y_1) = (k^2 x_1, k^2 y_1)$
(ii) $(x_1, y_1) + (x_2, y_2) = (x_1 + x_2, y_1 + y_2)$, $k.(x_1, y_1) = (|k|x_1, |k|y_1)$
Ans. (i) Not vector space (ii) Not a vector space

8. Prove that the set of all 2nd order square matrices is a vector space with respect to matrix addition and multiplication of a matrix by a number.

9. If a_1, a_2, a_3 are the fixed elements of a field F, then show that the set W of all ordered triads (x_1, x_2, x_3) of elements of F, such that $a_1 x_1 + a_2 x_2 + a_3 x_3 = 0$, is a subspace of $V_3(F)$.

10. Show that the set W of all ordered triads (x, y, z) such that $x - 3y + 4z = 0$, is a subspace of $V_3(R)$.

11. Prove that the set of all solutions (a, b, c) of the equation $a + b + 2c = 0$ is a subspace of the vector space $V_3(R)$.

12. Let $R^n = \{\alpha : \alpha = (a_1, a_2, ..., a_n)$, where a_i^s are in $R\}$.
Show that the set of all vectors a such that $a_1 \le 0$ is not a subspace of $R^n(R)$.
Also, show that the set of all $\alpha^{'s}$ such that a_2 is an integer is not a subspace of $R^n(R)$.

13. Let V be the vector space of all square $n \times n$ matrices over a field F and let $W = \{A \in V : AM = MA\}$ for a given matrix M, then show that W is a subspace of $V(F)$.

33.28 BASIS (*R.G.P.V., Bhopal, III Semester, Dec. 2006*)

Let V be a vector space. A collection of vectors $\alpha_1, \alpha_2, ...\alpha_r$ is said to form a basis of V if $\alpha_1, \alpha_2, ... \alpha_r$ are linearly independent and if they generate V.

Coordinate of a Vector

Let $V(F)$ be a finite dimensional vector space. Let $B = \alpha_1, \alpha_2, ... \alpha_n$ be ordered basis of V. Let $\alpha \in V$. Then there exists a unique n-tuple $(x_1, x_2, ... x_n)$ of scalars such that

$$\alpha = x_1 \alpha_1 + x_2 \alpha_2 + ... x_n \alpha_n$$

$(x_1, x_2, ... x_n)$ is called coordinates of the basis V.

33.29 DIMENSION OR RANK OF A VECTOR SPACE

(*R.G.P.V., Bhopal, III Semester, Dec. 2005*)

The number of vectors presents in a basis of a vector space V is called the dimension of V. It is denoted by dim (V).

Example 25. *Dimension of the vector space V_4 is 4, since the four vectors (1, 0, 0, 0), (0, 1, 0, 0), (0, 0, 1, 0), and (0, 0, 0, 1) form a basis of V_4.*

Example 26. *Dim $(V_n) = n$, since there are n number of vectors in a basis of V_n.*

Here, we are mainly concern with finite dimensional vector space. The dimension of vector space may be infinite.

Example 27. *Each set of $(n + 1)$ or more vectors of a finite dimensional vector space $V(F)$ of dimension n is :*

(i) linearly dependent *(ii) a basis of $V(F)$*

(iii) a subspace of $V(F)$ *(iv) linearly independent*

(*R.G.P.V., Bhopal, III Semester, Dec. 2007, 2006*)

Solution. Dimension of vector space $V(F)$ is n, therefore $V(F)$ may have at most n independent vectors. Here the number of vectors are $(n + 1)$, so they are linearly dependent.

Ans.

Example 28. *Show that the vectors* (1, 0, 0), (1, 1, 0), (1, 1, 1) form a basis for R^3.

(*R. G. P. V. Bhopal, III Semester, June 2007*)

Solution. Let $a_1, a_2, a_3 \in R$ be such that

$$a_1(1, 0, 0) + a_2(1, 1, 0) + a_3(1, 1, 1) = 0$$

$$(a_1 + a_2 + a_3) + (0 + a_2 + a_3) + (0 + 0 + a_3) = 0 \quad ...(1)$$

$$a_1 + a_2 + a_3 = 0$$

$$a_2 + a_3 = 0$$

$$a_3 = 0$$

The matrix of the coefficients of the equations is

$$A = \begin{bmatrix} 1 & 1 & 1 \\ 0 & 1 & 1 \\ 0 & 0 & 1 \end{bmatrix}$$

Rank $(A) = 3$

$a_3 = 0,\ a_2 = 0$ and $a_1 = 0$

The non-zero values of a_1, a_2, a_3 do not exist which can satisfy (1).

Thus, (1, 0, 0), (1, 1, 0) and (1, 1, 1) are linearly independent.

Hence, the set of given vectors form a basis of R^3. **Proved.**

Example 29. *Show that* $a_1 = (1, 0, 0)$, $a_2 = (0, 1, 0)$, $a_3 = (0, 0, 1)$ *form a basis of the vectors space* V_3.

Solution. Let a_1, a_2, a_3 be non-zero real numbers.

$$a_1\alpha_1 + a_2\alpha_2 + a_3\alpha_3 = 0$$

$$a_1 (1, 0, 0) + a_2 (0, 1, 0) + a_3 (0, 0, 1) = 0 \quad \text{... (1)}$$

$$(a_1,\ a_2, a_3) = 0$$

$$\Rightarrow \quad a_1 = 0, \quad a_2 = 0, \quad a_3 = 0$$

Thus non-zero values of a_1, a_2, a_3 do not exist which can satisfy (1).

Hence, the given system of vectors is linearly independent.

Let $\alpha = x_1 (1, 0, 0) + x_2 (0, 1, 0) + x_3 (0, 0, 1)$

$= x_1\,\alpha_1 + x_2\,\alpha_2 + x_3\,\alpha_3$

It shows that any vector space V_3 can be expressed as a linear combination of $\alpha_1, \alpha_2, \alpha_3$. So $\alpha_1, \alpha_2, \alpha_3$ are the generators.

Hence, $\alpha_1, \alpha_2, \alpha_3$ form basis of V. **Proved.**

Example 30. *Determine whether the following vectors form a basis of* R^3 *or not*

(1, 1, 2), (1, 2, 5), (5, 3, 4).

Solution. We know that dim $R^3 = 3$. Thus if the given set of vectors is linearly independent, then it will be a basis of R^3 otherwise not.

Now, for $a_1, a_2, a_3 \in R$

$$a_1 (1, 1, 2) + a_2 (1, 2, 5) + a_3 (5, 3, 4) = (0, 0, 0)$$

$$\Rightarrow \quad (a_1 + a_2 + 5a_3, a_1 + 2a_2 + 3a_3, 2a_1 + 5a_2 + 4a_3 = (0, 0, 0)$$

$$\therefore \quad a_1 + a_2 + 5a_3 = 0 \quad \text{... (1)}$$

$$a_1 + 2a_2 + 3a_3 = 0 \quad \text{... (2)}$$

$$2a_1 + 5a_2 + 4a_3 = 0 \quad \text{... (3)}$$

$\therefore$ The coefficient matrix of these equations is

$$A = \begin{bmatrix} 1 & 1 & 5 \\ 1 & 2 & 3 \\ 2 & 5 & 4 \end{bmatrix}$$

Here $|A| = 1 (8 - 15) - 1 (4 - 6) + 5 (5 - 4) = -7 + 2 + 5 = 0$

$$\text{Rank of } A \neq 3$$

The scalars a_1, a_2, a_3 are not all zero, therefore, the given set S of vectors is linearly dependent and hence the given set of vectors are not basis set. **Ans.**

Example 31. *Show that the vectors (2, 1, 4), (1, –1, 2), (3, 1, –2) form a basis for* R^3.

(R.G.P.V., Bhopal, III Semester, Dec. 2007)

Solution. Let $S = \{(2, 1, 4), (1, -1, 2), (3, 1, -2)\}$

Again, let $a_1, a_2, a_3 \in R$ be such that

$$a_1 (2, 1, 4) + a_2 (1, -1, 2) + a_3 (3, 1, -2) = 0$$

$\Rightarrow$ $$(2a_1 + a_2 + 3a_3, a_1 - a_2 + a_3, 4a_1 + 2a_2 - 2a_3) = (0, 0, 0)$$

$\therefore$ $$2a_1 + a_2 + 3a_3 = 0 \quad \ldots (1)$$

$$a_1 - a_2 + a_3 = 0 \quad \ldots (2)$$

$$4a_1 + 2a_2 - 2a_3 = 0 \quad \ldots (3)$$

$\therefore$ The coefficient matrix of these equations is

$$A = \begin{bmatrix} 2 & 1 & 3 \\ 1 & -1 & 1 \\ 4 & 2 & -2 \end{bmatrix}$$

Here $$|A| = 2(2-2) - 1(-2-4) + 3(2+4) = 24 \neq 0$$

$\therefore$ Rank $(A) = 3$, *i.e.*, the number of unknowns is 3.

Hence the only solution of these equations is $a_1 = 0, a_2 = 0, a_3 = 0$. Therefore, the set is linearly independent.

Also the dimension of vector space R^3 is 3. Hence any set of three linearly independent vectors is a basis for $\mathbf{R}^3$. **Proved.**

Therefore, the set $S = \{(2, 1, 4), (1, -1, 2), (3, 1, -2)\}$ is a basis for $\mathbf{R}^3$.

Example 32. *Show that the set $S = \{(1, 0, 0), (1, 1, 0) (1, 1, 1), (0, 1, 0)\}$ is not a basis set.*

Solution. Let $a_1, a_2, a_3, a_4 \in R$ be such that

$$a_1 (1, 0, 0) + a_2 (1, 1, 0) + a_3 (1, 1, 1) + a_4 (0, 1, 0) = 0$$

$\Rightarrow$ $$(a_1 + a_2 + a_3 + 0a_4,\ 0a_1 + a_2 + a_3 + a_4, 0a_1 + 0a_2 + a_3 + 0a_4) = (0, 0, 0)$$

$\Rightarrow$ $$a_1 + a_2 + a_3 = 0,\ a_2 + a_3 + a_4 = 0,\ \ a_3 = 0$$

The coefficient matrix of these equations are in the matrix form.

$$\text{A} = \begin{bmatrix} 1 & 1 & 1 & 0 \\ 0 & 1 & 1 & 1 \\ 0 & 0 & 1 & 0 \end{bmatrix}$$

Here, $$|A| = 1(1-0) = 1 \neq 0$$

$\Rightarrow$ $$R(A) = 3$$

$\therefore$ The scalars a_1, a_2, a_3, a_4 are not all zero, therefore, the given set S of vectors is linearly dependent and hence S is not a basis set. **Proved.**

Example 33. *Show that set : $S \{(1, 2, 1), (2, 1, 0), (1, -1, 2)\}$ forms a basis of V_3 (F)* (*R.G.P.V., Bhopal, III Semester, June 2008*)

Solution. Let $a_1, a_2, a_3 \in R$ be such that

$$a_1 (1, 2, 1) + a_2 (2, 1, 0) + a_3 (1, -1, 2) = 0 \quad \ldots (1)$$

$$(a_1 + 2a_2 + a_3) + (2a_1 + a_2 - a_3) + (a_1 + 2a_3) = 0$$

$$a_1 + 2a_2 + a_3 = 0$$

$$2a_1 + a_2 - a_3 = 0$$

$$a_1 + 0a_2 + 2a_3 = 0$$

The matrix of the coefficients of the above equations is

$$A = \begin{bmatrix} 1 & 2 & 1 \\ 2 & 1 & -1 \\ 1 & 0 & 2 \end{bmatrix}$$

Here, $|A| = 1(2-0) - 2(4+1) + 1(0-1)$

$$= 2 - 10 - 1 = -9 \neq 0$$

$R(A) = 3$

$$a_1 = 0,\ a_2 = 0,\ a_3 = 0,$$

The non-zero values of a_1, a_2, a_3 do not exist which can satisfy (1).

Thus, (1, 2, 1), (2, 1, 0), (1, –1, 2) are linearly independent.

Hence, S forms a basis of $\mathbf{R}^3$. **Proved.**

Example 34. *Prove that the vectors $\alpha_1 = (1, 0, -1)$, $\alpha_2 = (1, 2, 1)$ and $\alpha_3 = (0, -3, 2)$ form a basis of $V_3(R)$.*

Solution. Let $S = \{\alpha_1, \alpha_2, \alpha_3\}$ be the given set of vectors of $V_3(R)$ where $\alpha_1 = (1, 0, -1)$, $\alpha_2 = (1, 2, 1)$ and $\alpha_3 = (0, -3, 2)$.

(*i*) S is linearly independent in $V_3(R)$. Let $a_1, a_2, a_3 \in R$ be such that

$$a_1\alpha_1 + a_2\alpha_2 + a_3\alpha_3 = 0$$

$$\Rightarrow \quad a_1(1, 0, -1) + a_2(1, 2, 1) + a_3(0, -3, 2) = (0, 0, 0)$$

$$\Rightarrow \quad (a_1 + a_2 + 0a_3, 0a_1 + 2a_2 - 3a_3, -a_1 + a_2 + 2a_3) = (0,0,0)$$

$$\Rightarrow \quad a_1 + a_2 + 0a_3 = 0,$$

$$0a_1 + 2a_2 - 3a_3 = 0$$

$$-a_1 + a_2 + 2a_3 = 0.$$

The coefficient matrix of the above equations is

$$A = \begin{bmatrix} 1 & 1 & 0 \\ 0 & 2 & -3 \\ -1 & 1 & 2 \end{bmatrix}$$

Here $|A| = 1(4+3) - 1(0-3) + 0(0+2) = 10 \neq 0$

$\therefore$ Rank of $A = 3$ the number of unknown constants.

Hence, $a_1 = 0$, $a_2 = 0$, $a_3 = 0$ is the only solution of these equations.

Therefore, α_1, α_2 and α_3 are linearly independent in $V_3(R)$.

(*ii*) $L(S) = V_3(R)$. To prove this, it is just sufficient to show that $V_3(R) \subseteq L(S)$, because

$$S \subseteq V_3(R) \Rightarrow L(S) \subseteq V_3(R).$$

Let $(x, y, z) \in V_3(R)$ be arbitrary, where $x, y, z \in R$.

If possible, let $a_1, a_2, a_3 \in R$ be such that

$$(x, y, z) = a_1\alpha_1 + a_2\alpha_2 + a_3\alpha_3$$

$\Rightarrow \quad (x, y, z) = a_1\,(1, 0, -1) + a_2\,(1, 2, 1) + a_3\,(0, -3, 2)$

$\Rightarrow \quad (x, y, z) = (a_1 + a_2, 2a_2 - 3a_3, -a_1 + a_2 + 2a_3)$

$\Rightarrow \quad x = a_1 + a_2, \quad y = 2a_2 - 3a_3,$

$z = -a_1 + a_2 + 3a_3 \qquad \text{... (1)}$

Solving for a_1, a_2, a_3, we get

$$a_1 = \frac{7x - 2y - 3z}{10}, a_2 = \frac{3x + 2y + 3z}{10}, a_3 = \frac{x - y + z}{5}$$

Clearly $a_1, a_2, a_3 \in R$ satisfies the system of equations (1). Hence every element of V_3 (*R*) can be expressed as a linear combination of vectors $\alpha_1, \alpha_2, \alpha_3$.

$\therefore \quad L(S) = V_3(R)$

Therefore, *S* is a basis of V_3 (*R*). **Proved.**

Example 35. *Show that the following set is a basis of* V_4 *(R).*

$S = \{(1, 0, 0, 0), (1, 1, 0, 0), (1, 1, 1, 0), (1, 1, 1, 1)\}$

Solution. Let $a_1, a_2, a_3, a_4 \in R$ be such that

$a_1\,(1, 0, 0, 0) + a_2\,(1, 1, 0, 0) + a_3\,(1, 1, 1, 0) + a_4\,(1, 1, 1, 1) = 0$

$\Rightarrow \quad (a_1 + a_2 + a_3 + a_4, a_2 + a_3 + a_4, a_3 + a_4, a_4) = (0,0,0,0)$

$\Rightarrow \quad a_1 + a_2 + a_3 + a_4 = 0, a_2 + a_3 + a_4 = 0, a_3 + a_4 = 0, a_4 = 0$

The coefficient matrix *A* of these equations is

$$A = \begin{bmatrix} 1 & 1 & 1 & 1 \\ 0 & 1 & 1 & 1 \\ 0 & 0 & 1 & 1 \\ 0 & 0 & 0 & 1 \end{bmatrix}$$

$$A \sim \begin{bmatrix} 1 & 0 & 0 & 0 \\ 0 & 1 & 0 & 0 \\ 0 & 0 & 1 & 0 \\ 0 & 0 & 0 & 1 \end{bmatrix} \quad \begin{matrix} R_1 \to R_1 - R_2 \\ R_2 \to R_2 - R_3 \\ R_3 \to R_3 - R_4. \end{matrix}$$

$\Rightarrow$ Rank (*A*) = 4 = number of unknowns.

Hence the only solution of these equations is $a_1 = 0, a_2 = 0, a_3 = 0, a_4 = 0$. Therefore the given set *S* of vectors is linearly independent.

Again the dimension of V_4 is 4. Hence any set of 4 linearly independent vectors is a basis for V_4 (*R*).

Hence, *S* is a basis for V_4 (*R*). **Proved.**

Example 36. *Show that the following set S of* R^3 *form a basis for* R^3.

$S = \{(1, 2, 1), (2, 1, 0), (1, -1, 2)\}$.

Express each of the standard basis vectors e_i, $i = 1, 2, 3$ *as linear combination of the above basis vectors.*

Solution. We know that the set of the vectors {(1, 0, 0), (0, 1, 0), (0, 0, 1)} forms a basis for R^3. Therefore, dim $R^3 = 3$. Further, we know that any set of n linearly independent vectors of n-dimensional vector space $V(F)$ forms a basis for V, so that in order to show that S forms a basis for R^3 it is just sufficient to show that the set $S = \{(1, 2, 1), (2, 1, 0), (1, -1, 2)\}$ is linearly independent. Now,

$$a_1 (1, 2, 1) + a_2 (2, 1, 0) + a_3 (1, -1, 2) = (0, 0, 0)$$

$$\Rightarrow \quad (a_1 + 2a_2 + a_3, 2a_1 + a_2 - a_3, a_1 + 2a_3) = (0, 0, 0)$$

$$\therefore \quad a_1 + 2a_2 + a_3 = 0 \qquad \text{... (1)}$$

$$2a_1 + a_2 - a_3 = 0 \qquad \text{... (2)}$$

$$a_1 + 2a_3 = 0 \qquad \text{... (3)}$$

$\therefore$ The coefficient matrix of these equations is

$$A = \begin{bmatrix} 1 & 2 & 1 \\ 2 & 1 & -1 \\ 1 & 0 & 2 \end{bmatrix}$$

$$|A| = 1(2-0) - 2(4+1) + 1(0-1) = -9 \neq 0.$$

$\therefore$ Rank $(A) = 3 =$ the number of unknowns.

Hence the only solution of these equations is $a_1 = 0, a_2 = 0, a_3 = 0$. Therefore, the set S is *L.I.*

Hence S forms a basis for R^3.

Now, $\quad e_1 = (1, 0, 0) = a(1, 2, 1) + b(2, 1, 0) + c(1, -1, 2)$

for $a, b, c \in R$.

Thus, $a + 2b + c = 1, \quad 2a + b - c = 0, \quad a + 2c = 0.$

Solving, we get $\quad a = -\dfrac{2}{9}, b = \dfrac{5}{9}, c = \dfrac{1}{9}.$

$$\therefore \quad e_1 = -\frac{2}{9}(1,2,1) + \frac{5}{9}(2,1,0) + \frac{1}{9}(1, -1, 2),$$

Thus showing that e_1 can be expressed as linear combination of elements of S.

Similarly, $\quad e_2 = (0, 1, 0) = \dfrac{4}{9}(1, 2, 1) - \dfrac{1}{9}(2, 1, 0) - \dfrac{2}{9}(1, -1, 2)$

$$e_3 = (0, 0, 1) = \frac{1}{3}(1, 2, 1) - \frac{1}{3}(2, 1, 0) + \frac{1}{3}(1, -1, 2). \qquad \textbf{Ans.}$$

Example 37. *If $V_3(R)$ is a vector space then show that $S = \{(0, 1, -1), (1, 1, 0), (1, 0, 2)\}$ is a basis of V_3 and hence find the co-ordinates of vector $(1, 0, -1)$ with respect to this basis.* (*R.G.P.V., Bhopal, III Semester, Dec. 2005*)

Solution. Here, $S = \{(0, 1, -1), (1, 1, 0), (1, 0, 2)\}$. Let $a_1, a_2, a_3 \in R$ be such that

$$a_1 (0, 1, -1) + a_2 (1, 1, 0) + a_3 (1, 0, 2) = (0, 0, 0)$$

$$\Rightarrow \quad (a_2 + a_3, a_1 + a_2, -a_1 + 2a_3) = (0, 0, 0)$$

$$\Rightarrow \quad a_2 + a_3 = 0, \quad a_1 + a_2 = 0, \quad -a_1 + 2a_3 = 0$$

The coefficient matrix of these equations is

$$A = \begin{bmatrix} 0 & 1 & 1 \\ 1 & 1 & 0 \\ -1 & 0 & 2 \end{bmatrix}.$$

Here $$|A| = 0 - 1(2-0) + 1(0+1) = -1 \neq 0$$

$\therefore$ Rank $(A) = 3$ = number of unknowns. Hence the only solution of these equations is $a_1 = 0, a_2 = 0, a_3 = 0$. Therefore, S is linearly independent. Again the dimension of V_3 is 3. Hence any set of 3 linearly independent vectors is a basis of V_3. Hence the set S is a basis of $V_3\ (R)$.

Now to find the co-ordinates of the vector (1, 0, –1) with respect to the ordered basis S, let (x_1, x_2, x_3) be the co-ordinates of the vector (1, 0, –1) where $x_1, x_2, x_3 \in R$ are scalars such that

$$(1, 0, -1) = x_1 (0, 1, -1) + x_2 (1, 1, 0) + x_3 (1, 0, 2)$$

$$\Rightarrow \quad (1, 0, -1) = (x_2 + x_3, x_1 + x_2, -x_1 + 2x_3)$$

$$\Rightarrow \quad x_2 + x_3 = 1,\ x_1 + x_2 = 0,\ -x_1 + 2x_3 = -1$$

Solving for x_1, x_2 and x_3, we get $\quad x_1 = -3, x_2 = 3, x_3 = -2$

Hence, the co-ordinates of the vector (1, 0, –1) with respect to the basis S are (x_1, x_2, x_3) *i.e.*, (–3, 3, –2).

Ans.

Example 38. *Find the co-ordinate vector V = (3, 5, –2) relative to the basis of $e_1 = (1, 1, 1)$, $e_2 = (0, 2, 3)$, $e_3 = (0, 2, -1)$* (*R.G.P.V., Bhopal, III Semester, June, 2006*)

Solution. Let $(3, 5, -2) = x_1 (1, 1, 1) + x_2 (0, 2, 3) + x_3 (0, 2, -1)$

$$\Rightarrow \quad (3, 5, -2) = (x_1 + 0x_2 + 0x_3) + (x_1 + 2x_2 + 2x_3) + (x_1 + 3x_2 - x_3)$$

$$x_1 + 0x_2 + 0x_3 = 3$$

$$x_1 + 2x_2 + 2x_3 = 5$$

$$x_1 + 3x_2 - x_3 = -2$$

$$\begin{bmatrix} 1 & 0 & 0 \\ 1 & 2 & 2 \\ 1 & 3 & -1 \end{bmatrix} \begin{bmatrix} x_1 \\ x_2 \\ x_3 \end{bmatrix} = \begin{bmatrix} 3 \\ 5 \\ -2 \end{bmatrix}$$

$$\Rightarrow \quad \begin{bmatrix} 1 & 0 & 0 \\ 3 & 8 & 0 \\ 1 & 3 & -1 \end{bmatrix} \begin{bmatrix} x_1 \\ x_2 \\ x_3 \end{bmatrix} = \begin{bmatrix} 3 \\ 1 \\ -2 \end{bmatrix} \quad R_2 \to R_2 + 2R_3$$

$$x_1 = 3$$

$$3x_1 + 8x_2 = 1 \quad \Rightarrow \quad 3(3) + 8x_2 = 1 \Rightarrow x_2 = -1$$

$$x_1 + 3x_2 - x_3 = -2 \quad \Rightarrow 3 + 3(-1) - x_3 = -2 \Rightarrow x_3 = 2$$

Hence, the co-ordinates of the vector (3, 5, –2) with respect to the given basis are (3, –1, 2)

Ans.

Example 39. *If S = {$\alpha_1, \alpha_2, \alpha_3$} is a basis of the vector space V (F), then show that every element α of V can be uniquely expressed as a linear combination of} elements of S.*

Solution. Since $S = \{\alpha_1, \alpha_2, \alpha_3\}$ is a basis of the vector space $V(F)$, S satisfies the following conditions:

(*i*) S is linearly independent set in V;

(*ii*) $L(S) = V$,

Let α be any arbitrary element of V, then there exist $a_1, a_2, a_3 \in F$ such that

$$\alpha = a_1\alpha_1 + a_2\alpha_2 + a_3\alpha_3. \quad \text{... (1)}$$

If possible, let $b_1, b_2, b_3 \in F$, which gives the following second representation of α.

$$\alpha = b_1\alpha_1 + b_2\alpha_2 + b_3\alpha_3. \quad \text{... (2)}$$

From (1) and (2), we have

$$a_1\alpha_1 + a_2\alpha_2 + a_3\alpha_3 = b_1\alpha_1 + b_2\alpha_2 + b_3\alpha_3$$

$$\Rightarrow (a_1 - b_1)\alpha_1 + (a_2 - b_2)\alpha_2 + (a_3 - b_3)\alpha_3 = 0$$

$$\Rightarrow a_1 - b_1 = 0, \ a_2 - b_2 = 0, \ a_3 - b_3 = 0 \quad [\because \{\alpha_1, \alpha_2, \alpha_3,\} \text{ is linearly independent}]$$

$$\Rightarrow a_1 = b_1, \ a_2 = b_2, a_3 = b_3.$$

Hence, the representation of α as a linear combination of elements of S is unique. **Proved.**

EXERCISE 33.4

1. Show that the set $S = \{1, x, x^2, \ldots\ldots. x^n\}$ is a basis of the vector space $F[x]$ of all polynomials of degree at most n, over the real field R.
2. Show that the set, $S = \{(1, 0, 0), (0, 1, 0), (1, 1, 0), (1, 1, 1)\}$ spans the vector space $V_3(R)$, but is not a basis.
3. Show that the vectors $(1, i, 0)$, $(2i, 1, 1)$, $(0, 1 + i, 1 - i)$ form a basis of $V_3(C)$.
4. Prove that the set, $S\ \{a + ib,\ c + id\}$ is a basis set of $C(R)$, if and only if $(ad - bc) \neq 0$.
5. Let V be the vector space of all ordered pairs of complex numbers over the real field R. Show that the set $S = \{(1, 0), (0, 1), (i, 0), (0, i)\}$ is *a* basis of $V(R)$.
6. Let W be the subspace of $V_4(R)$, generated by the vectors $(1, -2, 5, -3)$, $(2, 3, 1, -4)$ and $(3, 8, -3, -5)$. Find a basis and the dimension of W.

   Extend the basis of W to a basis of the whole space $V_4(R)$.

   Hints. Let $a_1(1, -2, 5, -3) + a_2(2, 3, 1, -4) + a_3(3, 8, -3, -5) = (0, 0, 0, 0)$.

   Then, we get four eqations in a_1, a_2, a_3.

   Reduce the coefficient matrix of this system to Echelon form. The vectors formed by non-zero rows of this matrix form the basis of W.

   For extending the basis of W to basis of $V_4(R)$, we adjoin $(0, 0, 1, 0)$ and $(0, 0, 0, 1)$ with the basis of W, as these four vectors form an Echelon matrix.

   Ans. The basis of W is $\{(1, -2, 5, -3), (0, 7, -9, 2)\}$ and that of $V_4(R)$ is $\{(1, -2, 5, -3), (0, 7, -9, 2), (0, 0, 1, 0), (0, 0, 0, 1)\}$.
7. Let W_1 and W_2 be the subspaces of 4-tuples space $V_4(R)$, generated by the sets

   $S_1 = \{(1, 1, 0, -1), (1, 2, 3, 0), (2, 3, 3, -1)\}$

   and $S_2 = \{(1, 2, 2, -2), (2, 3, 2, -3), (1, 3, 4, -3)\}$ respectively. Show that:-

   (*i*) $\dim(W_1 + W_2) = 3$ and (*ii*) $\dim(W_1 \cap W_2) = 1$.

   Hints. (*i*) $W_1 + W_2 = L(S_1 \cup S_2)$. Now, proceed to find the basis as in prob. 6.

   (*ii*) Find $\dim W_1$ and dim. W_2 as in prob. 6 and use the result

   $\dim W_1 + \dim W_2 = \dim(W_1 + W_2) + \dim(W_1 \cap W_2)$.

8. Show that:

(*i*) dim. $F(F) = 1$, where F is a field.

Hint. Show that $\{1\}$ is a basis of $F(F)$.

(*ii*) dim $C(R) = 2$.

Hint. Show that $\{1, i\}$ is a basis.

(*iii*) dim $C^n(C) = n$ (*iv*) dim $C^n(R) = 2n$

9. Let V be the vector space of $m \times n$ matrices over a field F. Let $A_{ij} \in V$ be the matrix with 1 as the (i, j) th entry and 0 elsewhere. Show that $\{Aij : 1 \le i \le m \text{ and } 1 \le j \le n\}$ is a basis of V and hence, dim $V = mn$.

10. Find the co-ordinate vector of (3, 1, –4) relative to the basis

$S = \{(1, 1, 1), (0, 1, 1), (0, 0, 1)\}$ in $V_3(R)$. **Ans.** (3, –2, 5)

11. Let $B = \{\alpha_1, \alpha_2, \alpha_3\}$ be an ordered basis of $R^3(R)$; where $\alpha_1 = (1, 0, -1)$, $\alpha_2 = (1, 1, 1)$ and $\alpha_3 = (1, 0, 0)$. Obtain the co-ordinates of the vector (a, b, c) in the ordered basis B.

Ans. $(b - c, b, a - 2b + c)$

12. Let V be the vector space of polynomials with degree not exceeding two.

Show that the set $S = \{1, x - 1, x^2 - 2x + 1\}$ forms a basis of V.

Find the co-ordinate vector of $(2x^2 - 5x + 6)$ relative to the above basis.

Ans. (3, –1, 2)

13. Let V be the vector space of all polynomials with degree not exceeding two. For a fixed $t \in R$, let $g_1(x) = 1$, $g_2(x) = t$ and $g_3(x) = (x + t)^2$.

Prove that $\{g_1(x), g_2(x), g_3(x)\}$ is a basis of V and obtain the co-ordinates of $c_0 + c_1x + c_2x^2$ in this ordered basis. **Ans.** $(c_0 - c_1t + c_2t^2, c_1 - 2c_2t, c_2)$

CHAPTER 34

LINEAR TRANSFORMATIONS

34.1 LINEAR TRANSFORMATIONS

Let U and V be two vector spaces over the same field F. Then, a mapping T of U into V is called a linear transformation or a homomorphism of U into V, if

(*i*) $T(\alpha + \beta) = T(\alpha) + T(\beta)\ \forall \alpha, \beta \in U$ and (*ii*) $T(a\alpha) = a\,T(\alpha)\ \forall a \in F, \alpha \in U$.

The conditions (*i*) and (*ii*) above can be combined as

$$T(a\alpha + b\beta) = aT(\alpha) + bT(\beta)\ \forall a, b \in F,\ \forall \alpha, \beta \in U$$

A transformation of U into itself is called a linear operator.

A one-one linear transformation of U onto V is called an isomorphism. In case, there exist an isomorphism of U onto V, we say that U is isomorphic to V and we write, $U \cong V$.

(*i*) Zero Transformation. If $U(F)$ and $V(F)$ be two vector spaces over the same field F, then the mapping $\hat{0}: U \to V$ defined by $\hat{0}(x) = 0, \forall x \in U$ is said to be **zero transformation.** $\bar{0}$ **is called zero operator.**

(ii) Identity Transformation (or Identity operator). If $V(F)$ is a vector space, then the mapping $T: V \to V$ defined by $T(v) = v\ \forall v \in V$ is called an **identity transformation.** T is called identity operator.

Theorem 1. *To prove that zero operator is linear operator.*

Proof. Let $U(F)$ and $V(F)$ be two vector spaces over the same field F. Let $\hat{0}: U \to V$ defined by $\hat{0}(x) = \hat{0}\ \forall x \in U$ be the zero operator.

Let $\alpha, \beta \in U$ and $a, b \in F$, then $\hat{0}(a\alpha + a\beta) = 0 = 0 + 0 = a\hat{0}(\alpha) + b\hat{0}(\beta)$.

Hence, $\hat{0}$ is a linear operator. **Proved.**

Theorem 2. *To prove that identity operator is linear operator.*

Proof. Let T be the identity operator on $V(F)$. Then $T(x) = x,\ \forall x \in V$.

Let $\alpha, \beta \in V$ and $a, b \in F$, then $a\alpha + b\beta \in V$

Now, $T(a\alpha + b\beta) = a\alpha + b\beta$ {by definition of T}

$= aT(\alpha) + bT(\beta)$ [$\because \alpha = T(\alpha), \beta = T(\beta)$]

Hence, T is a linear operator. **Proved.**

Example 1. *Show that the translation mapping $f: V_2(R) \to V_2(R)$ defined by $f(x, y) = (x + 2, y + 3)$ is not linear.*

Solution. Here $\bar{0} = (0, 0)$ is the zero vector of V_2 (R). Thus, by definition of f, we have

$$f(0) = f(0, 0) = (0 + 2, 0 + 3) = (2, 3) \neq 0.$$

Since f does not map the zero vector onto the zero vector, hence f is not linear.

Proved.

Example 2. *Prove that:* $T : R^2 \to R^2$

$T(x_1, x_2) = (x_1, 0)$, *is a linear transformation.*

Solution. Let $a, b \in R$ and $x \equiv (x_1, x_2)$ and $y \equiv (y_1, y_2) \in R^2$

We see $T = (a\,x + b\,y) = T\,\{a(x_1, x_2) + b(y_1, y_2)\}$

$$= T\,\{(a\,x_1, a\,x_2) + (b\,y_1, b\,y_2)\}$$

$$= T\,\{(a\,x_1 + b\,y_1 ; a\,x_2 + b\,y_2)\}$$

$$= (ax_1 + by_1, 0)$$

$$= (ax_1, 0) + (by_1, 0)$$

$$= a(x_1, 0) + b(y_1, 0)$$

$$= aT(x_1) + bT(y_1).$$

So, T is a linear transformation. **Proved.**

Example 3. *Show that the mapping* $f : V_2\,(R) \to V_3\,(R)$ *defined by* $f(a, b) = (a, b, 0)$ *is a linear transformation.*

(*R.G.P.V. Bhopal, III Semester, Dec. 2007*)

Solution. Let $\alpha = (a_1, b_1)$, $\beta = (a_2, b_2) \in V_2\,(R)$

If $a, b, \in F$, then $f(a\alpha + b\beta) = f\,[a\,(a_1, b_1) + b\,(a_2, b_2)]$

$$= f\,[(aa_1, ab_1) + (ba_2, bb_2)]$$

$$= f\,(aa_1 + ba_2, ab_1 + bb_2)$$

$$= (aa_1 + ba_2, ab_1 + bb_2, 0)$$

$$= a\,(a_1, b_1, 0) + b\,(a_2, b_2, 0)$$

$$= a\,f(\alpha) + b\,f(\beta)$$

Hence, f is a linear transformation. **Proved.**

Example 4. *Show that the mapping* $f\!: V_3(R) \to V_2(R)$ *defined by* $f(a, b, c) = (c, a + b)$ *is a linear transformation.*

Solution. Let $\alpha = (a_1, b_1, c_1)$, $\beta = (a_2, b_2, c_2) \in V_3\,(R)$. If $a, b, c \in R$, then

$$f(a\alpha + b\beta) = f\,[a(a_1, b_1, c_1) + b(a_2, b_2, c_2)] = f\,[(aa_1 + ba_2, ab_1 + bb_2, ac_1 + bc_2)]$$

$$= (ac_1 + bc_2, aa_1 + ba_2 + ab_1 + bb_2) \qquad \text{[By definition of } f\,]$$

$$= (ac_1, aa_1 + ab_1) + (bc_2, ba_2 + bb_2) = a(c_1, a_1 + b_1) + b(c_2, a_2 + b_2)$$

$$= af(a_1, b_1, c_1) + bf(a_2, b_2, c_2) = af(\alpha) + bf(\beta)$$

Proved.

Hence, f is a linear transformation.

Example 5. *Show that the mapping* $f : V_2(R) \to V_2(R)$ *defined by* $f(x, y) = (x^3, y^3)$ *is not a linear transformation.*

Solution. Let $\alpha = (x_1, y_1)$, $\beta = (x_2, y_2) \in V_2\,(R)$, then

$$f(\alpha + \beta) = f\,[(x_1, y_1) + (x_2, y_2)] = f\,[(x_1 + x_2, y_1 + y_2)] = [(x_1 + x_2)^3, (y_1 + y_2)^3]$$

$$\neq \left(x_1^3 + x_2^3, y_1^3 + y_2^3\right)$$

$$\neq \left(x_1^3, y_1^3\right) + \left(x_2^3, y_2^3\right)$$

$$\neq f(\alpha) + f(\beta)$$

Hence, f is not a linear transformation. **Proved.**

Example 6. *Show that the mapping $f : V_3(R) \to V_2(R)$ defined by $f(a, b, c) = (a - b, a + c)$ is linear transfromation.* (*R.G.P.V., Bhopal, III Semester, Dec. 2006*)

Solution. Let $\alpha, \beta \in V_3$

$$\alpha = (a_1, b_1, c_1),\ \beta = (a_2, b_2, c_2)$$

$$\begin{aligned} f(\alpha + \beta) &= f[(a_1, b_1, c_1) + (a_2, b_2, c_2)] \\ &= f[(a_1 + a_2, b_1 + b_2, c_1 + c_2] \\ &= f[a_1 + a_2 - b_1 + b_2,\ \ a_1 + a_2, c_2 + c_2)] \\ &= f[a_1 - b_1, + a_2 + b_2 + a_1 + c_1, + a_2 + c_2)] \\ &= (a_1 - b_1, a_1 + c_1) + (a_2 + b_2, a_2 + c_2) \\ &= f[(a_1, b_1, c_1) + f(a_2, b_2, c_2) \\ &= f(\alpha) + f(\beta) \end{aligned}$$

For any real number

$$\begin{aligned} f(k, \alpha) &= f[k(a_1, b_1, c_1)] \\ &= f(ka_1, kb_1, kc_1) \\ &= (ka_1 - kb_1, ka_1 + kc_1) \\ &= k(a_1 - b_1, a_1 + c_1) \\ &= kf(a_1, b_1, c_1) \\ &= kf(\alpha) \end{aligned}$$

So, f is a linear transformation from V_3 to V_2. **Proved.**

Example 7. *Show that the function $T : V_2 \to V_2$ defined by $T(x, y) = (2x + 3y, 3x - 4y)$ is a linear mapping.*

Solution. Let $a, b \in F$ and $\alpha = (x_1, y_1), \beta = (x_2, y_2) \in V_2$,

then $T(a\alpha + b\beta) = T[a(x_1, y_1) + b(x_2, y_2)] = T(ax_1 + bx_2, ay_1 + by_2)$

$$\begin{aligned} &= [2(ax_1 + bx_2) + 3(ay_1 + by_2), 3(ax_1 + bx_2) - 4(ay_1 + by_2)] \\ &= (2ax_1 + 3ay_1, 3ax_1 - 4ay_1) + (2bx_2 + 3by_2, 3bx_2 - 4by_2) \\ &= a(2x_1 + 3y_1, 3x_1 - 4y_1) + b(2x_2 + 3y_2, 3x_2 - 4y_2) \\ &= aT(x_1, y_1) + bT(x_2, y_2) = aT(\alpha) + bT(\beta). \end{aligned}$$

Hence, T is a linear mapping. **Proved.**

Example 8. *Let $V(F)$ be the vector space of all $n \times n$ matrices over the field F and if $M \in V$ be given, prove that the mapping $f : V(F) \to V(F)$ defined by $f(A) = AM + MA$, $\forall A \in V$, is a linear transformation.*

Solution. Let $A, B \in V(F)$ and $a \in F$,

then $f(A + B) = (A + B)M + M(A + B) = AM + BM + MA + MB$

$= (AM + MA) + (BM + MB)$

[By associativity of matrix addition in V]

$= f(A) + f(B)$ [By definition of f]

and $f(aA) = (aA)M + M(aA) = a(AM + MA) = af(A)$

Hence, f is a linear transformation. **Proved.**

Example 9. *Let V be the linear space of $R \to R$ functions which have derivative. Let D (f) be mean derivative of f. Then show that $D : V \to V$ is a linear map.*

Solution. Let $f, g \in V$; $a, b \in R$ and $x \in R$,

$$
\begin{aligned}
[D(af + bg)](x) &= D[(af + bg)(x)] \\
&= D[a f(x) + bg(x)] \\
&= D[a f(x)] + D[bg(x)] \\
&= a D[f(x)] + b D[g(x)] \\
&= a [D(f)](x) + b [D(g)](x) \\
&= [a D(f)](x) + [b D(g)](x) \\
&= [\{a D(f)\} + \{b D(g)\}](x)
\end{aligned}
$$

So, $D(af + bg) = a D(f) + b D(g)$

Hence, $D : V \to V$ is a linear map. **Proved.**

Example 10. *The matrix of linear transformation $T : R \to R^2$ defined by $T(x) = (3x, 5x)$ with respect to a standard basis is :*

(*i*) $\begin{bmatrix} 5 \\ 3 \end{bmatrix}$ (*ii*) $\begin{bmatrix} 3 \\ 5 \end{bmatrix}$ (*iii*) [5, 3] (*iv*) [3, 5]

(*R.G.P.V. Bhopal, III Semester, Dec. 2007*)

Solution. The standard basis of R and R^2 are $B = \{e\}$ and $B' = \{e_1, e_2\}$ respectively; where $e = (1)$.

$e_1 = (1, 0)$, $e_2 = (0, 1)$. Thus, by definition of T, we have

$T(1) = (3, 5) = 3(1, 0) + 5(0, 1) = 3e_1 + 5e_2$

Hence, the coefficient matrix of T is [3, 5] and its transpose matrix is $\begin{bmatrix} 3 \\ 5 \end{bmatrix}$

$$\therefore \quad [T : B, B'] = \begin{bmatrix} 3 \\ 5 \end{bmatrix}$$

Hence, (*ii*) is correct answer. **Ans.**

34.2 MATRIX OF A LINEAR TRANSFORMATION

Consider the simultaneous equations given below:

$$
\begin{aligned}
2x_1 + 3x_2 - x_3 &= 2 \\
5x_1 - 6x_2 - 3x_3 &- 10 \\
x_1 + x_2 + x_3 &= 8
\end{aligned}
$$

The left hand side of the equations can be considered as the linear transformations of T

$$T\begin{bmatrix} x_1 \\ x_2 \\ x_3 \end{bmatrix} = \begin{bmatrix} 2x_1 + 3x_2 - x_3 \\ 5x_1 + 6x_2 - 3x_3 \\ x_1 + x_2 + x_3 \end{bmatrix} = \begin{bmatrix} 2 & 3 & -1 \\ 5 & 6 & -3 \\ 1 & 1 & 1 \end{bmatrix}\begin{bmatrix} x_1 \\ x_2 \\ x_3 \end{bmatrix}$$

we can write the formula $A : T_A(X) = AX$

In general for $m \times n$ matrix the transformation is $TA : R^n \to R^m$

such transformation is called matrix transformation.

For example. the matrix $\begin{bmatrix} 1 & 5 & 3 \\ 2 & 7 & 9 \\ 4 & 1 & 2 \end{bmatrix}$ gives matrix transformation

$$T\begin{bmatrix} x_1 \\ x_2 \\ x_3 \end{bmatrix} = \begin{bmatrix} x_1 + 5x_2 + 3x_3 \\ 2x_1 + 7x_2 + 9x_3 \\ 4x_1 + x_2 + 2x_3 \end{bmatrix} = \begin{bmatrix} 1 & 5 & 3 \\ 2 & 7 & 9 \\ 4 & 1 & 2 \end{bmatrix}\begin{bmatrix} x_1 \\ x_2 \\ x_3 \end{bmatrix}$$

This can also be written horizontally

$$T : R^3 \to R^3, [x_1, x_2, x_3] \to [x_1 + 5x_2 + 3x_3,\ \ 2x_1 + 7x_2 + 9x_3,\ \ 4x_1 + x_2 + 2x_3]$$

This transformation is not matrix transformation because it can not be expressed as $A\,X$ for constant matrix A.

Example 11. *The matrix of linear mapping $T : R^3 \to R^3$ given by $T(a, b, c) = (a, b, 0)$ relative to standard basis is*

(*i*) $\begin{bmatrix} 0 & 1 & 0 \\ 2 & 0 & 1 \\ 1 & 0 & 2 \end{bmatrix}$ (*ii*) $\begin{bmatrix} 2 & 0 & 1 \\ 1 & 2 & 0 \\ 2 & 1 & 0 \end{bmatrix}$ (*iii*) $\begin{bmatrix} 1 & 0 & 0 \\ 0 & 1 & 0 \\ 0 & 0 & 1 \end{bmatrix}$ (*iv*) $\begin{bmatrix} 2 & 2 & 0 \\ 0 & 2 & 0 \\ 0 & 0 & 2 \end{bmatrix}$.

(*R.G.P.V., Bhopal, III Semester, Dec. 2006*)

Solution. The standard basis of R^3 is $B = (e_1, e_2, e_3)$, where

$e_1 = (1, 0, 0)$, $e_2 = (0, 1, 0)$ and $e_3 = (0, 0, 1)$

Thus by definition of T, we have

$$T(e_1) = T(1, 0, 0) = (1, 0, 0) = 1\,(1, 0, 0) + 0\,(0, 1, 0) + 0\,(0, 0, 1)$$
$$= 1\,e_1 + 0e_2 + 0e_3$$
$$T(e_2) = T(0, 1, 0) = (0, 1, 0) = 0\,(1, 0, 0) + 1\,(0, 1, 0) + 0\,(0, 0, 1)$$
$$= 0e_1 + 1.e_2 + 0.e_3$$
$$T(e_3) = T(0, 0, 1) = (0, 0, 1) = 0\,(1, 0, 0) + 0\,(0, 1, 0) + 1\,(0, 0, 1)$$
$$= 0e_1 + 0e_2 + e_3$$

Hence, the coefficient matrix is $\begin{bmatrix} 1 & 0 & 0 \\ 0 & 1 & 0 \\ 0 & 0 & 1 \end{bmatrix}$ and its transpose

matrix is $\begin{bmatrix} 1 & 0 & 0 \\ 0 & 1 & 0 \\ 0 & 0 & 1 \end{bmatrix}$

$$[T, B] = \begin{bmatrix} 1 & 0 & 0 \\ 0 & 1 & 0 \\ 0 & 0 & 1 \end{bmatrix}$$

Ans. (iii) is correct

Example 12. *The standard basis for a* 2×2 *matrix*

$$B = \left\{\begin{bmatrix} 1 & 0 \\ 0 & 0 \end{bmatrix}, \begin{bmatrix} 0 & 0 \\ 1 & 0 \end{bmatrix}, \begin{bmatrix} 0 & 1 \\ 0 & 0 \end{bmatrix}, \begin{bmatrix} 0 & 0 \\ 0 & 1 \end{bmatrix}\right\},$$

and the basis $A = \left\{\begin{bmatrix} 1 & 0 \\ 0 & 0 \end{bmatrix}, \begin{bmatrix} 2 & 0 \\ -1 & 0 \end{bmatrix}, \begin{bmatrix} 0 & 1 \\ 0 & 1 \end{bmatrix}, \begin{bmatrix} -3 & 0 \\ 0 & 2 \end{bmatrix}\right\}$

(*i*) *Find the transition matrix from A to B.*

(*ii*) *Find the matrix that has coordinate vector*

$$[V]_A = (-8, 3, 5, -2)$$

Solution. (*i*) From A to B the transition matrix is

$$P = \left\{ \begin{bmatrix} 1 & 0 \\ 0 & 0 \end{bmatrix}_B \begin{bmatrix} 2 & 0 \\ -1 & 0 \end{bmatrix}_B \begin{bmatrix} 0 & 1 \\ 0 & 1 \end{bmatrix}_B \begin{bmatrix} -3 & 0 \\ 0 & 2 \end{bmatrix}_B \right\}$$

Now $$\begin{bmatrix} -3 & 0 \\ 0 & 2 \end{bmatrix} = c_1 \begin{bmatrix} 1 & 0 \\ 0 & 0 \end{bmatrix} + c_2 \begin{bmatrix} 0 & 0 \\ 1 & 0 \end{bmatrix} + c_3 \begin{bmatrix} 0 & 1 \\ 0 & 0 \end{bmatrix} + c_4 \begin{bmatrix} 0 & 0 \\ 0 & 1 \end{bmatrix}$$

$$= \begin{bmatrix} c_1 & 0 \\ 0 & 0 \end{bmatrix} + \begin{bmatrix} 0 & 0 \\ c_2 & 0 \end{bmatrix} + \begin{bmatrix} 0 & c_3 \\ 0 & 0 \end{bmatrix} + \begin{bmatrix} 0 & 0 \\ 0 & c_4 \end{bmatrix}$$

$$= \begin{bmatrix} c_1 + 0 + 0 + 0 & 0 + 0 + c_3 + 0 \\ 0 + c_2 + 0 + 0 & 0 + 0 + 0 + c_4 \end{bmatrix} = \begin{bmatrix} c_1 & c_3 \\ c_2 & c_4 \end{bmatrix}$$

$$\Rightarrow \qquad c_1 = -3, \quad c_2 = 0, \quad c_3 = 0, \quad c_4 = 2$$

Thus, $$\begin{bmatrix} -3 & 0 \\ 0 & 2 \end{bmatrix}_B = \begin{bmatrix} -3 \\ 0 \\ 0 \\ 2 \end{bmatrix}$$

Similarly, we have

$$\begin{bmatrix} 1 & 0 \\ 0 & 0 \end{bmatrix}_B = \begin{bmatrix} 1 \\ 0 \\ 0 \\ 0 \end{bmatrix}, \quad \begin{bmatrix} 0 & 0 \\ 1 & 0 \end{bmatrix}_B = \begin{bmatrix} 2 \\ 0 \\ -1 \\ 0 \end{bmatrix}, \quad \begin{bmatrix} 0 & 1 \\ 0 & 0 \end{bmatrix}_B = \begin{bmatrix} 0 \\ 1 \\ 0 \\ 1 \end{bmatrix}$$

Hence, $$P = \begin{bmatrix} 1 & 2 & 0 & -3 \\ 0 & 0 & 1 & 0 \\ 0 & -1 & 0 & 0 \\ 0 & 0 & 1 & 2 \end{bmatrix}$$ **Ans.**

(*ii*) The matrix that has the coordinate vector $[V]_4 = (-8, 5, 3. -2)$is given by

$$[V]_A = \begin{bmatrix} 1 & 2 & 0 & -3 \\ 0 & 0 & 1 & 0 \\ 0 & -1 & 0 & 0 \\ 0 & 0 & 1 & 2 \end{bmatrix} \begin{bmatrix} -8 \\ 3 \\ 5 \\ -2 \end{bmatrix}$$

$$= \begin{bmatrix} -8 + 6 + 0 + 6 \\ 0 + 0 + 5 + 0 \\ 0 - 3 + 0 + 0 \\ 0 + 0 + 5 - 4 \end{bmatrix} = \begin{bmatrix} 4 \\ -3 \\ 5 \\ 1 \end{bmatrix}$$

which is the coordinates of V *w.r.t.* standard basis.

Hence, $$V = \begin{bmatrix} 4 & 5 \\ -3 & 1 \end{bmatrix}$$ **Ans.**

34.3 CHANGE OF BASIS

Here we will illustrate the change in O basis by the following:

Let the coordinate of the point A be (4,5).

If two persons X and Y want to go from O to A.

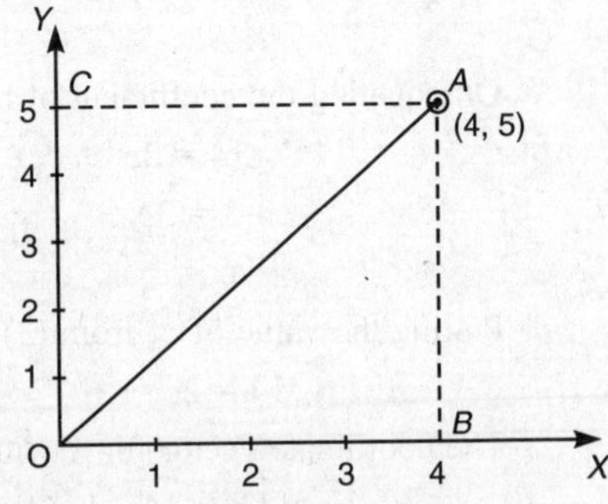

Person X starts from O and uses the path OB and BA accordingly he has to go first four steps along OX and then five steps parallel to y- axis to reach A.

Person Y starts from O and uses the path OC and then CA accordingly he has to go first 5-steps along OY-axis and then 4 steps parallel to X-axis to reach A.

Here both the persons reach A by using different paths.

The only difference is in change of order of paths. These paths are expressed in the standard basis of R^2.

$$S_1 = \{(1, 0), (0, 1)\}$$

$$S_2 = \{(0, 1), (1, 0)\}$$

Path 1 = (4, 5) = 4 (1, 0) + 5 (0, 1)

Path 2 = (4, 5) = 5 (0, 1) + 4 (1, 0)

In path 1 and 2 the coefficients are actually directions in terms of the standard basis of R^2.

Now, we have seen that to express the vectors order is important in a basis. Here the direction vectors or coordinate vectors are $\begin{bmatrix} 4 \\ 5 \end{bmatrix}$ and $\begin{bmatrix} 5 \\ 4 \end{bmatrix}$.

The vectors can be expressed in form of matrix as $\begin{bmatrix} 4 \\ 5 \end{bmatrix} = \begin{bmatrix} 1 & 0 \\ 0 & 1 \end{bmatrix} \begin{bmatrix} 4 \\ 5 \end{bmatrix}$

$$\begin{bmatrix} 4 \\ 5 \end{bmatrix} = \begin{bmatrix} 0 & 1 \\ 1 & 0 \end{bmatrix} \begin{bmatrix} 5 \\ 4 \end{bmatrix}$$

We can express any vector by using standard basis easily. But we have to describe the given vector in term of different basis.

Example 13. *Let $X = (4, 5)$ in R^2, find the coordinate vector for X with respect to the basis $A = \{(1, 1), (-1, 2)\}$*

Solution. $(4, 5) = C_1 (1, 1) + C_2 (-1, 2)$

$$= (C_1, C_1) + (-C_2, 2C_2)$$

$$= (C_1 - C_2, C_1 + 2C_2)$$

Here, $C_1 - C_2 = 4, \qquad C_1 + 2C_2 = 5$

On solving, we get $C_1 = \dfrac{13}{3}, \qquad C_2 = \dfrac{1}{3}$

The coordinate vector for X with respect to A is $[X]_A = \left(\dfrac{13}{3}, \dfrac{1}{3}\right)$ **Ans.**

Example 14. *Find the coordinate vector for* $X = 14 - 3x + 6x^2$ *with respect to the basis* $A = (2, 3x, x^2 + 1)$

Solution. Here, $14 - 3x + 6x^2 = c_1 (2) + c_2 (3x) + c_3 (x^2 + 1)$

$$= (2c_1 + c_3) + c_2 (3x) + c_3 x^2$$

On equating the coefficient of both sides, we get

$$14 = 2c_1 + c_3 \qquad \ldots(1)$$

$$-3 = 3c_2 \quad \Rightarrow \quad c^2 = -1 \qquad \ldots(2)$$

$$6 = c_3 \quad \Rightarrow \quad c_3 = 6 \qquad \ldots(3)$$

Putting the value of c_3 from (3) in (1), we get

$$14 = 2c_1 + 6 \quad \Rightarrow \quad c_1 = 4$$

The coordinate vector for X with respect to basis A is

$$[X]^A = [4, -1, 6]$$

Ans.

Example 15. *Let* $A = \{(1, 0, 1), (0, 1, 1), (1, 2, 4)\}$ *is a basis for* R^3. *Suppose* $(X)_A = (-1, 2, 4)$. *Find X.*

Solution. $X = (c_1e_1, \quad c_2e_2, \quad c_3e_3)$

$$X = -1(1, 0, 1), \quad 2(0, 1, 1), \quad 4(1, 2, 4) = (-1, 0, -1), \quad (0, 2, 2), \quad (4, 8, 16)$$

$$= (-1 + 0 + 4), \quad (0 + 2 + 8), \quad (-1 + 2 + 16) = (3, 10, 17)$$

Ans.

Example 16. *Let V be the vector space of all polynomials over the field R of all real numbers, then the mapping*

$$T : V \to V : T[p(x)] = \frac{d}{dx}[p(x)], \forall\, p(x) \in V$$

is a linear transformation, but not an isomorphism.

Solution. $\forall a, b \in R$ and $p_1(x), p_2(x) \in V$, we have

$$T[a.p_1(x) + b.p_2(x)] = \frac{d}{dx}[a.p_1(x) + b.p_2(x)]$$

$$= a.\frac{d}{dx}[p_1(x)] + b.\frac{d}{dx}[p_2(x)] = a \,.\, T[p_1(x)] + b.T[p_2(x)].$$

$\therefore$ T is *a linear transformation.*

However, T is not one-one, for if $p(x) \in V$, and $c_1, c_2 \in R$ such that $c_1 \neq c_2$, then

$$p(x) + c_1 \neq p(x) + c_2, \text{ but } T[p(x) + c_1] = T[p(x) + c_2] = \frac{d}{dx}[p(x)]$$

Hence, T is not an isomorphism.

Ans.

34.4 CO-DOMAIN OF A LINEAR TRANSFORMATION

If $T : A \to B$ is a transformation then the set A is called the domain and the set B is called the co-domain of T.

For example; R = set of all real numbers

C = set of all complex numbers

For example; $T(x) = 2x + 3i$. Then, for any element 6 in R, we get an element $= 2(6) + 3i = 12 + 3i$ in C.

$T(6) = 12 + 3i, \qquad T : R \to C$ is a transformation.

Here, R is the domain and C is the co-domain of T.

Range of Transformation or set of images

Let $T : A \to B$ be a transformation set of all images by T is a subset of B. This subset is called image set or range of T. It is denoted by $T(A)$.

For example; $T : R \to C$ defined by $T(x) = 3x + i$

$$T(R) = \{3x + i : x \text{ is real}\}$$

Kernel of a transformation

Ker $(f) = \{\alpha \in U : f(\alpha) = 0'$, where $0'$ is the zero of $V\}$

34.5 RANK AND NULLITY OF A LINEAR TRANSFORMATION

Let $T : V \to W$ be a Linear Transformation. We know that $T(V)$ is a subspace of the vector space. The dimension of this subspace $T(V)$ is called the rank of T.

Nullity of f = dim [ker (f)]

34.6 PRODUCT OF LINEAR TRANSFORMATIONS

Let U, V and W be three vector spaces over the same field F. Let f, g be linear transformation from U into V and V into W respectively. Then their product to be denoted by gf is the composite mapping from U into W, defined by $(gf)(\alpha) = g[f(\alpha)]\ \forall\ \alpha \in U$

34.7 PRODUCT OF TWO LINEAR TRANSFORMATION IS A LINEAR TRANSFORMATION

Since, for all $\alpha, \beta \in U$ and $a, b \in f$. Then

$$(gf)(a\alpha + b\beta) = g[f(a\alpha + b\beta)]$$

$$= g[a f(\alpha) + b f(\beta)] \qquad (f \text{ is linear})$$

$$= a\, g[f(\alpha)] + b\, g[f(\beta)]$$

$$= a\, gf(\alpha) + b\, gf(\beta)$$

Proved.

34.8 INVERTIBLE LINEAR TRANSFORMATION

Let U and V be the vector spaces and T be a linear transformation from U into V. Such that T is one-one and onto; then T is called *invertible*.

Example 17. *Show that the transformation $T : V_2 \to V_3$ such that $T(x, y) = (x + y, x - y, y)$ is a linear transformation.*

Find the range, rank null space and nullity of T.

Solution. (*i*) Let $\alpha = (x_1, y_1), \beta = (x_2, y_2) \in V_2$

$$T(\alpha) = T(x_1, y_1) = (x_1 + y_1, x_1 - y_1, y_1)$$

and $T(\beta) = T(x_2, y_2) = (x_2 + y_2, x_2 - y_2, y_2)$

$x, y \in R \Rightarrow x\alpha + y\beta \in V_2$

$$T(x\alpha + y\beta) = T[x(x_1, y_1) + y(x_2, y_2)] = T[x x_1 + y x_2, xy_1 + yy_2]$$
$$= xx_1 + yx_2 + xy_1 + yy_2, xx_1 + yx_2 - xy_1 - yy_2, xy_1 + yy_2$$
$$= x(x_1 + y_1) + y(x_2 + y_2), x(x_1 - y_1) + y(x_2 - y_2), xy_1 + yy_2$$
$$= x[x_1 + y_1, x_1 - y_1, y_1] + y[x_2 + y_2, x_2 - y_2, y_2]$$
$$= xT(\alpha) + yT(\beta)$$ **Proved.**

Hence, T is a linear transformation from V_2 to V_3.

(ii) Range of T

We have, $R(T) = \{\beta \in V_3 : \beta = T(\alpha), \alpha \in V_2\}$

Now, (1, 0), (0, 1) is the basis of V_2 also.

$T_1(1, 0) = (1 + 0, 1 - 0, 0) = (1, 1, 0)$

$T(0, 1) = (0 + 1, 0 - 1, 1) = (1, -1, 1)$

Range of $T \subset V_3$ generated by (1, 1, 0) and (1, –1, 1).

Now, vectors $(1, 1, 0), (1, -1, 1) \in V_3$ are linearly independent if $a, b \in R$. Then

$a(1, 1, 0) + b(1, -1, 1) = (0, 0, 0) \Rightarrow (a + b, a - b, b) = (0, 0, 0)$

$\Rightarrow a + b = 0,\ a - b = 0,\ b = 0 \Rightarrow a = 0,\ b = 0$

Hence {(1, 1, 0), (1, –1, 1)} is the basis of $R(T)$.

Hence rank of $T = \dim R(T) = 2$

Nullity of $T = \dim V_2(R) - \text{Rank}(T) = 2 - 2 = 0$

(iii) Null space of

$N(T) = \{\alpha \in V_2 : T(\alpha) = 0, \in V_3(R)\}$

$(x, y) \in N(T) \Rightarrow T(x, y) = (0, 0, 0)$

$(x + y, x - y, y) = (0, 0, 0) \Rightarrow x + y = 0, x - y = 0, y = 0.$

$\Rightarrow x = 0,\ y = 0.$

Null space of T consist of zero vector (0, 0) of $V_2(R)$.

$\Rightarrow$ Null space of T is the zero subspace of $V_2(R)$.

Nullity of $T = \dim. N(T) = 0.$ **Ans.**

Example 18. *If the matrix of a linear transformation T on V_3 (R) with respect to basis B = {(1, 0, 0), (0, 1, 0), (0, 0, 1)} is :*

$$\begin{bmatrix} 0 & 1 & 1 \\ 1 & 0 & -1 \\ -1 & -1 & 0 \end{bmatrix}$$

then what is the matrix of T relative to the basis

B ′ = {(0, 1, – 1), (1, – 1, 1), (– 1, 1, 0)}?

(*R.G.P.V., Bhopal, III Semester, Dec. 2007*)

Solution. The standard basis of V_3 (R) is

$$B = \{e_1 = (1, 0, 0),\ \ e_2 = (0, 1, 0),\ \ e_3 = (0, 0, 1)\} \quad ...(1)$$

$$\text{Also, given that } [T]_B = \begin{bmatrix} 0 & 1 & 1 \\ 1 & 0 & -1 \\ -1 & -1 & 0 \end{bmatrix} \quad ...(2)$$

$$\text{Now, transpose of matrix (2) = coefficient matrix} = \begin{bmatrix} 0 & 1 & -1 \\ 1 & 0 & -1 \\ 1 & -1 & 0 \end{bmatrix}$$

Hence,
$$\left.\begin{aligned} T(e_1) &= 0e_1 + 1e_2 + (-1)e_3 = (0, 1, -1) \\ T(e_2) &= 1e_1 + 0e_2 + (-1)e_3 = (1, 0, -1) \\ T(e_3) &= 1e_1 + (-1)e_2 + 0e_3 = (1, -1, 0) \end{aligned}\right\} \quad ...(3)$$

Now, if $(a, b, c) \in V_3\ (R)$, then we can write

$$\begin{aligned} (a, b, c) &= ae_1 + be_2 + ce_3 \\ \therefore \quad T(a, b, c) &= T[ae_1 + be_2 + ce_3] = aT(e_1) + bT(e_2) + cT(e_3) \\ &= a(0, 1, -1) + b(1, 0, -1) + c(1, -1, 0) \quad \text{[Using (3)]} \\ &= (b + c, a - c, -a - b) \quad ...(4) \end{aligned}$$

This is an explicit expression for T.

Now, let us find the matrix of T with respect to the basis B'.

We have,
$$\begin{aligned} T(e_1') &= T(0, 1, -1) = (1 - 1, 0 + 1, 0 - 1) = (0, 1, -1) \\ T(e_2') &= T(1, -1, 1) = (0, 0, 0) \\ T(e_3') &= T(-1, 1, 0) = (1, -1, 0) \end{aligned}$$

$$\begin{aligned} \text{Also, let } (a, b, c) &= xe'_1 + ye_2' + ze_3' \\ &= x\ (0, 1, -1) + y(1, -1, 1) + z(-1, 1, 0) \quad ...(5) \end{aligned}$$

$$\Rightarrow \quad (a, b, c) = (y - z, x - y + z, -x + y)$$

$$\Rightarrow \quad y - z = a,\ x - y + z = b,\ -x + y = c$$

$$\Rightarrow \quad x = a + b,\ y = a + b + c,\ z = b + c$$

Hence, from equation (5), we have

$$(a, b, c) = (a + b)\, e_1' + (a + b + c)\, e_2' + (b + c)\, e_3' \quad ...(6)$$

For $a = 0,\ b = 1,\ c = -1$, we get

$$T(e_1') = (0, 1, -1) = 1e_1' + 0e_2' + 0e_3' \quad ...(7)$$

For $a = b = c = 0$, we get

$$T(e_2') = (0, 0, 0) = 0e_1' + 0e_2' + 0e_3' \quad ...(8)$$

For $a = 1,\ b = -1,\ c = 0$, we get

$$T(e_3') = (1, -1, 0) = 0e_1' + 0e_2' + (-1)\, e_3' \quad ...(9)$$

From (7), (8) and (9), we get

$$\text{co-efficient matrix} = \begin{bmatrix} 1 & 0 & 0 \\ 0 & 0 & 0 \\ 0 & 0 & -1 \end{bmatrix}$$

$$[T]_{B'} = \text{transpose of the coefficient matrix} \begin{bmatrix} 1 & 0 & 0 \\ 0 & 0 & 0 \\ 0 & 0 & -1 \end{bmatrix}.$$ **Ans.**

34.9 SIMILARITY OF MATRICES

If A and B are square matrices of order n over the field F, then B is said to be similar to A, if there exists an $n \times n$ invertible square matrix P with elements in F is such that

$$B = P^{-1}AP$$

Theorem 3. The relation of similarity is an *equivalence relation* in the set of all $n \times n$ matrices over the field F.

Proof. If A and B are two $n \times n$ matrices over the field F, then B is said to be similar to A. If there exists an $n \times n$ invertible matrix P over the field F such that

$$B = P^{-1}AP$$

(*i*) **Reflexive :** A is similar to A because

$$A = I^{-1} AI$$

where I is $n \times n$ unit matrix over F.

(*ii*) **Symmetric:** If A is similar to B then there exists an $n \times n$ invertible matrix P over the field F.

$$A = P^{-1}BP \qquad \text{...(1)}$$

Pre multiplying by P and post multiplying by P^{-1} to equation (1), we get

$$PAP^{-1} = P(P^{-1}BP)P^{-1}$$
$$= I\,B\,I$$
$$= B$$
$$\Rightarrow \quad B = PAP^{-1} \qquad [(P^{-1})^{-1} = P]$$
$$B = (P^{-1})^{-1}AP^{-1}$$

Hence, B is similar to A.

(*iii*) **Transitive:** If A is similar to B and B is similar to C then

$$A = P^{-1}BP \qquad \text{...(2)}$$

and

$$B = Q^{-1}CQ \qquad \text{...(3)}$$

Putting the value of B in (2) from (3), we get

$$\mathrm{A} = P^{-1}(Q^{-1}CQ)P$$
$$= (\mathrm{P}^{-1}Q^{-1})C(QP)$$
$$= (QP)^{-1}C(QP) \qquad [(QP)^{-1} = P^{-1}Q^{-1}]$$

Thus, A is similar to C.

Hence, similarity is an equivalence relation on the set of $n \times n$ matrices over the field F.

34.10 SIMILARITY OF LINEAR TRANSFORMATIONS

Let $L(V, V)$ denote the set of all linear transformations on $V(F)$.

If $T_1, T_2 \in L(V, V)$, then T_2 is said to be similar to T_1 if there exists an invertible linear transformation T in $L(V, V)$ such that

$$T_2 = TT_1T^{-1}$$

Example 19. *Find the matrix representation of linear transformation T on V_3 (R) defines as $T(a, b, c) = (2b + c, a - 4b, 3a)$ corresponding to the basis $B = \{(1, 0, 0), (0, 1, 0), (0, 0, 1)\}$ (R. G. P. V. Bhopal; III Sem. Dec 2005)*

Solution. Let $T: V_3(R) \to V_3(R)$ be a linear transformation and V_3 (R) has a basis set $B = \{(1, 0, 0), (0, 1, 0), (0, 0, 1)\}$. Then

$$T(e_1) = T(1, 0, 0) = (2 \times 0 + 0, \quad 1 - 4(0), \quad 3 \times 1)$$
$$= (0, 1, 3) = 0(1, 0, 0) + 1(0, 1, 0) + 3(0, 0, 1)$$
$$T(e_2) = T(0,1,0) = (2\times1+0, \quad 0-4\times1, \quad 3\times0) = (2, -4, 0)$$
$$= 2(1, 0, 0) - 4(0, 1, 0) + 0(0, 0, 1)$$
$$T(e_3) = T(0,0,1) = (2\times0+1, \quad 0-4\times0, \quad 3\times0) = (1, 0, 0)$$
$$= 1(1, 0, 0) + 0(0, 1, 0) + 0(0, 0, 1)$$

$$[T : B] = \begin{bmatrix} 0 & 2 & 1 \\ 1 & -4 & 0 \\ 3 & 0 & 0 \end{bmatrix}$$ **Ans.**

Example 20. *Let T be the linear transformation operater on R^3 defined by $T(x, y, z) = (2x + z, x - 4y, 3x)$. Find the matrix representation of T in the basis $f_1 = (1, 0, 0), f_2 = (1, 1, 0), f_3 = (1, 1, 1)$. (Delhi University, April 2010)*

Solution. Let $T : R^3 \to R^3$ be a linear transformation and R^3 has a basis set

$$f_1 = (1, 0, 0), \; f_2 = (1, 1, 0), f_3 = (1, 1, 1)$$
$$T(f_1) = T(1, 0, 0) = (2, 1, 3) = 2(1, 0, 0) + 1(1, 1, 0) + 3(1, 1, 1)$$
$$T(f_2) = T(1, 1, 0) = (2, -3, 3) = 2(1, 0, 0) - 3(1, 1, 0) + 3(1, 1, 1)$$
$$T(f_3) \quad = T(1, 1, 1) = (3, -3, 3) = 3(1, 0, 0) - 3(1, 1, 0) + 3(1, 1, 1)$$

$$[T : B] = \begin{bmatrix} 2 & 2 & 3 \\ -1 & -3 & -3 \\ 3 & 3 & 3 \end{bmatrix}$$ **Ans.**

EXERCISE 34.1

1. Show that the mapping $T : V_2(R) \to V_2(R) : T(a, b) = (b, a)$ is an isomorphism.
2. Show that each one of the following mappings is a linear transformation :-
   (i) $T : V_3(R) \to V_2(R) : T(a, b, c) = (a - b, a - c)$;
   (ii) $T : V_2(R) \to V_2(R) : T(a, b) = (a + b, a)$;
   (iii) $T : V_3(R) \to V_3(R) : T(a, b, c) = (a - b + 2c, 2a + b, -a - 2b + 2c)$.
3. Let $f : R^3 \to R^2 : f(a, b, c) = (a, b) \; \forall \; (a, b, c) \in R^3$.
   Show that f is a linear transformation. Also, find ker (f).

Hint. $\ker(f) = \{(a, b, c) \in R^3 : f(a, b, c) = (0, 0)\}$
$= \{(a, b, c) \in R^3 : (a, b) = (0, 0)\}$
$= \{(0, 0, c) : c \in R\}$

4. Let $f : R^2 \to R^2 : f(a, b) = (a + b, a + b)\ \forall\ (a, b) \in R^2$.

Show that f is a linear transformation. Also, find the basis and dimensions of range space of f and the kernel of f.

Hint. The standard basis of R^2 is $\{(1, 0), (0, 1)\}$. The images of these vectors generate $R(f)$.

Now, $f(1, 0) = (1 + 0, 1 + 0) = (1, 1)$ and $f(0, 1) = (0 + 1, 0 + 1) = (1, 1)$.

$\therefore$ $\{(1, 1)\}$ is a basis of $R(f)$ and hence dim. $R(f) = 1$.

Also, $f(a, b) = (0, 0) \Rightarrow (a + b, a + b) = (0, 0)$

$\Rightarrow a + b = 0, a + b = 0.$

Taking $b = 1$, we have $a = -1$.

$\therefore$ $\{(-1, 1)\}$ is a basis of $\ker(f)$ and hence dim. $\ker(f) = 1$

5. Let U and V be two vector subspaces over the same field F. Show that a function $T : U \to V$ is linear transformation iff :

$T(au + v) = aT(u) + T(v)$ for all $u, v \in U$ and $a \in F$. *(R.G.P.V., Bhopal, III Semester, June 2007)*

6. Let $F[x]$ denote the set of all polynomials over the real field R. Let D, M and S be the linear operators on $F[x]$, defined by:-

$$D[p(x)] = \frac{d}{dx}[p(x)];\ M[p(x)] = x \cdot p(x) \text{ and } S[p(x)] = \int p(x)\,dx, \forall\ p(x) \in F[x].$$

Then, show that

(*i*) $DM \neq MD$ (ii) $DM - MD = I$ (iii) $DS = I \neq SD$.

7. (*i*) Show that the translation mapping

$f : R^2 \to R^2 : f(x, y) = (x + 2, y + 3)$ is not linear.

Hint. $f(0, 0) = (2, 3) \neq (0, 0)$.

(*ii*) Show that the mapping $f : R \to R : f(ab) = ab$ is not linear.

8. Show that transformation $T : R^2 \to R^3$ defined by

$T(a, b) = (a - b, b - a, -a).\ \forall\ a, b \in R.$

is a linear transformation from R^2 into R^3. Find the range, rank, null space and nullity of T.

Ans. Range $(T) = \{(1, -1, 0), (0, 0, 1)\}$

Rank $(T) = 2$

Nullity of $T = \{(0, 0)\}$

Null space of contains zero vector of R^2.

9. Prove that the system S consisting of n vectors

$e_1 = (1, 0, 0..... 0), e_2 = (0, 1, 0, 0, ...0),$

$e_n = (0, 0, ...1)$ is a basis of $V_n(F)$.

10. Find the matrix representation of linear transformation T on $V_3(R)$ defined as

$T(a, b, c) = (2b + c, a - 4b, 3a)$

$B = \{(1, 1, 1), (1, 1, 0), (1, 0, 0)\}$

Ans. $$[T : B] = \begin{bmatrix} 3 & 3 & 3 \\ -6 & -6 & -2 \\ 6 & 5 & -1 \end{bmatrix}$$

11. Find the matrix of the following linear maps with respect to the standard basis of R^n.

(*i*) $T: R^3 \to R^2$ defined by $T(x, y, z) = (2x - 4y + 9z, 5x + 3y - 2z)$

Ans. $[T : B, B'] = \begin{bmatrix} 2 & -4 & 9 \\ 5 & 3 & -2 \end{bmatrix}$

(*ii*) $T: R^3 \to R^3$ defined by $T(x, y, z) = (z, y + z, x + y + z)$ **Ans.** $[T : B] = \begin{bmatrix} 0 & 0 & 1 \\ 0 & 1 & 1 \\ 1 & 1 & 1 \end{bmatrix}$

12. If the matrix of a linear transformation T on $V_2(R)$ with respect to the basis A = [(1, 0), (0, 1)] is $\begin{bmatrix} 1 & 1 \\ 1 & 1 \end{bmatrix}$ then what is the matrix of T with respect to the ordered basis A = {(1, 1), (1, –1)} **Ans.**

$[T : A'] = \begin{bmatrix} 2 & 0 \\ 0 & 0 \end{bmatrix}$

13. Let T be a linear transformation on R^3 defined by

$T(x_1, x_2, x_3) = (3x_1 + x_3, -2x_1 + x_2, -x_1 + 2x_2 + 4x_3)$.

Then find the matrix of T with respect to standard basis of R^3.

Prove that T is invertible and find a formula for T^{-1}. **Ans.** $[T]_B = \begin{bmatrix} 3 & 0 & 1 \\ -2 & 1 & 0 \\ -1 & 2 & 4 \end{bmatrix}$,

$$T^{-1}(\alpha) = T^{-1}(x, y, z) = \frac{1}{9}(4x + 2y - z, \quad 8x + 13y - 2z, \quad -3x - 6y + 3z)$$

Choose the correct alternative:

14. The zero vector in the vector space R^3 is :

(*a*) (0, 1, 0) (*b*) (0, 0) (*c*) (0, 1) (*d*) (0, 0, 0)

(*R.G.P.V., Bhopal, III Semester, June 2007*) **Ans.** (*d*)

15. The vector space $V_3(R)$ is of dimension :

(*a*) 1 (*b*) 2 (*c*) 3 (*d*) 4

(*R.G.P.V., Bhopal, III Semester, June 2007*) **Ans.** (*c*)

BASIS OF NULL SPACE, ROW SPACE AND COLUMN SPACE

35.1 INTRODUCTION

Now we will discuss the fundamental subspaces associated with matrices.

35.2 ROW VECTORS

Here we have a $m \times n$ matrix.

$$A = \begin{bmatrix} a_{11} & a_{12} \ldots\ldots\ldots\ldots a_{1n} \\ a_{21} & a_{22} \ldots\ldots\ldots\ldots a_{2n} \\ \ldots\ldots\ldots\ldots & \ldots\ldots\ldots\ldots \\ a_{m1} & a_{m2} \ldots\ldots\ldots\ldots a_{mn} \end{bmatrix}$$

The elements a_{11}, a_{12} ...... are known as enteries. If a_{11}, a_{12} .... a_{1n} are real, then enteries are in R. The rows of A described as vectors in R^n are called *row vectors* or *row matrix* as

$$r_1 = (a_{11}, a_{12}, \text{------------ } a_{1n})$$
$$r_2 = (a_{21}, a_{22} \text{ ----------- } a_{2n})$$
$$\text{------------------------------------}$$
$$\text{------------------------------------}$$
$$r_m = (a_{m1}, a_{m2} \text{ ---------- } a_{mn})$$

35.3 COLUMN VECTORS

Column vectors of A are

$$C_1 = \begin{bmatrix} a_{11} \\ a_{21} \\ \ldots \\ \ldots \\ a_{m1} \end{bmatrix}, C_2 = \begin{bmatrix} a_{12} \\ a_{22} \\ \ldots \\ \ldots \\ a_{m2} \end{bmatrix} \text{ -------------- } C_n = \begin{bmatrix} a_{1n} \\ a_{2n} \\ \ldots \\ \ldots \\ a_{mn} \end{bmatrix}$$

Example 1. *Write the row vectors and column vectors for the matrix*

$$A = \begin{bmatrix} 2 & 4 & 5 & 0 \\ -1 & 0 & 3 & 1 \\ 2 & 1 & 6 & 4 \end{bmatrix}$$

Solution. The row vectors are

$r_1 = [2\ 4\ 5\ 0]$, $\quad r_2 = [-1\ 0\ 3\ 1]$, $\quad r_3 = [2\ 1\ 6\ 4]$

The column vectors are

$$C_1 = \begin{bmatrix} 2 \\ -1 \\ 2 \end{bmatrix}, C_2 = \begin{bmatrix} 4 \\ 0 \\ 1 \end{bmatrix}, C_3 = \begin{bmatrix} 5 \\ 3 \\ 6 \end{bmatrix}, C_4 = \begin{bmatrix} 0 \\ 1 \\ 4 \end{bmatrix}$$ **Ans**

35.4 ROW SPACE

If r_1, r_2, --------------- r_n are the rows of matrix A then subspace of R^n that is spanned by the row vectors is called the row space of A.

Row space of $A = [r_1, r_2$, -----------, $r_n]$

It is denoted by row (A).

35.5 COLUMN SPACE

If C_1, C_2, --------------- C_m are the columns of the matrix A, then the subspace of R^n that is spanned by the column vectors of A is called column space.

Column space of $A = \{C_1, C_2$, ------------- $C_n\}$

It is denoted by col (A).

35.6 NULL SPACE

The solution set of $AX = 0$ in R^n is called the null space of A. It is denoted by null (A).

OR

Consider a system $AX = 0$ of homogeneous linear equations, then the solutions of this system contribute a vector space, called the *null space* of A.

Example 2. *Determine the null space of each of the following matrices:*

(*a*) $A = \begin{bmatrix} -3 & 0 \\ 2 & -4 \end{bmatrix}$, $\quad$ (*b*) $B = \begin{bmatrix} 1 & -2 \\ -4 & 8 \end{bmatrix}$, $\quad$ (*c*) $0 = \begin{bmatrix} 0 & 0 \\ 0 & 0 \end{bmatrix}$

Solution. We solve the following equations to find out the null space of A.

(*a*) $$\begin{bmatrix} -3 & 0 \\ 2 & -4 \end{bmatrix}\begin{bmatrix} x_1 \\ x_2 \end{bmatrix} = \begin{bmatrix} 0 \\ 0 \end{bmatrix} \Rightarrow \begin{bmatrix} -3x_1 = 0 \Rightarrow x_1 = 0 \\ 2x_1 - 4x_2 = 0 \Rightarrow 0 - 4x_2 = 0 \Rightarrow x_2 = 0 \end{bmatrix}$$

$$\Rightarrow \quad x_1 = x_2 = 0$$

The single solution is $\begin{bmatrix} 0 \\ 0 \end{bmatrix}$ in terms of vectors from R^2. Hence the null space of A is $\{0\}$.

Ans.

(*b*) To find the null space of B, we solve the following equations

$$\begin{bmatrix} 1 & -2 \\ -4 & 8 \end{bmatrix}\begin{bmatrix} x_1 \\ x_2 \end{bmatrix} = \begin{bmatrix} 0 \\ 0 \end{bmatrix} \Rightarrow \begin{array}{l} x_1 - 2x_2 = 0 \\ -4x_1 + 8x_2 = 0 \end{array}$$

Let, $x_2 = k$ then $x_1 - 2k = 0 \Rightarrow x_1 = 2k$

The null space of B consists of all solutions of $BX = 0$

The null space of B in term of vector notation consists of all the vector $X = (x_1, x_2)$

The null space of B, form R^2 of the form $X = (2k, k) = k\,(2, 1)$ **Ans.**

(c) $$0 = \begin{bmatrix} 0 & 0 \\ 0 & 0 \end{bmatrix} \begin{bmatrix} x_1 \\ x_2 \end{bmatrix} = \begin{bmatrix} 0 \\ 0 \end{bmatrix}$$

As every vector X in R^2 will be a solution of the obove equations, so the null space of 0 is all of R^2. **Ans.**

35.7 DIMENSION OF A VECTOR SPACE

The number of vectors present in a basis of vector space V is called the dimension of V. It is denoted by dim (V).

35.8 NULLITY

The dimension of the null space of the matrix A is called the nullity of A and is denoted by nullity (A) or the number of free variables in the solution of $AX = 0$.

Example 3. *Determine a basis for the null space of the matrix*

$$A = \begin{bmatrix} 1 & 2 & -3 & 2 & -3 \\ 1 & -1 & 1 & -2 & 0 \\ 0 & 1 & -1 & 1 & -1 \\ 1 & 2 & 5 & -6 & -3 \end{bmatrix}$$

Solution. For finding null space, we have to solve the following system of equations:

$$\begin{aligned} x_1 + 2x_2 - 3x_3 + 2x_4 - 3x_5 &= 0 \\ x_1 - x_2 + x_3 - 2x_4 + 0x_5 &= 0 \\ 0x_1 + x_2 - x_3 + x_4 - x_5 &= 0 \\ x_1 + 2x_2 + 5x_3 - 6x_4 - 3x_5 &= 0 \end{aligned}$$

It can be easily verified that the solution of this system is

$$x_1 = x_4 + x_5$$
$$x_2 = x_5$$
$$x_3 = x_4$$

when x_4 and x_5 are arbitrary constants.

Thus, null space of A consists of vectors of the form:

$$X = \begin{bmatrix} x_1 \\ x_2 \\ x_3 \\ x_4 \\ x_5 \end{bmatrix} = \begin{bmatrix} x_4 + x_5 \\ x_5 \\ x_4 \\ x_4 \\ x_5 \end{bmatrix} = x_4 \begin{bmatrix} 1 \\ 0 \\ 1 \\ 1 \\ 0 \end{bmatrix} + x_5 \begin{bmatrix} 1 \\ 1 \\ 0 \\ 0 \\ 1 \end{bmatrix} = x_4 u + x_5 v \text{ where } u = \begin{bmatrix} 1 \\ 0 \\ 1 \\ 1 \\ 0 \end{bmatrix}, \; v = \begin{bmatrix} 1 \\ 1 \\ 0 \\ 0 \\ 1 \end{bmatrix}$$

Since x_4 and x_5 are arbitrary, we have null (A) = span (u, v): Moreover the following argument shows that u and v are linearly independent.

$$x_4 u + x_5 v = 0$$

The dimension of null space of $A = 2$ or Nullity $A = 2$

Another solution $X = 0$

$\Rightarrow$ All the coordinates of the solution $X = 0$

$\Rightarrow$ $x_4 = 0, x_5 = 0$.

Hence, u and v form a basis of null (A). **Ans.**

35.9 THEOREM 1. FOR BASIS OF ROW SPACE AND COLUMN SPACE OF ECHELON FORM MATRIX

Let A be a matrix in row-Echelon form. The row vectors containing leading 1 will form a basis for the row space of A.

The column vector that contain the leading 1 from the row vectors will form a basis for the column space of A.

Example 4. *Determine a basis and dimension for the row and column space of the matrix*

$$A = \begin{bmatrix} 1 & 4 & 5 & 0 & 2 & 3 \\ 0 & 1 & -1 & 2 & -4 & 5 \\ 0 & 0 & 0 & 1 & 3 & 1 \\ 0 & 0 & 0 & 0 & 1 & 2 \\ 0 & 0 & 0 & 0 & 0 & 0 \end{bmatrix}$$

Solution. The given matrix A is in the Echelon form. The basis of A is the row vectors containing leading term 1.

So the basis of the row space of A is

$r_1 = [1\ 4\ 5\ 0\ 2\ 3]$, $r_2 = [0\ 1\ -1\ 2\ -4\ 5]$, $r_3 = [0\ 0\ 0\ 1\ 3\ 1]$, $r_4 = [0\ 0\ 0\ 0\ 1\ 2]$

The dimension of the row space is 4. **Ans.**

Again the basis for the column space of A will be the columns that contain leading 1 from the row vectors.

So the basis for the column space of A is

$$C_1 = \begin{bmatrix} 1 \\ 0 \\ 0 \\ 0 \\ 0 \end{bmatrix}, C_2 = \begin{bmatrix} 4 \\ 1 \\ 0 \\ 0 \\ 0 \end{bmatrix}, \qquad C_4 = \begin{bmatrix} 0 \\ 2 \\ 1 \\ 0 \\ 0 \end{bmatrix}, C_5 = \begin{bmatrix} 2 \\ -4 \\ 3 \\ 1 \\ 0 \end{bmatrix},$$

The dimemsion of the column space is 4. **Ans.**

35.10 BASIS OF ROW SPACE OF A GENERAL MATRIX

If the matrix A is not in the row echelon form, then the given matrix is transformed in the row echelon form and apply the theorem given below.

Theorm 1. Let A be a general matrix and a row Echelon form matrix E is formed then the row space of A = Row space of E.

Example 5. *Find a basis for the row space of the matrix*

$$A = \begin{bmatrix} 1 & 3 & 2 & 4 & 1 \\ 0 & 0 & 2 & 2 & 0 \\ 2 & 6 & 2 & 6 & 2 \\ 3 & 9 & 1 & 10 & 6 \end{bmatrix}$$

Solution. Let us transform matrix A to echelon form matrix E.

$$\sim\begin{bmatrix}1&3&2&4&1\\0&0&2&2&0\\2&6&2&6&2\\3&9&1&10&6\end{bmatrix}\sim\begin{bmatrix}1&3&2&4&1\\0&0&2&2&0\\0&0&-2&-2&0\\0&0&-5&-2&3\end{bmatrix}\begin{matrix}\\\\R_3\to R_3-2R_1\\R_4\to R_4-3R_1\end{matrix}$$

$$\sim\begin{bmatrix}1&3&2&4&1\\0&0&2&2&0\\0&0&0&0&0\\0&0&-5&-2&3\end{bmatrix}R_3\to R_3+R_2\quad\sim\begin{bmatrix}1&3&2&4&1\\0&0&1&1&0\\0&0&-5&-2&3\\0&0&0&0&0\end{bmatrix}\begin{matrix}R_2\to\frac{1}{2}R_2\\R_4\leftrightarrow R_3\end{matrix}$$

$$\sim\begin{bmatrix}1&3&2&4&1\\0&0&1&1&0\\0&0&0&3&3\\0&0&0&0&0\end{bmatrix}R_3\to R_3+5R_2\sim\begin{bmatrix}1&3&2&4&1\\0&0&1&1&0\\0&0&0&1&1\\0&0&0&0&0\end{bmatrix}R_3\to\frac{1}{3}R_3\quad=E$$

We know that the row space of A = The row space of E. Therefore, basis for row space of A is

$r_1=\{1\ 3\ 2\ 4\ 1\},\ r_2=\{0\ 0\ 1\ 1\ 0\},\ r_3=\{0\ 0\ 0\ 1\ 1\}$ **Ans.**

Basis of column space of a general matrix

Therom 2. Let E be the echelon form matrix of a general matrix A and column vectors of E form a basis for the column space of E, then a set of corresponding column vectors form A be a basis for the column space of A.

Example 6. *Find a basis for the column space of the matrix A given in the previous example.*

Solution. From previous example, we have

$$E=\begin{bmatrix}1&3&2&4&1\\0&0&1&1&0\\0&0&0&1&1\\0&0&0&0&0\end{bmatrix}$$

We know that the basis for the column space of matrix E is

$$C_1=\begin{bmatrix}1\\0\\0\\0\end{bmatrix},C_3=\begin{bmatrix}2\\1\\0\\0\end{bmatrix},C_4=\begin{bmatrix}4\\1\\1\\0\end{bmatrix},C_5=\begin{bmatrix}1\\0\\1\\0\end{bmatrix}$$

From the theorem on the basis of column space, now we have a basis for column space of the given matrix A is the column number 1, 3, 4, 5.

This a basis of the column space of E is

$$C_1=\begin{bmatrix}1\\0\\2\\3\end{bmatrix},C_3=\begin{bmatrix}2\\2\\2\\1\end{bmatrix},C_4=\begin{bmatrix}4\\2\\6\\10\end{bmatrix},C_5=\begin{bmatrix}1\\0\\2\\6\end{bmatrix}$$ [Using Exp. 5] **Ans.**

35.11 METHOD II

(1) If E is the column echelon form of A then col (E) = col (A).

(2) The non zero columns in E are linearly independent and space col (B).

From above two statements, we can find a basis of col (A).

Non-zero column of a column echelon form is a basis of col (A).

Example 7. *Find the basis for the column space of the matrix*

$$A = \begin{bmatrix} 1 & 1 & 1 & 3 & 4 \\ 2 & 3 & 3 & 6 & 8 \\ -3 & 2 & 3 & -9 & -12 \\ 1 & 3 & 7 & 3 & 4 \end{bmatrix}$$

Solution. We have,

$$A = \begin{bmatrix} 1 & 1 & 1 & 3 & 4 \\ 2 & 3 & 3 & 6 & 8 \\ -3 & 2 & 3 & -9 & -12 \\ 1 & 3 & 7 & 3 & 4 \end{bmatrix}$$

$$= \begin{bmatrix} 1 & 0 & 0 & 0 & 0 \\ 2 & 1 & 1 & 0 & 0 \\ -3 & 5 & 6 & 0 & 0 \\ 1 & 2 & 6 & 0 & 0 \end{bmatrix} \begin{matrix} C_2 \to C_2 - C_1 \\ C_3 \to C_3 - C_1 \\ C_4 \to C_4 - 3C_1 \\ C_5 \to C_5 - 4C_1 \end{matrix}$$

$$= \begin{bmatrix} 1 & 0 & 0 & 0 & 0 \\ 2 & 1 & 0 & 0 & 0 \\ -3 & 5 & 1 & 0 & 0 \\ 1 & 2 & 4 & 0 & 0 \end{bmatrix} C_3 \to C_3 - C_2$$

$$= E$$

Col (A) = Col (E)

= Span $\{C_1, C_2, C_3\}$ [Zero columns does not effect span]

Here first three columns of E are linearly independent.

$C_1 + C_2 + C_3 = 0 \Rightarrow C_1 = C_2 = C_3 = 0$

Therefore, three columns of E $\begin{bmatrix} 1 \\ 2 \\ -3 \\ 1 \end{bmatrix}, \begin{bmatrix} 0 \\ 1 \\ 5 \\ 2 \end{bmatrix}, \begin{bmatrix} 0 \\ 0 \\ 1 \\ 4 \end{bmatrix}$

form a basis of col (A). **Ans.**

35.12 METHOD III.

Find the transpose of the matrix A and then determine a basis for the row space of the transposed matrix of A. This row space gives a basis for the column space of A.

Example 8. *Determine a basis for the column space of the matrix*

$$A = \begin{bmatrix} 1 & 1 & 1 & 3 & 4 \\ 2 & 3 & 2 & 6 & 8 \\ 4 & 7 & 4 & 12 & 16 \\ 5 & 11 & 6 & 15 & 20 \end{bmatrix}$$

Solution.

$$\text{The transpose of } A = A^T = \begin{bmatrix} 1 & 2 & 4 & 5 \\ 1 & 3 & 7 & 11 \\ 1 & 2 & 4 & 6 \\ 3 & 6 & 12 & 15 \\ 4 & 8 & 16 & 20 \end{bmatrix}$$

$$\text{Echelon form of } A^T = E \approx \begin{bmatrix} 1 & 2 & 4 & 5 \\ 0 & 1 & 3 & 6 \\ 0 & 0 & 0 & 1 \\ 0 & 0 & 0 & 0 \\ 0 & 0 & 0 & 0 \end{bmatrix} \begin{matrix} \\ R_2 \to R_2 - R_1 \\ R_3 \to R_3 - R_1 \\ R_4 \to R_4 - 3R_1 \\ R_5 \to R_5 - 4R_1 \end{matrix}$$

A basis of Row space of A^T is $r_1 = [1\ 2\ 4\ 5]$, $r_2 = [0\ 1\ 3\ 6]$, $r_3 = [0\ 0\ 0\ 1]$

A basis of column space of the given matrix A, $r_1 = [1\ 2\ 4\ 5]$, $r_2 = [0\ 1\ 3\ 6]$ and $r_3 = [0\ 0\ 0\ 1]$ **Ans.**

35.13 RANK OF A MATRIX

The row rank of a matrix is equal to the dimension of the row space of the matrix.

The column rank of matrix is equal to the dimension of the column space of the given matrix.

35.14 RANK NULLITY THEOREM

Consider a matrix A then rank (A) + null (A) = number of columns of A.

Example 9. *Determine a basis for the null space, row space, column space of the matrix*

$$A = \begin{bmatrix} 1 & 3 & 4 & -2 & -1 \\ 3 & 7 & 6 & 2 & 1 \\ 2 & 4 & 2 & 4 & 2 \\ 1 & 1 & -2 & 6 & 3 \end{bmatrix}$$

Also determine rank and nullity of the matrix.

Solution. We solve the following system of equations to find null space:

$$x_1 + 3x_2 + 4x_3 - 2x_4 - x_5 = 0 \quad \text{... (1)}$$

$$3x_1 + 7x_2 + 6x_3 + 2x_4 + x_5 = 0 \quad \text{... (2)}$$

$$2x_1 + 4x_2 + 2x_3 + 4x_4 + 2x_5 = 0 \quad \text{... (3)}$$

$$x_1 + x_2 - 2x_3 + 6x_4 + 3x_5 = 0 \quad \text{... (4)}$$

Equation (3) is the same as (2) – (1).

Equation (4) is the same as (2) – 2 (1).

Now, we have to solve equations (1) and (2) only. Hence, there are two equations and 5 unknowns. Let $x_3 = k_1$, $x_4 = k_2$ and $x_5 = k_3$.

Putting the values of x_3, x_4 and x_5 in (1) and (2), we get

$$x_1 + 3x_2 + 4k_1 - 2k_2 - k_3 = 0 \quad \text{... (5)}$$

$$3x_1 + 7x_2 + 6k_1 + 2k_2 + k_3 = 0 \quad \text{... (6)}$$

On solving (5) and (6), we get

$$x_1 = 5k_1 - 10k_2 - 5k_3$$

and $\quad x_2 = -3k_1 + 4k_2 + 2k_3$

The null space of A consists of the following vectors

$$X = \begin{bmatrix} x_1 \\ x_2 \\ x_3 \\ x_4 \\ x_5 \end{bmatrix} = \begin{bmatrix} 5k_1 - 10k_2 - 5k_3 \\ -3k_1 + 4k_2 + 2k_3 \\ k_1 \\ k_2 \\ k_3 \end{bmatrix} = k_1 \begin{bmatrix} 5 \\ -3 \\ 1 \\ 0 \\ 0 \end{bmatrix} + k_2 \begin{bmatrix} -10 \\ 4 \\ 0 \\ 1 \\ 0 \end{bmatrix} + k_3 \begin{bmatrix} -5 \\ 2 \\ 0 \\ 0 \\ 1 \end{bmatrix} = k_1 u + k_2 v + k_3 w$$

Here u, v and w form a basis of null (A). Hence, nullity (A) = 3, since null space has dimension 3.

Rank (A) = Number of column of A – nullity (A)

$= 5 - 3 = 2$

$$A = \begin{bmatrix} 1 & 3 & 4 & -2 & -1 \\ 0 & -2 & -6 & 8 & 4 \\ 0 & 0 & 0 & 0 & 0 \\ 0 & 0 & 0 & 0 & 0 \end{bmatrix} \begin{matrix} \\ R_2 \to R_2 - 3R_1 \\ R_3 \to R_3 - (R_2 - R_1) \\ R_4 \to R_4 - (R_2 - 2R_1) \end{matrix}$$

$$= \begin{bmatrix} 1 & 3 & 4 & -2 & -1 \\ 0 & 1 & 3 & -4 & -2 \\ 0 & 0 & 0 & 0 & 0 \\ 0 & 0 & 0 & 0 & 0 \end{bmatrix} \quad R_2 \to \frac{1}{-2} R_2$$

A basis for the row space of A is

$$r_1 = [1\ 3\ 4\ -2\ -1], \quad r_2 = [0\ 1\ 3\ -4\ -2]$$

A basis for the column space of A is

$$C_1 = \begin{bmatrix} 1 \\ 3 \\ 2 \\ 1 \end{bmatrix}, \quad C_2 = \begin{bmatrix} 3 \\ 7 \\ 4 \\ 1 \end{bmatrix}$$ **Ans.**

Example 10. *Find a basis for:*

$$\text{Span} \left\{ \begin{bmatrix} 1 & 1 \\ 2 & 1 \end{bmatrix} \begin{bmatrix} 1 & 0 \\ 7 & 4 \end{bmatrix} \begin{bmatrix} 3 & 3 \\ 7 & 5 \end{bmatrix} \begin{bmatrix} 2 & 2 \\ 4 & 2 \end{bmatrix} \begin{bmatrix} -3 & -3 \\ -6 & -3 \end{bmatrix} \right\}$$

Solution. We write matrix $\begin{bmatrix} a & b \\ c & d \end{bmatrix}$ as $\begin{bmatrix} a \\ b \\ c \\ d \end{bmatrix}$ to convert the problem to the column space.

$$\begin{bmatrix} 1 & 1 & 3 & 2 & -3 \\ 1 & 0 & 3 & 2 & -3 \\ 2 & 7 & 7 & 4 & -6 \\ 1 & 4 & 5 & 2 & -3 \end{bmatrix}$$

By column transformation, we get column Echelon from:

$$E = \begin{bmatrix} 1 & 0 & 0 & 0 & 0 \\ 1 & -1 & 0 & 0 & 0 \\ 2 & 5 & 1 & 0 & 0 \\ 1 & 3 & 2 & 0 & 0 \end{bmatrix} \begin{pmatrix} C_2 \to C_2 - C_1 \\ C_3 \to C_3 - 3C_1 \\ C_4 \to C_4 - 2C_1 \\ C_5 \to C_5 + 3C_1 \end{pmatrix}$$

Non zero column of E form a basis of col (A).

$$\begin{bmatrix} 1 & 0 & 0 \\ 1 & -1 & 0 \\ 2 & 5 & 1 \\ 1 & 3 & 2 \end{bmatrix}$$

Corresponding matrices $\begin{bmatrix} 1 & 1 \\ 2 & 1 \end{bmatrix}, \begin{bmatrix} 0 & -1 \\ 5 & 3 \end{bmatrix}$ and $\begin{bmatrix} 0 & 0 \\ 1 & 2 \end{bmatrix}$ are the basis of the span.

Ans.

EXERCISE 35.1

1. Determine a basis for the null space of the matrix $A = \begin{bmatrix} 1 & 1 & 0 \\ 0 & 1 & -1 \\ 1 & 0 & 1 \end{bmatrix}$ **Ans.** $\begin{bmatrix} 1 \\ 1 \\ -1 \end{bmatrix}$

2. Find a basis for the column space of the matrix $A = \begin{bmatrix} 1 & 2 & 3 \\ 2 & 4 & 5 \\ 3 & 5 & 6 \end{bmatrix}$

Ans. $r_1 = \begin{bmatrix} 1 \\ 0 \\ 0 \end{bmatrix}, r_2 = \begin{bmatrix} 0 \\ 1 \\ 0 \end{bmatrix}, r_3 = \begin{bmatrix} 0 \\ 0 \\ 1 \end{bmatrix}$

3. Find the rank and nullity of A and A^T

(*i*) $\{0\}$, (*ii*) $\begin{bmatrix} 1 & 2 & 3 \\ 4 & 5 & 6 \\ 7 & 8 & 9 \end{bmatrix}$, (*iii*) $\begin{bmatrix} 1 & 0 & 1 & 0 \\ 1 & 0 & 0 & 1 \\ 0 & 1 & 1 & 0 \end{bmatrix}$

Ans. (*i*) rank (A) = rank $(A^T) = 0$, nullity (A) = nullity $(A^T) = 1$

(*ii*) rank (A) = rank $(A^T) = 2$, nullity $(A) = 3 - 2 = 1$, nullity $(A^T) = 3 - 2 = 1$

(*iii*) rank (A) = rank $(A^T) = 3$, nullity $(A) = 4 - 3 = 1$, nullity $(A^T) = 3 - 3 = 0$

4. Determine a basis for the null space of the matrix

$$A = \begin{bmatrix} 1 & 1 & 3 & 3 \\ 0 & 2 & 2 & 4 \\ 1 & 0 & 2 & 1 \\ 1 & 1 & 3 & 3 \end{bmatrix}$$

Ans. $\begin{bmatrix} 1 \\ 2 \\ 0 \\ -1 \end{bmatrix}, \begin{bmatrix} 2 \\ 1 \\ -1 \\ 0 \end{bmatrix}$

5. Find basis for the row space and column space of $A = \begin{bmatrix} 1 & 2 & -1 & 4 \\ 2 & -1 & 3 & 3 \\ -1 & 3 & -4 & 1 \\ 0 & 1 & -1 & 1 \end{bmatrix}$

6. Find a basis for the column space of the matrix. $A = \begin{bmatrix} 1 & 2 & 5 & 0 \\ -4 & -6 & -13 & 1 \\ -1 & -2 & -4 & 0 \\ -2 & -3 & -7 & \frac{1}{2} \end{bmatrix}$

7. Find the basis for the column space of the matrix. $A = \begin{bmatrix} 1 & 2 & -1 & 2 \\ 2 & 5 & -3 & 4 \\ -2 & -4 & 2 & -1 \\ 3 & 6 & -2 & 6 \end{bmatrix}$

Ans. $r_1 = \begin{bmatrix} 1 \\ 2 \\ -2 \\ 3 \end{bmatrix}, r_2 = \begin{bmatrix} 0 \\ 1 \\ 0 \\ 0 \end{bmatrix}, r_3 = \begin{bmatrix} 0 \\ 1 \\ 0 \\ -1 \end{bmatrix}, r_4 = \begin{bmatrix} 0 \\ 0 \\ 1 \\ 0 \end{bmatrix}$

8. Find a basis for the null space of $A = \begin{bmatrix} 1 & 3 & 2 & 0 & 1 \\ -1 & -1 & -1 & 1 & 0 \\ 0 & 4 & 2 & 4 & 3 \\ 1 & 3 & 2 & -2 & 0 \end{bmatrix}$

Ans. nullity $(A) = 2$

9. Determine a basis for the row space of $A = \begin{bmatrix} 1 & 3 & -5 & 1 & 5 \\ 1 & 4 & -7 & 3 & -2 \\ 1 & 5 & -9 & 5 & -9 \\ 0 & 3 & -6 & 2 & -1 \end{bmatrix}$

Ans $r_1 = (1\ 3\ -5\ 1\ 5)$, $r_2 = (0\ 1\ -2\ 2\ -7)$, $r_3 = (0\ 0\ 0\ 1\ -5)$

10. Find a basis for the column space of $A = \begin{bmatrix} -1 & 2 & -1 & 5 & 6 \\ 4 & -4 & -4 & -12 & -8 \\ 2 & 0 & -6 & -2 & 4 \\ -3 & 1 & 7 & -2 & 12 \end{bmatrix}$

Ans $C_1 = \begin{bmatrix} -1 \\ 4 \\ 2 \\ -3 \end{bmatrix}, C_2 = \begin{bmatrix} 2 \\ -4 \\ 0 \\ 1 \end{bmatrix}, C_3 = \begin{bmatrix} 5 \\ -12 \\ -2 \\ -2 \end{bmatrix}$

REAL INNER PRODUCT SPACES

36.1 INTRODUCTION

In this chapter are will discuss about the idea of inner products in a real vector space. It is extremily fruitful as for as the applications of the theory to problems are concerned.

36.2 INNER PRODUCT SPACES

A vector space together with an inner product defined on it is called inner product space. We know the scalar product of vectors $\vec{a}$ and $\vec{b}$ in R^n we can define an inner product of two-column vectors a and b as $(a, b) = a^T.b$.

This definition can be extended to general real vector spaces by taking basic property of $\vec{a} \cdot \vec{b}$.

Let X and Y be two column vectors over the real field R,

$$X = \begin{pmatrix} x_1 \\ x_2 \\ \vdots \\ x_n \end{pmatrix} \text{ and } Y = \begin{pmatrix} y_1 \\ y_2 \\ \vdots \\ y_n \end{pmatrix}$$

$$(X, Y) = X^T Y = [x_1, x_2, ..., x_n] \begin{pmatrix} y_1 \\ y_2 \\ \vdots \\ y_n \end{pmatrix}$$

$$= x_1 y_1 + x_2 y_2 + \text{.............} + x_n y_n.$$

This inner product reduces to the ordinary dot product in R^2 and R^3. It is called the *standard inner product* for R^n.

Definition

A real vector space V is called a real inner product space if it has the following properties:

1. Symmetry. $(X, Y) = (Y, X)$

2. Additivity. $(X + Y, Z) = (X, Z) + (Y, Z)$

3. Linearity. $C(X, Y) = (C X, Y) = (X, CY)$

4. Positivity. $(X, X) \geq 0$ and $(X, X) = 0$ if and only if $X = 0$.

The length or norm of a vector X in V is defined by $\| X \| = \sqrt{(X, X)}$

A vector of norm 1 is called a *unit vector*.

Example 1. *Let* $u = \begin{pmatrix} 2 \\ 3 \\ 4 \end{pmatrix}$ *and* $v = \begin{pmatrix} 5 \\ -1 \\ 1 \end{pmatrix}$, *find their inner product and length of each.*

Solution. $(u, v) = u^T v$

$$= [2,\ 3,\ 4] \begin{bmatrix} 5 \\ -1 \\ 1 \end{bmatrix} = 2.5 + 3.(-1) + 4.1 = 10 - 3 + 4 = 11$$

$$\| u \|^2 = (u, u) = u^T u = [2\ 3\ 4] \begin{bmatrix} 2 \\ 3 \\ 4 \end{bmatrix} = 2.2 + 3.3 + 4.4 = 4 + 9 + 16 = 29$$

$$\| u \| = \sqrt{29}$$

Also, $\| v \|^2 = (v, v) = v^T v = [\,5\ -1\ 1\,] \begin{bmatrix} 5 \\ -1 \\ 1 \end{bmatrix} = 5.5 + (-1).(-1) + 1.\ 1 = 25 + 1 + 1 = 27$

$\Rightarrow \quad \| v \| = \sqrt{27} = 3\sqrt{3}$ **Ans.**

Example 2. *Prove that* $(X, Y) = (Y, X)$

Solution. If $X = \begin{bmatrix} x_1 \\ x_2 \\ \vdots \\ x_n \end{bmatrix}$ and $Y = \begin{bmatrix} y_1 \\ y_2 \\ \vdots \\ y_n \end{bmatrix}$

$$(X, Y) = X^T Y = [x_1\ x_2, \ldots\ldots, x_n] \begin{bmatrix} y_1 \\ y_2 \\ \vdots \\ y_n \end{bmatrix} = x_1 y_1 + x_2 y_2 + \ldots + x_n y_n \qquad \ldots (1)$$

$$(Y, X) = Y^T X = [y_1, y_2, \ldots\ldots, y_n] \begin{bmatrix} x_1 \\ x_2 \\ \vdots \\ x_n \end{bmatrix}$$

$$= y_1 x_1 + y_2 x_2 + \ldots. + y_n x_n = x_1 y_1 + x_2 y_2 + \ldots + x_n y_n \qquad \ldots (2)$$

From (1) and (2), we have $(X, Y) = (Y, X)$ **Proved.**

Example 3. *Prove that*

$$(X+Y, Z) = (X, Z) + (Y, Z)$$

Solution. Let $X = \begin{bmatrix} x_1 \\ x_2 \\ \vdots \\ x_3 \end{bmatrix}, Y = \begin{bmatrix} y_1 \\ y_2 \\ \vdots \\ y_n \end{bmatrix}$ and $Z = \begin{bmatrix} z_1 \\ z_2 \\ \vdots \\ z_n \end{bmatrix}$

$$(X+Y, Z) = (X+Y)^T Z = [x_1 + y_1, x_2 + y_2, \ldots\ldots, x_n + y_n] \begin{bmatrix} z_1 \\ z_2 \\ \vdots \\ z_n \end{bmatrix}$$

$$= (x_1 + y_1) z_1 + (x_2 + y_2) z_2 + \ldots\ldots + (x_n + y_n) z_n$$

$$= x_1 z_1 + y_1 z_1 + x_2 z_2 + y_2 z_2 + \ldots\ldots\ldots + x_n z_n + y_n z_n$$

$$= (x_1 z_1 + x_2 z_2 + \ldots\ldots x_n z_n) + (y_1 z_1 + y_2 z_2 + \ldots\ldots\ldots + y_n z_n)$$

$$= (x_1\ x_2 \ldots\ldots x_n) \begin{bmatrix} z_1 \\ z_2 \\ \vdots \\ z_3 \end{bmatrix} + [y_1\ y_2 \ldots\ldots y_n] \begin{bmatrix} z_1 \\ z_2 \\ \vdots \\ z_n \end{bmatrix}$$

$$= X^T Z + Y^T Z = (X, Z) + (Y, Z)$$ **Proved.**

Example 4. *Prove that:* $c(X, Y) = (cX, Y) = (X, cY)$

Solution. Let $X = \begin{bmatrix} x_1 \\ x_2 \\ \vdots \\ x_n \end{bmatrix}$ and $Y = \begin{bmatrix} y_1 \\ y_2 \\ \vdots \\ y_n \end{bmatrix}$

$$\therefore \quad c(X, Y) = cX^T Y = c\,[x_1, x_2, \ldots\ldots, x_n] \begin{bmatrix} y_1 \\ y_2 \\ \vdots \\ y_n \end{bmatrix}$$

$$= c(x_1 y_1 + x_2 y_2 + \ldots\ldots + x_n y_n) = c x_1\ y_1 + c x_2 y_2 + \ldots\ldots + c x_n\ y_n \quad \ldots (1)$$

$$= (c x_1,\ c x_2, \ldots\ldots\ldots\ldots\ldots\ldots, c x_n) \begin{bmatrix} y_1 \\ y_2 \\ \vdots \\ y_n \end{bmatrix} = c\, X^T Y \quad \ldots (2)$$

$$= (c\,X, Y)$$ **Proved.**

Again $c\,(X, Y) = cx_1 y_1 + c\,x_2 y_2 + \ldots\ldots + c\,x_n y_n$ [From (1)]

$$= x_1 c y_1 + x_2 c y_2 + \ldots\ldots\ldots\ldots\ldots\ldots\ldots\ldots + x_n\, c y_n$$

$$= [x_1, x_2, \ldots.. x_n] \begin{bmatrix} cy_1 \\ cy_2 \\ \vdots \\ cy_n \end{bmatrix}$$

$$= X^T cY \qquad \ldots (3)$$

$$= (X, cY)$$ **Proved.**

Hence, $c\,(X, Y) = (cX, Y) = (X, cY)$ [From (2) and (3)]

Example 5. *Prove that*

$(X, X) \geq 0$ *and* $(X, X) = 0$, *if and only if* $X = 0$

Solution. Let $X = \begin{bmatrix} x_1 \\ x_2 \\ \vdots \\ x_n \end{bmatrix}$

$$(X, X) = X^T X = [x_1, x_2, \ldots\ldots\ldots. x_n] \begin{bmatrix} x_1 \\ x_2 \\ \vdots \\ x_n \end{bmatrix}$$

$$= x_1 x_1 + x_2 x_2 + \ldots\ldots\ldots\ldots + x_n\, x_n = x_1^2 + x_2^2 + \ldots.. + x_n^2 \qquad \ldots (1)$$

≥ 0 $\qquad (\because x^2$ is always positive) **Proved.**

Again, from (1), we have

$$(X, X) = x_1^2 + x_2^2 + \ldots. + x_n^2 \qquad \ldots (2)$$

But $(X, X) = 0$ (given) ... (3)

From (2) and (3), we have

$$x_1^2 + x_2^2 + \ldots.. + x_n^2 = 0 \qquad \ldots (4)$$

Equation (4) is only possible if

$$x_1 = x_2 = \ldots.. = x_n = 0$$

$\Rightarrow \quad X = 0$ **Proved.**

36.3 ORTHOGONAL VECTORS (PERPENDICULAR VECTORS)

According to Cauchy Schwarz inequality

$$|(X, Y)| \leq \|X\|\ \|Y\|$$

Scalar product of two vectors

$$(X, Y) = \|X\|\ \|Y\| \cos\theta$$

Where θ is the angle between two vectors X and Y

If $\cos\theta = 0$, then $\qquad \left[\cos\theta = 0 \Rightarrow \theta = \frac{\pi}{2}\right]$

$$(X, Y) = 0$$

These two vectors X and Y are known as orthogonal vectors.

36.4 ORTHONORMAL VECTORS

Two vectors X and Y are said to be orthonormal if

(*i*) inner product of these vectors is zero, and

(*ii*) the norm of each vector X and Y is 1.

We know some properties involving orthogonality

(*i*) $0 \perp X, \quad \forall X \in V$

(*ii*) $X \perp X$, iff $X = 0$, where $X \in V$

(*iii*) $X \perp Y \Rightarrow Y \perp X$, for $X \in V$

(*iv*) $X \perp Y \Rightarrow \alpha X \perp Y$, for any $\alpha \in F$, where $X, Y \in V$.

Note: Every orthonormal set is orthogonal. But the converse is not true.

Example 6. *Let* $X_1 = (1, 2, -1)$, $X_2 = (2, 1, 4)$ and $X_3 = (3, -2, -1)$ in R^3 then.

(*i*) Show that they form an orthogonal set under the standard Euclidean inner product for R^3 but not an orthonormal set.

(*ii*) Convert them into a set of vectors that will form an orthonormal set of vectors under the standard Euclidean inner product for R^3.

Solution.

(*i*) $$(X_1, X_2) = (1, 2, -1)(2, 1, 4)^T = (1, 2, -1)\begin{bmatrix} 2 \\ 1 \\ 4 \end{bmatrix} = 2 + 2 - 4 = 0$$

$$(X_1, X_3) = (1, 2, -1)(3, -2, -1)^T = (1, 2, -1)\begin{bmatrix} 3 \\ -2 \\ -1 \end{bmatrix} = 3 - 4 + 1 = 0$$

$$(X_2, X_3) = (2, 1, 4)(3, -2, -1)^T = (2, 1, 4)\begin{bmatrix} 3 \\ -2 \\ -1 \end{bmatrix} = 6 - 2 - 4 = 0$$

Hence X_1, X_2, X_3 form an orthogonal set.

$$\| X_1 \| = \sqrt{1+4+1} = \sqrt{6} \neq 1$$

Thus they are not orthonormal.

(*ii*) $$\frac{X_1}{\| X_1 \|} = \frac{1}{\sqrt{1+4+1}}(1, 2, -1) = \left(\frac{1}{\sqrt{6}}, \frac{2}{\sqrt{6}}, \frac{-1}{\sqrt{6}}\right)$$

$$\frac{X_2}{\| X_2 \|} = \frac{1}{\sqrt{4+1+16}}(2, 1, 4) = \left(\frac{2}{\sqrt{21}}, \frac{1}{\sqrt{21}}, \frac{4}{\sqrt{21}}\right)$$

$$\frac{X_3}{\| X_3 \|} = \frac{1}{\sqrt{9+4+1}}(3, -2, -1) = \left(\frac{3}{\sqrt{14}}, \frac{-2}{\sqrt{14}}, \frac{-1}{\sqrt{14}}\right)$$

Therefore the orthonormal set is

$$\left(\frac{1}{\sqrt{6}}, \frac{2}{\sqrt{6}}, \frac{-1}{\sqrt{6}}\right), \left(\frac{2}{\sqrt{21}}, \frac{1}{\sqrt{21}}, \frac{4}{\sqrt{21}}\right) \text{ and } \left(\frac{3}{\sqrt{14}}, \frac{-2}{\sqrt{14}}, \frac{-1}{\sqrt{14}}\right).$$

36.5 GRAM-SCHMIDT ORTHOGONALISATION PROCESS

If we are given a set of linearly independent vectors

$X_1, X_2, \ldots X_n$, we can construct an orthogonal set of vectors from them as follows:

Let $\quad Y_1 = X_1$

$$Y_2 = X_2 - k_1 Y_1 \qquad \ldots (1)$$

Here $\quad Y_1, Y_2$ are orthogonal

$$\Rightarrow \quad (Y_1, Y_2) = 0 \qquad \ldots (2)$$

Putting the value of Y_2 in (2) from (1), we get

$$\Rightarrow \quad (Y_1, X_2 - k_1 Y_1) = 0 \quad \Rightarrow \quad (Y_1, X_2) - k_1(Y_1, Y_1) = 0$$

$$\Rightarrow \quad k_1 = \frac{(Y_1, X_2)}{(Y_1, Y_1)} \qquad \ldots (3)$$

Putting the value of k_1 from (3) in (1), we get

$$Y_2 = X_2 - \frac{(Y_1, X_2)}{(Y_1, Y_1)} Y_1$$

Again we take $Y_3 = X_3 - k_2 Y_1 - k_3 Y_2 \qquad \ldots (4)$

Since Y_3 is orthogonal to Y_1 and Y_2. Therefore $(Y_1, Y_3) = 0$

$$\Rightarrow \quad (Y_1, X_3 - k_2 Y_1 - k_3 Y_2) = 0 \qquad \text{[Using (4)]}$$

$$\Rightarrow \quad (Y_1, X_3) - k_2(Y_1, Y_1) - k_3(Y_1, Y_2) = 0 \qquad [\because (Y_1, Y_2) = 0]$$

$$(Y_1, X_3) - k_2 (Y_1, Y_1) = 0$$

$$\Rightarrow \quad k_2 = \frac{(Y_1, X_3)}{(Y_1, Y_1)} \quad \Rightarrow \quad k_2 = \frac{(Y_1, X_3)}{\| Y_1 \|^2} \qquad \ldots (5)$$

Again $\quad (Y_2, Y_3) = 0$

$$\Rightarrow \quad (Y_2, X_3 - k_2 Y_1 - k_3 Y_2) = 0 \qquad \text{[Using(4)]}$$

$$\Rightarrow \quad (Y_2, X_3) - k_2(Y_2, Y_1) - k_3(Y_2, Y_2) = 0$$

$$\Rightarrow \quad (Y_2, X_3) - k_3(Y_2, Y_2) = 0 \qquad [\because (Y_1, Y_2) = 0]$$

$$\Rightarrow \quad k_3 = \frac{(Y_2, X_3)}{(Y_2, Y_2)} \quad \Rightarrow \quad k_3 = \frac{(Y_2, X_3)}{\| Y_2 \|^2} \qquad \ldots (6)$$

Putting the values of k_2 and k_3 from (5) and (6) in (4), we get

$$Y_3 = X_3 - \frac{(Y_1, X_3)}{\| Y_1 \|^2} Y_1 - \frac{(Y_2, X_3)}{\| Y_2 \|^2} Y_2$$

$$\boxed{\begin{aligned} \boldsymbol{Y_1} &= \boldsymbol{X_1} \\ \boldsymbol{Y_2} &= \boldsymbol{X_2} - \frac{(\boldsymbol{X_2}, \boldsymbol{Y_1})}{\| \boldsymbol{Y_1} \|^2} \boldsymbol{Y_1} \\ \boldsymbol{Y_3} &= \boldsymbol{X_3} - \frac{(\boldsymbol{X_3}, \boldsymbol{Y_1})}{\| \boldsymbol{Y_1} \|^2} \boldsymbol{Y_1} - \frac{(\boldsymbol{X_3}, \boldsymbol{Y_2})}{\| \boldsymbol{Y_2} \|^2} \boldsymbol{Y_2} \end{aligned}}$$

Example 7. *Let R^3 have the Euclidian inner product. Use the Gram Schmidt process to transform the basis vector $u_1 = (1, 1, 1)$, $u_2 = (-1, 1, 0)$ and $u_3 = (1, 2, 1)$ into an orthogonal basis $\{v_1, v_2, v_3\}$.* *(Gujarat, II Semester, June 2009)*

Solution. Let $v_1 = u_1 = (1, 1, 1)$

Then the condition of orthogonality $v_1 \cdot u_2$ gives

$$\frac{v_1 \cdot u_2}{v_1 \cdot v_1} = \frac{(1,1,1).(-1,1,0)}{(1,1,1)(1,1,1)} = \frac{(1)(-1)+(1).(1)+(1).(0)}{1+1+1} = \frac{-1+1+0}{1+1+1} = 0$$

$$v_2 = u_2 - \frac{(u_2, v_1)}{(v_1, v_1)} v_1 = (-1,1,0) - \frac{(-1,1,0).(1,1,1)}{1^2+1^2+1^2}(1,1,1)$$

$$= (-1, 1, 0) - 0.\,(1, 1, 1) = (-1, 1, 0)$$

$$v_3 = u_3 - \frac{(u_3, v_1)}{\| v_1 \|^2} v_1 - \frac{(u_3, v_2)}{\| v_2 \|^2} v_2$$

$$= (1,2,1) - \frac{(1,2,1).(1,1,1)}{(1^2+1^2+1^2)}.(1,1,1) - \frac{(1,2,1)(-1,1,0)}{(-1)^2+1^2+0}(-1,1,0)$$

$$= (1, 2, 1) - \frac{1+2+1}{3}(1,1,1) - \frac{-1+2+0}{2}(-1, 1, 0)$$

$$= (1, 2, 1) - \frac{4}{3}(1, 1, 1) - \frac{1}{2}(-1, 1, 0)$$

$$= \left(1 - \frac{4}{3} + \frac{1}{2},\; 2 - \frac{4}{3} - \frac{1}{2},\; 1 - \frac{4}{3} - 0\right) = \left(\frac{1}{6}, \frac{1}{6}, -\frac{1}{3}\right) \quad \textbf{Ans.}$$

Example 8. *Construct an orthonormal set of vectors from the set*

$$X_1 = (1, 2, 1),\ X_2 = (2, 1, 4),\ and\ X_3 = (4, 5, 6)$$

Solution. Let $Y_1 = X_1 = (1, 2, 1)$ and $Y_2 = X_2 - k_1 Y_1$

then the condition of orthogonality Y_1, Y_2 gives

$$\left[\frac{(Y_1, X_2)}{(Y_1, Y_1)} = \frac{1.2+2.1+1.4}{1.1+2.2+1.1} = \frac{2+2+4}{1+4+1} = \frac{8}{6} = \frac{4}{3}\right]$$

$$\therefore \quad Y_2 = X_2 - \frac{(X_2, Y_1)}{(Y_1, Y_1)} Y_1 = (2,1,4) - \frac{4}{3}(1,2,1) \qquad \text{(Art. 36.5)}$$

$$= (2,1,4) - \left(\frac{4}{3}, \frac{8}{3}, \frac{4}{3}\right) = \left(\frac{2}{3}, \frac{-5}{3}, \frac{8}{3}\right) = \frac{1}{3}(2, -5, 8)$$

Y_3 can be calculated either by Art. 36.5 or by the following method.

Let $Y_3 = (a, b, c)$

Since, it is orthogonal to Y_1 and Y_2, therefore

$$Y_3.Y_1 = 0 \quad \Rightarrow (a, b, c).(1, 2, 1) = 0 \Rightarrow a + 2b + c = 0 \qquad \text{... (1)}$$

$$Y_3.Y_2 = 0 \quad \Rightarrow (a, b, c).\frac{1}{3}(2, -5, 8) = 0 \Rightarrow 2a - 5b + 8c = 0 \qquad \text{... (2)}$$

Solving (1) and (2) by cross multiplication method, we get

$$\frac{a}{16+5} = \frac{-b}{8-2} = \frac{c}{-5-4} \quad \Rightarrow \quad \frac{a}{7} = \frac{-b}{2} = \frac{c}{-3}$$

So, we take $Y_3 = (7, -2, -3)$

Now, $$\|Y_1\| = \sqrt{1^2 + 2^2 + 1^2} = \sqrt{6}$$

$$\|Y_2\| = \frac{1}{3}\sqrt{2^2 + 5^2 + 8^2} = \frac{1}{3}\sqrt{93}$$

$$\|Y_3\| = \sqrt{7^2 + 2^2 + 3^2} = \sqrt{62}$$

Therefore the orthonormal set is

$$\frac{Y_1}{\|Y_1\|}, \frac{Y_2}{\|Y_2\|}, \frac{Y_3}{\|Y_3\|} = \frac{(1, 2, 1)}{\sqrt{6}}, \frac{\frac{1}{3}(2, -5, 8)}{\frac{1}{3}\sqrt{93}}, \frac{(7, -2, -3)}{\sqrt{62}}$$

$$\left(\frac{1}{\sqrt{6}}, \frac{2}{\sqrt{6}}, \frac{1}{\sqrt{6}}\right), \left(\frac{2}{\sqrt{93}}, \frac{-5}{\sqrt{93}}, \frac{8}{\sqrt{93}}\right), \left(\frac{7}{\sqrt{62}}, \frac{-2}{\sqrt{62}}, \frac{-3}{\sqrt{62}}\right)$$ **Ans.**

Example 9. *Show that the vectors* $\begin{bmatrix} 2 \\ 4 \\ 3 \end{bmatrix}$ *and* $\begin{bmatrix} 5 \\ -4 \\ 2 \end{bmatrix}$ *are orthogonal.*

Solution. Let the given vector be X and Y respectively then $(X, Y) = X^T Y$

$$= [\,2,\ 4,\ 3\,]\begin{bmatrix} 5 \\ -4 \\ 2 \end{bmatrix} = 10 - 16 + 6 = 0$$

Hence, X and Y are orthogonal vectors. **Proved.**

36.6 SCHWARZ INEQUALITY

Prove that: $|(X, Y)| \le \|X\|\|Y\|$

Proof. If $X = 0$ or $Y = 0$, then $|(X, Y)| = 0 = \|X\|\|Y\|$

So, let us assume that $X \neq 0$ and $Y \neq 0$. Hence, $\|Y\| > 0$

Let $Z = \frac{Y}{\|Y\|}$. Then $Z \in V$, and $\|Z\| = 1$

Now, for any $\alpha \in F$, consider the norm of the vector $X - \alpha Z \in V$

$$\|X - \alpha Z\|^2 = (X - \alpha Z, X - \alpha Z)$$
$$= (X, X) - \alpha(Z, X) - \alpha(X, Z) + \alpha\alpha(Z, Z)$$
$$= \|X\|^2 - \alpha(X, Z) - \alpha(X, Z) + \alpha\alpha \qquad [\because (Z, Z) = 1]$$

Adding and subtracting(X, Z) (X, Z), we have

$$= \|X\|^2 - (X, Z)(X, Z) + (X, Z)(X, Z) - \alpha(X, Z) - \alpha(X, Z) + \alpha\alpha$$
$$= \|X\|^2 - |(X, Z)|^2 + (X, Z)\{(X, Z) - \alpha\} - \alpha\{(X, Z) - \alpha\}$$
$$= \|X\|^2 - |(X, Z)|^2 + \{(X, Z) - \alpha\}\{(X, Z) - \alpha\}$$
$$= \|X\|^2 - |(X, Z)|^2 + |(X, Z) - \alpha|^2$$

Now, $\|X_1 Z\|^2 \ge 0$ this means that

$$\|X\|^2 - |(X,Z)|^2 + |(X,Z) - \alpha|^2 \geq 0 \qquad \forall\, \alpha \in F$$

In particular, if we choose $\alpha = (X, Z)$, we get

$$0 \leq \|X\|^2 - |(X,Z)|^2$$

Hence, $|(X,Z)| \leq \|X\| \;\Rightarrow\; \left|\left(X, \frac{Y}{\|Y\|}\right)\right| \leq \|X\|$

$$\Leftrightarrow \frac{1}{\|Y\|}|(X,Y)| \leq \|X\| \Leftrightarrow |(X,Y)| \leq \|X\|\,\|Y\|$$ **Proved.**

Example 10. *Obtain an orthogonal basis for P_2, the space of all real polynomials of degree at most 2, the inner product being defined by $(P_1, P_2) = \int_0^1 P_1(t)\, P_2(t)\, dt$.*

Solution. $(1, t, t^2)$ is a basis for P_2.

From this we will obtain an orthogonal basis (w_1, w_2, w_3).

Now, $w_1 = 1$ and $(w_1, w_1) = \int_0^1 dt = [t]_0^1 = 1$

$$w_2 = t - \frac{(t, w_1)}{(w_1, w_1)} w_1 \qquad \text{(Art. 36.5)} \qquad ...(1)$$

Now, $(t, w_1) = \int_0^1 t\, dt = \left[\frac{t^2}{2}\right]_0^1 = \frac{1}{2}$

Putting $(t, w_1) = \frac{1}{2}$ and $(w_1, w_2) = 1$ in (1), we get

$$\Rightarrow \quad w_2 = t - \frac{1}{2}\, w_1 = t - \frac{1}{2}(1) = t - \frac{1}{2}$$

$$\therefore \quad (w_2, w_2) = \int_0^1 \left(t - \frac{1}{2}\right)^2 dt = \left[\frac{1}{3}\left(t - \frac{1}{2}\right)^3\right]_0^1 = \frac{1}{3}\left[\frac{1}{8} + \frac{1}{8}\right] = \frac{1}{12}$$

$$w_3 = t^2 - \frac{(t^2, w_2)}{(w_2, w_2)} w_2 - \frac{(t^2, w_1)}{(w_1, w_1)} w_1 \qquad \text{(Art. 36.5)}$$

$$= t^2 - \frac{(t^2, w_2)}{\frac{1}{12}}\left(t - \frac{1}{2}\right) - \frac{\frac{1}{3}}{1}(1) = t^2 - 12\left\{\frac{1}{12}\left(t - \frac{1}{2}\right)\right\} - \frac{1}{3}$$

$$= t^2 - \left(t - \frac{1}{2}\right) - \frac{1}{3} = t^2 - t + \frac{1}{6}$$

Also, $(w_3, w_3) = \int_0^1 \left(t^2 - t + \frac{1}{6}\right)^2 dt$

$$= \int_0^1 \left(t^4 + t^2 + \frac{1}{36} - 2t^3 + \frac{t^2}{3} - \frac{t}{3}\right) dt$$

$$= \int_0^1 \left(t^4 - 2t^3 + \frac{4t^2}{3} - \frac{t}{3} + \frac{1}{36}\right) dt$$

$$(t^2, w_2) = \int_0^1 t^2\left(t - \frac{1}{2}\right) dt = \int_0^1 \left(t^3 - \frac{t^2}{2}\right) dt = \left(\frac{t^4}{4} - \frac{t^3}{6}\right)_0^1 = \frac{1}{4} - \frac{1}{6} = \frac{1}{12}$$

$$= \left(\frac{t^5}{5} - \frac{t^4}{2} + \frac{4t^3}{9} - \frac{t^2}{6} + \frac{t}{36} \right)_0^1$$

$$= \frac{1}{5} - \frac{1}{2} + \frac{4}{9} - \frac{1}{6} + \frac{1}{36} = \frac{36 - 90 + 80 - 30 + 5}{180} = \frac{1}{180}$$

Thus, the orthogonal basis is $\left\{ \frac{W_1}{\|W_1\|}, \frac{W_2}{\|W_2\|}, \frac{W_3}{\|W_3\|} \right\}$

$$= \left\{ 1, \sqrt{12}\left(t - \frac{1}{2}\right), \sqrt{180}\left(t^2 - t + \frac{1}{6}\right) \right\}$$ **Ans.**

36.7 UNITARY TRANSFORMATION

A linear transformation, $Y = AX$, where A is Unitary (*i.e.*, A is such $A^\theta A = A A^\theta = I_n$), is called a *unitary transformation.*

Theorem 1. The necessary and sufficient candition for a linear transformation $Y = AX$ and $Vn(C)$ to preserve lengths is that A is unitary.

Proof. Let A be unitary, then, $A^\theta A = AA^\theta = I_n$

Now $Y = AX \Rightarrow Y^\theta = (AX)^\theta = X^\theta A^\theta$

$\Rightarrow \quad Y^\theta Y = X^\theta A^\theta (AX) = X^\theta (A^\theta A) X = X^\theta X \qquad [\because A^\theta A = I_n]$

$\Rightarrow \quad \|Y\|^2 = \|X\|^2 \Rightarrow \|Y\| = \|X\|$

Thus whenever A is unitary, then the length of the vectors are preserved.

Conversely, let the lengths be preserved, so that

$$\|X\| = \|Y\|$$

Now, $\quad \|X\|^2 = \|Y\|^2 \quad \Rightarrow \quad X^\theta X = Y^\theta Y$

$\Rightarrow \quad = (AX)^\theta (AX) = X^\theta (A^\theta A) X = I_n$

$\Rightarrow \quad X^\theta (A^\theta A) X - I_n = 0 \Rightarrow A^\theta A - I_n = 0 \Rightarrow A^\theta A = I_n$

Hence, A is unitary. **Proved.**

Theorem 2. Every unitary transformation $Y = AX$ preserves inner products.

Proof. Let $Y = AX$ be a unitary transformation. Then A is unitary and therefore, $A^\theta A = I_n$

Let $Y_1 = AX_1$ and $Y_2 = AX_2$ $[A^\theta A = I_n]$

$\therefore \quad Y_2 Y_1 = Y_2^\theta Y_1 = (AX_2)^\theta (AX_1)$

$= X_2^\theta (A^\theta A) X_1 = X_2^\theta X_1 = X_2 \cdot X_1 \; A^\theta A = I_n$

So, the given linear transformation preserves the inner products. **Proved.**

Theorem 3. If (V, (,)) is an inner product space and $X, Y \in V$, then

$$\|X + Y\| \le [\|X\| + \|Y\|] \qquad \text{(Triangle inequality)}$$

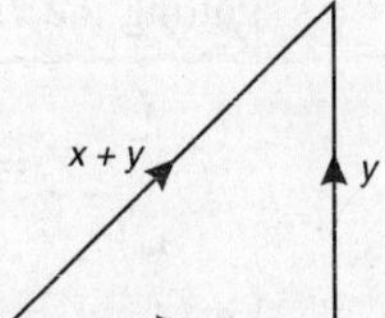

Proof. Now $\|X + Y\|^2 = (X + Y, X + Y)$

$= \|X\|^2 + (X, Y) + (Y, X) + \|Y\|^2$

$$= \|X\|^2 + (X,Y) + (X,Y) + \|Y\|^2$$

$$= \|X\|^2 + 2(X,Y) + \|Y\|^2$$

$$\leq \|X\|^2 + 2|(X,Y)| + \|Y\|^2 \quad [\because |(X,Y)| \geq (X,Y)]$$

$$\leq \|X\|^2 + 2\ \|X\|\|Y\| + \|Y\|^2 \quad \text{(By Sch. ineq.)}$$

$$\leq (\|X\| + \|Y\|)^2$$

Hence $\|X+Y\|^2 \leq (\|X\| + \|Y\|)^2$

Taking square root of both sides, we get

$$\|X+Y\| \leq \|X\| + \|Y\|$$ **Proved.**

Theorem 4. (Parallelogram law)

If X and Y are any two vectors in an inner product space, then

$$\|X+Y\|^2 + \|X-Y\|^2 = 2\left(\|X\|^2 + \|Y\|^2\right)$$

Proof. $\|X+Y\|^2 = (X+Y, X+Y)$

$$= (X, X+Y) + (Y, X+Y) \quad \text{(by linearity)}$$

$$= (X,X) + (X,Y) + (Y,X) + (Y,Y)$$

$$= \|X\|^2 + (X,Y) + (X,Y) + \|Y\|^2 \quad \text{(By symmetry of linear product)}$$

$$= \|X\|^2 + \|Y\|^2 + 2(X,Y) \quad \ldots (1)$$

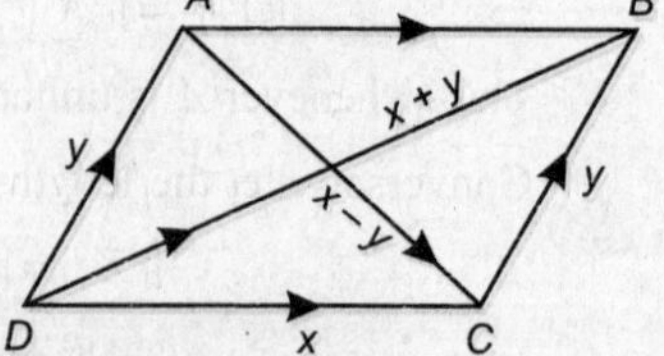

Similarly, $\|X-Y\|^2 = \|X\|^2 + \|Y\|^2 - 2(X,Y)$ ... (2)

Adding (1) and (2), we get

$$\|X+Y\|^2 + \|X-Y\|^2 = 2\left(\|X\|^2 + \|Y\|^2\right)$$ **Proved.**

Example 11. *Orthonormalize the set of linearly independent vectors* X_1 = (1, 0, 1, 1), X_2 = (–1, 0, –1, 1) *and* X_3 = (0, –1, 1, 1) *of* R^4 *with standard inner product.*

Solution. Let $Y_1 = X_1 = (1, 0, 1, 1)$

and $$Y_2 = X_2 - \frac{(X_2, Y_1)}{(Y_1, Y_1)} Y_1 \quad \text{(Art. 36.5)} \quad \ldots (1)$$

Let us calculate the condition of orthogonality $\frac{(X_2, Y_1)}{(Y_1, Y_1)} Y_1$

$$\frac{(Y_1, X_2)}{(Y_1, Y_1)} = \frac{[(1,0,1,1).(-1,0,-1,1)]}{[(1,0,1,1)(1,0,1,1)]} = \frac{1(-1)+0(0)+1(-1)+1(1)}{1\cdot 1+0\cdot 0+1\cdot 1+1\cdot 1}$$

$$= \frac{-1+0-1+1}{1+0+1+1} = \frac{-1}{3}$$

Putting the value of $\frac{(X_2, Y_1)}{(Y_1, Y_1)}$ in (1), we get

$$Y_2 = (-1, 0, -1, 1) + \frac{1}{3}(1, 0, 1, 1) = (-1, 0, -1, 1) + \left(\frac{1}{3}, 0, \frac{1}{3}, \frac{1}{3}\right) = \left(-\frac{2}{3}, 0, -\frac{2}{3}, \frac{4}{3}\right)$$

Let $$Y_3 = X_3 - \frac{(Y_1, X_3)}{\|Y_1\|^2} Y_1 - \frac{(Y_2, X_3)}{\|Y_2\|^2} Y_2$$

$$= (0,-1,1,1) - \frac{((1,0,1,1).(0,-1,1,1))}{((1,0,1,1),(1,0,1,1))}(1,0,1,1)$$

$$- \frac{\left(\left(\frac{-2}{3},0,\frac{-2}{3},\frac{4}{3}\right).(0,-1,1,1)\right)}{\left(\left(\frac{-2}{3},0,\frac{-2}{3},\frac{4}{3}\right).\left(\frac{-2}{3},0,\frac{-2}{3},\frac{4}{3}\right)\right)}\left(\frac{-2}{3},0,\frac{-2}{3},\frac{4}{3}\right)$$

$$= (0,-1,1,1) - \frac{0+0+1+1}{1+0+1+1}(1,0,1,1) - \frac{0+0-\frac{2}{3}+\frac{4}{3}}{\frac{4}{9}+0+\frac{4}{9}+\frac{16}{9}}\left(\frac{-2}{3},0,\frac{-2}{3},\frac{4}{3}\right)$$

$$= (0,-1,1,1) - \frac{2}{3}(1,0,1,1) - \frac{\frac{2}{3}}{\frac{24}{9}}\left(\frac{-2}{3},0,\frac{-2}{3},\frac{4}{3}\right)$$

$$= (0,-1,1,1) - \left(\frac{2}{3},0,\frac{2}{3},\frac{2}{3}\right) - \frac{1}{4}\left(\frac{-2}{3},0,\frac{-2}{3},\frac{4}{3}\right)$$

$$= (0,-1,1,1) + \left(\frac{-2}{3},0,\frac{-2}{3},\frac{-2}{3}\right) + \left(\frac{2}{12},0,\frac{2}{12},\frac{-1}{3}\right) = \left(\frac{-1}{2},-1,\frac{1}{2},0\right)$$

Thus, the required othogonal set is $\left\{(1,0,1,1),\left(\frac{-2}{3},0,\frac{-2}{3},\frac{4}{3}\right),\left(\frac{-1}{2},-1,\frac{1}{2},0\right)\right\}$ and the corresponding orthonormal set is

$$\left\{\frac{Y_1}{\|Y_1\|},\frac{Y_2}{\|Y_2\|},\frac{Y_3}{\|Y_3\|}\right\}$$

$$= \frac{(1,0,1,1)}{\sqrt{(1,0,1,1)\,(1,0,1,1)}},\ \frac{\left(-\frac{2}{3},0,-\frac{2}{3},\frac{4}{3}\right)}{\sqrt{\left(-\frac{2}{3},0,-\frac{2}{3},\frac{4}{3}\right)\left(-\frac{2}{3},0,\frac{-2}{3},\frac{4}{3}\right)}},\ \frac{\left(-\frac{1}{2},-1,\frac{1}{2},0\right)}{\sqrt{\left(-\frac{1}{2},-1,\frac{1}{2},0\right)\left(-\frac{1}{2},-1,\frac{1}{2},0\right)}}$$

$$= \left\{\frac{(1,0,1,1)}{\sqrt{3}},\ \frac{\left(-\frac{2}{3},0,-\frac{2}{3},\frac{4}{3}\right)}{\frac{2\sqrt{6}}{3}},\ \frac{\left(-\frac{1}{2},-1,\frac{1}{2},0\right)}{\sqrt{\frac{3}{2}}}\right\}$$

$$= \left\{\left(\frac{1}{\sqrt{3}},0,\frac{1}{\sqrt{3}},\frac{1}{\sqrt{3}}\right),\left(\frac{-1}{\sqrt{6}},0,\frac{-1}{\sqrt{6}},\frac{2}{\sqrt{6}}\right),\left(\frac{-1}{\sqrt{6}},\frac{-2}{\sqrt{6}},\frac{1}{\sqrt{6}},0\right)\right\}$$ **Ans.**

Example 12. *If $p = p(x) = p_0 + p_1x + p_2x^2$ and $q = q(x) = q_0 + q_1x + q_2x^2$, then the inner product is defined by*

$(p, q) = p_0q_0 + p_1q_1 + p_2q_2$ for the vectors

$$X_1 = 1+2x+3x^2,\ X_2 = 3+5x+5x^2,\ X_3 = 2+x+8x^2$$

Find the orthonormal vectors.

Solution. Let $Y_1 = X_1 = 1 + 2x + 3x^2$

$$Y_2 = X_2 - \frac{(Y_1 . X_2)}{\|Y_1\|^2}Y_1 \qquad \text{(Art. 36.5)}$$

$$= (3+5x+5x^2) - \frac{(1+2x+3x^2 \, . \, 3+5x+5x^2)}{||1+2x+3x^2||^2}(1+2x+3x^2)$$

$$= (3+5x+5x^2) - \frac{28}{14}(1+2x+3x^2)$$

$$\left| \begin{aligned} &(1+2x+3x^2 . 3+5x+5x^2) \\ &= [(1,2,3).(3,5,5)] \\ &= [3+10+15] = 28 \end{aligned} \right.$$

$$= (3+5x+5x^2) - (2+4x+6x^2) = (1+x-x^2)$$

$$\left| \begin{aligned} &(1+2x+3x^2 . 1+2x+3x^2) \\ &= [(1,2,3).(1,2,3)] \\ &= 1+4+9 = 14 \end{aligned} \right.$$

$$Y_3 = X_3 - \frac{(Y_1, X_3)}{||Y_1||^2} Y_1 - \frac{(Y_2, X_3)}{||Y_2||^2} Y_2 \quad \text{(Art. 36.5)}$$

$$Y_3 = (2+x+8x^2) - \frac{(1+2x+3x^2 . 2+x+8x^2)}{||1+2x+3x^2||^2}(1+2x+3x^2) - \frac{[(1+x-x^2).(2+x+8x^2)]}{||1+x-x^2||^2}(1+x-x^2)$$

$$\left| \begin{aligned} &(1+2x+3x^2, 2+x+8x^2) \\ &= [(1,2,3),(2,1,8)] \\ &= 2+2+24 = 28 \end{aligned} \right.$$

$$Y_3 = (2+x+8x^2) - \frac{28}{14}(1+2x+3x^2) - \frac{-5}{3}(1+x-x^2)$$

$$= (2+x+8x^2) - (2+4x+6x^2) + \left(\frac{5}{3} + \frac{5}{3}x - \frac{5}{3}x^2\right)$$

$$\left| \begin{aligned} &(1+x-x^2, 2+x+8x^2) \\ &= [(1,1,-1),(2,1,8)] \\ &= 2+1-8 = -5 \end{aligned} \right.$$

$$= \frac{5}{3} - \frac{4}{3}x + \frac{x^2}{3} = \frac{1}{3}(5-4x+x^2)$$

Thus, the required orthogonal set is

$$\left\{(1+2x+3x^2), (1+x-x^2), \frac{1}{3}(5-4x+x^2)\right\}$$ **Ans.**

And the corresponding orthonormal set is $\left\{\frac{Y_1}{||Y_1||}, \frac{Y_2}{||Y_2||}, \frac{Y_3}{||Y_3||}\right\}$

i.e.; $\left\{\frac{1}{\sqrt{14}}(1+2x+3x^2), \frac{1}{\sqrt{3}}(1+x-x^2), \frac{1}{3\sqrt{42}}(5-4x+x^2)\right\}$ **Ans.**

EXERCISE 36.1

1. Given that $X_1 = (2, -1, 0)$, $X_2 = (1, 0, -1)$ and $X_3 = (3, 7, -1)$ is a basis of R^3 and assuming that we are working with the standard Euclidean inner product, construct an orthonormal basis for R^3.

Ans. $\left\{\left(\frac{2}{\sqrt{5}}, \frac{-1}{\sqrt{5}}, 0\right), \left(\frac{1}{\sqrt{30}}, \frac{2}{\sqrt{30}}, \frac{-5}{\sqrt{30}}\right), \left(\frac{1}{\sqrt{6}}, \frac{2}{\sqrt{6}}, \frac{1}{\sqrt{6}}\right)\right\}$

2. Find an orthonormal basis for $V = \text{span}\ \{(1, -1, 0, 2), (3, 2, 1, 2), (-2, 1, 0, 1)\}$ in R^4 with standard inner product. **Ans.** $\left\{\frac{1}{\sqrt{6}}(1,-1,0,2), \frac{1}{\sqrt{498}}(13,17,6,2), \frac{1}{\sqrt{355572}}(-411,267,21,339)\right\}$

3. Find the value of k so that the following expression forms an inner product:
$(X, Y) = x_1y_1 - 3x_1y_2 - 3x_2y_1 + kx_2y_2$, where $X = (x_1, y_1)$ and $Y = (x_2, y_2)$ in R^2. **Ans.** $k > 9$.

4. Define the following with example:
(*a*) Inner product spaces

(*b*) Norm of a vector

(*c*) Cauchy schwarz inequality

(*d*) Orthogonal and orthonormal set of vectors

(*e*) Finite dimension of inner product space

(*f*) Unitary matrix

5. Prove that the set of vectors {(4, 1, –5), (2, –3, 1), (1, 1, 1)} is an orthogonal set of vectors in R^3 with standard inner product. Is it a basis of R^3. Give reasons.

6. Prove that the set of vectors {(2, –2, –1), (1, 2, –3), (2, 1, 2)} is orthogonal basis of V_3.

7. Obtain an orthonormal basis, with respect to the standard inner product, for the subspace of R^4 generated by (1, 0, 2, 0) and (1, 2, 3, 1). **Ans.** $\left\{\frac{1}{\sqrt{5}}(1,0,2,0), \sqrt{\frac{5}{26}}\left(\frac{-2}{5}, 2, \frac{1}{5}, 1\right)\right\}$

8. Let (*V*, (,)) be an inner product space. Then, for any *x*, *y*, *z* ∈ *V* and α, μ ∈ *R*, prove that

(*i*) $(\alpha x + \mu y, z) = \alpha(x, z) + \mu(y, z)$ (*ii*) $(x - y, z) = (x, z) - (y, z)$

(*iii*) $(x, z) = (y, z)\ \forall\ z \in V \Rightarrow x = y$

9. Show that $\|x\| - \|y\| \le \|x - y\|$ for *x*, *y* ∈ (V, (,)).

[**Hint:** Use the triangle inequality for *y* and (*x* – *y*)]

10. Obtain a vector *V* = (*x*, *y*, *z*) ∈ R^3 so that *V* is perpendicular to (1, 0, 0) as well as (–1, 2, 0), with respect to the standard inner product. **Ans** (0, 0, *Z*), *Z* ∈ *R*.

11. Consider the standard basis *B* = {e_1, ... , e_n}of R^n. Show that the set *R* = {$2e_1$, $2e_2$, ..., $2e_n$} is orthogonal but not orthonormal, with respect to the standard inner product.

12. If $p(x) = p_0 + p_1x + p_2x^2$ and $q(x) = q_0 + q_1x + q_2x^2$ and its inner product is

$(p, q) = p_0\,q_0 + p_1\,q_1 + p_2\,q_2$

for the basis $X_1 = 3 + 4x + 5x^2$, $X_2 = 9 + 12x + 5x^2$, $X_3 = 1 - 7x + 25x^2$.

Find the orthonormal basis: **Ans.** $\left(\frac{3}{\sqrt{50}} + \frac{4}{\sqrt{50}}x + \frac{5}{\sqrt{50}}x^2, \frac{3}{\sqrt{50}} + \frac{4}{\sqrt{50}}x - \frac{5}{\sqrt{50}}x^2, \frac{4}{5} - \frac{3}{5}x + 0x^2\right)$

36.8 ORTHOGONAL TRANSFORMATION

A transformation *Y* = *AX* is said to be orthogonal if its matrix is orthogonal.

Theorem. A linear transformation preserves lenghts, iff it preserves inner product.

Proof. Let the linear transformation is

$$Y = AX$$

Let X_1 and X_2 be any two vectors whose images are Y_1 and Y_2 respectively.

i.e.; $Y_1 = AX_1$

$Y_2 = AX_2$

Now, $\|X_1 + X_2\|^2 = \|X_1\|^2 + \|X_2\|^2 + 2X_1 . X_2$

$$\Rightarrow \quad X_1 . X_2 = \frac{1}{2}\left[\|X_1 + X_2\|^2 - \|X_1\|^2 - \|X_2\|^2\right]$$

Similarly, $Y_1 \cdot Y_2 = \frac{1}{2}\left[\|Y_1 + Y_2\|^2 - \|Y_1\|^2 - \|Y_2\|^2\right]$

From the above results, we can say that the given linear transformation preserves length if and only if it preserves the inner product.

Theorem 5. A linear transformation preserves lengths if and only if its matrix is orthogonal.

Proof. Let X_1 and X_2 be two vectors and Y_1, Y_2 be their images respectively

$$Y_1 = AX_1 \qquad (Y = AX \text{ given})$$

$$Y_2 = AX_2$$

$$(Y_1, Y_2) = Y_1^T\, Y_2 = (AX_1)^T\,(AX_2)$$

$$= X_1^T\, A^T\, AX_2$$

$$= X_1^{\ T} I X_2$$

$$= X_1^T X_2 \qquad (A^T A = 1 \text{ since } A \text{ is orthogonal})$$

$$= (X_1, X_2)$$

Thus, the linear transformation preserves the inner product and therefore it preserves length.

Conversely: If a linear transformation preserves the inner product if and only if it preserves length.

36.9 ORTHOGONAL PROJECTIONS

Let $\vec{a}$ and $\vec{b}$ be two vectors, then

$$a\,.\,b = (OA)(OB)(\cos\theta)$$

$$= (OA)(OB)\left(\frac{OC}{OA}\right) \quad \text{along } b = (OB)\,(OC)$$

$$\Rightarrow \quad \bar{a}\cdot\bar{b} = (\text{Length of } \bar{b}\,)\ (\text{Projection of } \bar{a} \text{ on } \bar{b}\,)$$

Projection of $\bar{a}$ along $\bar{b}$ $= \dfrac{\bar{a}\cdot\bar{b}}{|\,\bar{b}\,|}\hat{b} = \dfrac{\bar{a}\cdot\bar{b}}{|\,b\,|^2}\vec{b}$

Let a vector be Y_1 in R^n.

Let us decompose a vector X_2 in R^n as the sum of two vectors one $X_{2,P}$ is multiple of Y_1 and the other is Y_2 perpendicular to Y_1.

$$X_2 = X_{2,p} + Y_2$$

$$X_2 = \alpha Y_1 + Y_2 \qquad OA = \alpha Y_1$$

$$Y_2 = X_2 - \alpha Y_1 \qquad \text{...(1)}$$

Y_2 is orthogonal to Y_1.

$$\Rightarrow \quad Y_2\,.\,Y_1 = 0$$

$$\Rightarrow \quad (X_2 - \alpha Y_1)\,.\,Y_1 = 0 \quad [\text{From (1)}]$$

$$\Rightarrow \quad X_2 Y_1 - \alpha Y_1 . Y_1 = 0\text{d}$$

$$\Rightarrow \quad \alpha = \frac{X_2\, Y_1}{Y_1\,.\,Y_1} \qquad \text{... (2)}$$

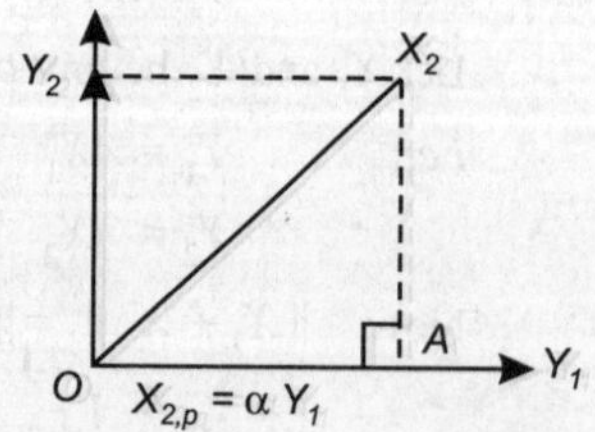

But $\quad X_{2,p} = \alpha Y_1 \qquad$ ... (3)

Putting the value of α from (2) in (3), we get

$$X_{2,p} = \frac{X_2 \cdot Y_1}{Y_1 \cdot Y_1} Y_1$$

The vector $X_{2,p}$ is called the orthogonal projection of X_2 on Y_1. And the vector Y_2 is called the other component of X_2 and perpendicular to Y_1.

36.10 DECOMPOSITION OF VECTOR INTO THE SUM OF TWO PROJECTIONS

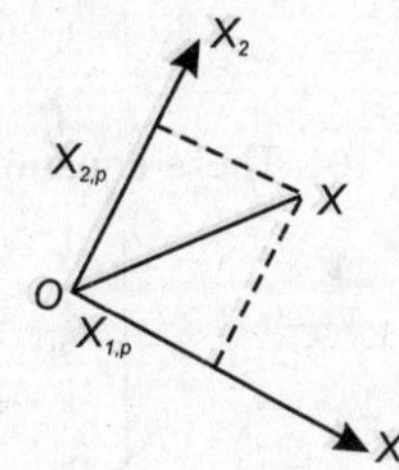

If $Z = R^n = \text{span}(X_2, X_1)$

X_1 and X_2 are orthogonal

$X = X_{1,p} + X_{2,p} =$ Projection of X on X_1 + Projection of X on X_2

$$X = \frac{X.X_1}{X_1.X_1} X_1 + \frac{X.X_2}{X_2.X_2} X_2 \qquad \text{(Where } X \text{ is any vector)}$$

36.11 ORTHOGONAL COMPONENT

Let Z be a subspace of R^n. The set of all vectors that are orthogonal to the vector X is the orthogonal component of Z and is denoted by Z^1.

Working Rule to find orthogonal components

Step 1. Find a matrix A having as row vectors generating set for Z.

Step 2. Find the null space of A by solving $AX = 0$

Example 13. *Find the orthogonal projection of X on*

$X = (4, -2)$ and $Y = (3, 5)$

Solution. Projection of X on $Y = \frac{X \cdot Y}{\|Y\|^2} \vec{Y}$

$$= \frac{(4,-2).(3,5)}{(3^2+5^2)}(3,5) = \frac{12-10}{34}(3,5) = \frac{1}{17}(3,5) = \left(\frac{3}{17}, \frac{5}{17}\right) \qquad \textbf{Ans.}$$

Example 14. *Find the orthogonal projection of X on Y_1 and Y_2 where*

$X = (2, 3)$

$Y_1 = (1, 2)$

$Y_2 = (4, 5)$.

Solution $X = \frac{X \cdot Y_1}{Y_1 . Y_1} Y_1 + \frac{X \cdot Y_2}{Y_2 . Y_2} Y_2 = \frac{(2,3).(1,2)}{(1^2+2^2)}(1,2) + \frac{(2,3).(4,5)}{(4^2+5^2)}(4,5)$

$$= \frac{2+6}{5}(1,2) + \frac{8+15}{41}(4,5) = \frac{8}{5}(1,2) + \frac{23}{41}(4,5)$$

$$= \left(\frac{8}{5}, \frac{16}{5}\right) + \left(\frac{92}{41}, \frac{115}{41}\right) = \left(\frac{788}{205}, \frac{1231}{205}\right) \qquad \textbf{Ans.}$$

36.12 LEAST SQUARE APPROXIMATION

Earlier we have solved consistent equations. But here we are going to find out the best approximate solution of the given inconsistent system of the equations.

Example 15. *Find the equation of a line that passes through the following points*

(2, 12), (–3, –20), (4, 20) and (–4, 10).

Solution Let the equation of the line passing through the given four points be

$$y = a + bx \qquad \dots (1)$$

Let the points A (2, 12), B (–3, –20), C (4, 20) and D (–4, –10) be on the line whose equation is (1).

$$a + 2b = 12$$
$$a - 3b = -20$$
$$a + 4b = 20$$
$$a - 4b = -10$$

These equations are written in the matrix form as

$$\begin{bmatrix} 1 & 2 \\ 1 & -3 \\ 1 & 4 \\ 1 & -4 \end{bmatrix} \begin{bmatrix} a \\ b \end{bmatrix} = \begin{bmatrix} 12 \\ -20 \\ 20 \\ -10 \end{bmatrix}$$

$$AX = B$$

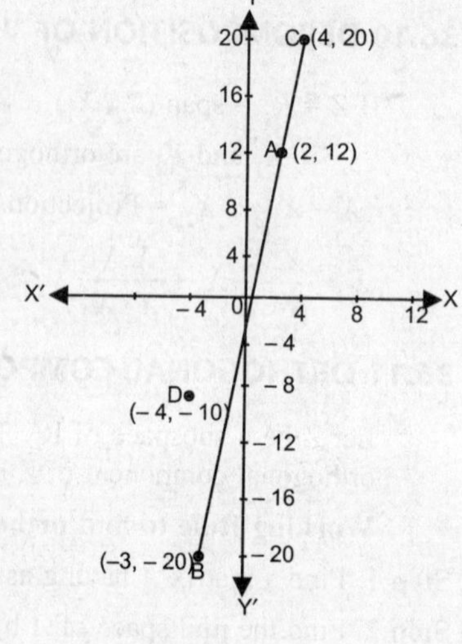

System of these equations is inconsistent and has no solution. In this case the best thing what we can do is to find a that makes AB as close as possible to B.

36.13 APPROXIMATION THEOREM

Let Z be a finite dimentional sub-space of inner product space V and X is any vector in V. The best approximation to X from Z is $\text{Proj}_z X$.

36.14 SOLUTION OF THE LEAST SQUARE PROBLEM

(i) Method

Let X_{ap} be the best approximate solution of the equation $AX = b$ ... (1)

Step 1. We find $\text{Proj}_{\text{col}(A)}\, b$.

Step 2. By the above theorem we know that $AX_{ap} = \text{Proj}_{\text{col}(A)} b$ ... (2)

Step 3. We solve equation (2) for X_{ap}.

***(ii)* Method:** The other method to find Least Square Solution.

Suppose the given system of equation is $AX = b$, whose least square solution we have to find.

Let the least square solution of $AX = b$ be X_{ap}.

Premultiplying $AX = b$ by A^T, we get

$$A^T AX = A^T b$$

$$\Rightarrow \quad X_{ap} = \left(A^T A\right)^{-1} A^T b.$$

($A^T A$ is invertible since columns of A, $m \times n$ matrix are linearly independent.)

Example 16. *Find the best approximate solution of the following inconsistent equations:*

$$2x_1 + x_2 = 4$$
$$3x_1 + 2x_2 = 7$$
$$x_1 - x_2 = 3$$

by least square method.

Solution We have,

$$2x_1 + x_2 = 4$$

$$3x_1 + 2x_2 = 7$$

$$x_1 - x_2 = 3$$

Equations in matrix form are

$$\begin{bmatrix} 2 & 1 \\ 3 & 2 \\ 1 & -1 \end{bmatrix}\begin{bmatrix} x_1 \\ x_2 \end{bmatrix} = \begin{bmatrix} 4 \\ 7 \\ 3 \end{bmatrix}$$

$$AX = b$$

Here, $A = \begin{bmatrix} 2 & 1 \\ 3 & 2 \\ 1 & -1 \end{bmatrix} \Rightarrow A^T = \begin{bmatrix} 2 & 3 & 1 \\ 1 & 2 & -1 \end{bmatrix}$

Now, $A^TAX = A^Tb$

$$\begin{bmatrix} 2 & 3 & 1 \\ 1 & 2 & -1 \end{bmatrix}\begin{bmatrix} 2 & 1 \\ 3 & 2 \\ 1 & -1 \end{bmatrix}\begin{bmatrix} x_1 \\ x_2 \end{bmatrix} = \begin{bmatrix} 2 & 3 & 1 \\ 1 & 2 & -1 \end{bmatrix}\begin{bmatrix} 4 \\ 7 \\ 3 \end{bmatrix}$$

$$\begin{bmatrix} 14 & 7 \\ 7 & 6 \end{bmatrix}\begin{bmatrix} x_1 \\ x_2 \end{bmatrix} = \begin{bmatrix} 32 \\ 15 \end{bmatrix}$$

$$\begin{bmatrix} x_1 \\ x_2 \end{bmatrix} = \begin{bmatrix} 14 & 7 \\ 7 & 6 \end{bmatrix}^{-1}\begin{bmatrix} 32 \\ 15 \end{bmatrix}$$

$$\Rightarrow \quad \begin{bmatrix} x_1 \\ x_2 \end{bmatrix} = \frac{1}{35}\begin{bmatrix} 6 & -7 \\ -7 & 14 \end{bmatrix}\begin{bmatrix} 32 \\ 15 \end{bmatrix}$$

$$\Rightarrow \quad \begin{bmatrix} x_1 \\ x_2 \end{bmatrix} = \frac{1}{35}\begin{bmatrix} 87 \\ -14 \end{bmatrix} \Rightarrow \begin{bmatrix} x_1 \\ x_2 \end{bmatrix} = \begin{bmatrix} \frac{87}{35} \\ \frac{-14}{35} \end{bmatrix}$$

Hence, $x_1 = \frac{87}{35}$ and $x_2 = \frac{-14}{35}$. **Ans.**

Example 17. *Find the errors for the solution of the previous example.*

Solution From the previous example we get the approximate solution as $x_1 = \frac{87}{35}$ and $x_2 = \frac{-14}{35}$.

1. $2x_1 + x_2 = 4$

$$\text{Error} = e_1 = 4 - 2x_1 - x_2 = 4 - 2\left(\frac{87}{35}\right) - \left(-\frac{14}{35}\right) = -\frac{20}{35} \qquad \left(\begin{array}{l} AX = b \\ \Rightarrow A^T AX = A^T b \end{array}\right)$$

2. Agrain $3x_1 + 2x_2 = 7$

$$\text{Error} = e_2 = 7 - 3x_1 - 2x_2 = 7 - 3\left(\frac{87}{35}\right) - 2\left(\frac{-14}{35}\right) = \frac{12}{35}$$

3. Again $x_1 - x_2 = 3$

$$\text{Error} = e_3 = 3 - x_1 + x_2 = 3 - \frac{87}{35} + \left(\frac{-14}{35}\right) = \frac{4}{35}$$

$$r^2 = (r_1)^2$$

$$e^2 = (e_1)^2 + (e_2)^2 + (e_3)^2$$

$$= \left(\frac{-20}{35}\right)^2 + \left(\frac{12}{35}\right)^2 + \left(\frac{4}{35}\right)^2 = \frac{1}{35^2}(400+144+16) = \frac{560}{1225} = \frac{16}{35}$$

$$\text{Error} = e = \sqrt{\frac{16}{35}} = 0.676$$ **Ans.**

Example 18. *Find the best approximate solution of the system of the following inconsistent equations:*

$$4x_1 + 3x_2 - 4x_3 = 3$$
$$x_1 + x_2 + x_3 = 3$$
$$2x_1 + x_2 + 6x_3 = 10$$
$$x_1 + x_2 - x_3 = 0$$

By least square method. Also compute error.

Solution. We have

$$4x_1 + 3x_2 - 4x_3 = 3$$
$$x_1 + x_2 + x_3 = 3$$
$$2x_1 + x_2 + 6x_3 = 10$$
$$x_1 + x_2 - x_3 = 0$$

Equations in matrix form are:

$$\begin{bmatrix} 4 & 3 & -4 \\ 1 & 1 & 1 \\ 2 & 1 & 6 \\ 1 & 1 & -1 \end{bmatrix} \begin{bmatrix} x_1 \\ x_2 \\ x_3 \end{bmatrix} = \begin{bmatrix} 3 \\ 3 \\ 10 \\ 0 \end{bmatrix}$$

$$AX = b$$

Here $A = \begin{bmatrix} 4 & 3 & -4 \\ 1 & 1 & 1 \\ 2 & 1 & 6 \\ 1 & 1 & -1 \end{bmatrix} \Rightarrow A^T = \begin{bmatrix} 4 & 1 & 2 & 1 \\ 3 & 1 & 1 & 1 \\ -4 & 1 & 6 & -1 \end{bmatrix}$

Now, $A^T AX = A^T b$

$$\begin{bmatrix} 4 & 1 & 2 & 1 \\ 3 & 1 & 1 & 1 \\ -4 & 1 & 6 & -1 \end{bmatrix} \begin{bmatrix} 4 & 3 & -4 \\ 1 & 1 & 1 \\ 2 & 1 & 6 \\ 1 & 1 & -1 \end{bmatrix} \begin{bmatrix} x_1 \\ x_2 \\ x_2 \end{bmatrix} = \begin{bmatrix} 4 & 1 & 2 & 1 \\ 3 & 1 & 1 & 1 \\ -4 & 1 & 6 & -1 \end{bmatrix} \begin{bmatrix} 3 \\ 3 \\ 10 \\ 0 \end{bmatrix}$$

$$\Rightarrow \quad \begin{bmatrix} 22 & 16 & -4 \\ 16 & 12 & -6 \\ -4 & -6 & 54 \end{bmatrix} \begin{bmatrix} x_1 \\ x_2 \\ x_3 \end{bmatrix} = \begin{bmatrix} 35 \\ 22 \\ 51 \end{bmatrix}$$

$$\Rightarrow \quad \begin{bmatrix} x_1 \\ x_2 \\ x_3 \end{bmatrix} = \begin{bmatrix} 22 & 16 & -4 \\ 16 & 12 & -6 \\ -4 & -6 & 54 \end{bmatrix}^{-1} \begin{bmatrix} 35 \\ 22 \\ 51 \end{bmatrix} = \frac{1}{216} \begin{bmatrix} 612 & -840 & -48 \\ -840 & 1172 & 68 \\ -48 & 68 & 8 \end{bmatrix} \begin{bmatrix} 35 \\ 22 \\ 51 \end{bmatrix}$$

$$= \frac{1}{216} \begin{bmatrix} 492 \\ -148 \\ 224 \end{bmatrix} = \begin{bmatrix} \frac{41}{18} \\ \frac{-37}{54} \\ \frac{28}{27} \end{bmatrix}$$

Hence, $x_1 = \frac{41}{18}, \quad x_2 = \frac{-37}{54}, \quad x_3 = \frac{28}{27}.$ **Ans.**

Computation of Error:

$$\text{Error} = e_1 = 3 - 4x_1 - 3x_2 + 4x_3 = 3 - 4\left(\frac{41}{18}\right) - 3\left(\frac{-37}{54}\right) - 4\left(\frac{28}{27}\right) = \frac{-443}{54}$$

$$\text{Error} = e_2 = 3 - x_1 - x_2 - x_3 = 3 - \frac{41}{18} - \left(\frac{-37}{54}\right) - \frac{28}{27} = \frac{10}{27}$$

$$\text{Error} = e_3 = 10 - 2x_1 - x_2 - 6x_3 = 10 - 2\left(\frac{41}{18}\right) - \left(\frac{-37}{54}\right) - 6\left(\frac{28}{27}\right) = \frac{-5}{54}$$

$$\text{Error} = e_4 = 0 - x_1 - x_2 + x_3 = 0 - \frac{41}{18} - \left(\frac{-37}{54}\right) + \frac{28}{27} = \frac{-5}{9}$$

$$e^2 = e_1^2 + e_2^2 + e_3^2 + e_4^2$$

$$= \left(\frac{-443}{54}\right)^2 + \left(\frac{10}{27}\right)^2 + \left(\frac{-5}{54}\right)^2 + \left(\frac{-5}{9}\right)^2$$

$$= 67.30 + 0.14 + 0.0086 + 0.31$$

$$= 67.7586$$

$$\text{Error} = e = \sqrt{67.7586} = 8.2316$$

Hence, error = 8.2316 **Ans.**

Example 19. *Obtain the equation of the best fit straight line by the method of least squares to the following data:*

x	1	2	3	4
y	2.4	3.8	3.6	4.0

(*Delhi University, April 2010*)

Solution. Let the equation of the line be

$$y = a + bx \qquad \text{... (1)}$$

As line (1) passes through (1, 2.4), (2, 3.8) (3, 3.6) and (4, 4.0)

$$\therefore \quad a + b = 2.4$$
$$a + 2b = 3.8$$
$$a + 3b = 3.6$$
$$a + 4b = 4$$

These equations can be written in matrix form as

$$\begin{bmatrix} 1 & 1 \\ 1 & 2 \\ 1 & 3 \\ 1 & 4 \end{bmatrix} \begin{bmatrix} a \\ b \end{bmatrix} = \begin{bmatrix} 2.4 \\ 3.8 \\ 3.6 \\ 4.0 \end{bmatrix}$$

$$A\,X = \mathrm{B}$$

where $A = \begin{bmatrix} 1 & 1 \\ 1 & 2 \\ 1 & 3 \\ 1 & 4 \end{bmatrix} \Rightarrow A^T = \begin{bmatrix} 1 & 1 & 1 & 1 \\ 1 & 2 & 3 & 4 \end{bmatrix}$

Now, $\quad A^{\mathrm{T}}AX = A^T B$

$$\begin{bmatrix} 1 & 1 & 1 & 1 \\ 1 & 2 & 3 & 4 \end{bmatrix} \begin{bmatrix} 1 & 1 \\ 1 & 2 \\ 1 & 3 \\ 1 & 4 \end{bmatrix} \begin{bmatrix} a \\ b \end{bmatrix} = \begin{bmatrix} 1 & 1 & 1 & 1 \\ 1 & 2 & 3 & 4 \end{bmatrix} \begin{bmatrix} 2.4 \\ 3.8 \\ 3.6 \\ 4.0 \end{bmatrix} \Rightarrow \begin{bmatrix} 4 & 10 \\ 10 & 30 \end{bmatrix} \begin{bmatrix} a \\ b \end{bmatrix} = \begin{bmatrix} 13.8 \\ 36.8 \end{bmatrix}$$

$$\Rightarrow \quad \begin{bmatrix} a \\ b \end{bmatrix} = \begin{bmatrix} 4 & 10 \\ 10 & 30 \end{bmatrix}^{-1} \begin{bmatrix} 13.8 \\ 36.8 \end{bmatrix} = \frac{1}{20} \begin{bmatrix} 30 & -10 \\ -10 & 4 \end{bmatrix} \begin{bmatrix} 13.8 \\ 36.8 \end{bmatrix}$$

$$= \frac{1}{20} \begin{bmatrix} 46 \\ 9.2 \end{bmatrix} = \begin{bmatrix} 2.3 \\ 0.46 \end{bmatrix}$$

$\Rightarrow \quad a = 2.3$ and $b = 0.46$

So, the line that best aproximates all the given four points is

$$y = 2.3 + 0.46x.$$

Ans.

Example 20. *Find the best approximate equation of the line that pass through the points* (2, 3), (3, 5), (4, 3) *and* (5, 6). *Use the method of the least square.*

Solution Let the best approximate equation of the line be

$$y = a + bx. \qquad \text{... (1)}$$

Since (1) passes through the points (2, 3), (3, 5), (4, 3) and (5, 6), so

$$3 = a + 2b$$
$$5 = a + 3b$$
$$3 = a + 4b$$
$$6 = a + 5b.$$

These equations are written in the matrix form as

$$\begin{bmatrix} 1 & 2 \\ 1 & 3 \\ 1 & 4 \\ 1 & 5 \end{bmatrix} \begin{bmatrix} a \\ b \end{bmatrix} = \begin{bmatrix} 3 \\ 5 \\ 3 \\ 6 \end{bmatrix}$$

$$AX = B$$

Where $$A = \begin{bmatrix} 1 & 2 \\ 1 & 3 \\ 1 & 4 \\ 1 & 5 \end{bmatrix}, A^T = \begin{bmatrix} 1 & 1 & 1 & 1 \\ 2 & 3 & 4 & 5 \end{bmatrix}$$

Now $$A^T AX = A^T B$$

$$\begin{bmatrix} 1 & 1 & 1 & 1 \\ 2 & 3 & 4 & 5 \end{bmatrix} \begin{bmatrix} 1 & 2 \\ 1 & 3 \\ 1 & 4 \\ 1 & 5 \end{bmatrix} \begin{bmatrix} a \\ b \end{bmatrix} = \begin{bmatrix} 1 & 1 & 1 & 1 \\ 2 & 3 & 4 & 5 \end{bmatrix} \begin{bmatrix} 3 \\ 5 \\ 3 \\ 6 \end{bmatrix}$$

$$\Rightarrow \quad \begin{bmatrix} 4 & 14 \\ 14 & 54 \end{bmatrix} \begin{bmatrix} a \\ b \end{bmatrix} = \begin{bmatrix} 17 \\ 63 \end{bmatrix}$$

$$\begin{bmatrix} a \\ b \end{bmatrix} = \begin{bmatrix} 4 & 14 \\ 14 & 54 \end{bmatrix}^{-1} \begin{bmatrix} 17 \\ 63 \end{bmatrix}$$

$$\begin{bmatrix} a \\ b \end{bmatrix} = \frac{1}{20} \begin{bmatrix} 54 & -14 \\ -14 & 4 \end{bmatrix} \begin{bmatrix} 17 \\ 63 \end{bmatrix} = \frac{1}{20} \begin{bmatrix} 36 \\ 14 \end{bmatrix} = \begin{bmatrix} \frac{36}{20} \\ \frac{14}{20} \end{bmatrix}$$

$$= \begin{bmatrix} 1.8 \\ 0.7 \end{bmatrix}$$

$$a = 1.8, b = 0.7$$

So the best approximate equation of the line is

$$y = 1.8 + 0.7x$$ **Ans.**

36.15 LINEAR TRANSFORMATION OF MATRICES

Let X and Y be two vectors such that

$$X = \begin{bmatrix} x_1 \\ x_2 \\ \vdots \\ x_n \end{bmatrix}, Y = \begin{bmatrix} y_1 \\ y_2 \\ \vdots \\ y_n \end{bmatrix}$$

$$y_1 = [\; a_{11}, a_{12}, ..., a_{1n} \;] \begin{bmatrix} x_1 \\ x_2 \\ \vdots \\ x_n \end{bmatrix}$$

$$y_1 = [\; a_{11}x_1 + a_{12}x_2 + ... + a_{1n}x_n \;]$$

Similarily $y_2 = [\ a_{21}x_1 + a_{22}x_2 + ... + a_{2n}x_n\ ]$

....

....

$y_n = [\ a_{n1}x_1 + a_{n2}x_2 + ... + a_{nn}x_n\ ]$

In short $Y = AX$

$$\text{Here}\quad A = \begin{bmatrix} a_{11}, & a_{12}, ... a_{1n} \\ a_{21}, & a_{22}, ... a_{2n} \\ & .... \\ & .... \\ a_{n1}, & a_{n2}, ..., a_{nn} \end{bmatrix}$$

Matrix A in $Y = AX$ is called the *matrix of transformation.*

$|A|$ is the *modulus* of transformation. Y is called the *image* of X.

If $|A| = 0$, then linear transformation is called *singular*.

If $|A| \neq 0$, then the linear transformation is known as non-singular or regular.

Now, $Y = AX$, Here vector X is transformed into Y.

$$\Rightarrow \quad X = A^{-1}Y$$

$$Z = BY \qquad \text{(Putting the value of } Y)$$

$$= B\,(AX)$$

$$= (BA)\,X$$

(BA) transforms X into Z.

36.16 ORTHOGONAL TRANSFORMATION OF MATRICES

If a transformation transforms $(x_1^2 + x_2^2 + ... + x_n^2)$ into $(y_1^2 + y_2^2 + ... + y_n^2)$ then this transformation is known as orthogonal transformation.

Matrix A is known as orthogonal matrix.

$$X'X = \begin{bmatrix} x_1, x_2, ..., x_n \end{bmatrix} \begin{bmatrix} x_1 \\ x_2 \\ \vdots \\ x_n \end{bmatrix} = x_1^2 + x_2^2 + x_3^2 + ... + x_n^2$$

Similarly, $YY' = y_1^2 + y_2^2 + ... + y_n^2$

$$= Y'Y$$

$$= (AX)'\,(AX)$$

$$= (X'A')\,(AX)$$

$$= X\,(A'A)X$$

If $AA' = 1$, then A is known as orthogonal matrix.

Example 21. *If* (2, –3) *and* (1, 1) *are transformed into* (4, 5) *and* (3, 1) *respectively then find the matrix of transformation.*

Solution. We know that

$$Y = AX$$

$$\begin{bmatrix} 4 \\ 5 \end{bmatrix} = \begin{bmatrix} a_{11} & a_{12} \\ a_{21} & a_{22} \end{bmatrix} \begin{bmatrix} 2 \\ -3 \end{bmatrix}$$

$$\Rightarrow \quad \begin{bmatrix} 4 \\ 5 \end{bmatrix} = \begin{bmatrix} 2a_{11} & -3a_{12} \\ 2a_{21} & -3a_{22} \end{bmatrix}$$

$$\Rightarrow \quad 2a_{11} - 3a_{12} = 4 \qquad \text{... (1)}$$

and $$2a_{21} - 3a_{22} = 5 \qquad \text{... (2)}$$

Again $$\begin{bmatrix} 3 \\ 1 \end{bmatrix} = \begin{bmatrix} a_{11} & a_{12} \\ a_{21} & a_{22} \end{bmatrix} \begin{bmatrix} 1 \\ 1 \end{bmatrix}$$

$$\Rightarrow \quad \begin{bmatrix} 3 \\ 1 \end{bmatrix} = \begin{bmatrix} a_{11} + a_{12} \\ a_{21} + a_{22} \end{bmatrix}$$

$$\Rightarrow \quad a_{11} + a_{12} = 3 \qquad \text{... (3)}$$

and $$a_{21} + a_{22} = 1 \qquad \text{... (4)}$$

Solving (1) and (3), we get

$$a_{11} = \frac{13}{5}, \qquad a_{12} = \frac{2}{5}$$

Again solving (2) and (4), we get

$$a_{21} = \frac{8}{5}, \qquad a_{22} = \frac{-3}{5}$$

Hence, $$A = \begin{bmatrix} a_{11} & a_{12} \\ a_{21} & a_{22} \end{bmatrix} = \begin{bmatrix} \frac{13}{5} & \frac{2}{5} \\ \frac{8}{5} & \frac{-3}{5} \end{bmatrix}$$ **Ans.**

Example 22. *If* $y_1 = 3x_1 + 4x_2, y_2 = 4x_1 - x_2$ *and* $z_1 = +6y_1 + 2y_2$ *and* $z_2 = 2y_1 - 3y_2$ *then show these transformation in matrix form and find the composite transform of each transform* z_1, z_2 *in terms of* x_1 *and* x_2.

Solution. Here, we have

$$y_1 = 3x_1 + 4x_2 \qquad \text{... (1)}$$

Again $$y_2 = 4x_1 - x_2 \qquad \text{... (2)}$$

From (1) and (2)

$$\begin{bmatrix} y_1 \\ y_2 \end{bmatrix} = \begin{bmatrix} 3x_1 + 4x_2 \\ 4x_1 - x_2 \end{bmatrix}$$

$$\Rightarrow \quad \begin{bmatrix} y_1 \\ y_2 \end{bmatrix} = \begin{bmatrix} 3 & 4 \\ 4 & -1 \end{bmatrix} \begin{bmatrix} x_1 \\ x_2 \end{bmatrix} \qquad \text{... (3)}$$

The other given transformations are:

$$z_1 = 6y_1 + 2y_2$$
$$z_2 = 2y_1 - 3y_2$$

$$\Rightarrow \quad \begin{bmatrix} z_1 \\ z_2 \end{bmatrix} = \begin{bmatrix} 6 & 2 \\ 2 & -3 \end{bmatrix} \begin{bmatrix} y_1 \\ y_2 \end{bmatrix} \qquad \dots (4)$$

Putting the value of $\begin{bmatrix} y_1 \\ y_2 \end{bmatrix}$ from (3) in (4), we get

$$\begin{bmatrix} z_1 \\ z_2 \end{bmatrix} = \begin{bmatrix} 6 & 2 \\ 2 & -3 \end{bmatrix} \begin{bmatrix} 3 & 4 \\ 4 & -1 \end{bmatrix} \begin{bmatrix} x_1 \\ x_2 \end{bmatrix} = \begin{bmatrix} 26 & 22 \\ -6 & 11 \end{bmatrix} \begin{bmatrix} x_1 \\ x_2 \end{bmatrix} = \begin{bmatrix} 26x_1 + 22x_2 \\ -6x_1 + 11x_2 \end{bmatrix}$$

$$\Rightarrow \quad z_1 = 26x_1 + 22x_2$$

and $\quad z_2 = -6x_1 + 11x_2$ **Ans.**

Example 23. *What are the conditions for a matrix to be orthogonal matrix.*

Solution. If A is any matrix, then its transpose matrix is A'.

For a matrix to be orthogonal conditions are:

(*i*) $AA' = I$

(*ii*) $|A| = 1$

Example 24. *If A is an orthogonal matrix, then show that* $|A| = 1$.

Solution. Since A is an orthogonal matrix, therefore

$$AA' = I$$

$$\Rightarrow \quad |AA'| = |I| \quad \Rightarrow \quad |A||A'| = 1$$

$$\Rightarrow \quad |A||A| = 1 \quad \Rightarrow \quad |A^2| = 1$$

$$\Rightarrow \quad |A| = 1$$ **Proved.**

Example 25. *Show that inverse of an orthogonal matrix is orthogonal.*

Solution. Here, we have orthogonal matrix A, therefore

$$AA' = \mathrm{I}$$

$$(\mathrm{AA'})^{-1} = (\mathrm{I})^{-1}$$

$$\Rightarrow \quad (A')^{-1} . A^{-1} = I \qquad [(AB)^{-1} = B^{-1} . A^{-1}]$$

$$\Rightarrow \quad (A^{-1})' . A^{-1} = \mathrm{I}$$

Hence A^{-1} is also orthogonal matrix. **Proved.**

Example 26. *Show that the transpose of an orthogonal matrix is also orthogonal.*

Solution. Here, we have orthogonal matrix A

$$\therefore \quad AA' = \mathrm{I}$$

$$\Rightarrow \quad (AA')' = I'$$

$\Rightarrow \quad (A')'A' = I \qquad [\because (AB)' = B'A']$

Hence, A' is also orthogonal matrix. **Proved.**

Example 27. *If A and B are two orthogonal matrices, show that AB is also orthogonal matrix.*

Solution. $AA'\ A'A = \text{I}$ as A is an orthogonal matrix.

$BB'\ B'B = \text{I}$ as B is an orthogonal matrix.

Now, $(AB)\,(AB)' = (AB)\,(B'A') \qquad [\because (AB)' = B'A']$

$= A(BB')A'$

$= AIA' \qquad [\because BB' = I]$

$= AA' = I \qquad ...(1)$

Similarly, $(AB)'\,(AB) = (B'A')(AB) \qquad [\because (AB)' = B'A']$

$= B'A'AB$

$= B'(A'A)B$

$= B'IB = B'B \qquad [\because A'A = I]$

$= \text{I} \qquad ...(2)$

From (1) and (2), AB is an orthogonal matrix. **Proved.**

Example 28. *Verify that*

$$A = \frac{1}{3}\begin{bmatrix} 1 & 2 & 2 \\ 2 & 1 & -2 \\ -2 & 2 & -1 \end{bmatrix} \text{ is orthogonal.}$$

Solution. We have,

$$A = \frac{1}{3}\begin{bmatrix} 1 & 2 & 2 \\ 2 & 1 & -2 \\ -2 & 2 & -1 \end{bmatrix} \Rightarrow A' = \frac{1}{3}\begin{bmatrix} 1 & 2 & -2 \\ 2 & 1 & 2 \\ 2 & -2 & -1 \end{bmatrix}$$

Now $$AA' = \frac{1}{9}\begin{bmatrix} 1 & 2 & 2 \\ 2 & 1 & -2 \\ -2 & 2 & -1 \end{bmatrix}\begin{bmatrix} 1 & 2 & -2 \\ 2 & 1 & 2 \\ 2 & -2 & -1 \end{bmatrix} = \frac{1}{9}\begin{bmatrix} 9 & 0 & 0 \\ 0 & 9 & 0 \\ 0 & 0 & 9 \end{bmatrix} = \begin{bmatrix} 1 & 0 & 0 \\ 0 & 1 & 0 \\ 0 & 0 & 1 \end{bmatrix}$$

$\Rightarrow \quad AA' = I$

$$A = \frac{1}{3}\begin{bmatrix} 1 & 2 & 2 \\ 2 & 1 & -2 \\ -2 & 2 & -1 \end{bmatrix} = \begin{bmatrix} \frac{1}{3} & \frac{2}{3} & \frac{2}{3} \\ \frac{2}{3} & \frac{1}{3} & \frac{-2}{3} \\ \frac{-2}{3} & \frac{2}{3} & \frac{-1}{3} \end{bmatrix}$$

$$|A| = \begin{vmatrix} \frac{1}{3} & \frac{2}{3} & \frac{2}{3} \\ \frac{2}{3} & \frac{1}{3} & -\frac{2}{3} \\ -\frac{2}{3} & \frac{2}{3} & -\frac{1}{3} \end{vmatrix}$$

$$= \frac{1}{3}\left(\frac{-1}{9}+\frac{4}{9}\right)-\frac{2}{3}\left(\frac{-2}{9}-\frac{4}{9}\right)+\frac{2}{3}\left(\frac{4}{9}+\frac{2}{9}\right) = \frac{1}{3}\left(\frac{3}{9}\right)+\frac{2}{3}\left(\frac{6}{9}\right)+\frac{2}{3}\left(\frac{6}{9}\right)$$

$$= \frac{1}{9}+\frac{4}{9}+\frac{4}{9}=\frac{9}{9}=(1)$$

Hence, matrix A is orthogonal. **Proved.**

Example 29. *Show that* $\begin{bmatrix} \cos\phi & 0 & \sin\phi \\ \sin\theta\sin\phi & \cos\theta & -\sin\theta\cos\phi \\ -\cos\theta\sin\phi & \sin\theta & \cos\theta\cos\phi \end{bmatrix}$ *is an orthogonal matrix.*

Solution. We have,

$$A = \begin{bmatrix} \cos\phi & 0 & \sin\phi \\ \sin\theta\sin\phi & \cos\theta & -\sin\theta\cos\phi \\ -\cos\theta\sin\phi & \sin\theta & \cos\theta\cos\phi \end{bmatrix} \Rightarrow A' = \begin{bmatrix} \cos\phi & \sin\theta\sin\phi & -\cos\theta\sin\phi \\ 0 & \cos\theta & \sin\theta \\ \sin\phi & -\sin\theta\cos\phi & \cos\theta\cos\phi \end{bmatrix}$$

Now,

$$AA' = \begin{bmatrix} \cos\phi & 0 & \sin\phi \\ \sin\theta\sin\phi & \cos\theta & -\sin\theta\cos\phi \\ -\cos\theta\sin\phi & \sin\theta & \cos\theta\cos\phi \end{bmatrix} \begin{bmatrix} \cos\phi & \sin\theta\sin\phi & -\cos\theta\sin\phi \\ 0 & \cos\theta & \sin\theta \\ \sin\phi & -\sin\theta\cos\phi & \cos\theta\cos\phi \end{bmatrix}$$

$$= \begin{bmatrix} \cos^2\phi+\sin^2\phi & \sin\theta\sin\phi\cos\phi - \sin\theta\cos\phi\sin\phi & -\cos\phi\cos\theta\sin\phi + \sin\phi\cos\theta\cos\phi \\ \sin\theta\sin\phi\cos\phi - \sin\theta\cos\phi\sin\phi & \sin^2\theta\sin^2\phi+\cos^2\theta + \sin^2\theta\cos^2\phi & -\sin\theta\ \cos\theta\ \sin^2\phi+\sin\theta\cos\theta - \sin\theta\cos\theta\cos^2\phi \\ -\cos\theta\sin\phi\cos\phi + \cos\theta\cos\phi\sin\phi & -\cos\theta\sin\theta\sin^2\phi+\sin\theta\cos\theta - \sin\theta\cos\theta\cos^2\phi & \cos^2\theta\sin^2\phi+\sin^2\theta + \cos^2\theta\cos^2\phi \end{bmatrix}$$

$$= \begin{bmatrix} 1 & 0 & 0 \\ 0 & \sin^2\theta(\sin^2\phi+\cos^2\phi)+\cos^2\theta & -\sin\theta\cos\theta(\sin^2\phi+\cos^2\phi)+\sin\theta\cos\theta \\ 0 & -\sin\theta\cos\theta(\sin^2\phi+\cos^2\phi)+\sin\theta\cos\theta & \cos^2\theta\,(\sin^2\phi+\cos^2\phi)\ +\sin^2\theta \end{bmatrix}$$

$$= \begin{bmatrix} 1 & 0 & 0 \\ 0 & \sin^2\theta+\cos^2\theta & -\sin\theta\cos\theta+\sin\theta\cos\theta \\ 0 & -\sin\theta\cos\theta+\sin\theta\cos\theta & \cos^2\theta+\sin^2\theta \end{bmatrix}$$

$$= \begin{bmatrix} 1 & 0 & 0 \\ 0 & 1 & 0 \\ 0 & 0 & 1 \end{bmatrix} = I$$

$$|A| = \begin{vmatrix} \cos\phi & 0 & \sin\phi \\ \sin\theta\sin\phi & \cos\theta & -\sin\theta\cos\phi \\ -\cos\theta\sin\phi & \sin\theta & \cos\theta\cos\phi \end{vmatrix}$$

$$= \cos\phi\left[\ \cos^2\theta\cos\phi+\sin^2\theta\cos\phi\ \right]+0+\sin\phi\left[\ \sin^2\theta\sin\phi+\cos^2\theta\sin\phi\ \right]$$

$$= \cos\phi\left[\ (\cos^2\theta+\sin^2\theta)\cos\phi\ \right]+\sin\phi\left[\ (\sin^2\theta+\cos^2\theta)\sin\phi\ \right]$$

$$= \cos\phi \cdot \cos\phi + \sin\phi \cdot \sin\phi = \cos^2\phi + \sin^2\phi = 1$$

Hence, matrix A is an orthogonal. **Proved.**

EXERCISE 36.2

1. Determine the values of α, β and γ when $\begin{bmatrix} 0 & 2\beta & \gamma \\ \alpha & \beta & -\gamma \\ \alpha & -\beta & \gamma \end{bmatrix}$ is orthogonal.

Ans. $\alpha = \pm\dfrac{1}{\sqrt{2}},\ \beta = \pm\dfrac{1}{\sqrt{6}},\ \gamma = \pm\dfrac{1}{\sqrt{3}}$

2. Verify whether the given matrix is orthogonal. $A = \dfrac{1}{3}\begin{bmatrix} 2 & 2 & 1 \\ 2 & -1 & 2 \\ -1 & 2 & 2 \end{bmatrix}$.

3. Show that $A = \begin{bmatrix} \cos\theta & 0 & \sin\theta \\ 0 & 1 & 0 \\ -\sin\theta & 0 & \cos\theta \end{bmatrix}$ is an orthogonal matrix.

4. Verify that the matrix $A = \begin{bmatrix} \frac{1}{\sqrt{3}} & \frac{1}{\sqrt{6}} & \frac{-1}{\sqrt{2}} \\ \frac{1}{\sqrt{3}} & \frac{-2}{\sqrt{6}} & 0 \\ \frac{1}{\sqrt{3}} & \frac{1}{\sqrt{6}} & \frac{1}{\sqrt{2}} \end{bmatrix}$ is orthogonal.

5. Show that BA is an orthogonal matrix, if A and B both are orthogonal matrices.

6. Find the least squares solution to the following system of equations:

$$\begin{bmatrix} 2 & -1 & 1 \\ 1 & -5 & 2 \\ -3 & 1 & -4 \\ 1 & -1 & 1 \end{bmatrix}\begin{bmatrix} x_1 \\ x_2 \\ x_3 \end{bmatrix} = \begin{bmatrix} -4 \\ 2 \\ 5 \\ -1 \end{bmatrix}$$

Ans. $x_1 = \dfrac{-18}{7},\ x_2 = \dfrac{-151}{210},\ x_3 = \dfrac{107}{210}$

7. Find the least squares solution of the inconsistent system $AX = b$, where

$A = \begin{bmatrix} 4 & 0 \\ 0 & 2 \\ 1 & 1 \end{bmatrix}$, $b = \begin{bmatrix} 2 \\ 0 \\ 11 \end{bmatrix}$. Also determine the least squares error. **Ans.** $\hat{x} = \begin{bmatrix} 1 \\ 2 \end{bmatrix}; \sqrt{84}$

8. By the method of least square find the best approximate equation of a parabola ($y = a + bx + cx^2$) passing through the points (2, 3),(3, 5), (4, 3) and (5, 6). **Ans.** $y = \dfrac{91}{20} - \dfrac{21}{20}x + \dfrac{x^2}{4}$

9. Find the least square solution of $AX = b$, $x - 6y = -1$, $x - 2y = 2$, $x + y = 1$, $x + 7y = 6$,

Ans. $x = 2,\ y = \dfrac{1}{2}$

10. Find the best approximate equation of the line ($y = a + bx$) that will pass through the points (–3, 70), (1, 21), (–7, 110), (5, –35). Use method of least squares and also compute the error.

Ans. $y = \dfrac{149}{5} - \dfrac{121}{10}x$, Error = 8.0125

37

CHAPTER

DETERMINANTS

37.1 INTRODUCTION

With the help of determinants, we can solve a system of simultaneous equations by Cramer's Rule. Determinants are also used in calculating inverse of a square matrix.

37.2 DETERMINANT AS ELIMINANT

Consider the following three equations having three unknowns, x, y and z.

$$a_1 x + b_1 y + c_1 z = 0 \qquad ...(1)$$

$$a_2 x + b_2 y + c_2 z = 0 \qquad ...(2)$$

$$a_3 x + b_3 y + c_3 z = 0 \qquad ...(3)$$

From (2) and (3) by cross-multiplication; we get

$$\frac{x}{b_2 c_3 - b_3 c_2} = \frac{y}{a_3 c_2 - a_2 c_3} = \frac{z}{a_2 b_3 - a_3 b_2} = k \text{ (say)}$$

$$x = (b_2 c_3 - b_3 c_2)\, k$$

$$y = (a_3 c_2 - a_2 c_3)\, k$$

and $$z = (a_2 b_3 - a_3 b_2)\, k$$

Substituting the values of x, y and z in (1), we get the eliminant

$$a_1 (b_2 c_3 - b_3 c_2)\, k + b_1 (a_3 c_2 - a_2 c_3)\, k + c_1 (a_2 b_3 - a_3 b_2)\, k = 0$$

or $$a_1 (b_2 c_3 - b_3 c_2) - b_1 (a_2 c_3 - a_3 c_2) + c_1 (a_2 b_3 - a_3 b_2) = 0 \qquad ...(4)$$

From (1), (2) and (3) by suppressing x, y, z the remaining can be written in the determinant as

$$\begin{vmatrix} a_1 & b_1 & c_1 \\ a_2 & b_2 & c_2 \\ a_3 & b_3 & c_3 \end{vmatrix} = 0 \qquad ...(5)$$

This is the determinant of third order.

As (4) and (5) both are the eliminant of the same equations.

$$\therefore \quad \begin{vmatrix} a_1 & b_1 & c_1 \\ a_2 & b_2 & c_2 \\ a_3 & b_3 & c_3 \end{vmatrix} = a_1 (b_2 c_3 - b_3 c_2) - b_1 (a_2 c_3 - a_3 c_2) + c_1 (a_2 b_3 - a_3 b_2) = 0$$

or $\begin{vmatrix} a_1 & b_1 & c_1 \\ a_2 & b_2 & c_2 \\ a_3 & b_3 & c_3 \end{vmatrix} = a_1 \begin{vmatrix} b_2 & c_2 \\ b_3 & c_3 \end{vmatrix} - b_1 \begin{vmatrix} a_2 & c_2 \\ a_3 & c_3 \end{vmatrix} + c_1 \begin{vmatrix} a_2 & b_2 \\ a_3 & b_3 \end{vmatrix}$

37.3. MINOR

The minor of an element is defined as a determinant obtained by deleting the row and column containing the element.

Thus the minors of a_1, b_1 and c_1 are respectively.

$$\begin{vmatrix} b_2 & c_2 \\ b_3 & c_3 \end{vmatrix}, \begin{vmatrix} a_2 & c_2 \\ a_3 & c_3 \end{vmatrix} \text{ and } \begin{vmatrix} a_2 & b_2 \\ a_3 & b_3 \end{vmatrix}$$

Thus $\begin{vmatrix} a_1 & b_1 & c_1 \\ a_2 & b_2 & c_2 \\ a_3 & b_3 & c_3 \end{vmatrix} = a_1$ (minor of a_1) $- b_1$ (minor of b_1) $+ c_1$ (minor of c_1).

37.4. COFACTOR

Cofactor = $(-1)^{r+c}$ Minor

where r is the number of rows of the element and c is the number of columns of the element.

The cofactor of any element of ith row and jth column is

$(-1)^{i+j}$ minor

Thus the cofactor of $a_1 = (-1)^{1+1} (b_2c_3 - b_3c_2) = +(b_2c_3 - b_3c_2)$

The cofactor of $b_1 = (-1)^{1+2} (a_2c_3 - a_3c_2) = -(a_2c_3 - a_3c_2)$

The cofactor of $c_1 = (-1)^{1+3} (a_2b_3 - a_3b_2) = +(a_2b_3 - a_3b_2)$

The determinant $= a_1$ (cofactor of a_1) $+ a_2$ (cofactor of a_2) $+ a_3$ (cofactor of a_3).

Example 1. *Write down the minors and co-factors of each element and also evaluate the determinant:*

$$\begin{vmatrix} 1 & 3 & -2 \\ 4 & -5 & 6 \\ 3 & 5 & 2 \end{vmatrix}$$

Solution. M_{11} = Minor of element (1) $= \begin{vmatrix} 1 & 3 & -2 \\ 4 & -5 & 6 \\ 3 & 5 & 2 \end{vmatrix}$

By eleminating the row and column of (1), the remaining is minor of (1)

$$= \begin{vmatrix} -5 & 6 \\ 5 & 2 \end{vmatrix} = (-5) \times 2 - (6 \times 5) = -10 - 30 = -40$$

Cofactor of element (1) $= A_{11} = (-1)^{1+1} M_{11} = (-1)^2 (-40) = -40$

M_{12} = Minor of element (3)

By eleminating the row and column of (3), we get

$$= \begin{vmatrix} 1 & 3 & -2 \\ 4 & -5 & 6 \\ 3 & 5 & 2 \end{vmatrix} = \begin{vmatrix} 4 & 6 \\ 3 & 2 \end{vmatrix} = (4 \times 2) - (3 \times 6) = 8 - 18 = -10$$

$\Rightarrow$ Cofactor of element (3) $= A_{12} = (-1)^{1+2} (-10) = 10$

M_{13} = Minor of element (–2)

$$= \begin{vmatrix} 1 & 3 & -2 \\ 4 & -5 & 6 \\ 3 & 5 & 2 \end{vmatrix} = \begin{vmatrix} 4 & -5 \\ 3 & 5 \end{vmatrix} = (4\times 5)-(-5)\times 3 = 20+15 = 35$$

$\Rightarrow$ Coafactor of element (–2) = $A_{13} = (-1)^{1+3} M_{13} = (-1)^4 (35) = 35$

M_{21} = Minor of element (4)

$$= \begin{vmatrix} 1 & 3 & -2 \\ 4 & -5 & 6 \\ 3 & 5 & 2 \end{vmatrix} = \begin{vmatrix} 3 & -2 \\ 5 & 2 \end{vmatrix} = (3\times 2)-(-2)\times 5 = 6+10 = 16$$

$\Rightarrow$ Coafactor of element (4) = $A_{21} = (-1)^{2+1} M_{21} = (-1)^{2+1} (16) = -16$

M_{22} = Minor of element (–5)

$$= \begin{vmatrix} 1 & 3 & -2 \\ 4 & -5 & 6 \\ 3 & 5 & 2 \end{vmatrix} = \begin{vmatrix} 1 & -2 \\ 3 & 2 \end{vmatrix} = (1\times 2)-(-2)\times 3 = 2+6 = 8$$

$\Rightarrow$ Coafactor of element (–5) = $A_{22} = (-1)^{2+2} M_{22} = (-1)^{2+2} (8) = (8)$

M_{23} = Minor of element (6)

$$= \begin{vmatrix} 1 & 3 & -2 \\ 4 & -5 & 6 \\ 3 & 5 & 2 \end{vmatrix} = \begin{vmatrix} 1 & 3 \\ 3 & 5 \end{vmatrix} = (1\times 5)-3\times 3 = 5-9 = -4$$

$\Rightarrow$ Coafactor of element (6) = $A_{23} = (-1)^{2+3} M_{23} = (-1)^{2+3} (-4) = 4$

M_{31} = Minor of element (3)

$$= \begin{vmatrix} 1 & 3 & -2 \\ 4 & -5 & 6 \\ 3 & 5 & 2 \end{vmatrix} = \begin{vmatrix} 3 & -2 \\ -5 & 6 \end{vmatrix} = (3\times 6)-(-2)\times(-5) = 18-10 = 8$$

$\Rightarrow$ Coafactor of element (3) = $A_{31} = (-1)^{3+1} M_{31} = (-1)^{3+1} (8) = 8$

M_{32} = Minor of element (5)

$$= \begin{vmatrix} 1 & 3 & -2 \\ 4 & -5 & 6 \\ 3 & 5 & 2 \end{vmatrix} = \begin{vmatrix} 1 & -2 \\ 4 & 6 \end{vmatrix} = (1\times 6)-(-2)\times 4 = 6+8 = 14$$

$\Rightarrow$ Coafactor of element (5) = $A_{32} = (-1)^{3+2} M_{32} = (-1)^{3+2} (14) = -14$

M_{33} = Minor of element (2)

$$= \begin{vmatrix} 1 & 3 & -2 \\ 4 & -5 & 6 \\ 3 & 5 & 2 \end{vmatrix} = \begin{vmatrix} 1 & 3 \\ 4 & -5 \end{vmatrix} = 1\times(-5)-(4\times 3) = -5-12 = -17$$

$\Rightarrow$ Coafactor of element (5) = $A_{33} = (-1)^{3+3} M_{33} = (-1)^{3+3} (-17) = -17$

Example 2. *Evaluate the determinants* :

$$\begin{vmatrix} 3 & -1 & -2 \\ 0 & 0 & -1 \\ 3 & -5 & 0 \end{vmatrix}$$

Solution. We have, two zero entries in the second row. So, expanding along 2nd row:

$$\begin{vmatrix} 3 & -1 & -2 \\ 0 & 0 & -1 \\ 3 & -5 & 0 \end{vmatrix} = -0\begin{vmatrix} -1 & -2 \\ -5 & 0 \end{vmatrix} + 0\begin{vmatrix} 3 & -2 \\ 3 & 0 \end{vmatrix} - (-1)\begin{vmatrix} 3 & -1 \\ 3 & -5 \end{vmatrix}$$

$$= -0 + 0 + 1(-15 + 3) = -12$$ **Ans.**

Example 3. *Prove that the determinant* $\begin{vmatrix} x & \sin\theta & \cos\theta \\ -\sin\theta & -x & 1 \\ \cos\theta & 1 & x \end{vmatrix}$ *is independent of* θ

Solution. We have,

$$\begin{vmatrix} x & \sin\theta & \cos\theta \\ -\sin\theta & -x & 1 \\ \cos\theta & 1 & x \end{vmatrix} = x\begin{vmatrix} -x & 1 \\ 1 & x \end{vmatrix} - \sin\theta\begin{vmatrix} -\sin\theta & 1 \\ \cos\theta & x \end{vmatrix} + \cos\theta\begin{vmatrix} -\sin\theta & -x \\ \cos\theta & 1 \end{vmatrix}$$

$$= x(-x^2 - 1) - \sin\theta(-x\sin\theta - \cos\theta) + \cos\theta(-\sin\theta + x\cos\theta)$$

$$= -x^3 - x + x\sin^2\theta + \sin\theta\cos\theta - \sin\theta\cos\theta + x\cos^2\theta$$

$$= -x^3 - x + x(\sin^2\theta + \cos^2\theta) = -x^3 - x + x$$

Thus, the determinant is independent of θ **Proved.**

Example 4. *Evaluate the determinant* $\begin{vmatrix} 1 & 0 & 4 \\ 3 & 5 & -1 \\ 0 & 1 & 2 \end{vmatrix}$.

(i) With the help of second row, *(ii) with the help of third column.*

Solution.

(i) $\begin{vmatrix} 1 & 0 & 4 \\ 3 & 5 & -1 \\ 0 & 1 & 2 \end{vmatrix} = 3 \times (\text{cofactor of } 3) + 5 \times (\text{cofactor of } 5) + (-1)(\text{cofactor of } -1)$.

$$= 3 \times (-1)^{2+1}\begin{vmatrix} 0 & 4 \\ 1 & 2 \end{vmatrix} + 5 \times (-1)^{2+2}\begin{vmatrix} 1 & 4 \\ 0 & 2 \end{vmatrix} + (-1) \times (-1)^{2+3}\begin{vmatrix} 1 & 0 \\ 0 & 1 \end{vmatrix}$$

$$= -3 \times (0 - 4) + 5(2 - 0) + (1 - 0) = 12 + 10 + 1 = 23$$ **Ans.**

(ii) $\begin{vmatrix} 1 & 0 & 4 \\ 3 & 5 & -1 \\ 0 & 1 & 2 \end{vmatrix} = 4 \times (\text{cofactor of } 4) + (-1)(\text{cofactor of } (-1)) + 2 \times (\text{cofactor of } 2)$

$$= 4 \times (-1)^{1+3}\begin{vmatrix} 3 & 5 \\ 0 & 1 \end{vmatrix} + (-1)(-1)^{2+3}\begin{vmatrix} 1 & 0 \\ 0 & 1 \end{vmatrix} + 2 \times (-1)^{3+3}\begin{vmatrix} 1 & 0 \\ 3 & 5 \end{vmatrix}$$

$$= 4 \times (3 - 0) + (1 - 0) + 2(5 - 0) = 12 + 1 + 10 = 23$$ **Ans.**

EXERCISE 37.1

Write the minors and co-factors of each element of the following determinants and also evaluate the determinant in each case :

1. $\begin{vmatrix} 42 & 1 & 6 \\ 28 & 7 & 4 \\ 14 & 3 & 2 \end{vmatrix}$

Ans. $M_{11} = 2$, $M_{12} = 0$, $M_{13} = -14$, $M_{21} = -16$, $M_{22} = 0$
$M_{23} = 112$, $M_{31} = -38$, $M_{32} = 0$, $M_{33} = 266$
$A_{11} = 2$, $A_{12} = 0$, $A_{13} = -14$, $A_{21} = 16$, $A_{22} = 0$
$A_{23} = -112$, $A_{31} = -38$, $A_{32} = 0$, $A_{33} = 266$, $|A| = 0$

2. $\begin{vmatrix} 1 & a & bc \\ 1 & b & ca \\ 1 & c & ab \end{vmatrix}$

Ans. $M_{11} = (ab^2 - ac^2)$, $M_{12} = (ab - ac)$, $M_{13} = (c - b)$, $M_{21} = a^2b - bc^2$
$M_{22} = (ab - bc)$, $M_{23} = (c - a)$, $M_{31} = (ca^2 - cb^2)$, $M_{32} = ca - bc, M_{33} = (b - a)$,
$A_{11} = (ab^2 - ac^2)$, $A_{12} = (ac - ab)$, $A_{13} = (c - b)$, $A_{21} = bc^2 - a^2b$
$A_{22} = (ab - bc)$, $A_{23} = (a - c)$, $A_{31} = (ca^2 - cb^2)$, $A_{32} = (bc - ca), A_{33} = (b - a)$
$|A| = (a - b)(b - c)(c - a)$.

Expand the following determinants:

3. $\begin{vmatrix} 2 & -3 & 4 \\ 5 & 1 & -6 \\ -7 & 8 & -9 \end{vmatrix}$ **Ans.** $|A| = 5$

4. $\begin{vmatrix} 5 & 0 & 7 \\ 8 & -6 & -4 \\ 2 & 3 & 9 \end{vmatrix}$ **Ans.** $|A| = 42$

5. $\begin{vmatrix} a & h & g \\ h & b & f \\ g & f & c \end{vmatrix}$ **Ans.** $|A| = abc + 2fgh - af^2 - bg^2 - ch^2$

Expand the following determinants by two methods:

(*i*) along the-third row. (*ii*) along the-third column.

6. $\begin{vmatrix} 1 & -3 & 2 \\ 4 & -1 & 2 \\ 3 & 5 & 2 \end{vmatrix}$ **Ans.** $|A| = 40$

7. $\begin{vmatrix} 3 & -2 & 4 \\ 1 & 2 & 1 \\ 0 & 1 & -1 \end{vmatrix}$ **Ans.** $|A| = -7$

8. $\begin{vmatrix} 2 & 3 & -2 \\ 1 & 2 & 3 \\ -2 & 1 & -3 \end{vmatrix}$ **Ans.** $|A| = -37$

37.5. PROPERTIES OF DETERMINANTS

Property (*i*). *The value of a determinant remains unaltered; if the rows are interchanged into columns (or the columns into rows).*

Verification. Let $\Delta = \begin{vmatrix} a_1 & a_2 & a_3 \\ b_1 & b_2 & b_3 \\ c_1 & c_2 & c_3 \end{vmatrix}$

Expanding along first row, we get

$$\Delta = a_1 \begin{vmatrix} b_2 & b_3 \\ c_2 & c_3 \end{vmatrix} - a_2 \begin{vmatrix} b_1 & b_3 \\ c_1 & c_3 \end{vmatrix} + a_3 \begin{vmatrix} b_1 & b_2 \\ c_1 & c_2 \end{vmatrix}$$

$$= a_1(b_2c_3 - b_3c_2) - a_2(b_1c_3 - b_3c_1) + a_3(b_1c_2 - b_2c_1) \qquad ...(1)$$

By interchanging the rows and columns of Δ, we get the determinant

$$\Delta_1 = \begin{vmatrix} a_1 & b_1 & c_1 \\ a_2 & b_2 & c_2 \\ a_3 & b_3 & c_3 \end{vmatrix}$$

$$= a_1 (b_2c_3 - b_3c_2) - b_1 (a_2c_3 - a_3c_2) + c_1 (a_2b_3 - a_3b_2)$$
$$= a_1b_2c_3 - a_1b_3c_2 - a_2b_1c_3 + a_3b_1c_2 + a_2b_3c_1 - a_3b_2c_1$$
$$= (a_1b_2c_3 - a_1b_3c_2) - (a_2b_1c_3 - a_2b_3c_1) + (a_3b_1c_2 - a_3b_2c_1)$$
$$= a_1 (b_2c_3 - b_3c_2) - a_2 (b_1c_3 - b_3c_1) + a_3 (b_1c_2 - b_2c_1) \quad ...(2)$$

From (1) and (2), we have

$$\Delta = \Delta_1.$$

It follows: The value of determinant remains unaltered, if the rows are interchanged into columns (or the columns into rows). **Proved.**

Property (*ii*). *If two rows (or two columns) of a determinant are interchanged, the sign of the value of the determinant changes.*

Verification. Let $\Delta = \begin{vmatrix} a_1 & a_2 & a_3 \\ b_1 & b_2 & b_3 \\ c_1 & c_2 & c_3 \end{vmatrix}$

$$= a_1 (b_2 c_3 - b_3 c_2) - a_2 (b_1 c_3 - b_3 c_1) + a_3 (b_1 c_2 - b_2 c_1) \quad ... (1)$$

Interchanging first and third rows, the new determinant obtained is given by

$$\Delta_1 = \begin{vmatrix} c_1 & c_2 & c_3 \\ b_1 & b_2 & b_3 \\ a_1 & a_2 & a_3 \end{vmatrix}$$

Expanding along third row, we get

$$\Delta_1 = -[a_1 (c_2 b_3 - b_2 c_3) - a_2 (c_1 b_3 - c_3 b_1) + a_3 (b_2 c_1 - b_1 c_2)]$$
$$= -[a_1 (b_2 c_3 - b_3 c_2) - a_2 (b_1 c_3 - b_3 c_1) + a_3 (b_1 c_2 - b_2 c_1)] \quad ... (2)$$

From (1) and (2), we have

$$\Delta_1 = \Delta$$

Hence, property (*ii*) is verified. **Proved.**

Example 5. *Verify property (ii) for* $\begin{vmatrix} 1 & 2 & 3 \\ 4 & 0 & 6 \\ 7 & 8 & 9 \end{vmatrix}$

Solution. Let $\Delta = \begin{vmatrix} 1 & 2 & 3 \\ 4 & 0 & 6 \\ 7 & 8 & 9 \end{vmatrix} = 1(0-48) - 2(36-42) + 3(32-0)$

$$= -48 + 12 + 96 = 60$$

Interchanging second the third rows, we have

$$\Delta_1 = \begin{vmatrix} 1 & 2 & 3 \\ 7 & 8 & 9 \\ 4 & 0 & 6 \end{vmatrix} = 1(48-0) - 2(42-36) + 3(0-32)$$

$$= 48 - 12 - 96 = -60$$

Thus, $\Delta_1 = \Delta$

Hence property (*ii*) is verified. **Verified.**

Property (*iii*). *If two rows (or columns) of a determinant are identical, the value of the determinant is zero.*

Verification. Let $\Delta = \begin{vmatrix} a_1 & b_1 & c_1 \\ a_1 & b_1 & c_1 \\ a_3 & b_3 & c_3 \end{vmatrix}$, so that the first two rows are identical.

By interchanging the first two rows, we get the same determinant D.

By property (*ii*), on interchanging the rows, the sign of the determinant changes.

or $\Delta = -\Delta$ or $2\Delta = 0$ or $\Delta = 0$ **Proved.**

Example 6. *Evaluate* : $\Delta = \begin{vmatrix} 2 & 3 & 4 \\ 2 & 3 & 4 \\ 1 & 5 & 6 \end{vmatrix}$

Solution. Expanding along first row, we have

$$\Delta = 2(18 - 20) - 3\,(12 - 4) + 4(10 - 3)$$
$$= 2 \times (-2) - 3\,(8) + 4\,(7) = -4 - 24 + 28 = 0$$

Here, R_1 and R_2 are identical. **Verified.**

Property (*iv*). *If the elements of any row (or column) of a determinant be each multiplied by the same number, the determinant is multiplied by that number.*

Verification. Let $\Delta = \begin{vmatrix} a_1 & b_1 & c_1 \\ a_2 & b_2 & c_2 \\ a_3 & b_3 & c_3 \end{vmatrix}$

and Δ_1 be the determinant obtained by multiplying the elements of the first row by k. Then

$$\Delta_1 = \begin{vmatrix} ka_1 & kb_1 & kc_1 \\ a_2 & b_2 & c_2 \\ a_3 & b_3 & c_3 \end{vmatrix}$$
$$= ka_1\,(b_2c_3 - b_3c_2) - kb_1\,(a_2c_3 - a_3c_2) + kc_1\,(a_2b_3 - a_3b_2)$$
$$= k\,[a_1\,(b_2c_3 - b_3c_2) - b_1\,(a_2c_3 - a_3c_2) + c_1\,(a_2b_3 - a_3b_2)]$$
$$= k\begin{vmatrix} a_1 & b_1 & c_1 \\ a_2 & b_2 & c_2 \\ a_3 & b_3 & c_3 \end{vmatrix} = k\,\Delta.$$

Hence, $\begin{vmatrix} ka_1 & kb_1 & kc_1 \\ a_2 & b_2 & c_2 \\ a_3 & b_3 & c_3 \end{vmatrix} = k\begin{vmatrix} a_1 & b_1 & c_1 \\ a_2 & b_2 & c_2 \\ a_3 & b_3 & c_3 \end{vmatrix}$ **Proved.**

Example 7. *Verify the property (iv) by*

$$\Delta = \begin{vmatrix} 2 & 5 & 8 \\ 3 & 7 & 1 \\ 2 & 0 & 2 \end{vmatrix}$$

Solution. $\Delta = \begin{vmatrix} 2 & 5 & 8 \\ 3 & 7 & 1 \\ 2 & 0 & 2 \end{vmatrix} = 2\,(14-0) - 5\,(6-2) + 8\,(0-14) = 28 - 20 - 112 = -104$

Multiplying the first column by 5, we get

$$\Delta_1 = \begin{vmatrix} 10 & 5 & 8 \\ 15 & 7 & 1 \\ 10 & 0 & 2 \end{vmatrix} = 10\,(14-0) - 5\,(30-10) + 8\,(0-70)$$

$$= 140 - 100 - 560 = -520 = 5\,(-104)$$

$$\Delta_1 = 5\Delta$$

Property (*iv*) is verified. **Verified.**

Property (*v*). *The value of the determinant remains unaltered if to the elements of one row (or column) be added any constant multiple of the corresponding elements of any other row (or column) respectively.*

Verification. Let
$$\Delta = \begin{vmatrix} a_1 & b_1 & c_1 \\ a_2 & b_2 & c_2 \\ a_3 & b_3 & c_3 \end{vmatrix}$$

On multiplying the second column by l and the third column by m and adding to the first column, we get

$$\Delta' = \begin{vmatrix} a_1 + lb_1 + mc_1 & b_1 & c_1 \\ a_2 + lb_2 + mc_2 & b_2 & c_2 \\ a_3 + lb_3 + mc_3 & b_3 & c_3 \end{vmatrix} = \begin{vmatrix} a_1 & b_1 & c_1 \\ a_2 & b_2 & c_2 \\ a_3 & b_3 & c_3 \end{vmatrix} + l \begin{vmatrix} b_1 & b_1 & c_1 \\ b_2 & b_2 & c_2 \\ b_3 & b_3 & c_3 \end{vmatrix} + m \begin{vmatrix} c_1 & b_1 & c_1 \\ c_2 & b_2 & c_2 \\ c_3 & b_3 & c_3 \end{vmatrix}$$

$$= \Delta + 0 + 0$$ (Since columns are identical)

$$= \Delta$$ **Proved.**

Example 8. *Verify the property (v) by*

$$\Delta = \begin{vmatrix} 1 & 2 & 4 \\ 3 & 2 & 5 \\ 0 & 4 & 6 \end{vmatrix}$$

Solution.
$$\Delta = \begin{vmatrix} 1 & 2 & 4 \\ 3 & 2 & 5 \\ 0 & 4 & 6 \end{vmatrix} = 1\,(12-20) - 2\,(18-0) + 4\,(12-0) = -8 - 36 + 48 = 4$$

On multiplying the second column by 5 and third column by 6 and adding to the first column, we get

$$\Delta_1 = \begin{vmatrix} 1+10+24 & 2 & 4 \\ 3+10+30 & 2 & 5 \\ 0+20+36 & 4 & 6 \end{vmatrix} = \begin{vmatrix} 35 & 2 & 4 \\ 43 & 2 & 5 \\ 56 & 4 & 6 \end{vmatrix} = 35\,(12-20) - 2\,(258-280) + 4\,(172-112)$$

$$= 35\,(-8) - 2\,(-22) + 4\,(60) = -280 + 44 + 240 = 284 - 280 = 4$$

$$\Delta_1 = \Delta$$ **Verified.**

Example 9. *Show that*

$$\Delta = \begin{vmatrix} b-c & c-a & a-b \\ c-a & a-b & b-c \\ a-b & b-c & c-a \end{vmatrix} = 0$$

Solution. Let
$$\Delta = \begin{vmatrix} b-c & c-a & a-b \\ c-a & a-b & b-c \\ a-b & b-c & c-a \end{vmatrix}$$

Applying $C_1 \to C_1 + C_2 + C_3$, we get

$$\Delta = \begin{vmatrix} 0 & c-a & a-b \\ 0 & a-b & b-c \\ 0 & b-c & c-a \end{vmatrix} = 0 \qquad [\because C_1 \text{ consists of all zeroes.}]$$

Proved.

Example 10. *Without expanding, evaluate the determinant*

$$\begin{vmatrix} \sin\alpha & \cos\alpha & \sin(\alpha+\delta) \\ \sin\beta & \cos\beta & \sin(\beta+\delta) \\ \sin\gamma & \cos\gamma & \sin(\gamma+\delta) \end{vmatrix}$$

Solution. Let
$$\Delta = \begin{vmatrix} \sin\alpha & \cos\alpha & \sin(\alpha+\delta) \\ \sin\beta & \cos\beta & \sin(\beta+\delta) \\ \sin\gamma & \cos\gamma & \sin(\gamma+\delta) \end{vmatrix}$$

$$\Rightarrow \quad \Delta = \begin{vmatrix} \sin\alpha & \cos\alpha & \sin\alpha\cos\delta + \cos\alpha\sin\delta \\ \sin\beta & \cos\beta & \sin\beta\cos\delta + \cos\beta\sin\delta \\ \sin\gamma & \cos\gamma & \sin\gamma\cos\delta + \cos\gamma\sin\delta \end{vmatrix}$$

$[\because \sin(A+B) = \sin A \cos B + \cos A \sin B]$

$$\Rightarrow \quad \Delta = \begin{vmatrix} \sin\alpha & \cos\alpha & 0 \\ \sin\beta & \cos\beta & 0 \\ \sin\gamma & \cos\gamma & 0 \end{vmatrix} \qquad [\text{Applying } C_3 \to C_3 - \cos\delta.\, C_1 - \sin\delta . C_2]$$

$$\Rightarrow \quad \Delta = 0 \qquad [\because C_3 \text{ consists of all zeroes.}]$$ **Ans.**

Example 11. *By using property of determinants prove that :*

$$\begin{vmatrix} 1 & x & x^2 \\ x^2 & 1 & x \\ x & x^2 & 1 \end{vmatrix} = (1-x^3)^2$$

Solution. L.H.S. $= \begin{vmatrix} 1 & x & x^2 \\ x^2 & 1 & x \\ x & x^2 & 1 \end{vmatrix} = \begin{vmatrix} 1+x+x^2 & x & x^2 \\ 1+x+x^2 & 1 & x \\ 1+x+x^2 & x^2 & 1 \end{vmatrix}$ [Applying $C_1 \to C_1 + C_2 + C_3$,]

$$= (1+x+x^2)\begin{vmatrix} 1 & x & x^2 \\ 1 & 1 & x \\ 1 & x^2 & 1 \end{vmatrix}$$

$$= (1+x+x^2)\begin{vmatrix} 1 & x & x^2 \\ 0 & 1-x & x-x^2 \\ 0 & x^2-x & 1-x^2 \end{vmatrix} \qquad [\text{Applying } R_2 \to R_2 - R_1 \text{ and } R_3 \to R_3 - R_1]$$

$= (1+x+x^2)\,(1)\,\{(1-x)\,(1-x^2) - (x^2-x)\,(x-x^2)\}$

$= (1+x+x^2)\,(1-x)^2\,\{1+x+x^2\} = \{(1-x)\,(1+x+x^2)\}^2 = (1-x^3)^2 =$ R.H.S.

Proved.

Example 12. *Using properties of determinants, prove that*

$$\begin{vmatrix} x & y & z \\ x^2 & y^2 & z^2 \\ x^3 & y^3 & z^3 \end{vmatrix} = x\,y\,z(x-y)(y-z)(z-x).$$

Solution. Let $\Delta = \begin{vmatrix} x & y & z \\ x^2 & y^2 & z^2 \\ x^3 & y^3 & z^3 \end{vmatrix} = xyz \begin{vmatrix} 1 & 1 & 1 \\ x & y & z \\ x^2 & y^2 & z^2 \end{vmatrix}$

Operate: $C_1 \to C_1 - C_2$; $C_2 \to C_2 - C_3$.

$$\Delta = xyz \begin{vmatrix} 0 & 0 & 1 \\ x-y & y-z & z \\ x^2-y^2 & y^2-z^2 & z^2 \end{vmatrix} = xyz \begin{vmatrix} x-y & y-z \\ x^2-y^2 & y^2-z^2 \end{vmatrix} \quad \text{(On expanding by } R_1\text{)}$$

$$= xyz(x-y)(y-z) \begin{vmatrix} 1 & 1 \\ x+y & y+z \end{vmatrix} = xyz(x-y)(y-z)(z-x).$$ **Proved.**

Example 13. *Using the properties of determinants, show that*

$$\begin{vmatrix} a+x & y & z \\ x & a+y & z \\ x & y & a+z \end{vmatrix} = a^2(a+x+y+z).$$

Solution. Let $\Delta = \begin{vmatrix} a+x & y & z \\ x & a+y & z \\ x & y & a+z \end{vmatrix}$

Operate: $C_1 \to C_1 + C_2 + C_3$

$$\Delta = \begin{vmatrix} a+x+y+z & y & z \\ a+x+y+z & a+y & z \\ a+x+y+z & y & a+z \end{vmatrix}$$

Taking $(a+x+y+z)$ common from Ist column, we get

$$= (a+x+y+z) \begin{vmatrix} 1 & y & z \\ 1 & a+y & z \\ 1 & y & a+z \end{vmatrix}$$

Operate: $R_2 \to R_2 - R_1, R_3 \to R_3 - R_1$

$$= (a+x+y+z) \begin{vmatrix} 1 & y & z \\ 0 & a & 0 \\ 0 & 0 & a \end{vmatrix} = (a+x+y+z) \times 1 \begin{vmatrix} a & 0 \\ 0 & a \end{vmatrix} \quad \text{[Expanding along } C_1\text{]}$$

$$= (a+x+y+z)\,a^2 = a^2(a+x+y+z)$$ **Proved.**

Example 14. *If w is the one of the imaginary cube roots of unity, find the value of the determinant:*

$$\begin{vmatrix} 1 & \omega & \omega^2 \\ \omega & \omega^2 & 1 \\ \omega^2 & 1 & \omega \end{vmatrix}$$

Solution. The given determinant $= \begin{vmatrix} 1 & \omega & \omega^2 \\ \omega & \omega^2 & 1 \\ \omega^2 & 1 & \omega \end{vmatrix}$

By $R_1 \to R_1 + R_2 + R_3$, we get

$$= \begin{vmatrix} 1+\omega+\omega^2 & 1+\omega+\omega^2 & 1+\omega+\omega^2 \\ \omega & \omega^2 & 1 \\ \omega^2 & 1 & \omega \end{vmatrix} = \begin{vmatrix} 0 & 0 & 0 \\ \omega & \omega^2 & 1 \\ \omega^2 & 1 & \omega \end{vmatrix} \qquad [\because 1+\omega+\omega^2 = 0]$$

$= 0$ (Since each entry in R_1 is zero.) **Ans.**

Example 15. *Without expanding the determinant, show that $(a + b + c)$ is a factor of the determinant* $\begin{vmatrix} a & b & c \\ b & c & a \\ c & a & b \end{vmatrix}$.

Solution. Let $\Delta = \begin{vmatrix} a & b & c \\ b & c & a \\ c & a & b \end{vmatrix}$ Operate: $C_1 \to C_1 + C_2 + C_3$

$$\Rightarrow \qquad \Delta = \begin{vmatrix} a+b+c & b & c \\ a+b+c & c & a \\ a+b+c & a & b \end{vmatrix} = (a+b+c)\begin{vmatrix} 1 & b & c \\ 1 & c & a \\ 1 & a & b \end{vmatrix}$$

$\Rightarrow$ $(a + b + c)$ is a factor of Δ. **Proved.**

Example 16. *Without expanding the determinant, prove that* $\begin{vmatrix} 1 & a & b+c \\ 1 & b & c+a \\ 1 & c & a+b \end{vmatrix} = 0.$

Solution. Let $\Delta = \begin{vmatrix} 1 & a & b+c \\ 1 & b & c+a \\ 1 & c & a+b \end{vmatrix}$

Operate: $C_3 \to C_3 + C_2$.

$$\therefore \qquad \Delta = \begin{vmatrix} 1 & a & a+b+c \\ 1 & b & a+b+c \\ 1 & c & a+b+c \end{vmatrix} = (a+b+c)\begin{vmatrix} 1 & a & 1 \\ 1 & b & 1 \\ 1 & c & 1 \end{vmatrix}$$

$= 0$ ($\because$ C_1 and C_3 are identical). **Proved.**

Example 17. *Without expanding the determinant, prove that*

$$\begin{vmatrix} \frac{1}{a} & a^2 & bc \\ \frac{1}{b} & b^2 & ca \\ \frac{1}{c} & c^2 & ab \end{vmatrix} = 0$$

Solution. Let $\Delta = \begin{vmatrix} \frac{1}{a} & a^2 & bc \\ \frac{1}{b} & b^2 & ca \\ \frac{1}{c} & c^2 & ab \end{vmatrix}$

Multiply R_1 by a, R_2 by b and R_3 by c.

$$\Delta = \frac{1}{abc}\begin{vmatrix} 1 & a^3 & abc \\ 1 & b^3 & abc \\ 1 & c^3 & abc \end{vmatrix} = \frac{1}{abc}.abc\begin{vmatrix} 1 & a^3 & 1 \\ 1 & b^3 & 1 \\ 1 & c^3 & 1 \end{vmatrix} = 1 \times 0 = 0.$$

(Since C_1 and C_3 are identical) **Proved.**

Example 18. *Using properties of determinants, prove that :*

$$\begin{vmatrix} 1 & a & a^3 \\ 1 & b & b^3 \\ 1 & c & c^3 \end{vmatrix} = (a-b)(b-c)(c-a)(a+b+c)$$

Solution. Let $\Delta = \begin{vmatrix} 1 & a & a^3 \\ 1 & b & b^3 \\ 1 & c & c^3 \end{vmatrix}$

Operate: $R_1 \to R_1 - R_2$; $R_2 \to R_2 - R_3$

$$\Delta = \begin{vmatrix} 0 & a-b & a^3-b^3 \\ 0 & b-c & b^3-c^3 \\ 1 & c & c^3 \end{vmatrix} = 1.\begin{vmatrix} a-b & a^3-b^3 \\ b-c & b^3-c^3 \end{vmatrix} \quad \text{(Expand along } C_1\text{)}$$

$$= (a-b)(b-c)\begin{vmatrix} 1 & a^2+ab+b^2 \\ 1 & b^2+bc+c^2 \end{vmatrix}$$

Operate: $R_1 \to R_1 - R_2$

$$\Delta = (a-b)(b-c)\begin{vmatrix} 0 & (a^2-c^2)+(ab-bc) \\ 1 & b^2+bc+c^2 \end{vmatrix}$$

$$= (a-b).(b-c).(-1)\,[(a^2-c^2)+b\,(a-c)]$$

$$= (a-b).(b-c)(c-a)\,(a+b+c).$$

Proved.

[**Note :** It can also be proved by factor Theorem easily]

Example 19. *Evaluate*

$$\begin{vmatrix} a-b-c & 2a & 2a \\ 2b & b-c-a & 2b \\ 2c & 2c & c-a-b \end{vmatrix}$$

Solution. By $R_1 \to R_1 + R_2 + R_3$, we get

$$\begin{vmatrix} a+b+c & a+b+c & a+b+c \\ 2b & b-c-a & 2b \\ 2c & 2c & c-a-b \end{vmatrix} = (a+b+c)\begin{vmatrix} 1 & 1 & 1 \\ 2b & b-c-a & 2b \\ 2c & 2c & c-a-b \end{vmatrix}$$

$$= (a+b+c)\begin{vmatrix} 1 & 0 & 0 \\ 2b & -(a+b+c) & 0 \\ 2c & 0 & -(a+b+c) \end{vmatrix} \begin{matrix} \\ C_2 \to C_2 - C_1 \\ C_3 \to C_3 - C_1 \end{matrix}$$

On expanding by first row $= (a + b + c)\,(a + b + c)^2 = (a + b + c)^3$. **Ans.**

Example 20. *By using properties of determinants prove that:*

$$\begin{vmatrix} 1+a^2-b^2 & 2ab & -2b \\ 2ab & 1-a^2+b^2 & 2a \\ 2b & -2a & 1-a^2-b^2 \end{vmatrix} = (1+a^2+b^2)^3$$

Solution. Let $\Delta = \begin{vmatrix} 1+a^2-b^2 & 2ab & -2b \\ 2ab & 1-a^2+b^2 & 2a \\ 2b & -2a & 1-a^2-b^2 \end{vmatrix}$

Applying $C_1 \to C_2 - bC_3$, $\quad C_2 \to C_2 - aC_3$, we get

$$\Delta = \begin{vmatrix} 1+a^2+b^2 & 0 & -2b \\ 0 & 1+a^2+b^2 & 2a \\ b\,(1+a^2+b^2) & -a\,(1+a^2+b^2) & 1-a^2-b^2 \end{vmatrix}$$

Taking $(1 + a^2 + b^2)$ common from C_1 and C_2, we get

$$\Delta = (1+a^2+b^2)^2 \begin{vmatrix} 1 & 0 & -2b \\ 0 & 1 & 2a \\ b & -a & 1-a^2-b^2 \end{vmatrix}$$

Applying $C_3 \to C_3 + 2bC_1$, we get

$$\Delta = (1+a^2+b^2)^2 \begin{vmatrix} 1 & 0 & 0 \\ 0 & 1 & 2a \\ b & -a & 1-a^2+b^2 \end{vmatrix}$$

Expanding along R_1, we get

$$\Delta = (1+a^2+b^2)^2 \begin{vmatrix} 1 & 2a \\ -a & 1-a^2+b^2 \end{vmatrix} = (1 + a^2 + b^2)^2\,(1 - a^2 + b^2 + 2a^2)$$

$$= (1 + a^2 + b^2)^3$$

Proved.

Example 21. *Prove that*

$$\begin{vmatrix} \alpha & \beta & \gamma \\ \alpha^2 & \beta^2 & \gamma^2 \\ \beta+\gamma & \gamma+\alpha & \alpha+\beta \end{vmatrix} = (\alpha-\beta)\,(\beta-\gamma)\,(\gamma-\alpha)\,(\alpha+\beta+\gamma)$$

Solution. Let $\Delta = \begin{vmatrix} \alpha & \beta & \gamma \\ \alpha^2 & \beta^2 & \gamma^2 \\ \beta+\gamma & \gamma+\alpha & \alpha+\beta \end{vmatrix}$.

Applying $R_3 \to R_1 + R_3$, we get

$$\Delta = \begin{vmatrix} \alpha & \beta & \gamma \\ \alpha^2 & \beta^2 & \gamma^2 \\ \alpha+\beta+\gamma & \alpha+\beta+\gamma & \alpha+\beta+\gamma \end{vmatrix}$$

$$= (\alpha+\beta+\gamma)\begin{vmatrix} \alpha & \beta & \gamma \\ \alpha^2 & \beta^2 & \gamma^2 \\ 1 & 1 & 1 \end{vmatrix} \quad \text{[Taking out } (\alpha+\beta+\gamma) \text{ common from } R_3]$$

$$= (\alpha+\beta+\gamma)\begin{vmatrix} \alpha & \beta-\alpha & \gamma-\alpha \\ \alpha^2 & \beta^2-\alpha^2 & \gamma^2-\alpha^2 \\ 1 & 0 & 0 \end{vmatrix} \quad \text{Applying } C_2 \to C_2 - C_1,\ C_3 \to C_3 - C_1$$

$$= (\alpha+\beta+\gamma)(\beta-\alpha)(\gamma-\alpha)\begin{vmatrix} \alpha & 1 & 1 \\ \alpha^2 & \beta+\alpha & \gamma+\alpha \\ 1 & 0 & 0 \end{vmatrix}$$

$$= (\alpha+\beta+\gamma)(\beta-\alpha)(\gamma-\alpha).1\begin{vmatrix} 1 & 1 \\ \beta+\alpha & \gamma+\alpha \end{vmatrix} \quad \text{[Expanding along } R_3]$$

$$= (\alpha+\beta+\gamma)(\beta-\alpha)(\gamma-\alpha)(\gamma+\alpha-\beta-\alpha)$$

$$= (\alpha+\beta+\gamma)(\beta-\gamma)(\gamma-\alpha)(\alpha-\beta)$$ **Proved.**

Example 22. *Prove that*

$$\begin{vmatrix} -a^2 & ab & ac \\ ba & -b^2 & bc \\ ac & bc & -c^2 \end{vmatrix} = 4\,a^2b^2c^2$$

Solution. Let $\Delta = \begin{vmatrix} -a^2 & ab & ac \\ ba & -b^2 & bc \\ ac & bc & -c^2 \end{vmatrix}$

Taking a, b, c common from C_1, C_2 and C_3 respectively, we get

$$\Delta = abc\begin{vmatrix} -a & a & a \\ b & -b & b \\ c & c & -c \end{vmatrix}$$

$$= a^2b^2c^2\begin{vmatrix} -1 & 1 & 1 \\ 1 & -1 & 1 \\ 1 & 1 & -1 \end{vmatrix} \quad \text{[Taking } a, b, c \text{ common from } R_1, R_2 \text{ and } R_3 \text{ respectively]}$$

$$= a^2b^2c^2\begin{vmatrix} -1 & 0 & 0 \\ 1 & 0 & 2 \\ 1 & 2 & 0 \end{vmatrix} \quad \text{[Applying } C_2 \to C_2 + C_1,\ C_3 \to C_3 + C_1]$$

$$= a^2b^2c^2(-1)\begin{vmatrix} 0 & 2 \\ 2 & 0 \end{vmatrix}$$

$$= a^2b^2c^2(-1)(0-4) = 4a^2b^2c^2 \quad \text{[Expanding along } R_1]$$ **Proved.**

Example 23. *Using properties of determinants, prove the following:*

$$\begin{vmatrix} a & b & c \\ a-b & b-c & c-a \\ b+c & c+a & a+b \end{vmatrix} = a^3+b^3+c^3-3abc.$$

Solution. We have,

$$\begin{vmatrix} a & b & c \\ a-b & b-c & c-a \\ b+c & c+a & a+b \end{vmatrix} = \begin{vmatrix} a+b+c & b & c \\ a-b+b-c+c-a & b-c & c-a \\ b+c+c+a+a+b & c+a & a+b \end{vmatrix}$$

By applying $C_1 \to C_1 + C_2 + C_3]$

$$= \begin{vmatrix} a+b+c & b & c \\ 0 & b-c & c-a \\ 2(a+b+c) & c+a & a+b \end{vmatrix} = \begin{vmatrix} a+b+c & b & c \\ 0 & b-c & c-a \\ 0 & c+a-2b & a+b-2c \end{vmatrix} \quad \text{By } R_3 \to R_3 - 2R_1$$

Expanding along C_1, we get

$$= (a+b+c)\{(b-c)(a+b-2c) - (c-a)(c+a-2b)\}$$
$$= (a+b+c)\{(ab+b^2-2bc-ac-bc+2c^2) - (c^2+ac-2bc-ac-a^2+2ab)\}$$
$$= (a+b+c)\{ab+b^2-2bc-ac-bc+2c^2-c^2-ac+2bc+ac+a^2-2ab\}$$
$$= (a+b+c)(a^2+b^2+c^2-ab-bc-ca) = a^3+b^3+c^3-3abc$$ **Proved.**

Example 24. *If* $\begin{vmatrix} a & a^2 & a^3-1 \\ b & b^2 & b^3-1 \\ c & c^2 & c^3-1 \end{vmatrix} = 0$, *prove that* $abc = 1$.

Solution. $\begin{vmatrix} a & a^2 & a^3-1 \\ b & b^2 & b^3-1 \\ c & c^2 & c^3-1 \end{vmatrix} = 0 \Rightarrow \begin{vmatrix} a & a^2 & a^3 \\ b & b^2 & b^3 \\ c & c^2 & c^3 \end{vmatrix} + \begin{vmatrix} a & a^2 & -1 \\ b & b^2 & -1 \\ c & c^2 & -1 \end{vmatrix} = 0$

$$\Rightarrow \quad abc\begin{vmatrix} 1 & a & a^2 \\ 1 & b & b^2 \\ 1 & c & c^2 \end{vmatrix} - \begin{vmatrix} a & a^2 & 1 \\ b & b^2 & 1 \\ c & c^2 & 1 \end{vmatrix} = 0$$

(Taking out common a, b, c from R_1, R_2 and R_3 from 1st determinant)

$$\Rightarrow \quad abc\begin{vmatrix} 1 & a & a^2 \\ 1 & b & b^2 \\ 1 & c & c^2 \end{vmatrix} - \begin{vmatrix} a & 1 & a^2 \\ b & 1 & b^2 \\ c & 1 & c^2 \end{vmatrix} = 0 \quad \text{(Interchanging } C_2 \text{ and } C_3)$$

$$\Rightarrow \quad abc\begin{vmatrix} 1 & a & a^2 \\ 1 & b & b^2 \\ 1 & c & c^2 \end{vmatrix} - \begin{vmatrix} 1 & a & a^2 \\ 1 & b & b^2 \\ 1 & c & c^2 \end{vmatrix} = 0$$

(Interchanging C_1 and C_2 of the second determinant)

$$\Rightarrow \quad (abc-1)\begin{vmatrix} 1 & a & a^2 \\ 1 & b & b^2 \\ 1 & c & c^2 \end{vmatrix} = 0 \quad \Rightarrow \quad abc - 1 = 0 \quad \Rightarrow abc = 1$$ **Proved.**

Example 25. *Show that* $\begin{vmatrix} b+c & c+a & a+b \\ q+r & r+p & p+q \\ y+z & z+x & x+y \end{vmatrix} = 2\begin{vmatrix} a & b & c \\ p & q & r \\ x & y & z \end{vmatrix}$

Solution. The above determinant can be expressed as the sum of 8 determinants as given below:

$$\begin{vmatrix} b+c & c+a & a+b \\ q+r & r+p & p+q \\ y+z & z+x & x+y \end{vmatrix} = \begin{vmatrix} b & c & a \\ q & r & p \\ y & z & x \end{vmatrix} + \begin{vmatrix} b & a & a \\ q & p & p \\ y & x & x \end{vmatrix} + \begin{vmatrix} b & c & b \\ q & r & q \\ y & z & y \end{vmatrix} + \begin{vmatrix} b & a & b \\ q & p & q \\ y & x & y \end{vmatrix}$$

$$+ \begin{vmatrix} c & c & a \\ r & r & p \\ z & z & x \end{vmatrix} + \begin{vmatrix} c & a & a \\ r & p & p \\ z & x & x \end{vmatrix} + \begin{vmatrix} c & c & b \\ r & r & q \\ z & z & y \end{vmatrix} + \begin{vmatrix} c & a & b \\ r & p & q \\ z & x & y \end{vmatrix}$$

$$= \begin{vmatrix} b & c & a \\ q & r & p \\ y & z & x \end{vmatrix} + 0+0+0+0+0+0+ \begin{vmatrix} c & a & b \\ r & p & q \\ z & x & y \end{vmatrix}$$

$$= (-1)^2 \begin{vmatrix} a & b & c \\ p & q & r \\ x & y & z \end{vmatrix} + (-1)^2 \begin{vmatrix} a & b & c \\ p & q & r \\ x & y & z \end{vmatrix} = 2 \begin{vmatrix} a & b & c \\ p & q & r \\ x & y & z \end{vmatrix}$$ **Proved.**

Example 26. *If a, b, c are in A.P; then find the determinant:*

$$\begin{vmatrix} x+2 & x+3 & x+2a \\ x+3 & x+4 & x+2b \\ x+4 & x+5 & x+2c \end{vmatrix}$$

Solution. Applying $R_1 \to R_1 + R_3 - R_2$ to the given determinant, we have

$$\begin{vmatrix} (x+2)+(x+4)-2(x+3) & (x+3)+(x+5)-2(x+4) & (x+2a)+(x+2c)-2(x+2b) \\ x+3 & x+4 & x+2b \\ x+4 & x+5 & x+2c \end{vmatrix}$$

$$= \begin{vmatrix} 0 & 0 & 2a+2c-4b \\ x+3 & x+4 & x+2b \\ x+4 & x+5 & x+2c \end{vmatrix}$$

$$= \begin{vmatrix} 0 & 0 & 0 \\ x+3 & x+4 & x+2b \\ x+4 & x+5 & x+2c \end{vmatrix} = 0 \qquad \left[\begin{array}{ll} \because & a, b, c \text{ are in A.P.} \\ \Rightarrow & 2b = a+c \\ \Rightarrow & 2a+2c = 4b \end{array}\right]$$

[$\because$ R_1 consists of all zeroes.] **Ans.**

Example 27. *Prove that*

$$\begin{vmatrix} 2\alpha & \alpha+\beta & \alpha+\gamma \\ \beta+\alpha & 2\beta & \beta+\gamma \\ \gamma+\alpha & \gamma+\beta & 2\gamma \end{vmatrix} = 0$$

Solution. Given determinant $= \begin{vmatrix} \alpha+\alpha & \alpha+\beta & \alpha+\gamma \\ \beta+\alpha & \beta+\beta & \beta+\gamma \\ \gamma+\alpha & \gamma+\beta & \gamma+\gamma \end{vmatrix}$

The above determinant can be expressed as the sum of 8 determinants.

Each of the 8 determinants has either two identical columns or identical rows.

$\therefore$ Each of the resulting determinant is zero. Hence the result. **Proved.**

Example 28. *Using properties of determinants, prove that:*

$$\begin{vmatrix} 3a & -a+b & -a+c \\ -b+a & 3b & -b+c \\ -c+a & -c+b & 3c \end{vmatrix} = 3(a+b+c)(ab+bc+ca)$$

Solution.

$$\begin{vmatrix} 3a & -a+b & -a+c \\ -b+a & 3b & -b+c \\ -c+a & -c+b & 3c \end{vmatrix} = \begin{vmatrix} a+b+c & -a+b & -a+c \\ a+b+c & 3b & -b+c \\ a+b+c & -c+b & 3c \end{vmatrix} \text{ (Applying } C_1 \to C_1 + C_2 + C_3)$$

$$= (a+b+c)\begin{vmatrix} 1 & -a+b & -a+c \\ 1 & 3b & -b+c \\ 1 & -c+b & 3c \end{vmatrix} = (a+b+c)\begin{vmatrix} 1 & -a+b & -a+c \\ 0 & a+2b & a-b \\ 0 & a-c & a+2c \end{vmatrix} \begin{matrix} \\ R_2 \to R_2 - R_1 \\ R_3 \to R_3 - R_1 \end{matrix}$$

$= (a + b + c) .1.\{(a + 2b)(a + 2c) - (a - c) (a - b)\}$ [Expanding along C_1]

$= (a + b + c) \{(a^2 + 2ac + 2ab + 4bc) - (a^2 - ab - ac + bc)\}$

$= (a + b + c) (3ab + 3bc + 3ca) = 3 (a + b + c) (ab + bc + ca)$ **Proved.**

Example 29. *Show that $x = -(a + b + c)$ is one root of the equation:*

$$\begin{vmatrix} x+a & b & c \\ b & x+c & a \\ c & a & x+b \end{vmatrix} = 0$$

and solve the equation completely.

Solution. By $C_1 \to C_1 + C_2 + C_3$, we get

$$\begin{vmatrix} x+a+b+c & b & c \\ x+a+b+c & x+c & a \\ x+a+b+c & a & x+b \end{vmatrix} = 0 \Rightarrow (x+a+b+c)\begin{vmatrix} 1 & b & c \\ 1 & x+c & a \\ 1 & a & x+b \end{vmatrix} = 0$$

$$\Rightarrow (x+a+b+c)\begin{vmatrix} 1 & b & c \\ 0 & x-b+c & a-c \\ 0 & a-b & x+b-c \end{vmatrix} = 0, R_2 \to R_2 - R_1;\ R_3 \to R_3 - R_1$$

On expanding by first column, we get

$(x + a + b + c) [(x - b + c) (x + b - c) - (a - b) (a - c)] = 0$

$\Rightarrow$ $(x + a + b + c) [x^2 - (b - c)^2 - (a^2 - ac - ab + bc)] = 0$

$\Rightarrow$ $(x + a + b + c) (x^2 - b^2 - c^2 + 2bc - a^2 + ac + ab - bc] = 0$

$\Rightarrow$ $(x + a + b + c) (x^2 - a^2 - b^2 - c^2 + ab + bc + ca) = 0$

Either $x + a + b + c = 0 \Rightarrow x = -(a + b + c)$

or $x^2 - a^2 - b^2 - c^2 + ab + bc + ca = 0$

$\Rightarrow$ $x = \pm\sqrt{a^2 + b^2 + c^2 - ab - bc - ca}$

Hence, $x = -(a + b + c)$ is one root of the given equation. **Proved.**

Example 30. *Find the value of*

$$\begin{vmatrix} (b+c)^2 & a^2 & a^2 \\ b^2 & (c+a)^2 & b^2 \\ c^2 & c^2 & (a+b)^2 \end{vmatrix}$$

Solution. By $C_1 \to C_1 - C_3, C_2 \to C_2, - C_3$, we get

$$\begin{vmatrix} (b+c)^2 - a^2 & a^2 - a^2 & a^2 \\ b^2 - b^2 & (c+a)^2 - b^2 & b^2 \\ c^2 - (a+b)^2 & c^2 - (a+b)^2 & (a+b)^2 \end{vmatrix}$$

$$= \begin{vmatrix} (a+b+c)(b+c-a) & 0 & a^2 \\ 0 & (a+b+c)(c+a-b) & b^2 \\ (a+b+c)(c-a-b) & (a+b+c)(c-a-b) & (a+b)^2 \end{vmatrix}$$

On taking out $(a + b + c)$ as common from 1st and 2nd columns, we get

$$= (a+b+c)^2 \begin{vmatrix} b+c-a & 0 & a^2 \\ 0 & c+a-b & b^2 \\ c-a-b & c-a-b & (a+b)^2 \end{vmatrix}$$

$$= (a+b+c)^2 \begin{vmatrix} -a+b+c & 0 & a^2 \\ 0 & a-b+c & b^2 \\ -2b & -2a & 2ab \end{vmatrix} R_3 \to R_3 - (R_1 + R_2)$$

$$= -2(a+b+c)^2 \begin{vmatrix} -a+b+c & 0 & a^2 \\ 0 & a-b+c & b^2 \\ b & a & -ab \end{vmatrix}$$

On expanding by first row, we get

$= -2\,(a+b+c)^2\,[(-a+b+c)\,\{-ab\,(a-b+c) - ab^2\} + a^2\,\{0 - b\,(a-b+c)\}]$

$= -2\,(a+b+c)^2\,[(-a+b+c)\,(-a^2b - abc) - a^2b\,(a-b+c)]$

$= -2ab\,(a+b+c)^2\,[(-a+b+c)\,(-a-c) - a\,(a-b+c)]$

$= -2ab\,(a+b+c)^2\,[a^2 + ac - ab - bc - ac - c^2 - a^2 + ab - ac]$

$= -2ab\,(a+b+c)^2\,(-bc - ac - c^2) = 2abc\,(a+b+c)^2\,(b+a+c)$

$= \quad 2abc\,(a+b+c)^3.$

Example 31. *Using properties of determinants, solve for x :*

$$\begin{vmatrix} a+x & a-x & a-x \\ a-x & a+x & a-x \\ a-x & a-x & a+x \end{vmatrix} = 0$$

Solution. Given that,

$$\begin{vmatrix} a+x & a-x & a-x \\ a-x & a+x & a-x \\ a-x & a-x & a+x \end{vmatrix} = 0$$

Applying $C_1 \to C_1 + C_2 + C_3$, we get

$$\begin{vmatrix} 3a-x & a-x & a-x \\ 3a-x & a+x & a-x \\ 3a-x & a-x & a+x \end{vmatrix} = 0$$

$$\Rightarrow \quad (3a-x) \begin{vmatrix} 1 & a-x & a-x \\ 1 & a+x & a-x \\ 1 & a-x & a+x \end{vmatrix} = 0$$

Now applying $R_2 \to R_2 - R_1$ and $R_3 \to R_3 - R_1$ we get

$$\Rightarrow \qquad (3a-x)\begin{vmatrix} 1 & a-x & a-x \\ 0 & 2x & 0 \\ 0 & 0 & 2x \end{vmatrix} = 0$$

Expanding along C_1, we get

$\Rightarrow (3a-x)(4x^2-0) = 0 \quad \Rightarrow \quad 4x^2(3a-x) = 0 \quad \Rightarrow$ If $4x^2 = 0$, then $x = 0$

$\Rightarrow$ If $3a - x = 0$, then $x = 3a$ Hence, $x = 0$ or $3a$ **Ans.**

Example 32. *Using properties of determinants, prove the following*

$$\begin{vmatrix} 1+a & 1 & 1 \\ 1 & 1+b & 1 \\ 1 & 1 & 1+c \end{vmatrix} = abc\left(\frac{1}{a}+\frac{1}{b}+\frac{1}{c}+1\right)$$

Solution. Let $$\Delta = \begin{vmatrix} 1+a & 1 & 1 \\ 1 & 1+b & 1 \\ 1 & 1 & 1+c \end{vmatrix}$$

Taking a, b and c common from R_1, R_2 and R_3 rows respectively.

$$\Delta = abc\begin{vmatrix} \frac{1+a}{a} & \frac{1}{a} & \frac{1}{a} \\ \frac{1}{b} & \frac{1+b}{b} & \frac{1}{b} \\ \frac{1}{c} & \frac{1}{c} & \frac{1+c}{c} \end{vmatrix} = abc\begin{vmatrix} 1+\frac{1}{a} & \frac{1}{a} & \frac{1}{a} \\ \frac{1}{b} & 1+\frac{1}{b} & \frac{1}{b} \\ \frac{1}{c} & \frac{1}{c} & 1+\frac{1}{c} \end{vmatrix}$$

$$= abc\begin{bmatrix} 1+\frac{1}{a}+\frac{1}{b}+\frac{1}{c} & 1+\frac{1}{a}+\frac{1}{b}+\frac{1}{c} & 1+\frac{1}{a}+\frac{1}{b}+\frac{1}{c} \\ \frac{1}{b} & 1+\frac{1}{b} & \frac{1}{b} \\ \frac{1}{c} & \frac{1}{c} & 1+\frac{1}{c} \end{bmatrix} R_1 \to R_1 + R_2 + R_3$$

Taking $\left(1+\frac{1}{a}+\frac{1}{b}+\frac{1}{c}\right)$ common from R_1, we get

$$\Delta = abc\left(1+\frac{1}{a}+\frac{1}{b}+\frac{1}{c}\right)\begin{vmatrix} 1 & 1 & 1 \\ \frac{1}{b} & 1+\frac{1}{b} & \frac{1}{b} \\ \frac{1}{c} & \frac{1}{c} & 1+\frac{1}{c} \end{vmatrix}$$

Operate : $C_2 \to C_2 - C_1$; $C_3 \to C_3 - C_1$ **Proved**

$$\Delta = abc\left(\frac{1}{a}+\frac{1}{b}+\frac{1}{c}+1\right)\begin{vmatrix} 1 & 0 & 0 \\ \frac{1}{b} & 1 & 0 \\ \frac{1}{c} & 0 & 1 \end{vmatrix} = abc\left(\frac{1}{a}+\frac{1}{b}+\frac{1}{c}+1\right) \text{ (On expanding by } R_1\text{)}.$$

Example 33. *Prove that :*

$$\begin{vmatrix} a & a^2 & bc \\ b & b^2 & ac \\ c & c^2 & ab \end{vmatrix} = (a-b)(b-c)(c-a)(ab+bc+ac).$$

Solution. Let
$$\Delta = \begin{vmatrix} a & a^2 & bc \\ b & b^2 & ac \\ c & c^2 & ab \end{vmatrix}$$

$$= \frac{1}{abc}\begin{vmatrix} a^2 & a^3 & abc \\ b^2 & b^3 & abc \\ c^2 & c^3 & abc \end{vmatrix} = \frac{1}{abc}.abc\begin{vmatrix} a^2 & a^3 & 1 \\ b^2 & b^3 & 1 \\ c^2 & c^3 & 1 \end{vmatrix}$$

$$\therefore \quad \Delta = \begin{vmatrix} a^2-b^2 & a^3-b^3 & 0 \\ b^2-c^2 & b^3-c^3 & 0 \\ c^2 & c^3 & 1 \end{vmatrix} \begin{matrix} R_1 \to R_1 - R_2 \\ R_2 \to R_2 - R_3 \\ \end{matrix}$$

$$= (a-b)(b-c)\begin{vmatrix} a+b & a^2+ab+b^2 & 0 \\ b+c & b^2+bc+c^2 & 0 \\ c^2 & c^3 & 1 \end{vmatrix}$$

Expand by C_3

$$= (a-b)(b-c).1\begin{vmatrix} a+b & a^2+ab+b^2 \\ b+c & b^2+bc+c^2 \end{vmatrix}$$

$$= (a-b)(b-c)\begin{vmatrix} a+b & a^2+ab+b^2 \\ c-a & bc+c^2-a^2-ab \end{vmatrix} R_2 \to R_2 - R_1$$

$$= (a-b)(b-c)\begin{vmatrix} a+b & a^2+ab+b^2 \\ c-a & b(c-a)+(c^2-a^2) \end{vmatrix}$$

$$= (a-b)(b-c)(c-a)\begin{vmatrix} a+b & a^2+ab+b^2 \\ 1 & b+c+a \end{vmatrix}$$

$$= (a-b)(b-c)(c-a)[(a+b)(a+b+c) - 1.(a^2+ab+b^2)]$$

$$= (a-b)(b-c)(c-a)(ab+bc+ac).$$ **Proved.**

EXERCISE 37.2

Expand the following determinants, using properties of the determinants :

1. $\begin{vmatrix} 1 & 3 & 7 \\ 4 & 9 & 1 \\ 2 & 7 & 6 \end{vmatrix}$ **Ans.** 51.

2. $\begin{vmatrix} x & a & a \\ a & x & a \\ a & a & x \end{vmatrix}$ **Ans.** $(x+2a)(x-a)^2$

3. Show that $\begin{vmatrix} 0 & x-a & x-b \\ x-a & 0 & x-c \\ x-b & x-c & 0 \end{vmatrix} = 2(x-a)(x-b)(x-c).$

4. $\begin{vmatrix} \frac{1}{a} & a & bc \\ \frac{1}{b} & b & ca \\ \frac{1}{c} & c & ab \end{vmatrix} = 0$

5. $\begin{vmatrix} x+y & y+z & z+x \\ z & x & y \\ 1 & 1 & 1 \end{vmatrix} = 0.$

6. $\begin{vmatrix} x+4 & 2x & 2x \\ 2x & x+4 & 2x \\ 2x & 2x & x+4 \end{vmatrix} = (5x+4)(4-x)^2$

7. $\begin{vmatrix} a+b+2c & a & b \\ c & b+c+2a & b \\ c & a & c+a+2b \end{vmatrix} = 2(a+b+c)^3$

8. $\begin{vmatrix} 1 & x+y & x^2+y^2 \\ 1 & y+z & y^2+z^2 \\ 1 & z+x & z^2+x^2 \end{vmatrix} = (x-y)(y-z)(z-x).$

9. $\begin{vmatrix} b^2c^2 & bc & b+c \\ c^2a^2 & ca & c+a \\ a^2b^2 & ab & a+b \end{vmatrix} = 0$

10. $\begin{vmatrix} 1 & a & a^2-bc \\ 1 & b & b^2-ca \\ 1 & c & c^2-ab \end{vmatrix} = 0.$

11. $\begin{vmatrix} a & a+b & a+b+c \\ 2a & 3a+2b & 4a+3b+2c \\ 3a & 6a+3b & 10a+6b+3c \end{vmatrix} = a^3.$

12. $\begin{vmatrix} 1 & a & bc \\ 1 & b & ca \\ 1 & c & ab \end{vmatrix} = (a-b)(b-c)(c-a).$

13. $\begin{vmatrix} 1 & a & a^2 \\ 1 & b & b^2 \\ 1 & c & c^2 \end{vmatrix} = \begin{vmatrix} 1 & a & bc \\ 1 & b & ca \\ 1 & c & ab \end{vmatrix}$

14. $\begin{vmatrix} a+b+c & -c & -b \\ -c & a+b+c & -a \\ -b & -a & a+b+c \end{vmatrix} = 2(a+b)(b+c)(c+a).$

15. $\begin{vmatrix} a^2 & bc & ac+c^2 \\ a^2+ab & b^2 & ac \\ ab & b^2+bc & c^2 \end{vmatrix} = 4a^2b^2c^2$

16. $\begin{vmatrix} 3a & -a+b & -a+c \\ -b+a & 3b & -b+c \\ -c+a & -c+b & 3c \end{vmatrix} = 3\,(a+b+c)\,(ab+bc+ca)$

37.6 FACTOR THEOREM

If the elements of a determinant are polynomials in a variable x and if the substitution $x = a$ makes two rows (or columns) identical then $(x - a)$ is a factor of the determinant.

When two rows are identical, the value of the determinant is zero. The expansion of a determinant being polynomial in x vanishes on putting $x = a$, then $x - a$ is its factor by the Remainder theorem.

Example 34. *Show that*

$$\begin{vmatrix} 1 & 1 & 1 \\ x & y & z \\ x^2 & y^2 & z^2 \end{vmatrix} = (x-y)(y-z)(z-x)1$$

Solution. If we put $x = y$, $y = z$ *and* $z = x$ then in each case two columns become identical and the determinant vanishes.

$\therefore$ $(x-y)$, $(y-z)$, and $(z-x)$ are the factors.

Since the determinant is of third degree, the other factor can be numerical only k (say).

$$\begin{vmatrix} 1 & 1 & 1 \\ x & y & z \\ x^2 & y^2 & z^2 \end{vmatrix} = k(x-y)(y-z)(z-x) \qquad \text{... (1)}$$

This leading term (product of the elements of the diagonal elements) L.H.S. of (1) is yz^2 and in the expansion of R.H.S. *i.e.* $k(x-y)(y-z)(z-x)$ we get kyz^2

Equating the coefficient of yz^2 on both sides of (1), we have

$$k = 1$$

Hence, the expansion $= (x-y)(y-z)(z-x)$. **Proved.**

Example 35. *Using properties of determinants, prove that*

$$\begin{vmatrix} x & x^2 & 1+px^3 \\ y & y^2 & 1+py^3 \\ z & z^2 & 1+pz^3 \end{vmatrix} = (1+pxyz)(x-y)(y-z)(z-x), \text{ where } p \text{ is any scalar.}$$

Solution. We have,

$$\begin{vmatrix} x & x^2 & 1+px^3 \\ y & y^2 & 1+py^3 \\ z & z^2 & 1+pz^3 \end{vmatrix} = \begin{vmatrix} x & x^2 & 1 \\ y & y^2 & 1 \\ z & z^2 & 1 \end{vmatrix} + \begin{vmatrix} x & x^2 & px^3 \\ y & y^2 & py^3 \\ z & z^2 & pz^3 \end{vmatrix} = \begin{vmatrix} x & x^2 & 1 \\ y & y^2 & 1 \\ z & z^2 & 1 \end{vmatrix} + pxyz \begin{vmatrix} 1 & x & x^2 \\ 1 & y & y^2 \\ 1 & z & z^2 \end{vmatrix}$$

$$= \begin{vmatrix} 1 & x & x^2 \\ 1 & y & y^2 \\ 1 & z & z^2 \end{vmatrix} + pxyz \begin{vmatrix} 1 & x & x^2 \\ 1 & y & y^2 \\ 1 & z & z^2 \end{vmatrix} = (1+pxyz) \begin{vmatrix} 1 & x & x^2 \\ 1 & y & y^2 \\ 1 & z & z^2 \end{vmatrix} \qquad \text{... (1)}$$

If we put $x = y$ then two rows become identical and the determinant vanishes.

$\Rightarrow$ $(x-y)$ is a factor.

If we put $y = z$ then two rows become identical and the determinant vanishes.

$\Rightarrow$ $(y-z)$ is a factor.

If we put $z = x$, then two rows become identical and the determinant vanishes.

$\Rightarrow$ $(z-x)$ is also a factor.

Since the determinant is of the third degree, the other factor can be numerical k,

$$\begin{vmatrix} 1 & x & x^2 \\ 1 & y & y^2 \\ 1 & z & z^2 \end{vmatrix} = k(x-y)(y-z)(z-x) \qquad \text{...(2)}$$

This leading term (product of the diagonal elements) L.H.S. of (2) is yz^2 and in the expansion of R.H.S. i.e., $k(x-y)(y-z)(z-x)$ we get kyz^2.

Equating the coefficients of yz^2, we have $k = 1$

$$\text{Hence, } \begin{vmatrix} 1 & x & x^2 \\ 1 & y & y^2 \\ 1 & z & z^2 \end{vmatrix} = (x-y)(y-z)(z-x) \qquad \text{... (3)}$$

From (1) and (3), we have

$$\text{The given determinant } = \begin{vmatrix} x & x^2 & 1+px^3 \\ y & y^2 & 1+py^3 \\ z & z^2 & 1+pz^3 \end{vmatrix} = (1+pxyz)(x-y)(y-z)(z-x) \qquad \textbf{Proved.}$$

Example 36. *Factorize*

$$\Delta = \begin{vmatrix} 1 & 1 & 1 \\ a^2 & b^2 & c^2 \\ a^3 & b^3 & c^3 \end{vmatrix}$$

Solution. Putting $a = b$, $C_1 = C_2$ and hence $\Delta = 0$.

$\therefore$ $a - b$ is a factor of Δ.

Similarly, $b - c$, $c - a$ are also factors of Δ.

$\therefore$ $(a - b)(b - c)(c - a)$ is a third degree factor of Δ which itself is of the fifth degree as is judged from the leading term b^2c^3.

$\therefore$ The remaining factor must be of the second degree. As Δ is symmetrical in a, b, c the remaining factor must, therefore, be of the form

$$k(a^2 + b^2 + c^2) + l(ab + bc + ca)$$

$$\therefore \qquad \Delta = (a-b)(b-c)(c-a)\{k(a^2+b^2+c^2) + l(ab+bc+ca)\}$$

If $k \neq 0$, we shall get terms like a^4b, b^4c etc. which do not occur in Δ. Hence, k must be zero.

$$\therefore \qquad \Delta = (a-b)(b-c)(c-a)\{0 + l(ab+bc+ca)\}$$

or $\qquad \Delta = l(a-b)(b-c)(c-a)(ab+bc+ca)$

The leading term in $\quad \Delta = b^2c^3$

The corresponding term on R.H.S. $= l\, b^2c^3$

$\therefore \qquad l = 1$

Hence, $\qquad \Delta = (a-b)(b-c)(c-a)(ab+bc+ca)$. **Ans.**

EXERCISE 37.3

1. Evaluate, without expanding

$$\begin{vmatrix} a & a^2 & 1+a^3 \\ b & b^2 & 1+b^3 \\ c & c^2 & 1+c^3 \end{vmatrix}$$ **Ans.** $(a-b)(b-c)(c-a)(1+abc)$

2. Solve the equation $\begin{vmatrix} x^3-a^3 & x^2 & x \\ b^3-a^3 & b^2 & b \\ c^3-a^3 & c^2 & c \end{vmatrix} = 0, b \neq c, c \neq 0, b \neq 0.$ **Ans.** $x = \dfrac{a^3}{bc}, x = b, x = c$

3. Without expanding, show that

$$\Delta = \begin{vmatrix} (a-x)^2 & (a-y)^2 & (a-z)^2 \\ (b-x)^2 & (b-y)^2 & (b-z)^2 \\ (c-x)^2 & (c-y)^2 & (c-z)^2 \end{vmatrix} = 2(a-b)(b-c)(c-a)(x-y)(y-z)(z-x).$$

4. Show (without expanding) that

$$\begin{vmatrix} bc & a^2 & a^2 \\ b^2 & ca & b^2 \\ c^2 & c^2 & ab \end{vmatrix} = \begin{vmatrix} bc & ab & ca \\ ab & ca & bc \\ ca & bc & ab \end{vmatrix}$$

$$= -\frac{1}{2}(ab+bc+ca)[(ab-bc)^2 + (bc-ca)^2 + (ca-ab)^2]$$

37.7 SPECIAL TYPES OF DETERMINANTS

***(i)* Ortho-symmetric Determinant.** If every element of the leading diagonal is the same and the equidistant elements from the diagonal are equal, then the determinant is said to be ortho-symmetric determinant.

$$\begin{vmatrix} a & h & g \\ h & a & f \\ g & f & a \end{vmatrix}$$

***(ii)* Skew-Symmetric Determinant.** If the elements of the leading diagonal are all zero and every other element is equal to its conjugate with sign changed, the determinant is said to be Skew-symmetric.

$$\begin{vmatrix} 0 & -a & -b \\ a & 0 & -c \\ b & c & 0 \end{vmatrix}$$

Property 1. *A Skew-symmetric determinant of odd order vanishes.*

Example 37. *Prove that*

$$= \begin{vmatrix} 0 & -a & -b \\ a & 0 & -c \\ b & c & 0 \end{vmatrix} = 0$$

Solution. Taking out (– 1) common from each of the three columns

$$\Delta = (-1)^3 \begin{vmatrix} 0 & a & b \\ -a & 0 & c \\ -b & -c & 0 \end{vmatrix}$$

Changing rows into columns

$$\Delta = (-1)^3 \begin{vmatrix} 0 & -a & -b \\ a & 0 & -c \\ b & c & 0 \end{vmatrix} = (-1)^3\ \Delta = -\ \Delta \quad \text{or} \quad 2\Delta = 0 \quad \text{or} \quad \Delta = 0$$

37.8 APPLICATION OF DETERMINANTS

Area of triangle. We know that the area of a triangle, whose vertices are (x_1, y_1), (x_2, y_2) and (x_3, y_3) is given by

$$\Delta = \frac{1}{2}\left[x_1(y_2 - y_3) - x_2(y_1 - y_3) + x_3(y_1 - y_2)\right]$$

$$= \frac{1}{2}\left[x_1 \begin{vmatrix} y_2 & 1 \\ y_3 & 1 \end{vmatrix} - x_2 \begin{vmatrix} y_1 & 1 \\ y_3 & 1 \end{vmatrix} + x_3 \begin{vmatrix} y_1 & 1 \\ y_2 & 1 \end{vmatrix}\right] = \frac{1}{2}\begin{vmatrix} x_1 & y_1 & 1 \\ x_2 & y_2 & 1 \\ x_3 & y_3 & 1 \end{vmatrix}$$

Note. Since area is always a positive quantity, therefore we always take the absolute value of the determinant for the area.

Condition of collinearity of three points. Let $A(x_1, y_1)$, $B(x_2, y_2)$ and $C(x_3, y_3)$ be three points. Then, *A*, *B*, *C* are collinear

$\Leftrightarrow$ area of triangle $ABC = 0$

$$\Leftrightarrow \frac{1}{2}\begin{vmatrix} x_1 & y_1 & 1 \\ x_2 & y_2 & 1 \\ x_3 & y_3 & 1 \end{vmatrix} = 0 \Leftrightarrow \begin{vmatrix} x_1 & y_1 & 1 \\ x_2 & y_2 & 1 \\ x_3 & y_3 & 1 \end{vmatrix} = 0$$ **Proved.**

Example 38. *Using determinants, find the area of the triangle with vertices (– 2, –3), (3, 2) and (–1, –8).*

Solution. The area of the given triangle $= \frac{1}{2}\begin{vmatrix} -2 & -3 & 1 \\ 3 & 2 & 1 \\ -1 & -8 & 1 \end{vmatrix}$

$$= \frac{1}{2}\begin{vmatrix} -2 & -3 & 1 \\ 5 & 5 & 0 \\ 1 & -5 & 0 \end{vmatrix} \quad \begin{matrix} R_2 \to R_2 - R_1 \\ R_3 \to R_3 - R_1 \end{matrix}$$

Expand by C_3 we get

$$= \frac{1}{2}.1.\begin{vmatrix} 5 & 5 \\ 1 & -5 \end{vmatrix} = \frac{1}{2}(-25-5) = \frac{|-30|}{2} = 15 \text{ sq. units}$$ **Ans.**

Example 39. *If area of triangle is 35 sq. units with vertices (2, –6), (5, 4) and (k, 4). Then find k.*

Solution. Let the vertices of triangle be A (2, –6), B (5, 4) and C (k, 4). Since the area of the triangle ABC is 35 sq. units, so we have

$$\frac{1}{2}\begin{vmatrix} 2 & -6 & 1 \\ 5 & 4 & 1 \\ k & 4 & 1 \end{vmatrix} = \pm 35$$

$$\Rightarrow \quad \frac{1}{2}\begin{vmatrix} 2 & -6 & 1 \\ 3 & 10 & 0 \\ k-2 & 10 & 0 \end{vmatrix} = \pm 35 \quad [\text{Applying } R_2 \to R_2 - R_1 \text{ and } R_3 \to R_3 - R_1]$$

$$\Rightarrow \quad \frac{1}{2}\begin{vmatrix} 3 & 10 \\ k-2 & 10 \end{vmatrix} = \pm 35 \Rightarrow \frac{1}{2}\{30 - 10(k-2)\} = \pm 35 \quad [\text{Exanding along } C_3]$$

$\Rightarrow \quad 30 - 10k + 20 = \pm 70$

$\Rightarrow \quad 10k = 50 \mp 70 \quad \Rightarrow \quad k = 12 \quad$ or $\quad k = -2$ **Ans.**

Example 40. *Show that points A(a, b + c), B(b, c + a), C(c, a+ b) are collinear.*

Solution. The area of the triangle formed by the given points:

$$= \frac{1}{2}\begin{vmatrix} a & b+c & 1 \\ b & c+a & 1 \\ c & a+b & 1 \end{vmatrix}$$

Operate: $R_2 \to R_2 - R_1, R_3 \to R_3 - R_1$

$$= \frac{1}{2}\begin{vmatrix} a & b+c & 1 \\ b-a & a-b & 0 \\ c-a & a-c & 0 \end{vmatrix} = \frac{1}{2}(1)\{(b-a)(a-c) - (a-b)(c-a)\} \quad [\text{Expanding along } C_3]$$

$$= \frac{1}{2}[ab - bc - a^2 + ac - ac + a^2 + bc - ab] = \frac{1}{2}[0] = 0$$

Hence, the given points are collinear. **Proved.**

Example 41. *Using determinants, show that the points (11, 7), (5, 5) and (– 1, 3) are collinear.*

Solution. The area of the triangle formed by the given points

$$= \frac{1}{2}\begin{vmatrix} 11 & 7 & 1 \\ 5 & 5 & 1 \\ -1 & 3 & 1 \end{vmatrix}$$

Operate: $R_1 \to R_1 - R_2$; $R_2 \to R_2 - R_3$

$$= \frac{1}{2}\begin{vmatrix} 6 & 2 & 0 \\ 6 & 2 & 0 \\ -1 & 3 & 1 \end{vmatrix} = \frac{1}{2}.0 = 0. \qquad (\because R_1 \text{ and } R_2 \text{ are identical})$$

Hence, the given points are collinear. **Proved.**

Example 42. *Using determinants, find the area of the triangle whose vertices are (1, 4) (2, 3) and (– 5, – 3). Are the given points collinear?*

Solution. Area of the required triangle

$$= \frac{1}{2}\begin{vmatrix} 1 & 4 & 1 \\ 2 & 3 & 1 \\ -5 & -3 & 1 \end{vmatrix} = \frac{1}{2}[1(3+3) - 4(2+5) + 1(-6+15)] = \frac{1}{2}[6 - 28 + 9] = \frac{13}{2} \neq 0$$

Hence, the given points are not collinear. **Ans.**

Example 43. *Find the equation of line joining A* (1, 2) *and B* (3, 6) *using determinants.*

Solution. Let *P* (*x*, *y*) be any point on *AB*. Then, area of triangle *ABP* is zero. So,

$$\frac{1}{2}\begin{vmatrix} 1 & 2 & 1 \\ 3 & 6 & 1 \\ x & y & 1 \end{vmatrix} = 0$$

$$\Rightarrow \frac{1}{2}\begin{vmatrix} 1 & 2 & 1 \\ 2 & 4 & 0 \\ x-1 & y-2 & 0 \end{vmatrix} = 0 \qquad [\text{Applying } R_2 \to R_2 - R_1, R_3 \to R_3 - R_1]$$

$$\Rightarrow \frac{1}{2}(1)\{2(y-2) - 4(x-1)\} = 0 \qquad [\text{Expanding along } C_3]$$

$$\Rightarrow y - 2 - 2x + 2 = 0 \Rightarrow \quad y = 2x$$ **Ans.**

EXERCISE 37.4

1. Using determinants, find the area of the triangle with vertices (2,– 7), (1, 3), (10, 8). **Ans.** Area = $\frac{95}{2}$
2. Using determinants, show that the points (3, 8), (– 4, 2) and (10, 14) are collinear.
3. Using determinants, find the area of the triangle whose vertices are (– 2, 4), (2, – 6) and (5, 4). Are the given points collinear ? **Ans.** Area = 35, not collinear
4. Using determinants, find the area of the triangle whose vertices are (– 1, – 3), (2, 4) and (3, – 1). Are the given points collinear ? **Ans.** Area = 11, not collinear
5. Using determinants, find the area of the triangle whose vertices are (1, – 1), (2, 4) and (– 3, 5). Are the given points collinear ? **Ans.** Area = 13, not collinear

6. Find the value of α, so that the points (1, – 5), (– 4, 5) and (α, 7) are collinear. **Ans.** α = – 5
7. Find the value of x, if the area of triangle is 35 square cms with vertices (x, 4), (2, – 6), (5, 4). **Ans.** $x = -2, 12$
8. Using determinants find the value of k, so that the points (k, 2 – 2k), (– k + 1, 2k) and (– 4 – k, 6 – 2 k) may be collinear. **Ans.** $k = -1, \frac{1}{2}$
9. If the points (x, – 2), (5, 2) and (8, 8) are collinear, find x using determinants. **Ans.** $x = 3$
10. If the points (3, – 2), (x, 2) and (8, 8) are collinear, find x using determinants. **Ans.** $x = 5$

37.9 SOLUTION OF SIMULTANEOUS LINEAR EQUATIONS BY DETERMINANTS (CRAMER'S RULE)

Let us solve the following equations.

$$a_1 x + b_1 y + c_1 z = d_1$$
$$a_2 x + b_2 y + c_2 z = d_2$$
$$a_3 x + b_3 y + c_3 z = d_3$$

Let
$$D = \begin{vmatrix} a_1 & b_1 & c_1 \\ a_2 & b_2 & c_2 \\ a_3 & b_3 & c_3 \end{vmatrix} \quad \text{or} \quad x\,D = \begin{vmatrix} a_1 x & b_1 & c_1 \\ a_2 x & b_2 & c_2 \\ a_3 x & b_3 & c_3 \end{vmatrix}$$

Multiplying the 2nd column by y and 3rd column by z and adding to the 1st column, we get

$$xD = \begin{vmatrix} a_1x + b_1y + c_1z & b_1 & c_1 \\ a_2x + b_2y + c_2z & b_2 & c_2 \\ a_3x + b_3y + c_3z & b_3 & c_3 \end{vmatrix}, \quad xD = \begin{vmatrix} d_1 & b_1 & c_1 \\ d_2 & b_2 & c_2 \\ d_3 & b_3 & c_3 \end{vmatrix} \Rightarrow x = \frac{\begin{vmatrix} d_1 & b_1 & c_1 \\ d_2 & b_2 & c_2 \\ d_3 & b_1 & c_3 \end{vmatrix}}{\begin{vmatrix} a_1 & b_1 & c_1 \\ a_2 & b_2 & c_2 \\ a_3 & b_3 & c_3 \end{vmatrix}} = \frac{D_1}{D}$$

Similarly,
$$y = \frac{D_2}{D} = \frac{\begin{vmatrix} a_1 & d_1 & c_1 \\ a_2 & d_2 & c_2 \\ a_3 & d_3 & c_3 \end{vmatrix}}{\begin{vmatrix} a_1 & b_1 & c_1 \\ a_2 & b_2 & c_2 \\ a_3 & b_3 & c_3 \end{vmatrix}} \Rightarrow z = \frac{D_3}{D} = \frac{\begin{vmatrix} a_1 & b_1 & d_1 \\ a_2 & b_2 & d_2 \\ a_3 & b_3 & d_3 \end{vmatrix}}{\begin{vmatrix} a_1 & b_1 & c_1 \\ a_2 & b_2 & c_2 \\ a_3 & b_3 & c_3 \end{vmatrix}}$$

Thus,
$$x = \frac{D_1}{D}, \quad y = \frac{D_2}{D}, \quad z = \frac{D_3}{D}$$ **Ans.**

Example 44. *Solve the following system of equations using Cramer's rule :*

$$5x - 7y + z = 11$$
$$6x - 8y - z = 15$$
$$3x + 2y - 6z = 7$$

Solution. The given equations are

$$5x - 7y + z = 11$$
$$6x - 8y - z = 15$$
$$3x + 2y - 6z = 7$$

Here, $D = \begin{vmatrix} 5 & -7 & 1 \\ 6 & -8 & -1 \\ 3 & 2 & -6 \end{vmatrix} = 5\,(48 + 2) + 7\,(-36 + 3) + 1\,(12 + 24) = 55$

$$D_1 = \begin{vmatrix} 11 & -7 & 1 \\ 15 & -8 & -1 \\ 7 & 2 & -6 \end{vmatrix} = 11\,(48 + 2) + 7\,(-90 + 7) + 1\,(30 + 56) = 55$$

$$D_2 = \begin{vmatrix} 5 & 11 & 1 \\ 6 & 15 & -1 \\ 3 & 7 & -6 \end{vmatrix} = 5\,(-90 + 7) - 11\,(-36 + 3) + 1\,(42 - 45) = -55$$

$$D_3 = \begin{vmatrix} 5 & -7 & 11 \\ 6 & -8 & 15 \\ 3 & 2 & 7 \end{vmatrix} = 5\,(-56 - 30) + 7\,(42 - 45) + 11\,(12 + 24) = -55$$

By Cramer's Rule

$$x = \frac{D_1}{D} = \frac{55}{55} = 1, \quad y = \frac{D_2}{D} = \frac{-55}{55} = -1, \quad z = \frac{D_3}{D} = \frac{-55}{55} = -1$$

Hence, $x = 1, y = -1, z = -1$ **Ans.**

Example 45. *Solve, by determinants, the following set of simultaneous equations:*

$$5x - 6y + 4z = 15$$

$$7x + 4y - 3z = 19$$

$$2x + y + 6z = 46$$

Solution. $D = \begin{vmatrix} 5 & -6 & 4 \\ 7 & 4 & -3 \\ 2 & 1 & 6 \end{vmatrix} = 419$ $\qquad D_1 = \begin{vmatrix} 15 & -6 & 4 \\ 19 & 4 & -3 \\ 46 & 1 & 6 \end{vmatrix} = 1257$

$$D_2 = \begin{vmatrix} 5 & 15 & 4 \\ 7 & 19 & -3 \\ 2 & 46 & 6 \end{vmatrix} = 1676 \qquad D_3 = \begin{vmatrix} 5 & -6 & 15 \\ 7 & 4 & 19 \\ 2 & 1 & 46 \end{vmatrix} = 2514$$

By Cramer's Rule

$$x = \frac{D_1}{D} = \frac{1257}{419} = 3 \qquad y = \frac{D_2}{D} = \frac{1676}{419} = 4. \qquad z = \frac{D_3}{D} = \frac{2514}{419} = 6.$$

Hence $x = 3, y = 4, z = 6$ **Ans.**

Example 46. *Solve, using Cramer's rule*

$$3x - 2y + 4z = 5$$

$$x + y + 3z = 2$$

$$-x + 2y - z = 1$$

Solution. $D = \begin{vmatrix} 3 & -2 & 4 \\ 1 & 1 & 3 \\ -1 & 2 & -1 \end{vmatrix} = -5$ $\quad D_1 = \begin{vmatrix} 5 & -2 & 4 \\ 2 & 1 & 3 \\ 1 & 2 & -1 \end{vmatrix} = -33$

$$D_2 = \begin{vmatrix} 3 & 5 & 4 \\ 1 & 2 & 3 \\ -1 & 1 & -1 \end{vmatrix} = -13 \qquad D_3 = \begin{vmatrix} 3 & -2 & 5 \\ 1 & 1 & 2 \\ -1 & 2 & 1 \end{vmatrix} = 12$$

By Cramer's Rule

$$x = \frac{D_1}{D} = \frac{-33}{-5} = \frac{33}{5} \qquad y = \frac{D_2}{D} = \frac{-13}{-5} = \frac{13}{5}$$

Hence, $z = \frac{D_3}{D} = \frac{12}{-5} = \frac{-12}{5}$ $\qquad x = \frac{33}{5}, \quad y = \frac{13}{5}, \quad z = \frac{-12}{5}$ **Ans.**

Example 47. *Solve the following system of equations by using determinants :*

$$x + y + z = 1$$
$$ax + by + cz = k$$
$$a^2x + b^2y + c^2z = k^2$$

Solution. We have,

$$D = \begin{vmatrix} 1 & 1 & 1 \\ a & b & c \\ a^2 & b^2 & c^2 \end{vmatrix} = \begin{vmatrix} 1 & 0 & 0 \\ a & b-a & c-a \\ a^2 & b^2-a^2 & c^2-a^2 \end{vmatrix} \text{[Applying } C_2 \to C_2 - C_1 \text{ and } C_3 \to C_3 - C_1]$$

$$= (b-a)(c-a)\begin{vmatrix} 1 & 0 & 0 \\ a & 1 & 1 \\ a^2 & b+a & c+a \end{vmatrix}$$

$$= (b-a)(c-a).1.\begin{vmatrix} 1 & 1 \\ b+a & c+a \end{vmatrix} \qquad \text{[Expanding along } R_1]$$

$$= (b-a)(c-a)(c+a-b-a) = (b-c)(c-a)(a-b) \qquad ...(1)$$

$$D_1 = \begin{vmatrix} 1 & 1 & 1 \\ k & b & c \\ k^2 & b^2 & c^2 \end{vmatrix} = (b-c)(c-k)(k-b) \qquad \text{[Replacing } a \text{ by } k \text{ in (1)]}$$

$$D_2 = \begin{vmatrix} 1 & 1 & 1 \\ a & k & c \\ a^2 & k^2 & c^2 \end{vmatrix} = (k-c)(c-a)(a-k) \qquad \text{[Replacing } b \text{ by } k \text{ in (1)]}$$

and $$D_3 = \begin{vmatrix} 1 & 1 & 1 \\ a & b & k \\ a^2 & b^2 & k^2 \end{vmatrix} = (a-b)(b-k)(k-a) \qquad \text{[Replacing } c \text{ by } k \text{ in (1)]}$$

$\therefore$ $$x = \frac{D_1}{D} = \frac{(b-c)(c-k)(k-b)}{(b-c)(c-a)(a-b)} = \frac{(c-k)(k-b)}{(c-a)(a-b)},$$

$$y = \frac{D_2}{D} = \frac{(k-c)(c-a)(a-k)}{(b-c)(c-a)(a-b)} = \frac{(k-c)(a-k)}{(b-c)(a-b)},$$

and $$z = \frac{D_3}{D} = \frac{(a-b)(b-k)(k-a)}{(a-b)(b-c)(c-a)} = \frac{(b-k)(k-a)}{(b-c)(c-a)}$$

Hence, $x = \dfrac{(c-k)(k-b)}{(c-a)(a-b)}$, $y = \dfrac{(k-c)(a-k)}{(b-c)(a-b)}$ and $z = \dfrac{(b-k)(k-a)}{(b-c)(c-a)}$ **Ans.**

EXERCISE 37.6

Using Cramer's Rule, solve the following system of equations :

1. $2x - 3y = 7$ **Ans.** $x = \frac{3}{5}, y = -\frac{29}{15}$

$x - 2y = 8$

2. $2x + y = 1$ **Ans.** $x = 2, y = -3$

$7x - 3y = 10$

3. $2x + 3y = 10$ **Ans.** $x = \frac{16}{3}, y = -\frac{2}{9}$

$3x + 2y = 5.$

4. $5x + 2y = 3$ **Ans.** $x = -1, \; y = 4$

$x + 6y = 4.$

5. $7x - 2y = -7$ **Ans.** $x = -3, y = -7$

$-3x + 5y = -7$

6. $x - 2y = 4$ **Ans.** $x = -6, \; y = -5$

$2x - y = 1.$

7. $x - 4y - z = 11$ **Ans.** $x = -1, \; y = -5, \; z = 8$

$2x - 5y + 2z = 39$

$-3x + 2y + z = 1.$

8. $x + 3y - 2z = 5$ **Ans.** $x = 1, \; y = 2, z = 1$

$2x + y + 4z = 8$

$6x + y - 3z = 5$

9. $x + 2y + 5z = 23$ **Ans.** $x = 4, \; y = 2, \; z = 3$

$3x + y + 4z = 26$

$6x + y + 7z = 47$

10. $x + y + z = 1$ **Ans.** $x = \frac{1}{3}, y = 1, z = -\frac{1}{3}$

$3x + 5y + 6z = 4$

$9x + 2y - 36z = 17$

11. $2y - z = 0$ **Ans.** $x = 5, y = -3, z = -6$

$x + 3y = -4$

$3x + 4y = 3$

12. $x + y + z = -1$ **Ans.** $x = 1, \; y = -1, \; z = -1$

$x + 2y + 3z = -4$

$x + 3y + 4z = -6$

13. $x + y + z = 1$ **Ans.** $x = \dfrac{(2-k)(3-k)}{2}$, $y = \dfrac{(1-k)(3-k)}{-1}$, $z = \dfrac{(1-k)(2-k)}{2}$

$x + 2y + 3z = k$

$1^2x + 2^2y + 3^2z = k^2$

14. Show that there are three real values of λ for which the equations:

$(a - \lambda)\, x + by + cz = 0$

$bx + (c - \lambda)\, y + az = 0$

$cx + ay + (b - \lambda)\, z = 0$

are simultaneously true, and that the product of these values of λ is $\begin{vmatrix} a & b & c \\ b & c & a \\ c & a & b \end{vmatrix}$

37.10 RULE FOR MULTIPLICATION OF TWO DETERMINANTS

Multiply the elements of the first row of Δ_1 with the corresponding elements of the first, the second and the third row of Δ_2 respectively.

Their respective sums form the elements of the first row of $\Delta_1\Delta_2$. Similarly multiply the elements of the second row of $\Delta_1$with the corresponding elements of first, second and third row of the Δ_2 to from the second row of $\Delta_1\Delta_2$ and so on.

Example 48. *Find the product*

$$\begin{vmatrix} a_1 & b_1 & c_1 \\ a_2 & b_2 & c_2 \\ a_3 & b_3 & c_3 \end{vmatrix} \times \begin{vmatrix} \alpha_1 & \beta_1 & \gamma_1 \\ \alpha_2 & \beta_2 & \gamma_2 \\ \alpha_3 & \beta_3 & \gamma_3 \end{vmatrix}$$

Solution. Product of the given determinants

$$= \begin{vmatrix} a_1\alpha_1 + b_1\beta_1 + c_1\gamma_1 & a_1\alpha_2 + b_1\beta_2 + c_1\gamma_2 & a_1\alpha_3 + b_1\beta_3 + c_1\gamma_3 \\ a_2\alpha_1 + b_2\beta_1 + c_2\gamma_1 & a_2\alpha_2 + b_2\beta_2 + c_2\gamma_2 & a_2\alpha_3 + b_2\beta_3 + c_2\gamma_3 \\ a_3\alpha_1 + b_3\beta_1 + c_3\gamma_1 & a_3\alpha_2 + b_3\beta_2 + c_3\gamma_2 & a_3\alpha_2 + b_3\beta_3 + c_3\gamma_3 \end{vmatrix}$$ **Ans.**

Example 49. *Find*

$$\begin{vmatrix} a & b & c \\ b & c & a \\ c & a & b \end{vmatrix} \times \begin{vmatrix} -a & c & b \\ -b & a & c \\ -c & b & a \end{vmatrix}$$ *and hence show that*

$$\begin{vmatrix} 2bc - a^2 & c^2 & b^2 \\ c^2 & 2ca - b^2 & a^2 \\ b^2 & a^2 & 2ab - c^2 \end{vmatrix} = (a^3 + b^3 + c^3 - 3abc)^2$$

Solution. Product of the given determinants

$$= \begin{vmatrix} -a^2 + bc + bc & -ab + ab + c^2 & -ac + b^2 + ac \\ -ab + c^2 + ab & -b^2 + ac + ac & -bc + bc + a^2 \\ -ca + ca + b^2 & -bc + a^2 + bc & -c^2 + ab + ab \end{vmatrix} = \begin{vmatrix} 2bc - a^2 & c^2 & b^2 \\ c^2 & 2ca - b^2 & a^2 \\ b^2 & a^2 & 2ab - c^2 \end{vmatrix}$$

$$\text{Now}\quad \begin{vmatrix} -a & c & b \\ -b & a & c \\ -c & b & a \end{vmatrix} = (-1)^2 \begin{vmatrix} a & b & c \\ b & c & a \\ c & a & b \end{vmatrix} = a(bc-a^2)-b(b^2-ac)+c(ab-c^2)$$

$= -(a^3 + b^3 + c^3 - 3abc)$

Product $= (a^3 + b^3 + c^3 - 3abc)^2$ **Proved.**

Example 50. *Prove that the determinant*

$$\begin{vmatrix} 2b_1+c_1 & c_1+3a_1 & 2a_1+3b_1 \\ 2b_2+c_2 & c_2+3a_2 & 2a_2+3b_2 \\ 2b_3+c_3 & c_3+3a_3 & 2a_3+3b_3 \end{vmatrix}$$

is a multiple of the determinant $\begin{vmatrix} a_1 & b_1 & c_1 \\ a_2 & b_2 & c_2 \\ a_3 & b_3 & c_3 \end{vmatrix}$ *and find the other factor.*

Solution. $$\begin{vmatrix} 2b_1+c_1 & c_1+3a_1 & 2a_1+3b_1 \\ 2b_2+c_2 & c_2+3a_2 & 2a_2+3b_2 \\ 2b_3+c_3 & c_3+3a_3 & 2a_3+3b_3 \end{vmatrix} = \begin{vmatrix} a_1 & b_1 & c_1 \\ a_2 & b_2 & c_2 \\ a_3 & b_3 & c_3 \end{vmatrix} \times \begin{vmatrix} 0 & 2 & 1 \\ 3 & 0 & 1 \\ 2 & 3 & 0 \end{vmatrix}$$ **Ans.**

Example 51. *Prove that* $$\begin{vmatrix} 1 & \cos(\beta-\alpha) & \cos(\gamma-\alpha) \\ \cos(\alpha-\beta) & 1 & \cos(\gamma-\beta) \\ \cos(\alpha-\gamma) & \cos(\beta-\gamma) & 1 \end{vmatrix} = 0$$

Solution. $$\begin{vmatrix} \cos\alpha & \sin\alpha & 0 \\ \cos\beta & \sin\beta & 0 \\ \cos\gamma & \sin\gamma & 0 \end{vmatrix} \times \begin{vmatrix} \cos\alpha & \sin\alpha & 0 \\ \cos\beta & \sin\beta & 0 \\ \cos\gamma & \sin\gamma & 0 \end{vmatrix} = 0$$

$$\Rightarrow \begin{vmatrix} \cos^2\alpha+\sin^2\alpha & \cos\alpha\cos\beta+\sin\alpha\sin\beta & \cos\alpha\cos\gamma+\sin\alpha\sin\gamma \\ \cos\beta\cos\alpha+\sin\beta\sin\alpha & \cos^2\beta+\sin^2\beta & \cos\beta\cos\gamma+\sin\beta\sin\gamma \\ \cos\gamma\cos\alpha+\sin\gamma\sin\alpha & \cos\gamma\cos\beta+\sin\gamma & \cos^2\gamma+\sin^2\gamma \end{vmatrix} = 0$$

$\Rightarrow$ The above determinant can be split into eight determinants and each determinant having identical column is zero. **Proved.**

ALGEBRA OF MATRICES

38.1 DEFINITION

Let us consider a set of simultaneous equations,

$$x + 2y + 3z + 5t = 0$$
$$4x + 2y + 5z + 7t = 0$$
$$3x + 4y + 2z + 6t = 0.$$

Now we write down the coefficients of x, y, z, t of the above equations and enclose them within brackets and then we get

$$A = \begin{bmatrix} 1 & 2 & 3 & 5 \\ 4 & 2 & 5 & 7 \\ 3 & 4 & 2 & 6 \end{bmatrix}$$

The above system of numbers, arranged in a rectangular array in rows and columns and bounded by the brackets, is called a matrix.

It has got 3 rows and 4 columns and in all $3 \times 4 = 12$ elements. It is termed as 3×4 matrix, to be read as [3 by 4 matrix]. In the double subscripts of an element, the first subscript determines the row and the second subscript determines the column in which the element lies, a_{ij} lies in the ith row and jth column.

38.2 VARIOUS TYPES OF MATRICES

(*i*) **Row Matrix.** If a matrix has only one row and any number of columns, it is called a *Row matrix, e.g.,*

$$[2\ 7\ 3\ 9]$$

(*b*) **Column Matrix.** A matrix, having one column and any number of rows, is called a *Column matrix, e.g.,* $\begin{bmatrix} 1 \\ 2 \\ 3 \end{bmatrix}$

(*c*) **Null Matrix or Zero Matrix.** Any matrix, in which all the elements are zeros, is called a *Zero matrix* or *Null matrix e.g.,*

$$\begin{bmatrix} 0 & 0 & 0 & 0 \\ 0 & 0 & 0 & 0 \end{bmatrix}$$

(*d*) **Square Matrix.** A matrix, in which the number of rows is equal to the number of columns, is called a square matrix *e.g.*,

$$\begin{bmatrix} 2 & 5 \\ 1 & 4 \end{bmatrix}$$

(*e*) **Diagonal Matrix.** A square matrix is called a diagonal matrix, if all its non-diagonal elements are zero *e.g.*,

$$\begin{bmatrix} 1 & 0 & 0 \\ 0 & 3 & 0 \\ 0 & 0 & 4 \end{bmatrix}$$

(*f*) **Scalar matrix.** A diagonal matrix in which all the diagonal elements are equal to a scalar, say (k) is called a scalar matrix.
For example;

$$\begin{bmatrix} 2 & 0 & 0 \\ 0 & 2 & 0 \\ 0 & 0 & 2 \end{bmatrix}, \begin{bmatrix} -6 & 0 & 0 & 0 \\ 0 & -6 & 0 & 0 \\ 0 & 0 & -6 & 0 \\ 0 & 0 & 0 & -6 \end{bmatrix}$$

i.e., $A = [a_{ij}]_{n \times n}$ is a scalar matrix if $a_{ij} = \begin{cases} 0, & \text{when } i \neq j \\ k, & \text{when } i = j \end{cases}$

(*g*) **Unit or Identity Matrix.** A square matrix is called a unit matrix if all the diagonal elements are unity and non-diagonal elements are zero *e.g.*,

$$\begin{bmatrix} 1 & 0 & 0 \\ 0 & 1 & 0 \\ 0 & 0 & 1 \end{bmatrix}, \begin{bmatrix} 1 & 0 \\ 0 & 1 \end{bmatrix}$$

(*h*) **Symmetric Matrix.** A square matrix will be called symmetric, if for all values of i and j, $a_{ij} = a_{ji}$ *i.e.*, $A' = A$

$$e.g., \begin{bmatrix} a & h & g \\ h & b & f \\ g & f & c \end{bmatrix}$$

(*i*) **Skew Symmetric Matrix.** A square matrix is called skew symmetric matrix, if
(1) $a_{ij} = -a_{ji}$ for all values of i and j, or $A' = -A$
(2) All diagonal elements are zero, *e.g.*,

$$\begin{bmatrix} 0 & -h & -g \\ h & 0 & -f \\ g & f & 0 \end{bmatrix}$$

(*j*) **Triangular Matrix.** (Echelon form) A square matrix, all of whose elements below the leading diagonal are zero, is called an *upper triangular matrix.* A square matrix, all of whose elements above the leading diagonal are zero, is called *a lower triangular matrix* *e.g.*,

$$\begin{bmatrix} 1 & 3 & 2 \\ 0 & 4 & 1 \\ 0 & 0 & 6 \end{bmatrix} \qquad \begin{bmatrix} 2 & 0 & 0 \\ 4 & 1 & 0 \\ 5 & 6 & 7 \end{bmatrix}$$

Upper triangular matrix   Lower triangular matrix

(*k*) **Transpose of a Matrix.** If in a given matrix A, we interchange the rows and the corresponding columns, the new matrix obtained is called the transpose of the matrix A and is denoted by A' or A^T *e.g.*,

$$A = \begin{bmatrix} 2 & 3 & 4 \\ 1 & 0 & 5 \\ 6 & 7 & 8 \end{bmatrix}, \ A' = \begin{bmatrix} 2 & 1 & 6 \\ 3 & 0 & 7 \\ 4 & 5 & 8 \end{bmatrix}$$

(*l*) **Orthogonal Matrix.** A square matrix A is called an orthogonal matrix if the product of the matrix A and the transpose matrix A' is an identity matrix *e.g.*,

$$A.\ A' = I$$

if $|A| = 1$, matrix A is proper.

(*m*) **Conjugate of a Matrix**

Let
$$A = \begin{bmatrix} 1+i & 2-3i & 4 \\ 7+2i & -i & 3-2i \end{bmatrix}$$

Conjugate of matrix A is $\bar{A}$

$$\bar{A} = \begin{bmatrix} 1-i & 2+3i & 4 \\ 7-2i & i & 3+2i \end{bmatrix}$$

(*n*) **Matrix A^θ.** Transpose of the conjugate of a matrix A is denoted by A^θ.

Let
$$A = \begin{bmatrix} 1+i & 2-3i & 4 \\ 7+2i & -i & 3-2i \end{bmatrix}$$

$$\bar{A} = \begin{bmatrix} 1-i & 2+3i & 4 \\ 7-2i & +i & 3+2i \end{bmatrix}$$

$$(\bar{A})' = \begin{bmatrix} 1-i & 7-2i \\ 2+3i & i \\ 4 & 3+2i \end{bmatrix}$$

$$A^\theta = \begin{bmatrix} 1-i & 7-2i \\ 2+3i & i \\ 4 & 3+2i \end{bmatrix}$$

(*o*) **Unitary Matrix.** A square matrix A is said to be unitary if

$$A^\theta A = I$$

e.g.
$$A = \begin{bmatrix} \frac{1+i}{2} & \frac{-1+i}{2} \\ \frac{1+i}{2} & \frac{1-i}{2} \end{bmatrix}, \quad A^\theta = \begin{bmatrix} \frac{1-i}{2} & \frac{1-i}{2} \\ \frac{-1-i}{2} & \frac{1+i}{2} \end{bmatrix}, \quad A \cdot A^\theta = I$$

(*p*) **Hermitian Matrix.** A square matrix $A = (a_{ij})$ is called Hermitian matrix, if every *i-jth* element of A is equal to conjugate complex *j-i*th element of A.

In other words, $a_{ij} = \bar{a}_{ji}$

e.g.
$$\begin{bmatrix} 1 & 2+3i & 3+i \\ 2-3i & 2 & 1-2i \\ 3-i & 1+2i & 5 \end{bmatrix}$$

Necessary and sufficient condition for a matrix A to be Hermitian is that $A = A^\theta$ *i.e.* conjugate transpose of A

$$\Rightarrow \quad A = (\bar{A})'.$$

(*q*) **Skew Hermitian Matrix.** A square matrix $A = (a_{ij})$ will be called a Skew Hermitian matrix if every *i-j*th element of A is equal to negative conjugate complex of *j-i*th element of A.

In other words, $a_{ij} = -\bar{a}_{ji}$

All the elements in the principal diagonal will be of the form

$$a_{ii} = -\bar{a}_{ii} \quad \text{or} \quad a_{ii} + \bar{a}_{ii} = 0$$

If $\quad a_{ii} = a + ib \quad$ then $\quad \bar{a}_{ii} = a - ib$

$$(a + ib) + (a - ib) = 0 \quad \Rightarrow \quad 2a = 0 \Rightarrow a = 0$$

So, a_{ii} is pure imaginary $\Rightarrow a_{ii} = 0$.

Hence, all the diagonal elements of a Skew Hermitian Matrix are either zeros or pure imaginary.

e.g. $$\begin{bmatrix} i & 2-3i & 4+5i \\ -(2+3i) & 0 & 2i \\ -(4-5i) & 2i & -3i \end{bmatrix}$$

The necessary and sufficient condition for a matrix A to be Skew Hermitian is that

$$A^{\theta} = -A$$

$$(\bar{A})' = -A$$

(*r*) **Idempotent Matrix.** A matrix, such that $A^2 = A$ is called Idempotent Matrix.

$$e.g.\ A = \begin{bmatrix} 2 & -2 & -4 \\ -1 & 3 & 4 \\ 1 & -2 & -3 \end{bmatrix},\ A^2 = \begin{bmatrix} 2 & -2 & -4 \\ -1 & 3 & 4 \\ 1 & -2 & -3 \end{bmatrix}\begin{bmatrix} 2 & -2 & -4 \\ -1 & 3 & 4 \\ 1 & -2 & -3 \end{bmatrix} = \begin{bmatrix} 2 & -2 & -4 \\ -1 & 3 & 4 \\ 1 & -2 & -3 \end{bmatrix} = A$$

(*s*) **Periodic Matrix.** A matrix A will be called a Periodic Matrix, if

$$A^{k+1} = A$$

where k is a +ve integer. If k is the least + ve integer, for which $A^{k+1} = A$, then k is said to be the period of A. If we choose $k = 1$, we get $A^2 = A$ and we call it to be idempotent matrix.

(*t*) **Nilpotent Matrix.** A matrix will be called a Nilpotent matrix, if $A^k = 0$ (null matrix) where k is a +ve integer ; if however k is the least +ve integer for which $A^k = 0$, then k is the *index* of the nilpotent matrix.

$$e.g.,\ A = \begin{bmatrix} ab & b^2 \\ -a^2 & -ab \end{bmatrix},\ A^2 = \begin{bmatrix} ab & b^2 \\ -a^2 & -ab \end{bmatrix}\begin{bmatrix} ab & b^2 \\ -a^2 & -ab \end{bmatrix} = \begin{bmatrix} 0 & 0 \\ 0 & 0 \end{bmatrix} = 0$$

A is nilpotent matrix whose index is 2.

(*u*) **Involuntary Matrix.** A matrix A will be called an Involuntary matrix, if $A^2 = I$ (unit matrix). Since $I^2 = I$ always $\quad \therefore$ Unit matrix is involuntary.

(*v*) **Equal Matrices.** Two matrices are said to be equal if

(*i*) They are of the same order.

(*ii*) The elements in the corresponding positions are equal.

Thus if $$A = \begin{bmatrix} 2 & 3 \\ 1 & 4 \end{bmatrix},\ B = \begin{bmatrix} 2 & 3 \\ 1 & 4 \end{bmatrix}$$

Here $$A = B$$

(*w*) **Singular Matrix.** If the determinant of the matrix is zero, then the matrix is known as singular matrix *e.g.* $A = \begin{bmatrix} 1 & 2 \\ 3 & 6 \end{bmatrix}$ is singular matrix, because $|A| = 6 - 6 = 0$.

Example 1. *Find the values of x, y, z and 'a' which satisfy the matrix equation.*

$$\begin{bmatrix} x+3 & 2y+x \\ z-1 & 4a-6 \end{bmatrix} = \begin{bmatrix} 0 & -7 \\ 3 & 2a \end{bmatrix}$$

Solution. As the given matrices are equal, so their corresponding elements are equal.

$$x + 3 = 0 \quad \Rightarrow \quad x = -3 \qquad ...(1)$$
$$2y + x = -7 \qquad ...(2)$$
$$z - 1 = 3 \quad \Rightarrow \quad z = 4 \qquad ...(3)$$
$$4a - 6 = 2a \quad \Rightarrow \quad a = 3 \qquad ...(4)$$

Putting the value of $x = -3$ from (1) into (2), we have

$$2y - 3 = -7 \quad \Rightarrow \quad y = -2$$

Hence, $x = -3,\ y = -2,\ z = 4,\ a = 3$ **Ans.**

38.3 ADDITION OF MATRICES

If A and B be two matrices of the same order, then their sum, $A + B$ is defined as the matrix, each element of which is the sum of the corresponding elements of A and B.

Thus if $A = \begin{bmatrix} 4 & 2 & 5 \\ 1 & 3 & -6 \end{bmatrix},\ B = \begin{bmatrix} 1 & 0 & 2 \\ 3 & 1 & 4 \end{bmatrix}$

then $A + B = \begin{bmatrix} 4+1 & 2+0 & 5+2 \\ 1+3 & 3+1 & -6+4 \end{bmatrix} = \begin{bmatrix} 5 & 2 & 7 \\ 4 & 4 & -2 \end{bmatrix}$

If $A = [a_{ij}],\ B = [b_{ij}]$ then $A + B = [a_{ij} + b_{ij}]$

Example 2. *Show that any square matrix can be expressed as the sum of two matrices, one symmetric and the other anti-symmetric.*

Solution. Let A be a given square matrix.

Then $A = \frac{1}{2}(A + A') + \frac{1}{2}(A - A')$

Now, $(A + A')' = A' + A = A + A'$.

$\therefore$ $A + A'$ is a symmetric matrix.

Also, $(A - A')' = A' - A = -(A - A')$

$\therefore$ $A - A'$ or $\frac{1}{2}(A - A')$ is an anti-symmetric matrix.

$$\boxed{A = \frac{1}{2}(A + A') + \frac{1}{2}(A - A')}$$

Square matrix = Symmetric matrix + Anti-symmetric matrix **Proved.**

Example 3. *Write matrix A given below as the sum of a symmetric and a skew symmetric matrix.*

$$A = \begin{pmatrix} 1 & 2 & 4 \\ -2 & 5 & 3 \\ -1 & 6 & 3 \end{pmatrix}$$

Solution. $A = \begin{bmatrix} 1 & 2 & 4 \\ -2 & 5 & 3 \\ -1 & 6 & 3 \end{bmatrix}$ On transposing, we get $A' = \begin{bmatrix} 1 & -2 & -1 \\ 2 & 5 & 6 \\ 4 & 3 & 3 \end{bmatrix}$

On adding A and A′, we have

$$A + A' = \begin{bmatrix} 1 & 2 & 4 \\ -2 & 5 & 3 \\ -1 & 6 & 3 \end{bmatrix} + \begin{bmatrix} 1 & -2 & -1 \\ 2 & 5 & 6 \\ 4 & 3 & 3 \end{bmatrix} = \begin{bmatrix} 2 & 0 & 3 \\ 0 & 10 & 9 \\ 3 & 9 & 6 \end{bmatrix} \qquad ...(1)$$

On subtracting A' from A, we get

$$A - A' = \begin{bmatrix} 1 & 2 & 4 \\ -2 & 5 & 3 \\ -1 & 6 & 3 \end{bmatrix} - \begin{bmatrix} 1 & -2 & -1 \\ 2 & 5 & 6 \\ 4 & 3 & 3 \end{bmatrix} = \begin{bmatrix} 0 & 4 & 5 \\ -4 & 0 & -3 \\ -5 & 3 & 0 \end{bmatrix} \quad ...(2)$$

On adding (1) and (2), we have

$$2A = \begin{bmatrix} 2 & 0 & 3 \\ 0 & 10 & 9 \\ 3 & 9 & 6 \end{bmatrix} + \begin{bmatrix} 0 & 4 & 5 \\ -4 & 0 & -3 \\ -5 & 3 & 0 \end{bmatrix}$$

$$A = \begin{bmatrix} 1 & 0 & \frac{3}{2} \\ 0 & 5 & \frac{9}{2} \\ \frac{3}{2} & \frac{9}{2} & 3 \end{bmatrix} + \begin{bmatrix} 0 & 2 & \frac{5}{2} \\ -2 & 0 & -\frac{3}{2} \\ \frac{-5}{2} & \frac{3}{2} & 0 \end{bmatrix}$$

A = [Symmetric matrix] + [Skew symmetric matrix.] **Ans.**

Example 4. *Express* $A = \begin{bmatrix} 1 & -2 & -3 \\ 3 & 0 & 5 \\ 5 & 6 & 1 \end{bmatrix}$ *as the sum of a lower triangular matrix and upper triangular matrix.*

Solution. Let $A = L + U$

$$\begin{bmatrix} 1 & -2 & -3 \\ 3 & 0 & 5 \\ 5 & 6 & 1 \end{bmatrix} = \begin{bmatrix} a & 0 & 0 \\ b & c & 0 \\ d & e & f \end{bmatrix} + \begin{bmatrix} 1 & p & q \\ 0 & 1 & r \\ 0 & 0 & 1 \end{bmatrix}$$

$$\begin{bmatrix} 1 & -2 & -3 \\ 3 & 0 & 5 \\ 5 & 6 & 1 \end{bmatrix} = \begin{bmatrix} a+1 & 0+p & 0+q \\ b+0 & c+1 & 0+r \\ d+0 & e+0 & f+1 \end{bmatrix}$$

Equating the corresponding elements on both the sides, we get

$$\begin{array}{lll} a+1 = 1 & p = -2 & q = -3 \\ b = 3 & c+1 = 0 & r = 5 \\ d = 5 & e = 6 & f+1 = 1 \end{array}$$

On solving these equations, we get

$$\begin{array}{lll} a = 0 & p = -2 & q = -3 \\ b = 3 & c = -1 & r = 5 \\ d = 5 & e = 6 & f = 0 \end{array}$$

Hence, $L = \begin{bmatrix} 0 & 0 & 0 \\ 3 & -1 & 0 \\ 5 & 6 & 0 \end{bmatrix}$ and $U = \begin{bmatrix} 1 & -2 & -3 \\ 0 & 1 & 5 \\ 0 & 0 & 1 \end{bmatrix}$ **Ans.**

38.4 PROPERTIES OF MATRIX ADDITION

Only matrices of the same order can be added or subtracted.

(*i*) **Commutative Law.** $A + B = B + A$.

(*ii*) **Associative law.** $A + (B + C) = (A + B) + C$.

38.5 SUBTRACTION OF MATRICES

The difference of two matrices is a matrix, each element of which is obtained by subtracting the elements of the second matrix from the corresponding element of the first.

$$A - B = [a_{ij} - b_{ij}]$$

Thus

$$\begin{bmatrix} 8 & 6 & 4 \\ 1 & 2 & 0 \end{bmatrix} - \begin{bmatrix} 3 & 5 & 1 \\ 7 & 6 & 2 \end{bmatrix}$$

$$= \begin{bmatrix} 8-3 & 6-5 & 4-1 \\ 1-7 & 2-6 & 0-2 \end{bmatrix} = \begin{bmatrix} 5 & 1 & 3 \\ -6 & -4 & -2 \end{bmatrix} \quad \textbf{Ans.}$$

38.6 SCALAR MULTIPLE OF A MATRIX

If a matrix is multiplied by a scalar quantity k, then each element is multiplied by k, *i.e.*

$$A = \begin{bmatrix} 2 & 3 & 4 \\ 4 & 5 & 6 \\ 6 & 7 & 9 \end{bmatrix}$$

$$3A = 3\begin{bmatrix} 2 & 3 & 4 \\ 4 & 5 & 6 \\ 6 & 7 & 9 \end{bmatrix} = \begin{bmatrix} 3\times 2 & 3\times 3 & 3\times 4 \\ 3\times 4 & 3\times 5 & 3\times 6 \\ 3\times 6 & 3\times 7 & 3\times 9 \end{bmatrix} = \begin{bmatrix} 6 & 9 & 12 \\ 12 & 15 & 18 \\ 18 & 21 & 27 \end{bmatrix}$$

EXERCISE 38.1

1. (*i*) If $A = \begin{bmatrix} -1 & 7 & 1 \\ 2 & 3 & 4 \\ 5 & 0 & 5 \end{bmatrix}$, represent it as A = B + C where B is a symmetric and C is a skew-symmetric matrix.

(*b*) Express $\begin{bmatrix} 1 & 2 & 0 \\ 3 & 7 & 1 \\ 5 & 9 & 3 \end{bmatrix}$ as a sum of symmetric and skew-symmetric matrix.

Ans. (*i*) $A = \begin{bmatrix} -1 & \frac{9}{2} & 3 \\ \frac{9}{2} & 3 & 2 \\ 3 & 2 & 5 \end{bmatrix} + \begin{bmatrix} 0 & \frac{5}{2} & -2 \\ \frac{-5}{2} & 0 & 2 \\ 2 & -2 & 0 \end{bmatrix}$ (*b*) $A = \begin{bmatrix} 1 & \frac{5}{2} & \frac{5}{2} \\ \frac{5}{2} & 7 & 5 \\ \frac{5}{2} & 5 & 3 \end{bmatrix} + \begin{bmatrix} 0 & -\frac{1}{2} & -\frac{5}{2} \\ \frac{1}{2} & 0 & -4 \\ \frac{5}{2} & 4 & 0 \end{bmatrix}$

2. Matrices A and B are such that

$3A - 2B = \begin{bmatrix} 2 & 1 \\ -2 & -1 \end{bmatrix}$ and $-4A + B = \begin{bmatrix} -1 & 2 \\ -4 & 3 \end{bmatrix}$

Find A and B. **Ans.** $A = \begin{bmatrix} 0 & -1 \\ 2 & -1 \end{bmatrix}$, $B = \begin{bmatrix} -1 & -2 \\ 4 & -1 \end{bmatrix}$

3. Given $3\begin{bmatrix} x & y \\ z & w \end{bmatrix} = \begin{bmatrix} x & 6 \\ -1 & 2w \end{bmatrix} + \begin{bmatrix} 4 & x+y \\ z+w & 3 \end{bmatrix}$

Find x, y, z and w. **Ans.** $x = 2,\ y = 4,\ z = 1,\ w = 3$

4. If $A = \begin{bmatrix} 0 & 2 & 0 \\ 1 & 0 & 3 \\ 1 & 1 & 2 \end{bmatrix}$, $B = \begin{bmatrix} 1 & 2 & 1 \\ 2 & 1 & 0 \\ 0 & 0 & 3 \end{bmatrix}$ **Ans.** (*i*) $= \begin{bmatrix} 3 & 10 & 3 \\ 8 & 3 & 6 \\ 2 & 2 & 13 \end{bmatrix}$, (*ii*) $= \begin{bmatrix} -4 & -2 & -4 \\ -5 & -4 & 9 \\ 3 & 3 & -6 \end{bmatrix}$

Find (*i*) $2A + 3B$ (*ii*) $3A - 4B$.

38.7 MULTIPLICATION

The product of two matrices A and B is only possible if the number of columns in A is equal to the number of rows in B.

Let $A = [a_{ij}]$ be an $m \times n$ matrix and $B = [b_{ij}]$ be an $n \times p$ matrix. Then the product AB of these matrices is an $m \times p$ matrix $C = [c_{ij}]$ where

$$c_{ij} = a_{i1}\, b_{1j} + a_{i2}\, b_{2j} + a_{i3}\, b_{3j} + ... + a_{in}\, b_{nj}$$

38.8 $(AB)' = B'A'$

If A and B are two matrices conformal for product AB, then show that $(AB)' = B'A'$, where dash represents transpose of a matrix.

Solution. Let $A = (a_{ij})$ be an $m \times n$ matrix and $B = (b_{ij})$ be $n \times p$ matrix.

Since AB is $m \times p$ matrix, $(AB)'$ is a $p \times m$ matrix.

Further B' *is* $p \times n$ matrix and A' an $n \times m$ matrix and therefore $B'\, A'$ is a $p \times m$ matrix.

Then $(AB)'$ and $B'\, A'$ are matrices of the same order.

Now the (j, i)th element of $(AB)' = (i, j)$th element of $(AB) = \sum_{k=1}^{n} a_{ik} b_{kj}$ ...(1)

Also the jth row of B' is b_{1j}, b_{2j} .... b_{nj}, and ith column of A' is $a_{i1}, a_{i2}, a_{i3}.... a_{in}$.

$\therefore$ (j, i)th element of $B'A' = \sum_{k=1}^{n} b_{kj} a_{ik}$ ...(2)

From (1) and (2), we have (j, i)th element of $(AB)' = (j, i)$ th element of $B'A'$.

As the matrices $(AB)'$ and $B'A'$ are of the same order and their corresponding elements are equal, we have $(AB)' = B'A'$. **Proved.**

38.9 PROPERTIES OF MATRIX MULTIPLICATION

1. Multiplication of matrices is not commutative.
   $AB \neq BA$
2. Matrix multiplication is associative, if conformability is assured.
   $A\,(BC) = (AB)\,C$
3. Matrix multiplication is distributive with respect to addition.
   $A\,(B + C) = AB + AC$
4. Multiplication of matrix A by unit matrix.
   $AI = IA = A$
5. Multiplicative inverse of a matrix exists if $|A| \neq 0$.
   $A\,.\,A^{-1} = A^{-1}\,.\,A = I$
6. If A is a square then $A \times A = A^2$, $A \times A \times A = A^3$.
7. $A^0 = I$
8. $I^n = I$, where n is positive integer.

Example 5. *If* $A = \begin{bmatrix} 0 & 1 & 2 \\ 1 & 2 & 3 \\ 2 & 3 & 4 \end{bmatrix}$ *and* $B = \begin{bmatrix} 1 & -2 \\ -1 & 0 \\ 2 & -1 \end{bmatrix}$

obtain the product AB and explain why BA is not defined.

Solution. The number of columns in A is 3 and the number of rows in B is also 3, therefore the product AB is defined.

$$AB = \begin{bmatrix} 0 & 1 & 2 \\ 1 & 2 & 3 \\ 2 & 3 & 4 \end{bmatrix} \begin{matrix} R_1 \\ R_2 \\ R_3 \end{matrix} \times \overset{\begin{matrix} C_1 & C_2 \end{matrix}}{\begin{bmatrix} 1 & -2 \\ -1 & 0 \\ 2 & -1 \end{bmatrix}} = \begin{bmatrix} R_1\,C_1 & R_1\,C_2 \\ R_2\,C_1 & R_2\,C_2 \\ R_3\,C_1 & R_3\,C_2 \end{bmatrix}$$

R_1, R_2, R_3 are rows of A and C_1, C_2 are columns of B.

$$= \begin{bmatrix} \boxed{0\ 1\ 2}\boxed{\begin{matrix} 1 \\ -1 \\ 2 \end{matrix}} & \boxed{0\ 1\ 2}\boxed{\begin{matrix} -2 \\ 0 \\ -1 \end{matrix}} \\ \boxed{1\ 2\ 3}\boxed{\begin{matrix} 1 \\ -1 \\ 2 \end{matrix}} & \boxed{1\ 2\ 3}\boxed{\begin{matrix} -2 \\ 0 \\ -1 \end{matrix}} \\ \boxed{2\ 3\ 4}\boxed{\begin{matrix} 1 \\ -1 \\ 2 \end{matrix}} & \boxed{2\ 3\ 4}\boxed{\begin{matrix} -2 \\ 0 \\ -1 \end{matrix}} \end{bmatrix}$$

For convenience of multiplication, we write the columns in horizontal rectangles.

$$= \begin{bmatrix} \boxed{0\ \ 1\ \ 2} & \boxed{0\ \ 1\ \ 2} \\ \boxed{1\ -1\ \ 2} & \boxed{-2\ \ 0\ -1} \\ \boxed{1\ \ 1\ \ 3} & \boxed{1\ \ 2\ \ 3} \\ \boxed{1\ -1\ \ 2} & \boxed{-2\ \ 0\ -1} \\ \boxed{2\ \ 3\ \ 4} & \boxed{2\ \ 3\ \ 4} \\ \boxed{1\ -1\ \ 2} & \boxed{-2\ \ 0\ -1} \end{bmatrix}$$

$$= \begin{bmatrix} 0\times 1+1\times(-1)+2\times 2 & 0\times(-2)+1\times 0+2\times(-1) \\ 1\times 1+2\times(-1)+3\times 2 & 1\times(-2)+2\times 0+3\times(-1) \\ 2\times 1+3\times(-1)+4\times 2 & 2\times(-2)+3\times 0+4\times(-1) \end{bmatrix}$$

$$= \begin{bmatrix} 0-1+4 & 0+0-2 \\ 1-2+6 & -2+0-3 \\ 2-3+8 & -4+0-4 \end{bmatrix} = \begin{bmatrix} 3 & -2 \\ 5 & -5 \\ 7 & -8 \end{bmatrix}$$ **Ans.**

Since, the number of columns of B is (2) ≠ the number of rows of A is 3, BA is not defined.

Example 6. *If* $A = \begin{bmatrix} 1 & -2 & 3 \\ 2 & 3 & -1 \\ -3 & 1 & 2 \end{bmatrix}$ *and* $B = \begin{bmatrix} 1 & 0 & 2 \\ 0 & 1 & 2 \\ 1 & 2 & 0 \end{bmatrix}$

from the products AB and BA, and show that AB ≠ BA.

Solution. Here,
$$AB = \begin{bmatrix} 1 & -2 & 3 \\ 2 & 3 & -1 \\ -3 & 1 & 2 \end{bmatrix} \begin{bmatrix} 1 & 0 & 2 \\ 0 & 1 & 2 \\ 1 & 2 & 0 \end{bmatrix}$$

$$= \begin{bmatrix} 1-0+3 & 0-2+6 & 2-4+0 \\ 2+0-1 & 0+3-2 & 4+6-0 \\ -3+0+2 & 0+1+4 & -6+2+0 \end{bmatrix} = \begin{bmatrix} 4 & 4 & -2 \\ 1 & 1 & 10 \\ -1 & 5 & -4 \end{bmatrix}$$

$$BA = \begin{bmatrix} 1 & 0 & 2 \\ 0 & 1 & 2 \\ 1 & 2 & 0 \end{bmatrix} \begin{bmatrix} 1 & -2 & 3 \\ 2 & 3 & -1 \\ -3 & 1 & 2 \end{bmatrix} = \begin{bmatrix} 1+0-6 & -2+0+2 & 3-0+4 \\ 0+2-6 & 0+3+2 & 0-1+4 \\ 1+4+0 & -2+6+0 & 3-2+0 \end{bmatrix} = \begin{bmatrix} -5 & 0 & 7 \\ -4 & 5 & 3 \\ 5 & 4 & 1 \end{bmatrix}$$

$AB \neq BA$ **Proved.**

Example 7. *If* $A = \begin{bmatrix} 1 & 2 \\ -2 & 3 \end{bmatrix}$, $B = \begin{bmatrix} 2 & 1 \\ 2 & 3 \end{bmatrix}$ *and* $C = \begin{bmatrix} -3 & 1 \\ 2 & 0 \end{bmatrix}$

Verify that $(AB)\,C = A\,(BC)$ *and* $A\,(B + C) = AB + AC$.

Solution. We have, $AB = \begin{bmatrix} 1 & 2 \\ -2 & 3 \end{bmatrix} \times \begin{bmatrix} 2 & 1 \\ 2 & 3 \end{bmatrix}$

$$= \begin{bmatrix} (1)(2)+(2)(2) & (1)(1)+(2)(3) \\ (-2)(2)+(3)(2) & (-2)(1)+(3)(3) \end{bmatrix} = \begin{bmatrix} 6 & 7 \\ 2 & 7 \end{bmatrix}$$

$$BC = \begin{bmatrix} 2 & 1 \\ 2 & 3 \end{bmatrix} \times \begin{bmatrix} -3 & 1 \\ 2 & 0 \end{bmatrix} = \begin{bmatrix} -6+2 & 2+0 \\ -6+6 & 2+0 \end{bmatrix} = \begin{bmatrix} -4 & 2 \\ 0 & 2 \end{bmatrix}$$

$$AC = \begin{bmatrix} 1 & 2 \\ -2 & 3 \end{bmatrix} \times \begin{bmatrix} -3 & 1 \\ 2 & 0 \end{bmatrix} = \begin{bmatrix} -3+4 & 1+0 \\ 6+6 & -2+0 \end{bmatrix} = \begin{bmatrix} 1 & 1 \\ 12 & -2 \end{bmatrix}$$

$$B + C = \begin{bmatrix} 2+(-3) & 1+1 \\ 2+2 & 3+0 \end{bmatrix} = \begin{bmatrix} -1 & 2 \\ 4 & 3 \end{bmatrix}$$

(*i*) $(AB)\,C = \begin{bmatrix} 6 & 7 \\ 2 & 7 \end{bmatrix} \times \begin{bmatrix} -3 & 1 \\ 2 & 0 \end{bmatrix} = \begin{bmatrix} -18+14 & 6+0 \\ -6+14 & 2+0 \end{bmatrix} = \begin{bmatrix} -4 & 6 \\ 8 & 2 \end{bmatrix}$ ...(1)

and $A\,(BC) = \begin{bmatrix} 1 & 2 \\ -2 & 3 \end{bmatrix} \begin{bmatrix} -4 & 2 \\ 0 & 2 \end{bmatrix} = \begin{bmatrix} -4+0 & 2+4 \\ 8+0 & -4+6 \end{bmatrix} = \begin{bmatrix} -4 & 6 \\ 8 & 2 \end{bmatrix}$ ...(2)

Thus from (1) and (2), we get

$(AB)\,C = A\,(BC)$ **Verified.**

(*ii*) $A\,(B + C) = \begin{bmatrix} 1 & 2 \\ -2 & 3 \end{bmatrix} \begin{bmatrix} -1 & 2 \\ 4 & 3 \end{bmatrix}$

$$= \begin{bmatrix} -1+8 & 2+6 \\ 2+12 & -4+9 \end{bmatrix} = \begin{bmatrix} 7 & 8 \\ 14 & 5 \end{bmatrix} \quad ...(3)$$

$$AB + AC = \begin{bmatrix} 6+1 & 7+1 \\ 2+12 & 7-2 \end{bmatrix} = \begin{bmatrix} 7 & 8 \\ 14 & 5 \end{bmatrix} \quad ...(4)$$

Thus from (3) and (4), we get

$A\,(B + C) = AB + AC$ **Verified.**

Example 8. *If* $A = \begin{pmatrix} 1 & 2 & 2 \\ 2 & 1 & 2 \\ 2 & 2 & 1 \end{pmatrix}$ *show that* $A^2 - 4A - 5I = 0$ *where I, 0 are the unit matrix and the null matrix of order 3 respectively. Use this result to find* A^{-1}.

Solution. Here, we have $A = \begin{bmatrix} 1 & 2 & 2 \\ 2 & 1 & 2 \\ 2 & 2 & 1 \end{bmatrix}$

$$A^2 = \begin{bmatrix} 1 & 2 & 2 \\ 2 & 1 & 2 \\ 2 & 2 & 1 \end{bmatrix}\begin{bmatrix} 1 & 2 & 2 \\ 2 & 1 & 2 \\ 2 & 2 & 1 \end{bmatrix} = \begin{bmatrix} 9 & 8 & 8 \\ 8 & 9 & 8 \\ 8 & 8 & 9 \end{bmatrix}$$

$$A^2 - 4A - 5I = \begin{bmatrix} 9 & 8 & 8 \\ 8 & 9 & 8 \\ 8 & 8 & 9 \end{bmatrix} - 4\begin{bmatrix} 1 & 2 & 2 \\ 2 & 1 & 2 \\ 2 & 2 & 1 \end{bmatrix} - 5\begin{bmatrix} 1 & 0 & 0 \\ 0 & 1 & 0 \\ 0 & 0 & 1 \end{bmatrix}$$

$$A^2 - 4A - 5I = \begin{bmatrix} 9-4-5 & 8-8-0 & 8-8-0 \\ 8-8-0 & 9-4-5 & 8-8-0 \\ 8-8-0 & 8-8-0 & 9-4-5 \end{bmatrix} = \begin{bmatrix} 0 & 0 & 0 \\ 0 & 0 & 0 \\ 0 & 0 & 0 \end{bmatrix}$$

$$A^2 - 4A - 5I = 0 \quad \Rightarrow \quad 5I = A^2 - 4A$$

Multiplying by A^{-1}, we get

$$5A^{-1} = A - 4I$$

$$= \begin{bmatrix} 1 & 2 & 2 \\ 2 & 1 & 2 \\ 2 & 2 & 1 \end{bmatrix} - 4\begin{bmatrix} 1 & 0 & 0 \\ 0 & 1 & 0 \\ 0 & 0 & 1 \end{bmatrix} = \begin{bmatrix} -3 & 2 & 2 \\ 2 & -3 & 2 \\ 2 & 2 & -3 \end{bmatrix}$$

$$A^{-1} = \frac{1}{5}\begin{bmatrix} -3 & 2 & 2 \\ 2 & -3 & 2 \\ 2 & 2 & -3 \end{bmatrix}$$ **Ans.**

Example 9. *Show by means of an example that in matrices $AB = 0$ does not necessarily mean that either $A = 0$ or $B = 0$, where 0 stands for the null matrix.*

Solution. Let $A = \begin{bmatrix} 1 & -1 & 1 \\ -3 & 2 & -1 \\ -2 & 1 & 0 \end{bmatrix}$, $B = \begin{bmatrix} 1 & 2 & 3 \\ 2 & 4 & 6 \\ 1 & 2 & 3 \end{bmatrix}$

$$AB = \begin{bmatrix} 1-2+1 & 2-4+2 & 3-6+3 \\ -3+4-1 & -6+8-2 & -9+12-3 \\ -2+2+0 & -4+4+0 & -6+6+0 \end{bmatrix} = \begin{bmatrix} 0 & 0 & 0 \\ 0 & 0 & 0 \\ 0 & 0 & 0 \end{bmatrix}$$

$$AB = 0.$$

But here neither $A = 0$ nor $B = 0$. **Proved.**

Example 10. *If $AB = AC$, it is not necessarily true that $B = C$ i.e. like ordinary algebra, the equal matrices in the identity cannot be cancelled.*

Solution. Let $AB = \begin{bmatrix} 1 & -3 & 2 \\ 2 & 1 & -3 \\ 4 & -3 & -1 \end{bmatrix}\begin{bmatrix} 1 & 4 & 1 & 0 \\ 2 & 1 & 1 & 1 \\ 1 & -2 & 1 & 2 \end{bmatrix} = \begin{bmatrix} -3 & -3 & 0 & 1 \\ 1 & 15 & 0 & -5 \\ -3 & 15 & 0 & -5 \end{bmatrix}$

$$AC = \begin{bmatrix} 1 & -3 & 2 \\ 2 & 1 & -3 \\ 4 & -3 & -1 \end{bmatrix}\begin{bmatrix} 2 & 1 & -1 & -2 \\ 3 & -2 & -1 & -1 \\ 2 & -5 & -1 & 0 \end{bmatrix} = \begin{bmatrix} -3 & -3 & 0 & 1 \\ 1 & 15 & 0 & -5 \\ -3 & 15 & 0 & -5 \end{bmatrix}$$ **Proved.**

Here, $AB = AC$. But $B \neq C$.

Example 11. *Represent each of the transformations*

$x_1 = 3y_1 + 2y_2,\quad y_1 = z_1 + 2z_2 \quad and \quad x_2 = -y_1 + 4y_2,\ y_2 = 3z_1$

by the use of matrices and find the composite transformation which expresses x_1, x_2 in terms of z_1, z_2.

Solution. The equations in the matrix form are

$$\begin{bmatrix} x_1 \\ x_2 \end{bmatrix} = \begin{bmatrix} 3 & 2 \\ -1 & 4 \end{bmatrix} \begin{bmatrix} y_1 \\ y_2 \end{bmatrix} \qquad ...(1)$$

$$\begin{bmatrix} y_1 \\ y_2 \end{bmatrix} = \begin{bmatrix} 1 & 2 \\ 3 & 0 \end{bmatrix} \begin{bmatrix} z_1 \\ z_2 \end{bmatrix} \qquad ...(2)$$

Substituting the values of y_1, y_2 in (1), we get

$$\begin{bmatrix} x_1 \\ x_2 \end{bmatrix} = \begin{bmatrix} 3 & 2 \\ -1 & 4 \end{bmatrix} \begin{bmatrix} 1 & 2 \\ 3 & 0 \end{bmatrix} \begin{bmatrix} z_1 \\ z_2 \end{bmatrix} = \begin{bmatrix} 9 & 6 \\ 11 & -2 \end{bmatrix} \begin{bmatrix} z_1 \\ z_2 \end{bmatrix} = \begin{bmatrix} 9z_1 + 6z_2 \\ 11z_1 - 2z_2 \end{bmatrix}$$

$$x_1 = 9z_1 + 6z_2,\quad x_2 = 11z_1 - 2z_2$$ **Ans.**

Example 12. *Prove that the product of two matrices*

$$\begin{bmatrix} \cos^2\theta & \cos\theta\sin\theta \\ \cos\theta\sin\theta & \sin^2\theta \end{bmatrix} \text{ and } \begin{bmatrix} \cos^2\phi & \cos\phi\sin\phi \\ \cos\phi\sin\phi & \sin^2\phi \end{bmatrix}$$

is zero when θ and ϕ differ by an odd multiple of $\frac{\pi}{2}$.

Solution. $= \begin{bmatrix} \cos^2\theta & \cos\theta\sin\theta \\ \cos\theta\sin\theta & \sin^2\theta \end{bmatrix} \times \begin{bmatrix} \cos^2\phi & \cos\phi\sin\phi \\ \cos\phi\sin\phi & \sin^2\phi \end{bmatrix}$

$$= \begin{bmatrix} \cos^2\theta\cos^2\phi + \cos\theta\sin\theta\cos\phi\sin\phi & \cos^2\theta\cos\phi\sin\phi + \cos\theta\sin\theta\sin^2\phi \\ \cos\theta\sin\theta\cos^2\phi + \sin^2\theta\cos\phi\sin\phi & \cos\theta\sin\theta\cos\phi\sin\phi + \sin^2\theta\sin^2\phi \end{bmatrix}$$

$$= \begin{bmatrix} \cos\theta\cos\phi(\cos\theta\cos\phi + \sin\theta\sin\phi) & \cos\theta\sin\phi(\cos\theta\cos\phi + \sin\theta\sin\phi) \\ \sin\theta\cos\phi(\cos\theta\cos\phi + \sin\theta\sin\phi) & \sin\theta\sin\phi(\cos\theta\cos\phi + \sin\theta\sin\phi) \end{bmatrix}$$

$$= \begin{bmatrix} \cos\theta\cos\phi\cos(\theta-\phi) & \cos\theta\sin\phi\cos(\theta-\phi) \\ \sin\theta\cos\phi\cos(\theta-\phi) & \sin\theta\sin\phi\cos(\theta-\phi) \end{bmatrix}$$

Given $\qquad \theta - \phi = (2n+1)\frac{\pi}{2}$

$$\cos(\theta-\phi) = \cos(2n+1)\frac{\pi}{2} = 0$$

$\therefore$ The product $= \begin{bmatrix} 0 & 0 \\ 0 & 0 \end{bmatrix} = 0.$ **Proved.**

Example 13. *Verify that*

$$A = \frac{1}{3}\begin{bmatrix} 1 & 2 & 2 \\ 2 & 1 & -2 \\ -2 & 2 & -1 \end{bmatrix} \text{ is orthogonal.}$$

Solution. $A = \frac{1}{3}\begin{bmatrix} 1 & 2 & 2 \\ 2 & 1 & -2 \\ -2 & 2 & -1 \end{bmatrix} \quad \therefore \quad A' = \frac{1}{3}\begin{bmatrix} 1 & 2 & -2 \\ 2 & 1 & 2 \\ 2 & -2 & -1 \end{bmatrix}$

$$AA' = \frac{1}{9}\begin{bmatrix} 1 & 2 & 2 \\ 2 & 1 & -2 \\ -2 & 2 & -1 \end{bmatrix}\begin{bmatrix} 1 & 2 & -2 \\ 2 & 1 & 2 \\ 2 & -2 & -1 \end{bmatrix} = \frac{1}{9}\begin{bmatrix} 9 & 0 & 0 \\ 0 & 9 & 0 \\ 0 & 0 & 9 \end{bmatrix}\begin{bmatrix} 1 & 0 & 0 \\ 0 & 1 & 0 \\ 0 & 0 & 1 \end{bmatrix} = I$$

Hence, A is an orthogonal matrix. **Verified.**

Example 14. *Determine the values of* α, β, γ *when*

$$\begin{bmatrix} 0 & 2\beta & \gamma \\ \alpha & \beta & -\gamma \\ \alpha & -\beta & \gamma \end{bmatrix} \text{ is orthogonal.}$$

Solution. Let $A = \begin{bmatrix} 0 & 2\beta & \gamma \\ \alpha & \beta & -\gamma \\ \alpha & -\beta & \gamma \end{bmatrix}$

On transposing A, we have

$$A' = \begin{bmatrix} 0 & \alpha & \alpha \\ 2\beta & \beta & -\beta \\ \gamma & -\gamma & \gamma \end{bmatrix}$$

If A is orthogonal, then $AA' = I$

$$\begin{bmatrix} 0 & 2\beta & \gamma \\ \alpha & \beta & -\gamma \\ \alpha & -\beta & \gamma \end{bmatrix}\begin{bmatrix} 0 & \alpha & \alpha \\ 2\beta & \beta & -\beta \\ \gamma & -\gamma & \gamma \end{bmatrix} = \begin{bmatrix} 1 & 0 & 0 \\ 0 & 1 & 0 \\ 0 & 0 & 1 \end{bmatrix}$$

$$\Rightarrow \begin{bmatrix} 4\beta^2+\gamma^2 & 2\beta^2-\gamma^2 & -2\beta^2+\gamma^2 \\ 2\beta^2-\gamma^2 & \alpha^2+\beta^2+\gamma^2 & \alpha^2-\beta^2-\gamma^2 \\ -2\beta^2+\gamma^2 & \alpha^2-\beta^2-\gamma^2 & \alpha^2+\beta^2+\gamma^2 \end{bmatrix} = \begin{bmatrix} 1 & 0 & 0 \\ 0 & 1 & 0 \\ 0 & 0 & 1 \end{bmatrix}$$

Equating the corresponding elements, we have

$$\left.\begin{array}{r} 4\beta^2+\gamma^2 = 1 \\ 2\beta^2-\gamma^2 = 0 \end{array}\right\} \Rightarrow \beta = \pm\frac{1}{\sqrt{6}}, \gamma = \pm\frac{1}{\sqrt{3}}$$

But $\alpha^2 + \beta^2 + \gamma^2 = 1$ as $\beta = \pm\frac{1}{\sqrt{6}}, \gamma = \pm\frac{1}{\sqrt{3}}, \alpha = \pm\frac{1}{\sqrt{2}}$ **Ans.**

Example 15. *Prove that*

$$(AB)^n = A^n \,.\, B^n, \text{ if } A \,.\, B = B \,.\, A$$

Solution.

$$(AB)^1 = AB = (A)\,.\,(B)$$
$$(AB)^2 = (AB)\,.\,(AB) = (ABA)\,.\,B = \{A\,(AB)\}\,.\,B$$
$$= (A^2B)\,.\,B = A^2\,(B\,.\,B) = A^2.B^2$$

Suppose that

$$(AB)^n = A^n\,.\,B^n$$
$$(AB)^{n+1} = (AB)^n\,.\,(AB) = (A^n\,.\,B^n)\,.\,(AB) = A^n\,.\,(B^nA)\,.\,B$$
$$= A^n\ (B^{n-1}\,.\,BA)\,.\,B = A^n\,.\,(B^{n-1}\,.\,AB)\,.\,B$$
$$= A^n\,.\,(B^{n-2}\,.\,B\,.\,AB)\,.\,B = A^n\,.\,(B^{n-2}\,.\,AB\,.\,B)\,.\,B$$
$$= A^n\,.\,(B^{n-2}\,.\,AB^2)\,.\,B, \text{ continuing the process } n \text{ times.}$$
$$= A^n\,.\,(A\,.\,B^n)\,.\,B = A^n\,.\,(A\,.\,B^{n+1}) = A^{n+1}\,.\,B^{n+1}$$

Hence, taking the above to be true for $n = n$, we have shown that it is true for $n = n + 1$ and also it was true for $n = 1, 2, ...$ so it is universally true. **Proved.**

EXERCISE 38.2

1. Compute AB, if

$$A = \begin{bmatrix} 1 & 2 & 3 \\ 4 & 5 & 6 \end{bmatrix} \text{ and } B = \begin{bmatrix} 2 & 5 & 3 \\ 3 & 6 & 4 \\ 4 & 7 & 5 \end{bmatrix}$$

Ans. $\begin{bmatrix} 20 & 38 & 26 \\ 47 & 92 & 62 \end{bmatrix}$

2. If $A = \begin{bmatrix} 1 & 3 & 0 \\ -1 & 2 & 1 \\ 0 & 0 & 2 \end{bmatrix}$, $B = \begin{bmatrix} 2 & 3 & 4 \\ 1 & 2 & 3 \\ -1 & 1 & 2 \end{bmatrix}$. From the product AB and BA. Show that $AB \neq BA$.

3. If $A = \begin{bmatrix} 0 & 1 & 0 \\ 0 & 0 & 1 \\ 0 & 0 & 0 \end{bmatrix}$, $B = \begin{bmatrix} 0 & 0 & 0 \\ 1 & 0 & 0 \\ 0 & 1 & 0 \end{bmatrix}$

(*i*) Calculate AB and BA. Hence evaluate $A^2 B + B^2 A$

(*ii*) Show that for any number k, $(A + kB^2)^3 = KI$, where I is the unit matrix.

4. If $A = \begin{bmatrix} 0 & 1 \\ -1 & 0 \end{bmatrix}$ choose α and β so that $(\alpha I + \beta A)^2 = A$ **Ans.** $\alpha = \beta = \pm \dfrac{1}{\sqrt{2}}$

5. Write the following transformation in matrix form :

$$x_1 = \frac{\sqrt{3}}{2} y_1 + \frac{1}{2} y_2 \ ; x_2 = -\frac{1}{2} y_1 + \frac{\sqrt{3}}{2} y_2$$

Hence, find the transformation in matrix form which expresses y_1, y_2 in terms of x_1, x_2.

Ans. $y_1 = \dfrac{\sqrt{3}}{2} x_1 - \dfrac{1}{2} x_2 , y_2 = \dfrac{1}{2} x_1 + \dfrac{\sqrt{3}}{2} x_2$

6. If $A = \begin{bmatrix} 0 & -\tan\dfrac{\alpha}{2} \\ \tan\dfrac{\alpha}{2} & 0 \end{bmatrix}$ and I is a unit matrix, show that $I + A = (I - A)\begin{bmatrix} \cos\alpha & -\sin\alpha \\ \sin\alpha & \cos\alpha \end{bmatrix}$

7. If $f(x) = x^3 - 20x + 8$, find $f(A)$ where $A = \begin{bmatrix} 1 & 1 & 3 \\ 1 & 3 & -3 \\ -2 & -4 & -4 \end{bmatrix}$ **Ans.** 0

8. Show that $\begin{bmatrix} \cos\theta & -\sin\theta \\ \sin\theta & \cos\theta \end{bmatrix} = \begin{bmatrix} 1 & -\tan\dfrac{\theta}{2} \\ \tan\dfrac{\theta}{2} & 1 \end{bmatrix} \begin{bmatrix} 1 & \tan\dfrac{\theta}{2} \\ -\tan\dfrac{\theta}{2} & 1 \end{bmatrix}^{-1}$

9. If $A = \begin{bmatrix} 3 & -3 & 4 \\ 2 & -3 & 4 \\ 0 & -1 & 1 \end{bmatrix}$ then show that $A^3 = A^{-1}$.

10. Verify whether the matrix $A = \dfrac{1}{3}\begin{bmatrix} 2 & 2 & 1 \\ 2 & -1 & 2 \\ -1 & 2 & 2 \end{bmatrix}$ is orthogonal.

11. If A and B are square matrices of the same order, explain in general

(*i*) $(A + B)^2 \neq A^2 + 2AB + B^2$ (*ii*) $(A - B)^2 \neq A^2 - 2AB + B^2$ (*iii*) $(A + B)(A - B) \neq A^2 - B^2$

38.10 ADJOINT OF A SQUARE MATRIX

Let the determinant of the square matrix A be $|A|$.

If $A = \begin{bmatrix} a_1 & a_2 & a_3 \\ b_1 & b_2 & b_3 \\ c_1 & c_2 & c_3 \end{bmatrix}$, Than $|A| = \begin{vmatrix} a_1 & a_2 & a_3 \\ b_1 & b_2 & b_3 \\ c_1 & c_2 & c_3 \end{vmatrix}$.

The matrix formed by the co-factors of the elements in

$$|A| \text{ is } \begin{vmatrix} A_1 & A_2 & A_3 \\ B_1 & B_2 & B_3 \\ C_1 & C_2 & C_3 \end{vmatrix}$$

where $A_1 = \begin{vmatrix} b_2 & b_3 \\ c_2 & c_3 \end{vmatrix} = b_2c_3 - b_3c_2,\qquad A_2 = -\begin{vmatrix} b_1 & b_3 \\ c_1 & c_3 \end{vmatrix} = -b_1c_3 + b_3c_1$

$$A_3 = \begin{vmatrix} b_1 & b_2 \\ c_1 & c_2 \end{vmatrix} = b_1c_2 - b_2c_1,\qquad B_1 = -\begin{vmatrix} a_2 & a_3 \\ c_2 & c_3 \end{vmatrix} = -a_2c_3 + a_3c_2$$

$$B_2 = \begin{vmatrix} a_1 & a_3 \\ c_1 & c_3 \end{vmatrix} = a_1c_3 - a_3c_1,\qquad B_3 = -\begin{vmatrix} a_1 & a_2 \\ c_1 & c_2 \end{vmatrix} = -a_1c_2 + a_2c_1$$

$$C_1 = \begin{vmatrix} a_2 & a_3 \\ b_2 & b_3 \end{vmatrix} = a_2b_3 - a_3b_2,\qquad C_2 = -\begin{vmatrix} a_1 & a_3 \\ b_1 & b_3 \end{vmatrix} = -a_1b_3 + a_3b_1$$

$$C_3 = \begin{vmatrix} a_1 & a_2 \\ b_1 & b_2 \end{vmatrix} = a_1b_2 - a_2b_1$$

Then the transpose of the matrix of co-factors

$$\begin{bmatrix} A_1 & B_1 & C_1 \\ A_2 & B_2 & C_2 \\ A_3 & B_3 & C_3 \end{bmatrix}$$

is called the adjoint of the matrix A and is written as adj A.

38.11 MATHEMATICAL INDUCTION

By mathematical induction we can prove results for all positive integers. If the result to be proved for the positive integer n then we apply the following method.

Working Rule:

Step 1. Verify the result for $n = 1$

Step 2. Assume the result to be true for $n = k$ and then prove that it is true for $n = k + 1$.

Explanation. By step 1, the result is true for $n = k = 1$

By step 2, the result is true for $n = k + 1 = 1 + 1 = 2$ $(k = 1)$

Again, the result is also true for $n = k + 1 = 2 + 1 = 3$ $(k = 2)$

Similarly, the result is also true for $n = k + 1 = 3 + 1 = 4$ $(k = 3)$

Hence, in this way the result is true for all positive integer n.

Example 16. *By mathematical induction,*

if $A = \begin{bmatrix} \cos\alpha & \sin\alpha \\ -\sin\alpha & \cos\alpha \end{bmatrix}$, *show that* $A^n = \begin{bmatrix} \cos n\alpha & \sin n\alpha \\ -\sin n\alpha & \cos n\alpha \end{bmatrix}$

Where n is a positive integer.

Solution. We prove the result by mathematical induction :

$$A^n = \begin{bmatrix} \cos n\alpha & \sin n\alpha \\ -\sin n\alpha & \cos n\alpha \end{bmatrix}$$

Let us verify the result for $n = 1$.

$$A^1 = \begin{bmatrix} \cos 1\alpha & \sin 1\alpha \\ -\sin 1\alpha & \cos 1\alpha \end{bmatrix} = \begin{bmatrix} \cos \alpha & \sin \alpha \\ -\sin \alpha & \cos \alpha \end{bmatrix} = A \qquad \text{[Given]}$$

The result is true when $n = 1$.

Let us assume that the result is true for any positive integer k.

$$A^k = \begin{bmatrix} \cos k\alpha & \sin k\alpha \\ -\sin k\alpha & \cos k\alpha \end{bmatrix}$$

Now, $$A^{k+1} = A^k . A = \begin{bmatrix} \cos k\alpha & \sin k\alpha \\ -\sin k\alpha & \cos k\alpha \end{bmatrix} \begin{bmatrix} \cos \alpha & \sin \alpha \\ -\sin \alpha & \cos \alpha \end{bmatrix}$$

$$= \begin{bmatrix} \cos k\alpha \cos\alpha - \sin k\alpha \sin\alpha & \cos k\alpha \sin\alpha + \sin k\alpha \cos\alpha \\ -\sin k\alpha \cos\alpha - \cos k\alpha \sin\alpha & -\sin k\alpha \sin\alpha + \cos k\alpha \cos\alpha \end{bmatrix}$$

$$= \begin{bmatrix} \cos(k\alpha + \alpha) & \sin(k\alpha + \alpha) \\ -\sin(k\alpha + \alpha) & \cos(k\alpha + \alpha) \end{bmatrix} = \begin{bmatrix} \cos(k+1)\alpha & \sin(k+1)\alpha \\ -\sin(k+1)\alpha & \cos(k+1)\alpha \end{bmatrix}$$

The result is true for $n = k + 1$.

Hence, by mathematical induction the result is true for all positive integer n. **Proved.**

Example 17. *Factorise the matrix* $A = \begin{bmatrix} 5 & -2 & 1 \\ 7 & 1 & -5 \\ 3 & 7 & 4 \end{bmatrix}$ *into the form LU, where L is lower triangular and U is upper triangular matrix.*

Solution. Let $A = LU$

$$\Rightarrow \begin{bmatrix} 5 & -2 & 1 \\ 7 & 1 & -5 \\ 3 & 7 & 4 \end{bmatrix} = \begin{bmatrix} l_{11} & 0 & 0 \\ l_{21} & l_{22} & 0 \\ l_{31} & l_{32} & l_{33} \end{bmatrix} \begin{bmatrix} 1 & u_{12} & u_{13} \\ 0 & 1 & u_{23} \\ 0 & 0 & 1 \end{bmatrix} \qquad \text{... (1)}$$

$$\Rightarrow \begin{bmatrix} 5 & -2 & 1 \\ 7 & 1 & -5 \\ 3 & 7 & 4 \end{bmatrix} = \begin{bmatrix} l_{11} & l_{11}u_{12} & l_{11}u_{13} \\ l_{21} & l_{21}u_{12} + l_{22} & l_{21}u_{13} + l_{22}u_{23} \\ l_{31} & l_{31}u_{12} + l_{32} & l_{31}u_{13} + l_{32}u_{23} + l_{33} \end{bmatrix}$$

Equating the corresponding elements of equal matrices, we get

$\Rightarrow l_{11} = 5 \qquad l_{11}u_{12} = -2 \qquad l_{11}u_{13} = 1$

$l_{21} = 7 \qquad l_{21}u_{12} + l_{22} = 1 \qquad l_{21}u_{13} + l_{22}u_{23} = -5$

$l_{31} = 3 \qquad l_{31}u_{12} + l_{32} = 7 \qquad l_{31}u_{13} + l_{32}u_{23} + l_{33} = 4$

Let us solve the above equations along first column.

$l_{11} = 5$

$l_{21} = 7$

$l_{31} = 3$

Let us solve along first row.

$$l_{11}u_{12} = -2 \qquad \Rightarrow 5\, u_{12} = -2 \qquad \Rightarrow u_{12} = -\frac{2}{5}$$

$$l_{11}u_{13} = 1 \qquad \Rightarrow 5\, u_{13} = 1 \qquad \Rightarrow u_{13} = \frac{1}{5}$$

Let us solve along second column.

$$l_{21}\, u_{12} + l_{22} = 1 \quad \Rightarrow 7\left(-\frac{2}{5}\right) + l_{22} = 1 \quad \Rightarrow l_{22} = 1 + \frac{14}{5} = \frac{19}{5}$$

$$l_{31}\, u_{12} + l_{32} = 7 \quad \Rightarrow 3\left(-\frac{2}{5}\right) + l_{32} = 7 \quad \Rightarrow l_{32} = 7 + \frac{6}{5} = \frac{41}{5}$$

Let us solve along second row,

$$l_{21}u_{13} + l_{22}u_{23} = -5 \qquad \Rightarrow 7\left(\frac{1}{5}\right) + \frac{19}{5}u_{23} = -5 \qquad \Rightarrow u_{23} = \left(-5 - \frac{7}{5}\right)\frac{5}{19} = -\frac{32}{19}$$

Let us solve along third column,

$$l_{31}u_{13} + l_{32}u_{23} + l_{33} = 4 \Rightarrow 3\left(\frac{1}{5}\right) + \left(\frac{41}{5}\right)\left(-\frac{32}{19}\right) + l_{33} = 4 \quad \Rightarrow \quad l_{33} = 4 - \frac{3}{5} + \frac{1312}{95} = \frac{327}{19}$$

Putting the values of $l_{11}, l_{21}, l_{22}, l_{31}, l_{32}, l_{33}, u_{12}, u_{13}, u_{23}$ in (1), we get

$$\begin{bmatrix} 5 & -2 & 1 \\ 7 & 1 & -5 \\ 3 & 7 & 4 \end{bmatrix} = \begin{bmatrix} 5 & 0 & 0 \\ 7 & \frac{19}{5} & 0 \\ 3 & \frac{41}{5} & \frac{327}{19} \end{bmatrix} \begin{bmatrix} 1 & -\frac{2}{5} & \frac{1}{5} \\ 0 & 1 & -\frac{32}{19} \\ 0 & 0 & 1 \end{bmatrix}$$ **Ans.**

38.12 PROPERTY OF ADJOINT MATRIX

The product of a matrix A and its adjoint is equal to unit matrix multiplied by the determinant A.

Proof. If A be a square matrix, then (Adjoint A) . A = A . (Adjoint A) = $|A|$. I

Let $$A = \begin{bmatrix} a_1 & a_2 & a_3 \\ b_1 & b_2 & b_3 \\ c_1 & c_2 & c_3 \end{bmatrix} \text{ and adj . } A = \begin{bmatrix} A_1 & B_1 & C_1 \\ A_2 & B_2 & C_2 \\ A_3 & B_3 & C_3 \end{bmatrix}$$

$$A\,.\,(\text{adj. } A) = \begin{bmatrix} a_1 & a_2 & a_3 \\ b_1 & b_2 & b_3 \\ c_1 & c_2 & c_3 \end{bmatrix} \times \begin{bmatrix} A_1 & B_1 & C_1 \\ A_2 & B_2 & C_2 \\ A_3 & B_3 & C_3 \end{bmatrix}$$

$$= \begin{bmatrix} a_1 A_1 + a_2 A_2 + a_3 A_3 & a_1 B_1 + a_2 B_2 + a_3 B_3 & a_1 C_1 + a_2 C_2 + a_3 C_3 \\ b_1 A_1 + b_2 A_2 + b_3 A_3 & b_1 B_1 + b_2 B_2 + b_3 B_3 & b_1 C_1 + b_2 C_2 + b_3 C_3 \\ c_1 A_1 + c_2 A_2 + c_3 A_3 & c_1 B_1 + c_2 B_2 + c_3 B_3 & c_1 C_1 + c_2 C_2 + c_3 C_3 \end{bmatrix}$$

$$= \begin{bmatrix} |A| & 0 & 0 \\ 0 & |A| & 0 \\ 0 & 0 & |A| \end{bmatrix} = |A| \begin{bmatrix} 1 & 0 & 0 \\ 0 & 1 & 0 \\ 0 & 0 & 1 \end{bmatrix} = |A|\, I$$

38.13 INVERSE OF A MATRIX

If A and B are two square matrices of the same order, such that

$$AB = BA = I \qquad (I = \text{unit matrix})$$

then B is called the inverse of A *i.e.* $B = A^{-1}$ and A is the inverse of B.

Condition for a square matrix A to possess an inverse is that matrix A is non-singular, i.e., $|A| \neq 0$

If A is a square matrix and B be its inverse, then $AB = I$

Taking determinant of both sides, we get

$$|AB| = |I| \text{ or } |A||B| = I$$

From this relation it is clear that $|A| \neq 0$

i.e. the matrix A is non-singular.

To find the inverse matrix with the help of adjoint matrix

We know that $A\,.\,(Adj.\,A) = |A|\,I$

$$\Rightarrow \quad A \cdot \frac{1}{|A|}(Adj.\,A) = I \qquad [\text{Provided } |A| \neq 0] \qquad ...(1)$$

and

$$A\,.\,A^{-1} = I \qquad ...(2)$$

From (1) and (2), we have

$$\therefore \quad \boxed{A^{-1} = \frac{1}{|A|}(Adj.\,A)}$$

Example 18. *If* $A = \begin{bmatrix} 3 & -3 & 4 \\ 2 & -3 & 4 \\ 0 & -1 & 1 \end{bmatrix}$, *find* A^{-1}.

Solution. $A = \begin{bmatrix} 3 & -3 & 4 \\ 2 & -3 & 4 \\ 0 & -1 & 1 \end{bmatrix}$

$$|A| = 3(-3+4) + 3(2-0) + 4(-2-0) = 3 + 6 - 8 = 1$$

The co-factors of elements of various rows of $|A|$ are

$$\begin{bmatrix} (-3+4) & (-2-0) & (-2-0) \\ (3-4) & (3-0) & (3-0) \\ (-12+12) & (-12+8) & (-9+6) \end{bmatrix}$$

Therefore, the matrix formed by the co-factors of $|A|$ is

$$\begin{bmatrix} 1 & -2 & -2 \\ -1 & 3 & 3 \\ 0 & -4 & -3 \end{bmatrix}, \quad Adj.A = \begin{bmatrix} 1 & -1 & 0 \\ -2 & 3 & -4 \\ -2 & 3 & -3 \end{bmatrix}$$

$$\therefore \quad A^{-1} = \frac{1}{|A|} Adj.A = \frac{1}{1}\begin{bmatrix} 1 & -1 & 0 \\ -2 & 3 & -4 \\ -2 & 3 & -3 \end{bmatrix} = \begin{bmatrix} 1 & -1 & 0 \\ -2 & 3 & -4 \\ -2 & 3 & -3 \end{bmatrix} \qquad \textbf{Ans.}$$

Example 19. *If* $A = \frac{1}{9}\begin{bmatrix} -8 & 1 & 4 \\ 4 & 4 & 7 \\ 1 & -8 & 4 \end{bmatrix}$, *prove that* $A^{-1} = A'$, *A' being the transpose of A.*

Solution. We have, $A = \frac{1}{9}\begin{bmatrix} -8 & 1 & 4 \\ 4 & 4 & 7 \\ 1 & -8 & 4 \end{bmatrix}, \quad A' = \frac{1}{9}\begin{bmatrix} -8 & 4 & 1 \\ 1 & 4 & -8 \\ 4 & 7 & 4 \end{bmatrix}$

$$AA' = \frac{1}{9}\begin{bmatrix} -8 & 1 & 4 \\ 4 & 4 & 7 \\ 1 & -8 & 4 \end{bmatrix} \cdot \frac{1}{9}\begin{bmatrix} -8 & 4 & 1 \\ 1 & 4 & -8 \\ 4 & 7 & 4 \end{bmatrix}$$

$$= \frac{1}{81}\begin{bmatrix} 64+1+16 & -32+4+28 & -8-8+16 \\ -32+4+28 & 16+16+49 & 4-32+28 \\ -8-8+16 & 4-32+28 & 1+64+16 \end{bmatrix}$$

$$= \frac{1}{81}\begin{bmatrix} 81 & 0 & 0 \\ 0 & 81 & 0 \\ 0 & 0 & 81 \end{bmatrix} = \begin{bmatrix} 1 & 0 & 0 \\ 0 & 1 & 0 \\ 0 & 0 & 1 \end{bmatrix} \text{ or } AA' = I$$

$$A' = A^{-1}$$ **Proved.**

Example 20. *If A and B are non-singular matrices of the same order then,*

$$(AB)^{-1} = B^{-1}.A^{-1}$$

Hence prove that $(A^{-1})^m = (A^m)^{-1}$ for any positive integer m.

Solution. We know that,

$$(AB) . (B^{-1} A^{-1}) = [(AB) B^{-1}] . A^{-1} = [A (BB^{-1}] . A^{-1}$$
$$= [AI] A^{-1} = A . A^{-1} = I$$

Also, $$B^{-1} A^{-1} . (AB) = B^{-1}[A^{-1} . (AB)] = B^{-1} [(A^{-1} A) . B]$$
$$= B^{-1} [I . B] = B^{-1} . B = I$$

By definition of the inverse of a matrix, $B^{-1} A^{-1}$ is inverse of AB.

$\Rightarrow$ $$B^{-1} A^{-1} = (AB)^{-1}$$ **Proved.**

$$(A^m)^{-1} = [A . A^{m-1}]^{-1} = (A^{m-1})^{-1} A^{-1}$$
$$= (A . A^{m-2})^{-1} . A^{-1} = [(A^{m-2})^{-1} . A^{-1}] . A^{-1} = (A^{m-2})^{-1} (A^{-1})^2$$
$$= (A . A^{m-3})^{-1} . (A^{-1})^2 = [(A^{m-3})^{-1} . A^{-1}] (A^{-1})^2 = (A^{m-3})^{-1} (A^{-1})^3$$
$$= A^{-1} (A^{-1})^{m-1} = (A^{-1})^m$$ **Proved.**

Example 21. *Find A satisfying the Matrix equation.*

$$\begin{bmatrix} 2 & 1 \\ 3 & 2 \end{bmatrix} A \begin{bmatrix} -3 & 2 \\ 5 & -3 \end{bmatrix} = \begin{bmatrix} -2 & 4 \\ 3 & -1 \end{bmatrix}$$

Solution. $$\begin{bmatrix} 2 & 1 \\ 3 & 2 \end{bmatrix} A \begin{bmatrix} -3 & 2 \\ 5 & -3 \end{bmatrix} = \begin{bmatrix} -2 & 4 \\ 3 & -1 \end{bmatrix}$$

Both sides of the equation are pre-multiplied by the inverse of $\begin{bmatrix} 2 & 1 \\ 3 & 2 \end{bmatrix}$ *i.e.,* $\begin{bmatrix} 2 & -1 \\ -3 & 2 \end{bmatrix}$

$$\begin{bmatrix} 2 & -1 \\ -3 & 2 \end{bmatrix}\begin{bmatrix} 2 & 1 \\ 3 & 2 \end{bmatrix} A \begin{bmatrix} -3 & 2 \\ 5 & -3 \end{bmatrix} = \begin{bmatrix} 2 & -1 \\ -3 & 2 \end{bmatrix}\begin{bmatrix} -2 & 4 \\ 3 & -1 \end{bmatrix}$$

$$\begin{bmatrix} 1 & 0 \\ 0 & 1 \end{bmatrix} A \begin{bmatrix} -3 & 2 \\ 5 & -3 \end{bmatrix} = \begin{bmatrix} -7 & 9 \\ 12 & -14 \end{bmatrix}$$

$$A \begin{bmatrix} -3 & 2 \\ 5 & -3 \end{bmatrix} = \begin{bmatrix} -7 & 9 \\ 12 & -14 \end{bmatrix}$$

Again both sides are post-multiplied by the inverse of $\begin{bmatrix} -3 & 2 \\ 5 & -3 \end{bmatrix}$ *i.e.* $\begin{bmatrix} 3 & 2 \\ 5 & 3 \end{bmatrix}$

$$A\begin{bmatrix} -3 & 2 \\ 5 & -3 \end{bmatrix}\begin{bmatrix} 3 & 2 \\ 5 & 3 \end{bmatrix} = \begin{bmatrix} -7 & 9 \\ 12 & -14 \end{bmatrix}\begin{bmatrix} 3 & 2 \\ 5 & 3 \end{bmatrix}$$

$$A\begin{bmatrix} 1 & 0 \\ 0 & 1 \end{bmatrix} = \begin{bmatrix} 24 & 13 \\ -34 & -18 \end{bmatrix} \Rightarrow A = \begin{bmatrix} 24 & 13 \\ -34 & -18 \end{bmatrix}$$ **Ans.**

EXERCISE 38.3

Find the adjoint and inverse of the following matrices: (1 - 3)

1. $\begin{bmatrix} 2 & 5 & 3 \\ 3 & 1 & 2 \\ 1 & 2 & 1 \end{bmatrix}$ **Ans.** $\frac{1}{4}\begin{bmatrix} -3 & 1 & 7 \\ -1 & -1 & 5 \\ 5 & 1 & -13 \end{bmatrix}$

2. $\begin{bmatrix} 1 & 1 & 2 \\ 1 & 9 & 3 \\ 1 & 4 & 2 \end{bmatrix}$ **Ans.** $-\frac{1}{3}\begin{bmatrix} 6 & 6 & -15 \\ 1 & 0 & -1 \\ -5 & -3 & 8 \end{bmatrix}$

3. $\begin{bmatrix} 1 & 0 & -1 \\ 3 & 4 & 5 \\ 0 & -6 & -7 \end{bmatrix}$ **Ans.** $\frac{1}{20}\begin{bmatrix} 2 & 6 & 4 \\ 21 & -7 & -8 \\ -18 & 6 & 4 \end{bmatrix}$

4. If $A = \begin{bmatrix} 3 & -4 \\ 1 & -1 \end{bmatrix}$, then show that $A^n = \begin{bmatrix} 1+2n & -4n \\ n & 1-2n \end{bmatrix}$

5. If $A = \begin{bmatrix} 1 & 1 & -2 \\ -1 & 2 & 1 \\ 0 & 1 & -1 \end{bmatrix}, P = \begin{bmatrix} 1 & 1 & 3 \\ 0 & 3 & 2 \\ 1 & 1 & 1 \end{bmatrix}$, show that $P^{-1}AP = \begin{bmatrix} -1 & 0 & 0 \\ 0 & 2 & 0 \\ 0 & 0 & 1 \end{bmatrix}$

6. If $A = \begin{bmatrix} 1 & 1 & 1 \\ 1 & 2 & 3 \\ 1 & 4 & 9 \end{bmatrix}, B = \begin{bmatrix} 2 & 5 & 3 \\ 3 & 1 & 2 \\ 1 & 2 & 1 \end{bmatrix}$, show that $(AB)^{-1} = B^{-1}A^{-1}$.

7. Given the matrix $A = \begin{bmatrix} 3 & 2 & 2 \\ 1 & 3 & 1 \\ 5 & 3 & 4 \end{bmatrix}$ compute det (A), A^{-1} and the matrix B such that $AB = \begin{bmatrix} 3 & 4 & 2 \\ 1 & 6 & 1 \\ 5 & 6 & 4 \end{bmatrix}$

Also compute BA. Is $AB = BA$?

Ans. $5, \frac{1}{5}\begin{bmatrix} 9 & -2 & -4 \\ 1 & 2 & -1 \\ -12 & 1 & 7 \end{bmatrix}. B = \begin{bmatrix} 1 & 0 & 0 \\ 0 & 2 & 0 \\ 0 & 0 & 1 \end{bmatrix}, AB \neq BA$

8. Find the condition of k such that the matrix

$A = \begin{bmatrix} 1 & 3 & 4 \\ 3 & k & 6 \\ -1 & 5 & 1 \end{bmatrix}$ has an inverse. Obtain A^{-1} for $k = 1$. **Ans.** $k \neq -\frac{3}{5}, A^{-1} = \frac{1}{8}\begin{bmatrix} -29 & 17 & 14 \\ -9 & 5 & 6 \\ 16 & -8 & -8 \end{bmatrix}$

9. Prove that $(A^{-1})^T = (A^T)^{-1}$.

10. If $A\begin{bmatrix} 0 & 1 \\ 2 & -1 \end{bmatrix} = \begin{bmatrix} 2 & 1 \\ -1 & 0 \end{bmatrix}$ where $A = \begin{bmatrix} a & b \\ c & d \end{bmatrix}$, then A is

(*a*) $\begin{bmatrix} 2 & 1 \\ 0 & 0 \end{bmatrix}$ (*b*) $\begin{bmatrix} 0 & 1 \\ 2 & -1 \end{bmatrix}$ (*c*) $\begin{bmatrix} 2 & 1 \\ -1 & 0 \end{bmatrix}$ (*d*) $\begin{bmatrix} 2 & 1 \\ -\frac{1}{2} & -\frac{1}{2} \end{bmatrix}$ (*AMIETE, June 2010*) **Ans.** (*d*)

38.14 ELEMENTARY TRANSFORMATIONS

Any one of the following operations on a matrix is called an elementary transformation.

1. Interchanging any two rows (or columns). This transformation is indicated by R_{ij}, if the ith and jth rows are interchanged.
2. Multiplication of the elements of any row R_i (or column) by a non-zero scalar quantity k is denoted by $(k.R_i)$.
3. Addition of constant multiplication of the elements of any row R_j to the corresponding elements of any other row R_j is denoted by $(R_i + kR_j)$.

If a matrix B is obtained from a matrix A by one or more E-operations, then B is said to be equivalent to A. The symbol ~ is used for equivalence.

i.e., $A \sim B$.

Example 22. *Reduce the following matrix to upper triangular form (Echelon form) :*

$$\begin{bmatrix} 1 & 2 & 3 \\ 2 & 5 & 7 \\ 3 & 1 & 2 \end{bmatrix}$$

Solution. *Upper triangular matrix.* If in a square matrix, all the elements below the principal diagonal are zero, the matrix is called an upper triangular matrix.

$$\begin{bmatrix} 1 & 2 & 3 \\ 2 & 5 & 7 \\ 3 & 1 & 2 \end{bmatrix} \sim \begin{bmatrix} 1 & 2 & 3 \\ 0 & 1 & 1 \\ 0 & -5 & -7 \end{bmatrix} \begin{matrix} R_2 \to R_2 - 2\,R_1 \\ R_3 \to R_3 - 3\,R_1 \end{matrix} \sim \begin{bmatrix} 1 & 2 & 3 \\ 0 & 1 & 1 \\ 0 & 0 & -2 \end{bmatrix} R_3 \to R_3 + 5\,R_2 \qquad \textbf{Ans.}$$

Example 23. *Transform* $\begin{bmatrix} 1 & 3 & 3 \\ 2 & 4 & 10 \\ 3 & 8 & 4 \end{bmatrix}$ *into a unit matrix.*

Solution.

$$\begin{bmatrix} 1 & 3 & 3 \\ 2 & 4 & 10 \\ 3 & 8 & 4 \end{bmatrix} \sim \begin{bmatrix} 1 & 3 & 3 \\ 0 & -2 & 4 \\ 0 & -1 & -5 \end{bmatrix} \begin{matrix} R_2 \to R_2 - 2\,R_1 \\ R_3 \to R_3 - 3\,R_1 \end{matrix}$$

$$\sim \begin{bmatrix} 1 & 3 & 3 \\ 0 & 1 & -2 \\ 0 & -1 & -5 \end{bmatrix} R_2 \to -\frac{1}{2} R_2 \sim \begin{bmatrix} 1 & 0 & 9 \\ 0 & 1 & -2 \\ 0 & 0 & -7 \end{bmatrix} \begin{matrix} R_1 \to R_1 - 3\,R_2 \\ \\ R_3 \to R_3 + R_2 \end{matrix}$$

$$\sim \begin{bmatrix} 1 & 0 & 9 \\ 0 & 1 & -2 \\ 0 & 0 & 1 \end{bmatrix} R_3 \to -\frac{1}{7} R_3 \sim \begin{bmatrix} 1 & 0 & 0 \\ 0 & 1 & 0 \\ 0 & 0 & 1 \end{bmatrix} \begin{matrix} R_1 \to R_1 - 9R_3 \\ R_2 \to R_2 + 2R_3 \\ \end{matrix}$$

38.15 ELEMENTARY MATRICES

A matrix obtained from a unit matrix by a single elementary transformation is called elementary matrix.

$$I = \begin{bmatrix} 1 & 0 & 0 \\ 0 & 1 & 0 \\ 0 & 0 & 1 \end{bmatrix}$$

Consider the matrix obtained by $R_2 + 3\,R_1$

$$\begin{bmatrix} 1 & 0 & 0 \\ 3 & 1 & 0 \\ 0 & 0 & 1 \end{bmatrix} \text{ is called the elementary matrix.}$$

38.16 THEOREM

Every elementary row transformation of a matrix can be affected by pre-multiplication with the corresponding elementary matrix.

Consider the matrix $$A = \begin{bmatrix} 2 & 3 & 4 \\ 5 & 6 & 7 \\ 3 & 5 & 9 \end{bmatrix}$$

Let us apply row transformation $R_3 + 4\,R_1$ and we get a matrix B.

$$B = \begin{bmatrix} 2 & 3 & 4 \\ 5 & 6 & 7 \\ 11 & 17 & 25 \end{bmatrix}$$

Now we shall show that pre-multiplication of A by corresponding elementary matrix $R_3 + 4\,R_1$ will give us B.

Now, if $I = \begin{bmatrix} 1 & 0 & 0 \\ 0 & 1 & 0 \\ 0 & 0 & 1 \end{bmatrix}$ then, Elementary matrix $= \begin{bmatrix} 1 & 0 & 0 \\ 0 & 1 & 0 \\ 4 & 0 & 1 \end{bmatrix} R_3 \to R_3 + 4R_1$

$$\therefore \quad \text{Elementary matrix} \times A = \begin{bmatrix} 1 & 0 & 0 \\ 0 & 1 & 0 \\ 4 & 0 & 1 \end{bmatrix} \times \begin{bmatrix} 2 & 3 & 4 \\ 5 & 6 & 7 \\ 3 & 5 & 9 \end{bmatrix} = \begin{bmatrix} 2 & 3 & 4 \\ 5 & 6 & 7 \\ 11 & 17 & 25 \end{bmatrix} = B$$

Similarly, we can show that every elementary column transformation of a matrix can be affected by post-multiplication with the corresponding elementary matrix.

38.17 TO COMPUTE THE INVERSE OF A MATRIX FROM ELEMENTARY MATRICES (Gauss-jordan Method)

If A is reduced to I by elementary transformation then

$$PA = I \qquad \text{where} \qquad P = P_n P_{n-1} \ldots P_2 P_1$$

$$\therefore \qquad P = A^{-1} \qquad\qquad = \text{Elementary matrix.}$$

Working rule. Write $A = IA$. Perform elementary row transformation on A of the left side and on I of the right hand side so that A is reduced to I and I of right hand side is reduced to P getting $I = PA$.

Then P is the inverse of A.

38.18 THE INVERSE OF A SYMMETRIC MATRIX

The elementary transformations are to be transformed so that the property of being symmetric is preserved. This requires that the transformations occur in pairs, a row transformation must be followed immediately by the same column transformation.

Example 24. *Find the inverse of the following matrix employing elementary transformations:*

$$\begin{bmatrix} 3 & -3 & 4 \\ 2 & -3 & 4 \\ 0 & -1 & 1 \end{bmatrix} \qquad \textit{(U.P., I Semester, Compartment 2002)}$$

Solution. The given matrix is $A = \begin{bmatrix} 3 & -3 & 4 \\ 2 & -3 & 4 \\ 0 & -1 & 1 \end{bmatrix}$

$$\begin{bmatrix} 3 & -3 & 4 \\ 2 & -3 & 4 \\ 0 & -1 & 1 \end{bmatrix} = \begin{bmatrix} 1 & 0 & 0 \\ 0 & 1 & 0 \\ 0 & 0 & 1 \end{bmatrix} A \Rightarrow \begin{bmatrix} 1 & -1 & \frac{4}{3} \\ 2 & -3 & 4 \\ 0 & -1 & 1 \end{bmatrix} = \begin{bmatrix} \frac{1}{3} & 0 & 0 \\ 0 & 1 & 0 \\ 0 & 0 & 1 \end{bmatrix} A \quad R_1 \to \frac{R_1}{3}$$

$$\Rightarrow \begin{bmatrix} 1 & -1 & \frac{4}{3} \\ 0 & -1 & \frac{4}{3} \\ 0 & -1 & 1 \end{bmatrix} = \begin{bmatrix} \frac{1}{3} & 0 & 0 \\ -\frac{2}{3} & 1 & 0 \\ 0 & 0 & 1 \end{bmatrix} A \quad R_2 \to R_2 - 2R_1 \qquad \Rightarrow \begin{bmatrix} 1 & -1 & \frac{4}{3} \\ 0 & 1 & -\frac{4}{3} \\ 0 & -1 & 1 \end{bmatrix} = \begin{bmatrix} \frac{1}{3} & 0 & 0 \\ \frac{2}{3} & -1 & 0 \\ 0 & 0 & 1 \end{bmatrix} A \quad R_2 \to -R_2$$

$$\Rightarrow \begin{bmatrix} 1 & -1 & \frac{4}{3} \\ 0 & 1 & -\frac{4}{3} \\ 0 & 0 & -\frac{1}{3} \end{bmatrix} = \begin{bmatrix} \frac{1}{3} & 0 & 0 \\ \frac{2}{3} & -1 & 0 \\ \frac{2}{3} & -1 & 1 \end{bmatrix} A \quad R_3 \to R_3 + R_2 \qquad \Rightarrow \begin{bmatrix} 1 & -1 & \frac{4}{3} \\ 0 & 1 & -\frac{4}{3} \\ 0 & 0 & 1 \end{bmatrix} = \begin{bmatrix} \frac{1}{3} & 0 & 0 \\ \frac{2}{3} & -1 & 0 \\ -2 & 3 & -3 \end{bmatrix} A \quad R_3 \to -3R_3$$

$$\Rightarrow \begin{bmatrix} 1 & -1 & 0 \\ 0 & 1 & 0 \\ 0 & 0 & 1 \end{bmatrix} = \begin{bmatrix} 3 & -4 & 4 \\ -2 & 3 & 4 \\ -2 & 3 & -3 \end{bmatrix} A \quad \begin{matrix} R_1 \to R_1 - \frac{4}{3}R_3 \\ R_2 \to R_2 + \frac{4}{3}R_3 \end{matrix}$$

$$\Rightarrow \begin{bmatrix} 1 & 0 & 0 \\ 0 & 1 & 0 \\ 0 & 0 & 1 \end{bmatrix} = \begin{bmatrix} 1 & -1 & 0 \\ -2 & 3 & -4 \\ -2 & 3 & -3 \end{bmatrix} A \quad R_1 \to R_1 + R_2$$

Hence, $$A^{-1} = \begin{bmatrix} 1 & -1 & 0 \\ -2 & 3 & -4 \\ -2 & 3 & -3 \end{bmatrix}$$ **Ans.**

Example 25. *Find the inverse of the following matrix by elementary row transformation.*

$$\begin{bmatrix} 0 & 1 & 2 \\ 1 & 2 & 3 \\ 3 & 1 & 1 \end{bmatrix}$$

Solution. Let $A = \begin{bmatrix} 0 & 1 & 2 \\ 1 & 2 & 3 \\ 3 & 1 & 1 \end{bmatrix}$

Elementary row transformation, which will reduce $A = IA$ to $I = PA$, then matrix P will be the inverse of matrix A.

$$\begin{bmatrix}0&1&2\\1&2&3\\3&1&1\end{bmatrix}=\begin{bmatrix}1&0&0\\0&1&0\\0&0&1\end{bmatrix}A \Rightarrow \begin{bmatrix}1&2&3\\0&1&2\\3&1&1\end{bmatrix}=\begin{bmatrix}0&1&0\\1&0&0\\0&0&1\end{bmatrix}A \quad R_1 \leftrightarrow R_2$$

$$\Rightarrow \begin{bmatrix}1&2&3\\0&1&2\\0&-5&-8\end{bmatrix}=\begin{bmatrix}0&1&0\\1&0&0\\0&-3&1\end{bmatrix}A \quad R_3 \to R_3-3R_1 \Rightarrow \begin{bmatrix}1&2&3\\0&1&2\\0&0&2\end{bmatrix}=\begin{bmatrix}0&1&0\\1&0&0\\5&-3&1\end{bmatrix}A \quad R_3 \to R_3+5R_2$$

$$\begin{bmatrix}1&2&3\\0&1&2\\0&0&1\end{bmatrix}=\begin{bmatrix}0&1&0\\1&0&0\\\frac{5}{2}&-\frac{3}{2}&\frac{1}{2}\end{bmatrix}A \quad R_3 \to \frac{1}{2}R_3 \Rightarrow \begin{bmatrix}1&2&0\\0&1&0\\0&0&1\end{bmatrix}=\begin{bmatrix}-\frac{15}{2}&\frac{11}{2}&-\frac{3}{2}\\-4&3&-1\\\frac{5}{2}&-\frac{3}{2}&\frac{1}{2}\end{bmatrix}A \quad R_1 \to R_1-3R_3,\ R_2 \to R_2-2R_3,$$

$$\Rightarrow \begin{bmatrix}1&0&0\\0&1&0\\0&0&1\end{bmatrix}=\begin{bmatrix}\frac{1}{2}&-\frac{1}{2}&\frac{1}{2}\\-4&3&-1\\\frac{5}{2}&-\frac{3}{2}&\frac{1}{2}\end{bmatrix}A \quad R_1 \to R_1-2R_2$$

$$I = PA \Rightarrow P = A^{-1}$$

$$A^{-1}=\begin{bmatrix}\frac{1}{2}&-\frac{1}{2}&\frac{1}{2}\\-4&3&-1\\\frac{5}{2}&\frac{-3}{2}&\frac{1}{2}\end{bmatrix}$$ **Ans.**

Example 26. *Find the inverse of the matrix A by applying elementary transformations*

$$\begin{bmatrix}0&2&1&3\\1&1&-1&-2\\1&2&0&1\\-1&1&2&6\end{bmatrix}.$$

Solution. Here, we have $A=\begin{bmatrix}0&2&1&3\\1&1&-1&-2\\1&2&0&1\\-1&1&2&6\end{bmatrix}$

Let
$$\begin{bmatrix}0&2&1&3\\1&1&-1&-2\\1&2&0&1\\-1&1&2&6\end{bmatrix}=\begin{bmatrix}1&0&0&0\\0&1&0&0\\0&0&1&0\\0&0&0&1\end{bmatrix}A$$

$$\begin{bmatrix}1&1&-1&-2\\0&2&1&3\\1&2&0&1\\-1&1&2&6\end{bmatrix}=\begin{bmatrix}0&1&0&0\\1&0&0&0\\0&0&1&0\\0&0&0&1\end{bmatrix}A \quad R_1 \leftrightarrow R_2$$

$$\begin{bmatrix} 1 & 1 & -1 & -2 \\ 0 & 2 & 1 & 3 \\ 0 & 1 & 1 & 3 \\ 0 & 2 & 1 & 4 \end{bmatrix} = \begin{bmatrix} 0 & 1 & 0 & 0 \\ 1 & 0 & 0 & 0 \\ 0 & -1 & 1 & 0 \\ 0 & 1 & 0 & 1 \end{bmatrix} A \quad \begin{matrix} \\ \\ R_3 \to R_3 - R_1 \\ R_4 \to R_4 + R_1 \end{matrix}$$

$$\begin{bmatrix} 1 & 1 & -1 & -2 \\ 0 & 1 & 1 & 3 \\ 0 & 2 & 1 & 3 \\ 0 & 2 & 1 & 4 \end{bmatrix} = \begin{bmatrix} 0 & 1 & 0 & 0 \\ 0 & -1 & 1 & 0 \\ 1 & 0 & 0 & 0 \\ 0 & 1 & 0 & 1 \end{bmatrix} A \quad R_3 \leftrightarrow R_2$$

$$\begin{bmatrix} 1 & 1 & -1 & -2 \\ 0 & 1 & 1 & 3 \\ 0 & 0 & -1 & -3 \\ 0 & 0 & -1 & -2 \end{bmatrix} = \begin{bmatrix} 0 & 1 & 0 & 0 \\ 0 & -1 & 1 & 0 \\ 1 & 2 & -2 & 0 \\ 0 & 3 & -2 & 1 \end{bmatrix} A \quad \begin{matrix} \\ \\ R_3 \to R_3 - 2R_2 \\ R_4 \to R_4 - 2R_2 \end{matrix}$$

$$\begin{bmatrix} 1 & 1 & -1 & -2 \\ 0 & 1 & 1 & 3 \\ 0 & 0 & -1 & -3 \\ 0 & 0 & 0 & 1 \end{bmatrix} = \begin{bmatrix} 0 & 1 & 0 & 0 \\ 0 & -1 & 1 & 0 \\ 1 & 2 & -2 & 0 \\ -1 & 1 & 0 & 1 \end{bmatrix} A \quad R_4 \to R_4 - R_3$$

$$\begin{bmatrix} 1 & 1 & -1 & -2 \\ 0 & 1 & 1 & 3 \\ 0 & 0 & 1 & 3 \\ 0 & 0 & 0 & 1 \end{bmatrix} = \begin{bmatrix} 0 & 1 & 0 & 0 \\ 0 & -1 & 1 & 0 \\ -1 & -2 & 2 & 0 \\ -1 & 1 & 0 & 1 \end{bmatrix} A \quad R_3 \to -R_3$$

$$\begin{bmatrix} 1 & 1 & -1 & 0 \\ 0 & 1 & 1 & 0 \\ 0 & 0 & 1 & 0 \\ 0 & 0 & 0 & 1 \end{bmatrix} = \begin{bmatrix} -2 & 3 & 0 & 2 \\ 3 & -4 & 1 & -3 \\ 2 & -5 & 2 & -3 \\ -1 & 1 & 0 & 1 \end{bmatrix} A \quad \begin{matrix} R_1 \to R_1 + 2R_4 \\ R_2 \to R_2 - 3R_4 \\ R_3 \to R_3 - 3R_4 \\ \\ \end{matrix}$$

$$\begin{bmatrix} 1 & 1 & 0 & 0 \\ 0 & 1 & 0 & 0 \\ 0 & 0 & 1 & 0 \\ 0 & 0 & 0 & 1 \end{bmatrix} = \begin{bmatrix} 0 & -2 & 2 & -1 \\ 1 & 1 & -1 & 0 \\ 2 & -5 & 2 & -3 \\ -1 & 1 & 0 & 1 \end{bmatrix} A \quad \begin{matrix} R_1 \to R_1 + R_3 \\ R_2 \to R_2 - R_3 \\ \\ \\ \end{matrix}$$

$$\begin{bmatrix} 1 & 0 & 0 & 0 \\ 0 & 1 & 0 & 0 \\ 0 & 0 & 1 & 0 \\ 0 & 0 & 0 & 1 \end{bmatrix} = \begin{bmatrix} -1 & -3 & 3 & -1 \\ 1 & 1 & -1 & 0 \\ 2 & -5 & 2 & -3 \\ -1 & 1 & 0 & 1 \end{bmatrix} A \quad R_1 \to R_1 - R_2$$

$$I = A^{-1} A$$

Hence, $$A^{-1} = \begin{bmatrix} -1 & -3 & 3 & -1 \\ 1 & 1 & -1 & 0 \\ 2 & -5 & 2 & -3 \\ -1 & 1 & 0 & 1 \end{bmatrix}$$ **Ans.**

EXERCISE 38.4

Reduce the matrices to triangular form:

1. $A = \begin{bmatrix} 1 & 2 & 3 \\ 2 & 5 & 7 \\ 3 & 1 & 2 \end{bmatrix}$ **Ans.** $\begin{bmatrix} 1 & 2 & 3 \\ 0 & 1 & 1 \\ 0 & 0 & -2 \end{bmatrix}$

2. $\begin{bmatrix} 3 & 1 & 4 \\ 1 & 2 & -5 \\ 0 & 1 & 5 \end{bmatrix}$ **Ans.** $\begin{bmatrix} 0 & 1 & 4 \\ 0 & 5 & -19 \\ 0 & 0 & 22 \end{bmatrix}$

Find the inverse of the following matrices:

3. $\begin{bmatrix} 1 & 3 & 3 \\ 1 & 4 & 3 \\ 1 & 3 & 4 \end{bmatrix}$ **Ans.** $\begin{bmatrix} 7 & -3 & -3 \\ -1 & 1 & 0 \\ -1 & 0 & 1 \end{bmatrix}$

4. $\begin{bmatrix} 1 & -1 & 1 \\ 4 & 1 & 0 \\ 8 & 1 & 1 \end{bmatrix}$ **Ans.** $\begin{bmatrix} 1 & 2 & -1 \\ -4 & -7 & 4 \\ -4 & -9 & 5 \end{bmatrix}$

Use elementary row operations to find inverse of

5. $A = \begin{bmatrix} 1 & 1 & 3 \\ 1 & 3 & -3 \\ -2 & -4 & -4 \end{bmatrix}$ **Ans.** $\frac{1}{4}\begin{bmatrix} 12 & 4 & 6 \\ -5 & -1 & -3 \\ -1 & -1 & -1 \end{bmatrix}$ *(AMIETE, June 2010)*

6. $\begin{bmatrix} 2 & 1 & -1 & 2 \\ 1 & 3 & 2 & -3 \\ -1 & 2 & 1 & -1 \\ 2 & -3 & -1 & 4 \end{bmatrix}$ **Ans.** $\frac{1}{18}\begin{bmatrix} 2 & 5 & -7 & 1 \\ 5 & -1 & 5 & -2 \\ -7 & 5 & 11 & 10 \\ 1 & -2 & 10 & 5 \end{bmatrix}$

7. $\begin{bmatrix} 1 & 2 & 3 & 1 \\ 1 & 3 & 3 & 2 \\ 2 & 4 & 3 & 3 \\ 1 & 1 & 1 & 1 \end{bmatrix}$ **Ans.** $\begin{bmatrix} 1 & -2 & 1 & 0 \\ 1 & -2 & 2 & -3 \\ 0 & 1 & -1 & 1 \\ -2 & 3 & -2 & 3 \end{bmatrix}$

8. $\begin{bmatrix} 2 & 1 & -1 & 2 \\ 1 & 3 & 2 & -3 \\ -1 & 2 & 1 & -1 \\ 2 & -3 & -1 & 4 \end{bmatrix}$ **Ans.** $\frac{1}{18}\begin{bmatrix} 2 & 5 & -7 & 1 \\ 5 & -1 & 5 & -2 \\ -7 & 5 & 11 & 10 \\ 1 & -2 & 10 & 5 \end{bmatrix}$

9. $\begin{bmatrix} 2 & -6 & -2 & -3 \\ 5 & -13 & -4 & -7 \\ -1 & 4 & 1 & 2 \\ 0 & 1 & 0 & 1 \end{bmatrix}$ **Ans.** $\begin{bmatrix} -2 & 1 & 0 & 1 \\ 1 & 0 & 2 & -1 \\ -4 & 1 & -3 & 1 \\ -1 & 0 & -2 & 2 \end{bmatrix}$

10. $\begin{bmatrix} 1 & 3 & 3 & 2 & 1 \\ 1 & 4 & 3 & 3 & -1 \\ 1 & 3 & 4 & 1 & 1 \\ 1 & 1 & 1 & 1 & -1 \\ 1 & -2 & -1 & 2 & 2 \end{bmatrix}$ **Ans.** $\frac{1}{15}\begin{bmatrix} 30 & -20 & -15 & 25 & -5 \\ 30 & -11 & -18 & 7 & -8 \\ -30 & 12 & 21 & -9 & 6 \\ -15 & 12 & 6 & -9 & 6 \\ 15 & -7 & -6 & -1 & -1 \end{bmatrix}$

11. If X, Y are non-singular matrices and $B = \begin{bmatrix} X & O \\ O & Y \end{bmatrix}$, show that $B^{-1} = \begin{bmatrix} X^{-1} & O \\ O & Y^{-1} \end{bmatrix}$ where O is a null matrix.

RANK OF MATRIX

39.1 RANK OF A MATRIX

The rank of a matrix is said to be r if

(*a*) It has at least one non-zero minor of order r.

(*b*) Every minor of A of order higher than r is zero.

Note: (*i*) Non-zero row is that row in which all the elements are not zero.

(*ii*) The rank of the product matrix AB of two matrices A and B is less than the rank of either of the matrices A and B.

(*iii*) Corresponding to every matrix A of rank r, there exist non-singular matrices P and Q such that
$$PAQ = \begin{bmatrix} I_r & 0 \\ 0 & 0 \end{bmatrix}$$

39.2 NORMAL FORM (CANONICAL FORM)

By performing elementary transformation, any non-zero matrix A can be reduced to one of the following four forms, called the Normal form of A :

(*i*) I_r (*ii*) $[I_r \ \ 0]$ (*iii*) $\begin{bmatrix} I_r \\ 0 \end{bmatrix}$ (*iv*) $\begin{bmatrix} I_r & 0 \\ 0 & 0 \end{bmatrix}$

The number r so obtained is called the rank of A and we write $\rho(A) = r$. The form $\begin{bmatrix} I_r & 0 \\ 0 & 0 \end{bmatrix}$ is called first canonical form of A. Since both row and column transformations may be used here, the element 1 of the first row obtained can be moved in the first column. Then both the first row and first column can be cleared of other non-zero elements. Similarly, the element 1 of the second row can be brought into the second column, and so on.

Example 1. *Find the rank of the following matrix by reducing it to normal form*
$$A = \begin{bmatrix} 1 & 2 & -1 & 3 \\ 4 & 1 & 2 & 1 \\ 3 & -1 & 1 & 2 \\ 1 & 2 & 0 & 1 \end{bmatrix}$$

Solution.
$$\begin{bmatrix} 1 & 2 & -1 & 3 \\ 4 & 1 & 2 & 1 \\ 3 & -1 & 1 & 2 \\ 1 & 2 & 0 & 1 \end{bmatrix} \sim \begin{bmatrix} 1 & 2 & -1 & 3 \\ 0 & -7 & 6 & -11 \\ 0 & -7 & 4 & -7 \\ 0 & 0 & 1 & -2 \end{bmatrix} \begin{matrix} \\ R_2 \to R_2 - 4R_1 \\ R_3 \to R_3 - 3R_1 \\ R_4 \to R_4 - R_1 \end{matrix}$$

$$\sim \begin{bmatrix} 1 & 2 & -1 & 3 \\ 0 & -7 & 6 & -11 \\ 0 & -7 & 4 & -7 \\ 0 & 0 & 1 & -2 \end{bmatrix} \sim \begin{bmatrix} 1 & 2 & -1 & 3 \\ 0 & -7 & 6 & -11 \\ 0 & 0 & -2 & 4 \\ 0 & 0 & 1 & -2 \end{bmatrix} R_3 \to R_3 - R_2$$

$$C_2 \to C_2 - 2\,C_1, C_3 \to C_3 + C_1, C_4 \to C_4 - 3C_1$$

$$\sim \begin{bmatrix} 1 & 0 & 0 & 0 \\ 0 & -7 & 6 & 11 \\ 0 & 0 & -2 & 4 \\ 0 & 0 & 1 & -2 \end{bmatrix} \sim \begin{bmatrix} 1 & 0 & 0 & 0 \\ 0 & -7 & 6 & 11 \\ 0 & 0 & -2 & 4 \\ 0 & 0 & 0 & 0 \end{bmatrix} R_4 \to R_4 + \frac{1}{2} R_3$$

$$C_3 \to C_3 + \frac{6}{7}\, C_2, C_4 \to C_4 - \frac{11}{7}\, C_2,$$

$$C_4 \to C_4 + 2C_3 \sim \begin{bmatrix} 1 & 0 & 0 & 0 \\ 0 & -7 & 0 & 0 \\ 0 & 0 & -2 & 0 \\ 0 & 0 & 0 & 0 \end{bmatrix} \sim \begin{bmatrix} 1 & 0 & 0 & 0 \\ 0 & 1 & 0 & 0 \\ 0 & 0 & 1 & 0 \\ 0 & 0 & 0 & 0 \end{bmatrix} \begin{matrix} R_2 \to -1/7\, R_2 \\ R_3 \to -1/2\, R_3 \end{matrix}$$

Rank of $A = 3$ **Ans.**

Example 2. *For which value of 'b' the rank of the matrix*

$$A = \begin{bmatrix} 1 & 5 & 4 \\ 0 & 3 & 2 \\ b & 13 & 10 \end{bmatrix} \text{ is 2,}$$ *(AMIETE, June 2009, U.P., I Semester, 2008)*

Solution. Here, we have

$$A = \begin{bmatrix} 1 & 5 & 4 \\ 0 & 3 & 2 \\ b & 13 & 10 \end{bmatrix} \sim \begin{bmatrix} 1 & 5 & 4 \\ 0 & 3 & 2 \\ 0 & 13-5b & 10-4b \end{bmatrix} R_3 \to R_3 - bR_1$$

$$\sim \begin{bmatrix} 1 & 0 & 0 \\ 0 & 3 & 2 \\ 0 & 13-5b & 10-4b \end{bmatrix} \begin{matrix} C_2 \to C_2 - 5C_1 \\ C_3 \to C_3 - 4C_1 \end{matrix} \sim \begin{bmatrix} 1 & 0 & 0 \\ 0 & 3 & 2 \\ 0 & 0 & \frac{2(2-b)}{3} \end{bmatrix} R_3 \to R_3 - \frac{13-5b}{3} R_2$$

If rank of A is 2, then $\frac{2(2-b)}{3}$ must be zero.

i.e; $\frac{2(2-b)}{3} = 0 \quad \Rightarrow 2 - b = 0 \quad \Rightarrow b = 2$ **Ans.**

Example 3. *Reduce the matrix to normal form and find its rank.*

$$\begin{bmatrix} 2 & 3 & 4 & 5 \\ 3 & 4 & 5 & 6 \\ 4 & 5 & 6 & 7 \\ 9 & 10 & 11 & 12 \end{bmatrix}$$

(R.G.P.V., Bhopal, I Sem. April 2009)

Solution. $\begin{bmatrix} 2 & 3 & 4 & 5 \\ 3 & 4 & 5 & 6 \\ 4 & 5 & 6 & 7 \\ 9 & 10 & 11 & 12 \end{bmatrix} \sim \begin{bmatrix} 2 & 3 & 4 & 5 \\ 0 & -\frac{1}{2} & -1 & \frac{-3}{2} \\ 0 & -1 & -2 & -3 \\ 0 & \frac{-7}{2} & -7 & \frac{-21}{2} \end{bmatrix} \begin{matrix} R_2 \to R_2 - \frac{3}{2}R_1 \\ R_3 \to R_3 - 2R_1 \\ R_4 \to R_4 - \frac{9}{2}R_1 \end{matrix}$

$$\sim \begin{bmatrix} 2 & 0 & 0 & 0 \\ 0 & -\frac{1}{2} & -1 & \frac{-3}{2} \\ 0 & -1 & -2 & -3 \\ 0 & \frac{-7}{2} & -7 & \frac{-21}{2} \end{bmatrix} \begin{matrix} C_2 \to C_2 - \frac{3}{2}C_1 \\ C_3 \to C_3 - 2\,C_1 \\ C_4 \to C_4 - \frac{5}{2}C_1 \end{matrix} \sim \begin{bmatrix} 1 & 0 & 0 & 0 \\ 0 & 1 & 2 & 3 \\ 0 & 1 & 2 & 3 \\ 0 & 1 & 2 & 3 \end{bmatrix} \begin{matrix} R_1 \to \frac{1}{2}R_1 \\ R_2 \to -2R_2 \\ R_3 \to -R_3 \\ R_4 \to \frac{-2}{7}R_4 \end{matrix}$$

$$\sim \begin{bmatrix} 1 & 0 & 0 & 0 \\ 0 & 1 & 2 & 3 \\ 0 & 0 & 0 & 0 \\ 0 & 0 & 0 & 0 \end{bmatrix} \begin{matrix} R_3 \to R_3 - R_2 \\ R_4 \to R_4 - R_2 \end{matrix} \sim \begin{bmatrix} 1 & 0 & 0 & 0 \\ 0 & 1 & 0 & 0 \\ 0 & 0 & 0 & 0 \\ 0 & 0 & 0 & 0 \end{bmatrix} \begin{matrix} C_3 \to C_3 - 2C_2 \\ C_4 \to C_4 - 3C_2 \end{matrix} = \begin{bmatrix} I_2 & 0 & 0 \\ 0 & 0 & 0 \\ 0 & 0 & 0 \end{bmatrix}$$

Hence its rank = 2 **Ans.**

Example 4. *Find the rank of the matrix.*

$$A = \begin{bmatrix} 1 & 3 & 4 & 2 \\ 2 & -1 & 3 & 2 \\ 3 & -5 & 2 & 2 \\ 6 & -3 & 8 & 6 \end{bmatrix}, \text{ by reducing it to normal form.}$$

(*Uttarakhand I semester, Dec 2006*)

Solution. We have, $A = \begin{bmatrix} 1 & 3 & 4 & 2 \\ 2 & -1 & 3 & 2 \\ 3 & -5 & 2 & 2 \\ 6 & -3 & 8 & 6 \end{bmatrix} = \begin{bmatrix} 1 & 3 & 4 & 2 \\ 0 & -7 & -5 & -2 \\ 0 & -14 & -10 & -4 \\ 0 & -21 & -16 & -6 \end{bmatrix} \begin{matrix} R_2 \to R_2 - 2R_1 \\ R_3 \to R_3 - 3R_1 \\ R_4 \to R_4 - 6R_1 \end{matrix}$

$$= \begin{bmatrix} 1 & 0 & 0 & 0 \\ 0 & -7 & -5 & -2 \\ 0 & -14 & -10 & -4 \\ 0 & -21 & -16 & -6 \end{bmatrix} \begin{matrix} C_2 \to C_2 - 3C_1 \\ C_3 \to C_3 - 4C_1 \\ C_4 \to C_4 - 2C_1 \end{matrix}$$

$$= \begin{bmatrix} 1 & 0 & 0 & 0 \\ 0 & -7 & -5 & -2 \\ 0 & 0 & 0 & 0 \\ 0 & 0 & -1 & 0 \end{bmatrix} \begin{matrix} R_3 \to R_3 - 2R_2 \\ R_4 \to R_4 - 3R_2 \end{matrix} = \begin{bmatrix} 1 & 0 & 0 & 0 \\ 0 & -7 & -5 & -2 \\ 0 & 0 & -1 & 0 \\ 0 & 0 & 0 & 0 \end{bmatrix} R_3 \leftrightarrow R_4$$

$$= \begin{bmatrix} 1 & 0 & 0 & 0 \\ 0 & -7 & 0 & 0 \\ 0 & 0 & -1 & 0 \\ 0 & 0 & 0 & 0 \end{bmatrix} \begin{matrix} C_3 \to C_3 - \frac{5}{7}C_2 \\ C_4 \to C_4 - \frac{2}{7}C_2 \end{matrix} = \begin{bmatrix} 1 & 0 & 0 & 0 \\ 0 & 1 & 0 & 0 \\ 0 & 0 & 1 & 0 \\ 0 & 0 & 0 & 0 \end{bmatrix} \begin{matrix} R_2 \to -\frac{1}{7}R_2 \\ R_3 \to -R_3 \end{matrix}$$

$$= \begin{bmatrix} I_3 & 0 \\ 0 & 0 \end{bmatrix} \text{ which is normal form.}$$

Hence, Rank (A) = 3. **Ans.**

Example 5. *Reduce the matrix A to its normal form, when*

$$A = \begin{bmatrix} 1 & 2 & -1 & 4 \\ 2 & 4 & 3 & 4 \\ 1 & 2 & 3 & 4 \\ -1 & -2 & 6 & -7 \end{bmatrix}$$

Hence, find the rank of A.

Solution. The given matrix is $A = \begin{bmatrix} 1 & 2 & -1 & 4 \\ 2 & 4 & 3 & 4 \\ 1 & 2 & 3 & 4 \\ -1 & -2 & 6 & -7 \end{bmatrix}$

$$\sim \begin{bmatrix} 1 & 2 & -1 & 4 \\ 0 & 0 & 5 & -4 \\ 0 & 0 & 4 & 0 \\ 0 & 0 & 5 & -3 \end{bmatrix} \begin{matrix} \\ R_2 \to R_2 - 2R_1 \\ R_3 \to R_3 - R_1 \\ R_4 \to R_4 + R_1 \end{matrix} \sim \begin{bmatrix} 1 & 0 & 0 & 0 \\ 0 & 0 & 5 & -4 \\ 0 & 0 & 4 & 0 \\ 0 & 0 & 5 & -3 \end{bmatrix} \begin{matrix} C_2 \to C_2 - 2C_1 \\ C_3 \to C_3 + C_1 \\ C_4 \to C_4 - 4C_1 \end{matrix}$$

$$\sim \begin{bmatrix} 1 & 0 & 0 & 0 \\ 0 & 5 & 0 & -4 \\ 0 & 4 & 0 & 0 \\ 0 & 5 & 0 & -3 \end{bmatrix} C_3 \leftrightarrow C_2 \sim \begin{bmatrix} 1 & 0 & 0 & 0 \\ 0 & 5 & 0 & -4 \\ 0 & 0 & 0 & \frac{16}{5} \\ 0 & 0 & 0 & 1 \end{bmatrix} \begin{matrix} \\ \\ R_3 \to R_3 - \frac{4}{5}R_2 \\ R_4 \to R_4 - R_2 \end{matrix}$$

$$\sim \begin{bmatrix} 1 & 0 & 0 & 0 \\ 0 & 5 & -4 & 0 \\ 0 & 0 & \frac{16}{5} & 0 \\ 0 & 0 & 1 & 0 \end{bmatrix} C_4 \leftrightarrow C_3 \sim \begin{bmatrix} 1 & 0 & 0 & 0 \\ 0 & 5 & 0 & 0 \\ 0 & 0 & \frac{16}{5} & 0 \\ 0 & 0 & 0 & 0 \end{bmatrix} \begin{matrix} \\ R_2 \to R_2 + \frac{5}{4}R_3 \\ \\ R_4 \to R_4 - \frac{5}{16}R_3 \end{matrix}$$

$$\sim \begin{bmatrix} 1 & 0 & 0 & 0 \\ 0 & 1 & 0 & 0 \\ 0 & 0 & 1 & 0 \\ 0 & 0 & 0 & 0 \end{bmatrix} \begin{matrix} R_2 \to 1/5\,R_2 \\ R_3 \to 5/16\,R_3 \end{matrix} \sim \begin{bmatrix} I_3 & 0 \\ 0 & 0 \end{bmatrix}$$

Which is the required normal form.

And since, the non-zero rows are 3 hence, the rank of the given matrix is 3. **Ans.**

Example 6. *Find non-singular matrices P, Q so that PAQ is a normal form where*

$$A = \begin{bmatrix} 2 & 1 & -3 & -6 \\ 3 & -3 & 1 & 2 \\ 1 & 1 & 1 & 2 \end{bmatrix}$$

and hence find its rank.

Solution. Order of A is 3×4

Total number of rows in $A = 3$; $\therefore$ Consider unit matrix I_3.

Total number of columns in $A = 4$

Hence, consider unit matrix I_4,

$$\therefore \qquad A_{3\times 4} = I_3\, A\, I_4$$

$$\begin{bmatrix} 2 & 1 & -3 & -6 \\ 3 & -3 & 1 & 2 \\ 1 & 1 & 1 & 2 \end{bmatrix} = \begin{bmatrix} 1 & 0 & 0 \\ 0 & 1 & 0 \\ 0 & 0 & 1 \end{bmatrix} A \begin{bmatrix} 1 & 0 & 0 & 0 \\ 0 & 1 & 0 & 0 \\ 0 & 0 & 1 & 0 \\ 0 & 0 & 0 & 1 \end{bmatrix}$$

$$\begin{bmatrix} 1 & 1 & 1 & 2 \\ 3 & -3 & 1 & 2 \\ 2 & 1 & -3 & -6 \end{bmatrix} = \begin{bmatrix} 0 & 0 & 1 \\ 0 & 1 & 0 \\ 1 & 0 & 0 \end{bmatrix} A \begin{bmatrix} 1 & 0 & 0 & 0 \\ 0 & 1 & 0 & 0 \\ 0 & 0 & 1 & 0 \\ 0 & 0 & 0 & 1 \end{bmatrix} R_1 \leftrightarrow R_3$$

$$\begin{bmatrix} 1 & 1 & 1 & 2 \\ 0 & -6 & -2 & -4 \\ 0 & -1 & -5 & -10 \end{bmatrix} = \begin{bmatrix} 0 & 0 & 1 \\ 0 & 1 & -3 \\ 1 & 0 & -2 \end{bmatrix} A \begin{bmatrix} 1 & 0 & 0 & 0 \\ 0 & 1 & 0 & 0 \\ 0 & 0 & 1 & 0 \\ 0 & 0 & 0 & 1 \end{bmatrix} \begin{matrix} R_2 \to R_2 - 3R_1 \\ R_3 \to R_3 - 2R_1 \end{matrix}$$

$$C_2 \to C_2 - C_1,\ C_3 \to C_3 - C_1,\ C_4 \to C_4 - 2C_1$$

$$\begin{bmatrix} 1 & 0 & 0 & 0 \\ 0 & -6 & -2 & -4 \\ 0 & -1 & -5 & -10 \end{bmatrix} = \begin{bmatrix} 0 & 0 & 1 \\ 0 & 1 & -3 \\ 1 & 0 & -2 \end{bmatrix} A \begin{bmatrix} 1 & -1 & -1 & -2 \\ 0 & 1 & 0 & 0 \\ 0 & 0 & 1 & 0 \\ 0 & 0 & 0 & 1 \end{bmatrix}$$

$$\begin{bmatrix} 1 & 0 & 0 & 0 \\ 0 & 6 & 2 & 4 \\ 0 & 1 & 5 & 10 \end{bmatrix} = \begin{bmatrix} 0 & 0 & 1 \\ 0 & -1 & 3 \\ -1 & 0 & 2 \end{bmatrix} A \begin{bmatrix} 1 & -1 & -1 & -2 \\ 0 & 1 & 0 & 0 \\ 0 & 0 & 1 & 0 \\ 0 & 0 & 0 & 1 \end{bmatrix} \begin{matrix} R_2 \to (-1)R_2 \\ R_3 \to (-1)R_3 \end{matrix}$$

$$\begin{bmatrix} 1 & 0 & 0 & 0 \\ 0 & 1 & 5 & 10 \\ 0 & 6 & 2 & 4 \end{bmatrix} = \begin{bmatrix} 0 & 0 & 1 \\ -1 & 0 & 2 \\ 0 & -1 & 3 \end{bmatrix} A \begin{bmatrix} 1 & -1 & -1 & -2 \\ 0 & 1 & 0 & 0 \\ 0 & 0 & 1 & 0 \\ 0 & 0 & 0 & 1 \end{bmatrix} R_2 \leftrightarrow R_3$$

$$\begin{bmatrix} 1 & 0 & 0 & 0 \\ 0 & 1 & 5 & 10 \\ 0 & 0 & -28 & -56 \end{bmatrix} = \begin{bmatrix} 0 & 0 & 1 \\ -1 & 0 & 2 \\ 6 & -1 & -9 \end{bmatrix} A \begin{bmatrix} 1 & -1 & -1 & -2 \\ 0 & 1 & 0 & 0 \\ 0 & 0 & 1 & 0 \\ 0 & 0 & 0 & 1 \end{bmatrix} R_3 \to R_3 - 6R_2$$

$$C_3 \to C_3 - 5C_2,\quad C_4 \to C_4 - 10C_2$$

$$\begin{bmatrix} 1 & 0 & 0 & 0 \\ 0 & 1 & 0 & 0 \\ 0 & 0 & -28 & -56 \end{bmatrix} = \begin{bmatrix} 0 & 0 & 1 \\ -1 & 0 & 2 \\ 6 & -1 & -9 \end{bmatrix} A \begin{bmatrix} 1 & -1 & 4 & 8 \\ 0 & 1 & -5 & -10 \\ 0 & 0 & 1 & 0 \\ 0 & 0 & 0 & 1 \end{bmatrix}$$

$$\begin{bmatrix} 1 & 0 & 0 & 0 \\ 0 & 1 & 0 & 0 \\ 0 & 0 & 1 & 2 \end{bmatrix} = \begin{bmatrix} 0 & 0 & 1 \\ -1 & 0 & 2 \\ -\frac{6}{28} & \frac{1}{28} & \frac{9}{28} \end{bmatrix} A \begin{bmatrix} 1 & -1 & 4 & 8 \\ 0 & 1 & -5 & -10 \\ 0 & 0 & 1 & 0 \\ 0 & 0 & 0 & 1 \end{bmatrix} R_3 \to -\frac{1}{28} R_3$$

$$\begin{bmatrix}1&0&0&0\\0&1&0&0\\0&0&1&0\end{bmatrix}=\begin{bmatrix}0&0&1\\-1&0&2\\-\frac{6}{28}&\frac{1}{28}&\frac{9}{28}\end{bmatrix}A\begin{bmatrix}1&-1&4&0\\0&1&-5&0\\0&0&1&-2\\0&0&0&1\end{bmatrix}\quad C_4\to C_4-2C_3$$

$$N=PAQ$$

$$P=\begin{bmatrix}0&0&1\\-1&0&2\\-\frac{3}{14}&\frac{1}{28}&\frac{9}{28}\end{bmatrix},\ Q=\begin{bmatrix}1&-1&4&0\\0&1&-5&0\\0&0&1&-2\\0&0&0&1\end{bmatrix}$$ **Ans.**

Note. P and Q are not unique.

Normal form of the given matrix is $\begin{bmatrix}1&0&0&0\\0&1&0&0\\0&0&1&0\end{bmatrix}$

The number of non zero rows in the normal matrix = 3

Hence Rank = 3 **Ans.**

Example 7. *If* $A=\begin{bmatrix}3&-3&4\\2&-3&4\\0&-1&1\end{bmatrix}$, *Find two non singular matrices P and Q such that PAQ = I. Hence find* A^{-1}.

Solution. $A_{3\times 3}=I_3\,A\,I_3$

$$\begin{bmatrix}3&-3&4\\2&-3&4\\0&-1&1\end{bmatrix}=\begin{bmatrix}1&0&0\\0&1&0\\0&0&1\end{bmatrix}A\begin{bmatrix}1&0&0\\0&1&0\\0&0&1\end{bmatrix}$$

$$\begin{bmatrix}1&0&0\\2&-3&4\\0&-1&1\end{bmatrix}=\begin{bmatrix}1&-1&0\\0&1&0\\0&0&1\end{bmatrix}A\begin{bmatrix}1&0&0\\0&1&0\\0&0&1\end{bmatrix}\quad R_1\to R_1-R_2$$

$$\begin{bmatrix}1&0&0\\0&-3&4\\0&-1&1\end{bmatrix}=\begin{bmatrix}1&-1&0\\-2&3&0\\0&0&1\end{bmatrix}A\begin{bmatrix}1&0&0\\0&1&0\\0&0&1\end{bmatrix}\quad R_2\to R_2-2R_1$$

$$\begin{bmatrix}1&0&0\\0&3&4\\0&1&1\end{bmatrix}=\begin{bmatrix}1&-1&0\\-2&3&0\\0&0&1\end{bmatrix}A\begin{bmatrix}1&0&0\\0&-1&0\\0&0&1\end{bmatrix}\quad C_2\to -C_2$$

$$\begin{bmatrix}1&0&0\\0&1&1\\0&3&4\end{bmatrix}=\begin{bmatrix}1&-1&0\\0&0&1\\-2&3&0\end{bmatrix}A\begin{bmatrix}1&0&0\\0&-1&0\\0&0&1\end{bmatrix}\quad R_2\leftrightarrow R_3$$

$$\begin{bmatrix}1&0&0\\0&1&1\\0&0&1\end{bmatrix}=\begin{bmatrix}1&-1&0\\0&0&1\\-2&3&-3\end{bmatrix}A\begin{bmatrix}1&0&0\\0&-1&0\\0&0&1\end{bmatrix}\quad R_3\to R_3-3R_2$$

$$\begin{bmatrix} 1 & 0 & 0 \\ 0 & 1 & 0 \\ 0 & 0 & 1 \end{bmatrix} = \begin{bmatrix} 1 & -1 & 0 \\ 0 & 0 & 1 \\ -2 & 3 & -3 \end{bmatrix} A \begin{bmatrix} 1 & 0 & 0 \\ 0 & -1 & 1 \\ 0 & 0 & 1 \end{bmatrix} C_3 \to C_3 - C_2$$

$$I_3 = PAQ$$

$$\left[\begin{aligned} I &= P\,A\,Q \\ P^{-1} &= A\,Q \\ P^{-1}Q^{-1} &= A \\ (P^{-1}\,Q^{-1})^{-1} &= A^{-1} \\ QP &= A^{-1} \end{aligned}\right]$$

$$A^{-1} = QP, \qquad A^{-1} = \begin{bmatrix} 1 & 0 & 0 \\ 0 & -1 & 1 \\ 0 & 0 & 1 \end{bmatrix} \begin{bmatrix} 1 & -1 & 0 \\ 0 & 0 & 1 \\ -2 & 3 & -3 \end{bmatrix}$$

$$\Rightarrow \qquad A^{-1} = \begin{bmatrix} 1 & -1 & 0 \\ -2 & 3 & -4 \\ -2 & 3 & -3 \end{bmatrix}$$ **Ans.**

39.3 RANK OF MATRIX BY TRIANGULAR FORM

Rank = Number of non-zero row in upper triangular matrix.

Note. Non-zero row is that row which does not contain all the elements as zero.

Example 8. *Find the rank of the matrix*

$$\begin{bmatrix} 1 & 2 & 3 & 2 \\ 2 & 3 & 5 & 1 \\ 1 & 3 & 4 & 5 \end{bmatrix}$$

Solution.

$$\begin{bmatrix} 1 & 2 & 3 & 2 \\ 2 & 3 & 5 & 1 \\ 1 & 3 & 4 & 5 \end{bmatrix} \sim \begin{bmatrix} 1 & 2 & 3 & 2 \\ 0 & -1 & -1 & -3 \\ 0 & 1 & 1 & 3 \end{bmatrix} \begin{matrix} \\ R_2 \to R_2 - 2R_1 \\ R_3 \to R_3 - R_1 \end{matrix}$$

$$\sim \begin{bmatrix} 1 & 2 & 3 & 2 \\ 0 & -1 & -1 & -3 \\ 0 & 0 & 0 & 0 \end{bmatrix} \begin{matrix} \\ \\ R_3 \to R_3 + R_2 \end{matrix}$$

Rank = Number of non zero rows = 2. **Ans.**

Example 9. *Find the rank of the matrix*

$$\begin{bmatrix} -1 & 2 & 3 & -2 \\ 2 & -5 & 1 & 2 \\ 3 & -8 & 5 & 2 \\ 5 & -12 & -1 & 6 \end{bmatrix}$$

Solution.

$$\begin{bmatrix} -1 & 2 & 3 & -2 \\ 2 & -5 & 1 & 2 \\ 3 & -8 & 5 & 2 \\ 5 & -12 & -1 & 6 \end{bmatrix} \sim \begin{bmatrix} -1 & 2 & 3 & -2 \\ 0 & -1 & 7 & -2 \\ 0 & -2 & 14 & -4 \\ 0 & -2 & 14 & -4 \end{bmatrix} \begin{matrix} \\ R_2 \to R_2 + 2\,R_1 \\ R_3 \to R_3 + 3\,R_1 \\ R_4 \to R_4 + 5\,R_1 \end{matrix}$$

$$\sim \begin{bmatrix} -1 & 2 & 3 & -2 \\ 0 & -1 & 7 & -2 \\ 0 & 0 & 0 & 0 \\ 0 & 0 & 0 & 0 \end{bmatrix} \begin{matrix} \\ R_3 \to R_3 - 2\,R_2 \\ R_4 \to R_4 - 2\,R_2 \\ \\ \end{matrix}$$

Here the 4th order and 3rd order minors are zero. But a minor of second order

$$\begin{vmatrix} 3 & -2 \\ 7 & -2 \end{vmatrix} = -6+14 = 8 \neq 0$$

Rank = Number of non-zero rows = 2. **Ans.**

Example 10. *Find the rank of matrix* $\begin{bmatrix} 2 & 3 & -2 & 4 \\ 3 & -2 & 1 & 2 \\ 3 & 2 & 3 & 4 \\ -2 & 4 & 0 & 5 \end{bmatrix}$ *(U.P., I Semester, Dec., 2006)*

Solution. Multiplying R_1 by $\frac{1}{2}$, we get 1 as pivotal element

$$\sim \begin{bmatrix} 1 & \frac{3}{2} & -1 & 2 \\ 3 & -2 & 1 & 2 \\ 3 & 2 & 3 & 4 \\ -2 & 4 & 0 & 5 \end{bmatrix} \sim \begin{bmatrix} 1 & \frac{3}{2} & -1 & 2 \\ 0 & \frac{-13}{2} & 4 & -4 \\ 0 & \frac{-5}{2} & 6 & -2 \\ 0 & 7 & -2 & 9 \end{bmatrix} \begin{matrix} R_2 \to R_2 - 3R_1 \\ R_3 \to R_3 - 3R_1 \\ R_4 \to R_4 + 2R_1 \end{matrix}$$

$$\sim \begin{bmatrix} 1 & 0 & 0 & 0 \\ 0 & \frac{-13}{2} & 4 & -4 \\ 0 & \frac{-5}{2} & 6 & -2 \\ 0 & 7 & -2 & 9 \end{bmatrix} \begin{matrix} C_2 \to C_2 \frac{-3}{2} C_1 \\ C_3 \to C_3 + C_1 \\ C_4 \to C_4 - 2C_1 \end{matrix}$$

$$\sim \begin{bmatrix} 1 & 0 & 0 & 0 \\ 0 & 1 & -\frac{8}{13} & \frac{8}{13} \\ 0 & 0 & \frac{58}{13} & -\frac{6}{13} \\ 0 & 0 & \frac{30}{13} & \frac{61}{13} \end{bmatrix} \begin{matrix} R_3 \to R_3 + \frac{5}{2} R_2 \\ R_4 \to R_4 - 7 R_2 \end{matrix} \sim \begin{bmatrix} 1 & 0 & 0 & 0 \\ 0 & 1 & 0 & 0 \\ 0 & 0 & \frac{58}{13} & -\frac{6}{13} \\ 0 & 0 & \frac{30}{13} & \frac{61}{13} \end{bmatrix} \begin{matrix} C_3 \to C_3 + \frac{8}{13} C_2 \\ C_4 \to C_4 - \frac{8}{13} C_2 \end{matrix}$$

$$\sim \begin{bmatrix} 1 & 0 & 0 & 0 \\ 0 & 1 & -\frac{8}{13} & \frac{8}{13} \\ 0 & 0 & \frac{58}{13} & -\frac{6}{13} \\ 0 & 0 & \frac{30}{13} & \frac{61}{13} \end{bmatrix} \begin{matrix} R_3 \to R_3 + \frac{5}{2} R_2 \\ R_4 \to R_4 - 7 R_2 \end{matrix} \sim \begin{bmatrix} 1 & 0 & 0 & 0 \\ 0 & 1 & 0 & 0 \\ 0 & 0 & \frac{58}{13} & -\frac{6}{13} \\ 0 & 0 & \frac{30}{13} & \frac{61}{13} \end{bmatrix} \begin{matrix} C_3 \to C_3 + \frac{8}{13} C_2 \\ C_4 \to C_4 - \frac{8}{13} C_2 \end{matrix}$$

$$\sim \begin{bmatrix} 1 & 0 & 0 & 0 \\ 0 & 1 & 0 & 0 \\ 0 & 0 & 1 & -\frac{3}{29} \\ 0 & 0 & \frac{30}{13} & \frac{61}{13} \end{bmatrix} R_3 \to \frac{13}{58} R_3 \sim \begin{bmatrix} 1 & 0 & 0 & 0 \\ 0 & 1 & 0 & 0 \\ 0 & 0 & 1 & -\frac{3}{29} \\ 0 & 0 & 0 & \frac{143}{29} \end{bmatrix} R_4 \to R_4 - \frac{30}{13} R_3$$

$$\sim\begin{bmatrix}1 & 0 & 0 & 0\\ 0 & 1 & 0 & 0\\ 0 & 0 & 1 & 0\\ 0 & 0 & 0 & \frac{143}{29}\end{bmatrix} C_4 \to C_4 + \frac{3}{29}C_3 \quad \sim\begin{bmatrix}1 & 0 & 0 & 0\\ 0 & 1 & 0 & 0\\ 0 & 0 & 1 & 0\\ 0 & 0 & 0 & 1\end{bmatrix} R_4 \to \frac{29}{143}R_4$$

$$\simeq I_4$$

Hence, the rank of the given matrix = 4 **Ans.**

Example 11. *Use elementary transformation to reduce the following matrix A to triangular from and hence find the rank of A.*

$$A = \begin{bmatrix}2 & 3 & -1 & -1\\ 1 & -1 & -2 & -4\\ 3 & 1 & 3 & -2\\ 6 & 3 & 0 & -7\end{bmatrix}$$

(R.G.P.V., Bhopal, June 2007, U.P., I Semester, Dec. 2005)

Solution. We have,

$$A = \begin{bmatrix}2 & 3 & -1 & -1\\ 1 & -1 & -2 & -4\\ 3 & 1 & 3 & -2\\ 6 & 3 & 0 & -7\end{bmatrix} \approx \begin{bmatrix}1 & -1 & -2 & -4\\ 2 & 3 & -1 & -1\\ 3 & 1 & 3 & -2\\ 6 & 3 & 0 & -7\end{bmatrix} R_1 \longleftrightarrow R_2$$

$$\approx \begin{bmatrix}1 & -1 & -2 & -4\\ 0 & 5 & 3 & 7\\ 0 & 4 & 9 & 10\\ 0 & 9 & 12 & 17\end{bmatrix} \begin{matrix} \\ R_2 \to R_2 - 2R_1\\ R_3 \to R_3 - 3R_1\\ R_4 \to R_4 - 6R_1\end{matrix} \approx \begin{bmatrix}1 & -1 & -2 & -4\\ 0 & 5 & 3 & 7\\ 0 & 0 & 33/5 & 22/5\\ 0 & 0 & 33/5 & 22/5\end{bmatrix} \begin{matrix} \\ \\ R_3 \to R_3 - 4/5\,R_2\\ R_4 \to R_4 - 9/5\,R_2\end{matrix}$$

$$\approx \begin{bmatrix}1 & -1 & -2 & -4\\ 0 & 5 & 3 & 7\\ 0 & 0 & 33/5 & 22/5\\ 0 & 0 & 0 & 0\end{bmatrix} R_4 \to R_4 - R_3$$

$R(A)$ = Number of non-zero rows.

$\Rightarrow \quad R(A) = 3$ **Ans.**

Example 12. *Prove that the points (x_1, y_1), (x_2, y_2) and (x_3, y_3) are collinear if and only if the rank of the matrix* $\begin{bmatrix}x_1 & y_1 & 1\\ x_2 & y_2 & 1\\ x_3 & y_3 & 1\end{bmatrix}$ *is less than three.*

Solution. Necessary condition.

Since the points (x_1, y_1), (x_2, y_2) and (x_3, y_3) are collinear. Therefore, the area of the triangle formed by these points is zero.

$$\therefore \quad \frac{1}{2}\begin{bmatrix} x_1 & y_1 & 1 \\ x_2 & y_2 & 1 \\ x_3 & y_3 & 1 \end{bmatrix} = 0 \qquad \Rightarrow \begin{bmatrix} x_1 & y_1 & 1 \\ x_2 & y_2 & 1 \\ x_3 & y_3 & 1 \end{bmatrix} = 0$$

$\Rightarrow$ The rank of matrix $\begin{bmatrix} x_1 & y_1 & 1 \\ x_2 & y_2 & 1 \\ x_3 & y_3 & 1 \end{bmatrix}$ is less than 3.

Given, the three points are collinear and we have proved that the rank of matrix is less than 3.

Hence, the condition is necessary.

Sufficient condition.

Given : The rank of the following matrix is less than 3.

$$\text{Rank of } \begin{bmatrix} x_1 & y_1 & 1 \\ x_2 & y_2 & 1 \\ x_3 & y_3 & 1 \end{bmatrix} \le 3 \quad \Rightarrow \begin{vmatrix} x_1 & y_1 & 1 \\ x_2 & y_2 & 1 \\ x_3 & y_3 & 1 \end{vmatrix} = 0 \quad \Rightarrow \quad \frac{1}{2}\begin{vmatrix} x_1 & y_1 & 1 \\ x_2 & y_2 & 1 \\ x_3 & y_3 & 1 \end{vmatrix} = 0$$

Thus, the three points (x_1, y_1), (x_2, y_2) and (x_3, y_3) are collinear. Given, the rank of matrix is less than 3 and we have proved that the points are collinear.

Hence, the condition is sufficient. **Proved.**

Theorem

The rank of the product matrix AB of two matrices *A* and *B* is less than the rank of the either of the matrices *A* and *B*.

Proof. Let r_1 and r_2 be the ranks of the matrices *A* and *B* respectively.

Since r_1 is the rank of the matrix *A*, therefore

$$A \sim \begin{bmatrix} Ir_1 & 0 \\ 0 & 0 \end{bmatrix} \qquad \text{... (1)}$$

Where Ir_1 is the unit matrix of order r_1 and contains r_1 rows.

Post multiplying (1) by *B*, we get

$$AB \sim \begin{bmatrix} Ir_1 & 0 \\ 0 & 0 \end{bmatrix} B$$

But $\begin{bmatrix} Ir_1 & 0 \\ 0 & 0 \end{bmatrix}$ B can have r_1 non-zero rows at the most.

$$\text{Rank of } AB = \text{Rank of } \begin{bmatrix} Ir_1 & 0 \\ 0 & 0 \end{bmatrix} B$$

$$\text{Rank of } AB = \text{Rank } \begin{bmatrix} Ir_1 & 0 \\ 0 & 0 \end{bmatrix} B \le r_1$$

Rank of $AB \le$ Rank of A

Similarly we can prove that,

Rank of $AB \le$ Rank of B. **Proved.**

EXERCISE 39.1

Find the rank of the following matrices:

1. $\begin{bmatrix} 1 & 2 & 3 \\ 2 & 4 & 7 \\ 3 & 6 & 10 \end{bmatrix}$ **Ans.** 2

2. $\begin{bmatrix} 1 & 2 & 1 \\ -1 & 0 & 2 \\ 2 & 1 & -3 \end{bmatrix}$ **Ans.** 3

3. $\begin{bmatrix} 0 & 1 & 2 & -2 \\ 4 & 0 & 2 & 6 \\ 2 & 1 & 3 & 1 \end{bmatrix}$ **Ans.** 2

4. $\begin{bmatrix} 2 & 4 & 3 & -2 \\ -3 & -2 & -1 & 4 \\ 6 & -1 & 7 & 2 \end{bmatrix}$ **Ans.** 3

5. $\begin{bmatrix} 3 & 4 & 1 & 1 \\ 2 & 4 & 3 & 6 \\ -1 & -2 & 6 & 4 \\ 1 & -1 & 2 & -3 \end{bmatrix}$ **Ans.** 4

6. $\begin{bmatrix} 1 & 4 & 3 & -2 & 1 \\ -2 & -3 & -1 & 4 & 3 \\ -1 & 6 & 7 & 2 & 9 \\ -3 & 3 & 6 & 6 & 12 \end{bmatrix}$ **Ans.** 2

Reduce the following matrices to Echelon form and find out the rank

7. $\begin{bmatrix} 1 & 1 & 2 \\ 1 & 2 & 2 \\ 2 & 2 & 3 \end{bmatrix}$ **Ans.** $\begin{bmatrix} 1 & 0 & 0 \\ 0 & 1 & 0 \\ 0 & 0 & 1 \end{bmatrix}$, Rank = 3

8. $\begin{bmatrix} 1 & 2 & 3 & 0 \\ 2 & 4 & 3 & 2 \\ 3 & 2 & 1 & 3 \\ 6 & 8 & 7 & 5 \end{bmatrix}$ **Ans.** $\begin{bmatrix} I_3 & 0 \\ 0 & 0 \end{bmatrix}$, Rank = 3

9. $\begin{bmatrix} 3 & 2 & 5 & 7 & 12 \\ 1 & 1 & 2 & 3 & 5 \\ 3 & 3 & 6 & 9 & 15 \end{bmatrix}$ **Ans.** $\begin{bmatrix} I_2 & 0 \\ 0 & 0 \end{bmatrix}$, Rank = 2

10. $\begin{bmatrix} 2 & -4 & 3 & 1 & 0 \\ 1 & -2 & 1 & -4 & 2 \\ 0 & 1 & -1 & 3 & 1 \\ 4 & -7 & 4 & -4 & 5 \end{bmatrix}$ **Ans.** $\begin{bmatrix} I_3 & 0 \\ 0 & 0 \end{bmatrix}$, Rank = 3

Using elementary transformations, reduce the following matrices to the canonical form (or row-reduced Echelon form):

11. $A = \begin{bmatrix} 0 & 0 & 0 & 0 & 0 \\ 0 & 1 & 2 & 3 & 4 \\ 0 & 2 & 3 & 4 & 1 \\ 0 & 3 & 4 & 1 & 2 \end{bmatrix}$

12. $A = \begin{bmatrix} 0 & 4 & -12 & 8 & 9 \\ 0 & 2 & -6 & 2 & 5 \\ 0 & 1 & -3 & 6 & 4 \\ 0 & -8 & 24 & 3 & 1 \end{bmatrix}$

Using elementary transformations, reduce the following matrices to the normal form:

13. $A = \begin{bmatrix} 1 & 2 & 0 & -1 \\ 3 & 4 & 1 & 2 \\ -2 & 3 & 2 & 5 \end{bmatrix}$ **Ans.** $A = \begin{bmatrix} 1 & 0 & 0 & 0 \\ 0 & 1 & 0 & 0 \\ 0 & 0 & 1 & 0 \end{bmatrix}$

14. $A=\begin{bmatrix}1 & 2 & 3 & 4\\3 & 4 & 1 & 2\\4 & 3 & 1 & 2\end{bmatrix}$ **Ans.** $A=\begin{bmatrix}1 & 0 & 0 & 0\\0 & 1 & 0 & 0\\0 & 0 & 1 & 0\end{bmatrix}$

Obtain a matrix N in the normal form equivalent to

15. $A=\begin{bmatrix}0 & 0 & 0 & 0 & 0\\0 & 4 & 5 & 0 & 0\\0 & 9 & 1 & -1 & 2\\0 & 10 & 0 & 1 & 11\end{bmatrix}$

Hence find non-singular matrices P and Q such that $PAQ = N$.

16. $\begin{bmatrix}1 & -3 & 1 & 2\\0 & 1 & 2 & 3\\3 & 4 & 1 & -2\end{bmatrix}$

Find the rank of the following matrix by reducing it into normal form:

17. $A=\begin{bmatrix}1 & 3 & 2 & 5 & 1\\2 & 2 & -1 & 6 & 3\\1 & 1 & 2 & 3 & -1\\0 & 2 & 5 & 2 & -3\end{bmatrix}$ **Ans.** 3

18. $A=\begin{bmatrix}1 & 2 & 3 & 1\\1 & 3 & 3 & 2\\2 & 4 & 3 & 3\\1 & 1 & 1 & 1\end{bmatrix}$ **Ans.** 4

Choose the correct answer:

19. Rank of matrix $A=\begin{bmatrix}1 & 2 & 3\\1 & 4 & 2\\2 & 6 & 5\end{bmatrix}$ is

(*a*) 0 (*b*) 1 (*c*) 3 (*d*) 2 (*AMIETE, June 2009*) **Ans.** (*d*)

CHAPTER

Consistency of Linear System of Equations and their Solution (Linear Dependence)

40.1 SOLUTION OF SIMULTANEOUS EQUATIONS

The matrix of the coefficients of x, y, z is reduced into Echelon form by elementary row transformations. At the end of the row transformation the value of z is calculated from the last equation and value of y and the value of x are calculated by the backward substitution.

Example 1. *Solve the following equations*

$$x - y + 2z = 3,\quad x + 2y + 3z = 5,\quad 3x - 4y - 5z = -13$$

Solution. In the matrix form, the equations are written in the following form.

$$\begin{bmatrix} 1 & -1 & 2 \\ 1 & 2 & 3 \\ 3 & -4 & -5 \end{bmatrix}\begin{bmatrix} x \\ y \\ z \end{bmatrix} = \begin{bmatrix} 3 \\ 5 \\ -13 \end{bmatrix} \text{ or } \begin{bmatrix} 1 & -1 & 2 \\ 0 & 3 & 1 \\ 0 & -1 & -11 \end{bmatrix}\begin{bmatrix} x \\ y \\ z \end{bmatrix} = \begin{bmatrix} 3 \\ 2 \\ -22 \end{bmatrix} \begin{matrix} R_2 \to R_2 - R_1 \\ R_3 \to R_3 - 3R_1 \end{matrix}$$

$$\begin{bmatrix} 1 & -1 & 2 \\ 0 & 3 & 1 \\ 0 & 0 & -\frac{32}{3} \end{bmatrix}\begin{bmatrix} x \\ y \\ z \end{bmatrix} = \begin{bmatrix} 3 \\ 2 \\ -\frac{64}{3} \end{bmatrix} \quad R_3 \to R_3 + \frac{1}{3}R_2$$

$$x - y + 2z = 3 \qquad \ldots(1)$$

$$3y + z = 2 \qquad \ldots(2)$$

$$\frac{-32}{3}z = \frac{-64}{3} \Rightarrow z = 2$$

Putting the value of z in (2), we get

$$3y + 2 = 2 \Rightarrow y = 0$$

Putting the value of y, z in (1), we get

$$x - 0 + 4 = 3 \Rightarrow x = -1$$

$$x = -1,\ y = 0,\ z = 2$$ **Ans.**

Example 2. *Find all the solutions of the system of equations*

$$x_1 + 2x_2 - x_3 = 1,\ 3x_1 - 2x_2 + 2x_3 = 2,\ 7x_1 - 2x_2 + 3x_3 = 5$$

Solution. $\begin{bmatrix} 1 & 2 & -1 \\ 3 & -2 & 2 \\ 7 & -2 & 3 \end{bmatrix}\begin{bmatrix} x_1 \\ x_2 \\ x_3 \end{bmatrix} = \begin{bmatrix} 1 \\ 2 \\ 5 \end{bmatrix}$

$$R_2 \to R_2 - 3R_1, \quad R_3 \to R_3 - 7R_1$$

$$\begin{bmatrix} 1 & 2 & -1 \\ 0 & -8 & 5 \\ 0 & -16 & 10 \end{bmatrix}\begin{bmatrix} x_1 \\ x_2 \\ x_3 \end{bmatrix} = \begin{bmatrix} 1 \\ -1 \\ -2 \end{bmatrix}, \begin{bmatrix} 1 & 2 & -1 \\ 0 & -8 & 5 \\ 0 & 0 & 0 \end{bmatrix}\begin{bmatrix} x_1 \\ x_2 \\ x_3 \end{bmatrix} = \begin{bmatrix} 1 \\ -1 \\ 0 \end{bmatrix} R_3 \to R_3 - 2R_2$$

$$x_1 + 2x_2 - x_3 = 1 \qquad ...(1)$$

$$-8x_2 + 5x_3 = -1 \qquad ...(2)$$

Let $x_3 = k$

Putting $x_3 = k$ in (2), we get

$$-8x_2 + 5k = -1 \Rightarrow x_2 = \frac{1}{8}(5k+1)$$

Substituting the values of x_3, x_1 in (1), we get

$$x_1 + \frac{1}{4}(5k+1) - k = 1$$

$$\therefore \quad x_1 = 1 + k - \frac{5k}{4} - \frac{1}{4} = -\frac{k}{4} + \frac{3}{4}$$

$$\therefore \quad x_1 = -\frac{k}{4} + \frac{3}{4}, \; x_2 = \frac{5k}{8} + \frac{1}{8}, \; x_3 = k$$

The equations have infinite solution. **Ans.**

Example 3. *Express the following system of equations in matrix form and solve them by the elimination method (Gauss Jordan Method)*

$$2x_1 + x_2 + 2x_3 + x_4 = 6$$
$$6x_1 - 6x_2 + 6x_3 + 12x_4 = 36$$
$$4x_1 + 3x_2 + 3x_3 - 3x_4 = -1$$
$$2x_1 + 2x_2 - x_3 + x_4 = 10$$

Solution. The equations are expressed in matrix form as

$$\begin{bmatrix} 2 & 1 & 2 & 1 \\ 6 & -6 & 6 & 12 \\ 4 & 3 & 3 & -3 \\ 2 & 2 & -1 & 1 \end{bmatrix}\begin{bmatrix} x_1 \\ x_2 \\ x_3 \\ x_4 \end{bmatrix} = \begin{bmatrix} 6 \\ 36 \\ -1 \\ 10 \end{bmatrix}$$

$$\Rightarrow \begin{bmatrix} 2 & 1 & 2 & 1 \\ 0 & -9 & 0 & 9 \\ 0 & 1 & -1 & -5 \\ 0 & 1 & -3 & 0 \end{bmatrix}\begin{bmatrix} x_1 \\ x_2 \\ x_3 \\ x_4 \end{bmatrix} = \begin{bmatrix} 6 \\ 18 \\ -13 \\ 4 \end{bmatrix} \begin{matrix} R_2 \to R_2 - 3R_1 \\ R_3 \to R_3 - 2R_1 \\ R_4 \to R_4 - R_1 \end{matrix}$$

$$\begin{bmatrix} 2 & 1 & 2 & 1 \\ 0 & 1 & 0 & -1 \\ 0 & 1 & -1 & -5 \\ 0 & 1 & -3 & 0 \end{bmatrix}\begin{bmatrix} x_1 \\ x_2 \\ x_3 \\ x_4 \end{bmatrix} = \begin{bmatrix} 6 \\ -2 \\ -13 \\ 4 \end{bmatrix} R_2 \to \frac{R_2}{-9}$$

$$\begin{bmatrix} 2 & 1 & 2 & 1 \\ 0 & 1 & 0 & -1 \\ 0 & 0 & -1 & -4 \\ 0 & 0 & -3 & 1 \end{bmatrix}\begin{bmatrix} x_1 \\ x_2 \\ x_3 \\ x_4 \end{bmatrix} = \begin{bmatrix} 6 \\ -2 \\ -11 \\ 6 \end{bmatrix} \begin{matrix} R_3 \to R_3 - R_2 \\ \\ R_4 \to R_4 - R_2 \end{matrix}$$

$$\begin{bmatrix} 2 & 1 & 2 & 1 \\ 0 & 1 & 0 & -1 \\ 0 & 0 & -1 & -4 \\ 0 & 0 & 0 & 13 \end{bmatrix} \begin{bmatrix} x_1 \\ x_2 \\ x_3 \\ x_4 \end{bmatrix} = \begin{bmatrix} 6 \\ -2 \\ -11 \\ 39 \end{bmatrix} \quad R_4 \to R_4 - 3R_3$$

$2x_1 + x_2 + 2x_3 + x_4 = 6$ ...(1)

$x_2 - x_4 = -2$ ...(2)

$-x_3 - 4x_4 = -11$ ...(3)

$13x_4 = 39 \Rightarrow x_4 = 3$

Putting the value of x_4 in (3), we get

$$-x_3 - 12 = -11 \Rightarrow x_3 = -1$$

Putting the value of x_4 in (2), we get

$$x_2 - 3 = -2 \Rightarrow x_2 = 1$$

Substituting the values of x_4, x_3 and x_2 in (1), we get

$$2x_1 + 1 - 2 + 3 = 6 \text{ or } 2x_1 = 4 \Rightarrow x_1 = 2$$

$\therefore \quad x_1 = 2, x_2 = 1, x_3 = -1, x_4 = 3$ **Ans.**

Example 4. *Find the general solution of the system of equations:*

$$3x_1 + 2x_3 + 2x_4 = 0$$
$$-x_1 + 7x_2 + 4x_3 + 9x_4 = 0$$
$$7x_1 - 7x_2 - 5x_4 = 0$$

Solution. The system of equations in the matrix form is expressed as

$$\begin{bmatrix} 3 & 0 & 2 & 2 \\ -1 & 7 & 4 & 9 \\ 7 & -7 & 0 & -5 \end{bmatrix} \begin{bmatrix} x_1 \\ x_2 \\ x_3 \\ x_4 \end{bmatrix} = \begin{bmatrix} 0 \\ 0 \\ 0 \end{bmatrix}$$

$$\begin{bmatrix} -1 & 7 & 4 & 9 \\ 3 & 0 & 2 & 2 \\ 7 & -7 & 0 & -5 \end{bmatrix} \begin{bmatrix} x_1 \\ x_2 \\ x_3 \\ x_4 \end{bmatrix} = \begin{bmatrix} 0 \\ 0 \\ 0 \end{bmatrix} \quad R_1 \leftrightarrow R_2$$

$$\begin{bmatrix} -1 & 7 & 4 & 9 \\ 0 & 21 & 14 & 29 \\ 0 & 42 & 28 & 58 \end{bmatrix} \begin{bmatrix} x_1 \\ x_2 \\ x_3 \\ x_4 \end{bmatrix} = \begin{bmatrix} 0 \\ 0 \\ 0 \end{bmatrix} \quad \begin{matrix} R_2 \to R_2 + 3R_1 \\ R_3 \to R_3 + 7R_1 \end{matrix}$$

$$\begin{bmatrix} -1 & 7 & 4 & 9 \\ 0 & 21 & 14 & 29 \\ 0 & 0 & 0 & 0 \end{bmatrix} \begin{bmatrix} x_1 \\ x_2 \\ x_3 \\ x_4 \end{bmatrix} = \begin{bmatrix} 0 \\ 0 \\ 0 \end{bmatrix} \quad R_3 \to R_3 - 2R_2$$

$-x_1 + 7x_2 + 4x_3 + 9x_4 = 0$ ...(1)

$21x_2 + 14x_3 + 29x_4 = 0$ ...(2)

Let $x_4 = a, x_3 = b$

From (2), $21\,x_2 + 14\,b + 29\,a = 0$ or $x_2 = -\dfrac{2b}{3} - \dfrac{29a}{21}$

From (1), $-x_1 + 7\left(-\dfrac{2b}{3} - \dfrac{29a}{21}\right) + 4b + 9a = 0$

$$x_1 = -\frac{2a}{3} - \frac{2b}{3}$$

$$x_1 = -\frac{2}{3}(a+b),\; x_2 = -\frac{1}{21}(29a+14b)$$

$$x_3 = b,\; x_4 = a$$

Ans.

40.2 TYPES OF LINEAR EQUATIONS

(1) Consistent. A system of equations is said to be *consistent,* if they have one or more solution *i.e.*

$$x + 2y = 4 \qquad\qquad x + 2y = 4$$
$$3x + 2y = 2 \qquad\qquad 3x + 6y = 12$$

Unique solution   Infinite solution

(2) Inconsistent. If a system of equation has no solution, it is said to be inconsistent *i.e.*

$$x + 2y = 4$$
$$3x + 6y = 5$$

40.3 CONSISTENCY OF A SYSTEM OF LINEAR EQUATIONS

$$a_{11}\,x_1 + a_{12}\,x_2 + \ldots a_{1n}\,x_n = b_1$$
$$a_{21}\,x_1 + a_{22}\,x_2 + \ldots a_{2n}\,x_n = b_2$$
$$\ldots\ldots\ldots\ldots\ldots\ldots\ldots\ldots\ldots$$
$$a_{m1}\,x_1 + a_{m2}\,x_2 + \ldots a_{mn}\,x_n = b_m$$

$$\Rightarrow \begin{bmatrix} a_{11} & a_{12} \ldots\ldots & a_{1n} \\ a_{21} & a_{22} \ldots\ldots & a_{2n} \\ \ldots\ldots & \ldots\ldots & \ldots\ldots \\ a_{m1} & a_{m2} \ldots\ldots & a_{mn} \end{bmatrix} \begin{bmatrix} x_1 \\ x_2 \\ \ldots \\ x_m \end{bmatrix} = \begin{bmatrix} b_1 \\ b_2 \\ \ldots \\ b_m \end{bmatrix} \quad \text{and } C = [A, B] = \begin{bmatrix} a_{11} & a_{12} \ldots\ldots & a_{1n} & b_1 \\ a_{21} & a_{22} \ldots\ldots & a_{2n} & b_2 \\ \ldots\ldots & \ldots\ldots & \ldots\ldots & \ldots \\ a_{m1} & a_{m2} \ldots\ldots & a_{mn} & b_m \end{bmatrix}$$

$$\Rightarrow \qquad AX = B$$

is called the **augmented** matrix.

$$\boxed{[A{:}B] = C}$$

***(a)* Consistent equations.** If Rank A = Rank C

(*i*) *Unique solution:* Rank A = Rank $C = n$ where n = number of unknown.

(*ii*) *Infinite solution:* Rank A = Rank $C = r,\; r < n$

***(b)* Inconsistent equations.** If Rank $A \neq$ Rank C.

In Brief:

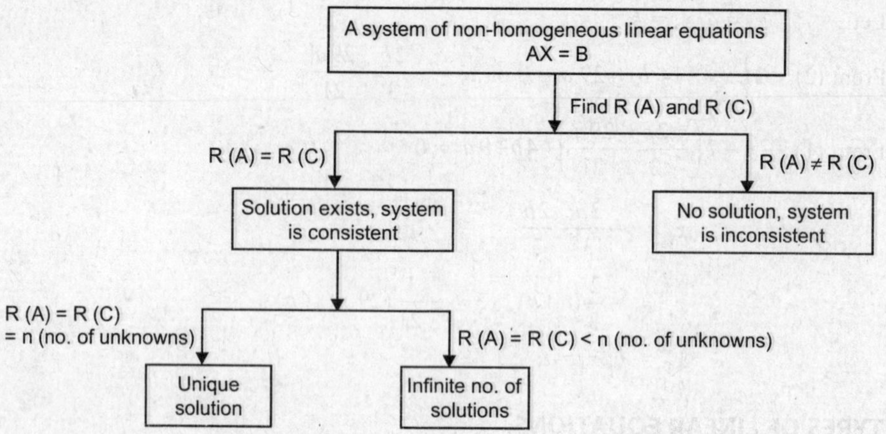

Example 5. *Show that the equations*

$$2x + 6y = -11,\ 6x + 20y - 6z = -3,\ 6y - 18z = -1$$

are not consistent.

Solution. Augmented matrix $C = [A, B]$

$$= \begin{bmatrix} 2 & 6 & 0 & : & -11 \\ 6 & 20 & -6 & : & -3 \\ 0 & 6 & -18 & : & -1 \end{bmatrix} \sim \begin{bmatrix} 2 & 6 & 0 & : & -11 \\ 0 & 2 & -6 & : & 30 \\ 0 & 6 & -18 & : & -1 \end{bmatrix} R_2 \to R_2 - 3R_1$$

$$\sim \begin{bmatrix} 2 & 6 & 0 & : & -11 \\ 0 & 2 & -6 & : & 30 \\ 0 & 0 & 0 & : & -91 \end{bmatrix} R_3 \to R_3 - 3R_2$$

The rank of C is 3 and the rank of A is 2.

Rank of $A \neq$ Rank of C. The equations are not consistent. **Ans.**

Example 6. *Test the consistency and hence solve the following set of equations.*

$$x_1 + 2x_2 + x_3 = 2$$
$$3x_1 + x_2 - 2x_3 = 1$$
$$4x_1 - 3x_2 - x_3 = 3$$
$$2x_1 + 4x_2 + 2x_3 = 4$$

Solution. The given set of equations is written in the matrix form:

$$\begin{bmatrix} 1 & 2 & 1 \\ 3 & 1 & -2 \\ 4 & -3 & -1 \\ 2 & 4 & 2 \end{bmatrix} \begin{bmatrix} x_1 \\ x_2 \\ x_3 \end{bmatrix} = \begin{bmatrix} 2 \\ 1 \\ 3 \\ 4 \end{bmatrix}$$

$$AX = B$$

Here, we have augmented matrix $C = [A, B] \sim \begin{bmatrix} 1 & 2 & 1 & 2 \\ 3 & 1 & -2 & 1 \\ 4 & -3 & -1 & 3 \\ 2 & 4 & 2 & 4 \end{bmatrix}$

$$\sim \begin{bmatrix} 1 & 2 & 1 & 2 \\ 0 & -5 & -5 & -5 \\ 0 & -11 & -5 & -5 \\ 0 & 0 & 0 & 0 \end{bmatrix} \begin{matrix} R_2 \to R_2 - 3R_1 \\ R_3 \to R_3 - 4R_1 \\ R_4 \to R_4 - 2R_1 \end{matrix} \sim \begin{bmatrix} 1 & 2 & 1 & 2 \\ 0 & 1 & 1 & 1 \\ 0 & -11 & -5 & -5 \\ 0 & 0 & 0 & 0 \end{bmatrix} R_2 \to -\frac{1}{5} R_2$$

$$\sim \begin{bmatrix} 1 & 2 & 1 & 2 \\ 0 & 1 & 1 & 1 \\ 0 & 0 & 6 & 6 \\ 0 & 0 & 0 & 0 \end{bmatrix} R_3 \to R_3 + 11 R_2 \sim \begin{bmatrix} 1 & 2 & 1 & 2 \\ 0 & 1 & 1 & 1 \\ 0 & 0 & 1 & 1 \\ 0 & 0 & 0 & 0 \end{bmatrix} R_3 \to \frac{1}{6} R_3$$

Number of non-zero rows= Rank of matrix.

$$\Rightarrow \quad R(C) = R(A) = 3$$

Hence, the given system is consistent and possesses a unique solution. In matrix form the system reduces to

$$\begin{bmatrix} 1 & 2 & 1 \\ 0 & 1 & 1 \\ 0 & 0 & 1 \\ 0 & 0 & 0 \end{bmatrix} \begin{bmatrix} x_1 \\ x_2 \\ x_3 \end{bmatrix} = \begin{bmatrix} 2 \\ 1 \\ 1 \\ 0 \end{bmatrix}$$

$$x_1 + 2x_2 + x_3 = 2 \qquad \text{...(1)}$$

$$x_2 + x_3 = 1 \qquad \text{...(2)}$$

$$x_3 = 1$$

From (2), $x_2 + 1 = 1 \Rightarrow x_2 = 0$

From (1), $x_1 + 0 + 1 = 2 \Rightarrow x_1 = 1$

Hence, $x_1 = 1, x_2 = 0$ and $x_3 = 1$ **Ans.**

Example 7. *Test for consistency and solve :*

$$5x + 3y + 7z = 4, \; 3x + 26y + 2z = 9, \; 7x + 2y + 10z = 5$$

Solution. The augmented matrix $C = [A, B]$ *(R.G. P.V. Bhopal I. Sem. April 2009-08)*

$$\begin{bmatrix} 5 & 3 & 7 & : & 4 \\ 3 & 26 & 2 & : & 9 \\ 7 & 2 & 10 & : & 5 \end{bmatrix} \sim \begin{bmatrix} 1 & \frac{3}{5} & \frac{7}{5} & : & \frac{4}{5} \\ 3 & 26 & 2 & : & 9 \\ 7 & 2 & 10 & : & 5 \end{bmatrix} R_1 \to \frac{1}{5} R_1$$

$$\sim \begin{bmatrix} 1 & \frac{3}{5} & \frac{7}{5} & : & \frac{4}{5} \\ 0 & \frac{121}{5} & -\frac{11}{5} & : & \frac{33}{5} \\ 0 & -\frac{11}{5} & \frac{1}{5} & : & -\frac{3}{5} \end{bmatrix} \begin{matrix} \\ R_2 \to R_2 - 3R_1 \\ R_3 \to R_3 - 7R_1 \end{matrix} \sim \begin{bmatrix} 1 & \frac{3}{5} & \frac{7}{5} & : & \frac{4}{5} \\ 0 & \frac{121}{5} & -\frac{11}{5} & : & \frac{33}{5} \\ 0 & 0 & 0 & : & 0 \end{bmatrix} R_3 \to R_3 + \frac{1}{11} R_2$$

Rank of $A = 2 =$ Rank of C

Hence, the equations are consistent. But the rank is less than 3 *i.e.* number of unknowns. So its solutions are infinite.

$$\begin{bmatrix} 1 & \frac{3}{5} & \frac{7}{5} \\ 0 & \frac{121}{5} & -\frac{11}{5} \\ 0 & 0 & 0 \end{bmatrix} \begin{bmatrix} x \\ y \\ z \end{bmatrix} 2 = \begin{bmatrix} \frac{4}{5} \\ \frac{33}{5} \\ 0 \end{bmatrix}$$

$$x+\frac{3}{5}y+\frac{7}{5}z = \frac{4}{5}$$

$$\frac{121}{5}y-\frac{11z}{5} = \frac{33}{5} \text{ or } 11y - z = 3$$

Let $z = k$ then $\quad 11y - k = 3$ or $y = \frac{3}{11}+\frac{k}{11}$

$$x+\frac{3}{5}\left[\frac{3}{11}+\frac{k}{11}\right]+\frac{7}{5}k = \frac{4}{5} \text{ or } x = -\frac{16}{11}k+\frac{7}{11}$$ **Ans.**

Example 8. *Test the consistency of following system of linear equations and hence find the solution.*

$$4x_1 - x_2 = 12$$
$$-x_1 + 5x_2 - 2x_3 = 0$$
$$-2x_2 + 4x_3 = -8$$

(U.P., I semester Dec. 2005)

Solution. The given equation in the matrix form is

$$\begin{bmatrix} 4 & -1 & 0 \\ -1 & 5 & -2 \\ 0 & -2 & 4 \end{bmatrix}\begin{bmatrix} x_1 \\ x_2 \\ x_3 \end{bmatrix} = \begin{bmatrix} 12 \\ 0 \\ -8 \end{bmatrix}$$

$$AX = B$$

where, $\quad A = \begin{bmatrix} 4 & -1 & 0 \\ -1 & 5 & -2 \\ 0 & -2 & 4 \end{bmatrix}, X = \begin{bmatrix} x_1 \\ x_2 \\ x_3 \end{bmatrix}, B = \begin{bmatrix} 12 \\ 0 \\ -8 \end{bmatrix}$

$$C = [A, B]$$

$$C = \begin{bmatrix} 4 & -1 & 0 & : & 12 \\ -1 & 5 & -2 & : & 0 \\ 0 & -2 & 4 & : & -8 \end{bmatrix} \sim \begin{bmatrix} -1 & 5 & -2 & : & 0 \\ 4 & -1 & 0 & : & 12 \\ 0 & -2 & 4 & : & -8 \end{bmatrix} R_1 \leftrightarrow R_2$$

$$\sim \begin{bmatrix} -1 & 5 & -2 & : & 0 \\ 0 & 19 & -8 & : & 12 \\ 0 & -2 & 4 & : & -8 \end{bmatrix} R_2 \to R_2 + 4R_1$$

$$\sim \begin{bmatrix} -1 & 5 & -2 & : & 0 \\ 0 & 19 & -8 & : & 12 \\ 0 & 0 & \frac{60}{19} & : & \frac{-128}{19} \end{bmatrix} R_3 \to R_3 + \frac{2}{19}R_2$$

Here, rank of A is 3 and Rank of C is also 3.

$$R(A) = R(C) = 3$$

Hence, the equations are consistent with unique solution.

$$\begin{bmatrix} -1 & 5 & -2 \\ 0 & 19 & -8 \\ 0 & 0 & \frac{60}{19} \end{bmatrix}\begin{bmatrix} x_1 \\ x_2 \\ x_3 \end{bmatrix} = \begin{bmatrix} 0 \\ 12 \\ \frac{-128}{19} \end{bmatrix}$$

$$-x_1 + 5x_2 - 2x_3 = 0 \quad ...(1)$$
$$19x_2 - 8x_3 = 12 \quad ...(2)$$
$$\frac{60}{19}x_3 = \frac{-128}{19} \Rightarrow x_3 = -\frac{128}{19}\times\frac{19}{60} \Rightarrow x_3 = \frac{-32}{15}$$

On putting the value of x_3 in (2), we get

$$19x_2 - 8\left(\frac{-32}{15}\right) = 12 \Rightarrow 19x_2 = 12 - \frac{256}{15} = \frac{-76}{15}$$

$$\Rightarrow \quad x_2 = \frac{-76}{15\times 19} = -\frac{4}{15}$$

On putting the values of x_2 and x_3 in (1), we get

$$-x_1 + 5\left(-\frac{4}{15}\right) - 2\left(\frac{-32}{15}\right) = 0$$

$$\Rightarrow \quad -x_1 = \frac{20}{15} - \frac{64}{15} = \frac{-44}{15} \Rightarrow x_1 = \frac{44}{15}$$

Hence, $\quad x_1 = \frac{44}{15}, x_2 = \frac{-4}{15}$ and $x_3 = \frac{-32}{15}$. **Ans.**

Example 9. *Test for consistency of the following system of equations and, if consistent, solve them.*

$$x_1 + 2x_2 - x_3 = 3$$
$$3x_1 - x_2 + 2x_3 = 1$$
$$2x_1 - 2x_2 + 3x_3 = 2$$
$$x_1 - x_2 + x_3 = -1$$

Solution. The augmented matrix $C = [A, B]$

$$\begin{bmatrix} 1 & 2 & -1 & : & 3 \\ 3 & -1 & 2 & : & 1 \\ 2 & -2 & 3 & : & 2 \\ 1 & -1 & 1 & : & -1 \end{bmatrix} \sim \begin{bmatrix} 1 & 2 & -1 & : & 3 \\ 0 & -7 & 5 & : & -8 \\ 0 & -6 & 5 & : & -4 \\ 0 & -3 & 2 & : & -4 \end{bmatrix} \begin{matrix} \\ R_2 \to R_2 - 3R_1 \\ R_3 \to R_3 - 2R_1 \\ R_4 \to R_4 - R_1 \end{matrix}$$

$$\sim \begin{bmatrix} 1 & 2 & -1 & : & 3 \\ 0 & -7 & 5 & : & -8 \\ 0 & 0 & \frac{5}{7} & : & \frac{20}{7} \\ 0 & 0 & \frac{-1}{7} & : & \frac{-4}{7} \end{bmatrix} \begin{matrix} R_3 \to R_3 - \frac{6}{7}R_2 \\ R_4 \to R_4 - \frac{3}{7}R_2 \end{matrix} \sim \begin{bmatrix} 1 & 2 & -1 & : & 3 \\ 0 & -7 & 5 & : & -8 \\ 0 & 0 & \frac{5}{7} & : & \frac{20}{7} \\ 0 & 0 & 0 & : & 0 \end{bmatrix} R_4 \to R_4 + \frac{1}{5}R_3$$

Rank of $C = 3 =$ Rank of A

Hence, the system of equations is consistent with unique solution.

Now, $$\begin{bmatrix} 1 & 2 & -1 \\ 0 & -7 & 5 \\ 0 & 0 & \frac{5}{7} \end{bmatrix}\begin{bmatrix} x_1 \\ x_2 \\ x_3 \end{bmatrix} = \begin{bmatrix} 3 \\ -8 \\ \frac{20}{7} \end{bmatrix}$$

$$x_1 + 2x_2 - x_3 = 3 \quad ...(1)$$
$$-7x_2 + 5x_3 = -8 \quad ...(2)$$

$$\frac{5}{7}x_3 = \frac{20}{7} \Rightarrow x_3 = 4$$

Form (2), $-7x_2 + 5 \times 4 = -8 \Rightarrow -7x_2 = -28 \quad \Rightarrow x_2 = 4$

Form (1), $x_1 + 2 \times 4 - 4 = 3 \quad \Rightarrow x_1 = 3 - 8 + 4 = -1$

Hence, $\quad x_1 = -1, \quad x_2 = 4, \quad x_3 = 4$ **Ans.**

Example 10. *Discuss the consistency of the following system of equations*

$$2x + 3y + 4z = 11, \quad x + 5y + 7z = 15, \quad 3x + 11y + 13z = 25.$$

If found consistent, solve it.

Solution. The augmented matrix $C = [A, B]$

$$\begin{bmatrix} 2 & 3 & 4 & 11 \\ 1 & 5 & 7 & 15 \\ 3 & 11 & 13 & 25 \end{bmatrix} \sim \begin{bmatrix} 1 & 5 & 7 & 15 \\ 2 & 3 & 4 & 11 \\ 3 & 11 & 13 & 25 \end{bmatrix} R_1 \leftrightarrow R_2$$

$R_2 \to R_2 - 2R_1, \; R_3 \to R_3 - 3R_1,$

$$\begin{bmatrix} 1 & 5 & 7 & 15 \\ 0 & -7 & -10 & -19 \\ 0 & -4 & -8 & -20 \end{bmatrix} \quad R_2 \to -\frac{1}{7}R_2, \; R_3 \to -\frac{1}{4}R_3,$$

$$\sim \begin{bmatrix} 1 & 5 & 7 & 15 \\ 0 & 1 & \frac{10}{7} & \frac{19}{7} \\ 0 & 1 & 2 & 5 \end{bmatrix} R_3 \to R_3 - R_2 \sim \begin{bmatrix} 1 & 5 & 7 & 15 \\ 0 & 1 & \frac{10}{7} & \frac{19}{7} \\ 0 & 0 & \frac{4}{7} & \frac{16}{7} \end{bmatrix}$$

Rank of $C = 3 =$ Rank of A

Hence, the system of equations is consistent with unique solution.

Now, $$\begin{bmatrix} 1 & 5 & 7 \\ 0 & 1 & \frac{10}{7} \\ 0 & 0 & \frac{4}{7} \end{bmatrix} \begin{bmatrix} x \\ y \\ z \end{bmatrix} = \begin{bmatrix} 15 \\ \frac{19}{7} \\ \frac{16}{7} \end{bmatrix}$$

$$\Rightarrow \qquad x + 5y + 7z = 15 \qquad \text{...(1)}$$

$$y + \frac{10z}{7} = \frac{19}{7} \qquad \text{...(2)}$$

$$\frac{4z}{7} = \frac{16}{7} \quad \Rightarrow \quad z = 4$$

From (2), $\quad y + \frac{10}{7} \times 4 = \frac{19}{7} \quad \Rightarrow \quad y = -3$

From (1), $\quad x + 5(-3) + 7(4) = 15 \quad \Rightarrow \quad x = 2$

$x = 2, \; y = -3, \; z = 4$ **Ans.**

Example 11. Test for the consistency of the following system of equations :

$$x_1 + 2x_2 + 3x_3 + 4x_4 = 5$$
$$6x_1 + 7x_2 + 8x_3 + 9x_4 = 10$$
$$11x_1 + 12x_2 + 13x_3 + 14x_4 = 15$$
$$16x_1 + 17x_2 + 18x_3 + 19x_4 = 20$$
$$21x_1 + 22x_2 + 23x_3 + 24x_4 = 25$$

Solution. The given equations are written in the matrix form.

$$\begin{bmatrix} 1 & 2 & 3 & 4 \\ 6 & 7 & 8 & 9 \\ 11 & 12 & 13 & 14 \\ 16 & 17 & 18 & 19 \\ 21 & 22 & 23 & 24 \end{bmatrix} \begin{bmatrix} x_1 \\ x_2 \\ x_3 \\ x_4 \end{bmatrix} = \begin{bmatrix} 5 \\ 10 \\ 15 \\ 20 \\ 25 \end{bmatrix}$$

$AX = B$

$$C = [A : B] = \begin{bmatrix} 1 & 2 & 3 & 4 & : & 5 \\ 6 & 7 & 8 & 9 & : & 10 \\ 11 & 12 & 13 & 14 & : & 15 \\ 16 & 17 & 18 & 19 & : & 20 \\ 21 & 22 & 23 & 24 & : & 25 \end{bmatrix} = \begin{bmatrix} 1 & 2 & 3 & 4 & 5 \\ 0 & 5 & 10 & 15 & 20 \\ 0 & 10 & 20 & 30 & 40 \\ 0 & 15 & 30 & 45 & 60 \\ 0 & 20 & 40 & 60 & 80 \end{bmatrix} \begin{matrix} \\ R_2 \to R_2 - 6R_1 \\ R_3 \to R_3 - 11R_1 \\ R_4 \to R_4 - 16R_1 \\ R_5 \to R_5 - 21R_1 \end{matrix}$$

$$= \begin{bmatrix} 1 & 2 & 3 & 4 & 5 \\ 0 & 5 & 10 & 15 & 20 \\ 0 & 0 & 0 & 0 & 0 \\ 0 & 0 & 0 & 0 & 0 \\ 0 & 0 & 0 & 0 & 0 \end{bmatrix} \begin{matrix} \\ \\ R_3 \to R_3 - 2R_2 \\ R_4 \to R_4 - 3R_2 \\ R_5 \to R_5 - 4R_2 \end{matrix}$$

Number of non zero rows is only 2.
So Rank (A) = Rank (C) = 2
Since Rank (A) = Rank (C) < Number of unknows.
The given system of equations is consistent and has infinite number of solutions. **Ans.**

Example 12. *For what values of k, the equations $x + y + z = 1$, $2x + y + 4z = k$, $4x + y + 10z = k^2$ has a solution?*

Solution. Here, we have

$$x + y + z = 1$$
$$2x + y + 4z = k$$
$$4x + y + 10z = k^2$$

$$\Rightarrow \quad \begin{bmatrix} 1 & 1 & 1 \\ 2 & 1 & 4 \\ 4 & 1 & 10 \end{bmatrix} \begin{bmatrix} x \\ y \\ z \end{bmatrix} = \begin{bmatrix} 1 \\ k \\ k^2 \end{bmatrix}$$

$$AX = B$$

$$C = [A : B] = \begin{bmatrix} 1 & 1 & 1 & : & 1 \\ 2 & 1 & 4 & : & k \\ 4 & 1 & 10 & : & k^2 \end{bmatrix} \sim \begin{bmatrix} 1 & 1 & 1 & : & 1 \\ 0 & -1 & 2 & : & k-2 \\ 0 & -3 & 6 & : & k^2-4 \end{bmatrix} \begin{matrix} \\ R_2 \to R_2 - 2R_1 \\ R_3 \to R_3 - 4R_1 \end{matrix}$$

$$\sim \begin{bmatrix} 1 & 1 & 1 & : & 1 \\ 0 & -1 & 2 & : & k-2 \\ 0 & 0 & 0 & : & k^2-3k+2 \end{bmatrix} R_3 \to R_3 - 3R_2 \qquad \Rightarrow \begin{bmatrix} 1 & 1 & 1 \\ 0 & -1 & 2 \\ 0 & 0 & 0 \end{bmatrix} \begin{bmatrix} x \\ y \\ z \end{bmatrix} = \begin{bmatrix} 1 \\ k-2 \\ k^2-3k+2 \end{bmatrix}$$

If the given system has solutions, then $R(A) = R(C)$ But $R(A) = 2$

$R(C) = 2$ if $k^2 - 3k + 2 = 0 \Rightarrow (k-1)(k-2) = 0 \Rightarrow k = 1, k = 2$

Case I. When $k = 1$, we have

$$x + y + z = 1 \quad ...(1)$$
$$-y + 2z = 1 - 2 = -1 \quad ...(2)$$

Let $z = \lambda$

Putting the value of $z = \lambda$ in (2) , we have

$$-y + 2\lambda = -1 \Rightarrow y = 2\lambda + 1$$

Putting the values of y and z in (1), we have

$$x + (2\lambda + 1) + \lambda = 1 \Rightarrow x = -3\lambda$$

Hence solution is $x = -3\lambda$
$y = 2\lambda + 1$
$z = \lambda$ (λ is an arbitray constant)

Case II. When $k = 2$, we have

$$x + y + z = 1 \quad ...(3)$$
$$-y + 2z = 4 - 6 + 2 \Rightarrow -y + 2z = 0 \quad ...(4)$$

Let $z = c$

Putting the value of $z = c$ in (4), we have

$$-y + 2c = 0 \Rightarrow y = 2c$$

Putting the values of y and z in (1), we have

$$x + 2c + c = 1 \Rightarrow x = -3c + 1$$

Hence the solution is

$x = 1 - 3c, y = 2c, z = c,$ where c is an arbitrary constant. **Ans.**

Example 13. *Investigate the values of* λ *and* μ *so that the equations:*

$$2x + 3y + 5z = 9$$
$$7x + 3y - 2z = 8$$
$$2x + 3y + \lambda z = \mu$$

have (i) no solution (ii) a unique solution
(iii) an infinite number of solutions. *(R.G.P.V. Bhopal, I Semester, June 2007)*

Solution. Here, we have,

$$2x + 3y + 5z = 9$$
$$7x + 3y - 2z = 8$$
$$2x + 3y + \lambda z = \mu$$

The above equations are written in the matrix form

$$\begin{bmatrix} 2 & 3 & 5 \\ 7 & 3 & -2 \\ 2 & 3 & \lambda \end{bmatrix} \begin{bmatrix} x \\ y \\ z \end{bmatrix} = \begin{bmatrix} 9 \\ 8 \\ \mu \end{bmatrix}$$
$$AX = B$$

$$C = [A : B] = \begin{bmatrix} 2 & 3 & 5 & : & 9 \\ 7 & 3 & -2 & : & 8 \\ 2 & 3 & \lambda & : & \mu \end{bmatrix} = \begin{bmatrix} 2 & 3 & 5 & : & 9 \\ 0 & -\frac{15}{2} & -\frac{39}{2} & : & -\frac{47}{2} \\ 0 & 0 & \lambda - 5 & : & \mu - 9 \end{bmatrix} \begin{matrix} \\ R_2 \to R_2 - \frac{7}{2} R_1 \\ R_3 \to R_3 - R_1 \end{matrix}$$

(i) **No solution.** Rank $(A) \neq$ Rank (C)

$\lambda - 5 = 0$ or $\lambda = 5$ and $\mu - 9 \neq 0 \quad \Rightarrow \quad \mu \neq 9$

(ii) **A unique solution.** Rank $(A) = R(C)$ = Number of unknowns

$\lambda - 5 \neq 0 \quad \Rightarrow \quad \lambda \neq 5$ and $\mu \neq 9$

(iii) **An infinite number of solutions.** Rank (A) = Rank $(C) = 2$

$\lambda - 5 = 0$ and $\mu - 9 = 0$

$\lambda = 5$ and $\mu = 9$ **Ans.**

Example 14. *Determine for what values of λ and μ the following equations have (i) no solution; (ii) a unique solution; (iii) infinite number of solutions.*

$$x + y + z = 6, \quad x + 2y + 3z = 10, \quad x + 2y + \lambda z = \mu$$

Solution. $\begin{bmatrix} 1 & 1 & 1 \\ 1 & 2 & 3 \\ 1 & 2 & \lambda \end{bmatrix} \begin{bmatrix} x \\ y \\ z \end{bmatrix} = \begin{bmatrix} 6 \\ 10 \\ \mu \end{bmatrix}$

$$AX = B$$

$$C = (A, B) = \begin{bmatrix} 1 & 1 & 1 & \cdot & 6 \\ 1 & 2 & 3 & \cdot & 10 \\ 1 & 2 & \lambda & \cdot & \mu \end{bmatrix} \sim \begin{bmatrix} 1 & 1 & 1 & \cdot & 6 \\ 0 & 1 & 2 & : & 4 \\ 0 & 1 & \lambda - 1 & \cdot & \mu - 6 \end{bmatrix} \begin{matrix} \\ R_2 \to R_2 - R_1 \\ R_3 \to R_3 - R_1 \end{matrix}$$

$$\sim \begin{bmatrix} 1 & 1 & 1 & \cdot & 6 \\ 0 & 1 & 2 & \cdot & 4 \\ 0 & 0 & \lambda - 3 & \cdot & \mu - 10 \end{bmatrix} R_3 \to R_3 - R_2$$

(*i*) There is no solution if $R(A) \neq R(C)$

i.e. $\lambda - 3 = 0$ or $\lambda = 3$ and $\mu - 10 \neq 0$ or $\mu \neq 10$

(*ii*) There is a unique solution if $R(A) = R(C) = 3$

i.e. $\lambda - 3 \neq 0$ or $\lambda \neq 3$, μ may have any value.

(*iii*) There are infinite solutions if $R(A) = R(C) = 2$

$\lambda - 3 = 0$ or $\lambda = 3$ and $\mu - 10 = 0$ or $\mu = 10$ **Ans.**

Example 15. *Find for what values of λ and μ the system of linear equations:*

$$x + y + z = 6$$
$$x + 2y + 5z = 10$$
$$2x + 3y + \lambda z = \mu$$

has (i) a unique solution (ii) no solution (iii) infinite solutions. Also find the solution for $\lambda = 2$ and $\mu = 8$.

(Uttarakhand, 1st semester, Dec. 2006)

Solution. $\begin{bmatrix} 1 & 1 & 1 \\ 1 & 2 & 5 \\ 2 & 3 & \lambda \end{bmatrix} \begin{bmatrix} x \\ y \\ z \end{bmatrix} = \begin{bmatrix} 6 \\ 10 \\ \mu \end{bmatrix}$

$$AX = B$$

$$C = (A, B) = \begin{bmatrix} 1 & 1 & 1 & : & 6 \\ 1 & 2 & 5 & : & 10 \\ 2 & 3 & \lambda & : & \mu \end{bmatrix} \sim \begin{bmatrix} 1 & 1 & 1 & : & 6 \\ 0 & 1 & 4 & : & 4 \\ 0 & 1 & \lambda - 2 & : & \mu - 12 \end{bmatrix} \begin{matrix} \\ R_2 \to R_2 - R_1 \\ R_3 \to R_3 - 2R_1 \end{matrix}$$

$$\sim \begin{bmatrix} 1 & 1 & 1 & : & 6 \\ 0 & 1 & 4 & : & 4 \\ 0 & 0 & \lambda - 6 & : & \mu - 16 \end{bmatrix} R_3 \to R_3 - R_2 \qquad \text{...(1)}$$

(*i*) **A unique solution**

If $R(A) = R(C) = 3$

then $\lambda - 6 \neq 0 \Rightarrow \lambda \neq 6$

(*ii*) **No solutions**

If $R(A) \neq R(C)$, then $R(A) = 2$ and $R(C) = 3$

$\lambda - 6 = 0 \Rightarrow \lambda = 6$ and $\mu - 16 \neq 0 \Rightarrow \mu \neq 16$

(*iii*) **Infinite solutions**

If $R(A) = R(C) = 2$

then $\lambda - 6 = 0$ and $\mu - 16 = 0$

$\Rightarrow \quad \lambda = 6$ and $\mu = 16$

(*iv*) Putting $\lambda = 2$ and $\mu = 8$ in (1), we get

$$\begin{bmatrix} 1 & 1 & 1 & : & 6 \\ 0 & 1 & 4 & : & 4 \\ 0 & 0 & -4 & : & -8 \end{bmatrix} \Rightarrow \begin{bmatrix} 1 & 1 & 1 \\ 0 & 1 & 4 \\ 0 & 0 & -4 \end{bmatrix} \begin{bmatrix} x \\ y \\ z \end{bmatrix} = \begin{bmatrix} 6 \\ 4 \\ -8 \end{bmatrix}$$

$$x + y + z = 6$$
$$y + 4z = 4$$
$$-4z = -8 \quad \Rightarrow \quad z = 2$$

Putting $z = 2$ in (3), we get

$$y + 8 = 4 \quad \Rightarrow \quad y = -4$$

Putting $y = -4$, $z = 2$ in (2), we get

$$x - 4 + 2 = 6 \quad \Rightarrow \quad x = 8$$

Hence, $x = 8,\ y = -4,\ z = 2$ **Ans.**

Example 16. *Show that the equations*

$$-2x + y + z = a$$
$$x - 2y + z = b$$
$$x + y - 2z = c$$

have no solution unless $a + b + c = 0$. In which case they have infinitely many solutions? Find these solutions when $a = 1$, $b = 1$ and $c = -2$.

Solution. Augmented matrix,

$$C = [A : B] = \begin{bmatrix} -2 & 1 & 1 & : & a \\ 1 & -2 & 1 & : & b \\ 1 & 1 & -2 & : & c \end{bmatrix} \qquad [\text{Rank } (A) = 2]$$

$R_1 \leftrightarrow R_3$

$$\sim \begin{bmatrix} \textcircled{1} & 1 & -2 & : & c \\ 1 & -2 & 1 & : & b \\ -2 & 1 & 1 & : & a \end{bmatrix} \sim \begin{bmatrix} 1 & \textcircled{1} & -2 & : & c \\ 0 & -3 & 3 & : & b-c \\ 0 & 3 & -3 & : & a+2c \end{bmatrix} \begin{matrix} \\ R_2 \to R_2 - R_1 \\ R_3 \to R_3 + 2R_1 \end{matrix}$$

$$\sim \begin{bmatrix} 1 & 1 & -2 & : & c \\ 0 & -3 & 3 & : & b-c \\ 0 & 0 & 0 & : & a+b+c \end{bmatrix} R_3 \to R_3 + R_2 \qquad \text{... (1)}$$

Case I. If $a + b + c \neq 0$

Rank of $C = 3$. But Rank of $A = 2$

$\Rightarrow \quad R(C) \neq R(A)$ where A is the coefficient matrix.

Hence, the system being inconsistent, have no solution.

Case II. If $a + b + c = 0$

Rank of $C = 2$ and $R(A) = 2$

$\Rightarrow \quad R(C) = R(A)$

Hence, the system has infinite number of solutions.

Case III. On putting $a = 1$, $b = 1$ and $c = -2$ in (1), we get

$$\begin{bmatrix} 1 & 1 & -2 & : & -2 \\ 0 & -3 & 3 & : & 3 \\ 0 & 0 & 0 & : & 0 \end{bmatrix}$$

$$x + y - 2z = -2 \qquad ...(2)$$
$$-3y + 3z = 3 \qquad ...(3)$$

Let $z = k$, k being an arbitrary constant.

From (3) $-3y + 3k = 3 \quad \Rightarrow \quad y = k - 1$

Putting $y = k - 1$ and $\quad z = k$ in (2), we get

$$x + (k-1) - 2k = -2 \quad \Rightarrow \quad x = k - 1$$

Hence, the solutions are $x = k - 1,\ y = k - 1,\quad z = k$ **Ans.**

Example 17. *Find for what values of k the set of equations*

$$2x - 3y + 6z - 5t = 3,\ y - 4z + t = 1,\ 4x - 5y + 8z - 9t = k$$

has (i) no solution (ii) infinite number of solutions.

Solution. The augmented matrix $C = [A, B]$

$$\begin{bmatrix} 2 & -3 & 6 & -5 & \cdot & 3 \\ 0 & 1 & -4 & 1 & \cdot & 1 \\ 4 & -5 & 8 & -9 & \cdot & k \end{bmatrix} \sim \begin{bmatrix} 2 & -3 & 6 & -5 & \cdot & 3 \\ 0 & 1 & -4 & 1 & \cdot & 1 \\ 0 & 1 & -4 & 1 & \cdot & k-6 \end{bmatrix} \quad R_3 \to R_3 - 2R_1$$

$$\sim \begin{bmatrix} 2 & -3 & 6 & -5 & \cdot & 3 \\ 0 & 1 & -4 & 1 & \cdot & 1 \\ 0 & 0 & 0 & 0 & \cdot & k-7 \end{bmatrix} \quad R_3 \to R_3 - R_2$$

(*i*) There is no solution if $R(A) \neq R(C)$

$$k - 7 \neq 0 \text{ or } k \neq 7,\ R(A) = 2 \text{ and } R(C) = 3.$$

(*ii*) There are infinite solutions if $R(A) = R(C) = 2$

$$k - 7 = 0 \Rightarrow k = 7 \qquad \textbf{Ans.}$$

$$\begin{bmatrix} 2 & -3 & 6 & -5 \\ 0 & 1 & -4 & 1 \end{bmatrix} \begin{bmatrix} x \\ y \\ z \\ t \end{bmatrix} = \begin{bmatrix} 3 \\ 1 \end{bmatrix}$$

$$2x - 3y + 6z - 5t = 3 \qquad ...(1)$$
$$y - 4z + t = 1 \qquad ...(2)$$

Let $\quad t = k_1$ and $z = k_2$.

From (2), $y - 4k_2 + k_1 = 1$ or $y = 1 + 4k_2 - k_1$

From (l), $2x - 3 - 12k_2 + 3k_1 + 6k_2 - 5k_1 = 3$

$\Rightarrow \quad 2x = 6 + 6k_2 + 2k_1 \quad \Rightarrow \quad x = 3 + 3k_2 + k_1$

$\quad y = 1 + 4k_2 - k_1 \quad \Rightarrow \quad z = k_2,\ t = k_1$ **Ans.**

40.4. HOMOGENEOUS EQUATIONS

For a system of homogeneous linear equations $AX = O$

(*i*) $X = O$ is always a solution. This solution in which each unknown has the value zero is called the **Null Solution** or the **Trivial solution.** Thus a homogeneous system is always consistent.

A system of homogeneous linear equations has either the trivial solution or an infinite number of solutions.

(*ii*) If $R(A)$ = number of unknowns, the system has only the trivial solution.
(*iii*) If $R(A)$ < number of unknowns, the system has an infinite number of non-trivial solutions.

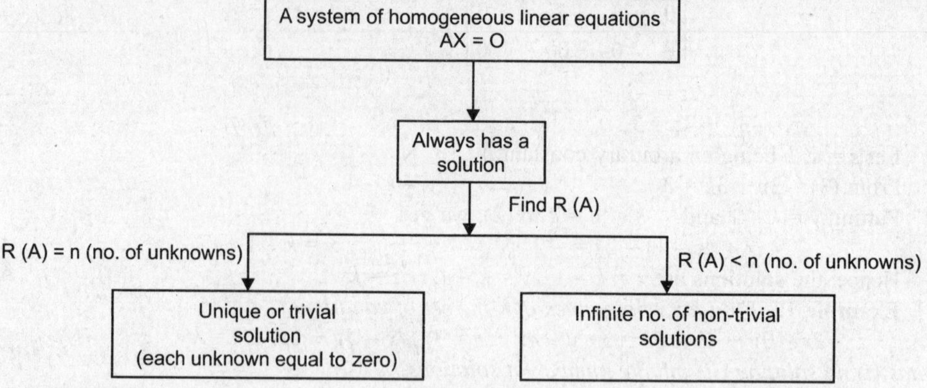

Example 18. *Determine 'b' such that the system of homogeneous equations*

$$2x + y + 2z = 0$$
$$x + y + 3z = 0$$
$$4x + 3y + bz = 0$$

has (i) Trivial solution
(ii) Non-Trivial solution. Find the Non-Trivial solution using matrix method.

(U.P., I Sem Dec 2008)

Solution. Here, we have

$$2x + y + 2z = 0$$
$$x + y + 3z = 0$$
$$4x + 3y + bz = 0$$

(*i*) **For trivial solution:** We know that $x = 0$, $y = 0$ and $z = 0$. So, b can have any value.
(*ii*) **For non-trivial solution:** The given equations are written in the matrix form as :

$$\begin{bmatrix} 2 & 1 & 2 \\ 1 & 1 & 3 \\ 4 & 3 & b \end{bmatrix}\begin{bmatrix} x \\ y \\ z \end{bmatrix} = \begin{bmatrix} 0 \\ 0 \\ 0 \end{bmatrix}$$

$$AX = B$$

$$R_1 \leftrightarrow R_2, \quad R_2 \to R_2 - 2R_1, R_3 \to R_3 - 4R_1, \quad R_3 \to R_3 - R_2$$

$$C = \begin{bmatrix} 2 & 1 & 2 & : 0 \\ 1 & 1 & 3 & : 0 \\ 4 & 3 & b & : 0 \end{bmatrix} \sim \begin{bmatrix} 1 & 1 & 3 & : 0 \\ 2 & 1 & 2 & : 0 \\ 4 & 3 & b & : 0 \end{bmatrix} \sim \begin{bmatrix} 1 & 1 & 3 & : 0 \\ 0 & -1 & -4 & : 0 \\ 0 & -1 & b-12 & : 0 \end{bmatrix} \sim \begin{bmatrix} 1 & 1 & 3 & : 0 \\ 0 & -1 & -4 & : 0 \\ 0 & 0 & b-8 & : 0 \end{bmatrix}$$

For non trivial solution or infinite solutions $R(C) = R(A) = 2 <$ Number of unknowns

$$b - 8 = 0, \quad b = 8$$

Ans.

Example 19. *Find the values of k such that the system of equations*

$$x + ky + 3z = 0, \quad 4x + 3y + kz = 0, \quad 2x + y + 2z = 0$$

has non-trivial solution.

Solution. The set of equations is written in the form of matrices

$$\begin{bmatrix} 1 & k & 3 \\ 4 & 3 & k \\ 2 & 1 & 2 \end{bmatrix}\begin{bmatrix} x \\ y \\ z \end{bmatrix} = \begin{bmatrix} 0 \\ 0 \\ 0 \end{bmatrix}, \qquad AX = B, \; C = [A : B] = \begin{bmatrix} 1 & k & 3 & : & 0 \\ 4 & 3 & k & : & 0 \\ 2 & 1 & 2 & : & 0 \end{bmatrix}$$

On interchanging first and third rows, we have

$$\begin{bmatrix} 2 & 1 & 2 & : & 0 \\ 4 & 3 & k & : & 0 \\ 1 & k & 3 & : & 0 \end{bmatrix}$$

$$R_2 \to R_2 - 2R_1,\quad R_3 \to R_3 - \frac{1}{2}R_1 \quad R_3 \to R_3 - \left(k - \frac{1}{2}\right)R_2$$

$$\sim \begin{bmatrix} 2 & 1 & 2 & : & 0 \\ 0 & 1 & k-4 & : & 0 \\ 0 & k - \frac{1}{2} & 2 & : & 0 \end{bmatrix} \sim \begin{bmatrix} 2 & 1 & 2 & : & 0 \\ 0 & 1 & k-4 & : & 0 \\ 0 & 0 & 2 - \left(k - \frac{1}{2}\right)(k-4) & : & 0 \end{bmatrix}$$

For a non-trivial solution or for infinite solution, $R(A) = R(C) = 2$

so $\quad 2 - \left(k - \frac{1}{2}\right)(k-4) = 0 \;\Rightarrow\; 2 - k^2 + 4k + \frac{k}{2} - 2 = 0$

$\Rightarrow\quad -k^2 + \frac{9}{2}k = 0 \;\Rightarrow\; k\left(-k + \frac{9}{2}\right) = 0 \;\Rightarrow\; k = \frac{9}{2}, k = 0$ **Ans.**

40.5 CRAMER'S RULE

$$a_1x + b_1y + c_1z = d_1$$
$$a_2x + b_2y + c_2z = d_2$$
$$a_3x + b_3y + c_3z = d_3$$

then $\quad D = \begin{vmatrix} a_1 & b_1 & c_1 \\ a_2 & b_2 & c_2 \\ a_3 & b_3 & c_3 \end{vmatrix}, \qquad D_1 = \begin{vmatrix} d_1 & b_1 & c_1 \\ d_2 & b_2 & c_2 \\ d_3 & b_3 & c_3 \end{vmatrix}$

$$D_2 = \begin{vmatrix} a_1 & d_1 & c_1 \\ a_2 & d_2 & c_2 \\ a_3 & d_3 & c_3 \end{vmatrix}, \qquad D_3 = \begin{vmatrix} a_1 & b_1 & d_1 \\ a_2 & b_2 & d_2 \\ a_3 & b_3 & d_3 \end{vmatrix}$$

$$x = \frac{D_1}{D},\quad y = \frac{D_2}{D},\quad z = \frac{D_3}{D}$$

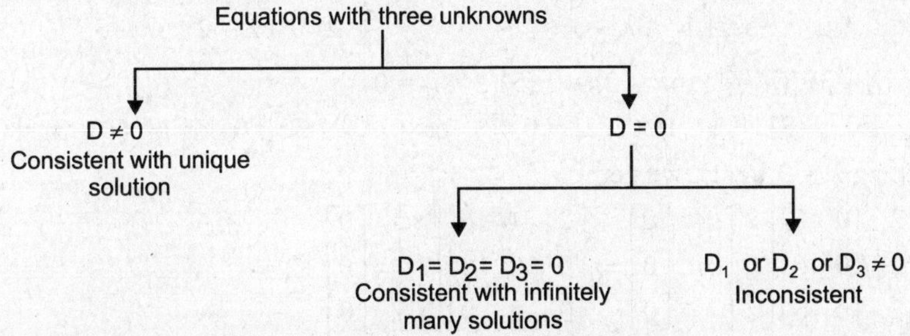

Example 20. *Show that the homogeneous system of equations*

$x + y\cos\gamma + z\cos\beta = 0;\quad x\cos\gamma + y + z\cos\alpha = 0;\quad x\cos\beta + y\cos\alpha + z = 0$

has non-trivial solution if $\alpha + \beta + \gamma = 0$.

Solution. If the system has only non-trivial solutions , then

$$\begin{vmatrix} 1 & \cos\gamma & \cos\beta \\ \cos\gamma & 1 & \cos\alpha \\ \cos\beta & \cos\alpha & 1 \end{vmatrix} = 0$$

$$\Rightarrow \quad 1 - \cos^2\alpha + \cos\gamma\,(\cos\alpha\cos\beta - \cos\gamma) + \cos\beta\,(\cos\gamma\cos\alpha - \cos\beta) = 0$$

$$\Rightarrow \quad \sin^2\alpha - \cos^2\beta - \cos^2\gamma + 2\cos\alpha\cos\beta\cos\gamma = 0$$

$$\Rightarrow \quad -(\cos^2\beta - \sin^2\alpha) - \cos^2\gamma + 2\cos\alpha\cos\beta\cos\gamma = 0$$

$$\Rightarrow \quad -\cos(\alpha+\beta)\cos(\beta-\alpha) - \cos^2\gamma + 2\cos\alpha\cos\beta\cos\gamma = 0$$

[if $\alpha + \beta + \gamma = 0$]

$$\Rightarrow \quad -\cos(-\gamma)\cos(\beta-\alpha) - \cos^2\gamma + 2\cos\alpha\cos\beta\cos\gamma = 0$$

$$\Rightarrow \quad -\cos\gamma\,[\cos(\beta-\alpha) + \cos(\beta+\alpha)] + 2\cos\alpha\cos\beta\cos\gamma = 0$$

$$\Rightarrow \quad -2\cos\beta\cos\alpha\cos\gamma + 2\cos\alpha\cos\beta\cos\gamma = 0$$

which is true.

Hence, the given homogeneous system of equations has non-trivial solution if $\alpha + \beta + \gamma = 0$.

Proved.

Example 21. *Find values of λ for which the following system of equations is consistent and has non-trivial solutions. Solve equations for all such values of λ.*

$$(\lambda - 1)x + (3\lambda + 1)y + 2\lambda z = 0$$

$$(\lambda - 1)x + (4\lambda - 2)y + (\lambda + 3)z = 0$$

$$2x + (3\lambda + 1)y + 3(\lambda - 1)z = 0$$

(*A.M.I.E.T.E., Summer 2010*)

Solution.

$$\begin{bmatrix} (\lambda-1) & (3\lambda+1) & 2\lambda \\ (\lambda-1) & (4\lambda-2) & (\lambda+3) \\ 2 & (3\lambda+1) & (3\lambda-3) \end{bmatrix} \begin{bmatrix} x \\ y \\ z \end{bmatrix} = \begin{bmatrix} 0 \\ 0 \\ 0 \end{bmatrix} \quad \ldots(1)$$

$$AX = 0$$

For infinite solutions, $|A| = 0$

$$\begin{vmatrix} \lambda-1 & 3\lambda+1 & 2\lambda \\ \lambda-1 & 4\lambda-2 & \lambda+3 \\ 2 & 3\lambda+1 & 3\lambda-3 \end{vmatrix} = 0,$$

$$\begin{vmatrix} 0 & -\lambda+3 & \lambda-3 \\ \lambda-1 & 4\lambda-2 & \lambda+3 \\ 2 & 3\lambda+1 & 3\lambda-3 \end{vmatrix} = 0, \; R_1 \to R_1 - R_2 \qquad \begin{vmatrix} 0 & 0 & \lambda-3 \\ \lambda-1 & 5\lambda+1 & \lambda+3 \\ 2 & 6\lambda-2 & 3\lambda-3 \end{vmatrix} = 0 \; C_2 \to C_2 + C_3$$

$$(\lambda - 3)\,[(\lambda - 1)(6\lambda - 2) - 2(5\lambda + 1)] = 0$$

$$[6\lambda^2 - 8\lambda + 2 - 10\lambda - 2] = 0 \quad \text{or} \quad 6\lambda^2 - 18\lambda = 0 \quad \text{or} \quad 6\lambda(\lambda - 3) = 0,\ \lambda = 3, 0$$

On putting $\lambda = 3$ in (1), we get

$$\begin{bmatrix} 2 & 10 & 6 \\ 2 & 10 & 6 \\ 2 & 10 & 6 \end{bmatrix} \begin{bmatrix} x \\ y \\ z \end{bmatrix} = \begin{bmatrix} 0 \\ 0 \\ 0 \end{bmatrix} \Rightarrow \begin{bmatrix} 2 & 10 & 6 \\ 0 & 0 & 0 \\ 0 & 0 & 0 \end{bmatrix} \begin{bmatrix} x \\ y \\ z \end{bmatrix} = \begin{bmatrix} 0 \\ 0 \\ 0 \end{bmatrix}$$

$$2x + 10y + 6z = 0 \Rightarrow x + 5y + 3z = 0$$

Let $\quad x = k_1,\ y = k_2,\ 3z = -k_1 - 5k_2 \Rightarrow z = \dfrac{-k_1}{3} - \dfrac{5k_2}{3}$ **Ans.**

EXERCISE 40.1

Test the consistency of the following equations and solve them if possible.

1. $3x + 3y + 2z = 1, \quad x + 2y = 4, \quad 10y + 3z = -2, \quad 2x - 3y - z = 5$

Ans. Consistent, $x = 2, \ y = 1, \ z = -4$

2. $x_1 - x_2 + x_3 - x_4 + x_5 = 1, \qquad 2x_1 - x_2 + 3x_3 + 4x_5 = 2,$
$3x_1 - 2x_2 + 2x_3 + x_4 + x_5 = 1, \qquad x_1 + x_3 + 2x_4 + x_5 = 0$

Ans. $x_1 = -3k_1 + k_2 - 1, x_2 = -3k_1 - 1, x_3 = k_1 - 2k_2 + 1, x_4 = k_1, x_4 = k_1, x_5 = k_2$

3. Find the value of k for which the following system of equations is consistent.

$3x_1 - 2x_2 + 2x_3 = 3, \ x_1 + kx_2 - 3x_3 = 0, \ 4x_1 + x_2 + 2x_3 = 7$ **Ans.** $k = \frac{1}{4}$

4. Find the value of λ for which the system of equations
$x + y + 4z = 1, x + 2y - 2z = 1, \ \lambda x + y + z = 1$
will have a unique solution. **Ans.** $\lambda \neq \frac{7}{10}$

5. Determine the values of a and b for which the system $\begin{bmatrix} 3 & -2 & 1 \\ 5 & -8 & 9 \\ 2 & 1 & a \end{bmatrix}\begin{bmatrix} x \\ y \\ z \end{bmatrix} = \begin{bmatrix} b \\ 3 \\ -1 \end{bmatrix}$

(*i*) has a unique solution, (*ii*) has no solution and, (*iii*) has infinitely many solutions.

Ans. (*i*) $a \neq -3$, (*ii*) $a = -3, \ b \neq \frac{1}{3}$, (*iii*) $a = -3, \ b = \frac{1}{3}$

6. Choose λ that makes the following system of linear equations consistent and find the general solution of the system for that λ.
$x + y - z + t = 2, \ 2y + 4z + 2t = 3, \ x + 2y + z + 2t = \lambda$

Ans. $\lambda = \frac{7}{2}, \ x = \frac{1}{2} + 3k_2, \ y = \frac{3}{2} - 2k_2 - k_1, \ z = k_2, \ t = k_1$

7. Show that the equations
$3x + 4y + 5z = a, \ 4x + 5y + 6z = b, \ 5x + 6y + 7z = c$
don't have a solution unless $a + c = 2b$.
Solve the equations when $a = b = c = -1$ **Ans.** $x = k + 1, \ y = -2k - 1, z = k$

8. Find the values of k, such that the system of equations
$4x_1 + 9x_2 + x_3 = 0, \ kx_1 + 3x_2 + kx_3 = 0, \ x_1 + 4x_2 + 2x_3 = 0$
has non-trivial solution. Hence, find the solution of the system.

Ans. $k = 1, \ x_1 = 2\lambda, \ x_2 = -\lambda, \ x_3 = \lambda$

9. Find values of λ for which the following system of equations has a non-trivial solution.
$3x_1 + x_2 - \lambda x_3 = 0, \ 2x_1 + 4x_2 + \lambda x_3 = 0, \ 8x_1 - 4x_2 - 6x_3 = 0$ **Ans.** $\lambda = 1$

10. Find value of λ so that the following system of homogeneous equations have exactly two linearly independent solutions
$\lambda x_1 - x_2 - x_3 = 0, \quad -x_1 + \lambda x_2 - x_3 = 0, \ -x_1 - x_2 + \lambda x_3 = 0,$ **Ans.** $\lambda = -1$

11. Find the values of k for which the following system of equations has a non-trivial solution.
$(3k - 8)x + 3y + 3z = 0, \ 3x + (3k - 8)y + 3z = 0, \ 3x + 3y + (3k - 8)z = 0$ (*AMIETE, June 2010*)

Ans. $k = \frac{2}{3}, \frac{11}{3}$

12. Solve the homogeneous system of equations :
$4x + 3y - z = 0, \ 3x + 4y + z = 0, \ x - y - 2z = 0, \ 5x + y - 4z = 0$ **Ans.** $x = k, \ y = -k, \ z = k$

13. If $A = \begin{bmatrix} -1 & 2 & 1 \\ 3 & -1 & 2 \\ 0 & 1 & \lambda \end{bmatrix}$ **Ans.** (*i*) $\lambda \neq 1$, (*ii*) $\lambda - 1$

find the values of λ for which equation $AX = 0$ has (*i*) a unique solution, (*ii*) more than one solution.

14. Show that the following system of equations:

$x + 2y - 2u = 0, \quad 2x - y - u = 0, \quad x + 2z - u = 0, \quad 4x - y + 3z - u = 0$

do not have a non-trivial solution.

15. Determine the values of λ and μ such that the following system has (*i*) no solution (*ii*) a unique solution (*iii*) infinite number of solutions:

$2x - 5y + 2z = 8, \quad 2x + 4y + 6z = 5, \quad x + 2y + \lambda z = \mu$

Ans. (*i*) $\lambda = 3,\ \mu \neq \frac{5}{2}$ (*ii*) $\lambda \neq 3$, (*iii*) $\lambda = 3,\ \mu = \frac{5}{2}$

16. Test the following system of equations for consistency. If possible, solve for non-trivial solutions.

$3x + 4y - z - 6t = 0,\ 2x + 3y + 2z - 3t = 0, \quad 2x + y - 14z - 9t = 0,\ x + 3y + 13z + 3t = 0$

Ans. $x = 11k_1 + 6k_2,\ y = -8k_1 - 3k_2,\ z = k_1\ \ t = k_2$

17. Given the following system of equations

$2x - 2y + 5z + 3z = 0,\ 4x - y + z + w = 0, \quad 3x - 2y + 3z + 4w = 0,\ x - 3y + 7z + 6w = 0$

Reduce the coefficient matrix A into Echelon form and find the rank utilising the property of rank, test the given system of equation for consistency and if possible find the solution of the given system.

Ans. $x = 5k,\ y = 36k,\ z = 7k,\ w = 9k$

18. Find the values of λ for which the equations

$(2 - \lambda)\, x + 2y + 3 = 0, \qquad 2x + (4 - \lambda)\, y + 7 = 0, \qquad 2x + 5y + (6 - \lambda) = 0$

are consistent and find the values of x and y corresponding to each of these values of λ.

Ans. $\lambda = 1, -1, 12.$

40.6 VECTORS

A n-tuple is a set of n similar things. If the place of every members of a set is fixed then it is called an *ordered* set. Any ordered n-tuple of numbers is called a *n-vector*. Thus the coordinates of a point in space is called 3-vector (x, y, z). The members of a set are called the components of a vector so x, y, z in a 3-vector are called components.

$x_1, x_2, x_3 .... x_n$ are the components of a n-vector $X = (x_1, x_2, x_3 .... x_n)$.

Each row of a matrix is a vector and each column of the matrix is also a vector.

40.7 LINEAR DEPENDENCE AND INDEPENDENCE OF VECTORS

Vectors (matrices) $X_1, X_2, ... X_n$ are said to be dependent if

(1) all the vectors (row or column matrices) are of the same order.

(2) n scalars $\lambda_1, \lambda_2, ... \lambda_n$ (not all zero) exist such that

$$\lambda_1 X_1 + \lambda_2 X_2 + \lambda_3 X_3 + ... + \lambda_n X_n = 0$$

Otherwise they are linearly independent.

Remember: If in a set of vectors, any vector of the set is the combination of the remaining vectors, then the vectors are called dependent vectors.

Example 22. *Show that the vector (1, 0, 0) (1, 1, 0) and (1, 1, 1) are linearly independent.*

(Delhi University, April 2010)

Solution. Let $X_1 = (1, 0, 0),\ X_2 = (1, 1, 0)$ and $X_3 = (1, 1, 1)$

consider the matrix equation

$$\lambda_1 X_1 + \lambda_2 X_2 + \lambda_3 X_3 = 0$$

$$\Rightarrow \lambda_1 (1, 0, 0) + \lambda_2 (1, 1, 0) + \lambda_3 (1, 1, 1) = 0$$

$$\Rightarrow \quad \lambda_1 + \lambda_2 + \lambda_3 = 0$$

$$\lambda_2 + \lambda_3 = 0$$

$$\lambda_3 = 0$$

This is the homogeneous system

$$\lambda_1 = 0,\ \lambda_2 = 0 \text{ and } \lambda_3 = 0$$

Thus non-zero values of $\lambda_1, \lambda_2, \lambda_3$ do not exist which can satisfy.

Hence by definition the given system of vectors is linearly independent. **Proved.**

Example 23. *Examine the following vectors for linear dependence and find the relation if it exists.*

$X_1 = (1, 2, 4), X_2 = (2, -1, 3), X_3 = (0, 1, 2), X_4 = (-3, 7, 2)$

Solution. Consider the matrix equation

$\lambda_1 X_1 + \lambda_2 X_2 + \lambda_3 X_3 + \lambda_4 X_4 = 0$

$\Rightarrow \lambda_1 (1, 2, 4) + \lambda_2 (2, -1, 3) + \lambda_3 (0, 1, 2) + \lambda_4 (-3, 7, 2) = 0$

$$\begin{aligned} \lambda_1 + 2\lambda_2 + 0\lambda_3 - 3\lambda_4 &= 0 \\ 2\lambda_1 - \lambda_2 + \lambda_3 + 7\lambda_4 &= 0 \\ 4\lambda_1 + 3\lambda_2 + 2\lambda_3 + 2\lambda_4 &= 0 \end{aligned}$$

This is the homogeneous system

$$\begin{bmatrix} 1 & 2 & 0 & -3 \\ 2 & -1 & 1 & 7 \\ 4 & 3 & 2 & 2 \end{bmatrix} \begin{bmatrix} \lambda_1 \\ \lambda_2 \\ \lambda_3 \\ \lambda_4 \end{bmatrix} = \begin{bmatrix} 0 \\ 0 \\ 0 \end{bmatrix} \quad \text{or } A\,\lambda = 0$$

$$\begin{bmatrix} 1 & 2 & 0 & -3 \\ 0 & -5 & 1 & 13 \\ 0 & -5 & 2 & 14 \end{bmatrix} \begin{bmatrix} \lambda_1 \\ \lambda_2 \\ \lambda_3 \\ \lambda_4 \end{bmatrix} = \begin{bmatrix} 0 \\ 0 \\ 0 \end{bmatrix} \qquad \begin{matrix} R_2 \to R_2 - 2R_1 \\ R_3 \to R_3 - 4R_1 \end{matrix}$$

$$\begin{bmatrix} 1 & 2 & 0 & -3 \\ 0 & -5 & 1 & 13 \\ 0 & 0 & 1 & 1 \end{bmatrix} \begin{bmatrix} \lambda_1 \\ \lambda_2 \\ \lambda_3 \\ \lambda_4 \end{bmatrix} = \begin{bmatrix} 0 \\ 0 \\ 0 \end{bmatrix} \qquad R_3 \to R_3 - R_2$$

$$\begin{aligned} \lambda_1 + 2\lambda_2 - 3\lambda_4 &= 0 \\ -5\lambda_2 + \lambda_3 + 13\lambda_4 &= 0 \\ \lambda_3 + \lambda_4 &= 0 \end{aligned}$$

Let $\quad \lambda_4 = t, \lambda_3 + t = 0, \lambda_3 = -t$

$$-5\lambda_2 - t + 13t = 0, \lambda_2 = \frac{12t}{5}$$

$$\lambda_1 + \frac{24t}{5} - 3t = 0 \text{ or } \lambda_1 = \frac{-9t}{5}$$

Hence, the given vectors are linearly dependent.

Substituting the values of λ in (1), we get

$$-\frac{9tX_1}{5} + \frac{12t}{5}X_2 - tX_3 + tX_4 = 0 \Rightarrow -\frac{9X_1}{5} + \frac{12X_2}{5} - X_3 + X_4 = 0$$

$\Rightarrow \quad 9X_1 - 12X_2 + 5X_3 - 5X_4 = 0$ **Ans.**

Example 24. *Define linear dependence and independence of vectors.*

Examine for linear dependence [1, 0, 2, 1], [3, 1, 2, 1], [4, 6, 2, –4], [–6, 0, –3, –4] and find the relation between them, if possible.

Solution. Consider the matrix equation

$\lambda_1 X_1 + \lambda_2 X_2 + \lambda_3 X_3 + \lambda_4 X_4 = 0$ ...(1)

$\lambda_1 (1, 0, 2, 1) + \lambda_2 (3, 1, 2, 1) + \lambda_3 (4, 6, 2, -4) + \lambda_4 (-6, 0, -3, -4) = 0$

$$\lambda_1 + 3\lambda_2 + 4\lambda_3 - 6\lambda_4 = 0$$
$$0\lambda_1 + \lambda_2 + 6\lambda_3 + 0\lambda_4 = 0$$
$$2\lambda_1 + 2\lambda_2 + 2\lambda_3 - 3\lambda_4 = 0$$
$$\lambda_1 + \lambda_2 - 4\lambda_3 - 4\lambda_4 = 0$$

$$\begin{bmatrix} 1 & 3 & 4 & -6 \\ 0 & 1 & 6 & 0 \\ 2 & 2 & 2 & -3 \\ 1 & 1 & -4 & -4 \end{bmatrix}\begin{bmatrix} \lambda_1 \\ \lambda_2 \\ \lambda_3 \\ \lambda_4 \end{bmatrix} = \begin{bmatrix} 0 \\ 0 \\ 0 \\ 0 \end{bmatrix}$$

$$\begin{bmatrix} 1 & 3 & 4 & -6 \\ 0 & 1 & 6 & 0 \\ 0 & -4 & -6 & 9 \\ 0 & -2 & -8 & 2 \end{bmatrix}\begin{bmatrix} \lambda_1 \\ \lambda_2 \\ \lambda_3 \\ \lambda_4 \end{bmatrix} = \begin{bmatrix} 0 \\ 0 \\ 0 \\ 0 \end{bmatrix} \quad \begin{matrix} R_3 \to R_3 - 2R_1 \\ R_4 \to R_4 - R_1 \end{matrix}$$

$$\begin{bmatrix} 1 & 3 & 4 & -6 \\ 0 & 1 & 6 & 0 \\ 0 & 0 & 18 & 9 \\ 0 & 0 & 4 & 2 \end{bmatrix}\begin{bmatrix} \lambda_1 \\ \lambda_2 \\ \lambda_3 \\ \lambda_4 \end{bmatrix} = \begin{bmatrix} 0 \\ 0 \\ 0 \\ 0 \end{bmatrix} \quad \begin{matrix} R_3 \to R_3 + 4R_2 \\ R_4 \to R_4 + 2R_2 \end{matrix}$$

$$\begin{bmatrix} 1 & 3 & 4 & -6 \\ 0 & 1 & 6 & 0 \\ 0 & 0 & 18 & 9 \\ 0 & 0 & 0 & 0 \end{bmatrix}\begin{bmatrix} \lambda_1 \\ \lambda_2 \\ \lambda_3 \\ \lambda_4 \end{bmatrix} = \begin{bmatrix} 0 \\ 0 \\ 0 \\ 0 \end{bmatrix} \quad R_4 \to R_4 - \frac{2}{9}R_3$$

$$\lambda_1 + 3\lambda_2 + 4\lambda_3 - 6\lambda_4 = 0$$
$$\lambda_2 + 6\lambda_3 = 0$$
$$18\lambda_3 + 9\lambda_4 = 0$$

Let $\lambda_4 = t, 18\lambda_3 + 9t = 0$ or $\lambda_3 = \frac{-t}{2}$

$$\lambda_2 - 3t = 0 \text{ or } \lambda_2 = 3t$$
$$\lambda_1 + 9t - 2t - 6t = 0$$
$$\lambda_1 = -t$$

Substituting the values of λ_1, λ_2, λ_3 and λ_4 in (1), we get

$$-tX_1 + 3tX_2 - \frac{t}{2}X_3 + tX_4 = 0 \text{ or } 2X_1 - 6X_2 + X_3 - 2X_4 = 0 \quad \textbf{Ans.}$$

Example 25. *Show that row vectors of the matrix*

$$\begin{bmatrix} 1 & 2 & -2 \\ -1 & 3 & 0 \\ 0 & -2 & 1 \end{bmatrix}$$ *are linearly independent.* (*U.P., I Sem, Dec 2009*)

Solution. Here, we have three vectors

$$X_1 = (1, 2, -2)'$$
$$X_2 = (-1, 3, 0)'$$
$$X_3 = (0, -2, 1)'$$

$$X_1 = \begin{bmatrix} 1 \\ 2 \\ -2 \end{bmatrix}, X_2 = \begin{bmatrix} -1 \\ 3 \\ 0 \end{bmatrix}, X_3 = \begin{bmatrix} 0 \\ -2 \\ 1 \end{bmatrix}$$

Consider the equation

$$\lambda_1 X_1 + \lambda_2 X_2 + \lambda_3 X_3 = 0 \qquad ...(1)$$

$$\lambda_1 \begin{bmatrix} 1 \\ 2 \\ -2 \end{bmatrix} + \lambda_2 \begin{bmatrix} -1 \\ 3 \\ 0 \end{bmatrix} + \lambda_3 \begin{bmatrix} 0 \\ -2 \\ 1 \end{bmatrix} = \begin{bmatrix} 0 \\ 0 \\ 0 \end{bmatrix}$$

$$\lambda_1 - \lambda_2 + 0\,\lambda_3 = 0$$
$$2\,\lambda_1 + 3\,\lambda_2 - 2\,\lambda_3 = 0$$
$$-2\,\lambda_1 + 0\,\lambda_2 + \lambda_3 = 0$$

which is the system of homogeneous equations

$$\begin{bmatrix} 1 & -1 & 0 \\ 2 & 3 & -2 \\ -2 & 0 & 1 \end{bmatrix} \begin{bmatrix} \lambda_1 \\ \lambda_2 \\ \lambda_3 \end{bmatrix} = \begin{bmatrix} 0 \\ 0 \\ 0 \end{bmatrix} \Rightarrow \begin{bmatrix} 1 & -1 & 0 \\ 0 & 5 & -2 \\ 0 & -2 & 1 \end{bmatrix} \begin{bmatrix} \lambda_1 \\ \lambda_2 \\ \lambda_3 \end{bmatrix} = \begin{bmatrix} 0 \\ 0 \\ 0 \end{bmatrix} \begin{matrix} \\ R_2 \to R_2 - 2R_1 \\ R_3 \to R_3 + 2R_1 \end{matrix}$$

$$\begin{bmatrix} 1 & -1 & 0 \\ 0 & 5 & -2 \\ 0 & 0 & \frac{1}{5} \end{bmatrix} \begin{bmatrix} \lambda_1 \\ \lambda_2 \\ \lambda_3 \end{bmatrix} = \begin{bmatrix} 0 \\ 0 \\ 0 \end{bmatrix} R_3 \to R_3 + \frac{2}{5} R_2$$

$$\lambda_1 - \lambda_2 = 0 \qquad ...(2)$$
$$5\lambda_2 - 2\lambda_3 = 0 \qquad ...(3)$$
$$\frac{1}{5}\lambda_3 = 0 \Rightarrow \lambda_3 = 0 \qquad ....(4)$$

Putting the value of λ_3 in (3), we get

$$5\,\lambda_2 - 2\,(0) = 0 \Rightarrow \lambda_2 = 0$$

Putting the value of λ_2 in (2) , we get

$$\lambda_1 - 0 = 0 \Rightarrow \lambda_1 = 0$$

Thus non zero values of λ_1, λ_2, λ_3 do not exist which can satisfy (1). Hence by definition the given system of vectors is linearly independent. **Proved.**

40.8 LINEARLY DEPENDENCE AND INDEPENDENCE OF VECTORS BY RANK METHOD

1. If the rank of the matrix of the given vectors is equal to number of vectors, then the vectors are linearly independent.
2. If the rank of the matrix of the given vectors is less than the number of vectors, then the vectors are linearly dependent.

Example 26. *Show using a matrix that the set of vectors*
$X = [1, 2, -3, 4]$, $Y = [3, -1, 2, 1]$, $Z = [1, -5, 8, -7]$ is linearly dependent.

Solution. Here, we have

$$X = [1, 2, -3, 4], \quad Y = [3, -1, 2, 1], \quad Z = [1, -5, 8, -7]$$

Let us form a matrix of the above vectors

$$\begin{bmatrix} 1 & 2 & -3 & 4 \\ 3 & -1 & 2 & 1 \\ 1 & -5 & 8 & -7 \end{bmatrix} \sim \begin{bmatrix} 1 & 2 & -3 & 4 \\ 0 & -7 & 11 & -11 \\ 0 & -7 & 11 & -11 \end{bmatrix} \begin{matrix} \\ R_2 \to R_2 - 3R_1 \\ R_3 \to R_3 - R_1 \end{matrix} \sim \begin{bmatrix} 1 & 2 & -3 & 4 \\ 0 & -7 & 11 & -11 \\ 0 & 0 & 0 & 0 \end{bmatrix} R_3 \to R_3 - R_2$$

Here the rank of the matrix = 2 < Number of vectors

Hence, vectors are linearly dependent. **Proved.**

Example 27. *Show using a matrix that the set of vectors : [2, 5, 2, –3], [3, 6, 5, 2], [4, 5, 14, 14], [5, 10, 8, 4] is linearly independent.*

Solution. Here, the given vectors are

[2, 5, 2, –3], [3, 6, 5, 2], [4, 5, 14, 14], [5, 10, 8, 4]

Let us form a matrix of the above vectors :

$$\begin{bmatrix} 2 & 5 & 2 & -3 \\ 3 & 6 & 5 & 2 \\ 4 & 5 & 14 & 14 \\ 5 & 10 & 8 & 4 \end{bmatrix} \sim \begin{bmatrix} 2 & 5 & 2 & -3 \\ 1 & 1 & 3 & 5 \\ 1 & -1 & 9 & 12 \\ 1 & 5 & -6 & -10 \end{bmatrix} \begin{matrix} \\ R_2 \to R_2 - R_1 \\ R_3 \to R_3 - R_2 \\ R_4 \to R_4 - R_3 \end{matrix}$$

$$\sim \begin{bmatrix} 1 & 1 & 3 & 5 \\ 2 & 5 & 2 & -3 \\ 1 & -1 & 9 & 12 \\ 1 & 5 & -6 & -10 \end{bmatrix} R_1 \leftrightarrow R_2 \sim \begin{bmatrix} 1 & 1 & 3 & 5 \\ 0 & 3 & -4 & -13 \\ 0 & -2 & 6 & 7 \\ 0 & 4 & -9 & -15 \end{bmatrix} \begin{matrix} \\ R_2 \to R_2 - 2R_1 \\ R_3 \to R_3 - R_1 \\ R_4 \to R_4 - R_1 \end{matrix}$$

$$\sim \begin{bmatrix} 1 & 1 & 3 & 5 \\ 0 & 3 & -4 & -13 \\ 0 & 0 & \frac{10}{3} & \frac{-5}{3} \\ 0 & 0 & \frac{-11}{3} & \frac{7}{3} \end{bmatrix} \begin{matrix} \\ \\ R_3 \to R_3 + \frac{2}{3}R_2 \\ R_4 \to R_4 - \frac{4}{3}R_2 \end{matrix} \sim \begin{bmatrix} 1 & 1 & 3 & 5 \\ 0 & 3 & -4 & -13 \\ 0 & 0 & \frac{10}{3} & \frac{-5}{3} \\ 0 & 0 & 0 & \frac{1}{2} \end{bmatrix} R_4 \to R_4 + \frac{11}{10}R_3$$

Here, the rank of the matrix = 4 = Number of vectors

Hence, the vectors are linearly independent. **Proved.**

EXERCISE 40.2

Examine the following system of vectors for linear dependence. If dependent, find the relation between them.

1. $X_1 = (1, -1, 1), X_2 = (2, 1, 1), X_3 = (3, 0, 2)$. **Ans.** Dependent, $X_1 + X_2 - X_3 = 0$
2. $X_1 = (1, 2, 3), X_2 = (2, -2, 6)$. **Ans.** Independent
3. $X_1 = (3, 1, -4), X_2 = (2, 2, -3), X_3 = (0, -4, 1)$. **Ans.** Dependent, $2X_1 - 3X_2 - X_3 = 0$
4. $X_1 = (1, 1, 1, 3), X_2 = (1, 2, 3, 4), X_3 = (2, 3, 4, 7)$. **Ans.** Dependent, $X_1 + X_2 - X_3 = 0$
5. $X_1 = (1, 1, -1, 1), X_2 = (1, -1, 2, -1), X_3 = (3, 1, 0, 1)$. **Ans.** Dependent, $2X_1 + X_2 - X_3 = 0$
6. $X_1 = (1, -1, 2, 0), X_2 = (2, 1, 1, 1), X_3 = (3, -1, 2, -1), X_4 = (3, 0, 3, 1)$.
   Ans. Dependent, $X_1 + X_2 - X_4 = 0$
7. Show that the column vectors of following matrix A are linearly independent:

   $$A = \begin{bmatrix} 1 & 0 & 0 \\ 6 & 2 & 1 \\ 4 & 3 & 2 \end{bmatrix}$$
8. Show that the vectors $x_1 = (2, 3, 1, -1)$, $x_2 = (2, 3, 1, -2)$, $x_3 = (4, 6, 2, 1)$ are linearly dependent. Express one of the vectors as linear combination of the others.
9. Find whether or not the following set of vectors are linearly dependent or independent:
   (*i*) (1, –2), (2, 1), (3, 2) (*ii*) (1, 1, 1, 1), (0, 1, 1, 1), (0, 0, 1, 1), (0, 0, 0, 1).
   Ans. (*i*) Dependent (*ii*) Independent

10. Show that the vectors $x_1 = (a_1, b_1)$, $x_2 = (a_2, b_2)$ are linearly dependent if $a_1 b_2 - a_2 b_1 = 0$.

40.9 ANOTHER METHOD (ADJOINT METHOD) TO SOLVE LINEAR EQUATIONS

Let the equations be

$$a_1 x + a_2 y + a_3 z = d_1$$
$$b_1 x + b_2 y + b_3 z = d_2$$
$$c_1 x + c_2 y + c_3 z = d_3$$

We write the above equations in the matrix form

$$\begin{bmatrix} a_1 x + a_2 y + a_3 z \\ b_1 x + b_2 y + b_3 z \\ c_1 x + c_2 y + c_3 z \end{bmatrix} = \begin{bmatrix} d_1 \\ d_2 \\ d_3 \end{bmatrix} \text{ or } \begin{bmatrix} a_1 & a_2 & a_3 \\ b_1 & b_2 & b_3 \\ c_1 & c_2 & c_3 \end{bmatrix} \begin{bmatrix} x \\ y \\ z \end{bmatrix} = \begin{bmatrix} d_1 \\ d_2 \\ d_3 \end{bmatrix}$$

$$AX = B \qquad \text{...(1)}$$

where $A = \begin{bmatrix} a_1 & a_2 & a_3 \\ b_1 & b_2 & b_3 \\ c_1 & c_2 & c_3 \end{bmatrix}, X = \begin{bmatrix} x \\ y \\ z \end{bmatrix}$ and $B = \begin{bmatrix} d_1 \\ d_2 \\ d_3 \end{bmatrix}$

Multiplying (1) by A^{-1}.

$$A^{-1} AX = A^{-1} B \text{ or } IX = A^{-1} B \text{ or } X = A^{-1} B.$$

Example 28. *Solve, with the help of matrices, the simultaneous equations*

$$x + y + z = 3, \quad x + 2y + 3z = 4, \quad x + 4y + 9z = 6$$

Solution. The given equations in the matrix form are written as below:

$$\begin{bmatrix} 1 & 1 & 1 \\ 1 & 2 & 3 \\ 1 & 4 & 9 \end{bmatrix} \begin{bmatrix} x \\ y \\ z \end{bmatrix} = \begin{bmatrix} 3 \\ 4 \\ 6 \end{bmatrix}$$

$$AX = B$$

where $A = \begin{bmatrix} 1 & 1 & 1 \\ 1 & 2 & 3 \\ 1 & 4 & 9 \end{bmatrix}, X = \begin{bmatrix} x \\ y \\ z \end{bmatrix}, B = \begin{bmatrix} 3 \\ 4 \\ 6 \end{bmatrix}$

Now we have to find out the A^{-1}.

$$|A| = 1 \times 6 + 1 \times (-6) + 1 \times 2 = 6 - 6 + 2 = 2$$

$$\text{Matrix of co-factors} = \begin{bmatrix} 6 & -6 & 2 \\ -5 & 8 & -3 \\ 1 & -2 & 1 \end{bmatrix}, \textit{Adjoint } A = \begin{bmatrix} 6 & -5 & 1 \\ -6 & 8 & -2 \\ 2 & -3 & 1 \end{bmatrix}$$

$$A^{-1} = \frac{1}{|A|} \textit{Adjoint } A = \frac{1}{2} \begin{bmatrix} 6 & -5 & 1 \\ -6 & 8 & -2 \\ 2 & -3 & 1 \end{bmatrix}$$

$$X = A^{-1} B = \frac{1}{2} \begin{bmatrix} 6 & -5 & 1 \\ -6 & 8 & -2 \\ 2 & -3 & 1 \end{bmatrix} \begin{bmatrix} 3 \\ 4 \\ 6 \end{bmatrix}$$

$$\Rightarrow \quad \begin{bmatrix} x \\ y \\ z \end{bmatrix} = \frac{1}{2} \begin{bmatrix} 18 - 20 + 6 \\ -18 + 32 - 12 \\ 6 - 12 + 6 \end{bmatrix} = \frac{1}{2} \begin{bmatrix} 4 \\ 2 \\ 0 \end{bmatrix} = \begin{bmatrix} 2 \\ 1 \\ 0 \end{bmatrix}$$

$$x = 2, y = 1, z = 0$$ **Ans.**

Example 29. *Given the matrices*

$$A \equiv \begin{bmatrix} 1 & 2 & 3 \\ 3 & -1 & 1 \\ 4 & 2 & 1 \end{bmatrix}, X \equiv \begin{bmatrix} x \\ y \\ z \end{bmatrix} \text{ and } C \equiv \begin{bmatrix} 1 \\ 2 \\ 3 \end{bmatrix}$$

Write down the linear equations given by $AX = C$ and solve for x, y, z by the matrix method.

Solution. $AX = C$

$$\begin{bmatrix} 1 & 2 & 3 \\ 3 & -1 & 1 \\ 4 & 2 & 1 \end{bmatrix}\begin{bmatrix} x \\ y \\ z \end{bmatrix} = \begin{bmatrix} 1 \\ 2 \\ 3 \end{bmatrix}$$

$$X = A^{-1} . C$$

$$\begin{bmatrix} x \\ y \\ z \end{bmatrix} = \begin{bmatrix} 1 & 2 & 3 \\ 3 & -1 & 1 \\ 4 & 2 & 1 \end{bmatrix}^{-1}\begin{bmatrix} 1 \\ 2 \\ 3 \end{bmatrix}$$

$$\text{Matrix of co-factors of } A = \begin{bmatrix} -3 & 1 & 10 \\ 4 & -11 & 6 \\ 5 & 8 & -7 \end{bmatrix}$$

$$|A| = 1(-3) + 2(1) + 3(10) = -3 + 2 + 30 = 29$$

$$\text{Adj. } A = \begin{bmatrix} -3 & 4 & 5 \\ 1 & -11 & 8 \\ 10 & 6 & -7 \end{bmatrix}$$

$$A^{-1} = \frac{1}{|A|} Adj. A = \frac{1}{29}\begin{bmatrix} -3 & 4 & 5 \\ 1 & -11 & 8 \\ 10 & 6 & -7 \end{bmatrix}.$$

$$X = A^{-1} C$$

$$\begin{bmatrix} x \\ y \\ z \end{bmatrix} = \frac{1}{29}\begin{bmatrix} -3 & 4 & 5 \\ 1 & -11 & 8 \\ 10 & 6 & -7 \end{bmatrix}\begin{bmatrix} 1 \\ 2 \\ 3 \end{bmatrix}$$

$$\begin{bmatrix} x \\ y \\ z \end{bmatrix} = \frac{1}{29}\begin{bmatrix} -3 & +8 & +15 \\ 1 & -22 & +24 \\ 10 & +12 & -21 \end{bmatrix} = \frac{1}{29}\begin{bmatrix} 20 \\ 3 \\ 1 \end{bmatrix} = \begin{bmatrix} \frac{20}{29} \\ \frac{3}{29} \\ \frac{1}{29} \end{bmatrix}$$

Hence, $x = \frac{20}{29}, y = \frac{3}{29}, z = \frac{1}{29}$ **Ans.**

Example 29. *By the method of matrix, inversion, solve the system.*

$$\begin{bmatrix} 1 & 1 & 1 \\ 2 & 5 & 7 \\ 2 & 1 & -1 \end{bmatrix}\begin{bmatrix} x & u \\ y & v \\ z & w \end{bmatrix} = \begin{bmatrix} 9 & 2 \\ 52 & 15 \\ 0 & -1 \end{bmatrix}$$

Solution. $\begin{bmatrix} 1 & 1 & 1 \\ 2 & 5 & 7 \\ 2 & 1 & -1 \end{bmatrix}\begin{bmatrix} x & u \\ y & v \\ z & w \end{bmatrix} = \begin{bmatrix} 9 & 2 \\ 52 & 15 \\ 0 & -1 \end{bmatrix}$

$$\begin{bmatrix} x & u \\ y & v \\ z & w \end{bmatrix} = \begin{bmatrix} 1 & 1 & 1 \\ 2 & 5 & 7 \\ 2 & 1 & -1 \end{bmatrix}^{-1}\begin{bmatrix} 9 & 2 \\ 52 & 15 \\ 0 & -1 \end{bmatrix}$$

$$= \frac{-1}{4}\begin{bmatrix} -12 & 2 & 2 \\ 16 & -3 & -5 \\ -8 & 1 & 3 \end{bmatrix}\begin{bmatrix} 9 & 2 \\ 52 & 15 \\ 0 & -1 \end{bmatrix} = \frac{-1}{4}\begin{bmatrix} -4 & 4 \\ -12 & -8 \\ -20 & -4 \end{bmatrix} = \begin{bmatrix} 1 & -1 \\ 3 & 2 \\ 5 & 1 \end{bmatrix}$$

$x = 1,\quad u = -1$

$y = 3,\quad v = 2$

$z = 5,\quad w = 1$ **Ans.**

EXERCISE 40.3

Solve the following equations

1. $3x + y + 2z = 3,\ 2x - 3y - z = -3,\ x + 2y + z = 4$ **Ans.** $x = 1,\ y = 2,\ z = -1$
2. $x + 2y + 3z = 1,\ 2x + 3y + 8z = 2,\ x + y + z = 3$ **Ans.** $x = \frac{9}{2}, y = -1, z = -\frac{1}{2}$
3. $4x + 2y - z = 9,\ x - y + 3z = -4,\ 2x + z = 1$ **Ans.** $x = 1,\ y = 2,\ z = -1$
4. $5x + 3y + 3z = 48,\ 2x + 6y - 3z = 18,\ 8x - 3y + 2z = 21$ **Ans.** $x = 3,\ y = 5,\ z = 6$
5. $x + y + z = 6,\ x - y + 2z = 5,\ 3x + y + z = 8$ **Ans.** $x = 1,\ y = 2,\ z = 3$
6. $x + 2y + 3z = 1,\ 3x - 2y + z = 2,\ 4x + 2y + z = 3$ **Ans.** $x = \frac{7}{10}, y = \frac{3}{40}, z = \frac{1}{20}$
7. $9x + 4y + 3z = -1,\ 5x + y + 2z = 1,\ 7x + 3y + 4z = 1$ **Ans.** $x = 0,\ y = -1,\ z = 1$
8. $x + y + z = 8,\ x - y + 2z = 6,\ 9x + 5y - 7z = 14$ **Ans.** $x = 5, y = \frac{5}{3}, z = \frac{4}{3}$
9. $3x + 2y + 4z = 7,\ 2x + y + z = 4,\ x + 3y + 5z = 2$ **Ans.** $x = \frac{9}{4}, y = -\frac{9}{8}, z = \frac{5}{8}$

40.10 PARTITIONING OF MATRICES

Sub matrix. A matrix obtained by deleting some of the rows and columns of a matrix A is said to be sub matrix.

For example, $A = \begin{bmatrix} 4 & 1 & 0 \\ 5 & 2 & 1 \\ 6 & 3 & 4 \end{bmatrix}$, then $\begin{bmatrix} 4 & 1 \\ 5 & 2 \end{bmatrix}, \begin{bmatrix} 5 & 2 \\ 6 & 3 \end{bmatrix}, \begin{bmatrix} 1 & 0 \\ 2 & 1 \end{bmatrix}$ are the sub matrices.

Partitioning: A matrix may be subdivided into sub matrices by drawing lines parallel to its rows and columns. These sub matrices may be considered as the elements of the original matrix.

For example, $$A = \begin{bmatrix} 2 & 1 & : & 0 & 4 & 1 \\ 1 & 0 & : & 2 & 3 & 4 \\ .... & .... & : & .... & .... & .... \\ 4 & 5 & : & 1 & 6 & 5 \end{bmatrix}$$

$$A_{11} = \begin{bmatrix} 2 & 1 \\ 1 & 0 \end{bmatrix}, \quad A_{12} = \begin{bmatrix} 0 & 4 & 1 \\ 2 & 3 & 4 \end{bmatrix}$$

$$A_{21} = [4 \;\; 5], \quad A_{22} = [1 \;\; 6 \;\; 5]$$

Then we may write $\quad A = \begin{bmatrix} A_{11} & A_{12} \\ A_{21} & A_{22} \end{bmatrix}$

So, the matrix is partitioned. The dotted lines divide the matrix into sub-matrices. A_{11}, A_{12}, A_{21}, A_{22} are the sub-matrices but behave like elements of the original matrix A. The matrix A can be partitioned in several ways.

Addition by submatrices: Let A and B be two matrices of the same order and are partitioned identically.

$$A = \begin{bmatrix} 2 & 3 & 4 & \vdots & 5 \\ 0 & 1 & 2 & \vdots & 3 \\ \cdots & \cdots & \cdots & \vdots & \cdots \\ 3 & 4 & 5 & \vdots & 6 \\ \cdots & \cdots & \cdots & \vdots & \cdots \\ 4 & 5 & 0 & \vdots & 1 \end{bmatrix} \qquad B = \begin{bmatrix} 3 & 1 & 4 & \vdots & 6 \\ 2 & 1 & 0 & \vdots & 4 \\ \cdots & \cdots & \cdots & \vdots & \cdots \\ 4 & 5 & 1 & \vdots & 2 \\ \cdots & \cdots & \cdots & \vdots & \cdots \\ 1 & 3 & 4 & \vdots & 5 \end{bmatrix}$$

For example;

$$A = \begin{bmatrix} A_{11} & A_{12} \\ A_{21} & A_{22} \\ A_{31} & A_{32} \end{bmatrix}, \quad B = \begin{bmatrix} B_{11} & B_{12} \\ B_{21} & B_{22} \\ B_{31} & B_{32} \end{bmatrix}$$

$$A = \begin{bmatrix} A_{11} + B_{11} & A_{12} + B_{12} \\ A_{21} + B_{21} & A_{22} + B_{22} \\ A_{31} + B_{31} & A_{32} + B_{32} \end{bmatrix}$$

40.11 MULTIPLICATION BY SUB-MATRICES

Two matrices A and B, which are conformable to the product AB are partitioned in such a way that the columns of A partitioned in the same way as the rows of B are partitioned. But the rows of A and columns of B can be partitioned in any way.

For example, Here A is a 3×4 matrix and B is 4×3 matrix.

$$A = \begin{bmatrix} 1 & 2 & 3 & \vdots & 4 \\ 0 & 1 & 2 & \vdots & 3 \\ 1 & 4 & 1 & \vdots & 2 \end{bmatrix} \text{ and } B = \begin{bmatrix} 4 & 5 & 6 \\ 3 & 2 & 1 \\ 1 & 0 & 4 \\ \cdots & \cdots & \cdots \\ 2 & 5 & 3 \end{bmatrix}$$

The partitioning of the columns of A is the same as the partitioning of the rows of B. Here, A is partitioned after third column, B has been partitioned after third row.

Example 31. *If C and D are two non-singular matrices, show that if*

$$A = \begin{bmatrix} C & 0 \\ 0 & D \end{bmatrix}, \quad \textit{then } A^{-1} = \begin{bmatrix} C^{-1} & 0 \\ 0 & D^{-1} \end{bmatrix}$$

Solution. Let $\quad A = \begin{bmatrix} E & F \\ G & H \end{bmatrix} \quad$ ...(1)

Then $$AA^{-1} = \begin{bmatrix} C & 0 \\ 0 & D \end{bmatrix}\begin{bmatrix} E & F \\ G & H \end{bmatrix} = \begin{bmatrix} CE+0G & CF+0H \\ 0E+DG & 0F+DH \end{bmatrix}$$

So that $$\begin{bmatrix} CE+0G & CF+0H \\ 0E+DG & 0F+DH \end{bmatrix} = \begin{bmatrix} I & 0 \\ 0 & I \end{bmatrix}$$

$$CE + 0G = I \Rightarrow CE = I$$

$$CF + 0H = 0 \Rightarrow CF = 0$$

$$0E + DG = 0 \Rightarrow DG = 0$$

$$OF + DH = I \Rightarrow DH = I$$

Since, C is non singular and $CF = 0$, $\therefore F = 0$

$$CE = I \Rightarrow E = C^{-1}$$

Similarly, D is non singular and $DG = 0 \Rightarrow G = 0$ and $DH = I \Rightarrow H = D^{-1}$

Putting these values in (1), we get

$$A^{-1} = \begin{bmatrix} C^{-1} & 0 \\ 0 & D^{-1} \end{bmatrix}$$ **Proved.**

Inverse By Partitioning: Let the matrix B be the inverse of the matrix A. Matrices A and B are partitioned as

$$A = \begin{bmatrix} A_{11} & A_{12} \\ A_{21} & A_{22} \end{bmatrix}, \quad B = \begin{bmatrix} B_{11} & B_{12} \\ B_{21} & B_{22} \end{bmatrix}$$

Since, $$AB = BA = I$$

$$\begin{bmatrix} A_{11} & A_{12} \\ A_{21} & A_{22} \end{bmatrix}\begin{bmatrix} B_{11} & B_{12} \\ B_{21} & B_{22} \end{bmatrix} = \begin{bmatrix} B_{11} & B_{12} \\ B_{21} & B_{22} \end{bmatrix}\begin{bmatrix} A_{11} & A_{12} \\ A_{21} & A_{22} \end{bmatrix} = \begin{bmatrix} I & 0 \\ 0 & I \end{bmatrix}$$

$$\begin{bmatrix} A_{11}B_{11} + A_{12}B_{21} & A_{11}B_{12} + A_{12}B_{22} \\ A_{21}B_{11} + A_{22}B_{21} & A_{21}B_{12} + A_{22}B_{22} \end{bmatrix} = \begin{bmatrix} B_{11}A_{11} + B_{12}A_{21} & B_{11}A_{12} + B_{12}A_{22} \\ B_{21}A_{11} + B_{22}A_{21} & B_{21}A_{12} + B_{22}A_{22} \end{bmatrix} = \begin{bmatrix} I & 0 \\ 0 & I \end{bmatrix}$$

Let us solve the equations for B_{11}, B_{12}, B_{21} and B_{22}.

Let, $$B_{22} = M^{-1}$$

From (2), $$B_{12} = -A_{11}^{-1}(A_{22}B_{22}) = -(A_{11}^{-1}A_{22})M^{-1}$$

From (3), $$B_{21} = -(B_{22}A_{21})A_{11}^{-1} = -M^{-1}(A_{21}A_{11}^{-1})$$

From (1), $$B_{11} = A_{11}^{-1} - A_{11}^{-1}(A_{12}B_{21}) = A_{11}^{-1} - (A_{11}^{-1}A_{12})B_{21}$$

$$= A_{11}^{-1} + (A_{11}^{-1}A_{12})M^{-1}(A_{21}A_{11}^{-1})$$

Here $$M = A_{22} - A_{21}(A_{11}^{-1}A_{22})$$

Note: A is usually taken of order $n - 1$.

Example 32. *Find the inverse of the following matrix by partitioning*

$$\begin{bmatrix} 1 & 3 & 3 \\ 1 & 4 & 3 \\ 1 & 3 & 4 \end{bmatrix}$$

Solution. Let the matrix be partitioned into four submatrices as follows:

Let $A = \begin{bmatrix} 1 & 3 & 3 \\ 1 & 4 & 3 \\ 1 & 3 & 4 \end{bmatrix}$

$A_{11} = \begin{bmatrix} 1 & 3 \\ 1 & 4 \end{bmatrix}; \ A_{12} = \begin{bmatrix} 3 \\ 3 \end{bmatrix}$

$A_{21} = [1 \quad 3]; \ A_{22} = [4]$

We have to find $A^{-1} = \begin{bmatrix} B_{11} & B_{12} \\ B_{21} & B_{22} \end{bmatrix}$ where

$$B_{11} = A_{11}^{-1} + (A_{11}^{-1} A_{12})(M^{-1})(A_{21} A_{11}^{-1})$$

$$B_{21} = -M^{-1}(A_{21} A_{11}^{-1})$$

$$B_{12} = -A_{11}^{-1} A_{12} M^{-1}; \ B_{22} = M^{-1}$$

and $M = A_{22} - A_{21}(A_{11}^{-1} A_{12})$

Now $\begin{bmatrix} 4 & -3 \\ -1 & 1 \end{bmatrix}$; $A_{11}^{-1} A_{12} = \begin{bmatrix} 4 & -3 \\ -1 & 1 \end{bmatrix} \begin{bmatrix} 3 \\ 3 \end{bmatrix} = \begin{bmatrix} 3 \\ 0 \end{bmatrix}$

$$A_{21} A_{11}^{-1} = [1 \quad 3] \begin{bmatrix} 4 & -3 \\ -1 & 1 \end{bmatrix} = [1 \ 0]$$

$$M = [4] - [1 - 3] \begin{bmatrix} 3 \\ 0 \end{bmatrix} = [4] - [3] = [1]$$

$$M^{-1} = [3]$$

$$\therefore \quad B_{11} = \begin{bmatrix} 4 & -3 \\ -1 & 1 \end{bmatrix} + \begin{bmatrix} 3 \\ 0 \end{bmatrix} [1 \ 0] = \begin{bmatrix} 4 & -3 \\ -1 & 1 \end{bmatrix} + \begin{bmatrix} 3 & 0 \\ 0 & 0 \end{bmatrix} \Rightarrow B_{11} = \begin{bmatrix} 7 & -3 \\ -1 & 1 \end{bmatrix}$$

$$B_{21} = -[1] \, [1 \ 0] = -[1 \ 0]$$

$$B_{12} = -\begin{bmatrix} 3 \\ 0 \end{bmatrix}$$

$$B_{22} = [1]$$

$$A^{-1} = \begin{bmatrix} B_{11} & B_{12} \\ B_{21} & B_{22} \end{bmatrix} = \begin{bmatrix} 7 & -3 & -3 \\ -1 & 1 & 0 \\ -1 & 0 & 1 \end{bmatrix}$$ **Ans.**

Example 33. *Find the inverse of* $A = \begin{bmatrix} 1 & 2 & 3 & 1 \\ 1 & 3 & 3 & 2 \\ 2 & 4 & 3 & 3 \\ 1 & 1 & 1 & 1 \end{bmatrix}$ *by partitioning.*

Solution. (*a*) Take $G_3 = \begin{bmatrix} 1 & 2 & 3 \\ 1 & 3 & 3 \\ 2 & 4 & 3 \end{bmatrix}$ and partition so that

$$A_{11} = \begin{bmatrix} 1 & 2 \\ 1 & 3 \end{bmatrix}, A_{12} = \begin{bmatrix} 3 \\ 3 \end{bmatrix}, A_{21} = [2 \ 4], \text{ and } A_{22} = [3]$$

Now, $A_{11}^{-1} = \begin{bmatrix} 3 & -2 \\ -1 & 1 \end{bmatrix}, A_{11}^{-1} A_{12} = \begin{bmatrix} 3 & -2 \\ -1 & 1 \end{bmatrix} \begin{bmatrix} 3 \\ 3 \end{bmatrix} = \begin{bmatrix} 3 \\ 0 \end{bmatrix},$

$$A_{21} A_{11}^{-1} = [2 \quad 4] \begin{bmatrix} 3 & -2 \\ -1 & 1 \end{bmatrix} = [2 \ 0]$$

$$M = A_{22} - A_{21}(A_{11}^{-1} A_{12}) = [3] - [2 \quad 4]\begin{bmatrix} 3 \\ 0 \end{bmatrix} = [-3] \text{ And } M^{-1} = [1/3]$$

Then $B_{11} = A_{11}^{-1} + (A_{11}^{-1} A_{12}) M^{-1}(A_{21} A_{11}^{-1}) = \begin{bmatrix} 3 & -2 \\ -1 & 1 \end{bmatrix} + \begin{bmatrix} 3 \\ 0 \end{bmatrix}\begin{bmatrix} -\frac{1}{3} \end{bmatrix}[2 \quad 0] = \begin{bmatrix} 3 & -2 \\ -1 & 1 \end{bmatrix} - \begin{bmatrix} 2 & 0 \\ 0 & 0 \end{bmatrix}$

$$= \frac{1}{3}\begin{bmatrix} 3 & -6 \\ -3 & 3 \end{bmatrix}$$

$$B_{12} = -(A_{11}^{-1} A_{12}) M^{-1} = \frac{1}{3}\begin{bmatrix} 3 \\ 0 \end{bmatrix}$$

$$B_{21} = -M^{-1}(A_{21} A_{11}^{-1}) = \frac{1}{3}[2 \quad 0]$$

$$B_{22} = M^{-1} = \begin{bmatrix} -\frac{1}{3} \end{bmatrix}$$

and $$G_3^{-1} = \begin{bmatrix} B_{11} & B_{12} \\ B_{21} & B_{22} \end{bmatrix} = \frac{1}{3}\begin{bmatrix} 3 & -6 & 3 \\ -3 & 3 & 0 \\ 2 & 0 & -1 \end{bmatrix}$$

(*b*) Partition A so that $A_{11} = \begin{bmatrix} 1 & 2 & 3 \\ 1 & 3 & 3 \\ 2 & 4 & 3 \end{bmatrix}$, $A_{12} = \begin{bmatrix} 1 \\ 2 \\ 3 \end{bmatrix}$, $A_{21} = [1 \ 1 \ 1]$, and $A_{22} = [1]$.

Now, $A_{11}^{-1} = \frac{1}{3}\begin{bmatrix} 3 & -6 & 3 \\ -3 & 3 & 0 \\ 2 & 0 & -1 \end{bmatrix}$, $A_{11}^{-1} A_{12} = \frac{1}{3}\begin{bmatrix} 0 \\ 3 \\ -1 \end{bmatrix}$, $A_{21} A_{11}^{-1} = \frac{1}{3}[2 \quad -3 \quad 2]$

$$M = [1] - [1 \quad 1 \quad 1]\left(\frac{1}{3}\right)\begin{bmatrix} 0 \\ 3 \\ -1 \end{bmatrix} = \begin{bmatrix} \frac{1}{3} \end{bmatrix}. \text{ and } M^{-1} = [3]$$

Then $$B_{11} = \frac{1}{3}\begin{bmatrix} 3 & -6 & 3 \\ -3 & 3 & 0 \\ 2 & 0 & -1 \end{bmatrix} + \frac{1}{3}\begin{bmatrix} 0 \\ 3 \\ -1 \end{bmatrix}[3]\frac{1}{3}[2 \ -3 \ 2]$$

$$= \frac{1}{3}\begin{bmatrix} 3 & -6 & 3 \\ -3 & 3 & 0 \\ 2 & 0 & -1 \end{bmatrix} + \frac{1}{3}\begin{bmatrix} 0 & 0 & 0 \\ 6 & -9 & 6 \\ -2 & 3 & -2 \end{bmatrix} = \begin{bmatrix} 1 & -2 & 1 \\ 1 & -2 & 2 \\ 0 & 1 & -1 \end{bmatrix}$$

$$B_{12} = \begin{bmatrix} 0 \\ -3 \\ 1 \end{bmatrix}, \quad B_{21} = [-2 \quad 3 \quad -2], \quad B_{22} = [3]$$

$$A^{-1} = \begin{bmatrix} B_{11} & B_{12} \\ B_{21} & B_{22} \end{bmatrix} = \begin{bmatrix} 1 & -2 & 1 & 0 \\ 1 & -2 & 2 & -3 \\ 0 & 1 & -1 & 1 \\ -2 & 3 & -2 & 3 \end{bmatrix}$$ **Ans.**

EXERCISE 40.4

1. Compute A + B using partitioning

$$A = \begin{bmatrix} 4 & 1 & 0 & 5 \\ 6 & 7 & 8 & 1 \\ 0 & 2 & 1 & 1 \\ 1 & 2 & 0 & 1 \end{bmatrix}, \; B = \begin{bmatrix} 3 & 2 & 1 & 1 \\ 1 & 0 & 1 & 1 \\ 2 & 1 & 2 & 1 \\ 0 & 1 & 2 & 3 \end{bmatrix}$$

Ans. $\begin{bmatrix} 7 & 3 & 1 & 6 \\ 7 & 7 & 9 & 2 \\ 2 & 3 & 3 & 2 \\ 1 & 3 & 2 & 4 \end{bmatrix}$

2. Compute AB using partitioning

$$A = \begin{bmatrix} 1 & 2 & 0 & 1 \\ 4 & 1 & 3 & 2 \\ 2 & 1 & 3 & 0 \end{bmatrix}, \; B = \begin{bmatrix} 2 & 3 & 1 \\ 0 & 1 & 4 \\ 4 & 1 & 2 \\ 2 & 1 & 2 \end{bmatrix}$$

Ans. $\begin{bmatrix} 4 & 6 & 11 \\ 24 & 18 & 18 \\ 16 & 10 & 12 \end{bmatrix}$

3. Find the inverse of $\begin{bmatrix} A & B \\ C & 0 \end{bmatrix}$

where B, C are non-singular. **Ans.** $\begin{bmatrix} 0 & C^{-1} \\ B^{-1} & -B^{-1}AC^{-1} \end{bmatrix}$

Find the inverse of the following metrices by partitioning:

4. $\begin{bmatrix} 2 & 1 & -1 \\ 1 & 3 & 2 \\ -1 & 2 & 1 \end{bmatrix}$ **Ans.** $\dfrac{1}{10}\begin{bmatrix} 1 & 3 & -5 \\ 3 & -1 & 5 \\ -5 & 5 & -5 \end{bmatrix}$

5. $\begin{bmatrix} 1 & 2 & -1 \\ -1 & 1 & 2 \\ 2 & -1 & 1 \end{bmatrix}$ **Ans.** $\dfrac{1}{14}\begin{bmatrix} 3 & -1 & 5 \\ 5 & 3 & -1 \\ -1 & 5 & 3 \end{bmatrix}$

6. $\begin{bmatrix} 2 & 3 & 4 \\ 4 & 3 & 1 \\ 1 & 2 & 4 \end{bmatrix}$ **Ans.** $\dfrac{1}{5}\begin{bmatrix} -10 & 4 & 9 \\ 15 & -4 & -14 \\ -5 & 1 & 6 \end{bmatrix}$

7. $\begin{bmatrix} 1 & 2 & 3 \\ 2 & 4 & 5 \\ 3 & 5 & 6 \end{bmatrix}$ **Ans.** $\begin{bmatrix} 1 & -3 & 2 \\ -3 & 3 & -1 \\ 2 & -1 & 0 \end{bmatrix}$

8. $\begin{bmatrix} 3 & 4 & 2 & 7 \\ 2 & 3 & 3 & 2 \\ 5 & 7 & 3 & 9 \\ 2 & 3 & 2 & 3 \end{bmatrix}$ **Ans.** $\dfrac{1}{2}\begin{bmatrix} -1 & 11 & 7 & -26 \\ -1 & -7 & -3 & 16 \\ 1 & 1 & -1 & 0 \\ 1 & -1 & -1 & 2 \end{bmatrix}$

Choose the correct answer:

9. If $3x + 2y + z = 0$, $x + 4y + z = 0$, $2x + y + 4z = 0$, be a system of equations then

(*i*) System is inconsistent

(*ii*) it has only trivial solution

(*iii*) it can be reduced to a single equation thus solution does not exist

(*iv*) Determinant of the coefficient matrix is zero. (*AMIETE, June 2010*)

CHAPTER

Eigen Values, Eigen Vector, Cayley Hamilton Theorem, Diagonalisation (Complex and Unitary Matrices)

41.1 INTRODUCTION

Eigen values and eigen vectors are used in the study of ordinary differential equations, analysing population growth and finding powers of matrices.

41.2 EIGEN VALUES

Let $\begin{bmatrix} a_{11} & a_{12} & a_{13} & \cdots & a_{1n} \\ a_{21} & a_{22} & a_{23} & \cdots & a_{2n} \\ a_{31} & a_{32} & a_{33} & \cdots & a_{3n} \\ \cdots & \cdots & \cdots & \cdots & \cdots \\ a_{n1} & a_{n2} & a_{n3} & \cdots & a_{nn} \end{bmatrix} \begin{bmatrix} x_1 \\ x_2 \\ x_3 \\ \vdots \\ x_n \end{bmatrix} = \begin{bmatrix} y_1 \\ y_2 \\ y_3 \\ \vdots \\ y_n \end{bmatrix}$

$$AX = Y \qquad ...(1)$$

Where A is the matrix, X is the column vector and Y is also column vector.

Here column vector X is transformed into the column vector Y by means of the square matrix A.

Let X be a such vector which transforms into λX by means of the transformation (1). Suppose the linear transformation $Y = AX$ transforms X into a scalar multiple of itself i.e. λX.

$$AX = Y = \lambda X$$

$$AX - \lambda IX = 0$$

$$(A - \lambda I)X = 0 \qquad ...(2)$$

Thus the unknown scalar λ is known as an eigen value of the matrix A and the corresponding non zero vector X as **eigen vector.**

The eigen values are also called characteristic values or proper values or latent values.

Let $A = \begin{bmatrix} 2 & 2 & 1 \\ 1 & 3 & 1 \\ 1 & 2 & 2 \end{bmatrix}$

$A - \lambda I = \begin{bmatrix} 2 & 2 & 1 \\ 1 & 3 & 1 \\ 1 & 2 & 2 \end{bmatrix} - \lambda \begin{bmatrix} 1 & 0 & 0 \\ 0 & 1 & 0 \\ 0 & 0 & 1 \end{bmatrix} = \begin{bmatrix} 2-\lambda & 2 & 1 \\ 1 & 3-\lambda & 1 \\ 1 & 2 & 2-\lambda \end{bmatrix}$ is characteristic matrix (*b*) **Characteristic Polynomial:** The determinant $| A - \lambda I |$ when expanded will give a polynomial, which we call as characteristic polynomial of matrix A.

For example; $\begin{vmatrix} 2-\lambda & 2 & 1 \\ 1 & 3-\lambda & 1 \\ 1 & 2 & 2-\lambda \end{vmatrix}$

$$= (2 - \lambda)(6 - 5\lambda + \lambda^2 - 2) - 2(2 - \lambda - 1) + 1(2 - 3 + \lambda)$$
$$= -\lambda^3 + 7\lambda^2 - 11\lambda + 5$$

(*c*) **Characteristic Equation:** The equation $| A - \lambda I| = 0$ is called the characteristic equation of the matrix A *e.g.*

$$\lambda^3 - 7\lambda^2 + 11\lambda - 5 = 0$$

(*d*) **Characteristic Roots or Eigen Values:** The roots of characteristic equation $| A - \lambda I | = 0$ are called characteristic roots of matrix A. *e.g.*

$$\lambda^3 - 7\lambda^2 + 11\lambda - 5 = 0$$

$$\Rightarrow \quad (\lambda - 1)(\lambda - 1)(\lambda - 5) = 0 \qquad \therefore \lambda = 1, 1, 5$$

Characteristic roots are 1, 1, 5.

Some Important Properties of Eigen Values

(AMIETE, Dec. 2009)

(1) Any square matrix A and its transpose A' have the same eigen values.

Note. The sum of the elements on the principal diagonal of a matrix is called the **trace** of the matrix.

(2) The sum of the eigen values of a matrix is equal to the **trace** of the matrix.

(3) The product of the eigen values of a matrix A is equal to the **determinant** of A.

(4) If $\lambda_1, \lambda_2, \ldots \lambda_n$ are the eigen values of A, then the eigen values of

(*i*) kA are $k\lambda_1, \; k\lambda_2, \; \ldots, \; k\lambda_n$ (*ii*) A^m are $\lambda_1^m, \lambda_2^m, \ldots, \lambda_n^m$

(*iii*) A^{-1} are $\frac{1}{\lambda_1}, \frac{1}{\lambda_2}, \ldots, \frac{1}{\lambda_n}$.

Example 1. *Find the characteristic roots of the matrix* $\begin{bmatrix} 6 & -2 & 2 \\ -2 & 3 & -1 \\ 2 & -1 & 3 \end{bmatrix}$

Solution. The characteristic equation of the given matrix is

$$\begin{vmatrix} 6-\lambda & -2 & 2 \\ -2 & 3-\lambda & -1 \\ 2 & -1 & 3-\lambda \end{vmatrix} = 0$$

$$\Rightarrow \quad (6 - \lambda)(9 - 6\lambda + \lambda^2 - 1) + 2(-6 + 2\lambda + 2) + 2(2 - 6 + 2\lambda) = 0$$

$$\Rightarrow \quad -\lambda^3 + 12\lambda^2 - 36\lambda + 32 = 0$$

By trial, $\lambda = 2$ is a root of this equation.

$$\Rightarrow \quad (\lambda - 2)(\lambda^2 - 10\lambda + 16) = 0 \Rightarrow (\lambda - 2)(\lambda - 2)(\lambda - 8) = 0$$

$\Rightarrow$ $\lambda = 2, 2, 8$ are the characteristic roots or Eigen values. **Ans.**

Example 2. *Find the eigen values of the matrix :*

$$\begin{bmatrix} 2 & 1 & 1 \\ 1 & 2 & 1 \\ 0 & 0 & 1 \end{bmatrix}$$

(R.G.P.V. Bhopal, I Semester, June 2007)

Solution. Let $A=\begin{bmatrix} 2 & 1 & 1 \\ 1 & 2 & 1 \\ 0 & 0 & 1 \end{bmatrix}$

The characteristic equation of A is

$$|A-\lambda I|=0$$

$$\begin{vmatrix} 2-\lambda & 1 & 1 \\ 1 & 2-\lambda & 1 \\ 0 & 0 & 1-\lambda \end{vmatrix}=0$$

Expanding the determinant with the help of third row, we have

$$\Rightarrow \quad (1-\lambda)\left[(2-\lambda)^2-1\right]=0 \quad \Rightarrow \quad (1-\lambda)\left(\lambda^2-4\lambda+4-1\right)=0$$

$$\Rightarrow \quad (1-\lambda)\left(\lambda^2-4\lambda+3\right)=0 \quad \Rightarrow \quad (1-\lambda)(\lambda-3)(\lambda-1)=0$$

$$\Rightarrow \quad \lambda=1,\ 1,\ 3$$

The eigen values of the given matrix are 1, 1 and 3. **Ans.**

Example 3. *The matrix A is defined as* $A=\begin{bmatrix} 1 & 2 & -3 \\ 0 & 3 & 2 \\ 0 & 0 & -2 \end{bmatrix}$

Find the eigen values of $3A^3+5A^2-6A+2I$.

Solution. $|A-\lambda I|=0$

$$\begin{vmatrix} 1-\lambda & 2 & -3 \\ 0 & 3-\lambda & 2 \\ 0 & 0 & -2-\lambda \end{vmatrix}=0$$

$\Rightarrow \quad (1-\lambda)(3-\lambda)(-2-\lambda)=0$ or $\lambda=1,\ 3,\ -2$

Eigen values of $A^3 = 1, 27, -8$; Eigen values of $A^2 = 1, 9, 4$

Eigen values of $A = 1, 3, -2$; Eigen values of $I = 1, 1, 1$

$\therefore$ Eigen values of $3A^3+5A^2-6A+2I$

First eigen value $= 3(1)^3+5(1)^3-6(1)+2(1) = 4$

Second eigen value $= 3(27)+5(9)-6(3)+2(1) = 110$

Third eigen value $= 3(-8)+5(4)-6(-2)+2(1) = 10$

Required eigen values are 4, 110, 10 **Ans.**

Example 4. *Show that for any square matrix A, the product of all the eigen values of A is equal to det (A), and the sum of all the eigen values of A is equal to the sum of the diagonal elements.*

Solution. Let $A=\begin{bmatrix} a_{11} & a_{12} & a_{13} \\ a_{21} & a_{22} & a_{23} \\ a_{31} & a_{32} & a_{33} \end{bmatrix}$, $|A-\lambda I| = \begin{vmatrix} a_{11}-\lambda & a_{12} & a_{13} \\ a_{21} & a_{22}-\lambda & a_{23} \\ a_{31} & a_{32} & a_{33}-\lambda \end{vmatrix}$

$|A-\lambda I|$

$= (a_{11}-\lambda)[(a_{22}-\lambda)(a_{33}-\lambda)-a_{32}a_{23}] - a_{12}[a_{21}(a_{33}-\lambda)-a_{31}a_{23}] +$

$a_{13}[a_{21}a_{32}-a_{31}(a_{22}-\lambda)]$

$$= (a_{11} - \lambda)\,[a_{22}\,a_{33} - (a_{22} + a_{33})\,\lambda + \lambda^2 - a_{32}\,a_{23}] - a_{12}\,[a_{21}\,a_{33} - a_{21}\,\lambda - a_{31}\,a_{23}] + a_{13}\,(a_{21}\,a_{32} - a_{31}a_{22} + a_{31}\,\lambda)$$

$$= a_{11}\,a_{22}\,a_{33} + (-\,a_{11}\,a_{22} - a_{11}\,a_{33})\,\lambda + a_{11}\,\lambda^2 - a_{11}\,a_{32}\,a_{23} + (-\,a_{22}\,a_{33} + a_{32}\,a_{23})\,\lambda + (a_{22} + a_{33})\,\lambda^2 - \lambda^3 - a_{12}\,a_{21}\,a_{33} + a_{12}\,a_{31}\,a_{23} + a_{12}\,a_{21}\,\lambda + a_{13}\,a_{21}\,a_{32} - a_{13}\,a_{31}\,a_{22} + a_{13}\,a_{31}\,\lambda$$

$$= -\,\lambda^3 + \lambda^2\,(a_{11} + a_{22} + a_{33}) + \lambda(-\,a_{11}\,a_{22} - a_{11}\,a_{33} + a_{12}\,a_{21} - a_{22}\,a_{33} + a_{23}\,a_{32} + a_{13}\,a_{31}) - [a_{11}\,(a_{22}\,a_{33} - a_{23}\,a_{32}) - a_{12}\,(a_{21}\,a_{33} - a_{23}\,a_{31}) + a_{13}\,(a_{21}\,a_{32} - a_{31}\,a_{22})] \quad ...(1)$$

If λ_1, λ_2, λ_3 be the roots of the equation (1) then

Sum of the roots $= \lambda_1 + \lambda_2 + \lambda_3 = a_{11} + a_{22} + a_{33}$ = Sum of the diagonal elements.

Product of the roots

$$= \lambda_1\lambda_2\lambda_3 = [a_{11}\,(a_{22}\,a_{33} - a_{23}\,a_{32}) - a_{12}\,(a_{21}\,a_{33} - a_{23}\,a_{31}) + a_{13}\,(a_{21}\,a_{32} - a_{31}\,a_{22})]$$

= Determinant A. **Proved.**

Example 5. *Let λ be an eigen value of a matrix A. Then prove that*

(*i*) $\lambda + k$ is an eigen value of $A + kI$

(*ii*) $k\lambda$ is an eigen value of kA. (*Gujarat, II Semester, June 2009*)

Solution. Here, A has eigen value λ. $\Rightarrow \; |A - \lambda\, I| = 0$ ... (1)

(*i*) Adding and subtracting kI from (1) we get

$$|A + kI - \lambda I - kI| = 0$$

$\Rightarrow \quad |(A + kI) - (\lambda + k)\, I| = 0 \Rightarrow \quad A + kI$ has $\lambda + k$ eigen value.

(*ii*) Multiplying (1), by k, we get

$$k\,|A - \lambda\, I| = 0 \qquad \Rightarrow \; |kA - k\lambda I| = 0$$

$\Rightarrow \quad kA$ has eigen value $k\lambda$. **Proved.**

Example 6. *If λ_1, λ_2, ... λ_n are the eigen values of A, find the eigen values of the martrix $(A - \lambda I)^2$.*

Solution. $(A - \lambda\, I)^2 = A^2 - 2\lambda\, AI + \lambda^2\, I^2 = A^2 - 2\,\lambda A + \lambda^2\, I$

Eigen values of A^2 are $\lambda_1^2, \; \lambda_2^2, \; \lambda_3^2 \; ... \; \lambda_n^2$

Eigen values of $2\lambda\, A$ are $2\lambda\, \lambda_1, \quad 2\lambda\, \lambda_2, \quad 2\lambda\, \lambda_3 \; ... \; 2\lambda\, \lambda_n$.

Eigen values of $\lambda^2\, I$ are λ^2.

$\therefore$ Eigen values of $A^2 - 2\lambda\, A + \lambda^2\, I$

$$\lambda_1^2 - 2\lambda\lambda_1 + \lambda^2, \quad \lambda_2^2 - 2\lambda\lambda_2 + \lambda^2, \quad \lambda_3^2 - 2\lambda\lambda_3 + \lambda^2 ......$$

$$\Rightarrow \quad (\lambda_1 - \lambda)^2, \quad (\lambda_2 - \lambda)^2, \quad (\lambda_3 - \lambda)^2, ... (\lambda_n - \lambda)^2$$ **Ans.**

Example 7. *Prove that a matrix A and its transpose A′ have the same characteristic roots.*

Solution. Characteristic equation of matrix A is

$$|A - \lambda I| = 0 \quad ... (1)$$

Characteristic equation of matrix A' is

$$|A' - \lambda I| = 0 \quad ...(2)$$

Clearly both (1) and (2) are same, as we know that

$$|A| = |A'|$$

i.e., a determinant remains unchanged when rows be changed into columns and columns into rows. **Proved.**

Example 8. *If A and P be square matrices of the same type and if P be invertible, show that the matrices A and $P^{-1}AP$ have the same characteristic roots.*

Solution. Let us put $B = P^{-1} AP$ and we will show that characteristic equations for both A and B are the same and hence they have the same characteristic roots.

$$B - \lambda I = P^{-1} AP - \lambda I = P^{-1} AP - P^{-1} \lambda IP = P^{-1} (A - \lambda I) P$$

$$\therefore \quad | B - \lambda I | = |P^{-1} (A - \lambda I) P | = | P^{-1} | |A - \lambda I| | P |$$

$$= |A - \lambda I | | P^{-1} | | P | = |A - \lambda I| | P^{-1}P|$$

$$= |A - \lambda I | | I | = | A - \lambda I| \text{ as } | I | = 1$$

Thus the matrices A and B have the same characteristic equations and hence the same characteristic roots. **Proved.**

Example 9. *If A and B be two square invertible matrices, then prove that AB and BA have the same characteristic roots.*

Solution. Now $\quad AB = IAB = B^{-1} B (AB) = B^{-1} (BA) B \quad$ ...(1)

But by Ex. 8, matrices BA and $B^{-1} (BA) B$ have same characteristic roots or matrices BA and AB by (1) have same characteristic roots. **Proved.**

Example 10. *If A and B be n rowed square matrices and if A be invertible, show that the matrices $A^{-1} B$ and BA^{-1} have the same characteristics roots.*

Solution. $\quad A^{-1} B = A^{-1} BI = A^{-1} B (A^{-1}A) = A^{-1} (BA^{-1}) A. \quad$ ...(1)

But by Ex. 8, matrices BA^{-1} and $A^{-1} (BA^{-1})A$ have same characteristic roots or matrices BA^{-1} and $A^{-1} B$ by (1) have same characteristic roots. **Proved.**

Example 11. *Show that 0 is a characteristic root of a matrix, if and only if, the matrix is singular.*

Solution. Characteristic equation of matrix A is given by

$$| A - \lambda I | = 0$$

If $\lambda = 0$, then from above it follows that $| A | = 0$ *i.e.* Matrix A is singular.

Again if Matrix A is singular *i.e.*, $| A | = 0$ then

$$| A - \lambda I | = 0 \Rightarrow | A | - \lambda | I | = 0, 0 - \lambda \cdot 1 = 0 \Rightarrow \lambda = 0.$$ **Proved.**

Example 12. *Show that characteristic roots of a triangular matrix are just the diagonal elements of the matrix.*

Solution. Let us consider the triangular matrix.

$$A = \begin{bmatrix} a_{11} & 0 & 0 & 0 \\ a_{21} & a_{22} & 0 & 0 \\ a_{31} & a_{32} & a_{33} & 0 \\ a_{41} & a_{42} & a_{43} & a_{44} \end{bmatrix}$$

Characteristic equation is $|A - \lambda I| = 0$

or $$\begin{vmatrix} a_{11} - \lambda & 0 & 0 & 0 \\ a_{21} & a_{22} - \lambda & 0 & 0 \\ a_{31} & a_{32} & a_{33} - \lambda & 0 \\ a_{41} & a_{42} & a_{43} & a_{44} - \lambda \end{vmatrix} = 0$$

On expansion it gives

$$(a_{11} - \lambda) (a_{22} - \lambda) (a_{33} - \lambda) (a_{44} - \lambda) = 0$$

$$\therefore \quad \lambda = a_{11}, \quad a_{22}, \quad a_{33}, \quad a_{44}$$

which are diagonal elements of matrix A. **Proved.**

Example 13. *If λ is an eigen value of an orthogonal matrix, then $\frac{1}{\lambda}$ is also eigen value.*

[**Hint:** $AA' = I$ if λ is the eigen value of A, then $\lambda^2 = 1, \quad \lambda = \frac{1}{\lambda}$]

Example 14. *Find the eigen values of the orthogonal matrix.*

$$B = \frac{1}{3}\begin{bmatrix} 1 & 2 & 2 \\ 2 & 1 & -2 \\ 2 & -2 & 1 \end{bmatrix}$$

Solution. The characteristic equation of

$$A = \begin{bmatrix} 1 & 2 & 2 \\ 2 & 1 & -2 \\ 2 & -2 & 1 \end{bmatrix} \text{ is } \begin{vmatrix} 1-\lambda & 2 & 2 \\ 2 & 1-\lambda & -2 \\ 2 & -2 & 1-\lambda \end{vmatrix} = 0$$

$$\Rightarrow \quad (1-\lambda)[(1-\lambda)(1-\lambda)-4]-2[2(1-\lambda)+4]+2[-4-2(1-\lambda)]=0$$

$$\Rightarrow \quad (1-\lambda)(1-2\lambda+\lambda^2-4)-2(2-2\lambda+4)+2(-4-2+2\lambda)=0$$

$$\Rightarrow \quad \lambda^3-3\lambda^2-9\lambda+27=0 \qquad \Rightarrow \qquad (\lambda-3)^2(\lambda+3)=0$$

The eigen values of A are 3, 3, –3, so the eigen values $B = \frac{1}{3}A$ are 1, 1, –1.

Note. If $\lambda = 1$ is an eigen value of B then its reciproc $\frac{1}{\lambda} = \frac{1}{1} = 1$ is also an eigen value of B. **Ans.**

EXERCISE 41.1

1. If λ be an eigen value of a non singular matrix A, show that $\frac{|A|}{\lambda}$ is an eigen value of matrix adj A.
2. There are infinitely many eigen vectors corresponding to a single eigen value.
3. Find the eigen values of the matrix $\begin{bmatrix} 2 & -3 & 1 \\ 3 & 1 & 3 \\ -5 & 2 & -4 \end{bmatrix}$ **Ans.** Eigen values are 0, + 1, –2
4. Find the eigen value $\begin{bmatrix} 4 & 2 & -2 \\ -5 & 3 & 2 \\ -2 & 4 & 1 \end{bmatrix}$. **Ans.** 1, 2, 5
5. Find the product of the eigen value of the matrix $\begin{bmatrix} 3 & -3 & 3 \\ 2 & 1 & 1 \\ 1 & 5 & 6 \end{bmatrix}$ **Ans.** 18
6. Find the sum of the eigen values of the matrix $\begin{bmatrix} 3 & 2 & 1 \\ 1 & 3 & 2 \\ 4 & 1 & 5 \end{bmatrix}$ **Ans.** 11
7. Find the eigen value of the inverse of the matrix $\begin{bmatrix} 4 & 6 & 6 \\ 1 & 3 & 2 \\ -1 & -4 & -3 \end{bmatrix}$ **Ans.** $-1, \frac{1}{4},$

8. Find the eigen value of the square of the matrix $\begin{bmatrix} 1 & 0 & -1 \\ 1 & 2 & 1 \\ 2 & 2 & 3 \end{bmatrix}$ **Ans.** 1, 4, 9

9. Find the eigen values of the matrix $\begin{bmatrix} 3 & 1 & 4 \\ 0 & 2 & 6 \\ 0 & 0 & 5 \end{bmatrix}^3$ **Ans.** 8, 27, 125

10. The sum and product of the eigen values of the matrix $A = \begin{bmatrix} 2 & 2 & 1 \\ 1 & 3 & 1 \\ 1 & 2 & 2 \end{bmatrix}$ are respectively

(*a*) 7 and 7 (*b*) 7 and 5 (*c*) 7 and 6 (*d*) 7 and 8 (*AMIETE, June 2010*) **Ans.** (*b*)

41.3 CAYLEY-HAMILTON THEOREM

Satement. Every square matrix satisfies its own characteristic equation.

If $|A - \lambda I| = (-1)^n \left(\lambda^n + a_1\lambda^{n-1} + a_2\lambda^{n-2} + \cdots + a_n\right)$ be the characteristic polynomial of $n \times n$ matrix $A = (a_{ij})$, then the matrix equation

$$X^n + a_1X^{n-1} + a_2X^{n-2} + \ldots + a_nI = 0 \text{ is satisfied by } X = A \text{ i.e}$$

$$A^n + a_1A^{n-1} + a_2A^{n-2} + \ldots + a_nI = 0$$

Proof. Since the elements of the matrix $A - \lambda I$ are at most of the first degree in λ, the elements of adj. $(A - \lambda I)$ are at most degree $(n - 1)$ in λ. Thus, adj. $(A - \lambda I)$ may be written as a matrix polynomial in λ, given by

$$Adj(A - \lambda I) = B_0\lambda^{n-1} + B_1\lambda^{n-2} + \ldots + B_{n-1}$$

where $B_0, B_1, \ldots, B_{n-1}$ are $n \times n$ matrices, their elements being polynomial in λ.

We know that $(A - \lambda I)\,Adj(A - \lambda I) = |A - \lambda I|\,I$

$$(A - \lambda I)\left(B_0\lambda^{n-1} + B_1\lambda^{n-2} + \ldots. + B_{n-1}\right) = (-1)^n\left(\lambda^n + a_1\lambda^{n-1} + \ldots + a_n\right)I$$

Equating coefficient of like power of λ on both sides, we get

$$-IB_0 = (-1)^n\, I$$

$$AB_0 - IB_1 = (-1)^n\, a_1I$$

$$AB_1 - IB_2 = (-1)^n\, a_2I$$

....

$$AB_{n-1} = (-1)^n\, a_nI$$

On multiplying the equation $A^n, A^{n-1}, \ldots, I$ respectively and adding, we obtain

$$0 = (-1)^n\left[A^n + a_1A^{n-1} + \ldots + a_nI\right]$$

Thus $\qquad A^n + a_1A^{n-1} + \ldots + a_nI = 0$

for example, Let A be square matrix and if $\lambda^3 - 2\lambda^2 + 3\lambda - 4 = 0$ ...(1)

be its characteristic equation, then according to Cayley Hamilton Theorem (1) is satisfied by A.

$$A^3 - 2A^2 + 3A - 4I = 0 \qquad ...(2)$$

We can find can A^{-1} from equation (2). On premultiplying equation (2) by A^{-1},we get

$$A^2 - 2A + 3I - 4A^{-1} = 0$$

$$A^{-1} = \frac{1}{4}[A^2 - 2A + 3I]$$

Example 15. *Verify Cayley-Hamilton theorem for the matrix*

$$A = \begin{pmatrix} 1 & 2 \\ 2 & -1 \end{pmatrix}$$ *and hence find* A^{-1}. (*U.P.,I Sem., Dec 2008*)

Solution. The characteristic equation of the matrix is $|A - \lambda I| = 0$

$$\begin{vmatrix} 1-\lambda & 2 \\ 2 & -1-\lambda \end{vmatrix} = 0$$

$$(1-\lambda)(-1-\lambda) - 4 = 0 \Rightarrow -1 + \lambda^2 - 4 = 0 \Rightarrow \lambda^2 - 5 = 0$$

By Cayley-Hamilton Theorem, $A^2 - 5I = 0$ ...(1)

Now, $A^2 = A.A = \begin{bmatrix} 1 & 2 \\ 2 & -1 \end{bmatrix}\begin{bmatrix} 1 & 2 \\ 2 & -1 \end{bmatrix} = \begin{bmatrix} 5 & 0 \\ 0 & 5 \end{bmatrix}$

$$A^2 - 5I = \begin{bmatrix} 5 & 0 \\ 0 & 5 \end{bmatrix} - 5\begin{bmatrix} 1 & 0 \\ 0 & 1 \end{bmatrix} = \begin{bmatrix} 5 & 0 \\ 0 & 5 \end{bmatrix} + \begin{bmatrix} -5 & 0 \\ 0 & -5 \end{bmatrix} + \begin{bmatrix} 0 & 0 \\ 0 & 0 \end{bmatrix} = 0 \qquad ....(2)$$

From (1) and (2), Cayley-Hamilton theorem is verified.

Again from (1), we have

$$A^2 - 5I = 0$$

Multiplying by A^{-1}, we get

$$A - 5A^{-1} = 0 \Rightarrow A^{-1} = \frac{1}{5}A \quad \Rightarrow A^{-1} = \frac{1}{5}\begin{bmatrix} 1 & 2 \\ 2 & -1 \end{bmatrix} = \begin{bmatrix} \frac{1}{5} & \frac{2}{5} \\ \frac{2}{5} & -\frac{1}{5} \end{bmatrix}$$ **Ans.**

Example 16. *Verify Cayley-Hamilton Theorem for the following matrix:*

$$A = \begin{bmatrix} 2 & -1 & 1 \\ -1 & 2 & -1 \\ 1 & -1 & 2 \end{bmatrix}$$

and use the theorem to find A^{-1}. (*Delhi University, April 2010*)

Solution. We have $A = \begin{bmatrix} 2 & -1 & 1 \\ -1 & 2 & -1 \\ 1 & -1 & 2 \end{bmatrix}$

Characteristic equation

$$|A - \lambda I| = 0$$

$$\begin{vmatrix} 2-\lambda & -1 & 1 \\ -1 & 2-\lambda & -1 \\ 1 & -1 & 2-\lambda \end{vmatrix} = 0$$

$$\Rightarrow (2-\lambda)(4 + \lambda^2 - 4\lambda - 1) + (\lambda - 2 + 1) + (1 + \lambda - 2) = 0$$

$\Rightarrow \quad 2\lambda^2 - 8\lambda + 6 - \lambda^3 + 4\lambda^2 - 3\lambda + 2\lambda - 2 = 0$

$\Rightarrow \quad -\lambda^3 + 6\lambda^2 - 9\lambda + 4 = 0$

$\Rightarrow \quad \lambda^3 - 6\lambda^2 + 9\lambda - 4 = 0$

By Cayley-Hamilton theorem

$$A^3 - 6A^2 + 9A - 4I = 0 \qquad \text{... (1)}$$

$$A^2 = \begin{bmatrix} 2 & -1 & 1 \\ -1 & 2 & -1 \\ 1 & -1 & 2 \end{bmatrix}\begin{bmatrix} 2 & -1 & 1 \\ -1 & 2 & -1 \\ 1 & -1 & 2 \end{bmatrix} = \begin{bmatrix} 6 & -5 & 5 \\ -5 & 6 & -5 \\ 5 & -5 & 6 \end{bmatrix}$$

$$A^3 = \begin{bmatrix} 6 & -5 & 5 \\ -5 & 6 & -5 \\ 5 & -5 & 6 \end{bmatrix}\begin{bmatrix} 2 & -1 & 1 \\ -1 & 2 & -1 \\ 1 & -1 & 2 \end{bmatrix} = \begin{bmatrix} 22 & -21 & 21 \\ -21 & 22 & -21 \\ 21 & -21 & 22 \end{bmatrix}$$

From equation (1), we get

$$\text{L.H.S.} = \begin{bmatrix} 22 & -21 & 21 \\ -21 & 22 & -21 \\ 21 & -21 & 22 \end{bmatrix} - 6\begin{bmatrix} 6 & -5 & 5 \\ -5 & 6 & -5 \\ 5 & -5 & 6 \end{bmatrix} + 9\begin{bmatrix} 2 & -1 & 1 \\ -1 & 2 & -1 \\ 1 & -1 & 2 \end{bmatrix} + \begin{bmatrix} -4 & 0 & 0 \\ 0 & -4 & 0 \\ 0 & 0 & -4 \end{bmatrix}$$

$$= \begin{bmatrix} 22-36+18-4 & -21+30-9 & 21-30+9 \\ -21+30-9 & 22-36+18-4 & -21+30-9 \\ 21-30+9 & -21+30-9 & 22-36+18-4 \end{bmatrix} = \begin{bmatrix} 0 & 0 & 0 \\ 0 & 0 & 0 \\ 0 & 0 & 0 \end{bmatrix}$$

$= \text{R.H.S.}$

Verify Cayley-Hamilton Theorem.

From (1), $\quad A^3 - 6A^2 + 9A - 4I = 0$

$\Rightarrow \quad A^2 - 6A + 9I - 4A^{-1} = 0$

$\Rightarrow \quad A^{-1} = \frac{1}{4}[A^2 - 6A + 9I]$

$$= \frac{1}{4}\left[\begin{bmatrix} 6 & -5 & 5 \\ -5 & 6 & -5 \\ 5 & -5 & 6 \end{bmatrix} - 6\begin{bmatrix} 2 & -1 & 1 \\ -1 & 2 & -1 \\ 1 & -1 & 2 \end{bmatrix} + \begin{bmatrix} 9 & 0 & 0 \\ 0 & 9 & 0 \\ 0 & 0 & 9 \end{bmatrix}\right]$$

$$= \frac{1}{4}\begin{bmatrix} 3 & 1 & -1 \\ 1 & 3 & 1 \\ -1 & 1 & 3 \end{bmatrix} \qquad \textbf{Ans.}$$

Example 17. *Find the characteristic equation of the matrix A.*

$$A = \begin{bmatrix} 4 & 3 & 1 \\ 2 & 1 & -2 \\ 1 & 2 & 1 \end{bmatrix}$$

Hence fiind A^{-1}. (*R.G.P.V., Bhopal, Feb. 2006*)

Solution Characteristic equation is

$$\begin{vmatrix} 4-\lambda & 3 & 1 \\ 2 & 1-\lambda & -2 \\ 1 & 2 & 1-\lambda \end{vmatrix} = 0$$

$$(4-\lambda)\left[1+\lambda^2-2\lambda+4\right]-3(2-2\lambda+2)+1\cdot(4-1+\lambda)=0$$

$$\Rightarrow \quad (4-\lambda)\left(\lambda^2-2\lambda+5\right)-3(-2\lambda+4)+(3+\lambda)=0$$

$$\Rightarrow \quad 4\lambda^2-8\lambda+20-\lambda^3+2\lambda^2-5\lambda+6\lambda-12+3+\lambda=0$$

$$\Rightarrow \quad -\lambda^3+6\lambda^2-6\lambda+11=0 \quad \text{or} \quad \lambda^3-6\lambda^2+6\lambda-11=0$$

By Cayley-Hamilton Theorem

$$A^3-6A^2+6A-11I=0 \qquad ...(1)$$

Multiplying (1) by A^{-1}, we get

$$A^2-6A+6I-11A^{-1}=0 \quad \text{or} \quad 11A^{-1}=A^2-6A+6I$$

$$11A^{-1}=\begin{bmatrix}4&3&1\\2&1&-2\\1&2&1\end{bmatrix}\begin{bmatrix}4&3&1\\2&1&-2\\1&2&1\end{bmatrix}-6\begin{bmatrix}4&3&1\\2&1&-2\\1&2&1\end{bmatrix}+6\begin{bmatrix}1&0&0\\0&1&0\\0&0&1\end{bmatrix}$$

$$\therefore \quad =\begin{bmatrix}23&17&-1\\8&3&-2\\9&7&-2\end{bmatrix}+\begin{bmatrix}-24&-18&-6\\-12&-6&12\\-6&-12&-6\end{bmatrix}+\begin{bmatrix}6&0&0\\0&6&0\\0&0&6\end{bmatrix}=\begin{bmatrix}5&-1&-7\\-4&3&10\\3&-5&-2\end{bmatrix} \quad \textbf{Ans.}$$

$$A^{-1}=\frac{1}{11}\begin{bmatrix}5&-1&-7\\-4&3&10\\3&-5&-2\end{bmatrix}$$

Example 18. *Find the characteristic equation of the matrix* $A=\begin{bmatrix}2&1&1\\0&1&0\\1&1&2\end{bmatrix}$

Verify Cayley Hamilton Theorem and hence prove that :

$$A^8-5A^7+7A^6-3A^5+A^4-5A^3+8A^2-2A+I=\begin{bmatrix}8&5&5\\0&3&0\\5&5&8\end{bmatrix}$$

(*Gujarat, II Semester, June 2009*)

Solution. Characteristic equation of the matrix A is

$$\begin{vmatrix}2-\lambda&1&1\\0&1-\lambda&0\\1&1&2-\lambda\end{vmatrix}=0$$

$$\Rightarrow \quad (2-\lambda)[(1-\lambda)(2-\lambda)]-1(0)+1(0-1+\lambda)=0$$

$$\Rightarrow \quad \lambda^3-5\lambda^2+7\lambda-3=0$$

According to Cayley-Hamilton Theorem

$$A^3-5A^2+7A-3I=0 \qquad ...(1)$$

We have to verify the equation (1).

$$A^2=\begin{bmatrix}2&1&1\\0&1&0\\1&1&2\end{bmatrix}\begin{bmatrix}2&1&1\\0&1&0\\1&1&2\end{bmatrix}=\begin{bmatrix}5&4&4\\0&1&0\\4&4&5\end{bmatrix}$$

$$A^3 = A^2.A = \begin{bmatrix} 2 & 1 & 1 \\ 0 & 1 & 0 \\ 1 & 1 & 2 \end{bmatrix}\begin{bmatrix} 5 & 4 & 4 \\ 0 & 1 & 0 \\ 4 & 4 & 5 \end{bmatrix} = \begin{bmatrix} 14 & 13 & 13 \\ 0 & 1 & 0 \\ 13 & 13 & 14 \end{bmatrix}$$

$$A^3 - 5A^2 + 7A - 3I = \begin{bmatrix} 14 & 13 & 13 \\ 0 & 1 & 0 \\ 13 & 13 & 14 \end{bmatrix} - 5\begin{bmatrix} 5 & 4 & 4 \\ 0 & 1 & 0 \\ 4 & 4 & 5 \end{bmatrix} + 7\begin{bmatrix} 2 & 1 & 1 \\ 0 & 1 & 0 \\ 1 & 1 & 2 \end{bmatrix} - 3\begin{bmatrix} 1 & 0 & 0 \\ 0 & 1 & 0 \\ 0 & 0 & 1 \end{bmatrix}$$

$$= \begin{bmatrix} 14-25+14-3 & 13-20+7+0 & 13-20+7+0 \\ 0+0+0+0 & 1-5+7-3 & 0-0+0-0 \\ 13-20+7+0 & 13-20+7-0 & 14-25+14-3 \end{bmatrix} = \begin{bmatrix} 0 & 0 & 0 \\ 0 & 0 & 0 \\ 0 & 0 & 0 \end{bmatrix} = 0$$

Hence Cayley Hamilton Theorem is verified.

Now, $A^8 - 5A^7 + 7A^6 - 3A^5 + A^4 - 5A^3 + 8A^2 - 2A + I$,

$$= A^5\left(A^3 - 5A^2 + 7A - 3I\right) + A\left(A^3 - 5A^2 + 7A - 3I\right) + A^2 + A + I$$

$$= A^5 \times O + A \times O + A^2 + A + I = A^2 + A + I$$

$$= \begin{bmatrix} 5 & 4 & 4 \\ 0 & 1 & 0 \\ 4 & 4 & 5 \end{bmatrix} + \begin{bmatrix} 2 & 1 & 1 \\ 0 & 1 & 0 \\ 1 & 1 & 2 \end{bmatrix} + \begin{bmatrix} 1 & 0 & 0 \\ 0 & 1 & 0 \\ 0 & 0 & 1 \end{bmatrix}$$

$$= \begin{bmatrix} 5+2+1 & 4+1+0 & 4+1+0 \\ 0+0+0 & 1+1+1 & 0+0+0 \\ 4+1+0 & 4+1+0 & 5+2+1 \end{bmatrix} = \begin{bmatrix} 8 & 5 & 5 \\ 0 & 3 & 0 \\ 5 & 5 & 8 \end{bmatrix}$$ **Proved.**

41.4 POWER OF MATRIX (by Cayley Hamilton Theorem)

Any positive integral power A^m of matrix A is linearly expressible in terms of those of lower degree, where m is a positive integer and n is the degree of characteristic equation such that $m > n$.

Example 19. *Find A^4 with the help of Cayley Hamilton Theorem, if*

$$A = \begin{bmatrix} 1 & 0 & -1 \\ 1 & 2 & 1 \\ 2 & 2 & 3 \end{bmatrix}$$

Solution. Here, we have $$A = \begin{bmatrix} 1 & 0 & -1 \\ 1 & 2 & 1 \\ 2 & 2 & 3 \end{bmatrix}$$

Characteristic equation of the matrix A is

$$\begin{vmatrix} 1-\lambda & 0 & -1 \\ 1 & 2-\lambda & 1 \\ 2 & 2 & 3-\lambda \end{vmatrix} = 0 \quad \Rightarrow \quad \lambda^3 - 6\lambda^2 - 11\lambda - 6 = 0$$

$$\Rightarrow \quad (\lambda-1)(\lambda-2)(\lambda-3) = 0$$

Eigen values of A are 1, 2, 3.

Let $\lambda^4 = (\lambda^3 - 6\lambda^2 - 11\lambda - 6)Q(\lambda) + (a\lambda^2 + b\lambda + c) = 0$ ...(1)

(where Q (λ) is quotient)

Put $\lambda = 1$ in (1), $(1)^4 = a + b + c \quad \Rightarrow \quad a + b + c = 1$ ...(2)

Put $\lambda = 2$ in (1), $(2)^4 = 4a + 2b + c \quad \Rightarrow \quad 4a + 2b + c = 16$ ... (3)

Put $\lambda = 3$ in (1), $(3)^4 = 9a + 3b + c \quad \Rightarrow \quad 9a + 3b + c = 81$ ... (4)

Solving (2), (3) and (4), we get

$a = 25, \quad b = -60, \quad c = 36$

Replacing λ by matrix A in (1), we get

$$A^4 = \left(A^3 - 6A^2 + 11A - 6\right)Q(A) + \left(aA^2 + bA + c\right)$$

$$= O + aA^2 + bA + cI$$

$$= 25\begin{bmatrix} 1 & 0 & -1 \\ 1 & 2 & 1 \\ 2 & 2 & 3 \end{bmatrix}\begin{bmatrix} 1 & 0 & -1 \\ 1 & 2 & 1 \\ 2 & 2 & 3 \end{bmatrix} + (-60)\begin{bmatrix} 1 & 0 & -1 \\ 1 & 2 & 1 \\ 2 & 2 & 3 \end{bmatrix} + 36\begin{bmatrix} 1 & 0 & 0 \\ 0 & 1 & 0 \\ 0 & 0 & 1 \end{bmatrix}$$

$$= \begin{bmatrix} -25 & -50 & -100 \\ 125 & 150 & 100 \\ 250 & 250 & 225 \end{bmatrix} + \begin{bmatrix} -60 & 0 & 60 \\ -60 & -120 & -60 \\ -120 & -120 & -180 \end{bmatrix} + \begin{bmatrix} 36 & 0 & 0 \\ 0 & 36 & 0 \\ 0 & 0 & 36 \end{bmatrix}$$

$$= \begin{bmatrix} -25-60+36 & -50+0+0 & -100+60+0 \\ 125-60+0 & 150-120+36 & 100-60+0 \\ 250-120+0 & 250-120+0 & 225-180+36 \end{bmatrix} = \begin{bmatrix} -49 & -50 & -40 \\ 65 & 66 & 40 \\ 130 & 130 & 81 \end{bmatrix}$$

EXERCISE 41.2

1. Find the characteristic polynomial of the matrix

$$A = \begin{bmatrix} 3 & 1 & 1 \\ -1 & 5 & -1 \\ 1 & -1 & 3 \end{bmatrix}$$

Verify Cayley-Hamilton Theorem for this matrix. Hence find A^{-1}. **Ans.** $A^{-1} = \dfrac{1}{20}\begin{bmatrix} 7 & -2 & -3 \\ 1 & 4 & 1 \\ -2 & 2 & 8 \end{bmatrix}$

2. Use Cayley-Hamilton Theorem to find the inverse of the matrix

$$\begin{bmatrix} \cos\theta & \sin\theta \\ -\sin\theta & \cos\theta \end{bmatrix}$$

Ans. $\begin{bmatrix} \cos\theta & -\sin\theta \\ \sin\theta & \cos\theta \end{bmatrix}$

3. Using Cayley-Hamilton Theorem, find A^{-1}, given that

$$A = \begin{bmatrix} 2 & -1 & 3 \\ 1 & 0 & 2 \\ 4 & -2 & 1 \end{bmatrix}$$

Ans. $-\dfrac{1}{5}\begin{bmatrix} 4 & -5 & -2 \\ 7 & -10 & -1 \\ -2 & 0 & 1 \end{bmatrix}$

4. Find the characteristic equation of the matrix

$$A = \begin{bmatrix} 1 & 3 & 7 \\ 4 & 2 & 3 \\ 1 & 2 & 1 \end{bmatrix}$$

and show that the equation is also satisfied by A. **Ans.** $\lambda^3 - 4\lambda^2 - 20\lambda - 35 = 0$.

5. Using, Cayley-Hamilton Theorem obtain the inverse of the matrix

$$\begin{bmatrix} 1 & 1 & 3 \\ 1 & 3 & -3 \\ -2 & -4 & -4 \end{bmatrix} \qquad \textbf{Ans. } \frac{1}{4}\begin{bmatrix} 12 & 4 & 6 \\ -5 & -1 & -3 \\ -1 & -1 & -1 \end{bmatrix}$$

6. Show that the matrix $A = \begin{bmatrix} 1 & -2 & 2 \\ 1 & 2 & 3 \\ 0 & -1 & 2 \end{bmatrix}$

satisfies its characteristic equation. Hence find A^{-1}. **Ans.** $\frac{1}{9}\begin{bmatrix} 7 & 2 & -10 \\ -2 & 2 & -1 \\ -1 & 1 & 4 \end{bmatrix}$

7. Verify Cayley-Hamilton Theorem for the matrix

$$A = \begin{bmatrix} 1 & 1 & 2 \\ 3 & 1 & 1 \\ 2 & 3 & 1 \end{bmatrix}$$

Hence evaluate A^{-1}. **Ans.** $\frac{1}{11}\begin{bmatrix} -2 & 5 & -1 \\ -1 & -3 & 5 \\ 7 & -1 & -2 \end{bmatrix}$

8. Verify Cayley-Hamilton theorem for the matrix

$$A = \begin{bmatrix} 1 & -2 & 3 \\ 2 & 4 & -2 \\ -1 & 1 & 2 \end{bmatrix}$$

9. Find adj. A by using Cayley-Hamilton thmeorem where A is given by

$$A = \begin{bmatrix} 1 & 2 & 1 \\ 0 & 1 & -1 \\ 3 & -1 & 1 \end{bmatrix}$$ *(R.G.P.V., Bhopal, April 2010)* **Ans.** $\begin{bmatrix} 0 & -3 & -3 \\ -3 & -2 & 1 \\ -3 & 7 & 1 \end{bmatrix}$

10. If a matrix $A = \begin{bmatrix} 1 & 0 & 0 \\ 0 & -1 & 0 \\ 1 & 0 & 1 \end{bmatrix}$, find the matrix A^{32}, using Cayley Hamilton Theorem. **Ans.** $\begin{bmatrix} 1 & 0 & 0 \\ 0 & 1 & 0 \\ 32 & 0 & 1 \end{bmatrix}$

41.5 CHARACTERISTIC VECTORS OR EIGEN VECTORS

As we have discussed in Art 41.2,

A column vector X is transformed into column vector Y by means of a square matrix A.

Now we want to multiply the column vector X by a scalar quantity λ so that we can find the same transformed column vector Y.

i.e., $AX = \lambda X$

X is known as eigen vector.

Example 20. *Show that the vector (1, 1, 2) is an eigen vector of the matrix*

$$A = \begin{bmatrix} 3 & 1 & -1 \\ 2 & 2 & -1 \\ 2 & 2 & 0 \end{bmatrix}$$ *corresponding to the eigen value 2.*

Solution. Let $X = (1, 1, 2)$.

Now, $$AX = \begin{bmatrix} 3 & 1 & -1 \\ 2 & 2 & -1 \\ 2 & 2 & 0 \end{bmatrix} \begin{bmatrix} 1 \\ 1 \\ 2 \end{bmatrix} = \begin{bmatrix} 3+1-2 \\ 2+2-2 \\ 2+2+0 \end{bmatrix} = \begin{bmatrix} 2 \\ 2 \\ 4 \end{bmatrix} = 2\begin{bmatrix} 1 \\ 1 \\ 2 \end{bmatrix} = 2X$$

Corresponding to each characteristic root λ, we have a corresponding non-zero vector X which satisfies the equation $[A - \lambda I]\, X = 0$.The non-zero vector X is called characteristic vector or Eigen vector.

41.6 PROPERTIES OF EIGEN VECTORS

1. The eigen vector X of a matrix A is not unique.

2. If $\lambda_1, \lambda_2, ..., \lambda_n$ be distinct eigen values of an $n \times n$ matrix then corresponding eigen vectors $X_1, X_2, ..., X_n$ form a linearly independent set.

3. If two or more eigen values are equal it may or may not be possible to get linearly independent eigen vectors corresponding to the equal roots.

4. Two eigen vectors X_1 and X_2 are called orthogonal vectors if $X_1' X_2 = 0$.

5. Eigen vectors of a symmetric matrix corresponding to different eigen values are orthogonal.

Normalised form of vectors. To find normalised form of $\begin{bmatrix} a \\ b \\ c \end{bmatrix}$, we divide each element by $\sqrt{a^2 + b^2 + c^2}$.

For example, normalised form of $\begin{bmatrix} 1 \\ 2 \\ 2 \end{bmatrix}$ is $\begin{bmatrix} 1/3 \\ 2/3 \\ 2/3 \end{bmatrix}$ $\left[\sqrt{1^2 + 2^2 + 2^2} = 3\right]$

41.7 ORTHOGONAL VECTORS

Two vectors X and Y are said to be orthogonal if $X_1^T X_2 = X_2^T X_1 = 0$.

Example 21. *Determine whether the eigen vectors of the matrix*

$$A = \begin{bmatrix} 1 & 0 & -1 \\ 1 & 2 & 1 \\ 2 & 2 & 3 \end{bmatrix}$$

are orthogonal.

Solution. Characteristic equation is

$$\begin{vmatrix} 1-\lambda & 0 & -1 \\ 1 & 2-\lambda & 1 \\ 2 & 2 & 3-\lambda \end{vmatrix} = 0$$

$\Rightarrow \quad (1-\lambda)[(2-\lambda)(3-\lambda) - 2] - 0 - 1[2 - 2(2-\lambda)] = 0$

$\Rightarrow \quad (1-\lambda)(6 - 5\lambda + \lambda^2 - 2) - (2 - 4 + 2\lambda) = 0 \Rightarrow (\lambda - 1)(\lambda^2 - 5\lambda + 4) + 2(\lambda - 1) = 0$

$\Rightarrow \quad (1-\lambda)(\lambda^2 - 5\lambda + 4) - 2(\lambda - 1) = 0 \qquad \Rightarrow (\lambda - 1)[\lambda^2 - 5\lambda + 4 + 2] = 0$

$\Rightarrow \quad (\lambda - 1)(\lambda^2 - 5\lambda + 6) = 0 \qquad \Rightarrow (\lambda - 1)(\lambda - 2)(\lambda - 3) = 0$

So, $\lambda = 1, 2, 3$ are three distinct eigen values of A.

For λ = 1

$$\begin{bmatrix} 1-\lambda & 0 & -1 \\ 1 & 2-\lambda & 1 \\ 2 & 2 & 3-\lambda \end{bmatrix}\begin{bmatrix} x_1 \\ x_2 \\ x_3 \end{bmatrix} = \begin{bmatrix} 0 \\ 0 \\ 0 \end{bmatrix}$$

$$\Rightarrow \begin{bmatrix} 1-1 & 0 & -1 \\ 1 & 2-1 & 1 \\ 2 & 2 & 3-1 \end{bmatrix}\begin{bmatrix} x_1 \\ x_2 \\ x_3 \end{bmatrix} = \begin{bmatrix} 0 \\ 0 \\ 0 \end{bmatrix} \Rightarrow \begin{bmatrix} 0 & 0 & -1 \\ 1 & 1 & 1 \\ 2 & 2 & 2 \end{bmatrix}\begin{bmatrix} x_1 \\ x_2 \\ x_3 \end{bmatrix} = \begin{bmatrix} 0 \\ 0 \\ 0 \end{bmatrix}$$

$$\Rightarrow \qquad x_3 = 0$$

$$x_1 + x_2 + x_3 = 0 \Rightarrow x_2 = -x_3 - x_1$$

Let $x_1 = k$ then $x_2 = 0 - k = -k$

$$X_1 = \begin{bmatrix} k \\ -k \\ 0 \end{bmatrix} \Rightarrow X_1 = k\begin{bmatrix} 1 \\ -1 \\ 0 \end{bmatrix}$$

For λ = 2

$$\begin{bmatrix} 1-2 & 0 & -1 \\ 1 & 2-2 & 1 \\ 2 & 2 & 3-2 \end{bmatrix}\begin{bmatrix} x_1 \\ x_2 \\ x_3 \end{bmatrix} = \begin{bmatrix} 0 \\ 0 \\ 0 \end{bmatrix} \Rightarrow \begin{bmatrix} -1 & 0 & -1 \\ 1 & 0 & 1 \\ 2 & 2 & 1 \end{bmatrix}\begin{bmatrix} x_1 \\ x_2 \\ x_3 \end{bmatrix} = \begin{bmatrix} 0 \\ 0 \\ 0 \end{bmatrix}$$

$$\Rightarrow \qquad x_1 + 0x_2 + x_3 = 0$$

$$2x_1 + 2x_2 + x_3 = 0$$

$$\frac{x_1}{0-2} = \frac{x_2}{2-1} = \frac{x_3}{2-0} = k$$

$$\Rightarrow \qquad x_1 = 2k, \quad x_2 = -k, \quad x_3 = -2k$$

$$X_2 = k\begin{bmatrix} 2 \\ -1 \\ -2 \end{bmatrix}$$

For λ = 3

$$\begin{bmatrix} 1-3 & 0 & -1 \\ 1 & 2-3 & 1 \\ 2 & 2 & 3-3 \end{bmatrix}\begin{bmatrix} x_1 \\ x_2 \\ x_3 \end{bmatrix} = \begin{bmatrix} 0 \\ 0 \\ 0 \end{bmatrix} \Rightarrow \begin{bmatrix} -2 & 0 & -1 \\ 1 & -1 & 1 \\ 2 & 2 & 0 \end{bmatrix}\begin{bmatrix} x_1 \\ x_2 \\ x_3 \end{bmatrix} = \begin{bmatrix} 0 \\ 0 \\ 0 \end{bmatrix}$$

$$\left.\begin{matrix} -2x_1 + 0x_2 - x_3 = 0 \\ x_1 - x_2 + x_3 = 0 \end{matrix}\right\} \Rightarrow \frac{x_1}{0-1} = \frac{x_2}{-1+2} = \frac{x_3}{2-0} = k$$

$$\Rightarrow \qquad x_1 = k, \quad x_2 = -k, \quad x_3 = -2k \; X_3 = \begin{bmatrix} x_1 \\ x_2 \\ x_3 \end{bmatrix} = \begin{bmatrix} k \\ -k \\ -2k \end{bmatrix} = k\begin{bmatrix} 1 \\ -1 \\ -2 \end{bmatrix}$$

$$X'_1 X_2 = [1, -1, \begin{bmatrix} 2 \\ -1 \\ -2 \end{bmatrix}] = 3, \quad X'_2 X_3 \; [2, -1, -2]\begin{bmatrix} 1 \\ -1 \\ -2 \end{bmatrix} = 7, = \; X'_3 X_1 = [1, -1, -2]\begin{bmatrix} 1 \\ -1 \\ 0 \end{bmatrix} = 2$$

Since $X_1^T X_2 = 3 \neq 0, \quad X_2^T X_3 = 7 \neq 0, \quad X_3^T X_1 = 2 \neq 0$

Thus, there are three distinct eigen vectors. So X_1, X_2, X_3 are not orthogonal eigen vectors.

41.8 NON-SYMMETRIC MATRICES WITH NON-REPEATED EIGEN VALUES

Example 22. *Show that if* $\lambda_1, \lambda_2, \ldots, \lambda_n$ *be the eigen values of the matrix A, then* A^n *has the eigen values* $\lambda_1^n, \lambda_2^n, \ldots \lambda_n^n$.

Solution. Let λ be an eigen value of the matrix A.

Therefore, $AX = \lambda X$ ...(1)

By premultiplying both sides of (1) by A^{n-1}, we get

$$A^{n-1}(AX) = A^{n-1}(\lambda X) \quad \Rightarrow \quad A^n X = \lambda(A^{n-1}X) \qquad \ldots (2)$$

But $A^2 X = A(AX) = A(\lambda X)$

$$= \lambda(AX) = \lambda(\lambda X) = \lambda^2 X \qquad \text{(From } (AX = \lambda X)$$

$$A^3 X = A(A^2 X) = \lambda(\lambda^2 X) = \lambda^3 X$$

Similarly, $A^4 X = \lambda^4 X$

....

....

$$A^n X = \lambda^n X$$

$\Rightarrow$ λ^n is an eigen value of A^n.

Hence, if $\lambda_1, \lambda_2, \ldots\ldots\lambda_n$ be the eigen values of A, then $\lambda_1^n, \lambda_2^n, \lambda_3^n, \ldots\ldots\ldots\ldots\lambda_n^n$ be the eigen values of A^n.

Example 23. *If* λ *be an eigen value of matrix A (non-zero matrix), show that* λ^{-1} *is an eigen value of* A^{-1}.

Solution. We hav λ, is an eigen value of matrix A.

$$AX = \lambda X \qquad \ldots (1)$$

where X is eigen vector

Premultiplying both sides of (1) by A^{-1}, we get

$$A^{-1}(AX) = A^{-1}(\lambda X) \quad \Rightarrow \quad (A^{-1}A)X = \lambda(A^{-1}X)$$

$$\Rightarrow \quad IX = \lambda(A^{-1}X) \quad \Rightarrow \quad X = \lambda(A^{-1}X)$$

$$\Rightarrow \quad \frac{1}{\lambda}X = A^{-1}X \quad \Rightarrow \quad A^{-1}X = \lambda^{-1}X$$

Henc λ^{-1}, is an eigen value of A^{-1}. **Proved.**

Example 24. *Find the eigen value and corresponding eigen vectors of the matrix*

$$A = \begin{pmatrix} -5 & 2 \\ 2 & -2 \end{pmatrix} \qquad \textit{(U.P.I Sem., Dec 2008)}$$

Solution. $|A - \lambda I| = 0$

$$\Rightarrow \begin{vmatrix} -5-\lambda & 2 \\ 2 & -2-\lambda \end{vmatrix} = 0 \Rightarrow (-5-\lambda)(-2-\lambda) - 4 = 0$$

$$\Rightarrow \lambda^2 + 7\lambda + 10 - 4 = 0 \Rightarrow \lambda^2 + 7\lambda + 6 = 0$$

$$(\lambda + 1)(\lambda + 6) = 0 \quad \Rightarrow \lambda = -1, -6$$

The eigen values of the given matrix are -1 and -6.

(*i*) When $\lambda = -1$, the corresponding eigen vectors are given by

$$\begin{bmatrix} -5+1 & 2 \\ 2 & -2+1 \end{bmatrix}\begin{bmatrix} x_1 \\ x_2 \end{bmatrix} = \begin{bmatrix} 0 \\ 0 \end{bmatrix} \Rightarrow \begin{bmatrix} -4 & 2 \\ 2 & -1 \end{bmatrix}\begin{bmatrix} x_1 \\ x_2 \end{bmatrix} = \begin{bmatrix} 0 \\ 0 \end{bmatrix}$$

$$\Rightarrow \quad 2x_1 - x_2 = 0 \Rightarrow x_1 = \frac{1}{2}x_2$$

Let $x_1 = k$, then $x_2 = 2k$, Hence, eigen vector $X_1 = \begin{bmatrix} k \\ 2k \end{bmatrix}$

(*ii*) When $\lambda = -6$, the corresponding eigen vectors are given by

$$\begin{bmatrix} -5+6 & 2 \\ 2 & -2+6 \end{bmatrix}\begin{bmatrix} x_1 \\ x_2 \end{bmatrix} = \begin{bmatrix} 0 \\ 0 \end{bmatrix} \Rightarrow \begin{bmatrix} 1 & 2 \\ 2 & 4 \end{bmatrix}\begin{bmatrix} x_1 \\ x_2 \end{bmatrix} = \begin{bmatrix} 0 \\ 0 \end{bmatrix}$$

$$\Rightarrow \quad x_1 + 2x_2 = 0 \qquad \Rightarrow x_1 = -2x_2$$

Let $x_1 = k_1$, then $x_2 = -\frac{1}{2}k_1$

Hence eigen vector $X_2 = \begin{bmatrix} k_1 \\ -\frac{k_1}{2} \end{bmatrix}$ or $\begin{bmatrix} 2k_1 \\ -k_1 \end{bmatrix}$

Hence eigen vectors are $\begin{bmatrix} k \\ 2k \end{bmatrix}$ and $\begin{bmatrix} 2k_1 \\ -k_1 \end{bmatrix}$ **Ans.**

Example 25. *Find the eigen values and eigen vectors of matrix* $A = \begin{bmatrix} 3 & 1 & 4 \\ 0 & 2 & 6 \\ 0 & 0 & 5 \end{bmatrix}$

(*AMIETE, June 2010, 2009*)

Solution. $|A - \lambda I| = \begin{vmatrix} 3-\lambda & 1 & 4 \\ 0 & 2-\lambda & 6 \\ 0 & 0 & 5-\lambda \end{vmatrix} = (3-\lambda)(2-\lambda)(5-\lambda)$

Hence the characteristic equation of matrix A is given by

$$|A - \lambda I| = 0 \quad \Rightarrow \quad (3-\lambda)(2-\lambda)(5-\lambda) = 0$$

$\therefore \quad \lambda = 2, 3, 5.$

Thus the eigen values of matrix A are 2, 3, 5.

The eigen vectors of the matrix A corresponding to the eigen value λ is given by the non-zero solution of the equati $(A - \lambda I)X = 0$.

or
$$\begin{bmatrix} 3-\lambda & 1 & 4 \\ 0 & 2-\lambda & 6 \\ 0 & 0 & 5-\lambda \end{bmatrix}\begin{bmatrix} x_1 \\ x_2 \\ x_3 \end{bmatrix} = \begin{bmatrix} 0 \\ 0 \\ 0 \end{bmatrix} \quad \text{... (1)}$$

When $\lambda = 2$, the corresponding eigen vector is given by

$$\begin{bmatrix} 3-2 & 1 & 4 \\ 0 & 2-2 & 6 \\ 0 & 0 & 5-2 \end{bmatrix}\begin{bmatrix} x_1 \\ x_2 \\ x_3 \end{bmatrix} = \begin{bmatrix} 0 \\ 0 \\ 0 \end{bmatrix}$$

$$\Rightarrow \quad \begin{bmatrix} 1 & 1 & 4 \\ 0 & 0 & 6 \\ 0 & 0 & 3 \end{bmatrix}\begin{bmatrix} x_1 \\ x_2 \\ x_3 \end{bmatrix} = \begin{bmatrix} 0 \\ 0 \\ 0 \end{bmatrix}$$

$$x_1 + x_2 + 4x_3 = 0$$

$$\Rightarrow \quad 0x_1 + 0x_2 + 6x_3 = 0$$

$$\frac{x_1}{6-0} = \frac{x_2}{0-6} = \frac{x_3}{0-0} = k \quad \Rightarrow \quad \frac{x_1}{1} = \frac{x_2}{-1} = \frac{x_3}{0} = k \quad \Rightarrow \quad x_1 = k,\ x_2 = -k,\ x_3 = 0$$

Hence $X_1 = \begin{bmatrix} k \\ -k \\ 0 \end{bmatrix} = k\begin{bmatrix} 1 \\ -1 \\ 0 \end{bmatrix}$ can be taken as an eigen vector of A corresponding to the eigen value $\lambda = 2$

When $\lambda = 3$, substituting in (1), the corresponding eigen vector is given by

$$\begin{bmatrix} 3-3 & 1 & 4 \\ 0 & 2-3 & 6 \\ 0 & 0 & 5-3 \end{bmatrix}\begin{bmatrix} x_1 \\ x_2 \\ x_3 \end{bmatrix} = \begin{bmatrix} 0 \\ 0 \\ 0 \end{bmatrix} \quad \Rightarrow \quad \begin{bmatrix} 0 & 1 & 4 \\ 0 & -1 & 6 \\ 0 & 0 & 2 \end{bmatrix}\begin{bmatrix} x_1 \\ x_2 \\ x_3 \end{bmatrix} = \begin{bmatrix} 0 \\ 0 \\ 0 \end{bmatrix}$$

$$0x_1 + x_2 + 4x_3 = 0$$

$$0x_1 - x_2 + 6x_3 = 0$$

$$\frac{x_1}{6+4} = \frac{x_2}{0-0} = \frac{x_3}{0-0} \quad \Rightarrow \quad \frac{x_1}{10} = \frac{x_2}{0} = \frac{x_3}{0} = \frac{k}{10}$$

$$x_1 = k,\ x_2 = 0,\ x_3 = 0$$

Hence, $X_2 = \begin{bmatrix} k \\ 0 \\ 0 \end{bmatrix} = k\begin{bmatrix} 1 \\ 0 \\ 0 \end{bmatrix}$ can be taken as an eigen vector of A corresponding to the eigen value $\lambda = 3$.

When $\lambda = 5$.

Again, when $\lambda = 5$, substituting in (1), the corresponding eigen vector is given by

$$\begin{bmatrix} 3-5 & 1 & 4 \\ 0 & 2-5 & 6 \\ 0 & 0 & 5-5 \end{bmatrix}\begin{bmatrix} x_1 \\ x_2 \\ x_3 \end{bmatrix} = \begin{bmatrix} 0 \\ 0 \\ 0 \end{bmatrix} \quad \Rightarrow \quad \begin{bmatrix} -2 & 1 & 4 \\ 0 & -3 & 6 \\ 0 & 0 & 0 \end{bmatrix}\begin{bmatrix} x_1 \\ x_2 \\ x_3 \end{bmatrix} = \begin{bmatrix} 0 \\ 0 \\ 0 \end{bmatrix}$$

$$-2x_1 + x_2 + 4x_3 = 0$$

$$-3x_2 + 6x_3 = 0$$

By cross-multiplication method, we have

$$\frac{x_1}{6+12} = \frac{x_2}{0+12} = \frac{x_3}{6-0} \quad \Rightarrow \quad \frac{x_1}{18} = \frac{x_2}{12} = \frac{x_3}{6} \quad \Rightarrow \quad \frac{x_1}{3} = \frac{x_2}{2} = \frac{x_3}{1} = k$$

$$x_1 = 3k, \quad x_2 = 2k, \quad x_3 = k$$

Hence, $X_3 = \begin{bmatrix} 3k \\ 2k \\ k \end{bmatrix} = k\begin{bmatrix} 3 \\ 2 \\ 1 \end{bmatrix}$ can be taken as an eigen vector of A corresponding to the eigen value $\lambda = 5$.

Ans.

EXERCISE 41.3

Non-symmetric matrix with different eigen values:

Find the eigen values and the corresponding eigen vectors for the following matrices:

1. $\begin{bmatrix} 4 & 2 & -2 \\ -5 & 3 & 2 \\ -2 & 4 & 1 \end{bmatrix}$ **Ans.** $1, 2, 5; \begin{bmatrix} 2 \\ 1 \\ 4 \end{bmatrix}, \begin{bmatrix} 1 \\ 1 \\ 2 \end{bmatrix}, \begin{bmatrix} 0 \\ 1 \\ 1 \end{bmatrix}$

2. $\begin{bmatrix} 2 & -2 & 3 \\ 1 & 1 & 1 \\ 1 & 3 & -1 \end{bmatrix}$ **Ans.** $-2, 1, 3; \begin{bmatrix} 11 \\ 1 \\ 14 \end{bmatrix}, \begin{bmatrix} -1 \\ 1 \\ 1 \end{bmatrix}, \begin{bmatrix} 1 \\ 1 \\ 1 \end{bmatrix}$

3. $\begin{bmatrix} -9 & 2 & 6 \\ 5 & 0 & -3 \\ -16 & 4 & 11 \end{bmatrix}$ **Ans.** $-1, 1, 2; \begin{bmatrix} 2 \\ -1 \\ 3 \end{bmatrix}, \begin{bmatrix} 1 \\ -1 \\ 2 \end{bmatrix}, \begin{bmatrix} 2 \\ -1 \\ 4 \end{bmatrix}$

4. $\begin{bmatrix} 4 & 6 & 6 \\ 1 & 3 & 2 \\ -1 & -4 & -3 \end{bmatrix}$ **Ans.** $-1, 1, 4; \begin{bmatrix} -6 \\ -2 \\ 7 \end{bmatrix}, \begin{bmatrix} 0 \\ 1 \\ -1 \end{bmatrix}, \begin{bmatrix} 3 \\ 1 \\ -1 \end{bmatrix}$

41.9 NON-SYMMETRIC MATRIX WITH REPEATED EIGEN VALUES

Example 26. *Find the eigen values and eigen vectors of the matrix:*

$$\begin{bmatrix} 2 & 1 & 1 \\ 1 & 2 & 1 \\ 0 & 0 & 1 \end{bmatrix}$$

Solution. We have, $A = \begin{bmatrix} 2 & 1 & 1 \\ 1 & 2 & 1 \\ 0 & 0 & 1 \end{bmatrix}$.

Characteristic equation of A is $|A - \lambda I| = 0$

$$\begin{vmatrix} 2-\lambda & 1 & 1 \\ 1 & 2-\lambda & 1 \\ 0 & 0 & 1-\lambda \end{vmatrix} = 0$$

On expanding the determinant by the third row, we get

$$\Rightarrow \quad (1-\lambda)\{(2-\lambda)(2-\lambda)-1\} = 0 \quad \Rightarrow \quad (1-\lambda)\{(2-\lambda)^2-1\} = 0$$

$$\Rightarrow \quad (1-\lambda)(2-\lambda+1)(2-\lambda-1) = 0 \quad \Rightarrow \quad (1-\lambda)(3-\lambda)(1-\lambda) = 0$$

$$\Rightarrow \quad \lambda = 1, 1, 3$$

when $\lambda = 1$

$$\begin{bmatrix} 2-1 & 1 & 1 \\ 1 & 2-1 & 1 \\ 0 & 0 & 1-1 \end{bmatrix}\begin{bmatrix} x \\ y \\ z \end{bmatrix} = \begin{bmatrix} 0 \\ 0 \\ 0 \end{bmatrix} \quad \Rightarrow \quad \begin{bmatrix} 1 & 1 & 1 \\ 1 & 1 & 1 \\ 0 & 0 & 0 \end{bmatrix}\begin{bmatrix} x \\ y \\ z \end{bmatrix} = \begin{bmatrix} 0 \\ 0 \\ 0 \end{bmatrix}$$

$$\begin{bmatrix} 1 & 1 & 1 \\ 0 & 0 & 0 \\ 0 & 0 & 0 \end{bmatrix}\begin{bmatrix} x \\ y \\ z \end{bmatrix} = \begin{bmatrix} 0 \\ 0 \\ 0 \end{bmatrix} R_2 \to R_2 - R_1 \quad \Rightarrow \quad x + y + z = 0$$

Let $x = k_1$ and $y = k_2$

$$k_1 + k_2 + z = 0 \quad \Rightarrow \quad z = -(k_1 + k_2)$$

$$X_1 = \begin{bmatrix} k_1 \\ k_2 \\ -(k_1+k_2) \end{bmatrix} = k\begin{bmatrix} 1 \\ 1 \\ -2 \end{bmatrix} \qquad \text{[If } k_1 = k_2 = k\text{]}$$

Again $\lambda = 1$, $X_2 = \begin{bmatrix} 1 \\ 0 \\ -1 \end{bmatrix}$ [Again if $k_1 = 1,\ k_2 = 0,\ -(k_1 + k_2) = -1$]

when $\lambda = 3$

$$\begin{bmatrix} 2-3 & 1 & 1 \\ 1 & 2-3 & 1 \\ 0 & 0 & 1-3 \end{bmatrix}\begin{bmatrix} x \\ y \\ z \end{bmatrix} = \begin{bmatrix} 0 \\ 0 \\ 0 \end{bmatrix} \quad \Rightarrow \quad \begin{bmatrix} -1 & 1 & 1 \\ 1 & -1 & 1 \\ 0 & 0 & -2 \end{bmatrix}\begin{bmatrix} x \\ y \\ z \end{bmatrix} = \begin{bmatrix} 0 \\ 0 \\ 0 \end{bmatrix}$$

$$\begin{bmatrix} -1 & 1 & 1 \\ 0 & 0 & 2 \\ 0 & 0 & -2 \end{bmatrix}\begin{bmatrix} x \\ y \\ z \end{bmatrix} = \begin{bmatrix} 0 \\ 0 \\ 0 \end{bmatrix} \quad R_2 \to R_2 + R_1$$

$$-x + y + z = 0$$

$$2z = 0 \quad \Rightarrow \quad z = 0$$

$$-x + y + 0 = 0 \quad \Rightarrow \quad x = y = k \text{ (say)}$$

$$X_3 = \begin{bmatrix} k \\ k \\ 0 \end{bmatrix} = k\begin{bmatrix} 1 \\ 1 \\ 0 \end{bmatrix}$$

Ans.

Example 27. *Find all the Eigen values and Eigen vectors of the matrix*

$$A = \begin{bmatrix} -2 & 2 & -3 \\ 2 & 1 & -6 \\ -1 & -2 & 0 \end{bmatrix}$$

(*AMIETE, Dec. 2009*)

Solution. Characteristic equation of A is

$$\begin{vmatrix} -2-\lambda & 2 & -3 \\ 2 & 1-\lambda & -6 \\ -1 & -2 & 0-\lambda \end{vmatrix} = 0$$

$$\Rightarrow \quad (-2-\lambda)[-\lambda+\lambda^2-12]-2(-2\lambda-6)-3(-4+1-\lambda)=0$$

$$\lambda^3+\lambda^2-21\lambda-45=0 \qquad \text{...(1)}$$

By trial: If $\lambda = -3$, then $-27 + 9 + 63 - 45 = 0$, so $(\lambda + 3)$ is one factor of (1).

The remaining factors are obtained on dividing equation (1) by $\lambda + 3$.

−3	1	1	−21	−45
		−3	6	45
	1	−2	−15	0

$\lambda^2 - 2\lambda - 15 = 0 \quad \Rightarrow \quad (\lambda - 5)(\lambda + 3) = 0$

$\Rightarrow \quad (\lambda+3)(\lambda+3)(\lambda-5)=0 \quad \Rightarrow \lambda = 5, -3, -3$

To find the eigen vectors for corresponding eigen values, we will consider the matrix equation

$$(A-\lambda I)X = 0 \qquad i.e., \begin{bmatrix} -2-\lambda & 2 & -3 \\ 2 & 1-\lambda & -6 \\ -1 & -2 & 0-\lambda \end{bmatrix}\begin{bmatrix} x \\ y \\ z \end{bmatrix} = \begin{bmatrix} 0 \\ 0 \\ 0 \end{bmatrix} \qquad \text{... (2)}$$

On putting $\lambda = 5$ in eq^n. (2), it becomes $\begin{bmatrix} -7 & 2 & -3 \\ 2 & -4 & -6 \\ -1 & -2 & -5 \end{bmatrix}\begin{bmatrix} x \\ y \\ z \end{bmatrix} = \begin{bmatrix} 0 \\ 0 \\ 0 \end{bmatrix}$

We have $\quad -7x + 2y - 3z = 0,$

$\quad 2x - 4y - 6z = 0$

$$\frac{x}{-12-12} = \frac{y}{-6-42} = \frac{z}{28-4} \quad \text{or} \quad \frac{x}{-24} = \frac{y}{-48} = \frac{z}{24} \quad \text{or} \quad \frac{x}{1} = \frac{y}{2} = \frac{z}{-1} = k$$

$$x = k, \quad y = 2k, \quad z = -k$$

Hence, the eigen vecter $X_1 = \begin{bmatrix} k \\ 2k \\ -k \end{bmatrix} = k\begin{bmatrix} 1 \\ 2 \\ -1 \end{bmatrix}$

Put $\lambda = -3$ in eq^n. (2), it becomes $\begin{bmatrix} 1 & 2 & -3 \\ 2 & 4 & -6 \\ -1 & -2 & 3 \end{bmatrix}\begin{bmatrix} x \\ y \\ z \end{bmatrix} = \begin{bmatrix} 0 \\ 0 \\ 0 \end{bmatrix}$

We have $\quad x + 2y - 3z = 0,$

$\quad 2x + 4y - 6z = 0,$

$\quad -x - 2y + 3z = 0$

Here first, second and third equations are the same.

Let $x = k_1$, $y = k_2$ then $z = \frac{1}{3}(k_1 + 2k_2)$

Hence, the eigen vector is $\begin{bmatrix} k_1 \\ k_2 \\ \frac{1}{3}(k_1+2k_2) \end{bmatrix}$

Let $k_1 = 0$, $k_2 = 3$, Hence $X_2 = \begin{bmatrix} 0 \\ 3 \\ 2 \end{bmatrix}$

Since the matrix is non-symmetric, the corresponding eigen vectors X_2 and X_3 must be linearly independent. This can be done by choosing

$$k_1 = 3, \; k_2 = 0, \text{ and Hence } X_3 = \begin{bmatrix} 3 \\ 0 \\ 1 \end{bmatrix}$$

Hence, $\quad X_1 = \begin{bmatrix} 1 \\ 2 \\ -1 \end{bmatrix}, \; X_2 = \begin{bmatrix} 0 \\ 3 \\ 2 \end{bmatrix}, \quad X_3 = \begin{bmatrix} 3 \\ 0 \\ 1 \end{bmatrix}.$ **Ans.**

Example 28. *Find the eigen values and eigen vectors of the following matrix*

$$A = \begin{bmatrix} 1 & -3 & 3 \\ 3 & -5 & 3 \\ 6 & -6 & 4 \end{bmatrix}$$ (*Delhi University, April 2010*)

Solution. We have $$A = \begin{bmatrix} 1 & -3 & 3 \\ 3 & -5 & 3 \\ 6 & -6 & 4 \end{bmatrix}$$

The characteristic equation of A is

$\Rightarrow \quad |A - \lambda I| = 0$

$$\begin{vmatrix} 1-\lambda & -3 & 3 \\ 3 & -5-\lambda & 3 \\ 6 & -6 & 4-\lambda \end{vmatrix} = 0$$

$\Rightarrow (1-\lambda)(\lambda^2 + \lambda - 20 + 18) + 3(12 - 3\lambda - 18) + 3(-18 + 30 + 6\lambda) = 0$

$\Rightarrow (1 - \lambda)(\lambda^2 + \lambda - 2) - 9\lambda - 18 + 18\lambda + 36 = 0$

$\Rightarrow \lambda^2 + \lambda - 2 - \lambda^3 - \lambda^2 + 2\lambda + 9\lambda + 18 = 0$

$\Rightarrow \lambda^3 - 12\lambda - 16 = 0 \quad \Rightarrow (\lambda - 4)(\lambda + 2)(\lambda + 2) = 0$

$\lambda = -2, -2, 4$

When $\lambda = -2$

$$[A + 2I]\, X_1 = 0$$

$$\begin{bmatrix} 3 & -3 & 3 \\ 3 & -3 & 3 \\ 6 & -6 & 6 \end{bmatrix} \begin{bmatrix} x_1 \\ x_2 \\ x_3 \end{bmatrix} = \begin{bmatrix} 0 \\ 0 \\ 0 \end{bmatrix}$$

$$3x_1 - 3x_2 + 3x_3 = 0$$

$$x_1 - x_2 + x_3 = 0$$

If $\quad x_1 = x_2 = 1, \quad x_3 = 0$

and $\quad x_1 = 1, \ x_2 = -1, \ x_3 = -2$

$$\therefore X_1 = \begin{bmatrix} 1 \\ 1 \\ 0 \end{bmatrix} \text{ and } X_2 = \begin{bmatrix} 1 \\ -1 \\ -2 \end{bmatrix}$$

When $\lambda = 4$

$$[A - 4I]\, X_3 = 0$$

$$\begin{bmatrix} -3 & -3 & 3 \\ 3 & -9 & 3 \\ 6 & -6 & 0 \end{bmatrix} \begin{bmatrix} x_1 \\ x_2 \\ x_3 \end{bmatrix} = \begin{bmatrix} 0 \\ 0 \\ 0 \end{bmatrix}$$

$\Rightarrow \quad -x_1 - x_2 + x_3 = 0$

and $\quad x_1 - 3x_2 + x_3 = 0$

$$\frac{x_1}{-1+3} = \frac{x_2}{1+1} = \frac{x_3}{3+1} \Rightarrow \frac{x_1}{2} = \frac{x_2}{2} = \frac{x_3}{4} \Rightarrow \frac{x_1}{1} = \frac{x_2}{1} = \frac{x_3}{2}$$

$$\therefore \quad X_3 = \begin{bmatrix} 1 \\ 1 \\ 2 \end{bmatrix}$$ **Ans.**

EXERCISE 41.4

Non-symmetric matrices with repeated eigen values

Find the eigen values and eigen vectors of the following matrices:

1. $\begin{bmatrix} 2 & -2 & 2 \\ 1 & 1 & 1 \\ 1 & 3 & -1 \end{bmatrix}$ **Ans.** $-2, 2, 2;$ $\begin{bmatrix} -4 \\ -1 \\ 7 \end{bmatrix}, \begin{bmatrix} 0 \\ 1 \\ 1 \end{bmatrix}$ **2.** $\begin{bmatrix} 2 & 2 & 1 \\ 1 & 3 & 1 \\ 1 & 2 & 2 \end{bmatrix}$ **Ans.** $1, 1, 5;$ $\begin{bmatrix} 0 \\ 1 \\ -2 \end{bmatrix}, \begin{bmatrix} 1 \\ 0 \\ -1 \end{bmatrix}$

3. $\begin{bmatrix} 2 & 1 & 1 \\ 2 & 3 & 2 \\ 3 & 3 & 4 \end{bmatrix}$ **Ans.** $1, 1, 7;$ $\begin{bmatrix} 0 \\ 1 \\ -1 \end{bmatrix}, \begin{bmatrix} 1 \\ 0 \\ -1 \end{bmatrix}, \begin{bmatrix} 1 \\ 2 \\ 3 \end{bmatrix}$ **4.** $\begin{bmatrix} -9 & 4 & 4 \\ -8 & 3 & 4 \\ -16 & 8 & 7 \end{bmatrix}$ **Ans.** $-1, -1, 3;$ $\begin{bmatrix} 0 \\ 1 \\ -1 \end{bmatrix}, \begin{bmatrix} 1 \\ 1 \\ 1 \end{bmatrix}, \begin{bmatrix} 1 \\ 1 \\ 2 \end{bmatrix}$

5. $\begin{bmatrix} 1 & 1 & 0 \\ 0 & 1 & 0 \\ 0 & 0 & 1 \end{bmatrix}$ *(AMIETE, Dec. 2010)* **Ans.** $1, 1, 1,$ $\begin{bmatrix} 1 \\ 0 \\ 1 \end{bmatrix}$

41.10 SYMMETRIC MATRICES WITH NON REPEATED EIGEN VALUES

Example 29. *Find the eigen values and the corresponding eigen vectors of the matrix*

$$\begin{bmatrix} -2 & 5 & 4 \\ 5 & 7 & 5 \\ 4 & 5 & -2 \end{bmatrix}$$

Solution. $|A - \lambda I| = 0$

$$\begin{vmatrix} -2-\lambda & 5 & 4 \\ 5 & 7-\lambda & 5 \\ 4 & 5 & -2-\lambda \end{vmatrix} = 0 \quad \Rightarrow \quad \lambda^3 - 3\lambda^2 - 90\lambda - 216 = 0$$

By trial: Take $\lambda = -3$, then $-27 - 27 + 270 - 216 = 0$

By synthetic division

$$\begin{array}{r|rrrr} -3 & 1 & -3 & -90 & -216 \\ & & -3 & 18 & 216 \\ \hline & 1 & -6 & -72 & 0 \end{array}$$

$$\lambda^2 - 6\lambda - 72 = 0 \;\Rightarrow\; (\lambda - 12)(\lambda + 6) = 0 \;\Rightarrow\; \lambda = -3, -6, \; 12$$

Matrix equation for eigen vectors $[A - \lambda I]\, X = 0$

$$\begin{bmatrix} -2-\lambda & 5 & 4 \\ 5 & 7-\lambda & 5 \\ 4 & 5 & -2-\lambda \end{bmatrix} \begin{bmatrix} x \\ y \\ z \end{bmatrix} = \begin{bmatrix} 0 \\ 0 \\ 0 \end{bmatrix} \qquad ...(1)$$

Eigen Vector

On putting $\lambda = -3$ in (1), it will become

$$\begin{bmatrix} 1 & 5 & 4 \\ 5 & 10 & 5 \\ 4 & 5 & 1 \end{bmatrix} \begin{bmatrix} x \\ y \\ z \end{bmatrix} = \begin{bmatrix} 0 \\ 0 \\ 0 \end{bmatrix} \quad \Rightarrow \quad \begin{cases} x + 5y + 4z = 0 \\ 5x + 10y + 5z = 0 \end{cases}$$

$$\frac{x}{25-40} = \frac{y}{5-20} = \frac{z}{10-25} \qquad \text{or} \qquad \frac{x}{1} = \frac{y}{-1} = \frac{z}{1}$$

Eigen vector $X_1 = \begin{bmatrix} 1 \\ -1 \\ 1 \end{bmatrix}$.

Eigen vector corresponding to eigen value $\lambda = 6$.

Equation (1) becomes

$$\begin{bmatrix} 4 & 5 & 4 \\ 5 & 13 & 5 \\ 4 & 5 & 4 \end{bmatrix}\begin{bmatrix} x \\ y \\ z \end{bmatrix} = \begin{bmatrix} 0 \\ 0 \\ 0 \end{bmatrix} \quad \text{or} \quad \begin{cases} 4x+5y+4z=0 \\ 5x+13y+5z=0 \end{cases}$$

$$\frac{x}{25-52} = \frac{y}{20-20} = \frac{z}{52-25} \quad \text{or} \quad \frac{x}{1} = \frac{y}{0} = \frac{z}{-1}$$

eigen vector $X_2 = \begin{bmatrix} 1 \\ 0 \\ -1 \end{bmatrix}$

Eigen vector corresponding to eigen value $\lambda = 12$.

Equation (1) becomes

$$\begin{bmatrix} -14 & 5 & 4 \\ 5 & -5 & 5 \\ 4 & 5 & -14 \end{bmatrix}\begin{bmatrix} x \\ y \\ z \end{bmatrix} = \begin{bmatrix} 0 \\ 0 \\ 0 \end{bmatrix} \quad \text{or} \quad \begin{cases} -14x+5y+4z=0 \\ 5x-5y+5z=0 \end{cases}$$

$$\frac{x}{25+20} = \frac{y}{20+70} = \frac{z}{70-25} \quad \text{or} \quad \frac{x}{1} = \frac{y}{2} = \frac{z}{1}$$

Eigen vector $X_3 = \begin{bmatrix} 1 \\ 2 \\ 1 \end{bmatrix}$ **Ans.**

EXERCISE 41.5

Symmetric matrices with non-repeated eigen values

Find the eigen values and eigen vectors of the following matrices:

1. $\begin{bmatrix} 5 & 0 & 1 \\ 0 & -2 & 0 \\ 1 & 0 & 5 \end{bmatrix}$ **Ans.** $-2, 4, 6;\ \begin{bmatrix} 0 \\ 1 \\ 0 \end{bmatrix}, \begin{bmatrix} 1 \\ 0 \\ -1 \end{bmatrix}, \begin{bmatrix} 1 \\ 0 \\ 1 \end{bmatrix}$

2. $\begin{bmatrix} 3 & -1 & 1 \\ -1 & 5 & -1 \\ 1 & -1 & 3 \end{bmatrix}$ **Ans.** $2, 3, 6;\ \begin{bmatrix} -1 \\ 0 \\ 1 \end{bmatrix}, \begin{bmatrix} 1 \\ 1 \\ 1 \end{bmatrix}, \begin{bmatrix} 1 \\ -2 \\ 1 \end{bmatrix}$

3. $\begin{bmatrix} 8 & -6 & 2 \\ -6 & 7 & -4 \\ 2 & -4 & 3 \end{bmatrix}$ (*U.P., I Semester, Jan 2011*) **Ans.** $0, 3, 15;\ \begin{bmatrix} 1 \\ 2 \\ 2 \end{bmatrix}, \begin{bmatrix} 2 \\ 1 \\ -2 \end{bmatrix}, \begin{bmatrix} 2 \\ -2 \\ 1 \end{bmatrix}$

4. $\begin{bmatrix} 2 & 4 & -6 \\ 4 & 2 & -6 \\ -6 & -6 & -15 \end{bmatrix}$ **Ans.** $-2, 9, -18;\ \begin{bmatrix} 1 \\ -1 \\ 0 \end{bmatrix}, \begin{bmatrix} 2 \\ 2 \\ -1 \end{bmatrix}, \begin{bmatrix} 1 \\ 1 \\ 4 \end{bmatrix}$

5. $\begin{bmatrix} 1 & 1 & 3 \\ 1 & 5 & 1 \\ 3 & 1 & 1 \end{bmatrix}$ **Ans.** $-2, 3, 6;\ \begin{bmatrix} -1 \\ 0 \\ 1 \end{bmatrix}, \begin{bmatrix} 1 \\ -1 \\ 1 \end{bmatrix}, \begin{bmatrix} 1 \\ 2 \\ 1 \end{bmatrix}$

41.11 SYMMETRIC MATRICES WITH REPEATED EIGEN VALUES

Example 30. *Find all the eigen values and eigen vectors of the matrix* $\begin{bmatrix} 2 & -1 & 1 \\ -1 & 2 & -1 \\ 1 & -1 & 2 \end{bmatrix}$

Solution. The characteristic equation $\begin{vmatrix} 2-\lambda & -1 & 1 \\ -1 & 2-\lambda & -1 \\ 1 & -1 & 2-\lambda \end{vmatrix} = 0$

$$\Rightarrow \quad (2-\lambda)[(2-\lambda)^2 - 1] + 1[-2+\lambda+1] + 1[1-2+\lambda] = 0$$

$$\Rightarrow \quad (2-\lambda)(4-4\lambda+\lambda^2-1) + (\lambda-1) + \lambda - 1 = 0$$

$$\Rightarrow \quad 8 - 8\lambda + 2\lambda^2 - 2 - 4\lambda + 4\lambda^2 - \lambda^3 + \lambda + 2\lambda - 2 = 0$$

$$\Rightarrow \quad -\lambda^3 + 6\lambda^2 - 9\lambda + 4 = 0$$

$$\Rightarrow \quad \lambda^3 - 6\lambda^2 + 9\lambda - 4 = 0 \qquad \text{... (1)}$$

On putting $\lambda = 1$ in (1), the equation (1) is satisfied. So $\lambda = -1$ is one factor of the equation (1). The other factor $(\lambda^2 - 5\lambda + 4)$ is got on dividing (1) $\lambda - 1$.

$$\Rightarrow \quad (\lambda-1)(\lambda^2 - 5\lambda + 4) = 0 \quad \text{or} \quad (\lambda-1)(\lambda-1)(\lambda-4) = 0 \quad \Rightarrow \quad \lambda = 1, 1, 4$$

The eigen values are 1, 1, 4.

When $\lambda = 4$ $\begin{pmatrix} 2-4 & -1 & 1 \\ -1 & 2-4 & -1 \\ 1 & -1 & 2-4 \end{pmatrix}\begin{pmatrix} x_1 \\ x_2 \\ x_3 \end{pmatrix} = \begin{bmatrix} 0 \\ 0 \\ 0 \end{bmatrix} \Rightarrow \begin{pmatrix} -2 & -1 & 1 \\ -1 & -2 & -1 \\ 1 & -1 & -2 \end{pmatrix}\begin{pmatrix} x_1 \\ x_2 \\ x_3 \end{pmatrix} = \begin{bmatrix} 0 \\ 0 \\ 0 \end{bmatrix}$

$$-2x_1 - x_2 + x_3 = 0$$

$$x_1 - x_2 - 2x_3 = 0$$

$$\Rightarrow \quad \frac{x_1}{2+1} = \frac{x_2}{1-4} = \frac{x_3}{2+1} \qquad \Rightarrow \qquad \frac{x_1}{1} = \frac{x_2}{-1} = \frac{x_3}{1} = k$$

$$\Rightarrow \quad x_1 = k, \quad x_2 = -k, \quad x_3 = k$$

$$X_1 = \begin{bmatrix} k \\ -k \\ k \end{bmatrix} = k\begin{bmatrix} 1 \\ -1 \\ 1 \end{bmatrix} \quad \text{or} \quad X_1 = \begin{bmatrix} 1 \\ -1 \\ 1 \end{bmatrix}$$

When $\lambda = 1$ $\begin{pmatrix} 2-1 & -1 & 1 \\ -1 & 2-1 & -1 \\ 1 & -1 & 2-1 \end{pmatrix}\begin{pmatrix} x_1 \\ x_2 \\ x_3 \end{pmatrix} = 0$

$$\Rightarrow \quad \begin{pmatrix} 1 & -1 & 1 \\ -1 & 1 & -1 \\ 1 & -1 & 1 \end{pmatrix}\begin{pmatrix} x_1 \\ x_2 \\ x_3 \end{pmatrix} = 0 \quad \Rightarrow \quad \begin{pmatrix} 1 & -1 & 1 \\ 0 & 0 & 0 \\ 0 & 0 & 0 \end{pmatrix}\begin{pmatrix} x_1 \\ x_2 \\ x_3 \end{pmatrix} = 0, \; \begin{matrix} R_2 \to R_2 + R_1 \\ R_3 \to R_3 - R_1 \end{matrix}$$

$$x_1 - x_2 + x_3 = 0$$

Let $x_1 = k_1$ and $x_2 = k_2$

$$k_1 - k_2 + x_3 = 0 \quad \text{or} \quad x_3 = k_2 - k_1$$

$$X_2 = \begin{bmatrix} k_1 \\ k_2 \\ k_2 - k_1 \end{bmatrix} \Rightarrow X_2 = \begin{bmatrix} 1 \\ 1 \\ 0 \end{bmatrix} \quad \begin{bmatrix} k_1 = 1 \\ k_2 = 1 \end{bmatrix}$$

Let $X_3 = \begin{bmatrix} l \\ m \\ n \end{bmatrix}$

As X_3 is orthogonal to X_1 since the given matrix is symmetric

$$[1, -1, 1]\begin{bmatrix} l \\ m \\ n \end{bmatrix} = 0 \quad \text{or} \quad l - m + n = 0 \qquad \ldots (2)$$

As X_3 is orthogonal to X_2 since the given matrix is symmetric

$$[1, 1, 0]\begin{bmatrix} l \\ m \\ n \end{bmatrix} = 0 \quad \text{or} \quad l + m + 0 = 0 \qquad \ldots (3)$$

Solving (2) and (3), we get $\dfrac{l}{0-1} = \dfrac{m}{1-0} = \dfrac{n}{1+1} \Rightarrow \dfrac{l}{-1} = \dfrac{m}{1} = \dfrac{n}{2}$

$$X_3 = \begin{bmatrix} -1 \\ 1 \\ 2 \end{bmatrix}$$ **Ans.**

EXERCISE 41.6

Symmetric matrices with repeated eigen values

Find the eigen values and the corresponding eigen vectors of the following matrices:

1. $\begin{bmatrix} 1 & 2 & 3 \\ 2 & 4 & 6 \\ 3 & 6 & 9 \end{bmatrix}$ **Ans.** 0, 0, 14; $\begin{bmatrix} -2 \\ 1 \\ 0 \end{bmatrix}, \begin{bmatrix} 3 \\ 6 \\ -5 \end{bmatrix}, \begin{bmatrix} 1 \\ 2 \\ 3 \end{bmatrix}$

2. $\begin{bmatrix} 2 & 0 & 1 \\ 0 & 3 & 0 \\ 1 & 0 & 2 \end{bmatrix}$ **Ans.** 1, 3, 3; $\begin{bmatrix} 1 \\ 0 \\ -1 \end{bmatrix}, \begin{bmatrix} 1 \\ 1 \\ 1 \end{bmatrix}, \begin{bmatrix} 1 \\ -2 \\ 1 \end{bmatrix}$

3. $\begin{bmatrix} 6 & -2 & 2 \\ -2 & 3 & -1 \\ 2 & -1 & 3 \end{bmatrix}$ **Ans.** 8, 2, 2; $\begin{bmatrix} 2 \\ -1 \\ 1 \end{bmatrix}, \begin{bmatrix} 1 \\ 0 \\ -2 \end{bmatrix}, \begin{bmatrix} 2 \\ 5 \\ 1 \end{bmatrix}$

4. $\begin{bmatrix} 6 & -3 & 3 \\ -3 & 6 & -3 \\ 3 & -3 & 6 \end{bmatrix}$ **Ans.** 3, 3, 12 $\begin{bmatrix} 0 \\ 0 \\ 0 \end{bmatrix}, \begin{bmatrix} 0 \\ 0 \\ 0 \end{bmatrix}, \begin{bmatrix} 1 \\ -1 \\ 1 \end{bmatrix}$

41.12 MATRIX HAVING ONLY ONE LINEARLY INDEPENDENT EIGEN VECTOR

Example 31. *Find the eigen values and eigen vectors of*

$$A = \begin{bmatrix} -3 & -7 & -5 \\ 2 & 4 & 3 \\ 1 & 2 & 2 \end{bmatrix}$$

has less than three linearly independent eigen vectors. It is possible to obtain a similarity transformation that will diagonalise this matrix.

Solution. The characteristic equation of the given matrix is

$$|A - \lambda I| = 0$$

$$\Rightarrow \quad \begin{vmatrix} -3-\lambda & -7 & -5 \\ 2 & 4-\lambda & 3 \\ 1 & 2 & 2-\lambda \end{vmatrix} = 0$$

$$\Rightarrow \quad (-3-\lambda)[(4-\lambda)(2-\lambda)-6]+7[2(2-\lambda)-3]-5[4-(4-\lambda)]=0$$

$$\Rightarrow \quad \lambda^3 - 3\lambda^2 + 3\lambda - 1 = 0 \quad \Rightarrow \quad (\lambda-1)^3 = 1 \quad \Rightarrow \quad \lambda = 1, 1, 1$$

Eigen values of the given matrix A are 1, 1, 1. Eigen vector when $\lambda = 1$

$$\begin{bmatrix} -3-1 & -7 & -5 \\ 2 & 4-1 & 3 \\ 1 & 2 & 2-1 \end{bmatrix}\begin{bmatrix} x_1 \\ x_2 \\ x_3 \end{bmatrix} = \begin{bmatrix} 0 \\ 0 \\ 0 \end{bmatrix} \quad \Rightarrow \quad \begin{bmatrix} -4 & -7 & -5 \\ 2 & 3 & 3 \\ 1 & 2 & 1 \end{bmatrix}\begin{bmatrix} x_1 \\ x_2 \\ x_3 \end{bmatrix} = \begin{bmatrix} 0 \\ 0 \\ 0 \end{bmatrix}$$

$$\Rightarrow \quad -4x_1 - 7x_2 - 5x_3 = 0 \qquad \text{... (1)}$$

$$2x_1 + 3x_2 + 3x_3 = 0 \qquad \text{... (2)}$$

$$\Rightarrow \quad \frac{x_1}{-21+15} = \frac{x_2}{-10+12} = \frac{x_3}{-12+14}$$

$$\Rightarrow \quad \frac{x_1}{-6} = \frac{x_2}{2} = \frac{x_3}{2} = k \quad \text{(say)}$$

Thus, $x_1 = -6k$, $x_2 = 2k$ and $x_3 = 2k$

$$X = \begin{bmatrix} x_1 \\ x_2 \\ x_3 \end{bmatrix} = \begin{bmatrix} -6k \\ 2k \\ 2k \end{bmatrix} = 2k\begin{bmatrix} -3 \\ 1 \\ 1 \end{bmatrix}$$

All the eigen vectors are same and hence linearly independent. **Ans.**

41.13 MATRIX HAVING ONLY TWO EIGEN VECTORS

Example 32. *Find the eigen values and eigen vectors of* $A = \begin{bmatrix} 3 & 10 & 5 \\ -2 & -3 & -4 \\ 3 & 5 & 7 \end{bmatrix}$ *has less than three linearly independent eigen vectors. Is it possible to obtain a similarity transformation that will diagonalise this matrix?*

Solution. The characteristic equation of the given matrix A is

$$|A - \lambda I| = 0$$

$$\begin{vmatrix} 3-\lambda & 10 & 5 \\ -2 & -3-\lambda & -4 \\ 3 & 5 & 7-\lambda \end{vmatrix} = 0$$

$$\Rightarrow \quad (3-\lambda)[(-3-\lambda)(7-\lambda)+20]-10[-2(7-\lambda)+12]+5[-10-3(-3-\lambda)]=0$$

$$\Rightarrow \quad (3-\lambda)[-21+3\lambda-7\lambda+\lambda^2+20]-10[-14+2\lambda+12]+5[-10+9+3\lambda]=0$$

$$\Rightarrow \quad (3-\lambda)(\lambda^2-4\lambda-1)-10(2\lambda-2)+5(3\lambda-1)=0$$

$$\Rightarrow \quad \lambda^3 - 7\lambda^2 + 16\lambda - 12 = 0 \quad \Rightarrow \quad (\lambda-3)(\lambda-2)(\lambda-2) = 0 \quad \Rightarrow \quad \lambda = 3, 2, 2$$

Eigen values of the given matrix A are 3, 2, 2.

Eigen vector, when $\lambda = 3$

$$\begin{bmatrix} 3-3 & 10 & 5 \\ -2 & -3-3 & -4 \\ 3 & 5 & 7-3 \end{bmatrix}\begin{bmatrix} x_1 \\ x_2 \\ x_3 \end{bmatrix} = \begin{bmatrix} 0 \\ 0 \\ 0 \end{bmatrix} \Rightarrow \begin{bmatrix} 0 & 10 & 5 \\ -2 & -6 & -4 \\ 3 & 5 & 4 \end{bmatrix}\begin{bmatrix} x_1 \\ x_2 \\ x_3 \end{bmatrix} = \begin{bmatrix} 0 \\ 0 \\ 0 \end{bmatrix}$$

$$-2x_1 - 6x_2 - 4x_3 = 0 \qquad \text{... (1)}$$

$$3x_1 + 5x_2 + 4x_3 = 0 \qquad \text{... (2)}$$

Solving (1) and (2) by cross multiplication method, we have

$$\frac{x_1}{-24+20} = \frac{x_2}{-12+8} = \frac{x_3}{-10+18}$$

$$\Rightarrow \qquad \frac{x_1}{-4} = \frac{x_2}{-4} = \frac{x_3}{8} = k \text{ (say)}$$

Thus, $x_1 = -4k$, $x_2 = -4k$ and $x_3 = 8k$

$$X = \begin{bmatrix} x_1 \\ x_2 \\ x_3 \end{bmatrix} = \begin{bmatrix} -4k \\ -4k \\ 8k \end{bmatrix} = 4k\begin{bmatrix} -1 \\ -1 \\ 2 \end{bmatrix}$$

Eigen vector when $\lambda = 2$

$$\begin{bmatrix} 3-2 & 10 & 5 \\ -2 & -3-2 & -4 \\ 3 & 5 & 7-2 \end{bmatrix}\begin{bmatrix} x_1 \\ x_2 \\ x_3 \end{bmatrix} = \begin{bmatrix} 0 \\ 0 \\ 0 \end{bmatrix} \Rightarrow \begin{bmatrix} 1 & 10 & 5 \\ -2 & -5 & -4 \\ 3 & 5 & 5 \end{bmatrix}\begin{bmatrix} x_1 \\ x_2 \\ x_3 \end{bmatrix} = \begin{bmatrix} 0 \\ 0 \\ 0 \end{bmatrix}$$

$$\Rightarrow \qquad x_1 + 10x_2 + 5x_3 = 0 \qquad \text{... (3)}$$

$$-2x_1 - 5x_2 - 4x_3 = 0 \qquad \text{... (4)}$$

Solving (3) and (4) by cross multiplication method, we have

$$\frac{x_1}{-40+25} = \frac{x_2}{-10+4} = \frac{x_3}{-5+20} \quad \Rightarrow \quad \frac{x_1}{-15} = \frac{x_2}{-6} = \frac{x_3}{15} = k \text{ (say)}$$

$$\Rightarrow \qquad x_1 = -15k, \qquad x_2 = -6k, \; x_3 = 15k$$

$$X = \begin{bmatrix} x_1 \\ x_2 \\ x_3 \end{bmatrix} = \begin{bmatrix} -15k \\ -6k \\ 15k \end{bmatrix} = 3k\begin{bmatrix} -5 \\ -2 \\ 5 \end{bmatrix}$$

We get one eigen vector corresponding to repeated root $\lambda_2 = 2 = \lambda_3$.

Eigen vectors corresponding to $\lambda_2 = 2 = \lambda_3$ are not linearly independent. Similarity transformation is not possible. **Ans.**

41.14 COMPLEX EIGEN VALUES

Example 33. *Show that if* $0 < \theta < \pi$, *then* $A = \begin{bmatrix} \cos\theta & -\sin\theta \\ \sin\theta & \cos\theta \end{bmatrix}$ *has no real eigen values and consequently no eigen vector.* *(Gujarat, II Semester, June 2009)*

Solution. The characteristic equation of A is $\begin{vmatrix} \cos\theta - \lambda & -\sin\theta \\ \sin\theta & \cos\theta - \lambda \end{vmatrix} = 0$

$\Rightarrow \quad (\cos\theta - \lambda)^2 + \sin^2\theta = 0$

$\Rightarrow \quad \cos^2\theta - 2\lambda\cos\theta + \lambda^2 + \sin^2\theta = 0$

$\Rightarrow \quad \lambda^2 - 2\lambda\cos\theta + 1 = 0$

$$\Rightarrow \quad \lambda = \frac{2\cos\theta \pm \sqrt{4\cos^2\theta - 4}}{2} = \frac{2\cos\theta \pm 2i\sqrt{1-\cos^2\theta}}{2} = \cos\theta \pm i\sin\theta$$

Hence, the given matrix A has no real eigen values and consequently no eigen vector. **Proved.**

Example 34. *If a matrix A is non-singular. Then $\lambda = 0$ is not its eigen value.*

Solution. Since matrix A is non-singular then $|A| \neq 0$

$$\Rightarrow \quad |A - 0I| \neq 0$$

Hence $\lambda = 0$ is not its eigen value. **Proved.**

41.15 ALGEBRAIC MULTIPLICITY

Algebraic multiplicity of an eigen value is the number of times of repetition of an eigen value.

It is denoted by $\text{mult}_a(\lambda)$.

For example, the eigen values of a matrix $\begin{bmatrix} -2 & 2 & -3 \\ 2 & 1 & -6 \\ -1 & -2 & 0 \end{bmatrix}$ are –3, –3, 5.

The $\text{mult}_a(-3) = 2$ and $\text{mult}_a(5) = 1$

41.16 GEOMETRIC MULTIPLICITY

Geometric multiplicity of an eigen value is the number of linearly independent eigen vectors corresponding to λ.

It is denoted by $\text{Mult}_g(\lambda)$

In example 27, two linearly independent eigen vectors corresponding to

$$\lambda = -3 \text{ are } \begin{bmatrix} 0 \\ 3 \\ 2 \end{bmatrix} \text{ and } \begin{bmatrix} 3 \\ 0 \\ 1 \end{bmatrix}.$$

so the $\text{mult}_g(-3) = 2$

And the eigen vector corresponding to $\lambda = 5$ is $\begin{bmatrix} 1 \\ 2 \\ -1 \end{bmatrix}$ so the $\text{mult}_g(5) = 1$.

41.17 REGULAR EIGEN VALUE

If the algebraic multiplicity and geometric multiplicity of an eigen value are equal, then the eigen value is called *regular*.

Example 35. *Find the algebraic multiplicity and geometric multiplicity of an eigen value of the matrix* $A = \begin{bmatrix} 3 & 10 & 5 \\ -2 & -3 & -4 \\ 3 & 5 & 7 \end{bmatrix}$ *and show geometric multiplicity cannot be greater than algebraic multiplicity.*

Solution. The characteristic equation of the given matrix is

$$\begin{vmatrix} 3-\lambda & 10 & 5 \\ -2 & -3-\lambda & -4 \\ 3 & 5 & 7-\lambda \end{vmatrix} = 0$$

$$\Rightarrow \qquad \lambda^3 - 7\lambda^2 + 16\lambda - 12 = 0$$

$$\Rightarrow \qquad (\lambda-2)(\lambda-2)(\lambda-3) = 0$$

$$\Rightarrow \qquad \lambda = 2, 2, 3$$

Therefore 2 is a multiple eigen value repeating 2 times. So Algebraic Multiplicity of 2 is 2.

$$\text{Mult}_a(2) = 2. \qquad \text{...(A)}$$

We shall find the eigen vector corresponding to the eigen value 2.

$$X = \begin{bmatrix} 3-2 & 10 & 5 \\ -2 & -3-2 & -4 \\ 3 & 5 & 7-2 \end{bmatrix}\begin{bmatrix} x_1 \\ x_2 \\ x_3 \end{bmatrix} = \begin{bmatrix} 0 \\ 0 \\ 0 \end{bmatrix}$$

$$\Rightarrow \qquad \begin{bmatrix} 1 & 10 & 5 \\ -2 & -5 & -4 \\ 3 & 5 & 5 \end{bmatrix}\begin{bmatrix} x_1 \\ x_2 \\ x_3 \end{bmatrix} = \begin{bmatrix} 0 \\ 0 \\ 0 \end{bmatrix}$$

$$x_1 + 10x_2 + 5x_3 = 0 \qquad \text{...(1)}$$

$$-2x_1 - 5x_2 - 4x_3 = 0 \qquad \text{...(2)}$$

Solving (1) and (2) by cross multiplication method, we have

$$\frac{x_1}{-40+25} = \frac{x_2}{-10+4} = \frac{x_3}{-5+20}$$

$$\Rightarrow \qquad \frac{x_1}{-15} = \frac{x_2}{-6} = \frac{x_3}{15} = k \text{ (say)}$$

Thus $x_1 = -15k$, $x_2 = -6k$, $x_3 = -15k$.

$$X = \begin{bmatrix} x_1 \\ x_2 \\ x_3 \end{bmatrix} = \begin{bmatrix} -15k \\ -6k \\ 15k \end{bmatrix} = 3k\begin{bmatrix} -5 \\ -2 \\ 5 \end{bmatrix}$$

Here the linearly independent eigen vector is 1.

So the, geometric multiplicity of eigen value 2 is 1

$$\text{Mult}_g(2) = 1 \qquad \text{...(B)}$$

Hence from (A) and (B) **Ans.**

Geometric multiplicity < Algebraic multiplicity

Notes: (1) If the values of x_1, x_2, x_3 are in terms of k (one independent value), then there is only one linearly independent eigen vector. So the geometric multiplicity is 1.

(2) If the values of x_1, x_2, x_3 are in terms of k_1, k_2 two independent values, then there are two linearly independent eigen vectors. So the geometric multiplicity is 2.

EXERCISE 41.7

From the following matrices; find eigen value, Algebraic multiplicity, Geometric multiplicity.

1. $\begin{bmatrix} -2 & -1 \\ 5 & 4 \end{bmatrix}$ **Ans.** $\lambda = -1$, $Mult_a(-1) = 1$, $Mult_g(-1) = 1$

$\lambda = 3$, $Mult_a(3) = 1$, $Mult_g(3) = 1$

2. $\begin{bmatrix} 1 & 1 & 1 \\ 0 & 1 & 1 \\ 0 & 0 & 1 \end{bmatrix}$ **Ans.** $\lambda = 1$, $Mult_a(1) = 3$, $Mult_g(1) = 1$

3. $\begin{bmatrix} 0 & 0 & 1 \\ 1 & 0 & -3 \\ 0 & 1 & 3 \end{bmatrix}$ **Ans.** $\lambda = 1$, $Mult_a(1) = 3$, $Mult_g(1) = 1$

4. $\begin{bmatrix} 1 & 2 & 3 \\ 0 & 2 & 3 \\ 0 & 0 & 2 \end{bmatrix}$ **Ans.** $\lambda = 2$, $Mult_a(2) = 2$, $Mult_g(2) = 1$

5. $\begin{bmatrix} 2 & 2 & 1 \\ 1 & 3 & 1 \\ 1 & 2 & 2 \end{bmatrix}$ **Ans.** $\lambda = 5$, $Mult_a(5) = 1$, $Mult_g(5) = 1$

6. $\begin{bmatrix} 1 & 2 & 2 \\ 0 & 2 & 1 \\ -1 & 2 & 2 \end{bmatrix}$ **Ans.** $\lambda = 1$, $Mult_a(1) = 2$, $Mult_g(1) = 1$

$\lambda = 2$, $Mult_a(2) = 2$, $Mult_g(2) = 1$

7. $\begin{bmatrix} 5 & 4 & -4 \\ 4 & 5 & -4 \\ -1 & -1 & 2 \end{bmatrix}$ **Ans.** $\lambda = 1$, $Mult_a(1) = 2$, $Mult_g(1) = 2$

$\lambda = 10$, $Mult_a(10) = 1$, $Mult_g(10) = 1$

8. $\begin{bmatrix} 1 & 0 & 0 & 0 \\ 0 & 1 & 0 & 0 \\ 0 & 0 & 1 & 1 \\ 0 & 0 & 0 & 1 \end{bmatrix}$ **Ans.** $\lambda = 1$, $Mult_a(1) = 4$, $Mult_g(1) = 3$

41.18 SIMILARITY TRANSFORMATION

Let A and B be two square matrices of order n. Then B is said to be similar to A if there exists a non-singular matrix P such that

$$B = P^{-1} AP \qquad ...(1)$$

Equation (1) is called a similar transformation.

41.19 DIAGONALISATION OF A MATRIX

Diagonalisation of a matrix A is the process of reduction of A to a diagonal form 'D'. If A is related to D by a similarity transformation such that $D = P^{-1} AP$ then A is reduced to the diagonal matrix D through *modal matrix P*. D is also called *spectral matrix* of A.

41.20 ORTHOGONAL TRANSFORMATION OF A SYMMETRIC MATRIX TO DIAGONAL FORM

Let A be a symmetric matrix, then

$$A \,.\, A' = \text{I} \qquad ...(1)$$

and
$$A \,.\, A^{-1} = I \qquad ...(2)$$

From (1) and (2), we have $\quad A^{-1} = A'$

We know that, diagonalisation transformation of a symmetric matrix is

$$P^{-1}AP = D$$

If we normalize each eigen vector and use them to form the normalized modal matirx N then N is an orthogonal matrix.

Then, $\quad N'\,AN = D$

Transforming A into D by means of the transformation $N'\,AN = D$ is called as orthogonal transformation.

Note. To normalize eigen vector divide each element of the vector by the square root of the sum of the squares of all the elements of the vector.

Example 36. *Show that similar matrices have same trace.* *(D. U. April 2010)*

Solution. As we know that similar matrices have eigen value.

Trace of matrices is sum of all eigen value. Hence similar matrices have same trace. **Proved.**

41.21 THEOREM ON DIAGONALIZATION OF A MATRIX

Theorem. *If a square matrix A of order n has n linearly independent eigen vectors, then a matrix P can be found such that $P^{-1}\,AP$ is a diagonal matrix.*

Proof. We shall prove the theorem for a matrix of order 3. The proof can be easily extended to matrices of higher order.

Let
$$A = \begin{bmatrix} a_1 & b_1 & c_1 \\ a_2 & b_2 & c_2 \\ a_3 & b_3 & c_3 \end{bmatrix}$$

and let λ_1, λ_2, λ_3 be its eigen values and X_1, X_2, X_3 the corresponding eigen vectors, where

$$X_1 = \begin{bmatrix} x_1 \\ y_1 \\ z_1 \end{bmatrix}, \qquad X_2 = \begin{bmatrix} x_2 \\ y_2 \\ z_2 \end{bmatrix}, \qquad X_3 = \begin{bmatrix} x_3 \\ y_3 \\ z_3 \end{bmatrix}$$

For the eigen value λ_1, the eigen vector is given by

$$\left.\begin{aligned} (a_1 - \lambda_1)x_1 + b_1y_1 + c_1z_1 &= 0 \\ a_2x_1 + (b_2 - \lambda_1)y_1 + c_2z_1 &= 0 \\ a_3x_1 + b_3y_1 + (c_3 - \lambda_1)z_1 &= 0 \end{aligned}\right\} \qquad ...(1)$$

$\therefore$ We have

$$\left.\begin{aligned} a_1\,x_1 + b_1\,y_1 + c_1\,z_1 &= \lambda_1\,x_1 \\ a_2\,x_1 + b_2\,y_1 + c_2\,z_1 &= \lambda_1\,y_1 \\ a_3\,x_1 + b_3\,y_1 + c_3\,z_1 &= \lambda_1\,z_1 \end{aligned}\right\} \qquad ...(2)$$

Similarly, for λ_2 and λ_3, we have

$$\left.\begin{aligned} a_1 x_2 + b_1 y_2 + c_1 z_2 &= \lambda_2 x_2 \\ a_2 x_2 + b_2 y_2 + c_2 z_2 &= \lambda_2 y_2 \\ a_3 x_2 + b_3 y_2 + c_3 z_2 &= \lambda_2 z_2 \end{aligned}\right\} \qquad ...(3)$$

and

$$\left.\begin{aligned} a_1 x_3 + b_1 y_3 + c_1 z_3 &= \lambda_3 x_3 \\ a_2 x_3 + b_2 y_3 + c_2 z_3 &= \lambda_3 y_3 \\ a_3 x_3 + b_3 y_3 + c_3 z_3 &= \lambda_3 z_3 \end{aligned}\right\} \qquad ...(4)$$

We consider the matrix $P = \begin{bmatrix} x_1 & x_2 & x_3 \\ y_1 & y_2 & y_3 \\ z_1 & z_2 & z_3 \end{bmatrix}$

whose columns are the eigen vectors of A.

Then $AP = \begin{bmatrix} a_1 & b_1 & c_1 \\ a_2 & b_2 & c_2 \\ a_3 & b_3 & c_3 \end{bmatrix} \begin{bmatrix} x_1 & x_2 & x_3 \\ y_1 & y_2 & y_3 \\ z_1 & z_2 & z_3 \end{bmatrix}$

$$= \begin{pmatrix} a_1 x_1 + b_1 y_1 + c_1 z_1 & a_1 x_2 + b_1 y_2 + c_1 z_2 & a_1 x_3 + b_1 y_3 + c_1 z_3 \\ a_2 x_1 + b_2 y_1 + c_2 z_1 & a_2 x_2 + b_2 y_2 + c_2 z_2 & a_2 x_3 + b_2 y_3 + c_2 z_3 \\ a_3 x_1 + b_3 y_1 + c_3 z_1 & a_3 x_2 + b_3 y_2 + c_3 z_2 & a_3 x_3 + b_3 y_3 + c_3 z_3 \end{pmatrix}$$

$$= \begin{pmatrix} \lambda_1 x_1 & \lambda_2 x_2 & \lambda_3 x_3 \\ \lambda_1 y_1 & \lambda_2 y_2 & \lambda_3 y_3 \\ \lambda_1 z_1 & \lambda_2 z_2 & \lambda_3 z_3 \end{pmatrix} \qquad \text{[Using results (2), (3) and (4)]}$$

$$= \begin{pmatrix} x_1 & x_2 & x_3 \\ y_1 & y_2 & y_3 \\ z_1 & z_2 & z_3 \end{pmatrix} \begin{pmatrix} \lambda_1 & 0 & 0 \\ 0 & \lambda_2 & 0 \\ 0 & 0 & \lambda_3 \end{pmatrix} = PD$$

where D is the Diagonal matrix $\begin{pmatrix} \lambda_1 & 0 & 0 \\ 0 & \lambda_2 & 0 \\ 0 & 0 & \lambda_3 \end{pmatrix}$.

$\therefore \qquad AP = PD$

$\Rightarrow \qquad P^{-1} AP = P^{-1} PD = D$

Notes 1. The square matrix P, which diagonalises A, is found by grouping the eigen vectors of A into square-matrix and the resulting diagonal matrix has the eigen values of A as its diagonal elements.

2. The transformation of a matrix A to $P^{-1} AP$ is known as a *similarity transformation.*

3. The reduction of A to a diagonal matrix is, obviously, a particular case of similarity transformation.

4. The matrix P which diagonalises A is called the *modal matrix* of A and the resulting diagonal matrix D is known as the *spectra matrix* of A.

Example 37. *Find the eigen values, eigen vectors the modal matrix and diagonalise the matrix given below.*

$$\begin{bmatrix} 1 & 0 & 0 \\ 0 & 3 & -1 \\ 0 & -1 & 3 \end{bmatrix}$$

Solution. The characteristic equation of the given matrix is

$$\begin{vmatrix} 1-\lambda & 0 & 0 \\ 0 & 3-\lambda & -1 \\ 0 & -1 & 3-\lambda \end{vmatrix} = 0$$

$$\Rightarrow \quad (1-\lambda)\{(3-\lambda)^2 - 1\} = 0 \quad \Rightarrow \quad (1-\lambda)(3-\lambda+1)(3-\lambda-1) = 0$$

$$\Rightarrow \quad (1-\lambda)(4-\lambda)(2-\lambda) = 0 \quad \Rightarrow \quad \lambda = 1, 2, 4$$

Eigen vectors

When $\lambda = 1$,

$$\begin{bmatrix} 0 & 0 & 0 \\ 0 & 2 & -1 \\ 0 & -1 & 2 \end{bmatrix}\begin{bmatrix} x_1 \\ x_2 \\ x_3 \end{bmatrix} = \begin{bmatrix} 0 \\ 0 \\ 0 \end{bmatrix} \sim \begin{bmatrix} 0 & 0 & 0 \\ 0 & 2 & -1 \\ 0 & 0 & \frac{3}{2} \end{bmatrix}\begin{bmatrix} x_1 \\ x_2 \\ x_3 \end{bmatrix} = \begin{bmatrix} 0 \\ 0 \\ 0 \end{bmatrix} \quad R_3 \to R_3 + \frac{1}{2}R_2$$

$$\Rightarrow \quad 2x_2 - x_3 = 0 \qquad \text{... (1)}$$

$$\frac{3}{2}x_3 = 0 \Rightarrow x_3 = 0 \qquad \text{... (2)}$$

Putting $x_3 = 0$ from (2) in (1), we get $2x_2 - 0 = 0 \Rightarrow x_2 = 0$

$$\text{Eigen Vector} = \begin{bmatrix} 1 \\ 0 \\ 0 \end{bmatrix}$$

When $\lambda = 2$,

$$\begin{bmatrix} -1 & 0 & 0 \\ 0 & 1 & -1 \\ 0 & -1 & 1 \end{bmatrix}\begin{bmatrix} x_1 \\ x_2 \\ x_3 \end{bmatrix} = \begin{bmatrix} 0 \\ 0 \\ 0 \end{bmatrix} \Rightarrow \begin{bmatrix} 1 & 0 & 0 \\ 0 & 1 & -1 \\ 0 & 0 & 0 \end{bmatrix}\begin{bmatrix} x_1 \\ x_2 \\ x_3 \end{bmatrix} = \begin{bmatrix} 0 \\ 0 \\ 0 \end{bmatrix} \quad \begin{matrix} R_1 \to -R_1 \\ R_3 \to R_3 + R_2 \end{matrix}$$

$$x_1 = 0$$

$$x_2 - x_3 = 0 \Rightarrow x_2 = x_3$$

$$\text{Eigen vector} = \begin{bmatrix} 0 \\ 1 \\ 1 \end{bmatrix}$$

When $\lambda = 4$,

$$\begin{bmatrix} -3 & 0 & 0 \\ 0 & -1 & -1 \\ 0 & -1 & -1 \end{bmatrix}\begin{bmatrix} x_1 \\ x_2 \\ x_3 \end{bmatrix} = \begin{bmatrix} 0 \\ 0 \\ 0 \end{bmatrix}$$

$$-3x_1 = 0$$

$$-x_2 - x_3 = 0$$

$$x_2 = -x_3$$

$$\text{Eigen Vector} = \begin{bmatrix} 0 \\ 1 \\ -1 \end{bmatrix}$$

Modal matrix $= \begin{bmatrix} 1 & 0 & 0 \\ 0 & 1 & 1 \\ 0 & 1 & -1 \end{bmatrix}$ **Ans.**

Let us diagonalise the given matrix:

$$P^{-1}AP = -\frac{1}{2}\begin{bmatrix} -2 & 0 & 0 \\ 0 & -1 & -1 \\ 0 & -1 & 1 \end{bmatrix}\begin{bmatrix} 1 & 0 & 0 \\ 0 & 3 & -1 \\ 0 & -1 & 3 \end{bmatrix}\begin{bmatrix} 1 & 0 & 0 \\ 0 & 1 & 1 \\ 0 & 1 & -1 \end{bmatrix}$$

$$= -\frac{1}{2}\begin{bmatrix} -2 & 0 & 0 \\ 0 & -1 & -1 \\ 0 & -1 & 1 \end{bmatrix}\begin{bmatrix} 1 & 0 & 0 \\ 0 & 2 & 4 \\ 0 & 2 & -4 \end{bmatrix} = -\frac{1}{2}\begin{bmatrix} -2 & 0 & 0 \\ 0 & -4 & 0 \\ 0 & 0 & -8 \end{bmatrix} = \begin{bmatrix} 1 & 0 & 0 \\ 0 & 2 & 0 \\ 0 & 0 & 4 \end{bmatrix}$$

Example 38. *Find a matrix P which diagonalizes the matrix*

$A = \begin{bmatrix} 4 & 1 \\ 2 & 3 \end{bmatrix}$, *verify* $P^{-1}AP = D$ *where D is the diagonal matrix.* (*U.P., I Semester, Dec. 2008*)

Solution. The characteristic equation of matrix A is

$$\begin{vmatrix} 4-\lambda & 1 \\ 2 & 3-\lambda \end{vmatrix} = 0 \quad \Rightarrow \quad (4-\lambda)(3-\lambda) - 2 = 0$$

$$\Rightarrow \lambda^2 - 7\lambda + 12 - 2 = 0 \quad \Rightarrow \quad \lambda^2 - 7\lambda + 10 = 0$$

$$\Rightarrow (\lambda - 2)(\lambda - 5) = 0 \quad \Rightarrow \quad \lambda = 2, \lambda = 5$$

Eigen values are 2 and 5.

(*i*) When $\lambda = 2$, eigen vectors are given by the matrix equation

$$\begin{bmatrix} 4-2 & 1 \\ 2 & 3-2 \end{bmatrix}\begin{bmatrix} x_1 \\ x_2 \end{bmatrix} = \begin{bmatrix} 0 \\ 0 \end{bmatrix} \Rightarrow \begin{bmatrix} 2 & 1 \\ 2 & 1 \end{bmatrix}\begin{bmatrix} x_1 \\ x_2 \end{bmatrix} = \begin{bmatrix} 0 \\ 0 \end{bmatrix}$$

$\Rightarrow$ $\qquad 2x_1 + x_2 = 0 \Rightarrow x_2 = -2x_1$

Let $\qquad x_1 = k, x_2 = -2k$

Hence, the eigen vector $\qquad X_1 = \begin{bmatrix} k \\ -2k \end{bmatrix}$ or $\begin{bmatrix} 1 \\ -2 \end{bmatrix}$

(ii) When $\lambda = 5$, eigen vectors are given by the matrix equation

$$\begin{bmatrix} 4-5 & 1 \\ 2 & 3-5 \end{bmatrix}\begin{bmatrix} x_1 \\ x_2 \end{bmatrix} = \begin{bmatrix} 0 \\ 0 \end{bmatrix} \Rightarrow \begin{bmatrix} -1 & 1 \\ 2 & -2 \end{bmatrix}\begin{bmatrix} x_1 \\ x_2 \end{bmatrix} = \begin{bmatrix} 0 \\ 0 \end{bmatrix}$$

$\Rightarrow$ $\qquad -x_1 + x_2 = 0 \Rightarrow x_1 = x_2$

Let $x_1 = k$, then $x_2 = k$

Hence, the eigen vector $\qquad X_2 = \begin{bmatrix} k \\ k \end{bmatrix}$ or $\begin{bmatrix} 1 \\ 1 \end{bmatrix}$

Modal matrix $\qquad P = \begin{bmatrix} 1 & 1 \\ -2 & 1 \end{bmatrix} \Rightarrow P^{-1} = \frac{1}{3}\begin{bmatrix} 1 & -1 \\ 2 & 1 \end{bmatrix}$

For diagonalization

$$D = P^{-1}AP = \frac{1}{3}\begin{bmatrix} 1 & -1 \\ 2 & 1 \end{bmatrix}\begin{bmatrix} 4 & 1 \\ 2 & 3 \end{bmatrix}\begin{bmatrix} 1 & 1 \\ -2 & 1 \end{bmatrix}$$

$$= \frac{1}{3}\begin{bmatrix} 1 & -1 \\ 2 & 1 \end{bmatrix}\begin{bmatrix} 2 & 5 \\ -4 & 5 \end{bmatrix} = \frac{1}{3}\begin{bmatrix} 6 & 0 \\ 0 & 15 \end{bmatrix} = \begin{bmatrix} 2 & 0 \\ 0 & 5 \end{bmatrix}$$ **Verified.**

Example 39. *Let* $A = \begin{bmatrix} 6 & -2 & 2 \\ -2 & 3 & -1 \\ 2 & -1 & 3 \end{bmatrix}$ *Find matrix P such that* $P^{-1}AP$ *is diagonal matrix.*

Solution. The characteristic equation of the matrix A is

$$\begin{vmatrix} 6-\lambda & -2 & 2 \\ -2 & 3-\lambda & -1 \\ 2 & -1 & 3-\lambda \end{vmatrix} = 0$$

$$\Rightarrow \quad (6-\lambda)[9+\lambda^2-6\lambda-1]+2[-6+2\lambda+2]+2[2-6+2\lambda]=0$$

$$\Rightarrow \quad (6-\lambda)(\lambda^2-6\lambda+8)-8+4\lambda-8+4\lambda=0$$

$$\Rightarrow \quad 6\lambda^2-36\lambda+48-\lambda^3+6\lambda^2-8\lambda-16+8\lambda=0$$

$$\Rightarrow \quad -\lambda^3+12\lambda^2-36\lambda+32=0 \quad \Rightarrow \quad \lambda^3-12\lambda^2+36\lambda-32=0$$

$$\Rightarrow \quad (\lambda-2)^2(\lambda-8)=0 \quad \Rightarrow \quad \lambda = 2, 2, 8$$

Eigen vector for $\lambda = 2$

$$\begin{bmatrix} 4 & -2 & 2 \\ -2 & 1 & -1 \\ 2 & -1 & 1 \end{bmatrix}\begin{bmatrix} x_1 \\ x_2 \\ x_3 \end{bmatrix} = \begin{bmatrix} 0 \\ 0 \\ 0 \end{bmatrix} R_1 \to R_1 + R_2 \quad \text{or} \quad \begin{bmatrix} 2 & -1 & 1 \\ -2 & 1 & -1 \\ 2 & -1 & 1 \end{bmatrix}\begin{bmatrix} x_1 \\ x_2 \\ x_3 \end{bmatrix} = \begin{bmatrix} 0 \\ 0 \\ 0 \end{bmatrix} \begin{matrix} \\ R_2 \to R_1 + R_2 \\ R_3 \to R_2 + R_3 \end{matrix}$$

$$\begin{bmatrix} 2 & -1 & 1 \\ 0 & 0 & 0 \\ 0 & 0 & 0 \end{bmatrix}\begin{bmatrix} x_1 \\ x_2 \\ x_3 \end{bmatrix} = \begin{bmatrix} 0 \\ 0 \\ 0 \end{bmatrix} \quad \text{or} \quad 2x_1 - x_2 + x_3 = 0$$

This equation is satisfied by $x_1 = 0, x_2 = 1, x_3 = 1$

$$X_1 = \begin{bmatrix} 0 \\ 1 \\ 1 \end{bmatrix}$$

and again $x_1 = 1, x_2 = 3, x_3 = 1.$

$$X_2 = \begin{bmatrix} 1 \\ 3 \\ 1 \end{bmatrix}$$

Eigen vector for $\lambda = 8$

$$\begin{bmatrix} -2 & -2 & 2 \\ -2 & -5 & -1 \\ 2 & -1 & -5 \end{bmatrix}\begin{bmatrix} x_1 \\ x_2 \\ x_3 \end{bmatrix} = \begin{bmatrix} 0 \\ 0 \\ 0 \end{bmatrix}$$

$$-2x_1 - 2x_2 + 2x_3 = 0$$

$$-2x_1 - 5x_2 - x_3 = 0$$

$$\frac{x_1}{2+10} = \frac{x_2}{-4-2} = \frac{x_3}{10-4} \Rightarrow \frac{x_1}{12} = \frac{x_2}{-6} = \frac{x_3}{6} \Rightarrow \frac{x_1}{2} = \frac{x_2}{-1} = \frac{x_3}{1}$$

$$X_3 = \begin{bmatrix} 2 \\ -1 \\ 1 \end{bmatrix}$$

$$\therefore \quad P = \begin{bmatrix} 0 & 1 & 2 \\ 1 & 3 & -1 \\ 1 & 1 & 1 \end{bmatrix}, \quad P^{-1} = -\frac{1}{6}\begin{bmatrix} 4 & 1 & -7 \\ -2 & -2 & 2 \\ -2 & 1 & -1 \end{bmatrix}$$

Now $P^{-1}AP = -\frac{1}{6}\begin{bmatrix} 4 & 1 & -7 \\ -2 & -2 & 2 \\ -2 & 1 & -1 \end{bmatrix}\begin{bmatrix} 6 & -2 & 2 \\ -2 & 3 & -1 \\ 2 & -1 & 3 \end{bmatrix}\begin{bmatrix} 0 & 1 & 2 \\ 1 & 3 & -1 \\ 1 & 1 & 1 \end{bmatrix} = \begin{bmatrix} 2 & 0 & 0 \\ 0 & 2 & 0 \\ 0 & 0 & 8 \end{bmatrix}$ **Ans.**

Example 40. *The matrix* $A = \begin{bmatrix} a & h \\ h & b \end{bmatrix}$ *is transformed to the diagonal form* $D = T^{-1}AT$, *where* $T = \begin{bmatrix} \cos\theta & \sin\theta \\ -\sin\theta & \cos\theta \end{bmatrix}$. *Find the value of* θ *which gives this diagonal transformation.*

Solution. $T = \begin{bmatrix} \cos\theta & \sin\theta \\ -\sin\theta & \cos\theta \end{bmatrix} \quad \therefore \quad T^{-1} = \begin{bmatrix} \cos\theta & -\sin\theta \\ \sin\theta & \cos\theta \end{bmatrix}$

Now $T^{-1}AT = \begin{bmatrix} \cos\theta & -\sin\theta \\ \sin\theta & \cos\theta \end{bmatrix}\begin{bmatrix} a & h \\ h & b \end{bmatrix}\begin{bmatrix} \cos\theta & \sin\theta \\ -\sin\theta & \cos\theta \end{bmatrix}$

$$= \begin{bmatrix} a\cos\theta - h\sin\theta & h\cos\theta - b\sin\theta \\ a\sin\theta + h\cos\theta & h\sin\theta + b\cos\theta \end{bmatrix}\begin{bmatrix} \cos\theta & \sin\theta \\ -\sin\theta & \cos\theta \end{bmatrix}$$

$$= \begin{bmatrix} a\cos^2\theta - 2h\sin\theta\cos\theta + b\sin^2\theta & (a-b)\sin\theta\cos\theta - h\sin^2\theta + h\cos^2\theta \\ (a-b)\sin\theta\cos\theta + h\cos^2\theta - h\sin^2\theta & a\sin^2\theta + 2h\sin\theta\cos\theta + b\cos^2\theta \end{bmatrix}$$

$$= \begin{bmatrix} a\cos^2\theta - h\sin 2\theta + b\sin^2\theta & (a-b)\sin\theta\cos\theta + h\cos 2\theta \\ (a-b)\sin\theta\cos\theta + h\cos 2\theta & a\sin^2\theta + h\sin 2\theta + b\cos^2\theta \end{bmatrix}$$

$$= \begin{bmatrix} d_1 & 0 \\ 0 & d_2 \end{bmatrix} \text{ being diagonal matrix}$$

$$\therefore \quad (a-b)\sin\theta\cos\theta + h\cos 2\theta = 0$$

$$\Rightarrow \quad \frac{a-b}{2}\sin 2\theta + h\cos 2\theta = 0 \quad \Rightarrow \quad \frac{a-b}{2}\sin 2\theta = -h\cos 2\theta$$

$$\Rightarrow \quad \tan 2\theta = \frac{2h}{b-a} \quad \Rightarrow \quad \theta = \frac{1}{2}\tan^{-1}\frac{2h}{b-a}$$ **Ans.**

EXERCISE 41.8

1. Find the matrix B which transforms the matrix

$A = \begin{bmatrix} 8 & -8 & -2 \\ 4 & -3 & -2 \\ 3 & -4 & 1 \end{bmatrix}$ to a diagonal matrix. **Ans.** $B = \begin{bmatrix} 4 & 3 & 2 \\ 3 & 2 & 1 \\ 2 & 1 & 1 \end{bmatrix}$

2. For the matrix $A = \begin{bmatrix} 4 & 1 & 0 \\ 1 & 4 & 1 \\ 0 & 1 & 4 \end{bmatrix}$, determine a matrix P such that $P^{-1}AP$ is diagonal matrix.

$$\textbf{Ans. } P = \begin{bmatrix} -1 & 1 & 1 \\ 0 & -\sqrt{2} & \sqrt{2} \\ 1 & 1 & 1 \end{bmatrix}$$

3. Determine the eigen values and the corresponding eigen vectors of the matrix $A = \begin{bmatrix} 5 & 7 & -5 \\ 0 & 4 & -1 \\ 2 & 8 & -3 \end{bmatrix}$

Hence find the matrix P such that $P^{-1}AP$ is diagonal matrix. **Ans.** $P = \begin{bmatrix} 2 & 1 & -1 \\ 1 & 1 & 1 \\ 3 & 2 & 1 \end{bmatrix}$

4. Reduce the following matrix A into a diagonal matrix

$$A = \begin{bmatrix} 8 & -6 & 2 \\ -6 & 7 & -4 \\ 2 & -4 & 3 \end{bmatrix} \qquad \textbf{Ans. } \begin{bmatrix} 0 & 0 & 0 \\ 0 & 3 & 0 \\ 0 & 0 & 15 \end{bmatrix}$$

41.22 POWERS OF A MATRIX (By diagonalisation)

We can obtain powers of a matrix by using diagonalisation.

We know that $D = P^{-1} AP$

Where A is the square matrix and P is a non-singular matrix.

$$D^2 = (P^{-1} AP)(P^{-1} AP) = P^{-1} A (P P^{-1}) AP = P^{-1} A^2 P$$

Similarly $D^3 = P^{-1} A^3 P$

In general $D^n = P^{-1} A^n P$ ...(1)

Pre-multiply (1) by P and post-multiply by P^{-1}

$$P D^n P^{-1} = P (P^{-1} A^n P) P^{-1}$$
$$= (P P^{-1}) A^n (P P^{-1})$$
$$= A^n$$

Procedure: (1) Find eigen values for a square matrix A.

(2) Find eigen vectors to get the modal matrix P.

(3) Find the diagonal matrix D, by the formula $D = P^{-1}$ AP

(4) Obtain A^n by the formula $A^n = P D^n P^{-1}$.

Example 41. *Find a matrix P which transform the matrix* $A = \begin{bmatrix} 1 & 0 & -1 \\ 1 & 2 & 1 \\ 2 & 2 & 3 \end{bmatrix}$ *to diagonal form. Hence* A^4.

Solution. Characteristic equation of the matrix A is

$$\begin{vmatrix} 1-\lambda & 0 & -1 \\ 1 & 2-\lambda & 1 \\ 2 & 2 & 3-\lambda \end{vmatrix} = 0 \quad \text{or} \quad \lambda^3 - 6\lambda^2 + 11\lambda - 6 = 0$$

$$\text{or} \quad (\lambda-1)(\lambda-2)(\lambda-3) = 0$$

$$\Rightarrow \lambda = 1, 2, 3$$

For $\lambda = 1$, eigen vector is given by

$$\begin{bmatrix} 1-1 & 0 & -1 \\ 1 & 2-1 & 1 \\ 2 & 2 & 3-1 \end{bmatrix} \begin{bmatrix} x_1 \\ x_2 \\ x_3 \end{bmatrix} = \begin{bmatrix} 0 \\ 0 \\ 0 \end{bmatrix} \Rightarrow \begin{bmatrix} 0 & 0 & -1 \\ 1 & 1 & 1 \\ 2 & 2 & 2 \end{bmatrix} \begin{bmatrix} x_1 \\ x_2 \\ x_3 \end{bmatrix} = \begin{bmatrix} 0 \\ 0 \\ 0 \end{bmatrix}$$

$$\left.\begin{array}{r} 0x_1+0x_2-x_3=0 \\ x_1+x_2+x_3=0 \end{array}\right] \Rightarrow \frac{x_1}{0+1}=\frac{x_2}{-1+0}=\frac{x_3}{0} \text{ or } x_1=1, x_2=-1, x_3=0$$

Eigen vector is [1, –1, 0].

For $\lambda = 2$, eigen vector is given by

$$\begin{bmatrix} 1-2 & 0 & -1 \\ 1 & 2-2 & 1 \\ 2 & 2 & 3-2 \end{bmatrix}\begin{bmatrix} x_1 \\ x_2 \\ x_3 \end{bmatrix}=\begin{bmatrix} 0 \\ 0 \\ 0 \end{bmatrix} \Rightarrow \begin{bmatrix} -1 & 0 & -1 \\ 1 & 0 & 1 \\ 2 & 2 & 1 \end{bmatrix}\begin{bmatrix} x_1 \\ x_2 \\ x_3 \end{bmatrix}=\begin{bmatrix} 0 \\ 0 \\ 0 \end{bmatrix}$$

$$\Rightarrow \begin{bmatrix} 0 & 0 & 0 \\ 1 & 0 & 1 \\ 2 & 2 & 1 \end{bmatrix}\begin{bmatrix} x_1 \\ x_2 \\ x_3 \end{bmatrix}=\begin{bmatrix} 0 \\ 0 \\ 0 \end{bmatrix} \quad R_1 \to R_1+R_2$$

$$\left.\begin{array}{r} x_1+0x_2+x_3=0 \\ 2x_1+2x_2+x_3=0 \end{array}\right]$$

$$\Rightarrow \frac{x_1}{0-2}=\frac{x_2}{2-1}=\frac{x_3}{2-0} \Rightarrow x_1=-2, \quad x_2=1, \quad x_3=2$$

Eigen vector is [–2, 1, 2].

For $\lambda = 3$, eigen vector is given by

$$\begin{bmatrix} 1-3 & 0 & -1 \\ 1 & 2-3 & 1 \\ 2 & 2 & 3-3 \end{bmatrix}\begin{bmatrix} x_1 \\ x_2 \\ x_3 \end{bmatrix}=\begin{bmatrix} 0 \\ 0 \\ 0 \end{bmatrix} \Rightarrow \begin{bmatrix} -2 & 0 & -1 \\ 1 & -1 & 1 \\ 2 & 2 & 0 \end{bmatrix}\begin{bmatrix} x_1 \\ x_2 \\ x_3 \end{bmatrix}=\begin{bmatrix} 0 \\ 0 \\ 0 \end{bmatrix}$$

$$\left.\begin{array}{r} -2x_1+0x_2-x_3=0 \\ x_1-x_2+x_3=0 \end{array}\right]$$

$$\Rightarrow \frac{x_1}{0-1}=\frac{x_2}{-1+2}=\frac{x_3}{2-0} \Rightarrow x_1=-1, \quad x_2=1, \quad x_3=2$$

Eigen vector is [–1, 1, 2].

$$\text{Modal matrix } P=\begin{bmatrix} 1 & -2 & -1 \\ -1 & 1 & 1 \\ 0 & 2 & 2 \end{bmatrix} \text{ and } P^{-1}=-\frac{1}{2}\begin{bmatrix} 0 & 2 & -1 \\ 2 & 2 & 0 \\ -2 & -2 & -1 \end{bmatrix}$$

$$\text{Now} \quad P^{-1}AP=\begin{bmatrix} 0 & -1 & \frac{1}{2} \\ -1 & -1 & 0 \\ 1 & 1 & \frac{1}{2} \end{bmatrix}\begin{bmatrix} 1 & 0 & -1 \\ 1 & 2 & 1 \\ 2 & 2 & 3 \end{bmatrix}\begin{bmatrix} 1 & -2 & -1 \\ -1 & 1 & 1 \\ 0 & 2 & 2 \end{bmatrix}=\begin{bmatrix} 1 & 0 & 0 \\ 0 & 2 & 0 \\ 0 & 0 & 3 \end{bmatrix}=D$$

$$A^4=PD^4P^{-1}=\begin{bmatrix} 1 & -2 & -1 \\ -1 & 1 & 1 \\ 0 & 2 & 2 \end{bmatrix}\begin{bmatrix} 1 & 0 & 0 \\ 0 & 16 & 0 \\ 0 & 0 & 81 \end{bmatrix}\begin{bmatrix} 0 & -1 & \frac{1}{2} \\ -1 & -1 & 0 \\ 1 & 1 & \frac{1}{2} \end{bmatrix}=\begin{bmatrix} -49 & -50 & -40 \\ 65 & 66 & 40 \\ 130 & 130 & 81 \end{bmatrix}$$ **Ans.**

EXERCISE 41.9

Find a matrix P which transforms the following matrices to diagonal form. Hence calculate the power matrix.

1. If $A=\begin{bmatrix}1 & 1 & 3\\ 1 & 5 & 1\\ 3 & 1 & 1\end{bmatrix}$, calculate A^4. **Ans.** $\begin{bmatrix}251 & 405 & 235\\ 405 & 891 & 405\\ 235 & 405 & 251\end{bmatrix}$

2. If $A=\begin{bmatrix}3 & -1 & 1\\ -1 & 5 & -1\\ 1 & -1 & 3\end{bmatrix}$, calculate A^4. **Ans.** $\begin{bmatrix}251 & -405 & 235\\ -405 & 891 & -405\\ 235 & -405 & 251\end{bmatrix}$

3. If $A=\begin{bmatrix}2 & -1 & 1\\ -1 & 2 & -1\\ 1 & -1 & 2\end{bmatrix}$, calculate A^6. **Ans.** $\begin{bmatrix}1366 & -1365 & 1365\\ -1365 & 1366 & -1365\\ 1365 & -1365 & 1366\end{bmatrix}$

4. If $A=\begin{bmatrix}1 & 1 & 1\\ 0 & 2 & 1\\ -4 & 4 & 3\end{bmatrix}$, calculate A^8. **Ans.** $\begin{bmatrix}-12099 & 12355 & 6305\\ -12100 & 12356 & 6305\\ -13120 & 13120 & 6561\end{bmatrix}$

5. Show that the matrix A is diagonalisable $A=\begin{bmatrix}3 & 1 & -1\\ -2 & 1 & 2\\ 0 & 1 & 2\end{bmatrix}$. If so obtain the matrix P such that P^{-1} AP is a diagonal matrix. *(AMIETE, June 2010)*

41.23 SYLVESTER THEOREM

Let $\quad P(A) = C_0 A^n + C_1 A^{n-1} + C_2 A^{n-2} + \ldots + C_{n-1} A + C_n I$

and $\quad |\lambda I - A| = f(\lambda)$ and Adjoint matrix of $[\lambda I - A] = [f(\lambda)]$

$$z(\lambda) = \frac{[f(\lambda)]}{f'(\lambda)} = \frac{\text{Adjoint matrix of } [\lambda I - A]}{f'(\lambda)}$$

Then according to Sylvester's theorem

$$P(A) = P(\lambda_1).\, Z(\lambda_1) + P(\lambda_2).\, Z(\lambda_2) + P(\lambda_3).\, Z(\lambda_3) + \ldots$$

$$= \sum_{r=1}^{n} P(\lambda_r).\, Z(\lambda_r)$$

Example 42. *If* $A=\begin{bmatrix}2 & 0\\ 0 & 1\end{bmatrix}$, *find* A^{100}.

Solution. $f(\lambda) = |\lambda I - A| = \begin{vmatrix}\lambda & 0\\ 0 & \lambda\end{vmatrix} - \begin{vmatrix}2 & 0\\ 0 & 1\end{vmatrix} = \begin{vmatrix}\lambda-2 & 0\\ 0 & \lambda-1\end{vmatrix} = 0$

$\Rightarrow \quad f(\lambda) = (\lambda-2)(\lambda-1) = 0$ or $\lambda = 1, \lambda = 2$

$f(\lambda) = \lambda^2 - 3\lambda + 2,\ f'(\lambda) = 2\lambda - 3$

$f'(2) = 4 - 3 = 1,\ f'(1) = 2 - 3 = -1$

$[f(\lambda)]$ = Adjoint matrix of the matrix $[\lambda I - A] = \begin{bmatrix}\lambda-1 & 0\\ 0 & \lambda-2\end{bmatrix}$

$$Z(\lambda_1) = Z(1) = \frac{[f(1)]}{f'(1)} = \frac{1}{-1}\begin{bmatrix}0 & 0\\ 0 & -1\end{bmatrix} = \begin{bmatrix}0 & 0\\ 0 & 1\end{bmatrix}$$

$$Z(\lambda_2) = Z(2) = \frac{[f(2)]}{f'(2)} = \frac{1}{1}\begin{bmatrix} 1 & 0 \\ 0 & 0 \end{bmatrix}$$

By Sylvester theorem $P(A) = P(\lambda_1).\, Z(\lambda_1) + P(\lambda_2).\, Z(\lambda_2)$

$$A^{100} = P(\lambda_1)\, Z(\lambda_1) + P(\lambda_2)\, Z(\lambda_2)$$

$$= \lambda_1^{100}\begin{bmatrix} 0 & 0 \\ 0 & 1 \end{bmatrix} + \lambda_2^{100}\begin{bmatrix} 1 & 0 \\ 0 & 0 \end{bmatrix} = 1^{100}\begin{bmatrix} 0 & 0 \\ 0 & 1 \end{bmatrix} + 2^{100}\begin{bmatrix} 1 & 0 \\ 0 & 0 \end{bmatrix}$$

$$= \begin{bmatrix} 0 & 0 \\ 0 & 1 \end{bmatrix} + \begin{bmatrix} 2^{100} & 0 \\ 0 & 0 \end{bmatrix} = \begin{bmatrix} 2^{100} & 0 \\ 0 & 1 \end{bmatrix}$$ **Ans.**

Example 43. *Find A^{10} for the matrix*

$$A = \begin{bmatrix} -1 & 3 \\ 1 & 1 \end{bmatrix}$$ (*D. U. April* 2010)

Solution. $f(\lambda) = |\lambda I - A| = \begin{vmatrix} \lambda & 0 \\ 0 & \lambda \end{vmatrix} - \begin{vmatrix} -1 & 3 \\ 1 & 1 \end{vmatrix} = \begin{vmatrix} \lambda+1 & -3 \\ -1 & -1+\lambda \end{vmatrix} = 0$

$\Rightarrow \quad f(\lambda) = \lambda^2 - 1 - 3 = 0 \quad \Rightarrow \lambda^2 = 4 \Rightarrow \lambda = 2, \lambda = -2$

$\Rightarrow \quad f(\lambda) = \lambda^2 - 4 \quad \Rightarrow f'(\lambda) = 2\lambda$

$f'(-2) = -4, \quad f'(2) = 4$

$[f(\lambda)] =$ Adjoint matrix of the matrix $[\lambda I - A] = \begin{bmatrix} -1+\lambda & -3 \\ -1 & 1+\lambda \end{bmatrix}$

$$Z(\lambda_1) = Z(-2) = \frac{[f(-2)]}{f'(-2)} = \frac{-1}{4}\begin{bmatrix} -3 & -3 \\ -1 & -1 \end{bmatrix} = \frac{1}{4}\begin{bmatrix} 3 & 3 \\ 1 & 1 \end{bmatrix}$$

$$Z(\lambda_2) = Z(2) = \frac{[f(2)]}{f'(2)} = \frac{1}{4}\begin{bmatrix} 1 & -3 \\ -1 & 3 \end{bmatrix}$$

By Sylvester theorem

$$P(A) = P(\lambda_1)\,.\,Z(\lambda_1) + P(\lambda_2)\,.\,Z(\lambda_2)$$

$$A^{10} = P(\lambda_1)\,.\,Z(\lambda_1) + P(\lambda_2)\,.\,Z(\lambda_2)$$

$$= \lambda_1^{10}\frac{1}{4}\begin{bmatrix} 3 & 3 \\ 1 & 1 \end{bmatrix} + \frac{\lambda_2^{10}}{4}\begin{bmatrix} 1 & -3 \\ -1 & 3 \end{bmatrix}$$

$$= \frac{(-2)^{10}}{4}\begin{bmatrix} 3 & 3 \\ 1 & 1 \end{bmatrix} + \frac{2^{10}}{4}\begin{bmatrix} 1 & -3 \\ -1 & 3 \end{bmatrix}$$

$$= \frac{2^{10}}{4}\begin{bmatrix} 3+1 & 3-3 \\ 1-1 & 1+3 \end{bmatrix} = 2^8\begin{bmatrix} 4 & 0 \\ 0 & 4 \end{bmatrix} = \begin{bmatrix} 2^{10} & 0 \\ 0 & 2^{10} \end{bmatrix}$$ **Ans.**

Example 44. *If* $A = \begin{bmatrix} 1 & 0 \\ 0 & 3 \end{bmatrix}$, *find A^{50}.*

Solution. $f(\lambda) = |\lambda I - A| = \begin{vmatrix} \lambda & 0 \\ 0 & \lambda \end{vmatrix} - \begin{vmatrix} 1 & 0 \\ 0 & 3 \end{vmatrix} = \begin{vmatrix} \lambda-1 & 0 \\ 0 & \lambda-3 \end{vmatrix} = 0$

$\Rightarrow \quad f(\lambda) = (\lambda - 1)(\lambda - 3) = 0$ or $\lambda = 1, \lambda = 3$

$$f(\lambda)=\lambda^2-4\lambda+3,\quad f'(\lambda)=2\lambda-4$$
$$f'(1)=2-4=-2,\quad f'(3)=6-4=2$$

$[f(\lambda)]$ = Adjoint matrix of the matrix $[\lambda I-A]=\begin{bmatrix}\lambda-3 & 0\\ 0 & \lambda-1\end{bmatrix}$

$$Z(\lambda_1)=Z(1)=\frac{[f(1)]}{f'(1)}=\frac{1}{-2}\begin{bmatrix}-2 & 0\\ 0 & 0\end{bmatrix}=\begin{bmatrix}1 & 0\\ 0 & 0\end{bmatrix}$$
$$Z(\lambda_2)=Z(3)=\frac{[f(3)]}{f'(3)}=\frac{1}{2}\begin{bmatrix}0 & 0\\ 0 & 2\end{bmatrix}=\begin{bmatrix}0 & 0\\ 0 & 1\end{bmatrix}$$

By Sylvester theorem $P(A)=P_1(\lambda_1).Z(\lambda_1)+P(\lambda_2).Z(\lambda_2)$

$$A^{50}=P(\lambda_1)Z(\lambda_1)+P(\lambda_2)Z(\lambda_2)=\lambda_1^{50}\begin{bmatrix}1 & 0\\ 0 & 0\end{bmatrix}+\lambda_2^{50}\begin{bmatrix}0 & 0\\ 0 & 1\end{bmatrix}$$
$$=1^{50}\begin{bmatrix}1 & 0\\ 0 & 0\end{bmatrix}+3^{50}\begin{bmatrix}0 & 0\\ 0 & 1\end{bmatrix}=\begin{bmatrix}1 & 0\\ 0 & 0\end{bmatrix}+\begin{bmatrix}0 & 0\\ 0 & 3^{50}\end{bmatrix}=\begin{bmatrix}1 & 0\\ 0 & 3^{50}\end{bmatrix}$$ **Ans.**

EXERCISE 41.10

Use sylvesters theorem in solving the following:

1. Verify Sylvesters theorem for A^3, where $A=\begin{bmatrix}1 & 4\\ 3 & 2\end{bmatrix}$

2. Given $A=\begin{bmatrix}1 & 0\\ 0 & 3\end{bmatrix}$, find A^{256}. **Ans.** $\begin{bmatrix}1 & 0\\ 0 & 3^{256}\end{bmatrix}$

3. Given $A=\begin{bmatrix}\lambda_1 & 0\\ 0 & \lambda_2\end{bmatrix}$, show that $e^A=\begin{bmatrix}e^{\lambda_1} & 0\\ 0 & e^{\lambda_2}\end{bmatrix}$

4. Given $A=\begin{bmatrix}-1 & 3\\ 1 & 1\end{bmatrix}$, show that 2 sin A = | sin 2 | A.

5. Prove that 3 tan A = A tan (3) where $A=\begin{bmatrix}-1 & 4\\ 2 & 1\end{bmatrix}$

6. Prove that $\sin^2 A+\cos^2 A=1$, where $A=\begin{bmatrix}1 & 2\\ -1 & 4\end{bmatrix}$

7. Given $A=\begin{bmatrix}1 & -2 & -3\\ 0 & 2 & 0\\ 0 & 0 & 3\end{bmatrix}$, find A^{-1}. **Ans.** $\begin{bmatrix}1 & 1 & 1\\ 0 & -\frac{1}{2} & 0\\ 0 & 0 & \frac{1}{3}\end{bmatrix}$

8. Given $A=\begin{bmatrix}1 & 20 & 0\\ -1 & 7 & 1\\ 3 & 0 & -2\end{bmatrix}$, find tan A.

Ans. $\frac{\tan 1}{2}\begin{bmatrix}-18 & 60 & 20\\ 0 & 0 & 0\\ -18 & 60 & 20\end{bmatrix}+\frac{\tan 2}{-1}\begin{bmatrix}-20 & 80 & 20\\ -1 & 4 & 1\\ -15 & 60 & 15\end{bmatrix}+\frac{\tan 3}{2}\begin{bmatrix}-20 & 100 & 20\\ -2 & 10 & 2\\ -12 & 60 & 12\end{bmatrix}$

41.24 COMPLEX MATRICES

Conjugate of a Complex Number

$z = x + i\,y$ is called a complex number where $\sqrt{-1} = i$, x, y are real numbers. $\overline{z} = x - iy$ is called the conjugate of the complex number z, e.g.,

Complex number	Conjugate number
$2 + 3i$	$2 - 3i$
$-4 - 5i$	$-4 + 5i$
$6i$	$-6i$
2	2

Conjugate of a matrix. The matrix formed by replacing the elements of a matrix by their respective conjugate numbers is called the conjugate of A and is denoted by $\overline{A}$.

$$A = (a_{ij})_{m\times n}, \text{ then } \overline{A} = (\overline{a}_{ij})_{m\times n}$$

Example

If $A = \begin{bmatrix} 3+4i & 2-i & 4 \\ i & 2 & -3i \end{bmatrix}$ then $\overline{A} = \begin{bmatrix} 3-4i & 2+i & 4 \\ -i & 2 & 3i \end{bmatrix}$

41.25 THEOREM

If A and B be two matrices and their conjugate matrices are $\overline{A}$ and $\overline{B}$ respectively, then

(*i*) $\overline{(\overline{A})} = A$ (*ii*) $\overline{(A+B)} = \overline{A}+\overline{B}$ (*iii*) $\overline{(kA)} = \overline{k}\,\overline{A}$ (*iv*) $\overline{(AB)} = \overline{A}\,\overline{B}$

Proof. Let $A = [a_{ij}]_{m\times n}$, then

$\overline{A} = [\overline{a}_{ij}]_{m\times n}$ where $\overline{a}_{ij}$ is the conjugate complex of a_{ij}.

The (i, j) th element of $\overline{(\overline{A})}$ = the conjugate complex of the (i, j)th element of $\overline{A}$

= the conjugate complex of $\overline{a}_{ij}$

= a_{ij} = the (i, j)th element of A.

Hence $\overline{(\overline{A})} = A$. **Proved.**

(*ii*) Let $A = [a_{ij}]_{m\times n}$ and $B = [b_{ij}]_{m\times n}$

$\overline{A} = [\overline{a}_{ij}]_{m\times n}$ and $\overline{B} = [\overline{b}_{ij}]_{m\times n}$

(i, j) th element of $\overline{(A+B)}$ = conjugate complex of (i, j) th element of $(A + B)$

= conjugate complex of $(a_{ij} + b_{ij})$

$= \overline{(a_{ij}+b_{ij})} = \overline{a}_{ij} + \overline{b}_{ij}$

= (i, j)th element of $\overline{A}$ + (i, j)th element of $\overline{B}$

= (i, j)th element of $(\overline{A}+\overline{B})$

Hence, $\overline{(A+B)} = \overline{A}+\overline{B}$ **Proved.**

(*iii*) Let $A = [a_{ij}]_{m\times n}$, let k be any complex number.

The (i, j)th element of $\overline{(kA)}$ = conjugate complex of the (i, j)th element of kA

$$= \text{conjugate complex of } ka_{ij}$$

$$= \overline{ka_{ij}} = \overline{k} \cdot \overline{a_{ij}}$$

$$= \overline{k} \cdot (i, j)\text{th element of } \overline{A} = (i, j)\text{th element of } \overline{k} \cdot \overline{A}$$

Hence, $\overline{kA} = \overline{k} \cdot \overline{A}$ **Proved.**

(*iv*) Let $A = [a_{ij}]_{m \times n}$, $B = [b_{ij}]_{n \times p}$

Then $\overline{A} = [\overline{a_{ij}}]_{m \times n}$, $\overline{B} = [\overline{b_{ij}}]_{n \times p}$

The (i, j)th element of $(\overline{AB})$ = conjugate complex of (i, j)th element of AB

$$= \text{conjugate complex of } \sum_{j=1}^{n} a_{ij}\, b_{jk} = \left(\sum_{j=1}^{n} \overline{a_{ij} b_{jk}}\right) = \sum_{j=1}^{n} \overline{a_{ij}} \cdot \overline{b_{jk}}$$

$$= (i, j)\text{th element of } \overline{A} \cdot \overline{B}$$

Hence, $(\overline{AB}) = \overline{A} \cdot \overline{B}$ **Proved.**

41.26 TRANSPOSE OF CONJUGATE OF A MATRIX

The transpose of a conjugate of a matrix A is denoted by

$$(\overline{A})' = A^{\theta}$$

The (i, j)th element of A^{θ} = (j, i)th element of $\overline{A}$

= conjugate complex of (j, i)th element of A.

Example 45. *If* $A = \begin{bmatrix} 2+3i & 1-2i & 2+4i \\ 3-4i & 4+3i & 2-6i \\ 5 & 5+6i & 3 \end{bmatrix}$, *find* A^{θ}

Solution. We have $A = \begin{bmatrix} 2+3i & 1-2i & 2+4i \\ 3-4i & 4+3i & 2-6i \\ 5 & 5+6i & 3 \end{bmatrix} \Rightarrow \overline{A} = \begin{bmatrix} 2-3i & 1+2i & 2-4i \\ 3+4i & 4-3i & 2+6i \\ 5 & 5-6i & 3 \end{bmatrix}$

$$A^{\theta} = (\overline{A})' = \begin{bmatrix} 2-3i & 3+4i & 5 \\ 1+2i & 4-3i & 5-6i \\ 2-4i & 2+6i & 3 \end{bmatrix}$$ **Ans.**

EXERCISE 41.11

1. If the matrix $A = \begin{bmatrix} 1+i & 3-5i \\ 2i & 5 \end{bmatrix}$, find (*i*) $\overline{A}$ (*ii*) $(\overline{A})'$ (*iii*) A^{θ} (*iv*) $(A^{\theta})^{\theta}$

Ans. (*i*) $\overline{A} = \begin{bmatrix} 1-i & 3+5i \\ -2i & 5 \end{bmatrix}$ (*ii*) $(\overline{A})' = \begin{bmatrix} 1-i & -2i \\ 3+5i & 5 \end{bmatrix}$

(*iii*) $A^{\theta} = \begin{bmatrix} 1-i & -2i \\ 3+5i & 5 \end{bmatrix}$ (*iv*) $(A^{\theta})^{\theta} = \begin{bmatrix} 1+i & 3-5i \\ 2i & 5 \end{bmatrix}$

41.27 HERMITIAN MATRIX

Definition. A square matrix $A = [a_{ij}]$ is said to be Hermitian if the (i, j)th element of A, *i.e.*,

$$a_{ij} = \overline{a}_{ji} \quad \text{for all } i \text{ and } j.$$

For example $\begin{bmatrix} 2 & 3+4i \\ 3-4i & 1 \end{bmatrix}, \begin{bmatrix} a & b-id \\ b+id & c \end{bmatrix}$

Hence all the elements of the principal diagonal are real.

A necessary and sufficient condition for a matrix A to be Hermitian is that $A = A^\theta$.

Example 46. *Prove that the following*

(i) $(A^\theta)^\theta = A$ *(ii)* $(A + B)^\theta = A^\theta + B^\theta$ *(iii)* $(kA)^\theta = \overline{k}\, A^\theta$ *(iv)* $(AB)^\theta = B^\theta \cdot A^\theta$

where A^θ and B^θ be the transposed conjugates of A and B respectively, A and B being conformable to multiplication.

Solution.

(i) $(A^\theta)^\theta = \overline{\left[\left((\overline{A})'\right)\right]}' = \overline{\overline{[A]}} = A \qquad \text{as } \left\{(\overline{A})'\right\}' = \overline{A}$

(ii) $(A+B)^\theta = (\overline{A+B})' = (\overline{A}+\overline{B})'$

$$= (\overline{A})' + (\overline{B})' = A^\theta + B^\theta$$

(iii) $(kA)^\theta = (\overline{kA})' = (\overline{k}\,\overline{A})' = \overline{k}(\overline{A})' = \overline{k}\, A^\theta$

(iv) $(AB)^\theta = (\overline{AB})' = (\overline{A}\cdot\overline{B})' = (\overline{B})'\cdot(\overline{A})' = B^\theta \cdot A^\theta$ **Proved.**

Example 47. *Prove that matrix* $A = \begin{bmatrix} 1 & 1-i & 2 \\ 1+i & 3 & i \\ 2 & -i & 0 \end{bmatrix}$ *is Hermitian.*

Solution. $\overline{A} = \begin{bmatrix} 1 & 1+i & 2 \\ 1-i & 3 & -i \\ 2 & i & 0 \end{bmatrix} \Rightarrow (\overline{A})' = \begin{bmatrix} 1 & 1-i & 2 \\ 1+i & 3 & i \\ 2 & -i & 0 \end{bmatrix}$

$A^\theta = A \quad \Rightarrow \quad A$ is Hermitian matrix. **Proved.**

Example 48. *Show that* $A = \begin{bmatrix} -i & 3+2i & -2-i \\ -3+2i & 0 & 3-4i \\ 2-i & -3-4i & -2i \end{bmatrix}$ *is Skew-Hermitian matrix.*

Solution. $\overline{A} = \begin{bmatrix} i & 3-2i & -2+i \\ -3-2i & 0 & 3+4i \\ 2+i & -3+4i & 2i \end{bmatrix}$

$$(\overline{A})' = \begin{bmatrix} i & -3-2i & 2+i \\ 3-2i & 0 & -3+4i \\ -2+i & 3+4i & 2i \end{bmatrix}$$

$$\Rightarrow \quad A^\theta = \begin{bmatrix} i & -3-2i & 2+i \\ 3-2i & 0 & -3+4i \\ -2+i & 3+4i & 2i \end{bmatrix} \qquad [\because A^\theta = (\overline{A})']$$

$$= -\begin{bmatrix} -i & 3+2i & -2-i \\ -3+2i & 0 & 3-4i \\ 2-i & -3-4i & -2i \end{bmatrix} = -A$$

$A^\theta = -A \Rightarrow$ A is Skew-Hermitian matrix. **Proved.**

Example 49. *Show that the matrix $B^\theta AB$ is Hermitian or Skew-Hermitian according as A is Hermitian or Skew-Hermitian.*

Solution. (*i*) Let A be Hermitian $\Rightarrow A^\theta = A$

Now
$$(B^\theta AB)^\theta = (AB)^\theta (B^\theta)^\theta$$
$$= B^\theta \cdot A^\theta \cdot B$$
$$= B^\theta \cdot A \cdot B \qquad (A^\theta = A)$$

Hence, $B^\theta AB$ is Hermitian.

(*ii*) Let A be Skew-Hermitian $\Rightarrow A^\theta = -A$

Now,
$$(B^\theta AB)^\theta = (AB)^\theta \cdot (B^\theta)^\theta$$
$$= B^\theta \cdot A^\theta \cdot B$$
$$= -B^\theta A \cdot B \qquad (A^\theta = -A)$$

Hence $B^\theta AB$ is Skew-Hermitian. **Proved.**

41.28 THE CHARACTERISTIC ROOTS OF A HERMITIAN MATRIX ARE ALL REAL

Solution.

We know that matrix A is Hermitian if

$$A^\theta = A \text{ i.e., where } A^\theta \ (\overline{A}') \text{ or } (\overline{A})'$$

Also
$$(\lambda A)^\theta = \overline{\lambda} A^\theta \text{ and } (AB)^\theta = B^\theta A^\theta.$$

If λ is a characteristic root of matrix A then $AX = \lambda X$. ...(1)

$\therefore \qquad (AX)^\theta = (\lambda X)^\theta \qquad$ or $\qquad X^\theta A^\theta \ \overline{\lambda} X^\theta.$

But A is Hermitian. $\therefore A^\theta = A$.

$\therefore \qquad X^\theta A = \overline{\lambda} X^\theta \qquad \therefore \ X^\theta AX = \overline{\lambda} X^\theta X$ ...(2)

Again from (1) $X^\theta AX = X^\theta \lambda X = \lambda X^\theta X$ ...(3)

Hence from (2) and (3) we conclude that $\overline{\lambda} = \lambda$ showing that λ is real.

Deduction 1. *From above we conclude that characteristic roots of real symmetric matrix are all real, as in this case, real symmetric matrix will be Hermitian.*

For symmetric, we know that $A' = A$. $(\overline{A}') = \overline{A}$

$\therefore \qquad$ or $\quad A^\theta = A \qquad\qquad \overline{A} = A$ as A is real. Rest as above.

41.29 SKEW-HERMITIAN MATRIX

Definition. A square matrix $A = (a_{ij})$ is said to be Skew-Hermitian matrix if the (i, j)th element of A is equal to the negative of the conjugate complex of the (j, i)th element of A, *i.e.*, $a_{ij} = -\overline{a}_{ji}$ for all i and j.

If A is a Skew-Hermitian matrix, then

$$a_{ii} = -\overline{a_{ii}}$$

$$a_{ii} + \overline{a_{ii}} = 0$$

Obviously, a_{ii} is either a pure imaginary number or must be zero.

For example $\begin{bmatrix} 0 & -3+4i \\ 3+4i & 0 \end{bmatrix}$ and $\begin{bmatrix} 0 & a-ib \\ -a-ib & 0 \end{bmatrix}$ are Skew-Hermitian matrixes.

A necessary and sufficient condition for a matrix A to be Skew-Hermitian is that $A^\theta = -A$.

Deduction 2. *Characteristic roots of a skew Hermitian matrix is either zero or a pure imaginary numbers.* *(D.U. III Sem. 2012, April 2010)*

If A is skew Hermitian, then iA is Hermitian.

Also λ be a characteristic root of A then $AX = \lambda X$.

$\therefore$
$$(i\,.A)\,X = (i\lambda)\,X.$$

Above shows that $i\lambda$ is characteristic root of matrix iA, which is Hermitian and hence $i\lambda$ should be real, which will be possible if λ is either pure imaginary or zero.

Example 50. *Show that every square matrix can be expressed as R + iS uniquely where R and S are Hermitian matrices.*

Solution. Let A be any square matrix. It can be rewritten as

$$A = \left\{\frac{1}{2}(A + A^\theta)\right\} + i\left\{\frac{1}{2i}(A - A^\theta)\right\} = R + iS$$

where $R = \frac{1}{2}(A + A^\theta),\ S = \frac{1}{2i}(A - A^\theta)$

Now we have to show that R and S are Hermitian matrices.

$$R^\theta = \frac{1}{2}(A + A^\theta)^\theta = \frac{1}{2}[A^\theta + (A^\theta)^\theta] = \frac{1}{2}(A^\theta + A) = \frac{1}{2}(A + A^\theta) = R$$

Thus R is Hermitian matrix.

Now, $$S^\theta = \left[\frac{1}{2i}(A - A^\theta)\right]^\theta = -\frac{1}{2i}(A - A^\theta)^\theta$$

$$= -\frac{1}{2i}[A^\theta - (A^\theta)^\theta] = \frac{-1}{2i}(A^\theta - A) = \frac{1}{2i}(A - A^\theta) = S$$

Thus S is a Hermitian matrix.

Hence $A = R + iS$, where R and S are Hermitian matrices.

Now, we have to show its **uniqueness**.

Let $A = P + iQ$ be another expression, where P and Q are Hermitian matrices, *i.e.*,

$$P^\theta = P,\ Q^\theta = Q$$

Then $$A^\theta = (P + iQ)^\theta = P^\theta + (iQ)^\theta = P^\theta - iQ^\theta = P - iQ$$

$$A = P + iQ \text{ and } A^\theta = P - iQ$$

$$P = \frac{1}{2}(A + A^\theta) = R \text{ and } Q = \frac{1}{2i}(A - A^\theta) = S$$

Hence $A = R + iS$ is the unique expression, where R and S are Hermitian matrices. **Proved.**

Example 51. *Express the matrix* $A\begin{bmatrix} i & 2-3i & 4+5i \\ 6+i & 0 & 4-5i \\ -i & 2-i & 2+i \end{bmatrix}$ *as a sum of Hermitian and Skew Hermitian matrix.* *(U.P.I Sem Dec, 2009)*

Solution. Here, we have

$$A = \begin{bmatrix} i & 2-3i & 4+5i \\ 6+i & 0 & 4-5i \\ -i & 2-i & 2+i \end{bmatrix} \quad \ldots (1)$$

$$\overline{A} = \begin{bmatrix} -i & 2+3i & 4-5i \\ 6-i & 0 & 4+5i \\ i & 2+i & 2-i \end{bmatrix}$$

$$(\overline{A})' = \begin{bmatrix} -i & 6-i & i \\ 2+3i & 0 & 2+i \\ 4-5i & 4+5i & 2-i \end{bmatrix}$$

$$A^{\theta} = \begin{bmatrix} -i & 6-i & i \\ 2+3i & 0 & 2+i \\ 4-5i & 4+5i & 2-i \end{bmatrix} \quad \ldots (2)$$

On adding (1) & (2), we get

$$A + A^{\theta} = \begin{bmatrix} 0 & 8-4i & 4+6i \\ 8+4i & 0 & 6-4i \\ 4-6i & 6+4i & 4 \end{bmatrix}$$

Let $$R = \frac{1}{2}[A + A^{\theta}] = \begin{bmatrix} 0 & 4-2i & 2+3i \\ 4+2i & 0 & 3-2i \\ 2-3i & 3+2i & 2 \end{bmatrix} \quad \ldots (3)$$

On subtracting (2) from (1), we get

$$A - A^{\theta} = \begin{bmatrix} 2i & -4-2i & 4+4i \\ 4-2i & 0 & 2-6i \\ -4+4i & -2-6i & 2i \end{bmatrix}$$

$$\frac{1}{2}(A - A^{\theta}) = \begin{bmatrix} i & -2-i & 2+2i \\ 2-i & 0 & 1-3i \\ -2+2i & -1-3i & i \end{bmatrix} \quad \ldots (4)$$

From (3) and (4), we have

$$A = \underbrace{\begin{bmatrix} 0 & 4-2i & 2+3i \\ 4+2i & 0 & 3-2i \\ 2-3i & 3+2i & 2 \end{bmatrix}}_{\text{Hermitian matrix}} + \underbrace{\begin{bmatrix} i & -2-i & 2+2i \\ 2-i & 0 & 1-3i \\ -2+2i & -1-3i & i \end{bmatrix}}_{\text{Skew-Hermitian matrix}}$$

Example 52. *Express the matrix* $A = \begin{bmatrix} 1+i & 2 & 5-5i \\ 2i & 2+i & 4+2i \\ -1+i & -4 & 7 \end{bmatrix}$ *as the sum of Hermitian matrix and Skew-Hermitian matrix.*

Solution. $A=\begin{bmatrix}1+i & 2 & 5-5i\\ 2i & 2+i & 4+2i\\ -1+i & -4 & 7\end{bmatrix} \Rightarrow \overline{A}=\begin{bmatrix}1-i & 2 & 5+5i\\ -2i & 2-i & 4-2i\\ -1-i & -4 & 7\end{bmatrix}$ ...(1)

$$(\overline{A})'=\begin{bmatrix}1-i & -2i & -1-i\\ 2 & 2-i & -4\\ 5+5i & 4-2i & 7\end{bmatrix} \Rightarrow A^{\theta}=\begin{bmatrix}1-i & -2i & -1-i\\ 2 & 2-i & -4\\ 5+5i & 4-2i & 7\end{bmatrix} \quad ...(2)$$

On adding (1) and (2), we get

$$A+A^{\theta}=\begin{bmatrix}2 & 2-2i & 4-6i\\ 2+2i & 4 & 2i\\ 4+6i & -2i & 14\end{bmatrix}$$

Let $$R=\frac{1}{2}(A+A^{\theta})=\begin{bmatrix}1 & 1-i & 2-3i\\ 1+i & 2 & i\\ 2+3i & -i & 7\end{bmatrix} \quad ...(3)$$

On subtracting (2) from (1), we get

$$A-A^{\theta}=\begin{bmatrix}2i & 2+2i & 6-4i\\ -2+2i & 2i & 8+2i\\ -6-4i & -8+2i & 0\end{bmatrix}$$

Let $$S\ \frac{1}{2}(A-A^{\theta})=\begin{bmatrix}i & 1+i & 3-2i\\ -1+i & i & 4+i\\ -3-2i & -4+i & 0\end{bmatrix} \quad ...(4)$$

From (3) and (4), we have

$$A=\underset{\text{Hermitian matrix}}{\begin{bmatrix}1 & 1-i & 2-3i\\ 1+i & 2 & i\\ 2+3i & -i & 7\end{bmatrix}}+\underset{\text{Skew-Hermitian matrix}}{\begin{bmatrix}i & 1+i & 3-2i\\ -1+i & i & 4+i\\ -3-2i & -4+i & 0\end{bmatrix}}$$ **Ans.**

Example 53. *For any square matrix, $AA^{\theta} = I$ show that $A^{\theta} = I$.*

Solution. $AA^{\theta} = I$ (given)

So A is invertible.

Let B be another matrix such that

$$AB = BA = I \quad ...(1)$$

Now $B = BI = B(AA^{\theta})\ (AA^{\theta} = I)$

$= (BA)\, A^{\theta}$

$= IA^{\theta} = A^{\theta}$ [Using (1)]

We know that $BA = I$ [From (1)]

Putting the value of B from (2) in (1), we get

$\Rightarrow$ $A^{\theta} A = I$ **Proved.**

41.30 CHARACTERISTIC ROOTS OF A SKEW-HERMITIAN MATRIX IS EITHER ZERO OR PURELY AN IMAGINARY NUMBER

Since A is a skew-Hermitian matrix:

$\therefore$ iA is Hermitian matrix.

Let λ be a characteristic root of A.

Then, $AX = \lambda X \Rightarrow (iA)X = (i\lambda)X$

$\Rightarrow$ $i\lambda$ is a characteristic root of matrix iA.

But $i\lambda$ is a characteristic root of Hermitian matrix.

Therefore, $i\lambda$ should be real.

Hence, λ is either zero or purely imaginary. **Proved.**

41.31 PERIODIC MATRIX

A square matrix is said to be periodic, if $A^{k+1} = A$, where k is a positive integer. If k is the least positive integer for which $A^{k+1} = A$, then A is said to be of period k.

41.32 IDEMPOTENT MATRIX

A square matrix is said to be idempotent provided $A^2 = A$.

41.33 PROVE THAT THE EIGEN VALUES OF AN IDEMPOTENT MATRIX ARE EITHER ZERO OR UNITY

(*R.G.P.V., Bhopal, I Semester, June* 2007)

Solution. Let A be an idempotent matrix

$\therefore$ $A^2 = A$

Let λ be a characteristic root of A and the corresponding vector be X. Hencce $X \neq 0$ and

$$AX = \lambda X \quad \text{...(1)}$$

$\Rightarrow$ $A(AX) = A(\lambda X) = \lambda(AX)$

$\Rightarrow$ $(AA)X = \lambda(\lambda X)$ $[\because$ From (1), $AX = \lambda X]$

$\Rightarrow$ $A^2X = \lambda^2 X$

$\Rightarrow$ $AX = \lambda^2 X$ $[\because A^2 = A]$

$\Rightarrow$ $\lambda X = \lambda^2 X$ [From (1) $AX = \lambda X$]

$\Rightarrow$ $(\lambda^2 - \lambda)X = 0$ $\Rightarrow$ $\lambda^2 - \lambda = 0$

$\Rightarrow$ $\lambda(\lambda - 1) = 0$ $\Rightarrow$ $\lambda^2 = 0, 1$ $[\because X \neq 0]$

Hence, the eigen values of an idempotent matrix are either zero or unity. **Proved.**

Example 54. *Determine all the idempotent diagonal matrices of order n.*

Solution. Let $A = \text{diag.}\ [d_1, d_2, d_3, \ldots d_n]$ be an idempotent matrix of order n.

Here, for the matrix 'A' to be idempotent $A^2 = A$

$$\begin{bmatrix} d_1 & 0 & 0........0 \\ 0 & d_2 & 0........0 \\ 0 & 0 & d_3.......0 \\ \hline 0 & 0 & 0........d_n \end{bmatrix}\begin{bmatrix} d_1 & 0 & 0........0 \\ 0 & d_2 & 0........0 \\ 0 & 0 & d_3.......0 \\ \hline 0 & 0 & 0........d_n \end{bmatrix} = \begin{bmatrix} d_1 & 0 & 0........0 \\ 0 & d_2 & 0........0 \\ 0 & 0 & d_3.......0 \\ \hline 0 & 0 & 0........d_n \end{bmatrix}$$

$$\Rightarrow \quad \begin{bmatrix} d_1^2 & 0 & 0........0 \\ 0 & d_2^2 & 0........0 \\ 0 & 0 & d_3^2.......0 \\ \hline 0 & 0 & 0........d_n^2 \end{bmatrix} = \begin{bmatrix} d_1 & 0 & 0........0 \\ 0 & d_2 & 0........0 \\ 0 & 0 & d_3.......0 \\ \hline 0 & 0 & 0........d_n \end{bmatrix} \therefore$$

$$d_1^2 = d_1; \quad d_2^2 = d_2........d_n^2 = d_n$$

i.e., $d_1 = 0, 1;\ d_2 = 0, 1;\ d_3 = 0, 1\ ....\ d_n = 0, 1.$

Hence diag. $[d_1, d_2, d_3 \dots d_n]$, is the required idempotent matrix where

$d_1 = d_2 = d_3 = \dots d_n = 0$ or 1. **Ans.**

EXERCISE 41.12

1. Which of the following matrices are Hermitian:

$(a)\ \begin{bmatrix} 1 & 2+i & 3-i \\ 2+i & 2 & 4-i \\ 3+i & 4+i & 3 \end{bmatrix}$ $\quad (b)\ \begin{bmatrix} 2i & 3 & 1 \\ 4 & -1 & 6 \\ 3 & 7 & 2i \end{bmatrix}$

$(c)\ \begin{bmatrix} 4 & 2-i & 5+2i \\ 2+i & 1 & 2-5i \\ 5-2i & 2+5i & 2 \end{bmatrix}$ $\quad (d)\ \begin{bmatrix} 0 & i & 3 \\ -7 & 0 & 5i \\ 3i & 1 & 0 \end{bmatrix}$ **Ans.** (*c*)

2. Which of the following matrices are Skew-Hermitian:

$(a)\ \begin{bmatrix} 2i & -3 & 4 \\ 3 & 3i & -5 \\ -4 & 5 & 4i \end{bmatrix}$ $\quad (b)\ \begin{bmatrix} 3i & -1 & 2 \\ 1 & 2i & -6 \\ 4 & 6 & -3i \end{bmatrix}$

$(c)\ \begin{bmatrix} 0 & 1-i & 2+3i \\ -1-i & 0 & 6i \\ -2+3i & 6i & 4i \end{bmatrix}$ $\quad (d)\ \begin{bmatrix} 1 & 3 & 7+i \\ 3i & -i & 6 \\ 7-i & 8 & 0 \end{bmatrix}$ **Ans.** (*a*), (*c*)

3. Give an example of a matrix which is Skew-symmetric but not Skew-Hermitian.

Ans. $\begin{bmatrix} 0 & 2+3i \\ -2-3i & 0 \end{bmatrix}$

4. If A be a Hermitian matrix, show that iA is Skew-Hermitian. Also show that if B be a Skew-Hermitian matrix, then iB must be Hermitian.

5. If A and B are Hermitian matrices, then show that $AB + BA$ is Hermitian and $AB - BA$ is Skew-Hermitian.

6. If A is any square matrix, show that $A + A^\theta$ is Hermitian.

7. If $H = \begin{bmatrix} 3 & 5+2i & -3 \\ 5-2i & 7 & 4i \\ -3 & -4i & 5 \end{bmatrix}$, show that H is a Hermitian matrix.

Verify that iH is a Skew-Hermitian matrix.

8. Show that for any complex square matrix A,

(*i*) $(A + A^*)$ is a Hermitian matrix, where $A^* = A^{-T}$

(*ii*) $(A - A^*)$ is Skew-Hermitian matrix.

(*iii*) A^*A^* and A^*A are Hermitian matrices.

9. Show that any complex square matrix can be uniquely expressed as the sum of a Hermitian matrix and a Skew-Hermitian matrix.

10. Express $A = \begin{bmatrix} i & 2-3i & 4+5i \\ 6+i & 0 & 4-5i \\ -i & 2-i & 2+i \end{bmatrix}$ as the sum of Hermitian and Skew-Hermitian matrices.

11. Prove that the latent roots of a Hermitian matrix are all real.

12. If $A = \begin{bmatrix} 2+i & 3 & -1+3i \\ -5 & i & 4-2i \end{bmatrix}$ show that AA^* is a Hermitian matrix; where A^* is the conjugate transpose of A. *(AMIETE, June 2010)*

41.34 UNITARY MATRIX

A square matrix A is said to be unitary matrix if

$$A \cdot A^\theta = A^\theta A = I$$

Example 55. *If A is a unitary matrix, show that A^T is also unitary.*

Solution. $A \cdot A^\theta = A^\theta A = I$, since A is a unitary matrix.

$$(AA^\theta)^\theta = (A^\theta A)^\theta = I^\theta \qquad (I^\theta = I)$$

$$(AA^\theta)^\theta = (A^\theta A)^\theta = I$$

$$(A^\theta)^\theta A^\theta = A^\theta (A^\theta)^\theta = I$$

$$AA^\theta = A^\theta A = I \qquad [\text{since } (A^\theta)^\theta = A]$$

$$(AA^\theta)^T = (A^\theta A)^T = (I)^T$$

$$(A^\theta)^T A^T = A^T (A^\theta)^T = I$$

$$(A^T)^\theta \cdot A^T = A^T (A^T)^\theta = I$$

Hence, A^T is a unitary matrix. **Proved.**

Example 56. *If A is a unitary matrix, show that A^{-1} is also unitary.* (*DU, III Sem. 2012*)

Solution. $AA^\theta = A^\theta A = I$, since A is a unitary matrix.

$$(AA^\theta)^{-1} = (A^\theta \cdot A)^{-1} = (I)^{-1} \qquad \text{taking inverse}$$

$$(A^\theta)^{-1} \cdot A^{-1} = A^{-1} (A^\theta)^{-1} = I$$

$$(A^{-1})^\theta \cdot A^{-1} = A^{-1} (A^{-1})^\theta = I$$

Hence, A^{-1} is a unitary matrix. **Proved.**

Example 57. *If A and B are two unitary matrices, show that AB is a unitary matrix.*

Solution. $A \cdot A^\theta = A^\theta A = I$ since A is a unitary matrix. ...(1)

Similarly, $$B \cdot B^\theta = B^\theta B = I \qquad ...(2)$$

Now, $$(AB)(AB)^\theta = (AB)(B^\theta \cdot A^\theta)$$

$$= A(BB^{\theta})\cdot A^{\theta}$$

$$= A I A^{\theta} \qquad \text{[From (2)]}$$

$$= AA^{\theta} = I \qquad \text{[From (1)]}$$

Again, $(AB)^{\theta}\cdot(AB) = (B^{\theta}\cdot A^{\theta})(AB)$

$$= B^{\theta}(A^{\theta}A)B \qquad \text{[From (1)]}$$

$$= B^{\theta} I B$$

$$= B^{\theta}B$$

$$= I \qquad \text{[From (2)]}$$

Hence, AB is a unitary matrix. **Proved.**

Example 58. *Prove that the matrix* $\frac{1}{\sqrt{3}}\begin{bmatrix}1 & 1+i\\ 1-i & -1\end{bmatrix}$ *is unitary.*

Solution. Let $A = \frac{1}{\sqrt{3}}\begin{bmatrix}1 & 1+i\\ 1-i & -1\end{bmatrix}$

$$A^{\theta} = \frac{1}{\sqrt{3}}\begin{bmatrix}1 & 1+i\\ 1-i & -1\end{bmatrix}$$

$$A^{\theta}\cdot A = \frac{1}{\sqrt{3}}\begin{bmatrix}1 & 1+i\\ 1-i & -1\end{bmatrix}\times\frac{1}{\sqrt{3}}\begin{bmatrix}1 & 1+i\\ 1-i & -1\end{bmatrix}$$

$$= \frac{1}{3}\begin{bmatrix}1+(1+1) & (1+i)-(1+i)\\ (1-i)-1(1-i) & (1+1)+1\end{bmatrix} = \frac{1}{3}\begin{bmatrix}3 & 0\\ 0 & 3\end{bmatrix} = \begin{bmatrix}1 & 0\\ 0 & 1\end{bmatrix} = I$$

Hence, A is a unitary matrix. **Proved.**

Example 59. *Show that the matrix* $A = \begin{bmatrix}\alpha+i\gamma & -\beta+i\delta\\ \beta+i\delta & \alpha-i\gamma\end{bmatrix}$ *is a unitary matrix if*

$$\alpha^2+\beta^2+\gamma^2+\delta^2 = 1$$ (*U.P., I Semester, Dec.* 2005)

Solution. We have, $A = \begin{bmatrix}\alpha+i\gamma & -\beta+i\delta\\ \beta+i\delta & \alpha-i\gamma\end{bmatrix}$

$$A^{\theta} = \begin{bmatrix}\alpha-i\gamma & \beta-i\delta\\ -\beta-i\delta & \alpha+i\gamma\end{bmatrix}$$

We know that, a square matrix A is said to be unitary if $A\,A^{\theta} = I$

$$\begin{bmatrix}\alpha+i\gamma & -\beta+i\delta\\ \beta+i\delta & \alpha-i\gamma\end{bmatrix}\begin{bmatrix}\alpha-i\gamma & \beta-i\delta\\ -\beta-i\delta & \alpha+i\gamma\end{bmatrix} = \begin{bmatrix}1 & 0\\ 0 & 1\end{bmatrix}$$

$$\begin{bmatrix}\alpha^2+\gamma^2+\beta^2+\delta^2 & \alpha\beta-i\alpha\delta+i\beta\gamma+\gamma\delta-\alpha\beta-i\beta\gamma+i\alpha\delta-\delta\gamma\\ \alpha\beta-i\beta\gamma+i\alpha\delta+\gamma\delta-\alpha\beta-i\alpha\delta+i\beta\gamma-\delta\gamma & \beta^2+\delta^2+\alpha^2+\gamma^2\end{bmatrix} = \begin{bmatrix}1 & 0\\ 0 & 1\end{bmatrix}$$

$$\Rightarrow \begin{bmatrix}\alpha^2+\beta^2+\gamma^2+\delta^2 & 0\\ 0 & \alpha^2+\beta^2+\gamma^2+\delta^2\end{bmatrix} = \begin{bmatrix}1 & 0\\ 0 & 1\end{bmatrix}$$

$$\Rightarrow \quad \alpha^2+\beta^2+\gamma^2+\delta^2 = 1$$ **Proved.**

Example 60. *Define a unitary matrix. If* $N=\begin{bmatrix}0 & 1+2i\\ -1+2i & 0\end{bmatrix}$ *is a matrix, then show that* $(I-N)(I+N)^{-1}$ *is a unitary matrix, where I is an identity matrix.*

(D. U. April 2010, U.P.,)

Solution. Unitary matrix: A square matrix 'A' is said to be unitary if $A^{\theta}A = I$, where $A^{\theta}=(\bar{A})^{T}$ and I is an identity matrix.

we have $N=\begin{bmatrix}0 & 1+2i\\ -1+2i & 0\end{bmatrix}$

$$I-N=\begin{bmatrix}1 & 0\\ 0 & 1\end{bmatrix}-\begin{bmatrix}0 & 1+2i\\ -1+2i & 0\end{bmatrix}=\begin{bmatrix}1 & -1-2i\\ 1-2i & 1\end{bmatrix} \quad ...(1)$$

Now we have to find $(I+N)^{-1}$

$$I+N=\begin{bmatrix}1 & 0\\ 0 & 1\end{bmatrix}+\begin{bmatrix}0 & 1+2i\\ -1+2i & 0\end{bmatrix}=\begin{bmatrix}1 & 1+2i\\ -1+2i & 1\end{bmatrix}$$

$$|I+N| = 1-(-1-4)=6$$

$$\text{Adj. }(I+N)=\begin{bmatrix}1 & -1-2i\\ 1-2i & 1\end{bmatrix}$$

$$(I+N)^{-1}=\frac{Adj(I+N)}{|I+N|}=\frac{1}{6}\begin{bmatrix}1 & -1-2i\\ 1-2i & 1\end{bmatrix} \quad ...(2)$$

For unitary matrix $A^{\theta}A=I$

From (1) and (2), we get

$$\therefore \quad (I-N)(I+N)^{-1}=\frac{1}{6}\begin{bmatrix}1 & -1-2i\\ 1-2i & 1\end{bmatrix}\begin{bmatrix}1 & -1-2i\\ 1-2i & 1\end{bmatrix}=\frac{1}{6}\begin{bmatrix}-4 & -2-4i\\ 2-4i & -4\end{bmatrix}=B \text{ (say)}$$

Now $\quad (\bar{B})^{T}=\frac{1}{6}\begin{bmatrix}-4 & 2+4i\\ -2+4i & -4\end{bmatrix}$

$$(\bar{B})^{T}B=\frac{1}{36}\begin{bmatrix}-4 & 2+4i\\ -2+4i & -4\end{bmatrix}\begin{bmatrix}-4 & -2-4i\\ 2-4i & -4\end{bmatrix}=\frac{1}{36}\begin{bmatrix}36 & 0\\ 0 & 36\end{bmatrix}=I.$$

Hence the result. **Proved.**

41.35. THE MODULUS OF EACH CHARACTERISTIC ROOT OF A UNITARY MATRIX IS UNITY.

(D. U. April 2010, U.P.,)

Solution. Suppose A is a unitary matrix. Then

$$A^{\theta}A=I.$$

Let λ be a characteristic root of A. Then

$$AX=\lambda X \quad ...(1)$$

Taking conjugate transpose of both sides of (1), we get

$$(AX)^{\theta}=\bar{\lambda}X^{\theta} \quad ...(2)$$

$$\Rightarrow \quad X^{\theta}A^{\theta}=\bar{\lambda}X^{\theta}$$

From (1) and (2), we have

$$\Rightarrow \quad (X^\theta A^\theta)(AX) = \bar{\lambda}\lambda X^\theta X$$

$$\Rightarrow \quad X^\theta (A^\theta A) X = \bar{\lambda}\lambda X^\theta X$$

$$\Rightarrow \quad X^\theta I X = \bar{\lambda}\lambda X^\theta X \qquad (\because A^\theta . A = I)$$

$$\Rightarrow \quad X^\theta X = \bar{\lambda}\lambda X^\theta X$$

$$\Rightarrow \quad X^\theta X(\bar{\lambda}\lambda - 1) = 0 \qquad ...(3)$$

Since $X^\theta X \neq 0$ therefore (3) gives

$\lambda\bar{\lambda} - 1 = 0.$ or $\lambda\bar{\lambda} = 1$ or $|\lambda|^2 = 1 \quad \Rightarrow \quad |\lambda| = 1$ **Proved.**

EXERCISE 41.13

1. Show that the matrix $A = \frac{1}{\sqrt{2}}\begin{bmatrix} 1 & i \\ -i & -1 \end{bmatrix}$ is unitary.

2. Prove that a real matrix is unitary if it is orthogonal.

3. Prove that the following matrix is unitary:

$$\begin{bmatrix} \frac{1}{2}(1+i) & \frac{1}{2}(-1+i) \\ \frac{1}{2}(1+i) & \frac{1}{2}(1-i) \end{bmatrix}$$

4. Show that $U = \frac{1}{\sqrt{3}}\begin{bmatrix} 1 & 1 & 1 \\ 1 & \omega & \omega^2 \\ 1 & \omega^2 & \omega \end{bmatrix}$ is a unitary matrix, where ω is the complex cube root of unity.

5. Prove that the latent roots of a unitary matrix have unit modulus.

6. Verify that the matrix

$$A = \frac{1}{2}\begin{bmatrix} 1+i & 1-i \\ 1-i & 1+i \end{bmatrix}$$

has eigen values with unit modulus.

Tick (✓) the correct answer:

7. If λ is an eigen value of the matrix 'M' then for the matrix $(M - \lambda I)$, which of the following statement (*s*) is/are coorrect ?

(*i*) Skew symmetric (*ii*) Non singular (*iii*) Singular (*iv*) None of these **Ans.** (*iii*)

(*U.P., I Sem. Dec. 2009*)

8. A square matrix A is idempotent if :

(*i*) $A' = A$ (*ii*) $A' = -A$ (*iii*) $A^2 = A$ (*iv*) $A^2 = I$ **Ans.** (*iii*)

(*R.G.P.V. Bhopal, I Semester June,* 2007)

9. If a square matrix U such tha $\overline{U}' = U^{-1}$ then U is

(*i*) Orthogonal (*ii*) Unitary (*iii*) Symmetric (*iv*) Hermitian **Ans.** (*ii*)

(*R.G.P.V. Bhopal, I Semester June,* 2007)

10. If λ is an eigen value of a non-singular matrix A then the eigen value of A^{-1} is

(*i*) $1/\lambda$ (*ii*) λ (*iii*) $-\lambda$ (*iv*) $-1/\lambda$

(*AMIETE, June 2010*) **Ans.** (*i*)

First Order Lagrange's Linear and Non-Linear Partial Differential Equations

42.1 PARTIAL DIFFERENTIAL EQUATIONS

Partial differential equations are those equations which contain partial derivatives, independent variables and dependent variables.

The independent variables will be denoted by x and y and the dependent variable by z. The partial differential coefficients are denoted as follows :

$$\frac{\partial z}{\partial x}=p, \qquad \frac{\partial z}{\partial y}=q. \qquad \frac{\partial^2 z}{\partial x^2}=r, \qquad \frac{\partial^2 z}{\partial x \partial y}=s, \qquad \frac{\partial^2 z}{\partial y^2}=t$$

42.2 ORDER

Order of a partial differential equation is the same as that of the order of the highest differential coefficient in it.

42.3 METHOD OF FORMING PARTIAL DIFFERENTIAL EQUATIONS

A partial differential equation is formed by two methods.

(*i*) By eliminating arbitrary constants.

(*ii*) By eliminating arbitrary functions.

(*i*) **Method of elimination of arbitrary constants**

Example 1. *Find the PDE of all sphere whose centre lie on z-axis and given by equations $x^2 + y^2 + (z - a)^2 = b^2$, a and b being constants.* (*U.P., II Semester, 2009*)

Solution. We have, $x^2 + y^2 + (z - a)^2 = b^2$ ... (1)

(1) contains two arbitrary constants a and b.

Differentiating (1) partially w.r.t. x, we get

$$2x+2(z-a)\frac{\partial z}{\partial x}=0$$

$\Rightarrow \qquad x + (z - a)\, p = 0$ ... (2)

Differentiating (1) partially w.r.t. y, we get

$$2y+2(z-a)\frac{\partial z}{\partial y}=0$$

$y + (z - a)\, q = 0$ ... (3)

Let us eliminate a from (2) and (3).

From (2) $$(z-a) = -\frac{x}{p}$$

Putting this value of $z - a$ in (3), we get

$$y - \frac{x}{p}q = 0$$

$\Rightarrow$ $$y\,p - x\,q = 0$$ **Ans.**

(*ii*) **Method of elimination of arbitrary functions**

Example 2. *Form the partial differential equation from*

$$z = f(x^2 - y^2)$$

Solution. We have, $$z = f(x^2 - y^2) \qquad \text{... (1)}$$

Differentiating (1) w.r.t. x and y, we get

$$p = \frac{\partial z}{\partial x} = f'(x^2 - y^2)\,2x \qquad \text{... (2)}$$

$$q = \frac{\partial z}{\partial y} = f'(x^2 - y^2)\,(-2y) \qquad \text{... (3)}$$

Dividing (2) by (3), we get $\dfrac{p}{q} = \dfrac{-x}{y} \quad \Rightarrow \quad py = -\,qx$

$\Rightarrow$ $$yp + xq = 0$$ **Ans.**

EXERCISE 42.1

Form the partial differential equations from:

1. $z = (x + a)(y + b)$ **Ans.** $pq = z$
2. $(x - h)^2 + (y - k)^2 + z^2 = a^2$ **Ans.** $z^2(p^2 + q^2 + 1) = a^2$
3. $2z = (a\,x + y)^2 + b$ **Ans.** $p\,x + q\,y = q^2$
4. $a\,x^2 + b\,y^2 + z^2 = 1$ **Ans.** $z\,(p\,x + q\,y) = z^2 - 1$
5. $x^2 + y^2 = (z - c)^2 \tan^2 \alpha$ **Ans.** $y\,p - x\,q = 0$
6. $z = f(x^2 + y^2)$ **Ans.** $yp - xq = 0$
7. $2z = \dfrac{x^2}{a^2} + \dfrac{y^2}{b^2}$ **Ans.** $2z = xp + yq$
8. $f(x + y + z, x^2 + y^2 + z^2) = 0$ **Ans.** $(y - z)\,p + (z - x)\,q = x - y$

42.4 SOLUTION OF EQUATION BY DIRECT INTEGRATION

Example 3. *Solve* $$\frac{\partial^3 z}{\partial x^2 \partial y} = \cos(2x + 3y)$$

Solution. We have, $$\frac{\partial^3 z}{\partial x^2 \partial y} = \cos(2x + 3y)$$

Integrating w.r.t. 'x', we get $$\frac{\partial^2 z}{\partial x \partial y} = \frac{1}{2}\sin(2x + 3y) + f(y)$$

Again, integrating w.r.t. x, we get $$\frac{\partial z}{\partial y} = -\frac{1}{4}\cos(2x + 3y) + \int f(x)\,dx + g(y)$$

$$= -\frac{1}{4}\cos(2x+3y)+f(y)+g(y)$$

Integrating w.r.t. 'y', we get $z = -\frac{1}{12}\sin(2x+3y)+x\int f(y)\,dy+\int g(y)\,dy$

$$z = -\frac{1}{12}\sin(2x+3y)+x\phi_1(y)+\phi_2(y)$$ **Ans.**

Example 4. *Solve* $\frac{\partial^2 z}{\partial x\,\partial y} = x^2 y$

subject to the conditions $z(x, 0) = x^2$ *and* $z(1, y) = \cos y$.

Solution. We have, $\frac{\partial}{\partial x}\left(\frac{\partial z}{\partial y}\right) = x^2 y$

On integrating w.r.t. x, we obtain

$$\frac{\partial z}{\partial y} = \frac{x^3}{3}y + f(y)$$

Integrating w.r.t. y, we obtain

$$z = \frac{x^3}{3}\cdot\frac{y^2}{2}+\int f(y)\,dy+g(x) \qquad \left[F(y) = \int f(y)\,dy\right]$$

$$\Rightarrow \qquad z = \frac{x^3 y^2}{6}+F(y)+g(x) \qquad \text{... (1)}$$

Condition 1 : Putting $z = x^2$ and $y = 0$ in (1), we get

$$x^2 = 0 + F(0) + g(x)$$

Putting the value of $g(x)$ in (1), we get

$$z = \frac{x^3 y^2}{6}+F(y)+x^2-F(0) \qquad \text{... (2)}$$

Condition 2 : $z(1, y) = \cos y$

Putting $x = 1$ and $z = \cos y$ in (2), we get

$$\cos y = \frac{y^2}{6}+F(y)+1-F(0) \Rightarrow F(y) = \cos y - \frac{1}{6}y^2 - 1 + F(0)$$

Putting the value of $F(y)$ in (2), we get

$$z = \frac{1}{6}x^3 y^2+\cos y-\frac{1}{6}y^2-1+F(0)+x^2-F(0)$$

$$\Rightarrow \qquad z = \frac{1}{6}x^3 y^2+\cos y-\frac{1}{6}y^2-1+x^2$$ **Ans.**

EXERCISE 42.2

Solve the following:

1. $\frac{\partial^2 z}{\partial x\partial y} = xy^2$ **Ans.** $z = \frac{x^2 y^3}{6}+f(y)+\phi(x)$
2. $\frac{\partial^2 z}{\partial x\partial y} = e^y\cos x$ **Ans.** $z = e^y\sin x+f(y)+\phi(x)$
3. $\frac{\partial^2 z}{\partial x\partial y} = \frac{y}{x}+2$ **Ans0.** $z = \frac{y^2}{2}\log x+2xy+f(y)+\phi(x)$

4. $\dfrac{\partial^2 z}{\partial x^2} = a^2 z$, when $x = 0$, $\dfrac{\partial z}{\partial x} = a \sin y$ and $\dfrac{\partial z}{\partial y} = 0$ **Ans.** $z = \sin x + e^y \cos x$

5. $\dfrac{\partial^2 z}{\partial x \partial y} = \sin x \sin y$ if $\dfrac{\partial z}{\partial y} = -2 \sin y$ and $z = 0$, when $x = 0$.

Choose the correct answer:

6. The solution of the partial differential equation $\dfrac{\partial^2 z}{\partial y^2} = \sin(xy)$ is

(*a*) $z = -x^{-2}\sin(xy) + yf(x) + g(x)$ (b) $z = -x^2 \sin(xy) - xf(x) + g(x)$
(*c*) $z = -y^2 \sin(xy) + yf(x) + g(x)$ (*d*) $z = -x^{-2}\sin(xy) + xf(x) + g(x)$

(*AMIETE, June 2009*) **Ans.** (*d*)

42.5 LAGRANGE'S LINEAR EQUATION IS AN EQUATION OF THE TYPE

$$Pp + Qq = R$$

where P, Q, R are the functions of x, y, z and $p = \dfrac{\partial z}{\partial x}$, $q = \dfrac{\partial z}{\partial y}$

Solution. $\qquad Pp + Qq = R \qquad$ ... (1)

This form of the equation is obtained by eliminating an arbitrary function f from

$$f(u, v) = 0 \qquad \text{... (2)}$$

where u, v are functions of x, y, z.

Differentiating (2) partially w.r.t. to x and y.

$$\frac{\partial f}{\partial u}\left(\frac{\partial u}{\partial x} + \frac{\partial u}{\partial z}\frac{\partial z}{\partial x}\right) + \frac{\partial f}{\partial v}\left(\frac{\partial v}{\partial x} + \frac{\partial v}{\partial z}\frac{\partial z}{\partial x}\right) = 0 \qquad \text{... (3)}$$

and

$$\frac{\partial f}{\partial u}\left(\frac{\partial u}{\partial y} + \frac{\partial u}{\partial z}\frac{\partial z}{\partial y}\right) + \frac{\partial f}{\partial v}\left(\frac{\partial v}{\partial y} + \frac{\partial v}{\partial z}\frac{\partial z}{\partial y}\right) = 0 \qquad \text{... (4)}$$

Let us eliminate $\dfrac{\partial f}{\partial u}$ and $\dfrac{\partial f}{\partial v}$ from (3) and (4).

From (3), $$\frac{\partial f}{\partial u}\left[\frac{\partial u}{\partial x} + \frac{\partial u}{\partial z}p\right] = -\frac{\partial f}{\partial v}\left[\frac{\partial v}{\partial x} + \frac{\partial v}{\partial z}p\right] \qquad \text{... (5)}$$

From (4), $$\frac{\partial f}{\partial u}\left[\frac{\partial u}{\partial y} + \frac{\partial u}{\partial z}q\right] = -\frac{\partial f}{\partial v}\left[\frac{\partial v}{\partial y} + \frac{\partial v}{\partial z}q\right] \qquad \text{... (6)}$$

Dividing (5) by (6), we get

$$\frac{\dfrac{\partial u}{\partial x} + \dfrac{\partial u}{\partial z}\cdot p}{\dfrac{\partial u}{\partial y} + \dfrac{\partial u}{\partial z}\cdot q} = \frac{\dfrac{\partial v}{\partial x} + \dfrac{\partial v}{\partial z}\cdot p}{\dfrac{\partial v}{\partial y} + \dfrac{\partial v}{\partial z}\cdot q}$$

$$\Rightarrow \quad \left[\frac{\partial u}{\partial x} + \frac{\partial u}{\partial z}\cdot p\right]\left[\frac{\partial v}{\partial y} + \frac{\partial v}{\partial z}\cdot q\right] = \left[\frac{\partial u}{\partial y} + \frac{\partial u}{\partial z}\cdot q\right]\left[\frac{\partial v}{\partial x} + \frac{\partial v}{\partial z}\cdot p\right]$$

$$\frac{\partial u}{\partial x}\times\frac{\partial v}{\partial y} + \frac{\partial u}{\partial x}\times\frac{\partial v}{\partial z}\cdot q + \frac{\partial u}{\partial z}\times p\times\frac{\partial v}{\partial y} + \frac{\partial u}{\partial z}\times\frac{\partial v}{\partial z}\cdot pq$$

$$= \frac{\partial u}{\partial y}\times\frac{\partial v}{\partial x}+\frac{\partial u}{\partial y}\times\frac{\partial v}{\partial z}p+\frac{\partial u}{\partial z}\cdot q\times\frac{\partial v}{\partial x}+\frac{\partial u}{\partial z}\times\frac{\partial v}{\partial z}\cdot pq$$

$$\Rightarrow \left[\frac{\partial u}{\partial y}\times\frac{\partial v}{\partial z}-\frac{\partial u}{\partial z}\times\frac{\partial v}{\partial y}\right]p+\left[\frac{\partial u}{\partial z}\times\frac{\partial v}{\partial x}-\frac{\partial u}{\partial x}\times\frac{\partial v}{\partial z}\right]q = \frac{\partial u}{\partial x}\times\frac{\partial v}{\partial y}-\frac{\partial u}{\partial y}\times\frac{\partial v}{\partial x} \quad \text{... (7)}$$

If (1) and (7) are the same, then the coefficients of p, q are equal.

$$P=\frac{\partial u}{\partial y}\times\frac{\partial v}{\partial z}-\frac{\partial u}{\partial z}\times\frac{\partial v}{\partial y}, \quad Q=\frac{\partial u}{\partial z}\times\frac{\partial v}{\partial x}-\frac{\partial u}{\partial x}\times\frac{\partial v}{\partial z} \quad \text{... (8)}$$

$$R=\frac{\partial u}{\partial x}\times\frac{\partial v}{\partial y}-\frac{\partial u}{\partial y}\times\frac{\partial v}{\partial x}$$

Now suppose $u = c_1$ and $v = c_2$ are two solutions, where c_1, c_2 are constants.

Differentiating $u = c_1$ and $v = c_2$

$$\frac{\partial u}{\partial x}dx+\frac{\partial u}{\partial y}dy+\frac{\partial u}{\partial z}dz = 0 \quad \text{... (9)}$$

and

$$\frac{\partial v}{\partial x}dx+\frac{\partial v}{\partial y}dy+\frac{\partial v}{\partial z}dz = 0 \quad \text{... (10)}$$

Solving (9) and (10), we get

$$\frac{dx}{\frac{\partial u}{\partial y}\times\frac{\partial v}{\partial z}-\frac{\partial u}{\partial z}\times\frac{\partial v}{\partial y}} = \frac{dy}{\frac{\partial u}{\partial z}\times\frac{\partial v}{\partial x}-\frac{\partial u}{\partial x}\times\frac{\partial v}{\partial z}} = \frac{dz}{\frac{\partial u}{\partial x}\times\frac{\partial v}{\partial y}-\frac{\partial u}{\partial y}\times\frac{\partial v}{\partial x}} \quad \text{... (11)}$$

From (8) and (11), we have $\frac{dx}{P}=\frac{dy}{Q}=\frac{dz}{R}$

Solutions of these equations are $u = c_1$ and $v = c_2$

$\therefore$ $f(u, v) = 0$ is the required solution of (1).

42.6 WORKING RULE TO SOLVE Pp + Qq = R

Step 1. Write down the auxiliary equations $\frac{dx}{P}=\frac{dy}{Q}=\frac{dz}{R}$

Step 2. Solve the above auxiliary equations.

Let the two solutions be $u = c_1$ and $v = c_2$.

Step 3. Then $f(u, v) = 0$ or $u = \phi(v)$ is the required solution of $Pp + Qq = R$.

Example 5. *Solve the following partial differential equation*

$$yq - xp = z. \quad \text{where } p=\frac{\partial z}{\partial x}, q=\frac{\partial z}{\partial y}.$$

Solution. We have, $y q - x p = z$

Here the auxiliary equations are

$$\frac{dx}{-x}=\frac{dy}{y}=\frac{dz}{z}$$

$$-\log x = \log y - \log a \quad \text{(From first two equations)}$$

$$x y = a \quad \text{... (1)}$$

$$\log y = \log z + \log b \quad \text{(From last two equations)}$$

$$\frac{y}{z}=b \qquad \text{... (2)}$$

From (1) and (2) we get the solution

$$f\left(xy, \frac{y}{z}\right)=0. \qquad \textbf{Ans.}$$

Example 6. *Solve $y^2p - xyq = x(z - 2y)$.*

Solution. We have, $y^2p - xyq = x(z - 2y)$

The auxiliary equations are $\dfrac{dx}{y^2}=\dfrac{dy}{-xy}=\dfrac{dz}{x(z-2y)}$ ... (1)

Considering first two members of the equations

$$\frac{dx}{y}=\frac{dy}{-x} \quad \Rightarrow \quad x\,dx = -y\,dy$$

Integrating $\quad \dfrac{x^2}{2}=-\dfrac{y^2}{2}+\dfrac{C_1}{2} \quad \Rightarrow \quad x^2 + y^2 = C_1$ ... (2)

From last two equations of (1), we have $-\dfrac{dy}{y}=\dfrac{dz}{z-2y}$

$\Rightarrow \quad -zdy + 2y\,dy = ydz \quad \Rightarrow \quad 2ydy = ydz + zdy$

On integration, we get $\quad y^2 = yz + C_2$

$$y^2 - yz = C_2 \qquad \text{... (3)}$$

From (2) and (3), we have $x^2 + y^2 = f(y^2 - yz)$ **Ans.**

Example 7. *Solve $(x^2 - yz)\,p + (y^2 - zx)\,q = z^2 - xy$* *(U.P., II Semester, 2010)*

Solution. $(x^2 - y\,z)\,p + (y^2 - z\,x)\,q = z^2 - x\,y$ ... (1)

The auxiliary equations are

$$\frac{dx}{x^2-yz}=\frac{dy}{y^2-zx}=\frac{dz}{z^2-xy}$$

$$\Rightarrow \quad \frac{dx-dy}{x^2-yz-y^2+zx}=\frac{dy-dz}{y^2-zx-z^2+xy}=\frac{dz-dx}{z^2-xy-x^2+yz}$$

$$\Rightarrow \quad \frac{dx-dy}{(x-y)(x+y+z)}=\frac{dy-dz}{(y-z)(x+y+z)}=\frac{dz-dx}{(z-x)(x+y+z)}$$

$$\Rightarrow \quad \frac{dx-dy}{x-y}=\frac{dy-dz}{y-z}=\frac{dz-dx}{z-x} \qquad \text{... (2)}$$

Integrating first two members of (2), we have

$$\log(x-y)=\log(y-z)+\log c_1$$

$$\log\frac{x-y}{y-z}=\log c_1 \quad \Rightarrow \quad \frac{x-y}{y-z}=c_1 \qquad \text{... (3)}$$

Similarly from last two members of (2), we have $\dfrac{y-z}{z-x}=c_2$ .... (4)

From (3) and (4), the required solution is

$$f\left[\frac{x-y}{y-z}, \frac{y-z}{z-x}\right]=0 \qquad \textbf{Ans.}$$

42.7 METHOD OF MULTIPLIERS

Let the auxiliary equations be $\dfrac{dx}{P}=\dfrac{dy}{Q}=\dfrac{dz}{R}$

l, m, n may be constants or functions of x, y, z then, we have

$$\frac{dx}{P}=\frac{dy}{Q}=\frac{dz}{R}=\frac{l\,dx+m\,dy+n\,dz}{lP+mQ+nR}$$

l, m, n are chosen in such a way that

$$lP+mQ+nR=0$$

Thus $\qquad ldx+m\,dy+n\,dz=0$

Solve this differential equation, if the solution is $u=c_1$.

Similarly, choose another set of multipliers (l_1, m_1, n_1) and if the second solution is $v=c_2$.

$\therefore$ Required solution is $f(u, v)=0$.

Example 8. *Solve*

$$(mz-ny)\frac{\partial z}{\partial x}+(nx-lz)\frac{\partial z}{\partial y}=ly-mx.$$

Solution. We have, $(mz-ny)\dfrac{\partial z}{\partial x}+(nx-lz)\dfrac{\partial z}{\partial y}=ly-mx$

Here, the auxiliary equations, are $\dfrac{dx}{mz-ny}=\dfrac{dy}{nx-lz}=\dfrac{dz}{ly-mx}$

Using multipliers x, y, z, we get

$$\text{each fraction}=\frac{x\,dx+y\,dy+z\,dz}{x(mz-ny)+y(nx-lz)+z(ly-mx)}=\frac{x\,dx+y\,dy+z\,dz}{0}$$

$\therefore \qquad x\,dx+y\,dy+z\,dz=0$

which on integration gives

$$x^2+y^2+z^2=c_1 \qquad \text{... (1)}$$

Again using multipliers, l, m, n, we get

$$\text{each fraction}=\frac{l\,dx+m\,dy+n\,dz}{l(mz-ny)+m(nx-lz)+n(ly-mx)}=\frac{l\,dx+m\,dy+n\,dz}{0}$$

$\therefore \qquad ldx+m\,dy+n\,dz=0$

which on integration gives

$$lx+my+nz=c_2 \qquad \text{... (2)}$$

Hence from (1) and (2), the required solution is

$$x^2+y^2+z^2=f(lx+my+nz)$$ **Ans.**

Example 9. *Solve the partial differential equation* $x(y^2+z)\,p-y(x^2+z)\,q=z(x^2-y^2)$ *where,* $p=\dfrac{\partial z}{\partial x}$ *and* $q=\dfrac{\partial z}{\partial y}$. *(U.P., II Semester, 2008)*

Solution. Lagrange's subsidiary equations are

$$\frac{dx}{x(y^2+z)}=\frac{dy}{-y(x^2+z)}=\frac{dz}{z(x^2-y^2)} \qquad \text{...(1)}$$

Using $x, y, -1$ as multipliers, we get

$$\text{each fraction}=\frac{x\,dx+y\,dy-dz}{x^2(y^2+z)-y^2(x^2+z)-z(x^2-y^2)}=\frac{x\,dx+y\,dy-dz}{0}$$

$\therefore \qquad x\,dx + y\,dy - dz = 0$

Integrating, we get

$$\frac{x^2}{2} + \frac{y^2}{2} - z = \frac{C_1}{2}$$

$$\Rightarrow \qquad x^2 + y^2 - 2z = C_1 \qquad \text{...(2)}$$

Again, using $\frac{1}{x}$, $\frac{1}{y}$ and $\frac{1}{z}$ as multipliers, we get

$$\text{each fraction} = \frac{\frac{1}{x}dx + \frac{1}{y}dy + \frac{1}{z}dz}{y^2 + z - x^2 - z + x^2 - y^2} = \frac{\frac{1}{x}dx + \frac{1}{y}dy + \frac{1}{z}dz}{0}$$

$$\therefore \qquad \frac{1}{x}dx + \frac{1}{y}dy + \frac{1}{z}dz = 0$$

Integrating, we get

$\log x + \log y + \log z = \log C_2 \quad \Rightarrow \quad xyz = C_2$ ...(3)

Hence the general solution is

$\phi(x^2 + y^2 - 2z, xyz) = 0$ **Ans.**

Example 10. *Find the general solution of*

$$x(z^2 - y^2)\frac{\partial z}{\partial x} + y(x^2 - z^2)\frac{\partial z}{\partial y} = z(y^2 - x^2)$$ *(AMIETE, Dec. 2009)*

Solution. $x(z^2 - y^2)\frac{\partial z}{\partial x} + y(x^2 - z^2)\frac{\partial z}{\partial y} = z(y^2 - x^2)$ ... (1)

$\therefore$ The auxiliary simultaneous equations are

$$\frac{dx}{x(z^2 - y^2)} = \frac{dy}{y(x^2 - z^2)} = \frac{dz}{z(y^2 - x^2)} \qquad \text{... (2)}$$

Using multipliers x, y, z, we get

$$\text{Each term of (2)} = \frac{x\,dx + y\,dy + z\,dz}{x^2(z^2 - y^2) + y^2(x^2 - z^2) + z^2(y^2 - x^2)} = \frac{x\,dx + y\,dy + z\,dz}{0}$$

$\therefore \qquad x\,dx + y\,dy + z\,dz = 0$ ... (3)

On integration $x^2 + y^2 + z^2 = C_1$

Again (2) can be written as

$$\frac{\frac{dx}{x}}{z^2 - y^2} = \frac{\frac{dy}{y}}{x^2 - z^2} = \frac{\frac{dz}{z}}{y^2 - x^2} = \frac{\frac{dx}{x} + \frac{dy}{y} + \frac{dz}{z}}{(z^2 - y^2) + (x^2 - z^2) + (y^2 - x^2)} \qquad \text{... (4)}$$

$$= \frac{\frac{dx}{x} + \frac{dy}{y} + \frac{dz}{z}}{0} \quad \Rightarrow \quad \frac{dx}{x} + \frac{dy}{y} + \frac{dz}{z} = 0$$

On integration, we get

$\Rightarrow \qquad \log x + \log y + \log z = \log C_2$

$\qquad \log x\,y\,z = \log C_2 \quad \Rightarrow \quad x\,y\,z = C_2$ .... (5)

From (3) and (5), the general solution is

$$xyz = f(x^2 + y^2 + z^2)$$ **Ans.**

Example 11. *Solve the partial differential equation*

$$\frac{y-z}{yz}p=\frac{z-x}{zx}q=\frac{x-y}{xy}$$ *(AMIETE, June 2010, 2009)*

Solution. We have, $\frac{y-z}{yz}p=\frac{z-x}{zx}q=\frac{x-y}{xy}$

Multiplying by xyz, we get

$$x(y-z)\,p+y(z-x)\,q=z(x-y)$$

$$\frac{dx}{x(y-z)}=\frac{dy}{y(z-x)}=\frac{dz}{z(x-y)}=\frac{dx+dy+dz}{x(y-z)+y(z-x)+z(x-y)} \quad \text{... (1)}$$

$$=\frac{dx+dy+dz}{0}$$

$\therefore$ $dx+dy+dz=0$

Which on integration gives

$$x+y+z=a \quad \text{...(2)}$$

Again (1) can be written as

$$\frac{\frac{dx}{x}}{y-z}=\frac{\frac{dy}{y}}{z-x}=\frac{\frac{dz}{z}}{x-y}=\frac{\frac{dx}{x}+\frac{dy}{y}+\frac{dz}{z}}{(y-z)+(z-x)+(x-y)}=\frac{\frac{dx}{x}+\frac{dy}{y}+\frac{dz}{z}}{0} \Rightarrow \frac{dx}{x}+\frac{dy}{y}+\frac{dz}{z}=0$$

On integration, we get

$$\log x+\log y+\log z=\log b \Rightarrow \log xyz=\log b \Rightarrow xyz=b \quad \text{... (3)}$$

From (2) and (3) the general solution is

$$xyz=f(x+y+z)$$ **Ans.**

Example 12. *Solve* $(x^2-y^2-z^2)\,p+2xy\,q=2xz$.

Solution. We have, $(x^2-y^2-z^2)\,p+2xy\,q=2xz$ ... (1)

Here the auxiliary equations are $\frac{dx}{x^2-y^2-z^2}=\frac{dy}{2xy}=\frac{dz}{2xz}$ ... (2)

From the last two members of (2), we have $\frac{dy}{y}=\frac{dz}{z}$

which on integration gives

$$\log y=\log z+\log a \Rightarrow \log\frac{y}{z}=\log a$$

$$\Rightarrow \frac{y}{z}=a \quad \text{... (3)}$$

Using multipliers $x, y, z,$ we have

$$\frac{dx}{x^2-y^2-z^2}=\frac{dy}{2xy}=\frac{dz}{2xz}=\frac{x\,dx+y\,dy+z\,dz}{x(x^2+y^2+z^2)}$$

$$\frac{2x\,dx+2y\,dy+2z\,dz}{(x^2+y^2+z^2)}=\frac{dz}{z}$$

which on integration gives

$$\log(x^2+y^2+z^2)=\log z+\log b$$

$$\frac{x^2+y^2+z^2}{z}=b \qquad \text{... (4)}$$

Hence from (3) and (4), the required solution is $x^2+y^2+z^2=z\,f\left(\frac{y}{z}\right)$ **Ans.**

Example 13. *Solve the differential equation*

$$x^2\frac{\partial z}{\partial x}+y^2\frac{\partial z}{\partial y}=(x+y)z$$

Solution. We have, $x^2\frac{\partial z}{\partial x}+y^2\frac{\partial z}{\partial y}=(x+y)z$ ... (1)

The auxiliary equations of (1) are

$$\frac{dx}{x^2}=\frac{dy}{y^2}=\frac{dz}{(x+y)z} \qquad \text{... (2)}$$

Take first two members of (2) and integrate them

$$-\frac{1}{x}=-\frac{1}{y}+c$$

$$\frac{1}{x}-\frac{1}{y}=c_1 \qquad \text{... (3)}$$

(2) can be written as

$$\frac{\frac{dx}{x}}{x}=\frac{\frac{dy}{y}}{y}=\frac{\frac{dz}{z}}{x+y}=\frac{\frac{dx}{x}+\frac{dy}{y}-\frac{dz}{z}}{(x+y)-(x+y)}$$

$$\Rightarrow \qquad \frac{dx}{x}+\frac{dy}{y}-\frac{dz}{z}=0$$

On integration, we get

$$\Rightarrow \qquad \log x+\log y-\log z=\log c_2$$

$$\Rightarrow \qquad \log\frac{xy}{z}=\log c_2 \quad \Rightarrow \quad \frac{xy}{z}=c_2 \qquad \text{... (4)}$$

From (3) and (4), we have $f\left[\frac{1}{x}-\frac{1}{y},\frac{xy}{z}\right]=0.$ **Ans.**

Example 14. *Find the general solution of* $x\frac{\partial z}{\partial x}+y\frac{\partial z}{\partial y}+t\frac{\partial z}{\partial t}=xyt$

Solution. The auxiliary equations are $\frac{dx}{x}=\frac{dy}{y}=\frac{dt}{t}=\frac{dz}{xyt}$ ... (1)

Taking the first two members and integrating, we get

$$\log x=\log y+\log a=\log ay$$

$$\Rightarrow \qquad x=ay, \qquad \text{i.e.} \qquad x/y=a \qquad \text{... (2)}$$

Similarly, from the 2nd and 3rd members $\frac{t}{y}=b$ ... (3)

Multiplying the equation (1) by xyt, we get

$$dz=\frac{tydx}{1}=\frac{txdy}{1}=\frac{xydt}{1}=\frac{tydx+txdy+xydt}{3}$$

Integrating, we get

$$z = \frac{1}{3}xyt + c \Rightarrow z - \frac{1}{3}xyt = c \quad \text{... (4)}$$

From (2), (3) and (4) the solution is $z - \frac{1}{3}xyt = f\left(\frac{y}{x}\right) + \phi\left(\frac{t}{y}\right)$ **Ans.**

Example 15. *Solve* $(y+z)\,p - (x+z)\,q = x - y$ *(AMIETE, June 2010)*

Solution. $(y+z)\,p - (x+z)\,q = x - y$ ... (1)

$\therefore$ The auxiliary equations are

$$\frac{dx}{y+z} = \frac{dy}{-(x+z)} = \frac{dz}{x-y} \quad \text{... (2)}$$

$$\Rightarrow \frac{dx}{y+z} = \frac{dy}{-(x+z)} = \frac{dz}{x-y} = \frac{dx+dy+dz}{y+z-(x+z)+x-y}$$

$$\Rightarrow \frac{dx}{y+z} = \frac{dy}{-(x+z)} = \frac{dz}{x-y} = \frac{dx+dy+dz}{0}$$

Thus, we have $dx + dy + dz = 0$

Which on integration gives $x + y + z = c_1$ ... (3)

Using multipliers $x, y, -z$ for (2), we get

$$\frac{dx}{y+z} = \frac{dy}{-(x+z)} = \frac{dz}{x-y} = \frac{x\,dx + y\,dy - z\,dz}{x(y+z) - y(x+z) - z(x-y)}$$

$$\Rightarrow \frac{dx}{y+z} = \frac{dy}{-(x+z)} = \frac{dz}{x-y} = \frac{x\,dx + y\,dy - z\,dz}{0}$$

Integrating $x\,dx + y\,dy - z\,dz = 0$, we get

$$\frac{x^2}{2} + \frac{y^2}{2} - \frac{z^2}{2} = c_2 \Rightarrow x^2 + y^2 - z^2 = 2c_2 \quad \text{... (4)}$$

From (3) and (4), we get the required solution $f(x+y+z,\, x^2+y^2-z^2) = 0$ **Ans.**

Example 16. *Solve* $z\,p + yq = x.$

Solution. The auxiliary equations are $\frac{dx}{z} = \frac{dy}{y} = \frac{dz}{x}$ ...(1)

(*i*) (*ii*) (*iii*)

From (*i*) and (*iii*)

$$\frac{dx}{z} = \frac{dz}{x} \Rightarrow x\,.\,dx = z\,.\,dz$$

$$\Rightarrow \frac{x^2}{2} = \frac{z^2}{2} - \frac{c_1}{2} \Rightarrow x^2 = z^2 - c_1 \quad \text{... (2)}$$

$$\Rightarrow z = \sqrt{x^2 + c_1}$$

Putting the value of z in (1), we get

$$\frac{dx}{\sqrt{x^2+c_1}} = \frac{dy}{y}, \quad \sinh^{-1}\frac{x}{\sqrt{c_1}} = \log y + c_2 \quad \text{... (3)}$$

From (2) and (3), the required solution is

$$f(z^2 - x^2) = \sinh^{-1}\frac{x}{\sqrt{c_1}} - \log y$$ **Ans.**

Example 17. *Solve* $px(z-2y^2)=(z-qy)(z-y^2-2x^3)$.

Solution. $px(z-2y^2)=(z-qy)(z-y^2-2x^3)$

$$\Rightarrow \quad px(z-2y^2)+qy(z-y^2-2x^3)=z(z-y^2-2x^3) \quad \text{... (1)}$$

Here the auxiliary equations are

$$\frac{dx}{x(z-2y^2)}=\frac{dy}{y(z-y^2-2x^3)}=\frac{dz}{z(z-y^2-2x^3)} \quad \text{... (2)}$$

From the last two members of (2), we have $\dfrac{dy}{y}=\dfrac{dz}{z}$

which gives on integration

$$\log y=\log z+\log a \quad \Rightarrow \quad y=az \quad \text{... (3)}$$

From the first and third members of (2), we have

$$\frac{dx}{x(z-2y^2)}=\frac{dz}{z(z-y^2-2x^3)}$$

$$\Rightarrow \quad \frac{dx}{x(z-2a^2z^2)}=\frac{dz}{z(z-a^2z^2-2x^3)} \quad \text{[Using (3), } y=az]$$

$$\frac{dx}{x(1-2a^2z)}=\frac{dz}{z-a^2z^2-2x^3}$$

$$\Rightarrow \quad z\,dx-a^2z^2dx-2x^3dx=x\,dz-2a^2\,x\,z\,dz$$

$$\Rightarrow \quad (x\,dz-z\,dx)-a^2(2\,x\,z\,dz-z^2dx)+2x^3\,dx=0$$

$$\Rightarrow \quad \frac{x\,dz-z\,dx}{x^2}-a^2\frac{(2xz\,dz-z^2dx)}{x^2}+2x\,dx=0$$

On integrating, we have $\dfrac{z}{x}-a^2\dfrac{z^2}{x}+x^2=b$ ... (4)

From (3) and (4), we have the required solution

$$\frac{y}{z}=f\left(\frac{z}{x}-\frac{a^2z^2}{x}+x^2\right)$$ **Ans.**

EXERCISE 42.3

Solve the following partial differential equations:

1. $p\tan x+q\tan y=\tan z$ **Ans.** $f\left(\dfrac{\sin x}{\sin y},\dfrac{\sin y}{\sin z}\right)=0$
2. $(y-z)p+(x+y)q=z-x$ **Ans.** $f(x+y+z,\ x^2+2yz)=0$
3. $(y+zx)p-(x+yz)q=x^2-y^2$ **Ans.** $f(x^2+y^2-z^2)=(x-y)^2-(z+1)^2$
4. $zx\dfrac{\partial z}{\partial x}-zy\dfrac{\partial z}{\partial y}=y^2-x^2$ **Ans.** $f(x^2+y^2+z^2,\ xy)=0$
5. $pz-qz=z^2+(x+y)^2$ **Ans.** $[z^2+(x+y)^2]\,e^{-2x}=f(x+y)$
6. $p+q+2xz=0$ **Ans.** $f(x-y)=x^2+\log z$
7. $x^2p+y^2q+z^2=0$ **Ans.** $f\left(\dfrac{1}{y}-\dfrac{1}{x},\dfrac{1}{y}+\dfrac{1}{z}\right)=0$
8. $(x^2+y^2)p+2xyq=(x+y)z$ **Ans.** $f\left(\dfrac{x+y}{z},\dfrac{2y}{x^2-y^2}\right)=0$

9. $\dfrac{\partial z}{\partial x}-2\dfrac{\partial z}{\partial y}=2x-e^y+1$ **Ans.** $f(2x+y)=z-\dfrac{(2x+1)^2}{4}-\dfrac{e^y}{2}$

10. $p + 3q = 5z + \tan(y - 3x)$ **Ans.** $f(y-3x)=\dfrac{e^{5x}}{5z+\tan(y-3x)}$

11. $xp - yq + x^2 - y^2 = 0$ **Ans.** $f(xy)=\dfrac{x^2}{2}+\dfrac{y^2}{2}+z$

12. $(x+y)\left(\dfrac{\partial z}{\partial x}+\dfrac{\partial z}{\partial y}\right)=z-1$ **Ans.** $f(x-y)=\dfrac{x+y}{(z-1)^2}$.

13. $(x^3+3xy^2)\dfrac{\partial z}{\partial x}+(y^3+3x^2y)\dfrac{\partial z}{\partial y}=2(x^2+y^2)z$ **Ans.** $f\left(\dfrac{xy}{z^2}\right)=\dfrac{4xy}{(x^2-y^2)^2}$

14. $(z^2 - 2yz - y^2)\,p + (xy + zx)\,q = xy - zx$ *(A.M.I.E.T.E., Dec. 2010)*

Ans. $f(z^2 - 2yz - y^2,\ x^2 + y^2 + z^2) = 0$

15. Find the solution of the equation $\dfrac{x\partial z}{\partial y}-\dfrac{y\partial z}{\partial x}=0$, which passes through the curve $z = 1$, $x^2 + y^2 = 4$

Ans. $f(x^2 + y^2 - 4,\ z - 1) = 0$

Indicate True or False for the following statements

16. With usual symbols, the *P.D.E.* $u_{xx} + u^2u_{yy} = f(xy)$ is non-linear in 'u' and is of second order.

(U.P., II Semester, 2009) **Ans.** True

42.8 PARTIAL DIFFERENTIAL EQUATIONS NON-LINEAR IN *p* AND *q*.

We give below the methods of solving non-linear partial differntial equations in certain standard form only.

Type I. Equation of the type $f(p, q) = 0$ i.e., equations containing p and q only.

Method. Let the required solution be

$$z = ax + by + c \quad \text{... (1)}$$

$$\frac{\partial z}{\partial x}=a,\ \frac{\partial z}{\partial y}=b.$$

On putting these values in $f(p, q) = 0$

we get $f(a, b) = 0$,

From this, find the value of b in terms of a and substitude the value of b in (1), that will be the required solution.

Example 18. *Solve* $p^2 + q^2 = 1$ ...(1)

Solution. Let

$$z = ax + by + c \quad \text{... (2)}$$

$$p=\frac{\partial z}{\partial x}=a \quad \text{and} \quad q=\frac{\partial z}{\partial y}=b$$

On substituting the values of p and q in (1), we have

$$\therefore \quad a^2 + b^2 = 1 \ \Rightarrow\ b=\sqrt{1-a^2}$$

Putting the value of b in (2), we get

$$z = ax + \sqrt{1-a^2}\ y + c$$ **Ans.**

Example 19. *Solve* $x^2p^2 + y^2q^2 = z^2$ *(MDU, May 2010, R.G.P.V. Bhopal, Feb. 2008)*

Solution. This equation can be transformed in the above type.

$$\frac{x^2}{z^2}p^2 + \frac{y^2}{z^2}q^2 = 1 \Rightarrow \left(\frac{x}{z}\frac{\partial z}{\partial x}\right)^2 + \left(\frac{y}{z}\frac{\partial z}{\partial y}\right)^2 = 1$$

$$\left(\frac{\frac{\partial z}{z}}{\frac{\partial x}{x}}\right)^2 + \left(\frac{\frac{\partial z}{z}}{\frac{\partial y}{y}}\right)^2 = 1 \qquad \dots (1)$$

Let $\frac{\partial z}{z} = \partial Z, \quad \frac{\partial x}{x} = \partial X, \quad \frac{\partial y}{y} = \partial Y$

$\therefore \quad \log z = Z, \quad \log x = X, \quad \log y = Y$

$\therefore$ (1) can be written as

$$\left(\frac{\partial Z}{\partial X}\right)^2 + \left(\frac{\partial Z}{\partial Y}\right)^2 = 1$$

or $\quad P^2 + Q^2 = 1 \qquad \dots (2)$

Let the required solution be

$$Z = aX + bY + c$$

$$P = \frac{\partial Z}{\partial X} = a, \quad Q = \frac{\partial Z}{\partial Y} = b$$

From (2) we have

$$a^2 + b^2 = 1 \Rightarrow b = \sqrt{1-a^2}$$

$\therefore \quad Z = aX + \sqrt{1-a^2}\,Y + c$

$\Rightarrow \quad \log z = a \log x + \sqrt{1-a^2}\,\log y + c$ **Ans.**

Example 20. *Solve* $(x+y)(p+q)^2 + (x-y)(p-q)^2 = 1$

(R.G.P.V. Bhopal, II Semester, June 2006)

Solution. We have,

$$(x+y)(p+q)^2 + (x-y)(p-q)^2 = 1 \qquad \dots (1)$$

Put $\quad x + y = U^2 \Rightarrow 2U\frac{\partial U}{\partial x} = 1$ and $2U\frac{\partial U}{\partial y} = 1$

$\Rightarrow \quad \frac{\partial U}{\partial x} = \frac{1}{2U}$ and $\frac{\partial U}{\partial y} = \frac{1}{2U}$

And $\quad x - y = V^2 \Rightarrow 2V\frac{\partial V}{\partial x} = 1$ and $2V\frac{\partial V}{\partial y} = -1 \Rightarrow \frac{\partial V}{\partial x} = \frac{1}{2V}$ and $\frac{\partial V}{\partial y} = -\frac{1}{2V}$

Also, $\quad p = \frac{\partial z}{\partial x} = \frac{\partial z}{\partial U}.\frac{\partial U}{\partial x} + \frac{\partial z}{\partial V}.\frac{\partial V}{\partial x} \Rightarrow p = \frac{1}{2U}.\frac{\partial z}{\partial U} + \frac{1}{2V}.\frac{\partial z}{\partial V} \qquad \dots (2)$

and $\quad q = \frac{\partial z}{\partial y} = \frac{\partial z}{\partial U}.\frac{\partial U}{\partial y} + \frac{\partial z}{\partial V}.\frac{\partial V}{\partial y} \Rightarrow q = \frac{1}{2U}.\frac{\partial z}{\partial U} - \frac{1}{2V}.\frac{\partial z}{\partial V} \qquad \dots (3)$

(2) + (3), $\quad p + q = \frac{1}{U}\frac{\partial z}{\partial U}, \qquad$ (2) – (3), $p - q = \frac{1}{V}\frac{\partial z}{\partial V}$

On putting the values of $(x+y)$, $(x-y)$, $(p+q)$, $(p-q)$ in (1), we get

$$U^2\left(\frac{1}{U}\frac{\partial z}{\partial U}\right)^2+V^2\left(\frac{1}{V}\frac{\partial z}{\partial V}\right)^2=1, \quad \left(\frac{\partial z}{\partial U}\right)^2+\left(\frac{\partial z}{\partial V}\right)^2=1$$

The complete integral is

$$z=aU+\sqrt{1-a^2}\,V+C \qquad [z=ax+by+c]$$

Hence,
$$z=a\sqrt{x+y}+\sqrt{1-a^2}\,\sqrt{x-y}+C \qquad \textbf{Ans.}$$

EXERCISE 42.4

Solve the following parital differential equations

1. $pq=1$ **Ans.** $z=ax+\frac{1}{a}y+c$ **2.** $\sqrt{p}+\sqrt{q}=1$ **Ans.** $z=ax+(1-\sqrt{a})^2y+c$

3. $p^2-q^2=1$ **Ans.** $z=ax+\sqrt{(a^2-1)}\,y+c$ **4.** $pq+p+q=0$ **Ans.** $z=ax-\frac{a}{1+a}y+c$

Type II. *Equation of the type*

$$z=px+qy+f(p,q)$$

Its solution is $z=ax+by+f(a,b)$

Example 21. *Solve* $z=px+qy+p^2+q^2$

Solution.
$$z=px+qy+p^2+q^2, \qquad [p=a,\ q=b]$$

Its solution is
$$z=ax+by+a^2+b^2 \qquad \textbf{Ans.}$$

Example 22. *Solve* $z=px+qy+2\sqrt{pq}$

Solution.
$$z=px+qy+2\sqrt{pq}$$

Its solution is
$$z=ax+by+2\sqrt{ab} \qquad \textbf{Ans.}$$

Type III. Equation of the type $f(z,p,q)=0$ *i.e.* equations not containing x and y.

Let z be a function of u where

$$u=x+ay.$$

$$\frac{\partial u}{\partial x}=1 \quad \text{and} \quad \frac{\partial u}{\partial y}=a$$

Then
$$p=\frac{\partial z}{\partial x}=\frac{\partial z}{\partial u}\cdot\frac{\partial u}{\partial x}=\frac{\partial z}{\partial u}\ ; \quad q=\frac{\partial z}{\partial y}=\frac{\partial z}{\partial u}\frac{\partial u}{\partial y}=\frac{\partial z}{\partial u}(a)$$

On putting the values of p and q in the given equation $f(z,p,q)=0$, it becomes $f\left(z,\frac{\partial z}{\partial u},a\frac{\partial z}{\partial u}\right)=0$

which is an ordinary differential equation of the first order.

Example 23. *Solve*

$$p\,(1+q)=qz \qquad \textit{(MDU, May 2005)}$$

Solution.
$$p\,(1+q)=qz \qquad \ldots(1)$$

Let
$$u=x+ay \ \Rightarrow\ \frac{\partial u}{\partial x}=1,\ \frac{\partial u}{\partial y}=a$$

$$p = \frac{\partial z}{\partial x} = \frac{\partial z}{\partial u}\frac{\partial u}{\partial x} = \frac{dz}{du} \quad \text{and} \quad q = \frac{\partial z}{\partial y} = \frac{\partial z}{\partial u}\frac{\partial u}{\partial y} = a\frac{dz}{du}$$

(1) becomes

$$\frac{dz}{du}\left(1 + a\frac{dz}{du}\right) = a\frac{dz}{du}z \quad \Rightarrow \quad 1 + a\frac{dz}{du} = az$$

$$a\frac{dz}{du} = az - 1 \quad \Rightarrow \quad \frac{dz}{du} = \frac{az-1}{a}$$

$$\Rightarrow \quad \frac{du}{dz} = \frac{a}{az-1} \quad \Rightarrow \quad du = \frac{adz}{az-1}$$

$$u = \log(az-1) + \log c$$

$$x + ay = \log c\,(az-1)$$ **Ans.**

Example 24. *Solve* $z^2(p^2 + q^2) = x^2 + y^2$, *where* $p = \frac{\partial z}{\partial x}$ *and* $q = \frac{\partial z}{\partial y}$

(R.G.P.V, Bhopal, Dec. 2010, June 2009, 2008, MDU, May, 2006)

Solution. $z^2(p^2 + q^2) = x^2 + y^2$

The equation can be written as:

$$\left(z\frac{\partial z}{\partial x}\right)^2 + \left(z\frac{\partial z}{\partial y}\right)^2 = x^2 + y^2 \quad \ldots (1)$$

Let $zdz = dZ$ so that $Z = \frac{1}{2}z^2$

Now, $\frac{\partial Z}{\partial x} = \frac{\partial Z}{\partial z}\frac{\partial z}{\partial x} = z\frac{\partial z}{\partial x}$ and $\frac{\partial Z}{\partial y} = \frac{\partial Z}{\partial z}\frac{\partial z}{\partial y} = z\frac{\partial z}{\partial y}$

Therefore, equation (1) becomes

$$\left(\frac{\partial Z}{\partial x}\right)^2 + \left(\frac{\partial Z}{\partial y}\right)^2 = x^2 + y^2$$

$$P^2 + Q^2 = x^2 + y^2, \text{ where } P = \frac{\partial Z}{\partial x} \text{ and } \frac{\partial Z}{\partial y} = Q$$

Or $P^2 - x^2 = y^2 - Q^2$ which is of the form $f_1(x, P) = f_2(y, Q)$

Let $P^2 - x^2 = y^2 - Q^2 = a$ then $P = \sqrt{x^2 + a}$ and $Q = \sqrt{y^2 - a}$

Substituting these values of P and Q in

$dZ = Pdx + Qdy$, we get

$$dZ = \sqrt{x^2 + a}\,dx + \sqrt{y^2 - a}\,dy$$

Integrating $Z = \frac{1}{2}x\sqrt{x^2 + a} + \frac{a}{2}\log\left(x + \sqrt{x^2 + a}\right) + \frac{1}{2}y\sqrt{y^2 - a} - \frac{a}{2}\log\left(y + \sqrt{y^2 - a}\right) + b$

$$\Rightarrow z^2 = x\sqrt{x^2 + a} + y\sqrt{y^2 - a} + a\log\frac{x + \sqrt{x^2 + a}}{y + \sqrt{y^2 - a}} + c \quad \left(\text{where } c = 2b,\ Z = \frac{1}{2}z^2\right)$$ **Ans.**

Example 25. *Solve $p(1 + q^2) = q(z - a)$.*

Solution. Let $u = x + by$

so that $$p = \frac{dz}{du} \quad \text{and} \quad q = b\frac{dz}{du}$$

Substituting these values of p and q in the given equation, we have

$$\frac{dz}{du}\left[1 + b^2\left(\frac{dz}{du}\right)^2\right] = b\frac{dz}{du}(z - a)$$

$$1 + b^2\left(\frac{dz}{du}\right)^2 = b(z - a) \Rightarrow b^2\left(\frac{dz}{du}\right)^2 = bz - ab - 1$$

$$\frac{dz}{du} = \frac{1}{b}\sqrt{bz - ab - 1}, \quad \int\frac{b\,dz}{\sqrt{bz - ab - 1}} = \int du + c$$

$$2\sqrt{bz - ab - 1} = u + c$$

$$4(bz - ab - 1) = (u + c)^2$$

$$4(bz - ab - 1) = (x + by + c)^2$$ **Ans.**

Example 26. *Solve $z^2(p^2x^2 + q^2) = 1$*

(*MDU, May 2007, R.G.P.V., Bhopal, II Semester, June 2005*)

Solution. $$z^2(p^2x^2 + q^2) = 1 \qquad ...(1)$$

$$\Rightarrow \quad z^2\left[\left(x\frac{\partial z}{\partial x}\right)^2 + \left(\frac{\partial z}{\partial y}\right)^2\right] = 1 \quad \Rightarrow \quad z^2\left[\left(\frac{\frac{\partial z}{\partial x}}{x}\right)^2 + \left(\frac{\partial z}{\partial y}\right)^2\right] = 1$$

$$\Rightarrow \quad z^2\left[\left(\frac{\partial z}{\partial X}\right)^2 + \left(\frac{\partial z}{\partial y}\right)^2\right] = 1 \qquad ...(2)$$

where $$\frac{\partial x}{x} = \partial X \Rightarrow \log x = X, \quad \text{Let } u = X + ay$$

$$\frac{\partial z}{\partial X} = \frac{dz}{du} \quad \text{and} \quad \frac{\partial z}{\partial y} = a\frac{dz}{du}$$

Then (2) becomes

$$z^2\left[\left(\frac{dz}{du}\right)^2 + \left(a\frac{dz}{du}\right)^2\right] = 1 \Rightarrow \left(\frac{dz}{du}\right)^2 + a^2\left(\frac{dz}{du}\right)^2 = \frac{1}{z^2}$$

$$\Rightarrow \quad \left(\frac{dz}{du}\right)^2 = \frac{1}{z^2(1 + a^2)} \Rightarrow \frac{dz}{du} = \frac{1}{z\sqrt{1 + a^2}} \Rightarrow z\,dz = \frac{du}{\sqrt{1 + a^2}}$$

$$\Rightarrow \quad \int z\,dz = \int\frac{du}{\sqrt{1 + a^2}} + c \Rightarrow \frac{z^2}{2} = \frac{u}{\sqrt{1 + a^2}} + c$$

$$\sqrt{1 + a^2}\,\frac{z^2}{2} = u + c\sqrt{1 + a^2} = X + ay + c\sqrt{1 + a^2}$$

$$\Rightarrow \quad \sqrt{1 + a^2}\,\frac{z^2}{2} = \log x + ay + c\sqrt{1 + a^2}$$ **Ans.**

EXERCISE 42.5

Solve the following partial differential equations:

1. $z^2(p^2+q^2+1)=1$ **Ans.** $(1-z^2)^{\frac{1}{2}}=-\dfrac{x+ay}{\sqrt{1+a^2}}+c$

2. $1+q^2=q(z-a)$ **Ans.** $\dfrac{x+by}{b}+\dfrac{1}{4}(z-a)^2=\dfrac{1}{4}(z-a)\sqrt{(z-a)^2-2^2}+y\cosh^{-1}\left(\dfrac{z-a}{2}\right)$

3. $x^2p^2+y^2q^2=z$ **Ans.** $2\sqrt{z}=\dfrac{\log x+a\log y}{\sqrt{1+a^2}}+c$

Type IV. Equation of the type

$$f_1(x,p)=f_2(y,q)$$

In these equations, z is absent and the terms containing x and p can be written on one side and the terms containing y and q can be written on the other side.

Method. Let $f_1(x,p)=f_2(y,q)=a$

$f_1(x,p)=a$, solve it for p.

$f_2(y,q)=a$, solve it for q.

Since $$dz=\frac{\partial z}{\partial x}dx+\frac{\partial z}{\partial y}dy \Rightarrow dz=p\,dx+q\,dy$$

$$\Rightarrow \quad dz=F_1(x)\,dx+F_2(y)\,dy \Rightarrow z=\int F_1(x)dx+\int F_2(y)dy+c$$

Example 27. *Solve* $p-x^2=q+y^2$

Solution. $p-x^2=q+y^2=c$ (say)

i.e. $p=x^2+c$ and $q=c-y^2$

Putting these values of p and q in

$$dz=p\,dx+q\,dy=(x^2+c)\,dx+(c-y^2)\,dy$$

$$z=\left(\frac{x^3}{3}+c\,x\right)+\left(cy-\frac{y^3}{3}\right)+c_1$$ **Ans.**

Example 28. *Solve* $p^2+q^2=z^2(x+y)$

Solution $$p^2+q^2=z^2(x+y) \Rightarrow \left(\frac{p}{z}\right)^2+\left(\frac{q}{z}\right)^2=(x+y)$$

$$\Rightarrow \left(\frac{1}{z}\frac{\partial z}{\partial x}\right)^2+\left(\frac{1}{z}\frac{\partial z}{\partial y}\right)^2=x+y \Rightarrow \left(\frac{\frac{\partial z}{z}}{\partial x}\right)^2+\left(\frac{\frac{\partial z}{z}}{\partial y}\right)^2=x+y$$

$$\Rightarrow \left(\frac{\partial Z}{\partial x}\right)^2+\left(\frac{\partial Z}{\partial y}\right)^2=x+y \Rightarrow \frac{\partial z}{z}=\partial Z \text{ or } \log z=Z$$

$$\Rightarrow P^2+Q^2=x+y \Rightarrow P^2-x=y-Q^2=a$$

$$P^2-x=a \Rightarrow P=\sqrt{a+x}$$

$$y-Q^2=a \Rightarrow Q=\sqrt{y-a}$$

Therefore, the equation $$dZ=\frac{\partial Z}{\partial x}dx+\frac{\partial Z}{\partial y}dy$$

$$dZ = P\,dx + Q\,dy$$

gives
$$dZ = \sqrt{a+x}\,dx + \sqrt{y-a}\,dy$$

$$Z = \int \sqrt{a+x}\,dx + \int \sqrt{y-a}\,dy + c$$

$\Rightarrow$
$$\log z = \frac{2}{3}(a+x)^{\frac{3}{2}} + \frac{2}{3}(y-a)^{\frac{3}{2}} + c$$ **Ans.**

EXERCISE 42.6

Solve the following partial differential equations:

1. $q - p + x - y = 0$ (*MDU, May 2010*) **Ans.** $2z = (x+a)^2 + (y+a)^2 + b$
2. $\sqrt{p} + \sqrt{q} = 2x$ **Ans.** $z = \frac{1}{6}(2x-a)^3 + a^2y + b$
3. $q = xp + p^2$ **Ans.** $z = -\frac{x^2}{4} + \left\{\frac{x\sqrt{x^2+4a}}{4} + a\log\left(x + \sqrt{x^2+4a}\right)\right\} + ay + b$
4. $z(p^2 - q^2) = x - y$ **Ans.** $z^{3/2} = (x+a)^{3/2} + (y+a)^{3/2} + b$
5. $p^2 - q^2 = x - y$ (*MDU, May 2010*) **Ans.** $z = \frac{2}{3}(x+c)^{3/2} + \frac{2}{3}(y+c)^{3/2} + b$
6. $\frac{y^2 z}{x}p + xzq = y^2$ **Ans.** $\phi\,(x^3 - y^3, x^2 - z^2) = 0$

Tick ✓ the correct answer.

7. The partial differential equation from $z = (a+x)^2 + y$ is

   (*i*) $z = \frac{1}{4}\left(\frac{\partial z}{\partial x}\right)^2 + y$ (*ii*) $z = \frac{1}{4}\left(\frac{\partial z}{\partial y}\right)^2 + y$ (*iii*) $z = \left(\frac{\partial z}{\partial x}\right)^2 + y$ (*iv*) $z = \left(\frac{\partial z}{\partial y}\right)^2 + y$ **Ans.** (*i*)

8. The solution of $xp + yq = z$ is

   (*i*) $f(x, y) = 0$ (*ii*) $f\left(\frac{x}{y}, \frac{y}{z}\right) = 0$ (*iii*) $f(xy, yz) = 0$ (*iv*) $f(x^2, y^2) = 0$ **Ans.** (*ii*)

9. The solution of $p + q = z$ is

   (*i*) $f(x + y, y + \log z) = 0$ (*ii*) $f(xy, y\log z) = 0$

   (*iii*) $f(x - y, y - \log z) = 0$ (*iv*) None of these **Ans.** (*iii*)

10. The solution of $(y - z)p + (z - x)q = x - y$ is

    (*i*) $f(x + y + z) = xyz$ (*ii*) $f(x^2 + y^2 + z^2) = xyz$

    (*iii*) $f(x^2 + y^2 + z^2, x^2y^2z^2) = 0$ (*iv*) $f(x + y + z) = x^2 + y^2 z^2$ **Ans.** (*iv*)

42.9 CHARPIT'S METHOD

General method for solving partial differential equation with two independent variables.

Solution. Let the general partial differential equation be

$$f(x, y, z, p, q) = 0 \quad \text{... (1)}$$

Since z depends on x, y, we have

$$dz = \frac{\partial z}{\partial x}dx + \frac{\partial z}{\partial y}dy$$

$$dz = p\,dx + q\,dy \quad \text{... (2)}$$

The main thing in Charpit's method is to find another relation between the variables x, y, z and p, q. Let the relation be

$$\phi(x, y, z, p, q) = 0 \qquad \text{... (3)}$$

On solving (1) and (3), we get the values of p and q.

These values of p and q when substituted in (2), it becomes integrable.

To determine ϕ, (1) and (3) are differentiated w.r.t. x and y giving

$$\left.\begin{aligned}\frac{\partial f}{\partial x}+\frac{\partial f}{\partial z}p+\frac{\partial f}{\partial p}\frac{\partial p}{\partial x}+\frac{\partial f}{\partial q}\frac{\partial q}{\partial x}=0\\ \frac{\partial \phi}{\partial x}+\frac{\partial \phi}{\partial z}p+\frac{\partial \phi}{\partial p}\frac{\partial p}{\partial x}+\frac{\partial \phi}{\partial q}\frac{\partial q}{\partial x}=0\end{aligned}\right\}\quad \text{w.r.t. } x, \text{ (First pair)}$$

$$\left.\begin{aligned}\frac{\partial f}{\partial y}+\frac{\partial f}{\partial z}p+\frac{\partial f}{\partial p}\frac{\partial p}{\partial y}+\frac{\partial f}{\partial q}\frac{\partial q}{\partial y}=0\\ \frac{\partial \phi}{\partial y}+\frac{\partial \phi}{\partial z}p+\frac{\partial \phi}{\partial p}\frac{\partial p}{\partial y}+\frac{\partial \phi}{\partial q}\frac{\partial q}{\partial y}=0\end{aligned}\right\}\quad \text{w.r.t. } y, \text{ (Second pair)}$$

Eliminating $\frac{\partial p}{\partial x}$ between the equation of first pair, we have

$$-\frac{\partial p}{\partial y}=\frac{\frac{\partial f}{\partial x}+\frac{\partial f}{\partial z}p+\frac{\partial f}{\partial q}\frac{\partial q}{\partial x}}{\frac{\partial f}{\partial p}}=\frac{\frac{\partial \phi}{\partial x}+\frac{\partial \phi}{\partial z}p+\frac{\partial \phi}{\partial q}\frac{\partial q}{\partial x}}{\frac{\partial \phi}{\partial p}}$$

$$\Rightarrow\quad \left(\frac{\partial f}{\partial x}\frac{\partial \phi}{\partial p}-\frac{\partial \phi}{\partial x}\frac{\partial f}{\partial p}\right)+p\left(\frac{\partial f}{\partial z}\frac{\partial \phi}{\partial p}-\frac{\partial \phi}{\partial z}\frac{\partial f}{\partial p}\right)+\frac{\partial q}{\partial x}\left(\frac{\partial f}{\partial q}\frac{\partial \phi}{\partial p}-\frac{\partial \phi}{\partial q}\frac{\partial f}{\partial p}\right)=0 \qquad \text{... (4)}$$

On eliminating $\frac{\partial q}{\partial y}$ between the equations of second pair, we have

$$\left(\frac{\partial f}{\partial y}\frac{\partial \phi}{\partial q}-\frac{\partial \phi}{\partial y}\frac{\partial f}{\partial q}\right)+q\left(\frac{\partial f}{\partial z}\frac{\partial \phi}{\partial q}-\frac{\partial \phi}{\partial z}\frac{\partial f}{\partial q}\right)+\frac{\partial p}{\partial y}\left(\frac{\partial f}{\partial p}\frac{\partial \phi}{\partial q}-\frac{\partial \phi}{\partial p}\frac{\partial f}{\partial q}\right)=0 \qquad \text{... (5)}$$

Adding (4) and (5) and keeping in view the relation $\frac{\partial p}{\partial y}=\frac{\partial^2 z}{\partial x\,\partial y}=\frac{\partial q}{\partial x}$, the terms of the last brackets of (4) and (5) cancel. On rearranging, we get

$$\frac{\partial \phi}{\partial f}\left(\frac{\partial f}{\partial x}+p\frac{\partial f}{\partial z}\right)+\frac{\partial \phi}{\partial q}\left(\frac{\partial f}{\partial y}+q\frac{\partial f}{\partial z}\right)+\frac{\partial \phi}{\partial z}\left(-p\frac{\partial f}{\partial p}-q\frac{\partial f}{\partial q}\right)+\left(-\frac{\partial f}{\partial p}\right)\frac{\partial \phi}{\partial x}+\left(-\frac{\partial f}{\partial q}\right)\frac{\partial \phi}{\partial y}=0$$

$$\Rightarrow\left(-\frac{\partial f}{\partial q}\right)\left(\frac{\partial \phi}{\partial x}\right)+\left(-\frac{\partial f}{\partial q}\right)\frac{\partial \phi}{\partial y}+\left(-p\frac{\partial f}{\partial p}-q\frac{\partial f}{\partial q}\right)\frac{\partial \phi}{\partial z}+\left(\frac{\partial f}{\partial x}+p\frac{\partial f}{\partial z}\right)\frac{\partial \phi}{\partial p}+\left(\frac{\partial f}{\partial y}+q\frac{\partial f}{\partial z}\right)\frac{\partial \phi}{\partial q}=0$$

... (6)

Equation (6) is a Lagrange's linear equation of the first order with x, y, z, p, q as independent variables and ϕ as dependent variable. Its subsidiary equations are

$$\frac{dx}{-\frac{\partial f}{\partial p}}=\frac{dy}{-\frac{\partial f}{\partial q}}=\frac{dz}{-p\frac{\partial f}{\partial p}-q\frac{\partial f}{\partial q}}=\frac{dp}{\frac{\partial f}{\partial x}+p\frac{\partial f}{\partial z}}=\frac{dq}{\frac{\partial f}{\partial y}+q\frac{\partial f}{\partial z}}=\frac{\partial \phi}{0} \qquad \text{... (7)}$$

(Commit to memory)

Any of the integrals of (7) satisfies (6). Such an integral involving p or q or both may be taken as assumed relation (3). However, we should choose the simplest integral involving p and q derived from (7). This relation and equation (1) gives the values p and q. The values of p and q are substituted in (2). On integration new equation (2) gives the solution of (1).

Example 29. *Solve $px + qy = pq$*

Solution. $f(x, y, z, p, q) = 0$ is $px + qy - pq = 0$ ... (1)

$$\frac{\partial f}{\partial x} = p, \quad \frac{\partial f}{\partial y} = q, \quad \frac{\partial f}{\partial z} = 0, \quad \frac{\partial f}{\partial p} = x - q, \quad \frac{\partial f}{\partial q} = y - p$$

Charpit's equations are

$$\frac{dx}{-\frac{\partial f}{\partial p}} = \frac{dy}{-\frac{\partial f}{\partial q}} = \frac{dz}{-p\frac{\partial f}{\partial p} - q\frac{\partial f}{\partial q}} = \frac{dp}{\frac{\partial f}{\partial x} + p\frac{\partial f}{\partial z}} = \frac{dq}{\frac{\partial f}{\partial y} + q\frac{\partial f}{\partial z}} = \frac{d\phi}{0}$$

$$\frac{dx}{-(x-q)} = \frac{dy}{-(y-p)} = \frac{dz}{-p(x-q) - q(y-p)} = \frac{dp}{p} = \frac{dq}{q} = \frac{d\phi}{0}$$

We have to choose the simplest integral involving p and q

$$\Rightarrow \qquad \frac{dp}{p} = \frac{dq}{q} \quad \Rightarrow \quad \log p = \log q + \log a \;\Rightarrow\; p = aq$$

Putting for p in the given equation (1), we get

$$q\,(ax + y) = aq^2 \qquad \therefore \; q = \frac{y + ax}{a}$$

$$\therefore \qquad p = aq = y + ax$$

Now $\qquad dz = pdx + qdy$ ... (2)

Putting for p and q in (2), we get

$$dz = (y + ax)\,dx + \frac{y + ax}{a}\,dy$$

$$adz = (y + ax)\,dx + (y + ax)\,dy$$

$$adz = (y + ax)\,(adx + dy)$$

Integrating $\qquad az = \dfrac{(y + ax)^2}{2} + b$ **Ans.**

Example 30. *Solve $(p^2 + q^2)y = qz$* ... (1)

(MDU, Dec. 2007, R.G.P.V. Bhopal, II Semester, June 2007)

Solution. $f(x, y, z, p, q) = 0$ is $(p^2 + q^2)y - qz = 0$

$$\frac{\partial f}{\partial x} = 0, \frac{\partial f}{\partial y} = p^2 + q^2, \frac{\partial f}{\partial z} = -q, \frac{\partial f}{\partial p} = 2py, \frac{\partial f}{\partial q} = 2qy - z$$

Now, Charpit's equations are

$$\frac{dx}{-\frac{\partial f}{\partial p}} = \frac{dy}{-\frac{\partial f}{\partial q}} = \frac{dz}{-p\frac{\partial f}{\partial p} - q\frac{\partial f}{\partial q}} = \frac{dp}{\frac{\partial f}{\partial x} + p\frac{\partial f}{\partial z}} = \frac{dq}{\frac{\partial f}{\partial y} + q\frac{\partial f}{\partial z}}$$

$$\Rightarrow \qquad \frac{dx}{-2py} = \frac{dy}{-2q + z} = \frac{dz}{-2p^2y - 2q^2y + qz} = \frac{dp}{-pq} = \frac{dq}{p^2 + q^2 - q^2} = \frac{d\phi}{0}$$

We have to choose the simplest integral involving p and q

$$\frac{dp}{-pq} = \frac{dq}{p^2} \Rightarrow -\frac{dp}{q} = \frac{dq}{p} \Rightarrow pdp + qdq = 0$$

Integrating $\qquad p^2 + q^2 = a^2 \qquad$ (say)

Putting for $p^2 + q^2$ in the equation (1), we get

$$a^2 y = qz \Rightarrow q = \frac{a^2 y}{z} \text{ so } p = \sqrt{a^2 - q^2} = \sqrt{a^2 - \frac{a^4 y^2}{z^2}}$$

$$\Rightarrow \qquad p = \frac{a}{z}\sqrt{z^2 - a^2 y^2}$$

Now $\qquad dz = pdx + qdy$

Putting for p and q in (2), we get

$$dz = \frac{a}{z}\sqrt{z^2 - a^2 y^2}\, dx + \frac{a^2 y}{z} dy \Rightarrow zdz = a\sqrt{z^2 - a^2 y^2}\, dx + a^2 ydy$$

$$\frac{zdz - a^2 ydy}{\sqrt{z^2 - a^2 y^2}} = a.dx$$

Integrating, we get

$$\frac{1}{2}\frac{2}{1}\sqrt{z^2 - a^2 y^2} = ax + b \qquad \text{(Put } z^2 - a^2y^2 = t\text{)}$$

On squaring, $\qquad z^2 - a^2 y = (ax + b)^2$ **Ans.**

Example 31. *Solve* $2z + p^2 + qy + 2y^2 = 0$

(MDU, Dec. 2006, R.G.P.V. Bhopal, II Semester, June 2006)

Solution. Here, we have

Let $\qquad f = 2z + p^2 + qy + 2y^2 = 0 \qquad$ ... (1)

$$\Rightarrow \qquad \frac{\partial f}{\partial x} = 0, \frac{\partial f}{\partial z} = 2, \text{ and } \frac{\partial f}{\partial p} = 2p \qquad \text{... (2)}$$

Charpit's equations are

$$\frac{dx}{-\frac{\partial f}{\partial p}} = \frac{dy}{-\frac{\partial f}{\partial q}} = \frac{dz}{-p\frac{\partial f}{\partial p} - q\frac{\partial f}{\partial q}} = \frac{dp}{\frac{\partial f}{\partial x} + p\frac{\partial f}{\partial z}} = \frac{dq}{\frac{\partial f}{\partial y} + q\frac{\partial f}{\partial z}} = \frac{d\phi}{0}$$

Here, we take $\qquad \dfrac{dp}{\frac{\partial f}{\partial x} + p\frac{\partial f}{\partial z}} = \dfrac{dx}{-\frac{\partial f}{\partial p}} \qquad$ ... (3)

Putting the values of $\frac{\partial f}{\partial x}, \frac{\partial f}{\partial z}$ and $\frac{\partial f}{\partial p}$ from (2) in (3), we get

$$\frac{dp}{0 + 2p} = \frac{dx}{-2p}$$

$$dp + dx = 0 \Rightarrow p + x = a \Rightarrow p = a - x \qquad \text{... (4)}$$

On putting $p = a - x$ in (1), we get

$$2z + (a - x)^2 + qy + 2y^2 = 0$$

$$q = -\frac{2z + (a - x)^2 + 2y^2}{y} \qquad \text{... (5)}$$

Putting for p from (4) and for q from (5) in $dz = pdx + qdy$, we get

$$dz = (a-x)\,dx - \frac{2z+(a-x)^2+2y^2}{y}\,dy$$

$$\Rightarrow \quad dz-(a-x)dx + y\,dy = -\frac{[2z+(a-x)^2+y^2]}{y}\,dy$$

$$\frac{2[dz-(a-x)dx+ydy]}{2z+(a-x)^2+y^2} = -\frac{2}{y}dy$$

In the L.H.S. the numerator is the differential coefficient of denominator.

Hence integrating, we get

$$\log\{2z+(a-x)^2+y^2\} = \log y^{-2} + \log b \quad \Rightarrow \quad 2z+(a-x)^2+y^2 = \frac{b}{y^2}$$

$$\therefore \qquad 2z = \frac{b}{y^2} - (a-x)^2 - y^2$$ **Ans.**

Example 32. *Solve by using the charpit's method*

$$p^2 + qy = z$$ (*MDU, Dec., 2005*)

Solution. $f(x, y, z, p, q) = 0$ is $p^2 + qy - z = 0$

$$\frac{\partial f}{\partial x} = 0, \quad \frac{\partial f}{\partial y} = q, \quad \frac{\partial f}{\partial z} = -1, \quad \frac{\partial f}{\partial p} = 2p, \quad \frac{\partial f}{\partial q} = y$$

The subsidiary equations are

$$\frac{dx}{2p} = \frac{dy}{y} = \frac{dz}{2p^2+qy} = \frac{-dp}{-p} = \frac{-dp}{q+q(-1)}$$

$$\frac{dx}{2p} = \frac{dy}{y} = \frac{dz}{2p^2+qy} = \frac{dp}{p} = \frac{dq}{0}$$

$$dq = 0 \quad \Rightarrow \quad q = a \quad \text{and} \quad p = \sqrt{z-ay}$$

Substitute the values of p and q in $dz = pdx + qdy$, we get

$$dz = \sqrt{z-ay}\,dx + ady$$

$$\frac{dz - ady}{\sqrt{z-ay}} = dx$$

On integration $$\frac{\sqrt{z-ay}}{\frac{1}{2}} = x+b, \quad 2\sqrt{z-ay} = x+b$$

$$4(z-ay) = (x+b)^2 \quad \Rightarrow \quad z - ay = \frac{(x+b)^2}{4}$$

$$\boxed{z = ay + \frac{x^2+b^2+2xb}{4}}$$ **Ans.**

Example 33. *Solve* $2(z + xp + yq) = yp^2$ (*MDU, Dec., 2008, RGPV, II Sem., Feb., 2006*)

Solution. Here, we have

$$f = 2(z + xp + yq) - yp^2$$

Forming the auxiliary equations

$$\frac{dx}{2x-2yp} = \frac{dy}{2y} = \frac{dz}{2xp-2yp^2+2qy} = \frac{dp}{-(2p+2p)} = \frac{dq}{-(2q-p^2+2q)}$$

$$\Rightarrow \quad \frac{dx}{x-yp} = \frac{dy}{y} = \frac{dz}{xp-yp^2+yq} = \frac{dp}{-2p} = \frac{dq}{-\left(2q-\frac{p^2}{2}\right)}$$

Using second & fourth

$$\frac{dy}{y} = \frac{dp}{-2p} \Rightarrow \log y = -\frac{1}{2}\log p \Rightarrow -2\log y = \log p \Rightarrow p = ay^{-2} = \frac{a}{y^2}$$

Substituting the value of p in given PDE, we get

$$2z + 2x\left(\frac{a}{y^2}\right) + 2yq = y\left(\frac{a}{y^2}\right)^2$$

$$\Rightarrow \quad 2yq = y\left(\frac{a}{y^2}\right)^2 - 2z - 2x\left(\frac{a}{y^2}\right) \Rightarrow q = \frac{a^2}{2y^4} - \frac{z}{y} - \frac{ax}{y^3}$$

Now, $$dz = p\,dx + q\,dy$$

$$\Rightarrow \quad dz = \frac{a}{y^2}dx + \left(\frac{a^2}{2y^4} - \frac{z}{y} - \frac{ax}{y^3}\right)dy$$

Regrouping the forms, we get

$$\left(\frac{ydz + zdy}{y}\right) = \left(\frac{ay\,dx - ax\,dy}{y^3}\right) + \frac{a^2}{2y^4}dy$$

Multiplying throughout by y, we get

$$y\,dz + z\,dy = a\frac{y\,dx - x\,dy}{y^2} + \frac{a^2}{2y^3}dy$$

$$d(yz) = ad\left(\frac{x}{y}\right) + \frac{a^2}{2}\frac{dy}{y^3}$$

On integration, we get

$$yz = a\frac{x}{y} + \frac{a^2}{2}\left(\frac{1}{-2y^2}\right) + b \Rightarrow z = \frac{ax}{y^2} - \frac{a^2}{4y^3} + \frac{b}{y} \quad \textbf{Ans.}$$

EXERCISE 42.7

Solve the following partial differential equations:

1. $z = p.q$ **Ans.** $2\sqrt{az} = ax + y + \sqrt{ab}$
2. $(p + q)(px + qy) - 1 = 0$ **Ans.** $z\sqrt{(1+a)} = 2\sqrt{(ax+y)} + b$
3. $z = px + qy + p^2 + q^2$ **Ans.** $z = ax + by + a^2 + b^2$
4. $z = p^2x + q^2y$ *(MDU, May, 2005)* **Ans.** $(1+a)z = [\sqrt{ax} + \sqrt{(b+y)}]^2$
5. $z = pq\,xy$ *(MDU, May, 2008)* **Ans.** $z = ax^b y^{1/b}$
6. $pxy + pq + qy = yz$ **Ans.** $\log(z - ax) = y - a\log(a + y) + b$
7. $q + xp = p^2$ **Ans.** $z = axe^{-y} - \frac{1}{2}a^2e^{-2y} + b$
8. $q = px^2 + q^2$ **Ans.** $z = a\log x + \left(\frac{1 \pm \sqrt{1-4a}}{2}\right)y + b$
9. $2xz - px^2 - 2qxy + pq = 0$ *(MDU, Dec., 2009, Rajasthan, 2006)* **Ans.** $z = ay + b(x^2 - a)$

Linear and Non-linear Partial Differential Equations with Constant Coefficients of 2nd Order

43.1 LINEAR HOMOGENEOUS PARTIAL DIFFERENTIAL EQUATIONS OF nTH ORDER WITH CONSTANT COEFFICIENTS

An equation of the type

$$a_0 \frac{\partial^n z}{\partial x^n} + a_1 \frac{\partial^n z}{\partial x^{n-1} \partial y} + ... + a_n \frac{\partial^n z}{\partial y^n} = F(x, y) \qquad ... (1)$$

is called a homogeneous linear partial differential equation of nth order with constant coefficients. It is called homogeneous because all the terms contain derivatives of the same order.

Putting $\frac{\partial}{\partial x} = D$ and $\frac{\partial}{\partial y} = D'$ in (1), we get

$$(a_0 D^n + a_1 D^{n-1} D' + ... + a_n D'^n) z = F(x, y) \quad \Rightarrow \quad f(D, D') z = F(x, y)$$

43.2 RULES FOR FINDING THE COMPLEMENTARY FUNCTION

Consider the equation

$$a_0 \frac{\partial^2 z}{\partial x^2} + a_1 \frac{\partial^2 z}{\partial x \partial y} + a_2 \frac{\partial^2 z}{\partial y^2} = 0 \Rightarrow \qquad (a_0 D^2 + a_1 DD' + a_2 D'^2) z = 0$$

Step 1 : Put $\qquad D = m$ and $D' = 1$

$$a_0 m^2 + a_1 m + a_2 = 0$$

This is the auxiliary equation.

Step 2 : Solve the auxiliary equation.

Case 1. If the roots of the auxiliary equation are real and different; say m_1, m_2.

Then $\qquad$ C.F. $= f_1(y + m_1 x) + f_2(y + m_2 x)$

Theory: $(D - m_1 D')(D - m_2 D') z = 0 \qquad ...(1)$

(1) will be satisfied by the solution of

$$(D - m_2 D') z = 0 \quad \Rightarrow \quad p - m_2 q = 0 \qquad ...(2)$$

This is a Lagrange's linear equation. Its subsidiary equations are

$$\frac{dx}{1} = \frac{dy}{-m_2} = \frac{dz}{0} \quad \Rightarrow \quad \frac{x}{1} = \frac{y}{-m_2} \quad \Rightarrow \quad y + m_2 x = C_1 \text{ and}$$

$$\frac{x}{1} = \frac{z}{0} \quad \Rightarrow \quad z = C_2$$

$\therefore$ Solution of (2) is $\qquad z = f_2(y + m_2 x)$

Similarly the solution of $(D - m_1 D')z = 0$ is

$$z = f_1(y + m_1 x)$$

Hence the complete solution of (1) is

$$z = f_1(y + m_1 x) + f_2(y + m_2 x)$$

Case 2. If the roots are equal; say m

Then $\qquad C.F. = f_1(y + mx) + xf_2(y + mx)$

Theory:

$$(D - mD')(D - mD')z = 0 \qquad ...(1)$$

Let $\qquad (D - mD')z = u \qquad ...(2)$

(1) becomes $\qquad (D - mD')z = 0 \qquad ...(3)$

Solution of (3) is $\qquad u = f(y + mx)$

(2) becomes $\qquad (D - mD')z = f(y + mx) \quad \Rightarrow p - mq = f(y + mx)$

This is Lagrange's equation and its subsidiary equations are

$$\frac{dx}{1} = \frac{dy}{-m} = \frac{dz}{f(y+mx)}$$

$$(i) \quad (ii) \quad (iii)$$

From (*i*) and (*ii*), $y + mx = C_1$ and $dz = f(y + mx)dx$

$$dz = f(C_1)dx \quad \Rightarrow \quad z = f(y+mx).x + f_1(y+mx)$$

S. No.	*Roots of A.E.*	*C.F.*
1.	m_1, m_2, m_3 (different)	$f_1(y+m_1x) + f_2(y+m_2x) + f_3(y+m_3x)$
2.	m_1, m_2, m_3 $\begin{bmatrix} m_2 = m_1 \\ m_3 \neq m_1 \end{bmatrix}$	$f_1(y+m_1x) + x\, f_2(y+m_1x) + f_3(y+m_3x)$
3.	m_1, m_2, m_3 $m_1 = m_2 = m_3$	$f_1(y+m_1x) + x\, f_2(y+m_1x) + x^2 f_3(y+m_1x)$

Example 1. *Solve* $(D^3 - 4D^2D' + 3DD'^2)z = 0.$

Solution. $\qquad (D^3 - 4D^2D' + 3DD'^2)z = 0. \qquad [D = m, D' = 1]$

Its auxiliary equation is $m^3 - 4m^2 + 3m = 0$

$$m(m^2 - 4m + 3) = 0$$

$$m(m-1)(m-3) = 0 \qquad \Rightarrow \qquad m = 0, 1, 3$$

The required solution is

$$z = f_1(y) + f_2(y+x) + f_3(y+3x) \qquad \textbf{Ans.}$$

Example 2. *Solve* $\dfrac{\partial^2 z}{\partial x^2} - 4\dfrac{\partial^2 z}{\partial x \partial y} + 4\dfrac{\partial^2 z}{\partial y^2} = 0.$

Solution. $(D^2 - 4DD' + 4D'^2)z = 0$

Its auxiliary equation is $m^2 - 4m + 4 = 0 \qquad [D = m, D' = 1]$

$\Rightarrow \qquad (m-2)^2 = 0 \Rightarrow \qquad m = 2, 2$

The required solution is

$$z = f_1(y + 2x) + xf_2(y + 2x)$$ **Ans.**

Example 3. *Solve the linear partial differential equation*

$$\frac{\partial^4 z}{\partial x^4} - 2\frac{\partial^4 z}{\partial x^3 \partial y} + 2\frac{\partial^4 z}{\partial x \partial y^3} - \frac{\partial^4 z}{\partial y^4} = 0.$$

Solution. Here, we have

$$(D^4 - 2D^3D' + 2DD'^3 - D'^4)\, z = 0, \text{ where } D \equiv \frac{\partial}{\partial x} \text{ and } D' \equiv \frac{\partial}{\partial y}$$

Auxiliary equation is

$$m^4 - 2m^3 + 2m - 1 = 0$$

$$m^3 (m - 1) - m^2 (m - 1) - m (m - 1) + 1 (m - 1) = 0$$

$$\Rightarrow \quad (m^3 - m^2 - m + 1)(m - 1) = 0$$

$$\Rightarrow \quad (m + 1)(m - 1)^3 = 0$$

$$\Rightarrow \quad m = -1, 1, 1, 1$$

$$\therefore \quad z = f_1(y - x) + f_2(y + x) + x f_3(y + x) + x^2 f_4(y + x)$$ **Ans.**

Example 4. *Solve the linear partial differential equation* $\dfrac{\partial^4 z}{\partial x^4} + \dfrac{\partial^4 z}{\partial y^4} = 0.$

Solution. $(D^4 + D'^4)\, z = 0$

Auxiliary equation is $m^4 + 1 = 0$

$$\Rightarrow \quad m^4 + 1 + 2m^2 = 2m^2$$

$$\Rightarrow \quad (m^2 + 1)^2 - (m\sqrt{2})^2 = 0$$

$$\Rightarrow \quad \left(m^2 + \sqrt{2}m + 1\right)\left(m^2 - \sqrt{2}m + 1\right) = 0$$

so that $m^2 + \sqrt{2}m + 1 = 0$ or $m^2 - \sqrt{2}m + 1 = 0$

$$\Rightarrow \quad m = \frac{-1+i}{\sqrt{2}} \text{ and } \frac{1+i}{\sqrt{2}} \text{ and } \frac{-1-i}{\sqrt{2}}, \frac{1-i}{\sqrt{2}},$$

Hence
$$z = f_1\left\{y + \left(\frac{-1+i}{\sqrt{2}}\right)x\right\} + f_2\left\{y + \left(\frac{1+i}{\sqrt{2}}\right)x\right\}$$

$$+ f_3\left\{y + \left(\frac{-1-i}{\sqrt{2}}\right)x\right\} + f_4\left\{y + \left(\frac{1-i}{\sqrt{2}}\right)x\right\}$$ **Ans.**

EXERCISE 43.1

Solve the following equations:

1. $\dfrac{\partial^2 z}{\partial x^2} + \dfrac{4\partial^2 z}{\partial x \partial y} - 5\dfrac{\partial^2 z}{\partial y^2} = 0$ **Ans.** $z = f_1(y + x) + f_2(y - 5x)$

2. $2\dfrac{\partial^2 z}{\partial x^2} + 5\dfrac{\partial^2 z}{\partial x \partial y} + 2\dfrac{\partial^2 z}{\partial y^2} = 0$ **Ans.** $z = f_1(2y - x) + f_2(y - 2x)$

3. $(D^3 - 6D^2D' + 11DD'^2 - 6D'^3)z = 0$ **Ans.** $z = f_1(y + x) + f_2(y + 2x) + f_3(y + 3x)$

4. $\dfrac{\partial^2 z}{\partial x^2} - 2\dfrac{\partial^2 z}{\partial x \partial y} + \dfrac{\partial^2 z}{\partial y^2} = 0$ **Ans.** $z = f_1(y + x) + x f_2(y + x)$

5. $(D^3 - 6D^2D' + 12DD'^2 - 8D'^3)z = 0$ **Ans.** $z = f_1(y + 2x) + x f_2(y + 2x) + x^2 f_3(y + 2x)$

6. $\dfrac{\partial^4 z}{\partial x^4} - \dfrac{\partial^4 z}{\partial y^4} = 0$ **Ans.** $z = f_1(y + x) + f_2(y - x) + f_3(y + ix) + f_4(y - ix)$

7. $\frac{\partial^3 z}{\partial x^3} - 4\frac{\partial^3 z}{\partial x^2 \partial y} + 4\frac{\partial^3 z}{\partial x \partial y^2} = 0$ **Ans.** $z = f_1(y) + f_2(y + 2x) + xf_3(y + 2x)$

8. $\frac{\partial^3 z}{\partial x^3} - 7\frac{\partial^3 z}{\partial x \partial y^2} + 6\frac{\partial^3 z}{\partial y^3} = 0$ **Ans.** $z = f_1(y + x) + f_2(y + 2x) + f_3(y - 3x)$

9. $\frac{\partial^4 z}{\partial x^4} + \frac{\partial^4 z}{\partial y^4} + 2\frac{\partial^4 z}{\partial x^2 \partial y^2} = 0$ **Ans.** $z = f_1(y - x) + x f_2(y - 2x) + f_3(y + x) + xf_4(y + x)$

10. $\frac{\partial^2 z}{\partial x^2} - \frac{\partial^2 z}{\partial x \partial y} - 6\frac{\partial^2 z}{\partial y^2} = 0$ **Ans.** $z = f(y + 3x) + f_2(y - 2x)$

43.3 GENERAL RULES FOR FINDING THE PARTICULAR INTEGRAL

Given partial differential equation is

$$f(D, D')\, z = F(x, y)$$

$$\text{P.I.} = \frac{1}{f(D, D')} F(x, y)$$

If f (D, D′) is a homogeneous function of D and D′ of degree n and R.H.S. function $\phi(ax + by)$, $e^{(ax + by)}$, $ax + by$, $\sin(ax + by)$. Then the particular integral

$$\text{P.I.} = \frac{1}{F(D, D')} \phi(ax + by)$$

P.I. of $\frac{1}{F(D, D')} F(x, y) = \frac{1}{F(a, b)} \iiint \cdots \int \phi(u)\, du\, du \ldots du$ (n times), where $u = ax + by$

GENERAL RULE

Integrate $\phi(u)$ w.r.t. u, n times and after integration replace u by $ax + by$.

Case of failure:

To find P.I.

The given equation is

$$F(D, D')\, z = \phi(ax + by) \text{ and } F(a, b) = 0$$

Procedure: Let $F(D, D')$ is a homogeneous function of degree n.

Differentiating $F(D, D')$ partially w.r.t D and multiply L.H.S. by x, we get

$$x \frac{1}{\frac{\partial}{\partial D} F(D, D')} \phi(ax + by)$$

If F (a, b) is again zero.

Differentiate it second time and multiply by x to get $x^2 \frac{1}{\frac{\partial^2}{\partial D^2} F(D, D')} \phi(ax + by)$

If $F(a, b) \neq 0$ then stop.

If $F(a, b) = 0$, then repeat the above procedure. After m times differentiating and multiplying x by m times, we get

$$x^m \frac{1}{\frac{\partial^m}{\partial D^m} F(D, D')} \phi(ax + by)$$

Let $F(a, b) \neq 0$, then $\left[\text{Put } \dfrac{\partial^m}{\partial D^m} F(D, D') = \psi(D, D')\right]$

$$\text{P.I.} = x^m \frac{1}{(a, b)} \phi(ax + by)$$

SHORT FORMULAE:

We can find P.I. of function on the R.H.S. of the form e^{ax+by}, sin $(ax + by)$, cos $(ax + by)$ either by general formula or by the short formula given below:

(*i*) When $F(x, y) = e^{ax+by}$

$$\text{P.I.} = \frac{1}{f(D, D')} e^{ax+by} = \frac{e^{ax+by}}{f(a, b)} \qquad [\text{Put } D = a, D' = b]$$

Theory, $De^{ax+by} = a e^{ax+by}$, $D' e^{ax+by} = b e^{ax+by}$

$$D^2 e^{ax+by} = a^2 e^{ax+by}, \; DD' e^{ax+by} = ab e^{ax+by}, \; D'^2 e^{ax+by} = b^2 e^{ax+by}$$

$\therefore$ $(D^2 + k_1 DD' + k_2 D'^2) e^{ax+by} = (a^2 + k_1 ab + k_2 b^2) e^{ax+by}$

i.e., $\boxed{f(D, D') e^{ax+by} = f(a, b) e^{ax+by}}$

$\Rightarrow$ $\dfrac{1}{f(D, D')} f(D, D') e^{ax+by} = \dfrac{1}{f(D, D')} f(a, b) e^{ax+by}$

$\Rightarrow$ $\boxed{\dfrac{1}{f(D, D')} e^{ax+by} = \dfrac{1}{f(a, b)} e^{ax+by}}$

(*ii*) When $F(x, y) = \sin(ax + by)$ or $F(x, y) = \cos(ax + by)$

$$\text{P.I.} = \frac{1}{f(D^2, DD', D'^2)} \sin(ax + by) = \frac{1}{f(-a^2, -ab, -b^2)} \sin(ax + by)$$

Put $D^2 = -a^2, \quad DD' = -ab, \quad D'^2 = -b^2$

$$\text{P.I.} = \frac{1}{f(D^2, DD', D'^2)} \cos(ax + by) = \frac{1}{f(-a^2, -ab, -b^2)} \cos(ax + by)$$

Theory.

$$D^2 \sin(ax + by) = -a^2 \sin(ax + by)$$

$$DD' \sin(ax + by) = -ab \sin(ax + by)$$

$$D'^2 \sin(ax + by) = -b^2 \sin(ax + by)$$

$$f(D^2, DD', D'^2) \sin(ax + by) = f(-a^2, -ab, -b^2) \sin(ax + by)$$

$\Rightarrow$ $\dfrac{1}{f(D^2, DD', D'^2)} f(D^2, DD', D'^2) \sin(ax + by)$

$$= \frac{1}{f(D^2, DD', D'^2)} f(-a^2, -ab, -b^2) \sin(ax + by)$$

$$\boxed{\frac{1}{f(-a^2, -ab, -b^2)} \sin(ax + by) = \frac{1}{f(D^2, DD', D'^2)} \sin(ax + by)}$$

Case 1. When R.H.S. $= e^{ax+by}$

Example 5. *Solve:* $\dfrac{\partial^3 z}{\partial x^3} - 3\dfrac{\partial^3 z}{\partial x^2 \partial y} + 4\dfrac{\partial^3 z}{\partial y^3} = e^{x+2y}$ (*U.P. II Semester, June 2007*)

Solution. $$\frac{\partial^3 z}{\partial x^3}-3\frac{\partial^3 z}{\partial x^2 \partial y}+4\frac{\partial^3 z}{\partial y^3}=e^{x+2y}$$

Given equation in symbolic form is

$$(D^3-3D^2D'+4D'^3)z=e^{x+2y}$$

Its A.E. is $m^3-3m^2+4=0 \quad \Rightarrow \quad m=-1, 2, 2.$

$\therefore$ $$\text{C.F.} = f_1(y-x)+f_2(y+2x)+x f_3(y+2x)$$

$$\text{P.I.} = \frac{1}{D^3-3D^2D'+4D'^3}e^{x+2y}$$

Put $D=1, D'=2$ $$= \frac{1}{1-6+32}e^{x+2y}=\frac{e^{x+2y}}{27}$$

Hence, complete solution is

$$z=f_1(y-x)+f_2(y+2x)+x f_3(y+2x)+\frac{e^{x+2y}}{27}$$ **Ans.**

EXERCISE 43.2

Solve the following equations:

1. $\frac{\partial^2 z}{\partial x^2}-\frac{\partial^2 z}{\partial y^2}=e^{x+2y}$ **Ans.** $z=f_1(y+x)+f_2(y-x)-\frac{e^{x+2y}}{3}$

2. $\frac{\partial^2 z}{\partial x^2}-5\frac{\partial^2 z}{\partial x\partial y}+6\frac{\partial^2 z}{\partial y^2}=e^{x+y}$ **Ans.** $z=f_1(y+2x)+f_2(y+3x)+\frac{1}{2}e^{x+y}$

3. $\frac{\partial^2 z}{\partial x^2}-4\frac{\partial^2 z}{\partial x\partial y}+4\frac{\partial^2 z}{\partial y^2}=e^{2x+y}$ **Ans.** $z=f_1(y+2x)+x f_2(y+2x)+\frac{x^2}{2}e^{2x+y}$

4. $\frac{\partial^2 z}{\partial x^2}-7\frac{\partial^2 z}{\partial x\partial y}+12\frac{\partial^2 z}{\partial y^2}=e^{x-y}$ **Ans.** $z=f_1(y+3x)+f_2(y+4x)+\frac{1}{20}e^{x-y}$

5. $\frac{\partial^3 z}{\partial x^3}-2\frac{\partial^3 z}{\partial x^2\partial y}=2e^{2x-y}$ **Ans.** $z=f_1(y)+x f_2(y)+f_3(y+2x)+\frac{1}{8}e^{2x-y}$

6. $(D^2-2DD'+D'^2)z=e^{x+2y}$ **Ans.** $z=f_1(y+x)+x f_2(y+x)+e^{x+2y}$

7. $\frac{\partial^2 z}{\partial x^2}-\frac{\partial^2 z}{\partial y^2}-2\frac{\partial z}{\partial x}+2\frac{\partial z}{\partial y}=e^{2x+3y}$ **Ans.** $z=f_1(y+x)+e^{2x}f_2(y-x)-\frac{1}{3}e^{2x+3y}$

8. $\frac{\partial^2 z}{\partial x^2}-5\frac{\partial^2 z}{\partial x\partial y}+6\frac{\partial^2 z}{\partial y^2}=\exp(3x-2y)$ **Ans.** $z=f_1(y+2x)+f_2(y+3x)+\frac{1}{63}e^{3x-2y}$

Case II. When R.H.S. = $\sin(ax+by)$ or $\cos(ax+by)$

Example 6. *Solve the linear partial differential equation*

$$\frac{\partial^2 z}{\partial x^2}+2\frac{\partial^2 z}{\partial x\,\partial y}+\frac{\partial^2 z}{\partial y^2}=\sin(2x+3y)$$ *(U.P. II Semester Summer 2006)*

Solution. We have,

$$\frac{\partial^2 z}{\partial x^2}+2\frac{\partial^2 z}{\partial x\,\partial y}+\frac{\partial^2 z}{\partial y^2}=\sin(2x+3y)$$

$$(D^2+2DD'+D'^2)z=\sin(2x+3y) \quad \text{where } D=\frac{\partial}{\partial x} \text{ and } D'=\frac{\partial}{\partial y}$$

Put $D=m, \quad D'=1$

The auxiliary equation is

$$m^2 + 2m + 1 = 0$$

$$\Rightarrow \quad (m+1)^2 = 0 \Rightarrow m = -1, -1 \quad \Rightarrow \quad \text{C.F.} = f_1(y-x) + x\, f_2(y-x)$$

$$\text{P.I.} = \frac{1}{D^2 + 2DD' + D'^2} \sin(2x+3y) = \frac{1}{-4+2(-6)-9} \sin(2x+3y)$$

$$= \frac{1}{-25} \sin(2x+3y) \qquad \begin{bmatrix} D^2 = -2^2 = -4 \\ D'^2 = -3^2 = -9 \\ DD' = -2\times 3 = -6 \end{bmatrix}$$

Hence, the complete solution is

$$z = C.F. + P.I.$$

$$\Rightarrow \quad z = f_1(y-x) + x\, f_2(y-x) + \frac{1}{-25} \sin(2x+3y)$$

$$\Rightarrow \quad z = f_1(y-x) + x\, f_2(y-x) - \frac{1}{25} \sin(2x+3y) \qquad \textbf{Ans.}$$

Example 7. *Solve* $(D+1)(D+D'-1)z = \sin(x+2y)$ *(U.P. II Semester, 2010)*

Solution. $\text{C.F.} = e^{-x}\phi(y) + e^{x}\phi_2(y-x)$

$$\text{P.I.} = \frac{1}{(D+1)(D+D'-1)} \sin(x+2y) = \frac{1}{D^2 + DD' + D' - 1} \sin(x+2y)$$

$$= \frac{1}{-1+(-2)+D'-1} \sin(x+2y) = \frac{1}{D'-4} \sin(x+2y)$$

$$= \frac{D'+4}{(D'^2-16)} \sin(x+2y) = \frac{D'+4}{(-4-16)} \sin(x+2y)$$

$$= -\frac{1}{20}(D'+4)\sin(x+2y) = -\frac{1}{20}[D'\sin(x+2y) + 4\sin(x+2y)]$$

$$= -\frac{1}{20}[2\cos(x+2y) + 4\sin(x+2y)]$$

Hence, the solution is

$$z = e^{-x}\phi(y) + e^{x}\phi_2(y-x) - \frac{1}{10}[\cos(x+2y) + 2\sin(x+2y)] \qquad \textbf{Ans.}$$

Example 8. *Solve* $\dfrac{\partial^2 z}{\partial x^2} - \dfrac{\partial^2 z}{\partial x \partial y} = \sin x \cos 2y$ *(U.P., II Semester, June, 2010, 2008)*

Solution. We have, $\dfrac{\partial^2 z}{\partial x^2} - \dfrac{\partial^2 z}{\partial x \partial y} = \sin x \cos 2y$

The given equation can be written in the form

$$(D^2 - DD')z = \sin x \cos 2y \quad \text{where } D = \frac{\partial}{\partial x},\ D' = \frac{\partial}{\partial y}$$

Writing $D = m$ and $D' = 1$, the auxiliary equation is

$$m^2 - m = 0 \quad \Rightarrow \quad m(m-1) = 0 \quad \Rightarrow \quad m = 0, 1$$

$$\therefore \quad \text{C.F.} = f_1(y) + f_2(y+x)$$

$$\text{P.I.} = \frac{1}{D^2 - DD'}\sin x\cos 2y = \frac{1}{D^2 - DD'}\frac{1}{2}[\sin(x+2y)+\sin(x-2y)]$$

$$= \frac{1}{2}\frac{1}{D^2 - DD'}\sin(x+2y) + \frac{1}{2}\frac{1}{D^2 - DD'}\sin(x-2y)$$

Put $D^2 = -1, DD' = -2$ in the first integral and $D^2 = -1, DD' = 2$ in the second integral.

$$= \frac{1}{2}\left[\frac{\sin(x+2y)}{-1-(-2)}\right] + \frac{1}{2}\left[\frac{\sin(x-2y)}{-1-(2)}\right] = \frac{1}{2}\sin(x+2y) - \frac{1}{6}\sin(x-2y)$$

Hence the complete solution is z = C.F. + P.I.

i.e., $$z = f_1(y) + f_2(y+x) + \frac{1}{2}\sin(x+2y) - \frac{1}{6}\sin(x-2y)$$ **Ans.**

Example 9. *Solve the partial differential equation:*

$$\frac{\partial^2 z}{\partial x^2} - 3\frac{\partial^2 z}{\partial x\,\partial y} + 2\frac{\partial^2 z}{\partial y^2} = e^{2x-y} + e^{x+y} + \cos(x+2y)$$ (*U.P. II Semester Summer 2006*)

Solution. Given equation is

$$\frac{\partial^2 z}{\partial x^2} - 3\frac{\partial^2 z}{\partial x\,\partial y} + 2\frac{\partial^2 z}{\partial y^2} = e^{2x-y} + e^{x+y} + \cos(x+2y)$$

Given equation can be written as :

$$(D^2 - 3DD' + 2D'^2)\,z = e^{2x-y} + e^{x+y} + \cos(x+2y)$$

The auxiliary equation is

$$m^2 - 3m + 2 = 0 \Rightarrow m^2 - 2m - m + 2 = 0$$

$$\Rightarrow m(m-2) - 1(m-2) = 0 \Rightarrow (m-1)(m-2) = 0 \Rightarrow m = 1, 2$$

Hence, C.F. $= \phi_1(y+x) + \phi_2(y+2x)$

Now $$\text{P.I.} = \frac{1}{(D-D')(D-2D')}\{e^{2x-y} + e^{x+y} + \cos(x+2y)\}$$

$$= \frac{1}{(D-D')(D-2D')}e^{2x-y} + \frac{1}{(D-D')(D-2D')}e^{x+y} + \frac{1}{(D-D')(D-2D')}\cos(x+2y)$$

$$= I_1 + I_2 + I_3$$

Let $$I_1 = \frac{1}{(D-D')(D-2D')}e^{2x-y}$$ (Replacing D by 2 and D' by – 1.)

$$= \frac{1}{(2+1)(2+2)}e^{2x-y} = \frac{1}{12}e^{2x-y}$$

Now, $$I_2 = \frac{1}{(D-D')(D-2D')}e^{x+y},$$ (Replacing D by 1 and D' by 1)

$$= \frac{1}{(D-D')(-1)}e^{x+y} = -\frac{1}{(D-D')}e^{x+y} = -x\frac{1}{1}e^{x+y} = -x\,e^{x+y}$$

Now, $$I_3 = \frac{1}{(D-D')(D-2D')}\cos(x+2y) = \frac{1}{D^2 - 3DD' + 2D'^2}\cos(x+2y)$$

$$= \frac{1}{-1-3(-2)+2(-4)}\cos(x+2y)$$ (Replacing D^2 by –1; DD' by –2; D'^2 by –4)

$$= \frac{1}{-1+6-8}\cos(x+2y) = -\frac{1}{3}\cos(x+2y)$$

$$\text{P.I.} = I_1 + I_2 + I_3$$

Thus required $P.I. = \frac{1}{12}e^{2x-y} - xe^{x+y} - \frac{1}{3}\cos(x+2y)$

Hence, the complete solution is

$$Z = C.F. + P.I.$$

$$= \phi_1(y+x) + \phi_2(y+2x) + \frac{1}{12}e^{2x-y} - xe^{x+y} - \frac{1}{3}\cos(x+2y) \quad \textbf{Ans.}$$

Example 10. *Solve the P.D.E.* $\frac{\partial^2 z}{\partial x^2} - 2\frac{\partial^2 z}{\partial x \partial y} + \frac{\partial^2 z}{\partial y^2} = \sin x.$ (*U.P II Semester 2009*)

Solution. Here, we have

$$\frac{\partial^2 z}{\partial x^2} - 2\frac{\partial^2 z}{\partial x \partial y} + \frac{\partial^2 z}{\partial y^2} = \sin x$$

$$\Rightarrow \quad (D^2 - 2DD' + D'^2)z = \sin x$$

Its auxiliary equation is $(m^2 - 2m + 1) = 0 \quad \Rightarrow \quad (m-1)^2 = 0 \Rightarrow m = 1, 1$

$$\text{C.F.} = f_1(x+y) + x f_2(x+y)$$

$$\text{P.I.} = \frac{1}{D^2 - 2DD' + D'^2}\sin x = \left[\frac{1}{-1-0+0}\right]\sin x = -\sin x \quad \begin{bmatrix} D=1 \\ D'=0 \end{bmatrix}$$

Hence, the complete solution is

$$z = \text{C.F.} + \text{P.I.} = f_1(x+y) + x f_2(x+y) - \sin x \quad \textbf{Ans.}$$

EXERCISE 43.3

Solve the following equations:

1. $[2D^2 - 5DD' + 2D'^2]z = 5\sin(2x+y)$ **Ans.** $z = f_1(y+2x) + f_2(2y+x) - \frac{5}{3}x\cos(2x+y)$

2. $\frac{\partial^2 z}{\partial x^2} - \frac{\partial^2 z}{\partial x \partial y} = \cos(x+2y)$ **Ans.** $z = f_1(y) + f_2(y+x) + \cos(x+2y)$

3. $(D^2 - DD')z = \cos x \cos 2y$ **Ans.** $z = f_1(y) + f_2(y+x) + \frac{1}{2}\cos(x+2y) - \frac{1}{6}\cos(x-2y)$

4. $r - 2s = \sin x \cos 2y$ **Ans.** $z = f_1(y) + f_2(y+2x) + \frac{1}{15}(\sin x \cos 2y) + 4\sin 2y \cos x$

5. $(D^2 + D'^2)z = \cos mx . \cos ny$ **Ans.** $z = f_1(y+ix) + f_2(y-ix) - \frac{\cos mx . \cos xy}{(m^2+n^2)}$

6. $(D^3 - 7DD'^2 - 6D'^3)z = \sin(x+2y) + e^{3x+y}$

Ans. $z = f_1(y-x) + f_2(y-2x) + f_3(y+3x) - \frac{1}{75}\cos(x+2y) + \frac{x}{20}e^{3x+y}$

7. $(D^2 - DD')z = \cos 2y(\sin x + \cos x)$

Ans. $z = f_1(y) + f_2(y+x) + \frac{1}{2}[\sin(x+2y) + \cos(x+2y)] - \frac{1}{6}[\sin(x-2y) + \cos(x-2y)]$

Case III. When R.H.S. = $\phi\,(ax + by)$ polynomial

Example 11. *Solve* $\frac{\partial^2 z}{\partial x^2}+\frac{\partial^2 z}{\partial x \partial y}-6\frac{\partial^2 z}{\partial y^2}=x+y.$

Solution. With $D=\frac{\partial}{\partial x}$, $D'=\frac{\partial}{\partial y}$, the given equation can be written in the form

$$(D^2+DD'-6D'^2)z=x+y$$

Writing $D = m$ and $D' = 1$, the auxiliary equation is $m^2 + m - 6 = 0$

$\Rightarrow \quad (m+3)(m-2)=0 \Rightarrow m=-3, 2$

$\therefore \quad \text{C.F.} = f_1(y-3x)+f_2(y+2x)$

$\therefore \quad P.I. = \frac{1}{D^2+DD'-6D'^2}(x+y) = \frac{1}{(1)^2+(1)(1)-6(1)^2}\iint u\,du\,du$ [where $u = x + y$]

$$=\frac{1}{-4}\frac{u^3}{6}=-\frac{u^3}{24}=\frac{(x+y)^3}{-24}$$

The complete solution is $z = f_1(y-3x)+f_2(y+2x)-\frac{(x+y)^3}{24}$ **Ans.**

Example 12. *Solve* : $(D^2 + 2DD' - 8D'^2)z = \sqrt{2x+3y}$

Solution. Here, we have

$$(D^2+2DD'-8D'^2)z=\sqrt{2x+3y}$$

$A.E.$ is $m^2 + 2m - 8 = 0 \Rightarrow (m+4)(m-2) = 0 \Rightarrow m = 2, m = -4$

$$C.F. = f_1(y+2x)+f_1(y-4x)$$

$$P.I. = \frac{1}{D^2+2DD'-8D'^2}\sqrt{2x+3y} = \frac{1}{D^2+2DD'-8D'^2}(2x+3y)^{\frac{1}{2}}$$

$$=\frac{1}{(2)^2+2(2)(3)-8(3)^2}\iint u^{\frac{1}{2}}du\,du, \text{ where } u = 2x+3y$$

$$=\frac{1}{-56}\frac{u^{5/2}}{\frac{3}{2}\cdot\frac{5}{2}} = -\frac{1}{56}\left[\frac{4}{15}(2x+3y)^{\frac{5}{2}}\right] = -\frac{1}{210}(2x+3y)^{\frac{5}{2}}$$

Hence, the complete solution $= C.F. + P.I.$

$$= f_1(y+2x)+f_2(y-4x)-\frac{1}{210}(2x+3y)^{\frac{5}{2}}$$ **Ans.**

Example 13. *Solve* $(D^3 - 3D^2D' - 4DD'^2 + 12D'^3)z = \sin(y+2x)$.

Solution. Here, we have

$(D^3 - 3D^2D' - 4DD'^2 + 12D'^3)z = 0$ ...(1)

Putting $D = m$ and $D' = 1$ in (1) ; we get

A.E. is $m^3 - 3m^2 - 4m + 12 = 0$

$\Rightarrow \quad m^2(m-3) - 4(m-3) = 0$

$\Rightarrow \quad (m^2-4)(m-3) = 0 \Rightarrow m = \pm 2, 3$

$\therefore \quad \text{C.F.} = f_1(y+2x)+f_2(y-2x)+f_3(y+3x)$

$$P.I. = \frac{1}{D^3-3D^2D'-4DD'^2+12D'^3}\sin(y+2x)$$

$$= x \frac{1}{3D^2 - 6DD' - 4D'^2} \sin(y+2x)$$

$$= \frac{x}{-3(2)^2 + 6(2)(1) + 4(1)^2} \sin(y+2x)$$

$$= \frac{x}{4} \sin(y+2x)$$ **Ans.**

Complete solution is $Z = C.F. + P.I. = F_1(y + x) + F_2(y - 2x) F_3(y + 3x) + \frac{x}{4} \sin(y + 2x)$

Example 14. *Solve : $(4D^2 - 4DD' + D'^2)z = 16 \log(x + 2y)$*

Solution. Here, we have $(4D^2 - 4DD' + D'^2)z = 16 \log(x + 2y)$

Auxiliary equation is

$$4m^2 - 4m + 1 = 0 \quad \Rightarrow \quad (2m - 1)^2 = 0 \quad \Rightarrow m = \frac{1}{2}, \frac{1}{2}$$

$$C.F. = f_1\left(y + \frac{x}{2}\right) + x\, f_2\left(y + \frac{x}{2}\right)$$

$$\text{P.I.} = \frac{1}{4D^2 - 4DD' + D'^2} 16 \log(x + 2y)$$

$$= \frac{1}{4(1)^2 - 4(1)(2) + (2)^2} 16 \int\int \log u\, du\, du, \text{ where } u = x + 2y \qquad \text{(case of failure)}$$

$$= x \frac{1}{8D - 4D'} 16 \int \log u\, du = x \frac{1}{8(1) - 4(2)} 16 \log u \qquad \text{(Case of failure)}$$

$$= 16x^2 \left(\frac{1}{8}\right) \log u = 16 \frac{x^2}{8} \log(x + 2y) = 2x^2 \log(x + 2y)$$

The complete solution = $C.F. + P.I.$

$$= f_1\left(y + \frac{x}{2}\right) + x f_2\left(y + \frac{x}{2}\right) + 2x^2 \log(x + 2y)$$ **Ans.**

EXERCISE 43.4

Solve the following equations

1. $\frac{\partial^2 z}{\partial x^2} - \frac{\partial^2 z}{\partial y^2} = x - y$ **Ans.** $z = f_1(y - x) + f_2(y + x) + \frac{x^3}{6} - \frac{x^2 y}{2}$

2. $\frac{\partial^2 z}{\partial x^2} + 3\frac{\partial^2 z}{\partial x \partial y} - 4\frac{\partial^2 z}{\partial y^2} = x + \sin y$ **Ans.** $z = f_1(y + x) + f_2(y - 4x) + \frac{x^3}{6} + \frac{1}{4}\sin y$

3. $\frac{\partial^3 z}{\partial x^2 \partial y} - 2\frac{\partial^3 z}{\partial x \partial y^2} + \frac{\partial^3 z}{\partial y^3} = \frac{1}{x^2}$ **Ans.** $z = f_1(x) + f_2(y + x) + x f_3(y + x) - y \log x$

4. $(D^3 - 4D^2D' + 5DD'^2 - 2D'^3)z = e^{y+2x} + (y + x)^{\frac{1}{2}}$

 Ans. $z = f_1(y + x) + x f_2(y + x) + f_3(y + 2x) + x e^{y+x} - \frac{x^2}{3}(y + x)^{\frac{3}{2}}$

5. $\frac{\partial^3 z}{\partial x^3} - 4\frac{\partial^3 z}{\partial x^2 \partial y} + 4\frac{\partial^3 z}{\partial x \partial y^2} = 2\sin(3x + 2y)$ **Ans.** $z = f_1(y + 2x) + f_2(2y + x) - \frac{5}{3} x \cos(2x + y)$

6. $\frac{\partial^2 z}{\partial x^2} - 4\frac{\partial^2 z}{\partial x \partial y} + 3\frac{\partial^2 z}{\partial y^2} = \sqrt{x+3y}$ **Ans.** $z = f_1(y+x) + f_2(y+3x) + \frac{1}{60}(x+3y)^{\frac{5}{2}}$

Case IV. *When* $F(x, y) = x^m y^n$

$$P.I. = \frac{1}{f(D, D')} x^m y^n = [f(D, D')]^{-1} x^m y^n$$

(*a*) If $m > n$, then $\frac{1}{f(D, D')}$ is expanded in the powers of $\frac{D'}{D}$.

(*b*) If $m < n$, then $\frac{1}{f(D, D')}$ is expanded in the powers of $\frac{D}{D'}$.

Example 15. *Solve :* $(D^2 + D'^2)z = x^2y^2$

Solution. Here, we have

$$(D^2 + D'^2)z = x^2y^2$$

Putting $D = m$ and $D' = 1$, we get the A.E. as

$$m^2 + 1 = 0 \Rightarrow m^2 = -1 \Rightarrow m = \pm i.$$

$$\therefore \quad C.F. = f_1(y + ix) + f_2(y - ix)$$

$$P.I. = \frac{1}{D^2 + D'^2}(x^2y^2) = \frac{1}{D^2} \cdot \frac{1}{\left(1 + \frac{D'^2}{D^2}\right)}(x^2y^2) = \frac{1}{D^2}\left(1 + \frac{D'^2}{D^2}\right)^{-1}(x^2y^2)$$

$$= \frac{1}{D^2}\left(1 - \frac{D'^2}{D^2}\right)(x^2y^2) = \frac{1}{D^2}(x^2y^2) - \frac{D'^2}{D^4}(x^2y^2) = \frac{x^4}{12}y^2 - \frac{1}{D^4}(2x^2)$$

$$= \frac{x^4}{12}y^2 - 2\cdot\frac{x^6}{3\cdot4\cdot5\cdot6} = \frac{1}{180}(15x^4y^2 - x^6)$$

Thus, the complete solution is

$$z = f_1(y + ix) + f_2(y - ix) + \frac{1}{180}(15x^4y^2 - x^6)$$ **Ans.**

Example 16. *Solve :* $\frac{\partial^3 z}{\partial x^3} - \frac{\partial^3 z}{\partial y^3} = x^3y^3$

Solution. Here, we have $\frac{\partial^3 z}{\partial x^3} - \frac{\partial^3 z}{\partial y^3} = x^3y^3 \Rightarrow (D^3 - D'^3)z = x^3y^3$

Putting $D = m$ and $D' = 1$ in above, we have

$$A.E. \text{ is } m^3 - 1 = 0 \Rightarrow m = 1, w, w^2$$

Where w is one of the cube roots of unity.

$$\therefore \quad C.F. = f_1(y + x) + f_2(y + wx) + f_3(y + w^2x)$$

$$P.I. = \frac{1}{D^3 - D'^3}(x^3y^3) = \frac{1}{D^3\left(1 - \frac{D'^3}{D^3}\right)}x^3y^3$$

$$= \frac{1}{D^3}\cdot\left(1 - \frac{D'^3}{D^3}\right)^{-1}(x^3y^3) = \frac{1}{D^3}\left(1 + \frac{D'^3}{D^3}\right)(x^3y^3)$$

$$= \frac{1}{D^3}\left[x^3y^3 + \frac{1}{D^3}D'^3(x^3y^3)\right] = \frac{1}{D^3}\left[x^3y^3 + \frac{1}{D^3}(6x^3)\right]$$

$$= \frac{1}{D^3}(x^3y^3) + \frac{1}{D^6}(6x^3) = \frac{x^6y^3}{6\cdot5\cdot4} + \frac{6x^9}{4\cdot5\cdot6\cdot7\cdot8\cdot9} = \frac{x^6y^3}{120} + \frac{x^9}{10080}$$

Hence, the complete solution is

$$z = C.F. + P.I. = f_1(y+x) + f_2(y+wx) + f_3(y+w^2x) + \frac{x^6y^3}{120} + \frac{x^9}{10080}$$ **Ans.**

Example 17. *Solve* $\frac{\partial^2 z}{\partial x^2} + 2\frac{\partial^2 z}{\partial x\,\partial y} + \frac{\partial^2 z}{\partial y^2} = x^2 + xy + y^2$

Solution. Here, we have $\frac{\partial^2 z}{\partial x^2} + 2\frac{\partial^2 z}{\partial x\,\partial y} + \frac{\partial^2 z}{\partial y^2} = x^2 + xy + y^2$

A.E. is $m^2 + 2m + 1 = 0 \Rightarrow (m+1)^2 = 0 \quad \Rightarrow \quad m = -1, -1$

$$C.F. = f_1(y-x) + xf_2(y-x)$$

$$P.I. = \frac{1}{D^2 + 2DD' + D'^2}(x^2+xy+y^2) = \frac{1}{D^2}\frac{1}{\left(1+\frac{2D'}{D}+\frac{D'^2}{D^2}\right)}(x^2+xy+y^2)$$

$$= \frac{1}{D^2}\left(1+\frac{2D'}{D}+\frac{D'^2}{D^2}\right)^{-1}(x^2+xy+y^2)$$

$$= \frac{1}{D^2}\left(1-\frac{2D'}{D}-\frac{D'^2}{D^2}+\frac{4D'^2}{D^2}+.....\right)(x^2+xy+y^2)$$

$$= \left(\frac{1}{D^2}-\frac{2D'}{D^3}+\frac{3D'^2}{D^4}\right)(x^2+xy+y^2)$$

$$= \frac{1}{D^2}(x^2+xy+y^2) - \frac{2D'}{D^3}(x^2+xy+y^2) + \frac{3D'^2}{D^4}(x^2+xy+y^2)$$

$$= \left(\frac{x^4}{12}+\frac{x^3y}{6}+\frac{x^2y^2}{2}\right) - \left(\frac{2}{D^3}\right)(x+2y) + \frac{3}{D^4}(2)$$

$$= \frac{x^4}{12}+\frac{x^3y}{6}+\frac{x^2y^2}{2}-\frac{x^4}{12}-\frac{2x^3y}{3}+\frac{6x^4}{2\cdot3\cdot4}$$

$$= \frac{x^4}{12}+\frac{x^3y}{6}+\frac{x^2y^2}{2}-\frac{x^4}{12}-\frac{2x^3y}{3}+\frac{x^4}{4} = \frac{x^4}{4}-\frac{1}{2}x^3y+\frac{x^2y^2}{2}$$

Hence, the complete solution is

$$z = f_1(y-x) + xf_2(y-x) + \frac{x^4}{4} - \frac{x^3y}{2} + \frac{x^2y^2}{2}$$ **Ans.**

EXERCISE 43.5

Solve the following equations :

1. $\frac{\partial^2 z}{\partial x^2} + 3\frac{\partial^2 z}{\partial x\,\partial y} + 2\frac{\partial^2 z}{\partial y^2} = 12xy$ **Ans.** $z = f_1(y-x) + f_2(y-2x) + 2x^3y - \frac{3x^4}{2}$
2. $\frac{\partial^2 z}{\partial x^2} - \frac{\partial^2 z}{\partial x\partial y} - 6\frac{\partial^2 z}{\partial y^2} = xy$ **Ans.** $z = f_1(y-2x) + f_2(y+3x) + \frac{x^3y}{6} + \frac{x^4}{24}$
3. $(D^3 - 3D^2D')z = x^2y$ **Ans.** $z = \phi_1(y+x) + \phi_2(y-x) + \frac{1}{12}e^{2x-y} - xe^{x+y} - \frac{1}{3}\cos(x+2y)$
4. $\frac{\partial^3 z}{\partial x^3} - 2\frac{\partial^3 z}{\partial x^2\partial y} = 2e^{2x} + 3x^2y$ **Ans.** $z = f_1(y) + xf_2(y) + f_3(y+2x) + \frac{1}{60}(15e^{2x} + 3x^5y + x^6)$
5. $(D^2 - 6DD' + 9D'^2)z = 12x^2 + 36xy$ **Ans.** $z = f_1(y+3x) + xf_2(y+3x) + 6x^3y + 10x^4$

6. $(D^2 - D'^2 + D + 3D' - 2)z = e^{x-y} - x^2y$

Ans. $z = e^{-2x} f_1(y+x) + e^x f_2(y-x) - \frac{1}{4}e^{x-y} + \frac{1}{2}\left(x^2y + xy + \frac{3}{2}x^2 + \frac{3y}{2} + 3x + \frac{21}{4}\right)$

43.4 P.I. OF ANY FUNCTION

If the function on the R.H.S. of the P.D.E. is not of the form, given in previous cases, then

$$\text{P.I.} = \frac{1}{F(D, D')}\phi(x, y)$$

$F(D, D')$ is factorized to get

$$F(D, D') = (D - m_1D')(D - m_2D') \ldots\ldots (D - m_nD')$$

$$\text{P.I.} = \frac{1}{(D - m_1D)(D - m_2D') \ldots (D - m_nD')}\phi(x, y)$$

Let us consider

$$\text{P.I.} = \frac{1}{D - m_1D'}\phi(x, y) \qquad \text{(Taking only one term)}$$

$$\Rightarrow \qquad p - m_1q = \phi(x, y)$$

Subsidiary equations are (Lagrange's equations)

$$\frac{dx}{1} = \frac{dy}{-m_1} = \frac{dz}{\phi(x, y)}$$

From the first two

$$dy + m_1dx = 0 \qquad \Rightarrow \qquad y + m_1x = c$$

From the first and last equations, we get

$$dz = \phi(x, y)dx = \phi(x, c - m_1x)\,dx$$

$$\Rightarrow \qquad z = \int \phi(x, c - m_1x)dx$$

$$\boxed{\text{P.I.} = \frac{1}{D - mD'}F(x, y) = \int \phi(x, c - mx)dx}$$

where c is replaced by $y + mx$ after integration.

Similarly we repeat the above method to get *P.I.*

Case V. When R.H.S. = Any function

Example 18. *Solve* $(D^2 - DD' - 2D'^2)z = (y-1)e^x$

Solution. $(D^2 - DD' - 2D'^2)z = (y-1)e^x$

A.E. is $\qquad m^2 - m - 2 = 0$

$$(m-2)(m+1) = 0 \Rightarrow m = 2, -1$$

$$\text{C.F.} = f_1(y + 2x) + f_2(y - x)$$

$$\text{P.I.} = \frac{1}{D^2 - DD' - 2D'^2}(y-1)\,e^x$$

$$= \frac{1}{(D+D')(D-2D')}(y-1)e^x = \frac{1}{D+D'}\int[(c-2x-1)e^x dx] \qquad [\text{Put } y = c - 2x]$$

$$= \frac{1}{D+D'}[(c-2x-1)e^x + 2e^x] = \frac{1}{D+D'}[ce^x - 2xe^x + e^x] \qquad [\text{Put c} = y + 2x]$$

$$= \frac{1}{D+D'}[(y+2x)e^x - 2xe^x + e^x] = \frac{1}{D+D'}[ye^x + e^x]$$

$$= \int [(c+x)e^x + e^x]dx \qquad \text{[Put } y = c + x]$$

$$= (c+x)e^x - e^x + e^x \qquad \text{[Put } c = y - x]$$

$$= ce^x + xe^x = (y-x)e^x + xe^x = ye^x$$

Hence, complete solution is

$$\therefore \quad z = f_1(y+2x) + f_2(y-x) + ye^x \qquad \textbf{Ans.}$$

Example 19. *Solve* $\frac{\partial^2 z}{\partial x^2} + \frac{\partial^2 z}{\partial x \partial y} - 6\frac{\partial^2 z}{\partial y^2} = y \cos x$.

(R.G.P.V., Bhopal, June 2009, Feb. 2008, June 2006)

Solution. We have, $\frac{\partial^2 z}{\partial x^2} + \frac{\partial^2 z}{\partial x \partial y} - 6\frac{\partial^2 z}{\partial y^2} = y\cos x$

$$\Rightarrow \quad (D^2 + DD' - 6D'^2)\, z = y \cos x$$

Its auxiliary equation is

$$m^2 + m - 6 = 0 \quad \Rightarrow \quad (m+3)(m-2) = 0 \quad \Rightarrow \quad m = 2, -3$$

$$\therefore \qquad \text{C.F.} = f_1(y+2x) + f_2(y-3x)$$

$$\text{P.I.} = \frac{1}{D^2 + DD' - 6D'^2} y\cos x = \frac{1}{(D-2D')(D+3D')} y\cos x$$

$$= \frac{1}{D-2D'}\int (c+3x)\cos x\, dx \qquad \text{[Put } y = c + 3x]$$

$$= \frac{1}{D-2D'}[(c+3x)\sin x + 3\cos x] = \frac{1}{D-2D'}[y \sin x + 3\cos x] \qquad \text{[Put } c + 3x = y]$$

$$= \int [(c-2x)\sin x + 3\cos x]dx \qquad \text{[Put } y = c - 2x]$$

$$= (c-2x)(-\cos x) - 2\sin x + 3\sin x = -y\cos x + \sin x \qquad \text{[Put } c - 2x = y]$$

Hence, the complete solution is

$$z = f_1(y+2x) + f_2(y-3x) + \sin x - y\cos x \qquad \textbf{Ans.}$$

Example. 20. *Solve:* $(D^2 + D\,D' - 6\,D'^2)\, z = x^2 \sin(x+y)$

Solution. Here, we have

$$(D^2 + D\,D' - 6\,D'^2)\, z = x^2 \sin(x+y)$$

Putting $D = m$ and $D' = 1$, we have

A.E. is $m^2 + m - 6 = 0 \Rightarrow (m+3)(m-2) = 0 \Rightarrow m = 2, -3$

$$\text{C.F.} = f_1(y+2x) + f_2(y-3x)$$

$$\text{P.I.} = \frac{1}{D^2 + D\,D' - 6\,D'^2}\left[x^2 \sin(x+y)\right] = \frac{1}{(D-2D')(D+3D')}\left[x^2 \sin(x+y)\right].$$

$$\text{Let } \frac{1}{D+3D'}[x^2 \sin(x+y)] = u \Rightarrow (D+3D')\, u = x^2 \sin(x+y)$$

$$u = \int x^2 \sin(x+c+3x)\, dx = \int x^2 \sin(4x+c)\, dx \qquad [y = c + 3x]$$

$$= x^2\left(\frac{-\cos(4x+c)}{4}\right) - (2x)\left(\frac{-\sin(4x+c)}{16}\right) + 2\frac{\cos(4x+c)}{64}$$

$$= \left[\frac{-x^2}{4} + \frac{1}{32}\right] \cos(4x + c) + \frac{x}{8} \sin(4x + c) \qquad ...(1)$$

On eliminating c, we put $c = y - 3x$ in (1) and get

$$u = \left(\frac{-x^2}{4} + \frac{1}{32}\right) \cos(4x + y - 3x) + \frac{x}{8} \sin(4x + y - 3x)$$

$$= \left(\frac{-x^2}{4} + \frac{1}{32}\right) \cos(x + y) + \frac{x}{8} \sin(x + y)$$

$$\text{P.I.} = \frac{1}{(D - 2D')(D + 3D')} x^2 \sin(x + y) = \frac{1}{(D - 2D')} u \qquad (y = c - 2x)$$

$$= \frac{1}{D - 2D'} \left[\left(\frac{-x^2}{4} + \frac{1}{32}\right) \cos(x + y) + \frac{x}{8} \sin(x + y)\right]$$

$$= \int \left[\left(\frac{-x^2}{4} + \frac{1}{32}\right) \cos(x + c - 2x) + \frac{x}{8} \sin(x + c - 2x)\right] dx \qquad (y = c - 2x)$$

$$= \int \left[\left(\frac{-x^2}{4} + \frac{1}{32}\right) \cos(c - x) + \frac{x}{8} \sin(c - x)\right] dx$$

$$= \left(\frac{-x^2}{4} + \frac{1}{32}\right) \{-\sin(c - x)\} - \left(\frac{-x}{2}\right) [-\cos(c - x)]$$

$$+ \left(\frac{-1}{2}\right) \sin(c - x) + \frac{x}{8} \cos(c - x) - \frac{1}{8} [-\sin(c - x)]$$

$$= \left(\frac{x^2}{4} - \frac{1}{32} - \frac{1}{2} + \frac{1}{8}\right) \sin(c - x) + \left(\frac{-x}{2} + \frac{x}{8}\right) \cos(c - x)$$

$$= \left(\frac{x^2}{4} - \frac{13}{32}\right) \sin(c - x) - \frac{3x}{8} \cos(c - x) \qquad ...(2)(c = 2x + y)$$

On eliminating c, we put $c = 2x + y$ in (2) and get

$$\text{P.I.} = \left(\frac{x^2}{4} - \frac{13}{32}\right) \sin(2x + y - x) - \frac{3x}{8} \cos(2x + y - x)$$

$$= \left(\frac{x^2}{4} - \frac{13}{32}\right) \sin(x + y) - \frac{3x}{8} \cos(x + y)$$

The complete solution = *C.F.* + *P.I.*

$$z = f_1(y + 2x) + f_2(y - 3x) + \left(\frac{x^2}{4} - \frac{13}{32}\right) \sin(x + y) - \frac{3x}{8} \cos(x + y) \textbf{ Ans.}$$

Example 21. *Solve* : $(D^2 - 4D'^2) = \frac{4x}{y^2} - \frac{y}{x^2}$

Solution. Here, we have

$$(D^2 - 4D'^2) z = \frac{4x}{y^2} - \frac{y}{x^2} \qquad ...(1)$$

A.E. is $\quad m^2 - 4 = 0 \;\Rightarrow\; (m + 2)(m - 2) = 0 \;\Rightarrow\; m = 2, -2$

$$\text{C.F.} = f_1(y+2x) + f_2(y-2x)$$

$$P.I. = \frac{1}{D^2 - 4D'^2}\left(\frac{4x}{y^2} - \frac{y}{x^2}\right) = \frac{1}{(D+2D')(D-2D')}\left(\frac{4x}{y^2} - \frac{y}{x^2}\right)$$

Let $u = \frac{1}{D-2D'}\left(\frac{4x}{y^2} - \frac{y}{x^2}\right) \Rightarrow (D-2D')\,u = \left(\frac{4x}{y^2} - \frac{y}{x^2}\right)$

$$\Rightarrow \quad u = \int\left[\frac{4x}{(c-2x)^2} - \frac{c-2x}{x^2}\right]dx \qquad (y = c-2x)$$

$$= \int\left[\frac{-2(c-2x)+2c}{(c-2x)^2} - \frac{c}{x^2} + \frac{2}{x}\right]dx = \int\left[\frac{-2}{(c-2x)} + \frac{2c}{(c-2x)^2} - \frac{c}{x^2} + \frac{2}{x}\right]dx$$

$$= \log(c-2x) + \frac{c}{c-2x} + \frac{c}{x} + 2\log x$$

On eliminating c, replace c *by* $2x+y$ and have

$$u = \log(2x+y-2x) + \frac{2x+y}{2x+y-2x} + \frac{2x+y}{x} + 2\log x$$

$$= \log y + \frac{2x+y}{y} + 2 + \frac{y}{x} + 2\log x \qquad ...(2)$$

$$\text{Now P.I.} = \frac{1}{(D+2D')(D-2D')}\left(\frac{4x}{y^2} - \frac{y}{x^2}\right) = \frac{1}{(D+2D')}u$$

$$= \frac{1}{D+2D'}\left[\log y + \frac{2x+y}{y} + 2 + \frac{y}{x} + 2\log x\right] \qquad \text{[using (2)]}$$

$$= \int\left[\log(c+2x) + \frac{2x+c+2x}{c+2x} + 2 + \frac{c+2x}{x} + 2\log x\right]dx \qquad [y = c+2x]$$

$$= \int\left[\log(c+2x) + \frac{2x}{c+2x} + 1 + 2 + \frac{c}{x} + 2 + 2\log x\right]dx$$

$$= \int \log(c+2x) + \frac{2x+c-c}{2x+c} + 5 + \frac{c}{x} + 2\log x]\,dx$$

$$= \int \log(c+2x) + 1 - \frac{c}{2x+c} + 5 + \frac{c}{x} + 1\cdot\log x^2\Big]\,dx$$

$$= \Big[x\log(c+2x) - \int x\frac{1}{c+2x}2\,dx + 6x - \frac{c}{2}\log(c+2x)$$

$$+ c\log x + x\log x^2 - \int \frac{2x}{x^2}.x\,dx\Big]$$

$$= x\log(c+2x) - \int \frac{c+2x-c}{c+2x}dx + 6x - \frac{c}{2}\log(c+2x)$$

$$+ c\log x + x\log x^2 - \int 2\,dx$$

$$= x\log(c+2x) - x + \frac{c}{2}\log(c+2x) + 6x - \frac{c}{2}\log(c+2x)$$

$$+ c\log x + x\log x^2 - 2x$$

$$= x \log (c + 2x) + 3x + c \log x + x \log x^2$$

On eliminating c, replacing c by $y - 2x$ and have

$$= x \log (y - 2x + 2x) + 3x + (y - 2x) \log x + x \log x^2$$

$$= x \log y + 3x + y \log x - 2x \log x + x \log x^2$$

$$= x \log y + 3\, x + y \log x - x \log x^2 + x \log x^2$$

$$= x \log y + 3x + y \log x$$

Hence, the complete solution is

$y = C.F + P.I. = f_1 (y + 2x) + f_2 (y - 2x) + x \log y + 3x + y \log x$ **Ans.**

Example 22. *Solve :* $[D^3 + D^2 D' - D D'^2 - D'^3]\, z = e^x \cos 2\, y$

Solution. We have

$$[D^3 + D^2 D' - D D'^2 - D'^3]\, z = e^x \cos 2y$$

A.E. is $\quad m^3 + m^2 - m - 1 = 0 \Rightarrow (m + 1)^2 (m - 1) = 0 \quad \Rightarrow \quad m = 1, -1, -1$

C.F. $= f_1 (y + x) + f_2 (y - x) + x f_3 (y - x)$

$$\text{P. I.} = \frac{1}{D^3 + D^2 D' - D D'^2 - D'^3} e^x \cos 2y = \frac{1}{(D + D')^2 (D - D')} e^x \cos 2y$$

Let $\quad u = \dfrac{1}{D - D'} e^x \cos 2y \qquad (y = c - x)$

$$= \int e^x \cos 2 (c - x)\, dx = \frac{e^x}{1+4} [\cos 2 (c - x) - 2 \sin 2 (c - x)] \qquad \text{... (1)}$$

On eliminating c, replace c by $x + y$ in (1), and have

$$u = \frac{e^x}{5} [\cos 2 (x + y - x) - 2 \sin 2 (x + y - x)] = \frac{e^x}{5} [\cos 2y - 2 \sin 2y]$$

Now $\left[\dfrac{1}{(D + D')(D - D')}\right] e^x \cos 2y = \dfrac{1}{(D + D')} u$

$$= \frac{1}{D + D'} \left[\frac{e^x}{5} (\cos 2y - 2 \sin 2y)\right] \qquad (y = c + x)$$

$$= \int \left[\frac{e^x}{5} \{\cos (2\, c + 2x) - 2 \sin (2c + 2x)\}\right] dx$$

$$= \int \frac{e^x}{5} \cos (2c + 2x)\, dx - 2 \int \frac{e^x}{5} \sin (2c + 2x)\, dx$$

$$= \frac{e^x}{5\,(1+4)} [\cos (2c + 2x) + 2 \sin (2c + 2x)] - \frac{2\, e^x}{5\,(1+4)} [\sin (2c + 2x) - 2 \cos (2c + 2x)]$$

$$= \frac{e^x}{25} [\cos (2c + 2x) + 2 \sin (2c + 2x) - 2 \sin (2c + 2x) + 4 \cos (2c + 2x)]$$

$$= \frac{e^x}{25} [5 \cos (2c + 2x)] = \frac{e^x}{5} \cos (2c + 2x)$$

On eliminating c, replace c by $y - x$ and have

$$= \frac{e^x}{5}\cos(2y - 2x + 2x) = \frac{e^x}{5}\cos 2y$$

$$\text{P.I.} = \frac{1}{(D+D')}\left(\frac{1}{D+D'}\frac{1}{D-D'}\right)(e^x \cos 2y) = \frac{1}{D+D'}\frac{e^x}{5}\cos 2y \qquad (y = c + x)$$

$$= \int \frac{e^x}{5}\cos 2(c+x)\,dx = \frac{e^x}{5(1+4)}[\cos 2(c+x) + 2\sin 2(c+x)]$$

On eliminating c, replace c by $(y - x)$ and get

$$= \frac{e^x}{25}[\cos 2(y-x+x) + 2\sin 2(y-x+x)] = \frac{e^x}{25}[\cos 2y + 2\sin 2y]$$

The complete solution *is z = C.F. + P.I.*

$$= f_1(y+x) + f_2(y-x) + x\, f_3(y-x) + \frac{e^x}{25}(\cos 2y + 2\sin 2y)] \qquad \textbf{Ans.}$$

EXERCISE 43.6

Solve the following equations:

1. $(D-D')(D+2D')z = (y+1)e^x$ **Ans.** $z = f_1(y+x) + f_2(y-2x) + ye^x$
2. $\frac{\partial^2 z}{\partial x^2} - \frac{\partial^2 z}{\partial y^2} = \tan^3 x \tan y - \tan x \tan^3 y$ **Ans.** $z = f_1(y+x) + f_2(x-y) + \frac{1}{2}\tan x \tan y$
3. $(D^2 - DD' - 2D'^2)z = (2x^2 + xy - y^2)\sin xy - \cos xy$ **Ans.** $z = f_1(y+2x) + f_2(y-x) + \sin xy$
4. $\frac{\partial^2 z}{\partial x^2} + \frac{\partial^2 z}{\partial x\,\partial y} - 2\frac{\partial^2 z}{\partial y^2} = (y-1)e^x$ **Ans.** $z = f_1(y+x) + f_2(y-2x) + (y-2)e^x$
5. $(D^2 - 6DD' + 9D'^2)z = 12x^2 + 36xy$ **Ans.** $z = f_1(y+3x) + x f_2(y+3x) + 10x^4 + 6x^3 y$

43.5 NON-HOMOGENEOUS LINEAR EQUATIONS

The linear differential equations which are not homogeneous are called Non-homogeneous Linear Equations.

For example, $3\frac{\partial^2 z}{\partial x^2} + 2\frac{\partial^2 z}{\partial x \partial y} + 4\frac{\partial^2 z}{\partial y^2} + 5\frac{\partial z}{\partial x} + \frac{\partial z}{\partial y} + z = 0$

$$f(D, D') = f_1(x, y)$$

Its solution, $z = \text{C.F.} + \text{P.I.}$

Complementary Function: Let the non-homogeneous equation be

$$(D - mD' - a)z = 0 \quad \text{or} \quad \frac{\partial z}{\partial x} - m\frac{\partial z}{\partial y} - az = 0$$

$$p - mq = az$$

The Lagrange's subsidiary equations are $\frac{dx}{1} = \frac{dy}{-m} = \frac{dz}{az}$

From first two relations, we have $-mdx = dy$

$$dy + mdx = 0 \Rightarrow \quad y + mx = c_1 \qquad \text{... (1)}$$

and from first and third relation, we have

$$dx = \frac{dz}{az} \Rightarrow x = \frac{1}{a}\log z + c_2 \Rightarrow z = c_3 e^{ax} \qquad \text{... (2)}$$

From (1) and (2), we have $z = e^{ax}\phi(y+mx)$

Similarly the solution of $(D - mD' - a)^2 z = 0$ is

$$z = e^{ax}\phi_1(y+mx) + x e^{ax}\,\phi_2(y+mx)$$

43.6 IF THE EQUATION IS OF THE FORM

$$(\alpha D + \beta D' + \gamma)\, z = 0 \quad \Rightarrow \quad \alpha p + \beta q = -\gamma z$$

It is of Lagrange's form.

Lagrange's subsidiary equations are $\dfrac{dx}{\alpha} = \dfrac{dy}{\beta} = \dfrac{dz}{-\gamma z}$

From first two, we have $\alpha y - \beta x = C_1$

From first and last, we have $\dfrac{dz}{z} = -\dfrac{\gamma}{\alpha}\, dx$

$$\Rightarrow \quad \log z = -\frac{\gamma}{\alpha}x + \log C_2 \quad \Rightarrow \quad z = C_2 e^{-\frac{\gamma}{\alpha}x} = \phi(C_1)\, e^{-\frac{\gamma}{a}x} \quad \Rightarrow \quad z = e^{-\frac{\gamma}{\alpha}x}\, \varphi(\alpha y - \beta x)$$

where φ is an arbitrary function.

Example 23. *Solve $(D + D' - 2)(D + 4D' - 3)\, z = 0$*

Solution. The equation can be rewritten as

$$\{D - (-D') - 2\{D - (-4D') - 3\}\, z = 0$$

Hence the solution is $z = e^{2x}\,\phi_1(y - x) + e^{3x}\,\phi_2(y - 4x)$ **Ans.**

Example 24. *Solve $(D + 3D' + 4)^2\, z = 0$*

Solution. The equation is rewritten as $[D - (-3D') - (-4^2)]\, z = 0$

Hence the solution is $z = e^{-4x}\phi_1(y - 3x) + x e^{-4x}\phi_2(y - 3x)$ **Ans.**

Example 25. *Solve $r + 2s + t + 2p + 2q + z = 0$*

Solution. The equation is rewritten as

$$(D^2 + 2DD' + D'^2 + 2D + 2D' + 1)z = 0$$

$$\Rightarrow \quad [(D + D')^2 + 2(D + D') + 1]z = 0$$

$$\Rightarrow \quad (D + D' + 1)^2\, z = 0$$

$$\Rightarrow \quad [D - (-D') - (-1)]^2\, z = 0$$

Hence the solution is

$$z = e^{-x}\phi_1 y(y - x) + x e^{-x}\phi_2(y - x) \qquad \textbf{Ans.}$$

Example 26. *Solve $r - t + p - q = 0$*

Solution. The equation is rewritten as

$$(D^2 - D'^2 + D - D')\, z = 0$$

$$\Rightarrow \quad [(D - D')(D + D') + 1\,(D - D')]\, z = 0$$

$$\Rightarrow \quad (D - D')(D + D' + 1)\, z = 0$$

Hence the solution is

$$z = \phi_1(y + x) + e^{-x}\phi_2(y - x) \qquad \textbf{Ans.}$$

EXERCISE 43.7

Solve the following equations:

1. $(D-D')(D+D'-3)z=0$ **Ans.** $z=\phi(y+x)+e^{3x}\phi_2(y-x)$
2. $(D-D'-1)(D-D'-2)z=0$ **Ans.** $z=e^{x}\phi_1(y+x)+e^{2x}\phi_2(y+x)$
3. $(D+D'-1)(D+2D'-2)z=0$ **Ans.** $z=e^{x}\phi_1(y-x)+e^{2x}\phi_2(y-2x)$
4. $(D^2+DD'+D'-1)z=0$ **Ans.** $z=e^{-x}\phi_1(y)+e^{x}\phi_2(y-x)$
5. $(D^2-DD'-2D'^2+2D+2D')z=0$ **Ans.** $z=\phi_1(y-x)+e^{-2x}\phi_2(y+2x)$
6. $[D^2-D'^2+D+3D'-2]z=0$ **Ans.** $z=e^{-2x}\phi_1(y+x)+e^{x}\phi_2(y-x)$
7. $(D^2-a^2D'^2+2ab\,D+2ab\,D')z=0$ **Ans.** $z=\phi_1(y-ax)+e^{-2abx}\phi_2(y+ax)$
8. $t+s+q=0$ **Ans.** $z=\phi_1(x)+e^{-x}\phi_2(y-x)$
9. $(D+D'-1)(D+2D'-2)z=0$ **Ans.** $z=e^{x}\phi_1(y-x)+e^{3x}\phi_2(y-2x)$
10. $(D-2D'+5)^2z=0$ **Ans.** $z=e^{-5x}\phi_1(y+2x)+xe^{-5x}\phi_2(y+2x)$

Particular Integral

Case I. $$\frac{1}{F(D,D')}e^{ax+by}=\frac{1}{F(a,b)}e^{ax+by}$$

Example 27. *Solve* $(D-D'-2)(D-D'-3)z=e^{3x-2y}$

Solution. The complementary function is

$$\text{C.F.}=e^{2x}\phi_1(y+x)+e^{3x}\phi_2(y+x).$$

$$\text{P.I.}=\frac{1}{(D-D'-2)(D-D'-3)}e^{3x-2y}=\frac{1}{[3-(-2)-2][3-(-2)-3]}e^{3x-2y}=\frac{1}{6}e^{3x-2y}$$

Hence, the complete solution is

$$z=e^{2x}\phi_1(y+x)+e^{3x}\phi_2(y+x)+\frac{1}{6}e^{3x-2y}$$ **Ans.**

Case II. $$\frac{1}{F(D^2,DD',D'^2)}\sin(ax+by)=\frac{1}{F(-a^2,-ab,-b^2)}\sin(ax+by)$$

Case III. $$\frac{1}{F(D,D')}x^my^n=[F(D,D')^{-1}]x^my^n$$

Example 28. *Solve* $[D^2-D'^2+D+3D'-2]z=x^2y$

Solution. $(D-D'+2)(D+D'-1)z=0$

$$\text{C.F.}=e^{-2x}\phi_1(y+x)+e^{x}\phi_2(y-x)$$

$$\text{P.I.}=\frac{1}{(D-D'+2)(D+D'-1)}x^2y$$

$$=\frac{1}{D^2-D'^2+D+3D'-2}x^2y=-\frac{1}{2}\left[\frac{1}{1-\frac{3D'}{2}-\frac{D}{2}+\frac{D'^2}{2}-\frac{D^2}{2}}\right]x^2y$$

$$=-\frac{1}{2}[1-\frac{1}{2}(3D'+D-D'^2+D^2)]^{-1}x^2y$$

$$= -\frac{1}{2}[1+\frac{1}{2}(3D'+D-D'^2+D^2)+\frac{1}{4}(3D'+D-D'^2+D^2)^2$$

$$+\frac{1}{8}(3D'+D-D'^2+D^2)^3+.........]x^2y$$

$$= -\frac{1}{2}[1+\frac{1}{2}(3D'+D-D'^2+D^2)+\frac{1}{4}(9D'^2+D^2+6DD'+6D^2D')+\frac{1}{8}(9D^2D')+....]x^2y$$

$$= -\frac{1}{2}[x^2y+\frac{1}{2}(3x^2+2xy-0+2y)+\frac{1}{4}(0+2y+12x+12)+\frac{1}{8}(18)]$$

$$= -\frac{1}{2}\left[x^2y+\frac{3x^2}{2}+xy+y+\frac{y}{2}+3x+3+\frac{9}{4}\right]=-\frac{1}{2}\left[x^2y+\frac{3x^2}{2}+xy+\frac{3y}{2}+3x+\frac{21}{4}\right]$$

Hence, the complete solution is

$$z=e^{-2x}\phi_1(y+x)+e^x\phi_2(y-x)-\frac{1}{2}\left(x^2y+\frac{3x^2}{2}+xy+\frac{3y}{2}+3x+\frac{21}{4}\right)$$ **Ans.**

Case IV. $\frac{1}{F(D,D')}[e^{ax+by}\phi(x,y)]=e^{ax+by}\frac{1}{F(D+a,D'+b)}\phi(x,y)$

Example 29. *Solve* $(D-3D'-2)^2z=2e^{2x}\sin(y+3x)$

Solution. A.E. is $(D-3D'-2)^2=0$

$$\text{C.F.} = e^{2x}\phi_1(y+3x)+xe^{2x}\phi_2(y+3x) \qquad [D'=1, D=3]$$

$$\text{P.I.} = \frac{1}{(D-3D'-2)^2}2e^{2x}.\sin(y+3x)$$

$$= 2e^{2x}\frac{1}{(D+2-3D'^2-2)^2}\sin(y+3x)=2e^{2x}\frac{1}{(D-3D'^2)^2}\sin(y+3x)$$

$$= 2e^{2x}\frac{1}{D^2+9D'^2-6DD'}\sin(y+3x)$$

$$[D^2+9D'^2-6DD'=9+9-6(3)=0]$$

$$= 2e^{2x}.x\frac{1}{2(D-3D'^2)}\sin(y+3x) \qquad \text{(As denominator becomes zero)}$$

$$= 2x^2e^{2x}\frac{1}{2}\sin(y+3x) \qquad \text{(Again differentiate)}$$

$$= x^2e^{2x}\sin(y+3x)$$

Hence, the complete solution is

$$z=e^{2x}\phi_1(y+3x)+xe^{2x}\phi_2(y+3x)+x^2e^{2x}\sin(y+3x)$$ **Ans.**

Example 30. *Solve* $(D^2+DD'-6D'^2)z=x^2\sin(x+y)$

Solution. $(D^2+DD'-6D'^2)z=x^2\sin(x+y)$

For complementary function

$$(D^2+DD'-6D'^2)=0 \text{ or } (D-2D')(D+3D')=0$$

$$\text{C.F.} = \phi_1(y+2x)+\phi_2(y-3x)$$

$$\text{P.I.} = \frac{1}{D^2 + DD' - 6D'^2} x^2 \sin(x+y)$$

$$= \text{Imaginary part of } \frac{1}{D^2 + DD' - 6D'^2} x^2 [\cos(x+y) + i\sin(x+y)]$$

$$= \text{Imaginary part of } \frac{1}{D^2 + DD' - 6D'^2} x^2 e^{i(x+y)}$$

$$= \text{Imaginary part of } e^{iy} \frac{1}{D^2 + Di - 6(i)^2} x^2 e^{ix}$$

$$= \text{Imaginary part of } e^{i(x+y)} \frac{1}{(D+i)^2 + (D+i)i + 6} x^2$$

$$= \text{Imaginary part of } e^{i(x+y)} \frac{1}{D^2 + 3iD + 4} x^2$$

$$= \text{Imaginary part of } \frac{e^{i(x+y)}}{4} \frac{1}{1 + \frac{3iD}{4} + \frac{D^2}{4}} x^2$$

$$= \text{Imaginary part of } \frac{e^{i(x+y)}}{4} \left[1 + \frac{3iD}{4} + \frac{D^2}{4}\right]^{-1} x^2$$

$$= \text{Imaginary part of } \frac{e^{i(x+y)}}{4} \left[1 - \frac{3iD}{4} - \frac{D^2}{4} - \frac{9D^2}{16} \ldots\right] x^2$$

$$= \text{Imaginary part of } \frac{e^{i(x+y)}}{4} \left[x^2 - \frac{3ix}{2} - \frac{2}{4} - \frac{9}{16}(2)\right]$$

$$= \text{Imaginary part of } \frac{1}{4}[\cos(x+y) + i\sin(x+y)] \left[x^2 - \frac{3ix}{2} - \frac{13}{8}\right]$$

$$= \frac{1}{4}\left[\sin(x+y)\left(x^2 - \frac{13}{8}\right) - \frac{3}{2} x\cos(x+y)\right]$$

$$= \frac{1}{4}\sin(x+y)\left(x^2 - \frac{13}{8}\right) - \frac{3x}{8}\cos(x+y)$$

Hence, the complete solution is

$$z = \phi_1(y+2x) + \phi_2(y-3x) + \frac{1}{4}\sin(x+y)\left(x^2 - \frac{13}{8}\right) - \frac{3x}{8}\cos(x+y) \quad \textbf{Ans.}$$

EXERCISE 43.8

Solve the following equations:

1. $[D^2 + 2DD' + D'^2 - 2D - 2D']z = 0$ **Ans.** $z = \phi_1(x-y) + e^{2x}\phi_2(x-y)$

2. $(D^2 - D'^2 - 3D + 3D')z = e^{x-2y}$ **Ans.** $z = \phi_1(y+x) + e^{3x}\phi_2(y-x) - \frac{1}{12}e^{x-2y}$

3. $(D - D' - 1)(D + D' - 2)z = e^{2x-y}$ **Ans.** $z = e^x\phi_1(x+y) + e^{2x}\phi_2(y-x) - \frac{1}{2}e^{2x-y}$

4. $(D^2 - D'^2 - 3D + 3D')z = e^{x+2y}$ **Ans.** $z = \phi_1(y+x) + e^{3x}\phi_2(x-y) - xe^{x+2y}$

5. $(D+D')(D+D'-2)z = \sin(x+2y)$

$$\textbf{Ans. } z = \phi_1(y-x) + e^{2x}\phi_2(y-x) + \frac{1}{117}[6\cos(x+2y) - 9\sin(x+2y)]$$

6. $(D^2 - DD' - 2D)z = \cos(3x+4y)$ **Ans.** $z = \phi_1(y) + e^{2x}\phi_2(y+x) + \frac{1}{15}[\cos(3x+4y) - 2\sin(3x+4y)]$

7. $(DD' + D - D' - 1)z = xy$ **Ans.** $z = e^{-x}\phi_1(x) + e^{x}\phi_2(y) - (xy + y - x - 1)$

8. $(D + D' - 1)(D + 2D' - 3)z = 4 + 3x + 6y$ **Ans.** $z = e^{x}\phi_1(x-y) + e^{3x}\phi_2(2x-y) + 6 + x + 2y$

9. $D(D + D' - 1)(D + 3D' - 2)z = x^2 - 4xy + 2y^2$

$$\textbf{Ans. } z = \phi_1(y) + e^{x}\phi_2(x-y) + e^{2x}\phi_3(3x-y) + \frac{1}{2}\left[\frac{x^3}{3} - 2x^2y + 2xy^2 - \frac{7}{2}x^2 + 4xy + \frac{x}{2}\right]$$

10. $(D - D' + 2)(D + D' - 1)z = e^{x-y} - x^2y$

$$\textbf{Ans. } z = e^{-2x}\phi_1(x+y) + e^{x}\phi_2(x-y) - \frac{e^{x-y}}{4} + \frac{1}{2}\left[x^2y + xy + \frac{3x^2}{2} + \frac{3}{2}y + 3x + \frac{21}{4}\right]$$

11. $(D^2 - DD' - 2D'^2 + 2D' + 2D)z = e^{2x+3y} + \sin(2x+y) + xy$

$$\textbf{Ans. } z = \phi_1(x-y) + e^{x}\phi_2(2x+y) - \frac{1}{10}e^{2x+3y} - \frac{1}{6}\cos(2x+y) + \frac{x}{24}(6xy - 6y + 9x - 2x^2 - 12)$$

12. $\dfrac{\partial^2 z}{\partial x^2} - \dfrac{\partial^2 z}{\partial y^2} - 3\dfrac{\partial z}{\partial x} + 3\dfrac{\partial z}{\partial y} = xy + e^{x+2y}$ *(Uttarakhand, June 2009)*

$$\textbf{Ans. } z = f_1(y+x) + e^{3x}f_2(y-x) - \frac{1}{3}\left(\frac{x^2y}{2} + \frac{x^3}{6} + \frac{x^2}{3} + \frac{xy}{3} + \frac{2x}{9}\right) - xe^{x+2y}$$

43.7 MONGE'S METHOD

Let the equation be

$$Rr + Ss + Tt = V \qquad \text{... (1)}$$

where R, S, T, V are functions of x, y, z, p and q.

$$r = \frac{\partial^2 f}{\partial x^2}, \quad s = \frac{\partial^2 f}{\partial x \partial y}, \quad t = \frac{\partial^2 f}{\partial y^2}$$

We have $$dp = \frac{\partial p}{\partial x}dx + \frac{\partial p}{\partial y}dy = rdx + sdy \qquad \text{... (2)}$$

and $$dq = \frac{\partial q}{\partial x}dx + \frac{\partial q}{\partial y}dy = sdx + tdy \qquad \text{... (3)}$$

From (2) and (3), $$r = \frac{dp - sdy}{dx} \text{ and } t = \frac{dq - sdx}{dy}$$

Putting for r and t in (1), we get

$$R\left(\frac{dp - sdy}{dx}\right) + Ss + T\left(\frac{dq - sdx}{dy}\right) = V$$

$$\Rightarrow \quad Rdpdy + Tdqdx - Vdxdy - s(Rdy^2 - Sdxdy + Tdx^2) = 0 \qquad \text{... (4)}$$

Equation (4) is satisfied if

$$Rdpdy + Tdqdx - Vdxdy = 0 \quad \text{... (5)}$$

$$Rdy^2 - Sdxdy + Tdx^2 = 0 \quad \text{.... (6)}$$

Equations (5) and (6) are called **Monge's equations.**

Since (6) can be factorised into two equations.

$$dy - m_1\,dx = 0 \text{ and } dy - m_2 dx = 0$$

Now combine $dy - m_1 dx = 0$ and equation (5). If need be, we may also use the relation $dz = p.dx + q.dy$ while solving (5) and (6). The solution leads to two integrals

$$u(x, y, z, p, q) = a \text{ and } v(x, y, z, p, q) = b$$

Then we get a relation between u and v.

$$V = f_1(u) \quad \text{... (7)}$$

Equation (7) is further integrated by methods of first order equations.

Note. If the intermediate solution is of the form $Pr + Qq = R$, then we use lagrange's equation.

Example 31. *Solve* $r = a^2t$.

Solution. We have $dr = rdx + sdy$ and $dq = sdx + rdy$ which gives

$$r = \frac{dp - sdy}{dx} \text{ and } t = \frac{dq - sdx}{dy}$$

Putting these values of r and t in $r = a^2t$, we get

$$\frac{dp - sdy}{dx} = a^2\frac{dq - sdx}{dy} \quad \Rightarrow \quad dp\,dy - a^2\,dx\,dq - s\,(dy^2 - a^2\,dx^2) = 0$$

Thus, the Monges' equations are $dpdy - a^2dxdq = 0$ ... (1)

$$dy^2 - a^2dx^2 = 0 \quad \text{... (2)}$$

(2) can be resolved into factors

$$dy - adx = 0 \quad \text{... (3)}$$

and $$dy + adx = 0 \quad \text{... (4)}$$

Combining (3) with (1), we get

$$dp\,(adx) - a^2\,dx\,dq = 0 \text{ or } dp - adq = 0 \quad \text{... (5)}$$

(3) and (5) on integration give respectively

and $$\left.\begin{array}{r} y - ax = A \\ p - aq = B \end{array}\right] \Rightarrow p - aq = f_1(y - ax) \quad \text{... (6)}$$

Similarly combining (4) and (1)

$$p + aq = f_2(y + ax) \quad \text{... (7)}$$

Adding and subtracting (6) and (7), we get

$$p = \frac{1}{2}[f_1(y - ax) + f_2(y + ax)],\; q = \frac{1}{2a}[f_2(y + ax) - f_1(y - ax)]$$

Substituting these values in $dz = pdx + qdy$

$$dz = \frac{1}{2}[f_1(y - ax) + f_2(y + ax)]dx + \frac{1}{2a}[f_2(y + ax) - f_1(y - ax)]dy$$

$$dz = \frac{1}{2a}(dy + adx) f_2(y + ax) - \frac{1}{2a}(dy - adx) f_1(y - ax)$$

Integrating, $z = \frac{1}{2a}\phi_1(y + ax) - \frac{1}{2a}\phi_2(y - ax)$

$\Rightarrow$ $z = F_1(y + ax) + F_2(y - ax)$ **Ans.**

Example 32. *Solve $r - t\cos^2 x + p\tan x = 0$*

Solution. $r = \frac{dp - sdy}{dx}$ and $t = \frac{dq - sdx}{dy}$

Putting for r and t in the given equation, we get

$$\frac{dp - sdy}{dx} - \frac{dq - sdx}{dy}\cos^2 x + p\tan x = 0$$

$$\Rightarrow \quad dp\,dy - sdy^2 - dx\,dp\cos^2 x + sdx^2\cos^2 x + p\,dx\,dy\tan x = 0$$

$$\Rightarrow \quad dp\,dy - dx\,dq\cos^2 x + p\,dx\,dy\tan x - s(dy^2 - dx^2\cos^2 x) = 0$$

Monge's equations are

$$dp\,dy - dx\,dq\cos^2 x + p\,dx\,dy\tan x = 0 \quad \text{...(1)}$$

$$dy^2 - dx^2\cos^2 x = 0 \quad \text{... (2)}$$

Eq. (2) is factorised $(dy + dx\cos x)(dy - dx\cos x) = 0$

$$dy - dx\cos x = 0 \quad \text{... (3)}$$

$$dy + dx\cos x = 0 \quad \text{... (4)}$$

Integrating (3) and (4), we get

$$y - \sin x = A \quad \text{... (5)}$$

$$y + \sin x = B. \quad \text{... (6)}$$

Combining (3) and (1), we get

$$dp - dq.\cos x + p\tan x dx = 0$$

$$\Rightarrow \quad (dp\sec x + p\sec x\tan x\,dx) - dq = 0$$

Integrating $p\sec x - q = B$ ... (7)

Combining (5) and (7), we have

$$p\sec x - q = f_1(y - \sin x) \quad \text{... (8)}$$

In combining (6) and (7), we get

$$p\sec x + q = f_2(y + \sin x) \quad \text{... (9)}$$

From (8) and (9)

$$p = \frac{1}{2}\cos x[f_1(y - \sin x) + f_2(y + \sin x)] \text{ and } q = \frac{1}{2}[f_2(y + \sin x) - f_1(y - \sin x)]$$

Putting for p and q in $dz = pax + qdy$, we get

$$dz = \frac{1}{2}\cos x[f_1(y - \sin x) + f_2(y + \sin x)]dx + \frac{1}{2}[f_2(y + \sin x) - f_1(y - \sin x)]dy$$

$$\Rightarrow \quad dz = \frac{1}{2}f_2(y + \sin x)[dy + \cos x dx] - \frac{1}{2}f_1(y - \sin x)[dy - \cos x\,dx]$$

Integrating we get $z = \frac{1}{2}F_2(y + \sin x) + F_1(y - \sin x)$ **Ans.**

43.8 CLASSIFICATION OF PARTIAL DIFFERENTIAL EQUATIONS

(U.P., II Semester, June 2007)

In practical problems, the following types of equations are generally used :

(i) *Wave equation*: $\dfrac{\partial^2 u}{\partial t^2} = c^2 \dfrac{\partial^2 u}{\partial x^2}$

(ii) *One-dimensional heat flow*: $\dfrac{\partial u}{\partial t} = c^2 \dfrac{\partial^2 u}{\partial y^2}$

(iii) *Two-dimensional heat flow*: $\dfrac{\partial^2 u}{\partial x^2} + \dfrac{\partial^2 u}{\partial y^2} = 0$

(iv) *Radio equations*: $-\dfrac{\partial V}{\partial x} = L\dfrac{\partial I}{\partial t}, -\dfrac{\partial I}{\partial x} = C\dfrac{\partial V}{\partial t}$

Consider the Equation. $A\dfrac{\partial^2 u}{\partial x^2} + B\dfrac{\partial^2 u}{\partial x \partial y} + C\dfrac{\partial^2 u}{\partial y^2} + F(x, y, u, p, q) = 0$ ...(1)

Where A, B, C may be constants or functions of x and y. Now the equation (1) is

1. Parabolic; if $B^2 - 4AC = 0$

2. Elliptic; if $B^2 - 4AC < 0$

3. Hyperbolic; if $B^2 - 4AC > 0$

1. Parabolic Equation. The one-dimensional heat conduction equation

$\dfrac{\partial u}{\partial t} = C^2 \dfrac{\partial^2 u}{\partial x^2}$ is an example of parabolic partial differential equation e.g.

One dimensional heat flow equation $\dfrac{\partial u}{\partial t} = \dfrac{\partial^2 u}{\partial x^2}$ is parabolic type.

2. Elliptic Equations

The following are the examples of elliptic equations.

Two dimensional heat flow equation in steady state given by $\dfrac{\partial^2 z}{\partial x^2} + \dfrac{\partial^2 z}{\partial y^2} = 0$ is elliptic in nature.

(i) Laplace equation: $\dfrac{\partial^2 u}{\partial x^2} + \dfrac{\partial^2 u}{\partial y^2} = 0$

(ii) Poisson equation: $\dfrac{\partial^2 u}{\partial x^2} + \dfrac{\partial^2 u}{\partial y^2} = f(x, y)$

Laplace equation arrises in study-state flow and potential problems.

Poisson equations arrises in fluid mechanics electricity and magnetism and torsion problems.

3. Hyperbolic Equations

The wave equation $\dfrac{\partial^2 u}{\partial t^2} = c^2 \dfrac{\partial^2 u}{\partial x^2}$

is the simplest example of hyperbolic partial differential equation.

Remark 1. If A, B, C in (1) are constants, then nature of equation (1) will be the same for all values of x and y.

Remark 2. If A, B, C are functions of x and y in (1), then nature of equation (1) will not for all values of x and y.

Example 33. *Classify the following equations.*

(*a*) $2\frac{\partial^2 u}{\partial x^2}+4\frac{\partial^2 u}{\partial x \partial y}+3\frac{\partial^2 u}{\partial y^2}=0$

(*b*) $\frac{\partial^2 u}{\partial x^2}+4\frac{\partial^2 u}{\partial x \partial y}+4\frac{\partial^2 u}{\partial y^2}=0$ (*c*) $2\frac{\partial^2 u}{\partial x^2}+6\frac{\partial^2 u}{\partial x \partial y}+3\frac{\partial^2 u}{\partial y^2}=0$

Solution. (*a*) $A = 2, B = 4, \quad C = 3$

$B^2 - 4AC = (4)^2 - 4(2)(3) < 0$ **Ans.** Elliptic

(*b*) $A = 1, B = 4, \quad C = 4$

$B^2 - 4AC = (4)^2 - 4(1)(4) = 0$ **Ans.** Parabolic

(*c*) $A = 2, B = 6, C = 3$

$B^2 - 4AC = (6)^2 - 4(2)(3) = +12 > 0$ **Ans.** Hyperbolic

Example 34. *Determine whether the following equations are hyperbolic, parabolic and elliptic ?*

(*a*) $x^2\frac{\partial^2 u}{\partial t^2}-\frac{\partial^2 u}{\partial x^2}=u$ (*b*) $t\frac{\partial^2 u}{\partial t^2}+2\frac{\partial^2 u}{\partial x \partial t}+x\frac{\partial^2 u}{\partial x^2}+\frac{\partial u}{\partial x}=0$

(*c*) $x\frac{\partial^2 u}{\partial t^2}+t\frac{\partial^2 u}{\partial x \partial t}+\frac{\partial^2 u}{\partial t^2}=0$

Solution. (*a*) Here $A = x^2, \; B = 0, \; C = -1$

Now, $B^2 - 4AC = (0)^2 - 4x^2(-1) = 4x^2$

$\therefore$ It is hyperbolic if $4x^2 > 0$ *i.e.*, $x > 0$

parabolic if $4x^2 = 0$ *i.e.*, $x = 0$

Since $4x^2$ being a square, cannot be negative hence it cannot be elliptic.

(*b*) Here $A = t, \; B = 2, C = x$

Now, $B^2 - 4AC = 4 - 4tx$

It is hyperbolic if $4 - 4tx > 0$ *i.e.*, $tx < 1$

elliptic if $4 - 4tx < 0$ *i.e.*, $tx > 1$

and parabolic if $4 - 4tx = 0$ *i.e.*, $tx = 1$.

(*c*) Here $A = x, \quad B = t, \quad C = 1$

Now, $B^2 - 4AC = (t)^2 - 4(x)(1) = t^2 - 4x$

$\therefore$ It is hyperbolic if $t^2 - 4x > 0$ *i.e.*, $t^2 > 4x$

elliptic if $t^2 - 4x < 0$ *i.e.*, $t^2 < 4x$

and parabolic if $t^2 - 4x = 0$ *i.e.*, $t^2 = 4x$. **Ans.**

Example 35. *Classify the partial differential equations*

$$\frac{\partial^2 u}{\partial t^2}+t\frac{\partial^2 u}{\partial x \partial t}+x\frac{\partial^2 u}{\partial x^2}+2\frac{\partial u}{\partial t}+\frac{\partial u}{\partial x}+6u=0$$

Solution. We have,

$$\frac{\partial^2 u}{\partial t^2}+t\frac{\partial^2 u}{\partial x\partial t}+x\frac{\partial^2 u}{\partial x^2}+2\frac{\partial u}{\partial t}+\frac{\partial u}{\partial x}+6u=0$$

Here $A = 1, B = t, C = x$

Now, $B^2 - 4AC = t^2 - 4(1)(x) = t^2 - 4x$

The equation is elliptic if $t^2 - 4x < 0.$

The equation is parabolic if $t^2 - 4x = 0.$

The equation is hyperbolic if $t^2 - 4x > 0.$ **Ans.**

Example 36. *Classify the partial differential equation*

$$x^2\frac{\partial^2 u}{\partial t^2}+3\frac{\partial^2 u}{\partial x\partial t}+x\frac{\partial^2 u}{\partial x^2}+17\frac{\partial u}{\partial t}=100u$$

Solution. We have,

$$x^2\frac{\partial^2 u}{\partial t^2}+3\frac{\partial^2 u}{\partial x\partial t}+x\frac{\partial^2 u}{\partial x^2}+17\frac{\partial u}{\partial t}=100u$$

Here $A = x^2, B = 3, C = x$

Now, $B^2 - 4AC = (3)^2 - 4x^2.x = 9 - 4x^3$

The equation is elliptic if $9 - 4x^3 < 0.$

The equation is parabolic if $9 - 4x^3 = 0.$

The equation is hyperbolic if $9 - 4x^3 > 0.$ **Ans.**

Example 37. *Show that the equation.*

$$\frac{\partial^2 u}{\partial t^2}=c^2\frac{\partial^2 u}{\partial x^2}$$ *is hyperbolic.*

Solution. We have,

$$\frac{\partial^2 u}{\partial t^2}-c^2\frac{\partial^2 u}{\partial x^2}=0.$$

Here $A = 1, B = 0, C = -c^2$

Now, $B^2 - 4AC = (0)^2 - 4(1)(-c^2) = 4c^2$

Hence, $B^2 - 4AC > 0$

Thus, the given equation is hyperbolic. **Proved.**

Example 38. *Classify the following differential equation as to type in the second quadrant of xy-plane* $\sqrt{y^2+x^2}\,u_{xx}+2(x-y)\,u_{xy}+\sqrt{y^2+x^2}\,u_{yy}=0$

Solution. Here $A=\sqrt{y^2+x^2},\quad B=2(x-y),\quad C=\sqrt{y^2+x^2}$

None $B^2 - 4AC = 4(x-y)^2 - 4(y^2+x^2) = -8xy$

In second quadrant, y is positive while x is – ve.

$\Rightarrow$ $B^2 - 4AC = +\text{ ve} > 0$

Hence, differential equation is hyperbolic. **Ans.**

Example 39. *Match the column for the items of the Left side to that of right side :*

A second order P.D.E. in the function 'u' of two independent variables x, y given with usual symbols $Au_{xx} + Bu_{xy} + Cu_{yy} + F(u) = 0$ *then*

(i) Hyperbolic	*(a)* $B^2 - 4AC = 0$
(ii) Parabolic	*(b)* $B^2 - 4AC < 0$
(iii) Elliptic	*(c)* $B^2 - 4AC > 0$
(iv) Not classifies	*(d)* $A = B = C = 0$

(U.P., II Semester, June 2009)

Solution.

(*i*) Hyperbolic	(*c*) $B^2 - 4AC > 0$
(*ii*) Parabolic	(*a*) $B^2 - 4AC = 0$
(*iii*) Elliptic	(*b*) $B^2 - 4AC < 0$
(*iv*) Not classifies	(*d*) $A = B = C = 0$

Ans.

EXERCISE 43.9

Classify the following partial differential equations:

1. $9\frac{\partial^2 u}{\partial x^2} - 6\frac{\partial^2 u}{\partial x \partial t} + \frac{\partial^2 u}{\partial t^2} = 0$ **Ans.** Parabolic

2. $3\frac{\partial^2 z}{\partial x^2} + 2\frac{\partial^2 z}{\partial x \partial y} + 4\frac{\partial^2 z}{\partial y^2} = 0$ **Ans.** Elliptic

3. $2\frac{\partial^2 z}{\partial x^2} - 6\frac{\partial^2 z}{\partial x \partial y} + 3\frac{\partial^2 z}{\partial y^2} = 0$ **Ans.** Hyperbolic

4. $t\frac{\partial^2 u}{\partial t^2} + 3\frac{\partial^2 u}{\partial x \partial t} + x\frac{\partial^2 u}{\partial x^2} + 17\frac{\partial u}{\partial x} = 0$

   Ans. Hyperbolic if $xt < \frac{9}{4}$, parabolic if $xt = \frac{9}{4}$, elliptic if $xt > \frac{9}{4}$

5. $\frac{\partial^2 z}{\partial x^2} = \frac{\partial x}{\partial y}$ **Ans.** Parabolic

6. $\frac{\partial^2 z}{\partial x^2} = \frac{\partial^2 z}{\partial y^2}$ **Ans.** Hyperbolic

7. $\frac{\partial^2 z}{\partial x^2} + \frac{\partial^2 z}{\partial y^2} = 0$ *(U.P., II Semester, 2010)* **Ans.** Elliptic

Choose the correct answer:

8. The complementary function of $r - 7s + 6t = e^{x+y}$ is :

   (*i*) $f_1(y - x) + f_2(y - 6x)$ (*ii*) $f_1(y + x) + f_2(y + 6x)$

   (*iii*) $f_1(y + 2x) + f_2(y - 2x)$ (*iv*) $f_1(y + 3x) + f_2(y - 4x)$ **Ans.** (*ii*)

   (R.G.P.V Bhopal, II Semester Feb. 2006)

APPLICATIONS OF PARTIAL DIFFERENTIAL EQUATIONS

44.1 INTRODUCTION

In applied mathematics, the partial differential equations generally arise from the mathematical formulation of physical problems. Subject to certain given conditions, called boundary conditions solving such an equation is known as solving a boundary value problem.

The method of solution of such equations differs from that used in the case of ordinary differential equations. We first find out the general solution of the ordinary differential equation and determine the particular solution with the help of given conditions. Here, from the start, we try to find particular solutions of the partial differential equations which satisfy all the boundary conditions. Method of separation of variables is employed to solve the applied partial differential equation.

44.2 METHOD OF SEPARATION OF VARIABLES

(U.P., II Semester, June 2007)

In this method, we assume that the dependent variable is the product of two functions, each of which involves only one of the independent variables. So two ordinary differential equations are formed.

Example 1. *Applying the method of separation of variables techniques, find the solution to the P.D.E.*

$$3u_x + 2u_y = 0 \text{ ..., where } u_x = \frac{\partial u}{\partial x}, u_y = \frac{\partial u}{\partial y}.$$

Solution. Here we have

$$\frac{3\,\partial u}{\partial x} + \frac{2\,\partial u}{\partial y} = 0 \quad \text{...(1)}$$

Let $$u = X(x)\,Y(y) \quad \text{...(2)}$$

Where X is a function of x only and Y is a function of y only.

On differentiating (2) partially w.r.t. x, we get

$$\frac{\partial u}{\partial x} = \frac{\partial X}{\partial x} \cdot Y \quad \text{...(3)}$$

On differentiating (2) partially w.r.t. y, we get

$$\frac{\partial u}{\partial y} = X.\frac{\partial Y}{\partial Y} \qquad ...(4)$$

Putting the values of $\frac{\partial u}{\partial x}$ and $\frac{\partial u}{\partial y}$ from (3) and (4) in (1), we get

$$3\frac{\partial X}{\partial x}.Y + 2X\frac{\partial Y}{\partial y} = 0 \qquad ...(5)$$

Dividing (5) by XY, we get

$$\frac{3}{X}\frac{\partial X}{\partial x} + \frac{2}{Y}\frac{\partial Y}{\partial y} = 0 \qquad \left[\begin{array}{l}\text{R.H.S is constant for L.H.S,}\\ \text{So we take both equations}\\ \text{are equal to } k \text{ (constant)}\end{array}\right]$$

$$\Rightarrow \quad \frac{3}{X}\frac{\partial X}{\partial x} = -\frac{2}{Y}\frac{\partial Y}{\partial y} = k \quad \Rightarrow \quad \frac{3}{X}\frac{\partial X}{\partial x} = k \text{ and } -\frac{2}{Y}\frac{\partial Y}{\partial y} = k$$

$$\Rightarrow \quad \frac{\partial X}{X} = \frac{k}{3}\partial x \text{ and } \frac{\partial Y}{Y} = -\frac{k}{2}\partial y \quad \Rightarrow \quad \log X = \frac{k}{3}x + c_1 \text{ and } \log Y = \frac{-k}{2}y + c_2$$

$$\Rightarrow \quad X = e^{\frac{k}{3}x + c_1} \text{ and } Y = e^{-\frac{k}{2}y + c_2}$$

Putting the values of X and Y in (2), we get

$$u = e^{\frac{k}{3}x + c_1}\, e^{\frac{-k}{2}y + c_2} = e^{k\left(\frac{x}{3} - \frac{y}{2}\right) + c_1 + c_2} = e^{k\left(\frac{x}{3} - \frac{y}{2}\right)} . e^{c_1 + c_2}$$

Hence $$u = A\, e^{k\left(\frac{x}{3} - \frac{y}{2}\right)} \qquad [\text{where } A = e^{c_1 + c_2}]$$ **Ans.**

Example 2. *Solve the following equation* $\frac{\partial^2 z}{\partial x^2} - 2\frac{\partial z}{\partial x} + \frac{\partial z}{\partial y} = 0$ *by the method of separation of variables.* *(AMIETE, June 2009, U.P., II Semester, Summer 2009, 2005)*

Solution. Given equation is

$$\frac{\partial^2 z}{\partial x^2} - 2\frac{\partial z}{\partial x} + \frac{\partial z}{\partial y} = 0 \qquad ... (1)$$

Let $$z = X(x)\, Y(y) \qquad ... (2)$$

where X is a function of x only and Y is a function of y only.

$$\frac{\partial z}{\partial x} = Y\frac{dX}{dx}, \qquad \frac{\partial^2 z}{\partial x^2} = Y\frac{d^2 X}{dx^2}$$

$$\frac{\partial z}{\partial y} = X\frac{dY}{dy}$$

Putting all values in equation (1), we get $Y\frac{d^2 X}{dx^2} - 2Y\frac{dX}{dx} + X\frac{dY}{dy} = 0$

Dividing by XY, we have $\frac{1}{X}\frac{d^2 X}{dx^2} - \frac{2}{X}\frac{dX}{dx} + \frac{1}{Y}\frac{dY}{dy} = 0$

Separating the variables, we have $\frac{1}{X}\frac{d^2 X}{dx^2} - \frac{2}{X}\frac{dX}{dx} = -\frac{1}{Y}\frac{dY}{dy} = K$ *(let)* where K is a constant.

$$\frac{1}{X}\frac{d^2X}{dx^2}-\frac{2}{X}\frac{dX}{dx}=K \qquad\qquad -\frac{1}{Y}\frac{dY}{dy}=K$$

$$\frac{d^2X}{dx^2}-2\frac{dX}{dx}=KX \qquad\qquad \frac{dY}{dy}+KY=0$$

$\Rightarrow$ $(D^2-2D-K)X=0$ $\qquad$ $(D+K)Y=0$

A. E. is $m^2-2m-K=0$ $\qquad$ A.E. is $m+K=0 \Rightarrow m=-K$

$\Rightarrow$ $m=\dfrac{2\pm\sqrt{4+4K}}{2}$ $\qquad$ $\Rightarrow Y=C_3e^{-Ky}$ ...(4)

$\Rightarrow$ $m=1\pm\sqrt{1+K}$

Thus $X=C_1e^{(1+\sqrt{1+K})x}+C_2e^{(1-\sqrt{1+K})x}$ ...(3)

Putting the values of X and Y from (3) and (4) in (2), we get

$$z=\left\{C_1e^{(1+\sqrt{1+K})x}+C_2e^{(1+\sqrt{1+K})x}\right\}C_3e^{-Ky}$$ **Ans.**

Example 3. *Using the method of separation of variables, solve*

$$\frac{\partial u}{\partial x}=2\frac{\partial u}{\partial t}+u$$

where $u(x,0)=6e^{-3x}$ *(U.P. II Semester summer 2006)*

Solution. $\dfrac{\partial u}{\partial x}=2\dfrac{\partial u}{\partial t}+u$ ... (1)

Let $u=X(x)\cdot T(t)$ ... (2)

where X is a function of x only and T is a function of t only.

Putting the value of u in (1), we get

$$\frac{\partial(X.T)}{\partial x}=2\frac{\partial}{\partial t}(X.T)+X.T, \qquad T\frac{dX}{dx}=2X\frac{dT}{dt}+X.T$$

On separating the variables, we get

$$\frac{1}{X}\frac{dX}{dx}=\frac{2}{T}\frac{dT}{dt}+1=C$$ [On dividing by XT]

$$\frac{1}{X}\frac{dX}{dx}=C \qquad\qquad \frac{2}{T}\frac{dT}{dt}+1=C$$

$\Rightarrow$ $\dfrac{dX}{dx}=CX$ $\qquad$ $\Rightarrow$ $\dfrac{dT}{dt}+\dfrac{T}{2}=\dfrac{CT}{2}$

$\Rightarrow$ $DX-CX=0$ $\qquad$ $\Rightarrow$ $DT-\left(\dfrac{C}{2}-\dfrac{1}{2}\right)T=0$

$\Rightarrow$ $(D-C)X=0$ $\qquad$ A.E. is $m-\left(\dfrac{C}{2}-\dfrac{1}{2}\right)=0$

A.E. is $m-C=0 \Rightarrow m=C$ $\qquad$ $\Rightarrow m=\dfrac{1}{2}(C-1) \Rightarrow T=be^{\frac{1}{2}(c-1)t}$

$\Rightarrow$ $X=ae^{cx}$

Putting the values of X and T in (2), we have

$$u = ae^{cx} . be^{\frac{1}{2}(c-1)t}$$

$$\Rightarrow \quad u = abe^{cx + \frac{1}{2}(c-1)t} \quad ...(3)$$

On putting $t = 0$ and $u = 6e^{-3x}$ in (3), we get

$$6e^{-3x} = abe^{cx} \quad \Rightarrow ab = 6 \text{ and } c = -3$$

Putting the values of ab and c in (3), we have

$$u = 6e^{-3x-2t}$$ **Ans.**

which is the required solution.

Example 4. *Solve the following equation by the method of separation of variables*

$$\frac{\partial^2 u}{\partial x \, \partial t} = e^{-t} \cos x$$

given that $u = 0$ *when* $t = 0$ *and* $\frac{\partial u}{\partial t} = 0$ *when* $x = 0$. *[U.P. II Semester, (SUM) 2008]*

Solution. Let $u = XT$ ...(1)

where X is a function of x only and T is a function of t only.

Then, $$\frac{\partial u}{\partial t} = \frac{\partial}{\partial t}(XT) = X\frac{dT}{dt}$$

$$\therefore \quad \frac{\partial^2 u}{\partial x \partial t} = \frac{\partial}{\partial x}\left(X\frac{dT}{dt}\right) = \frac{dT}{dt}.\frac{dX}{dx} \quad ...(2)$$

Substituting the value $\frac{\partial^2 u}{\partial x \partial t}$ from (2) in the given equation, we get

$$\frac{dT}{dt}\frac{dX}{dx} = e^{-t} \cos x$$

Separating the variables, we get

$$e^t \frac{dT}{dt} = \frac{\cos x}{\left(\frac{dX}{dx}\right)} = -p^2 \text{ (say)} \quad ...(3)$$

Now, $$e^t \frac{dT}{dt} = -p^2$$

$$\Rightarrow \quad dT = -p^2 e^{-t} \, dt$$

On integration, we get

$$T = p^2 e^{-t} + c_1 \quad ...(4)$$

Also, $$\frac{dX}{dx} = -\frac{1}{p^2}\cos x$$

$$dX = -\frac{1}{p^2}\cos x \, dx$$

On integration, we get

$$X = -\frac{1}{p^2}\sin x + c_2 \quad ...(5)$$

Putting the values of X and T from (4) and (5) in (1), we get

$$u = XT = \left(-\frac{1}{p^2}\sin x + c_2\right)(p^2 e^{-t} + c_1) \quad ...(6)$$

On putting $u = 0$ and $t = 0$ in (6), we get

$$0 = \left(-\frac{1}{p^2}\sin x + c_2\right)(p^2 + c_1)$$

$$\Rightarrow \quad p^2 + c_1 = 0 \Rightarrow c_1 = -p^2$$

Differentiating (6) w.r.t. "t", we get

$$\frac{\partial u}{\partial t} = \left(-\frac{1}{p^2}\sin x + c_2\right)(-p^2 e^{-t}) \quad ...(7)$$

Putting $\frac{\partial u}{\partial t} = 0$ when $x = 0$ in (7), we get

$$0 = c_2\,(-p^2 e^{-t})$$

$$\Rightarrow \quad c_2 = 0$$

Substituting the values of $c_1 = -p^2$ and $c_2 = 0$ in (6), we get

$$u = -\frac{1}{p^2}\sin x\,(p^2 e^{-t} - p^2)$$

$$= (1 - e^{-t})\sin x$$ **Ans.**

Example 5. *Solve the P.D.E. by separation of variables method,*

$$u_{xx} = u_y + 2u,\ u(0, y) = 0$$

$$\frac{\partial}{\partial x} u(0, y) = 1 + e^{-3y}.$$ *(U.P. II Semester, 2010, 2009)*

Solution. Let $u = XY$ ...(1)

where X is a function of x only and Y is a function of y only.

On differentiating, we get

$$\frac{\partial u}{\partial y} = \frac{\partial}{\partial y}(XY) = X\frac{dY}{dy} = XY'$$

and $\frac{\partial u}{\partial x} = \frac{\partial}{\partial x}(XY) = Y\frac{dX}{dx}$ and $\frac{\partial^2 u}{\partial x^2} = \frac{\partial}{\partial x}\left[Y\frac{dX}{dx}\right] = Y\frac{d^2X}{dx^2}$

Putting the values of u_{xx} and u_y and u in the given equation, we get

$$YX'' = XY' + 2XY$$

On separating the variables, we get

$$\frac{X''}{X} = \frac{Y' + 2Y}{Y}$$

$$\Rightarrow \quad \frac{X''}{X} = \frac{Y'}{Y} + 2 = k \text{ (say)} \quad ...(2)$$

(*i*)	$\frac{X''}{X} = k$	$\frac{Y'}{Y} + 2 = k$
$\Rightarrow$	$X'' - kX = 0$	$\Rightarrow \frac{Y'}{Y} = k - 2$
A.E. is	$m^2 - k = 0 \Rightarrow m = \pm\sqrt{k}$	$\Rightarrow \frac{dY}{Y} = (k-2)\,dy$
$\therefore$	$\Rightarrow X = C_1 e^{\sqrt{k}x} + C_2 e^{-\sqrt{k}x}$	$\Rightarrow \log Y = (k-2)\,y + \log C_3$
$\therefore$		$\Rightarrow Y = C_3 e^{(k-2)y}$

On putting the values of X and Y in (1), we get

$$u = (C_1 e^{\sqrt{kx}} + C_2 e^{-\sqrt{kx}})\, C_3 e^{(k-2)y} \qquad ...(3)$$

On putting $x = 0$ and $u = 0$ in (3), we get

$$0 = (C_1 + C_2)\, C_3 e^{(k-2)y} \qquad (C_3 e^{(k-2)y} \neq 0) \qquad ...(4)$$

$$\Rightarrow \qquad C_1 + C_2 = 0 \Rightarrow C_2 = -C_1$$

On putting $C_2 = -C_1$ in (3), we get

$$u = \Sigma C_1 C_3 (e^{\sqrt{kx}} - e^{-\sqrt{kx}})\, e^{(k-2)y} \qquad ...(5)$$

On differentiating (5), w.r.t. x, we get

$$\frac{\partial u}{\partial x} = \Sigma C_1 C_3 \sqrt{k}\, (e^{\sqrt{kx}} + e^{-\sqrt{kx}})\, e^{(k-2)y} \qquad ...(6)$$

On putting $x = 0$ and $\frac{\partial u}{\partial x} = 1 + e^{-3y}$ in (6), we get

$$1 + e^{-3y} = \Sigma C_1 C_3 \sqrt{k}\, (2)\, e^{(k-2)y} = \sum_{n=1}^{\infty} b_n e^{(k-2)y} \qquad [2c_1 c_3 = b_n]$$

$$1 + e^{-3y} = \Sigma C_1 C_3 \sqrt{k}\, (2)\, e^{(k-2)y} = b_1 e^{(k-2)y} + b_2 e^{(k-2)y} + b_3 e^{(k-2)y} + ...$$

Comparing the coefficients, we get

$$b_1 = 1,\ k - 2 = 0 \Rightarrow k = 2$$

Again comparing, we get

$$2C_1 C_3 \sqrt{k} = 1 \Rightarrow C_1 C_3 = \frac{1}{2\sqrt{k}} = \frac{1}{2\sqrt{2}} \qquad (\because k = 2)$$

Again comparing b_3 on both the sides, we get

$$b_3 = -1,\ k - 2 = -3 \Rightarrow k = -1$$

Again comparing b_3, we get

$$2C_1 C_3 \sqrt{k} = 1 \Rightarrow C_1 C_3 = \frac{1}{2\sqrt{k}} = \frac{1}{2\sqrt{-1}} = \frac{1}{2i} \qquad (\because k = -1)$$

Hence, $u(x,y) = \dfrac{1}{2\sqrt{2}}(e^{\sqrt{2}x} - e^{-\sqrt{2}x}) + \dfrac{1}{2i}(e^{ix} - e^{-ix})\, e^{-3y}$

$$u(x,y) = \frac{1}{\sqrt{2}} \sinh \sqrt{2}\, x + e^{-3y} \sin x.$$ **Ans.**

EXERCISE 44.1

Using the method of separation of variables, find the solution of the following equations

1. $2x\dfrac{\partial z}{\partial x} - 3y\dfrac{\partial z}{\partial y} = 0$ **Ans.** $z = cx^{\frac{k}{2}} y^{\frac{k}{3}}$

2. $\dfrac{\partial u}{\partial x} + u = \dfrac{\partial u}{\partial t}$ if $u = 4e^{-3x}$ when $t = 0$ **Ans.** $u = 4e^{-3x-2t}$

3. $4\dfrac{\partial u}{\partial x} + \dfrac{\partial u}{\partial y} = 3u$ and $u = e^{-5y}$ when $x = 0$ (*AMIETE, June 2010*) **Ans.** $u = e^{2x-5t}$

4. $4\dfrac{\partial u}{\partial t} + \dfrac{\partial u}{\partial x} = 3u,\ u = 3e^{-x} - e^{-5x}$ at $t = 0$ **Ans.** $u = 3e^{t-x} - e^{2t-5x}$

5. $3\dfrac{\partial u}{\partial x} + 2\dfrac{\partial u}{\partial y} = 0;\ u(x, 0) = 4e^{-x}$ **Ans.** $u = 4e^{-x+3/2y}$

6. $\dfrac{\partial u}{\partial x}+\dfrac{\partial u}{\partial y}=2(x+y)u$ *(AMIETE, June 2010)* **Ans.** $u=ce^{x^2+y^2+k(x-y)}$

7. $\dfrac{\partial u}{\partial t}=4\dfrac{\partial^2 u}{\partial x^2}$ if $u(x,0)=4x-\dfrac{1}{2}x^2$ **Ans.** $u=\left(4x-\dfrac{x^2}{2}\right)e^{-p^2 t}$

8. $\dfrac{\partial u}{\partial t}=\dfrac{\partial^2 u}{\partial x^2}$ if $u(x, 0) = \sin \pi x$ **Ans.** $u=\sin \pi x . e^{-pl}$

9. $\dfrac{\partial^2 u}{\partial x^2}=\dfrac{\partial u}{\partial t}$ if $u(x,0)=x^2(25-x^2)$ **Ans.** $u=x^2(25-x^2)e^{-p^2 t}$

10. $\dfrac{\partial^2 z}{\partial x^2}-2\dfrac{\partial z}{\partial x}-\dfrac{\partial z}{\partial y}=0$ **Ans.** $z=c_1 e^{[1+\sqrt{(1+p)}]x+p^2 y}+c_2 e^{[1-\sqrt{1+P}]x+p^2 y}$

11. $\dfrac{\partial^2 u}{\partial t^2}=\dfrac{\partial^2 u}{\partial x^2}$ If $u(x,0)=\dfrac{1}{2}x(1-x)$ **Ans.** $u=\dfrac{x}{2}(1-x)\cos pt+c_2 \sin pt(c_3 \cos px+c_4 \sin px)$

12. $16\dfrac{\partial^2 u}{\partial x^2}=\dfrac{\partial^2 u}{\partial t^2}$ if $u(x,0)=x^2(5-x)$ **Ans.** $u=x^2(5-x)\cos pt+c_4 \sin pt\left(c_1 \cos\dfrac{px}{4}+c_2 \sin\dfrac{px}{4}\right)$

13. $\dfrac{\partial u}{\partial x}=2\dfrac{\partial u}{\partial y}+u,\ u(x, 0)=4e^{-3x}$ **Ans.** $u=4e^{-(3x+2y)}$

14. $\dfrac{\partial u}{\partial x}-2\dfrac{\partial u}{\partial y}=u$, given that $u(x,0)=3e^{-5x}+2e^{-3x}$

Ans. $u=3e^{-5x-3y}+2e^{-3x-2y}$

44.3 EQUATION OF VIBRATING STRING

(AMIETE, June 2010, U.P. II Semester, June 2007)

Let us consider small transverse vibrations of an elastic string of length l, which is stretched and then fixed at its two ends. Now we will study the transverse vibration of the string when no external forces act on it. Take an end of the string as the origin and the string in the equilibrium position as the x-axis and the line through the origin and perpendicular to the x-axis as the y-axis. We make the following assumptions:

1. The motion takes place entirely in one plane. This plane is chosen as the *xy plane*.
2. In this plane, each particle of the string moves in a direction perpendicular to the equilibrium position of the string.
3. The tension T caused by stretching the string before fixing it at the end points is constant at all times at all points of the deflected string.
4. The tension T is very large compared with the weight of the string and hence the gravitational force may be neglected.
5. The effect of friction is negligible.
6. The string is perfectly flexible. It can transmit only tension but not bending or shearing forces.
7. The slope of the deflection curve is small at all points and at all times.

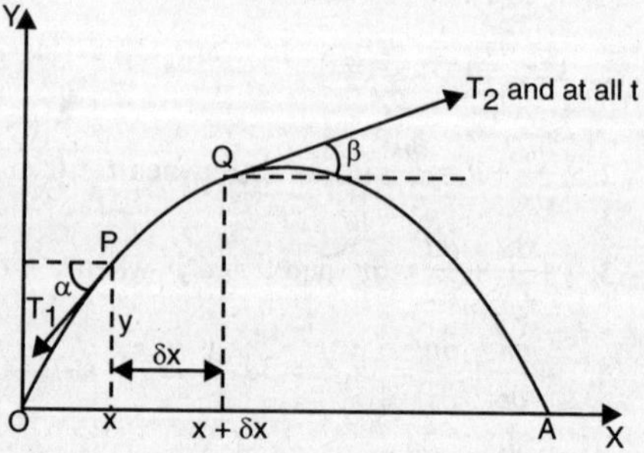

When the string is in motion in the xy-plane, the displacement y of any point of the string is a function of x and time t. Let $P(x, y)$ and $Q(x + \delta x, y + \delta y)$ be two neighbouring points on the string. Let α and β be the inclinations made by the tangents at P and Q respectively with the x-axis. The tension T_1 at P and tension T_2 at Q are tangential. Let m be the mass per unit length of the string which is homogeneous. Consider the motion of the infinitesimal element PQ of the string. The vertical component of the force to which this element is subjected to is

$$m\delta s\frac{\partial^2 y}{\partial t^2} = T_2 \sin\beta - T_1 \sin\alpha \qquad ...(1)$$

Since, there is no motion in the horizontal direction,

$$T_1 \cos\alpha = T_2 \cos\beta = T \text{ (constant)} \qquad ...(2)$$

Dividing (1) by T, we get

$$\frac{m\delta s\dfrac{\partial^2 y}{\partial t^2}}{T} = \frac{T_2 \sin\beta}{T} - \frac{T_1 \sin\alpha}{T}$$

$$\frac{m\delta s}{T}\frac{\partial^2 y}{\partial t^2} = \frac{T_2 \sin\beta}{T_2 \cos\beta} - \frac{T_1 \sin\alpha}{T_1 \cos\alpha} \qquad \begin{bmatrix} T = T_1 \cos\alpha \\ T = T_2 \cos\beta \end{bmatrix}$$

$$\Rightarrow \quad \frac{m\delta s}{T}\frac{\partial^2 y}{\partial t^2} = \tan\beta - \tan\alpha \qquad \Rightarrow \quad \frac{\partial^2 y}{\partial t^2} = \frac{T}{m\delta s}(\tan\beta - \tan\alpha)$$

Slope $\tan\alpha = \left(\dfrac{\partial y}{\partial x}\right)_x$ at P and slope $\tan\beta = \left(\dfrac{\partial y}{\partial x}\right)_{x+\delta x}$ at Q.

$$\frac{\partial^2 y}{\partial t^2} = \frac{T}{m\delta x}\left[\left(\frac{\partial y}{\partial x}\right)_{x+\delta x} - \left(\frac{\partial y}{\partial x}\right)_x\right] \quad \Rightarrow \quad \frac{\partial^2 y}{\partial t^2} = \frac{T}{m}\left[\frac{\left(\dfrac{\partial y}{\partial x}\right)_{x+\delta x} - \left(\dfrac{\partial y}{\partial x}\right)_x}{\delta x}\right]$$

On taking limit as $\delta x \to 0$, we have

$$\frac{\partial^2 y}{\partial t^2} = \frac{T}{m}\frac{\partial^2 y}{\partial x^2}, \qquad \boxed{\frac{\partial^2 y}{\partial t^2} = a^2\frac{\partial^2 y}{\partial x^2}} \qquad \left(\text{Take } a^2 = \frac{T}{m}\right)$$

Note. The partial differential equation is known as one-dimensional wave equation, because the motion is only in vertical direction (Transverse vibration) not in horizontal direction.

Boundary conditions.

At O, $x = 0$ and $y = 0$, $\dfrac{\partial y}{\partial t} = 0$ as $t = 0$ *At A*, $x = l$ and $y = 0$,

Example 6. *Solve completely the equation* $\dfrac{\partial^2 y}{\partial t^2} = c^2\dfrac{\partial^2 y}{\partial x^2}$, *representing the vibrations of a string of length l, fixed at both ends, given that* $y(0, t) = 0$, $y(l, t) = 0$; $y(x, 0) = f(x)$ *and* $\dfrac{\partial}{\partial t}y(x, 0) = 0$, $0 < x < l$. *(U.P. II Semester summer 2005)*

Solution. Here,

$$\frac{\partial^2 y}{\partial t^2} = c^2\frac{\partial^2 y}{\partial x^2} \qquad ...(1)$$

Let

$$y = X(x)\,T(t) \qquad ...(2)$$

$$\frac{\partial^2 y}{\partial t^2} = X\frac{d^2 T}{dt^2}$$

$$\frac{\partial^2 y}{\partial x^2} = T\frac{d^2 X}{dx^2}$$

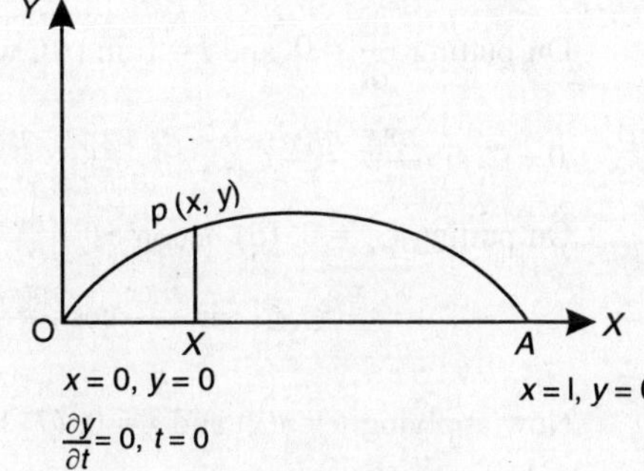

Equation (1) becomes $X\dfrac{d^2T}{dt^2}=c^2T\dfrac{d^2X}{dx^2}$

Separating the variables,

$$\frac{1}{X}\frac{d^2X}{dx^2}=\frac{1}{c^2T}\frac{d^2T}{dt^2}=-p^2\text{ (let)}$$

If $\dfrac{1}{X}\dfrac{d^2X}{dx^2}=-p^2$

$\Rightarrow\quad \dfrac{d^2X}{dx^2}=-Xp^2$

$\Rightarrow\quad \dfrac{d^2X}{dx^2}+Xp^2=0$

$\Rightarrow\quad (D^2+p^2)X=0$

A.E. is $m^2+p^2=0 \quad\Rightarrow\quad m=\pm pi$

$X=(C_1\cos px+C_2\sin px)$

If $\dfrac{1}{c^2T}\dfrac{\partial^2T}{\partial t^2}=-p^2$

$\Rightarrow\quad (D^2+p^2c^2)\,T=0$

A.E. is $m^2+p^2c^2=0$

$\Rightarrow\quad m^2=-p^2c^2 \quad\Rightarrow\quad m=\pm pci$

$T=(C_3\cos pct+C_4\sin pct)$

Putting the values of X and T in (2), we get

$$y=(C_1\cos px+C_2\sin px)\,(C_3\cos pct+C_4\sin pct) \quad ..(3)$$

Now applying the boundary condition

$x=0$ and $y=0$

Putting these values in (3), we get

$$0=C_1\,(C_3\cos pct+C_4\sin pct) \quad\Rightarrow\quad C_1=0$$

Equation (3) becomes, $\quad y=C_2\sin px\,(C_3\cos pct+C_4\sin pct) \quad ..(4)$

Putting $x=l$ and $y=0$ in (4), we get

$$0=C_2\sin pl\,(C_3\cos pct+C_4\sin pct) \Rightarrow \sin pl=0=\sin n\pi$$

$$\Rightarrow pl=n\pi \Rightarrow \quad p=\frac{n\pi}{l}$$

On putting $p=\dfrac{n\pi}{l}$, (4) becomes

$$y=C_2\sin\frac{n\pi x}{l}\left(C_3\cos\frac{n\pi ct}{l}+C_4\sin\frac{n\pi ct}{l}\right) \quad ...(5)$$

On differentiating (5) w.r.t. t, we get

$$\frac{\partial y}{\partial t}=C_2\sin\frac{n\pi x}{l}\left(-\frac{n\pi c}{l}C_3\sin\frac{n\pi ct}{l}+\frac{n\pi c}{l}C_4\cos\frac{n\pi c}{l}t\right) \quad ...(6)$$

On putting $\dfrac{\partial y}{\partial t}=0$ and $t=0$ in (6), we get

$$0=C_2\sin\frac{n\pi x}{l}.\frac{n\pi c}{l}.C_4 \quad\Rightarrow\quad C_4=0$$

On putting $C_4=0$, (5) becomes

$$y=C_2C_3\,\sin\frac{n\pi x}{l}\cos\frac{n\pi ct}{l} \quad [\text{let } C_2\,C_3=b_n] \quad ...(7)$$

Now applying $y=f(x)$ and $t=0$, (7) becomes $f(x)=b_n\sin\dfrac{n\pi x}{l}$

C_2C_3 can be calculated using Fourier sine series as $b_n = \frac{2}{l}\int_0^l f(x)\sin\frac{n\pi x}{l}dx$ ...(8)

Thus, required solution for the given equation is

$$y = b_n \sin\frac{n\pi x}{l}\cos\frac{n\pi ct}{l}$$

where b_n is given by equation (8). **Ans.**

Example 7. *A tightly stretched string with fixed end points x =0 and x = π is initially at rest in its equilibrium position. If it is set vibrating by giving to each of its points an initial velocity.*

$$\left(\frac{\partial y}{\partial t}\right)_{t=0} = 0.03 \sin x - 0.04 \sin 3x$$

then find the displacement y (x,t) at any point of string at any time t.

Solution. Here we have equation for vibration of a string $\frac{\partial^2 y}{\partial t^2} = c^2\frac{\partial^2 y}{\partial x^2}$

Its solution is

$$y(x,t) = (c_1 \cos pct + c_2 \sin pct)(c_3 \cos px + c_4 \sin px) \quad ...(1)$$

[See Example 6 on page 1193]

Putting $x = 0$ and $y = 0$ in (1), we get

$$0 = (c_1 \cos pct + c_2 \sin pct)(c_3)$$

$$\Rightarrow \quad c_3 = 0$$

Putting $c_3 = 0$ in (1), we have $y = (c_1 \cos pct + c_2 \sin pct)(c_4 \sin px)$ ...(2)

Putting $x = \pi$ and $y = 0$ in (2), we get

$$0 = (c_1 \cos pct + c_2 \sin pct)\, c_4 \sin p\pi$$

$$\Rightarrow \quad 0 = \sin p\pi$$

and $\sin n\pi = \sin p\pi \quad \Rightarrow \quad p = n$

Putting $p = n$ in (2), we get

$$y = (c_1 \cos nct + c_2 \sin nct)(c_4 \sin nx) \quad ...(3)$$

Putting $t = 0$, $y = 0$ in (3), we get

$$0 = (c_1 + 0)\, c_4 \sin nx \quad c_4 \sin nx \neq 0 \text{ since } \sin nx \text{ is a part of } X)$$

$$\Rightarrow \quad c_1 = 0$$

On putting $c_1 = 0$ in (3), we get

$$y = (c_2 \sin nct)(c_4 \sin nx)$$

General equation is $(c_2\, c_4 = b_n)$

$$y = \sum_{n=1}^{\infty} b_n \sin nct \sin nx \quad ...(4)$$

Differentiating (4) w.r.t. 't', we get

$$\frac{\partial y}{\partial t} = \sum_{n=1}^{\infty} b_n nc \cos nct \sin nx \quad ...(5)$$

On putting $\frac{\partial y}{\partial t} = 0.03 \sin x - 0.04 \sin 3x$ and $t = 0$ in (5) we get

$$0.03 \sin x - 0.04 \sin 3x = \sum_{n=1}^{\infty} b_n nc \sin nx$$

$\Rightarrow$ $0.03 \sin x - 0.04 \sin 3x = cb_1 \sin x + 2cb_2 \sin 2x + 3c_3b_3 \sin 3x + ...$

Comparing the coefficient, we get

$$0.03 = cb_1 \Rightarrow b_1 = \frac{0.03}{c}$$

$$0 = 2cb_2 \Rightarrow b_2 = 0$$

$$-0.04 = 3c\, b_3 \Rightarrow b_3 = \frac{-0.04}{3c} = \frac{-0.0133}{c}$$

Putting the values of b_1, b_2, b_3 ... in (4), we get

$$y = \frac{0.03}{c} \sin ct \sin x - \frac{0.0133}{c} \sin 3ct \sin 3x$$

Hence $$y = \frac{1}{c}[0.03 \sin ct \sin x - 0.0133 \sin 3x]$$ **Ans.**

Example 8. *A string is stretched and fastened to two points l apart. Motion is started by displacing the string in the form $y = a\sin\frac{\pi x}{l}$ from which it is released at a time $t = 0$. Show that the displacement of any point at a distance x from one end at time t is given by*

$$y(x, t) = a\sin\left(\frac{\pi x}{l}\right)\cos\left(\frac{\pi ct}{l}\right)$$

(A.M.I.E.T.E., U.P., 2009)

Solution. Consider an elastic string tightly stretched between two points O and A. Let O be the origin and OA as x-axis. On giving a small displacement to the string, perpendicular to its length (parallel to the y-axis). Let y be the displacement at the point P (x, y) at any time. The wave equation.

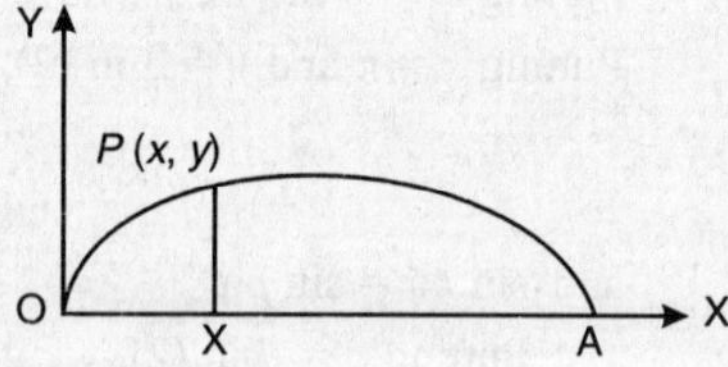

$$\frac{\partial^2 y}{\partial t^2} = c^2 \frac{\partial^2 y}{\partial x^2}$$

where c is a constant. The vibration of the string is given by:

$$\frac{\partial^2 y}{\partial t^2} = c^2 \frac{\partial^2 y}{\partial x^2} \quad ... (1)$$

As the end points of the string are fixed, for all time,

$$y(0, t) = 0 \quad ... (2)$$

and $$y(l, t) = 0 \quad ... (3)$$

Since the initial transverse velocity of any point of the string is zero, therefore,

$$\left(\frac{\partial y}{\partial t}\right)_{t=0} = 0 \quad ... (4)$$

Also $$y(x,0) = a\sin\frac{\pi x}{l} \quad ... (5)$$

Now we have to solve (1), subject to the above boundary conditions. Since the vibration of the string is periodic, therefore, the solution of (1) is of the form

$$y(x, t) = (C_1 \cos px + C_2 \sin px)(C_3 \cos cpt + C_4 \sin cpt) \quad ... (6)$$

On putting $x = 0$ and $y = 0$ in (6), we get

$$0 = C_1(C_3 \cos c\,pt + C_4 \sin c\,pt) \Rightarrow C_1 = 0$$

On putting $C_1 = 0$ in (6), we get

$$y(x, t) = C_2 \sin px (C_3 \cos c\,pt + C_4 \sin c\,pt) \quad \text{... (7)}$$

On differentiating (7) w.r.t. t, we get

$$\frac{\partial y}{\partial t} = C_2 \sin px[C_3(-cp \sin c\,pt) + C_4(cp \cos c\,pt)] \quad \text{...(8)}$$

On putting $\frac{\partial y}{\partial t} = 0$ and $t = 0$ in (8), we get

$$0 = C_2 \sin px(C_4 cp) \quad \Rightarrow \quad C_2 C_4 cp = 0$$

If $\quad C_2 = 0$, (7) will lead to the trivial solution $y(x, t) = 0$

Thus, the only possibility is that $C_4 = 0$

On putting $C_4 = 0$ in (7), we get

$$y(x, t) = C_2 C_3 \sin px \cos c\,pt \quad \text{...(9)}$$

On putting $x = l$ and $y = 0$ in (9), we get

$$0 = C_2\ C_3 \sin pl \cos cpt, \text{ for all } t.$$

Since C_2 and $C_3 \neq 0$, we have $\sin pl = \sin n\pi$.

$\therefore \ pl = n\pi \Rightarrow \quad p = \frac{n\pi}{l}$, where n is an integer.

On putting the value of p in (9), we get

$$y(x, t) = C_2 C_3 \sin\frac{n\pi x}{l} \cos\frac{n\pi ct}{l} \quad \text{... (10)}$$

On putting $t = 0$ and $y = a \sin\frac{\pi x}{l}$ in (10), we get

$$a \sin\frac{\pi x}{l} = C_2 C_3 \sin\frac{n\pi x}{l}$$

which will be satisfied by taking $C_2\ C_3 = a$ and $n = 1$

Hence the required solution is

$$y(x, t) = a \sin\frac{\pi x}{l} \cos\frac{\pi ct}{l}$$ **Proved.**

Example 9. *The vibrations of an elastic string is governed by the partial differential equation*

$$\frac{\partial^2 u}{\partial t^2} = \frac{\partial^2 u}{\partial x^2}$$

The length of the string is π and the ends are fixed. The initial velocity is zero and the initial deflection is $u(x, 0) = 2 (\sin x + \sin 3x)$. Find the deflection $u(x, t)$ of the vibrating string for $t > 0$.

Solution. $\quad \frac{\partial^2 u}{\partial t^2} = \frac{\partial^2 u}{\partial x^2}$

Solution of the given differential equation is

$$\Rightarrow \quad u = (c_1 \cos pt + c_2 \sin pt)(c_3 \cos px + c_4 \sin px) \quad \text{... (1)}$$

On putting $x = 0$, $u = 0$ in (1), we get

$$0 = (c_1 \cos pt + c_2 \sin pt)\, c_3 \quad \Rightarrow \quad c_3 = 0$$

On putting $c_3 = 0$ in (1), it reduces

$$u = (c_1 \cos pt + c_2 \sin pt) c_4 \sin px \qquad \text{... (2)}$$

On putting $x = \pi$ and $u = 0$ in (2), we have

$$0 = (c_1 \cos pt + c_2 \sin pt) c_4 \sin p\pi$$

$$\sin p\pi = 0 = \sin n\pi \qquad n = 1, 2, 3, 4$$

$$\therefore \qquad p\pi = n\pi \quad \text{or} \quad p = n$$

On substituting the value of p in (2), we get

$$u = (c_1 \cos nt + c_2 \sin nt) c_4 \sin nx \qquad \text{... (3)}$$

On differentiating (3) w.r.t. "t", we get

$$\frac{\partial u}{\partial} = (-c_1 n \sin nt + c_2 n \cos nt)\, c_4 \sin nx \qquad \text{...(4)}$$

On putting $\frac{\partial u}{\partial} = 0,\ t = 0$ in (4), we have

$$0 = (c_2 n)\,(c_4 \sin nx) \Rightarrow c_2 = 0$$

On putting $c_2 = 0$, (3) becomes

$$u = (c_1 \cos nt)(c_4 \sin nx)$$

$$u = c_1 c_4 \cos nt \sin nx = \Sigma b_n \cos nt \sin nx \quad [\because b_n = c_1 c_4] \qquad \text{...(5)}$$

On putting $u = 2(\sin x + \sin 3x)$ and $t = 0$ in (5), we have

$$2(\sin x + \sin 3x) = b_1 \sin x + b_2 \sin 2x + b_3 \sin 3x + ...$$

Comparing the coefficient of sin on both side, we have

$$b_1 = 2,\ b_2 = 0 \text{ and } b_3 = 2$$

On substituting the values of b_1, b_2, b_3 in (5), we get

$$u = 2\,[\cos t \sin x + \cos 3t \sin 3x]$$

Ans.

Example 10. *A tightly stretched string with fixed end points $x = 0$ and $x = l$ is initially in a position given by $y = y_0 \sin^3\left(\frac{\pi x}{l}\right)$. If it is released from rest from this position, find the displacement $y\ (x, t)$.*

Solution. Let the equation to the vibrating string be

$$\frac{\partial^2 y}{\partial t^2} = c^2 \frac{\partial^2 y}{\partial x^2} \qquad \text{... (1)}$$

Here the initial conditions are

$$y(0, t) = 0,\ y(l, t) = 0$$

$$\frac{\partial y}{\partial t} = 0 \text{ at } t = 0,\ y(x, 0) = y_0 \sin^3 \frac{\pi x}{l} = \frac{y_0}{4}\left[3 \sin \frac{\pi x}{l} - \sin \frac{3\pi x}{l}\right]$$

The solution of (1) is of the form

$$y = (c_1 \cos px + c_2 \sin px)(c_3 \cos pct + c_4 \sin pct) \qquad \text{... (2)}$$

On putting $x = 0$ and $y = 0$ in (2), we get

$$0 = c_1\,(c_3 \cos pct + c_4 \sin pct) \qquad \Rightarrow c_1 = 0$$

On putting $c_1 = 0$ in (2), we get

$$y = c_2 \sin px (c_3 \cos pct + c_4 \sin pct) \qquad \text{... (3)}$$

On putting $x = l$ and $y = 0$ in (3), we get

$$0 = c_2 \sin pl(c_3 \cos pct + c_4 \sin pct)$$

$\therefore$ $\sin pl = 0 = \sin n\pi$ or $pl = n\pi$, or $p = \frac{n\pi}{l}$, where $n = 0, 1, 2, 3.....$

On putting the value of p in (3), we get

$$y = c_2 \sin\frac{n\pi x}{l}\left(c_3 \cos\frac{n\pi ct}{l} + c_4 \sin\frac{n\pi ct}{l}\right) \qquad \text{... (4)}$$

On differentiating (4), w.r.t. t, we get

$$\frac{\partial y}{\partial t} = c_2 \sin\frac{n\pi x}{l}\left(-\frac{n\pi c}{l} c_3 \sin\frac{n\pi c}{l} t + c_4 \frac{n\pi c}{l} \cos\frac{n\pi ct}{l}\right) \qquad \text{...(5)}$$

On putting $\frac{\partial y}{\partial t} = 0$ and $t = 0$ in (5), we have

$$0 = c_2 \sin\frac{n\pi x}{l} c_4 \frac{n\pi c}{l} \quad \Rightarrow \quad c_4 = 0$$

On putting $c_4 = 0$ in (4), we get

$$y = c_2 c_3 \sin\frac{n\pi x}{l} \cos\frac{n\pi ct}{l}$$

$$y = b_n \sin\frac{n\pi x}{l} \cos\frac{n\pi c}{l} t \qquad (b_n = c_2 c_3)$$

General solution is

$$y = \sum_{n=1}^{\infty} b_n \sin\frac{n\pi x}{l} \cos\frac{n\pi c}{l} t \qquad \text{... (6)}$$

On putting $t = 0$ and $y = \frac{y_0}{4}\left[3\sin\frac{\pi x}{l} - \sin\frac{3\pi x}{l}\right]$ in (6), we have

$$\frac{y_0}{4}\left(3\sin\frac{\pi x}{l} - \sin\frac{3\pi x}{l}\right) = \sum_{n=1}^{\infty} b_n \sin\frac{n\pi x}{l}$$

$$\Rightarrow \quad \frac{y_0}{4}\left(3\sin\frac{\pi x}{l} - \sin\frac{3\pi x}{l}\right) = b_1 \sin\frac{\pi x}{l} + b_2 \sin\frac{2\pi x}{l} + b_3 \sin\frac{3\pi x}{l} + \ldots$$

On equating the coefficients of sin on both the sides, we get

$$b_1 = \frac{3y_0}{4}, \quad b_3 = -\frac{y_0}{4}$$

and all others b's are zero.

Hence (5) becomes

$$y = \frac{y_0}{4}\left(3\sin\frac{\pi x}{l}\cos\frac{c\pi t}{l} - \sin\frac{3\pi x}{l}\cos\frac{3c\pi t}{l}\right) \qquad \textbf{Ans.}$$

Example 11. *A string is stretched and fastened to two points l apart. Motion is started by displacing the string into the form $y = k(lx - x^2)$ from which it is released at time $t = 0$. Find the displacement of any point on the string at a distance of x from one end at time t.*

Solution. The vibration of the string is given by

$$\frac{\partial^2 y}{\partial t^2} = c^2 \frac{\partial^2 y}{\partial x^2} \quad \text{... (1)}$$

As the end points of the string are fixed for all time,

$$y(0, t) = 0 \quad \text{... (2)}$$

and $$y(l, t) = 0 \quad \text{... (3)}$$

since the initial transverse velocity of any point of the string is zero, therefore,

$$\left(\frac{\partial y}{\partial t}\right)_{t=0} = 0 \quad \text{... (4)}$$

and $$y(x, 0) = k(lx - x^2) \quad \text{... (5)}$$

Solution of (1) is $$y = (c_1 \cos px + c_2 \sin px)(c_3 \cos c\,pt + c_4 \sin cpt) \quad \text{... (6)}$$

On putting $x = 0$ and $y = 0$ in (6), we get

$$0 = c_1 (c_3 \cos c\,pt + c_4 \sin cpt) \quad \Rightarrow \quad c_1 = 0$$

On putting $c_1 = 0$ in (6), we get

$$y = c_2 \sin px(c_3 \cos cpt + c_4 \sin cpt) \quad \text{... (7)}$$

On differentiating (7) w.r.t. t, we get

$$\frac{\partial y}{\partial t} = c_2 \sin px(-c_3 cp \sin cpt + c_4 cp \cos cpt) \quad \text{...(8)}$$

On putting $\left(\frac{\partial y}{\partial t}\right) = 0$ and $t = 0$ in (8), we get

$$0 = c_2 \sin px\,(c_4 cp) \quad \Rightarrow \quad c_4 = 0 \text{ since } c_2 \neq 0$$

On putting $c_4 = 0$ in (7), we get

$$y = c_2 \sin px(c_3 \cos cpt)$$

$$y = c_2 c_3 \sin px \cos cpt \quad \text{... (9)}$$

On putting $x = l$ and $y = 0$ in equation (9), we get

$$0 = c_2 c_3 \sin pl \cos cpt \text{ or } 0 = \sin pl$$

$$\Rightarrow \quad \sin n\pi = \sin pl \text{ or } pl = n\pi,\ p = \frac{n\pi}{l} \quad \text{where } n = 1, 2, 3......$$

On putting $p = \frac{n\pi}{l}$, equation (9) becomes

$$\Rightarrow \quad y = b_n \sin\frac{n\pi x}{l} \cos\frac{n\pi c}{l} t \quad [c_2 c_3 = b_n]$$

We can have any number of solutions by taking different integral values of n and the complete solution will be the sum of these solutions. Thus,

$$y(x, t) = \sum_{n=1}^{\infty} b_n \sin\frac{n\pi x}{l} \cos\frac{n\pi c}{l} t \quad \text{... (10)}$$

On putting $t = 0$ and $y = k(lx - x^2)$in (10), we get

$$k(l\,x - x^2) = \sum_{n=1}^{\infty} b_n \sin\frac{n\pi x}{l} \quad \text{... (11)}$$

Now it is clear that (11) represents the expansion of $f(x)$ in the form of a Fourier sine series and consequently

$$b_n = \frac{2}{l}\int_0^l f(x)\sin\frac{n\pi x}{l}dx \quad \text{... (12)}$$

$$b_n = \frac{2}{l}\int_0^l k\,(lx - x^2)\sin\frac{n\pi x}{l}dx$$

$$= \frac{2k}{l}\left[(lx-x^2)\left(-\cos\frac{n\pi x}{l}\right)\frac{l}{n\pi} - (l-2x)\left(-\sin\frac{n\pi x}{l}\right)\frac{l^2}{n^2\pi^2} + (-2)\left(\cos\frac{n\pi x}{l}\right)\frac{l^3}{n^3\pi^3}\right]_0^l$$

$$= \frac{2k}{l}\left[(-1)^{n+1}\frac{2l^3}{n^3\pi^3} + \frac{2l^3}{n^3\pi^3}\right] = \frac{8l^2k}{n^3\pi^3} \text{ when } n \text{ is odd.}$$

$= 0,$ when n is even

Putting the value of b_n in (10), we get

$$y = \sum_{n=1}^{\infty}\frac{8l^2k}{n^3\pi^3}\sin\frac{n\pi x}{l}\cos\frac{n\pi c}{l}t, \text{ when } n \text{ is odd.} \quad \textbf{Ans.}$$

Example 12. *A string is stretched between the fixed points (0, 0) and (l, 0) and released at rest from the initial deflection given by*

$$f(x) = \begin{cases} \dfrac{2kx}{l} & \text{when } 0 < x < \dfrac{l}{2} \\ \dfrac{2k}{l}(l-x) & \text{when } \dfrac{l}{2} < x < l \end{cases}$$

Find the deflection of the string at any time t.

Solution. As we have done in the example 11 (see equation 10)

$$y(x, t) = \sum_{n=1}^{\infty} b_n \cos\frac{n\pi ct}{l}\sin\frac{n\pi x}{l} \quad \text{... (1)}$$

$$b_n = \frac{2}{l}\int_0^l f(x)\sin\frac{n\pi x}{l}dx$$

$$b_n = \frac{2}{l}\left[\int_0^{l/2}\frac{2kx}{l}\sin\frac{n\pi x}{l}dx + \int_{\frac{l}{2}}^{l}\frac{2k}{l}(l-x)\sin\frac{n\pi x}{l}dx\right]$$

$$= \frac{4k}{l^2}\left[x\left(-\frac{l}{n\pi}\cos\frac{n\pi x}{l}\right) - 1\left(-\frac{l^2}{n\pi^2}\sin\frac{n\pi x}{l}\right)\right]_0^{\frac{l}{2}}$$

$$+ \frac{4k}{l^2}\left[(l-x)\left(-\frac{l}{n\pi}\cos\frac{n\pi x}{l}\right) - (-1)\left(\frac{-l^2}{n^2\pi^2}\sin\frac{n\pi x}{l}\right)\right]_{\frac{l}{2}}^{l}$$

$$= \frac{4k}{l^2}\left[-\frac{l}{2}\frac{l}{n\pi}\cos\frac{n\pi}{2} + \frac{l^2}{n^2\pi^2}\sin\frac{n\pi}{2}\right] + \frac{4k}{l^2}\left[\frac{l}{2}\frac{l}{n\pi}\cos\frac{n\pi}{2} + \frac{l^2}{n^2\pi^2}\sin\frac{n\pi}{2}\right]$$

$$= \frac{8k}{l^2}\frac{l^2}{n^2\pi^2}\sin\frac{n\pi}{2} = \frac{8k}{n^2\pi^2}\sin\frac{n\pi}{2}$$

On putting the value of b_n in (1), we get

$$y = \sum_{n=1}^{\infty}\frac{8k}{n^2\pi^2}\sin\frac{n\pi}{2}\cos\frac{n\pi ct}{l}\sin\frac{n\pi x}{l} \quad \textbf{Ans.}$$

Example 13. *A taut string of length 2l is fastened at both ends. The mid point of the string is taken to a height b and then released from the rest in that position. Find the displacement of the string.*

Solution. Taking an end as origin, the boundary conditions are

$$y(0, t) = 0 \quad t \geq 0 \qquad \text{...(1)}$$

$$y(2l, t) = 0, \quad t \geq 0 \qquad \text{... (2)}$$

$$\left(\frac{\partial y}{\partial t}\right)_{t=0} = 0,$$

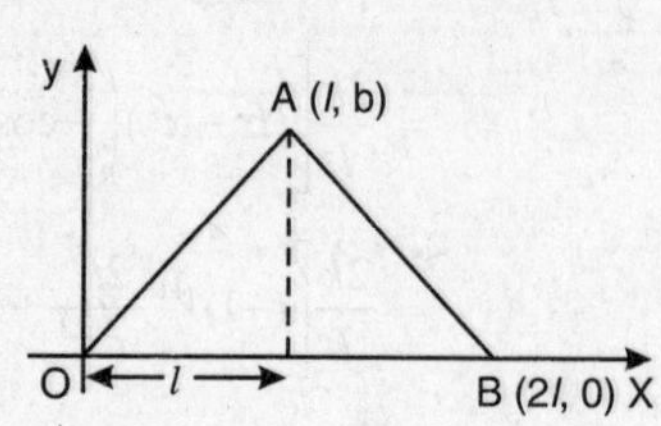

OA passes through $(0, 0)$ and (l, b)

$$\text{slope} = \frac{b}{l}x, \quad \Rightarrow \quad y = \frac{bx}{l}, \qquad 0 \leq x \leq l$$

Equation of OA is $y = -\frac{b}{l}(x - 2l),\ l \leq x \leq 2l$ ...(3) $\left[y - y_1 = \frac{y_2 - y_1}{x_2 - x_1}(x - x_1)\right]$

$\left(\text{Equation of } AB,\ y - 0 = \frac{b - 0}{l - 2l}(x - 2l)\right)$

Equation of the vibrating string is $\frac{\partial^2 y}{\partial t^2} = c^2 \frac{\partial^2 y}{\partial x^2}$ ... (4)

Starting with the solution of (4), we get

$$y(x, t) = (c_1 \cos px + c_2 \sin px)(c_3 \cos pct + c_4 \sin pct) \qquad \text{... (5)}$$

On putting $x = 0$, $y = 0$ in (5), we get

$$0 = c_1\ (c_3 \cos pct + c_4 \sin pct) \Rightarrow c_1 = 0$$

On putting $c_1 = 0$ in (5), we get

$$y(x, t) = c_2 \sin px (c_3 \cos pct + c_4 \sin pct) \qquad \text{... (6)}$$

On putting $y = 0, x = 2l$ in (6), we get

$$0 = c_2 \sin p(2l)\ (c_3 \cos pct + c_4 \sin pct) \Rightarrow \sin 2pl = 0 = \sin n\pi \Rightarrow 2pl = n\pi,\ p = \frac{n\pi}{2l}$$

Substituting the value of p in (6), we have

$$y(x, t) = c_2 \sin \frac{n\pi}{2l} x \left(c_3 \cos \frac{n\pi ct}{2l} + c_4 \sin \frac{n\pi ct}{2l} \right) \qquad \text{... (7)}$$

On differentiating (7) w.r.t., t, we get

$$\frac{\partial y}{\partial t} = c_2 \sin \frac{n\pi x}{2l} \left[-c_3 \frac{n\pi c}{2l} \sin \frac{n\pi ct}{2l} + c_4 \frac{n\pi c}{2l} \cos \frac{n\pi ct}{2l} \right] \qquad \text{... (8)}$$

Putting $\left(\frac{\partial y}{\partial t}\right) = 0$, $t = 0$ in (8), we get

$$0 = c_2 \sin \frac{n\pi x}{2l} \left[0 + c_4 \frac{n\pi c}{2l} \right] \Rightarrow c_4 = 0$$

Putting $c_4 = 0$ in (7), we have

$$y(x, t) = \left(c_2 \sin \frac{n\pi x}{2l} \right) \left(c_3 \cos \frac{n\pi ct}{2l} \right)$$

General solution is

$$y(x,t)=\sum_{n=1}^{\infty} b_n \sin\frac{n\pi x}{2l}\cos\frac{n\pi ct}{2l} \qquad (c_2c_3=b_n) \qquad \text{... (9)}$$

Equation of OA is $y=\frac{bx}{l},\ 0\le x\le l$

Equation of AB is $y=\frac{-b}{l}(x-2l),\ l\le x\le 2l$

$$\Rightarrow \quad f(x)=\begin{cases}\frac{bx}{l}, & 0\le x\le l\\ \frac{-b}{l}(x-2l), & l\le x\le 2l\end{cases}$$

$$b_n=\frac{2}{2l}\int_0^{2l} f(x)\sin\frac{n\pi x}{2l}dx=\frac{1}{l}\int_0^{l} f(x)\sin\frac{n\pi x}{2l}dx+\frac{1}{l}\int_l^{2l} f(x)\sin\frac{n\pi x}{2l}dx$$

$$=\frac{1}{l}\int_0^{l}\frac{bx}{l}\sin\frac{n\pi x}{2l}dx+\frac{1}{l}\int_l^{2l}\frac{-b}{l}(x-2l)\sin\frac{n\pi x}{2l}dx$$

$$=\frac{1}{l}\left[\frac{bx}{l}\frac{2l}{n\pi}\left(-\cos\frac{n\pi x}{2l}\right)-\frac{b}{l}\left(-\frac{4l^2}{n^2\pi^2}\sin\frac{n\pi x}{2l}\right)\right]_0^l$$

$$+\frac{1}{l}\left[\frac{-b}{l}(x-2l)\left(-\frac{2l}{n\pi}\cos\frac{n\pi x}{2l}\right)-\left(-\frac{b}{l}\right)-\frac{4l^2}{n^2\pi^2}\sin\frac{n\pi x}{2l}\right]_l^{2l}$$

$$=\frac{b}{l^2}\left[\frac{-2l^2}{n\pi}\cos\frac{n\pi}{2}+\frac{4l^2}{n^2\pi^2}\sin\frac{n\pi}{2}+\frac{2l^2}{n\pi}\cos\frac{n\pi}{2}+\frac{4l^2}{n^2\pi^2}\sin\frac{n\pi}{2}\right]$$

$$=\frac{b}{l^2}\left[\frac{8l^2}{n^2\pi^2}\sin\frac{n\pi}{2}\right]=\frac{8b}{n^2\pi^2}\sin\frac{n\pi}{2}=0 \text{ for } n \text{ even}=\frac{8b}{n^2\pi^2}\sin\frac{n\pi}{2} \text{ for } n \text{ odd.}$$

Substituting the value of b_n in (9), we get

$$y(x,t)=\frac{8b}{\pi^2}\sum_{n=1}^{\infty}\frac{1}{(2n-1)^2}\sin(2n-1)\frac{\pi}{2}\sin\frac{(2n-1)\pi x}{2l}\cos\frac{(2n-1)\pi ct}{2l} \qquad \textbf{Ans.}$$

Example 14. *The points of trisection of a string are pulled aside through the same distance on opposite sides of the position of equilibrium and the string is released from rest. Derive an expression for the displacement of the string at subsequent time and show that the mid-point of the string always remains at rest.*

Solution. Let the string OA (l) be trisected at B and C. Initially the string is held in the form OB' $C'A$, where $BB'=CC'=a$

The equation of vibrating string is

$$\frac{\partial^2 y}{\partial t^2}=c^2\frac{\partial^2 y}{\partial x^2} \qquad \text{...(1)}$$

Boundary conditions are

$$y(0,t)=0,\ y(l,t)=0,\ \left(\frac{\partial y}{\partial t}\right)_{t=0}=0$$

Fourth condition is that at $t=0$, the string rests in the form $OB'C'A$.

Equation of OB' is $y=\frac{a}{l/3}x \Rightarrow y=\frac{3a}{l}x$ $\qquad (y=mx)$

Equation of $B'C'$ is $y-a=\frac{a+a}{\frac{l}{3}-\frac{2l}{3}}\left(x-\frac{l}{3}\right)$ $\qquad \left[y-y_1=\frac{y_2-y_1}{x_2-x_1}(x-x_1)\right]$

$$i.e.,\ y-a=\frac{2a}{\frac{-l}{3}}\left(x-\frac{l}{3}\right) \Rightarrow y-a=-\frac{6a}{l}\left(x-\frac{l}{3}\right)$$

$$\Rightarrow \quad y=a-\frac{6ax}{l}+2a \quad \Rightarrow \quad y=3a-\frac{6ax}{l} \quad \Rightarrow \quad y=\frac{3a}{l}(l-2x)$$

Equation of $C'A$ is $y-0=\dfrac{-a-0}{\frac{2l}{3}-l}(x-l) \qquad \Rightarrow \qquad y=\dfrac{3a}{l}(x-l)$

Hence fourth condition is

$$f(x)=\begin{cases}\frac{3a}{l}x, & 0\le x\le \frac{l}{3}\\ \frac{3a}{l}(l-2x), & \frac{l}{3}\le x\le \frac{2l}{3}\\ \frac{3a}{l}(x-l), & \frac{2l}{3}\le x\le l\end{cases}$$

General solution of (1) is

$$y(x,t)=\sum_{n=1}^{\infty} b_n \sin\frac{n\pi x}{l}\cos\frac{n\pi ct}{l} \qquad \text{... (2)}$$

On putting $t = 0$ in (2), we get

$$y(x,0)=\sum_{n=1}^{\infty} b_n \sin\frac{n\pi x}{l}$$

We have to find the value of b_n by Fourier half range formula.

$$b_n=\frac{2}{l}\int_0^l f(x)\sin\frac{n\pi x}{l}dx$$

$$=\frac{2}{l}\left[\int_0^{\frac{l}{3}}\frac{3ax}{l}\sin\frac{n\pi x}{l}dx+\int_{\frac{l}{3}}^{\frac{2l}{3}}\frac{3a}{l}(l-2x)\sin\frac{n\pi x}{l}dx+\int_{\frac{2l}{3}}^{l}\frac{3a}{l}(x-l)\sin\frac{n\pi x}{l}dx\right]$$

$$=\frac{6a}{l^2}\left[x\left(-\frac{l}{n\pi}\cos\frac{n\pi x}{l}\right)-(1)\left(-\frac{l^2}{n^2\pi^2}\sin\frac{n\pi x}{l}\right)\right]_0^{\frac{l}{3}}+\frac{6a}{l^2}\left[(l-2x)\left(\frac{-l}{n\pi}\cos\frac{n\pi x}{l}\right)\right.$$

$$\left.-(-2)\left(\frac{-l^2}{n^2\pi^2}\sin\frac{n\pi x}{l}\right)\right]_{\frac{l}{3}}^{\frac{2l}{3}}+\frac{6a}{l^2}\left[(x-l)\left(\frac{-l}{n\pi}\cos\frac{n\pi x}{l}\right)-(1)\left(\frac{-l^2}{n^2\pi^2}\sin\frac{n\pi x}{l}\right)\right]_{\frac{2l}{3}}^{l}$$

$$=\frac{6a}{l^2}\left[\left(-\frac{l^2}{3n\pi}\cos\frac{n\pi}{3}+\frac{l^2}{n^2\pi^2}\sin\frac{n\pi}{3}\right)+\frac{l^2}{3n\pi}\cos\frac{2n\pi}{3}-\frac{2l^2}{n^2\pi^2}\sin\frac{2n\pi}{3}\right.$$

$$\left.+\frac{l^2}{3n\pi}\cos\frac{n\pi}{3}+\frac{2l^2}{n^2\pi^2}\sin\frac{n\pi}{3}-\left(\frac{l^2}{3n\pi}\cos\frac{2n\pi}{3}+\frac{l^2}{n^2\pi^2}\sin\frac{2n\pi}{3}\right)\right]$$

$$=\frac{6a}{l^2}.\frac{3l^2}{n^2\pi^2}\left(\sin\frac{n\pi}{3}-\sin\frac{2n\pi}{3}\right)$$

$$=\frac{18a}{n^2\pi^2}\sin\frac{n\pi}{3}[1+(-1)^n] \qquad \left[\sin\frac{2n\pi}{3}=\sin\left(n\pi-\frac{n\pi}{3}\right)=-(-1)^n\sin\frac{n\pi}{3}\right]$$

$= 0, \qquad$ when n is odd

$$= \frac{36a}{n^2\pi^2}\sin\frac{n\pi}{3}, \quad \text{when } n \text{ is even}$$

On substituting the value of b_n in (2), we get

$$y(x,t) = \sum_{n=2}^{\infty} \frac{36a}{n^2\pi^2}\sin\frac{n\pi}{3}\sin\frac{n\pi x}{l}\cos\frac{n\pi ct}{l} \quad \text{...(3) } (n \text{ is even}) \textbf{ Ans.}$$

Putting $x = \frac{l}{2}$ in equation (3), we get

$$y\left(\frac{l}{2}, t\right) = \frac{36a}{n^2\pi^2}\sin\frac{n\pi}{3}\sin\frac{n\pi}{2}\cos\frac{n\pi ct}{l} = 0 \quad \left[\begin{array}{c}\sin\frac{n\pi}{2} = 0 \\ \text{as } n \text{ is even}\end{array}\right]$$

Hence, mid-point of the string is always at rest. **Proved.**

44.4 SOLUTION OF WAVE EQUATION BY D'ALEMBERT'S METHOD

$$\frac{\partial^2 y}{\partial t^2} = c^2 \frac{\partial^2 y}{\partial x^2} \quad \text{...(1)}$$

Let us introduce the two new independent variables $u = x + ct$, $v = x - ct$

So that y becomes a function of u and v. Then,

$$\frac{\partial y}{\partial x} = \frac{\partial y}{\partial u}\frac{\partial u}{\partial x} + \frac{\partial y}{\partial v}\frac{\partial v}{\partial x} = \frac{\partial y}{\partial u}(1) + \frac{\partial y}{\partial v}(1) = \frac{\partial y}{\partial u} + \frac{\partial y}{\partial v}$$

$$\frac{\partial}{\partial x} \equiv \frac{\partial}{\partial u} + \frac{\partial}{\partial v}$$

$$\frac{\partial^2 y}{\partial x^2} = \frac{\partial}{\partial x}\left(\frac{\partial y}{\partial x}\right) = \left(\frac{\partial}{\partial u} + \frac{\partial}{\partial v}\right)\left(\frac{\partial y}{\partial u} + \frac{\partial y}{\partial v}\right)$$

$$= \frac{\partial}{\partial u}\left(\frac{\partial y}{\partial u} + \frac{\partial y}{\partial v}\right) + \frac{\partial}{\partial v}\left(\frac{\partial y}{\partial u} + \frac{\partial y}{\partial v}\right) = \frac{\partial^2 y}{\partial u^2} + 2\frac{\partial^2 y}{\partial u \partial v} + \frac{\partial^2 y}{\partial v^2} \quad \text{... (2)}$$

$$\frac{\partial y}{\partial t} = \frac{\partial y}{\partial u}\frac{\partial u}{\partial t} + \frac{\partial y}{\partial v}\frac{\partial v}{\partial t} = \frac{\partial y}{\partial u}c + \frac{\partial y}{\partial v}(-c) = c\left(\frac{\partial y}{\partial u} - \frac{\partial y}{\partial v}\right) \quad \left[\because \frac{\partial u}{\partial t} = c, \ \frac{\partial v}{\partial t} = -c\right]$$

$$\frac{\partial}{\partial t} = c\left(\frac{\partial}{\partial u} - \frac{\partial}{\partial v}\right)$$

$$\frac{\partial^2 y}{\partial t^2} = \frac{\partial}{\partial t}\left(\frac{\partial y}{\partial t}\right) = c\left(\frac{\partial}{\partial u} - \frac{\partial}{\partial v}\right)c\left(\frac{\partial y}{\partial u} - \frac{\partial y}{\partial v}\right) = c^2\left(\frac{\partial^2 y}{\partial u^2} - 2\frac{\partial^2 y}{\partial u \partial v} + \frac{\partial^2 y}{\partial v^2}\right) \quad \text{... (3)}$$

Substituting the values of $\frac{\partial^2 y}{\partial x^2}$ and $\frac{\partial^2 y}{\partial t^2}$ from (2) and (3) in (1), we get

$$c^2\left(\frac{\partial^2 y}{\partial u^2} - 2\frac{\partial^2 y}{\partial u \partial v} + \frac{\partial^2 y}{\partial v^2}\right) = c^2\left(\frac{\partial^2 y}{\partial u^2} + 2\frac{\partial^2 y}{\partial u \partial v} + \frac{\partial^2 y}{\partial v^2}\right) \Rightarrow \frac{\partial^2 y}{\partial u \partial v} = 0 \quad \text{... (4)}$$

Integrating (4) *w.r.t.* v, we get $\quad \frac{\partial y}{\partial u} = f(u) \quad$ ... (5)

where $f(u)$ is constant in respect of v. Again integrating (5) *w.r.t.* 'u' we get

$$y = \int f(u)\,du + \psi(v)$$

where $\psi(v)$ is constant in respect of u

$$y = \phi(u) + \psi(v) \qquad \text{where } \phi(u) = \int f(u)\,du$$

$$\Rightarrow \quad y(x,t) = \phi(x+ct) + \psi(x-ct) \qquad ...(6)$$

This is D'Alemberts solution of wave equation (1)

To determine ϕ, ψ, let us apply initial conditions, $y(x,0) = f(x)$ and $\frac{\partial y}{\partial t} = 0$ when $t = 0$.

Differentiating (6) *w.r.t.* "t", we get

$$\frac{\partial y}{\partial t} = c\phi'(x+ct) - c\psi'(x-ct) \qquad ...(7)$$

Putting $\frac{\partial y}{\partial t} = 0$, and $t = 0$ in (7) we get $0 = c\phi'(x) - c\psi'(x)$

$$\Rightarrow \quad \phi'(x) = \psi'(x) \text{ or } \phi(x) = \psi(x) + b$$

Again substituting $y = f(x)$ and $t = 0$ in (6), we get

$$f(x) = \phi(x) + \psi(x) \text{ or } f(x) = [\psi(x) + b] + \psi(x)$$

$$\Rightarrow \quad f(x) = 2\psi(x) + b \qquad ...(8)$$

So that $\psi(x) = \frac{1}{2}[f(x) - b]$ and $\phi(x) = \frac{1}{2}[f(x) + b]$ [Using 8)]

On putting the values of $\phi(x + ct)$ and $y(x - ct)$ in (6), we get

$$y(x,t) = \frac{1}{2}[f(x+ct) + b] + \frac{1}{2}[f(x-ct) - b]$$

$$\Rightarrow \quad y(x,t) = \frac{1}{2}[f(x+ct) + f(x-ct)] \qquad \textbf{Ans.}$$

EXERCISE 44.2

1. A tightly stretched string with fixed end points $x = 0$ and $x = l$ is initially at rest in its equilibrium position. If it is set vibrating by giving to each of its points a velocity $\lambda x(l - x)$, find the displacement of the string at any distance x from one end at any time t.

   Ans. $y = \frac{8\lambda l^3}{c\pi^2} \sum_{(n=1)}^{\infty} \frac{1}{(2n-1)^4} \sin\frac{(2n-1)\pi x}{l} . \sin\frac{(2n-1)\pi ct}{l}$

2. A tightly stretched string of length l fastened at both ends, is disturbed from the position of equilibrium imparting to each of its points an initial velocity of magnitude $f(x)$. Show that the solution of the problem.

$$u(x,t) = \frac{2}{l}\sum_{n=1}^{\infty}\left[\int_0^l f(x)\sin\frac{n\pi}{l}x\,dx\right]\sin\frac{n\pi x}{l}\cos\frac{n\pi at}{l}$$

3. Find the solution of the equation $\frac{\partial^2 y}{\partial x^2} = \frac{1}{c^2}\frac{\partial^2 y}{\partial t^2}$ subject to the boundary conditions

   $y(0, t) = 0$, $y(l, t) = 0$, $y(x, 0) = \phi(x)$, $\frac{\partial y}{\partial t}(x, 0) = \psi(x)$ **Ans.** $y = \phi(x).\cos\frac{n\pi ct}{l} + \frac{l\psi(x)}{n\pi c}.\frac{\sin\frac{n\pi ct}{l}}{\sin\frac{n\pi x}{l}}$

4. The vibration of an elastic string of length l are governed by the one-dimensional wave equation $\frac{\partial^2 u}{\partial t^2} = c^2\frac{\partial^2 u}{\partial x^2}$. The string is fixed at the ends.

$u(0, t) = 0 = u(l, t)$ for all t. The initial deflection is $u(x,0)=\begin{cases} x, & 0<x<\dfrac{l}{2} \\ l-x, & \dfrac{l}{2}\le x\le l \end{cases}$

and the initial velocity is zero. Find the deflection of the string at any instant of time.

Ans. $\dfrac{4l}{\pi^2}\sum_{n=1}^{\infty}\dfrac{\sin\dfrac{n\pi}{2}}{n^2}\sin\dfrac{n\pi x}{l}\cos\dfrac{n\pi ct}{l}$

5. A tightly stretched violin string of length l is fixed at both ends and is plucked at $x=\dfrac{l}{3}$ and assumes initially the shape of a triangle of height a. Find the displacement y at any distance x and any time t after the string is released from rest. **Ans.** $y(x,t)=\dfrac{9a}{\pi^2}\sum_{1}^{\infty}\dfrac{1}{n^2}\sin\dfrac{n\pi}{3}\cos\dfrac{n\pi ct}{l}\sin\dfrac{n\pi x}{l}$

44.5 ONE DIMENSIONAL HEAT FLOW

In this article, we shall consider the flow of heat and the accompanying variation of temperature with position and with time in conducting solids the following empirical laws are taken as the basis of investigation.

1. Heat flows from higher to lower temperature.
2. The amount of heat required to produce a given temperature change in a body is proportional to the mass of the body and to the temperature change. This constant of proportionality is known as the specific heat (c) of the conducting material.
3. The rate at which heat flows through an area is proportional to the area and to the temperature gradient normal to the area. This constant of proportionality is known as the thermal conductivity (k) of the material.

Consider a bar or rod of homogeneous material of density ρ(gr/cm³) and having a constant cross-sectional area A (cm²). We suppose that the sides of the bar are insulated and the loss of heat from the sides by conduction or radiations is negligible.

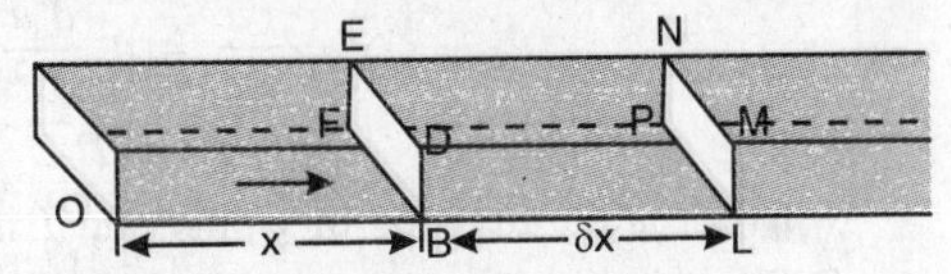

Take an end of the bar as the origin and the direction of heat flow as the positive x-axis.

Let c be the specific heat and k the thermal conductivity of the material.

Consider an element between two parallel sections $BDEF$ and $LMNP$ at distances x and $x + \delta x$ from the origin O, the sections being perpendicular to the x-axis. The mass of the element $= A\,\rho\,\delta\,x$.

Let $u\,(x, t)$ be the temperature at a distance x at time t. By the second law stated above, the rate of increase of heat in the element $= A\,\rho\,\delta\,x\,c\dfrac{\partial u}{\partial t}$. If R_1 and R_2 are respectively the rates (cal/sec) of inflow and outflow, for the sections $x = x$ and $x = x + \delta x$, then

$$R_1 = -kA\left(\frac{\partial u}{\partial x}\right)_x$$

and $R_2 = -kA\left(\dfrac{\partial u}{\partial x}\right)_{x+\delta x}$ the negative sign being due to the fact that heat flows from higher to lower temperature. $\left(i.e., \dfrac{\partial u}{\partial x}\text{ is negative}\right)$

Equating the rates of increase of heat from the two empirical laws.

$$A\rho c\delta x.\frac{\partial u}{\partial t} = R_1 - R_2 = kA\left[\left(\frac{\partial u}{\partial x}\right)_{x+\delta x} - \left(\frac{\partial u}{\partial x}\right)_x\right]$$

$$\therefore \qquad \frac{\partial u}{\partial t} = \frac{k}{\rho c}\left[\frac{\left(\frac{\partial u}{\partial x}\right)_{x+\delta x} - \left(\frac{\partial u}{\partial x}\right)_x}{\delta x}\right]$$

Taking the limit as $\delta x \to 0$ *i.e.* when $x + \delta x \to x$.

$$\frac{\partial u}{\partial t} = \frac{k}{\rho c}\underset{\delta x \to 0}{\text{Lt}}\left[\frac{\left(\frac{\partial u}{\partial x}\right)_{x+\delta x} - \left(\frac{\partial u}{\partial x}\right)_x}{\delta x}\right] = \frac{k}{\rho c}\frac{\partial}{\partial x}\left(\frac{\partial u}{\partial x}\right)$$

i.e.,

$$\frac{\partial u}{\partial t} = \frac{k}{\rho c}\frac{\partial^2 u}{\partial x^2}$$

$\frac{k}{pc}$ is called the *diffusivity* (cm^2/sec.) of the substance. If we denote it by c^2, the above equation takes the form

$$\frac{\partial u}{\partial t} = c^2\frac{\partial^2 u}{\partial x^2}$$

Example 15. *By method of separation of variables solve the P.D.E.*

$$\frac{\partial u}{\partial t} = c^2\frac{\partial^2 u}{\partial x^2}$$

Solution. Here, we have

$$\frac{\partial u}{\partial t} = c^2\frac{\partial^2 u}{\partial x^2} \qquad ...(1)$$

Let

$$u = XT \qquad ...(2)$$

where X is the function of x only and T is the function of t only.

Differentiating (2) partially w.r.t. t, we get

$$\frac{\partial u}{\partial t} = \frac{\partial}{\partial t}(XT) = X\frac{dT}{dt}$$

and

$$\frac{\partial^2 u}{\partial x^2} = c^2\frac{\partial^2}{\partial x^2}(XT) = \frac{\partial}{\partial x}\left(\frac{\partial}{\partial x}XT\right) = \frac{\partial}{\partial x}\left(T\frac{dX}{dx}\right) = T\frac{d^2 X}{dx^2}$$

Putting the values of $\frac{\partial u}{\partial t}$ and $\frac{\partial^2 u}{\partial x^2}$ in (1), we get

$$X\frac{dT}{dt} = c^2 T\frac{d^2 X}{dx^2}$$

Separating the variables, we get

$$\frac{1}{c^2}\frac{\frac{dT}{dt}}{T} = \frac{\frac{d^2 X}{dx^2}}{X}$$

L.H.S. is constant for x so we take $\frac{1}{c^2}\frac{\frac{dT}{dt}}{T} = -p^2$

R.H.S is constant for t so we take $\dfrac{\frac{d^2X}{dx^2}}{X} = -p^2$

$$\therefore \quad \frac{1}{c^2}\frac{\frac{dT}{dt}}{T} = \frac{\frac{d^2X}{dx^2}}{X} = -p^2$$

$$\frac{1}{c^2}\frac{\frac{dT}{dt}}{T} = -p^2$$

$$\frac{dT}{dt} = -p^2c^2T$$

A.E. is $\quad m = -p^2c^2$

$\Rightarrow \quad T = c_1 e^{-c^2p^2t}$

$$\frac{\frac{d^2X}{dx^2}}{X} = -p^2$$

$$D^2X = -p^2X$$

A.E. is $m^2 = -p^2$

$\Rightarrow \quad m = \pm\, ip$

$X \;= c_2 \cos px + c_3 \sin px$

Putting the values of T and X in (2), we get

$$u = c_1 e^{-c^2p^2t}\,(c_2 \cos px + c_3 \sin px)$$ **Ans.**

Example 16. *Find the temperature in a bar of length 2 whose ends kept at zero and lateral surface insulated if the initial temperature is sin* $\dfrac{\pi x}{2} + 3\sin\dfrac{5\pi x}{2}$.

[*U.P. II Sem., 2007; U.P. II Semester, 2009*]

Solution. Here, we have

One-dimensional heat flow equation is

$$\frac{\partial u}{\partial t} = c^2\frac{\partial^2 u}{\partial x^2}$$

Its solution is

$$u(x,t) = (c_1 \cos px + c_2 \sin px)c_3 e^{-c^2p^2t} \qquad \text{...(1)}$$

Putting $x = 0$, $u = 0$ in (1), we get

$$0 = c_1c_3e^{-c^2p^2t} \qquad \text{(given)}$$

$$\Rightarrow \qquad c_1 = 0$$

Putting $c_1 = 0$ in (1), we get

$$u = c_2c_3 \sin px\, e^{-c^2p^2t} \qquad \text{... (2)}$$

Putting $x = 2$, $u = 0$ in (2), we get (given)

$$0 = c_2c_3 \sin 2p\, e^{-c^2p^2t}$$

$$\Rightarrow \qquad \sin 2p = 0 = \sin n\pi$$

$$\Rightarrow \qquad p = \frac{n\pi}{2},\, n \in \text{I}$$

Putting the value of p in (2), we get

$$u = b_n \sin\frac{n\pi x}{2}\, e^{-\frac{n^2\pi^2c^2t}{4}} \qquad (c_2c_3 = b_n)$$

$$u = \sum_{n=1}^{\infty} b_n \sin\frac{n\pi x}{2}\, e^{-\frac{c^2\pi^2c^2t}{4}} \qquad \text{... (3) (General solution)}$$

Putting $t = 0$ and $u = \sin\frac{\pi x}{2} + 3\sin\frac{5\pi x}{2}$ in (3), we get

$$\sin\left(\frac{\pi x}{2}\right) + 3\sin\left(\frac{5\pi x}{2}\right) = \sum_{n=1}^{\infty} b_n \sin\frac{n\pi x}{2}$$

$$= b_1 \sin\left(\frac{\pi x}{2}\right) + b_2 \sin\left(\frac{2\pi x}{2}\right) + \ldots + b_5 \sin\left(\frac{5\pi x}{2}\right) + \ldots \quad \ldots (4)$$

On equating the coefficients of both sides, we get

$$b_1 = 1 \text{ and } b_5 = 3$$

On putting the values of b_1 and b_5 in (4), we get

Hence, $$u = \sin\left(\frac{\pi x}{2}\right) e^{-\pi^2 c^2 t/4} + 3\sin\left(\frac{5\pi x}{2}\right) e^{-25\pi^2 c^2 t/4}.$$ **Ans.**

Example 17. *Solve the equation* $\frac{\partial u}{\partial t} = \frac{\partial^2 u}{\partial x^2}$ *with boundary condition u(x, 0) = 3 sin nπx, u(0, t), u(l, t) = 0 where 0 < x < l.*

Solution. Here, we have $\frac{\partial u}{\partial t} = \frac{\partial^2 u}{\partial x^2}$

Its solution is $u = c_1 e^{-p^2 t}(c_2 \cos px + c_3 \sin px)$ ... (1)

Putting $x = 0$, and $u = 0$ in (1), we get (given)

$$0 = c_1 c_2 e^{-p^2 t} \Rightarrow c_2 = 0$$

Putting the value of c_2 in (1), we get $u = c_1 c_3 e^{-p^2 t} \sin px$ ... (2)

Putting $x = l$ and $u = 0$ in (2), we get (given)

$$0 = c_1 c_3 e^{-p^2 t} \sin pl$$

$$\Rightarrow \sin pl = 0 = \sin n\pi \Rightarrow pl = n\pi \Rightarrow p = \frac{n\pi}{l}$$

Putting the value of p in (2), we get

$$u(x, t) = c_1 c_3 e^{-\frac{n^2\pi^2}{l}t} \sin\frac{n\pi x}{l} = b_n e^{-\frac{n^2\pi^2}{l^2}t} \sin\frac{n\pi x}{l} \quad (b_n = c_1 c_3)$$

$$u = \sum_{n=1}^{\infty} b_n e^{-(n^2\pi^2 t/l^2)} \sin\frac{n\pi x}{l} \quad \ldots (3) \text{ (general solution)}$$

On putting $t = 0$ and $u = 3\sin n\pi x$ in (3), we get

$$3 \sin n\pi x = \sum_{n=1}^{\infty} b_n \sin\frac{n\pi x}{l}. \quad \ldots (4)$$

Equating the coefficients, we get $b_n = 3$, $l = 1$.

On putting the value of b_n and l in (3), we get $u = 3\sum_{n=1}^{\infty} b_n e^{-n^2\pi^2 t} \sin n\pi x$ **Ans.**

Example 18. *A rod of length l with insulated sides is initially at a uniform temperature u. Its ends are suddenly cooled to 0°C and are kept at that temperature. Prove that the temperature function u(x, t) is given by*

$$u(x, t) = \sum_{n=1}^{\infty} b_n \sin\frac{n\pi x}{l} . e^{-\frac{c^2\pi^2 n^2 t}{l^2}}$$

where b_n *is determined from the equation.*

$$b_n = \frac{2}{l}\int_0^l u_0(x) \sin\frac{n\pi x}{l} dx$$

Solution. Let heat flow along a bar of uniform cross-section, in the direction perpendicular to the cross-section. Take one end of the bar as origin and the direction of heat flow is along x-axis.

Let the temperature of the bar at any time t at a point x distance from the origin be $u\ (x, t)$.

Then the equation of one dimensional heat flow is $\frac{\partial u}{\partial t} = c^2 \frac{\partial^2 u}{\partial x^2}$ ...(1)

Let us assume that $u = XT$, where X is a function of x alone and T that of t alone.

$$\therefore \quad \frac{\partial u}{\partial t} = X\frac{dT}{dt} \quad \text{and} \quad \frac{\partial^2 u}{\partial x^2} = T\frac{d^2 X}{dx^2}$$

Substituting these values in (1), we get $X\frac{dT}{dt} = c^2 T \frac{d^2 X}{dx^2}$

$$\Rightarrow \quad \frac{1}{c^2 T}\frac{dT}{dt} = \frac{1}{X}\frac{d^2 X}{dx^2} = -p^2 \text{ (constant)} \quad ...(2)$$

$\frac{1}{c^2 T}\frac{dT}{dt} = -p^2$	$\frac{1}{X}\frac{d^2 X}{dx^2} = -p^2$
$\frac{dT}{dt} + p^2 c^2 T = 0$	$\frac{d^2 X}{dx^2} + p^2 X = 0$
$\Rightarrow \quad DT + p^2c^2T = 0$	$\Rightarrow \quad D^2X + p^2X = 0$
$\Rightarrow \quad (D + p^2c^2)T = 0$	$\Rightarrow \quad (D^2 + p^2)\ X = 0$
A.E. is $\quad m + p^2c^2 = 0$	A.E. is $\quad m^2 + p^2 = 0$
$\Rightarrow \quad m = -p^2c^2$	$\Rightarrow \quad m = \pm\ ip$
$\Rightarrow \quad T = c_1\ e^{-p^2c^2t}$	$\Rightarrow \quad X = c_2 \cos px + c_3 \sin px$

$$\therefore\ u = XT \quad \Rightarrow \quad u = c_1 e^{-p^2c^2t}(c_2 \cos px + c_3 \sin px) \quad ...\ (3)$$

Putting $x = 0$, $u = 0$ in (3), we get

$$0 = c_1 e^{-p^2c^2t}(c_2) \Rightarrow c_2 = 0 \quad [\text{since } c_1 \neq 0]$$

(3) becomes $\quad u = c_1 e^{-p^2c^2t} c_3 \sin px$ ... (4)

Again putting $x = l$, $u = 0$ in (4), we get

$$0 = c_1 e^{-p^2c^2t} c_3 \sin pl \Rightarrow \sin pl = 0 = \sin n\pi \Rightarrow pl = n\pi$$

$$\Rightarrow \quad p = \frac{n\pi}{l}, \ n \text{ is any integer}$$

Hence (4) becomes $\quad u = c_1 c_3 e^{\frac{-n^2c^2\pi^2}{l^2}t} \sin\frac{n\pi x}{l} = b_n e^{-\frac{n^2c^2\pi^2 t}{l^2}} \sin\frac{n\pi}{l}x, \quad [b_n = c_1c_3]$

This equation satisfies the given conditions for all integral values of n. Hence taking $n = 1, 2, 3, \ldots$, *the most general solution is*

$$u = \sum_{n=1}^{\infty} b_n e^{\frac{-n^2c^2\pi^2 t}{l^2}} \sin\frac{n\pi}{l}x$$

$$b_n = \frac{2}{l}\int_0^l u_0 \sin\frac{n\pi x}{l}dx = \frac{2}{l}u_0\left[-\frac{l}{n\pi}\cos\frac{n\pi x}{l}\right]_0^l$$

$$= \frac{2}{l}u_0\left[-\frac{l}{n\pi}(\cos n\pi - \cos 0)\right] = -\frac{2u_0}{n\pi}\left[(-1)^n - 1\right]$$

$$\Rightarrow \qquad b_n = \begin{cases} 0, & \text{if } n \text{ is even} \\ \dfrac{4u_0}{n\pi}, & \text{if } n \text{ is odd} \end{cases}$$

Hence the temperature function

$$u(x,t) = \frac{4u_0}{\pi}\sum_{n=1,3,5\ldots}^{\infty}\frac{1}{n}\sin\frac{n\pi x}{l}e^{-\frac{c^2n^2\pi^2 t}{l^2}}$$

$$= \frac{4u_0}{\pi}\sum_{n=1}^{\infty}\frac{1}{2n-1}\sin\frac{(2n-1)\pi x}{l}e^{-\frac{c^2(2n-1)^2\pi^2 t}{l^2}}$$

Example 19. *Determine the solution of one dimensional heat equation*

$$\frac{\partial u}{\partial t} = c^2\frac{\partial^2 u}{\partial x^2}$$

Subject to the boundary conditions u (0, t) = 0, u (l, t) = 0 (t > 0) and the initial condition u(x, 0) = x, l being the length of the bar. *(U.P. II Semester Summer 2006)*

Solution. $$\frac{\partial u}{\partial t} = c^2\frac{\partial^2 u}{\partial x^2} \qquad \ldots (1)$$

Boundary conditions are

$$u(0, t) = 0$$
$$u(l, t) = 0 \qquad (t > 0)$$
$$u(x, 0) = x$$

On solving (1), we get

$$u = c_1 e^{-p^2c^2t}(c_2\cos px + c_3\sin px) \qquad \ldots(2)$$

Putting $x = 0$ and $u = 0$ in (2), we get

$$0 = c_1 e^{-p^2c^2t}(c_2) \quad \Rightarrow \quad c_2 = 0$$

Putting $c_2 = 0$ in (2), we get

$$u = c_1 e^{-p^2c^2t}c_3\sin px \qquad \ldots(3)$$

Again putting $x = l$, $u = 0$ in (3), we get

$$0 = c_1 e^{-p^2c^2t}c_3\sin pl$$

$$\Rightarrow \qquad \sin pl = 0 = \sin n\pi \qquad \Rightarrow \qquad pl = n\pi$$

$$\Rightarrow \qquad p = \frac{n\pi}{l}, \; n \text{ is any integer}$$

Hence, (3) becomes $u = c_1 c_3 e^{-\frac{n^2c^2\pi^2}{l^2}t} \sin\frac{n\pi x}{l} = b_n e^{\frac{n^2c^2\pi^2 t}{l^2}} \sin\frac{n\pi}{l}x$ ...(4)

On putting $t = 0$ and $u = x$ in (4), we get

$$x = b_n \sin\frac{n\pi}{l}x$$

General solution is

$$x = \sum_{n=1}^{\infty} b_n \sin\frac{n\pi}{l}x$$

Now, $b_n = \frac{2}{l}\int_0^l x\sin\frac{n\pi x}{l}dx = \frac{2}{l}\left[x\cdot\frac{l}{n\pi}\left(-\cos\frac{n\pi x}{l}\right)-(1)\left(\frac{-l^2}{n^2\pi^2}\sin\frac{n\pi x}{l}\right)\right]_0^l$

$$= \frac{2}{l}\left[\left(l\cdot\frac{l}{n\pi}(-\cos n\pi)+\frac{l^2}{n^2\pi^2}\sin n\pi\right)-0\right] = \frac{2}{l}\left[-\frac{l^2}{n\pi}(-1)^n\right] = (-1)^{n+1}\frac{2l}{n\pi}$$

Putting the value of b_n in (4), we get

$$u = \frac{2l}{\pi}\sum_{n=1}^{\infty}\frac{(-1)^{n+1}}{n}\sin\frac{n\pi x}{l}e^{-\frac{n^2c^2\pi^2 t}{l^2}}$$ **Ans.**

Example 20. *An insulated rod of length l has its ends A and B maintained at 0°C and 100°C respectively until steady state conditions prevail. If B is suddenly reduced to 0°C and maintained at 0°C, find the temperature at a distance x from A at time t.*

(U.P. II Semester Summer, 2005)

Solution. The initial temperature of the rod can be written as:

$$u(x,t) = 0 + \frac{100}{l}x = \frac{100}{l}x$$

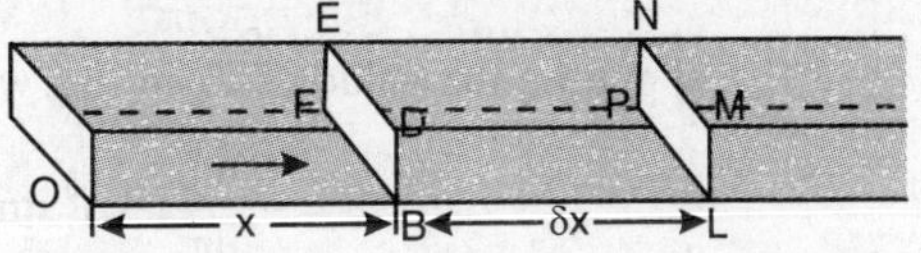

While in steady state, the temperature distribution can be written as

$$u(x,t) = 0 + \frac{0}{l}x = 0$$

To find u in the intermediate period, calculating time from the instant when the end temperature were changed.

$$u = u_1(x) + u_2(x)$$

where $u_2(x)$ is temperature after a sufficient long time and $u_1(x, t)$ is the transient temperature distribution tending to zero as $t \to \infty$ Hence $u_2(x) = 0$.

Also $u_1(x, t)$ satisfies one dimensional heat flow

$$C^2\frac{\partial^2 u}{\partial x^2} = \frac{\partial u}{\partial t}$$

Thus $$u = (C_1\cos px + C_2\sin px)e^{-c^2p^2t} \quad \text{... (1)}$$

On putting $x = 0$, $u = 0$ in (1), we get

$$0 = C_1 e^{-p^2c^2t} \quad \Rightarrow \quad C_1 = 0$$

On putting $C_1 = 0$ in (1), we get

$$u = C_2 \sin px\, e^{-c^2p^2t} \quad \text{... (2)}$$

On putting $x = l$, $u = 0$ in (2), we get

$$0 = C_2 \sin ple^{-p^2c^2t} \qquad \Rightarrow \qquad \sin pl = 0 = \sin n\pi$$

$$\Rightarrow \qquad pl = n\pi \quad \Rightarrow \quad p = \frac{n\pi}{l}$$

On putting the value of p in (2), we get

$$u = C_2 \sin \frac{n\pi x}{l} e^{-\frac{n^2\pi^2c^2}{l^2}t} \qquad \text{... (3)}$$

On putting $t = 0$, $u = \frac{100}{l}x$ in (3), we get

$$\frac{100}{l}x = C_2 \sin \frac{n\pi x}{l}$$

$$C_2 = \frac{2}{l}\int_0^l \frac{100}{l}x \cdot \sin\frac{n\pi x}{l}dx$$

$$C_2 = \frac{200}{l^2}\int_0^l x \sin\frac{n\pi x}{l}dx$$

$$C_2 = \frac{200}{l^2}\left[-\frac{xl}{n\pi}\cos\frac{n\pi x}{l} - (-1)\frac{l^2}{n^2\pi^2}\sin\frac{n\pi x}{l}\right]_0^l$$

$$C_2 = \frac{200}{l^2}\left[\frac{-l^2}{n\pi}\cos n\pi\right]$$

$$\Rightarrow \qquad C_2 = -\frac{200}{n\pi}(-1)^n$$

On putting the value of C_2 in (3), we get

$$u = -\frac{200}{n\pi}(-1)^n \sin\frac{n\pi x}{l} e^{-\frac{n^2\pi^2c^2}{l^2}t}$$

$$\Rightarrow \qquad u = (-1)^{n+1} \cdot \frac{200}{n\pi} \sin\frac{n\pi x}{l} e^{-\frac{n^2\pi^2c^2}{l^2}t}$$ **Ans.**

Example 21. *The ends A and B of a rod 20 cm long have the temperatures at 30° C and at 80°C until steady state prevails. The temperature of the ends are changed to 40°C and 60°C respectively. Find the temperature distribution in the rod at time t.*

Solution. The initial temperature distribution in the rod is

$$u_1(x,t) = 30 + \frac{80-30}{20}x = 30 + \frac{50}{20}x \quad i.e., \; u_1(x,t) = 30 + \frac{5}{2}x$$

and the final distribution (*i.e.* in steady state) is

$$u_2(x) = 40 + \frac{60-40}{20}x = 40 + \frac{20}{20}x = 40 + x$$

To get u in the intermediate period, reckoning time from the instant when the end temperature were changed, we assumed

$$u = u_1(x, t) + u_2(x)$$

where $u_2(x)$ is the steady state temperature distribution in the rod (*i.e.,* temperature after a sufficiently long time) and $u_1(x, t)$ is the transient temperature distribution which tends to zero as t increases.

Thus $\qquad u_2(x) = 40 + x$

Now $u_1(x. t)$ satisfies the one-dimensional heat-flow equation $c^2 \dfrac{\partial^2 u}{\partial x^2} = \dfrac{\partial u}{\partial t}$

Hence its solution is $(a_k \cos kx + b_k \sin kx)\ e^{-c^2k^2t}$
Hence u is of the form

$$u = 40 + x + \Sigma(a_k \cos kx + b_k \sin kx) e^{-c^2k^2t}$$

Since $\qquad u = 40°$ when $x = 0$ and $u = 60°$ when $x = 20$, we get

$$a_k = 0, \text{ and } k = \frac{n\pi}{20}$$

Hence
$$u = 40 + x + \sum_{n=1}^{\infty} b_n \sin\frac{n\pi x}{20}\, e^{-c^2\left(\frac{n\pi}{20}\right)^2 t} \qquad ...(1)$$

Using the initial conditions *i.e.,*

$$u = 30 + \frac{5}{2}x \text{ when } t = 0, \text{ we get}$$

$$30 + \frac{5}{2}x = 40 + x + \Sigma b_n \sin\frac{n\pi}{20}x \quad \Rightarrow \quad \frac{3}{2}x - 10 = \Sigma b_n \sin\frac{n\pi x}{20}$$

Hence
$$b_n = \frac{2}{20}\int_0^{20}\left(\frac{3}{2}x - 10\right)\sin\frac{n\pi x}{20}\,dx$$

$$\Rightarrow \qquad = \frac{1}{10}\left[\left(\frac{3x}{2} - 10\right)\left(-\frac{20}{n\pi}\cos\frac{n\pi x}{20}\right) - \frac{3}{2}\left(-\frac{400}{n^2\pi^2}\sin\frac{n\pi x}{20}\right)\right]_0^{20}$$

$$= \frac{1}{10}\left[-20\left(\frac{20}{n\pi}\right)(-1)^n - (-10)\left(-\frac{20}{n\pi}\right)\right] = -\frac{20}{n\pi}[2(-1)^n + 1]$$

Putting this value of b_n in (1), we get

$$\therefore \qquad u = 40 + x - \frac{20}{\pi}\sum\left[\left(\frac{2(-1)^n + 1}{n}\right)\sin\frac{n\pi x}{20}.e^{-\left(\frac{cn\pi}{20}\right)^2 t}\right] \qquad \textbf{Ans.}$$

EXERCISE 44.3

1. The equation $\dfrac{\partial u}{\partial t} = \mu\dfrac{\partial^2 u}{\partial x^2}$ refers to the conduction of heat along a bar, without radiation, show that if $u = Ae^{-gx}\sin(nt - gx)$, where A, g and n are positive constants, then $g = \sqrt{\dfrac{n}{2\mu}}$

2. Solve $\dfrac{\partial u}{\partial t} = a^2\dfrac{\partial^2 u}{\partial x^2}$ under the conditions
$u'(0, t) = 0 \quad t > 0, \qquad u'(\pi, t) = 0, \qquad u(x, 0) = x^2, 0 < x < \pi$

Ans. $u(x,t) = \dfrac{\pi^2}{3} + 4\displaystyle\sum_{n=1}^{\infty}\frac{(-1)^n}{n^2}\cos nx e^{-a^2n^2t}$

3. A uniform rod of length 'a' whose surface is thermally insulated is initially at temperature $\theta = \theta_0$. At time $t = 0$, one end is suddenly cooled to $\theta = 0$ and subsequently maintained at this temperature, the other end remains thermally insulated. Find the temperature $\theta(x, t)$.

Ans. $\theta(x,t) = \dfrac{4\theta_0}{\pi}\displaystyle\sum_{n=1}^{\infty}\frac{\sin\dfrac{(2n+1)\pi x}{2a}}{2n+1}\, e^{-\frac{(2n+1)^2\pi^2c^2t}{4a^2}}$

4. Solve $\dfrac{\partial U}{\partial t} = a^2 \dfrac{\partial^2 U}{\partial x^2}$ under the conditions

(*i*) $U \neq \infty$ if $t \to \infty$; (*ii*) $U(0, t) = U(\pi, t) = 0$; (*iii*) $U(x, 0) = \pi x - x^2$

Ans. $u = \dfrac{8}{\pi} \sum_{n=1}^{\infty} \dfrac{\sin(2n-1)x}{(2n-1)^3} e^{-a^2(2n-1)^2 t}$

5. The temperature distribution in a bar of length π, which is perfectly insulated at the ends $x = 0$ and $x = \pi$ is governed by the partial differential equation $\dfrac{\partial u}{\partial t} = \dfrac{\partial^2 u}{\partial x^2}$. Assuming the initial temperature as $u(x, 0) = f(x) = \cos 2x$, find the temperature distribution at any instant of time. **Ans.** $u = e^{-4t} \cos 2x$

6. The heat flow in a bar of length 10 cm of homogeneous material is governed by the partial differential equation $u_t = c^2 u_{xx}$. The ends of the bar are kept at temperature 0°C, and the initial temperature is $f(x) = x(10 - x)$. Find the temperature in the bar at any instant of time.

Ans. $u(x, t) = \dfrac{800}{\pi^3} \sum_{n=1}^{\infty} \dfrac{1}{(2n-1)^3} \sin \dfrac{(2n-1)\pi x}{10} e^{-\left[\frac{(2n-1)^2 \pi^2 c^2 t}{100}\right]}$

7. Find the temperature $u(x, t)$ in a bar of length π which is perfectly insulated everywhere including the ends $x = 0$ and $x = \pi$. This leads to the conditions $\dfrac{\partial u}{\partial x}(0, t) = 0, \dfrac{\partial u}{\partial x}(\pi, t) = 0$. Further the initial conditions are as given below:

$u(x, 0) = \begin{cases} x, & 0 < x < \pi/2 \\ \pi x, & \pi/2 \le x < \pi \end{cases}$ Find the solution by the separation of variables.

44.6 TWO DIMENSIONAL HEAT FLOW

When the heat-flow is along curves instead of along straight lines, all the curves lying in parallel planes, then the flow is called two-dimensional. Let us consider now the flow of heat in a metal-plate in the *xoy* plane. Let the plate be of uniform thickness *h*, density ρ, thermal conductivity *k* and the specific heat *C*. Since the flow is two dimensional, the temperature at any point of the plate is independent of the *z*-coordinate. The heat-flow lies in the *xoy* plane and is zero along the direction normal to the *xoy* plane.

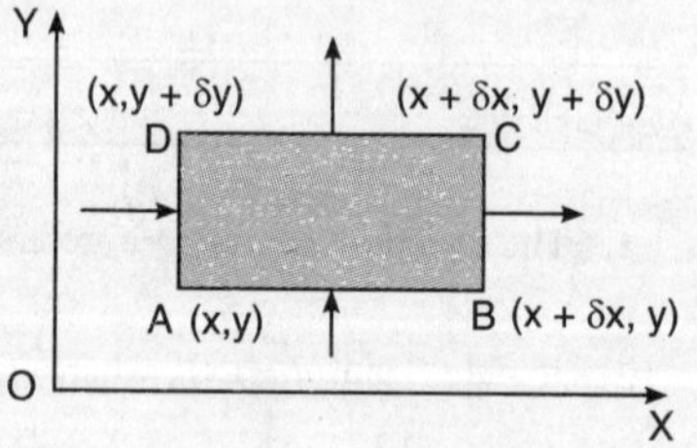

Now, consider a rectangular element *ABCD* of the plate with sides δx and δy, the edges being parallel to the coordinates axes, as shown in the figure, Then the quantity of heat *entering* the element *ABCD* per sec. through the surface *AB* is

Y
(x,y + δy)
D
(x + δx, y + δy)
C
A (x,y)
B (x + δx, y)
O
X

$$= -k\left(\frac{\partial u}{\partial y}\right)_y \delta x.h.$$

Similarly the quantity of heat entering the element *ABCD* per sec. through the surface *AD* is

$$= -k\left(\frac{\partial u}{\partial x}\right)_x \delta y.h.$$

The amount of heat which flows out through the surfaces *BC* and *CD* are

$$-k\left(\frac{\partial u}{\partial x}\right)_{x+\delta x} .\delta y.h \text{ and } -k\left(\frac{\partial u}{\partial y}\right)_{y+\delta y} .\delta x.h \text{ respectively.}$$

Therefore the total gain of heat by the rectangular element *ABCD* per sec.

= inflow – outflow.

$$= kh\left[\left\{\left(\frac{\partial u}{\partial x}\right)_{x+\delta x} - \left(\frac{\partial u}{\partial x}\right)_x\right\}\delta y + \left\{\left(\frac{\partial u}{\partial y}\right)_{y+\delta y} - \left(\frac{\partial u}{\partial y}\right)_y\right\}\delta x\right]$$

$$= kh\delta x.\delta y\left[\frac{\left(\frac{\partial u}{\partial x}\right)_{x+\delta x} - \left(\frac{\partial u}{\partial x}\right)_x}{\delta x} + \frac{\left(\frac{\partial u}{\partial y}\right)_{y+\delta y} - \left(\frac{\partial u}{\partial y}\right)_y}{\delta y}\right] \quad ...(1)$$

The rate of gain of heat by the element *ABCD* is also given by

$$\rho\delta x.\delta y.h.c.\frac{\partial u}{\partial t} \quad ...(2)$$

Equating the two-expressions for gain of heat per sec. from (1) and (2), we have

$$\rho\delta x.\delta y.h.c.\frac{\partial u}{\partial t} = hk\delta x\delta y\left[\frac{\left(\frac{\partial u}{\partial x}\right)_{x+\delta x} - \left(\frac{\partial u}{\partial x}\right)_x}{\delta x} + \frac{\left(\frac{\partial u}{\partial y}\right)_{y+\delta y} - \left(\frac{\partial u}{\partial y}\right)_y}{\delta y}\right]$$

i.e.,
$$\frac{\partial u}{\partial t} = \frac{k}{\rho c}\left[\frac{\left(\frac{\partial u}{\partial x}\right)_{x+\delta x} - \left(\frac{\partial u}{\partial x}\right)_x}{\delta x} + \frac{\left(\frac{\partial u}{\partial y}\right)_{y+\delta y} - \left(\frac{\partial u}{\partial y}\right)_y}{\delta y}\right]$$

Taking the limit as $\delta x \to 0$, $\delta y \to 0$, the above equation reduces to $\frac{\partial u}{\partial t} = \frac{k}{\rho c}\left(\frac{\partial^2 u}{\partial x^2} + \frac{\partial^2 u}{\partial y^2}\right)$

Putting $\alpha^2 = \frac{k}{\rho c}$ as before, the equation becomes,

$$\frac{\partial u}{\partial t} = \alpha^2\left(\frac{\partial^2 u}{\partial x^2} + \frac{\partial^2 u}{\partial y^2}\right) \quad ...(3)$$

The equation (3) gives the temperature distribution of the plate in the transit state.

In the steady-state, u is independent of t, so that $\frac{\partial u}{\partial t} = 0$. Hence the temperature distribution of the plate in the steady-state is $\frac{\partial^2 u}{\partial x^2} + \frac{\partial^2 u}{\partial y^2} = 0.$

Example 22. *Solve* $\frac{\partial^2 u}{\partial x^2} + \frac{\partial^2 u}{\partial y^2} = 0$ *which satisfies the conditions*

$$u(0, y) = u(l, y) = u(x, 0) = 0$$

and
$$u(x, a) = \sin\frac{n\pi x}{l}$$

Solution. Consider the heat flow in a metal plate of uniform thickness, in the directions parallel to length and breadth of the plate. There is no heat flow along the normal to the plane of the rectangle.

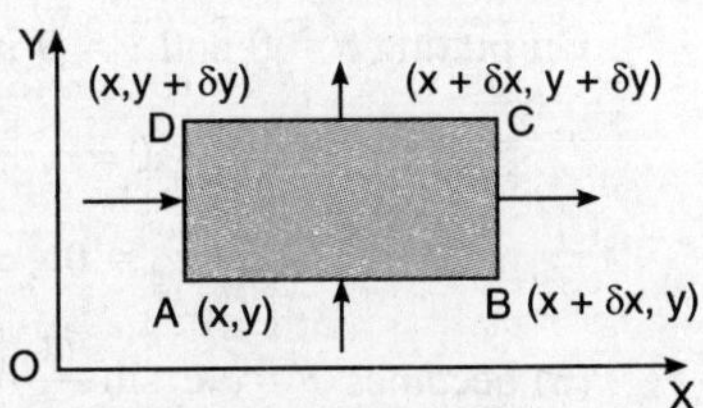

Let $u(x, y)$ be the temperature at any point (x, y) of the plate at time t is given by

$$\frac{\partial u}{\partial t} = c^2\left(\frac{\partial^2 u}{\partial x^2} + \frac{\partial^2 u}{\partial y^2}\right) \qquad \text{... (1)}$$

In the steady state, u does not change with t.

$$\therefore \qquad \frac{\partial u}{\partial t} = 0$$

(1) becomes $\frac{\partial^2 u}{\partial x^2} + \frac{\partial^2 u}{\partial y^2} = 0$

This is called Laplace equation in two dimensions.

Let $\qquad u = X(x) \,.\, Y(y) \qquad$ ...(2)

Putting the values of $\frac{\partial^2 u}{\partial x^2}$ and $\frac{\partial^2 u}{\partial y^2}$ in (1), we have

$$X''Y + XY'' = 0 \qquad \text{...(3)}$$

$$\Rightarrow \qquad \frac{X''}{X} = -\frac{Y''}{Y} = -p^2 \text{ (say)}$$

$D^2X = -p^2X$	$D^2Y = p^2Y$
$\Rightarrow \quad D^2X + p^2X = 0$	$\Rightarrow D^2Y - p^2Y = 0]$
$\Rightarrow \quad (D^2 + p^2)X = 0$	$\Rightarrow (D^2 - p^2)Y = 0$
$A.E.$ is $\quad m^2 + p^2 = 0$	$A.E.$ is $m^2 - p^2 = 0$
$\Rightarrow \quad m^2 = -p^2$	$\Rightarrow m^2 = p^2$
$\Rightarrow \quad m = \pm\, ip$	$\Rightarrow m = \pm\, p$
$X = c_1 \cos px + c_2 \sin px$	$Y = c_3 e^{py} + c_4 e^{-py}$

Putting the values of X and Y in (2), we have

$$u = (c_1 \cos px + c_2 \sin px)(c_3 e^{py} + c_4 e^{-py}) \qquad \text{... (4)}$$

Putting $x = 0$, $u = 0$ in (4), we have

$$0 = c_1 (c_3 e^{py} + c_4 e^{-py})$$

$$\therefore \qquad c_1 = 0$$

(4) is reduced to $u = c_2 \sin px (c_3 e^{py} + c_4 e^{-py}) \qquad$ ... (5)

On putting $x = l$, $u = 0$ in (5), we have

$$0 = c_2 \sin pl (c_3 e^{py} + c_4 e^{-py})$$

$$c_2 \neq 0 \qquad \therefore \sin pl = 0 = \sin n\pi \;\Rightarrow\; pl = n\pi \;\Rightarrow\; p = \frac{n\pi}{l}$$

Now (5) becomes $\quad u = c_2 \sin\frac{n\pi x}{l}\left(c_3 e^{\frac{n\pi y}{l}} + c_4 e^{\frac{-n\pi y}{l}}\right) \qquad$ ... (6)

On putting $u = 0$ and $y = 0$ in (6), we have

$$0 = c_2 \sin\frac{n\pi x}{l}(c_3 + c_4)$$

$$\Rightarrow \qquad c_3 + c_4 = 0 \;\Rightarrow\; c_3 = -c_4$$

(6) becomes $u = c_2 c_3 \sin\frac{n\pi x}{l}\left(e^{\frac{n\pi y}{l}} - e^{\frac{-n\pi y}{l}}\right) \qquad$ ... (7)

On putting $y = a$ and $u = \sin\frac{n\pi x}{l}$ in (7), we have

$$\sin\frac{n\pi x}{l} = c_2 c_3 \sin\frac{n\pi x}{l}\left(e^{\frac{n\pi a}{l}} - e^{\frac{-n\pi a}{l}}\right) \quad i.e. \quad c_2 c_3 = \frac{1}{e^{\frac{n\pi a}{l}} - e^{\frac{-n\pi a}{l}}}$$

Putting this value in (7), we have

$$u = \sin\frac{n\pi x}{l}\frac{e^{\frac{n\pi y}{l}} - e^{\frac{-n\pi y}{l}}}{e^{\frac{n\pi a}{l}} - e^{\frac{-n\pi a}{l}}} \quad \text{or} \quad u = \sin\frac{n\pi x}{l}\frac{\sinh\frac{n\pi y}{l}}{\sinh\frac{n\pi a}{l}} \qquad \textbf{Ans.}$$

Example 23. *A rectangular plate with insulated surfaces is 10 cm wide and so long compared to its width that it may be considered infinite in length without introducing an appreciable error. If the temperature along the short edge y = 0 is given by*

$$u(x_1 0) = \begin{cases} 20x, & 0 < x \le 5 \\ 20\,(10 - x), & 5 < x < 10 \end{cases}$$

while the two long edges x = 0 and x = 10 as well as the other short edges are kept at 0°C. Find the steady state temperature at any point (x, y) of the plate.

Solution. In the steady state, the temperature $u,(x, y)$ at any point $p\ (x, y)$ satisfy the equation

$$\frac{\partial^2 u}{\partial x^2} + \frac{\partial^2 u}{\partial y^2} = 0 \qquad \text{... (1)}$$

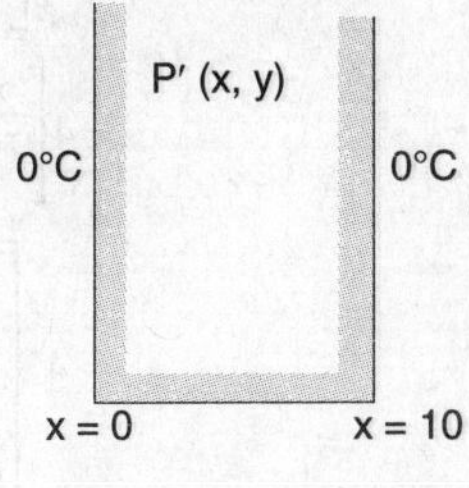

The boundary conditions are

$u\ (0, y) = 0$ for all values of y ... (2)

$u\ (10, y) = 0$ for all values of y ... (3)

$u(x, \infty) = 0$ for all values of x ... (4)

$$u(x_1 0) = \begin{cases} 20x, & 0 < x \le 5 \\ 20\,(10 - x), & 5 < x < 10 \end{cases} \qquad \text{... (5)}$$

Now three possible solutions of (1) are

$$u = (C_1 e^{px} + C_2 e^{-px})(C_3 \cos py + C_4 \sin py) \qquad \text{... (6)}$$

$$u = (C_5 \cos px + C_6 \sin px)(C_7 e^{py} + C_8 e^{-py}) \qquad \text{... (7)}$$

$$u = (C_9 x + C_{10})(C_{11} y + C_{12}) \qquad \text{... (8)}$$

Of these, we have to choose that solution which is consistent with the physical nature of the problem. The solution (6) and (8) cannot satisfy the condition (2), (3) and (4). Thus, only possible solution is (7) *i.e.,* of the form.

$$u(x, y) = (C_1 \cos px + C_2 \sin px)(C_3 e^{py} + C_4 e^{-py}) \qquad \text{... (9)}$$

By (2), $u(0, y) = C_1(C_3 e^{py} + C_4 e^{-py}) = 0$ for all values of y

$\therefore \quad C_1 = 0$

$\therefore$ (9) reduces to $u(x, y) = C_2 \sin px (C_3 e^{py} + C_4 e^{-py})$ ... (10)

By (3), $u\ (10, y) = C_2 \sin 10p\ (C_3\ e^{py} + C_4\ e^{-py}) = 0,\ C_2 \ne 0$

$$\therefore \quad \sin 10\,p = 0 = \sin n\pi \Rightarrow 10p = n\pi \Rightarrow p = \frac{n\pi}{10}$$

Also to satisfy the condition (4) *i.e.* $u = 0$ as $y \to \infty$

$$C_3 = 0$$

Hence (10) takes the form $u(x, y) = C_2 C_4 \sin px\, e^{-py}$

or $u(x,y) = b_n \sin px.e^{-py}$ where $b_n = C_2 C_4$

$\therefore$ The most general solution that satisfies (2), (3) and (4) is of the form

$$u(x,y) = \sum_{n=1}^{\infty} b_n \sin px\, e^{-py} \qquad \text{... (11)}$$

Putting $y = 0$, $u(x,0) = \sum_{n=1}^{\infty} b_n \sin px$, where $p = \dfrac{n\pi}{10}$

This requires the expansion of u in Fourier series in the interval $x = 0$ and $x = 5$ and from $x = 5$ to $x = 10$.

$$b_n = \frac{2}{10}\int_0^5 20x \sin px dx + \frac{2}{10}\int_5^{10} 20(10-x)\sin px\, dx$$

$$b_n = 4\int_0^5 x \sin px\, dx + 4\int_5^{10} (10-x)\sin px\, dx$$

$$= 4\left[x\left(\frac{-\cos px}{p}\right) - (1)\left(\frac{-\sin px}{p^2}\right)\right]_0^5 + 4\left[(10-x)\left(\frac{-\cos px}{p}\right) - (-1)\left(\frac{-\sin px}{p^2}\right)\right]_5^{10}$$

$$= 4\left[\frac{-5\cos 5p}{p} + \frac{\sin 5p}{p^2}\right] + 4\left[0 - \frac{\sin 10p}{p^2} + \frac{5\cos 5p}{p} + \frac{\sin 5p}{p^2}\right]$$

$$= 4\left[\frac{2\sin 5p}{p^2} - \frac{\sin 10p}{p^2}\right] \qquad \left(p = \frac{n\pi}{10}\right)$$

$$= 4\left[\frac{2\sin 5.\frac{n\pi}{10}}{\frac{n^2\pi^2}{100}} - \frac{\sin 10.\frac{n\pi}{10}}{\frac{n^2\pi^2}{100}}\right] = \frac{800}{n^2\pi^2}\sin\frac{n\pi}{2} - \frac{400}{n^2\pi^2}\sin n\pi$$

$$= \frac{800}{n^2\pi^2}\sin\frac{n\pi}{2} = 0, \text{ if } n \text{ is even.}$$

$$= \pm\frac{800}{n^2\pi^2}, \qquad \text{if } n \text{ is odd.}$$

$$\Rightarrow \quad b_n = \frac{(-1)^{n+1}800}{(2n-1)^2\pi^2}$$

On putting the value of b_n in (11) the temperature at any point (x, y) is given by

$$u(x,y) = \frac{800}{\pi^2}\sum_{n=1}^{\infty}\frac{(-1)^{n+1}}{(2n-1)^2}\sin\frac{(2n-1)\pi x}{10}\, e^{-\frac{(2n-1)\pi y}{10}} \qquad \textbf{Ans.}$$

Example 24. *A thin rectangular plate whose surface is impervious to heat flow has t = 0 an arbitrary distribution of temperature f (x, y). Its four edges x = 0, x = a, y = 0, y = b are kept at zero temperature. Determine the temperature at a point of the plate as t increases.*

Solution. The partial differential equation of two dimensional heat conduction problem is

$$\frac{\partial^2 u}{\partial x^2} + \frac{\partial^2 u}{\partial y^2} = \frac{1}{c^2}\frac{\partial u}{\partial t} \qquad \text{... (1)}$$

The boundary conditions are $u(0, y) = u(a, y) = u(x, 0) = u(x, b)$ and the initial condition is $u(x, y) = f(x, y)$

Let the solution be $\qquad u = X.Y.T.$

On putting the values of the derivatives in (1), we get

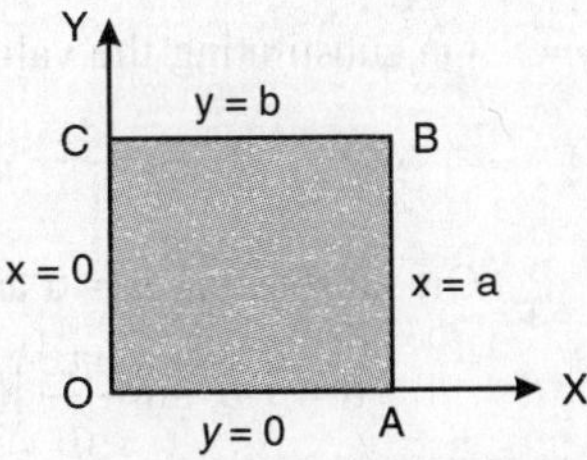

$$YT\frac{d^2X}{dx^2} + XT\frac{d^2Y}{dy^2} = \frac{XY}{c^2}\frac{dT}{dt}$$

Separating the variables, we get, (dividing by XYT)

$$\frac{1}{X}\frac{d^2X}{dx^2} + \frac{1}{Y}\frac{d^2Y}{dy^2} = \frac{1}{C^2T}\frac{dT}{dt}$$

Since X is function of independent variable x alone, Y of y alone and T of t alone, there are three possibilities.

1. $\frac{1}{X}\frac{d^2X}{dx^2} = 0, \quad \frac{1}{Y}\frac{d^2Y}{dy^2} = 0, \quad \frac{1}{c^2T}\frac{dT}{dt} = 0$

2. $\frac{1}{X}\frac{d^2X}{dx^2} = K_1^2, \quad \frac{1}{Y}\frac{d^2Y}{dy^2} = K_2^2, \quad \frac{1}{c^2T}\frac{dT}{dt} = K^2$

3. $\frac{1}{X}\frac{d^2X}{dx^2} = -K_1^2, \quad \frac{1}{Y}\frac{d^2Y}{dy^2} = -K_2^2, \quad \frac{1}{c^2T}\frac{dT}{dt} = -K^2$

$$K^2 = K_1^2 + K_2^2 \qquad \text{...(A)}$$

Out of these three solutions, we have to choose that solution which is consistent with the physical nature of the problem. Accordingly third solution is accepted here.

Then $\qquad Y = A\cos K_1x + B\sin K_1x$

$$T = Ee^{-C^2kt}$$

$$u = XYT$$

$$u = (A\cos K_1x + B\sin K_1x)(C\cos K_2y + D\sin K_2y).Ee^{-C^2kt} \qquad \text{...(1)}$$

Now we apply boundary conditions

On putting $u = 0$ and $x = 0$ in (1), we get

$$0 = A(C\cos K_2y + D\sin K_2y)Ee^{-C^2kt} \Rightarrow A = 0$$

On putting the value of A in (1), it reduces to

$$u = (B\sin K_1x)(C\cos K_2y + D\sin K_2y).Ee^{-C^2kt}$$

$$\Rightarrow \qquad u = (B_1\sin K_1x)(C\cos K_2y + D\sin K_2y).e^{-C^2kt} \qquad \text{...(2)}$$

$$(B_1 = B.E)$$

On putting $u = 0$ and $x = a$ in (2), we get

$$0 = (B_1\sin K_1a)(C\cos K_2y + D\sin K_2y).e^{-C^2kt}$$

$$\Rightarrow \qquad 0 = B_1\sin K_1a \Rightarrow \sin K_1a = 0 = \sin n\pi \Rightarrow K_1a = n\pi \Rightarrow K_1 = \frac{n\pi}{a}$$

On putting the value of K_1 in (2), we have

$$u = \left(B_1\sin\frac{n\pi}{a}x\right)(C\cos K_2y + D\sin K_2y).e^{-C^2kt} \qquad \text{...(3)}$$

On putting $u = 0, y = 0$ in (3), we obtain

$$0 = \left(B_1 \sin\frac{n\pi}{a}x\right)(C)e^{-C^2kt} \quad \Rightarrow \quad C = 0$$

On substituting the value of C in (3), we have

$$u = \left(B_1 \sin\frac{n\pi x}{a}\right)(D \sin K_2 y)e^{-C^2kt} \qquad \text{... (4)}$$

On substituting $u = 0$ and $y = b$ in (4), we have

$$0 = \left(B_1 \sin\frac{n\pi x}{a}\right)(D \sin K_2 b).e^{-C^2kt} \quad \Rightarrow \sin K_2 b = 0 = \sin m\pi \Rightarrow K_2 = \frac{m\pi}{b}$$

On putting the value of K_2 in (4), we have

$$u = \left(B_1 \sin\frac{n\pi x}{a}\right)\left(D \sin\frac{m\pi y}{b}\right).e^{-C^2kt}$$

$$u = A_{mn} \sin\frac{n\pi x}{a} \sin\frac{m\pi y}{b}.e^{-C^2kt}, \qquad \text{... (5) } (B_1 D = A_{mn})$$

But from (A), $K^2 = K_1^2 + K_2^2 = \pi^2\left(\frac{n^2}{a^2} + \frac{m^2}{b^2}\right)$

$$\Rightarrow \qquad K_{mn}^2 = \pi^2\left(\frac{n^2}{a^2} + \frac{m^2}{b^2}\right)$$

By using K_{mn}, (5) becomes

$$u = \sum_{n=1}^{\infty}\sum_{m=1}^{\infty} A_{mn} \sin\frac{n\pi x}{a} \sin\frac{m\pi y}{b} e^{-C^2 k_{mn} t}$$

On applying the initial condition $u = f(x, y)$, $t = 0$

$$f(x, y) = \sum_{n=1}^{\infty}\sum_{m=1}^{\infty} A_{mn} \sin\frac{n\pi x}{a} \sin\frac{m\pi y}{b}$$

Which is the double Fourier sine series of $f(x, y)$

Where $\qquad A_{mn} = \frac{2}{a}\frac{2}{b}\int_{x=0}^{a}\int_{y=0}^{b} \sin\frac{n\pi x}{a} \sin\frac{m\pi y}{b}.f(x, y)\,dx\,dy$ **Ans.**

Example 25. *A rectangular plate with insulated surface is 8 cm wide and so long compared to its width that it may be considered infinite in length without introducing an appreciable. If the tempearature along one short edge y = 0 is given by*

$$u(x, 0) = 100 \sin\frac{\pi x}{8}, 0 < x < 8$$

While the two long edges x = 0 and x = 8 as well as the other short edge are kept at 0°C, show that the steady state temperature at any plate is given by

$$u(x, y) = 100 e^{-\frac{\pi y}{8}} \sin\frac{\pi x}{8}.$$

Soution. Here, we have

Two dimensional heat flow equation in steady state is given by

$$\frac{\partial^2 u}{\partial x^2} + \frac{\partial^2 u}{\partial y^2} = 0$$

Its solution is $u(x, y) = (c_1 \cos px + c_2 \sin px)(c_3e^{py} + c_4e^{-py})$ ...(1)

Putting $x = 0$ and $u = 0$ in (1), we get (given)

$$0 = c_1(c_3e^{py} + c_4e^{-py})$$

$$\Rightarrow \quad c_1 = 0$$

Putting the value of c_1 in (1), we get

$$u = c_2 \sin px(c_3e^{py} + c_4e^{-py}) \quad ...(2)$$

Again putting $x = 8$ and $u = 0$ in (2), we get (given)

$$0 = c_2 \sin 8p(c_3e^{py} + c_4e^{-py})$$

$$\Rightarrow \quad \sin 8p = 0 = \sin n\pi \Rightarrow p = \frac{n\pi}{8} \; (n \in I)$$

Putting the value of p in (2), we get

$$u(x, y) = c_2 \sin \frac{n\pi x}{8}\left(c_3e^{\frac{n\pi y}{8}} + c_4e^{-\frac{n\pi y}{8}}\right) \quad ...(3)$$

Putting $y = \infty$ and $u = 0$ in (3), we get (given)

$$0 = c_2 \sin \frac{n\pi x}{8} c_3e^{\frac{n\pi y}{8}}$$

$$\Rightarrow c_3 = 0$$

Putting $c_3 = 0$ in (3), we get

$$u = c_2c_4 \sin \frac{n\pi x}{8} e^{-\frac{n\pi y}{8}}$$

$$u = b_n \sin \frac{n\pi x}{8} e^{-\frac{n\pi y}{8}} \quad ...(4)$$

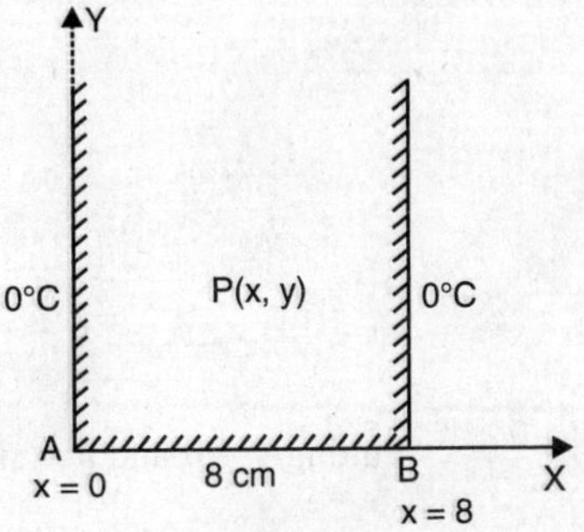

Putting $y = 0$ and $u = 100 \sin \frac{\pi x}{8}$ in (4), we get (given)

$$100 \sin \frac{\pi x}{8} = b_n \sin \frac{n\pi x}{8}$$

$$\Rightarrow b_n = 100,$$

Putting the value of b_n in (4), we get

$$u = 100 \sin\left(\frac{\pi x}{8}\right)e^{-\frac{\pi y}{8}}$$ **Ans.**

Example 26. *Solve* $\frac{\partial^2 u}{\partial x^2} + \frac{\partial^2 u}{\partial y^2} = 0$*,* $0 < x < \pi$*,* $0 < y < \pi$*, which satisfies the conditions :*

$u(0, y) = u(\pi, y) = u(x, \pi) = 0$ *and* $u(x, 0) = \sin^2 x$.

(GBTU 2011)

Solution. Here, we have

$$\frac{\partial^2 u}{\partial x^2} + \frac{\partial^2 u}{\partial y^2} = 0$$

Its solution is $u(x, y) = (c_1 \cos px + c_2 \sin px)(c_3e^{py} + c_4e^{-py})$ ... (1)

Putting $x = 0$ and $u = 0$ in (1), we get (given)

$$0 = c_1(c_3e^{py} + c_4e^{-py})$$

$$\Rightarrow c_1 = 0$$

Putting the value of c_1 in (1), we get

$$u(x, y) = c_2 \sin px(c_3 e^{py} + c_4 e^{-py}) \quad \text{... (2)}$$

Putting $x = \pi$ and $u = 0$ in (2), we get (given)

$$0 = c_2 \sin p\pi(c_3 e^{py} + c_4 e^{-py})$$

$$\Rightarrow \quad \sin p\pi = 0 = \sin n\pi$$

$$\Rightarrow \quad p = n$$

Putting $p = n$ in (2), we get $u = c_2 \sin nx(c_3 e^{ny} + c_4 e^{-ny})$

$$= \sin nx(Ae^{ny} + Be^{-ny}) \quad \text{... (3) } (c_2c_3 = A \text{ and } c_2c_4 = B.)$$

Putting $u = 0$ and $y = \pi$ in (3), we get (given)

$$0 = \sin nx(Ae^{n\pi} + Be^{-n\pi})$$

$$\Rightarrow \quad 0 = Ae^{n\pi} + Be^{-n\pi}$$

$$\Rightarrow \quad Ae^{n\pi} = -Be^{-n\pi} = -\frac{1}{2}b \quad \Rightarrow A = -\frac{1}{2}be^{-n\pi} \text{ and } B = \frac{1}{2}be^{n\pi}$$

On putting the values of A and B in (3), we get

$$u = \sin nx\left[-\frac{1}{2}be^{-n\pi}e^{ny} + \frac{1}{2}be^{n\pi}e^{-ny}\right]$$

$$= \frac{1}{2}b\,[e^{n(\pi-y)} - e^{-n(\pi-y)}]\sin nx$$

$$= b \sinh n\,(\pi - y) \sin nx.$$

$$u = \sum_{n=1}^{\infty} b_n \sinh n(\pi - y)\sin nx \quad \text{... (4) (General solution)}$$

Putting $y = 0$ and $u = \sin^2 x$ in (4), we get (given)

$$\sin^2 x = \sum_{n=1}^{\infty} b_n \sinh n\pi \sin nx$$

Here

$$b_n \sinh n\pi = \frac{2}{\pi}\int_0^{\pi} \sin^2 x \sin nx\,dx = \frac{1}{\pi}\int_0^{\pi}(1 - \cos 2x)\sin nx\,dx$$

$$= \frac{1}{\pi}\int_0^{\pi}\left[\sin nx - \frac{1}{2}\{\sin(n+2)x + \sin(n-2)x\}\right]dx$$

$$= \frac{1}{\pi}\left[\frac{-\cos nx}{n} + \frac{\cos(n+2)x}{2(n+2)} + \frac{\cos(n-2)x}{2(n-2)}\right]_0^{\pi}$$

$$= \frac{1}{\pi}\left[\frac{-\cos n\pi}{n} + \frac{\cos(n+2)\pi}{2(n+2)} + \frac{\cos(n-2)\pi}{2(n-2)} + \frac{1}{n} - \frac{1}{2(n+2)} - \frac{1}{2(n-2)}\right]$$

$$= \frac{1}{\pi}\left[\left(\frac{-(-1)^n}{n} + \frac{(-1)^{n+2}}{2(n+2)} + \frac{(-1)^{n-2}}{2(n-2)}\right) + \left(\frac{1}{n} - \frac{1}{2(n+2)} - \frac{1}{2(n-2)}\right)\right]$$

$$= \frac{1}{2\pi}\left[\left(\frac{1}{n+2} + \frac{1}{n-2} - \frac{2}{n}\right)\{(-1)^n - 1\}\right], \text{ where } n \neq 2$$

$$b_n \sinh n\pi = \frac{-8}{\pi n\,(n^2 - 4)}, \text{ when } n \text{ is odd} \quad \text{... (5)}$$

$$b_n \sinh n\pi = 0, \text{ when } n \text{ is even and} \neq 2$$

Now, we have to find out $b_2 \sinh 2\pi$.

$$b_2 \sinh 2\pi = \frac{2}{\pi}\int_0^{\pi} \sin^2 x \sin 2x \, dx$$

$$= \frac{1}{\pi}\int_0^{\pi} (1-\cos 2x)\sin 2x \, dx = \frac{1}{\pi}\int_0^{\pi}\left(\sin 2x - \frac{1}{2}\sin 4x\right)dx = \left(\frac{-\cos 2x}{2} + \frac{1}{8}\cos 4x\right)_0^{\pi} = 0$$

$$\Rightarrow \qquad b_2 = 0 \qquad \text{...(6)}$$

On putting the value of b_n from (5) in (4), we get

$$u = \frac{-8}{\pi}\sum_{n=1,3,5,\ldots}^{\infty} \frac{\sin nx \sinh n(\pi - y)}{n(n^2-4)\sinh n\pi} \qquad \textbf{Ans.}$$

EXERCISE 44.4

1. Find by the method of separation of variables, a particular solution of the equation

$$\frac{\partial^2 u}{\partial x^2} + \frac{\partial^2 u}{\partial y^2} = 0$$

that tends to zero as x tends to infinity and is equal to 2 cos y when $x = 0$. **Ans.** $u = 2e^{-x}\cos y$

2. Solve the equation: $u_{xx} + u_{yy} = 0$

$$u(0, y) = u(\pi, y) = 0 \text{ for all } y,$$
$$u(x, 0) = k, \quad 0 < x < \pi$$
$$\lim_{y \to \infty} u(x, y) = 0, \quad 0 < x < \pi$$

Ans. $u(x, y) = \sum_{n=1}^{\infty} b_n \sin nx e^{-ny}, \ k = \sum_{n=1}^{\infty} b_n \sin nx$

3. A infinitely long uniform plate is bounded by two parallel edges and an end at right angles to them. The breadth is π. This end is maintained at a temperature u_0 at all points and other edges are at zero temperature. Determine the temperature at any point of the plate in the steady state.

Ans. $u(x, y) = \frac{4u_0}{\pi}\left[e^{-y}\sin x + \frac{1}{3}e^{-3y}\sin 3x + \frac{1}{5}e^{-5y}\sin 5x + \ldots\right]$

4. Solve $\frac{\partial^2 V}{\partial x^2} + \frac{\partial^2 V}{\partial y^2} = 0$, given that

(*i*) $V = 0$ when $x = 0$ and $x = c$; (*ii*) $V \to 0$ as $y \to \infty$ (*iii*) $V = V_0$ when $y = 0$.

Ans. $V(x, y) = \sum_{n=1}^{\infty} b_n \sin\frac{n\pi x}{c} e^{-\frac{n\pi y}{c}}, V_0 = \sum_{n=1}^{\infty} b_n \sin\frac{n\pi x}{c}$

5. The steady state temperature distribution in a thin plate bounded by the lines $x = 0$, $x = a$, $y = 0$ and $y = \infty$, is governed by the partial differential equation

$$\frac{\partial^2 u}{\partial x^2} + \frac{\partial^2 u}{\partial y^2} = 0.$$

Obtain the steady state temperature distribution under the conditions

$u(0, y) = 0, \quad u(a, y) = 0, \quad u(x, \infty) = 0$

$u(x, 0) = x, \quad 0 \le x \le a/2$

$\lim 4(x, y) = a - x, \ a/2 \le x \le a$ **Ans.** $u(x, y) = \sum_{n=1}^{\infty} \frac{4a}{n^2\pi^2}\sin\frac{n\pi}{2}\sin\frac{n\pi x}{a} e^{-\frac{\pi y}{a}}$

6. An infinitely long plane uniform plate is bounded by two parallel edges and an end at right angles to them. The breadth is π. This end is maintained at temperature u_0 at all points and the other edges are at zero temperature. Determine the temperature at any point of the plate in the steady state.

Ans. $u(x, y) = \frac{4u_0}{\pi}\sum_{n=1}^{\infty}\frac{1}{n}\sin x e^{-ny}$.

44.7 EQUATION OF HEAT FLOW IN TWO DIMENSIONS IN POLAR COORDINATES

Equation of heat flow in two dimension in polar coordinates.

Consider a sheet of conducting material of uniform density ρ, uniform thickness h, thermal conductivity k and specific heat c. Let O, the pole and OX, the initial line, be taken on the plane of sheet. As we are dealing with two dimensional heat-flow, the temperature function $u(r, \theta, t)$ at point (r, θ) of the plate is a function of r, θ and time t.

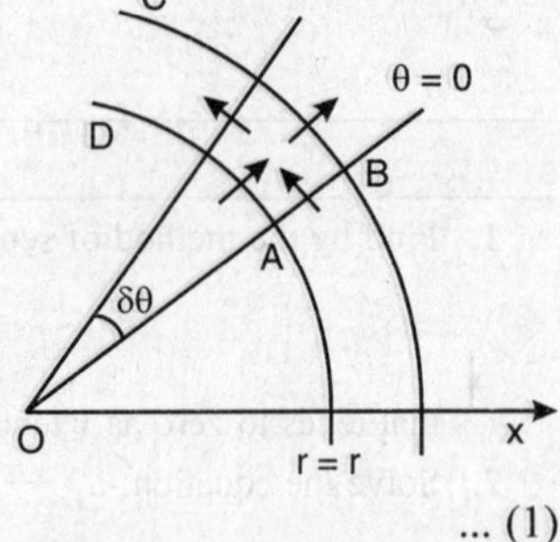

Consider an element of the sheet included between the circles $r = r$, $r = r + \delta r$ and the straight lines $\theta = \theta$ and $\theta = \theta + \delta\theta$ through the pole. Heat-flow directions are assumed to be positive in the positive directions associated with r and θ. The mass of the element $ABCD = \rho h(r\delta r \rho\theta)$

Let δu denote the temperature rise in the element during a short interval of time δt succeeding the time t.

∴ Rate of increase of heat content in the element

$$= \underset{\delta t \to 0}{\text{Lt}}\ \rho h(r\delta r\delta\theta)c\frac{\delta u}{\delta t} = \rho hcr\delta r\delta\theta\frac{\partial u}{\partial t} \quad \text{... (1)}$$

If R_1, R_2, R_3, R_4 are the roots of flow of heat across the sides of the element through the edges AB, CD, AD and BC respectively, then

$$R_1 = -k\left(\frac{1}{r}\frac{\partial u}{\partial\theta}\right)_\theta h\delta r, \qquad R_2 = -k\left(\frac{1}{r}\frac{\partial u}{\partial\theta}\right)_{\theta+\delta\theta} h\delta r$$

$$R_3 = -k\left(\frac{\partial u}{\partial r}\right)_r hr\delta\theta, \qquad R_4 = -k\left(\frac{\partial u}{\partial r}\right)_{r+\delta r} h(r+\delta r)\delta\theta$$

The rate of increase of heat in the element $= R_1 - R_2 + R_3 - R_4$ which is equal to the expression in (1)

$$\therefore\ \rho hcr\,\delta r\,\delta\theta\frac{\partial u}{\partial t} = kh\left[\left\{\left(\frac{1}{r}\frac{\partial u}{\partial\theta}\right)_{\theta+\delta\theta} - \left(\frac{1}{r}\frac{\partial u}{\partial\theta}\right)_\theta\right\}\delta r + \left\{\left(\frac{\partial u}{\partial r}\right)_{r+\delta r} - \left(\frac{\partial u}{\partial r}\right)_r\right\} r\delta\theta + \delta r\delta\theta\left(\frac{\partial u}{\partial r}\right)_{r+\delta r}\right]$$

Dividing by $\delta r.\delta\theta.hr\rho c$,

$$\frac{\partial u}{\partial t} = \frac{k}{\rho c}\left[\frac{1}{r^2}\frac{\left\{\left(\frac{\partial u}{\partial\theta}\right)_{\theta+\delta\theta} - \left(\frac{\partial u}{\partial\theta}\right)_\theta\right\}}{\delta\theta} + \frac{\left(\frac{\partial u}{\partial r}\right)_{r+\delta r} - \left(\frac{\partial u}{\partial r}\right)_r}{\delta r} + \frac{1}{r}\left(\frac{\partial u}{\partial r}\right)_{r+\delta r}\right]$$

Taking the limit as $\delta\theta, \delta r \to 0$,

$$\frac{\partial u}{\partial t} = \frac{k}{\rho c}\left[\frac{1}{r^2}\frac{\partial^2 u}{\partial\theta^2} + \frac{\partial^2 u}{\partial r^2} + \frac{1}{r}\frac{\partial u}{\partial r}\right]$$

Therefore the equation of heat-flow in polar coordinates is

$$\frac{\partial^2 u}{\partial r^2} + \frac{1}{r}\frac{\partial u}{\partial r} + \frac{1}{r^2}\frac{\partial^2 u}{\partial\theta^2} = \frac{1}{\alpha^2}\frac{\partial u}{\partial t} \qquad \text{where } \alpha^2 = \frac{k}{\rho c}$$

In steady-state, $\frac{\partial u}{\partial t} = 0$. Hence the equation of heat-flow in steady-state is

$$\nabla^2 u \equiv \frac{\partial^2 u}{\partial r^2} + \frac{1}{r}\frac{\partial u}{\partial r} + \frac{1}{r^2}\frac{\partial^2 u}{\partial\theta^2} = 0$$

Example 27. *Solve* $\dfrac{\partial^2 u}{\partial r^2}+\dfrac{1}{r}\dfrac{\partial u}{\partial r}+\dfrac{1}{r^2}\dfrac{\partial^2 u}{\partial \theta^2}=0$

by the method of separation of variables.

Solution. $\dfrac{\partial^2 u}{\partial r^2}+\dfrac{1}{r}\dfrac{\partial u}{\partial r}+\dfrac{1}{r^2}\dfrac{\partial^2 u}{\partial \theta^2}=0$ or $r^2\dfrac{\partial^2 u}{\partial r^2}+r\dfrac{\partial u}{\partial r}+\dfrac{\partial^2 u}{\partial \theta^2}=0$ ... (1)

Let $u = R(r)\,.\,T(\theta)$

$$\frac{\partial u}{\partial r}=\frac{dR}{dr}.T(\theta) \text{ and } \frac{\partial^2 u}{\partial r^2}=\frac{d^2R}{dr^2}.T(\theta)$$

$$\frac{\partial u}{\partial \theta}=R(r).\frac{dT}{d\theta} \text{ and } \frac{\partial^2 u}{\partial \theta^2}=R(r).\frac{d^2T}{d\theta^2}$$

Putting the values of $\dfrac{\partial^2 u}{\partial r^2}, \dfrac{\partial u}{\partial r}$ and $\dfrac{\partial^2 u}{\partial \theta^2}$ in (1), we get

$$r^2.\frac{d^2R}{dr^2}.T(\theta)+r\frac{dR}{dr}.T(\theta)+R(r)\frac{d^2T}{d\theta^2}=0$$

$$\left(r^2\frac{d^2R}{dr^2}+r.\frac{dR}{dr}\right).T+R\frac{d^2T}{d\theta^2}=0$$

$$\frac{r^2\dfrac{d^2R}{dr^2}+r.\dfrac{dR}{dr}}{R}=-\frac{1}{T}\frac{d^2T}{d\theta^2}=h \text{ (say)} \qquad ...(2)$$

$$r^2\frac{d^2R}{dr^2}+r.\frac{dR}{dr}-hR=0 \qquad\qquad \frac{d^2T}{d\theta^2}+hT=0$$

Put $r = e^z$ $\qquad$ $(D^2 + h)\,T = 0$

$[D(D-1)+D-h]R=0$ $\qquad$ A.E. is $m^2 + h = 0$ ...(4)

A.E. is $m^2 - h = 0$ ...(3) $\qquad$ $\Rightarrow m=\pm i\sqrt{h}$

$\Rightarrow \quad m=\pm\sqrt{h}$ $\qquad$ $T=c_3\cos(\sqrt{h}\theta)+c_4\sin(\sqrt{h}\theta)$

$$R=c_1e^{\sqrt{h}z}+c_2e^{-\sqrt{h}z}$$

$$R=c_1r^{\sqrt{h}}+c_2r^{-\sqrt{h}}$$

$$u=(c_1r^{\sqrt{h}}+c_2r^{-\sqrt{h}})[c_3\cos(\sqrt{h}\theta)+c_4\sin(\sqrt{h}\theta)] \qquad ...(5)$$

Case 1. If $h = k^2$

(2) becomes $u=(c_1r^k+c_2r^{-k})\,[c_3\cos(k\theta)+c_4\sin(k\theta)]$ ...(6)

Case 2. If $h = 0$

On putting $h = 0$ in (3), we get

$m^2 = 0 \Rightarrow m = 0, 0$

$R = (c_5 + Zc_6) \Rightarrow R = [c_5 + (\log r)\, c_6]$

Again on putting $h = 0$ in (4), we get

$m^2 = 0 \Rightarrow m = 0, 0$

$T = c_7 + \theta c_8$

But $u = R.T. = (c_5 + c_6 \log r)\,(c_7 + \theta c_8)$

Case 3. If $h = -p^2$

On putting $h = -p^2$ in (3), we get

$$m^2 + p^2 = 0 \Rightarrow m = \pm ip$$

$$R = c_9 \cos p \log r + c_{10} \sin (p \log r)$$

Again on putting $h = -p^2$ in (4), we get

$$m^2 - p^2 = 0 \Rightarrow m = \pm p$$

$$T = c_{11}e^{p\theta} + c_{12}e^{-p\theta}$$

$$u = RT = c_9 \cos (p \log r) + c_{10} \sin (p \log r)\,(c_{11}e^{p\theta} + c_{12}e^{-p\theta})$$

Then there are three possible solutions

$$u = (c_1 r^k + c_2 r^{-k})[c_3 \cos(k\theta) + c_4 \sin(k\theta)] = (c_5 + c_6 \log r)(c_7 + c_8\theta)$$

$$= [c_9 \cos(p \log r) + c_{10} \sin(p \log r)][c_{11}e^{p\theta} + c_{12}e^{-p\theta}]$$

On putting $c_3 = a \cos \alpha$ and $c_4 = -a \sin \alpha$ in (6), we get

$$u = (c_1 r^k + c_2 r^{-k})\,(a \cos \alpha \cos k\theta - a \sin \alpha \sin k\theta)$$

$$= [(c_1 r^k + c_2 r^{-k})\,(a \cos (k\theta + \alpha)]$$

$$= (Ar^k + Br^{-k}) \cos (k\theta + \alpha)$$ **Ans.**

Example 28. *Solve* $\frac{\partial^2 V}{\partial r^2} + \frac{1}{r}\frac{\partial V}{\partial r} + \frac{1}{r^2}\frac{\partial^2 V}{\partial \theta^2} = 0$ *with boundary conditions,*

(*i*) V is finite when $r \to 0$

(*ii*) $V = \Sigma c_n \cos n\theta$ on $r = a$ (*U.P. II Sem 2010*)

Solution. Solution of the given equation is as equation (7) in above example, we have

$$V = \Sigma(A_n r^n + B_n r^{-n}) \cos (\theta + \alpha) \quad ...(1)$$

Case I. On putting $r = a$, $V = \Sigma c_n \cos n\theta$ in (1), we get

$$\Sigma c_n \cos n\theta = \Sigma(A_n a^n + B_n a^{-n}) \cos (n\theta + \alpha) \quad ...(2)$$

On equating the constant terms both the sides of (2), we get

$$\Rightarrow \quad c_n = A_n a^n + B_n a^{-n} \quad ...(3)$$

Case II. When $r = 0$ and V is finite

$$\Rightarrow \quad B_n = 0$$

On putting $B_n = 0$ in (2), we get

$$\Sigma c_n \cos n\theta = \Sigma A_n a^n \cos (\theta + \alpha) \quad ...(4)$$

Comparing the constant terms of (4) on both sides, we get

$$c_n = A_n a^n \Rightarrow A_n = \frac{c_n}{a^n} \quad (\alpha = 0)$$

On putting $A_n = \frac{c_n}{a^n}$, $B_n = 0$ and $\alpha = 0$ in (1), we get

$$V = \sum\left(\frac{c_n}{a^n} r^n + 0\right) \cos (n\theta + 0) \Rightarrow V = \sum c_n \left(\frac{r}{a}\right)^n \cos n\theta$$ **Ans.**

Example 29. *The diameter of a semicircular plate of radius a is kept at 0°C and the temperature at the semicircular boundary is T°C. Find the steady state temperature in the plate.*

Solution. Let the centre O of the semicircular plate be the pole and the bounding diameter be as the initial line. Let $u(r, \theta)$ be the steady state temperature at any point $P(r, \theta)$ and u satisfies the equation

$$r^2\frac{\partial^2 u}{\partial r^2}+r\frac{\partial u}{\partial r}+\frac{\partial^2 u}{\partial \theta^2}=0 \qquad ...(1)$$

The boundary conditions are

(*i*) $u(r, 0) = 0 \qquad 0 \le r \le a$

(*ii*) $u(r, \pi) = 0 \qquad 0 \le r \le a$

(*iii*) $u(a, \theta) = T$.

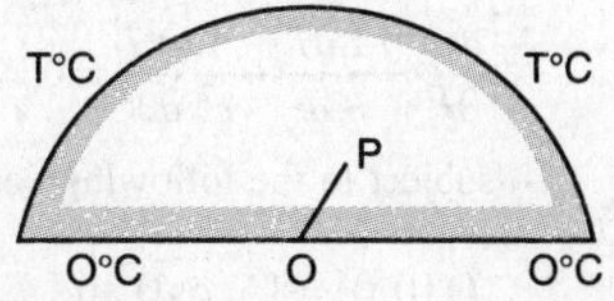

From conditions (*ii*) and (*iii*), we have $u \to 0$ as $r \to 0$. Hence the appropriate solution of (1) is as solved in example 27.

$$u=(c_1 r^p+c_2 r^{-p})(c_3\cos p\theta+c_4\sin p\theta) \qquad ...(2)$$

Putting $u(r, 0) = 0$ in (2), we get

$$0=(c_1 r^p+c_2 r^{-p})\, c_3 \;\Rightarrow\; c_3=0$$

On putting $c_3 = 0$ in (2), we get

$$u=(c_1 r^p+c_2 r^{-p})\, c_4 \sin p\theta \qquad ...(3)$$

Putting $u(r, \pi) = 0$ in (3), we get

$$0=(c_1 r^p+c_2 r^{-p})c_4\sin p\pi \Rightarrow \sin p\pi=0=\sin n\pi \;\Rightarrow\; p\pi=n\pi \Rightarrow p=n$$

On putting $p = n$, (3) becomes

$$u=(c_1 r^n+c_2 r^{-n})c_4\sin n\theta \qquad ... (4)$$

Since, $\qquad u = 0$ when $r = 0$

$$0 = c_2$$

(4) becomes, $u=c_1c_4r^n\sin n\theta$

The most general solution of (1) is

$$u(r,\theta)=\sum_{n=1}^{\infty} b_n r^n \sin n\theta \qquad (c_1\, c_2 = b_n) \qquad ... (5)$$

Putting $r = a$ and $u = T$ in (5), we have

$$T=\sum_{n=1}^{\infty} b_n\, a^n \sin n\theta$$

By Fourier half range series, we get

$$b_n a^n=\frac{2}{\pi}\int_0^{\pi} T\sin n\theta\, d\theta=\frac{2}{\pi}T\left(\frac{-\cos n\theta}{n}\right)_0^{\pi}=\frac{2T}{n\pi}[-(-1)^n+1]$$

$b_n a^n = 0$, when n is even,

$b_n a^n = \dfrac{4T}{n\pi}$, when n is odd.

$$\Rightarrow \qquad b_n=\frac{4T}{n\pi a^n}$$

Hence, (5) becomes

$$u(r,\theta)=\frac{4T}{\pi}\left[\frac{r/a}{1}\sin\theta+\frac{(r/a)^3}{3}\sin 3\theta+\frac{(r/a)^5}{5}\sin 5\theta+...\right] \qquad \textbf{Ans.}$$

EXERCISE 44.5

1. Solve the steady-state temperature equation

$$\frac{\partial^2 T}{\partial r^2}+\frac{1}{r}\frac{\partial T}{\partial r}+\frac{1}{r^2}\frac{\partial^2 T}{\partial \theta^2}=0;\quad 10\le r\le 20,\quad 0\le\theta\le 2\pi$$

subject to the following conditions:

$T(10,\theta)=15\cos\theta$ and $T(20,\theta)=30\sin\theta$ **Ans.** $T(r,\theta)=\frac{4T}{\pi}\left[\frac{r}{a}\sin\theta+\frac{1}{3}\left(\frac{r}{a}\right)^3\sin 3\theta+....\right]$

2. A semi-circular plate of radius a has its circumference kept at temperature $u(a,\theta)=k\theta(\pi-\theta)$ while the boundary diameter is kept at zero temperature. Find the steady state temperature distribution $u(r, \theta)$ of the plate assuming the lateral surfaces of the plate to be insulated.

Ans. $u(r,\theta)=\frac{8k}{\pi}\sum_1^{\infty}\left(\frac{r}{a}\right)^{2n-1}\frac{\sin(2n-1)\theta}{(2n-1)^3}$

3. Find the steady state temperature in a circular plate of radius a which has one half of its circumference at 0°C and the other half at 60°C.

Ans. $u(r,\theta)=50-\frac{200}{n}\sum_{n=1}^{\infty}\frac{1}{2n-1}\left(\frac{r}{a}\right)^{2n-1}\sin(2n-1)\theta.$

44.8 TRANSMISSION LINE EQUATIONS

$$\frac{\partial^2 V}{\partial x^2}=RC\frac{\partial V}{\partial t}$$

$$\frac{\partial^2 i}{\partial x^2}=RC\frac{\partial i}{\partial t}\quad\text{are called telegraph equations,}$$

where V = potential, i = current, C = capacitance, L = inductance

$$\frac{\partial^2 V}{\partial x^2}=LC\frac{\partial^2 V}{\partial t^2}$$

$$\frac{\partial^2 i}{\partial x^2}=LC\frac{\partial^2 i}{\partial t^2}$$

are called radio equations.

Example 30. *Find the current i and voltage v in a transmission line of length l, t seconds after the ends are suddenly grounded, given that i (x, 0) =* i_0, *v (x, 0) =* $v_0\sin\left(\frac{\pi x}{l}\right)$ *and that R and G are negligible.*

Solution. $$\frac{\partial^2 v}{\partial x^2}=LC\frac{\partial^2 v}{\partial t^2}$$

Let $v = XT$ where X and T are the functions of x and t respectively.

$$\frac{\partial^2 v}{\partial x^2}=T\frac{d^2 X}{dx^2}\quad\text{and}\quad\frac{\partial^2 v}{\partial t^2}=X\frac{d^2 T}{dt^2}$$

$$T\frac{d^2 X}{dx^2}=LCX\frac{d^2 T}{dt^2}$$

$$\frac{\frac{d^2 X}{dx^2}}{X}=LC\frac{\frac{d^2 T}{dt^2}}{T}=-p^2\text{ say}$$

Since the initial conditions suggest the values of v and i are periodic functions,

$\therefore \qquad X = c_1 \cos px + c_2 \sin px$

$$T = c_3 \cos \frac{pt}{\sqrt{LC}} + c_4 \sin \frac{pt}{\sqrt{LC}}$$

$$\Rightarrow \qquad v = (c_1 \cos px + c_2 \sin px)\left(c_3 \cos \frac{pt}{\sqrt{LC}} + c_4 \sin \frac{pt}{\sqrt{LC}}\right) \qquad ...(1)$$

When $\qquad t = 0,\ v = v_0 \sin \frac{\pi x}{l}$

$$v_0 \sin \frac{\pi x}{l} = (c_1 \cos px + c_2 \sin px) c_3 \qquad ...(2)$$

On equating the coefficients, we get

$$c_1 c_3 = 0 \Rightarrow c_1 = 0 \quad \text{and} \quad c_2 c_3 = v_0,\ p = \frac{\pi}{l}$$

(1) becomes

$$v = \sin \frac{\pi x}{l}\left[v_0 \cos \frac{\pi t}{l\sqrt{LC}} + c_2 c_4 \sin \frac{\pi t}{l\sqrt{LC}}\right] \qquad ...(3)$$

Now when $\qquad t = 0,\ i = i_0$ (constant)

Hence, $\qquad \frac{\partial i}{\partial x} = 0$

$$\frac{\partial i}{\partial x} = -c \frac{\partial v}{\partial t} \qquad \therefore \qquad \frac{\partial v}{\partial t} = 0 \text{ when } t = 0$$

Now $\qquad \frac{\partial v}{\partial t} = \sin \frac{\pi x}{l}\left(\frac{\pi}{l\sqrt{LC}}\right)\left[-v_0 \sin \frac{\pi t}{l\sqrt{LC}} + c_2 c_4 \cos \frac{\pi t}{l\sqrt{LC}}\right] \qquad ...(4)$

On putting $\frac{\partial v}{\partial t} = 0$ and $t = 0$ in (4), we get $\quad c_2\, c_4 = 0 \Rightarrow c_4 = 0$

(3) is reduced to $\qquad v = v_0 \sin \frac{\pi x}{l} \cos \frac{\pi t}{l\sqrt{LC}}$

$$\frac{\partial v}{\partial x} = \frac{\pi}{l} v_0 \cos \frac{\pi x}{l} \cos \frac{\pi t}{l\sqrt{LC}} = -L \frac{\partial i}{\partial t} \qquad ...(5)$$

$$\frac{\partial v}{\partial t} = -\frac{v_0 \pi}{l\sqrt{LC}} \sin \frac{\pi x}{l} \sin \frac{\pi t}{l\sqrt{LC}} = \frac{-1}{C} \frac{\partial t}{\partial x} \qquad ...(6)$$

Integrating (5) and (6), we get

$$i = -v_0 \sqrt{\frac{C}{L}} \cos \frac{\pi x}{l} \sin \frac{\pi t}{l\sqrt{LC}} + f(x)$$

$$i = -v_0 \sqrt{\frac{C}{L}} \cos \frac{\pi x}{l} \sin \frac{\pi t}{l\sqrt{LC}} + F(t)$$

$\therefore \quad f(x)$ and $F(t)$ must be constant only, since $i = i_0$ when $t = 0$

$\therefore \qquad$ constant $= i_0 = f(x)$

Hence, $\qquad i = i_0 - v_0 \sqrt{\frac{C}{L}} \cos \frac{\pi x}{l} \sin \frac{\pi t}{l\sqrt{LC}}$. **Ans.**

EXERCISE 44.6

1. A transmission line 1,000 miles long is initially under steady state condition with potential 1,300 volts at the sending end ($x = 0$) and 1,200 volts at the receiving end ($x = 1000$). The terminal end of the line is suddenly grounded but the potential at the source is kept at 1,300 volts. Assuming the inductance and leakage to be negligible, find the potential $v(x, t)$, if it satisfies the equation $v_t = (1/RC)v_{xx}$.

Ans. $v(x,y) = \sum_1^{\infty} b_n \sin nx\, e^{-ny}$ and $k = \sum_1^{\infty} b_n \sin nx$

2. Obtain a solution of the telegraph equation

$$\frac{\partial^2 e}{\partial x^2} = RC\frac{\partial e}{\partial t}$$

suitable for the case when e decays with the time and when there is a steady fall of potential from e_0 to 0 along the line of length l initially and the sending end is suddenly earthed.

Ans. $e(x,t) = \frac{2e_0}{\pi}\sum_{n=1}^{\infty} \frac{1}{n}\sin\frac{n\pi x}{l} e^{\frac{-n^2\pi^2 l}{CRl^2}}$

44.9 LAPLACE EQUATION

Laplace equation is used to solve engineering problems and its theory is called potential theory. Its solutions are known as harmonic functions.

Example 31. *Solve the Laplace equation* $\frac{\partial^2 u}{\partial x^2} + \frac{\partial^2 u}{\partial y^2} = 0$ *in a rectangle in the xy-plane with* $u(x, 0) = 0$, $u(x, b) = 0$, $u(0, y)$ *and* $u(a, y) = f(y)$ *parallel to y-axis.* [U.P.T.U. 2010, 2008]

Solution. Here, the Laplace equation is

$$\frac{\partial^2 u}{\partial x^2} + \frac{\partial^2 u}{\partial y^2} = 0 \quad ...(1)$$

Let $\qquad u = XY$

Where X is a function of x only Y is a function of y only.

$$\frac{\partial u}{\partial x} = \frac{\partial (XY)}{\partial x} = Y\frac{\partial X}{\partial x}$$

$$\frac{\partial^2 u}{\partial x^2} = \frac{\partial}{\partial x}\left(\frac{\partial u}{\partial x}\right) = \frac{\partial}{\partial x}\left(Y\frac{\partial X}{\partial x}\right) = Y\frac{\partial^2 X}{\partial x^2}$$

$$\frac{\partial u}{\partial y} = \frac{\partial (XY)}{\partial y} = X\frac{\partial Y}{\partial y}$$

$$\frac{\partial^2 u}{\partial y^2} = \frac{\partial}{\partial y}\left(\frac{\partial}{\partial y}(XY)\right) = \frac{\partial}{\partial y}\left(X\frac{\partial Y}{\partial y}\right) = X\frac{\partial^2 Y}{\partial^2 y}$$

Putting the values of $\frac{\partial^2 u}{\partial x^2}$ and $\frac{\partial^2 u}{\partial y^2}$ in (1), we get

$$Y\frac{\partial^2 u}{\partial x^2} + X\frac{\partial^2 u}{\partial y^2} = 0 \text{ or } YX'' + XY'' = 0$$

On dividing by XY and separating the variables, we get

$$\frac{Y''}{Y} = -\frac{X''}{X}$$

Case I. If $\dfrac{Y''}{Y} = -\dfrac{X''}{X} = p^2$(say)

$$\frac{Y''}{Y} = p^2$$
$$Y'' - p^2Y = 0$$
A. E. is $m^2 - p^2 = 0 \Rightarrow m = \pm p$

$Y = c_1e^{py} + c_2e^{-py}$ ...(2)

$$-\frac{X''}{X} = p^2$$
$\Rightarrow$ $X'' + p^2X = 0$

Auxiliary equation is
$$m^2 + p^2 = 0 \Rightarrow m = \pm ip$$
$X = c_3 \cos px + c_4 \sin px$ ...(3)

Putting $y = 0$ and $Y = 0$ in (2), we get
$$c_1 + c_2 = 0 \Rightarrow c_2 = -c_1$$
On putting $y = b$ and Y = 0, in (2) we get
$$c_1e^{pb} + c_2e^{-pb} = 0$$
$$c_1(e^{pb} - e^{-pb}) = 0$$
$$c_1 = 0 \qquad [c_1 = c_2 = 0]$$
$\Rightarrow$ $Y = 0$

But $u = XY$ $[Y = 0]$

$= X(0) = 0$, which is not possible

Case II. If $\dfrac{Y''}{Y} = -\dfrac{X''}{X} = 0$ (say)

$$\frac{Y''}{Y} = 0$$
$Y'' = 0 \Rightarrow Y = c_5 + c_6y$ ...(4)

Putting $y = 0$, $Y = 0$ in (4), we get
$$0 = c_5$$
Putting $y = b$, $Y = 0$ in (4), we get
$$0 = c_6b \Rightarrow c_6 = 0$$
$\Rightarrow$ $Y = 0$
$$u = XY$$

$$-\frac{X''}{X} = 0$$
$X'' = 0 \Rightarrow X = c_7 + c_8x$

$[Y = 0]$

$= X(0) = 0$, which is not possible.

Case III. If $\dfrac{Y''}{Y} = -\dfrac{X''}{X} = -p^2$

(*i*) $\dfrac{Y''}{Y} = -p^2$

$\Rightarrow$ $Y'' + p^2Y = 0$

A.E. is $m^2 + p^2 = 0 \Rightarrow m = \pm pi$

$Y = c_9 \cos py + c_{10} \sin py$ ...(5)

On putting $y = 0$, $Y = 0$ in (5), we get

$\Rightarrow$ $c_9 = 0$

$$-\frac{X''}{X} = -p^2$$
$\Rightarrow$ $X'' - p^2X = 0$

A.E. is $m^2 - p^2 = 0 \Rightarrow m = \pm p$
$$X = c_{11}\, e^{px} + c_{12}e^{-px}$$

Putting $y = b$ and $Y = 0$ in (5), we get

$$0 = c_9 \cos pb + c_{10} \sin pb.$$

$$c_{10} \sin pb = 0 \qquad (\because c_9 = 0)$$

$$\Rightarrow \qquad \sin pb = 0$$

$$\text{and } \sin n\pi = 0$$

$$\Rightarrow \qquad pb = n\pi \Rightarrow p = \frac{n\pi}{b}$$

Putting $c_9 = 0,\ p = \frac{n\pi}{b}$ in (5), we get

$$Y = c_{10} \sin \frac{n\pi y}{b}$$

$$u = XY$$

$$u = c_{10} \sin \frac{n\pi y}{b}\left(c_{11} e^{\frac{n\pi x}{b}} + c_{12} e^{-\frac{n\pi x}{b}}\right) \qquad \text{... (6)}$$

Putting $x = 0$ and $u = 0$ in (6), we get

$$0 = c_{10} \sin \frac{n\pi y}{b}(c_{11} + c_{12})$$

$$c_{11} + c_{12} = 0 \Rightarrow c_{12} = -c_{11}$$

Putting the value of $c_{12} = -c_{11}$ in (6), we get

$$u = c_{10} c_{11} \sin \frac{n\pi y}{b}\left(e^{\frac{n\pi x}{b}} - e^{-\frac{n\pi x}{b}}\right)$$

$$u = 2c_{10} c_{11} \sin \frac{n\pi y}{b} \sinh \frac{n\pi x}{b}$$

$$u = b_n \sin \frac{n\pi y}{b} \sinh \frac{n\pi x}{b} \qquad (b_n = 2\, c_{10}\, c_{11})$$

$$u = \sum_{n=1}^{\infty} b_n \sin \frac{n\pi y}{b} \sinh \frac{n\pi x}{b} \qquad \text{... (7) (General solution)}$$

Putting $x = a$ and $u = f(y)$ in (7), we get

$$u = f(y) = \sum_{n=1}^{\infty} b_n \sin \frac{n\pi y}{b} \sinh \frac{n\pi x}{b}$$ **Ans.**

We also know that

$$\sinh \frac{n\pi a}{b} = \frac{2}{b}\int_0^b f(y) \sin \frac{n\pi y}{b}\, dy$$

$$b_n = \frac{2}{b \sinh\left(\frac{n\pi a}{b}\right)}\int_0^b f(y) \sin \frac{n\pi y}{b}\, dy.$$

Example 32. *Find the deflection u(x, y, t) of the square membrane with a = b = c = 1, if the initial velocity is zero and the initial deflection f(x, y) = A sin πx 2πy.*

(Q.Bank U.P.T.U. 2001)

Solution. Here, the equation of the vibrations of the square membrane

$$\frac{\partial^2 u}{\partial t^2} = c^2\left(\frac{\partial^2 u}{\partial x^2} + \frac{\partial^2 u}{\partial y^2}\right) \qquad \text{...(1)}$$

Let $\qquad u = XYT \qquad$ ... (2)

Where X is function of x only, Y is a function of y only, and T is a function of t only.

$$\frac{\partial^2 u}{\partial t^2} = \frac{\partial^2}{\partial t^2}(XYT) = XY\frac{d^2T}{dt^2}$$

$$\frac{\partial^2 u}{\partial x^2} = \frac{\partial^2}{\partial x^2}(XYT) = YT\frac{d^2X}{dx^2}$$

$$\frac{\partial^2 u}{\partial y^2} = \frac{\partial^2}{\partial y^2}(XYT) = XT\frac{d^2Y}{dy^2}$$

On putting the values of $\frac{\partial^2 u}{\partial t^2}, \frac{\partial^2 u}{\partial x^2}$ and $\frac{\partial^2 u}{\partial y^2}$ in (1), we get

$$XY\frac{d^2T}{dy^2} = c^2\left(YT\frac{d^2X}{dx^2} + XT\frac{d^2Y}{dy^2}\right)$$

On dividing by $c^2\, XYT$, we get

$$\frac{\frac{d^2T}{dt^2}}{c^2T} = \frac{\frac{d^2X}{dx^2}}{X} + \frac{\frac{d^2Y}{dy^2}}{Y}$$

This will be true only when each member is a constant. Choosing suitably, we have

$$\frac{\frac{d^2X}{dx^2}}{X} = -k^2, \quad \frac{\frac{d^2Y}{dy^2}}{Y} = -l^2 \quad \Rightarrow \quad \frac{\frac{d^2T}{dt^2}}{c^2T} = -(k^2+l^2)$$

$\frac{d^2X}{dx^2} + k^2X = 0$	$\frac{d^2Y}{dy^2} + l^2Y = 0$	$\frac{d^2T}{dt^2} + (k^2+l^2)c^2T = 0$
$A.E.\ is\ m^2 + k^2 = 0$	$A.E.\ is\ m^2 + l^2 = 0$	$A.E.\ is\ m^2 + (k^2+l^2)c^2 = 0$
$\Rightarrow m = ik, m = -ik$	$\Rightarrow m = il, m = -il$	$m = i\sqrt{k^2+l^2}\ c, m = -i\sqrt{k^2+l^2}\ c$
$X = c_1\cos kx + c_2\sin kx$	$Y = c_3\cos ly + c_4\sin ly$	$T = c_5\cos\sqrt{k^2+l^2}\,ct + c_6\sin\sqrt{k^2+l^2}\,ct.$

Putting the values of X, Y and T in (2), we get

$$u = (c_1\cos kx + c_2\sin kx)(c_3\cos ly + c_4\sin ly)[c_5\cos\sqrt{k^2+l^2}\ ct + c_6\sin\sqrt{k^2+l^2}\ ct] \quad ...(3)$$

On putting $x = 0$ and $u = 0$ in (3), we get (given)

$$0 = c_1(c_3\cos ly + c_4\sin ly)[c_5\cos\sqrt{k^2+l^2}\ ct + c_6\sin\sqrt{k^2+l^2}\ ct] \quad ...(4)$$

$$\Rightarrow \quad c_1 = 0$$

On putting $c_1 = 0$ in (3), we get (given)

$$u = c_2\sin kx(c_3\cos ly + c_4\sin ly)(c_5\cos\sqrt{k^2+l^2}\ ct + c_6\sin\sqrt{k^2+l^2}\ ct) \quad ...(5)$$

Putting $x = 1$ and $u = 0$ in (5), we get (given)

$$u = c_2\sin m\pi x(c_3\cos ly + c_4\sin ly)(c_5\cos\sqrt{m^2\pi^2+l^2}\ ct + c_6\sin\sqrt{m^2\pi^2+l^2}\ ct) \quad ...(6)$$

Putting $y = 0$, $u = 0$ in (7), we get

$$0 = c_2\sin m\pi x.\, c_3\,(c_5\cos\sqrt{m^2\pi^2+l^2}\ ct + c_6\sin\sqrt{m^2\pi^2+l^2}\ ct) \quad ...(7)$$

$$\Rightarrow \quad c_3 = 0$$

Putting the value of c_3 in (7), we get

$$u = c_2 c_4 \sin m\pi x \sin ly \left(c_5 \cos \sqrt{m^2\pi^2 + l^2}\ ct + c_6 \sin \sqrt{m^2\pi^2 + l^2} + ct\right) \qquad ...(8)$$

Putting $y = 1$ and $u = 0$ in (9), we get

$$0 = c_2 c_4 \sin m\pi x \sin l \left(c_5 \cos \sqrt{m^2\pi^2 + l^2}\ ct + c_6 \sin \sqrt{m^2\pi^2 + l^2}\ ct\right) \qquad ...(9)$$

$$\Rightarrow \quad \sin l = 0 = \sin n\pi \quad \Rightarrow \quad l = n\pi$$

Putting the value $l = n\pi$ in (9), we get

$$u = c_2 c_4 \sin m\pi x \sin n\pi y \left(c_5 \cos \sqrt{m^2\pi^2 + n^2\pi^2}\ ct + c_6 \sin \sqrt{m^2\pi^2 + m^2\pi^2 + n^2\pi^2}\ ct\right) \ ...(10)$$

Putting $c_2c_4c_5 = A_{mn}$, $c_2c_4c_6 = B_{mn}$ and $p = \sqrt{m^2 + n^2}$ in (10), we get the general solution as

$$u(x, y, t) = \sum_{m=1}^{\infty}\sum_{n=1}^{\infty} \sin m\pi x \sin n\pi y \,(A_{mn} \cos pt + B_{mn} \sin pt) \qquad \textbf{Ans.}$$

EXERCISE 44.7

Indicate True or False for the following :

1. The small transverse vibrations of a string are governed by one dimensional heat equation $y_t = a^2 y_{xx}$. (True/False) *(U.P. II Semester, 2009)* **Ans.** False

2. Two dimensional steady state heat flow is given by Laplace's equation $u_t = a^2(u_{xx} + u_{tt})$. (True/False) *(U.P. II Semester, 2009)* **Ans.** False

3. $\dfrac{\partial^2 y}{\partial t^2} = c^2 \dfrac{\partial^2 y}{\partial x^2}$ is a two-dimensional wave equation. **Ans.** True

4. Radio equations are $V_{xx} = LCV_n$ and $I_{xx} = LCI_{tt}$ **Ans.** True

5. The small transverse vibrations of a string are $y_t^2 = a^2 y_{xx}$. **Ans.** False

44.10 VIBRATING MEMBRANE

Now we will discuss the equation of thevibrations of a tightly stretched membrane. Such as membrane of drum. Let T be the tension per unit length and m be the mass per unit area.

Now we want to discuss the forces acting on an element $\delta x \delta y$ of the membrance. Forces $T\delta x$ and $T\delta y$ act on the edges along the tangent to the membrane. Let the small displacement perpendicular to xy-plane be μ.

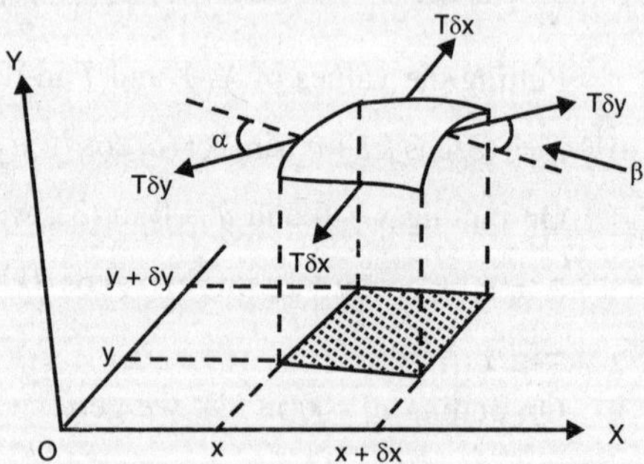

$T\delta y$ on the opposite edges of length δy make angles α and β to the horizontal.

Thus Vertical component

$$= T\delta y \sin \beta - T\delta y \sin \alpha$$

$$= T\delta y (\tan \beta - \tan \alpha).$$

$$[\tan \alpha = \sin \alpha, \text{ if } \alpha \text{ is very small}]$$

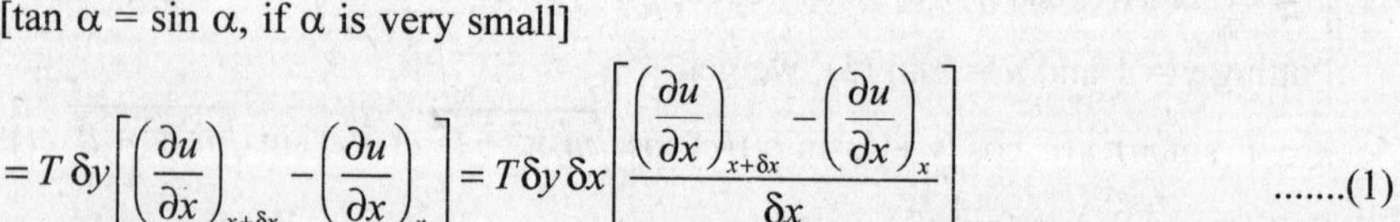

$$= T\,\delta y\left[\left(\frac{\partial u}{\partial x}\right)_{x+\delta x} - \left(\frac{\partial u}{\partial x}\right)_x\right] = T\delta y\,\delta x\left[\frac{\left(\frac{\partial u}{\partial x}\right)_{x+\delta x} - \left(\frac{\partial u}{\partial x}\right)_x}{\delta x}\right] \qquad ....(1)$$

$$= T\,\delta x\,\delta y \frac{\partial^2 u}{\partial x^2} \text{ upto a first order of approximation.}$$

Another vertical component of the force $T\delta x$ acting on the edges of the length δx.

$$= T\,\delta x \frac{\partial^2 u}{\partial y^2}\delta y \qquad ...(2)$$

Equation of the motion of the element is

$$\frac{d^2X}{dx^2} + k^2X = 0, \quad \frac{d^2Y}{dy^2} + l^2Y = 0$$

$$m\,\delta x\,\delta y \frac{\partial^2 u}{\partial t^2} = T\left[\frac{\partial^2 u}{\partial x^2} + \frac{\partial^2 u}{\partial y}\right]\delta x\,\delta y$$

$$\frac{\partial^2 u}{\partial t^2} = C^2\left[\frac{\partial^2 u}{\partial x^2} + \frac{\partial^2 u}{\partial y^2}\right] \text{ where } C^2 = \frac{T}{m} \qquad ...(3)$$

Where m is the mass per unit area of the membrane.

44.11 SOLUTION OF THE EQUATION OF THE VIBRATING MEMBRANE (RECTANGULAR MEMBRANE)

Let u be the solution of equation (3) and we assume that

$$u = X(x)\,.\,Y(y)\,.\,T(t) \qquad ...(4)$$

On differentiating (4), we have

$$\frac{\partial^2 u}{\partial x^2} = Y \cdot T \frac{d^2X}{dx^2}$$

$$\frac{\partial^2 u}{\partial y^2} = XT \frac{d^2Y}{dy^2}$$

$$\frac{\partial^2 u}{\partial t^2} = XY \frac{d^2T}{dt^2}$$

Substituting these values in (3), we get

$$XY\frac{d^2T}{dt^2} = C^2\left[\frac{d^2X}{dx^2}YT + XT\frac{d^2Y}{dy^2}\right]$$

Dividing by $XYTc^2$, we get $\frac{1}{C^2}\frac{d^2T}{dt^2} = \left[\frac{1}{X}\frac{d^2X}{dx^2} + \frac{1}{Y}\frac{d^2Y}{dy^2}\right]$ ...(5)

Each member is constant $\frac{d^2X}{dx^2} + k^2X = 0, \quad \frac{d^2Y}{dy^2} + l^2Y = 0$

Putting these values in (5), we get $\frac{d^2T}{dt^2} + (k^2 + l^2)c^2T = 0$

$\frac{d^2X}{dx^2} + k^2X = 0$	$\frac{d^2Y}{dy^2} + l^2Y = 0$	$\frac{d^2T}{dt^2} + (k^2 + l^2)c^2T = 0$
$(D^2 + k^2)\,X = 0$	$(D^2 + l^2)\,Y = 0$	$[D^2 + (k^2 + l^2)\,c^2]T = 0$
$m^2 + k^2 = 0$	$m^2 + l^2 = 0$	$m^2 + [k^2 + l^2]c^2 = 0$
$m = \pm\, ik$	$m = \pm il$	$m = iC\sqrt{k^2 + l^2}$
$X = c_1 \cos kx + c_2 \sin kx$	$Y = c_3 \cos ly + c_4 \sin ly$	$T = c_5 \cos c\sqrt{k^2 + l^2}$ $+ c_6 \sin c\sqrt{k^2 + l^2}$

Putting the values of X,Y,T in (4), we get

$$u(x, y, t) = (c_1 \cos kx + c_2 \sin kx)(c_3 \cos ly + c_4 \sin ly)$$

$$\left[c_5 \cos\sqrt{(k^2+l^2)}ct + c_6 \sin\sqrt{(k^2+l^2)}ct\right] \qquad \text{...(6)}$$

Suppose if the membrane is rectangular and stretched between the lines

$$x = 0, x = a, y = 0, y = b.$$

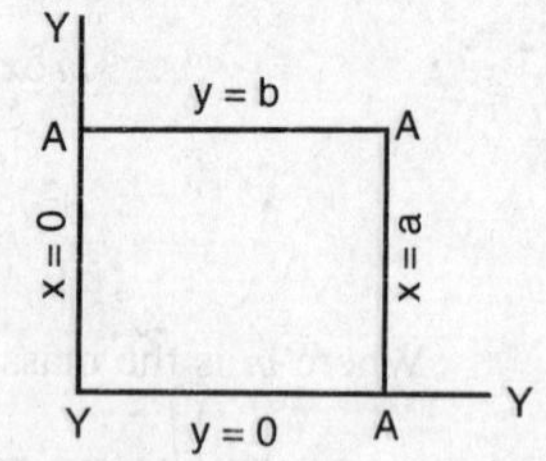

Then the boundary conditions are

(*i*) $u = 0$, when $x = 0$

(*ii*) $u = 0$, when $x = a$

(*iii*) $u = 0$, when $y = 0$

(*iv*) $u = 0$, when $y = b$ for all t.

On putting $x = 0$, $u = 0$ in (6), we get

$$0 = c_1(c_3 \cos ly + c_4 \sin ly)\left[c_5 \cos\sqrt{(k^2+l^2)}\,ct + c_6 \sin\sqrt{(k^2+l^2)}\,ct\right]$$

$$c_1 = 0$$

On putting $c_1 = 0$, (6) becomes

$$u = c_2 \sin kx(c_3 \cos ly + c_4 \sin ly)\left[c_5 \cos x\sqrt{(k^2+l^2)}ct + c_6 \sin\sqrt{(k^2+l^2)}ct\right] \qquad \text{...(7)}$$

On putting $u = 0$ and $x = a$ in (7), we get

$$0 = c_2 \sin ka(c_3 \cos ly + c_4 \sin ly)\left[c_5 \cos\sqrt{(k^2+l^2)}\,ct + c_6 \sin\sqrt{(k^2+l^2)}\,ct\right]$$

$$\Rightarrow \quad \sin ka = 0 \quad \Rightarrow \sin ka = \sin m\pi \quad \Rightarrow k = \frac{m\pi}{a}$$

On putting the value of k in (7), we get

$$u = c_2 \sin\frac{m\pi x}{a}[c_3 \cos ly + c_4 \sin ly]\left[c_5 \cos x\sqrt{\frac{m^2\pi^2}{a^2}+l^2}\cdot ct + c_6 \sin\sqrt{\frac{m^2\pi^2}{a^2}+l^2}\cdot ct\right] \qquad \text{...(8)}$$

Similarly, applying the conditions (*iii*) and (*iv*) in (8), we get

$$c_3 = 0 \text{ or } l = \frac{n\pi}{b}, \text{ where } n \text{ is an integer.}$$

(8) becomes,

$$u = c_2c_4 \sin\frac{m\pi x}{a}\cdot\sin\frac{n\pi y}{b}\left(c_5 \cos x\sqrt{\frac{m^2\pi^2}{a^2}+\frac{n^2\pi^2}{b^2}}\,ct + c_6 \sin\sqrt{\frac{m^2\pi^2}{a^2}+\frac{n^2\pi^2}{b^2}}\,ct\right) \qquad \text{...(9)}$$

Let $\qquad p = c\pi\sqrt{\dfrac{m^2}{a^2}+\dfrac{n^2}{b^2}}$

Now, $u = c_2c_4 \sin\dfrac{m\pi x}{a}\sin\dfrac{n\pi y}{b}(c_5 \cos pt + c_6 \sin pt)$ [From (9)]

These are the solution of wave equation (1) and are called eigen functions. Choosing the constants c_2 and c_4 so that $c_2c_4 = 1$, we can write the general solution of (1) as

$$u = \sum_{m=1}^{\infty}\sum_{n=1}^{\infty} \sin\frac{m\pi x}{a}\sin\frac{n\pi y}{b}(\mathrm{A}_{mn} \cos pt + \mathrm{B}_{mn} \sin pt) \qquad \text{...(10)}$$

Differentiating (10) with respect to 't', we get

$$\frac{\partial u}{\partial t}=\sum_{m=1}^{\infty}\sum_{n=1}^{\infty}\sin\frac{m\pi x}{a}\sin\frac{n\pi y}{b}\left(-pA_{mn}\sin pt+pB_{mn}\cos pt\right) \quad ...(11)$$

On putting $\frac{\partial u}{\partial t}=0$ when $t = 0$ in (11), we get

$$0=\sum_{m=1}^{\infty}\sum_{n=1}^{\infty}\sin\frac{m\pi x}{a}\sin\frac{n\pi y}{b}\left(pB_{mn}\cos pt\right) \Rightarrow B_{mn}=0$$

Also using the condition : $u = f(x, y)$ when $t = 0$, we get

$$f(x,y)=\sum_{m=1}^{\infty}\sum_{n=1}^{\infty}A_{mn}\sin\frac{m\pi x}{a}\sin\frac{n\pi y}{b}$$

This is a double Fourier series. Multiplying both sides by $\sin\frac{m\pi x}{a}\sin\frac{n\pi y}{b}$ and integrating from $x = 0$ to $x = a$ and $y = 0$ to $y = b$, every term on the right except one, all become zero. Thus, we get

$$\int_0^a\int_0^b f(x,y)A_{mn}\sin\frac{m\pi x}{a}\sin\frac{n\pi y}{b}\,dydx=\frac{ab}{4}A_{mn}$$

It is the generalised Euler's formula and gives the coefficients in the solution.

On putting the value of A_{mn}, we get the solution of (1),

Example 33. *Find the deflection u (x, y, t) of the square membrane with a = b = c = 1, if the initial velocity is zero and the initial deflection f(x, y) = A sin πx sin 2πy.*

Solution : The equation of the vibration of the square membrane is

$$\frac{\partial^2 u}{\partial t^2}=c^2\left(\frac{\partial^2 u}{\partial x^2}+\frac{\partial^2 u}{\partial y^2}\right) \quad ...(1)$$

Boundary conditions are,

$u(0, y, t) = 0,\ u(1, y, t) = 0$

$u(x, 0, t) = 0,\ u(x, 1, t) = 0$

the initial conditions are

$$u(x,y,0)=f(x,y)=A\sin\pi x\sin 2\pi y \text{ and } \left(\frac{\partial u}{\partial t}\right)_{t=0}=0$$

Let, u be the solution of (1) where $u = XYT$ and X is function of x only, Y is a function of y only, and T is a function of t only.

On differentiating u partially with respect to t, x, y, we get

$$\frac{\partial^2 u}{\partial t^2}=\frac{\partial^2}{\partial t^2}(XYT)=XY\frac{d^2T}{dt^2}$$

$$\frac{\partial^2 u}{\partial x^2}=\frac{\partial^2}{\partial x^2}(XYT)=YT\frac{d^2X}{dx^2}$$

$$\frac{\partial^2 u}{\partial y^2}=\frac{\partial^2}{\partial y^2}(XYT)=XT\frac{d^2Y}{dy^2}$$

On putting the derivatives in (1), we get $XY\frac{d^2T}{dt^2}=c^2\left(YT\frac{d^2X}{dx^2}+XT\frac{d^2Y}{dy^2}\right)$

On dividing by $XYTc^2$, we get $\frac{1}{c^2}\frac{1}{T}\frac{d^2T}{dt^2}=\left(\frac{1}{X}\frac{d^2X}{dx^2}+\frac{1}{Y}\frac{d^2Y}{dy^2}\right)$ ...(3)

Equation (3) holds good when each member is a constant. *i.e.* l, and k are constants.

$$\frac{1}{X}\frac{d^2X}{dx^2}=l^2$$

$$\frac{d^2X}{dx^2}+l^2X=0,$$

$$\frac{d^2X}{dx^2}+l^2X=0$$

$$(D^2+l^2)X=0$$

$$\Rightarrow\quad m^2+l^2=0$$

$$m=\pm il$$

$$\frac{1}{Y}\frac{d^2Y}{dy^2}=-k^2$$

$$\frac{d^2Y}{dy^2}+k^2Y=0 \text{ and}$$

$$\frac{d^2Y}{dy^2}+k^2Y=0$$

$$(D^2+k^2)Y=0$$

$$m^2+k^2=0$$

$$\Rightarrow\quad m=\pm ik$$

$$\frac{1}{T}\frac{d^2T}{dt^2}=(l^2+k^2)$$

$$\frac{d^2T}{dt^2}+(l^2+k^2)c^2T=0$$

$$\frac{d^2T}{dt^2}+(l^2+k^2)c^2T=0$$

$$(D^2+(l^2+k^2)t^2=0$$

$$\Rightarrow\quad m=\pm ic\sqrt{l^2+k^2}$$

$$T=c_5\cos\sqrt{l^2+k^2}\,ct+c_6\sin\sqrt{l^2+k^2}\,ct$$

$X=c_1\cos kx+c_2\sin kx$, $Y=x_3\cos ly+c_4\sin ly$

Putting the values of X, Y and T in (2), we get

$$u=(c_1\cos kx+c_2\sin kx)(c_3\cos ly+c_4\sin ly)\left[c_5\cos\sqrt{l^2+k^2}\,ct+c_6\sin\sqrt{l^2+k^2}\,ct\right] \quad ...(4)$$

On putting $x=0$, $u=0$ in (4), we get

$$0=c_1\left(c_3\cos ly+c_4\sin ly\right)\left(c_5\cos\sqrt{l^2+k^2}\,ct+c_6\sin\sqrt{l^2+k^2}\,ct\right)c_1=0 \quad ...(5)$$

On putting $c_1=0$ in (5), we get

$$u=c_2\sin kx\left(c_3\cos ly+c_4\sin ly\right)\left(c_5\cos\sqrt{l^2+k^2}\,ct+c_6\sin\sqrt{l^2+k^2}\,ct\right) \quad ...(6)$$

On putting $x=1$ and $u=0$ in (6), we get

$$0=c_2\sin k\left(c_3\cos ly+c_4\sin ly\right)\left(c_5\cos\sqrt{l^2+k^2}\,ct+c_6\sin\sqrt{l^2+k^2}\,ct\right)$$

$\Rightarrow\quad \sin l=0 \Rightarrow\quad \sin l=\sin m\pi \quad\Rightarrow l=m\pi$

Putting the value of $l=m\pi$ in (6), we get

$$u=c_2\sin m\pi x\left(c_3\cos ly+c_4\sin ly\right)\left(c_5\cos\sqrt{k^2+l^2}\,ct+c_6\sin\sqrt{k^2+l^2}\,ct\right) \quad ...(7)$$

Putting $y=0$ and $u=0$ in (7), we get

$$0=c_2\sin m\pi x\cdot c_3\left(c_5\cos\sqrt{k^2+l^2}\,ct+c_6\sin\sqrt{k^2+l^2}\cdot ct\right)$$

$\Rightarrow\quad c_3=0$

On putting $c_3=0$ in (7), we get

$$u=c_2c_4\sin m\pi x\sin ly\left(c_5\cos\sqrt{k^2+l^2}\cdot ct+c_6\sin\sqrt{k^2+l^2}\cdot ct\right) \quad ...(8)$$

Now on putting $y=1$ and $u=0$ in (8), we get

$$u=c_2c_4\sin m\pi x\sin l\left(c_5\cos\sqrt{k^2+l^2}\cdot ct+c_6\sin\sqrt{k^2+l^2}\cdot ct\right)$$

$\Rightarrow\quad \sin l=0=\sin n\pi \quad\Rightarrow\quad l=n\pi$

Putting the value of $l=n\pi$ in (8), we get

$$u=c_2c_4\sin m\pi x\sin n\pi y\left(c_5\cos\sqrt{m^2\pi^2+n^2\pi^2}\cdot ct+c_6\sin\sqrt{m^2\pi^2+n^2\pi^2}\cdot ct\right) \quad ...(9)$$

If we put $p = \pi c\sqrt{m^2 + n^2}$ in (9), we get

$$u = \sin m\pi x \sin n\pi y \left(A_{mn} \cos pt + B_{mn} \sin pt\right)$$

The general equation is

$$u = \sum_{m=1}^{\infty}\sum_{n=1}^{\infty} \sin m\pi x \sin n\pi y \left(A_{mn} \cos pt + B_{mn} \sin pt\right) \qquad \text{...(10)}$$

Differentiating (10) w.r.t 't', we get

$$\frac{\partial u}{\partial t} = \sum_{m=1}^{\infty}\sum_{n=1}^{\infty} \sin m\pi x \sin n\pi y \left(-p\,A_{mn} \sin pt + p\,B_{mn} \cos pt\right) \qquad \text{...(11)}$$

On putting $\frac{\partial u}{\partial t} = 0$ and $t = 0$ in (11), we get

$$0 = \sum_{m=1}^{\infty}\sum_{n=1}^{\infty} \sin m\pi x \sin n\pi y \left[pB_{mn}\right] \qquad \text{...(12)}$$

$$\Rightarrow \qquad B_{mn} = 0$$

On putting $u = a \sin \pi x \sin 2\pi y$ and $t = 0$ in (12), we get

$$A \sin \pi x \sin 2\pi y = \sum_{n=1}^{\infty}\sum_{m=1}^{\infty} A_{mn} \sin m\pi x \sin n\pi y$$

$$A_{mn} = \frac{2}{1} \times \frac{2}{1} \int_0^1\int_0^1 A_{mn} \sin \pi x \sin 2\pi y \sin m\pi x \sin n\pi y\, dx\, dy$$

$$\Rightarrow \qquad Am_1 = Am_3 = Am_5 = \ldots\ldots\ldots\ldots = 0$$

But, $\quad Am_2 = 4A\int_0^1\int_0^1 \sin \pi x \sin m\pi x \sin^2 2\pi y\, dx\, dy$ (if n is odd)

$$= 2A\int_0^1\int_0^1 \sin \pi x \sin m\pi x \left(1 - \cos 4\pi y\right) dx\, dy$$

$$= 2A\int_0^1 \sin \pi x \sin m\pi x \left(y - \frac{\sin 4\pi y}{4\pi}\right)_0^1 dx = 2A\int_0^1 \sin \pi x \sin m\pi x\, dx$$

$$\Rightarrow \qquad A_{22} = A_{32} = A_{42} = \ldots\ldots\ldots\ldots = 0$$

$$A_{12} = 2A\int_0^1 \sin^2 n\pi\, dx = A\int_0^1 \left(1 - \cos 2\pi x\right) dx = A\left(x - \frac{\sin 2\pi x}{2\pi}\right)_0^1 = A$$

$$u = A \sin \pi x \sin 2\pi y \cos pt$$

where $\quad p = \pi c\sqrt{m^2 + n^2} = \pi(1)\sqrt{1+4} = \pi\sqrt{5}.$

$$u = A \cos \pi\sqrt{5}t \sin \pi x \sin 2\pi y.$$ **Ans.**

Example 34. *Find the deflection u(x, y, t) of a rectangular membrane, $0 \le x \le a$, $0 \le y \le b$ given that is entire its entire boundary is fixed, initial velocity is zero (starts from rest) and initial deflection $f(x, y) = kxy\,(a - x)\,(b - y)$.*

Solution. We know from the Article 44.3, equation (10) that

$$u = \sum_{m=1}^{\infty}\sum_{n=1}^{\infty} \sin \frac{m\pi x}{a} \sin \frac{n\pi y}{b} \left(A_{mn} \cos pt + B_{mn} \sin pt\right) \qquad \text{...(1)}$$

On differentiating (1), we get

$$\frac{\partial u}{\partial t} = \sum\sum \sin \frac{m\pi x}{a} \sin \frac{n\pi y}{b} \left(-PA_{mn} \sin pt + B_{mn} P \cos pt\right) \qquad \text{...(2)}$$

On putting initial velocity $\frac{\partial u}{\partial t} = 0$, $t = 0$ in (2), we have

$$0 = \sum\sum \sin\frac{m\pi x}{a}\sin\frac{n\pi y}{b}(\text{PB}_{mn}\sin pt)$$

$$\Rightarrow \qquad \text{B}_{mn} = 0$$

Putting $\text{B}_{mn} = 0$ in (1) we have

$$u = \text{A}_{mn}\sum\sum \sin\frac{m\pi x}{a}\sin\frac{n\pi y}{b}\cos pt \qquad ...(3)$$

The membrane starts from rest with initial condition is

$u = kxy\,(a - x)\,(b - y)$, when $t = 0$

$$kxy(a-x)(b-y) = \text{A}_{mn}\sum\sum \sin\frac{m\pi x}{a}\sin\frac{n\pi y}{b}$$

$$\text{A}_{mn} = \frac{4}{ab}\int_0^a\int_0^b f(x,y)\sin\frac{m\pi x}{a}\sin\frac{n\pi y}{b}\,dy\,dx$$

$$= \frac{4}{ab}\int_0^a\int_0^b kxy(a-x)(b-y)\sin\frac{m\pi x}{a}\sin\frac{n\pi y}{b}\,dy\,dx$$

$$\frac{a\,b\,A_{mn}}{4k} = \int_0^a\left(x(a-x)\sin\frac{m\pi x}{a}dx\right)\times\left[\int_0^a y(b-y)\sin\frac{n\pi y}{b}dy\right]$$

$$= I_1 \times I_2 \qquad ...(4)$$

$$\text{I}_1 = \int_0^a x(a-x)\sin\frac{m\pi x}{a}dx$$

$$= \int_0^a x(a-x)\frac{a}{mn}\left(-\cos\frac{m\pi n}{a}\right) - (a-2x)\frac{a^2}{m^2\pi^2}\left(-\sin\frac{m\pi x}{a}\right) + (-2)\frac{a^3}{m^3\pi^3}\cos\left(\frac{m\pi x}{a}\right)\Bigg]_0^a$$

$$= 0 - 0 - \frac{2a^3}{m^3\pi^3}(\cos m\pi - 1)$$

Similarly, $\qquad I_2 = -\frac{2b^3}{n^3\pi^3}(\cos n\pi - 1)$

Putting the values of I_1 and I_2, we get

$$\frac{ab}{4k}\text{A}_{mn} = \left[-\frac{2a^3}{m^3\pi^3}(\cos m\pi - 1)\right]\left[-\frac{2b^3}{n^3\pi^3}(\cos n\pi - 1)\right]$$

$$\text{A}_{mn} = \frac{4k}{ab}\left[-\frac{2a^3}{m^3\pi^3}\{(-1)^m - 1\}\right]\left[-\frac{2b^3}{n^3\pi^3}((-1)^n - 1)\right] = \frac{16a^3b^3k}{m^3n^3\pi^6}\left[(-1)^m - 1\right]\left[(-1)^n - 1\right]$$

$$= \left[\begin{array}{ll}\frac{64a^2b^2k}{\pi^3 n^3 m^3}, & \text{when } m \text{ and } n \text{ are odd.}\\ 0, & \text{otherwise}\end{array}\right.$$

The required deflection $= u\,(x, y, t)$

$$= \left[\frac{64a^2bk}{\pi}\sum_{m=1}^{\infty}\sum_{n=1}^{\infty}\frac{1}{m^3}\sin\frac{m\pi x}{a}\right)\times\frac{1}{n^3}\sin\frac{n\pi y}{b}, \text{ where } k^2\pi^2\left[\frac{m^2}{a^2} + \frac{n^2}{b^2}\right] \qquad \textbf{Ans.}$$

INTEGRAL TRANSFORMS

45.1 INTRODUCTION

Integral transforms are used in the solution of partial differential equations. The choice of a particular transform to be used for the solution of a differential equations depends upon the nature of the boundary conditions of the equation and the facility with which the transform $F(s)$ can be converted to give $f(x)$.

45.2 INTEGRAL TRANSFORMS

The integral transform $F(s)$ of a function $f(x)$ with the kernel $k(s, x)$ is defined as

$$I[f(x)] = F(s) = \int_a^b f(x)\, k(s,x)\, dx.$$

For example

1. Laplace transform with the kernel $k(s, x) = e^{-sx}$

$$L[f(x)] = F(s) = \int_0^\infty f(x).e^{-sx} dx$$

2. Fourier Complex transform with the kernel $k(s, x) = e^{isx}$

$$F[f(x)] = F(s) = \frac{1}{\sqrt{2\pi}} \int_{-\infty}^{\infty} f(x)\, e^{isx} dx.$$

$$f(x) = \frac{1}{\sqrt{2\pi}} \int_{-\infty}^{\infty} F(s)\, e^{-isx} ds \qquad \text{(Inversion formula)}$$

3. Fourier Sine transform with the kernel $k(s, x) = \sin sx$

$$F_s[f(x)] = F_s(s) = \sqrt{\frac{2}{\pi}} \int_0^\infty f(x) \sin sx.dx$$

$$f(x) = \sqrt{\frac{2}{\pi}} \int_0^\infty F_s(s) \sin sx\, ds \qquad \text{(Inversion formula)}$$

4. Fourier Cosine transform with the kernel $k(s, x) = \cos sx$

$$F_c[f(x)] = F_c(s) = \sqrt{\frac{2}{\pi}} \int_0^\infty f(x) \cos sx.dx$$

$$f(x) = \sqrt{\frac{2}{\pi}} \int_0^\infty F_c(s) \cos sx\, ds \qquad \text{(Inversion formula)}$$

5. Hankel Transform with the kernel $(k, s) = x J_n(sx)$

$$H[f(x)] = F(s) = \int_0^{\infty} f(x).x J_n(sx)\,dx$$

$$f(x) = \frac{1}{\sqrt{2\pi}}\int_{-\infty}^{\infty} F(s)e^{-isx}ds \qquad \text{(Inversion formula)}$$

6. Hilbert Transform with the kernel $k(s,x) = \dfrac{1}{s-x}$

$$F(s) = \frac{1}{\pi}\int_{-\infty}^{\infty}\frac{f(x)}{s-x}dx$$

$$f(x) = \frac{-1}{\pi}\int_{-\infty}^{\infty}\frac{F(s)}{s-x}ds \qquad \text{(Inversion formula)}$$

7. Mellin transform with the kernel $k(s,x) = x^{s-1}$

$$M[f(x)] = F(s) = \int_0^{\infty} f(x).x^{s-1}dx.$$

The students have already done "Laplace transform" and also learnt to solve the ordinary differential equations by using Laplace transforms.

Integral transforms are used in solving the partial differential equation with boundary conditions.

List of Formulae of Fourier Integrals

1. Fourier Integral for $f(x)$ is $\quad f(x) = \dfrac{1}{\pi}\displaystyle\int_0^{\infty}\int_{-\infty}^{\infty} f(t)\cos u(t-x)\,du\,dt$

2. Fourier Sine Integral for $f(x)$ is $\quad f(x) = \dfrac{2}{\pi}\displaystyle\int_0^{\infty}\int_0^{\infty} f(t)\sin ut\,\sin ux\,du\,dt$

3. Fourier Cosine Integral for $f(x)$ is $f(x) = \dfrac{2}{\pi}\displaystyle\int_0^{\infty}\int_0^{\infty} f(t)\cos ut\,\cos ux\,du\,dt$

45.3 FOURIER INTEGRAL THEOREM

It states that $f(x) = \dfrac{1}{\pi}\displaystyle\int_0^{\infty}\int_{-\infty}^{\infty} f(t)\cos u(t-x)\,dt\,du$

Proof. We know that Fourier series of a function $f(x)$ in $(-c, c)$ is given by

$$f(x) = \frac{a_0}{2} + \sum_{n=1}^{\infty} a_n\cos\frac{n\pi x}{c} + \sum_{n=1}^{\infty} b_n\sin\frac{n\pi x}{c} \qquad \text{... (1)}$$

where a_0, a_n and b_n are given by

$$a_0 = \frac{1}{c}\int_{-c}^{c} f(t)dt,$$

$$a_n = \frac{1}{c}\int_{-c}^{c} f(t)\cos\frac{n\pi t}{c}dt$$

$$b_n = \frac{1}{c}\int_{-c}^{c} f(t)\sin\frac{n\pi t}{c}dt$$

Substituting the values of a_0 a_n and b_n in (1), we get

$$f(x) = \frac{1}{2c}\int_{-c}^{c} f(t)\,dt + \sum_{n=1}^{\infty}\frac{1}{c}\int_{-c}^{c} f(t)\cos\frac{n\pi t}{c}\cos\frac{n\pi x}{c}dt + \sum_{n=1}^{\infty}\frac{1}{c}\int_{-c}^{c} f(t)\sin\frac{n\pi t}{c}\sin\frac{n\pi x}{c}dt$$

$$= \frac{1}{2c}\int_{-c}^{c} f(t)\,dt + \sum_{n=1}^{\infty} \frac{1}{c}\int_{-c}^{c} f(t)\left[\cos\frac{n\pi t}{c}\cos\frac{n\pi x}{c} + \sin\frac{n\pi t}{c}\sin\frac{n\pi x}{c}\right]dt$$

$$= \frac{1}{2c}\int_{-c}^{c} f(t)\,dt + \sum_{n=1}^{\infty} \frac{1}{c}\int_{-c}^{c} f(t)\cos\left(\frac{n\pi t}{c} - \frac{n\pi x}{c}\right)dt$$

$$= \frac{1}{2c}\int_{-c}^{c} f(t)\,dt + \sum_{n=1}^{\infty} \frac{1}{c}\int_{-c}^{c} f(t)\cos\frac{n\pi}{c}(t-x)\,dt$$

$$= \frac{1}{2c}\int_{-c}^{c} f(t)\left\{1 + 2\sum_{n=1}^{\infty}\cos\frac{n\pi}{c}(t-x)\right\}dt \qquad \text{... (2)}$$

Since cosine functions are even functions *i.e.*, $\cos(-\theta) = \cos\theta$ the expression

$$1 + 2\sum_{n=1}^{\infty}\cos\frac{n\pi}{c}(t-x) = \sum_{n=-\infty}^{\infty}\cos\frac{n\pi}{c}(t-x)$$

Therefore, (2) becomes

$$f(x) = \frac{1}{2c}\int_{-c}^{c} f(t)\left\{\sum_{-\infty}^{\infty}\cos\frac{n\pi}{c}(t-x)\right\}dt$$

$$= \frac{1}{2\pi}\int_{-c}^{c} f(t)\left\{\frac{\pi}{c}\sum_{-\infty}^{\infty}\cos\frac{n\pi}{c}(t-x)\right\}dt \qquad \text{... (3)}$$

Let us now assume that c increases indefinitely, so that we may write $\frac{n\pi}{c} = u$ and $\frac{\pi}{c} = du$.

This assumption gives

$$\lim_{c\to\infty}\left\{\frac{\pi}{c}\sum_{-\infty}^{\infty}\cos\frac{n\pi}{c}(t-x)\right\} = \int_{-\infty}^{\infty}\cos u(t-x)\,du$$

$$= 2\int_{0}^{\infty}\cos u(t-x)\,du \qquad \text{... (4)}$$

Substituting in (3) from (4), we obtain

$$f(x) = \frac{1}{2\pi}\int_{-\infty}^{\infty} f(t)\left\{2\int_{0}^{\infty}\cos u(t-x)\,du\right\}dt \qquad \text{...(5)}$$

Thus $\boxed{f(x) = \frac{1}{\pi}\int_{0}^{\infty}\int_{-\infty}^{\infty} f(t)\cos u(t-x)\,du\,dt}$ **Proved.**

Note. We have assumed the following conditions on $f(x)$.

(*i*) $f(x)$ is defined as single-valued except at finite points in $(-c, c)$.

(*ii*) $f(x)$ is periodic outside $(-c, c)$ with period $2c$.

(*iii*) $f(x)$ and $f'(x)$ are sectionally continuous in $(-c, c)$.

(*iv*) $\int_{-\infty}^{\infty} |f(x)|\,dx$ converges, *i.e.*, $f(x)$ is absolutely integrable in $(-\infty, \infty)$.

45.4 FOURIER SINE AND COSINE INTEGRALS

$$f(x) = \frac{2}{\pi}\int_{0}^{\infty}\sin ux\,du\int_{0}^{\infty} f(t)\sin ut\,dt \qquad \textbf{(Fourier Sine Integral)}$$

$$f(x) = \frac{2}{\pi}\int_{0}^{\infty}\cos ux\,du\int_{0}^{\infty} f(t)\cos ut\,dt \qquad \textbf{(Fourier Cosine Integral)}$$

Proof. We know that,

$$\cos u(t-x) = \cos(ut - ux)$$

$$\Rightarrow \quad \cos u(t-x) = \cos ut \cos ux + \sin ut \sin ux$$

Then equation (5) of article 45.3, can be written as

$$f(x) = \frac{1}{\pi}\int_0^\infty \int_{-\infty}^\infty f(t)\,(\cos ut \cos ux + \sin ut \sin ux)\,du\,dt$$

$$\Rightarrow \quad f(x) = \frac{1}{\pi}\int_0^\infty \int_{-\infty}^\infty f(t)\,\cos ut\,\cos ux\,du\,dt + \frac{1}{\pi}\int_0^\infty \int_{-\infty}^\infty f(t)\,\sin ut\,\sin ux\,du\,dt \quad \text{... (6)}$$

Case 1. When $f(t)$ is odd.

$\therefore$ $f(t)\cos ut$ is odd hence $\int_0^\infty \int_{-\infty}^\infty f(t)\cos ut\,\cos ux\,du\,dt = 0$

$$\left[\begin{array}{ll} \text{For odd function,} & \int_{-a}^a f(x)dx = 0 \\ \text{For even function,} & \int_{-a}^a f(x)\,dx = 2\int_0^a f(x)\,dx \end{array}\right]$$

From (6), we have

$$\therefore \quad \boxed{f(x) = \frac{2}{\pi}\int_0^\infty \sin ux\,du \int_0^\infty f(t)\,\sin ut\,dt} \quad \text{... (7)}$$

The relation (7) is called **Fourier sine integral**.

Case 2. When $f(t)$ is even.

$\therefore$ $f(t)\sin ut$ is odd hence $\int_0^\infty \int_{-\infty}^\infty f(t)\sin ut\,\sin ux\,du\,dt = 0$

$\therefore$ $f(t)\cos ut$ is even.

From (6), we have

$$\boxed{f(x) = \frac{2}{\pi}\int_0^\infty \cos ux\,du \int_0^\infty f(t)\cos ut\,dt} \quad \text{... (8)}$$

The relation (8) is known as **Fourier cosine integral.**

45.5 FOURIER'S COMPLEX INTEGRAL

$$f(x) = \frac{1}{2\pi}\int_{-\infty}^\infty e^{-iux}\,du \int_{-\infty}^\infty f(t)\,e^{iut}\,dt$$

Proof. We know that $\int_{-a}^a f(x)\,dx = 0$ if $f(x)$ is odd function.

$$\therefore \quad \int_{-\infty}^\infty \sin u(t-x)\,du = 0 \qquad \text{[since } \sin u\,(t-x) \text{ is odd.]}$$

Obviously we have

$$\frac{1}{2\pi}\int_{-\infty}^\infty f(t)\,dt \int_{-\infty}^\infty \sin u\,(t-x)\,du = 0$$

$$\Rightarrow \quad \frac{i}{2\pi}\int_{-\infty}^\infty f(t)\,dt \int_{-\infty}^\infty \sin u\,(t-x)\,du = 0 \quad \text{(Multiplying by } i) \quad \text{... (9)}$$

On adding (5) and (9), we have

$$f(x) = \frac{1}{2\pi}\int_{-\infty}^\infty \int_{-\infty}^\infty f(t)\,\cos u\,(t-x)\,du\,dt + \frac{i}{2\pi}\int_{-\infty}^\infty f(t)\,dt \int_{-\infty}^\infty \sin u\,(t-x)\,du$$

$$= \frac{1}{2\pi}\int_{-\infty}^{\infty} f(t)\,dt \int_{-\infty}^{\infty} [\cos u\,(t-x) + i \sin u\,(t-x)]\,du$$

$$= \frac{1}{2\pi}\int_{-\infty}^{\infty} f(t)\,dt \int_{-\infty}^{\infty} e^{iu(t-x)}\,du$$

$$\Rightarrow \quad \boxed{f(x) = \frac{1}{2\pi}\int_{-\infty}^{\infty} e^{-iux}\,du \int_{-\infty}^{\infty} f(t)\,e^{iut}\,dt} \qquad \text{... (10)}$$

Relation (10) is called **Fourier's Complex Integral.**

Some Useful Results

1. $\boxed{\int_0^{\infty} \frac{\sin ax}{x}\,dx = \frac{\pi}{2} \qquad (a > 0)}$

2. $\boxed{\int_0^{\infty} e^{-x^2}\,dx = \frac{\sqrt{\pi}}{2}}$

3. $\boxed{\int_{-\infty}^{\infty} \frac{\sin mx}{(x-b)^2 + a^2}\,dx = \frac{\pi}{a} e^{-am} \sin bm, \qquad (m > 0)}$

Examples based on Fourier integral

Example 1. *Express the function*

$$f(x) = \begin{cases} 1 & \text{when} \quad |x| \le 1 \\ 0 & \text{when} \quad |x| > 1 \end{cases}$$

as a Fourier integral. Hence evaluate

$$\int_0^{\infty} \frac{\sin\lambda \cos\lambda x}{\lambda}\,d\lambda$$ *(U.P. & Uttarakhand, III Semester 2008, Dec. 2004)*

Solution. The Fourier Integral for $f(x)$ is $\frac{1}{\pi}\int_0^{\infty}\int_{-\infty}^{\infty} f(t) \cos u\,(t-x)\,du\,dt$

On replacing u by λ, we have

$$f(x) = \frac{1}{\pi}\int_0^{\infty}\int_{-\infty}^{\infty} f(t) \cos\lambda\,(t-x)\,dt\,d\lambda$$

$$= \frac{1}{\pi}\int_0^{\infty}\int_{-1}^{1} \cos\lambda\,(t-x)\,dt\,d\lambda \qquad \text{(Since } f(t) = 1)$$

$$= \frac{1}{\pi}\int_0^{\infty}\left[\frac{\sin\lambda\,(t-x)}{\lambda}\right]_{-1}^{1} d\lambda = \frac{1}{\pi}\int_0^{\infty}\left[\frac{\sin\lambda\,(1-x)}{\lambda} - \frac{\sin\lambda(-1-x)}{\lambda}\right] d\lambda$$

$$= \frac{1}{\pi}\int_0^{\infty} \frac{\sin\lambda\,(1-x) + \sin\lambda\,(1+x)}{\lambda}\,d\lambda \qquad \left[\sin C + \sin D = 2 \sin\frac{C+D}{2} \cos\frac{C-D}{2}\right]$$

$$= \frac{1}{\pi}\int_0^{\infty} \frac{2\sin\dfrac{[\lambda(1-x)+\lambda(1+x)]}{2}\cos\dfrac{[\lambda(1-x)-\lambda(1+x)]}{2}}{\lambda}$$

Thus, $$f(x) = \frac{2}{\pi}\int_0^{\infty} \frac{\sin\lambda \cos\lambda x}{\lambda}\,d\lambda$$ **Ans.**

$$\Rightarrow \quad \int_0^{\infty} \frac{\sin\lambda \cos\lambda x}{\lambda}\,d\lambda = \frac{\pi}{2} f(x)$$

$$\Rightarrow \quad \int_0^\infty \frac{\sin\lambda\cos\lambda x}{\lambda}d\lambda = \begin{cases} \frac{\pi}{2}\times 1 = \frac{\pi}{2} & \text{for } |x|<1 \\ \frac{\pi}{2}\times 0 = 0 & \text{for } |x|>1 \end{cases} \quad \left[\begin{array}{l} f(x)=1 \text{ for } |x|\le 1 \\ f(x)=0 \text{ for } |x|>1 \end{array}\right.$$

For $|x| = 1$, which is a point of discontinuity of $f(x)$, value of integral $= \frac{\pi/2+0}{2} = \frac{\pi}{4}$ **Ans.**

Example 2. *Find the Fourier Sine integral for*

$$f(x) = e^{-\beta x} \qquad \beta > 0 \qquad \textit{(U.P. III Semester Comp., 2004)}$$

hence show that $\frac{\pi}{2}e^{-\beta x} = \int_0^\infty \frac{\lambda \sin \lambda x}{\beta^2+\lambda^2}d\lambda$

Solution. The Fourier Sine Integral of $f(x)$ is

$$f(x) = \frac{2}{\pi}\int_0^\infty \sin ux\, du \int_0^\infty f(t)\sin ut\, dt$$

On replacing u by λ, we get

$$f(x) = \frac{2}{\pi}\int_0^\infty \sin\lambda x\, d\lambda \int_0^\infty f(t)\sin\lambda t\, dt \qquad \text{... (1)}$$

Putting the value of $f(x)$ in (1), we get

$$e^{-\beta x} = \frac{2}{\pi}\int_0^\infty \sin\lambda x\, d\lambda \int_0^\infty e^{-\beta t}\sin\lambda t\, dt \qquad \left[\int e^{ax}\sin bx\, dx = \frac{e^{ax}}{a^2+b^2}(a\sin bx - b\cos bx)\right]$$

$$= \frac{2}{\pi}\int_0^\infty \sin\lambda x\, d\lambda \left[\frac{e^{-\beta t}}{\beta^2+\lambda^2}(-\beta\sin\lambda t - \lambda\cos\lambda t)\right]_0^\infty = \frac{2}{\pi}\int_0^\infty \sin\lambda x\, d\lambda \left[0 + \frac{\lambda}{\beta^2+\lambda^2}\right]$$

$$e^{-\beta x} = \frac{2}{\pi}\int_0^\infty \frac{\lambda\sin\lambda x}{\beta^2+\lambda^2}d\lambda \quad \text{or} \quad \frac{\pi}{2}e^{-\beta x} = \int_0^\infty \frac{\lambda\sin\lambda x}{\beta^2+\lambda^2}d\lambda. \qquad \textbf{Proved.}$$

Example 3. *Using Fourier Cosine Integral representation of an appropriate function, show that*

$$\int_0^\infty \frac{\cos wx}{k^2+w^2}dw = \frac{\pi e^{-kx}}{2k}, \; x > 0,\; k > 0$$

Solution. We know that Fourier Cosine Integral is $f(x) = \frac{2}{\pi}\int_0^\infty \cos u\, x\, du \int_0^\infty f(t)\cos u\, t\, dt$

Putting the value of $f(t)$ and replacing u by w, we get

$$e^{-kx} = \frac{2}{\pi}\int_0^\infty \cos wx\, dw \int_0^\infty e^{-kt}\cos wt\, dt \qquad \left[f(t) = e^{-kt}\right]$$

$$= \frac{2}{\pi}\int_0^\infty \cos wx\, dw \left[\frac{e^{-kt}}{k^2+w^2}\{-k\cos wt + w\sin wt\}\right]_0^\infty$$

$$\left\{\int e^{ax}\cos bx\, dx = \frac{e^{ax}}{a^2+b^2}[a\cos bx + b\sin bx]\right\}$$

$$e^{-kx} = \frac{2}{\pi}\int_0^\infty \cos wx\, dw \left[0 + \frac{k}{k^2+w^2}\right]$$

$$e^{-kx} = \frac{2k}{\pi}\int_0^\infty \frac{\cos wx\, dw}{k^2+w^2} \quad \Rightarrow \quad \int_0^\infty \frac{\cos wx}{k^2+w^2}dw = \frac{\pi e^{-kx}}{2k} \qquad \textbf{Proved.}$$

Example 4. *Using Fourier integral representation, show that*

$$\int_0^{\infty} \frac{\cos(\lambda x)}{(1+\lambda^2)}\, d\lambda = \frac{\pi}{2} e^{-x}, (x>0)$$ *(D.U., April 2010, U.P. & Uttarakhand, III Semester 2008)*

Solution. Fourier cosine integral is

$$F(x) = \frac{2}{\pi}\int_0^{\infty} \cos\lambda x \int_0^{\infty} F(t)\cos\lambda t\, dt\, d\lambda \;\Rightarrow\; e^{-x} = \frac{2}{\pi}\int_0^{\infty}\cos\lambda x\int_0^{\infty} e^{-t}\cos\lambda t\, dt\, d\lambda$$

[Put $F(x) = e^{-x}$]

$$e^{-x} = \frac{2}{\pi}\int_0^{\infty}\cos\lambda x\left[\frac{e^{-t}}{1+\lambda^2}(-\cos\lambda t + \lambda\sin\lambda t)\right]_0^{\infty} d\lambda$$

$$e^{-x} = \frac{2}{\pi}\int_0^{\infty}\cos\lambda x\,.\left(\frac{1}{1+\lambda^2}\right)d\lambda \;\Rightarrow\; \int_0^{\infty}\frac{\cos\lambda x}{1+\lambda^2}\,d\lambda = \frac{\pi}{2}e^{-x}, x\geq 0$$ **Proved.**

Example 5. *Using Fourier integral representation, show that*

$$\int_0^{\infty}\frac{\cos|x|\omega + \omega\sin|x|\omega}{1+\omega^2}\,d\omega = \begin{cases} 0, & \text{if } x<0 \\ \pi/2, & \text{if } x=0. \\ \pi e^{-x}, & \text{if } x>0 \end{cases}$$

Solution. Putting $k = 1$ in example 4, we have

$$\int_0^{\infty}\frac{\cos x\omega}{1+\omega^2}\,d\omega = \frac{\pi}{2}e^{-x}, x>0 \qquad \text{... (1)}$$

Putting $\beta = 1$ and $\lambda = \omega$ in example 2, we have

$$\int_0^{\infty}\frac{\omega\sin x\omega}{1+\omega^2}\,d\omega = \frac{\pi}{2}e^{-x}, x>0 \qquad \text{... (2)}$$

Adding (1) and (2), we get

$$\int_0^{\infty}\frac{\cos x\omega + \omega\sin x\omega}{1+\omega^2}\,d\omega = \frac{\pi}{2}e^{-x} + \frac{\pi}{2}e^{-x} \qquad \text{... (3)}$$

Case I. When $x < 0 \;\Rightarrow\; |x|$ is replaced by $-x$ in (3), we have

$$\int_0^{\infty}\frac{\cos(-x\omega)+\omega\sin(-x\omega)}{1+\omega^2}\,d\omega = \int_0^{\infty}\frac{\cos(x\omega)-\omega\sin(x\omega)}{1+\omega^2}\,d\omega = \frac{\pi}{2}e^{-x} - \frac{\pi}{2}e^{-x} = 0$$

Case II. When $x = 0$, L.H.S. of (3) becomes

$$\int_0^{\infty}\frac{\cos 0\omega + \omega\sin 0\omega}{1+\omega^2}\,d\omega = \int_0^{\infty}\frac{d\omega}{1+\omega^2} = \left(\tan^{-1}\omega\right)_0^{\infty} = \frac{\pi}{2}$$

Case III. When $x > 0$, (3) is

$$\int_0^{\infty}\frac{\cos x\omega + \omega\sin x\omega}{1+\omega^2}\,d\omega = \frac{\pi}{2}e^{-x} + \frac{\pi}{2}e^{-x}$$

$$= \pi e^{-x}.$$

From case I, case II and case III, we get

$$\int_0^{\infty}\frac{\cos x\omega + \omega\sin x\omega}{1+\omega^2}\,d\omega = \begin{cases} 0, & \text{if } x<0 \\ \dfrac{\pi}{2}, & \text{if } x=0 \\ \pi e^{-x}, & \text{if } x>0 \end{cases}$$ **Proved.**

Example 6. *Find the complex form of the Fourier integral representation of*

$$f(x) = \begin{cases} e^{-kx}, & x > 0 \text{ and } k > 0 \\ 0, & \text{otherwise} \end{cases}$$

(U.P. III semester Dec. 2005; U.P. III Semester Dec. 2004)

Solution. We have, $f(x) = \begin{cases} e^{-kx}, & x > 0 \text{ and } k > 0 \\ 0, & \text{otherwise} \end{cases}$

We know that the complex form of Fourier integral representation of $f(x)$ is given

$$f(x) = \frac{1}{2\pi}\int_{-\infty}^{\infty} e^{-iux}\, du \int_{-\infty}^{\infty} f(t)\, e^{iut}\, dt$$

On replacing u by λ, we have

$$f(x) = \frac{1}{2\pi}\int_{-\infty}^{\infty} e^{-i\lambda x}\, d\lambda \int_{-\infty}^{\infty} f(t)\, e^{i\lambda t}\, dt \quad \text{... (1)}$$

On putting the value of $f(t)$ in (1), we get

$$f(x) = \frac{1}{2\pi}\int_{-\infty}^{\infty} e^{-i\lambda x}\, d\lambda \left[\int_0^{\infty} e^{-kt}\, e^{i\lambda t}\, dt\right] = \frac{1}{2\pi}\int_{-\infty}^{\infty} e^{-i\lambda x}\, d\lambda \left[\int_0^{\infty} e^{-(k-i\lambda)t}\, dt\right]$$

$$= \frac{1}{2\pi}\int_{-\infty}^{\infty} e^{-i\lambda x}\, d\lambda \left[\frac{e^{-(k-i\lambda)t}}{-(k-i\lambda)}\right]_0^{\infty} = \frac{1}{2\pi}\int_{-\infty}^{\infty} e^{-i\lambda x}\, d\lambda \left[0 - \frac{1}{-k+i\lambda}\right] = \frac{1}{2\pi}\int_{-\infty}^{\infty} \frac{e^{-i\lambda x}}{k - i\lambda}\, d\lambda.$$

EXERCISE 45.1

1. Find the Fourier Sine Integral representation of

$$f(x) = \begin{cases} 0, & 0 < x < 1 \\ k, & 1 < x < 2 \\ 0, & x > 2 \end{cases} \quad \text{where } k \text{ is constant.}$$

Ans. $f(x) = \dfrac{2k}{\pi}\displaystyle\int_0^{\infty}\left(\frac{\cos\lambda - \cos 2\lambda}{\lambda}\right)\sin\lambda x\, d\lambda$

2. Find Fourier Sine Integral representation of

$$f(x) = x^2, \quad 0 \le x \le 1.$$

Ans. $f(x) = \dfrac{2}{\pi}\displaystyle\int_0^{\infty}\left[\left(\frac{-1}{\lambda} + \frac{2}{\lambda^3}\right)\cos\lambda + \frac{2\sin\lambda}{\lambda^2} - \frac{2}{\lambda^3}\right]\sin\lambda x\, d\lambda$

3. Find Fourier Sine Integral of

$$f(x) = \begin{cases} x & \text{for } 0 < x < 1 \\ -2 - x & \text{for } 1 < x < 2. \\ 0 & \text{for } x > 2 \end{cases}$$

Ans. $f(x) = \dfrac{2}{\pi}\displaystyle\int_0^{\infty}\left[\frac{4\sin u}{u} - \frac{4\sin \partial u}{u^2} + \frac{2\cos u}{u} - \frac{\cos 2u}{u^2} - \frac{1}{u^2}\right]\sin x\, dx$

4. Find Fourier Cosine Integral representation of

$$f(x) = \begin{cases} \sin x, & 0 \le x \le \pi \\ 0, & x > \pi \end{cases}$$

Ans. $f(x) = -\dfrac{2}{\pi}\displaystyle\int_0^{\infty}\left(\frac{1 + \cos\lambda\pi}{\lambda^2 - 1}\right)\cos\lambda x\, d\lambda$

5. Find Fourier Cosine Integral representation of

$$f(x) = \begin{cases} x, & 0 \le x \le 2 \\ 0, & \text{otherwise} \end{cases}$$

Ans. $f(x) = \dfrac{2}{\pi}\displaystyle\int_0^{\infty}\left(\frac{2\sin 2\lambda}{\lambda} + \frac{\cos 2\lambda - 1}{\lambda^2}\right)\cos\lambda x\, d\lambda$

6. Find the Fourier Cosine Integral of the function e^{-ax}. Hence show that

$$\int_0^{\infty}\frac{\cos x}{\lambda^2 + 1^2}\, d\lambda = \frac{\pi}{2}e^{-x}, \quad x \ge 0$$

Ans. $\dfrac{2a}{\pi}\displaystyle\int_0^{\infty}\frac{\cos ax}{\lambda^2 + a^2}\, d\lambda.$

7. Express $f(x) = \begin{cases} 1 & \text{for } 0 \le x \le \pi \\ 0 & \text{for } x > \pi \end{cases}$ as a Fourier Sine Integral and hence evaluate

$\int_0^\infty \frac{1-\cos\pi\lambda}{\lambda} \sin\lambda x \, d\lambda$ **Ans.** $\frac{\pi}{4}$

8. Find the Fourier Integral representation of the function

$$f(x) = \begin{cases} 1 & \text{for } |x|<1 \\ 0 & \text{for } |x|>1 \end{cases}$$

Evaluate $\int_0^\infty \frac{\cos ux \sin u}{u} du$ at $x = 1$ and $\int_0^\infty \frac{\sin u}{u} du$ **Ans.** $\frac{2}{\pi}\int_0^\infty \frac{\sin u \cos ux}{u} dx$ $\frac{\pi}{4}$, $\frac{\pi}{2}$

(U.P. III Semester, 2008)

9. Using Fourier Integral representation, show that

$$\int_0^\infty \frac{\lambda \sin\lambda x}{(\lambda^2+\alpha^2)(\lambda^2+\beta^2)} d\lambda = \frac{\pi}{2}\frac{(e^{-\alpha x} - e^{-\beta x})}{(\beta^2-\alpha^2)}$$ **Ans.** $\frac{6}{\pi}\int_0^\infty \frac{\lambda\sin\lambda x}{(\lambda^2+1)(\lambda^2+4)} d\lambda$

Hence find the Fourier Sine integral representation of $e^{-x} - e^{-2x}$.

10. If $f(x) = \begin{Bmatrix} 0, & x<0 \\ x, & 0 \le x \le \pi \\ 0, & x > \pi \end{Bmatrix}$ then show that

$$f(x) = \frac{1}{\pi\lambda^2}\int_0^\infty [\lambda\pi \sin\lambda(\pi - x) + \cos\lambda(\pi - x) - \cos\lambda x] \, d\lambda$$

11. Using Fourier Integral formula, prove that $\int_0^\infty \left(\frac{\lambda^2+2}{\lambda^2+4}\right)\cos\lambda x \, d\lambda = \frac{\pi}{2} e^{-x} \cos x$, if $x > 0$.

12. Using Fourier Integral method, prove that $\int_0^\infty \left(\frac{\sin\pi\lambda}{1-\lambda^2}\right)\sin\lambda x \, d\lambda = \begin{cases} \frac{\pi}{2}\sin x, & \text{if } 0 \le x \le \pi \\ 0, & \text{if } x > \pi \end{cases}$

13. Solve the integral equation

$\int_0^\infty f(x) \cos\lambda x \, dx = e^{-\lambda}$ **Ans.** $f(x) = \frac{2}{\pi(1+x^2)}$

45.6 FOURIER TRANSFORMS

We have done in Article 45.5 that *(U.P. III Semester, Dec. 2005)*

$$f(x) = \frac{1}{2\pi}\int_{-\infty}^{\infty} e^{-iux} du \int_{-\infty}^{\infty} f(t) e^{iut} dt$$

$$= \frac{1}{2\pi}\int_{-\infty}^{\infty} e^{-isx} ds \int_{-\infty}^{\infty} f(t)e^{ist} dt = \frac{1}{\sqrt{2\pi}}\int_{-\infty}^{\infty} e^{-isx} ds\left[\frac{1}{\sqrt{2\pi}}\int_{-\infty}^{\infty} f(t)e^{ist} dt\right] \quad (u = s) \ldots (1)$$

Putting $\frac{1}{\sqrt{2\pi}}\int_{-\infty}^{\infty} f(t)e^{ist} dt = F(s)$ in (1), we get ... (2)

$$\Rightarrow \quad f(x) = \frac{1}{\sqrt{2\pi}}\int_{-\infty}^{\infty} e^{-isx} . F(s) ds \quad \ldots (3)$$

In (2), $F(s)$ is called the **Fourier transform** of $f(x)$.
In (3), $f(x)$ is called the **inverse Fourier transform** of $F(s)$.

Note: For reasons of symmetry, we multiply both $f(x)$ and $F(s)$ by $\sqrt{\frac{1}{2\pi}}$ instead of having the factor $\frac{1}{2\pi}$ in only one function. Thus, we obtain the definition of Fourier transform as

$$F(s) = F[f(x)] = \frac{1}{\sqrt{2\pi}}\int_{-\infty}^{\infty} f(t)e^{ist}\,dt$$

$$f(x) = \frac{1}{\sqrt{2\pi}}\int_{-\infty}^{\infty} F(s)e^{-isx}\,ds$$

45.7 FOURIER SINE AND COSINE TRANSFORMS

Fourier Sine Transform

From equation (7) of Article 45.4 we know that

$$f(x) = \frac{2}{\pi}\int_0^{\infty} \sin sx\,ds\int_0^{\infty} f(t)\sin st\,dt = \sqrt{\frac{2}{\pi}}\int_0^{\infty} \sin sx\,ds\left[\sqrt{\frac{2}{\pi}}\int_0^{\infty} f(t)\sin st\,dt\right] \quad (s = u)$$

$$= \sqrt{\frac{2}{\pi}}\int_0^{\infty} \sin sx\,ds\,F(s)$$

$$F(s) = F_s[f(x)] = \sqrt{\frac{2}{\pi}}\int_0^{\infty} f(t)\sin st\,dt \quad \text{... (1)}$$

$$f(x) = \sqrt{\frac{2}{\pi}}\int_0^{\infty} F(s)\sin sx\,ds \quad \text{... (2)}$$

In equation (1), $F(s)$ is called **Fourier Sine transform** of $f(x)$.

In equation (2), $f(x)$ is called the **Inverse Fourier Sine transform** of $F(s)$.

Fourier Cosine Transform

From equation (8) of Article 45.4, we have

$$f(x) = \sqrt{\frac{2}{\pi}}\int_0^{\infty} \cos sx\,ds\left[\sqrt{\frac{2}{\pi}}\int_0^{\infty} f(t)\cos st\,dt\right] = \sqrt{\frac{2}{\pi}}\int_0^{\infty} \cos sx\,ds\,F(s)$$

$$F(s) = F_c[f(x)] = \sqrt{\frac{2}{\pi}}\int_0^{\infty} f(t)\cos st\,dt \quad \text{... (3)}$$

$$f(x) = \sqrt{\frac{2}{\pi}}\int_0^{\infty} \cos sx\,F(s)\,ds \quad \text{... (4)}$$

In equation (3), $F(s)$ is called **Fourier Cosine transform** of $f(x)$.

In equation (4), $f(x)$ is called the **Inverse Fourier Cosine transform of** $F(s)$.

Examples based on Fourier Transform.

Example 7. *Find the Fourier transform of*

$$f(x) = \begin{cases} 1 \text{ for } & |x| < a \\ 0 \text{ for } & |x| > a \end{cases}$$

Solution. The Fourier transform of a function $f(x)$ is given by

$$F[f(x)] = \frac{1}{\sqrt{2\pi}}\int_{-\infty}^{\infty} f(x).e^{isx}\,dx \quad \text{... (1)}$$

Substituting the value of $f(x)$ in (1), we get

$$F[f(x)] = \frac{1}{\sqrt{2\pi}}\int_{-a}^{a} 1.e^{isx}\,dx = \left[\frac{e^{isx}}{is}\right]_{-a}^{a} = \frac{1}{\sqrt{2\pi}}\frac{1}{(is)}\left[e^{ias} - e^{-ias}\right]$$

$$= \frac{1}{\sqrt{2\pi}} \frac{2}{s} \cdot \frac{e^{ias} - e^{-ias}}{2i} = \frac{1}{\sqrt{2\pi}} \frac{2 \sin sa}{s} = \sqrt{\frac{2}{\pi}} \frac{\sin sa}{s}$$ **Ans.**

Example 8. *Find the Fourier transform of* *(Uttarakhand, III Semester, 2009)*

$$f(x) = \begin{cases} 1 - x^2 & \text{if } |x| \le 1. \\ 0 & \text{if } |x| > 1. \end{cases}$$

and use it to evaluate $\int_0^{\infty} \left(\frac{x \cos x - \sin x}{x^3} \right) \cos \frac{x}{2} dx.$

Solution. We have $f(x) = \begin{cases} 1 - x^2, & -1 < x < 1 \\ 0 \quad , & |x| > 1 \end{cases}$

The Fourier transform of a function $f(x)$ is given by

$$F[f(x)] = \frac{1}{\sqrt{2\pi}} \int_{-\infty}^{\infty} f(x) . e^{isx} . dx \qquad \dots (1)$$

Substituting the values of $f(x)$ in (1), we get

$$F[f(x)] = \frac{1}{\sqrt{2\pi}} \int_{-1}^{1} (1 - x^2) e^{isx} dx$$

Integrating by parts, we get $\left[\int [uv] = uv_1 - u'v_2 + u''v_3 \dots \right]$

$$F[f(x)] = \frac{1}{\sqrt{2\pi}} \left[(1 - x^2) \frac{e^{isx}}{is} - (-2x) \frac{e^{isx}}{(is)^2} + (-2) \frac{e^{isx}}{(is)^3} \right]_{-1}^{1}$$

$$\Rightarrow \qquad F(s) = \frac{1}{\sqrt{2\pi}} \left[-2 \frac{e^{is}}{s^2} + 2 \frac{e^{is}}{is^3} - 2 \frac{e^{-is}}{s^2} - \frac{2e^{-is}}{is^3} \right]$$

$$= \frac{1}{\sqrt{2\pi}} \left[-\frac{2}{s^2} (e^{is} + e^{-is}) + \frac{2}{is^3} (e^{is} - e^{-is}) \right]$$

$$= \frac{1}{\sqrt{2\pi}} \left[-\frac{2}{s^2} (2 \cos s) + \frac{2}{is^3} (2i \sin s) \right] = \frac{1}{\sqrt{2\pi}} \frac{4}{s^3} [-s \cos s + \sin s]$$

Ans.

By inversion formula for Fourier transform

$$f(x) = \frac{1}{\sqrt{2\pi}} \int_{-\infty}^{\infty} F(s) e^{-isx} dx \qquad \dots (2)$$

Putting the value of $F(s)$ in (2), we get

$$f(x) = \frac{1}{\sqrt{2\pi}} \int_{-\infty}^{\infty} \frac{1}{\sqrt{2\pi}} \frac{4}{s^3} (\sin s - s \cos s) \, e^{-isx} ds$$

$$= \frac{2}{\pi} \int_{-\infty}^{\infty} \frac{1}{s^3} (\sin s - s \cos s) (\cos sx - i \sin sx) \, ds$$

$$= \frac{2}{\pi} \left[\int_{-\infty}^{\infty} \frac{1}{s^3} \sin s \cos sx \, ds - \int_{-\infty}^{\infty} \frac{i}{s^3} \sin s \sin sx \, ds - \int_{-\infty}^{\infty} \frac{1}{s^3} s \cos s \cos sx \, ds \right.$$

$$\left. + i \int_{-\infty}^{\infty} \frac{s}{s^3} \cos s \sin sx \, ds \right]$$

[I, III, functions are even and II, IV functions are odd]

$$= \frac{4}{\pi}\left[\int_0^\infty \frac{1}{s^3}\sin s\cos sx\, ds - 0 - \int_0^\infty \frac{s}{s^3}\cos s\cos sx\, ds + 0\right]$$

$$= \frac{4}{\pi}\int_0^\infty \left(\frac{\sin s - s\cos s}{s^3}\right)\cos sx\, ds$$

Putting $x = \frac{1}{2}$, we get

$$f\left(\frac{1}{2}\right) = \frac{4}{\pi}\int_0^\infty \left(\frac{\sin s - s\cos s}{s^3}\right)\cos\frac{s}{2}\, ds \qquad \left[f\left(\frac{1}{2}\right) = 1 - \left(\frac{1}{2}\right)^2 = \frac{3}{4}\right]$$

$$\frac{3}{4} = -\frac{4}{\pi}\int_0^\infty \left(\frac{s\cos s - \sin s}{s^3}\right)\cos\frac{s}{2}\, ds$$

$$-\frac{3\pi}{16} = \int_0^\infty \left(\frac{s\cos s - \sin s}{s^3}\right)\cos\frac{s}{2}\, ds$$

Hence, $\int_0^\infty \left(\frac{s\cos s - \sin s}{s^3}\right)\cos\frac{s}{2}\, ds = -\frac{3\pi}{16}$ **Ans.**

Example 9. *Find the Fourier transform of* e^{-ax^2}, where $a > 0$.

(*U.P. III Semester, Dec. 2002*)

Solution. The Fourier transform of $f(x)$:

$$F\{f(x)\} = \frac{1}{\sqrt{2\pi}}\int_{-\infty}^{\infty} f(x)e^{isx}\, dx \qquad \text{... (1)}$$

Putting the value of $f(x)$ in (1), we get

$$F\left\{e^{-ax^2}\right\} = \frac{1}{\sqrt{2\pi}}\int_{-\infty}^{\infty} e^{-ax^2} e^{isx}\, dx = \frac{1}{\sqrt{2\pi}}\int_{-\infty}^{\infty} e^{-x^2a - \frac{s^2}{4a} + isx + \frac{s^2}{4a}}\, dx$$

$$= \frac{1}{\sqrt{2\pi}}\int_{-\infty}^{\infty} e^{-\left(x\sqrt{a} - \frac{is}{2\sqrt{a}}\right)^2} e^{-\frac{s^2}{4a}}\, dx$$

Putting $x\sqrt{a} - \frac{is}{2\sqrt{a}} = y$, $dx = \frac{dy}{\sqrt{a}}$, we get

$$F\left\{e^{-ax^2}\right\} = \frac{e^{-s^2/4a}}{\sqrt{2\pi}}\int_{-\infty}^{\infty} e^{-y^2}\frac{dy}{\sqrt{a}}.$$

$$= \frac{e^{-s^2/4a}}{\sqrt{2\pi}} \times \frac{\sqrt{\pi}}{\sqrt{a}} \qquad \left(\text{Since } \int_{-\infty}^{\infty} e^{-y^2} dy = \sqrt{\pi}\right)$$

$$F\left\{e^{-ax^2}\right\} = \frac{e^{-s^2/4a}}{\sqrt{2a}}$$ **Ans.**

Example 10. *Find the Fourier transform of the function*

$$f(x) = \begin{cases} \frac{1}{2\in}, & |x| \le \in \\ 0, & x > \in \end{cases}$$

Solution. We have, $f(x) = \begin{cases} \dfrac{1}{2\in}, & |x| \le \in \\ 0, & x > \in \end{cases}$

The Fourier transform of the function $f(x)$ is given by

$$F(s) = \frac{1}{\sqrt{2\pi}} \int_{-\infty}^{\infty} f(x) e^{isx} dx \qquad ...(1)$$

Substituting the value of $f(x)$ in (1), we get

$$F[f(x)] = \frac{1}{\sqrt{2\pi}} \int_{-\in}^{\in} \frac{1}{2\in} e^{isx}\, dx = \frac{1}{\sqrt{2\pi}} \cdot \frac{1}{2\in} \int_{-\in}^{\in} (\cos sx + i \sin sx)\, dx$$

$$= \frac{1}{\sqrt{2\pi}} \cdot \frac{1}{2\in} \left[\int_{-\in}^{\in} \cos sx\, dx + i \int_{-\in}^{\in} \sin sx\, dx \right] = \frac{1}{\sqrt{2\pi}} \frac{2}{2\in} \left[\int_0^{\in} \cos sx\, dx + 0 \right]$$

[First function is even and second is odd]

$$= \frac{1}{\sqrt{2\pi}} \cdot \frac{1}{\in} \left[\frac{\sin sx}{s} \right]_0^{\in} = \frac{1}{\sqrt{2\pi}} \cdot \frac{1}{\in} \left[\frac{\sin s\in - \sin 0}{s} \right] = \frac{1}{\sqrt{2\pi}} \frac{\sin s\in}{s\in}$$ **Ans.**

Example 11. *Find the Fourier transform of function*

$$f(t) = \begin{cases} t, & for\, |t| < a \\ 0, & for\, |t| > a \end{cases}$$

Solution. We have,

$$f(t) = \begin{cases} t, & for\, |t| < a \\ 0, & for\, |t| > a \end{cases}$$

The Fourier Transform of the function $f(t)$ is given by

$$F[f(t)] = \frac{1}{\sqrt{2\pi}} \int_{-\infty}^{\infty} f(t) e^{ist} dt \qquad ...(1)$$

Substituting the value of $f(t)$ in (1), we get

$$F[f(t)] = \frac{1}{\sqrt{2\pi}} \int_{-a}^{a} t\, e^{ist} dt = \frac{1}{\sqrt{2\pi}} \int_{-a}^{a} t(\cos st + i \sin st)\, dt$$

$$= \frac{1}{\sqrt{2\pi}} \left[0 + 2 \int_0^a it \sin st\, dt \right]$$ [First is odd and second is even]

$$= \frac{2i}{\sqrt{2\pi}} \left[\left\{ t \left(\frac{-\cos st}{s} \right) \right\}_0^a - \int_0^a 1 \left(\frac{-\cos st}{s} \right) dt \right] = \frac{2i}{\sqrt{2\pi}} \left[-\frac{a}{s} \cos as + \frac{1}{s} \left[\frac{\sin st}{s} \right]_0^a \right]$$

$$= \frac{2i}{\sqrt{2\pi}} \left[-\frac{a}{s} \cos as + \frac{1}{s^2} \sin as \right] = \frac{2i}{\sqrt{2\pi}} \frac{1}{s^2} [\sin sa - as \cos as]$$ **Ans.**

Example 12. *Find the Fourier transform of the function shown in the adjoining figure.*

Solution. Here, $f(t) = \begin{cases} 2, \text{ for } -1 < t < 1 \\ 1, \text{ for } -2 < t < -1 \\ 1, \text{ for } 1 < t < 2 \end{cases}$

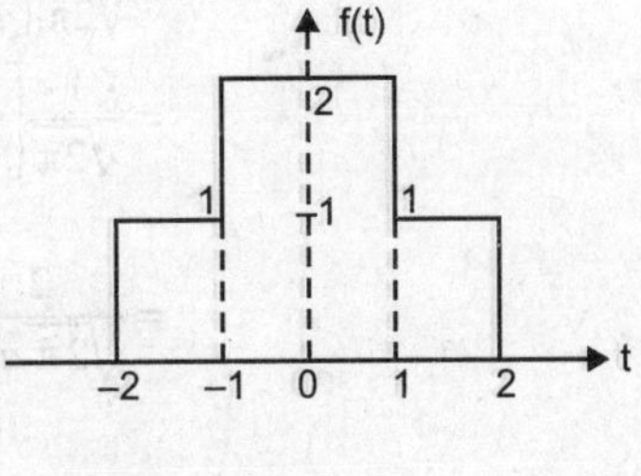

The Fourier transform of the function $f(t)$ is given by

$$F[f(t)] = \frac{1}{\sqrt{2\pi}} \int_{-\infty}^{\infty} f(t)\, e^{ist}\, dt \qquad ...(1)$$

Putting the value of $f(t)$ in (1), we get

$$= \frac{1}{\sqrt{2\pi}}\left[\int_{-1}^{1} 2e^{ist}\,dt + \int_{-2}^{-1} 1\cdot e^{ist}\,dt + \int_{1}^{2} 1\cdot e^{ist}\,dt\right] = \frac{1}{\sqrt{2\pi}}\left[2\left(\frac{e^{ist}}{is}\right)_{-1}^{1} + \left(\frac{e^{ist}}{is}\right)_{-2}^{-1} + \left(\frac{e^{ist}}{is}\right)_{1}^{2}\right]$$

$$= \frac{1}{\sqrt{2\pi}}\left[2\left(\frac{e^{is}-e^{-is}}{is}\right) + \left(\frac{e^{-is}-e^{-2is}}{is}\right) + \left(\frac{e^{2is}-e^{is}}{is}\right)\right]$$

$$= \frac{1}{\sqrt{2\pi}s}\left[2\left(\frac{e^{is}-e^{-is}}{i}\right) + \left(\frac{e^{2is}-e^{-2is}}{i}\right) - \left(\frac{e^{is}-e^{-is}}{i}\right)\right]$$

$$= \frac{1}{\sqrt{2\pi}s}\left[\left(\frac{e^{is}-e^{-is}}{i}\right) + \left(\frac{e^{2is}-e^{-2is}}{i}\right)\right]$$

$$= \frac{1}{\sqrt{2\pi}s}\,[2\sin s + 2\sin 2s] = \frac{1}{\sqrt{2\pi}s}\,[2\sin s + 4\sin s\cos s]$$

$$= \frac{2}{\sqrt{2\pi}s}\sin s\,(1+2\cos s) = \sqrt{\frac{2}{\pi}}\cdot\frac{\sin s}{s}\,(1+2\cos s)$$ **Ans.**

Example 13. *Find the Fourier transform of the function*

$$f(x) = \begin{cases} 1+\dfrac{x}{a}, & (-a<x<0) \\ 1-\dfrac{x}{a}, & 0<x<a \\ 0, & \text{otherwise} \end{cases}$$

(*U.P., III Semester, Summer 2002*)

Solution. Fourier transform of $f(x)$ is given by

$$F[f(x)] = \frac{1}{\sqrt{2\pi}}\int_{-\infty}^{\infty} f(x)\cdot e^{isx}\,dx \qquad \text{... (1)}$$

Putting the value of $f(x)$ in (1), we get

$$F[f(x)] = \frac{1}{\sqrt{2\pi}}\int_{-a}^{0}\left(1+\frac{x}{a}\right)e^{isx}\,dx + \frac{1}{\sqrt{2\pi}}\int_{0}^{a}\left(1-\frac{x}{a}\right)e^{isx}\,dx$$

$$= \frac{1}{\sqrt{2\pi}}\left[\left(1+\frac{x}{a}\right)\times\frac{e^{isx}}{is} - \left(\frac{1}{a}\right)\frac{e^{isx}}{-s^2}\right]_{-a}^{0} + \frac{1}{\sqrt{2\pi}}\left[\left(1-\frac{x}{a}\right)\frac{e^{isx}}{is} - \left(-\frac{1}{a}\right)\frac{e^{isx}}{-s^2}\right]_{0}^{a}$$

$$= \frac{1}{\sqrt{2\pi}}\left[\frac{1}{is} + \frac{1}{a}\cdot\frac{1}{s^2} + \frac{1}{a}\frac{e^{-isa}}{-s^2}\right] + \frac{1}{\sqrt{2\pi}}\left[\frac{1}{a}\cdot\frac{e^{isa}}{-s^2} - \frac{1}{is} + \frac{1}{a}\frac{1}{s^2}\right]$$

$$= \frac{1}{\sqrt{2\pi}}\left[\frac{2}{as^2} + \frac{1}{-as^2}\left(e^{isa}+e^{-isa}\right)\right]$$

$$= \frac{1}{\sqrt{2\pi}}\left[\frac{2}{as^2} - \frac{2}{as^2}\cos sa\right] = \frac{1}{\sqrt{2\pi}}\,\frac{2}{as^2}\,[1-\cos as]$$

$$= \frac{2}{\sqrt{2\pi}\,as^2}\,2\sin^2\frac{as}{2} = \frac{2\sqrt{2}\,\sin^2\dfrac{as}{2}}{\sqrt{\pi}\,as^2}$$ **Ans.**

Examples based on Fourier Sine Transform:

Example 14. *Find Fourier Sine transform of* $\frac{1}{x}$.

Solution. Here, $f(x) = \frac{1}{x}$

$$F_s[f(x)] = \sqrt{\frac{2}{\pi}} \int_0^\infty f(x) \sin sx \, dx \quad \text{... (1)}$$

Putting the value of $f(x)$ in (1), we get

$$F_s\left(\frac{1}{x}\right) = \sqrt{\frac{2}{\pi}} \int_0^\infty \frac{\sin sx}{x} dx = \sqrt{\frac{2}{\pi}} \int_0^\infty \frac{\sin\theta}{\frac{\theta}{s}} \frac{d\theta}{s} \qquad \text{Putting } s\,x = \theta \text{ so that } s\,dx = d\theta$$

$$= \sqrt{\frac{2}{\pi}} \int_0^\infty \frac{\sin\theta}{\theta} d\theta = \sqrt{\frac{2}{\pi}} \left(\frac{\pi}{2}\right)$$ [Some useful result I on page 1208] **Ans.**

$$= \sqrt{\frac{\pi}{2}}$$

Example 15. *Find the Fourier Sine Transform of* e^{-ax}.

Solution. Here, $f(x) = e^{-ax}$

The Fourier sine transform of $f(x)$:

$$F_s[f(x)] = \sqrt{\frac{2}{\pi}} \int_0^\infty f(x) \sin sx \, dx \quad \text{... (1)}$$

On putting the value of $f(x)$ in (1), we get

$$F_s[e^{-ax}] = \sqrt{\frac{2}{\pi}} \int_0^\infty e^{-ax} \sin sx \, dx$$

On integrating by parts, we get

$$F_s[e^{-ax}] = \sqrt{\frac{2}{\pi}} \left[\frac{e^{-ax}}{a^2+s^2}[-a \sin sx - s \cos sx]\right]_0^\infty \left[\int e^{ax} \sin bx \, dx = \frac{e^{ax}}{a^2+b^2}(a \sin bx - b \cos bx)\right]$$

$$= \sqrt{\frac{2}{\pi}} \left[0 - \frac{1}{a^2+s^2}(-s)\right] = \sqrt{\frac{2}{\pi}} \left(\frac{s}{a^2+s^2}\right) \quad \textbf{Ans.}$$

Examples based on Fourier Cosine transform

Example 16. *Find the Fourier Cosine Transform of* $f(x) = e^{-ax}$.

Solution. The Fourier Cosine Transform is

$$F_c[f(x)] = \sqrt{\frac{2}{\pi}} \int_0^\infty f(x) \cos sx \, dx \quad \text{... (1)}$$

Putting the value of $f(x)$ in (1), we get

$$F_c[e^{-ax}] = \sqrt{\frac{2}{\pi}} \int_0^\infty e^{-ax} \cos sx \, dx \qquad \left[\int e^{ax} \cos bx \, dx = \frac{e^{ax}}{a^2+b^2}(a \cos bx + b \sin bx)\right]$$

$$= \sqrt{\frac{2}{\pi}} \left[\frac{e^{-ax}}{a^2+s^2}\{-a \cos sx + s \sin sx\}\right]_0^\infty = \sqrt{\frac{2}{\pi}} \left[0 + \frac{a}{a^2+s^2}\right] = \sqrt{\frac{2}{\pi}} \frac{a}{a^2+s^2} \quad \textbf{Ans.}$$

Example 17. *Find the Fourier Cosine Transform of* $f(x) = 5e^{-2x} + 2e^{-5x}$

Solution. The Fourier Cosine Transform of $f(x)$ is given by

$$F_c\{f(x)\} = \sqrt{\frac{2}{\pi}} \int_0^\infty f(x) \cos sx \, dx \qquad \text{... (1)}$$

Putting the value of $f(x)$ in (1), we get $F_c\{f(x)\} = \int_0^\infty (5e^{-2x} + 2e^{-5x}) \cos sx \, dx$

$$= 5\int_0^\infty e^{-2x} \cos sx \, dx + 2\int_0^\infty e^{-5x} \cos sx \, dx$$

$$= 5\left[\frac{e^{-2x}}{(-2)^2 + s^2}(-2\cos sx + s\sin sx)\right]_0^\infty + 2\left[\frac{e^{-5x}}{(-5)^2 + s^2}(-5\cos sx + s\sin sx)\right]_0^\infty$$

$$= 5\left[0 - \frac{1}{4+s^2}(-2)\right] + 2\left[0 - \frac{1}{25+s^2}(-5)\right] = 5\left(\frac{2}{s^2+4}\right) + 2\left(\frac{5}{s^2+25}\right)$$

$$= 10\left(\frac{1}{s^2+4} + \frac{1}{s^2+25}\right) \qquad \textbf{Ans.}$$

Example 18. *Obtain Fourier Cosine Transfrom of*

$$f(x) = \begin{cases} x, & \text{for } 0<x<1 \\ 2-x, & \text{for } 1<x<2 \\ 0, & \text{for } x>2. \end{cases} \qquad \textit{(U.P., III Semester Dec. 2002)}$$

Solution. Fourier Cosine Transform

$$F_c\{f(x)\} = \sqrt{\frac{2}{\pi}} \int_0^\infty f(x) \cos sx \, dx \qquad \text{... (1)}$$

Putting the value of $f(x)$ in (1), we get

$$F_c[f(x)] = \sqrt{\frac{2}{\pi}}\left[\int_0^1 x\cos sx\,dx + \int_1^2 (2-x)\cos sx\,dx + \int_2^\infty 0.\cos sx\,dx\right]$$

$$= \sqrt{\frac{2}{\pi}}\left[\left\{x\frac{\sin sx}{s} - \left(-\frac{\cos sx}{s^2}\right)\right\}_0^1 + \left\{(2-x)\frac{\sin sx}{s} - (-1)\left(-\frac{\cos sx}{s^2}\right)\right\}_1^2\right] + 0$$

$$= \sqrt{\frac{2}{\pi}}\left[\left\{\left(\frac{\sin s}{s} + \frac{\cos s}{s^2}\right) - \frac{1}{s^2}\right\} + \left\{\left(-\frac{\cos 2s}{s^2}\right) - \left(\frac{\sin s}{s} - \frac{\cos s}{s^2}\right)\right\}\right]$$

$$= \sqrt{\frac{2}{\pi}}\left[\frac{2\cos s}{s^2} - \frac{1}{s^2} - \frac{\cos 2s}{s^2}\right] = \sqrt{\frac{2}{\pi}}\left[\frac{2\cos s - 1 - (2\cos^2 s - 1)}{s^2}\right]$$

$$= \sqrt{\frac{2}{\pi}}\left[\frac{2\cos s - 1 - 2\cos^2 s + 1}{s^2}\right] = \sqrt{\frac{2}{\pi}}\,\frac{2\cos s(1-\cos s)}{s^2} \qquad \textbf{Ans.}$$

Examples based on Fourier Sine and Fourier Cosine Transform

Example 19. *Find Fourier Sine and Cosine Transform of (a)* x^{n-1}. *(b)* $\frac{1}{\sqrt{x}}$.

(U.P. III Semester (Comp.) 2004)

Solution. Here, $f(x) = x^{n-1}$

The Fourier Sine Transform of $f(x)$:

$$F_s[f(x)] = \sqrt{\frac{2}{\pi}} \int_0^\infty f(x) \sin sx \, dx \qquad \ldots (1)$$

Putting the value of $f(x)$ in (1), we get

$$(a) \quad F_s(x^{n-1}) = \sqrt{\frac{2}{\pi}} \int_0^\infty \sin sx \cdot x^{n-1} \, dx \qquad \ldots (2)$$

The Fourier Cosine Transform of $f(x)$:

$$F_c[f(x)] = \sqrt{\frac{2}{\pi}} \int_0^\infty f(x) \cos sx \, dx \qquad \ldots (3)$$

Putting the value of $f(x)$ in (3), we get

$$F_c(x^{n-1}) = \sqrt{\frac{2}{\pi}} \int_0^\infty \cos sx \cdot x^{n-1} dx \qquad \ldots (4)$$

Multiplying (2) by i and adding to (4), we have

$$F_c(x^{n-1}) + iF_s(x^{n-1}) = \sqrt{\frac{2}{\pi}} \int_0^\infty (\cos sx + i \sin sx) x^{n-1} dx = \sqrt{\frac{2}{\pi}} \int_0^\infty e^{isx} x^{n-1} dx$$

On putting $isx = -t$ so that $isdx = -dt$, we get

$$= \sqrt{\frac{2}{\pi}} \int_0^\infty e^{-t} \left(-\frac{t}{is}\right)^{n-1} \left(-\frac{dt}{is}\right) = \sqrt{\frac{2}{\pi}} \frac{1}{(is)^n} (-1)^n \int_0^\infty e^{-t} t^{n-1} dt$$

$$= \sqrt{\frac{2}{\pi}} \frac{(i)^{2n}}{(i)^n s^n} \Gamma n \qquad = \sqrt{\frac{2}{\pi}} \frac{(i)^n}{s^n} \Gamma n \qquad \left[\int_0^\infty e^{-t} t^{n-1} dt = \Gamma n\right]$$

$$= \sqrt{\frac{2}{\pi}} \frac{\left(\cos\frac{\pi}{2} + i \sin\frac{\pi}{2}\right)^n \Gamma n}{s^n} = \sqrt{\frac{2}{\pi}} \frac{\left(\cos\frac{n\pi}{2} + i \sin\frac{n\pi}{2}\right) \Gamma n}{s^n}$$

Equating real and imaginary parts, we get

$$F_c(x^{n-1}) = \sqrt{\frac{2}{\pi}} \frac{\Gamma n}{s^n} \cos\frac{n\pi}{2} \qquad \ldots (5)$$

$$F_s(x^{n-1}) = \sqrt{\frac{2}{\pi}} \frac{\Gamma n}{s^n} \sin\frac{n\pi}{2} \qquad \ldots (6) \quad \textbf{Ans.}$$

(*b*) Putting $n = \frac{1}{2}$ in (5), we get $F_c\left(\frac{1}{\sqrt{x}}\right) = \sqrt{\frac{2}{\pi}} \frac{\Gamma\frac{1}{2}}{s^{\frac{1}{2}}} \cos\frac{\pi}{4} = \sqrt{\frac{2}{\pi}} \frac{\sqrt{\pi}}{\sqrt{s}} \frac{1}{\sqrt{2}} = \frac{1}{\sqrt{s}}$

Putting $n = \frac{1}{2}$ in (6), we get $F_s\left(\frac{1}{\sqrt{x}}\right) = \sqrt{\frac{2}{\pi}} \frac{\Gamma\frac{1}{2}}{s^{\frac{1}{2}}} \sin\frac{\pi}{4} = \sqrt{\frac{2}{\pi}} \frac{\sqrt{\pi}}{\sqrt{s}} \frac{1}{\sqrt{2}} = \frac{1}{\sqrt{s}}$ **Ans.**

Second Method.

Example 20. *Find the Fourier Sine Transform of*

$$f(x) = \frac{e^{-ax}}{x}. \qquad \textit{(U.P. III Semester 2008)}$$

Solution. The Sine Transform of the function $f(x)$ is given by

$$F(s)=\sqrt{\frac{2}{\pi}}\int_0^\infty f(x)\ \sin sx\, dx \qquad \text{... (1)}$$

Substituting the value of $f(x)$ in (1), we get

$$F(s)=\sqrt{\frac{2}{\pi}}\int_0^\infty \frac{e^{-ax}}{x}\sin sx\, dx$$

Differentiating both sides w.r.t. 's', we get

$$\frac{d}{ds}[F(s)]=\sqrt{\frac{2}{\pi}}\int_0^\infty \frac{e^{-ax}}{x}(x\cos sx)\,dx$$

$$=\sqrt{\frac{2}{\pi}}\int_0^\infty e^{-ax}\cos sx\, dx=\sqrt{\frac{2}{\pi}}\left[\frac{e^{-ax}}{a^2+s^2}(-a\cos sx+s\sin sx)\right]_0^\infty$$

$$=\sqrt{\frac{2}{\pi}}\frac{a}{s^2+a^2} \qquad \text{... (2)}$$

Integrating (2) w.r.t. 's', we get

$$F(s)=\sqrt{\frac{2}{\pi}}\int\frac{a}{s^2+a^2}ds = \sqrt{\frac{2}{\pi}}\tan^{-1}\frac{s}{a}+C \qquad \text{... (3)}$$

For $s=0$, $\quad F(s)=0$

Putting $s=0$, $F(s)=0$ in (3), we get

$$0=0+C \quad \text{or} \quad C=0$$

On putting the value of C in (3), we get

$$F(s)=\sqrt{\frac{2}{\pi}}\tan^{-1}\frac{s}{a}$$

Ans.

Example 21. *Find Fourier Cosine Transform of* $e^{-a^2x^2}$ *and hence evaluate Fourier Sine Transform of* $xe^{-a^2x^2}$.

Solution. *Here,* $f(x) = e^{-a^2x^2}$

The Fourier Cosine Transform of $f(x)$:

$$F_c[f(x)]=\sqrt{\frac{2}{\pi}}\int_0^\infty f(x)\cos sx\, dx \qquad \text{... (1)}$$

Putting the value of $f(x)$ in (1), we get

$$F_c(e^{-a^2x^2})=\sqrt{\frac{2}{\pi}}\int_0^\infty e^{-a^2x^2}\cdot\cos sx\, dx$$

$$=\text{Real part of }\sqrt{\frac{2}{\pi}}\int_0^\infty e^{-a^2x^2}\cdot e^{isx}\,dx \qquad [\cos sx+i\sin sx=e^{isx}]$$

$$=\text{Real part of }\sqrt{\frac{2}{\pi}}\int_0^\infty e^{-a^2x^2+isx}dx=\frac{1}{a\sqrt{2}}e^{-\frac{s^2}{4a^2}}$$ (See Example 9 on page 1215)

We know that

$$F_s[(x f(x)] = -\frac{d}{ds}F_c f(x)$$ (See Example 28 page1225)

$$F_s(xe^{-a^2x^2}) = -\frac{d}{ds}F_c(e^{-a^2x^2}) \qquad \left[\because f(x) = e^{-a^2x^2}\right]$$

$$= -\frac{d}{ds}\frac{1}{a\sqrt{2}}e^{-\frac{s^2}{4a^2}} = \frac{1}{a\sqrt{2}}e^{-\frac{s^2}{4a^2}}\frac{s}{2a^2} = \frac{s}{2\sqrt{2}\,a^3}e^{-\frac{s^2}{4a^2}}$$ **Ans.**

Example 22. *Find Fourier Cosine Transform of* $\frac{1}{1+x^2}$ *and hence find Fourier Sine Transform of* $\frac{x}{1+x^2}$. *(Uttrakhand, III Semester, June 2009; U.P. III Semester Dec. 2004)*

Solution. Here, $f(x) = \frac{1}{1+x^2}$

The Fourier Cosine Transform of $f(x)$:

$$F_c[f(x)] = \sqrt{\frac{2}{\pi}}\int_0^\infty f(x)\cos sx\,dx \qquad \text{... (1)}$$

Putting the value of $f(x)$ in (1), we get

$$F_c(s) = \sqrt{\frac{2}{\pi}}\int_0^\infty \frac{1}{1+x^2}\cos sx\,dx = I \text{ (say)} \qquad \text{...(2)}$$

Differentiating w.r.t. 's', we get

$$\frac{dI}{ds} = \sqrt{\frac{2}{\pi}}\int_0^\infty \frac{1}{1+x^2}\left(\frac{d}{ds}\cos sx\right)dx$$

$$\Rightarrow \quad \frac{dI}{ds} = \sqrt{\frac{2}{\pi}}\int_0^\infty \frac{-x\sin sx}{1+x^2}dx = -\sqrt{\frac{2}{\pi}}\int_0^\infty \frac{x^2\sin sx}{x(1+x^2)}dx$$

$$= -\sqrt{\frac{2}{\pi}}\int_0^\infty \frac{(1+x^2-1)\sin sx}{x(1+x^2)}dx$$

$$= -\sqrt{\frac{2}{\pi}}\int_0^\infty \frac{(1+x^2)\sin sx}{x(1+x^2)}dx + \sqrt{\frac{2}{\pi}}\int_0^\infty \frac{\sin sx}{x(1+x^2)}dx$$

$$= -\sqrt{\frac{2}{\pi}}\int_0^\infty \frac{\sin sx}{x}dx + \sqrt{\frac{2}{\pi}}\int_0^\infty \frac{\sin sx}{x(1+x^2)}dx \qquad \left[\int_0^\infty \frac{\sin sx}{x}dx = \frac{\pi}{2}\right]$$

$$\Rightarrow \quad \frac{dI}{ds} = -\sqrt{\frac{2}{\pi}}.\frac{\pi}{2} + \sqrt{\frac{2}{\pi}}\int_0^\infty \frac{\sin sx}{x(1+x^2)}dx \qquad \text{...(3)]}$$

Again differentiating, we get

$$\Rightarrow \quad \frac{d^2I}{ds^2} = 0 + \sqrt{\frac{2}{\pi}}\int_0^\infty \frac{\frac{d}{ds}\sin sx}{x(1+x^2)}dx$$

$$\Rightarrow \quad \frac{d^2I}{ds^2} = \sqrt{\frac{2}{\pi}}\int_0^\infty \frac{x\cos sx}{x(1+x^2)}dx = \sqrt{\frac{2}{\pi}}\int_0^\infty \frac{\cos sx}{1+x^2}dx \qquad \text{[From (2)]}$$

$$\Rightarrow \quad \frac{d^2 I}{ds^2} - I = 0 \Rightarrow \quad D^2 I - I = 0 \Rightarrow \quad (D^2 - 1)\, I = 0$$

A.E. is $m^2 - 1 = 0 \Rightarrow m = \pm 1$

C.F. $= C_1 e^s + C_2 e^{-s}$ and P.I. $= 0$

$$\Rightarrow \qquad I = C_1 e^s + C_2 e^{-s} \qquad ...(4)$$

$$\therefore \qquad \frac{dI}{ds} = C_1 e^s - C_2 e^{-s} \qquad ...(5)$$

Putting $s = 0$ in (2), we get

$$I = \sqrt{\frac{2}{\pi}} \int_0^\infty \frac{1}{1+x^2}\, dx = \sqrt{\frac{2}{\pi}} \left[\tan^{-1} x\right]_0^\infty = \sqrt{\frac{2}{\pi}} \left[\tan^{-1} \infty - 0\right] = \sqrt{\frac{2}{\pi}} \left(\frac{\pi}{2}\right) = \sqrt{\frac{\pi}{2}} \qquad ...(6)$$

Again putting s = 0 in (3), we get $\frac{dI}{ds} = -\sqrt{\frac{\pi}{2}}$ ...(7)

Putting $s = 0$ and equating (4) and (6) ; (5) and (7), we get

$$C_1 + C_2 = \sqrt{\frac{\pi}{2}} \text{ and } C_1 - C_2 = -\sqrt{\frac{\pi}{2}}$$

On solving, we get $\qquad C_1 = 0, \qquad C_2 = \sqrt{\frac{\pi}{2}}$

Putting the values of C_1 and C_2 in (4), we get

$$I = \sqrt{\frac{\pi}{2}}\, e^{-s}$$

$$\Rightarrow \qquad \sqrt{\frac{2}{\pi}} \int_0^\infty \frac{\cos sx}{1+x^2}\, dx = \sqrt{\frac{\pi}{2}}\, e^{-s} \qquad \left[I = \sqrt{\frac{2}{\pi}} \int_0^\infty \frac{\cos sx}{1+x^2}\, dx\right]$$

Differentiating w.r.t. s, we get

$$\sqrt{\frac{2}{\pi}} \int_0^\infty \frac{-x \sin sx}{1+x^2}\, dx = -\sqrt{\frac{\pi}{2}}\, e^{-s}$$

$$\Rightarrow \qquad \sqrt{\frac{2}{\pi}} \int_0^\infty \frac{x \sin sx}{1+x^2}\, dx = \sqrt{\frac{\pi}{2}}\, e^{-s}$$

$$\Rightarrow \qquad \int_0^\infty \frac{x \sin sx}{1+x^2}\, dx = \frac{\pi}{2}\, e^{-s}$$

Ans.

Example 23. *Taking the function* $f(x) = \begin{cases} 1, & 0 < x < \pi \\ 0, & x > \pi \end{cases}$

show that $\int_0^\infty \left(\frac{1 - \cos s\pi}{s}\right) . \sin sx\, ds = \begin{cases} \frac{\pi}{2}, & 0 < x < \pi \\ 0, & x > \pi \end{cases}$.

Solution. We have, $f(x) = \begin{cases} 1, & 0 < x < \pi \\ 0, & x > \pi \end{cases}$

The Fourier Sine Transform of $f(x)$:

$$F_s[f(x)] = \sqrt{\frac{2}{\pi}} \int_0^\infty f(x) \sin sx\, dx \qquad ...\ (1)$$

Putting the value of $f(x)$ in (1), we get

$$= \sqrt{\frac{2}{\pi}}\int_0^{\pi} 1 \cdot \sin sx\, dx + \sqrt{\frac{2}{\pi}}\int_{\pi}^{\infty} 0.\sin sx\, dx = \sqrt{\frac{2}{\pi}}\left[-\frac{\cos sx}{s}\right]_0^{\pi} = \sqrt{\frac{2}{\pi}}\left(\frac{1-\cos s\pi}{s}\right)$$

By inverse formula for Fourier sine transform

$$\sqrt{\frac{2}{\pi}}\int_0^{\infty}\sqrt{\frac{2}{\pi}}\left(\frac{1-\cos s\pi}{s}\right)\sin sx\, ds = \begin{cases} 1, & 0<x<\pi \\ 0, & x>\pi \end{cases}$$

$$\Rightarrow \quad \frac{2}{\pi}\int_0^{\infty}\left(\frac{1-\cos s\pi}{s}\right)\sin sx\, ds = \begin{cases} 1, & 0<x<\pi \\ 0, & x>\pi \end{cases}$$

$$\Rightarrow \quad \int_0^{\infty}\left(\frac{1-\cos s\pi}{s}\right)\sin sx\, ds = \begin{cases} \frac{\pi}{2}, & 0<x<\pi \\ 0, & x>\pi \end{cases}$$ **Ans.**

Example 24. *Find the Fourier sine transform of* $e^{-|x|}$.

Hence evaluate $\int_0^{\infty}\frac{x\sin mx}{1+x^2}dx$ *(U.P. III Semester, Dec. 2003)*

Solution. In the interval $(0, \infty)$, x is always positive therefore $e^{-|x|} = e^{-x}$

Now, Fourier sine transform of e^{-x} is given by

$$F_s\{e^{-x}\} = \sqrt{\frac{2}{\pi}}\int_0^{\infty} e^{-x}\sin sx\, dx = \sqrt{\frac{2}{\pi}}\left[\frac{e^{-x}}{(-1)^2+s^2}(-\sin sx - s\cos sx)\right]_0^{\infty}$$

$$= \sqrt{\frac{2}{\pi}}\frac{1}{1+s^2}(0+s) = \sqrt{\frac{2}{\pi}}\left(\frac{s}{1+s^2}\right) = F(s) \qquad \dots (1)$$

Now, the inverse sine transform of F (s), is e^{-x}. Using inverse formula for the Sine Transform, we get

$$e^{-x} = \sqrt{\frac{2}{\pi}}\int_0^{\infty} F(s)\sin sx\, ds = \sqrt{\frac{2}{\pi}}\int_0^{\infty}\sqrt{\frac{2}{\pi}}\left(\frac{s}{1+s^2}\right)\sin sx\, ds \qquad \text{[Using (1)]}$$

Replacing x by m, we get

$$e^{-m} = \frac{2}{\pi}\int_0^{\infty}\frac{s}{1+s^2}\sin ms\, ds$$

Replacing s by x, we get

$$e^{-m} = \frac{2}{\pi}\int_0^{\infty}\frac{x\sin mx}{1+x^2}dx \qquad \left[\int_a^b f(s)ds = \int_a^b f(x)dx\right]$$

Hence, we get $\int_0^{\infty}\frac{x\sin mx}{1+x^2}dx = \frac{\pi}{2}e^{-m}$. **Ans.**

Example 25. *Find the function whose Sine Transform is* $\frac{e^{-as}}{s}$.

Solution. Here, $F_s\,[f(x)] = \frac{e^{-as}}{s}$

The inverse Fourier Sine Transform of $F_s[f(x)]$ or $F_s(s)$:

$$f(x)=\sqrt{\frac{2}{\pi}}\int_0^\infty F(s)\,.\,\sin sx\,ds \qquad \ldots (1)$$

Putting the value of $F(s)$ in (1), we get $f(x)=\sqrt{\frac{2}{\pi}}\int_0^\infty \frac{e^{-as}}{s}\sin sx\,dx$

Differentiating w.r.t 'x' we get

$$\frac{df}{dx}=\sqrt{\frac{2}{\pi}}\int_0^\infty \frac{e^{-as}}{s}\frac{d}{dx}(\sin sx)\,ds=\sqrt{\frac{2}{\pi}}\int_0^\infty \frac{e^{-as}}{s}(s\cos sx)\,ds$$

$$=\sqrt{\frac{2}{\pi}}\int_0^\infty e^{-as}\cos sx\,ds=\sqrt{\frac{2}{\pi}}\left[\frac{e^{-as}}{a^2+x^2}(-a\cos sx+x\sin sx)\right]_0^\infty$$

$$=\sqrt{\frac{2}{\pi}}\left[0-\frac{1}{a^2+x^2}(-a)\right]=\sqrt{\frac{2}{\pi}}\frac{a}{a^2+x^2} \qquad \ldots (2)$$

Integrating both sides, of (2) w.r.t. 'x', we get

$$f(x)=\left(\sqrt{\frac{2}{\pi}}\,a\right)\frac{1}{a}\tan^{-1}\frac{x}{a}+C \qquad =\sqrt{\frac{2}{\pi}}\,.\tan^{-1}\frac{x}{a}+C \qquad \ldots (3)$$

On putting $x = 0$ and $f(0) = 0$ in (2), we get

$$f(0)=0+C \qquad \Rightarrow 0=0+C \qquad \Rightarrow C=0$$

On putting the value of C in (3), we get

$$f(x)=\sqrt{\frac{2}{\pi}}\tan^{-1}\frac{x}{a}$$

$$F_s^{-1}\left(\frac{e^{-as}}{s}\right)=\sqrt{\frac{2}{\pi}}\tan^{-1}\frac{x}{a} \qquad \ldots (3)$$

On substituting $a = 0$ in (3), we have

$$F^{-1}\left(\frac{1}{s}\right)=\sqrt{\frac{2}{\pi}}\,.\,\frac{\pi}{2}=\sqrt{\frac{\pi}{2}}$$ **Ans.**

Example 26. *Find the Fourier Cosine transform of*

$$f(x)=\begin{cases} x & \text{for } \; 0<x<\frac{1}{2} \\ 1-x & \text{for } \; \frac{1}{2}<x<1 \\ 0 & \text{for } \; x>1. \end{cases}$$

Write the inverse transform.

Solution. The Fourier Cosine Transform of $f(x)$:

$$F_c\,[f(x)]=\sqrt{\frac{2}{\pi}}\int_0^\infty f(x)\cdot\cos sx\,dx \qquad \ldots (1)$$

Putting the value of $f(x)$ in (1), we get

$$=\sqrt{\frac{2}{\pi}}\int_0^{1/2} x\cos sx\,dx+\sqrt{\frac{2}{\pi}}\int_{1/2}^1 (1-x)\cos sx\,dx$$

$$= \sqrt{\frac{2}{\pi}}\left[x\frac{\sin sx}{s} - \left(\frac{-\cos sx}{s^2}\right)\right]_0^{1/2} + \sqrt{\frac{2}{\pi}}\left[(1-x)\frac{\sin sx}{s} - (-1)\frac{(-\cos sx)}{s^2}\right]_{1/2}^{1}$$

$$= \sqrt{\frac{2}{\pi}}\left[\frac{1}{2}\frac{\sin\frac{s}{2}}{s} + \frac{\cos\frac{s}{2}}{s^2} - \frac{1}{s^2}\right] + \sqrt{\frac{2}{\pi}}\left[-\frac{\cos s}{s^2} - \frac{1}{2}\frac{\sin\frac{s}{2}}{s} + \frac{\cos\frac{s}{2}}{s^2}\right]$$

$$= \sqrt{\frac{2}{\pi}}\left[-\frac{\cos s}{s^2} + \frac{2\cos s/2}{s^2} - \frac{1}{s^2}\right]$$ **Ans.**

Example 27. *Solve the integral equation*

$$\int_0^\infty f(x)\cos sx\, dx = \begin{cases} 1-s, & 0 \le s \le 1 \\ 0, & s > 1 \end{cases}$$

Hence prove that $\int_0^\infty \frac{\sin^2 t}{t^2}dt = \frac{\pi}{2}$. *(U.P. III Semester (SUM) 2004)*

Solution. $F_c(s) = \sqrt{\frac{2}{\pi}} \int_0^\infty f(x)\cos sx\, dx$ and $F_c(s) = \begin{cases} 1-s, & 0 \le s \le 1 \\ 0, & s > 1 \end{cases}$

$\therefore$ By inversion formula for Fourier cosine transform, we have

$$f_c(x) = \sqrt{\frac{2}{\pi}}\int_{-\infty}^{\infty} F_c(s)\cos sx\, ds \qquad ...(1)$$

Putting the value of $F_c(s)$ in (1), we get

$$f(x) = \sqrt{\frac{2}{\pi}}\left[\int_0^1 (1-s)\cos sx\, ds + \int_1^\infty 0.\cos sx\, ds\right] = \sqrt{\frac{2}{\pi}}\int_0^1 (1-s)\cos sx\, ds$$

$$= \sqrt{\frac{2}{\pi}}\left[(1-s)\cdot\frac{\sin sx}{x} - (-1)\cdot\frac{-\cos sx}{x^2}\right]_0^1 = \sqrt{\frac{2}{\pi}}\left[0 - \frac{\cos x}{x^2} + \frac{1}{x^2}\right] = \sqrt{\frac{2}{\pi}}\left[\frac{1-\cos x}{x^2}\right] \qquad ...(2)$$

Deduction. Since $\sqrt{\frac{2}{\pi}} \int_0^\infty f(x)\cos sx\, dx = \begin{cases} 1-s, & 0 \le s \le 1 \\ 0, & s > 1 \end{cases}$ where $f(x) = \sqrt{\frac{2}{\pi}}\left[\frac{1-\cos x}{x^2}\right]$

From (1) and (2), we have

$$\sqrt{\frac{2}{\pi}}\int_0^\infty \sqrt{\frac{2}{\pi}}\left(\frac{1-\cos x}{x^2}\right)\cos sx\, dx = \begin{cases} 1-s, & 0 \le s \le 1 \\ 0, & s > 1 \end{cases} \qquad ... (3)$$

On putting $s = 0$ in (3), we have

$$\frac{2}{\pi}\int_0^\infty \frac{1-\cos x}{x^2}dx = 1 \qquad \Rightarrow \int_0^\infty \frac{2\sin^2\frac{x}{2}}{x^2}dx = \frac{\pi}{2}$$

Putting $x = 2t$ so that $dx = 2dt$, we get $\int_0^\infty \frac{\sin^2 t}{t^2}dt = \frac{\pi}{2}$. **Proved.**

Example 28. *Show that*

(a) $F_s[x\, f(x)] = -\frac{d}{ds}F_c(s)$ *(b)* $F_c[x\, f(x)] = \frac{d}{ds}F_s(s)$

and hence find Fourier Cosine and Sine Transform of $x\, e^{-ax}$.

Solution. (*a*) The Fourier Cosine Transform of $f(x)$

$$F_c\,[f(x)] = \sqrt{\frac{2}{\pi}}\int_0^\infty f(x)\cos sx\,dx \qquad \text{... (1)}$$

Differentiating (1), w.r.t. 's', we get $\frac{d}{ds}F_c[f(x)] = -\sqrt{\frac{2}{\pi}}\int_0^\infty x\,f(x)\sin sx\,dx$

$$\frac{d}{ds}F_C(s) = -F_s\{x\,f(x)\} \qquad \text{... (2)}$$

(*b*) The Fourier Sine Transform of $f(x)$

$$F_s\,[f(x)] = \sqrt{\frac{2}{\pi}}\int_0^\infty f(x)\sin sx\,dx$$

Differentiating w.r.t 's' we get

$$\frac{d}{ds}F_s[f(x)] = \sqrt{\frac{2}{\pi}}\int_0^\infty f(x)\frac{d}{ds}(\sin sx)\,dx = \sqrt{\frac{2}{\pi}}\int_0^\infty x\,f(x)\cos sx\,dx$$

$$\frac{d}{ds}F_s(s) = F_c\{x\,f(x)\} \qquad \text{... (3)}$$

(*c*) $\qquad F_c\,\{x\,f(x)\} = \frac{d}{ds}F_s\,[f(x)]$ [From (3)]

$$\Rightarrow \quad F_c(xe^{-ax}) = \frac{d}{ds}F_s(e^{-ax})$$

$$= \frac{d}{ds}\left[\sqrt{\frac{2}{\pi}}\frac{s}{a^2+s^2}\right]$$ (Using example 15 on page 1217)

$$= \sqrt{\frac{2}{\pi}}\frac{(a^2+s^2)-s(2s)}{(a^2+s^2)^2} = \sqrt{\frac{2}{\pi}}\frac{a^2-s^2}{(a^2+s^2)^2}$$

(*d*) $\qquad F_s\{x\,f(x)\} = \frac{-d}{ds}F_c\{f(x)\}$ [From (2)]

$$F_s(xe^{-ax}) = -\frac{d}{ds}F_c(e^{-ax}) = -\frac{d}{ds}\left(\sqrt{\frac{2}{\pi}}\frac{a}{a^2+s^2}\right) = \sqrt{\frac{2}{\pi}}\frac{2as}{(a^2+s^2)^2}$$ **Ans.**

Example 29. *If* $F_c(s) = \frac{1}{2}\tan^{-1}\left(\frac{2}{s^2}\right)$, *then find* $f(x)$.

Solution. $\tan^{-1}\left(\frac{2}{s^2}\right) = \tan^{-1}\left[\frac{2}{(s^2-1)+1}\right] = \tan^{-1}\left[\frac{(s+1)-(s-1)}{1+(s+1)(s-1)}\right]$

$$= \tan^{-1}\left(\frac{\frac{1}{s-1}-\frac{1}{s+1}}{1+\left(\frac{1}{s+1}\right)\left(\frac{1}{s-1}\right)}\right) = \tan^{-1}\left(\frac{1}{s-1}\right)-\tan^{-1}\left(\frac{1}{s+1}\right)$$

The inversion Fourier Cosine formula is

$$f(x) = \sqrt{\frac{2}{\pi}}\int_0^\infty F(s)\cos sx\,dx \qquad \text{... (1)}$$

Putting the value of F (s) in (1), we get

$$f(x) = \sqrt{\frac{2}{\pi}}\int_0^\infty \frac{1}{2}\tan^{-1}\left(\frac{2}{s^2}\right)\cos sx\, ds = \sqrt{\frac{1}{2\pi}}\int_0^\infty \left\{\tan^{-1}\left(\frac{1}{s-1}\right) - \tan^{-1}\left(\frac{1}{s+1}\right)\right\}\cos sx\, ds$$

$$= \sqrt{\frac{1}{2\pi}}\left[\int_0^\infty \tan^{-1}\left(\frac{1}{s-1}\right)\cos sx\, ds - \int_0^\infty \tan^{-1}\left(\frac{1}{s+1}\right)\cos sx\, ds\right]$$

$$= I_1 - I_2 = I\ \text{(say)} \qquad \ldots(2)$$

Where $I_1 = \sqrt{\frac{1}{2\pi}}\left[\left\{\tan^{-1}\left(\frac{1}{s-1}\right)\cdot\frac{\sin sx}{x}\right\}_0^\infty - \int_0^\infty \frac{-1}{(s-1)^2}\cdot\frac{1}{\left\{1+\frac{1}{(s-1)^2}\right\}}\cdot\frac{\sin sx}{x}ds\right]$

$$I_1 = 0 + \frac{1}{x}\sqrt{\frac{1}{2\pi}}\int_0^\infty \frac{\sin sx}{(s-1)^2+1}ds$$

Similarly, $$I_2 = \frac{1}{x}\sqrt{\frac{1}{2\pi}}\int_0^\infty \frac{\sin sx}{(s+1)^2+1}ds$$

On putting the values of I_1 and I_2 in (2), we get

$$I = \frac{1}{x}\sqrt{\frac{1}{2\pi}}\int_0^\infty \left\{\frac{\sin sx}{(s-1)^2+1} - \frac{\sin sx}{(s+1)^2+1}\right\}ds$$

$$= \frac{1}{x}\sqrt{\frac{1}{2\pi}}\,\pi\left[-\pi e^{-x}\sin x - e^{-x}\sin(-x)\right]$$ [See result 3 on page 1247]

$$\Rightarrow \quad f(x) = \frac{1}{x}\sqrt{\frac{1}{2\pi}}\,\pi\left[e^{-x}\sin x + e^{-x}\sin x\right] = \frac{\pi}{x}\sqrt{\frac{2}{\pi}}e^{-x}\sin x$$

$$= \sqrt{\frac{\pi}{2}}\left[\frac{e^{-x}\sin x}{x}\right]$$ **Ans.**

Example 30. *By finding the Fourier Transform of f (x) =* $e^{-a^2x^2}$, $a > 0$, *show that the transform of* $e^{-\frac{x^2}{2}}$ is $e^{-\frac{s^2}{2}}$.

Solution. The Fourier Transform of $f(x)$ is

$$F_c[f(x)] = \sqrt{\frac{2}{\pi}}\int_0^\infty f(x)\cos sx\, dx \qquad \ldots(1)$$

Putting the value of $f(x)$ in (1), we get

$$F\{e^{-a^2x^2}\} = \sqrt{\frac{2}{\pi}}\int_{-\infty}^\infty e^{-a^2x^2}e^{isx}dx = \sqrt{\frac{2}{\pi}}\int_{-\infty}^\infty e^{-a^2x^2+isx+\frac{s^2}{4a^2}}e^{\frac{-s^2}{4a^2}}dx$$

$$= \sqrt{\frac{2}{\pi}}\int_{-\infty}^\infty e^{-\left[(ax)^2-\left(\frac{is}{2a}\right)^2+2(ax)\left(\frac{is}{2a}\right)\right]}\cdot e^{-\frac{s^2}{4a^2}}dx = \sqrt{\frac{2}{\pi}}\int_{-\infty}^\infty e^{-\left(ax-\frac{is}{2a}\right)^2}e^{-\frac{s^2}{4a^2}}dx$$

$$= e^{-\frac{s^2}{4a^2}}\cdot\sqrt{\frac{2}{\pi}}\int_{-\infty}^\infty e^{-t^2}\frac{dt}{a}$$ $\left[\begin{array}{l}\text{Putting } ax-\frac{is}{2a}=t \text{ and } a>0\\ \text{so that } a\,dx = dt \Rightarrow dx = \frac{dt}{a}\end{array}\right]$

$$= e^{-\frac{s^2}{4a^2}} \cdot \frac{1}{a}\sqrt{\frac{2}{\pi}} \times \frac{\sqrt{\pi}}{2} \qquad \left[\text{since } \int_{-0}^{\infty} e^{-t^2}\, dt = \frac{\sqrt{\pi}}{2}\right]$$

$$= \frac{1}{\sqrt{2}a} e^{-\frac{s^2}{4a^2}}$$

Putting $a = \frac{1}{\sqrt{2}}$, we get $F\left\{e^{-\frac{x^2}{2}}\right\} = e^{-\frac{s^2}{2}}$

Thus, $e^{-\frac{x^2}{2}}$ is self-reciprocal. **Proved.**

EXERCISE 45.2

1. Find the Fourier Transform of $f(x)$ if

$$f(x) = \begin{cases} x, & |x| \le a \\ 0, & |x| > a \end{cases}$$

Ans. $\frac{1}{\sqrt{2\pi}} \frac{2i}{s^2}(as \cos as - \sin as)$

2. Show that the Fourier Transform of

$$f(x) = \begin{cases} a - |x| & \textit{for } |x| < a \\ 0 & \textit{for } |x| > a > 0 \end{cases}$$

is $\sqrt{\frac{2}{\pi}}\left(\frac{1-\cos as}{s^2}\right)$.

Hence show that $\int_0^{\infty}\left(\frac{\sin t}{t}\right)^2 dt = \frac{\pi}{2}$

3. Show that the Fourier Transform of

$$f(x) = \begin{cases} \frac{\sqrt{2\pi}}{2a} & \text{for } |x| \le a \\ 0 & \text{for } |x| > a \end{cases}$$

is $\frac{\sin sa}{sa}$

4. Find Fourier Transform of $e^{-a|x|}$ if $a > 0$ and $x > 0$. **Ans.** $\frac{2a}{a^2+s^2}$

5. Find Fourier Transform of $\frac{1}{\sqrt{|x|}}$.

6. Find the Fourier Transform of $f(x) = \begin{cases} e^{ikx}, & a < x < b \\ 0, & x < a \text{ and } x > b \end{cases}$ **Ans.** $\frac{i}{\sqrt{2\pi}(k+s)}\left[e^{i(k+s)a} - e^{i(k+s)b}\right]$

7. Show that the Fourier Transform of

$$f(x) = \begin{cases} 0 & \text{for } x < \alpha \\ 1 & \text{for } \alpha < x < \beta \\ 0 & \text{for } x > \beta \end{cases}$$

is $\frac{1}{\sqrt{2\pi}} \frac{2i}{s^2}(as \cos as - \sin as)$

8. If $F(s)$ is the Fourier Transform of $f(x)$, prove that

$$f[e^{iax} f(x)] = F(s + a)$$

9. Find Fourier transform of

$$F(x)=\begin{cases}x^2, & |x|<a\\ 0, & |x|>a\end{cases}.$$ **Ans.** $\left(\frac{2a^2}{s}-\frac{4}{s^3}\right)\sin as+\frac{4a}{s^2}\cos as$

10. Show that the Fourier Sine Transform of $f(x)=\begin{cases}x & \text{for } 0<x<1\\ 2-x & \text{for } 1<x<2\\ 0 & \text{for } x>2\end{cases}$ is $\frac{2\sin s(1-\cos s)}{s^2}$.

11. Show that the Fourier Sine Transform of $\frac{x}{1+x^2}$ is $\sqrt{\frac{\pi}{2}}\,as\,e^{-as}$.

12. Find Fourier Sine Transform of

$$f(x)=\frac{1}{x(x^2+a^2)}$$ *(U.P.T.U. 2001)* **Ans.** $\frac{\pi}{2a^2}(1-e^{-as})$

13. Find Fourier Sine Transform of $xe^{-x^2/2}$. **Ans.** $\frac{1}{2}se^{-s^2/2}$

14. Find Fourier Sine Transform of $\frac{e^{-ax}-e^{-bx}}{x}$. **Ans.** $\sqrt{\frac{2}{\pi}}\left(\tan^{-1}\frac{s}{a}-\tan^{-1}\frac{s}{b}\right)$

15. Find Fourier Sine transform of $\frac{\cosh ax}{\sinh \pi x}$. **Ans.** $\frac{\sinh s}{2(\cos a+\cosh s)}$

16. Find the Fourier Cosine Transform of $\frac{e^{ax}+e^{-ax}}{e^{\pi x}-e^{-\pi x}}$. **Ans.** $\cos\frac{a}{2}\left(\frac{e^{\frac{s}{2}}+e^{-\frac{s}{2}}}{2\cos sa+e^{s}+e^{-s}}\right)$

17. Find the Fourier Sine and Cosine Transform of $ae^{-\alpha x}+be^{-\beta x}$, $\alpha, \beta>0$

Ans. $\frac{as}{s^2+\alpha^2}+\frac{bs}{s^2+\beta^2},\ \frac{a\alpha}{s^2+\alpha^2}+\frac{b\beta}{s^2+\beta^2}$

18. If $F(s)$ is the Fourier Transform of $f(x)$, prove that $F[x^n f(x)]=(-i)^n\frac{d^n}{ds^n}\{F(s)\}$

[Hint: See Art. 45.8 Properties of Fourier Transform given below]

19. Find the function $f(x)$ if its Cosine Transform is

$$F_c(s)=\begin{cases}\frac{1}{2\pi}\left(a-\frac{s}{2}\right), & s<2a\\ 0, & s\geq 2a\end{cases}$$ **Ans.** $\frac{2\sin^2 ax}{\pi^2x^2}$

20. Find $f(x)$ if its Fourier Sine Transform is $\frac{s}{1+s^2}$ **Ans.** e^{-x}

21. Find $f(x)$ if its Fourier Sine Transform is $e^{-\pi s}$ **Ans.** $\frac{2}{\pi}\left(\frac{s}{1+s^2}\right)$

22. Find $f(x)$ if its Fourier Sine Transform is $\frac{\pi}{2}$. **Ans.** $\frac{1}{s}$

23. Find $f(x)$ if its Fourier Sine Transform is $(2\pi s)^{\frac{1}{2}}$. **Ans.** $\frac{1}{x\sqrt{x}}$

24. Find $f(x)$ if its Fourier Sine Transform is $\begin{cases}\sin s, & 0<s<\pi\\ 0, & s\geq\pi\end{cases}$ **Ans.** $\frac{2}{\pi}\cdot\frac{\sin \pi x}{(1-x^2)}$

25. Find $f(x)$ whose Fourier Sine Transform is se^{-as}. **Ans.** $\frac{2}{\pi}\frac{\sin 2\theta}{a^2+x^2}$, where $\tan\theta=\frac{x}{a}$

45.8 PROPERTIES OF FOURIER TRANSFORMS

(1) **LINEAR PROPERTY.** If $F_1(s)$ and $F_2(s)$ are Fourier transforms of $f_1(x)$ and $f_2(x)$ respectively, then

$$F[af_1(x)+b\,f_2(x)]=a\,F_1(s)+b\,F_2(s)$$

where a and b are constants.

We know that $$F_1(s)=\frac{1}{\sqrt{2\pi}}\int_{-\infty}^{\infty}e^{isx}\cdot f_1(x)\,dx$$

and $$F_2(s)=\frac{1}{\sqrt{2\pi}}\int_{-\infty}^{\infty}e^{isx}f_2(x)\,dx$$

$$F[af_1(x)+b\,f_2(x)]=\frac{1}{\sqrt{2\pi}}\int_{-\infty}^{\infty}e^{isx}\left[af_1(x)+b\,f_2(x)\right]dx$$

$$=a\frac{1}{\sqrt{2\pi}}\int_{-\infty}^{\infty}e^{isx}f_1(x)\,dx+b\int_{-\infty}^{\infty}e^{isx}f_2(x)\,dx=a\,F_1(s)+b\,F_2(s)$$ **Proved.**

(2) **CHANGE OF SCALE PROPERTY**

If $F(s)$ is the complex Fourier transform of $f(x)$, then $F\{f(ax)\}=\frac{1}{a}F\left(\frac{s}{a}\right)$

Proof. We know that $F(s)=\frac{1}{\sqrt{2\pi}}\int_{-\infty}^{\infty}e^{isx}f(x)\,dx$

$$F\{f(ax)\}=\frac{1}{\sqrt{2\pi}}\int_{-\infty}^{\infty}e^{isx}f(ax)\,dx \qquad \left[\text{Put } ax=t \Rightarrow dx=\frac{dt}{a}\right]$$

$$=\frac{1}{\sqrt{2\pi}}\int_{-\infty}^{\infty}e^{is\frac{t}{a}}f(t)\frac{dt}{a}=\frac{1}{a}\frac{1}{\sqrt{2\pi}}\int_{-\infty}^{\infty}e^{i\left(\frac{s}{a}\right)t}f(t)\,dt=\frac{1}{a}F\left(\frac{s}{a}\right)$$ **Proved.**

(3) **SHIFTING PROPERTY**

If $F(s)$ is the complex Fourier transform of $f(x)$, then

$$F\{f(x-a)\}=e^{isa}F(s)$$

Proof. $$F(s)=\int_{-\infty}^{\infty}e^{isx}f(x)\,dx$$

$$F\{f(x-a)\}=\frac{1}{\sqrt{2\pi}}\int_{-\infty}^{\infty}e^{isx}f(x-a)\,dx \qquad [\text{Put } x-a=t, \text{ so that } dx=dt]$$

$$=\frac{1}{\sqrt{2\pi}}\int_{-\infty}^{\infty}e^{is(t+a)}f(t)\,dt=\frac{e^{isa}}{\sqrt{2\pi}}\int_{-\infty}^{\infty}e^{ist}f(t)\,dt=e^{isa}F(s)$$

Proved.

(4) $F\left\{e^{iax}f(x)\right\}=F(s+a)$

Proof. $$F\{f(x)e^{iax}\}=\frac{1}{\sqrt{2\pi}}\int_{-\infty}^{\infty}e^{iax}f(x)e^{isx}dx=\frac{1}{\sqrt{2\pi}}\int_{-\infty}^{\infty}e^{i(s+a)x}f(x)\,dx$$

$$=F(s+a)$$ **Proved.**

(5) **MODULATION THEOREM** [*U.P. III semester Dec. 2005*]

If $F(s)$ is the complex Fourier transform of $f(x)$, then

$$F\{f(x)\cos ax\}=\frac{1}{2}[F(s+a)+F(s-a)]$$

Proof. We know that F (s) $= \frac{1}{\sqrt{2\pi}}\int_{-\infty}^{\infty} e^{isx} f(x)\,dx$

$$F\ \{f(x)\cos ax\} = \frac{1}{\sqrt{2\pi}}\int_{-\infty}^{\infty} e^{isx} f(x)\cos ax\ dx = \frac{1}{\sqrt{2\pi}}\int_{-\infty}^{\infty} e^{isx} f(x)\frac{e^{iax}+e^{-iax}}{2}dx$$

$$= \frac{1}{2}\frac{1}{\sqrt{2\pi}}\int_{-\infty}^{\infty} e^{isx} f(x) e^{iax} dx + \frac{1}{2}\frac{1}{\sqrt{2\pi}}\int_{-\infty}^{\infty} e^{isx} f(x)\ e^{-iax} dx$$

$$= \frac{1}{2}\frac{1}{\sqrt{2\pi}}\int_{-\infty}^{\infty} e^{i(s+a)x} f(x)\,dx + \frac{1}{2}\frac{1}{\sqrt{2\pi}}\int_{-\infty}^{\infty} e^{i(s-a)x} f(x)dx$$

$$= \frac{1}{2}F(s+a) + \frac{1}{2}F(s-a) = \frac{1}{2}[F(s+a)+F(s-a)]$$ **Proved.**

(6) If **$F\ \{f(x)\} = -F(s)$, then $F\ \{x^n f(x)\}$** $= (-i)^n \frac{d^n}{ds^n}F(s).$

Proof. We know that $F(s) = \frac{1}{\sqrt{2\pi}}\int_{-\infty}^{\infty} e^{isx} f(x)\,dx$ ... (1)

Differentiating (1) w.r.t. s both sides, n times, we get

$$\frac{d^n F(s)}{ds^n} = \frac{1}{\sqrt{2\pi}}\int_{-\infty}^{\infty} (ix)^n e^{isx} f(x)\,dx = (i)^n \frac{1}{\sqrt{2\pi}}\int_{-\infty}^{\infty} x^n e^{isx}\cdot f(x)\cdot dx = (i)^n\ F(x^n\ f(x))$$

$$F(x^n f(x)) = (-i)^n \frac{d^n\{F(s)\}}{ds^n}$$

(7) **$F\{f'(x)\} = -$ is $F(s)$ if $f(x) \to 0$ as $x \to \pm\infty$**

Proof. $F\{f'(x)\}1 = \frac{1}{\sqrt{2\pi}}\int_{-\infty}^{\infty} e^{isx} f'(x)\,dx = \frac{1}{\sqrt{2\pi}}\int_{-\infty}^{\infty} e^{isx} d\{f(x)\}\,dx$

$$= \frac{1}{\sqrt{2\pi}}\left[\left\{e^{isx} f(x)\right\}_{-\infty}^{\infty} - is\int_{-\infty}^{\infty} e^{isx} f(x)\,dx\right] = \frac{1}{\sqrt{2\pi}}\left[0 - is\int_{-\infty}^{\infty} e^{isx}\ f(x)\,dx\right]$$

$$= -is\ F(s)$$ **Proved.**

(8) $\boldsymbol{F\left\{\int_a^x f(x)\,dx\right\} = \frac{F(s)}{(-is)}}$

Proof. Let $f_1(x) = \int_a^x f(x)\,dx \quad\Rightarrow\quad f_1'(x) = f(x)$

$$F\{f'(x)\} = (-is)F_1(s) = (-is)F\{f_1(x)\} = -is\,F\left\{\int_a^x f(x)\,dx\right\}$$

$$F\left\{\int_a^x f(x)\,dx\right\} = \frac{1}{(-is)}F\{f_1'(x)\} = \frac{1}{(-is)}F\{f(x)\} = \frac{F(s)}{(-is)}$$ **Proved.**

Note. $F_s(s)$ and $F_c(s)$ are Fourier Sine and Cosine transforms of $f(x)$ respectively.

Properties.

1. $F_s\{af(x)+bg(x)\} = aF_s\{f(x)\} + b\,F_s\{g(x)\}$
2. $F_c\{af(x)+bg(x)\} = aF_c\{f\{x\}\} + b\,F_c\{g(x)\}$
3. $F_s\{f(ax)\} = \frac{1}{a}F_s\left(\frac{s}{a}\right)$
4. $F_c\{f(ax)\} = \frac{1}{a}F_c\left(\frac{s}{a}\right)$

5. $F_s[f(x)\sin ax] = \frac{1}{2}[F_c(s-a) - F_c(s+a)]$

6. $F_c\{f(x)\sin ax\} = \frac{1}{2}[F_s(s+a) - F_s(s-a)]$

7. $F_s\{f(x)\cos ax\} = \frac{1}{2}[F_s(s+a) + F_s(s-a)]$

Proof of (5) : $F_s\ \{f(x)\ \sin ax\} = \int_0^\infty f(x)\ \sin ax\ .\ \sin sx\ dx$

$$= \frac{1}{2}\int_0^\infty f(x)\{\cos(s-a)x - \cos(s+a)x\}\,dx$$

$$= \frac{1}{2}\left[\int_0^\infty f(x)\cos(s-a)x\,dx - \int_{0.}^\infty f(x)\cos(s+a)x\,dx\right]$$

$$= \frac{1}{2}[F_c(s-a) - F_c(s+a)]$$ **Proved.**

45.9 CONVOLUTION

The Convolution of two functions $f(x)$ and $g(x)$ is defined as

$$f(x)*g(x) = \int_{-\infty}^{\infty} f(u)\,g(x-u)\,du$$

Convolution Theorem on Fourier Transform *(U.P., III Semester, Dec. 2006)*

The Fourier transform of the convolution of $f(x)$ and $g(x)$ is the product of their Fourier transforms, *i.e.,*

$$F[f(x)*g(x)] = F[f(x)]\cdot F[g(x)]$$

Proof. We know that $f(x)*g(x) = \frac{1}{\sqrt{2\pi}}\int_{-\infty}^{\infty} f(u)\cdot g(x-u)\,du$ ... (1)

Taking Fourier transform of both sides of (1), we have

$$F[f(x)*g(x)] = F\left[\int_{-\infty}^{\infty} f(u)\cdot g(x-u)\,du\right]$$

$$= \frac{1}{\sqrt{2\pi}}\int_{-\infty}^{\infty}\left[\frac{1}{\sqrt{2\pi}}\int_{-\infty}^{\infty} f(u)\cdot g(x-u)\,du\right]e^{isx}dx$$

$$= \frac{1}{\sqrt{2\pi}}\int_{-\infty}^{\infty} f(u)\,du\,\frac{1}{\sqrt{2\pi}}\left\{\int_{-\infty}^{\infty} g(x-u)e^{isx}dx\right\} = \frac{1}{\sqrt{2\pi}}\int_{-\infty}^{\infty} f(u)\cdot du\cdot F\{g(x-u)\}$$

$$= \frac{1}{\sqrt{2\pi}}\int_{-\infty}^{\infty} f(u)du\cdot e^{ius}G(s)$$ (Using shifting property)

$$= G(s)\cdot\frac{1}{\sqrt{2\pi}}\int_{-\infty}^{\infty} f(u)e^{ius}du = G(s)\,.\,F(s) = F(s)\,.\,G(s)$$ **Proved.**

By inversion

$$F^{-1}\{F(s)\cdot G(s)\} = f*g = F^{-1}\{F(s)\}*F^{-1}\{G(s)\}$$

Question: Verify the above statement to find the Fourier Inverse Transform of e^{-as} sin bs.

(U.P., III Semester, Dec. 2006)

45.10 PARSEVAL'S IDENTITY FOR FOURIER TRANSFORMS

(*U.P. III Semester Dec. 2005*)

If the Fourier transforfm of $f(x)$ and $g(x)$ be $F(s)$ and $G(s)$ respectively, then

(*i*) $\int_{-\infty}^{\infty} F(x)\overline{G}(s)\,ds = \int_{-\infty}^{\infty} f(x)\overline{g}(x)\,dx$

where $\overline{G}(s)$ is the complex conjugate of $G(s)$ and $\overline{g}(x)$ is the complex conjugate of $g(x)$

(*ii*) $\int_{-\infty}^{\infty} |F(s)|^2\,ds = \int_{-\infty}^{\infty} |f(x)|^2\,dx$

Proof. (*i*) $\int_{-\infty}^{\infty} [f(x)\overline{g}(x)dx = \int_{-\infty}^{\infty} f(x)\left[\frac{1}{\sqrt{2\pi}}\int_{-\infty}^{\infty} \overline{G}(s)e^{isx}ds\right]dx$

Since $$\overline{g}(x) = \frac{1}{\sqrt{2\pi}}\int_{-\infty}^{\infty} \overline{G}(s)e^{isx}ds$$

$$\int_{-\infty}^{\infty} f(x)\overline{g}(x)dx = \frac{1}{\sqrt{2\pi}}\int_{-\infty}^{\infty} \overline{G}(s)ds\cdot\int_{-\infty}^{\infty} f(x)e^{isx}dx$$

$$\left[\text{since } \frac{1}{\sqrt{2\pi}}\int_{-\infty}^{\infty} f(x)e^{isx}dx = F(s)\right] \text{ Fourier Transform}$$

$$= \int_{-\infty}^{\infty} \overline{G}(s)F(s)\,ds \qquad \text{... (1)}$$

Putting $g(x) = f(x)$ in (1), we get

$$\int_{-\infty}^{\infty} F(s)\cdot\overline{F}(s) = \int_{-\infty}^{\infty} f(x)\cdot\overline{f}(x)\,dx$$

$$\Rightarrow \quad \int_{-\infty}^{\infty} |F(s)|^2\,ds = \int_{-\infty}^{\infty} |f(x)|^2\,dx$$ **Proved.**

45.11 PARSEVAL'S IDENTITY FOR COSINE TRANSFORM

(*i*) $\frac{2}{\pi}\int_0^{\infty} F_C(s)\cdot G_C(s)\,ds = \int_0^{\infty} f(x)\cdot g(x)\,dx$ (*ii*) $\frac{2}{\pi}\int_0^{\infty} |F_c(s)|^2\,ds = \int_0^{\infty} |f(x)|^2\,dx$

45.12 PARSEVAL'S IDENTITY FOR SINE TRANSFORM

(*i*) $\frac{2}{\pi}\int_0^{\infty} F_s(s)\cdot G_s(s)\,ds = \int_0^{\infty} f(x)\cdot g(x)\,dx$ (*ii*) $\frac{2}{\pi}\int_0^{\infty} |F_s(s)|^2\,ds = \int_0^{\infty} |f(x)|^2\,dx$

Example 31. *Using Parseval's identity, show that*

$$\int_0^{\infty} \frac{x^2dx}{(x^2+1)^2} = \frac{\pi}{4}$$

Solution. Let $f(x) = \frac{x}{x^2+1}$ so that $F_s(s) = \frac{\pi}{2}e^{(-s)}$

By Parseval's identity for sine transformation

$$\frac{2}{\pi}\int_0^{\infty} |F_s(s)|^2\,ds = \int_0^{\infty} |f(x)|^2\,dx$$

$$\int_0^{\infty} \left|\frac{x}{x^2+1}\right|^2 dx = \frac{2}{\pi}\int_0^{\infty} \left|\frac{\pi}{2}e^{-s}\right|^2 ds$$

$$=\left(\frac{2}{\pi}\right)\left(\frac{\pi^2}{4}\right)\int_0^\infty |e^{-2s}|\,ds=\frac{\pi}{2}\left[\frac{e^{-2s}}{-2}\right]_0^\infty=\frac{\pi}{2}\left[0+\frac{1}{2}\right]=\frac{\pi}{4}$$ **Proved.**

Example 32. *Using Parseval's identity, prove that*

$$\int_0^\infty \frac{dt}{(a^2+t^2)(b^2+t^2)}=\frac{\pi}{2ab(a+b)}$$

Solution. Let $f(x)=e^{-ax},\ g(x)=e^{-bx}$

Then $$F_c(s)=\frac{a}{a^2+s^2},\ G_c(s)=\frac{b}{b^2+s^2}$$

By Parseval's identity for Fourier cosine transformation

$$\frac{2}{\pi}\int_0^\infty F_c(s)\,G_c(s)\,ds=\int_0^\infty f(x)\cdot g(x)\,dx \quad \text{... (1)}$$

On substituting the values of $F_c(s)$, $G_c(s)$, $f(x)$ and $g(x)$ in (1), we get

$$\frac{2}{\pi}\int_0^\infty\left(\frac{a}{a^2+s^2}\right)\left(\frac{b}{b^2+s^2}\right)ds=\int_0^\infty e^{-ax}\cdot e^{-bx}\,dx$$

$$\Rightarrow\quad \frac{2}{\pi}\int_0^\infty \frac{ab}{(a^2+s^2)}\frac{ds}{(b^2+s^2)}=\int_0^\infty e^{-(a+b)x}\,dx=\left[\frac{e^{-(a+b)x}}{-(a+b)}\right]_0^\infty=\left[0+\frac{1}{a+b}\right]$$

$$\Rightarrow\quad \int_0^\infty \frac{ds}{(a^2+s^2)(b^2+s^2)}=\frac{\pi}{2ab}\frac{1}{a+b}$$

$$\int_0^\infty \frac{dt}{(a^2+t^2)(b^2+t^2)}=\frac{\pi}{2ab(a+b)}$$ **Proved.**

Example 33. *Using Parseval's identity, prove* $\int_0^\infty\left(\frac{\sin t}{t}\right)^2 dt=\frac{\pi}{2}.$

Solution. We know that

if $$f(x)=\begin{cases}1 & \text{for } |x|< a\\ 0 & \text{for } |x|>a>0\end{cases}$$

then $$F(s)=\sqrt{\frac{2}{\pi}}\frac{\sin as}{s}$$

Using Parseval's identity

$$\int_{-\infty}^\infty |f(t)|^2\,dt=\int_{-\infty}^\infty |F(s)|^2\,ds$$

$$\therefore\quad \int_{-a}^a (1)^2\,dt=\int_{-\infty}^\infty \frac{2}{\pi}\left(\frac{\sin as}{s}\right)^2 ds$$

$$2a=\frac{2}{\pi}\int_{-\infty}^\infty\left(\frac{\sin as}{s}\right)^2 ds$$

Putting $as=t$, we get $$2a=\frac{2}{\pi}\int_{-\infty}^\infty\left(\frac{\sin t}{\frac{t}{a}}\right)^2\frac{dt}{a}\qquad\left(\because\ a\,ds=dt\Rightarrow ds=\frac{dt}{a}\right)$$

$$a\pi = a\int_{-\infty}^{\infty}\left(\frac{\sin t}{t}\right)^2 dt \quad \Rightarrow \quad \pi = \int_{-\infty}^{\infty}\left(\frac{\sin t}{t}\right)^2 dt$$

$$\Rightarrow \quad \int_0^{\infty}\left(\frac{\sin t}{t}\right)^2 dt = \frac{\pi}{2}$$ **Proved.**

Exampe 34. *Find the Fourier transform of*

$$f(x) = \begin{cases} 1 - |x| & \text{if } |x| < 1 \\ 0 & \text{if } |x| > 1 \end{cases}$$

and hence find the value of $\int_0^{\infty} \frac{sin^4 t}{t^4} dt$.

Solution. $F\{f(x)\} = \frac{1}{\sqrt{2\pi}}\int_{-\infty}^{\infty}(1-|x|)e^{isx}dx$

$$= \frac{1}{\sqrt{2\pi}}\int_{-\infty}^{\infty}(1-|x|)(\cos sx + i\sin sx)\,dx$$

$$= \frac{1}{\sqrt{2\pi}}\int_{-1}^{1}(1-|x|)\cos sx\,dx + \frac{i}{\sqrt{2\pi}}\int_{-1}^{1}(1-|x|)\sin sx\,dx$$

(Even function) (odd function)

$$= \frac{2}{\sqrt{2\pi}}\int_0^1 (1-x)\cos sx\,dx + 0 = \sqrt{\frac{2}{\pi}}\left[\left\{(1-x)\frac{\sin sx}{s}\right\}_0^1 + \int_0^1 \frac{\sin sx}{s}dx\right]$$

$$= \sqrt{\frac{2}{\pi}}\left[0 + \left\{\frac{-\cos sx}{s^2}\right\}_0^1\right] = \sqrt{\frac{2}{\pi}}\left(\frac{1-\cos s}{s^2}\right)$$

Using Parseval's identity, we get $\int_{-\infty}^{\infty} |F(s)|^2\,ds = \int_{-\infty}^{\infty} |f(t)|^2\,dt$

$$\frac{2}{\pi}\int_{-\infty}^{\infty}\frac{(1-\cos s)^2}{s^4}ds = \int_{-1}^{+1}(1-|x|)^2 dx$$

$$\frac{4}{\pi}\int_0^{\infty}\frac{\left(1-1+2\sin^2\frac{s}{2}\right)^2}{s^4}ds = \int_{-1}^{+1}(1+x^2-2|x|)dx$$

(L.H.S. is even function, $2x$ on R.H.S. is odd function)

$$\frac{16}{\pi}\int_0^{\infty}\frac{\sin^4\frac{s}{2}}{s^4}ds = 2\int_0^1(1+x^2-2x)dx = 2\left(x+\frac{x^3}{3}-x^2\right)_0^1 = \frac{2}{3}$$

Putting $\frac{s}{2} = x$, we get

$$\frac{16}{\pi}\int_0^{\infty}\frac{\sin^4 x}{16x^4}2\,dx = \frac{2}{3} \quad \Rightarrow \quad \int_0^{\infty}\frac{\sin^4 x}{x^4}dx = \frac{\pi}{6}$$ **Ans.**

Example 35. *Solve for f(x) from the integral equation*

$$\int_0^{\infty} f(x)\cos sx\,dx = e^{-s}$$

Solution. $\int_0^\infty f(x)\cos sx\,dx = e^{-s}$ ... (1)

Multiplying (1) by $\sqrt{\frac{2}{\pi}}$, we get $\sqrt{\frac{2}{\pi}}\int_0^\infty f(x)\cos sx\,dx = \sqrt{\frac{2}{\pi}}\,e^{-s}$

$$F_c\{f(x)\} = \sqrt{\frac{2}{\pi}}\,e^{-s}$$

$$f(x) = F_c^{-1}\left[\sqrt{\frac{2}{\pi}}e^{-s}\right] = \sqrt{\frac{2}{\pi}}\left[\sqrt{\frac{2}{\pi}}\int_0^\infty e^{-s}\cos sx\,ds\right]$$

$$= \frac{2}{\pi}\int_0^\infty e^{-s}\cos sx\,ds = \frac{2}{\pi}\left[\frac{e^{-s}}{1+x^2}\{-\cos sx + x\sin sx\}\right]_0^\infty = \frac{2}{\pi}\frac{1}{1+x^2}$$ **Ans.**

Example 36. *Solve for f (x) from the integral equation*

$$\int_0^\infty f(x)\sin sx\,dx = \begin{cases} 1 \text{ for } 0 \le s < 1 \\ 2 \text{ for } 1 \le s < 2 \\ 0 \text{ for } \quad s \ge 2 \end{cases}$$

Solution. Multiplying by $\sqrt{\frac{2}{\pi}}$ both sides of the given equation, we get

$$\sqrt{\frac{2}{\pi}}\int_0^\infty f(x)\sin sx\,dx = \begin{cases} \sqrt{\frac{2}{\pi}} \text{ for } 0 \le s < 1 \\ 2\sqrt{\frac{2}{\pi}} \text{ for } 1 \le s < 2 \\ 0 \text{ for } s \ge 2 \end{cases}$$

$$F_s[f(x)] = \begin{cases} \sqrt{\frac{2}{\pi}} \text{ for } 0 \le s < 1 \\ 2\sqrt{\frac{2}{\pi}} \text{ for } 1 \le s < 2 \\ 0 \text{ for } s \ge 2 \end{cases}$$

$$f(x) = F_s^{-1} \quad \text{(R.H.S.)} = \sqrt{\frac{2}{\pi}}\int_0^1 \sqrt{\frac{2}{\pi}}\sin sx\,ds + \sqrt{\frac{2}{\pi}}\int_1^2 2\sqrt{\frac{2}{\pi}}\sin sx\,ds$$

$$= \frac{2}{\pi}\left[\frac{-\cos sx}{x}\right]_0^1 + \frac{4}{\pi}\left[\frac{-\cos sx}{x}\right]_1^2 = \frac{2}{\pi}\left(\frac{1-\cos x}{x}\right) + \frac{4}{\pi}\left(\frac{\cos x - \cos 2x}{x}\right)$$

$$= \frac{2}{\pi x}[1 - \cos x + 2\cos x - 2\cos 2x]$$

$$f(x) = \frac{2}{\pi x}(1 + \cos x - 2\cos 2x)$$ **Ans.**

Example 37. *Prove (i) $F\{f^n(x)\} = (-is)^n F(s)$*

(ii) Hence solve for f (x) if $\int_{-\infty}^{\infty} f(t)e^{|x-t|}dt = \phi(x)$ *is known.*

Proof. (*i*) See property (7) on page 1231 $F\{f'(x)\} = -is\,F(s)$

Similarly $$F\{f''(x)\} = (-is)^2\,F(s)$$

$$F\{f'''(x)\} = (-is)^3\,F(s)$$

Using integration by parts successively and making assumptions that $f, f', \ldots, f^{(n-1)} \to 0$ as $f(x) \to \pm\infty$.

$$F\{f^n(x)\} = \frac{1}{\sqrt{2\pi}}\int_{-\infty}^{\infty} e^{isx}\frac{d^n}{dx^n}f(x)\cdot dx \quad = (-is)^n F(s).$$

(*ii*) $$\frac{1}{\sqrt{2\pi}}\phi(x) = \frac{1}{\sqrt{2\pi}}\int_{-\infty}^{\infty} f(t)e^{|x-t|}dt, \text{ from the given equation } = f(x) * e^{-|x|}$$

By convolution theorem,

$$\frac{1}{\sqrt{2\pi}}\bar{\phi}(s) = F(s)\cdot\sqrt{\frac{2}{\pi}}\frac{1}{1+s^2}$$

$$F(s) = \frac{1}{2}(1+s^2)\bar{\phi}(s) \quad = \frac{1}{2}[\bar{\phi}(s) - (-is)^2\bar{\phi}(s)]$$

$\therefore$ $$f(x) = \frac{1}{2}\phi(x) - \frac{1}{2}\phi''(x) \text{ using the result derived in } (i)$$

EXERCISE 45.3

Using Parseval's identity

1. Evaluate $\int_0^\infty \left(\frac{1-\cos x}{x}\right)^2 dx$ **Ans.** $\frac{\pi}{2}$

2. Prove that $\int_0^\infty \frac{\sin at}{t(a^2+t^2)}dt = \frac{\pi}{2}\frac{1-e^{-a^2}}{a^2}$

45.13 FOURIER TRANSFORM OF DERIVATIVES

We have already seen that,

$$F\{f^n(x)\} = (-is)^n\,F(s)$$

(*i*) $\therefore$ $F\left(\frac{\partial^2 u}{\partial x^2}\right) = (-is)^2\,F\{u(x)\} = -s^2\bar{u}$ where $\bar{u}$ is Fourier transform of u w.r.t. x.

(*ii*) $$F_c\{f'(x)\} = -\sqrt{\frac{2}{\pi}}f(o) + s\,F_s(s)$$

$$\text{L.H.S.} = \sqrt{\frac{2}{\pi}}\int_0^\infty f'(x)\cdot\cos sx\,dx = \sqrt{\frac{2}{\pi}}\int_0^\infty \cos sx\,d\{f(x)\}$$

$$= \sqrt{\frac{2}{\pi}}\left[\{f(x)\cos sx\}_0^\infty + s\int_0^\infty f(x)\sin sx\,dx\right]$$

$$= s\,F_s(s) - \sqrt{\frac{2}{\pi}}f(0) \text{ assuming } f(x) \to 0 \text{ as } x \to \infty$$

(*iii*) $$F_s\{f'(x)\} = \sqrt{\frac{2}{\pi}}\int_0^\infty \sin sx\,d\,[f(x)] = \sqrt{\frac{2}{\pi}}\left[\{f(x)\sin sx\}_0^\infty - s\int_0^\infty f(x)\cos sx\,dx\right]$$

$$= -s F_c(s)$$

(iv) $F_c\{f''(x)\} = \sqrt{\dfrac{2}{\pi}} \int_0^\infty \cos sx\, d\,[f'(x)] = \sqrt{\dfrac{2}{\pi}} \left[\{f'(x) \cos sx\}_0^\infty + s \int_0^\infty f'(x) \sin sx\, dx \right]$

$$= -\sqrt{\frac{2}{\pi}} f'(0) + s F_s\{f'(x)\} = -s^2 F_c(s) - \sqrt{\frac{2}{\pi}} f'(0) \text{ assuming } f(x),\ f'(x) \to 0 \text{ as } x \to \infty$$

(v) $F_s\{f''\}(x) = \sqrt{\dfrac{2}{\pi}} \left[\int_0^\infty \sin sx\, d[f'(x)] \right] = \sqrt{\dfrac{2}{\pi}} \left[(f'(x) \sin sx)_0^\infty - s \int_0^\infty f'(x) \cos sx\, dx \right]$

$$= -s F_c\{f'(x)\} = -s \left[s F_s(s) - \sqrt{\frac{2}{\pi}} f(0) \right]$$

$$= -s^2 F_s(s) + \sqrt{\frac{2}{\pi}}\, s f(0) \text{ assuming } f(x),\ f'(x) \to 0 \text{ as } x \to \infty.$$

45.14 RELATIONSHIP BETWEEN FOURIER AND LAPLACE TRANSFORMS

Consider

$$f(t) = \begin{cases} e^{-st} g(t) \text{ for } t > 0 \\ 0 \quad \text{for } t < 0 \end{cases} \qquad \text{... (1)}$$

Then the Fourier transform of $f(t)$ is given by

$$F\{f(t)\} = \frac{1}{\sqrt{2\pi}} \int_{-\infty}^{\infty} e^{ist} f(t)\, dt = \frac{1}{\sqrt{2\pi}} \int_0^\infty e^{ist} f(t)\, dt$$

$$= \frac{1}{\sqrt{2\pi}} \int_0^\infty e^{(is-x)t} g(t)\, dt = \frac{1}{\sqrt{2\pi}} \int_0^\infty e^{-pt} g(t)\, dt \text{ where } p = x - is = \frac{1}{\sqrt{2\pi}} L\{g(t)\}$$

$\therefore$ Fourier transform of $f(t) = \dfrac{1}{\sqrt{2\pi}} \times$ Laplace transform of $g(t)$ defined by (1).

45.15 SOLUTION OF BOUNDARY VALUE PROBLEMS BY USING INTEGRAL TRANSFORM

Solution of heat conduction problems by Laplace transform.

Example 38. *Use Fourier sine transform to solve the equation*

$$\frac{\partial u}{\partial t} = k \frac{\partial^2 u}{\partial x^2}$$

under the conditions

(i) $u(0, t) = 0$ *(ii)* $u(x, 0) = e^{-x}$

(iii) $u(x, t)$ *is bounded.* *(U.P. III Semester, Dec. 2006)*

Solution. The given equation is

$$\frac{\partial u}{\partial t} = k \frac{\partial^2 u}{\partial x^2} \qquad \text{... (1)}$$

Taking Fourier sine transform of both the sides, we get

$$\int_0^\infty \frac{\partial u}{\partial t} \sin sx\, dx = k \int_0^\infty \frac{\partial^2 u}{\partial x^2} \sin sx\, dx$$

$$\frac{d\bar{u}}{dt} = k\left[s(u)_{x=0} - s^2\,\bar{u}\right]$$

$$\Rightarrow \quad \frac{d\bar{u}}{dt} + ks^2\,\bar{u} = 0 \qquad \left[\text{where } \bar{u} = \int_0^\infty u \sin sx\, dx\right]$$

$$\Rightarrow \quad D\bar{u} + ks^2\,\bar{u} = 0$$

$$\Rightarrow \quad (D + ks^2)\,\bar{u} = 0$$

A.E. is $\quad D + ks^2 = 0 \;\Rightarrow\; D = -ks^2$

$$\text{C.F.} = c_1\, e^{-ks^2 t}$$

$$\bar{u} = c_1\, e^{-ks^2 t} \qquad \ldots (2)$$

At $t = 0$, $\quad (\bar{u})_{t=0} = \int_0^\infty (u)_{t=0} \sin sx\, dx = \int_0^\infty e^{-x} \sin sx\, dx$

$$= \frac{e^{-s}}{1+s^2}\left[-\sin sx - s\cos sx\right]_0^\infty \qquad \left[\int e^{ax} \sin bx\, dx = \frac{e^{ax}}{a^2+b^2}(a \sin bx - b\cos bx)\right]$$

$$= \frac{s}{1+s^2} \qquad \ldots (3)$$

From (2), $\quad (\bar{u})_{t=0} = c_1 \qquad \ldots (4)$

From (3) and (4), we have

$$c_1 = \frac{s}{1+s^2}$$

From (2), $\quad \bar{u} = \dfrac{s}{1+s^2} e^{-ks^2 t}$

Taking inverse Fourier sine transform, we get

$$u(x, t) = \sqrt{\frac{2}{\pi}} \int_0^\infty \sqrt{\frac{2}{\pi}}\, \frac{s}{1+s^2}\, e^{-ks^2 t} \sin sx\, ds.$$

$$\Rightarrow \quad u(x, t) = \frac{2}{\pi}\int_0^\infty \frac{s}{1+s^2} e^{-ks^2 t} \sin sx\, ds \qquad \textbf{Ans.}$$

Example 39. *Solve the equation* $\dfrac{\partial u}{\partial t} = \dfrac{\partial^2 u}{\partial x^2}, \qquad x>0, t>0$

subject to the conditions:

(*i*) $u = 0$, *when* $x = 0$, $t > 0$ $\qquad$ (*ii*) $u = \begin{cases} 1, & 0 \le x < 1 \\ 0 & x > 1 \end{cases}$, *when* $t = 0$

(*iii*) u (x, t) *is bounded.* (*U.P., III Semester Dec. 2003*)

Solution. In view of the initial condition (*i*), we apply Fourier Sine Transform to both sides of the given equation.

$$\int_0^\infty \frac{\partial u}{\partial t} \sin sx\, dx = \int_0^\infty \frac{\partial^2 u}{\partial x^2} \sin sx\, dx$$

$$\Rightarrow \quad \frac{\partial}{\partial t}\int_0^\infty u \sin sx\, dx = -s^2\,\bar{u}(s) + su(0)$$

(See Article 45.13 on page 1236)

On putting $u = 0$ and $x = 0$, we get

$$\Rightarrow \qquad \frac{\partial \bar{u}}{\partial t} + s^2\bar{u} = 0, \qquad \Rightarrow \qquad \left(\frac{\partial}{\partial t} + s^2\right)\bar{u} = 0$$

Its solution *i.e.* complementary function is $\bar{u}(s,t) = Ae^{-s^2 t}$ ... (1)

Since we have assumed

$$\bar{u} = \bar{u}(s,t) = \int_0^\infty u(x,t)\sin sx\,dx, \quad \Rightarrow \quad \bar{u}(s,0) = \int_0^\infty u(x,0)\sin sx\,dx$$

$$\Rightarrow \qquad \bar{u}(s,0) = \int_0^1 1.\sin sx\,dx = \left[\frac{-\cos sx}{s}\right]_0^1 = \frac{1-\cos s}{s} \qquad [u(x,0)=1] \quad \text{... (2)}$$

From (1) and (2), we get $A = \bar{u}(s,0) = \dfrac{1-\cos s}{s}$

On putting the value of A in (1), we get

Hence, $\bar{u}(s,t) = \dfrac{1-\cos s}{s}\,e^{-s^2 t}$

Now applying the Inverse Fourier Sine Transform, we have

$$u(x,t) = \frac{2}{\pi}\int_0^\infty \left(\frac{1-\cos s}{s}\right).\,e^{-s^2 t}\,ds,$$

which is the required solution. **Ans.**

Solution of wave equation by Laplace transform

Example 40. *An infinitely long string having one end at x = 0 is initialy at rest along x-axis. The end x = 0 is given a transverse displacement f(t), when t > 0. Find the displacement of any point of the string at any time.*

Solution. Let $y(x, t)$ be the displacement, then wave equation is

$$\frac{\partial^2 y}{\partial x^2} = c^2\frac{\partial^2 y}{\partial t^2} \qquad \text{... (1)}$$

subject to the conditions

$$y(x,0) = 0 \quad \text{... (2)} \qquad \frac{\partial y}{\partial t}(x,0) = 0 \quad \text{... (3)}$$

$$y(0,t) = f(t) \quad \text{... (4)} \qquad y(x,t) \text{ is bounded} \quad \text{... (5)}$$

On taking Laplace transform of (1), we have

$$L\left(\frac{\partial^2 y}{\partial x^2}\right) = c^2 L\frac{\partial^2 y}{\partial t^2}$$

$$s^2\bar{y} - sy(x,0) - \frac{\partial y}{\partial t}(x,0) = c^2\frac{d^2\bar{y}}{dx^2} \qquad \text{... (6)}$$

On putting $y(x, 0) = 0$, $\dfrac{\partial y}{\partial t}(x, 0) = 0$ in (6) get

$$s^2\,\bar{y} = c^2\frac{d^2\bar{y}}{dx^2} \quad \text{or} \quad \frac{d^2\bar{y}}{dx^2} = \left(\frac{s}{c}\right)^2\bar{y} \qquad \text{... (7)}$$

Laplace transform of (4), $\bar{y}(0,s) = \bar{f}(s)$ at $x = 0$ ... (8)

On solving (7), we get $\bar{y} = Ae^{\frac{sx}{c}} + Be^{-\frac{sx}{c}}$ ... (9)

According to condition (5), y is finite at $x \to \infty$, this gives $A = 0$.
Equation (9) becomes

$$\bar{y} = Be^{-\frac{sx}{c}} \qquad \text{... (10)}$$

Putting the value of $\bar{y}(0, s) = \bar{f}(s)$ at $x = 0$ in (10), we get $\bar{f}(s) = B$

Thus (10) becomes $\quad y = \bar{f}(s) \cdot e^{-\frac{sx}{c}}$

To get y from $\bar{y}$, we use complex inversion formula

$$y = \frac{1}{2\pi i}\int_{a-i\infty}^{a+i\infty} e^{\left(t-\frac{x}{c}\right)s} - f(s)\,ds$$

Hence $\quad y = f\left(t - \frac{x}{c}\right)$ **Ans.**

Solution of Transmission Lines equations by Laplace Transformations.

Example 41. *A semi-infinite transmission line, of negligible inductance and leakage per unit length has its voltage and current equal to zero. A constant voltage v_0 is applied at the sending end ($x = 0$) at $t = 0$. Find the voltage and current at any point ($x > 0$) at any instant.*

Solution. Let v and i be the voltage and current at any point x and at any time t.

$$\left.\begin{aligned} -\frac{\partial v}{\partial x} &= Ri + L\frac{\partial i}{\partial t} \\ -\frac{\partial i}{\partial x} &= c\frac{\partial v}{\partial t} + GV \end{aligned}\right]$$

On putting $L = 0$, $G = 0$ in above equations, we get

$$\frac{\partial v}{\partial x} = -Ri \qquad \text{... (1)}$$

$$\frac{\partial i}{\partial x} = -c\frac{\partial v}{\partial t} \qquad \text{... (2)}$$

Conditions are $v(x, 0) = 0$ ... (3)

$i(x, 0) = 0$ ... (4)

$v(0, t) = v_0$ ... (5)

$v(x, t)$ finite for all x and t. ... (6)

Applying Laplace transform of (1) and (2), we get

$$\frac{d\bar{v}}{dx} = -R\bar{i} \qquad \text{... (7)}$$

$$\frac{di}{dx} = -c(s\bar{v} - v) \text{ and } v(x, 0) = 0 \text{ or } \frac{di}{dx} = -cs_{\bar{v}} \qquad \text{... (8)}$$

Differentiating (7) w.r.t. 'x', we get $\quad \frac{d^2\bar{v}}{dx^2} = -R\frac{di}{dx}$

$$\Rightarrow \quad \frac{d^2\bar{v}}{dx^2} = -R(-cs\bar{v}) \qquad \left(\frac{d\bar{v}}{dx} = -cs\bar{v}\right)$$

$$\Rightarrow \quad \frac{d^2\bar{v}}{dx^2} = Rcs\bar{v} \Rightarrow \frac{d^2\bar{v}}{dx^2} - Rcs\bar{v} = 0$$

$$\Rightarrow \quad (D^2 - Rcs)\bar{v} = 0 \Rightarrow D = \pm\sqrt{Rcs}) \qquad \text{... (9)}$$

Laplace transform of (5) is $\bar{v}(0, s) = \frac{v_0}{s}$ ... (10)

And Laplace transform of (6) is $\overline{v}(x,s)$ remains finite as $x \to \infty$ ... (11)

Equation (9) is an ordinary differential equation and its solution is

$$\overline{v} = Ae^{\sqrt{Rcs}\,x} + Be^{-\sqrt{Rcs}\,x} \quad .. (12)$$

As $x \to \infty, \overline{v}$ remains finite only when $A = 0$.

So (12) becomes $\overline{v} = Be^{-\sqrt{Rcs}\,x}$ ... (13)

Putting $\overline{v} = \frac{v_0}{s}$ in (13), we get $\frac{v_0}{s} = Be^{-\sqrt{Rcsx}}$ ...(14)

Putting $x = 0$ in (14), we get $\frac{v_0}{s} = B$

Substituting the value of B in (13) we have

$$\overline{v} = \frac{v_0}{s} e^{-\sqrt{Rcs}\,x}$$

On applying inversion transform, we get

$$v = v_0 \, L^{-1}\left[\frac{e^{-\sqrt{Rcs}\,x}}{s}\right] = v_0 \, erf\,c\left[\frac{x\sqrt{Rc}}{2\sqrt{t}}\right]$$

$$v = v_0 \frac{x\sqrt{Rc}}{2\sqrt{\pi}} \int_0^t u^{-\frac{3}{2}} e^{-\left(\frac{Rcx^2}{4u}\right)} \cdot du \quad ... (15)$$

From (1) $$i = \frac{-1}{R}\frac{\partial v}{\partial x} \quad ... (16)$$

On differentiating (15), we get $\frac{\partial v}{\partial x} = \frac{v_0 \, x\sqrt{Rc}}{2\sqrt{\pi}} t^{-\frac{3}{2}} e^{-\frac{Rcx^2}{4t}}$

Putting the value of $\frac{\partial v}{\partial x}$ in (16), we obtain $i = \frac{v_0 x}{2\sqrt{\pi}}\sqrt{\frac{c}{R}} \cdot t^{-\frac{3}{2}} e^{-\frac{Rcx^2}{4t}}$ **Ans.**

Solution of partial differential Equations by Fourier Transform

Example 42. *Solve* $\frac{\partial^2 u}{\partial t^2} = \alpha^2 \frac{\partial^2 u}{\partial x^2}$, $\quad -\infty < x < \infty, \quad t \geq 0$

with conditions u (x, 0) = f(x),

$\frac{\partial u}{\partial t}$ *(x, 0) = g (x) and assuming u,* $\frac{\partial u}{\partial x} \to 0$ *as* $x \to \pm\infty$.

Solution. Taking Fourier transform on both sides of the differential equation,

$\frac{d^2\overline{u}}{dt^2} = \alpha^2(-s^2\overline{u})$ where $\overline{u}$ is Fourier transform of u with respect to x.

$$\frac{d^2\overline{u}}{dt^2} + \alpha^2 s^2 \overline{u} = 0$$

Auxiliary equation is $D^2 + \alpha^2 s^2 = 0 \Rightarrow D = \pm i\alpha s$

$$\therefore \quad \overline{u}(s,t) = Ae^{i\alpha st} + Be^{-i\alpha st} \quad ... (1)$$

Since $$u\,(x,0) = f(x) \text{ and } \frac{\partial u}{\partial t}(x,0) = g(x),$$

$$\overline{u}(s,0) = F(s) \text{ and } \frac{d\overline{u}}{dt}(s,0) = G(s) \text{ on taking transform.}$$

Using these condition in (1), we get

$$\bar{u}(s,0) = A + B = F(s) \quad \text{... (2)}$$

$$\frac{d\bar{u}}{dt}(s,0) = i\alpha s(A-B) = G(s) \quad \text{...(3)}$$

Solving (2) and (3), we get

$$A = \frac{1}{2}\left[F(s) + \frac{G(s)}{i\alpha s}\right]$$

$$B = \frac{1}{2}\left[F(s) - \frac{G(s)}{i\alpha s}\right]$$

Using these values in (1), we get

$$\bar{u}(s,t) = \frac{1}{2}\left[F(s) + \frac{G(s)}{i\alpha s}\right]e^{i\alpha st} + \frac{1}{2}\left[F(s) - \frac{G(s)}{i\alpha s}\right]e^{-i\alpha st} \quad \text{... (4)}$$

By inversion theorem, (4) reduces to

$$u(x,t) = \frac{1}{2}\left[f(x-\alpha t) - \frac{1}{\alpha}\int_{\alpha}^{x-\alpha t} g(\theta)\,d\theta\right] + \frac{1}{2}\left[f(x+\alpha t) + \frac{1}{\alpha}\int_{\alpha}^{x+\alpha t} g(\theta)\,d\theta\right]$$

Using the result

$$F\left(\int_{\alpha}^{x} f(t)\,dt\right) = \frac{F(s)}{(-is)}$$

Ans.

45.16 FOURIER TRANSFORMS OF PARTIAL DERIVATIVE OF A FUNCTION

$$F_f\left[\frac{\partial^2 u}{\partial x^2}\right] = -s^2 F(u) \quad \text{where } F(u) \text{ is Fourier transform of } u \text{ w.r.t. } x.$$

$$F_s\left[\frac{\partial^2 u}{\partial x^2}\right] = s(u)_{x=0} - s^2 F_s(u) \quad \text{(sine transform)}$$

$$F_c\left[\frac{\partial^2 u}{\partial x^2}\right] = -\left[\frac{\partial u}{\partial x}\right]_{x=0} - s^2 F_c(u) \quad \text{(cosine transform)}$$

Proof. Let $F[u(x,t)]$ be the Fourier transform of the function $u(x,t)$, *i.e.*

$$F[u(x,t)] = \int_{-\infty}^{\infty} e^{isx} u(x,t)\,dx$$

The Fourier transform of $\frac{\partial^2 u}{\partial x^2}$ is given by $F\left[\frac{\partial^2 u}{\partial x^2}\right] = \int_{-\infty}^{\infty} e^{isx}\frac{\partial^2 u}{\partial x^2}\,dx.$

Integrating by parts, we have

$$F\left[\frac{\partial^2 u}{\partial x^2}\right] = \left[e^{isx}\frac{\partial u}{\partial x} - \int ise^{isx}\frac{\partial u}{\partial x}\,dx\right]_{-\infty}^{\infty}$$

$$= \left[e^{isx}\frac{\partial u}{\partial x} - ise^{isx}u + \int (is)^2 e^{isx}u\,dx\right]_{-\infty}^{\infty} \quad \text{Again integrating}$$

$$= \left[0 - 0 - s^2\int_{-\infty}^{\infty} e^{isx}.u\,dx\right] \quad \left[u = 0, \frac{\partial u}{\partial x} = 0 \text{ when } x \to \infty\right]$$

Thus $$F\left[\frac{\partial^2 u}{\partial x^2}\right] = -s^2 F[u(x,t)]$$

Similarly, the Fourier sine transform of $\frac{\partial^2 u}{\partial x^2}$ is given by

$$F_s\left[\frac{\partial^2 u}{\partial x^2}\right] = \int_0^\infty \frac{\partial^2 u}{\partial x^2} \sin sx\, dx$$

$$\Rightarrow \quad F_s\left[\frac{\partial^2 u}{\partial x^2}\right] = s[u(x,t)]_{x=0} - s^2 F_s[u(s,t)] \qquad \text{(sine transform)}$$

and $$F_c\left[\frac{\partial^2 u}{\partial x^2}\right] = -\left[\frac{\partial u}{\partial x}\right]_{x=0} - s^2 F_c[u(s,t)] \qquad \text{(cosine transform)}$$

45.17 APPLICATIONS TO SIMPLE HEAT TRANSFER EQUATIONS

Solution of heat conduction problems by Fourier sine Transforms

Example 43. *Solve the equation*

$$\frac{\partial u}{\partial t} = \frac{\partial^2 u}{\partial x^2}, \qquad x > 0,\ t > 0$$

subject to the conditions

(i) $u = 0$ *when* $x = 0,\ t > 0$ *(ii)* $u = \begin{cases} 1, & 0 < x < 1 \\ 0, & x \geq 1 \end{cases}$ *when* $t = 0$

(iii) $u(x, t)$ *is bounded.*

(**Note.** If u at $x = 0$ is given, take Fourier sine transform and if $\frac{\partial u}{\partial x}$ at $x = 0$ is given, use Fourier cosine transform.)

Solution. In view of the initial conditions, we apply Fourier sine transform

$$\int_0^\infty \frac{\partial u}{\partial t} \sin sx\, dx = \int_0^\infty \frac{\partial^2 u}{\partial x^2} \sin sx\, dx \qquad \left[\frac{\partial u}{\partial t} = \frac{\partial^2 u}{\partial x^2}\right]$$

$$\frac{\partial}{\partial t}\int_0^\infty u \sin sx\, dx = -s^2\,\bar{u}(s) + s\,u(0) \qquad u = 0 \text{ when } x = 0$$

$$\frac{\partial \bar{u}}{\partial t} = -s^2 \bar{u} \quad \text{or} \quad \frac{\partial \bar{u}}{\partial t} + s^2 \bar{u} = 0 \Rightarrow (D + s^2)\bar{u} = 0$$

A.E. is $m + s^2 = 0 \quad \Rightarrow \quad m = -s^2$

$$\therefore \quad \bar{u} = Ae^{-s^2 t} \qquad \text{... (1)}$$

$$\bar{u} = \bar{u}(s,t) = \int_0^\infty u(x,t) \sin sx\, dx$$

$$\bar{u} = \bar{u}(s, 0) = \int_0^\infty u(x, 0) \sin sx\, dx$$

$$\bar{u}(s, 0) = \int_0^\infty 1 \cdot \sin sx\, dx = \left[\frac{-\cos sx}{s}\right]_0^1 = \frac{1 - \cos s}{s} \qquad \text{... (2)}$$

From (2) putting the value of $\bar{u}$ $(s, 0)$ in (1), we get $\frac{1 - \cos s}{s} = A$

$\therefore \quad \bar{u} = \frac{1-\cos s}{s} e^{-s^2 t} \quad \Rightarrow \quad u = \frac{2}{\pi}\int_0^\infty \left(\frac{1-\cos s}{s}\right) e^{-s^2 t} ds$ **Ans.**

Example 44. *Solve* $\frac{\partial u}{dt} = k\frac{\partial^2 u}{\partial x^2}$ *for* $x \geq 0, t \geq 0$ *under the given conditions* $u = u_0$ *at* $x = 0$, $t > 0$ *with initial condition* $u(x, 0) = 0, x \geq 0$

Solution. Taking Fourier sine transforms

$$F_s\left(\frac{\partial u}{\partial t}\right) = F_s\left(k\frac{\partial^2 u}{\partial x^2}\right)$$

$$\frac{d\bar{u}}{dt} = k\left[-s^2\bar{u} + \sqrt{\frac{2}{\pi}}\, s\, u(0,t)\right]$$

$$= -ks^2\bar{u} + \sqrt{\frac{2}{\pi}}\, k\, s\, u_0 \text{ where } \bar{u} \text{ is the Fourier sine transform of } u.$$

$$\frac{d\bar{u}}{dt} + ks^2\bar{u} = \sqrt{\frac{2}{\pi}}\, s\, k\, u_0$$

This is linear in $\bar{u}$.

$$\therefore \quad \bar{u}\, e^{ks^2 t} = \sqrt{\frac{2}{\pi}} k u_0 \int s\, e^{ks^2 t} dt = \sqrt{\frac{2}{\pi}}\frac{u_0}{s} e^{ks^2 t} + c \quad \text{... (1)}$$

Since, $u(x, 0) = 0$, $\bar{u}(s, 0) = 0$. Using this in (1), we have

$$0 = \sqrt{\frac{2}{\pi}}\frac{u_0}{s} + c \quad \Rightarrow \quad c = -\sqrt{\frac{2}{\pi}}\frac{u_0}{s}$$

$$e^{ks^2 t} = \sqrt{\frac{2}{\pi}}\frac{u_0}{s}\left(e^{ks^2 t} - 1\right) \quad \Rightarrow \quad \bar{u} = \sqrt{\frac{2}{\pi}}\frac{u_0}{s}\left(1 - e^{ks^2 t}\right)$$

By inversion theorem, we have

$$u(x,t) = \frac{2u_0}{\pi}\int_0^\infty \left(\frac{1-e^{ks^2 t}}{s}\right)\sin sx\, ds.$$ **Ans.**

Example 45. *Solve* $\frac{\partial u}{\partial t} = k\frac{\partial^2 u}{\partial x^2}$ *for* $0 \leq x < \infty, t > 0$ *given the conditions*

(i) $u(x, 0) = 0$ *for* $x \geq 0$ *(ii)* $\frac{\partial u}{\partial x}(0,t) = -a$ *(constant)*

(iii) $u(x, t)$ *is bounded.*

Solution. In this problem, $\frac{\partial u}{\partial x}$ at $x = 0$ is given. Hence, take Fourier cosine transform on both sides of the given equation.

$$F_c\left(\frac{\partial u}{\partial t}\right) = F_c\left(k\frac{\partial^2 u}{\partial x^2}\right)$$

$$\frac{d\bar{u}}{dt} = k\left(-s^2\bar{u} - \sqrt{\frac{2}{\pi}}\cdot\frac{\partial u}{\partial x}(0, t)\right) = -ks^2\bar{u} + \sqrt{\frac{2}{\pi}}ka \quad \text{[Using condition (ii)]}$$

$$\frac{d\bar{u}}{dt} + ks^2\bar{u} = \sqrt{\frac{2}{\pi}}ka$$

This is linear in $\bar{u}$. Therefore, solving

$$\bar{u}e^{ks^2t} = \int \sqrt{\frac{2}{\pi}}\, ka\, e^{ks^2t}\, dt = \sqrt{\frac{2}{\pi}}\, ka \frac{e^{ks^2t}}{ks^2} + c$$

$$\bar{u}(s,t) = \sqrt{\frac{2}{\pi}} \frac{a}{s^2} + c\, e^{-ks^2t} \qquad \text{... (1)}$$

Since $\quad u(x, 0) = 0 \quad$ for $\quad x \geq 0$

$$\bar{u}(s, 0) = 0.$$

Using this in (1), we get

$$\bar{u}(s,0) = c + \sqrt{\frac{2}{\pi}} \frac{a}{s^2} = 0$$

$$\therefore \qquad c = -\sqrt{\frac{2}{\pi}} \frac{a}{s^2}$$

Substituting this in (1), we have $\bar{u}(s, t) = \sqrt{\frac{2}{\pi}} \frac{a}{s^2}\left(1 - e^{-ks^2t}\right)$

By inversion theorem, we have $u(x,t) = \frac{2}{\pi} \cdot a \int_0^\infty \frac{1 - e^{-ks^2t}}{s^2} \cos sx\, ds.$ **Ans.**

EXERCISE 45.4

1. Use Fourier transform to solve the equation $\frac{\partial u}{\partial t} = \frac{\partial^2 u}{\partial x^2}$

under the conditions $\quad u = 0$ at $x = 0$

$$u = \begin{cases} 1, & 0 < x < 1 \\ 0, & x \geq 1 \end{cases} \quad \text{when } t = 0$$

and u is bounded. **Ans.** $u = \frac{2}{\pi} \int_\pi^\infty \frac{1 - \cos s}{s} e^{-s^2 t} \sin sx\, ds$

2. Use Fourier sine transform to solve the equation $\frac{\partial u}{\partial t} = 2 \frac{\partial^2 u}{\partial x^2}$

Under the conditions $u(0, t) = 0$, $u(x, 0) = e^{-x}$, $u(x, t)$ is bounded.

Ans. $u(x,t) = \frac{2}{\pi} \int_0^\infty \frac{s}{s^2 + 1} e^{-2s^2 t} \sin sx\, ds$

3. A tightly stretched string with fixed end points $x = b$ and $x = c$ is initially in a position given by $y = b \sin\left(\frac{\pi x}{c}\right)$. It is released from rest in this position. Show by the method of Laplace transform that the displacement y at any distance x from one end and at any time t is given by

$$y = b \sin \frac{\pi x}{c} \cos \frac{\pi q}{c} t.$$

and y satisfies the equation $\frac{\partial^2 y}{\partial x^2} = a^2 \frac{\partial^2 y}{\partial t^2}$

4. A string is stretched tightly between $x = 0$ and $x = l$ and both ends are given displacement $y = a \sin pt$ perpendicular to the string. If the string satisfies the differential equation

$$\frac{\partial^2 y}{\partial x^2} = \frac{1}{c^2} \frac{\partial^2 y}{\partial t^2}$$

Show that the oscillations of the string are given by

$$y = a \sec \frac{Pl}{2c} \cos\left(\frac{Px}{c} - \frac{Pl}{2c}\right) \sin pt.$$

5. An infinite cable with resistance R ohms/km, capacitance C Farads/km, and negligible inductance and leakage is subjected to constant E.M.F. E_0 at the home end at time $t = 0$. Using the operational method show that the entering current at any subsequent time t is

$$I(t) = E_0\left(\frac{C}{\pi Rt}\right)^{1/2}$$

6. Solve the equation for high voltage semi-infinite line with the following initial and boundary conditions

$v(x, t) = 0$ and $i(x, 0) = 0$, $v(0, t) = v_0\, u(t)$, $v(x, t)$ is finite as $x \to \infty$

Ans. $v = v_0 u[t - x\sqrt{LC}]$, for $x \le \dfrac{t}{\sqrt{LC}}$ and

$v = 0$ for $x > \dfrac{t}{\sqrt{LC}}$

7. Solve $\dfrac{\partial u}{\partial t} = \dfrac{\partial^2 u}{\partial x^2}$ if

(*i*) $\dfrac{\partial u}{\partial x}(0,t) = 0$ for $t > 0$. (*ii*) $u(x, 0) = \begin{cases} x & 0 \le x \le 0 \\ 0 & x > 1 \end{cases}$

(*iii*) and $u(x, t)$ is bounded for $x > 0$, $t > 0$ **Ans.** $\left[u(x,t) = \dfrac{2}{\pi}\displaystyle\int_0^\infty \left(\frac{\sin s}{s} + \frac{\cos s - 1}{s^2}\right) e^{-s^2 t} \cos sx\, ds\right]$

45.18 FINITE FOURIER TRANSFORMS

Let $f(x)$ denote a function which is sectionally continuous over the range $(0, l)$. Then the **Finite Fourier sine transform** of $f(x)$ on this interval is defined as

$$F_s(p) = \bar{f}_s(P) = \int_0^l f(x) \sin\frac{p\pi x}{l}\, dx$$

where p is an integer (Instead of s, we take p as a parameter)

Inversion formula for sine transform

If $\bar{f}_s(p) = F_s(p)$ is the finite Fourier sine transform of $f(x)$ in $(0, l)$ then the inversion formula for sine transform is

$$f(x) = \frac{2}{l}\sum_{p=l}^{\infty} \bar{f}_s(P) \sin\frac{p\pi x}{l}$$

Proof. For the given function $f(x)$ in $(0, l)$, if we find the half range Fourier sine series, we get.

$$f(x) = \sum_{n=1}^{\infty} b_n \sin\frac{n\pi x}{l} \quad \text{... (1)}$$

where $$b_n = \frac{2}{l}\int_0^l f(x) \sin\frac{n\pi x}{l}\, dx$$

$$\therefore \quad b_p = \frac{2}{l}\int_0^l f(x) \sin\frac{p\pi x}{l}\, dx = \frac{2}{l}\bar{f}_s(p) \text{ by definition}$$

Substituting in (1), we get $f(x) = \dfrac{2}{l}\displaystyle\sum_{p=1}^{\infty} f_s(p) \sin\frac{p\pi x}{l}$

Finite Fourier Cosine Transform

Let $f(x)$ denote a sectionally continuous function in $(0, l)$.

Then the Finite Fourier cosine transform of $f(x)$ over $(0, l)$ is defined as

$$F_c(p) = \bar{f}_c(p) = \int_0^l f(x) \cos \frac{p\pi x}{l} dx \quad \text{where } p \text{ is an integer.}$$

Inversion formula for cosine transform

If $\bar{f}_c\ (P)$ is the finite Fourier cosine transform of $f(x)$ in $(0, l)$, then the inversion formula for cosine transform is

$$f(x) = \frac{1}{l}\bar{f}_c(0) + \frac{2}{l}\sum_{p=l}^{\infty} \bar{f}_c(p) \cos \frac{p\pi x}{l}$$

where
$$\bar{f}_c(0) = \int_0^l f(x)\, dx.$$

Proof. If we find half range Fourier cosine series for $f(x)$ in $(0, l)$, we obtain.

$$f(x) = \frac{a_0}{2} + \sum_{n=1}^{\infty} a_n \cos \frac{n\pi x}{l} \qquad \text{... (2)}$$

where
$$a_n = \frac{2}{l}\int_0^l f(x) \cos \frac{n\pi x}{l} dx$$

$\therefore$
$$a_p = \frac{2}{l}\bar{f}_c(p)$$

$$a_0 = \frac{2}{l}\int_0^l f(x)\, dx = \frac{2}{l}\bar{f}_c(0).$$

Substituting in (2), we get

$$f(x) = \frac{1}{l} f_c(0) + \frac{2}{l}\sum_{p=1}^{\infty} f_c(p) \cos \frac{p\pi x}{l}$$

Example 46. *Find the finite Fourier sine and cosine transforms of*

(*i*) $f(x) = 1$ in $(0, \pi)$

(*ii*) $f(x) = x$ in $(0, l)$

(*iii*) $f(x) = x^2$ in $(0, l)$

(*iv*) $f(x) = 1$ in $0 < x < \pi/2$

$\qquad = -1$ in $\pi/2 < x < \pi$

(*v*) $f(x) = x^3$ in $(0, l)$

(*vi*) $f(x) = e^{ax}$ in $(0, l)$

Solution.

(*i*) $\bar{f}_s\ (p) = F_s(1) = \int_0^{\pi} 1.\sin \frac{p\pi x}{\pi} dx = \left(-\frac{\cos px}{p}\right)_0^{\pi} = \frac{1-\cos p\pi}{p}$ [If $p \neq 0$]

$$\bar{f}_c(p) = \int_0^{\pi} 1 \cdot \cos px\, dx = \left(\frac{\sin px}{p}\right)_0^{\pi} = \frac{1}{p}(0-0) = 0$$

$(ii)\quad \overline{f}_s(p) = F_s(p) = \int_0^l x \sin\frac{p\pi x}{l}dx$

$$= \left[(x)\left(\frac{-\cos\frac{p\pi x}{l}}{\frac{p\pi}{l}}\right) - (1)\left(-\frac{\sin\frac{p\pi x}{l}}{\frac{p^2\pi^2}{l^2}}\right)\right]_0^l = \frac{-l}{p\pi}(l\cos p\pi)$$

$$= \frac{-l^2}{p\pi}(-1)^P \qquad [\text{If } p \neq 0]$$

$\overline{f}_c(p) = F_c(x) = \int_0^l x\cos\frac{p\pi x}{l}dx$

$$= \left[(x)\left(\frac{\sin\frac{p\pi x}{l}}{\frac{p\pi}{l}}\right) - (1)\left(-\frac{\cos\frac{p\pi x}{l}}{\frac{p^2\pi^2}{l^2}}\right)\right]_0^l = \frac{l^2}{p^2\pi^2}[(-1)^p - 1] \quad [\text{If } p \neq 0]$$

$(iii)\quad \overline{f}_s(p) = F_s(x^2) = \int_0^l x^2 \sin\frac{p\pi x}{l}dx$

$$= \left[(x^2)\left(-\frac{\cos\frac{p\pi x}{l}}{\frac{p\pi}{l}}\right) - (2x)\left(-\frac{\sin\frac{p\pi x}{l}}{\frac{p^2\pi^2}{l^2}}\right) + (2)\left(\frac{\cos\frac{p\pi x}{l}}{\frac{p^3\pi^3}{l^3}}\right)\right]_0^l$$

$$= \frac{-l^3}{p\pi}(-1)^p + \frac{2l^3}{p^3\pi^3}\left[(-1)^p - 1\right] \qquad [\text{If } p \neq 0]$$

$\overline{f}_c(p) = \int_0^l (x^2)\cos\frac{p\pi x}{l}dx$

$$= \left[(x^2)\left(\frac{\sin\frac{p\pi x}{l}}{\frac{p\pi}{l}}\right) - (2x)\left(-\frac{\cos\frac{p\pi x}{l}}{\frac{p^2\pi^2}{l^2}}\right) + (2)\left(-\frac{\sin\frac{p\pi x}{l}}{\frac{p^3\pi^3}{l^3}}\right)\right]_0^l = \frac{2l^3}{p^2\pi^2}(-1)^p \quad [\text{If } p \neq 0]$$

$(iv)\quad F_s\{f(x)\} = \int_0^{\pi/2} \sin px\,dx + \int_{\pi/2}^{\pi}(-1)\sin px\,dx$

$$= \left(-\frac{\cos px}{p}\right)_0^{\pi/2} + \left(\frac{\cos px}{p}\right)_{\pi/2}^{\pi} = -\frac{1}{p}\left(\cos\frac{p\pi}{2} - 1\right) + \frac{1}{p}\left(\cos p\pi - \cos\frac{p\pi}{2}\right)$$

$$= -\frac{1}{p}\left(\cos p\pi - 2\cos\frac{p\pi}{2} - 1\right) \qquad [\text{If } p \neq 0]$$

$$F_c(f(x)) = \int_0^{\pi/2} \cos px\,dx - \int_{\pi/2}^{\pi} \cos px\,dx$$

$$= \left(\frac{\sin px}{p}\right)_0^{\pi/2} - \left(\frac{\sin px}{p}\right)_{\pi/2}^{\pi} = \frac{2}{p}\sin\frac{p\pi}{2} \qquad [\text{If } p \neq 0]$$

(v) $$F_s(x^3) = \int_0^l x^3 \sin\frac{p\pi x}{l}\,dx$$

$$= \left[(x^3)\left(-\frac{\cos\frac{p\pi x}{l}}{\frac{p\pi}{l}}\right) - (3x^2)\left(\frac{\sin\frac{p\pi x}{l}}{\frac{p^2\pi^2}{l^2}}\right) + (6x)\left(\frac{\cos\frac{p\pi x}{l}}{\frac{p^3\pi^3}{l^3}}\right) - (6)\left(\frac{\sin\frac{p\pi x}{l}}{\frac{p^4\pi^4}{l^4}}\right)\right]_0^l$$

$$= -\frac{l^4}{p\pi}(-1)^p + \frac{6\,l^4}{p^3\pi^3}(-1)^p \qquad [\text{If } p \neq 0]$$

$$F_c(x^3) = \int_0^l x^3 \cos\frac{p\pi x}{l}\,dx$$

$$= \left[(x^3)\left(\frac{\sin\frac{p\pi x}{l}}{\frac{p\pi}{l}}\right) - (3x^2)\left(-\frac{\cos\frac{p\pi x}{l}}{\frac{p^2\pi^2}{l^2}}\right) + (6x)\left(\frac{-\sin\frac{p\pi x}{l}}{\frac{p^3\pi^3}{l^3}}\right) - (6)\left(\frac{\cos\frac{p\pi x}{l}}{\frac{p^4\pi^4}{l^4}}\right)\right]_0^l$$

$$= \frac{3l^4}{\pi^2 p^2}(-1)^p - \frac{6l^4}{p^4\pi^4}[(-1)^p - 1] \qquad [\text{If } p \neq 0]$$

(vi) $$F_s(e^{ax}) = \int_0^l e^{ax}\sin\frac{p\pi x}{l}\,dx$$

$$= \left\{\frac{e^{ax}}{a^2 + \frac{p^2\pi^2}{l^2}}\left[a\sin\frac{p\pi x}{l} - \frac{p\pi}{l}\cos\frac{p\pi x}{l}\right]\right\}_0^l = \frac{e^{al}}{a^2 + \frac{p^2\pi^2}{l^2}}\cdot\left(-\frac{p\pi}{l}(-1)^p\right) + \frac{1}{a^2 + \frac{p^2\pi^2}{l^2}}\left(\frac{p\pi}{l}\right)$$

$$F_c(e^{ax}) = \int_0^l e^{ax}\cos\frac{p\pi x}{l}\,dx$$

$$F_c(e^{ax}) = \left\{\frac{e^{ax}}{a^2 + \frac{p^2\pi^2}{l^2}}\left[a\cos\frac{p\pi x}{l} + \frac{p\pi}{l}\sin\frac{p\pi x}{l}\right]\right\}_0^l = \frac{e^{al}}{a^2 + \frac{p^2\pi^2}{l^2}}a(-1)^p - \frac{1}{a^2 + \frac{p^2\pi^2}{l^2}}(a) \quad \textbf{Ans.}$$

Example 47. *Find f (x) if its finite Fourier sine transform is* $\frac{2\pi}{p^3}(-1)^{p-1}$ *for p = 1, 2, ...,*

0 < x < π.

Solution. By inversion Theorem, we have

$$f(x) = \frac{2}{\pi}\sum_{p=1}^{\infty}\frac{2\pi}{p^3}(-1)^{p-1}\sin px = 4\sum_{p=1}^{\infty}\frac{(-1)^{p-1}}{p^3}\sin px \qquad \textbf{Ans.}$$

Example 48. *Find f (x) if its finite Fourier sine transform is given by*

(*i*) $F_s(p)=\dfrac{1-\cos p\pi}{p^2\pi^2}$ *for p = 1, 2, 3, ... and 0 < x < π*

(*ii*) $F_s(p)=\dfrac{16(-1)^{p-1}}{p^3}$ *for p = 1, 2, 3, ... and 0 < x < 8*

(*iii*) $F_s(p)=\dfrac{\cos\dfrac{2\pi p}{3}}{(2p+1)^2}$ *for p = 1, 2, 3, ... and 0 < x < 1.*

Solution. By inversion theorem, we have

(*i*) $f(x)=\dfrac{2}{\pi}\sum_{p=1}^{\infty}\left(\dfrac{1-\cos p\pi}{p^2\pi^2}\right).\sin px=\dfrac{2}{\pi^3}\sum_{p=1}^{\infty}\left(\dfrac{1-\cos p\pi}{p^2}\right).\sin px$

(*ii*) $f(x)=\dfrac{2}{l}\sum_{p=1}^{\infty}F_s(p)\sin\left(\dfrac{p\pi x}{l}\right)=\dfrac{2}{8}\sum_{p=1}^{\infty}\dfrac{16(-1)^{p-1}}{p^3}\sin\left(\dfrac{p\pi x}{8}\right)$ [Since $l = 8$]

$$=4\sum_{p=1}^{\infty}\frac{(-1)^{p-1}}{p^3}\sin\left(\frac{p\pi x}{8}\right)$$

(*iii*) $f(x)=\dfrac{2}{l}\sum_{p=1}^{\infty}F_s(p)\sin\left(\dfrac{p\pi x}{l}\right)=2\sum_{p=1}^{\infty}\dfrac{\cos\left(\dfrac{2\pi p}{3}\right)}{(2p+1)^2}\sin(p\pi x)$ [Since $l = 1$] **Ans.**

Example 49. *Find f (x) if its finite Fourier cosine transform is*

(*i*) $F_c(p)=\dfrac{1}{2p}\sin\left(\dfrac{p\pi}{2}\right)$ *for p = 1, 2, 3, ...*

$=\dfrac{\pi}{4}$ *for p = 0 given 0 < x < 2π*

(*ii*) $F_c(p)=\dfrac{6\sin\dfrac{p\pi}{2}-\cos p\pi}{(2p+1)\pi}$ *for p = 1, 2, 3, ...*

$=\dfrac{2}{\pi}$ *for p = 0 given 0 < x < 4*

(*iii*) $F_c(p)=\dfrac{\cos\left(\dfrac{2p\pi}{3}\right)}{(2p+1)^2}$ *for p = 1, 2, 3, ...*

$= 1$ *for p = 0 given 0 < x < 1*

Solution. By inversion theorem, we have

$$f(x)=\frac{1}{l}F_c(0)+\frac{2}{l}\sum_{p=1}^{\infty}F_c(p).\cos\frac{p\pi x}{l}.$$

(*i*) Here $F_c(0)=\pi/4$ and $l=2\pi$

$\therefore \qquad f(x)=\frac{1}{2\pi}\left(\frac{\pi}{4}\right)+\frac{2}{2\pi}\sum_{p=1}^{\infty}\frac{1}{2p}\sin\left(\frac{p\pi}{2}\right)\cos\left(\frac{p\pi x}{2\pi}\right)$

$$=\frac{1}{8}+\frac{1}{2\pi}\sum_{p=1}^{\infty}\frac{1}{p}\sin\left(\frac{p\pi}{2}\right)\cos\left(\frac{px}{2}\right)$$

(*ii*) Here $\qquad F_c(0)=\frac{2}{\pi}$ and $l = 4$

$$f(x)=\frac{1}{4}\left(\frac{2}{\pi}\right)+\frac{2}{4}\sum_{p=1}^{\infty}\frac{\left(6\sin\frac{p\pi}{2}-\cos p\pi\right)}{(2p+1)\pi}\cos\left(\frac{p\pi x}{4}\right)$$

$$=\frac{1}{2\pi}+\frac{1}{2\pi}\sum_{p=1}^{\infty}\frac{\left(6\sin\frac{p\pi}{2}-\cos p\pi\right)}{(2p+1)}.\cos\left(\frac{p\pi x}{4}\right)$$

(*iii*) Here $\qquad F_c(0) = 1,\ l = 1$

$\therefore \qquad f(x)=\frac{1}{1}+\frac{2}{1}\sum_{p=1}^{\infty}\frac{1}{(2p+1)^2}\cos\left(\frac{2p\pi}{3}\right).\cos(p\pi x)$

$$=1+2\sum_{p=1}^{\infty}\frac{\cos\left(\frac{2p\pi}{3}\right)}{(2p+1)^2}\cos(p\pi x)$$ **Ans.**

Example 50. *Find the finite Fourier sine transform of f (x) = 1 in (0, π). Use the inversion theorem and find Fourier series for f (x) = 1 in (0, π). Hence prove*

(*i*) $1-\frac{1}{3}+\frac{1}{5}-\frac{1}{7}+....=\pi/4$ $\qquad$ (*ii*) $\frac{1}{1^2}+\frac{1}{3^2}+\frac{1}{5^2}+....=\pi^2/8$

Solution. $\qquad F_s(1)=\int_0^{\pi}1.\sin\left(\frac{p\pi x}{\pi}\right)dx=\left[-\frac{\cos px}{p}\right]_0^{\pi}$

$$\bar{f}_s(p)=\frac{1-\cos p\pi}{p}\text{ if p}\neq 0$$

By inversion theorem,

$$f(x)=\frac{2}{l}\sum_{p=1}^{\infty}F_s\ (p)\sin\frac{p\pi x}{l}$$

$$1=\frac{2}{\pi}\sum_{p=1}^{\infty}\frac{1-(-1)^p}{p}.\sin px \qquad [\text{Since } l=\pi]$$

$$1=\frac{4}{\pi}\left[\sin x+\frac{1}{3}\sin 3x+\frac{1}{5}\sin 5x+...\right] \qquad [p\text{ is odd}] \qquad .... (1)$$

This is the half range Fourier sine series for $f(x) = 1$ in $(0, \pi)$ getting $x = \pi/2$.

On putting $x=\frac{\pi}{2}$ in (1), we get $\frac{4}{\pi}\left[1-\frac{1}{3}+\frac{1}{5}-\frac{1}{7}+....\right]=1$

$$\therefore \qquad 1-\frac{1}{3}+\frac{1}{5}-\frac{1}{7}+\ldots.=\frac{\pi}{4}$$

In the half Fourier sine series $l_n=\frac{4}{\pi}.\frac{1}{n}$ for n odd

By using Parseval's Theorem

$$\text{(Range)}\left[\frac{1}{2}\sum b_n^2\right]=\int_0^{\pi}(1)^2\,dx$$

$$\pi\left[\frac{1}{2}.\frac{16}{\pi^2}\sum_{n=1.3.5....}^{\pi}\frac{1}{n^2}\right]=\pi \qquad i.e., \quad \sum_{n=1.3.5}^{\infty}\frac{1}{n^2}=\frac{\pi^2}{8} \qquad \textbf{Ans.}$$

$$(ii) \qquad \frac{1}{1^2}+\frac{1}{3^2}+\frac{1}{5^2}+\ldots\ldots\ldots=\frac{\pi^2}{8}$$

EXERCISE 45.5

Find the finite Fourier sine and cosine transforms of

1. $f(x) = 2x$ in (0, 4) **Ans.** $F_s(s)=\begin{cases}\frac{32}{s\pi}(1-\cos s\pi), & s\neq 0\\ 0, & s=0\end{cases}$, $F_c(s)=\begin{cases}\frac{32}{s^2\pi^2}(\cos s\pi-1), & s\neq 0\\ 16, & s=0\end{cases}$

2. $f(x) = x$ in $(0, \pi)$ **Ans.** $F_s(s)=\begin{cases}\frac{\pi}{s}(-1)^{s+1}, & s\neq 0\\ 0, & s=0\end{cases}$, $F_c(s)=\begin{cases}\frac{(-1)^s-1}{s^2}, & s\neq 0\\ \frac{\pi^2}{2}, & s=0\end{cases}$

3. $f(x) = \cos ax$ in $(0, \pi)$ Ans. $F_s(s)=\frac{s}{s^2-a^2}[1-(-1)^s\cos a\pi]$

4. $f(x) = 1-\frac{x}{\pi}$ in $(0, \pi)$ Ans. $F_s(s)=\frac{1}{s}$

5. $f(x)=\begin{cases}x \text{ in } (0,\pi/2)\\ \pi-x \text{ in } (\pi/2,\pi)\end{cases}$ **Ans.** $F_s(s)=\frac{2}{s^2}\sin\frac{s\pi}{2}$

6. $f(x)=e^{-ax}$ in $(0, l)$ **Ans.** $F_s(e^{-ax})=\frac{e^{-al}}{a^2+\frac{s^2a^2}{l^2}}\left(-\frac{s\pi}{l}(-1)^s\right)+\frac{1}{a^2+\frac{s^2\pi^2}{l^2}}\left(\frac{s\pi}{l}\right)$

$$F_c(e^{-ax})=\frac{ae^{-al}}{a^2+\frac{s^2\pi^2}{l^2}}(-1)^{s+1}+\frac{a}{a^2+\frac{s^2\pi^2}{l^2}}$$

7. $f(x)=\frac{\pi}{3}-x+\frac{x^2}{2\pi}$ in $(0, \pi)$ **Ans.** $F_s(s)=\frac{\pi(-1)^p}{6p}+\frac{1(-1)^p}{\pi p^3}+\frac{\pi}{3p}-\frac{1}{\pi p^3}$, $F_c[f(x)]=\frac{1}{s^2}$

8. Find finite Fourier cosine transform of $\left(1-\frac{x}{\pi}\right)^2, 0<x<\pi$. **Ans.** $F_c(s)=\begin{cases}\frac{2}{\pi s^2}, & s\neq 0\\ \frac{\pi}{3}, & s=0\end{cases}$

9. Find $f(x)$ if $\bar{f}_c(s)=\begin{cases}\dfrac{\sin\left(\frac{s\pi}{2}\right)}{2s}, & s=1,2,3\ldots\\ \frac{\pi}{4} & s=0 \text{ given } 0<x<2x.\end{cases}$ **Ans.** $\frac{1}{8}+\frac{1}{\pi}\sum_{s=1}^{\infty}\frac{\sin\frac{s\pi}{2}}{2s}\cos\frac{sx}{2}$

45.19 FINITE FOURIER SINE AND COSINE TRANSFORMS OF DERIVATIVES

Using the definition and the integration by parts, we can easily prove the following results. For $0 \le x \le l$

(*i*) $$F_s(f'(x)) = -\frac{p\pi}{l}\bar{f}_c(p)$$

(*ii*) $$F_c\{f'(x)\} = f(l)(-1)^p - f(0) + \frac{p\pi}{l}\bar{f}_s(p)$$

(*iii*) $$F_s\{f''(x)\} = -\frac{p^2\pi^2}{l^2}\bar{f}_s(p) + \frac{p\pi}{l}\left[f(0) - (-1)^p f(l)\right]$$

(*iv*) $$F_c\{f''(x)\} = -\frac{p^2\pi^2}{l^2}\bar{f}_c(p) + f'(l)(-1)^p - f'(0)$$

Proof: (*i*) $$F_s(f'(x)) = \int_0^l f'(x)\sin\frac{p\pi x}{l}dx = \int_0^l \sin\frac{p\pi x}{l}\cdot d\{f(x)\}$$

$$= \left(f(x)\sin\frac{p\pi x}{l}\right)_0^l - \int_0^l f(x)\frac{p\pi}{l}.\cos\frac{p\pi x}{l}\,dx$$

$$= -\frac{p\pi}{l}\bar{f}_c(p) \qquad \text{... (1)}$$

(*ii*) $$F_c\{f'(x)\} = \int_0^l f'(x)\cos\frac{p\pi x}{l}dx = \left(f(x)\cos\frac{p\pi x}{l}\right)_0^l - \int_0^l f(x)\cdot\left(-\frac{p\pi}{l}\sin\frac{p\pi x}{l}\right)dx$$

$$= (-1)^P f(l) f(0) + \frac{p\pi}{l}\bar{f}_s(p) \qquad \text{... (2)}$$

(*iii*) $$F_s[f''(x)] = \int_0^l \sin\frac{p\pi x}{l}d[f'(x)]$$

$$= \left(f'(x)\sin\frac{p\pi x}{l}\right)_0^l - \frac{p\pi}{l}\int_0^l f'(x)\cos\frac{p\pi x}{l}ds$$

$$= -\frac{p\pi}{l}\left[(-1)^p f(l) - f(0) + \frac{p\pi}{l}\bar{f}_s(p)\right] \qquad \text{[Using (2)]}$$

$$= -\frac{p^2\pi^2}{l^2}\bar{f}_s(p) + \frac{p\pi}{l}\left[f(0) - (-1)^p f(l)\right] \qquad \text{... (3)}$$

(*iv*) $$F_c\{f''(x)\} = \int_0^l \cos\frac{p\pi x}{l}d[f'(x)]$$

$$= \left[f'(x)\cos\frac{p\pi x}{l}\right]_0^l + \frac{p\pi}{l}\int_0^l f'(x)\sin\frac{p\pi x}{l}dx \qquad \text{[Using (1)]}$$

$$= (-1)^P f'(l) - f'(0) + \frac{p\pi}{l}\left[-\frac{p\pi}{l}\bar{f}_c(p)\right] \qquad \text{... (4)}$$

$$= -\frac{p^2\pi^2}{l^2}\bar{f}_c(p) + f'(l)(-1)^p - f'(0)$$

Note. if $u = u(x, t)$ then

$$F_s\left[\frac{\partial u}{\partial x}\right]=\frac{-p\pi}{l}F_c(u)$$

$$F_c\left[\frac{\partial u}{\partial x}\right]=\frac{p\pi}{l}F_s(u)-u(0,t)+(-1)^p u(l,t)$$

$$F_s\left[\frac{\partial^2 u}{\partial x^2}\right]=\frac{p^2\pi^2}{l^2}F_s(u)+\frac{p\pi}{l}[u(0,t)-(-1)^p u(l,t)]$$

$$F_c\left[\frac{\partial^2 u}{\partial x^2}\right]=-\frac{p^2\pi^2}{l^2}F_c(u)+\frac{\partial u}{\partial x}(l,t)\cos p\pi-\frac{\partial u}{\partial x}(0,\ t)$$

Example 51. *Using finite Fourier transform, solve*

$$\frac{\partial u}{\partial t}=\frac{\partial^2 u}{\partial x^2},\ \textit{given } u\ (0,\ t)=0 \textit{ and } u\ (4,\ t)=0$$

and $u\ (x,\ 0)=2x$ where $0<x<4,\ t>0$.

Solution. Since $u\ (0,\ t)$ given, take finite Fourier sine transform.

$$\int_0^4 \frac{\partial u}{\partial t}\sin\frac{p\pi x}{4}dx=\int_0^4\frac{\partial^2 u}{\partial x^2}\sin\frac{p\pi x}{4}dx$$

$$\frac{d\bar{u}_s}{dt}=F_s\left(\frac{\partial^2 u}{\partial x^2}\right)=-\frac{p^2\pi^2}{16}\bar{u}_s+\frac{p\pi}{4}[u\ (0,t)-(-1)^p u(4,t)]$$

$$=-\frac{p^2\pi^2}{16}\bar{u}_s \text{ using } u(0,\ t)=0,\ u(4,\ t)=0$$

$$\frac{d\,\bar{u}_s}{\bar{u}_s}=-\frac{p^2\pi^2}{16}dt$$

Integrating $$\log\ \bar{u}_s=-\frac{p^2\pi^2}{16}t+c$$

$$\bar{u}_s=Ae^{-\frac{\pi^2 p^2}{16}t} \qquad \text{... (1)}$$

Since $u(x,\ 0)=2x$

$$\bar{u}_s\,(p,\ 0)=\int_0^4(2x)\ \sin\left(\frac{p\pi x}{4}\right)dx=\left[(2x)\left(-\frac{4}{p\pi}\cos\left(\frac{p\pi x}{4}\right)+(2)\frac{16}{p^2\pi^2}\sin\left(\frac{p\pi x}{4}\right)\right)\right]_0^4$$

$$=-\frac{32}{p\pi}\cos p\pi \qquad \text{... (2)}$$

Using (2) in (1), we get $\bar{u}_s\ (p,\ 0)=A=-\dfrac{32}{p\pi}\cos p\pi\ =\ -\dfrac{32}{p\pi}(-1)^p$

Substituting in (1), we get

$$\therefore \qquad \bar{u}_s=-\frac{32}{p\pi}(-1)^p\ e^{-\frac{p^2\pi^2}{16}t}$$

By inversion Theorem, we get

$$u(x, t) = \frac{2}{4}\sum_{p=1}^{\infty} \frac{32}{p\pi}(-1)^{p+1} e^{-\frac{p^2\pi^2}{16}t} \sin\left(\frac{p\pi x}{4}\right) \qquad \textbf{Ans.}$$

Example 52. *Solve* $\frac{\partial u}{\partial t} = \frac{\partial^2 u}{\partial x^2}$, $\quad 0 < x < 6, t > 0$

given $\quad \frac{\partial u}{\partial x}(0, t) = 0, \frac{\partial u}{\partial x}(6, t) = 0$ *and* $u(x, 0) = 2x$.

Solution. Since $\frac{\partial u}{\partial x}(0, t)$ is given, use finite Fourier cosine transform of

$$\frac{\partial u}{\partial t} = \frac{\partial^2 u}{\partial x^2}$$

$$\int_0^6 \frac{\partial u}{\partial t} \cos\frac{p\pi x}{6} dx = \int_0^6 \frac{\partial^2 u}{\partial x^2} \cos\frac{p\pi x}{6} dx$$

$$\frac{d\bar{u}_c}{dt} = -\frac{p^2\pi^2}{36}\bar{u}_c + \frac{\partial u}{\partial x}(6, t)\cos p\pi - \frac{\partial u}{\partial x}(0, t) = -\frac{p^2\pi^2}{36}\bar{u}_c$$

$$\Rightarrow \qquad \frac{d\bar{u}_c}{\bar{u}_c} = -\frac{p^2\pi^2}{36}dt \qquad \Rightarrow \qquad \log \bar{u}_c = -\frac{p^2\pi^2}{36}t + c$$

$$\bar{u}_c = A\, e^{-\frac{p^2\pi^2}{36}t} \qquad \text{... (1)}$$

$u(x, 0) = 2x.$

$\therefore$ At $t = 0$

$$\bar{u}_c(p, 0) = \int_0^6 (2x) \cos\frac{p\pi x}{6} dx = \frac{72}{p^2\pi^2}(\cos p\pi - 1) \qquad \text{... (2)}$$

Using this in (1), we get

$$\bar{u}_c(p, 0) = A = \frac{72}{p^2\pi^2}(\cos p\pi - 1)$$

Substituting in (1), we get

$$\bar{u}_c(p, t) = \frac{72}{p^2\pi^2}(\cos p\pi)$$

By inversion theorem, we get

$$u(x, t) = \frac{1}{l}\bar{f}_c(0) + \frac{2}{l}\sum_{p=1}^{\infty} \bar{f}_c(p)\cos\left(\frac{p\pi x}{l}\right)$$

$$= \frac{1}{6}\int_0^6 (2x)\, dx + \frac{2}{6}\sum_{p=1}^{\infty} \frac{72}{p^2\pi^2}(\cos p\pi - 1)\, e^{-\frac{p^2\pi^2}{36}t}.\cos\left(\frac{p\pi x}{6}\right)$$

$$= 6 + \frac{24}{\pi^2}\sum_{p=1}^{\infty} \frac{(\cos p\pi - 1)}{p^2} e^{-\frac{p^2\pi^2}{36}t} \cos\left(\frac{p\pi x}{6}\right) \qquad \textbf{Ans.}$$

Example 53. *Solve* $\frac{\partial u}{\partial t} = 2\frac{\partial^2 u}{\partial x^2}$, $0 < x < 4, t > 0$

given $u(0, t) = 0$; $u(4, t) = 0$; $u(x, 0) = 3 \sin \pi x - 2 \sin 5\pi x$.

Solution. Since $u(0, t)$ is given, take finite Fourier sine transform. The equation becomes (as in example 51 on page 1253).

$$\frac{d\bar{u}_c}{dt} = 2\left[-\frac{p^2\pi^2}{16}\bar{u}_s + \frac{p\pi}{4}\{u(0,t) - (-1)^p u(4,t)\}\right] = -\frac{p^2\pi^2}{8}\bar{u}_s$$

Solving, we get $\bar{u}_s = Ae^{-\frac{p^2\pi^2}{8}t}$ ... (1)

$$u(x, 0) = 3\sin \pi x - 2\sin 5\pi x$$

Taking sine Transform, $\bar{u}_s(p, 0) = \int_0^4 (3\sin \pi x - 2\sin 5\pi x)\sin\frac{p\pi x}{4}dx$

$$= 0, \text{ if } p \neq 4 \text{ or } p \neq 20.$$

If $p = 4$, $\bar{u}_s(4, 0) = 6$

If $p = 20$, $\bar{u}_s(20, 0) = -4$

$$u(x, t) = \frac{2}{4}\sum_{p=1}^{\infty} \bar{u}_s(p, t)\sin\left(\frac{p\pi x}{4}\right)$$

$$= \frac{1}{2}[6e^{-\frac{p^2\pi^2}{8}t}\sin \pi x - 4e^{-\frac{p^2\pi^4}{8}t}\sin 5\pi x]$$

where p in the first term is 4 and p in the second term is 20

$$= 3e^{-2\pi^2 t}\sin \pi x - 2e^{-50\pi^2 t}\sin 5\pi x.$$ **Ans.**

EXERCISE 45.6

1. Solve $\frac{\partial u}{\partial t} = \frac{\partial^2 u}{\partial x^2}$, $0 < x < 6, t > 0$ given that $u(0, t) = 0 = u(6, t)$ and $u(x, 0) = \begin{cases} 1 \text{ for } 0 < x < 3 \\ 0 \text{ for } 3 < x < 6 \end{cases}$

Ans. $u(x,t) = \frac{2}{\pi}\sum_{p=1}^{\pi}\left(\frac{1 - \cos\frac{p\pi}{2}}{p}\right)e^{-\frac{p^2\pi^2 t}{36}}\sin\left(\frac{p\pi x}{6}\right)$

2. Solve $\frac{\partial v}{\partial t} = \frac{\partial^2 v}{\partial x^2}$ subject to conditions $v(0, t) = 1$, $v(\pi, t) = 3$

$v(x, 0) = 1$ for $0 < x < \pi$, $t > 0$ **Ans.** $v(x, t) = \frac{4}{\pi}\sum_{p=1}^{\infty}\frac{\cos p\pi}{p}e^{-p^2 t}\sin px + 1 + \frac{2x}{\pi}$

3. Solve $\frac{\partial \theta}{\partial t} = \frac{\partial^2 \theta}{\partial x^2}$

given $\theta(0, t) = 0$, $\theta(\pi, t) = 0$, $\theta(x, 0) = 2x$ for $0 < x < \pi, t > 0$

Ans. $\theta(x, t) = 4\sum_{n=1}^{\infty}\frac{1}{n}(-1)^{n+1}e^{-n^2 t}\sin nx$

TABLE

S.No.	Function $f(x)$	Fourier Sine Transform $F_s(s)$
1.	$\begin{cases} 1, & 0 < x < b \\ 0, & x > b \end{cases}$	$\dfrac{1-\cos bs}{s}$
2.	x^{-1}	$\dfrac{\pi}{2}$
3.	$\dfrac{x}{x^2+b^2}$	$\dfrac{\pi}{2}e^{-bs}$
4.	e^{-bx}	$\dfrac{s}{s^2+b^2}$
5.	$x^{n-1}e^{-bx}$	$\dfrac{\Gamma(n)\sin(n\tan^{-1}s/b)}{(s^2+b^2)^{n/2}}$
6.	xe^{-bx^2}	$\dfrac{\sqrt{\pi}}{4b^{3/2}}se^{-s^2/4b}$
7.	$x^{-1/2}$	$\sqrt{\dfrac{\pi}{2s}}$
8.	x^{-n}	$\dfrac{\pi s^{n-1}\csc(n\pi/2)}{2\Gamma(n)}, \quad 0 < n < 2$
9.	$\dfrac{\sin bx}{x}$	$\dfrac{1}{2}\text{In}\left(\dfrac{s+b}{s-b}\right)$
10.	$\dfrac{\sin bx}{x^2}$	$\begin{cases} \pi s/2, & s < b \\ \pi b/2, & s > b \end{cases}$
11.	$\dfrac{\cos bx}{x}$	$\begin{cases} 0, & s < b \\ \pi/4, & s = b \\ \pi/2, & s > b \end{cases}$
12.	$\tan^{-1}(x/b)$	$\dfrac{\pi}{2s}e^{-bs}$
13.	$\csc bx$	$\dfrac{\pi}{2b}\tanh\dfrac{\pi s}{2b}$
14.	$\dfrac{1}{e^{2x}-1}$	$\dfrac{\pi}{4}\cot h\left(\dfrac{\pi s}{2}\right)-\dfrac{1}{2s}$

15.	$\begin{cases} 1, & 0<x<b \\ 0, & x>b \end{cases}$	$\frac{\sin bs}{s}$ Fourier Cosine Transform
16.	$\frac{1}{x^2+b^2}$	$\frac{\pi e^{-bs}}{2b}$
17.	e^{-bx}	$\frac{b}{s^2+b^2}$
18.	$x^{n-1}\, e^{-bx}$	$\frac{\Gamma(n)\cos(n\tan^{-1} s/b)}{(s^2+b^2)^{n/2}}$
19.	xe^{-bx^2}	$\frac{1}{2}\sqrt{\frac{\pi}{b}}e^{-\frac{s}{4b}}$
20.	$\ln\left(\frac{x^2+b^2}{x^2+c^2}\right)$	$\frac{e^{-cs}-e^{-bs}}{\pi s}$
21.	$\frac{\sin bx}{x}$	$\begin{cases} \pi/2, & s<b \\ \pi/4, & s=b \\ 0, & s>b \end{cases}$
22.	$\sin bx^2$	$\frac{\sqrt{\pi}}{8b}\left(\cos\frac{s^2}{4b}-\sin\frac{s^2}{4b}\right)$
23.	$\cos bx^2$	$\sqrt{\frac{\pi}{8b}}\left(\cos\frac{s^2}{4b}+\sin\frac{s^2}{4b}\right)$
24.	$\operatorname{sech} bx$	$\frac{\pi}{2b}\operatorname{sech}\frac{s\pi}{2b}$
25.	$\frac{\cosh(\sqrt{\pi}x/2)}{\cosh(\sqrt{\pi}\,x)}$	$\frac{\sqrt{\pi}}{2}\frac{\cosh(\sqrt{\pi}s/2)}{\cosh(\sqrt{\pi}\ s)}$
26.	$\frac{e^{-b\sqrt{x}}}{\sqrt{x}}$	$\sqrt{\frac{\pi}{2s}}\{\cos(2b\sqrt{s})-\sin(2b\sqrt{s})\}$

LAPLACE TRANSFORM

46.1 INTRODUCTION

Laplace transforms help in solving the differential equations with boundary values without finding the general solution and the values of the arbitrary constants.

46.2 LAPLACE TRANSFORM

Definition. Let $f(t)$ be function defined for all positive values of t, then

$$F(s) = \int_0^\infty e^{-st} f(t)\, dt$$

provided the integral exists, is called the **Laplace Transform** of $f(t)$. It is denoted as

$$\boxed{L[f(t)] = F(s) = \int_0^\infty e^{-st} f(t)\, dt}$$

46.3 IMPORTANT FORMULAE

1. $L(1) = \dfrac{1}{s}$

2. $L(t^n) = \dfrac{n!}{s^{n+1}}$, when $n = 0, 1, 2, 3...$

3. $L(e^{at}) = \dfrac{1}{s-a}$ $\quad (s > a)$

4. $L(\cosh at) = \dfrac{s}{s^2 - a^2}$ $\quad (s^2 > a^2)$

5. $L(\sinh at) = \dfrac{a}{s^2 - a^2}$ $\quad (s^2 > a^2)$

6. $L(\sin at) = \dfrac{a}{s^2 + a^2}$ $\quad (s > 0)$

7. $L(\cos at) = \dfrac{s}{s^2 + a^2}$ $\quad (s > 0)$

1. $\boxed{L(1) = \dfrac{1}{s}}$

Proof. $L(1) = \int_0^\infty 1 \cdot e^{-st} dt = \left[\dfrac{e^{-st}}{-s}\right]_0^\infty = -\dfrac{1}{s}\left[\dfrac{1}{e^{st}}\right]_0^\infty = -\dfrac{1}{s}[0-1] = \dfrac{1}{s}$

Hence $L(1) = \dfrac{1}{s}$ **Proved.**

2. $\boxed{L(t^n)=\dfrac{n!}{s^{n+1}}}$ where n and s are positive.

Proof. $$L(t^n) = \int_0^\infty e^{-st} t^n dt$$

Putting $st = x$ or $t = \dfrac{x}{s}$ or $dt = \dfrac{dx}{s}$

Thus, we have $L(t^n) = \int_0^\infty e^{-x}\left(\dfrac{x}{s}\right)^n \dfrac{dx}{s} \Rightarrow L(t^n) = \dfrac{1}{s^{n+1}}\int_0^\infty e^{-x}.x^n\,dx$

$$\Rightarrow L(t^n) = \frac{\overline{|n+1}}{s^{n+1}} \Rightarrow L(t^n) = \frac{n!}{s^{n+1}} \quad \left[\begin{matrix}\overline{|n+1} = \int_0^\infty e^{-x}.x^n\,dx \\ \text{and} \quad \overline{|n+1} = n!\end{matrix}\right]$$ **Proved.**

3. $\boxed{L(e^{at})=\dfrac{1}{s-a}}$, where $s > a$

Proof. $L(e^{at}) = \int_0^\infty e^{-st}.e^{at} dt = \int_0^\infty e^{-st+at}.dt$

$$= \int_0^\infty e^{(-s+a)t}.dt = \int_0^\infty e^{-(s-a)t}.dt = \left[\frac{e^{-(s-a)t}}{-(s-a)}\right]_0^\infty = -\frac{1}{s-a}\left[\frac{1}{e^{(s-a)t}}\right]_0^\infty$$

$$= \frac{-1}{(s-a)}(0-1) = \frac{1}{s-a}$$ **Proved.**

4. $\boxed{L(\cosh at)=\dfrac{s}{s^2-a^2}}$

Proof. $L(\cosh at) = L\left[\dfrac{e^{at}+e^{-at}}{2}\right]$ $\left(\therefore \cosh at = \dfrac{e^{at}+e^{-at}}{2}\right)$

$$= \frac{1}{2}L(e^{at}) + \frac{1}{2}L(e^{-at}) = \frac{1}{2}\left[\frac{1}{s-a}+\frac{1}{s+a}\right] \quad \left[L(e^{at}) = \frac{1}{s-a}\right]$$

$$= \frac{1}{2}\left[\frac{s+a+s-a}{s^2-a^2}\right] = \frac{s}{s^2-a^2}$$ **Proved.**

5. $\boxed{L(\sinh at)=\dfrac{a}{s^2-a^2}}$

Proof. $L(\sinh at) = L\left[\dfrac{1}{2}(e^{at}-e^{-at})\right]$

$$= \frac{1}{2}[L(e^{at}) - L(e^{-at})] = \frac{1}{2}\left[\frac{1}{s-a}-\frac{1}{s+a}\right] = \frac{1}{2}\left[\frac{s+a-s+a}{s^2-a^2}\right]$$

$$= \frac{a}{s^2-a^2}$$ **Proved.**

6. $\boxed{L(\sin at)=\dfrac{a}{s^2+a^2}}$

Proof. $L(\sin at) = L\left[\dfrac{e^{iat}-e^{-iat}}{2i}\right]$ $\quad\left[\because \sin at = \dfrac{e^{iat}-e^{-iat}}{2i}\right]$

$$= \frac{1}{2i}\left[L\left(e^{iat}-e^{-iat}\right)\right] = \frac{1}{2i}\left[L(e^{iat})-L(e^{-iat})\right]$$

$$= \frac{1}{2i}\left[\frac{1}{s-ia}-\frac{1}{s+ia}\right] = \frac{1}{2i}\frac{s+ia-s+ia}{s^2+a^2} = \frac{1}{2i}\frac{2ia}{s^2+a^2} = \frac{a}{s^2+a^2}$$ **Proved.**

7. $\boxed{\mathbf{L(\cos at) = \dfrac{s}{s^2+a^2}}}$

Proof. $L(\cos at) = L\left(\dfrac{e^{iat}+e^{-iat}}{2}\right)$ $\quad\left[\because \cos at = \dfrac{e^{iat}+e^{-iat}}{2}\right]$

$$= \frac{1}{2}[L(e^{iat}+e^{-iat})] = \frac{1}{2}[L(e^{iat})+L(e^{-iat})] = \frac{1}{2}\left[\frac{1}{s-ia}+\frac{1}{s+ia}\right] = \frac{1}{2}\frac{s+ia+s-ia}{s^2+a^2}$$

$$= \frac{s}{s^2+a^2}$$ **Proved.**

Example 1. *Find the Laplace transform of f(t) defined as*

$$f(t) = \begin{cases} \dfrac{t}{k}, & \text{when } 0<t<k \\ 1, & \text{when } t>k \end{cases}$$

Solution. $L\,[f(t)] = \displaystyle\int_0^k \frac{t}{k} e^{-st} dt + \int_k^\infty 1.e^{-st} dt = \frac{1}{k}\left[\left(t\frac{e^{-st}}{-s}\right)_0^k - \int_0^k \frac{e^{-st}}{-s} dt\right] + \left[\frac{e^{-st}}{-s}\right]_k^\infty$

$$= \frac{1}{k}\left[\frac{ke^{-ks}}{-s} - \left(\frac{e^{-st}}{s^2}\right)_0^k\right] + \frac{e^{-ks}}{s} = \frac{1}{k}\left[\frac{ke^{-ks}}{-s} - \frac{e^{-sk}}{s^2} + \frac{1}{s^2}\right] + \frac{e^{-ks}}{s}$$

$$= -\frac{e^{-sk}}{s} - \frac{1}{k}\frac{e^{-ks}}{s^2} + \frac{1}{k}\frac{1}{s^2} + \frac{e^{-ks}}{s} = \frac{1}{ks^2}[-e^{-ks}+1]$$ **Ans.**

Example 2. *Find the Laplace transform of the function* $f(t) = \begin{cases} t-1, & 1<t<2 \\ 3-t, & 2<t<3 \end{cases}$.

(*U.P., II Semester, 2009*)

Solution. The given function is periodic with period 3.

$$L\,[f(t)] = \int_1^3 f(t)\, e^{-st} dt$$

$$= \left[\int_1^2 (t-1)\, e^{-st} dt + \int_2^3 (3-t)\, e^{-st} dt\right]$$

$$= \left[\int_1^2 (t-1)\, e^{-st} dt + \int_2^3 (3-t)\, e^{-st} dt\right]$$

$$= \left[\left\{(t-1)\frac{e^{-st}}{-s} - \frac{e^{-st}}{(-s)^2}\right\}_1^2 + \left\{(3-t)\frac{e^{-st}}{-s} + \frac{e^{-st}}{(-s)^2}\right\}_2^3\right]$$

$$= \left[\left\{\frac{e^{-2s}}{-s} - \frac{e^{-2s}}{s^2} + \frac{e^{-s}}{s^2}\right\} + \left\{\frac{e^{-3s}}{s^2} - \frac{e^{-2s}}{-s} - \frac{e^{-2s}}{s^2}\right\}\right]$$

$$= \left[-\frac{e^{-2s}}{s} - \frac{e^{-2s}}{s^2} + \frac{e^{-s}}{s^2} + \frac{e^{-3s}}{s^2} + \frac{e^{-2s}}{s} - \frac{e^{-2s}}{s^2} \right]$$

$$= \left[\frac{1}{s^2}(-e^{-2s} + e^{-s} + e^{-3s} - e^{-2s}) \right] = \frac{1}{s^2}[e^{-s} - 2e^{-2s} + e^{-3s}] \qquad \textbf{Ans.}$$

Example 3. *Find the Laplace transform of* $F(t) = \begin{cases} 1, & 0 \le t < 1 \\ t, & 1 \le t < 2 \\ t^2, & 2 \le t < \infty \end{cases}$ *(Q. Bank U.P. 2001)*

Solution. Here, we have $F(t) = \begin{cases} 1, & 0 \le t < 1 \\ t, & 1 \le t < 2 \\ t^2, & 2 \le t < \infty \end{cases}$

$$L[F(t)] = \int_0^\infty e^{-st} \cdot F(t)dt = \int_0^1 e^{-st}dt + \int_1^2 t\, e^{-st}dt + \int_2^\infty t^2 e^{-st}dt$$

$$= \left(\frac{e^{-st}}{-s} \right)_0^1 + \left(t\frac{e^{-st}}{-s} - \frac{e^{-st}}{s^2} \right)_1^2 + \left(t^2 \frac{e^{-st}}{-s} \right)_2^\infty - \int_2^\infty 2t \cdot \frac{e^{-st}}{-s} dt$$

$$= \left(\frac{1 - e^{-s}}{s} \right) + \left(\frac{-2}{s} e^{-2s} - \frac{e^{-2s}}{s^2} \right) - \left(\frac{e^{-s}}{-s} - \frac{e^{-s}}{s^2} \right) + \frac{4}{s} e^{-2s} + \frac{2}{s} \int_2^\infty t\, e^{-st}\, dt$$

$$= \frac{1}{s} + \frac{2}{s} e^{-2s} + \frac{e^{-s}}{s^2} - \frac{e^{-2s}}{s^2} + \frac{2}{s}\left[\left(t\frac{e^{-st}}{-s} \right)_2^\infty - \int_2^\infty 1 \cdot \frac{e^{-st}}{-s} dt \right]$$

$$= \frac{1}{s} + \frac{2}{s} e^{-2s} + \frac{e^{-s}}{s^2} - \frac{e^{-2s}}{s^2} + \frac{2}{s}\left[\frac{2}{s} e^{-2s} + \frac{1}{s} \left(\frac{e^{-st}}{-s} \right)_2^\infty \right]$$

$$= \frac{1}{s} + \frac{2}{s} e^{-2s} + \frac{e^{-s}}{s^2} + \frac{3}{s^2} e^{-2s} + \frac{2}{s^3} e^{-2s}. \qquad \textbf{Ans.}$$

Example 4. *Find the Laplace transform of* $f(t) = \begin{cases} t^2, & 0 < t < 2 \\ t - 1, & 2 < t < 3 \\ 7, & t > 3 \end{cases}$

(U.P, II Semester, June 2007)

Solution. $L[f(t)] = \int_0^\infty e^{-st} f(t)dt = \int_0^2 t^2 e^{-st}dt + \int_2^3 (t-1)e^{-st}dt + \int_3^\infty 7e^{-st}dt$

$$\left[\int I\, II = I\, II_1 - I'\, II_{11} + I''\, II_{111} - \ldots \right]$$

$$= \left[t^2 \left(\frac{e^{-st}}{(-s)} \right) - 2t \frac{e^{-st}}{(-s)^2} + 2\frac{e^{-st}}{(-s)^3} \right]_0^2 + \left[(t-1)\left(\frac{e^{-st}}{(-s)} \right) - \frac{e^{-st}}{(-s)^2} \right]_2^3 + 7\left[\frac{e^{-st}}{-s} \right]_3^\infty$$

$$= \left[-4\left(\frac{e^{-2s}}{s} \right) - 4\left(\frac{e^{-2s}}{s^2} \right) - 2\left(\frac{e^{-2s}}{s^3} \right) + \frac{2}{s^3} \right] + \left[2\left(\frac{e^{-3s}}{-s} \right) - \frac{e^{-3s}}{s^2} + \frac{e^{-2s}}{s} + \frac{e^{-2s}}{s^2} \right] + 7\left(0 + \frac{e^{-3s}}{s} \right)$$

$$= \frac{2}{s^3} + e^{-2s}\left[-\frac{4}{s} - \frac{4}{s^2} - \frac{2}{s^3} \right] + e^{-3s}\left[-\frac{2}{s} - \frac{1}{s^2} \right] + e^{-2s}\left[\frac{1}{s} + \frac{1}{s^2} \right] + e^{-3s}\left[\frac{7}{s} \right]$$

$$= \frac{2}{s^3} + e^{-2s}\left[-\frac{4}{s} - \frac{4}{s^2} - \frac{2}{s^3} + \frac{1}{s} + \frac{1}{s^2}\right] + e^{-3s}\left[-\frac{2}{s} - \frac{1}{s^2} + \frac{7}{s}\right]$$

$$= \frac{2}{s^3} + e^{-2s}\left[-\frac{3}{s} - \frac{3}{s^2} - \frac{2}{s^3}\right] + e^{-3s}\left[\frac{5}{s} - \frac{1}{s^2}\right] = \frac{2}{s^3} - \frac{e^{-2s}}{s^3}(2+3s+3s^2) + \frac{e^{-3s}}{s^2}(5s-1) \quad \textbf{Ans.}$$

Example 5. *Find the Laplace transform of (1 + sin 2t).*

Solution. Laplace transform of (1 + sin 2*t*)

$$= \int_0^\infty e^{-st}(1+\sin 2t)\,dt = \int_0^\infty e^{-st}\left(1 + \frac{e^{2it} - e^{-2it}}{2i}\right)dt$$

$$= \frac{1}{2i}\int_0^\infty [2ie^{-st} + e^{(-s+2i)t} - e^{(-s-2i)t}]dt = \frac{1}{2i}\left[\frac{2ie^{-st}}{-s} + \frac{e^{(-s+2i)t}}{-s+2i} - \frac{e^{(-s-2i)t}}{-s-2i}\right]_0^\infty$$

$$= \frac{1}{2i}\left[\left(0 + \frac{2i}{s}\right) + \frac{1}{-s+2i}(0-1) - \frac{1}{-s-2i}(0-1)\right]$$

$$= \frac{1}{2i}\left[\frac{2i}{s} + \frac{1}{s-2i} - \frac{1}{s+2i}\right] = \frac{1}{2}\left[\frac{2}{s} + \frac{4}{s^2+4}\right] = \frac{1}{s} + \frac{2}{s^2+4} \quad \textbf{Ans.}$$

Alternate Method

$$L(1 + \sin 2t) = L(1) + L \sin 2t = \frac{1}{s} + \frac{2}{s^2+4} \quad \textbf{Ans.}$$

46.4 PROPERTIES OF LAPLACE TRANSFORMS

$$L[af_1(t) + bf_2(t)] = a\,L[f_1(t)] + b\,L[f_2(t)]$$

Proof. $L[af_1(t) + bf_2(t)] = \int_0^\infty e^{-st}[af_1(t) + bf_2(t)]\,dt$

$$= a\int_0^\infty e^{-st} f_1(t)dt + b\int_0^\infty e^{-st} f_2(t)dt$$

$$= a\,L[f_1(t)] + b\,L[f_2(t)] \quad \textbf{Proved.}$$

46.5 CHANGE OF SCALE PROPERTY

If $L\{f(t)\} = F(s)$ then $\boxed{L\{f(at)\} = \frac{1}{a}F\left(\frac{s}{a}\right)}$

Proof. $L\{f(at)\} = \int_0^\infty e^{-st} f(at)dt = \int_0^\infty e^{-\left(\frac{s}{a}\right)u} f(u)\frac{du}{a}$ $\left[\text{Put } at = u \Rightarrow dt = \frac{du}{a}\right]$

$$= \frac{1}{a}\int_0^\infty e^{-\left(\frac{s}{a}\right)u} f(u)du = \frac{1}{a}\int_0^\infty e^{-\left(\frac{s}{a}\right)t} f(t)dt$$

$$= \frac{1}{a}\int_0^\infty e^{-St} f(t)dt = \frac{1}{a}L\{f(t)\} = \frac{1}{a}F(S) \quad \left[\text{Put } S = \frac{s}{a}\right]$$

$$= \frac{1}{a}F\left(\frac{s}{a}\right) \quad \textbf{Proved.}$$

Example 6. *If* $L\{J_0(\sqrt{t})\} = \dfrac{e^{-\frac{1}{4s}}}{s}$, *find* $L\{J_0(2\sqrt{t})\}$.

Solution. Here, we have

$$L\{J_0(\sqrt{t})\} = \frac{e^{-\frac{1}{4s}}}{s}$$

By change of scale property,

$$L\{J_0(\sqrt{4t})\} = \frac{1}{4}\cdot\left\{\frac{e^{-\frac{1}{4(s/4)}}}{(s/4)}\right\}$$

$$\Rightarrow \quad L\{J_0(2\sqrt{t})\} = \frac{1}{s}e^{-1/s}$$ **Ans.**

(2) First Shifting Theorem. If L $\{f(t)\}$ = F(s), then

$$\boxed{\mathbf{L}[e^{at}f(t)] = \mathbf{F}(s-a)}$$

Proof.

$$L[e^{at}f(t)] = \int_0^\infty e^{-st}.e^{at}f(t)dt = \int_0^\infty e^{-(s-a)t}f(t)dt$$

$$= \int_0^\infty e^{-rt}f(t)dt \qquad \text{where } r = s-a$$

$$= F(r) = F(s-a)$$ **Proved.**

With the help of this property, we can have the following important results :

1. $L(e^{at}t^n) = \dfrac{n!}{(s-a)^{n+1}}$

2. $L(e^{at}\cosh bt) = \dfrac{s-a}{(s-a)^2-b^2}$ **3.** $L(e^{at}\sinh bt) = \dfrac{b}{(s-a)^2-b^2}$

4. $L(e^{at}\sin bt) = \dfrac{b}{(s-a)^2+b^2}$ **5.** $L(e^{at}\cos bt) = \dfrac{s-a}{(s-a)^2+b^2}$

46.6 HEAVISIDE'S SHIFTING THEOREM (Second Translation Property)

If L $\{f(t)\} = F(s)$ and $g(t) = \begin{cases} f(t-a), & t>a \\ 0, & 0<t<a \end{cases}$ then prove that

$$\boxed{L\{g(t)\} = e^{-as}F(s)}$$ *(U.P. II Semester, Summer 2006)*

Proof. $L\{g(t)\} = \int_0^\infty e^{-st}g(t)dt$

$$= \int_0^a e^{-st}g(t)dt + \int_a^\infty e^{-st}g(t)dt \qquad [g(t) = 0, \text{ when } 0<t<a]$$

$$= 0 + \int_a^\infty e^{-st}.f(t-a)dt = \int_a^\infty e^{-st}f(t-a)dt \qquad [\text{Put } t-a = u \Rightarrow dt = du]$$

$$= \int_0^\infty e^{-s(u+a)}f(u)du = e^{-sa}\int_0^\infty e^{-su}f(u)du = e^{-as}\int_0^\infty e^{-st}f(t)\,dt$$ **Proved.**

$$\boxed{\mathbf{L}\{g(t)\} = e^{-as}\mathbf{F}(s)}$$

Example 7. *Find the Laplace transform of $\cos^2 t$.*

Solution. We know that $\cos 2t = 2\cos^2 t - 1$

$$\cos^2 t = \frac{1}{2}[\cos 2t + 1]$$

$$L(\cos^2 t) = L\left[\frac{1}{2}(\cos 2t+1)\right] = \frac{1}{2}[L(\cos 2t)+L(1)]$$

$$= \frac{1}{2}\left[\frac{s}{s^2+(2)^2}+\frac{1}{s}\right] = \frac{1}{2}\left[\frac{s}{s^2+4}+\frac{1}{s}\right]$$ **Ans.**

Example 8. *If* $L(\cos^2 t)\ \frac{s^2+2}{s(s^2+4)}$, *find* $L(\cos^2 at)$. *(U.P., II Semester, Summer 2006)*

Solution. We have, $L(\cos^2 t) = \frac{s^2+2}{s(s^2+4)}$

By change of scale property, we have

$$L(\cos^2 at)\ \frac{1}{a}\cdot\frac{\left(\frac{s}{a}\right)^2+2}{\frac{s}{a}\left[\left(\frac{s}{a}\right)^2+4\right]} = \frac{1}{a}\left[\frac{s^2+2a^2}{\frac{s}{a}(s^2+4a^2)}\right]\ \frac{s^2+2a^2}{s(s^2+4a^2)}$$ **Ans.**

Example 9. *Find the Laplace transform of* $t^{-\frac{1}{2}}$.

Solution. We know that $L(t^n)\ \frac{\overline{|n+1}}{s^{n+1}}$

Put $n\ -\frac{1}{2}$, $L(t^{-1/2}) = \frac{\overline{\left|-\frac{1}{2}+1\right.}}{s^{-1/2+1}} = \frac{\overline{\left|\frac{1}{2}\right.}}{\sqrt{s}} = \frac{\sqrt{\pi}}{\sqrt{s}}$, where $\overline{\left|\frac{1}{2}\right.} = \sqrt{\pi}$ **Ans.**

Example 10. *Find the Laplace transform of 2 sin 2t cos 4t.*

Solution. We have

$$f(t) = 2\sin 2t\cos 4 = \sin(2t+4t) + \sin(2t-4t) = \sin 6t - \sin 2t$$

$$L\,f(t) = L(\sin 6t) - L(\sin 2t) = \frac{6}{s^2+36} - \frac{2}{s^2+4}$$ **Ans.**

Example 11. *Find the Laplace transform of $4\sin^3 t$.*

Solution. We have

$$f(t) = 4\sin^3 t = 3\sin t - \sin 3t \qquad [\sin 3t = 3\sin t - 4\sin^3 t]$$

$$L\,f(t) = 3L\sin t - L\sin 3t = \frac{3}{s^2+1} - \frac{3}{s^2+9}$$ **Ans.**

Example 12. *Find the Laplace transform of 4 cosh 2t sin 4t*

Solution. We have

$$f(t) = 4\cosh 2t\sin 4t = 4\left(\frac{e^{2t}+e^{-2t}}{2}\right)\left(\frac{e^{4it}-e^{-4it}}{2i}\right)$$

$$= \frac{1}{i}\left[e^{(2+4i)t} - e^{(2-4i)t} + e^{(-2+4i)t} - e^{(-2-4i)t}\right]$$

$$L\,[f(t)] = -i\left[L(e^{(2+4i)t}) - L(e^{(2-4i)t}) + L(e^{(-2+4i)t}) - L(e^{(-2-4i)t})\right]$$

$$= -i\left[\frac{1}{s-2-4i} - \frac{1}{s-2+4i} + \frac{1}{s+2-4i} - \frac{1}{s+2+4i}\right]$$

$$= -i\left[\left(\frac{1}{s-2-4i} - \frac{1}{s+2+4i}\right) - \left(\frac{1}{s-2+4i} - \frac{1}{s+2-4i}\right)\right]$$

$$= -i\left[\frac{4+8i}{s^2-(2+4i)^2} - \frac{4-8i}{s^2-(2-4i)^2}\right] \quad \textbf{Ans.}$$

EXERCISE 46.1

Find the Laplace transforms of the following:

1. $t + t^2 + t^3$ **Ans.** $\frac{1}{s^2} + \frac{2}{s^3} + \frac{6}{s^4}$

2. $\sin t \cos t$ **Ans.** $\frac{1}{s^2+4}$

3 $t^{7/2} e^{5t}$ *(M.D.U. Dec. 2009)* **Ans.** $\frac{105\sqrt{\pi}}{16\,(s-5)^{9/2}}$

4. $\sin^3 2t$ **Ans.** $\frac{48}{(s^2+4)(s^2+36)}$

5 $e^{-t}\cos^2 t$ **Ans.** $\frac{1}{2s+2} + \frac{s+1}{2s^2+4s+10}$

6. $\sin 2t \cos 3t$ **Ans.** $\frac{2(s^2-5)}{(s^2+1)(s^2+25)}$

7. $\sin 2t \sin 3t$ **Ans.** $\frac{12s}{(s^2+1)(s^2+25)}$

8. $\cos at \sinh at$ **Ans.** $\frac{1}{2}\left[\frac{s-a}{(s-a)^2+a^2} - \frac{s+a}{(s+a)^2+a^2}\right]$

9. $\sinh^3 t$ **Ans.** $\frac{6}{(s^2-1)(s^2-9)}$

10. $\cos t \cos 2t$ **Ans.** $\frac{s(s^2+5)}{(s^2+1)(s^2+9)}$

11. $\cosh at \sin at$ **Ans.** $\frac{a(s^2+2a^2)}{s^4+4a^4}$

12 $f(t) = \begin{cases} \cos\left(t - \frac{2\pi}{3}\right), & t > \frac{2\pi}{3} \\ 0, & t < \frac{2\pi}{3} \end{cases}$ **Ans.** $e^{\frac{-2\pi s}{3}} \frac{s}{s^2+1}$

46.7 EXISTENCE THEOREM

According to this theorem $\int_0^\infty e^{-st} f(t)\,dt$ exists is $\int_0^\lambda e^{-st} f(t)dt$ can actually be evaluated and its limit a $\lambda \to \infty$ exists.

Otherwise we may use the following theorem:

If $f(t)$ is continuous an $\lim_{t\to\infty}\left[e^{-at} f(t)\right]$ is finite, then Laplace transform of $f(t)$ *i.e* $\int_0^\infty e^{-st} f(t)dt$ exists for $s > a$.

It should however, be kept in mind that the above foresaid conditions are sufficient but not necessary.

For example $L\left(\frac{1}{\sqrt{t}}\right)$ exists though $\frac{1}{\sqrt{t}}$ is infinite at $t = 0$. Similarly a function $f(t)$ for which $\lim_{t\to\infty}\left[e^{-at} f(t)\right]$ is finite and having a finite discontinuity will have a Laplace transform of $s > a$.

46.8 LAPLACE TRANSFORM OF THE DERIVATIVE OF *f(t)* (*D.U. April 2010*)

$$\boldsymbol{L[f'(t)] = sL[f(t)] - f(0)} \qquad \text{where} \qquad L[f(t)] = F(s).$$

Proof $L[f'(t)] = \int_0^\infty e^{-st} f'(t)dt$

Integrating by parts, we get

$$L[f'(t)] = \left[e^{-st} f(t)\right]_0^\infty - \int_0^\infty (-se^{-st}) f(t)\, dt$$

$$= -f(0) + s\int_0^\infty e^{-st} f(t)dt \qquad (e^{-st} f(t) = 0, \quad \text{when } t = \infty)$$

$$= -f(0) + s\, L\,[f(t)]$$

$$\boxed{\boldsymbol{L[f'(t)] = s\,L\,[f(t)] - f(0).}}$$ **Proved.**

Note. Roughly, Laplace transform of **derivative** of $f(t)$ corresponds to **multiplication** of the Laplace transform of $f(t)$ by s.

46.9 LAPLACE TRANSFORM OF DERIVATIVE OF ORDER *n* (*M.D.U. Dec. 2009*)

$$L[f^n(t)] = s^n L[f(t)] - s^{n-1} f(0) - s^{n-2} f'(0) - s^{n-3} f''(0) - \dots - f^{n-1}(0)$$

Proof. We have already proved in Article 46.8 that

$$L[f'(t)] = s\,L[f(t)] - f(0) \qquad \text{...(1)}$$

Replacing $f(t)$ by $f'(t)$ and $f'(t)$ by $f''(t)$ in (1), we get

$$L[f''(t)] = s\,L[f'(t)] - f'(0) \qquad \text{...(2)}$$

Putting the value of $L[f'(t)]$ from (1) in (2), we have

$$L[f''(t)] = s\,[s\,L[f(t)] - f(0)] - f'(0)$$

$$\Rightarrow \qquad L[f''(t)] = s^2 L[f(t)] - sf(0) - f'(0)$$

Similarly $\qquad L[f'''(t)] = s^3 L[f(t)] - s^2 f(0) - s\,f'(0) - f''(0)$

$$L[f^{iv}(t)] = s^4 L[f(t)] - s^3 f(0) - s^2 f'(0) - s\,f''(0) - f'''(0)$$

$$\boxed{\boldsymbol{L[f^n(t)] = s^n L[f(t)] - s^{n-1} f(0) - s^{n-2} f'(0) - s^{n-3} f''(0) + \dots\dots - f^{n-1}(0)}}$$

Example 13. *Given* $L\left(2\sqrt{\frac{t}{\pi}}\right) = \frac{1}{s^{3/2}}$, *show that* $L\left(\frac{1}{\sqrt{\pi t}}\right) = \frac{1}{\sqrt{s}}$. (*U.P., II Semester, 2005*)

Solution. Let $F(t) = 2\sqrt{\frac{t}{\pi}} \quad \Rightarrow \quad F'(t) = \frac{1}{\sqrt{\pi t}}$. Also $F(0) = 0$

$$\therefore \quad L\{F'(t)\} = s\,L\{F(t)\} - F(0) = s\frac{1}{s^{3/2}} - 0$$

$$\therefore \quad L\left(\frac{1}{\sqrt{\pi t}}\right) = \frac{1}{\sqrt{s}}.$$ **Ans.**

Example 14. *Find the Laplace transform of* $\sin\sqrt{t}$ *; Hence find* $L\left(\frac{\cos\sqrt{t}}{\sqrt{t}}\right)$

Solution. $\sin\sqrt{t} = \sqrt{t} - \frac{(\sqrt{t})^3}{3!} + \frac{(\sqrt{t})^5}{5!} - \ldots\ldots$

$$= t^{1/2} - \frac{t^{3/2}}{3!} + \frac{t^{5/3}}{5!} - \ldots\ldots \qquad \left[\sin x = x - \frac{x^3}{3!} + \frac{x^5}{5!} - \ldots\ldots\right]$$

$$\therefore \quad L(\sin\sqrt{t}) = L\left(t^{1/2} - \frac{t^{3/2}}{3!} + \frac{t^{5/2}}{5!} - \ldots\ldots\right) \quad \frac{\Gamma 3/2}{s^{3/2}} - \frac{\Gamma 5/2}{3!s^{5/2}} + \frac{\Gamma 7/2}{5!s^{7/2}} - \ldots\ldots$$

$$= \frac{\sqrt{\pi}}{2s^{3/2}}\left\{1 - \left(\frac{1}{2^2 s}\right) + \frac{1}{2!}\left(\frac{1}{2^2 s}\right)^2 - \ldots\ldots\right\}$$

$$= \frac{\sqrt{\pi}}{2s^{3/2}}\left\{1 - \left(\frac{1}{2^2 s}\right) + \frac{1}{2!}\left(\frac{1}{2^2 s}\right)^2 - \ldots\ldots\right\} \qquad \left[e^x = 1 + \frac{x}{1!} + \frac{x^2}{2!} + \frac{x^3}{3!} + \ldots\ldots\right]$$

$$\Rightarrow \quad L(\sin\sqrt{t}) = \frac{\sqrt{\pi}}{2s^{3/2}} e^{-(1/4s)}$$

No $L\left[\frac{d}{dt}(\sin\sqrt{t})\right], = s\,L(\sin\sqrt{t}) - 0 \qquad \left[\because F(0) = 0 \text{ and } L\left[\frac{d}{dt}[F(t)]\right] = sF(s)\right]$

$$L\left(\frac{\cos\sqrt{t}}{2\sqrt{t}}\right) = \frac{\sqrt{\pi}}{2\sqrt{s}} e^{-\left(\frac{1}{4s}\right)} \Rightarrow L\left(\frac{\cos\sqrt{t}}{\sqrt{t}}\right) = \frac{\sqrt{\pi}}{\sqrt{s}} e^{-\left(\frac{1}{4s}\right)}$$ **Ans.**

46.10 LAPLACE TRANSFORM OF INTEGRAL OF *f(t)*

$$\boxed{L\left[\int_0^t f(t)dt\right] = \frac{1}{s}F(s)} \qquad \text{where} \quad L[f(t)] = F(s)$$

Proof. Let $\phi(t) = \int_0^t f(t)\,dt$ and $\phi(0) = 0$ the $\phi'(t) = f(t)$

We know the formula of Laplace transforms of $\phi'(t)$ *i.e.*

$$L[\phi'(t)] = sL[\phi(t)] - \phi(0)$$

$$\Rightarrow \quad L[\phi'(t)] = sL[\phi(t)] \qquad [\phi(0) = 0]$$

$$\Rightarrow \quad L[\phi(t)] = \frac{1}{s}L[\phi'(t)]$$

Putting the values of $\phi(t)$ and $\phi'(t)$, we get

$$L\left[\int_0^t f(t)dt\right] = \frac{1}{s}L[f(t)] \Rightarrow \boxed{\mathbf{L}\left[\int_0^t f(t)\,dt\right] = \frac{1}{s}F(s)}$$ **Proved.**

Note. (1) Laplace transform of **Integral** of $f(t)$ corresponds to the division of the Laplace transform of $f(t)$ by s.

(2) $\int_0^t f(t)\,dt = L^{-1}\left[\frac{1}{s}F(s)\right]$

46.11 LAPLACE TRANSFORM OF *t.f(t)* (Multiplication by *t*)

If $L[f(t)] = F(s)$, then

$$L[t^n f(t)] = (-1)^n \frac{d^n}{ds^n}[F(s)].$$ (*U.P., II Semester, Summer 2005*)

Proof. $$L[f(t)] = F(s) = \int_0^\infty e^{-st} f(t)dt \quad ...(1)$$

Differentiating (1) w.r.t. 's', we get

$$\frac{d}{ds}[F(s)] = \frac{d}{ds}\left[\int_0^\infty e^{-st} f(t)dt\right] = \int_0^\infty \frac{\partial}{\partial s}(e^{-st}) f(t)dt$$

$$= \int_0^\infty (-t e^{-st}).f(t)dt = \int_0^\infty e^{-st}[-t.f(t)]\,dt$$

$$= L[-t\,f(t)] \quad \Rightarrow \quad L[t\,f(t)] = (-1)^1 \frac{d}{ds}[F(s)]$$

Similarly, $L\left[t^2 f(t)\right] = (-1)^2 \frac{d^2}{ds^2}[F(s)]$

$$L\left[t^3 f(t)\right] = (-1)^3 \frac{d^3}{ds^3}[F(s)]$$

$$\boxed{L\left[t^n f(t)\right] = (-1)^n \frac{d^n}{ds^n}[F(s)]}$$ **Proved.**

46.12 INITIAL AND FINAL VALUE THEOREMS

(*a*) Initial Value Theorem $L\{f(t)\} = F(s)$

$\Rightarrow$ $$\lim_{t \to 0} f(t) = \operatorname{Lim}_{s \to \infty}[sF(s)]$$, provided the limit exists.

Proof. $$L\{f'(t)\} = s\,L\{f(t)\} - f(0)$$

$\Rightarrow$ $$\int_0^\infty e^{-st} f'(t)dt = s\,F(s) - f(0)$$

$\Rightarrow$ $$\operatorname{Lim}_{s \to \infty} \int_0^\infty e^{-st} f'(t)dt = \operatorname{Lim}_{s \to \infty}[sF(s) - f(0)]$$

$\Rightarrow$ $$\operatorname{Lim}_{s \to \infty}[s\,F(s)] = f(0) + \int_0^\infty \left(\operatorname{Lim}_{s \to \infty} e^{-st}\right) f'(t)dt$$

$$= f(0) + \int_0^\infty (0)\,f'(t)\,dt \qquad (\because \operatorname{Lim}_{s \to \infty} e^{-st} = 0)$$

$$= f(0) + 0 = f(0) = \operatorname{Lim}_{t \to 0} f(t)$$

(*b*) Final Value Theorem. $L\{f(t)\} = F(s)$

$\Rightarrow$ $$\operatorname{Lim}_{t \to \infty} f(t) = \operatorname{Lim}_{s \to 0}[sF(s)]$$, provided the limits exist.

Proof $L\{f'(t)\} = sL\{f(t)\} - f(0) \quad \Rightarrow \quad \int_0^\infty e^{-st} f'(t)dt = sF(s) - f(0)$

$\Rightarrow$ $$\operatorname{Lim}_{s \to 0} \int_0^\infty e^{-st} f'(t)\,dt = \operatorname{Lim}_{s \to 0}[sF(s) - f(0)]$$

$$\Rightarrow \qquad \lim_{s\to 0}[s\,F(s)-f(0)]=\lim_{s\to 0}\int_0^\infty e^{-st}f'(t)dt$$

$$\Rightarrow \qquad \lim_{s\to 0}[s\,F(s)]-f(0)=\lim_{s\to 0}\int_0^\infty e^{-st}f'(t)dt$$

$$\Rightarrow \qquad \lim_{s\to 0}\left[sF(s)\right]-f(0)=\int_0^\infty \lim_{s\to 0} e^{-st}f'(t)dt=\int_0^\infty (1)\,f'(t)dt \qquad \left[\because \lim_{s\to 0} e^{-st}=1\right]$$

$$\boxed{\lim_{s\to 0}\left[sF(s)\right]=\lim_{t\to\infty} f(t)}$$

Example 15. *If* $L\{F(t)\}=\dfrac{1}{s(s+\beta)}$ *then, find* $\lim\limits_{t\to\infty} F(t)$

Solution. By final-value theorem,

$$\lim_{t\to\infty} F(t) = \lim_{s\to 0} sL\{F(t)\}=\lim_{s\to 0}\frac{s}{s\,(s+\beta)}=\lim_{s\to 0}\frac{1}{(s+\beta)}=\frac{1}{\beta} \qquad \textbf{Ans.}$$

46.13. EXPONENTIAL INTEGRAL FUNCTION

$$\int_t^\infty\left(\frac{e^{-x}}{x}\right)dx$$

Let $\quad f(t)=\int_t^\infty \dfrac{e^{-x}}{x}dx$

$$\Rightarrow \qquad f'(t)=-\frac{e^{-t}}{t} \qquad \Rightarrow \qquad tf'(t)=-e^{-t} \quad \text{[Here – ve sign appears due to lower limit]}$$

Taking Laplace Transform of $tf'(t)$, we get $L\{t\,f'(t)\}=L\{-e^{-t}\}=-L\{e^{-t}\}$

$$\Rightarrow \qquad -\frac{d}{ds}\,[sF(s)-f(0)]=-\frac{1}{s+1}$$

$$\Rightarrow \qquad \frac{d}{ds}\,[s\,F(s)]=\frac{1}{s+1} \qquad \left[\because f(0)=\text{constant} \therefore \frac{d}{ds}f(0)=0\right]$$

Integrating both the sides, we get

$$s\,F\,(s) = \log\,(s+1) + C \qquad ...(1)$$

Now, by final value theorem, we have

$$\lim_{s\to 0} sF(s)=\lim_{t\to\infty} f(t) \qquad ...(2)$$

Hence $\lim\limits_{s\to 0}\left[sF(s)\right]=\lim\limits_{s\to 0}[\log(s+1)+C]=0+C=C \qquad ...(3)$

Also, $$\lim_{t\to\infty}\left[f(t)\right]=\lim_{t\to\infty}\int_t^\infty\left(\frac{e^{-x}}{x}\right)dx=0 \qquad ...(4)$$

Putting the values of $\lim\limits_{x\to 0}[s\,F(s)]$ and $\lim\limits_{t\to\infty}\left[f(t)\right]$ from (3) and (4) in (2), we get

$$C=0.$$

Hence from (1) $sF(s)=\log(s+1)\Rightarrow F(s)=\left\{\dfrac{\log(s+1)}{s}\right\}$

$$\Rightarrow \qquad \boxed{L\int_t^{\infty}\left(\frac{e^{-x}}{x}\right)dx = \left[\frac{\log(s+1)}{s}\right]}$$

Example 16. *Find the Laplace Transform of t sin at.*

Solution. $\mathrm{L}(t \sin at) = \mathrm{L}\left(t\frac{e^{iat}-e^{-iat}}{2i}\right) = \frac{1}{2i}[\mathrm{L}(t.e^{iat}) - \mathrm{L}(te^{-iat})]$

$$= \frac{1}{2i}\left[-\frac{d}{ds}\frac{1}{s-ia} + \frac{d}{ds}\frac{1}{s+ia}\right] = \frac{1}{2i}\left[\frac{1}{(s-ia)^2} - \frac{1}{(s+ia)^2}\right] = \frac{1}{2i}\left[\frac{(s+ia)^2-(s-ia)^2}{(s-ia)^2(s+ia)^2}\right]$$

$$= \frac{1}{2i}\frac{(s^2+2ias-a^2)-(s^2-2ias-a^2)}{(s^2+a^2)^2} = \frac{1}{2i}\frac{4ias}{(s^2+a^2)^2} = \frac{2as}{(s^2+a^2)^2} \qquad \textbf{Ans.}$$

Example 17. *Find the Laplace transform of t sinh at.*

Solution. $\mathrm{L}(\sinh at) = \frac{a}{s^2-a^2}$

$$\mathrm{L}[t \sinh at] = -\frac{d}{ds}\left(\frac{a}{s^2-a^2}\right) = \frac{2as}{(s^2-a^2)^2} \qquad \textbf{Ans.}$$

Example 18. *Find the Laplace transform of t^2 cos at*

Solution. $L(\cos at) = \frac{s}{s^2+a^2}$

$$L(t^2 \cos at) = (-1)^2\frac{d^2}{ds^2}\left[\frac{s}{s^2+a^2}\right] = \frac{d}{ds}\frac{(s^2+a^2).1-s(2s)}{(s^2+a^2)^2} = \frac{d}{ds}\frac{a^2-s^2}{(s^2+a^2)^2}$$

$$= \frac{(s^2+a^2)^2(-2s)-(a^2-s^2).2(s^2+a^2)(2s)}{(s^2+a^2)^4} = \frac{(s^2+a^2)(-2s)-(a^2-s^2)4s}{(s^2+a^2)^3}$$

$$= \frac{-2s^3-2a^2s-4a^2s+4s^3}{(s^2+a^2)^3} = \frac{2s(s^2-3a^2)}{(s^2+a^2)^3} \qquad \textbf{Ans.}$$

Example 19. *Obtain the Laplace transform of t^2e^t . sin 4t.*

(U.P. II Semester, Summer 2002, Uttarakhand II Sem. 2010)

Solution. $\mathrm{L}(\sin 4t) = \frac{4}{s^2+16}$,

$$L(e^t \sin 4t) = \frac{4}{(s-1)^2+16}$$

$$L(t\,e^t \sin 4t) = -\frac{d}{ds}\left(\frac{4}{s^2-2s+17}\right) = \frac{4(2s-2)}{(s^2-2s+17)^2}$$

$$L(t^2e^t \sin 4t) = -\frac{d}{ds}\left(\frac{4(2s-2)}{(s^2-2s+17)^2}\right)$$

$$= -4\frac{(s^2-2s+17)^2 2-(2s-2)2(s^2-2s+17)(2s-2)}{(s^2-2s+17)^4}$$

$$= 4\frac{(s^2-2s+17)\,2-2(2s-2)^2}{(s^2-2s+17)^3} = \frac{-4(2s^2-4s+34-8s^2+16s-8)}{(s^2-2s+17)^3}$$

$$= \frac{-4(-6s^2+12s+26)}{(s^2-2s+17)^3} = \frac{8[3s^2-6s-13]}{(s^2-2s+17)^3}$$ **Ans.**

Example 20. *Find the Laplace transform of the function*

$f(t) = te^{-t}\sin 2t$ *(U.P. II Semester, Summer 2002)*

Solution. $L[\sin 2t] = \dfrac{2}{s^2+4}$

$$L\left[e^{-t}\sin 2t\right] = \frac{2}{(s+1)^2+4} = F(s) \quad \text{(say)}$$

$$L(te^{-1}\sin 2t) = -F'(s) = -\frac{d}{ds}\left[\frac{2}{(s+1)^2+4}\right] = \frac{2\cdot 2(s+1)}{\left[(s+1)^2+4\right]^2} = \frac{4(s+1)}{\left[(s+1)^2+4\right]^2}$$ **Ans.**

EXERCISE 46.2

Find the Laplace transforms of the following :

1. te^{at} **Ans.** $\dfrac{1}{(s-a)^2}$
2. $t\cosh at$ **Ans.** $\dfrac{s^2+a^2}{(s^2-a^2)^2}$
3. $t\cos t$ **Ans.** $\dfrac{s^2-1}{(s^2+1)^2}$
4. $t\cosh t$ **Ans.** $\dfrac{s^2+1}{(s^2-1)^2}$
5. $t^2\sin t$ **Ans.** $\dfrac{2(3s^2-1)}{(s^2+1)^3}$
6. t^3e^{-3t} **Ans.** $\dfrac{6}{(s+3)^4}$
7. $t\sin^2 3t$ **Ans.** $\dfrac{1}{2}\left[\dfrac{1}{s^2}-\dfrac{s^2-36}{(s^2+36)^2}\right]$
8. $te^{at}\sin at$ **Ans.** $\dfrac{2a(s-a)}{(s^2-2as+2a^2)^2}$
9. $te^{-t}\cosh t$ **Ans.** $\dfrac{s^2+2s+2}{(s^2+2s)^2}$
10. $t^2e^{-2}t\cos t$ **Ans.** $\dfrac{2(s^3+6s^2+9s+2)}{(s^2+4s+5)^3}$
11. $\int_0^t e^{-2t}t\sin^3 t\,dt$ **Ans.** $\dfrac{3(s+2)}{2s}\left[\dfrac{1}{\left[(s+2)^2+9\right]^2}-\dfrac{1}{\left[(s+2)^2+1\right]^2}\right]$
12. If $f(t)$ is continuous, except for an ordinary discontinuity at $t = a$, $(a > 0)$ as given in the figure, then show that
$L\left[f'(t)\right] = s\left[f(t)\right] - f(0) - e^{-as}\left[f(a+0)-f(a-0)\right]$
(U.P. II Semester 2003)

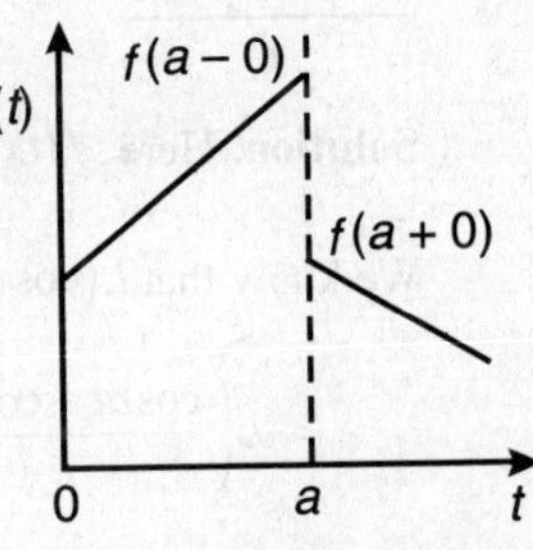

13. Pick the correct statement for final value theorem of Laplace transform:
(*i*) $\underset{t\to 0}{\text{Lt}}\, f(t) = \underset{s\to\infty}{\text{Lt}}\, sF(s)$ (*ii*) $\underset{t\to\infty}{\text{Lt}}\, f(t) = \underset{s\to 0}{\text{Lt}}\, sF(s)$
(U.P. II Semester 2010) **Ans.** (*ii*)

46.14 LAPLACE TRANSFORM OF $\frac{1}{t}f(t)$ (Division by t) *(D.U. April 2010)*

If $L[f(t)] = F(s)$, the $L\left[\dfrac{1}{t}f(t)\right] = \int_s^\infty F(s)\,ds$ *(U.P. II Semester Summer, 2007, 2005)*

Proof. We know that $L[f(t)] = F(s)$ or $F(s) = \int_0^\infty e^{-st} f(t)\,dt$ ...(1)

Integrating (1) w.r.t. 's', we have

$$\int_s^\infty F(s)ds \int_s^\infty\left[\int_0^\infty e^{-st} f(t)dt\right]ds = \int_0^\infty f(t)\left[\int_s^\infty e^{-st}ds\right]dt = \int_0^\infty f(t)\left[\frac{e^{-st}}{-t}\right]_s^\infty dt$$

$$\int_0^\infty \frac{-f(t)}{t}[e^{-st}]_s^\infty dt = \int_0^\infty \frac{-f(t)}{t}[0-e^{-st}]dt = \int_0^\infty e^{-st}\left\{\frac{1}{t}.f(t)\right\}dt = L\left[\frac{1}{t}f(t)\right]$$

$$\Rightarrow \quad \boxed{L\left[\frac{1}{t}f(t)\right] = \int_s^\infty F(s)ds}$$ **Proved.**

Cor. $L^{-1}\int_s^\infty F(s)\,ds = \frac{1}{t}f(t)$

Example 21. *Find the Laplace transform of* $\frac{\sin 2t}{t}$

Solution. $L(\sin 2t) = \frac{2}{s^2+4}$

$$L\left(\frac{\sin 2t}{t}\right) = \int_s^\infty \frac{2}{s^2+4}ds = 2.\frac{1}{2}\left[\tan^{-1}\frac{s}{2}\right]_s^\infty = \left[\tan^{-1}\infty - \tan^{-1}\frac{s}{2}\right] = \frac{\pi}{2} - \tan^{-1}\frac{s}{2}$$

$$= \cot^{-1}\frac{s}{2}$$ **Ans.**

Example 22. *Find the Laplace transform of f(t)* $= \frac{\sin at}{t}dt$

(M.D.U., Dec. 2009, U.P., II Semester, Summer 2005)

Solution. $L(\sin at) = \frac{a}{s^2+a^2}$

$$L\left(\frac{\sin at}{t}\right) = \int_s^\infty \frac{a}{s^2+a^2}ds = \left[\tan^{-1}\frac{s}{a}\right]_s^\infty = \frac{\pi}{2} - \tan^{-1}\frac{s}{a} = \cot^{-1}\frac{s}{a}$$

Hence, $L\left[\frac{\sin at}{t}dt\right] = \frac{1}{s}\cot^{-1}\frac{s}{a}$ **Ans.**

Example 23. *Find the Laplace transform of :*

$\frac{\cos at - \cos bt}{t}$ *(Uttarakhand, II Semester, June 2007, U.P., II Semester, 2004)*

Solution. Here $f(t) = \frac{\cos at - \cos bt}{t}$

We know that $L(\cos at - \cos bt) = L(\cos at) - L(\cos bt) = \frac{s}{s^2+a^2} - \frac{s}{s^2+b^2}$

$$L\left(\frac{\cos at - \cos bt}{t}\right) = \int_s^\infty\left(\frac{s}{s^2+a^2} - \frac{s}{s^2+b^2}\right)ds = \left[\frac{1}{2}\log(s^2+a^2) - \frac{1}{2}\log(s^2+b^2)\right]_s^\infty$$

$$= \frac{1}{2}\left[\log\frac{s^2+a^2}{s^2+b^2}\right]_s^\infty = \frac{1}{2}\left[\log\frac{1+\frac{a^2}{s^2}}{1+\frac{b^2}{s^2}}\right]_s^\infty = \frac{1}{2}\log 1 - \frac{1}{2}\log\frac{1+\frac{a^2}{s^2}}{1+\frac{b^2}{s^2}} = 0 - \frac{1}{2}\log\frac{s^2+a^2}{s^2+b^2} \quad [\log 1 = 0]$$

$$\frac{1}{2}\log\frac{s^2+b^2}{s^2+a^2}$$ **Ans.**

Example 24. *If* $f(t)=\dfrac{e^{at}-\cos bt}{t}$, *find the Laplace transform of f (t).*

(U.P. II Semester, Summer 2003)

Solution $f(t)=\dfrac{e^{at}-\cos bt}{t}=\dfrac{e^{at}}{t}-\dfrac{\cos bt}{t}$

We know that $L(e^{at}-\cos bt)=\left(\dfrac{1}{s-a}-\dfrac{s}{s^2+b^2}\right)$

$$\therefore \quad L\left(\frac{e^{at}-\cos bt}{t}\right)=\int_s^\infty\left(\frac{1}{s-a}-\frac{s}{s^2+b^2}\right)ds=\left[\log(s-a)-\frac{1}{2}\log(s^2+b^2)\right]_s^\infty$$

$$=\left[\frac{2\log(s-a)-\log(s^2+b^2)}{2}\right]_s^\infty=\frac{1}{2}\left[\log(s-a)^2-\log(s^2+b^2)\right]_s^\infty$$

$$=\frac{1}{2}\left[\log\frac{(s-a)^2}{s^2+b^2}\right]_s^\infty=\frac{1}{2}\left[\log\left\{\frac{\left(1-\frac{a}{s}\right)^2}{1+\frac{b^2}{s^2}}\right\}\right]_s^\infty$$

$$=\frac{1}{2}\left[0-\log\frac{\left(1-\frac{a}{s}\right)^2}{\left(1+\frac{b^2}{s^2}\right)}\right]=\frac{1}{2}\left[\log\frac{s^2+b^2}{(s-a)^2}\right] \qquad \textbf{Ans.}$$

Example 25. *Find the Laplace transform of* $\dfrac{1-\cos t}{t^2}$.

Solution. $L(1-\cos t)=L(1)-L(\cos t)=\dfrac{1}{s}-\dfrac{s}{s^2+1}$

$$L\left[\frac{1-\cos t}{t}\right]=\int_s^\infty\left(\frac{1}{s}-\frac{s}{s^2+1}\right)ds=\left[\log s-\frac{1}{2}\log\left(s^2+1\right)\right]_s^\infty$$

$$=\frac{1}{2}\left[\log s^2-\log(s^2+1)\right]_s^\infty=\frac{1}{2}\left[\log\frac{s^2}{s^2+1}\right]_s^\infty$$

$$=\frac{1}{2}\left[\log\frac{1}{\left(1+\frac{1}{s^2}\right)}\right]_s^\infty=\frac{1}{2}\left[0-\log\frac{s^2}{s^2+1}\right]=-\frac{1}{2}\log\frac{s^2}{s^2+1}$$

Again, $L\left[\dfrac{1-\cos t}{t^2}\right]=-\dfrac{1}{2}\displaystyle\int_s^\infty\log\frac{s^2}{s^2+1}ds=-\frac{1}{2}\int_s^\infty\left(\log\frac{s^2}{s^2+1}.1\right)ds$

Integrating by parts, we have,

$$=-\frac{1}{2}\left[\log\frac{s^2}{s^2+1}.s-\int\frac{s^2+1}{s^2}\frac{(s^2+1)2s-s^2(2s)}{(s^2+1)^2}.sds\right]_s^\infty$$

$$= -\frac{1}{2}\left[s\log\frac{s^2}{s^2+1} - 2\int\frac{1}{s^2+1}ds\right]_s^\infty = -\frac{1}{2}\left[s\log\frac{s^2}{s^2+1} - 2\tan^{-1}s\right]_s^\infty$$

$$= -\frac{1}{2}\left[0 - 2\left(\frac{\pi}{2}\right) - s\log\frac{s^2}{s^2+1} + 2\tan^{-1}s\right] = -\frac{1}{2}\left[-\pi - s\log\frac{s^2}{s^2+1} + 2\tan^{-1}s\right]$$

$$= \frac{\pi}{2} + \frac{s}{2}\log\frac{s^2}{s^2+1} - \tan^{-1}s = \left(\frac{\pi}{2} - \tan^{-1}s\right) + \frac{s}{2}\log\frac{s^2}{s^2+1} = \cot^{-1}s + \frac{s}{2}\log\frac{s^2}{s^2+1} \quad \textbf{Ans.}$$

Example 26. *Evaluat* $L\left[e^{-4t}\frac{\sin 3t}{t}\right]$

Solution. $L[\sin 3t] = \dfrac{3}{s^2+3^2}$

$$\Rightarrow \quad L\left[\frac{\sin 3t}{t}\right] = \int_s^\infty \frac{3}{s^2+9}ds = \left[\frac{3}{3}\tan^{-1}\frac{s}{3}\right]_s^\infty = \left[\tan^{-1}\frac{s}{3}\right]_s^\infty = \frac{\pi}{2} - \tan^{-1}\frac{s}{3} = \cot^{-1}\frac{s}{3}$$

$$L\left[e^{-4t}\frac{\sin 3t}{t}\right] = \cot^{-1}\frac{s+4}{3} = \tan^{-1}\frac{3}{s+4} \quad \textbf{Ans.}$$

EXERCISE 46.3

Find Laplace transform of the following:

1. $\frac{1}{t}(1-e^t)$ **Ans.** $\log\frac{s-1}{s}$
2. $\frac{1}{t}(e^{-at}-e^{-bt})$ **Ans.** $\log\frac{s+b}{s+a}$
3. $\frac{1}{t}(1-\cos at)$ **Ans.** $-\frac{1}{2}\log\frac{s^2}{s^2+a^2}$
4. $\frac{1}{t}\sin^2 t$ **Ans.** $\frac{1}{4}\log\frac{s^2+4}{s^2}$
5. $\frac{1}{t}\sinh t$ **Ans.** $-\frac{1}{2}\log\frac{s-1}{s+1}$
6. $\frac{1}{t}(e^{-t}\sin t)$ **Ans.** $\cot^{-1}(s+1)$
7. $\frac{1}{t}(1-\cos t)$ **Ans.** $\frac{1}{2}[\log(s^2+1)-\log s^2]$
8. $\int_0^\infty \frac{1}{t}e^{-2t}\sin t\,dt$ **Ans.** $\frac{1}{s}\cot^{-1}(s+2)$
9. $\int_0^\infty \frac{e^{-t}-e^{-3t}}{t}dt$ **Ans.** $\log 3$

46.15 LAPLACE TRANSFORM OF ERROR FUNCTION

Example 27. *Find* $\left\{erf\sqrt{t}\right\}$ *and hence prove that*

$$L\left\{t.erf\,2\sqrt{t}\right\} = \frac{3s+8}{s^2(s+4)^{3/2}}$$

(U.P. II Semester, Summer 2001)

Solution. We know that $erf\sqrt{t} = \dfrac{2}{\sqrt{\pi}}\displaystyle\int_0^{\sqrt{t}} e^{-x^2}dx$

$$= \frac{2}{\sqrt{\pi}}\int_0^{\sqrt{t}}\left(1 - x^2 + \frac{x^4}{2!} - \frac{x^6}{3!} + \ldots\right)dx = \frac{2}{\sqrt{\pi}}\left[x - \frac{x^3}{3} + \frac{x^5}{10} - \frac{x^7}{42}\ldots\right]_0^{\sqrt{t}}$$

$$= \frac{2}{\sqrt{\pi}}\left[\sqrt{t} - \frac{t^{3/2}}{3} + \frac{t^{5/2}}{10} - \frac{t^{7/2}}{42} + \ldots\right]$$

$$L\{erf\sqrt{t}\} = \frac{2}{\sqrt{\pi}}\left[\frac{\left|\frac{3}{2}\right.}{s^{3/2}}-\frac{\left|\frac{5}{2}\right.}{3s^{5/2}}+\frac{\left|\frac{7}{2}\right.}{10s^{7/2}}-\frac{\left|\frac{9}{2}\right.}{42s^{9/2}}+\dots\right]$$

$$= \frac{2}{\sqrt{\pi}}\left[\frac{\frac{1}{2}\left|\frac{1}{2}\right.}{s^{3/2}}-\frac{\frac{3}{2}.\frac{1}{2}\left|\frac{1}{2}\right.}{3s^{5/2}}+\frac{\frac{5}{2}.\frac{3}{2}.\frac{1}{2}\left|\frac{1}{2}\right.}{10s^{7/2}}-\frac{\frac{7}{2}.\frac{5}{2}.\frac{3}{2}.\frac{1}{2}\left|\frac{1}{2}\right.}{42s^{9/2}}+\dots\right] \qquad \left[\therefore \left|\frac{1}{2}\right. = \sqrt{\pi}\right]$$

$$= \frac{1}{s^{3/2}}-\frac{1}{2}\frac{1}{s^{5/2}}+\frac{1.3}{2.4}\frac{1}{s^{7/2}}-\frac{1}{2}.\frac{3}{4}.\frac{5}{6}\frac{1}{s^{9/2}}+\dots$$

$$= \frac{1}{s^{3/2}}\left[1-\frac{1}{2}.\frac{1}{s}+\frac{1}{2}.\frac{3}{4}.\frac{1}{s^2}-\frac{1}{2}.\frac{3}{4}.\frac{5}{6}.\frac{1}{s^3}+\dots\right]$$

$$= \frac{1}{s^{3/2}}\left[1-\frac{1}{2}.\frac{1}{s}+\frac{\left(-\frac{1}{2}\right)\left\{-\frac{3}{2}\right\}}{2!}\frac{1}{s^2}+\frac{\left(-\frac{1}{2}\right)\left(-\frac{3}{2}\right)\left(-\frac{5}{2}\right)}{3!}\frac{1}{s^3}+\dots\right]$$

$$= \frac{1}{s^{3/2}}\left[1+\frac{1}{s}\right]^{-\frac{1}{2}} = \frac{1}{s^{3/2}}\left[\frac{s}{s+1}\right]^{\frac{1}{2}} = \frac{1}{s\sqrt{(s+1)}} \qquad \textbf{Ans.}$$

Now $\quad L\left\{erf\left(2\sqrt{t}\right)\right\} = L\left\{erf\sqrt{4t}\right\} = \frac{1}{4}\frac{1}{\frac{s}{4}\sqrt{\frac{s}{4}+1}} = \frac{2}{s\sqrt{s+4}}$

$$L\left\{t.erf\left(2\sqrt{t}\right)\right\} = -\frac{d}{ds}\frac{2}{\sqrt{s^3+4s^2}} = -2\left(-\frac{1}{2}\right)\left[s^3+4s^2\right]^{-\frac{3}{2}}(3s^2+8s)$$

$$= \frac{3s^2+8s}{(s^3+4s^2)^{3/2}} = \frac{s(3s+8)}{s^3(s+4)^{3/2}} = \frac{3s+8}{s^2(s+4)^{3/2}} \qquad \textbf{Proved.}$$

46.16 COMPLEMENTARY ERROR FUNCTION

This function is defined by

$$erf_c\left(\sqrt{t}\right) = 1-erf\left(\sqrt{t}\right) = 1-\frac{2}{\sqrt{\pi}}\int_0^{\sqrt{t}} e^{-x^2}\,dx$$

Now, $L\left\{erf_c\left(\sqrt{t}\right)\right\} = L\left\{1-\frac{2}{\sqrt{\pi}}\int_0^{\sqrt{t}} e^{-x^2}\,dx\right\} \quad L(1)-\frac{2}{\sqrt{\pi}}L\left\{\int_0^{\sqrt{t}} e^{-x^2}\,dx\right\} = \frac{1}{s}-\frac{1}{s\sqrt{s+1}}$

$$\frac{\sqrt{(s+1)}-1}{s\sqrt{(s+1)}} = \frac{\left\{\sqrt{(s+1)}-1\right\}\left\{\sqrt{(s+1)}+1\right\}}{s\sqrt{(1+s)}\left\{\sqrt{(s+1)}+1\right\}}$$

$$= \frac{s+1-1}{s\sqrt{s+1}\,(\sqrt{s+1}+1)} = \frac{1}{\sqrt{(s+1)}\left\{\sqrt{(s+1)}+1\right\}}$$

$$\boxed{\therefore L\left[erf_c\,(\sqrt{t})\right] = \frac{1}{\sqrt{s+1}\left\{\sqrt{s+1}+1\right\}}} \qquad \textbf{Ans.}$$

46.17 UNIT STEP FUNCTION

With the help of unit step functions, we can find the inverse transform of functions, which cannot be determined with previous methods.

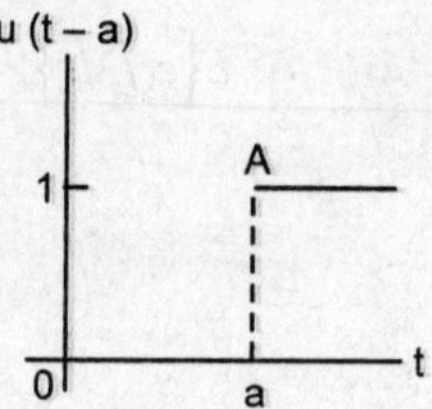

The unit step function $u(t-a)$ is defined as follows:

$$u(t-a)=\begin{cases}0, \text{ when } t<a\\ 1, \text{ when } t\geq a\end{cases}\qquad \text{where } a\geq 0$$

46.18 LAPLACE TRANSFORM OF UNIT FUNCTION

$$L[u(t-a)]=\frac{e^{-as}}{s}$$

Proof. $L\left[u(t-a)\right]=\int_0^\infty e^{-st}u(t-a)dt\;=\int_0^a e^{-st}.0dt+\int_a^\infty e^{-st}.1\,dt=0+\left[\frac{e^{-st}}{-s}\right]_a^\infty$

$$L[u(t-a)]=\frac{e^{-as}}{s}$$

Proved.

Example 28. *Express the following function in terms of unit step functions and find its Laplace transform:*

$$f(t)=\begin{cases}8, & t<2\\ 6, & t\geq 2\end{cases}$$

Solution. $f(t)=\begin{cases}8+0, & t<2\\ 8-2, & t\geq 2\end{cases}=8+\begin{cases}0, & t<2\\ -2, & t\geq 2\end{cases}=8+(-2)\begin{cases}0, & t<2\\ 1, & t\geq 2\end{cases}=8-2u(t-2)$

$$L\{f(t)\}=8L(1)-2Lu(t-2)=\frac{8}{s}-2\frac{e^{-2s}}{s}$$

Ans.

Example 29. *Draw the graph of* $u(t-a)-u(t-b)$.

Solution. As in Art 46.17 the graph of $u(t–a)$ is a straight line parallel to t-axis from A to ∞.

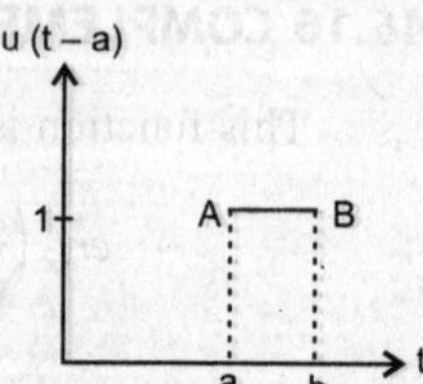

Similarly, the graph of $u(t-b)$ is a straight line parallel to t-axis from B to ∞.

Hence, the graph of $u(t-a)$ is AB. **Ans.**

46.19 SECOND SHIFTING THEOREM

If $L[f(t)]=F(s)$, then $\boxed{L[f(t-a).u(t-a)]=e^{-as}F(s)}$

Proof $L[f(t-a).u(t-a)]=\int_0^\infty e^{-st}\left[f(t-a).u(t-a)\right]dt$

$$=\int_0^a e^{-st}f(t-a).0\,dt+\int_a^\infty e^{-st}f(t-a)(t)dt\;=\int_a^\infty e^{-st}f(t-a)dt$$

$$=\int_0^\infty e^{-s(u+a)}f(u)du,\quad \text{where } u=t-a$$

$$=e^{-sa}\int_0^\infty e^{-su}.f(u)du=e^{-sa}F(s)$$

Proved.

Example 30. *Express the following function in terms of unit step function and find its Laplace transform:*

$$f(t)=\begin{cases}E, & a<t<b\\ 0, & t\geq b\end{cases}$$

Solution.

$$f(t)=E\begin{cases}1, & a<t<b\\ 0, & t\geq b\end{cases} \qquad \left[L[f(t-a).u(t-a)]=e^{-as}F(s)\right]$$

$$L\{f(t)\}=E\left[\frac{e^{-as}}{s}-\frac{e^{-bs}}{s}\right] \qquad \textbf{Ans.}$$

Example 31. *Express the following function in terms of unit step function :*

$$f(t)=\begin{cases}t-1, & 1<t<2\\ 3-t, & 2<t<3\end{cases}$$

and find its Laplace transform. *(U.P.; II Semester, 2009)*

Solution.

$$f(t)=\begin{cases}t-1, & 1<t<2\\ 3-t, & 2<t<3\end{cases}$$

$$=(t-1)\left[u(t-1)-u(t-2)\right]+(3-t)\left[u(t-2)-u(t-3)\right]$$

$$=(t-1)u(t-1)-(t-1)u(t-2)+(3-t)u(t-2)+(t-3)u(t-3)$$

$$=(t-1)u(t-1)-2(t-2)u(t-2)+(t-3)u(t-3)$$

$$=e^{-s}L(t)-2e^{-2s}L(t)-e^{-3s}L(t)$$

$$\left[L[f(t-a).u(t-a)]=e^{-as}F(s)\right]$$

$$L[f(t)]=\frac{e^{s}}{s^2}-2\frac{e^{-2s}}{s^2}+\frac{e^{-3s}}{s^2} \qquad \textbf{Ans.}$$

Example 32. *Find L{F(t)} if*

$$F(t)=\begin{cases}\sin\left(t-\frac{\pi}{3}\right), & t>\frac{\pi}{3}\\ 0, & t<\frac{\pi}{3}\end{cases}$$

Solution.

$$L\{F(t)\}=e^{-s\frac{\pi}{3}}L(\sin t) \qquad \left[\because a=\frac{\pi}{3}\right]$$

$$=e^{-s\frac{\pi}{3}}.\frac{1}{s^2+1} \qquad \text{(Using second shifting property)} \ \textbf{Ans.}$$

46.20 THEOREM $L[f(t)u(t-a)]\ \ e^{-as}L[f(t+a)]$

Proof. $L[f(t).u(t-a)]=\int_0^\infty e^{-st}[f(t).u(t-a)]dt$

$$=\int_0^a e^{-st}[f(t).u(t-a)]dt+\int_a^\infty e^{-st}.[f(t).u(t-a)]dt=0+\int_a^\infty e^{-st}.f(t)(1)dt$$

$$=\int_a^\infty e^{-s(y+a)}.f(y+a)dy=e^{-as}\int_a^\infty e^{-sy}.f(y+a)\,dy \qquad (t-a=y)$$

$$e^{-as}\int_a^{\infty} e^{-st} f(t+a)dt = e^{-as} L[f(t+a)]$$ **Proved.**

Example 33. *Find the Laplace transform of $t^2\, u\,(t-3)$.*

Solution. $t^2.u(t-3) = \left[(t-3)^2 + 6(t-3) + 9\right]u(t-3)$

$$= (t-3)^2.u(t-3) + 6(t-3).u(t-3) + 9u(t-3)$$

$$L[t^2 .\, u\,(t-3)] = L[(t-3)^2.u(t-3)] + 6L[(t-3).u(t-3)] + 9L[u(t-3)]$$

$$= e^{-3s}\left[\frac{2}{s^3} + \frac{6}{s^2} + \frac{9}{s}\right]$$

Aliter $L[t^2u(t-3)] = e^{-3s}L(t+3)^2 = e^{-3s}L\left[t^2 + 6t + 9\right] = e^{-3s}\left[\frac{2}{s^3} + \frac{6}{s^2} + \frac{9}{s}\right]$ **Ans.**

Example 34. *Find the Laplace transform of $e^{-2t}u_{\pi}(t)$ where*

$$u_{\pi}(t) = \begin{cases} 0; & t < \pi \\ 1; & t > \pi \end{cases}$$

Solution $u_{\pi}(t) = \begin{cases} 0; & t < \pi \\ 1; & t > \pi \end{cases}$

$$u_{\pi}(t) = u\,(t-\pi)$$

$$L\,[u_{\pi}\,(t)] = L\,[u\,(t-\pi) = \frac{e^{-\pi s}}{s}$$

$$\Rightarrow \quad L\left[e^{-2t}u_{\pi}(t)\right] = \frac{e^{-\pi(s+2)}}{s+2}$$ **Ans.**

Example 35. *Express the following function in terms of unit step function and find its Laplace transform* $f(t) = f(t) = \begin{cases} 0, & 0 < t < 1 \\ t-1, & 1 < t < 2 \\ 1, & 2 < t \end{cases}$ *(U.P. II Semester, Summer 2002)*

Solution. The above function shown in the figure is expressed in algebraic for

$$f(t) = \begin{cases} 0, & 0 < t < 1 \\ t-1, & 1 < t < 2 \\ 1, & 2 < t \end{cases}$$

f (t)
1
1
O 1 2 t

$$f(t) = (t-1)\left[u(t-1) - u(t-2)\right] + u(t-2)$$

$$= (t-1)u(t-1) - u(t-2)\{t-1-1\} = (t-1)u(t-1) - (t-2)u(t-2)$$

$$Lf(t) = L(t-1)u(t-1) - L(t-2)u(t-2)$$

$$= \frac{e^{-s}}{s^2} - \frac{e^{-2s}}{s^2}$$ **Ans.**

Example 36. *Represent $f(t) = \sin 2t$, $2\pi < t < 4\pi$ and $f(t) = 0$ otherwise, in terms of unit step function and then find its Laplace transform.*

Solution. $f(t) = \begin{cases} \sin 2t, & 2\pi < t < 4\pi \\ 0, & \text{otherwise} \end{cases}$

$$f(t) = \sin 2t\,[u(t-2\pi) - u(t-4\pi)]$$

$$L[f(t)] = L[\sin 2t\,.\,u\,(t-2\pi)] - L[\sin 2t\,.\,u(t-4\pi)]$$

$$= e^{-2\pi s} L \sin 2(t+2\pi) - e^{-4\pi s} L \sin 2(t+4\pi)$$

$$= e^{-2\pi s} L \sin 2t - e^{-4\pi s} L \sin(2t)$$

$$= e^{-2\pi s} \frac{2}{s^2+4} - e^{-4\pi s} \frac{2}{s^2+4} \quad (e^{-2\pi s} - e^{-4\pi s}) \frac{2}{s^2+4}$$ **Ans.**

Example 37. *A function f (t) obeys the equation f (t) +* $\int_0^t f(t)\ dt$ *= cosh 2t*

Find the Laplace transform of f (t). (*U.P. II Semester Summer* 2006)

Solution. We have, $f(t) + 2\int_0^t f(t)\ dt = \cosh 2t$

Taking Laplace transformation of both the sides, we ge

$$L\{f(t)\} + 2L\int_0^t f(t)\,dt = L(\cosh 2t) \quad \Rightarrow \quad F(s) + 2.\frac{1}{s}F(s) = \frac{s}{s^2-4}$$

$$\Rightarrow \quad F(s)\left\{1 + \frac{2}{s}\right\} = \frac{s}{s^2-4} \quad \Rightarrow \quad F(s)\left\{\frac{s+2}{s}\right\} = \frac{s}{s^2-4}$$

$$\Rightarrow \quad F(s) = \left(\frac{s}{s^2-4}\right)\left(\frac{s}{s+2}\right) \quad \Rightarrow \quad F(s) = \frac{s^2}{(s^2-4)(s+2)}$$ **Ans.**

EXERCISE 46.4

Find the Laplace transform of the following:

1. $f(t) = \begin{cases} t-1, & 1 < t < 2 \\ 0, & \text{otherwise} \end{cases}$ **Ans.** $\frac{e^{-s} - e^{-2s}}{s^2} - \frac{e^{-2s}}{s}$

2. $e^t u(t-1)$ **Ans.** $\frac{e^{-(s-1)}}{s-1}$

3. $\frac{1-e^{2t}}{t} + tu(t) + \cosh t.\cos t$ **Ans.** $\log\frac{s-2}{s} + \frac{1}{s^2} + \frac{s^3}{s^4+4}$

4. $t^2 u(t-2)$ **Ans.** $\frac{e^{-2s}}{s^3}(4s^2 + 4s + 2)$

5. $\sin t\ u\ (t-4)$ **Ans.** $\frac{e^{-4s}}{s^2+1}[\cos 4 + s\sin 4]$

6. $f(t) = K(t-2)[u(t-2) - u(t-3)]$ **Ans.** $\frac{K}{s^2}\left[e^{-2s} - (s+1)e^{-3s}\right]$

7. $f(t) = K\frac{\sin \pi t}{T}[u(t-2T) - u(t-3T)]$ **Ans.** $\frac{K\pi T}{s^2T^2+\pi^2}(e^{-2sT} - e^{-3sT})$

Express the following in terms of unit step functions and obtain Laplace transforms.

8. $f(t) = \begin{cases} t, & 0 < t < 2 \\ 0, & 2 < t \end{cases}$ **Ans.** $u(t) - u(t-2), \frac{1-(2s+1)e^{-2s}}{s^2}$

9. $f(t) = \begin{cases} \sin t, & 0 < t < \pi \\ t, & t > \pi \end{cases}$ **Ans.** $\frac{1+e^{-\pi s}}{s^2+1} + \frac{e^{-\pi s}(\pi s+1)}{s^2}$

10. $f(t) = \begin{cases} 4, & 0 < t < 1 \\ -2, & 0 < t < 3 \\ 5, & t > 3 \end{cases}$ **Ans.** $\frac{4 - 6e^{-s} + 7e^{-3s}}{s}$

46.21. PERIODIC FUNCTIONS

Let $f(t)$ be a periodic function with period T, then

$$L[f(t)] = \frac{\int_0^T e^{-st} f(t)\,dt}{1-e^{-sT}}$$

Proof. $L[f(t)] = \int_0^\infty e^{-st} f(t)dt \qquad \int_0^T e^{-st} f(t)dt + \int_T^{2T} e^{-st} f(t)dt + \int_{2T}^{3T} e^{-st} f(t)dt + ...$

Substituting $t = u + T$ in second integral and $t = u + 2T$ in third integral, and so on.

$$L[f(t)] = \int_0^T e^{-st} f(t)dt + \int_0^T e^{-s(u+T)} f(u+T)du + \int_0^T e^{-s(u+2T)} f(u+2T)du + ...$$

$$= \int_0^T e^{-st} f(t)dt + e^{-sT}\int_0^T e^{-su} f(u)du + e^{-2sT}\int_0^T e^{-su} f(u)du + ...$$

$$[f(u) = f(u+T) = f(u+2T) = f(u+3T) = ...]$$

$$= \int_0^T e^{-st} f(t)dt + e^{-sT}\int_0^T e^{-st} f(t)dt + e^{-2sT}\int_0^T e^{-st} f(t)dt + ...$$

$$= \left[1 + e^{-sT} + e^{-2sT} + e^{-3sT} + ...\right]\int_0^T e^{-st} f(t)dt \qquad \left[1 + a + a^2 + a^3 + ... = \frac{1}{1-a}\right]$$

$$= \frac{1}{1-e^{-sT}}\int_0^T e^{-st} f(t)dt.$$

Proved.

Example 38. *Find the Laplace transform of the waveform*

$$f(t) = \left(\frac{2t}{3}\right), 0 \le t \le 3.$$

Solution. $L[f(t)] = \frac{1}{1-e^{-sT}}\int_0^T e^{-st} f(t)dt$

$$L\left[\frac{2t}{3}\right] = \frac{1}{1-e^{-3s}}\int_0^3 e^{-st}\left(\frac{2}{3}t\right)dt = \frac{1}{1-e^{-3s}}\frac{2}{3}\left[\frac{te^{-st}}{-s} - (1)\frac{e^{-st}}{s^2}\right]_0^3$$

$$= \frac{2}{3}\frac{1}{1-e^{-3s}}\left[\frac{3e^{-3s}}{-s} - \frac{e^{-3s}}{s^2} + \frac{1}{s^2}\right] = \frac{2}{3}.\frac{1}{1-e^{-3s}}\left[\frac{3e^{-3s}}{-s} + \frac{1-e^{-3s}}{s^2}\right]$$

$$= \frac{2e^{-3s}}{-s(1-e^{-3s})} + \frac{2}{3s^2}$$

Ans.

Example 39. *Draw the graph and find the Laplace transform of the triangular wave function of period 2 C given by* $f(t) = \begin{cases} t, & 0 < t \le C \\ 2C - t, & C < t < 2C \end{cases}$

(Uttarakhand, II Semester, June 2007)

Solution. Period $= 2C = T$

Laplace transform of periodic function $f(t)$

$$L\{f(t)\} = \frac{\int_0^T e^{-st} f(t)dt}{1-e^{-sT}}$$

$$L\{f(t)\} = \frac{1}{1-e^{-2Cs}}\int_0^{2C} e^{-st} f(t)dt \qquad (T = 2c)$$

On putting the values of $f(t)$, we get

$$L[f(t)]=\frac{1}{1-e^{-2Cs}}\left[\int_0^C e^{-st}t\,dt+\int_C^{2C}e^{-st}(2C-t)\,dt\right]$$

$$=\frac{1}{1-e^{-2Cs}}\left[\left\{\frac{te^{-st}}{-s}-1.\frac{e^{-st}}{(-s)^2}\right\}_0^C+\left\{(2C-t)\frac{e^{-st}}{(-s)}-(-1)\frac{e^{-st}}{(-s)^2}\right\}_C^{2C}\right]$$

$$=\frac{1}{1-e^{-2Cs}}\left[\left\{\frac{C.e^{-Cs}}{-s}-\frac{e^{-Cs}}{(-s)^2}-0+\frac{1}{s^2}\right\}+\left\{(2C-2C)\frac{e^{-2Cs}}{(-s)}+\frac{e^{-2Cs}}{s^2}-\left((2C-C)\frac{e^{-Cs}}{-s}+\frac{e^{-Cs}}{s^2}\right)\right\}\right]$$

$$=\frac{1}{1-e^{-2Cs}}\left\{-\frac{Ce^{-Cs}}{s}-\frac{e^{-Cs}}{s^2}+\frac{1}{s^2}+\frac{e^{-2Cs}}{s^2}+\frac{Ce^{-Cs}}{s}-\frac{e^{-Cs}}{s^2}\right\}$$

$$=\frac{1}{1-e^{-2Cs}}\left\{\frac{1}{s^2}(1-2e^{-Cs}+e^{-2Cs})\right\}=\frac{\left(1-e^{-Cs}\right)^2}{s^2\left(1+e^{-Cs}\right)\left(1-e^{-Cs}\right)}=\frac{1-e^{-Cs}}{s^2\left(1+e^{-Cs}\right)}\quad\textbf{Ans.}$$

Example 40. *Draw the graph of the periodic function*

$$f(t)=\begin{cases}t, & 0<t<\pi\\ \pi-t, & \pi<t<2\pi\end{cases}$$

and find its Laplace transform. *(U.P. Second Semester, 2003)*

Solution. Period = $2\pi = T$

Laplace transform of Periodic functions

$$L\{f(t)\}=\frac{\int_0^T e^{-st}f(t)\,dt}{1-e^{-sT}}$$

$$=\frac{1}{1-e^{-2\pi s}}\int_0^{2\pi}e^{-st}f(t)\,dt=\frac{1}{1-e^{-2\pi s}}\left[\int_0^{\pi}e^{-st}t\,dt+\int_{\pi}^{2\pi}e^{-st}(\pi-t)\,dt\right]$$

$$=\frac{1}{1-e^{-2\pi s}}\left\{\frac{t.e^{-st}}{-s}-1.\frac{e^{-st}}{(-s)^2}\right\}_0^{\pi}+\left\{(\pi-t)\frac{e^{-st}}{(-s)}-(-1)\frac{e^{-st}}{(-s)^2}\right\}_{\pi}^{2\pi}$$

$$=\frac{1}{1-e^{-2\pi s}}\left[\left\{\frac{\pi.e^{-\pi s}}{-s}-\frac{e^{-\pi s}}{(-s)^2}-0+\frac{1}{s^2}\right\}+\left\{(\pi-2\pi)\frac{e^{-2\pi s}}{-s}+\frac{e^{-2\pi s}}{s^2}-\left((\pi-\pi)\frac{e^{-\pi s}}{-s}+\frac{e^{-\pi s}}{s^2}\right)\right\}\right]$$

$$=\frac{1}{1-e^{-2\pi s}}\left\{-\frac{\pi e^{-\pi s}}{s}-\frac{e^{-\pi s}}{s^2}+\frac{1}{s^2}+\pi\frac{e^{-2\pi s}}{s}+\frac{e^{-2\pi s}}{s^2}-0-\frac{e^{-\pi s}}{s^2}\right\}$$

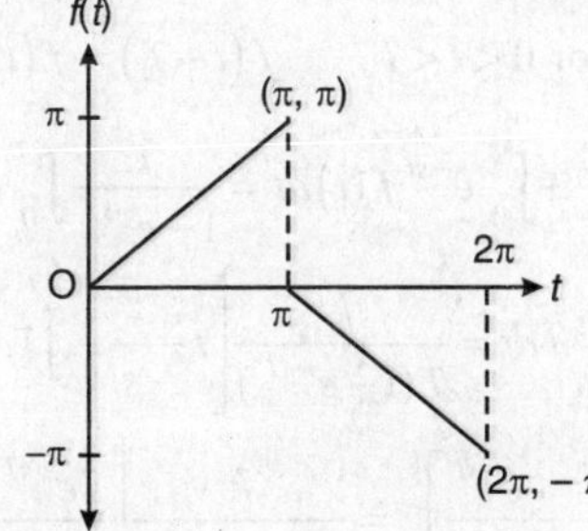

$$=\frac{1}{1-e^{-2\pi s}}\left\{-\frac{\pi}{s}e^{-\pi s}+\frac{\pi}{s}e^{-2\pi s}+\frac{1}{s^2}-\frac{1}{s^2}.e^{-\pi s}+\frac{1}{s^2}e^{-2\pi s}-\frac{e^{-\pi s}}{s^2}\right\}$$

$$= \frac{1}{1-e^{-2\pi s}}\left[\frac{\pi}{s}\left(e^{-2\pi s}-e^{-\pi s}\right)+\frac{1}{s^2}\left(1+e^{-2\pi s}-2e^{-\pi s}\right)\right] = \frac{-\pi s e^{-\pi s}\left(1-e^{-\pi s}\right)+\left(1-e^{-\pi s}\right)^2}{s^2\left(1+e^{-\pi s}\right)\left(1-e^{-\pi s}\right)}$$

$$= \frac{-\pi s e^{-\pi s}+1-e^{-\pi s}}{s^2\left(1+e^{-\pi s}\right)}$$ **Ans.**

Example 41. *Find the Laplace transform of the function (Half wave rectifier)*

$$f(t) = \begin{cases} \sin \omega t & \text{for } 0 < t < \dfrac{\pi}{\omega} \\ 0 & \text{for } \dfrac{\pi}{\omega} < t < \dfrac{2\pi}{\omega}. \end{cases}$$ *(U.P. II Semester, 2010, Summer 2002)*

Solution. $L[f(t)] \quad \dfrac{1}{1-e^{-sT}}\displaystyle\int_0^T e^{-st} f(t)\,dt$

$$= \frac{1}{1-e^{-\frac{2\pi s}{\omega}}}\int_0^{2\pi/\omega} e^{-st} f(t)\,dt \qquad \left[\begin{array}{l} f(t) \text{ is a periodic function} \\ T = \dfrac{2\pi}{\omega} \end{array}\right]$$

$$= \frac{1}{1-e^{-\frac{2\pi s}{\omega}}}\left[\int_0^{\pi/\omega} e^{-st}\sin \omega t\,dt + \int_{\pi/\omega}^{2\pi/\omega} e^{-st}\times 0\times dt\right]$$

$$= \frac{1}{1-e^{-\frac{2\pi s}{\omega}}}\int_0^{\pi/\omega} e^{-st}\sin\ \omega t\ dt \qquad \left[\int e^{ax}\sin bx\,dx = e^{ax}\frac{(a\sin bx - b\cos bx)}{a^2+b^2}\right]$$

$$L\,[f(t)] = \frac{1}{1-e^{-\frac{2\pi s}{\omega}}}\left[\frac{e^{-st}(-s\sin\omega t-\omega\cos\omega t)}{s^2+\omega^2}\right]_0^{\pi/\omega}$$

$$= \frac{1}{1-e^{-\frac{2\pi s}{\omega}}}\left[\frac{\omega e^{-\frac{\pi}{\omega}s}+\omega}{s^2+\omega^2}\right] = \frac{\omega\left[1+e^{-\frac{\pi s}{\omega}}\right]}{(s^2+\omega^2)\left[1-e^{-\frac{2\pi s}{\omega}}\right]} = \frac{\omega\left[1+e^{-\frac{\pi}{\omega}}\right]}{(s^2+\omega^2)\left(1-e^{-\frac{\pi}{\omega}}\right)\left(1+e^{-\frac{\pi}{\omega}}\right)}$$

$$= \frac{\omega}{(s^2+\omega^2)\left[1-e^{-\frac{\pi s}{\omega}}\right]}$$ **Ans.**

Example 42. *Find the Laplace Transform of the Periodic function (saw tooth wave)*

$$f(t) = \frac{kt}{T} \text{ for } 0 < t < T, \qquad f(t+T) = f(t)$$

Solution. $L[f(t)] = \dfrac{1}{1-e^{-sT}}\displaystyle\int_0^T e^{-st} f(t)\,dt = \dfrac{1}{1-e^{-sT}}\displaystyle\int_0^T e^{-st}\frac{kt}{T}\,dt$

$$= \frac{1}{1-e^{-sT}}\frac{k}{T}\int_0^T e^{-st}.t\,dt = \frac{k}{T(1-e^{-sT})}\left[t\frac{e^{-st}}{-s}-\int 1.\frac{e^{-st}}{-s}dt\right]_0^T \quad \text{Integrating by parts}$$

$$= \frac{k}{T(1-e^{-sT})}\left[\frac{te^{-st}}{-s}-\frac{e^{-st}}{s^2}\right]_0^T = \frac{k}{T(1-e^{-sT})}\left[\frac{Te^{-sT}}{-s}-\frac{e^{-sT}}{s^2}+\frac{1}{s^2}\right]$$

$$= \frac{k}{T(1-e^{-sT})}\left[\frac{Te^{-sT}}{-s}+\frac{1}{s^2}(1-e^{-sT})\right] = -\frac{ke^{-sT}}{s(1-e^{-sT})}+\frac{k}{Ts^2}$$ **Ans.**

Example 43. *Obtain Laplace transform of rectangular wave given*

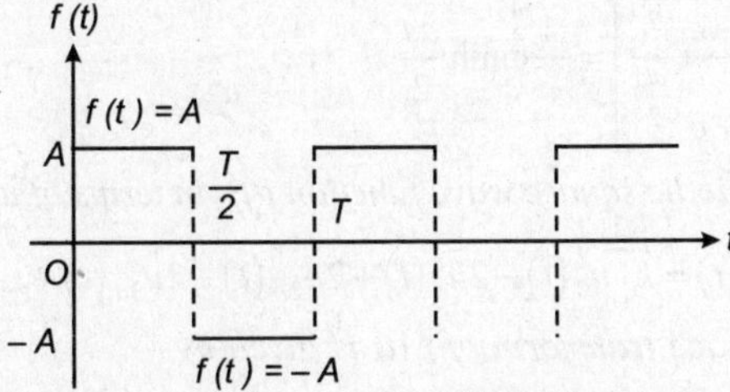

Solution. We know that Laplace transform of a periodic function *i.e.*,

$$Lf(t)=\frac{\int_0^T e^{-st}f(t)dt}{1-e^{-sT}}=\frac{\int_0^{\frac{T}{2}}e^{-st}A\,dt+\int_{\frac{T}{2}}^{T}e^{-st}(-A)dt}{1-e^{-sT}}$$

$$=A\frac{\left[\frac{e^{-st}}{-s}\right]_0^{\frac{T}{2}}-\left[\frac{e^{-st}}{-s}\right]_{\frac{T}{2}}^{T}}{1-e^{-sT}}=\frac{A}{1-e^{-sT}}\left[-\frac{e^{-\frac{sT}{2}}}{s}+\frac{1}{s}+\frac{e^{-sT}}{s}-\frac{e^{-\frac{sT}{2}}}{s}\right]$$

$$=\frac{A}{s(1-e^{-sT})}\left[1-2e^{-\frac{sT}{2}}+e^{-sT}\right]=\frac{A}{s(1-e^{-sT})}\left[1-e^{-\frac{sT}{2}}\right]^2$$

$$=\frac{A\left[1-e^{-\frac{sT}{2}}\right]^2}{s\left(1+e^{\frac{-sT}{2}}\right)\left(1-e^{-\frac{sT}{2}}\right)}=\frac{A\left(1-e^{-\frac{sT}{2}}\right)}{s\left(1+e^{-\frac{sT}{2}}\right)}=\frac{A\left(e^{\frac{sT}{4}}-e^{-\frac{sT}{4}}\right)}{s\left(e^{\frac{sT}{4}}+e^{-\frac{sT}{4}}\right)}=\frac{A}{s}\tanh\frac{sT}{4}$$ **Ans.**

Example 44. *Draw the graph of the following periodic function and find its Laplace transform:*

$$f(t)=\begin{cases}t & \text{for } 0<t\le a\\ 2a-t & \text{for } a<t<2a\end{cases}$$

(*U.P. II Semester, Summer 2002*)

Solution. The given function is periodic with period $2a$.

$$\therefore\quad L[f(t)]=\frac{1}{1-e^{-2as}}\int_0^{2a}f(t)e^{-st}dt$$

$$=\frac{1}{1-e^{-2as}}\left[\int_0^a f(t)e^{-st}dt+\int_a^{2a}f(t)e^{-st}dt\right]$$

$$=\frac{1}{1-e^{-2as}}\left[\int_0^a te^{-st}dt+\int_a^{2a}(2a-t)e^{-st}dt\right]$$

$$=\frac{1}{1-e^{-2as}}\left[\left\{t\frac{e^{-st}}{-s}-\frac{e^{-st}}{(-s)^2}\right\}_0^a+\left\{(2a-t)\frac{e^{-st}}{-s}+\frac{e^{-st}}{(-s)^2}\right\}_a^{2a}\right]$$

$$=\frac{1}{1-e^{-2as}}\left[-\frac{ae^{-as}}{s}-\frac{e^{-as}}{s^2}+\frac{1}{s^2}+\frac{e^{-2as}}{s^2}+\frac{ae^{-as}}{s}-\frac{e^{-as}}{s^2}\right]=\frac{1}{1-e^{-2as}}\left[\frac{1}{s^2}+\frac{e^{-2as}}{s^2}-2\frac{e^{-as}}{s^2}\right]$$

$$\frac{1}{s^2}\frac{1}{(1-e^{-2as})}\left(1+e^{-2as}-2e^{-as}\right)=\frac{1}{s^2}\frac{(1-e^{-as})^2}{(1+e^{-as})(1-e^{-as})}=\frac{1}{s^2}\left[\frac{1-e^{-as}}{1+e^{-as}}\right]$$

$$= \frac{1}{s^2}\left[\frac{e^{\frac{as}{2}} - e^{-\frac{as}{2}}}{e^{\frac{as}{2}} + e^{-\frac{as}{2}}}\right] = \frac{1}{s^2}\tanh\frac{as}{2}$$ **Ans.**

Example 45. *A periodic square wave function f(t), in terms of unit step functions, is written as*

$$f(t) = k[u_0(t) - 2u_a(t) + 2u_{2a}(t) - 2u_{3a}(t) + ...]$$

Show that the Laplace transform of f (t) is given by

$$L[f(t)] = \frac{k}{s}\tanh\left(\frac{as}{2}\right)$$

Solution. $f(t) = k[u_0(t) - 2u_a(t) + 2u_{2a}(t) - 2u_{3a}(t) + ...]$

$$f(t) = k\,[u\,(t-0) - 2u\,(t-a) + 2u\,(t-2a) - 2u\,(t-3a) + ...]$$

$$L\,[f(t)] = k\,[L\,u\,(t-0) - 2\,L\,u\,(t-a) + 2\,L\,u\,(t-2a) - 2\,L\,u\,(t-3a) + ...]$$

$$= k\left[\frac{1}{s} - 2\frac{e^{-as}}{s} + 2\frac{e^{-2as}}{s} - 2\frac{e^{-3as}}{s} + ...\right] = \frac{k}{s}\left[1 - 2e^{-as} + 2e^{-2as} - 2e^{-3as} + ...\right]$$

$$= \frac{k}{s}\left[1 - 2(e^{-as} - e^{-2as} + e^{-3as} - ...)\right] = \frac{k}{s}\left[1 - 2\frac{e^{-as}}{1+e^{-as}}\right] = \frac{k}{s}\left[\frac{1 + e^{-as} - 2e^{-as}}{1 + e^{-as}}\right]$$

$$= \frac{k}{s}\left[\frac{1 - e^{-as}}{1 + e^{-as}}\right] = \frac{k}{s}\left[\frac{e^{\frac{as}{2}} - e^{\frac{-as}{2}}}{e^{\frac{as}{2}} + e^{-\frac{as}{2}}}\right] = \frac{k}{s}\tan\text{h}\,\frac{as}{2}$$ **Ans.**

EXERCISE 46.5

1. Find the Laplace transform of the periodic function

$f(t) = e^t$ for $0\ t < 2\pi$ **Ans.** $\dfrac{e^{2(1-s)\pi} - 1}{(1-s)\left(1 - e^{-2\pi s}\right)}$

2. Obtain Laplace transform of full wave rectified sine wave given by

$f(t) = \sin \omega t,\ 0 < t < \dfrac{\pi}{\omega}$ **Ans.** $\dfrac{\omega}{(s^2+\omega^2)}\coth\dfrac{\pi s}{2\omega}$

3. Find the Laplace transform of the staircase function

$f(t) = kn,\ np < t < (n+1)p,\ n = 0, 1, 2, 3$ **Ans.** $\dfrac{ke^{ps}}{s\left(1 - e^{-ps}\right)}$

Find Laplace transform of the following:

4. $f(t) = t^2,\ 0 < t < 2,\ f(t+2) = f(t)$ **Ans.** $\dfrac{2 - e^{-2s} - 4se^{-2s} - 4s^2e^{-2s}}{s^3(1 - e^{-2s})}$

5. $f(t) = \begin{cases} 1, & 0 \le t \le \dfrac{a}{2} \\ -1, & \dfrac{a}{2} \le t < a \end{cases}$ *(U.P. II Semester, 2004)* **Ans.** $\dfrac{1}{s}\tanh\dfrac{as}{4}$

6. $f(t) = \begin{cases} \cos \omega t, & 0 < t < \dfrac{\pi}{\omega} \\ 0, & \dfrac{\pi}{\omega} < t < \dfrac{2\pi}{\omega} \end{cases}$ **Ans.** $\dfrac{s}{(s^2+\omega^2)\left(1 - e^{-\frac{\pi s}{\omega}}\right)}$

7. $f(t)=\begin{cases} t, & 0<t<1 \\ 0, & 1<t<2 \end{cases}\quad f(t+2)=f(t)$ **Ans.** $\dfrac{1-e^{-s}(s+1)}{s^2\left(1-e^{-2s}\right)}$

8. $f(t)=\begin{cases} \dfrac{2t}{T}, & 0\le t\le \dfrac{T}{2} \\ \dfrac{2}{T}(T-t), & \dfrac{T}{2}\le t\le T \end{cases}\quad f(t+T)=f(t)$ **Ans.** $\dfrac{2}{Ts^2}\tanh\dfrac{sT}{4}-\dfrac{1}{s\left(e^{\frac{sT}{2}}+1\right)}$

46.22 CONVOLUTION THEOREM

If $L[f_1(t)] = F_1(s)$ and $L[f_2(t)] = F_2(s)$

then $L\left\{\int_0^t f_1(x)f_2(t-x)\,dx\right\}=F_1(s)\cdot F_2(s)$

or $$\boldsymbol{L^{-1}\left(F_1(s).F_2(s)\right)=\int_0^t f_1(x)f_2(t-x)\,dx}$$

Proof. We have

$$L\left\{\int_0^\infty f_1(x)f_2(t-x)\,dx\right\}=\int_0^\infty e^{-st}\left[\int_0^t f_1(x)f_2(t-x)\,dx\right]dt \qquad \text{(By Definition)}$$

where the double integral is taken over the infinite region in the first quadrant lying between the lines $x = 0$ and $x = t$.

Here first we are integrating w.r.t. "x", within limits $x = 0$ and $x = t$, and then we will integrate w.r.t. "t" with limits $t = 0$ and $t = \infty$

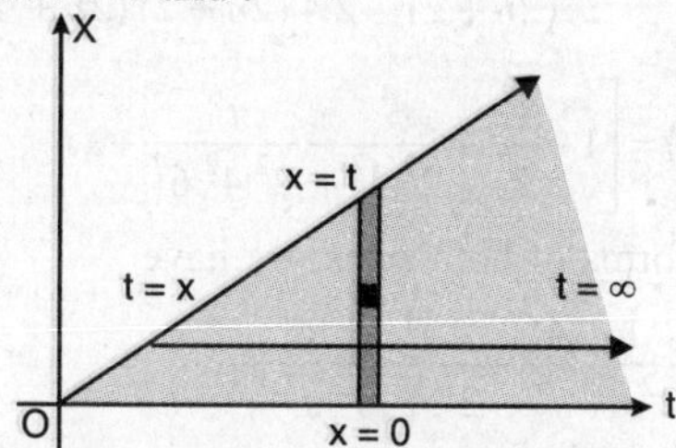

On changing the order of integration first we integrate w.r.t. "t" with limits $t = x$ and $t = \infty$ and then w.r.t. "x" with limits $x = 0$ and $x = \infty$

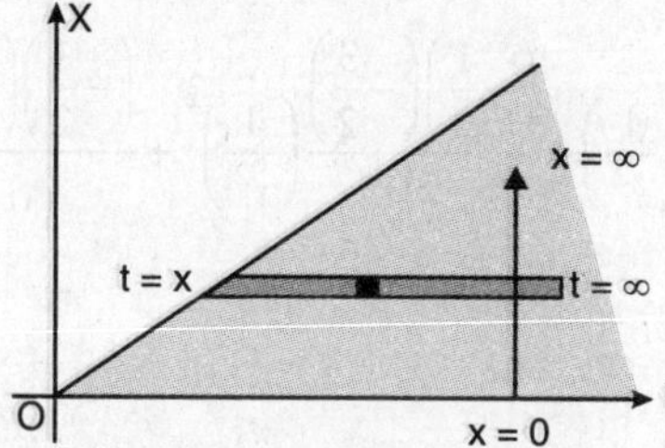

On changing the order of integration, the integral becomes

$$\int_0^\infty dx\left[\int_x^\infty e^{-st} f_1(x).f_2(t-x)\,dt\right]$$

$$=\int_0^\infty dx\left[\int_x^\infty e^{-s(t-x+x)} f_1.(x).f_2(t-x)\,dt\right]$$

$$=\int_0^\infty dx\left[\int_x^\infty e^{-s(t-x)}.e^{-sx} f_1(x).f_2(t-x)\,dt\right]$$

$$= \int_0^\infty e^{-sx} f_1(x)\, dx \left[\int_x^\infty e^{-s(t-x)} f_2(t-x)\, dt\right] = \int_0^\infty e^{-sx} f_1(x)\, dx \left[\int_0^\infty e^{-sz} f_2(z)\, dz\right]$$

[Put $t - x = z \Rightarrow dt = dz$]

$$\int_0^\infty e^{-sx} f_1(x) dx \int_0^\infty e^{-sz} f_2(z) dz,$$ Lower limit $x - x = z \Rightarrow z = 0$]

$$\int_0^\infty e^{-sx} f_1(x) F_2(s) dx = \left[\int_0^\infty e^{-sx} f_1(x) dx\right] F_2(s) = F_1(s)\, F_2(s)$$ **Proved.**

Example 46. *Find the Laplace transform of* $\int_0^t e^x . \sin(t-x)\, dx$

Solution. By Convolution Theorem

$$L \int_0^t f_1(x)\, f_2(t-x)\, dx = F_1(s) . F_2(s)$$

$$\Rightarrow \quad L \int_0^t e^x . \sin(t-x)\, dx = L(e^x) . L \sin t = \frac{1}{s-1} \frac{1}{s^2+1} = \frac{1}{(s-1)(s^2+1)}$$ **Ans.**

Note. Convolution Theorem is generally used to find Inverse Laplace transform of the product of two functions, discussed in the next chapter.

46.23 LAPLACE TRANSFORM OF BESSEL FUNCTIONS $J_0(x)$ and $J_1(x)$

We know that

$$J_n(x) = \frac{x^2}{2^n \Gamma(n+1)}\left[1 - \frac{x^2}{2.(2n+2)} + \frac{x^4}{2.4\,(2n+2)\,(2n+4)} - .....\right]$$

$$J_0(t) = \left[1 - \frac{t^2}{2^2} + \frac{t^4}{2^2.4^2} - \frac{t^6}{2^2.4^2.6^2} + ...\right]$$

Taking Laplace transforms of both sides, we have

$$LJ_0(t) = \frac{1}{s} - \frac{1}{2^2}.\frac{2!}{s^3} + \frac{1}{2^2.4^2}\frac{4!}{s^5} - \frac{1}{2^2.4^2.6^2}\frac{6!}{s^7} + ...$$

$$= \frac{1}{s}\left[1 - \frac{1}{2}\left(\frac{1}{s^2}\right) + \frac{1.3}{2.4}\left(\frac{1}{s^4}\right) - \frac{1.3.5}{2.4.6}\left(\frac{1}{s^6}\right) + ...\right]$$

$$= \frac{1}{s}\left[1 + \left(-\frac{1}{2}\right)\left(\frac{1}{s^2}\right) + \frac{\left(-\frac{1}{2}\right)\left(-\frac{3}{2}\right)}{2!}\left(\frac{1}{s^2}\right)^2 + \frac{\left(-\frac{1}{2}\right)\left(-\frac{3}{2}\right)\left(-\frac{5}{2}\right)}{3!}\left(\frac{1}{s^2}\right)^3 + ...\right]$$

$$= \frac{1}{s}\left[1 + \frac{1}{s^2}\right]^{-\frac{1}{2}}$$ (By Binomial theorem)

$$= \frac{1}{s}\left[\frac{s^2+1}{s^2}\right]^{-\frac{1}{2}} = \frac{1}{s}\left[\frac{s^2}{s^2+1}\right]^{\frac{1}{2}} = \frac{1}{\sqrt{s^2+1}}$$ ... (1) **Ans.**

We know that $Lf(at) = \frac{1}{a} F\left(\frac{s}{a}\right)$

$$LJ_0(at) = \frac{1}{a}\frac{1}{\sqrt{\frac{s^2}{a^2}+1}} = \frac{1}{\sqrt{s^2+a^2}} \qquad \text{[From (1)]}$$

$$LJ_1(x) = -LJ_0'(x) = -\left[s\,LJ_0(x) - J_0(0)\right] = -\left[s.\frac{1}{\sqrt{s^2+1}} - 1\right] = 1 - \frac{s}{\sqrt{s^2+1}} \qquad \textbf{Ans.}$$

EXERCISE 46.6

Find the Laplace transform of the following:

1. $e^{ax} J_0(bx)$ **Ans.** $\frac{1}{\sqrt{s^2+2as+a^2+b^2}}$

2. $x J_0(ax)$ **Ans.** $\frac{s}{(s^2+a^2)^{3/2}}$

3. $x J_1(x)$ **Ans.** $\frac{1}{(s^2+1)^{3/2}}$

46.24 EVALUATION OF INTEGRALS

We can evaluate number of integrals having lower limit 0 and upper limit ∞ by the help of Laplace transform.

Example 47. *Evaluate* $\int_0^\infty te^{-3t}\sin t\,dt$

Solution. $\int_0^\infty te^{-3t}\sin t\,dt = \int_0^\infty te^{-st}\sin t\,dt \qquad (s = 3)$

$$L(t\sin t) = -\frac{d}{ds}\left(\frac{1}{s^2+1}\right) = \frac{2s}{\left(s^2+1\right)^2}$$

Putting $s = 3$, we get $\frac{2\times 3}{(3^2+1)^2} = \frac{6}{100} = \frac{3}{50}$ **Ans.**

Example 48. *Evaluate* $\int_0^\infty \frac{e^{-t}\sin t}{t}dt$ and $\int_0^\infty \frac{\sin t}{t}dt$ *(U.P., II Semester, 2009)*

Solution. $\int_0^\infty \frac{e^{-t}\sin t}{t}dt = \int_0^\infty e^{-st}\frac{\sin t}{t}dt \qquad (s = 1)$

$$= L\left[\frac{\sin t}{t}\right] = \int_s^\infty \frac{1}{s^2+1}ds = \left[\tan^{-1}s\right]_s^\infty = \frac{\pi}{2} - \tan^{-1}s \qquad ...(1)$$

$$= \frac{\pi}{2} - \tan^{-1}(1) = \frac{\pi}{2} - \frac{\pi}{4} = \frac{\pi}{4} \qquad (s = 1) \qquad \textbf{Ans.}$$

On putting $s = 0$ in (1), we get $\int_0^\infty \frac{\sin t}{t}dt = \frac{\pi}{2} - \tan^{-1}(0) = \frac{\pi}{2}$ **Ans.**

Example 49. *Evaluate* $\int_0^\infty \frac{e^{-at}-e^{-bt}}{t}dt$ *(GBTU, 2011)*

Solution. $\int_0^\infty e^{-st}(e^{-at}-e^{-bt})dt = L(e^{-at}-e^{-bt}) = L(e^{-at}) - L(e^{-bt}) = \left(\frac{1}{s+a} - \frac{1}{s+b}\right)$

$$\int_0^\infty e^{-st}\left(\frac{e^{-at}-e^{-bt}}{t}\right)dt = L\frac{1}{t}(e^{-at}-e^{-bt}) = \int_s^\infty\left(\frac{1}{s+a} - \frac{1}{s+b}\right)ds$$

$$= \left[\log(s+a) - \log(s+b)\right]_s^\infty$$

$$= \left[\log\left(\frac{s+a}{s+b}\right)\right]_s^\infty = \left[\log \frac{1+\frac{a}{s}}{1+\frac{b}{s}}\right]_s^\infty = \left[\log 1 - \log\frac{s+a}{s+b}\right] = \log\frac{s+b}{s+a}$$

Putting $s = 0$ in above, we get $\int_0^\infty \left(\frac{e^{-at} - e^{-bt}}{t}\right) dt = \log\left(\frac{b}{a}\right)$ **Ans.**

Example 50. *Show that* $\int_0^\infty t^3 e^{-t} \sin t\, dt = 0$

Solution. $\{t^3 \sin t\} = (-1)^3 \frac{d^3}{ds^3} L\{\sin t\}$

$$\Rightarrow \quad \int_0^\infty e^{-st} t^3 \sin t\, dt = \frac{-d^3}{ds^3} \frac{1}{s^2+1}$$

$\left[\text{This is G.P. } \text{Sum} = \frac{a}{1-r}\right]$

$$= -\frac{d^2}{ds^2}\left[-\frac{2s}{(s^2+1)^2}\right] = \frac{d}{ds}\left[\frac{(s^2+1)^2(2) - 2s\,[2(s^2+1)]\,(2s)}{(s^2+1)^4}\right]$$

$$= \frac{d}{ds}\left[\frac{2(s^2+1) - 8s^2}{(s^2+1)^3}\right] = \frac{d}{ds}\left[\frac{-6s^2+2}{(s^2+1)^3}\right]$$

$$= \frac{(s^2+1)^3(-12s) - (-6s^2+2)\,3\,(s^2+1)^2(2s)}{(s^2+1)^6}$$

$$= \frac{(s^2+1)(-12s) - (-6s^2+2)\,6s}{(s^2+1)^4} = \frac{-12s^3 - 12s + 36s^3 - 12s}{(s^2+1)^4}$$

$$\int_0^\infty e^{-st} t^3 \sin t\, dt = \frac{24s^3 - 24s}{(s^2+1)^4} = \frac{24s\,(s^2-1)}{(s^2+1)^4} \qquad \text{... (1)}$$

Putting $s = 1$ in (1), we get $\int_0^\infty e^{-t} t^3 \sin t\, dt = 0$ **Ans.**

Example 51. *Evaluat* $\int_0^\infty t^2 e^{3t} \sin^2 t\, dt.$

Solution. We have $\sin^2 t = \frac{1}{2}(1 - \cos 2t)$

$$\Rightarrow \quad L\{\sin^2 t\} = \frac{1}{2}[L\{1\} - L\{\cos 2t\}] = \frac{1}{2}\left[\frac{1}{s} - \frac{s}{s^2+4}\right]$$

$$\Rightarrow \quad L\left[t^2 \sin^2 t\right] = (-1)^2 \cdot \frac{d^2}{ds^2}\left[\frac{1}{2}\left\{\frac{1}{s} - \frac{s}{s^2+4}\right\}\right]$$

$$\Rightarrow \quad \int_0^\infty e^{-st}.t^2 \sin^2 t\, dt = \frac{1}{2}\frac{d}{ds}\left[\frac{d}{ds}\left\{\frac{1}{s} - \frac{s}{s^2+4}\right\}\right] = \frac{1}{2}\frac{d}{ds}\left[-\frac{1}{s^2} - \frac{(s^2+4)(1) - s\,(2s)}{(s^2+4)^2}\right]$$

$$= \frac{1}{2}\frac{d}{ds}\left[-\frac{1}{s^2} - \frac{-s^2+4}{(s^2+4)^2}\right] = \frac{1}{2}\left[\frac{2}{s^3} - \frac{(s^2+4)^2(-2s) - (-s^2+4)\,2\,(s^2+4)(2s)}{(s^2+4)^4}\right]$$

$$= \frac{1}{2}\left[\frac{2}{s^3} - \frac{(s^2+4)(-2s) - (-s^2+4)\,4s}{(s^2+4)^3}\right] \qquad \text{... (1)}$$

Putting the value of $s = -3$ in (1), we get

$$\int_0^\infty e^{3t}\, t^2 \sin^2 t\, dt = \frac{1}{2}\left[\frac{2}{-27} - \frac{(13)6-(-5)(-12)}{(9+4)^3}\right]$$

$$= -\frac{1}{27} - \frac{9}{(13)^3} = \frac{-2197-243}{59319} = \frac{-2440}{59319}$$ **Ans.**

EXERCISE 46.7

Evaluate the following by using Laplace Transform:

1. $\int_0^\infty t e^{-4t} \sin t\, dt$ **Ans** $\frac{8}{289}$

2. $\int_0^\infty \frac{e^{-2t} \sinh t \sin t}{t} dt$ **Ans** $\frac{1}{2}\tan^{-1}\frac{1}{2}$

3. $\int_0^\infty \frac{\sin^2 t}{t^2} dt$ **Ans.** $i\frac{5}{2}$

4. $\int_0^\infty \frac{e^{-t} - e^{-4t}}{t} dt$ **Ans.** log 4

46.25 FORMULATION OF LAPLACE TRANSFORM

S.No.	$f(t)$	$F(s)$
1.	e^{at}	$\frac{1}{s-a}$
2.	t^n	$\frac{\overline{\lvert n+1}}{s^{n+1}}$ or $\frac{n!}{s^{n+1}}$
3.	sin *at*	$\frac{a}{s^2+a^2}$
4.	cos *at*	$\frac{s}{s^2+a^2}$
5.	sinh *at*	$\frac{a}{s^2-a^2}$
6.	cosh *at*	$\frac{s}{s^2-a^2}$
7.	$u(t-a)$	$\frac{e^{-as}}{s}$
8.	$\delta(t-a)$	e^{-as}
9.	e^{bt} sin *at*	$\frac{a}{(s-b)^2+a^2}$
10.	e^{bt} cos *at*	$\frac{s-b}{(s-b)^2+a^2}$

11	$\frac{t}{2a}\sin at$	$\frac{s}{(s^2+a^2)^2}$
12.	$t\cos at$	$\frac{s^2-a^2}{(s^2+a^2)^2}$
13.	$\frac{1}{2a^3}(\sin at - at\cos at)$	$\frac{1}{(s^2+a^2)^2}$
14.	$\frac{1}{2a}(\sin at + at\cos at)$	$\frac{s^2}{(s^2+a^2)^2}$

46.26 PROPERTIES OF LAPLACE TRANSFORM

S.No.	*Property*	*f(t)*	*F (s)*
1.	Scaling	$f(at)$	$\frac{1}{a}F\left(\frac{s}{a}\right),\quad a>0$
		$\frac{df(t)}{dt}$	$s\,F(s)-f(0),\quad s>0$
2.	Derivative	$\frac{d^2 f(t)}{dt^2}$	$s^2 F(s) - sf(0) - f'(0),\quad s>0$
		$\frac{d^3 f(t)}{dt^3}$	$s^3F(s) - s^2f(0) - s\,f'(0) - f''(0),\ s>0$
3.	Integral	$\int_0^t f(t)\,dt$	$\frac{1}{s}F(s)\,,\ s>0$
4.	Initial Value	$\lim_{t\to 0} f(t)$	$\lim_{s\to\infty} sF(s)$
5.	Final Value	$\lim_{t\to\infty} f(t)$	$\lim_{s\to\infty} sF(s)$
6.	First shifting	$e^{-at}f(t)$	$F(s+a)$
7.	Second shifting	$f(t)\,u(t-a)$	$e^{-a}\,L\,f(t+a)$
8.	Multiplication by *t*	$t\,f(t)$	$-\frac{d}{ds}F(s)$
9.	Multiplication by t^n	$t^n f(t)$	$(-1)^n\frac{d^n}{ds^n}F(s)$
10.	Division by *t*	$\frac{1}{t}f(t)$	$\int_s^\infty F(s)\,ds$

11.	Periodic function	$f(t)$	$\dfrac{\int_0^T e^{-st} f(t)}{1-e^{-st}}$ $f(t+T)=f(t)$
12.	Convolution	$f(t) * g(t)$	$F(s)\, G(s)$

OBJECTIVE TYPE QUESTIONS

Choose the correct alternative :

1. Laplace transform of $t^3 e^{-3t}$ is :

(i) $\dfrac{7}{(s+4)^3}$ (ii) $\dfrac{s}{(s+3)^3}$ (iii) $\dfrac{6}{(s+3)^4}$ (iv) $\dfrac{2}{(s+6)^3}$ **Ans.** (iii)

(R.G.P.V., Bhopal, II Semester, Feb. 2006)

2. Laplace transform of $e^{-2t} \sin 4t$ is :

(i) $\dfrac{2}{s^2+4s+20}$ (ii) $\dfrac{s-2}{s^2+4s+20}$ (iii) $\dfrac{s-4}{s^2+4s+20}$ (iv) $\dfrac{4}{s^2+4s+20}$ **Ans.** (iv)

(R.G.P.V., Bhopal, II Semester, June 2007)

3. If $L\{F(t)\}\ \bar{f}(s)$, the $L\left\{\int_0^t F(x)\,dx\right\}$ is:

(i) $\int_0^s \bar{f}(s)\,ds$ (ii) $\int_0^s \frac{1}{s}\bar{f}(s)\,ds$ (iii) $\frac{1}{s}\bar{f}(s)$ (iv) $s\,\bar{f}(s)$ **Ans.** (iii)

(R.G.P.V., Bhopal, II Semester, June 2006)

4. If $L\{F(t)\}\ \bar{f}(s)$, then $L\{t\,F(t)\}$ is:

(i) $\bar{f}'(s)$ (ii) $-\bar{f}'(s)$ (iii) $\bar{f}'(s)+\bar{f}(s)$ (iv) $s\bar{f}'(s)+\bar{f}(s)$ **Ans.** (ii)

(R.G.P.V., Bhopal, II Semester, June 2006)

5. Laplace transform of $\dfrac{\cos at - \cos bt}{t}$ is

(i) $\log\dfrac{s^2+b^2}{s^2+a^2}$ (ii) $\dfrac{1}{2}\log\dfrac{s^2+a^2}{s^2+b^2}$ (iii) $\dfrac{1}{2}\log\dfrac{s^2+b^2}{s^2+a^2}$ (iv) $\log\dfrac{s+b}{s+a}$ **Ans.** (iii)

(R.G.P.V., Bhopal, II Semester, Feb. 2006, 2005)

6. The Laplace transform of the function

$F(t) \begin{cases} 1, & 0 \le t < 2 \\ -1, & 2 \le t < 4 \end{cases}$, $f(t+4) = f(t)$ is given as,

(i) $\dfrac{1-e^{-2s}}{s(1+e^{-2s})}$ (ii) $\dfrac{1+e^{-2s}}{s(1+e^{-2s})}$ (iii) 0 (iv) $\dfrac{s+1}{s-1}$ **Ans.** (i)

(U.P., II Semester, 2009)

Fill in the blank of the following question:

7. The Laplace transform of $\int_0^t\int_0^t\int_0^t \cos au\; du\; du\; du$ is given as ...... *[U.P.T.U. (SUM) 2009]*

Ans. $\dfrac{1}{s^2(s^2+a^2)}$

INVERSE LAPLACE TRANSFORMS

(SOLUTION OF DIFFERENTIAL EQUATIONS)

47.1 INVERSE LAPLACE TRANSFORMS

If $F(s)$ is the Laplace Transform of a function $f(t)$, then $f(t)$ is known as Inverse Laplace Transform. Now we will discuss how to find $f(t)$ when $F(s)$ is given.

If $L[f(t)] = F(s)$, then $L^{-1}[F(s)] = f(t)$, where L^{-1} is called the Inverse Laplace Transform operator.

From the application point of view, the Inverse Laplace Transform is very useful.

Inverse Laplace Transform is used in solving differential equations without finding the general solution and arbitrary constants.

47.2 IMPORTANT FORMULAE

1. $L^{-1}\left(\frac{1}{s}\right) = 1$

2. $L^{-1}\frac{1}{s^n} = \frac{t^{n-1}}{(n-1)!}$

3. $L^{-1}\frac{1}{s-a} = e^{at}$

4. $L^{-1}\frac{s}{s^2-a^2} = \cosh at$ (*R.G.P.V., Bhopal, Dec. 2007*)

5. $L^{-1}\frac{1}{s^2-a^2} = \frac{1}{a}\sinh at$

6. $L^{-1}\frac{1}{s^2+a^2} = \frac{1}{a}\sin at$

7. $L^{-1}\frac{s}{s^2+a^2} = \cos at$

8. $L^{-1}F(s-a) = e^{at}f(t)$

9. $L^{-1}\frac{1}{(s-a)^2+b^2} = \frac{1}{b}e^{at}\sin bt$

10. $L^{-1}\frac{s-a}{(s-a)^2+b^2} = e^{at}\cos bt$

11. $L^{-1}\frac{1}{(s-a)^2-b^2} = \frac{1}{b}e^{at}\sinh bt$

12. $L^{-1}\frac{s-a}{(s-a)^2-b^2} = e^{at}\cosh bt$

13. $L^{-1}\frac{1}{(s^2+a^2)^2} = \frac{1}{2a^3}(\sin at - at\cos at)$

14. $L^{-1}\frac{s}{(s^2+a^2)^2} = \frac{1}{2a}t\sin at$

15. $L^{-1}\frac{s^2-a^2}{(s^2+a^2)^2} = t\cos at$

16. $L^{-1}(1) = s(t)$

17. $L^{-1}\frac{s^2}{(s^2+a^2)^2} = \frac{1}{2a}[\sin at + at\cos at]$

18. $L^{-1}\left\{\frac{1}{t}F(t)\right\} = \int_0^t f(t)\,dt$

Example 1. *Show that* $\frac{1}{s^{1/2}} = L\left[\frac{1}{\sqrt{\pi t}}\right]$. *(U.P., II Semester, Summer 2005)*

Solution. We have to show that $\frac{1}{s^{1/2}} = L\left[\frac{1}{\sqrt{\pi t}}\right]$.

Now, $$L^{-1}\left\{\frac{1}{s^n}\right\} = \frac{t^{n-1}}{(n-1)!} = \frac{t^{n-1}}{\overline{|n}}$$

So $$L^{-1}\left\{\frac{1}{s^{1/2}}\right\} = \frac{t^{\frac{1}{2}-1}}{\overline{|1/2}} = \frac{t^{-1/2}}{\overline{|1/2}} = \frac{t^{-1/2}}{\sqrt{\pi}}$$

$\Rightarrow$ $$L^{-1}\left\{\frac{1}{s^{1/2}}\right\} = \frac{1}{\sqrt{\pi t}} \quad \Rightarrow \quad \frac{1}{s^{1/2}} = L\left[\frac{1}{\sqrt{\pi t}}\right]$$ **Proved.**

Example 2. *Find the inverse Laplace Transform of the following:*

(*i*) $\frac{1}{s-2}$ (*ii*) $\frac{1}{s^2-9}$ (*iii*) $\frac{1}{s^2+25}$

(*iv*) $\frac{s}{s^2+9}$ (*v*) $\frac{1}{(s-2)^2+1}$ (*vi*) $\frac{s-1}{(s-1)^2+4}$

(*vii*) $\frac{1}{(s+3)^2-4}$ (*viii*) $\frac{s+2}{(s+2)^2-25}$

Solution.

(*i*) $L^{-1}\frac{1}{s-2} = e^{2t}$

(*ii*) $L^{-1}\frac{1}{s^2-9} = L^{-1}\frac{1}{3}\cdot\frac{3}{s^2-(3)^2} = \frac{1}{3}\sinh 3t$

(*iii*) $L^{-1}\frac{1}{s^2+25} = \frac{1}{5}\frac{5}{s^2+(5)^2} = \frac{1}{5}\sin 5t$

(*iv*) $L^{-1}\frac{s}{s^2+9} = \frac{s}{s^2+(3)^2} = \cos 3t$

(*v*) $L^{-1}\frac{1}{(s-2)^2+1} = e^{2t}\sin t$

(*vi*) $L^{-1}\frac{s-1}{(s-1)^2+(2)^2} = e^{t}\cos 2t$

(*vii*) $L^{-1}\frac{1}{(s+3)^2-4} = \frac{1}{2}\frac{2}{(s+3)^2-(2)^2} = \frac{1}{2}e^{-3t}\sinh 2t$

(*viii*) $L^{-1}\frac{s+2}{(s+2)^2-25} = L^{-1}\frac{(s+2)}{(s+2)^2-(5)^2} = e^{-2t}\cosh 5t$ **Ans.**

Example 3. *Find* $L^{-1}\frac{s^2+2s+6}{s^3}$ *(M.D.U. 2010)*

Solution. Here, we have

$$L^{-1}\frac{s^2+2s+6}{s^3} = L^{-1}\left[\frac{1}{s}+\frac{2}{s^2}+\frac{6}{s^3}\right] = 1+\frac{2t}{1!}+\frac{6}{2!}t^2$$

$$= 1 + 2t + 3t^2$$ **Ans.**

Example 4. *Find Inverse Laplace Transform of*

(*a*) $\left\{\frac{6}{2s-3}-\frac{3+4s}{9s^2-16}+\frac{8-6s}{16s^2+9}\right\}$ (*b*) $\frac{2s-5}{9s^2-25}$

(*c*) $\frac{s-2}{6s^2+20}$

Solution.

$$(a)\ L^{-1}\left\{\frac{6}{2s-3}-\frac{3}{9s^2-16}-\frac{4s}{9s^2-16}+\frac{8}{16s^2+9}-\frac{6s}{16s^2+9}\right\}$$

$$= L^{-1}\left\{\frac{3}{s-\frac{3}{2}}-\frac{\frac{1}{3}}{s^2-\left(\frac{4}{3}\right)^2}-\frac{\frac{4}{9}s}{s^2-\left(\frac{4}{3}\right)^2}+\frac{\frac{1}{2}}{s^2+\left(\frac{3}{4}\right)^2}-\frac{\frac{3}{8}s}{s^2+\left(\frac{3}{4}\right)^2}\right\}$$

$$= L^{-1}\left\{\frac{3}{s-\frac{3}{2}}-\frac{1}{4}\frac{\frac{4}{3}}{s^2-\left(\frac{4}{3}\right)^2}-\frac{4}{9}\frac{s}{s^2-\left(\frac{4}{3}\right)^2}+\frac{2}{3}\frac{\frac{3}{4}}{s^2+\left(\frac{3}{4}\right)^2}-\frac{3}{8}\frac{s}{s^2+\left(\frac{3}{4}\right)^2}\right\}$$

$$= 3\,e^{\frac{3}{2}t}-\frac{1}{4}\sinh\frac{4}{3}t-\frac{4}{9}\cosh\frac{4}{3}t+\frac{2}{3}\sin\frac{3}{4}t-\frac{3}{8}\cos\frac{3}{4}t$$ **Ans.**

$$(b)\ L^{-1}\frac{2s-5}{9s^2-25}=L^{-1}\left[\frac{2s}{9s^2-25}-\frac{5}{9s^2-25}\right]=L^{-1}\left[\frac{2s}{9\left[s^2-\left(\frac{5}{3}\right)^2\right]}-\frac{5}{9\left[s^2-\left(\frac{5}{3}\right)^2\right]}\right]$$

$$=\frac{2}{9}\cosh\frac{5}{3}t-\frac{1}{3}L^{-1}\left(\frac{\frac{5}{3}}{s^2-\left(\frac{5}{3}\right)^2}\right)=\frac{2}{9}\cosh\frac{5t}{3}-\frac{1}{3}\sin\frac{5t}{3}$$ **Ans.**

$$(c)\ L^{-1}\frac{s-2}{6s^2+20}=L^{-1}\frac{s}{6s^2+20}-L^{-1}\frac{2}{6s^2+20}=\frac{1}{6}L^{-1}\frac{s}{s^2+\frac{10}{3}}-\frac{1}{3}L^{-1}\frac{1}{s^2+\frac{10}{3}}$$

$$=\frac{1}{6}L^{-1}\frac{s}{s^2+\frac{10}{3}}-\frac{1}{3}\times\sqrt{\frac{3}{10}}\,L^{-1}\frac{\sqrt{\frac{10}{3}}}{s^2+\frac{10}{3}}=\frac{1}{6}\cos\sqrt{\frac{10}{3}}t-\frac{1}{\sqrt{30}}\sin\sqrt{\frac{10}{3}}t$$ **Ans.**

Example 5. *Find the inverse Laplace transform of following function:*

$\dfrac{14s+10}{49s^2+28s+13}$. *[U.P., II Semester, 2007]*

Solution. The given function can be written as

$$\frac{14s+10}{49s^2+28s+13}=\frac{14s+10}{(7s+2)^2+9}=\frac{14\left(s+\frac{2}{7}\right)+6}{49\left(s+\frac{2}{7}\right)^2+9}$$

$$\therefore\quad L^{-1}\left(\frac{14\,s+10}{49\,s^2+28\,s+13}\right)=L^{-1}\left[\frac{14\left(s+\frac{2}{7}\right)+6}{49\left(s+\frac{2}{7}\right)^2+9}\right]=e^{-\frac{2t}{7}}L^{-1}\left(\frac{14\,s+6}{49\,s^2+9}\right)$$

$$= e^{-\frac{2}{7}t} L^{-1} \frac{14}{49}\left(\frac{s+\frac{6}{14}}{s^2+\frac{9}{49}}\right) = e^{-\frac{2t}{7}}\left[\frac{14}{49} L^{-1}\left(\frac{s}{s^2+\frac{9}{49}}\right)+\left(\frac{14}{49}\right)\left(\frac{6}{14}\right) L^{-1}\left(\frac{1}{s^2+\frac{9}{49}}\right)\right]$$

$$= e^{-\frac{2t}{7}}\left[\frac{2}{7}\cos\frac{3}{7}t+\frac{6}{49}.\frac{7}{3}\sin\frac{3}{7}t\right]$$

$$= \frac{2}{7} e^{-\frac{2t}{7}}\left(\cos\frac{3}{7}t+\sin\frac{3}{7}t\right)$$ **Ans.**

EXERCISE 47.1

Find the Inverse Laplace Transform of the following:

1. $\frac{3s-8}{4s^2+25}$ **Ans.** $\frac{3}{4}\cos\frac{5t}{2}-\frac{4}{5}\sin\frac{5t}{2}$

2. $\frac{3(s^2-2)^2}{2s^5}$ **Ans.** $\frac{3}{2}-3t^2+\frac{1}{2}t^4$

3. $\frac{2s-5}{4s^2+25}+\frac{4s-18}{9-s^2}$ **Ans.** $\frac{1}{2}\left(\cos\frac{5t}{2}-\sin\frac{5t}{2}\right)-4\cosh 3t+6\sinh 3t$

4. $\frac{5s-10}{9s^2-16}$ **Ans.** $\frac{5}{9}\cosh\frac{4}{3}t-\frac{5}{6}\sinh\frac{4}{3}t$

5. $\frac{1}{4s}+\frac{16}{1-s^2}$ **Ans.** $\frac{1}{4}-16\sinh t$

6. $L^{-1}\left\{\frac{1}{s^n}\right\}$ exist only when the value of n is :

   (*i*) Negative integer (*ii*) Positive integer

   (*iii*) Zero (*iv*) None of these **Ans.** (*ii*) (*U.P. II Semester, 2010*)

47.3 MULTIPLICATION BY *S*

$$L^{-1}[s\,F(s)] = \frac{d}{dt} f(t) + f(0)\,\delta(t)$$

Example 6. *Find the Inverse Laplace Transform of* (*i*) $\frac{s}{s^2+1}$ (*ii*) $\frac{s}{4s^2-25}$ (*iii*) $\frac{3s}{2s+9}$

Solution.

(*i*) $L^{-1}\frac{1}{s^2+1} = \sin t$

$L^{-1}\frac{s}{s^2+1} = \frac{d}{dt}(\sin t) + \sin(0)\,\delta(t) = \cos t$ **Ans.**

(*ii*) $L^{-1}\frac{1}{4s^2-25} = \frac{1}{4}L^{-1}\frac{1}{s^2-\frac{25}{4}} = \frac{1}{4}.\frac{2}{5}L^{-1}\frac{\frac{5}{2}}{s^2-\left(\frac{5}{2}\right)^2} = \frac{1}{10}\sinh\frac{5}{2}t$

$$L^{-1}\frac{s}{4s^2-25}=\frac{1}{10}\frac{d}{dt}\sinh\frac{5}{2}t+\frac{1}{10}\sinh\frac{5}{2}(0)\,\delta(t)$$

$$=\frac{1}{10}\left(\frac{5}{2}\right)\cosh\frac{5}{2}t=\frac{1}{4}\cosh\frac{5}{2}t \quad \textbf{Ans.}$$

(iii) $$L^{-1}\frac{3}{2s+9}=\frac{3}{2}L^{-1}\frac{1}{s+\frac{9}{2}}=\frac{3}{2}e^{-\frac{9}{2}t}$$

$$L^{-1}\frac{3s}{2s+9}=\frac{3}{2}\frac{d}{dt}\left(e^{-\frac{9}{2}t}\right)+\frac{3}{2}e^{-\frac{9}{2}(0)}\delta(t)=\frac{3}{2}\left(-\frac{9}{2}\right)e^{-\frac{9}{2}t}+\frac{3}{2}=-\frac{27}{4}e^{-\frac{9}{2}t}+\frac{3}{2} \quad \textbf{Ans.}$$

EXERCISE 47.2

Find the Inverse Laplace Transform of the following:

1. $\frac{s}{s+5}$ **Ans.** $1-5e^{-5t}$
2. $\frac{2s}{3s+6}$ **Ans.** $-\frac{4}{3}e^{-2t+\frac{2}{3}}$
3. $\frac{s}{2s^2-1}$ **Ans.** $\frac{1}{2}\cosh\frac{t}{2}$
4. $\frac{s^2}{s^2+a^2}$ **Ans.** $-a\sin at+1$
5. $\frac{s^2+4}{s^2+9}$ **Ans.** $-\frac{5}{3}\sin 3t+1$

47.4 DIVISION BY s (MULTIPLICATION BY $\frac{1}{s}$)

$$L^{-1}\left[\frac{F(s)}{s}\right]=\int_0^t\left[L^{-1}[F(s)]\right]dt=\int_0^t f(t)\,dt$$

Example 7. *Find the Inverse Laplace Transform of*

(i) $\frac{1}{s(s+a)}$ (ii) $\frac{1}{s(s^2+1)}$ (iii) $\frac{s^2+3}{s(s^2+9)}$

Solution.

(i) $$L^{-1}\left(\frac{1}{s+a}\right)=e^{-at}$$

$$L^{-1}\left[\frac{1}{s(s+a)}\right]=\int_0^t L^{-1}\left(\frac{1}{s+a}\right)dt=\int_0^t e^{-at}\,dt=\left[\frac{e^{-at}}{-a}\right]_0^t$$

$$=\frac{e^{-at}}{-a}+\frac{1}{a}=\frac{1}{a}\left[1-e^{-at}\right] \quad \textbf{Ans.}$$

(ii) $$L^{-1}\frac{1}{s^2+1}=\sin t$$

$$L^{-1}\frac{1}{s}\left(\frac{1}{s^2+1}\right)=\int_0^t L^{-1}\left(\frac{1}{s^2+1}\right)dt=\int_0^t \sin t\,dt=[-\cos t]_0^t=-\cos t+1 \quad \textbf{Ans.}$$

(iii) $$L^{-1}\frac{s^2+3}{s(s^2+9)}=L^{-1}\left[\frac{s^2+9-6}{s(s^2+9)}\right]=L^{-1}\left[\frac{1}{s}-\frac{6}{s(s^2+9)}\right]$$

$$=1-2\int_0^t \sin 3t\,dt=1+2\times\frac{1}{3}[\cos 3t]_0^t=1+\frac{2}{3}\cos 3t-\frac{2}{3}$$

$$=\frac{2}{3}\cos 3t+\frac{1}{3}=\frac{1}{3}[2\cos 3t+1] \quad \textbf{Ans.}$$

EXERCISE 47.3

Find the Inverse Laplace Transform of the following:

1. $\dfrac{1}{2s(s-3)}$ **Ans.** $\dfrac{1}{2}\left[\dfrac{e^{3t}}{3}-1\right]$ **2.** $\dfrac{1}{s(s+2)}$ **Ans.** $\dfrac{1-e^{-2t}}{2}$

3. $\dfrac{1}{s(s^2-16)}$ **Ans.** $\dfrac{1}{16}[\cosh 4t-1]$ **4.** $\dfrac{1}{s(s^2+a^2)}$ **Ans.** $\dfrac{1-\cos at}{a^2}$

5. $\dfrac{s^2+2}{s(s^2+4)}$ **Ans.** $\cos^2 t$ **6.** $\dfrac{1}{s^2(s+1)}$ **Ans.** $t-1+e^{-t}$

7. $\dfrac{1}{s^3(s^2+1)}$ **Ans.** $\dfrac{t^2}{2}+\cos t-1$

47.5 FIRST SHIFTING PROPERTY

If $L^{-1}F(s)=f(t)$, then $\boxed{\boldsymbol{L^{-1}F(s+a)=e^{-at}L^{-1}[F(s)]}}$

Example 8. *Find the Inverse Laplace Transform of*

$$(i)\ \frac{1}{(s+2)^5} \qquad (ii)\ \frac{s}{s^2+4s+13} \qquad (iii)\ \frac{1}{9s^2+6s+1}$$

Solution.

$$(i)\ L^{-1}\frac{1}{s^5}=\frac{t^4}{4!} \text{ then } L^{-1}\frac{1}{(s+2)^5}=e^{-2t}.\frac{t^4}{4!}$$ **Ans.**

$$(ii)\ L^{-1}\left(\frac{s}{s^2+4s+13}\right)=L^{-1}\frac{s+2-2}{(s+2)^2+(3)^2}=L^{-1}\frac{s+2}{(s+2)^2+(3)^2}-L^{-1}\frac{2}{(s+2)^2+(3)^3}$$

$$=e^{-2t}L^{-1}\frac{s}{s^2+3^2}-e^{-2t}L^{-1}\frac{2}{3}\left(\frac{3}{s^2+3^2}\right)=e^{-2t}\cos 3t-\frac{2}{3}e^{-2t}\sin 3t$$ **Ans.**

$$(iii)\ L^{-1}\frac{1}{9s^2+6s+1}=L^{-1}\frac{1}{(3s+1)^2}=\frac{1}{9}L^{-1}\frac{1}{\left(s+\frac{1}{3}\right)^2}=\frac{1}{9}e^{-t/3}L^{-1}\frac{1}{s^2}$$

$$=\frac{1}{9}e^{-t/3}\,t=\frac{t\,e^{-t/3}}{9}$$ **Ans.**

Example 9. *Find the Inverse Laplace Transform of* $\dfrac{s+1}{s^2-6s+25}$

(*U.P. II Semester Summer* 2010)

Solution. $$L^{-1}\left(\frac{s+1}{s^2+6s+25}\right)=L^{-1}\left[\frac{s+1}{(s-3)^2+(4)^2}\right]=L^{-1}\left[\frac{s-3+4}{(s-3)^2+(4)^2}\right]$$

Let $$=L^{-1}\left[\frac{s-3}{(s-3)^2+(4)^2}\right]+L^{-1}\left[\frac{4}{(s-3)^2+(4)^2}\right]$$

$$=e^{3t}\cos 4t+\frac{1}{4}e^{3t}\sin 4t.$$ **Ans.**

EXERCISE 47.4

Obtain the Inverse Laplace Transform of the following:

1. $\dfrac{s+8}{s^2+4s+5}$ **Ans.** $e^{-2t}(\cos t + 6\sin t)$

2. $\dfrac{s}{(s+3)^2+4}$ **Ans.** $e^{-3t}(\cos 2t - 1.5\sin 2t)$

3. $\dfrac{s}{(s+7)^4}$ **Ans.** $e^{-7t}\dfrac{t^2}{6}(3-7t)$

4. $\dfrac{s+2}{s^2-2s-8}$ **Ans.** $e^{-t}(\cosh 3t + \sinh 3t)$

5. $\dfrac{s}{s^2+6s+25}$ **Ans.** $e^{-3t}\left[\cos 4t - \dfrac{3}{4}\sin 4t\right]$

6. $\dfrac{1}{2(s-1)^2+32}$ **Ans.** $\dfrac{e^t}{8}\sin 4t$

7. $\dfrac{s-4}{4(s-3)^2+16}$ **Ans.** $\dfrac{1}{4}e^{3t}\cos 2t - \dfrac{1}{8}e^{3t}\sin 2t$

47.6 SECOND SHIFTING PROPERTY

$$\boxed{L^{-1}\left[e^{-as}F(s)\right] = f(t-a)\,u(t-a)}$$

Example 10. *Obtain Inverse Laplace Transform of*

(*i*) $\dfrac{e^{-\pi s}}{(s+3)}$ (*ii*) $\dfrac{e^{-s}}{(s+1)^3}$

Solution.

(*i*) $L^{-1}\dfrac{1}{s+3} = e^{-3t}, \quad L^{-1}\dfrac{e^{-\pi s}}{s+3} = e^{-3(t-\pi)}\,u(t-\pi)$ **Ans.**

(*ii*) $L^{-1}\dfrac{1}{s^3} = \dfrac{t^2}{2!} \quad \Rightarrow \quad L^{-1}\dfrac{1}{(s+1)^3} = e^{-t}\dfrac{t^2}{2!}$

$$L^{-1}\frac{e^{-s}}{(s+1)^3} = e^{-(t-1)}\cdot\frac{(t-1)^2}{2!}\cdot u(t-1)$$ **Ans.**

Example 11. *Evaluate*

$$L^{-1}\left[\frac{e^{-s}-3e^{-3s}}{s^2}\right]$$

Solution. $L^{-1}\left[\dfrac{e^{-s}-3e^{-3s}}{s^2}\right] = L^{-1}\left[\dfrac{e^{-s}}{s^2} - \dfrac{3e^{-3s}}{s^2}\right]$ ...(1)

We know that $L[u(t-a)] = \dfrac{e^{-as}}{s}$

and $L[(t-a)\,u(t-a)] = \dfrac{e^{-as}}{s^2}$

Using these results in (1), we get

$$\therefore \quad L^{-1}\left[\frac{e^{-s}-3e^{-3s}}{s^2}\right] = (t-1)\,u(t-1) - 3(t-3)\,u(t-3)$$ **Ans.**

Example 12. *Find the Inverse Laplace Transform of* $\dfrac{se^{-s/2}+\pi e^{-s}}{s^2+\pi^2}$ *in terms of unit step functions.*

Solution.
$$L^{-1}\frac{\pi}{s^2+\pi^2}=\sin \pi t$$

$$L^{-1}\left[e^{-s}\frac{\pi}{s^2+\pi^2}\right]=\sin\pi\,(t-1).u(t-1)=-\sin(\pi t).u(t-1) \qquad ...(1)$$

and
$$L^{-1}\frac{s}{s^2+\pi^2}=\cos \pi t$$

$$L^{-1}\left[e^{-s/2}\frac{s}{s^2+\pi^2}\right]=\cos\pi\left(t-\frac{1}{2}\right).u\left(t-\frac{1}{2}\right)=\sin\pi t\,.u\left(t-\frac{1}{2}\right) \qquad ...(2)$$

On adding (1) and (2), we get

$$L^{-1}\left[\frac{e^{-s/2}\,s+e^{-s}.\pi}{s^2+\pi^2}\right]=\sin(\pi t).u\left(t-\frac{1}{2}\right)-\sin(\pi t).u(t-1)$$

$$=\sin\pi t\left[u\left(t-\frac{1}{2}\right)-u(t-1)\right]$$ **Ans.**

Example 13. *Find the value of* $L^{-1}\left\{\dfrac{1}{(s^2+a^2)^2}\right\}$

Solution.
$$\frac{1}{(s^2+a^2)^2}=\frac{1}{s}.\frac{s}{(s^2+a^2)^2}=-\frac{1}{2s}\frac{d}{ds}\left(\frac{1}{s^2+a^2}\right)$$

$$\Rightarrow\quad L^{-1}\left\{\frac{1}{(s^2+a^2)^2}\right\}=L^{-1}\left\{-\frac{1}{2s}\frac{d}{ds}\left(\frac{1}{s^2+a^2}\right)\right\}$$

$$\Rightarrow\quad -\frac{1}{2s}\left\{-t\frac{1}{a}\sin at\right\}=\frac{1}{2a}\frac{1}{s}\{t.\sin at\}=\frac{1}{2a}\int_0^t t\sin at\,dt$$

$$=\frac{1}{2a}\left[t\left(\frac{-\cos at}{a}\right)-\int\frac{-\cos at}{a}dt\right]_0^t=\frac{1}{2a}\left[-\frac{t}{a}\cos at+\frac{\sin at}{a^2}\right]_0^t$$

$$=\frac{1}{2a^3}(\sin at-at\cos at)$$ **Ans.**

EXERCISE 47.5

Obtain Inverse Laplace Transform of the following:

1. $\dfrac{e^{-s}}{(s+2)^3}$ **Ans.** $e^{-(t-2)}\dfrac{(t-2)^2}{2}u(t-2)$

2. $\dfrac{e^{-2s}}{(s+1)(s^2+2s+2)}$ **Ans.** $e^{-(t-2)}\,|1-\cos(t-2)|\,u(t-2)$

3. $\dfrac{e^{-s}}{\sqrt{s+1}}$ **Ans.** $\dfrac{e^{-(t-1)}}{\sqrt{\pi(t-1)}}.u(t-1)$

4. $\dfrac{e^{-\frac{\pi}{2}s}+e^{-\frac{3\pi}{2}s}}{s^2+1}$ **Ans.** $\cot t\left[u\left(t-\frac{3\pi}{2}\right)-u\left(t-\frac{\pi}{2}\right)\right]$

5. $\dfrac{e^{-4s}(s+2)}{s^2+4s+5}$ **Ans.** $e^{-2(t-u)}\cos(t-u)\,u(t-4)$

6. $\dfrac{e^{-as}}{s^2}$ **Ans.** $f(t)=\begin{cases} t-a, \text{ where } t>a \\ 0, \quad \text{ where } t<0 \end{cases}$

7. $\dfrac{e^{-\pi s}}{s^2+1}$ **Ans.** $-\sin t.\ u(t-\pi)$

47.7 INVERSE LAPLACE TRANSFORMS OF DERIVATIVES

$$L^{-1}\left[\frac{d}{ds}F(s)\right]=-t\,L^{-1}[F(s)]=-t\,f(t) \quad \Rightarrow \quad \boxed{L^{-1}[F(s)]=-\frac{1}{t}L^{-1}\left[\frac{d}{ds}F(s)\right]}$$

Example 14. *Find* $L^{-1}\left\{\log\dfrac{s+1}{s-1}\right\}$. *(Uttarakhand, II Semester, June 2010, 2009, 2007)*

Solution. $L^{-1}\left\{\log\left(\dfrac{s+1}{s-1}\right)\right\}=-\dfrac{1}{t}L^{-1}\left[\dfrac{d}{ds}\log\left(\dfrac{s+1}{s-1}\right)\right]$

$$=-\frac{1}{t}L^{-1}\left[\frac{d}{ds}\log(s+1)-\frac{d}{ds}\log(s-1)\right]=-\frac{1}{t}L^{-1}\left[\frac{1}{s+1}-\frac{1}{s-1}\right]$$

$$=-\frac{1}{t}\left[e^{-t}-e^{t}\right]=\frac{1}{t}\left[e^{t}-e^{-t}\right]$$ **Ans.**

Example 15. *Find the Inverse Laplace Transform of* $F(s)=\log\dfrac{s+a}{s+b}$

Solution. $L^{-1}\log\left(\dfrac{s+a}{s+b}\right)=-\dfrac{1}{t}L^{-1}\left[\dfrac{d}{ds}\log\dfrac{s+a}{s+b}\right]$

$$=-\frac{1}{t}L^{-1}\left[\frac{d}{ds}\log(s+a)-\frac{d}{ds}\log(s+b)\right]=-\frac{1}{t}L^{-1}\left[\frac{1}{s+a}-\frac{1}{s+b}\right]$$

$$=-\frac{1}{t}\left[e^{-at}-e^{-bt}\right]=\frac{1}{t}\left(e^{-bt}-e^{-at}\right)$$ **Ans.**

Example 16. *Obtain the Inverse Laplace Transform of* $\log\dfrac{s^2-1}{s^2}$.

Solution. $L^{-1}\left[\log\dfrac{s^2-1}{s^2}\right]=-\dfrac{1}{t}L^{-1}\left[\dfrac{d}{ds}\log\dfrac{s^2-1}{s^2}\right]$

$$=-\frac{1}{t}L^{-1}\left[\frac{d}{ds}\left\{\log(s^2-1)-2\log s\right\}\right]=-\frac{1}{t}L^{-1}\left[\frac{2s}{s^2-1}-\frac{2}{s}\right]=-\frac{1}{t}[2\cosh t-2]$$

$$=\frac{2}{t}[1-\cosh t]$$ **Ans.**

Example 17. *Find the function whose Laplace transform is* $\log\left(1+\frac{1}{s}\right)$.

(U.P., II Semester, June 2007)

Solution. $L^{-1}\left[\log\left(1+\frac{1}{s}\right)\right]=\frac{1}{t}L^{-1}\left[\frac{d}{ds}\log\left(\frac{s+1}{s}\right)\right]$

$$=-\frac{1}{t}L^{-1}\left[\left(\frac{s}{s+1}\right)\left(-\frac{1}{s^2}\right)\right]=-\frac{1}{t}L^{-1}\left[-\frac{1}{s\,(s+1)}\right]$$

$$=-\frac{1}{t}L^{-1}\left[\frac{1}{s+1}-\frac{1}{s}\right] \qquad \text{(Partial fraction)}$$

$$=-\frac{1}{t}[e^{-t}-1]=\frac{1}{t}[1-e^{-t}] \text{ **Ans.**}$$

Example 18. *Find the inverse Laplace transform of* $\tan^{-1}\left(\frac{2}{s^2}\right)$ *(Q. Bank U.P.T.U. 2001)*

Solution. Here, we have $L^{-1}\left[\tan^{-1}\left(\frac{2}{s^2}\right)\right]=-\frac{1}{t}L^{-1}\left[\frac{d}{ds}\tan^{-1}\frac{2}{s^2}\right]$

$$=-\frac{1}{t}L^{-1}\left[\frac{1}{1+\frac{4}{s^4}}\left(\frac{4}{s^3}\right)\right]=-\frac{1}{t}L^{-1}\left[\frac{s^4}{s^4+4}\left(-\frac{4}{s^3}\right)\right]=\frac{1}{t}L^{-1}\left[\frac{4s}{s^4+4}\right]$$

$$=\frac{4}{t}L^{-1}\left[\frac{s}{s^4+4}\right]=\frac{4}{t}L^{-1}\left[\frac{s}{(s^2+2s+2)\,(s^2-2s+2)}\right]$$

$$=\frac{4}{t}L^{-1}\left[-\frac{1}{4}\frac{1}{(s^2+2s+2)}+\frac{1}{4}\frac{1}{(s^2-2s+2)}\right] \qquad \left(\text{By partial fraction}\right)$$

$$=\frac{1}{t}L^{-1}\left[-\frac{1}{(s^2+2s+2)}+\frac{1}{(s^2-2s+2)}\right]=\frac{1}{t}L^{-1}\left[-\frac{1}{(s+1)^2+1}+\frac{1}{(s-1)^2+1}\right]$$

$$=\frac{1}{t}\left[-e^{-t}\sin t+e^{t}\sin t\right]=\frac{\sin t}{t}\left[e^{t}-e^{-t}\right] \qquad \textbf{Ans.}$$

Example 19. *Find Inverse Laplace Transform of* $\tan^{-1}\frac{1}{s}$. *(M.D.U., 2010)*

Solution. $L^{-1}\left(\tan^{-1}\frac{1}{s}\right)=-\frac{1}{t}L^{-1}\left[\frac{d}{ds}\tan^{-1}\frac{1}{s}\right]$

$$=-\frac{1}{t}L^{-1}\left[\frac{1}{1+\frac{1}{s^2}}\left(-\frac{1}{s^2}\right)\right]=\frac{1}{t}L^{-1}\left[\frac{1}{1+s^2}\right]=\frac{\sin t}{t} \qquad \textbf{Ans.}$$

Example 20. *Find* $L^{-1}[\tan^{-1}(1+s)]$ *(M.D.U. 2010)*

Solution. $L^{-1}\left[\tan^{-1}(1+s)\right]=-\frac{1}{t}L^{-1}\left[\frac{d}{ds}\tan^{-1}(1+s)\right]$

$$=-\frac{1}{t}L^{-1}\left[\frac{1}{1+(s+1)^2}\right]=-\frac{1}{t}L^{-1}\left[\frac{1}{(s+1)^2+1}\right]$$

$$=-\frac{1}{t}e^{-t}\sin t \qquad \textbf{Ans.}$$

Example 21. *Find the inverse Laplace transform of*

$$\cot^{-1}\left(\frac{s}{2}\right)$$

Solution.

Let $\quad L^{-1}\left[\cot^{-1}\left(\frac{s}{2}\right)\right] = f(t) \quad \Rightarrow \quad L^{-1}\left[\frac{d}{ds}\cot^{-1}\left(\frac{s}{2}\right)\right] = -t\,f(t)$

$\Rightarrow \quad L^{-1}\left[\frac{-1}{1+\frac{s^2}{4}}\,\frac{1}{2}\right] = -t\,f(t) \Rightarrow \quad L^{-1}\left[\frac{2}{s^2+2}\right] = t\,f(t)$

$\Rightarrow \quad \sin 2t = t\,f(t) \quad \Rightarrow \quad f(t) = \frac{1}{t}\sin 2t$ **Ans.**

Example 22. *Obtain the Inverse Laplace Transform of* $cot^{-1}\left(\frac{s+3}{2}\right)$

(*U. P., II Semester, Summer* 2002)

Solution. We know that $L^{-1}\left[F(s)\right] = -\frac{1}{t}L^{-1}\left[\frac{d}{ds}F(s)\right]$

$$\therefore \quad L^{-1}\left[\cot^{-1}\left(\frac{s+3}{2}\right)\right] = -\frac{1}{t}L^{-1}\left[\frac{d}{ds}\cot^{-1}\left(\frac{s+3}{2}\right)\right]$$

$$= -\frac{1}{t}L^{-1}\left\{\frac{-\frac{1}{2}}{1+\left(\frac{s+3}{2}\right)^2}\right\} = \frac{1}{2t}L^{-1}\left\{\frac{4}{4+(s+3)^2}\right\}$$

$$= \frac{1}{t}L^{-1}\left\{\frac{2}{2^2+(s+3)^2}\right\} = \frac{1}{t}e^{-3t}\,L^{-1}\left(\frac{2}{2^2+s^2}\right)$$

$$= \frac{e^{-3t}}{t}\sin 2t$$ **Ans.**

Example 23. *Find the inverse Laplace transform of* $\frac{2as}{(s^2+a^2)^2}$

Solution. $\quad L^{-1}\left(\frac{a}{s^2+a^2}\right) = \sin at$

$$L^{-1}\left[\frac{d}{ds}\left\{\frac{a}{s^2+a^2}\right\}\right] = -t\sin at \Rightarrow L^{-1}\left\{\frac{-2as}{(s^2+a^2)^2}\right\} = -t\sin at$$

$\Rightarrow \quad L^{-1}\left\{\frac{2as}{(s^2+a^2)^2}\right\} = t\sin at$ **Ans.**

Example 24. *Find the inverse Laplace transform of* $\dfrac{s^2-a^2}{(s^2+a^2)^2}$

Solution. We know that

$$L^{-1}\left\{\frac{s}{s^2+a^2}\right\} = \cos at$$

$$\therefore \quad L^{-1}\left[\frac{d}{ds}\left\{\frac{s}{s^2+a^2}\right\}\right] = -t\cos at$$

$$\Rightarrow \quad L^{-1}\left\{\frac{(s^2+a^2).1-s(2s)}{(s^2+a^2)^2}\right\} = -t\cos at \quad \Rightarrow \quad L^{-1}\left[\frac{a^2-s^2}{(s^2+a^2)^2}\right] = -t\cos at$$

$$\therefore \quad L^{-1}\left[\frac{s^2-a^2}{(s^2+a^2)^2}\right] = t\cos at$$ **Ans.**

EXERCISE 47.6

Obtain Inverse Laplace Transform of the following:

1. $\log\left(1+\dfrac{\omega^2}{s^2}\right)$ **Ans.** $-\dfrac{2}{t}\cos\omega t+2$

2. $\log\left(1+\dfrac{1}{s^2}\right)$ **Ans.** $\dfrac{2}{t}[1-\cos\omega t]$

3. $\dfrac{s}{1+s^2+s^4}$ **Ans.** $\dfrac{2}{\sqrt{3}}\sin\dfrac{\sqrt{3}}{2}t\sinh\dfrac{t}{2}$

4. $\dfrac{s}{(s^2+a^2)^2}$ **Ans.** $\dfrac{t\sin at}{2a}$

5. S $\log\dfrac{s}{\sqrt{s^2+1}}+\cot^{-1}s$ **Ans.** $\dfrac{1-\cos t}{t^2}$

6. $\dfrac{1}{2}\log\left\{\dfrac{s^2+b^2}{(s-a)^2}\right\}$ **Ans.** $\dfrac{e^{-at}-\cos bt}{t}$

47.8 INVERSE LAPLACE TRANSFORM OF INTEGRALS

$$L^{-1}\left[\int_s^\infty F(s)\,ds\right] = \frac{f(t)}{t} = \frac{1}{t}L^{-1}[F(s)] \quad \text{or} \quad \boxed{L^{-1}[F(s)] = t\,L^{-1}\left[\int_s^\infty F(s)\,ds\right]}$$

Example 25. *Obtain* $L^{-1}\dfrac{2s}{(s^2+1)^2}$

Solution. $L^{-1}\dfrac{2s}{(s^2+1)^2} = t\,L^{-1}\displaystyle\int_s^\infty \frac{2s\,ds}{(s^2+1)^2} = t\,L^{-1}\left[-\frac{1}{s^2+1}\right]_s^\infty = t\,L^{-1}\left[-0+\frac{1}{s^2+1}\right]$

$$= t\sin t$$ **Ans.**

47.9 PARTIAL FRACTIONS METHOD

Example 26. *Find the Inverse Laplace Transform of* $\dfrac{1}{s^2-5s+6}$.

Solution. Let us convert the given function into partial fractions.

$$L^{-1}\left[\frac{1}{s^2-5s+6}\right] = L^{-1}\left[\frac{1}{s-3}-\frac{1}{s-2}\right]$$

$$= L^{-1}\left(\frac{1}{s-3}\right) - L^{-1}\left(\frac{1}{s-2}\right) = e^{3t}-e^{2t}$$ **Ans.**

Example 27. *Find the Inverse Laplace Transform of* $\dfrac{s^3}{s^4-a^4}$ *(Q.Bank U.P., 2001)*

Solution. $L^{-1}\left(\dfrac{s^3}{s^4-a^4}\right)=L^{-1}\left[s\left\{\dfrac{s^2}{(s^2-a^2)(s^2+a^2)}\right\}\right]=L^{-1}\left[\dfrac{s}{2}\left(\dfrac{1}{s^2-a^2}+\dfrac{1}{s^2+a^2}\right)\right]$

$$=\frac{1}{2}L^{-1}\left[\frac{s}{s^2-a^2}+\frac{s}{s^2+a^2}\right] \quad \text{(By partial fractions)}$$

$$\frac{1}{2}(\cosh at+\cos at)$$ **Ans.**

Example 28. *Find the inverse Laplace transform of*

$$\frac{s+4}{s(s-1)(s^2+4)}$$ *(Q. Bank U.P. 2001)*

Solution. Let us first resolve $\dfrac{s+4}{s(s-1)(s^2+4)}$

$$\frac{s+4}{s(s-1)(s^2+4)}\equiv\frac{A}{s}+\frac{B}{s-1}+\frac{Cs+D}{s^2+4}$$

$$=\frac{s+4}{s(s-1)(s^2+4)}\equiv\frac{A}{s}+\frac{B}{s-1}+\frac{Cs+D}{s^2+4}$$

$$s+4\equiv A(s-1)(s^2+4)+Bs(s^2+4)+(Cs+D)\,s(s-1)$$

Putting $s = 0$, we get $4 = -4A$ $\Rightarrow A=-1$

Putting $s = 1$, we get $5 = B.1.\ (1 + 4)$ $\Rightarrow B=1$

Equating the coefficients of s^3 on both sides of (1), we get

$$1 = 4A + 4B - D \Rightarrow 1 = -4 + 4 - D \Rightarrow D = -1.$$

On putting the values of A, B, C, D in (1), we get

$$\frac{s+4}{s(s-1)(s^2+4)}=-\frac{1}{s}+\frac{1}{s-1}-\frac{1}{s^2+4}$$

$$L^{-1}\left[\frac{s+4}{s(s-1)(s^2+4)}\right]=L^{-1}\left[-\frac{1}{s}+\frac{1}{s-1}-\frac{1}{s^2+4}\right]$$

$$-L^{-1}\left(\frac{1}{s}\right)+L^{-1}\left(\frac{1}{s-1}\right)-\frac{1}{2}L^{-1}\left(\frac{2}{s^2+2^2}\right)=-1+e^t-\frac{1}{2}\sin 2t.$$ **Ans.**

Example 29. *Find the inverse Laplace transform of* $\dfrac{1}{s^4+4}$

[U.P., II Semester, (SUM) 2007]

Solution. Here, we have

$$s^4 + 4 = (s^2 + 2)^2 - (2s)^2 = (s^2 - 2s + 2)\,(s^2 + 2s + 2)$$

$$\frac{1}{s^4+4}=\frac{1}{(s^2-2s+2)\,(s^2+2s+2)}$$

$$=\frac{1}{4s}\left[\frac{1}{s^2-2s+2}-\frac{1}{s^2+2s+2}\right] \quad \left(\text{By partial fractions}\right) \quad ...(1)$$

Now, $$L^{-1}\left(\frac{1}{s^2-2s+2}\right)=L^{-1}\left[\frac{1}{(s-1)^2+1}\right]=e^t\sin t$$

and $$L^{-1}\left(\frac{1}{s^2+2s+2}\right)=L^{-1}\left[\frac{1}{(s+1)^2+1}\right]=e^{-t}\sin t$$

$\therefore$ $$\frac{1}{4}L^{-1}\left(\frac{1}{s^2-2s+2}-\frac{1}{s^2+2s+2}\right)=\frac{1}{4}(e^t-e^{-t})\sin t$$

Hence, $$L^{-1}\left[\frac{1}{4s}\left(\frac{1}{s^2-2s+2}-\frac{1}{s^2+2s+2}\right)\right]=\frac{1}{4}\int_0^t(e^t-e^{-t})\sin t\,dt$$

$$\Rightarrow\quad L^{-1}\left(\frac{1}{s^2+4}\right)=\frac{1}{4}\left[\frac{e^t}{2}(\sin t-\cos t)-\frac{e^{-t}}{2}(-\sin t-\cos t)\right]$$

$$=\frac{1}{4}\left[\sin t\left(\frac{e^t+e^{-t}}{2}\right)-\cos t\left(\frac{e^t-e^{-t}}{2}\right)\right]$$

$$\Rightarrow\quad L^{-1}\left(\frac{1}{s^2+4}\right)=\frac{1}{4}[\sin t\cosh t-\cos t\sinh t]$$ **Ans.**

Example 30. *Find the inverse Laplace transform of* $\frac{s}{s^4+4a^4}$

Solution. $$s^4+4a^4=(s^2+2a^2)^2-(2as)^2=(s^2-2as+2a^2)(s^2+2as+2a^2)$$
$$=\{(s-a)^2+a^2\}\{(s+a)^2+a^2\}$$

$$\frac{s}{s^4+4a^4}=\frac{s}{\{(s-a)^2+a^2\}\{(s+a)^2+a^2\}}$$

$$=\frac{1}{4a}\left[\frac{1}{(s-a)^2+a^2}-\frac{1}{(s+a)^2+a^2}\right]$$ (By partial fraction)

$\therefore$ $$L^{-1}\left(\frac{s}{s^4+4a^4}\right)=\frac{1}{4a}\left[L^{-1}\left\{\frac{1}{(s-a)^2+a^2}\right\}-L^{-1}\left\{\frac{1}{(s+a)^2+a^2}\right\}\right]$$

$$=\frac{1}{4a}\left[\frac{1}{a}e^{at}\sin at-e^{-at}\frac{1}{a}\sin at\right]$$

$$=\frac{1}{2a^2}\sin at\left(\frac{e^{at}-e^{-at}}{2}\right)=\frac{1}{2a^2}\sin at\sinh at.$$ **Ans.**

Example 31. *Find the Inverse Laplace Transform of* $\frac{e^{-cs}}{s^2(s+a)}$*, c > 0.*

Solution. We have,

$$L^{-1}\left[\frac{e^{-cs}}{s^2(s+a)}\right]=L^{-1}\left[-\frac{e^{-cs}}{a^2s}+\frac{e^{-cs}}{as^2}+\frac{e^{-cs}}{a^2(s+a)}\right]$$ (By Partial fractions)

$$=L^{-1}\left[\left(\frac{-1}{a^2}\frac{e^{-cs}}{s}\right)+\left(\frac{1}{a}\right)\frac{e^{-cs}}{s^2}+\left(\frac{1}{a^2}\right)\frac{e^{-c(s+a)}}{e^{-ca}(s+a)}\right]$$

$$=-\frac{1}{a^2}u(t-c)+\frac{1}{a}(t-c)u(t-c)+\frac{1}{a^2e^{-ca}}e^{at}u(t-c)$$

$$= u(t-c)\left[\frac{-1}{a^2}+\frac{1}{a}(t-c)+\frac{1}{a^2}e^{a(c+t)}\right]$$, where $u(t-c)$ is unit step function. **Ans.**

Example 32. *Find the Inverse Laplace Transform of* $\dfrac{5s+3}{(s-1)(s^2+2s+5)}$

(*U.P. II Semester, Summer* 2005)

Solution. $L^{-1}\left\{\dfrac{5s+3}{(s-1)(s^2+2s+5)}\right\}$

Let $$\frac{5s+3}{(s-1)(s^2+2s+5)}=\frac{A}{s-1}+\frac{Bs+C}{S^2+2s+5}$$

$$5s+3 = A(s^2+2s+5)+(Bs+C)(s-1)$$

$$5s+3 = s^2(A+B)+s(2A-B+C)+(5A-C)$$

Comparing the coefficients of s^2, s and constant, we get

$$A+B=0 \quad \text{...(1)}$$

$$2A-B+C=5 \quad \text{...(2)}$$

$$5A-C=3 \quad \text{...(3)}$$

On adding equations (1) and (2), we have $3A+C=5$ ...(4)

Adding equations (3) and (4), we get $8A=8 \Rightarrow A=1$

Putting $A=1$, $C=2$ in (2), we get $B=1$

Thus $$\frac{5s+3}{(s-1)(s^2+2s+5)}=\frac{1}{s-1}+\frac{-s+2}{s^2+2s+5}=\frac{1}{s-1}\ \frac{s-2}{(s+1)^2+2^2}$$

$$=\frac{1}{s-1}-\frac{s+1}{(s+1)^2+2^2}+\frac{3}{(s+1)^2+2^2}$$

$$L^{-1}\left\{\frac{5s+3}{(s-1)(s^2+2s+5)}\right\}=L^{-1}\left\{\frac{1}{s-1}\right\}+L^{-1}\left\{\frac{3}{(S+1)^2+2^2}\right\}-L^{-1}\left\{\frac{s+1}{(s+1)^2-2^2}\right\}$$

$$=e^t+3e^{-t}L^{-1}\left\{\frac{1}{s^2+2^2}\right\}-e^{-t}L^{-1}\left\{\frac{s}{s^2+2^2}\right\}$$

$$=e^t+3e^{-t}.\frac{1}{2}\sin 2t-e^{-t}\cos 2t$$

Example 33. *Find the Inverse Laplace Transform of* $\dfrac{s^2}{(s^2+a^2)(s^2+b^2)}$.

Solution. Let us convert the given function into partial fractions.

$$L^{-1}\left[\frac{s^2}{(s^2+a^2)(s^2+b^2)}\right]=L^{-1}\left[\frac{a^2}{a^2-b^2}.\frac{1}{s^2+a^2}-\frac{b^2}{a^2-b^2}.\frac{1}{s^2+b^2}\right]$$

$$=\frac{1}{a^2-b^2}L^{-1}\left[\frac{a^2}{s^2+a^2}-\frac{b^2}{s^2+b^2}\right]=\frac{1}{a^2-b^2}\left[a^2\left(\frac{1}{a}\sin at\right)-b^2\left(\frac{1}{b}\sin bt\right)\right]$$

$$=\frac{1}{a^2-b^2}[\,a\sin at-b\sin bt\,]$$ **Ans.**

Note: This question is also solved by using the Convolution Theorem as an example 37.

EXERCISE 47.7

Find the Inverse Laplace Transforms of the follwoing by partial fractions method:

1. $\dfrac{1}{s^2-7s+12}$ **Ans.** $e^{4t}-e^{3t}$

2. $\dfrac{s+2}{s^2-4s+13}$ **Ans.** $e^{2t}\cos 3t+\dfrac{4}{3}e^{2t}\sin 3t$

3. $\dfrac{3s+1}{(s-1)(s^2+1)}$ **Ans.** $e^t-2\cos t+\sin t$

4. $\dfrac{11s^2-2s+5}{2s^3-3s^2-3s+2}$ **Ans.** $2e^{-t}+5e^{2t}-\dfrac{3}{2}e^{t/2}$

5. $\dfrac{2s^2-6s+5}{(s-1)(s-2)(s-3)}$ **Ans.** $\dfrac{1}{2}e^t-e^{2t}+\dfrac{5}{2}e^{3t}$

6. $\dfrac{s-4}{(s-4)^2+9}$ **Ans.** $e^{4t}\cos 3t$

7. $\dfrac{16}{(s^2+2s+5)^2}$ **Ans.** $e^{-t}(\sin 2t-2t\cos 2t)$

8. $\dfrac{1}{(s+1)(s^2+2s+2)}$ **Ans.** $e^{-t}(1-\cos t)$

9. $\dfrac{1}{(s-2)(s^2+1)}$ **Ans.** $\dfrac{1}{5}e^{2t}-\dfrac{1}{5}\cos t-\dfrac{2}{5}\sin t$

10. $\dfrac{s^2-6s+7}{(s^2-4s+5)^2}$ **Ans.** $te^{2t}\{\cos t-\sin t\}$

47.10 INVERSE LAPLACE TRANSFORM BY CONVOLUTION

$$L\left\{\int_0^t f_1(x)*f_2(t-x)\,dx\right\}=F_1(s)\,.\,F_2(s) \text{ or } \boxed{\int_0^t f_1(x)\,.\,f_2(t-x)\,dx=L^{-1}[F_1(s)\,.\,F_2(s)]}$$

Example 34. *Use convolution theorem to evaluate:*

$$L^{-1}\left\{\frac{s}{(s^2+4)^2}\right\}$$

Solution.
$$\frac{s}{(s^2+4)^2}=\frac{1}{s^2+4}\cdot\frac{s}{s^2+4}$$

Let
$$F_1(s)=\frac{1}{s^2+4} \text{ and } F_2(s)=\frac{s}{s^2+4}$$

and
$$L^{-1}[F_1(s)]=L^{-1}\left(\frac{1}{s^2+4}\right)=\frac{1}{2}\sin 2t$$

and
$$L^{-1}[F_2(s)]=L^{-1}\left(\frac{s}{s^2+4}\right)=\cos 2t$$

According to Convolution Theorem

$$L^{-1}[F_1(s)\,.\,F_2(s)]=\int_0^t f_1(x).f_2(t-x)dx=\int_0^t\frac{1}{2}\sin 2x\cos 2(t-x)dx$$

$$=\frac{1}{4}\int_0^t[\sin(2x+2t-2x)+\sin(2x-2t+2x)]\,dx=\frac{1}{4}\int_0^t[\sin 2t+\sin(4x-2t)]dx$$

$$=\frac{1}{4}\left[x\sin 2t-\frac{1}{4}\cos(4x-2t)\right]_0^t=\frac{1}{4}\left[t\sin 2t-\frac{1}{4}\cos(4t-2t)+\frac{1}{4}\cos(-2t)\right]$$

$$=\frac{1}{4}\left[t\sin 2t-\frac{1}{4}\cos 2t+\frac{1}{4}\cos 2t\right]=\frac{1}{4}t\sin 2t$$ **Ans.**

Example 35. *Use convolution theorem to find the inverse of the function* $\dfrac{1}{(s^2+a^2)^2}$.

Solution. We know that *(U.P., II Semester, 2009)*

$$L^{-1}\left[\frac{1}{s^2+a^2}\right] = \frac{1}{a}\sin at$$

Hence by convolution theorem

$$L^{-1}\frac{1}{(s^2+a^2)}\frac{1}{(s^2+a^2)} = \int_0^t \frac{1}{a}\sin ax \frac{1}{a}\sin a(t-x)\,dx$$

$$= \frac{1}{a^2}\int_0^t \frac{1}{2}[\cos(ax-at+ax)-\cos(ax+at-ax)]\,dx \quad \left\{\sin A\sin B = \frac{1}{2}[\cos(A-B)-\cos(A+B)]\right\}$$

$$= \frac{1}{2a^2}\int_0^t [\cos(2ax-at)-\cos at]\,dx = \frac{1}{2a^2}\left[\frac{1}{2a}\sin(2ax-at) - x\cos at\right]_0^t$$

$$= \frac{1}{2a^2}\left[\frac{1}{2a}\sin(2at-at) - t\cos at - \frac{1}{2a}\sin(-at)\right] = \frac{1}{2a^2}\left[\frac{1}{2a}\sin at - t\cos at + \frac{1}{2a}\sin at\right]$$

$$= \frac{1}{2a^2}\left[\frac{2}{2a}\sin at - t\cos at\right] = \frac{1}{2a^3}[\sin at - at\cos at]$$ **Ans.**

Example 36. *State convolution theorem and hence find*

$$L^{-1}\left\{\frac{1}{(s+2)^2(s-2)}\right\}$$ *(Uttarakhand, II Semester, June 2007)*

Solution. Convolution Theorem (See Art 47.10 on page 1349).

Let $L\{f_1(t)\} = F_1(s)$ and Let $L\{f_2(t)\} = F_2(s)$

$$F_1(s) = \frac{1}{(s+2)^2} \quad \text{and} \quad F_2(s) = \frac{1}{s-2}$$

$$f_1(t) = L^{-1}\left[\frac{1}{(s+2)^2}\right] = t\,e^{-2t}$$

$$f_2(t) = L^{-1}\left[\frac{1}{(s-2)}\right] = e^{2t}$$

According to Convolution Theorem

$$L^{-1}[F_1(s).F_2(s)] = \int_0^t f_1(x) f_2(t-x)\,dx$$

$$L^{-1}\left[\frac{1}{(s+2)^2(s-2)}\right] = \int_0^t x\,e^{-2x}\cdot e^{2(t-x)}dx = \int_0^t x\,e^{2t-4x}\,dx$$

$$= \left[x\frac{e^{2t-4x}}{-4} - \int 1.\frac{e^{(2t-4x)}}{-4}dx\right]_0^t = \left[-\frac{x}{4}e^{2t-4x} + \frac{1}{4}\left\{\frac{e^{(2t-4x)}}{-4}\right\}\right]_0^t$$

$$= \frac{-t}{4}e^{2t-4t} - \frac{1}{16}e^{2t-4t} + \frac{1}{16}e^{2t} = \frac{-t}{4}e^{-2t} - \frac{1}{16}e^{-2t} + \frac{1}{16}e^{2t}$$

$$= \frac{e^{2t}}{16} - \frac{1}{16}e^{-2t}[4t+1]$$ **Ans.**

Example 37. *Using the Convolution Theorem find*

$$L^{-1}\left\{\frac{s^2}{(s^2+a^2)(s^2+b^2)}\right\},\ a \neq b.$$

(M.D.U., 2009, U.P. II Semester Summer 2006)

Solution. We have, $L(\cos at) = \dfrac{s}{s^2+a^2}$ and $L(\cos bt) = \dfrac{s}{s^2+b^2}$

Hence, by the convolution theorem

$$L\left\{\int_0^t \cos ax \cos b(t-x)\,dx\right\} = \frac{s^2}{(s^2+a^2)(s^2+b^2)}$$

Therefore,

$$L^{-1}\left\{\frac{s^2}{(s^2+a^2)(s^2+b^2)}\right\} = \int_0^t \cos ax \cos b(t-x)\,dx$$

$$= \frac{1}{2}\int_0^t \{\cos(ax+bt-bx) + \cos(ax-bt+bx)\}\,dx$$

$$= \frac{1}{2}\int_0^t \cos[(a-b)x+bt]\,dx + \frac{1}{2}\int_0^t \cos[(a+b)x-bt]\,dx$$

$$= \left[\frac{\sin[(a-b)x+bt]}{2(a-b)}\right]_0^t + \left[\frac{\sin[(a+b)x-bt]}{2(a+b)}\right]_0^t = \frac{\sin at - \sin bt}{2(a-b)} + \frac{\sin at + \sin bt}{2(a+b)}$$

$$= \frac{a\sin at - b\sin bt}{a^2-b^2}$$ **Ans.**

Example 38. *Evaluate* $L^{-1}\left\{\dfrac{s}{(s^2+1)(s^2+4)}\right\}$

Solution. We know that $L^{-1}\dfrac{s}{s^2+1} = \cos t$ and $L^{-1}\dfrac{2}{s^2+2^2} = \sin 2t$

$$L^{-1}\left(\frac{s}{(s^2+1)(s^2+4)}\right) = \frac{1}{2}L^{-1}\left[\left(\frac{s}{s^2+1}\right)\left(\frac{2}{s^2+4}\right)\right]$$

$$= \frac{1}{2}\int_0^t \sin 2x \cos(t-x)\,dx$$ [By Convolution Th.]

$$= \int_0^t \sin x\cos x\,\{\cos t\cos x + \sin t\sin x\}\,dx = \int_0^t \left[\sin x\cos^2 x\cos t + \sin^2 x\cos x\sin t\right]dx$$

$$= \left[-\frac{\cos^3 x}{3}\cos t + \frac{\sin^3 x}{3}\sin t\right]_0^t = -\frac{\cos^4 t}{3} + \frac{\sin^4 t}{3} + \frac{\cos t}{3} = \frac{1}{3}\left[\sin^4 t - \cos^4 t\right] + \frac{\cos t}{3}$$

$$= \frac{1}{3}(\sin^2 t + \cos^2 t)(\sin^2 t - \cos^2 t) + \frac{\cos t}{3} = \frac{1}{3}(\sin^2 t - \cos^2 t) + \frac{\cos t}{3} = -\frac{1}{3}\cos 2t + \frac{\cos t}{3}$$

$$= \frac{1}{3}(\cos t - \cos 2t)$$ **Ans.**

Example 39. *Obtain* $L^{-1}\dfrac{1}{s\,(s^2+a^2)}$.

Solution. $L^{-1}\dfrac{1}{s}=1$ and $L^{-1}\dfrac{1}{s^2+a^2}=\dfrac{\sin at}{a}$.

$$L^{-1}\{F_1(s)\,.\,F_2(s)\}=\int_0^t f_1(t)\,f_2(t-x)\,dx \qquad \text{(Convolution Theorem)}$$

Hence by the Convolution Theorem

$$L^{-1}\left[\frac{1}{s}\cdot\frac{1}{s^2+a^2}\right]=\int_0^t \frac{\sin a\,(t-x)}{a}\,dx=\left[\frac{-\cos(at-ax)}{-a^2}\right]_0^t$$

$$=\frac{1}{a^2}[\,1-\cos at\,] \qquad \textbf{Ans.}$$

Example 40. *Using Convolution Theorem, prove that*

$$L^{-1}\left[\frac{1}{s^3(s^2+1)}\right]=\frac{t^2}{2}+\cos t-1$$ (*U.P., II Semester, Summer* 2005)

Solution. We know that,

$$L^{-1}\left\{\frac{1}{s^3}\right\}=\frac{t^2}{2\,!}$$

$$L^{-1}\left\{\frac{1}{s^2+1}\right\}=\sin t$$

Using Convolution Theorem,

$$L^{-1}\left\{\frac{1}{s^3\,(s^2+1)}\right\}=\int_0^t \frac{(t-x)^2}{2\,!}\sin x\,dx$$

$$=\frac{1}{2}\int_0^t \left(t^2+x^2-2tx\right)\sin x\,dx=\frac{1}{2}\left[\left(t^2+x^2-2tx\right)(-\cos x)-\int(2x-2t)(-\cos x)\,dx\right]_0^t$$

$$=\frac{1}{2}\left[(t^2+x^2-2tx)(-\cos x)+2\int(x-t)\cos x\,dx\right]_0^t$$

$$=\frac{1}{2}\left[(t^2+x^2-2tx)(-\cos x)+2\,(x-t)\sin x+2\cos x\right]_0^t$$

$$=\frac{1}{2}\left[(t^2+t^2-2t^2)(-\cos t)+0+2\cos t+t^2\cos 0-2\cos 0\right]$$

$$=\frac{1}{2}[\,2\cos t+t^2-2\,]=\cos t+\frac{t^2}{2}-1=\frac{t^2}{2}+\cos t-1 \qquad \textbf{Ans.}$$

EXERCISE 47.8

Obtain the Inverse Laplace Transform of the following by convolution theorem:

1. $\dfrac{s^2}{(s^2+a^2)^2}$ **Ans.** $\dfrac{1}{2}t\cos at+\dfrac{1}{2a}\sin at$

2. $\dfrac{1}{(s^2+1)^3}$ **Ans.** $\dfrac{1}{8}\left[\left(3-t^2\right)\sin t-3t\cos t\right]$

3. $\dfrac{s}{(s^2+a^2)^2}$ **Ans.** $\dfrac{t\sin at}{2a}$

4. $\dfrac{1}{s^2(s^2-a^2)}$ **Ans.** $\dfrac{1}{a^3}[\,-at+\sinh at\,]$

5. $\dfrac{1}{(s+1)(s^2+1)}$ **Ans.** $\dfrac{1}{2}(\cos t-\sin t-e^{-t})$

47.11 HEAVISIDE INVERSE FORMULA OF $\frac{F(s)}{G(s)}$

If $F(s)$ and $G(s)$ be two polynomials in S. The degree of F(s) is less than that of $G(s)$. Let $\alpha_1, \alpha_2, \alpha_3, \ldots \alpha_n$ be n roots of the equation $G(s) = 0$

Inverse Laplace formula of $\frac{F(s)}{G(s)}$ is given by

$$L^{-1}\left\{\frac{F(s)}{G(s)}\right\} = \sum_{i=1}^{n} \frac{F(\alpha_i)}{G'(\alpha_i)} e^{\alpha_i t}$$

Example 41. *Find* $L^{-1}\left\{\frac{2s^2+5s-4}{s^3+s^2-2s}\right\}$.

Solution. Let $\quad F(s) = 2s^2 + 5s - 4$

and $\quad G(s) = s^3 + s^2 - 2s = s(s^2 + s - 2) = s(s+2)(s-1)$

$G'(s) = 3s^2 + 2s - 2$

$G(s) = 0$ has three roots, 0, 1, – 2

$\Rightarrow \quad \alpha_1 = 0, \quad \alpha_2 = 1, \quad \alpha_3 = -2$

By Heaviside Inverse formula

$$L^{-1}\left\{\frac{F(s)}{G(s)}\right\} = \sum_{i=1}^{n} \frac{F(\alpha_i)}{G'(\alpha_i)} e^{\alpha_i t}$$

$$= \left\{\frac{F(\alpha_1)}{G'(\alpha_1)}\right\} e^{t\alpha_1} + \frac{F(\alpha_2)}{G'(\alpha_2)} e^{t\alpha_2} + \frac{F(\alpha_3)}{G'(\alpha_3)} e^{t\alpha_3} = \frac{F(0)}{G'(0)} e^0 + \frac{F(1)}{G'(1)} e^t + \frac{F(-2)}{G'(-2)} e^{-2t}$$

$$= \frac{-4}{-2} e^0 + \frac{3}{3} e^t + \frac{(-6)}{(6)} e^{-2t} = 2 + e^t - e^{-2t}$$ **Ans.**

Example 42. *Find* $L^{-1}\left[\frac{2s^2-6s+5}{s^3-6s^2+11s-6}\right]$ *(U.P. II Semester,* 2004)

Solution. Let $\quad F(s) = 2s^2 - 6s + 5$

$G(s) = s^3 - 6s^2 + 11s - 6 = (s-1)(s-2)(s-3)$

$G(s) = 0$ has three roots, 1, 2, 3.

$\Rightarrow \quad \alpha_1 = 1, \quad \alpha_2 = 2, \quad \alpha_3 = 3$

$G'(s) = 3s^2 - 12s + 11$

By Heaviside Inverse formula, we have $L^{-1}\left\{\frac{F(s)}{G(s)}\right\} = \sum_{i=1}^{n} \frac{F(\alpha_i)}{G'(\alpha_i)} e^{t\alpha_i}$

$$L^{-1}\left\{\frac{2s^2-6s+5}{s^3-6s^2+11s-6}\right\} = \frac{F(\alpha_1)}{G'(\alpha_1)} e^{t\alpha_1} + \frac{F(\alpha_2)}{G'(\alpha_2)} e^{t\alpha_2} + \frac{F(\alpha_3)}{G'(\alpha_3)} e^{t\alpha_3}$$

$$= \frac{F(1)}{G'(1)} e^t + \frac{F(2)}{G'(2)} e^{2t} + \frac{F(3)}{G'(3)} e^{3t} = \frac{(1)}{(2)} e^t + \frac{(1)}{(-1)} e^{2t} + \frac{(5)}{(2)} e^{3t} = \frac{1}{2} e^t - e^{2t} + \frac{5}{2} e^{3t}$$ **Ans.**

EXERCISE 47.9

Using Heaviside expansion formula, find the Inverse Laplace Transform of the following :

1. $\dfrac{s-1}{s^2+3s+2}$ **Ans.** $-2e^{-t}+3e^{-2t}$ **2.** $\dfrac{s}{(s-1)(s-2)(s-3)}$ **Ans.** $\dfrac{1}{2}e^{t}-2e^{2t}+\dfrac{3}{2}e^{3t}$

3. $\dfrac{2s+3}{(s-2)(s-3)(s-4)}$ **Ans.** $\dfrac{7}{2}e^{2t}-9e^{3t}+\dfrac{11}{2}e^{4t}$ **4.** $\dfrac{11s^2-2s+5}{2s^3-3s^2-3s+2}$ **Ans.** $2e^{-t}+5\cdot e^{2t}-\dfrac{3}{2}e^{\frac{t}{2}}$

47.12 SOLUTION OF DIFFERENTIAL EQUATIONS BY LAPLACE TRANSFORMS

Ordinary linear differential equations with constant coefficients can be easily solved by the Laplace Transform method, without finding the general solution and the arbitrary constants. The method will be clear from the following examples:

Example 43. *Solve the following equation by Laplace transform*

$$y''' - 2y'' + 5y' = 0\ ;\ y = 0,\quad y' = 1 \text{ at } t = 0 \text{ and } y = 1 \text{ at } t = \frac{\pi}{8}.$$

Solution. Here, we have

$$y''' - 2y'' + 5y' = 0 \qquad ...(1)$$

Taking Laplace transform on both sides of (1), we get

$$L(y''') - 2L(y'') + 5L(y') = L\,(0)$$

$$\Rightarrow s^3\bar{y} - s^2 y\,(0) - sy'\,(0) - y''\,(0) - 2[s^2\bar{y} - sy\,(0) - y'(0)] + 5[s\bar{y} - y(0)] = 0$$

$$\Rightarrow (s^3 - 2s^2 + 5s)\,\bar{y} - s - k + 2 = 0 \qquad [\text{Let } y''\,(0) = k \text{ (say)}]$$

$$\Rightarrow \bar{y} = \frac{(k-2)+s}{s\,(s^2-2s+5)}$$

$$= \left(\frac{k-2}{5}\right)\left(\frac{1}{s} - \frac{s-2}{s^2-2s+5}\right) + \frac{1}{s^2-2s+5}$$

$$= \left(\frac{k-2}{5}\right)\frac{1}{s} - \left(\frac{k-2}{5}\right)\left\{\frac{(s-1)-1}{(s-1)^2+4}\right\} + \frac{1}{(s-1)^2+4}$$

$$\Rightarrow \bar{y} = \left(\frac{k-2}{5}\right)\frac{1}{s} - \left(\frac{k-2}{5}\right)\left\{\frac{(s-1)}{(s-1)^2+4}\right\} + \left(\frac{k+3}{10}\right).\left\{\frac{2}{(s-1)^2+4}\right\} \qquad ...(3)$$

Taking Inverse Laplace transform on both sides of (3), we get

$$y = \left(\frac{k-2}{5}\right) - \left(\frac{k-2}{5}\right)e^t \cos 2t + \left(\frac{k+3}{10}\right)e^t \sin 2t \qquad ...(4)$$

Puttting $y\left(\frac{\pi}{8}\right) = 1$, we get $1 = \left(\frac{k-2}{5}\right) - \left(\frac{k-2}{5}\right)e^{\pi/8}.\frac{1}{\sqrt{2}} + \left(\frac{k+3}{10}\right)e^{\pi/8}.\frac{1}{\sqrt{2}}$ ...(5)

$$\Rightarrow \qquad k = 7$$

On putting the value of k in (4), we get (on simplification)

Hence required solution is

$$y = 1 + e^t(\sin 2t - \cos 2t)$$ **Ans.**

Example 44. *Using Laplace transforms, find the solution of the initial value problem*

$$y'' - 4y' + 4y = 64 \sin 2t$$
$$y(0) = 0, \; y'(0) = 1.$$

Solution. Here, we have $y'' - 4y' + 4y = 64 \sin 2t$ ... (1)

$$y(0) = 0, \; y'(0) = 1.$$

Taking Laplace transform of both sides of (1), we have

$$[s^2 \bar{y} - sy(0) - y'(0)] - 4[s\bar{y} - y(0)] + 4\bar{y} = \frac{64 \times 2}{s^2 + 4} \quad \text{... (2)}$$

On putting the values of $y(0)$ and $y'(0)$ in (2), we get

$$s^2 \bar{y} - 1 - 4s\bar{y} + 4\bar{y} = \frac{128}{s^2 + 4}$$

$$\Rightarrow \quad (s^2 - 4s + 4)\bar{y} = 1 + \frac{128}{s^2 + 4}, \quad \Rightarrow \quad (s-2)^2 \bar{y} = 1 + \frac{128}{s^2 + 4}$$

$$\Rightarrow \quad \bar{y} = \frac{1}{(s-2)^2} + \frac{128}{(s-2)^2 (s^2 + 4)} = \frac{1}{(s-2)^2} - \frac{8}{s-2} + \frac{16}{(s-2)^2} + \frac{8s}{s^2 + 4}$$

$$y = L^{-1}\left[-\frac{8}{s-2} + \frac{17}{(s-2)^2} + \frac{8s}{s^2 + 4}\right]$$

$$y = -8e^{2t} + 17t\,e^{2t} + 8 \cos 2t$$ **Ans.**

Example 45. *Using Laplace transforms, find the solution of the initial value problem*

$$y'' + 9y = 6 \cos 3t$$
$$y(0) = 2, \; y'(0) = 0$$

(*U. P. II Semester Summer* 2006)

Solution. We have, $y'' + 9y = 6 \cos 3t$ ... (1)

$$y(0) = 2, \; y'(0) = 0$$

Taking Laplace transform of (1), we get

$$[s^2 \bar{y} - sy(0) - y'(0)] + 9\bar{y} = 6\frac{s}{s^2 + 9} \quad \text{... (2)}$$

Putting the values of $y(0)$ and $y'(0)$ in (2), we have

$$s^2 \bar{y} - 2s + 9\bar{y} = \frac{6s}{s^2 + 9}$$

$$\Rightarrow \quad (s^2 + 9)\bar{y} = 2s + \frac{6s}{s^2 + 9} \quad \Rightarrow \quad \bar{y} = \frac{2s}{s^2 + 9} + \frac{6s}{(s^2 + 9)^2}$$

$$y = L^{-1}\frac{2s}{s^2 + 9} + L^{-1}\frac{6s}{(s^2 + 9)^2} = 2 \cos 3t + L^{-1}\frac{d}{ds}\left[\frac{-3}{(s^2 + 9)}\right]$$

$$= 2 \cos 3t - t \sin 3t$$ **Ans.**

Example 46. *Using Laplace transformation solve the following differential equation:*

$$\frac{d^2 x}{dt^2} + 9x = \cos 2t, \quad \text{if } x(0) = 1, \quad x\left(\frac{\pi}{2}\right) = -1$$

Solution. $\dfrac{d^2 x}{dt^2} + 9x = \cos 2t$ ... (1)

Taking Laplace transform of both the sides of (1), we get

$$L\frac{d^2 x}{dt^2} + 9\,L\,x = L\cos 2t$$

$$\Rightarrow \quad s^2\,\bar{x} - sx(0) - x'(0) + 9\,\bar{x} = \frac{s}{s^2+4} \quad \text{... (2)}$$

On putting $x(0) = 1$ in (2), we get

$$s^2\,\bar{x} - s + 9\,\bar{x} - x'(0) = \frac{s}{s^2+4}$$

$$(s^2+9)\,\bar{x} = s + \frac{s}{s^2+4} + x'(0) = \frac{s(s^2+4)+s}{s^2+4} + x'(0) = \frac{s^3+5s}{s^2+4} + x'(0)$$

$$\Rightarrow \quad \bar{x} = \frac{(s^3+5s)}{(s^2+4)(s^2+9)} + \frac{x'(0)}{s^2+9} = \frac{1}{5}\frac{s}{s^2+4} + \frac{4}{5}\frac{s}{s^2+9} + \frac{x'(0)}{s^2+9}$$

Taking the Inverse Laplace Transform, we get

$$x(t) = \frac{1}{5}L^{-1}\frac{s}{s^2+4} + \frac{4}{5}L^{-1}\frac{s}{s^2+9} + L^{-1}\frac{x'(0)}{s^2+9}$$

$$x(t) = \frac{1}{5}\cos 2t + \frac{4}{5}\cos 3t + \frac{x'(0)\sin 3t}{3} \quad \text{... (3)}$$

On putting $x\left(\frac{\pi}{2}\right) = -1$ in (3), we get

$$-1 = -\frac{1}{5} + 0 - \frac{x'(0)}{3} \quad \Rightarrow \quad x'(0) = \frac{12}{5}$$

On putting the value of $x'(0)$ in (3), we get

$$x = \frac{1}{5}\cos 2t + \frac{4}{5}\cos 3t + \frac{12}{5}\frac{\sin 3t}{3} = \frac{1}{5}[\cos 2t + 4\cos 3t + 4\sin 3t] \quad \textbf{Ans.}$$

Example 47. *Solve, using Laplace transform method*

$$y(0) = -2,\ y'(0) = 8;\quad y'' + 4y' + 4y = 6e^{-t}$$ *(U.P., II Semester, 2007)*

Solution. Here, we have

$$y'' + 4y' + 4y = 6e^{-t} \quad \text{...(1)}$$

Taking Laplace transform on both sides of (1), we get

$$L(y'') + 4L(y') + 4L(y) = 6L(e^{-t})$$

$$\Rightarrow [s^2\bar{y} - sy(0) - y'(0)] + 4[s\,\bar{y} - y(0)] + 4\bar{y} = \frac{6}{s+1} \quad \text{... (2)} \quad [\text{Here } \bar{y} = L(y)]$$

Putting the values of $y(0) = -2$ and $y'(0) = 8$ in (2), we get

$$(s^2 + 4s + 4)\ \bar{y} + 2s - 8 + 8 = \frac{6}{s+1}$$

$$\Rightarrow \quad (s^2 + 4s + 4)\ \bar{y} + 2s = \frac{6}{s+1}$$

$$(s+2)^2\ \bar{y} + 2s = \frac{6}{s+1}$$

$$\bar{y} = \frac{6}{(s+1)(s+2)^2} - \frac{2s}{(s+2)^2} = 6\left[\frac{1}{s+1} - \frac{1}{s+2} - \frac{1}{(s+2)^2}\right] - \frac{2\{(s+2)-2\}}{(s+2)^2}$$

$$= \frac{6}{s+1} - \frac{6}{s+2} - \frac{6}{(s+2)^2} - \frac{2}{s+2} + \frac{4}{(s+2)^2} = \frac{6}{s+1} - \frac{8}{s+2} - \frac{2}{(s+2)^2} \quad ...(3)$$

Taking inverse Laplace transform on both sides of (3), we get

$$y = 6e^{-t} - 8e^{-2t} - 2t\, e^{-2t}$$ **Ans.**

Example 48. *Solve the following differential equation using Laplace transform*

$$\frac{d^3y}{dt^3} - 3\frac{d^2y}{dt^2} + 3\frac{dy}{dt} - y = t^2 e^t$$

where $y(0) = 1, \left(\frac{dy}{dt}\right)_{t=0} = 0, \left(\frac{d^2y}{dt^2}\right)_{t=0} = -2$ *(U. P., II Semester, (SUM) 2008)*

Solution. Here we have equation

$$y''' - 3y'' + 3y' - y = t^2 e^t \quad ...(1)$$

Taking Laplace transform on both side of equation (1), we get

$$L(y''') - 3L(y'') + 3L(y') - L(y) = L(t^2 e^t)$$

$$\Rightarrow [s^3\bar{y} - s^2y(0) - sy'(0) - y''(0)] - 3[s^2\bar{y} - sy(0) - y'(0)] + 3[s\bar{y} - y(0)] - \bar{y} = \frac{2}{(s-1)^3}$$

$$...(2) \quad [\text{where } \bar{y} = L(y)]$$

Putting the values of $y(0)$, $y'(0)$ and y'' at $x = 0$ in (2), we get

$$\Rightarrow (s^3\bar{y} - s^2 + 2) - 3(s^2\bar{y} - s) + 3(s\bar{y} - 1) - \bar{y} = \frac{2}{(s-1)^3}$$

$$\Rightarrow \quad (s^3 - 3s^2 + 3s - 1)\,\bar{y} - s^2 + 3s - 1 = \frac{2}{(s-1)^3}$$

$$\Rightarrow \quad (s-1)^3\,\bar{y} = s^2 - 3s + 1 + \frac{2}{(s-1)^3}$$

$$\Rightarrow \quad \bar{y} = \frac{(s-1)^2}{(s-1)^3} - \frac{s}{(s-1)^3} + \frac{2}{(s-1)^6} = \frac{1}{s-1} - \frac{(s-1)+1}{(s-1)^3} + \frac{2}{(s-1)^6}$$

$$= \frac{1}{s-1} - \frac{1}{(s-1)^2} - \frac{1}{(s-1)^3} + \frac{2}{(s-1)^6} \quad ...(3)$$

Taking inverse Laplace transform on both sides of equation (3), we get

$$y = e^t - te^t - \frac{t^2}{2}e^t + \frac{t^5 e^t}{60} = \left(1 - t - \frac{t^2}{2} + \frac{t^5}{60}\right)e^t$$ **Ans.**

Example 49. *Solve* $\frac{d^2y}{dx^2} + 2\frac{dy}{dx} + 5y = e^{-x}\sin x$ *where* $y(0) = 0$, $y'(0) = 1$.

Solution. Here, we have $\frac{d^2y}{dx^2} + 2\frac{dy}{dx} + 5y = e^{-x}\sin x$

Taking the Laplace Transform of both the sides, we get

$$[s^2\bar{y} - sy(0) - \bar{y}(0)] + 2[s\bar{y} - y(0)] + 5\bar{y} = L(e^{-x}\sin x)$$

$$[s^2\,\bar{y} - sy(0) - y'(0)] + 2[s\,\bar{y} - y(0)] + 5\,\bar{y} = \frac{1}{(s+1)^2+1} \qquad \ldots (1)$$

On substituting the values of y (0) and y' (0) in (1), we get

$$(s^2\,\bar{y} - 1) + 2(s\,\bar{y}) + 5\bar{y} = \frac{1}{s^2+2s+2}$$

$$(s^2+2s+5)\,\bar{y} = 1 + \frac{1}{s^2+2s+2} = \frac{s^2+2s+3}{s^2+2s+2}$$

$$\bar{y} = \frac{s^2+2s+3}{(s^2+2s+5)\,(s^2+2s+2)}$$

On resolving the R.H.S. into partial fractions, we get

$$\bar{y} = \frac{2}{3}\,\frac{1}{s^2+2s+5} + \frac{1}{3}\,\frac{1}{s^2+2s+2}$$

On inversion, we obtain

$$y = \frac{2}{3}\,L^{-1}\,\frac{1}{s^2+2s+5} + \frac{1}{3}\,L^{-1}\,\frac{1}{s^2+2s+2}$$

$$\Rightarrow \qquad y = \frac{1}{3}\,L^{-1}\,\frac{2}{(s+1)^2+(2)^2} + \frac{1}{3}\,L^{-1}\,\frac{1}{(s+1)^2+(1)^2}$$

$$\Rightarrow \qquad y = \frac{1}{3}\,e^{-x}\sin 2\,x + \frac{1}{3}\,e^{-x}\sin x \qquad \Rightarrow \qquad y = \frac{1}{3}\,e^{-x}\,(\sin x + \sin 2\,x) \qquad \textbf{Ans.}$$

Example 50. *Solve the equation by the transform method:*

$$\frac{dy}{dt} + 2y + \int_0^t y\,dt = \sin t\,, \qquad y(0) = 0, \quad y'(0) = 1$$

Solution. Taking Laplace transform of the given equation, we get

$$[s\,\bar{y} - y(0)] + 2\,\bar{y} + \frac{\bar{y}}{s} = \frac{1}{s^2+1} \qquad \ldots (1) \qquad \left[\because L\left\{\int_0^t y\,dt = \frac{\bar{y}}{s}\right\}\right]$$

On substituting the values of y (0) and y' (0) in (1), we get

$$[s\,\bar{y} - 1] + 2\,\bar{y} + \frac{\bar{y}}{s} = \frac{1}{s^2+1}$$

$$\bar{y}\left(s + 2 + \frac{1}{s}\right) = 1 + \frac{1}{s^2+1} \qquad [\because y(0) = 1]$$

$$\frac{1}{s}\,\bar{y}\,(s^2+2s+1) = \frac{s^2+1+1}{s^2+1}$$

$$\frac{1}{s}\,\bar{y}\,(s+1)^2 = \frac{s^2+2}{s^2+1} \qquad \Rightarrow \qquad \bar{y}\,(s+1)^2 = \frac{s^3+2s}{s^2+1}$$

$$\Rightarrow \qquad \bar{y} = \frac{s^3+2s}{(s+1)^2\,(s^2+1)} = \frac{1}{s+1} - \frac{3}{2\,(s+1)^2} + \frac{1}{2\,(s^2+1)} \qquad \text{[By partial fractions]}$$

Taking Inverse Laplace Transform, we have

$$y = e^{-t} - \frac{3}{2}\,t\,e^{-t} + \frac{1}{2}\sin t \qquad \textbf{Ans.}$$

Example 51. *Solve the following differential equation using Laplace transform:*

$$x\frac{d^2y}{dx^2}+\frac{dy}{dx}+xy=0, \qquad \textit{given } y(0)=2,\ y'(0)=0.$$

(*Uttarakhand, II Semester, June* 2007)

Solution. Taking Laplace transform on both sides of the given equation, we have

$$L[xy'']+L[y']+L[xy]=L(0)$$

$$\Rightarrow \quad -\frac{d}{ds}L(y'')+L(y')-\frac{d}{ds}L(y)=0$$

$$\Rightarrow \quad -\frac{d}{ds}\{s^2\bar{y}-sy(0)-y'(0)\}+\{s\bar{y}-y(0)\}-\frac{d\bar{y}}{ds}=0 \qquad \dots (1)$$

Putting $y(0)=2$ and $y'(0)=0$ in (1), we get

$$-\frac{d}{ds}\{s^2\bar{y}-2s-0\}+\{s\bar{y}-2\}-\frac{d\bar{y}}{ds}=0$$

$$\Rightarrow \quad -s^2\frac{d\bar{y}}{ds}-(2s)\bar{y}+2+s\bar{y}-2-\frac{d\bar{y}}{ds}=0$$

$$\Rightarrow \quad (s^2+1)\frac{d\bar{y}}{ds}+s\bar{y}=0 \qquad \dots (2)$$

Separating the variables, we have

$$\frac{d\bar{y}}{\bar{y}}+\frac{s\,ds}{s^2+1}=0 \qquad \dots (3)$$

On integrating, we have

$$\log\bar{y}+\frac{1}{2}\log(s^2+1)=\log C$$

$$\Rightarrow \quad \log\bar{y}=\log C-\log\sqrt{s^2+1} \quad \Rightarrow \quad \log\bar{y}=\log\frac{C}{\sqrt{s^2+1}}$$

$$\Rightarrow \quad \bar{y}=\frac{C}{\sqrt{s^2+1}} \qquad \dots (4)$$

Taking Inverse Laplace Transform, we get

$$y=L^{-1}\left[\frac{C}{\sqrt{s^2+1}}\right]$$

$$\Rightarrow \quad y=C\,J_0(x) \qquad \dots (5)$$ [See Art. 46.22 on page 1285]

At $x=0$ $\quad y(0)=C\,J_0(0) \qquad \dots (6)$

Putting $y(0)=2$ and $J_0(0)=1$ in (6), we get

$$2=C(1) \quad \Rightarrow \quad C=2$$

On putting the value of C in (5), we get

$$y=2\,J_0(x)$$ **Ans.**

Example 52. *Using Laplace transforms, find the solution of the initial value problem*

$$y''+9y=9\,u(t-3), \qquad y(0)=y'(0)=0$$

where $u(t-3)$ *is the unit step functions.*

Solution. We have, $\quad y''+9y=9\,u(t-3) \qquad \dots (1)$

Taking Laplace transform of (1), we have

$$s^2\,\bar{y} - sy(0) - y'(0) + 9\bar{y} = 9\,\frac{e^{-3s}}{s} \qquad \text{... (2)}$$

Putting the values of y (0) = 0 and y' (0) = 0 in (2), we get

$$s^2\,\bar{y} + 9\bar{y} = 9\,\frac{e^{-3s}}{s}$$

$$(s^2+9)\,\bar{y} = 9\,\frac{e^{-3s}}{s}$$

$$\bar{y} = \frac{9\,e^{-3s}}{s\,(s^2+9)} \quad \Rightarrow \quad y = L^{-1}\,\frac{9\,e^{-3s}}{s\,(s^2+9)} \qquad \text{... (3)}$$

We know that $\quad L^{-1}\,\dfrac{3}{s^2+9} = \sin 3t$

$$\Rightarrow \qquad 3\,L^{-1}\,\frac{3}{s\,(s^2+9)} = 3\int_0^t \sin 3t\,dt = -[\cos 3t]_0^t = 1 - \cos 3t \qquad \text{... (4)}$$

[Using second shifting theorem]

Using (4), we get the inverse of (3)

$$y = [1 - \cos 3\,(t-3)]\,u\,(t-3) \qquad \textbf{Ans.}$$

Example 53. *Solve, by the method of Laplace transform, the differential equation*

$(D^2 + n^2)\,x = a \sin (nt + \alpha),$

$x = Dx = 0$ *at* $t = 0.$ *(U.P., II Semester, Summer, 2010)*

Solution. Taking Laplace transform of the given differential equation, we get

$$[s^2\,\bar{x} - sx(0) - x'(0)] + n^2\,\bar{x} = a\,L \sin (nt + \alpha)$$

$$= a\,L\,[\sin n\,t \cos \alpha + \cos n\,t \sin \alpha] = a\left[\frac{n}{s^2+n^2}\cos\alpha + \frac{s}{s^2+n^2}\sin\alpha\right]$$

$$= \frac{a\,[n\cos\alpha + s\sin\alpha]}{s^2+n^2} \qquad \text{... (1)}$$

Putting x (0) = x' (0) = 0 in (1), we get

$$s^2\,\bar{x} + n^2\,\bar{x} = a\left[\frac{n\cos\alpha + s\sin\alpha}{s^2+n^2}\right] \quad \Rightarrow \quad (s^2+n^2)\,\bar{x} = \frac{a\,[\cos\alpha + s\sin\alpha]}{s^2+n^2}$$

$$\Rightarrow \qquad \bar{x} = \frac{a\,n\cos\alpha + as\sin\alpha}{(s^2+n^2)^2}$$

$$\Rightarrow\tau \qquad \bar{x} = a\cos\alpha\,.\,\frac{n}{(s^2+n^2)^2} + a\sin\alpha\,.\,\frac{s}{(s^2+n^2)^2} \qquad \text{... (2)}$$

Taking the Inverse Laplace Transform of (2), we get

$$x = a\cos\alpha\,L^{-1}\left[\frac{n}{(s^2+n^2)^2}\right] + (a\sin\alpha)\,L^{-1}\left[\frac{s}{(s^2+n^2)^2}\right] \qquad \text{... (3)}$$

Let us find out the inverse of the term on R.H.S.

But $\quad L^{-1}\left\{\dfrac{2s}{(s^2+n^2)^2}\right\} = L^{-1}\,\dfrac{d}{ds}\,\dfrac{1}{s^2+n^2} = \dfrac{t}{n}\sin nt \qquad \left[-\dfrac{d}{ds}[F(s)] = t\,f(t)\right]$

$$\Rightarrow \quad L^{-1}\left\{\frac{s}{(s^2+n^2)^2}\right\}=\frac{t}{2n}\sin nt \qquad \dots (4)$$

Again $L^{-1}\frac{1}{s}\left\{\frac{s}{(s^2+n^2)^2}\right\}=\frac{1}{2n}\int_0^t t\sin nt\,dt \qquad \left[\frac{F(s)}{s}=\int_0^t f(t)\,dt\right]$

$$L^{-1}\left[\frac{1}{(s^2+n^2)^2}\right]=\frac{1}{2n}\left[t\left(\frac{-\cos nt}{n}\right)+\frac{1}{n^2}\sin nt\right]_0^t \qquad \dots (5)$$

Putting the values of $L^{-1}\left[\frac{s}{(s^2+n^2)^2}\right]$ from (4) and $L^{-1}\left[\frac{n}{(s^2+n^2)^2}\right]$ from (5) in (3), we get

$$x=(a\cos\alpha)\frac{1}{2n^2}[\sin nt-nt\cos nt]+(a\sin\alpha)\frac{t}{2n}\sin nt$$

$$=\frac{a}{2n^2}[\cos\alpha\sin nt-nt\cos\alpha\cos nt+nt\sin\alpha\sin nt]$$

$$=\frac{a}{2n^2}[\cos\alpha\sin nt-nt(\cos nt\cos\alpha-\sin nt\sin\alpha)]=\frac{a}{2n^2}[\sin nt\cos\alpha-nt\cos(nt+\alpha)]$$

Which is the required solution. **Ans.**

Example 54. *Solve* $t\,y''+2\,y'+t\,y=\cos t$ *if* $y(0)=1,\; y'(0)=0$.

Solution. We have, $t\,y''+2\,y'+t\,y=\cos t$

$$\Rightarrow \quad L\{t\,y''\}+L\{2y'\}+L\{t\,y\}=L\{\cos t\}$$

$$\Rightarrow \quad -\frac{d}{ds}L\{y''\}+2L\{y'\}-\frac{d}{ds}L\{y\}=L\{\cos t\}$$

$$\Rightarrow \quad -\frac{d}{ds}[s^2\bar{y}-sy(0)-y'(0)]+2[s\bar{y}-y(0)]-\frac{d}{ds}\bar{y}=\frac{s}{s^2+1}$$

$$\Rightarrow \quad -\frac{d}{ds}[s^2\bar{y}-s]+2s\bar{y}-2-\frac{d}{ds}\bar{y}=\frac{s}{s^2+1}$$

$$\Rightarrow \quad -2s\bar{y}-s^2\frac{d\bar{y}}{ds}+1+2s\bar{y}-2-\frac{d\bar{y}}{ds}=\frac{s}{s^2+1}$$

$$\Rightarrow \quad (s^2+1)\frac{d\bar{y}}{ds}+1=\frac{-s}{s^2+1} \quad \Rightarrow \quad \frac{d\bar{y}}{ds}=\frac{-s}{(s^2+1)^2}-\frac{1}{s^2+1}$$

Taking Inverse Laplace Transform, we get

$$L^{-1}\frac{d}{ds}(\bar{y})=L^{-1}\left[\frac{-s}{(s^2+1)^2}-\frac{1}{s^2+1}\right] \qquad \left[L^{-1}\frac{d}{ds}F(s)=(-1)^1t^1f(t)\right]$$

$$\Rightarrow \quad (-1)^1t^1\,y=-\frac{1}{2}L^{-1}\left\{\frac{2s}{(s^2+1)^2}\right\}-L^{-1}\left\{\frac{1}{s^2+1}\right\}=-\frac{1}{2}t\sin t-\sin t$$

$$\Rightarrow \quad y=\frac{1}{2}\left(1+\frac{2}{t}\right)\sin t$$ **Ans.**

Example 55. *Solve* $[\,tD^2+(1-2t)D-2]\,y=0$, *where* $y(0)=1, y'(0)=2$.

(*M.D.U., 2010*)

Solution. Here, $tD^2y+(1-2t)Dy-2y=0 \quad \Rightarrow \quad ty''+y'-2ty'-2y=0$

Taking Laplace transform of given differential equation, we get

$$L(t\,y'') + L\,(y') - 2\,L\,(t\,y') - 2\,L\,(y) = 0$$

$$\Rightarrow \quad -\frac{d}{ds}L\{\,y''\,\}+L\{\,y'\,\}+2\frac{d}{ds}L\,(y')-2\,L\,(y)=0$$

$$\Rightarrow \quad -\frac{d}{ds}[\,s^2\,\bar{y}-sy\,(0)-y'\,(0)\,]+[\,s\,\bar{y}-y\,(0)\,]+2\,\frac{d}{ds}[\,s\,\bar{y}-y\,(0)\,]-2\bar{y}=0$$

Putting the values of $y(0)$ and $y'\,(0)$, we get

$$-\frac{d}{ds}(\,s^2\,\bar{y}-s-2\,)+(\,s\,\bar{y}-1\,)+2\frac{d}{ds}(s\,\bar{y}-1)-2\,\bar{y}=0 \qquad [\because\ y(0)=1,\ y'\,(0)=2]$$

$$\Rightarrow \quad -s^2\frac{d\,\bar{y}}{ds}-2s\,\bar{y}+1+s\,\bar{y}-1+2\left(s\frac{d\,\bar{y}}{ds}+\bar{y}\right)-2\,\bar{y}=0$$

$$\Rightarrow \quad -(\,s^2-2s\,)\frac{d\,\bar{y}}{ds}-s\,\bar{y}=0$$

$$\Rightarrow \quad -\frac{d\,\bar{y}}{\bar{y}}-\frac{1}{s-2}ds=0 \qquad \textit{(Separating the variables)}$$

$$\Rightarrow \quad \int\frac{d\,\bar{y}}{\bar{y}}+\int\frac{ds}{s-2}=0 \quad \Rightarrow \quad \log\bar{y}+\log(\,s-2\,)=\log C \quad \Rightarrow \quad \log\bar{y}\,(\,s-2\,)=\log C$$

$$\Rightarrow \quad \bar{y}\,(\,s-2\,)=C \quad \Rightarrow \quad \bar{y}=\frac{C}{s-2} \quad \Rightarrow \quad y=C\,L^{-1}\left\{\frac{1}{s-2}\right\} \Rightarrow y=C\,e^{2t} \qquad \ldots (1)$$

$$y\,(0)=C\,e^0 \qquad \ldots (2)$$

Putting $y\,(0) = 1$ in (2), we get $1 = C\,e^0 \quad \Rightarrow \quad C = 1.$

Putting $C = 1$ in (1), we get $y = e^{2t}$

This is the required solution. **Ans.**

Example 56. *Using Laplace transform, solve the following differential equation: -*

$$y'' + 2t\,y' - y = t$$

when $y\,(0) = 0$ *and* $y'\,(0) = 1$

Solution. We have, $y'' + 2t\,y' - y = t$ ... (1)

Taking Laplace transform of (1), we get

$$[\,s^2\,\bar{y}-sy\,(0)-y'\,(0)\,]-2\frac{d}{ds}[\,s\,\bar{y}-y\,(0)\,]-\bar{y}=\frac{1}{s^2} \qquad \ldots (2)$$

On putting $y\,(0) = 0$ and $y'\,(0) = 1$ in (2), we get

$$(s^2\,\bar{y}-1)-2\frac{d}{ds}(s\,\bar{y}-0)-\bar{y}=\frac{1}{s^2}$$

$$\Rightarrow \quad (s^2\,\bar{y}-1)-2\,\bar{y}-2s\frac{d\,\bar{y}}{ds}-\bar{y}=\frac{1}{s^2} \quad \Rightarrow \quad -2s\frac{d\,\bar{y}}{ds}+(\,s^2-3\,)\,\bar{y}=\frac{1}{s^2}+1=\frac{1+s^2}{s^2}$$

$$\Rightarrow \quad \frac{d\,\bar{y}}{ds}-\frac{s^2-3}{2s}\,\bar{y}=\frac{1+s^2}{-2s^3} \quad \Rightarrow \quad \frac{d\,\bar{y}}{ds}-\left(\frac{s}{2}-\frac{3}{2s}\right)\bar{y}=-\frac{1}{2s^3}-\frac{1}{2s} \qquad \ldots (3)$$

Thus (3) is a linear differential equation.

$$\text{I.F.} = e^{\frac{1}{2}\int\left(\frac{3}{s}-s\right)ds} = e^{\frac{1}{2}\left(3\log s-\frac{s^2}{2}\right)} = e^{-\frac{s^2}{4}}\,.\,s^{\frac{3}{2}}$$

Solution of differential equation (3) is

$$\bar{y}\, e^{-\frac{s^2}{4}} . s^{\frac{3}{2}} = -\frac{1}{2}\int \left(\frac{1}{s^3}+\frac{1}{s}\right) s^{\frac{3}{2}} . e^{-\frac{s^2}{4}}\, ds = -\frac{1}{2}\int \left(\sqrt{s}+\frac{1}{s^{\frac{3}{2}}}\right) e^{-\frac{s^2}{4}}\, ds$$

Put $s^2 = 4z \quad \Rightarrow \quad s = 2\sqrt{z} \quad$ so that $\quad ds = \dfrac{dz}{\sqrt{z}}$

$$\bar{y}\, s^{\frac{3}{2}} . e^{-\frac{s^2}{4}} = -\frac{1}{2}\int \left(\sqrt{2}\, z^{\frac{1}{4}} + \frac{1}{2\sqrt{2}}\, z^{-\frac{3}{4}}\right) e^{-z}\, \frac{dz}{\sqrt{z}}$$

$$= -\frac{1}{\sqrt{2}}\int \left(z^{-\frac{1}{4}} + \frac{1}{4}\, z^{-\frac{5}{4}}\right) e^{-z}\, dz = -\frac{1}{\sqrt{2}}\int t^{-\frac{1}{4}}\, e^{-z}\, dz - \frac{1}{4\sqrt{2}}\int z^{-\frac{5}{4}}\, e^{-z}\, dz$$

$$= -\frac{1}{\sqrt{2}}\left[z^{-\frac{1}{4}}\, \frac{e^{-z}}{-1} + \int \left(-\frac{1}{4}\right) z^{-\frac{5}{4}}\, e^{-z}\, dz\right] + \frac{1}{4\sqrt{2}}\int z^{-\frac{5}{4}}\, e^{-z}\, dz + C$$

$$= \frac{1}{\sqrt{2}}\, e^{-z}\, z^{-\frac{1}{4}} z + C = \frac{1}{\sqrt{2}}\, e^{-\frac{s^2}{4}} \left(\frac{s^2}{4}\right)^{-\frac{1}{4}} + C = \frac{1}{\sqrt{5}}\, e^{-\frac{5^2}{4}} + C \Rightarrow \bar{y} = \frac{1}{s^2} + C$$

(Particular case)

$$\Rightarrow \quad \bar{y} = \frac{1}{s^2} + C \quad \Rightarrow \quad y = L^{-1}\left(\frac{1}{s^2} + C\right) = t + C$$

$y = t$ (c must vanish if $\bar{y}$ is a transform since $\bar{y} \to 0$ as $\to \infty$) **Ans.**

Example 57. *A particle moves in a line so that its displacement x from a fixed point O at any time t, is given by*

$$\frac{d^2x}{dt^2} + 4\frac{dx}{dt} + 5x = 80 \sin 5t$$

Using Laplace transform, find its displacement at any time t if initially particle is at rest at x = 0 *(U.P., II Semester 2009)*

Solution. Here, we have

$$\frac{d^2x}{dt^2} + 4\frac{dx}{dt} + 5x = 80 \sin 5t, \; x(0) = 0, \; x'(0) = 0 \qquad ...(1)$$

Taking Laplace transform of both sides of (1), we get

$$[s^2\bar{x} - sx(0) - x'(0)] + 4[s\bar{x} - x(0)] + 5\bar{x} = L[80 \sin 5t]$$

$$[s^2\bar{x} - 0 - 0] + 4s\bar{x} + 5\bar{x} = 80\left(\frac{5}{s^2+25}\right), \quad \Rightarrow \quad (s^2 + 4s + 5)\,\bar{x} = \frac{400}{s^2+25}$$

$$\bar{x} = \left(\frac{1}{s^2+4s+5}\right)\left(\frac{400}{s^2+25}\right)$$

$$\bar{x} = \frac{2s+18}{s^2+4s+5} - \frac{2s+10}{s^2+25} \qquad \text{[By Partial fraction]}$$

$$= \frac{2(s+2)+14}{(s+2)^2+1} - \frac{2s}{s^2+(5)^2} - \frac{10}{s^2+(5)^2}$$

$$= \frac{2(s+2)}{(s+2)^2+1} + \frac{14}{(s+2)^2+1} - \frac{2s}{s^2+(5)^2} - \frac{10}{s^2+(5)^2}$$

$$\Rightarrow \quad x = 2\,L^{-1}\frac{(s+2)}{(s+2)^2+1} + 14\,L^{-1}\frac{1}{(s+2)^2+1} - 2L^{-1}\frac{s}{s^2+(5)^2} - 2L^{-1}\frac{5}{s^2+(5)^2}$$

$$= 2e^{-2t}\cos t + 14e^{-2t}\sin t - 2\cos 5t - 2\sin 5t$$

$$= 2e^{-2t}(\cos t + 7\sin t) - 2(\cos 5t + \sin 5t)$$ **Ans.**

47.13 ELECTRIC CIRCUIT

Consider an electric circuit consisting of a resistance R, inductance L, a condensor of capacitor C and electromotive power of voltage E in a series. A switch is also connected in the circuit. Here,

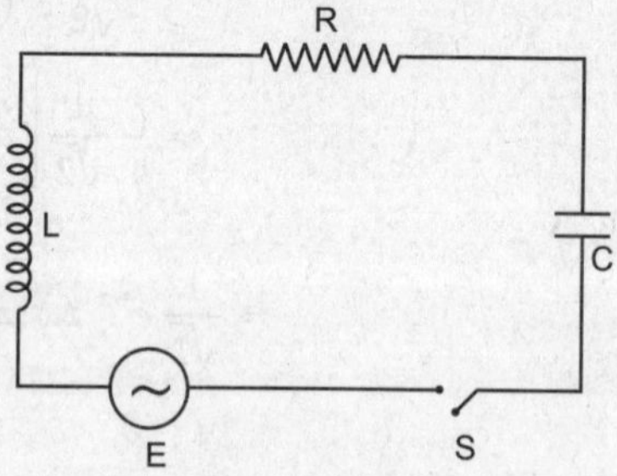

$$i = \frac{dq}{dt}$$

Voltage developed by $Ri\,,\, L\frac{di}{dt}$ and $\frac{q}{C}$

By Kirchhoff low $\quad \boxed{L\frac{di}{dt} + Ri + \frac{q}{C} = E}$

Example 58. *A resistance R in series with inductance L is connected with e.m.f. E (t). The current i is given by*

$$L\frac{di}{dt} + Ri = E$$

If the switch is connected at t = 0 and disconnected at t = a, find the current i in terms of t.

Solution. Conditions under which current i flows are

$$E(t) = \begin{cases} E, & 0 < t < a \\ 0, & t > a \end{cases} \qquad [\, i = 0 \text{ at } t = 0]$$

Given equation is $L\frac{di}{dt} + Ri = E$ ... (1)

Taking Laplace transform of (1), we get

$$L[\,s\,\bar{i} - i(0)\,] + R\,\bar{i} = \int_0^\infty e^{-st}\,E\,dt$$

$$L\,s\,\bar{i} + R\,\bar{i} = \int_0^\infty e^{-st}\,E\,dt \qquad [\,i(0) = 0]$$

$$(Ls+R)\,\bar{i} = \int_0^\infty e^{-st}\,.\,E\,dt = \int_0^a e^{-st}E\,dt + \int_a^\infty e^{-st}\,E\,dt$$

$$= E\left[\frac{e^{-st}}{-s}\right]_0^a + 0 = \frac{E}{s}[1 - e^{-as}\,] = \frac{E}{s} - \frac{E}{s}e^{-as}$$

$$\Rightarrow \qquad \bar{i} = \frac{E}{s\,(Ls+R)} - \frac{Ee^{-as}}{s\,(Ls+R)}$$

Taking Inverse Laplace Transform, we obtain

$$i = \text{Inverse Lap.}\left[\frac{E}{s\,(Ls+R)}\right] - \text{Inverse Lap.}\left[\frac{E\,e^{-as}}{s\,(Ls+R)}\right] \qquad \text{... (2)}$$

Now we have to find the value of Inverse Lap. $\left[\frac{E}{s(Ls+R)}\right]$

$$\text{Inverse Lap}.\left[\frac{E}{s(Ls+R)}\right]=\frac{E}{L}\text{ Inverse Lap}.\left[\frac{1}{s\left(s+\frac{R}{L}\right)}\right]$$

$$=\frac{E}{L}\frac{L}{R}\text{ Inverse Lap}.\left[\frac{1}{s}-\frac{1}{s+\frac{R}{L}}\right]=\frac{E}{R}\left[1-e^{-\frac{R}{L}t}\right]$$

(Resolving into partial fractions)

and $$\text{Inverse Lap}.\left[\frac{E\,e^{-as}}{s(Ls+R)}\right]=\frac{E}{R}\left[1-e^{-\frac{R}{L}(t-a)}\right]u(t-a)$$

(By the Second Shifting Theorem)

On substituting the values of the inverse transforms in (2), we get

$$i=\frac{E}{R}\left[1-e^{-\frac{R}{L}t}\right]-\frac{E}{R}\left[1-e^{-\frac{R}{L}(t-a)}\right]u(t-a)$$

Hence $$i=\frac{E}{R}\left[1-e^{-\frac{R}{L}t}\right]\text{ for } 0<t<a, \qquad [u(t-a)=0]$$

$$i=\frac{E}{R}\left[1-e^{-\frac{R}{L}t}\right]-\frac{E}{R}\left[1-e^{-\frac{R}{L}(t-a)}\right]\quad \text{for } t>a\ [u(t-a)=1]$$

$$=\frac{E}{R}\left[e^{-\frac{R}{L}(t-a)}-e^{-\frac{R}{L}t}\right]=\frac{E}{R}e^{-\frac{R}{L}t}\left[e^{\frac{Ra}{L}}-1\right] \qquad \textbf{Ans.}$$

Example 59. *Voltage Ee^{-at} is applied at $t=0$ to a circuit of inductance L and resistance R. Show that the current at time t is* $\frac{E}{R-aL}(e^{-at}-e^{-Rt/L})$. [*U.P., II Semester, (SUM) 2007*]

Solution. We know that

$$L\frac{dI}{dt}+RI=Ee^{-at}, \qquad \text{...(1)}$$

where $I(0)=0$

Taking Laplace transform of both sides of (1), we get

$$L[s\bar{I}-I(0)]+R\bar{I}=\frac{E}{s+a} \qquad \text{...(2)}$$

Putting I (0) = 0 in (2), we get

$$(Ls+R)\bar{I}=\frac{E}{s+a}$$

$$\Rightarrow \quad \bar{I}=\frac{E}{(s+a)(Ls+R)}=\frac{E}{R-aL}\left(\frac{1}{s+a}-\frac{1}{Ls+R}\right)$$

$$=\frac{E}{R-aL}\left(\frac{1}{s+a}-\frac{1}{s+R/L}\right) \qquad \text{...(3)}$$

Taking the Inverse Laplace transform of both sides of (3), we get

$$I = \frac{E}{R - aL} L^{-1}\left\{\frac{1}{s+a} - \frac{1}{s+R/L}\right\} = \frac{E}{R-aL}[e^{-at} - e^{-Rt/L}] \qquad \textbf{Ans.}$$

Example 60. *Using the Laplace transform, find the current i (t) in the LC - circuit. Assuming L = 1 henry, C = 1 farad, zero initial current and charge on the capacitor, and*

$v(t) = t$, *when* $0 < t < 1$

$= 0$ *otherwise.*

i C L V(t)

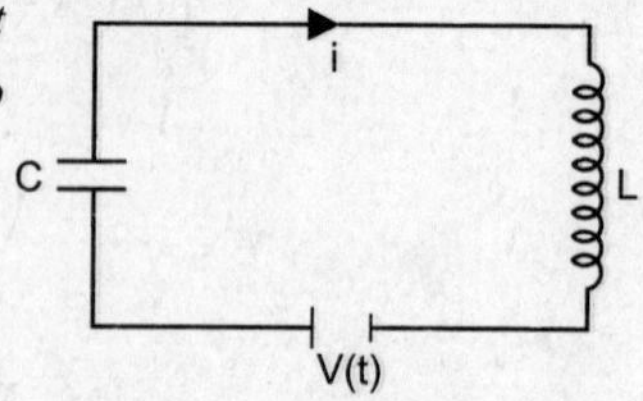

Solution. The differential equation for L and C circuit is given by

$$L\frac{d^2q}{dt^2} + \frac{q}{C} = E \qquad \text{... (1)}$$

Putting $L = 1$, $C = 1$, $E = v(t)$ in (1), we get

$$\frac{d^2q}{dt^2} + q = v(t) \qquad \text{... (2)}$$

Taking Laplace Transform of (2), we have

$$s^2\,\bar{q} - sq(0) - q'(0) + \bar{q} = \int_0^\infty v(t)\,e^{-st}\,dt$$

Substituting $q(0) = 0$, and $q'(0) = 0$, we get

$$s^2\,\bar{q} + \bar{q} = \int_0^1 te^{-st}\,dt + \int_1^\infty 0\,e^{-st}\,dt$$

$$(s^2+1)\,\bar{q} = \left[t\frac{e^{-st}}{-s}\right]_0^1 - \int_0^1 \frac{e^{-st}}{-s}\,dt = \frac{e^{-s}}{-s} - \left[\frac{e^{-st}}{s^2}\right]_0^1 = -\frac{e^{-s}}{s} - \frac{e^{-s}}{s^2} + \frac{1}{s^2}$$

$$\bar{q} = \frac{1}{s^2+1}\left[-\frac{e^{-s}}{s} - \frac{e^{-s}}{s^2} + \frac{1}{s^2}\right]$$

$$\bar{q} = \frac{-e^{-s}}{s(s^2+1)} - \frac{e^{-s}}{s^2(s^2+1)} + \frac{1}{s^2(s^2+1)}$$

Taking Inverse Laplace Transform, we get

$$q = \text{Inverse Lap.}\frac{-e^{-s}}{s(s^2+1)} - \text{Inverse Lap.}\frac{e^{-s}}{s^2(s^2+1)} + \text{Inverse Lap.}\frac{1}{s^2(s^2+1)} \qquad \text{... (3)}$$

We know that

$$\text{Inverse Lap. } L^{-1}[e^{-as}F(s)] = f(t-a)\,u(t-a)$$

$$\text{Inverse Lap.}\frac{1}{s(s^2+1)} = \int_0^t \sin t\,dt = [-\cos t]_0^t = 1 - \cos t \qquad \text{... (4)}$$

$$\text{Inverse Lap.}\frac{1}{s^2(s^2+1)} = \int_0^t (1-\cos t)\,dt = t - \sin t \qquad \text{... (5)}$$

In view of this, we have

$$\text{Inverse Lap.}\left[\frac{-e^{-s}}{s(s^2+1)}\right] = -[1 - \cos(t-1)]\,u(t-1) \qquad \text{[From (4)]}$$

$$\text{Inverse Lap. } L^{-1}\frac{e^{-s}}{s^2(s^2+1)} = [(t-1)-\sin(t-1)]\,u(t-1) \qquad \text{[From (5)]}$$

Putting the above values in (3), we get

$$q = -[1-\cos(t-1)]\,u(t-1) - [(t-1)-\sin(t-1)]\,u(t-1) + t - \sin t \qquad \textbf{Ans.}$$

EXERCISE 47.10

Solve the following differential equations:

1. $\dfrac{d^2y}{dx^2}+y=0$ where $y=1$ and $\dfrac{dy}{dx}=-1$ at $x=0$. **Ans.** $y=\cos x-\sin x$

2. $\dfrac{d^2y}{dx^2}+2\dfrac{dy}{dx}+5y=0$,where $y=2$, $\dfrac{dy}{dx}=-4$ at $x=0$. **Ans.** $y=e^{-x}(2\cos 2x-\sin 2x)$

3. $\dfrac{d^3y}{dx^3}+2\dfrac{d^2y}{dx^2}-\dfrac{dy}{dx}-2y=0$, given $y=\dfrac{dy}{dx}=0$, $\dfrac{d^2y}{dx^2}=6$ at $x=0$.

Ans. $y=e^{x}-3e^{-x}+2e^{-2x}$

4. $\dfrac{d^2y}{dx^2}+y=3\cos 2x$, where $y=\dfrac{dy}{dx}=0$ at $x=0$. **Ans.** $y=\cos x-\cos 2x$.

5. $\dfrac{d^2y}{dx^2}+\dfrac{dy}{dx}-2y=1-2x$, given $y=0$, $\dfrac{dy}{dx}=4$ at $x=0$. **Ans.** $y=e^{x}-e^{-2x}+x$

6. $\dfrac{d^2y}{dx^2}-3\dfrac{dy}{dx}+2y=4e^{2x}$, given $y=-3$, and $\dfrac{dy}{dx}=5$ at $x=0$ **Ans.** $y=-7e^{x}+4e^{2x}+4xe^{2x}$

7. $\dfrac{d^2y}{dx^2}-3\dfrac{dy}{dx}+2y=4x+e^{2x}$, where y = 1, $\dfrac{dy}{dx}=-1$ at $x=0$. **Ans.** $y=3+2x+\dfrac{1}{2}e^{3x}-2e^{2x}-\dfrac{1}{2}e^{x}$.

8. $\dfrac{d^3y}{dx^3}+2\dfrac{d^2y}{dx^2}-\dfrac{dy}{dx}-2y=0$, where y = 1, $\dfrac{dy}{dx}=2$, $\dfrac{d^2y}{dx^2}=2$ at $x=0$. **Ans.** $y=\dfrac{5}{3}e^{x}-e^{-x}+\dfrac{1}{2}e^{-2x}$

9. $(D^2-D-2)\,x=20\sin 2t$, $x_0=-1$, $x_1=2$ **Ans.** $x=2e^{2t}-4e^{-t}+\cos 2t-3\sin 2t$

10. $(D^3+D^2)\,x=6t^2+4$, $x(0)=0$, $x'(0)=2$, $x''(0)=0$ **Ans.** $x=\dfrac{1}{2}t^4-2t^3+8t^2-16t+16-16e^{-t}$

11. $\dfrac{d^2x}{dt^2}-2\dfrac{dx}{dt}+x=e^{t}$, where $x(0)=2$, $\dfrac{dx}{dt}=-1$ at $t=0$ **Ans.** $x=2e^{t}-3te^{t}+\dfrac{1}{2}t^2e^{t}$

12. $y''+2y'+y=te^{-t}$ if $y(0)=1$, $y'(0)=-2$. **Ans.** $y=\left(1-t+\dfrac{t^3}{6}\right)e^{-t}$.

13. $\dfrac{d^2y}{dx^2}+y=x\cos 2x$, where $y=\dfrac{dy}{dx}=0$ at $x=0$. **Ans.** $y=\dfrac{4}{9}\sin 2x-\dfrac{5}{9}\sin x-\dfrac{x}{3}\cos 2x$

14. $\dfrac{d^3y}{dx^3}-3\dfrac{d^2y}{dx^2}+3\dfrac{dy}{dx}-y=x^2e^{2x}$, where $y=1$, $\dfrac{dy}{dx}=0$, $\dfrac{d^2y}{dx^2}=-2$ at $x=0$.

Ans. $y=e^{2x}(x^2-6x+12)-e^{x}(15x^2+7x+11)$.

15. $y''+4y'+3y=t$, $t>0$; given that $y(0)=0$ and $y'(0)=1$. **Ans.** $y=-\dfrac{4}{9}+\dfrac{t}{6}+e^{-t}-\dfrac{5}{9}e^{-3t}$

47.14 SOLUTION OF SIMULTANEOUS DIFFERENTIAL EQUATIONS BY LAPLACE TRANSFORMS

Simultaneous differential equations can also be solved by Laplace Transform method.

Example 61. *Solve* $\frac{dx}{dt} + y = 0$ *and* $\frac{dy}{dt} - x = 0$ *under the condition* $x(0) = 1$, $y(0) = 0$.

Solution. We have, $x' + y = 0$ ... (1)

and $y' - x = 0$ ... (2)

Taking the Laplace transform of (1) and (2), we get

$$[s\bar{x} - x(0)] + \bar{y} = 0 \quad \text{... (3)}$$

$$[s\bar{y} - y(0)] - \bar{x} = 0 \quad \text{... (4)}$$

On substituting the values of $x(0)$ and $y(0)$ in (3) and (4), we get

$$s\bar{x} - 1 + \bar{y} = 0 \quad \text{... (5)}$$

$$s\bar{y} - \bar{x} = 0 \quad \text{... (6)}$$

Solving (5) and (6) for $\bar{x}$ and $\bar{y}$, we get

$$\bar{x} = \frac{s}{s^2+1}, \quad \bar{y} = \frac{1}{s^2+1},$$

On inversion, we obtain $x = L^{-1}\left(\frac{s}{s^2+1}\right)$, $y = L^{-1}\left(\frac{1}{s^2+1}\right)$,

$x = \cos t$, $y = \sin t$ **Ans.**

Example 62. *Using Laplace Transform solve the following simultaneous equations:*

$$\frac{dx}{dt} + x + y = 0$$

$$\frac{dy}{dt} + 4x + y = 0$$

given $x(0) = y(0) = 1$ *(D.U. April 2010)*

Solution. Here, we have

$$\frac{dx}{dt} + x + y = 0 \quad \text{... (1)}$$

$$\frac{dy}{dt} + 4x + y = 0 \quad \text{... (2)}$$

Taking Laplace transform of (1) on both sides, we get

$$s\bar{x} - x(0) + \bar{x} + \bar{y} = 0 \qquad [\text{where } \bar{x} = (x) \text{ and } y = 2(y)]$$

$$(s+1)\bar{x} + \bar{y} = 1 \quad \text{... (3)}$$

Again, taking Laplace transform of (2) on both sides, we get

$$s\bar{y} - y(0) + 4\bar{x} + \bar{y} = 0$$

$$\Rightarrow \quad (s+1)\bar{y} + 4\bar{x} = 1 \quad \text{... (4)} \; [\because y(0) = 1]$$

Multiplying (3) by $(s + 1)$ and subtracting (4), from (3), we gets

$$(s+1)^2\bar{x} - 4\bar{x} = s$$

$$\bar{x} = \frac{s}{(s+1)^2 - 4} = \frac{s+1}{(s+1)^2 - 4} - \frac{1}{(s+1)^2 - 4} \qquad ... (5)$$

Taking inverse Laplace transform of (5), we get

$$x = e^{-t} \cos h2t - \frac{1}{2} e^{-t} \sin h2t \qquad ... (6)$$

Again, multiplying (3) by 4 and (4) by (σ + 1) then subtract, we get

$$4\bar{y} - (s+1)^2 \bar{y} = -\delta - 3$$

$$\bar{y} = \frac{s+3}{(s+1)^2} = \frac{s+1}{(s+1)^2 - 4} + \frac{2}{(s+1)^2 - 4} \qquad ... (7)$$

Again, taking inverse Laplace transform of (7), we get

$$y = e^{-t} \cos h\, 2t + e^{-t} \sin h\, 2t \qquad ... (8)$$

(8) and (6) are solution of differential equation. **Ans.**

Example 63. *Solve the following simultaneous differential equations by Laplace transform*

$$3\frac{dx}{dt} - y = 2t, \quad \frac{dx}{dt} + \frac{dy}{dt} - y = 0$$

with the conditions $x(0) = y(0) = 0.$ [*U.P., II Semester, (SUM) 2008*]

Solution. Here, we have

$$3\frac{dx}{dt} - y = 2t, \qquad ...(1)$$

and
$$\frac{dx}{dt} + \frac{dy}{dt} - y = 0 \qquad ...(2)$$

Taking Laplace transform on both sides of equation (1), we get

$$3L(x') - L(y) = L(2t)$$

$$3[s\bar{x} - x(0)] - \bar{y} = \frac{2}{s^2} \qquad \text{[where } L(x) = \bar{x} \text{ and } L(y) = \bar{y}\text{]}$$

$$3s\bar{x} - \bar{y} = \frac{2}{s^2} \qquad ...(3)$$

Again taking Laplace transform on both sides of equation (2), we get

$$L(x') + L(y') - L(y) = L(0)$$

$$\Rightarrow \quad [s\bar{x} - x(0)] + [s\bar{y} - y(0)] - \bar{y} = 0 \quad \Rightarrow \quad s\bar{x} + (s-1)\bar{y} = 0 \qquad ...(4)$$

Multiplying equation (4) by 3, we get

$$3s\bar{x} + 3(s-1)\bar{y} = 0 \qquad ...(5)$$

Subtracting equation (3) from (5), we get

$$(3s - 2)\bar{y} = -\frac{2}{s^2}$$

$$\Rightarrow \quad \bar{y} = -\frac{2}{s^2(3s-2)} = \frac{1}{s^2} + \frac{3}{2s} - \frac{3}{2\left(s - \frac{2}{3}\right)}$$

Taking inverse Laplace transform on both sides, we get

$$y = t + \frac{3}{2} - \frac{3}{2} e^{\frac{2t}{3}} \quad \text{...(6)}$$

Substituting $\bar{y}$ in (3), we get

$$3s\bar{x} - \frac{1}{s^2} - \frac{3}{2s} + \frac{3}{2\left(s - \frac{2}{3}\right)} = \frac{2}{s^2} \quad \Rightarrow \quad 3s\bar{x} = \frac{3}{s^2} + \frac{3}{2s} - \frac{3}{2\left(s - \frac{2}{3}\right)}$$

$$\Rightarrow \quad \bar{x} = \frac{1}{s^3} + \frac{1}{2s^2} - \frac{1}{2s\left(s - \frac{2}{3}\right)} \quad \Rightarrow \quad x = \frac{1}{s^3} + \frac{1}{2s^2} - \frac{3}{4}\left(\frac{1}{s - \frac{2}{3}} - \frac{1}{s}\right)$$

Taking inverse Laplace transform on both sides, we get

$$x = \frac{t^2}{2} + \frac{t}{2} - \frac{3}{4} e^{\frac{2t}{3}} + \frac{3}{4} \quad \text{...(7)}$$

Equation (6) and (7) when taken together, give the complete solution. **Ans.**

Example 64. *Solve the simultaneous equations:*

$$\frac{dx}{dt} - y = e^t,$$

$$\frac{dy}{dt} + x = \sin t, \text{ given } x(0) = 1,\ y(0) = 0,$$ (*U.P.,II Semester, Summer* 2006)

Solution. $\frac{dx}{dt} - y = e^t$ ... (1) $\qquad \frac{dy}{dt} + x = \sin t$ ... (2)

Taking Laplace transform of (1), we get

$$[\, s\,\bar{x} - x(0)\,] - \bar{y} = \frac{1}{s-1}$$

i.e., $$s\,\bar{x} - 1 - \bar{y} = \frac{1}{s-1}$$ $[\because x(0) = 1]$

$$\bar{x} - \bar{y} = 1 + \frac{1}{s-1}$$

$$s\,\bar{x} - \bar{y} = \frac{s}{s-1} \quad \text{... (3)}$$

Taking Laplace Transform of (2), we get

$$[\, s\bar{y} - y(0)\,] + \bar{x} = \frac{1}{s^2+1}$$ $[y(0) = 0)$

$$\bar{x} + s\,\bar{y} = \frac{1}{s^2+1} \quad \text{... (4)}$$

Solving (3) and (4) for $\bar{x}$ and $\bar{y}$, we have

$$\bar{x} = \frac{s^2}{(s-1)(s^2+1)} + \frac{1}{(s^2+1)^2} = \frac{1}{2}\left[\frac{1}{s-1} + \frac{s}{s^2+1} + \frac{1}{s^2+1}\right] + \frac{1}{(s^2+1)^2} \quad \text{... (5)}$$

and $$\bar{y} = \frac{s}{(s^2+1)^2} - \frac{s}{(s-1)(s^2+1)} = \frac{s}{(s^2+1)^2} - \frac{1}{2}\left[\frac{1}{s-1} - \frac{s}{s^2+1} + \frac{1}{s^2+1}\right] \quad \text{... (6)}$$

Taking Inverse Laplace Transform of (5), we get

$$x = \frac{1}{2} L^{-1}\left[\frac{1}{s-1} + \frac{s}{s^2+1} + \frac{1}{s^2+1}\right] + L^{-1}\left[\frac{1}{(s^2+1)^2}\right]$$

$$= \frac{1}{2}[\,e^t + \cos t + \sin t\,] + \frac{1}{2}(\sin t - t\cos t)$$

$$\left[\because L^{-1}\left[\frac{1}{(s^2+a^2)^2}\right] = \frac{1}{2a^2}(\sin at - at\cos at)\right]$$

$$= \frac{1}{2}[e^t + \cos t + 2\sin t - t\cos t]$$

Taking Inverse Laplace Transform of (6), we get

$$y = L^{-1}\left[\frac{s}{(s^2+1)^2}\right] - \frac{1}{2} L^{-1}\left[\frac{1}{s-1} - \frac{s}{s^2+1} + \frac{1}{s^2+1}\right]$$

$$= \frac{1}{2} t\sin t - \frac{1}{2}[e^t - \cos t + \sin t] \qquad \left[\because L^{-1}\left[\frac{s}{(s^2+a^2)^2}\right] = \frac{1}{2a^2} t\sin at\right]$$

$$= \frac{1}{2}[t\sin t - e^t + \cos t - \sin t]$$

Hence, $x = \frac{1}{2}(e^t + \cos t + 2\sin t - t\cos t)$

$y = \frac{1}{2}(t\sin t - e^t + \cos t - \sin t)$ **Ans.**

Example 65. *Use Laplace transform to solve:*

$$\frac{dx}{dt} + y = \sin t, \quad \frac{dy}{dt} + x = \cos t$$

given that $x = 2$, $y = 0$ at $t = 0$.

Solution. Here, we have

$$\frac{dx}{dt} + y = \sin t \qquad ...(1)$$

$$\frac{dy}{dt} + x = \cos t \qquad ...(2)$$

Taking Laplace transform of (1) and (2), we get

$$s\overline{x} - x(0) + \overline{y} = \frac{1}{s^2+1} \quad \Rightarrow \quad s\overline{x} + \overline{y} = \frac{1}{s^2+1} + 2 \qquad ...(3)$$

and $s\overline{y} - y(0) + \overline{x} = \frac{s}{s^2+1} \quad \Rightarrow \quad \overline{x} + s\overline{y} = \frac{s}{s^2+1}$ ...(4)

Solving (3) and (4) for $\overline{x}$ an $\overline{y}$, we get

$$\overline{x} = \frac{2s}{s^2-1} \text{ and } \overline{y} = \frac{1}{1+s^2} + \frac{2}{1-s^2}$$

$$\overline{x} = \frac{1}{s+1} + \frac{1}{s-1} \qquad ...(5)$$

and $\overline{y} = \frac{1}{1+s^2} + \frac{1}{s+1} - \frac{1}{s-1}$ ... (6) (By partial fractions)

Taking Inverse Laplace transform on both sides of (5) and (6), we get

$$x = e^{-t} + e^{t}$$

and $$y = \sin t + e^{-t} - e^{t}$$ **Ans.**

Example 66. *The co-ordinates (x, y) of a particle moving along a plane curve at any time t are given by*

$$\frac{dy}{dt} + 2x = \sin 2t, \quad \frac{dx}{dt} - 2y = \cos 2t; \ (t > 0)$$

It is given that at t = 0, x = 1 and y = 0. Show using transforms that the particle moves along the curve $4x^2 + 4xy + 5y^2 = 4$.

Solution. Here, we have

$$\left[\begin{array}{l} \dfrac{dy}{dt} + 2x = \sin 2t \\ \dfrac{dx}{dt} - 2y = \cos 2t \end{array}\right. \Rightarrow \left[\begin{array}{ll} 2x + Dy = \sin 2t & \quad ...(1) \\ Dx - 2y = \cos 2t, & \quad ...(2) \end{array}\right.$$

Taking Laplace transform of (1) on both sides, we get

$$2\bar{x} + s\bar{y} - y(0) = \frac{2}{s^2+4} \qquad \text{where } \bar{x} = L(x) \text{ and } \bar{y} = L(y)$$

$$2\bar{x} + s\bar{y} = \frac{2}{s^2+4} \qquad ...(3) \ [\because y(0) = 0]$$

Again, taking Laplace transform of equation, (2) on both sides, we get

$$s\bar{x} - x(0) - 2\bar{y} = \frac{s}{s^2+4}, \text{ where } \bar{x} = L(x) \text{ and } \bar{y} = L(y)$$

$$\Rightarrow \qquad s\bar{x} - 2\bar{y} = \frac{s}{s^2+4} + 1 \qquad ...(4) \because x(0) = 0]$$

Multiplying equation (3) by 2 and equation (4) by s and then adding, we get

$$s\bar{x} + s^2\bar{x} = \frac{4}{s^2+4} + \frac{s^2}{s^2+4} + s$$

$$(4+s^2)\bar{x} = 1 + s$$

$$\bar{x} = \frac{1+s}{4+s^2} = \frac{1}{s^2+4} + \frac{s}{s^2+4} \qquad ...(5)$$

Taking Inverse Laplace transform of (5), we get

$$x = \frac{1}{2}\sin 2t + \cos 2t \qquad ...(6)$$

Again, multiplying (3) by s and (4) by 2 then subtracting equation (6) from (3), we get

$$s^2\bar{y} + 4\bar{y} = \frac{2s}{s^2+4} - \frac{2s}{s^2+4} - 2$$

$$\Rightarrow \qquad \bar{y} = \frac{-2}{s^2+4} \qquad ...(7)$$

Taking Inverse Laplace transform of (7), we get $y = -\sin 2t$

Now, $$4x^2 = 4\left[\frac{1}{4}\sin^2 2t + \cos^2 2t + \sin 2t \cos 2t\right]$$

$$5y^2 = 5\sin^2 2t$$

$$4xy = 4\left[\left(\frac{1}{2}\sin 2t + \cos 2t\right).(-\sin 2t)\right]$$

$$= -(2\sin^2 2t + 4\sin 2t\cos 2t)$$

$$\therefore \quad 4x^2 + 5y^2 + 4xy = 4\sin^2 2t + 4\cos^2 2t = 4$$ **Ans.**

Example 67. *Solve the following simultaneous differential equations by Laplace transform*

$$\frac{dx}{dt} + 4\frac{dy}{dt} - y = 0; \frac{dx}{dt} + 2y = e^{-t}$$

with conditions $x(0) = y(0) = 0.$ [*U.P., II Semester, 2008*]

Solution. Here, we have

$$\frac{dx}{dt} + 4\frac{dy}{dt} - y = 0 \quad ...(1)$$

and $$\frac{dx}{dt} + 2y = e^{-t} \quad ...(2)$$

Taking Laplace transform on both sides of equation (1), we get

$$L(x') + 4L(y') - L(y) = L(0)$$

$$\Rightarrow \quad s\overline{x} - x(0) + 4[s\overline{y} - y(0)] - \overline{y} = 0$$

$$\Rightarrow \quad s\overline{x} + (4s-1)\overline{y} = 0 \quad ...(3)$$

Again, taking Laplace transform on both sides of equation (2), we get

$$L(x') + 2L(y) = L(e^{-t})$$

$$\Rightarrow \quad s\overline{x} - x(0) + 2\overline{y} = \frac{1}{s+1} \quad \Rightarrow \quad s\overline{x} + 2\overline{y} = \frac{1}{s+1} \quad ...(4)$$

Subtracting (4) from (3), we get

$$(4s-3)\overline{y} = -\frac{1}{s+1}$$

$$\overline{y} = -\frac{1}{(s+1)(4s-3)} = -\frac{1}{7}\left(\frac{-1}{s+1} + \frac{1}{s-3/4}\right) = \frac{1}{7}\left(\frac{1}{s+1} - \frac{1}{s-3/4}\right) \quad ...(5)$$

Taking inverse Laplace transform on both sides of (5), we get

$$y = \frac{1}{7}\left(e^{-t} - e^{\frac{3t}{4}}\right) \quad ...(6)$$

Substituting $\overline{y}$ in (4), we get

$$s\overline{x} + \frac{2}{7}\left(\frac{1}{s+1} - \frac{1}{s-3/4}\right) = \frac{1}{s+1}$$

$$\Rightarrow \quad s\overline{x} = \frac{5}{7(s+1)} + \frac{2}{7(s-3/4)}$$

$$\Rightarrow \quad \overline{x} = \frac{5}{7s(s+1)} + \frac{2}{7s(s-3/4)} = \frac{5}{7}\left(\frac{1}{s} - \frac{1}{s+1}\right) + \frac{8}{21}\left(\frac{1}{s-3/4} - \frac{1}{s}\right)$$

$$= \frac{1}{3s} - \frac{5}{7(s+1)} + \frac{8}{21(s-3/4)} \qquad \text{...(7)}$$

Taking Inverse Laplace transform on both sides of (7), we get

$$x = \frac{1}{3} - \frac{5}{7}e^{-t} + \frac{8}{21}e^{\frac{3t}{4}}$$ **Ans.**

Example 68. *Using Laplace Transformation, solve*

$$(D-2)\,x - (D+1)\,y = 6\,e^{3t} \qquad \text{... (1)}$$
$$(2D-3)\,x + (D-3)\,y = 6e^{3t}$$

Given $x = 3$, $y = 0$ *when* $t = 0$.

Solution. Taking Laplace transformation of the given equations, we get

$$\begin{bmatrix} L\,Dx - 2\,Lx - LDy - Ly = 6\,Le^{3t} \\ 2L\,Dx - 3Lx + LDy - 3Ly = 6\,Le^{3t} \end{bmatrix}$$

$$\Rightarrow \begin{bmatrix} s\,\bar{x} - x(0) - 2\,\bar{x} - s\bar{y} + y(0) - \bar{y} = 6\dfrac{1}{s-3} \\ 2s\,\bar{x} - 2x(0) - 3\,\bar{x} + s\,\bar{y} - y(0) - 3\,\bar{y} = \dfrac{6}{s-3} \end{bmatrix}$$

$$\Rightarrow \begin{bmatrix} (s-2)\,\bar{x} - (s+1)\,\bar{y} - 3 = \dfrac{6}{s-3} \\ (2s-3)\,\bar{x} + (s-3)\,\bar{y} - 6 = \dfrac{6}{s-3} \end{bmatrix} \Rightarrow \begin{bmatrix} (s-2)\,\bar{x} - (s+1)\,\bar{y} = \dfrac{3s-3}{s-3} \\ (2s-3)\,\bar{x} + (s-3)\,\bar{y} = \dfrac{6s-12}{s-3} \end{bmatrix}$$

$$\Rightarrow \begin{bmatrix} (s-3)(s-2)\,\bar{x} - (s-3)(s+1)\,\bar{y} = 3s-3 \\ (s+1)(2s-3)\,\bar{x} + (s+1)(s-3)\,\bar{y} = \dfrac{(s+1)(6s-12)}{s-3} \end{bmatrix} \text{ on adding, we get}$$

$$(3s^2 - 6s + 3)\,x = 3(s-1) + \frac{6(s^2 - s - 2)}{s-3} \quad \Rightarrow \quad \bar{x} = \frac{3(s-1)}{3(s-1)^2} + \frac{6(s^2 - s - 2)}{3(s-1)^2(s-3)}$$

$$x = L^{-1}\left[\frac{1}{s-1} + \frac{2}{(s-1)^2} + \frac{2}{s-3}\right] = e^t + 2t\,e^t + 2e^{3t}$$

Putting the value of x in (1), we get

$$(D-2)(e^t + 2te^t + 2e^{3t}) - (D+1)\,y = 6e^{3t}$$

$$\Rightarrow \quad e^t + 2te^t + 2e^t + 6e^{3t} - 2e^t - 4te^t - 4e^{3t} - (D+1)\,y = 6e^{3t}$$

$$\Rightarrow \quad (D+1)\,y = e^t - 2te^t - 4e^{3t} \qquad \text{... (2)}$$

Taking Laplace transform of (2), we get

$$s\,\bar{y} - y(0) + \bar{y} = \frac{1}{s-1} - \frac{2}{(s-1)^2} - \frac{4}{s-3}$$

$$\Rightarrow \quad (s+1)\,\bar{y} = \frac{1}{s-1} - \frac{2}{(s-1)^2} - \frac{4}{s-3} \qquad [\because y(0) = 0]$$

$$\Rightarrow \quad \bar{y} = \frac{1}{s^2 - 1} - \frac{2}{(s+1)(s-1)^2} - \frac{4}{(s+1)(s-3)}$$

$$= \frac{1}{s^2-1} - \frac{\frac{1}{2}}{s+1} + \frac{\frac{1}{2}}{s-1} - \frac{1}{(s-1)^2} + \frac{1}{s+1} - \frac{1}{s-3}$$

$$\Rightarrow \qquad \bar{y} = \frac{1}{s^2-1} + \frac{\frac{1}{2}}{s+1} + \frac{\frac{1}{2}}{s-1} - \frac{1}{(s-1)^2} - \frac{1}{s-3}$$

$$\Rightarrow \qquad y = L^{-1}\left[\frac{1}{s^2-1} + \frac{1}{2}\frac{1}{s+1} + \frac{1}{2}\frac{1}{s-1} - \frac{1}{s-3} - \frac{1}{(s-1)^2}\right]$$

$$\Rightarrow \qquad y = \sinh t + \frac{1}{2}e^{-t} + \frac{1}{2}e^{t} - e^{3t} - te^{t}$$

$$y = \sinh t + \cosh t - e^{3t} - te^{t}$$ **Ans.**

Example 69. *Solve the simultaneous equations*

$$(D^2 - 3)x - 4y = 0$$
$$x + (D^2 + 1)y = 0$$

for $t > 0$, given that $x = y = \frac{dy}{dt} = 0$ and $\frac{dx}{dt} = 2$ at $t = 0$.

Solution. Here, we have

$$(D^2 - 3)x - 4y = 0 \quad ...(1)$$
$$x + (D^2 + 1)y = 0 \quad ...(2)$$

Taking Laplace transform of (1) and (2), we get

$$s^2\bar{x} - sx(0) - x'(0) - 3\bar{x} - 4\bar{y} = 0$$

i.e., $$(s^2 - 3)\bar{x} - 4\bar{y} = 2 \quad ...(3)[\because x(0) = 0, x'(0) = 2]$$

and $\bar{x} + s^2\bar{y} - sy(0) - y'(0) + \bar{y} = 0$

i.e., $$\bar{x} + (s^2 + 1)\bar{y} = 0 \quad ...(4)[\because y(0) = 0, y'(0) = 0]$$

Solving (3) and (4) for $\bar{x}$ and $\bar{y}$, we get

$$\bar{x} = \frac{2(s^2+1)}{(s^2-1)^2} = \frac{1}{(s-1)^2} + \frac{1}{(s+1)^2} \quad ...(5)$$

and $$\bar{y} - \frac{2}{(s^2-1)^2} = -\frac{1}{2}\left[\frac{1}{s+1} - \frac{1}{s-1} - \frac{1}{(s+1)^2} + \frac{1}{(s-1)^2}\right] \quad ...(6)$$

Taking Inverse Laplace transform of both sides of (5) and (6), we get

$$x = L^{-1}\left[\frac{1}{(s-1)^2} + \frac{1}{(s+1)^2}\right] = te^{t} + te^{-t} = 2t\left(\frac{e^{t} + e^{-t}}{2}\right) = 2t\cosh t$$

and $$y = -\frac{1}{2}L^{-1}\left(\frac{1}{s+1} - \frac{1}{s-1} - \frac{1}{(s+1)^2} + \frac{1}{(s-1)^2}\right)$$

$$= -\frac{1}{2}(e^{-t} - e^{t} - te^{-t} + te^{t}) = \frac{e^{t} - e^{-t}}{2} - t\left(\frac{e^{t} - e^{-t}}{2}\right) = (1-t)\sinh t$$

Hence, $$x = 2t\cosh t, \; y = (1-t)\sinh t.$$ **Ans.**

EXERCISE 47.11

Solve the following:

1. $\frac{dx}{dt}+4y=0, \frac{dy}{dt}-9x=0$. Given $x = 2$ and $y = 1$ at $t = 0$.

Ans. $x=-\frac{2}{3}\sin 6t+2\cos 6t, y=\cos 6t+3\sin 6t$

2. $4\frac{dy}{dt}+\frac{dx}{dt}+3y=0,\ 3\frac{dx}{dt}+2x+\frac{dy}{dt}=1$ under the condition $x = y = 0$ at $t = 0$.

Ans. $x=\frac{1}{2}-\frac{1}{5}e^{-t}-\frac{3}{10}e^{-\frac{6}{11}t}, y=\frac{1}{5}e^{-t}-\frac{1}{5}e^{-\frac{6}{11}t}$

3. $\frac{dx}{dt}+5x-2y=t, \frac{dy}{dt}+2x+y=0$ being given $x = y = 0$ when $t = 0$.

Ans. $x=-\frac{1}{27}(1+6t)e^{-3t}+\frac{1}{27}(1+3t), y=-\frac{2}{27}(2+3t)e^{-3t}-\frac{2t}{9}+\frac{4}{27}$

4. The currents i_1 and i_2 in mesh are given by the differential equations:

$$\frac{di_1}{dt}-\omega i_2=a\cos pt, \quad \frac{di_2}{dt}+\omega i_1=a\sin pt$$

Find the currents i_1 and i_2 by Laplace transform, if $i_1 = i_2 = 0$ at $t = 0$.

Ans. $i_1=\frac{a}{p+\omega}(\sin\omega t+\sin pt), i_2=\frac{a}{p+\omega}(\cos\omega t-\cos pt)$

47.15 SOLUTION OF PARTIAL DIFFERENTIAL EQUATION BY LAPLACE TRANSFORM

Example 70. *Solve the differential equation using Laplace transform method:*

$$\frac{\partial y}{\partial t}=3\frac{\partial^2 y}{\partial t^2} \quad \textit{where} \quad y\left(\frac{\pi}{2},t\right)=0, \left(\frac{\partial y}{\partial x}\right)_{x=0}=0 \quad \textit{and } y(x,0)=30\cos 5x.$$

(U.P., II Semester Summer 2005)

Solution. Given equation is

$$\frac{\partial y}{\partial t}=3\frac{\partial^2 y}{\partial t^2}$$

Taking Laplace transform of both sides, we get

$$sL\{y\}-y(x,0)=3\frac{d^2}{dx^2}L\{y\}, \qquad [\text{Let } L\{y\}=\bar{y}]$$

$$s\,\bar{y}-y(x,0)=3\frac{d^2\bar{y}}{dx^2}$$

$$3\frac{d^2\bar{y}}{dx^2}-s\,\bar{y}=-30\cos 5x$$

$$\left(D^2-\frac{s}{3}\right)\bar{y}=-10\cos 5x$$

A.E. is $m^2 - \frac{s}{3} = 0 \qquad \Rightarrow \qquad m = \pm\sqrt{\frac{s}{3}}$

$$C.F. = C_1\, e^{x\sqrt{s/3}} + C_2\, e^{-x\sqrt{s/3}}$$

$$P.I. = \frac{1}{\left(D^2 - \frac{s}{3}\right)}(-10 \cos 5x)$$

$$P.I. = \frac{30 \cos 5x}{75 + s}$$

Thus $$\bar{y} = C_1\, e^{-x\sqrt{s/3}} + C_2\, e^{x\sqrt{s/3}} + \frac{30 \cos 5x}{75 + s} \qquad \dots (1)$$

$$\frac{\partial y}{\partial x} = 0 \quad \text{when} \quad x = 0$$

$$\Rightarrow \quad L\left\{\frac{\partial y}{\partial x}\right\} = 0 \quad \text{at} \quad x = 0$$

$$\Rightarrow \quad \frac{d\bar{y}}{dx} = 0 \quad \text{at} \quad x = 0$$

Again, $$y\left(\frac{\pi}{2}, t\right) = 0 \Rightarrow L\left\{y\left(\frac{\pi}{2}, t\right)\right\} = 0 \Rightarrow \bar{y}\left(\frac{\pi}{2}, s\right) = 0$$

$$\frac{d\bar{y}}{dx} = \sqrt{\frac{s}{3}}\left[Ae^{x\sqrt{s/3}} - Be^{-x\sqrt{s/3}} - \frac{150 \sin 5x}{75 + s}\right]$$

Putting $$\frac{d\bar{y}}{dx} = 0 \quad \text{at} \quad x = 0$$

$$0 = \sqrt{\frac{s}{3}}[A - B] \Rightarrow A = B$$

Equation (1) becomes,

$$\bar{y} = A\left[e^{x\sqrt{s/3}} + e^{-x\sqrt{s/3}}\right] + \frac{30 \cos 5x}{75 + s} \qquad \dots (2)$$

Subjecting this to the condition

$$\bar{y}\left(\frac{\pi}{2}, s\right) = 0$$

$$0 = A\left[e^{\frac{\pi}{2}\sqrt{s/3}} + e^{-\frac{\pi}{2}\sqrt{s/3}}\right] + \frac{30 \cos (5\pi/2)}{75 + s}$$

$$24 \cosh\left[\frac{\pi}{2}\sqrt{\frac{s}{3}}\right] = 0 \qquad \Rightarrow \qquad A = 0$$

From equation (2), $\bar{y} = \frac{30 \cos 5x}{75 + s}$, taking inverse Laplace, we get

$$y = 30\, e^{-75\, t} \cos 5x$$ **Ans.**

EXERCISE 47.12

Solve the following differential equations using Laplace transform method:

1. $\dfrac{\partial u}{\partial t}=\dfrac{\partial^2 u}{\partial x^2}$ if $u\,(x\,,\,0)=\sin\pi x$ **Ans.** $u = \sin\pi x.\ e^{-p2t}$

2. $\dfrac{\partial u}{\partial t}=4\dfrac{\partial^2 u}{\partial x^2}$ if $u\,(x,0)=4x-\dfrac{1}{2}x^2$ **Ans.** $u=\left(4x-\dfrac{x^2}{2}\right)e^{-p^2 t}$

3. $\dfrac{\partial^2 u}{\partial t^2}=\dfrac{\partial^2 u}{\partial x^2}$ if $u(x,0)=\dfrac{1}{2}x\,(1-x)$ **Ans.** $u=\dfrac{x}{2}(1-x)\cos pt+C_2\sin pt\,(C_3\cos px+C_4\sin px)$

4. $16\dfrac{\partial^2 u}{\partial t^2}=\dfrac{\partial^2 u}{\partial x^2}$ if $u(x,0)=x^2\,(5-x)$ **Ans.** $u=x^2\,(5-x)\cos pt+C_4\sin pt\left(C_1\cos\dfrac{px}{4}+C_2\sin\dfrac{px}{4}\right)$

5. $\dfrac{\partial u}{\partial t}=\dfrac{\partial^2 u}{\partial x^2}$ if $u\,(x,\,0)=\begin{cases}2x, & \text{when } 0\le x\le\dfrac{1}{2}\\[2mm] 2(1-x), & \text{when } \dfrac{1}{2}\le x\le 1\end{cases}$

Dirac-Delta Function

48.1 DEFINITION

The function δ (x) which is zero everywhere except at $x = 0$, and tends to infinity in such a manner that

$$\int_{-\infty}^{+\infty} \delta(x)\, dx = 1$$

$$\delta(x) = \begin{cases} 0, & if\ x \neq 0 \\ \infty, & if\ x = 0 \end{cases}$$

48.2 IMPULSE FUNCTION

When a large force acts for a short time, then the product of the force and the time is called impulse in applied mechanics. The unit impulse function is the limiting function.

$$\delta(t) = \begin{bmatrix} \frac{1}{\varepsilon}, & 0 < t < \varepsilon \\ 0, & \text{otherwise} \end{bmatrix}$$

The value of the function (height of the strip in the figure) becomes infinite as $\varepsilon \to 0$ and the area of the rectangle is unity.

(1) The Unit Impulse function is defined as follow:

$$\delta(t) = \begin{cases} \infty & \text{for } t = 0 \\ 0 & \text{for } t \neq 0 \end{cases}$$

and $$\int_0^{\infty} \delta(t).dt = 1$$ [Area of strip = 1]

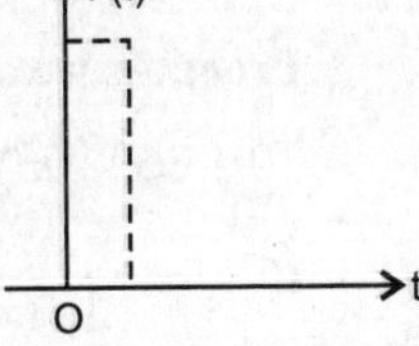

For example Hammerring the iron bar.

48.3 PROPERTIES

(1) $$\boxed{\int_0^{\infty} f(t)\,\delta(t)\,dt = f(0)}$$

If $f(x)$ is defined at $x = 0$, then the action of δ(s) to $f(x)$ is $f(0)$. This is expressed by an analytic expression.

$$\int_{0}^{+\infty} f(x)\,\delta(x)dx = f(0)$$

Here, L.H.S. can depend only on value of $f(x)$ very close to origin.

If we shift the singularity at $x = a$, then we can transform equation to

$$\int_{-\infty}^{+\infty} f(x)\,\delta(x-a)dx = f(a)$$

Symbolically, $f(x)\,\delta(x-a) = f(a)\,\delta(x-a)$

Example 1. *Evaluate* $\int_{-\infty}^{\infty} e^{-5t}\,\delta(t-2)dt.$

Solution. $\int_{-\infty}^{\infty} e^{-5t}\,\delta(t-2)\,dt = e^{-5\times 2} = e^{-10}$

(2) $\boxed{\int_{-\infty}^{\infty} f(t)\delta'(t-a)dt = -f'(a)}$

Proof. $\int_{-\infty}^{\infty} f(t)\delta'(t-a)dt = [f(t).\delta(t-a)]_{-\infty}^{\infty} - \int_{-\infty}^{\infty} f'(t)\,\delta(t-a)dt$

$$= 0 - 0 - f'(a) = -f'(a)$$

(3) $\delta(x)$ is an even function of x.

$$\delta(-x) = \delta(x)$$

(4) Put $f(x) = x$ in $f(x)\,\delta(x-a) = f(a)\,\delta(x-a)$

$$\therefore \quad x\,\delta(x) = 0$$

(5) $$\delta(ax) = \frac{1}{|a|}\delta(x)$$

(6) $$\int \delta(a-x)\,\delta(x-b)dx = \delta(a-b)$$

(7) Symmetry property

$$\delta(x-a) = \delta(a-x)$$

(8) $$\delta(x^2-a^2) = \frac{\delta(x-a)+\delta(x+a)}{2|a|}$$

Proof. As we observed that it vanishes at $x = a$ and $x = -a$ and hence it has two components;

$$\delta(x^2-a^2) = \delta[(x-a)(x+a)] = \frac{1}{|x+a|}\delta(x-a) + \frac{1}{|x-a|}\delta(x+a) \qquad \text{[Property 5]}$$

$$\int f(x)\,\delta(x^2-a^2)dx = \int \frac{1}{|x+a|} f(x)\,\delta(x-a)dx + \int \frac{1}{|x-a|} f(x)\,\delta(x+a)dx$$

$$= \frac{1}{|2a|} f(a) + \frac{1}{|2a|} f(-a)$$

$$= \frac{1}{|2a|}\left[\int f(x)\,\delta(x-a)dx + \int f(x)\,\delta(x-a)dx\right]$$

$$\Rightarrow \quad \delta(x^2-a^2) = \frac{1}{|2a|}[\delta(x-a)+\delta(x+a)]$$ **Proved.**

48.4 LAPLACE TRANSFORM

We know that

$$\int_0^{\infty} f(t)\,\delta(t-a) = f(a)$$

Replacing $f(t)$ by e^{-st} in above equation, we get

$$\int_0^{\infty} e^{-st}(t-a)dt = [e^{-st}]_{t=a} = e^{-sa}$$

$$\therefore \quad L[\delta(t-a)] = e^{-sa}$$

If $a = 0$, then $L[\delta(t)] = 1$.

Example 2. *Find the Laplace transform of $t^3\,\delta(t-4)$.*

Solution. $$L[t^3\,\delta\,(t-4)] = \int_0^{\infty} e^{-st}\,t^3\,\delta(t-4)dt = 4^3 e^{-4s}$$ **Ans.**

Example 3. *Find the Laplace transform of $e^{-4t}\,\delta\,(t-3)$.*

Solution. $$L[e^{-4t}\,\delta(t-3)] = \int_0^{\infty} e^{-st}\,e^{-4t}\,\delta(t-3)dt = e^{-3(s+4)}$$ **Ans.**

48.5 FOURIER TRANSFORM

We know that $$\int_{-\infty}^{+\infty} f(t)\,\delta(t-a)dt = f(a)$$

Replacing $f(t)$ by e^{-ist} in above equation, we get

$$\int_{-\infty}^{+\infty} e^{-ist}\,\delta(t-a)dt = e^{-isa}$$

$$\therefore \quad F.T\,[\delta(t-a)] = \frac{e^{-isa}}{2\pi}.$$ If $a = 0$, then $F.T\,[\delta(t)] = \dfrac{1}{2\pi}$.

ECERCISE 48.1

Evaluate the following:

1. $\int_0^{\infty} e^{-3t}\,\delta(t-4)dt$ **Ans.** e^{-12}

2. $\int_{-\infty}^{\infty} \sin 2t\,\delta\left(t-\frac{\pi}{4}\right)dt$ **Ans.** 1

3. $\int_{-\infty}^{\infty} e^{-3t}\,\delta'(t-2)dt$ **Ans.** $3e^{-6}$

Find the Laplace transform of:

4. $\dfrac{\delta(t-4)}{t}$ **Ans.** $\dfrac{e^{-4s}}{4}$

5. $\cos t \log t\,\delta(t-\pi)$ **Ans.** $-e^{-\pi s}\log \pi$

Find fourier transform of

6. $f(x) = \begin{cases} 1/\epsilon, |x| \le \epsilon \\ 0, |x| > t \end{cases}$ **Ans.** $\sqrt{\dfrac{2}{\pi}}$

Taking the limit as $\epsilon \to 0$ and dicuss the result.

49

CHAPTER

TENSOR ANALYSIS

49.1 INTRODUCTION

Scalars are specified by magnitude only, ***vectors*** have magnitude as well as direction.

Tensors are associated with magnitude and two or more directions. For example, the stress of an elastic solid at a point depends upon two directions. One of the directions is given by the normal to the area, while the other is that of the force on it.

Tensors are similar to vectors. A vector can be specified by its components (Magnitude and direction). A tensor can be specified only by its components which depend upon the system of reference. The components of the same tensor will be different for two different sets of axes with different orientations.

Tensors analysis is suitable for mathematical formulation of natural laws in forms which are invariant with respect to different frames of reference. That is why Einstein used tensors for the formulation of his Theory of Relativity.

49.2 CO-ORDINATE TRANSFORMATION

If we have two systems of rectangular co-ordinate axes *OX, OY, OZ; OX', or, OZ'*; having same origin such that the direction cosines of the lines *OX', OY', OZ'* relative to the system *OX YZ* are $l_1, m_1, n_1; l_2, m_2, m_2, n_2; l_3, m_3, n_3$

Two equivalent systems of transformation equations express x', y', z' in terms of x, y, z and vice versa.

$$\left.\begin{aligned} x' &= l_1 x + m_1 y + n_1 z \\ y' &= l_2 x + m_2 y + n_2 z \\ z' &= l_3 x + m_3 y + n_3 z \end{aligned}\right\} \quad \dots (1)$$

$$\left.\begin{aligned} x &= l_1 x' + l_2 y' + l_3 z' \\ y &= m_1 x' + m_2 y' + m_3 z' \\ z &= n_1 x' + n_2 y' + n_3 z' \end{aligned}\right\} \quad \dots (2)$$

where (x', y', z') and (x, y, z) are co-ordinates of point P relative to two systems of co-ordinate axes. System of transformation equation shown above in (1) and (2) can be written as

	x	y	z
x'	l_1	m_1	n_1
y'	l_2	m_2	n_2
z'	l_3	m_3	n_3

49.3 SUMMATION CONVENTION

The sum of the following $a_1 x_1 + a_2 x_2 + a_3 x_3 + ... + a_n x_n$ ...(1)

can be written in brief as $\sum_{i=1}^{i=n} a_i x_i$ ... (2)

More simplified and compact notation for (2) used by Einstein is $a_i x^i$. ...(3)

In (3) we have omitted Σ-sign.

$$a_i x^i = a_1 x^1 + a_2 x^2 + a_3 x_3^3 + ...a_n x^n$$

We write $x_1, x_2, x_3, ... x_n$ as $x^1, x^2, x^3, ... x^n$ in tensor analysis. These superscripts do not stand for powers of x but indicate different symbols. The power of x^i is written as $(x^i)^2, (x^i)^3 ...$

Example 1. *Write out $a_{rs} x^s = b_r (r, s = 1, 2, 3 .... n)$ in full:*

Solution. $a_{rs} x^s = b_r$

$a_{1s} x^s + a_{2s} x^s + a_{3s} x^s + ... + a_{ns} x^s = b_1 + b_2 + b_3 + ... + b_n$ (r occurs 1 to n)

$$(a_{11} x^1 + a_{12} x^2 + a_{13} x^3 + ... a_{1n} x^n) + (a_{21} x^1 + a_{22} x^2 + a_{23} x^3 + ... + a_{2n} x^n)$$
$$+ (a_{31} x^1 + a_{32} x^2 + a_{33} x^3 + ... +....a_{3n} x^n) + ... = b_1 + b_2 + b_3 +....+ b_n$$

Example 2. *If $f = f(x^1, x^2, x^3, ... x^n)$ then show that $df = \frac{\partial f}{dx^i} dx^i$*

Solution. $df = \frac{\partial f}{dx^1} dx^1 + \frac{\partial f}{dx^2} dx^2 + ... + \frac{\partial f}{dx^n} dx^n = \frac{\partial f}{\partial x^i} dx^i$ **Proved.**

49.4 SUMMATION OF CO-ORDINATES

The equations of co-ordinates can be written in very compact form in terms of summation convention. We write (x_1, x_2, x_3) and $(\bar{x}_1, \bar{x}_2, \bar{x}_3)$ instead of (x, y, z) and (x', y', z') and denote the co-ordinate axes as OX_1, OX_2, OX_3 and $O\bar{X}_1, O\bar{X}_2, O\bar{X}_3$. Also we denote $x_i, \bar{x}_j$ as the co-ordinates of a point P relative to the two systems of axes; where $i = 1, 2, 3, j = 1, 2, 3$.

Let l_{ij} denote the cosines of the angles between $OX_i, O\bar{X}_j$. In general $l_{ij} \neq l_{ji}$

The equation of co-ordinate transformation can be written as

$$\left.\begin{aligned} \bar{x}_1 &= l_{11} x_1 + l_{21} x_2 + l_{31} x_3 \\ \bar{x}_2 &= l_{12} x_1 + l_{22} x_2 + l_{32} x_3 \\ \bar{x}_3 &= l_{13} x_1 + l_{23} x_2 + l_{33} x_3 \end{aligned}\right\} \quad ...(1a)$$

$$\left.\begin{aligned} x_1 &= l_{11} \bar{x}_1 + l_{12} \bar{x}_2 + l_{13} \bar{x}_3 \\ x_2 &= l_{21} \bar{x}_1 + l_{22} \bar{x}_2 + l_{23} \bar{x}_3 \\ x_3 &= l_{31} \bar{x}_1 + l_{32} \bar{x}_2 + l_{33} \bar{x}_3 \end{aligned}\right\} \quad ...(1b)$$

These equations of co-ordinate transformation can be represented by means of a table form such that

	x_1	x_2	x_3
$\overline{x}_1$	l_{11}	l_{21}	l_{31}
$\overline{x}_2$	l_{12}	l_{22}	l_{32}
$\overline{x}_3$	l_{13}	l_{23}	l_{33}

X_3
$\overline{X}_3$
(l_{31}, l_{32}, l_{33})
(l_{13}, l_{23}, l_{33})
(l_{12}, l_{22}, l_{32})
$\overline{X}_2$
X_2
O
(l_{21}, l_{22}, l_{23})
(l_{11}, l_{12}, l_{13})
X_1
(l_{11}, l_{21}, l_{31})
$\overline{X}_1$

Adopting summation on convention *i.e.*,

$$a_{11} + a_{22} + a_{33} = a_{ij}$$

$a_{ip}\, b_{iq} = a_{1p} b_{1q} + a_{2p}\, b_{2q} + a_{3p} b_{3q}$ we re-write above equations as

$$\overline{x}_1 = l_{i1}x_i \qquad x_1 = l_{1j}\overline{x}_j$$

$$\overline{x}_2 = l_{i2}x_i \qquad x_2 = l_{2j}\overline{x}_j$$

$$\overline{x}_3 = l_{i3}x_i \qquad x_3 = l_{3j}\overline{x}_j$$

We can re-write these equations in single equation in the form.

$$\overline{x}_j = l_{ij}x_i,\ x_i = l_{ij}\overline{x}_j$$

which are complete equivalents of the equations of co-ordinate transformation from either system to another.

49.5 RELATION BETWEEN THE DIRECTION COSINES OF THREE MUTUALLY PERPENDICULAR STRAIGHT LINES

The direction cosines of any three mutually perpendicular straight lines $O\overline{X}_1$, $O\overline{X}_2$, $O\overline{X}_3$ relative to the system OX_1, OX_2, $0X_3$ are $l_{11}, l_{21}, l_{31}, l_{12}, l_{22}, l_{32}, l_{23}, l_{33}$.

The relation between these direction cosines are

$$l_{11}l_{11} + l_{21}l_{21} + l_{31}l_{31} = l_{j1}\, l_{j1} = 1,\ l_{12}l_{12} + l_{22}l_{22} + l_{32}l_{32} = l_{j2}l_{j2} = 1$$

$$l_{13}l_{13} + l_{23}l_{23} + l_{33}l_{33} = l_{j3}\, l_{j3} = 1.$$

Similarly,

$$l_{11}l_{12} + l_{21}l_{22} + l_{31}l_{32} = l_{j1}\ l_{j2} = 0,\ l_{12}l_{13} + l_{22}l_{23} + l_{32}l_{33} = l_{j2}l_{j3} = 0$$

$$l_{13}l_{11} + l_{23}l_{21} + l_{33}l_{31} = l_{j3}\ l_{j1} = 0$$

Finally, we can write these equations by means of a single equation as

$$l_{ij}l_{kj} = \begin{cases} 1, \text{ when } i = k \\ 0, \text{ when } i \neq k \end{cases} \quad \text{or} \quad \delta_{ik} = \begin{cases} 1, \text{ when } i = k \\ 0, \text{ when } i \neq k \end{cases}$$

where δ_{ik} is the kronecker delta.

or $$\delta_{ik} = l_{ij}l_{kj}$$

Now, we know that $\overline{x}_j = l_{ij}x_i$

Multiplying both sides by l_{jk} then

or $$l_{jk}\overline{x}_j = l_{ij}l_{jk}x_x \quad \Rightarrow \quad l_{jk}\overline{x}_j = \delta_{ik}x_i$$

putting $i = k$ *i.e.*, $\delta_{ik} = 1$ when $i = k$

$$\delta_{kk}x_k = l_{jk}\overline{x}_j \Rightarrow x_k = l_{jk}\overline{x}_j$$

49.6 TRANSFORMATION OF VELOCITY COMPONENTS ON CHANGE FROM ONE SYSTEM OF RECTANGULAR AXES TO ANOTHER

We know that with the help of parallelogram law of velocities, that any given velocity can be represented by means of its three components along three mutually perpendicular lines and the three components characterise velocity completely. The components change as we pass from one system of mutually perpendicular lines to another.

Let OX_1, OX_2, OX_3 and $O\overline{X}_1$, $O\overline{X}_2$, $O\overline{X}_3$ are two systems of rectangular axes and suppose that $l_i, \overline{l}_j$ are the direction cosines of the line of action of the velocity and v, denote the magnitude of the velocity. Then

$$v_i = l\,v,\ \overline{v}_j = v\overline{l}_j \qquad ...(1)$$

where v_i and $\overline{v}_j$, denotes the components of velocity relative to the two systems of axes.

By the equation of co-ordinate transformation, we have

$$\overline{l}_j = l_{ij}l_i,\ l_i = l_{ij}\overline{l}_j \qquad ...(2)$$

From (1) and (2), we have

$$\frac{\overline{v}_j}{v} = l_{ij}\frac{v_i}{v},\ \frac{v_i}{v} = l_{ij}\frac{\overline{v}_j}{v}. \qquad i.e.,\ \overline{v}_j = l_{ij}v_i$$

Thus we see equation of transformation of velocity components are same as for the transformation of co-ordinate of points.

49.7 RANK OF A TENSOR

The rank of a tensor is the number (without counting an index which appears once as a subscript) of indices in the symbol representing a tensor. For example

Tensor	*Symbol*	*Rank*
Scalar	A	zero
Contravariant Tensor	B^i	1
Covariant Tensor	C_k	1
Covariant Tensor	D_y	2
Mixed Tensor	E^{il}_{jkl}	3

In an n-dimensional space, the number of components of a tensor of rank r is n^r.

49.8 FIRST ORDER TENSORS

Definition. Any entity representable by a set of three numbers (called components) relatively to a system of rectangular axes is called first order tensors, if its components a_r, a_j relatively to any two systems of rectangular axes OX_1, OX_2, OX_3, $O\overline{X}_1$, $O\overline{X}_2$, $O\overline{X}_3$ are connected by the relation,

$$\overline{a}_j = l_{ij}a_i \qquad ...(3)$$

$$\Rightarrow \qquad a_i = l_{ij}\,\overline{a}_j$$

l_{ij} being cosines of angle between OX_i and $O\overline{X}_j$. A tensor of first order is also called a *vector.*

Note. Consider any two tensors of first order and let $a_i, b_j, \overline{a}_p, \overline{b}_q$; be the components of the same relatively to two different systems of axes, we have

$$\overline{a}_p = l_{ip}a_i,\ \overline{b}_q = l_{jq}b_j$$

where l_{ip} and l_{jq} have their usual meanings. This gives

$$\overline{a}_p\overline{b}_q = l_{ip}a_i l_{jq}b_j = l_{ip}l_{jq}a_i b_j \quad \text{... (1)}$$

The R.H.S. of (1) denotes the sum of 9 terms obtained by giving all possible pair of values to the dummy suffixes i,j so that each components of $\overline{a}_p\overline{b}_q$ is expressed as a linear combination of nine components of the set a_i, b_j, the coefficient being dependent only upon the positions of the two systems of axes relative to each other and not on the components of the sets $\overline{a}_p\overline{b}_q$, $a_i b_j$

49.9 SECOND ORDER TENSORS

Definition. Any entity representanle by a two suffixes set relatively to a system of rectangular axes is called a second order tensor, if the sets $a_{ij}, \overline{a}_{pq}$ representing the entity relative to any two systems of rectangular axes OX_1, OX_2, OX_3, $O\overline{X}_1, O\overline{X}_2, O\overline{X}_3$ are connected by relation.

$$\overline{a}_{pq} = l_{ip}\, l_{jp}\, a_{ij}$$

49.10 TENSORS OF ANY ORDER

Definition. Any entity representable by a set with *m*, suffixes relatively to a system of rectangular co-ordinate axes is called a tensor of order *m*, if the set a_{ijkl} ...... , $\overline{a}_{pqrs}$ .... representing the entity relatively to any two systems of rectangular axes OX_1, OX_2, OX_3, $O\overline{X}_1, O\overline{X}_2, O\overline{X}_3$ are connected by the relation

$$\overline{a}_{pqrs}\ldots\ldots = l_{ip}l_{jq}l_{kr}l_{ls}\ldots a_{ijkl}\ldots$$

We say that a_{ijkl} ..... are the components of tensor relatively to the rectangular system of axes OX_1, OX_2, OX_3.

49.11 TENSOR OF ZERO ORDER

Definition. Any entity representable by a single number such that the same number represents the entity irrespective of any underlying system of axes is called a tensor of order zero. A tensor of order zero is also called a *scalar*.

49.12 ALGEBRAIC OPERATIONS ON TENSORS

Theorem. If a_{ijkl} ...., b_{ijkl} .... are two tensors of the same order then

$$c_{ijkl}\ldots = a_{ijkl}\ldots + b_{ijkl}\ldots$$

is a tensor of the same order.

Proof. Let a_{ijkl} ...., b_{ijki} ..... and $\overline{a}_{pqrs}$ ....., $\overline{b}_{pqrs}$ ... be the components of the given tensors relatively to two systems OX_1, OX_2, OX_3, $O\overline{X}_1, O\overline{X}_2, O\overline{X}_3$.

We write

$$c_{ijkl}\ldots = a_{ijkl}\ldots + b_{ijkl}\ldots$$

$$\overline{c}_{pqrs}\ldots\ldots = \overline{a}_{pqrs}\ldots\ldots + \overline{b}_{pqrs}\ldots\ldots$$

Let l_{ij} denote the cosine of the angle between OX_i and OX_j.

$$\overline{c}_{pqrs}\ldots\ldots = l_{ip}l_{iq}l_{kr}\ldots\ldots c_{ijkl}\ldots\ldots \quad \text{...(1)}$$

As a_{ijkl} ..... and b_{ijkl} ...... are tensors, we have

$$\overline{a}_{pqrs}\ldots\ldots = l_{ip}l_{jq}l_{kr}l_{ls}\ldots\ldots a_{ijkl}\ldots\ldots \quad \text{...(2)}$$

$$\overline{b}_{pqrs}\ldots\ldots = l_{ip}l_{jq}l_{kr}l_{ls}\ldots\ldots b_{ijkl}\ldots\ldots \qquad \ldots(3)$$

Adding (2) and (3), we obtain (1).

Hence the theorem

Similarly, we can show for difference

$$\overline{a}_{pqrs}\ldots\ldots - \overline{b}_{pqrs}\ldots\ldots = l_{ip}l_{jq}l_{kr}l_{ls}\ldots\ldots\left[a_{ijkl}\ldots - b_{ijkl}\ldots\right]$$

$$\overline{d}_{pqrs}\ldots\ldots = l_{ip}l_{jq}l_{kr}l_{ls}\ldots\ldots d_{ijkl}\ldots\ldots$$

Proved.

49.13 PRODUCT OF TWO TENSORS

Theorem. If a_{ijkl} ...., b_{pqrs} ....be two tensors of order α and β respectively, then

$c_{ijkl....pqrs}\ldots = a_{ijkl}\ldots b_{pqrs}\ldots$ is a tensor of order α + β.

Proof. Let a_{ijkl} ....., b_{pqrs} ...... and $\overline{a}_{i_1j_1k_1l_1}\ldots,\overline{b}_{p_1j_1k_1l_1\ldots},\ \overline{b}_{p_1q_1r_1s_1}\ldots$ be the components of given tensor relatively to two systems OX_1, OX_2, OX_3, $O\overline{X}_1, O\overline{X}_2, O\overline{X}_3$ we write

$$c_{ijkl\ldots pqrs} = a_{ijkl}\ldots b_{pqrs}$$

$$\overline{c}_{i_1j_1k_1l_1\ldots\ldots p_1q_1r_1s_1}\ldots = \overline{a}_{i_1j_1k_1l_1}\ldots\overline{b}_{p_1q_1r_1s_1}\ldots$$

Let l_{ij} be the direction cosines of the angle between OX_i and $O\overline{X}_j$, then

$$\overline{c}_{i_1j_1k_1l_1\ldots\ldots p_1q_1}\ldots = l_{ii_1}l_{jj_1}\ldots l_{pp_1}l_{qq_1}\ldots c_{ijk\ldots pqr\ldots\ldots} \qquad \ldots(1)$$

As a_{ijkl} ..... and b_{pqrs} ...... are tensors we have

$$\overline{a}_{i_1j_1k_1l_1}\ldots = l_{ii_1}l_{jj_1}l_{kk_1}\ldots a_{ijkl}\ldots \qquad \ldots(2)$$

$$\overline{b}_{p_1q_1r_1s_1}\ldots = l_{pp_1}l_{qq_1}l_{rr_1}\ldots b_{pqrs}\ldots \qquad \ldots(3)$$

Multiplying (2) and (3) we get (1). The new tensor obtained is called product of the tensors.

Proved.

49.14 QUOTIENT LAW OF TENSORS

Thorem. If there be an entity representable by a multisuffix set a_{ij} relatively to any given system of rectangular axes and if $a_{ij}b_i$ is a vector, where b_i is any arbitrary vector whatsoever then a_{ij} is a tensor of order two.

Proof. $a_{ij}b_i = c_j$ so that c_j is a vector. Let $a_{ij}b_i$, c_j and $\overline{a}_{pq}, \overline{b}_p, \overline{c}_q$ be the components of the given entity and two vectors relatively to two systems of axes OX_1, OX_2, OX_3, $O\overline{X}_1, O\overline{X}_2, O\overline{X}_3$, then we have

$$a_{ij}b_i = c_j \qquad \ldots(1)$$

$$\overline{a}_{pq}\overline{b}_p = \overline{c}_q \qquad \ldots(2)$$

Also, b_i, c_j being vectors, we have

$$\overline{c}_q = l_{jq}c_j \qquad \ldots(3)$$

$$b_i = l_{ip}\overline{b}_p \qquad \ldots(4)$$

From these, we have $\overline{a}_{pq}\overline{b}_p = \overline{c}_q = l_{jq}c_j - l_{jq}a_{ij}b_i = l_{jq}a_{ij}l_{ip}\overline{b}_p = l_{ip}l_{jq}a_{ij}\overline{b}_p$

i.e., $\left(\overline{a}_{pq} - l_{ip}l_{iq}a_{ij}\right)\overline{b}_p = 0$

As the vector $\overline{b}_p$ is arbitrary, we consider three vectors whose components relatively to $O\overline{X}_1, O\overline{X}_2, O\overline{X}_3$ are 1, 0, 0 ; 0, 1, 0 ; 0, 0, 1.

For these vectors, we have from (5)

$$\overline{a}_{1q} - l_{i1} l_{jq} a_{ij} = 0, \qquad \overline{a}_{2q} - l_{i2} l_{jq} a_{ij} = 0, \qquad \overline{a}_{3q} - l_{i3} l_{jq} a_{ij} = 0$$

These are equivalent to

$$\overline{a}_{pq} - l_{ip} l_{jq} a_{ij} = 0$$

i.e.,
$$\overline{a}_{pq} = l_{ip} l_{jq} a_{ij}, \qquad \text{[This shows that } a_{ij} \text{ is of second order]}$$

so that the components of the given entity obey the tensorial transformation laws. Hence the result. **Proved.**

49.15 CONTRACTION THEOREM

(D.U. April 2010)

Theorem. If $a_{ijkl......}$ is a tensor of order *m,* then the set obtained on identifying any two suffixes is a tensor of order $(m-2)$.

Proof. Let $a_{ijkl.....}$, $\overline{a}_{pqrs.....}$, be the components of the given tensor relatively to two coordinate systems of axes OX_1, OX_2, OX_3, $O\overline{X}_1, O\overline{X}_2, O\overline{X}_3$, so that we have,

$$\overline{a}_{pqrs.....} = l_{ip} l_{jq} l_{kr} .... a_{ijkl.....} \qquad ...(1)$$

Let us identify q and s then,

$$\overline{a}_{pqrs.....} = l_{ip} l_{js} l_{kr} l_{ls} .... a_{ijkl.....}$$

$$\overline{a}_{pr.....} = l_{ip} l_{kr} \delta_{jl} .... a_{ijkl.....} \qquad \left[\begin{matrix} \because \delta_{jl} = 0, & j \neq l \\ \delta_{jl} = 1, & j = l \end{matrix}\right]$$

$$= l_{ip} l_{kr} .... a_{ilkl....} = l_{ip} l_{kr} .... a_{ik....}$$

This shows that the order of the tensor reduces by two.

Hence the theorem. **Proved.**

49.16 SYMMETRIC AND ANTISYMMETRIC TENSORS

If $\quad A^k_{rs} = A^k_{sr} \quad$ or $\quad \left(A^{r,s}_k = A^{s,r}_k\right)$

then A^k_{rs} (or A^k_{sr}) are said to be symmetric tensors.

If $\quad B^k_{rs} = -B^k_{sr} \quad$ or $\quad \left(B^{rs}_k = -B^{sr}_k\right)$

then B^k_{rs} (or B^{sr}_k) are known as antisymmetric tensors.

The symmetric (or antisymmetric) property is conserved under a transform of co-ordinates.

49.17 SYMMETRIC AND SKEW SYMMETRIC TENSORS

Invariance of the symmetric and skew-symmetric character of the sets of components of tensors

Theorem. Show that if $a_{ijkl.....}$ is symmetric (skew-symmetric) in any two suffixes, then so is also $\overline{a}_{pqrs.....}$ in the same suffixes.

Proof. Let $a_{ijkl.....}$, $\overline{a}_{pqrs.....}$ be the components of a tensor respectively to two systems of axes OX_1, OX_2, OX_3, $O\overline{X}_1, O\overline{X}_2, O\overline{X}_3$. Then we have

$$\bar{a}_{pqrs}\ldots.. = l_{ip}l_{jq}l_{kr}l_{ls}\ldots a_{ijkl}\ldots.. \qquad \ldots(1)$$

Now, suppose that a_{ijkl}...... is symmetric in the second and fourth suffixes. Interchanging q and s on the two sides of (1) we obtain

$$\bar{a}_{psrq}\ldots.. = l_{ip}l_{js}l_{kr}l_{lq}\ldots a_{ijkl}\ldots.. \qquad \ldots(2)$$

As j and l are dummy, we can interchange them. Then interchanging j and l on the R.H.S. of (2) we get

$$\bar{a}_{psrq}\ldots.. = l_{ip}l_{ls}l_{kr}l_{jq}\ldots a_{ijkl}\ldots.. \qquad \ldots(3)$$

$$\bar{a}_{psrq}\ldots.. = l_{ip}l_{jq}l_{kr}l_{ls}\ldots a_{ijkl}\ldots..$$

The set a_{ijkl}.........is symmetric in the second and fourth suffixes.

Now from (1) and (3) we have

$$\bar{a}_{pqrs}\ldots.. = \bar{a}_{psrq}\ldots...$$

Hence the result. **Proved.**

Definition. A tensor is said to be symmetric (skew-symmetric) in any two suffixes if its components relatively to every co-ordinate system are symmetric (skew-symmetric) in the two suffixes, in question.

A tensor is said to be symmetric (skew-symmetric) if it is so in every pair of suffixes, *e.g.*,

If u_i, v_j be any two vectors then the two second order tensors $u_iv_j + u_jv_i$, $u_iv_j - u_jv_i$ are respectively symmetric and skew-symmetric.

49.18 THEOREM

Every second order tensor can be expressed as the sum of a symmetric and a skew-symmetric tensor.

Proof. Let a_{ij} be any tensor of order 2. Now,

$$\bar{a}_{pq} = l_{ip}l_{jq}a_{ij} = l_{jp}l_{iq}a_{ji} \qquad \ldots(1)$$

where we have interchanged the two dummy suffixes i and j. Then (1) shows that a_{ij} is also a tensor of order two.

Now, $$a_{ij} = \frac{1}{2}\left[a_{ij} + a_{ji}\right] + \frac{1}{2}\left[a_{ij} - a_{ji}\right]$$

$$= \text{symmetric} + \text{skew-symmetric}$$

Thus a_{ij} is the sum of symmetric and skew symmetric tensors. **Proved.**

49.19 A FUNDAMENTAL PROPERTY OF TENSORS

Theorem. If the components of a tensor relatively to any one system of co-ordinate axes are all zero, then the components relatively to every system of co-ordinate axes are also zero.

Proof. Consider a tensor whose components relatively to the systems of axes OX_1, OX_2, OX_3, $O\bar{X}_1, O\bar{X}_2, O\bar{X}_3$ are a_{ijkl} ...., $\bar{a}_{pqrs}$ ......and let a_{ijkl} ....= 0 for every system of values of i, j, k, l...... we have

$$\bar{a}_{pqrs}\ldots.. = l_{ip}l_{jq}\,l_{kr}\,l_{ls}\ldots a_{ijkl}\ldots.. = 0$$

for every system of values of p, q, r, s..... **Proved.**

49.20 ZERO TENSOR

Def. A tensor whose components relatively to one co-ordinate system and, also relatively to every co-ordinate system are all zero is known as zero tensor.

A zero tensor of every order is denoted by 0.

EXERCISE 49.1

1. Write the following using summation convention:

(a) $(x^1)^1 + (x^1)^2 + (x^1)^3 \ldots\ldots (x^1)^n$ **Ans.** $(x^1)^i$

(b) $(x^1)^2 + (x^2)^2 + (x^3)^2 + \ldots (x^n)^2$ **Ans.** $(x^i)^2$

(c) $\frac{df}{dt} = \frac{\partial f}{\partial x^1}\frac{dx^1}{dt} + \frac{\partial f}{\partial x^2}\frac{dx^2}{dt} + \ldots\ldots + \frac{\partial f}{\partial x^n}\frac{dx^n}{dt}$ **Ans.** $\frac{df}{dt} = \frac{\partial f}{\partial x^i}\frac{dx^i}{dt}$

2. Write out all the tensor in $S = a_{ij}x^i x^j$ taking $n = 3$.

Ans. $S = (a_{11}x^1x^1 + a_{12}x^1x^2 + a_{13}x^1x^3) + (a_{21}x^2x^1 + a_{22}x^2x^2 + a_{23}x^2x^3) + (a_{31}x^3x^1 + a_{32}x^3x^2 + a_{33}x^3x^3)$

3. Write the tensor contained in x^{pq}, x_{qr} if $n = 2$

Ans. $(x^{11} + x^{21})x_{11} + (x^{11} + x^{21})x_{12} + (x^{12} + x^{22})x_{21} + (x^{12} + x^{22})x_{22}$

4. How many equations in a four dimensional space are represented by $R^{\alpha}_{\beta rp} = 0$ **Ans.** 8

5. Show that every tensor can be expressed as the sum of two tensors, one of which is symmetric and the other skew-symmetric in a pair of covariant or contravariant indices.

6. Show that the symmetric (or antisymmetric) property of a tensor is conserved under a transformation of co-ordinates.

7. If A^1 and B_j, are components of a contravariant and covariant tensor of rank one, then show that $C^i_j = A^i B_j$ are the components of a mixed tensor of rank 2.

8. Write down the laws of transformation for the tensors A^{ij}_k and B^{ij}_{klm}

Ans. $\bar{A}^{ij}_k = \frac{\partial \bar{x}^i}{\partial x^p}\frac{\partial \bar{x}^j}{\partial x^q}\frac{\partial x^r}{\partial \bar{x}^k}A^{pq}_r$, $\bar{B}^{ij}_{klm} = \frac{\partial \bar{x}^i}{\partial x^p}\frac{\partial \bar{x}^j}{\partial x^q}\frac{\partial x^r}{\partial \bar{x}^k}\frac{\partial x^s}{\partial \bar{x}^l}\frac{\partial x^t}{\partial \bar{x}^m}B^{rq}_{rst}$

9. Evaluate (a) $\delta^i_j\,\delta^j_k$ (b) $\delta^i_j\delta^j_k\,\delta^k_l$ **Ans.** (a)δ^i_k(b) δ^i_j

10. Show that the velocity of a fluid at any point is a contravariant tensor of rank one.

49.21 TWO SPECIAL TENSORS

1. Alternate tensor

Consider an abstract entity of order 3 and dimension 3 such that its components relatively to every system of co-ordinate axes are the same and given by $\in_{ijk}$ where

$$\in_{ijk} = \begin{cases} 0 & \text{if any two of } ijk \text{ are equal} \\ 1 & \text{if } ijk \text{ is a cyclic permutation } 1, 2, 3 \\ -1 & \text{if } ijk \text{ is an anti cyclic permutation } 1, 2, 3 \end{cases}$$

Thus for unequal values of the suffixes, we have

$$\in_{123} = \in_{312} = \in_{231} = 1, \in_{132} = \in_{213} = \in_{321} = -1$$

Let OX_1, OX_2, OX_3, $O\bar{X}_1, O\bar{X}_2, O\bar{X}_3$ be two systems of rectangular axes. We want to show that $\in_{ijk}$ is a tensor of order three. Consider, now expression

$$l_{ip}\,l_{jq}\,l_{kr} \in_{ijk} \qquad \ldots(1)$$

For any given system of values $p, q, r,$ the expression (1) consists of a sum of $3^3 = 27$ terms of which 6 only are non-zero, for the other 21 terms corresponds to a case when atleast two of i, j, k are equal. The expression (1) can be written as in the form of determinant

$$\begin{vmatrix} l_{1p} & l_{2p} & l_{3p} \\ l_{1q} & l_{2q} & l_{3q} \\ l_{1r} & l_{2r} & l_{3r} \end{vmatrix}$$

From properties of determinants,

$$\text{Above determnant} = \begin{cases} 0 & \text{if any two of } p, q, r \text{ have equal value.} \\ 1 & \text{if } p, q, r \text{ is a cyclic permutation of } 1, 2, 3 \\ -1 & \text{if } p, q, r \text{ is a non cyclic permutation of } 1, 2, 3 \end{cases}$$

Thus we see that the components of the given entity in any two systems of rectangular axes satisfy the tensorial transformation equations so that the entity is a tensor. This tensor is known as *Alternate tensor.* Thus, we see alternate tensor is same as skew-symmetric tensor. $\in_{ijk}$, always denote the alternate tensor.

49.22 KRONECKER TENSOR

The symbol δ_l^k kronecker delta is defined as

$$\delta_l^k = 0 \text{ when } k \neq l$$

$$\delta_l^k = 0 \text{ when } k = l$$

It means $\delta_1^1 = \delta_2^2 = \ldots.. = \delta_n^n = 1$ and $\delta_1^2 = \delta_1^3 = \delta_2^1 = \delta_2^3 = \ldots\ldots = 0$

In general $A_{ij}\delta_k^i = A_{i1}\delta_k^1 = A_{i2}\delta_k^2 + \ldots. + A_{ik}\delta_k^k + \ldots. + A_{1n}\delta_k^n$

$$= 0 + 0 + \ldots. + A_{ik}(1) + \ldots.. + 0 = A_{ik}$$

Example 3. *If A^{ij} are the cofactors of a^{ij} in a determinant Δ of order three, then show that*

$$a_{ij}A^{kj} = \Delta\delta_i^k$$

Solution. We know that

$$a_{11}A^{11} + a_{12}A^{12} + a_{13}A^{13} = \Delta \quad \ldots (1)$$

$$a_{11}A^{21} + a_{12}A^{22} + a_{13}A^{23} = 0 \quad \ldots(2)$$

$$a_{11}A^{31} + a_{12}A^{32} + a_{13}A^{33} = 0 \quad \ldots(3)$$

These three equations can be written in brief as

$$a_{ij}A^{1j} = \Delta \quad \ldots (4) \qquad a_{ij}A^{2j} = 0 \quad \ldots (5) \qquad a_{ij}A^{3j} = 0 \quad \ldots(6)$$

Using kronecker delta, equations (4), (5), (6) can be combined into a single equation:

$$a_{ij}A^{kj} = \Delta\delta_1^k \quad \ldots (7)$$

Similarly six more equations are given by

$$a_{2j}A^{kj} = \Delta\delta_2^k \quad \ldots. (8) \quad \text{and} \quad a_{3j}A^{kj} = \Delta\delta_3^k \quad \ldots (9)$$

Equations (7), (8), (9) can be written as a single equation.

$$a_{ij}A^{kj} = \Delta\delta_i^k \quad \ldots (10)$$

All the nine equations of the determinant are included in one equation (10).

49.23 ISOTROPIC TENSOR

A tensor which has the same set of components relatively to every system of co-ordinate axes is called an *Isotropic tensor.*

49.24 RELATION BETWEEN ALTERNATE AND KRONECKER TENSOR

Prove that $\in_{ijm}\in_{klm} = \delta_{ik}\delta_{jl} - \delta_{il}\delta_{jk}$. Here each side is a tensor of order 4 so that tensor equality is equivalent to set of 81 scalar equality. We have to prove that

$$\in_{ij1}\in_{kl1} + \in_{ij2}\in_{kl2} + \in_{ij3}\in_{kl3} = \delta_{ik}\delta_{jl} - \delta_{il}\delta_{jk}$$

Proof: Case I. When $i = j$ or $k=l$. There will be 45 such equations and for all these equations L .H. S. = 0 = R.H.S.

Case II. If the pair (i, j) such that $i \neq j$ is different from the pair (l, k), $l \neq k$ we see that there will be 24 such scalar equations for which L.H.S. = 0 = R.H.S.

Ex. $(i, j) = (1, 2), (j, k) = (1, 3), (3, 1), (2, 3), (3, 2)$

$(i, j) = (2, 1), (j, k) = (1, 3), (3, 1), (2, 3), (3, 2).$

Case III. Thus we are left to consider the possibility when i, j and k, l take the pairs of values,

$$(1, 2); (1, 3); (2, 3); (2, 1); (3, 1); (3, 2).$$

Consider the first case we have

$i =1, j = 2, k= 1, l = 2; i=1, j = 2, k = 2, l = 1; i = 2, j= 1, k = 1, l = 2; i=2, j = 1, k = 2, l = 1.$

Each pair of (i, j) *i.e* (1, 2) gives two scalar equations. Thus 6 pairs of (i, j) give 12 such scalar equations. In these cases we have

L.H.S. = 1 = R.H.S., L.H.S. = –1 = R.H.S.

L.H.S. = – 1 = R.H.S., L.H.S. = 1 = R.H.S.

This result is also true for other five cases. Hence we have the result.

Example 4. *Prove that* $\in_{ilm} \in_{jlm} = 2\delta_{ij}$

Proof. We know that $\in_{ilm} \in_{jlm} = 2\delta_{ij}\, 2\delta_{ik} - 2\delta_{ik}\, 2\delta_{ij}$

Taking $k = l$ we get

$$\in_{ilm} \in_{jlm} = 2\delta_{ij}\, 2\delta_{ll} - 2\delta_{il}\, 2\delta_{ij}$$

Now $$\delta_{ll} = \delta_{11} + \delta_{22} + \delta_{33} = 1 + 1 + 1 = 3$$

$$2\delta_{ij}\, 2\delta_{lk} = 2\delta_{ij} \qquad \because \quad \delta_{ij}\, 2\delta_{im} = a_{jm}$$

$\therefore$ $$\in_{ilm} \in_{jlm} = 3\delta_{ij} - \delta_{ij} = 2\,\delta_{ij}$$ **Proved.**

Example 5. *Prove that* $\in_{ilm} \in_{jlm} = 6$

Proof. $\in_{ilm} \in_{jlm} = 2\delta_{ij}\, 2\delta_{lk} - 2\delta_{ik}\, 2\delta_{lj}$

Taking $k = l$, we get $\in_{ilm} \in_{jlm} = 2\delta_{ij}$

Taking $i = j$ (Contraction)

$$\in_{ilm} \in_{jlm} = 2\delta_{ii} = 2 \times 3 = 6$$ **Proved.**

49.25 MATRICES AND TENSORS OF FIRST AND SECOND ORDER

Consider any vector. Its components a_i relatively to any system of axes may be written in the form of a row or a column matrix as

$$[a_1 a_2 a_3] \quad \text{or} \quad \begin{bmatrix} a_1 \\ a_2 \\ a_3 \end{bmatrix}$$

We shall be writing $a_i = [a_1 a_2 a_3]$ or $a_i = \begin{bmatrix} a_1 \\ a_2 \\ a_3 \end{bmatrix}$

Now consider second order tensor. Its components a_{ij} relatively to any system of rectangular axes can be written as the form of matrix such that a_{ij} occurs at the intersection of the ith row and j^{th} column. Thus we shall write

$$\begin{bmatrix} a_{ij} \end{bmatrix} = \begin{bmatrix} a_{11} & a_{12} & a_{13} \\ a_{21} & a_{22} & a_{23} \\ a_{31} & a_{32} & a_{33} \end{bmatrix}$$

A matrix obtained by interchanging rows and columns of a given matrix is called the transpose of the same. Transpose of $[a_{ij}]$ will be denoted by $[a_{ij}]'$.

Sum of two matrices of the same type is the matrix whose elements are the sums of the corresponding elements of two matrices.

49.26 SCALAR AND VECTOR PRODUCTS OF TWO VECTORS

Def. 1. Scalar product. The scalar $u_i\ v_i$ is called the scalar product of the two vectors u_i, v_j. Thus the scalar product $= u_1 v_1 + u_2\ v_2 + u_3 v_3$.

Def. 2. Vector product. The vector $E_{ijk} u_i v_j$ is called vector product of two vectors u_i, v_j taken in this order. Components of these vectors are $u_2\ v_3 = u_3 v_2$, $u_3 v_1 = u_1 v_3$, $u_1 v_2 = u_2\ v_1$.

49.27 THE THREE SCALAR INVARIANTS OF A SECOND ORDER TENSOR

I. a_{ii} or $a_{11} + a_{22} + a_{33}$

II. $\frac{1}{2}(a_{ii}a_{jj} - a_{ij}a_{ji})$ or $\quad a_{11}a_{22} + a_{22}a_{33} + a_{33}a_{11} - a_{12}a_{21} - a_{23}a_{32} - a_{31}a_{13}$

III. $|a_{ij}|$ or $\begin{vmatrix} a_{11} & a_{12} & a_{13} \\ a_{21} & a_{22} & a_{23} \\ a_{31} & a_{32} & a_{33} \end{vmatrix}$

Proof. I. Identifying i, j we see that a_{ii} is scalar. Thus is invariant.

$$a_{ii} = a_{11} + a_{22} + a_{33} \qquad \text{...(1)}$$

II. Consider now the tensor of 4th order, $a_{ij}a_{pq}$. Identifying i with q and j with p we see that $a_{ij}a_{ji}$ is scalar.

Thus $\quad a_{ij}a_{ji} = (a_{11})^2 + (a_{22})^2 + (a_{33})^2 + 2a_{12}a_{21} + 2a_{23}a_{32} + 2a_{31}a_{13} \qquad \text{... (2)}$

is invariant. Subtracting(2) from square of (1) and dividing by 2. We establish invariance of II.

III. If a_{ij}, $\overline{a}_{pq}$ denote the components of tensor relatively to any two co-ordinate systems of axes, then in the usual notation, we have

$$\overline{a}_{pq} = l_{ip} l_{jq} a_{ij}$$

$$|\overline{a}_{pq}| = |l_{ip}||l_{jq}||a_{ij}|$$

since $\quad |l_{ip}| = |l_{jq}|$

$\therefore \quad |\overline{a}_{pq}| = |l_{ip}|^2 |a_{ij}| \quad$ but $\quad |l_{ip}|^2 = 1$

$\therefore \quad |\overline{a}_{pq}| = |a_{ij}|$

Hence it is an invariance.

49.28 SINGULAR AND NON-SINGULAR TENSORS OF SECOND ORDER

A tensor of second order is said to be singular or non-singular according as its determinant is zero or non zero.

49.29 RECIPROCAL OF A NON-SINGULAR TENSOR

Suppose a_{ij} be a second order tensor such that $|a_{ij}| \neq 0$

Lemma 1. We form another matrix

$$A_{ij} = \frac{\text{Cofactor of } a_{ij} \text{ in the determinant } a_{ij}}{|a_{ij}|}$$

Now, by theory of determinants, we know

$$A_{ki} a_{ij} = \delta_{kj} \qquad \text{...(1)}$$

We shall now show that A_{ij} is a second order tensor, we can not do so, by using Quotient law, from equation (1) since a_{ij} is not an arbitrary tensor. Let c_j be an arbitrary vector, then

$$c_j a_{ij} = d_i \quad \text{... (2)}$$

So that d_i is also a vector. We shall prove that this is an arbitrary vector. Now (2) is equivalent to a set of 3 linear equations in the components of c_j and as the determinant of $a_{ij} \neq 0$, we may assign any arbitrary values to d_i and the resulting equations can be uniquely solved for the components of c_j. Thus d_i is an arbitrary vector. We now have

$$d_i A_{ki} = a_{ij} c_j A_{ki} = A_{ki} a_{ij} c_j = \delta_{kj} c_j = c_k$$

$$c_k = d_i a_{ki} \quad \text{...(3)}$$

Therefore by quotient law A_{ki} is a second order tensor.

Lemma 2. $e_{ij} = \dfrac{\text{Cofactor of } A_{ij} \text{ in the determinant } A_{ij}}{\left|A_{ij}\right|}$

We know from the theory of determinants.

$$\left|A_{ij}\right|\left|a_{ij}\right| = 1 \qquad \text{But} \qquad \left|a_{ij}\right| \neq 0$$

Hence determinant $\left|A_{ij}\right| \neq 0$

We shall now show that $e_{ij} = a_{ij}$

$$e_{ki} A_{ij} = \delta_{kj}$$

Take inner product with a_{ji}

$$e_{ki} A_{ij} a_{jl} = \delta_{kj} a_{jl}$$

$$e_{ki} \delta_{il} = a_{kl}$$

$$e_{kl} = a_{kl}$$

Def. Two second order non-singular tensors a_{ij} and A_{ij} are said to be conjugate (or reciprocal) tensors if they satisfy the equation

$$A_{ki} a_{ij} = \delta_{kj}$$

49.30 EIGEN VALUES AND EIGEN VECTORS OF A TENSOR OF SECOND ORDER

Def. A scalar, λ, is called an eigen value of second order tensor a_{ij}, if there exists a non-zero vector x, such that $a_{ij} x_j = \lambda x_i$. This equation is equivalent to

$$a_{ij} x_j = \lambda \delta_{ij} x_j$$

or

$$\left(a_{ij} - \lambda \delta_{ij}\right) x_j = 0 \quad \text{... (1)}$$

since $x_j \neq 0$, Hence $\left|a_{ij} - \lambda \delta_{ij}\right| = 0$ ...(2)

This is a necessary condition for λ, to be eigen value. Eq. (2) is cubic eq. in λ and therefore in general will give us three eigen values may not all be distinct corresponding to the tensor a_{ij}.

Consider now any system of co-ordinate axes OX_1, OX_2, OX_3, and let a_{ij} be the component of the given tensor in this system. Consider now a vector x_j, whose components relatively to OX_1, OX_2, OX_3 are given on solving (1). As the components of x_i are not zero relatively to one system OX_1, OX_2, OX_3, this vector can be zero vector *i.e.* its components relatively to any system of axes can not all be zero.

The tensor eq. (1) being true for one system OX_1, OX_2, OX_3 will be true for every system of axes.

Thus we see that every second order tensor possesses three eigen values, not necessarily all distinct. These eigen values are the roots of the cubic $\left|a_{ij} - \lambda \delta_{ij}\right| = 0$ in λ. Also to each eigen value corresponds an eigen vector. The vector x_i corresponding to eigen value λ is called an eigen vector.

49.31 THEOREM

Orthogonality of eigen vectors corresponding to distinct eigen values of a symmetric second order tensor.

Proof. Let a_{ij} be a symmetric second order tensor, and let x_i and y_i be the eigen vectors corresponding to the distinct eigen values λ_1 and λ_2 $(\lambda_1 \neq \lambda_2)$, we have

$$a_{ij}\, x_j = \lambda_1\, x_i \qquad \text{...(1)}$$

$$a_{ij}\, y_j = \lambda_2\, y_i \qquad \text{...(2)}$$

Now,

$$\lambda_1 x_i\, y_i = a_{ij}\, x_j y_i$$

$$= a_{ji}\, x_j\, y_i \qquad [\because a_{ij} = a_{ji}]$$

$$= a_{ij}\, y_j x_i \qquad \text{(Interchanging dummy indices)}$$

$$\lambda_1 x_i\, y_i = a_{ij}\, x_j y_i$$

$$\therefore \quad \lambda_1 x_i\, y_i = \lambda_2\, x_j y_i \quad \text{or} \quad (\lambda_1 - \lambda_2)\,(x_i y_i) = 0$$

since $\lambda_1 - \lambda_2 \neq 0 \;\therefore\; x_i y_i = 0$

Thus x_i and y_i are orthogonal *i.e*. the eigen vectors are orthogonal.

49.32 REALITY OF THE EIGEN VALUES

Theorem. The eigen values of symmetric second order tensor are real

Proof. Let λ be any eigen value so that we have a relation

$$a_{ij} x_j = \lambda\, x_i \qquad \text{...(1)}$$

Here the components of x_j cannot be assumed to be all real. Taking complex conjugate (denoted by bar) in (1), we get

$$\bar{a}_{ij}.\bar{x}_j = \bar{\lambda}\bar{x}_i \qquad \left[\because a_{ij} \text{ is symmetric } \therefore a_{ij} = a_{ji}\right.$$

$$a_{ij}\bar{x}_j = \bar{\lambda}\bar{x}_i \qquad \left.\bar{a}_{ij} = a_{ij} \text{ (all elements are real)}\right.$$

Take inner product by x_i

$$a_{ij}\bar{x}_j x_i = \bar{\lambda}\bar{x}_i x_i$$

$$\bar{\lambda}\{\bar{x}_i x_i\} = a_{ji}(\bar{x}_j x_i) \qquad \text{by symmetry}$$

$$= a_{ij}\bar{x}_i x_j \qquad \text{[interchanging dummy indices]}$$

$$= a_{ij} x_i \bar{x}_j \qquad [\bar{x}_i x_i = \text{real}]$$

$$= \bar{a}_{ij} x_i \bar{x}_j = \text{real} \qquad [\because (a - ib)\,(a + ib) = a^2 + b^2 \text{ which is real}]$$

This shows that the right hand side is real. Hence $\bar{\lambda}$ is real. Thus λ is real *i.e.* eigen values are real. **Proved.**

49.33 ASSOCIATION OF A SKEW SYMMETRIC TENSORS OF ORDER TWO AND VECTORS

We associate the skew symmetric tensor of order two.

$$a_{ij} = \in_{ijk}\, a_k \qquad \text{... (1)}$$

The tensor a_{ij} is skew symmetric for

$$a_{ji} = \in_{jlk}\, a_k = -\in_{ijk}\, a_k = -a_{ij}$$

The relation (1) is equivalent to statements

$$a_{23} = a_1,\, a_{32} = -a_1;\, a_{31} = a_2,\, a_{13} = -a_2;\, a_{12} = a_3,\, a_{21} = -a_3;\; a_{11} = 0,\, a_{22} = 0;\, a_{33} = 0.$$

On the inner multiplication with $\in_{ijm}$ we obtain from (1)

$$\in_{ijm} a_{ij} = \in_{ijm} \in_{ijk} a_k \qquad \in_{ijk} \in_{pqk} = \delta_{ip}\delta_{jq} - \delta_{iq}\delta_{jp}$$

$$= 2\delta_{mk} a_k \qquad \in_{ijk} \in_{pjk} = \delta_{ip}\delta_{jj} - \delta_{jp}\delta_{ij}$$

$$= 2a_m \text{ when } k = m \qquad = 3\delta_{ip} - \delta_{ip} = 2\delta_{ip}$$

Hence $a_m = \frac{1}{2} \in_{ijm} a_{ij}$

This shows that association is one-one.

49.34 TENSOR FIELDS

A tensor field or a tensor point function is said to be defined when there is given a law which associates to each point of given region of space a tensor of the same order. Thus a tensor field a_{ij}...... of any order is defined if the components a_{ij}...... are functions of x_1, x_2, x_3.

49.35 GRADIENT OF TENSOR FIELDS:GRADIENT OF A SCALAR FUNCTION.

Let u be a scalar point function so that there is a value of u associated with each point of a given region of space. Thus if OX_1, OX_2, OX_3 and $O\overline{X}_1$, $O\overline{X}_2$, $O\overline{X}_3$ be any two systems, then u is a function of x_i and $\overline{x}_p$ which are co-ordinates of any point P relatively to the two systems of axes. For any point P, x_i, $\overline{x}_p$ are different but the values of u are same. Consider now two sets of first order

$$\frac{\partial u}{\partial x_i}, \frac{\partial u}{\partial \overline{x}_p} \qquad \text{we have } \frac{\partial u}{\partial x_i} = \frac{\partial u}{\partial \overline{x}_p}\frac{\partial \overline{x}_p}{\partial x_i}$$

We know that $\overline{x}_p = l_{ip} x_i$

$$\therefore \quad \frac{\partial \overline{x}_p}{\partial x_i} = l_{ip} \qquad \therefore \quad \frac{\partial u}{\partial x_i} = l_{ip}\frac{\partial u}{\partial \overline{x}_p}$$

Thus we see that $\frac{\partial u}{\partial x_i}$ is a tensor of order one *i.e.* a vector. This is usually denoted by *u, i,*

grad $u = u, i$

If components $\frac{\partial u}{\partial x_i}$ and $\frac{\partial u}{\partial \overline{x}_p}$ relatively to two systems of axes OX_1, OX_2, OX_3 and $O\overline{X}_1$, $O\overline{X}_2$, $O\overline{X}_3$ obey the tensorial transformation law. This vector is called the *gradient of scalar u.*

49.36 GRADIENT OF VECTOR

(Delhi University, April 2010)

Consider now any tensor field u_i of order one. If u_i, $\overline{u}_p$ be the components relatively to two systems of axes OX_1, $0X_2$, OX_3, $O\overline{X}_1$, $O\overline{X}_2$, $O\overline{X}_3$ we have $\overline{u}_p = l_{ip} u_i$

$$\therefore \quad \frac{\partial \overline{u}_p}{\partial \overline{x}_j} = l_{ip}\frac{\partial u_i}{\partial \overline{x}_j} = l_{ip}\frac{\partial u_i}{dx_k}\cdot\frac{\partial x_k}{\partial \overline{x}_j} = l_{ip} l_{kj}\frac{\partial u_i}{\partial x_k} \qquad \left[\begin{array}{l} x_k = l_{kj}\overline{x}_j \\ \frac{\partial x_k}{\partial \overline{x}_j} = l_{kj} \end{array}\right.$$

We see $\frac{\partial u_i}{\partial x_k}$ is a tensor of second order. It is denoted by symbol $u_{i,j}$ and is called the *gradient of* $u_{i,j}$.

49.37 DIVERGENCE OF VECTOR POINT FUNCTION

The scalar of the gradient of a vector point function is called the divergence of the point function. Thus if u_i is a vector point function so that

$$u_{i,j} = \frac{\partial u_i}{\partial x_j} \text{ is its gradient,then } u_{i,i} = \frac{\partial u_i}{\partial x_i} = \frac{\partial u_1}{\partial x_1} + \frac{\partial u_2}{\partial x_2} + \frac{\partial u_3}{\partial x_3} \text{ is called div } u_i$$

$$\text{div } u_i = u_{i,i}$$

49.38 CURL OF A VECTOR POINT FUNCTION

The vector of the gradient of a vector point function is called the curl of the point function.

Thus if u_i is a vector point function so that $u_{i,j} = \dfrac{\partial u_i}{\partial x_j}$ is its gradient, then the vector of a tensor

i.e. the vector $\in_{jik} u_{i,j}$ is called the curl of u_i denoted by the symbol curl id, $u_i = \in_{jik} u_{i,j}$.

Example 6. *Prove the following results*

(*i*) grad $(\phi\psi) = \phi$ grad $\psi + \psi$ grad ϕ (*ii*) grad $(\vec{f}.\vec{g}) = \vec{f} \times \text{curl } \vec{g} + \vec{g} \times \text{curl } \vec{f} + \vec{f}.\nabla\vec{g} + \vec{g}.\nabla\vec{f}$

(*iii*) div $(\phi\vec{f}) = \phi$ div $\vec{f} + \vec{f}$ grad ϕ (*iv*) div $(\vec{f} \times \vec{g}) = \vec{g}.$ curl $\vec{f} - \vec{f}.$curl $\vec{g}$

(*v*) curl $(\phi\vec{f}) = \text{grad}\,\phi \times \vec{f} + \phi \text{ curl } \vec{f}$ (*vi*) curl $(\vec{f} \times \vec{g}) = \vec{f} \,\text{div}\, \vec{g} - \vec{g} \text{ div } \vec{f} + \vec{g}.\nabla\vec{f} - \vec{f}.\nabla\vec{g}$

Proof. (*i*) grad $(\phi\psi) = (\phi\psi)_i = \phi\psi_i + \psi\phi_i = \phi$ grad $\psi + \psi$ grad ϕ.

(*ii*) grad $(\vec{f}.\vec{g}) = (f_i g_i)_{,j}$

$$= f g_{i,j} + g f_{i,j} \quad ...(1)$$

Now $\vec{f} \times \text{curl } \vec{g} = \in_{pkm} f_p \in_{jik} g_{i,j} = -\in_{pmk} \in_{jik} f_p g_{i,j}$

$$= -\left[\delta_{pj}\delta_{mi} - \delta_{pi}\delta_{mj}\right] f_p g_{i,j} = -\delta_{pj}\delta_{mi} f_p g_{i,j} + \delta_{pi}\delta_{mj} f_p g_{i,j}$$

Identifying *p, j* and *m, i* in first part and *p, i* and *m, j* in second part, we get.

$$\vec{f} \times \text{curl}\,\vec{g} = -\delta_{pp}\delta_{mm} f_p g_{m,p} + \delta_{pp}\delta_{mm} f_p g_{p,m} = -f_p g_{m,p} + f_p g_{p,m}$$

$$= -\vec{f}.\nabla\vec{g} + f_p g_{p,m} \quad ...(2)$$

similarly, $\vec{g} \times \text{curl } \vec{f} = -g.\nabla\vec{f} + g_p f_{p,m}$ ...(3)

Substituting the value of $f_p g_{p,m} g_p f_{p,m}$ from (2) and (3) into (1), we get

$$\text{grad}(\vec{f}.\vec{g}) = \vec{g} \times \text{curl}\,\vec{f} + \vec{g}.\nabla\vec{f} + \vec{f} \times \text{curl}\,\vec{g} + \vec{f}.\nabla\vec{g}$$

or

$$\text{grad}(\vec{f}.\vec{g}) = \vec{g} \times \text{curl}\,\vec{f} + \vec{f} \times \text{curl } \vec{g} + g.\nabla\vec{f} + \vec{f}.\nabla\vec{g}$$

(*iii*) $\text{div}(\phi\vec{f}) = (\phi f_i)_{,i} = \phi f_{i,i} + \phi_{,i} f_i = \phi\,\text{div}\vec{f} + f_i\phi_{,i} = \phi\,\text{div}\,\vec{f} + \vec{f}.\text{grad}\,\phi$

(*iv*) $\text{div}(\vec{f} \times \vec{g}) = (\in_{ijk} f_i g_j)_{,k} = \in_{ijk} \{f_i g_{j,k} + g_j f_{i,k}\} = \in_{ijk} f_i g_{j,k} + \in_{ijk} g_j f_{i,k}$

$$= -\in_{kji} f_i g_{j,k} + \{\in_{kij} g_j f_{i,k}\} = -f_i \in_{kji} g_{j,k} + g_j \in_{kij} f_{i,k}$$

$$= -\vec{f}.(\text{curl}\,\vec{g}) + \vec{g}.(\text{curl}\,\vec{f}) = \vec{g}.(\text{curl}\,\vec{f}) - \vec{f}.(\text{curl}\,\vec{g})$$

(*v*) $\text{curl}(\phi\vec{f}) = \in_{jik} (\phi f_i)_{,j} = \in_{jik} \phi f_{i,j} + \in_{jik} \phi_{,j} f_i$

$$= \phi \in_{jik} f_{i,j} + \in_{jik} f_i \phi_{,j} = \phi(\text{curl } \vec{f}) + (\text{grad } \phi) \times \vec{f}$$

(*vi*) $\quad$ $\text{curl}\,(\vec f \times \vec g) = \epsilon_{mkn}\,(\epsilon_{ijk}\, f_i g_j)_{,m} = -\,\epsilon_{mnk}\,\epsilon_{ijk}\,[f_i g_{j,m} + g_j f_{i,m}]$

$$= -[\delta_{mi}\delta_{nj} - \delta_{mj}\delta_{nj}][f_i g_{j,m} + g_j f_{i,m}] = -[\delta_{mi}\delta_{nj} f_i g_{j,m} - \delta_{mj}\delta_{ni} f_i g_{j,m} + \delta_{mi}\delta_{nj} g_j f_{i,m} - \delta_{mj}\delta_{ni} g_j f_{i,m}]$$

$$= -\delta_{mm}\delta_{nn} f_m g_{n,m} + \delta_{mm}\delta_{nn} f_n g_{m,m} - \delta_{mm}\delta_{nn} g_n f_{m,m} + \delta_{mm}\delta_{nn} g_m f_{n,m}$$

$$= -f_m g_{n,m} + f_n g_{m,m} - g_n f_{m,m} + g_m f_{n,m} = -\vec f.\nabla\vec g + (\text{div}\,\vec g)\vec f - (\text{div}\,\vec f)\vec g + \nabla\vec f.\vec g$$

$$= \vec f\,\text{div}\vec g - \vec g\,\text{div}\,\vec f + \vec g.\nabla\vec f - \vec f.\nabla\vec g$$

Proved.

49.39 SECOND ORDER DIFFERENTIAL OPERATORS

(*i*) $\text{div (grad}\,\phi) = \nabla^2\phi$ $\quad$ (*ii*) $\text{curl (grad}\,\phi) = 0$

(*iii*) $\text{div (curl}\,\vec f\,) = 0$ $\quad$ (*iv*) $\text{grad (div}\,\vec f\,) = \text{curl (curl}\,\vec f\,) + \nabla^2\vec f$

Proof. (I) $\text{div (grad}\,\phi) = (\phi_{,i})_{,i} = \phi_{,ii} = \nabla^2\phi$

(*ii*) $\text{curl (grad}\,\phi) = \epsilon_{jik}\,(\phi_{,i})_{,j} = \epsilon_{jik}\,\phi_{,ij} = I$, say ...(1)

$$\phi_{,ij}\ \phi_{,ji} \qquad \left[\frac{\partial^2\phi}{\partial x\partial y} = \frac{\partial^2\phi}{\partial y\partial x}\right]$$

Now, $\quad I = \epsilon_{jik}\,\phi_{,ji} = \epsilon_{ijk}\,\phi_{,ij}$ ...(2)

From (1) and (2) $\quad 2I = (\epsilon_{jik} + \epsilon_{ijk})\phi_{,ij}$

$$2I = (-\,\epsilon_{ijk} + \epsilon_{ijk})\phi_{,ij} = 0$$

Hence, $I = 0$ or $\text{curl (grad}\,\phi) = 0$

(*iii*) $\quad \text{div (curl}\,\vec f) = (\epsilon_{jik}\; f_{i,j})_{,k} = \epsilon_{jik}\; f_{i,jk} = I$ (say) ... (1)

Because $\quad f_{i,jk} = f_{i,kj}$

Then $\quad I = \epsilon_{jik}\; f_{i,kj} = \epsilon_{kij}\; f_{i,jk}$ ...(2)

From (1) and (2) $\quad 2I = (\epsilon_{jik} + \epsilon_{kij})f_{i,jk} = (\epsilon_{jik} - \epsilon_{jik})f_{i,jk} = 0$

Hence, $\quad I = 0$ or $\text{div (curl}\,\vec f\,) = 0$

(*iv*) $\quad \text{grad (div}\,\vec f\,) = (f_{i,i})_{,j} = f_{i,ij}$ ... (1)

$$\text{curl (curl}\,\vec f = \epsilon_{mkn}\,(\epsilon_{jik}\; f_{i,j})_m = \epsilon_{nmk}\,\epsilon_{jik}\; f_{i,jm} = (\delta_{nj}\delta_{mi} - \delta_{ni}\delta_{mj})\,f_{i,jm}$$

$$= (\delta_{nn}\delta_{mm} f_{m,nm} - \delta_{nn}\delta_{mm} f_{n,mm}) = f_{m,nm} - f_{n,mm}$$

$$= f_{m,mn} - f_{n,mm} = \text{grad(div}\,\vec f) = \nabla^2\vec f \qquad \text{[From (1)]}$$

Thus, $\quad \text{grad(div}\,\vec f) = \text{curl (curl}\vec f) + \nabla^2\vec f$

49.40 TENSORIAL FORM OF GAUSS'S AND STOKE'S THEOREM

Gauss's divergence theorem.

If $\vec F$, is a continuously differentiable vector point function and *S* is a closed surface enclosing a region *V*, then

$$\oint_s \vec F.\hat n ds = \int_V \text{div}\,\vec F\, dv \qquad ...(1)$$

where $\hat\eta$ is a unit vector $\oint_s \vec F_i\, n_i ds = \int_V F_{i,i} dV = \int \frac{\partial F_i}{\partial x_i} dV$

49.41 STOKE'S THEOREM

If $\vec{F}$ is any continuously differentiable vector point function and S is a surface bounded by a curve c, then

$$\oint_c \vec{F}.dr = \int_S \text{curl}\,\vec{F}.\hat{\eta}\,ds \qquad ...(2)$$

where $\hat{\eta}$ is a unit vector $\quad \oint_c \vec{F}_i.dr_i = \int_S (\in_{jik} F_{i,j})n_k ds$

Example 7. *By means of divergence theorem of Gauss's, show that*

$$\oint_S \in_{qpi} n_p \in_{ijk} a_j x_k ds = 2a_q V$$

where V is the volume enclosed by the surface S, having the outward drawn normal n. The position vector to any point in V is x_i and a_i, *is an arbitrary constant vector.*

Proof. $\quad$ L.H.S. $= \oint_S \in_{ipq} \in_{ijk} n_p a_j x_k ds$

$$= \oint_S (\delta_{qj}\delta_{pk} - \delta_{qk}\delta_{pj}) n_p a_j x_k ds = \oint_S n_k a_q x_k ds - \oint_S n_j a_j x_q ds$$

$$= a_q \oint_S n_k x_k ds - a_j \oint_S n_j x_q ds = a_q \oint_V \frac{\partial x_k}{\partial x_k} dv - a_j \oint_V \frac{\partial x_q}{\partial x_j} dv = a_q \delta_{kk} v - a_j \delta_{qj} v$$

$$= 3a_q v - a_q v = 2a_q v$$

Proved.

Example 8. *If* $\vec{q} = \vec{w} \times \vec{r}$, *show that* $2\vec{w} = \nabla \times \vec{q}$ *using the index notation. The vector* $\vec{w}$ *is a constant.*

Solution. $\quad q_k = \in_{ijk} w_i x_j \qquad$ (given)

$$\left[\nabla \times \vec{q}\right]_m = \in_{lkm} q_{k,l} = \in_{lkm} \in_{ijk} (w_i x_j)_{,l}$$

$$\left[\nabla \times \vec{q}\right]_m = \in_{mlk} \in_{ijk} (w_i x_{j,l} + x_j w_{l,l}) = \in_{mlk} \in_{ijk} w_i x_{j,l}$$

Since $\vec{w}$ is a constant vector

$$\therefore \quad w_{i,l} = 0 = (\delta_{mi}\delta_{ij} - \delta_{mj}\delta_{li}) w_i x_{j,l} = w_m x_{l,l} - w_l x_{m,l} = w_m \delta_{ll} - w_i \delta_{ml} = 3w_m - w_m$$

$$\left[\nabla \times \vec{q}\right]_m w = 2w_m$$

Hence $\quad \nabla \times \vec{q} = 2\vec{w}.$

Proved.

49.42 RELATION BETWEEN ALTERNATE AND KRONECKER TENSOR

(Delhi University, April 2010)

$$\in_{ijk} \in_{lmn} = \begin{vmatrix} l_{1i} & l_{2i} & l_{3i} \\ l_{1j} & l_{2j} & l_{3j} \\ l_{1k} & l_{2k} & l_{3k} \end{vmatrix} \times \begin{vmatrix} l_{1l} & l_{2l} & l_{3l} \\ l_{1m} & l_{2m} & l_{3m} \\ l_{1n} & l_{2n} & l_{3n} \end{vmatrix} = \begin{vmatrix} \delta_{il} & \delta_{im} & \delta_{in} \\ \delta_{jl} & \delta_{jm} & \delta_{jn} \\ \delta_{kl} & \delta_{km} & \delta_{kn} \end{vmatrix}$$

Identifying k and l, we get

$$\in_{ijk} \in_{kmn} = \begin{vmatrix} \delta_{ik} & \delta_{im} & \delta_{in} \\ \delta_{jk} & \delta_{jm} & \delta_{jn} \\ \delta_{kk} & \delta_{km} & \delta_{kn} \end{vmatrix} = \begin{vmatrix} \delta_{ik} & \delta_{im} & \delta_{in} \\ \delta_{jk} & \delta_{jm} & \delta_{jn} \\ 3 & \delta_{km} & \delta_{kn} \end{vmatrix}$$

$$\delta_{kk} = \delta_{11} + \delta_{22} + \delta_{33} = 1 + 1 + 1 = 3.$$

Expanding the determinant, we have

$$\in_{ijk} \in_{kmn} = \delta_{ik}(\delta_{jm}\delta_{kn} - \delta_{jn}\delta_{km}) + \delta_{im}(3\delta_{jn} - \delta_{jk}\delta_{kn}) + \delta_{in}(\delta_{jk}\delta_{km} - 3\delta_{jm})$$

$$= \delta_{jm}\delta_{in} - \delta_{jn}\delta_{im} + 3\delta_{im}\delta_{jn} - \delta_{im}\delta_{jn} + \delta_{in}\delta_{jm} - 3\delta_{in}\delta_{jm}$$

$$\in_{ijk} \in_{kmn} = \delta_{im}\delta_{jn} - \delta_{in}\delta_{jm}$$

49.43 THE THREE SCALAR INVARIANTS OF A SECOND ORDER TENSOR

Let a_{ij} be a second order tensor

(*i*) a_{ii}

Proof. $\bar{a}_{pq} = l_{ip}\, l_{jq}\, a_{ij}$

Identifying p and q, we have $\bar{a}_{pp} = l_{ip}\, l_{jp}\, a_{ij}$

$\bar{a}_{pp} = \delta_{ij}\, a_{ij} = a_{ii}$. Hence a_{ii} is an invariant.

(*ii*) $\frac{1}{2}(a_{ii}\, a_{jj} - a_{ij}\, a_{ij})$

We know that a_{ii} and a_{jj} are invariants. Now we have to show that $a_{ij} a_{ji}$ is also an invariant. Then $(a_{ii}\, a_{jj} - a_{ij}\, a_{ij})$ will also be invariant.

Let $\bar{a}_{pq} = l_{ip}\, l_{jp}\, a_{ij}$ and $\bar{a}_{rs} = l_{mr}\, l_{ns}\, a_{mn}$

Now consider the tensor of 4th order $\bar{a}_{pq}\, \bar{a}_{rs} = l_{ip}\, l_{jq}\, l_{mr}\, l_{ns}\, a_{ij} a_{mn}$

First identifying r and q and then identifying p and s we have

$$\bar{a}_{pq}\, \bar{a}_{qp} = l_{is}\, l_{jr}\, l_{mr}\, l_{ns}\, a_{ij}\, a_{mn} = \delta_{in}\, \delta_{jm}\, a_{ij}\, a_{mn} = a_{ij}\, a_{ji}$$

Hence $a_{ij}\, a_{ji}$ is an invariant. Therefore $\frac{1}{2}(a_{ii}\, a_{jj} - a_{ij}\, a_{ji})$ is invariant

(*iii*) $|a_{ij}|$ **Proof.** $\bar{a}_{pq} = l_{ip}\, l_{jq}\, a_{ij}$ $\Rightarrow$ $|\bar{a}_{pq}| = |l_{ip}|\, |l_{jq}|\, |a_{ij}|$

We know by the property of determinants $|l_{ip}|\, |l_{jq}| = 1$ $\Rightarrow |\bar{a}_{pq}| = |a_{ij}|$

Hence $|a_{ij}|$ is an invariant.

49.44 TENSOR ANALYSIS

Example 9. *What is a mixed tensor of second rank? Prove that $\delta_q^{\,p}$ is a mixed tensor of the second rank.*

Solution. The N^2 quantities $A_s^{\,q}$ are called components of a mixed tensor of the second rank if

$$\bar{A}_r^{\,p} = \frac{\partial \bar{x}^p}{\partial x^q} \frac{\partial x^s}{\partial \bar{x}^r} A_s^{\,q}$$

Now, if $\delta_s^{\,q}$ defined by $\delta_s^{\,q} = \begin{cases} 0 & \text{if } q \neq s \\ 1 & \text{if } q = s \end{cases}$

is a mixed tensor of second rank, it must transform according to the rule $\bar{\delta}_k^i = \frac{\partial \bar{x}^i}{\partial x^p} \frac{\partial x^q}{\partial \bar{x}^k} \delta_q^p$

The right side equals $\frac{\partial \bar{x}^j}{\partial x^p} \frac{\partial x^p}{\partial \bar{x}^k} = \delta_k^j$

since $\bar{\delta}_k^j = \delta_k^j = 1$ if $j = k$, and 0 if $j \neq k$, if follows that δ_r^p is a mixed tensor of rank two.

Example 10. *Evaluate (i)* $\delta^p_q A^{qr}_s$ *(ii)* $\delta^p_q \delta^p_r$

Solution. (*i*) $\delta^p_q A^{qr}_s = \delta^p_q A^{pr}_s = A^{pr}_s$ (*ii*) $\delta^p_q \delta^p_r = \delta^{\ p}_p \delta^p_r = \delta^p_r (\because \delta^{\ p}_q = 1)$ **Ans.**

Example 11. *Show that every tensor can be expressed as the sum of two tensors one of which is symmetric and the other skew-symmetric in a pair of covariant or contravariant indices.*

Solution. Consider the tensor B^{pq}, we have

$$B^{pq} = \frac{1}{2}(B^{pq} + B^{qp}) + \frac{1}{2}(B^{pq} - B^{qp})$$

But $$R^{pq} = \frac{1}{2}(B^{pq} + B^{qp}) = R^{qp}$$ is symmetric and

$$S^{pq} = \frac{1}{2}(B^{pq} - B^{qp}) = R^{qp}$$ is skew-symmetric.

Thus B^{pq} = symm tensor + skew-symm tensor.

By similar reasoning the result is seen to be true for any tensor. **Proved.**

Example 12. *What is contraction as applied to tensors? Prove that the contraction of the tensor* A^p_q *is a scalar or invariant.*

Solution. *Contraction.* If one contravariant and one covariant index of a tensor are set equal, the result indicates that a summation over the equal indices is to be taken according to the summation convention. This resulting sum is a tensor of rank two less than that of the original tensor. The process is called contraction. For example, in the tensor of rank 3, B^{mp}_q, set $p = q$ we get $B_q^{\ mp} = C^m$, a tensor of rank 1.

To prove that contraction of A^p_q is a scalar or invariant.

we have $$\bar{A}^j_k = \frac{\partial \bar{x}^j}{\partial x^p}\frac{\partial x^q}{\partial \bar{x}^k} A^p_q$$

putting $j = k$, $$\bar{A}^j_j = \frac{\partial \bar{x}^j}{\partial x^p}\frac{\partial x^q}{\partial \bar{x}^j} A^p_q = \delta^q_{\ p} A^p_q = A^p_p$$

Then $\bar{A}^j_j = A^p_q$ and it follows that $A_p^{\ p}$ must be an invariant. Since A^p_q is a tensor of rank two and contraction with respect to a single index lowers the rank by two. Therefore, an invariant is a tensor of rank zero. **Proved.**

Example 13. *A covariant tensor has components xy,* $2y - z^2$*, xz in rectangular co-ordinates. Find its covariant components in spherical co-ordinates.*

Solution. Let A_j denote the covariant component in rectangular co-ordinates

$$x^1 = x, x^2 = y, x^3 = z.$$

Then
$$A_1 = xy = x^1 x^2$$
$$A_2 = 2y - z^2 = 2x^2 - (x^3)^2$$
$$A_3 = xz = x^1 x^3$$

Let $\bar{A}_k$ denote the covariant component in spherical co-ordinates $\bar{x}^1 = r, \bar{x}^2 = \theta, \bar{x}^3 = \phi$

Then $$\bar{A}_k = \frac{\partial \bar{x}^j}{\partial \bar{x}^k} A_j \quad \text{... (1)}$$

In spherical co-ordinates

$$x = r \sin\theta \cos\phi$$

$\Rightarrow$ $$x^1 = \bar{x}^1 \sin \bar{x}^2 \cos \bar{x}^3 \quad \text{...(2)}$$

$$y = r \sin\theta \sin\phi$$

$$\Rightarrow \qquad x^2 = \bar{x}^1 \sin \bar{x}^2 \sin \bar{x}^3 \qquad ...(3)$$

$$z = r\cos\theta$$

$$\Rightarrow \qquad x^3 = \bar{x}^1 \cos \bar{x}^2 \qquad ...(4)$$

Therefore equation (1) yields the covariant component.

$$\bar{A}_1 = \frac{\partial x^1}{\partial \bar{x}^1} A_1 + \frac{\partial x^2}{\partial \bar{x}^1} A_2 + \frac{\partial x^3}{\partial \bar{x}^1} A_3$$

$$= (\sin \bar{x}^2 \;\cos \bar{x}^3)\,(x^1 x^2) + (\sin \bar{x}^2 \;\sin \bar{x}^3) \times [(2x^2 - (x^3)^2)] + (\cos \bar{x}^2)\,(x^1 x^3)$$
$$= (\sin\theta\cos\phi))\,(r^2 \sin^2\theta \sin\phi\cos\phi) + (\sin\theta\sin\phi)\,(2r\sin\theta\sin\phi - r^2\cos^2\theta)$$
$$+ (\cos\theta)\,(r^2 \sin\theta\cos\theta$$

$\cos\phi)$

$$\bar{A}_2 = \frac{\partial x^1}{\partial \bar{x}^2} A_1 + \frac{\partial x^2}{\partial \bar{x}^2} A_2 + \frac{\partial x^3}{\partial \bar{x}^2} A_3$$

$$= \left(\bar{x}^1 \cos\bar{x}^2 \cos \bar{x}^3\right)(x^1 x^2) + \left(\bar{x}^1 \cos\bar{x}^2 \sin \bar{x}^3\right)[(2x^2 - (x^3)^2)] + \bar{x}^1(-\sin\bar{x}^2)\,(x^1 x^2)$$

or $\bar{A}_2 = (r\cos\theta\cos\phi)\,(r^2\sin^2\theta\sin\phi\cos\phi) + (r\cos\theta\sin\phi)\,(2r\sin\theta\sin\phi - r^2\cos^2 0)$

$$+ (-r\sin\theta)\,(r^2\sin\theta\cos\theta\cos\phi) \quad \bar{A}_3 = \frac{\partial x^1}{\partial \bar{x}^3} A_1 + \frac{\partial x^2}{\partial \bar{x}^3} A_2 + \frac{\partial x^3}{\partial \bar{x}^3} A_3$$

$= (-r\sin\theta\sin\phi)\,(r^2\sin^2\theta\;\sin\phi\cos\phi) + (r\sin\theta\cos\phi)\,(2r\sin\theta\sin\phi - r^2\cos^2\theta)$. **Ans.**

Example 14. *Define symmetric and skew-symmetric tensors. Prove that a symmetric tensor of rank two has at most* $\frac{N(N+1)}{2}$ *different components in N-dimensional space* V_N.

Solution. Symmetric Tensor. A tensor is called symmetric with respect to two contravariant or two covariant indices if its components remain unaltered upon interchange of the indices.

Thus if $A_{qs}^{mpr} = A_{qs}^{pmr}$, the tensor is symmetric in *m* and *p*. If a tensor is symmetric with respect to any two contravariant and any two covariant indices, it is called symmetric.

Skew-symmetric. A tensor is called skew-symmetric with respect to two contravariant or two covariant indices if its component change sign upon interchange of the indices. Thus, if $A_{qs}^{mpr} = -A_{qs}^{pmr}$ the tensor is skew symmetric in *m* and *p*. If a tensor is skew-symmetric with respect to any two contravariant and any two covariant indices it is called skew-symmetric.

Let A_{pq} be a tensor of rank 2. The number of its all components in V_N is N^2.

The components of A_{pq} are

$$\begin{matrix} A_{11} & A_{12} & A_{13} & \cdots\cdots & A_{1N} \\ A_{21} & A_{22} & A_{23} & \cdots\cdots & A_{2N} \\ \cdots & \cdots & \cdots & \cdots & \cdots \\ A_{N1} & A_{N2} & A_{N3} & \cdots\cdots & A_{NN} \end{matrix}$$

There are N independent components of the form

$$A_{11}, A_{22}, A_{33} \;\cdots\cdots\; A_{NN}$$

Hence number of components of the form A_{12}, A_{23}, A_{34}, ...in which there are distinct subscripts will be $N^2 - N$. But these component are symmetric. *i.e.*, $A_{12} = A_{21}$ etc.

Hence number of different component of this form are $\frac{1}{2}(N^2 - N)$

$\therefore$ Total number of different components are

$$= \frac{1}{2}\,(N^2 - N) + N = \frac{N^2}{2} + \frac{N}{2} = \frac{N(N+1)}{2}$$ **Proved.**

Example 15. *Define a metric or fundamental tensor. Determine the components of the fundamental tensor in cylindrical co-ordinates.*

Solution. Metric or Fundamental Tensor.

In rectangular coordinates (x, y, z) the differential of arc length ds is obtained from $ds^2 = dx^2 + dy^2 + dz^2$. By transforming to general curvilinear co-ordinates this becomes

$$ds^2 = \sum_{p=1}^{3} \sum_{q=1}^{3} g_{pq}\, du_p\, du_q$$

Such spaces are called three-dimensional Euclidean spaces. We define the line element ds in this space to be given by the quadratic form, called the metric form or metric,

$$ds^2 = \sum_{p=1}^{N} \sum_{q=1}^{N} g_{pq}\, dx^p\, dx^q \qquad ...(1)$$

$$\Rightarrow \quad ds^2 = g_{pq}\, dx^p\, dx^q$$

The quantities g_{pq} are the components of a covariant tensor of rank 2 called the *metric* tensor or *fundamental* tensor.

We know that $ds^2 = dx^2 + dy^2 + dz^2$

In cylindrical co-ordinates,

$$x = r\cos\theta,\ y = r\sin\theta,\ z = z$$

$$\therefore \quad dx = -r\sin\theta\, d\theta + \cos\theta\, dr$$

$$dy = r\cos\theta\, d\theta + \sin\theta\, dr$$

$$dz = dz$$

$$ds^2 = dx^2 + dy^2 + dz^2$$

$$= (-r\sin\theta\, d\theta + \cos\theta\, dr)^2 + (r\cos\theta\, d\theta + \sin\theta\, dr)^2 + (dz)^2$$

$$\Rightarrow \quad ds^2 = (dr)^2 + r^2\,(d\theta)^2 + (dz)^2$$

Also metric is given by

$$ds^2 = g_{pq}\, dx^p\, dx^q \qquad ... (2)$$

If $\quad x^1 = r,\ x^2 = \theta,\ x^3 = z$

then comparing (1) & (2), we have

$$g_{11} = 1,\ g_{22} = r^2,\ g_{33} = 1,\ g_{12} = g_{21} = 0,\ g_{13} = g_{31} = 0,\ g_{23} = g_{32} = 0.$$

Metric tensor is given by $\begin{pmatrix} g_{11} & g_{12} & g_{13} \\ g_{21} & g_{22} & g_{23} \\ g_{31} & g_{32} & g_{33} \end{pmatrix}$

Metric tensor in cylindrical co-ordinates $= \begin{pmatrix} 1 & 0 & 0 \\ 0 & r^2 & 0 \\ 0 & 0 & 1 \end{pmatrix}$ **Ans.**

Example 16. *Define what is meant by invariant? Show that the contraction of the outer product of the tensors A^p and B_q is an invariant.*

Solution. Scalar or Invariant. Suppose ϕ is a function of the co-ordinates x^k, and let ϕ denote the functional value under a transformation to a new set of co-ordinates $\overline{x}^k$. Then ϕ is called a *scalar* or *invariant* with respect to the co-ordinate transformation if $\phi = \overline{\phi}$.

Since A^P and B_q are tensors.

$$\overline{A}^j = \frac{\partial \overline{x}^j}{\partial x^p} A^p, \overline{B}_k = \frac{\partial x^q}{\partial \overline{x}^k} B_q$$

Then $$\overline{A}^j\, \overline{B}_k = \frac{\partial \overline{x}^j}{\partial x^p} \frac{\partial x^q}{\partial \overline{x}^k} A^p B_q$$

By contraction (putting $j = k$)

$$\overline{A}^j\, \overline{B}_j = \frac{\partial \overline{x}^j}{\partial x^p} \frac{\partial x^q}{\partial \overline{x}^j} A^p B_q = \delta^q_p A^p B_q = A^p B_p$$ **Proved.**

and so $A^P B_p$ is an invariant.

Example 17. *What do you understand by associated tensors ?*

Solution. Associated Tensors. Given a tensor we can derive other tensors by raising or lowering indices. For example, given the tensor A_{pq} we obtain by raising the index p, the tensor $A^p_{\cdot q}$, the dot indicating the original position of the moved index. By raising the index q also we obtain A^{pq}. We shall often write A^{pq}. These derived tensors can be obtained by forming inner products of the given tensor with the metric tensor g_{pq} or its conjugate g^{pq}. Thus, for example

$$A^p_q = g^{rq} A_{rq}, \; A^{pq} = g^{rP} g^{sq} A_{rs}$$

All tensors obtained from a given tensor by forming inner products with the metric tensor and its conjugate are called *associated tensors* of the given tensor. For example; A^m and A_m are associated tensors, the first are contravariant and the second covariant components. The relation between them is given by

$$A_p = g^{pq} A^q \quad \Rightarrow A^P = g^{pq} A_q.$$

For rectangular co-ordinates $g_{pq} = 1$ if $p = q$, and 0 if $p \neq q$, so that $A_p = A^P$, which explains why no distinction was made between contravariant and covariant components of a vector for rectangular co-ordinates.

49.45 CONJUGATE OR RECIPROCAL TENSORS

Let $g = |g_{pq}|$ denote the determinant with elements g_{pq} and suppose $g \neq 0$. Define g^{pq} by

$$g^{pq} = \frac{\text{cofactor of } g_{pq}}{g}$$

Then g^{pq} is a symmetric contravariant tensor of rank two called the conjugate or Reciprocal tensor of g_{pq}.

Also $$g^{pq}\, g_{rq} = \delta^p_r$$

49.46 CHRISTOFFEL SYMBOLS

The symbols $$[pq,r] = \frac{1}{2}\left(\frac{\partial g_{pr}}{\partial x^q} + \frac{\partial g_{qr}}{\partial x^p} - \frac{\partial g_{pq}}{\partial x^r}\right); \left\{ \begin{matrix} s \\ pq \end{matrix} \right\} = g^{sr}[pq,r]$$

are called the Christoffel symbols of the first and second kind respectively.

Example 18. *Prove that* $[pq, r] = g_{rs} \begin{Bmatrix} s \\ pq \end{Bmatrix}$

Solution. $g_{ks} \begin{Bmatrix} s \\ pq \end{Bmatrix} = g_{ks}\, g^{sr}[pq,r] = \delta_k^r[pq, r] = [pq, k]$

$$\Rightarrow \qquad [pq, k] = g_{ks} \begin{Bmatrix} s \\ pq \end{Bmatrix}$$

ie.
$$[pq, r] = g_{rs} \begin{Bmatrix} s \\ pq \end{Bmatrix}$$
Proved.

Example 19. *Prove that (i) [pq, r] = [qp, r] (ii)* $\begin{Bmatrix} s \\ pq \end{Bmatrix} = \begin{Bmatrix} s \\ qp \end{Bmatrix}$.

Solution. (*i*)
$$[pq,r] = \frac{1}{2}\left(\frac{\partial g_{pr}}{\partial x^q} + \frac{\partial g_{qr}}{\partial x^p} - \frac{\partial g_{pq}}{\partial x^r}\right)$$
$$= \frac{1}{2}\left(\frac{\partial g_{qr}}{\partial x^p} + \frac{\partial g_{pr}}{\partial x^q} - \frac{\partial g_{qp}}{\partial x^r}\right)$$
$$[pq, r] = [qp, r]$$

(*ii*) $\begin{Bmatrix} s \\ pq \end{Bmatrix} = g^{sr}\,[pq,r] = g^{sr}[qp,r] = \begin{Bmatrix} s \\ qp \end{Bmatrix}$ **Proved.**

Example 20. *Prove that* $\dfrac{\partial g_{pq}}{\partial x^m} = [pm, q] + [qm, p]$

Solution. $[pm, q] + [qm, p]$

$$= \frac{1}{2}\left(\frac{\partial g_{pq}}{\partial x^m} + \frac{\partial g_{mq}}{\partial x^p} - \frac{\partial g_{pm}}{\partial x^q}\right) + \frac{1}{2}\left(\frac{\partial g_{qp}}{\partial x^m} + \frac{\partial g_{mp}}{\partial x^q} - \frac{\partial g_{qm}}{\partial x^p}\right) = \frac{1}{2}\frac{\partial g_{pq}}{\partial x^m} + \frac{1}{2}\frac{\partial g_{qp}}{\partial x^m} = \frac{\partial g_{pq}}{\partial x^m}$$
Proved.

Example 21. *Prove that* $\dfrac{\partial g^{pq}}{\partial x^m} = -g^{pn}\begin{Bmatrix} q \\ mn \end{Bmatrix} - g^{qn}\begin{Bmatrix} p \\ mn \end{Bmatrix}$

Solution.
$$\frac{\partial}{\partial x^m}(g^{jk} g_{ij}) = \frac{\partial}{\partial x^m}\left(\delta_i^k\right) = 0$$

Then
$$g^{jk}\frac{\partial g_{ij}}{\partial x^m} + \frac{\partial g^{jk}}{\partial x^m} g_{ij} = 0 \quad \Rightarrow \quad g^{ij}\frac{\partial g^{jk}}{\partial x^m} = -g^{jk}\frac{\partial g_{ij}}{\partial x^m}$$

Multiplying by g^{ir}

i.e.
$$\delta_j^r \frac{\partial g^{jk}}{\partial x^m} = -g^{ir} g^{jk}[im, j] + [jm, i]$$

$$\Rightarrow \qquad \frac{\partial g^{rk}}{\partial x^m} = -g^{ir}\begin{Bmatrix} k \\ im \end{Bmatrix} - g^{jk}\begin{Bmatrix} r \\ jm \end{Bmatrix}$$

Replace r, k, i, j by p, q, n, n, we get $\dfrac{\partial g^{pq}}{\partial x^m} = -g^{pn}\begin{Bmatrix} q \\ mn \end{Bmatrix} - g^{qn}\begin{Bmatrix} p \\ mn \end{Bmatrix}$ **Proved.**

Example 22. *Derive transformation laws for the christoffel symbols of the first and the second kind.*

Solution. Since $\bar{g}_{jk} = \frac{\partial x^p}{\partial \bar{x}^j}\frac{\partial x^q}{\partial \bar{x}^k} g_{pq}$

$$\therefore \quad \frac{\partial \bar{g}_{jk}}{\partial \bar{x}^m} = \frac{\partial x^p}{\partial \bar{x}^j}\frac{\partial x^q}{\partial \bar{x}^k}\frac{\partial g_{pq}}{\partial x^r}\frac{\partial x^r}{\partial \bar{x}^m} + \frac{\partial x^p}{\partial \bar{x}^j}\frac{\partial^2 x^q}{\partial \bar{x}^m \partial \bar{x}^k} g_{pq} + \frac{\partial^2 x^p}{\partial \bar{x}^m \partial \bar{x}^j}\frac{\partial x^q}{\partial \bar{x}^k} g_{pq} \quad ...(1)$$

By cyclic permutation of indices *n, k, m* and *p, q, r*

$$\frac{\partial \bar{g}_{km}}{\partial \bar{x}^j} = \frac{\partial x^q}{\partial \bar{x}^k}\frac{\partial x^r}{\partial \bar{x}^m}\frac{\partial g_{qr}}{\partial x^p}\frac{\partial x^p}{\partial \bar{x}^j} + \frac{\partial x^q}{\partial \bar{x}^k}\frac{\partial^2 x^r}{\partial \bar{x}^j \partial \bar{x}^m} g_{qr} + \frac{\partial^2 x^q}{\partial \bar{x}^j \partial \bar{x}^k}\frac{\partial x^r}{\partial \bar{x}^m} g_{pq} \quad ...(2)$$

$$\frac{\partial \bar{g}_{mj}}{\partial \bar{x}^k} = \frac{\partial x^r}{\partial \bar{x}^m}\frac{\partial x^p}{\partial \bar{x}^j}\frac{\partial g_{rp}}{\partial x^q}\frac{\partial x^q}{\partial \bar{x}^k} + \frac{\partial x^r}{\partial \bar{x}^m}\frac{\partial^2 x^p}{\partial \bar{x}^k \partial \bar{x}^j} g_{rp} + \frac{\partial^2 x^r}{\partial \bar{x}^k \partial \bar{x}^m}\frac{\partial x^p}{\partial \bar{x}^j} g_{rp} \quad ...(3)$$

Subtracting (1) from the sum of (2) and (3) and multiplying by $\frac{1}{2}$, we obtain on using the definition of the Christoffel symbols of the first kind,

$$\overline{[jk,m]} = \frac{\partial x^p}{\partial \bar{x}^j}\frac{\partial x^q}{\partial \bar{x}^k}\frac{\partial x^r}{\partial \bar{x}^m}[pq,r] + \frac{\partial^2 x^p}{\partial \bar{x}^j \partial \bar{x}^k}\frac{\partial x^q}{\partial \bar{x}^m} g_{pq} \quad ... (4) \textbf{ Ans.}$$

49.47 TRANSFORMATION LAW FOR SECOND KIND

Multiplying (4) by $\bar{g}^{nm}$

$$\bar{g}^{nm} = \frac{\partial \bar{x}^n}{\partial x^s}\frac{\partial \bar{x}^m}{\partial x^t} g^{st} \qquad \text{we get}$$

$$\bar{g}^{nm}\overline{[jk,m]} = \frac{\partial x^p}{\partial \bar{x}^j}\frac{\partial x^q}{\partial \bar{x}^k}\frac{\partial x^r}{\partial \bar{x}^m}\frac{\partial \bar{x}^n}{\partial x^s}\frac{\partial \bar{x}^m}{\partial x^t} g^{st}[pq,r] + \frac{\partial^2 x^p}{\partial \bar{x}^j \partial \bar{x}^k}\frac{\partial x^q}{\partial \bar{x}^m}\frac{\partial \bar{x}^n}{\partial x^s}\frac{\partial \bar{x}^m}{\partial x^t} g^{st} g_{pq}$$

Then $$\overline{\begin{Bmatrix} n \\ jk \end{Bmatrix}} = \frac{\partial x^p}{\partial \bar{x}^j}\frac{\partial x^q}{\partial \bar{x}^k}\frac{\partial \bar{x}^n}{\partial x^s}\delta_t^r g^{st}[pq,r] + \frac{\partial^2 x^p}{\partial \bar{x}^j \partial \bar{x}^k}\frac{\partial \bar{x}^n}{\partial x^s}\delta_t^q g^{st} g_{pq}$$

$$\overline{\begin{Bmatrix} n \\ jk \end{Bmatrix}} = \frac{\partial x^p}{\partial \bar{x}^j}\frac{\partial x^q}{\partial \bar{x}^k}\frac{\partial \bar{x}^n}{\partial x^s}\begin{Bmatrix} s \\ pq \end{Bmatrix} + \frac{\partial^2 x^p}{\partial \bar{x}^j \partial \bar{x}^k}\frac{\partial \bar{x}^n}{\partial x^p} \quad ... (5)$$

(4) and (5) are required transformation laws.

Example 23. *If* $(ds)^2 = r^2 (d\theta)^2 + r^2 \sin^2\theta \, (d\phi)^2$, *find the value of*

(*a*) [22, 1] (*b*) [12, 2] (*c*) [1, 22] (*d*) [2, 12].

Solution. $(ds)^2 = r^2 (d\theta)^2 + r^2 \sin^2\theta \, (d\phi)^2$

$$g_{11} = r^2, \; g_{22} = r^2 \sin^2\theta, \; g_{12} = 0 = g_{21}$$

$$g = \begin{vmatrix} g_{11} & g_{12} \\ g_{21} & g_{22} \end{vmatrix} = \begin{vmatrix} r^2 & 0 \\ 0 & r^2\sin^2\theta \end{vmatrix} = r^4 \sin^2\theta$$

$$g^{11} = \frac{\text{cofactor of } g_{11}}{g} = \frac{r^2\sin^2\theta}{r^4\sin^2\theta} = \frac{1}{r^2}$$

$$g^{22} = \frac{\text{cofactor of } g_{22}}{g} = \frac{r^2}{r^4\sin^2\theta} = \frac{1}{r^2\sin^2\theta}$$

The christoffel symbols of first kind are

$$[ij,k]=\frac{1}{2}\left[\frac{\partial g_{jk}}{\partial x^i}+\frac{\partial g_{ik}}{\partial x^j}-\frac{\partial g_{ij}}{\partial x^k}\right]$$

(*a*) $$[22,1]=\frac{1}{2}\left[\frac{\partial g_{21}}{\partial x^2}+\frac{\partial g_{21}}{\partial x^2}-\frac{\partial g_{22}}{\partial x^1}\right]=\frac{1}{2}\left[\frac{\partial(0)}{\partial\phi}+\frac{\partial(0)}{\partial\phi}-\frac{\partial(r^2\sin^2\theta)}{\partial\theta}\right]$$

$$= r^2 \sin\theta\cos\theta$$

(*b*) $$[12,2]=\frac{1}{2}\left[\frac{\partial g_{22}}{\partial x^1}+\frac{\partial g_{12}}{\partial x^2}-\frac{\partial g_{12}}{\partial x^2}\right]=\frac{1}{2}\left[\frac{\partial(r^2\sin^2\theta)}{\partial\theta}+\frac{\partial(0)}{\partial\phi}-\frac{\partial(0)}{\partial\phi}\right]$$

$$= r^2 \sin\theta\cos\theta$$

(*c*) The christoffel symbols of second kind are

$$\begin{Bmatrix} k \\ ij \end{Bmatrix} = g^{kl}[ij,l]$$

$$\begin{Bmatrix} 1 \\ 22 \end{Bmatrix} = g^{ll}[22,l] = g^{11}[22,1]\, g^{12}[22,2]$$

$$= \frac{1}{r^2}[-r^2\sin\theta\cos\theta]+0 \quad (g^{12}=0) = -\sin\theta\cos\theta$$

(*d*) $$\begin{Bmatrix} 2 \\ 12 \end{Bmatrix} = g^{2l}[22,l] = g^{21}[12,1]+g^{22}[12,2]$$

$$= 0+\frac{1}{r^2\sin^2\theta}[r^2\sin\theta\cos\theta]=\frac{\cos\theta}{\sin\theta}=\cot\theta = r^4\sin\theta\cos\theta$$ **Ans.**

Example 24. *If* $(ds)^2 = (dr)^2 + r^2(d\theta)^2 + r^2\sin^2\theta\,(d\phi)^2$, *find the value of*

(*a*) [22, 1], (*b*) [33, 1], (*c*) [13, 3], (*d*) [23, 3],

(*e*) $\begin{Bmatrix} 1 \\ 22 \end{Bmatrix}$, (*f*) $\begin{Bmatrix} 1 \\ 33 \end{Bmatrix}$, (*g*) $\begin{Bmatrix} 3 \\ 13 \end{Bmatrix}$, (*h*) $\begin{Bmatrix} 3 \\ 23 \end{Bmatrix}$

Solution. $(ds)^2 = (dr)^2 + r^2(d\theta)^2 + r^2\sin^2\theta\,(d\phi)^2$

$x_1 = r,\ x_2 = \theta,\ x_3 = \phi$

$g_{11} = 1,\ g_{22} = r^2,\ g_{33} = r^2\sin\theta$

$g_{12} = 0 = g_{13} = \ldots$

$$g = \begin{vmatrix} g_{11} & g_{12} & g_{13} \\ g_{21} & g_{22} & g_{23} \\ g_{31} & g_{32} & g_{33} \end{vmatrix} = \begin{bmatrix} 1 & 0 & 0 \\ 0 & r^2 & 0 \\ 0 & 0 & r^2\sin^2\theta \end{bmatrix} = r^4\sin^2\theta$$

$$g^{11} = \frac{\text{cofactor of } g_{11}}{g} = \frac{\begin{vmatrix} r^2 & 0 \\ 0 & r^2\sin^2\theta \end{vmatrix}}{r^4\sin^2\theta} = \frac{r^4\sin^2\theta}{r^4\sin^2\theta} = 1$$

$$g^{22} = \frac{\text{cofactor of } g_{22}}{g} = \frac{\begin{vmatrix} 1 & 0 \\ 0 & r^2\sin^2\theta \end{vmatrix}}{r^4\sin^2\theta} = \frac{r^2\sin^2\theta}{r^4\sin^2\theta} = \frac{1}{r^2}$$

$$g^{33} = \frac{\text{cofactor of } g_{33}}{g} = \frac{\begin{vmatrix} 1 & 0 \\ 0 & r^2 \end{vmatrix}}{r^4\sin^2\theta} = \frac{r^2}{r^4\sin^2\theta} = \frac{1}{r^2\sin^2\theta}$$

The christoffel symbols of the first kind are

$$[ij,k]=\frac{1}{2}\left[\frac{\partial g_{jk}}{\partial x^i}+\frac{\partial g_{ik}}{\partial x^i}-\frac{\partial g_{ij}}{\partial x^k}\right]$$

(*a*) $$[22,1]=\frac{1}{2}\left[\frac{\partial g_{21}}{\partial x^2}+\frac{\partial g_{21}}{\partial x^2}-\frac{\partial g_{22}}{\partial x^1}\right]=\frac{1}{2}\left[\frac{\partial(0)}{\partial\theta}+\frac{\partial(0)}{\partial\theta}-\frac{\partial(r^2)}{\partial r}\right]=\frac{1}{2}(-2r)=-r$$

(*b*) $$[33,1]=\frac{1}{2}\left[\frac{\partial g_{31}}{\partial x^3}+\frac{\partial g_{31}}{\partial x^3}-\frac{\partial g_{33}}{\partial x^1}\right]$$

$$=\frac{1}{2}\left[\frac{\partial(0)}{\partial\phi}+\frac{\partial(0)}{\partial\phi}-\frac{\partial(r^2\sin^2\theta)}{\partial r}\right]=\frac{1}{2}(-2r\sin^2\theta)=-r\sin^2\theta$$

(*c*) $$[13,3]=\frac{1}{2}\left[\frac{\partial g_{33}}{\partial x^1}+\frac{\partial g_{13}}{\partial x^3}-\frac{\partial g_{13}}{\partial x^3}\right]=\frac{1}{2}\frac{\partial g_{33}}{\partial x^1}=\frac{1}{2}\left[\frac{\partial(r^2\sin^2\theta)}{\partial r}\right]=r\sin^2\theta$$

(*d*) $$[23,3]=\frac{1}{2}\left[\frac{\partial g_{33}}{\partial x^2}+\frac{\partial g_{23}}{\partial x^3}-\frac{\partial g_{23}}{\partial x^3}\right]=\frac{1}{2}\frac{\partial g_{33}}{\partial x^2}=\frac{1}{2}\frac{\partial}{\partial\theta}(r^2\sin^2\theta)=r^2\sin\theta\cos\theta$$

The christoffel symbols of the second kind are

$$\begin{Bmatrix}k\\ij\end{Bmatrix}=g^{kl}[ij,l]$$

(*e*) $$\begin{Bmatrix}1\\22\end{Bmatrix}=g^{1l}[22,l]=g^{11}[22,1]+g^{12}[22,2]+g^{13}[22,3]=(1)(-r)+0+0=-r$$

(*f*) $$\begin{Bmatrix}1\\33\end{Bmatrix}=g^{1l}[33,l]=g^{11}[33,1]+g^{12}[33,2]+g^{13}[33,3]=(1)(-r\sin^2\theta)+0+0=-r\sin^2\theta$$

(*g*) $$\begin{Bmatrix}3\\13\end{Bmatrix}=g^{3l}[13,l]=g^{31}[13,1]+g^{32}[13,2]+g^{33}[13,3]$$

$$=0\,(r\sin^2\theta)+0\,[0]+\frac{1}{r^2\sin^2\theta}[r\sin^2\theta]=\frac{1}{r}$$

(*h*) $$\begin{Bmatrix}3\\23\end{Bmatrix}=g^{3l}[23,l]=g^{31}[23,1]+g^{32}[23,2]+g^{33}[23,3]$$

$$=0\,[0]+0\,[0]+\frac{1}{r^2\sin^2\theta}(r\sin\theta\cos\theta)=\cot\theta$$ **Ans.**

Example 25. *Prove that* $\begin{Bmatrix}p\\pq\end{Bmatrix}=\frac{\partial}{\partial x^q}\log\sqrt{g}$

Solution. $g=g_{jk}\,G(j,k)$ (Sum over k only)

where $G(j,k)$ is the cofactor of g_{jk} in the determinant $g=|g_{jk}|\neq 0$ since $G(j,k)$ does not contain g_{jk} explicitly,

$$\frac{\partial g}{\partial g_{jr}}=G(j,r)$$

Then, summing over j and r

$$\frac{\partial g}{\partial x^m} = \frac{\partial g}{\partial g_{jr}} \frac{\partial g_{jr}}{\partial x^m} = G(j,r)\frac{\partial g_{jr}}{\partial x^m} = g\, g^{jr} \frac{\partial g_{jr}}{\partial x^m} = g\, g^{jr}\,([jm,r]+[rm,j])$$

$$\frac{\partial g}{\partial x^m} = g\left(\begin{Bmatrix} j \\ jm \end{Bmatrix} + \begin{Bmatrix} r \\ rm \end{Bmatrix}\right) = 2g\begin{Bmatrix} j \\ jm \end{Bmatrix}$$

Thus $$\frac{1}{2g}\frac{\partial g}{\partial x^m} = \begin{Bmatrix} j \\ jm \end{Bmatrix} \quad \text{or} \quad \begin{Bmatrix} j \\ jm \end{Bmatrix} = \frac{\partial g}{\partial x^m}\log\sqrt{g}$$

Replacing j by p and m by q, $\begin{Bmatrix} p \\ pq \end{Bmatrix} = \dfrac{\partial}{\partial x^q}\log\sqrt{g}$ **Proved.**

49.48 CONTRAVARIANT, COVARIANT AND MIXED TENSOR

If A_i be a set of n functions of the co-ordinates $x^1, x^2, x^3 \ldots x^n (x^i)$. They are transformed in another system of co-ordinates $x^{-1}, x^{-2}, x^{-3} \ldots x^n$ according to $\overline{A}_i = \dfrac{\partial x^j}{\partial \overline{x}^i} A_j$

A_i are called the components of a covariant tensor.

If $\phi\,(x^1, x^2 \ldots x^n)$ be a scalar functions, then $\dfrac{\partial \phi}{\partial \overline{x}^i} = \dfrac{\partial \phi}{\partial x^1}\dfrac{\partial x^1}{\partial \overline{x}^i} + \dfrac{\partial \phi}{\partial x^2}\dfrac{\partial x^2}{\partial \overline{x}^i} + \ldots + \dfrac{\partial \phi}{\partial x^n}\dfrac{\partial x^n}{\partial \overline{x}^i}$ ... (1)

then $\dfrac{\partial \phi}{\partial x^1}, \dfrac{\partial \phi}{\partial x^2} \ldots \dfrac{\partial \phi}{\partial x^n}$ are the components of a covariant vector.

Since x is a function of x^i (*i.e.*, $x^1, x^2, \ldots x^n$)

so $$d\overline{x}^i = \frac{\partial \overline{x}^i}{\partial x^1}dx^1 + \frac{\partial \overline{x}^i}{\partial x^2}dx^2 + \ldots + \frac{\partial \overline{x}^i}{\partial x^n}dx^n \quad \ldots (2)$$

$$= \frac{\partial \overline{x}^i}{\partial x^1}dx^1$$

On comparing (1) and (2) we can say that $dx^1, dx^2, \ldots dx^n$ is an example of a contravariant tensor.

If q and s varry from 1 to n, then A^{qs} will be n^2 functions.

If N^2 quantities A^{qs} in a co-ordinate system $(x^1, x^2, \ldots x^N)$ are related to N^2 other quantities $\overline{A}^{pr}$ in another system $(\overline{x}^1, \overline{x}^2, \ldots \overline{x}^N)$ by the transformation equations

$$\overline{A}^{pr} = \sum_{s=1}^{N}\sum_{q=1}^{N}\frac{\partial \overline{x}^p}{\partial x^q}\frac{\partial \overline{x}^r}{\partial x^s}A^{qs} \quad \Rightarrow \quad \overline{A}^{pr} = \frac{\partial \overline{x}^p}{\partial x^q}\frac{\partial \overline{x}^r}{\partial x^s}A^{qs}$$

they are called *contravariant* components of a tensor of the second rank.

If the transformation law is $\overline{A}^{pr} = \dfrac{\partial x^q}{\partial \overline{x}^p}\dfrac{\partial x^s}{\partial \overline{x}^r}A_{qs}$,

then quantities A_{qs} are called components of *covariant* tensor of second rank.

The N^2 quantities A_s^q are called components of a *mixed* tensor of second rank if

$$\overline{A}_r^p = \frac{\partial \overline{x}^p}{\partial x^q}\frac{\partial x^s}{\partial \overline{x}^r}A_s^q$$

EXERCISE 49.2

1. If A^i are the components of an absolute contravariant tensor of rank one, show that $\frac{\partial A_i}{\partial x_j}$ are the components of a mixed tensor.

2. If A^{ji} and A_{ij} are reciprocal symmetric tensors and x_i are the components of a covariant tensor of rank one, show that $A_{ij}\ x^i x^j, = A^{ij}\ x_i\ x_j$ where $x^i = A^{i\alpha} x_\alpha$.

3. If the components of a tensor are zero in one co-ordinate system, then prove that the components are zero in all co-ordinate systems.

4. Show that the expression $A\ (i,j,\ k)$ is a tensor if its inner product with an arbitrary tensor B_k^{jl} is a tensor.

5. A^{ij} is a contravariant tensor and B_i a covariant tensor. Show that $A^{ij}\ B_k$ is a tensor of rank three, but $A^{ij}\ B_j$ is a tensor of rank one.

6. If g_{ij} denotes the components of a covariant tensor of rank two, show that the product $g_{ij} dx^i\ dx^j$ is an invariant scalar.

7. Find g and g^{ij} corresponding to the metric $ds^2 = 5\ (dx^1)^2 + 3\ (dx^2)^2 + 4\ (dx^3)^2 - 6\ dx^1\ dx^2 + 4dx^2\ dx^3$.

Ans. $g = 4$, $g^{11} = 2$, $g^{22} = 5$, $g^{33} = 1.5$, $g^{12} = 3$, $g^{23} = -2.5$, $g^{13} = -1.5$

8. Find the values of g and g^{ij}, if

$$ds^2 = \frac{dr^2}{1 - \frac{r^2}{R^2}} + r^2 (d\theta^2 + \sin^2\theta d\phi^2) \text{, where R is constant}$$

Ans. $g = \dfrac{r^4 \sin^2\theta}{1 - \frac{r^2}{R^2}}; g^{11} = 1 - \dfrac{r^2}{R^2}, g^{22} = \dfrac{1}{r^2}, g^{33} = \dfrac{1}{r^2 \sin^2\theta}, g^{ij} = 0\ (i \neq j)$

9. Prove that the angle $\theta_{12}, \theta_{23}, \theta_{31}$ between the co-ordinate curves in a three dimensional co-ordinate system are given by

$$\cos\theta_{12} = \frac{g_{12}}{\sqrt{g_{11} g_{22}}}, \cos\theta_{23} = \frac{g_{23}}{\sqrt{g_{22}\, g_{33}}}, \cos\theta_{31} = \frac{g_{31}}{\sqrt{g_{33}\, g_{11}}}$$

10. Prove that for an orthogonal co-ordinate system

(*a*) $g_{12} = g_{23} = g_{31} = 0$ (*b*) $g^{11} = \dfrac{1}{g_{11}}, g^{22} = \dfrac{1}{g_{22}}, g^{33} = \dfrac{1}{g_{33}}$

11. Surface of a sphere is a two dimensional Riemannian space. Find its fundamental metric tensor. If a be the fixed radius of the sphere.

Ans. $g_{11} = a^2$, $g_{22} = a^2 \sin^2\theta$, $g = a^4 \sin^2\theta$

$$g^{11} = \frac{1}{a^2}, g^{22} = \frac{1}{a^2 \sin^2\theta}, g^{12} = 0 = g^{21}.$$

INDEX

A

B

C

D

E

F

G

H

I

L

M

N

O

U

V

W